10TH EDITION

& TOPLEY WILSON'S
MICROBIOLOGY & MICROBIAL INFECTIONS

PARASITOLOGY

10TH EDITION

TOPLEY & WILSON'S

MICROBIOLOGY & MICROBIAL INFECTIONS

Topley & Wilson's Microbiology and Microbial Infections has grown from one to eight volumes since first published in 1929, reflecting the ever-increasing breadth and depth of knowledge in each of the areas covered. This tenth edition continues the tradition of providing the most comprehensive reference to microorganisms and the resulting infectious diseases currently available. It forms a unique resource, with each volume including examples of the best writing and research in the fields of virology, bacteriology, medical mycology, parasitology, and immunology from around the globe.

www.topleyandwilson.com

VIROLOGY Volumes 1 and 2
Edited by Brian W.J. Mahy and Volker ter Meulen
Volume 1 ISBN 0 340 88561 0; Volume 2 ISBN 0 340 88562 9; 2 volume set ISBN 0 340 88563 7

BACTERIOLOGY Volumes 1 and 2
Edited by S. Peter Borriello, Patrick R Murray, and Guido Funke
Volume 1 ISBN 0 340 88564 5; Volume 2 ISBN 0 340 88565 3; 2 volume set ISBN 0 340 88566 1

MEDICAL MYCOLOGY
Edited by William G. Merz and Roderick J. Hay
ISBN 0 340 88567 X

PARASITOLOGY
Edited by F.E.G. Cox, Derek Wakelin, Stephen H. Gillespie, and Dickson D. Despommier
ISBN 0 340 88568 8

IMMUNOLOGY
Edited by Stephan H.E. Kaufmann and Michael W. Steward
ISBN 0 340 88569 6

Cumulative index
ISBN 0 340 88570 X

8 volume set plus CD-ROM
ISBN 0 340 80912 4

CD-ROM only
ISBN 0 340 88560 2

For a full list of contents, please see the *Complete table of contents* on page 883

10TH EDITION

TOPLEY & WILSON'S

MICROBIOLOGY & MICROBIAL INFECTIONS

PARASITOLOGY

EDITED BY

F.E.G. Cox PHD DSc
Senior Visiting Research Fellow
Department of Infectious and Tropical Diseases
London School of Hygiene and Tropical Medicine, London, UK

Derek Wakelin PHD DSc FRCPATH
Professor Emeritus, University of Nottingham, Nottingham, UK

Stephen H. Gillespie MD FRCP (EDIN) FRCPATH
Professor of Medical Microbiology, and Regional Microbiologist
Centre for Medical Microbiology,
Royal Free and University College Medical School, London, UK

Dickson D. Despommier PHD
Professor of Public Health and Microbiology
Department of Environmental Health Sciences
Columbia University, New York, NY, USA

 Hodder Arnold

A MEMBER OF THE HODDER HEADLINE GROUP

 ASM PRESS

First published in Great Britain in 1929
Second edition 1936
Third edition 1946
Fourth edition 1955
Fifth edition 1964
Sixth edition 1975
Seventh edition 1983 and 1984
Eighth edition 1990
Ninth edition 1998
This tenth edition published in 2005 by
Hodder Arnold, an imprint of Hodder Education and a member of the Hodder Headline Group,
338 Euston Road, London NW1 3BH

http://www.hoddereducation.com

Distributed in the United States of America by ASM Press, the book publishing division of the American Society for Microbiology
1752 N Street, N.W. Washington, D.C. 20036, USA

Hodder Headline's policy is to use papers that are natural, renewable and recyclable products and made from wood grown in sustainable forests. The logging and manufacturing processes are expected to conform to the environmental regulations of the country of origin.

Whilst the advice and information in this book are believed to be true and accurate at the date of going to press, neither the author[s] nor the publisher can accept any legal responsibility or liability for any errors or omissions that may be made. In particular (but without limiting the generality of the preceding disclaimer) every effort has been made to check drug dosages; however it is still possible that errors have been missed. Furthermore, dosage schedules are constantly being revised and new side-effects recognized. For these reasons the reader is strongly urged to consult the drug companies' printed instructions before administering any of the drugs recommended in this book.

British Library Cataloguing in Publication Data
A catalogue record for this book is available from the British Library

Library of Congress Cataloging-in-Publication Data
A catalog record for this book is available from the Library of Congress

This volume only ISBN-10 0 340 885 688   ISBN-13 978 0 340 885 680
Complete set and CD-ROM ISBN-10   0 340 80912 4   ISBN-13   978 0 340 80912 9
Indian edition ISBN-10   0 340 88559 9   ISBN-13   978 0 340 88559 8

1 2 3 4 5 6 7 8 9 10

Commissioning Editor: Serena Bureau / Joanna Koster
Development Editor: Layla Vandenberg
Project Editor: Zelah Pengilley
Production Controller: Deborah Smith
Index: Merrall-Ross International Ltd.
Cover Designer: Sarah Rees

Cover image: Toxocara roundworm. Jackie Lewin, EM Unit, Royal Free Hospital / Science Photo Library

Typeset in 9/11 Times New Roman by Lucid Digital, Salisbury, UK
Printed and bound in Italy

What do you think about this book? Or any other Hodder Arnold title? Please send your comments to www.hoddereducation.com

Contents

Contributors

John P. Ackers MA DPHIL MSc
Professor of Postgraduate Education in Public Health
Department of Infectious and Tropical Diseases
London School of Hygiene and Tropical Medicine
London, UK

Roy M. Anderson
Department of Infectious Disease Epidemiology
Imperial College School of Medicine
London, UK

Jørn Andreassen
Department of Population Biology
Biological Institute, University of Copenhagen
Copenhagen, Denmark

R.W. Ashford BSc PHD DSc
Liverpool School of Tropical Medicine
Liverpool, UK

Subash Babu MBBS PHD
Helminth Immunology Section
Laboratory of Parasitic Diseases
National Institutes of Health
Bethesda, MD, USA

Maria-Gloria Basáñez MSc PHD FRES
Senior Lecturer and MSc Course Organiser
Department of Infectious Disease Epidemiology
Imperial College School of Medicine
London, UK

Paul A. Bates BA PHD
Senior Lecturer in Medical Parasitology
Liverpool School of Tropical Medicine
Liverpool, UK

Samuel J. Black BSc PHD
Head of Department
Department of Veterinary & Animal Sciences
University of Massachusetts
Amherst, MA, USA

Janette E. Bradley BSc PHD
School of Biology, University of Nottingham
Nottingham, UK

Donald E. Burgess PHD
Collection Scientist, Protistology
American Type Culture Collection
Manassas, VA, USA

Michael Cappello MD
Associate Professor
Departments of Pediatrics and
Epidemiology and Public Health
Yale School of Medicine
New Haven, CT, USA

F.E.G. Cox PHD DSc
Senior Visiting Research Fellow
Department of Infectious and Tropical Diseases,
London School of Hygiene and Tropical Medicine
London, UK

Philip S. Craig
Biosciences Research Institute
School of Environment and Life Sciences
University of Salford
Salford, UK

Alan Curry BSc PHD
Head of Unit, Electron Microscopy Unit
Manchester Royal Infirmary
Manchester, UK

Dickson D. Despommier
Professor of Public Health and Microbiology
Department of Environmental Health Sciences
Mailman School of Public Health
Columbia University
New York, NY, USA

J.P. Dubey BVSC&AH MVSc PHD
Senior Scientist
United States Department of Agriculture
Beltsville Agricultural Research Centre
MD, USA

Martha Espinosa-Cantellano MD DSc
Center for Research and Advanced Studies
Mexico City, Mexico

Ana Flisser
Directora de Investigacion,
Hospital General 'Dr Manuel Gea Gonzalez'; and
SSA Investigadora Departamento de Microbiologia
y Parasitologia
Facultad de Medicina, UNAM
Mexico DF, Mexico

Lynne S. Garcia MS MT FAAM
Director, LSG and Associates
Santa Monica, CA, USA

David I. Gibson BSc PhD DSc
Head of Division, Parasitic Worms Division
Department of Zoology
The Natural History Museum
London, UK

Herbert M. Gilles MD DSc FRCP FFPH DTM&H
Professor Emeritus
Liverpool School of Tropical Medicine
University of Liverpool
Liverpool, UK

Stephen H. Gillespie MD FRCP(Edin) FRCPath
Professor of Medical Microbiology; and
Regional Microbiologist
Centre for Medical Microbiology
Royal Free and University College Medical School
London, UK

Melissa R. Haswell-Elkins PhD
Senior Lecturer, Indigenous Health and
Indigenous Stream Co-ordinator
Australian Integrated Mental Health Initiative
North Queensland Health Equalities Promotion Unit
School of Population Health
University of Queensland, Cairns
Queensland, Australia

Celia V. Holland PhD
Associate Professor and Head of Department
Department of Zoology, Trinity College
Dublin, Ireland

Marcel Hommel MD PhD
Professor of Tropical Medicine
Liverpool School of Tropical Medicine
Liverpool, UK

Peter J. Hotez MD PhD
Professor and Chair
Department of Microbiology and
Tropical Medicine
The George Washington University
Washington, DC, USA

David T. John MSPH PhD
Professor of Microbiology/Parasitology; and
Associate Dean for Basic Sciences and Graduate Studies
Oklahoma State University, Center for Health Studies
Tulsa, OK, USA

M. Paul Kelly
Senior Lecturer (Wellcome Fellow)
Barts and the London School of Medicine and Dentistry
Adult and Paediatric Gastroenterology
Digestive Disease Research Centre
London, UK

Somei Kojima MD PhD
Professor Emeritus
The University of Tokyo, Tokyo, Japan; and
Deputy Director, Center for Medical Science
International University of Health and Welfare
Otawara, Tochigi Prefecture, Japan

Laetitia M. Kortbeek MD
Medical Microbiologist
Diagnostic Laboratory for Infectious Diseases
and Perinatal Screening (LIS)
National Institute of Public Health
and the Environment, Bilthoven
The Netherlands

Jaroslav Kulda RNDr, PhD
Professor of Parasitology
Department of Parasitology, Faculty of Science
Charles University in Prague
Prague, Czech Republic

Ralph Lainson OBE FRS AFTWAS BSc PhD DSc
Department of Parasitology
Instituto Evandro Chagas
Belém, Pará, Brazil

Francisco J. López-Antuñano
Instituto Nacional de Salud Publica
Colonia Santa Maria Ahuacatitlan
Morelos, Mexico

Andrew S. MacDonald BSc PhD
MRC Research Fellow
Institute of Immunology and Infection Research
University of Edinburgh
Edinburgh, UK

Adolfo Martínez-Palomo MD DSc
Emeritus Professor
Center for Research and Advanced Studies
Mexico City, Mexico

Vincent McDonald PHD
Reader in Gastroenterology
Centre for Gastroenterology
Institute of Cell and Molecular Science
Barts and the London School of Medicine and Dentistry
Queen Mary College, University of London
London, UK

Heinz Mehlhorn
Department of Parasitology
University of Duesseldorf
Duesseldorf, Germany

Michael A. Miles MSC PHD DSC FRCPATH
Professor of Medical Protozoology
Department of Infectious and Tropical Diseases
London School of Hygiene and Tropical Medicine
London, UK

David H. Molyneux MA PHD DSC
Director, Lymphatic Filariasis Support Centre
Liverpool School of Tropical Medicine
Liverpool, UK

Ralph Muller BSC PHD DSc FIBIOL
Honorary Senior Lecturer
Department of Infectious and Tropical Diseases
London School of Hygiene and Tropical Medicine
London, UK

Eva Nohýnková RNDR PHD
Head, Department of Tropical Medicine
Faculty Hospital Bulovka
Charles University in Prague
Prague, Czech Republic

Thomas B. Nutman MD
Head, Helminth Immunology Section
Laboratory of Parasitic Diseases
National Institutes of Health
Bethesda, MD, USA

Elena Pinelli MSC PHD
Senior Scientist
Diagnostic Laboratory for Infections Diseases and
Perinatal Screening (LIS)
National Institute of Public Health
and the Environment (RIVM)
Bilthoven, The Netherlands

Gabriel A. Schmunis
Communicable Diseases Unit,
Pan American Health Organization
Washington, DC, USA

John Richard Seed PHD
Professor, Department of Epidemiology
School of Public Health
University of North Carolina
Chapel Hill, NC, USA

Jeffrey J. Shaw OBE DSC PHD DAP&E
Parasitology Department
Biomedical Sciences Institute
São Paulo University, São Paulo, Brazil

Andrew Spielman BS SCD MA(Hon)
Department of Immunology and Infectious Diseases
Harvard School of Public Health
Boston, MA, USA

Sam R. Telford III SCD
Associate Professor of Infectious Diseases
Tufts University School of Veterinary Medicine
North Grafton, MA, USA; and
Visiting Scientist in Immunology and
Infectious Diseases
Harvard School of Public Health
Boston, MA, USA

Joke W.B. van der Giessen DVM PHD
Senior Scientist, Parasitic Zoonoses,
Microbiological Laboratory for Health Protection
National Institute of Public Health and the
Environment (RIVM)
Bilthoven, The Netherlands

Derek Wakelin PHD DSC FRCPATH
Professor Emeritus, University of Nottingham
Nottingham, UK

James A.G. Whitworth MD FRCP MFPH DTM&H
Head of International Activities
Wellcome Trust
London, UK

Jeffrey J. Windsor MSC CSCI FIBMS
Senior Biomedical Scientist
National Public Health Service for Wales
Microbiology Aberystwyth, Bronglais Hospital
Aberystwyth, Ceredigion, Wales, UK

Kentaro Yoshimura DVM PHD
Professor Emeritus
Akita University School of Medicine
Akita, Japan

Viqar Zaman DSC FRCPATH
Honorary Visiting Professor
Department of Pathology & Microbiology
The Aga Khan University
Karachi, Pakistan

Preface

Much of this new volume will look familiar but there have been a number of important changes since Parasitology first appeared as a new volume in the 9th edition of Topley and Wilson in 1998. The complexity of the subject has necessitated the appointment of some new editors. Frank Cox has retained responsibility for the general and protozoal diseases chapters. Stephen Gillespie, with his particular expertise with intestinal protozoan infections, clinical parasitology, and diagnosis, has replaced Julius Kreier who made such an important contribution to the success of the previous volume. Derek Wakelin has been joined by Dickson Despommier to share responsibility for the helminth infections. We have been very fortunate in being able to retain almost all of the original team of authors and in being able to recruit some additional authors with internationally recognized expertise in particular areas where there have been significant changes as outlined below.

Our better understanding of the taxonomy of the protozoa has necessitated some important changes to this volume. *Pneumocystis*, long regarded as a protozoan, is now firmly established as a fungus and has been removed from this volume. The Microsporidia constitute an important group that has had a chequered history and until recently has always been classified with the Protozoa. It is now clear that the microsporidians are definitely fungi but are retained in this volume on the grounds of history and continuity. The removal of *Pneumocystis* has allowed us to bring in a new chapter on *Blastocystis*, a parasite whose taxonomic status is still uncertain but one that has become increasingly prevalent and important. There have been a number of changes to the content of some chapters. The chapter on Intestinal amebae has been replaced by two, one on *Entamoeba histolytica* infections and one on Other intestinal amebae. Similarly, the chapter on Trichomonads and related flagellates has been replaced by one on *Trichomonas* infections and one on Intestinal flagellates. Our increasing understanding of the intestinal Coccidia has made it necessary to reorganize and redistribute the content of the three previous chapters. The single chapter on Schistosomes has been divided: the Asian schistosomes and African and American schistosomes are now covered in separate chapters. *Toxocara* and *Trichinella* now have a chapter each and the former

chapter now has a wider coverage. Finally, from a practical point of view, there is a new chapter on Diagnosis.

Parasitic infections continue to present a major challenge to the health and well-being of millions of people across the world, particularly those living in the poorest regions, mainly in the tropics and subtropics. Global unrest, floods and famines, the migrations of huge populations, and the almost unstoppable spread of HIV infections have added to already serious problems not only in terms of increased parasite burdens but also in terms of the interactions between parasites and other infectious agents. Despite all this, however, progress is being made in the containment of many parasitic infections and there is real optimism about the possibilities of eliminating or eradicating some of the diseases that have plagued humans for centuries. Hopes for the future revolve around two quite disparate themes; the application of the potential of molecular biology and the systematic and rational use of well-tried methods of control. These themes run through the chapters in this volume.

The main objective of the application of the techniques of molecular biology in the context of the control of infections is the determination of the genetic make up of the host, parasite, and vector, This should enable us to understand why particular individuals become infected or ill, to design chemotherapeutic and immunological strategies to prevent or ameliorate disease, and to identify features of vectors that might be susceptible to attack. The genomes of a number of important parasites have already been determined and these include the malaria parasite, *Plasmodium falciparum*, the trypanosomes *Trypanosoma brucei* and *T. cruzi*, *Leishmania major*, *Toxoplasma gondii*, *Cryptosporidium parvum*, the schistosomes *Schistosoma mansoni* and *S. japonicum*, and the filarial worm *Brugia malayi*. The genome of the malaria mosquito vector, *Anopheles gambiae*, has also been determined and progress is being made towards the determination of the genomes of other parasites and vectors. So far, however, no direct applications of knowledge have been possible but this information is now being evaluated in the context of drug and vaccine development.

Very little progress has been made in the development of new cheap, safe, and effective drugs against parasitic infections. Most progress has been made in the field of malaria where resistance against chloroquine, pyrimethamine, and other drugs has made the discovery of new remedies imperative. The Chinese plant-derived drug artemisinin and its derivatives hold out most hope for the immediate future particularly in combination with well-tried but no longer effective drugs. Other combination therapies are being increasingly used and there are a number of new drugs at various stages of development. A new drug, Miltefosine, has proved to be useful against leishmaniasis but, unfortunately, there has been little progress in the development of drugs against other important protozoan diseases and the treatment of Chagas disease and African trypanosomiasis has hardly moved on for over half a century. There has been very little progress on the development of new drugs against helminth infections, praziquantel and ivermectin have continued to be invaluable for the treatment of schistosomiasis and microfilarial infections respectively but there is still a requirement for a good macrofilaricide. The almost certain prospect of drug resistance has heightened the requirement for new drugs to replace even those that are currently effective.

The most important development has been the realization that we cannot wait for the promises of molecular biology to be fulfilled and the systematic application of well-established methods of control such as anti-vector measures and the mass administration of drugs is producing dividends. Emphasis has moved from eradication (i.e. reduction to zero of the world-wide incidence of infection) to more realistic aims of the elimination of disease or infection in particular geographical areas. The success of this kind of approach has been very encouraging. The interruption of vector transmission of *Trypanosoma cruzi* in endemic countries in South America is on target for 2010; the interruption of transmission and the elimination of lymphatic filariasis as a public health problem is on target for 2020; and the elimination of onchocerciasis as a public health problem was achieved in 2002 in West Africa and is on target for 2007 in the remaining endemic regions of the world. The WHO's Roll Back Malaria programme now relies heavily on the use of bednets impregnated with insecticides which have proved to be very effective in reducing transmission and the incidence of malaria where they have been used in a systematic way. Any optimism must be moderated with the realization that the best made plans and predictions often go astray; guinea worm should have been eradicated in 1995 but this date was postponed to 2004 and now 2009 almost entirely due to civil unrest in the Sudan.

Parasitic infections are not confined to humans but exact a massive toll on domesticated animals. Parasites still remain one of the major constraints on the development of the poorer parts of the world not only by directly affecting those infected but also destroying the animals required for food thus creating a vicious circle in which those that are infected are malnourished and those that are malnourished fare worse when infected. Studies on the parasites of humans have led to insights into diseases of domesticated animals and vice versa so this volume is only able to present part of much larger interrelated fields of study. It is hoped that the chapters in this volume will provide an insight into the human diseases caused by parasites, what their affects are and ways in which they can be controlled not only now but also in the future.

F.E.G. Cox
Derek Wakelin
Stephen H. Gillespie
Dickson D. Despommier
London, Nottingham and New York
May 2005

Preface to the 9th edition

The inclusion of a volume devoted to parasitic infections is a new initiative for Topley and Wilson. Strictly speaking, the term parasite can be applied to any infectious agent but, by convention, is generally restricted to infections caused by protozoa and helminths and excludes the viruses and prokaryotic organisms traditionally covered in these volumes. The grouping together of protozoa and helminths is not entirely satisfactory because, from the point of view of systematics, they have little in common except their eukaryotic nature, the former being unicellular and the latter multicellular. In fact, infections caused by protozoa have more in common with viral and bacterial infections than with those caused by helminths; modern epidemiological concepts therefore tend to group protozoa with viruses and bacteria as microparasites; characteristically, small organisms that multiply within their vertebrate hosts; and regard helminths as macroparasites that are characteristically large and usually do not do so. Nevertheless, parasitology is a well established and inclusive discipline: parasitology textbooks, courses, societies and international meetings all continue to embrace both protozoa and helminths, a tradition followed in this volume.

Until relatively recently, parasitic infections in humans were regarded mainly as exotic problems of concern only to those living in the tropics and subtropics, but in 1976 WHO drew attention to the magnitude of the problem that these diseases cause and listed five parasitic infections; malaria, trypanosomiasis, leishmaniasis, schistosomiasis and filariasis; among six diseases presenting the greatest challenges in the developing world (the sixth being leprosy).

The impact of parasitic diseases in developing countries is enormous, but they are no longer confined to such areas. The rapid spread of quick and affordable international travel has meant that parasitic infections are frequently imported into more temperate regions by immigrants and returning travelers; a number of parasites are now world wide in their distribution and many infections once thought to be harmless are now known to be life-threatening in immunocompromised individuals.

The chapters that follow summarize what is known about the most important parasitic diseases of humans, some caused by one species and some, like leishmaniasis, by several. Many of these parasites use a vector or one or more intermediate hosts in their complex life cycles and these are also described. Of particular interest is the fact that parasites have developed many ways to survive in hostile environments. This adaptability includes the capacity to evade the immune response: thus, most parasitic infections are of long duration and cause chronic diseases; many helminth infections last for two to ten years; and some protozoan infections persist for the lifetimes of their hosts. Partly because of the ability of parasites to evade the immune response, there is currently no vaccine against any parasitic infection in humans. Parasites also have the capacity to develop resistance to the few effective drugs available, resistance to antimalarials being the most spectacular, presenting problems also common in bacterial infections. Furthermore, attempts to control parasites by attacking their vectors have been hampered by the development of insecticide resistance. Parasitic infections, therefore, present enormous challenges, many of which are outlined in the chapters that follow, but there have been successes: guinea worm has all but been eliminated by the provision of clean water and onchocerciasis in Africa is one of the few infections that have almost totally succumbed to chemotherapeutic intervention. On the other hand, the news is not all good and several parasites are included in the growing list of those causing emerging infections, particularly in people infected with HIV or undergoing immunosuppressive therapy. Prominent among these are the microsporidians, several species of which have been recovered only from immunocompromised patients. Toxoplasmosis and cryptosporidiosis are other examples of parasitic diseases encountered mainly in immunosuppressed individuals as is pneumocystosis, the causative agent of which, Pneumocystis carinii, is included in this volume for completeness because, although now clearly established as a fungus (see Volume 4, Chapter 30) it has in the past been classified with the Protozoa and is often included in the protozoological literature.

The overall objective of this volume is to enhance, update and strengthen the literature familiar to those working with tropical diseases and to introduce parasitic infections to those in all parts of the world who are less well versed in their importance.

F.E.G. Cox
Julius P. Kreir
Derek Wakelin
1998

Abbreviations

A+T	adenine and thymidine
AB	alcian blue
ABR	annual biting rate
ADCC	antibody-dependent cellular cytotoxicity; or antibody dependent cell-mediated cytotoxicity
ADCI	antibody-dependent cellular inhibition
ADCL	anergic diffuse cutaneous leishmaniasis
ADH	alcohol dehydrogenase
AdoMet	S-adenosylmethionine
ADP	adenosine diphosphate
AE	alveolar echinococcosis
AIDS	acquired immune deficiency syndrome
AIM	dichlorodiphenyltrichloroethane
ALDH	aldehyde dehydrogenase
ALS	antilymphocyte serum
AMA	apical membrane antigen
AMP	adenosine monophosphate
APC	antigen-presenting cell
APOC	African Programme for Onchocerciasis Control
ARDS	adult respiratory distress syndrome
AT	adenine, thymine
ATP	adenosine triphosphate; or annual transmission potential
ATPase	adenosine triphosphatase
AV	atrioventricular
AVL	American visceral leishmaniasis
BCG	bacille Calmette-Guérin
BDCL	borderline disseminated cutaneous leishmaniasis
BHC	benzene hexachloride
BMT	bone marrow transplant
BSE	bovine spongiform encephalopathy
Bti	*Bacillus thuringiensis israeliensis* H–14 serotype toxin
CAA	circulating anodic antigen
CAEP	ceramide aminoethyl phosphonate
CAT	computer-assisted tomography or computerized axial tomography
CATT	card agglutination test for trypanosomiasis
CCA	circulating cathodic antigen
CDC	Centers for Disease Control and Prevention
CDD	community directed distributors
CDP	cytidine diphosphate

CDTI	community-directed treatment with ivermectin
CE	cystic echinococcosis
CFT	complement fixation text
CIE	counterimmunoelectrophoresis
CM	cerebral malaria
CMFL	community microfilarial load
CMP	cytidine monophosphate
CNS	central nervous system
COI	cytochrome C oxidase I
COP	circum oval precipitin
COWP	cryptosporidial oocyst wall protein
CQR	Chloroquine-resistant
CQS	chloroquine-sensitive
CRD	Cross-reacting determinant
CSF	cerebrospinal fluid
CSL	circumsporozoite-like antigen
CSO	civil society organizations
CSP	circumsporozoite protein
CT	computed tomography; or covert toxocariasis
CTP	cytidine triphosphate
DALY	disability-adjusted life years
DAT	direct agglutination test
DCL	Diffuse (more correctly 'disseminated') cutaneous leishmaniasis
DDT	dichlorodiphenyltrichloroethane
DEET	diethyltoluamide
DHFR	dihydrofolate reductase
DHPS	dihydropteroate synthase
DS	double sandwich
DT	Dye test
DTH	delayed-type hypersensitivity
EANMAT	East African Network for Monitoring Anti-malarial Treatment
EBA	erythrocyte binding antigen
ECG	eletrocardiograph or eletrocardiogram
ECM	extracellular matrix components
EDTA	ethylenediaminetetraacetic acid
efl-α	elongation factor 1-alpha
EGF	epidermal growth factor
EGTA	ethyleneglycol-bis(b-aminoethylether)-N,N,N′,N′-tetraacetic acid
EIA	enzyme immunoassay

ELISA	enzyme-linked immunosorbent assay		**IHA**	indirect hemagglutination assay
EMP	Embden-Meyerhof-Parnas; or erythrocyte membrane protein		**IHAT**	indirect hemagglutination test
			IL	interleukin
EN	endemic normal		**IMDM**	Iscove's modified Dulbecco's medium
ENT	ear, nose and throat		**iNOS**	inducible nitric oxide synthase
epg	eggs per gram		**IPT**	Intermittent presumptive treatment
EPI	expanded program of immunization		**ISAGA**	immunosorbent agglutination assay
EPM	equine protozoal encephalomyelitis		**ISG**	invariable surface glycoprotein
ER	endoplasmic reticulum		**ITFDE**	International Task Force for Disease Eradication
ERCP	endoscopic retrograde cholangio-pancreatography		**ITN**	insecticide-treated nets
E-S	excretory-secretory		**ITS**	internal transcribed spacer
ESAG	expression site associated genes			
EST	expressed sequence tag		**KAP**	knowledge, attitudes, and practice
ESV	encystation-specific vesicles		**kDNA**	kinetoplast DNA
			KIVI	kit for in vitro isolation
FAD	flavin adenine dinucleotide		**KLH**	keyhole limpet hemocyanin
FILCO	Filariasis Control Movement			
FITC	fluorescein isothiocyanate		**LAH**	left anterior hemiblock
FML	fucose−mannose ligand		**LCL**	localized forms of cutaneous leishmaniasis
			LDH	lactate dehydrogenasc
G6PD	glucose-6-phosphate dehydrogenase		**LPG**	lipophosphoglycan
GABA	γ-aminobutyric acid		**LPR**	lipoprotein receptor-related
GAE	granulomatous amebic encephalitis		**LR**	leishmaniasis recidivans
GAPDH	glyceraldehyde-3-phosphate dehydrogenase		**LS-A**	liver-stage antigens
GBP	glycophorin-binding protein		**LT-α**	lymphotoxin-α
GIS	geographic information system			
GM-CSF	granulocyte−macrophage colony-stimulating factor		**MALT**	mucosa-associated lymphoid tissue
			MAP	multiple antigenic peptides
Gp	glycoprotein		**MAT**	modified agglutination test
GPELF	Global Program to Eliminate Filariasis		**MBR**	monthly biting rate
GPI	glycosylphosphatidylinositol		**MCP-1**	monocyte chemoattractant protein-1
GR	glutathione reductase		**MDCK**	Madin−Darby canine kidney
gRNA	guide RNA		**medRNA**	mini-exon derived RNA
GSH	glutathione		**MHC**	major histocompatibility complex
GSSG	glutathione reductase substrate glutathione disulphide		**MIF**	macrophage inhibition factor; or migration inhibitory factor
GST	glutathione S-transferase		**MIM**	Multilateral Initiative on Malaria
			MIP-1α	macrophage inflammatory protein 1α
H&E	hematoxylin and eosin		**MMV**	Medicines for Malaria Venture
HAART	highly active antiretroviral therapy		**MOH**	Ministry of Health
HIV	human immunodeficieny virus		**MR**	magnetic resonance
HM	hexose monophosphate		**MRI**	magnetic resonance imaging
HSPG	heparan sulfate protein glycoconjugates		**MSF**	Médecins Sans Frontières
HTLV	human T-cell lymphotropic virus		**MSP**	merozoite surface protein
HTLV-1	human T-lymphocyte virus type 1		**MTOC**	microtubule organizing center
			MTP	monthly transmission potential
IBS	irritable bowel syndrome			
ICAM-1	intercellular adhesion molecule-1		**NACM**	*Naegleria* ameba cytopathogenic material
IFA	immunofluorescence assay; or indirect fluor-escent antibody assay		**NADPH**	nicotinamide adenine dinucleotide phosphate (reduced)
IFAT	indirect fluorescent antibody test		**NANP**	asparagine−alanine−asparagine−proline
IFN	interferon		**NCL**	neotropical cutaneous leishmaniasis
IFN-α	interferon-alpha		**NFκB**	nuclear factor kappa B
IFN-β	interferon-beta		**NGDO**	nongovernmental development organization
IFN-γ	interferon gamma		**NGO**	nongovernmental organization
Ig	immunoglobuin		**NK**	natural killer

NMR	nuclear magnetic resonance		**RT-PCR**	reverse transcription polymerase chain reaction
NNN	Novy, MacNeal, Nicolle			
NO	nitric oxide		**SAF**	sodium-acetate acetic-acid formalin
NVDP	asparagine−valine−aspartate−proline		**SAG**	surface antigen
			SAPA	shared acute phase antigen
OCP	Onchocerciasis Control Programme		**SCF**	Save the Children Fund
OD	optical denisty		**SCID**	severe combined immunodeficiency
ODC	ornithine decarboxylase		**SDS PAGE**	sodium dodecylsulfate-polyacrylamide gel electrophoresis
ODRS	oxygen-derived reactive species		**SEA**	soluble egg antigen
OEPA	Onchocerciasis Elimination Program for the Americas		**SEM**	scanning electron microscope
OLM	ocular lava migrans		**sER**	smooth endoplasmic reticulum
ONCHOSIM	onchocerciasis simulation model		**SERA**	serine rich antigen
			SERP	serine-rich protein
*p*ABA	*para*-amino benzoic acid		**SICA**	schizont-infected cell agglutination antigen
PAIR	percutaneous aspiration–injection–re-aspiration		**SIV**	simian immunodeficiency virus
PAM	primary amebic meningoencephalitis		**s.l.**	*sensu lato*
PAS	periodic acid-Schiff		**snRNAs**	small nuclear RNA
PBMC	peripheral blood mononuclear cells		**SOD**	superoxide dismutase
PCR	polymerase chain reaction		**SPOV**	sporophorous vesicle
6-PGD	6-phosphate gluconate dehydrogenase		**SRA**	serum resistance associated
PGM	phosphoglucomutase		**SREHP**	serine-rich *E. histolytica* protein
PID	pelvic inflammatory disease		**SSU**	small subunit
PKDL	post kala-azar dermal leishmaniasis		**SSUrRNA**	small subunit ribosomal RNA
pLDH	parasite lactate dehydrogenase		**STAT**	signal transduction activator of transcription
PMN	polymorphonuclear neutrophil		**SWAP**	sector-wide approaches to financing; or soluble worm antigen preparation
POP	persistent organic pollutants			
PPDK	pyruvate phosphate dikinase			
PPi-PFK	pyrophosphate-dependent phosphofructokinase		**TB**	tuberculosis
PPi	inorganic pyrophosphate		**TBA**	traditional birth attendants
PPP	public−private partnerships		**TBF**	thick blood film
PSG	promastigote secretory gel		**Tc**	cytotoxic T cell
PSP	promastigote surface protease		**TCA**	tricarboxylic acid
PTM	peritrophic matrix		**TCNA**	*T. cruzi* neuraminidase
PV	parasitophorous vacuole		**TCR**	T cell receptor
PVA	polyvinyl alcohol		**TEM**	transmission electron microscopy
PVM	parasitophorous vacuole membrane		**TGF-β**	transforming growth factor beta
			Th	T helper [cell]
QBC	quantitative buffy coat		**Th1**	T helper 1
			Th2	T helper 2
RAPD	random amplification of polymorphic DNA		**TLTF**	trypanosome-derived lymphocyte triggering factor
RAPLOA	rapid assessment of prevalence of eye worm		**TMAF**	trypanosomal macrophage activating factor
RBBB	right bundle branch block		**TMN**	tubular membranous network
RBM	Roll Back Malaria		**TNF**	tumor necrosis factor
rDNA	ribosomal DNA		**TNF-α**	tumor necrosis factor-alpha
REMO	rapid epidemiological mapping of onchocerciasis		**TNF-β**	tumor necrosis factor-beta
RFMS	rapid field assessment process		**TNFR**	TNF receptor
rER	rough endoplasmic reticulum		**TPE**	Tropical pulmonary eosinophilia
RESA	ring-infected erythrocyte surface antigen		**TPI**	Triose-phosphate isomerase
RFLP	restriction fragment length polymorphism		**TPx-1**	thioredoxin peroxidase-1
RIA	radioimmunoassay		**Tr**	T-regulatory
RK	rabbit kidney; or receptor kinase		**TR**	trypanothione reductase
RNAi	RNA inhibition		**TRAP**	thrombospondin related adhesion protein
ROI	reactive oxygen intermediate		**TS**	thymidylate synthase
rSJPM	recombinant *S. japonicum* paramyosin			

T[SH]₂	dihydrotrypanothione		**VCAM-1**	vascular cell adhesion molecule-1
T[S]₂	trypanothione disulfide		**VCRC**	Vector Control Research Centre
			VHDL	very high density lipoprotein
UMP	uridine monophosphate		**VHW**	village health worker
UNICEF	United Nations'(International) Children's		**VLM**	visceral larva migrans
	(Emergency) Fund		**VSG**	variable surface glycoprotein
US	ultrasonography		**VSP**	variant-specific surface protein
UV	ultra-violet			
			WB	Western blot
VAT	variable antigen type		**WHO**	World Health Organization

PART I

GENERAL PARASITOLOGY

History of human parasitology

F.E.G. COX

The history of parasitology embraces three interconnected fields: zoology, with its interests in parasites for their own sakes, veterinary medicine, and human medicine, particularly tropical medicine. During our relatively short history on Earth we humans have acquired about 300 species of helminth worms and over 70 species of protozoa (Ashford and Crewe 2003) of which a small proportion cause some of the most important diseases in the world, particularly in the tropics. In this chapter it is only possible to deal with those parasites that are the most common and which cause the most important diseases.

HUMAN EVOLUTION AND PARASITIC INFECTIONS

Human evolution and the acquisition of parasitic infections cannot be separated and, in order to understand the history of parasitology, we have to delve deeply into the past. It is now widely accepted that our species, *Homo sapiens*, emerged somewhere in Africa, but we do not know exactly where or when; opinions on this subject are extreme. Molecular evidence dates our origins to somewhere between 150 000 years ago in eastern Africa (Tishkoff et al. 2001) and 60 000 years ago in the northern part of southern Africa (Wells 2002). The details are not really relevant to us here, but what is important is that most authorities agree that our early ancestors spread from Africa throughout the world in several waves (Templeton 2002) until, at the end of the last Ice Age, about 11 000 years ago, humans had migrated to, and inhabited, virtually the whole of the Earth. During these migrations humans were not alone

and brought with them various parasites as 'heirlooms,' inherited from primate ancestors in Africa, and 'souvenirs,' acquired from animals encountered during these migrations. Agriculture, the practice of domesticating animals, and the development of settlements and cities increased the opportunities for the transmission of infections between animals and humans and between humans. Later, the opening up of trade routes brought parasitic infections to new geographical regions, for example, colonization and the slave trade brought new parasites to the New World from the Old World (Desowitz 1997).

For many years the distribution of parasites in the past has been largely a matter of speculation, but we have now begun to learn a lot about the past history of parasitic infections from studies of archeological artifacts such as the presence of helminth eggs or protozoan cysts in coprolites (fossilized or desiccated feces) and naturally or artificially preserved bodies. From such studies a new science, palaeoparasitology, has emerged and some examples of these discoveries will be discussed later.

Written records

Direct evidence concerning our knowledge of parasites in the past is fragmentary, so we have had to rely heavily on the written record. The first written records date from a period of Egyptian medicine between 3000 and 400 BC, particularly the papyri of 1500 BC discovered by Georg Ebers at Thebes (Bryan 1930). Many diseases, specifically fevers, are described in detail in the writings of Greek physicians between 800 and 300 BC, the best known being the collected works known as the *Corpus*

Hippocratorum (see Jones and Whithington 1948–1953). Physicians from other civilizations also made significant contributions to our understanding of parasitic diseases and there are extensive writings from China (3000–300 BC), India (2500 – 200 BC), Rome 700 BC – 400 AD, and the Arab Empire (latter part of the first millennium). Two important Arabic medical works that include a great deal of information about diseases caused by parasites are *Al-hawi* by Rhazes (850–923) and *Al canon fi al tib* by Avicenna (980–1037).

Religious and superstitious beliefs in the Dark and Middle Ages held back medical progress in Europe, and it was not until the Renaissance that scientific and medical revolution freed scientists from the chains of superstition and religion and unleashed a period of spectacular investigation that gradually led to the great discoveries of the 19th century. These include the renunciation of the theory of spontaneous generation and the development of the 'germ theory', by Louis Pasteur; the demonstration that diseases are caused by bacteria, also by Pasteur; the discovery of viruses, by Pierre Emile Roux; the introduction of methods of preventing diseases caused by microorganisms, by Robert Koch; and the incrimination of vectors in the transmission of parasites, by Patrick Manson.

Tropical diseases

Most commentators date the origin of tropical diseases as a distinct branch of medicine to 1898 when Sir Patrick Manson, now universally regarded as the 'father of tropical medicine,' published the first edition of his *Tropical Diseases* (Manson 1898). The important tropical diseases described by Manson included a number caused by parasites – malaria, sleeping sickness, amoebic dysentery, kala azar, and bilharzia – but Manson did not consider parasitology to be a distinct branch of tropical medicine.

Michael Warboys has suggested that 'the dominant tradition of the history of tropical diseases has been to celebrate the discoveries of the etiologies of the classic group of vector-borne parasitic diseases in the period 1870–1920' (Warboys 1993). There is, however, much more to the history of tropical diseases than this and it is now clear that the work of those who investigated parasitic diseases in the centuries before Manson should not be dismissed as irrelevant to our understanding of these diseases. Nor should the discoveries of the 20th-century workers be regarded as mere finishing touches to previous knowledge. Scientific progress during any period of history depends on the availability of techniques and of individuals capable of using and interpreting the information that becomes available. The remarkable thing about the last decades of the 19th century was the coming together of so many new ideas and personalities.

Parasitic diseases have always held a fascination for travelers, clinicians, and scientists. Some, such as sleeping sickness, are unique to the tropics; others, such as malaria, are of particular importance in the tropics but also extend into more temperate regions; still others, such as hookworm, have worldwide distribution (either now or in the past). The present distribution of parasitic diseases tends to reflect the success of hygiene and control measures in the more developed parts of the world rather than any clear geographical or climatic restriction. Thus, many parasites that were once widespread are now found in only the warmer and poorer parts of the world and are considered as 'exotic' diseases when imported into the cooler northern and southern latitudes where they were once prevalent. The term 'tropical diseases' has now become one of convenience rather than a definitive description

The history of parasitology has been well served in the scientific literature, particularly in works devoted to tropical medicine. As well as the publications already mentioned there is a considerable amount of information about the history of parasitology in the more general works by Scott (1939), Ackernecht (1965), Brothwell and Sandison (1967), Campbell et al. (1992), Chernin (1977), Cox (1996), Mack (1991), Norman (1991), Ranger and Slack (1992), and Ransford (1983). More specifically, there are a number of publications dedicated to the history of parasitology, including those of Cook (2001), Cox (2002, 2004a), Foster (1965), Garnham (1970), Hoeppli (1956, 1959), Kean et al. (1978), Moore (1976), and Warboys (1983). The most comprehensive work on the history of any aspect of parasitology, one that will be impossible to emulate, is David Grove's *A history of human helminthology* (Grove 1990), which contains over 800 pages of detailed accounts of all the discoveries in human helminthology and allied fields. *The Wellcome Trust illustrated history of tropical diseases* (Cox 1996) contains chapters on most of the important human parasitic infections. Foster's *A history of parasitology* (Foster 1965) covers some, but not all, important parasitic infections. In addition, there are a number of important monographs mainly relating to the associations between parasitic infections and human endeavor and welfare, and these will be mentioned in the appropriate places.

Helminth worms, being large and easily recognizable, and common parasites of domesticated and wild animals, very soon attracted the attention of zoologists who, schooled in the concepts of comparative zoology, treated parasites in the same way as any other group of animals and tended to ignore the fact that they also caused disease. Veterinarians and clinicians, on the other hand, were quick to note the implications of these infections and to draw parallels between parasitic infections in domesticated animals and humans and discoveries in the field of zoology. To this day, helminthology is usually regarded as a branch of zoology. Protozoa, being microscopic, had to await the development of the microscope before they were first recognized; the history of the protozoa, therefore, parallels that of other microorganisms and has largely been the province of medicine rather than zoology. There is thus a dichotomy that is not always apparent in that the interest of most helminthologists has centered on the worms themselves and, to a lesser extent, the diseases they cause whereas, in contrast, scientists interested in parasitic protozoa have tended to focus predominantly on diseases. There are other obvious reasons for this dichotomy; for example, the fact that helminths are multicellular organisms and can be examined outside their hosts has facilitated physiological and biochemical studies, whereas isolating protozoa from their host cells has proved to be very difficult. These differences in approach have had a marked effect on our understanding and interpretation of the history of parasitology. Although discoveries about the helminths and protozoa originally tended to run on parallel paths, these paths converged at the end of the 19th century when Patrick Manson introduced the medical and veterinary world to the concept of vector-borne parasitic diseases and brought entomologists into the field of parasitology. Thereafter, research and teaching institutions have traditionally considered protozoology, helminthology, and entomology as three interrelated fields.

The history of parasitology embraces some of the most exciting discoveries ever made in medicine, and these findings have had an impact well outside their immediate field. For example, the discovery of the vector-borne nature of the filarial worms influenced ideas about the transmission of malaria which, in turn, led to the discovery of the mosquito-borne transmission of yellow fever, the whole field of arboviruses, and the implication of other vectors in trypanosomiasis, relapsing fever, and plague. Similarly, the discovery that trypanosomes caused diseases in both animals and humans opened up a new field of zoonoses (infections common to animals and humans) and the study of reservoir hosts for a number of infections.

Many of the great discoveries in parasitology were made over a relatively short period, in the 30 years between 1885 and 1915, when major advances were also being made in our understanding of other infectious diseases. The great personalities of this period made discoveries in a number of fields, and their findings and ideas fed off one another. The names of Pasteur, Koch, Bruce, and Manson occur time and time again in the history of parasitology and microbiology.

The history of parasitic diseases can be considered as a number of phases: the recognition of the disease; the discovery of the organism involved; the connection between the organism and the disease; the consolidation of all these strands; and the application of this knowledge to control schemes.

So vast is the field of human parasitology, with its many and far-reaching discoveries, that it is not possible to do justice to the whole subject and in the rest of this chapter only the most significant aspects will be discussed. No attempt will be made to discuss the parasites or the diseases they cause other than briefly, and the reader is referred to other chapters in this volume to provide the necessary background information and details.

HELMINTHS AND HELMINTH INFECTIONS

Discovery of the organisms

Given the ubiquitous nature and large size of the helminth worms, it would be surprising if our earliest ancestors had not been aware of the more common types; evidence for this assumption comes from the fact that in primitive tribes in Sarawak and North Borneo most people are aware of their intestinal roundworms and tapeworms (Hoeppli 1959). Much has been made of quotations from the Bible on the subject of helminth worms, but these are often open to several interpretations; the only conclusion that can be safely drawn is that those who contributed to the Bible were aware of the existence of parasitic worms (i.e. we cannot conclude that they possessed any detailed knowledge of the subject). With the first written records by physicians we are on firmer ground. The Egyptian medical papyrus usually referred to as the *Papyrus Ebers*, dating from around 1500 BC, refers to intestinal worms and these records are backed by the discovery of calcified helminth eggs in mummies dating from 1200 BC. The Greeks, Aristotle (384 – 322 BC) in particular, were familiar with a range of parasites from humans and also domesticated and wild animals. Greek and Roman physicians, including Hippocrates (460 – 375 BC), Celsus (25 BC – AD 50), and Galen (129 – 200 AD), were familiar with the human roundworms, *Ascaris lumbricoides* and *Enterobius vermicularis*, and tapeworms. In fact, the terms 'roundworm' (helmins strongyle) and 'ribbon worm' (helmins taenia) can be traced back to the earliest Greek records (Grove 1990). Paulus Aegineta (625–690) clearly described *Ascaris*, *Enterobius*, and tapeworms, and gave good clinical descriptions of the infections they caused. With the decline and fall of the Roman Empire, the study of medicine switched to Arabic physicians. Avicenna recognized not only *Ascaris*, *Enterobius*, and tapeworms but also the Guinea worm, *Dracunculus medinensis*, a worm that had been recorded in parts of the Arab world, particularly around the Red Sea, for over 1 000 years.

The medical literature of the Middle Ages is very limited, but the existence of parasitic worms was widely recorded and, in many cases, these were regarded as the possible causes of a number of diseases. Unfortunately, many of these 'worms' were fictitious, as were the symptoms ascribed to their presence, for example toothache and heart attacks (Hoeppli 1959). On the other hand, Chinese physicians believed that humans should harbor at least three worms to be in good health, and this belief that worms were beneficial also extended into Europe (Foster 1965).

The history of helminthology really took off in the 17th and 18th centuries following the re-emergence of science and scholarship during the Renaissance. In his *Systema Naturae*, Linnaeus described six helminth worms: *Ascaris lumbricoides*, *Ascaris vermicularis* (= *Enterobius vermicularis*), *Gordius medinensis* (= *Dracunculus medinensis*), *Fasciola hepatica*, *Taenia solium*, and *Taenia lata* (= *Diphyllobothrium latum*) (Linnaeus 1758). Thereafter there was a gradual accretion of new species; nine species were identified by 1782, with the addition of *Trichuris trichiura*, *Taenia saginata*, and *Echinococcus granulosus*. This number remained virtually unchanged in various texts published between 1800 and 1819 but was subsequently extended by the inclusion of *Trichinella spiralis* and *Loa loa* by 1845; *Ancylostoma duodenale*, *Schistosoma mansoni*, and *Hymenolepis nana* by 1855; and *Strongyloides stercoralis*, *Wuchereria bancrofti*, *Clonorchis sinensis*, *Fasciolopsis buski*, *Paragonimus westermani*, *Heterophyes heterophyes*, *Schistosoma haematobium*, and *Dipylidium caninum* between 1885 and 1890 (see Grove 1990).

By the beginning of the 20th century, 28 species had been recorded in humans, a number that has grown to >300 species including accidental and very rare records (Coombs and Crompton 1991; Ashford and Crewe 2003). Of these, about 80 can be regarded as fairly common and about 20 very common. The description of a new species of helminth worm in humans is now very unusual, but advances in biochemistry and molecular biology permit the identification of minute but significant differences between parasites and this is likely to lead to the creation of additional species and subspecies from within the current list.

Our understanding of parasitology could not really progress until the theory of spontaneous generation had been disproved. Doubts had been raised about the validity of this concept in the 17th century, but it still persisted until the definitive experiments of Pasteur at the end of the 19th century. A number of helminthologists were among those who had begun to doubt the concept, including Francesco Redi and Edward Tyson who, in the 17th century, went so far as to describe sexual reproduction in worms. Studies on the biology and life cycles of helminth worms effectively destroyed the theory of spontaneous generation (see Farley 1977) well before Pasteur delivered its deathblow and promulgated the 'germ theory' of infection.

Table 1.1 lists some of the most important events in our knowledge and understanding of parasitic worms. Note the massive activity that occurred at the end of the 19th and beginning of the 20th centuries.

ASCARIS AND ASCARIASIS

Ascaris lumbricoides, the large roundworm, is one of the six worms listed by Linnaeus and its name has remained unchanged ever since. Human infections are worldwide, with >250 million people infected. The presence of this worm, which is often voided in the feces, is very obvious and its existence has been recorded from Egyptian medical papyri of 1500 BC, the works of Hippocrates in the 5th century BC (see Jones and Whithington 1948–1953), Chinese writings of the 2nd and 3rd centuries BC, and the texts of Roman and Arabic physicians from the 1st to 10th centuries AD (see Grove 1990; Goodwin 1996b). Direct evidence of the antiquity of human infections comes from the discovery of eggs of *Ascaris* in human coprolites dated 227 BC from Peru (Horne 1985), in Egyptian mummies from 1938–1600 BC (Cockburn et al. 1998) and in Chinese mummies from 1368–1644 BC (Dexian et al. 1981). This worm is so common that it is impossible to say who discovered it, but the first scientific accounts of its anatomy were by Edward Tyson (1683) and Francesco Redi (1684). The mode of infection by the ingestion of eggs was established by Casimir Davaine (1862) and Giovanni Grassi (1881) who infected himself and later found eggs in his feces. The whole life cycle, including the migration of the larval stages around the body, was worked out in 1922 by Shimesu Koino who also infected himself and deduced the life cycle when he found eggs in his sputum (Koino 1922).

TRICHINELLA AND TRICHINOSIS

Trichinosis, also known as trichinellosis and trichina infection, is caused by the nematode worm *Trichinella spiralis* and is usually acquired by eating infected pork. Foster (1965) considers this to be one of the most interesting parasitic infections and more likely to be responsible for some religious traditions of avoiding pork than the possibility of tapeworm infection (see below). Although the association with pigs had been recognized since the earliest times, the encysted larvae in the muscle were not seen until 1821 and even then their presence was not associated with disease. The story of the association between the worm and human infection is described in full by Foster (1965) and Grove (1990) and can only be summarized here. It revolves around the discovery of the encysted larvae of the worm in humans in 1835 by James Paget, a medical student at St Bartholomew's Hospital in London; this discovery led to intensive investigations by some of the world's most important parasitologists including Rudolf Leuckart, Rudolph Virchow, Thomas

Table 1.1 *Important events in the history of our knowledge and understanding of parasitic worms.*

Date	Event
c. 5000 BC	Hookworm ova in a human (?) coprolite from Brazil
c. 2500 BC	*Paragonimus* ova in human coprolites from Chile
c. 2330 BC	*Ascaris* ova coprolites from Peru
c. 2000 BC	*Taenia* ova in an Egyptian mummy
c. 1500 BC	*Papyrus Ebers* (Egypt): reference to roundworms (*Ascaris lumbricoides*), threadworms (*Enterobius vermicularis*), and tapeworms (*Taenia* spp.)
1250–1000 BC	*Schistosoma haematobium* eggs in Egyptian mummies
1300–1234 BC	Biblical references to *Dracunculus medinensis* in Red Sea region
c. 800 BC	*Ancylostoma* worms in a Peruvian mummy
c. 600–700 BC	Records of *Dracunculus medinensis* worms from Mesopotamia
c. 400 BC	*Corpus Hippocraticorum* (Aphorisms and Epidemics): references to roundworms (*Ascaris lumbricoides*), threadworms (*Enterobius vermicularis*, *Echinococcus granulosus*, and hydatid disease
500–400 BC	Hookworm eggs in coprolites from Brazil
c. 384–322 BC	Aristotle refers to cysticerci in tongues of pigs and distinguishes between large and flat worms (*Taenia*), cylindrical worms (*Ascaris*), and thin worms (*Enterobius*)
c. 300 BC	Chinese descriptions of threadworms, tapeworms, hookworms, and hookworm disease
200–300 BC	*Chinese Book of Plain Questions* mentions *Ascaris*
167 BC	*Clonorchis* ova from China
AD 20	Celsus recognizes tapeworms taenia, tinea, and taeniola (*Taenia* spp.), vermes cucurbitini (tapeworm proglottids), 'hailstones' (cysticerci) and roundworms, lumbrici teretes (*Ascaris lumbricoides*)
AD 27–39	Pliny describes tapeworms in Egyptians (presumably *T. saginata* because pork was not eaten) and *Tinea rotunda* (Ascaris)
130–200	Galen recognizes three types of worms: roundworms (*Ascaris lumbricoides*), threadworms (*Enterobius vermicularis*), and tapeworms (*Taenia* spp.) and also cysticerci in livers of slaughtered animals
c. 900	Hookworms in a pre-Columbian Andean mummy
c. 1000	Avicenna (Ibn Sina) refer to 'four kinds of worms'
c. 1340	Johannes Actuarius (a Byzantine physician) describes *Trichocephalus trichiuris* (=*T. trichiura*).
1379	Jehan de Brie refers to *Fasciola hepatica* from sheep
1592	Dunus discovers *Diphyllobothrium latum*
1602	Plater distinguishes betweem *Diphyllobothrium* and *Taenia*
1634	Herbert describes *Dracunculus medinensis* worms in the leg
1674	Welsch makes a detailed study of *Dracunculus medinensis* and dracunculiasis
1683	Tyson describes the anatomy of *Ascaris lumbricoides*
1683	Tyson discovers the scolex of *Taenia solium*
1684	Redi describes helminths from domestic and wild animals and arranges them according to their hosts
1684	Redi recognizes parasitic nature of *Echinococcus* cysts
1700	Andry publishes *De la génération des vers dans le corps de l'homme*, the first textbook on medical parasitology, in which he suggests that worms are derived from 'seeds'. He also draws *Taenia saginata*, but calls it *T. solium*, and describes *Diphyllobothrium latum*
1766	Palas realizes that *Echinococcus* cysts are tapeworm larvae
1770	Mongin describes *Loa loa* and the clinical features of loiasis
1793	Chisholm observes transmission of *Dracunculus medinensis*
1808–10	Rudolphi publishes his text *Entozoorum, sive verminum intestinalium, historia naturalis*, regarded as the first textbook on helminthology.
1819	Bremser describes *Enterobius vermicularius*
1819	Rudolphi describes the larvae of *Dracunculus medinensis*
1833	Hilton finds but does not describe *Trichinella spiralis* in human muscle
1835	Paget discovers and Owen reports on the larvae of *Trichinella spiralis* in human muscle
1843	Dubini recognizes the hookworm *Ancylostoma duodenale*
1843	Gruby and Delafond find microfilariae in blood of dogs
1843	Busk discovers *Fasciolopsis buski*
1846	Leidy describes *Trichinella spiralis* in pig muscle
1847	Fujinami describes Katayama disease caused by *Schistosoma japonicum*
1850	Bilharz discovers *Schistosoma haematobium*
1853	von Siebold describes *Echinococcus granulosus* tapeworms in dogs fed with hydatid cysts from sheep

(Continued over)

Table 1.1 *Important events in the history of our knowledge and understanding of parasitic worms. (Continued)*

Date	Event
1855	Küchenmeister demonstrates that *Taenia solium* infections are acquired by eating pork
1859	Virchow finds adult worms of *Trichinella spiralis* in a dog fed on infected muscle
1863	Bastian gives a detailed description of *Dracunculus medinensis*
1863	Demarquay finds microfilariae of *Wuchereria bancrofti* in fluid from a scrotal tumor
1863	Naunyn infects dogs with *Echinococcus* hydatid cysts from a human
1863	Cobbold suggests that snails might be the intermediate hosts of schistosomes
1864–9	Cobbold's book *Entozoa* is published
1864	Harley suggests that snails might be the intermediate hosts of schistosomes
1867	Leuckart describes life cycle of *Echinococcus granulosus*
1868	Wucherer finds microfilariae of *Wuchereria bancrofti* in the urine of a patient with hematuria
1869	Fedchenko finds *Dracunculus medinensis* in *Cyclops*
1872	Lewis sees microfilariae of *Wuchereria bancrofti* in human blood
1874	McConnell describes *Clonorchis sinensis*
1875	O'Neill discovers microfilariae of *Onchocerca volvulus* in skin nodules
1876	Normand discovers larvae and adults of *Strongyloides stercoralis*
1876	Joseph Bancroft finds and recognizes adult *Wuchereria bancrofti* worms
1877	Manson demonstrates that *Wuchereria bancrofti* is transmitted by mosquitoes
1877	Bavay describes *Strongyloides stercoralis* larvae and adult worms
1877	Perroncito establishes the bovine source of *Taenia saginata* infections
1879	Perroncito realizes that conditions in mines favor the development of hookworm larvae
1879	Ringer discovers *Paragonimus westermani*
1881	Manson suggests that *Paragonimus westermani* develops in a snail
1881	Leuckart and Thomas independently describe the life cycle of *Fasciola hepatica*
1881	Braun finds larvae of *Diphyllobothrium latum* in fish and induces infections in dogs
1893	Manson reports on the finding of adult *Onchocerca volvulus* worms
1890	Mackenzie finds larvae of *Loa loa*
1895	Argyll-Robertson suggests that *Loa loa* is transmitted by blood-sucking insects
1900	Low, acting on a suggestion by Bancroft, demonstrates microfilariae in mosquito mouthparts and confirms transmission through bite of mosquito
1904	Katsurada describes *Schistosoma japonicum* from a cat
1904	Fujinami finds *Schistosoma japonicum* in a human
1905	Looss elucidates the life cycle of *Ancylostoma duodenale*
1912	Leiper demonstrates transmission of *Loa loa* by flies of the genus *Chrysops*
1913	Turkhud completes the life cycle of *Dracunculus medinensis*
1914	Miyairi and Suzuki describe the life cycle of *Schistosoma japonicum* in snails
1914	Leiper distinguishes between *Schistosoma mansoni* and *S. haematobium* and identifies their snail intermediate hosts
1915	Kobayashi realizes that *Clonorchis sinensis* uses a fish host
1917	Janicki and Rosen independently implicate copepods and fish in the life cycle of *Diphyllobothrium latum*
1922	Koino describes the life cycle of *Ascaris lumbricoides*
1926	Blacklock elucidates the life cycle of *Onchocerca volvulus* in blackflies
1927	Lichtenstein and Brug recognize *Brugia malayi*

References are given in the text and in Grove (1990). Note that the dates of particular discoveries do not always coincide with the dates of publications. This is because in the 19th century and earlier, scientists often made their findings known at meetings or in letters before the publication of the definitive paper. In some cases the reference refers to a later and fuller report. Some of the earlier dates are very approximate.

Cobbold, Richard Owen, and Gottlob Küchenmeister. Although the discovery was made by Paget, much of the credit has been attributed to Owen who not only wrote the definitive report (Owen 1835) but also played down Paget's role in the discovery. The adult worm itself was discovered by Rudolph Virchow and Friedrich Zenker and it was the latter who finally recognized the clinical significance of the infection and realized that humans became infected by eating uncooked pork (see Grove 1990; Bundy and Michael 1996).

HOOKWORMS AND HOOKWORM INFECTION

Human hookworm infections caused by *Ancylostoma duodenale* and *Necator americanus*, the former originating in Asia and the latter in Africa, have been associated with humans for over 5 000 years (Hoeppli 1959). Hookworms reached America before the 5th century BC and larvae have been found in coprolites dating from this period (Ferreira et al. 1980). Hookworm disease has been recognized since the earliest records began and the Egyptian papyrus of 1500 BC, the works of Hippocrates in the 5th century BC, and Avicenna in the 10th century all give accurate descriptions of the disease; its history has been reviewed by Foster (1965), Grove (1990), and Ball (1996). Adult *A. duodenale* worms were first recognized by Angelo Dubini in 1843 (Dubini 1843) and credit for the discovery that humans become infected through the skin goes to Arthur Looss who accidentally infected himself and realized that the infection begins when larvae in the soil penetrate the skin, usually of the foot (Looss 1898).

The connection between the presence of worms and hookworm disease was established in 1854 by Wilhelm Griesinger (Griesinger 1854). The clinical signs of hookworm infection – anemia, greenish-yellow pallor, and lassitude – had been observed by Hippocrates and his followers and many early physicians (see Grove 1990) and associated with miners, but it was not until 1879 that Edoardo Perroncito, working with miners in the St Gothard tunnel, realized that the conditions in mines favor the development of the larval stages (Perroncito 1880). The importance of hookworm infections in the USA in the early years of the 20th century led to the establishment of a number of Schools of Public Health and eventually to the creation of the World Health Organization (Ettling 1990).

STRONGYLOIDES AND STRONGYLOIDIASIS

Strongyloidiasis is caused by two species of intestinal nematodes, of which *Strongyloides stercoralis* is the more common and important species. The life cycle involves both parasitic and free-living generations. Parthenogenic female worms lay eggs that hatch within the host and the larvae that are passed out in the feces adopt a free-living existence in the soil from which they penetrate the skin of a new host. Alternatively, the larvae can mature to the infective stage in feces on the skin and reinfect the host through the skin, or mature to the infective stage without leaving the gut and penetrate the gut wall. It is not surprising, given the absence of eggs and confusion with other free-living species of nematodes, that *S. stercoralis* was not recognized until 1876 when both the larvae and the disease, strongyloidiasis, were discovered by Louis Normand who later also found the adult worms and sent them for identification to Arthur Bavay who realized what they were and described them (Bavay 1877). In 1883 Rudolf Leuckart described the alternation of generations involving both parasitic and free-living phases (Leuckart 1883). The discovery that infection with *Strongyloides* occurred through the skin was made by Paul Van Durme based on the work of Looss

mentioned above, who had shown that *A. duodenale* infects its host in this way (Van Durme 1901–1902). However, Van Durme was probably working with the other species, *S. fuelleborni* (see Grove 1990), and the mode of infection by *S. stercoralis* was established by Looss who infected himself by putting larvae on his skin and later detecting larvae in his feces (Looss 1905). The phenomenon of autoinfection was described by Friedrich Fülleborn, working with dogs in Hamburg in 1914 (Fülleborn 1914). For nearly a century *S. stercoralis* received little attention, until it was realized that it could cause disseminated infections in immunosuppressed patients, including those infected with the human T-lymphocyte virus type 1 (HTLV-1); although at one time thought to be implicated, *S. stercoralis* is no longer regarded as a major concomitant of acquired immunodeficiency syndrome (AIDS) (see Gutierrez 2000).

LYMPHATIC FILARIASIS

Lymphatic filariasis is caused by infection with the nematode worms *Wuchereria bancrofti*, *Brugia malayi*, and *B. timori*, which are transmitted by mosquitoes. The discovery of the life cycle by Patrick Manson in 1877 is widely regarded as the most significant discovery in tropical medicine, with implications that reached far beyond helminthology into such diverse areas as malaria and the arboviruses. The massive swellings of the limbs associated with lymphatic filariasis cannot have gone unnoticed by our early ancestors, and descriptions of what was almost certainly this disease date back to 2000 BC in Egypt and to Nok sculptures of the Sudan and West Africa from 500 AD (see Grove 1990; Nelson 1996). The larval stages of the filarial worms, microfilariae, were first seen in the blood of dogs in 1843 (Gruby and Delafond 1843) and in humans by Demarquay (1863). This latter discovery attracted the attention of a number of eminent parasitologists, including Wilhelm Greisinger, Theodor Bilharz, and Otto Wucherer but it was a relatively unknown Scottish clinician, Timothy Lewis, who realized that the worms were associated with filariasis (Lewis 1872). The mode of transmission remained a mystery until Patrick Manson discovered that the mosquito was involved (Manson 1878).

The story of Manson's discoveries has been told many times (see Manson-Bahr 1962; Foster 1965; Service and Wilmott 1978; Chernin 1983; Grove 1990; Eldridge 1992; Nelson 1996). Manson, then working in Amoy in China, found microfilariae in the blood of dogs and humans; he hypothesized that these parasites in the blood might be transmitted by bloodsucking insects. Accordingly, he fed mosquitoes on the blood of his gardener, who was harboring the parasites, and subsequently was able to detect larval stages in the mosquitoes (Manson 1878). Manson, probably prompted by Fedchenko's discovery of the life cycle of *Dracunculus medinensis* (see below), mistakenly believed that the parasites escaped from the mosquito into water and that human infections were acquired through contact with contaminated water. The actual mode of transmission, through the bite of a mosquito, was established only when suggestions by the Australian parasitologist, Thomas Bancroft, were followed up by Manson's assistant George Carmichael Low (Low 1900). Manson was instrumental in the foundation of the London School of Hygiene and Tropical Medicine in 1899 and the Royal Society of Tropical Medicine and Hygiene in London in 1907.

ONCHOCERCA AND ONCHOCERCIASIS

The history of onchocerciasis, or river blindness, caused by the nematode worm *Onchocerca volvulus*, is a relatively short one and there are few reliable early records. This is partly because there are so many causes of blindness in the tropics with which this condition can be confused. Accounts of the history of onchocerciasis are given by Grove (1990) and Muller (1996b). The skin lesions associated with this infection were well known in West Africa and called 'craw craw' by the time John O'Neill first identified nematode larvae in the lesions of those suffering from this condition in 1875 (O'Neill 1875). Adult worms were not seen until some years later when Leuckart received specimens from a German doctor in Ghana and sent them to Manson who, with due acknowledgements, described them in *Davidson's hygiene and diseases of warm climates* (Manson 1893).

For a number of years, this parasite was considered to be a rare curiosity until the connection between the worm and the skin lesions was made by Rodolfo Robles in Central America (Robles 1917). Robles was also the first person to suggest that blackflies might be involved in the transmission of the infection but it was not until 1923–1926 that Donald Breadalbane Blacklock demonstrated conclusively that blackflies were indeed the vectors (Blacklock 1926). In 1928, there came the first report of microfilariae in histological sections of the eyes of blind individuals (Ochoterena 1928) and between 1928 and 1935 a flurry of activity clearly established the role of *Onchocerca* infection as a cause of blindness (see Grove 1990). The 20th century has been one of triumph over this disease, first with the use of insecticides to control the blackfly larvae and later with the development and free distribution of a very effective drug, ivermectin, making the eradication of onchocerciasis a real possibility.

LOIASIS

Loa loa, the nematode that causes loiasis, is a large worm that occasionally passes through the eye and must have been recognized from the earliest times. There are few early records but there is no doubt that Huighen van Linschoten became aware of this worm during the course of his researches into Guinea worm at the end of the 16th century (van Linschoten 1610). The adult worm was first seen by a French surgeon, Mongin, in 1770 (Mongin 1770) and the microfilariae were recognized in 1890 by Stephen Mackenzie who referred his material to Patrick Manson (see Manson 1891). The mode of transmission remained a mystery until 1895 when, basing his ideas on Manson's earlier work on lymphatic filariasis, Douglas Argyll-Robertson predicted that the vector would be a bloodsucking insect but failed to find any parasites in his suspected insect hosts (Argyll-Robertson 1895). In 1912 Robert Leiper threw the list of possible vectors wide open and conclusively demonstrated that the vectors were actually tabanid flies belonging to the genus *Chrysops* (Leiper 1913). One of the signs of loiasis is the presence of swellings, known as Calabar swellings, particularly on the arms and legs. Argyll-Robertson suspected that they were caused by the worm and this was eventually accepted by Patrick Manson who was unsure as to whether the swellings were caused by the blockage of lymphatics or as a result of a pathological reaction to some product of the worm (Manson 1910).

DRACUNCULUS AND DRACUNCULIASIS (GUINEA WORM DISEASE)

Dracunculiasis, caused by the nematode *Dracunculus medinensis*, is the only parasitic infection of which a good description is given in the Bible; most observers agree that the 'fiery serpents' that afflicted the children of Israel in the region of the Red Sea between 1300 and 1234 BC were dracunculus worms (The Bible, Numbers 21, Verse 6; Foster 1965). There are also numerous records of this worm in second millennium papyri from Egypt; writings from Mesopotamia in the 7th century BC; and later records by Greek, Roman, and Arabic physicians (see Grove 1990; Tayeh 1996). One of the best descriptions is that by Avicenna in his book *Al canon fe al tib*. Avicenna recognized that this was a worm and not a rotten vein, as had previously been maintained (hence the old name 'Medina vein'). The presence of the worm in the leg was formally recognized by Thomas Herbert in 1634 (Herbert 1634). Throughout the 18th and early 19th centuries the conviction strengthened that dracunculiasis was caused by the worm *D. medinensis* (listed as *Gordius medinensis* by Linnaeus in 1758) and that infection was acquired through drinking water as suggested by the Dutch navigator Huighen van Linschoten in 1610 (van Linschoten 1610). On the other hand, George Busk (Busk 1846) and H. J. Carter (Carter 1855) argued that the infection could be acquired through the skin. The actual life cycle was unravelled by a Russian scientist, Aleksej Fedchenko, who suspected that water fleas, *Cyclops*, were involved and that infection was acquired by accidentally consuming these invertebrates, but was unable to induce infections in experimental animals by feeding them infected water fleas (Fedchenko 1870). It was not until 1905 that Robert Leiper demonstrated that *Dracunculus* larvae could survive in digestive juices (Leiper 1906) and the whole life cycle was finally elaborated by Dyneshvar Turkhud in 1913 (Turkhud 1914). The 20th century was one of considerable success in controlling this infection, which is now well on the way to being eradicated.

SCHISTOSOMES AND SCHISTOSOMIASIS

Human schistosomiasis, or bilharzia, affects between 200 and 300 million people and is caused by several species of *Schistosoma* each giving rise to a characteristic form of the disease. Calcified eggs of *Schistosoma haematobium*, the cause of urinary schistosomiasis, have been found in Egyptian mummies dating from 1250 to 1000 BC (Contis and David 1996) providing direct evidence that these schistosomes were present in ancient Egypt. The most obvious sign of urinary schistosomiasis is blood in the urine, and this cannot have gone unobserved by the ancient Egyptians. Inevitably there have been numerous attempts to find descriptions of this condition in the medical papyri and many historians have described the disease as 'aaa' recorded in over 50 early papyri, including the *Papyrus Ebers*. In some papyri, 'aaa' occurs together with the initial hieroglyph interpreted as a penis discharging blood, suggesting schistomiasis haematobia (Ebbell 1937). This interpretation is widely quoted in historical and parasitological textbooks but no passages from the papyri link 'aaa' with the bladder or urine and the discharge from the penis might equally represent semen and not blood. This subject is discussed by Nunn and Tapp (2000), who abandon 'aaa' as a possible ancient Egyptian word for schistosomiasis. However, as schistosomiasis was almost certainly common and widespread in ancient Egypt, it is curious that the Egyptians did not have a word for it unless it was so common that it was ignored. It should be mentioned here that there have been a number of other suggestions as to what 'aaa' might be, including hookworm infection.

The worm itself, *S. haematobium*, was first described by Theodor Bilharz in 1851 and the popular name for all forms of schistosomiasis, 'bilharzia,' now honours his discovery (Bilharz and von Siebold 1852–53). The connection with the urinary disease was made a year later by Bilharz working with Wilhelm Griesinger (Bilharz 1853). The search for the intermediate stages in the life cycle of *S. haematobium* took a long time. A number of experienced parasitologists, including Arthur Looss, Prospero Sonsino, and Thomas Cobbold, working at the end of the 19th century, all failed to infect snails and it was not until 1915 that Leiper demonstrated the complete life cycle in the snail host (Leiper 1916).

Our knowledge of the history of intestinal schistosomiasis, caused by *S. mansoni*, dates back to conclusions reached by Manson in 1902 that there were two species of *Schistosoma* in humans (Manson 1902). At the time this was not universally accepted and it was Leiper who firmly established the existence of *S. mansoni* as a separate species in 1915 (Leiper 1916).

The third important form of schistosomiasis, Katayama disease, caused by *S. japonicum*, was first recognized by Akira Fujinami in 1847 in a report that did not become available until 1909 (Fujinami and Nakamura 1909). The worm itself was discovered by Fujiro Katsurada in 1904 (Katsurada 1904). Development in the snail host was described by Miyairi and Suzuki (1913) 2 years before Leiper independently described the life cycle of *S. haematobium*.

The history of such an important disease as schistosomiasis involves a great number of observations, events, and individuals, and a detailed account of the history is given by Grove (1990). There are shorter accounts by Hoeppli (1959), Foster (1965), and Goodwin (1996a). A bibliography covering the early history of schistosomiasis is given by Warren (1973) and there is an excellent account of schistosomiasis in the context of British and American imperialism by Farley (1991).

LIVER AND LUNG FLUKES

Several flatworms or flukes, probably over 100 as adults or larvae, infect humans, of which the most important are *Paragonimus westermani*, causing paragonimiasis; *Clonorchis sinensis*, causing clonorchiasis; and *Opisthorchis* spp., causing opisthorchiasis. Virtually all the important discoveries about the parasites themselves were made over a period of less than 45 years, between 1874 and 1918, and all were based on observations that had been made on other parasitic flukes such as *Fasciola hepatica* in sheep and others of zoological rather than medical interest. The various discoveries were made by a large number of people and often reported in obscure publications, and no attempt will be made here to list the individual achievements; for this the reader is referred to Grove (1990) and Muller (1996a).

The histories of these infections as diseases begin with the discovery of the worms and continue with the elaboration of the various life cycles. The adult worm of *P. westermani* was discovered in the lungs of a human in Taiwan in 1879 by B.S. Ringer who wrote to Manson who later described his find (see Manson 1881). The presence of eggs in the sputum was recognized independently by Erwin von Baelz in 1880 (von Baelz 1880) and by Manson in 1881 who, having read Ringer's report, realized the connection with the worms he had seen and postulated that a snail might act as an intermediate host (see Manson 1881). A number of Japanese workers, including Koan Nakagawa, Sadamu Yokogawa, Harujiro Kobayashi, and Keinosuke Miyairi, reported on the whole life cycle in the snail *Semisulcospira* between 1916 and 1922 (see Grove 1990).

The human liver fluke, *C. sinensis*, was first recognized by James McConnell in 1875 (McConnell 1875). The whole life cycle remained a puzzle until Kobayashi discovered that there was a second intermediate host, a fish (Kobayashi 1915), and the snail host was discovered by Masatomo Muto in 1918 (Muto 1918). Kobayashi's discovery that the second intermediate host could be an important food fish had an immediate impact on the control of this infection.

The first records of *Opisthorchis* adult worms in humans were made by Konstantin Wingradoff (1892) and the snail and fish hosts were described by Hans Vogel (1934).

CESTODES AND CESTODIASIS

Humans harbor about 55 species of tapeworms, about 40 adult and 15 larval stages. The most important are the adults of *Diphyllobothrium latum* (the broad tapeworm), *Taenia solium* (the pork tapeworm), *T. saginata* (the beef tapeworm), and the larval stages of the dog tapeworm, *Echinococcus granulosus*. The large size of the adult worms and their common occurrence attracted the attention of all the major medical writers including the authors of the *Papyrus Ebers* and the *Corpus Hippocratorum* and physicians such as Celsus and Avicenna.

It is generally held that the broad tapeworm *D. latum* was first recognized by Felix Plater (sometimes spelled Platter) at the beginning of the 17th century (see Foster 1965; Grove 1990). By the middle of the 18th century it was apparent that individuals harboring *D. latum* were those whose diet was mainly fish, but it was not until the life cycles of other tapeworms of zoological interest had been elaborated that further progress became possible. The existence of three hosts in the life cycle of *D. latum* (human, fish, and copepod) confused the issue until 1917 when Janicki and Rosen independently worked out the whole of the life cycle (Janicki and Rosen 1917). Plater had already provided an excellent description of the disease, and the only significant subsequent discovery was made in 1948 when it was realized that this worm has an affinity for vitamin B_{12}, thus accounting for the pernicious anemia associated with this infection (see von Bonsdorff 1977).

The scientific study of the taeniid tapeworms can be traced to the work of Edward Tyson who, in the late 17th century, studied not only the tapeworms of humans but also those of dogs and other animals. This gave him the opportunity to recognize the 'head' (scolex) of a tapeworm for the first time, to describe the anatomy and physiology of the adult worms, and to lay the foundations for our knowledge of the biology of the taeniid tapeworms of humans (Tyson 1683). Although the distinction between broad and taeniid tapeworms had been recognized by the 17th century, the existence of two species of *Taenia* in humans presented problems because, even to the trained eye, the distinctions between *T. solium* and *T. saginata* are not obvious. This led to confusion between the two worms until a century

after the work of Tyson when, in 1782, Johann Goeze noted the similarities between the heads of tapeworms in humans and pigs (Goeze 1782). It was not until 1855 that Gottlob Küchenmeister demonstrated experimentally that eating infected pork gave rise to tapeworms in humans (Küchenmeister 1855) although the presence of cysts in pork had been noted throughout recorded history and is mentioned by Aristotle in the 4th century BC, Hippocrates, and Galen (see Foster 1965; Grove 1990). It has been suggested that, even before these discoveries, people had long been aware of the dangers of infected pork and that this accounted for the Jewish and Islamic prohibitions on the consumption of pork (Küchenmeister 1857) but there is no real evidence for this and the same arguments have been put forward for *Trichinella spiralis*.

The history of our knowledge of *T. saginata* parallels that of *T. solium*, and Goeze and Küchenmeister are credited with recognizing the differences between the two species of *Taenia* in humans. Several years after the implication of pigs as the intermediate hosts of *T. solium*, J.H. Oliver noted tapeworm infections in individuals who had eaten 'measly' beef (Oliver 1871) and the role of cattle as intermediate hosts of *T. saginata* was confirmed in a more controlled experiment in 1877 (Perroncito 1877). Together with observations on the pork tapeworm, these discoveries had a massive impact on the control of tapeworm infections in humans, as they led to a restriction of the amount of infected meat consumed by humans.

Humans are also host to two important kinds of larval tapeworm, cysticerci of the pork tapeworm *T. solium* and hydatid cysts of the dog tapeworm *E. granulosus*. The demonstration of the life cycle of *T. solium* threw new light on the nature of cysticercosis, and it was soon apparent that humans could become infected with the larval stages of *T. solium* when they ingested eggs. This made humans, in effect, dead-end intermediate hosts for this parasite, which also used humans as its definitive hosts. Although, for ethical reasons, the conclusive experiments could not be carried out, by the middle of the 19th century various experiments with animals and observations on humans had established without doubt that cysticercosis was caused by the ingestion of the eggs of *T. solium* (Küchenmeister 1860).

Infections with the larval stages of the dog tapeworm *E. granulosus* cause hydatid disease in humans. These cysts have been recognized from the earliest times and there are accurate descriptions by Hippocrates and Galen among others (see Grove 1990) but it was not until 1684 that Redi recognized the parasitic nature of these cysts (Redi 1684). Pierre Pallas was the first to suggest that these cysts were the larval stages of tapeworms (Pallas 1766) and in 1853 Carl von Siebold demonstrated that echinococcus cysts from sheep gave rise to adult tapeworms when fed to dogs (von Siebold 1853). The whole story was completed in

1863 when Bernhard Naunyn found adult tapeworms in dogs fed with hydatid cysts from a human (Naunyn 1863).

PROTOZOA AND PROTOZOAL INFECTIONS

Discovery of the organisms

Because of their small size, it was not possible to see the unicellular eukaryotic organisms we know as protozoa until the invention of the microscope and Antony van Leeuwenhoek's observations towards the end of the 17th century (Dobell 1960). All commentators agree that the first person to see a parasitic protozoan was Leeuwenhoek, who observed *Giardia* in his own stools. His written descriptions and his clinical observations leave little doubt that this flagellate is indeed the organism that he saw (Dobell 1960). Leeuwenhoek also described, very accurately, a number of other protozoa including flagellates and ciliates in frogs. The protozoa were recognized as a phylum of the animal kingdom a century later (Müller 1786) but it was not until over 140 years after Leeuwenhoek's observations that other parasitic protozoa were described and these were the large and conspicuous gregarines of insects (Dufour 1828), the beloved introduction to parasitic protozoology of generations of parasitologists. Among the conspicuous parasites in the blood are the trypanosomes, and the first to be recognized was from the blood of a fish (Valentin 1841); it is interesting to note that it was more than 50 years before similar parasites were seen in humans. The first associations between blood parasites and disease were those recorded by Griffiths Evans, who found trypanosomes in the blood of horses suffering from surra (Evans 1881). David Bruce found similar organisms in cattle, suffering from nagana, and dogs in 1894–7 (see Bruce 1915). Amebae are also conspicuous protozoa and those in humans were first seen in 1849 by Gros who recognized the nonpathogenic *Entamoeba gingivalis* from the mouth (see Dobell 1919); the pathogenic *E. histolytica* was recognized for the first time by Alexandrovitch Lösch in 1875 (Lösch 1875).

The detection of protozoa living in red blood cells was not possible until the development of adequate staining techniques; the first to be seen were malaria parasites by Charles Laveran in 1880 (Laveran 1880); *Babesia* in cattle by Theobald Smith and Frederick Kilbourne in 1893 (Smith and Kilbourne 1893); and *Theileria* in cattle by Charles Lounsbury in 1904 (Lounsbury 1904). Leishmania parasites were discovered independently by P.F. Borovsky in 1898, William Leishman in 1903, Charles Donovan in 1903, and James Homer Wright in 1903. The first leishmania parasites associated with New World leishmaniasis were discovered by Gaspar Vianna in 1911.

The remaining important protozoan parasites of humans were not detected until the beginning of the 20th century when, in a flurry of activity, *Trypanosoma cruzi* was identified as the cause of the condition that is now called Chagas disease (Chagas 1909). In a remarkable discovery that was to have much wider implications, *Toxoplasma gondii*, a parasite with the widest known host range including humans, was found in an obscure mammal, the gundi, by Charles Nicolle and Louis Manceaux in 1909.

Table 1.2 lists some of the most important events in our knowledge and understanding of parasitic protozoa. Note that the considerable activity that occurred during the second half of the 19th century mirrors what was also being done in the field of helminthology.

AMOEBAE AND AMOEBIASIS

Humans harbor nine species of intestinal amoebae of which only one, *Entamoeba histolytica*, is a pathogen. *E. histolytica* causes two forms of the same disease: amoebic dysentery, resulting from the invasion of the gut wall, and hepatic amoebiasis caused by extraintestinal amoebae in the liver. There is circumstantial evidence that both forms of the disease were recognized from the earliest times, but there are so many causes of both dysentery and liver disease that these records are open to other interpretations (see Bray 1996). The history of amoebiasis has been described by Dobell (1919), Foster (1965) and Bray (1996). James Annersley (1828) is credited with the first accurate descriptions of both forms of the disease and was the first to suggest the connection between them. From the middle of the 19th century, several workers observed amoebae in the stools of patients with amoebic dysentery but credit for the discovery of *E. histolytica* is now given to Alexandrovitch Lösch (Lösch 1875). The association between the amoeba and liver abscesses was finally determined shortly afterwards when Stephan Kartulis demonstrated amoebae in both intestine and liver (Kartulis 1886). Many workers did not believe that *E. histolytica* could cause disease, as all intestinal amoebae were generally thought to be harmless. It was therefore suggested that humans might harbor two morphologically similar amoebae, one pathogenic and one not (see Dobell 1919). This situation has only been resolved comparatively recently using biochemical techniques that clearly show that there are indeed two species: *E. histolytica*, which can cause disease, and *E. dispar*, which cannot (see Chapter 8, Classification and introduction to the parasitic protozoa).

GIARDIA AND GIARDIASIS

Giardia holds a special place in the affection of all protozoologists because the parasite that causes giardiasis, *Giardia duodenalis* (also known as *G. intestinalis* or *G. lamblia*) was the first parasitic protozoan ever to

Table 1.2 *Important events in our knowledge of parasitic protozoa.*

Date	Event
1681	Leeuwenhoek discovers *Giardia duodenalis* in his own feces
1773	Muller describes *Cercaria tenax* (*Trichomonas tenax*) from human mouth
1786	Müller recognizes Protozoa as a phylum of the Kingdom Animalia
1841	Valentin sees trypanosomes in the blood of fish
1848	Gros describes *Entamoeba gingivalis* from a human mouth
1853–4	Devaine describes *Cercomonas hominis* (*Trichomonas* and *Chilomastix*) from human feces
1857	Nägli describes a microsporidian from silkworms
1859	Lambl describes *Giardia duodenalis*
1870	Lewis describes *Entamoeba coli* from a human
1875	Lösch describes *Entamoeba histolytica*
1878	Lewis describes *Trypanosoma lewisi* from a rat
1879	Leuckart creates the phylum Sporozoa
1880	Laveran sees the malarial parasite *Plasmodium* in human blood
1881	Evans finds trypanosomes in the blood of a horse suffering from surra
1885	Cunningham sees *Leishmania* parasites in cutaneous lesions
1887	Metchnikov recognizes the coccidial nature of the malaria parasites
1891	Nepveu sees African trypanosomes in human blood
1893	Smith and Kilborne describe *Babesia* from cattle and incriminate ticks in its transmission
1894	Bruce sees trypanosomes in the blood of cattle and suggests a connection with tsetse flies
1897	Ross demonstrates the mosquito transmission of malaria parasites
1898	Bignami, Grassi, and Bastianelli describe mosquito stages of human malaria parasites
1898	Borovsky sees *Leishmania* parasites in a cutaneous lesion
1902	Dutton finds *Trypanosoma brucei gambiense* in human blood
1903	Wright describes *Leishmania* from cutaneous lesions
1903	Leishman and Donovan independently describe *Leishmania* from kala azar patients
1904	Bruce suggests that tsetse flies transmit African trypanosomes
1908	Nicolle and Manceaux and Splendore independently discover *Toxoplasma gondii*
1909–12	Chagas discovers *Trypanosoma cruzi*, its role in South American trypanosomiasis and transmission by bugs
1909	Kleine describes the cyclical transmission of trypanosomes
1910	Stevens and Fantham recognize *Trypanosoma brucei gambiense*
1911	Vianna describes *Leishmania braziliensis* in the New World
1912	Brumpt describes the transmission of *Trypanosoma cruzi* via the feces of bugs
1912	Tyzzer describes *Cryptosporidium* from a mouse
1921	Sergent and Sergent show that *Leishmania* parasites are transmitted by sandflies
1937	Wolf and Cowan implicate *Toxoplasma gondii* in human disease
1941	Adler and Ber demonstrate that leishmaniasis is transmitted through the bite of a sandfly
1948	Shortt and Garnham describe liver stages of malaria parasites
1959	Matsubayashi and colleagues describe a human infection with the microsporidian, *Encephalitozoon*
1976	Nime and colleagues and Meisel and colleagues independently describe human infections with *Cryptosporidium*
1969	Hutchison and colleagues describe life cycle of *Toxoplasma gondii*
1979	Ashford finds *Cyclospora* in a human
1982	Krotoski and colleagues discover *Plasmodium* hypnozoites

References are given in the text. Note that the dates of particular discoveries do not always coincide with the dates of publications. This is because in the 19th century and earlier scientists often made their findings known at meetings or in letters before the publication of the definitive paper. In some cases the reference refers to a later and fuller report.

be seen. An account of the history of giardiasis is given by Farthing (1996).

The first good accounts and illustrations of *Giardia* were by Vilém Lambl (1859), but the parasite received little attention until the war of 1914–1918 when troops returning to the UK with diarrhea were found to have, in their feces, *Giardia* cysts which caused similar infec-

tions in laboratory animals (Fantham and Porter 1916). In 1921 the distinguished British protozoologist Clifford Dobell suggested that *Giardia* could be a serious pathogen (Dobell 1921) and in 1926 Reginald Miller observed diarrhea and malabsorption in children infected with *Giardia* (Miller 1926). It still took some time before the association between infection with

Giardia and disease became widely recognized, following the detailed studies of Robert Rendtorff who produced unambiguous evidence linking the parasite with disease (Rendtorff 1954).

In the 300 years since *Giardia* was first discovered it has become recognized as a common and serious pathogen worldwide. It is still not known how many species infect humans and what role, if any, is played by reservoir hosts in the epidemiology of the infection.

AFRICAN TRYPANOSOMES AND SLEEPING SICKNESS

The story of African sleeping sickness is told briefly by Hoare (1972) and in more detail by Foster (1965), Nash (1969), Lyons (1992), Williams (1996), and Cox (2004b). The early records of African trypanosomiasis or sleeping sickness are vague. The first definitive accounts are those given by an English naval surgeon, John Atkins, in 1721 (see Atkins 1734) and Thomas Winterbottom, who coined the term 'negro lethargy,' in 1803 (Winterbottom 1803). Many explanations were put forward as causes of sleeping sickness, but the real basis of the disease was not forthcoming until Pasteur had established the germ theory towards the end of the 19th century. The first clues came from observations on surra, a horse disease, and nagana, a wasting disease of cattle. Although trypanosomes had been seen in the blood of fishes, frogs, and mammals several years before, it was not until 1881 that Griffith Evans found trypanosomes in the blood of horses and camels suffering from surra and suggested that the parasites might be the cause of the condition (Evans 1881).

In 1894, David Bruce, a British army surgeon, was sent to Zululand (now part of South Africa) to investigate an outbreak of nagana in cattle. Given the scientific climate of the period, Bruce suspected a bacterial cause but instead found trypanosomes in the blood of diseased cattle; he later demonstrated that these caused nagana in cattle, horses, and dogs. Bruce also noticed that infected cattle had spent some time in the fly-infested 'tsetse belt' and that the disease was similar to that in humans suffering from the 'fly disease' of hunters (Bruce 1895, 1915). Bruce also showed that the infection was acquired from tsetse flies, but mistakenly thought that transmission was purely mechanical.

It was later discovered that the trypanosomes that cause nagana and surra are not the same as those that cause human sleeping sickness. Although organisms that were certainly trypanosomes had been seen in human blood by Gustave Nepveu in 1891 (see Nepveu 1898) it was not until 1902 that Everett Dutton found the trypanosome that causes Gambian or chronic sleeping sickness (*Trypanosoma brucei gambiense*) in humans (Dutton 1902). In 1910 *T. b. rhodesiense*, the cause of Rhodesian or acute sleeping sickness, was described by J.W.W Stephens and Harold Fantham (Stephens and Fantham 1910). The role of the tsetse fly in the transmission of sleeping sickness remained controversial until Friedrich Kleine, a colleague of Robert Koch, demonstrated cyclical transmission, but he mistakenly thought that there was a sexual stage in the life cycle (Kleine 1909).

The persistence of trypanosomes in the blood and the existence of successive waves of parasitemia have attracted the attention of all those who have worked with trypanosomiasis. This phenomenon was described in detail by Ronald Ross and David Thomson in 1911 (Ross and Thompson 1911). The underlying mechanism and the way in which the parasite evades the immune response, now called 'antigenic variation,' were not elaborated until the work of Keith Vickerman in 1969 (Vickerman and Luckins 1969); this discovery precipitated a vast amount of interest in trypanosomes as cells and not only as causes of disease. Other 20th-century investigations have been concerned with refining the early discoveries and establishing the epidemiology of African sleeping sickness, leading to the development of methods for the control of both the human and animal forms of trypanosomiasis and the development of new and effective drugs.

SOUTH AMERICAN TRYPANOSOMIASIS: CHAGAS DISEASE

The earliest records of Chagas disease in South America are from 2 500-year-old mummies that show clear signs of the destructive nature of the disease (Rothammer et al. 1985). There are also a number of possible early written records, but the signs and symptoms of Chagas disease are so vague that it is difficult to know whether or not many of the conditions attributed to this infection are really valid. The early history of Chagas disease is described by Guerra (1970) and Miles (1996) and is passed over here because, from a scientific and medical viewpoint, the history really begins with a remarkable series of discoveries that were made between 1907 and 1912 by Carlos Chagas. Chagas not only discovered the trypanosome that causes the disease but also demonstrated its life cycle in the bugs that transmit it and described the disease which, although it affected millions of people, had until then remained enigmatic. Chagas's discoveries have been extensively described elsewhere (Wenyon 1926; Scott 1939; Guerra 1970; Lewinsohn 1979; Kean et al. 1978; Leonard 1990; Miles 1996). Briefly, Chagas, who was then in charge of an antimalaria campaign in Brazil, noticed that the bugs that infested the poorly constructed houses harbored flagellated protozoa; these could experimentally infect monkeys and guineapigs, in the blood of which a new trypanosome, *Trypanosoma cruzi*, was subsequently detected (Chagas 1909). Chagas suspected that these bugs might also transmit the parasite to humans and confirmed this when he found the trypanosomes in the blood of children suffering from an acute febrile condition (Chagas 1911). In 1912, Chagas demonstrated that *T. cruzi* was maintained in a number of reservoir hosts (Chagas 1912) thus completing in less than 5 years a cycle of discovery that had taken many years for earlier workers in other fields. The disease that affects some 18 million people now commemorates his name.

One thing that Chagas did not get right was the actual mode of transmission by the bug, as he thought that the trypanosomes were transmitted via the bite of the insect; it was left to Emile Brumpt to demonstrate that transmission was via the fecal route (Brumpt 1912). The links between infection with *T. cruzi* and the various signs of Chagas disease, such as megacolon, megaesophagus, and cardiac failure, were not determined until the work of Fritz Koberle in the 1960s (Koberle 1968). Exactly how the damage to heart and nerves is caused and whether or not there is an autoimmune component are still controversial issues.

LEISHMANIA AND LEISHMANIASIS

Leishmaniasis, caused by several species of *Leishmania*, is transmitted by sandflies and occurs in various forms in both the Old World and New World. Human leishmaniasis was, and still is in some cases, acquired accidentally from naturally infected wild animals. The conspicuous cutaneous lesions have been the subject of numerous records dating back to 2500–1000 BC (see Manson-Bahr 1996). Detailed descriptions of the Old World forms are given by the Arab physicians including Avicenna in the 10th century, and missionaries were well aware of this disease in the New World in the 16th century (see Lainson 1996).

The history of Old World leishmaniasis is described by Hoare (1938), Garnham (1987), and Manson-Bahr (1996). The discovery of the parasite responsible for the Old World cutaneous disease has been a matter of some controversy. There is no doubt that the organism was actually seen in 1885 by David Cunningham (Cunningham 1885) and in 1898 by a Russian military surgeon, P.F. Borovsky (see Hoare 1938), but credit for its discovery is usually given to James Homer Wright (Wright 1903). Credit for the discovery of *Leishmania donovani*, the parasite that causes visceral leishmaniasis (kala azar), goes to William Leishman and Charles Donovan who independently discovered the parasite in the tissues of patients suffering from this condition (Donovan 1903; Leishman 1903). Borovsky's discoveries were unknown to Wright and to Leishman and Donovan, but the names of the two latter workers are commemorated in the popular designation of the intracellular parasites as Leishman–Donovan bodies.

The search for a vector was a long one, partly because of the small size of the sandfly vectors. Although enough epidemiological evidence to incriminate sandflies belonging to the genus *Phlebotomus* had accumulated by the mid 1910s, it was not until 1921 that the experimental proof of transmission to humans from a sandfly was demonstrated by the Sergent brothers, Edouard and Etienne (Sergent et al. 1921). Infection through the bite of a sandfly was not finally demonstrated until 1941 (Adler and Ber 1941).

Leishmaniasis also occurs in the New World; the disfiguring conditions caused have been recognized in sculptures since the 5th century and in the writings of the Spanish missionaries in the 16th century (see Lainson 1996). Until 1911, it was thought that the New World forms of leishmaniasis were the same as those of the Old World, but in that year Gaspar Vianna found that the parasites in South America differed from those in Africa and India and created a new species, *Leishmania braziliensis* (Vianna 1911). Since then a number of other species, causing cutaneous and mucocutaneous leishmaniasis, unique to the New World have been described suggesting that this is where they emerged independently (see Chapter 17, New World leishmaniasis). There is, however, some controversy about the status of New World visceral leishmaniasis which was once thought to be caused by *L. donovani*, a species imported from the Old World possibly with the Spanish military and missionaries. This parasite has long been regarded as a separate species, *L. chagasi*, named in honor of the great Brazilian parasitologist Carlos Chagas, but the validity of *L. chagasi* as a New World species has been challenged by biochemical and molecular studies and it is now thought to be *L. infantum*, a species related to *L. donovani*, and thus of Old World origin (Maurício et al. 2000).

Following the discovery of the sandfly transmission of Old World leishmaniasis, the vectors in the Old World were also thought to belong to the same genus, *Phlebotomus*, but in 1922 it was discovered that the genus involved was actually *Lutzomyia*. Over the last two decades the complex pattern of species of parasite, vector, reservoir host, and disease has been painstakingly elaborated by Ralph Lainson and his colleagues (see Lainson 1996 and Chapter 17, New World leishmaniasis).

MALARIA

Malaria ranks among the most important infectious diseases in the world. Its history extends into antiquity and has been reviewed many times by, for example, Garnham (1966), Harrison (1978), Bruce-Chwatt (1985, 1988), McGregor (1996), Nye and Gibson (1997), Dobson (1999), and Poser and Bruyn (1999). Malaria almost certainly evolved with humans during our evolution from our primate ancestors; similar parasites are common in monkeys and apes. The disease probably originated in Africa and spread with human migrations, first throughout the tropics, subtropics, and temperate regions of the Old World and then to the New World with explorers, missionaries, and slaves. The periodic fevers of malaria are characteristic and are mentioned in the records of every civilized society from China in 2700 BC, through the writings of Greek, Roman, Assyrian, Indian, Arabic, and European physicians up to the 19th century. The earliest detailed accounts are those of Hippocrates in the 5th century BC, and for the next 700 years much of what we know about malaria relates to the disease in Greece, Italy, and throughout the Roman Empire. Thereafter, references to malaria become commonplace in Europe and elsewhere.

The science of malariology could not take off until the end of the 19th century, with the establishment of the germ theory and the birth of microbiology; a start was then made in earnest to discover the cause of the disease that was threatening many parts of the European empires. The story of the discovery of the malaria parasite and its mode of transmission is one of the most exciting in the history of infectious diseases and, like all good stories, is full of triumphs, disappointments, and rivalries (see McGregor 1996).

In 1880, a French army surgeon, Charles Laveran, looking for a bacterial cause of malaria, found and described malarial parasites in human blood and the parasite was immediately causally linked with the disease (Laveran 1880). The mode of transmission was not at all clear and numerous suggestions were put forward, none of which stood up to any detailed analysis. The actual incrimination of the mosquito as a vector was largely due to the intuition of Patrick Manson. In 1877 Manson had demonstrated the mosquito transmission of lymphatic filariasis (Manson 1878), in 1893 Smith and Kilbourne had shown that ticks transmit piroplasms (Smith and Kilbourne 1893), and in 1894 Bruce had implicated tsetse flies in the transmission of African trypanosomiasis (Bruce 1895). Thus, it was evident to Manson that a vector might be involved in the transmission of malaria and he postulated that it could be a mosquito (Manson 1894). Manson was unable to undertake this investigation himself, and secured the services of Ronald Ross, an army surgeon working in India. After several false starts using the wrong kinds of mosquito, *Culex* and *Aedes* instead of *Anopheles*, Ross, who had already demonstrated that bird malaria parasites developed in mosquitoes, found developing parasites in a mosquito fed on the blood of a patient suffering from malaria (Ross 1897). The rest of the life cycle in the mosquito was worked out by the Italian scientists Amico Bignami, Battista Grassi, and Giovanni Bastianelli (see Bastianelli and Bignami 1900; Fantini 1999). There is considerable controversy about the respective roles of Ross and the Italian workers in the discovery of the life cycle of the malaria parasite in the mosquito, and the arguments have been well rehearsed elsewhere. Nevertheless, most commentators agree that it was Ross who incriminated the mosquito as the vector of malaria parasites in birds and that he and the Italian workers established that this was also the case for the human malaria parasites independently at about the same time (see Dobson 1999; Fantini 1999).

The life cycle in humans, however, remained incompletely understood and nobody knew where the parasites developed during the first 10 days after infection, as they could not be found in the blood. In part this was due to too much reliance on a mistake by the influential German scientist Fritz Schaudinn who, in 1903, described the direct penetration of red blood cells by the infective sporozoites injected by the mosquito (Schaudinn 1903). The question was not resolved until 1947 when Henry Shortt and Cyril Garnham, working in London, showed that a phase of division in the liver preceded the development of parasites in the blood (Shortt and Garnham 1948); thus the whole life cycle had taken over 50 years to elucidate. There remained one further problem; what caused the long prepatent period between infection and the appearance and reappearance of parasites in the blood seen in some temperate strains of *P. vivax*? This led to the discovery in 1982 of dormant exoerythrocytic stages, hypnozoites, by Wojciech Krotoski, working with Garnham's team (Krotoski et al. 1982).

The last 50 years have been dominated by the control of the malaria parasite and its vectors by drugs and insecticides and the search for a vaccine. For a fascinating and amusing insight into the problems, solutions, intrigues, and personalities involved, the reader is referred to the popular book *The malaria capers* by Robert Desowitz (Desowitz 1991).

TOXOPLASMA, TOXOPLASMOSIS, AND INFECTIONS CAUSED BY RELATED ORGANISMS

The history of toxoplasmosis is a relatively recent one despite the fact that the organism that causes the disease, *Toxoplasma gondii*, is the most common human parasite world wide and one that infects the widest range of hosts (see Chapter 21, Toxoplasmosis). The history of toxoplasmosis has largely escaped the attention of the major reference books, but the basic events are well covered by Dubey and Beattie (1988) and Moulin (1993). *T. gondii* was discovered, largely by accident, by Charles Nicolle while searching for a reservoir host of *Leishmania* in a North African rodent, *Ctenodactylus gondi* (Nicolle and Manceaux 1909). At about the same time, Alfonso Splendore, working in São Paulo, discovered the same parasite in rabbits (Splendore 1909) and subsequently there have been numerous records from mammals and birds both in the wild and in captivity. The association with human disease was not made until 1937 when Arne Wolf and David Cowen reported on a congenital infection (Wolf and Cowen 1937). This report stimulated a vast amount of research, quickly leading to the knowledge that *T. gondii* is actually a very common parasite of humans but rarely causes disease, and that the parasite can cross the placenta and damage the fetus. The life cycle of *T. gondii* remained elusive until 1970 when William Hutchison and his colleagues demonstrated that this parasite is a stage in the life cycle of a common intestinal coccidian of cats (Hutchison et al. 1970), an observation quickly confirmed by other workers who had been working on this problem at the same time (see Cox 2002).

The parasitological significance of this discovery was that until then it had been assumed that the intestinal coccidians of vertebrates had only one host. The discovery of the *T. gondii* life cycle initiated a massive search for similar phases in the life cycles of other coccidian parasites. As a result, a number of protozoa that had not been properly identified could be classified as stages in the life cycle of poorly understood coccidians that had occasionally been encountered in the intestine and tissues of humans and other animals and in which transmission depends on predator–prey relationships (see Chapter 22, *Sarcocystis*). The species that infect humans are *Sarcocystis hominis*, acquired from cattle, and *S. suihominis*, acquired from pigs, and a complex of species collectively known as *S. lindemanni*. There is a detailed account of the nature and discovery of these species by Tadros and Laarman (1982).

CRYPTOSPORIDIUM, CYCLOSPORA, AND ISOSPORA

Humans also harbor three other species of coccidia, *Cryptosporidium parvum*, *Cyclospora cayetanensis*, and *Isospora belli*, once regarded as rare and accidental curiosities but now identified as possible pathogens in AIDS patients and immunocompromised individuals. All have simple single-host life cycles. *Cryptosporidium parvum* was discovered in 1912 by the American parasitologist Edward Ernest Tyzzer in the gastric glands of laboratory mice in which he had previously found another species, *C. muris* (Tyzzer 1912). The first sporadic human cases were recorded independently in 1976 by Nime and colleagues (Nime et al. 1976) and Meisel and colleagues (Meisel et al. 1976) but from 1981 onwards numerous new cases began to be recognized in AIDS patients. *C. parvum* is not very host-specific, and the source of these infections is probably drinking water contaminated with cattle feces. *Cryptosporidium* infections are now known to be very common and have caused a number of epidemics. There are histories of human cryptosporidiosis by McDonald (1996) and in Dubey et al. (1990).

Cyclospora cayetanensis, another coccidian that is mainly associated with AIDS, was discovered in 1979 by the English parasitologist Richard Ashford who found an unidentified coccidian in patients in Papua New Guinea (Ashford 1979), but it received little attention until 1986 when it was found again in the stools of patients with HIV (Soave et al. 1986). In 1993, this parasite was formally named *Cyclospora cayetanensis* (Ortega et al. 1993) and since then it has been identified as the cause of a number of outbreaks of diarrhea and fatigue in both immunocompetent and immunosup-

pressed individuals but the reservoir host, if any, is not known.

Isospora belli, the last of this group of parasites, was discovered in 1915 (Woodcock 1915) and is another coccidian frequently found in asymptomatic immunocompetent individuals but is associated with diarrhea in AIDS patients.

MICROSPORIDIANS

Microsporidians are common spore-forming parasites of vertebrates closely related to the Fungi (Cavalier-Smith 1998) but traditionally regarded as parasites. During the 19th century microsporidians attracted considerable attention, mainly as parasites of invertebrates, following the discovery of spores of *Nosma bombycis* by Nägeli in 1857 while investigating an outbreak of a disease called pébrine in the silkworm *Bombyx mori* (Nägeli 1857) and subsequent detailed investigations by Louis Pasteur in 1870 (see Vallery-Radot 1924). In the last decade of the 19th century, and throughout the 20th century, increasing numbers of species of microsporidians were discovered in fish, amphibians, reptiles, birds, and mammals (see Canning and Lom 1986). There have also been several sporadic reports of what might have been human microsporidial infections, but it is difficult to interpret the various structures that have been described as spores. The first human case was probably that of 'Encephalitozoon chagasi' in a newborn baby in 1927 (Torres 1927) but this cannot be confirmed because the original material has been lost. The parasite named *Encephalitozoon hominis* by Wolf and Cowen (1937) is now known to have been *Toxoplasma gondii*. The first authenticated record of human microsporidiosis was in 1959 when Hisakichi Matsubayashi and his colleagues in Japan found an *Encephalitozoon* sp. in a boy with convulsions (Matsubayashi et al. 1959). Thereafter, there have been increasing numbers of reports of cases of microsporidian infections in humans. In 1985 a new microsporidian, *Enterocytozoon bieneusi*, was found in an AIDS patient (Desportes et al. 1985) since when about 14 species, belonging to seven genera, have been associated with fulminating infections in immunodepressed patients, some with less serious infections in immunocompetent individuals (Canning 2001), and the number of cases, particularly in AIDS patients, continues to rise. Very little is known about the transmission and epidemiology of the microsporidians. There is a detailed account of the microsporidians in Chapter 25, Microsporidiosis.

SUMMARY AND CONCLUSIONS

The history of parasitology is a fascinating one, and parasites have been the subject of some of the most exciting discoveries in the field of medicine. In this chapter, it has not been possible to do more than merely touch on the major events, each one of which has represented the culmination of years of observation, conjecture, and experimentation. The personalities involved have been among the most eminent of their generations, but the cooperation and competition that existed between them would require a book on its own. The history of parasitology to a large extent reflects the availability of new concepts and techniques; thus each discovery has been the product of its own time and has been possible only because particular individuals have been willing and able to exploit new knowledge as it became accessible. There is more to the history of parasitology than this, because most of the great discoveries were made by individuals who were not looking in the same direction as everybody else. The conviction that diseases arose from decay and the air we breathe held back the discovery of the microbial causes of diseases despite the fact that, with hindsight, all the clues had been there. Had the helminthologists been able to persuade others not only that worms do not arise de novo, but also that they cause disease, the germ theory would have emerged two centuries earlier. The germ theory, with its emphasis on the search for bacterial causes of disease, held back the discovery of the protozoal causes of disease; in another context, it also inhibited a logical approach to nutritional disorders. Those who broke through the barriers of preconceived ideas, no matter how fashionable, were the ones who made the greatest discoveries. The history of 20th-century medicine will pinpoint the mistakes made in the study of AIDS and bovine spongiform encephalopathy (BSE) and will, almost certainly, tell us that our approaches to such subjects as vaccines and drugs were misplaced. For the present, it is widely assumed that all the important discoveries have been made; moreover, the climate of research at the beginning of the 21st century favors large teams working on relatively small projects. It is difficult to imagine that any future discoveries in the field of parasitology will engender the excitement of many of those outlined in this chapter, but there will certainly be new inventions and developments. There will, for example, be new drugs and new vaccines; new species will be described, and molecular techniques will be used to unravel details about both the parasites and the diseases they cause. These developments will bring fame and fortune, membership of learned societies, and possibly Nobel Prizes, but the most important achievements will probably go unsung. The campaign for the eradication of Guinea worm is reaching its final stages, the eradication of river blindness is well on its way, there is now an ambitious program to eradicate lymphatic filariasis and, in many parts of the world, malaria is being controlled by the use of insecticide-impregnated bed nets. These events, which will probably not be mentioned in any publications other than specialized reports, will be much more important to those suffering from these diseases than any other more spectacular discoveries or advances. The future history of parasitology will be, and should be, written in terms of the development of the underprivileged parts of the world that have borne the burden of these diseases for far too long.

REFERENCES

Ackernecht, E.H. 1965. *History and geography of the most important diseases*. New York: Hufner.

Adler, S. and Ber, M. 1941. The transmission of *Leishmania tropica* by the bite of *Phlebotomus papatasi*. *Indian J Med Res*, **122**, 803–9.

Annersley, J. 1828. *Researches into the causes, nature and treatment of the more prevalent diseases of warm climates generally*. London: Longman, Rees, Orme, Brown and Green.

Argyll-Robertson, D.M. 1895. Case of *Filaria loa* in which the parasite was removed from under the conjunctiva. *Trans Ophthalmol Soc*, **15**, 137–67.

Ashford, R.W. 1979. Ocurrence of an undescribed coccidian in man. *Ann Trop Med Parasitol*, **73**, 497–500.

Ashford, R.W. and Crewe, W. 2003. *The parasites of Homo sapiens*, 2nd edn. London: Taylor and Francis.

Atkins, J. 1734. *The navy surgeon or a practical system of surgery*. London: Caesar Ward and Richard Chandler.

Baelz, E.O. 1880. Parasiticidal haemoptysis (anonymous abstract). *Lancet*, **ii**, 548–9.

Ball, P.A.J. 1996. Hookworm disease. In: Cox, F.E.G. (ed.), *The Wellcome Trust illustrated history of tropical diseases*. London: Wellcome Trust, 318–25.

Bastianelli, G. and Bignami, A.P. 1900. Malaria and mosquitoes. *Lancet*, **i**, 79–83.

Bavay, A. 1877. Sur l'anguillule intestinale (*Anguillula intestinalis*), nouveau ver nématode trouvé par le Dr. Normand chez les malades atteints de diarrhée de Cochinchine. *C R Acad Sci*, **84**, 266–8.

Bilharz, T. 1853. Fernere mittheilungen über *Distomun haematobium*. *Z Wiss Zool*, **4**, 454–6.

Bilharz, T. and von Siebold, C.T. 1852-1853. Ein Beitrag zur Helminhographia humana, aus briefichen Mittheilungen des Dr. Bilharz in Cairo, nenst Bemerkungen von Prof. C.Th. von Siebold in Breslau. *Z Wiss Zool*, **4**, 53–76.

Blacklock, B. 1926. The development of *Onchocerca volvulus* in *Simulium damnosum*. *Ann Trop Med Parasitol*, **20**, 1–48.

Bray, R.S. 1996. Amoebiasis. In: Cox, F.E.G. (ed.), *The Wellcome Trust illustrated history of tropical diseases*. London: Wellcome Trust, 170–7.

Brothwell, D. and Sandison, A.T. (eds) 1967. *Diseases in antiquity*. Springfield, Illinois: C. C. Thomas.

Bruce, D. 1895. *Preliminary report on the tsetse fly disease or nagana in Zululand*. Durban: Bennett and David.

Bruce, D. 1915. Croonian lectures. *Br Med J*, **i**, 1073–8.

Bruce-Chwatt, L.J. 1985. *Essential malariology*, 2nd edn. London: Heinemann, 1–11.

Bruce-Chwatt, L.J. 1988. History of malaria from prehistory to eradication. In: Wernsdorfer, W.H. and McGregor, I. (eds), *Malaria: principles and practice of malariology*, vol. 1. . Edinburgh: Churchill Livingstone, 1–59.

Brumpt, E. 1912. Le *Trypanosoma cruzi* évolué chez *Conorhinus megistus*, , *Cimex lectularis*, *Cimex boueti*, *Ornithodorus moubata*. Cycle évolutif de ce parasite. *Bull Soc Pathol Exot*, **5**, 360–7.

Bryan, C.P. 1930. *The papyrus Ebers*. London: Geoffrey Bles, (translated from the German).

Bundy, D.A.P. and Michael, E. 1996. Trichinosis. In: Cox, F.E.G. (ed.), *The Wellcome Trust illustrated history of tropical diseases*. London: Wellcome Trust, 310–17.

Busk, G. 1846. Observations on the structure and nature of the *Filaria medinensis* or Guinea worm. *Trans Microsc Soc*, **2**, 65–80.

Campbell, S., Hall, B. and Klausner, D. (eds) 1992. *Health, disease and healing in medieval culture*. London: Macmillan.

Canning, E.U. 2001. Microsporidia. In: Gillespie, S. and Pearson, R.D. (eds), *Principles and practice of clinical parasitology*. London: Wiley, 171–95.

Canning, E.U. and Lom, J. 1986. *The microsporidia of vertebrates*. London: Academic Press.

Carter, H.J. 1855. Notes on dracunculus in the island of Bombay. *Trans Med Phys Soc Bombay*, **2**, 45–6.

Cavalier-Smith, T. 1998. A revised six-kingdom system of life. *Biol Rev*, **73**, 203–66.

Chagas, C. 1909. Nova tripanosomiase humana. Estudos sobre e morfologia e o ciclo evolutivo do *Schizotrypanum cruzi* n.gen. n.sp. agente etiologico de nova entidade morbida do homem. *Mem Inst Oswaldo Cruz*, **1**, 159–218.

Chagas, C. 1911. Nova entidade morbida do homem. Rezumo general de estudios etiologicos e clinicos. *Mem Inst Oswaldo Cruz*, **3**, 276–94.

Chagas, C. 1912. Sobre un trypanosomo do tatu, *Tatusia novemcincta*, transmittido pela *Triatoma geniculata* Latr. (1811), Possibilidade do ser o tatu um depositario do *Trypanosoma cruzi* no mundo exterior. *Brasil Médico*, **26**, 305–6.

Chernin, E. 1977. Milestones in the history of tropical medicine and hygiene. *Am J Trop Med Hyg*, **26**, 1053–104.

Chernin, J. 1983. Sir Patrick Manson's studies on the transmission and biology of filariasis. *Rev Infect Dis*, **5**, 148–66.

Cockburn, A., Cockburn, E. and Reyman, T.A. (eds) 1998. *Mummies, disease and ancient cultures*, 2nd edn. Cambridge: Cambridge University Press.

Contis, G. and David, A.R. 1996. The epidemiology of bilharzia in ancient Egypt: 5000 years of schistosomiasis. *Parasitol Today*, **12**, 253–5.

Cook, G.C. 2001. History of parasitology. In: Gillespie, S. and Pearson, R.D. (eds), *Principles and practice of clinical parasitology*. London: John Wiley.

Coombs, I. and Crompton, D.W.T. 1991. *A guide to human helminthology*. London: Taylor and Francis.

Cox, F.E.G. (ed.) 1996. *The Wellcome Trust illustrated history of tropical diseases*. London: Wellcome Trust.

Cox, F.E.G. 2002. History of human parasitology. *Clin Microbiol Rev*, **15**, 595–612.

Cox, F.E.G. 2004a. History of human parasitic diseases. *Infect Dis Clin N Am*, **18**, 171–88.

Cox, F.E.G. 2004b. History of sleeping sickness (African trypanosomiasis). *Infect Dis Clin N Am*, **18**, 231–45.

Cunningham, D.D. 1885. On the presence of peculiar parasitic organisms in the tissue of a specimen of Delhi boil. *Scientific Memoirs by Officers of the Medical and Sanitary Departments of the Government of India*, **1**, 21–31.

Davaine, C.J. 1862. Nouvelles researches sur le développment de la propogation de l'ascaride lombricoide et du trichocéphale de l'homme. *C R Acad Sci*, **18**, 665–7.

Demarquay, J.N. 1863. Sur une tumeur des bourses contenant un liquide laiteux (galactocèle de Vidal) et renferment des petits entres vermiformes que l'on peut considerée comme des helminthes hematoides a l'état d'embryon. *Gaz Med Paris*, **18**, 665–7.

Desowitz, R.S. 1991. *The malaria capers*. New York: Norton.

Desowitz, R.S. 1997. *Tropical diseases from 50 000 BC to 2500 AD*. London: Harper Collins.

Desportes, L., Le Charpentier, Y., et al. 1985. Occurrence of a new microsporidian: *Enterocytozoon bieneusi* n.g., n.sp. in the enterocytes of a human patient with AIDS. *J Protozool*, **23**, 250–4.

Dexian, W., Wenyuan, Y., et al. 1981. Parasitological investigation on the ancient corpse of the Western Han Dynesty unearthed from tomb no. 168 on Phoenix Hill in Jiangling County. *Acta Acad Med Wuhan*, **1**, 16–23.

Dobell, C. 1919. *The amoebae living in man*. London: John Bale Sons and Danielsson.

Dobell, C. 1921. A report on the occurrence of intestinal parasites in the inhabitants of Britain with special reference to Entamoeba histolytica. *Medical Research Council Special Report Series No. 59*. London: His Majesty's Stationery Office.

Dobell, C. 1960. *Antony van Leeuwenhoek and his 'little animals'*. New York: Dover Publications.

Dobson, M.J. 1999. The malaria centenary. *Parasitologia*, **41**, 21–32.

Donovan, C. 1903. The etiology of the heterogeneous fevers in India. *Br Med J*, **ii**, 1401.

Dubey, J.P. and Beattie, C.P. 1988. *Toxoplasmosis of animals and man*. Boca Raton: CRC Press.

Dubey, J.P., Speer, C.A. and Fayer, R. (eds) 1990. *Cryptosporidiosis of man and animals*. Boca Raton: CRC Press.

Dubini, A. 1843. Nuovo verme intestinal umano (*Agchylostoma duodenale*) constituente un sesto genere dei nematoidea propri dell'uomo. *Ann Univ Med Milano*, **106**, 5–13.

Dufour, L. 1828. Note sur la grégarine, nouveau genre de ver qui vit en troupeau dans les intestines de divers insectes. *Ann Sci Nat Paris*, **13**, 366–9.

Dutton, J.E. 1902. Preliminary note upon a trypanosome occurring in the blood of man. *Thompson Yates Lab Rep*, **4**, 455–68.

Ebbell, B. 1937. *The papyrus Ebers*. London: Oxford University Press.

Eldridge, B.F. 1992. Patrick Manson and the discovery age of vector biology. *J Am Mosquito Control Assoc*, **8**, 215–20.

Ettling, J. 1990. The role of the Rockefeller Foundation in hookworm research and control. In: Schad, G. and Warren, K.S. (eds), *Hookworm disease: current studies and new directions*. London: Taylor and Francis, 3–14.

Evans, G. 1881. On a horse disease in India known as 'Surra' probably due to a haematozoon. *Vet J Ann Comp Pathol*, **13**, 1–10, 82–8, 180–200, 326–33.

Fantham, H.B. and Porter, A. 1916. The pathogenicity of *Giardia* (*Lamblia*) *intestinalis* to men and experimental animals. *Br Med J*, **ii**, 139–41.

Fantini, B. 1999. The concept of specificity and the Italian contribution to the discovery of the malaria transmission cycle. *Parasitologia*, **41**, 39–47.

Farley, J. 1977. *The spontaneous generation controversy*. Baltimore: Johns Hopkins University Press.

Farley, J. 1991. *Bilharzia. A history of imperial tropical medicine*. Cambridge: Cambridge University Press.

Farthing, M.J.G. 1996. Giardiasis. In: Cox, F.E.G. (ed.), *The Wellcome Trust illustrated history of tropical diseases*. London: Wellcome Trust, 248–55.

Fedchenko, A.P. 1870. Concerning the structure and reproduction of the Guinea Worm *Filaria medinensis* (translated from the Russian). *Am J Trop Med Hyg*, **20**, 1971, 511–23.

Ferreira, L.F., Araujo, A.G.E. and Confalonieri, U.E.C. 1980. The finding of eggs and larvae of parasitic helminths in archaeological material from Unai, Minas Gerais, Brazil. *Trans R Soc Trop Med Hyg*, **174**, 798–800.

Foster, W.D. 1965. *A history of parasitology*. Edinburgh: Livingstone.

Fujinami, A. and Nakamura, A. 1909. [The mode of transmission of Katayama disease of Hiroshima Prefecture. Japanese schistosomiasis, the development of the causative worm and the disease in animals caused by it] (in Japanese). *Hiroshima Iji Geppo*, **132**, 324–41.

Fülleborn, F. 1914. Untersuchungen über den Infektionsweg bei *Strongyloides* und *Ankylostomum* und die Biologie dieser Parasiten. *Arch Schiffs- Tropen-Hyg*, **18**, 26–80.

Garnham, P.C.C. 1966. *Malaria parasites and other Haemosporidia*. Oxford: Blackwell Scientific Publications.

Garnham, P.C.C. 1970. *Progress in parasitology*. London: Athlone Press.

Garnham, P.C.C. 1987. Introduction. In: Peters, W. and Killick-Kendrick, R. (eds), *The leishmaniases in biology and medicine*. vol. 1. London: Academic Press, xiii–xv.

Goeze, J.A.E. 1782. *Versuch einer Naturgeschichte der Eingeweidewürmer thierischer Körper*. Blankenberg: P. Pape.

Goodwin, L. 1996a. Schistosomiasis. In: Cox, F.E.G. (ed.), *The Wellcome Trust illustrated history of tropical diseases*. London: Wellcome Trust, 264–73.

Goodwin, L. 1996b. Ascariasis. In: Cox, F.E.G. (ed.), *The Wellcome Trust illustrated history of tropical diseases*. London: Wellcome Trust, 326–31.

Grassi, B. 1881. Noto interno ad alcuni parassiti dell'uomo III. Interno all'Ascaris lumbricoides. *Gaz Osp Milano*, **2**, 432.

Griesinger, W. 1854. Klinische und anatomische Beobachtungen über die Krankheiten von Egypten. *Arch Physiol*, **13**, 528–75.

Grove, D.I. 1990. *A history of human helminthology*. Wallingford: CAB International.

Gruby, D. and Delafond, H.M. 1843. Note sur une altération vermineuse de sang d'un chien determiné par un grand nombre d'hematozoaires du genre filaire. *C R Acad Sci*, **16**, 325–35.

Guerra, F. 1970. American trypanosomiasis. An historical and human lesson. *J Trop Med Hyg*, **72**, 83–118.

Gutierrez, Y. 2000. *Diagnostic pathology of parasitic infections*, 2nd edn. New York: Oxford University Press.

Harrison, G. 1978. *Mosquitoes, malaria and man*. London: John Murray.

Herbert, T. 1634. *A relation of some yeares travaile into Afrique, Asia, Indies*. Amsterdam: Da Capo. (Also published as *Some years travels into Africa and Asia*, London, 1677).

Hoare, C.A. 1938. Early discoveries regarding the parasites of oriental sore. *Trans R Soc Trop Med Hyg*, **32**, 67–92.

Hoare, C.A. 1972. *The trypanosomes of mammals*. Oxford: Blackwell Scientific Publications, 3–5.

Hoeppli, R. 1956. The knowledge of parasites and parasitic infections from ancient times to the 17th century. *Exp Parasitol*, **5**, 398–419.

Hoeppli, R. 1959. *Parasites and parasitic infections in early science and medicine*. Singapore: University of Malaya Press.

Horne, P.D. 1985. A review of the evidence for human endoparasitism in the pre-Columbian New World through the study of coprolites. *J Archaeol Sci*, **12**, 299–310.

Hutchison, W.M., Dunachie, J.F., et al. 1970. Coccidian-like nature of *Toxoplasma gondii*. *Br Med J*, **i**, 142–4.

Janicki, C. and Rosen, F. 1917. Le cycle évolutif du *Bothriocephalus latus* L. *Bull Soc Sci Nat Neuchâtel*, **42**, 19–21.

Jones, W.H. and Whithington, E.T. 1948–53. *Works of Hippocrates*, 4 vols. London: Loeb Classical Library.

Kartulis, S. 1886. Zur aetiologie der dysenterie in Aegyptien. *Arch Pathol Anat Physiol*, **105**, 521–31.

Katsurada, F. 1904. [The etiology of a parasitic disease]. *Iji Shinbun*, **669**, 1325-32. (In Japanese, translated as *Schistosoma japonicum*, a new human parasite which gives rise to an endemic disease in different parts of Japan. *J Trop Med Hyg*, **8**, 108-11: 1905).

Kean, B.H., Mott, K.E. and Russell, A.J. (eds) 1978. *Tropical medicine and parasitology: classic investigations*, 2 vols. Ithaca: Cornell University Press.

Kleine, F.K. 1909. Positiv Infektionversuche mit *Trypanosoma brucei* durch *Glossina palpalis*. *Deutsch Med Wochenschr*, **35**, 469–70.

Kobayashi, H. 1915. On the life history and morphology of *Clonorchis sinensis*. *Zentralbl Bakteriol Parasitenk*, **75**, 299–317.

Koberle, F. 1968. Chagas' disease and Chagas' disease syndrome: the pathology of American trypanosomiasis. *Adv Parasitol*, **6**, 63–116.

Koino, S. 1922. Experimental infection of the human body with ascarides. *Jap Med World*, **15**, 317–20.

Krotoski, W.A., Collins, W.E. and Bray, R.S. 1982. Demonstration of hypnozoites in sporozoite-transmitted *Plasmodium vivax* infection. *Am J Trop Med Hyg*, **31**, 1291–3.

Küchenmeister, F. 1855. Offenes Sendschreiben an die k.k. Gesellschaft der Aertze zu Wien. Experimenteller Nachweis dass *Cysticercus cellulosae* innerhalb des menschlichen Damarkanales sich in *Taenia solium* umwandelt. *Wien Med Wochenschr*, **5**, 1–4.

Küchenmeister, F. 1857. *Animal and vegetable parasites*. London: Sydenham Society, Translated from the German by Edwin Lankaster.

Küchenmeister, F. 1860. Erneuter Versuch der Umwandlung des *Cysticercus cellulosae* in *Taenia solium hominis*. *Deut Klin Berlin*, **12**, 187–9.

Lainson, R. 1996. New World leishmaniasis. In: Cox, F.E.G. (ed.), *The Wellcome Trust illustrated history of tropical diseases*. London: Wellcome Trust, 218–29.

Lambl, V. 1859. Microscopische untersuchungen der darmexcrete. Beitrag zur pathologisches des darmes und zur diagnostik am krankenbette. *Vierteljahrsschr Prak Heilk (Prague)*, **61**, 1–57.

Laveran, A. 1880. Note sur un nouveau parasite trouvé dans le sang de plusieurs malades atteints de fièvre palustre. *Bull Acad Med Paris*, **9**, 1235–6.

Leiper, R.T. 1906. The influence of acid on Guinea worm larvae encysted in *Cyclops*. *Br Med J*, **i**, 19–20.

Leiper, R.T. 1913. Report to the Advisory Committee of the Tropical Diseases Research Fund Colonial Office London. *Trop Dis Bull*, **2**, 195–6.

Leiper, R.T. 1916. On the relation between the terminal-spined and lateral-spined eggs of bilharzia. *Br Med J*, **i**, 411.

Leishman, W.B. 1903. On the possibility of the occurrence of trypanosomiasis in India. *Br Med J*, **i**, 1252–4.

Leonard, J. 1990. Carlos Chagas, health pioneer of the Brazilian backlands. *Bull Pan Am Health Org*, **24**, 226–39.

Leuckart, R. 1883. Ueber die Lebensgeschichte der sogenannten *Anguillula stercoralis* und deren Beziehungen zu der sogenannten *Anguillula strongloides*. Bericht über die Verhandlungen der königlich sachsischen. *Ges Wiss Leipzig. Math-Phys Kl (1882)*, **34**, 1882, 84–107.

Lewinsohn, R. 1979. The discovery of *Trypanosoma cruzi* and of American trypanosomiasis (footnote to the history of Chagas' disease). *Trans R Soc Trop Med Hyg*, **73**, 513–23.

Lewis, T.R. 1872. On a haematozoon inhabiting human blood, its relation to chyluria and other diseases. *8th Annual Report of the Sanitary Commissioners of the Government of India*. Calcutta: Government Printing House, 241–66.

Linschoten, J.H. van. 1610. *Histoire de la navigaation de JHL*. Amsterdam.

Linnaeus, C. 1758. *Systema Naturae, sive regina tria naturae systematice proposita por classes, ordines, genera, species cum characteribus differentiis synonymis, locis*, 10th edn. Holmiae [Stockholm]: L. Salvi.

Looss, A. 1898. Zur Lebensgeschichte des *Ankylostoma duodenale*. *Zentralbl Bakteriol Parasitenk*, **24**, 483–8.

Looss, A. 1905. Die Wanderung der Ancylostomum- und Strongyloides-Larven von der haut nach dem Darm. *C R 6ième Congr Internat Zool. Berne*, **1904**, 225–33.

Lösch, F.A. 1875. Massive development of amebas in the large intestine (translated from the Russian). *Am J Trop Med Hyg*, **24**, 1975, 383–92.

Lounsbury, C.P. 1904. Transmission of African Coast fever. *Agr J Cape of Good Hope*, **24**, 428–32.

Low, G.C. 1900. A recent observation on *Filaria nocturna* in *Culex*, probable mode of infection in man. *Br Med J*, **i**, 1456–7.

Lyons, M. 1992. *A colonial disease: a social history of sleeping sickness in Northern Zaire 1900–1940*. Cambridge: Cambridge University Press.

Mack, A. 1991. *In time of plague: the history and social consequences of lethal epidemic disease*. New York: New York University Press.

Manson, P. 1878. On the development of *Filaria sanguis hominis* and on the mosquito considered as a nurse. *J Linn Soc (Zool)*, **14**, 304–11.

Manson, P. 1881. Distoma ringeri. *Med Times Gaz, London*, **2**, 8–9.

Manson, P. 1891. The *Filaria sanguinis hominis major* and *minor*, two new species of haematozoa. *Lancet*, **i**, 4–8.

Manson, P. 1893. Diseases of the skin in tropical climates. In: Davidson, A.H. (ed.), *Hygiene and diseases of warm climates*. London: Young J. Pentland, 928–95.

Manson, P. 1894. On the nature and significance of crescentic and flagellated bodies in malarial blood. *Br Med J*, **ii**, 1306–8.

Manson, P. 1898. *Tropical diseases*. London: Cassell.

Manson, P. 1902. Report of a case of bilharzia from the West Indies. *Br Med J*, **ii**, 1894–5.

Manson, P. 1910. On the nature and origin of Calabar swellings. *Trans R Soc Trop Med Hyg*, **3**, 244–51.

Manson-Bahr, P. 1962. *Patrick Manson: the father of tropical medicine*. London: Thomas Nelson.

Manson-Bahr, P.E.C. 1996. Old World leishmaniasis. In: Cox, F.E.G. (ed.), *The Wellcome Trust illustrated history of tropical diseases*. London: Wellcome Trust, 206–17.

Matsubayashi, H., Kioke, T., et al. 1959. A case of *Encephalitozoon*-like body infection in man. *Arch Pathol*, **67**, 181–7.

Maurício, I.L., Stothard, J.R. and Miles, M.A. 2000. The strange case of *Leishmania chagasi*. *Parasitol Today*, **16**, 30-2188-18, 9.

Meisel, J.L., Perera, D.G., et al. 1976. Overwhelming wasting diarrhoea associated with a *Cryptosporidium* in an immunodepressed patient. *Gastroenterology*, **70**, 1156–60.

McConnell, J.F. 1875. Remarks on the anatomy and pathological relations of a new species of liver-fluke. *Lancet*, **ii**, 271–4.

McDonald, V. 1996. : Cryptosporidiosis. In: Cox, F.E.G. (ed.), *The Wellcome Trust illustrated history of tropical diseases*. London: Wellcome Trust, 257–65.

McGregor, I. 1996. Malaria. In: Cox, F.E.G. (ed.), *The Wellcome Trust illustrated history of tropical diseases*. London: Wellcome Trust, 230–47.

Miles, M.A. 1996. New World trypanosomiasis. In: Cox, F.E.G. (ed.), *The Wellcome Trust illustrated history of tropical diseases*. London: Wellcome Trust, 192–205.

Miller, R. 1926. Lambliasis as a cause of chronic enteritis in children. *Arch Dis Child*, **1**, 93–8.

Miyairi, K. and Suzuki, M. 1913. [On the development of Schistosoma japonicum]. *Tokyo Iji Shinshi*, **1836**, 1–5, (In Japanese, partly translated from a German translation in Kean et al. 1978, 540–5).

Mongin 1770. Sur un ver trouvé sous la conjunctive à Maribarou, ïsle Saint-Dominique. *J Med Chirurg Pharm*, **32**, 338–9.

Moore, D.V. 1976. Fifty years of American parasitology. *J Parasitol*, **62**, 498–514.

Moulin, A.M. 1993. Historical introduction: the Institut Pasteur's contribution. *Res Immunol*, **144**, 8–13.

Müller, O.F. 1786. *Animalcula infusoria fluviatilia et marina*. Copenhagen: Hauniae.

Muller, R. 1996a. Liver and lung flukes. In: Cox, F.E.G. (ed.), *The Wellcome Trust illustrated history of tropical diseases*. London: Wellcome Trust, 274–85.

Muller, R. 1996b. Onchocerciasis. In: Cox, F.E.G. (ed.), *The Wellcome Trust illustrated history of tropical diseases*. London: Wellcome Trust, 304–9.

Muto, M. 1918. [On the primary intermediate host of *Clonorchis sinensis*]. *Chuo Igakkai Zassh*, **25**, 49–52, (In Japanese, partly translated in Kean et al. 1978, 557–60).

Nägeli 1857. Ueber die neue Krankheit der Seidenraupe und verwandte Organismen. *Versamml Deut Naturforsch Aert*, **15**, 760–1.

Nash, T.A.M. 1969. *Africa's bane. The tsetse fly*. London: Collins.

Naunyn, B. 1863. Ueber die zu *Echinococcus hominis* gehörige täen. *Arch Anat Physiol Wiss Med*, **1863**, 412–16.

Nelson, G. 1996. Lymphatic filariasis. In: Cox, F.E.G. (ed.), *The Wellcome Trust illustrated history of tropical diseases*. London: Wellcome Trust, 294–303.

Nepveu, G. 1898. Sur un trypanosome dans le sang de l'homme. *C R Soc Biol (Marseilles)*, **5**, 1172–4.

Nicolle, C. and Manceaux, L. 1909. Sur un protozoaire nouveau du gondi. *Arch Inst Pasteur*, **2**, 97–103.

Nime, R.A., Burek, J.D., et al. 1976. Acute enterocolitis in a human being infected with the protozoan *Cryptosporidium*. *Gastroenterology*, **70**, 592–8.

Norman, J.M. 1991. *Morton's medical bibliography. An annotated check list of texts illustrating the history of medicine. Garrison and Morton*, 5th edn. Aldershot: Scolar Press.

Nunn, J.F. and Tapp, E. 2000. Tropical diseases in ancient Egypt. *Trans R Soc Trop Med Hyg*, **94**, 147–53.

Nye, E.R. and Gibson, M.E. 1997. *Ronald Ross: malariologist and polymath*. London: Macmillan.

Ochoterena, B. 1928. Contribucíon para el conocimento de la onchocercosis en Mexico. *Arbeiten über Tropenkrankheiten. Festschrift Bernhard Nocht*, 386–9.

Oliver, J.H. 1871. *Seventh Annual Report of the Sanitary Commissioner of the Government of India*. Calcutta: Government Printing House, 82–3.

O'Neill, J. 1875. On the presence of a filaria in 'craw craw'. *Lancet*, **i**, 265–6.

Ortega, Y.R., Sterling, C.R., et al. 1993. *Cyclospora* species-a newÂprotozoan pathogen of humans. *N Engl J Med*, **328**, 1308–12.

Owen, R. 1835. Description of a microsopic entozoon infesting the muscles of the human body. *London Med Gaz*, **16**, 125–7.

Pallas, P.S. 1766. *Miscellanea zoologica: Quibus novae imprimus atque obscurae animalium species. Describuntur et observationibus iconbusque illustrantur*. Hagae: Petrum van Cleff.

Perroncito, E. 1877. On the tenacity of life of the cysticercus in the flesh of oxen and on the rapid development of the corresponding *Taenia mediocanellata* in the human body. *The Veterinarian*, **50**, 817–18.

Perroncito, E. 1880. Helminthological observations upon the endemic disease developed among the labourers in the Tunnel of Mount St Gothard. *J Queckett Microsc Club*, **6**, 141–8.

Poser, C.M. and Bruyn, G.W. 1999. *An illustrated history of malaria*. New York: Parthenon.

Ranger, T. and Slack, P. (eds) 1992. *Epidemics and ideas*. Cambridge: Cambridge University Press.

Ransford, O. 1983. *Bid the sickness cease: disease in the history of black Africa*. London: John Murray.

Redi, F. 1684. *Osservazione intorno agli animali viventi che si trouvano negli animali viventi*. Florence: Pietro Martini.

Rendtorff, R.C. 1954. The experimental transmission of human protozoan parasites. II *Giardia lamblia* cysts given in capsules. *J Hyg*, **59**, 209–20.

Robles, R. 1917. Enfermedad neuva en Guatemala. *Juventid Med*, **17**, 97–115.

Ross, R. 1897. On some peculiar pigmented cells found in two mosquitoes fed on malarial blood. *Br Med J*, **ii**, 1736–88.

Ross, R. and Thompson, D. 1911. A case of sleeping sickness studied by precise enumerative methods: further observations. *Ann Trop Med Parasitol*, **4**, 395–415.

Rothammer, F., Allison, M., et al. 1985. Chagas disease in pre-Colombian South America. *Am J Phys Anthropol*, **68**, 355–6.

Schaudinn, F. 1903. Studien über krankheitserregende Protozoen II. Plasmodium vivax (Grassi et Feletti) der Erreger des Tertianfiebers beim Menschen. *Arb Kaiserl Gesundheitsamte*, **19**, 169–250.

Scott, H.H. 1939. *A history of tropical medicine*, 2 vols. London: Edward Arnold.

Sergent, E., Sergent, Et., et al. 1921. Transmission du clou de Biskra par le phlebotome *Phlebotomus papatasi* Scop. *C R Acad Sci*, **73**, 1030–2.

Service, M.W. and Wilmott, S. (eds) 1978. *Medical entomology centenary symposium proceedings*. London: Royal Society of Tropical Medicine and Hygiene.

Shortt, H.E. and Garnham, P.C.C. 1948. Pre-erythrocytic stages in mammalian malaria parasites. *Nature (London)*, **161**, 126.

Smith, T. and Kilbourne, F.C. 1893. Investigations into the nature, causation and prevention of Texas or Southern fever. *Bull Bureau Anim Ind US Dept Agric, Washington*, **1**, 177–304.

Soave, R., Dubey, J.P., et al. 1986. A new intestinal pathogen? *Clin Res*, **34**, 533A.

Splendore, A. 1909. Sur un nouveau protozoaire parasite du lapin, deuxième note préliminaire. *Bull Soc Pathol Exot*, **2**, 462–5.

Stephens, J.W.W. and Fantham, H.B. 1910. On the peculiar morphology of a trypanosome from a case of sleeping sickness and the possibility of its being a new species (*T. rhodesiense*). *Proc Roy Soc London (B)*, **83**, 28–33.

Tadros, W. and Laarman, J.J. 1982. Current concepts on the biology, evolution and taxonomy of tissue cyst forming Eimeriid coccidia. *Adv Parasitol*, **20**, 293–468.

Tayeh, A. 1996. Dracunculiasis. In: Cox, F.E.G. (ed.), *The Wellcome Trust illustrated history of tropical diseases*. London: Wellcome Trust, 286–303.

Templeton, A.R. 2002. Out of Africa again and again. *Nature*, **416**, 45–51.

Tishkoff, S.A., Pakstis, A.J., et al. 2001. Short tandem-repeat polymorphism/alu haplotype variation at the PLAT locus: implications for modern human origins. *Am J Hum Genet*, **67**, 901–25.

Torres, C.M. 1927. Sur une nouvelle maladie de l'homme, characterisée par la présence d'une parasite intracellulaire, très proche de *Toxoplasma* et de l'*Encephalitozoon*, dans le tissu musculaire cardique, les muscles du squelette, le tissu cellulaire sous-cutane et le tissu nerveux. *C R Soc Biol*, **97**, 1778–81.

Turkhud, D.A. 1914. *Report of the Bombay Bacteriological Laboratory for the Year 1913*. Bombay: Government of India Central Press, 14–16.

Tyson, E. 1683. Lumbricus teres, or some anatomical observations on the round worm bred in human bodies. *Phil Trans Roy Soc London*, **13**, 153–61.

Tyzzer, E.E. 1912. *Cryptosporidium parvum* (sp. nov.), a coccidium found in the small intestine of the common mouse. *Arch Protistenk*, **26**, 394–413.

Valentin, G.G. 1841. Ueber ein Entozoon im Blut von *Salmo fario*. *Muller's Archives für 1841*, 435–6.

Vallery-Radot, R. 1924. *The Life of Pasteur*. New York: McClure Phillips, (translated from the French).

Van Durme, P. 1901–1902. Quelques notes sur les embryons de 'Strongyloides intestinalis' et leur pénétration par le peau. *Thompson Yates Lab Rep*, **4**, 471–4.

Vianna, G. 1911. Sobre uma nova espécie de *Leishmania*. *Brasil Med*, **25**, 411.

Vickerman, K. and Luckins, A.G. 1969. Localization of variable antigens in the surface coat of *Trypanosoma brucei* using ferritin conjugated antibody. *Nature (London)*, **224**, 1125–6.

Vogel, H. 1934. Der Entwicklungszyklus von *Opisthorchis felineus* (Riv.) nebst Bemerkungen über Systematik und Epidemiologie. *Zoologica*, **33**, 1–103.

von Bonsdorff, G. 1977. *Diplyllobothriasis in man*. London: Academic Press.

von Siebold, C.T. 1853. Ueber die Verwandlung der *Echinococcus*-brut in Taenien. *Z Wiss Zool*, **4**, 409–25.

Warboys, M. 1983. The emergence and early development of parasitology. In: Warren, K.S. and Bowers, J.Z. (eds), *Parasitology: a global perspective*. New York: Springer-Verlag, 1–18.

Warboys, M. 1993. Tropical diseases. In: Bynum, W.F. and Porter, R. (eds), *Encyclopedia of the history of medicine*, vol. 1. . London and New York: Routledge, 512–36.

Warren, K.S. 1973. *Schistosomiasis. The evolution of a medical literature*. Cambridge, Massachusetts: MIT Press.

Wells, S. 2002. *The journey of man: a genetic odyssey*. London: Allan Lane.

Wenyon, C.M. 1926. *Protozoology. A manual for medical men, veterinarians and zoologists*, 2 vols. . London: Ballière and Co.

Williams, B.I. 1996. African trypanosomiasis. In: Cox, F.E.G. (ed.), *The Wellcome Trust illustrated history of tropical diseases*. London: Wellcome Trust, 178–91.

Winogradoff, K. 1892. [On a new species of distomum [*Distomum sibiricum*] in the human liver] (abstracted from the Russian). *Zentralbl Allgemeine Patholog Pathol Anat*, **3**, 910–11.

Winterbottom, T.M. 1843. *An account of the native Africans in the neighbourhood of Sierra Leone to which is added an account of the present state of medicine among them*, vol. 2. London: C. Whittingham, 29–31.

Wolf, A. and Cowen, D. 1937. Granulomatous encephalomyelitis due to an encephalitozoon (encephalitozoic encephomyelitis). A new protozoan disease of man. *Bull Neurol Inst New York*, **6**, 306–71.

Woodcock, H.M. 1915. Notes on the protozoan parasites in the excreta. *Br Med J*, **ii**, 704.

Wright, J.H. 1903. Protozoa in a case of tropical ulcer. *J Med Res*, **10**, 472–82.

2

World-wide importance of parasites

GABRIEL A. SCHMUNIS AND FRANCISCO J. LÓPEZ-ANTUÑANO

PARASITIC DISEASES AND GLOBAL HEALTH

Introduction

This chapter describes the prevalence and overall distribution of parasitic infections in the world, the burden they constitute to the human population, and the costs they represent to individuals and society.

In 1993, life expectancy in the least developed countries of the world was 43 years, compared with 73 years in developed countries. Although life expectancy was projected to increase to 79 years by the year 2000 in developed countries, it was expected to remain less than 60 years in 45 developing countries (WHO 1995a). About 51 million people died in the world in 1993, 39 million of them in the developing countries. Worldwide, communicable diseases were responsible for c. 40 percent of the total number of deaths (20 million). In contrast, 80 percent (16 million) of deaths in developing countries were due to infectious and parasitic diseases (WHO 1995a). Five years later, communicable diseases, including bacterial, viral, parasitic, and zoonotic diseases, still caused 13.3 million (25 percent) of the 53.9 million deaths that occurred in 1998. In that year, communicable diseases were the main cause of death among children aged 0–4 years (63 percent of all deaths), and of premature deaths (48 percent of all deaths) in ages 0–44 years. They were also the main cause of death (45 percent of all deaths) in low-income countries such as those from South East Asia and Africa (WHO 1999). Six diseases were the leading infectious disease killers: acute respiratory infections (including pneumonia and influenza), acquired immune deficiency syndrome (AIDS), diarrheal diseases, tuberculosis, malaria, and measles (WHO 1999). Even in 2001, 19 percent of the 56.554 million deaths that occurred in the world were caused by infectious or parasitic diseases. Lower respiratory tract infection was the main killer (3.871 million deaths), followed by HIV/AIDS (2.866 million), diarrheal diseases (2.001 million), tuberculosis (1.644 million), malaria (1.124 million), and measles (745 thousand deaths) (WHO 2002).

Communicable diseases occur worldwide. Poverty and its accompanying features (lack of sanitation, malnutrition, illiteracy, and overcrowding) all contribute to most of the parasitic diseases of public health importance. Although infectious and parasitic diseases are more significant causes of morbidity and mortality among the poor, no-one is exempt from infection. Internal and external parasites are among the most common pathogens worldwide and anyone can be exposed to them at home or when they travel.

The burden of parasitic diseases

A comprehensive study by the World Bank (1993a) attempted to measure the burden that diseases, as a group or individually, represent in different regions of the world. The measurement took into account both losses from premature death (defined as the difference between the actual age of death and the age at which death would have occurred in a population with a low mortality) and the effects of losses caused by weakened health. The effects of the loss of healthy life were quantified for the diseases listed in the 109 categories included in the International Classification of Diseases (WHO 1977) and for approximately 95 percent of possible causes of disability (World Bank 1993a). To evaluate losses from illness, the incidence of cases by age, sex and demographic region was estimated and the number of years of healthy life lost was obtained by multiplying the expected duration of the disease by a severity weight that estimated the severity of the disability in comparison with loss of life caused by the illness (World Bank 1993a). The global burden of disease estimated in this manner was quantified using a unit called the disability adjusted life year (DALY), one DALY being defined as the loss of 1 year of healthy life to disease. The relative global burden of a specific disease was estimated by comparing the DALYs lost from different diseases or groups of diseases. In many areas of the world, communicable diseases continue to impose a significant burden that is greater in developing countries than in countries with established market economies. Injuries also cause greater losses in developing countries than in developed countries (Figure 2.1).

Parasitic diseases contribute heavily to the burden produced by communicable diseases, their contribution being comparable with such common diseases as tuberculosis, sexually transmitted diseases, diseases preven-

table by vaccination, and acute respiratory infections. Malaria is the parasitic disease responsible for the greatest burden worldwide; in Africa its burden is almost as great as that produced by acute respiratory infections. Worm infections are also a significant burden in Africa and some other regions (Figure 2.2) (World Bank 1993b).

The burden of infectious diseases, including parasitic infections, is borne mainly by the poorest people in the poorest countries of the world. Parasitic diseases are thus contributing to inequalities that exist both within and between societies (Evans and Jamison 1994). In Figure 2.3, the DALYs lost as a result of infectious diseases of microbial and viral origin are compared with the losses caused by parasitic diseases in developing countries and in countries with established market economies. Although the burdens produced by diseases of viral or microbial origin are the highest, the burden produced by parasitic diseases is also great (World Bank 1993b). The effects and burdens produced by parasitic diseases vary greatly from region to region. Different parasitic diseases are responsible for the bulk of the burden in different geographical areas; for example, malaria and schistosomiasis are the diseases that cause the highest burden globally, but Chagas disease produces the highest burden in Latin America and the Caribbean (Figure 2.4).

The burden of different communicable diseases also varies with age (Table 2.1). Among parasitic diseases in 1990, malaria produced the highest burden in children under 5 years of age, and intestinal worms produced the highest burden in children aged 5–14 years (Table 2.1). The situation remained the same in 2001: independent of age, malaria was still the parasitic disease that produced the highest burden among all parasitic diseases (Table 2.2) (WHO 2002).

Disease burdens vary not only with age and geographical region but also among demographic groups. Not surprisingly, a significant proportion of the deaths caused by communicable diseases occurs in developing countries (Tables 2.3 and 2.4) and the individuals at greatest risk are those living in rural areas and in urban squatter settlements. Such people are usually poor and lack adequate housing, safe water supplies, and good waste-disposal systems. Workers involved in mining and timber exploitation are also often exposed to parasitic diseases as an occupational hazard (Rosenfield et al. 1984).

Population movement and parasitic diseases

Population movements have helped to raise awareness of parasites as a worldwide public health problem. In the early 1990s, 500 million people crossed international borders by air, 18–20 million refugees were generated from conflicts and 20–30 million people were displaced

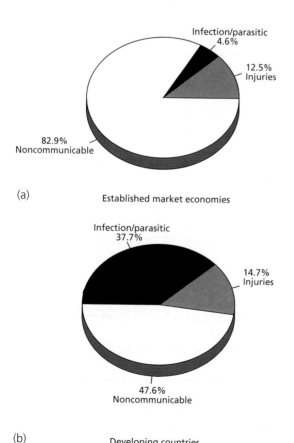

(a) Established market economies

(b) Developing countries

Figure 2.1 Percentage distribution of DALYs lost by overall cause in (a) countries with established market economies and (b) developing countries (data from World Bank 1993b).

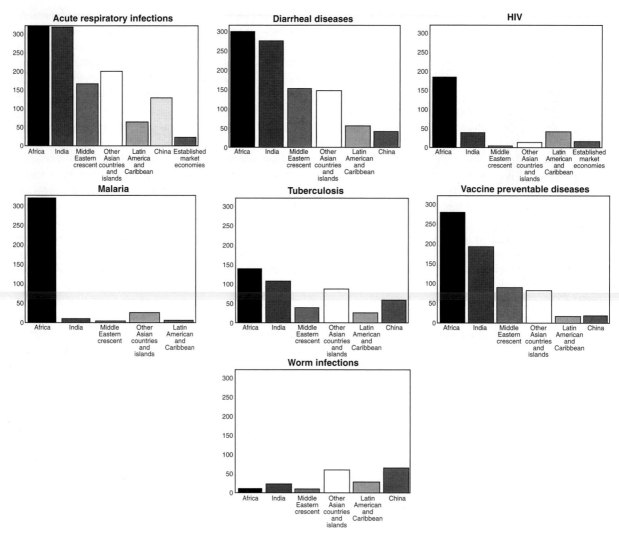

Figure 2.2 *Distribution of DALYs lost by selected communicable diseases by region, 1990 (hundreds of thousands of DALYs lost) (data from World Bank 1993b).*

for other reasons (Wilson 1995; WHO 1995b). Of the 5.2 billion people in the world, 1 in every 134 was forced to leave their home in 1992 (WHO 1995b) and an estimated 70 million people, mostly from developing countries, work in countries to which they have migrated (Wilson 1995). Many of these people have carried parasites to their new homes, sometimes requiring treatment themselves and sometimes serving as sources of infection to others. As a result of these population shifts, medical personnel in many advanced countries have gained personal experience of parasitic diseases for the first time.

The USA exemplifies the problems faced by the countries with immigrants with intestinal parasites. It is estimated that more than 600 000 immigrants enter the USA each year from countries where intestinal parasites are endemic: Asia, the Middle East, sub-Saharan Africa, Eastern Europe, Latin America, and the Caribbean. At entry, people with parasitic infections may be asymptomatic, and stool examination is not sensitive enough as a screening method for detecting all of those infected.

There is, however, a broad-spectrum antiparasitic drug, albendazole, which is safe and effective in eradicating many intestinal worms; it would be possible to screen all immigrants for parasites and to treat those infected or to treat all immigrants presumptively (400 mg of albendazole per day for 5 days). Another possibility would be to do nothing unless the immigrants have symptoms (watchful waiting). Estimates of the costs of these three types of intervention were made, and effectiveness was expressed both in terms of the cost of treatment per DALY averted and in terms of the cost per hospitalization averted. Presumptive treatment of all immigrants at risk for parasitosis, when compared with watchful waiting, would avert at least 870 DALYs, prevent at least 33 deaths and 374 hospitalizations, and save at least US$4.2 million per year. Screening compared with watchful waiting would cost US$159 236 per DALY averted. It was therefore concluded that presumptive treatment of all immigrants at risk of parasitosis would save lives and money, while universal screening and treatment only of those with positive stools,

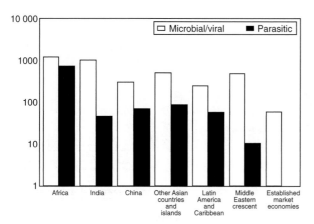

Figure 2.3 *Burden of disease caused by parasites (including malaria, African and American trypanosomiasis, schistosomiasis, leishmaniasis, filariasis, onchocerciasis, ascariasis, trichuriasis, and hookworm) and by microbial and viral infections (including tuberculosis, sexually transmitted diseases, HIV, diarrheal diseases, pertussis, poliomyelitis, diphtheria, measles, tetanus, meningitis, hepatitis, and acute respiratory infections) by region, 1990 (hundreds of thousands of DALYs lost) (data from World Bank 1993b).*

would save lives but would be less cost-effective (Muennig et al. 1999).

In addition to international migration, many people also migrate between regions of a country; in the last six decades, economic hardship in the rural areas of developing countries has stimulated migration to urban areas within the same country. As a result, c. 45 percent of the world's population now live in cities, and diseases that were traditionally considered to be 'rural' have spread into the cities, usually affecting the most deprived populations. This phenomenon has brought many people within the orbit of public health services for the first time and diseases previously not noticed have now made themselves apparent. It has also increased demands on already overburdened public health services. From their bridgeheads in the slums, the parasites may spread to other urban areas, reaching new hosts through vectors, through fecal contamination of soil, food, and water, and through transfused blood.

Transfusion-acquired malaria (especially *Plasmodium vivax* malaria) and Chagas disease pose particular problems as they are both easily transmitted by blood transfusion. Other parasites, such as the African trypanosomes (particularly *Trypanosoma gambiense*), *Babesia microti*, and the filarial worms such as *Brugia malayi*, *Loa loa*, *Wuchereria bancrofti*, *Mansonella ozzardi*, and *M. perstans*, may also be transmitted by blood transfusion but fortunately this mode of transmission is rare even in the endemic countries. It is not surprising that malaria has resulted from blood transfusion as plasmodia can remain viable in refrigerated blood for up to 10 days. In developing countries, transfusion malaria is most common in people who have received blood sold by low-income donors whereas in developed countries, transfusion malaria results from the migration of carriers

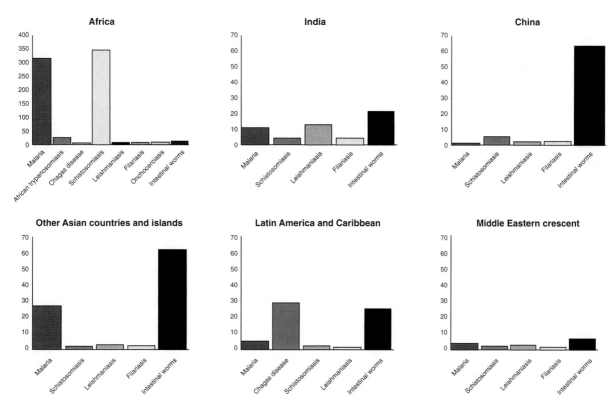

Figure 2.4 *Burden of disease caused by various parasitic diseases by regions, 1990 (hundreds of thousands of DALYs lost) (data from World Bank 1993b).*

Table 2.1 *Main causes of disease burden in children from developing countries, 1990*

Disease	DALYs lost (millions) ⩽4 years old	%	Disease	DALYs lost (millions) 5–14 years old	%
Acute respiratory infections	93.2	18	Intestinal helminths[b]	16.8	11.8
Diarrheal diseases	82.5	16	Vaccine preventable diseases	11.8	8.3
Vaccine preventable diseases[a]	55.1	10.6	Acute respiratory infections	10.4	7.36
Malaria	24.3	4.7	Diarrheal diseases	9.3	6.5
Tuberculosis	2.6	0.5	Tuberculosis	6.8	4.85
			Malaria	6.5	4.57
Total DALYs lost[c] (millions)	518.0			142.0	

Adapted from World Bank 1993b
a) Pertussis, polio, measles and tetanus; b) *Ascaris* and *Trichuris*; c) Communicable, noncommunicable and injury DALYs.

from endemic areas. To avoid such infections, blood donation is not permitted in some countries from those who have had malaria or who have visited or lived in malarious areas. Individuals with positive serology for *T. cruzi*, the causative agent of Chagas disease, have been found among the 7 million or more legal immigrants from South and Central America to the USA. Their presence poses a risk of transmission of Chagas disease by blood transfusion. A similar risk exists in Europe where more than 250 000 immigrants from Central and South America now live (Schmunis 1994). Transfusion-transmitted *T. cruzi* infection has been reported in

Table 2.2 *DALYs lost (millions) due to nine 'highest worldwide burden of disease' communicable diseases (all ages) 2001*

Acute respiratory infections	HIV/AIDS	Diarrheal diseases	Malaria	TB	Measles	Tropical diseases[a]	Pertussis	Sexually transmitted diseases
94	88	62	42	36	26	13	12	12

WHO 2002
a) African trypanosomiasis, Chagas disease, schistosomiasis, leishmaniasis, lymphatic filariasis, and onchocerciasis

Table 2.3 *Deaths by cause and demographic group according to types of country and age group, 1990*

Disease	Developing countries Deaths (thousands) ⩽4 years	Developing countries Deaths (thousands) ⩾5 years	Formerly socialist economies of Europe and established market economies Deaths (thousands) ⩽4 years	Formerly socialist economies of Europe and established market economies Deaths (thousands) ⩾5 years	World
Acute respiratory infections	2 710	1 274	21	309	4 314
Diarrheal diseases	2 474	392	4	3	2 873
Tuberculosis	71	1 907	0	38	2 016
Malaria	633	294	0	0	926
African trypanosomiasis	5	51	0	0	55
Chagas disease	0	23	0	0	23
Schistosomiasis	1	37	0	0	38
Leishmaniasis	7	46	0	0	54
Onchocerciasis	0	29	0	0	30
Ascariasis	0	13	0	0	0
Trichuriasis	0	10	0	0	9
Hookworm	0	6	0	0	0
Lymphatic filariasis	0	0	0	0	0

Data from World Bank 1993b

Table 2.4 *Annual deaths by selected infectious and parasitic disease*

Selected infectious or parasitic disease	Deaths (thousands)				
	1990[a]	1993[b]	1995[c]	1999[d]	2001[e]
Acute respiratory infections (upper and lower)	4 314	4110[f]	4 416	4 010	3 947
HIV/AIDS		700	1 063	2 673	2 866
Diarrheal diseases	2 873	3 010[f]	3 115	2 213	2 001
Tuberculosis	2 016	2 709	3 072	1 669	1 644
Malaria	926	000	1 423	1 086	1 124
Diseases preventable by vaccination		1 679	2 100	1 554	1 318
Sexually transmitted diseases				178	189
Meningitis		245	35[g]	124	173
Hepatitis		933[g]	1 156[h]	171[h]	127[i]
African trypanosomiasis	55	55	20	66	50
Chagas disease	23	45	45	21	13
Leishmaniasis	54	197	20	57	59
Schistosomiasis	38	200	80	14	15
Onchocerciasis	30	35	47	0	0
Ascariasis	13	60	60	3	4
Hookworm	6	90	65	7	4
Trichuriasis	9		10	2	2
Total world deaths		51 000	51 882	55 965	56 554

a) World Bank 1993b; b) WHO 1995a; c) WHO 1996; d) WHO 2000a; e) WHO 2002; f) Children under 5 years; g) Meningococcal; h) HBV; i) HBV and HCV.

Canada, the USA, and Spain (Villalba et al. 1992; Schmunis 1991). Infection may occur in nonendemic areas by means other than transfusion; for example, infection in a migrant was the cause of a case of congenital *T. cruzi* infection in Sweden.

As air travel has become more common, imported malaria has been seen with increasing frequency in Canada, the Caribbean, Europe, Oceania, and the USA and it has been stated that patients with malaria are now more often seen in temperate London (England) than in tropical Jamaica (West Indies) (Alleyne 1983). In the UK, 8 353 cases of imported malaria were diagnosed between 1987 and 1992 (WHO 1999) and more than 1 000 cases of imported malaria are detected yearly in the USA. For travelers in most areas of Asia, the species of *Plasmodium* responsible for malaria is usually *P. vivax*, whereas for travelers in Africa and some parts of Asia it is *P. falciparum*, and either or both species may occur in travelers in the Americas. There is a constant risk that travelers infected with plasmodia who return to areas where there are *Anopheles* mosquitoes susceptible to infection may start local transmission (Maldonado et al. 1986; Phillips-Howard 1991; López-Antuñano and Schmunis 1993). Malaria has also been reported in individuals living in nonmalarious areas who have never left that area and have never had a blood transfusion. Most cases have been acquired by people living near or passing through airports and have probably originated in airplane-transported infected mosquitoes.

Food-borne trematodes, such as *Opisthorchis* spp., are responsible for diseases in migrants from South East

Asia and the eastern Pacific, and a high prevalence has been reported in refugees in camps in Cambodia and Laos and in Thai immigrant workers in Kuwait and China. Pulmonary paragonimiasis has been diagnosed in refugees from South East Asia residing in France and in the USA. Liver flukes have been found in immigrants from Asia to Europe and the USA (WHO 1995b). Ethiopian refugees in Israel also harbor liver flukes. The round worm *Strongyloides stercoralis* and the protozoa *Giardia duodenalis* and *Entamoeba histolytica* are among the parasites often acquired by migrants and by travelers to developing countries.

NON-HELMINTH PARASITES

African trypanosomiasis (sleeping sickness)

Trypanosomiasis, caused by *T. brucei gambiense* and *T. brucei rhodesiense*, is endemic in rural areas of 36 countries of sub-Saharan Africa, particularly in West, Central, and East Africa in areas where the tsetse fly is present. Sleeping sickness caused by the trypanosomes has eliminated entire communities in Africa. Outbreaks of sleeping sickness produced by *T. b. gambiense* continue to occur in Angola, Congo, Sudan, Uganda, and Zaire and *T. b. rhodesiense* is a serious threat in Tanzania and Mozambique. Sporadic cases of sleeping sickness have also been reported from some southern African countries (WHO 1986, 1994a). Even tourists to certain areas of Africa can be affected, as shown by recent cases in UK travelers returning from East Africa (Moore et al. 2002). It is estimated that over 55 million people in 36 countries are at risk of acquiring sleeping sickness and that 250 000–300 000 people are infected. The yearly incidence is difficult to ascertain because surveillance is poor in the c. 200 confirmed foci of endemic sleeping sickness. The number of cases reported annually varies from 20 000 to 25 000 (WHO 1986, 1994a), and the estimated annual mortality 20 000 to 70 000 from 1990 to 2001. (World Bank, 1993b; WHO 1995a, 1996, 2000a, 2002).

The impact of the animal trypanosome disease, nagana, is also great, depriving much of Africa of milk, meat, and draught oxen for plowing and transport. The cost of these problems is more than US$4.5 million per year (Kabayo 2002).

Projects for economic development, if planned without proper consideration for their effects on the environment and on health, can cause unexpected health problems. Development projects for producing electricity and improving agriculture may change the local ecology and facilitate the spread of trypanosomiasis and other parasitic infections. For instance, they may encourage people to migrate to areas where their contact with parasites and their vectors is increased. Dams may displace villages, forcing people into tsetse-infested grasslands. Extensive lake margins may encourage growth of vegetation in which tsetse flies can breed, increasing the risk of bites among people who come to collect water. Such exposures may cause recrudescence of infection and even bring about full-blown epidemics (WHO 1986). Political upheavals and droughts may also stimulate life-style changes, many of which facilitate the spread of diseases. One of the most common consequences of civil disorder and of drought is the mass movement of people. While in transit, they are vulnerable to disease and many may become infected if they migrate into or across trypanosomiasis endemic areas.

If untreated, both forms of African trypanosomiasis are fatal, but there are important differences between the two forms of the disease. *T. b. gambiense* infections may run a protracted course of several years and infected individuals may serve as sources of infection if they migrate to areas where susceptible hosts and vectors are present. The disease caused by *T .b. rhodesiense* in East Africa causes death in weeks or (at most) a few months and, therefore, carriers pose less of a threat to individuals in uninfected areas. Melarsoprol has been used for the treatment of African trypanosomiasis for decades and eflornithine, which is much more expensive but has fewer side-effects and improves the overall cure rate, has been introduced for treatment more recently. An economic study in Uganda indicated that melarsoprol treatment is associated with an incremental cost per life and DALY saved of US$209 and US$8, respectively. Each additional life saved by switching from melarsoprol alone to a combination of melarsoprol and eflornithine would cost an extra US$1 033 per life saved, and an extra US$40.9 per DALY gained. Shifting from this second alternative to treatment of all patients with eflornithine would lead to an incremental cost per life saved of US$4 444 and an incremental cost of US$166.8 per DALY gained (Politi et al. 1995).

Programs for control of sleeping sickness are based on a combination of active and passive medical surveillance and on vector control. Costs include those incurred as a result of death and loss of productivity due to illness and those incurred as a result of the cost of treatment and disease control. The estimated cost from loss of income due to premature death from sleeping sickness is US$615–761 per person, varying in different areas and among different segments of society; the cost per individual is higher for those with a higher income. In 1984–85, the cost of surveillance was estimated at US$0.79–1.39 per person depending on the strategy used. The cost of treatment, including hospitalization, was US$35, US$88 and US$133 per person for treatment with pentamidine, suramin, and melarsoprol, respectively. Costs for prevention using screens and traps varied from US$2.30 to 11.50 per person. Ground-spraying costs were estimated at US$200–1000 per km. The cost per person protected by screens and traps depended on the number of people served per screen or trap and on the density of the population in the area under surveillance (WHO 1986).

American trypanosomiasis (Chagas disease)

Chagas disease occurs frequently in South and Central America, from Mexico in the north to Argentina and Chile in the south. Most of the people infected with *T. cruzi* in the Americas live in poor rural or semi-urban areas. Infected individuals have been detected in all Central and South American countries, including a few in French Guiana, Guyana, and Surinam. The only Spanish-speaking countries in the western Hemisphere where human *T. cruzi* infections have not been found are Cuba and the Dominican Republic. American trypanosomiasis also occurs as a zoonotic disease in the USA, where its range extends to Northern California in the west and to Maryland in the east.

Chagas disease is a major public health problem in Latin America. About 16–18 million people were considered to be infected and 90 million others exposed to the risk of infection in the 1980s. More recent data suggest that there are 11 million infected individuals (Table 2.5) and that 2–3 million have the cardiac and/or hollow visceral lesions that characterize the chronic stage of the disease. Many infected individuals show few or no clinical symptoms of disease and thus may not seek treatment. Infected individuals who are not treated

Table 2.5 *Infection by Trypanosoma cruzi in the Americas based on data for 1980–85 or later estimates*

Country	Endemic area (thousands km²)	% total country area	Estimated population at risk		Estimated infected population		Estimated yearly incidence
			Number (thousands)[a]	% total population	Number (thousands)	% total population	
Argentina	1 946	70	6 900	23	2 333	7.2	b
Bolivia	1 300	100	2 834	55	1 134	22.2	86 676
Brazil	3 615	42	41 054	32	1 961	1.21	b
Chile	350		1 800	15	142	1.06	b
Colombia	200	18	3 000	10	900–1 300	3.3–4.02	39 162–31 330
Costa Rica		...	1 112	45	130	5.3	4 030–3 320
Ecuador	100	35	3 823	41	30–450	0.34 to 4.09	7 488–13 365
El Salvador		...	21	43	322	6.9	10 048
Guatemala		...	4 022	52	730	9.8	30 076–28 387
Honduras		...	1 824	42	300	7.4	9 891–11 490
Mexico		...	...	...	152–540	0.18–0.63	14 2880
Nicaragua		...	...	...	67	1.82	5 016–2 660
Panama		...	898	42	220	10.6	7 130–5 346
Paraguay		...	1 475	45	397	11.59	14 680
Peru	120	9	6 766	34	643	3.47	19 072–24 320
Uruguay	125	71	975	33	37–51	1.25–1.59	b
Venezuela	697	76	11 392	68	800–1 200	7.42–4.14	17 9703–22 960

Data from Schmunis 1994, 2000

a) Total population the year the study was made or average population for the years the study was made (Schmunis 1994). Incidence data from 1990 and 1997 (Hayes and Schofield 1990; Schofield and Dujardin 1997);
b) Incidence data from these countries are not included because the beneficial effect of control programs is not considered in the calculation (Hayes and Schofield 1990).
... No data.

early after infection or in the early chronic stage of the infection will remain infected throughout their lives. The majority of cases of human infection occur in rural areas, particularly where houses are colonized by the triatome bugs that transmit the infection. The extent of vector transmission in the human population is related to socioeconomic status; poor people living in inadequate housing being most commonly infected. The infection persists wherever living conditions permit intimate contact between the triatome bugs and the human host. Prevalence in the general population is 0.34 percent in Ecuador, 1.25 percent in Uruguay, 11.5 percent in Paraguay, and 22.2 percent in Bolivia. Incidence estimates range from 3 320–4 030 for Costa Rica to 22 960–179 703 for Venezuela (Table 2.5) (Hayes and Schofield 1990; Schmunis 1994; Schofield and Dujardin 1997).

Transfusion is the second most common mode of acquiring the infection. Infection rates in blood donors vary from 6 to 12 percent in some areas of Argentina, Brazil and Chile and from 2 percent in Honduras to 48 percent in Bolivia. In 1997, six countries had more than 99 percent screening coverage for *T. cruzi*, but seven countries still showed risk of transfusion-transmitted infection that year (Schmunis et al. 2001). The risk of receiving an infected blood unit and acquiring a transfusion-transmitted infection has been reduced with time due to improvements in screening coverage (Schmunis 1999)

Congenital transmission, the third most common route of acquiring the infection, occurs in at least 3–5 percent of babies born to infected mothers (Schmunis 1994).

Chagas disease is a debilitating and incapacitating chronic condition that develops in 20 percent of those infected with *T. cruzi*, killing from 23 000 (World Bank 1993b) to 45 000 (Moncayo 1993) people every year in the early 1990s. Recent estimates suggest that mortality has decreased to 13 000 per year (WHO 2002). Chagasic cardiopathy was ranked third as a cause of disability in a rural area of Brazil where *T. cruzi* infection is endemic (Dias et al. 1985) and was also the main cause of early retirement in the region (Lopes and Chapadeiro 1986). In another endemic area of Brazil, 4–9 percent of all incapacity benefits received by individuals 30–50 years old were attributed to the effects of Chagas disease (Zicker and Zicker 1985). The number of productive years lost because of the disease among individuals 15–64 years of age in endemic areas of Brazil was estimated to be 2.275 and 1.369 per 100 000 population for males and females, respectively (Pereira 1984). Using data from the World Bank (1993b) study and from other sources, it is possible to compare the burden and general costs of Chagas disease with those of other diseases (World Bank 1993b). The disease burdens produced in 1990 by malaria (35 700 000 DALYs) and schistosomiasis (4 500 000 DALYs) world-wide are higher than that produced by Chagas disease (2 740 000 DALYs),

but the Chagas disease burden is higher than those produced by leishmaniasis, African trypanosomiasis, leprosy, filariasis, and onchocerciasis (Figure 2.4). In Latin America, the burden of Chagas disease exceeds that of the so-called tropical diseases (i.e. malaria, schistosomiasis, and leishmaniasis) and intestinal worms. In Latin America, tropical diseases together produced a disease burden that was only about one-quarter of that caused by Chagas disease. The disease burden caused by Chagas disease is, moreover, the fourth highest among all infectious diseases that occur in the region. Only acute respiratory infections, diarrhea, and AIDS produced higher disease burdens. The situation has improved since 1990, and the burden of Chagas disease is now considered to be 676 000 DALYs (WHO 2002).

As the affected population is mostly at a low socioeconomic level, the cost of treatment is usually provided by the State. In Brazil in 1987, the estimated cost for pacemakers and surgery to correct enlarged viscera was US$250 million and the cost of lost labor by the 75 000 chagasic workers represented a loss of another US$625 million (Dias 1987). These costs did not include consultation, care, and supportive treatment, costs for which amounted to another US$1 000 per patient per year (Schofield and Dias 1991), or disability payments which, in one Brazilian State in 1987, were US$400 000 (Dias 1987). In Argentina, treatment of an acute case cost US$591.80; treatment of an asymptomatic case cost US$ 174.49; and treatment of one with cardiac lesions cost US$603.62 (del Rey et al. 1995). In the same country, the cost for 30 months of treatment of 128 patients who had cardiac lesions was US$350 000 (Evequoz 1993). If all patients from Bolivia who needed treatment were treated, that country would spend US$215 000 each year for treatment of patients with acute Chagas disease; US$21 million for treatment of chronic disease and US$186 000 for treatment of congenital disease. In 1994, direct costs to Bolivia as a result of deaths from the disease were US$343 000 per year and annual indirect costs of morbidity and mortality were estimated at US$43.8 and 57.5 million, respectively (Human Development Ministry 1994). In Chile, the yearly cost of the disease was estimated at US$37 million in addition to the cost of pacemakers and surgery for implantation of pacemakers (Apt 1991).

In addition to the costs of treatment, there are also costs associated with the control of vectors responsible for Chagas disease. The annual cost for vector control through insecticide spraying, housing improvement, and health education in countries infested with the vector *T. infestans* is large: US$100 000 for Uruguay; US$300 000 for Chile; US$18 million for Argentina, and US$25 million for Brazil.

Brazil spent US$516 682 000 in prevention and control activities, mostly for vector control, from 1975 to 1995. In 1978, it was estimated that there were 3 573 000 individuals infected with *T. cruzi*, 3.1 percent of the population; by 1995 the seropositivity rate had decreased to 1.3 percent, 1 961 000 infected individuals. Between 1975 and 1995, vector control prevented 277 000 new infections and 88 000 deaths while the expenditures prevented (expected benefits) were US$847 million. The loss of 1 620 000 DALYs, 41 percent from averted deaths and 59 percent from averted disability (Akhavan 2000), was also prevented. These data indicate the benefit of vector control. Sustained vector-control activities were responsible for the interruption of *T. cruzi* vectoral transmission in Uruguay in 1997, Chile in 1999, and in nine of the 11 endemic states of Brazil in 2000–02.

Leishmaniasis

Leishmaniasis is considered to be endemic in Africa, in most countries of the Americas and Asia, and in many countries of Europe. Most cases of cutaneous leishmaniasis (more than 90 percent) occur in Afghanistan, Brazil, Iran, Peru, Saudi Arabia, and Syria. More than 90 percent of visceral leishmaniasis cases occur in Bangladesh, India, Nepal, and Sudan (WHO 1994b). In all its forms, there are 12–13 million cases of leishmaniasis worldwide; the annual incidence is 600 000 new clinical cases, and 350 million people are at risk of infection. The development of new territories, explosive migration, rapid urbanization and the associated environmental changes, are expanding the areas where human cutaneous leishmaniasis commonly occurs. Leishmaniasis is transmitted by sandflies and the clinical manifestations of infection vary depending on the species of *Leishmania* involved (WHO 1990a).

Cutaneous leishmaniasis is characterized by skin ulcers that heal slowly and leave scars. The species causing cutaneous leishmaniasis in the Old World are *L. tropica*, *L. major*, and *L. aethiopica*; the latter is also the etiological agent of diffuse leishmaniasis. In the New World, cutaneous leishmaniasis is mainly caused by *L. braziliensis*, *L. mexicana*, *L. amazonensis*, *L. venezuelensis*, *L. panamensis*, and *L. guyanensis*. *L. braziliensis* and *L. panamensis* are mainly responsible for mucocutaneous leishmaniasis, a condition in which cutaneous lesions spread to the mucosa of the mouth and nose and *L. mexicana* and related forms are responsible for diffuse cutaneous leishmaniasis in the New World (WHO 1990a; Rey 1991).

Cutaneous leishmaniasis is not usually life threatening, but clinical visceral leishmaniasis (kala azar) is often fatal if untreated. Visceral leishmaniasis is caused by *L. donovani* and *L. infantum* in the Old World and *L. infantum chagasi* in the New World. Although most infections are asymptomatic or subclinical, those associated with malnourishment often result in severe disease. In the Indian subcontinent, 400 000 new cases are estimated to have occurred each year with a case fatality rate of 5–7 percent. Large epidemics of kala azar

occurred in India in 1978, 1982, and 1987 (WHO 1990a). An epidemic in Southern Sudan was first recognized in mid 1988 and, by 1993, 300 000–400 000 people were at risk of infection and 40 000 were thought to have died. In some villages, the population decreased by 30–40 percent as a consequence of the epidemic (WHO 1993a). Overall it is estimated that visceral leishmaniasis kills at least 50 000 people per year.

Prevention of cutaneous leishmaniasis is based on:

- case detection and treatment
- the destruction of sandflies and their breeding places
- the prevention of contact between humans and sandflies.

In addition to these measures, the canine reservoir must also be eliminated to prevent visceral leishmaniasis in the Mediterranean and American regions. Treatment costs (drug only) for cutaneous and mucocutaneous leishmaniasis vary from US$60 to 70 per person, in different regions. Control costs for visceral leishmaniasis are US$100 per person. When other related costs are included, the cost of each case of visceral leishmaniasis amounts to US$250. When treatment with amphotericin is needed (because of failure of other drugs to bring about cure) the cost of each case increases by US$45 (WHO 1990a, 1993a).

An epidemic of visceral leishmaniasis in Sudan provided an opportunity to obtain accurate data on excess mortality, the costs of intervention measures and their effects, thus making it possible to express the cost-effectiveness of intervention as the cost per DALY averted. The cost-effectiveness ratio of US$18.40 per DALY places the treatment of visceral leishmaniasis in Sudan among health interventions considered 'very good value for money' (i.e. interventions of less than US$25 per DALY) (Griekspoor et al. 1999).

Malaria

Malaria is the most important protozoal disease affecting humans. People suffering from malaria mostly live in poverty in rural areas where *Anopheles* mosquitoes are present. Over 40 percent of the world's population remains exposed to varying degrees of risk from malaria and one or more types of malaria occur in 90 regions of the world (WHO 1994c, d, e). The global incidence of malaria is c. 120 million clinical cases each year, the parasite being carried by 300–400 million people. Countries in tropical Africa account for more than 80 percent of all clinical cases and more than 90 percent of all carriers (WHO 1992a, 1995a). The reporting of incidence figures for large regions tends to mask trends within countries and among countries in a region. The malaria situation has improved in some countries but has deteriorated in others. In India, the reduction in reported malaria cases from 9 million to 5 million between 1976 and 1984 was largely due to a return to levels that had occurred before a peak just prior to 1976.

Some areas are particularly susceptible to epidemics of malaria. These include the 'frontier areas' in South East Asia and South America that have experienced substantial population increases and are now being rapidly developed. In the highly endemic areas of Africa, there has been little change in levels of infection, but epidemics associated with weather and population movement have occurred in areas of low endemicity (WHO 1992a). Worldwide distribution of malaria and major problem areas are shown in Figure 2.5 (WHO 1993b). Of the 5.3 million cases reported to the World Health Organization in 1992 (excluding the African Region), 75 percent were concentrated in six countries: Afghanistan, Brazil, Colombia, India, Sri Lanka, and Vietnam (WHO 1994c). Within these countries, malaria is concentrated in certain areas (WHO 1992a). In 1994, 27 million cases were reported from Africa, 1.1 million from the Americas, 3.5 million from South East Asia, 2.1 million from the western Pacific, 321 000 from the eastern Mediterranean Region, and 91 000 from Europe (WHO 1997a).

Severe and fatal malaria infections are usually caused by *P. falciparum*, the predominant species of malaria in tropical Africa *P. falciparum* is less common in the rest of the world. For certain areas, including 'frontier areas' and areas with civil war or other conflicts, illegal trade and mass movements of refugees, malaria mortality is probably greatly underestimated (WHO 1994c). For example, 1 428 deaths from malaria were reported from the Americas in 1986 but the annual malaria mortality for the Brazilian Amazon region alone (a 'frontier area') was between 6 000 and 10 000. Implementation of early diagnosis and prompt treatment through strengthening the local health services, has drastically decreased mortality by falciparum malaria in the Amazon Region of South America.

The vast majority of deaths from malaria occur in Africa, where the death toll so caused is now estimated to be at least 1 million per year in children (WHO 1993b). Between 1970 and 1975, malaria was responsible for 20–30 percent of infant mortality in Kenya and Nigeria and, in the Gambia, mortality from malaria was 6.3 per 1 000 per year in infants and 10.7 per 1 000 per year in children under 4 years of age (WHO 1992a).

Resistance of *P. falciparum* to the antimalarial drug chloroquine was detected almost simultaneously in Brazil, Colombia, Venezuela, and Thailand in the late 1950s and the 1960s. Several countries have now reported the widespread presence of *P. falciparum* strains resistant not only to chloroquine but also to amodiaquine, pyrimethamine in combination with long-acting sulfonamides, and to other drugs. The degree of resistance varies from country to country (López-Antuñano and Schmunis 1993). Resistance to *P. vivax*

to chloroquine has been shown in Indonesia, Papua New Guinea, Myanmar, and Vanuatu (WHO 1997a).

P. falciparum has a high prevalence in areas that border jungles where the population is very mobile owing to expansion of agricultural and mining activities. In South America, these areas also have a high frequency of infection with *P. falciparum* strains resistant to antimalarial drugs. In such areas, migration favors contact between carriers and others who are susceptible to the infection. Exposure of large numbers of susceptible people to infected vectors gives rise to epidemics with high rates of morbidity and mortality. In these areas, the breakdown of the social structure, the absence of health service facilities, and the resultant failure of diagnosis and treatment of infected people adds to the risk of infection and death from malaria.

In many of the nations that are promoting national development and experiencing expanding populations, new roads are opened into jungle areas that are undergoing agricultural development and from which natural resources are being extracted. One of the detrimental consequences of such development is an increase in the incidence of malaria, mainly because of internal migrations. The destinations of the migrants are often determined by roads planned by development officers in remote cities without any thought for public health considerations. The migrants use the new roads to penetrate into jungle areas where disease vectors abound; here they often construct temporary dwellings that lack even minimum sanitary facilities. These areas often lack basic health and education programs as well as political and administrative organization, leaving the impoverished migrants vulnerable to disease. Some types of development program are particularly likely to increase the risk of malaria; for example, flood-based irrigation systems used for growing rice and banana crops. Poorly built roads, dams, and irrigation canals can also produce anopheline breeding sites, particularly during long periods of rain.

For most countries, the cost of malaria has been enormous. In the USA, it was estimated to be US$500 million in 1938 (Williams 1938). In Greece prior to the 1940s, the annual costs of malaria were huge: US$12 million for the loss of working days and earnings (using the minimum wage for the calculations), US$3.5 million for treatment and US$7.5 million for deaths (Livadas and Athanassatos 1963). In India, annual costs were US$50 million, US$60 million, and US$340 million, respectively (Sinton 1938). In the 1950s, lost wages amounted to US$60 million per year in Indonesia and US$14 million per year in the Philippines, where treatment costs were US$25 million per year and costs due to deaths were US$15 million per year (Wernsdorfer and Wernsdorfer 1988). The cost of deaths associated with malaria was estimated at US$157 million in Mexico for the period 1949–1963 (Wernsdorfer and Wernsdorfer 1988). In Paraguay, the economic potential

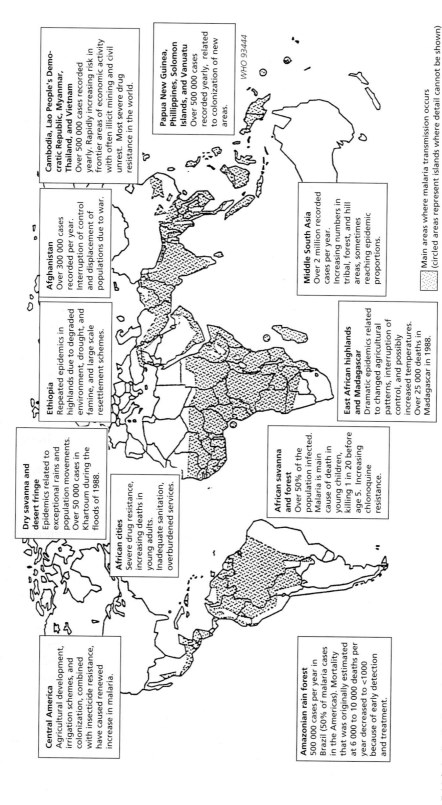

Central America
Agricultural development, irrigation schemes, and colonization, combined with insecticide resistance, have caused renewed increase in malaria.

Dry savanna and desert fringe
Epidemics related to exceptional rains and population movements. Over 50 000 cases in Khartoum during the floods of 1988.

African cities
Severe drug resistance, increasing deaths in young adults. Inadequate sanitation, overburdened services.

Ethiopia
Repeated epidemics in highlands due to degraded environment, drought, and famine, and large scale resettlement schemes.

Afghanistan
Over 300 000 cases recorded per year. Interruption of control and displacement of populations due to war.

Cambodia, Lao People's Demo-cratic Republic, Myanmar, Thailand, and Vietnam
Over 500 000 cases recorded yearly. Rapidly increasing risk in frontier areas of economic activity with often illicit mining and civil unrest. Most severe drug resistance in the world.

Papua New Guinea, Phillippines, Solomon Islands, and Vanuatu
Over 500 000 cases recorded yearly, related to colonization of new areas.

Amazonian rain forest
500 000 cases per year in Brazil (50% of malaria cases in the Americas). Mortality that was originally estimated at 6 000 to 10 000 deaths per year decreased to <1000 because of early detection and treatment.

African savanna and forest
Over 50% of the population infected. Malaria is main cause of death in young children, killing 1 in 20 before age 5. Increasing chlonoquine resistance.

East African highlands and Madagascar
Dramatic epidemics related to changed agricultural patterns, interruption of control, and possibly increased temperatures. Over 25 000 deaths in Madagascar in 1988.

Middle South Asia
Over 2 million recorded cases per year. Increasing numbers in tribal, forest, and hill areas, sometimes reaching epidemic proportions.

WHO 93444

Main areas where malaria transmission occurs
(circled areas represent islands where detail cannot be shown)

Figure 2.5 Malaria distribution and problem areas (reproduced with permission from WHO 1993b).

of a family with malaria was 74–86 percent of the potential of families without the disease (Conly 1975). In Peru in 1998, there were 77 000 cases of malaria with 43 deaths and the cost to the Peruvian economy was US$37.9 million that year, 1.5 percent of the total health expenditure. The government cost was US$ 9.3 million, 1 percent of the public health expenditure, The costs to families, who are mostly poor, amounted to US$26.6 million, averaging US$475 per family. The cost to families per malaria case was US$95, each family having an average of five cases (Ministry of Health 2000).

High malaria prevalence with its concomitant high social and economic costs interferes with the productivity and stability of agricultural settlements (Sawyer 1993). In Africa, the estimated average time lost by adults due to one episode of malaria were 6.2 days, 4.3 days, and 4 days in Sudan, Sri Lanka, and Burkina Faso respectively; in Kenya, Malawi, Nigeria, and Rwanda, on the other hand, it was 2–3 days (WHO 1999). In 1985 in Africa, a malaria episode in a person cost US$1.80 directly and US$9.80 indirectly. Another study showed that the cost of providing treatment was US$0.21 per child and US$0.63 per adult. The direct expenditures incurred as a result of malaria by low-income and high-income households during a year in Malawi were US$19.13 and 19.84, respectively and the indirect costs were US$2.13 and 20.61, respectively (Ettling et al. 1994). In the sub-Saharan countries only, the direct cost of malaria is estimated to be US$500 million annually (WHO 1999). Thus, it is not surprising that direct and indirect costs of malaria in Africa as a whole were estimated as US$800 million in 1987 (WHO 1993b) and were predicted to reach US$1.8 billion by 1995 (WHO 1995a).

The high costs caused by malaria are not limited to Africa. In Thailand in 1952, at the beginning of the intensification of the antimalarial program, the loss in agricultural production caused by malaria was US$1.52–1.90 million. Fourteen years later, this figure had decreased to US$0.30–0.37 million per year (Kühner 1971). The cost of the malaria control program was US$17 million in 1980 (Kaewsonthi and Harding 1984). Resistance to DDT and other insecticides and to chloroquine and other drugs has significantly increased the costs associated with malaria prevention and control.

In areas where the population is relatively stable, malaria can often be controlled by vector source reduction and the protection of households from vectors by various methods including indoor spraying with residual insecticides and insecticide-impregnated bednets. For migrant populations, the best approach for prevention and control is promotion of self-protection; for example, using screened tents, insecticide-impregnated bednets, and insect repellents. The efficacy and relative cost-effectiveness of insecticide-treated bednets for the control of malaria in children under 5 years of age have

recently been demonstrated by several large-scale trials (WHO 2000b). However, it has been suggested that long-term use of insecticide-treated bednets in areas of high transmission could lead to mortality rebound in later childhood. This would reduce the cost-effectiveness of the intervention and, in the extreme, could lead to negative overall effects (Coleman et al. 1999).

In all situations, early detection and treatment will decrease morbidity and prevent mortality. Selection of the appropriate control measures requires assessment of the risk factors present in the area. In most cases, it is necessary to use several control measures in an integrated fashion (López-Antuñano and Schmunis 1993).

Among infants, severe anemia and malaria are often associated. The cost and effectiveness of control measures to prevent both anemia and malaria have been evaluated based on data from the expanded program on immunization (EPI) and not on the malaria control program. For severe anemia, and from the perspective of the health provider, the cost-effectiveness ratios were, respectively, US$8, 9, and 21 per DALY for malaria chemoprophylaxis with Deltaprim (a combination of 3.125 mg pyrimethamine and 25 mg dapsone) + iron, Deltaprim alone, or iron supplementation alone. For malaria prevention, Deltaprim + iron cost US$9.7 per DALY and Deltaprim alone cost US$10.2 per DALY. The cost-effectiveness ratios ranged from US$9 to 26 for severe anemia prevention and from US$11 to 12 for the prevention of clinical malaria. These measures were highly cost-effective, as defined by the World Bank's proposed threshold of less than US$25 per DALY for comparative assessments. Furthermore, all the preventive interventions were less costly than current malaria and anemia control strategies that rely on clinical case management. This economic analysis supports the inclusion of both malaria chemoprophylaxis and iron supplementation delivered through the EPI as part of the control strategies for these major killers of infants in parts of sub-Saharan Africa (Alonzo Gonzalez et al. 2000).

The success of malaria control programs has been uneven in some countries of the Americas. Such is the case in Brazil: malaria transmission was controlled elsewhere in the country by 1980 but not in the Amazon Basin where cases increased steadily until 1989 to almost half a million a year, and the coefficient of mortality quadrupled during the period 1977–1988. A reorganization of program activities from 1989 to 1996 facilitated a change towards earlier and more aggressive case treatment and more concentrated vector-control measures. The epidemic stopped expanding during 1990–1991 and reversed during 1992–1996. The total cost of the program from 1989 through mid-1996 was US$616 million: US$526 million for prevention and US$90 million for treatment. Compared to what would have happened in the absence of the program, nearly 2 million cases of malaria and 231 000 deaths were prevented; the lives saved were almost equally due to preventing infection and to case treatment. Converting the savings in lives and morbidity into DALYs yields almost 9 million DALYs, 5.1 million from treatment and 3.9 million from prevention. Nearly all the gain came from controlling deaths and therefore from controlling *P. falciparum*. The overall cost-effectiveness was US$2 672 per life saved or US$69 per DALY, which is low compared to most previous estimates and compares favorably with many other disease-control interventions. Contrary to much previous experience, case treatment appears to be more cost-effective than vector control, particularly where *P. falciparum* is prevalent and unfocussed insecticide spraying is relatively ineffective. Halting the epidemic, by better targeted vector control and emphasizing treatment, paid off in much reduced mortality from malaria and in significantly lower costs per life saved (Akhavan et al. 1999).

In the Americas, good records of the cost of malaria control programs have been kept for several years and these indicate a continuous increase in the funds spent on antimalarial activities. Unfortunately, funding from international organizations and bilateral agencies has steadily decreased (Figure 2.6) (López-Antuñano and Schmunis 1993; PAHO 1994, 1995). Table 2.6 shows the overall budget for the health sector, the budget devoted to public health and the budget for malaria control programs in countries of the Americas in 1993 (PAHO 1994). From 1996 to 1999, expenditures by the 21 American countries with active control programs were US$263 million from the national budget and US$23 million from external contributions or loans. In 2000, countries spent US$105 million on malaria; US$97 million from the national budget and US$8.2 million from external contributions or loans (Organizacion Panamericana de la Salud 2000). Despite high expenditures for malaria control, the problem of malaria in the Americas remains unsolved.

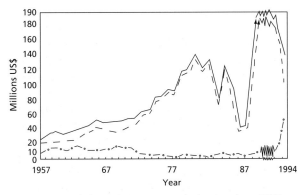

Figure 2.6 *Funds for malaria programs in the Americas, 1957–93. Total funds ——; funds provided by local governments - - -; funds provided by international or bilateral aid agencies -.-. Data from López-Antuñano and Schmunis 1993; PAHO 1994, 1995.*

Intestinal protozoa

Intestinal protozoa are the etiological agents of several widespread parasitic diseases, the most common of which are caused by *Entamoeba histolytica* and *Giardia duodenalis* (also known as *G. lamblia*).

Amebiasis, caused by *E. histolytica*, is a cosmopolitan infection transmitted by the fecal–oral route. Its greatest impact is in Africa and Asia. In Africa, Egypt, Morocco and countries located between 10°N and 10°S are severely affected. Prevalence is high in Asia, particularly in Bangladesh, Myanmar, China, India, Iraq, the Republic of Korea, and Vietnam. The amebae in these countries are highly pathogenic. Amebiasis is also a problem in Mexico and other Latin-American countries. On the other hand, in Europe and the USA, amebic infection is often asymptomatic or benign in spite of prevalence rates as high as 2–5 percent (WHO 1981). In 1984, it was estimated that 500 million people worldwide were infected with *E. histolytica* and that 40–50 million individuals had clinical symptoms of amebiasis, including diarrhea, dysentery, and liver disease. Worldwide amebiasis causes 40 000–100 000 deaths per year (WHO 1981, 1987, 1993c) and in 1996 was responsible for 48 000 cases (incidence) and 70 000 deaths (WHO 1997b).

The highest prevalence of *G. duodenalis* occurs in the tropics and subtropics where sanitation is poor. Transmission may be direct by ingestion of feces or indirect through ingestion of contaminated water or food. Travelers to tropical Africa, Mexico, Russia, South East Asia, southern Asia, and western South America are at a high risk of acquiring giardiasis (Wolfe 1992). Giardiasis may infect 200 million people worldwide and may produce symptoms in 500 000 individuals every year (Walsh and Warren 1979; WHO 1987, 1993c, 1997b). In developing countries, *G. duodenalis* is one of the first pathogens to infect infants and peak prevalence rates of 15–20 percent occur in children under 10 years old (Hill 1993). In developed countries, outbreaks frequently occur in child-care settings (Thompson 1994) and in adults who drink contaminated water. From 1977 to 1988, giardiasis was the leading cause of outbreaks of water-borne disease in the USA, causing an estimated 4 600 hospital admissions annually (Lengerich et al. 1994). Homosexual males constitute a special risk group. *Giardia duodenalis* may cause acute or chronic diarrhea, steatorrhea, loose stools, malabsorption of fat (Benenson 1990), and growth retardation and malnutrition in children (Farthing et al. 1986). Its major impact is on children under 3 years old, the undernourished, and the immunocompromised. It is also a cause of morbidity in adults (Farthing 1993).

Prevention of amebiasis and giardiasis is based on the implementation of good personal hygiene, appropriate disposal of human waste, and adequate handling and treatment of water supplies. The combination of filtra-

Table 2.6 *Funds assigned to the health sector, public health, and malaria prevention and control programs in the Americas 1993*

Country (geographical sub-region)	Budget for health sector (US$)[a]	Budget for public health (US$)[a]	% assigned for public health	Budget for malaria programs (US$)[a]	% of health sector	Loans or grants for malaria programs (US$)[a]
Mexico	13 376 794 032	1 494 709 452	11.17	28 441 613	0.21	...
Belize	9 717 172	1 655 934	17.04	477 919	4.92	100 000
Costa Rica	1 166 108 757	50 479 687	4.33	1 714 017	0.15	344 310
El Salvador	887 956 536	84 883 721	9.56	1 220 930	0.14	1 023 255
Guatemala	134 204 319	93 324 434	69.54	2 434 719	1.81	166 985
Honduras	805 829 807	83 904 586	10.41	2 016 013	0.25	283 072
Nicaragua	...	...	...	...	...	301 647
Panama	344 163 190	172 081 595	50.00	3 719 976	1.08	71 000
Haiti	973 600	...	...	20 000	2.05	250 000
Dominican Republic	...	...	...	599 334	...	517 815
French Guiana	...	...	...	538 535	...	...
Guyana	...	...	...	91 973	...	...
Brazil	...	1 130 947 000	...	97 124 000	...	5 500 000
Bolivia	96 404 722	35 846 246	37.18	187 066	0.19	...
Colombia	...	738 829 332	...	13 524 381	...	...
Ecuador	204 632 124	100 307 839	4.90	4 963 244	0.24	...
Peru	5 380 095 982	...	...	...	...	...
Venezuela	10 735 841 686	851 385 818	7.93	6 976 914	0.06	4 600 000
Argentina	...	...	...	1 826 000	...	...
Paraguay	225 233 859 410	83 547 789 813	37.09	6 405 522	0.003	...
Total	260 218 581 338	88 386 145 456	33.97	172 282 156	0.066	13 158 084

Reproduced from PAHO 1994 with permission.
a) conversion based on United National exchange rates at 31 December 1993.
... No data.

tion and chlorination usually eliminates these organism from water, but higher concentrations of chlorine and longer exposure times are needed than those required for the elimination of bacteria. Boiling is an effective method of purifying drinking water for personal use.

There are few estimates of the cost of amebiasis, but one calculation suggested that in Mexico in 1984, expenditure for the treatment of invasive amebiasis was 1.6 percent of the budget of the Health Ministry (WHO 1987). In the developing countries, the costs associated with giardiasis are difficult to estimate because there are no hard data on the number of cases of diarrhea induced by giardiasis. An estimate made on the basis of the number of admissions to hospitals for giardiasis in the USA suggests direct costs of US$3–5 million annually.

THE HELMINTH WORMS

Dracunculiasis (Guinea worm disease)

Dracunculiasis occurs in 16 sub-Saharan countries in Africa, as well as in India and Pakistan. It afflicts the ten African countries considered to be among the least developed countries of the world, where the gross domestic product ranges from US$450 to 550 per inhabitant per year. The number of cases reported in the 1980s worldwide was 5–10 million and this decreased to 3 million in 1992, to 229 773 in 1993, and to 164 973 in 1994. Niger, Nigeria, and Sudan accounted for two-thirds of all cases from Africa (WHO 1993c, 1995c). During 1998, there were almost 72 000 cases reported from Africa, the most affected countries being Ghana, Nigeria, and Sudan (WHO 1999).

Dracunculiasis causes disability as a result of the migration of the parasite through the host. This migration results in localized pain, mainly in the areas around the joints. When the worm emerges through the skin, the pain at the site is severe and is accompanied by general symptoms of fever, joint pain, nausea, and vomiting. There is a painful edema followed by a blister and then an ulcer at the site of penetration. Extraction of the worm may cause disability for 2–4 weeks.

The success of the eradication campaign now underway in many countries is based on the provision of potable drinking water through the construction of safe wells. In infected areas, prevention is implemented through health education programs that convey the message that drinking contaminated water is the origin of the disease, that individuals with open blisters and ulcers should not be permitted to come into contact with any source of drinking water, and that water must be treated (by boiling, chlorinating or filtering) to eliminate the larvae of the crustaceans that act as the intermediate hosts. In Nigeria, control of the intermediate host costs US$3.00 per household for water filters and US$1 per capita for cement ring wells. This amounts to

c. 30 percent of the estimated potential losses suffered in a season due to incapacity caused by the disease (Guyatt and Evans 1992).

Filariasis

About 750 million people in 76 countries live in areas of endemic filariasis, an estimated 90 million of them are infected with filariae. Of these, 72.8 million are infected with *Wuchereria bancrofti* and 5.8 million are infected with a variety of species of *Brugia* (WHO 1993c). There are 300 000 individuals infected with filariae in seven countries of the Americas; 48.6 million in eight countries of South East Asia (4.8 million infected with *Brugia*); 3.9 million in 14 countries or territories of the western Pacific (1 million infected with *Brugia*); and 25.6 million in 40 countries in Africa (WHO 1992b). Almost 40 million people worldwide suffer from lymphedema, elephantiasis or hydrocoele caused by bancroftian filariasis, 14 million of them in Africa and 21 million in South East Asia. Worldwide, *Brugia* is believed to cause lymphedema and elephantiasis in 2.8 million people of whom 1 900 000 are in South East Asia and 900 000 are in the western Pacific (WHO 1994f).

Clinical manifestations of bancroftian and brugian filariasis vary widely in different geographical areas from asymptomatic microfilaremia and malaise to acute symptoms characterized by adenolymphagitis with fever, hydrocoele, lymphedema, elephantiasis, and chyluria, the last four being characteristic of chronic bancroftian filariasis (WHO 1992b). In India, it is estimated that an infected person suffers 4.47 acute attacks per year if they have bancroftian filariasis and 2.20 acute attacks per year if they have brugian filariasis. Each attack has an average duration of 4 days and the numbers of working days lost per year per person are 17.43 and 8.91, for bancroftian and brugian filariasis, respectively (WHO 1992b).

The vectors of filariasis are *Culex*, *Anopheles*, *Aedes*, and *Mansonia* mosquitoes. Slum dwellers with poor housing and sanitation are at the highest risk of bancroftian filariasis transmitted by *C. quinquefasciatus* whereas the rural poor suffer more from filariasis transmitted by other mosquitoes. Control measures are directed towards reduction of vector density and interruption of contact between human and vector by the application of a combination of chemical or biological control measures, environmental management, and personal protection measures (WHO 1992b). Treatment is another factor in control and involves annual administration of a single dose of diethylcarbamazine (DEC) or a combination of DEC and ivermectin (WHO 1994g). Ivermectin is the drug of choice for both individual treatment and mass-treatment programs for the control of onchocerciasis (Ottesen 1994) and a single dose can also be used for treatment or as a control measure for lymphatic filariasis (Brown et al. 2000). The combinations of DEC plus ivermectin or ivermectin plus albendazole caused a decrease in microfilaremia greater than that achieved with either drug alone (Ismail et al. 1998; Beach et al. 1999). Thus, combinations of DEC or ivermectin with albendazole could be effective for mass treatment of lymphatic filariasis and at the same time would decrease the burden produced by soil-transmitted helminth infections. Control costs vary with the scope of coverage and the strategy or combination of strategies used. Control costs are US$1.8 million in China; US$1 million in India and Thailand; US$500 000 in Egypt and Malaysia; and from US$27 000 to 300 000 in French Polynesia, Papua New Guinea, and Sri Lanka (WHO 1994g).

Between 75 and 122.9 million people are at risk of onchocerciasis (river blindness), which is endemic in 28 countries in Africa and six countries in the Americas. In Africa, 17.7 million people are infected with *Onchocerca volvulus*, of whom 500 000 have damaged vision and 267 000 are blind. In the Americas, 140 000 people are infected with *Onchocerca* of whom 750 are blind (WHO 1995d). Many individuals infected with *O. volvulus* are asymptomatic. Symptoms of acute dermal or ocular infection are rare clinical manifestations of chronic established infection, the most severe consequences of which are reduction of the visual field and blindness. The impact of the disease is seen mainly in sub-Saharan countries where, in some areas, up to 40 percent of adults may have severe visual impairment. This results in an inability of the family to support itself and a decrease in agricultural output that leads to unstable communities (WHO 1995d).

Control activities were originally based on control of the vector of onchocerciasis, the blackfly *Simulium*, but are now based on yearly administration of single doses of ivermectin to those at risk. Evaluation of a community-directed, ivermectin-treatment program in Uganda indicated that desired treatment coverage of the population eligible to take the drug, was achieved in 42.6–51.0 percent of the communities. The annual cost per person treated with ivermectin was higher in the districts with small populations (fewer than 15 000) than in those with large populations (more than 40 000), US$0.40 as opposed to US$0.10 or less (Katabarwa et al. 1999). The costs of ivermectin distribution vary between US$0.10–5 per person per year (WHO 1995d). Implementation of control measures has been successful in many areas and has resulted in interruption of transmission in 11 countries of West Africa. Levels of infection in children have been significantly reduced and the numbers of infected people decreased from 1 million in 1975 to 10 000 in 1992. Since the inception of the prevention program, 100 000 people have avoided blindness.

Food-borne trematode infections

Food-borne trematode infections have their highest prevalence in South East Asia and in the western Pacific; they are also present in areas of Africa, the Americas, and Europe. The geographical areas involved are expanding as are the populations at risk as a result of improvements in transportation that favor population movements and trade. It is estimated that food-borne trematode infections affect 40 million people throughout the world (WHO 1995b). The food-borne parasites of greatest public health importance are *Fasciola*, *Clonorchis*, *Paragonimus*, and *Opisthorchis*.

Human fascioliasis, caused by *F. hepatica* and *F. gigantica*, has been reported from Bolivia, China, Ecuador, Egypt, Iran, Peru, Portugal, and Spain; and 2.4 million people are infected. Clonorchiasis has been observed in China, Hong Kong, Macao, Republic of Korea, Russia, and Vietnam and estimates suggest that 7 million people are infected. Opisthorchiasis, caused by *O. viverrini* and *O. felinus*, occurs in Kazakhstan, Laos, Russia, Thailand, and Ukraine; 10 million people are infected. Of less importance from the public health viewpoint are the intestinal flukes that cause echinostomiasis, fasciolopsiasis, heterophyiasis, metagonimiasis, and nanophyetiasis. These infections are found in China, Egypt, Japan, Republic of Korea, Russia, and Thailand and c. 1.2 million people are infected (WHO 1995b).

The epidemiology of these diseases is determined by ecological and environmental factors. Aquatic or semi-terrestrial snails are the intermediate hosts and the reservoir hosts are fish, crustaceans or mammals. Human infection occurs from ingestion of raw or uncooked fish, shellfish or aquatic plants that harbor the infective metacercariae (Figure 2.7). In some areas, industrial production of cultured fish and shellfish may be related to the spread of food-borne trematodes (Benenson 1990; WHO 1995b).

To prevent human *Fasciola* infection, infection of livestock must also be controlled, as humans are only accidental hosts. The consumption of watercress or other aquatic plants from areas where sheep or cattle graze should be avoided, as should be the use of animal feces for fertilizing water plants. Mollusk hosts should be eliminated when possible. The best method for prevention of infection with *Clonorchis* and *Opisthorchis* is to avoid eating raw fish and to use sanitary methods for disposal of feces. Infection with these parasites is serious as they not only produce liver disease but also increase the risk of cholangiocarcinoma. *P. westermani* and several other related species cause lung fluke disease. These parasites have freshwater crabs as their intermediate hosts and prevention is effected by thoroughly cooking crustacea, by disposal of sputum and feces in a sanitary manner and by snail control when feasible (Benenson 1990). In migrants to developed nations, lung fluke infection is often misdiagnosed as tuberculosis. Programs for control of food-borne trematodes should be multisectoral, integrating the activities related to human health, food safety, aquaculture, agriculture, and education (WHO 1995b).

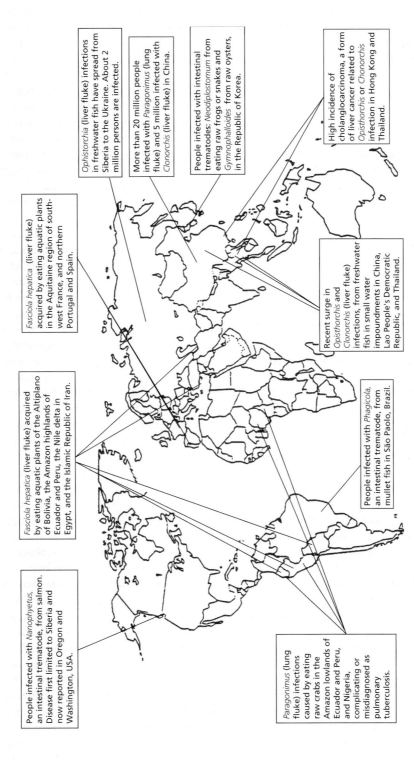

People infected with *Nanophyetus*, an intestinal trematode, from salmon. Disease first limited to Siberia and now reported in Oregon and Washington, USA.

Fasciola hepatica (liver fluke) acquired by eating aquatic plants of the Altiplano of Bolivia, the Amazon highlands of Ecuador and Peru, the Nile delta in Egypt, and the Islamic Republic of Iran.

Fasciola hepatica (liver fluke) acquired by eating aquatic plants in the Aquitaine region of south-west France, and northern Portugal and Spain.

Ophistorchia (liver fluke) infections in freshwater fish have spread from Siberia to the Ukraine. About 2 million persons are infected.

More than 20 million people infected with *Paragonimus* (lung fluke) and 5 million infected with *Clonorchis* (liver fluke) in China.

People infected with intestinal trematodes: *Neodiplostomum* from eating raw frogs or snakes and *Gymnophalloides* from raw oysters, in the Republic of Korea.

High incidence of cholanglocarcinoma, a form of liver cancer related to *Opisthorchis* or *Chonorchis* infection in Hong Kong and Thailand.

Recent surge in *Opisthorchis* and *Clonorchis* (liver fluke) infections, from freshwater fish in small water impoundments in China, Lao People's Democratic Republic, and Thailand.

People infected with *Phagicola*, an intestinal trematode, from mullet fish in São Paolo, Brazil.

Paragonimus (lung fluke) infections caused by eating raw crabs in the Amazon lowlands of Ecuador and Peru, and Nigeria, complicating or misdiagnosed as pulmonary tuberculosis.

Figure 2.7 Food-borne trematode infections: the global distribution is changing with the environment and with changes in human behavior (reproduced with permission from WHO 1995b).

The economic burden of these diseases stems from a number of direct and indirect costs associated with the infections, including morbidity, loss of productivity, absenteeism, and cost of healthcare in infected people as well as losses in agricultural and aquacultural enterprises. The wages lost annually in Thailand as a result of opisthorchiasis in those aged 15–60 years are estimated at US$60 million and the cost of medical care is US$19.4 million. In addition, the cost of control for a 6-year period (1988–1993), with a coverage of 3 million people, was US$8.3 million, including costs for fecal examinations, drugs, and the promotion and training of personnel for community participation in control programs. In the Republic of Korea, control programs provide medical examinations and treatment for 15 million people and cost an estimated US$2.5 million per year (WHO 1995b).

Intestinal worms: *Ascaris*, hookworm, and *Trichuris*

Ascaris lumbricoides infections have been reported during the last few years from more than 150 of the 208 states and countries of the world. Hookworms and the whipworm, *Trichuris trichura*, are as widely distributed as *Ascaris* (Bundy and Cooper 1989; WHO 1987). These parasites are soil-transmitted helminths of global importance. Strongyloidosis has a patchy global distribution and is less prevalent than ascariasis. *Ascaris* has high prevalence in the subtropical and tropical areas of Africa, Asia, and the Latin-American countries. Although common in humid parts of the world, *Trichuris* is also found in temperate climates. Hookworm infection has almost been eradicated from Europe and the USA. The hookworm *Necator americanus* occurs in the Americas, equatorial Africa, southern Asia, South East Asia, Polynesia, and Australia. *Ancylostoma duodenale*, another hookworm, is more common in Africa and in northern and south-western Asia. The ranges of both species of hookworm (and of other parasites of global importance) often overlap.

During 1977–1978, 800–1 000 million people were infected with *Ascaris* and, of these, 1 million had symptoms; hookworm infected 700–900 million people, of whom 1.5 million had symptoms; and *Trichuris* affected 500 million people of whom 100 000 had symptoms. The number of deaths due annually to *Ascaris* and hookworms in this period were approximately 20 000 and 50 000, respectively (WHO 1993c). Subsequently, it was reported that *Ascaris* and *Trichuris* cause clinical symptoms in 214 and 133 million people, respectively, and that hookworm causes clinical symptoms in 96 million (WHO 1995a). More recent estimates for *Ascaris* are 1 400–1 500 million infected, 120–220 million cases (Albonico et al. 1999) and 4 000 deaths (WHO 2002). An estimated 1 049 million people harbor *Trichuris* (Crompton 1999); 114 million of them are preschool

children and 233 million are children aged 5–14 years (Chan 1997). It is estimated that 27 million school-age children are infected in sub-Saharan Africa, 7 million in India, 42 million in China, 70 million in the rest of Asia, 39 million in Latin America and the Caribbean, and 18 million in the Middle-Eastern Crescent (Stephenson et al. 2000). Deaths attributed to *Trichuris* and hookworm were 2 000 and 4 000, respectively (WHO 2002).

The impact of these parasites on public health is related to the severity of the infection; heavy infections cause malabsorption and large blood and protein losses that are compounded in the context of poor nutritional status. Heavily infected people are also at risk of developing complications such as intestinal obstruction or anemia. The impact of these parasitic diseases is difficult to evaluate, as they may cause effects such as retarded physical growth, retarded development of cognitive skills, low participation in education, and poor performance (Evans and Jamison 1994). None of these effects is easy to isolate and measure routinely. Although there is a relationship between parasitic infections and human malnutrition, it is sometimes difficult to determine which is the primary factor because there are similarities in the socioeconomic conditions in which these conditions both occur (Crompton and Nesheim 1982).

Ascaris is mainly prevalent in young people (i.e. children aged 6–12 years). *Trichuris* infections peak in early life and then generally remain steady, whereas hookworm prevalence peaks in adolescence. Heavy worm burdens of *Ascaris* and *Trichuris* are found in children of primary school age (WHO 1981, 1987).

In many communities, 15 percent of the infected population harbor more than 60 percent of the worms. As morbidity is usually related to intensity of infection, heavily infected individuals have a greater risk of developing disease symptoms, and they also cause the most contamination of the environment (Anderson 1986). In addition, individuals with a heavy worm burden are more likely to re-acquire heavy infection after treatment and to suffer from polyparasitism (Anderson and Medley 1985; Bundy and Cooper 1989). The growth and nutrition of children are adversely affected by *Ascaris* and *Trichuris* infection and there are remarkable improvements in appetite, food intake, digestion, absorption, and growth in infected children who receive treatment (Cooper et al. 1990; Stephenson et al. 1989). Hookworm infection often causes iron-deficiency anemia, which may reduce productivity in affected individuals. The infection and the resulting anemia may impair mental development in children, negatively influencing their cognitive performance. Infection may also complicate pregnancy (Stephenson 1984; Prescott and Jancloes 1984; WHO 1987).

The cost of infection by intestinal worms is huge. In the late 1970s, US$4.4 million was annually lost in Kenya in the form of nutrients wasted as food unabsorbed by worm-infested people. In the same country in 1976, the cost incurred by the government for treatment of parasitic intestinal infections was US$339 000 (costs for examining patients and for drugs) and the cost incurred by affected families US$394 000 (cost for drugs, lost wages, and for transportation). These figures seem to be low as 88 804 patients were admitted to hospitals in Kenya that year as a result of ascariasis (Stephenson 1984; WHO 1987).

Programs for prevention of infection with intestinal worms must be based on the use of several strategies, focusing on improvements in basic sanitation, waste disposal, health education, and health services. In the short term, immediate action can be taken by administering the drug albendazole, which decreases morbidity and mortality. One strategy is to select the target population and then treat only those who are infected. This approach requires screening the entire population. It has been estimated that in Rwanda, application of this strategy would cost US$600 000 in salaries, excluding costs for the collection and handling of samples. Treatment of infected individuals (about two-thirds of the population for round worms and one-third for flat worms) would cost US$64 000 per year. An alternative program using chemotherapy (piperazine and niclosamide) on all individuals has been estimated to cost US$440 000 per year (De Schaepdryver 1984). Improving basic environmental sanitation and treatment facilities would be considerably more expensive than either of these policies. Control of these parasites requires a massive effort to raise the standard of living of the affected population if the benefits of any treatment programs are to have a lasting effect.

Schistosomiasis

Schistosomiasis is the most important and prevalent of the water-borne parasitic diseases. It is a major health risk in the rural areas of 74 developing countries. Five species of *Schistosoma* affect humans:

1 *S. mansoni* occurs in 36 countries in Africa, nine in the Americas, and seven in the eastern Mediterranean
2 *S. japonicum* is found in China, Indonesia, the Philippines, and Thailand
3 *S. mekongi* occurs in Cambodia and Laos
4 *S. intercalatum* is present in five countries in Central Africa
5 *S. haematobium* is prevalent in 51 eastern Mediterranean and African countries and in India and Turkey.

The first four of these species cause intestinal schistosomiasis and *S. haematobium* causes urinary schistosomiasis. It is estimated that more than 200 million people residing in rural and agricultural areas are infected with schistosomes and that 500–600 million people are exposed to infection. The estimated mortality was about 200 000 per year (WHO 1985, 1993d, 1994h) but has now decreased to c. 17 000 (WHO 2002).

The prevalence of schistosomiasis within countries varies widely: in some countries, the entire population is exposed; in others, there are only scattered foci of disease. Morbidity also varies widely, the foci of clinical disease usually being highly localized. The incidence of clinical schistosomiasis may be significantly higher in countries where the prevalence of the infection is greater than 40 percent (WHO 1985). In the past 10 years, the incidence and prevalence of schistosome infection have decreased in some areas but the infection has become more widespread in others. Large hydroelectric and agricultural irrigation projects (some involving the creation of artificial water reservoirs) and associated population movements are considered to be responsible for many of the increases.

Rates of infection with schistosomes are higher for men than for women because cultural and social biases in behavior and occupation cause men to be more exposed to contaminated water. Of the individuals infected with *S. haemotobium,* 60–70 percent are aged 5–14 years and this group is also the most heavily infected group in areas where *S. haemotobium* is endemic. Hematuria is present in 80 percent of infected children. The prevalence and intensity of urinary disease is lower in older age groups. In contrast, *Schistosoma japonicum* infection has no typical age prevalence and intensity distribution (WHO 1985, 1993d). In areas endemic for *S. mansoni,* the highest prevalence of infection is in those aged 10–24 years, but the prevalence in older groups may also be high. Usually, the most heavily infected individuals are those aged 10–14 years. A significant proportion of these individuals (those with an egg count of fewer than 800 eggs per gram of feces) have hepatomegaly and splenomegaly (WHO 1985, 1993d). Poverty, ignorance, substandard housing, and poor sanitation are the main factors that maintain these endemic foci.

Several studies have been made on the public-health impact of schistosomiasis. In these studies, the following parameters have been assessed: school performance, physical fitness, growth, productivity, and earnings. Although results have not been conclusive (Tanner 1989) there is consensus that in north-eastern Brazil, Egypt and Sudan there is a severe reduction of the work capacity of the rural population as a consequence of schistosomiasis (WHO 1994h).

The economic impact of schistosomiasis is partly a result of the large number of working days lost per infected person per year. This loss is 4.4 days and 40 days per year per person for those infected with *S. haematobium* and *S. japonicum,* respectively. About 10 percent of people infected with *S. mansoni* develop severe symptoms. The number of days of life lost because of schistosomiasis is c. 600–1 000 per case

(Walsh and Warren 1979). The global burden caused by schistosomiasis was 4 500 000 DALYs in 1990 (World Bank 1993b), but decreased to 1 760 000 DALYs in 2001 (WHO 2002).

The annual per capita cost for control of schistosomiasis depends on the nature of the program used. In Central Liberia, focal application of a molluscicide (niclosamide) was combined with mass chemotherapy. A three-dose metrifonate mass treatment was applied in one area with *S. haematobium* infections only, concurrent metrifonate and niridazole mass treatment in one dose was applied in another area with *S. haematobium* infections only, while in another area with both *S. haematobium* and *S. mansoni*, praziquantel (1 dose) was used. The costs per capita protected were US$3.33 for metrifonate, US$1.53 for metrifonate and niridazole combined, and US$1.67 for praziquantel. These figures do not include costs of parasitological examinations (Saladin et al. 1983). In the case of Saint Lucia, the costs were US$1.10 per person per year for chemotherapy, US$3.70 for snail control, and US$4.50 for a clean water supply (US Congress 1985). Costs per protected person were estimated at US$0.70–3.10 per year in three African countries. This amount may seem little, but it should be taken into account that total health expenditures per person in these countries varied from US$1 to US$3 per person per year (Gryseels 1989). Case detection of infected individuals in Sudan cost US$0.03 per capita per year, just for operating the health services. In Brazil, the costs, including laboratory diagnosis, consultation, and drugs, were US$3.73 (Guyatt and Evans 1992). The introduction of specialized programs, instead of using the general health services for control of schistosomiasis, increased the annual cost per person from US$1.50–6.53 (WHO 1993d). A schistosomiasis control program for control of *S. haematobium* or *S. mansoni* in a rural area with a population of 100 000 will have a 5-year cost of US$3.70–25.91 per capita. The variation is determined by the prevalence, intensity and rate of infection in the population (Rohde 1989).

Miscellaneous worms of regional importance

In addition to the parasites of global importance, there are others of regional or national relevance. These are the roundworms *Capillaria philippinensis*, *Enterobius vermicularis*, and *S. stercoralis*, the cestodes *Hymenolepis nana*, *Taenia saginata*, *T. solium*, and *Diphyllobothrium latum* and the trematode *Fasciolopsis buski* (WHO 1987). It has been estimated that there are 80–100 million cases of *Strongyloides*, 400 million of *Enterobius* and 78 million of *Taenia* worldwide (WHO 1990b; Warren et al. 1993). The most widespread of these worms is *S. stercoralis* which is prevalent in tropical and subtropical areas and is a severe problem in South America and the Caribbean. *Enterobius* is cosmopolitan,

but human infections are most common in temperate regions and it is often found in developed countries in the northern hemisphere (WHO 1981, 1987). Morbidity is usually low. Taeniasis, caused by *T. saginata*, is also cosmopolitan but is most prevalent in African countries south of the Sahara, eastern Mediterranean countries, and in parts of the former Soviet Union. Lower prevalences are found in Europe, India and southern Asia, Japan, the Philippines, and much of Latin America. The lowest prevalences are found in Australia, Canada, and the USA (WHO 1979). *T. solium* is mainly restricted to poor socioeconomic areas of central and southern Africa, Mexico, Central and South America, and southern Asia. Sporadic cases are found in southern Europe (WHO 1979). Human cysticercosis, caused by the larval stages of *T. solium* in tissues, is far more important as a public health problem than is human taeniasis (i.e. the adult form in the intestine). The highest prevalences of human cysticercosis are found in some areas of Africa, Asia, and South America (WHO 1981, 1987). In Mexico, hospitalization and treatment costs for humans with neurocysticercosis was US$17 million annually (Murrell 1991).

EXPANDING THREATS

New diseases may emerge as a result of changes in ecology, human demographics, and human behavior. Factors such as international travel and commerce, introduction of new technology and industry, microbial adaptation and change, breakdown of public health measures, and social disorder all affect human health (Institute of Medicine 1992; Morse 1995). As a result, several parasites, such as *Anisakis*, *Babesia*, *Cryptosporidium*, *Cyclospora*, *Giardia*, microsporidians, *Plasmodium*, *S. stercolaris*, and *Toxoplasma gondii* may be poised for an increase in importance (Institute of Medicine 1992; Wurtz 1994). This list could also be extended to include a number of other parasites. Some of these parasites (i.e. *Giardia*, *Plasmodium*, *Strongyloides*, and *Schistosoma*) and the factors that favor their emergence or reemergence, have been mentioned before. For example, schistosomiasis has increased in incidence in many areas after the construction of dams. A brief review of the others follows.

Anisakis is a common parasite of marine mammals and fish and the etiological agent of herring worm disease (anisakiasis) in humans, who are incidental hosts. Humans are infected by larvae present in undercooked fish, squid or octopus. The larvae localize in the gastrointestinal tract and then migrate upwards, attaching to the oropharynx. Many cases have been observed in Japan, where it is customary to consume raw fish. Infected fish can be transported around the world and the incidence of infection is increasing because of the growing trend in the consumption of raw fish. Cases have been reported in the Americas, Asia,

Europe, and the South Pacific (WHO 1987; Benenson 1990; Institute of Medicine 1992). Prevention is effected by not eating raw fish.

Babesia microti, a protozoan transmitted by nymphs of the *Ixodes* tick, infects red blood cells, and is carried mainly by deer mice. Human babesiosis is usually a result of infection by *B. microti* but other species of *Babesia* have also been implicated. Babesiosis in humans has been observed in Mexico, Europe, and the USA. The emergence of cases in the USA is linked to reforestation and a resultant increase in the deer population and the number of deer mice. Preventive measures include the elimination of deer mice in the vicinity of human habitations and the use of tick repellents (Benenson 1990).

The coccidian *Cryptosporidium parvum* was recognized as a human pathogen less than 30 years ago. Cryptosporidia live intracellularly in the gut cells of humans and domestic animals such as cattle. Transmission is by the fecal–oral route with direct person-to-person and animal-to-person transmission and indirect water-borne transmission. The parasite is found worldwide and its importance as a human pathogen was recognized as a result of several small outbreaks that were either water-borne or food-borne and one massive outbreak associated with drinking water (the latter affected more than 400 000 people in Wisconsin, USA). In immunocompetent individuals, the infection may be asymptomatic or it may produce an acute or persistent diarrhea. Individuals immunosuppressed because of infection with human immunodeficiency virus (HIV) or from other causes have been the most severely affected, experiencing cryptosporidiosis as an aggressively opportunistic infection. Prevention should be achieved by improving surveillance of water purification plants to detect malfunction and by applying high standards of personal hygiene. Appropriate measures include the sanitary disposal of human feces and animal excreta. Chemical disinfection of water is ineffective against cryptosporidia because the oocysts (the infective stage) are very resistant to chlorine (Benenson 1990; Fraser 1994; Colley 1995).

Described in 1979, *Cyclospora cayetanensis*, another coccidia, is now recognized as producing long-lasting diarrhea in humans in North, Central and South America, and the Caribbean, as well as in South East Asia and Eastern Europe. It is transmitted through contaminated water or food, but is highly unlikely to be transmitted from person to person because oocysts require days or weeks to sporulate and become infectious (Ortega et al. 1993; Wurtz 1994). An outbreak of c. 850 cases occurred in Canada and USA (Centers for Disease Control 1996). *C. cayetanensis* may be the origin of persistent diarrhea in immunocompromised patients with AIDS. Prevention is similar to that of the other food- or water-borne diseases.

Microsporidia are normally parasites of animals other than humans. Four genera of microsporidia, *Encephalitozoon*, *Enterocytozoon*, *Nosema*, and *Trachipleistophora*, have been recognized as human pathogens worldwide, usually affecting the immunosuppressed, including those with HIV. *Encephalitozoon* has been reported to cause systemic infection with central nervous system and kidney involvement; *Nosema* is considered to be the cause of keratitis; and *Trachipleistophora* has been recognized as an agent of myositis. *Enterocytozoon*, which causes the majority of microsporidial infections in humans, produces chronic diarrhea in AIDS patients. The mode of transmission of microsporidia to humans is unknown (Desportes et al. 1985; Canning and wHollister 1987; Shadduck and Greeley 1989; Centers for Disease Control 1990; Orenstein et al. 1990; Cali et al. 1991; Madi et al. 1991; Molina et al. 1993; Pol et al. 1993).

Toxoplasma gondii is a coccidian parasite of cats that produces disease opportunistically in immunosuppressed humans on all continents. Infection is also common in immunocompetent individuals, but most of them are asymptomatic. Infection occurs as a result of eating infected raw or undercooked meat from the intermediate hosts (sheep, swine, chicken, goats, cattle or birds) or by ingestion of food or water contaminated with oocysts from cat feces. Transplacental infection occurs when pregnant women acquire the infection during pregnancy. Prevention is based on personal hygiene, avoiding eating raw or undercooked meat, and avoiding contact with cat feces. Patients with AIDS must receive prophylactic treatment to prevent disease caused by *Toxoplasma* (Benenson 1990; Wong and Remington 1994). Human toxoplasmosis in the USA is estimated to be an annual economic/public health burden of more than US$400 million. (Murrell 1991).

CONCLUSION

Infectious disease must be viewed as a global problem, a problem exacerbated by dynamic ecological changes brought about by technological, socioeconomic, environmental, and demographic changes, as well as by microbial change and adaptation (Institute of Medicine 1992). The mobility of modern humans and the evolution of parasites (including drug-resistant strains) mean that parasitic diseases are a threat in both developing and developed regions.

Parasitic infections cause human suffering and economic losses worldwide but the burden is greatest in underdeveloped countries. Millions of people are at risk of infection, disease, and death from protozoan and metazoan parasites. These parasites cause social problems and their effects are increased by the conditions in disordered societies. Their presence hampers educational, political, and economic development, imposing a tremendous burden on the already precarious health services in the poor areas of the world. This situation constitutes a serious public health concern and

Table 2.7 *Global estimates of parasitic infections and disease burden*

Infection	Infected population[a]	Number of cases[a]	Population at risk[a]	Burden of disease (thousands of DALYs)[b]
African trypanosomes[c]	250–300*	250–300*	50–55[†]	1 598
American trypanosomes[d, e]	16–18[†]	3.2–3.6[†]	60[†]	649
Amebiasis[f]	500[†]	40–50[†]		
Ascariasis[g]	785–1 500[†]	120-220[†]		1 181
Clonorchiasis[h]	7[†]		289.3[†]	
Dracunculiasis[i]	164*		100[†]	
Enterobiasis[j]	400[†]			
Fascioliasis[h]	2.4[†]		180[†]	
Lymphatic filariasis[k]	78.6–90[†]	5[†, l]	751[†]	5 644
Giardiasis[m]	200[†]	500*		
Hookworms[n]	750–1 000[†]	1.5–96[†]		1 825
Intestinal flukes[h]	1.3[a]			
Leishmaniasis[o]	12–13[†]	600*	350[†]	2 357
Malaria[p]	500[†]	250–450[†]	2 400[†]	42 280
Onchocerciasis[q]	17.5[†]	270[*, r]	75–80[†]	987
Opisthorchiasis[h]	10.3[†]		63.5[†]	
Paragonimiasis[h]	20.5[†]		194.8[†]	
Schistosomiasis[s]	200[†]		400–500[†]	1 760
Strongyloidosis[j]	80–100[†]			
Taeniasis[t]	70[†]			
Trichuriasis[u]	750–1 049[†]	133[†]		1 649

a) Numbers in thousands (*) and millions ([†]); b) WHO 2002; c) WHO 1994d, 1995a; d) Schmunis 1994; e) 400 000 cases of heart and hollow viscera disease annually (WHO 1995a); f) WHO 1987, 1993c, 1997b; g) WHO 1990b, 1993c, 1995a; Albonico et al. 1999; h) WHO 1995b; i) WHO 1995c; j) WHO 1990b; k) WHO 1992b, 1993c, 1994f; l) disabled; m) WHO 1987, 1993c; n) WHO 1993c, 1995a; o) WHO 1990a, 1993c, 1995a; p) WHO 1993b, 1995a; q) WHO 1995c; r) Blind; s) WHO 1985, 1993d, 1994h; t) Warren et al. 1993; u) WHO 1990b, 1993c, 1995a; Crompton 1999; Stephenson et al. 2000.

inhibits the healthy development of individuals and social groups (Tables 2.4 and 2.7).

In developed countries, worms and protozoa present increased threats because of population movement, trade, and changing life-styles. Implementation of prevention and control strategies requires a systematic multidisciplinary and multisectoral approach.

For many diseases, there is sufficient knowledge to develop successful control programs. The following factors are vital:

- knowledge of the means of spread of the parasite and of any reservoirs
- a means of detecting both clinically apparent and asymptomatic infection
- a system of treatment or a method of interrupting spread.

In addition, the political will to implement control programs is also necessary but is often lacking in societies that lack cohesion and where social and economic chaos prevail. Meanwhile, failure to use classic public health measures while waiting for a 'miraculous' system of cure will retard progress. Because of the complexity and cost of controlling infectious diseases, the worldwide burden caused by parasitic infections is only likely to decrease in the context of both social and economic development.

REFERENCES

Akhavan, D. 2000. *Analise de custo-efetividade do programa de controle da doenca de Chagas no Brasil*. Brasilia: Organizacao Pan-Americana da Saude/Organizacao Mundial da Saude, pp. 7–9.

Akhavan, D., Musgrove, P., et al. 1999. Cost-effective malaria control in Brazil. Cost-effectiveness of a Malaria Control Program in the Amazon Basin of Brazil, 1988–1996. *Soc Sci Med*, **49**, 1385–99.

Albonico, M., Crompton, D.W.T. and Savioli, L. 1999. Control strategies for human intestinal nematode infections. *Adv Parasitol*, **42**, 278–341.

Alleyne, G.A.O. 1983. What is the role of institutions in the developing world? In: Simpson, T.W., Strickland, G.T. and Mercer, M.A. (eds), *New developments in tropical medicine*, Vol. 2. . Washington, DC: Ntl Council Intl Hlth, 9–14.

Alonzo, M., Gonzalez, C., et al. 2000. Cost-effectiveness of iron supplementation and malaria chemoprophylaxis in the prevention of anemia and malaria among Tanzanian infants. *Bull World Health Organ*, **78**, 97–107.

Anderson, R.M. 1986. The population dynamics and epidemiology of intestinal nematode infections. *Trans R Soc Trop Med Hyg*, **80**, 706–18.

Anderson, R.M. and Medley, G.F. 1985. Community control of helminth infections of man by mass and selective therapy. *Parasitology*, **90**, 629–60.

Apt, W.B. 1991. *Aspectos clínicos de la enfermedad de Chagas en Chile y sus repercusiones económicas, Programa de actividades*. Taller sobre erradicación o control de la enfermedad de Chagas en Chile, Santiago, 8.

Beach, M.J., Streit, T.G. and Addiss, D.G. 1999. Assessment of combined ivermectin and albendazole for treatment of intestinal helminth and *Wuchereria bancrofti* infections in Haitian schoolchildren. *Am J Trop Med Hyg*, **60**, 479–86.

Benenson, A.S. *Control of communicable diseases in man*, 15th edn. Washington, DC: Pub Hlth Ass, pp. 96, 97, 112, 163, 183, 440.

Brown, K.R., Ricci, F.M. and Ottesen, E.A. 2000. Ivermectin: effectiveness in lymphatic filariasis. *Parasitology*, **121**, Suppl, S133–S46.

Bundy, D.A.P. and Cooper, E.S. 1989. Human trichuris and trichuriasis. *Adv Parasitol*, **29**, 107–73.

Cali, A., Meisler, D.M., et al. 1991. Corneal microsporidiosis in a patient with AIDS. *Am J Trop Med Hyg*, **44**, 463–8.

Canning, E.U. and Hollinster, W.S. 1987. Microsporidia of mammals – widespread pathogens or opportunistic curiosities? *Parasitol Today*, **3**, 262–73.

Centers for Disease Control, 1990. Microsporidian keratoconjunctivitis in patients with AIDS. *Morbid Mortal Weekly Rep*, **39**, 188–9.

Centers for Disease Control, 1996. Update: outbreaks of Cyclospora cayetanensis infection – United States and Canada, 6. *Morbid Mortal Weekly Rep*, **199**, 611–12.

Chan, M.S. 1997. The global burden of intestinal nematode infection – fifty years on. *Parasitol Today*, **13**, 438–43.

Coleman, P.G., Goodman, C.A. and Mills, A. 1999. Rebound mortality and the cost-effectiveness of malaria control: potential impact of increased mortality in late childhood following the introduction of insecticide treated nets. *Trop Med Int Health*, **4**, 175–86.

Colley, D.G. 1995. Waterborne cryptosporidiosis threat addressed. *Emerg Infect Dis*, **1**, 67–8.

Conly, G.N. 1975. *The impact of malaria on economic development: a case study,* Sci Pub 297. Washington, DC: Pan American Health Organization.

Cooper, E.S., Bundy, D.A., et al. 1990. Growth suppression in the trichuris dysentery syndrome. *Eur J Clin Nutr*, **44**, 285–91.

Crompton, D.W.T. 1999. How much human helminthiasis is there in the world? *J Parasitol*, **85**, 397–403.

Crompton, D.W.T. and Nesheim, M.C. 1982. Nutritional science and parasitology: a case for collaboration. *BioScience*, **32**, 677–80.

De Schaepdryver, A. 1984. Costs of training and maintenance of expert man-power vs costs of drugs. Priorities in the field of helminthic diseases in developing countries. *Soc Sci Med*, **19**, 1113–16.

del Rey, E.C., Basombrio, M.A. and Rojas, C.L. 1995. Beneficios brutos de la prevencion del mal de Chagas. *Castañares*, **4**, 5–73.

Desportes, I., Le Charpentier, Y., et al. 1985. Occurrence of a new microsporidian: *Enterocytozoon bieneusi* n. g. n.sp., in the enterocytes of human patient with AIDS. *J Protozool*, **32**, 250–4.

Dias, J.C.P. 1987. Control of Chagas' disease in Brazil. *Parasitol Today*, **3**, 336–41.

Dias, J.C.P., Loyola, C.C.P. and Brener, S. 1985. Doença de Chagas em Minas Gerais: Situação actual e perspectivas. *Rev Bras Malariol Doencas Trop*, **37**, 7–28.

Ettling, M., McFarland, D.A., et al. 1994. Economic impact of malaria in Malawian households. *Trop Med Parasitol*, **45**, 74–9.

Evans, D.B. and Jamison, D.T. 1994. Economics and the argument for parasitic disease control. *Science*, **264**, 1866–7.

Evequoz, M.C. 1993. *Evaluación de la miocardiopatia Chagásica. Grados II y III. Estimación de costos* (Tesis). Universidad Nacional de Córdoba, Facultad de Medicina, Escuela de Salud Pública, Córdoba, Argentina.

Farthing, M.J.G. 1993. Diarrheal disease: Current concepts and future challenges. Pathogenesis of giardiasis. *Trans R Soc Trop Med Hyg*, **3**, 17–21.

Farthing, M.J.G., Mata, L., et al. 1986. Natural history of *Giardia* infection in infants and children in rural Guatemala and its impact on physical growth. *Am J Clin Nutr*, **43**, 393–405.

Fraser, D. 1994. Epidemiology of *Giardia lamblia* and *Cryptosporidium* infections in childhood. *Isr J Med Sci*, **30**, 356–61.

Griekspoor, A., Sondorp, E. and Vos, T. 1999. Cost-effectiveness analysis of humanitarian relief interventions: visceral leishmaniasis treatment in the Sudan. *Hlth Policy Plan*, **14**, 70–6.

Gryseels, B. 1989. The relevance of schistosomiasis for public health. *Trop Med Parasitol*, **40**, 134–42.

Guyatt, H.L. and Evans, D. 1992. Economic considerations for helminth control. *Parasitol Today*, **18**, 397–402.

Hayes, R.J. and Schofield, C. 1990. Estimación de las tasas de incidencia de infecciones y parasitosis crónicas a partir de la prevalencia: la enfermedad de chagas en América Latina. *Bol Sanit Panam*, **108**, 308–16.

Hill, D.R. 1993. Giardiasis. Issues in diagnosis and management. *Infect Dis Clin North Am*, **7**, 503–25.

Human Development Ministry, 1994. *Chagas in Bolivia*. La Paz, Bolivia: US Agency International Development, 81–90.

Institute of Medicine, 1992. *Emerging infections*. Washington, DC: Natural Academy Press.

Ismail, M.M., Jayakodi, R.L., et al. 1998. Efficacy of a single dose combination of albendazole, ivermectin, and diethilcarbamizine for the treatment of bancroftian filariasis. *Trans Roy Soc Trop Med Hyg*, **92**, 94–7.

Kabayo, J.P. 2002. Aiming to eliminate tsetse from Africa. *Trends Parasitol*, **18**, 473–5.

Kaewsonthi, S. and Harding, A. 1984. Cost and performance of malaria surveillance in Thailand. *Soc Sci Med*, **19**, 1081–97.

Katabarwa, M., Mutabazi, D. and Richards, F. Jnr. 1999. The community-directed, ivermectin-treatment program for onchocerciasis control in Uganda – an evaluative study (1993–1997). *Ann Trop Med Parasitol*, **93**, 727–35.

Kühner, A. 1971. The impact of public health programs on economic development. *Int J Health Serv*, **1**, 285–92.

Lengerich, E.J., Adiss, D.G. and Juranek, D.D. 1994. Severe giardiasis in the United States. *Clin Infect Dis*, **18**, 760–3.

Livadas, G.A. and Athanassatos, D. 1963. The economic benefits of malaria eradication in Greece. *Riv Malariol*, **42**, 177–87.

Lopes, E.R. and Chapadeiro, E. 1986. Doenca de Chagas no Triangulo Mineiro. *Rev Goiana Med*, **32**, 109–13.

López-Antuñano, F.J. and Schmunis, G.A. 1993. Plasmodia of humans. In: Kreier, J.P. (ed.), *Parasitic protozoa*, 2nd edn. San Diego: Academic Press, 135–266.

Madi, K., Trajman, A., et al. 1991. Jejunal biopsy in HIV-infected patients. *J Acquired Immune Defic Syndr*, **4**, 930–7.

Maldonado, Y.A., Nahlen, B.L., et al. 1986. Transmission of *Plasmodium vivax* malaria in San Diego County California. *Am J Trop Med Hyg*, **42**, 3–9.

Ministry of Health, 2000. The economic impact of malaria in Peru. Lima, Peru: MINSA-Peru, p. 126.

Molina, J.M., Sarfati, C., et al. 1993. Intestinal microsporidiosis in human immunodeficiency virus-infected patients with chronic unexplained diarrhea: prevalence and clinical and biological features. *J Infect Dis*, **167**, 217–21.

Moncayo, A. 1993. Chagas' disease. In *Tropical Diseases Research, Progress 1991–92*. Eleventh program report of the UNDP/World Bank/WHO Special Program for Research and Training in Tropical Diseases. Geneva: World Health Organization, 67–75.

Moore, D.A., Edwards, M. and Escombe, R. 2002. African trypanosomiasis in travelers returning to the United Kingdom. *Emerg Infect Dis*, **8**, 74–6.

Morse, S.M. 1995. Factors in the emergence of infectious diseases. *Emerg Infect Dis*, **1**, 7–15.

Muennig, P., Pallin, D., et al. 1999. The cost-effectiveness of strategies for the treatment of intestinal parasites in immigrants. *N Engl J Med*, **340**, 773–9.

Murrell, K.D. 1991. Economic losses resulting from food-borne parasitic zoonoses. *Southeast Asian J Trop Med Public Health*, **22**, Suppl, S377–81.

Orenstein, J., Chiang, J., et al. 1990. Intestinal microsporidiosis as a cause of diarrhea in human immunodeficiency virus-infected patients: a report of 20 cases. *Hum Pathol*, **21**, 475–81.

Organizacion Panamericana de la Salud, 2000. Vigilancia de las enfermeades infecciosas emergentes en los paises amazonicos. *Rev Patol Trop*, **31**, Suppl, 2–76.

Ortega, Y.R., Sterling, C., et al. 1993. *Cyclospora* species – a new protozoan pathogen of humans. *N Engl J Med*, **328**, 1308–12.

Ottesen, E.A. 1994. The human filariasis: new understandings, new therapeutic strategies. *Curr Opin Inf Dis*, **7**, 500–58.

PAHO. 1994. *Status of malaria programs in the Americas*, XLII Report, CSP24/INF/2. Washington, DC: Pan American Health Organization.

PAHO. 1995. *Status of malaria programs in the Americas*, XLIII Report, CD38/INF/2. Washington, DC: Pan American Health Organization.

Pereira, M.G. 1984. Caracteristicas da mortalidade urbana por Doença de Chagas Distrito Federal, Brasil. *Bol Sanit Panam*, **104**, 213–20.

Phillips-Howard, P. 1991. Travelers beware. *World Health*, **Sep–Oct**, 24–5.

Pol, S., Romana, C.A. and Richard, S. 1993. Microsporidia infection in patients with the human immunodeficiency virus and unexplained cholangitis. *N Engl J Med*, **328**, 95–9.

Politi, C., Carrin, G., et al. 1995. Cost-effectiveness analysis of alternative treatments of African gambiense trypanosomiasis in Uganda. *Health Econ*, **4**, 273–87.

Prescott, N. and Jancloes, M.F. 1984. Selected economic issues in helminth control. *Soc Sci Med*, **19**, 1060–84.

Rey, L. 1991. *Parasitologia*, 2nd edn. Rio de Janeiro: Guanabara Koogan, pp. 182-215.

Rohde, R. 1989. Schistosomiasis control: an estimation of costs. *Trop Med Parasitol*, **40**, 240–4.

Rosenfield, P.L., Golladay, F. and Davidson, R.K. 1984. The economics of parasitic diseases: research priorities. *Soc Sci Med*, **19**, 1117–26.

Saladin, B., Saladin, K., et al. 1983. A pilot control trial of schistosomiasis in central Liberia by mass chemotherapy of target populations, combined with focal application of molluscicide. *Acta Trop*, **40**, 271–95.

Sawyer, D. 1993. Economic and social consequences of malaria in new colonization projects in Brazil. *Soc Sci Med*, **37**, 1131–6.

Schmunis, G.A. 1991. *Trypanosoma cruzi*, the etiologic agent of chagas' disease: status in the blood supply in endemic and non endemic countries. *Transfusion*, **31**, 547–57.

Schmunis, G.A. 1994. American trypanosomiasis as a public health problem, *Chagas disease and the nervous system*, Sci Pub 547. Washington, DC: Pan American Health Organization, 3–29.

Schmunis, G.A. 1999. Prevention of transfusional *Trypanosoma cruzi* infection in Latin America. *Mem Inst Owaldo Cruz*, **94**, 93–101.

Schmunis, G.A. 2000. A tripanossomiase Americana e seu impacto na saude publica das Americas. In: Andrade, Z, Barral-Neto, M. and Brener, Z. (eds), *Trypanosoma cruzi e doenca de Chagas*, 2nd edn. Rio de Janeiro: Guanabara Koogan, 2–15.

Schmunis, G.A., Zicker, F., et al. 2001. Safety of blood supply for infectious diseases in Latin American countries, 1994–1997. *Am J Trop Med Hyg*, **65**, 924–30.

Schofield, C.J. and Dias, J.C.P. 1991. A cost benefit analysis of Chagas' disease control. *Mem Inst Oswaldo Cruz*, **86**, 285–95.

Schofield, C.J. and Dujardin, J.P. 1997. Chagas disease vector control in Central America. *Parasitol Today*, **13**, 141–4.

Shadduck, J.A. and Greeley, E. 1989. Microsporidia and human infections. *Clin Microbiol Rev*, **2**, 158–69.

Sinton, J.A. 1938. What malaria cost India. *Govt India Hlth Bull*, **26**.

Stephenson, L.S. 1984. Methods to evaluate nutritional and economic implications of ascaris infections. *Soc Sci Med*, **19**, 1061–5.

Stephenson, L.S., Holland, C.V. and Cooper, E.S. 2000. The public health significance of *Trichuris trichiura*. *Parasitology*, **121**, Suppl, S73–95.

Stephenson, L.S., Kinoti, S.N., et al. 1989. Treatment with a single dose of albendazole improves growth of Kenyan school children with hookworm, *Trichuris trichura* and *Ascaris lumbricoides* infections. *Am J Trop Med Hyg*, **41**, 78–87.

Tanner, M. 1989. Evaluation of public health impact of schistosomiasis. *Trop Med Parasitol*, **40**, 143–8.

Thompson, S.C. 1994. *Giardia lamblia* in children and the child care setting: A review of the literature. *J Pediatr Child Health*, **30**, 202–9.

US Congress. 1985. *Status of biomedical research and related technology for tropical diseases*. Washington, DC: US Congress office of technological assessment OTA H-258, US Govt Print Off, 101–2.

Villalba, R., Fornes, G., et al. 1992. Acute Chagas disease in a recipient of a bone marrow transplant in Spain: case report. *Clin Infect Dis*, **14**, 594–5.

Walsh, J.A. and Warren, K.S. 1979. Selective primary health care. An interim strategy for disease control in developing countries. *N Engl J Med*, **301**, 967–74.

Warren, K.S., Bundy, D.A.P., et al. 1993. Helminth infections. In: Jamison, D.T., Mosley, W.H., et al. (eds), *Disease control priorities in developing countries*. Oxford: Oxford Medical Publications, 131–60.

Wernsdorfer, G and Wernsdorfer, W.H. In: Wernsdorfer, W.H. and McGregor, I. (eds), *Social and economic implications of malaria and its control*. , *Principles and practices of malariology*, Vol. 2. . New York: Churchill Livingstone, 1421–71.

WHO, 1977. *Manual of the international statistical classification of diseases injuries and causes of death*. Geneva: World Health Organization.

WHO, 1979. *Parasitic zoonosis*, WHO Tech Rep Ser 637. Geneva: World Health Organization.

WHO, 1981. *Intestinal protozoan and helminthic infections*, WHO Tech Rep Ser 666. Geneva: World Health Organization.

WHO, 1985. *The control of schistosomiasis*, WHO Tech Rep Ser 728. Geneva: World Health Organization.

WHO, 1986. *Epidemiology and control of African trypanosomiasis*, WHO Tech Rep Ser 739. Geneva: World Health Organization.

WHO, 1987. *Prevention and control of intestinal parasitic infection*, WHO Tech Rep Ser 749. Geneva: World Health Organization.

WHO, 1990a. *Control of the leishmaniases*, WHO Tech Rep Ser 793. Geneva: World Health Organization.

WHO, 1990b. *Informal consultation on intestinal helminth infections*, WHO /CDS/IPI/90.1. Geneva: World Health Organization.

WHO, 1992a. World malaria situation in 1990, Part I. *WHO Wkly Epidemiol Rec*, **67**, 161–168.

WHO, 1992b. *Lymphatic filariasis: the disease and its control*, WHO Tech Rep Ser 821. Geneva: World Health Organization.

WHO, 1993a. Leishmaniasis epidemic in Southern Sudan. *WHO Weekly Epid Rec*, **68**, 41–2.

WHO, 1993b. *A global strategy for malaria control*. Geneva: World Health Organization.

WHO, 1993c. Global health situation IV. Selected infectious and parasitic diseases due to identified organisms. *WHO Wkly Epidemiol Rec*, **68**, 43–4.

WHO, 1993d. *The control of schistosomiasis*, WHO Tech Rep Ser 830. Geneva: World Health Organization.

WHO, 1994a. *Control of tropical diseases, 1. Progress report*, WHO CTD/MIP/94.4. Geneva: World Health Organization, 41–4.

WHO, 1994b. *Control of tropical diseases 1. Progress report*. WHO CTD/MIP/94.4. Geneva: World Health Organization, 33–5.

WHO, 1994c. World malaria situation in 1992. *WHO Wkly Epidem Rec*, **42**, 309–14.

WHO, 1994d. World malaria situation in 1992. *WHO Wkly Epidem Rec*, **43**, 317–21.

WHO, 1994e. World malaria situation in 1992. *WHO Wkly Epidem Rec*, **44**, 325–30.

WHO, 1994f. *Control of tropical diseases, 1. Progress report*, WHO CTD/MIP/94.4. Geneva: World Health Organization, 27–31.

WHO, 1994g. *Lymphatic filariasis infection and disease: control strategies*, WHO TDR/CTD/FIL/ Penang/94.1. Penang: World Health Organization.

WHO, 1994h. *Control of tropical diseases, 1. Progress report*, WHO CTD/MIP/94.4. Geneva: World Health Organization, 23–6.

WHO, 1995a. *The world health report. Bridging the gap. Executive summary*. Geneva: World Health Organization, 1–5.

WHO, 1995b. *Control of food borne trematode infections*, WHO Tech Rep Ser 849. Geneva: World Health Organization.

WHO, 1995c. Dracunculiasis. *WHO Wkly Epidemiol Rec*, **70**, 125–32.

WHO, 1995d. *Onchocerciasis and its control*, WHO Tech Rep Ser 852. Geneva: World Health Organization.

WHO, 1996. *The world health report 1996*. Geneva: World Health Organization, 24.

WHO, 1997a. World malaria situation 1994. *WHO Wkly Epidemiol Rec*, **72**, 269–74.

WHO, 1997b. *The world health report 1997*. Geneva: World Health Organization, Geneva, 15.

WHO, 1999. *Report on infectious diseases. Removing obstacles to healthy development*. Geneva: World Health Organization.

WHO, 2000a. *The world health report 2000*. Geneva: World Health Organization, 164.

WHO, 2000b. *WHO expert committee on malaria*, WHO tech Rep Ser 892. Geneva: World Health Organization.

WHO, 2002. *The world health report 2002*. Geneva: World Health Organization, 186–92.

Williams, L.L. 1938. Economic importance of malaria control. *New Jersey Mosq Extermin Ass*, **25**, 148–51.

Wilson, M.E. 1995. Travel and the emergence of infectious diseases. *Emerg Infect Dis*, **1**, 39–46.

Wolfe, M.S. 1992. Giardiasis. *Clin Microbiol Rev*, **5**, 93–100.

Wong, S.V. and Remington, J.S. 1994. Toxoplasmosis and the setting of AIDS. In: Broder, S., Merigan, T.C. and Bolognesi, D. (eds), *Textbook of AIDS medicine*. Baltimore: Williams and Wilkins, 233–257.

World Bank. 1993a. *Investing in health. World development indicators*. World Development Report 1993a. Oxford: Oxford University Press, 25–36.

World Bank, 1993b. *Investing in health. World development indicators*. World Development Report 1993b. Oxford: Oxford University Press, 213–25.

Wurtz, R. 1994. *Cyclospora*: a newly identified intestinal pathogen of humans. *Clin Infect Dis*, **18**, 620–3.

Zicker, F. and Zicker, E.M. 1985. Beneficios providenciarios por incapacidade como indicador de morbidade. Estudo da doenca de Chagas em Goias. *Rev Goiana Med*, **31**, 125–36.

3

Epidemiology of parasitic infections

ROY M. ANDERSON

INTRODUCTION

Epidemiology is the study of patterns of infection and associated disease within populations or defined communities. It is interdisciplinary in character, drawing on concepts and methods from a wide variety of biomedical fields including molecular biology, immunology, genetics, population biology, statistics, and mathematics, plus the specific fields associated with a particular infection, such as virology, bacteriology, parasitology, and vector biology. It is a quantitative field of medicine relying on accurate surveillance and measurement of infection plus disease, and dependent on statistical methods for hypothesis testing, parameter measurement, and description, plus mathematical techniques for the provision of a theoretical template to facilitate the interpretation of observed pattern or experimental outcome. A sound and detailed knowledge of the biology of the host and infectious agent, however, is an essential prerequisite for the successful application of such quantitative tools.

The past decade has witnessed significant changes in the manner in which epidemiological research is conducted, largely arising from the availability of new methods of measurement or detection of infection or disease linked to rapid advances in the disciplines of molecular biology and immunology. One example is the use of polymerase chain reaction (PCR) methods to detect the presence of very low concentrations of a particular pathogen within the human host. Such techniques enable quantitative measures to be made of parasite load or burdens within infected patients, in addition to the more precise determination of the prevalence (proportion of hosts infected) of infection. New immunological tools greatly facilitate the detection of current or past infection with non-invasive sample collection procedures (such as saliva-based assays for antibodies specific to viral, bacterial, and parasite infections) being of particular value in field-based studies.

Advances in other areas have also made very important contributions to epidemiology, particularly in the field of genetics. Molecular methods and the increased availability of genome sequence data for both pathogen and host open new avenues for the study of the determinants of pathogenesis and the coevolution of the parasite and its human or vector host. Increasingly, linkage or segregation analyses are beginning to find associations between the occurrence of disease, or persistent infection, and particular genotypes within human populations. Molecular or genetic epidemiology will undoubtedly be a major growth area in the coming decade. Associated population genetic studies can define the frequency of predisposition to infection or disease within defined populations.

There has also been a growing realization amongst researchers in this area that conventional statistical approaches to epidemiology, which have been so successful in improving understanding of non-infectious disease problems, are less appropriate for the study of the population level consequences of infection and immunity. Increasingly, concepts and methods that have evolved in evolutionary biology, ecology, and population genetics are being used to study the population biology, transmission dynamics, and evolution of infectious diseases within human communities. Such approaches recognize that the interaction between parasite and host is a very dynamic one, in terms of transmission and persistence, and of the evolution of both organisms (with that of the pathogen having the potential to take place much more rapidly than that of the human host).

Observed patterns of infection and disease in human communities are determined by the interplay between the variables that control susceptibility to infection, its typical course, and the likelihood of serious disease within an individual, and the variables that control the rate of transmission between people. The former include innate susceptibility, the distribution of latent and infectious periods plus the duration of acquired immunity, while the latter include behavioral, social, environmental, and demographic factors. Researchers are typically concerned with the recondite biological and clinical details that make each infection unique. This chapter focuses on the principles that determine observed epidemiological patterns, independent of the type of infectious agent under study. Historically, work on infectious disease is often compartmentalized into distinct areas based on the taxonomy of the infectious agent, namely, virology, bacteriology, and parasitology. However, the principles governing the spread, persistence, and evolution of all infectious agents are the same. A simple example is provided by the notion of reproductive or transmission success. If we consider the host as the basic unit of study, then the reproductive number of an infection (R_0) is the average number of secondary cases of infection generated by one primary case in a susceptible host population (Anderson and May 1991). The magnitude of this parameter reflects the potential for transmission and evolution, and the degree to which control measures must inhibit transmission if an infection is to be eradicated in a given population.

The real world is of course replete with complications – economic and social as well as biological – and it is easy to lose sight of generalities when grappling with the details that ultimately make each infectious agent unique. However, it is often the case in epidemiology that the influence of a few factors dominates the generation of observed pattern. Understanding the key principles that underpin spread and persistence helps in the identification of these factors.

This chapter highlights the key processes that influence the transmission, persistence, and control of protozoan and helminth parasites with a focus on malaria, intestinal nematodes, schistosome flukes, and the filarial worms. Bear in mind, however, that the principles are equally relevant to the study of viral and bacterial infections (Anderson and May 1991).

BASIC CONCEPTS

Microparasites and macroparasites

A distinction is made between microparasites and macroparasites to cut across conventional taxonomic lines in order to focus on the population biology of the infectious agent.

MICROPARASITES

Microparasites may be thought of as those agents which have direct reproduction – usually at very high rates – within the host (Anderson and May 1991). They tend to be characterized by small size and a short generation time. Recovery from infection is usually associated with acquired immunity against reinfection with the same strain of parasite for some time, and often for life. For many viruses, bacteria, and protozoa, however, immunity may be highly strain specific although a degree of cross-protection may arise via shared antigens. Although there are important exceptions, the duration of infection is typically short relative to the expected life span of the human host. This feature, combined with acquired immunity, means that microparasite infections are typically of a transient nature in individual hosts. Most viral, bacterial, and (in a less certain manner) many protozoan parasites fall broadly into the microparasitic category.

For such infectious agents, it makes sense to stratify the host population into a few classes namely: susceptibles, infecteds but latent (i.e. non-infectious), infecteds and infectious, and immune individuals. A reasonable operational definition of a microparasite is an organism whose population biology can, to a sensible first approximation, be described by such a compartmental model. An essential feature of the framework is that no account is taken of the degree or severity of the infection (i.e. the abundance of the infectious agent within the host). The reality of heterogeneity between individuals with respect to genetic background, behavior, and environment is replaced by the abstraction of some average 'infected' or 'immune' individual.

It is often necessary to distinguish between infection and disease. The period between the point of infection and the appearance of symptoms of disease is termed the incubation period. The duration of symptoms of disease is not necessarily synchronous with the period during which an infected host is infectious to susceptible individuals. Furthermore, the host may be infected but not infectious. The period from the point of infection to the beginning of the state of infectiousness is termed the latent period. For some infections such as malaria, intermittent bouts of infectiousness may occur. For others, such as *Human immunodeficiency virus* (HIV), the degree of infectiousness may vary widely throughout the incubation period. The sum of the latent and infectious periods is referred to as the average generation time of the infection. The term average refers to the fact that latent, infectious, and incubation periods are often very variable between individuals due to a variety of factors such as host and parasite genetic background and the size of the infecting inoculum of the infectious agent.

Most epidemiological and demographic parameters – human birth and death rates, disease-induced death rate, recovery rates, rate of loss of immunity – can be measured directly by appropriate observational studies. The transmission rate, however, combines many biological, social, and environmental factors, and is thus rarely amenable to direct measurement. The best way is to infer this rate indirectly via longitudinal or cross-sectional cohort based studies to directly measure the incidence of infection.

MACROPARASITES

Macroparasites may be thought of as those having no direct reproduction within the definitive host. This category embraces helminths and arthropods, where transmission stages produced within or on the host pass to the exterior to complete development to the next stage of often complex life cycles which may involve more than one host (e.g. the schistosome flukes and the filarial worms). Macroparasites are typically larger than microparasites and have much longer generation times, often being an appreciable fraction of the human host's life span. When an immune response is elicited, it usually depends on the past and present number of parasites harbored, and tends to be of a relatively short duration once parasites are removed (e.g. by chemotherapy) from the host. Thus, macroparasitic infections are typically persistent in character, with hosts continually being reinfected. The evocation of an immune response and the pathology induced by infection both depend on the burden of parasites harbored by the host. This means that to quantify levels of infection and morbidity in the population it is necessary to measure the distribution between individuals of parasite numbers per host. As shown in Figure 3.1, parasite numbers are rarely uniformly or randomly distributed. Typically, the distributions are highly heterogeneous with most people harboring few parasites and a few harboring many. Compartmental models based on divisions into suscep-

tible and infected persons are therefore inappropriate descriptions of the epidemiology of macroparasitic infections. The appropriate framework for study is one that records the prevalence (fraction infected) of infection, the average intensity of infection and the distribution of parasite numbers within the human population.

The division into microparasites and macroparasites is rather crude. Many infectious agents are not easily forced into this dichotomous scheme. Many protozoan parasites such as the *Plasmodium* species that induce the disease malaria may, to a good approximation, have their epidemiology described by a microparasitic compartmental framework (i.e. prevalence of infecteds, susceptibles, and immunes), although their patterns of persistence within the human population, with individuals being repeatedly infected, are characteristic of macroparasites. However, the dichotomy serves as a useful starting point for highlighting the population biology and epidemiology, as opposed to emphasizing conventional taxonomic categorizations.

Basic reproductive number of infection, R_0

For a microparasite, the basic or case reproductive rate, R_0, is defined as the average number of secondary cases of infection produced by one primary case in a susceptible population. When a microparasitic infection first invades a human community, the fraction susceptible decreases over time. Eventually, some sort of equilibrium state is reached (endemic infection) which may involve oscillatory fluctuations in prevalence or incidence of a seasonal or longer term nature. At this equilibrium (or oscillating equilibrium) the rate at which susceptible infants or children are infected is exactly balanced by the rate at which new susceptibles are born into the community and each infection will on average produce exactly one secondary case. That is, at equilibrium, the effective reproductive rate $R = 1$. If we assume that the human community mixes homogeneously (independent of age, sex, residence location, etc.), the effective reproductive rate R is equal to the basic reproductive rate, R_0, discounted by x, the fraction of the human community that is susceptible, R, is equal to $R_0 x$. Given that at equilibrium $R = 1$, then

$$R_0 = 1/x \tag{3.1}$$

where x is the fraction susceptible at equilibrium. This simple relationship provides a method of estimating the value of R_0 from serological or other data on age-specific susceptibility.

For a macroparasite, R_0 is the average number of female offspring (in the case of a dioecious parasite), produced throughout the lifetime of a mature female parasite, which themselves achieve reproductive maturity in the absence of density-dependent constraints. In the absence of these constraints, which include competition

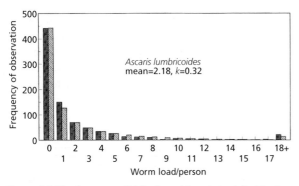

Figure 3.1 *The frequency distribution of* Ascaris lumbricoides *in a fishing community in Pulicat, Tamil Nadu, India (after Elkins et al. 1986). The negative binomial probability distribution provides a good empirical description of the observed pattern with mean = 2.8 worms per person and* k = 0.32. *Dark bars, observed value; hatched bars, expected value*

for space or other resources within the human host and acquired immunity dependent on parasite load, parasite numbers within the population would grow exponentially provided $R_0 > 1$. In practice, macroparasite populations tend to be rather stable through time and a variety of density-dependent factors act to generate such stability.

For most common viral and bacterial infections and some protozoa (e.g. *Toxoplasma*), serological tools are available to establish whether or not a person has acquired and recovered from infection at some time in their past. Age-stratified cross-sectional serological surveys or longitudinal cohort studies provide the best information for estimating the magnitude of R_0 in a given community. A diagrammatic serological profile is shown in Figure 3.2. The magnitude of R_0 can either be estimated from this profile by calculating the fraction of the total population susceptible to infection or equally simply from the following expression:

$$R_0 = (L - M)/(A - M) \tag{3.2}$$

where L is life expectancy, M is the average duration of maternal antibody-derived protection (typically 6 months for many common viral infections such as measles), and A is the average age at infection. The key quantity is A, which can be estimated directly from the serological profile (Grenfell and Anderson 1985). If the value of A is small, R_0 is large, and the infection has high transmission efficiency, while, conversely, if A is large, R_0 is small, and the infection has low reproductive success. Transmission success of a given parasite can vary widely in value between different locations and at different times of the year.

Serological surveys provide a wealth of information about the transmission dynamics of infectious agents. They not only provide information on the values of M (duration of maternally derived protection) and A (average age at infection), but they can also guide

decisions on the optimum age at which to vaccinate and on the fraction who should be immunized to block transmission. Carrying out large-scale surveys to include infants, children, and adults can be difficult if blood samples are required to extract serum for the evaluation of the presence or absence of antibodies specific to particular antigens. Techniques based on the testing of saliva for secreted antibodies (Perry et al. 1993) have been developed for a variety of infectious agents and these provide epidemiologists with a powerful tool for the study of transmission. Serum or saliva samples must be collected at random within the community, taking due note of important stratifications within the population such as age, sex, ethnic background, residence location, etc. However, problems can arise if the duration of measurable antibody production following infection is not lifelong. This is thought to be the case for many bacterial and protozoan infections which makes serological surveys of less value in determining past and current patterns of transmission. We require more sophisticated techniques to detect immunological markers of past infection with antigenically variable parasitic organisms. In general, however, the collection of serological data, finely stratified according to age, is of vital importance in any quantitative assessment of transmission and the potential impact of control measures. Ideally, age-stratified cross-sectional surveys should be conducted each year, such that a profile of herd immunity can be constructed prior to and post the introduction of any intervention program.

The measurement and interpretation of the magnitude of R_0 can be made complicated by genetic variability within the parasite population when such variability is reflected by different phenotypic properties such as transmissibility or antigenic characteristics (Gupta and Day 1994). Consider a protozoan that induces lifelong immunity in those who recover from infection. In such cases R_0 is inversely related to the average age at infection, A, by a simple relationship

$$R_0 = L/A \tag{3.3}$$

where L is human life expectancy. However, lifelong immunity is rarely observed in parasitic protozoa due largely to the existence of many distinct antigenic strains of a particular parasite species. In such cases, recovery from infection by one strain may or may not confer a degree of resistance to another strain depending on their degree of 'antigenic' relatedness. Immunity may then only develop as a consequence of exposure to many different antigenic types or strains circulating in a given locality. This appears to be the case for the malaria parasite *Plasmodium falciparum*. Young children in hyperendemic malaria zones may experience many clinical attacks of malaria per year (Greenwood et al. 1991). Although older children develop a functional but non-sterilizing immunity manifest as a reduction in clinical episodes, a substantial reduction in parasite load is only

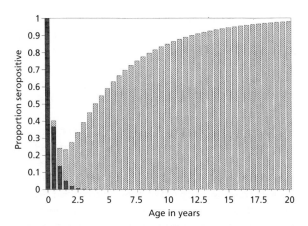

Figure 3.2 *Diagrammatic representation of a cross-sectional seroepidemiological survey, stratified by age, for the presence of antibodies specific to parasite antigens. Dark bars, maternal antibodies; hatched bars, antibodies from infection*

observed in adults after a long period of exposure to the genetical diverse population of parasites. If the delay in the development of immunity is a direct consequence of the antigenic diversity within the parasite population (as opposed to parasite evasion or modulation of the host's immunological attack) then we would expect to see a slower rise by age of immunological (seroepidemiology) evidence of exposure to a particular strain, but a much faster rise in the exposure to 'malaria', where the latter is defined as the experience of any one of several strains. When n strains are independently circulating in a defined population, the average age on first exposure to malaria, A_n, is

$$A_n = 1 / \sum_{i=1}^{n} R_{0i} \qquad (3.4)$$

Here it assumed that strain-specific immunity is life-long and R_{0i} defines the case reproductive number of strain i. This equation makes clear that a low average age at first infection (often the case for *P. falciparum*) can arise when the transmissibilities of each of many strains circulating is low. These conclusions are well supported by field observation on the rise with age in the fraction of a population in Madang, Papua New Guinea who had experienced infection with five antigenically distinct strains of *P. falciparum* (Gupta et al. 1994b) (Figure 3.3).

The relationship between prevalence and intensity of infection

We have briefly noted that many sources of heterogeneity influence the distribution of parasites within a human community (Figure 3.1). The precise degree of heterogeneity, measured either by a statistic such as the variance/mean ratio (V/M) of parasite load per host, or

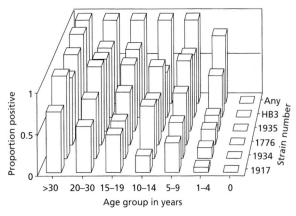

Figure 3.3 *Age-stratified serological survey for presence of past infection by a series of distinct isolates or strains (serotypes) of* Plasmodium falciparum *from a community in Papua New Guinea (data from Gupta et al. 1994b). The isolates differ in parasite induced erythrocyte surface antigens (PIESA). The proportion exposed to any type is recorded at the back of the graph (= Any)*

the aggregation parameter, k (which varies inversely with the degree of aggregation) of the negative binomial probability distribution (which provides a good empirical model of many observed distributions), determines the relationship between the prevalence and mean intensity of infection (Figure 3.4a). As the degree of contagion or aggregation increases, the prevalence reaches a plateau, well below 100 percent, as the mean parasite load rises. For a given species of parasite, observed patterns of heterogeneity are often remarkably constant independent of the study community or geographical region (Anderson and May 1985, see Figure 3.4b). For helminth parasites in particular, worm load is directly correlated with the likelihood of symptoms of disease. As such parasite aggregation determines the prevailing burden of disease in a given population for a fixed mean worm load, the greater the degree of parasite aggregation, the smaller the number of individuals in the population showing symptoms of disease. However, for extreme aggregation the severity of disease will be greater in those showing symptoms due to their very high worm loads.

Heterogeneity and predisposition to infection

The epidemiological measures of prevalence and average intensity of infection are summary statistics of the frequency distribution of parasite numbers per host (Figure 3.1). The form of these distributions is of great significance to the burden of disease in a population and the regulating constraints acting on the net rate of parasite transmission. For helminths and protozoa, the distributions are highly aggregated or contagious in form such that most individuals harbor few parasites and a few individuals harbor many. In other words, the variance in parasite load is much greater in value than the mean.

Parasite aggregation may arise as a consequence of a wide variety of factors, either acting alone or concomitantly (Anderson and Gordon 1982). These include heterogeneity in exposure to infection (due to social, environmental, or behavioral factors, or to aggregation in the spatial distribution of infective stages or infected intermediate hosts), differences in susceptibility to infection (due to genetic or nutritional factors, or to varying past experience of infection), or to variability in parasite survival within different individuals (due to genetic, immunological, or nutritional factors). Which factors are of major importance as determinants of observed patterns is often difficult to ascertain in particular settings. Heterogeneity in the human behavioral patterns associated with exposure to infection is often very important as is well illustrated by water contact patterns in relation to *Schistosoma mansoni* transmission (Fulford et al. 1996) (see Figure 3.6, see p. 56). However, host genetic background is undoubtedly of greater impor-

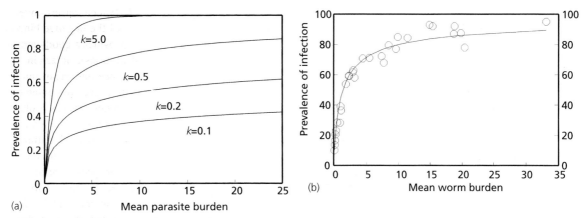

Figure 3.4 (a) *The relationship between the prevalence and mean intensity of infection predicted by the negative binomial distribution for different values of the aggregation parameter* k *(where* k → 0 *reflects extreme aggregation and* k → 5 *reflects a random distribution of parasite numbers per host).* **(b)** *A comparison of the observed and expected (on the basis of the negative binomial model with* k = 0.586*) relationship between the prevalence and intensity of infection with* Ascaris lumbricoides *in different human communities. The circles represent observed values; the continuous line is the expected value (after Guyatt et al. 1990).*

tance as well illustrated by the practice of using inbred strains of rodent hosts in the experimental study of many parasitic infections in order to reduce variability in the typical course of infection.

A growing number of studies are pointing to genetic factors in human communities as key determinants of susceptibility to infection and disease. For example, severe clinical disease in schistosomiasis is typically the consequence of heavy infection, the occurrence of which is determined largely by the susceptibility or resistence of individuals (Wilkins et al. 1987). The intensity of infection in a particular study site in Brazil has been shown to be influenced by a major gene leading to the hypothesis that worm burden is largely controlled by the genetic background of the human host concomitant with the degree of exposure to infection (Abel et al. 1991). The gene, referred to as *sm1*, is located on chromosome 5, and the region in which segregation analysis indicates it lies contains several candidate genes that encode immunological molecules that are known to play important roles in the acquisition of resistance to *S. mansoni* (Marquet et al. 1996).

The discovery of a clear genetic marker for predisposition to heavy infection in the study of Marquet et al. (1996) in the case of schistosomes may point the way to a better understanding in general of predisposition to helminth infection and associated disease. Epidemiological studies based on the quantification of worm load, chemotherapeutic treatment, and the monitoring of patterns of re-infection post treatment, have provided firm evidence of predisposition to heavy infection for a variety of parasites including *Ascaris lumbroides* (Elkins et al. 1986), *Necator americanus* (Schad and Anderson 1985), *Trichuris trichiura* (Bundy et al. 1987), and *Schistosoma mansoni* (Bensted-Smith et al. 1987). Undoubtedly, however, genetic factors act concomitantly with the host behaviors to determine parasite load.

Aggregated distributions of parasite numbers per host are also of importance for protozoans such as the malarial parasites. Observed heterogeneities include the distribution of oocysts in mosquito populations, the number of distinct antigenic types or strains per human host or per mosquito, and the frequency distribution of mosquito bites per person (Paul et al. 1995; Dye and Hasibeder 1986; Anderson 1994). If there is significant heterogeneity in exposure/susceptibility to infection then in epidemic situations, the fraction of the population infected will depend critically on the degree of heterogeneity in exposure. This point is illustrated in Figure 3.5 which records the relationship between the fraction infected in an epidemic and the magnitude of transmission success of the parasite (R_0), as a function of the degree of heterogeneity in the rate of exposure to infection defined by the coefficient of variation (where cv = 0

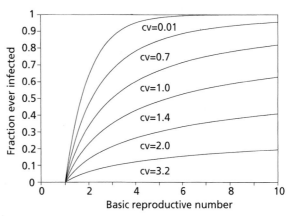

Figure 3.5 *The predicted relationship between the fraction of a population infected by a microparasite over the course of an epidemic in a virgin population and the degree of heterogeneity in the human behaviors that dictate exposure to infection as measured by the coefficient of variation (cv) where cv = 0 reflects a random distribution (see Anderson and May 1991).*

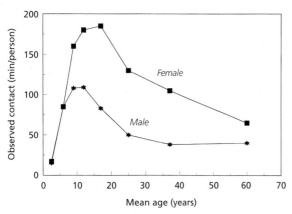

Figure 3.6 *Recorded water contact rates, stratified by age and sex in a Kenyan community where* Schistosoma mansoni *is endemic. The vertical axis records the arithmetic mean duration of water contact per individual (Kintengei community (1990–1992), after Fulford et al. 1996).*

reflects a random distribution of exposure) of the frequency distribution. An example of variation in exposure to infection is shown in Figure 3.6.

Irrespective of the generative mechanisms of such patterns, aggregation of parasites within human communities has important implications for the epidemiology and control of these infections. Firstly, it enhances the likelihood of an individual parasite finding a mate of the opposite sex (Figure 3.7). It therefore enhances the net reproductive rate of the total parasite population. Second, it increases the net regulatory impact of density-dependent constraints on parasite establishment, survival, and reproduction (often immunologically mediated). Third, it results in severe symptoms of disease typically occurring in a relatively small fraction of the total population. Finally, it has important implications for the design of control programs (i.e. chemotherapy can be focused on those predisposed to heavy infection).

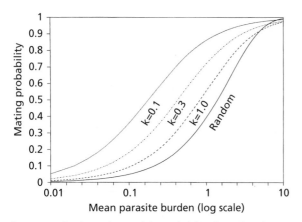

Figure 3.7 *The dependence of the probability that a female worm is mated, on the mean worm burden per person, for various assumptions concerning the distribution of parasite numbers per host for a monogamous parasite. The distributions are the negative binomial (= aggregated) with various k values and the Poisson (= random) distribution (after May 1977).*

If host genetic background has a major influence on immunocompetence, however, as suggested by work on *S. mansoni* (Marquet et al. 1996), then work towards vaccine development must take account of the fact that the aim of immunization must be to protect those least able to acquire immunity.

TRANSMISSION DYNAMICS

Observed epidemiological pattern is generated by the population or transmission dynamics of the parasite within a defined community. Before turning to the key processes influencing transmission, a brief comment is made on observed patterns of prevalence and intensity of infection.

Cross-sectional surveys of prevalence and intensity of infection

A variety of methods may be employed to assess the manner in which the prevalence or intensity of infection changes over time or between different strata (such as age groups) of the population, depending on the type of infectious agent under study. These range from molecular and immunological methods to the direct counting of either parasite eggs in fecal material (intestinal nematodes) or infected red blood cells in blood samples (i.e. for *Plasmodium*). The majority of published studies are cross-sectional and stratified by age owing to the long period of study required to monitor longitudinal changes in a cohort of people over their average life span.

As illustrated in Figures 3.8–3.10, changes in both prevalence and intensity with age (= time) are typically convex in form with the latter showing a greater degree of convexity than the former. Changes in prevalence as individuals age are a poor reflection of average parasite load due to the high degree of heterogeneity in the frequency distribution of parasite numbers per person. Convex patterns of change with age may arise either as a result of age-related changes in exposure to infection (due to behavioral factors) (see Figure 3.6) and/or increased resistance to infection (where acquired immunity may influence establishment or parasite survival once in the host) in older individuals with considerable past experience of infection. With respect to convexity in prevalence an additional factor may be of importance. A decay in prevalence in older age groups could arise, independent of any change in the average worm load, if the degree of parasite aggregation increases with age. Much evidence supports this notion and it may arise due to heterogeneity in genetic make-up within the host population where some individuals develop a good degree of immunity after limited exposure while others remain predisposed to heavy infection. In general, convexity is a common feature for all of the major

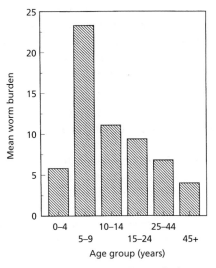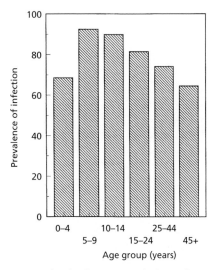

Figure 3.8 *Cross-sectional survey of the prevalence of infection and mean worm burden in a community in southern India infected with the intestinal nematode* Ascaris lumbricoides *(after Elkins et al. 1986)*

parasitic infections and may arise from ecological (= behavioral) or immunological processes, or a combination of both.

Studies in which age-related changes in exposure have been quantified (i.e. water contact patterns in the case of schistosomes, vector biting rates for malaria and the filarial worms) suggest that a combination of both factors acts to determine the observed convex profiles. A good indication of the significance of the slow build up of immunity with repeated exposure as individuals age is provided by comparative studies of the degree of convexity as a function of the overall net intensity of transmission in a defined community. For intestinal nematodes (Anderson and May 1991), schistosomes (Anderson 1987), and malaria (Boyd 1949), the degree of convexity is greater in areas of high transmission intensity and less in areas of low average exposure (Figure 3.11). The issue of why immunity to parasite

infection builds up slowly as individuals age is discussed later in this chapter.

In summary, observed patterns of infection with parasitic organisms tend to reveal three key points, namely, heterogeneity in parasite burdens per person, predisposition to heavy or light infection, and convexity in the manner prevalence or intensity change as individuals age.

Life cycles and reproductive success

Microparasite and macroparasite life or developmental cycles may be direct or indirect. In the latter case one or more intermediate hosts may be involved, such as the molluskan intermediate host of the schistosomes and the mosquito vector of the malaria parasites. The transmission potential of a parasite with a complex life cycle is determined by many distinct population-determining

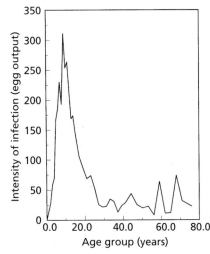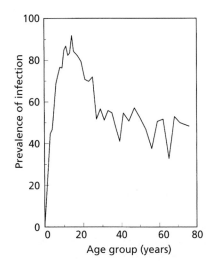

Figure 3.9 *Cross-sectional survey of the prevalence of infection and the mean intensity of infection (based on fecal egg counts) in a community in Tanzania infected with* Schistosoma haematobium *(after Bradley and McCullough 1973)*

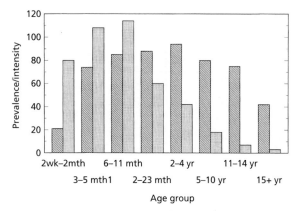

Figure 3.10 *Cross-sectional survey of the prevalence (dark bars) and intensity (lighter bars) of* Plasmodium falciparum *(after Davidson and Draper 1953)*

rate processes that influence the many distinct developmental stages. Examples of the population and rate processes involved in some direct (intestinal nematodes) and indirect cycles are portrayed in Figures 3.12–3.14. The overall transmission or reproductive success is best defined by the basic reproductive number, R_0, of the parasite. For a direct life cycle intestinal nematode with a free-living larval or egg stage as the transmission stage (Figure 3.12), the magnitude of R_0 in a human community of size N is defined as follows (see Anderson and May 1991):

$$R_0 = [s\lambda\beta Nd_1d_2]/[(\mu + \mu_1)(\mu_2 + \beta N)] \qquad (3.5)$$

The individual parameters are defined in Table 3.1. Provided $R_0 > 1$, the parasite population will persist in the host population. However, unlike many directly transmitted viral or bacterial infections, the critical host density for persistence of most protozoa and helminths is typically very low. Infections such as intestinal nematodes or malaria parasites were therefore able to persist endemically in small hunter-gatherer societies in areas such as the Amazon basin or Papua New Guinea (Tyrrell 1977) prior to the encroachment of civilization into remote rain forest habitats.

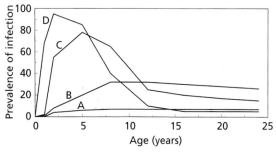

Figure 3.11 *Prevalence of acute malaria infection versus age in years in populations with differing levels of endemicity. A, low endemicity; B, moderate endemicity; C, high endemicity; D, hyperendemicity (after Boyd 1949)*

In the absence of regulating constraints on population growth, the parasite population growth rate Λ is given approximately by:

$$\Lambda = (R_0 - 1)/(\mu^{-1} + \mu_1^{-1}). \qquad (3.6)$$

This expression makes clear the role of net reproductive success (R_0) and adult parasite life expectancy μ_1^{-1} as important determinants of re-infection rates, or parasite population recovery, after the cessation of control measures. Populations of short-lived parasites (i.e. malaria) will bounce back rapidly to pre-controlled levels once control measures cease, while long-lived species may take many years to recover (Table 3.2).

For microparasites with indirect life cycles (Figure 3.14) such as the *Plasmodium* species, the magnitude of R_0 is determined by rate processes acting on the human and vector populations. For malaria R_0 is defined as follows (see Aron and May 1982):

$$R_0 = [ma^2bc/\mu_2\gamma]exp(-\mu_1T_1 - \mu T_2) \qquad (3.7)$$

where m is the ratio of female mosquitoes per human host, a is the rate of biting on humans by a single mosquito (the 'human biting rate' defined as the number of bites per unit of time), b is the proportion of infectious bites that produce a patent infection, γ^{-1} is the average duration of infectiousness over the course of infection in the human host, c is the proportion of bites by a susceptible mosquito on infected people that produce a patent infection in the vector, μ_1^{-1} is mosquito life expectancy, μ_2^{-1} is human life expectancy, T_2 is the latent period in the mosquito (before it becomes infectious to humans), and T_1 is the latent period in humans (before gametocytes appear) (Table 3.3). In practice, the effective time the insect vector has to transmit is much shorter than its life expectancy due to the fact that the latent period of infection in the vector is often long by comparison with life expectancy (Table 3.3). This is the reason why the prevalence of infection in vectors is often very low (a few percent or less), since few survive long enough for the infection to mature. If $R_0 > 1$, the infection is endemic in the human population.

Vector densities often fluctuate widely over the different seasons of a year due to the influence of environmental factors, such as rainfall on breeding success and life expectancy. The magnitude of m in eq. (5) may therefore fluctuate widely such that at certain times of the year $R_0 > 1$, whilst at others (the dry season) it falls below unity in value. In such circumstances the endemic persistence of malaria depends on the survival of the infection in the human host. Areas of low transmission are much more sensitive to seasonal changes in mosquito or vector density.

More generally, the longest lived stage in the parasite's life cycle plays a key role in ensuring the long persistence of the organism. Control measures aimed at

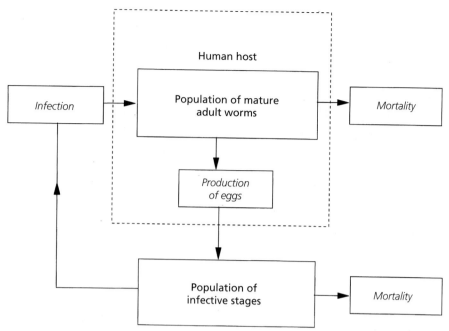

Figure 3.12 *Flow chart of the principal population and rate processes involved in the life cycle of a directly transmitted intestinal nematode (macroparasite)*

different stages in the cycle (e.g. insecticides acting on the vector or chemotherapeutic agents acting on the infection in the human host) may have similar effects on transmission success, but very different effects on the ability of the parasite population to recover rapidly post the cessation or interruption of control measures.

Regulating constraints on parasite population growth

A characteristic of many parasitic infections is their stability or robustness to perturbation, whether induced by environmental changes or control interventions.

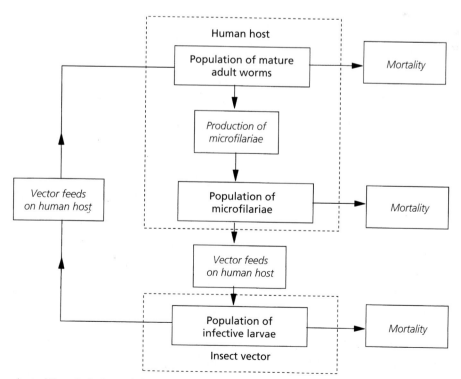

Figure 3.13 *Flow chart of the principal populations and rate processes involved in the life cycle of an indirectly transmitted filarial nematode (macroparasite)*

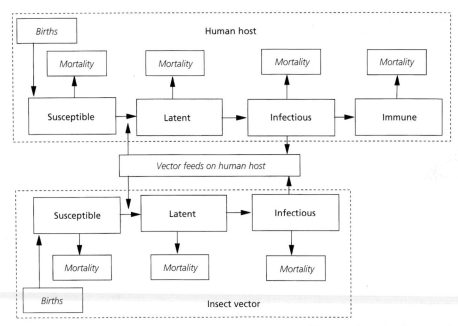

Figure 3.14 *Flow chart of the principal population and rate processes involved in the life cycle of a malarial parasite (vector transmitted microparasite)*

Helminth infections are particularly remarkable in this sense, where after the cessation of mass chemotherapy, parasite abundance typically returns, over a time span of a year or so, to pre-control levels. In part this is a consequence of their transmission potential (e.g. very high fecundity) and in part due to the long life expectancy of the mature worms in the human host (Table 3.2). In the absence of density-dependent constraints on population growth, however, provided $R_0 > 1$, their abundance would grow exponentially. As originally noted by Malthus in the context of human population, resource limitation must eventually regulate the rate of growth until some sort of equilibrium is attained. This may be oscillating in character but for most parasites, if we ignore short-term seasonal fluctuations, endemic states appear to be stable and non-oscillatory. For parasites occupying an environment created by another living organism, the constraints on growth are many and varied and act on establishment, fecundity, and survival. An example is presented in Figure 3.15, where the fecundity of *S. mansoni* is plotted as a function of the intensity of infection. They may arise via competition for limiting resources such as food or space or, more commonly, as a result of the non-linear action of the host's immunological responses. The non-linear nature of such responses is reflected by modest or little response at very low parasite densities, and strong responses at high densities. At very high parasite (or antigen) densities, in some cases the effectiveness of the immune response decays as depicted in Figure 3.16, where T cell proliferation in response to different concentrations of *P. falciparum* antigens is plotted. This may be due to modulation by the parasite, the accumulation of parasite

Table 3.1 *The rate processes that determine the magnitude of the basic reproductive number, R_0, of a direct life cycle intestinal nematode (see Anderson and May 1991)*

Parameter	Definition
s	Sex rate of adult worms
λ	Per capita fecundity of female worms
β	Per capita transmission rate of infective stages to the human host
d_1	Proportion of adult worms that attain sexual maturity
d_2	Proportion of transmission stages that survive to the infective stage
μ^{-1}	Life expectancy of the human host
μ_1^{-1}	Life expectancy of adult worms
μ_2^{-1}	Life expectancy of infective larvae
N	Human population size

Table 3.2 *Helminth parasite life expectancy in the human host (see Anderson and May 1985 for source references)*

Parasite	Life expectancy (years)[a]
Enterobius vermicularis	<1
Trichuris trichiura	1–2
Ascaris lumbricoides	1–2
Necator americanus	2–3
Ancylostoma duodenale	2–3
Schistosoma mansoni	3–6
Schistosoma haematobium	3–6
Wuchereria bancrofti	3–5
Onchocerca volvulus	8–10

a) Rough approximations, owing to the practical difficulties inherent in estimation.

Table 3.3 *Latent periods and vector life expectancies (see Anderson and May 1991 for source references)*

Parasite	Latent period (days)	Vector	Life expectancy (days)
Plasmodium falciparum	11 days (24°C)	*Anopheles gambiae*	8–15
Trypanosoma brucei	15–35	*Glossina morsitans*	28–32
Yellow fever virus	10–12	*Aedes aegypti*	8–14
Onchocerca volvulus	14	*Simulium damnosum*	10–20
Wuchereria bancrofti	13–15	*Anopheles funestus*	10–15
Schistosoma mansoni	20–42 (22°C)	*Biomphalaria glabrata*	7–42

secreted toxins or some sort of malfunctioning of the host induced regulatory constraints on the immune response.

Immunity to parasitic infections

One of the major features of the epidemiology of parasite infections is evidence for repeated infection as individuals age in areas of endemic exposure despite abundant evidence of immune recognition and attack. This is a puzzle of both practical and fundamental significance. Protozoa and helminths have complex genomes and elaborate developmental cycles, and have evolved a wide array of mechanisms to evade or modulate immunological attack (Maizels et al. 1993 and Chapter 4, Immunology and immunopathology of human parasitic infections).

As mentioned earlier, the observed aggregated distributions of parasite number per host may reflect differences in innate susceptibility to infection or in the ability to mount effective immunological responses. Work on the population dynamics of the interaction between subsets of T helper cells (Th1 and Th2 subsets) suggests that early exposure to infection in infancy (or even in the womb via maternal experience) may in part determine predisposition to heavy or light infection where high exposure generates tolerance with elevated Th2 responses concomitant with high helminth loads

(Schweitzer and Anderson 1992). The quantitative details of the host's ability to recognize and respond to particular parasite antigens (influenced by the genetic make-up of the host) is of importance in determining whether the host is persistently susceptible or develops protective immunity.

To tease out the factors that control observed patterns of infection and immunity in communities, epidemiological study must turn more and more to detailed longitudinal studies of changes in parasite numbers and immunological responses specific to parasite antigens. For the blood-dwelling schistosome flukes in humans (*S. mansoni* and *S. haematobium*), circulating specific and non-specific serum IgE correlates both positively with accumulated past experience of infection and negatively with re-infection rates after drug treatment (Hagan et al. 1987). For the gut-dwelling nematode *Trichuris trichiuria*, comparison of age-dependent isotope responses and the age-profile of worm burden in two endemic communities with very different levels of transmission suggests that parasite-specific secretory IgA responses best reflect accumulated past experience of infection, whereas serum IgG isotopes simply correlate positively with current worm burden (Needham et al. 1994) (Figure 3.17).

How specific immunological responses to protozoa and helminths operate to generate resistance (effector mechanisms) is poorly understood. Some information can be drawn from epidemiological studies via associa-

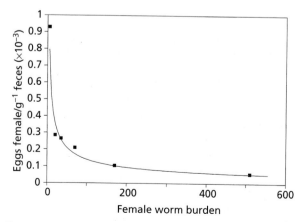

Figure 3.15 *Density-dependent fecundity in* Schistosoma mansoni *based on autopsy data collected by Cheever (1968) (see Medley and Anderson 1985)*

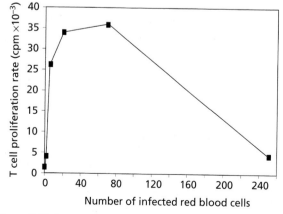

Figure 3.16 *Dose–response curve of the influence of antigen concentration (number of* P. falciparum *infected red blood cells) on T-cell proliferation rates (after Jones et al. 1990)*

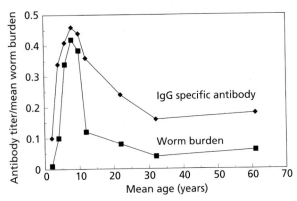

Figure 3.17 *Age-related changes in the burden of* Trichuris trichiura *and IgG antibodies specific to parasite antigens in a community in St Lucia, West Indies (after Needham et al. 1994)*

tion of parasite load with different types of response, or by following immune responses following drug treatment and during re-infection. A general pattern seems to be that individual responses mounted to protozoan and helminths rarely operate in isolation and resistance is typically a multi-component response (see Chapter 4, Immunology and immunopathology of human parasitic infections).

The many and exciting opportunities presented by advances in molecular biology for the study of immunity to parasites are parallelled by the challenge of an ever-expanding descriptive literature. At present our under-standing of how immunity (as reflected by resistance to infection and increased rates of parasite clearance from the host) works for any single protozoan or helminth parasite of humans is limited despite abundant descriptive study (see Chapter 4, Immunology and immuno-pathology of human parasitic infections). The population ecology of the immune system is a very non-linear world and future research will need to emphasize the measure-ment and quantification of the many rate parameters that control the humoral and cellular responses to para-site antigens, and the functional dependencies between parasite and immune system variables that determine the dynamics of infection both within the host and in the community of hosts. To achieve this for the major infec-tions of human communities will require more detailed longitudinal studies that combine conventional measures of parasite burden with immunological and molecular studies of parasite genetic diversity and the dynamics of immune responses under repeated exposure to infection.

Antigen diversity and strain structure in parasite populations

The maintenance of antigenic diversity is a key strategy adopted by many infectious agents to evade immunolo-gical attack in the host and to facilitate the invasion of hosts who have prior experience of infection. Advances in molecular biology have provided many new tools for

the study of genetic diversity within parasite populations and these have stimulated a shift in emphasis in epide-miological research towards a more evolutionary approach to the study of infection and immunity. Evolu-tion is at the core of the relationship between host and pathogen. In seeking explanations of observed epide-miological pattern, in the search for new drugs and vaccines, or indeed in the study of immunological responses to infection, the enormous potential of viruses, bacteria, protozoa, and helminths to evolve rapidly (either via mutation or recombination events) on time-scales much shorter than the generation time of the human host, is often inadequately appreciated. In many cases, particularly for microparasites, such evolution may take place on a timescale faster than that required to mount an effective immunological response. Genetic variation in microparasites, via antigenic change, is often a central pillar of the pathogen's strategy to persist in the host in the face of immunological attack. Persistence enhances the likelihood of transmission to new hosts in the population. This strategy is particularly common amongst persistent viral or protozoan infections, such as HIV or *Plasmodium* or *Trypanosoma* species, where the immune system acts as the main selective force driving the emergence and reproductive success of 'escape' anti-genic variants. One spin-off of the rapidly expanding databases on the antigenic diversity of important human pathogens (sequence databases), has been the develop-ment of statistical techniques to examine relatedness between different parasite isolates via the construction of phylogenetic trees (Harvey 1996). These techniques often enable inferences to be drawn (about who acquired infection from whom) via sequence data from parasite isolates from different patients. In the coming decade, sequence information on genetic variation within pathogen populations will increasingly be linked to phenotype, as mirrored by reproductive success of the parasite or the pathogenesis induced by different genetic variants.

At the between-host level, antigenic variation for both the microparasites and macroparasites may permit repe-ated infection of individual hosts as is evident in *P. falci-parum* infection in humans. The slow build up of cross-reactive (across parasite strains) immunological responses after repeated exposure to different antigenic strains will influence the duration of infection, the infec-tiousness of a person and in some cases the likelihood of the occurrence of disease. In many cases the delay in the development of immunity to parasitic infection may be a consequence of the antigenic diversity of the parasite in which protective immunity only develops after exposure to many different antigenic types or strains circulating in a given locality. The antigenic genetic constitution of a parasite population may vary from one locality to another (Gupta and Day 1994).

Where many strains persist in a given community the question arises of how this diversity is maintained. A

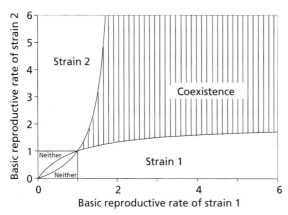

Figure 3.18 *The influence of the degree of cross-immunity (c = 0, total cross-immunity, c = 1, no cross-immunity, with c set at 0.5) on the pattern of coexistence of two strains of a microparasite as a function of their respective transmission success as measured by the basic reproductive number, R_0 (after Gupta et al. 1994a)*

simple view might be that the strain with the greatest reproductive success would, over time, competitively displace other strains. However, recent work shows that this is not necessarily the case, provided a degree of cross-immunity builds up to antigens shared by all strains. In the case of complete cross-immunity, competitive exclusion occurs, while in the case of no cross-immunity (unlikely in reality), all strains can persist since they circulate independently in the host population (Gupta et al. 1994a). For intermediate cases (probably the norm), coexistence is possible provided the reproductive successes (the R_0) of the various strains do not differ too widely in magnitude (Figure 3.18). Long-term studies of temporal and spatial trends in the prevalence of different parasite strains (e.g. *P. falciparum*, African trypanosomes, and some helminth species) are in progress, but as yet little is understood about longitudinal trends in parasite diversity within defined human communities (Anderson 1994; Snow et al. 1993). What has become clear, however, is that superinfection of an infected host by a new strain appears common for many of the microparasitic organisms, including HIV and malaria (Paul et al. 1995; Robertson et al. 1995).

The observation that many distinct strains persist in human communities over long periods of time raises a further question concerning the role of genetic exchange in the maintenance of parasite strain structure. Many parasites have the opportunity for the exchange of genetic material because of, for example, sexual processes in the case of *P. falciparum*, coinfection of the same host cell for HIV, or transformation in bacteria. The persistence of strain structure plus the frequent opportunity for genetic exchange appears paradoxical, because it would seem that the former would be lost rapidly as a result of genetic recombination. This issue has recently been resolved by theoretical and observational studies which reveal that dominant polymorphic

antigenic determinants (that is, those that elicit the most effective immune responses) will be organized into non-overlapping combinations of antigenic determinants as a result of selection by the host's immune system, thereby defining a set of distinctive independently transmitted strains (Gupta et al. 1996). Dominant polymorphic antigen determinants will be in linkage disequilibrium, despite frequent genetic exchange, even though they may be encoded by several unlinked genes. By contrast, weaker polymorphic determinants within the same parasite population will be in linkage equilibrium. This suggests that the detection of non-random association between epitope regions of parasite antigens can be employed as a strategy for identifying the dominant polymorphic antigens. To date, analyses of linkage disequilibrium patterns within infectious agent populations have, for the most part, only addressed the question of clonality (Walliker 1989; Tibayrenc et al. 1990).

The population dynamics and genetics of a parasite population that is structured into strains by a dominant immune response may be influenced at other levels by weaker responses to both conserved and other polymorphic determinants. For example, the typical age distribution of *P. falciparum* prevalence (see Figure 3.10) suggests that although the variant surface antigens (VSA) (which undergo antigenic variation within the host) categorize the population into strains (each with a given repertoire of expressed antigens), some degree of infection blocking immunity to all strains is eventually established after repeated exposure to a conserved antigen (Gupta and Day 1994).

More generally, these and other studies reveal the importance of a better understanding of the population genetics of parasite and host populations for the interpretation of epidemiological patterns. Molecular epidemiological studies that address the question of the linkage between genetic diversity in the parasite and the incidence of infection and associated disease are likely to be a major focus in the epidemiology of infectious diseases in the coming decade. The occurrence of disease in infected patients may often be linked to the subtle interplay between the genetic background of the host and that of the invading pathogen.

CONTROL OF INFECTION AND DISEASE

At present, the options for the community-based control of the major tropical parasitic infections are either chemotherapy (e.g. for malaria and the helminth infections) or interventions to prevent exposure to infection (e.g. impregnated bed nets to prevent contact with biting vectors, safe water supplies, and good sanitation to reduce contact with the transmission stages of intestinal nematodes). Vaccines are not available for any of the key parasitic infections and early trial results with candidate products have been disappointing. To date, no effective vaccine has been developed for an antigenically variable infectious agent that provides long protection duration, irrespective of whether the target agent is a virus, bacterium or protozoan. In research on candidate vaccines, most attention is directed at achieving high efficacy. However, duration of protection is of equal significance in terms of potential effectiveness within a community-based immunization program.

For both chemotherapy and interventions to reduce the likelihood of transmission, the difficulty of the task of achieving long-term control is related to the magnitude of the transmission intensity in a defined community as measured by R_0. To block or prevent transmission, the magnitude of R_0 must be reduced to less than unity in value. As control measures are introduced, and the value of R_0 progressively declines over time, significant changes in the prevalence of infection are only likely to occur as the critical point of $R_0 = 1$ is approached (Figure 3.19).

For microparasite infections, community-based chemotherapy or chemoprophylaxis (in the case of malaria) is expensive and beyond the reach of most communities in poor tropical regions. In addition, treatment tends to select for resistant strains of the parasite, as well illustrated by malaria. In many regions of the world, but particularly in South-East Asia, multi-drug resistant strains of *P. falciparum* are a major cause of morbidity and mortality. Some take a pessimistic view that with the decline in the rate of development of new antimalarials (because of drug development costs and the endemism of the infection in poor developing, as opposed to wealthy developed countries) the battle to control malaria by community-based drug use is likely to be futile in the longer term. As such, many see the future in terms of vaccines although current progress does not promote optimism for rapid advances in this area. The main hope at present lies in an integrated approach involving vector control by insecticides, the promotion of community-wide impregnated bed net use, and drug treatment to prevent mortality (see Chapter 5,

Control of parasites, parasitic infections and parasitic diseases).

Epidemiological study of the emergence and persistence of drug resistant parasites has been limited in scope and scale to date. Detailed long-term longitudinal studies (ideally cohort based) are required in defined areas to measure the frequency of the genes that determine resistance to defined drugs. In parallel with such molecular epidemiological studies, quantitative information must be acquired on patterns of drug use to define the intensity of selection that induces the observed change in resistance frequency. Much more thought must be given to defining optimum methods for the use of a number of different drugs, both for the individual patient and for the community, to maximize the effective life expectancy of a drug and minimize the rate of evolution of resistance.

A more encouraging picture emerges for the helminth infections where community-wide use of anthelmintics (often targeted at the child age groups in which the intensity of infection and associated morbidity is greatest) has been very effective in parasite control. To reduce R_0 to less than unity in value, the proportion of the population that must be treated per unit of time (i.e. monthly or yearly) must exceed a critical value, g, where

$$g = 1 - exp[(1 - R_0)/A]/h \qquad (3.8)$$

Here, A is the life expectancy of the parasite in the human host and h is the efficacy of the drug (Anderson and May 1991). To take a specific example, namely *Ascaris lumbricoides* for R_0 values around 3 and with $A = 1$ year, a drug of 95 percent efficacy would have to be administered to more than 91 percent of the population each year to eradicate the infection (Figure 3.20). However, if morbidity control is the aim, a much more targeted program of treatment of heavy infection in schoolchildren would suffice. Recent studies show that additional benefits accrue from the treatment of heavy infection, such as improved rates of growth in children,

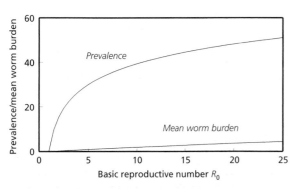

Figure 3.19 *The relationship between the prevalence and intensity of a macroparasitic infection as a function of the magnitude of transmission success as measured by* R_0*. Below the point* $R_0 = 1$*, the parasite is unable to persist in the host population.*

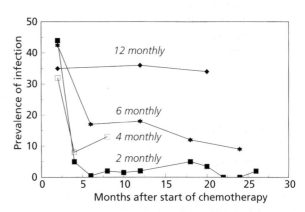

Figure 3.20 *The influence of different monthly intervals between rounds of mass chemotherapy on the prevalence of* Ascaris lumbricoides *in a study population in Korea (after Seo and Chai 1980)*

improved cognitive function, and a net reduction in transmission in the community as a whole (Nokes and Bundy 1994). A major practical advantage of targeting treatment to children is their accessibility during school attendance. The frequency of treatment following the introduction of control will depend on rates of re-infection as defined by the magnitude of R_0. For long-lived species with limited transmission success, such as the schistosome and filarial worms, intervals of a few to many years may suffice (Butterworth et al. 1991). Conversely, for the short-lived worms such as the intestinal parasites, yearly or more frequent treatment is required (Thein-Hliang et al. 1984).

Long-term control of all parasites in areas of endemic infection will require continual intervention or chemotherapeutic treatment unless the intensity of control is sufficient to eradicate the parasite. For the important infections in tropical regions, eradication is difficult to envisage given limited resources for public health measures. Vaccines are ideally required and their development must remain a research priority in the coming years (see Chapter 5, Control of parasites, parasitic infections and parasitic diseases).

The scientific assessment of the effectiveness of a given intervention program in both epidemiological and economic terms, is often poor despite large expenditure in the implementation of a specific program. More rigorous approaches to intervention study design and evaluation are required that meld quantitative epidemiology study pre- and post-intervention, with assessments of cost and benefit. Cost–benefit analyses of different control options must be based on a framework that captures the interaction between the transmission dynamics of the parasite and the intensity of the control measure. This relationship is invariably non-linear and hence intuition alone will not suffice to accurately deduce the potential impact of a given policy. The key problem, however, is often in the design of the intervention program. Detailed statistical analyses are required to determine the correct sample size (whether in units of people or communities) required to detect a defined level of change in the incidence of infection or disease following intervention.

The practical aspects of the control of parasitic diseases are discussed in more detail in Chapter 5, Control of parasites, parasitic infections, and parasitic diseases.

CONCLUSIONS

The interdisciplinary character of epidemiological research implies that advances in a wide variety of disciplines will influence our understanding of the processes that control the spread and persistence of infectious agents in human communities in the coming years. These include those in molecular biology, immunology, mathematics, genetics, information handling, storage,

and retrieval. Modern desktop computers plus associated software provide a striking illustration of how advances in one field of science can influence progress in epidemiological study. Database construction and handling for a large cohort intervention study is now a simple task, as are comparative studies of sequence data from different patients or pathogen isolates, or detailed statistical analyses seeking associations between host and parasite variables.

However, in parallel with such technical advances, whether in data management or the measurement plus detection of infection or immunity, there is an associated need to broaden the conceptual template on which epidemiological study is based. Until recently, much epidemiological research on infectious agents ignored the very dynamic nature of the interaction between host and parasite populations, both with respect to the dynamics of transmission and evolution. A failure to take account of population biology and evolution may even influence the more practical end of epidemiology, namely, cost–benefit analyses of different control interventions. Such analyses must be based on measures of the impact of control on both transmission success and parasite evolution.

Direct and indirect benefits arise from vaccinating or drug treating individual patients. The direct benefits to the patient are obvious, but the indirect benefits relate to the removal of a host that can potentially pass infection to others. As such, net benefit to the community includes an element related to the reduction in transmission success of the parasite. Drug use may promote the rate of spread of drug resistant parasite strains and hence reduce the period over which a given drug is an effective treatment.

A key problem in all epidemiological studies is the interpretation of heterogeneity in patterns of infection and disease within human communities. It is in this area that recent advances in biological research, particularly at the molecular genetic level, are likely to promote the greatest change in epidemiological research. Genetic diversity in host and pathogen populations is now measurable on the scale required for population-based comparisons and linkage analyses. There seems little doubt that such work will begin to provide answers to such fundamental questions as to why some are predisposed to specific infections and associated disease, while others are not.

REFERENCES

Abel, L., Demenai, F., et al. 1991. Evidence for the segregation of a major gene in human susceptibility/resistance to infection by *Schistosoma mansoni. Am J Hum Genet*, **48**, 959–70.

Anderson, R.M. 1987. Determinants of infection in human schistosomiasis. In: Mahmoud, A.F. (ed.), *Baillière's Clinics in tropical medicine and communicable diseases, Schistosomiasis*, vol. 2. London: Baillère Tindall, 278–300.

Anderson, R.M. 1994. The Croonian Lecture. Populations, infectious disease and immunity: a very nonlinear world. *Phil Trans R Soc Lond B*, **346**, 457–505.

Anderson, R.M. and Gordon, D.M. 1982. Processes influencing the distribution of parasite numbers within host populations with special emphasis on parasite-induced host mortalities. *Parasitology*, **85**, 373–98.

Anderson, R.M. and May, R.M. 1985. Helminth infections of humans: mathematical models, population dynamics and control. *Adv Parasitol*, **24**, 1–101.

Anderson, R.M. and May, R.M. 1991. *Infectious diseases of humans; dynamics and control*. Oxford: Oxford University Press.

Aron, J.L. and May, R.M. 1982. The population dynamics of malaria. In: Anderson, R.M. (ed.), *The population dynamics of infectious diseases: theory and applications*. London: Chapman & Hall, 139–79, Chapter 5.

Bensted-Smith, R., Anderson, R.M., et al. 1987. Evidence for predisposition of individual patients to reinfection with *Schistosoma mansoni* after treatment. *Trans R Soc Trop Med Hyg*, **81**, 651–6.

Boyd, M.F. 1949. *Malariology*. Philadelphia: Saunders.

Bradley, D.J. and McCullough, F.S. 1973. Egg output stability and the epidemiology of *Schistosoma haematobium*. Part II. An analysis of the epidemiology of endemic *S. haemotobium*. *Trans R Soc Trop Med Hyg*, **67**, 491–500.

Bundy, D.A.P., Cooper, E.S., et al. 1987. Predisposition to *Trichuris trichiura* infection in humans. *Epidemiol Infect*, **98**, 65–72.

Butterworth, A.E., Sturrock, R.F., et al. 1991. Comparison of different chemotherapy strategies against *Schistosoma mansoni* in Machakos District, Kenya: effects on human infection and morbidity. *Parasitology*, **103**, 339–55.

Cheever, A.W. 1968. A quantitative post-mortem study of *Schistosoma mansoni* in man. *Am J Trop Med Hyg*, **17**, 38–64.

Davidson, G. and Draper, C.C. 1953. Field study of some of the basic factors concerned in the transmission of malaria. *Trans R Soc Trop Med Hyg*, **47**, 522–35.

Dye, C.M. and Hasibeder, H.G. 1986. Population dynamics of mosquito-borne disease: effects of flies which bite some people more than others. *Trans R Soc Trop Med Hyg*, **83**, 69–77.

Elkins, D., Haswell-Elkins, M., et al. 1986. The epidemiology and control of intestinal helminths in the Publicat Lake region of Southern India. I. Study design and pre- and post-treatment on *Ascaris lumbricoides* infection. *Trans R Soc Trop Med Hyg*, **80**, 774–92.

Fulford, A.J.C., Ouma, J.H., et al. 1996. Water contact observations in Kenyan communities endemic for schistosomiasis: methodology and patterns of behaviour. *Parasitology*, **113**, 223–4.

Grenfell, B.T. and Anderson, R.M. 1985. The estimation of age-related rates of infection from case notifications and serological data. *J Hyg (Camb)*, **95**, 419–36.

Greenwood, B.M., Marsh, K. and Snow, R. 1991. Why do some African children develop severe malaria? *Parasitol Today*, **7**, 227–8.

Gupta, S.D. and Day, K.P. 1994. A strain theory of malarial transmission. *Parasitol Today*, **10**, 476–81.

Gupta, S., Swinton, J. and Anderson, R.M. 1994a. Theoretical studies of the effects of heterogeneity in the parasite population on the transmission dynamics of malaria. *Proc R Soc London B*, **256**, 231–8.

Gupta, S., Trenholme, K., et al. 1994b. Antigenic diversity and the transmission dynamics of *P. falciparum*. *Science*, **263**, 961–3.

Gupta, S., Maiden, M.C.J., et al. 1996. The maintenance of strain structure in populations of recombining infectious agents. *Nature Medicine*, **2**, 437–42.

Guyatt, H.L., Bundy, D.A.P., et al. 1990. The relationship between the frequency distribution of *Ascaris lumbricoides* and the prevalence and intensity of infection in human communities. *Parasitology*, **101**, 139–43.

Hagan, P., Blumenthal, U.J., et al. 1987. Resistance to infection with *Schistosoma haematobium* in Gambian children: analysis of their immune response. *Trans R Soc Trop Med Hyg*, **81**, 938–46.

Harvey, P.H. 1996. Phylogenies for ecologists. *J Anim Ecol*, **65**, 255–63.

Jones, K.R., Hickling, J.K., et al. 1990. Polyclonal in vitro proliferative responses from non-immune donors to *Plasmodium falciparum* malaria antigens require UCHL+ (memory) T cells. *Eur J Immunol*, **20**, 307–15.

Maizels, R.M., Bundy, D.A.P., et al. 1993. Immunological modulation and evasion by helminth parasites in human populations. *Nature (London)*, **365**, 686–805.

Marquet, S., Abel, I., et al. 1996. Genetic control of a locus controlling the intensity of infection by *Schistosoma mansoni* on chromosome 5q31-q33. *Nat Genetics*, **14**, 181–4.

May, R.M. 1977. Togetherness among schistosomes: its effects on the dynamics of the infection. *Math Biosci*, **35**, 301–43.

Medley, G. and Anderson, R.M. 1985. Density-dependent fecundity in *Schistosoma mansoni* infections in man. *Trans R Soc Trop Med Hyg*, **79**, 532–4.

Needham, C.S., Lillywhite, J.E., et al. 1994. Temporal changes in *Trichuris trichiura* infection intensity and serum isotype responses in children. *Parasitology*, **109**, 197–200.

Nokes, C.B. and Bundy, D.A.P. 1994. Does helminth infection affect mental processing and educational achievement? *Parasitol Today*, **10**, 14–18.

Paul, R.E.L., Packer, M.J., et al. 1995. Mating patterns in malaria parasite populations of Papua New Guinea. *Science*, **269**, 1709–11.

Perry, K.R., Brown, D.W.G., et al. 1993. Detection of measles, mumps and rubella antibodies in saliva using capture radioimmunoassay. *J Med Virol*, **40**, 235–40.

Robertson, D.L., Sharp, P.M., et al. 1995. Recombination in HIV-1. *Nature*, **374**, 124–6.

Schad, G.A. and Anderson, R.M. 1985. Predisposition to hookworm infection in man. *Science*, **228**, 1537–40.

Schweitzer, N. and Anderson, R.M. 1992. The regulation of immunological responses to parasitic infections and the development of tolerance. *Proc R Soc Lond B*, **247**, 107–12.

Seo, B. and Chai, J. 1980. Comparative efficacy of various internal mass treatment of *Ascaris lumbricoides* infection in Korea. *Korean J Parasitol*, **18**, 145–51.

Snow, R.W., Schellenberg, J.R.M.A., et al. 1993. Periodicity and space-time clustering of severe childhood malaria on the coast of Kenya. *Trans R Soc Trop Med Hyg*, **87**, 386–90.

Thein-Hliang, Saur, J. et al. 1984. Epidemiology and transmission dynamics of *Ascaris lumbricoides* in Okpo, rural Burma. *Trans R Soc Trop Med Hyg*, **78**, 497–504.

Tibayrenc, M., Kjellberg, F. and Ayala, F.J.A. 1990. A clonal theory of parasitic protozoa:the population structure of *Entamoeba, Giardia, Leishmania, Naegleria, Plasmodium, Trichomonas* and *Trypanosoma* and their medical significance. *Proc Natl Acad Sci USA*, **87**, 2414–18.

Tyrrell, D.A.J. 1977. Aspects of infection in isolated communities. *Infectious diseases in isolated communities*. London: CIBA Foundation, 137–53.

Walliker, D. 1989. Genetic recombination in malaria parasites. *Exp Parasitol*, **69**, 303–9.

Wilkins, H.A., Blumenthal, U.J., et al. 1987. Resistance to infection after treatment of urinary schistomisasis. *Trans R Soc Trop Med Hyg*, **81**, 29–35.

4

Immunology and immunopathology of human parasitic infections

F.E.G. COX AND DEREK WAKELIN

INTRODUCTION

The human immune system is concerned with defense against invading organisms and the removal of malignant cells. The role of the immune system is to deliver an appropriate immune response at the right time and place and it is not surprising, therefore, that different infectious agents should elicit different immune responses. In order to understand the nature of infectious agents, particularly parasites, it is necessary to distinguish between microparasites and macroparasites (see Chapter 3, Epidemiology of parasitic infections). Microparasites are small and multiply within their vertebrate host, often inside cells, thus posing an immediate threat unless contained by an appropriate immune response. Viruses, bacteria, and protozoa are microparasites, typically inducing an infection in which the host becomes infected, then experiences a latent period during which the parasites multiply. This is followed by a period of disease and discomfort during which the immune response brings the infection under control and, finally, a gradual recovery and long-term resistance to reinfection. Macroparasites (i.e. helminths) are large and most do not multiply within their vertebrate host and, therefore, do not present an immediate threat after initial infection. However, the host must protect itself from overwhelming infections and reinvasion by infective stages and can only do so by eliciting an appropriate immune response. Immune responses to protozoa (microparasites) and helminths (macroparasites) are therefore very different from one another.

Although protozoa and helminths are very different in terms of the immune responses they elicit, the infections they cause have a great deal in common, usually being long-lasting and, over a period of years, inducing immunopathological changes that may be more dangerous than the infection itself. In short, the parasite invades the host, is recognized as foreign and elicits an apparently appropriate immune response. However, parasites have all evolved mechanisms for evading the immune response and the net result is a situation in which the infection is only partially controlled. As the parasite may not be totally eliminated, the host continues to mount increasingly complex immune responses until the parasite kills the host, the host overcomes the infection or there is an uneasy compromise. During the course of all infections, parasites die or are killed and parasite molecules deposited on host cells may elicit autoimmune responses which, in turn, contribute to the pathology of the infection. Some parasites can avoid eliciting an immune response by mimicking host molecules which, if effective, can be very successful but, if unsuccessful, can initiate an autoimmune reaction. The majority of parasitic infections also elicit powerful inflammatory responses that may alter, render nonfunctional or even destroy host tissues.

Virtually all parasitic infections are long-term and chronic, and it was once thought that there were no effective, protective immune responses against parasitic infections in humans. It is now clear, however, that some form of immunity is the rule. The evidence for this comes from a number of sources:

- in endemic areas, the majority of people experience clinical immunity to parasitic infections and the prevalence of infection falls with age while immunological parameters increase
- the uneasy compromise that exists after recovery can be broken, for example by immunosuppressive drugs or concomitant infections
- in every experimental model investigated, immunity can be demonstrated and there is no reason why humans should be different.

IMMUNE RESPONSES TO PARASITIC INFECTIONS

The overall immune response

The immune response to any parasite, as with any other infectious agent, begins when the host becomes infected and the parasite is recognized by the cells of the immune system. In order to initiate an immune response, molecules of the parasite must come into contact with the cells of the immune system and, if this does not happen, no immune response will be initiated. Many parasites invade host cells and tissues and thus come into direct contact with the immune system; most intestinal parasites pass through host tissues before reaching their final location. Even for intestinal parasites that do not breach the integrity of gut wall, such as the protozoan flagellate *Giardia duodenalis*, there is the potential for direct antigen uptake across the mucosa and the possibility of evoking an immune response. Once a parasite is recognized as foreign, the next series of events is initiated. The events involved in recognition are complex and much of what we know about them is concerned with peptide antigens. This chapter contains no detailed discussion of carbohydrate antigens although these play an important role in many antiparasite responses. In the simplest situation, the first event that occurs is when parasite peptide antigens are processed by antigen-presenting cells (APC) and transported to the surface of the cell in combination with specific major histocompatibility complex (MHC) molecules. The whole complex, consisting of the MHC molecule and the peptide, is then recognized by antigen receptors on helper T lymphocytes and the immune response proper is initiated. These early steps are known as antigen processing and presentation and these events determine the eventual outcome of the immune response. A number of cell types, particularly macrophages and dendritic cells, can act as APCs. In many parasitic infections, macrophages are involved and, as the antigen on the surface of the

APC binds to the receptor on the helper T lymphocyte, the macrophage releases T-cell-activating molecules known as cytokines. In response, the helper T cells produce other cytokines that, in their turn, activate the effector cells of the immune system. The specific target of the immune attack is the antigen that initiated the immune response in the first place but parasites, particularly helminths, are very large targets and the immune system has to recognize a large number of different surface antigens. Not all of these are involved in protection, and some are even able to deflect the immune response away from more vulnerable targets. Much research has been devoted to the identification and characterization of protective antigens in order to try to understand how the immune response to parasites works and how to target vaccines to relevant molecules. Considerable effort has also been put into identifying antigens that are irrelevant to protection, or even counter-protective, or are involved in the evasion of the immune response and the pathology of the disease.

Although the immune responses to parasites are complex, the individual components are relatively simple and there is nothing unique or special about immunity to parasites except for the degree of complexity. In this context, it is convenient to consider the protective immune response against infectious organisms as having three arms:

1 cytotoxic T (Tc) cells
2 activated macrophages
3 antibody.

The Tc cells plus activated macrophages constitute what is conventionally called cell-mediated or antibody-independent immunity, while antibody constitutes humoral or antibody-dependent immunity (Figure 4.1). The Tc cells engage target cells expressing antigens in combination with class I MHC molecules and bring about their cytotoxic activity by creating pores in the membrane of the target cell, which then ruptures. Tc cells are frequently involved in viral infections but are seldom implicated in parasitic infections. Natural killer (NK) cells constitute a heterogeneous collection of lymphocytes that are neither B nor T cells, are not specific to a particular antigen and do not involve MHC class I molecules. NK cells seem to be able to kill target cells directly by forming pores in the target cell membrane, which ruptures as in Tc cell killing. NK cells are being increasingly implicated in protection against parasitic infections, not necessarily as effector cells but as the sources of important regulatory molecules.

Activated macrophages, macrophages that have been sensitized and triggered by signals such as lipopolysaccharide and interferon-gamma (IFN-γ) are among the most important cells involved in protection against parasites and can kill target cells either intracellularly or extracellularly. If it is small, the target cell can be

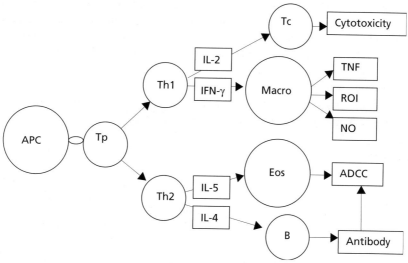

Figure 4.1 *Diagrammatic representation of the main features of the overall immune response. Antigen is recognized and processed by APCs and presented to precursor T (Tp) cells that carry receptors for the antigen. Depending on the nature of the antigen, and how it is presented, the Tp cells differentiate into Th1 or Th2 cells. Th1 cells produce a number of cytokines including IL-2 and INF-γ. IL-2 activates Tc cells that kill target cells carrying the original antigen in an MHC restricted manner. INF-γ activates resting macrophages (Macro) making them more phagocytic and causing them to release a variety of molecules, including TNF-α, ROI, and NO. ROI and NO are involved in the intracellular killing of ingested parasites or the destruction of parasites in close proximity to the activated macrophage. Th2 cells release a different set of cytokines, including IL-4 and IL-5. IL-4 is involved in B cell differentiation and activation and the release of antibodies specific to the antigen. IL-5 activates eosinophils (Eos). Antibodies may act in cooperation with cells such as eosinophils and macrophages by acting as a bridge bringing the activated cell and the parasite together and facilitating the release of toxic molecules directly onto the surface of the parasite, a process known as ADCC.*

phagocytosed and destroyed in the phagolysosome via a number of pathways including:

- proteolytic enzyme activity
- reactive oxygen intermediates (ROI) (superoxide, hydrogen peroxide and hydroxyl radicals) that destroy the target cell though a process of lipid peroxidation
- nitric oxide (NO) that reacts with the superoxide anion to produce highly toxic peroxynitrite and hydroxyl radicals.

Macrophages also produce tumor necrosis factor-alpha (TNF-α), which has a synergistic effect with NO. These molecules, among other things, interfere with membrane integrity, DNA structure and mitochondrial function and bring about the destruction of, or damage to, target cells. One of the key features of the activation of macrophages is that the various toxic molecules produced are released not only into the phagolysosome but also extracellularly and are thus able to destroy cells in the close vicinity of the activated macrophage. Macrophages are, therefore, able to kill large cells or multicellular organisms, especially if the macrophage is closely bound to the target via, for example, an antibody bridge as described below. These toxic molecules can also damage host cells, and their release in an inappropriate place or in excess is a major cause of immunopathology. TNF-α, for example, is highly toxic and can cause cachexia, hemodynamic dysfunction, neurotoxicity, acidosis, and

renal failure; IFN-γ can cause tachycardia, leukopenia, and malaise, and can enhance the effects of TNF-α. The roles of TNF-α, oxygen-derived free radicals, NO, and other molecules in pathogenesis are reviewed by Clark and Cowden (2003).

The most important cells with direct effector functions in immune responses are Tc cells, activated macrophages and B lymphocytes (Figure 4.1). Each of these cells has its specific role in the immune response and can act concurrently or sequentially, with or without other cells, to bring about the immobilization or destruction of the invading organism. Other cells commonly seen in response to parasitic infections, particularly helminths, include eosinophils (Figure 4.1). Eosinophilia is a characteristic of a number of helminth infections but the role of these cells in human infections is enigmatic. Much of what we know about the part played by eosinophils in parasitic infections has been derived from histological studies but the elucidation of the role of interleukin (IL)-5, the key eosinophil growth and development factor, has made it possible to examine the role of these cells in more detail. It is now clear that eosinophils do play an important role in resistance to a range of helminth infections in mice, rats, dogs, primates, and humans, and that they are involved in the killing of the larval, but not the adult, stages of these parasites (Meeusen and Balic 2000).

The humoral immune response is concerned with B cell products, i.e. immunoglobulin (Ig) antibodies. The four major classes, IgA, IgG, IgM, and IgE, occur in different situations and are effective against parasites in different ways although there is some sharing of functions. Antibodies are directly effective against invading organisms in four main ways: neutralization, agglutination, complement activation, and facilitated opsonization. IgA occurs as two subclasses, IgA1 and IgA2, each of which can occur in a monomeric, dimeric or, occasionally, polymeric form. IgA is essentially an antibody found in secretions where it protects the mucous surfaces of the gut, respiratory tract and urinogenital tract; it also occurs in breast milk. Its main functions are the neutralization of bacterial toxins and the inhibition of the invasion of the mucosa, either by agglutinating the infectious agents or blocking attachment to host cells. IgG and IgM are found in the serum and extra-vascular spaces. IgM, a pentameric immunoglobulin that occurs as a single subclass, is the first antibody produced in response to antigenic stimulation; it is very good at agglutinating microorganisms with repeated epitopes and at activating complement via the classic pathway. IgG occurs as four subclasses (IgG1, IgG2, IgG3, and IgG4), and usually replaces IgM; it is the most common antibody in secondary immune responses. All subclasses can agglutinate microorganisms. IgG1, IgG2, and IgG3 can activate complement, IgG3 being the most effective. IgG1 and IgG4 can pass across the placenta, and IgG1 and IgG3 can bind to macrophages and neutrophils via their Fc portions. IgE has an affinity for mast cells, basophils and eosinophils and, when bound to an antigen, causes cross-linking, resulting in the release of pharmacologically active substances from the cells to which it is bound.

Antibodies seldom work in isolation from other components of the immune system. IgG and IgM fix complement and initiate complement-mediated lysis of the target cell. IgG and IgE bind to macrophages, eosinophils, and mast cells, via the Fc portion, and to the target cell by the Fab portion, and thus facilitate the destruction of the target cell by toxic molecules released by the effector cell, a process known as antibody dependent cell-mediated cytotoxicity (ADCC).

The immune responses evoked during infections with parasites are not unusual and are essentially the same as those involved in immunity to other infectious organisms. What is different is the complexity of the immune response in reaction to diverse antigens expressed by different stages of complicated life cycles often in different locations in the host. This leads to a plethora of antibody-independent and antibody-dependent responses which, over a period of time, can give rise to misdirected immune responses leading to pathological changes instead of protection. For example, the production of TNF-α in malaria is beneficial but in excess is detrimental; in schistosomiasis potentially protective responses lead to the formation of harmful granulomas around nonthreatening eggs deposited in tissues. For these reasons, current interest in immunity to parasites is being concentrated on an understanding of the overall control of the immune response which is brought about by the interactions of the various cytokines involved in a cytokine network.

The cytokine network in parasitic infections

Cytokines are low molecular weight protein mediators that are involved in the growth, differentiation and repair of cells and occupy a central position in the regulation of the immune response. Since their discovery in the 1960s, the study of cytokines has revolutionized understanding of both the control of the immune response and the nature of the pathology associated with diseases caused by infectious organisms. This has become a massive field and is covered in more detail in Chapter 15, Cytokines in the Immunology volume of this series and, in the context of infectious diseases, in Kotb and Calandra (2003). The cytokine network acts as a series of signals between the cells of the immune system. Many cytokines are growth or activation factors and without them the immune system could not function. An understanding of the cytokine network has permitted immunologists to unravel the interactions that come into play during an immune response and to consider individual components separately before considering them in the context of the response as a whole. Four main groups of cytokines are involved in the immune response: interleukins, interferons, transforming growth factors, and tumor necrosis factors (Tables 4.1 and 4.2). All of these have been intensively studied, particularly the interleukins; about 30 interleukins have been described but there is a considerable degree of overlap in structure and function, particularly concerning the 'over-16s', giving rise to some confusion and concern (Eberl 2002). Among the most important cytokines in the context of immunity to infectious diseases are IL-1–6, 8–10, 12, 13, and 15–18, interferon-alpha (IFN-α), interferon-beta (INF-β) and IFN-γ, TNF-α and tumor necrosis factor-beta (TNF-β). Other cytokines involved, directly or indirectly, in infectious diseases include a group known as chemokines (including IL-8) that are responsible for the induction and migration of subsets of leukocytes. Many cytokines occur in more than one form, e.g. IL-1a and IL-1b, with subtly different functions. Cytokines are produced by many cell types and act on many cell types, including the ones that produce them, in some cases promoting cell activity and in other cases inhibiting it. Thus, there is a network of cytokine activity that can be very difficult to unravel. Nevertheless, it is possible to distinguish two distinct pathways of cytokine activity:

1 the pro-inflammatory pathway (mainly involving IL-1, IL-6, IL-12, IL-15, IL-18, IFN-γ and TNF-α) that leads to cell-mediated immune responses

Table 4.1 Interleukins involved in immunity to parasitic infections

Cytokine	Important source cell(s)	Important target cell(s)	Important action(s)	Role in immunity to parasitic infections
IL-1	Macrophages and virtually all nucleated cells	Macrophages, T cells, B. cells, eosinophils, neutrophils	Lymphocyte activating factor. Stimulation of immune response, increase in production of other cytokines	Involved in all parasitic infections
IL-2	Th1 cells	Th and Tc cells	T cell growth factor. Growth and proliferation of T cells, induction of cytotoxic T cells	Cytotoxic killing of liver stages of *Plasmodium*
IL-3	T cells	Hemopoietic and other cells	Multicolony stimulating factor. Hemopoiesis	Involved in all parasitic infections
IL-4	Th2 cells	B cells, macrophages, mast cells, and other cell types	Proliferation of B cells, direction of immune response to Th2 pole, immunoglobulin switching, inhibition of inflammation	Involved in antibody production in all parasitic infections. Production of IgE. Expulsion of gastro-intestinal nematodes, Inhibition of cell mediated immunity in leishmaniasis
IL-5	T cells, mast cells, eosinophils	Eosinophils, B cells	Eosinophil differentiation factor. Production of eosinophils	Eosinophilia in helminth infections, ADCC killing of helminth larval stages
IL-6	T cells, monocytes, macrophages, and other cell types	B cells, myeloid cells, and other cell types	B cell stimulatory factor. B cell growth and differentiation	
IL-7	Bone marrow cells	Pre-B and pre-T cells	Lymphopoietin. Growth of pre-T and pre-B cells	
IL-8	Monocytes and macrophages	Neutrophils	Neutrophil chemoattraction	
IL-9	Th2 cells	T cells, mast cells	T cell growth factor. Synergizes with IL-4 in antibody production	
IL-10	Th2 cells, monocytes, macrophages	Th1 cells, NK cells, macrophages	Immunoregulation. Enhancement of Th2 responses, inhibition of Th1 cell and macrophage function, inhibition of IL-12, IL-13 and IFN-γ	Inmunity in leishmaniasis. Downregulation of immunity in filariasis
IL-11	Bone marrow cells	Pre-B cells	Promotion of pro-inflammatory cytokines IL-1, IL-6 and NO	
IL-12	Macrophages	T cells, NK cells	NK cell stimulatory factor. Promotion of Th1 responses, inhibition of IL-4 and IL-10	Enhancement of immunity to intracellular protozoa, e.g. *Leishmania, Trypanosoma cruzi, Toxoplasma*
IL-13	Th2 cells, other T cells, eosinophils, mast cells	B cells, monocytes	Proliferation of B cells, promotion of Th2 differentiation, inhibition of Th1 activities	Expulsion of gastro-intestinal nematodes
IL-14	T cells	B cells	Proliferation of B cells, inhibition of antibody secretion	
IL-15	Monocytes	T cells, NK cells	T cell growth factor. Promotion of Th1 cell activities, enhances cytotoxic activities of T cells and NK cells	
IL-16	T cells	T cells	Chemoattraction	
IL-17	T cells		Initiation of inflammatory responses	
IL-18	Activated macrophages	Macrophages	Induction and enhancement of IFN-γ and NK cell cytotoxicity	Regulation of mast cell and Th2 cytokines in trichinellosis

There is a considerable degree of cross-function among the post-16 cytokines (IL-17 to IL-27) (see Eberl 2002). Most of these cytokines have not been thoroughly investigated in parasitic infections.

Table 4.2 *Properties of important cytokines in immunity to parasitic infections*

Th1 cytokines	Th2 cytokines	Pro-inflammatory cytokines	Counter-inflammatory cytokines	Cytokines that can lead to pathology[b]
IFN-γ[a]	IL-4[a]	IL-12	IL-4	IL-1
IL-2	IL-5[a]	IL-15	IL-10	IL-6
IL-3	IL-3	IL-18	TGF-β	IL-8
TNF-α	IL-13	IFN-γ		IL-12
TNF-β	IL-6			TNF-α
GM-CSF	IL-10			MIF
	TGF-β			

a) These are the most important cytokines.
b) For example, increased vascular permeability, tissue damage, circulatory collapse, multi-organ failure etc.

2 a pathway (involving IL-3, IL-4, IL-5 and IL-13) that leads to the activation of the humoral immune response, the production of antibodies and the activation of inflammatory cells.

These distinctions are not absolute and some cytokines can have contrasting functions, for example IL-6 is a pro-inflammatory cytokine that can also indirectly damp down such responses through its ability to control the production of acute phase proteins. In all infections there are well-regulated pathways of cytokine activity that lead to protection but these are nearly always accompanied by counter-protective pathways that damp down potentially damaging side-effects (Figure 4.2). Many cell-mediated immune responses, for example, are characterized by the production of dangerous molecules such as TNF-α and NO that can cause pathology but are modulated by other cytokines such as IL-4, IL-10, and IL-13.

From the moment an immunologically intact host is infected with a parasite or any other infectious agent, the host begins to mount an appropriate protective immune response. The key cells are the so-called T helper (Th) lymphocytes characterized by the possession of a CD4 surface marker. At first, these are uncommitted but gradually differentiate into Th1 and Th2 cells, each characterized by the cytokines they produce, until eventually they become fully differentiated; when this happens, they are mutually exclusive. The Th1 cells produce T1 cytokines that drive the immune response towards antibody-independent immune responses, in particular IL-2 that drives the immune response towards the production of CD8 Tc cells and IFN-γ that drives the response towards the activation of macrophages. Th2 cells, on the other hand, produce IL-4, IL-5, IL-10, and IL-13 that lead to the activation of B cells, and the subsequent production of antibody, and to the proliferation and differentiation of eosinophils. A third subset of Th cells, Th3 or T-regulatory (Tr) cells produce cytokines such as IL-10 and transforming growth factor-beta (TGF-β) that down-regulate the activities of other Th cells and, therefore depress or even suppress the immune response. In short, the Th1 responses represent the cell-mediated arm of the immune response and the Th2 responses represent the humoral arm. Tc cells are ideally suited for the destruction of virus-infected cells, IFN-γ-activated macrophages are involved in the killing of intracellular pathogens and the antibody produced by B cells is most effective against extracellular pathogens such as helminth worms. The role of Th1 and Th2 cells in a number of infectious diseases is well discussed in the various contributions in Romagnani (1996) and Kotb and Calandra (2003). Parasites are no different from other pathogens in that they inevitably induce some kind of immune response except that Tc cells are less involved in parasitic infections than in viral infections. In general, it is widely accepted that protective Th1 responses predominate in infections caused by intracellular protozoa whereas Th2 responses are more important in immunity to extracellular helminth infections. With reference to specific examples, toxoplasmosis (like tuberculosis) is highly polarized towards the Th1 pole, leishmaniasis (and leprosy) tend towards the Th1 pole, malaria tends towards the Th2 pole and helminth

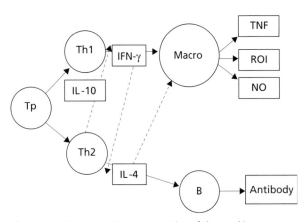

Figure 4.2 *Diagrammatic representation of the cytokine cross-regulation of the immune response. IFN-γ, as well as having a positive role in the activation of macrophages (Th1 response) has a negative effect on Th2 responses. Similarly, the Th2 cell products IL-4 and IL-10 inhibit Th1 responses. Positive pathways are here indicated by solid lines and negative pathways by broken lines.*

infections are polarized to the Th2 pole (Garraud et al. 2003). In addition, the mutual exclusivity mentioned above frequently results in extreme polarization in which one arm of the immune response is protective and the other counter-protective. However, these are generalizations and the details of each individual immune response can differ from time to time or from stage to stage of an infection. It is also important to bear in mind that the Th1/Th2 paradigm is not a rigid set of rules and that there may be switches in the patterns of Th1 and Th2 cytokines during the course of an infection (Allen and Maizels 1997).

Th cell subsets and the cytokine network were originally identified and defined in mice (Mossman and Coffman 1989) but virtually every component has since been shown to apply to humans as well (Romagnani 1991); parasitic infections provided the first evidence for human Th cell dichotomy. Currently, extrapolations are made from mouse to human and are tested before any definite assumptions are made.

The polarization of T cells towards cell-mediated or antibody-mediated responses does not depend entirely on the infectious agent involved but can be modulated by pre-existing factors including the cytokines present locally. For example, the presence of IL-12 or IFN-γ drives the immune response towards the Th1 pole whereas IL-4 drives it towards the Th2 pole (Ma et al. 1997). Initial or subsequent polarization involves the interaction of a number of regulatory cytokines some of which act as growth and differentiation factors. What is important here is that these cytokines, and also effector molecules, act nonspecifically. It is, therefore, possible for one infectious agent to be caught up in the cytokine network generated in response to another acquired previously, concurrently or subsequently (see Cox 2001).

Another important factor in determining the outcome of an infection is whether or not the established infection is inducing a Th1 or a Th2 response. The Th1 immunological milieu involves a number of molecules and the cells that produce them, in particular, NK cells, IL-12, and IFN-γ. The initiation of a new immune response in such a situation is gradually forced towards the Th1 pole and, if the superimposed agent is controlled by Th2 responses, it is at an advantage in such a situation. The converse applies if IL-4 predominates in the immunological milieu at the time of the second infection.

By the beginning of the 1990s, a general pattern emerged in which it appeared that, in the majority of protozoal infections in experimental models, Th1 cells were protective and Th2 cells were counter-protective, whereas in helminth infections the opposite was the case, Th2 cells were protective and Th1 cells counter-protective (Cox and Liew 1992). It is now clear that the interplay between cytokines in the immune response to parasites is not as simple as it once seemed to be and

can vary over the duration of the infection, at different sites in the host, between different species or strains of hosts and between different strains of parasite (Allen and Maizels 1997). Nevertheless, the Th1/Th2 dichotomy is a useful concept when attempting to unravel the complex immune responses that occur in parasitic infections; this can be illustrated by reference to NO, a reactive gas with many functions that has assumed a central role in our understanding of the immunology and immunopathology of parasitic infections. Its role in cell signaling has been known for a long time but it is only in the past decade that its role in immunity to infection has been understood (Clark and Rockett 1996). There are two important isoforms of NO: constitutive and inducible. Constitutive NO is found in a variety of cells. Inducible NO is produced by macrophages under the control of the enzyme inducible nitric oxide synthase (iNOS). NO is a powerful antimicrobial compound and is capable of killing a range of organisms including bacteria, protozoa, and helminth worms. It acts on iron–sulfur clusters in macromolecules and represses DNA synthesis and iron-containing enzymes, including those involved in respiration. Nitric oxide synthase is up-regulated by IFN-γ and TNF-α and is down-regulated by IL-4, IL-5, and IL-10. NO is, therefore, a key molecule in the immune response to parasitic infections and its production and activity are favored by Th1 responses. However, NO is also potentially dangerous and is inhibited by Th2 responses (James 1995).

It is important to understand the cytokine system in parasitic infections because of the complexity of the immune responses involved when coping with highly evolved and antigenically diverse pathogens, and in order to interpret events leading to protection on one hand and pathology on the other. Interpretations from only a few years ago of the immune response to parasites now seem very simplistic and new discoveries have called for a reinterpretation of many of the facts observed. On the other hand, new discoveries have enabled us to understand much more clearly many of the phenomena previously observed. In the remainder of this chapter, these points are illustrated by reference to the most important parasitic infections in terms of human parasitic diseases.

IMMUNITY TO PROTOZOA

Characteristics of protozoa in relation to the immune response

Unlike most viral and bacterial infections, protozoal infections tend to be chronic and to be associated with inappropriate immune responses, immunodepression to superimposed infections and immunopathological damage. This pattern has parallels with the patterns found in mycobacterial infections, leprosy, and tuberculosis, in which most individuals usually develop some immunity to the clinical disease while never actually eliminating the infection altogether. The immune response patterns to these infections are also seen in bacterial infections, such as listeriosis, and in sepsis. Leprosy, tuberculosis, and many protozoal infections present similar problems and the study of protozoa is now providing important clues as to what is happening in other infections (see Kotb and Calandra 2003). Our increasing knowledge of immune responses to protozoa is also beginning to explain the immunopathology associated with such infections because long-term chronic infections continually stimulate the immune system, which responds inappropriately and can eventually begin to attack the body itself. It is now clear that an understanding of the roles of cytokines is central to understanding the protective immune response, immunopathology and the possibility of developing vaccines. In this context, the use of inbred strains of laboratory mice has been invaluable and the ease with which genes can be deleted or inserted makes it possible to dissect these immune responses in great detail, to identify the roles of particular cytokines, and to extrapolate from mice to humans.

Infections with *Leishmania* species in inbred strains of mice have become the paradigms for studies on T cell subsets and the cytokine network as a whole. Early investigations using *L. major* in mice revealed that there is antagonism between the products of Th1 cells (IFN-γ) and Th2 cells (IL-4 and IL-10), the former leading to protection and the latter to counter-protection. It was later shown that the effector molecule involved in parasite killing was NO. These early observations subsequently proved to be very simplistic but provided a basis for more detailed investigations of leishmaniasis and of other parasitic protozoa. *Toxoplasma gondii* is also a parasite of macrophages and, like *Leishmania*, is controlled by IFN-γ-activated macrophages and the production of NO, TNF-α synergizes with IFN-γ in this killing. Malaria is a complex infection and immunity involves both Th1 and Th2 responses; parasite killing involves both antibodies and NO, which is also responsible for much of the pathology associated with severe malaria. Studies on cytokines have now extended to all the protozoa that infect humans, and research in this area is probably one of the most intense and productive areas of parasite immunology. It is probably true to say that in most, if not all, infections either the Th1 or the Th2 response is protective whereas the other is counterprotective and can lead to pathological changes or switching off the protective immune response.

Protozoa are microparasites that multiply within their vertebrate hosts by dividing at a genetically determined rate; thus, no matter how small the initial infective dose, the infection will eventually overwhelm the host unless it is controlled. The only ways in which this rate of multiplication can be curtailed are exhaustion of the cells available for infection or intervention of an immune response. In protozoal infections, some kind of acquired immunity is the rule and the main reason why this does not eliminate the infection is because all protozoan parasites have evolved ways of evading the consequences of the immune responses that they have evoked (Zambrano-Villa et al. 2002). Protozoal infections, therefore, tend to be long and chronic. The general pattern is, typically, a latent period during which few or no parasites can be detected, a phase of logarithmic increase, a crisis during which the infection is brought under control, and a rapid or gradual decline leading to a chronic infection with the possibility of subsequent recrudescences and long-term immunopathological damage.

All protozoal infections are accompanied by a number of immunological changes including the production of specific antibodies, usually IgM followed by IgG, and cell-mediated responses that can be measured in various ways. It is often not appreciated that many of these immunological responses have little or nothing to do with protection. All protozoa are antigenically complex and are, therefore, recognized as foreign so a variety of immune responses is inevitably initiated. These responses are often either due to irrelevant antigens or to relevant antigens that the parasite is able to use as part of its repertoire of immune evasion techniques. For example, antibodies to the dominant antigen of the malaria sporozoite deflect the immune response away from more vulnerable targets; in leishmaniasis, antibodies merely represent the stimulation of a counterprotective Th2 response.

In summary, protozoal infections are accompanied by changes in immunological parameters that, in some cases, mirror recovery but do not necessarily represent a protective mechanism. Unfortunately, a vast amount of the literature on the immunology of protozoal infections is devoted to these 'will-o'-the-wisps' and this has tended to obscure and confuse valid and relevant observations.

The importance of protective immune responses to protozoa has been highlighted by the consequences of co-infections with human immunodeficieny virus (HIV). Protozoal infections such as cryptosporidiosis, cyclosporiasis, and also microsporidiosis, once thought to be rare and harmless, are now recognized as concomitants of acquired immune deficiency syndrome (AIDS) with major pathological potential (see Chapters 20, Intestinal coccidia: cryptosporidiosis, isosporiasis, cyclosporiasis, and 25, Microsporidiosis). HIV is known to adversely affect the immune response to a number of other protozoal infections including leishmaniasis, toxoplasmosis, and Chagas disease (Ambroise-Thomas 2001), but does not seem to affect the outcome of malaria or African trypanosomiasis. The occurrence of such protozoal infections in immunosuppressed and immunocompromised patients has become an important aspect in the management of AIDS patients. Thus, in many parts of the world, an understanding of the immune response to parasitic protozoa has taken on a new significance.

Immunity to intestinal protozoa

The gastrointestinal tract serves as a habitat for a number of organisms including viruses, bacteria, protozoa and helminth worms. In normal, healthy individuals the mucus layer of the gastrointestinal tract passively protects the underlying cells from invasion. In addition, there is a specialized mucosal immune system consisting of mucosa-associated lymphoid tissue (MALT), intra-epithelial lymphocytes, macrophages and T and B lymphocytes in the lamina propria together with secreted noncomplement fixing immunoglobulin, IgA. Singly or together, these elements provide the first line of defense against invading organisms that gain entry to the body via the alimentary tract. Thus, the immunology of the gut can be considered as a discrete component of the overall immune system. Once an organism has broached the intestinal defenses, it effectively no longer occupies the gastrointestinal compartment of the body and becomes subject to attack by other components of the immune system. It is necessary to consider immunity to intestinal protozoa against this background. In this context, it is important to note that many experimental studies have involved the administration of parasite antigens by routes other than the intestinal one; thus, the immune responses generated have not necessarily been appropriate. The literature on immune responses to intestinal protozoa has been confused by results of such experiments and many of the conclusions drawn probably have little validity. This section, therefore, stresses what is known about immunity to intestinal protozoa in humans and under-emphasizes the contributions made by animal and in vitro studies. The comprehensive reviews available will be cited.

INTESTINAL AMEBAE

Entamoeba histolytica is the only ameba of importance about which anything concerning the immunology is known. There is some controversy as to whether or not all strains of this parasite are pathogenic (see Chapter 9, Amebiasis: *Entamoeba histolytica* infections) but, in thecontext of this chapter it is assumed that they are not. Most strains of *E. histolytica* are nonpathogenic and liveas harmless commensals in the large intestine where they do not appear to elicit any immune response. Pathogenic forms can invade the mucosa and submucosa, causing deep ulcers, and may be carried in the bloodstream to the liver, where they cause large abscesses. Immune responses, indicated by the production of specific antibodies, are elicited at various

stages after invasion but there is no real evidence that recovery from infection induces any protection against subsequent infections nor, at the population level, is there any age-dependent diminution in prevalence that can be ascribed to the acquisition of immunity. In patients with invasive amebiasis, there is a transient secretory immune response followed by the production of serum antibodies (Pérez-Montfort and Kretschmer 1990). The secretory immunoglobulin, IgA, is probably the most important antibody in the gut. Mucosal IgA antibodies directed against a galactose lectin correlate with the development of resistance in children (Haque et al. 2002) but this resistance is very short-lived (Stanley 2001). There is no correlation between serum antibodies, predominantly IgG, and the severity of the infection or recovery. It is not clear whether the apparent absence of acquired immunity is due to some defect in the immune system or to the fact that the amebae have evolved ways of evading the humoral immune response. Evasive mechanisms include the capacity to degrade IgG and secretory IgA, to resist complement lysis and to redistribute and shed surface antigens (see Zambrano-Villa et al. 2002). In some but not all individuals with invasive amebiasis, there is a transient cell-mediated immune response as detected by skin tests, but it is not at all clear what role cellular immune responses play. The epithelial cells that the amebae invade can produce a variety of chemokines and cytokines as well as mediators such as NO so it is possible that cell-mediated immune mechanisms could play some role in protection. Cell-mediated responses are depressed during *E. histolytica* infections; in vitro studies show that amebae can lyse target cells, including those of the immune system, by inserting ion channels called amebapores into their membranes. *E. histolytica* can also promote the production of the Th1 cytokines, IL-4 and IL-10, and block the production of IL-18 (Que et al. 2003), two mechanisms known to inhibit cell-mediated immune responses. In conclusion, there is some evidence that humans mount an effective immune response against *E. histolytica*, that this involves secretory mucosal IgA, and that any immunity is transient. However, the fact that recurrences of amebic ulcers are rare could suggest that there is some kind of longer-lasting immune response; this warrants some attention. Various aspects of amebiasis, including immunology, are reviewed by Stanley (2001), Petri (2002), and Haque et al. (2002) (see also Chapter 9, Amebiasis: *Entamoeba histolytica* infections).

There have been a number of attempts to develop vaccines against *E. histolytica* including DNA vaccines incorporating genes for a galactose lectin; this vaccine elicits a cell-mediated immune response (Gaucher and Chadee 2002). The possibilities and problems associated with developing vaccines against amebiasis are discussed by Miller-Sims and Petri (2002).

INTESTINAL FLAGELLATES

Giardia duodenalis has been long considered to be a harmless commensal of the small intestine but increasing evidence that it can cause malabsorption and severe diarrhea has stimulated interest in immune responses to this parasite (see Chapter 12, Giardiasis). A number of experimental studies with *G. muris* in mice clearly demonstrate that there is acquired immunity and protection against subsequent challenge and immunity is genetically controlled. In mice, the key molecule is secretory mucosal IgA. Mice can also be infected with the human parasite *G. duodenalis* and, in this model, there is an IL-4-dependent, Th2-type protective response involving the production of IgA (Bienz et al. 2003; Zhou et al. 2003). In humans, the evidence for acquired immunity is more circumstantial but comes from a number of sources. There is usually spontaneous clearance and resistance to reinfection; those living in endemic areas are less affected than visitors, young children are more affected than older ones or adults, and immunodepressed individuals tend to experience long chronic infections (reviewed by Janoff and Smith 1990; Nash 1994; Kulda and Nohynkova 1995; Faubert 1996; see also Chapter 12, Giardiasis). All the available evidence suggests that the effector mechanism or mechanisms involve secretory IgA but the case is by no means proven. IgA is the most characteristic *Giardia*-specific antibody in patients with acute giardiasis and individuals with hypogammaglobulemia tend to experience chronic infections, but there is little direct evidence of any IgA antiparasitic effects. On the other hand, specific antibodies are found in human milk and saliva; breast-fed infants in endemic areas seem to acquire some degree of protection (Walterspiel et al. 1994). Other antibodies may also be involved, for example, current infections are accompanied by high levels of specific IgM but this is quickly replaced by IgA in acute infections and after re-exposure. Overall, it seems that immunity to giardiasis is similar to immunity to a number of bacterial infections in which secretory IgA in breast milk protects the new born infant until it can produce its own antibodies. IgA, which cannot fix complement, presumably adversely affects the parasites indirectly by preventing attachment to the villi. *G. duodenalis* can undergo antigenic variation although the actual mechanism is not understood (Nash 1997). The variant specific surface antigens constitute a family of unusual proteins and it is more likely that they have a biological role, possibly protecting them against intestinal proteases, than any immune evasive mechanism (Nash 2002). There have been reports of giardiasis in AIDS patients, but *Giardia* is not considered to be a major concomitant of HIV infections (Janoff et al. 1988). There is no vaccine against giardiasis and most observers do not believe that one is necessary. The immunoprophylaxis and immunotherapy of giardiasis is reviewed by Olson et al. (2002).

Although the human intestine is parasitized by a number of other flagellates including *Chilomastix mesnili*, *Dientamoeba fragilis*, and *Trichomonas hominis*, these are not normally regarded as important pathogens and virtually nothing is known about any immunological responses to them.

INTESTINAL COCCIDIANS

Little is known about the immune responses of humans to any of the three intestinal coccidians, *Cryptosporidium parvum*, *Isospora belli*, or *Cyclospora cayetanensis*. Cryptosporidiosis is the only important disease caused by any of these parasites and has only been recognized as such during the last 20 years (reviewed by Martins and Guerrant, 1995; see also Chapter 20, Intestinal coccidia: cryptosporidiosis, isosporiasis, cyclosporiasis). In healthy individuals, the infection is an unpleasant but self-limiting one, lasting less than a month and characterized by mild to severe watery diarrhea. In AIDS patients, the infections are more persistent, the symptoms much more severe and death may result (Adal et al. 1995). This suggests that normally there is immune involvement and specific antibodies have been detected in the serum of infected individuals; characteristically, IgM is produced first and then replaced by IgG, levels of which begin to decline after about a year. Nothing is known about the possible mechanisms or relevance of these immunological responses.

INTESTINAL CILIATES

Balantidium coli, the only intestinal ciliate of humans, is comparatively rare. Nothing is known about the immune responses to this ciliate and, by analogy with *Entamoeba histolytica*, it is unlikely that there is any reaction unless the mucosa and submucosa are invaded.

Immunity to protozoa inhabiting the urinogenital tract

The only protozoan that inhabits the human urinogenital tract is *Trichomonas vaginalis*, which parasitizes both men and women. The nature of the infection and the disease caused in men is poorly documented, but vaginitis caused by this parasite is an important sexually transmitted disease in women, in whom it persists for long periods unless treated (reviewed by Honigberg and Burgess 1994; see also Chapter 13, *Trichomonas* infections). There are no really appropriate animal models of human trichomoniasis and, although experimental infections in guinea pigs and mice have provided some information (e.g. complement-dependent and complement-independent killing in animals given intra-peritoneal, intradermal or subcutaneous injections), little that is really relevant can be derived from these observations (Ackers 1989). Similarly, observations on *T. foetus* in cattle have been largely irrelevant to human trichomoniasis. In humans, secretory IgA antibodies to *T. vaginalis* have been detected in vaginal secretions. Antibodies produced in response to infection diminish with time and do not appear to confer any protection against reinfection. Epidemiological evidence, derived from many studies in sexually transmitted disease clinics, suggests that infections with *T. vaginalis* tend to be persistent, that reinfection frequently occurs after cure and that the prevalence of the infection does not plateau or fall off in the older age groups. All this suggests that there is little or no acquired immunity and that specific immune responses play an insignificant role in limiting this infection. The reasons why there appears to be no immunity to *T. vaginalis* are not clear, although it has been suggested that this parasite has considerable antigenic diversity; whether or not it can also undergo antigenic variation is an open question (Alderete 1983).

Immunity to macrophage-inhabiting protozoa

INTRODUCTION

Three groups of parasitic protozoa inhabit macrophages at some stage during their life cycles: *Leishmania* spp., *Toxoplasma gondii*, and *Trypanosoma cruzi*. Each has evolved its own way of evading the destructive mechanisms of the macrophage: leishmanias survive in the fused phagolysosome; *T. gondii* prevents phagosome–lysolysosome fusion and *T. cruzi* escapes from the phagosome into the macrophage cytoplasm. In addition, all these parasites also possess enzymes that counteract the various toxic molecules present in the phagolysosome. Because of their ability to avoid destruction by macrophages, they all escape attack by one of the main arms of the immune system and this enables them to give rise to long chronic infections. On the other hand, none of these parasites is able to survive in IFN-γ activated macrophages, thus they live on a knife edge by inhabiting cells that can sustain them but which also have the capacity to destroy them. *Leishmania* spp. are totally dependent on macrophages whereas *T. gondii* and *T. cruzi* are also able to parasitize other cell types. This section deals with immunity to *Leishmania* spp. and *Toxoplasma gondii*. *T. cruzi* is discussed later under Immunity to tissue-inhabiting protozoa.

Leishmania spp.

About 20 species of *Leishmania* are known to infect humans in the New and Old Worlds and the number is gradually increasing. Leishmaniasis can be conveniently classified into three forms, cutaneous, mucocutaneous, and visceral, but these categories are not absolute (see Chapters 16, Old World leishmaniasis and 17, New World leishmaniasis). As these parasites are very common and as the majority of those infected (particularly with the cutaneous forms) do recover, albeit slowly, the immunology has been intensively studied, mainly with the ultimate aim of producing a vaccine. *L. major*, the causative agent of Old World cutaneous leishmaniasis, and *L. donovani*, the causative agent of visceral leishmaniasis, have been the most studied species, mainly because of the ease with which they can be maintained in genetically characterized laboratory mice. In particular, certain strains of mice (e.g. C57Bl) are relatively resistant to infection and develop small self-healing cutaneous lesions whereas others (e.g. BALB/c) are more susceptible and develop overwhelming visceralizing infections (reviewed by Liew 1990; Alexander and Russell 1992; Liew and O'Donnell 1993; Bogdan and Röllinghoff 1998, 1999; Tacchini-Cottier et al. 2003). Developments in molecular biology, particularly the use of knock-out mice, have made it possible to dissect the immune response to *L. major* in considerable detail and the *L. major*–mouse model has become a paradigm for all leishmanial infections. *L. major* infections in resistant mice tend to resemble most *L. major* infections in humans in the New World and *L. major* infections in susceptible mice resemble the condition known as diffuse cutaneous leishmaniasis (DCL). Models of visceral leishmaniasis, caused by *L. donovani*, and infections with the New World species have not been so extensively investigated but, in general, extrapolations from mice and hamsters to the human situation have tended to confirm the universal nature of the immune response involved. Immunity to leishmaniasis is cell-mediated and involves Th1 cytokines. Leishmanial parasites survive and multiply within normal resting macrophages but are killed by activated macrophages through a mechanism involving NO. NK cells and the cytokines IL-12, IL-18, TNF-α, and TNF inducers play major roles in this process by up-regulating the IFN-γ-induced NO killing. In essence, the protective immune response in leishmaniasis is a classical Th1-type cell-mediated response. However, this protective response is counteracted by an equally effective Th2 response involving IL-4, IL-10, IL-13, and TGF-β. IL-4 and IL-13 counteract the macrophage-activating effects of IFN-γ, IL-10 down-regulates macrophage activity, and TGF-β blocks IFN-γ-activated macrophage activity. The net result is that NO production is inhibited and parasite killing reduced. Resistant strains of mice tend to have high levels of IFN-γ and low levels of IL-4 and IL-10, whereas the opposite is the case in susceptible strains. The overall control of the immune response, therefore, depends on the activation of one of two subsets of T lymphocytes: activation of Th1 cells leads to the production of IFN-γ and protection whereas activation of Th2 cells produces IL-4 and IL-10 and is counter-protective (see Reiner and Locksley 1995; Scott 1996). It is now clear, however, that this is not the whole story; the actual control of immunity to leishmaniasis is much more complicated and the role of IL-12 and the interplay between IL-12 and IL-4 are crucial to the outcome of the infection (Scott and Farrell 1998). It is also important to note that in mice infected with *L. donovani*, the dichotomy between Th1 and Th2 responses is not nearly so clear cut (Kaye 1995) possibly indicating important differences in antigen processing and presentation in the skin and in the viscera (Moll 1993; Solbach and Laskay 1995, 2000; Ritter and Korner 2002). It is also significant that although IL-12 enhances the production of chemokines important in protection against *L. major* infections (Zalph and Scott 2003), in *L. mexicana* infections in mice there is an alternative pathway leading to the production of IFN-γ that is independent of IL-12 (Buxbaum et al. 2002). Overall, it is now clear that although the general patterns of immunity are similar, there are significant differences between the actual mechanisms involved with different species of *Leishmania* and to some extent this may account for differences in the pathologies associated with different species.

Although the distinctions between the Th1 and Th2 subsets are not so clear cut as in mice, this pattern is also true for human leishmaniasis including visceral leishmaniasis (Kemp et al. 1996). In patients with fulminating *L. donovani* infections, IL-4 and IL-10 levels tend to be high and IFN-γ levels low; in cutaneous infections (in which recovery is normally the rule) IFN-γ levels are high, and in chronic mucocutaneous infections IL-4 levels are high. All these facts are in accordance with well-established clinical observations such as the development of strong delayed type hypersensitivity (DTH) indicating macrophage activation, which correlates with recovery from cutaneous leishmaniasis; diffuse cutaneous leishmaniasis is characterized by the absence of DTH; and high levels of nonprotective IgM and IgG antibodies (resulting from Th2 cell activity) are seen in visceral leishmaniasis (see Blackwell 1993). In localized mucocutaneous leishmaniasis, Th1 cytokines predominate whereas in the more destructive forms there is a mixture of Th1 and Th2 responses, although the Th2 cytokine, IL-4, tends to predominate in accordance with the well-established paradigm (Pirmez et al. 1993). In humans, the nature of the immune response is also determined by numerous extraneous factors such as the species or strain of parasite involved, the genetic makeup of the host or host population, the number of bites by infected sandflies, and the number of parasites injected.

There have been a number of experimental vaccines developed against leishmaniasis and there have been some encouraging results in humans (Greenblatt 1988; Convit et al. 1989; Modabber 1995; Handman 1997). Vaccines that have been used or considered for use include attenuated strains, killed parasites, parasite extracts or metabolic products and, more recently, DNA vaccines. The use of relatively crude vaccines is still going on. Vaccines based on autoclaved *L. major* and *L. infantum* plus bacille Calmette–Guérin (BCG) have been successful in protecting dogs against these infections (Mohebali et al. 1999) and multiple injections of autoclaved *L. major*, with or without BCG, elicit immune responses that suggest that BCG could boost the immune response in humans (Alimohammadian et al. 2002). The possibility of using such vaccines against *L. donovani* in humans is discussed by Dube and Srivastava (2002). Two kinds of vaccine, one using genetically attenuated *L. major* and the other using autoclaved parasites, produced encouraging results in monkeys; both protocols are safe in humans but require further improvement before vaccine development can be contemplated (Amaral et al. 2002). A better characterized vaccine using five *L. infantum* antigenic determinants plus BCG gave 90 percent protection in dogs (Molano et al. 2003). These promising results have encouraged others to develop more advanced vaccines such as recombinant (de Carvalho et al., 2003) or DNA vaccines in mice infected with *L. major* (López-Fuertes et al. 2002; Méndez et al. 2002) or *L. donovani* (Sukumaran et al. 2003). Currently there are many leads but no effective and safe commercially available vaccines against leishmaniasis.

Toxoplasma gondii

Toxoplasma gondii is a common parasite of cats, which can be transmitted to virtually all warm-blooded animals (Dubey and Beattie 1988; see also Chapter 21, Toxoplasmosis). The parasites infect macrophages in which they survive by preventing phagosome–lysosome fusion; they undergo rapid division to form merozoites known as tachyzoites. The tachyzoites escape from the host cell and invade new cells. After a short time, this rapid division slows down and the parasites transform into slowly dividing bradyzoites, which remain dormant in cysts in various tissues, including the brain, where they may die and become calcified. If the infected individual is subsequently immunocompromised in some way, the dormant parasites in the cysts may become activated and give rise to a fulminating infection. In humans, most infections are benign or inapparent but if the parasite crosses the placenta it can seriously damage the fetus. *Toxoplasma gondii* infections are also now becoming serious concomitants of AIDS (Ambroise-Thomas and Pelloux 1993). The most obvious features of human *T. gondii* infections are that individuals acquire the infection relatively early in life and after recovering from it are immune to reinfection. Acute infections are characterized by specific IgM and chronic or recovered infections are characterized by high levels of specific IgG. *T. gondii* infections in all mammals except cats seem to be very similar so there is no major problem in extrapolating from studies in laboratory animals to humans. Much of what is known about immunity to toxoplasmosis has been derived from studies in mice mainly because of the ease with which various components of the immune response can be manipulated in these animals (Gazzinelli et al. 1993; Alexander and Hunter 1998; Wille and Hunter 2003). There have also been numerous parallel studies using sheep, as toxoplasmosis is a serious veterinary problem associated with abortions. The immune response varies throughout the infections. First, there is the innate immune response during which the infected resting macrophages produce small amounts of IL-1, IL-12, IL-18 and TNF-α. These cytokines stimulate NK cells to produce IFN-γ which activates the infected macrophage and brings about NO-mediated parasite killing. This innate immunity is, however, of limited efficacy and duration and the infection then moves into its acute phase during which the infected macrophages produce IL-12 that stimulates both CD4 and CD8 lymphocytes to produce IFN-γ which, in turn, stimulates the production of NO in the infected cell and subsequent parasite killing. During the chronic phase of the infection, CD8 cells seem to be the main source of IFN-γ and the infection is tolerated at a low level until the parasites die and become calcified. If, however, an individual is immunocompromised in some way during the chronic phase of the infection, there is a rapid and acute recrudescence of the parasitemia, which can be fatal. In addition to the Th1-type immune response, there is evidence of some Th2-type activity during which tachyzoites are coated with antibody and are either prevented from invading host macrophages or are unable to prevent phagosome–lysosome fusion if invasion is successful. As pointed out above, antibody levels are indicative of acute or chronic infections.

The key molecule in immunity to toxoplasmosis is NO, which kills the target cell by interfering with DNA and metabolic pathways. NO is a very toxic molecule and over-production can cause pathological damage to host cells and tissues and can also cause immunosuppression. In toxoplasmosis, the pathology is manifested in toxoplasmic encephalitis and retinopathies. The host, however, is able to moderate these adverse effects by producing IL-6, IL-10, and TGF-β, which inhibit the activity of IFN-γ and hence the production of NO. However, TGF-β also promotes the replication of the parasite (Nagineni et al. 2002). The roles of IL-6 and IL-10 are equivocal in that IL-6 actually promotes parasite proliferation which, with IL-10, can contribute to toxoplasmic encephalitis. IFN-γ plays a central role in toxoplasmosis as it not only induces microbicidal activity but also promotes cyst formation and prevents cyst rupture. There are numerous possible sources of IFN-γ, both T cell and non-T cell, and at least five mechanisms whereby IFN-γ-activated macrophages can affect the parasite – including the production of NO and ROI and

limiting the supply of iron to the parasite (Suzuki 2002). The immune responses to *T. gondii* are very complex; immunity to infection and the deleterious effects of inflammation are two sides of the same coin and it is essential for the host to maintain a balance between the various cytokines. In order to understand the mechanisms involved in the control of the cytokine network, recent research has moved away from the cytokines themselves to the signaling pathways and the ways in which *T. gondii* can disrupt macrophage function. *T. gondii* induces the activation of two transcription factors, the signal transduction activator of transcription (STAT)-1 and nuclear factor kappa B (NFκB), but blocks their translocation so that the macrophage cannot then produce IL-12 or TNF-α (Denkers et al. 2003).

It is not easy to make direct extrapolations from mice to humans but the available evidence suggests that the mechanisms of immunity in humans are similar to those in mice; for example, IFN-γ inhibits replication of the tachyzoites in macrophages from infected AIDS patients (Delemarre et al. 1994). In vitro studies using human cells have also confirmed murine studies. Using *T. gondii*-infected human myelomonocytic cells, Belloni et al. (2003) have confirmed that TNF-α and IFN-γ act synergistically and have also shown that TNF-α production is inhibited by the parasite and that this is caused by the shedding of TNF-α receptors. In retinochorditis, one of the most common manifestations of toxoplasmosis, the parasite actually inhibits the production of TGF-β suggesting that the interplay between *T. gondii* and retinal cells may play an important role in this condition (Nagineni et al. 2002). The role of NO in neurodegeneration is also controversial. NO is known to damage neuronal cells but this rarely happens during *T. gondii* infections, possibly because the parasite is able to inhibit the production of NO by microglia through a pathway involving prostaglandin-E and IL-10 (Rozenfeld et al. 2003). Whatever the mechanism, immunity is clearly very important in humans because the majority of the world's population will be infected at some time and that those that are immunoincompetent (e.g. the fetus) or immunosuppressed are at particular risk from a fulminating infection. Toxoplasmosis is now regarded as the most common cause of focal central nervous system infection in AIDS patients (Mariuz et al. 1994).

There is no vaccine against toxoplasmosis in humans. Commercial vaccines based on attenuated strains of *T. gondii* have been developed for use in sheep (Buxton 1993), but such a vaccine would be unacceptable for human use. Our understanding of the cytokine network and the development of DNA vaccines have encouraged the possibility of vaccines that might be acceptable for human use and there have been some successes in mice immunized with a plasmid containing a *T. gondii* surface antigen and IL-2 (Chen et al. 2002). Also being developed are DNA vaccines designed to activate CD8 cells (Scorza et al. 2003) and DNA vaccines designed to stimulate Th1-type responses and the production of IFN-γ (Fachado et al. 2003). However, a vaccine suitable for use in humans is a long way away.

Immunity to blood-inhabiting protozoa

INTRODUCTION

Four important groups of protozoan parasites inhabit the human bloodstream:

1 the African trypanosomes *Trypanosoma brucei gambiense* and *T. b. rhodesiense*
2 the American trypanosome *Trypanosoma cruzi* (but only transiently, it is best regarded as a tissue parasite)
3 the malaria parasites *Plasmodium falciparum, P. malariae, P. ovale,* and *P. vivax*
4 in rare and accidental infections, *Babesia* spp.

The bloodstream is a dangerous place for a parasite to live as there is continual and intimate contact with the cells and molecules of the immune system, particularly IgM and IgG antibodies and phagocytic cells. The ability to survive in what should be a very hostile environment has attracted the attention of numerous scientists. African trypanosomes have been particularly intensively studied in laboratory animals and this activity has generated a vast literature much of it, unfortunately, largely irrelevant to the human disease. In this section, only those aspects of the immune response directly concerned with human infections are discussed.

AFRICAN TRYPANOSOMES

The African trypanosomes that infect humans are *Trypanosoma brucei gambiense* and *T. b. rhodesiense*, both related to *T. b. brucei*, a trypanosome of wildlife and domesticated animals that does not infect humans (see Chapter 18, African trypanosomiasis). Little is known about the acquisition of immunity to trypanosomes in humans, but there is epidemiological evidence that immunity to *T. b. gambiense* does develop, albeit slowly, in human populations (Khonde et al. 1995). *T. b. brucei* has been widely used as a model for both the human and animal disease, largely because it is the only subspecies that is easy to maintain in laboratory mice and because of its similarities to the human forms. Infections with all three subspecies are similar; the parasites live and multiply in the blood and infections are characterized by successive waves of parasitemia representing different antigenic variants. Each trypanosome is covered with a thick glycoprotein coat, the variant surface glycoprotein (VSG). There is an almost limitless repertoire of VSGs, each known as a variable antigen type (VAT), that arise in 1 in 10^6 cells in laboratory syringe-passaged populations and 1 in 100 cells in field isolates. The normal pattern of events is that the parasites multiply in the blood, are destroyed by an antibody-mediated immune response specific to the major VAT, and are replaced by another population of parasites carrying a new VAT (Barry 1997). Each wave of parasitemia is accompanied by an increase in antibody specific to the VAT that preceded it and is responsible for the removal of most of the trypanosome population leaving a minor population that gives rise to the next wave of parasitemia; subsequent waves appear every 5–10 days. The antibody class involved in the destruction of the trypanosomes is IgM, which facilitates agglutination, complement-mediated lysis, and phagocytosis by macrophages and neutrophils. Later in the infection, VSG-specific IgG antibodies are involved in clearing trypanosomes from the tissues (see Sternberg 1998). The destruction of the trypanosomes, however, results in the release of a number of toxic molecules and internal and external antigens that adhere to or accumulate in various host tissues resulting in complex-mediated damage and resultant immunopathological changes which have been well documented in animal studies but have been little investigated in humans.

There is no evidence of protective cell-mediated immunity in African trypanosomiasis but there is some limited evidence that NO, which has an alternative role (see below), can kill trypanosomes possibly only in combination with TNF-α (Mansfield et al. 2001).

Trypanosomes are also enmeshed in the cytokine network and molecules involved in the immune response such as (TGF-β) and IFN-γ can act directly or indirectly on parasites; IFN-γ acts as a growth factor for *Trypanosoma brucei* (see Mansfield et al. 2001). Trypanosomes can also secrete or excrete products that affect cells of the immune system, for example trypanosome-derived lymphocyte triggering factor (TLTF) and trypanosomal macrophage activating factor (TMAF). TLTF induces lymphocytes to produce IFN-γ, which activates macrophages and promotes trypanosome growth, TMAF also stimulates macrophage activity (see Sternberg 1998; Hamadien et al. 2000). The production of TLTF has been demonstrated in the human trypanosomes *T. b. gambiense, T. b. rhodesiense* (Bakhiet et al. 1996). All the African trypanosomes are, therefore, potentially caught up in a series of interactions in which trypanosome-derived factors activate both T lymphocytes and macrophages. The net effect is that lymphocyte produced IFN-γ enhances trypanosome growth while other trypanosome-derived molecules induce macrophages to produce molecules that inhibit lymphocyte activation (Figure 4.3).

Trypanosome infections are accompanied by severe immunodepression to superimposed antigens and concurrent infections even though their hosts are capable of mounting effective specific immune response against trypanosome antigens. Although this phenomenon has been most extensively studied in animals, there is evidence that it also occurs in patients with trypanosomiasis (Greenwood et al. 1973). The underlying basis of

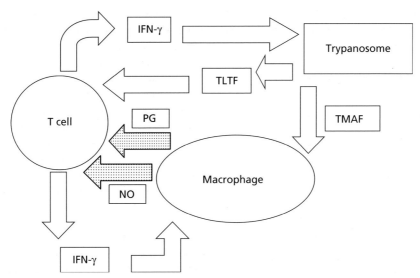

Figure 4.3 *Diagrammatic representation showing some of the interactions that exist between African trypanosomes and the cytokine network. Trypanosomes produce a number of factors including TMAF and TLTF that stimulate macrophages and lymphocytes respectively. IFN-γ, a product of stimulated lymphocytes, can act as a trypanosome growth factor but prostaglandins (PG) and NO produced by macrophages inhibit lymphocyte activity and hence the production of IFN-γ. Trypanosomes are therefore enmeshed in a network of cytokines and effector molecules that both stimulate and inhibit their growth and development. Positive signals are here indicated by open arrows and inhibitory ones by stippled arrows. (From Cox (2001) with permission from Cambridge University Press)*

this immunodepression is not at all clear but it is thought that impaired B-cell responses may result from overactivation and clonal depletion of B cells from lymph nodes during the course of infection. T-cell activity is also suppressed and this is thought to result from the development of immunosuppressive macrophages and the production of NO and prostaglandins.

The whole field of the immunology of trypanosome infections is a very controversial one and the traditional view that anti-VSG antibody is entirely responsible for the remission of the infection, albeit temporarily, has been challenged by several workers including Mansfield (1994), Mansfield et al. (2001), and de Baetselier (1996) based on in vitro observations and studies in mice. Mansfield suggests that macrophage-derived NO affects the survival of the trypanosomes and also inhibits T-cell function, leading to immunodepression. de Baetselier maintains that the trypanosomes induce CD8 T lymphocytes to produce IFN-γ and macrophages to produce TNF-α and that these synergize to produce toxic molecules that both adversely affect the trypanosomes and cause immunodepression. It is now clear that the traditional picture of antibody-mediated immune responses eliminating variant specific antigens and clearing successive parasitemias is not the whole picture (Mansfield et al. 2001). It is also not at all clear what the situation is in humans, whether or not there is any immunodepression and, if so, if it has any affect with respect to intercurrent infections such as HIV.

MALARIA PARASITES

Malaria is generally recognized as the most important parasitic disease, both in terms of mortality and morbidity (see Chapter 24, Malaria). Humans harbor four species of malaria parasite, *Plasmodium falciparum*, *P. vivax*, *P. malariae*, and *P. ovale*, in order of clinical importance. *P. falciparum* is the most intensively studied parasite and, unless otherwise specified, is the only one considered here. The control of malaria has largely depended on the combined use of insecticides and drugs but resistance has made these ineffective and attention has turned to the possibility of developing a vaccine. An important spin-off of this work has been the realization that much of the pathology of malaria is closely associated with immunity, in particular the involvement of cytokines. Research into immunity and immunopathology has resulted in a massive literature reaching tens of thousands of papers, mostly based on rodent models and in vitro systems, much of it, particularly many of the extrapolations made, irrelevant or peripheral to the human situation (Cox 1988). Until recently, our understanding of the ways in which humans react to infection with human malaria parasite has remained a mystery, but success in cloning the *Plasmodium* genome and the application of molecular techniques directly to the human malaria parasites is beginning to pay dividends. However, there are a number of things that we do not understand and our ignorance of these areas hinders our ability to develop vaccines or ways of ameliorating the pathology associated with the disease.

The starting point in our understanding of immunity to malaria is that from epidemiological studies it is known that:

- malaria infections are long-lived, 1–2 years in the case *P. falciparum* and 3–50 years for *P. malariae*

- individuals can be reinfected after natural recovery or cure
- there is a gradual build up of immunity over a period of many years
- any immunity fades quickly
- immunity is largely strain-specific
- children born to immune mothers are themselves resistant to that particular strain of malaria.

In summary, immunity to malaria can be considered to be the rule, although it is often incomplete and may take many years and numerous exposures to the bite of infected mosquitoes to develop. Thus, there is evidence of some immunity and the real challenge is to try to understand the mechanisms involved in order to develop effective vaccines.

In order to understand the immunology of malaria, it is necessary to describe the life cycle (see Chapter 24, Malaria for further details). The life cycle of *P. falciparum* begins when the infective stages, sporozoites, are injected by a mosquito directly into a capillary in the skin. The sporozoites circulate in the blood for 30–45 min before they enter liver cells where they undergo a phase of pre-erythrocytic schizogony involving repeated nuclear divisions resulting in the production of about 30 000 uninucleate merozoites after 6–7 days. The merozoites flood into the blood and actively attach to and invade red blood cells within about 20–30 seconds and initiate the erythrocytic stage of the life cycle (see Bannister and Mitchell 2003). Inside the circulating red blood cell, the merozoite feeds on hemoglobin and transforms into a uninucleate feeding stage called a trophozoite; the nucleus then begins to divide and, in *P. falciparum* infections but not in other species, the infected cells disappear from the circulation and adhere to endothelial cells of various internal organs. This process is known as sequestration and is very important in the pathology of the disease. Nuclear division occurs until a schizont containing about 16 merozoites is formed; the merozoites break out of the cell and invade new cells, a process that is repeated almost indefinitely. The liberation of parasites from the red cell is accompanied by the release of a number of toxic molecules, some of host origin and some of parasite origin, that are responsible for some of the pathology associated with malaria. Some young merozoites do not develop into schizonts but into male and female gametocytes that circulate in the peripheral blood until they are taken up by a mosquito when it feeds. Within the mosquito, the gametocytes mature into male and female gametes, fertilization occurs and a zygote called an ookinete is formed. Finally, there is another phase of multiplication resulting in the formation of sporozoites that migrate to the salivary glands and are injected when the mosquito feeds on a new host. The life cycles of the other malaria parasites of humans differ in the length of the pre-erythrocytic and erythrocytic phases, the numbers of merozoites produced and the absence of sequestration. Some strains of *P. vivax* have a persistent dormant pre-erythrocytic phase.

Each stage in this complex life cycle is antigenically distinct and a potential target for immune attack. The malaria parasite is most susceptible when the sporozoite is circulating in the blood, when merozoites are liberated from the liver into the bloodstream before they invade the red blood cells, and when merozoites have been released from red blood cells. The sporozoite is the obvious first target for immune attack. It is the first stage in the infection and the parasites are free in the blood for up to 30–45 min. The sporozoite possesses an immunodominant 40–60 kDa protein surface coat, the circumsporozoite protein (CSP), characterized by a repeat region consisting of four amino acids repeated 37 times, and three or four copies of a smaller repeat dispersed throughout the molecule. In *P. falciparum*, the amino acids of the dominant repeat are asparagine–alanine–asparagine–proline (NANP), and the minor repeat is asparagine–valine–aspartate–proline (NVDP). Other malaria parasites have similar repeat regions but with different amino acid sequences. The CSP is very antigenic and elicits a strong antibody response and has, therefore, been a favorite for vaccination studies. When exposed to antibody, however, the surface molecules are cross-linked and the sporozoite sheds the coat and escapes; thus, it is now thought that the repeat sequence may be a mechanism whereby the parasite evades the immune response by putting up a powerful 'smoke screen' that deflects the response away from more important targets. The repeat region of the CSP is not the only possible target; there are a number of other sporozoite antigens from nonrepeat regions and also those involved in the penetration of the hepatocytes. One of the most important is the thrombospondin related adhesion protein (TRAP) that is involved in sporozoite motility (Sultan et al. 1997).

It was previously thought that, once in the liver, the parasite was safe from immune attack but, in murine malaria models, there is now evidence that there is a Tc cell response to the early stages in hepatocytes and that what is recognized are parasite antigens in the presence of a hepatocyte MHC class I molecule (Tsuji and Zavala 2003). Although we do not know what happens in humans, there is evidence that in West Africa the outcome of infection is controlled by the MHC class I molecules, thus there has been increased interest in antigens associated with the liver stages. The best studied pre-erythrocytic-stage antigens are the liver-stage antigens (LS-A), LS-A1, and LS-A3 but at least 15 more have been identified some of which have some homologies with antigens of other stages in the life cycle (Grüner et al. 2003).

The erythrocytic stages, particularly the merozoites, have received the most attention. They are the easiest to study, they are responsible for much of the disease and they are obvious targets for attack. A vast amount of information about these stages has been accumulated and many of the antigens involved have been characterized and cloned. The best studied antigen is the merozoite surface protein (MSP)-1, found in all the malaria parasites studied to date; it is a glycoprotein with a molecular weight of c. 190–195 kDa. Experimentally, antibodies against MSP-1 block red blood cell invasion, thus this could be a major target for immune attack and for exploitation as a vaccine. Unfortunately, there is considerable diversity in MSP-1 between isolates, and immunity to one isolate does not confer immunity to another. Other *P. falciparum* antigens of importance are apical membrane antigen (AMA)-1, erythrocyte binding antigen (EBA)-175, serine rich antigen (SERA), and ring-infected erythrocyte surface antigen (RESA), all of which are involved in binding to or invading red blood cells. The erythrocyte membrane protein (EMP), PfEMP-1, found on the surface of late-infected red cells is involved in sequestration and undergoes antigenic variation and is, therefore, thought to be in part responsible for the recrudescences that occur in falciparum malaria (Gatton et al. 2003). The actual role of PfEMP-1 is unclear but it is thought that sequestration prevents infected erythrocytes from circulating through the spleen thus protecting them from the killing mechanisms that appear to operate there. Antibodies against PfEMP-1 prevent sequestration and facilitate parasite destruction in the spleen.

The antigens associated with the sexual stages would seem to be unlikely targets for immune attack but, in fact, they are very important. There are a number of antigens associated with the gametocytes, the male and female gametes and the zygote and, unlike most of the other antigens of malaria parasites, they are very susceptible to antibody attack. This only happens when the gametocytes are taken up by a mosquito together with antibodies in the serum. When the gametocytes escape from the infected red cell they are recognized by the antibodies and are either killed or inactivated. In either case, the infection in the mosquito cannot continue. The antigens of most interest are: a 250 kDa antigen on the gametocytes, gametes and zygote; a 25 kDa antigen on the gametes and zygote; and a 48 kDa antigen on the zygote and oocyst.

The targets for immune attack are, therefore, numerous and complex and the immune responses involved must also be correspondingly numerous and complex. Virtually all Th1 and Th2 components of the immune system are involved at some stage during an infection. Although many details are missing, we now have a very good idea of the overall immune response during the blood stages of a malaria infection. The first time an individual becomes infected with malaria, parasite antigens stimulate NK and T cells to produce low levels of IFN-γ that activate macrophages to produce TNF-α, ROI, and NO. All three molecules have been implicated in the killing of intra-erythrocytic parasites but the strongest evidence supports the role of NO as the key molecule in this process (see Clark and Cowden 2003). Whatever the actual mechanism, these non-antibody molecules, possibly in combination, bring the infection under control with few clinical symptoms. Where the destruction of parasites occurs is not known but it does not seem to be in the circulating blood: the most likely site is the spleen and evidence for this comes the observation that individuals who have had their spleens removed often have difficulty in controlling the parasitemia. It is possible that the flow of blood is slowed down in the spleen and the infected erythrocytes come into close contact with macrophages lining the blood vessels thus causing the release of pro-inflammatory cytokines and NO that damage the intra-erythrocytic parasites while infected cells and free merozoites are phagocytosed.

On exposure to a subsequent infection, the primed T cells produce larger amounts of IFN-γ that induces macrophages to produce excessive amounts of the pro-inflammatory cytokines, IL-12, IL-18, and TNF-α, leading to the symptoms associated with severe malaria. In individuals who recover, the immune system switches to the production of antibodies, mainly IgG2 and IgG3, which impede red blood cell invasion, inhibit parasite multiplication, enhance phagocytosis of merozoites and infected cells, and neutralize toxins (Garraud and Perraut 2003). At the same time, Tr cells produce anti-inflammatory cytokines, IL-10 and TGF-β, which down-regulate the production of TNF-α by macrophages and bring about amelioration of the symptoms of severe malaria. It is important to note that the different roles of Th1 and Th2 cytokines do overlap, for example IL-4, IFN-γ, and TGF-β are involved in the production of IgG1, IgG2a, and IgGb respectively (Stavnezer 1996).

The erythrocytic stages are not the only targets for immune attack; the early stages in the liver are other possible targets. The completion of the *Plasmodium* genome project has resulted in the description of a plethora of liver-stage antigens (see Grüner et al. 2003; Taylor-Robinson 2003). Immunity to liver stages is very difficult to investigate in humans but the only antigen expressed specifically by infected hepatocytes is LSA-1, thought to be the target for Tc cells or NK cells (Schmieg et al. 2003). Interest in the sporozoite antigens has also not waned, although attention has shifted towards the possibility of incorporating such antigens into multicomponent vaccines. Similarly, antigens from the sexual stages are now being considered as possible components of multistage vaccines (see below).

The complexity of the immune response means that inevitably some of the reactions will be adverse ones and the main manifestations of severe malaria, metabolic acidosis, cerebral malaria and anemia correlate with circulating levels of the pro-inflammatory mediator TNF-α. There is now incontrovertible evidence that severe malaria is caused, in part at least, by NO elicited during the protective immune response (Clark and Rockett 1996). The arguments that have gradually led to this conclusion are complex; further information is to be found in the comprehensive review by Clark and Cowden (2003). It is now also clear that in cases of severe malaria, NO as evidenced by the presence of iNOS, is present in many tissues of the body suggesting that the disease is a multi-organ one (Clark et al. 2003). The realization that protective immune responses may have deleterious side-effects has serious implications for the development of a vaccine against malaria.

The development of a vaccine must take into consideration what stage of the parasite life cycle to attack and what kind of antigens to use. The life cycle of the malaria parasite is very complex and includes a number of stages that are biochemically and immunologically distinct. Immune responses against all these stages have been detected but immunity against one does not protect against the next. In effect, malaria infections consist of a sequence of different infections, not just one.

The antigens are complex: each stage possesses a large number of different antigens some of which are protective and some of which are not. In addition, malaria parasites have evolved a number of ways to evade the immune response so some of the antigens identified could be counter-protective. Research on malaria vaccines has paralleled or, in some cases, led research on viral and bacterial vaccines and the emphasis has changed with our increasing understanding of the immune response as a whole. The traditional approaches have been to identify and characterize potentially protective surface antigens and to use these as:

- subunit antigens on their own
- as a basis for synthetic vaccines
- as recombinant molecules expressed in a suitable vector.

All three methods have been tried with varying success over the past two decades and this subject has been extensively reviewed (Hoffman 1996; Facer and Tanner 1997; Good et al. 1998; Miller and Hoffman 1998; Ballou et al. 1999; Anders and Saul 2000; Plebanski and Hill 2000). More recently, there has been a switch to the possibility of DNA vaccines and prime-boost methods of immunization whereby the host is first immunized with a DNA vaccine expressing the requisite antigens and subsequently with a recombinant vector expressing the same antigens.

Sporozoites are obvious targets, being the first stages to enter the body and circulating in the blood for a short time before entering the liver; a vaccine against the sporozoite should prevent infection. The first vaccines against malaria consisted of irradiated sporozoites (Clyde et al. 1973) and it soon became clear that the immune response was directed against the NANP repeat region of the major surface protein of the sporozoite, CSP. The NANP repeat was then used as a basis for recombinant vaccines expressed in *E. coli* (Ballou et al. 1987) and for synthetic vaccines (Herrington et al. 1987) in a number of trials with potentially promising results – about 20–30 percent of volunteers were protected. However, these trials also revealed that there was little correlation between antibody levels and protection. It is now widely accepted that the sporozoite repeat antigen is not a good vaccine candidate because if even one sporozoite escaped into the liver an infection would be initiated. In addition, the epidemiological evidence does not support the concept that a sporozoite vaccine would be effective. Returning to vaccines based on irradiated sporozoites, it was found that high levels of irradiation prevented sporozoites from invading liver cells and that exposure to such sporozoites was not protective, so attention moved on to sporozoite antigens involved in hepatocyte invasion. A promising recombinant vaccine called RTS,S and consisting of part of the CSP fused to the hepatitis B surface antigen was developed in the Gambia with excellent results (over 70 percent efficacy); unfortunately, this protection faded after about 15 weeks (Bojang et al. 2001). Modifications of this vaccine are now being developed for possible clinical trials. All these vaccine trials were based on eliciting antibody responses but attention has now turned towards T-cell responses (Taylor-Robinson 2003). It was once thought that there was no immune response against the liver stages but it is now apparent that there are immune responses in the liver that block the invasion of the hepatocytes and probably kill the early stages of the parasite within the hepatocytes (Aidoo and Udhayakumar 2000). Several liver-stage antigens are now recognized, particularly LSA-1 and LSA-3, as possible targets for immune attack and there have been several attempts to incorporate them into multicomponent vaccines. The most promising developments are concerned with DNA vaccines and the current front runner is a vaccine based on a string of 20 T- and B-cell epitopes derived from *P. falciparum* pre-erythrocytic-stage antigens fused with the TRAP molecule plus the modified vaccinia virus Ankara (MVA) (Moorthy and Hill 2002). This DNA ME-TRAP vaccine is now undergoing clinical trials with different prime-boost regimens (McConkey et al. 2003). Similar vaccines with the fowlpox virus FP9 substituted for MVA and the possibility of boosting the response to RTS,S are also at various phases of development and clinical trial (Moorthy and Hill 2002).

Blood stages are obvious vaccine targets because there are so many of them and because they cause the disease. A number of potential antigens have been recognized, either associated with the infected red cell, intra-erythrocytic parasites, merozoites or secreted soluble antigens. The prime targets are the antigens involved in adhesion and red blood cell invasion. A vaccine against the adhesion molecule, PfEMP-1, would prevent or decrease the pathology associated with sequestration but, because of the ability of PfEMP-1 to undergo antigenic variation, it is not a suitable vaccine candidate. Attention has, therefore, focused on antigens involved in red cell invasion. A recombinant form of MSP-1 has been assessed in Phase I and Phase II (safety and efficacy) trials but was found to produce adverse hypersensitivity effects (Keitel et al. 1999). However, MSP-1 together with other antigens, is being considered as a component of multicomponent vaccines incorporating the genes for antigens from various stages of the life cycle, including the sporozoite, liver, erythrocytic and sexual stages (Holder 1999; Shi et al. 1999). A number of multicomponent vaccines have been developed of which the most extensively studied is one based on three blood-stage antigens (MSP-1, MSP-2, and RESA), which has produced some encouraging results in field trials in Papua New Guinea (Genton et al. 2003).

Currently, the only vaccine that has been subject to extensive field trials is SPf66, a synthetic vaccine consisting of a polymer of three merozoite antigens: Pf83 (an 83 kDa peptide representing part of MSP-1), Pf55, and Pf35. These antigens, none of which has been identified with any of the known major malaria antigens, are linked trogether by the sporozoite antigen NANP (Valero et al. 1993). In a number of extensive trials in Colombia, Ecuador, and Venezuela, the results obtained indicated that the vaccine was safe and immunogenic and, although it did not prevent malaria, it reduced the number of malaria episodes. It has also undergone extensive trials in natural populations of children in Africa and adults in Thailand but, unfortunately, it did not prevent malaria in children in trials in Tanzania (Beck et al. 1997) or The Gambia (Bojang et al. 1998). SPf66 is the only malaria vaccine that has so far been approved by the WHO but has now been discontinued in its original form.

One of the possible vaccine components is the sexual stage in the mosquito (Targett 1995). At first, this might seem to be a very strange target for attack but it is both logical and feasible. The principle is to immunize individuals with sexual-stage antigens to raise antibodies that inactivate the sexual stages when taken up in the serum by a mosquito. This has been called the transmission blocking or 'altruistic' vaccine because it does not protect the person immunized although it does prevent transmission. Pfs25 and Pfs28, antigens on the surface of the malaria zygote, are the most favored candidates for such a vaccine and a recombinant fusion protein, TBV25-28, containing these two antigens has been developed and considered for clinical trials (Gozar et al. 2001). Up to now, the idea of such a vaccine would have been unthinkable but there is no reason why such a transmission blocking antigen should not be added to another vaccine such as SPf66.

Numerous other possibilities are currently being pursued and it will probably be necessary to have a multipronged approach to the development of a malaria vaccine. It may not be necessary to produce sterile immunity but it might be possible to reduce the severity of the infection. It may, therefore, be possible to consider novel approaches such as vaccinating against the disease rather than the infection and there is some evidence that it is possible to immunize laboratory animals against the toxins produced by malaria parasites (Schofield et al. 2002).

The development of a malaria vaccine has been a long and frustrating search with many false starts and unachievable promises. The future, however, is brighter and depends on scientific co-operation rather than competition which is now forthcoming (Medaglini and Hoeveler 2003; www.niaid.nih.gov/dmid/malaria/malariavac.htm).

BABESIAL INFECTIONS

Babesiosis, although common in wild and domesticated animals, is a rare accidental infection in humans (see Chapter 23, Babesiosis of humans). Little is known about immune responses in humans except that immunologically intact individuals are susceptible to infection with the rodent babesia *Babesia microti*. Immunity seems to be similar to that seen in malaria in that it involves the activation of macrophages and NK cells and the production of NO (Aguilar-Deflin et al. 2003). Such infections are accompanied by raised specific antibody levels, even in the presence of persisting parasitemias, a phenomenon common in many protozoal infections. Babesiosis caused by *B. microti* is more serious in immunologically compromised patients and the European form, caused by *B. divergens*, has been reported only from splenectomized individuals, suggesting that an intact immune system is required to keep this infection under control (see Telford et al. 1993). Although nothing is known about the ways in which *Babesia* spp. survive in humans, it is known that there are a number of ways in which these parasites survive in other hosts; these include antigenic variation, cytoadhesion to avoid immune attack (as in malaria infections), binding of host proteins to infected red blood cells and immunodepression (Allred 2003). The persistence of babesial infections and the ability to sequester suggest that the pathology of babesiosis might resemble that seen in malarial infections (see Clark and Cowden 2003).

Immunity to tissue-inhabiting protozoa

Several parasitic protozoa occur in human tissues, including the following:

- the ubiquitous coccidian *Toxoplasma gondii*, in a variety of nucleated cells, including macrophages, and the brain
- *Trypanosoma cruzi* in muscle and nerve cells
- *Sarcocystis* spp. in muscle
- microsporidians in various tissues including the brain

● opportunistic amebae in the brain.

Of these, *T. gondii*, already discussed, and *T. cruzi* are the most important and, partly because of the ease with which they can be maintained in laboratory animals, have been the most intensively studied. The remaining tissue-inhabiting protozoa cause long, chronic and usually harmless infections but these may become fulminating in immunologically compromised individuals such as those co-infected with HIV.

CHAGAS DISEASE (NEW WORLD TRYPANOSOMIASIS)

Chagas disease, caused by the trypanosome *Trypanosoma cruzi*, is confined to South and Central America (see WHO 2002; Chapter 19, New World trypanosomiasis). *T. cruzi* infects more than 150 species of mammals and most of the research done with this parasite has been carried out in rodents, particularly mice, in which the pattern of immunity that emerges is complex and is influenced by the genetic make-up of both the parasite and host. Chagas disease is one of the most insidious of all parasitic diseases. The infection is transmitted by blood-sucking bugs and is typically acquired in childhood; the victim may experience little more than a localized swelling and a transient fever but, in some individuals, possibly as many as 20–40 percent, the infection may be acute and life threatening. Adults who become infected tend to develop a life-long chronic infection which can result in death many years later. The trypanosomes divide at the site of the bite and circulate briefly in the blood before entering macrophages, in which they develop into nonflagellated amastigotes and survive by escaping from the phagolysosome and multiplying in the cytoplasm of the cell. From there, they escape and invade other macrophages or a variety of other nucleated cell types, including muscle cells (particularly cardiac muscle) and nerve cells. There is a strong immune response in individuals who survive an acute infection but this is only partially effective and does not eradicate all the parasites; the infection, therefore, persists throughout the life of the host. During the course of the infection, the host mounts specific antibody responses involving all subclasses of immunoglobulins. As in most other infections, there is an early rise in IgM followed by increasing levels of IgG, which persist throughout the infection and are, therefore, useful for diagnosis.

The use of genetically modified mice has made it possible to examine the immune response to *T. cruzi* in some detail (Tarleton et al. 2000). In mice, high levels of IgG2a, IgM, and IgE persist for about 13 weeks after which IgG2a and IgM levels begin to decline but, together with IgG1 and IgG2b, remain elevated throughout infection (Rowland et al. 1992). Studies in mice have also provided clues about cell-mediated immune responses to *T. cruzi* (reviewed by Takle and

Snary 1993; Kierszenbaum and Sztein 1994; Kirchoff 1994). Protective immunity to *T. cruzi* is T-cell dependent and cell-mediated (Wrightsman et al. 2002). The main defense mechanism is provided by macrophages activated by TNF-α, IFN-γ, and IL-12 resulting in the production of NO, which is responsible for the killing of intracellular parasites (see Machado et al. 2000; Talvani et al. 2003). However, this activity can be inhibited by IL-4, IL-10, and TGF-β (Munoz-Fernandez et al. 1992; Gazzinelli et al. 1992; DosReis 1997; Reed 1998). There is no clear-cut Th1 and Th2 polarity in *T. cruzi* infections and it appears that Th1 cells are required for the control of infection and that Th2 cells may be involved in the persistent parasitemia and severity of disease (Kumar and Tarleton 2001).

The long-term persistence of parasitemia is a characteristic feature of Chagas disease and it is clear that *T. cruzi* survives for the whole lifetime of patients with chronic Chagas disease as evidenced by the fact that, at autopsy, parasites have been detected in heart muscle (Brener 1994), smooth muscle and adrenal glands (Teixeira et al. 1997). The parasite can only survive by evading the immune response and this is partially achieved inactively by forsaking macrophages and invading nonphagocytic cells. *T. cruzi* also evades the immune response actively in a number of ways. Antibody attack is prevented in two ways: the inhibition of antibody binding, and the inhibition of complement activation. Antibody binding is inhibited by the overproduction of nonprotective IgM that interferes with the binding of potentially protective IgG; complement activation is inhibited by means of a surface glycoprotein that binds C3 (the third component of complement) and thus inhibits both the classical and alternative complement pathways (Beucher et al. 2003). *T. cruzi* also employs a variety of immune evasion mechanisms to enable it to avoid cell-mediated immune responses (Zambrano-Villa et al. 2002). The production of IL-12 and TNF-α (essential for parasite killing) is inhibited by a parasite surface antigen, AgC10, that binds to the surfaces of macrophages (de Diago et al., 1997); another parasite molecule promotes the production of IL-10 and TGF-β (Sztein et al. 1990) while also inhibiting the production of IL-12 (Silva et al. 1998). Immunosuppression is also characteristic of *T. cruzi* infections and this is brought about, in part, by the induction of a population of suppressor cells (Tarleton 1998). In the context of immune evasion, it might be to the advantage of the parasite to mimic host antigens but one possible outcome of the possession of shared antigens is autoimmunity, which some authorities consider to be characteristic of this infection.

Chagas disease is life-long and usually chronic and is accompanied by the gradual destruction of the infected cells, giving rise to cardiac failure and loss of control of the digestive system later in life, possibly 20–30 years later (Tarleton 2001). Whether or not Chagas disease is an autoimmune disease is a matter of controversy (Leon and Engman 2001), but the histopathological picture does suggest an autoimmune phenomenon. Sequestered antigens released when infected cells are destroyed and uninfected cells onto which *T. cruzi* antigens have been adsorbed are potential causes of autoimmunity. In experimental animals, autoantibodies appear after nerve cells have been destroyed, antibodies to rat ganglia cross-react with *T. cruzi* and anti-*T. cruzi* antibodies cross-react with mammalian tissues. Comparable observations have been made in patients with Chagas disease, for example the C-terminus of the *T. cruzi* ribosomal P protein shares epitopes with host cells (Mahler et al. 2001) and myosin shares epitopes with the *T. cruzi* B-13 antigen with evidence of some association with chronic chagasic cardiopathy (Cunha-Neto and Kalil 2001). There is further information on the pathology of Chagas disease in Chapter 19, New World trypanosomiasis.

In summary, immunity to *T. cruzi* involves some of the most complex immunological responses encountered in any infection involving antigens, host cells, chemokines and cytokines (see Teixeira et al. 2002); the interplay between immunity, immune evasion, and autoimmunity is likely to take many years to elucidate. In the meantime, although some experimental vaccines have been developed, the possibility of a protective vaccine that does not cause any autoimmunity against this important human pathogen seems remote.

MICROSPORIDIANS

Microsporidiosis, caused by at least eight species belonging to five genera (see Chapter 25, Microsporidiosis) is a rare but increasingly recognized condition in immunologically compromised patients, particularly those suffering from AIDS (Bryan 1995; Deraedt and Molina 1995). As the condition is virtually unknown in immunologically competent individuals, there must be some immune response that keeps these infections under control, but definite evidence for this, and its mechanisms, is lacking. Apart from the epidemiological evidence, most of our knowledge comes from animal and in vitro studies, which might not be relevant to what happens in humans. It is clear that perturbation of T-cell function is a major cause of immunological failure leading to microsporidiosis, but as T cells are central to so many immunological responses, it is difficult to draw firm conclusions about the nature of natural or acquired immunity to microsporidiosis in humans.

SARCOCYSTIS AND RELATED ORGANISMS

Little is known of these rare infections in humans (described in Chapter 22, *Sarcocystis*) in terms of the immune responses that they evoke in their natural hosts, laboratory animals or humans. All form thick-walled cysts with few signs of any immunological reaction or immunopathology.

OPPORTUNISTIC AMEBAE

Although these free-living amebae (belonging to the genera *Acanthamoeba*, *Balamuthia*, and *Naegleria*) are sometimes grouped with the intestinal amebae, it is convenient to consider them briefly here. As there have been relatively few human infections, there is very little information available and it is not known whether or not individuals mount a protective immune response. If they do, it is not clear whether this protects against subsequent infection (which, in any case, would be very unlikely). Virtually all that is known is that specific antibodies have been detected in the sera of infected individuals (see Chapter 11, Opportunistic amebae). Information that has accrued from observations in animal models has been reviewed by John (1993).

IMMUNITY TO HELMINTHS

Characteristics of helminths in relation to the immune response

Helminths (worm parasites) confront the immune system with problems that are quite different from those posed by protozoa. These problems are largely the consequences of greater size and structural complexity, but are compounded by three additional factors:

1 worms have the ability to move actively through the body of the host
2 as part of their life cycles, many undergo sequential developmental changes in the host, during which their structure and antigenic make-up may change dramatically
3 their bodies are covered by layers that are more complex and less vulnerable to immune attack than conventional plasma membranes.

Thus, although helminths occupy many of the sites in the body that are exploited by protozoa (other than the truly intracellular), the immune mechanisms required to destroy them in or remove them from these sites must often be qualitatively different. Unlike protozoa, the majority of helminths do not replicate within the host. In these circumstances, an increase in the worm burden carried (and the associated danger of pathology) occurs only when the host is repeatedly infected. Under conditions of natural infection, immune responses are, therefore, most important in preventing or reducing reinfection. There are, however, some helminths in which the numbers of parasite stages present in the body do increase after infection. These include:

- tapeworms, such as *Taenia solium* and *Echinococcus*, in which an initial infection may produce large numbers of larval stages
- the blood flukes (*Schistosoma*), in which eggs become trapped in tissues

- the nematodes, *Strongyloides* and *Trichinella*, in which larvae accumulate in the body.

In such cases, the development of the immune response can be considered analogous to that which occurs in protozoan infections.

Adult stages of helminths range in length from a few millimeters to several meters. Larval stages may be less than 1 mm in length, but even the smallest larva is too large to be phagocytosed or lysed by a single cell; cytotoxic activity, therefore, requires multiple interactions between the parasite surface and inflammatory cells. Such interactions can be mediated by both complement and antibody, ADCC being a common response to invasion by worms.

In addition to being large, helminths are complex multicellular organisms with well developed organs and tissues. When alive, they present the immune system with a diverse range of antigens, both on their surfaces and in excretory and secretory products. When they die, many more antigens may be released. Although immune responses are made to the majority of these antigens, they may have little or no effect on the parasite for some or all of the following reasons:

- the molecules concerned may not be essential for worm survival
- in the living worm, these molecules may be inaccessible to effector mechanisms
- the response to released antigens may take place at a considerable distance from the parasite
- the parasite may actively move away from the immune response generated at a particular site
- the antigens may elicit inappropriate immune responses
- the parasite may evade, inactivate or down-regulate immune effectors.

In general then, although there are important exceptions involving larval stages, worms are likely to be much less easily controlled by immediate interactions with immune effector mechanisms than are protozoa. Binding of antibodies or surface activation of complement rarely kills worms directly; effective immunity is more often the indirect result of immunologically specific recognition events. Among the more important of these are ADCC, immune-mediated inflammation, and antibody-mediated inhibition of enzyme activity.

The result of ADCC reactions is the adherence of large numbers of cells and the release onto the surface of a variety of enzymes and short-range mediators. These may harm the parasite directly, either by destroying the integrity of the surface layer or by causing damage to deeper tissues. The effectiveness of such attack is influenced by the nature of the surface available. For example, whereas flukes such as schistosomes have relatively delicate cytoplasmic surfaces, nematodes have tougher collagenous cuticles that provide them with considerable protection against external damage. Larval nematodes may be killed by ADCC, but adults are often unaffected. Even the theoretically more vulnerable schistosomes have multiple evasion strategies that result in their surfaces being difficult to damage. In addition, effective ADCC requires not only that cells bind to the target, but also that the cells maintain contact, which is sometimes difficult when the target is an active worm.

With the important exceptions of tapeworms and acanthocephalans, most helminths feed actively and digest their food source enzymatically, either within their own intestines or externally before uptake. The enzymes used in these processes can be important targets for the host immune response, as are those used by worms to penetrate the host and to move through its tissues. The interaction of antibodies with target parasite enzymes can inhibit their activity, which can prevent successful invasion, migration, growth, and development and may result in worm death. Indeed, the complex nutritional, environmental, and behavioral requirements of worms make any host-mediated interference with normal function a potential means of increasing resistance.

By mechanisms that are not fully understood, the range of immune responses elicited by infections with helminths typically show a number of important differences from those generated by protozoan infections. Both in humans (Yazdanbakhsh et al. 2001, 2002) and experimental rodents (Gause et al. 2003), helminth antigens elicit responses that are predominantly mediated through cytokines released from lymphocytes of the Th2 subset. Prominent among these responses are the increased production of reaginic antibody (IgE), mast cells and eosinophils. Immediate hypersensitivity reactions, in which these components are involved, are therefore characteristic accompaniments of worm infections, but rare in protozoa. The inflammatory consequences of such reactions can be very effective forms of host defense and provide substantial resistance to infection. They result in structural, functional, and biochemical changes in the tissues surrounding the worms and lead to the release of powerful biologically active mediators. If antigenic stimulation and inflammation persist, worms may become the focus of cellular accumulations that trap them and that may ultimately kill them. As well as endangering the parasite, such chronic reactions may also cause severe host pathology. In a number of infections (e.g. schistosomes, filarial nematodes, larval cestodes), the parasites release factors that lead to down-regulation of the immune response, IL-10 playing an important modulatory role. In this way, parasites protect themselves from the host response and prolong their survival.

One of the most interesting experimental models of helminth infection is that of the intestinal nematode *Trichuris muris* in mice. As with the protozoan

Leishmania major, there is a reciprocal relationship between the Th subset response and immunity but in the case of *T. muris*, Th2 activity leads to resistance and Th1 to continuing susceptibility – i.e. the opposite of what happens in *L. major* infection. The dependence of the host response to *T. muris* on specific cytokines is elegantly demonstrated by the fact that the phenotypes of genetically susceptible and resistant mice can be reversed by injection of the appropriate cytokines or anticytokine reagents (Else et al. 1994). Although these 'helminth-type' and 'protozoan-type' patterns are clear cut in certain experimental models, in the majority of species of both groups the interrelationship between Th1 and Th2 responses appears to be more complex.

It should, therefore, be clear that although protozoa and helminths are grouped together as 'parasites' and both are powerfully immunogenic to the hosts, if they are to be successfully controlled by immune responses rather different strategies are required and quite distinct mechanisms may be involved.

Immunity to intestinal helminths

The characteristics of the intestine as an environment for parasites have already been described (see Immunity to intestinal protozoa). Despite the many defenses that can act there, the intestine is one of the commonest environments for parasitic worms and it is colonized by representatives of all the major groups: digeneans (flukes), cestodes (tapeworms), and nematodes (roundworms). Relatively little is known of immune responses to intestinal flukes and tapeworms in humans. More is known about nematodes, but most of this information concerns responses that can be measured peripherally, i.e. antibodies, proliferative and cytokine responses of circulating lymphocytes, delayed hypersensitivity and inflammatory responses. For an insight into responses operative at the intestinal level, we are heavily dependent on data obtained from work with experimental models.

INTESTINAL NEMATODES

A very large number of nematode species live in or infect via the intestine. Collectively, they are among the commonest of all parasites, infecting an estimated one-quarter of the world's population (see Chapter 34, Gastrointestinal nematodes – *Ascaris* hookworm, *Trichuris*, and *Enterobius*). The major intestinal nematodes of humans (*Ascaris*, hookworms, *Strongyloides*, and *Trichuris*) all mature in the intestine, but reach that site in a variety of ways. *Trichuris* infections are established by ingestion of eggs and the worms develop directly in the large intestine. *Ascaris* also infects orally, but after hatching, the larval stages penetrate the intestinal mucosa and migrate via the liver and lungs before returning to the small intestine to mature. The larvae of

hookworms and *Strongyloides* characteristically infect through the skin and then migrate to the small intestine. In addition to the intestine itself, immunity to worms with migratory stages may, therefore, operate at several sites in the body. The mechanisms likely to be effective are determined by the characteristics of each organ system involved. Immune and inflammatory responses in parenteral sites such as the skin, liver, and lungs share a number of common features. Worms moving through these sites are likely to be in intimate contact with potential effectors, such as antibodies (IgM, IgG), complement, biologically active mediators and cytotoxic cells, and can therefore be damaged directly. In contrast, worms living in the intestine itself may have a quite different relationship with the immune system. Not only does the intestine have distinctive immune mechanisms, but unless worms penetrate into the mucosa they are unlikely to be accessible to some of the effector mechanisms that operate in the tissues. Those effectors that may affect intestinal worms include the antibody isotypes that are secreted into the lumen (IgA and IgM) or leak from the mucosa (IgG), and mediators such as amines, prostaglandins, leukotrienes and reactive oxygen and nitrogen metabolites that are produced by inflammatory cells. Intestinal worms, like all parasites, are wholly dependent on the physical and physicochemical characteristics of their environment, and these conditions may change dramatically when the intestine is inflamed as a result of infection. Such changes can themselves contribute to immunity by making the local habitats of the parasites unsuitable for their continued survival. As already mentioned (see The cytokine network in parasitic infections, and Characteristics of helminths in relation to the immune response), worm infections selectively elicit Th2 subset responses and the cytokines produced by these cells generate powerful responses, components of which (e.g. IgE, mast cells, and eosinophils) are known to be potent mediators of intestinal inflammation.

Experimental systems in laboratory animals show that intestinal worms can be very effectively controlled by host immune responses (Gause et al. 2003). It is much less clear whether or not intestinal worms in humans are similarly controlled. Infections are not only very common, they are often persistent. Individuals may acquire worms early in life, retain worms for long periods, and be easily reinfected. Epidemiological studies indicate that although the prevalence of infections often remains high as people age, the intensity of infection (numbers of worms) can decline significantly (Bundy 1995). This may indicate the operation of protective immune responses, although it is obvious that behavioral changes will also be involved, as transmission of all the major intestinal worms is dependent on some degree of contact with fecal or fecally contaminated material. Cross-sectional analysis of infected populations shows that the parasite burden is aggregated into relatively few individuals, most having small numbers of worms. Again, this may indicate an effective immunity or may reflect behavioral influences. It is well established that individuals are predisposed to a particular level of infection, i.e. they reacquire worm burdens after chemotherapy that are similar to those existing before treatment. In some, this predisposition results in heavy infections (i.e. the individuals are susceptible); in others, only light infections are acquired (i.e. they appear resistant). Predisposition suggests the operation of genetically determined levels of immunity to infection, and evidence supporting this interpretation is now beginning to appear (Williams-Blangero et al. 2002).

Knowledge of immune responses to intestinal nematodes of humans is much less detailed than that of responses to filarial worms and schistosomes (see Immunity to filarial nematodes, and Schistosomes). In general, correlative immunoepidemiological studies show that serological responses correlate positively with intensity and duration of infection, but certain isotypes appear to be associated with protective immunity. These include IgA responses in *Trichuris* infections (Bundy and Medley 1992) and IgE responses in hookworm (*Necator*) and *Trichuris* infections (Pritchard et al. 1995; Faulkner et al. 2002). The functional roles of antiworm antibodies are uncertain. ADCC responses may be generated against stages migrating through tissues, but the adults in the intestine are large and well protected organisms. Antibodies may interfere with feeding activities, particularly where, as in hookworms and possibly *Trichuris*, these depend on the worm releasing enzymes into the host's tissues. Antibodies may also reduce the effectiveness of worm secretions that interfere with the operation of host defense mechanisms, such as anti-enzymes, antioxidants and immunomodulators (Pritchard 1995). Recent studies have begun to analyze cellular and cytokine responses (Cooper et al. 2000; Geiger et al. 2002; Turner et al. 2002). So far, these have not shown the same clear-cut correlation of protective immunity with T2 cytokines that has emerged from work in animal models (see below). Geiger et al. (2002), for example, found that parasite-free individuals showed higher T1 cytokine production (TNF-α, IL-12, IFN-γ) than infected individuals, whereas T2 cytokines were present equally in parasite-free and infected patients. Both this group and Turner et al. (2002) recorded production of IL-10 in infected individuals, suggesting that chronic infections may be associated with activity of Th3 or Tr cells.

Experimental studies in rodents have given detailed pictures of the nature and operation of immune responses against intestinal nematodes. These may operate during the course of primary infections to expel worms from the intestine. On reinfection, protective responses operate more quickly and may prevent the establishment of infection almost completely. In all cases, immunity is Th cell-dependent, and Th2 cells play the major role, polarization of the response to this subset being critical for protection (Gause et al. 2003). In manipulative experiments, it has been shown that IFN-γ can prevent, and IL-4 together with IL-13 can promote, the normal expression of immunity; IL-10 production is also important to ensure a Th2 bias. The roles of the co-stimulatory and signaling membrane molecules that regulate T-cell activity and define whether host responses are protective or otherwise (e.g. B7.1-B7.2, CD40L, STAT 4/6 and chemokine ligands) have now been fully defined (Gause et al. 2003; deSchoolmeester et al. 2003). Analysis has also extended to the role of intracellular signaling pathways (Artis et al. 2002).

Responses to infection are accompanied by acute inflammatory changes that result in structural and functional modifications of the intestinal mucosa, and in some species (e.g. *Nippostrongylus*, *Strongyloides*, *Trichinella*) these act as important effector mechanisms, (Artis and Grencis 2001). Although IgE antibodies and eosinophilia accompany the intestinal inflammation induced by infection in rodents they do not seem to have a significant role, and deletion of these responses by using antibodies against the relevant cytokines does not affect the expression of immunity. In contrast, mast cells do play an important role against certain species (Helmby and Grencis 2002). In other parasites (e.g. *Trichuris*) antiworm IgA and IgG1 antibodies seem to be important, probably by interfering with important metabolic or behavioral activities. In worms that migrate through the tissues to reach the intestine (*Nippostrongylus*, *Strongyloides*), immunity acts against the intestinal stages during a primary infection, but against tissue stages in subsequent infections, larvae being trapped in inflammatory foci in the lungs or the skin. Similar mechanisms appear to act against migrating larval hookworms in immune mice (Girod et al. 2003).

Experimental models have not only allowed the study of protective responses, they have also shed light on situations in which immunity does not operate or operates at best ineffectively. Some nematodes in mice (like nematodes in humans) establish chronic, long-term infections. With one species (*Heligmosomoides polygyrus*), this is due to immunomodulatory influences exerted by adult worms, which effectively suppress the potentially protective responses elicited by the larval stages. Immunomodulation is associated with the release of particular molecules in the secretions of adult worms. Full expression of this immune suppression is under strong host genetic influences, which determine the speed and nature of antiparasite responses. *Trichuris muris* can also establish chronic infections in certain strains of inbred mice. In most strains, infection elicits Th2 responses that result in effective immunity and worms are lost before they can become mature. In certain strains, the initial Th2 response is switched to a Th1 response and the host is then not only unable to eliminate the existing infection but remains susceptible to subsequent infections as well, a situation that has many parallels with human *Trichuris* infections (Artis and Grencis 2001). Chronic worm infections are often associated with immunopathological changes to the intestine, which almost certainly reflect the effects of persisting, inappropriate immune responses.

Immunity to filarial nematodes

Filarial nematodes (members of the Filarioidea) are important human parasites in many subtropical and tropical countries (see Chapters 38, Lymphatic filariasis, and 39, Onchocerciasis). All are transmitted by insect vectors, which act as intermediate hosts allowing development of larvae to the infective stage. These larvae are transmitted when the insect bites, the initial stages of infection, therefore, involve the skin and dermal tissues. In some species, the adults remain in the superficial layers of the body but in others, the adult worms live within much deeper tissues; both are, therefore, theoretically susceptible to direct interaction with immune effectors such as antibody, complement and cytotoxic cells. There are several species of filariae that infect humans, many of which are of low pathogenicity or of only local significance. The most important species are those in which the adult worm lives in the lymphatic tissues (the genera *Wuchereria* and *Brugia*) and those in which the adults live in the skin (*Onchocerca*). In the former, adult females liberate embryos (microfilaria larvae) that circulate in the blood; female *Onchocerca* release microfilaria into the skin. Biting insects pick up these larvae while feeding, the larvae undergo molts to reach the infective third stage, and these then re-enter the human at the next blood meal. In the lymphatic species, pathology is caused by the adult stages, whereas in *Onchocerca* microfilaria are the primary cause of pathology.

It is characteristic of all filarial infections that they are long-lived; worms survive for many years and release millions of microfilaria. Many individuals in areas where these infections are endemic continue to accumulate worms over long periods. These two features suggest that there is little or no immunity, but more detailed epidemiological studies suggest otherwise. Even in areas where the overall prevalence is high, some individuals are parasitologically negative, show no pathology but do have antiparasite immune responses, indicating exposure to infection. The majority show signs of infection, of pathology, or both. Some are heavily infected, with many circulating microfilaria, but may have relatively little overt pathology; some have severe pathology but may have few or no microfilaria in the blood; others may have low microfilaremia but show abnormal pathology indicative of hypersensitivity reactions. Collectively, these data show that filarial infections do stimulate immune responses; in some individuals these responses appear to be fully protective, but in others, immune responses contribute to the development of pathologic changes. This spectrum of responsiveness reflects a variety of patterns of Th cell and cytokine activity as well as parasite-induced immune suppression.

Analysis of protective responses to filarial nematodes has been carried out in vitro, using human cells and sera, and in vivo, using a variety of animal models. It has been shown in several cases that immunity can effectively operate through ADCC reactions directed against antigens present on the cuticular surface. The filarial cuticle is relatively delicate, compared with that of the intestinal nematodes for example, and may play a significant role in the uptake of nutrients. Antigens on the cuticle are known to be targets for antibodies, particularly IgG and IgE isotypes, and these facilitate adherence of cells such as eosinophils, macrophages, and neutrophils. Under experimental conditions, ADCC reactions can kill all stages, from microfilaria to adult; under conditions of natural infection it is most likely that a primary target is the incoming infective larva. Although the cuticle appears to be a major target for immune responses, other protective antigens have also been identified. Antigens derived from internal tissues, including muscle proteins, can elicit protective responses when used to immunize experimental animals.

Immunity to filaria is T-cell dependent and involves the activity of Th cells. In humans, putative immunity correlates with production of T-cell dependent antibody isotypes (IgE and IgG3), and these isotypes are also required for effective ADCC. Production of eosinophils, a contributor to cytotoxicity, is also T dependent. Studies in experimental animals have confirmed the need for Th cell activity, and shown that in some cases the Th2 subset is functionally important, IL-4 being a key cytokine (Lawrence 1996; Volkmann et al. 2001). There are some interesting exceptions to these conclusions (e.g. Le Goff et al. 2002) and these raise the perennial problem as to how representative of human responses are those measured in experimental animals. However, reduced IL-4 production is often found in heavily infected individuals but present in putative immunes. A factor complicating interpretation of immune responses to filarial worms is the demonstration that they harbor symbiotic *Wohlbachia* bacteria (Taylor 2002). These play an important role in worm development and fertility; if the bacteria are removed by antibiotic treatment worms may die. *Wohlbachia*, however, releases a lipopolysaccharide-like endotoxin that can affect both the nature of the immune response and the consequential immunopathological consequences.

Some of the most interesting data concerning immune responses to filarial infections in humans are those that attempt to correlate patient status with particular components of the immune system (Nutman 2001). It is now widely accepted that, both for lymphatic filariasis and onchocerciasis, there are individuals in populations in endemic areas who neither develop patent infections nor show any signs of diseases, yet are immunologically positive in the sense of having antiparasite antibodies and antigen-specific T-cell activity. Such individuals are often referred to as 'endemic normals', although the use of this term is debatable. The fact that these individuals remain parasitologically negative despite exposure to infected vectors implies effective control of incoming infections and it is thought that this control operates against the infective larval stages. Experimental data from *Brugia* infections in cats, as well as data from rodent models, support this interpretation; indeed there is good evidence that irradiated infective larvae can induce an effective immunity when used as a vaccine and that this immunity is Th2-dependent (Bancroft et al. 1994).

The suggestion that some individuals do develop immunity to filarial infections raises the question of why the majority apparently do not, despite evidence for immune responses to the parasites. Those who show parasitological or pathological evidence of infection can be divided into a number of categories (Freedman 1998; Nutman 2001). In lymphatic filariasis, the extremes are patients who are microfilaremic and asymptomatic, and those who are microfilaria negative but show obstructive pathology. Another category shows atypical allergic hyper-responsiveness (tropical pulmonary eosinophilia). The lymphocytes of those with high microfilaremia often appear to be unresponsive to parasite antigens when tested in vitro (as is also the case with cells from patients with heavy skin burdens of the microfilaria of *Onchocerca*). There is evidence that altered T-cell-subset activity (including Tr cells) and cytokine production are contributory factors (Maizels et al. 1995; Sato-guina et al. 2002; Steel and Nutman 2003) but released parasite products themselves can be directly immuno-suppressive (Allen and MacDonald 1998; Pfaff et al. 2002). The existence of phosphorylcholine on filarial molecules is associated with altered immune responsiveness, mediated through IL-10 (Houston et al. 2000) and IL-10 is increasingly recognized as an important modulator of the antiparasite response leading to an altered balance between Th1 and Th2 activity. One consequence of this is the high levels of circulating antiworm IgG4 and IgE. These isotypes are found in both microfilaremic and elephantiasis patients but in the former, the ratio between the isotypes is biased to IgG4 and in the latter, the bias is towards IgE. The high levels of IgG4 may inhibit allergic responsiveness in the microfilaremic patients, whereas high IgE may promote pathological responses leading to obstructive pathology. IgE may, therefore, be important in protection against larval stages, through mediation of ADCC, but it may also contribute to immunopathology via hypersensitivity reactions to adult worms. Other IgG isotypes are low in microfilaremic individuals but more prominent in the elephantiasis group, in whom they may again contribute to pathology.

The T-cell anergy seen in patients heavily infected with lymphatic filarias or with *Onchocerca* appears similarly to be due to down-regulation of T-cell responsiveness by the cytokine IL-10 (Mahanty et al. 1997; Doetze et al. 2000). This is released from both Th2 cells, monocytes and macrophages, but predominantly the latter. Microfilaremic patients show high IL-10 production and low antigen-specific T-cell proliferation; symptomatic patients produce little IL-10 and their T cells proliferate well when exposed to parasite antigen. These results suggest that Th1 responses are somehow involved in antiparasite activity, which may be both beneficial, giving immunity against larvae, and harmful, via inflammatory responses to adult worms. IL-4 secretion appears to be linked with that of IL-5, lymphatic filariasis being characterized by high peripheral eosinophilia. In filarial infections there is, therefore, a complex, parasite-induced interplay between Th subsets that results in modulation of antibody and cellular responsiveness and determines both the parasitological and pathological outcomes of infection (Figure 4.4).

Immunity to flukes

Several species of digenean Platyhelminthes are parasitic in humans. They can be divided into four categories, based on their locations within the body:

1 the lung flukes (*Paragonimus*)
2 liver flukes (e.g. *Clonorchis/Opisthorchis*)
3 intestinal flukes (many genera)
4 blood flukes (*Schistosoma*).

Both lung and liver flukes can be highly pathogenic, with infections resulting in severe tissue damage (see Chapter 31, Lung and liver flukes). Serological and cellular responses to these parasites are well described, but are not easily correlated with protective immunity. Little is known of immune responses to the human intestinal flukes, although infections of mice with *Echinostoma* have shown that an effective immunity can operate. The most intensively studied flukes are the schistosomes.

SCHISTOSOMES

Adult schistosomes live in the blood vessels of the host. Infections are initiated by penetration through the skin of larval stages that have developed in aquatic snails. To reach their final location, the larvae (schistosomula) migrate from the skin via the lungs to the liver, where they develop into mature male and female worms. These form permanent pairs, and then move to the mesenteric veins around the intestine or the vesical veins around the bladder. Once there, the worms can survive for prolonged periods (possibly years) during which time the female releases large numbers of eggs. These must leave the body and hatch in water so that larvae can find and

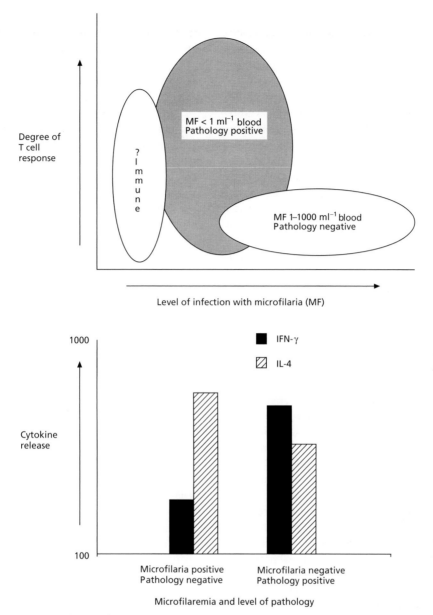

Figure 4.4 *Diagrammatic representation of the variations in T cell responsiveness and Th cell subset activity in patients infected with the lymphatic filaria* Brugia malayi. *T cell responses, reflected in level of proliferation when stimulated in vitro with antigen, are much higher in individuals without obvious infection (putatively immune) and in patients showing obstructive pathology than in those showing heavy microfilaremia but little pathology. Cytokine profiles, measured by in vitro cytokine production, indicate that heavy microfilaremia is associated with a predominant Th2 response (i.e. IL-4 production), whereas pathology in the absence of microfilaremia is associated with a predominant Th1 response (i.e. IFN-γ is the major cytokine). Data from Maizels et al. 1995*

penetrate a suitable snail. Eggs develop in the host's body and the contained larvae release enzymes that facilitate their passage out of blood vessels, through tissues into the lumen of the gut or bladder. Release from the body therefore occurs with feces or urine. There are three major species infecting humans: *Schistosoma japonicum*, *S. mansoni* (both in mesenteric veins), and *S. haematobium* (vesical veins).

Schistosomiasis is a multifaceted, immunopathological disease, but the most important components are the hypersensitivity responses initiated by eggs that are trapped in tissues (see Chapter 28, Schistosomes:

general). Granulomata form around the eggs, particularly those in the liver, and cause considerable local tissue damage. By blocking blood flow, granulomata can cause portal hypertension and serious vascular abnormalities, some of which (e.g. esophageal varices) may be life threatening.

Schistosome infections are common wherever climatic and environmental conditions favor snail development and where socioeconomic conditions allow fecal and urinary contamination of water that is used for drinking, bathing, washing or working. Infections are acquired early in life and are long-lasting; the disease is most serious in older children and young adults. Initial infections appear to generate little or no protective immunity, but the intensity of infection declines with age and it is now accepted that this decline has an immunological as well as a behavioral component.

Experimental models have primarily used *S. mansoni*, which infects a variety of rodents and primates – work with other species is covered in Chapter 30, Schistosomes: Asian. Primate studies defined the phenomenon of 'concomitant immunity' – the host is immune to reinfection while continuing to harbor the adult worms from an initial infection. Concomitant immunity was interpreted as showing that resistance mechanisms operated against the early schistosomula stage, but that adult worms were unaffected. In essence, this view still holds although there is now evidence that adults are also affected by immunity (Pearce and MacDonald 2002). Early studies also focused on effector mechanisms that resulted in damage to the tegumental surface which, in the larval stages at least, appeared vulnerable to direct damage by complement-mediated mechanisms and damage brought about by ADCC. The ability of adult worms (and indeed later larval stages) to survive in the face of responses capable of destroying early schistosomula depends on intrinsic changes in the tegument that render it less easily damaged, and the acquisition of host molecules (including glycolipids, immunoglobulin, and Fc receptors) that provide an immunological 'disguise'. In addition, schistosomes have a number of active evasion strategies that interfere with host immune mechanisms.

Experimental models, both in vivo and in vitro, indicate that there are multiple mechanisms through which resistance can be expressed. There is still no clear consensus about the mechanisms underlying protective immunity, although it seems clear that ADCC can be an important component of protective immunity. Several antibody isotypes and cell populations can participate in this response; eosinophils are major killer cells and IgG, IgE, and IgA are important isotypes. Interactions between parasites and macrophages may also be important, both in terms of ADCC and through the release of mediators capable of damaging the worms. A major difficulty, which has still to be resolved, is the relevance of data from animal experiments and from in vitro studies to the immune responses likely to occur in humans living in endemic areas (Capron 1998).

It is clear that schistosome infections elicit powerful CD4 T cell-dependent responses, the balance between Th1 and Th2 components being critical to the outcome (Hoffman et al. 2002). Data from experimental models show that the initial response is Th1 biased, switching to a Th2-biased response when egg laying begins. This switch is mediated by changes in local dendritic cells and co-stimulatory molecules as well as an altered cytokine environment in which IL-4 plays an important role (Gause et al. 2003) and is triggered by egg-associated

carbohydrate antigens. These have specific glycan determinants that appear selectively to boost Th2 responses both to themselves and to bystander antigens (Okano et al. 2001). The antigens that can initiate protective immunity include molecules present at the surface of the tegument, the targets for ADCC as well as internal antigens. The latter include structural proteins, e.g. paramyosin, as well as a number with enzyme activity, e.g. glutathione S-transferase (GST) and triosephosphate isomerase. Several antigens have been defined and produced as recombinant or synthetic molecules and have been exploited as vaccine candidates (Bergquist et al. 2002). Trials have been carried out in rodents, bovines, and primates, with some degree of success achieved, largely in terms of reducing worm egg production and thus reducing egg-induced pathology. Clinical trials have been carried out using GST as the major antigen (Capron et al. 2001). However, not everyone is convinced about the need for and feasibility of a vaccine (Gryseels 2000) and this topic has been fiercely debated (Hagan et al. 2000).

The demonstration that immunity does play a part in determining the age-related decline in infection seen in endemic areas has come from detailed population surveys; these have monitored levels of reinfection after chemotherapy, as well as degrees of water contact. They have shown that individuals may effectively resist reinfection despite frequent contact with infected waters. Immunological studies on these populations have identified significant correlations between protective immunity and levels of antiworm IgE, but have also shown that other isotypes (particularly IgM, IgG2, and IgG4) can act as blocking antibodies and interfere with the expression of immunity (Hoffman et al. 2002). In young children (under 10 years) blocking antibodies predominate; in older children, protective IgE predominates. It is assumed that IgE may act in ADCC and that blocking antibodies interfere with the optimal operation of this protective mechanism.

Invasion and development of schistosomes are associated with distinct immunopathological phases. Penetration of the skin by cercariae causes a dermatitis, which may be severe in individuals sensitized by prior infection. The developmental stages may also be associated with allergic symptoms. This is particularly true in the case of *S. japonicum* infection in which fever, eosinophilia, lymphadenopathy, splenomegaly, and intestinal disturbances form the syndrome known as 'Katayama fever'. The hypersensitivity responses that lead to granuloma formation around trapped eggs have been studied in great detail, particularly in *S. mansoni* infections. Granuloma formation is the result of a complex series of cellular interactions involving T and B cells, antigen-presenting cells, eosinophils, and macrophages These interactions are coordinated through adhesion molecules, co-stimulatory molecules, chemokines, and cytokines (Sandor et al. 2003). Both Th1 and Th2 cells contribute to granuloma formation, but down-regulation of Th2 activity occurs as infection progresses and this results in modulation of granuloma formation – IL-10 can play an important role in this process (Sadler et al. 2003). Granulomas eventually become surrounded by fibrous material, which helps to localize the lesion, but if excessive can lead to severe pathology. A balance between Th1 and Th2 responses appears to be critical: if either dominates, pathological levels of fibrosis may result. In the mouse, severe fibrosis appears to be Th2-mediated whereas in humans, a Th1 bias has been implicated (Fallon et al. 2000a; Hoffman et al. 2002; Pearce and MacDonald 2002), although IFN-γ has been associated with protection and TNF-α with aggravation of this disease (Henri et al. 2002). The severity of fibrosis in infected hosts is strongly influenced by genetic factors (Dessein et al. 1999).

S. mansoni and S. japonicum eggs leave the host's body via the intestine, which may become extensively damaged as tissues become sensitized to the eggs. Work in mice has shown that the enteropathy associated with this phase of infection is aggravated by Th2-dominated responses (Fallon et al. 2000b). In S. haematobium, the eggs of which leave via the bladder, inflammatory responses may produce severe pathological changes. There is a definite, but unexplained, association between such changes and the development of bladder cancers.

Immunity to tapeworms

Humans may act as hosts for adult stages of a number of tapeworm species. Some of these (e.g. Taenia saginata and T. solium) occur only in humans, others (e.g. Diphyllobothrium latum) occur in other hosts as well. Infections with adult tapeworms are acquired, in almost all cases, by ingestion of larval stages present in the bodies of intermediate hosts; for example, in the case of T. saginata by ingesting cysticerci present in infected beef and in the case of D. latum by ingesting plerocercoids present in infected fish. One species, Hymenolepis nana, has a unique life cycle in which development from the egg to the adult can take place within a single host. This species can, therefore, be locally quite common particularly in children. Very little is known of immune responses to adult tapeworms in humans; experimental studies with H. nana and other hymenolepids in rodents have shown that protective responses can act in the intestine against the adult stage.

LARVAL TAPEWORMS

In general, adult cestodes cause relatively little pathology, whereas larval tapeworms (metacestodes) can be highly pathogenic. In most cases, larvae are acquired by the accidental ingestion of eggs released from adult tapeworms in the intestines of other host species or from larval stages carried in transport hosts used for food. Larvae of the pork tapeworm T. solium, however, can develop from eggs released by adults in the intestine of the same or other humans.

The larval infections of greatest clinical importance are those caused by T. solium (cysticercosis), Echinococcus granulosus (hydatid disease), and E. multilocularis (alveolar echinococcosis) (see Chapter 33, Larval cestodes). In all of these, the metacestodes form cyst-like structures in host tissues. Cysticercosis occurs wherever T. solium infections are common in the natural intermediate host, the pig. If infected pig meat is eaten raw or undercooked, the larval stages (cysticerci) become activated in the intestine, attach to the mucosa, and transform into the adult worm. When mature, the adult begins to shed segments (proglottids) that pass out with feces. As these decay, they release infective eggs that may contaminate water or vegetables, which then become a source of human infection. It is also thought that eggs can hatch directly in the intestine of a host harboring an adult worm (autoinfection). By whatever route infection is achieved, eggs hatch, then larvae penetrate the intestinal wall and migrate via the bloodstream. Cysticerci can form in the muscles, where they are relatively harmless, and in the central nervous system (CNS), where they can cause severe symptoms.

Larvae of Echinococcus spp. are acquired by accidental ingestion of eggs deposited in the feces of dogs or wild carnivores such as foxes. The eggs hatch in the small intestine, the larvae penetrate the mucosa and migrate via the blood, and develop in internal organs such as the lungs, liver, peritoneal cavity, and CNS. Larvae of E. granulosus grow into large fluid-filled cysts (hydatid cysts), whereas larvae of E. multilocularis form a pseudo-malignant mass of proliferating vesicular structures.

The immunological relationships between larval tapeworms and their hosts are complex and not well understood. Metacestodes often elicit little inflammation, they can persist for long periods and some can metastasize. All of these characteristics imply minimal or ineffective immune responses, yet it is quite clear that infections are immunogenic. There are many descriptions of cellular and serological responses to metacestodes in humans, from which some correlations with protective immunity can be drawn. There has been considerable work on infections in animal models, from which have come more precise insights into determinants of resistance and susceptibility. Immune responses can be considered as operative early or late in infection. The term 'early' describes responses against the hexacanth larva or onchosphere that is released from the egg and which undertakes the initial migration into the tissues; 'late' describes responses against the subsequent developing cystic metacestode stages.

In the early stages of larval development, the tapeworms have an unprotected external layer called the tegument. They are, therefore, susceptible to damage by complement-fixing antibodies (e.g. IgG1 and IgG2 isotypes in mice; IgG2a in rats) and by complement- and antibody-mediated cellular cytotoxicity. Experimental studies show that host immunity can destroy larvae as they cross the mucosa or after they penetrate into other organs, but killing does not occur during an initial infection and the larvae become successfully established; early immunity is, therefore, most effective against reinfection. As the larvae develop, they acquire an external protective layer formed of both host and parasite components. They also produce sulfated polysaccharides and proteoglycans that interfere with host complement activation. In consequence, the larvae become insusceptible to the responses that can damage earlier life-cycle stages. Late immunity appears to operate exclusively through cytotoxic mechanisms, though these have little significant effect on larvae when they are fully developed. Larval tapeworm infections, like schistosome infections, therefore, show concomitant immunity, the host harboring fully developed primary infection larvae while being resistant to reinfection.

There is considerable evidence that larval tapeworms exert a variety of immunosuppressive effects on the host. For example, larval secretions (specifically glycoconjugates) inhibit T-cell proliferative responses and IL-2 production, prevent the accumulation of inflammatory cells, and interfere with normal macrophage functions (Dematteis et al. 2001; Rodriguez-Sosa et al. 2002). Infections, particularly with *E. granulosus*, generate powerful hypersensitivity responses, which can be life threatening if there is release of cyst fluid during surgical removal procedures.

Studies with *Echinococcus* spp. in experimental rodents and in humans are beginning to identify some of the key immunological events that may determine the outcome of infection (Gottstein and Felleisen 1995). As with many infections, there are striking genetically determined differences in the susceptibility and resistance of different strains of mice to this parasite. Mice that have greater resistance show responses that seem to indicate a dominant influence of Th1 cells, and the granulomata that develop around the mass of larval tissue contain a high proportion of CD8 T cells. In contrast, more susceptible mice, in which parasite growth is unchecked, have Th2-dominated responses and a high proportion of CD4 T cells in the granulomata. It is suggestive that similar data have come from studies of humans in endemic areas, some of whom, though susceptible to infection, seem to be resistant to disease. In these individuals, CD8 T cell numbers in the granulomata are high and peripheral lymphocytes secrete large amounts of the Th2 cytokine IL-5. Thus, Th1- and Th2-mediated mechanisms play a crucial role in the complex relationship between parasite and host (Vuitton 2003) and manipulation of the balance between the two may have clinical benefits (Godot et al. 2003).

There is no immediate prospect of immunoprophylaxis against larval tapeworm infections in humans, but the development of a recombinant vaccine against the larval stages of *Taenia ovis* in sheep has stimulated similar studies on vaccines against *Echinococcus* and *T. solium* cysticercosis (Lightowlers et al. 1996; Lightowlers 2003). Interestingly, one candidate antigen is paramyosin, also used in vaccine studies with schistosomes and filarial nematodes (Vazquez-Talavera et al. 2001). The most important practical application of immunological studies at present is the development of improved immunodiagnosis; larval tapeworm infections being difficult to diagnose definitively, even with modern scanning techniques.

In contrast to our current knowledge about immunity to intestinal nematodes, relatively little is known about responses to intestinal tapeworms. The human species (*Taenia*, *Diphyllobothrium*) appear to generate little or no protective immunity, although hosts do make immune responses (see Chapter 32, Intestinal tapeworms). There is a considerable body of work using hymenolepid tapeworms in experimental rodents and this has given a significant body of data (Andreassen et al. 1999). Many aspects of the intestinal response in these models (e.g. mastocytosis, goblet cell hyperplasia, eosinophil infiltration, IgE) resemble those seen in nematodes (Ishih and Uchikawa 2000; Starke and Oaks 2001), but in the rat (the natural host) these are not linked to worm loss. In the mouse, this parasite is rejected and recent work shows that the mechanisms involved, as with nematodes in mice, are Th2-dependent (McKay and Khan 2003).

RESPONSES TO PARASITES MAY INTERACT WITH OTHER DISEASES

The powerful immune responses induced by parasites may be expected to influence the susceptibility of infected individuals to other diseases. It has been known for many years that the presence of one parasite may alter responses to a second, but in recent years there has been increased interest in broader aspects of parasite–disease interactions, in part triggered by the hygiene hypothesis. This states that early exposure to the antigens of infectious agents and environmental organisms can affect the balance of Th cell responses made subsequently, and thus influence susceptibility to allergic and autoimmune diseases (Rook 2000). Particular attention has been paid to the possible consequences of infections with intestinal helminths because of the ability of such infections to polarize towards Th2-dominated responses. It has been proposed that, because these worms are widespread in children in the developing world, they may be an important factor underlying the reduced prevalence of autoimmune and (paradoxically) allergic diseases in these regions (Cooper 2002) and there is some experimental evidence to support this (e.g. Bashir et al. 2002). However, the position regarding the influence of parasites on autoimmune disease seem less clear cut (Agersberg et al. 2001). Evidence also suggests that infections with other worms such as schistosomes and *Onchocerca* can modulate responses to mycobacteria, responses to which are a powerful component of the hygiene hypothesis (Sewell et al. 2002; Stewart et al. 1999). Experimental data suggest that the strong Th2-biased responses associated with intestinal worms may alter susceptibility to auto-immune inflammatory diseases such as colitis and Crohn's disease (Elliot et al. 2000; Khan et al. 2002).

Worm infections have been associated with reduced responses to vaccines, although this may be more a reflection of an immune suppression (Cooper et al. 2001). They have also been linked to increased susceptibility to HIV and human T-cell lymphotropic virus (HTLV) infection, although the evidence here is sometimes contradictory (Ayash-Rashkovsky et al. 2002; Satoh et al. 2002; Wolday et al. 2002). A reciprocal effect of HIV infection and increased susceptibility to reinfection with *Schistosoma mansoni* has been described (Karanja et al. 2002). Recently, there has been interest in the interactions between intestinal worm infections and malaria. In endemic areas, where humans are commonly co-infected, epidemiological evidence has proposed both an increased incidence of *P. falciparum* and protection against cerebral malaria (Nacher et al. 2001, 2002).

The immune responses to infection with protozoa also influence infections with other infectious agents including viruses, bacteria, helminth worms, and other protozoa. There are numerous examples from experimental infections (see Cox 2001) but there is very little direct evidence that these interactions are important in humans, largely because of the difficulties inherent in carrying out the necessary definitive studies. The best-known example is Burkitt's lymphoma; the causative agent of this childhood lymphoma, the Epstein–Barr virus, is common all over the world but only causes tumors in areas where malaria is endemic (De The 1985) presumably as a consequence of malaria-induced immunodepression. There have been sporadic reports of enhanced bacterial infections in children infected with malaria (Walsh et al. 2000) and it has been known for some time that it is difficult to vaccinate children with malaria against tetanus, typhoid, or bacterial meningitis (Williamson and Greenwood, 1978). Acquired and innate immunodeficiencies also have an adverse outcome on infections with parasitic protozoa and the interactions between HIV and protozoa have been extensively studied. HIV infections are associated with enhanced *Toxoplasma gondii*, *Cryptosporidium parvum*, *Cyclospora cayetanensis*, *Isospora belli*, *Leishmania* spp., and *Trypanosoma cruzi* infections (see Ambroise-Thomas 2001). In all these cases, the parasitic infection is enhanced, and may even be fatal, in individuals infected with HIV. The most intensively studied interaction is the HIV/*Leishmania* combination, in which both organisms parasitize macrophages: HIV enhances the protozoan infection and vice versa, and the actual outcome of the dual infection is determined by a complex and changing balance between a number of different cytokines (Wolday et al. 1999). Interestingly there is very little evidence that HIV infections are enhanced in individuals infected with malaria although the reverse does seem to occur (Chandramohan and Greenwood 1998).

Our understanding of the Th1/Th2 dichotomy that has run as a thread throughout this chapter has opened up new prospects for the control of parasitic infections but it is clear that much more needs to be known about what happens in humans, particularly in those co-infected with another infectious agent, before any immunologically based preventative or therapeutic measures can be implemented.

REFERENCES

Ackers, J.P. 1989. Immunologic aspects of human trichomoniasis. In: Honigberg, B.M. (ed.), *Trichomonads parasitic in humans*. New York: Springer Verlag, 36–52.

Adal, K.A., Sterling, C.R. and Guerrant, R.L. 1995. *Cryptosporidium* and related species. In: Blaser, M.J., et al. (eds), *Infections of the gastrointestinal tract*. New York: Raven Press, 1107–28.

Agersberg, S.S., Garza, K.M. and Tung, K.S. 2001. Intestinal parasitisim terminates self-tolerance and enhances neonatal induction of autoimmune disease and memory. *Eur J Immunol*, **31**, 851–9.

Aguilar-Deflin, I., Wettstein, P.J. and Persing, D.H. 2003. Resistance to acute babesiosis is associated with interleukin 12 and gamma interferon-mediated responses and requires macrophages and natural killer cells. *Infect Immun*, **71**, 2002–8.

Aidoo, M. and Udhayakumar, V. 2000. Field studies of cytotoxic T-lymphocytes in malaria infections: implications for malaria vaccine development. *Parasitol Today*, **16**, 50–6.

Alderete, J.F. 1983. Antigen analysis of several pathogenic strains of *Trichomonas vaginalis*. *Infect Immun*, **39**, 1041–7.

Alexander, J. and Hunter, C.A. 1998. Immunoregulation during toxoplasmosis. In: Liew, F.Y. and Cox, F.E.G. (eds), *Immunology of intracellular parasitism*. Basel: Karger, 81–102.

Alexander, J. and Russell, D.G. 1992. The interaction of *Leishmania* with macrophages. *Adv Parasitol*, **31**, 175–254.

Alimohammadian, M.H., Khamesipour, A., et al. 2002. The role of BCG in human immune responses induced by multiple injections of autoclaved *Leishmania major* as a candidate vaccine against leishmaniasis. *Vaccine*, **21**, 174–80.

Allen, J.E. and MacDonald, A.S. 1998. Profound suppression of cellular proliferation mediated by the secretions of nematodes. *Parasite Immunol*, **20**, 241–7.

Allen, J.E. and Maizels, R.M. 1997. Th1-Th2: reliable paradigm or dangerous dogma? *Immunol Today*, **18**, 387–92.

Allred, D.R. 2003. Babesiosis: persistence in the face of adversity. *Trends Parasitol*, **19**, 51–5.

Amaral, V.F., Teva, A., et al. 2002. Study of the safety, immunogenicity and efficacy of attenuated and killed *Leishmania* (*Leishmania*) *major* vaccines in a rhesus monkey (*Macaca mulatta*) model of the human disease. *Mem Inst Oswaldo Cruz*, **97**, 1041–8.

Ambroise-Thomas, P. 2001. Parasitic diseases and immunodeficiencies. *Parasitology*, **122**, Supplement, S65–71.

Ambroise-Thomas, P. and Pelloux, H. 1993. Toxoplasmosis – congenital and in immunocompromised patients: a parallel. *Parasitol Today*, **9**, 61–3.

Anders, R.F. and Saul, A. 2000. Malaria vaccines. *Parasitol Today*, **16**, 444–7.

Andreassen, J., Bennet-Jenkins, E.M. and Bryant, C. 1999. Immunology and biochemistry of *Hymenolepis diminuta*. *Adv Parasitol*, **42**, 223–75.

Artis, D. and Grencis, R.K. 2001. T-helper cell cytokine responses during intestinal nematode infection, induction, regulation and effector function. In: Kennedy, M.W. and Harnett, W. (eds), *Parasitic nematodes*. Wallingford: CAB International, 331–71.

Artis, D., Shapira, S., et al. 2002. Differential requirement for NF-kappa B family members in control of helminth infection and intestinal inflammation. *J Immunol*, **169**, 4481–7.

Ayash-Rashkovsky, M., Weisman, Z., et al. 2002. Generation of Th1 immune responses to inactivated gp120-depleted HIV-1 in mice with a dominant Th2 biased immune profile via correction of immunostimulatory oligonucleotides – relevance to AIDS vaccine in developing countries. *Vaccine*, **20**, 2684–92.

Bakhiet, M.P., Olsson, J., et al. 1996. Human and rodent interferon-gamma as a growth factor for *Trypanosoma brucei*. *Eur J Immunol*, **26**, 1359–64.

Ballou, W.R., Hoffman, S.L., et al. 1987. Safety and efficacy of a recombinant *Plasmodium falciparum* sporozoite vaccine. *Lancet*, **1**, 1277–81.

Ballou, W.R., Kester, K.E., et al. 1999. Malaria vaccines, triumphs or tribulations? *Parassitologia*, **41**, 403–8.

Bancroft, A.J., Grencis, R.K., et al. 1994. The role of CD4 cells in protective immunity to *Brugia pahangi*. *Parasite Immunol*, **16**, 385–7.

Bannister, L.H. and Mitchell, G.F. 2003. The ins and outs and roundabouts of malaria. *Trends Parasitol*, **19**, 209–13.

Barry, J.D. 1997. The relative significance of mechanisms of antigenic variation in African trypanosomes. *Parasitol Today*, **13**, 212–18.

Bashir, M.E., Andersen, P., et al. 2002. An enteric helminth infection protects against an allergice response to dietary antigen. *J Immunol*, **169**, 3284–92.

Beck, H.P., Felger, I.H., et al. 1997. Analysis of multiple *Plasmodium falciparum* infections in Tanzanian children during phase III trial of the malaria vaccine SPf66. *J Infect Dis*, **175**, 921–6.

Belloni, A., Villena, I., et al. 2003. Regulation of tumor necrosis factor alpha and its specific receptors during *Toxoplasma gondii* infection in human monocytic cells. *Parasitol Res*, **89**, 207–13.

Bergquist, R., Al-Sherbiny, M., et al. 2002. Blueprint for schistosomiasis vaccine development. *Acta Trop*, **82**, 183–92.

Beucher, M., Meira, W.S., et al. 2003. Expression and purification of functional, recombinant *Trypanosoma cruzi* complement regulatory protein. *Protein Expr Purif*, **27**, 19–26.

Bienz, M., Dai, W.-J., et al. 2003. Interleukin-6 deficient mice are highly susceptible to *Giardia lamblia* infection but exhibit normal intestinal immunoglobulin A responses against the parasite. *Infect Immun*, **71**, 1569–73.

Blackwell, J.M. 1993. Immunology of leishmaniasis. In: Lachmann, P.J., Peters, K., et al. (eds), *Clinical aspects of allergy*, 5th edn. Oxford: Blackwell Scientific Publications, 1575–97.

Bogdan, C. and Röllinghoff, M. 1998. The immune response to *Leishmania*: mechanisms of parasite control and evasion. *Int J Parasitol*, **28**, 121–34.

Bogdan, C. and Röllinghoff, M. 1999. How do protozoan parasites survive inside macrophages? *Parasitol Today*, **15**, 22–8.

Bojang, K.A., Obaro, S.K., et al. 1998. An efficacy trial of the malaria vaccine SPf66 in Gambian infants – second year of follow-up. *Vaccine*, **16**, 62–7.

Bojang, K.A., Milligan, P.J., et al. 2001. Randomized, double blind, controlled trial of efficacy of RTS,S/AS02 malaria vaccine against *P. falciparum* infection in semi-immune adult men in The Gambia. *Lancet*, **358**, 1927–34.

Brener, Z. 1994. The pathogenesis of Chagas' disease, an overview of current theories. In: *Chagas' disease and the nervous system*. Sci Pub 547. Washington, DC: Pan American Health Organization.

Bryan, R.T. 1995. Microsporidiosis as an AIDS-related opportunistic infection. *Clin Infect Dis*, **21**, Suppl. 1, S62–5.

Bundy, D.A.P. 1995. Epidemiology and transmission of intestinal helminths. In: Farthing, M.J.G., Keusch, G.T. and Wakelin, D. (eds), *Enteric infection*, Vol. 2. London: Chapman and Hall Medical, 5–24.

Bundy, D.A.P. and Medley, G.F. 1992. Immunoepidemiology of human geohelminthiasis, ecological and immunological determinants of worm burden. *Parasitology*, **104**, S105–19.

Buxbaum, L.U., Uzonna, J.E., et al. 2002. Control of New World cutaneous leishmaniasis is IL-12 independent but STAT 4 dependent. *Eur J Immunol*, **32**, 3206–15.

Buxton, D. 1993. Toxoplasmosis: the first commercial vaccine. *Parasitol Today*, **9**, 335–7.

Capron, A. 1998. Schistosomiasis: forty year's war on the worm. *Parasitol Today*, **14**, 379–84.

Capron, A., Capron, M. and Riveau, G. 2001. Vaccine strategies against schistosomiasis, from concepts to clinical trials. *Br Med Bull*, **62**, 139–48.

Chandramohan, D. and Greenwood, B.M. 1998. Is there an interaction between human immunodeficiency virus and *Plasmodium falciparum*? *Int J Epidemiol*, **27**, 296–301.

Chen, G.-J, Chen, H-F, et al. 2002. Protective effect of DNA-mediated immunization with a combination of SAG1 and IL-2 gene adjuvant against infection with *Toxoplasma gondii* in mice. *Chin Med J (Beijing)*, **115**, 1448–52.

Clark, I.A. and Cowden, W.B. 2003. The pathophysiology of falciparum malaria. *Pharmacol Ther*, **99**, 221–60.

Clark, I.A. and Rockett, K.A. 1996. Nitric oxide and parasitic disease. *Adv Parasitol*, **37**, 1–56.

Clark, I.A., Awburn, M.M., et al. 2003. Tissue distribution of macrophage inhibitory factor and inducible nitric oxide synthase in falciparum malaria and sepsis in African children. *Malaria J*, **2**, 6.

Clyde, D.F., Most, H., et al. 1973. Immunization of man against sporozoite-induced falciparum malaria. *Am J Med Sci*, **266**, 169–77.

Convit, J., Castellanos, P., et al. 1989. Immunotherapy of localized, intermediate, and diffuse forms of American cutaneous leishmaniasis. *J Infect Dis*, **160**, 104–15.

Cooper, P.J. 2002. Can intestinal helminth infections (geohelminths) affect the development and expression of asthma and allergic disease? *Clin Exp Immunol*, **128**, 398–404.

Cooper, P.J., Chico, M.E., et al. 2000. Human infection with *Ascaris lumbricoides* is associated with a polarized cytokine response. *J Infect Dis*, **182**, 1207–13.

Cooper, P.J., Chico, M., et al. 2001. Human infection with *Ascaris lumbricoides* is associated with suppression of the interleukin-2 response to recombinant cholera toxin-B subunit following vaccination with the live oral cholera vaccine CVD 103-HgR. *Infect Immun*, **69**, 1574–80.

Cox, F.E.G. 1988. Major animal models in malaria research: rodent. In: Wernsdorfer, W.H. and McGregor, I. (eds), *Malaria: principles and practice of malariology*, Vol. 2. London: Heinemann, 1503–43.

Cox, F.E.G. 2001. Concomitant infections, parasites and immune responses. *Parasitology*, **122**, Suppl, S23–38.

Cox, F.E.G. and Liew, F.Y. 1992. T-cell subsets and cytokines in parasitic infections. *Immunol Today*, **13**, 445–8.

Cunha-Neto, E. and Kalil, J. 2001. Heart-infiltrating and peripheral T cells in the pathogenesis of human Chagas' disease cardiomyopathy. *Autoimmunity*, **34**, 187–92.

de Baetselier, P. 1996. Mechanisms underlying trypanosome-induced T-cell immunosuppression. In: Mustafa, A.S., Al-Attiyah, R.J., et al. (eds), *T-cell subsets and cytokine interplay in infectious diseases*. Basel: Karger, 124–39.

de Carvalho, L.P., Soto, M., et al. 2003. Characterization of the immune response to *Leishmania infantum* recombinant antigens. *Microbes Infect*, **5**, 7–12.

de Diago, J., Punzón, C., et al. 1997. Alteration of macrophage function by a *Trypanosoma cruzi* membrane mucin. *J Immunol*, **159**, 4983–9.

Delemarre, F.G.A., Stevenhagen, A. and Kroon, F.P. 1994. Effect of IFN-gamma on the proliferation of *Toxoplasma gondii* in monocytes and monocyte-derived macrophages from AIDS patients. *Immunology*, **83**, 646–50.

Dematteis, S., Pirotto, F., et al. 2001. Modulation of the cellular response by a carbohydrate rich fraction from *Echinococcus granulosus* protoscoleces in infected or immunized BALB/c mice. *Parasite Immunol*, **23**, 1–9.

Denkers, E.Y., Kim, L. and Butcher, B.A. 2003. In the belly of the beast; subversion of macrophage proinflammatory signaling cascades during *Toxoplasma gondii* infection. *Cell Microbiol*, **5**, 75–83.

Deraedt, S. and Molina, J.M. 1995. Les microsporidioses en pathologie humaine. *Méd Mal Infect*, **25**, 570–6.

deSchoolmeester, M.L., Little, M.C., et al. 2003. Absence of CC chemokine ligand 2 results in an altered Th1/Th2 cytokine balance and failure to expel *Trichuris muris* infection. *J Immunol*, **170**, 4693–700.

Dessein, A.J., Marquet, S., et al. 1999. Infection and disease in human schistosomiasis mansoni are under distinct major gene control. *Microbes Infect*, **1**, 561–7.

De The, G. 1985. Epstein-Barr virus and Burkitt's lymphoma worldwide: the causal relationship revisited. In Lenoir, G.M., O'Connor, G.T. and Olweny, C.L.M. (eds), Burkitt's lymphoma: a human cancer model. Sci Pub 60, 165-176. Lyon: IARC.

Doetze, A., Satoguina, J., et al. 2000. Antigen-specific cellular hyporesponsiveness in a chronic human helminth infection is mediated by T(h)3/T(r)1-type cytokine and transforming growth factor-beta but not by a T(h)1 to T(h)2 shift. *Int Immunol*, **12**, 623–30.

DosReis, G.A. 1997. Cell-mediated immunity in experimental *Trypanosoma cruzi* infection. *Parasitol Today*, **13**, 335–42.

Dube, A. and Srivastava, B. 2002. Kala-azar, the immunological consequences and feasibility of vaccination. *J Immunol Immunopathol*, **4**, 122–35.

Dubey, J.P. and Beattie, C.P. (eds). 1988. *Toxoplasmosis of animals and man*. Boca Raton: CRC Press.

Eberl, M. 2002. Don't count your interleukins before they've hatched. *Trends Immunol*, **23**, 341–2.

Elliot, D.E., Urban, J.F., et al. 2000. Does failure to acquire helminth parasites predispose to Crohn's disease? *FASEB J*, **14**, 1848–55.

Else, K.J., Finkelman, F.D., et al. 1994. Cytokine-mediated regulation of chronic intestinal helminth infection. *J Exp Med*, **179**, 347–51.

Facer, C.A. and Tanner, M. 1997. Clinical trials of malaria vaccines: progress and prospects. *Adv Parasitol*, **39**, 2–68.

Fachado, A., Rodriguez, A., et al. 2003. Protective effect of naked DNA vaccine cocktail vaccine against lethal toxoplasmosis in mice. *Vaccine*, **21**, 1327–35.

Fallon, P.G., Richardson, E.J., et al. 2000a. Schistosome infection of transgenic mice defines distinct and contrasting pathogenic roles for IL-4 and IL-13: IL-13 is a profibrinolytic agent. *J Immunol*, **164**, 2585–91.

Fallon, P.G., Smith, P., et al. 2000b. Expression of interleukin-9 leads to Th2 cytokine dominated responses and and fatal enteropathy in mice with chronic *Schistosoma mansoni* infections. *Infect Immun*, **68**, 6005–11.

Faubert, G.M. 1996. The immune response to *Giardia*. *Parasitol Today*, **12**, 140–5.

Faulkner, H., Turner, J., et al. 2002. Age- and infection intensity-dependent cytokine and antibody production in human trichuriasis, the importance of IgE. *J Infect Dis*, **185**, 665–72.

Freedman, D.O. 1998. Immune dynamics in the pathogenesis of human lymphatic filariasis. *Parasitol Today*, **14**, 229–34.

Garraud, O. and Perraut, R. 2003. Malaria-specific antibody subclasses in immune individuals: a key source of information for vaccine design. *Trends Immunol*, **24**, 30–5.

Garraud, O., Perraut, R., et al. 2003. Class and subclass selection in parasite-specific antibody responses. *Trends Parasitol*, **19**, 300–4.

Gatton, M.L., Peters, J.M., et al. 2003. Switching rates of *Plasmodium falciparum var* genes: faster than we thought? *Trends Parasitol*, **19**, 202–7.

Gaucher, D. and Chadee, K. 2002. Construction and immunogenicity of a codon-optimized *Entamoeba histolytica* Gal-lectin DNA vaccine. *Vaccine*, **20**, 3244–53.

Gause, W.C., Urban, J.F. and Stadecker, M.J. 2003. The immune response to parasitic helminths: insights from murine models. *Trends Immunol*, **24**, 269–77.

Gazzinelli, R.T., Denkers, E.Y. and Sher, A. 1993. Host resistance to *Toxoplasma gondii*: a model for studying the selective induction of cell-mediated immunity by intracellular parasites. *Infect Ag Dis*, **2**, 139–50.

Gazzinelli, R.T., Oswald, I.P., et al. 1992. The microbicidal activity of interferon-gamma treated macrophages against *Trypanosoma cruzi* involves a L-arginine dependent mechanism inhibitable by interleukin-10 and transforming growth factor-β. *Eur J Immunol*, **22**, 2501–6.

Genton, B., Anders, R.F., et al. 2003. The malaria vaccine development program in Papua New Guinea. *Trends Parasitol*, **19**, 264–70.

Geiger, S.M., Massara, C.L., et al. 2002. Cellular responses and cytokine profiles in *Ascaris lumbricoides* and *Trichuris trichiura* infected patients. *Parasite Immunol*, **24**, 499–509.

Girod, N., Brown, A., et al. 2003. Successful vaccination of BALB/c mice against human hookworm (*Necator americanus*), the immunological phenotype of the protective response. *Int J Parasitol*, **33**, 71–80.

Godot, V., Harraga, S., et al. 2003. IFN alpha-2a protects mice against a helminth infection of the liver and modulates immune responses. *Gastroenterology*, **124**, 1441–50.

Good, M.F., Kaslow, D.C. and Miller, L.H. 1998. Pathways and strategies for developing a malaria blood-stage vaccine. *Ann Rev Immunol*, **16**, 57–87.

Gottstein, B. and Felleisen, R. 1995. Protective immune mechanisms against the metacestode of *Echinococcus multilocularis*. *Parasitol Today*, **11**, 320–6.

Gozar, M.M., Muratova, O., et al. 2001. *Plasmodium falciparum*, immunogenicity of alum-adsorbed clinical grade TBV25-28, a yeast-secreted malaria transmission-blocking vaccine candidate. *Exp Parasitol*, **97**, 61–9.

Greenblatt, C.L. 1988. Cutaneous leishmaniasis: the prospects for a killed vaccine. *Parasitol Today*, **4**, 53–5.

Greenwood, B.M., Whittle, H.C. and Molyneux, D.H. 1973. Immunosuppression of Gambian trypanosomiasis. *Trans R Soc Trop Med Hyg*, **67**, 846–50.

Grüner, A.C., Snounou, G., et al. 2003. Pre-erythrocytic antigens of *Plasmodium falciparum*: from rags to riches. *Trends Parasitol*, **19**, 74–8.

Gryseels, B. 2000. Schistosomiasis vaccines: a devil's advocate view. *Parasitol Today*, **16**, 46–8.

Hagan, P., Doenhoff, M.J., et al. 2000. Schistosomiasis vaccines: a response to a devil's advocate view. *Parasitol Today*, **16**, 322–3.

Hamadien, M., Bakhiet, M. and Harris, R.A. 2000. Interferon-γ induces secretion of lymphocyte triggering factor via tyrosine protein kinases. *Parasitology*, **120**, 281–7.

Handman, E. 1997. Leishmania vaccines: old and new. *Parasitol Today*, **13**, 236–8.

Haque, R., Duggal, P. and Ali, I.M. 2002. Innate and acquired resistance to amebiasis in Bangladeshi children. *J Infect Dis*, **186**, 547–52.

Helmby, H. and Grencis, R.K. 2002. IL-18 regulates intestinal mastocytosis and Th2 cytokine production independently of IFN-gamma during *Trichinella spiralis* infection. *J Immunol*, **169**, 2553–60.

Henri, S., Chevillard, C., et al. 2002. Cytokine regulation of periportal fibrosis in humans infected with *Schistosoma mansoni*, IFN-gamma is associated with protection against fibrosis and TNF-alpha with aggravation of disease. *J Immunol*, **169**, 929–36.

Herrington, D.A., Clyde, D.F., et al. 1987. Safety and immunogenicity in man of a synthetic peptide malaria vaccine against *Plasmodium falciparum* sporozoites. *Nature*, **328**, 257–9.

Hoffman, K.F., Wynn, T.A. and Dunne, D.W. 2002. Cytokine-mediated host responses during schistosome infections, walking the fine line between immunological control and immunopathology. *Adv Parasitol*, **52**, 265–307.

Hoffman, S.L. 1996. *Malaria vaccine development*. Washington, DC: American Society for Microbiology.

Holder, A. 1999. Malaria vaccines. *Proc Natl Acad Sci USA*, **96**, 1167–9.

Honigberg, B.M. and Burgess, D.E. 1994. Trichomonads of importance in human medicine including *Dientamoeba fragilis*. In: Kreier, J.P. (ed.), *Parasitic protozoa*, Vol. 9, 2nd edn. San Diego: Academic Press, 1–109.

Houston, K.M., Wilson, E.H., et al. 2000. Presence of phosphorylcholine on a filarial nematode protein influences immunoglobulin G subclass response to the molecule by an interleukin-10 dependent mechanisms. *Infect Immun*, **68**, 5466–8.

Ishih, A. and Uchikawa, R. 2000. Immunoglobulin E and mast cell responses are related to worm biomass not expulsion of *Hymenolepis diminuta* during low dose infection in rats. *Parasite Immunol*, **22**, 561–6.

James, S.L. 1995. Role of nitric oxide in parasitic infections. *Microbiol Rev*, **59**, 533–47.

Janoff, E.N. and Smith, P.D. 1990. The role of immunity in *Giardia* infections. In: Meyer, E.A. (ed.), *Giardiasis*. Amsterdam: Elsevier, 215–37.

Janoff, E.N., Smith, P.D. and Blaser, M.J. 1988. Acute antibody responses to *Giardia lamblia* are depressed in patients with AIDS. *J Infect Dis*, **157**, 798–804.

John, D.T. 1993. Opportunistically pathogenic free-living amebae. In: Kreier, J.P. (ed.), *Parasitic protozoa*, Vol. 3, 2nd edn. San Diego: Academic Press, 143–246.

Karanja, D.M., Hightower, A.W., et al. 2002. Resistance to reinfection with *Schistosoma mansoni* in occupationally exposed adults and effect of HIV-1 co-infection on susceptibility to schistosomiasis, a longitudinal study. *Lancet*, **360**, 592–6.

Kaye, P.M. 1995. Costimulation and the regulation of antimicrobial immunity. *Immunol Today*, **16**, 423–7.

Keitel, W.A., Kester, K.E., et al. 1999. Phase I trial of two recombinant vaccines containing the 19 kDa carboxy terminal fragment of *Plasmodium falciparum* merozoite surface protein 1 (msp-1) and T helper epitopes of tetanus toxoid. *Vaccine*, **18**, 531–9.

Kemp, M., Theander, T.G. and Kharazmi, A. 1996. The contrasting roles of CD4+ T cells in intracellular infections in humans: leishmaniasis as an example. *Immunol Today*, **17**, 13–16.

Khan, W.I., Blennerhasset, P.A., et al. 2002. Intestinal nematode infection ameliorates experimental colitis in mice. *Infect Immun*, **70**, 5931–7.

Khonde, N., Pepin, J., et al. 1995. Epidemiological evidence for immunity following *Trypanosoma brucei gambiense* sleeping sickness. *Trans R SocTrop Med Hyg*, **89**, 607–11.

Kierszenbaum, F. and Sztein, M.B. 1994. Chagas' disease (American trypanosomiasis). In: Kierszenbaum, F. (ed.), *Parasitic infections and the immune system*. San Diego: Academic Press, 53–85.

Kirchoff, L.V. 1994. American trypanosomiasis (Chagas' disease) and African trypanosomiasis (Sleeping sickness). *Curr Opin Immunol*, **7**, 542–6.

Kotb, M. and Calandra, T. (eds) 2003. *Cytokines and chemokines in infectious diseases handbook*. Totowa, NJ: Humana Press.

Kulda, J. and Nohynkova, E. 1995. Giardia in humans and animals. In: Kreier, J.P. (ed.), *Parasitic protozoa*, Vol. 10, 2nd edn. San Diego: Academic Press, 225–422.

Kumar, S. and Tarleton, R.L. 2001. Antigen-specific Th1 but not Th2 cells provide protection from lethal *Trypanosoma cruzi* infection in mice. *J Immunol*, **166**, 4956–603.

Lawrence, R.A. 1996. Lymphatic filariasis: what mice can tell us. *Parasitology Today*, **12**, 267–71.

Le Goff, L., Lamb, T.J., et al. 2002. IL-4 is required to prevent filarial development in resistant but not susceptible strains of mice. *Int J Parasitol*, **32**, 1277–84.

Leon, J.S. and Engman, D.M. 2001. Autoimmunity in Chagas heart disease. *Int J Parasitol*, **31**, 555–61.

Liew, F.Y. 1990. Regulation of cell-mediated immunity in leishmaniasis. *Curr Top Microbiol Immunol*, **155**, 54–64.

Liew, F.Y. and O'Donnell, C.A. 1993. Immunology of leishmaniasis. *Adv Parasitol*, **32**, 161–81.

Lightowlers, M.W. 2003. Vaccines for the prevention of cysticercosis. *Acta Trop*, **87**, 129–35.

Lightowlers, M.W., Lawrence, S.B., et al. 1996. Vaccination against hydatidosis using a defined recombinant antigen. *Parasite Immunol*, **18**, 457–62.

López-Fuertes, L., Pérez-Jiménez, E., et al. 2002. DNA vaccination with linear minimalistic (MIDGE) vectors confers protection against *Leishmania major* infections in mice. *Vaccine*, **21**, 247–57.

Ma, X., Aste-Amezaga, et al. 1997. Immunomodulatory functions and molecular regulation of IL-12. In: Aldorini, A. (ed.) *IL-12, Chemical Immunology* 68, Basel: Karger, pp. 1-22.

Machado, C.R., Camargos, E.R., et al. 2000. Cardiac autonomic denervation in congestive heart failure, comparison of Chagas' heart disease with other dilated cardiomyopathy. *Hum Path*, **31**, 3–10.

Mahanty, S., Ravichandran, M., et al. 1997. Regulation of parasite antigen-driven immune responses by interleukin-10 (IL-10) and IL-12 in lymphatic filariasis. *Infect Immun*, **65**, 1742–7.

Mahler, E., Sepulveda, P., et al. 2001. A monoclonal antibody against the immunodominant epitope of the ribosomal P2beta protein of *Trypanosoma cruzi* interacts with the human beta 1-adrenergic receptor. *Eur J Immunol*, **31**, 2210–16.

Maizels, R.M., Sartono, E., et al. 1995. T-cell activation and the balance of antibody isotypes in human lymphatic filariasis. *Parasitol Today*, **11**, 50–6.

Mansfield, J.M. 1994. T-cell responses to the trypanosome variant surface glycoprotein: a new paradigm? *Parasitol Today*, **10**, 267–70.

Mansfield, J.M., Davis, T.H. and Dubois, M.E. 2001. Immunobiology of African trypanosomiasis, new paradigms, newer questions. In: Black, S.J. and Seed, J.R. (eds), *World class parasites. The African trypanosomes*, Vol. 1. Boston: Kluwer Academic Publishers, 79–96.

Mariuz, P., Bosler, E.M. and Luft, B.J. 1994. Toxoplasmosis in individuals with AIDS. *Infect Dis Clin North Am*, **8**, 365–81.

Martins, C.A.P. and Guerrant, R.L. 1995. *Cryptosporidium* and cryptosporidiosis. *Parasitol Today*, **11**, 434–5.

McConkey, S.J., Reece, W.H.H., et al. 2003. Enhanced T-cell immunogenicity of plasmid DNA vaccines boosted by recombinant modified vaccinia virus Ankara in humans. *Nature Med*, **9**, 729–35.

McKay, D.M. and Khan, W.I. 2003. STAT-6 is an absolute requirement for murine rejection of *Hymenolepis diminuta*. *J Parasitol*, **89**, 188–9.

Medaglini, D. and Hoeveler, A. 2003. The European research effort for HIV/AIDS, malaria and tuberculosis. *Vaccine*, **21**, Suppl. 2, S116–20.

Meeusen, E.N.T. and Balic, A. 2000. Do eosinophils have a role in the killing of helminth parasites? *Parasitol Today*, **16**, 95–101.

Méndez, S., Belkaid, Y., et al. 2002. Optimization of DNA vaccination against cutaneous leishmaniasis. *Vaccine*, **210**, 3702–8.

Miller, L.H. and Hoffman, S.L. 1998. Research towards vaccines against malaria. *Nat Med*, **4**, 520–4.

Miller-Sims, V.C. and Petri, W.A. 2002. Opportunities and obstacles in developing a vaccine for *Entamoeba histolytica*. *Curr Opin Immunol*, **14**, 549–52.

Modabber, F. 1995. Vaccines against leishmaniasis. *Ann Trop Med Parasitol*, **89**, 83–8.

Mohebali, M., Fallah, E., et al. 1999. Field trial of autoclaved leishmania vaccines for control of canine visceral leishmaniasis in Meshkin-Sharh. *Archives of Razi Institute*, **50**, 87–92.

Molano, I., Garcia Alonso, M., et al. 2003. A *Leishmania infantum* multi-component antigen protein mixed with live BCG confers protection to dogs experimentally infected with *L. infantum*. *Vet Immunol Immunopathol*, **92**, 1–13.

Moll, H. 1993. Epidermal Langerhans cells are critical for immunoregulation of cutaneous leishmaniasis. *Immunol Today*, **14**, 383–7.

Moorthy, V. and Hill, A.V.S. 2002. Malaria vaccines. *Br Med Bull*, **62**, 59–72.

Mossman, T.R. and Coffman, R.L. 1989. Heterogeneity of cytokine secretion patterns and functions of helper T-cells. *Adv Immunol*, **46**, 111–47.

Munoz-Fernandez, M.A., Fernandez, M.A. and Fresno, M. 1992. Activation of human macrophages for the killing of intracellular *Trypanosoma cruzi* by TNF-alpha and IFN-gamma through a nitric oxide dependent mechanism. *Immunol Lett*, **33**, 35–40.

Nacher, M., Gay, F., et al. 2001. *Ascaris lumbricoides* infection is associated with protection from cerebral malaria. *Parasite Immunol*, **22**, 107–13.

Nacher, M., Singhasivanon, P., et al. 2002. Intestinal helminth infections are associated with increased incidence of *Plasmodium falciparum* malaria in Thailand. *J Parasitol*, **88**, 55–8.

Nagineni, C.N., Detrick, B. and Hooks, J.J. 2002. Transforming growth factor-β expression in human retinal pigment epithelial cells is enhanced by *Toxoplasma gondii*, a possible role in immunopathogenesis of retinochorditis. *Clin Exp Immunol*, **128**, 372–8.

Nash, T.E. 1994. Immunology, the role of the parasite. In: Thompson, R.C.A., Reynoldson, J.A. and Lymbery, A.J. (eds), *Giardia, from molecules to disease*. Wallingford: CAB International, 139–54.

Nash, T.E. 1997. Antigenic variation in *Giardia lamblia* and the host's immune response. *Philos Trans R Soc Lond B Biol Sci*, **352**, 1369–75.

Nash, T.E. 2002. Surface antigenic variation in *Giardia lamblia*. *Mol Microbiol*, **45**, 585–90.

Nutman, T.B. 2001. Lymphatic filariasis, new insights and prospects for control. *Curr Opin Infect Dis*, **14**, 539–46.

Okano, M., Satoskar, A.R., et al. 2001. Lacto-N-fucopentose III found on *Schistosoma mansoni* egg antigens functions as adjuvant for proteinss by inducing Th2-type responses. *J Immunol*, **167**, 442–50.

Olson, M.E., Ceri, H. and Morck, D.W. 2002. *Giardia* immunoprpphylaxis and immunotherapy. In: Olson, B.E., Olson, M.E. and Wallis, P.M. (eds), *Giardia, the cosmopolitan parasite*. Wallingford: CABI Publishing, 139–55.

Pearce, E.J. and MacDonald, A.S. 2002. The immunobiology of schistosomiasis. *Nature Rev Immunol*, **2**, 499–511.

Pérez-Montfort, R. and Kretschmer, R.R. 1990. Humoral immune responses. In: Kretschmer, R.R. (ed.), *Amebiasis, infection and disease by Entamoeba histolytica*. Boca Raton: CRC Press, 91–103.

Petri, W.A. 2002. Pathogenesis of amebiasis. *Curr Opin Microbiol*, **5**, 443–7.

Pfaff, A.W., Schulz-Key, H., et al. 2002. *Litomosoides sigmodontis* cystatin acts as an immunomodulator during experimental filariasis. *Int J Parasitol*, **32**, 171–8.

Pirmez, C., Yamamura, M., et al. 1993. Cytokine patterns in the pathogenesis of human leishmaniasis. *J Clin Invest*, **91**, 1390–5.

Plebanski, M. and Hill, A.V.S. 2000. The immunology of malaria infection. *Curr Opin Immunol*, **12**, 43–1.

Pritchard, D.I. 1995. The survival strategies of hookworms. *Parasitol Today*, **11**, 255–9.

Pritchard, D.I., Quinnell, R.J. and Walsh, E.A. 1995. Immunity in humans to *Necator americanus*, IgE, parasite weight and fecundity. *Parasite Immunol*, **17**, 71–5.

Que, X.C., Kim, S.H., et al. 2003. A surface amebic cysteine proteinase inactivates interleukin-18. *Infect Immun*, **71**, 1274–80.

Reed, S.G. 1998. Immunology of *Trypanosoma cruzi* infections. In: Liew, F.Y. and Cox, F.E.G. (eds), *Immunology of intracellular parasitism*. Basel: Karger, 124–43.

Reiner, S.L. and Locksley, R.M. 1995. The regulation of immunity to *Leishmania major*. *Ann Rev Immunol*, **13**, 151–77.

Ritter, U. and Korner, H. 2002. Divergent expression of inflammatory dermal chemokines in cutaneous leishmaniasis. *Parasite Immunol*, **24**, 295–301.

Rodriguez-Sosa, M., Satoskar, A.R., et al. 2002. Chronic helminth infection induces alternatively activated macrophages expressing high levels of CCR5 with low interleukin-12 production and T biasing ability. *Infect Immun*, **70**, 3656–64.

Romagnani, S. 1991. Human Th1 and Th2 subsets: doubts no more. *Immunol Today*, **12**, 256–7.

Romagnani, S. (ed.) 1996. *Th1 and Th2 cells in health and disease*. Basel: Karger.

Rook, G.A. 2000. Clean living increases more than just atopic disease. *Immunol Today*, **21**, 118–20.

Rowland, E.C., Mikhail, K.S. and McCormick, T.S. 1992. Isotype determination of anti-*Trypanosoma cruzi* antibody in murine Chagas' disease. *J Parasitol*, **78**, 557–61.

Rozenfeld, C., Martinez, R., et al. 2003. Soluble factors released by *Toxoplasma gondii*-infected astrocytes down-modulate nitric oxide production by gamma interferon-activated microglia and prevent neuronal degeneration. *Infect Immun*, **71**, 2047–57.

Sadler, C.H., Rutitzky, L.I., et al. 2003. IL-10 is crucial for the transition from acute to chronic disease state during infection of mice with *Schistosoma mansoni*. *Eur J Immunol*, **33**, 888.

Sandor, M., Weinstock, J.V. and Wynn, T.A. 2003. Granulomas in schistosome and mycobacterial infections, a model of local immune responses. *Trends Immunol*, **24**, 44–52.

Satoguina, J., Mempel, M., et al. 2002. Antigen-specific T regulatory-1 cells are associated with immunosuppression in a chronic helminth infection (onchocerciasis). *Microbes Infect*, **13**, 1291–300.

Satoh, M., Toma, H., et al. 2002. Involvement of IL-2/IL-2R system activation by parasite antigen in polyclonal expansion of CD4(+)25(+) HTLV-1-infected T-cells in human carriers of both HTLV-1 and *S. stercoralis*. *Oncogene*, **21**, 2466–75.

Schmieg, J., Gonzalez-Aseguinolaza, G. and Tsuji, M. 2003. The role of natural killer T cells and other T cell subsets against infection by the pre-erythrocytic stages of malaria parasites. *Microbes Infect*, **5**, 499–506.

Schofield, L., Hewitt, M.C., et al. 2002. Synthetic GPI as a candidate anti-toxic vaccine in a model of malaria. *Nature*, **418**, 785–9.

Scorza, T., D'Souza, S., et al. 2003. A GRA1 DNA vaccine primes cytotoxic CD8[+] T cells to control acute *Toxoplasma gondii* infection. *Infect Immun*, **71**, 309–16.

Scott, P. 1996. Th cell development and regulation in experimental cutaneous leishmaniasis. In: Romagnani, S. (ed.) *Th1 and Th2 cells in health and disease*. 98–114. Basel: Karger.

Scott, P. and Farrell, J.P. 1998. Experimental cutaneous leishmaniasis, induction and regulation of T cells following infection of mice with *Leishmania major*. In: Liew, F.Y. and Cox, F.E.G. (eds), *Immunology of intracellular parasitism*. Basel: Karger, 60–80.

Sewell, D.L., Reinke, E.K., et al. 2002. Immunoregulation of CNS autoimmunity by helminth and mycobacterial infections. *Immunol Lett*, **82**, 101–10.

Shi, Y.P., Hasnain, S.E., et al. 1999. Immunogenicity and in vitro protective efficacy of a recombinant multistage *Plasmodium falciparum* candidate vaccine. *Proc Natl Acad Sci USA*, **96**, 1615–20.

Silva, J.S., Aliberti, J.C.S., et al. 1998. The role of IL-12 in experimental *Trypanosoma cruzi* infection. *Braz J Med Biol Res*, **31**, 111–15.

Solbach, W. and Laskay, T. 1995. *Leishmania major* infection, the overture. *Parasitol Today*, **11**, 394–7.

Solbach, W. and Laskay, T. 2000. The host response to *Leishmania* infection. *Adv Immunol*, **74**, 275–317.

Stanley, S.L. 2001. Protective immunity to amebiasis, new insights and new challenges. *J Infect Dis*, **184**, 504–6.

Starke, W.A. and Oaks, J.A. 2001. Ileal mucosal mast cell, eosinophil, and goblet cell populations during *Hymenolepis diminuta* infection of the rat. *J Parasitol*, **87**, 1222–5.

Stavnezer, J. 1996. Antibody class switching. *Adv Immunol*, **61**, 79–146.

Steel, C. and Nutman, T.B. 2003. CTLA-4 in filarial infections, implications for a role in diminished T cell reactivity. *J Immunol*, **170**, 1930–8.

Sternberg, J.M. 1998. Immunobiology of African trypanosomiasis. In: Liew, F.Y. and Cox, F.E.G. (eds), *Immunology of intracellular parasitism*. Basel: Karger, 144–62.

Stewart, G.R., Bonssinesq, M., et al. 1999. Onchocerciasis modulates the immune response to mycobacterial antigens. *Clin Exp Immunol*, **117**, 517–23.

Sukumaran, B., Tewary, P., et al. 2003. Vaccination with DNA encoding ORFF antigen confers protective immunity in mice infected with *Leishmania donovani*. *Vaccine*, **21**, 1292–9.

Sultan, A.A., Thathy, V., et al. 1997. TRAP is necessary for gliding motion and infectivity of plasmodium sporozoites. *Cell*, **90**, 511–12.

Suzuki, Y. 2002. Immunopathogenesis of cerebral toxoplasmosis. *J Infect Dis*, **186**, S234–40.

Sztein, M.B., Cuna, W.R. and Kierszenbaum, F. 1990. *Trypanosoma cruzi* inhibits the expression of CD3, CD4, CD8 and IL2R by mitogen activated and cytotoxic human lymphocytes. *J Immunol*, **144**, 3558–62.

Tacchini-Cottier, F., Milon, G. and Louis, J.A. 2003. Th1 and Th2 cytokines in leishmaniasis. In: Kotb, M. and Calandra, T. (eds), *Cytokines and chemokines in infectious diseases handbook*. Totowa, NJ: Humana Press, 245–58.

Takle, G.B. and Snary, D. 1993. South American trypanosomiasis (Chagas' disease). In: Warren, K.S. (ed.), *Immunology and molecular biology of parasitic infections*. London: Blackwell Scientific Publications, 213–36.

Talvani, A., Machado, F.S., et al. 2003. Leukotriene B(4) induces nitric oxide synthesis in *Trypanosoma cruzi*-infected murine macrophages and mediates resistance to infection. *Infect Immun*, **70**, 4247–53.

Targett, G.A.T. 1995. Malaria – advances in vaccines. *Curr Opin Infect Dis*, **8**, 322–7.

Tarleton, R.L. 1998. *Trypanosoma cruzi*-induced suppression of IL-2 production. Evidence for a role for suppressor cells. *J Immunol*, **140**, 2769–73.

Tarleton, R.L. 2001. Parasite persistence in the aetiology of Chagas disease. *Int J Parasitol*, **31**, 550–4.

Tarleton, R.L., Grusby, M.J. and Zhang, L. 2000. Increased susceptibility of Stat4-deficient and enhanced resistance in Stat6-deficient mice to infection with *Trypanosoma cruzi*. *J Immunol*, **165**, 1520–5.

Taylor, M.J. 2002. *Wolbachia* endosymbiotic bacteria of filarial nematodes. A new insight into disease pathogenesis and control. *Arch Med Res*, **33**, 422–4.

Taylor-Robinson, A.W. 2003. Immunity to liver stage malaria, considerations for vaccine design. *Immunol Res*, **27**, 53–70.

Teixeira, M.M., Gazzinelli, R.T. and Silva, J.S. 2002. Chemokines, inflammation and *Typanosoma cruzi* infection. *Trends Parasitol*, **18**, 262–5.

Teixeira, V.P.A., Hial, V., et al. 1997. Correlation between adrenal central vein parasitism and heart fibrosis in chronic chagasic myocarditis. *Am J Trop Med Hyg*, **56**, 177–80.

Telford, S.R., Gorenflot, A., et al. 1993. Babesial infections in humans and wildlife. In: Kreier, J.P. (ed.), *Parasitic protozoa*, Vol. 5, 2nd edn. San Diego: Academic Press, 1–47.

Tsuji, M. and Zavala, F. 2003. T cells as mediators of protective immunity against liver stages of *Plasmodium*. *Trends Parasitol*, **19**, 88–93.

Turner, J., Faulkner, H., et al. 2002. A comparison of cellular and humoral immune responses to trichuroid derived antigens in human trichuriasis. *Parasite Immunol*, **24**, 83–93.

Valero, M.V., Amador, R., et al. 1993. Vaccination with SPf66, a chemically synthesised vaccine, against *Plasmodium falciparum* in Colombia. *Lancet*, **341**, 705–10.

Vazquez-Talavera, J., Solis, C.F., et al. 2001. Characterization and protective potential of the immune response to *Taenia solium* paramyosin in a murine model of cysticercosis. *Infect Immun*, **69**, 5412–16.

Volkmann, I., Saeftel, M., et al. 2001. Interleukin-4 is essential for the control of microfilariae in murine infection with the filaria *Litomosoides sigmodontis*. *Infect Immun*, **69**, 2950–61.

Vuitton, D.A. 2003. The ambiguous role of immunity in echinococcosis, protection of the host or of the parasite? *Acta Trop*, **85**, 119–32.

Walsh, A.L., Phiri, A.J., et al. 2000. Bacteraemia in febrile Malawian children: clinical and microbiologic features. *Pediatr Infect Dis J*, **19**, 312–18.

Walterspiel, J.N., Morrow, A.L., et al. 1994. Secretory anti-*Giardia lamblia* antibodies in human milk, protective effects against diarrhoea. *Pediatrics*, **93**, 28–31.

WHO, 2002. *Control of Chagas disease*, Tech Rep Ser 905. Geneva: World Health Organization.

Wille, U. and Hunter, C.A. 2003. Cytokines in the regulation of innate and adaptive immunity to *Toxoplasma gondii*. In: Kotb, M. and Calandra, T. (eds), *Cytokines and chemokines in infectious diseases handbook*. Totowa, NJ: Humana Press, 259–81.

Williams-Blangero, S., VandeBerg, J.L., et al. 2002. Genes on chromosomes 1 and 13 have significant effects on *Ascaris* infection. *Proc Natl Acad Sci USA*, **99**, 5533–8.

Williamson, W.A. and Grenwood, B.M. 1978. Impairment of the immune response to vaccination after severe malaria. *Lancet* **1**, 1329.

Wolday, D., Berhe, N., et al. 1999. Leishmania-HIV interaction: immunopathogenic mechanisms. *Parasitol Today*, **15**, 182–7.

Wolday, D., Mayaan, S., et al. 2002. Treatment of intestinal worms is associated with decreased HIV plasma viral load. *J Acquir Immune Defic Syndr*, **31**, 56–62.

Wrightsman, R.A., Luhrs, K.A., et al. 2002. Paraflagellar rod protein-specific CD8+ cytotoxic T lymphocytes target *Trypanosoma cruzi*-infected host cells. *Parasite Immunol*, **24**, 401–12.

Yazdanbakhsh, M., Kremser, P.G. and van Ree, R. 2002. Allergy, parasites and the hygiene hypothesis. *Science*, **296**, 490–4.

Yazdanbakhsh, M., van den Bigggelaar, A. and Maizels, R.M. 2001. Th2 responses without atopy: immunoregulation in chronic helminth infections and reduced allergic disease. *Trends Immunol*, **22**, 372–7.

Zalph, C. and Scott, P. 2003. Interleukin-12 regulates chemokine gene expression during the early immune response to *Leishmania major*. *Infect Immun*, **71**, 1587–9.

Zambrano-Villa, S., Rosales-Borjas, D., et al. 2002. How protozoan parasites evade the immune response. *Trends Parasitol*, **18**, 272–8.

Zhou, P., Li, E.Q., et al. 2003. Role of interleukin-6 in the control of acute and chronic *Giardia lamblia* in mice. *Infect Immun*, **71**, 1566–8.

5

Control of parasites, parasitic infections, and parasitic diseases

DAVID H. MOLYNEUX

INTRODUCTION AND CONTEXT

Since the publication of the 9th edition of Topley and Wilson in 1998, the landscape of global health has changed considerably. These changes have been driven by a series of initiatives that address broader development issues. These initiatives are as listed below.

1 Changed political awareness of international health issues with an increasing political commitment to poverty alleviation as a core component of agreed International Development Targets and Millennium Development Goals.

2 Changes in international organization policy and focus generated by changed leadership. In particular, the initiatives of the World Health Organization (WHO) in recognizing the need for public—private partnerships for health development; in defining priorities; and in launching initiatives such as Roll Back Malaria and the Tobacco-Free Initiative.

3 The establishment of the Global Fund for human immunodeficiency virus (HIV)/acquired immune deficieny syndrome (AIDS), tuberculosis (TB) and malaria following the initiative of the UN Secretary General. There is awareness that with the relentless problems of these diseases, even a fund of approximately US$2 billion will have limited impact since the estimated needs of HIV/AIDS, TB, and malaria come to an annual figure of c. US$10 - billion.

4 The new approach to donor-funding of health in poor countries where sector-wide approaches to financing (SWAP) (Cassels 1997) have been developed to enable a more coordinated approach to financing. This prevents donors from influencing, via special projects, the overall national health policy and plans. This so-called 'basket' funding recognizes that ownership rests with the country; that donors all contribute to the 'basket'; that, once committed, control of resources is lost; that priorities are established through policy dialogue; that partnership relations are strengthened. The SWAPs approach is also layered onto the increasing decentralization of national budgets to district-level management in many countries. However, the approaches of SWAPs to discourage project-specific or disease-specific funding appears at odds with the approaches of the Global Fund where disease-specific projects and programs are applied for.

5 Over recent years, large international nongovernmental development organizations (NGDO) have become increasingly active in disease control implementation and policy. Médecins Sans Frontières (MSF), Save the Children Fund (SCF), and Oxfam have been vociferous on issues of equity and access to drugs and in criticizing 'vertical' programs and drug donation programs. Notwithstanding policy papers introduced by NGDOs, MSF has advocated the establishment of a drugs for neglected disease initiative which, while necessary and laudable, must recog-

nize that in many circumstances such drugs require delivery through disease-specific methodologies and disease control is of necessity through approaches that are antipathetic to perceived policy wisdom of SWAPs and decentralization (e.g. trypanosomiasis or leishmaniasis).

6 There has been increasing recognition that infectious diseases are more prevalent and inflict a greater burden of disease on the poorest quintile of the population. The poorest 20 percent would benefit proportionately more if there was a pro-poor focus in tackling infectious diseases compared with other health interventions (Gwatkin et al. 1999).

7 There has been, in parallel with the recognition of the need for increased partnership, an expansion of public−private partnerships in health. These are summarized by Widdus (2001) who identifies some 70+ such initiatives. The diversity of objectives, financing, governance, and management of these alliances prevents any significant generalizations about best practice and how lessons can be learned as most of the alliances/partnerships are disease- or intervention-specific. The nature of some of these partnerships has been examined in more detail by Reich (2002), where some of the different models and experiences are discussed. Many of these have been developed over the past decade and have been funded by extensive long-term commitments from new donors, drug donations, long-term commitment from the private sector, and a recognition that relatively cheap interventions sustained by health systems can bring long-term health benefits. The establishment of the Bill and Melinda Gates Foundation as a key player in global health has greatly enhanced the opportunities for both research and intervention in parasitic infections. This Foundation, recognizing the importance of alliances that were embedded in its approach to program support, commissioned the study 'Building *Successful* Global Health Alliances', which outlines the parameters, criteria, and experiences of such alliances and creates a template through which to measure success (McKinsey 2002).

8 The introduction in 1993, following the World Development Report (World Bank 1993), of the concept of global burden of disease has been further expanded in a series of publications. Such data has enabled an assessment of likely change in global disease burden between 1990 and 2020. These projections suggest that as a proportion of global disease burden, only malaria remains a significant burden as a parasitic infection. Malaria falls in global burden importance from a ranking of 11th to a predicted 26th over this period. The majority of the projected changes in global burden are in the increased burden of cerebrovascular events, depressive illness, conflict-related conditions, road traffic accidents, and cancers. The disability-adjusted life years (DALY) burden of para-

sitic disease is projected to remain largely stable while the surge of noncommunicable disease burden due to the epidemiological transition, diet, and lifestyle change associated with urbanization, substance abuse, environmental degradation, population growth, and increased conflict are projected to be proportionally greater contributions. The DALY burden and overall public health importances of major parasitic disease are shown in Table 5.1.

9 The impact of nonhealth-sector roles, their financing and policies (e.g. education, agriculture, transport, natural resources) within government will have significant impact on health outcomes. Of particular importance will be levels of investment and achievement of education targets in an attempt to provide universal primary education by 2015; the importance of increasing the proportion of females in primary education will be of particular significance in achieving targets of improving maternal, child, and infant mortality figures. The complexities of the interactions and interrelationships are summarized in Figure 5.1

10 The interest in and, indeed, fear of emerging diseases (defined as either new infections of humans or re-emerging infections (where a rapid increase in incidence is seen of an existing infection or in a new geographical area) has been an important element in debate on health policy as epidemics of West Nile fever, Lyme disease, and Hantavirus have been recognized in the USA. Considerable additional resources have become available for research on such emerging agents. Recent quantitative analysis of the risk of emergence allied to the nature of the organism's mode of transmission and source have been provided by Taylor et al. (2001). They note that viruses, bacteria, and protozoa are more likely to emerge than macroparasites (e.g. helminths); that c. 75 percent of emergent organisms are from zoonotic sources and that emergences are independent of the mode of transmission. It should be noted, however, that despite the emphasis in some circles of the importance of such agents, they are not predicted to play a significant role in the global burden of disease estimates as a proportion of global DALYs. Such conclusions are based on the current definition of species. However, the capacity now exists to identify species complexes/groups and a level of intraspecies variation is becoming apparent as a result of molecular analyses. The new discipline molecular epidemiology, which has been applied to vectors and causative parasites, clearly suggests that the absolute numbers of genetically distinct parasites and vectors, irrespective of subspecific variation, is much greater than hitherto recognized, emphasizing both the degree of biodiversity in microorganisms and its importance in the strategies of parasite and vector control.

Table 5.1 *Overview of public health importance of parasitic infection (see also Remme et al. 2002)*

Disease / Parasite	Population at risk (millions)	No. of endemic countries	No. of infected (millions)	Estimated deaths (humans × 1000)	DALYs female	DALYs male	Total DALYs
Malaria	2 000	90	300–500	1 080	182.3	17.5	357.3
Leishmaniasis	350	82	12	41	12	8.6	20.6
Lymphatic filariasis	750	65	119	No direct mortality	5.6	2.9	7.5
Guinea worm disease	140	18	c. 0.12	No direct mortality			
Onchocerciasis	122	34	17.6	No direct mortality	3.7	2.7	6.4
African trypanosomiasis	50	36	0.02–0.30	50	9	8.8	17.8
Chagas disease	90	19	16	21	14.8	12.6	27.4
Schistosomiasis	500–600	74	200	11	29.9	15.4	45.3
Ascaris infection			1 000		53.8	51.4	105.2
Trichuris infection			900		32.2	30.9	63.1
Hookworm infection			500		5.8	5.6	11.4
Entamoeba infection			500	40–100			
Giardiasis			200				
Taeniasis	40		15				
Neurocysticercosis			50	50			
Food-borne trematodes			500				
Fascioliasis	180.25	8	2.39				
Clonorchiasis	289.26	6	7				
Opisthorchiasis	63.60	5	10.30				
Paragonimiasis	194.80	5	20.60				
Other intestinal flukes		6	1.28				

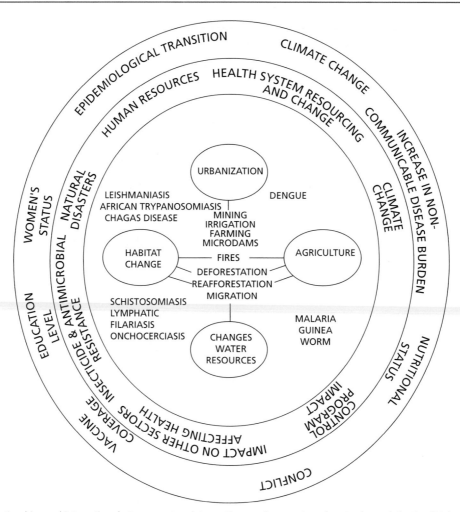

Figure 5.1 *Relationships and interactions between sectors intervention, environment, and vector-borne infection (Molyneux 2001)*

11 The suggestions that global climate change will have a widespread impact on health as mean temperatures rise over the next decades have provoked studies on the projected change in distribution of vector-borne infections. It is generally agreed (from different climate models) that the mean rise of temperature over the next 100 years will be of the order of 2–4°C. The impact of these changes, particularly on the distribution of *Plasmodium falciparum* malaria has been projected by various groups (IPCC 2001; Rogers and Randolph 2000; Hay et al. 2002; Patz et al. 2002) although little consensus is available (Patz et al., 2002). In addition, the role of El-Niño events has been studied and identified with changes in epidemic patterns in different regions of several vector-borne infections (dengue in Indonesia, malaria in most of Africa, Colombia, and India).

12 In many countries, the resources allocated to health remain disproportionately small compared with other sectors. The recent Macroeconomics and Health Report (WHO 2001) has made a significant contribution to the recognition that investment in health is critical to furtherance of human development and that significant returns on investment in health are obtained in the development process. Ill health is both a cause and consequence of poverty.

13 Civil unrest and disturbances, conflict and consequent migration, social disruption, and disruption of infrastructure including health services have had profound consequences for many populations over the last decade, particularly when associated with the end of the cold war. The impact of these events is summarized by Murray et al. (2002) and Supplement 1 to *Lancet* **360** (December 2002). The study of global burden of disease (Murray and Lopez 1996) suggests that the trend in disease burden associated with conflict will continue to rise to 2020. Molyneux (1997) summarized the impact of conflict on parasite infections; Table 5.2 updates available information on parasite infections in relation to conflict.

Epidemics may also be provoked by ecological, climatic, and environmental change; urbanization; human population movement resulting from civil unrest and conflict; reduced surveillance; and drug or insecticide resistance. Policymakers recommend that health systems are restructured to include a generalized 'horizontal' pattern of healthcare, insurance systems, and

Table 5.2 *Conflict related change in parasitic diseases*

Disease	Organism/location	Change
African trypanosomiasis (sleeping sickness)	*Trypanosoma brucei gambiense*	Epidemics in Democratic Republic of Congo (DRC) and Angloa over recent decades, associated with destruction/disruption of health services (Ekwanzala et al. 1996). Epidemics spread in north-western Uganda following conflict-related migration from Sudan.
	T. b. rhodesiense	Disruption of cotton and coffee production in Busoga during the Amin regime resulted in spread of *Lantana*, providing breeding sites for *Glossina fuscipes* and initiating transmission of acute sleeping sickness in peri-domestic environments, with cattle acting as reservoir hosts; restocking following cattle raiding induced epidemics following importations of infected cattle.
Visceral leishmaniasis	*Leishmania donovani*	Epidemics in southern Sudan: a changed ecological situation associated with an increase in *Phlebotomus orientalis* populations in maturing *Acacia / Balanites* woodland initially provoked the epidemics, which were left largely uncontrolled because of civil war, migration of infected populations, scarcity of treatment centers, and availability of drugs (Ashford and Thomson 1991; Seaman et al. 1992)
Cutaneous leishmaniasis	*Leishmania tropic*	Resurgence of *L. tropica* in Afghanistan (Kabul) following an increase in urban population density of non-immunes because of conflict (Ashford et al. 1992).
	Leishmania major	Movement of populations to Khartoum because of conflict and drought, and establishment of transmission among peri-urban reservoir of non-immunes living in shanties.
Malaria	Refugee camps in Africa	Refugee populations settled at lower altitude, in sites with relatively high rainfall; refugees with inadequate immunity or exposed to different strains of *Plasmodium falciparum*; absence of drugs and no control of *Anopheles gambiae*.
	Refugee camps in Afghanistan and Pakistan	Malaria epidemics in camps (controlled by spraying tents to control *A. stephensi* and *A. culicifacies*), exacerbated by increase in prevalence of *Plasmodium falciparum*.
	Cambodia	Mass deportation of urban non-immunes to forced labor in rice fields and forests; conscription for construction of defenses in border areas where multi-drug-resistant malaria commonly occurs.
Onchocerciasis	Sierra Leone	Weekly aerial larviciding against *Simulium* suspended because of security problems and local conflict. Levels of transmission consequently increased.
	Guinea Bissau, Sudan, Sierra Leone	Civil unrest prevented distribution of ivermectin with consequent resurgence in incidence.
	DRC, Central African Republic	Suspension or reduction of ivermectin distribution because of conflict.
	Liberia and Angola	Planning of program for community-directed distribution of ivermectin was retarded by collapse of national structures; remaining (passive) distribution by nongovernmental donors.
Dracunculiasis	Ghana	Increase in reported cases 1 year after local conflict failed to contain cases from previous year.
	Sudan	Civil war prevents adequate case-finding, case-containment, water-supply control, and filter distribution.

user charges, and decentralization of management to district level (or equivalent). Such restructuring reduces the ability of the system to respond to factors that lead to epidemics. It must be borne in mind that so called 'vertical' parasitic disease control activities (e.g. onchocerciasis control and the lymphatic filariasis, Chagas' disease, and Guinea worm eradication programs) have been remarkably successful, onchocerciasis control being judged particularly cost effective (Benton and Skinner 1990). Because of the complexity of the biological systems inherent in parasitic infections, particularly those that are vector-borne, such diseases are not easily amenable to control by a strictly 'horizontal' health system approach.

This chapter draws together information on the control of parasites, parasitic infections, and the diseases caused by parasites. The title distinguishes between the control of an organism (parasite, vector or ectoparasite) at the level of the individual and at the level of the community. An example of the paradox (individual vs. community) is the conflict between the treatment of individual patients compared with the need to control or even eradicate disease. It is important to emphasize these distinctions as they are frequently ignored by policy makers, health workers, and scientists seeking to develop new approaches and tools for control.

The changing health environment has already been emphasized, but it is important to recognize that approaches to control are dependent on an accurate knowledge of the problem. This requires biological, medical, and epidemiological inputs to define the etiology (causative organisms), the vectors (if vector-borne), the parameters and mode of transmission (vector-borne, water-borne, aerosol, oro–fecal, venereal). Systems of surveillance, monitoring and evaluation are required to define prevalence and trends of infection and disease. Without such fundamental information, a control strategy cannot be appropriately designed and implemented. Biological information needs to be supplemented by consideration of issues such as:

- logistics
- cost-effectiveness
- potential for integration within existing programs
- past successes or failures
- input from governmental sectors other than health (e.g. agriculture, forestry, education, water, other natural resources, wildlife)

- acceptability of an intervention to the target communities
- potential for ecological damage
- priority rating afforded by the Ministry of Health (MOH)
- availability of human resources for implementation
- potential of research to provide improved products within a particular timescale.

The overall framework of approaches towards parasitic disease control is summarized in Figure 5.2.

CONCEPTS OF CONTROL AND ERADICATION

A distinction must be maintained between the terms 'control' and 'eradication'; the latter is often used inappropriately and it should be employed with caution. In 1988, the International Task Force for Disease Eradication (ITFDE) was formed to evaluate systematically the potential for eradication of candidate diseases and to identify specific barriers to eradication. The criteria used to asses the feasibility of eradication are provided in

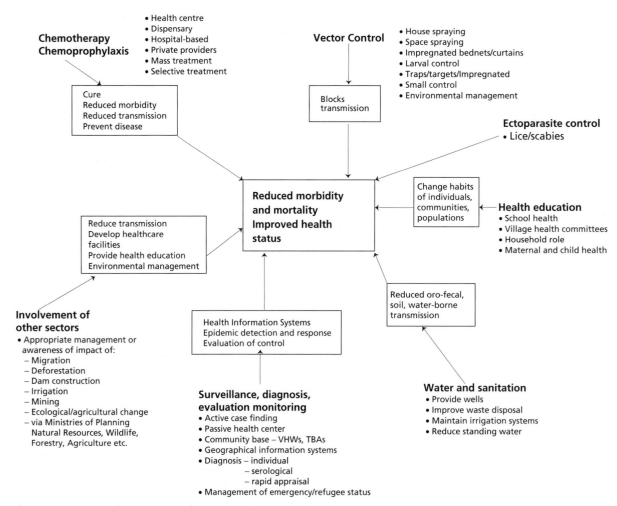

Figure 5.2 *Framework for the control of parasitic diseases*

Table 5.3 *Criteria for assessing eradicability of diseases or conditions (Dowdle and Hopkins 1998)*

General category	Specific criteria	Examples
Scientific feasibility	Epidemiologic vulnerability	Existence of nonhuman reservoir; ease of spread; natural cyclical decline in prevalence; naturally induced immunity; ease of diagnosis; and duration of any relapse potential.
	Effective, practical intervention available	Vaccine or other primary preventive, curative treatment, and means of eliminating vector. Ideally, intervention should be effective, safe, inexpensive, long lasting, and easily deployed.
	Demonstrated feasibility of elimination	Documented elimination from island or other geographic unit.
Political will / Popular support	Perceived burden of the disease	Extent, deaths, other effects; true burden may not be perceived; the reverse of benefits expected to accrue from eradication; relevance to rich and poor countries.
	Expected cost of elimination or eradication	Especially in relation to perceived burden from the disease.
	Synergy of eradication efforts with other interventions	Potential for added benefits or savings or spin off effects.
	Necessity for elimination rather than control.	

Table 5.3. The ITFDE has recently been reconstituted to evaluate the present situation as new developments have taken place over the last decade. The ITFDE defined eradication as 'reduction of the worldwide incidence of a disease to zero as a result of deliberate efforts obviating the necessity for further control measures.' The original ITFDE reviewed more than 90 diseases, 30 of them in depth, and concluded that dracunculiasis, rubella, poliomyelitis, mumps, lymphatic filariasis, and cysticercosis could probably be eradicated using existing technology. The term 'elimination' is increasingly being used to replace the term 'eradication,' which should be only used in a global context. The Dahlem conference held in Berlin in 1997 (Dowdle and Hopkins 1998; WHO 1998) also considered these issues in some detail and introduced the term 'extinction' to classify an organism that no longer exists (in contrast to the causative organism of smallpox, which has been eradicated as a cause of disease but still exists in stocks retained in secure laboratories). 'Elimination' is now regarded as meaning removal of an organism from a defined geographical region (local eradication) which creates problems for quantification of achievement towards the goal. The accepted position is that the disease is not eradicated but no longer requires ongoing investment in control and is maintained at a level where the problem is no longer a significant health burden. A new concept, 'elimination of a disease as a public health problem' has been introduced through World Assembly Resolutions. The definitions used in this chapter are from Dowdle and Hopkins (1998) and are as follows:

- Control: Reduction of disease incidence, prevalence, morbidity or mortality to a locally acceptable level as a result of deliberate efforts; continued intervention measures are required to maintain the reduction.

- Elimination of disease: Reduction to zero of the incidence of a specified disease in a defined geographical area as a result of deliberate efforts; continued intervention measures are required.
- Elimination of infection: Reduction to zero of the incidence of infection caused by a specified agent in a defined geographical area as a result of deliberate efforts; continued measures to prevent re-establishment of transmission are required.
- Eradication: Permanent reduction to zero of the worldwide incidence of infection caused by a specific agent as a result of deliberate efforts; intervention measures are no longer needed.
- Extinction: The specific infectious agent no longer exists in nature or the laboratory

The classic eradication program was that of smallpox, which achieved its target in 1977. To date, no parasitic disease has been eradicated, although attempts to eradicate Guinea worm are underway, through the program Global 2000 (Hopkins et al. 2002). Nevertheless, successful local eradication (correctly, elimination) has been achieved in some restricted geographical or epidemiological situations:

- onchocerciasis has been eliminated from several parts of Kenya and from the Nile at Jinja in Uganda by using the insecticide dichlorodiphenyltrichloroethane (DDT) to remove the local vectors (*Simulium neavei* and *S. damnosum*, respectively)
- the Onchocerciasis Control Programme (OCP) in West Africa has achieved the same goal by targeting particular cytoforms of the *S. damnosum* complex using aerial application of insecticides.

Local elimination has also been achieved as detailed below:

- the malaria vector *Anopheles gambiae* from Brazil in the late 1930s by Soper using larviciding measures and house spraying with pyrethrum, a success repeated in the early 1940s after the same species had been introduced into Egypt
- *Glossina palpalis*, the tsetse fly, the vector of human trypanosomiasis was eliminated from the island of Principe in 1905 by trapping flies using sticky back packs on plantation workers
- animal trypanosomiasis has been eliminated from parts of North East Nigeria by ground spraying of tsetse resting sites with persistent doses of DDT
- *Aedes aegypti*, the vector of yellow fever, has been eliminated from parts of Central and South America
- local antimosquito spraying has eliminated lymphatic filariasis from the Solomon Islands with no evidence of any resurgence over a 20-year period
- filariasis due to *Brugia malayi* has been eliminated from Sri Lanka through selective treatment with the drug diethylcarbamazine (DEC), antilarval measures (removal of host plants by herbiciding) and environmental improvements
- chemotherapeutic approaches have eliminated filariasis due to *Wuchereria bancrofti* from Japan, South Korea, and Taiwan in Asia, and Suriname and Trinidad and Tobago in the Americas (WHO 1994)
- filariasis has also been eliminated as a public health problem in large areas of China
- a long-term 'elimination' program has been successful against hydatid disease in Iceland
- malaria was eliminated from Sardinia by DDT spraying as well as in other marginal areas of distribution.

One noticeable feature of these successes is that many examples refer to islands or isolated populations or areas where the parasite is at the edge of its geographical range. Clearly, the advantages of isolation and a greater ability to control animal or human population movements are important. Elimination or global eradication of any disease is difficult to achieve and costs increase per case detected, controlled or averted as the end-point is reached.

However, the high cost of eradication or local elimination programs may be justified as these costs are time limited, whereas disease control implies a long-term commitment. Any control program must be cost-effective and should reduce the target disease to a level at which costs are sustainable by the local community (or by public or private healthcare systems). Control seeks to bring the problems to a level at which the disease is no longer of public health importance with morbidity at an acceptable level within the community, an absence of mortality and, if appropriate, greatly reduced levels of disability. To translate the level of control achieved to eradication or elimination status requires *either* a vastly increased cost per case treated or prevented which, for financial and ecological reasons, may never be feasible, *or* the development of a more effective intervention.

PUBLIC–PRIVATE PARTNERSHIPS IN DISEASE CONTROL

There have been significant developments of public–private partnerships (PPP) for disease control over recent years. These have, in part, been generated by the new policies of the Director-General of WHO, Dr. Gro-Harlem Brundtland, following her election in 1998 but also by the recognition that such partnerships represent a response to the changing environment in international health and that no single organization can undertake or be responsible for all aspects of health and its delivery. Recent PPPs in parasite control include:

- African Programme for Onchocerciasis Control (APOC)
- OCP (1974–2002)
- Onchocerciasis Elimination Programme in the Americas (OEPA)
- Roll Back Malaria
- Multilateral Initiative on Malaria
- Medicines for Malaria Venture
- Global Allliance to Eliminate Lymphatic Filariasis
- Global School Health Initiative
- Partnership for Parasite Control
- Global Alliance for African Human Trypanosomiasis
- Guinea Worm Eradication Programme (also known as 'Global 2000').

Each of these PPPs has different characteristics (e.g. governance, financing, time frame, links with the private sector) and different disease-specific objectives. Widdus (2001), Buse and Walt (2000a, b), Reich (2002) and Buse and Waxman (2001) provide a detailed analysis of the underlying principles of these alliances as well as case studies of functionality.

COMPONENTS OF CONTROL

Components of control are listed under the following headings:

- Situation analysis
- Definition of objectives and strategy
- Options and responsibilities at different levels of health system
- Planning and resourcing
- Evaluation and monitoring
- Implementation and integration of selected methods of control.

Situation analysis

- Perform desk study of published and unpublished reports to assess problems in the context of country, region, and district.

- Acquire information on prevalence and incidence.
- Appraise the validity of information.
- Evaluate current epidemiological situation by passive surveillance at health centers or by questionnaires (e.g. using the postal system).
- Observe changes over time and predict future change.
- Define the structure of health services and their existing capacity, human resources available, and needs for training and capacity building.
- Establish the priority afforded to the disease by the government, the MOH, the district management teams, and the communities.
- Establish links to other sectors or organizations in planning for control (e.g. other ministries, development organizations, nongovernmental organizations (NGO).
- Assess the influence of other activities such as development projects on planned programs.
- Carry out spot surveillance of local prevalence, vectors and, if applicable, animal reservoirs.
- Use rapid assessment methodologies (e.g. for schistosomiasis, onchocerciasis, filariasis, or loiasis).
- Assess the available methods for prediction of epidemics using remote sensing or climate prediction available to other sectors (e.g. agriculture, natural resources, environment).
- Establish a national task force composed of various stakeholder groups to address the problem.

Definition of objectives and strategy/options and responsibilities at different levels of health system

- Analyse cost-effectiveness of different control approaches and options.
- Select appropriate methodology and define control requirements.
- Establish an inventory of personnel and facilities (including estimation of training needs and requirements for equipment and drugs).
- Establish feasibility in the context of other health needs.
- Contrast epidemic ('firefighting') problems for which vertical, rapid intervention is necessary with endemic situations for which a long-term approach and integration are required (Table 5.4).
- Establish emergency response capacity to address predicted epidemic risk.

Planning and resourcing

- Define the expected contribution from the government.
- Develop national plan.
- Evaluate targeted approaches to donors in the context of donor priorities and prevailing national policy.
- Define appropriate timeframes for implementation of plans.

Table 5.4 *Role of different levels of the health system in parasitic disease control*

Level	Role
Community	Identification of suspects/patients
	Follow-up of patients
	Coordination of any appropriate vector-control activities (e.g. bednet distribution to vulnerable groups/reimpregnation)
	Facilitation of cooperation, local logistics
	Communication by village health committees
District	Passive detection and treatment
	Parasitological/serological diagnosis
	Treatment and minimum clinical care
	Follow-up of microscopy
Regional	Active surveillance
	Confirmatory diagnosis
	Data collection
	Technical supervision of vector control
	Distribution of reagents and materials for vector control
Ministry/country level	Situation analysis/policy position
	National strategy and plan
	Establish stakeholder group/national task force
	Financing
	Training needs and responsibility
	Health education
	Distribution of technical information, equipment, drugs, materials
	Purchase of equipment and supplies
	Human resource management

- Define the relationship of the action to overall health plans and budgets.
- Establish links with appropriate international reference centers for technical support; control of an epidemic may merit application for emergency status to provide rapid funding (e.g. requests for therapeutic drugs and insecticides from international aid agencies and NGDOs).
- Establish drug supply line following identification of sources; initiate quality assurance mechanisms; define tax status of drugs (e.g. donated products).
- Define role of the nongovernment sector (e.g. private providers, NGOs) in control policy.
- Ensure adequate information exchange about control policy between different bodies and individuals involved.
- Undertake knowledge, attitudes, and practice (KAP) studies as a basis to inform approaches to social mobilization strategies.
- Train personnel (including management) through courses, instruction of trainers, educational materials, and health education programs.
- Assess community acceptability and the perceived priority of any involvement that will require resource input from the communities (e.g. role and views of village health workers (VHW), volunteers, traditional birth attendants (TBA), community leaders, school teachers).
- Define the management structure of the program and its relationship with existing management structures.
- Assess capacity available (managerial, financial, technical) and ensure capacity building is embedded in planning.

Evaluation and monitoring

- Assess progress towards objectives (prevalence distribution, vector status).
- Establish sentinel site / baseline data in defined units.
- Define appropriate methods for epidemiological evaluation, e.g. parasitological, serological, and vector sampling methods.
- Carry out longitudinal surveys or spot surveys at indicator villages.
- Adjust the program in the light of results.
- Establish process indicators at national and subnational level.

Examples of the value of assessing longitudinal trends in disease control are illustrated in Figures 5.3a–e, 5.4a, b, p. 116, and 5.5, p. 117. These show trends in parameters that measure the impact of implementation of selected methods of control on onchocerciasis, Chagas' disease, and Guinea worm either by measures that indicate change of incidence or by directly measuring vector populations and hence transmission since the program began.

Implementation and integration of selected methods of control

CHEMOTHERAPY AND CHEMOPROPHYLAXIS

- Assess the availability and quality of drugs and the distribution system.
- Establish relationship between national bodies, donation programs, and NGDO community to define operational relationships (e.g. onchocerciasis, lymphatic filariasis, African trypanosomiasis, schistosomiasis, trachoma programs).
- Assess or monitor drug resistance (e.g. East African network for antimalarial drug resistance).
- Assess the role of private providers and control of quality and price (e.g. malaria drug policy).
- Utilize other systems for distribution (e.g. schools, agricultural extension workers, other health or government workers, NGOs, committees).

VECTOR AND RESERVOIR CONTROL

- Assess availability, cost, and appropriateness of insecticides.
- Assess availability of skills to monitor insecticide resistance.
- Assess availability and effectiveness of alternative chemicals.
- Assess capacity for management of the control program.
- Establish relationships with other sectors to provide support for environmental control measures.
- Assess acceptability and feasibility of reservoir control.
- Assess environmental acceptability of interventions.
- Assess personal protection (e.g. bednets, sustainability of a bednet program/retreatment modalities).
- Establish policy in relation to bednet distribution (e.g. vulnerable groups, social marketing).
- Investigate opportunities for integration if appropriate (e.g. malaria and lymphatic filariasis in Africa; dengue and filariasis in the Pacific; leishmaniasis, Chagas' disease, and malaria via bednets in Latin America).

ENVIRONMENTAL MANAGEMENT

- Ensure effective linkages between health and other sectors.
- Assess potential impact on other diseases.

IMMUNOPROPHYLAXIS

- Assess availability of vaccines.
- Assess capacity to manage the program and efficacy of cold chain.

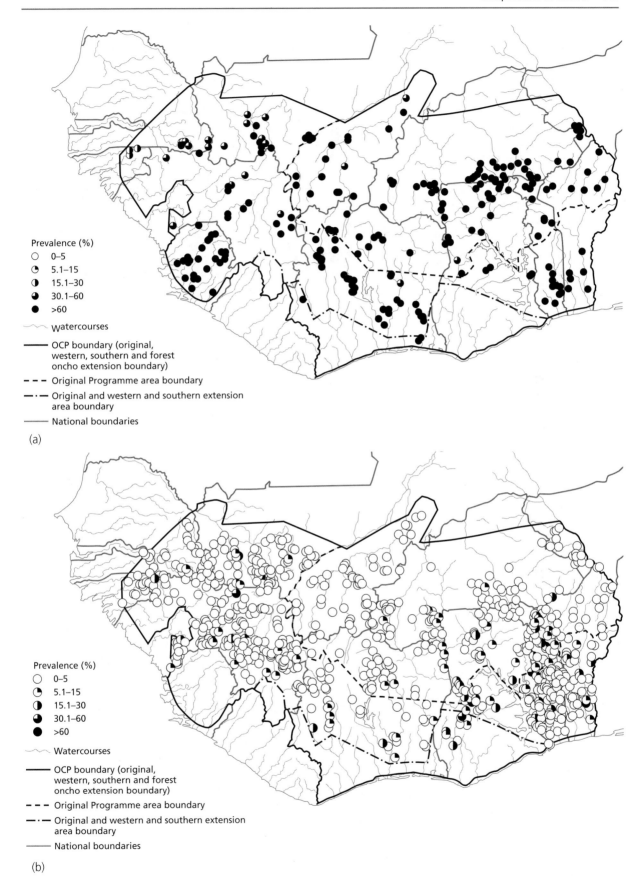

Figure 5.3 *Maps showing epidemiological changes in the prevalence and transmission of* Onchocerca volvulus *in countries of the OCP 1994–2001.* **(a)** *Epidemiological situation of villages evaluated before the control.* **(b)** *Prevalence of microfilariae in villages evaluated 2000–01 in OCP (Continued over)*

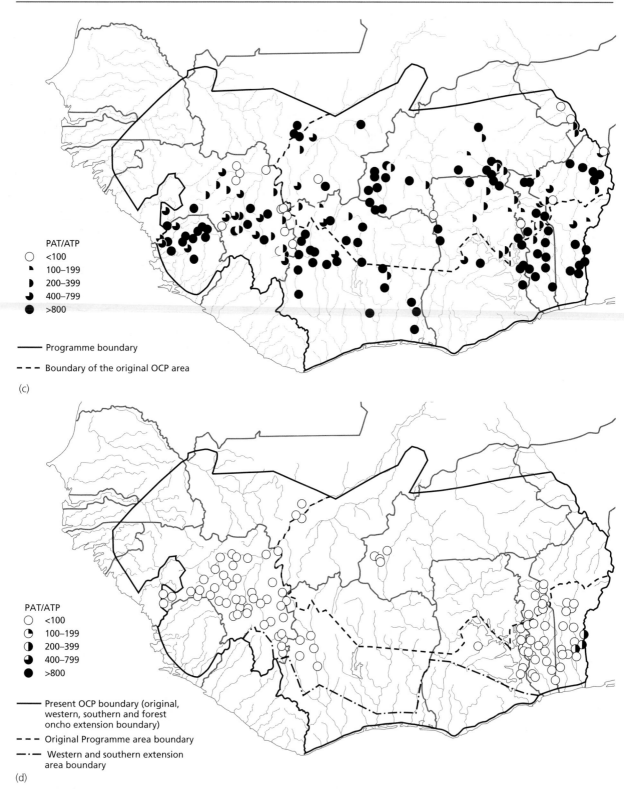

Figure 5.3 *Maps showing epidemiological changes in the prevalence and transmission of* Onchocerca volvulus *in countries of the OCP 1994–2001 (Continued).* **(c)** *Annual transmission potential of* Onchocerca volvulus *in West Africa, pre-control 1974.* **(d)** *Annual transmission potential of* Onchocerca volvulus *in West Africa, post-control 2001 (Continued over)*

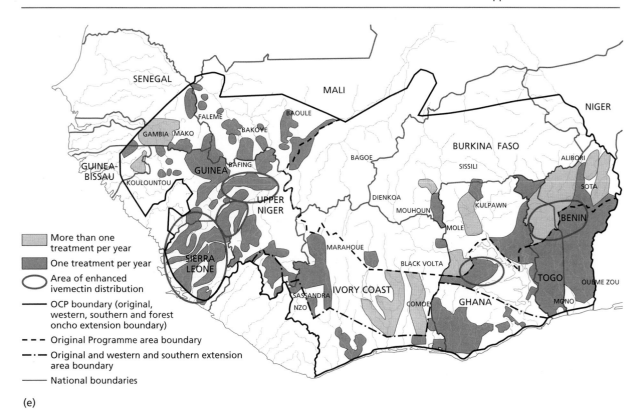

Figure 5.3 *Maps showing epidemiological changes in the prevalence and transmission of* Onchocerca volvulus *in countries of the OCP 1994–2001 (Continued).* **(e)** *Distribution of (ivermectin) in OCP area. Note the extensive distribution in areas formerly in extension areas with 'core area' free of onchocerciasis except in restricted area.*

- Establish links to existing programs (e.g. expanded program of immunization (EPI) to utilize the cold chain).

HEALTH EDUCATION / SOCIAL MOBILIZATION

- Use media resources, including radio, television, and video.
- Use posters and drama sessions oriented around the local environment and traditions.
- Ensure participation of teachers, local leaders, health workers, local medical practitioners, religious leaders.
- Ensure link to KAP and feedback to inform social mobilization strategy.

STRATIFICATION OF PARASITIC DISEASES IN RELATION TO ELIMINATION AND CONTROL

Control programs often involve specific approaches to arrest the transmission of infection (e.g. via vector control) or to prevent or cure a disease. Although such programs have been successful in the past, integrated approaches are now recognized as being more appropriate for reducing prevalence and incidence. This is important if the strategy is aimed at alleviation of a disease problem in a community or population rather than in an individual. Integrated control is based on coordinated planning and detailed knowledge from many different areas: scientific, technical, intersectoral,

financing, and managerial. An approach termed 'stratification' has been used in malaria control; it means that the strategy is modified according to different epidemiological situations (WHO 1993a). Malaria stratification has been taken a step further by those with particular interests in different environments and geographical regions, a process known as 'micro-stratification' (Rubio-Palis and Zimmerman 1997). While stratification has been most widely discussed with reference to malaria (Table 5.5, p. 131), the process is equally applicable to other parasitic diseases: leishmaniasis (WHO 1990), onchocerciasis, filariasis (WHO 1992a), schistosomiasis, and African trypanosomiasis. Tables 5.5–5.10, p. 131–141 illustrate the concept of stratification in the planning of control in selected parasitic diseases.

APPROACHES TO CONTROL

Control of a parasitic infection can be focused on the individual, with a view to alleviating pain, reducing disability or avoiding death, while at the same time reducing the parasite load and transmission within a community. Such an approach will be less cost-effective than larger-scale control programs that employ methods such as vector control, reservoir-host control or mass drug distribution. Large-scale measures have a public health objective that, while reducing individual suffering, also reduces community morbidity and mortality. This

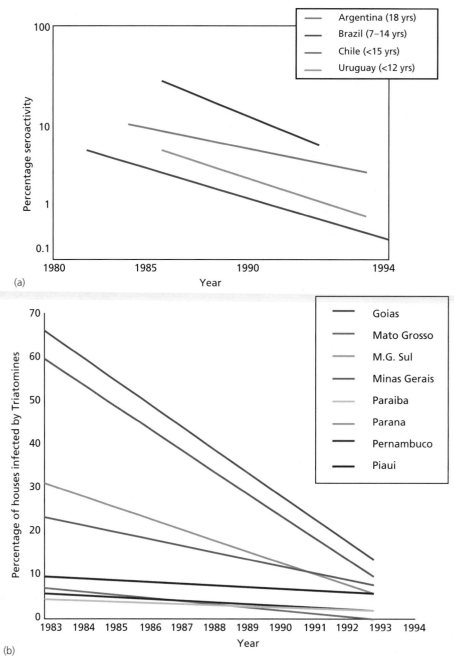

Figure 5.4 *Southern Cone Initiative.* **(a)** *Elimination of transmission: incidence of Chagas disease 1980–94 (Moncayo 1999).* **(b)** *Interruption of transmission: Brazil, endemic states 1983–93 (Gadelha 1993).*

provides socioeconomic benefits through improved agricultural productivity, improved cognitive function, better nutrition as a result of increased agricultural output of a more varied diet, and enhanced population mobility allowing for additional earning opportunities. Control of animal parasitic diseases also has benefits for human populations through increased protein availability and higher income from the sale of higher-quality livestock enhancing both local and national economies. Parasitic disease control programs vary in scale but they have generally been targeted at two different types of disease situation:

1 endemic disease in which long-term chronic infections have persisted in communities (e.g. onchocerciasis, hydatid disease (*Echinococcus* infection), schistosomiasis, Guinea worm (dracunculiasis), Chagas disease, and filariasis

2 epidemic disease where rapid intervention is required to prevent widespread morbidity and mortality.

Epidemics are frequently predictable but if health facilities are ill-equipped or nonexistent, high mortality may occur before control can be instigated. Recent examples of parasitic disease epidemics are described in Table 5.11, p. 142.

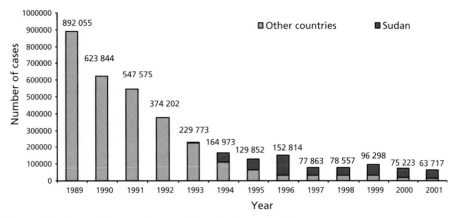

Figure 5.5 *Declining trend in dracunculiasis prevalence 1989–2001 (WHO 2002b)*

Anthropogenic and environmental changes frequently result in epidemics of parasitic disease. Several reviews have identified the primary drivers of these changes (Molyneux 1997, 2003a, b) as listed below:

- Movements of non-immune populations in areas where transmission occurs. Such movements may be of an organized nature (e.g. mobilization of the workforce in Brazil to exploit forest resources has resulted in malaria epidemics). Alternatively, they may occur without formal organization (e.g. movements of workers involved in mining for gold or gems in the Amazon and South East Asia).
- Climatic changes (e.g. temperature change is considered to be a cause of highland malaria in Kenya and Ethiopia). Unusual levels of rainfall following periods of drought result in epidemics of malaria in East and South Africa.
- Urbanization results in populations being exposed to new organisms as vectors establish in new habitats and peridomestic reservoirs act as sources of infection. Health services are grossly inadequate or nonexistent and service providers are often only NGDOs or faith groups. The increasing impact of urbanization on various different forms of leishmaniasis has been highlighted by WHO (2002a): for zoonotic cutaneous leishmaniasis (*L. guyanensis* and *L. brasiliensis*) in Brazil, Colombia, and Venezuela; for anthroponotic cutaneous leishmaniasis (*L. tropica*) in cities of south west Asia (Afghanistan, Iran, Syria, and Iraq); for zoonotic visceral disease (*L. infantumi*) in north-east Brazil where drought-induced migration to cities in which peri-urban shanty settlements provide conditions for transmission between humans and dog reservoirs via *Lu. longipalpis*.
- Changes in vegetation such as the development in Uganda of thickets of *Lantana*, which provided a habitat for *Glossina fuscipes* thus provoking epidemics of Rhodesian sleeping sickness. Another example is deforestation, which has resulted in exposure to leishmaniasis in the Amazon (Walsh et al. 1993).

- Development projects, particularly those involving water resource development, frequently exacerbate the health problems of the local or incoming population (Birley 1995; Hunter et al. 1993).
- Conflict, civil unrest, and associated population disruption have profound impacts on parasitic diseases and epidemics are frequently associated with such events. Disease-causing organisms with a capacity for rapid adaptation and reproduction that are associated with vectors having characteristics as generalists are more prone to create health problemsin conflict environments (Molyneux 1997). Tables 5.2 and 5.15, p. 145, summarize recent examples of changes in parasitic-disease epidemiology associated with conflict and characteristics of generalist vectors.
- Agricultural development projects, particularly those linked to irrigated agriculture and development of monocultures are associated with changed patterns ofinsect-borne infection, especially malaria (Ijumba and Lindsay 2001), Japanese encephalitis, leishmaniasis.

Common themes which operate in the above settings and are associated with changing vector-borne diseases include:

- epidemics associated with generalist vectors (characteristics of generalist vectors are listed below)
- animal reservoirs or mixing vessels associated as food sources with generalist vectors
- animal reservoirs that may be domestic, wild or intensively reared
- reduced biodiversity (often associated with the impact of hydrological change and deforestation/reforestation), which encourages expansion of adaptable generalist vectors and reservoirs
- ratio of *P. vivax*: *P. falciparum* changing with increasing *P. falciparum* (see also Table 5.15, p. 145)
- extractive activities (uncontrolled) generating the development of antimalaria resistance

- hydrological impacts over variable time-frames
     malaria and Japanese encephalitis – acute
     schistosomiasis / dracunculiasis – medium
     filariasis – chronic
- deforestation/reforestation impacts via
     behavior of humans, animal reservoirs, and vectors
     edge/interface effects/fragmentation patterns
     degree and type of reforestation
     loss of biodiversity
     loss of forest causing elimination of vector species
     common pattern of change occurring within different vector complexes.

Characteristics of generalist vectors include:

- wide geographical distribution
- species complexes or species groups
- capacity to feed on a range of available hosts
- ability to exploit peridomestic and peri-urban settings
- abiltiy to exploit new pre-imaginal habitats
- efficient vectors with high vectoral capacity
- no transovarial transmission.

Control of animal reservoir hosts

Many parasitic diseases are zoonoses, defined as 'those diseases and infections (the agents of) which are naturally transmitted between (other) vertebrate animals and man' (WHO, 1979). A list of recognized parasitic zoonoses is provided by the WHO (1979). Ostfeld and Keesing (2000) provide an updated list of vector-borne infections of potential public health importance, while a recent analysis of all emergent and re-emergent infections (Taylor et al. 2001) has identified that 75 percent of emerging pathogens are zoonotic and that such organisms are more than twice as likely to emerge as nonzoonotic ones. However, viruses and protozoa are more likely to emerge than macroparasites such as the helminths. The important zoonoses for which reservoir control can have a cost-effective impact are discussed below. The presence of an animal reservoir host may be a major impediment to control of a disease, particularly if the habits and habitats of the animal hosts prevent intervention either on the grounds of practicality or for reasons such as protected status of the host (e.g. primates or endangered species). The International Commission for Disease Eradication recognizes that the existence of a reservoir host precludes the likelihood of eradication or elimination of the infection (see Table 5.3).

LEISHMANIASIS

Among the most important examples of control programs targeted against animal reservoir hosts was that aimed at the great gerbil *Rhombomys opimus*, the reservoir host of *Leishmania major* in the former Soviet Union; a variety of techniques were employed to eliminate this rodent in the central Asian republics of

Uzbekistan, Kazakhstan, and Turkmenia. Burrow systems identified by aerial or ground surveys were destroyed by deep plowing. Alternatively, poisonous baits of zinc phosphide mixed with wheat and vegetable oil were introduced into every third or fourth hole. The effects of the zinc phosphide were enhanced by prior application of the anticoagulant, dicoumuarol, 5–7 days earlier. Because of its high toxicity, zinc phosphide had to be inserted directly into the rodent burrows. Elimination of *R. opimus* can also be achieved by flooding burrows in irrigation schemes. Reinvasion can be prevented by canal construction and agricultural development.

The fat sand rat *Psammomys obesus* (reservoir of *L. major* in the Near East and North Africa) has not been effectively controlled as it is not granivorous and because anticoagulants are too expensive for large-scale use. The other reservoir hosts of *L. major* (*Meriones* spp. and *Tatera* spp.) are not easily controlled. Large-scale control of hyraxes (reservoirs of *L. aethiopica*) needs to be organized at a local level using shooting and trapping.

The animal reservoirs of visceral leishmaniasis (*L. infantum* and *L. i. chagasi*) are domestic and wild canids. The instigation of dog control for rabies control in China has reportedly reduced the incidence of visceral leishmaniasis to almost zero. Elimination of stray dogs is justified for many reasons and although shooting and poisoning are effective they are not acceptable to some communities. In some countries (e.g. France) screening of domestic dogs for *Leishmania* spp. by serological examination in parallel with clinical examination, permits infected animals to be identified and either destroyed or treated. Treatment is not, however, entirely effective and an alternative approach is to administer a single prophylactic treatment shortly before the peak transmission season. Control of canine leishmaniasis could be integrated with rabies control by:

- registration of dogs
- regular evaluation and surveillance
- serological diagnosis
- mobilization and motivation of the dog-owning community.

However, a new approach has been the use of insecticide (deltamethrin)-impregnated collars on domestic dogs; this reduced the proportion of sandflies that took blood meals and survived by over 90 percent for at least 8 months after collars were applied (Killick-Kendrick et al. 1997). The importance of controlling canine leishmaniasis is underlined by the finding that human *L. infantum* infection is more widespread in southern Europe than was previously thought and by the fact that visceral leishmaniasis is an HIV-associated disease. In South America, and to a lesser extent in Europe, the presence of wild canid reservoirs necessitates a continuous commitment to control.

There is limited opportunity to control the wide range of reservoir hosts of New World cutaneous leishmaniasis. The WHO (1990) suggests that control of the opossum (*Didelphis marsupialis*) could be achieved by using baited pitfall traps in urban and peri-urban environments or in disturbed primary forest; this would result in control of *L. guyanensis*. *T. cruzi* might also be tackled by the same measures in peri-urban situations, but the cost-effectiveness of such measures remains to be validated. The WHO suggests an environmental management approach combining primary forest clearance with insecticide application, in order to create a vector-free and reservoir-free zone around villages. The transmission of *L. guyanensis* has been greatly reduced by these measures in French Guyana.

HYDATID DISEASE

Dogs and other canids are the definitive hosts of *Echinococcus granulosus*. Several countries have instigated dog control and surveillance and chemotherapeutic treatment (with praziquantel) to eliminate adult worms or to reduce access of dogs to larval stages in offal. Major hydatid control campaigns have been introduced in a number of countries including Australia, Iceland, New Zealand, Cyprus, Argentina, and Uruguay. The major measures applied are:

- preventing dogs from gaining access to raw offal at abattoirs and farms or to dead animals in the field
- reducing parasite loads by culling dog populations in combination with mass treatment.

The application of such measures resulted in a rapid reduction of cystic infections in livestock in Cyprus, where over 30 000 dogs were eliminated over a 2-year period. A parallel health education program was also introduced to inform the public about appropriate practices for feeding dogs and for slaughtering animals. A particular problem in hydatid control is that dog feces are used for specific purposes in some areas of the world; for instance, the Turkana of northern Kenya use dog feces to lubricate necklaces, and such feces are used for leather curing in Lebanon. Further problems arise from the inappropriate disposal of animal or even human corpses, and from close relationships between humans and dogs (e.g. 'nurse dogs' in Kenya, or family pets in England). In such cultures, the elimination of dogs cannot be envisaged.

Alveolar hydatid disease (caused by *E. multilocularis*) is a problem in mountainous areas of Europe, northern USA, Siberia, and China, where the fox is the main definitive host and larval (cystic) stages are found in microtine rodents (usually the genus *Microtus*). Health-education campaigns in Europe have successfully discouraged the consumption of wild fruits, particularly blueberries, which can be contaminated with fox feces. In China, reduction of the canine population has resulted in reduced transmission to humans; however, changes in land use associated with loss of forest has increased incidence of alveolar hydatid disease in other areas.

TAENIASIS, CYSTICERCOSIS, AND NEUROCYSTICERCOSIS

Cysticercosis of the central nervous system (neurocysticercosis) is caused by the larval stage (cysticerci) of the pork tapeworm *Taenia solium*. Human beings act as definitive hosts and swine as intermediate hosts. Pigs become infected when they ingest human feces containing *T. solium* eggs, which develop in the muscle and brain into cysticerci. When people eat the undercooked pork of an infected animal, they develop tapeworm infection. Human beings can, however, become infected by directly ingesting *T. solium* eggs from human feces. These eggs then develop into cysticerci, which migrate mostly into muscle (causing cysticercosis) and into the central nervous system where they can cause seizures and other neurological symptoms of neurocysticeriosis. Human cysticercosis infections are acquired by the feco–oral route in areas with poor hygiene and sanitation.

Neurocysticercosis has serious morbidity and is also a leading cause of epilepsy, which has profound social, physical, and psychological consequences. The diagnosis of neurocysticercosis involves the interpretation of nonspecific clinical manifestations, such as seizures, often with characteristic findings on computer-assisted tomography (CAT) scans or nuclear magnetic resonance (NMR) imaging of the brain, and the use of specific serological tests. Diagnostic criteria based on objective data (clinical, imaging, immunological, and epidemiological) have been proposed for different levels of the healthcare system but are not generally used in areas endemic for the disease. Lack of awareness and differences in quality and availability of medical services mean a lack of reporting. Human carriers of *T. solium* are diagnosed by the detection of proglottids or eggs in feces, or by more sensitive methods such as the detection of *Taenia* antigens in stools or specific antibodies in serum.

Human cysticercosis is a disease associated with poverty in areas where people eat pork and traditional pig husbandry is practiced. It is endemic in the Andean area of South America, Brazil, Central America and Mexico, China, the Indian subcontinent, South-East Asia, and subSaharan Africa. The spread of the disease is facilitated by poor hygiene, inadequate sanitation, and the use of untreated or partially treated wastewater in agriculture.

To control cysticercosis, the following intervention strategies are available (Lightowlers 1999):

- Case management, reporting, and surveillance. People with neurocysticercosis usually present with nonspecific neurological symptoms such as epilepsy, for which adequate case management should be available in health services. Such management requires a consensus on standardized criteria and guidelines for early differential diagnosis in peripheral healthcare structures (with emphasis on resource-poor areas) as well as guidelines for possible treatment or referral to the next level of the healthcare system. Better surveillance and reporting would lead to a more accurate understanding of the extent of the problem and to the identification of foci of transmission.

• Identification and treatment of individuals who are direct sources of contagion (human carriers of adult tapeworm) and their close contacts, combined with hygiene education and better sanitation, will interrupt or reduce the cycle of direct person-to-person transmission, an approach that has been successfully applied to other contagious diseases.

• Universal or selected treatment with praziquantel (10 mg/kg body weight) has significantly reduced the prevalence of human taeniasis in areas where T. solium infection is endemic (Mexico). To limit reinfection of humans by the intermediate hosts, treatment should be accompanied by veterinary sanitary measures, meat inspection, improvement of pig husbandry, and treatment of infected animals. Single-dose therapeutic agents, for example oxfendazole, have recently become available, with no deleterious effects on pigs or on the meat product. Animal vaccines are under development.

• Long-term success is more likely when anthelminthic chemotherapy programs are integrated into a wider intersectoral approach to increase public awareness and hygiene practices. Additional measures to sustain the impact of specific interventions include the provision of clean water and sanitation, and health education. Improvement in living conditions, legislation, modernization of swine husbandry, and improvement of meat inspection, have reduced transmission in many industrialized countries.

The International Task Force for Disease Eradication declared *Taenia solium* a potentially eradicable parasite, for the following reasons (CDC 1993):

• the life cycle requires humans as definitive hosts
• tapeworm infections in humans are the only source of infection for pigs, the natural intermediate host
• the transmission of infection from pigs to human beings can be controlled
• no reservoir for infection exists in wildlife.

The strategic use of anthelminthics against adult worms in people and the larval parasite in swine, combined with health education and regulation of pig slaughter, will interrupt transmission, but this approach has yet to be proven in any well-monitored intervention trials.

Taeniasis and cysticercosis do not lead to sudden, large-scale international epidemics and do not seem to constitute an appropriate subject for international notification although Roman et al. (2000) argue that neurocysticercosis should be made an international reportable disease. As with other potentially eradicable diseases, significant progress has been made over the past decade in developing tools, refining strategies, and defining the public health problems. Sotelo (2003) argues that eradication remains an attainable goal, the achievement of which requires more advocacy, resources, national commitment, and global structures. Health ministries should be encouraged to set up surveillance and reporting systems, and adopt an active approach towards prevention and control of these diseases.

Community participation in parasitic disease control

The drive towards primary healthcare following the Alma-Ata declaration of 1978 provoked a greater degree of involvement of communities in healthcare through:

• the use of community leaders to support various programs
• the identification of personnel to undertake health activities on a voluntary basis
• emphasis on the importance of such activities in community well being.

The topic of community participation has been reviewed by Curtis (1991) who provides a series of examples in vector-borne disease control. MacCormack (1991) provides an insight into the underlying principles of sustainable vector control in a community context and reviews the factors that influence success and failure. She emphasizes that much of the success achieved in small pilot programs has depended on particular characteristics such as:

• leadership
• a responsive, well motivated, well educated community
• incentives from agencies and insecticide manufacturers
• ease of communication.

Following initial success, there is a danger that a 'hot' project will fall into a steady state as enthusiasm and donor support wane while the project life-cycle faces inevitable problems. The scaling-up of pilot projects to national ones within a primary healthcare context presents additional challenges. For instance, the community may be affected by the replacement of local leaders with national bureaucracy. In establishing a functional link between the communities and the health systems, each group must be trained to understand the social role on the one hand and technical skills on the other. Communities' local knowledge about insects should be exploited to aid in vector control. Appropriate control methods, and the importance of maintaining them, must then be clearly explained to all those involved at the local level.

It must also be established whether unpaid community labor can be sustained over time; although it has been achieved in pilot programs, doubts exist about longer term sustainability (Walt 1988). Much is likely to depend on the community structure and its relationship with those in authority, who are perceived as those most likely to benefit. If, for example, a cost-recovery system operates, the volunteers are less able to collect fees from their social superiors. Professional interaction between technicians and volunteers can also fuel conflicts based on perceptions about status.

The outcome of community participation in any project depends on the numerous complex social interactions existing within the community environment. The interaction between weak and strong groups, and the impact of participation on such group relationships, are of critical importance (Antia 1988). It is valuable to define the boundaries of the community involved because individuals tend to identify with a particular locale; this is despite the inherent social instability of most villages, which results from factors such as migration, schooling, and marriage. For practical reasons, the community is usually defined by a geographical boundary such as an urban neighborhood or an agricultural village while nomadic groups represent a mobile community.

Communities differ in how they function and are stratified; for example, they may be democratic, autocratic or under military control. In a democratic environment, obtaining consensus may be a slow process, but the likelihood of sustainability is high. MacCormack concludes that community participation in vector control will be sustainable only if the assessment of the costs-to-benefits ratio takes account of 'opportunity costs' (the value of activities people would undertake if they had not committed themselves to a particular control activity). Sustainability will be enhanced if the following conditions apply: activities are linked to community priorities; skills training enhances community well being; and preventative work links to curative or care outcomes in the primary healthcare setting, or can produce income (Rajagopalan et al. 1987).

Dedicated control programs often achieve good results, but they are often part of a research project. Therefore, coverage of the population in need is relatively small, but delivery costs are high, as are recurrent costs as a proportion of overall costs. This leaves little financial input for operational expenses. In situations in which the frequency of the intervention is limited (e.g. once a year) the commitment of dedicated resources is difficult to justify.

Community-based treatments are usually better targeted and tend to involve volunteers, TBAs, and primary healthcare workers. Increasingly, other types of groups are also becoming involved; examples include women's groups, faith groups, civil society organizations (CSO), and NGDOs. The NGDO community has become increasingly involved in onchocerciasis control. Over the past decade, as the programs in Africa and the Americas have expanded using the donated drug Mectizan® (ivermectin), the momentum for NDGO involvement came from the organizations committed to blindness control that recognized the value of ivermectin as a tool for reducing morbidity associated with onchocercal eye disease (Drameh et al. 2002). At present, NGDOs provide some 25 percent of the resources required for national onchocerciasis programs, and 12 international NGDOs are active in some 20 countries in Africa through the APOC and the countries from the former OCP. The key element of the approach to control is community-directed treatment with ivermectin (CDTI), which is regarded as the key driver in ensuring sustainability of this program. The progress in the APOC program is documented in *Annals of Tropical Medicine and Parasitology*, volume 96, 2002, Supplement 1.

CDTI CHALLENGES

Amazigo et al. (2002) review the challenges presented by CDTI strategies. This approach is based on the principle of community participation but also ensures empowerment by allowing communities to decide who should be the community directed distributors (CDD) and how to plan ivermectin distribution (e.g. dates, location model of distribution). The replacement of the community-based treatment system with the community-directed approach has been encouraged as the latter is likely to be more sustainable, provides community ownership and empowerment and reduces costs to the health system. CDTI enables communities to organize distribution in line with cultural norms and organizational structures – such as kinship and clan structures in Uganda (Katabarwa et al. 2000), while stimulating basic healthcare infrastructure in remote areas (Hopkins 1998). Amazigo et al. (2002) articulate challenges for CDTI into managerial, technical, and sociopolitical headings. These can be summarized as shown below.

Managerial

- Develop drug collection mechanisms/including issues of delay and wastage.
- Integrate with health services including use of CDTI in other health problems and sharing infrastructure and personnel.
- Strengthen infrastructure at all levels to manage and sustain CDTI.

Technical

- Achieve and maintain 65 percent coverage of the total population within all villages (e.g. achieve

100 percent geographic coverall of meso- and hyper-endemic villages).

- Address refusals caused by perceptions of adverse reactions.
- Conduct an accurate census using a standard formula for calculating such a census.
- Establish commitment to self-monitoring especially by communities and use the results to improve projects.
- Undertake operational research that identifies problems and apply results.

Sociopolitical

- Increase recognition of role of CDDs and thereby prevent attrition.
- Determine and provide appropriate incentives and motivation for CDDs.
- Involve women as CDDs and in community decisive making.
- Increase involvement and build capacity of local NGDOs.
- Plan and achieve post APOC financial support.
- Achieve balance between cost-recovery strategies compared with guaranteed affordability at all levels; reduce costs.
- Develop advocacy to gain support from political leaders.

The experience that the CDTI concept has provided to date regarding community involvement in onchocerciasis control in Africa gives valuable insights into the link between communities and health systems, which are applicable and generalizable to other programs and have wider policy implications (e.g. cost recovery). However, CDTI may not be applicable in other settings where a high degree of coverage is needed and where health services themselves are better resourced than in rural Africa or in urban settings or conflict situations. The APOC has experienced the value of and problems presented by CDTI in conflict environments where the necessity of convincing communities to engage in onchocerciasis control is a priority: identifying partners prepared to work in insecure areas, and population movement. The needs of CDTI (i.e. high coverage, need for census, selection of CDDs, drug logistics, and monitoring) are difficult to achieve during conflict and civil unrest. Treatment can be given on specific days organized by governments such as independence days, Head of State's birthday, local festivals (e.g. to celebrate harvest or onset of rain). The use of such occasions may reduce the need for intense monitoring to ensure high coverage and compliance. There are no dedicated staff costs, administrative costs are absorbed by the system, transport costs are minimal, compliance is high, and costs of monitoring and evaluation are low. These arguments have been developed to justify such approaches to

lymphatic filariasis control in Papua New Guinea. Treatment with DEC was simplified by creating tablets based on a calculation of the average weight of adults and children; the appropriate amount of drug was formulated into one blue tablet for adults and one red tablet for children. This eliminated the need for scales for weighing, allowed the treatment to be administered by untrained personnel and improved compliance (by using a single tablet). The estimated cost per treatment was US$0.8 per child and US$1.2 per adult (not including costs for transportation, health-promotion materials, monitoring, and coordination). In seeking to develop community treatment of any disease, several issues (e.g. devising reporting forms for illiterate communities, drug collection points) need to be studied through locally organized operational research.

Because of the pressures on health systems, it is necessary to use personnel resources at the level of the community to participate in various phases of control activities. This is particularly relevant in the maintenance phase of programs when the community has seen benefits from intervention phases; this promotes confidence between the program participants, the communities themselves, and in health systems. Parasitic disease control programs are increasingly becoming dependent on the involvement of human resources without specific technical knowledge or with only limited training.

In the Chagas' disease control program, Garcia-Zapata and Marsden (1986) have described the value of community involvement in establishing vigilance over the presence of vector bugs following the attack phase. Bug information posts were linked to schools because it was recognized that children were sensitive to the renewed presence of triatomine bugs. Reappearance of bugs is the basis for initiating selective spraying at the local level. If given adequate information, local isolated farmers could also act as notifiers of bug presence. Simple devices for bug monitoring have been developed in Argentina. Similarly, successes in programs to control onchocerciasis, lymphatic filariasis, and Guinea worm have been due to the commitment of communities.

CONTROL OF BRUGIAN FILARIASIS

A focus of filariasis caused by *Brugia malayi* existed in the Indian coastal state of Kerala, in the Cherthala region of Alleppey. There was a high prevalence of the disease with serious clinical manifestations of elephantiasis. The vectors in the area were *Mansonia* mosquitoes that breed on certain water plants (*Pistia*, *Eichhornia*, and *Salvinia*), to which their larvae attach for respiration. The local people believed that vegetation in ponds improved the quality of the water, which was used for domestic purposes. Also, the water plants are used as green manure on coconut plantations (Rajagopalan et al.

1987) and, therefore, the community were reluctant to use chemical larvicides or herbicides.

An extensive education and awareness program was initiated via the media, targeting various sectors of the community. A group of sociologists was used to pass on the necessary technology and to create interest. The Filariasis Patients' Association, headed by a former school teacher, developed a school health-education program. This campaign was targeted towards the female population of Cherthala, who were often not chosen as brides because of filariasis; the effectiveness of the campaign was enhanced by a high level of literacy among the local women. The main control measure was the removal of water plants. The community first had to be persuaded that alternative green manure fertilizers were adequate and that ponds could be cleared without financial loss. As no government organization was available to supervise the weed removal, this was done by the people themselves. To provide motivation, the community was encouraged to culture fish that could be sustained only in weed-free ponds. Species of fast-growing edible fish were purchased from the State Fisheries Department and distributed free to those who agreed to remove weed from their ponds. The economic benefits derived from the fish motivated the community to keep the ponds free of weed, thereby removing the vector habitat.

The program is organized by the peoples' Filariasis Control Movement (FILCO), which supports the following variety of approaches to ensure that the whole community is involved:

- in six secondary schools, Student Filariasis Control Clubs run programs to detect and treat filariasis
- all educational institutions contribute voluntary labor to improve the environment
- the National Cadet Corps and National Service Scheme provide services for filariasis control
- the interns of medical colleges are involved in clinical and diagnostic night camps
- the Departments of Fisheries and Agriculture popularize fish culture and alternative fertilizers.

To publicize the message, hoardings depicting filariasis patients were erected in areas where they would have most impact (markets, hospitals, bus stations, railway stations). The slogan employed was 'Remove weeds and protect yourselves and your future generations from filariasis.' These messages were reinforced by the distribution of films depicting the suffering of a filariasis patients.

In parallel with environmental measures to reduce vector levels, screening and treatment of cases, previously under government control, are now undertaken by voluntary organizations. For instance, 75 filariasis detection and treatment centers are now screening the endemic area and positive individuals are being treated with DEC.

Brugian filariasis appears to have been eliminated from Sri Lanka over the last two decades as a result of selective chemotherapy using DEC and vector control of *Mansonia* host plants using herbicides. Extensive epidemiological investigations have indicated that *Brugia malayi* is no longer present in previously endemic areas.

GUINEA WORM ERADICATION

Guinea worm infection (dracunculiasis) is the subject of a global eradication campaign, initiated in 1980. The International Commission for the Certification of the Eradication of Dracunculiasis was established by the WHO in the mid 1990s to certify the absence of transmission in member states who applied for such status. Prior to such certification, a process of field visits by members of the Commission must be undertaken to validate the information submitted by the country. A pre-certification period of three years is necessary to be sure that transmission has ceased in previously endemic areas. A reward system is also introduced to ensure suspect cases are thoroughly investigated, and posters plus health education messages via media are used to promote the campaign. To date, elimination of transmission has been achieved in India and Pakistan, and both countries have been certified free of transmission. Much of the success in India can be attributed to the development of safe water supplies during the International Water Decade with the drive to ensure that every village had access to safe water. However, the disease remains endemic in 13 African countries while four countries in Africa (Cameroon, Chad, Kenya, and Senegal) and Yemen are in pre-certification status. While there has been a significant reduction in incidence in many countries, Sudan remains the country with the biggest problem due to civil unrest and population movement over the last two decades; Sudan accounts for around 75 percent of all global cases (Hopkins and Withers 2002). The impact of Guinea worm disease is concentrated at harvest and planting times when infected individuals are incapacitated and unable to farm or to attend school for up to 3 months at a time. The infection is painful, and secondary infections often result when the worms emerge through the skin of the extremities. As there are no effective drugs or vaccines, Guinea worm must be eradicated through a series of measures that require community involvement – filtering drinking water; treating bodies of water with the insecticide temephos to kill copepods, providing safe water, and case isolation (containment) to prevent contamination of water sources. The proportion of cases contained is an important measure of the success of the surveillance system, of the involvement of communities, and as a predictor of the likelihood of cases being detected the following season. In the pre-certification stage, it may be necessary to introduce rumor registers and provide rewards when detected cases are confirmed. Progress in the eradication program is shown in Figure 5.5.

The eradication strategy is based on the establishment of a national program in each affected country to survey every affected village, to estimate numbers of cases, and to initiate plans. It is then necessary to establish village-based surveillance, using VHWs to report cases on a monthly basis. Health education encourages the use of cloth filters and the prevention of contamination of water sources. Some villages can also be targeted for provision of safe water while the establishment of a community-based surveillance system (with monthly supervision) might be broadened to cover other major endemic conditions. *Dracunculus*-free certification can be issued to clear areas. To date, there has been considerable success with this approach and incidence has decreased dramatically. VHWs are directly involved in measures that benefit health and increase agricultural productivity and school attendance. Such success increases confidence within communities and motivates them to become involved in other self-sustaining health interventions. A new intervention using portable plastic drinking straws with a filter inserted has been introduced in Sudan to prevent ingestions of copepods.

Community surveillance methods for eradication require the establishment of village development committees whose members are responsible for case notification, health education, distribution and maintenance of filters, and patient care. In Mauritania, this role has been assigned to females who undertake the roles of water gathering and distribution. A potential weakness is that village eradication activities are dependent on monthly supervision from the health personnel at a sub-district level. Each village requires monthly visits, usually integrated with the extended vaccination program or other health programs. Dracunculiasis can be eradicated but needs considerable community commitment from VHWs; in Burkina Faso and Mali, these workers are rewarded by distribution of cloth in recognition of their contribution to the program.

COMMUNITY MALARIA CONTROL IN PONDICHERRY, INDIA

In the coastal areas of Pondicherry, South India, malaria has been controlled by a combination of approaches organized by the Vector Control Research Centre (VCRC). These approaches have achieved good control of the vector *A. subpictus* and also produced income for the local population. Vector control was achieved by removing the algae *Enteromorpha* from coastal lagoons where they sheltered mosquito larvae from fish and other predators. The algae were collected in the pre-monsoon season and provided the raw material for producing paper to be sold by villagers. The lagoon became a focus of mosquito-breeding during the dry season and the villagers were persuaded to deepen one of the ponds in order to allow the other to drain into it,

the mud from the deepened pond being used to fill low-lying areas. The deeper pond was then used for prawn culture, which provided extra income for the village.

Control of vectors was also necessary in other breeding sites, particularly the earthenware pots in which coconut husks are retted to make coir. Control was achieved by introducing larvae of the mosquito *Toxorhynchites* which predate larvae of vector mosquitoes. The water filled pits where *Casuarina* trees are grown also act as breeding sites and so these were stocked with the larvivorous fish *Gambusia*. This action was encouraged by using science clubs in schools to stock the pits.

Involving the community initially faced resistance from villagers who were divided among themselves and inherently suspicious of officialdom. These communities placed a higher priority on services such as water, electricity, and roads rather than vector control, except in situations in which malaria epidemics occurred. Only after the VCRC had intervened and asked government departments to provide electricity and a water supply did the villagers have enough confidence to allow vector control projects to commence. The success of this program, as that of filariasis control, was largely due to the initiative of active locally based indigenous scientists (VCRC 1992).

CONTROL OF PARASITIC INFECTIONS THROUGH SCHOOL-AGE CHILDREN

Population increases and investments in education have increased the global population of school-age children. This group is a particularly important entry point for improving health status by using education to create norms of healthy behavior and lifestyle. The school also provides access to a population with a high prevalence of helminth infection in whom mass drug distribution can rapidly reduce morbidity, enhance nutritional status, and improve cognition and hence school performance. Urbanization has increased enrolment into schools, but also promotes conditions such as lack of adequate waste disposal, overcrowding, and the absence of clean water, which may encourage increased transmission of helminth diseases.

A WHO pamphlet (WHO 1995a) estimated that of a global total of 1.2 billion school-age children, 700 million were registered in school with 400 million attending each day, and enrolment still increasing. The target parasitic diseases susceptible to control are schistosomiasis, *Ascaris lumbricoides* infection, *Trichuris trichiura* infection, hookworm infection (*Necator americanus* and *Ancylostoma duodenale*), and infection by the food-borne trematodes (*Clonorchis*, *Opisthorchis*, and *Paragonimus*). Many school-age children have mixed infections, providing justification for mass interventions based on effective single doses of drugs such as the benzimidazoles (mebendazole and albendazole) and praziquantel; these eliminate intestinal helminths and schistosomes.

The epidemiological characteristics of each particular situation require investigation; prevalence and intensity vary even in adjacent areas, as do the pattern and rate of reinfection. It is assumed that reinfection will occur and require retreatment, although emphasizing the importance of personal hygiene should reduce the rate of reinfection. There is thus a need for field studies to ascertain whether intervention for both schistosomiasis and intestinal helminths is warranted. Rapid assessment of hematuria is used to evaluate levels of morbidity due to *S. haematobium* while other approaches are being tested for *S. mansoni* and intestinal helminths.

Drugs for the treatment of intestinal helminths and schistosomes, and their treatment schedules and efficacy, are described by WHO (1995b). Albendazole and praziquantel have now been used in combination; combined drug delivery offers advantages in terms of lower cost and simplification of therapy. This results in improved compliance and improved drug efficacy by potentiation of the therapeutic effect. Trials have provided no evidence of adverse effects of combined treatment with praziquantel and albendazole. Phase I clinical trials showed a four-fold increase in the bioavailability of albendazole in the presence of praziquantel, while the bioavailability of praziquantel was unaffected by the administration of albendazole. Studies on the safety and efficacy of combined treatment have shown no increase in adverse effects and no reduction in efficacy.

The WHO emphasizes the importance of targeting chemotherapy for intestinal helminths at school-age children, as they harbor the most intense infections of *Ascaris*, *Trichiuris*, and *Schistosoma* although intensities of the different worm infections vary considerably between individuals. Treatment of this group achieves the maximum return in terms of reducing morbidity, and schools provide the most accessible entry point for maximum coverage. The WHO also stresses the value of targeting schoolchildren for oral antischistosomal drugs; programs should seek to provide complete coverage of the school-age population. Such approaches could be combined with other helminth control and also with other health-oriented activities such as immunization, nutritional programs, maternal and child health activities, and general health education as part of the school curriculum. Retreatment should not be undertaken for schistosomiasis more than once a year. Epidemiological studies need to be established to evaluate the criteria for retreatment frequency in each area.

The integration of helminth control into the education sector not only fulfils the immediate objective of infection control, but also brings broader health issues to the attention of a group who can strongly influence the behavior of future generations. WHO policy promotes school health programs, with de-worming as an entry point and with an emphasis on personal hygiene. The integration of school-based and community-based chemotherapy into the primary healthcare system will improve coverage and promote optimal retreatment schedules. Another potential benefit may be gained through training personnel at the periphery, who will develop activities based on disease-specific interventions to reduce prevalence and intensity of infection; this will reduce morbidity through coverage of schools and other entry points. While this approach to infection control will undoubtedly improve the health of schoolchildren and, to some extent, contribute to lower transmission rates over time, greater impact will be achieved if health education, improved water supplies, sanitation, and environmental management are integrated into the activities involved in community development.

Role of vector control in disease control

INSECT CONTROL

The role of insects in the transmission of parasitic infections has been recognized since Manson discovered that filaria parasites were transmitted by mosquitoes. From the latter half of the last century onwards, recognition followed that numerous other infectious diseases – sleeping sickness, malaria, Chagas disease, and onchocerciasis as well as various viral infections – were also insect-borne or tick-borne. Knowledge of biology of the vector (particularly of breeding sites, habitat of pre-adult stages, adult resting sites, biting behavior and host preferences, migration capacity and longevity) is a key factor in assessing whether or not appropriate control can be directed at the vector. Early attempts to target malaria by controlling mosquitoes depended on knowledge of larval stages and breeding sites. The removal of breeding sites (source reduction) was the only approach available to reduce vector populations; it was achieved by applying oil and copper compounds and by the rigorous removal or destruction of larval habitats by mechanical methods (WHO 1995c). Tsetse flies were controlled by trapping or destroying savanna and riverine habitats of *Glossina morsitans* and *G. palpalis* group flies, respectively. In some countries, game animals (the favored hosts of *Glossina*) were eliminated by shooting prior to the introduction of cattle. Removal of game animals rapidly reduced the tsetse population in large areas of South Africa, Uganda, and Zimbabwe, providing additional land for ranching.

The advent of chemical insecticides in the 1940s provided new tools for vector control that became incorporated into programs designed to control, if not eradicate, numerous diseases. DDT, initially the most widely used of these insecticides, was (together with chloroquine) a major tool in the campaign for malaria eradication launched by the WHO in 1955. In parallel, DDT was also being utilized in:

- tsetse-control programs for both human and animal trypanosomiasis
- control of *Simulium* (blackfly) larvae to control river blindness
- the successful elimination of *S. damnosum* in Uganda.

Meanwhile, another chlorinated hydrocarbon, benzene hexachloride (BHC), also known as lindane, was successfully used against triatomine bugs in Chagas disease control.

Recognition of the environmental side-effects of chlorinated hydrocarbons, through direct toxicity on nontarget insects and bio-accumulation, led to a suspension of their widespread use and a drive to utilize other chemical classes of insecticide. This was accelerated by the realization that resistance of mosquitoes to DDT was posing problems for malaria control, particularly in indoor house-spraying programs. Attempts were made by the environmental lobby to include DDT in the international treaty to ban persistent organic pollutants (POP); however, the public health utility of DDT has been recognized through the efforts of the WHO which, in the light of impending resistance to synthetic pyrethroids, advocated for its continued acceptability for indoor house-spraying against *Anopheles* vectors of malaria. Hence, DDT can still used in some public health control programs for indoor house-spraying; however, caution is needed and its use should be monitored and the potential for side-effects studied (Curtis 1994). While pyrethroids such as lambacyhalothrin are widely advocated for indoor house-spraying, the major emphasis in malaria vector control now is the drive to increase use, coverage and retreatment of permethrin-treated bednets as part of the Roll Back Malaria strategy. Ambitious targets have been set for the coverage of vulnerable groups (pregnant women, children under 5) but significant issues remain to be resolved in respect of payment for these insecticide-treated nets, subsidy for vulnerable groups, methods for ensuring re-treatment regularly takes place, the role of permanently impregnated nets, and how sustained coverage through the commercial sector is to be achieved given the policy that nets should be available free.

There are, however, many examples where vector control has had a profound public health impact. Outside the domain of parasitic disease, the elimination of yellow fever by source reduction of *Aedes aegypti* in Central America illustrates what can be achieved through intensive and well-organized campaigns of vector control. The elimination of *A. gambiae* from Brazil, following its establishment from West Africa in the 1930s, represented not only an important public health achievement but also a major contribution to the development of Brazil, given the efficiency of *A. gambiae* as a vector of malaria in Africa.

Currently, intensive programs are underway to eliminate *Triatoma infestans* (the vector of *Trypanosoma cruzi*) by domiciliary spraying using the synthetic pyrethroids (deltamethrin and lambacyhalothrin) in the Southern Cone countries of South America. This has had a major impact on transmission, as determined by standardized serological procedures and community-based bug-monitoring systems in Brazil, Argentina, Paraguay, Chile, and Uruguay. Figure 5.4(a–d) demonstrates the success of the program in terms of impact on transmission by incidence and by levels of house infestation. Applied research has been geared towards providing additional methodologies to support government-organized control schemes such as fumigant canisters, insecticidal paints, and simple bug-monitoring devices for surveillance of reinvasion. The Chagas disease control programs do not benefit from having an effective chemotherapeutic agent. This has extended the anticipated time for reducing Chagas disease as a public health problem because no drug or vaccine is likely to become available in the foreseeable future. Hence, the current approach of intensive vector control supported by national governments is the only feasible one (WHO 1991) with screening for infective bloodmeals.

To achieve the objectives of the OCP, vector control has been the core strategy in 11 countries in West Africa. The program, initiated in 1974, was based on larviciding the breeding sites of *S. damnosum* complex blackflies, initially in six West African countries. The strategy was based on the need to break the life cycle for the lifetime of the adult *Onchocerca volvulus* which, in 1974, was considered to be 18–20 years. At that time, only one insecticide (the organophosphate, temephos) had the appropriate chemical characteristics to satisfy the environmental prerequisite of avoiding long-term damage to the riverine ecology. Two major problems faced the program early in its existence: migration of savanna blackflies from outside the program area, and resistance of some blackfly populations to temephos. These problems were solved by extending the area of the program westwards and southwards and by engaging in a research program that rapidly developed alternative insecticides. The extensive activities of the OCP and its management and structures are described by Samba (1994), Molyneux (1995), and Boatin et al. (1997). At present, seven larvicides from different chemical classes are in operational use They are used on a rotational basis in accordance with parameters such as cost-effectiveness, environmental side-effects, river discharge rates, carry distance of insecticide, and risk of resistance development. The seven insecticides in current use (Hougard et al. 1993; Yameogo et al. 2004) are:

- *Bacillus thuringiensis israeliensis* H–14 serotype toxin (Bti)
- temephos, phoxim, and pyraclofos (all organophosphates)
- carbosulfan (a carbamate)
- permethrin (a pyrethroid)

● etofenprox also known as vectron (a pseudopyrethroid).

While vector control has been successful in breaking the onchocerciasis transmission cycle, the necessary duration of vector control has been reduced by the widespread use of ivermectin; this became available for use as a microfilaricide suitable for use in communities in 1986 (Remme 1995). Epidemiological studies indicate that the duration of adult worm life is 14 years and that distribution of ivermectin in hyperendemic areas on an annual basis rapidly reduces microfilarial loads and also has a rapid impact on ocular morbidity. Hence, combined vector control and chemotherapy reduce the duration of control to a 12-year period although ivermectin alone, at least in the African endemic areas, appears not to prevent transmission (Winnen et al. 2002).

Increasing numbers of displaced persons and refugees pose special problems for organizations responsible for management of disease. Intervention through vector control is an important component, as there is great potential for rapid disease outbreaks in crowded, unsanitary environments that support traumatized, malnourished, displaced individuals who are exposed to vector-borne infection or ectoparasites. Epidemics are sometimes most effectively controlled through vector control. Thomson (1995) provides details of how such epidemics should be controlled, with specific reference to relief organizations. The conflict-associated parasitic disease problems are shown in Table 5.2.

SNAIL CONTROL

The elimination of snails for the control of schistosomiasis remains an important component of control activities and three approaches have been used: chemical, environmental, and biological. Projects in several countries have shown that chemical control using molluscicides (e.g. niclosamide) in combination with other methods can reduce or eliminate transmission. Chemical molluscicides must be environmentally acceptable and must not produce adverse side-effects if they enter the food chain. In addition, they must be cost-effective, snail-specific, and easily applied. During the 1970s, snail control, either by chemical methods or through environmental management (weeding, irrigation-canal construction, and maintenance), was a key control component. The advent of safe and effective chemotherapy for schistosomiasis (oxamniquine, praziquantel) capable of reducing morbidity through population-based campaigns has shifted emphasis away from mollusciciding. Snail control is now considered as one of several approaches in integrated morbidity control, contributing to reduction in transmission at times of peak transmission. Target levels of reduction of snail populations should exceed 95 percent and must be maintained throughout the main season of transmission. Molluscicides can be used to destroy snails in breeding sites in transmission foci, reducing transmission in recreational areas such as lakes. They are also useful in the following ways (McCullough 1992):

● as a means of reinforcing community involvement
● to reduce transmission in areas where there is a special risk (e.g. fishing villages in the Volta Lake, Ghana)
● to eliminate newly introduced snail populations
● for total elimination of snails in isolated focal transmission sites (e.g. oases in North Africa)
● to prevent the establishment of dense populations in irrigation schemes.

Biological control of the snail hosts of schistosomes has been attempted by introducing other snail genera or species to out-compete the natural hosts but the impact of this approach on schistosomiasis morbidity has been limited.

Population-based chemotherapy, combined with local and seasonal use of molluscicides, will be the central features of future schistosomiasis control in high priority foci. Mollusciciding must be planned effectively and better strategies and delivery systems are required to improve cost-effectiveness. New, inexpensive but effective synthetic molluscicides are needed in addition to bayluscide, the only compound currently commercially available (WHO 1993b).

Role of chemotherapy and chemoprophylaxis

Drugs are the mainstay for controlling many diseases, but antiparasitic drugs present a considerable challenge for researchers interested in rational drug development. Investment by multinational drug companies in new agents for use in the tropics and the developing world is limited because the markets themselves determine the research and investment strategies of the pharmaceutical industry.

Parasite infections pose a variety of other problems for chemotherapeutic control:

● the organisms rapidly develop drug resistance
● effective compounds have a narrow therapeutic index and hence side-effects are common
● parasites can invade specific sites of the body making drug delivery difficult
● helminth parasites can be relatively large and require proportionally more drug than microscopic organisms
● compounds may be difficult to obtain when required and manufactured only infrequently, particularly those compounds used for treatment of diseases that affect relatively small numbers of individuals
● some currently used compounds contain toxic elements such as arsenic and antimony

- many compounds must be administered under medical supervision, require hospitalization, and are neither accessible nor affordable by individuals.

Despite these problems, some compounds are highly effective for controlling parasitic infections and for the treatment of individuals. New drugs for human diseases have also originated from alternative sources over the last decade. For instance, veterinary research produced ivermectin, used for treating onchocerciasis, lymphatic filariasis and scabies; ivermectin was originally marketed for the treatment of helminth infections of animals. Eflornithine, a drug that can cure late stage chronic sleeping sickness, was originally developed as an anti-tumor drug. The artemisine derivative qinghaosu, produced from the Chinese herb *Artemisia* and used as a cure for fever in China for 2 000 years, is now widely available for malaria treatment. Its availability is often uncontrolled, which will encourage rapid development of resistance. In addition, as with many well-used drugs, fake products are becoming widely available. This is of concern because the drug has valuable curative properties and, therefore, should only be introduced in appropriate circumstances in relation to the current drug-resistance spectrum of malaria parasites in each region. New formulations of existing drugs have been developed, such as encapsulated liposomes that provide slow release of a drug; a slow-release formulation of amphotericin B (AmBisome) has been developed to reduce toxicity and enhance efficacy in leishmaniasis treatment.

A distinction must be made between drug treatment initiated to alleviate individual suffering, and mass treatment strategies that seek not only to improve the well being of individuals, but also that of communities. This latter objective is achieved by ensuring that no parasites can be transmitted by vectors, or that no parasites remain to reinfect the population. Drugs must, therefore, be regarded as having several functions in control. Gutteridge (1993) provides a summary of the drugs available for parasitic diseases and the rationale for their use. Tables 5.12, 5.13 and 5.14 summarize the drugs available for treatment of the diseases discussed in this chapter.

A major problem in many parasitic diseases is that drug resistance frequently arises. Management of drug resistance in parasitic disease is difficult as there are few alternative compounds available. Strategies for reducing the likelihood of the development of drug resistance in malaria and helminth diseases are well documented and include the following:

- appropriate dosage regimes
- control of drug quality and distribution (in malaria)
- incorporation of vector control to reduce the development or spread of resistance.

Mechanisms responsible for development of resistance (Gutteridge 1993) include:

- the ability of the organism to metabolize the drug to an inactive form
- changes in the permeability of the organism so that the drug is no longer taken up
- development of alternative biochemical pathways
- increased levels of target enzyme production
- a change in the biochemical target so that the drug cannot bind as well as it did to the original compound.

The ability of microorganisms to change and adapt rapidly will clearly represent a continuing problem and drug resistance is unlikely to be avoidable. Hence, any disease control strategy based on chemotherapy should anticipate such a problem and should seek to reduce the rate at which resistance develops, while also promoting strategies for management of resistance if it occurs.

In many developing countries, drugs are available through many sources. From the perspective of the consumer, the availability of drugs at a health facility is a major indicator of quality of care. In the poorest households, up to 50 percent of expenditure on health is on drugs, many of variable quality (counterfeit, expired). In the developing world, pharmaceutical expenditure accounts for 20–30 percent of recurrent costs; it is second only to personnel costs (World Bank 1994). Supply of drugs for the control of parasites is usually organized in the public sector through government-formulated essential drug lists. The WHO essential drug list (WHO 1992b) and program includes drugs for the treatment of infection with parasites. The concept of an essential drug list is that such drugs should be available at all times, in adequate amounts, and in the appropriate dosage forms (WHO 1992b). Several important drugs used for parasite disease control are often unavailable, despite being specified on essential drug lists.

While governments have a responsibility to provide essential drugs to different levels of the health service, private providers play a major role in many countries where private expenditure on drugs considerably exceeds public expenditure. In Africa, c. 70 percent of the population has no regular access to essential drugs and drug stocks often run out as a result of problems with management, logistics or finances. This is particularly true in peri-urban and urban centers, even to the extent that hospitals may run out of drugs. Drug shortages are less frequent in commercial facilities and such supplies are more readily available in urban centers, but the perception of higher quality product availability in the private sector allows for prices to be inflated. Private suppliers are also frequently not subject to quality control. In the case of antimalarials, which are among the most frequently required drugs, there is over-pricing and supply is irregular while fake and adulterated products are promoted by importers. Antimalarials are often supplied without diagnostic confirmation.

The problems of drug supply, cost (generic versus brand name), quality control, distribution, irrational prescribing, noncompliance, and availability have been highlighted by the World Bank (1993, 1994). These problems potentially apply to all drugs, and awareness of them and the ways to circumvent them must be addressed if chemotherapy or chemoprophylaxis is to be a more effective part of disease control. Other problems may arise if vertical programs involving donated drug distribution are developed; for example, the distribution of ivermectin for onchocerciasis and albendazole for filariasis. Such an approach may conflict with the development of revolving funds to sustain drug supplies at primary healthcare centers, as might distribution for control of schistosomiasis or intestinal helminths targeted through school health programs. In the increasing number of emergency situations, the agencies involved need to be aware of the potential needs of the affected populations as part of their response planning. In this context, a predictive approach to supplying drugs (and insecticides) is necessary and must be based on a knowledge of the ecology of the area and the likely disease problems.

The use of reliable, systematic reviews of evidence of effectiveness to inform policy is becoming recognized as a vital contribution to enabling resources to be used appropriately. The Cochrane Collaboration has developed a series of systematic reviews of randomized control trials that provide a reliable assessment of the effectiveness of various healthcare interventions. This approach has been promoted because traditional reviews are unsystematic and neither respect scientific principles nor control for bias and random error. The approach involves worldwide partners and is designed to build on enthusiasm for the process, to minimize duplication, to avoid bias, to maintain an electronic database, and to ensure wide access in order to make the information available to decision makers. A list of the Cochrane reviews and protocols can be found at www.cochrane.org/cochrane/revabstr/g070index.htm.

The Parasitic Diseases Group is currently compiling a register of published and unpublished trials in parasitic diseases, identified by hand-searching relevant journals, using editors in China, Chile, the USA, and the UK. The review topics include malaria chemoprophylaxis in pregnancy, the use of artemisine derivatives in malaria, chemotherapy of filariasis, and the treatment of neurocysticercosis and giardiasis.

Role of diagnosis in control

Diagnosis of an infection or disease is a fundamental concept of curative medicine. In some parasitic infections, identification of more than simply the presence of the organism is essential if the appropriate treatment is to be given. For example, in sleeping sickness it is important to confirm the stage of the disease, as well as the presence of the parasite; this is because the drug-based treatment for late stage disease (melarsoprol) can be fatal. Hence, treatment is not initiated without confirmation of central nervous system involvement. In other diseases with high prevalence of infection in which mass treatment campaigns are in operation, diagnosis of the individual is not important. Diagnosis of malaria can only be confirmed by microscopical diagnosis; this is rarely undertaken in many settings in Africa. Clinic-based presumption diagnosis is often not malaria and this fact, together with first-line treatment-seeking behavior outside the formal public health sector, means few patients are correctly diagnosed. This results in overconsumption of antimalarials with unwanted consequences for drug resistance as well as increased cost to poor people.

In some diseases, such as schistosomiasis, onchocerciasis, and lymphatic filariasis, for which the overall prevalence is high, it is important that available resources are concentrated on those populations most in need. Therefore, the concept of rapid appraisal of disease burden in various communities has developed so that interventions can be better targeted. Ngoumou et al. (1994) described such an approach to the rapid mapping of hyperendemic onchocerciasis using nodule frequency, which correlates closely with community microfilarial load, as a means of targeting ivermectin distribution. The frequency of hematuria can be used as an indicator for the prevalence of schistosomiasis in school children. Recently, the rapid assessment of prevalence of eye worm (RAPLOA) technique, which uses the local name for the tropical eye-worm symptoms with a photograph of the worm passing through the eye or the presence of Calabar swelling, has been developed in communities in southern Nigeria and Cameroon. Recognition of the condition is valuable in prediction of endemicity.

There is, therefore, a trend away from 'gold standard' parasitological diagnostic techniques when disease-control programs are underway. This does not diminish the importance of diagnosis of the individual infection in need of specific treatment. The need to maintain diagnostic parasitological skills remains, and is particularly relevant, in the provision of diagnostic staff for microscopic diagnosis of malaria. Identification of parasites and differential diagnosis of *Plasmodium* species are important not only in determining treatment but also in saving unnecessary expenditure on drugs by defining whether or not a fever is of malarial origin. Despite the continuing need for skilled microscopists for malaria diagnosis, the amount of training provided is often inadequate, as are the quality of microscopes and staining materials available.

Serological diagnosis for screening populations has been used for surveillance of sleeping sickness. An antibody detection test, the card agglutination test for trypanosomiasis (CATT), and antigen detection tests are used to screen populations. In those suspected of infection, parasites are detected by examination of blood or lymph by direct or concentration techniques. Similarly, a direct agglutination test (DAT) has been developed to identify suspected cases of visceral leishmaniasis; this should be followed by a parasitological diagnosis, usually by the examination of bone marrow aspirate or splenic biopsy. These techniques are painful and risky, respectively, and not applicable to large numbers of cases.

Diagnostic developments include tests to detect antigens in urine – dip-stick techniques based on enzyme-linked immunosorbent assay (ELISA) systems and using color change as an indicator of the presence of antigens. In the case of urinary schistosomiasis, a color test is used to indicate the presence of blood.

Lymphatic filariasis due to *Wuchereria bancrofti* (and *Brugia malayi*) poses particular problems for diagnosis as the parasites appear in the blood only at night, when it is often difficult to take blood smears. Methods for detection of parasite antigen (Weil et al. 1997) or DNA represent more effective ways of obtaining information about the endemic status of filariasis and of monitoring the impact of control. Nevertheless, while such testing is being evaluated, parallel confirmatory night-time blood samples also need to be taken for monitoring and evaluation purposes at sentinel sites in control programs. Evaluation of lymphatic filariasis programs would also benefit from the availability of a test to detect antibodies resulting from exposure to infective larvae (L3) as a measure of effectiveness of transmission control. The importance of the continuing of monitoring and evaluation throughout a program allows programs to be determined by different methodologies (Figure 5.3(a–c) and Figure 5.4(a, b)). It is important to define the role of different age groups in the population in order to focus on incidence following interventions as a true measure of the impact and efficacy of the intervention while recognizing the sensitivity and specificity of any test as well as the sampling frame involved and the sensitivity of the expected population size to census error in defining real incidence prevalence or coverage if an arbitrary figure of success is selected.

There is also a need to develop an alternative diagnostic test to detect onchocerciasis in areas where the disease has been controlled, as skin snipping is no longer acceptable and both prevalence and microfilarial load are low. Here, it is important to determine if transmission is occurring and, although this is undertaken by examining blackflies, this may not be feasible. The ideal requirement for onchocerciasis serological diagnosis would be a test in humans to detect the pre-patent developing adult worm stage of the infection when few if any microfilaria have reached the skin.

The role of diagnosis in the control of parasites varies depending on how the question is approached; for instance, the following factors may need to be compared:

- treatment at the level of the individual for cure versus treatment at the level of the community for control
- safe drug versus toxic drug as a life-saving option
- expensive intervention versus cost-effective intervention
- cheap drug versus free drug
- monitoring control versus obtaining baseline data on prevalence pre-intervention.

The appropriate diagnostic approach depends on the particular situation.

Control of intestinal parasites

The impact on morbidity of intestinal helminths and, to a lesser extent, protozoal infections due to *Entamoeba* and *Giardia*, has been highlighted by the recognition that their public health impact was previously grossly underestimated (Table 5.1). There is now a body of evidence suggesting that intestinal helminths (*Ascaris*, *Trichuris*, and hookworms) have significant effects on growth, nutritional status, and school performance in children. The opportunity to provide alleviation of worm burdens through single dose, effective, inexpensive drugs (albendazole, mebendazole) is currently being exploited. This approach provides direct benefits as the worm burden is removed, reducing morbidity, and also possibly reducing the rate of transmission. Repeat treatment ensures that burdens are maintained below levels associated with morbidity, preventing protein–energy malnutrition and iron-deficiency anemia.

It is recommended that treatment for intestinal helminth infections is administered without prior screening as prevalence is high. The school-based approach is one of a range of strategies that can be used for helminth control at the community level; it can be integrated into existing systems such as maternal and child healthcare, family planning, water supply, and sanitation, and should be reinforced through health education. In areas where schistosomiasis is endemic, control approaches can be combined, reducing costs, improving compliance, and facilitating the logistics of control. In the longer term, health education, improved environmental sanitation, and safe water supplies are the components essential for the reduction of morbidity caused by intestinal parasites transmitted by the oro–fecal route or through the skin. In school-based programs, it is vital to train teachers about prevention and control, and these topics should be included in the curriculum.

There are continuing plans in some countries for major parasitic disease control programs. China has commenced a national de-worming program for a population of 200 million children; in Mauritius, the World Food Programme is including de-worming in a school meals initiative.

Table 5.5 *Stratification of malaria epidemiology in relation to approaches to control (after WHO 1993a)*

Malaria type (geographical distribution)	Epidemiological characteristics	Operational	Action required Disease management	Prevention
Savanna malaria. Sub-saharan Africa, Papua New Guinea.	*Plasmodium falciparum* dominant parasite. Perennial transmission but varying seasonality depending on distance from equator. Mortality and morbidity in young children and pregnant women. Drug resistance increasing problem.	Limited coverage by health services. Malaria control programs inadequate. Deficiency of appropriate capacity. Recognize interactions between HIV/AIDS and malaria treatment, control and policy. Increase access of poor to effective service	Increased interest of government, NGO, and private sector in malaria disease. Strengthen capacity for management of severe and complicated malaria. Improve diagnostic services, particularly in view of increased drug costs. Introduce appropriate national drug policies.	Implement impregnated bednets and curtains. Chemoprophylaxis for pregnant women unless precluded by drug resistance.
Malaria of highland or desert fringe (Africa, South East Asia highland; Sahel, South Africa, and South Pacific).	Risks of epidemics due to climate, conflict, changing agriculture practice or migration patterns. Drug resistance an increasing problem	Health services not available or limited. Terrain and distance present obstacles to control. Poor management of malaria where not historically a problem.	Establish epidemic preparedness, strategies, and policies. Active case detection and treatment of fever may be justified. Health services must be made aware of risk potential.	Vector control might reduce or eliminate transmission, particularly indoor residual spraying
Malaria of plains and valleys outside Africa (Central America, China, Indian sub-continent)	Variable, mainly moderate transmission. *P. vivax* may dominate. Strong seasonal variation. Drug resistance established.	Vector control not effective. Inadequate disease management. Health services inadequate; private services available.	Ensure responsibility of malaria control with general health services. Strengthen epidemiological systems.	Improve vector control if cost-effective. Environmental management might have impact hence intersectoral links necessary. Bednets can be useful (China).
Agricultural development projects. All malarious areas.	Increased transmission sometimes associated with irrigation. Seasonal malaria outbreaks in non-immune immigrant workers. Predictability of impact of change difficult to assess.	Insecticide resistance in cotton-growing areas. Finance for malaria control more likely to be available.	Establish services for treatment or strengthen existing ones.	Environmental management should be considered. Site and construct habitation appropriately. Impregnated bednets for labour force. Larvivorous fish in rice growing areas. House spraying and chemoprophylaxis appropriate for work force. Personal protection.

(Continued over)

Table 5.5 *Stratification of malaria epidemiology in relation to approaches to control (after WHO 1993a) (Continued)*

Malaria type (geographical distribution)	Epidemiological characteristics	Action required		
		Operational	Disease management	Prevention
Urban and peri-urban malaria (Africa, South America, South Asia).	Transmission and immunity variable. Increasing drug resistance.	Relatively good access to health services. Antimalarials available from different sources. High human population density. Breeding sites identifiable.	Standardize and harmonize treatment practices. Introduce drug quality and control systems. Facilities specific for malaria treatment required.	Personal protection and prevention. Impregnated bednets.
Malaria of forest and forest fringes South East Asia, South America.	Focally intense transmission. Occupational risk groups (e.g. mining). Severe multi-drug resistance. Non-immune migrant labour at high risk.	Health services inadequate or absent. No or limited social organization in communities. Variety of drugs sold via different outlets. Vector-control efficacy doubtful.	Treatment needs to be continually adjusted depending on experience of response to drug regimes. Establish communication between medical staff involved in treatment. Information, education, and communication of the private sector providers.	
War zone malaria. Conflict/ refugee emergency malaria.	Displacement of parasite-carrying population or non-immunes. Environmental degradation increases mosquito breeding.	Disruption of vector control. Limited curative services available. Drug distribution only via humanitarian assistance. Drugs provided may be inappropriate.	Awareness by responsible authorities of malaria risk. Drugs for treatment available via emergency programs must be appropriate to drug resistance pattern prevailing. Establish treatment facilities.	Refugees and soldiers may be protected by personal protection/chemoprophylaxis for vulnerable groups. Environmental measures and chemical control in refugee camps.

Table 5.6 *Stratification of epidemiology of African and South American trypanosomiasis in relation to control*

Disease and geographical locality	Epidemiological characteristics	Operational approach	Disease management	Required prevention
Rhodesiense sleeping sickness. East and Central Africa. *T. b. rhodesiense* Vector: *Glossina morsitans, S. pallidipes. G. fuscipes* in Uganda and Kenya epidemics	Usually zoonosis with range of game and domestic animal reservoirs. Transmission associated with entry into areas where savanna flies feed on game animals. Vector control inappropriate, occupation-associated transmission (e.g. honey gatherers, fishermen, poachers). Epidemics associated with human-to-human transmission by *G. fuscipes* with domestic cattle as reservoir hosts.	Acute disease. Passive case detection in local health facilities by routine diagnostic techniques. Need for active case detection and treatment centers in epidemics.	Accurate disease stage diagnosis; availability of drugs; recognition of toxicity of arsenicals and provision of treatment and care for serious adverse events	Avoidance of high risk areas. No vectors or reservoir control directly targeted but animal trypanosome control by vector control may have impact. Treatment of domestic livestock with trypanocides may reduce human cases. Epidemics stemmed by active case detection and vector control to reduce transmission (traps and targets)
Gambiense sleeping sickness. West and Central Africa. *T. b. gambiense.* Vector: riverine flies, *G. palpalis* group	Person-to-person transmission via riverine flies. Animal reservoirs of limited importance. Site associated transmission around high humidity areas.	Active case detection followed by treatment. Diagnosis by low parasitemias insensitive hence serological tests of value for screening. Vector control to arrest epidemic and halt transmission.	Provide diagnostic facilities at primary health care (PHC) and district level. Ensure drug availability. Maintain surveillance to detect early cases facilitating easier treatment. Provide adequate medical support for treatment and resuscitation in view of risk of side-effects to melarsoprol.	Vector control by community (traps); residual spraying of resting sites. Maintain surveillance to reduce epidemic risk. Prophylaxis with pentamidine no longer employed.
Chagas Disease. South and Central America. *T. cruzi.* Vector: triatomine bugs *Triatoma infestans* in Southern Cone	Zoonosis transmitted by intradomiciliary bugs. High vector infection rate. Many reservoir hosts. Acute disease in children but chronic in adults.	Diagnosis of chronic disease difficult and no effective remedy if diagnosed. Acute disease treatable but drugs expensive and toxic.	Limited efficacy of chemotherapy provides poor prognosis with death through cardiomyopathy. Transfusion disease averted by control of blood via trypanocidal additive (gentian violet).	Vector control by indoor house spraying or other insecticidal approaches – paints, fumigant cans. Improved housing provides beneficial long-term effect.

Table 5.7 *Stratification of epidemiology and control of leishmaniasis*

Disease and geographical locality	Epidemiological characteristics	Operational approach	Disease management	Prevention
Old World visceral leishmaniasis. Indian sub-continent. *L. donovani*. Vector: *P. argentipes*.	Anthroponotic subclinical cases frequent. No animal reservoir. Endemic with periodic epidemics.	Capacity to diagnose is limited. Antileishmanial drugs expensive, toxic, and not easily available. Inadequate reporting and case follow-up. Parasitological diagnosis to be available at district level with serological examination at PHC level if serodiagnostic capacity exists.	Improve diagnostics. Active case detection to reduce human reservoir.	Possible use of impregnated materials or residual insecticide application in epidemics to reduce transmission by sandflies.
East Africa. *L. donovani*. Vectors: *P. orientalis*; *P. martini*.	Reservoir hosts (rodents, carnivores) likely but importance unknown. Suggested association with termite hills in Kenya. Epidemic in S. Sudan associated with movement of population and development of *Acacia/Balanites* woodland.	Passive case detection and treatment. Annual incidence requires to be checked by active surveillance. S. Sudan epidemic places major demands on availability of Pentostam worldwide.	Improved diagnostics and drug availability.	Reservoir or vector control not feasible.
Visceral leishmaniasis with canine reservoir or assumed canine reservoir. North Africa, Middle East, Central and South America. *L. infantum*. Vectors: *Larroussius* subgenus; *P. perfiliewi*, *P. perniciosus*, *P. ariasi*. New World *Lutzomyia longipalpis*.	Zoonotic disease with foxes and dogs as reservoir hosts. Endemic over wide area; epidemics are infrequent. Sporadic cases over wide area with subclinical cases likely to be common. Found more frequently in children .	HIV-associated increase in cases in southern Europe through intravenous drug users. Vector control using insecticide impregnated collars on dogs. Diagnosis and treatment of dogs in endemic areas of Mediterranean and S. America has little impact. Reservoir control useful but has limited impact where feral reservoir hosts exist.	Ensure a minimum capacity for passive case detection and treatment.	Serological monitoring and treatment of dogs in Mediterranean. Use of insecticide impregnated collars on dogs.

(Continued over)

Table 5.7 *Stratification of epidemiology and control of leishmaniasis (Continued)*

Disease and geographical locality	Epidemiological characteristics	Operational approach	Disease management	Prevention
Anthroponotic cutaneous leishmaniasis. Old World urban centers. *L. tropica.* Vector: *P. sergenti.*	Predominantly found in densely populated settlements. Person-to-person transmission by *P. sergenti.* Incidental hosts not important. Transmission seasonal after sandfly peak.	Passive case detection as a minimum to reduce human reservoir. Integrate any vector control activities (house spraying, impregnated nets, and targets) with malaria control (but rarely sympatric).	Ensure awareness of disease; transmission, diagnosis, and treatment. Ensure drug availability.	Human reservoir control by active case detection. Vector control by residual insecticides in houses where cases found. Check susceptibility of vector to insecticide. Malaria control by house-spraying has impact as do bednets; use of impregnated materials feasible.
Zoonotic cutaneous leishmaniasis. Arid rural areas of Old World. *L. major.* Vectors: *P. papatasi, P. dubosqi.*	Reservoir (rodent)/sandfly transmission systems: *Rhombomys/P. papatasi* (Asia). *Psammomys/ P. papatasi* (Middle East/ N. Africa). *Arvicanthis/ P. dubosqi* (sub-Saharan Africa). Settlements around rodent colonies at risk; also workers (e.g. construction) who enter arid environment with rodent colonies.	Passive case detection and treatment with drugs. Live vaccines used in military situations. Rodent control/ environmental control possibly effective. Self-healing lesions with total immunity, but high level of disfigurement.	Awareness at health facilities. Availability of drugs.	Leishmanization has been used but 2.5 percent have large nonhealing lesions. Vector control not a cost-effective option. Environmental measures to prevent rodent establishment (e.g. plowing, poisons, flooding).
Zoonotic cutaneous leishmaniasis. East African highlands. *L. aethiopica.* Vectors: *P. longipes, P. pedifer.*	Association with hyraxes. Human cases associated with the proximity to hyrax colonies.	Case detection. Epidemiological knowledge of high-risk areas by health services.	Awareness at PHC level of clinical presentation; availability of antileishmanials.	Hyrax eradication possible locally. Vector control ineffective.
'Uta' cutaneous leishmaniasis. Andean highlands/western Cordilleras. *L. peruviana.* Vector: *Lu. peruensis.*	Putative reservoir hosts are rodents (*Akodon* and *Phyllotis*) and dogs. Seasonal transmission. Lesions predominantly in children below school age.	Passive surveillance. Case monitoring.	PHC awareness and case reporting.	Value of vector control is unknown except where proven peridomestic transmission.
Cutaneous leishmaniasis in Northern South America. *L. guyanensis* Vector: *Lu. umbratilis* (possibly *Lu. whitmani* and *Lu. anduzei*).	Forest edentates (Sloths *Choleopus* and anteater *Tamandua*). Reservoir hosts have high infection rate. *Didelphis* (oppossum) secondary peridomestic reservoir hosts where ecology has altered.	Humans infected by encroaching into forest or by establishment of habitation at forest edge.	Case detection and treatment. Availability of antileishmanials.	Vector control of limited impact. Clearing forest up to 300m from habitation reduces incidence.

(Continued over)

Table 5.7 *Stratification of epidemiology and control of leishmaniasis (Continued)*

Disease and geographical locality	Epidemiological characteristics	Operational approach	Disease management	Prevention
Cutaneous leishmaniasis in Central America/Northern South America. *L. panamensis.* Vectors: *Lu. trapidoi, Lu. gomezi, Lu. ylephiletor, Lu. panamensis.*	Also associated with forest penetration or association with secondary forest. Variety of reservoir hosts: edentates, primates, dogs, and rodents.	Passive case detection; awareness of risk for those entering or living in forest.	Case detection and treatment. Availability of antileishmanials.	Clearing of forest around villages reduces transmission. Vector and/or reservoir control not practicable.
Cutaneous and mucocutaneous leishmaniasis in Central and South America. *L. braziliensis.* Vectors: *Lu. wellcomei, Lu. intermedia, Lu. whitmani.*	Infection associated with forest activities and clearing land. Perennial transmission with high incidence in at-risk groups. Reservoir host not proven and many suspected vectors. In urban/peri-urban environment dogs, horses and pigs have been incriminated as reservoir hosts.	Espundia (mucocutaneous lesions) develop after varying periods following self-healing primary lesions.	Passive case detection; availability of drugs. Essential care of espundia cases to reduce sequelae of lesions. Commercially available vaccine probably not efficacious.	No vector or reservoir control applicable in forest but vector control possible in peri-urban situations, personal protection might help if affordable.
Cutaneous leishmaniasis in South and Central America. *L. mexicana* complex. Vectors: *Lu. olmeca, Lu. flaviscutellata.*	Variety of ecological situations – wet forest in Igapo, Brazil; dry forest in Yucatan; fringe forest in Dominican Republic. Rodent and marsupial reservoir hosts; highly dispersed and diverse.	Associated with penetration into forest environments.	Passive case detection, PHC, awareness, and availability of drugs.	Reservoir and vector control not possible. Personal protection in forest; reduce risk by not sleeping in forest. Health education.

Table 5.8 *Stratification of epidemiology and control of onchocerciasis*

Geographical locality	Epidemiological characteristics	Operational approach	Disease management	Prevention
Africa South of Sahara. Former OCP countries.				
1 Original areas: Benin, Burkina Faso, Cote d'Ivoire, Ghana, Mali, Niger, Togo.	1 Savannah woodland of northern tropics. Hyper-endemic villages close to rivers. Historically associated with high blindness rate resulting in depopulation. Major migratory vectors in savanna are *S. damnosum* and *S. sirbanum*. Annual transmission potential of more than 100 was associated with high blindness risk. Disease controlled by vector control since 1974 except in limited areas of Togo (Oti basin)	1 Vector control by weekly aerial larviciding for 14 years eliminates adult worms in human reservoir. Ivermectin twice per year for foci with inadequate vector control.	1 Vector control applied due to lack of availability of community-based drug until 1986. Thereafter, ivermectin for morbidity control to reduce community microfilarial load (CMFL) and arrest development of ocular lesions.	1–6 Maintain surveillance system for recrudescence using tests for early detection of transmission. Institute ivermectin if appropriate. Establish appropriate surveillance via integration into public health systems. Possible detection of early transmission by detection of infective larvae in blackflies. Maintenance of ivermectin distribution. Ensure high and sustainable delivery via community-directed treatment assisted by NGOs.
2 Extension areas: southern areas of Ghana, Togo, Benin; western extension to Guinea, Guinea Bissau, Sierra Leone, Senegal.	2 In forest areas, less blinding form of *O. volvulus* transmitted by forest vectors *S. soubrense, S. yahense, S. sanctipauli, S. squamosam, S. leonense* (Sierra Leone), which are less efficient and nonmigratory or less migratory.	2 Combined vector control and Ivermectin for 12 years projected to reduce disease to nonsignificant level. Ivermectin alone in areas of low CMFL (<10) where no invasion of *Simulium* threatens original OCP area.	2 Ivermectin throughout for extended periods up to 20 years as ivermectin does not interrupt transmission.	
3 Non-OCP countries: East of Benin to Sudan.		3 Ivermectin through community-based distribution systems.	3 Impact of ivermectin on skin disease and other manifestations of *O. volvulus* infection to be assessed.	
4 West and Equatorial Africa: Guinea-Bissau to Gabon; Rain forest.		4 Ivermectin through community-based distribution systems.		
5 Zaire Basin.		5 Ivermectin through community-based distribution systems.		
6 East Africa highlands: Ethiopia to Southern Malawi.		6 Focal vector control could achieve eradication of *S. damnosum* or *S. neavei*. Ivermectin through community-based distribution systems.		

(Continued over)

Table 5.8 *Stratification of epidemiology and control of onchocerciasis (Continued)*

Geographical locality	Epidemiological characteristics	Operational approach	Disease management	Prevention
Central America				
Guatemala; Mexico in well defined foci. Cross-border movement of migrant workers may spread the disease.	Anthrophilic *S. ochraceum* in highland foci; *S. metallicum* more zoophilic also involved. Inefficient as vector due to buccal armature: but compensated by vector abundance.	Control by ivermectin distribution via national programs through twice-yearly distribution.	Ivermectin-based.	Ivermectin distribution over years will reduce morbidity of skin and anterior segment eye disease
South America				
1 Brazil, S. Venezuela	1 Primary vectors: *S. oyapockense* and *S. guianense*. Secondary vector: *S. exiguum*.	Vector control not feasible. Integration of ivermectin with other programs.	Ivermectin-based.	Ivermectin distribution over years will reduce morbidity of skin and anterior segment eye disease.
2 N. Venezuela	2 Primary vector: *S. metallicum*. Secondary vector: *S. exiguum*.			
3 Ecuador, Columbia.	3 Primary vector: *S. exiguum*.			
Yemen				
Distributed along Wadis.	Vector: member of the *S. damnosum* complex (*S. rasyani*)	Ivermectin-based control initiated.	Ivermectin.	Ivermectin control embarked on.

Table 5.9 *Stratification of epidemiology and control of lymphatic filariasis (LF) (see Molyneux and Zagaria 2002)*

Geographical locality	Epidemiological characteristics	Operational approach	Disease management	Prevention
Tropical America: N E Brazil, Guyana, Surinam, French Guyana, Haiti, Dominican Republic, Costa Rica, Trinidad and Tobago.	Focal transmission by *Culex quinquefasciatus*. Varying levels of endemicity: Guyana endemic and about to launch program based on DEC salt. Haiti and Dominican Republic using DEC + albendazole. Surinam, Trinidad and Tobago, and Costa Rica have eliminated public health problem by DEC selective treatment	Limited control programs organized through Ministry of Health (in Brazil SUCAM). Urban mosquito control difficult. Strategy based on annual DEC + albendazole treatment of all eligible population. Introduction of home care for lymphedema	Opportunities for introduction of new approaches. Treatment using topical antibiotics, daily washing with soap and water of affected limbs thorough interdigital drying, limb elevation, and breathing. Single annual dose DEC + albendazole donated by GlaxoSmithKline	Very limited control implemented except in Recife. Potential for all appropriate measures to reduce transmission by single annual treatment of DEC + albendazole or DEC salt.
Tropical Africa: broad transmission zone in sub-Saharan Africa.	Urban filariasis *Culex quinquefasciatus* transmitted in E. Africa and Madagascar. Filariasis transmitted by *A. gambiae* complex and *A. funestus* in E. Africa and W. Africa. Mapping ongoing elsewhere to define implementation units for national control programs	Up-scaling control programs; nine countries active. Strategy based on ivermectin + albendazole where onchocerciasis and filariasis are co-endemic. DEC + albendazole elsewhere (Kenya, Zanzibar and Comoros).	Disease problems have been underestimated; hydrocoele surgery significant burden on district hospitals (30 percent of surgical interventions). Increase the development of home-based lymphedema management.	Pilot projects in Zanzibar using polystyrene beads for larval *Culex* control; impregnated bednets and, in urban areas, *Bacillus sphaericus* larval toxins. Development linkages and evaluate bednet programs on LF as part of malaria control.
Middle-East: Egypt	Urban and peri-urban transmission in Cairo and Nile delta. Resurgence since control relaxed in 1965 resulting in prevalence of c. 20 percent. *Culex molestus* and *C. quinquefasciatus* vectors.	DEC + albendazole campaigns active in all endemic villages in the Nile Delta (at-risk population 2.4 million) three rounds of drug distribution completed by end 2002. Vector control limited.	As above	Vector control likely to have limited impact; chemotherapeutic preventive approach will reduce transmission
Indian sub-continent: South Asia, India, Sri Lanka, Bangladesh, Nepal, Vietnam, Thailand, Myanmar.	*C. quinquefasciatus*, *Anopheles* species. Urban and semi-urban with high clinical morbidity particularly in India.	Variable depending on country and resources available. Chemotherapy via DEC + albendazole or DEC fortified salt. Mosquito control of limited impact.	As above. Surgery for serious deformities widely practiced in India, especially for hydrocoele, to reduce morbidity. Heavy cost burden on surgical services.	Mosquito control measures of limited impact. Define impact of DEC + albendazole on morbidity

(Continued over)

Table 5.9 *Stratification of epidemiology and control of lymphatic filariasisLF (see Molyneux and Zagaria 2002) (Continued)*

Geographical locality	Epidemiological characteristics	Operational approach	Disease management	Prevention
Pacific rim and islands	Aedes vector on Pacific islands. *Anopheles punctulatus* in Papua New Guinea, Vanuatu. *C. quinquefasciatus* in China. In the Solomon Islands *W. bancrofti* eliminated by malaria vector control in 1970s.	DEC + albendazole once a year for at least 5 years	As above.	Vector control difficult although some success in trials with *B. sphaericus* biocides. DEC salt and DEC + albendazole. Bednets contribute to reducing transmission in *Anopheles* settings (e.g. Vanuatu). Solomon Islands remain LF-free after elimination by indoor spraying of DDT in antimalaria campaign in 1980s
South Asia, India, Malaysia, Indonesia, China, Philippines, Vietnam. *Brugia malayi, Brugia timori*	*Mansonia, Anopheles.* In some areas animal reservoirs exist (cats and *Presbytis*, leaf monkeys).	Community participation in environmental vector control (weeding and fish culture) and chemotherapeutic control in India; use of DEC + ivermectin and DEC salt in same manner as for *Wuchereria bancrofti* filariasis.	As above. Treatment using topical antibiotics, hygiene of infected limbs in Kerala, India to ameliorate acute episodes.	Chemical control against *Mansonia* difficult due to variable adult behavior and larval association with weeds (dependence on environmental control).

Table 5.10 *Stratification of epidemiology and control of schistosomiasis*

Geographical locality	Epidemiological features	Operational approach	Disease and management	Prevention
Africa, South America, and Caribbean *Schistosoma mansoni, S. haematobium* widespread in Africa. *S. intercalatum* is limited in West and Central Africa. *S. mansoni* in South America and Caribbean	*S. mansoni/Biomphalaria* transmitted. *S. haematobium/Bulinus* transmitted. *S. mansoni* has peak prevalence in 10–24-year-olds; heaviest parasite loads 10–14 age group. *S. haematobium* peak prevalence and intensity 10–14-year-olds.	Linkage to other sectors in planning prevention and control via appropriate water-resource management. Population movement to be monitored in context of potential disease impact after exposure to new water-related development. Development of link to schools for chemotherapeutic control and for health education. Assessment of potential for mollusciciding.	Can result in significant morbidity and mortality, placing considerable burden on curative services. Severity of symptoms generally depends on intensity of infection. *S. mansoni* symptoms associated with hepatic, splenic, and intestinal systems. *S. haematobium* pathology concentrated in genital and urinary systems and lower intestine. Use of hematuria as rapid assessment tool for intervention priority.	Primary prevention is reduce access to contaminated water via improved management of water resources; involvement of other sectors in planning development projects. Chemotherapy assists in reducing infection reservoir; snail control may supplement control operations and be planned into irrigation schemes or impoundments. Health education via schools and community health providers. Maintenance of drug availability.
Middle East/North Africa *S. mansoni* and *S. haematobium.*			Praziquantel can reverse symptoms of severe disease. Carcinomas associated with schistosomal disease but precise etiology not established.	
South East Asia, China, Philippines, Thailand *S. japonicum, S. mekongi* in Kampuchea Laos.	*S. japonicum* transmitted by amphibious *Oncomelania* with many animal hosts (cattle, dogs, rats, pigs). *S. japonicum* has no typical age prevalence or intensity; variable depending on epidemiology. *S. mekongi* transmitted by *Neotricula* snails (dog reservoir).	National programs in China developed via communist structures in integrating chemotherapy and snail control. In Philippines, approaches to control through snail control, case detection and treatment, environmental sanitation, and health education. Redirection with increased emphasis with availability of praziquantel.	*S. japonicum* disease similar to *S. mansoni* with most severe disease between ages 2 and 40. Praziquantel effective in treatment.	

Operational phases of control operations: Phase 1 – planning data gathering; planning; resource allocation. Phase 2 – attack/intervention. Phase 3 – maintenance.

Table 5.11 *Recent epidemics of parasitic diseases, causes, impact and control*

Disease and geographical location	Putative cause	Epidemiological impact	Control problems and approaches
Malaria			
Madagascar, Highland malaria in Ethiopia, Kenya.	Change in agricultural patterns, interruption of control, environmental degradation (changing larval habitats), resettlement schemes, temperature change, and increased rainfall following period of drought. Non-immune population.	Over 25 000 deaths in Madagascar in 1988.	Institute epidemic preparedness policy. Ensure availability of appropriate drugs.
Forest-fringe malaria in South East Asia, South America.	Penetration into forest for gem and gold mining. Non-immune population exposed.	Epidemics in N.E. Kenya in 1999 following El Niño rains	Establish vector-control capacity and acquire insecticides.
Refugee camps.	New breeding sites; non-immune populations move to lower altitudes (e.g. Rwanda).		Inform and support local health structures and NGDOs.
Leishmaniasis			
Visceral leishmaniasis in South Sudan.	Ecological changes provide enhanced breeding sites for *P. orientalis* in *Acacia Balanites* woodland.	5 000 cases/year; 100 000 at risk – 75 percent mortality in children reported.	NGO-initiated treatment centers. Provision of emergency drugs. Limited diagnostic facilities. High incidence of post-kala azar dermal leishmanoid PKDL. Treatment established. Availability of drugs problematic.
Cutaneous leishmaniasis in Khartoum.	Population movement and breakdown of health services in conflict zone.	Outbreaks in Khartoum with 100 000 cases due to *L. major*.	Chemotherapy; vector control; rodent control.
Cutaneous leishmaniasis in Kabul, Afghanistan.	Lack of diagnostic capacity, drugs, and means of delivery.	New urban foci of transmission in *P. sergenti*. 4 000 cases in urban areas and 6 000 from provinces.	Active case detection and treatment; insecticide treatment of houses in cities; health education units.
Trypanosomiasis			
T. b. rhodesiense epidemic in Busoga, Uganda.	Change in ecology due to encroachment of *Lantana* following civil disturbance and change agricultural practices.	Up to 8 000 cases/year in 1980.	Surveillance and treatment. Establishment of treatment centers. Vector control via impregnated traps.
T. b. gambiense in Zaire–South Sudan, Angola	Breakdown of surveillance through disruption of health services.	Increased incidence.	Increase surveillance and treatment but coverage limited, drugs not available.
Schistosomiasis			
S. haematobium and *S. mansoni* epidemics increase associated with water-impoundment and rice irrigation throughout Africa.	Non-immune migrant population exposed for first time. *Bulinus senegalensis* breeds in irrigation systems; increase in snail populations. *Biomphalaria* increases. Major dam projects (Aswan, Egypt; Diama, Senegal and Akosombo, Ghana) are a well-recognized cause but micro-dams in West Africa provide additional and extensive new problems. War and refugee migration.	Increased incidence, high morbidity in non-immune populations.	Increase availability of chemotherapy. Target if possible through schools. Snail control or environmental management if feasible. Health education.

Table 5.12 *Widely used tools for parasitic disease control – diagnostics*

Disease	Diagnostic tools		Community assessment
	Personal diagnostic test (direct)	Personal diagnostic test (indirect)	
American trypanosomiasis		Serodiagnostic test using synthetic peptides.	Agglutination test for blood bank screening.
Dracunculiasis	Direct observation of worm emergence.		Use of village volunteers to identify cases and contain them.
Food-borne trematode infections	Kato technique (i.e. cellophane-fecal thick smear).	Intradermal screening for paragonimiasis.	
Leishmaniasis	Cutaneous: parasitological diagnosis by smear, culture. Visceral: parasitological diagnosis by bone-marrow puncture, spleen aspiration.	ELISA immunofluorescence. DAT.	
Lymphatic filariasis	Night blood films for nocturnally periodic forms; day blood films for diurnally periodic forms.	Assays to detect circulating filarial antigen by immunochromatographic test. 'Brugia Rapid' has similar characteristics for *Brugia*.	Hydrocoele survey; questionnaires to district hospitals for records; community assessment of lymphoedema rates.
Taeniasis, cysticercosis, neurocysticercosis	Fecal smear.	Coproantigen detection; CAT scan; NMR scan.	High prevalence of late-onset epilepsy; local pig husbandry, meat inspection, contaminated carcass detection.
Loa loa	Day blood films, presence of tropical eye worm or Calabar swelling.	No immunological methods available.	RAPLOA; use of remote sensing to determine areas of risk of high prevalence.
Malaria	Light-microscopic examination of Giemsa-stained blood films.	See Chapter 24, Malaria.	
Onchocerciasis	Skin snips.	DEC patch test as methodology for rapid diagnosis (mini Mazzotti).	Rapid assessment of nodule prevalence.
Schistosomiasis	Kato technique for: *S. intercalatum, S. japonicum,* and *S. mansoni.*	Quantitative urine filtration technique for *S. haematobium.*	Direct observation for gross hematuria, indicative of heavy *S. haematobium* infection in children in endemic areas. Detection of microhematuria by reagent strips (*S. haematobium*).
Sleeping sickness	Blood film. CSF examination. Bone-marrow aspiration.	Lymph gland palpation. CATT. Capillary tube centrifugation.	

Table 5.13 *Widely used tools for parasitic disease control – chemotherapy*

Disease	Treatment tools	
American trypanosomiasis	Acute phase: nifurtimox.	Blood bank transmission control: crystal violet/sodium ascorbate to kill parasites in infected blood.
Dracunculiasis	Treatment aimed only at superinfections.	
Food-borne trematode infections	Praziquantel.	
Leishmaniasis	Pentavalent antimonials.	
Lymphatic filariasis	DEC + albendazole outside areas of onchocerciasis endemicity.	Amphotericin B. Ivermectin + albendazole in onchocerciasis-endemic areas.
Malaria	Sporontocidal: proguanil, pyrimethamine, atovaquone.	Severe disease: quinine (+ tetracycline where needed). Artemether where quinine resistance.
	Tissue schizontocidal: primaquine, proguanil, tetracycline.	
	Gametocytocidal: primaquine, quinine, mefloquine, chloroquine, amodiaquine.	
	Blood schizontocidal: quinine, mefloquine, halofantin, chloroquine, amodiaquine, atovaquone, artemisinins, pyronaridine, tetracycline.	
Onchocerciasis	Ivermectin (single dose mass administration once a year). In the Americas, twice yearly treatment given.	
Schistosomiasis	Praziquantel.	Metrifonate. Oxamniquine.
Sleeping sickness	Early stage: pentamidine (*T. b. gambiense*). Suramin (*T. b. rhodesiense*).	Advanced stage: melarsoprol for both forms. Eflornithine if available for melarsoprol-resistant cases of *T. b. gambiense*. For arsenical-resistant cases of *T. b. gambiense* (mainly): eflornithine; nifurtimox has been produced for drug-resistant cases on humanitarian grounds.
Taeniasis, cysticercosis, neurocysticercosis	Praziquantel for *T. solium*.	Praziquental given as mass chemotherapy can reduce prevalence of cysticercosis.

Table 5.14 *Widely used tools for parasitic disease control – vector control*

Disease	Vector control tools		
American trypanosomiasis	Indoor residual spraying with pyrethroids.	Pyrethroid paint.	Fumigant canisters with pyrethroids.
Dracunculiasis	Temephos.		
Food-borne trematode infections	Environmental management, including land reclamation and drainage.	Niclosamide (molluscicide).	Biological control (natural predators).
Leishmaniasis	Residual insecticide spraying (DDT, pyrethroids).	Insecticide-treated nets have impact where transmission is in houses (Middle East *L. tropica* urban foci and in South America).	Insecticide-impregnated collars for dogs, which are reservoir hosts of *L. infantum* (Gavgani et al. 2002).
Lymphatic filariasis	Impregnated bednets in areas where disease is transmitted by *Anopheles*.	Polystyrene beads (if feasible) in settings where disease is transmitted by *Culex*; integrated control in Pacific where dengue control for *Aedes* is implemented.	Pyrethroid-impregnated bednets and curtains.
Malaria	Pyrethroid-impregnated bednets and curtains.	Indoor residual spraying (DDT, organophosphates, carbamates, pyrethroids).	Environmental management.
Onchocerciasis	Larviciding though aerial spraying (in rotational use: *B.t* H-14, temephos, pyraclofos, pyrethroids).	Ground larviciding employed in some settings where access is easy; *Simulium neavei* eliminated from several areas of East Africa by larvicides; deforestation reduces *S. neavei* populations.	
Schistosomiasis	Niclosamide (molluscicide).	Environmental management of irrigation systems.	
Sleeping sickness	Pyrethroid-impregnated screens.	Traps; pyrethroid-impregnated and non-impregnated.	

Table 5.15 *Examples of changes in* Plasmodium falciparum *:* P. vivax *ratios associated with anthropogenic change (conflict, irrigation, mining)*

Location	Projected changes	Reference
Tajikistan	Health systems disruption, conflict migration, chloroquine resistance in *P. falciparum*.	Pitt et al. (1998)
Afghanistan/Pakistan	Chloroquine resistance in *P. falciparum*.	Rowland and Nosten (2001)
Sri Lanka, Mahaweli	Irrigation on large scale.	Amerasinghe and Indrajith, 1994
Thar Desert, Rajasthan, India	Irrigation on large scale. Establishment of *Anopheles culcifacies* as efficient *P. falciparum* vector and dominance over *An. stephensi* (a poor vector).	Tyagi and Chaudhary (1997)
Amazonia, Brazil	New breeding sites for efficient vectors *An. darlingi* through mining, deforestation, road building.	Marques (1987)

REFERENCES

Amazigo, U.V., Brieger, W.R., et al. 2002. The challenges of community-directed treatment with ivermectin (CDTI) within the African Programme for Onchocerciasis Control (APOC). *Ann Trop Med Parasitol*, **96**, Supplement, S41–58.

Amerasinghe, F.P. and Indrajith, N.G. 1994. Postirrigation breeding patterns of surface water mosquitoes in the Mahaweli Project, Sri Lanka, and comparisons with preceding developmental phases. *J Med Entomol*, **31**, 516–23.

Antia, N.H. 1988. The Mandwa Project: an experiment in community participation. *Int J Health Serv*, **18**, 153–64.

Ashford, R.W., Kohestany, K.A. and Karimzad, M.A. 1992. Cutaneous leishmaniasis in Kabul: observations on a 'prolonged epidemic'. *Ann Trop Med Parasitol*, **86**, 361–71.

Ashford, R.W. and Thomson, M.C. 1991. Visceral leishmaniasis in Sudan. A delayed development disaster? *Ann Trop Med Parasitol*, **85**, 571–2.

Benton, B. and Skinner, E. 1990. Cost benefits of onchocerciasis control. *Acta Leiden*, **59**, 405–11.

Birley, M.H. 1995. *The health impact assessment of development projects*. London: HMSO.

Boatin, B., Molyneux, D.H., et al. 1997. Patterns of epidemiology and control of onchocerciasis in west Africa. *J Helminthol*, **71**, 91–101.

Buse, K. and Walt, G. 2000a. Global public–private partnerships: Part I – A new development in health? *Bull World Health Organ*, **78**, 4, 549–61.

Buse, K. and Walt, G. 2000b. Global public–private partnerships: Part II – What are the issues for global governance? *Bull World Health Organ*, **78**, 5, 699–709.

Buse, K. and Waxman, A. 2001. Public–private health partnerships: a strategy for WHO. *Bull World Health Org*, **79**, 748–54.

Cassels, A. 1997. *A guide to sector-wide approaches for health development: concepts, issues and working arrangements*, WHO/ARA/97.12. Geneva: World Health Organization.

CDC, 1993. Recommendations of the International Task Force for Disease Eradication. *MMWR*, **42**, 1–38.

Curtis, C.F. (ed.) 1991. *Control of disease vectors in the community*. London: Wolfe.

Curtis, C.F. 1994. Should DDT continue to be recommended for malaria vector control? *Med Vet Entomol*, **8**, 107–12.

Dowdle, W.R. and Hopkins, D.R. (eds) 1998. *The eradication of infectious diseases*. New York: John Wiley & Sons.

Drameh, P.S., Richards, F.O., et al. 2002. Ten years of NGDO action against river blindness. *Trends Parasitol*, **18**, 378–80.

Ekwanzala, M., Pepin, J., et al. 1996. In the heart of darkness; sleeping sickness in Zaire. *Lancet*, **348**, 1427–30.

Gadelha, M.C.A. 1993. Plan for the eradication of *Triatoma infestans* in Brazil (translated). *Rev Soc Bras Med Trop*, **26**, 27–32.

Garcia-Zapata, M.T.A. and Marsden, P.D. 1986. Chagas disease. In: Gilles, H.M. (ed.), *Clinics in tropical medicine and communicable diseases*. London: W.B. Saunders, 557–85.

Gavgani, A.S., Hodjati, M.H., et al. 2002. Effect of insecticide-impregnated dog collars on incidence of zoonotic visceral leishmaniasis in Iranian children: a matched-cluster randomized trial. *Lancet*, **360**, 374–9.

Gutteridge, W.E. 1993. Chemotherapy. In: Cox, F.E.G. (ed.), *Modern parasitology*, 2nd edn. Oxford: Blackwell Scientific Publications, 219–42.

Gwatkin, D.R., Guillot, M. and Heuvelin, P. 1999. The burden of disease among the global poor. *Lancet*, **354**, 586–9.

Hay, S.I., Cox, J., et al. 2002. Climate change and the resurgence of malaria in the East African highlands. *Nature*, **415**, 905–9.

Hopkins, A. 1998. Partnerships and distribution of Mectizan. Distribution of ivermectin in countries at war (translated). *Santé*, **8**, 72–4.

Hopkins, D.R., Ruiz-Tiben, E., et al. 2002. Dracunculiasis eradication: and now. Sudan. *Am J Trop Med Hyg*, **67**, 415–22.

Hopkins, D.R. and Withers, P.C. 2002. Sudan's war and eradication of dracunculiasis. *Lancet*, **360**, Supplement, S21–2.

Hougard, J.-M., Poudiougo, P., et al. 1993. Criteria for the selection of larvicides by the Onchocerciasis Control Programme in West Africa. *Ann Trop Med Parasitol*, **5**, 435–42.

Hunter, H.J., Rey, L., et al. 1993. *Parasitic diseases in water resources development. The need for intersectoral negotiations*. Geneva: World Health Organization.

Ijumba, J.N. and Lindsay, S.W. 2001. Impact of irrigation on malaria in Africa: paddies paradox. *Med Vet Entomol*, **15**, 1–11.

IPCC. 2001. *Climate change: synthesis report*. Watson, R.T. and Core Writing Team (eds). Cambridge: Cambridge University Press.

Katabarwa, N.M., Richards, F.O. and Ndyomugyenyi, R. 2000. In rural Ugandan communities the traditional kinship/clan system is vital to the success and sustainment of the African Programme for Onchocerciasis Control. *Ann Trop Med Parasitol*, **94**, 485–95.

Killick-Kendrick, R., Tang, Y., et al. 1997. Phlebotomine sandflies of Kenya (Diptera: Psychodidae). V. *Phlebotomus* (*Paraphlebotomus*) *mireillae* n.sp. *Ann Trop Med Parasitol*, **91**, 417–28.

Lightowlers, M.W. 1999. Eradication of *Taenia solium* cysticercosis: a role for vaccination of pigs. *Int J Parasitol*, **29**, 811–17.

MacCormack, C.P. 1991. Appropriate vector control in primary health care. In: Curtis, C.F. (ed.), *Control of disease vectors in the community*. London: Wolfe, 221–7.

Marques, A.C. 1987. Human migration and the spread of malaria in Brazil. *Parasitol Today*, **3**, 166–70.

McCullough, F.S. 1992. *The use of mollusciciding in schistosomiasis control*. Geneva: World Health Organization.

McKinsey and Co, Developing *successful* global health alliances. Seattle, WA: Bill and Melinda Gates Foundation.

Molyneux, D.H. 1995. Onchocerciasis control in West Africa: Current status and future of Onchocerciasis Control Programme. *Parasitol Today*, **11**, 399–402.

Molyneux, D.H. 1997. Patterns of change in vector-borne diseases. *Ann Trop Med Parasitol*, **91**, 827–39.

Molyneux, D.H. 2001. Sterile insect release and trypanosomiasis control: a plea for realism. *Trends Parasitol*, **17**, 413–14.

Molyneux, D.H. 2003a. Drivers of emergent parasitic diseases. *J Parasitol*, **89**, Suppl, S3–S13.

Molyneux, D.H. 2003b. Common themes in changing vector-borne disease scenarios. *Trans R Soc Med Hyg*, **97**, 129–32.

Molyneux, D.H. and Zagaria, N. 2002. Lymphatic filariasis elimination: progress in global programme development. *Ann Trop Med Parasitol*, **96**, Supplement 2, S15–40.

Moncayo, A. 1999. Progress towards the interuption of tansmission of Chagas disease. *Mem Inst Oswaldo Cruz*, **94**, Suppl. 1, S401–4.

Murray, C.J. and Lopez, A. *Global burden of diseases and injuries*, Vol. 1. Geneva: World Health Organization.

Murray, C.J.L., King, G., et al. 2002. Armed conflict as a public health problem. *Br Med J*, **324**, 346–9.

Ngoumou, P., Walsh, J.F. and Mace, J.-M. 1994. A rapid mapping technique for the prevalence and distribution of onchocerciasis: a Cameroon case study. *Ann Trop Med Parasitol*, **88**, 463–74.

Ostfeld, R.S. and Keesing, F. 2000. Pulsed resources and community dynamics of consumers in terrestrial ecosystems. *Trends Ecol Evol*, **15**, 232–7.

Patz, J.A., Hulme, M., et al. 2002. Climate change: Regional warming and malaria resurgence. *Nature*, **420**, 627–8.

Pitt, S., Pearcy, B.E., et al. 1998. War in Tajikistan and re-emergence of *Plasmodium falciparum*. *Lancet*, **35**, 2, 1279.

Reich, M.R. 2002. *Public–private: partnerships for public health*. London: Harvard University Press.

Rajagopalan, P.K., Paniker, K.N. and Das, P.K. 1987. Control of malaria and filariasis in South India. *Parasitol Today*, **3**, 233–41.

Remme, J.H.F. 1995. The African Programme for Onchocerciasis Control: preparing to launch. *Parasitol Today*, **11**, 403–6.

Remme, J.H.F., Blas, E., et al. 2002. Strategic emphases for tropical diseases research: a TDR perspective. *Trends Parasitol*, **18**, 421–6.

Rogers, D.J. and Randolph, S.E. 2000. The global spread of malaria in a future, warmer world. *Science*, **289**, 1763–6.

Roman, G., Sotelo, J., et al. 2000. A proposal to declare neurocysticercosis an international reportable disease. *Bull World Health Org*, **78**, 399–406.

Rowland, M. and Nosten, F. 2001. Malaria epidemiology and control in refugee camps and complex emergencies. *Ann Trop Med Parasitol*, **95**, 741–54.

Rubio-Palis, Y. and Zimmerman, R.H. 1997. Ecoregional classification of malaria vectors in the neotropics. *J Med Entom*, **34**, 499–510.

Samba, E.M. 1994. *The Onchocerciasis Control Programme in West Africa. An example of effective public health management*. Geneva: World Health Organization.

Seaman, J., Ashford, R.W., et al. 1992. Visceral leishmaniasis in southern Sudan: status of healthy villagers in epidemic conditions. *Ann Trop Med Parasitol*, **86**, 481–6.

Sotelo, J. 2003. Neurocysticercosis: eradication of cysticercosis is an attainable goal. *Br Med J*, **326**, 511–12.

Taylor, L.H., Latham, S.M. and Woolhouse, M.E.J. 2001. Risk factors for human disease emergence. *Philos Trans R Soc Lond B Biol Sci*, **356**, 983–9.

Thomson, M. 1995. *Disease prevention through vector control: guidelines for relief organisations*. UK and Ireland: Oxfam.

Tyagi, B.K. and Chaudhary, R.C. 1997. Outbreak of falciparum malaria in the Thar Desert (India), with particular emphasis on physiographic changes brought about by extensive canalization and their impact on vector density and dissemination. *J Arid Environ*, **36**, 541–55.

VCRC, 1992. *Control of Brugian filariasis*. Pondicherry: Misc Publ. Vector Control Research Centre.

Walsh, J.F., Molyneux, D.H. and Birley, M.H. 1993. Deforestation: effects on vector borne disease. *Parasitology*, **106**, 855–75.

Walt, G. 1988. CHWs: are national programmes in crisis? *Health Pol Plan*, **3**, 1–21.

Weil, G., Lammie, P.J. and Weiss, N. 1997. The ICT filariasis test: a rapid-format antigen test for diagnosis of bancroftian filariasis. *Parasitol Today*, **13**, 401–4.

Widdus, R. 2001. Public–private partnerships for health: their main targets, their diversity, and their directions. *Bull World Health Organ*, **79**, 713–20.

Winnen, M., Plaisier, A.P., et al. 2002. Can ivermectin mass treatments eliminate onchocerciasis in Africa? *Bull World Health Organ*, **80**, 384–91.

WHO, 1979. *Parasitic zoonoses*, WHO Tech Rep Ser 637. Geneva: World Health Organization.

WHO, 1990. *Control of the leishmaniases*, WHO Tech Rep Ser 793. Geneva: World Health Organization.

WHO, 1991. *Control of Chagas diseases*, WHO Tech Rep Ser 811. Geneva: World Health Organization.

WHO, 1992a. *Lymphatic filariasis. The disease and its control.* WHO Tech Rep Ser 721. Geneva: World Health Organization.

WHO, 1992b. *The use of essential drugs. Model list of essential drugs.* WHO Tech Rep Ser 825. Geneva: World Health Organization.

WHO, 1993a. *A global strategy for malaria control.* Geneva: World Health Organization.

WHO, 1993b. *The control of schistosomiasis*, WHO Tech Rep Ser 830. Geneva: World Health Organization.

WHO, 1994. *Lymphatic filariasis infection and disease control strategies. Report of a consultative meeting. Penang Malaysia.* Division of Control of Tropical Diseases (CTD) and UNDP/World Bank/WHO Special Programme for Research and Training in Tropical Diseases (TDR). TDR/CTD/FIL/PENANG. Geneva: World Health Organization.

WHO, 1995a. *Health of school children. Treatment of intestinal helminths and schistosomiasis*, WHO/SCHISTO/95.112. Geneva: World Health Organization.

WHO, 1995b. *WHO model prescribing information. Drugs used in parasitic diseases*, Vol. 2. Geneva: World Health Organization.

WHO, 1995c. *Vector control for malaria and other mosquito-borne diseases*, WHO Tech Rep Ser 857. Geneva: World Health Organization.

WHO, 1998. Global disease elimination and eradication as public health strategies. *Bull World Health Organ*, **76**, Suppl 2.

WHO, 2001. *Report of the commission on macroeconomics and health*, Chair Jeffrey D. Sachs. Executive summary. Geneva: World Health Organization.

WHO, 2002a. Urbanization: an increasing risk factor for leishmaniasis. *Wkly Epidemiol Rec*, **77**, 365–70.

WHO, 2002b. Dracunculiasis eradication. *Wkly Epidemiol Rec*, **77**, 141–52.

World Bank, 1993. *World development report. Investing in health.* Oxford: Oxford University Press.

World Bank, 1994. *Development in practice. Better health in Africa.* Washington: The World Bank.

Yameogo, L., Resh, V. and Molyneux, D.H. 2004. The control or river blindness in West Africa: A case history of biodiversity in a disease control programme. *Ecohealth*, in press.

6

Diagnosis of parasitic infections

STEPHEN H. GILLESPIE

INTRODUCTION

The ubiquity of human parasites is such that their diagnosis should be of importance to all involved in the management of patients with infection. Yet this branch of diagnostic microbiology is often strangely neglected. Parasite diagnosis frequently poses unique problems requiring novel solutions. The diversity of agents requires, on the part of the diagnostician, an encyclopedic knowledge covering pathogens that are imported and genuinely rare. Also, it requires the application of the full range of microbiological diagnostic methods. The majority of parasite diagnoses are made in a laboratory whose major interest is bacteriology but there are a number of important differences in the approach required for success. This chapter outlines the factors necessary for successful diagnosis of parasitic infections.

Life-cycle and natural history

Humans often form only one part of a complex parasitic life cycle. It is important to understand the way in which the organisms interact with the host if optimal diagnosis is to be obtained. For example, *Toxocara canis* is the ascarid parasite of canids and is unable to complete its life cycle in humans and other paratenic hosts (Gillespie et al. 1993). The diagnosis can be made in canids by examining stools for the presence of the characteristic eggs (Habluetzel et al. 2003). In contrast, in paratenic hosts such as the human, this approach will fail because parasite development is blocked at the L2 larval stage and adults cannot be found in the stool; it is thus essential to diagnose this infection by serological means.

Giardia has a simple life cycle which may only involve humans (and potentially an animal reservoir). The diagnosis is made simply by examining stools for the presence of *Giardia* cysts. However, a proportion of cases may not be diagnosed by this means and symptoms consistent with giardiasis may persist. In this instance, the fact that the protozoa preferentially infect the duodenal–jejunal junction means that this site can be sampled directly, either by duodenoscopy and biopsy or by the string test (see String test below).

Knowledge of the life cycle and natural history can indicate when to sample. For example, patients who have been exposed to schistosomiasis by swimming in contaminated water may present soon after their return from an endemic country. They may already have the symptoms and signs of Katayama fever. However, it would be futile to attempt to make a diagnosis by microscopic examination of stool or urine as it is likely that adults will not have matured and begun to deposit eggs. Later in the infection, the excretion of eggs may not be regular: in *Schistosoma haematobium* infection, eggs are preferentially excreted around mid-day. To obtain maximum yield, urine should be collected at this time. Similar considerations apply to the diagnosis of filarial infections where these parasites have become adapted to the circadian rhythms of the host. In areas where mosquitoes are night-biting, the maximum diagnostic yield can be obtained by sampling between midnight and 2 a.m. (Simonsen et al. 1997).

The pathogenic mechanisms of the parasite may affect the choice of specimen. *Plasmodium falciparum* produces knobs on the surface of the red cell that make it adhere to capillary endothelium, a point of considerable importance in the pathogenesis of cerebral malaria (Waller et al. 1999). It also has importance for laboratory diagnosis as capillary blood has a higher diagnostic yield than the venous blood traditionally sampled by doctors in developed countries. The microfilariae of *Onchocerca volvulus* can be found in the skin, but their distribution is not uniform. Knowledge of the normal distribution of microfilariae and the variations in different geographical locations will increase the diagnostic yield and reduce the number of (painful) pinch biopsies required to make a diagnosis (Scheiber et al., 1976).

Resources

The world is ill-divided, especially when it comes to infectious diseases. The poorest members of the world community have a disproportionate number of the world's parasites. Thus, those with the greatest need of parasite diagnosis have the least economic resource to acquire it (World Bank 1993). This financial fact of life is one that those charged with parasite diagnosis must bear in mind in delivering diagnostic services. It means that resources should be targeted on the most important clinical problems. For example, concentration techniques increase the yield of positive results for protozoal cysts and helminth eggs, but the cost of reagents and the expense of purchase and maintenance of a centrifuge may mean that simple wet preparations should be employed (Ramsay et al. 1991). Such compromises may not be as damaging as a superficial assessment might suggest. For individuals living in an endemic area and constantly at risk of acquiring infection, sensitive diagnosis may result in many treatment courses where the balance of benefit between parasite clearance and drug adverse events is uncertain.

The same considerations must be made in relation to technology. A range of diagnostic approaches can be applied to protozoa and helminths, but many of these methods are too expensive to be applied in the areas where the infections are most common.

Objectives of parasitic diagnosis

PATIENTS FROM NON-ENDEMIC AREAS

Many individuals now travel to countries where tropical parasites are endemic. Returning travelers will have some clear diagnostic demands. They may be very concerned about the risk of tropical parasites, no doubt enhanced by lurid stories from hardened expatriates. Their objective in presenting for diagnosis is to exclude the risk of an infection which is not yet causing symptoms (MacLean and Libman 1998). It may be that such a disease is evolving but the traveler is presenting before the infection has had time to become patent. Moreover, as infections in travelers are often light, the disease may present in a more subtle way than that described in the textbooks of parasitology and tropical medicine.

Travelers often lack the immunity that inhabitants of endemic countries possess. This may have a significant impact on the natural history of infection and a change in the speed at which diagnosis is required. *P. falciparum* is rapidly progressive in immunologically naïve subjects. Thus, when malaria is suspected in such a patient, the diagnosis of malaria should be made rapidly.

Patients from an endemic country with *Entamoeba* cysts in their stool rarely need further investigation. However, for a patient living in a country where this disease is rare, it would be useful if a species-specific diagnosis could be made. It is normal practice to attempt to eradicate cyst passage in returning travelers. To prevent unnecessary treatment in patients with *Entamoeba dispar* infection, a species-specific diagnosis is required (Sanchez-Guillen et al. 2002).

PATIENTS IN ENDEMIC AREAS

Most health facilities in endemic areas experience a severe shortage of resources for medical services. These scarce resources are often disproportionately directed towards clinical services. This frequently means that laboratories charged with the task of making a diagnosis lack the necessary resources to do so. Hospital managers often do not recognize that resources spent in obtaining an accurate diagnosis may produce savings in medical and pharmacy costs. Faced with this, it is essential that laboratory managers choose wisely which diseases to target and which methods to employ. Many patients have had to travel long distances to access healthcare. This means that, where possible, methods should be employed that provide a rapid diagnosis. The patients are now often required to make a financial contribution towards the cost of their laboratory costs and this may also serve to limit the scope of testing. These considerations do not necessarily mean that a poor quality service is the inevitable result. Although recent diagnostic developments have achieved similar sensitivity and specificity as a thin blood film examined by a trained microscopist, the latter remains the optimum diagnostic method for most parts of the world (Humar et al. 1997). Also, although limited resources probably limit stool examination to a direct saline wet preparation, this method will reliably diagnose the individuals that are most heavily infected with helminths and are thus most likely to benefit from treatment (Ramsay et al. 1991).

DIAGNOSIS FOR PUBLIC HEALTH

The approach to diagnosis required for public health purposes is often very different from that for routine curative services. In many instances, the questions posed are quite dissimilar and this has an impact on the methods employed. For example, for a patient who is thought to have schistosomiasis, the identification of a single egg in stool or urine is all that is required to initiate therapy. However, in a schistosomiasis control program, although this sort of result will be useful in quantifying the number of patients with infection, it is more likely that it will be necessary to quantify the worm burden of people living in infected communities. Thus, for *S. mansoni* infections, it would be necessary to perform a Kato-Katz investigation that would give a useful estimate of the number of eggs per gram, and thus the overall worm burden (Katz et al. 1972). To characterize the nature of hookworm infection, culture is necessary but this would have no impact on the management of the patient's illness (Harada and Mori 1955). In addition, parasite field studies often need to process specimens from a very large number of patients in a short time and methods must be adapted to this requirement.

Serology has limited value in the diagnosis of parasitic infections in endemic areas, but serological techniques are often employed in public health surveys as a way of rapidly determining the number of patients who have current or past infection with a particular pathogen.

In determining the epidemiology or epizoology of parasitic infections, it may be necessary to employ complex typing techniques to characterize pathogens identified in different communities. For example, the identification of a species of *Leishmania* might be sought so that its method of transmission could be better understood for control purposes. To achieve this, microsatellite DNA analysis and multilocus enzyme electrophoresis can be employed (Banuls et al. 2002; Barker 2002). Such techniques are not required for diagnosis in routine practice.

DIAGNOSIS FOR RESEARCH

Good-quality field research of parasitic infection should ask a very specific question for which diagnostic methods may be required. For example, if the nature of drug-resistant malaria is under investigation, it might be necessary to deploy DNA amplification techniques to determine the presence of various drug-resistance genes (Dorsey et al. 2001).

SPECIMENS

Any tissue or body fluid can be used for the diagnosis of parasitic infections. However, it is essential that the right specimen is obtained for diagnosis and in the right condition. It is pointless to spend time searching a stool for the presence of amebic trophozoites unless the stool is both fluid and fresh: amebic trophozoites rapidly lose motility as the temperature falls. Although females of *Trichinella spiralis* are found in the intestine and produce symptoms of diarrhea, the diagnosis of trichinosis is made by serology or by the examination of a muscle biopsy (see Chapter 37, *Trichinella*).

Stool samples

A minimum of three samples of stool is necessary to make a diagnosis of intestinal infection. As excretion of parasite eggs and cysts can be intermittent, it is preferable that specimens be sent on different days. To identify the trophozoite stage of *Entamoeba* and *Giardia*, freshly passed stools should be examined. If this is not possible, it may be necessary to fix the specimen for later examination. It may not be possible to make a diagnosis of amebiasis from a stool sample but motile trophozoites may be observed in scrapings from intestinal ulcers visualized on sigmoidoscopy.

Duodenal/jejunal biopsy or Enterotest may be needed to sample the jejunal fluid to diagnose giardiasis (Goldsmid and Davies 1978). A summary of the parasites that can be diagnosed using intestinal samples is listed in Table 6.1. Each of the modalities of laboratory diagnosis may need to be employed.

MACROSCOPIC EXAMINATION

It is essential to note the consistency of the stool to be examined. This can guide the technologist as to whether or not there is purpose in trying to identify trophozoites

Table 6.1 *Methods and pathogens that can be identified in intestinal specimens using these methods*

Sample	Method	Examples of pathogens identified
Stool	Microscopic examination of unstained and stained preparations	Protozoa: *Giardia lamblia*, *Entamoeba histolytica*, *Cryptosporidium parvum*, *Isospora belli*, Microsporidia Helminths: *Trichuris trichiura*, hookworms, *Ascaris lumbricoides*, *Schistosoma* spp.
	EIA, nucleic acid amplification	*Cryptosporidium parvum*, *E. histolytica*, *G. lamblia*
	Culture	*Strongyloides stercoralis*, speciation of hookworm
Adhesive tape sample, anal swab	Direct microscopy	*Enterobius vermicularis*
String test, duodenal fluid examination	Microscopy, EIA PCR	*G. lamblia*, *Cryptosporidium parvum*, *S. stercoralis*
Sigmoidoscopy	Direct examination and ulcer scrapings	*E. histolytica*
Proctoscopy	Rectal biopsy	*Schistosoma mansoni*

or cysts. Also, a negative result is often more easily explained to a clinician suspecting diarrheal disease if the fact that the stool is formed is noted. The presence of proglottids can be confirmed by macroscopic examination of the stool and sometimes adult ascarid worms can be found in stools.

MICROSCOPIC EXAMINATION

Direct saline wet preparations have the virtue of speed and cheapness. Unfortunately, this method of examination is less sensitive than the concentration techniques. Direct saline preparations are required if the trophozoite stages are to be seen. Quantification of the number of eggs present in stools may be achieved by variants of the Kato-Katz method (Katz et al. 1972).

Staining of stools can improve the diagnostic yield by making the target pathogen more easy to identify. Modified acid fast preparations or the use of UV illumination for the examination of auramine-stained preparations make the identification of *Cryptosporidium* and *Isospora* much easier (see Chapter 20, Intestinal coccidia: Cryptosporidiosis, isosporiasis, cyclosporiasis). In the same way, modified trichrome, Starey-Warthin, and calcofluor white may assist the diagnosis of microsporidiosis (see Chapter 25, Microsporidiasis). Rapid specific diagnosis can be obtained by the use of direct immunofluorescence (Kuczynska et al. 2003). For the diagnosis of microsporidiosis, direct immunofluorescence is probably essential. These techniques have proved of particular value for the diagnosis and quantification of intestinal protozoa in water supplies.

To improve the sensitivity of diagnosis, stools should be concentrated. A wide variety of methods is reported in the literature. Many of these methhods require expensive reagents and equipment such as a centrifuge, but others have been adapted for resource-poor communities (Ramsay et al. 1991). Different techniques are required for different pathogens and the optimal method should be employed for any specific target pathogen in, for example, a research study. To overcome the need to employ more than one concentration method, many laboratories choose to use an all-purpose method that gives good all-round performance. The formol–ether method and its variants are widely used as they provide good concentration of protozoal cysts and helminth eggs while also removing fecal debris (Allen and Ridley 1970). Commercially prepared centrifugation tubes can make this technique easier to apply.

IMMUNOLOGICAL AND MOLECULAR DIAGNOSTICS

Enzyme immunoassays can also be used to process stool samples. These methods have the advantage that they can be automated. Commercially produced diagnostic tests are now available for most intestinal protozoan pathogens (Garcia et al. 2000). In the same way, primer pairs for the main intestinal protozoa are now available and provide enhanced sensitivity. With the appearance of real-time polymerase chain reaction (PCR) equipment, it is now possible to make a rapid specific diagnosis of intestinal protozoal infections using these techniques (Bialek et al. 2002).

CULTURE METHODS

One of the major differences between parasitology and most other branches of microbiology is the fact that culture methods are rarely used. However, there are a limited number of situations where cultures of parasites are essential for a complete diagnosis. *Strongyloides* larvae are difficult to identify in stools, and an improved yield can be obtained by culturing stools. Cultures can reveal the presence of larvae when their numbers are low (Arakaki et al. 1990). Culture is also essential to permit species identification of hookworms (Harada and Mori 1955).

The viability of schistosome eggs can be determined by microscopic examination to see the cilia of the 'flame' cell. Alternatively, it is possible to hatch the miracidium from the eggs – the ultimate proof of viability. Fresh stool or urine specimens are placed in ten volumes of unchlorinated water (e.g. spring water) in an Erlenmeyer flask. The flask is covered and the sediment examined after 24 hours for the presence of miracidia.

SIGMOIDOSCOPY

The value of ulcer scrapings for diagnosis of amebiasis has already been discussed. In addition to this, rectal biopsy samples obtained at sigmoidoscopy or proctoscopy can be valuable in the diagnosis of schistosomiasis, especially in travelers when the number of eggs present in stools may be small (Harries et al. 1986).

STRING TEST

This technique depends on a length of nylon 'string' with a weighted end. This is swallowed and travels to the upper part of the jejunum. After a period of time, the string is pulled up and the fluid expressed can be examined. Bile-staining of the string can be read as proof that the string has traveled as far as the jejunum. Commercial versions of this device are available and the method can be used to make a diagnosis of giardiasis, strongyloidiasis, and cryptosporidiosis (Beal et al. 1970).

PINWORM

Pinworms, *Enterobius vermicularis*, lay their eggs on the skin around the anus. The optimum diagnostic yield is obtained from the perianal skin. Eggs can be harvested by using adhesive cellulose tape and then examined microscopically. Several commercial kits are available that make the diagnosis of pinworm more easy (Wagner and Eby 1983).

Urogenital samples

MACROSCOPIC EXAMINATION

Urine should be examined macroscopically where overt hematuria associated with urinary schistosomiasis can be observed. In cases of lymphatic filariasis, abnormal connections between the thoracic duct and ureters can permit leakage of lymph into the urine. This phenomenon, chyluria, is demonstrated by the excretion of milky urine after the patient has taken a fatty meal.

MICROSCOPIC EXAMINATION

Eggs of *Schistosoma haematobium* are mainly excreted into the urine were they can be demonstrated by microscopic examination. It is possible to improve the sensitivity of the method by expressing a larger volume of urine through a micropore membrane. Eggs are trapped in the membrane and can be seen under the microscope. An estimation of the worm burden can be obtained for research purposes by filtering a standard volume, usually 10 ml (Peters et al. 1976).

The diagnosis of vaginitis or urethritis caused by *Trichomonas vaginalis* is usually made by examination of vaginal discharge fluid or urethral pus. A simple saline wet preparation may permit the observation of the characteristic motile protozoa (see Figure 13.1, p. 256). Diagnosis can be made more sensitive by the use of direct immunofluoresence and culture (Draper et al. 1993).

CULTURE

Urine from patients infected with *S. haematobium* can be incubated overnight and the sediment observed for the presence of viable miracidia. The diagnosis of trichomoniasis can be enhanced by culture of vaginal and urethral secretions (see Chapter 13, *Trichomonas* infections).

Blood

Blood is a vital specimen for the diagnosis of parasitic infections because many organism have a blood stage as part of their life cycle. These are summarized in Table 6.2. It is essential that the specimens are taken at the right time. Malaria parasites are characteristically detectable in the blood during a fever and for some hours afterward. When a patient presents with fever and a history of residence or travel in an endemic area, it is essential to take a blood film for malaria immediately; however, a minimum of three samples should be obtained when the patient is febrile. Failure to follow this rule may result in a false negative diagnosis. In cases of lymphatic filariasis, there is diurnal variation in the presence microfilariae in blood. For this reason, it is essential to obtain blood samples at the correct time and to take the samples into bottles containing ethylenediaminetetraacetic acid (EDTA) (Simonsen et al. 1997).

CONCENTRATION TECHNIQUES

Many parasites may be present in blood in low numbers, which makes concentration techniques essential. The simplest of these is the thick film: red cells stacked on top each other are lysed by the hypotonic Field's stain – this allows a large number of red cells to be scanned quickly for the presence of the characteristic ring-shaped merozoites of malaria parasites. Centrifugation and fluorescence microscopy can be used to enhance the sensitivity of diagnosis. Using fluorescent antibodies makes this technique more sensitive for the diagnosis of *P. falciparum* and *P. vivax* (Forney et al. 2003). Buffy-coat concentration may permit the diagnosis of *Leishmania donovani* and African trypanosomes (Woo 1970). An improved concentration technique for African trypanosomes is to centrifuge blood through a DEAE column (Lumsden et al. 1979). To concentrate microfilariae, a measured volume of blood can be filtered through a micropore membrane (Bell 1975; Desowitz et al. 1973). The membrane is then stained and examined microscopically to enable species identification. This approach has the dual advantage of increasing the sensitivity of the technique while permitting the numbers of organisms present to be counted.

CULTURE

There are few culture techniques for blood samples because most of the organisms require the intervention of intermediate vectors. However, specimens of blood and buffy coat can be cultured for *Leishmania* on media such as Schneider's insect medium. After two weeks, promastigotes can be visualized. The sensitivity of blood culture for *Trypanosoma cruzi* can be improved by

Table 6.2 *Protozoan and helminth parasites that can be identified in blood*

Specimen	Technique	Pathogens
Whole or EDTA blood	Thick and thin films	*Plasmodium* spp., *Leishmania donovani* and *L. infantum*, *Trypanosoma brucei rhodesiense* and *T. b . gambiense*. *Babesia* spp., various microfilaria
Buffy coat	Stained film	*Leishmania* spp., *Trypanosoma* spp., microfilariae
EDTA blood	DEAE concentration	Trypanosomes
	Membrane filtration	Microfilariae

xenodiagnosis. Uninfected triatomid bugs are permitted to feed on a suspected patient for approximately 30 minutes. The bugs are then kept and sacrificed monthly for 3 months. The guts are dissected and examined microscopically for the presence of trypanosomes.

Biopsy specimens

In many instances, the life-cycle of the parasite means that the only effective means of diagnosis is biopsy. For example, *Leishmania donovani* can be obtained from concentrated buffy coat cells but this is less sensitive than biopsy of bone marrow or splenic fluid where the organisms are concentrated. For many parasitic infections, biopsy is the only practical means of making the diagnosis (e.g. cutaneous leishmaniasis or onchocerciasis). A list of the most common parasitic infections diagnosed by biopsy is listed in Table 6.3.

SKIN

Punch biopsies or 'skin snips' are the investigation of choice for onchocerciasis. Specimens should be obtained from areas of skin that exhibit the characteristic dry 'chicken skin' appearance that has been associated with infection. In addition to any suspect areas, samples should be taken from the shoulder area, from the buttocks and from the calves even when the skin appears normal. The biopsies are then placed in saline for transport to the laboratory. The microfilariae start to emerge after 30 minutes to 1 hour and can be identified by their characteristic movement. In addition, stained preparations permit the identification of species. The diagnostic yield can be increased by giving the patient a small dose of diethylcarbamazine, which stimulates the migration of the microfilariae.

The diagnosis of cutaneous leishmaniasis depends on a biopsy of the suspected ulcer. Material can be obtained for microscopy or culture using a slit skin smear. A touch preparation from a biopsy of the ulcer edge can be made. Alternatively, the biopsy material can be cultured in Schneider's insect medium for 2 weeks.

MUSCLE

Trichenella spiralis larvae are located in the striated muscle. Diagnosis can be made by muscle biopsy and then pressing the material between two slides for examination under the microscope.

BRAIN

Biopsy of the brain may form part of the diagnostic process for a number of parasitic infections where various stages may be found in the brain. Biopsy may be performed to diagnose the etiology of ring-enhancing lesions in the brain demonstrated by computed tomography (CT) scans or magnetic resonance imaging (MRI). This situation often arises in patients with late stage human immunodeficieny virus (HIV) infection with low CD4 cell counts. The differential diagnosis includes lymphoma and cerebral toxoplasmosis. Other parasites that may be observed in brain biopsy include tissue stages of *Taenia solium*, *Echinococcus multilocularis*, *Entamoeba histolytica*, and eggs of various schistosomes.

IMMUNOLOGICAL DIAGNOSIS

Antibody detection

Immunological diagnosis is an attractive approach to parasitic diagnosis but has only established itself in a limited number of diagnostic areas. This is not surprising since the immune response to protozoan and metazoan pathogens is often complex and many of these organisms have evolved methods to subvert the immune response.

Table 6.3 *Summary of techniques and pathogens identified with biopsy specimens*

Specimen	Technique	Pathogen
Skin	Slit skin smear	*Leishmania* spp.
	Pinch biopsy	*Onchocerca volvulus*
Brain	Microscopic examination of stained smears, culture, EIA PCR	*Toxoplasma gondii*, *Naegleria fowleri*, *Acantamoeba*, microsporidia, cysticercosis, *Echinococcus* spp.,
Muscle	Direct microscopy	*Trichinella spiralis*
Eye	Microscopy of stained samples	Microsporidia, *Acanthamoeba*
Vitreous or aqueous	EIA	*Toxocara canis*
Lymph node	Microscopic examination of stained material	*Trypanosoma* spp., *Toxoplasma gondii*, *Leishmania donovani*
Liver	Microscopic examination of stained preparation	*Toxoplasma*, *Naegleria*, *Entamoeba histolytica*, Microsporidia, *Opisthorchis sinensis*
	Examination of cyst fluid by microscopy and EIA	*Echinococcus granulosus*
Bone marrow	Microscopy of stained material and culture	*Leishmania donovani* and other species

Patients in many tropical countries have raised concentrations of polyclonal, low-affinity antibody that tends to produce false positive results in many immunological tests. Also, patients in endemic countries have life-long exposure to pathogens and many have serological evidence of previous infection making it impossible to identify a new infection.

TOXOPLASMOSIS

Toxoplasmosis is one disease for which immunological diagnosis is the method of choice. Although the organism is relatively easy to culture for sufficiently sensitive diagnosis, animal inoculation is required and, therefore, this approach to investigation is limited to the processing of precious specimens. Evidence of recent toxoplasmosis can be demonstrated using the Sabin-Feldman dye test or one of many commercial assays to detect immunoglobulin (Ig)M (Sabin and Feldman 1948; Evans and Ho-Yen 2000). The age of the infection can also be determined by measuring the affinity of serum IgA.

AMEBIASIS

Antibodies to E. histolytica can be detected by a variety of techniques including indirect immunofluorescence assay (IFA) and indirect hemagglutination assay (IHA). These titers are often negative in patients with intestinal disease but are elevated in patients with invasive infection. Acute diagnosis of amebic abscess can be assisted by the use of the cellulose acetate precipitin test, which becomes positive much earlier than the IFA and IHA.

MIGRATING WORMS

Most of the organisms responsible for larval migration are unable to complete their life cycle in the human host. As a consequence, it may not be possible to identify these organisms by conventional phenotypic methods. An example of such a parasite is Toxocara canis; the diagnosis is made by detection of antibodies specific to the excretory secretory antigens (de Savigny et al., 1979). Other examples include Trichinella, dog hookworm infection, and gnathosomiasis.

RETURNING TRAVELERS

Although of little value in endemic countries, many serological tests are of diagnostic value in returning travelers because the time frame of their exposure to the infective agents is known. In addition, the degree of the infection is frequently light, which makes diagnosis by conventional morphological methods more difficult. As travelers are unlikely to have been exposed to organisms such as schistosomiasis or filariasis previously, the detection of antibodies to such parasites could be diagnostic. Alternatively, because many travelers seek medical help to exclude tropical infections, a negative result may be seen as even more valuable in that it allows the patient to be reassured.

EVALUATION OF SPACE-OCCUPYING LESIONS

Detection of antibodies to T. solium, Echinococcus granulosis and Echinococcus multilocularis may assist the difficult diagnosis of these infections. Several enzyme immunoassays (EIA) have been described (Feldman et al. 1990).

Antigen detection

Direct detection of parasite antigens, as opposed to the detection of antibody responses, has found a regular place in diagnosis.

MALARIA

A simple dipstick presentation kit is available for the rapid diagnosis of malaria. Although the cost of this kit may restrict its application in resource-poor countries, it does provide a rapid and reliable diagnosis of P. falciparum and P. vivax malaria without the need for a microscope or EIA reader. It is probably a little less sensitive than the results obtained by an expert microscopist. It is indicated for the rapid diagnosis of malaria in laboratories that do not process many specimens for malaria. It would allow a rapid reliable diagnosis that could be confirmed by a reference facility.

INTESTINAL PROTOZOA

Commercially produced kits with a IFA or EIA format are available for the diagnosis of Giardia lamblia, Cryptosporidium parvum, Microsporidia, and Entamoeba histolytica infection. The latter method is especially valuable as E. histolytica and E. dispar cannot be distinguished under the light microscope. These methods appear to be at least as sensitive as light microscopy and, if the EIA format is employed, also permit large numbers of specimens to be processed (Ong et al. 1996).

MOLECULAR DIAGNOSIS

Although many molecular diagnostic methods have been described, molecular diagnostics have little part to play in diagnosis for the majority of parasitic infections. However, they do have considerable application in public health microbiology and in research when it is necessary not only to identify the presence of a pathogen but also to determine its subtype. Molecular methods can also be used to investigate the mechanisms of resistance to antiparasitic drugs. This has proved especially valuable for malaria and for intestinal nematode infections.

REFERENCES

Allen, A.V.H. and Ridley, D.S. 1970. Further observations on the formol-ether concentration technique for fecal parasites. *J Clin Pathol*, **23**, 545.

Arakaki, T.J., Iwanaga, M., et al. 1990. Efficacy of agar-plate culture in detection of *Strongyloides stercoralis* infection. *Parasitology*, **76**, 425.

Banuls, A.-L., Hide, M. and Tibayrenc, M. 2002. Evolutionary genetics and molecular diagnosis of *Leishmania* species. *Trans R Soc Trop Med Hyg*, **96**, Suppl. 1, 9–13.

Barker, G.C. 2002. Microsatellite DNA: a tool for population genetic analysis. *Trans R Soc Trop Med Hyg*, **96**, Suppl. 1, 21–4.

Beal, C.B., Viens, P. and Grant, R.H. 1970. A new technique for sampling the duodenal contents: demonstration of upper bowel pathogens. *Am J Trop Med Hyg*, **19**, 349–52.

Bell, D.R. 1975. Diagnosis of parasitic diseases by filtration. *Ann Soc Belg Med Trop*, **55**, 489–96.

Bialek, R., Binder, N., et al. 2002. Comparison of fluorescence, antigen and PCR assays to detect *Cryptosporidium parvum* in fecal specimens. *Diagn Microbiol Infect Dis*, **43**, 283–8.

Desowitz, R.S., Southgate, B.A. and Mataika, J.U. 1973. Studies on filariasis in the Pacific. 3. Comparative efficacy of the stained blood-film, counting-chamber and membrane-filtration techniques for the diagnosis of *Wucheria bancrofti* microfilaraemia in untreated patients in areas of low endemicity. *Southeast Asian J Trop Med Public Health*, **4**, 329–5.

de Savigny, D.H., Voller, A. and Woodruff, A.W. 1979. Toxocariasis serological diagnosis by enzyme immunoassay. *J Clin Pathol*, **32**, 284–8.

Dorsey, G., Kamya, M.R., et al. 2001. Polymorphisms in the *Plasmodium falciparum* pfcrt and *pfmdr*-1 genes and clinical response to chloroquine in Kampala, Uganda. *J Infect Dis*, **183**, 1417–20.

Draper, D., Parker, R., et al. 1993. Detection of *Trichomonas vaginalis* in pregnant women with the InPouch TV culture system. *J Clin Microbiol*, **31**, 1016–18.

Evans, R. and Ho-Yen, D.O. 2000. Evidence-based diagnosis of *Toxoplasma* infection. *Eur J Clin Microbiol Infect Dis*, **19**, 829–33.

Feldman, M., Plancarte, A., et al. 1990. Comparison of two assays (EIA and EITB) and two samples (saliva and serum) for the diagnosis of neurocysticercosis. *Trans R Soc Trop Med Hyg*, **84**, 559–62.

Forney, J.R., Wongsrichanalai, C., et al. 2003. Devices for rapid diagnosis of Malaria: evaluation of prototype assays that detect *Plasmodium falciparum* histidine-rich protein 2 and a *Plasmodium vivax*-specific antigen. *Clin Microbiol*, **41**, 2358–66.

Garcia, L.S., Shimizu, R.Y. and Bernard, C.N. 2000. Detection of *Giardia lamblia*, *Entamoeba histolytica*/*Entamoeba dispar* and *Cryptosporidium parvum* antigens in human fecal specimens using the triage parasite panel enzyme immunoassay. *J Clin Microbiol*, **38**, 3337–40.

Gillespie, S.H., Bidwell, D., et al. 1993. Diagnosis of human toxocariasis by antigen capture enzyme linked immunosorbent assay. *J Clin Pathol*, **46**, 551–4.

Goldsmid, J.M. and Davies, N. 1978. Diagnosis of parasitic infections of the small intestine by the Enterotest duodenal capsule. *Med J Aust*, **1**, 519–20.

Habluetzel, A., Traldi, G., et al. 2003. An estimation of *Toxocara canis* prevalence in dogs, environmental egg contamination and risk of human infection in the Marche region of Italy. *Vet Parasitol*, **113**, 243–52.

Harada, Y. and Mori, O. 1955. A new method for culturing hookworms. *Yonago Acta Med*, **1**, 177–9.

Harries, A.D., Fryatt, R., et al. 1986. Schistosomiasis in expatriates returning to Britain from the tropics: a controlled study. *Lancet*, **330**, 86.

Humar, A., Ohrt, C., et al. 1996. Parasight F test compared with the polymerase chain reaction and microscopy for the diagnosis of *Plasmodium falciparum* malaria in travelers. *Am J Trop Med Hyg*, **56**, 44–8.

Katz, N., Chaves, A. and Pellegrino, J. 1972. A simple device for quantitative stool thick-smear technique in *Schistosomiasis mansoni*. *Rev Inst Med Trop Sao Paulo*, **14**, 397–400.

Kuczynska, E., Boyer, D.G. and Shelton, D.R. 2003. Comparison of immunofluorescence assay and immunomagnetic electrochemiluminescence in detection of *Cryptosporidium parvum* oocysts in karst water samples. *J Microbiol Methods*, **53**, 17–26.

Lumsden, W.H.R., Kimber, C.R., et al. 1979. *Trypanosoma brucei*: Miniature anion-exchange centrifugation technique for detection of low parasitaemias: Adaptation for field use. *Trans R Soc Trop Med Hyg*, **73**, 312.

MacLean, J.D. and Libman, M. 1998. Screening returning travelers. *Infect Dis Clin North Am*, **12**, 431–43.

Ong, S.J., Cheng, M.Y., et al. 1996. Use of the ProSpecT microplate enzyme immunoassay for the detection of pathogenic and non-pathogenic *Entamoeba histolytica* in fecal specimens. *Trans R Soc Trop Med Hyg*, **90**, 248–9.

Peters, P.A., Warren, K.S. and Mahmoud, A.F.F. 1976. Rapid, accurate quantification of schistosome eggs via nuclepore filters. *J Parasitol*, **62**, 154.

Ramsay, A., Gillespie, S.H., et al. 1991. A field evaluation of the formol detergent method for concentrating fecal parasites. *J Trop Med Hyg*, **94**, 210–13.

Sabin, A.B. and Feldman, H.A. 1948. Dyes as microchemical indicators of a new immunity phenomenon affecting a protozoan parasite (*Toxoplasma*). *Science*, **108**, 660.

Sanchez-Guillen, Mdel. C., Perez-Fuentes, R., et al. 2002. Differentiation of *Entamoeba histolytica*/*Entamoeba dispar* by PCR and their correlation with humoral and cellular immunity in individuals with clinical variants of amoebiasis. *Am J Trop Med Hyg*, **66**, 731–7.

Scheiber, P., Braun-Munzinger, R.A. and Southgate, B.A. 1976. A new technique for the determination of microfilariael densities in onchocerciasis. *Bull World Health Organ*, **53**, 130–3.

Simonsen, P.E., Niemann, L. and Meyrowitsch, D.W. 1997. *Wuchereria bancrofti* in Tanzania: microfilariael periodicity and effect of blood sampling time on microfilariael intensities. *Trop Med Int Health*, **2**, 153–8.

Wagner, E.D. and Eby, W.C. 1983. Pinworm prevalence in California elementary school children, and diagnostic methods. *Am J Trop Med Hyg*, **32**, 998–1001.

Waller, K.L., Cooke, B.M., et al. 1999. Mapping the binding domains involved in the interaction between the *Plasmodium falciparum* knob-associated histidine-rich protein (KAHRP) and the cytoadherence ligand *P. falciparum* erythrocyte membrane protein 1 (PfEMP1). *J Biol Chem*, **274**, 23808–13.

Woo, P.K.T. 1970. The haematocrit centrifuge technique for the diagnosis of African trypanosomiasis. *Acta Trop*, **27**, 384.

World Bank, 1993. *Investing in health*. Oxford: World Bank.

PART II

PROTOZOA

7

Cellular organization of parasitic protozoa

HEINZ MEHLHORN

DEFINITION OF THE CELL

The organic entities on Earth exist in the following forms: various types of cells; short sequences of stable proteins (prions); extracellular genomes (viroids, or naked RNA molecules); and protein capsules, or capsids, containing relatively short molecules of DNA or RNA often surrounded by an additional membrane (bacteriophages and viruses). Prions, viroids, phages, and viruses lack their own metabolism and the ability to reproduce independently, whereas all types of cells are true living systems with metabolic ability and the capacity for independent reproduction. All cells share the following common attributes:

- They are enclosed by a cell membrane.
- Their systems of reproduction use DNA for information storage and RNA for directing cellular organization.
- Their genomes may undergo accidental change (i.e. mutate).
- They can use chemical bond energy or light energy to run their metabolic systems.
- They can detect and respond to environmental signals and can receive, recognize, and transmit signals and impulses.

Cells may also be motile and, in the case of eukaryotes, may have flowing cytoplasm (Mehlhorn and Ruthmann 1992; Alberts et al. 2002; Hausmann et al. 2003).

Two basic types of true cells are distinguishable: prokaryotes and eukaryotes. There are no transitional forms in existence today, and thus these two forms are quite distinct. Prokaryotes always occur as functionally single cells with no specialization, whereas eukaryotes may consist of a single cell (e.g. Protozoa) or may be multicellular organisms made up of differentiated (specialized) cells (e.g. Metazoa). If prokaryotic organisms (such as mycoplasms and bacteria) do aggregate, they occur as chains or clusters of unspecialized cells. In contrast, eukaryotes can consist of many cells functioning in a highly integrated fashion. Some significant differences between the cellular components of eukaryotes and prokaryotes are listed in Table 7.1.

Prokaryotic and eukaryotic cells represent the units of life. It was recognized by Virchow in 1855 that they are the smallest units capable of maintaining the continuity of life or, as he expressed it in Latin, 'Omnis cellula e cellula' (every cell derives from a cell).

UNICELLULAR EUKARYOTES

The higher protists (those that resemble plants, fungi, and animals) are all composed of eukaryotic cells. Eukaryotes may be unicellular in all their developmental stages, as are the protists, or unicellularity may be limited to certain developmental stages, such as the generative forms (the oocytes and spermatids) of plants and animals. Protists have followed many evolutionary

Table 7.1 *Differences between prokaryotes and eukaryotes*

Attributes	Prokaryotes	Eukaryotes
Cell nucleus		+
DNA		
Amount	Low (up to 1.4×10^{-2} pg/cell)	High (up to 1.6×10^{-2} – 96 pg/cell)
Organization	Circular	Linear (chromosomes) plus circular elements
Recombination	Conjugation	Meiosis and syngamy
Introns		+
Cell division		
Speed	Quick (20 min)	Slow (hours)
Mode	By formation of septa	By mitosis and cytokinesis
Organelles		
Ribosome type (subunits)	70S (30S + 50S)	80S (40S + 60S)
Membrane-bound organelles (mitochondria, plastids, Golgi, etc.)		+
Microtubules		+
Membrane-bound flagella ($9 \times 2 \times 2$ pattern)		+
Use of actomyosin for movement		+
Endo- and exocytotic activity (i.e. movement)		+

+ = present

paths in the development towards multicellularity, and the following protists represent intermediate steps in this process:

- protist forms with many nuclei per cell
- stable, double nucleated forms, such as *Giardia*
- forms with cell aggregation including some green algae, like *Volvox*
- forms with chains of dividing organisms, e.g. microsporidians
- forms that have multicellular stages, e.g. Myxozoa.

Note that although the Myxozoa, or myxosporidians, have long been classified with the protists, and are mentioned here, it is now accepted that they should be classified with the Metazoa (see Cavalier-Smith 1998). Even highly differentiated Metazoa retain vestiges of their unicellular origin, as shown by their development from unicellular eggs, some of which may develop even if they are not fertilized. They also have the ability to reconstruct their whole bodies from a single cell, as do the sponges (phylum Porifera).

The group termed protists is comprised exclusively of unicellular eukaryotic organisms that may be phototrophic, autotrophic, or heterotrophic. The Protozoa are named after the Greek term for 'first' (*proto*) and 'animal' (*zoon*). They are heterotrophic, lacking the ability to use light and inorganic materials to obtain energy and to synthesize structural components. Therefore, they must obtain preformed organic compounds, and on this basis may be considered to be animals. Apart from a few sedentary species, most Protozoa are motile. Because they have difficulty in retaining water,

partly because of their small size, most live in aquatic (or at least moist) environments. Although the majority of protozoans are free living, many species are mutualists, commensals, or true parasites. Some are highly pathogenic to their plant or vertebrate hosts and hence are relevant to veterinary and human medicine and agriculture (Mehlhorn 2002).

GENERAL MORPHOLOGY OF PARASITIC PROTOZOA

The classification of the Protozoa remains in flux, as new data continue to be obtained (see Chapter 8, Classification and introduction to the parasitic protozoa and Mehlhorn 2002). Some of the attributes of the protozoans are, however, beyond dispute. For example, they are all organized according to the basic pattern of the eukaryotic cell (Figure 7.1), the same type being found in all metazoan cells.

Eukaryotic cells consist of a membrane-bound cytoplasm containing one or more nuclei and various organelles that are also often membrane bound, their compartments and membranes acting as sites where reaction processes can occur (Alberts et al. 2002).

Cell membrane

The parasitic protozoans are surrounded by a membrane 4–10 nm thick (the unit membrane or plasmalemma) (Figures 7.2 and 7.3). In living cells this membrane always forms a closed sac or vesicle.

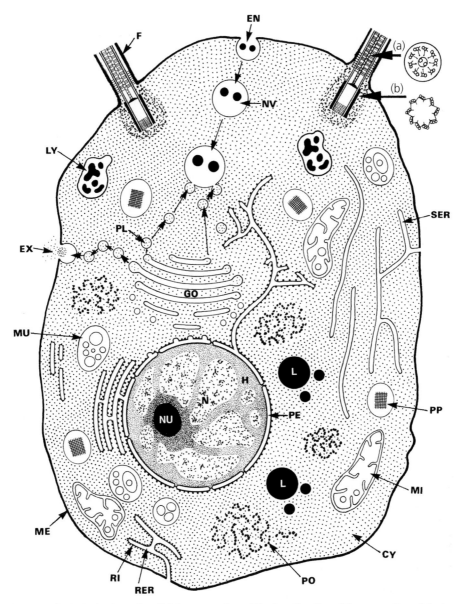

Figure 7.1 *Fine structure of a typical eukaryotic cell,* **(a)** *cross-section of the free flagellum:* **(b)** *cross-section of the basal body. B, basal body; CT, cytoplasm; EN, endocytosis; EX, exocytosis; F, flagellum; GO, Golgi apparatus; H, heterochromatin; L, lipid; LY, secondary lysosome (phagolysosome); ME, cell membrane (plus surface coat); MI, mitochondrion; MU, multivesicular body; N, nucleus; NU, nucleolus; NV, food vacuole (endocytotic vacuole); PE, perinuclear space; PL, primary lysosomes; PO, polyribosomes (chains); PP, peroxisome with protein crystal; RER, rough endoplasmic reticulum; RI, ribosome; SER, smooth endoplasmatic reticulum.*

The membrane is composed of species-specific amounts of various proteins and a double layer of lipids (Figure 7.2). It is physiologically and morphologically asymmetric, presenting a 'p face' (directed towards the cytoplasm) and an external 'e face.' High-magnification electron microscopy reveals that it is trilaminar (composed of three layers) (Figure 7.3). It is semipermeable, i.e. only certain types of molecules may cross it (Gennis 1989).

Several models have been created to explain how this biomembrane functions. The most widely accepted is the fluid mosaic model of Singer and Nicolson (1972) in which it is proposed that the membrane consists of a relatively stable double layer of lipid molecules within which proteins float like icebergs. At least some of the proteins may be structured to form pores or channels (Figure 7.2). The membranes expand by additive inclusion of vesicles formed inside the cytoplasm and shrink by formation of endocytotic vesicles (Figure 7.1) (Neupert and Lill 1992).

There are several systems for transporting substances through the cell membrane into the cytoplasm. Passage may occur by permeation (nonmediated transport), a process that is dependent on concentration gradients. Active transport (mediated transport) using motile carriers may also occur. In this system a protein binds

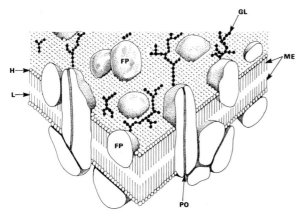

Figure 7.2 *Cell membrane of a typical eukaryotic cell following the fluid mosaic model of Singer and Nicolson (1972). Proteins (FP) float in a membrane that consists of a double layer of lipid molecules each with a hydrophilic component (H) and two lipophilic layers (L). The proteins, which may form pores (PO), often anchor glycoproteins, and together form the glycocalyx (GL).*

the molecule to be transported and then the complex moves actively from one side of the membrane to the other. Movement depends on changes in the electric charge that are linked with the binding and release of the transported molecule. The fixed pore (Figure 7.2) is a protein structure that stretches through the membrane. The molecules to be transported pass through the space, or channel, formed between the subunits of the pore. All forms of active transport require energy, which is derived from various metabolic reactions.

In addition to the direct transport of molecules through the cell membrane, other mechanisms exist for internalization of materials by cells. For example, relatively large organic materials may be internalized by the formation of endocytotic vesicles (Figure 7.2), a process termed pinocytosis for the uptake of liquids and phagocytosis for the uptake of solid particles. Within the vacuoles the material is degraded and then transported into the cytoplasm by mechanisms similar to those used by the cell membrane. Alternatively, it may be released

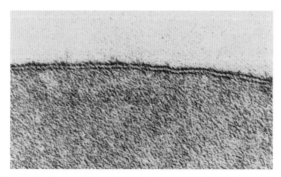

Figure 7.3 *Transmission electron micrograph of a section through the membrane of an erythrocyte showing the three layers of the membrane (× 100 000).*

by dissolution of the membrane. Some protozoans have developed special places for the uptake of food, called cytostomes. Many cells contain clathrin-coated pits that are involved in the receptor-mediated endocytosis of macromolecules, but our knowledge of coated pits in Protozoa remains limited (Morgan et al. 2002), although clathrin-dependent endocytosis has been seen in trypanosomes.

The cell membranes of both Protozoa and Metazoa are involved in many other tasks apart from transport functions, for example cell recognition. When cells come into contact with each other, recognition occurs via special receptors in the membranes that perceive chemical signals given off by other cells. Cell membranes also participate in excretion and secretion, these processes occurring by mechanisms similar to those for the intake of materials (e.g. bloodstream trypanosomes excrete trypanopains and oligopeptidases in order to cross the blood–brain barrier; Londsdale-Eccles and Grab 2002). Membrane-bound structures exist that mediate the joining of cells. These are needed for fusion of gametes and similar structures and are also found at places where flagella are attached to the surfaces of protozoa. The undulating membranes and recurrent flagella of trypanosomes and trichomonads are joined to the cell by means of such cell junctions (Figure 7.4; see also Figure 7.29a). Cell junctions also play a role in attachment of many parasites to the surfaces of host cells. For example, trypanosomes use junctions for attachment to the vector's intestine.

Pellicle and cytostomes

Many protozoans have more than one limiting membrane for at least some of their developmental stages (Table 7.2). In some groups the main cell membrane is underlined by one or more 'inner' membranes that are often derived from the endoplasmic reticulum (Table 7.2, Figure 7.5).

Like the outer membranes, these inner membranes are species specific and possess distinct inner and outer surfaces. The characteristics of the inner membranes have been revealed by the methods of freeze fracture and negative staining (Figure 7.6). The single outer membrane and the membranous complexes often have subpellicular microtubules (c. 25 nm in diameter) beneath them.

These shape-stabilizing cell boundary complexes are called pellicles (Figures 7.4 and 7.5). The motile stages of trypanosomes and *Leishmania* species form a pellicle that consists of one membrane plus underlying microtubules (Figure 7.7), whereas the motile stages of coccidians (sporozoites, merozoites, ookinetes, kinetes) have a pellicle consisting of three membranes plus underlying microtubules (Figure 7.6).

In ciliates the pellicle is composed of an outer membrane, a system of alveolar sacs, longitudinal micro-

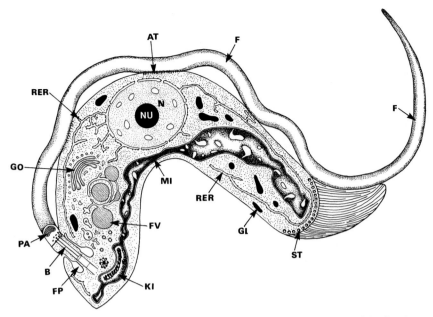

Figure 7.4 *Longitudinal section through a trypomastigote of* Trypanosoma *sp. AT, attaching zone of the flagellum; B, basal body; F, flagellum; FP, flagellar pocket (with cytostomal activity); FV, food vacuole; GL, glycosomes; GO, Golgi apparatus; KI, kinetoplast (containing DNA filaments); MI, mitochondrion; N, nucleus; NU, nucleolus; PA, paraxial rod; RER, rough endoplasmic reticulum; ST, subpellicular microtubules.*

Table 7.2 *Types of limiting membranes of some parasitic protozoans*

Species	Stage	Single membrane	Two or more membranes	Pellicle with subpellicular microtubules	Cyst wall
Trichomonas vaginalis	Trophozoite	+			
Giardia lamblia	Trophozoite	+		Ventral side	
	Cyst	+			+
Trypanosoma brucei group	Epi- and trypomastigotes			+	
Leishmania spp.	A-, promastigotes			+	
Entamoeba histolytica	Magna-form	+			
	Minuta-form	+			
	Cyst	+			+
Pneumocystis carinii	Trophozoite	+			
	Cyst	+			+
Eimeria spp.	Sporozoites			+	
	Merozoites			+	
	Oocysts				+
	Sporocysts	+			+
	Meronts	+			
	Male gametes	+			
	Female gametes of some species	+	+		
Toxoplasma gondii	Sporozoites			+	
	Merozoites			+	
	Oocysts	+			+
	Sporocysts	+			+
	Meronts	+			
	Male gametes	+			
	Gamonts	+			
	Female gametes		+		

+ = present

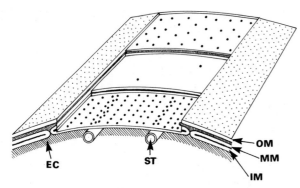

Figure 7.5 *Three-layered pellicle of the invasive stage of a sporozoan as revealed by freeze etching. Note that each of the membranes has unique intramembranous particles. EC, ectoplasm; IM, inner membrane; MM, middle membrane; OM, outer membrane; ST, subpellicular microtubules.*

tubules, and kinetodesmal fibres. This complex pellicle produces a stable base to hold rows of cilia (see Figure 7.18).

In the free-living euglenids the plasmalemma is reinforced by longitudinal microtubules and by a dense underlying epiplasm. The number of subpellicular microtubules is often stable within a species and may thus be used for species definition. For example, the merozoites of the tissue cyst-forming Coccidia (e.g. *Sarcocystis, Toxoplasma, Besnoitia, Frenkelia*) always have 22 microtubules and various *Eimeria* species have 24, 26, 28, 30, or 32 microtubules, respectively. In some parasites, such as the trypanosomatids, the number of subpellicular microtubules varies with the size of the individual parasite form, i.e. a long slender form with a small diameter has fewer microtubules than a form of the same species with a larger diameter (Figure 7.7).

The subpellicular microtubules may run from one pole to the other, as they do in gregarines and trypanosomes, or they may be restricted to limited portions of the pellicle. In ciliates, for example, they are restricted to the front half of the cell and in the coccidian motile stages they are localized to the front two thirds of the cell (Figure 7.8).

In the motile stages of the Coccidia the microtubules are anchored to an anterior polar ring. At the attachment point there is an interruption of the two inner membranes (Figure 7.9). If two or more polar rings are present (species specific), it is the outer one that is connected to the microtubules. The subpellicular microtubules are usually kept in contact with the inner surface of the pellicular membrane by means of side arms. The exact role of the microtubules in movement has not yet been identified.

The membrane of the eukaryotic cell can take up material by various methods including permeation, active transport and endocytosis. Endocytosis (see Figure 7.1) may occur at any point on the cell membrane or may occur only at predisposed places. For example, in *Giardia* trophozoites, endocytosis occurs only in the dorsal region, the region opposite the ventral sucker; in trypanosomes, endocytosis occurs in the flagellar pocket (Morgan et al. 2002) and in sporozoans it is found at small cytostomes called micropores (Figure 7.10) (Mehlhorn 2002).

Large endocytic elements (cytostomes or 'cell mouths') are characteristic of many ciliates (see Figure 7.31e) (Hausmann et al. 2003).

Cytostomes are reinforced by various structures (Figure 7.10), for instance by bundles of microtubules in trypanosomes and ciliates. After the phagosomes enter the cytoplasm the contents are digested and resorption of the necessary molecules occurs. The residue may be voided to the outside or stored as an inclusion. The process of excretion, called exocytosis, may occur anywhere on the surface or, as in ciliates, at a specialized place called the 'cell anus' or cytopyge. Exocytosis is a process similar to endocytosis but in reverse.

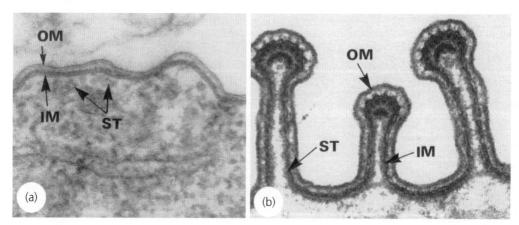

Figure 7.6 *Transmission electron micrographs of sections through the three-layered pellicle of a coccidian **(a)** and a gregarine **(b)**. Note that the two inner membranes appear mainly as a thick layer. IM, two inner membranes; OM, outer membrane; ST, subpellicular microtubule (×70 000).*

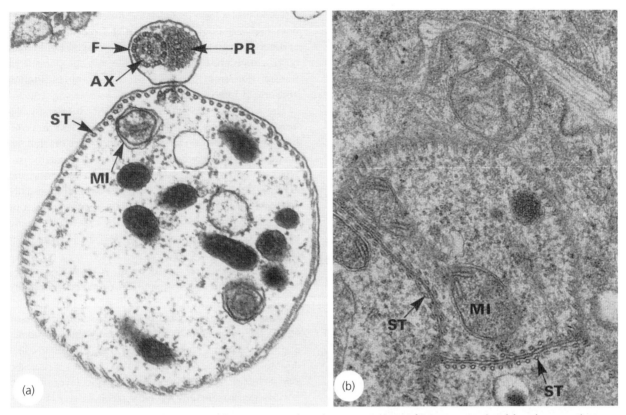

Figure 7.7 *Transmission electron micrographs of cross-sections through a trypomastigote of* Trypanosoma vivax **(a)** *and an amastigote of* Leishmania donovani **(b)**. *Note that in both cases the pellicle consists of a single cell membrane plus subpellicular microtubules (ST). AX, axoneme; F, flagellum; MI, mitochondrion; PR, paraxial rod (×35 000).*

Cyst wall

Many parasitic protozoans (Table 7.2) are capable of undergoing encystation, which involves formation of a cyst wall either outside or inside the cell membrane. This cyst wall may be single layered or multilayered.

Walls formed outside the plasmalemma are produced by exocytosis of materials. Cyst walls have two main

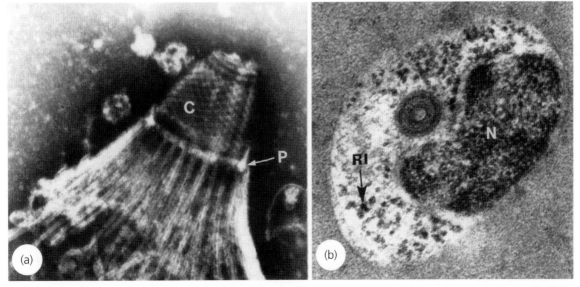

Figure 7.8 *Transmission electron micrographs:* **(a)** *Negatively stained apical pole of a merozoite of* Sarcocystis. *Note the protruded conoid (C), and the polar ring (P), to which the subpellicular microtubules are attached (×50 000).* **(b)** *Cross-section of a micropore (in an erythrocytic stage of* Theileria annulata, *the agent of Mediterranean coast fever of cattle). N, nucleus; RI, ribosome (×35 000).*

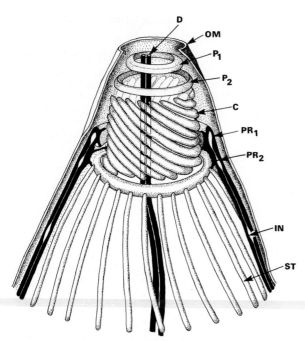

Figure 7.9 *Coccidian merozoite showing the apical pole and its protruded conoid. This structure is not present in hemosporidians and piroplasms. C, conoid consisting of microtubules; D, ductules of the rhoptries; IN, inclusion between the two inner membranes of the pellicle; OM, outer membrane of the three-layered pellicle; P1, P2, preconoidal rings; PR1, PR2, polar rings (the posterior one is connected with the subpellicular microtubules); ST, subpellicular microtubule.*

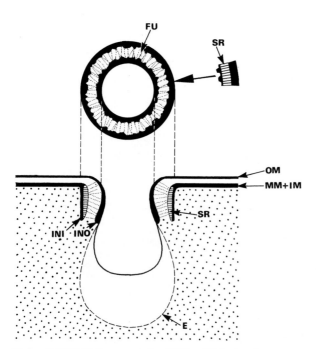

Figure 7.10 *A sporozoan micropore in longitudinal and tangential sections. E, enlargement during food uptake; FU, filamentous elements; IM, inner membrane of the pellicle; INI, interruption of the inner membrane (forms of the outer ring in tangential section); INO, dense material along the inner ring (invaginated outer membrane); MM, middle membrane; OM, outer membrane of the pellicle; SR, dense material between the inner and middle membranes.*

tasks: to protect the organism against unfavorable environmental conditions when passing from one host to another and to create spaces for reorganization and nuclear division. Cyst walls may also aid the parasite in transmission from one host to another by facilitating attachment to host-cell surfaces.

The cyst wall has one layer in cysts of *Entamoeba histolytica*, *Giardia* species, and some ciliates, such as the fish parasite *Ichthyophthirius multifiliis* and the human ciliate parasite *Balantidium coli*. There are usually two types of cysts in the sages of Coccidia in feces: the oocysts and the sporocysts. The oocyst wall is usually double-layered, but in a few species it may have four layers. It is formed by two types of cyst wall-forming bodies. The sporocyst wall usually has only a single layer (Figure 7.11).

The chemical composition of cyst walls varies according to the species, although proteins are usually the basic component. The cyst walls of *E. histolytica* and *G. lamblia* contain proteins which are keratin-like or elastin-like albuminoids composed of lysine, histidine, arginine, tyrosine, glutamic acid, and glycine. The double-layered oocyst wall of the sporozoans is periodic acid–Schiff (PAS) positive. The outermost layer of these oocysts consists mainly of fatty alcohols (e.g. hexacosanol), some phospholipids, and fatty acids. It has no carbohydrates or proteins. The inner layer is composed of glycoproteins and contains most of the carbohydrates found in the oocyst wall. These carbohydrates are composed of mannose, galactose, glucose, and hexosamine. The oocyst wall is highly resistant to the passage of potassium dichromate, sodium hypochloride, sulfuric acid, and sodium hydroxide and these chemicals are therefore used in storage and cleaning of oocysts. The oocyst wall is permeable to oxygen, carbon dioxide, ammonia, methylbromide, carbon disulfide, and various organic solvents. It is highly susceptible to mechanical pressure and therefore may easily be ruptured by shearing forces. Thus mechanical rupture of oocysts in the gizzard of the avian host is likely to be the normal method of excystation of avian coccidians. The micropyle (a preformed opening in some species) is probably used as rupture point to facilitate the escape of sporocysts. The sporocysts in the oocyst are bound by a double-layered wall, the outer layer of which is relatively smooth. The sporocysts of eimerians have an opening for exit of the sporozoites. This is closed by the Stieda body, which can be dissolved by trypsin. The sporocysts of *Isospora* and the tissue cyst-forming Coccidia (*Sarcocystis*, *Toxoplasma*) have sutures on which the excystation fluids act causing collapse of the sporocyst wall (Figure 7.11g).

The cyst walls of Myxozoa and Microspora are at least double-walled in most species (Figure 7.12) and the layers can be resolved by electron microscopy. In Myxozoa the wall consists of valves that open at preformed sites to release the infectious sporoplasm.

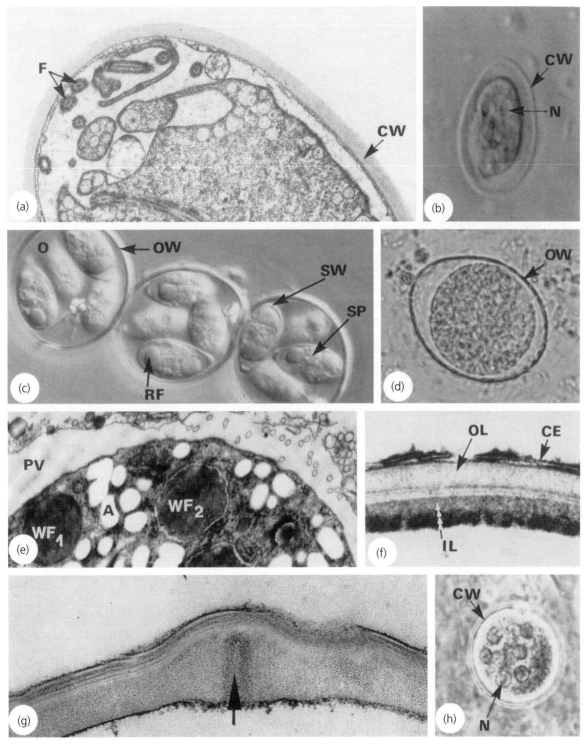

Figure 7.11 *Light (b, c, d, h) and electron micrographs (a, e, f, g) of cysts and cyst walls. (a, b)* Giardia lamblia. *CW, cyst wall; F, flagellum; N, nucleus. (c)* Eimeria *oocysts (0) containing sporocysts (SP). OW. oocyst wall; RF, refractile body of sporozoites; SW, sporocyst wall. (d)* Isospora *sp., unsporulated oocysts. OW, oocyst wall. (e)* Eimeria maxima; *macrogamete with wall forming bodies I and 2 (WFI, WF2). A, amylopectin; PV, parasitophorous vacuole. (1) Oocyst wall of* Eimeria *and* Isospora *spp. CE, cell membrane; IL, OL, inner and outer layer of oocyst wall. (g) Sporocyst wall in* Toxoplasma *and* Sarcocystis *spp., with the pre-formed suture (arrow) that ruptures during the excystation process. (h)* Entamoeba coli; *eight-nucleated cyst. CW, cyst wall; N, nucleus.*

The Microspora have developed a hollow tube that is protruded from the surface of the wall. It penetrates into a host cell and the infectious sporoplasm is passed through it. The exospore layer in Microspora is proteinaceous and is 15–100 nm thick, depending on the species. The endospore layer is chitinous and

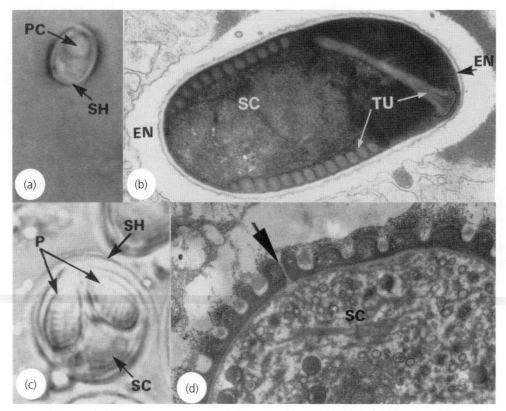

Figure 7.12 *Light (a, c) and electron micrographs (b, d) of Microspora (a, b) and Myxozoa (c, d). Note that the microsporidian wall (Nosema sp.) is entirely closed, while that of Myxozoa (d) (Myxosoma sp.) has two valves that leave a small opening (arrow). EN, endospore; EX, exospore; P, polar capsule with solid filaments; PC, polar capsule with a hollow tubule; SH, shell; SC, sporoplasm; TU, tubule.*

c. 150–200 nm thick. The spores are gram-positive (i.e. stain reddish purple with the Gram stain), a fact that is of diagnostic value. They are also stained light blue by Giemsa.

Surface coat

The plasma membrane of eukaryotic cells is strikingly asymmetric (Figure 7.2). The outer and inner layers are clearly delineated and the polypeptides on each surface (Figures 7.13 and 7.14) are distinct. Glycolipids and glycoproteins are present only on the external surface. The peripheral layer is rich in carbohydrate and is called the glycocalyx or surface coat. Molecules of glycosylphosphatidylinositol (GPI) serve as parasite-specific anchors for numerous surface proteins. The thickness of this layer varies with the species and with the developmental stage of the organism. Not only is the surface coat composed of glycoproteins and glycolipids, but various glycoproteins and proteoglycans (acid mucopolysaccharides) may also be adsorbed to it (Figure 7.13).

The surface coat may be a rather delicate coating, a mass of delicate filaments or a thick mat. Whatever its structure, it has several functions in the life of the organism: (1) it acts as a mechanical or chemical barrier; (2) it plays a role in recognition and adhesion to other

cells; (3) it contains enzymes that act on substances in the environment; and (4) it contains molecules that can act as antigens and thus plays an important role in immunological processes.

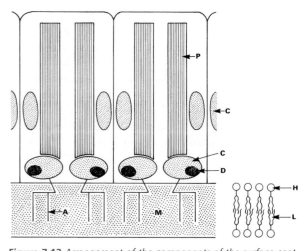

Figure 7.13 *Arrangement of the components of the surface coat along the surface of a trypanosome. A, anchor of a variant surface glycoprotein molecule; C, carbohydrate; D, cross-reacting determinant; H, hydrophilic component of the VSG molecule; L, lipophilic component; M, membrane consisting of two layers of lipid molecules; P, protein.*

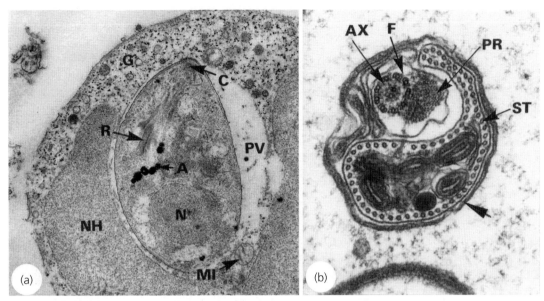

Figure 7.14 *Transmission electron micrographs of the surface coat:* **(a)** Toxoplasma gondii; *the tachyzoite within a host cell vacuole (PV) shows a slight positive Thièry reaction along its surface (arrow). Note the presence of amylopectin (A) granules in the parasite and of glycogen (G) in the host cell. C, conoid; MI, mitochondrion; N, nucleus; NH, nucleus of the host cell; R, rhoptry (×7 000).* **(b)** Trypanosoma vivax. *Cross-section through a trypomastigote stage showing the discharge of the surface coat (arrow). AX, axoneme; F, flagellum; PR, paraxial rod; ST, subpellicular microtubule (×25 000).*

The surface coat may change its composition as the parasite develops from stage to stage. In addition, many parasitic Protozoa, in particular the trypanosomes, have developed sophisticated methods of antigenic variation. This may be achieved by selective activation of different genes at different times. Organisms of the *Trypanosoma brucei* group have up to 1 000 genes that may become activated during the production of variant surface glycoproteins (VSGs). This selective activation results in changes in the variable antigen types (VATs) displayed and hinders the host defense against these blood-inhabiting flagellates. *Plasmodium* may also display variant antigen types, but there are fewer variants than those displayed by trypanosomes. Such different surface antigens (SAGs) are found in many of the coccidians, too (Lyons et al. 2002). The action of these genes results in antigenic variation and the production of immunologically different strains of protozoans; this explains why most antiprotozoal vaccines provide only limited protection, restricted to certain localities. The development of potent vaccines against protozoan parasites depends on the discovery of species-specific antigens with invariant epitopes that are accessible to the immune system during the parasite's life. In addition, they must be essential to the parasite's survival, possibly playing an important role in cell recognition, adhesion, immune evasion, metabolism, or cell invasion.

Cytoplasm

The cytoplasm in protozoa is generally divided into two zones: the peripheral, electron-lucent ectoplasm (hyaloplasm) and the denser central endoplasm. The endoplasm contains the cell organelles and the nucleus. This differentiation is particularly prominent in the amoebae and in the gregarines (Figures 7.15 and 7.16) but is not apparent in all species.

In some species, endoplasm and ectoplasm cannot even be distinguished by electron microscopy. Microgametes of most sporozoans, apart from those of the piroplasms, have a very reduced cytoplasm. They are comprised mainly of flagella, a mitochondrion, and a nucleus (Figure 7.17) (Mehlhorn 2002; Ferguson 2002).

The cytoplasm of most cells has a high viscosity and stability and a prominent cytoskeleton. Contractile elements of this cytoskeleton are responsible for the cytoplasmic flow and these include actin filaments with a diameter of c. 6 nm. These filaments are composed of a double chain of globular bodies. Cytoskeletons may also contain 10 nm filaments or intermediate filaments which,

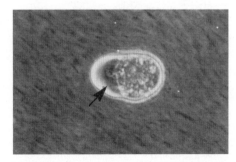

Figure 7.15 *Light micrograph of a trophozoite of* Entamoeba histolytica. *Note the presence of hyaline ectoplasm in the pseudopodium (arrow) and along the outer surface (×1 000).*

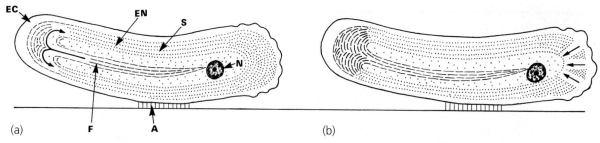

Figure 7.16 *Amebic movement brought about by a steady forward flux of the central liquid endoplasm and its transformation into a more stable form as it moves posteriorly and laterally. The origin of the forces needed for such movement is explained by either apical* **(a)** *or posterior* **(b)** *contractions (arrows). A, zone of adhesion to the surface; EC, hyalinic ectoplasm; EN, dense, granulomatous endoplasm; F, fluid portion of endoplasm; N, nucleus; S, stable, nonfluid portion of endoplasm.*

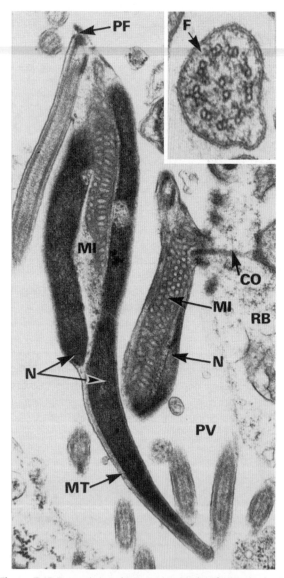

Figure 7.17 *Transmission electron micrographs of a longitudinal section through a coccidian microgamete (Sarcocystis sp.) and a cross-section through a flagellum (inset). CO, connection of the nucleus to the microgamont; F, flagellum; MI, mitochondrion; MT, microtubule; N, nucleus; PF, perforatorium (apical pole); PV, parasitophorous vacuole; RB, residual body (×25 000; inset ×100 000).*

to date, have been found only in the cells of higher vertebrates. The various types of filaments are organized into the microtubules of the cytoskeleton. The tubules have an outer diameter of 25 nm and an inner diameter of 15 nm. They are composed of protofilaments that are visible in cross-section. The protofilaments consist of α and β tubulin elements in a helical arrangement. The microtubules are polymerized at particular points called microtubule organizing centers (MTOCs). These centers exist at centromeres, centrioles, and at certain places in membranes. They also occur as constituents of flagella and cilia (see Figures 7.28 and 7.29).

The processes that bring about motility of the cytoplasm are well documented for metazoan cells, but for protozoans they are still poorly understood (Figure 7.17). In all cases, it is probably the aggregation of actin with myosin and tropomyosin to form an actomyosin complex that leads to movement, as proposed in the early sliding filament model of Huxley and Hanson (1954). This ATP-dependent system may produce relatively rapid movements, as occur in amoeba (20 μm s^{-1}) and in the sporozoites and merozoites of sporozoans. In addition to the cytoskeletal system, most parasitic protozoans have developed unique skeletal elements composed of combinations of the usual cytoskeleton elements. These structures include the following:

- subpellicular microtubules of trypanosomes and the motile stages of the Coccidia (see Figures 7.4 and 7.5)
- kinetodesmal fibrils of ciliates (Figure 7.18a)
- bundles of cytoplasmic microtubules in gamonts of piroplasms (Figure 7.18b)
- combined microtubules and filaments observed in the ventral disc of the diplomonadids (Figure 7.18c)
- crystalloid protein densifications that occur below the membrane of giardial trophozoites, at the apex of eimerian microgametes (Figure 7.18d) and also in the gamonts of piroplasms (Figure 7.18b)
- axostyles and pelta, occurring prominently in the trichomonads and consisting of one or more parallel rows of microtubules (Figure 7.19)
- costa and parabasal filaments, the filamentous, sometimes striated elements (Figure 7.19) that line the

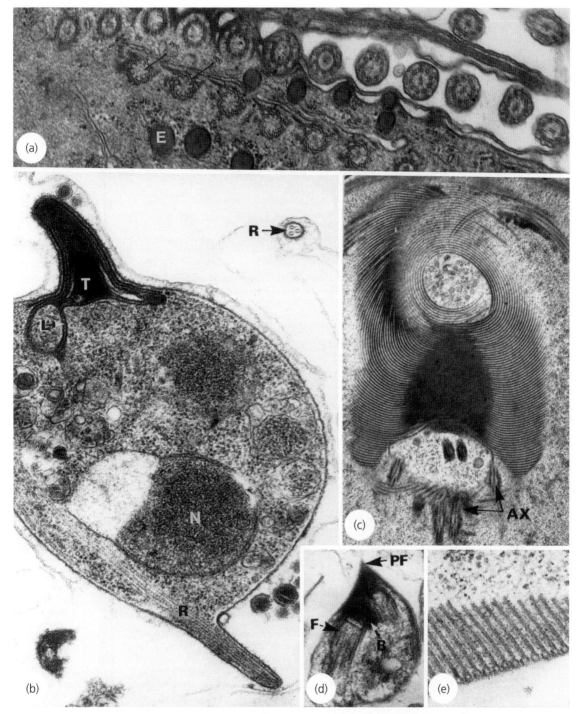

Figure 7.18 *Transmission electron micrographs:* **(a)** *Cross-section through the base of a cilium of Ichthyophthirius multifiliis showing interconnecting microtubules (arrows). E, Extrusomes (×30 000).* **(b)** *Babesia canis, longitudinal section through a microgamont. Note the thorn-like dense apex (A), the microtubule-containing ray (R), and the labyrinthine structure (L) at the base of the thorn (T). N, nucleus (×25 000).* **(c)** *Tangential section through the ventral disc of Giardia lamblia. Note the helically arranged rows of material. AX, axoneme (×20 000).* **(d)** *Perforatorium of the microgamete of eimerians, a structure needed for entering the macrogamete. B, basal body; F, flagellum; PF, perforatorium.* **(e)** *Cross section through the disc of Giardia lamblia, showing microtubules attached to filaments (×95 000).*

recurrent flagellum or the Golgi apparatus in trichomonads

- paraxial rods, which consist of a network of microfilaments that run along the axonemal microtubules of the flagella of Kinetoplastida (Figures 7.4, 7.7, and 7.29a)

- conoids, which are found in the motile stages of some Coccidia, such as in the genera *Eimeria, Sarcocystis,*

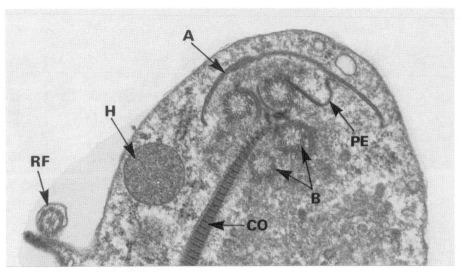

Figure 7.19 *Transmission electron micrographs of the apical pole of* Trichomonas vaginalis *showing the line of microtubules making up the axostyle (a), the basal bodies of the four anterior free flagella (b) and the costa (c) (×15 000). H, hydrogenosome; PE, pelta; RF, recurrent flagellum running along a surface fold.*

and *Toxoplasma* (see Figures 7.8 and 7.9a), but which are always absent from the hemosporidians (e.g. genera *Plasmodium*, *Theileria*, *Babesia*).

The conoid (Figure 7.9a) is a hollow truncated cone composed of spiralling microtubules 25 nm in diameter.

Two accessory structures, the conoidal or preconoidal rings, form an integral part of the conoid and they are connected with each other by a canopy-like membrane. During cell penetration the conoid protrudes through the anterior polar ring system. The conoid is apparently involved in penetration of host cells.

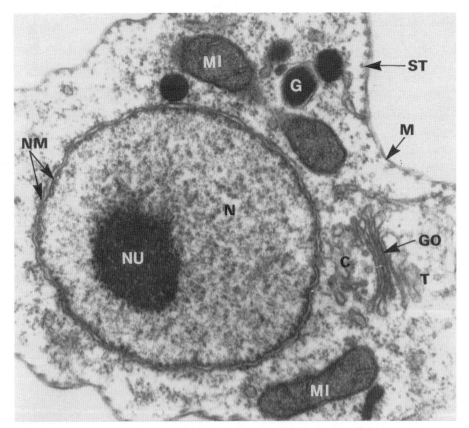

Figure 7.20 *Transmission electron micrograph of a developing stage of* Crithidia *sp. The section is in the region of the nucleus (×25 000). GO, Golgi apparatus; M, cell membrane MI, mitochondrion; N, nucleus; NM, nuclear pore; NU, nucleolus; ST, subpellicular microtubules; T, trans side of the Golgi apparatus.*

CYTOPLASMIC INCLUSIONS

The endoplasm contains a variety of organelles and other structures within which metabolic processes occur.

Nucleus

Protozoa possess at least one well-developed nucleus that is usually spherical to ovoid (Figure 7.20) and enclosed by a double-layered membrane containing pores (Figures 7.1 and 7.21). The pores are c. 50–70 nm in diameter with a central opening made up of eight subunits.

When viewed by light microscopy, the nuclei may appear vesicular or compact. The karyoplasm is composed of structural and enigmatic proteins: the chromatin (DNA) which may be organized into chromosomes; and the nucleolus. No membranes exist within the nucleus. There is usually only one nucleolus but in some species, and in certain developmental stages of other species, there may be several. The nucleolus is the site of synthesis of the large RNAs (28S, 18S, and 5.8S) and of the precursors of the ribosomes. It has two zones; one is composed of filaments 5–8 nm wide and the other is composed of granules 15 nm wide. The granular zone is situated at the periphery of the nucleolus. During nuclear division the nucleoli arc often dissolved and the chromosomes may condense and become visible (e.g. in ciliates) or they may remain stretched out and invisible

(e.g. in the eimerians and the hemosporidians). The nuclear membranes are retained in the protozoans during division. Separation of the chromosomes is carried out by a spindle apparatus, the appearance, arrangement, and placement of which is species specific. The apparatus always consists of microtubules and filaments.

In ciliates, two morphologically distinct nuclei occur: a generative micronucleus and a somatic macronucleus (see Figure 7.31e). In other parasitic species, and in some specific stages such as the meronts, sporonts, and gamonts of the coccidia and hemosporidia, there may be many nuclei, all of which may have a similar appearance. Morphologically similar nuclei may have similar functions or they may have different functions, as in the Myxosporea. During binary division the karyoplasm is usually completely distributed between the two daughter nuclei. In some cases, e.g. during the formation of microgametes in some coccidians, only one functional nucleus is produced and a part of the nucleus is left behind (see Figure 7.32c, e).

Mitochondria and related structures

Mitochondria contain the enzymes for oxidative phosphorylation and the tricarboxylic acid cycle, and are bounded by two membranes. Most parasitic Protozoa have one of three basic types of mitochondria, distinguished by characteristic infoldings of the inner

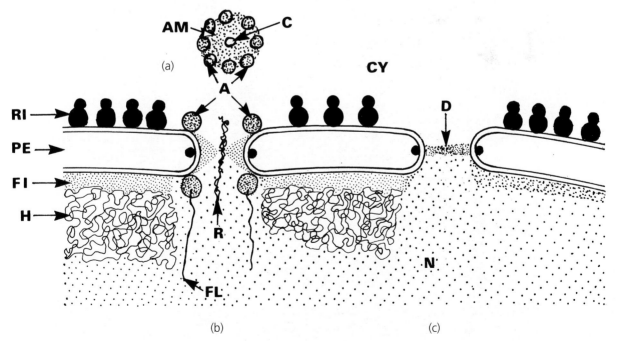

Figure 7.21 Nuclear pores (diameter c. 80 nm): **(a)** Surface view showing the eight outer peripheral annuli. **(b)** Longitudinal section of an active pore. **(c)** Longitudinal section of an inactive pore. A, annular element; AM, amorphous material; C, central channel (diameter 15–20 nm) also called central granulum when RNA passes through; CY, cytoplasm; D, diaphragm; Fl, filamentous layer; FL, filaments connected at inner annular elements; H, heterochromatin; N, nucleoplasm (karyoplasm); PE, perinuclear spore; R, RNA; RI, ribosome (with subunits).

membrane, which may be tubular, sack-like, or cristae-like (Figure 7.22).

Mitochondria reproduce by division and are therefore said to be semiautonomous organelles. This form of reproduction is possible because mitochondria have their own DNA. In most mitochondria this DNA is composed of two circular strands arranged in a supercoil. Some species have a single, large mitochondrion that contains more DNA, e.g. species of *Trypanosoma* and *Leishmania*. These organisms have 5 percent of their DNA in a single structure called the kinetoplast (Figures 7.4 and 7.22a) which is located close to the basal body of the flagellum. Kinetoplastid flagellates have no infoldings in this region of the mitochondria, providing space for the thousands of minicircles (0.3–0.8 μm in length) and the few maxicircles (9–11 μm in length) that make up their mitochondrial DNA; this region stains with Giemsa solution and is visible as a deep purple dot. Only the DNA of the maxicircles seems to be transcribed. During cell division the kinetoplast is always reproduced before nuclear division occurs. Trichomonads, some amoeba (e.g. *Entamoeba histolytica*), and Microspora have no mitochondria, but some amoeba contain symbiotic bacteria that may function as mitochondria. The trichomonads, which are anaerobic, have microbodies called hydrogenosomes. These spherical organelles have two closely attached membranes surrounding a granular matrix (Figure 7.19). The enzyme system of these bodies differs from that of mitochondria, as they metabolize pyruvate from glycolysis into acetate, carbon dioxide, and hydrogen. In ciliates, hydrogenosomes with more easily visible double membranes are present, as well as mitochondria.

The motile stages of sporozoans (i.e. merozoites and sporozoites, Figures 7.23 and 7.24a) have a Golgi-adjunct body (also called a double-walled organelle) close to the nucleus (Scholtyseck and Mehlhorn 1972).

In several species and genera (e.g. *T. gondii*, *Sarcocystis*, and *Plasmodium*) this organelle contains genes of the photosystem II and resembles a plastid, or a remnant of one, in terms of shape and activity (Hackstein et al. 1995). This body is now called the apicoplast and has been found to contain a third genome (in addition to that of the nucleus and mitochondria).

Endoplasmic reticulum, ribosomes, and golgi apparatus

The endoplasmic reticulum (ER) is a large system of tubes and sacs that runs throughout the cell. It connects the nuclear space with the cell interior and with the cell surface (Figures 7.1 and 7.25). There are two types of ER: rough endoplasmic reticulum (rER) and smooth endoplasmic reticulum (sER).

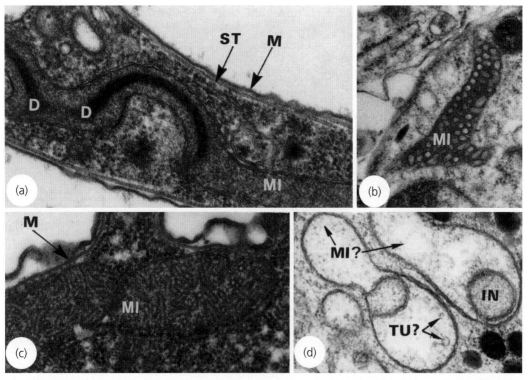

Figure 7.22 *Transmission electron micrographs of mitochondria or similar structures:* **(a)** Trypanosoma vivax, *kinetoplast in division (mitochondrion with cristae) (×25 000).* **(b)** Sarcocystis *microgamont with a mitochondrion of the sacculus type (×25 000).* **(c)** Ichthyophthirius multifiliis; *a trophozoite with a mitochondrion of the tubular type (×40 000).* **(d)** Theileria sp.; *mitochondria-like structures within kinetes. D, DNA filaments; IN, invagination; M, cell membrane; MI, mitochonidrion; ST, subpellicular microtubule; TU, tubule (×30 000).*

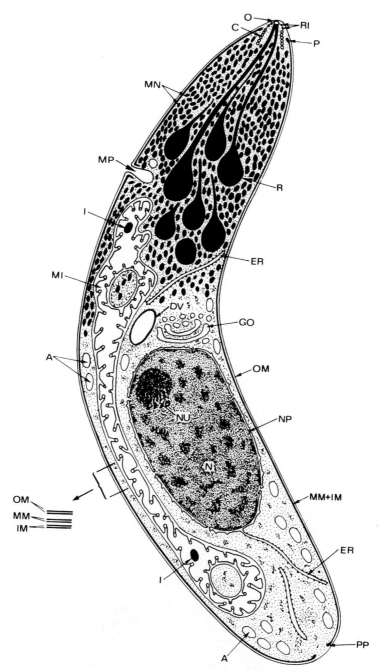

Figure 7.23 *Longitudinal section through a typical coccidian merozoite. A, amylopectin; C, conoid; D, dense, spherical bodies; DW, double-walled vesicle; ER, endoplasmic reticulum; GO, Golgi apparatus; I, dense inclusion; IM, inner membrane; MI, mitochondrion; MM, middle membrane; MN, micronemes; MP, micropore; N, nucleus; NP, nuclear pore; NU, nucleolus (karyosome); O, opening of the conoid; OM, outer membrane of pellicle; P, anterior polar ring; PP, posterior polar ring; R, rhoptries; RI ring-like elements of the conoidal canopy*

The rER is characterized by the presence of ribosomes (Figures 7.1 and 7.25) along its outer surface, whereas these are lacking in the sER. The rER and the sER may be interconnected.

The ribosomes consist of RNA and protein. The ribosomal RNA represents about four-fifths of the total cellular RNA. There are several types of ribosomes, depending on the nature of their subunits (Table 7.1; Figure 7.20); these subunits can be released by appro-

priate treatment of the ribosomal proteins. Several ribosomes often align in a chain to form polyribosomes, or polysomes (Figures 7.1 and 7.25), where they are collectively active in protein synthesis. Ribosomes occur along the surface of the rER, either singly or as polysomes. They also occur in the cytoplasm and inside the mitochondria. In Protozoa, ribosomes have a diameter of c. 30 nm and are composed of two subunits with sedimentation characteristics of 60S and 40S. The intact ribosome is of the 80S type. Mitochondrial ribosomes are of the 70S type or, in ciliates, 80S. The mitochondrial ribosomes are similar to those of prokaryotes (Bielka 1982) (see Table 7.1).

The Golgi apparatus is a flat sac with swollen regions (Figures 7.1 and 7.26), from which vesicles are formed (Farquhar and Palade 1981). Several Golgi apparatuses together form dictyosomes. These organelles are always situated close to the rER and the two organelles act together as functional units in the production and transport of membrane and in the formation and transport of all types of macromolecular proteins and lipids (Gal and

Raikhel 1993). The Golgi apparatus is never covered by ribosomes. Although many parasitic protozoans have prominent dictyosomes, certain developmental stages of protozoans have only a single Golgi apparatus which is difficult to detect in electron micrographs. The principal function of the Golgi apparatus is the transport of secretions and excretions formed on the rER. These reach the Golgi apparatus at its cis side and are transported within vacuoles that are cut off at its periphery (the trans region) (Figure 7.26). Endocytotic vesicles such as food vacuoles also come into contact with the Golgi apparatus and the rER system at points where they fuse

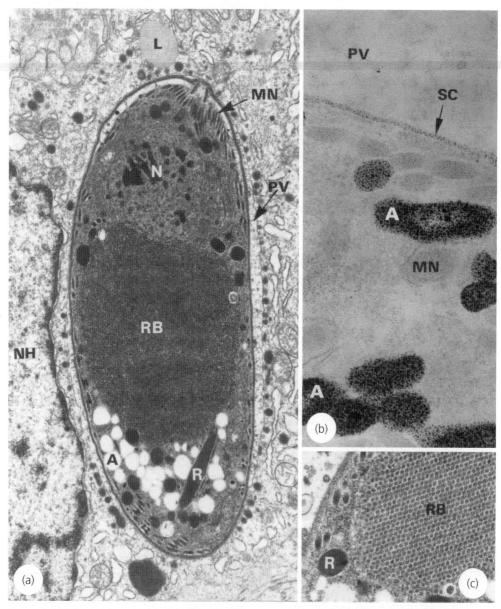

Figure 7.24 *Transmission electron micrograph of storage materials in coccidians:* **(a)** *Longitudinal section of a sporozoite of* Cystoisospora felis. *A, amylopectin; L, lipid; MN, micronemes; N, nucleus; NH, nucleus of the host cell; PV, parasitophorous vacuole; RB, refractile body (consisting of protein granules); RH, rhoptry; SC, surface coat (×12 000).* **(b)** *The Thièry reaction shows polysaccharides (e.g. amylopectin) within granules (A) and along the surface (SC) of a meiotic (×65 000).* **(c)** *Cross-section of C.* felis *showing structure of refractile body. RB, refractile body; R, rhoptry (×30 000).*

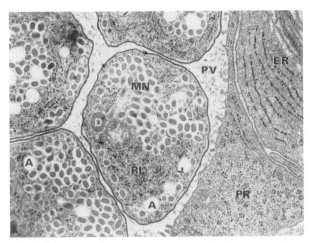

Figure 7.25 *Transmission electron micrograph of a cross-section through a mass of merozoites of* Eimeria. *The cytoplasm of the host cell adjacent to the parasitophorous vacuole (PV) shows the rough endoplasmic reticulum (ER) and many polyribosomes (PR). A, amylopectin; D, dense body; MN, micronemes; RI, ribosomes (×30 000).*

with enzyme-containing vacuoles produced by the Golgi apparatus; this is how digestive enzymes and other products enter the endocytotic vacuoles.

Lysosomes

Lysosomes are vesicles measuring 0.2–0.5 μm, bound by a single membrane. They are derived from the sER and are formed and released from the secretion side of the Golgi apparatus (the trans side). They contain enzymes such as phosphatases, proteinases, lipases, and nucleases, and have an internal pH of 4–5. When first released they are called primary lysosomes. After fusion with the

endocytotic vesicles their enzymes become active and the vesicle is then called a secondary lysosome. In these secondary lysosomes, or phagolysosomes, the ingested food is dispersed (Figures 7.1 and 7.26). Another type of secondary lysosome is the autolysosome, which is involved in the disintegration of cellular waste material, thus providing the function of disposal of debris. In *Leishmania* species so-called megasomes (reaching the size of the nucleus) may occur as special type of lysomal vacuole.

Microbodies

Microbodies are ubiquitous in eukaryotic cells. Several types of microbodies have been described (de Souza 2002):

- Hydrogenosomes (Benchimol 1999) are mostly spherical and occur in trichomonads. They reach a diameter of about 0.5–1 μm, are bound by two closely attached membranes, and reproduce by binary fission. However, they do not contain nucleic acid. Thus their proteins are synthesized in free ribosomes and post-translationally inserted into the organelle. As shown by biochemical methods, hydrogenosomes produce ATP and the molecular hydrogen that gives them their name.

- Peroxisomes (called glycosomes in trypanosomatids). These spherical, dense-appearing organelles are membrane-bound, randomly distributed throughout the cell, and reach a size of 0.7 μm diameter. Their name comes from the fact that they usually contain enzymes which use molecular oxygen to remove hydrogen atoms from organic substrate via an oxidative reaction, thus producing hydrogen peroxide

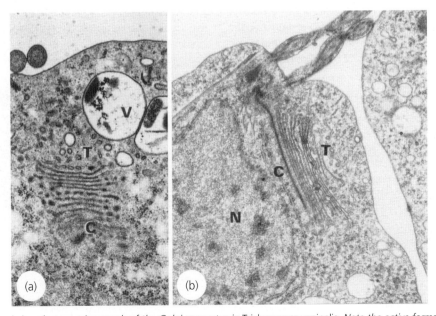

Figure 7.26 *Transmission electron micrograph of the Golgi apparatus in* Trichomonas vaginalis. *Note the active forms* **(a)** *and the inactive ones* **(b)** *(×20 000). C, cis side of Golgi; N, nucleus; T, trans side of Golgi; V, food vacuole.*

(H_2O_2). Furthermore, the peroxisomes contain catalase, which utilizes the H_2O_2 to oxidize a variety of other substrates by the so-called peroxidative reaction. In the case of the peroxisomes of trypanosomatids it has been clearly shown that they contain glycolytic enzymes being involved in the conversion of glucose to 3-phosphoglycerate (Opperdoes 1987). These bodies (originally named glycosomes) do not possess a genome and are thus encoded by nuclear genes as are the other peroxisomes. Reproduction occurs via binary fission.

- Acidocalcisomes (formerly designated in light microscopy as 'volutin granules') occur in large numbers in trypanosomes as spherical organelles of about 200 nm in diameter in all regions of the cell. They are limited by a 6-nm thick membrane and are involved in the storage and exchange of calcium, magnesium, and zinc ions (Docampo and Moreno 2001).

- Glyoxisomes of many protozoans process lipids for energy production. The reservosomes of *Trypanosoma cruzi* are spherical, reaching diameters of 0.7 μm and include all macromolecules ingested during the endocytic processes in the flagellar pocket. Also cruzipain, the major cysteine proteinase in the cell, is accumulated with reservosomes.

Rhoptries, micronemes, and dense bodies

The osmiophilic rhoptries, micronemes, and dense bodies are characteristic of the motile stages of sporozoans. They are usually located in the area between the nucleus and the apical pole. Rhoptries vary considerably in number (2–20) and in shape, appearing club-shaped (*Eimeria*), teardrop shaped (*Plasmodium*), or elongated (*Sarcocystis*, *Toxoplasma*). The anterior neck portions of rhoptries are narrow and duct-like and are found at the extreme tip of the cell running through the conoid if one is present (Figures 7.8 and 7.24). Rhoptries release

enzymes (Rhop 1–20) during the penetration process, after which they appear partly empty with a sponge-like interior. Micronemes are small rod-like structures 50–90 × 300–600 nm with rounded ends. These structures usually occupy the anterior regions of the motile stages of the sporozoans and are often arranged in bundles (Figures 7.24 and 7.27). The function of micronemes is not completely understood, although it is known that they contain at least seven proteins, which are used during adhesion of the parasite to the host cell as is the case in the so-called circumsporozoite protein (CSP) of *Plasmodium* species. Dense bodies (0.2 μm) are found at the anterior part of penetrating coccidian stages. At least six different proteins (GRA 1–6) are released during penetration into the parasitophorous vacuole (as defense against the digestive enzymes of the host cell?). They disappear during the reproductive process in meronts, macrogamonts, and microgamonts.

Wall-forming bodies

Parasites that produce walled cysts, such as amoebae, *Giardia*, coccidians, *Balantidium*, Myxozoa, and Microspora, develop wall-forming bodies of various types (see Figures 7.11e and 7.12). In amoebae, diplomonadids, Microspora, and Myxozoa, the contents of the wall-forming bodies fuse outside the cell after being excreted by exocytosis. This fused material forms an external cyst wall (see Figure 7.12). In macrogametes of coccidia the wall-forming bodies fuse in the region immediately below the cell membrane, thus producing an internal cyst wall (see Figure 7.11).

One or two different types of wall-forming body may occur in coccidia of the various genera. For example, in *Eimeria* and *Isospora* the macrogametes have two types of wall-forming bodies: (1) electron-dense bodies that give rise to the outer layer of the oocyst wall; and (2) sponge-like bodies that fuse to produce the inner layer of the oocyst wall. The entire oocyst wall is produced

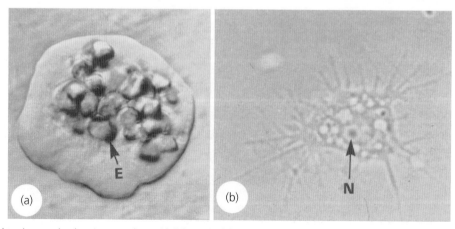

Figure 7.27 *Light micrographs showing amoebae with lobopodia* **(a)** *and filopodia (h):* **(a)** Entamoeba histolytica, *magna form. E, erythrocyte (×2 000).* **(b)** Acanthamoeba castellanii. *N, nucleus (×1 200).*

inside the cell membrane (see Figure 7.11e). The oocysts of *Sarcocystis* and the sporocysts of all coccidians are bound by a smooth wall that is formed by fusion of a single type of electron-dense wall-forming body (see Figure 7.11g). The cyst walls of all stages of Coccidia provide good protection against adverse environmental conditions. Whether they are single or double walled, the oocysts and sporocysts of all *Eimeria* have openings, whereas the sporocysts of *Isospora* and of the tissue cyst-forming coccidia have sutures in their walls (Figure 7.12d).

Vacuoles and storage elements

All parasitic protozoa have various types of vacuoles, the largest of which are the food vacuoles (see Figure 7.1); these disintegrate during digestion. Vacuoles that serve as storage elements contain crystalloid proteins (Figure 7.24a), lipids, or carbohydrates (Figure 7.24c). Lipid-containing vacuoles appear slightly grey in electron micrographs and are present mainly in resting stages such as cysts. Large amounts of protein are found in the refractile bodies of sporozoan sporozoites (Figure 7.24a) and in *Plasmodium* ookinetes. Carbohydrates are generally stored in the form of granules of glycogen or amylopectin (Figures 7.24a and 7.25). Amylopectin-containing granules appear as brilliant white areas in electron micrographs of gregarines, coccidians, and some endoparasitic ciliates. These granules are scattered throughout the cytoplasm of sporozoans of all developmental stages except microgametes, and are particularly numerous in the cytoplasm of macrogametes and oocysts (see Figure 7.11e). Glycogen (see Figure 7.14a) is present as small randomly distributed granules in the cytoplasm of *Tritrichomonas foetus* and as a large mass in the cytoplasm near the nucleus in *Entamoeba* or *Iodamoeba* cysts.

Locomotor systems

All protozoa are motile in at least one stage of their life cycle. The different species have developed distinct locomotor systems during their evolution.

Some protozoans use pseudopodia as locomotory organs. These structures, which may occur as thick lobopodia or as fine filopodia (Figure 7.27), are produced by the cytoplasmic movement mediated by the activity of the calcium-regulated actomyosin complexes. The contraction of filaments of actomyosin causes pressure on the cytoplasm at the posterior pole and this initiates a forward flow of cytoplasm at the apical pole. At the apical pole the amoeboid movement is enhanced by a transformation of the local cytoplasm from a stable gel form to a more liquid sol form (see Figure 7.16). Pseudopodia are most prominent in the various types of amoeba but may also be found in motile metazoan cells such as leucocytes.

Many protozoans use flagella or cilia as locomotor organs (Hausmann et al. 2003). These structures are constructed according to a common plan. Flagella are longer and generally less numerous than cilia, but their basic structures are similar (Figures 7.1, 7.28, and 7.29).

Both types of organelles are c. 0.2–0.4 μm in diameter and both possess an axoneme (an arrangement of nine pairs of outer microtubules and a single pair of central microtubules) which is anchored to a basal body (kinetosome) that resides inside the cortical cytoplasm of the cell (Figures 7.1 and 7.29d). The basal bodies are similar to centrioles in having nine sets of three microtubules arranged in a ring-like pattern. The basal bodies may be connected to filamentous elements such as the kinetodesmal filaments of cilia. Several species of Protozoa have a rod-like structure consisting of a network of protein filaments inside the flagellum (Figures 7.4, 7.7, and 7.29a). These rods lie beside the axoneme, probably adding to the thickness and stability of the flagellum. This enhancement may be particularly important for protozoans living in viscous media such as blood or intestinal fluids. Although flagella and cilia are constructed according to a general blueprint, in different protozoans the paired outer microtubules in the flagella and cilia may differ in shape (Figure 7.28). The A tubule, which is furnished with two dynein arms, typically possesses 13 protofilaments, whereas the B tubule has only nine protofilaments and shares three to four of the protofilaments of the A tubule. The dynein arms act as enzymes for breaking down ATP and are proposed to represent the motor system in the 'gliding filament' theory. This theory postulates that the movement of flagella and cilia is initiated by the gliding of microtubules along each other using the dynein arms as linking elements. The central regions of flagella are stabilized by spike-like elements

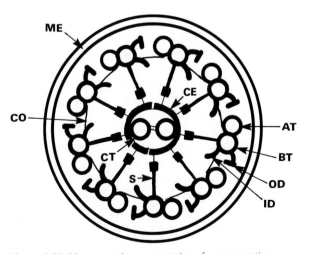

Figure 7.28 *Diagrammatic representation of a cross-section through a cilium or flagellum. AT, A tubules (consisting of 13 subunits); BT, B tubules; CF, central sheath; CO, connection between the outer pairs of microtubules; CT, central tubule; ID, inner dynein arm; ME, cell membrane; OD, outer dynein arm; S, spike.*

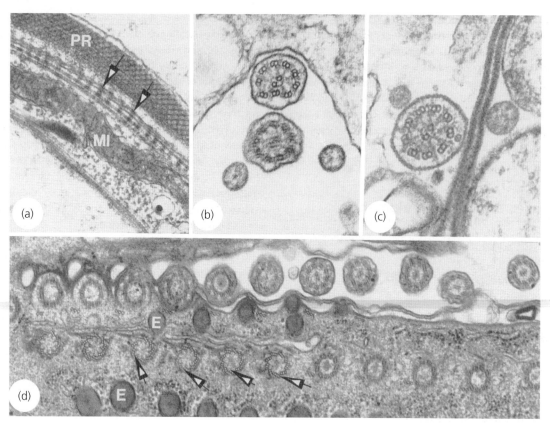

Figure 7.29 *Transmission electron micrographs of flagella* **(a–c)** *and cilia* **(d)**. **(a)** Trypanosoma vivax, *longitudinal section of a flagellum, attached to the surface by semidesmosomes (arrows) (×23 000).* **(b, c)** *Cross-section through the flagella of the gametes of* Eimeria *and* Sarcocystis *species showing the typical 9 × 2 + 2 pattern (b, ×25 000; c, ×40 000).* **(d)** *Cross-section through the basal bodies of cilia showing the typical 9 × 3 tubule arrangement (arrows). E, extrusomes; MI, mitochondrion; PR, paraxial rod of the flagellum.*

(Figure 7.28) in addition to the rod mentioned above. In the trichomonads and in the trypomastigote stages of trypanosomes, a flagellum may be connected to the cell surface by desmosomes (Figure 7.29). The recurrent flagella are attached in this manner and when the attached flagellum pulls the plasmalemma away from the body, the undulating membrane is created. The recurrent flagella never run inside the cytoplasm, but the axonemes of *Giardia* trophozoites run inside the cytoplasm for several micrometers (Figure 7.18c).

The invasive stages of sporozoans, i.e. the merozoites and sporozoites, have three types of movement: gliding, twisting, and bending. Only the first of these leads to active displacement of the organisms; the other two only change the direction of movement. The gliding form of movement is extremely rare in eukaryotic cells. It is temperature sensitive and cytochalasin B sensitive, the latter property suggesting the participation of actin in the process. The gliding movement may be related to the capping phenomenon in sporozoans. In capping, the organisms aggregate materials on their surfaces and move them towards the posterior pole, from where they release them into the surroundings. A parasite floating in a liquid could move forward using this type of action. The most studied of the capping phenomena is the circumsporozoite reaction of *P. falciparum* sporozoites.

Surface interactions and penetration

Parasites living inside host cells use a variety of means, either active or passive, to enter the host cell. In active processes the parasite's motility mechanisms play a role, whereas in passive processes the host cell internalizes the parasite by a phagocytic mechanism in the same way that food particles are ingested. In all cases, however, entry occurs after an initial recognition step in which some structure on the parasite surface reacts with and binds to some structure on the host-cell surface.

Different types of invasion processes are used by different intracellular pathogens. Some, such as the Microspora, penetrate the host-cell plasmalemma to enter the host-cell cytoplasm, whereas members of *Mycoplasma* and viruses invade by membrane fusion. Many parasites enter their host cells by phagocytosis. After entry by phagocytosis, different parasites follow different paths: the *Leishmania* develop in phagolysosomes; the mycobacteria inhibit fusion of the lysosomes with the phagosomes; and *T. cruzi*, some species of *Leishmania*, and all schizonts of *Theileria* (Piroplasmea) escape from the phagosome into the host-cell cytoplasm.

Some phagocytic processes are quite complex. Some organisms, such as malaria parasites, members of the genera *Eimeria*, *Sarcocystis*, and *Toxoplasma* actively invade by forming a moving junction between the host membrane and the parasite. After completion of the process, the parasite rests in a parasitophorous vacuole bound by the invaginated host-cell membrane. This parasitophorous vacuole membrane may disintegrate after the parasite has penetrated; for instance, stages of *Babesia* and *Theileria* are located directly in the cytoplasm of their host cells (red blood cells) shortly after invasion is completed (Figure 7.30a, b).

MODES OF REPRODUCTION

In parasitic protozoans, cell division occurs during both vegetative and sexual development (in which it occurs during the formation of gametes). Cell division may occur in production of gametes of both sexes, as in the gregarines, or only during production of the male gamonts, as in the Coccidia.

Binary fission

The most basic type of multiplication is binary fission (Figure 7.31). This process produces two daughter cells after duplication of the organelles of the mother cell. The axis of cell division is a characteristic feature of the various groups within the Protozoa, and different types of division can be distinguished: irregular, amoeba-like, longitudinal, oblique, and transverse. In microsporidians (e.g. *Nosema*) binary fissions occur after nuclear divisions in an irregular fashion, the cytoplasm being divided irregularly and different sizes of daughter cells being formed. After freeing, however, the daughter cells all develop a similar shape. In rhizopods such as *Entamoeba* and *Acanthamoeba*, cell division is irregular with respect to the cytoplasm but the constriction always occurs perpendicular to the spindle axis; this is called ameba-like division (Figure 7.31a). In flagellates such as the diplomonads, trichomonads, and trypanosomatids, and in sporozoans such as the Coccidia, the axis of cell division runs longitudinally and the process is therefore called longitudinal division. In flagellates the polarity of the division axis is determined by the initial axis of duplication of the basal bodies of the flagella (Figure 7.31b). Unusual binary fissions occur in a few species of sporozoans. For instance, the process by which the merozoites of piroplasms bud from the parent cell, one at a time, could be considered to be an irregular form of binary fission. The processes of endodyogeny in those Coccidia that form tissue cysts is a very peculiar form of longitudinal division (Figure 7.31d). In this process, two daughter cells are formed within a mother cell, arching over and parallel to the axis of the dividing nucleus. The daughter cells produce the inner two pellicular membranes de novo from the membranes of the endoplasmic reticulum and take over the outer membrane of the mother cell's pellicle. The two inner membranes of the mother cell disintegrate. In trichomonads, the axis (Figure 7.31c) of cell division is longitudinal at the beginning of the process when the flagellar basal bodies and the nucleus are reduplicated. Later, the divided nuclei move into positions opposite each other and as a result an oblique, or even a transverse, cytoplasmic division occurs.

In opalinid flagellates, which are characterized by oblique rows of 'cilia,' binary fission occurs during the sexual and asexual developmental phases (Figure 7.31f). The axis of the cytoplasmic division initially runs longitudinally, but soon becomes parallel to the oblique rows of cilia. This is called oblique division. It is intermediate between the longitudinal fission of flagellates and the transverse division of ciliates.

In ciliates such as *Balantidium* and *Ichthyophthirius*, binary fission starts with the duplication of the cytostomal kinetosomes (basal bodies) and is followed by the division of the macronuclei and micronuclei. The axis of cytoplasmic division runs perpendicular to the axis of the nuclear spindles. Cell constriction may occur simultaneously with nuclear division or there may be a time lag. This process is also called oblique division, or, as the division proceeds across the rows of the cilia, it is also described as homothetogenic fission (Figure 7.31e).

Rosette-like multiplication

If division of the cytoplasm is incomplete or delayed after nuclear division, the pair of newly formed nuclei may divide again before cytoplasmic division occurs. This may result in the simultaneous development of more than two offspring. In some cases the offspring remain attached at their posterior poles for some time, forming a mass that may appear as a rosette in stained preparations. Studies of this process by light microscopy were originally misinterpreted and the process was considered to be a multiple division. Rosettes of this type are frequently found in cultures of trichomonads and trypanosomes. They are also relatively common in preparations of *T. gondii* collected during the acute phase of infection, when endodyogenies are often repeated inside parasitophorous vacuoles of macrophages and reticuloendothelial cells.

Multiple divisions

Multiple divisions are characteristic of amoebae, sporozoans, and Microspora, but are rarely observed in the other groups of parasitic protozoans. The formation of daughter cells by multiple divisions may be initiated by three main types of mother cells (Figure 7.32a–f).

A multinuclear type of mother cell may initiate multiple divisions. In this case, daughter cell formation starts at the end of a phase of repeated nuclear division in the mother cell. In this way, multinucleate plasmodia

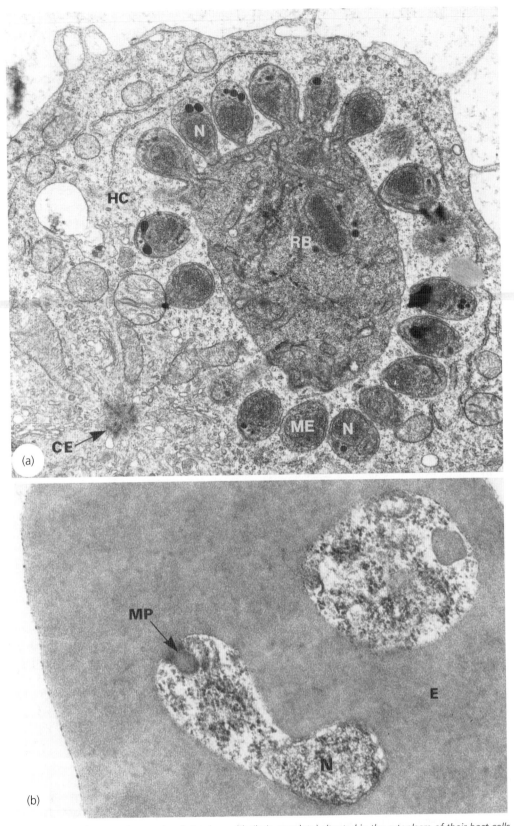

Figure 7.30 *Transmission electron micrographs of piroplasms* (Theileria annulata) *situated in the cytoplasm of their host cells.*
(a) *A meront within a lymphocyte. The host cell was in the process of dividing; note the centriole (CE) and the spindle (×12 000).*
(b) *A slender trophozoite and a spherical gamont in a bovine erythrocyte. CE, centriole; E, erythrocyte; HC, host cell cytoplasm; ME, merozoite, MP, micropore; N, nucleus; RB, residual body (×40 000).*

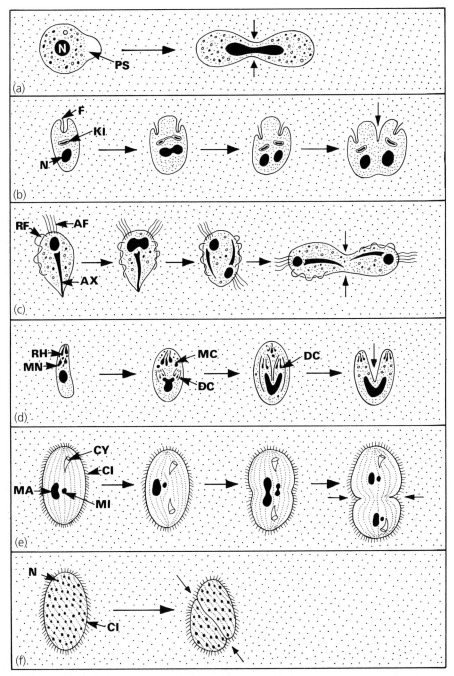

Figure 7.31 *Different types of binary fission (invaginations are shown by small arrows):* **(a)** *Amoeba type; division without fixed axis.* **(b)** *Trypanosomatid type (here* Leishmania*): longitudinal division.* **(c)** *Trichomonad type (here* T. vaginalis*): note that there is a longitudinal division only at the beginning.* **(d)** *Endodyogeny of tissue cyst forming* Coccidia *(e.g.* Toxoplasma, Sarcocystis*): inner development of daughter cells.* **(e)** *Ciliata type (e.g.* Balantidium*): cross-division.* **(f)** *Opalinata type: oblique division. AF, anterior free flagellum; AX, axostyle; B, basal body of flagellum; CI, cilium; CY, cytopharynx; DC, daughter cell; F, short flagellum in a pocket; KI, kinetoplast; MA, macronucleus; MC, mother cell; MI, micronucleus; MN, micronemes; N, nucleus; PS, pale pseudopodium; RF, recurrent flagellum; RH, rhoptries.*

are produced before cytoplasmic division starts. The occurrence of such multinucleate forms led to the naming of the genus *Plasmodium*. Multinucleate stages are also found in the life cycles of *E. histolytica*, in meronts and sporonts of gregarines, and in some stages of some coccidians. They occur in all gamonts of gregarines, but only in the microgamonts of some Coccidia.

The cytoplasm of the plasmodia may or may not be divided, but when it is divided the pattern is always species specific. In eimerian and theilerian meronts and in the sporonts of hemosporidians and of some piroplasms, daughter cell formation closely follows the last nuclear division. In eimerian microgamonts, a microgamete is formed around each nucleus and an electro-

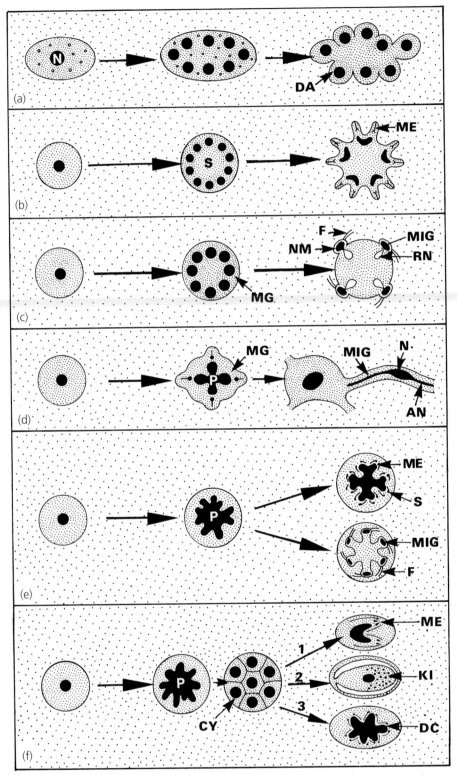

Figure 7.32 *Diagrammatic representation of multiple division:* **(a)** Ameba *sp.: formation of vegetative stages after excystation.* **(b)** *Formation of merozoites by meronts of species of* Eimeria, Plasmodium, Theileria, *etc.* **(c)** *Formation of microgametes, e.g.* Eimeria *spp.* **(d)** *Formation of microgametes of* Plasmodium *spp. and other hemosporidia (e.g. exflagellation).* **(e)** Sarcocystis *sp.: formation of merozoites by meronts (1) and of microgametes by microgamonts (2).* **(f)** *Formation of cytomeres. These may each develop: (1) two merozoites (some* Eimeria *spp.); (2) a single kinete (*Babesia *spp.); (3) many parasitic stages in the life cycle of various coccidians, i.e. merozoites of* Globidium *spp. or sporozoites of species of* Plasmodium, Babesia, *and* Theileria. *AN, axoneme (flagellum of microgamete); CT, cytomere; DA, daughter amoeba; DC, daughter cell; F, flagellum; KI, kinete; ME, merozoite anlage; MG, microgamont; MIG, microgamete; N, nucleus; NM, nucleus of microgamete (dense part); PN, polymorphous, multilobulated nucleus; RN, residual nucleus (light part); S, schizont.*

nlucent remnant of the mother cell's karyoplasm is left behind. Only occasionally do the nuclei of the mother cell divide after cytoplasmic division starts and therefore it is rare for two microgametes to originate from a single nucleus of the microgamont.

A uninuclear type of mother cell may initiate multiple division. In this case, daughter cell formation starts without a preceding phase of nuclear division, but the nucleus of the parasite has generally grown considerably and developed a lobulate surface before daughter cell formation occurs. When division of this giant nucleus begins, daughter cell formation occurs simultaneously. The daughter cells incorporate portions of the giant nucleus, the chromosomal components of which have been increased by a preceding endomitosis. This process of nuclear splitting occurs during the merogony of some cyst-forming coccidians and has been described as endo-polygeny. The microgamonts of some species of *Sarco-cystis* and of *Plasmodium* produce their microgametes in this way. A similar process of simultaneous nuclear splitting and daughter-cell formation is found in sporonts of some piroplasms during the production of sporozoites.

There is another type of multiple division that is inter-mediate between these two types. In these cases, a large mother cell with a lobated giant nucleus is divided into numerous uninuclear cytomeres. The single limiting membrane of these cytomeres originates from the surface membrane of the original cell or from the endo-plasmic reticulum of the mother cell. Such cytomeres are regularly formed in meronts of some species of *Eimeria* such as *E. tenella*, in some species of *Globidium* and in sporonts of all species of *Babesia, Theileria, Plasmo-dium,* and *Hepatozoon*. The uninuclear cytomeres ulti-mately produce the daughter cells. They give rise to two merozoites in *Eimeria*, to a single kinete in *Babesia* and *Theileria*, to many merozoites in *Globidium*, and to sporozoites in *Plasmodium, Babesia,* and *Theileria*. Cytomeres merge in Myxozoa.

Additional information concerning the composition, metabolism, genetics, and molecular biology of the cell can be found in de Puytorac et al. (1987), Alberts et al. (2002), Darnell et al. (1994), Mehlhorn (2002), and Hausmann et al. (2003).

CONCLUSION

This chapter illustrates that the Protozoa are a hetero-geneous group, with wide variations in morphology and life cycle. The parasitic protozoans are probably poly-phyletic in origin and this is also probably true of the Eukaryota (see Chapter 8, Classification and introduc-tion to the parasitic protozoa). There is no good reason to assume that the Eukaryota are of a single evolu-tionary line, but rather they may have arisen by muta-tion and probably symbiosis from various prokaryotic predecessors at more than one time.

There are only a few features that are shared by all parasitic protozoans; they are all basically unicellular organisms and they are all eukaryotic. Based on their common unicellular eukaryotic plan, a vast array of forms has developed producing a wide variety of organisms. These organisms have successfully inhabited a great range of ecological niches, some of which, to our detriment, are located in the human or animal body (Mehlhorn 2002).

REFERENCES

Alberts, B., Johnson, B., et al. 2002. *Molecular biology of the cell*, 4th edn. New York: Garland.

Benchimol, M. 1999. The hydrogenosome. *Acta Microsc*, **8**, 1–22.

Bielka, H. (ed.) 1982. *The eukaryotic ribosome*. Heidelberg: Springer.

Cavalier-Smith, T. 1998. A revised six-kingdom system of life. *Biol Rev*, **73**, 203–66.

Darnell, J., Lodish, H. and Baltimore, D. 1994. *Molekulare Zellbiologie*. Berlin: DeGruyter.

De Puytorac, P., Grain, J. and Mignot, J.P. 1987. *Précis de protistologie*. Paris: Editions Boubée.

De Souza, W. 2002. Special organelles of some pathogenic protozoa. *Parasitol Res*, **88**, 1013–25.

Docampo, R. and Moreno, S.N.J. 2001. The acidocalcisome. *Mol Biochem Parasitol*, **33**, 151–9.

Farquhar, M.G. and Palade, G.E. 1981. The Golgi apparatus – from artifact to center stage. *J Cell Biol*, **91**, 775–1035.

Ferguson, D.J.P. 2002. *Toxoplasma* and sex: essential or optional extra? *Trends Parasitol*, **18**, 355–9.

Gal, S. and Raikhel, V.V. 1993. Protein sorting in the endomembrane system of plant cells. *Curr Opin Cell Biol*, **5**, 636–40.

Gennis, R.B. 1989. *Biomembranes: Molecular structure and function*. New York: Springer.

Hackstein, J.H.P., Mackenstedt, U., et al. 1995. Parasitic apicomplexans harbor a chlorphyll-D1 complex. *Parasitol Res*, **81**, 207–16.

Hausmann, K., Hülsmann, N. and Renate Radek, R. 2003. *Protistology*, 3rd edn. Stuttgart: Schweizerbart'sche Verlagsbuchhandlung.

Huxley, H.E. and Hanson, J. 1954. Changes in the cross striations of muscle during contraction and stretch and their structural interpretation. *Nature*, **173**, 873–6.

Londsdale-Eccles, J.D. and Grab, D.J. 2002. *Trypanosoma* hydrolases and the blood-brain barrier. *Trends Parasitol*, **18**, 17–19.

Lyons, R.E., McLeod, R. and Roberts, C.W. 2002. *Toxoplasma gondii*: tachyzoite-bradyzoite intercoversion. *Trends Parasitol*, **18**, 198–203.

Mehlhorn, H. (ed.) 2002. *Encyclopedic references in parasitology*. Heidelberg: Springer.

Mehlhorn, H. and Ruthmann, A. 1992. *Allgemeine Protozoologie*. Jena: Fischer.

Morgan, G.W., Hall, B.S., et al. 2002. The kinetoplastida endocytic apparatus. part 1: a dynamic system for nutrition and evasion of host defences. *Trends Parasitol*, **18**, 491–6.

Neupert, W. and Lill, R. 1992. *New comprehensive biochemistry: membrane biogenesis and protein targeting*, Vol. 22. . Amsterdam: Elsevier.

Opperdoes, F.R. 1987. Compartmentalization of carbohydrate metabolism in trypanosomes. *Annu Rev Microbiol*, **41**, 127–51.

Scholtyseck, E. and Mehlhorn, H. 1972. Electron microscopic observations change the system of protozoans. *Naturwiss Rundschau*, **10**, 420–7.

Singer, S.J. and Nicolson, G.I. 1972. The fluid mosaic model of the structure of cell membranes. *Science*, **175**, 720–31.

8

Classification and introduction to the parasitic protozoa

F.E.G. COX

There are >200 000 named species of protozoa, of which nearly 10 000 are parasitic in invertebrates and in almost every species of vertebrate. It is therefore not surprising that humans should act as hosts to protozoa, but what is surprising is that we should harbor so many. During our short evolutionary history we have acquired >70 species belonging to >30 genera as heirlooms from our primate ancestors or souvenirs from the animals with which we have come in contact. Many of these protozoa are relatively harmless, but a few cause some of the most important diseases of tropical countries such as malaria, sleeping sickness, Chagas disease, and leishmaniasis, which together threaten over one quarter of the population of the world. Many also occur in temperate regions and some are increasingly being recognized and implicated as major pathogens in immunocompromised patients, particularly those suffering from HIV infections or undergoing immunosuppressive therapy.

From a practical viewpoint it is convenient to regard the protozoa as organisms that lie somewhere between the prokaryotic and higher eukaryotic organisms, sharing some of the characteristics of each. They are small, mostly 1–150 μm, the parasitic forms tending towards the lower end of this range, have short generation times, high rates of reproduction and a tendency to induce immunity to reinfection in those hosts that survive; features of infections with microparasites such as bacteria. On the other hand, protozoa are undoubtedly eukaryotic cells with organelles and metabolic pathways similar to those of their hosts.

Structurally, each protozoan is the equivalent of a single metazoan cell with its plasma membrane, nucleus, nuclear membrane, chromosomes, endoplasmic reticulum and, in most cases, mitochondria, Golgi body, ribosomes and various specialized structures adapted to meet particular needs (see Chapter 7, Cellular organization of parasitic protozoa). Parasitic protozoa are not simple or degenerate forms and their particular adaptations frequently include complex life cycles and specialized ways of entering their hosts and maintaining themselves once they have gained access. Their nutrition, physiology, and biochemistry are largely geared to the parasitic habit and are specialized rather than degenerate. Sexual reproduction occurs in some protozoa and, in the parasitic forms, is particularly important in the sporozoans in which it provides for apparently limitless variation and adaptability.

NATURE OF THE PARASITIC PROTOZOA

The definition of Protozoa as single-celled eukaryotic organisms is very simplistic and the group has always been an enigmatic one, largely because of the small size of the cells, the relative lack of morphological features, and the absence of any meaningful evolutionary history. In order to understand the background to the currently used classifications it is necessary to consider briefly the history of the subject, the changes that have occurred, and the reasons for these changes. It is generally agreed that the single-celled organisms that we now call protozoa were first recognized in 1676 by Antony van Leeuwenhoek, who described them as 'little animals' or 'animalcula' (Dobell 1960). The term Protozoa, meaning 'first animals', was introduced by Goldfuss in 1818 (Goldfuss 1818) and has been in use in a modified form ever since, although Goldfuss included a number of metazoans such as sponges and bryozoans within the group. The term 'single-celled animals' came into common usage after the botanist Matthias Jacob Schleiden and the zoologist Theodor Schwann had eluci- dated their cell theory in 1838 (see Cole 1926). By the middle of the 19th century, most biologists had realized that single-celled organisms included some that had greater affinities with plants than with animals and, in order to avoid the preconceptions inherent in the use of the term Protozoa, Hogg introduced the term 'Proto- ctista' to embrace those forms that had animal, plant, or no clear affinities (Hogg 1860). Haeckel also considered that all unicellular forms represented a primitive group for which he coined the term 'Protista' (Haeckel 1876). In 1938, Copeland grouped together all single-celled organisms that did not have obvious animal or plant affi- nities and used Haeckel's term Protista, which he later changed to Hogg's Protoctista on grounds of priority (Copeland 1938). All three terms, Protozoa, Protista, and Protoctista, are currently in use but this is not an appropriate place to discuss or to comment on the validity of these terms, the background of which is discussed in more detail by Lipscomb (1991) and Scamardella (1999). The term Protozoa is now almost universally used by parasitologists.

PARASITIC PROTOZOA AND PARASITIC INFECTIONS

Before attempting to classify the parasitic protozoa, it is necessary to introduce them and to comment briefly on the infections they cause. Traditionally, the parasitic protozoa of humans are listed in what used to be regarded as their taxonomic order but, as will be explained later, this is no longer appropriate. Never- theless, listing them according to their normally accepted groupings is a convenient and familiar starting place.

Parasitic amoebae

Several species of amoebae are common in humans in most parts of the world but only one, *Entamoeba histo- lytica*, is an important pathogen. *E. histolytica* invades the small intestine and colon and may be carried to the liver and other parts of the body causing amoebic dysen- tery or liver abscesses. Four other intestinal amoebae are commonly found all over the world: *E. dispar*, once regarded as a nonpathogenic race of *E. histolytica*, resembles the pathogenic form; *E. coli*, the most common amoeba of humans, is a harmless commensal, as are *Endolimax nana* and *Iodamoeba butschlii*. Humans are occasionally infected with other intestinal amoebae, *E. hartmanni*, *E. moshkovskii*, and *E. chattoni*. *E. gingivalis* occurs in the mouth and is sometimes asso- ciated with infected gums.

BLASTOCYSTIS

Cysts of *Blastocystis hominis* are increasingly being recognized in human feces. At one time this protozoan was thought to be a simple contaminant of fecal samples but it is now known to infect the intestine where it occurs in several forms including amoeboid and cystic but not flagellated forms. *B. hominis*, which occupies an unique position in the classification of the protozoa, was not considered in the 9th edition of Topley and Wilson.

Facultative amoebae and flagellates of humans

Several free-living amoebae occasionally infect humans, sometimes with fatal results. Species of *Acanthamoeba* can cause upper respiratory tract infections in immuno- compromised individuals. *Balamuthia mandrillaris* has been found in the brain. *Naegleria fowleri* and other *Naegleria* species, which are flagellates with amoeboid phases in their life cycles, have been implicated in primary meningoencephalitis in otherwise healthy indivi- duals.

INTESTINAL AND RELATED FLAGELLATES

A number of flagellates occur in the alimentary canals of humans. Large infestations can build up, but the infec- tions are seldom harmful although some may cause gastrointestinal disorders. Flagellates similar to those in the intestine can occur in other parts of the body, such as the urogenital system, and these may cause more serious infections. Eight species of flagellate are ubiqui- tous and common parasites of the human gastro- intestinal tract or urogenital system. Few do any real harm, but some occasionally give rise to unpleasant symptoms which can usually be easily treated. The most important by far is *Giardia duodenalis*, also known as *G. lamblia*, *G. intestinalis*, or *Lamblia intestinalis*. Three species of trichomonads are common in all parts of the world: *Trichomonas vaginalis* in the vagina and urethra of women and in the urethra, seminal vesicles, and pros- tate of men; *T. tenax* in the mouth; and *Penta- trichomonas hominis* in the cecum and large intestine. *Enteromonas hominis* is cosmopolitan and harmless.

Chilomastix mesnili is rare and harmless. *Dientamoeba fragilis* is transmitted through the eggs of the pinworm *Enterobius vermicularis* and may cause diarrhea.

TRYPANOSOMES OF HUMANS IN SOUTH AMERICA

Trypanosoma cruzi occurs in South and Central America and is infective to c. 100–150 species of wild and domesticated mammals. The vectors are bugs belonging to the family *Reduviidae*. In the human host, the flagellated trypanosomes enter various cells, particularly macrophages, muscle cells, and nerve cells, where they round up and multiply as nonflagellated amastigote forms. The disease, which can last for the lifetime of the host, is called Chagas disease and takes various forms depending on where the amastigotes develop, the most serious consequences being cardiac failure due to parasites in the heart muscles or the loss of the nervous control of the alimentary canal due to parasites in the nervous system. *T. rangeli*, which is usually harmless, occurs in humans, primates, cats, and dogs in Central and South America where it is transmitted by bugs and, although the parasite is quite different from *T. cruzi*, the two are occasionally confused.

TRYPANOSOMES OF HUMANS IN AFRICA

The African trypanosomes are found in the blood and are transmitted from host to host by tsetse flies belonging to the genus *Glossina*. The only form found is the flagellated form and, unlike the New World species, there are no amastigote forms. Two forms (subspecies) infect humans; *Trypanosoma brucei gambiense* and *T. b. rhodesiense*, the former in West and Central Africa where it causes chronic sleeping sickness and the latter in the savannah of East Africa where it causes acute sleeping sickness. If the trypanosomes cross the blood–brain barrier they can cause coma and death. Closely related species occur in horses, cattle, and game animals.

LEISHMANIA SPECIES

Leishmania species, which are closely related to the trypanosomes, are transmitted by sandflies and cause serious diseases in humans. The parasites, which infect macrophages, occur only in the nonflagellated amastigote form. The typical infection is a long-lasting or nonhealing cutaneous lesion but in many species, and in particular individuals, the parasites may invade subcutaneous or deeper tissues, causing hideous and permanent disfiguration. The most serious disease, kala azar, involves the macrophages of organs such as the liver and can result in multiorgan system failure. Leishmaniasis is now known to be caused by a complex of about 20 species. As the morphology of all these parasites is similar, identification tends to be based on isoenzyme and DNA techniques. In the Old World, the main species causing cutaneous leishmaniasis are *L. tropica* and *L. major* and the species causing visceral leishmaniasis are *L. donovani* and *L. infantum*. In the New World cutaneous and mucocutaneous leishmaniasis are caused by at least 13 distinct species of *Leishmania* and visceral leishmaniasis is caused by *L. infantum* (at one time thought to be a separate species, *L. chagasi*, and now probably best referred to as *L. infantum chagasi*).

COCCIDIA: *TOXOPLASMA* AND RELATED COCCIDIA

Coccidia are common parasites of the epithelial cells of the gut of vertebrates. The most common species in humans is *Toxoplasma gondii*, a parasite of felids with a very wide range of intermediate hosts, including humans. In cats and other felids the life cycle is confined to the gut, but if the infective cysts passed in the feces are ingested by other warm blooded animals, multiplication occurs in various cells throughout the body. Infections are normally symptomless, but in the fetus or in immunosuppressed patients they may be very serious and they may occasionally cause ocular damage in healthy individuals. Humans also act as definitive hosts to two intestinal species related to *Toxoplasma*, *Sarcocystis hominis*, and *S. suihominis*, the respective intermediate hosts of which are pigs and cattle. Humans can also act as accidental, dead-end, intermediate hosts to several other species of *Sarcocystis*, cysts of which occur in the muscles. *Isospora belli* is another coccidian with a life cycle similar to that of *Toxoplasma* in cats but which occurs in the intestine of humans in which it is very rare.

CRYPTOSPORIDIUM

Coccidians belonging to the genus *Cryptosporidium* are relatively common parasites in the intestinal and respiratory tracts of mammals, birds, and reptiles. *C. parvum* causes gastrointestinal disorders in cattle, sheep, and humans. The infection is unpleasant but not normally dangerous in humans except in immunocompromised individuals, such as acquired immunodeficiency syndrome (AIDS) sufferers, in whom it can be fatal.

CYCLOSPORA

Cyclospora cayetanensis is an intestinal coccidian that causes a self-limiting infection with diarrhea and malabsorption. First recorded in 1986, it has since been found in both immunologically competent and immunocompromised patients in many parts of the world. The actual route of infection is unknown, although water and meat have been implicated. It is also not clear whether or not there is any other host apart from humans.

MALARIA PARASITES

The malaria parasites belong to the same phylum as the coccidians but to a different group, the hemosporidians, members of which, as the name implies, are parasitic in the erythrocytes of vertebrates. The malaria parasites of mammals all belong to the genus *Plasmodium* and are transmitted by female mosquitoes of the genus

Anopheles. Human malaria, one of the most important diseases in the world with >500 million people at risk in tropical and subtropical regions, especially Africa, is caused by four species of *Plasmodium*: *P. falciparum*, *P. vivax*, *P. ovale*, and *P. malariae*. The disease is characterized by periodic fevers. *P. falciparum* causes malignant tertian malaria and is the most common and serious of all the forms of malaria, often causing cerebral damage resulting in death.

PIROPLASMS

The piroplasms are parasites of the erythrocytes of vertebrates and the vectors are ticks. *Babesia* species live in the blood of vertebrates and can cause serious disease, babesiosis. *B. microti* is common in wild rodents and occasionally infects humans in North America and sporadically elsewhere but the infections are not usually fatal. In Europe, splenectomized humans are sometimes infected with *B. divergens*, a cattle species, and the infection may be fatal.

CILIATES

Parasitic ciliates occur in most groups of vertebrates and invertebrates but only one species, *Balantidium coli*, infects humans. *Balantidium coli* is a common parasite of pigs in all parts of the world and has also been recorded in a number of other mammals including rats, dogs, camels, monkeys, apes, and c. 1 000 humans. The ciliate lives in the lumen of the large intestine and may invade the gut wall, where it produces ulcers resembling those caused by *Entamoeba histolytica*, although most cases are asymptomatic.

MICROSPORIDIANS

There are some 700 species of microsporidians in nearly all groups of vertebrates and invertebrates, particularly fish and arthropods. All microsporidians possess an inherent ability to multiply in their vertebrate hosts but this seems to be curtailed in humans, except in those individuals whose immune system is compromised. The first human case of microsporidiosis was recorded in 1959 and since then several new genera and species have been recorded, mainly in AIDS sufferers and others undergoing immunosuppressive therapy. Microsporidiosis is extremely rare in humans. *Encephalitozoon cuniculi*, a parasite of peritoneal macrophages that also spreads to other parts of the body including the brain, occurs in rodents, rabbits, dogs, and other carnivores and primates and there are increasing numbers of records from humans. Of the other species, few have been recorded from any animals although these must almost certainly be the sources of human infection. *Encephalitozoon hellem* parasitizes the cornea and conjunctiva and has been recorded in the viscera. *Encephalitozoon intestinalis* is an intestinal form that also occurs in a variety of cells in different parts of the body.

Enterocytozoon bieneusi is also an intestinal form associated with chronic diarrhea in AIDS patients. *Vittaforma corneae* infects the cornea. *Trachipleistophora hominis* has been recorded from the cornea and skeletal muscle of AIDS patients. The microsporidians are now thought to be more closely related to the fungi than the protozoa, but are retained in this volume because this is where they have been traditionally placed.

PNEUMOCYSTIS

The taxonomic position of *Pneumocystis* has long been uncertain but it is now clearly acknowledged to be a fungus (see Cavalier-Smith 2001). *Pneumocystis carinii*, which is normally a harmless parasite in the lungs but may be fatal in immunocompromised people, has previously been regarded as a protozoan and was included in this volume in the 9th edition of Topley and Wilson. In the present edition it is considered with the other fungi in the Medical Mycology volume, Chapter 37, *Pneumocystis* pnuemonia.

KINGDOMS IN DISPUTE

In his 1758 classification of all living things, Carl Linnaeus recognized two kingdoms, Animalia and Plantae. By the middle of the 19th century, Hogg (1860) and Haeckel (1876) had argued for the creation of an additional kingdom to accommodate single-celled organisms. Haeckel is often credited with the concept of three kingdoms of living things: Animalia (animals), Plantae (plants), and Protista (protozoa), which was expanded to four kingdoms, Animalia, Plantae, Protoctista, and Mychota (prokaryote organisms) by Copeland (1938). With the removal of the fungi from the plant kingdom the erection of a fifth kingdom, Fungi, was necessary and the viruses were added by Jahn and Jahn (1949) who recognized six kingdoms: Archetista (viruses), Monera (bacteria and blue green algae), Metazoa (multicellular animals), Metaphyta (multicellular plants), Fungi (fungi), and Protista (unicellular organisms). This classification did not recognize the fact that the single-celled eukaryotic organisms were not a homogeneous assemblage and that there were unicellular members in the animal and plant kingdoms that could not logically be separated from the Metazoa and Metaphyta. In other words, there was an argument for the creation of groupings such as Protozoa and Protophyta within the animal and plant kingdoms. Nevertheless, Jahn and Jahn's classification became widely adopted. With the removal of the viruses, now considered to be nonliving, there emerged what is now known as Whittaker's five-kingdom classification: Monera (prokaryotes), Animalia, Plantae, Fungi, and Protista (Whittaker 1969). Whittaker's five-kingdom scheme has stood the test of time fairly well but, as studies at the biochemical and molecular levels identified an increasing number of problems with the conventional ideas about single-celled organisms, it became necessary to adopt a rather more radical approach such as that set out by Cavalier-Smith (1993) and largely adopted by Corliss (1994). Conventionally, since the time of Linnaeus, the term 'kingdom' has been used as the highest category in the taxonomic hierarchy but Corliss, echoing the views of many protozoologists, accepted that the 'lower' eukaryotes could no longer be accommodated within a single kingdom and advocated the use of a superior category, the 'Empire,' containing six eukaryote kingdoms: Archezoa, Protozoa, Chromista (unicellular plant-like organisms), Plantae, Fungi, and Animalia. The first three contain only single-celled organisms, but a number of single-celled organisms also occur in the kingdom Plantae, a few in the kingdom Fungi, but none in the kingdom Animalia. One problem was that in this classification the kingdom Protozoa was not equivalent to any previous grouping with that name. In 1998, Cavalier-Smith came down in favor of six kingdoms – Bacteria, Protozoa, Chromista, Plantae, Fungi, and Animalia – effectively separating the protozoa from both the plant and animal kingdoms (Cavalier-Smith 1998). This broad classification will be adopted in this chapter.

INSTABILITY IN THE PHYLA AND CLASSES

When the Animalia was recognized as a kingdom, the Protozoa could be classified as a phylum and the subordinate groups as classes. However, when the status of the Protozoa was raised to that of a kingdom, the subordinate groups automatically became phyla and certain groups that had previously enjoyed ordinal status became elevated to classes or even phyla. This caused a great deal of confusion and requires some explanation. Goldfuss (1818) recognized three great groups of protozoa (amoebae, flagellates, and ciliates) on the basis of their mode of locomotion and this number was increased to four by the addition of the sporozoans by Bütschli (1883–5). These four groups persisted in various guises until comparatively recently, as the Rhizopoda or Sarcodina (amoebae, protozoa that move by means of pseudopodia), Mastigophora (flagellates, protozoa that move by means of flagella), Ciliophora or Ciliata (ciliates, protozoa that move by means of cilia), and Sporozoa (sporozoans, spore-forming protozoa that do not have any obvious means of locomotion). The Sporozoa sometimes included microsporidians and myxosporidians, often grouped together as cnidosporidians. This is essentially the traditional classification used in classical works by Calkins (1901), Craig (1926), Wenyon (1926), Kudo (1954), and Manwell (1961). By the beginning of the 1960s, with the availability of increasingly sophisticated ways of studying protozoa, it was becoming clear that this classification was unworkable and in 1964 the Society of Protozoologists published a revised classification (Honigberg et al. 1964). This grouped the Sarcodina and Mastigophora together as the Sarcomastigophora and removed the myxosporidians and microsporidians from the Sporozoa, with which they had been traditionally classified. One particularly controversial action was to remove the piroplasms from the Sporozoa and place them among the Sarcomastigophora, something that was ignored by virtually all parasitologists. The Society of Protozoologists' 1964 classification, with slight modifications (e.g. the replacement of the piroplasms among the Sporozoa) was widely accepted and appeared in standard textbooks such as those by Baker (1969) and Levine (1973). Subsequent changes involving *Toxoplasma* and related organisms were gradually incorporated into the 1964 scheme and a consensus classification, such as that of Baker (1977), gradually emerged.

This consensus did not last long, and rapid and major developments in our understanding of the protozoa necessitated a new classification. In 1980, the Society of Protozoologists published its second classification (Levine et al. 1980). The working party responsible for this classification recognized that the protozoa did not represent an assemblage of primitive organisms but embraced a number of diverse and distinct groups, and recognized seven phyla: Sarcomastigophora, Apicomplexa (essentially equivalent to the Sporozoa), Ciliophora, Microspora, Myxozoa, Ascetospora, and Labyrinthomorpha. Nevertheless, the affinities with the traditional classification remained clear and a modification which took into account only the parasitic forms was set out by Cox (1981) and used in information retrieval systems and parasitology textbooks such as Cox (1982), Kreier and Baker (1987), and Mehlhorn (1988). Like its predecessors, this scheme failed to keep up with developments at the ultrastructural, molecular, and biochemical levels and by the beginning of the 1990s Sleigh (1991) had decided to ignore the concept of phyla and simply recognize four 'groups': the flagellated protozoa, the amoeboid protozoa, the ciliated protozoa, and the sporozoans. This approach, also adopted by Cox (1991, 1993), has the merit of simplicity and clear links with the traditional classification, but is merely a classification of convenience and does not stand up to rigorous analysis at either the evolutionary or molecular level.

GENERA AND SPECIES: OASES OF STABILITY

The names of the genera and species of protozoa that infect humans have remained remarkably consistent throughout the upheavals that have characterized the classification of the protozoa over the past few years. The main changes have been the addition of the coccidians *Cryptosporidium* and *Cyclospora*, largely as concomitants of HIV infections, *Balamuthia* as an accidental infection, and *Blastocystis*. In addition, previously unidentified tissue parasites have now been identified as *Sarcocystis* spp. Another important change has been the growth in the number of species in the genus *Leishmania*. More recently we have seen the removal of the enigmatic species *Pneumocystis carinii*, now classified among the fungi, and the microsporidians *Brachiola*, *Encephalitozoon*, *Enterocytozoon*, *Microsporidium*, *Nosema*, *Septata*, and *Trachipleistophora*, which now rightly belong with the fungi but are retained in this volume for the sake of continuity.

The increase in the number of parasitic protozoa recorded from humans and the accumulation of knowledge about their biology has resulted in the creation of taxonomic and other groupings at the subgenus and subspecies levels, largely as a result of an unwillingness to tamper with well-established genera and species. As the parasitic protozoa have traditionally been classified with the animal kingdom, the International Code of Zoological Nomenclature applies to the naming of the subordinate groups, subgenera, and subspecies. However, the rules have not been applied rigorously and several apparent anomalies have arisen and the use of subgenera, species, and subspecies across the whole of the parasitic protozoa has not been uniform.

Among the amoebae, the only controversy relates to the most important parasite, *Entamoeba histolytica*. For many years it has been known that this parasite exists in two forms, one pathogenic and one nonpathogenic. This has resulted in some workers suggesting that the pathogenic form sensu strictu should be called *E. histolytica* and the nonpathogenic form something different, for example *E. dispar* as suggested by Brumpt (1925). Until relatively recently, most scientists were content to use one name, *E. histolytica*, but isoenzyme and other biochemical and molecular studies have established that there are real differences between the pathogenic and nonpathogenic forms and this has resulted in the reintroduction of *E. dispar* for the nonpathogenic form. Unfortunately, things are not that simple, and there have been reports that some of the biochemical criteria used are not stable (Spice and Ackers 1992). Nevertheless, it seems sensible to use the two specific names for which there is some precedent. Another apparently harmless species resembling *E. histolytica* is *E. hartmanni*, and there have been occasional reports of infections with another similar parasite, *E. moshkovskii*. There is currently some stability in the classification of amoebae belonging to the genus *Entamoeba*, but there may be other revisions on the way.

Giardia duodenalis is another intestinal parasite that causes a range of symptoms and what actually constitutes this species remains to be resolved, mainly because of the genetic variability that exists even within cloned lines (Thompson et al. 1993). There is little agreement about what the species name should actually be: *G. intestinalis* and *G. duodenalis* tend to be used interchangeably in western Europe and Australia, *G. lamblia* is used in the USA, and *Lamblia intestinalis* is sometimes used in eastern Europe. Electrophoretic analysis

suggests that the species in humans is morphologically *G. duodenalis*, which parasitizes a number of mammals, but that it could be afforded specific status as *G. intestinalis* on grounds of host specificity (Mayrhofer et al. 1995).

The genus *Leishmania* has received the most attention. Originally there were two Old World species, the cutaneous form *L. tropica* and the visceral form *L. donovani*. *L. tropica* was subsequently differentiated as two varieties, *L. tropica* var. *minor* and *L. tropica* var. *major*, and later subspecies, *L. t. minor* and *L. t. major*, which have now been raised to species level as *L. tropica* and *L. major*. Two subspecies of *L. donovani*, *L. d. donovani* and *L. d. infantum*, have now been raised to species level, as has *L. donovani* var. *archibaldi*. It was originally thought that the leishmanias in the New World were varieties of the Old World forms and the cutaneous form was called *L. tropica* var. *mexicana*, later changed to *L. mexicana*, and the mucocutaneous form was called *L. tropica* var. *braziliensis*, subsequently *L. braziliensis*. The New World visceral form of *L. donovani* subsequently became *L. d. chagasi* and then *L. chagasi*, but is now thought to be the same as the Old World species *L. infantum* (Maurício et al. 2000) which, in this volume, is referred to as a subspecies, *L. infantum chagasi*. The cutaneous forms have undergone massive revisions as more and more subspecies or species have been discovered. Initially, each new discovery was assigned either to *L. mexicana* or to *L. braziliensis* as a subspecies, for example *L. b. braziliensis*, but current opinion favors the use of species. Recognition of the fact that these species could and should be grouped as the 'mexicana complex' and the 'braziliensis complex' led to the creation of two subgenera; *Leishmania*, which embraces all the Old World species and members of the 'mexicana complex' and *Viannia*, which includes the 'braziliensis complex.' A detailed account of this classification and its background is given by Lainson and Shaw (1987). It should be pointed out here that the classification of the New World leishmanias is now based on morphological, behavioral, geographical, clinical, biochemical, and molecular criteria (see also Chapter 17, New World leishmaniasis).

Subgenera have also been used for the classification of the trypanosomes, *Trypanosoma* spp. Large numbers of trypanosomes parasitize all groups of vertebrates and, in an attempt to put some kind of order into what was becoming an increasingly heterogeneous genus, Hoare, in 1966, proposed the creation of a number of subgenera as a framework within which to classify the trypanosomes of mammals (see Hoare 1972). These subgenera were placed in two major groups; Stercoraria, containing *Megatrypanum*, *Herpetosoma*, and *Schizotrypanum*, and Salivaria, containing *Duttonella*, *Nannomonas*, *Pycnomonas*, and *Trypanozoon*. The New World trypanosome of humans, *T. cruzi*, is classified in the subgenus *Schizotrypanum* and *T. rangeli* is placed in the subgenus

Herpetosoma, whereas the Old World forms, *T. brucei gambiense* and *T. b. rhodesiense*, are placed in the subgenus *Trypanozoon* (see for example Lumsden and Evans 1976). However, the use of *Trypanosoma (Schizotrypanum) cruzi* and *Trypanosoma (Trypanozoon) brucei* is clumsy and the practice has now been all but abandoned.

The classification and nomenclature of the human trypanosomes is still not entirely satisfactory at the level of genus and species. There are real differences between *T. cruzi* and *T. brucei* and, although it would seem logical to classify them as separate genera, this cannot be done in isolation. Any changes would result in serious consequences for the classification of all the other trypanosomes of vertebrates and great disruption to information retrieval systems if one of the two generic names were to change.

The malaria parasites belonging to the genus *Plasmodium* have also been classified into subgenera largely on the same grounds as the trypanosomes; the numbers recorded in all groups of higher vertebrates, and a perceived need to bring some order into an increasingly complex situation. Nine subgenera have been proposed: *Plasmodium*, *Vinckeia*, *Laverania*, *Haemamoeba*, *Giovannolaia*, *Noyella*, *Huffia*, *Sauramoeba*, and *Carinia*, of which three occur in mammals, four in birds, and two in lizards (Garnham 1966). The human species, *Plasmodium falciparum*, was placed in the subgenus *Laverania* whereas the others, *P. malariae*, *P. ovale*, and *P. vivax*, were placed in the subgenus *Plasmodium*. As far as human parasites are concerned, the use of subgenera served very little purpose and has fallen into disuse.

CLASSIFICATION BELOW THE SPECIES LEVEL

There is a perceived need for some sort of category below that of species to differentiate between genetically distinct forms of parasites that produce markedly different diseases in humans, and the use of subspecies has been common among protozoologists. In the genus *Leishmania*, in which it was most widely used, this practice has fallen into disuse and there has been an incremental drift upwards towards separate species status. This tendency has been resisted by those working with the African trypanosomes and *Trypanosoma cruzi*, the causative agent of Chagas disease in South America. There is no doubt, however, that there are many different forms of *T. cruzi* and that their epidemiology and the diseases they produce are significantly different but, although there seems to be a de facto case for the consideration of subspecies, there has been little enthusiasm for such a move.

The only consistent use of subspecies is in the case of the African trypanosomes. Human sleeping sickness in Africa is caused by two different trypanosomes both closely related to a natural parasite of equines and game, *Trypanosoma brucei*. The two human forms have sometimes been recognized as species, *T. gambiense* and *T. rhodesiense* (Levine 1973) but, using morphological, clinical, epidemiological, biochemical, and molecular criteria, there can be no doubt about the close relationship between the human and the animal forms and the logical conclusion is that the animal form should be classified as *Trypanosoma brucei brucei* and the human forms as *T. brucei gambiense* and *T. brucei rhodesiense*. This approach has been adopted by most workers in the field and by information retrieval services. The widespread use of these subspecies was partly responsible for the demise of the use of subgenera because most authors and editors did not want to be bothered with names like *Trypanosoma (Trypanozoon) brucei gambiense*.

Toxoplasma gondii is another parasite that seems to exist in two forms, a virulent one and an avirulent one, and it is also very interesting to note that a single strain of *T. gondii* probably accounts for all the virulent forms worldwide (Sibley and Boothroyd 1992). There are, however, no serious moves to create any subspecies.

MOLECULAR AND BIOCHEMICAL APPROACHES TO THE CLASSIFICATION OF THE PARASITIC PROTOZOA

Fortunately there are a number of alternative approaches other than morphology and life cycles to the classification of protozoa, including the application of the techniques of biochemistry and molecular biology. Such techniques are increasingly being used to resolve phylogenetic, and consequently taxonomic, questions. The most widely used techniques can be roughly divided into those that are enzymatic or genetic. This topic is too extensive to be discussed in detail here and there is a useful review by Barta (2001), a comprehensive phylogenetic tree of parasites by de Meeûs and Renaud (2002), and accounts of particular applications in a number of papers in a symposium volume mainly devoted to leishmaniasis (Alvar and Baker 2002).

The first widely used technique was the use of isoenzyme profiles, which proved to be extremely useful tools for distinguishing between apparently identical parasites. Isoenzymes are enzymes that perform the same functions but have different mobilities in electric fields and can thus be distinguished very easily; populations of parasites with identical isoenzyme patterns are called zymodemes. This technique has been successfully used to distinguish the various species of *Leishmania* and the results obtained have correlated well with epidemiological, clinical, and molecular findings. Isoenzymes have also been used to distinguish between the subspecies of African trypanosomes and pathogenic and nonpathogenic forms of *Entamoeba histolytica* and *Toxoplasma gondii*, and have also been used extensively in studies on *Giardia duodenalis* and *Cryptosporidium parvum*.

Useful as they may be, isoenzyme techniques have the major disadvantage of being reliable only for distinguishing between organisms that are closely related (Richardson et al. 1986) and DNA and RNA technology is increasingly being used both for the diagnosis of parasitic infections and for resolving taxonomic and phylogenetic problems. Briefly, both DNA and RNA can be used to determine evolutionary distances, as nucleotide sequences tend to diverge over time and to evolve at a more predictable rate than do morphological characters. Differences in nucleotide sequences can, therefore, be compared and the more similar the sequences in two organisms the more likely it is that they are related; large numbers of such studies can be used to create realistic phylogenetic trees. Different kinds of RNA, particularly small nuclear RNA (16S and 18S SnRNA) and small subunit ribosomal RNA (SSUrRNA) have been extensively used for taxonomic and phylogenetic investigations and, although data are currently based on very few taxa, these techniques have been very useful. Johnson and Baverstock (1989) were the first to attempt to produce a comprehensive phylogenetic tree of the protozoa with special reference to the parasitic forms, using data derived from SSUrRNA. RNA sequences have confirmed the differences between pathogenic and nonpathogenic strains of *E. histolytica* determined by isoenzyme studies (Petri et al. 1993). Similar studies have also identified similarities between the Old World trypanosome *T. cruzi* and New World *T. brucei* and have distanced both from *Leishmania* species, thus confirming the current classification of the kinetoplastid flagellates (Maslov and Simpson 1995). SSUrRNA are the most commonly used sequences employed in systematic studies and, as far as the protozoa are concerned, the validity of such studies has been confirmed by comparing results obtained with SSUrRNA and those obtained using amino acid sequences of protein-encoding genes (Baldauf et al. 2000).

Ribosomal RNA studies using 18S rRNA have also shown that *Toxoplasma* and *Sarcocystis* are monophyletic and thus derived from a common ancestor (Ellis et al. 1995).

DNA methodology has also proved to be very useful. DNA probes have been extensively used for studies on *Leishmania* species both for diagnosis and for determining relationships. Initial studies that distinguished between the New World forms *L. mexicana* and *L. braziliensis* have now been extended to all species and, coupled with isoenzyme studies, have been largely responsible for the present classification of the genus. The development of the polymerase chain reaction (PCR) has revolutionized the use of DNA techniques in parasitology and has been used to confirm the existence of virulent and avirulent strains of *T. gondii* (Guo and Johnson 1995). Other widely used techniques include restriction fragment length polymorphism (RFLP) and random amplification of polymorphic DNA (RAPD). One example of the application of such techniques has been the establishment of the fact that *Leishmania chagasi* in the New World is the same as *L. infantum* in the New World (Maurício et al. 2000).

Increasingly, biochemical and molecular criteria are being used for the identification of new species and it is now unlikely that any future descriptions of protozoa lacking good morphological characteristics will be published and accepted unless the descriptions are soundly based at the molecular and genetic levels.

No matter how carefully any one particular species is characterized, this is of little use to any overall classification, which needs to take into account the characteristics of a diversity of organisms. One of the most powerful tools available for the study of evolution is parsimony analysis (Sober 1988; Stewart 1993). Essentially what this means is that the simplest explanation consistent with the available data should be used to reach conclusions about phylogenetic relationships. The concept is not a new one and many biologists are familiar with it in another guise, that of Occam's razor, which holds that entities should not be multiplied beyond necessity, i.e. where there is a simple explanation there is no need to look for a complex one. Parsimony analysis relies on the availability of raw data and, in the phylogenetic studies with which we are concerned here, these include morphology, life cycles, and molecular data. The ideal is to create taxa that can be ranked in a hierarchical order that reflects the underlying phylogeny. This approach has obvious advantages but also less obvious disadvantages: two taxa may possess characters that are only superficially similar; too much emphasis may be placed on too few examples; and the rates of morphological or molecular change may not be the same in the different groups under consideration. The most satisfactory examples of parsimony analysis are where traditional and molecular approaches produce the same overall pattern, but these may present the greatest difficulties if the various criteria used (e.g. morphological and molecular) are both based on the same invalid assumptions. On the other hand, where different approaches produce conflicting results it is possible that in seeking an explanation for the discrepancies a better understanding of the relationships will emerge.

CLASSIFICATION OF THE PARASITIC PROTOZOA OF HUMANS

In recent years, the availability of a number of molecular markers has made it possible to analyze relationships between protozoans and to draw conclusions that would not have been possible using morphological characters alone. So far, there has been remarkable agreement between the different approaches and, in most cases, molecular techniques have confirmed the more traditional findings; in other cases they have clarified long-standing controversies. In the long run, it is likely that molecular classifications of the protozoa will be the norm. What most parasitologists require, however, is a convenient, familiar, and stable framework within which to work and communicate, as most are concerned with only relatively few species of protozoa and those interested in the parasitic protozoa of humans need to be familiar with only around 20 genera and 50 species. It is therefore important that any classification should reflect modern thinking about the classification of the protozoa as a whole, while retaining sufficient traditional material to permit easy reference to past papers, textbooks, and information retrieval systems. There have been a number of attempts to do this and it is convenient to consider three classifications, those of the 1980s, 1994, and 2000s. The 1980s classifications are largely based on the Society of Protozoologists' second classification (Levine et al. 1980), the 1994 classifications are represented by the interim classification of Corliss (1994), and the 2000s classification is based on the comprehensive system drawn up by Cavalier-Smith (1993) and subsequent modifications. The 1980 and 1994 systems are set out below in order to set the scene for the more recent classification and to provide some continuity between the different systems several of which are still in use.

Traditional 1980s classification

Below is an outline version of a traditional 1980s classification. This classification is provided simply as a background for further discussion and to focus on the central problem of protozoan classification; the need to satisfy protozoologists on one hand and the need to be useful to parasitologists, especially medical parasitologists, on the other. The detailed classification by Lee et al. (1985) is typical of those used in the 1980s.

> Kingdom Animalia
> Subkingdom Protozoa
> > Phylum Sarcomastigophora (the amoebae and flagellates)
> > Subphylum Mastigophora. Intestinal, tissue, and blood-dwelling flagellates, e.g. *Giardia, Chilomastix, Trichomonas, Dientamoeba, Leishmania, Trypanosoma*
> > Subphylum Sarcodina. Obligate and facultative amoebae, e.g. *Entamoeba, Iodamoeba, Endolimax, Acanthamoeba, Naegleria*
> > Phylum Apicomplexa (the sporozoans)
> > > Class Sporozoea
> > > Subclass Coccidia
> > > > Order Eucoccidiida
> > > > Suborder Eimeriina, e.g. *Isospora, Sarcocystis, Toxoplasma, Cryptosporidium*
> > > > Suborder Haemosporina, e.g. *Plasmodium*
> > > Subclass Piroplasmea, e.g. *Babesia*
> > Phylum Microspora, e.g. *Encephalitozoon, Nosema*
> > Phylum Ciliophora, e.g. *Balantidium*

1994 'Interim user-friendly classification'

In 1994, in order to meet various and sometimes conflicting requirements, Corliss proposed a 'utilitarian and user-friendly classification' which took into account the numerous advances that had been made since the 1980 classifications while maintaining much that was traditional (Corliss 1994). In this classification, Corliss recognizes six kingdoms with the majority of single-celled organisms generally accepted as protozoa in two of them, the Archezoa and Protozoa. In the Archezoa there are three phyla, two of which contain parasites, and in the Protozoa there are 14 phyla, six of which contain parasites. This classification, which is outlined below, was accepted by a majority of protozoologists and was used in the 9th edition of Topley and Wilson.

> Empire Eukaryota
> > Kingdom Archezoa
> > > Phylum Metamonada
> > > > Class Trepomonadea
> > > > Order Diplomonadida, e.g. *Giardia*
> > > > Order Enteromonadida, e.g. *Enteromonas*
> > > > Class Retortamonadea
> > > > Order Retortamonadida, e.g. *Chilomastix, Retortamonas*
> > > Phylum Microspora
> > > > Class Microsporea
> > > > Order Microsporida, e.g. *Encephalitozoon, Enterocytozoon, Nosema*
> > Kingdom Protozoa
> > > Phylum Percolozoa
> > > > Class Heterolobosea
> > > > Order Schizopyrenida, e.g. *Naegleria*
> > > Phylum Parabasala
> > > > Class Trichomonadea
> > > > Order Trichomonadida, e.g. *Dientamoeba, Trichomonas*
> > > Phylum Euglenozoa
> > > > Class Kinetoplastidea
> > > > Order Trypanosomatida, e.g. *Leishmania, Trypanosoma*
> > > Phylum Ciliophora
> > > > Class Litostomatea
> > > > Order Trichomonatida, e.g. *Balantidium*
> > > Phylum Apicomplexa (Sporozoa)
> > > > Class Coccidea
> > > > Order Eimeriida e.g. *Cryptosporidium, Cyclospora, Isospora, Sarcocystis, Toxoplasma*
> > > > Class Haematozoea
> > > > Order Haemosporida, e.g. *Plasmodium*
> > > > Order Piroplasmida, e.g. *Babesia*
> > > Phylum Rhizopoda
> > > > Class Lobosea
> > > > Order Acanthopodida, e.g. *Acanthamoeba, Balamuthia*
> > > > Class Entamoebidea

Order Euamoebida, e.g. *Endolimax, Entamoeba, Iodamoeba*
Kingdom Chromista
Kingdom Plantae (the plant kingdom sensu strictu)
Kingdom Fungi (the fungi sensu strictu)
Kingdom Animalia (the animal kingdom sensu strictu)

This classification retained many of the familiar features of more traditional classifications and served as a transitional phase between the older classifications and the newer ones outlined below.

2000s Classification

Although the Corliss classification served the protozoological community well, subsequent developments in molecular biology have provided us with a much clearer understanding of phylogenetic relationships involving the single-celled organisms, to such an extent that the status of the various groups of protozoa has had to be reappraised. Comparison of SSUrRNA and protein sequences (Sogin and Silberman 1998; Doolittle 1999; Baldauf et al. 2000) have made it possible to arrange organisms within groups based on evolutionary distances. Such studies by themselves, however, do not necessarily reflect hierarchical arrangements. An alternative approach has been to consider the megaevolution of eukaryotic organisms, particularly the acquisition of prokaryotic organelles, for example mitochondria and chloroplasts, by a process of symbiogenesis,. Such studies, coupled with the realization that organelles can have been not only acquired but also modified, multiplied, or lost, have made it possible to arrange all living organisms within a realistic and evolutionarily and taxonomically sound overall system (Cavalier-Smith 1998). It is also clear that RNA and protein sequences broadly reflect the hierarchical systems based on evolutionary principles, thus providing, for the first time, a system of classification of the protozoa that is comparable with systems devised for higher eukaryotes. This scheme will therefore be adopted here and the major elements are easily recognizable as those in earlier classifications. In this classification there are six kingdoms – Bacteria, Protozoa, Animalia, Fungi, Plantae, and Chromista – of which the Protozoa constitute the basal eukaryotic kingdom. This system re-establishes the validity of the kingdom Protozoa with 13 phyla of which seven contain parasites that infect humans, but has necessitated the redistribution of some taxa, including some parasites of humans.

KINGDOM PROTOZOA. Unicellular eukaryotic phagotrophic, nonphotosynthetic organisms without cell walls
Subkingdom 1 Archezoa. Single-celled eukaryotic organisms exhibiting various prokaryotic features in their ribosomes and tRNA; lacking plastids, mitochondria, Golgi bodies, and cytoplasmic inclusions, e.g. hydrogenosomes and peroxisomes.

Phylum Metamonada. Unicellular intestinal flagellates with two, four or eight (occasionally more) flagella.
Class Trepomonadea. Intestinal flagellates; one or two karyomastigonts each with one to four flagella; no contractile axostyle; few cell-surface cortical microtubules.
Order Diplomonadida: *Giardia duodenalis*
Order Enteromonadida: *Enteromonas hominis*
Class Retortamonadea. Intestinal flagellates; cortical microtubules over entire body surface
Order Retortamonadida *Chilomastix mesnili, Retortamonas intestinalis*
Phylum Parabasalia. Unicellular flagellates with one or more nuclei and numerous flagella; characteristic complex parabasal body equivalent to Golgi body; no mitochondria.
Class Trichomonadea. Intestinal and related flagellates; typically four to six flagella, non-contractile axostyle.
Order Trichomonadida: *Dientamoeba fragilis, Trichomonas vaginalis, T. tenax, Pentatrichomonas hominis*
Subkingdom 2 Neozoa. Single-celled eukaryotic organisms typically possessing plastids, mitochondria, Golgi bodies, and cytoplasmic inclusions including hydrogenosomes and peroxisomes.
Phylum Percolozoa. Unicellular, nonpigmented organisms typically possessing one to four flagella, mitochondria and peroxisomes but lacking Golgi bodies.
Class Heterolobosea (flagellated amoebae). Amoeboflagellates; trophic form amoeboid; temporary flagellated phase in life cycle.
Order Schizopyrenida: *Naegleria fowleri*
Phylum Euglenozoa. Unicellular flagellates with one to four flagella, possess Golgi body and mitochondria.
Class Kinetoplastea (kinetoplastid flagellates). Unicellular flagellates with one to two flagella; prominent kinetoplast (DNA-containing body) within a single mitochondrion.
Order Trypanosomatida: *Leishmania donovani, L. infantum, L. major, L. tropica, L. braziliensis, L. mexicana, L. aethiopica, L. amazonensis, L. colombiensis, L. garnhami, L. guyanensis, L. lainsoni, L. lindenbergi, L. naiffi, L. panamensis, L. peruviana, L. pifanoi, L. shawi, Trypanosoma cruzi, T. brucei gambiense, T. brucei rhodesiense, T. rangeli*
Phylum Amoebozoa. Unicellular nonflagellated organisms with pseudopodia used for both feeding and locomotion.
Class Amoebaea. Amoebae with lobose pseudopodia and mitochondria; typically free-living but facultative parasites.

Order Acanthopodida: *Acanthamoeba castellani*, *Acanthamoeba* spp., *Balamuthia mandrillaris*

Class Entamoebidea. Amoebae with lobose pseudopodia; secondary loss of mitochondria, peroxisomes or hydrogenosomes; obligate parasites.

Order Euamoebida: *Entamoeba histolytica*, *E. coli*, *E. dispar*, *E. hartmanni*, *E. gingivalis*, *E. moshkovskii*, *E. chattoni* (=*E. polecki*), *Endolimax nana*, *Iodamoeba buetschlii*

Phylum Sporozoa (sporozoans). Unicellular organisms possessing at some stage an apical complex composed of polar rings, rhoptries, micronemes, and typically a conoid; elaborate life cycles involving a sexual process; all parasites.

Class Coccidea. Sexual stages small and intracellular usually in epithelial cells of vertebrate host; resistant oocyst.

Order Eimeriida: *Cryptosporidium parvum*, *Toxoplasma gondii*, *Cyclospora cayetanensis*, *Isospora belli*, *Sarcocystis hominis*, *S. suihominis*

Order Piroplasmida: *Babesia microti*, *B. divergens*, *B. gibsoni*

Order Haemosporida: *Plasmodium falciparum*, *P. malariae*, *P. ovale*, *P. vivax*

Phylum Ciliophora (ciliates). Unicellular organisms characterized by possession of many cilia used for locomotion and with complex oral ciliature used for feeding; two kinds of nuclei, one or more polyploid macronuclei and one or more diploid micronuclei.

Class Litostomatea. Mainly free-living ciliates with inconspicuous oral ciliature.

Order Vestibuliferida: *Balantidium coli*

KINGDOM CHROMISTA. Unicellular photosynthetic filamentous or colonial, organisms (in part 'algae'); some with secondary loss of plastids.

Subkingdom Chromobiota

Phylum Bigyra

Class Blastocystea: *Blastocystis hominis*

KINGDOM FUNGI. Eukaryotic heterotrophic organisms lacking plastids but possessing cell walls containing chitin and β-glycans.

Phylum Microspora (Microsporidians)

Class Microsporea: *Encephalitozoon cuniculi*, *E. hellem*, *E. intestinalis*, *Enterocytozoon bieneusi*, *Nosema ocularum*, *N. corneum*, *Brachiola connori*, *B. vesicularum*, *B. algerae*, *Microsporidium ceylonensis*, *M. africanum*, *Vittaforma corneae*, *Trachipleistophora hominis*, *T. anthropophthera*, *Pleistophora ronneafiei*.

KINGDOMS AND SUBKINGDOMS

The most important change from the 1980 and 1994 classification is that the Protozoa becomes a kingdom in its own right and is no longer regarded as a subkingdom of the kingdom Animalia as in the 1980s classification. The kingdom Protozoa now embraces the 1994 kingdoms Archezoa and Protozoa and it is now accepted that the kingdom Protozoa is the basal eukaryotic kingdom and has equal status with the other eukaryotic kingdoms, Chromista, Fungi, Plantae, and Animalia. As far as parasitologists are concerned this move is to be welcomed as it brings together most of the single-celled organisms traditionally regarded as protozoa within a single higher taxon.

ARCHEZOA

In the Corliss 1994 classification it was recognized that the flagellated protozoa, far from being classified in the same phylum as the amoebae as in the 1980s classification, did not constitute a coherent group and needed to be separated in some logical way. A number of amitochondrial flagellates and amoebae, together with the microsporidians, were placed in a distinct primitive kingdom, the Archezoa, the affinities of which were at that time obscure. In the present classification, the subkingdom Archezoa retains the phylum Metamonada and has added to it the phylum Parabasalia previously classified in the 1994 kingdom Protozoa. However, it is now apparent that the phylum Microspora has close affinities with the Fungi and should be removed from the subkingdom Archezoa and kingdom Protozoa (see below).

FLAGELLATED ARCHEZOANS: *GIARDIA*, *ENTEROMONAS*, *CHILOMASTIX* AND *RETORTAMONAS*, *DIENTAMOEBA*, *TRICHOMONAS*, AND *PENTATRICHOMONAS*

The flagellated species *Giardia duodenalis*, *Enteromonas hominis*, *Chilomastix mesnili*, and *Retortamonas intestinalis* (parasites of the human intestine) have been traditionally classified with a number of other flagellates within the subphylum or phylum Mastigophora. In this classification, as in the 1994 classification, they have been placed in the phylum Metamodada in what is now the subkingdom Archezoa. Most protozoologists, largely basing their evidence on ribosomal RNA, now regard *Giardia* as very primitive and consider that it represents a very early stage of eukaryote evolution and that in any overall evolutionarily sound scheme it should be placed in some basal, pivotal position. Parallels between *Giardia* and the other genera listed suggest that they should all be grouped together in the same phylum and that *Giardia* and *Enteromonas* should be in one class and *Chilomastix* and *Retortamonas* in another. The 1994 classification separated these flagellates from members of the Trichomonadida, containing *Dientamoeba fragilis* (parasitic in the intestine), two species of *Trichomonas*, *T. vaginalis* and *T. tenax*, and *Pentatrichomonas hominis* (parasitic respectively in the urogenital system, mouth, and intestine). *Dientamoeba* has no flagellum and is permanently amoeboid and was thus once classified with the amoebae, but there is now no doubt that it is a flagellate and should be classified with *Trichomonas*. Corliss (1994) pointed out that some protozoologists believed that the whole of the phylum Parabasala (now Parabasalia), which contains the Trichomonadida, should be classified with the Archezoa, bringing all these flagellated protozoa of humans closer together; in the present classification this has been done and they are assigned to the phylum Parasbasalia in the subkingdom Archezoa.

AMOEBOFLAGELLATES: *NAEGLERIA*

These freshwater forms are occasional accidental parasites of humans. In previous classifications they have been classified with other amoebae such as *Entamoeba* and *Acanthamoeba*, or with the flagellated protozoa. There is now no doubt that these organisms should be classified with a number of other free-living flagellated protozoa in a separate phylum, Percolozoa, as in the 1994 and present classifications.

AMOEBAE: *ACANTHAMOEBA, ENDOLIMAX, ENTAMOEBA*, AND *IODAMOEBA*

The arguments for keeping all these amoebae together in one phylum, Rhizopoda in the 1994 classification and Amoebozoa in the present one, are sound. There are also sound arguments for placing *Acanthamoeba* spp. (accidental parasites of humans) in one class, Amoebaea, and the obligate species, *Entamoeba histolytica*, in another, Entamoebidea. The class Amoebaea is equivalent to Lobosea in the 1994 classification. This classification is essentially a traditional one and should be acceptable to parasitologists.

BLOOD AND TISSUE FLAGELLATES: *LEISHMANIA* AND *TRYPANOSOMA*

These important genera include *Trypanosoma brucei gambiense*, *T. b. rhodesiense*, *T. cruzi*, and several species of *Leishmania*. They have long been recognized as comprising a distinct assemblage which has been given several names, ranging from the order Kinetoplastida to the phylum Kinetoplasta. In this classification, they are placed close to their free-living relatives, in the phylum Euglenozoa, which should be satisfactory for both parasitologists and protozoologists.

CILIATED PROTOZOA: *BALANTIDIUM*

The phylum Ciliophora contains ciliated protozoa that all possess distinctive patterns of ciliary distribution and can thus be classified satisfactorily on morphological grounds. *Balantidium coli* is the only ciliate that infects humans and, although in the past its taxonomic position has been uncertain, it is now accepted by most protozoologists that it is correctly classified in the phylum Ciliophora and class Litostomatea. *B. coli* is not an important parasite and is readily recognized, so its classification is of little interest to most parasitologists. However, what is of interest is the fact that the ciliates are now thought to be more closely related to the sporozoans than was previously believed when the protozoa were divided between the four major groups, amoebae, flagellates, sporozoans, and ciliates.

SPOROZOANS

Sporozoa is a traditional term applied to a large group of sexually reproducing, spore-forming protozoa, most of which have recognizably similar morphology at the electron microscope level and similar life cycles. This group originally contained a heterogeneous collection of spore-forming single-celled organisms including, at one time, the microsporidians, here classified in the kingdom Fungi. In the 1970s, a new parasite genus, *Perkinsus*, containing parasites of marine molluscs that were quite different from any of the other sporozoans, was allocated to this group and the phylum Sporozoa was downgraded to a class and the name of the phylum changed to Apicomplexa. The phylum Apicomplexa containing two classes Sporozoea, identical with the taxon Sporozoa established by Leuckart (1879), and Perkinsea containing the genus *Perkinsus* (see Levine et al. 1980). It is now clear that *Perkinsus* spp. do not belong with the sporozoans and should be classified close to, or with, the dinoflagellates (Siddall et al. 1997; Herrán et al. 2000) therefore the taxon Apicomplexa is redundant and should not be used. The traditional sporozoans (that is, all except the members of the class Perkinsea) possess sufficient common morphological features and life cycles, supported by studies at the molecular level, to justify inclusion in a single phylum. In the classification advocated here, Sporozoa replaces the phylum Apicomplexa and many parasitologists will be pleased to see the return of the phylum Sporozoa, which has both priority and common usage. References to Apicomplexa or apicomplexans should now revert to Sporozoa or sporozoans, the original names.

COCCIDIANS: *ISOSPORA, SARCOCYSTIS*, AND *TOXOPLASMA*

Members of this group are either parasites that undergo the whole of their life cycle in a single host, typically in the epithelial cells of the gut, or divide a similar cycle between two hosts. Until the early 1970s these parasites were known only from miscellaneous stages in their hosts but since then they have been classified as a tight assemblage on the grounds of morphology, life cycles, and molecular similarities. The species recognized in humans are *Isospora belli*, *Sarcocystis hominis*, *S. suihominis*, and *Toxoplasma gondii*. Although there may be some controversy about their exact identity and relationships, there is no doubt that in this classification they are satisfactorily accommodated in the phylum Sporozoa, Class Coccidea, and Order Eimeriida.

HEMOSPORIDANS: *PLASMODIUM* AND *BABESIA*

These parasites occur in the blood of their vertebrate hosts and have been extensively studied, largely because members of the genus *Plasmodium* cause malaria and members of the genus *Babesia* cause important diseases in cattle. In the 1994 classification, the class Haematozoea contains two orders; Haemosporida, containing members of the genus *Plasmodium*, and Piroplasmida, containing members of the genus *Babesia* which are rare and accidental parasites of humans. There can be little dispute about the classification of the Haemosporida, which is conventional and well accepted. The piroplasms have had a more checkered history, originally being described as sporozoans as long ago as 1889 (Smith and Kilborne 1893), regarded as a suborder by Wenyon (1926), transferred to the amoebae (Honigberg et al. 1964), elevated to the level of a subclass of the class Sporozoa (Levine et al. 1980) and to a class (Cox 1991). In the present classification the redundant class Haematozoa disappears and brings the malaria parasites and piroplasms together as equivalent orders. However, superficial similarities might mask fundamental differences and these may in time be sufficient to justify further separation of the two groups, either as subclasses or classes.

BLASTOCYSTIS

Blastocystis hominis has had a checkered taxonomic history since its discovery nearly 100 years ago and has been classified with the fungi, plants, flagellates, amoebae, and sporozoans. Molecular techniques have now established that it belongs close to the sporozoans and ciliates but in a separate kingdom, Chromista (Silberman et al. 1996; Cavalier-Smith 1998; Arisue et al. 2002). *Blastocystis* is the only parasitic member of this kingdom and is included in this volume for convenience and because of its affinities with other protozoa.

MICROSPORIDIANS: *BRACHIOLA, ENCEPHALITOZOON, ENTEROCYTOZOON, MICROSPORIDIUM, NOSEMA, PLEISTOPHORA, TRACHIPLEISTOPHORA,* AND *VITTAFORMA*

These tissue-inhabiting parasites have also had a checkered history, at one time being classified with the Sporozoa, later in a group with the myxosporidians and in the 1980s and 1994 classifications in a phylum of their own, the Microspora. As these parasites have become increasingly recognized as significant concomitants of HIV infections it is important that they should not be neglected on some classificatory sideline, and they are now thought to be more closely related to fungi than to protozoa (Hirt et al. 1999; Cavalier-Smith 2001). Accordingly they are now classified in the eukaryotic kingdom Fungi but are retained in this volume because they are traditionally considered together with the protozoa. It should be noted that *Pneumocystis carinii* was formerly considered to be a protozoan but has now been formally assigned to the Fungi, and this will probably also be the case with the microsporidians in due course.

REFERENCES

Alvar, J. and Baker, J.R. (eds) 2002. Molecular tools for epidemiological studies and diagnosis of leishmaniasis and selected other parasitic diseases. *Trans R Soc Trop Med Hyg*, **96** (suppl 1), S1/1-250.

Arisue, N., Hashimoto, T., et al. 2002. Phylogenetic position of *Blastocystis hominis* and of stramenopiles inferred from multiple molecular sequence data. *J Eukaryot Microbiol*, **49**, 42–53.

Baker, J.R. 1969. *Parasitic protozoa*. London: Hutchinson.

Baker, J.R. 1977. Systematics of parasitic protozoa. In: Kreier, J.P. (ed.), *Parasitic protozoa*. vol. 1. New York: Academic Press, 35–56.

Baldauf, S.L., Roger, A.J., et al. 2000. A kingdom-level phylogeny of eukaryotes based on combined protein data. *Science*, **290**, 972–6.

Barta, J.R. 2001. Molecular approaches for inferring evolutionary relationships among protistan parasites. *Vet Parasitol*, **101**, 175–86.

Brumpt, E. 1925. Etude sommaire de l' *Entamoeba dispar* n.sp. Amibe et kystes quadrinuclées parasite de l'homme. *Bull Acad Méd (Paris)*, **94**, 943–52.

Bütschli, O. 1883–5. Protozoa, In Bronn H.G. (ed.), *Bronn's Klassen und Ordnungen des Thier-Reichs*, vol. 1, Leipzig: C. F. Winter, 617–784, 785–864, 865–1088.

Calkins, G.N. 1901. *The protozoa*. London: Macmillan.

Cavalier-Smith, T. 1993. Kingdom Protozoa and its 18 phyla. *Microbiol Rev*, **57**, 953–94.

Cavalier-Smith, T. 1998. A revised six-kingdom system of life. *Biol Rev*, **73**, 203–66.

Cavalier-Smith, T. 2001. What are fungi? In: McLauchlin, D.J., McLauchlin, E.G. and Lemke, P.A. (eds), *Mycota VII, Part A, Systematics and evolution*. Berlin: Springer-Verlag, 1–37.

Cole, F.J. 1926. *The history of protozoology*. London: University of London Press.

Copeland, H.F. 1938. The kingdoms of organisms. *Quart Rev Biol*, **13**, 383–420.

Corliss, J.O. 1994. An interim utilitarian ('user-friendly') hierarchical classification and characterization of the protists. *Acta Protozool*, **33**, 1–51.

Cox, F.E.G. 1981. A new classification of parasitic protozoa. *Protozool Abstr*, **5**, 9–14.

Cox, F.E.G. (ed.) 1982. *Modern parasitology*. Oxford: Blackwell Scientific Publications, 3–4.

Cox, F.E.G. 1991. Systematics of parasitic protozoa. In: Kreier, J.P. and Baker, J.R. (eds), *Parasitic protozoa*, vol. 1. 2nd edn. San Diego: Academic Press, 55–80.

Cox, F.E.G. (ed.) 1993. *Modern parasitology*, 2nd edn. Oxford: Blackwell Scientific Publications, 1–2.

Craig, C.F. 1926. *A manual of the parasitic protozoa of man*. London: J.P. Lippincott, 1–2.

de la Herrán, R., Garrido-Ramos, M.A., et al. 2000. Molecular characterization of the ribosomal RNA gene region of *Perkinsus atlanticus*: its use in phylogenetic analysis and as a target for molecular diagnosis. *Parasitology*, **120**, 345–53.

de Meeûs, T. and Renaud, F. 2002. Parasites within the new phylogeny of eukaryotes. *Trends Parasitol*, **18**, 247–51.

Dobell, C. 1960. *Antony van Leeuwenhoek and his 'little animals'*. New York: Dover, 112–13.

Doolittle, W.F. 1999. Phylogenetic classification and the universal tree. *Science*, **284**, 2124–8.

Ellis, J.T., Luton, K., et al. 1995. Phylogenetic relationships between *Toxoplasma* and *Sarcocystis* deduced from a comparison of 18S rDNA sequences. *Parasitology*, **110**, 521–8.

Garnham, P.C.C. 1966. *Malaria parasites and other Haemosporidia*. Oxford: Blackwell Scientific Publications.

Goldfuss, G.A. 1818. Uber die Entwicklungsstufen des Thieres. *Nurenberg*, **18**, 21.

Guo, Z.-G. and Johnson, A.M. 1995. Genetic characterization of *Toxoplasma gondii* strains by random amplified polymorphic DNA polymerase chain reaction. *Parasitology*, **111**, 127–32.

Haeckel, E. 1876. *The history of creation: or the development of the earth and its inhabitants by the action of natural causes*. New York: Appleton.

Hirt, R.P., Logsdon, J.M., et al. 1999. Microsporidia are related to Fungi: evidence from the largest subunit of RNA polymerase II and other proteins. *Proc Nat Acad Sci USA*, **96**, 580–5.

Hoare, C.A. 1972. *The trypanosomes of mammals*. Oxford: Blackwell Scientific Publications.

Hogg, J. 1860. On the distinctions of a plant and an animal and on a fourth kingdom of nature. *Edinburgh New Philosoph J*, **12**, 216–25.

Honigberg, B.M., Balamuth, W., et al. 1964. A revised classification of the phylum Protozoa. *J Protozool*, **11**, 7–20.

Jahn, T.L. and Jahn, F.F. 1949. *How to know the Protozoa*. Dubuque, Iowa: William Brown.

Johnson, A.M. and Baverstock, P.R. 1989. Rapid ribosomal RNA sequencing and the phylogenetic analysis of protests. *Parasitol Today*, **5**, 102–5.

Kreier, J.P. and Baker, J.R. 1987. *Parasitic Protozoa*. Boston: Allen and Unwin.

Kudo, R.R. 1954. *Protozoology*, 4th edn. Springfield, Illinois: C.C. Thomas.

Lainson, R. and Shaw, J.J. 1987. Evolution, classification and geographical distribution. In: Peters, W. and Killick-Kendrick, R. (eds), *The leishmaniases in biology and medicine*. vol. 1. London: Academic Press, 1–120.

Lee, J.J., Hutner, S.H. and Bovee, E.C. 1985. *An illustrated guide to the Protozoa*. Lawrence, Kansas: Society of Protozoologists.

Leuckart, L. 1879. *Allgemeine Naturgeschichter der Parasiten*. Leipzig: C.F. Winter.

Levine, N.D. 1973. *Protozoan parasites of domestic animals and of man*, 2nd edn. Minneapolis: Burgess.

Levine, N.D., Corliss, J.O., et al. 1980. A newly revised classification of the Protozoa. *J Protozool*, **27**, 37–58.

Lipscomb, D. 1991. Broad classification. The Kingdoms and the Protozoa. In: Kreier, J.P. and Baker, J.R. (eds), *Parasitic protozoa*, vol. 1. 2nd edn. San Diego: Academic Press, 81–127.

Lumsden, W.H.R. and Evans, D.A. (eds) 1976. *Biology of the Kinetoplastida*. vol. 1. London: Academic Press.

Manwell, R.D. 1961. *Introduction to protozoology*. London: Edward Arnold.

Maslov, D.A. and Simpson, L. 1995. Evolution of parasitism in kinetoplastic protozoa. *Parasitol Today*, **11**, 30–2.

Maurício, I.L., Stothard, J.R. and Miles, M.A. 2000. The strange case of *Leishmania chagasi*. *Parasitol Today*, **16**, 188–9.

Mayrhofer, G., Andrews, R.H., Ey, P.L. and Chilton, N.B. 1995. Division of *Giardia* isolates from humans into two genetically distinct assemblages by electrophoretic analysis of enzymes encoded at 27 loci and comparison with *Giardia muris*. *Parasitology*, **111**, 11–17.

Mehlhorn, H. 1988. *Parasitology in focus: Facts and trends*. Berlin: Springer-Verlag.

Petri, W.A., Clark, C.G., et al. 1993. International seminar on amebiasis. *Parasitol Today*, **9**, 73–6.

Richardson, B.J., Adams, M. and Baverstock, P.R. 1986. *Allozyme electrophoresis*. New York: Academic Press.

Scamardella, J.M. 1999. Not plants or animals: a brief history of the origin of the kingdoms Protozoa, Protista and Protoctista. *Int Microbiol*, **2**, 207–16.

Sibley, L.D. and Boothroyd, J.C. 1992. Virulent strains of *Toxoplasma gondii* comprise a single clonal lineage. *Nature (London)*, **359**, 82–5.

Siddall, M.E., Reece, K.S., et al. 1997. 'Total evidence' refutes the inclusion of *Perkinsus* species in the phylum Apicomplexa. *Parasitology*, **115**, 165–76.

Silberman, J.D., Sogin, M.L., et al. 1996. Human parasite finds taxonomic home. *Nature (London)*, **380**, 398.

Sleigh, M.A. 1991. The nature of protozoa. In: Kreier, J.P. and Baker, J.R. (eds), *Parasitic protozoa*, vol. 1. 2nd edn. San Diego: Academic Press, 1–53.

Smith, T. and Kilborne, F.L. 1893. Investigations into the nature, causation and prevention of Texas or Southern cattle fever. *Bull Bureau Anim Indust*, **1**, 177–304.

Sober, E. 1988. *Reconstructing the past. Parsimony, evolution and inference*. Cambridge, Maryland: MIT Press.

Sogin, M.L. and Silberman, J.D. 1998. Evolution of the protists and protistan parasites from the perspective of molecular systematics. *Int J Parasitol*, **28**, 11–20.

Spice, W.M. and Ackers, J.P. 1992. The amoeba enigma. *Parasitol Today*, **8**, 402–6.

Stewart, C.B. 1993. The powers and pitfalls of parsimony. *Nature (London)*, **361**, 603–7.

Thompson, R.C.A., Reynoldson, J.A. and Lymbery, A.J. 1993. Giardia – from molecules to disease and beyond. *Parasitol Today*, **9**, 313–15.

Wenyon, C.M. 1926. *Protozoology. A manual for medical men, veterinarians and zoologists*, vols 1 and 2. London: Ballière Tindall, and Cox.

Whittaker, R.H. 1969. New concepts of kingdoms of organisms. *Science*, **163**, 150–60.

9

Amebiasis: *Entamoeba histolytica* infections

ADOLFO MARTÍNEZ-PALOMO AND MARTHA ESPINOSA-CANTELLANO

INTRODUCTION

There are six species of amebae commonly found in the human gastrointestinal tract. Of these, four belong to the genus *Entamoeba*: *E. histolytica*, *E. hartmanni*, *E. coli*, and the recently redescribed *E. dispar*. The rest represent separate genera: *Endolimax nana* and *Iodamoeba buetschlii*. *Dientamoeba fragilis*, long considered another intestinal amebae has now been shown to be an aberrant trichomonad. Only *E. histolytica* is of medical importance, since it is the causative agent of amebiasis, a potentially fatal disease. Other amebae are of interest mainly because their trophozoites may be difficult to distinguish by light microscopy from trophozoites of *E. histolytica*. *E. dispar*, however, deserves special mention because it is morphologically indistinguishable from *E. histolytica* and accounts for most asymptomatic infections previously assigned to *E. histolytica*.

Because of its clinical importance, most of this chapter is devoted to *E. histolytica*, although reference to the non-invasive *E. dispar* is necessarily included in those sections where differentiation between the two species is relevant. At the end of the chapter, brief consideration is given to the other intestinal amebae of humans, as well as to *E. moshkovskii* (the 'Laredo-type' that can be found in humans), *E. polecki* from pigs, *E. invadens*

from reptiles, and *E. chattoni* from monkeys. Additional information may be obtained from Martínez-Palomo (1982, 1993), Espinosa-Cantellano and Martínez-Palomo (2000), Haque et al. (2003), Stanley (2003), and Stauffer and Ravdin (2003).

For the last 70 years, two of the most puzzling aspects of the biology of *E. histolytica* were the unexplained variability of its pathogenic potential and the restriction of human invasive amebiasis to certain geographical areas despite the worldwide distribution of the parasite. The main debate centered on the question of whether there were one or two species of *E. histolytica*. In 1925, the French parasitologist Brumpt proposed that invasive amebiasis is produced by a species of ameba with a worldwide distribution that is biologically distinct but morphologically similar to nonpathogenic amebae (Brumpt 1925). Apart from some isolated studies carried out by Simic, there was nothing to refute or confirm this hypothesis for almost 50 years (Simic 1931). However, in the 1970s, differences in surface properties were found between strains of *E. histolytica* isolated from carriers and those obtained from patients with invasive amebiasis (Martínez-Palomo et al. 1973; Trissl et al. 1977, 1978). Starting in 1978, Sargeaunt and colleagues applied an isoenzyme technique to thousands of isolates of amebae obtained from several continents (Sargeaunt et al. 1982). This technique is based on the analysis of band patterns obtained after gel electrophoresis of the enzymes hexokinase and phosphoglucomutase. It revealed that invasive amebiasis is produced by strains that have characteristic isoenzyme patterns (zymodemes) distinct from those obtained from amebae harbored by most carriers. Moreover, monoclonal antibodies specifically recognized amebae belonging to pathogenic zymodemes (Strachan et al. 1988; Petri et al. 1990; Tachibana et al. 1990), and molecular probes were able to distinguish between pathogenic and nonpathogenic isoenzyme patterns (Tannich et al. 1991a).

The application of molecular biology to the field finally resolved the problem, and it is now accepted that there are two species: *E. histolytica*, previously known as pathogenic or invasive *E. histolytica*, and *E. dispar* formerly denominated as nonpathogenic or noninvasive *E. histolytica*. For a comprehensive review of the history of the debate, as well as the scientific evidence that supports the conclusion of the distinct natures of *E. histolytica* and *E. dispar*, refer to Spice and Ackers (1992), Diamond and Clark (1993), and Clark and Diamond (1993).

CLASSIFICATION

According to the Committee on Systematics and Evolution of the Society of Protozoologists, the *Entamoeba* are grouped in the family Endamoebidae, order Amoebida, subclass Gymnamoebia, class Lobosea, superclass Rhizopoda, subphylum Sarcodina, phylum Sarcomastigophora (Levine et al. 1980). However, in the 15 years since the last taxonomic scheme for the Protozoa was adopted, new data on the ultrastructure and on the molecular biology of numerous species have rendered it obsolete, and have indicated the necessity for revision of the classification of Protozoa (Cox, 1992; Corliss 1994; see also Chapter 8, Classification and introduction to the parasitic protozoa). In Corliss's proposed classification, the amebae are all placed in the phylum Rhizopoda, class Entamoebidea. Within this class, they are classified in the order Endamoebida and the family Endamoebidae. *Dientamoeba* is now considered a flagellate.

The factors currently considered to be useful for differentiation of *Entamoeba* from other amebae have been summarized by Neal (1988) and include intrinsic characteristics such as morphology, type of nuclear division, type of movement, type of physiology, antigenic nature, nature of DNA, variability of isoenzymes, susceptibility to drugs, and extrinsic characteristics such as host specificity, virulence factors, behavior in laboratory hosts, and clinical effects. Riboprinting has also been applied to distinguish different *Entamoeba* species from other intestinal protozoa.

STRUCTURE AND LIFE CYCLE

The trophozoite

The motile form of *E. histolytica*, the trophozoite (Figure 9.1), is a highly dynamic and pleomorphic cell, the form and motility of which are strongly affected by changes in temperature, pH, osmolarity, and redox potential. Actively motile amebae are elongated in form, with protruding lobopodia and a trailing uroid, whereas resting trophozoites tend to be spherical. The diameter of the cell varies between 10 and 60 μm, not only due to the pleomorphism of the parasite, but also to the feeding conditions: amebae obtained directly from intestinal or liver lesions are generally larger (20–40 μm) than those found in nondysenteric stools or in cultures (10–30 μm). The cell surface has numerous circular openings of diameter 0.2–0.4 μm that correspond to the mouths of micropinocytic vesicles. These openings are absent from the large protruding stomas of macropinocytic channels (2–6 μm in diameter) and from lobopodia. The uroid, when present, appears as a tail formed of irregular folds of the membrane and filiform processes called filopodia. Filopodia can also be found in other regions of the cell

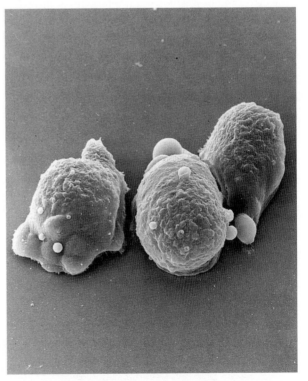

Figure 9.1 *Scanning electron micrograph of* E. histolytica *cultured trophozoites. Note the rough appearance of the cell surface, except in the regions of pseudopod extension. The parasite on the left shows a trailing uroid opposite to the pseudopod formations and abundant filopodia at the site of membrane attachment. A phagocytic opening of the membrane is clearly evident in the trophozoite located at the center of the micrograph.*

surface and are more frequently observed in amebae in monoxenic cultures or in contact with epithelial tissues (Martínez-Palomo 1982).

THE PLASMA MEMBRANE

The plasma membrane of *E. histolytica* is c. 10 nm thick and is covered by a uniform surface coat composed mainly of glycoproteins (Figure 9.2). The binding of the lectin concanavalin A to the surface of the trophozoite suggests a high content of sugar residues (i.e. mannose, glucose). In actively motile trophozoites, a thin layer of material is deposited on the substrate, leaving a trail of microexudate with cytochemical properties similar to those of the surface coat (Martínez-Palomo 1982).

Interaction of the trophozoite plasma membrane with specific ligands induces a dramatic redistribution of surface components that accumulate at the uroid and are later released into the medium. This capping of surface molecules, suggested as a mechanism of evasion of the humoral immune response, occurs through a sliding mechanism that involves both actin and myosin and is regulated by calmodulin and a myosin light-chain kinase (Espinosa-Cantellano and Martínez-Palomo 1994).

THE CYTOPLASM

The cytoplasm of *E. histolytica* trophozoites is characterized by the absence of most of the differentiated organelles found in other eukaryotic cells (i.e. mitochondria, Golgi apparatus, rough endoplasmic reticulum, centrioles, and microtubules). Instead, the cytoplasm contains abundant vacuoles of extremely variable size, with diameters 0.5–9.0 μm (Figure 9.3). Some of these vacuoles have been identified using biochemical and ultrastructural techniques. They include phagocytic vacuoles, macropinocytic and micropinocytic vacuoles, lysosomes, residual bodies, and autophagic vacuoles. Food vacuoles can be observed filled with starch or bacteria in amebae in xenic and monoxenic cultures. Amebae recovered from dysenteric stools usually contain ingested red blood cells; ingestion of red blood cells has traditionally been considered the best evidence

of the invasive nature of the parasite. The lysosomes of amebae differ from those of other eukaryotes in that the contained enzymes are bound to the lysosomal membrane rather than being free in the vacuolar compartment (Martínez-Palomo 1986). Our knowledge of this complex vacuolar system is mainly of the vacuoles involved in endocytic processes (Batista et al. 2000). The nature and functions of other vacuoles remain to be determined.

Although no structures resembling classical mitochondria are present in the amebic cytoplasm, the *E. histolytica* genome encodes molecular chaperone cpn60 and pyridine nucleotide transhydrogenase (Yu and Samuelson 1994; Clark and Roger 1995) that are associated with mitochondria in other eukaryotes. Cpn60 has been localized in one or two small (1–2 μm) vacuoles in the cytoplasm of *E. histolytica*. Targeting of cpn60 to this compartment is dependent on short signal peptides that can be replaced with mitochondrial targeting sequences from other protozoa. The crypton or mitosome seems to be a mitochondrial remnant whose function remains unknown.

A lattice of tubules and vesicles superficially resembling smooth endoplasmic reticulum can be found, although very seldom, in the cytoplasm of *E. histolytica* trophozoites. The lattice is made up of extremely thin tubules approximately 20 nm in diameter, forming irregular whorls or parallel arrays. In contrast to the 10 nm thick plasma and vacuolar membranes, the membrane enclosing these tubules is only 6 nm thick (Martínez-Palomo 1982).

Ribosomes appear to be mostly ordered in helical arrays approximately 300 nm long and 40 nm in diameter. In cysts and resting cultured trophozoites, the helices aggregate in large crystalline inclusions that can be several micrometers in length, forming the classic chromatoid body with a hexagonal packing pattern. Whether ribosomes in helices and in chromatoid bodies are functionally mature ribosomes or ribosomal precursors remains to be established.

Despite the striking motility and plasticity of the trophozoite and the important role that the cytoskeleton

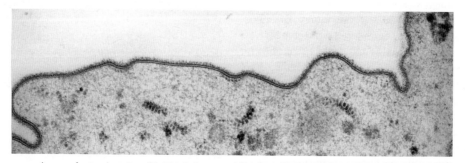

Figure 9.2 *Plasma membrane of a trophozoite of* E. histolytica *seen in a transmission electron micrograph of a cryofixed, cryosubstituted sample. A thick surface coat made of fibrilar material is clearly revealed with these techniques. The surface components here shown are usually lost during ordinary techniques used for fixation and embedding. The section was counterstained with only uranyl acetate and lead citrate.*

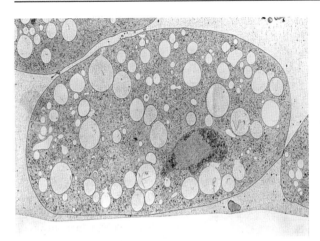

Figure 9.3 *Transmission electron micrograph of an* E. histolytica *trophozoite in culture. The ultrastructural appearance of the trophozoite is mainly characterized by the lack of mitochondria, a Golgi system, endoplasmic reticulum and cytoplasmic microtubules. Cryofixation and cryosubstitution of the sample provide a better preservation of cytoplasmic components than do standard procedures.*

plays in the invasion process, little was known about the structural organization of the cytoskeleton until recently. Prior to improvements in microscopy, only scarce microfilaments resembling actin (7 nm in diameter) were visible by transmission electron microscopy. The development of cryofixation and cryosubstitution allows much better preservation of biological materials and produces fewer artifacts than chemical fixation. Filaments with the appearance and size of actin have been visualized massed below the plasma membrane and in pseudopodial extensions and phagocytic stoma. Moreover, filaments with the size and appearance of myosin, which had only been observed in immuno-fluoresence preparations, have been observed with transmission electron microscopy. These studies have confirmed the absence of microtubules in both the cytoplasm and nuclei of nondividing amebae. Micro-tubules are known to be present in the nuclei of dividing trophozoites (González-Robles and Martínez-Palomo 1992).

Finding cylindrical particles in the cytoplasm of *E. histolytica* generated considerable interest. These particles are found in trophozoites obtained from such diverse sources as colonic exudates from patients with acute amebic colitis, human liver lesions, and axenic and polyxenic cultures. These particles are commonly arranged in rosettes; in thin sections of amebae one or two rosette conglomerates are usually found surrounding a finely granular specialized area of the cytoplasm. Each conglomerate is c. 1 μm in diameter and is composed of 9–30 cylindrical bodies. The bodies vary in size and may be up to 250 nm long and 90 nm wide. They are surrounded by a membrane that is 7 nm thick. They tend to be bullet-shaped, although occasionally they appear rounded at both ends. The

morphological similarity of the cylindrical particles to rhabdoviruses, and the cytochemical detection of a dense RNA-containing core in the rosette arrangements of the former have led to speculation of a possible viral nature. There is, however, no conclusive evidence to date about the nature and biological significance of these structures.

THE NUCLEUS

The nucleus of *E. histolytica* is 4–7 μm in diameter. The nuclear membrane is double and is interrupted by numerous nuclear pores c. 65 nm in diameter. The membrane is lined by a thin, uniform layer of granules that gives the nucleus the appearance of a ring in optical section. Chromatin clumps are usually uniform in size and evenly distributed inside the nuclear membrane although in some cells the chromatin appears concentrated on one side as a crescentic mass. The karyosome or endosome is a small, spherical mass c. 0.5 μm in diameter and located in the central part of the nucleus. Intranuclear bodies 0.2–1.0 μm in diameter are frequently observed but their nature and function are unknown. Nuclear division proceeds without dissolution of the nuclear membrane and involves the participation of thick microtubule spindles, but the precise mechanism is unknown. Fluorescence microscopy with the spreading technique has estimated that the number of chromosomes is about 30–50, which agrees with the results obtained from pulse-field gel electrophoresis (Willhoeft and Tannich 2000).

The cyst

Cysts have been studied far less extensively than trophozoites, mainly due to our inability to induce encystation in axenic cultures. Thus, most studies on *Entamoeba* cysts have been carried out on *E. invadens*, a parasite of reptiles that can be made to encyst in culture (reviewed in Eichinger 2001). *E. histolytica* cysts are round or slightly oval hyaline bodies 8–20 μm in diameter. They are surrounded by a refractile wall which is 125–150 nm thick and apparently composed of fibrillar material. This forms a tight mesh and may give rise to several lamellae (Figure 9.4). In *E. invadens*, the cyst wall has been shown to contain chitin. The plasma membrane frequently has deep invaginations; polyribosomes and vacuoles containing dense fibrogranular material can be observed close to its cytoplasmic face. Food vacuoles tend to disappear as the cyst matures. Staining with iron-hematoxylin makes the cytoplasm appear vacuolated with numerous glycogen deposits that decrease in size and number as the cyst matures. Chromatoid bodies, which are aggregated ribosomes, can be identified inside the cytoplasm as rod-shaped structures with blunt or rounded ends. Iodine stains allow the clear visualization of 1–4 small nuclei.

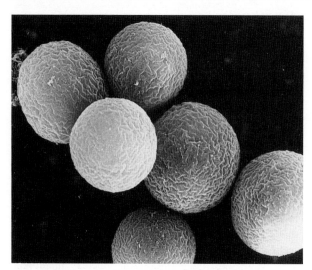

Figure 9.4 *Scanning electron micrograph of* E. histolytica *cysts recovered from feces of a human carrier. The cyst wall allows the parasite to survive in a relatively hostile environment outside the host.*

The life cycle

Trophozoites dwell in the colon, where they multiply and encyst typically producing four-nucleated cysts. These appear in the formed stools of carriers as round or slightly oval hyaline bodies with a refractive wall. When a cyst is ingested, the cyst wall is dissolved in the upper gastrointestinal tract and the parasite excysts in the terminal ileum, eventually giving rise to eight uninucleated trophozoites. Cysts do not develop within tissues, but the invasive form of the parasite, the trophozoite, can penetrate the intestinal mucosa and disseminate to other organs. Trophozoites are short-lived outside the body and do not survive passage through the upper gastrointestinal tract. In contrast, cysts may remain viable in a humid environment and stay infective for several days.

BIOCHEMISTRY, MOLECULAR BIOLOGY, AND GENETICS

Biochemistry

METABOLISM

The metabolism of the parasite is puzzling since *E. histolytica* is a facultative aerobe with peculiar glycolytic enzymes that are also found in other amitochondriate eukaryotes like *Naegleria*, and in certain bacteria. Glucose and, to a lesser extent, galactose are the main sources of energy. The uptake of glucose occurs via a specific transport system that provides c. 100 times the amount incorporated by endocytosis. This transport system is the rate-limiting step in glucose consumption. Glycogen is the main form of storage.

Glucose-6-phosphate is degraded to pyruvate via the Embden-Meyerhof pathway. A unique aspect of glycolysis in *Entamoeba* is the utilization of inorganic pyrophosphate (PPi), generally considered an end-product of metabolism, as an energy source in several glycolytic reactions. The genes coding for several of these enzymes have been cloned, including pyrophosphate-dependent phosphofructokinase (PPi-PFK) (Huang et al. 1995) and pyruvate phosphate dikinase (PPDK) (Bruchhaus and Tannich 1993; Saavedra-Lira and Pérez-Montfort 1996). In addition, the primary structure of enolase, which catalyses the conversion of 2-phosphoglycerate to phosphoenol pyruvate, has also been reported (Beanan and Bailey 1995).

The end-products of pyruvate degradation depend on the degree of anaerobiosis: under aerobic conditions (5 percent oxygen concentration) acetate, ethanol, and carbon dioxide are formed, whereas in an anaerobic environment, only ethanol and carbon dioxide are produced. The enzymatic activities involved in this process include a pyruvate synthase, aldehyde dehydrogenase (ALDH) and both NAD+-linked and NADP+-linked alcohol dehydrogenase (ADH). The genes encoding an NADP+-dependentADH and ALDH have been cloned (Kumar et al. 1992; Zhang et al. 1994). The presence of a multifunctional NAD+-dependent acetaldehyde/alcohol dehydrogenase, previously found only in anaerobic and facultative anaerobic bacteria, has been reported (Bruchhaus and Tannich 1994; Yang et al. 1994). It has been suggested that during aerobiosis, electrons are transferred from reduced substrates via a succession of carriers (including flavins, ferredoxin, other FeS proteins and ubiquinone) to molecular oxygen, which is reduced to water (Weinbach 1981; Huber et al. 1987; Ellis et al. 1994). The final electron acceptor under anaerobic conditions is not known.

Little is known about nucleic acid, lipid, and protein metabolism (Reeves 1984; McLaughlin and Aley 1985). *E. histolytica* is a purine auxotroph but can apparently synthesize pyrimidines. The parasite seems to be able to synthesize some of its lipids, including cholesterol and isoprenoids. At least three phospholipase activities (A1, A2, and L1) have been detected. Enzymes involved in inositol metabolism have been identified (Lohia et al. 1999), as have some of the early enzymes participating in the biosynthesis of glycosylated proteins (Vargas-Rodríguez et al. 1998; Villagómez-Castro et al. 1998).

E. histolytica is thought to be unique among eukaryotes in that it does not contain glutathione or glutathione-dependent enzymes (Fahey et al. 1984). In other organisms, the primary function of glutathione is to protect against oxygen toxicity. High concentrations of cysteine and other thiols have been identified in *Entamoeba*, which could carry out the functions of glutathione and its dependent enzymes (Fairlamb 1989). In addition, the presence of a disulphide oxidoreductase that binds flavin adenine dinucleotide (FAD) as a cofactor and can reduce alkyl hydroperoxides and hydrogen peroxide through NADH or NADPH has been suggested as a means of protection against damage by active oxidizing agents (Bruchhaus and Tannich 1995).

PLASMA MEMBRANE AND SURFACE MOLECULES

The biochemical characterization of cellular membranes, and particularly of the plasma membrane, is important for the understanding of host–parasite interactions. Progress in this field has been hampered by the presence of potent proteases, the unavailablity of suitable enzyme markers in the parasite, and the continuous turnover of surface membranes by endocytosis.

Membrane lipids have been studied both in whole extracts of trophozoites and in isolated plasma and internal membrane fractions (Aley et al. 1980). Cholesterol is enriched in the plasma membrane fraction to a molar ratio of 0.87 with phospholipid, a finding consistent with previous reports in other types of cells. Phosphatidylethanolamine predominates over phosphatidylcholine in the plasma membrane, although the latter is the most abundant lipid in internal vesicles. An unusual phospholipid, ceramide aminoethyl phosphonate (CAEP) has been demonstrated in internal vesicles, but is mostly concentrated in the plasma membrane, where it accounts for 35 percent of the total lipid content (Aley et al. 1980). This may have biological importance, as this compound is resistant to hydrolysis and may protect the parasite against the action of its own phospholipase.

Isolated plasma membranes with surface glycoproteins stabilized by concanavalin A treatment yielded a total of 12 glycoproteins ranging from 12 to 200 kDa in molecular weight. These were identified by radiolabeling and autoradiography (Aley et al. 1980). When Percoll density gradients were used to separate these glycoproteins after they had been labeled by lactoperoxidase iodination, 16 bands were identified, nine of which comigrated with ^{125}I-concanavalin A-labeled peptides, a result suggesting they had a glycoprotein nature. The discrepancy in the number of glycoproteins found by these two procedures may be due to the action of amebic proteases activated differently by various lysis conditions. Such proteases have been shown to affect the electrophoretic profile of whole lysates (Espinosa-Cantellano and Martínez-Palomo 1991).

Several surface molecules that mediate adhesion to cells or intestinal mucus are recognized by sera from patients cured of invasive amebiasis. They include:

- a 260 kDa N-acetyl-D-galactosamine inhibitable lectin formed of two subunits, one of 170 kDa and one of 35 kDa (Petri et al. 1989, 1990)
- an N-acetylglucosamine inhibitable lectin of 220 kDa (Rosales-Encina et al. 1987)
- a 112 kDa surface adhesin (Arroyo and Orozco 1987).

Antibodies raised against these proteins partially inhibit adhesion and phagocytosis of target cells in vitro, suggesting that the proteins participate in amebic adherence. The 112 kDa adhesin seems to be composed of two subunits, an adhesin and a protease encoded by adjacent genes (García-Rivera et al. 1999). In addition, E. histolytica expresses two lipophosphoglycan (LPG)-like molecules that may also participate in adhesion (Isibasi et al. 1982; Moody-Haupt et al. 2000). Conflicting reports on whether E. dispar synthesizes LPG have been published, one group having identified only one of the two molecules (Moody et al. 1998), while the other did not find any LPG even after using different methods of extraction in several isolates (Bhattacharya et al. 2000). Other surface antigens identified by sera from patients include peptides of 30, 96, and 125 kDa, a serine rich E. histolytica protein and a lipopeptidophosphoglycan (reviewed in Espinosa-Cantellano and Martínez-Palomo 1991).

Among the enzymes identified in the plasma membrane of E. histolytica are Ca2+-adenosine triphosphatase (ATPase), phospholipase A, neuraminidase and a metallocollagenase thought to play an important role during parasite invasion (reviewed in Espinosa-Cantellano and Martínez-Palomo 1991). Cysteine proteases are the most abundant proteases and have also been implicated in the pathogenicity of E. histolytica trophozoites (Keene et al. 1990). Four major cysteine proteases have been reported:

1 a 56 kDa cysteine protease (Keene et al. 1986)
2 a 26–29 kDa histolysin (Luaces and Barret 1988)
3 a 22–27 kDa amebapain (Scholze et al. 1992)
4 a 16 kDa cathepsin B-like protease (Lushbaugh et al. 1985).

Three different cysteine protease genes have been described. Protein sequences of purified enzymes suggest that one gene encodes amebapain and a second one histolysin. The amebic cysteine proteases are secreted enzymes and are, therefore, mainly located in the internal vesicles but some have been identified on the surface of the trophozoite. The identity of the higher molecular mass proteases was uncertain, but recent reports have started to answer this question. Cysteine proteases of 35 and 48 kDa were identified as products of EhCP2 and EhCP1, respectively, after purification and partial N-terminal sequencing (Hellberg et al. 2000). In addition, a novel 60 kDa cysteine protease, probably involved in erythrophagocytosis (Spinella et al. 1999) and a 50 kDa protease forming a tight complex with a 75 kDa adhesin (García-Rivera et al. 1999) have been reported in the parasite.

LYTIC FACTORS

E. histolytica derives its name from its ability to lyse virtually every tissue in the human body and in the

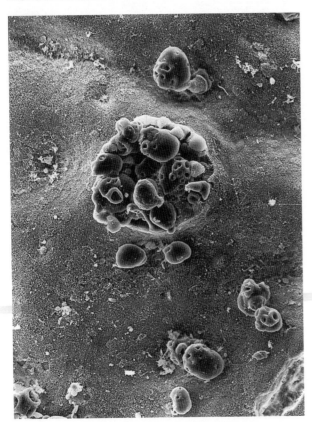

Figure 9.5 *Scanning electron micrograph of guinea pig cecal mucosa. A microulceration has been experimentally produced by trophozoites of* E. histolytica. *This lesion exemplifies the earliest stage in the development of intestinal invasive amebiasis.*

bodies of experimental animals (Figure 9.5). Lysis of target cells involves contact-dependent as well as contact-independent mechanisms. Initial attachment of the trophozoite occurs via the amebic adhesins described above. It has been suggested that, once attached, the parasite releases an active peptide, the amebapore, which is capable of inserting ion channels into liposomes and which possesses cytolytic and bactericidal activities (Leippe et al. 1994; Bruhn and Leippe 1999). *E. dispar* also possess channel-forming proteins that have been called disparpores, but seem to have different activities (Nickel et al. 1999). During the process of tissue invasion, degradation of extracellular matrix components (ECM) possibly involves specific receptors. A receptor common for fibronectin and laminin has been identified in trophozoites of *E. histolytica* (Talamás-Rohana et al. 1994). The actual degradation of ECM occurs as a result of action by the collagenase and secreted cysteine proteases that have been shown to be active against a wide variety of ECM proteins, including collagen, fibronectin, laminin, and some proteoglycans. Interestingly, the expression of cysteine protease genes has been found to be 10–100-fold higher in pathogenic *E. histolytica* than in nonpathogenic *E. dispar* strains (Tannich et al. 1991b).

Molecular biology and genetics

The total DNA content is 0.5 pg per cell, all of which is concentrated in the nucleus; there are no DNA-containing organelles (Byers 1986). Molecular analysis indicates that there is a complex karyotype with 31–35 chromosomal bands, ranging in size from 0.3 to 2.2 megabases and 14 linkage groups each with a ploidy of at least 4 (Willhoeft and Tannich 2000).

In most other unicellular eukaryotes, rDNA is present intrachromosomally. In contrast, rDNA in *Entamoeba* seems to exist exclusively as extrachromosomal circular molecules (Bhattacharya et al. 1989; Huber et al. 1989). This peculiar characteristic is also shared by *Naegleria*.

Various circular DNA molecules that are not circular rDNA have been detected by Southern blot analysis (Dhar et al. 1995). The fact that such circular DNA exists in the nucleus is supported by the identification (through hybridization with various circular DNA clones) of protein-coding regions in exonuclease-resistant DNA (Lioutas et al. 1995). The origin and function of these circular DNA elements remains to be established but their presence could explain, at least partially, the large variations in number and size of the karyotype of the parasite. The first *E. histolytica* gene was cloned in 1987 (Edman et al. 1987; Huber et al. 1987) and since then, numerous genes have been identified, their sequences analyzed, and their products characterized. An *E. histolytica* Genome Project has rapidly advanced. The genome is highly repetitive and introns are found in 15 percent of genes encoding proteins (Mann 2002).

Despite impressive advances in our understanding of the molecular biology of *Entamoeba*, many questions in the field are far from being resolved. *E. histolytica* has been successfully transfected and this allows studies on gene regulation, on the biological significance of various molecules for the survival of the parasite and on the mechanisms by which it produces disease (Nickel and Tannich 1994; Purdy et al. 1994).

CLINICAL ASPECTS

The term 'amebiasis' includes all cases of human infection with *E. histolytica*, but only a proportion of cyst-releasing individuals have symptoms due to the penetration of the parasite into the tissues, a condition known as 'invasive amebiasis.' The large group of infected, asymptomatic individuals, previously described as having 'luminal amebiasis,' is now thought to be composed mainly of *E. dispar* carriers.

Invasive amebiasis is a potentially fatal condition and there is a relatively high mortality among patients with severe forms of the disease. On a global scale, it ranks second after malaria as a cause of death among people with parasitic infections produced by protozoa (Walsh 1984).

CLINICAL MANIFESTATIONS

An epidemic of amebic dysentery that occurred in Chicago in 1933 provided an opportunity for studying the incubation period of the disease. Contamination of the water supply of two large hotels in the area was responsible for this outbreak, which resulted in around 1 400 cases of amebic dysentery. Although the complete history could not be obtained from all patients, reliable data were available from 391 cases. In these, the incubation periods ranged from 1 to 19 weeks, divided as follows: less than 1 week (6.7 percent), 1–4 weeks (59 percent), 4–8 weeks (24.7 percent), 9–13 weeks (7.4 percent), and more than 13 weeks (2.2 percent). Severe infections tended to have a shorter incubation period, most of them being reported within 1–6 weeks after initial exposure (US Treasury Department 1936).

Depending on the affected organ, the clinical manifestations of amebiasis are intestinal or extraintestinal. Both localizations can occur at the same time, but they are usually manifested separately.

Intestinal amebiasis

There are four clinical forms of invasive intestinal amebiasis, all of which are generally acute:

1 dysentery or bloody diarrhea
2 fulminating colitis
3 amebic appendicitis
4 ameboma of the colon.

The first of these is considered to be relatively benign, but the other three are severe forms of the disease and require prompt medical attention.

AMEBIC DYSENTERY

Dysenteric and diarrheic syndromes account for 90 percent of the cases of invasive intestinal amebiasis. Their various clinical manifestations depend on where the lesions are located within the rectosigmoid or higher regions of the colon. In people with this form of the disease, rectosigmoidoscopy may reveal superficial ulcerations extending over limited areas of the terminal portion of the large intestine. Patients with dysentery have an average of 3–5 mucosanguineous evacuations per day with moderate colic pain preceding discharge, they also have rectal tenesmus. In patients with bloody diarrhea, evacuations are also few, but the stools are composed of liquid fecal material stained with blood and, although there is moderate colic pain, there is no rectal tenesmus. Fever and systemic manifestations are generally absent. These syndromes constitute classic ambulatory dysentery and can easily be distinguished from diseases of bacterial origin. The clinical course is moderate and symptoms disappear rapidly with treatment; spontaneous remissions are occasionally observed after several days (Sepúlveda and Treviño-García Manzo 1986; Martínez-Palomo and Ruíz-Palacios 1990).

FULMINATING AMEBIC COLITIS

In contrast to the dysenteric syndrome, fulminating amebic colitis is an extremely severe, rapidly evolving clinical condition with necrotic ulcerous lesions extending over large areas, even the entire colon. These can affect all layers of the intestinal wall. Evacuations, preceded by intense colic pain, are frequent (20 or more in 24 h) and consist of fecal material mixed with blood or occasionally blood alone. Rectal tenesmus tends to be constant and acute. Systemic manifestations include abdominal discomfort, anorexia, and nausea. High fever (39–40°C) is usually present, accompanied by a weak, rapid pulse and low blood pressure. The patient suffers from dehydration and prostration, and may even develop shock. Peritonitis is a common complication due to the perforation of the intestinal wall (Sepúlveda and Treviño-García Manzo 1986; Martínez-Palomo and Ruíz-Palacios 1990).

AMEBIC APPENDICITIS

The symptoms of this condition are similar to those of bacterial appendicitis: acute pain and rigidity in the lower right quadrant of the abdomen, fever, tachycardia, and nausea. In more than two-thirds of cases of amebic appendicitis, patients have ulcerous lesions of the cecum. In these cases, diarrhea, often bloody, is also present (Sepúlveda and Treviño-García Manzo 1986; Martínez-Palomo and Ruíz-Palacios 1990).

AMEBOMA

Amebomas are pseudotumoral lesions, whose formation is associated with necrosis, inflammation and edema of the mucosa, and submucosa of the colon. Amebomas always co-exist with amebic ulcerations. These are generally single, but occasionally multiple, masses, usually found in the vertical segments of the large intestine: the cecum, the rectosigmoid region, and the hepatic and splenic angles of the colon. The condition is usually acute with dysentery or bloody diarrhea, abdominal pain, and a palpable mass in the corresponding area of the abdomen. If the lesion is located in the rectosigmoid region, it can be identified by endoscopy (Sepúlveda and Treviño-García Manzo 1986; Martínez-Palomo and Ruíz-Palacios 1990).

Extraintestinal amebiasis

E. histolytica can infect almost every organ of the body, including the liver, brain, lung, and skin. By far the most frequent form of extra-intestinal amebiasis is the amebic liver abscess which, due to its great clinical and epidemiological importance, is the most extensively discussed here.

Thoracic complications of amebiasis, including pleuropulmonary, pericardial, and mediastinal manifestations, are secondary to liver abscess. Involvement of the central nervous system is rare and lesions are single or multiple small areas of softening in the left cerebral hemisphere. Cutaneous amebiasis is usually a complication of intestinal amebiasis with dysentery. It may occur in the perianal region, or it may appear in the skin surrounding a fistula in cases of liver abscesses that involve the abdominal wall.

AMEBIC LIVER ABSCESS

This condition results from the migration of *E. histolytica* trophozoites from the colon to the liver via the portal circulation. The time lapse between penetration of the mucosa of the large intestine and damage to the hepatic parenchyma is unknown. It has been observed clinically and confirmed in a large series of autopsies that amebic colitis is found in only one-third of cases of hepatic abscess. Amebic liver abscesses have been reported in patients of all ages, but predominate in adults between 20–60 years. It has a marked preference for the right lobe of the liver and it is at least three times more frequent in males than in females. Interestingly, it is 10 times more common in adults than in children (Sepúlveda and Treviño-García Manzo, 1986).

The signs and symptoms of amebic hepatic abscess vary but, in general, the onset is abrupt with pain in the right hypochondrium radiating towards the right shoulder and scapular area. The pain usually increases with deep breathing, coughing, and while stepping on the right foot during walking. When the abscess is localized to the right lobe, symptoms include irritative cough, sometimes productive, and a pleuritic type of pain. Abscesses in the upper left lobe can cause epigastric, sometimes dyspneic pain, at times spreading to the base of the neck and to one or both shoulders. A sharp increase in pain centered in the precordial region strongly suggests that the abscess has penetrated into the pericardial space. This complication is nearly always fatal due to cardiac tamponade. Localization of the abscess in the vicinity of the diaphragm can lead to perforation of the pleura, causing pleurisy or empyema. If the bronchi become involved, pus may be vomited leading to clinical improvement. If the lung is involved, there may be pulmonary consolidation and abscess formation. When located in the inferior part of the liver, the abscess can penetrate into the peritoneal cavity and into neighboring organs (Sepúlveda 1970).

Fever between 38 and 40°C is found in all patients with amebic liver abscess. The patient commonly has chills and profuse sweating in the afternoon and at night. Other symptoms include anorexia, nausea, vomiting, and, of course, diarrhea (with or without blood), and dysentery. On physical examination, the cardinal sign of amebic liver abscess is painful hepatomegaly. Digital

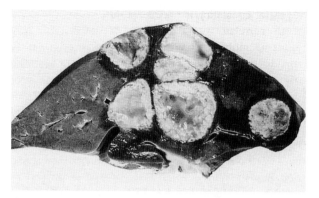

Figure 9.6 *Macroscopic appearance of amebic liver abscesses. When viewed microscopically, these lesions represent areas of liquefactive necrosis, rather than true abscesses.*

pressure and fist percussion often produce intense pain in the liver region. On palpation, the liver is soft and smooth, in contrast to the rough hard irregular character of the liver in patients with cirrhosis and hepatocarcinoma (Sepúlveda 1970). Jaundice is present in 8 percent of the patients that respond well to treatment. When jaundice is severe, multiple abscesses should be suspected (Figure 9.6). Ascites and hepatic coma occur rarely, but prognosis is poor in these patients.

IMMUNOLOGY

Humoral immune response

The time between infection with *E. histolytica* and the appearance of local antibody responses remains unknown. Coproantibodies have been found by indirect hemagglutination assay (IHA) in c. 80 percent of patients with amebic dysentery, compared to 2 percent of healthy controls and 4 percent of patients with non-amebic intestinal parasitic infections; this figure was obtained at the time when patients came in for treatment, but 3 weeks later it fell to 55 percent. This coincided with an increase in serum antibodies suggesting that the humoral immune reaction in people with invasive intestinal amebiasis is initiated by a short and transient local secretory response, followed by an increase in systemic antibodies (reviewed in Pérez-Montfort and Kretschmer 1990). Immunoglobulin (Ig)A-type anti-*E. histolytica* antibodies have also been found in human milk, colostrum, and saliva, and in the bile of intracecally immunized rats. The protective role of secretory IgA (and IgE) in amebiasis has not been established. In fact, *E. histolytica* trophozoites have been shown to degrade secretory IgA (Kelsall and Ravdin 1993).

Circulating antibodies to *E. histolytica* can be detected as early as 1 week after the onset of symptoms in humans and experimental animals. All immunoglobulin classes are involved, but there seems to be a predominance of IgG. In contrast to findings in animals, the levels of circulating antibodies in humans do not necessarily correlate with the severity of the disease. High titers of antibodies tend to appear early in the disease and to persist after invasive amebiasis is cured and in patients whose subclinical amebic infection is controlled (Knobloch and Mannweiler 1983). Antibody levels fall in different people at different rates following treatment. The detection of antibody in a patient depends on the sensitivity of the test used. Thus, if antibodies are measured by the sensitive techniques of IHA or by enzyme-linked immunosorbent assay (ELISA), antibodies can be detected more than 3 years after an invasive amebic episode in the absence of any recurrent infection (reviewed in Kretschmer 1986).

Complement may also be involved in limiting amebiasis, a role supported by the observation that experimental animals decomplemented with cobra venom factor are significantly more susceptible to development of amebic liver abscess than are normal animals. Also, the addition of cobra venom factor to normal human serum decreases the lysis of complement sensitive strains (Reed and Gigli 1990). Moreover, amebae are capable of activating complement through the classical and alternative pathways, even in the absence of antibody (Calderón and Schreiber 1985). This activation is lethal in vitro to the non-invasive *E. dispar*, and to *E. histolytica* strains that have been axenized and have partially lost their virulence. However, fresh isolates from patients with colitis or amebic liver abscesses are resistant to complement-mediated lysis (Reed et al. 1983) and repetitive treatment with human serum may induce complement resistance of axenic *E. histolytica* in culture (Calderón and Tovar 1986) suggesting that some amebae have developed a mechanism not yet understood for evasion of complement lysis. These observations may explain the conflicting reports of complement levels in humans and in experimental animals with invasive amebiasis: some suggest decreased levels of complement whereas others suggest elevated levels.

The deleterious effects of antiamebic antibodies on in vitro cultured trophozoites have been studied extensively. Immune sera can inhibit both erythrophagocytosis by *E. histolytica* and the cytotoxic effect of amebae on cells of cultured lines. At high concentrations, immune serum produces rapid lysis of trophozoites (reviewed in Kretschmer 1986). In apparent contrast to these observations, amebae may survive exposure to antibody in some circumstances. Incubation of amebae with immune serum induces the rapid redistribution of surface-bound antigens to the uroid of the cell and the caps thus formed can be either released or endocytosed. This process is both temperature and pH dependent, occurs through a sliding mechanism that involves both actin and myosin, and is regulated by calmodulin and a myosin light-chain kinase (Espinosa-Cantellano and Martínez-Palomo 1994). The parasite, which has thus shed the antibody–antigen complexes, apparently emerges undamaged and free of potentially harmful antibodies.

Taken together, these observations suggest that although a rapid humoral immune response is mounted by the host upon invasion by *E. histolytica*, the parasite has developed efficient evasion mechanisms that include resistance to complement and a mechanism for capping and shedding surface antigens. It is, therefore, considered that humoral antibodies are not protective against *E. histolytica*, a conclusion further supported by the observation of high rates of reinfection in persons with elevated antibody titers and by the detection of high levels of antibodies in patients with symptomatic amebiasis.

Cellular immune response

Even though the basic components of a local cellular immune response, such as mononuclear phagocytes and lymphocytes, are regularly present in early intestinal amebic lesions, their role in the establishment or prevention of invasive infection is not well understood. Evidence of systemic cell-mediated immunity has been confirmed in vivo through delayed hypersensitivity skin reactions, and in vitro through lymphokine (macrophage inhibition factor (MIF)), blastogenic response, leukocyte adherence inhibition, and lymphocytotoxic assays. Many patients, however, fail to react to delayed skin tests with amebic antigens during the early stages of the disease, apparently due to a state of specific unresponsiveness. Such unresponsiveness has been detected by in vitro assays, which show a transient failure to produce MIF, and by a reduction of lymphocyte cytotoxicity in patients in early stages of untreated hepatic amebiasis (reviewed in Kretschmer 1986). Thus tissue invasion by *E. histolytica* must be preceded by and be associated with some degree of T-cell suppression as a result of either selection or induction of suppressor T cells. Supporting the selective mechanism is the observation of increased susceptibility to invasive amebiasis of patients and experimental animals immunosuppressed by mechanisms mediated by suppressor T cells. The coincidence of malnutrition and amebic liver abscess in over 90 percent of clinical and autopsy cases and the significantly increased HLA-DR3 antigen levels found in Mexican patients with amebic liver abscess also suggest that T-cell mediated suppression is a factor in invasive amebiasis. The inductive proposal, on the other hand, is based on the observation that cell-free extracts of the parasite can exhaust, and thus suppress, the cellular immune response of the host (Kretschmer 1986).

The existence of an early transient anergy state does not necessarily preclude a protective role for cell-mediated immunity against invasive amebiasis. In fact, a state of cell-mediated immunity may be responsible for the rarity of recurrences of amebic liver abscesses in humans: 0.04 percent recurrences versus 0.2 percent amebic liver abscess per initial infection per year calculated in Mexico City (Kretschmer 1986), although a study of 2 013 cases of amebic liver abscess in Vietnam did not support the concept of a long-lasting immunity against recurrent liver abscess (Blesmann et al. 2002).

The existence of a partial cellular, rather than humoral, immunity against extraintestinal amebiasis is based on the following findings:

- immunization studies and studies of the effect of immunosuppression of experimental animals on infection
- the results of a few studies of passive transfer of immunity with cells
- the cytolytic effect of activated lymphocytes and macrophages (irrespective of the presence of antibodies or complement) against *E. histolytica*.

This conclusion is further supported by studies on severe combined immunodeficient (SCID) mice, in which 100 percent developed liver abscesses when challenged intrahepatically with *E. histolytica* trophozoites (Cieslak et al. 1992). The actual defense strategy seems to depend heavily on macrophages; depressing macrophage function with silica or antimacrophage serum increases the development of experimental amebiasis, whereas enhancing macrophage function with bacille Calmette-Guérin (BCG) decreases such development. Moreover, congenitally athymic nu/nu mice (which are devoid of T lymphocytes) develop amebic liver abscesses or intestinal amebic disease only after macrophage blockade with silica. *E. histolytica* may possess mechanisms to evade or modify the action of macrophages. The parasite has been found to release a small molecular weight factor that can inhibit the locomotion of monocytes, but not of polymorphonuclear cells (reviewed in Kretschmer 1986).

PATHOLOGY

A detailed description of the pathological changes in amebiasis is beyond the objectives of this chapter, but a brief summary of the main characteristics of intestinal and hepatic lesions is given. For a comprehensive review of the pathology of the disease, refer to Pérez-Tamayo (1986).

Invasion of the colonic and cecal mucosa begins in the interglandular epithelium. Cell infiltration around invading amebae leads to rapid lysis of inflammatory cells and tissue necrosis (Figure 9.7); thus, acute inflammatory cells are seldom found in biopsy samples or in scrapings of rectal mucosal lesions. Ulcerations may deepen and progress under the mucosa to form typical flask ulcers. These extend into the submucosa producing abundant microhemorrhages, the presence of which explains the finding of hematophagous amebae in stool specimens and in rectal scrapings. Such ameba are the best indication of the amebic nature of a case of

Figure 9.7 *Surface view of the mucosa of the colon in a fatal case of amebic colitis. Large and confluent ulcerations have destroyed most of the mucosal layers of the large intestine.*

dysentery or bloody diarrhea. The ulcers are initially superficial, with hyperemic borders, a necrotic base, and normal mucosa between the sites of invasion. Further progression causes the loss of mucosa and submucosa; the lesion can extend into the muscle layers and eventually cause rupture of the serosa.

Complications of intestinal amebiasis include perforation, direct extension to the skin and hematogenous dissemination, mainly to the liver. The presence and extent of liver involvement bears no relationship to the degree of intestinal amebiasis, and these conditions do not necessarily coincide. The early stages of hepatic amebic invasion have not been studied in humans. In experimental animals, inoculation of *E. histolytica* trophozoites into the portal vein produces multiple foci of infection in the liver with neutrophil accumulation around parasites, followed by focal necrosis and granulomatous infiltration. As the lesions grow in size, the granulomas gradually become necrotic and coalesce, necrotic tissue occupying progressively larger portions of the liver. Hepatocytes close to the early lesions degenerate and become necrotic, but direct contact of hepatocytes with amebae is rarely observed. The lesion can eventually consist of large areas of liquefied necrotic material surrounded by a thin capsule of fibrous appearance (Tsutsumi et al. 1984).

DIAGNOSIS

Traditionally, the diagnosis of invasive intestinal amebiasis has been based on the microscopic identification of *E. histolytica* trophozoites in rectal smears or recently evacuated stools and on the results of rectosigmoidoscopy. However, the microscopic detection of amebae has several drawbacks. The procedure is tedious and time consuming; it requires a skilled technician; the sensitivity is relatively low; several samples must be taken to detect cyst passers, and fresh samples are needed for the detection of trophozoites. Trophozoites are most likely to be found in material obtained during rectosigmoidoscopy such as the bloody mucus and in the yellowish exudate covering the mucosal ulcerations. The finding of motile trophozoites containing ingested red blood cells confirms the diagnosis of amebic infection in patients with the clinical gastrointestinal symptoms. Problems arise when cysts alone are identified in stools of healthy or diarrheic individuals, as the cysts may be those of the nonpathogenic *E. dispar*, which are morphologically indistinguishable from cysts of the pathogenic *E. histolytica* (Martínez-Palomo 1993).

The standard for the differentiation of *E. histolytica* from *E. dispar* was, until recently, the determination of their isoenzyme pattern by gel electrophoresis. This procedure is expensive, slow, and requires the cultivation of amebae from feces, which is not routinely applicable. New diagnostic methods are now based on the detection in stool of *E. histolytica*-specific antigen or DNA and by the presence of antiamebic antibodies in serum (reviewed in Tanyuksel and Petri 2003).

Monoclonal antibodies found by ELISA and immunofluorescence assays of fecal samples have been tested. Although the ELISA was reported to have high sensitivity and specificity in a study of 701 samples, it could not discriminate between pathogenic *E. histolytica* and nonpathogenic *E. dispar* (del Muro et al. 1987). The immunofluoresence test, on the other hand, proved to be much faster than the test based on isoenzyme determination, but still required cultivation of the organism, and detected only the motile form of the parasite (Strachan et al. 1988). An ELISA based on monoclonal antibodies directed against cross-reactive and specific *E. histolytica* epitopes of the amebic galactose adhesin is now available commercially. The authors reported that the test gives 93 percent specificity and 96 percent sensitivity in the detection of infection with pathogenic *E. histolytica* (Haque et al. 1994).

Molecular probes provide another approach for the detection of the parasite in stools. A study with a specific probe of 123 stool samples from patients in Mexico City revealed a 93 percent specificity and 100 percent sensitivity for detection of amebae in stools, although it could not distinguish between amebae with pathogenic and nonpathogenic isoenzyme patterns (Samuelson et al. 1989). A study carried out in the south-eastern state of Chiapas in Mexico, using polymerase chain reaction (PCR) to amplify specific pathogenic and nonpathogenic samples resulted in 98 percent specificity and 96 percent sensitivity for the identification of *E. histolytica* in stools of 201 randomly selected individuals (Acuña-Soto et al. 1993).

A comparison of the ELISA and PCR methods to differentiate *E. histolytica* from *E. dispar* in stool samples revealed that the ELISA test is less sensitive (Gonin and Trudel 2003). The specificity of PCR was confirmed in a study of a large number of isolates from both species obtained in three different continents (Zaki et al. 2002). In addition, a closed-tube, real-time PCR assay has given improved results over traditional PCR methods (Blessmann et al. 2002).

The diagnosis of invasive intestinal amebiasis may be confirmed by endoscopy. In patients with benign dysentery, examination by rectosigmoidoscopy reveals small superficial ulcerations of a linear or oval shape, covered by a yellowish exudate and surrounded by a normal or hyperemic mucosa. In patients with fulminating colitis, the ulcers are large and tend to be confluent, often with a necrotic appearance. The intervening mucosa shows intense inflammation and signs of hemorrhage. Endoscopy should be carried out with great care in these cases, as there is a risk of intestinal perforation. If an ameboma is situated in the terminal portion of the colon, it can be visualized by rectosigmoidoscopy. If it is located elsewhere, it is necessary to use colonoscopy or radiography and a barium enema to identify the pseudotumoral mass, which is often surrounded by ulcerations (Sepúlveda 1970).

The diagnosis of amebic liver abscess is sometimes difficult. In endemic areas or when there is a history of travel to such places, amebic abscess should always be suspected in patients with spiking fever, weight loss, and abdominal pain in the upper right quadrant or epigastrium, as well as in patients with tenderness in the liver area. The presence of leukocytosis, a high level of alkaline phosphatase and an elevated right diaphragm (visible in a chest X-ray) suggests the presence of an hepatic abscess. The diagnosis is confirmed by ultrasonography or by computerized axial tomography (CAT) scans. The latter is the most precise method for identifying hepatic abscesses, particularly when they are small, and it is of great value in the differential diagnosis of other focal lesions of the liver (Sepúlveda and Treviño-García Manzo 1986).

Serological tests for antiamebic antibodies are positive in c. 75 percent of patients with invasive colonic amebiasis and in more than 90 percent of patients with amebic liver abscesses. Circulating antibodies have been detected by virtually all known serologic tests, including tests based on immunofluorescent antibodies (IFA), IHA, radioimmunoassay (RIA), counterimmunoelectrophoresis (CIE), and ELISA. The latter is the most sensitive (it gives no false negatives in cases of amebic liver abscess), it is reasonably specific (3.6 percent false positives in controls living in endemic areas), and remains positive longest (more than 3 years in the absence of recurrent infection) (reviewed in Kretschmer 1986). The IFA test also deserves special mention because it is much simpler than all other tests, and when carried out in conjunction with IHA gives a 100 percent positivity in cases of amebic liver abscess. The Center for Disease Control in Atlanta, Georgia, has chosen the IHA as its standard serologic reference test for amebiasis. Titers below 1:256 are considered nonspecific. Because of its relative simplicity and efficiency, the CIE is particularly well suited for epidemiologic surveys.

Although serological techniques are useful in the diagnosis of invasive intestinal amebiasis and amebic liver abscess and as a tool for epidemiological studies of the disease, they do not aid in the diagnosis of simple intestinal infection. In endemic areas, the high prevalence of antiamebic antibodies in the general population reduces the usefulness of serologic tests for diagnosis and other tests must be performed before establishing the diagnosis of invasive amebiasis. Attempts have been made to discriminate serologically between present and past invasive amebiasis, using antibodies to specific antigens in sera of:

- patients with active or cured amebic liver abscesses
- individuals who are asymptomatic cyst carriers (Ximénez et al. 1993).

An antigen-detection ELISA in serum has proved to be useful (Haque et al. 2003).

EPIDEMIOLOGY AND CONTROL

Infection by *E. histolytica* is ubiquitous, but the highest incidence is usually found in communities with poor socioeconomic conditions and inadequate sanitation. In such communities, the infection is typically endemic. According to the World Health Organization, approximately 500 million people (not including those in the People's Republic of China) are infected with a 10 percent annual morbidity index (Walsh and Martínez-Palomo 1986), and up to 110 000 deaths annually can be attributed to complications of the disease (WHO, 1985). However, the actual frequency of amebiasis due to *E. histolytica* has been difficult to establish because there is a tendency to overestimate it in endemic areas, where cases of dysentery or bloody diarrhea are often misdiagnosed as amebiasis. In nonendemic areas with a low incidence of the disease, there is a tendency to overlook the presence of *E. histolytica* in stools.

In areas with high levels of poverty, illiteracy, and overcrowding, plus inadequate and contaminated water supplies and poor sanitation, direct fecal–oral transmission occurs frequently and the endemic character of the disease is maintained. Although difficult to implement, programs for the control of the disease require the improvement of environmental sanitation, health education, and early detection and treatment of infected people.

EPIDEMIOLOGY

Serological surveys

Antiamebic antibodies are relatively long-lasting in the serum of recovered patients and although antibodies cannot be used to discriminate between persons with recent and past infection, or between those infected with *E. histolytica* and *E. dispar*, detection of antibodies in the population has proved to be a valuable tool in epidemiological prevalence studies. A national seroepidemiological survey was undertaken in all 32 states of Mexico, with a probability-based sampling of almost 68 000 individuals who were representative of the population in age, socioeconomic status, and urban or rural origin. Antibodies to *E. histolytica* were detected by IHA assays in 8.41 percent of the population, ranging from 9.80 percent in the South Pacific region to 6.26 percent in the north-eastern parts, with a higher prevalence in rural (9.89 percent) than in urban (7.93 percent) areas. Frequency of presence of antibodies increased during the first decade of life, reaching the highest prevalence in children aged between 5–9 years. Antibody was detected more frequently in women (9.34 percent) than in men (7.09 percent). There was a clear correlation between high seroprevalence, low socioeconomic and educational levels, and inadequate housing conditions, although positive samples were found in all groups (Caballero-Salcedo et al. 1994).

A previous national serological survey of nearly 20 000 serum samples from 46 Mexican urban regions detected an average frequency of amebic seropositivity of 5.95 percent (Gutiérrez et al. 1976). The differences between these studies do not necessarily reflect an increase in seroprevalence in the Mexican population, but may be due to differences in the populations studied (the latter including only urban areas) and the methodology applied (IHA versus CIE).

Human susceptibility

The main reservoir of *E. histolytica* is human, although morphologically similar amebae may be found in primates, dogs, and cats. The parasite has been transmitted to various mammalian species, but it is doubtful that species other than humans serve as significant reservoirs of the parasite. Human susceptibility to infection appears to be widespread, but most individuals harboring the parasite do not develop disease. It is now known that most of these asymptomatic carriers are infected with the noninvasive *E. dispar* although *E. histolytica* has been positively identified in some asymptomatic persons. In a semi-rural area south of Durban, South Africa, 1 percent of apparently healthy individuals were carriers of pathogenic *E. histolytica* and a 1-year follow-up revealed that 90 percent of these carriers remained asymptomatic and underwent spontaneous cure, the remaining 10 percent developed amebic colitis (Ghadirian and Jackson 1987). The factors involved in the development of disease are unknown, but probably include host as well as parasite characteristics.

Frequent luminal amebic infections have been reported in male homosexual populations in North America, northern Europe, and Japan. The spread in this group is associated with specific sexual practices; prevalence rates are as high as 32 percent and most reported cases are asymptomatic or present diffuse symptoms. *E. dispar* has been most frequently identified, except in Japan, where pathogenic *E. histolytica* is not uncommon (Takeuchi et al. 1990).

Incidence of invasive amebiasis

There are a few epidemiological reports of invasive intestinal amebiasis that are accompanied by microscopic identification of *E. histolytica* trophozoites and rectosigmoidoscopy studies showing intestinal ulcerative lesions; these reports provide divergent data. The frequency of invasive intestinal amebiasis has been reported to range from 2.2 percent to 16.2 percent in persons with acute diarrhea or dysentery, and the case-fatality rate to range from 0.5 percent in uncomplicated cases to 40.2 percent in persons with amebic dysentery complicated by peritonitis – these figures are drawn from Mexico, Nigeria, Venezuela, and South Africa (reviewed in Muñoz 1986). Although a large number of cases were included in these studies, the populations sampled were biased by including only those admitted to hospitals and healthcare centers.

The frequency of amebic liver abscess is a reliable measure of rates of liver infection, as liver lesions can be identified clinically, in the laboratory, or through post-mortem studies. Liver abscess occurs in c. 2 percent of adult amebiasis patients in endemic areas (Sepúlveda 1982). Elsdon-Dew tabulated the geographical distribution of persons with amebic liver abscesses and concluded that the condition is prevalent in West and South East Africa, South East Asia, Mexico, and the western portion of South America (Elsdon-Dew 1968). Amebic hepatic abscess is at least ten times more common in adults than in children. It predominates in adults aged between 20 and 60 years and within this group, three times as many men suffer from this condition as women. The reasons for the notable differences in the frequency and severity of invasive amebiasis according to age and sex are unknown.

The differentiation between *E. histolytica* and *E. dispar* in areas of high endemicity has already provided relevant epidemiologic information in Mexico (Sánchez-Guillén et al. 2002), Côte d'Ivoire (Heckendorn et al. 2002), Ecuador (Gatti et al. 2002), and Vietnam (Blessmann et al. 2002)

CHEMOTHERAPY

The use of amebicides in the treatment of patients with amebiasis has contributed greatly to reducing the morbidity and mortality of the disease. Antiamebic drugs may be classified into three groups:

1 luminal amebicides
2 tissue amebicides
3 mixed amebicides.

The most frequently used amebicides with luminal action are di-iodohydroxyquin (650 mg orally three times daily for 20 days), diloxanide furoate (500 mg orally three times daily for 10 days) and paromomycin (500 mg orally three times daily for 5–10 days).

Amebicides effective in both tissues and the intestinal lumen include metronidazole and the nitroimidazole derivatives tinidazole and ornidazole. In addition to being active in both tissues and the intestinal lumen, these drugs have the advantage that they are given orally and are most effective therapeutically. They are reasonably well tolerated and despite reported carcinogenic effect in rodents and their mutagenic potential in bacteria, no such effect has been detected in humans. They are, therefore, the drugs of choice in the treatment of invasive amebiasis in spite of some unpleasant, but not serious side-effects. Metrodinazole is given orally 500–800 mg three times a day for 5 days to people with invasive intestinal amebiasis and for 5–10 days to persons with liver abscesses. Tinidazole or ornidazole are administered orally 2 g once a day for 1–3 days. For further details on the chemotherapy of amebiasis, refer to Martínez-Palomo and Ruíz-Palacios (1990).

There is no drug resistance in amebiasis. Those few reports of failed treatment refer to advanced cases in which most of the liver parenchyma was replaced by the amebic abscess and treatment was administered too late. In vitro studies performed with fresh isolates from symptomatic patients and asymptomatic carriers, as well as axenically cultured pathogenic amebae, confirmed drug sensitivity in all cases (Burchard and Mirelman 1988).

Two severe forms of invasive intestinal amebiasis require surgery in addition to chemotherapy: toxic megacolon and amebic appendicitis. Amebic liver abscesses should be treated by chemotherapy; percutaneous or surgical drainage is indicated in cases of imminent rupture.

One of the most striking features of invasive amebiasis, whether intestinal, hepatic or cutaneous, is that after adequate treatment amebic lesions heal without scarring. There is no apparent explanation of this peculiar aspect of the pathology of the disease.

VACCINATION

Evidence for the existence of acquired immunity against amebiasis is essentially based on two sets of observations. The first is the widely held knowledge that recurrence of amebic liver abscess is exceedingly rare. Although no recent data are available, estimates in the early 1970s in Mexico City indicated that 0.04 percent of patients recovered from an amebic liver abscess had recurrences during a 7-year follow-up period (versus a prevalence of 0.1–0.2 percent of amebic liver abscesses per year in the general population) (reviewed in Kretschmer 1986). The second set of observations supporting acquired immunity is drawn from studies on experimental animals that are refractory to amebic hepatic reinvasion after spontaneous or induced recovery. SCID mice can be partially protected by passive immunization procedures (Cieslak et al. 1992) and partial protection of intact mice has also been obtained through immunization with crude and fractionated preparations of *E. histolytica* and with live trophozoites. Local and systemic antibody responses have been reported to occur in experimental animals after immunization with fixed trophozoites and with an immunogenic peptide fused to the cholera toxin B subunit (Moreno-Fierros et al. 1995; Zhang et al. 1995).

There is also evidence against the development of immunity to invasive amebiasis. Recurrence of amebic hepatic abscesses was documented in a study in Vietnam (Blesmann et al. 2002). The rise in morbidity and mortality due to *E. histolytica* with increasing age, for example, casts some doubt on the concept of effective acquired protective immunity. In addition, it has been observed that not all antibodies are protective, as passive immunization with purified antiamebic immunoglobulins fails to protect hamsters against intrahepatic challenge with *E. histolytica* (Campos-Rodríguez et al. 1995). It has, therefore, been suggested that induction of protective immunity depends on the recognition by antibody of specific molecules. Recombinant peptides of three *E. histolytica* antigens show promise for prevention of experimental amebic liver abscess (reviewed in Stanley 1996): the serine-rich *E. histolytica* protein (SREHP), the 170 kDa subunit of N-acetyl-D-galactosamine inhibitable lectin, the 29 kDa cysteine-rich antigen. Much work still needs to be undertaken before an antiamebic vaccine is produced.

INTEGRATED CONTROL

Transmission of amebiasis may be accomplished through a variety of mechanisms. Asymptomatic carriers passing large numbers of cysts in their stools are important sources of infection, particularly if they are engaged in the preparation and handling of food. The cysts may remain viable and infective for a few days in feces, but are killed by desiccation and temperatures greater than 68°C, so boiled water is safe. The amount of chlorine normally used to purify water is insufficient to kill cysts; higher levels of chlorine are effective, but the water thus treated must be dechlorinated before use. Houseflies and cockroaches ingest cysts present in feces and can pass them from their guts following periods as long as 24 hours.

Fecal contamination of springs, unprotected shallow wells, and streams may occur as a result of surface runoff. Contamination may also result from discharge of sewage into rivers. Occasionally, siphonage of sewage into the water supply system has been responsible for outbreaks of infection. Freshening of vegetables and fruit with contaminated water and using human excreta as fertilizer may produce heavy contamination, a particularly serious problem with vegetables and fruit that are usually eaten raw (Walsh and Martínez-Palomo 1986).

The basic means of preventing amebic infection is the improvement of living conditions and education in countries where invasive amebiasis is prevalent. The main targets should be improvements in sanitation of the environment, detection and treatment of infected persons, and health education.

The most effective preventive measure is the adequate disposal of human feces through proper drainage systems or the use of septic tanks. Purified water should be distributed through pipelines to avoid contamination. In areas where amebic infection is common, drinking water should be boiled for several minutes, or filtered or chlorinated with higher levels of chlorine that those used to eliminate bacterial contamination. People should be instructed to clean and freshen vegetables carefully with uncontaminated running water, because treatment with iodine, chlorine, or silver solutions gives unreliable results. Food handlers should be periodically checked for intestinal infection and treated if found positive. Houseflies and cockroaches should be controlled and food adequately protected from them.

Cases of invasive amebiasis require prompt chemotherapy and it was previously recommended that all asymptomatic carriers of *Entamoeba* be treated. With the development of methods to differentiate between *E. histolytica* and *E. dispar*, *E. histolytica* carriers may be identified and should be treated, whereas *E. dispar* carriers should be monitored closely for the possibility of superinfection with *E. histolytica*. Whether such a course of action will be practicable for general control of amebiasis remains to be seen.

As part of health education of the population, hygienic practices such as hand washing after defecation and before eating, boiling of drinking water, and avoiding the consumption of raw vegetables and exposed food should be constantly reiterated in schools and healthcare units, and through periodic campaigns in the mass media.

Health personnel should be given training to improve the accuracy of their examination of stools. Doctors should be constantly reminded of the problem of amebiasis and informed of advances in diagnostic and therapeutic procedures; their active participation in prevention programs should be encouraged.

REFERENCES

Acuña-Soto, R., Samuelson, J., et al. 1993. Application of the polymerase chain reaction to the epidemiology of pathogenic and nonpathogenic *Entamoeba histolytica*. *Am J Trop Med Hyg*, **48**, 58–70.

Aley, S.B., Scott, W.A. and Cohn, Z.A. 1980. Plasma membrane of *Entamoeba histolytica*. *J Exp Med*, **152**, 391–404.

Arroyo, R. and Orozco, E. 1987. Localization and identification of *Entamoeba histolytica* adhesin. *Mol Biochem Parasitol*, **23**, 151–8.

Batista, E.J.O., de Menezes Feitosa, L.F. and de Souza, W. 2000. The endocytic pathway in *Entamoeba histolytica*. *Parasitol Res*, **86**, 881–90.

Beanan, M.J. and Bailey, G.B. 1995. The primary structure of an *Entamoeba histolytica* enolase. *Mol Biochem Parasitol*, **69**, 119–21.

Bhattacharya, S., Bhattacharya, A., et al. 1989. Circular DNA of *Entamoeba histolytica* encodes ribosomal RNA. *J Protozool*, **36**, 455–9.

Bhattacharya, A., Arya, R., et al. 2000. Absence of lipophosphoglycan-like glycoconjugates in *Entamoeba dispar*. *Parasitol*, **120**, 31–5.

Blesmann, J., Van Linh, P., et al. 2002. Epidemiology of amebiasis in a region of high incidence of liver abscess in central Vietnam. *Am J Top Med Hyg*, **66**, 578–83.

Bruchhaus, I. and Tannich, E. 1993. Primary structure of the pyruvate phosphate dikinase in *Entamoeba histolytica*. *Mol Biochem Parasitol*, **62**, 153–6.

Bruchhaus, I. and Tannich, E. 1994. Induction of an iron-containing superoxide dismutase in *Entamoeba histolytica* by a superoxide anion-generating system or by iron chelation. *Mol Biochem Parasitol*, **67**, 281–8.

Bruchhaus, I. and Tannich, E. 1995. Identification of an *Entamoeba histolytica* gene encoding a protein homologous to prokaryotic disulphide oxidoreductases. *Mol Biochem Parasitol*, **70**, 187–91.

Bruhn, H. and Leippe, M. 1999. Comparative modeling of amoebapores and granulysin based on the NK-lysin structure-structural and functional implications. *Biol Chem*, **380**, 1001–7.

Brumpt, E. 1925. Etude sommaire de l' '*Entamoeba dispar*' n sp Amibe à kystes quadrinucléés, parasite de l'homme. *Bull Acad Méd (Paris)*, **94**, 943–52.

Burchard, G.D. and Mirelman, D. 1988. *Entamoeba histolytica*: virulence potential and sensitivity to metronidazole and emetine of four isolates possessing nonpathogenic zymodemes. *Exp Parasitol*, **66**, 231–42.

Byers, T.J. 1986. Molecular biology of DNA in *Acanthamoeba*, *Amoeba*, *Entamoeba* and *Naegleria*. *Int Rev Cytol*, **99**, 311–41.

Caballero-Salcedo, A., Viveros-Rogel, M., et al. 1994. Seroepidemiology of amebiasis in Mexico. *Am J Trop Med Hyg*, **50**, 412–19.

Campos-Rodríguez, R., Shibayama-Salas, M., et al. 1995. Passive immunization during experimental amebic liver abscess development. *Parasitol Res*, **81**, 86–8.

Calderón, J. and Schreiber, R.D. 1985. Activation of the alternative and classical complement pathways by *Entamoeba histolytica*. *Infect Immun*, **50**, 560–5.

Calderón, J. and Tovar, R. 1986. Loss of susceptibility to complement lysis in *Entamoeba histolytica* HM1 by treatment with human sera. *Immunology*, **58**, 467–71.

Cieslak, P.R., Virgin, H.W. and Stanley, S.L. 1992. A severe combined immunodeficient (SCID) mouse model for infection with *Entamoeba histolytica*. *J Exp Med*, **176**, 1605–9.

Clark, C.G. and Diamond, L.S. 1993. *Entamoeba histolytica*: an explanation for the reported conversion of 'nonpathogenic' amebae to the 'pathogenic' form. *Exp Parasitol*, **77**, 456–60.

Clark, C.G. and Roger, A.J. 1995. Direct evidence for secondary loss of mitochondria in *Entamoeba histolytica*. *Proc Natl Acad Sci USA*, **92**, 6518–21.

Cox, F.E.G. 1992. Systematics of parasitic protozoa. In: Kreier, J.P. and Baker, J.R. (eds), *Parasitic protozoa*, Vol. 1. . San Diego: Academic Press, 55–80.

Corliss, J.O. 1994. An interim utilitarian ('user friendly') hierarchical classification and characterization of the protists. *Acta Protozool*, **33**, 1–51.

del Muro, R., Oliva, A., et al. 1987. Diagnosis of *Entamoeba histolytica* in feces by ELISA. *J Clin Lab Anal*, **1**, 322–5.

Dhar, S.K., Choudhury, N.R., et al. 1995. A multitude of circular DNAs exist in the nucleus of *Entamoeba histolytica*. *Mol Biochem Parasitol*, **70**, 203–6.

Diamond, L.S. and Clark, C.G. 1993. A redescription of *Entamoeba histolytica* Schaudinn, 1903 (emended Walker 1911) separating it from *Entamoeba dispar* Brumpt, 1925. *J Euk Microbiol*, **40**, 340–4.

Edman, U., Meza, Y. and Agabian, N. 1987. Genomic and cDNA actin sequences from a virulent strain of *Entamoeba histolytica*. *Proc Natl Acad Sci USA*, **84**, 3024–8.

Eichinger, D. 2001. Encystation in parasitic protozoa. *Curr Opin Microbiol*, **4**, 421–6.

Ellis, J.E., Setchell, K.D.R. and Kaneshiro, E.S. 1994. Detection of ubiquinone in parasitic and free-living protozoa, including species devoid of mitochondria. *Mol Biochem Parasitol*, **65**, 213–24.

Elsdon-Dew, R. 1968. The epidemiology of amoebiasis. *Adv Parasitol*, **6**, 1–62.

Espinosa-Cantellano, M. and Martínez-Palomo, A. 1991. The plasma membrane of *Entamoeba histolytica*: structure and dynamics. *Biol Cell*, **72**, 189–200.

Espinosa-Cantellano, M. and Martínez-Palomo, A. 1994. *Entamoeba histolytica*: Mechanism of surface receptor capping. *Exp Parasitol*, **79**, 424–35.

Espinosa-Cantellano, M. and Martínez-Palomo, A. 2000. Pathogenesis of intestinal amebiasis: from molecules to disease. *Clin Microbiol Rev*, **13**, 318–31.

Fahey, R.C., Newton, G.L., et al. 1984. *Entamoeba histolytica*: a eukaryote without glutathione metabolism. *Science*, **224**, 70–2.

Fairlamb, A.H. 1989. Novel biochemical pathways in parasitic protozoa. *Parasitol*, **99**, S93–S112.

García-Rivera, G., Rodríguez, M.A., et al. 1999. Entamoeba histolytica: a novel cysteine protease and an adhesin form the 112 kDa surface protein. *Mol Microbiol*, **33**, 556–68.

Gatti, S., Swierczynski, G., et al. 2002. Amebic infections due to *Entamoeba histolytica-Entamoeba dispar* complex: a study of the incidence in a remote rural area of Ecuador. *Am J Trop Med Hyg*, **67**, 123–7.

Ghadirian, V. and Jackson, T.F.H.G. 1987. A longitudinal study of asymptomatic carriers of pathogenic zymodemes of *Entamoeba histolytica*. *S Afr Med J*, **72**, 669–72.

Gonin, P. and Trudel, L. 2003. Detection and differentiation of *Entamoeba histolytica* and *Entamoeba dispar* isolates in clinical samples by PCR and enzyme-linked immunosorbent assay. *J Clin Microbiol*, **41**, 237–41.

González-Robles, A. and Martínez-Palomo, A. 1992. The fine structure of *Entamoeba histolytica* processed by cryo-fixation and cryo-substitution. *Arch Med Res*, **23**, 73–6.

Gutiérrez, G., Ludlow, A., et al. 1976. Encuesta serológica nacional, II, Investigación de anticuerpos contra *Entamoeba histolytica* en la República Mexicana. In: Sepúlveda, B. and Diamond, L.S. (eds),

Proceedings of the international conference on amebiasis. Mexico: Instituto Mexicano del Seguro Social, 599–608.

Haque, R., Neville, L.M., et al. 1994. Detection of *Entamoeba histolytica* and *E. dispar* directly in stool. *Am J Trop Med Hyg*, **50**, 595–6.

Haque, R., Huston, C.D., et al. 2003. Amebiasis. *New Eng Med J*, **348**, 1565–73.

Heckendorn, F., N'Goan, E.K., et al. 2002. Species-specific field testing of *Entamoeba* spp. in an area of high endemicity. *Trans R Soc Trop Med Hyg*, **96**, 521–8.

Hellberg, A., Leippe, M. and Bruchhaus, I. 2000. Two major 'higher molecular mass proteinases' of *Entamoeba histolytica* are identified as cysteine proteinases 1 and 2. *Mol Biochem Parasitol*, **105**, 305–9.

Huang, M., Albach, R.A., et al. 1995. Cloning and sequencing a putative pyrophosphate-dependent phosphofructokinase gene from *Entamoeba histolytica*. *Biochim Biophys Acta*, **120**, 215–17.

Huber, M., Garfinkel, L., et al. 1987. Entamoeba histolytica: cloning and characterization of actin cDNA. *Mol Biochem Parasitol*, **24**, 227–35.

Huber, M., Koller, B., et al. 1989. *Entamoeba histolytica* ribosomal RNA genes are carried on palindromic circular DNA molecules. *Mol Biochem Parasitol*, **32**, 285–96.

Isibasi, A., Santa-Cruz, M., et al. 1982. Localización en los trofozoítos de *Entamoeba histolytica* de una lipopeptidofosfoglicana extraída por fenol-agua de la cepa HK-9. *Arch Invest Méd (Mex)*, **13**, Suppl 3, 57–62.

Keene, W.E., Hidalgo, M.E. and Orozco, E. 1990. *Entamoeba histolytica*: correlation of the cytopathic effect of virulent trophozoites with secretion of a cysteine proteinase. *Exp Parasitol*, **71**, 199–206.

Keene, W.E., Petitt, M.G., et al. 1986. The major neutral proteinase of *Entamoeba histolytica*. *J Exp Med*, **163**, 536–49.

Kelsall, B.L. and Ravdin, J.I. 1993. Degradation of human IgA by *Entamoeba histolytica*. *J Infect Dis*, **168**, 1319–22.

Knobloch, J. and Mannweiler, E. 1983. Development and persistence of antibodies to *Entamoeba histolytica* in patients with amebic liver abscess. *Am J Trop Med Hyg*, **32**, 727–32.

Kretschmer, R.R. 1986. Immunology of amebiasis. In: Martínez-Palomo, A. (ed.), *Amebiasis*. Amsterdam: Elsevier, 95–167.

Kumar, S., Tripathi, L.M. and Sagar, P. 1992. Oxido-reductive functions of *Entamoeba histolytica* in relation to virulence. *Ann Trop Med Parasitol*, **86**, 239–48.

Leippe, M., Andra, J., et al. 1994. Amoebapores, a family of membranolytic peptides from cytoplasmic granules of *Entamoeba histolytica*: isolation, primary structure, and pore formation in bacterial cytoplasmic membrane. *Mol Microbiol*, **14**, 895–904.

Levine, N.D., Corliss, J.O., et al. 1980. A newly revised classification of the protozoa. *J Protozool*, **27**, 37–58.

Lioutas, C., Schmetz, C. and Tannich, E. 1995. Identification of various circular DNA molecules in *Entamoeba histolytica*. *Exp Parasitol*, **80**, 349–52.

Lohia, A., Hait, N.C. and Majunder, A.L. 1999. L-myo-inositol 1-phosphate synthase from *Entamoeba histolytica*. *Mol Biochem Parasitol*, **98**, 67–79.

Luaces, A.L. and Barret, A.J. 1988. Affinity purification and biochemical characterization of histolysin, the major cysteine proteinase of *Entamoeba histolytica*. *Biochem J*, **250**, 903–9.

Lushbaugh, W.B., Hofbauer, A.F. and Pittman, F.E. 1985. *Entamoeba histolytica*: purification of cathepsin B. *Exp Parasitol*, **59**, 328–36.

Mann, B.J. 2002. *Entamoeba histolytica* Genome Project: an update. *Trends Parasitol*, **18**, 147–8.

Martínez-Palomo, A. 1982. *The biology of Entamoeba histolytica*. Chichester: John Wiley & Sons.

Martínez-Palomo, A. 1986. Biology of *Entamoeba histolytica*. In: Martínez-Palomo, A. (ed.), *Amebiasis*. Amsterdam: Elsevier, 11–43.

Martínez-Palomo, A. 1993. Parasitic amebas of the intestinal tract. In: Kreier, J.P. and Baker, J.R. (eds), *Parasitic protozoa*, Vol. 3. . San Diego: Academic Press, 65–141.

Martínez-Palomo, A. and Ruíz-Palacios, G. 1990. Amebiasis. In: Warren, K.S. and Mahmoud, A.A.F. (eds), *Tropical and geographical medicine*. New York: McGraw Hill, 327–44.

Martínez-Palomo, A., González-Robles, A. and de la Torre, M. 1973. Selective agglutination of pathogenic strains of *Entamoeba histolytica* induced by Concanavalin A. *Nature New Biol*, **245**, 186–7.

McLaughlin, J. and Aley, S. 1985. The biochemistry and functional morphology of the *Entamoeba*. *J Protozool*, **32**, 221–40.

Moody, S., Becker, S., et al. 1998. Identification of significant variation in the composition of lipophosphoglycan-like molecules of *E. histolytica* and *E. dispar*. *J Eukaryot Microbiol*, **45**, Suppl, 9S–12S.

Moody-Haupt, S., Patterson, J.H., et al. 2000. The major surface antigens of *Entamoeba histolytica* trophozoites are GPI-anchored proteophosphoglycans. *J Mol Biol*, **24**, 409–20.

Moreno-Fierros, L., Domínguez-Robles, M.C. and Enríquez-Rincón, F. 1995. *Entamoeba histolytica*: Induction and isotype analysis of antibody producing cell responses in Peyer's patches and spleen after local and systemic immunization in male and female mice. *Exp Parasitol*, **80**, 541–9.

Muñoz, O. 1986. Epidemiology of amebiasis. In: Martínez-Palomo, A. (ed.), *Amebiasis*. Amsterdam: Elsevier, 213–39.

Neal, R.A. 1988. Phylogeny: the relationship of *Entamoeba histolytica* to morphologically similar amebae of the four-nucleate cyst group. In: Ravdin, J.I. (ed.), *Amebiasis: human infection by* Entamoeba histolytica. New York: Wiley, 13–26.

Nickel, R. and Tannich, E. 1994. Transfection and transient expression of chloramphenicol acetyltransferase gene in the protozoan parasite *Entamoeba histolytica*. *Proc Natl Acad Sci USA*, **91**, 7095–8.

Nickel, R., Ott, C., et al. 1999. Pore-forming peptides of *Entamoeba dispar*: Similarity and divergence to amoebapores in structure, expression and activity. *Eur J Biochem*, **265**, 1002–7.

Pérez-Montfort, R. and Kretschmer, R.R. 1990. Humoral immune responses. In: Kretschmer, R.R. (ed.), *Amebiasis: infection and disease by Entamoeba histolytica*. Boca Raton: CRC Press, 91–103.

Pérez-Tamayo, R. 1986. Pathology of amebiasis. In: Martínez-Palomo, A. (ed.), *Amebiasis*. Amsterdam: Elsevier, 45–94.

Petri, W.A., Chapman, M.D., et al. 1989. Subunit structure of the galactose and N-acetyl-D-galactosamine-inhibitable adherence lectin of *Entamoeba histolytica*. *J Biol Chem*, **264**, 3007–12.

Petri, W.A., Jackson, T.F.H.G., et al. 1990. Pathogenic and nonpathogenic strains of *Entamoeba histolytica* can be differentiated by monoclonal antibodies to the galactose-specific adherence lectin. *Infect Immun*, **58**, 1802–6.

Purdy, J.E., Mann, B.J., et al. 1994. Transient transfection of the enteric parasite *Entamoeba histolytica* and expression of firefly luciferase. *Proc Natl Acad Sci USA*, **91**, 7099–103.

Reed, S.L. and Gigli, I. 1990. Lysis of complement-sensitive *Entamoeba histolytica* by activated terminal complement components. *J Clin Invest*, **86**, 1815–22.

Reed, S.L., Sargeaunt, P.G. and Braude, A.L. 1983. Resistance to lysis by human serum of pathogenic *E. histolytica*. *Trans R Soc Trop Med Hyg*, **77**, 248–53.

Reeves, R.E. 1984. Metabolism of *Entamoeba histolytica* Schaudinn, 1903. In: Baker, J.R. and Muller, R. (eds), *Advances in parasitology*. London: Academic Press, 105–42.

Rosales-Encina, J.L., Meza, I., et al. 1987. Isolation of a 220 kDa protein with lectin properties from a virulent strain of *Entamoeba histolytica*. *J Infect Dis*, **156**, 790–7.

Saavedra-Lira, E. and Pérez-Montfort, R. 1996. Energy production in *Entamoeba histolytica*: new perspectives in rational drug design. *Arch Med Res*, **27**, 257–64.

Samuelson, J., Acuña-Soto, R., et al. 1989. DNA hybridization probe for clinical diagnosis of *Entamoeba histolytica*. *J Clin Microbiol*, **27**, 671–6.

Sánchez-Guillén, M.C., Pérez-Fuentes, R., et al. 2002. Differentiation of *Entamoeba histolytica/Entamoeba dispar* by PCR and correlation with humoral and cellular immunity in individuals with clinical amoebiasis. *Am J Trop Med Hyg*, **66**, 731–7.

Sargeaunt, P.G., Jackson, T.F.H.G. and Simjee, A. 1982. Biochemical homogeneity of *Entamoeba histolytica* isolates, especially those from liver abscess. *Lancet*, **1**, 1386–8.

Scholze, H., Löhden-Bendinger, U., et al. 1992. Subcellular distribution of amebapain, the major cysteine proteinase of *Entamoeba histolytica*. *Arch Med Res*, **23**, 105–8.

Sepúlveda, B. 1970. La amibiasis invasora por *Entamoeba histolytica*. *Gac Méd Méx*, **100**, 201–54.

Sepúlveda, B. 1982. Amebiasis: host-pathogen biology. *Rev Infect Dis*, **4**, 836–42.

Sepúlveda, B. and Treviño-García Manzo, N. 1986. Clinical manifestations and diagnosis of amebiasis. In: Martínez-Palomo, A. (ed.), *Amebiasis*. Amsterdam: Elsevier, 169–88.

Simic, T. 1931. Infection expérimentale de l'homme par *Entamoeba dispar* Brumpt. *Ann Parasitol Hum Compar*, **9**, 385–91.

Spinella, S., Levavasseur, E., et al. 1999. Purification and biochemical characterization of a novel cysteine protease of *Entamoeba histolytica*. *Eur J Biochem*, **266**, 170–80.

Spice, W.M. and Ackers, J.P. 1992. The amoeba enigma. *Parasitol Today*, **8**, 402–6.

Stanley, S.L. 1996. Progress towards an amebiasis vaccine. *Parasitol Today*, **12**, 7–14.

Stanley, S.L. 2003. Amoebiasis. *Lancet*, **311**, 1025–34.

Stauffer, W. and Ravdin, J.I. 2003. *Entamoeba histolytica*: an update. *Curr Opin Infect Dis*, **16**, 479–85.

Strachan, W.D., Spice, W.M., et al. 1988. Immunological differentiation of pathogenic and non-pathogenic isolates of *Entamoeba histolytica*. *Lancet*, **1**, 561–3.

Tachibana, H., Kobayashi, S., et al. 1990. Identification of a pathogenic isolate-specific 30,000-Mr antigen of *Entamoeba histolytica* by using a monoclonal antibody. *Infect Immun*, **58**, 955–60.

Takeuchi, T., Miyahira, Y., et al. 1990. High seropositivity for *Entamoeba histolytica* infection in Japanese homosexual men. *Trans R Soc Trop Med Hyg*, **84**, 250–1.

Talamás-Rohana, P., Hernández, V.I. and Rosales-Encina, J.L. 1994. A b1 integrin-like molecule in *Entamoeba histolytica*. *Trans R Soc Trop Med Hyg*, **88**, 596–9.

Tannich, E., Scholze, H., et al. 1991a. Homologous cysteine proteinases of pathogenic and nonpathogenic *Entamoeba histolytica*: differences in structure and expression. *J Biol Chem*, **266**, 4798–803.

Tannich, E., Bruchhaus, I., et al. 1991b. Pathogenic and nonpathogenic *Entamoeba histolytica*: identification and molecular cloning of an iron-containing superoxide dismutase. *Mol Biochem Parasitol*, **49**, 61–72.

Tanyuksel, M. and Petri, W.A. 2003. Laboratory diagnosis of amebiasis. *Clin Microbiol Rev*, **16**, 713–29.

Trissl, D., Martínez-Palomo, A., et al. 1977. Surface properties related to concanavalin A-induced agglutination A comparative study of several *Entamoeba* strains. *J Exp Med*, **145**, 652–65.

Trissl, D., Martínez-Palomo, A., et al. 1978. Surface properties of *Entamoeba*: increased rates of human erythrocyte phagocytosis in pathogenic strains. *J Exp Med*, **148**, 1137–45.

Tsutsumi, V., Mena, R. and Martínez-Palomo, A. 1984. Cellular bases of experimental liver abscess formation. *Am J Pathol*, **130**, 112–19.

US Treasury Dept and Public Health Service. 1936. *Epidemic amebic dysentery. The Chicago outbreak of 1933*. Bulletin 166. Washington, DC: National Institutes of Health.

Vargas-Rodríguez, L., Villagómez-Castro, J., et al. 1998. Identification and characterization of early reactions of asparagine-linked oligosaccharide assembly in *Entamoeba histolytica*. *Int J Parasitol*, **28**, 1333–40.

Villagómez-Castro, J.C., Calvo-Méndez, C., et al. 1998. *Entamoeba histolytica*: solubilization and biochemical characterization of dolichol phosphate mannose synthase, an essential enzyme in glycoprotein biosynthesis. *Exp Parasitol*, **88**, 111–20.

Walsh, J. 1984. Estimating the burden of illness in the tropics. In: Warren, K.S. and Mahmoud, A. (eds), *Tropical and geographical medicine*. New York: McGraw-Hill, 1073–85.

Walsh, J. and Martínez-Palomo, A. 1986. Control of amebiasis. In: Martínez-Palomo, A. (ed.), *Amebiasis*. Amsterdam: Elsevier, 241–60.

Weinbach, E.C. 1981. Biochemistry of enteric parasitic protozoa. *Trends Biochem Sci*, **6**, 254–7.

WHO, 1985. Amoebiasis and its control. *Bull World Health Organ*, **63**, 417–26.

Willhoeft, U. and Tannich, E. 2000. Fluorescence microscopy and fluorescence in situ hybridization of *Entamoeba histolytica* nuclei to analyze mitosis and the localization of repetitive DNA. *Mol Biochem Parasitol*, **105**, 291–6.

Ximénez, C., Leyva, O., et al. 1993. *Entamoeba histolytica*: antibody response to recent and past invasive events. *Ann Trop Med Parasitol*, **87**, 31–9.

Yang, W., Li, E., et al. 1994. *Entamoeba histolytica* has an alcohol dehydrogenase homologous to the multifunctional adhE gene product of *Escherichia coli*. *Mol Biochem Parasitol*, **64**, 253–60.

Yu, Y. and Samuelson, J. 1994. Primary structure of an *Entamoeba histolytica* nicotinamide nucleotide transhydrogenase. *Mol Biochem Parasitol*, **68**, 323–8.

Zaki, M., Meelu, P., et al. 2002. Simultaneous differentiation and typing of *Entamoeba histolytica* and *Entamoeba dispar*. *J Clin Microbiol*, **40**, 1271–6.

Zhang, T., Li, E. and Stanley, S.L. 1995. Oral immunization with the dodecapeptide repeat of the serine-rich *Entamoeba histolytica* protein (SREHP) fused to the cholera toxin B subunit induces a mucosal and systemic anti-SREHP antibody response. *Infect Immun*, **63**, 1349–55.

Zhang, W.-W., Shen, P.-S. and Descoteaux, S. 1994. Cloning and expression of an NADP+-dependent aldehyde dehydrogenase of *Entamoeba histolytica*. *Mol Biochem Parasitol*, **63**, 157–61.

10

Other intestinal amebae

JOHN P. ACKERS

INTRODUCTION AND SCOPE

The indubitably pathogenic protozoa that inhabit the human intestine have, quite rightly, received the bulk of our attention, both in research terms and in this volume (see Chapters 9, 12, 13, 14, and 15). Nevertheless, interms of numbers of persons infected, these pathogenic species are dwarfed by the unobtrusive, largely nonpathogenic and little studied organisms that are the subject of this chapter: *Entamoeba coli, E. gingivalis, E. moshkovskii, E. chattoni, E. polecki, Endolimax nana,* and *Iodamoeba bütschlii.*

In other chapters, the classification, immunology, epidemiology, and any progress towards the development of a vaccine are dealt with separately for each organism. So little is known about these aspects of most of the species considered here, however, that it would be tedious to repeat our ignorance for each one. These aspects are, therefore, considered briefly together before moving on to consider the clinical features of infection by each of these organisms.

CLASSIFICATION

Classically, all the *Entamoeba* are placed in one of the two great branches of the phylum Sarcomastigophora – the subphylum Sarcodina ('ameba'); and within this subphylum successively in the superclass Rhizopoda, class Lobosea, subclass Gymnamoebia, order Amoebida, and sub-order Tubulina (Levine et al. 1980) as are *Endolimax* and *Iodamoeba*. The overall classification of the protozoa is unsatisfactory and that of ameboid

organisms perhaps particularly so. However, in the case of the organisms included in this chapter, molecular taxonomy seems to confirm that they are more similar to each other (albeit only distantly related) than to any other major group of eukaryotes and may legitimately be placed in the family Entamoebidae (Silberman et al. 1999). Figure 10.1 shows that the number of nuclei in the mature cyst, long one of the bases of microscopical diagnosis, is a taxonomically valid character (note, however, that *I. bütschlii* was not included in this analysis).

None of these organisms possess mitochondria and it had been speculated that they and some other anaerobic protozoa were examples of surviving pre-mitochondrial eukaryotes ('Archezoa'). Enough genes of mitochondrial origin have now been observed in all the well-studied members of this group to make it virtually certain that they are descended from organisms that did possess this organelle and although no investigation has been made of the organisms described here, it seems very unlikely that the same is not true of them as well.

IMMUNOLOGY

Essentially, nothing is known of human immune responses to these parasites, although their ubiquity strongly suggests that protective immunity develops only after multiple infections, if at all. This is supported by some data from the Gambia (Bray and Harris 1977), which shows that the percentage of cyst-positive feces increases with age for all the gut protozoa studied (which included *E. coli, Endolimax nana,* and *I. bütschlii*)

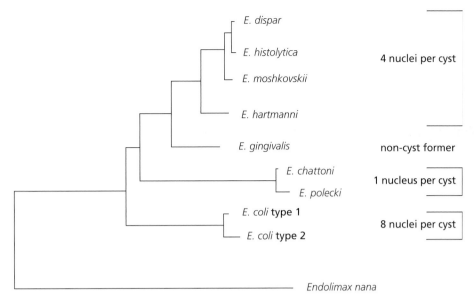

Figure 10.1 *Phylogenetic tree based on small sub-unit ribosomal RNA sequences, showing the relationship between the* Entamoeba *spp. discussed in this chapter and the number of nuclei in the mature cyst. (Courtesy of Dr Graham Clark, London School of Hygiene & Tropical Medicine; adapted from Silberman et al. 1999)*

except for *Giardia intestinalis*. Importantly, there is no evidence that any of these organisms can become clinically significant even in the presence of severe immunosuppression or clinical acquired immune deficiency syndrome (AIDS). It need hardly be added that there is no current work directed towards developing a vaccine against any of these organisms.

BIOCHEMISTRY AND MOLECULAR BIOLOGY

Again, there is virtually nothing known about the biochemistry of any of these organisms. All may be cultivated (with difficulty in some cases) in complex media in the presence of bacteria (Sargeaunt and Williams 1982; Diamond 1983) but only *E. moshkovskii* can currently be grown axenically (Diamond et al. 1995) and this has greatly hindered studies of metabolic pathways. The very limited amount of published molecular biological data suggests that these organisms may well share the peculiarities (such as ribosomal RNA genes carried on circular episomes) of *E. histolytica*. A comparative *Entamoeba* genome sequencing project, which includes *E. moshkovskii*, is underway at the Sanger Institute (www.sanger.ac.uk/Projects/Comp_Entamoeba/) and this will greatly increase our knowledge of at least this species.

EPIDEMIOLOGY AND LIFE-CYCLES

With the exception of *E. gingivalis* (see below), infection with all of these organisms is assumed to be acquired by

oral ingestion of fecally contaminated food or drink. Although no human volunteer studies have been carried out to prove this, the epidemiology of the infections and their close similarities to *E. histolytica/dispar*, where this has been demonstrated, support this. The life cycle is simple: excystment occurs, possibly as with *E. histolytica* in the small bowel, and the organisms then live as trophozoites in the lumen of the large bowel. There is no evidence that any of these organisms can invade the mucosa while the host is alive although they may do so rapidly after death (Vogel et al. 1996). Responding to unknown stimuli, trophozoites round up and secrete a resistant wall, becoming the cysts which, when shed in the feces, are responsible for transmitting the infection. Nuclear division frequently occurs following encystment to produce the characteristic numbers of nuclei so important in microscopical diagnosis. Although morphologically similar cysts are found in the feces of many wild and domestic animals, with the further exception of *E. chattoni* and *E. polecki* (see below), there is no evidence that animal reservoirs are important in the epidemiology of human infections (but also no evidence that they are not).

In a fascinating study, Garrido-Gonzalez et al. (2002) measured the cyst output from a number of infected persons. The results are complex but, in summary are thus:

- cyst excretion varies greatly between patients but a small group of persons produce a majority of the cysts
- cyst excretion varies widely over time and tends to occur in bursts

- in mixed infections cyst excretion by both species is affected
- different parasites produce different average numbers of cysts with *Endolimax nana > E. coli > E. histolytica/dispar* .

These results, which are supported by earlier and largely forgotten studies, clearly have implications for both diagnosis and prevention strategies.

DIAGNOSIS

In anything other than research settings, diagnosis of all these organisms except *E. gingivalis* is by microscopical examination of feces and is based on cyst morphology. Detailed protocols may be obtained from standard texts (Garcia and Bruckner 1997) or from the Centers for Disease Control and Prevention (CDC) (www.dpd.cdc.gov/dpdx/HTML/DiagnosticProcedures.htm) but, in outline, specimens should be stained with iodine and, if possible, a cyst concentrate should also be examined. Measurement of cysts is essential for accurate diagnosis, along with nuclear number and morphology.

Since it produces no cysts, diagnosis of *E. gingivalis* relies on trophozoite morphology; however, this is not distinctive (see below). As in all such cases, the trophozoites need either to be examined very shortly after the specimen is obtained or fixed immediately before being stained and examined at leisure.

Cultivation of these organisms, although not usually feasible routinely, has two advantages – it provides a (probably modest) increase in sensitivity and, more importantly, it allows much more accurate species diagnosis to be made based on methods such as isoenzymes, riboprinting and polymerase chain reaction (PCR) although this is exclusively carried out in a research setting. Protocols that allow the extraction of DNA directly from fecal samples and its use for PCR-based diagnosis of *E. histolytica* and *E. dispar* are now in routine use and could certainly be adapted for the nonpathogenic ameba if there were any demand for this.

CLINICAL ASPECTS

All of these organisms are regarded as nonpathogenic (occasional reports to the contrary are given below) and the advice from CDC is that persons infected with *E. coli, E. polecki, Endolimax nana* or *I. bütschlii* should not be treated (www.cdc.gov/ncidod/dpd/parasites/amebae/factsht_amebae.htm). Their presence is, however, evidence that close contact with fecally contaminated material has occurred and might prompt a more thorough search for pathogenic species. There seems no reason to doubt that the luminal amebicides such as paromomycin and diloxanide furoate would be effective if used. As would be anticipated, metronidazole appears to be effective in eliminating oral infection with *E. gingi-*

valis in the small number of reported cases where this was attempted (Sefer et al. 1989; el Azzouni and el Badry, 1994).

Clinical manifestations

ENTAMOEBA COLI

Entamoeba coli competes with *Endolimax nana* for the title of the world's commonest intestinal protozoan, with average incidence figures of 27 percent in a large number of older surveys (Levine 1973) and up to 40 percent reported in some populations; even in 1987, 4 percent of routine fecal samples taken in the US contained this parasite (Kappus et al. 1994). Apparently identical cysts have been found in the feces of many primates and, rarely, of other mammals. Typical cysts (Figure 10.2) are easy to identify being large (15–25 μm in diameter) and having eight smallish nuclei of one-sixth to one-eighth the cyst diameter. Peripheral chromatin is less regularly spread around the nuclear membrane than in *E. histolytica* cysts; chromidial bars, if present, have splintered rather than rounded ends. Occasionally, cysts may contain fewer than eight, or up to 16 nuclei; cyst size may be quite variable (Garrido-Gonzalez et al. 2002) and those 15 μm or less in size may be harder to distinguish from *E. histolytica/dispar*.

Trophozoites are usually 20–25 μm long with granular cytoplasm containing ingested bacteria, yeasts and other debris, but very rarely erythrocytes; in infected humans, trrophozoites are found in the cecum and colon. The single nucleus contains a large karyosome; blunt pseudopodia are extended in a sluggish, randomly directed manner. Although easily cultivated in bacteria-

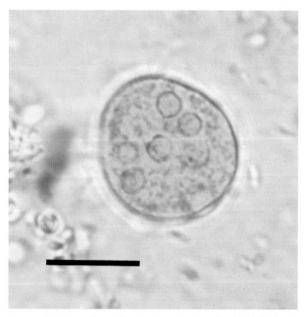

Figure 10.2 *Cyst of* E. coli, *wet preparation stained with iodine. Bar = 10 μm.*

containing media, very little work has been devoted to this common parasite and almost nothing is known of its cell biology. After being cultured, it may be distinguished from *E. histolytica* by isoenzyme analysis (Sargeaunt and Williams 1978), several monoclonal antibodies and riboprinting (Clark and Diamond 1992).

E. coli is normally regarded as wholly harmless, although two isolated reports (Wahlgren 1991; Corcoran et al. 1991) have suggested that it might be an occasional cause of diarrhea. In this, as in so many other cases, the evidence consists of symptomatic patients in whom no other pathogen could be identified and in whom treatment resulted in clinical cure accompanied by loss of the infection. There are a number of weaknesses with this kind of argument – however thoroughly cases of diarrhea are studied, in about 25 percent of them no pathogen can be demonstrated, and drugs such as metronidazole have a very broad spectrum of activity against a wide variety of gastrointestinal organisms. However, while *E. coli* clearly does not produce illness in the majority of those who carry it, that is not say that it cannot be pathogenic in some.

ENTAMOEBA GINGIVALIS

E. gingivalis is unusual among the *Entamoeba* in two respects – first, it inhabits the mouth rather than the large bowel; secondly, no cyst of *E. gingivalis* has ever been found. Transmission is presumably by direct mouth-to-mouth contact and, since the trophozoite is never exposed to desiccation or other unfavorable conditions, a resistant cyst is not needed. No selection pressure to maintain the ability to encyst therefore exists and so, like most redundant functions, it has probably been lost (Clark and Diamond 1997).

Trophozoites of this species may be recovered from the mouths of a significant percentage of patients with periodontal disease (Levine 1973; Favoreto and Machado 1995) and, although it is normally regarded as a harmless commensal that finds an ideal home among diseased gums (Pomes et al. 2000), some believe that it has a directly pathogenic role. Frequency of isolation appears to increase with age, possibly reflecting deteriorating condition of the gums and *E. gingivalis* has been implicated in the severe periodontal disease suffered by some human immunodeficiency virus (HIV)-positive patients (Lucht et al. 1998) and may even spread outside the oral cavity following radiotherapy (Perez Jaffe et al. 1998). *E. gingivalis* has also been recovered from the vaginas of a small number of women using intra-uterine devices and who have concomitant bacterial infections (deMoraes Ruehsen et al. 1980; Clark and Diamond 1992). In these cases, *E. gingivalisis* is presumed to have been acquired by orogenital contact.

The trophozoite of *E. gingivalis* (Figure 10.3) is about 10–20 μm long with a typical *Entamoeba* nucleus. A number of xenic media for its cultivation exist and it

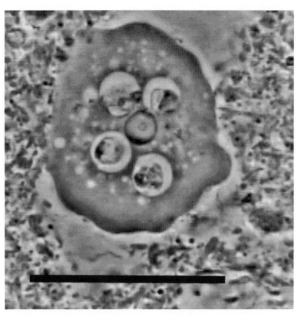

Figure 10.3 *Trophozoite of* E. gingivalis *in scrapings from a case of severe periodontal disease. Bar = 10 μm. (Courtesy of Dr Mark Bonner, Institut International de parodontie, Quebec)*

may be distinguished from other species by isoenzyme patterns, riboprinting, and PCR (Kikuta et al. 1996) By light microscopy, the trophozoites are not distinguishable from *E. histolytica* except in that they are frequently seen to have ingested both erythrocytes and leukocytes whereas *E. histolytica* when isolated from patients (but not in vitro) appears to ingest only the former (Dao 1985).

ENTAMOEBA MOSHKOVSKII

This organism was first isolated from the sewers of Moscow in 1941, but has subsequently been recovered from the same source in many parts of the world (Levine 1973). Its presence in sewers was a great mystery because no human excretors had ever been found. However, it is now known to be ubiquitous in both fresh and brackish anoxic environments. It may be grown in axenic culture and its optimum growth temperature is about 24°C.

Up to 1998, seven isolates of organisms (originally assumed to be *E. histolytica*) with the unusual property of being able to grow at room temperature and in hypotonic media had been recovered from asymptomatic human carriers. These organisms, referred to as atypical, *E. histolytica*-like or 'Laredo-type' (after the original isolate) are of low pathogenicity in laboratory animals, resistant to a number of drugs and differ antigenically from classical *E. histolytica* or *E. dispar* (Goldman 1969). In 1980, Sargeaunt et al. showed that the isoenzyme patterns of five such isolates were all very similar to that of two *E. moshkovskii* strains; Sargeaunt et al. (1982) and Clark and Diamond (1991) proved that the 'Laredo-type' and *E. moshkovskii* are the same species.

Both trophozoites and cysts of *E. moshkovskii* resemble those of *E. dispar*. The active trophozoites are usually 11–13 μm long; the cysts spherical, 7–17 μm in diameter and containing a large glycogen vacuole when fresh (Figure 10.4). The mature cysts contain four nuclei; chromidial bodies, if present, have rounded ends (Levine 1973). Unless they are cultured and their isoenzymes or other defining properties are tested, the organisms are likely to be identified as *E. dispar*. Epidemiologically, *E. moshkovskii* is presumably either a free-living species that is an occasional human parasite (apparently, it may be adapted to grow well at 37°C (Lachance 1959)) or a true human parasite which, because of its tolerance of low temperature and hypotonicity, can persist indefinitely outside its host(s). The latter possibility has been rendered much more likely by the recent and wholly unexpected finding that over 20 percent of a group of pre-school children in Bangladesh were infected with *E. moshkovskii* (Ali et al. 2003). The nested PCR diagnostic procedure developed by these authors should make it much easier to see if this situation exists in other populations.

One other Laredo-type human isolate – *E. ecuadoriensis* – is known; it possesses the same culture tolerance as *E. moshkovskii* but by both isoenzyme analysis (Sargeaunt et al. 1980) and riboprinting (Clark and Diamond 1991) is clearly distinct from all other known species of *Entamoeba*.

ENTAMOEBA POLECKI AND ENTAMOEBA CHATTONI

Entamoeba polecki (for which some authors prefer the name *Entamoeba suis* (Levine 1973) is a common parasite of the cecum and colon of pigs, where it apparently causes little or no pathology. The trophozoites may be 10–25 μm long (although they usually measure 12–18 μm) with vacuolated cytoplasm; the cyst (Figure 10.5) is usually 9–15 μm in diameter with a single nucleus even when mature. The nuclear endosome is usually quite large; glycogen granules and chromidial bodies may or may not be present (Noble and Noble 1952). The nucleus is 25–33 percent of the diameter of the whole cyst, compared with the relatively larger (33–50 percent) nucleus of the immature, uninucleate cyst of *E. histolytica*. However, the differences are subtle and, indeed, the parasites grown from one such specimen were isoenzymically indistinguishable from *E. histolytica* (Sargeaunt et al. 1980).

Entamoeba chattoni is an equally common parasite of wild primates (Levine 1973; Jackson et al. 1990), which is also found in captive animals. The cyst (Figure 10.6) is 6–18 μm in diameter, usually containing only one nucleus with a very variable appearance.

Both species have occasionally been reported from human feces; *E. polecki* most often in SE Asia (Desowitz and Barnish 1986; Muller et al. 1987; Barnish and Ashford 1989) or in refugees from that region (Chaker et al. 1982; DeGirolami and Kimber 1983), although it has also been reported from France (Masure et al. 1980) and Venezuela (Chacin Bonilla et al. 1992). No definite pathology has been linked to this organism. A small number of human infections with *E. chattoni* have also been described, mostly in those with occupational exposure to primates (Burrows and Klink 1955; Sargeaunt et al. 1992); some patients had diarrhea but whether or not this was due to the ameba is not clear.

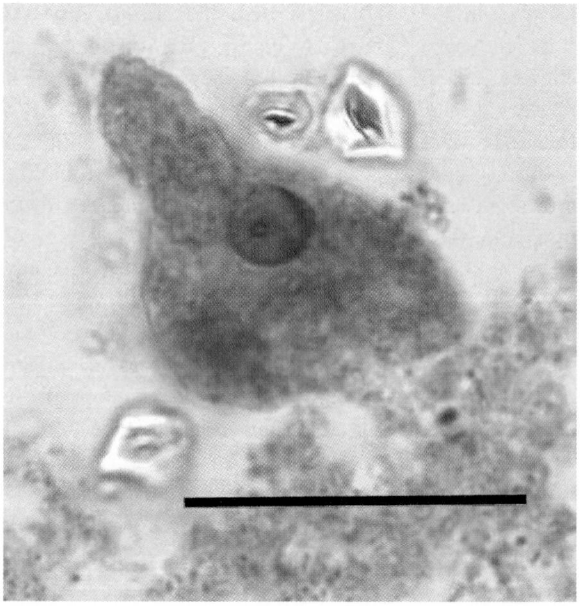

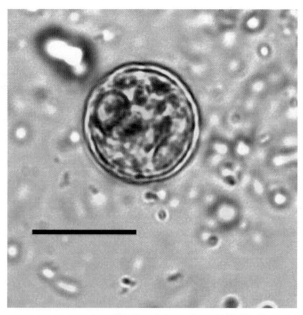

Figure 10.4 *Trophozoite of* E. moshkovskii *in culture. Bar = 10 μm. (Courtesy of John Williams, London School of Hygiene & Tropical Medicine)*

Figure 10.5 *Cyst of* E. polecki, *wet preparation stained with iodine. Bar = 10 μm. Note that these cysts and those of* E. chattoni *(Figure 10.6) are not morphologically distinguishable. (Courtesy of John Williams, London School of Hygiene & Tropical Medicine)*

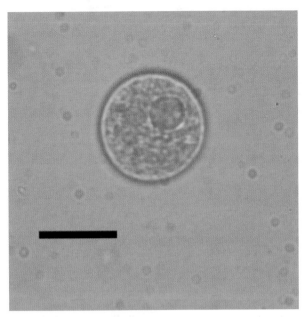

Figure 10.6 *Cyst of* E. chattoni, *wet preparation stained with iodine. Bar = 10 μm. (Courtesy of Dr Jaco Verweij, Centrum voor Infectieziekten, Afdeling Parasitologie, Universiteit Leiden)*

Although human infections with *E. polecki* and *E. chattoni* seem to be linked to exposure to different reservoir hosts, the cysts of both species are microscopically indistinguishable (Burrows 1959). Recently, therefore, species-specific PCR primers were developed to distinguish reliably between the two (Verweij et al. 2001). A most surprising result to emerge from this work was that, of 12 human fecal samples obtained from the Netherlands that were microscopically positive for uninucleate *Entamoeba* cysts, only one was infected with *E. polecki* and two with *E. chattoni*. The remainder were infected with two distinct *Entamoeba* species which, while clearly clustering with *E. polecki* and *E. chattoni*, were identical with neither. These results suggest that other related *Entamoeba* species can also infect humans; the authors suggest that, for the present, all such uninucleate cysts should be reported as '*E. polecki*-like.'

ENDOLIMAX NANA

Endolimax nana is a small, frequently reported ameba that is diagnosed in humans about as often as *E. coli* (Levine 1973; Kappus et al. 1994); similar if not identical parasites are also found in primates and pigs (Levine 1973). Both cysts and trophozoites possess a characteristic limax nuclear morphology that must be recognized to make an accurate diagnosis of fecal specimens. Limax nuclei have a large, irregularly-shaped mass of chromatin (the karyosome or endosome), which may be either centrally or eccentrically placed but there is no peripheral chromatin, so the nuclear membrane is all but invisible. The ring-shaped nucleus typical of *Entamoeba* and *Giardia* species is not seen; instead, wet prepara-

tions stained with iodine reveal the nuclei as dark, soft-edged specks which 'blink' into and out of view as the microscope is focused up and down.

Endolimax nana cysts may be ellipsoidal, spherical or irregular with a longest dimension of 6–10 μm, although it is usually less than 10 μm (Figure 10.7). They are thin-walled and almost always contain four limax-type nuclei, never more – smaller, thick-walled bodies with more than four dark nuclei are of fungal origin. Trophozoites are usually 8–10 μm long with a single limax nucleus and granular cytoplasm; they move sluggishly by extending blunt pseudopodia. The organism may be cultivated in bacteria-containing media and an iso-enzyme pattern has been published (Sargeaunt and Williams 1979); otherwise nothing is known of the metabolism.

This organism is normally regarded as nonpathogenic although there are isolated reports of gastrointestinal symptoms in AIDS patients for which no other cause could be found (Rolston et al. 1986; Peters et al. 1987). Reports of *Endolimax nana* being responsible for urticaria (Veraldi et al. 1991) and rheumatism (Burnstein and Liakos 1983; Alarcon Segovia and Abud Mendoza 1985) have yet to be confirmed.

IODAMOEBA BÜTSCHLII

Trophozoites and, above all, cysts of *I. bütschlii* contain glycogen that is colored deep brown in the usual iodine-stained wet preparations and gives the genus its name. Trophozoites are usually 12–15 μm long with a single nucleus containing a large karyosome and granules visible in permanent stained preparations; the cytoplasm may be coarse with ingested bacteria and glycogen-

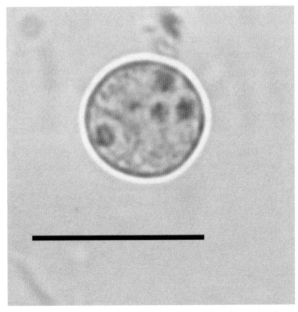

Figure 10.7 *Cyst of* Endolimax nana, *wet preparation stained with iodine. Bar = 10 μm. (Courtesy of John Williams, London School of Hygiene & Tropical Medicine)*

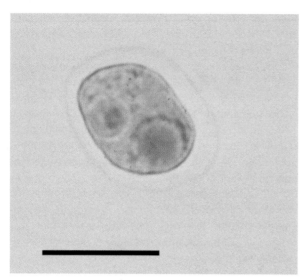

Figure 10.8 *Cyst of I.* bütschlii, *wet preparation stained with iodine. Bar = 10 μm. (Courtesy of John Williams, London School of Hygiene & Tropical Medicine)*

containing vacuoles. Cysts (Figure 10.8) are usually 9–12 μm across, often round or oval but not unusually very irregular with a single, limax-type nucleus containing a large central or eccentric karyosome. The glycogen vacuole is large and compact with a clearly defined edge; it stains deeply with iodine when the cysts are fresh but may disappear if they are stored for more than a few days. Distinction from other species is normally easy but may be difficult if immature, uninucleate *E. histolytica*/*E. dispar* cysts are present; these often contain glycogen stores, which will be used up as they mature, but this glycogen is usually present as a more diffuse, less deeply staining mass.

I. bütschlii is not rare in human feces but is usually reported in only a few percent of fecal samples from developing countries or from male homosexuals; identical cysts are found in the feces of several primates and the organism is (or was) very common in pigs all over the world. All such parasites are assigned to the same species but no evidence of human infections from animal reservoirs has been presented. The isoenzyme profile of an isolate has been published (Sargeaunt and Williams 1979).

I. bütschlii is not normally regarded as a pathogen (a disseminated infection and cerebral granuloma attributed to it were probably caused by an *Acanthamoeba* species, see Chapter 11, Opportunistic amebae) but has occasionally been proposed as a cause of diarrhea (Andrew 1947; Rolston et al. 1986).

REFERENCES

Alarcon Segovia, D. and Abud Mendoza, C. 1985. Parasitic rheumatism by *Endolimax nana*. Objections. *J Rheumatol*, **12**, 184–5.

Ali, I.K.M., Hossain, M.B, et al. 2003. *Entamoeba moshkovskii* infections in children in Bangladesh. *Emerg Infect Dis*, **9**, 580–4.

Andrew, R.R. 1947. Discussion on amoebiasis. *Trans R Soc Trop Med Hyg*, **41**, 88–9.

Barnish, G. and Ashford, R.W. 1989. Occasional parasitic infections of man in Papua New Guinea and Irian Jaya (New Guinea). *Ann Trop Med Parasitol*, **83**, 121–35.

Bray, R.S. and Harris, W.G. 1977. The epidemiology of infection with *Entamoeba histolytica* in the Gambia, West Africa. *Trans R Soc Trop Med Hyg*, **71**, 401–7.

Burnstein, S.L. and Liakos, S. 1983. Parasitic rheumatism presenting as rheumatoid arthritis. *J Rheumatol*, **10**, 514–15.

Burrows, R.B. 1959. Morphological identification of *Entamoeba hartmanni* and *Entamoeba polecki* from *Entamoeba histolytica*. *Am J Trop Med Hyg*, **8**, 583–9.

Burrows, R.B. and Klink, G.K. 1955. *Entamoeba polecki* infections in man. *Am J Hyg*, **62**, 156–67.

Chacin Bonilla, L., Bonilla, E., et al. 1992. Prevalence of *Entamoeba histolytica* and other intestinal parasites in a community from Maracaibo, Venezuela. *Ann Trop Med Parasitol*, **86**, 373–80.

Chaker, E., Kremer, M. and Kien, T.T. 1982. Presence of *Entamoeba polecki* in 14 refugees from South-East Asia. *Bull Soc Pathol Exot Filial*, **75**, 484–90.

Clark, C.G. and Diamond, L.S. 1991. The Laredo strain and other 'Entamoeba histolytica-like' amoebae are *Entamoeba moshkovskii*. *Mol Biochem Parasitol*, **46**, 11–18.

Clark, C.G. and Diamond, L.S. 1992. Colonization of the uterus by the oral protozoan *Entamoeba gingivalis*. *Am J Trop Med Hyg*, **46**, 158–60.

Clark, C.G. and Diamond, L.S. 1997. Intraspecific variation and phylogenetic relationships in the genus *Entamoeba* as revealed by riboprinting. *J Eukaryot Microbiol*, **44**, 142–54.

Corcoran, G.D., O'Connell, B., et al. 1991. *Entamoeba coli* as possible cause of diarrhoea. *Lancet*, **338**, 254.

Dao, A.H. 1985. Entamoeba gingivalis in sputum smears. *Acta Cytol*, **29**, 632–3.

DeGirolami, P.C. and Kimber, J. 1983. Intestinal parasites among Southeast Asian refugees in Massachusetts. *Am J Clin Pathol*, **79**, 502–4.

DeMoraes Ruehsen, M., McNeill, R.E., et al. 1980. Amebae resembling *Entamoeba gingivalis* in the genital tracts of IUD users. *Acta Cytol*, **24**, 413–20.

Desowitz, R.S. and Barnish, G. 1986. *Entamoeba polecki* and other intestinal protozoa in Papua New Guinea Highland children. *Ann Trop Med Parasitol*, **80**, 399–402.

Diamond, L.S. 1983. Lumen dwelling protozoa: *Entamoeba*, trichomonads and *Giardia*. In: Jensen, J.B. (ed.), *In vitro cultivation of protozoan parasites*. Boca Raton: CRC Press, 65–109.

Diamond, L.S., Clark, C.G. and Cunnick, C.C. 1995. YI-S, a casein-free medium for axenic cultivation of *Entamoeba histolytica*, related *Entamoeba*, *Giardia intestinalis* and *Trichomonas vaginalis*. *J Eukaryot Microbiol*, **42**, 277–8.

el Azzouni, M.Z. and el Badry, A.M. 1994. Frequency of *Entamoeba gingivalis* among periodontal and patients under chemotherapy. *J Egypt Soc Parasitol*, **24**, 649–55.

Favoreto, S. Jr and Machado, M.I. 1995. Incidence, morphology and diagnostic studies of *Entamoeba gingivalis*, Gros, 1849 (in Portuguese). *Rev Soc Bras Med Trop*, **28**, 379–87.

Garcia, L.S. and Bruckner, D.A. Diagnostic medical parasitology. Washington D.C: ASM Press.

Garrido-Gonzalez, E., Zurabian, R. and Acuna-Soto, R. 2002. Cyst production and transmission of *Entamoeba* and *Endolimax*. *Trans R Soc Trop Med Hyg*, **96**, 119–23.

Goldman, M. 1969. *Entamoeba histolytica*-like amoebae occurring in man. *Bull World Health Organ*, **40**, 355–64.

Jackson, T.F.H.G., Sargeaunt, P.G., et al. 1990. *Entamoeba histolytica*: naturally occurring infections in baboons. *Archivos de Investigacion Medica*, **21**, Suppl 1, 153–6.

Kappus, K.D., Lundgren, R.G.J., et al. 1994. Intestinal parasitism in the United States: update on a continuing problem. *Am J Trop Med Hyg*, **50**, 705–13.

Kikuta, N., Yamamoto, A. and Goto, N. 1996. Detection and identification of *Entamoeba gingivalis* by specific amplification of rRNA gene. *Can J Microbiol*, **42**, 1248–51.

Lachance, P.J. 1959. A Canadian strain of *Entamoeba moshkovskii* Chalaia 1941. *Can J Zool*, **37**, 415–17.

Levine, N.D. 1973. *The Amoebae*. Protozoan parasites of domestic animals and of man. Minneapolis: Burgess Publishing Company.

Levine, N.D., Corliss, J.O., et al. 1980. A newly revised classification of the protozoa. *J Protozool*, **27**, 37–58.

Lucht, E., Evengard, B., et al. 1998. *Entamoeba gingivalis* in human immunodeficiency virus type 1-infected patients with periodontal disease. *Clin Infect Dis*, **27**, 471–3.

Masure, O., Boles, J.M., et al. 1980. A case of human infection by *Entamoeba polecki* in France, with a review of literature. *Bull Soc Pathol Exot Filial*, **73**, 451–7.

Muller, R., Lillywhite, J., et al. 1987. Human cysticercosis and intestinal parasitism amongst the Ekari people of Irian Jaya. *J Trop Med Hyg*, **90**, 291–6.

Noble, G.A. and Noble, E.R. 1952. *Entamoeba* in farm mammals. *J Parasitol*, **38**, 571–95.

Perez Jaffe, L., Katz, R. and Gupta, P.K. 1998. *Entamoeba gingivalis* identified in a left upper neck nodule by fine-needle aspiration: a case report. *Diagn Cytopathol*, **18**, 458–61.

Peters, C., Kocka, F.E., et al. 1987. High carriage of *Endolimax nana* in diarrhoeal specimens from homosexual men. *Lett Appl Microbiol*, **5**, 65–6.

Pomes, C.E., Bretz, W.A., et al. 2000. Risk indicators for periodontal diseases in Guatemalan adolescents. *Braz Dent J*, **11**, 49–57.

Rolston, K.V.I., Hoy, J. and Mansell, P.W.A. 1986. Diarrhea caused by 'nonpathogenic amoebae' in patients with AIDS. *N Engl J Med*, **315**, 192.

Sargeaunt, P.G. and Williams, J.E. 1978. Electrophoretic isoenzyme patterns of *Entamoeba histolytica* and *Entamoeba coli*. *Trans R Soc Trop Med Hyg*, **72**, 164–6.

Sargeaunt, P.G. and Williams, J.E. 1979. Electrophoretic isoenzyme patterns of the pathogenic and nonpathogenic intestinal amoebae of man. *Trans R Soc Trop Med Hyg*, **73**, 225–7.

Sargeaunt, P.G. and Williams, J.E. 1982. The morphology in culture of the intestinal amoebae of man. *Trans R Soc Trop Med Hyg*, **76**, 465–72.

Sargeaunt, P.G., Patrick, S. and O'Keeffe, D. 1992. Human infections of *Entamoeba chattoni* masquerade as *Entamoeba histolytica*. *Trans R Soc Trop Med Hyg*, **86**, 633–4.

Sargeaunt, P.G., Williams, J.E., et al. 1980. A comparative study of *Entamoeba histolytica* (NIH:200, HK9, etc.), '*E. histolytica*-like' and other morphologically identical amoebae using isoenzyme electrophoresis. *Trans R Soc Trop Med Hyg*, **74**, 469–74.

Sargeaunt, P.G., Williams, J.E., et al. 1982. A review of isoenzyme characterization of *Entamoeba histolytica* with particular reference to pathogenic and nonpathogenic stocks isolated in Mexico. *Arch Invest Med*, **13**, 89–94.

Sefer, M., Boanchis, A.I., et al. 1989. Periodontal diseases with *Entamoeba gingivalis*. *Rev Chir Oncol Radiol ORL Oftalmol Stomatol Ser Stomatol*, **36**, 279–85.

Silberman, J.D., Clark, C.G., et al. 1999. Phylogeny of the genera *Entamoeba* and *Endolimax* as deduced from small-subunit ribosomal RNA sequences. *Mol Biol Evol*, **16**, 1740–51.

Veraldi, S., Schianchi Veraldi, R. and Gasparini, G. 1991. Urticaria probably caused by *Endolimax nana*. *Int J Dermatol*, **30**, 376.

Verweij, J.J., Polderman, A.M. and Clark, C.G. 2001. Genetic variation among human isolates of uninucleated cyst-producing *Entamoeba* species. *J Clin Microbiol*, **39**, 1644–6.

Vogel, P., Zaucha, G., et al. 1996. Rapid postmortem invasion of cecal mucosa of macaques by nonpathogenic *Entamoeba chattoni*. *Am J Trop Med Hygi*, **55**, 595–602.

Wahlgren, M. 1991. *Entamoeba coli* as cause of diarrhea? *Lancet*, **337**, 675.

11

Opportunistic amebae

DAVID T. JOHN

INTRODUCTION

Free-living amebae of the genera *Naegleria*, *Acanthamoeba*, and *Balamuthia* can cause disease in humans and other animals. Normally, they live as phagotrophs in aquatic habitats where they feed on bacteria but as opportunists, they may produce serious infection of the central nervous system (CNS) and the eye. The term amphizoic (from Greek 'amphi' meaning 'on both sides') has been proposed to describe the ability of these amebae to live in two worlds, as free-living organisms and as endoparasites (Page 1974).

Naegleria fowleri is responsible for a rapidly fatal infection involving the CNS and called primary amebic meningoencephalitis (PAM). Infection occurs most often in healthy young people who have a recent history of swimming in freshwater. Several species of *Acanthamoeba* cause disease. *Acanthamoeba* may produce a chronic CNS infection known as granulomatous amebic encephalitis (GAE) or an eye infection referred to as *Acanthamoeba* keratitis. Human infections originally attributed to *Hartmannella*, another free-living ameba, were actually caused by *Acanthamoeba*. *Balamuthia mandrillaris*, a leptomyxid ameba (Visvesvara et al. 1993), causes a chronic CNS infection similar to that produced by *Acanthamoeba*, also termed GAE. Additional reports have described eye infections caused by *Hartmannella* and *Vahlkampfia* (Aitken et al. 1996; Inoue et al. 1998), a fatal CNS infection caused by *Hartmannella vermiformis* (Centeno et al. 1996) and a nonfatal encephalitis caused by *Sappinia diploidea* (Gelman et al. 2001).

CLASSIFICATION

The classification of the Protozoa is in a state of flux and, following the precedents set out elsewhere in this volume (see Chapter 8, Classification and introduction to the parasitic protozoa), the scheme proposed by Corliss (1994) and widely adopted by scientists working with free-living protozoa is used in this chapter. On the basis of the traditional classification published by the Society of Protozoologists (Lee et al. 1985), all the opportunistic amebae were classified in the phylum Sarcomastigophora and subphylum Sarcodina, which contained two classes, class Lobosea and class Acarpomyxea. Class Lobosea contained two orders: order Amoebida (family Acanthamoebidae, e.g. *Acanthamoeba*) and order Schizopyrenida (family Valkampfidae, e.g. *Naegleria*). Class Acarpomyxea contained the order Leptomyxida (family Leptomyxidae, e.g. *Balamuthia*). In Corliss' classification both *Acanthamoeba* and *Balamuthia* are classified in the phylum Rhizopoda, class Lobosea, and *Naegleria* has been moved to the phylum Percolozoa and placed in the class Heterolobosea and order Schizopyrenida.

STRUCTURE AND LIFE-CYCLES

Morphology

The nuclei of the opportunistic amebae are characterized by a large central nucleolus or karyosome, and a nuclear membrane without chromatin granules. These features are especially useful when examining histological sections as they readily distinguish the opportunistic amebae from *Entamoeba histolytica*, the most important parasitic ameba of humans.

NAEGLERIA

The trophozoites of *N. fowleri* are known as limax amebae (from the Latin for 'slug'). These amebae are elongate and move in a directional manner by eruptive, blunt pseudopodia, called lobopodia. Actively moving amebae average about 22 μm in length (range 15–30 μm); inactive, rounded forms range from 9 to 15 μm in diameter (Carter 1970).

Trophozoites of pathogenic *Naegleria* have distinctive phagocytic structures known as amebostomes (Figure 11.1) (John et al. 1985). Amebostomes are used for engulfment and vary in number, depending on the strain. By transmission electron microscopy, amebostomes appear to be densely granular in contrast to the highly vacuolated body of the ameba. Amebostomes are visible by light microscopy but appear as thick-walled vacuoles. Similar food cups or phagocytic stomata occur on *E. histolytica* (Gonzalez-Robles and Martinez-Palomo 1983).

Reproduction in *Naegleria* is by simple binary fission of the trophozoite. Nuclear division is promitotic, which means that the nucleolus and the nuclear membrane persist during nuclear division (karyokinesis). The nucleolus elongates, forming a dumbbell-shaped structure, and divides into two polar masses, or nucleoli. During this process, the nuclear membrane remains intact (Page 1988).

When *Naegleria* are suspended in distilled water or non-nutrient buffer, they transform into temporary flagellated forms. The typical *N. fowleri* flagellate is a bluntly elongate cigar- or pear-shaped cell with two flagella emerging from beneath the anterior rostrum (Figure 11.2). Of the different species of *Naegleria*, *N. fowleri* flagellates are the most uniform, generally having two flagella (John et al. 1991).

The cysts of *N. fowleri* (Figure 11.3) are spherical, often clumped closely together, and are 7–15 μm in diameter (Carter 1970; Page 1988). Examination of cysts by electron microscopy reveals an average of fewer than two mucoid-plugged pores or ostioles per cyst and a relatively thin cyst wall (Schuster 1975), a feature that makes *N. fowleri* cysts susceptible to desiccation.

ACANTHAMOEBA

A feature that readily distinguishes the amebae of *Acanthamoeba* from those of *Naegleria* is the presence of acanthopodia (from Greek 'acanth-' meaing 'spine or thorn'), tapering spike-like pseudopodia (Figure 11.4). In contrast to *Naegleria*, which has rapid, directional locomotion, *Acanthamoeba* moves slowly on a broad front without direction. Trophozoites of *Acanthamoeba* are larger than those of *Naegleria* and average about 24–56 μm in length (Lewis and Sawyer 1979). Nuclear division in *Acanthamoeba* is metamitotic (i.e. the nucleolus and the nuclear membrane disintegrate during early karyokinesis), a pattern similar to that of dividing metazoan cells (Page 1988).

Considerable variation in cyst morphology occurs among the different species of *Acanthamoeba*, resulting in the naming of new species. Page (1967) recognized four species of *Acanthamoeba*; by 1976 he had recorded seven. On the basis of morphology and isoenzyme analysis, De Jonckheere (1987) identified 17 species of *Acanthamoeba*, of which seven have been associated with human infection. The cysts of *Acanthamoeba* are double-walled and, therefore, quite resistant in the environment. The cyst wall is made up of an outer, wrinkled or rippled ectocyst and an inner endocyst. The encysted ameba conforms to the shape of the endocyst.

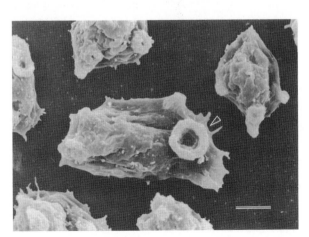

Figure 11.1 *Scanning electron micrograph (SEM) of a* Naegleria fowleri *ameba with a single amebostome (arrowhead). Bar = 5 μm*

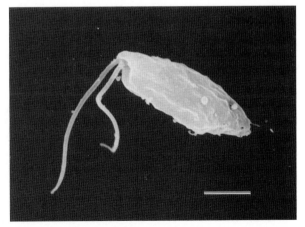

Figure 11.2 *SEM of a* Naegleria fowleri *flagellate with two flagella emerging from beneath the anterior rostrum. Bar = 3 μm*

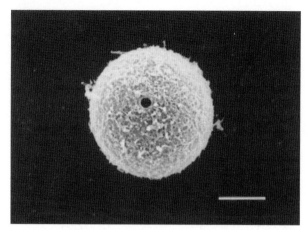

Figure 11.3 *SEM of a* Naegleria fowleri *cyst with a single pore or ostiole. Bar = 3 µm*

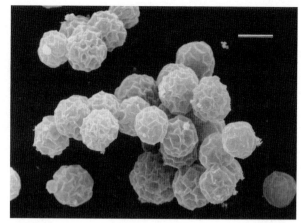

Figure 11.5 *SEM of pathogenic* Acanthamoeba polyphaga *cysts. Bar = 10 µm*

The characteristic wrinkled appearance of *Acanthamoeba* cysts is readily seen in culture (Figure 11.5) and in histological sections (see Figure 11.11).

BALAMUTHIA

The trophozoites of *B. mandrillaris* are mostly irregular or branching in shape (Figure 11.6), with some limax forms. Their length ranges from 12 to 60 µm (mean c. 30 µm) (Visvesvara et al. 1993). These amebae exhibit little directional motility, although a spider-like walking movement may be seen with amebae grown in Vero cell culture. Nuclear division is metamitotic, as in *Acanthamoeba*. Cysts are irregularly round and are 6–30 µm in diameter (mean 15 µm) with two walls. The inner cyst wall is thin and spherical and the outer cyst wall is thick and wavy or wrinkled (Figure 11.6) much like the cysts of *Acanthamoeba* (Visvesvara et al. 1993).

Life-cycles

Naegleria, *Acanthamoeba*, and leptomyxid amebae are distributed worldwide in freshwater and soil. *Acanth-*

amoeba and some leptomyxids are found in marine environments as well. Life-cycles are simple; a feeding trophozoite, or ameba, a resting cyst and, in *Naegleria*, a transient flagellate.

NAEGLERIA

The life-cycle of *N. fowleri* is illustrated in Figure 11.7. The term 'ameboflagellate' is used to describe amebae that can transform into flagellates. When *Naegleria* amebae are placed in a non-nutrient medium, such as distilled water or buffer, they differentiate into transient, nonfeeding, nondividing flagellates that, after a time, revert back to amebae. Amebae also encyst when conditions are appropriate and, later, excyst in a favorable environment.

The invasive stage of *N. fowleri* is the ameba and infection is acquired by exposure of the nasal passages to the amebae in fresh water. The amebae invade the nasal mucosa, the cribriform plate and the olfactory bulbs of the brain. It is likely that flagellates or cysts of

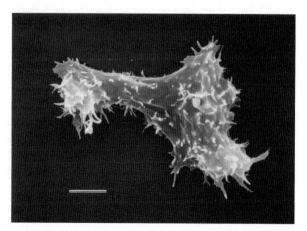

Figure 11.4 *SEM of a pathogenic* Acanthamoeba castellanii *ameba displaying numerous acanthopodia. Bar = 5 µm*

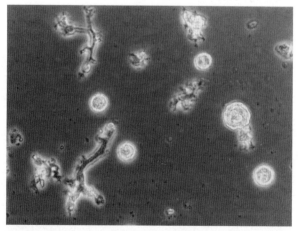

Figure 11.6 *Phase-contrast micrograph of* Balamuthia mandrillaris *amebae and cysts. (Courtesy Dr Govinda S. Visvesvara, Centers for Disease Control and Prevention, Atlanta, Georgia)*

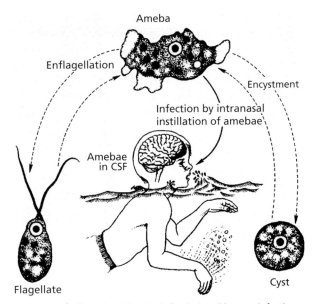

Figure 11.7 *Life cycle of* Naegleria fowleri *and human infection*

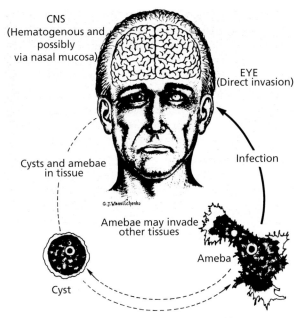

Figure 11.8 *Life cycle of* Acanthamoeba *spp. and human infection*

N. fowleri could enter the nose of a swimmer as readily as amebae. However, flagellates would revert quickly to amebae or the amebae could escape from cysts, the point being that amebae are the invading organisms. Flagellates or cysts of *N. fowleri* have never been found in tissue or cerebrospinal fluid.

ACANTHAMOEBA **AND** *BALAMUTHIA*

Figure 11.8 illustrates the life-cycle of *Acanthamoeba* and shows human involvement. The free-living cycle of the ameba and cyst is also reflected in human infection, in which both amebae and cysts are seen in tissue, in contrast to naeglerial infection in which only amebae occur. *Acanthamoeba* seems to be truly amphizoic in all respects. Amebae and cysts of *Balamuthia* are also found in tissue.

Human infection by *Acanthamoeba* involves the CNS, the eye and other organs. Although amebae of pathogenic *Acanthamoeba* are able to invade the nasal mucosa and cause fatal CNS disease in experimental animals, this is not thought to be the usual route of invasion in human infection. Invasion of the CNS seems to be by way of the circulation, the amebae originating from a primary focus elsewhere in the body, possibly the lower respiratory tract, ulcers of the skin or mucosa or other wounds. The modes of invasion by leptomyxid amebae are thought to be the same. GAE tends to occur in persons who are debilitated, chronically ill or immunocompromised. *Balamuthia* infections occur in both immunosuppressed and immunocompetent persons. In contrast, *Acanthamoeba* keratitis usually occurs in healthy individuals and infection is by direct invasion of the cornea through trauma to the eye or the wearing of contaminated contact lenses.

BIOCHEMISTRY, MOLECULAR BIOLOGY, AND GENETICS

Most research on the cell and molecular biology of free-living amebae has been with nonpathogenic rather than pathogenic species. Nonetheless, some observations can be made and comparisons drawn between the two groups.

Cell differentiation

Cell differentiation is the process by which amebae become cysts or flagellates and, of course, then become amebae again. *Acanthamoeba*, *Balamuthia*, and *Naegleria* all produce cysts, but only *Naegleria* produces flagellates. Thus, free-living amebae provide useful models for studying the developmental biology and molecular mechanisms controlling morphogenesis and differentiation in eukaryotic cells.

ENCYSTMENT

Encystment is an adaptive mechanism that enables the organism to survive conditions that would kill the ameba. Factors thought to induce cyst formation include starvation, drying and various chemicals. Excystment is the process by which amebae exit from their protective cysts. It is generally held that unfavorable conditions induce encystment and that favorable conditions stimulate excystment.

There are no definitive studies that describe the conditions for encystment and excystment in *N. fowleri*. Chemical factors that induce encystment in *A. castellanii* include inhibitors of DNA synthesis – fluorodeoxyuridine, mitomycin C, trenimon (Neff and Neff 1972) and hydroxyurea (Rudick 1971); the mitochondrial inhibitors – diminazine (berenil), ethidium bromide, erythromycin, and chloramphenicol (Akins and Byers 1980); and acetate and glucose starvation (Byers et al. 1980). Medium from encysting cultures of *A. castellanii* contains an extracellular encystment-enhancing activity that stimulates cyst formation in early log phase cultures (Akins and Byers 1980). Encystment-enhancing activity is required for encystment in cultures induced by diminazine and by glucose starvation, but not when it is induced by total deprivation of nutrients (Akins et al. 1985).

ENFLAGELLATION

Although spontaneous enflagellation occurs in cultures of *Naegleria*, especially exponential phase cultures of *N. fowleri*, synchronized enflagellation may be achieved by suspending amebae in distilled water or non-nutrient buffer. Enflagellation in *Naegleria* apparently occurs not as an obligatory phase in the life cycle but as a response to changes in the environment. Factors affecting enflagellation in *N. fowleri* include: nutrient depletion, temperature, phase of growth, and culture agitation. Maximum enflagellation in axenically grown *N. fowleri* occurs 4–5 hours after suspension in buffer (Cable and John 1986). For axenically grown nonpathogenic *N. gruberi*, maximum enflagellation occurs somewhat sooner, c. 1.5 hours after suspension in buffer (Fulton 1977).

Cyclohexamide and actinomycin D completely prevent enflagellation in *N. fowleri* when added before enflagellation commences but, when added after initiation of enflagellation, both prevent further differentiation and cause existing flagellates to revert to amebae (Woodworth et al. 1982a). The ultrastructure of *N. fowleri* flagellates is that of a typical eukaryotic protist. There is a distinct nuclear membrane and a prominent nucleolus, numerous vacuoles and cytoplasmic inclusions, pleomorphic mitochondria and some rough endoplasmic reticulum. Basal bodies, rootlets, and flagella are formed quickly after an initial lag of 90 minutes (Patterson et al. 1981).

Macromolecular composition

Changes in cell composition of *N. fowleri* are related to culture age. For shaken axenic cultures, average cell mass remains constant during logarithmic growth at 150 pg per ameba but decreases by 30 percent during the stationary phase at 96 hours. During logarithmic growth, 80–85 percent of the cell dry mass is protein (120 pg per ameba) (Weik and John 1978). Cell dry mass and protein of *N. fowleri* are c. 70 percent of values reported for *N. gruberi* (Weik and John 1977). The majority of *N. fowleri* polypeptides have molecular masses within the range of 20–60 kDa (Woodworth et al. 1982b). During logarithmic and stationary phases of growth of *N. fowleri*, carbohydrate content averages 15 pg per ameba and RNA is c. 18 pg per ameba (Weik and John 1978). By comparison, *N. gruberi* carbohydrate averages c. 35 pg per ameba and the RNA content is 8 pg per ameba (Weik and John 1977). The more than two-fold higher RNA value for the smaller pathogenic *N. fowleri* reflects different biosynthetic capabilities and maintenance of a larger ribosome complement. Total DNA in *N. fowleri* is 0.2 pg per ameba during logarithmic growth; it doubles during transition from exponential phase to stationary phase and then gradually decreases almost to initial levels. The peak in DNA content corresponds to an increase in the average number of nuclei per ameba; nuclear number then decreases as cells enter the stationary phase (Weik and John 1978). In *N. gruberi*, the total DNA content of 0.2 pg per cell increases by 50 percent during exponential growth (Weik and John 1977).

RESPIRATORY METABOLISM

As an opportunistic pathogen, *N. fowleri* lives in the brain, an oxygen-rich environment, and would thus be expected to have an aerobic metabolism. The use of the synonym '*N. aerobia*' (Singh and Das 1970) recognized the aerobic nature of the organism, in contrast to the anaerobic nature of strictly parasitic amebae. Unlike *E. histolytica*, an anaerobic parasite that lacks mitochondria, *N. fowleri* lives in aerobic aqueous environments and has many mitochondria.

Whole-cell respiration rates were measured polarographically throughout the growth cycle of *N. fowleri* (Weik and John 1979a). In agitated cultures, amebae consume 30 ng atoms O per minute per mg of cell protein during logarithmic growth. Under similar conditions, *N. gruberi* amebae consume 80 ng atoms O per minute per mg of cell protein (Weik and John 1979b). The lower oxygen consumption, and presumably oxygen requirement, of *N. fowleri* probably explains the presence of the pathogen in heated waters in which dissolved oxygen concentrations are substantially reduced. The respiratory rate gradually declines during the stationary phase. The reduction in respiratory rate may involve respiratory control as increases in respiratory rate did not occur despite the addition of oxygen (Weik and John 1979a). The respiratory process of isolated *N. fowleri* mitochondria is similar to that of classic mammalian cell mitochondria. Oxidation is coupled to phosphorylation (adenosine triphosphate (ATP) formation) as shown by the two- to three-fold increase in respiration on addition of a phosphate acceptor or an uncoupling agent. The spectra of oxidized and dithionite-reduced mitochondria show distinct absorption bands of flavins and c-type, b-type and a-type cytochromes (Weik and John 1979a).

Genetics

It has generally been assumed that reproduction in *Acanthamoeba* and *Naegleria* is asexual and that sexuality, or genetic exchange, does not occur. In the past, genetic studies have been hampered because of difficulty in visualizing the very small, numerous chromosomes of these amebae, which may number 16 for *Naegleria* (De Jonckheere 1989) and 80 for *Acanthamoeba* (Volkonsky 1931). Fulton (1970) reported that one strain of *N. gruberi* had twice the ploidy of another strain, based on cell and nuclear volume and DNA content. Cariou and Pernin (1987) and Pernin et al. (1992) provided evidence for diploidy and genetic recombination in *N. lovaniensis*. De Jonckheere (1989) compared the electrophoretic karyotypes of all the species of *Naegleria* and concluded that, although karyotype analysis cannot be used to identify *Naegleria* species, it is useful for studying gene localization and genetic exchange. No genetic studies on the opportunistic amebae have been published. Clark and Cross (1987) described a self-replicating ribosomal DNA plasmid in *N. gruberi*, making it the third eukaryotic genus to have a nuclear plasmid DNA. Additionally, they sequenced the small-subunit rRNA gene of *N. gruberi* and showed that circular rRNA genes are a general feature of schizopyrenid (ameboflagellate) amebae, including, presumably, *N. fowleri* (Clark and Cross 1988).

CLINICAL ASPECTS

Clinical manifestations

PAM

PAM typically occurs in healthy children or young adults with a recent history of swimming in freshwater. The disease is rapidly fatal, usually resulting in death within 72 hours of the onset of symptoms. Infection follows entry of water containing amebae or flagellates into the nose. It has been suggested that inhalation of cysts (e.g. during dust storms) could lead to infection (Lawande et al. 1979). From the nasal mucosa, the amebae penetrate the cribriform plate and travel along the olfactory nerves to the brain. They invade the olfactory bulbs and then spread to the more posterior regions of the brain. Within the brain, the amebae provoke inflammation and cause extensive destruction of tissue (Carter 1970; Martinez 1985).

The clinical course is dramatic. Symptoms begin with severe frontal headache, fever (39–40°C) and anorexia followed by nausea, vomiting, and signs of meningeal irritation, frequently evidenced by a positive Kernig's sign. Involvement of the olfactory lobes may cause disturbances in smell or taste and may be noted early in the course of the disease. Visual disturbance may occur.

The patient may experience confusion, irritability, and restlessness and may become irrational before lapsing into a coma. Generalized seizures may also occur. In order of frequency of occurrence, the more important symptoms are (Carter 1970; Martinez 1985):

- headache
- anorexia
- nausea
- vomiting
- fever
- neck stiffness.

GAE

GAE usually occurs in debilitated or chronically ill persons, some of whom may be undergoing immunosuppressive therapy. Infection by *Balamuthia* may occur in the immunocompetent. The underlying conditions reported in GAE are Hodgkin's disease, systemic lupus erythematosus, diabetes mellitus, glucose-6-phosphate dehydrogenase deficiency, alcoholism, and AIDS. Some of the victims of GAE are not debilitated or immunocompromised, but are otherwise healthy individuals. GAE is a disease not as well defined as that caused by *N. fowleri*. The course of infection is subacute or chronic, lasting weeks, months or even years, and is characterized by focal granulomatous lesions of the brain. The onset of GAE, unlike that of PAM, is insidious with a prolonged clinical course (Carter et al. 1981; Martinez 1987).

Acanthamoeba infection probably occurs through the lower respiratory tract or through ulcers of the skin or mucosa. Invasion into the CNS is by hematogenous spread from the primary focus of infection. As there are no lymphatic channels in the brain, invasion of the brain must be via the bloodstream (Martinez 1987). Even though some *Acanthamoeba* isolates are able to produce a CNS infection after intranasal instillation in mice, there is no proof that similar invasion occurs in the human disease. The portal of entry and spread in *Balamuthia* infection is believed to be the same as that for *Acanthamoeba*. The incubation period is not known, but probably lasts weeks or months and, during the prolonged clinical course, single or multiple space-occupying lesions develop. An altered mental state is a prominent feature in GAE. Headache, seizures, and neck stiffness occur in about half of the cases. Nausea and vomiting may also be noted (Martinez 1987).

ACANTHAMOEBA KERATITIS

Acanthamoeba keratitis is a chronic infection of the cornea caused by several species of *Acanthamoeba* including *A. castellanii*, *A. culbertsoni*, *A. hatchetti*, *A. polyphaga*, and *A. rhysodes* and infections are being diagnosed with increasing frequency (Stehr-Green et al. 1989). Infection is by direct contact of the cornea with amebae, which may be introduced through minor corneal trauma or by exposure to contaminated water or contact lenses. The wearing of contact lenses and the use of home-made saline solutions are important risk factors. Saline solutions contaminated with protein residues from contact lenses promote the growth of bacteria and yeast which, in turn, are a source of food for the amebae. Amebae attach to the contact lenses stored in contaminated solutions and are transferred to the eye when lenses are placed over the cornea. Amebae become established as part of the conjunctival flora and may invade the corneal stroma through a break in the epithelium or through the intact epithelium. They produce an infection that progresses to *Acanthamoeba* keratitis, which usually develops over a period of weeks to months, and is characterized by:

- severe ocular pain (often out of proportion to the degree of inflammation)
- affected vision
- a stromal infiltrate that is frequently ring-shaped and composed predominantly of neutrophils.

Acanthamoeba keratitis is a serious ocular infection which, if not properly managed, can lead to loss of vision and even loss of the eye.

IMMUNOLOGY

The factors responsible for susceptibility and innate resistance to infection by pathogenic free-living amebae are undefined. Relatively few human infections have occurred, even though large numbers of individuals must have been exposed to amebae. Most cases of PAM occur in otherwise healthy children or young adults. *Acanthamoeba* keratitis also generally occurs in healthy individuals. In contrast, GAE usually occurs in persons who are chronically ill, debilitated or immunosuppressed, although *Balamuthia* infections may occur in the immunocompetent.

Antibodies to opportunistic amebae have been detected in human sera and include agglutinating antibodies against pathogenic and nonpathogenic *Naegleria* (Marciano-Cabral et al. 1987; Reilly et al. 1983). The agglutinating activity was specific for each species, indicating exposure to each of the *Naegleria* species. Antibodies to both pathogenic and nonpathogenic *Acanthamoeba* and *Naegleria* were detected in normal human sera by indirect fluorescent antibody testing of 93 serum samples. All were positive, with titers ranging from 1:5 to 1:80 (Cursons et al. 1980). An avidin-biotin horseradish peroxidase assay has been used to detect antibodies to pathogenic and nonpathogenic *Naegleria* in samples of human serum (Dubray et al. 1987). Antibodies were detected in 88 percent of sera from 115 hospital patients. Antibodies were identified as immunoglobulin (Ig)G and IgM, with IgG antibody titers ranging from 1:20 to 1:640. Radioimmunoassay detected antibody to *N. fowleri* in a pool of normal human sera

(Tew et al. 1977). Antibody titers were nearly nine times greater against intracellular antigens than against cell surface antigens. The apparent widespread occurrence of antibodies to pathogenic free-living amebae in human sera may reflect the global distribution of these organisms. Alternatively, it may represent cross-reacting antibodies to antigens that have yet to be identified.

Serum from a victim of PAM in New Zealand was reported to have a low level of serum IgA, although IgG and IgM levels were normal (Cursons et al. 1979). In contrast, normal serum levels of IgA were found in a child who died of naeglerial infection in England (Cain et al. 1979). Because serum IgA levels may not be an accurate reflection of secretory IgA concentrations, both patients may have had deficient levels of secretory IgA. As *N. fowleri* invades the nasal mucosa, it seems reasonable to suggest that secretory IgA may play a role in protection. Except for the serological surveys described above, all other information on immune responses to pathogenic free-living amebae has come from experimental laboratory studies, mostly with animals (reviewed by John 1993).

PATHOLOGY

Pathogenicity

Pathogenicity is the ability of a microorganism to produce disease, whereas cytopathogenicity is the ability to produce pathologic change in cells or a cytopathic effect in vitro in cultured cells. The proposed mechanisms of cytopathogenicity for *N. fowleri* include phagocytosis, release of cytolytic substances and the presence of a biologically active component, *Naegleria* ameba cytopathogenic material (NACM).

Phagocytosis is a basic function of amebae. It causes the destruction of cells, whether in cell culture or in human tissue and is, therefore, intimately involved in pathogenesis. Brown (1979) called the piecemeal engulfment of mouse embryo cells by *N. fowleri* trogocytosis (from the Greek for 'to nibble'). It is now known that trogocytosis is accomplished by amebostomes (John et al. 1985) that the amebae use to engulf particles of various sizes, including cultivated mammalian cells. In addition to phagocytosis and trogocytosis, *N. fowleri* seems to injure cultivated mammalian cells by another contact-mediated means. Marciano-Cabral et al. (1990) reported cytolytic factors on the membranes of amebae that lyse B103 rat nerve cells on contact. Lysis is followed by engulfment of cellular debris. Cytopathic activity was enhanced by the divalent cations calcium or magnesium.

Various phospholipases (phospholipase A, lysophospholipase, sphingomyelinase) have been identified in culture media in which *N. fowleri* is grown (Cursons et al. 1978; Hysmith and Franson 1982). The phospholipase activities from cultures of pathogenic *Naegleria* are much greater than from cultures of nonpathogenic *Naegleria*. Phospholipases have also been reported to be present in media in which *A. culbertsoni* is grown. The amounts are greater in pathogenic *A. culbertsoni* cultures than in those of nonpathogenic *A. castellanii* (Cursons et al. 1978). Other enzymes, some with a possible role in pathogenesis, occur in *N. fowleri* extracts. These are various hydrolases, including acid phosphatase, several glycosidases, and elastase.

A cytopathic effect has been attributed to NACM, obtained from lysed amebae (Dunnebacke and Dixon 1989). NACM is isolated from lysates of *Naegleria* by centrifugation, filtration, and lyophilization and has been obtained from *N. fowleri*, *N. gruberi*, and *N. jadini*, but not from *Acanthamoeba*. NACM kills cells of a variety of avian and mammalian cell lines. After cultures are inoculated, there is a long latent period (4–10 days) followed by a short period (under 24 hours) during which the monolayer is destroyed. The cytopathic effect has been maintained in cell cultures through nine serial passages. NACM is a protein with a molecular mass of 36 000 kDa and an isoelectric point of pH 4.2. Monoclonal antibodies to NACM prevent its cytopathic activity. Fluorescent staining shows NACM to be located at the tips of the ameba's pseudopodia and in the peripheral cytoplasm. It forms ring-shaped structures resembling amebostomes (Dunnebacke and Dixon 1989).

PAM

The gross pathologic findings in PAM are remarkably constant. The cerebral hemispheres are usually edematous and swollen, meninges are diffusely hyperemic with a slight purulent exudate and the cortex contains many focal superficial hemorrhages. There is severe involvement of the olfactory bulbs, with hemorrhage, necrosis, and purulent exudate (Carter 1972; Martinez 1985).

Microscopic examination reveals many amebae in the subarachnoid and perivascular spaces. Presumably, the perivascular spaces provide a path of migration for the amebae, and the blood vessels supply the oxygen needed by these aerobic organisms. Small numbers of amebae are found clustered within the brain tissue and in the purulent exudate of the meninges and brain substance. Within the exudate, some amebae may be seen engulfed by macrophages. Many amebae contain phagocytosed cellular debris and erythrocytes. The purulent exudate contains numerous polymorphonuclear and mononuclear leukocytes (Carter 1972; Martinez 1985). Figure 11.9 shows *N. fowleri* amebae with engulfed erythrocytes in brain tissue; cysts do not occur in tissue.

The cortical gray matter is a preferred site for the ameba's development. Consequently, severe involvement occurs in the cerebral hemispheres, cerebellum, brain stem, and upper portions of the spinal cord. Encephalitis may be a result of light amebic invasion and inflammation or massive invasion with purulent,

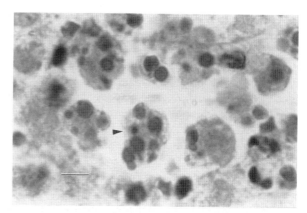

Figure 11.9 *Histological section of brain with* Naegleria fowleri *amebae containing engulfed erythrocytes (H&E stain). Prominent nucleolus visible within nucleus of one ameba (arrowhead). Bar = 10 μm*

hemorrhagic necrosis. Typically, the olfactory bulbs are extensively invaded, with hemorrhage and an inflammatory exudate; the involvement here is greater than in other areas of the brain. Infection of the central nervous system with *N. fowleri* may be described best as an acute, hemorrhagic, necrotizing meningoencephalitis (Carter 1972; Martinez 1985). Focal demyelination in the white matter of the brain and spinal cord may occur (Chang 1979; Duma et al. 1971). Curiously, demyelination may occur in the absence of amebae or cellular infiltrate. Chang (1979) suggests that demyelination may be caused by a phospholipolytic enzyme or enzyme-like substance produced by actively growing amebae present in the adjacent gray matter.

GAE

In contrast to naeglerial infection, which is characterized by a diffuse meningoencephalitis, *Acanthamoeba* and *Balamuthia* CNS disease is a focal granulomatous encephalitis. Martinez (1980) gave a summary of the neuropathological features for 15 patients with GAE. In affected areas, the leptomeninges contain a moderate amount of purulent exudate. The cerebral hemispheres show moderate or severe edema with foci of softened tissue and associated hemorrhagic necrosis. Lesions are usually multifocal and more posterior, including the upper portion of the spinal cord. The olfactory bulbs are not usually involved. Lesions of the CNS in GAE are characterized by necrosis with hemorrhagic foci and localized leptomeningitis. The chronic inflammatory exudate over the cortex comprises mostly mononuclear cells with a few polymorphonuclear leukocytes. The brain substance may have a prominent granulomatous reaction with foreign-body giant cells, which are never seen in naeglerial infection. The multinucleated giant cells may not be present in immunosuppressed patients (Carter et al. 1981; Martinez 1987).

Amebae reach the brain via the bloodstream, invasion of the CNS is thus centrifugal, from the deeper tissues toward the brain surface. Trophozoites and cysts (Figures 11.10 and 11.11) occur in most infected tissues and around blood vessels. *Acanthamoeba* and *Balamuthia* reach the CNS by hematogenous spread from a primary focus of infection elsewhere in the body, most probably the skin, mucosa or lungs. Within the infected primary tissues, there occurs a chronic granulomatous reaction like that seen in the brain, with multinucleated giant cells, trophozoites, and cysts. Similar lesions have been described from other tissues including prostate, thyroid, uterus, and pancreas, probably resulting from hematogenous dissemination of amebae from the primary focus in the skin or lungs, or possibly even a secondary CNS lesion (Martinez 1987).

ACANTHAMOEBA KERATITIS

Ocular infections with *Acanthamoeba* are characterized by chronic progressive ulcerative keratitis (Cohen et al. 1985). During early corneal infection, there may be pseudodendritic figures in the epithelium or just beneath the epithelium in the anterior stroma (Johns et al. 1987). In advanced cases of *Acanthamoeba* keratitis, there may be a marked stromal infiltrate and necrosis. The whitish inflammatory infiltrate, often appearing ring-shaped around the corneal ulcer, consists mainly of polymorphonuclear leukocytes and macrophages, with a few lymphocytes (Mathers et al. 1987). Although granulomatous inflammation has been described in *Acanthamoeba* keratitis, in most of the reports, neutrophils (not lymphocytes) are the predominant infiltrating cells. Corneal ulceration may progress to perforation (Lindquist et al. 1988).

DIAGNOSIS

Laboratory diagnosis

The laboratory diagnosis of infection by the opportunistic amebae depends on the recovery and identification of amebae in cerebrospinal fluid (CSF), brain tissue or

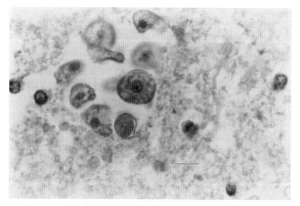

Figure 11.10 *Histological section of brain with* Acanthamoeba castellanii *amebae (H&E stain). Bar = 10 μm*

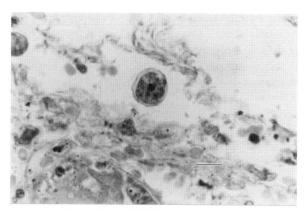

Figure 11.11 *Histological section of brain with* Acanthamoeba castellanii *cyst exhibiting typical wrinkled ectocyst (H&E stain). Bar = 10 μm*

corneal scrapings. Amebae in a clinical specimen may also be cultivated on non-nutrient agar spread with gram-negative bacteria and later transferred to liquid medium with antibiotics for axenic growth.

PAM

The diagnosis of PAM is made by microscopic identification of living or stained amebae in CSF; motile amebae are readily seen in simple wet-mount preparations. Amebae can be distinguished from other cells by their limax shape and progressive movement. It is not necessary to warm the slide as amebae remain fully active at room temperature. Refrigeration of the spinal fluid is not recommended because this may kill the amebae.

Spinal fluid smears may be stained with Wright or Giemsa stains. The bacterial Gram stain is of little value because heat fixing destroys the amebae, which causes them to stain poorly and appear as degenerating cells. Giemsa- or Wright-stained amebae have considerable amounts of sky-blue cytoplasm and relatively small, delicate, pink nuclei. Mononuclear leukocytes, on the other hand, have large purplish nuclei with only a small amount of sky-blue cytoplasm. In cytospin preparations of CSF, the amebae tend to be rounded and flattened, without pseudopodia. Occasionally, enlarged teardrop-shaped food vacuoles appear to radiate from the nucleus but this is an artefact induced by the cytospin procedure (Benson et al. 1985). Amebae may be cultivated by placing some of the CSF on non-nutrient agar (1.5 percent) spread with a lawn of washed *Esherichia coli* or *Enterobacter aerogenes* and incubated at 37°C. The amebae grow on the moist agar surface and use the bacteria as food, producing plaques as they clear the bacteria.

Clinically, PAM closely resembles fulminating bacterial meningitis, and the laboratory findings are also similar. The CSF is purulent or sanguinopurulent, with leukocyte counts (predominantly neutrophils) ranging from a few hundred to more than 20 000 cells mm^{-3}.

Spinal fluid glucose levels are low and protein content is generally increased. Typically, Gram-stained smears and cultures of spinal fluid are negative for bacteria (Carter 1972; Martinez 1985).

GAE

The laboratory diagnosis of GAE is made by identifying amebic forms of *Acanthamoeba* or *Balamuthia* in the CSF, or amebae and cysts in brain tissue. Whereas *N. fowleri* is readily cultured from CSF, *Acanthamoeba* and *Balamuthia* are not. *Acanthamoeba* has only rarely been isolated from patients with GAE. *A. culbertsoni* and *A. rhysodes* have been cultured from CSF, and *Acanthamoeba* sp. and *A. palestinensis* have been cultured from aspirated and biopsied brain material, respectively. Leptomyxid amebae have been cultivated twice from biopsy material (Visvesvara et al. 1990; Gordon et al. 1992).

As with *N. fowleri*, *Acanthamoeba* may be cultured on non-nutrient agar spread with washed *E. coli* or *E. aerogenes*. *Balamuthia* must be grown in tissue culture, preferably Vero cell (Visvesvara et al. 1993). *Acanthamoeba* and *Balamuthia* do not have a flagellate stage but amebae are identified by their small spiky acanthopodia or branching shape, respectively, and cysts are readily identified by their distinctive double-walled wrinkled appearance. Species identification may be made by using the indirect fluorescent antibody technique (IFAT) and specific antisera against *Acanthamoeba* spp. or *B. mandrillaris*. The species of *Acanthamoeba* identified most frequently from cases of GAE have been *A. castellanii* and *A. culbertsoni*.

ACANTHAMOEBA KERATITIS

Acanthamoeba keratitis is diagnosed by identifying amebae cultured from corneal scrapings or by histological examination of infected corneal tissue. As in GAE, *Acanthamoeba* may be cultured from corneal scrapings on non-nutrient agar spread with gram-negative bacteria. Cultures of corneal material should be incubated at 30°C rather than 37°C. Of the five species of *Acanthamoeba* identified as capable of causing eye infections, only *A. hatchetti* has not been cultured from clinical material. Species identification is based on indirect immunofluorescent antibody staining. The two species most frequently identified in *Acanthamoeba* keratitis have been *A. castellanii* and *A. polyphaga*. *A. castellanii* is the species that has most often been identified in cases of *Acanthamoeba* GAE and ocular infection.

Rapid diagnosis of *Acanthamoeba* keratitis may be made by identifying amebae or cysts in corneal scrapings using procedures for Giemsa staining, calcofluor white staining (Wilhelmus et al. 1986), and IFAT (Epstein et al. 1986). The calcofluor white procedure and IFAT both require the use of fluorescence microscopy and IFAT also requires an antiserum to *Acanthamoeba*.

Histopathological preparations of corneal tissue may be stained using the conventional hematoxylin and eosin (H&E) procedure or by the more specialized staining procedures of Heidenhain's hematoxylin, Gomori's chromium hematoxylin, periodic acid–Schiff, Bauer chromic acid-Schiff, and silver methenamine (McClellan et al. 1988). The special staining techniques are useful for demonstrating the presence of cysts in corneal tissue. IFAT and calcofluor white staining also may be used (Silvany et al. 1987).

Herpes simplex keratitis is the disease most commonly mistaken for *Acanthamoeba* keratitis (Johns et al. 1987; Mannis et al. 1986; Moore and McCulley 1989). The single most consistent clinical symptom of *Acanthamoeba* keratitis is severe ocular pain, which is not characteristic of an infection limited to the cornea and generally not present in persons with herpes simplex keratitis. Additional distinguishing features of *Acanthamoeba* keratitis include a history of direct exposure to soil or water, wearing contact lenses, scleritis and failure of cultures from the inflamed eye to reveal bacteria, fungi or viruses (Mannis et al. 1986).

EPIDEMIOLOGY

Literature review

Human infection by *Naegleria* and *Acanthamoeba* has been reported worldwide, as has the environmental isolation of these opportunistic amebae (reviewed by John 1993). Although leptomyxid amebae have been identified as a cause of human disease (Visvesvara et al. 1990; Lowichik et al. 1995) and have been cultivated twice from tissue (Visvesvara et al. 1990; Gordon et al. 1992), the environmental isolation of pathogenic leptomyxid amebae has been reported only once (John and Howard 1995).

PAM

Most of the reports of PAM have been from the developed rather than the developing nations, probably because of greater awareness rather than greater incidence. Australia, Czechoslovakia, and the USA have reported 75 percent of all cases of PAM. In the USA, most reported cases have been from the coastal states of Virginia, Florida, and Texas, accounting for 67 percent of the cases.

The majority of patients with naeglerial infection have had a history of recent swimming in freshwater during hot summer weather. In Richmond, Virginia, infection in 14 of 16 cases was probably acquired in two man-made lakes located within a few miles of each other (Callicott Jnr 1968; Duma et al. 1971; dos Santos 1970). Over a 3-year period, 16 young people died in Czechoslovakia after swimming in the same heated and chlorinated indoor swimming pool (Červa et al. 1968). Similar fatal cases have been reported following swimming in:

- swimming pools in Belgium, England, and New Zealand
- hot springs in California and New Zealand
- lakes in Arkansas, Florida, Missouri, Nevada, South Carolina, and Texas
- streams in Belgium, Mississippi, and New Zealand
- an irrigation ditch in Mexico.

Individual references for the above cases may be found in John (1993).

Infection has not always been acquired by swimming. The South Australian and northern Nigerian cases occurred in arid regions where swimming is unusual; the proposed means of infection were bathing and face washing (Anderson and Jamieson 1972; Lawande et al. 1980) or inhalation of dust-borne cysts (Lawande et al. 1979).

N. fowleri has been isolated from a variety of environmental sources in Africa, Asia, Australia, New Zealand, Europe, and North and South America, with 33 references describing 211 isolates (for specific references, see John 1993; John and Howard 1995). *N. fowleri* is more readily isolated from warm, clean water than from warm, organically enriched water. The average water temperature has been c. 30°C (range 15–45°C). Although laboratory animals may succumb to naeglerial infection, there is no evidence for an animal reservoir of *N. fowleri*.

GAE

There have been even fewer cases of *Acanthamoeba* CNS infection than of *Naegleria* infection. More than half of these have been reported from the USA. Unlike naeglerial infection, GAE is not associated with swimming. Infections often occur in persons who are debilitated or immunosuppressed including patients with acquired immunodeficiency syndrome (AIDS). In one of the AIDS patients, the clinical features were more like those of PAM than of GAE, presumably because of immunosuppression (Wiley et al. 1987). With the current increase in the number of immunosuppressed persons, it is likely that there will be a corresponding increase in disseminated *Acanthamoeba* infections.

GAE caused by the leptomyxid, *B. mandrillaris*, was first described in a pregnant mandrill baboon that died of meningoencephalitis at the San Diego Zoo (Visvesvara et al. 1990). Amebae were cultivated from brain tissue and used to prepare immune serum for indirect immunofluorescence assays; these revealed 16 human cases of leptomyxid infection originally identified as *Acanthamoeba* infections. Leptomyxid infections were also identified in a sheep, a horse, and a gorilla. Since the original description, leptomyxid infections continue to be reported in immunocompetent and immunosuppressed humans, including patients with AIDS.

At the time of writing, the world literature contains 26 references that describe the environmental isolation of 144 pathogenic *Acanthamoeba* isolates (for specific references see John 1993; John and Howard 1995). These have been recovered from a variety of sources at somewhat lower average temperatures than for the isolation of *N. fowleri* (below 25°C). Only one report describes the environmental recovery of pathogenic leptomyxid isolates and these do not appear to be *Balamuthia* (John and Howard 1995). As with *N. fowleri*, there is no evidence for an animal reservoir of pathogenic *Acanthamoeba* or *Balamuthia*.

ACANTHAMOEBA KERATITIS

The first cases of *Acanthamoeba* keratitis were reported from the UK in 1974 (Nagington and Watson 1974) and the USA in 1975 (Jones et al. 1975) and were associated with trauma to the eye or exposure to contaminated water. Since 1985, there has been a dramatic increase in the number of cases associated with the wearing of contact lenses; the greatest increase occurred in wearers of soft contact lenses (Stehr-Green et al. 1989). The continued increase in the number of contact lens wearers, and of those not properly caring for their lenses, will undoubtedly result in an increase in the incidence of *Acanthamoeba* keratitis, mainly in the developed nations where the wearing of contact lenses is widespread.

CHEMOTHERAPY

Overview

Although *Acanthamoeba* keratitis may be treated with antimicrobial agents, virtually all cases of PAM and GAE have been fatal because there is no effective treatment.

PAM

At present, there is no satisfactory treatment for PAM. The antibiotics used to treat bacterial meningitis are ineffective in naeglerial infection, as are the antiamebic drugs. Amphotericin B, a drug of considerable toxicity, is the antinaeglerial agent for which there is evidence of clinical effectiveness. The four known survivors of PAM, children from Australia (Anderson and Jamieson 1972), the UK (Apley et al. 1970), India (Pan and Ghosh 1971), and the USA (Seidel et al. 1982), were treated with amphotericin B, given intravenously and intrathecally. The patient in the USA was also given parenteral miconazole and oral rifampicin (Seidel et al. 1982). In experimental infections, tetracycline acts synergistically with amphotericin B to protect mice (Thong et al. 1979). Goswick and Brenner (2003) describe the effective treatment of experimental PAM in mice with azithromycin. Treatment was initiated 72 hours after infection and was continued for 5 days, with protection being 100 percent and greater than for amphotericin B-treated mice. Interestingly, the patient who recovered from *Sappinia diploidea* encephalitis also was treated with azithromycin (Gelman et al. 2001).

Amphotericin B is administered intravenously at high doses; 1–15 mg/kg of body weight daily for 3 days and then 1 mg/kg per day for 6 days. Additionally, amphotericin B may be given intrathecally and miconazole is administered intravenously (Carter 1972; Seidel et al. 1982). Amphotericin B is a polyene compound that acts on the plasma membrane, disrupting its selective permeability and causing leakage of cellular components. On exposure to amphotericin B, amebae round up and fail to form pseudopodia. Membrane-related changes, evident by electron microscopy, include enhanced nuclear plasticity, increased amounts of smooth and rough endoplasmic reticulum, decreased food vacuole formation and production of blebs on the plasma membrane (Schuster and Rechthand 1975).

GAE

As with naeglerial infection, there is no satisfactory treatment for GAE, partly because most cases have been diagnosed after death and there has not been adequate opportunity to evaluate therapeutic regimens. There are three reports of persons having recovered from *Acanthamoeba* CNS infection. A 7-year-old girl with a single *Acanthamoeba*-induced granulomatous brain tumor recovered following total excision of the mass and treatment with ketoconazole. *A. palestinensis* was cultured from the brain biopsy material (Ofori-Kwakye et al. 1986). The second report involved a 40-year-old man with *Acanthamoeba* meningitis who recovered following treatment with penicillin and chloramphenicol. *A. culbertsoni* was repeatedly cultured from the patient's CSF (Lalitha et al. 1985). The third case, for whom complete recovery cannot be claimed because the patient returned home and was not followed-up, was a 30-year-old man with chronic meningoencephalitis who was treated with sulfamethazine and from whose CSF *A. rhysodes* was cultured (Cleland et al. 1982).

Because the cysts of *Acanthamoeba* and *Balamuthia* form in tissues, a potentially effective drug for GAE must be able to destroy cysts as well as amebae. Otherwise, a possible relapse could occur after the course of treatment has ended.

ACANTHAMOEBA KERATITIS

Most of the earlier cases of *Acanthamoeba* keratitis required corneal transplants in order to manage the disease. Even so, there were reported instances of surgical enucleation. With present therapies, *Acanthamoeba* keratitis can be managed by medical treatment alone if infection is identified soon enough (Moore and McCulley 1989). The first successful medical cure of *Acanthamoeba* keratitis was reported by Wright et al.

(1985) and involved the use of a combination of dibromopropamidine and propamidine isethionate ointment and drops and neomycin drops. The success of this treatment has been confirmed by others. Signs of toxicity of propamidine and dibromopropamidine have been reported in one patient (Yeoh et al. 1987), but when treatment was discontinued, there seemed to be a recurrence of the *Acanthamoeba* keratitis.

In addition to topical propamidine, other successful treatment regimens have used:

- topical miconazole and systemic ketoconazole (Wilhelmus et al. 1986)
- topical miconazole and neosporin with epithelial debridement (Lindquist et al. 1988)
- topical clotrimazole (Driebe Jnr et al. 1988)
- oral itraconazole with topical miconazole and surgical debridement (Ishibashi et al. 1990)
- topical polyhexamethylene biguanide (Larkin et al. 1992).

PREVENTION AND CONTROL

Vaccination

There is no vaccine to protect against infection by the opportunistic amebae. Numerous attempts have been made to immunize mice against infection by *N. fowleri* or *Acanthamoeba*, without producing solid immunity (for details of the studies, see John (1993)).

Integrated control

Because of the relationship of swimming to naeglerial infection, many swimming areas have been subjected to intense investigation. Although *N. fowleri* has been isolated from many such areas, not all sampling efforts have yielded *N. fowleri*. Obviously, there are factors that favor the development of *N. fowleri* in swimming areas, including a warm temperature, the presence of an adequate food supply, insufficient residual free chlorine, minimal competition from other protozoans and probably optimal pH and oxygen levels. With the present limited understanding of the ecology of *N. fowleri*, practical measures for the prevention and control of the infection include education of the public, awareness within the medical community and adequate chlorination of public water supplies, including swimming facilities. Adequate chlorination requires a continuous free residual chlorine level of 0.5 mg per liter of water (Derreumaux et al. 1974) and remains the single most effective disinfectant system for controlling opportunistic amebae in public waters. This level of chlorination has effectively controlled the *N. fowleri* problem in the public water supplies of South Australia (Dorsch et al. 1983). The risk of acquiring naeglerial infection through swimming in Florida's freshwater lakes is estimated at c. 1 in 2.6 million exposures (Wellings 1977). Considering the millions of persons who swim outdoors each summer, it is truly remarkable that there are not more cases of PAM.

Factors that have been associated with *Acanthamoeba* keratitis in contact lens wearers are:

- using nonsterile, home-made saline
- disinfecting lenses less frequently than recommended
- wearing lenses while swimming (Stehr-Green et al. 1987).

Contact lens wearers should closely follow the manufacturer's recommendations for wear, care, and disinfection of lenses. Home-made saline solutions remain an important risk factor associated with *Acanthamoeba* keratitis. Contamination of the contact lens care system with bacteria or fungi encourages the survival and growth of *Acanthamoeba*. Amebae, including *A. polyphaga* and *A. hatchetti*, have been isolated from laboratory eyewash stations, especially stations containing reservoirs (Bier and Sawyer 1990; Tyndall et al. 1987). Eyewash stations with reservoirs should be flushed weekly, otherwise they potentially present a health hazard to users, particularly those wearing contact lenses.

With the expected increase in the number of contact lens wearers, there will undoubtedly be a corresponding increase in the number of cases of *Acanthamoeba* keratitis. As immunosuppression becomes more widespread (not just because of AIDS but also because of organ transplantation, cancer chemotherapy, congenitally acquired immunodeficiency, and suppression resulting from the indiscriminate release of toxic chemicals and carcinogens into the environment), the possibility of more CNS and disseminated *Acanthamoeba* infections also increases.

REFERENCES

Aitken, D., Hay, J., et al. 1996. Amebic keratitis in a wearer of disposable contact lenses due to a mixed *Vahlkampfia* and *Hartmannella* infection. *Ophthalmol*, **103**, 485–94.

Akins, R.A. and Byers, T.J. 1980. Differentiation promoting factors induced in *Acanthamoeba* by inhibitors of mitochondrial macromolecule synthesis. *Dev Biol*, **78**, 126–40.

Akins, R.A., Gozs, S.M. and Byers, T.J. 1985. Factors regulating the encystment enhancing activity (EEA) of *Acanthamoeba castellanii*. *J Gen Microbiol*, **131**, 2609–17.

Anderson, K. and Jamieson, A. 1972. Primary amoebic meningoencephalitis. *Lancet*, **1**, 902–3.

Apley, J., Clarke, S.K.R., et al. 1970. Primary amoebic meningoencephalitis in Britain. *Br Med J*, **1**, 596–9.

Benson, R.L., Ansbacher, L., et al. 1985. Cerebrospinal fluid centrifuge analysis in primary amebic meningoencephalitis due to *Naegleria fowleri*. *Arch Pathol Lab Med*, **109**, 668–71.

Bier, J.W. and Sawyer, T.K. 1990. Amoebae isolated from laboratory eyewash stations. *Curr Microbiol*, **20**, 349–50.

Brown, T. 1979. Observations by immunofluorescence microscopy and electron microscopy on the cytopathogenicity of *Naegleria fowleri* in mouse embryo-cell cultures. *J Med Microbiol*, **12**, 363–71.

Byers, T.J., Akins, R.A., et al. 1980. Rapid growth of *Acanthamoeba* in defined media; induction of encystment by glucose-acetate starvation. *J Protozool*, **27**, 216–19.

Cable, B.L. and John, D.T. 1986. Conditions for maximum enflagellation in *Naegleria fowleri*. *J Protozool*, **33**, 467–72.

Cain, A.R.R., Mann, P.G. and Warhurst, D.C. 1979. IgA and primary amoebic meningoencephalitis. *Lancet*, **1**, 441.

Callicott Jnr, J.H. 1968. Amoebic meningoencephalitis due to free-living amebas of the *Hartmannella* (*Acanthamoeba*)-*Naegleria* group. *Am J Clin Pathol*, **49**, 84–91.

Cariou, M.L. and Pernin, P. 1987. First evidence for diploidy and genetic recombination in free-living amoebae of the genus *Naegleria* on the basis of electrophoretic variation. *Genetics*, **115**, 265–70.

Carter, R.F. 1970. Description of a *Naegleria* sp. isolated from two cases of primary amoebic meningo-encephalitis, and of the experimental pathological changes induced by it. *J Pathol*, **100**, 217–44.

Carter, R.F. 1972. Primary amoebic meningo-encephalitis. An appraisal of present knowledge. *Trans R Soc Trop Med Hyg*, **66**, 193–213.

Carter, R.F., Cullity, G.J., et al. 1981. A fatal case of meningo-encephalitis due to a free-living amoeba of uncertain identity – probably *Acanthamoeba* sp. *Pathology*, **13**, 51–68.

Centeno, M., Rivera, F., et al. 1996. *Hartmannella vermiformis* isolated from the cerebrospinal fluid of a young male patient with meningoencephalitis and bronchopneumonia. *Arch Med Res*, **27**, 579–86.

Červa, L., Novak, K. and Culbertson, C.G. 1968. An outbreak of acute, fatal amebic meningoencephalitis. *Am J Epidemiol*, **88**, 436–44.

Chang, S.L. 1979. Pathogenesis of pathogenic *Naegleria* amoeba. *Folia Parasitol (Praha)*, **26**, 195–200.

Clark, C.G. and Cross, G.A.M. 1987. rRNA genes of *Naegleria gruberi* are carried exclusively on a 14-kilobase-pair plasmid. *Mol Cell Biol*, **7**, 3027–31.

Clark, C.G. and Cross, G.A.M. 1988. Circular ribosomal RNA genes are a general feature of schizopyrenid amoebae. *J Protozool*, **35**, 326–9.

Cleland, P.G., Lawande, R.V., et al. 1982. Chronic amebic meningoencephalitis. *Arch Neurol*, **39**, 56–7.

Cohen, E.J., Buchanan, H.W., et al. 1985. Diagnosis and management of *Acanthamoeba* keratitis. *Am J Ophthalmol*, **100**, 389–95.

Corliss, J.O. 1994. An interim 'user-friendly' hierarchical classification and characterization of the protists. *Acta Protozool*, **33**, 1–51.

Cursons, R.T.M., Brown, T.J. and Keys, E.A. 1978. Virulence of pathogenic free-living amebae. *J Parasitol*, **64**, 744–5.

Cursons, R.T.M., Keys, E.A., et al. 1979. IgA and primary amoebic meningoencephalitis. *Lancet*, **1**, 223–4.

Cursons, R.T.M., Brown, T.J., et al. 1980. Immunity to pathogenic free-living amoebae: role of humoral antibody. *Infect Immun*, **29**, 401–7.

De Jonckheere, J.F. 1987. Taxonomy. In: Rondanelli, E.G. (ed.), *Amphizoic amoebae: human pathology*. Padua: Piccin, 25–48.

De Jonckheere, J.F. 1989. Variation of electrophoretic karyotypes among *Naegleria* spp.. *Parasitol Res*, **76**, 55–62.

Derreumaux, A.L., Jadin, J.B., et al. 1974. Action du chlore sur les amibes de l'eau. *Ann Soc Belge Med Trop*, **54**, 415–18.

dos Santos, J.G. 1970. Fatal primary amebic meningoencephalitis: a retrospective study in Richmond, Virginia. *Am J Clin Pathol*, **54**, 737–42.

Dorsch, M.M., Cameron, A.S. and Robinson, B.S. 1983. The epidemiology and control of primary amoebic meningoencephalitis with particular reference to South Australia. *Trans R Soc Trop Med Hyg*, **77**, 372–7.

Driebe Jnr, W.T., Stern, G.A., et al. 1988. *Acanthamoeba* keratitis. Potential role for topical clotrimazole in combination therapy. *Arch Ophthalmol*, **106**, 1196–201.

Dubray, B.L., Wilhelm, W.E. and Jennings, B.R. 1987. Serology of *Naegleria fowleri* and *Naegleria lovaniensis* in a hospital survey. *J Protozool*, **34**, 322–7.

Duma, R.J., Rosenblum, W.I., et al. 1971. Primary amoebic meningoencephalitis caused by *Naegleria*. Two new cases, response to amphotericin B, and a review. *Ann Intern Med*, **74**, 923–31.

Dunnebacke, T.H. and Dixon, J.S. 1989. NACM, a cytopathogen from *Naegleria* ameba: purification, production of monoclonal antibody, and immunoreactive material in NACM-treated vertebrate cell cultures. *J Cell Sci*, **93**, 391–401.

Epstein, R.J., Wilson, L.A., et al. 1986. Rapid diagnosis of *Acanthamoeba* keratitis from corneal scraping using indirect fluorescent antibody staining. *Arch Ophthalmol*, **104**, 1318–21.

Fulton, C. 1970. Amebo-flagellates as research partners: the laboratory biology of *Naegleria* and *Tetramitus*. *Methods Cell Physiol*, **4**, 341–476.

Fulton, C. 1977. Cell differentiation in *Naegleria gruberi*. *Annu Rev Microbiol*, **31**, 597–629.

Gelman, B.B., Rauf, S.J., et al. 2001. Amoebic encephalitis due to *Sappinia diploidea*. *JAMA*, **285**, 2450–1.

Gonzalez-Robles, A. and Martinez-Palomo, A. 1983. Scanning electron microscopy of attached trophozoites of pathogenic *Entamoeba histolytica*. *JAMA*, **30**, 692–700.

Gordon, S.M., Steinberg, J.P., et al. 1992. Culture isolation of *Acanthamoeba* species and leptomyxid amebas from patients with amebic meningoencephalitis; including two patients with AIDS. *Clin Infect Dis*, **15**, 1024–30.

Goswick, S.M. and Brenner, G.M. 2003. Activities of azithromycin and amphotericin B against *Naegleria fowleri* in vitro and in a mouse model of primary amebic meningoencephalitis. *Antimicrob Agents Chemother*, **47**, 524–8.

Hysmith, R.M. and Franson, R.C. 1982. Degradation of human myelin phospholipids by phospholipase-enriched culture media of pathogenic *Naegleria fowleri*. *Biochem Biophys Acta*, **712**, 698–701.

Inoue, T., Asari, S., et al. 1998. *Acanthamoeba* Kerititis with symbiosis of *Hartmannella* ameba. *Am J Ophthalmol*, **125**, 721–3.

Ishibashi, Y., Matsumoto, Y., et al. 1990. Oral itraconazole and topical miconzaole with debridement for *Acanthamoeba* keratitis. *Am J Ophthalmol*, **109**, 121–6.

John, DT 1993. Opportunistically pathogenic free-living amebae. In: Kreier, J.P. and Baker, J.R. (eds), *Parasitic protozoa*, Vol. 3, 2nd edn. San Diego: Academic Press, 143–246.

John, D.T. and Howard, M.J. 1995. Seasonal distribution of pathogenic free-living amebae in Oklahoma waters. *Parasitol Res*, **81**, 193–201.

John, D.T., Cole Jnr, T.B. and Bruner, R.A. 1985. Amebostomes of *Naegleria fowleri*. *J Protozool*, **32**, 12–19.

John, D.T., Cole Jnr, T.B. and John, R.A. 1991. Flagella number among *Naegleria* flagellates. *Folia Parasitol (Praha)*, **38**, 289–95.

Johns, K.J., O'Day, D.M., et al. 1987. Herpes simplex masquerade syndrome: *Acanthamoeba* keratitis. *Curr Eye Res*, **6**, 207–12.

Jones, D.B., Visvesvara, G.S. and Robinson, N.M. 1975. *Acanthamoeba polyphaga* keratitis and *Acanthamoeba* uveitis associated with fatal meningoencephalitis. *Trans Ophthalmol Soc UK*, **95**, 221–32.

Lalitha, M.K., Anandi, V., et al. 1985. Isolation of *Acanthamoeba culbertsoni* from a patient with meningitis. *J Clin Microbiol*, **21**, 666–7.

Larkin, D.F.P., Kilvington, S. and Dart, J.K.G. 1992. Treatment of *Acanthamoeba* keratitis with polyhexamethylene biguanide. *Ophthalmology*, **99**, 185–92.

Lawande, R.V., Abraham, S.N., et al. 1979. Recovery of soil amebas from the nasal passages of children during the dusty harmattan period in Zaria. *Am J Clin Pathol*, **71**, 201–3.

Lawande, R.V., MacFarlane, J.T, et al. 1980. A case of primary amebic meningoencephalitis in a Nigerian farmer. *Am J Trop Med Hyg*, **29**, 21–5.

Lee, J.J., Hutner, S.H. and Bovee, E.C. 1985. *An illustrated guide to the protozoa*. Lawrence, Kansas: Allen Press.

Lewis, E.J. and Sawyer, T.K. 1979. *Acanthamoeba tubiashi* n. sp., a new species of fresh-water Amoebida (Acanthamoebidae). *Trans Am Micros Soc*, **98**, 543–9.

Lindquist, T.D., Sher, N.A. and Doughman, D.J. 1988. Clinical signs and medical therapy of early *Acanthamoeba* keratitis. *Arch Ophthalmol*, **106**, 73–7.

Lowichik, A., Rollins, N., et al. 1995. Leptomyxid amebic meningoencephalitis mimicking brain stem glioma. *Am J Neuroradiol*, **16**, 926–9.

Mannis, M.J., Tamaru, R., et al. 1986. *Acanthamoeba* sclerokeratitis. *Arch Ophthalmol*, **104**, 1313–17.

Marciano-Cabral, F., Cline, M.L. and Bradley, S.G. 1987. Specificity of antibodies from human sera for *Naegleria* species. *J Clin Microbiol*, **25**, 692–7.

Marciano-Cabral, F., Zoghby, K.L. and Bradley, S.G. 1990. Cytopathic action of *Naegleria fowleri* amoebae on rat neuroblastoma target cells. *J Protozool*, **37**, 138–44.

Martinez, A.J. 1980. Is *Acanthamoeba* encephalitis an opportunistic infection? *Neurology*, **30**, 567–74.

Martinez, A.J. 1985. *Free-living amebas: natural history, prevention, diagnosis, pathology, and treatment of disease*. Boca Raton: CRC Press.

Martinez, A.J. 1987. Clinical manifestations of free-living amebic infections. In: Rondanelli, E.G. (ed.), *Amphizoic amoebae: human pathology*. Padua: Piccin, 161–77.

Mathers, W., Stevens Jnr, G., et al. 1987. Immunopathology and electron microscopy of *Acanthamoeba* keratitis. *Am J Ophthalmol*, **103**, 626–35.

McClellan, K.A., Kappagoda, N.K., et al. 1988. Microbiological and histopathological confirmation of acanthamebic keratitis. *Pathology*, **20**, 70–3.

Moore, M.B. and McCulley, J.P. 1989. *Acanthamoeba* keratitis associated with contact lenses: six consecutive cases of successful management. *Br J Ophthalmol*, **73**, 271–5.

Nagington, J. and Watson, P.F. 1974. Amoebic infection of the eye. *Lancet*, **2**, 1537–40.

Neff, R.J. and Neff, R.H. 1972. Induction of differentiation in *Acanthamoeba* by inhibitors. *CR Trav Lab Carlsberg*, **39**, 111–68.

Ofori-Kwakye, S.K., Sidebottom, D.G., et al. 1986. Granulomatous brain tumor caused by *Acanthamoeba*. *J Neurosurg*, **64**, 505–9.

Page, F.C. 1967. Re-definition of the genus *Acanthamoeba* with description of three species. *J Protozool*, **14**, 709–24.

Page, F.C. 1974. *Rosculus ithacus* Hawes, 1963 (Amoebida, Flabellulidae) and the amphizoic tendency in amoebae. *Acta Protozool*, **13**, 143–54.

Page, F.C. 1988. *A new key to freshwater and soil gymnamoebae*. Ambleside, England: Freshwater Biol Assoc.

Pan, N.R. and Ghosh, T.N. 1971. Primary amoebic meningoencephalitis in two Indian children. *J Indian Med Assoc*, **56**, 134–7.

Patterson, M., Woodworth, T.W., et al. 1981. Ultrastructure of *Naegleria fowleri* enflagellation. *J Bacteriol*, **147**, 217–26.

Pernin, P., Ataya, A. and Cariou, M.L. 1992. Genetic structure of natural populations of the free-living amoeba, *Naegleria lovaniensis*. Evidence for sexual reproduction. *Heredity*, **68**, 173–81.

Reilly, M.F., Marciano-Cabral, F., et al. 1983. Agglutination of *Naegleria fowleri* and *Naegleria gruberi* by antibodies in human serum. *J Clin Microbiol*, **17**, 576–81.

Rudick, V.L. 1971. Relationships between nucleic acid synthetic patterns and encystment in aging unagitated cultures of *Acanthamoeba castellanii*. *J Cell Biol*, **49**, 498–506.

Schuster, F.L. 1975. Ultrastructure of cysts of *Naegleria* spp: a comparative study. *J Protozool*, **22**, 352–9.

Schuster, F.L. and Rechthand, E. 1975. In vitro effects of amphotericin B on growth and ultrastructure of the amoeboflagellates *Naegleria gruberi* and *Naegleria fowleri*. *Antimicrob Agents Chemother*, **8**, 591–605.

Seidel, J.S., Harmatz, P., et al. 1982. Successful treatment of primary amebic meningoencephalitis. *N Engl J Med*, **306**, 346–8.

Silvany, R.E., Luckenbach, M.A. and Moore, M.B. 1987. The rapid detection of *Acanthamoeba* in paraffin-embedded sections of corneal tissue with calcofluor white. *Arch Ophthalmol*, **105**, 1366–7.

Singh, B.N. and Das, S.R. 1970. Studies on pathogenic and non-pathogenic small free-living amoebae and the bearing of nuclear division on the classification of the order Amoebida. *Phil Trans R Soc Lond B Biol Sci*, **259**, 435–76.

Stehr-Green, J.K., Bailey, T.M. and Visvesvara, G.S. 1989. The epidemiology of *Acanthamoeba* keratitis in the United States. *Am J Ophthalmol*, **107**, 331–6.

Stehr-Green, J.K., Bailey, T.M., et al. 1987. *Acanthamoeba* keratitis in soft contact lens wearers. *J Am Med Assoc*, **258**, 57–60.

Tew, J.G., Burmeister, J., et al. 1977. A radioimmunoassay for human antibody specific for microbial antigens. *J Immunol Methods*, **14**, 231–41.

Thong, Y.H., Rowan-Kelly, B. and Ferrante, A. 1979. Delayed treatment of experimental amoebic meningo-encephalitis with amphotericin B and tetracycline. *Trans R Soc Trop Med Hyg*, **73**, 336–7.

Tyndall, R.L., Lyle, M.M. and Ironside, K.S. 1987. The presence of free-living amoebae in portable and stationary eye wash stations. *Am Ind Hyg Assoc J*, **48**, 933–4.

Visvesvara, G.S., Schuster, F.L. and Martinez, A.J. 1993. *Balamuthia mandrillaris*, n.g., n.sp., agent of amebic meningoencephalitis in humans and other animals. *J Euk Microbiol*, **40**, 504–14.

Visvesvara, G.S., Martinez, A.J., et al. 1990. Leptomyxid ameba, a new agent of amebic meningoencephalitis in humans and animals. *J Clin Microbiol*, **28**, 2750–6.

Volkonsky, M. 1931. *Hartmannella castellanii* Douglas et classification des Hartmannelles (Hartmanelliane nov. subfam., *Acanthamoeba* nov. gen., *Glaeseria* nov. gen.). *Arch Zool Exp Gén*, **72**, 317–39.

Weik, R.R. and John, D.T. 1977. Cell size, macromolecular composition and O_2 consumption during agitated cultivation of *Naegleria gruberi*. *J Protozool*, **24**, 196–200.

Weik, R.R. and John, D.T. 1978. Macromolecular composition and nuclear number during growth of *Naegleria fowleri*. *J Parasitol*, **64**, 746–7.

Weik, R.R. and John, D.T. 1979a. Cell and mitochondria respiration of *Naegleria fowleri*. *J Parasitol*, **65**, 700–8.

Weik, R.R. and John, D.T. 1979b. Preparation and properties of mitochondria from *Naegleria gruberi*. *J Protozool*, **26**, 311–18.

Wellings, F.M. 1977. Amoebic meningoencephalitis. *J Fla Med Assoc*, **64**, 327–8.

Wiley, C.A., Safrin, R.E., et al. 1987. *Acanthamoeba* meningoencephalitis in a patient with AIDS'. *J Infect Dis*, **155**, 130–3.

Wilhelmus, K.R., Osato, M.S., et al. 1986. Rapid diagnosis of *Acanthamoeba* keratitis using calcofluor white. *Arch Ophthalmol*, **104**, 1309–12.

Woodworth, T.W., Keefe, W.E. and Bradley, S.G. 1982a. Characterization of proteins in flagellates and growing amebae of *Naegleria fowleri*. *J Bacteriol*, **150**, 1366–74.

Woodworth, T.W., Keefe, W.E. and Bradley, S.G. 1982b. Characterization of the proteins of *Naegleria fowleri*: relationships between subunit size and charge. *J Protozool*, **29**, 246–51.

Wright, P., Warhurst, D. and Jones, B.R. 1985. *Acanthamoeba* keratitis successfully treated medically. *Br J Ophthalmol*, **69**, 778–82.

Yeoh, R., Warhurst, D.C. and Falcon, M.G. 1987. *Acanthamoeba* keratitis. *Br J Ophthalmol*, **71**, 500–3.

12

Giardiasis

LYNNE S. GARCIA

INTRODUCTION

Giardia was first seen by Leeuwenhoek in his own stool specimens in 1681, but was not described until 1859 by Lambl. In the early 1930s, Clifford Dobell translated a number of Leeuwenhoek's works from Dutch and Latin. In the account of his own giardiasis infection, Leeuwenhoek writes (Dobell 1920):

> All the particles aforesaid (description of stool debris) lay in a clear transparent medium, wherein I have sometimes also seen animalcules a-moving very prettily (trophozoites); some of 'em a bit bigger, others a bit less, than a blood-globule, but all of one and the same make. Their bodies were somewhat longer than broad, and their belly which was flatlike, furnisht with sundry little paws, wherewith they made such a stir in the clear medium and among the globules, that you might e'en fancy you saw a pissabed (woodlouse or sow-bug) running up against a wall; and albeit they made a quick motion with their paws, yet for all that they made but slow progress.

However, there is no evidence that either Leeuwenhoek or Lambl saw or recognized the cyst forms of the organism; they were first noted by Grassi in 1879 and were thought to be coccidia. Only later did Grassi associate the cyst form with the flagellated trophozoite form (Meyer 1990). The history of giardiasis is given in Chapter 1, History of human parasitology.

Giardia lamblia is worldwide in distribution, apparently more prevalent in children than in adults, and more common in warm climates than in cool ones. It is the most commonly diagnosed flagellate in the intestinal tract, and it may be the most commonly diagnosed intestinal protozoan in some areas of the world with an estimated 2.8×10^8 cases per year (Lane and Lloyd 2002).

CLASSIFICATION

Although various criteria have been used to differentiate species of *Giardia*, including host specificity, various body dimensions, and variations in structure, there is still considerable debate over the appropriate classification and nomenclature of this group of organisms. Based on work by Filice (1952) on structural variations, three groups have been proposed: amphibian *Giardia* (represented by *G. agilis*), the muris group from rodents and birds (represented by *G. muris*), and the intestinalis group from a variety of mammals (including humans), as well as birds and reptiles (represented by *G. duodenalis*). Despite disagreement concerning the names *intestinalis* and *lamblia*, both continue to be used to describe this organism. Meyer prefers to use *Giardia duodenalis*, which is often followed by the name of the animal from which the organism was obtained (Meyer 1990).

Ten strains from human- and animal-source *G. duodenalis* were evaluated using an isoelectric focusing technique. Banding patterns obtained from trophozoite total cell proteins demonstrated both similarities and differences. These findings confirm the heterogeneity of this group of *Giardia* spp., from animal and human hosts and from hosts within the same geographical area (Isaac-Renton et al. 1988; Ey et al. 1996; McIntyre et al. 2000; Monis et al. 1999; Sulaiman et al. 2003, 2004).

In a recent study from Australia, ribosomal RNA sequencing revealed differences between the genotypes of *Giardia* isolates recovered from humans and dogs living in the same locality. Results from the study suggest that zoonotic transmission of *Giardia* infections between humans and dogs does not occur frequently in the particular communities studied. The dog-associated sequences have not been reported before, suggesting that there may be a new *G. duodenalis* subgroup (Hopkins et al. 1997). However, in a recent study in India, *G. duodenalis* isolates recovered from humans and dogs living in the same locality were characterized at three different loci; the SSU-rDNA, elongation factor 1-alpha (ef1-α), and triose phosphate isomerase (*tpi*) gene. Analysis of the *tpi* gene placed canine *Giardia* isolates within the genetic groupings of human isolates (Assemblages A and B). Epidemiological data showing a highly significant association between the prevalence of *Giardia* in humans and the presence of a *Giardia*-positive dog in the same household provided further evidence for zoonotic transmission (Traub et al. 2004).

Molecular classification tools have been very valuable in delineating the pathogenesis and host range of *Giardia* isolates obtained from humans and a variety of other mammals. Molecular typing indicates that most animal parasites are not infective to humans but those that are can be genotypically classified into assemblage A or B (Lane and Lloyd 2002). In one study, assemblage A isolates were solely detected in patients with intermittent diarrheal complaints, while assemblage B isolates were present in patients with persistent diarrheal complaints (Homan and Mank 2001). Restriction fragment length polymorphism (RFLP) analysis has distinguished two groups (designated groups I and II) within assemblage A (Amar et al. 2002). Currently, assemblages A (groups 1 and 2), and B (groups 3 and 4) contain human pathogens, while C through G are found in other mammals (Adam 2001). The phylogeny of the organism has not been finalized, but there is a growing opinion that *Giardia* may not be an ancient eukaryote, but that it may be derived from a more complex mitochondria-containing protozoan (Lane and Lloyd 2002).

Within the USA and other areas throughout the world, the term *Giardia lamblia* has been commonly used for many years and refers to those organisms found in humans, as well as other mammals. In some geographic areas, this designation tends to eliminate any confusion, since the majority of health workers are used to this name and continue to report the presence of the organism using it. Others feel the designation *G. duodenalis* is taxonomically correct and refers to the many 'species' containing the 'duodenalis' double medial bodies. As long as both laboratory staff and physicians use the same species designation, the choice is optional.

STRUCTURE AND LIFE CYCLE

The life cycle is illustrated in Figure 12.1 and is similar to the simple, direct life cycle of most intestinal protozoa. Both the trophozoite and cyst are included in the life cycle of *G. lamblia*. Trophozoites divide by means of longitudinal binary fission producing two daughter trophozoites. The trophozoites are the intestinal dwelling stages and attach to the epithelium of the host villi by means of the ventral disc. The attachment is substantial and results in disc impression prints when the organism detaches from the surface of the epithelium. Trophozoites may remain attached or may detach from the mucosal surface. Since the epithelial surface sloughs off the tip of the villus every 72 h, the trophozoites apparently detach at that time.

There are several theories on possible mechanisms for attachment by the ventral disk; they include microtubule mediation, hydrodynamic action, contractile protein activity, and the interaction of lectins with surface bound sugars. Based on conflicting information, the attachment process probably depends on multiple mechanisms. A study by Katelaris and colleagues indicates that attachment to enterocyte-like differentiated Caco-2 cells is primarily by cytoskeletal mechanisms that can be inhibited by interfering with contractile filaments and microtubules; attachment by mannose binding lectin also seems to mediate binding (Katelaris et al. 1995).

The most common location of the organisms is in the crypts within the duodenum. For reasons that are not known, cyst formation takes place as the organisms move down through the colon.

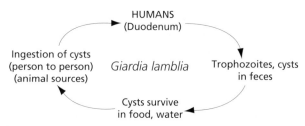

Figure 12.1 *Life cycle of* Giardia lamblia. *(Adapted from Garcia 2001)*

The *Giardia* encystation process can be divided into three parts:

1 stimulus for encystation
2 synthesis and intracellular transport of cyst wall components
3 assembly of the extracellular cyst wall.

It has been suggested that a cholesterol-regulated process similar to that in mammalian cells occurs during the induction of encystation in *Giardia*. The formation of the cyst wall during encystation is characterized by the appearance of dense encystation-specific vesicles (ESV). These ESVs transport cyst wall components to the plasma membrane of the encysting cell and release their contents to the cell exterior during cyst wall formation. Information also suggests that the cyst wall contains only two cyst proteins; these proteins appear to be common components of the cyst walls of a variety of intestinal parasitic protozoa.

The *Giardia* cyst wall is composed of an outer filamentous portion and an inner membranous portion. The process begins with the appearance of cyst wall antigens that appear on small protrusions of the trophozoite surface; these eventually form cap-like structures and may cover the entire surface of the trophozoite. Based on the filamentous nature of the cyst wall and its constant thickness, it may form from successive layers of cyst wall materials. The cyst wall has also been found to be somewhat flexible, but rigid enough to maintain its shape after the excysted trophozoite has left the cyst (Lujan et al. 1997).

Based on results by Halliday et al. (1995). *Giardia* appears to take up conjugated bile salts by active and passive transport mechanisms like the mammalian ileum. As conjugated bile salts are known to promote encystation, these uptake mechanisms may play an important role in the survival of the cyst stage and ultimate completion of the life cycle of the parasite (Das et al. 1997; Healy 1990). Cholesterol starvation may also play a role in stimulating encystation (Lujan et al. 1996). The trophozoites retract the flagella into the axonemes, the cytoplasm becomes condensed, and the cyst wall is secreted (Erlandsen et al. 1996). As the cyst matures, the internal structures are doubled so that when excystation occurs, the cytoplasm divides to produce two trophozoites. Excystation would normally occur in the duodenum or appropriate culture medium (Bingham and Meyer 1979). If kept cool and moist, cysts can remain viable for several months, but they cannot survive if moisture is lacking.

At the light microscopic level and seen on permanent stained fecal smears stained with trichrome or iron-hematoxylin stains, the trophozoite is usually described as teardrop-shaped from the front with the posterior end being pointed (Figures 12.2 and 12.3). Examined from the side, the trophozoite resembles the curved portion of a spoon. The concave portion is the area of the sucking

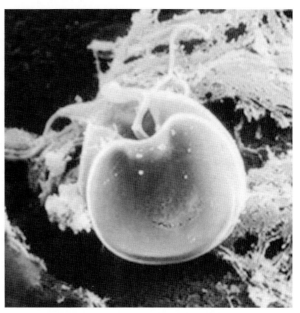

Figure 12.2 *Scanning electron micrograph of* Giardia lamblia *trophozoite on the mucosal surface. (Courtesy of Marietta Voge) (Garcia 2001)*

disc (Figure 12.2). There are four pairs of flagella, two nuclei, two axonemes, and two slightly curved bodies called the median bodies. The trophozoites usually measure 10–20 µm in length and 5–15 µm in width (Figure 12.3 and Table 12.1)

The cysts may be either round or oval and contain four nuclei, axonemes, and median bodies (Figure 12.3). Often, some cysts appear to be shrunk or distorted and one may see two halos, one around the cyst wall itself and one inside the cyst wall around the shrunken organism. The halo effect around the outside of the cyst is particularly visible on permanent stained smears. Cysts normally measure 11–14 µm in length and 7–10 µm in width (Figure 12.3 and Table 12.2).

BIOCHEMISTRY, MOLECULAR BIOLOGY, AND GENETICS

Organisms in the genus *Giardia* are described as aero-tolerant anaerobes that use substrate-level phosphorylation, iron-sulfur protein- and flavoprotein-mediated electron transport for the production of energy (Jarroll and Lindmark 1990). *Giardia* lacks mitochondria and microbodies, but contains lysosome-like organelles that accumulate ferritin and stain positively for acid phosphatase and aryl sulfatase (Upcroft and Upcroft 1998).

In the presence of exogenous glucose, intact *Giardia* trophozoites produce ethanol and acetate as organic end-products: aerobically, six times more acetate than ethanol is produced; anaerobically, twice as much ethanol as acetate is produced. In both situations, carbon dioxide is the only gaseous end-product that has been detected. Based on the pattern of carbon dioxide

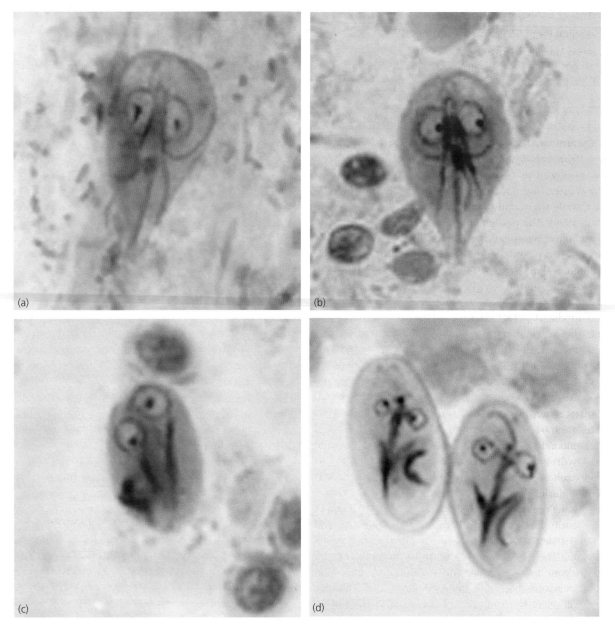

Figure 12.3 Giardia lamblia *(a, b) trophozoites stained with Wheatley's trichrome;* **(c)** *cyst stained with Wheatley's trichrome;* **(d)** *cysts stained with iron hematoxylin.*

production, glucose is metabolized using the Embden-Meyerhof-Parnas (EMP) pathway and the hexose monophosphate (HM) pathway. A number of energy metabolism enzymes have been identified and are reviewed by Jarroll and Lindmark (1990).

Carbohydrate and energy metabolism studies have also been conducted using the cyst stage; enzyme data confirm that specific activities occur regardless of the life-cycle stage involved. Studies also show that the cyst stage respires at c. 10–20 percent of the rate observed in trophozoites; respiration is not stimulated by exogenous glucose, but by exogenous ethanol. It is also interesting to note that cyst respiration decreases as water temperature decreases. These findings may explain increased cyst viability with decreased storage temperature.

In summary, a proposed energy metabolic pathway for *Giardia* indicates oxidation of endogenous and exogenous glucose incompletely to ethanol, acetate, and carbon dioxide. Energy is produced through substrate-level phosphorylation; iron-sulfur proteins and flavins are involved in electron transport. There is no evidence for oxidative phosphorylation or a cytochrome-mediated electron transport system (Jarroll and Lindmark 1990).

The fact that *Giardia* shares many metabolic attributes of the anaerobic bacteria provides a wide range of drug targets for future consideration. An example is the utilization of the amino acid arginine as an energy source, using the arginine dihydrolase pathway, which occurs in a number of anaerobic bacteria (Brown et al. 1998).

Table 12.1 Giardia lamblia *trophozoite morphology*

Organism stage	Shape and size	Motility	Nuclei	Flagella	Other features
Giardia lamblia trophozoite	Pear-shaped, 10–20 μm; width 5–15 μm	'Falling leaf' motility may be difficult to see if organism in mucus	Two nuclei Not visible in unstained mounts	Four lateral, two ventral, two caudal	Sucking disc occupying $^1/_2$–$^3/_4$ of ventral surface pear-shaped front view; spoon-shaped side view

Culture studies indicate that *Giardia* trophozoites have the capacity to synthesize isoprenoid lipids de novo, as well as to incorporate preformed lipids; also, steryl esters are abundant and increase during encystation, while triglycerides are low (Ellis et al. 1996; Gibson et al. 1999). The phospholipids present include phosphatidylcholine, phosphatidylethanolamine, phosphatidylglycerol, phosphatidylinositol, phosphatidylserine, and sphingomyelin. Neutral lipids present include sterols, mono-, di-, and triacylglycerides. Fatty acids can be incorporated into phospho- and neutral lipids (Jarroll and Lindmark 1990). Increased fatty acid unsaturation and the accumulation of storage lipids are consistent with the parasite differentiation into a cyst stage that can survive outside the host.

In the trophozoite, nucleic acid requirements are met by the salvage of pyrimidines, purines, and their nucleosides. Various carrier-mediated mechanisms have been proposed.

In contrast to trophozoites of *Entamoeba* and the trichomonads, *Giardia* has very few carbohydrate-splitting hydrolases. It has also been found that *Giardia* trophozoites exhibit increased levels of chitin synthetase activity during encystment (Jarroll and Lindmark 1990).

Although antigenic analysis shows promise in helping to identify and classify *Giardia*, both common and different antigens have been recovered in axenic cultures. The use of monoclonal antibodies in the study of antigens has proved useful; technical methods also involve quantitation, an important requirement in this type of analysis. Another potential problem involves antigenic variation; a single *Giardia* trophozoite can give rise in vivo, as well as in vitro, to trophozoites with varying surface antigens. Immune pressure may play a role in antigenic variation, but the speed with which antigenic changes take place after infection suggests that other mechanisms may also be involved. The antigenic profile of *Giardia* in culture may also change if the trophozoites harbor endosymbionts; the presence of these organisms may influence both antigenic and chemotaxonomic determinations (Meyer 1990).

Giardia lamblia is known to undergo surface antigenic variation. The antigens involved belong to a group of variant-specific surface proteins (VSP) that are unique cysteine-rich zinc finger proteins. Selection by immune-mediated processes is a possibility, especially because switching occurs at the same time that humoral responses are first detected. The purpose of antigenic variation may be the presentation of a wide variety of VSPs to the various hosts, increasing the chances of successful initial infection or reinfection (Nash 1997). Since VSPs are resistant to the effects of intestinal proteases, they may allow the organism to survive in the protease-rich small intestine (Nash 2002). Thus, the two major hypotheses concerning antigenic variation are evasion of the host immune system and allowing the parasites to survive in different intestinal environments; certainly these hypotheses are not mutually exclusive (Adam 2001).

CLINICAL ASPECTS

From available data, it appears the incubation time for giardiasis is c. 12–20 days. Because the acute stage usually lasts only a few days, giardiasis may not be recognized as the cause of the symptoms observed, which may mimic acute viral enteritis, bacillary dysentery, bacterial or other food poisonings, acute intestinal amebiasis or 'traveler's diarrhea' (toxigenic *E. coli*). However, the type of diarrhea plus the lack of blood, mucus, and cellular exudate is consistent with giardiasis.

Clinical manifestations

Nausea, anorexia, malaise, low-grade fever, and chills may accompany onset. There may be a sudden onset of

Table 12.2 Giardia lamblia *cyst morphology*

Organism stage	Shape and size	Motility	Nuclei	Flagella	Other features
Giardia lamblia cyst	Oval, ellipsoidal or may appear round 8–19 μm; usual range 11–14 μm; width, 7–10 μm	Nonmotile	Four nuclei – not distinct in unstained preparations; usually located at one end	Flagella not visible	Longitudinal fibers in cysts may be visible in unstained preparations; deep staining median bodies usually lie across the longitudinal fibers; there is often shrinkage and the cytoplasm pulls away from the cyst wall; there may also be a 'halo' effect around the outside of the cyst wall due to shrinkage caused by dehydrating reagents

explosive, watery, foul-smelling diarrhea with flatulence and abdominal distention. Other symptoms include epigastric pain and cramping. There is increased fat and mucus in the stool, but no blood. Weight loss often accompanies these symptoms. The acute infection usually resolves spontaneously, although in some patients, particularly children, the acute symptoms may last for months. Acute giardiasis must be differentiated from other acute viral, bacterial, and protozoal agents.

The acute phase is often followed by a subacute or chronic phase. Symptoms in these patients include recurrent, brief episodes of loose, foul stools; there may be increased distention and foul flatus. Between mushy stools, the patient may have normal stools or may be constipated. Abdominal discomfort continues to include marked distention and belching with a rotten egg taste. Chronic disease must be differentiated from amebiasis and other intestinal parasites such as *Dientamoeba fragilis*, *Cryptosporidium parvum*, *Cyclospora cayetanensis*, *Isospora belli*, microsporidia, *Strongyloides stercoralis*, and from inflammatory bowel disease and irritable colon (see also Chapters 20, Intestinal coccidia: cryptosporidiosis, isosporiasis, cyclosporiasis, 25, Microsporidiosis, and 35, *Strongyloides* and *capillaria*). Based on symptoms such as upper intestinal discomfort, heartburn and belching, giardiasis must also be differentiated from duodenal ulcer, hiatus hernia, gallbladder, and pancreatic disease.

Various types of malabsorption have been described including steatorrhea, disaccharidase (lactase, xylase) deficiency, Vitamin B_{12} malabsorption, hypocarotenemia, low serum folate, and protein-losing enteropathy. Lactose intolerance may persist after effective therapy, particularly in patients from ethnic groups with a predisposition for lactase deficiency. This should be considered prior to retreatment, especially if the patient continues to have diarrhea, but post-treatment negative stool specimens.

Although there is speculation that the organisms coating the mucosal lining may act to prevent fat absorption, this does not completely explain why the uptake of other substances normally absorbed at other intestinal levels is prevented (Tandon et al. 1977). Occasionally the gall bladder may also be involved, causing gall-bladder colic and jaundice. *G. lamblia* has also been identified from bronchoalveolar lavage fluid (Stevens and Vermeire 1981) and the urinary tract (Meyers et al. 1977). Reasons for such a dramatic variation in host susceptibility are not understood. Host factors such as nutritional status and both systemic and mucosal immunity may also play a role.

IMMUNOLOGY

Recent studies document antigenic variation with surface antigen changes during human infections with *Giardia*, and although the biological importance of this work is not clear, it suggests that this may provide a mechanism enabling the organism to escape the host's immune response (Nash et al. 1990). Also one would suspect that the more rapid the rate of change, the more likely that a chronic infection would be seen. Certain surface antigens may allow the organisms to survive better in the intestinal tract and might not be immunologically selected (Nash 1997).

Although patients with symptomatic giardiasis usually have no underlying abnormality of serum immunoglobulins, a high incidence of giardiasis has been shown to occur in patients with immunodeficiency syndromes, particularly in common variable hypogammaglobulinemia (Ament et al. 1973). Giardiasis was found to be the most common cause of diarrhea in these patients and was associated with mild to severe villus atrophy. Successful treatment of giardiasis led to symptomatic cure and improvement in mucosal abnormalities, with the exception of nodular lymphoid hyperplasia.

Acute (polymorphonuclear leukocyte and eosinophils) and chronic inflammatory cells have been found in the lamina propria together with increased epithelial mitoses – findings that revert to normal after therapy. The presence of these inflammatory cells has been linked to epithelial cell damage. The range of patient response is from normal to almost complete villous atrophy. Crypt mitosis is also increased based on the degree of villous damage, however, it has also been reported that mucosal changes return to normal within several days of treatment and that patients with giardiasis have reduced surface area compared with uninfected individuals.

Although no definitive pattern has been identified, it appears that lamina propria plasma cells producing immunoglobulin (Ig)M and IgE increase, whereas local production of IgA seems to be suppressed. More severe villous damage has also been reported in patients with diminished gamma globulins.

During the acute phase of giardiasis, jejunal active antigen uptake is increased, leading to delayed recruitment of mucosal and connective mast cells. These changes may influence the increased incidence of hypersensitivity reactions and allergic disease associated with *Giardia* infection (Hardin et al. 1997). Severe malabsorption has also been linked with isolated levothyroxine malabsorption, leading to severe hypothyroidism (Seppel et al. 1996) and secondary impairment of pancreatic function (Carroccio et al. 1997). In both cases, therapy with metronidazole led to complete remission of symptoms.

PATHOLOGY

Although the organisms in the crypts of the duodenal mucosa may reach very high densities, they might not cause any pathology. The organisms feed on the mucous secretions and do not penetrate the mucosa as *G. muris* does in mice where trophozoites have been recovered in the deeper tissues (Owen et al. 1981). Although some organisms have been seen in biopsy material inside the intestinal mucosa, others have been seen attached only to the epithelium. In one in vitro study using cultured epithelial cells, the organisms showed no toxic or invasive effect (Chavez et al. 1986). Another study by Chen and colleagues has demonstrated a *G. duodenalis* cysteine-rich surface protein (CRP136) that has 57 percent homology with the gene encoding the precursor of the sarafotoxins, a group of snake toxins from the burrowing adder known to cause symptoms similar to those of humans acutely infected with *Giardia* (Chen et al. 1995). Thus, CRP136 represents the first evidence for a potential *Giardia* toxin. In symptomatic cases, there may also be irritation of the mucosal lining, increased mucus secretion, and dehydration (Erlandsen 1974; Peterson 1957). In a recent study, it appears that *G. lamblia* may cause intestinal pathophysiology by disrupting tight junctional ZO-1 and increasing epithelial permeability. Apical administration of epidermal growth factor (EGF) prevents these abnormalities and reduces epithelial colonization by the live parasites (Buret et al. 2002).

With the advent of acquired immune deficiency syndrome (AIDS), there was speculation that *G. lamblia* might be an important pathogen in this group, but clinical findings to date have not confirmed this fear (Meyer 1990; Smith et al. 1988). Although giardiasis has certainly been reported in AIDS patients, clinical experience does not suggest a more pathogenic role for *G. lamblia* in these patients than in other non-AIDS individuals.

Isoenzyme studies, primarily designed to assist in organism identification and classification, have also provided additional information regarding pathogenicity, implication in waterborne outbreaks, and human disease. In one study where isoenzyme patterns of 32 isolates of *Giardia* obtained from both humans and animals were examined, there was no obvious correlation between clinical symptoms and isoenzyme patterns. Isolates from asymptomatic individuals were found in the same zymodemes (isoenzyme groups) as isolates from symptomatic hosts. This study also confirmed previous observations regarding genetic heterogeneity and demonstrated significant differences between isolates from within a single region and other widely separated geographic locations (Proctor et al. 1989). A study from Australia provided data on two *Giardia* demes derived from children with similar chronic symptoms both of which appear to be pathogenic (Upcroft et al. 1995).

DIAGNOSIS

Generally, routine laboratory test results are normal, including hematology; eosinophilia is rare. Malabsorption of fat, glucose, lactose, xylose, carotene, folic acid, and vitamin B_{12} is occasionally present in some patients. Diagnosis of giardiasis is usually based on identification of the organisms from stool or duodenal aspirate specimens (Table 12.3).

Routine stool examinations are normally recommended for the recovery and identification of intestinal protozoa. In the case of *G. lamblia*, because they are attached so securely to the mucosa by means of the sucking disk, even a series of five to six stools may be examined without recovering the organisms. These parasites also tend to be passed in the stool on a cyclical basis. The EnteroTest capsule may be helpful in recovering the organisms, as may duodenal aspirates. Although cysts can often be identified on the wet stool preparation, many infections may be missed without the examination of a permanent stained smear (Collins et al. 1978; NCCLS 1997; Garcia 2001). If material from the string test (EnteroTest) or mucus from a duodenal aspirate are submitted, they should be examined as a wet preparation for motility; motility may be represented by nothing more than a slight flutter of the flagella because the organism will be caught up in the mucus. After diagnosis, the rest of the positive material can be preserved and processed for permanent staining. Experience in the clinical laboratory setting has shown

Table 12.3 *Laboratory diagnosis of giardiasis – key points*

Laboratory diagnosis
Even if a series of three stool specimens (ova and parasite examinations) are submitted and examined correctly, the organisms may not be recovered and identified.
Motility on wet preparations may be difficult to see because the organisms may be caught up in mucus.
Any examination for parasites in stool specimens must include the use of a permanent stained smear (even on formed stool).
Duodenal drainage either alone or in combination with the EnteroTest capsule may be very helpful in organism recovery. This technique does not take the place of the ova and parasite examination.
Immunodiagnostic procedures have been found to be more sensitive than the routine stool methods (ova and parasite examination) and are now being used routinely by some laboratories. These procedures are normally performed on order; if negative and the patient remains symptomatic, additional testing (O&P examination, special stains for the coccidia and microsporidia) can be ordered. Other considerations such as cost, training, equipment availability, batching of tests, and number of requests also play a role in test selection.

that the examination of duodenal fluid after a series of negative stool examinations rarely confirms a positive infection. This may be due to parasite location or small numbers. Because of these considerations, the examination of intestinal fluid should not replace routine stool examinations, all of which should include the use of permanent stained smears (Garcia 2001).

Procedures have also been developed using the enzyme-linked immunosorbent assay (ELISA) to detect *Giardia* antigen in feces. The ELISA is at least as sensitive as microscopic wet examinations (Addis et al. 1991; Nash et al. 1987). A fluorescent method using monoclonal antibodies has also proved extremely sensitive and specific in detecting *Giardia* in fecal specimens (Garcia et al. 1992). Many of these newer methods are being used to screen patients suspected of having giardiasis or those who may be involved in an outbreak situation (Garcia and Shimizu 1997; Aldeen et al. 1998).

A number of studies have compared the newer immunoassay products, with excellent and comparable results. Several new products are available in a cartridge format (Figure 12.4). Some of these use an immunochromatographic strip-based detection system for the detection of *Giardia lamblia* and *Cryptosporidium parvum*, which can be used with fresh, frozen and/or formalinized fecal specimens (Garcia and Shimizu 2000; Katanik et al. 2001; Garcia et al. 2003). A third cartridge system uses enzyme immunoassay for the detection of the *Entamoeba histolytica/E. dispar* group, *Giardia lamblia*, and *Cryptosporidium parvum*, which requires fresh or frozen fecal specimens (Garcia et al. 2000). Results have demonstrated that ELISA for detection of copro-antigens in a single stool sample may be almost as sensitive for the detection of giardiasis as repeated microscopy on two sequential stool samples (Mank et al. 1997).

In a study of nine different immunoassay kits for the detection of *Giardia lamblia*, the sensitivity of ELISA ranged from 94 to 99 percent; the specificity was 100 percent. All direct fluorescence assay results were in agreement with both 100 percent specificity and 100 percent sensitivity; a total of 60 specimens positive for *Giardia*, 40 specimens positive for both *Giardia* and *Cryptosporidium*, and 50 negative fecal specimens were tested. In addition to specificity and sensitivity, factors such as cost, simplicity, ease of interpretation of results, equipment, available personnel, and number of tests ordered are also important considerations prior to method selection (Garcia and Shimizu 1997).

In another study of nine commercially available ELISAs for detection of *Giardia lamblia* in fecal specimens, all except one were found to be rapid, sensitive, and specific. Sensitivities and specificities ranged from 88.6 to 100 and 99.3 to 100 percent, respectively (Aldeen et al. 1998).

Occasionally, after multiple stool examinations and examination of intestinal fluid have been negative, a small bowel biopsy may confirm the suspect diagnosis of giardiasis. It is recommended that the biopsy be taken from the area of the duodenojejunal junction; but biopsies from multiple duodenal and jejunal sites are preferred. Fluid smears and touch preparations can be air dried, fixed in methanol, and stained with Giemsa stain. The organisms appear somewhat purple and the epithelial cells appear pink. Routine histological procedures should also be performed with very careful screening of the material; trophozoites are very difficult to see and may be present in very few of the sections. Normally, the trophozoites are more likely to be seen attached to the microvillous border within the crypts.

Radiological findings from the small intestine may be normal in giardiasis, but abnormalities can occur and are usually seen in the duodenum or jejunum. Findings may include thickened mucosal folds, increased secretions, and a pattern of edema and segmentation, none of which is diagnostic for giardiasis. Approximately 20 percent of patients with giardiasis may exhibit these findings after routine barium examination. It is always recommended that the stools be collected for routine examinations prior to the patient receiving barium as organisms may be difficult to identify in the stool for several days to more than a week afterwards. Because giardiasis may not produce any symptoms at all, demonstration of the organism in symptomatic patients may not rule out alternative diagnoses such as peptic ulcer, celiac disease, strongyloidiasis, and possibly carcinoma.

Unfortunately, serodiagnostic procedures for giardiasis do not yet fulfill the criteria necessary for wide clinical use, particularly since they may indicate either past or present infection. In contrast, the detection of antigen in a stool or visual identification of the organisms using immunodiagnostic reagents indicates current infection. With the increase in numbers of *Giardia* infections and awareness of particular situations such as nursery school settings, additional detection assays have been developed as rapid and reliable immunodiagnostic procedures (Keystone et al. 1978; Craft 1982; Sealy and Schuman 1983; Addis et al. 1992; Garcia and Shimizu 1997; Aldeen et al. 1998; Garcia and Shimizu 2000; Garcia et al. 2000; Katanik et al. 2001; Garcia et al. 2003).

Recent studies using indirect fluorescent assay (IFA) to measure anti-*Giardia* humoral responses demonstrated that more than 34 percent of asymptomatic patients have a titer equal to or less than 1:500, while more than 29 percent of the symptomatic patients have a titer of 1:8000 or higher. The circulating antiparasite IgM and IgA, measured by ELISA, were significantly higher in symptomatic patients and correlated with higher cyst output (Soliman et al. 1998).

The secretory immune response to membrane antigens during *Giardia lamblia* infection in humans may be important in future studies in the diagnosis of giardiasis; these antigens were not recognized by saliva samples from healthy individuals (Rosalesborjas et al. 1998).

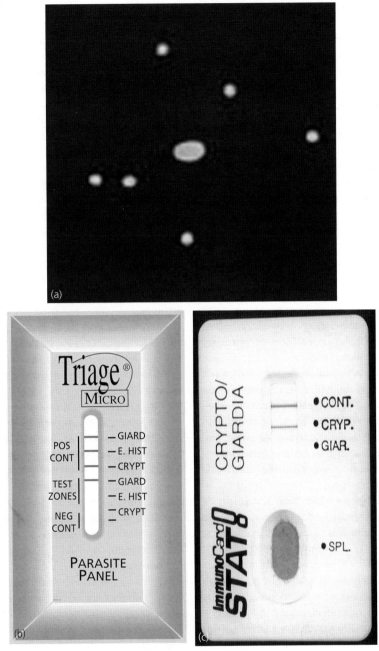

Figure 12.4 *Fecal immunoassays for the detection of Giardia lamblia:* **(a)** *Merifluor* Cryptosporidium/Giardia *combination FA reagent (Meridian Bioscience, Cinncinnati, OH) for the detection of* Giardia lamblia *cycsts (large oval structure) and* Cryptosporidium parvum *(small round objects) —can be used with preserved stools;* **(b)** *Triage® Paasite Panel (BIOSITE Diagnostics, San Diego, CA) cartridge format for the detection of* Giardia lamblia, Cryptosporidium parvum, *and the* Entamoeba histolytica/E. dispar *group – requires fresh stools;* **(c)** *immunochromatographic cartidge format (Meridian Bioscience, Cinncinnati, OH) for the detection of* Giardia lamblia *and* Cryptosporidium parvum.

EPIDEMIOLOGY

Giardiasis is one of the most common intestinal parasitic infections in humans and is distributed worldwide in developed and developing countries. There are, however, differences in the numbers of infections, not only between countries but also within geographic regions. Infections seem to be more common in children than adults, and other social, environmental, climatic, and economic factors also play a role in disease prevalence. Susceptibility to infection with *Giardia* is influenced by sex, age, environmental conditions, socioeconomic conditions, occupation, nutritional status, gastric acidity, and overall host immune status.

Transmission occurs through ingestion of as few as ten viable cysts. Infection can be acquired from food or water and directly person-to-person by the oro-fecal route. Often there are outbreaks due to poor sanitation facilities or breakdowns as evidenced by travelers and campers (Kettis and Magnius 1973; Knaus 1974; Lopez et al. 1978; Moore et al. 1969). There has also been an increase in the prevalence of giardiasis in the male homosexual population, probably due to anal and oral sexual practices (Phillips et al. 1981; Schmerin et al. 1978).

Geographical distribution varies considerably worldwide with reported prevalence figures ranging from a low of 5 percent in Indonesia to 43 percent in the Seychelles; in the US, the figure is c. 7.4 percent (Islam 1990). When reviewing reported prevalence figures, it is always important to remember that data varies considerably, depending on the sensitivity and specificity of the diagnostic methods used. In areas of the world in which *Giardia* is endemic, the majority of infections in children occur in those under 10 years of age. A decrease in infections seen in children older than 10 years may reflect acquired immunity and behavioral changes that decrease potential environmental exposure to the organism. High population density and overcrowding probably contribute to higher prevalences of giardiasis. In low-income populations residing on the United States–Mexico border with no sewerage and water services, households with children under 5 years of age were 1.3 times at risk for *Giardia* infection when compared with *Cryptosporidium*. These findings highlight the differences in the transmission of these two pathogens and emphasize the need for interventions related to hygiene, water supply, and sanitation (Redlinger et al. 2002).

Although seasonal patterns have been identified for some infectious diseases, limited information is available for giardiasis. Some data suggest an association with the cooler, wetter months of the year; this is not surprising if one considers the issue of environmental conditions advantageous to cyst survival.

It has been documented that certain occupations may place an individual at risk of infection and these include sewage and irrigation workers who may become exposed to infective cysts. In situations where young children are grouped together (e.g. nursery schools), there may be an increased incidence of exposure and subsequent infection of both children and staff.

The possible association between gastric acidity and giardiasis has been debated for some time; decreased gastric acid production may predispose to infection with *Giardia*. It is thought that normal gastric acidity acts as a barrier to the establishment of an infection; patients who have had a gastrectomy are prone to infection with *Giardia*. Although achlorhydria is associated with blood group A, and some evidence suggests that this group is more susceptible to giardiasis, subsequent evidence has not confirmed this association. Since reduction in gastric acid also occurs as a result of malnutrition, these factors may be linked and, as a group, increase the susceptibility to infection with this organism. This link between

malnutrition and giardiasis may also be explained by the impairment of the host's immune system.

The issue of breast milk in modifying infection with *Giardia* has also been discussed (Islam 1990). Lower incidence of giardiasis in children up to and including 6 months of age may be related to an association with breast-feeding and some protection against infection through secretory IgA. However, lower incidence may also be related to decreased exposure to *Giardia* in breast-fed infants.

Giardiasis is one of the more common causes of traveler's diarrhea and has been recorded from all parts of the world. It has also been speculated that visitors to areas endemic for *Giardia* are more likely to present with symptoms than individuals who live in the area; this difference is probably due to the development of immunity from prior, and possible continued, exposure to the organism (Soliman et al. 1998; Isaac-Renton et al. 1999).

In the USA, there have been a number of outbreaks attributed to either resort or municipal water supplies in Oregon, Colorado, Utah, Washington, New Hampshire, and New York (Craun 1990; Kirner 1978; Shaw et al. 1977; Wright et al. 1977). Most waterborne outbreaks have occurred in water systems using surface water sources (Dykes et al. 1980). High rates of infection have also been reported from hikers and campers who drank stream water: because some of these areas are remote from human habitation, infected wild animals, especially beavers, are suspected as being a possible source of infection (Dykes et al. 1980). In addition to beavers and muskrats, *Giardia* infects a number of different animals, including dogs, cats, cattle, sheep, pigs, goats, gerbils, rats, and gorillas; however, the role of some of these species in zoonotic giardiasis is uncertain (Halliday et al. 1995; Karanis et al. 1996; Graczyk et al. 2002; van Keulen et al. 2002).

Use of immunomagnetic beads to separate *Giardia* cysts from environmental surface waters, followed by DNA release and polymerase chain reaction (PCR) amplification of the target giardin gene has been used to improve the reliability of the detection of this parasite with improved sensitivity (Mahbubani et al. 1998). Another approach involves continuous separation channel centrifugation, which can simultaneously concentrate multiple pathogens as small as one micron with high and reproducible efficiency in a variety of water matrices (Borchardt and Spencer 2002). Unfortunately, cryptosporidial oocysts and *Giardia* cysts have also been found in bottled water in Brazil (Franco and Cantusio Neto 2002). It is now obvious that a large portion of the total organism loads in watercourses and in drinking water reservoirs result from rainfall and extreme runoff events. Thus, regular water samples are inadequate for accurately estimating the microbial contamination of watercourse systems. The procedures for raw-water surveillance within the context of multiple-barrier protection and risk assessment must include sampling during extreme runoff situations (Kistemann et al. 2002). A new dye, Beljian red, has been shown to discriminate *Giardia* cysts seeded into water samples from those naturally present in the sample; this approach should be useful in determining the specific recovery of protozoan parasites from environmental samples (Ferrari et al. 2003).

The phagocytic capacity of hemocytes of the Asian freshwater clam, *Corbicula fluminea*, for *Giardia* cysts may be another useful mechanism for indicating contamination of waste waters and agricultural drainage with these parasites (Graczyk et al. 1997).

Although control efforts have emphasized water treatment issues, a recent study suggests that a number of species of rotifers ingested varying numbers of *Giardia* cysts, which were retained within the rotifers' bodies throughout the observation period. This type of ingestion may represent another means of reducing water contamination (Trout et al. 2002).

With the recently recognized efficacy of ultra-violet (UV) irradiation against *Cryptosporidium* oocysts, the inactivation of *Giardia* cysts was also investigated, as was cyst ability to repair UV-induced damage following typical drinking water and wastewater doses of 160 and 400 JM(−2). The infectivity reduction of *Giardia* cysts at these UV doses remained unchanged after exposure to repair conditions. Thus, UV disinfection at practical doses achieves appreciable (greater than 4 log) inactivation of cysts in water with no evidence of DNA repair leading to infectivity recovery (Linden et al. 2002).

During the past few years, giardiasis has received much publicity. With increased travel, there has been a definite increase in symptomatic giardiasis within the USA (Addis et al. 1992). Various surveys show infection rates of 2–15 percent in various parts of the world.

CHEMOTHERAPY

If giardiasis is diagnosed, the patient should be treated; however, the therapy is controversial, depending on geographical location in the world. In the majority of cases, giardiasis can be eliminated with the use of quinacrine, but its side-effects include: nausea, vomiting, headache, and dizziness. Vertigo, excessive sweating, fever, pruritis, corneal edema, myalgias, and insomnia have also been reported.

Metronidazole is also very effective and is approved by the World Health Organization (WHO, 1995) but is listed as the second drug of choice by the Centers for Disease Control in the United States because of potential carcinogenicity in rats and mutagenic changes in bacteria. Although these changes have never been demonstrated in humans and the issue is still debated, metronidazole is not recommended for pregnant women. The most frequent side-effects are gastrointestinal and include: nausea, vomiting, diarrhea, and crampy abdominal pain. Additional complaints may include metallic

taste, headache, dizziness, drowsiness, lassitude, paresthesias, urticaria, and pruritis. Alcohol should be avoided when taking this medication. Although standard therapy usually cures the infection, some immunocompromised patients, including patients with AIDS, as well as some immunocompetent patients, have giardiasis that is refractory to therapy. Therapy with a combination of quinacrine and metronidazole has been reported to be effective (Nash et al. 2001).

Tinidazole has also been used and has proved more effective than metronidazole as a single dose (Jokipii and Jokipii 1979). Furazolidone is often used for treating children and paromomycin has been suggested for use in pregnancy (Davidson 1990).

In the absence of a parasitological diagnosis, the treatment of suspected giardiasis is a common question with no clear-cut answer. The approach depends on the alternatives and the degree of suspicion of giardiasis, both of which will vary among patients and physicians. It is not recommended that treatment be given without good parasitological evidence particularly since relief of symptoms does not allow a retrospective diagnosis of giardiasis; the most commonly used drug, metronidazole, also targets other organisms besides *Giardia*.

The third question involves treatment of asymptomatic patients. Generally, it is recommended that all cases of proven giardiasis be treated because the infection may cause subclinical malabsorption, symptoms are often periodic and may appear later, and a carrier is a potential source of infection for others. Certainly, in areas of the world where infection rates are extremely high, as well as the prospect of reinfection, the benefit per cost ratio also has to be examined.

INTEGRATED CONTROL

Because of the potential for wild animal, and possibly other domestic animal, reservoir hosts, other measures as well as personal hygiene and improved sanitary measures have to be considered (Brightman and Slonka 1979; Hewlett et al. 1982). Iodine has been recommended as an effective disinfectant for drinking water, but it must be used according to the appropriate directions (Jarroll et al. 1980; Jarroll et al. 1981; Zemlyn et al. 1981). Filtration systems have also been recommended, although they have certain drawbacks, such as clogging. Water-treatment systems tend to be unavailable in developing countries, many of which do not have piped water available to the general population. Boiling the water is effective, but carries with it the cost of fuel, another consideration in many parts of the world. Improved personal hygiene, routine hand-washing, better preparation and storage of food and water, control of insects that may come in contact with infected stools and then contaminate food or water, and treatment of symptomatic and asymptomatic individuals would all lead to a decrease in the overall numbers of infections with this parasite.

Although the potential for a vaccine has been discussed, prospects are poor for the development of actual vaccine reagents in the near future. Control measures will continue to be centered on environmental and personal hygiene issues.

REFERENCES

Adam, R.D. 2001. Biology of *Giardia lamblia*. *Clin Microbiol Rev*, **14**, 447–75.

Addis, D.G., Mathews, H.M., et al. 1991. Evaluation of a commercially available enzyme-linked immunosorbent assay for *Giardia lamblia* antigen in stool. *J Clin Microbiol*, **29**, 1137–42.

Addis, D.G., Davis, J.P., et al. 1992. Epidemiology of giardiasis in Wisconsin: increasing incidence of reported cases and unexplained seasonal trends. *Am J Trop Med Hyg*, **47**, 13–19.

Aldeen, W.E., Carroll, K., et al. 1998. Comparison of nine commercially available enzyme-linked immunosorbent assays for detection of *Giardia lamblia* in fecal specimens. *J Clin Microbiol*, **36**, 1338–40.

Amar, C.F., Dear, P.H., et al. 2002. Sensitive PCR-restriction fragment length polymorphism assay for detection and genotyping of *Giardia duodenalis* in human feces. *J Clin Microbiol*, **40**, 446–52.

Ament, M.E., Ochs, H.D. and David, S.D. 1973. Structure and function of the gastrointestinal tract in primary immunodeficiency syndromes: a study of 39 patients. *Medicine (Baltimore)*, **52**, 224–48.

Bingham, A.K. and Meyer, E.A. 1979. *Giardia* encystation can be induced in vitro in acidic solutions. *Nature (London)*, **277**, 301–2.

Borchardt, M.A. and Spencer, S.K. 2002. Concentration of *Cryptosporidium*, microsporidia and other waterborne pathogens by continuous separation channel centrifugation. *J Appl Microbiol*, **92**, 649–56.

Brightman II, A.H. and Slonka, G.F. 1979. A review of five clinical cases of giardiasis in cats. *J Am Anim Hosp Assoc*, **12**, 492–7.

Brown, D.M., Upcroft, J.A., et al. 1998. Anaerobic bacterial metabolism in the ancient eukaryote *Giardia duodenalis*. *Int J Parasitol*, **28**, 149–64.

Buret, A.G., Mitchell, K., et al. 2002. *Giardia lamblia* disrupts tight junctional ZO-1 and increases permeability in non-transformed human small intestinal epithelial monolayers: effects of epidermal growth factor. *Parasitology*, **125**, 11–19.

Carroccio, A., Montalto, G., et al. 1997. Secondary impairment of pancreatic function as a cause of severe malabsorption in intestinal giardiasis: A case report. *Am J Trop Med Hyg*, **56**, 599–602.

Chavez, B., Knaippe, F., et al. 1986. *Giardia lamblia*: Electrophysiology and ultrastructure of cytopathology in cultured epithelial cells. *Exp Parasitol*, **61**, 379–89.

Chen, N., Upcroft, J.A. and Upcroft, P. 1995. A *Giardia duodenalis* gene encoding a protein with multiple repeats of a toxin homologue. *Parasitology*, **111**, 423–31.

Collins, J.P., Keller, K.F. and Brown, L. 1978. 'Ghost' forms of *Giardia lamblia* cysts initially misdiagnosed as *Isospora*. *Am J Trop Med Hyg*, **27**, 334–5.

Craft, J.C. 1982. *Giardia* and giardiasis in children. *Pediatr Infect Dis*, **1**, 196–211.

Craun, G.F. 1990. Waterborne giardiasis. In: Meyer, E.A. (ed.), *Giardiasis*. Amsterdam: Elsevier, 267–93.

Das, S., Schteingart, C.D., et al. 1997. *Giardia lamblia*: Evidence for carrier-mediated uptake and release of conjugated bile acids. *Exp Parasitol*, **87**, 133–41.

Davidson, R.A. 1990. Treatment of giardiasis: the North American perspective. In: Meyer, E.A. (ed.), *Giardiasis*. Amsterdam: Elsevier, 325–53.

Dobell, C. 1920. The discovery of the intestinal protozoa of man. *Proc R Soc Med*, **13**, 1–15.

Dykes, A.C., Juranek, D.D., et al. 1980. Municipal waterborne giardiasis. An epidemiologic investigation. *Ann Intern Med*, **93**, 165–70.

Ellis, J.E., Wyder, M.A., et al. 1996. Changes in lipid composition during in vitro encystation and fatty acid desaturase activity of *Giardia lamblia*. *Mol Biochem Parasitol*, **81**, 13–25.

Erlandsen, S.L. 1974. Scanning electron microscopy of intestinal giardiasis: Lesions of the microvillous border of villus epithelial cells produced by trophozoites of *Giardia*. In: Johari, O. (ed.), *Scanning electron microscopy*. Chicago: IIT Research Institute, 775–82.

Erlandsen, S.L., Macechko, P.T., et al. 1996. Formation of the *Giardia* cyst wall: Studies on extracellular assembly using immunogold labeling and high resolution field emission SEM. *J Eukaryotic Microbiol*, **43**, 416–29.

Ey, P.L., Bruderer, T., et al. 1996. Comparison of genetic groups determined by molecular and immunological analyses of *Giardia* isolated from animals and humans in Switzerland and Australia. *Parasitol Res*, **82**, 52–60.

Ferrari, B.C., Attfield, P.V., et al. 2003. Application of the novel fluorescent dye Beljian red to the differentiation of *Giardia* cysts . *J Microbiol Methods*, **52**, 133–5.

Filice, F.P. 1952. Studies on the cytology and life history of a *Giardia* from the laboratory rat. *Univ Calif Publ Zool*, **47**, 53–146.

Franco, R. and Cantusio Neto, R. 2002. Occurrence of cryptosporidial oocysts and *Giardia* cysts in bottled mineral water commercialized in the city of Campinas, state of Sao Paulo, Brazil. *Mem Inst Oswaldo Cruz*, **97**, 205–7.

Garcia, L.S. 2001. *Diagnostic medical parasitology*, 4th edn. Washington, DC: American Society for Microbiology Press.

Garcia, L.S. and Shimizu, R.Y. 1997. Evaluation of nine immunoassay kits (enzyme immunoassay and direct fluorescence) for detection of *Giardia lamblia* and *Cryptosporidium parvum* in human fecal specimens. *J Clin Microbiol*, **35**, 1526–9.

Garcia, L.S. and Shimizu, R.Y. 2000. Detection of *Giardia lamblia* and *Cryptosporidium parvum* antigens in human fecal specimens using the ColorPAC combination rapid solid-phase qualitative immunochromatographic assay. *J Clin Microbiol*, **38**, 1267–8.

Garcia, L.S., Shimizu, R.Y. and Bernard, C.N. 2000. Detection of *Giardia lamblia*, *Entamoeba histolytica/Entamoeba dispar* and *Cryptosporidium parvum* antigens in human fecal specimens using the Triage Parasite Panel enzyme immunoassay. *J Clin Microbiol*, **38**, 3337–40.

Garcia, L.S., Shum, A.C. and Bruckner, D.A. 1992. Evaluation of a new monoclonal antibody combination reagent for direct fluorescence detection of *Giardia* cysts and *Cryptosporidium* oocysts in human fecal specimens. *J Clin Microbiol*, **30**, 3255–7.

Gaxrcia, L.S., Shimizu, R.Y., et al. 2003. Detection of *Giardia lamblia* and *Cryptosporidium parvum* antigens in human fecal specimens by rapid solid-phase qualitative immunochromatography. *J Clin Microbiol*, **41**, 209–12.

Gibson, G.R., Ramirez, D., et al. 1999. *Giardia lamblia*: Incorporation of free and conjugated fatty acids into glycerol-based phospholipids. *Exp Parasitol*, **92**, 1–11.

Graczyk, T.K., Cranfield, M.R. and Conn, D.B. 1997. In vitro phagocytosis of *Giardia duodenalis* cysts by hemocytes of the Asian freshwater clam *Corbicula fluminea*. *Parasitol Res*, **83**, 743–5.

Graczyk, T.K., Bosco-Nizeyi, J., et al. 2002. Anthropozoonotic *Giardia duodenalis* genotype (assemblage) A infections in habitats of free-ranging human-habituated gorillas, Uganda. *J Parasitol*, **88**, 905–9.

Halliday, C.E.W., Inge, P.M.G. and Farthing, M.J.G. 1995. Characterization of bile salt uptake by *Gardia lamblia*. *Int J Parastiol*, **25**, 1089–97.

Hardin, J.A., Buret, A.G., et al. 1997. Mast cell hyperplasia and increased macromolecular uptake in an animal model of giardiasis. *J Parasitol*, **83**, 908–12.

Healy, G.R. 1990. Giardiasis in perspective: the evidence of animals as a source of human *Giardia* infections. In: Meyer, E.A. (ed.), *Giardiasis*. Amsterdam: Elsevier, 305–13.

Hewlett, E.L., Andrews, J.S., et al. 1982. Experimental infection in mongrel dogs with *Giardia lamblia* cysts and cultured trophozoites. *J Infect Dis*, **145**, 89–93.

Homan, W.L. and Mank, T.G. 2001. Human giardiasis: genotype linked differences in clinical symptomatology. *Int J Parasitol*, **31**, 822–6.

Hopkins, R.M., Meloni, B.P., et al. 1997. Ribosomal RNA sequencing reveals differences between the genotypes of *Giardia* isolates recovered from humans and dogs living in the same locality. *J Parasitol*, **83**, 44–51.

Isaac-Renton, J.L., Byrne, S.K. and Prameya, R. 1988. Isoelectric focusing of ten strains of *Giardia duodenalis*. *J Parasitol*, **74**, 1054–6.

Isaac-Renton, J., Blatherwick, J., et al. 1999. Epidemic and endemic seroprevalence of antibodies to *Cryptosporidium* and *Giardia* in residents of three communities with different drinking water supplies. *Am J Trop Med Hyg*, **60**, 578–83.

Islam, A. 1990. Giardiasis in developing countries. In: Meyer, E.A. (ed.), *Giardiasis*. Amsterdam: Elsevier, 235–66.

Jarroll, E.L. and Lindmark, D.G. 1990. *Giardia* metabolism. In: Meyer, E.A. (ed.), *Giardiasis*. Amsterdam: Elsevier, 61–76.

Jarroll, E.L., Bingham, A.K. and Meyer, E.A. 1980. *Giardia* cyst destruction: Effectiveness of six small-quantity water disinfection methods. *Am J Trop Med Hyg*, **29**, 8–11.

Jarroll, E.L., Bingham, A.K. and Meyer, E.A. 1981. Effect of chlorine on *Giardia lamblia* cyst viability . *Appl Environ Microbiol*, **41**, 483–7.

Jokipii, L. and Jokipii, A.M.M. 1979. Single-dose metronidazole and tinidazole as therapy for giardiasis: success rates, side effects, and drug absorption and elimination. *J Infect Dis*, **140**, 984–8.

Karanis, P., Opiela, K., et al. 1996. Possible contamination of surface waters with *Giardia* spp. through muskrats. *Zentralbl Bakteriol*, **284**, 302–6.

Katanik, M.T., Schneider, S.K., et al. 2001. Evaluation of ColorPAC *Giardia/Cryptosporidium* rapid assay and ProSpecT *Giardia/Cryptosporidium* microplate assay for detection of *Giardia* and *Cryptosporidium* in fecal specimens . *J Clin Microbiol*, **39**, 4523–5.

Katelaris, P.H., Naeem, A. and Farthing, M.J.G. 1995. Attachment of *Giardia lamblia* trophozoites to a cultured human intestinal cell line. *Gut*, **37**, 512–18.

Kettis, A.A. and Magnius, L. 1973. *Giardia lamblia* infection in a group of students after a visit to Leningrad in March 1970. *Scand J Infect Dis*, **5**, 289–92.

Keystone, J.S., Karjden, S. and Warren, M.R. 1978. Person-to-person transmission of *Giardia lamblia* in day-care nurseries. *Can Med Assoc J*, **119**, 242–4.

Kirner, J.C. 1978. Waterborne outbreak of giardiasis in Camas, Washington. *J Am Waterworks Assoc*, **January**, 35–40.

Kistemann, T., Classen, T., et al. 2002. Microbial load of drinking water reservoir tributaries during extreme rainfall and runoff . *Appl Environ Microbiol*, **68**, 2188–97.

Knaus, W.A. 1974. Reassurance about Russian giardiasis. *N Engl J Med*, **291**, 156.

Lane, S. and Lloyd, D. 2002. Current trends in research into the waterborne parasite *Giardia*. *Crit Rev Microbiol*, **28**, 123–47.

Linden, K.G., Shin, G.A., et al. 2002. UV disinfection of *Giardia lamblia* cysts in water. *Environ Sci Technol*, **36**, 2519–22.

Lopez, C.E., Juranek, D.D., et al. 1978. Giardiasis in American travelers to Madeira Island, Portugal. *Am J Trop Med Hyg*, **27**, 1128–32.

Lujan, H.D., Mowatt, M.R. and Nash, T.E. 1997. Mechanisms of *Giardia lamblia* differentiation into cysts. *Microbiol Mol Biol Rev*, **61**, 294–304.

Lujan, H.D., Mowatt, M.R., et al. 1996. Cholesterol starvation induces differentiation of the intestinal parasite *Giardia lamblia*. *Proc Nat Acad Sci USA*, **93**, 7628–33.

Mahbubani, M.H., Schaefer, F.W., et al. 1998. Detection of *Giardia* in environmental waters by immuno-PCR amplification methods. *Curr Microbiol*, **36**, 107–13.

Mank, T.G., Zaat, J.O.M., et al. 1997. Sensitivity of microscopy versus enzyme immunoassay in the laboratory diagnosis of giardiasis. *Eur J Clin Microbiol Infect Dis*, **16**, 615–10.

McIntyre, L., Hoang, L., et al. 2000. Evaluation of molecular techniques to biotype *Giardia duodenalis* collected during an outbreak. *J Parasitol*, **86**, 172–7.

Meyer, E.A. 1990. Taxonomy and nomenclature. In: Meyer, E.A. (ed.), *Giardiasis*. Amsterdam: Elsevier, 51–60.

Meyers, J.D., Kuharic, H.A. and Holmes, K.K. 1977. *Giardia lamblia* infection in homosexual men. *Br J Vener*, **53**, 54–5.

Monis, P.T., Andrews, R.H., et al. 1999. Molecular systematics of the parasitic protozoan *Giardia intestinalis*. *Mol Biol Evol*, **16**, 1135–1144.

Moore, G.T., Cross, W.M., et al. 1969. Epidemic giardiasis at a ski resort. *N Engl J Med*, **281**, 402–7.

Nash, T.E. 1997. Antigenic variation in *Giardia lamblia* and the host's immune response. *Philos Trans R Soc Lond B Biol Sci*, **352**, 1369–75.

Nash, T.E. 2002. Surface antigenic variation in *Giardia lamblia*. *Mol Microbiol*, **45**, 585–90.

Nash, T.E., Herrington, D.A. and Levine, M.M. 1987. Usefulness of an enzyme-linked immunosorbent assay for detection of *Giardia* antigen in feces. *J Clin Microbiol*, **25**, 1169–71.

Nash, T.E., Herrington, D.A., et al. 1990. Antigenic variation of *Giardia lamblia* in experimental human infections. *J Immunol*, **144**, 4362–9.

Nash, T.E., Ohl, C.A., et al. 2001. Treatment of patients with refractory giardiasis. *Clin Infect Dis*, **33**, 22–8.

NCCLS, 1997. *Procedures for the recovery and identification of parasites from the intestinal tract. Approved guideline, M28-A*. Villanova, PA: National Committee for Clinical Laboratory Standards.

Owen, R.L., Allen, C.L. and Stevens, D.P. 1981. Phagocytosis of *Giardia muris* by macrophages in Peyer's patch epithelium in mice. *Infect Immun*, **33**, 591–601.

Peterson, J.M. 1957. Intestinal changes in *Giardia lamblia* infestation. *Am J Roentgenol*, **77**, 670–7.

Phillips, S.C., Mildran, D., et al. 1981. Sexual transmission of enteric protozoa and helminths in a venereal-disease-clinic population. *N Engl J Med*, **305**, 603–6.

Proctor, E.M., Isaac-Renton, J.L., et al. 1989. Isoenzyme analysis of human and animal isolates of *Giardia duodenalis* from British Columbia, Canada. *Am J Trop Med Hyg*, **41**, 411–15.

Redlinger, T., Corella-Barud, V., et al. 2002. Hyperendemic *Cryptosporidium* and *Giardia* in households lacking municipal sewer and water on the United States–Mexico border. *Am J Trop Med Hyg*, **66**, 794–8.

Rosalesborjas, D.M., Diazrivadeneyra, J., et al. 1998. Secretory immune response to membrane antigens during *Giardia lamblia* infection in humans. *Infect Immun*, **66**, 756–9.

Schmerin, M.J., Jones, T.C. and Klein, H. 1978. Giardiasis: Association with homosexuality. *Ann Intern Med*, **88**, 801–3.

Sealy, D.P. and Schuman, S.H. 1983. Endemic giardiasis and day care. *Pediatrics*, **72**, 154–8.

Seppel, T., Rose, F. and Schlaghecke, R. 1996. Chronic intestinal giardiasis with isolated levothyroxine malabsorption as reason for severe hypothyroidism---Implications for localization of thyroid hormone absorption in the gut. *Exp Clin Endocrinol Diabetes*, **104**, 180–2.

Shaw, P.K., Brodsky, R.E., et al. 1977. A community wide outbreak of giardiasis with evidence of transmission by a municipal water supply. *Ann Intern Med*, **87**, 426–32.

Smith, P.D., Lane, H.C., et al. 1988. Intestinal infections in patients with the acquired immunodeficiency syndrome (AIDS). *Ann Intern Med*, **108**, 328–33.

Soliman, M.M., Taghi-Kilani, R., et al. 1998. Comparison of serum antibody responses to *Giardia lamblia* of symptomatic and asymptomatic patients. *Am J Trop Med Hyg*, **58**, 232–9.

Stevens, W.J. and Vermeire, P.A. 1981. *Giardia lamblia* in bronchoalveolar lavage fluid. *Thorax*, **36**, 875.

Sulaiman, I.M., Fayer, R., et al. 2003. Triosephosphate isomerase gene characterization and potential zoonotic transmission of *Giardia duodenalis*. *Emerg Infect Dis*, **11**, 1444–52.

Sulaiman, I.M., Jiang, J., et al. 2004. Distribution of *Gardia duodenalis* genotypes and subgenotypes in raw urban wastewater in Milwaukee, Wisconsin. *Appl Environ Microbiol*, **70**, 3776–80.

Tandon, B.N., Tandon, R.K., et al. 1977. Mechanism of malabsorption in giardiasis: A study of bacterial flora and bile salt deconjugation in upper jejunum. *Gut*, **18**, 176–81.

Traub, R.J., Morris, P.T., et al. 2004. Epidemiological and molecular evidence supports the zoonotic transmission of *Giardia* among humans and dogs living in the same community. *Parasitology*, **128**, 253–62.

Trout, J.M., Walsh, E.J. and Fayer, R. 2002. Rotifers ingest *Giardia* cysts. *J Parasitol*, **88**, 1038–40.

Upcroft, J. and Upcroft, P. 1998. My favorite cell: *Giardia. Bioessays*, **20**, 256–63.

Upcroft, J.A., Boreham, P.F.L., et al. 1995. Biological and genetic analysis of a longitudinal collection of *Giardia* samples derived from humans. *Acta Tropica*, **60**, 35–46.

van Keulen, H., Macechko, P.T., et al. 2002. Presence of human *Giardia* in domestic, farm and wild animals, and environmental samples suggests a zoonotic potential for giardiasis. *Vet Parasitol*, **108**, 97–107.

WHO, 1995. *Drugs used in parasitic diseases*, 2nd edn. Geneva: World Health Organization.

Wright, R.A., Spencer, H.C., et al. 1977. Giardiasis in Colorado: An epidemiologic study. *Am J Epidemiol*, **105**, 330–6.

Zemlyn, S., Wilson, W.W. and Hillweg, P.A. 1981. A caution on iodine water purification. *West J Med*, **135**, 166–7.

13

Trichomonas infections

DONALD E. BURGESS

INTRODUCTION

Of the three to five trichomonad species found in humans, only *Trichomonas vaginalis* regularly displays the capacity for producing disease. Human trichomoniasis persists and is essentially cosmopolitan in its occurrence despite the availability of effective chemotherapeutic agents and little evidence of the development of significant drug resistance. A sexually transmitted disease, trichomoniasis caused by *T. vaginalis* ranges from a mild or inapparent infection to one causing a chronic and substantial level of inflammation in the reproductive tract of women and significant urethritis in men (Krieger et al. 1993a). Rare cases of perinatal infection of females born to infected mothers (Sobel 1992; Smith et al. 2002; Szarka et al. 2002) and respiratory trichomonad infections (Walzer et al. 1978) have been reported, the latter possibly attributable to trichomonads other than *T. vaginalis*.

CLASSIFICATION

T. vaginalis is a parasitic protozoan and is classified as follows:

Class Trichomonadea, Kirby, 1947
    Order Trichomonadida, Kirby, 1947 emend. Honigberg, 1974. There are typically 4–6 flagella, which may be recurved, free or attached to an undulating membrane (when present). There are usually no true cysts.
        Family Trichomonadidae, Wenyon, 1926. Characterized by possession of: a cytostome; 3–5 free flagella (one flagellum on the margin of an undulating membrane); and an axostyle protruding through the posterior of the cell.
            Genus Trichomonas, Donné, 1837. Trophozoites that typically possess: 4 free flagella (one along the outer margin of the undulating membrane); a costa at the base of the undulating membrane; and an axostyle. No cysts are present.

STRUCTURE AND LIFE CYCLE

The shape of *T. vaginalis* is typically pyriform in culture, although ameboid shapes are evident in parasites adhering to mammalian cells (Nielsen and Nielsen 1975). Light- and electron-microscopic studies (Honigberg and King 1964; Nielsen 1975), summarized by Honigberg and Burgess (1994), indicate that *T. vaginalis* is c. 9.7×7 μm and that nondividing organisms have four anterior flagella. In addition to the flagella (four anterior and one recurrent) the undulating membrane and the costa also originate in the kinetosomal complex at the anterior of the parasite. Internal organelles include a prominent nucleus and a rigid structure, the axostyle, that runs through the cell from the anterior to posterior end (Figure 13.1).

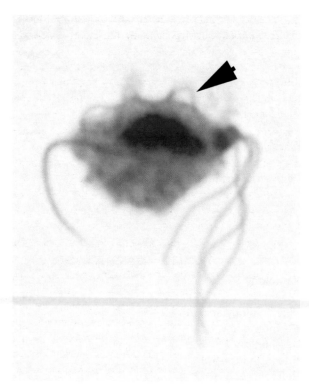

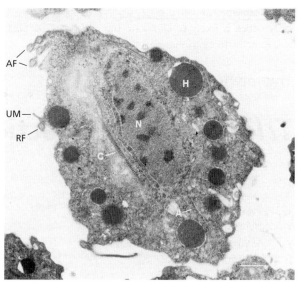

Figure 13.2 *Transmission electron micrograph of* Trichomonas vaginalis *from culture showing: axostyle (A), costa (C), hydrogenosome (H), nucleus (N), anterior flagella (AF), recurrent flagellum (R), undulating membrane (UM). Bar = 1 μm. (Courtesy of A. Blixt, VMBL)*

Figure 13.1 Trichomonas vaginalis *from culture, stained with Giemsa's stain, showing axostyle (left) and undulating membrane (arrow). Four distinct anterior flagella are visible (right). Bar = 10 μm. (Courtesy of M. Tighe, VMBL)*

A respiratory organelle, the hydrogenosome (Lindmark and Müller 1973), is present and appears by light microscopy as a paraxostylar and paracostal chromatic granule. The hydrogenosome is seen as an osmiophilic, dense granule by electron microscopy (Nielsen 1975, 1976) (Figure 13.2).

The life-cycle of *T. vaginalis* is simple in that the trophozoite is transmitted through coitus and no cyst form is known. The trophozoite divides by binary fission giving rise to a population in the urogenital tract of humans.

BIOCHEMISTRY, MOLECULAR BIOLOGY, AND GENETICS

Anaerobic metabolism produces most of the parasite's energy. Earlier work on the biochemistry of *T. vaginalis* (reviewed in Shorb 1964; von Brand 1973; Honigberg 1978) indicated that several sugars (e.g. maltose, glucose) are fermented by the parasite through a glycolytic pathway. Subsequent studies revealed that transport of glucose across the cell membrane occurred by facilitated diffusion (ter Kuile and Muller 1992). The metabolism of *T. vaginalis* differs from that of eukaryotic cells with mitochondria in three major areas:

1 the importance of inorganic pyrophosphates in carbohydrate catabolism
2 the importance of sulfur proteins in metabolism
3 the disposal of electrons by the formation of molecular hydrogen (Searle and Müller 1991; Müller 1992).

Trichomonads lack both cytochrome-mediated electron transport and the associated electron transport-linked phosphorylation, as well as the familiar mitochondrion that houses this system in other eukaryotic cells. Instead, the site of fermentative carbohydrate metabolism is the hydrogenosome (Lindmark and Müller 1973; Müller 1992), an organelle thought to have arisen by endosymbiosis with an anaerobic bacterium, or conversion of mitochondria (Lahti et al. 1992). Hydrogenosomes isolated from *T. vaginalis* have an NADH/ferredoxin oxidoreductase activity and these organelles ferment pyruvate to acetate, malate, molecular hydrogen and cxarbon dioxide. Pyruvate is also converted to acetyl-CoA via ferridoxin–oxidoreductase activity (Johnson et al. 1990; Land et al. 2001). This fermentation is dependent on adenosine diphosphate (ADP), inorganic phosphate, magnesium and succinate, and carbon dioxide-dependent carboxylation of pyruvate to malate by malate dehydrogenase. Detergent disruption (Triton X100) stops pyruvate-dependent formation of molecular hydrogen (this effect is overcome by adding exogenous ferredoxin), indicating that the structural integrity of the hydrogenosome is crucial to its effective functioning (Steinbuchel and Müller 1986).

The mechanism of development of anaerobic resistance to metronidazole also is controlled by hydrogenosomes in that metronidazole competes for hydrogen ions as an electron acceptor. In metronidazole-resistant *T. vaginalis* the expression levels of the hydrogenosomal enzymes (pyruvate:ferredoxin oxidoreductase, ferridoxin, malic enzyme, and hydrogenase) are reduced dramatically which is likely to eliminate the ability of the parasite to activate metronidazole (Land et al. 2001).

Hydrolytic enzymes

Proteinases of *T. vaginalis* have been known since the 1970s and detailed investigations of these enzymes have been made during the last 10 years. Cysteine proteinases seem to be particularly prevalent, as dithiothreital stimulates the activity of several proteinases from *T. vaginalis*. These range from 20 to 110 kDa (Lockwood et al. 1987), the lower molecular weight proteinases being released from the cell (Lockwood et al. 1984, 1987, 1988; North et al. 1990). The release of proteinases from *T. vaginalis* is of interest in the context of its ability to elicit varying degrees of inflammation and the possible role of these enzymes in pathogenesis. There are reports of cell-detaching factors released by the parasite into the growth medium (Garber et al. 1989; Lushbaugh et al. 1989) some of which are reported to have trypsin-like activity. These factors are active on human and hamster cells, causing them to detach and round up. Release of cell-detaching factors and proteinases from *T. vaginalis* clearly implies that these parasite products could degrade proteins such as laminin, vitronectin and other components of the extracellular matrix, thus effecting the release of host cells from tissue.

The substrate specificity and structure of some of the proteinases of *T. vaginalis* are now being determined. Parasite proteinases were not inhibited by pepstatin, phenyl methyl sulphonyl fluoride or EDTA, but they were inhibited by iodoacetic acid, antipain, leupeptin and N-a-p-tosyl-L-lysine chloromethyl ketone. Lower molecular sizes of proteinases were released from the parasite; proteinases of 25, 27, and 34 kDa specifically hydrolyzed synthetic substrates with arginine–arginine residues whereas other proteinases had activity over a wide substrate range (North et al. 1990). Recent cloning studies of cysteine proteinase genes of *T. vaginalis* indicate that at least four distinct genes are present and that they have considerable homology (up to 45 percent identity) with cysteine proteinase genes of *Dictyostelium discoideum*. Three of the proteinase genes were present as single copies, whereas the other was present as a multiple copy gene. The amino acid sequence data indicated that these proteinases were of the L/cathepsin H/ papain type (Mallinson et al. 1994).

Nucleic acids

Nucleic acid synthesis in trichomonads differs markedly from that in mammalian cells, as trichomonads do not synthesize either purines or pyrimidines de novo. *T. vaginalis* does not incorporate purine-ring precursors into nucleic acids but rather directly salvages purine nucleosides through adenosine and guanosine kinases. Neither inosine nor hypoxanthine are incorporated and no interconversion between adenylate and guanylate occurs (Heyworth et al. 1982). Trichomonads are also unable to synthesize pyrimidines from aspartate, orotate and bicarbonate, the typical de novo pathway. Neither thymidylate synthetase nor dihydrofolate reductase activity has been found. Conversion of cytidine by cytidine phosphotransferase and nucleotide kinases to cytidine monophosphate (CMP), cytidine diphosphate (CDP) and cytidine triphosphate (CTP) allows *T. vaginalis* to salvage exogenous cytidine, uridine, uracil and, to some extent, thymidine into its nucleotide pool (Wang and Cheng 1984). Cytidine is converted to uridine by cytidine deaminase, whereas uracil is converted to uridine by uridine phosphorylase before conversion into nucleotides. Nucleosides (but not nucleotides) are transported across the cell membrane of *T. vaginalis*. The nucleosides adenosine, guanosine and uridine are transported across the cell membrane by a two-carrier, facilitated transport mechanism at a rate apparently sufficient to sustain growth (Harris et al. 1988). Intracellularly, nucleosides are rapidly converted to nucleotides so that transport is not affected by any appreciable build up of nucleosides. Neither bases nor D-ribose affect uptake of nucleosides, suggesting that the carriers operating in *T. vaginalis* are similar to those in other parasitic protozoa such as *Plasmodium berghei* (Hansen et al. 1980) and *Leishmania donovani* (Aronow et al. 1987).

CLINICAL ASPECTS

T. vaginalis is site-specific, usually surviving in only the urogenital tract of humans although rare cases of pneumonia in newborns have been reported (Smith et al. 2002; Szarka et al. 2002). Approximately 10 percent of vulvovaginitis is due to infection with *T. vaginalis* (Sobel 1992). In men, clinical features of the infection may include urethral discharge and inflammation of the tissues lining the reproductive/urinary tract (Krieger et al. 1993a), but it is more usual for men to display no symptoms.

A female patient infected with *T. vaginalis* may have no obvious clinical signs or may suffer vaginitis of varying severity. Asymptomatic patients can also carry the infection and remain at risk of developing symptomatic disease at a later date (Krieger et al. 1988). Clinical features that may develop in 50–90 percent of infected women include: purulent vaginal discharge, pruritus, and dyspareunia. Leukorrhea and dysuria are occasionally seen. Vaginal discharge is observed in 50–75 percent of diagnosed women and is considered malodorous in only c. 10 percent of these patients. Although the discharge has often been described as yellow–green and frothy, only c. 8 percent of patients produce a discharge characterized as frothy and c. 59 percent produce a discharge characterized as purulent. Diffuse vulvar erythema and copious vaginal discharge are often present. Punctate hemorrhages of the cervix may result in a strawberry-like appearance, apparent by inspection in only 1–2 percent of patients but apparent in 45 percent of patients examined by colposcopy (Wølner-Hanssen et al. 1989). Pruritus, particularly vulvar itching, is a frequent symptom (25–50 percent of patients) and may be severe (Sobel 1992).

IMMUNOLOGY

Some of the earliest investigations into immunity centered on reports that sera obtained from a variety of mammals can lyse *T. vaginalis* and agglutinate the parasite. Although earlier investigators considered lysis to be dependent on antibody action, *T. vaginalis* is now known to activate complement by the alternative pathway (Gillin and Sher 1981). An extensive study of normal human, bovine, sheep, horse, swine, and dog sera against *T. vaginalis* (Reisenhofer 1963, summarized by Honigberg 1970) showed that these sera had low (equal to or less than 1:64) but consistently detected agglutinating activities. Current understanding of immunity to *T. vaginalis* is unsatisfactory and it is not clear whether induced immunity is protective and if so, how. Although there is evidence that protection may be achieved by immunization of laboratory animals, strong protective immunity does not seem to follow natural infection in humans. Immunosuppressed persons, such as those with acquired immune deficiency syndrome (AIDS), do not develop fulminating *T. vaginalis* infections despite the high prevalence of the parasite in the general population. A recent study of patients infected with *T. vaginalis* and human immunodeficiency virus (HIV) indicated no evidence of increased levels or longevity of parasite infection in these patients compared to patients with *T. vaginalis* but without HIV (Cu-Uvin et al. 2002). These observations may indicate that innate immunity involving chemotaxis and subsequent influx of neutrophils are much more important than acquired immunity in controlling infections with *T. vaginalis* since neutrophils are often the most numerous leukocytes present in sites of infection.

Innate immunity

Innate immune responses to *T. vaginalis* are well documented and may be important in controlling infection as well as in inflammation during infection. Both macrophages and polymorphonuclear neutrophils (PMN) have been shown to be capable of killing *T. vaginalis* in vitro (Rein et al. 1980) and PMNs have been reported to be attracted to a factor(s) secreted by live *T. vaginalis* (Mason and Forman 1980). Leukocytic vaginal discharge was previously considered to be an important clinical finding expected in women infected with *T. vaginalis*, but more recent studies indicate no significant difference between the numbers of infected and uninfected women with vaginal discharge (Fouts and Kraus 1980). Even so, recruitment of macrophages and PMNs to the site of infection (by the sensitized T lymphocytes and the lymphokines they produce, or by parasite-derived factors that are directly chemotactic for PMNs) could be important elements in protective immunity and the pathological changes present in persons with trichomoniasis.

Acquired immunity

Early experimental studies on acquired immune responses to *T. vaginalis* examined various immunizations of laboratory animals and evaluated antibody from patients with *T. vaginalis* infection. Guinea pigs given intraperitoneal inoculations of vaginal material containing trichomonads produced antibody responses, as did rabbits given formalin-killed parasites (Riedmuller 1932; Tokura 1935). In mice, intramuscular inoculation induced protection against challenge infection that lasted at least 15 weeks (Kelly and Schnitzer 1952; Schnitzer and Kelly 1953). Intramuscular injection of *T. vaginalis* protected up to 100 percent of mice against an intraperitoneal challenge, but much less protection was afforded against a subcutaneous challenge (Kelly et al. 1954). The protective effect of sera from infected humans was shown in experiments in which human sera were transferred to mice inoculated intraperitoneally with *T. vaginalis*. Similar mouse inoculation experiments with human sera or rabbit hyperimmune sera indicated that these antibodies were both protective. Levels of protective antibodies in the sera of patients quickly diminish following elimination of infection (Teras and Nigesen 1969; reviewed in Honigberg 1970). The conclusion from these and subsequent observations (Sobel 1992) is that although an immune response to *T. vaginalis* occurs in infected patients and produces circulating antibody against the parasite, the response is short-lived and does not provide protection against re-infection.

Antibody responses during infection with *T. vaginalis* have been evaluated in numerous surveys by several immunoassays (summarized in Honigberg and Burgess 1994). Although some infected patients do not have detectable anti-*T. vaginalis* antibodies, most display secreted and circulating (serum) antibodies. Secretory antibody levels in the reproductive tract seem to increase in women infected with *T. vaginalis* and parasite-specific antibodies occur in the urogenital tract of persons harboring *T. vaginalis*. Immunoglobulin IgA antibody, specific for *T. vaginalis*, was present in vaginal secretions of 76 percent of 29 infected women and 42 percent of 19 apparently uninfected women with higher levels in the infected patients (Ackers et al. 1975). Subsequent reports have verified the presence of parasite-specific antibody in vaginal secretions. Street et al. (1982) reported that 73.2 percent of infected women had anti-*T. vaginalis* IgG or IgA antibodies, compared to 41 percent of uninfected women. Additional studies on a women with proven *T. vaginalis* trichomoniasis have confirmed these findings (Romia and Othman 1991; Alderete et al. 1991a).

Early evidence that a cellular immune response is elicited in patients infected with *T. vaginalis* is based on the observation that such patients develop positive delayed-type hypersensitivity (DTH) reactions to whole parasite antigen preparations (Adler and Sadowsky 1947; Aburel et al. 1963). Later it was shown that blood lymphocytes from individuals with active infections proliferated in vitro when stimulated with whole cell antigen preparations; lymphocytes from uninfected persons gave minimal responses (Yano et al., 1983a, 1983b; Mason and Patterson 1985). Similar results have been obtained in early studies of DTH reactions (summarized in Honigberg and Burgess 1994) and lymphocyte proliferation responses in mice (Mason and Gwanzura 1988). These results, together with the findings in patients, indicate that systemic sensitization of T lymphocytes can occur during infection.

Antigens

Early immunological studies of *T. vaginalis* indicated that there were antigenic differences between geographically distinct strains of this parasite. In a series of studies in the 1960s, four serotypes (TLR, TN, TRT, and TR) were described by the use of cross-agglutination and complement fixation methods. These serotypes had common and unique antigens and were present in a large geographical area of central and eastern Europe (reviewed by Honigberg 1970). In another study, agglutination, complement fixation and fluorescent antibody methods were used to demonstrate three serotypes of *T. vaginalis* in one geographical area of the former USSR (Andreeva and Mihov 1976). This report was among the first to describe molecular differences between strains of *T. vaginalis* and to attempt to correlate these with serological type; on the basis of the electrophoretic patterns of ten strains of *T. vaginalis*, five different antigenic groups were detectable. Since this initial work, numerous antigens of *T. vaginalis* have been described, some being shared between strains and others being strain-specific.

Subsequently, antigens of *T. vaginalis* have been partially characterized including limited information about function. In one study, 20 immunogenic polypeptides, ranging from 20 to 200 kDa, were detected with rabbit antisera (Alderete 1983) many of which were determined to be on the parasite surface (Alderete et al. 1985). Additional studies using monoclonal antibodies prepared against one clone of *T. vaginalis*, showed that these monoclonal antibodies reacted with surface epitopes on four of nine distinct parasite clones (Torian et al. 1984; Connely et al. 1985). Western blot analysis of reactions indicated the presence of both conserved and heterogeneous epitopes and antigens (Torian et al. 1988).

Some isolates of *T. vaginalis* possess a 270 kDa surface antigen (P270) that is strongly expressed, whereas other isolates express little P270 on their surface (Alderete et al. 1986a, b). Expression of higher amounts of P270 correlates with lower *T. vaginalis* cytotoxicity toward HeLa cells, suggesting that the degree of expression of P270 may be a useful virulence marker. A 6-amino acid epitope of P270 (DREGRD) binds experimental antibodies specific for P270, as well as a high percentage of the antibody in serum from infected patients, suggesting that P270 could be an immunodominant antigen (Alderete and Neale 1989; Dailey and Alderete 1991). Additional *T. vaginalis* antigens have been detected, including a 230 kDa surface antigen that binds to vaginal antibodies present in infected patients (Alderete et al. 1991a) and proteinases that elicit serum and secreted antibodies (Alderete et al. 1991b; Bozner et al. 1992). Four antigenic surface molecules have also been implicated in the adhesion of *T. vaginalis* to vaginal epithelial cells, their expression being upregulated during attachment to host cells (Arroyo et al. 1993). Antibodies to these molecules protect target cells from parasite-mediated cytotoxicity (Arroyo et al. 1992) suggesting that anti-adhesion immune responses could be important in protecting against the pathogenic effects of *T. vaginalis*.

PATHOLOGY

Although *T. vaginalis* was first described by Donné (1836), its role as a cause of vaginitis was not recognized for another 80 years (Höhne 1916). The early report of Hogue (1943) on the pathogenic effects of *T. vaginalis* on cells in cultures suggested that the parasite could directly damage host cells. The clinical features, including pathogenicity, of human urogenital trichomonads have been reviewed (Jírovec and Petru 1968; Brown 1972; Catterall 1972; Honigberg 1990; Honigberg and Burgess 1994). Trichomoniasis is a frequently encountered, sexually transmitted disease (Lossick and Kent 1991) and a common cause of vaginitis and exocervicitis (Heine and McGregor 1993; Sobel 1992). Children born to women with trichomoniasis may suffer from low birth weight (discussed by Heine and McGregor 1993; Gibbs et al. 1992; Cotch et al. 1991), probably as a result of tissue damage from the cytopathic effect of this parasite on mammalian cells (Alderete and Perlman 1983; Krieger et al., 1985a).

Histological findings

Histological findings in women with trichomonad cervicitis have been described (Koss and Wolinska 1959); up to one-third of biopsies of female patients infected with *T. vaginalis* show no histological changes. Patients displaying changes had increased vascularity of the squamous epithelium and the presence of the trichomonads was often accompanied by distension of blood vessels within papillae. The 'strawberry cervix' (Lang and Ludmir 1961; Fouts and Kraus 1980) appearance of the exocervix is due to the local extravasation of blood resulting in petechiae (rather than to ulceration) and is not usually due to inflammation. There is more pronounced dilation of epithelial vessels in patients with trichomonad cervicitis compared with those with nontrichomonad-induced cervicitis. Edema of the squamous epithelium is accompanied by separation of epithelial cells from each other, inflammation of the squamous epithelium, and desquamation. Perinuclear haloes may develop that can be confused with the koilocytotic atypia caused by human papilloma virus infection (Quinn and Holmes 1984); these are often confined to the basal layers of the epithelium. A proportion of epithelial cells display a variety of other abnormalities including enlarged nuclei, hyperchromasia and binucleation. Pyknotic nuclei are evident in the damaged epithelium indicating necrosis and a purulent exudate often coats the epithelial surface; such findings are rare in nontrichomonad cervicitis. Biopsy material from 11 women infected with *T. vaginalis* was examined by Nielsen and Nielsen (1975) in an electron microscope study. Generally, there was evidence of chronic nonspecific inflammation with subepithelial infiltration by neutrophils and lymphocytes. Three of the patients had cervical erosion; the lumens of the glands were packed with neutrophils. In cases of more severe inflammation, neutrophils were present in deeper layers of the epithelium, the neutrophils near the surface being arranged in lacunae. *T. vaginalis* was occasionally found on the surface of the stratified squamous epithelium in clusters situated in areas of shallow depression in the epithelial surface. The trichomonads were often seen on the vaginal surface in a dense mantle, but were not attached to the epithelium in observations in seven patients. Ameboid trichomonads attached only to necrotic epithelial cells with the undulating membrane on the surface away from the substrate. In the areas of contact between parasites and epithelial cells, there was interdigitation of cytoplasmic projections from the parasite and host cell, which is evidence of intimate contact between the parasite and host during human infection. In women with florid vaginitis, the secretion typically contains numerous motile trichomonads, thus rendering diagnosis reasonably easy. In women with latent trichomoniasis, in which the parasite has retreated into the cervical region and in which bizarre and atypically hyperplastic cells may be present in smears, the trichomonads are more difficult to find. In the diffuse or patchy inflammatory secretion, the trichomonads may not be motile. *T. vaginalis* is often difficult to recognize in routinely fixed and stained smears (such as those stained with hematoxylin and eosin). Antibody-based immunohistochemical procedures have been developed that compare favorably with wet mount methods (Kreiger et al. 1988).

Mechanism of pathogenicity

Studies of the cytopathic effects of *T. vaginalis* on a variety of cell lines in vitro provide support for a cytopathic mechanism in which both parasite–host cell contact and soluble factors play a role. It was previously known that efficient killing of several target cell lines by *T. vaginalis* required close approximation or contact between parasite and target (Krieger et al. 1985a; Alderete and Garza 1985, 1988). Evidence that parasite adhesion to target cells is important for induction of damage of targets has come from experiments in which various treatments, (Arroyo and Alderete 1989), low temperature (4°C) (Alderete and Garza 1985), cytochalasin D (Krieger et al. 1985a), trypsin (Alderete and Garza 1985), or cysteine proteinase inhibitors lowered adhesion of parasites to targets. These studies suggested that proteinaceous surface structures synthesized by the parasite played a role in adhesion. Four antigenic surface molecules have specifically been implicated in the adhesion of *T. vaginalis* to host cells (Enbring and Alderete 1992) and parasites in contact with HeLa cells or vaginal epithelial cells increase their expression of these adhesion molecules (Arroyo et al. 1993). Although these reports indicate that intimate association of *T. vaginalis* to targets leads to efficient killing in these experimental systems, they do not prove the absence of one (or more) soluble mediators responsible for host cell damage. There are several reports of soluble factors produced by *T. vaginalis* that are active against nucleated host cells (Garber et al. 1989; Lushbaugh et al. 1989; Pindak et al. 1986). Although most of the cell-detachment factors remain uncharacterized there is a recent report of a soluble lytic factor with phospholipase activity (Lubick and Burgess 2004). When homogeneous preparations of such molecules become available, the precise structure and function of such factors will be established and a better understanding of the mechanisms of pathogenesis of trichomoniasis at the molecular level will be possible.

DIAGNOSIS

Microscopy and culture procedures are the most widely used methods for the detection of *T. vaginalis* in men and women. The most commonly employed procedures are:

- direct microscopic examination of fresh (stained or unstained) material in wet mounts, accomplished with the aid of bright-field, dark-field, or phase-contrast microscopy

- microscopic examination of fixed and stained preparations, usually smears, stained with Giemsa's stain
- cultivation employing a variety of media.

A combination of the wet mount and culture procedures is both efficient and cost-effective in detection of T. vaginalis-positive cases among both women and men. Reports of an improved culture device, InPouch™ (BioMed Diagnostics Inc.), indicate that it compares favorably with standard culture procedures in a clinical setting (Draper et al. 1993; Levi et al. 1997; Barenfanger et al., 2002).

Immunological methods have not been used widely for diagnosis of T. vaginalis infection although the relative specificity and sensitivity of individual techniques has been improved recently. In a comparison of several methods, direct detection of T. vaginalis using fluorescein-labeled monoclonal antibody compared favorably with wet mounts, staining of fixed material and culture (Krieger et al. 1988). Immunohistochemical methods detected the related trichomonad, *Tritrichomonas foetus*, in sections of formalin-fixed tissue from the reproductive tract of infected cattle more easily than standard microscopic methods (Burgess and Knoblock 1988; Rhyan et al. 1995).

Antibody and antigen detection

Several types of enzyme-linked immunosorbent assay (ELISA) (Engvall and Perlman 1971) have been developed for T. vaginalis, either to measure patient antibodies or to detect antigens of T. vaginalis in clinical samples. In one example, whole-cell antigen preparations of T. vaginalis (Street et al. 1982) together with aqueous extracts (Alderete 1984) were used as antigens to detect antibodies in serum and vaginal secretions of patients. Affinity-purified, rabbit antibodies to T. vaginalis have been used as both capture and detection antibodies in a sandwich ELISA to detect antigens of T. vaginalis (Watt et al. 1986). This assay had a sensitivity of 77 percent, a specificity of 100 percent, and was more sensitive than microscopic examination of wet mounts, but less sensitive than culture methods for detection of infection. Detection of T. vaginalis antigens in fluids on vaginal swabs by ELISA had a sensitivity of 93.2 percent and a specificity of 97.5 percent in a study of 44 culture-positive patients in a study group of 482 (Yule et al. 1987). Employing a monoclonal anti-T. vaginalis antibody as a capture and detection antibody, a sensitivity of 89 percent and a specificity of 97 percent have been achieved (Lisi et al. 1988).

Molecular methods

Nucleic acid hybridization methods for detection of T. vaginalis (Rubino et al. 1991) have sensitivity and specificity as good as culture methods. Riley et al. (1992) developed a polymerase chain reaction (PCR) method, using primers that produced a 102. bp product, termed A6p. They used the method to test 24 isolates of T. vaginalis, all of which reacted positively and similar results have been obtained with another PCR method using a different target (Kengne et al. 1994). In both cases, PCR procedures produced no false positive reactions with template DNA from sources such as human cells, other parasitic protozoa and viruses. A PCR assay using primers based on beta-tubulin sequences of T. vaginalis detected 22 of 23 culture-positive samples (Madico et al. 1998). Also, PCR methods were able to detect a higher prevalence in males than was detected by culture methods (Schwebke and Lawing 2002) and positive PCR results have recently been suggested as diagnostically useful where trichomoniasis may be undertreated (Wendel et al. 2002).

EPIDEMIOLOGY AND CONTROL

Trichomoniasis persists throughout the world, despite the availability of effective chemotherapeutic agents and seemingly adequate awareness of the disease on the part of physicians, public health workers and the general population (Schwebke 2002). In recent years, there has been a focus on T. vaginalis, due partly to heightened awareness of sexually transmitted diseases generally, as well as the fact that people with T. vaginalis infections seem to be at higher risk of HIV infection than are uninfected persons.

Infections with T. vaginalis are quite common; the prevalence of T. vaginalis in nonselected female populations is probably 5–20 percent. Most estimates of prevalence are based on data from selected patient populations (e.g. those attending sexually transmitted disease clinics); such estimates are unlikely to be representative of the population at large. Even so, the data indicate that there has been a slow decline in the prevalence of T. vaginalis infections over the past 20 years. In 1972, more than 180 million women and men were estimated to be infected with T. vaginalis worldwide (Brown 1972), including c. 2.5 million women in the USA and 1 million women in the UK (Catterall 1972). T. vaginalis infection prevalence in American women is estimated at 3 million cases annually (Rein 1990) and some investigators consider trichomoniasis to be the most common sexually transmitted protozoan disease (Levine 1991) and perhaps even the most common sexually transmitted disease (Hammill 1989). Although estimates in nonselected male populations are not as certain as for women, the prevalence in men is probably c. 50–60 percent of that in women (Rein and Müller 1989). One study noted a prevalence of 11 percent in men attending an STD clinic (Krieger et al. 1993b) which is comparable to the 17 percent recently detected in a similar study (Schwebke and Lawing 2002).

CHEMOTHERAPY

Treatment of vaginal trichomoniasis relies on the 5-nitroimidazole compounds: metronidazole, tinidazole, and ornidazole. Metronidazole is the only one of these currently recommended by the WHO (Schwebke 1995). Tinidazole with metronidazole have comparable efficacies when used in a single-dose regimen (Gabriel et al. 1982). Metronidazole acts by interfering with DNA synthesis, apparently as the result of the production of a transient cytotoxic intermediate form (Lossick 1990). Daily treatment for 7 days or treatment by a single 2 g dose gave equivalently high cure rates (more than 90 percent) (Lossick 1982). The daily treatment regimen consists of oral administration of 250 mg of metronidazole three times daily for 7 days, producing a cure rate of c. 95 percent (Lossick 1990; Sobel 1992). Although the multiple-dose regimen is efficacious for the patient even without treatment of the sexual partner, it suffers from a degree of noncompliance. The single-dose regimen is particularly effective if used with simultaneous treatment of sexual partners, but it does not protect against the relatively frequent event of prompt reinfection by untreated sexual partners. Because of the multifocal nature of infection with *T. vaginalis*, systemic (rather than local) treatment is preferred. In fact, endogenous reinfection by parasites harbored in the urethra and periurethral glands is known to occur when local treatment is attempted (Rein and Muller 1989).

VACCINATION

There is no vaccine currently available for use against *T. vaginalis*. The search for vaccines against protozoa is complicated by their ability to evade and modulate host immune responses. *T. vaginalis* may be able to evade the immune response as it exists largely outside the host in the lumen of the reproductive tract. Effector mechanisms such as secretory antibody (sIgA) may prove to be crucial in protection against the cytotoxic effects of this parasite on host cells. Many basic questions concerning the immune response to *T. vaginalis* remain unanswered and such lack of knowledge impedes vaccine development. For example, we do not know the mechanisms by which natural infections are eliminated, nor which target antigens elicit protective immunity. We do not even know if an effective protective immunity develops during natural infection. The absence of an appropriate laboratory animal model for trichomoniasis has delayed progress in answering these questions. The natural history of infection by *T. vaginalis* in humans suggests that adaptive immune responses during infection do not play a significant role in controlling infections. Indeed, the most prevalent leukocyte response in the reproductive tract during infection is by neutrophils and chemotaxis of neutrophils by components of *T. vaginalis* has

been reported (Mason and Forman 1980). These findings, together with the pattern of frequent reinfection, suggest that effective immunological memory may not be established in most infected persons. Other explanations for the absence of protective immunity after infection are the presence of a considerable number of distinct antigenic types of *T. vaginalis* (Krieger et al. 1985b), or the occurrence of some form of antigenic drift over time, as has been observed with certain surface antigens (Alderete et al. 1986b). A combination of these phenomena could produce a re-infection pattern similar to that seen with the rhinoviruses that cause the common cold. If such drifts occur, new antigenic profiles in the parasite population would occur so rapidly that the host could not effectively respond to the ever-changing epitopes being displayed.

INTEGRATED CONTROL

Control of trichomoniasis due to *T. vaginalis* should include:

- thorough examination to determine infection status of the patient
- appropriate treatment of the patient and sexual partner(s)
- continued surveillance by public health agencies to provide updated prevalence estimates.

Current knowledge of the immune response to *T. vaginalis* is insufficient to predict whether the development of a vaccine represents a realistic goal. A rational vaccine development strategy should include research toward understanding the roles of certain surface antigens and secreted antigenic molecules (e.g. proteinases) and identification of new target antigens.

REFERENCES

Aburel, E., Zervos, G., et al. 1963. Immunological and therapeutic investigations in vaginal trichomoniasis. *Rum Med Rev*, **7**, 13–19.

Ackers, J.P., Lumsden, W.H.R., et al. 1975. Anti-trichomonal antibody in the vaginal secretions of women infected with *T. vaginalis*. *Br J Vener Dis*, **51**, 319–23.

Adler, S. and Sadowsky, A. 1947. Intradermal reaction in trichomonad infection. *Lancet*, **252**, 1, 867–8.

Alderete, J.F. 1983. Identification of immunogenic and antibody-binding membrane proteins of pathogenic *Trichomonas vaginalis*. *Infect Immun*, **40**, 284–91.

Alderete, J.F. 1984. Enzyme-linked immunosorbent asay for detection of antibody to *Trichomonas vaginalis*: Use of whole cells and aqueous extracts as antigens. *Br J Vener Dis*, **60**, 164–70.

Alderete, J.F. and Garza, G.E. 1985. Specific nature of *Trichomonas vaginalis* parasitism of host cell surfaces. *Infect Immun*, **50**, 701–8.

Alderete, J.F. and Garza, G.E. 1988. Identification and properties of *Trichomonas vaginalis* proteins involved in cytadherence. *Infect Immun*, **56**, 28–33.

Alderete, J.F. and Neale, K.A. 1989. Relatedness of a major immunogen in *Trichomonas vaginalis* isolates. *Infect Immun*, **50**, 1845–53.

Alderete, J.F. and Perlman, E. 1983. Pathogenic *Trichomonas vaginalis* cytotoxicity to cell culture monolayers. *Br J Vener Dis*, **60**, 99–105.

Alderete, J.F., Suprun-Brown, L., et al. 1985. Heterogeneity of *Trichomonas vaginalis* and subpopulations with sera of patients and experimentally infected mice. *Infect Immun*, **49**, 463–8.

Alderete, J.F., Suprun-Brown, L. and Kasmala, L. 1986a. Monoclonal antibody to a major surface glycoprotein immunogen differentiates isolates and subpopulations of *Trichomonas vaginalis*. *Infect Immun*, **52**, 70–5.

Alderete, J.F., Kasmala, L., et al. 1986b. Phenotypic variation and diversity among *Trichomonas vaginalis* isolates and correlation of phenotype with trichomonal virulence determinants. *Infect Immun*, **53**, 285–93.

Alderete, J.F., Newton, E., et al. 1991a. Vaginal antibody of patients with trichomoniasis is to a prominent surface immunogen of *Trichomonas vaginalis*. *Genitourin Med*, **67**, 220–5.

Alderete, J.F., Newton, E., et al. 1991b. Antibody in sera of patients infected with *Trichomonas vaginalis* is to trichomonad proteinases. *Genitourin Med*, **67**, 331–4.

Andreeva, N. and Mihov, L. 1976. Electrophoretic studies of the water soluble proteins from ten local strains of *Trichomonas vaginalis*. *Folia Med (Prague)*, **18**, 67–73.

Aronow, B., Kaur, K., et al. 1987. Two high affinity nucleoside transporters in *Leishmania donovani*. *Molec Biochem Parasitol*, **22**, 29–37.

Arroyo, R. and Alderete, J.F. 1989. *Trichomonas vaginalis* surface proteinase activity is necessary for parasite adherence to epithelial cells. *Infect Immun*, **57**, 2991–7.

Arroyo, R., Enbring, J. and Alderete, J.F. 1992. Molecular basis of host epithelial cell recognition by *Trichomonas vaginalis*. *Mol Microbiol*, **6**, 853–62.

Arroyo, R., Gonzalez-Robles, A., et al. 1993. Signaling of *Trichomonas vaginalis* for amoeboid transformation and adhesion synthesis follows cytoadherence. *Molec Microbiol*, **7**, 299–309.

Barenfanger, J., Drake, C. and Hanson, C. 2002. Timing of inoculation of the pouch makes no difference in the increased detection of *Trichomonas vaginalis* by the InPouch TV method. *J Clin Microbiol*, **40**, 1387–9.

Bozner, P., Gombosova, A., et al. 1992. Proteinases of *Trichomonas vaginalis*: antibody response in patients with urogenital trichomoniasis. . *Parasitology*, **105**, 387–91.

Brown, M.T. 1972. Trichomoniasis. *Practitioner*, **209**, 639–44.

Burgess, D.E. and Knoblock, K.F. 1988. Identification of *Tritrichomonas foetus* in sections of bovine placental tissue with monoclonal antibodies. *J Parasitol*, **75**, 977–80.

Catterall, R.D. 1972. Trichomonal infections of the genital tract. *Med Clin North Am*, **56**, 1203–9.

Connely, R.J., Torian, B.E. and Stibbs, H.H. 1985. Identification of a surface antigen of *Trichomonas vaginalis*. *Infect Immun*, **49**, 270–4.

Cotch, M.F., Pastorek, J.G., et al. 1991. Demographic and behavioral predictors of *Trichomonas vaginalis* infection among pregnant women. *Obstet. Gynecol*, **78**, 1087–92.

Cu-Uvin, S., Ko, H., et al. 2002. Prevalence, incidence, and persistence or recurrence of trichomoniasis among human immunodeficiency virus (HIV)-positive women and among HIV-negative women at high risk for HIV infection. *Clin Infect Dis*, **34**, 1406–11.

Dailey, D.C. and Alderete, J.F. 1991. The phenotypically variable surface protein of *Trichomonas vaginalis* has a single, tandemly repeated immunodominant epitope. *Infect Immun*, **59**, 2083–8.

Donné, A. 1836. Animalcules observés dansles matières purulents et le produit des sécrétions des organes genitaux de l'home et de la femme. *CR Hebd Seances Acad Sc*, **3**, 385–6.

Draper, D., Parker, R., et al. 1993. Detection of *Trichomonas vaginalis* in pregnant women with the InPouch TV culture system. *J Clin Microbiol*, **31**, 1016–18.

Enbring, J. and Alderete, J.F. 1992. Molecular basis of host epithelial cell recognition by Trichomonas vaginalis. *Molec Microbiol*, **6**, 853–62.

Engvall, E. and Perlmann, P. 1971. Enzyme-linked immunosorbent assay (ELISA) quantitative assay of immunoglobulin G. *Immunochemistry*, **8**, 871–9.

Fouts, A.C. and Kraus, S.J. 1980. *Trichomonas vaginalis*: Re-evaluation of its clinical presentation and laboratory diagnosis. *J Infect Dis*, **141**, 137–43.

Gabriel, G., Robertson, E. and Thin, R.N. 1982. Single dose treatment of trichomoniasis. *J Int Med Res*, **10**, 129–30.

Garber, G.E., Lemchuk-Favel, L.T. and Bowie, W.R. 1989. Isolation of a cell-detaching factor of *Trichomonas vaginalis*. *J Clin Microbiol*, **27**, 1548–53.

Gibbs, R.S., Romero, R., et al. 1992. A review of premature birth and subclinincal infection. *Amer J Obstet Gynecol*, **166**, 1515–28.

Gillin, F.D. and Sher, A. 1981. Activation of the alternative complement pathway by *Trichomonas vaginalis*. *Infect Immun*, **34**, 268–73.

Hammill, H.A. 1989. *Trichomonas vaginalis*. *Obstet Gynecol Clin North Am*, **16**, 531–40.

Hansen, B.E., Sleeman, H.K. and Pappas, P.W. 1980. Purine base and nucleoside uptake in *Plasmodium berghei* and host erythrocytes. *J Parasitol*, **66**, 205–12.

Harris, D.I., Beechey, R.B., et al. 1988. Nucleoside uptake by *Trichomonas vaginalis*. *Molec Biochem Parasitol*, **29**, 105–16.

Heine, P. and McGregor, J.A. 1993. *Trichomonas vaginalis*: A re-emerging pathogen. *Clin Obstet Gynecol*, **36**, 137–44.

Heyworth, P.G., Gutteridge, W.E. and Ginger, C.D. 1982. Purine metabolism in *Trichomonas vaginalis*. *FEBS Lett*, **141**, 106–10.

Hogue, M.J. 1943. The effect of *Trichomonas vaginalis* on tissue culture cells. *Am J Hyg*, **37**, 142–52.

Höhne, O. 1916. *Trichomonas vaginalis* als häufiger Erreger einer typischen colpitis purulenta. *Centralbaltt Gynaekol*, **40**, 4–15.

Honigberg, B.M. 1970. Trichomonads. In: Jackson, G.J., Herman, R. and Singer, L. (eds), *Immunity to parasitic animals*, Vol. 2. . New York, NY: Appleton, 469–550.

Honigberg, B.M. 1978. Trichomonads of importance in human medicine. In: Kreier, J.P. (ed.), *Parasitic protozoa*, Vol. 2. . New York, NY: Academic Press, 275–454.

Honigberg, B.M. (ed.) 1990. *Trichomonads parasitic to humans*. New York, NY: Springer-Verlag.

Honigberg, B.M. and Burgess, D.E. 1994. Trichomonads of importance in human medicine including *Dientamoeba fragilis*. In: Kreier, J.P. (ed.), *Parasitic protozoa*, Vol. 9. . New York, NY: Academic Press, 1–109.

Honigberg, B.M. and King, V.M. 1964. Structure of *Trichomonas vaginalis* Donné. *J Parasitol*, **50**, 345–64.

Jírovec, O. and Petru, M. 1968. *Trichomonas vaginalis* and Trichomoniasis. *Adv Parasitol*, **6**, 117–88.

Johnson, P.J., d'Oliveira, C.E., et al. 1990. Molecular analysis of the hydrogenosomal ferredoxin of the anaerobic protist *Trichomonas vaginalis*. *Proc Natl Acad Sci USA*, **87**, 6097–101.

Kelly, D.R. and Schnitzer, R.J. 1952. Experimental studies on trichomoniasis II. Immunity to reinfection in *T. vaginalis* infections of mice. *J Immunol*, **69**, 337–42.

Kelly, D.R., Schumacher, A. and Schnitzer, R.J. 1954. Experimental studies in trichomoniasis III. Influence of the site of the immunizing infection with *Trichomonas vaginalis* on the immunity of mice to homologous reinfection by different routes. *J Immunol*, **73**, 40–3.

Kengne, P., Veas, F., et al. 1994. *Trichomonas vaginalis*: repeated DNA target for highly sensitive and specific polymerase chain reaction diagnosis. *Cell Molec Biol (Noisy-le-grand)*, **40**, 819–31.

Koss, L.G. and Wolinska, W.H. 1959. *Trichomonas vaginalis* cervicitis and its relationship to cervical cancer. *Cancer*, **12**, 117–19.

Krieger, J.N., Ravdin, J.I. and Rein, M.F. 1985a. Contact-dependent cytopathogenic mechanisms of *Trichomonas vaginalis*. *Infect Immun*, **50**, 778–86.

Krieger, J.N., Holmes, K.K., et al. 1985b. Geographic variation among isolates of *Trichomonas vaginalis*: Demonstration of antigen heterogenieity by using monoclonal antibodies and the indirect immunofluorescence technique. *J Infect Dis*, **152**, 979–84.

Krieger, J.N., Tam, M.R., et al. 1988. Diagnosis of trichomoniasis. Comparison of conventional wet-mount examination with cytologic studies, cultures, and monoclonal antibody staining of direct specimens. *J Am Med Assoc*, **259**, 1223–7.

Krieger, J.N., Jenny, C.V., et al. 1993a. Clinical manifestations of trichomoniasis in men. *Ann Intern. Med*, **118**, 844–9.

Krieger, J.N., Verdon, N.S., et al. 1993b. Natural history of urogenital trichomoniasis in men. *J Urol*, **149**, 1455–8.

Land, K.M., Clemens, D.L. and Johnson, P.J. 2001. Loss of multiple hydrogenosomal proteins associated with organelle metabolism and high-level drug resistance in trichomonads. *Exp Parasitol*, **97**, 102–10.

Lang, W.R. and Ludmir, A. 1961. A pathognomonic colposcopic sign of *Trichomonas vaginalis* aginitis. *Acta Cytol*, **5**, 390–2.

Lahti, C.J., d'Oliveira, C.E. and Johnson, P.J. 1992. Beta-succinyl-coenzyme A synthetase from *Trichomonas vaginalis* is a soluble hydrogenosomal protein with an amino-terminal sequence that resembles mitochondrial presequences. *J Bacteriol*, **174**, 6822–30.

Levi, M.H., Torres, J., et al. 1997. Comparison of the InPouch TV culture system and Diamond's modified medium for the detection of *Trichomonas vaginalis*. *J Clin Microbiol*, **35**, 3308–10.

Levine, G.I. 1991. Sexually transmitted parasitic diseases. *Prim Care*, **18**, 101–28.

Lindmark, D.G. and Müller, M. 1973. Hydrogenosome, a cytoplasmic organelle of the anaerobic flagellate *Tritrichomonas foetus*, and its role in pyruvate metabolism. *J Biol Chem*, **235**, 7724–8.

Lisi, P.J., Dondero, R.S., et al. 1988. Monoclonal-antibody-based enzyme-linked immunosorbent assay for *Trichomonas vaginalis*. *J Clin Microbiol*, **26**, 1684–6.

Lockwood, B.C., North, M.J. and Coombs, G.H. 1984. *Trichomonas vaginalis*, *Tritrichomonas foetus* and *Trichomitus batrachorum*: Comparative proteolytic activity. *Exp Parasitol*, **58**, 245–53.

Lockwood, B.C., North, M.J.S., et al. 1987. The use of a highly sensitive electrophoretic method to compare the proteinases of trichomonads. *Mol Biochem Parasitol*, **24**, 89–95.

Lockwood, B.C., North, M.J. and Coombs, G.H. 1988. The release of hydrolases from *Trichomonas vaginalis* and *Tritrichomonas foetus*. *Mol Biochem Parasitol*, **30**, 135–42.

Lossick, J.G. 1982. Treatment of *Trichomonas vaginalis* infections. *Rev Infect Dis*, **4**, S801–18.

Lossick, J.G. 1990. Treatment of sexually transmitted vaginosis/vaginitis. *Rev Infect Dis*, **12**, Suppl, S665–89.

Lossick, J.G. and Kent, H.L. 1991. Trichomoniasis: Trends in diagnosis and management. *Amer J Obstet Gynecol*, **165**, 1217–22.

Lubick, K.J. and Burgess, D.E. 2004. Purification and analysis of a phospholipase A2-like lytic factor of *Trichomonas vaginalis*. *Infect Immun*, **72**, 1284–90.

Lushbaugh, W.B., Turner, A.C., et al. 1989. Characterization of a secreted cytoactive factor from *Trichomonas vaginalis*. *Am J Trop Med Hyg*, **41**, 18–28.

Madico, G., Quinn, T.C., et al. 1998. Diagnosis of *Trichomonas vaginalis* infection by PCR using vaginal swab samples. *J Clin Microbiol*, **36**, 3205–10.

Mallinson, D.J., Lockwood, B.C., et al. 1994. Identification and molecular cloning of four cysteine proteinase genes from the pathogenic protozoan *Trichomonas vaginalis*. *Microbiol*, **140**, 2725–35.

Mason, P.R. and Forman, L. 1980. In vitro attraction of polymorphonuclear leukocytes by *Trichomonas vaginalis*. *J Parasitol*, **66**, 888–92.

Mason, P.R. and Gwanzura, L. 1988. Mouse spleen cell responses to trichomonal antigens in experimental *Trichomonas vaginalis* infection. *J Parasitol*, **74**, 93–7.

Mason, P.R. and Patterson, B.A. 1985. Proliferative response of human lymphocytes to secretory and cellular antigens of *Trichomonas vaginalis*. *J Parasitol*, **71**, 265–8.

Müller, M. 1992. Energy metabolism of ancestral eukaryotes: A hypothesis based on the biochemistry of amitochondriate parasitic protests. *Biosystem*, **28**, 33–40.

Nielsen, M.H. 1975. The ultrastructure of *Trichomonas vaginalis* Donné before and after transfer from vaginal secretion to Diamond's medium. *Acta Pathol Microbiol Scand Sect B*, **83**, 581–9.

Nielsen, M.H. 1976. In vitro effect of metronidazole on the ultrastructure of *Trichomonas vaginalis* Donné. *Acta Pathol Microbiol Scand Sect B*, **84**, 93–100.

Nielsen, M.H. and Nielsen, R. 1975. Electron microscopy of *Trichomonas vaginalis* Donné: interaction with vaginal epithelium in human trichomoniasis. *Acta Pathol Microbiol Scand Sect B*, **83B**, 305–20.

North, M.J., Robertson, C.D. and Coombs, G.H. 1990. The specificity of trichomonad proteinases analysed using fluorogenic substrates and specific inhibitors. *Mol Biochem Parasitol*, **39**, 183–94.

Pindak, F.F., Gardiner, W.A. and Pindak, M.M. 1986. Growth and cytopathogenicity of *Trichomonas vaginalis* in tissue cultures. *J Clin Microbiol*, **23**, 672–5.

Quinn, T.C. and Holmes, K.K. 1984. Trichomoniasis. In: Warren, K.S. and Mahmoud, A. (eds), *Tropical and geographical medicine*. New York, NY: McGraw Hill, 335–41.

Riedmüller, L. 1932. Zur Frage der ätiologischen Bedeutung der bei Pyometra und sporadischen Abortus des Rindes gefundenen Trichomonaden. *Schweiz Arch Tierheilkd*, **74**, 343–51.

Rein, M.F. 1990. Vaginitis. In: Rivlin, M.E., Morrison, J.C. and Bates, G.W. (eds), *Manual of clinical problems in obstetrics and gynecology*. Boston, MA: Little, Brown and Co, 287–95.

Rein, M.F. and Müller, M. 1989. *Trichomonas vaginalis* and trichomoniasis. In: Holmes, K.K., Mardt, D.A., et al. (eds), *Sexually transmitted diseases*. New York: McGraw Hill, 481–92.

Rein, M.F., Sullivan, J.A. and Mandell, G.L. 1980. Trichomonacidal activity of human polymorphonuclear neutrophils: killing by disruption and fragmentation. *Infect Dis*, **142**, 575–85.

Reisenhofer, U. 1963. Über die Beeinflussung von *Trichomonas vaginalis* durch verschiedene. *Sera Arch Hyg Bakteriol*, **146**, 628–35.

Rhyan, J.C., Wilson, K.L., et al. 1995. Immunohistochemical detection of *Tritrichomonas foetus* in formalin-fixed, paraffin-embedded sections of bovine placenta and fetal lung. *J Vet Diagn Invest*, **7**, 98–101.

Riley, D.E., Roberts, M.C., et al. 1992. Development of a polymerase chain reaction-based diagnosis of *Trichomonas vaginalis*. *J Clin Microbiol*, **30**, 465–72.

Romia, S.A. and Othman, T.A. 1991. Detection of antitrichomonal antibodies in sera and cervical secretions in trichomoniasis. *J Egypt Soc Parasitol*, **21**, 373–81.

Rubino, S., Muresu, R., et al. 1991. Molecular probe for identification of *Trichomonas vaginalis* DNA. *J Clin Microbiol*, **29**, 702–6.

Schnitzer, R.J. and Kelly, D.R. 1953. Short persistence of *Trichomonas vaginalis* in reinfected immune mice. *Proc Soc Exp Biol Med*, **82**, 404–6.

Schwebke, J.R. 1995. Metronidazole: utilization in the obstetric and gynecologic patient. *Sex Transm Dis*, **22**, 370–6.

Schwebke, J.R. 2002. Update on trichomoniasis. *Sex Transm Infect*, **78**, 378–9.

Schwebke, J.R. and Lawing, L.F. 2002. Improved Detection by DNA Amplification of *Trichomonas vaginalis* in males. *J Clin Microbiol*, **40**, 3681–3.

Searle, S.M. and Müller, M. 1991. Inorganic pyrophosphatase of Trichomonas vaginalis. *Molec Biochem Parasitol*, **44**, 91–6.

Shorb, M.S. 1964. The physiology of trichomonads. In: Hutner, S.H. and Lwoff, A. (eds), *Biochemistry and physiology of protozoa*, Vol. 3. . New York, NY: Academic Press, 383–457.

Sobel, J.D. 1992. Vulvovaginitis. *Dermatol Clin*, **10**, 339–59.

Smith, L.M., Wang, M., et al. 2002. *Trichomonas vaginalis* infection in a premature newborn. *J Perinatol*, **22**, 502–3.

Steinbuchel, A. and Müller, M. 1986. Anaerobic pyruvate metabolism of *Tritrichomonas foetus* and *Trichomonas vaginalis* hydrogenosomes. *Mol Biochem Parasitol*, **20**, 57–65.

Street, D.A., Taylor-Robinson, D., et al. 1982. Evaluation of an enzyme-linked immunosorbent assay for the detection of antibody to *Trichomonas vaginalis* in sera and vaginal secretions. *Br J Vener Dis*, **58**, 330–3.

Szarka, K., Temesvari, P., et al. 2002. Neonatal pneumonia caused by *Trichomonas vaginalis*. *Acta Microbiol Immunol Hung*, **49**, 15–19.

ter Kuile, B.H. and Müller, M. 1992. Interaction between facilitated diffusion of glucose across the plasma membrane and its metabolism in *Trichomonas vaginalis*. *FEMS Microbiol Lett*, **110**, 27–31.

Teras, J.K. and Nigesen, U. 1969. On the protective effect of blood sera of persons infected with *Trichomonas vaginalis* Donné. *Wiad Parazytol*, **15**, 481–3.

Tokura, N. 1935. Biologische und immunologische Untersuchungen über die memschenparasitären Trichomonaden. *Igaku Kenkyu*, **9**, 1–13.

Torian, B.E., Connelly, R.J., et al. 1984. Specific and common antigens of *Trichomonas vaginalis* detected by monoclonal antibodies. *Infect Immun*, **43**, 270–5.

Torian, B.E., Connelly, R.J., et al. 1988. Antigenic heterogeneity in the 115,000 Mr major surface antigen of *Trichomonas vaginalis*. *J Protozol*, **35**, 273–80.

von Brand, T. 1973. *Biochemistry of parasites*. New York, NY: Academic Press.

Walzer, P.D., Rutherford, I. and East, R. 1978. Empyema with *Trichomonas* species. *Amer Rev Resp Dis*, **118**, 415–18.

Wang, C.C. and Cheng, H.-W. 1984. Salvage of pyrimidine nucleosides by *Trichomonas vaginalis*. *Molec Biochem Parasitol*, **10**, 171–84.

Watt, R.M., Philip, A., et al. 1986. Rapid assay for immunological detection of *Trichomonas vaginalis*. *J Clin Microbiol*, **24**, 790–5.

Wendel, K.A., Erbelding, E.J., et al. 2002. *Trichomonas vaginalis* polymerase chain reaction compared with standard diagnostic and therapeutic protocols for detection and treatment of vaginal trichomoniasis. *Clin Infect Dis*, **35**, 576–80.

Wølner-Hanssen, P., Krieger, J.N., et al. 1989. Clinical manifestations of vaginal trichomoniasis. *J Am Med Assoc*, **264**, 571–6.

Yano, A., Yui, K., et al. 1983a. Immune response to *Trichomonas vaginalis* IV. Immunochemical and immunobiological analysis of T. vaginalis antigen. *Int Archiv Aller Appl Immunol*, **72**, 150–7.

Yano, A., Asoai, F., et al. 1983b. Antigen-specific proliferation responses on peripheral blood lymphocytes to *Trichomonas vaginalis* antigen in patients with *Trichomonas vaginalis*. *J Clin Microbiol*, **17**, 175–80.

Yule, A., Gellan, M.C., et al. 1987. Detection of *Trichomonas vaginalis* antigen in women by enzyme immunoassay. *J Clin Pathol*, **40**, 566–8.

14

Dientamoeba fragilis and other intestinal flagellates

JAROSLAV KULDA AND EVA NOHÝNKOVÁ

INTRODUCTION

The digestive tract of humans can serve as habitat for several species of flagellates belonging to protozoan orders Trichomonadida, Diplomonadida, and Retortamonadida. Besides *Giardia* and the trichomonads discussed in Chapters 12, Giardiasis and 13, *Trichomonas* infections, these are *Dientamoeba fragilis*, *Enteromonas hominis*, *Chilomastix mesnili*, and *Retortamonas intestinalis*. All are confined to the large intestine, the habitat with low digestive activity and an abundance of the intestinal flora. Although diverse in morphology and subcellular organization, these flagellates are united by their ability to live and multiply under anaerobic or microaerophilic conditions. All seem to depend on associated bacterial biocenosis for both nutrition and adjustment of proper physicochemical conditions. However, their metabolism remains unknown since they are rarely obtained in a condition suitable for biochemical experiments. Also, genomic studies are scarce and limited to sequence analyses of rRNA genes for phylogenetic investigations. The life cycles of most of these organisms are simple and direct and do not involve sexual processes.

Typically, these organisms alternate the proliferative trophozoite, multiplying by binary fission in the host intestine, and the resting resistant cyst that is discharged with feces and serves to transmit the infection. *Dientamoeba* does not produce cysts and is an exception from this generalization. Infection with the cyst-producing flagellates is acquired through contaminated food and water or by direct fecal–oral transmission. Just one of the species discussed (*D. fragilis*) is an enteropathogen

of clinical importance; the remaining species are harmless commensals. Some pathogenic potential was ascribed to *Chilomastix* but has not been proved. An increase in the flagellate population is often secondary to a pathological condition of another etiology, indicating merely that the intestinal biocenosis is out of balance. Still, even the nonpathogenic intestinal flagellates deserve consideration both from the diagnostic standpoint and as indicators of a poor standard of personal hygiene and sanitation.

DIENTAMOEBA FRAGILIS

Classification

Dientamoeba fragilis Jepps and Dobell, 1918 is an unusual flagellate with no flagella and was originally classified as an ameba. However, further morphological studies by Dobell (1940) revealed its relationship to *Histomonas meleagridis*, a turkey pathogen related to the trichomonads. Dobell noticed the presence of centrodesmus (centrodesmose), a fiber connecting the nuclei of the binucleate forms of *Dientamoeba* and later identified as the extranuclear spindle characteristic for the mitosis of Trichomonadida. Trichomonad affinities were unequivocally proven by the electron microscopic studies of Camp et al. (1974) and further confirmed by molecular phylogenies based on a complete sequence of the small subunit rRNA gene (Silberman et al. 1996). Biological, clinical, and diagnostic aspects of *Dientamoeba* have been recently reviewed (Johnson et al. 2004).

Structure and life cycle

Dientamoeba occurs in a trophozoite stage only and does not encyst. Low resistance of trophozoites to unfavorable environmental conditions, failure to infect humans and primates with cultured trophozoites (Dobell 1940), and the close relationship with *Histomonas* led to the assumption that *Dientamoeba*, like *Histomonas*, can develop in a helminth vector. Burrows and Swerdlow (1956) proposed the pinworm *Enterobius vermicularis* as a probable vector and pinworms eggs as a vehicle for *Dientamoeba* transmission. Uninucleate ameboid cells have been observed in pinworm eggs (Burrows and Swerdlow 1956; Ockert 1990) but convincing direct evidence of their identity with *Dientamoeba* is lacking.

Trophozoites of *Dientamoeba* usually measure 7–12 μm in diameter although they can vary within a wider range (3.5–22 μm). They are uninucleate or binucleate with predominance of binucleate forms (60–80 percent) resulting from protracted telophase of the parasite mitosis. Nuclei are rounded with a central group of chromatin granules and delicate nuclear membrane without adjacent peripheral chromatin (Figure 14.1). The extranuclear mitotic spindle can be seen in properly stained binucleate forms as a fibril extending between the nuclei (Figures 14.1 and 14.2c). *Dientamoeba* produces blunt pseudopodia and feeds by phagocytosis of bacteria or other particles such as rice starch in culture (Figure 14.2a, b).

Electron microscopy shows a profoundly reduced but recognizable pattern of the trichomonad cell (Camp et al. 1974), with typical association of striated parabasal fibrils and Golgi dyctiosomes into the parabasal apparatus, presence of the extranuclear mitotic spindle, and double-membrane-bound organelles corresponding to hydrogenosomes. *Dientamoeba* lacks kinetosomes, microtubular structures such as axostyle and pelta, and

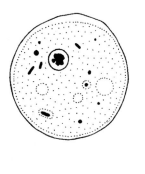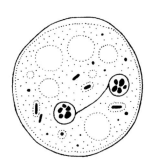

Figure 14.1 Dientamoeba fragilis: *uninucleate and binucleate trophozoites. Note the typical arrangements of chromatin granules within the nuclei and a fiber representing the extranuclear mitotic spindle (paradesmose) connecting the nuclei in the binucleate form. Based on iron-hematoxylin stained specimens. Bar = 10 μm.*

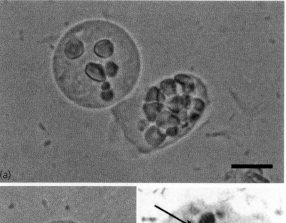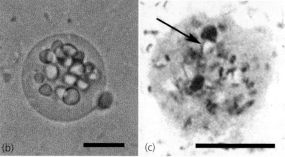

Figure 14.2 Dientamoeba fragilis: **(a, b)** *fresh mount from culture, trophozoites with ingested rice starch grains;* **(c)** *binucleate trophozoite with extranuclear spindle (arrow). Giemsa–Romanowski stain. Bar = 10 μm.*

any type of centriole. Microtubule organizing centers of the mitotic spindle are attached at the inner side of the base of parabasal fibrils that assume polar position in the mitotic cells. The cytoplasm contains glycogen granules and numerous food vacuoles (secondary lysosomes) with ingested particles in different phases of digestion. Mitochondria are absent.

Biochemistry and molecular biology

No experimental data are available on the biochemistry of *Dientamoeba*. The presence of hydrogenosomes and susceptibility to metronidazole suggest that the pyruvate-metabolizing pathway linked to ferredoxin-mediated electron transport known to operate in the hydrogenosome of trichomonads, is also functional in *Dientamoeba*.

The gene for small subunit ribosomal RNA (SSU rRNA) was cloned and sequenced for phylogenetic studies (Silberman et al. 1996) and fragments of this gene obtained by polymerase chain reaction amplification with trichomonad-specific primers, were used for riboprinting of *Dientamoeba* isolates (Johnson and Clark 2000). The genotyping of *Dientamoeba* based on direct amplification from human stool samples was reported by (Peek et al. 2004).

Clinical aspects

Despite controversies about the primary role of *D. fragilis* in the etiology of gastrointestinal disorders, sufficient evidence is available to demonstrate that the organism is pathogenic (reviewed by Yang and Scholten 1977; Ockert 1990; Windsor and Johnson 1999). Symptomatic infections with *Dientamoeba* occur in both children and adults. The most frequent symptoms are abdominal pain, intermittent diarrhea, flatulence, and fatigue, sometimes accompanied by nausea, vomiting, and anorexia. The disease usually has a chronic, recurrent character and often is associated with a peripheral eosinophilia that is more common in children than in adults (Spencer et al. 1979, 1982). Acute diarrhea or cases presenting as eosinophilic or ulcerative colitis have also been observed (Cuffari et al. 1998; Shein and Gelb 1983). Disappearance of symptoms after elimination of *Dientamoeba* by chemotherapy from patients where *D. fragilis* was the only pathogen detected, have been reported repeatedly (Kean and Malloch 1966; Spencer et al. 1979; Ockert 1990). *Dientamoeba* can cause diarrhea in those with acquired immune deficiency syndrome (AIDS). Although the prevalence of *Dientamoeba* infection is higher in groups that are positive for human immunodeficieny virus (HIV) than in groups not infected with HIV (Mendez et al. 1994), there is no conclusive evidence that the pathogenic effect is more severe in HIV-positive patients than it is in HIV-negative individuals (Lainson and da Silva 1999).

Immunology

No systematic study on the immune response to the infection with *D. fragilis* has been published so far. Peripheral eosinophilia has been reported repeatedly, but its significance is not known. Chan et al. (1996) employed indirect immunofluoresence assay in a seroprevalence study of *Dientamoeba* infection in children and confirmed by immunoblotting with representative sera that the dominant antibody response is directed against a 39 kDa protein of *D. fragilis*.

Pathology

Histopathologic studies on appendices removed from young patients infected with *Dientamoeba* showed changes in the intestinal wall that varied from suppurative appendicitis to lymphoid hyperplasia and marked fibrosis in connective tissue of the submucosa (Burrows et al. 1954). The authors concluded that *D. fragilis* is noninvasive and assumed that irritation caused by the protracted presence of the parasite can induce low-grade inflammatory changes in the appendix that ultimately lead to fibrosis. In contrast, Shein and Gelb (1983) observed by sigmoidoscopy an invasive ulcerative process with numerous punctiform ulcers in the intestine of a woman infected with *Dientamoeba*. Flat ulcers and both acute and chronic inflammatory changes were demonstrated on biopsy but *Dientamoeba* was absent in aspirates from the areas of ulceration. According to Ockert (1990) inflammatory changes of the mucosa of rectum and sigmoid region of colon were relatively frequent in patients infected with *D. fragilis*, usually displaying a picture of low-grade chronic proctitis.

Diagnosis

Infection by *D. fragilis* should be considered in differential diagnosis in patients with abdominal pain, diarrhea, flatulence, vomiting, and unexplained eosinophilia. Unfortunately, awareness of the infection is low and methods used in many laboratories are unsuitable for the detection of this parasite. Examination of permanently stained smears prepared from fresh stools is mandatory for successful laboratory diagnosis of *Dientamoeba*; cultivation significantly increases parasite detection. It is essential that the fecal specimens are preserved and fixed or inoculated in suitable media soon after defecation, as survival time of *Dientamoeba* trophozoites is short. The collected specimens must not be refrigerated. Routine coprological concentration techniques are useless because *D. fragilis* does not form cysts, and trophozoites are destroyed during the procedure. Chan et al. (1993) reported employment of indirect immunofluoresence to facilitate detection of *D. fragilis* in stools.

Schaudin/polyvinyl alcohol fixative/preservative followed by iron hematoxylin or trichrome stain or a non-mercury-based alternative EcoFix followed by EcoStain (Garcia and Shimizu 1998), as recommended for differentiation of intestinal amebae, are suitable methods for microscopic diagnosis of *Dientamoeba*. The main diagnostic characters are nuclear morphology and presence of binucleate cells. The extranuclear spindle can be recognized occasionally and, if visible, can support the diagnosis unequivocally.

Diagnostic cultivation can be performed in the simple biphasic medium of Dobell and Laidlaw or in the more complex Robinson's medium (Clark and Diamond 2002). The Dobell–Laidlaw biphasic medium is essentially a slant of coagulated horse serum overlaid with Ringer's solution containing native egg white and supplemented with rice starch powder. By using this medium, Ockert (1990) reported an increase in detection efficiency in a large (over 1000) random sample of people in Halle, Germany from c. 2 percent diagnosed from stained fecal smears alone to c. 39 percent diagnosed from a combination of microscopy and cultivation. *Dientamoeba* usually multiplies in a primary culture and often survives for several subcultures, but successful establishment of a long-term culture is rare. Robinson's medium may facilitate establishment of cultures as it is designed to exert partial control over the concomitant bacterial flora. It is preconditioned by inoculation of a standard strain of *Escherichia coli* and unwanted associates are suppressed by erythromycin added to the medium at each subculture.

Epidemiology

DISTRIBUTION

Infections with *D. fragilis* occur worldwide (Windsor and Johnson 1999) with higher prevalence in children than in adults, and in women than in men. Although some published data are difficult to compare due to different sensitivities of the detection methods used, it is apparent that the infection is relatively frequent. In studies based on large samples of stools examined for parasites over 2–5 years, the incidence of the infection varied from 4 to 30 percent (Yang and Scholten 1977; Ockert 1990; Windsor et al. 1998). Prevalence studies in Germany and the USA (Ockert 1990; Grendon et al. 1995) showed the highest frequency of infection in children aged 6–9 years (47 percent), with variations between 10 and 82 percent observed in individual communities of children in Germany (Ockert 1990). Higher infection rates are usually linked to crowded conditions and poor standards of personal hygiene. Reports on the highest prevalence in adults (up to 50 percent) concern inmates of certain mental institutions and other closed or semicommunal groups (Wenrich 1944; Windsor and Johnson 1999). Interestingly, infections by the cyst-forming intestinal protozoa are most prevalent in a different age group (3–5-year-olds) and co-infections with *Dientamoeba* occur less frequently than those among other intestinal flagellates and amebae. Reported high prevalence and differences in age and sex distribution of subjects infected with *Dientamoeba* and other intestinal protozoa suggest that factors involved in *D. fragilis* transmission are unique to this species.

TRANSMISSION

The transmission of *Dientamoeba* is not completely resolved. The relatively high prevalence of the parasite is in sharp contrast to the absence of a resistant cyst stage and the fragility of the trophozoite, which does not survive in water (Wenrich 1944), and its viability in feces decreases rapidly. Moreover, the trophozoites were found to die in a solution simulating gastric juice (Yang and Scholten 1977), and attempts to infect a human volunteer and primates with trophozoites from culture given per os were unsuccessful (Dobell 1940). Thus, transmission by fecally contaminated food or water as well as direct fecal–oral transmission is unlikely.

In analogy with the closely related poultry pathogen *Histomonas meleagridis*, dissemination of *D. fragilis* infection by eggs of a helminth vector was proposed (Dobell 1940) and indirect evidence has been accumulated that the pinworm *Enterobius vermicularis* might be the hypothetical vector (Burrows and Swerdlow 1956; Yang and Scholten 1977; Ockert 1990). Association of infections with *D. fragilis* and pinworm has been observed at histological examination of human appendices (Burrows and Swerdlow 1956; Ockert 1990; Červa

et al. 1991) as well as by routine examination for parasites, if proper detection methods were used (Yang and Scholten 1977; Ockert 1990). Statistical treatment confirmed that the frequency of these co-infections is significantly higher than would be expected at random, independent distribution of the parasite. Also, the increased prevalence of *Dientamoeba* in young people and women reflects a similar distribution of pinworm infections. Moreover, Ockert (1990) performed successful experimental self-infections with *D. fragilis* by ingesting washed pinworm eggs obtained from children infected with both *Enterobius* and *Dientamoeba*. There is also a report on *Dientamoeba* infection that followed accidental acquisition of enterobiosis by a person collecting perianal samples for pinworm diagnosis (Burrows and Swerdlow 1956).

Burrows and Swerdlow (1956) observed ameboid cells resembling mononuclear *Dientamoeba* trophozoites in eggs of pinworm females from subjects infected with *D. fragilis*, but not in worms from patients free from *Dientamoeba* infection. Ockert (1990) confirmed the presence of ameboid cells in the lumen of pinworm eggs and inside larvae encapsulated in the eggs. He further showed that these objects and *Dientamoeba* trophozoites from culture have the same value for isoelectric point after methylene blue staining. However, unequivocal identification of *Dientamoeba* in pinworms is lacking and attempts to isolate this parasite from pinworm eggs or larvae in culture were unsuccessful (Burrows and Swerdlow 1956; Yang and Scholten 1977; Ockert 1990).

Chemotherapy

Symptomatic *Dientamoeba* infections have been successfully treated with iodoquinol, metronidazole, secnidazole, chloramphenicol, tetracycline, and paromomycin (see Johnson et al. 2004 for references). Many authors consider iodoquinol (di-iodohydroxyquinoline) as the drug of choice for adults. Metronidazole and paromomycin have been used with success in children (Preiss et al. 1991; Cuffari et al. 1998).

It is difficult to recommend control measures against *Dientamoeba* infection without full knowledge of the factors involved in its transmission. However, strong indications for pinworm involvement suggest that good personal hygiene, such as that recommended for the prevention of enterobiosis, may be appropriate. Concurrent treatment for both *Enterobius* and *Dientamoeba* should also be considered when a patient infected with *Dientamoeba* belongs to a community where pinworm infections have been diagnosed.

Conclusions

Dientamoeba is a neglected pathogen that deserves higher awareness from diagnostic laboratories, clinicians and researchers. The infection rate of *Dientamoeba* all over the world appears to be higher than originally thought, but successful diagnosis depends heavily on proper collection and processing of fecal specimens. The enteropathogenic potential of *Dientamoeba* for both children and adults has been documented and efficient chemotherapy is available. Therefore, it would be desirable to consider *D. fragilis* infection more consistently in differential diagnosis of gastrointestinal disorders and laboratories should adapt their diagnostic protocols to ensure dependable diagnosis of the parasite. The unsolved problems of the parasite life cycle and transmission should be revisited, as new techniques are now available to face this challenge.

ENTEROMONAS HOMINIS

Classification

Enteromonas is the type genus of the suborder Enteromonadina, which comprises uninucleate members of the protozoan order Diplomonadida. An ultrastructural homology of enteromonads with higher Diplomonadida such as *Hexamita* is evident (Brugerolle 1975). By topology of cell structures and organelles, *Enteromonas* resembles one half of a typical diplomonad. An evolutionary route from simple enteromonad-like forms to organisms with the double karyomastigont characteristic of higher Diplomonadida has been proposed and supported by a phylogenetic analysis based on ultrastructural characters (Siddal et al. 1992).

Structure and life cycle

Enteromonas hominis da Fonseca, 1915 is a small, cyst forming flagellate, inhabiting the cecum and other parts of the large intestine of humans and primates (Dobell 1935). The broadly oval trophozoites measure approximately 4–8 μm by 3–5 μm and possess three anterior flagella and one recurrent flagellum attached along two-thirds of the body length and continuing thereafter with a free portion. The adherent part of the recurrent flagellum is accompanied by a darkly staining fibril. The oval nucleus is located in the anterior part of the body, with its longer axis perpendicular to the longitudinal axis of the cell (Figures 14.3a and 14.4). The cysts are oval or subspherical with one to four nuclei; four nucleate cysts are considered mature.

Electron microscopy (Brugerolle 1975) revealed the presence of a narrow, shallow cytostomal groove on the ventral surface of the cell that follows the attached portion of recurrent flagellum. The lateral margins of the groove are elevated and each reinforced by a microtubular band descending from the kinetosomal area. The recurrent flagellum is lodged inside the cytosomal depression. *Enteromonas* feeds on bacteria that are endocytosed in the bottom of the cytostomal groove all along its length; cytopharyngeal tube is not developed. The cytoplasm contains numerous food vacuoles and rough endoplasmic reticulum, but mitochondria, Golgi apparatus, and microbodies are absent.

Diagnosis

Cultivation on biphasic media of the Dobell-Laidlaw type is the best method for diagnosis of *Enteromonas hominis*. Identification of this organism in stools may be difficult even on permanently stained smears processed by routine techniques. The protargol silver stain can help with accurate determination (Figure 14.4).

Epidemiology

DISTRIBUTION

Infections with *Enteromonas hominis* occur worldwide, but their incidence is usually low. An increased prevalence in a population (up to 10 percent) indicates poor hygienic standards.

POTENTIAL PATHOGENICITY AND TREATMENT

Enteromonas is generally considered to be a harmless commensal and treatment is not recommended. However, occasional cases of symptomatic infections in the absence of detectable pathogens have been reported (Spriegel et al. 1989; Goldberg 1990). As symptoms resolved when parasites were eliminated by treatment with metronidazole, the authors advocate use of this drug in similar cases.

CHILOMASTIX MESNILI

Classification

The genus *Chilomastix* belongs to the protozoan order Retortamonadida. Besides the human parasite, the genus includes other species that inhabit the intestines of numerous vertebrate and some invertebrate hosts. A free-living species has also been described.

Structure and life cycle

Chilomastix mesnili (Wenyon, 1910) Alexeieff, 1912 is a cyst-forming flagellate living in the cecum and colon of humans. The trophozoites are pyriform with a rounded or flattened anterior end and a tapered posterior projection (Figures 14.3b and 14.5). They measure 14 × 6 μm on average, ranging from 6–20 μm in length and 4–7 μm in width. *Chilomastix* possesses three anterior flagella and one recurrent flagellum confined to the cytostomal cavity. The cytostome forms a large cleft in the dorsal cell surface, extending from the anterior end for half the body length. Two cytostomal fibrils support

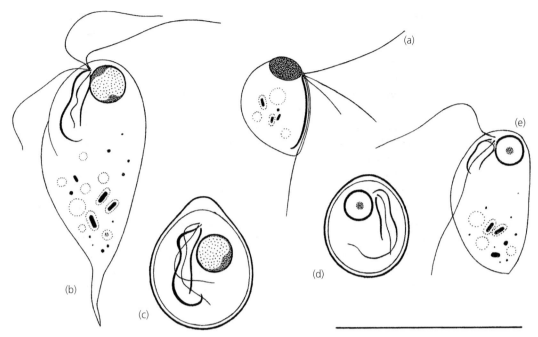

Figure 14.3 *Flagellates from the large intestine of humans, semidiagrammatic representation based on protargol stained specimens:* **(a)** Enteromonas hominis *trophozoite;* **(b, c)** Chilomastix mesnili: **(b)** *trophozoite,* **(c)** *cyst;* **(d, e)** Retortamonas intestinalis: **(d)** *cyst,* **(e)** *trophozoite. Bar = 10 μm.*

the edges of the cytostome. The fibril associated with the right cytostomal lip is prominent, characteristically curved, and turned round the posterior end of the cleft toward the tube of cytopharynx. The fibril of the left lip is straight and less conspicuous. The spherical nucleus is located dorsally at the anterior margin of the cell. The cysts are lemon-shaped with a broadly rounded posterior end and narrowed anteriorly. They measure 6–10 by 4–6 μm and contain a single nucleus, internalized flagella and both cytostomal fibrils (Figure 14.5).

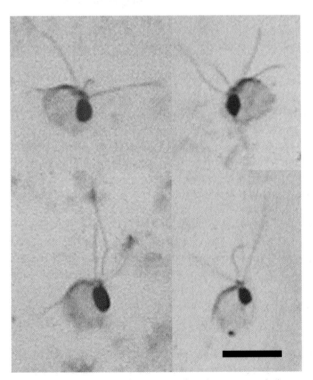

Figure 14.4 Enteromonas hominis *trophozoites. Protargol silver strain. Bar = 5 μm.*

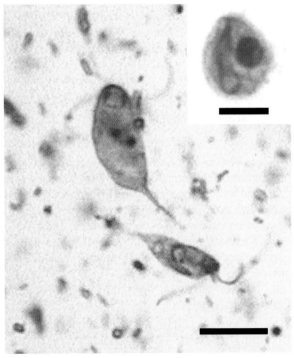

Figure 14.5 Chilomastix mesnili *trophozoites. Protargol silver stain. Bar = 10 μm. Insert:* Chilomastix mesnili *cyst. Iron-hematoxylin stain. Bar = 5 μm.*

Electron microscopy revealed details of components and the assembly of cytoskeletal structures associated with kinetosomes, cytostome, and cytopharynx, and showed that the cell is completely surrounded by a corset of subpellicular microtubules (Brugerolle 1973). The trophozoite cytoplasm contains numerous food vacuoles and rough endoplasmic reticulum but lacks mitochondria, Golgi apparatus, and any type of micro-body-like organelles. Endocytosis of bacterial food takes place in a microtubule-free area of the cytopharynx.

Potential pathogenicity

It is generally accepted that *C. mesnili* is nonpathogenic and its association with diarrhea does not reflect a causal relationship. Therefore, no specific therapy is recommended. Occasional case reports ascribing pathogenic potential to this organism are not very convincing, excepting a single well-documented study. Červa and Větrovská (1958) investigated the etiology of periodic outbreaks of watery diarrhea in a community of institutionalized children aged 1–3 years and concluded that *Chilomastix* was responsible for the disease. Affected children passed four to ten stools in a day; diarrhea was painless but caused a moderate dehydration, decreased weight gain or loss of weight. Infectious origin of the outbreaks was evident and linked to unsatisfactory hygienic conditions in common bathrooms. No bacterial pathogens were found despite repeated examination and symptoms were insensitive to antibacterial chemotherapy. When *Chilomastix* was detected as a potential agent of the infection, the children were treated with carbasone. The treatment resulted in the elimination of parasites and disappearance of symptoms, and marked improvement in weight gain followed. These observations suggest that the possibility of *Chilomastix* occasionally acting as moderate pathogen for small children should not be entirely ignored.

Diagnosis

Trophozoites can be found in soft or liquid stools, cysts predominate in the formed stools. Both trophozoites and cysts can be easily recognized on permanently stained smears by their characteristic morphology – in particular, by the presence of cytostome with the prominent, characteristically curved cytostomal fibril. Cultivation on biphasic media such as Dobell–Laidlaw medium can increase diagnostic recovery.

Epidemiology

DISTRIBUTION AND TRANSMISSION

Distribution of *Chilomastix mesnilli* in humans is cosmopolitan with incidence ranging from 0.2 to 1 percent in standard surveys. No marked differences can be seen in the results of surveys performed in Germany, USA, Mexico, Chile, Indonesia, and Australia. Increased prevalence of the infection has been reported in specific groups with a history of diarrhea; for example, in US troops deployed in Egypt (11 percent, Oyofo et al. 1997), closed communities of institutionalized children (40 percent, Červa and Větrovská 1958) or selected groups of diarrheic patients (12.3 percent, Felsenfeld and Young 1946). No significant differences have been observed in the infestation rate of *Chilomastix* in HIV-positive and HIV-negative groups (Mendez et al. 1994). The species of *Chilomastix* occurring in monkeys and pigs is probably identical with *C. mesnili* as indicated by morphology and by successful experimental transmissions of the human flagellate to these animals. However, natural infections of animals are infrequent; about 1 percent of pigs have been found infected in surveys performed in Europe and the USA.

Interhuman transmission of the *C. mesnili* infection doubtless is of primary importance. Infection of humans by cysts of animal origin cannot be ruled out, but has not been documented. Good sanitation and personal hygiene can prevent the infection.

RETORTAMONAS INTESTINALIS

Classification

Retortamonas is the type genus of the protozoan order Retortamonadida. The genus includes numerous species inhabiting the intestines of vertebrate and invertebrate hosts. A phylogenetic analysis based on sequences of the gene for SSU rRNA (Silberman et al. 2002) showed close relationship of retortamonads to Diplomonadida.

Structure and life cycle

Retortamonas intestinalis (Wenyon and O'Connor, 1917) Wenrich, 1932 is a rare inhabitant of the human cecum. It is a small cyst-forming flagellate with trophozoites of an elongated pyriform shape, measuring 4–8 by 3–4 µm and equipped with a cytostome (Figure 14.3e). The cytostomal cleft of oval shape is supported by two lateral fibrils of unequal length and covered anteriorly by a membranous roof. In routinely stained specimens, the margin of the roof together with both lateral fibrils may appear as a single fibril rounding the anterior edge of the cytostome and extending posteriorly along each side. A tube-like cytopharynx extends from the cytostomal groove into the cell interior and ingested bacteria are endocytosed at its end. *Retortamonas* possesses two flagella; the anterior flagellum is free, the recurrent flagellum passes through the cytostomal groove and after emerging from it continues as a relatively long free portion. The spherical nucleus is located at the anterior margin of the cell. Mitochondria, Golgi apparatus, and microbody-like organelles are absent. Arrangements of the *Retortamonas* cytoskeleton and other ultrastructural features of this parasite were described by Brugerolle (1977). The cysts are ovoid with a narrowed anterior end; they measure 4.5–7 by 3–4.5 μm and contain a single nucleus, internalized flagella, and cytostomal fibrils (Figure 14.3d).

Potential pathogenicity

Although the parasite has also been detected in diarrheic stools, there is no evidence of its pathogenicity. Therefore, no treatment is indicated. Metronidazole has been successfully used to clear the parasite from an unusual localization in the pancreatic juice of a patient with choledocholithiasis (Kawamura et al. 2001).

Diagnosis

Identification of trophozoites and cysts of *Retortamonas* on permanently stained smears stained by standard methods may be difficult but is possible. Small size, presence of the cytostome, and the spherical, apically localized nucleus can help in recognition of the organism even if the flagella are not clearly visible. Protargol silver stain can be used for exact determination. The best method for diagnosis is cultivation on biphasic media for intestinal protozoa and subsequent examination of cultured flagellates by phase contrast microscopy or in fixed and stained preparations.

Epidemiology

DISTRIBUTION AND TRANSMISSION

Infections of humans with *R. intestinalis* are infrequent, with a low prevalence even in groups living under conditions of poor personal hygiene and unsatisfactory sanitation (up to 2 percent). Autochthonous infections have been reported from all continents except Europe. Infection is alimentary, transmitted by cysts disseminated by fecal contamination. Retortamonads of similar morphology live in a variety of mammals, but zoonotic transmission is unlikely. The morphologically identical monkey parasite is considered to be conspecific with *R. intestinalis*, but attempts to infect monkeys by the human flagellate failed (Dobell 1935).

REFERENCES

Brugerolle, G. 1973. Etude ultrastructurale du trophozoite et du kyste chez le genre *Chilomastix* Alexeieff, 1910 (Zoomastigophora, Retortamonadida Grassé, 1952). *J Protozool*, **20**, 574–85.

Brugerolle, G. 1975. Etude ultrastructurale du genre *Enteromonas* da Fonseca, 1915 (Zoomastigophorea). *J Protozool*, **22**, 468–75.

Brugerolle, G. 1977. Ultrastructure du genre *Retortamonas* Grassi, 1879 (Zoomastigophorea, Retortamonadida Wenrich, 1931). *Protistologica*, **13**, 233–40.

Burrows, R.B., Swerdlow, M.A., et al. 1954. Pathology of *Dientamoeba fragilis* infections of the appendix. *J Trop Med Hyg*, **3**, 1033–9.

Burrows, R.B. and Swerdlow, M.A. 1956. *Enterobius vermicularis* as a probable vector of *Dientamoeba fragilis*. *Am J Trop Med Hyg*, **5**, 258–65.

Camp, R.R., Mattern, C.F.T and Honigberg, B.M. 1974. Study of *Dientamoeba fragilis* Jepps and Dobell, I. Electron microscopic observations of binucleate stages. II. Taxonomic position and revision of the genus. *J Protozool*, **21**, 69–82.

Chan, F.T.H., Guan, M.X. and Mackenzie, A.M.R. 1993. Application of indirect immunofluorescence to detection of *Dientamoeba fragilis* trophozoites in fecal specimens. *J Clin Microbiol*, **31**, 1710–14.

Chan, F., Steward, N., et al. 1996. Prevalence of *Dientamoeba fragilis* antibodies in children and recognition of a 39 kDa immunodominant protein antigen of the organism. *Eur J Clin Microbiol Infect Dis*, **15**, 950–4.

Clark, C.G. and Diamond, L.S. 2002. Methods for cultivation of luminal parasitic protists of clinical importance. *Clin Microbiol Rev*, **15**, 329–41.

Cuffari, C., Oligni, L. and Seiman, E.G. 1998. *Dientamoeba fragilis* masquerading as allergic colitis. *J Pediat Gastroenterol Nutr*, **26**, 16–20.

Červa, L. and Větrovská, G. 1958. Towards the question of pathogenic character of the flagellate *Chilomastix mesnili*. *Cesk Epid Mikrobiol Immunol*, **7**, 126–85.

Červa, L., Schrottenbaum, M. and Kliment, V. 1991. Intestinal parasites: a study of human appendices. *Folia Parasitol*, **38**, 5–9.

Dobell, C. 1935. Researches on the intestinal protozoa of monkeys and man. VII. On the *Enteromonas* of macaques and *Embadomonas intestinalis*. *Parasitology*, **27**, 564–92.

Dobell, C. 1940. Researches on the intestinal protozoa of monkeys and man. X. The life-history of *Dientamoeba fragilis*: Observations, experiments and speculations. *Parasitology*, **32**, 417–61.

Felsenfeld, O. and Young, V.M. 1946. The correlation of intestinal protozoa and enteric microorganisms of known and doubtfull pathogenicity. *Am J Dig Dis*, **13**, 233–4.

Garcia, L.S. and Shimizu, R.Y. 1998. Evaluation of intestinal protozoan morphology in human fecal specimens preserved on EcoFix: Comparison of Wheatley's Trichrome stain and EcoStain. *J Clin Microbiol*, **36**, 1974–6.

Goldberg, J. 1990. *Enteromonas hominis* incidence and diarrhea. *Am J Gastroenterol*, **85**, 480.

Grendon, J.H., Digiacomo, R.F. and Frost, F.J. 1995. Descriptive features of *Dientamoeba fragilis* infections. *J Trop Med Hyg*, **98**, 309–15.

Johnson, E.H., Windsor, J.J. and Clark, C.G. 2004. Emerging from obscurity: biological, clinical and diagnostic aspects of *Dientamoeba fragilis*. *Clin Microbiol Rev*, **17**, 553–70.

Johnson, J. and Clark, C.G. 2000. Cryptic genetic diversity in *Dientamoeba fragilis*. *J Clin Microbiol*, **38**, 4653–4.

Kawamura, O., Kon, Y., et al. 2001. *Retortamonas intestinalis* in the pancreatic juice of a patient with small nodular lesions of the main pancreatic duct. *Gastrointest Endoscopy*, **53**, 508–10.

Kean, B.H. and Malloch, C.L. 1966. The neglected ameba: *Dientamoeba fragilis*. A report of 100 'pure' infections. *Am J Dig Dis*, **11**, 735–46.

Lainson, R. and da Silva, B.A.M. 1999. Intestinal parasites of some HIV/seropositive individuals in North Brazil, with particular reference to *Isosopora belli* Wenyon, 1923 and *Dientamoeba fragilis* Jepps & Dobell, 1918. *Mem Inst Oswaldo Cruz*, **94**, 611–13.

Mendez, O.C., Szmulewicz, G., et al. 1994. Comparison between enteroparasite infestation rate in HIV-positive and negative populations. *Med-Buenos Aires*, **54**, 307–10.

Ockert, G. 1990. Symptomatology, pathology, epidemiology, and diagnosis of *Dientamoeba fragilis*. In: Honigberg, B.M. (ed.), *Trichomonads parasitic in humans*. New York: Springer, 293–306.

Oyofo, B.A., Peruski, L.F., et al. 1997. Enteropathogens associated with diarrhea among military personnel during Operation Bright Star 96, in Alexandria, Egypt . *Military Med*, **162**, 396–400.

Peek, R., Reedeker, F.R. and van Gool, T. 2004. Direct amplification and genotyping of *Dientamoeba fragilis* from human stool specimens. *J Clin Microbiol*, **42**, 831–5.

Preiss, U., Ockert, G., et al. 1991. On the clinical importance of *Dientamoeba fragilis* infections in childhood. *J Hyg Epidemiol Microbiol Immunol*, **35**, 27–34.

Siddal, M.E., Hong, H. and Desser, S.S. 1992. Phylogenetic analysis of the Diplomonadida (Wenyon, 1926) Brugerolle, 1975 – Evidence for heterochrony in protozoa and against *Giardia lamblia* as a missing link. *J Protozool*, **39**, 361–7.

Shein, R. and Gelb, A. 1983. Colitis due to *Dientamoeba fragilis*. *Int J Gastroenterol*, **78**, 634–6.

Silberman, J.D., Clark, C.G. and Sogin, M.L. 1996. *Dientamoeba fragilis* shares a recent common evolutionary history with the trichomonads. *Mol Biochem Parasitol*, **76**, 311–14.

Silberman, J.D., Simpson, A.G.B., et al. 2002. Retortamonad flagellates are closely related to diplomonads – implications for the history of mitochondrial function in eukaryote evolution. *Mol Biol Evol*, **19**, 777–86.

Spencer, M.J., Garcia, L.S. and Chapin, M.R. 1979. *Dientamoeba fragilis*: An intestinal pathogen in children? *Am J Dis Child*, **133**, 390–3.

Spencer, M.J., Chapin, M.R. and Garcia, L.S. 1982. *Dientamoeba fragilis*: a gastrointestinal protozoan infection in adults. *Am J Gastroenterol*, **77**, 565–9.

Spriegel, J.R., Saag, K.G. and Tsang, T.K. 1989. Infectious diarrhea secondary to *Enteromonas hominis*. *Am J Gastroenterol*, **84**, 1313–14.

Wenrich, D.H. 1944. Studies on *Dientamoeba fragilis* (Protozoa). IV. Further observations, with an outline of present-day knowledge of this species. *J Parasitol*, **30**, 322–38.

Windsor, J.J. and Johnson, E.H. 1999. *Dientamoeba fragilis*: the unflagellated human flagellate. *Br J Biomed Sci*, **56**, 293–306.

Windsor, J.J., Rafay, A.M., et al. 1998. Incidence of *Dientamoeba fragilis* in faecal samples submitted for routine microbiological analysis. *Br J Biomed Sci*, **55**, 172–5.

Yang, J. and Scholten, T.H. 1977. *Dientamoeba fragilis*: a review with notes on epidemiology, pathogenicity, mode of transmission, and diagnosis. *Am J Trop Med Hyg*, **26**, 16–22.

15

Balantidium coli

VIQAR ZAMAN AND F.E.G. COX

HISTORICAL INTRODUCTION

Balantidium coli is the only ciliate known to infect humans and can cause dysentery. Asymptomatic infections are not uncommon, but actual disease is rare and fewer than 1 000 cases of balantidial dysentery have ever been reported. In contrast, infection in pigs is extremely common and it is generally believed that pigs act as the main reservoir for human infection. The subject has been reviewed by Zaman (1993).

Historically, the genus *Balantidium* was first recognized by Claparède and Lachmann (1858–61), who found these ciliates in the rectum of frogs. Since then *Balantidium* species have been found in a variety of animals (Corliss 1979). In 1856 Petr Henrik Malmsten, a Swedish physician, found a ciliate in the dysenteric feces of two Swedish patients which, after wide consultation with Swedish, Norwegian, and Irish colleagues, he called *Paramoecium coli* (Malmsten 1857). This ciliate was indisputably *Balantidium coli* and in 1861 Rudolf Leuckart discovered a similar organism in the large intestine of a pig (Leuckart 1861). Shortly afterwards, Stein pointed out that the human parasite described by Malmsten and the pig ciliate described by Leuckart were morphologically identical and named them both *Balantidium coli* (Stein 1863). Several sightings of other ciliates from humans were made in the 19th and early 20th centuries (see Dobell and O'Connor 1921; Wenyon 1926) but these are no longer regarded as valid. Some early writers have suggested that Leeuwenhoek actually saw this ciliate and was even suffering from balantidial dysentery, suggestions that Dobell (1920, 1932) robustly dismisses. For many years there was some controversy concerning whether or not the species in humans, and pigs are the same and whether *B. coli* from pigs can infect humans, but there now seems to be no doubt that both of these statements are true (see Dobell and O'Connor 1921).

CLASSIFICATION

The genus *Balantidium* belongs to the phylum Ciliophora, class Litostomatea, order Vestibuliferida, and family Balantidiidae (Lee et al. 1985). McDonald (1922) proposed that the pig and human species should be separated as *B. suis* and *B. coli*. This separation has not been accepted, and the parasites in the two hosts continue to be regarded as a single species by most authors.

STRUCTURE AND LIFE CYCLE

Structure

B. coli has a trophozoite and a cyst stage. The trophozoite is actively motile and is the invasive stage; the cyst is the resistant form and the infective stage. The length of the trophozoite varies from 30 to 300 μm and the width from 30 to 100 μm (Sargeaunt 1971). The body shape is also variable, but is generally ovoid and slightly flattened on one side (Figures 15.1 and 15.2). At the

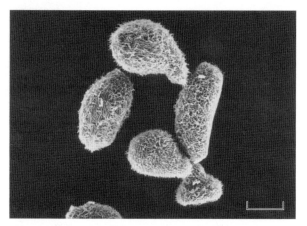

Figure 15.1 B. coli *trophozoites, showing variability in size and shape. SEM; bar = 30 μm*

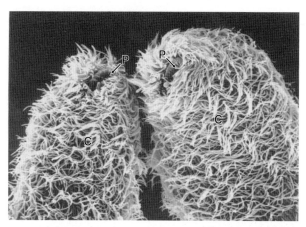

Figure 15.3 B. coli *trophozoites, showing peristome (P) and cilia (C) covering the body. SEM; bar = 3 μm*

anterior end of the trophozoite there is a funnel-shaped depression or mouth through which food is ingested. The mouth consists of a peristome, cytopharynx, and cytosome. The food passes from the peristome to the elongated cytopharynx and then to the cytosome. The cytopharynx, not visible when the parasite is filled with food, is often small but sometimes reaches about half the length of the parasite. The peristome appears as a rigid structure with striations when observed with the aid of a scanning electron microscope (SEM) (Figures 15.3 and 15.4).

The cilia, the organs of locomotion, cover the whole parasite and are embedded in the pellicle in longitudinal rows known as kineties. The number of kineties varies from 36 to 106 (Krascheninnikow 1962). The ciliary movement can be easily observed with the aid of a light microscope and the structure of the cilia is readily seen by SEM (Figures 15.3 and 15.4). The peristomal cilia are larger than those on the body (the somatic cilia). The peristomal cilia are used for propelling food into the cytopharynx (Figure 15.5). The food

particles on being ingested become surrounded by a vacuolar membrane and digestion takes place inside the vacuoles. *B. coli* is capable of ingesting a variety of food particles, such as bacteria, starch grains, red blood cells, fat droplets, etc. The parasite has an excretory opening at the posterior end known as the cytopyge, which is circular and much smaller than the peristome (Figure 15.6).

There are two contractile vacuoles which may lie side by side or one above the other. They are easily visible to an observer using a light microscope with interference or phase-contrast capability (Figure 15.7). These vacuoles are responsible for maintaining the proper osmotic pressure in the cell by drawing excess water from the cytoplasm and ejecting it to the exterior.

The trophozoite has two nuclei which are clearly visible in stained preparations (Figure 15.8). The macronucleus, which is large and situated near the middle of the body, may be spherical, curved, elongate, or kidney shaped (Figure 15.9) and is enclosed in a nuclear membrane from which it can be removed intact by

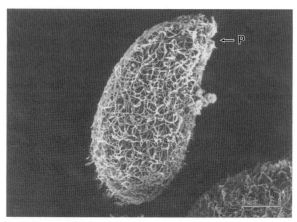

Figure 15.2 B. coli *trophozoite, showing peristome (P) at the anterior end. SEM; bar = 15 μm*

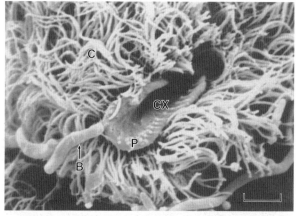

Figure 15.4 B. coli *trophozoite, showing peristome (P) with striations. A bacterium (B) is lying close to the peristome. CX, cytopharynx. SEM; bar = 1.5 μm*

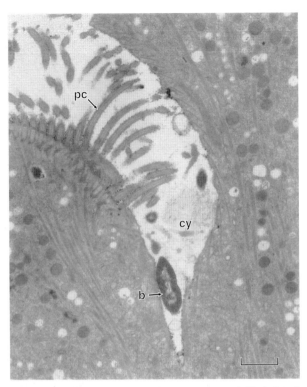

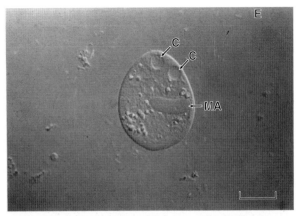

Figure 15.7 B. coli *trophozoite. Live organism showing macronucleus (MA) and 2 contractile vacuoles (C). Interference contrast; bar = 6 μm*

Figure 15.5 B. coli *trophozoite, showing peristomal cilia (pc) and cytopharynx (cy). A bacterium (b) is lying inside the cytopharynx. TEM; bar = 0.8 μm*

and the movement ceases. The encysted stage, like the trophozoite, has a macronucleus and a micronucleus.

Life cycle

The parasite is transmitted by the oral–fecal route and the cyst is the infective stage. Excystation probably occurs in the small intestine and multiplication occurs in the large intestine.

Multiplication is by binary fission and the earliest indication of it is the elongation of the organism. Elongation is followed by the formation of a transverse structure through the middle of the body. The body gradually begins to constrict and finally separates into two daughter individuals (Figure 15.14). The ciliary activity continues during this process; the anterior cell develops a new excretory pore and the posterior cell a new mouth.

Sexual union (syngamy) is an important aspect of this parasite's life cycle and occurs by a process of conjugation, in which two cells come in contact with each other at their anterior ends and exchange nuclear material

breaking the cell membrane (Figure 15.10). At the ultrastructural level numerous small nucleoli are visible inside the macronucleus (Figure 15.11). The single micronucleus is small and lies in close proximity to the macronucleus (Figures 15.8 and 15.9).

The cyst of *B. coli* is spherical to ovoid with a diameter of 40–60 μm (Figure 15.12). The cyst wall is thick and transparent. In electron microscopic images it is 400–500 nm thick (Figures 15.11 and 15.13). The parasite is visible inside the cyst and, in newly formed cysts, shows movement but, as the cyst matures, the cilia are absorbed

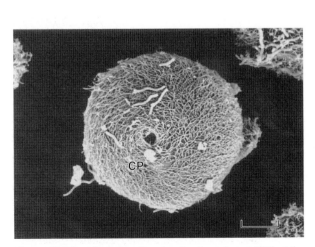

Figure 15.6 B. coli *trophozoite, showing the cytopyge (CP) at the posterior end. SEM; bar = 1.5 μm*

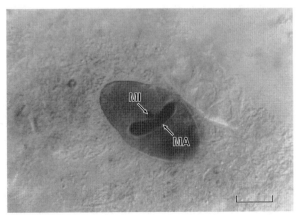

Figure 15.8 B. coli *trophozoite, showing macronucleus (MA) and micronucleus (MI). Trichrome stain; bar = 6 μm*

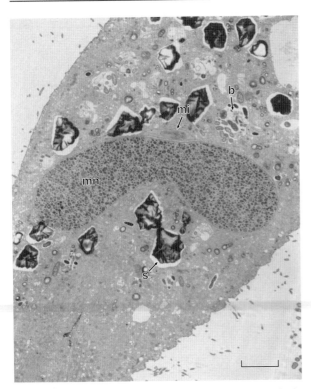

Figure 15.9 B. coli *trophozoite, showing the macronucleus (mn) and a small spindle-shaped micronucleus (mi). Ingested bacteria (b) and starch (s) are seen in the cytoplasm. TEM; bar = 3 μm*

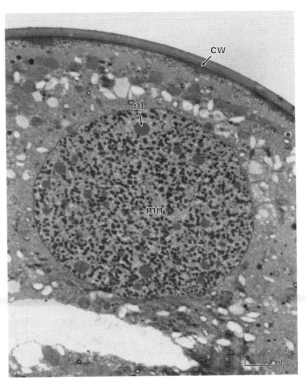

Figure 15.11 B. coli *cyst, showing the cyst wall (cw), macronucleus (mn) and nucleoli (nl). TEM; bar = 1 μm*

(Figure 15.15). Conjugation lasts for a few moments, after which the cells detach. There is no increase in numbers as a result of conjugation.

In the infected animal the parasite may be passed in the feces as a trophozoite or a cyst. The trophozoite does not encyst outside the body and disintegrates. The passed cyst survives and may contaminate food and water and may then be passed to other animals or humans. Pig-to-pig transmission is very common, and virtually 100 percent infection occurs in some piggeries where hygienic conditions are poor.

BIOCHEMISTRY

Little is known about the biochemistry of *B. coli*. Agosin and von Brand (1953) found that the ciliate is capable of consuming considerable amounts of oxygen despite the fact that it normally lives in the large intestine where little if any oxygen is available. The parasite prefers anaerobic conditions, however, and in the absence of oxygen it produces a large amount of carbon dioxide. Carbohydrates are the main source of energy. Templis and Lysenko (1957) found that *B. coli* can produce hyaluronidase, which probably helps it in its invasion of

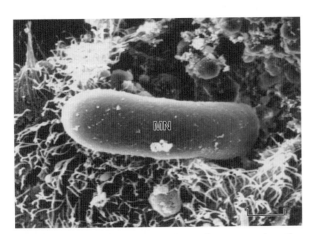

Figure 15.10 B. coli *trophozoite, showing the macronucleus after disruption of the cell membrane. SEM; bar = 3 μm*

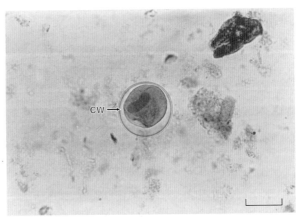

Figure 15.12 B. coli *cyst. The transparent cyst wall (CW) and the macronucleus are visible. Trichrome stain; bar = 15 μm*

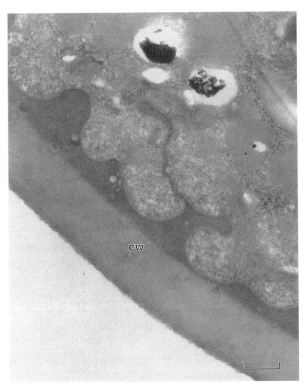

Figure 15.13 B. coli *cyst, showing the cyst wall (CW) at high magnification. TEM; bar = 0.2 μm*

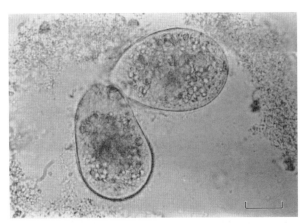

Figure 15.15 B. coli *undergoing conjugation. Attachment occurs at the anterior end. Live preparation; bar = 15 μm*

Chronic disease

Patients with chronic disease have periods in which frequent bowel movements alternate with periods of constipation. The feces are mucoid and rarely bloody. The organism is not easily seen in the feces and repeated stool examination may be necessary for its detection.

Acute disease

In acute disease, patients have diarrhea and the feces contain a great deal of mucus and blood (Castro et al. 1983). The clinical presentation is identical to that of acute amoebic dysentery. There may be fever and other intestinal symptoms such as anorexia, nausea, epigastric pain, vomiting, and intestinal colic. This may lead to severe dehydration, and sometimes to renal insufficiency. There is usually pain and tenderness in the cecal region. The symptoms may mimic those of appendicitis. In a majority of patients recovery occurs in 3–4 days even without treatment. In some patients, especially immunocompromised and malnourished ones, death may occur due to extensive destruction of the large intestine, involvement of the appendix, perforation, peritonitis, dehydration, and renal failure. In patients with acute infection, extraintestinal involvement such as liver abscess formation, pleuritis, and pneumonia may occur.

Immunology

There is no information about any immune responses to infection with *B. coli*. There is some evidence of age-acquired immunity in rhesus macaques where the parasite counts decrease significantly with age (Knezevich 1998). In some studies only children have been found to be infected (Devera et al. 1999), suggesting the acquisition of immunity, but this could be due to other factors such as changes in behavior. There has been a single

the host tissues by dissolution of the intracellular ground substance.

CLINICAL ASPECTS

Asymptomatic carrier condition

Asymptomatic carriers harbor the parasites and continue to pass cysts in their feces. They are responsible for spreading infection, especially in insanitary institutional environments, such as long-stay psychiatric units.

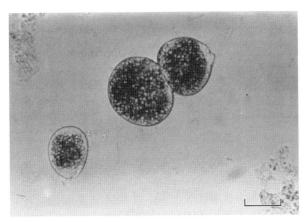

Figure 15.14 B. coli *undergoing binary fission. Live preparation; bar = 15 μm*

record of balantidial dysentery in an individual with HIV infection, and the authors suggest that the switch from asymptomatic infection to clinical dysentery might be due to immunosuppression resulting from the viral infection (Clyti et al. 1998). In this context it is interesting to note that some rhesus macaques infected with the simian immunodeficiency virus (SIV) were found to be infected with *B. coli* although the cause of diarrhea in these animals was primarily *Escherichia coli* infection (Mansfield et al. 2001). The rarity of severe human infections and the ease with which they can be cured mean that there is no need for a vaccine. The presence of conspicuous and easily identifiable cysts means that there is no need for an immunodiagnostic test.

Pathology

The gross pathologic appearance of the large intestine in patients with *Balantidium* infection is similar to that in patients with amoebiasis. On rare occasions the terminal ileum may also be infected. The gross changes consist of multiple ulcers with necrotic bases and undermined edges. The intervening mucosa may or may not be inflamed. On microscopic examination, parasites are frequently seen in clusters in the submucosa or at the bases of crypts. They can easily be recognized because of the presence of the macronucleus which stains deeply with hematoxylin and eosin (Figure 15.16). The cellular response is mainly lymphocytic, with some plasma cells being present. Neutrophils are few unless there is a superimposed bacterial infection. Sometimes the parasites may invade the regional lymph nodes and then they may be detected inside the lymphatic tissues. Extraintestinal balantidial infections are rare; there has been one report of *B. coli* in the bladder (Maleky 1998), but this report must be regarded with caution given our knowledge of ciliocytophthoria mentioned below.

The factors that determine whether or not *B. coli* invades the gut wall are unclear. However, in pigs, there is evidence that the presence of *Campylobacter coli*

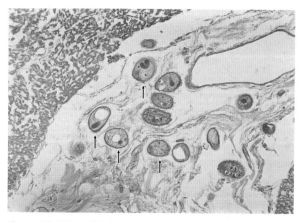

Figure 15.16 B. coli *in intestine. Many trophozoites (arrows) are lying in the submucosa. Hematoxylin and eosin; bar = 30 μm*

facilitates the invasion of the mucosa and both organisms seem to participate in the colitis associated with the infection (Hosino et al. 1999).

Diagnosis

Diagnosis of balantidiasis on clinical grounds is very difficult because of the nonspecific nature of the infection and possible confusion with other causes of diarrhea. Diagnosis is therefore best based on fecal examination, which reveals mainly trophozoites in acutely infected patients and cysts in patients with chronic infections. Diagnosis can also be made by the examination of biopsy specimens taken with the aid of a sigmoidoscope or by examination of scrapings of an ulcer. *Balantidium* can be cultured in all the media that support the growth of *Entamoeba histolytica* (Clark and Diamond 2002). However, culture is rarely necessary as the parasites are more easily detected in feces by microscopy and in tissues on histological examination (see Orihel and Ash 1995; Gutierrez 2000). One possible cause of confusion is the phenomenon of ciliocytophthoria resulting in normal ciliated cells appearing in unusual circumstances and there have been cases where such specimens have been tentatively identified as *B. coli* (Hadziyannis et al. 2000). These authors advocate the use of the Diff-Quik stain for the diagnosis of ciliocytophthoria.

EPIDEMIOLOGY

B. coli is worldwide in its distribution, particularly in the tropical and subtropical regions of the world including South and Central America, the Philippines, Iran, Central Asia, Papua New Guinea, and some Pacific islands. Although this ciliate is widely regarded as a common parasite of pigs that is transmissible to humans, it actually has a much broader host range and numerous surveys have shown that it occurs in a number of mammal species, both wild and domesticated. In a major survey of 297 Old World primates in Kenya, *B. coli* was detected in 6 out of 7 species with an overall prevalence of 24.9 percent (Muriuki et al. 1998). Also in Kenya, *B. coli* was found in 40 percent of 315 fecal samples from captive and wild-trapped baboons, vervets, and Sykes monkeys (Munene et al. 1998). In the New World *B. coli* was prevalent in 141 rhesus macaques in a free-ranging population (Knezevich 1998). *B. coli* is also common in camels, for example 11.9 percent of 260 fecal samples examined by Tekle and Abebe (2001) in Ethiopia and in young camels suffering from diarrhea in Bahrain (Abubakr et al. 2000). Buffaloes are also known to be infected with *B. coli*, and Patil et al. (1998) have recorded a 17 percent prevalence in 82 buffaloes suffering from diarrhea and dysentery in India. In a major survey of 567 mammalian species belonging to 17 families in Japan *B. coli* was found in 6 species of primates and artiodactyles but not in rodents, cats, or dogs (Nakauchi 1999). However, there have been occasional records of *B. coli* in dogs (Uptal Das 1999). Although there is little doubt that pigs are the main potential reservoirs of human infections, transmission to humans varies from place to place. In a survey of pigs at a Danish research farm, 57 percent of suckling pigs and 100 percent of pigs >4 weeks old were found to be infected with *B. coli*, presumably representative of this major pig-producing country as a whole, but no case of human balantidiasis has been recorded in Denmark (Hindsbo et al. 2000). In contrast, in a rural community in Venezuela, *B. coli* was detected in 33.3 percent of pigs and 12 percent of children, a situation that the authors attribute to deficient sanitation and a low socio-economic level (Devera et al. 1999). In a larger study of 2 124 children in northern Bolivia, Esteban et al. (1998) found that the overall prevalence of *B. coli* was 1.2 percent, again attributed to contact with pigs. In China the parasite appears to be endemic in Yunnan province, with infection rates of up to 4.24 percent in some villages (Yang et al. 1995). However, the highest prevalence (up to 20 percent) has been reported in the mountain districts of West Irian (Indonesia) where there is a close association between humans and pigs. McCarey (1952) has reviewed the prevalence of balantidiasis in Iran – an endemic area of unusual interest, as there is no pig farming there because of the Moslem prohibition on eating pork. The transmission was from human to human and the animal host, if any was not known; the possibility of infection acquired from camels was not considered as it was not known at that time that these acted as potential reservoir hosts.

There have been occasional sporadic outbreaks of balantidiasis in humans. A single outbreak of epidemic proportions was reported from the Pacific island of Truk (Walzer et al. 1973). It involved 110 persons in a short period of time and occurred as a result of contamination of the water supply by pig feces during a typhoon. Balantidiasis has been frequently observed in psychiatric units in the USA and many other countries. Here the transmission occurs by human-to-human contact and is due to a lack of hygienic conditions and to coprophagy.

CHEMOTHERAPY

A number of drugs, including metronidazole, secnidazole, tinidazole, tetracycline, trimethoprim plus sulfamethoxazole, sulfaguanidine, and sulfadimidine, are effective against *B. coli* in vitro (Choudhury et al. 1998), so there is no shortage of potentially useful drugs to treat humans or pigs. At present tetracycline and metronidazole are the drugs of choice for humans. Tetracycline, 500 mg, 4 times a day for 10 days, is recommended although clearance generally occurs in 2–3 days (WHO 1995). Alternatively, metronidazole, 750 mg, 3

times a day for 5–7 days, can be used but, according to WHO, results are less consistent (WHO 1995). No relapse of the infection after treatment and no resistance to these antibiotics have been reported. Treatment should be given to carriers in institutions to prevent the spread of infection to susceptible patients.

INTEGRATED CONTROL

Control consists of hygienic rearing of pigs and preventing the human–pig contact which can lead to human infections. Improved hygiene in psychiatric institutions will prevent human-to-human transmission in these settings, and treatment of humans shedding cysts will prevent human-to-human transmission. The relative lack of host specificity and close contact with other potential reservoir hosts, such as camels and buffalo, suggests that infection with *B. coli* is an ever-present problem and also that individuals working with or coming into contact with primates are at risk.

REFERENCES

Abubakr, M.I., Nayel, M.N., et al. 2000. Prevalence of gastrointestinal parasites in young camels in Bahrain. *Rev Elevage Méd Vét Pays Trop*, **53**, 267–71.
Agosin, M. and von Brand, T. 1953. Studies on the respiratory metabolism of *Balantidium coli*. *J Infect Dis*, **93**, 101–6.
Castro, J., Vazquez-Iglesias, J.L. and Arnal-Monreal, F. 1983. Dysentery caused by *Balantidium coli* – Report of two cases. *Endoscopy*, **15**, 272–4.
Choudhury, R., Sarmah, P.C. and Borkakoty, M.R. 1998. In vitro effect of some drugs on the growth and viability of Balantidium coli. *J Vet Parasitol*, **12**, 98–102.
Claparède, J. and Lachmann, J. 1858–1861. Etudes sur les infusoires et les rhizopodes. *Mém Inst Nat Genevois*, **5**, 1–260, **6**, 261–482; **7**, 1–291.
Clark, G.C. and Diamond, L.S. 2002. Methods for the cultivation of luminal protists of clinical importance. *Clin Microbiol Rev*, **15**, 329–41.
Clyti, E., Aznar, C., et al. 1998. Un cas de co-infection par *Balantidium coli* et VIH en Guyane Francaise. *Bull Soc Pathol Exot*, **91**, 309–11.
Corliss, J.O. 1979. *The ciliated protozoa*, 2nd edn. Oxford: Pergamon Press.
Devera, R., Requena, I., et al. 1999. Balantidiasis en una comunidad rural del Estado Boliver, Venezuela. *Bol Chil Parasitol*, **54**, 7–12.
Dobell, C. 1920. The discovery of the intestinal protozoa of man. *Proc Roy Soc Med*, **13**, 1–15.
Dobell, C. 1932. *Antony Van Leeuwenhoek and his 'little animals'*. London: John Bale, Sons & Danielsson.
Dobell, C. and O'Connor, F.W. 1921. *The intestinal protozoa of man*. London: John Bale Sons & Danielsson.
Esteban, J.G., Aguirre, C., et al. 1998. Balantidiasis in Aymara children from northern Bolivian Altiplano. *Am J Trop Med Hyg*, **59**, 922–7.
Gutierrez, Y. 2000. *Diagnostic pathology of parasitic infections*, 2nd edn. New York: Oxford University Press, 263–70.
Hadziyannis, E., Yen-Lieberman, B., et al. 2000. Ciliocytophthoria in clinical virology. *Arch Pathol Lab Med*, **124**, 1220–3.
Hindsbo, O., Nielsen, C.V., et al. 2000. Age-dependent occurrence of the intestinal ciliate *Balantidium coli* in pigs at a Danish research farm. *Acta Vet Scand*, **41**, 79–83.
Hosino, M., Sasagawa, G. and Tosaka, Y. 1999. Influence of *Balantidium* infection on colitis in pigs. *J Vet Med, Japan*, **52**, 287–91.

Knezevich, M. 1998. Geophagy as a therapeutic mediator of endoparasitism in a free-ranging group of rhesus monkeys (*Macaca mulatta*). *Am J Primatol*, **44**, 71–82.

Krascheninnikow, S. 1962. Variability in number of kineties in *Balantidium coli*. *J Parasitol*, **48**, 192.

Lee, J.J., Hutner, S.H. and Bovee, E.C. (eds) 1985. *An illustrated guide to the protozoa*. Lawrence, Kansas: Society of Protozoologists.

Leuckart, R. 1861. Über *Paramecium coli* Malmsten. *Arch Naturegesch*, **27**, 81.

Malmsten, P.H. 1857. Infusorien als intestinal-thiere beim Menschen. *Virchows Arch Path Anatomie Physiol*, **12**, 302–9.

McCarey, A.G. 1952. Balantidiasis in South Persia. *Br Med J*, **i**, 629–31.

McDonald, J.D. 1922. On *Balantidium coli* and *B. suis* (sp. nov.). *Univ Calif Publ Zool*, **20**, 243–6.

Maleky, F. 1998. Case report of *Balantidium coli* in human from south of Tehran, Iran. *Indian J Med Sci*, **52**, 201–2.

Mansfield, K.G., Kuei Chin, Lin, et al. 2001. Identification of enteropathogenic *Escherichia coli* in simian immunodeficiency virus-infected infant and adult rhesus monkeys. *J Clin Microbiol*, **39**, 971–6.

Munene, E., Otsyula, M., et al. 1998. Heminth and protozoan gastrointestinal tract parasites in captive and wild-trapped African non-human primates. *Vet Parasitol*, **78**, 195–201.

Muriuki, S.M.K., Murugu, R.K., et al. 1998. Some gastro-intestinal parasites of zoonotic (public health) importance commonly observed in Old World non-human primates in Kenya. *Acta Trop*, **71**, 73–82.

Nakauchi, K. 1999. The prevalence of *Balantidium coli* infection in fifty-six mammalian species. *J Vet Med Sci*, **61**, 63–5.

Orihel, T.C. and Ash, L.R. 1995. *Parasites in human tissue*. Chicago: American Society of Clinical Pathologists, 25–7.

Patil, N.A., Udupa, K.G., et al. 1998. Prevalence, clinical symptoms and treatment of balantidiosis in buffaloes. *Buffalo Bull*, **17**, 61, 64.

Sargeaunt, P.G. 1971. The size range of *Balantidium coli*. *Trans Roy Soc Trop Med Hyg*, **65**, 428.

Stein, F. 1863. Über das *Paramoecium* (?) *coli* Malmst. *Amt Bericht Versamml Deuts Naturforsch Aertze*, **37**, 165.

Tekle, T. and Abebe, G. 2001. Trypanosomiasis and helminthoses: major health problems of camels (*Camelus dromedarius*) in the southern rangelands of Borena, Ethiopia. *J Camel Practice Res*, **8**, 39–42.

Templis, C.H. and Lysenko, M.G. 1957. The production of hyaluronidase by *Balantidium coli*. *Exp Parasitol*, **6**, 31–6.

Uptal Das, 1999. A case of *Balantidium coli* infection in a dog. *Indian Vet J*, **76**, 174.

Walzer, P.D., Judson, F.M., et al. 1973. Balantidiasis outbreak in Truk. *Am J Trop Med Hyg*, **22**, 33–41.

Wenyon, C.M. 1926. *Protozoology*. London: Baillière, Tindall and Cox, 1201–10.

WHO, 1995. *Drugs used in parasitic diseases*, 2nd edn. Geneva: World Health Organization, 55.

Yang, Yuezhong, Zeng, Li, et al. 1995. Diarrhoea in piglets and monkeys experimentally infected with *Balantidium coli* isolated from human faeces. *J Trop Med Hyg*, **98**, 69–72.

Zaman, V. 1993. *Balantidium coli*. In: Kreier, J.P. and Baker, J.R. (eds), *Parasitic protozoa*, 2nd edn. vol. 3. New York: Academic Press, 43–60.

16

Old World leishmaniasis

PAUL A. BATES AND R.W. ASHFORD

INTRODUCTION

The leishmaniases are a group of human diseases that afflict people in many tropical and sub-tropical regions. They are caused by parasitic protozoa of the genus *Leishmania* and are transmitted to humans by the bite of female phlebotomine sandflies, small blood-feeding insects. Leishmaniasis is not a single entity but a collection of diseases, each with its own clinical manifestations and epidemiology. However, the basic pattern of the life cycle is well conserved among members of the genus, and each species of *Leishmania* tends to cause a certain type of disease within a specific epidemiological context. Therefore, *Leishmania* spp. are most conveniently studied by first considering their general properties before dealing with those specific to individual species; this is the approach adopted here. Five species of *Leishmania* are the medically important agents of Old World leishmaniasis: *L. major*, *L. tropica*, *L. aethiopica*, *L. donovani*, and *L. infantum* (Figures 16.1 and 16.2). The first three of these are predominantly agents of cutaneous leishmaniasis, an infection which is limited to the skin, and the last two are predominantly agents of visceral leishmaniasis, an infection of the liver and spleen that can be fatal. In addition to these five well-established species, a number of other named species can be found in the literature on Old World leishmaniasis. Some have subsequently turned out to be very close, if not identical, to known species; others have not

been examined in sufficient detail to reach a firm conclusion with regard to their identity.

CLASSIFICATION

The genus *Leishmania*

Leishmania is one of several genera within the family Trypanosomatidae (class Kinetoplastea, order Trypanosomatida) and, therefore, shares certain properties with other members of this family described elsewhere in this volume (see Chapter 18, African trypanosomiasis; Chapter 19, New World trypanosomiasis). These common properties include: the possession of a kinetoplast (a unique form of mitochondrial DNA); a single flagellum arising from a flagellar pocket; and an alternation between arthropod and mammalian hosts during the life cycle.

The features that uniquely identify the genus *Leishmania* are the nature of its vertebrate and invertebrate hosts and the developmental cycles, and morphological stages found within each. The vertebrate hosts are all mammals in which the parasites reside within the phagolysosomal system of mononuclear phagocytic cells, typically macrophages (Mauel 1996; Antoine et al. 1998; Rittig and Bogdan 2000). This is a unique location for a eukaryotic parasite and is only used by a handful of prokaryotic parasites (Garcia-del Portillo and Finlay

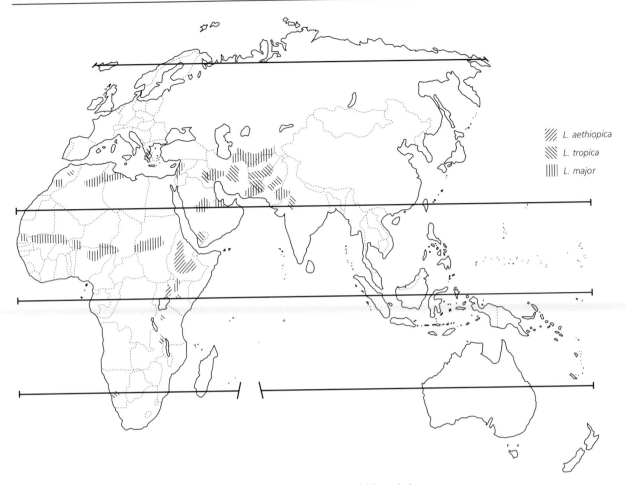

Figure 16.1 *Geographical distribution of the agents of Old World cutaneous leishmaniasis*

1995). The parasite gains entry to this intracellular residence via host cell phagocytosis. The invertebrate hosts
are all phlebotomine sandflies of two genera; *Phlebotomus* in the Old World and *Lutzomyia* in the New
World (see Chapter 17, New World leishmaniasis).
These are small, hairy, dipteran flies of the family
Psychodidae, in which only the females feed on blood
and transmit disease. In the sandfly, the parasites are
extracellular; development occurs exclusively in the gut
and transmission is via the mouthparts during blood
feeding (reviewed by Schlein 1993; Walters 1993; Bates
1994a; Sacks and Kamhawi 2001).

Various taxonomic distinctions have been proposed
at the subgeneric, species, and subspecies levels, but at
present there is no generally accepted classification that
incorporates all of these elements. Here, we adopt a
conservative pragmatic approach, essentially following
Lainson and Shaw (1987), but without any subgeneric
or subspecific qualifiers. In *Leishmania* the biological
species concept of a population of potentially interbreeding individuals cannot be applied, as the genus
seems to be largely, if not exclusively, asexual in its
mode of reproduction (Victoir and Dujardin 2002). The
ideal classification for an asexual organism at the level
of the 'species' combines the maximum biological

homogeneity within each species, but separates biologically different organisms. Among the techniques available for the classification of *Leishmania*, isoenzyme
analysis has been found to describe strains at a level that
allows the construction of phenograms and cladograms
at an ideal resolution (Rioux et al. 1990). Using a panel
of c. 12 enzymes, it is possible to identify and distinguish most, but not all of the currently accepted species
(Jamjoom et al. 2004). The group of strains that share a
given pattern of enzyme electrophoretic mobilities are
known as a 'zymodeme'. A given species may contain a
number of zymodemes indicating subpopulations or
subspecies. There has been an enormous accumulation
of data using this method; it has largely confirmed what
was known or expected biologically but has also
provided valuable new insights, such as the fact that
cutaneous leishmaniasis in Europe is caused by
L. infantum. An example of a cladogram constructed
using isoenzyme data is shown in Figure 16.3. The main
disadvantage of isoenzyme analysis is the lack of standardization between laboratories; the importance of the
strictest laboratory protocols to prevent mixing of
cultures cannot be overemphasized.

Of the other techniques that have been used, the resolution of restriction fragment length polymorphisms

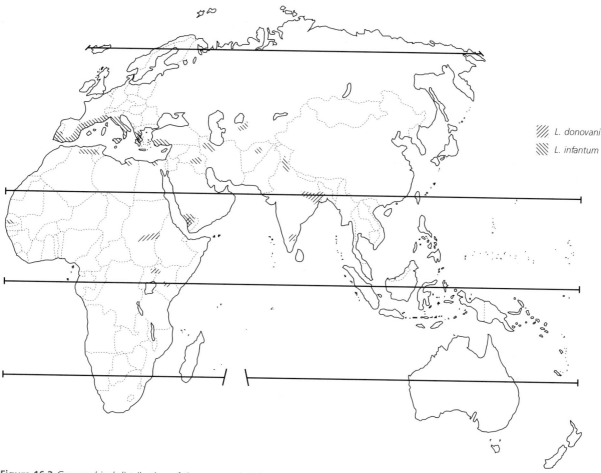

Figure 16.2 *Geographical distribution of the agents of Old World visceral leishmaniasis*

//// *L. donovani*

\\\\ *L. infantum*

(RFLP) and karyotype analysis has been found to be too fine except for studies at the population genetic level. Microsatellite DNA analysis is the latest technique being used with *Leishmania* and seems well suited to resolving taxonomic issues at the infraspecific level (Bulle et al. 2002; Jamjoom et al. 2002). DNA and ribosomal RNA sequence information hold much promise and have given valuable results at the generic level and above, but have not been sufficiently widely used yet to contribute to classification at the species level. Monoclonal antibody and DNA probes, with or without polymerase chain reaction (PCR) amplification, are of value in identification of stocks that have been well classified, but provide insufficient information to be of value in constructing a classification. A disadvantage of most current methods is the requirement to culture the parasite in question before identification can be made.

Of the five species considered in this chapter, all are distinct from those found in the New World (see Chapter 17, New World leishmaniasis) with one important exception, *L. infantum*. Molecular analysis reveals this to be indistinguishable from *L. chagasi* found in the New World (Mauricio et al. 1999, 2000). Because the former was named first both parasites should be referred to as *L. infantum*. We favor the explanation that

'*L. chagasi*' is, in fact, *L. infantum* that was introduced into the New World in post-Columbian times via the domestic dog (Momen et al. 1993; Courtenay et al. 2002). While others may disagree with this interpretation, it is certain that these two parasites are very closely related. The remaining New World parasites have rather different biology, clinical manifestations, and epidemiology to those of the Old World and are dealt with separately in Chapter 17, New World leishmaniasis. The main features of the Old World species are summarized in Table 16.1.

STRUCTURE AND LIFE CYCLE

The parasite exists in two main morphological forms, 'amastigotes' and 'promastigotes', found in the mammalian and sandfly hosts, respectively. Amastigotes are ovoid, non-motile, intracellular stages; promastigotes are elongated, motile, extracellular stages. The form introduced into the skin of the mammalian host is a promastigote, but this soon transforms into an amastigote, and the parasite remains in this form for the duration of the mammalian phase of the life cycle. Similarly, the form taken up by the sandfly is an amastigote, but this soon transforms into a promastigote. The basic life cycle of

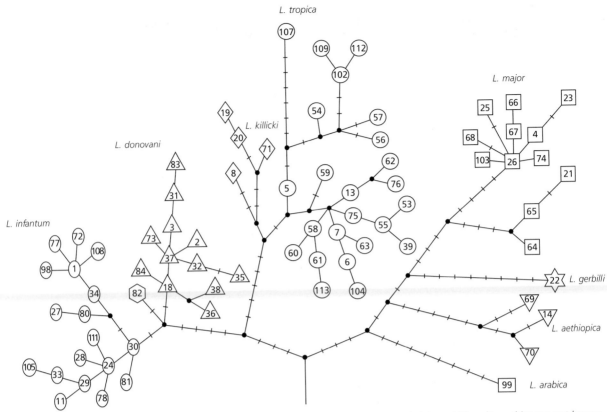

Figure 16.3 *Cladogram of Old World* Leishmania *zymodemes showing grouping into 'species'.* L. gerbilli *and* L. arabica *are not known from humans. (Modified from Rioux et al. 1990)*

Leishmania is illustrated in Figure 16.4 (reviewed by Molyneux and Killick-Kendrick 1987).

Amastigote structure

Amastigotes are ovoid cells of 3–5 μm length on the main axis and, therefore, lie at the lower limit of size described for eukaryotic cells. Because of their small size, little internal structure can be discerned in stained preparations at the light microscope level beyond the central round or oval nucleus and adjacent but smaller round or rod-shaped kinetoplast. At the ultrastructural level, more detail is revealed (Figure 16.5). The surface membrane is a conventional unit membrane under which lies a corset of microtubules, serving as a form of cytoskeleton. These are closely spaced in parallel rows

Table 16.1 *Summary of the main features of Old World* Leishmania *species*

Species	Disease in humans	Geographical distribution	Important mammalian hosts	Important sandfly hosts
L. major	Rural, zoonotic, cutaneous leishmaniasis, oriental sore	North Africa, Sahel of Africa, Central and West Asia	Great gerbil *Rhombomys opimus*, fat sand rat *Psammomys obesus*	*Phlebotomus papatasi, P. dubosqi, P. salehi*
L. tropica	Urban, anthroponotic cutaneous leishmaniasis, oriental sore	Central and West Asia	Humans	*Phlebotomus sergenti*
L. aethiopica	Cutaneous leishmaniasis, diffuse cutaneous leishmaniasis	Ethiopia, Kenya	Rock hyraxes *Heterohyrax brucei* and *Procavia* spp.	*Phlebotomus longipes, P. pedifer*
L. donovani	Visceral leishmaniasis, kala-azar, post kala-azar dermal leishmaniasis	Indian subcontinent, East Africa	Humans	*Phlebotomus argentipes, P. orientalis, P. martini*
L. infantum	Infantile visceral leishmaniasis	Mediterranean basin, Central and West Asia	Domestic dog	*Phlebotomus ariasi, P. perniciosus*

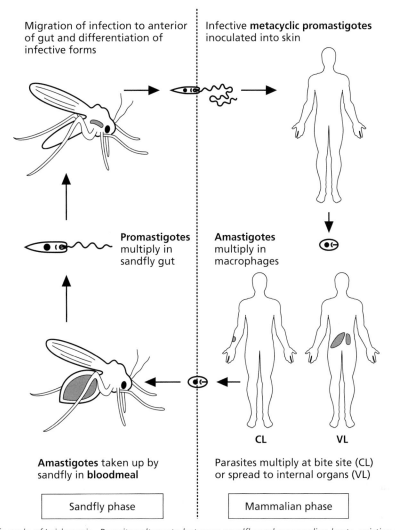

Migration of infection to anterior of gut and differentiation of infective forms

Infective **metacyclic promastigotes** inoculated into skin

Promastigotes multiply in sandfly gut

Amastigotes multiply in macrophages

CL   VL

Amastigotes taken up by sandfly in **bloodmeal**

Parasites multiply at bite site (CL) or spread to internal organs (VL)

Sandfly phase

Mammalian phase

Figure 16.4 *General life-cycle of* Leishmania. *Parasites alternate between sandfly and mammalian hosts, existing as promastigote and amastigote forms, respectively. Mammalian infection results in cutaneous leishmaniasis (CL) or visceral leishmaniasis (VL).*

which, together with the small size, make the amastigote a very robust eukaryotic cell. One unfortunate consequence of this feature is that it has not proved possible to develop subcellular fractionation techniques for amastigotes that will disrupt the surface membrane without destroying the organelles contained within. On the positive side, fairly vigorous homogenization techniques can be employed for the isolation of amastigotes from infected tissue without destroying their viability or structural integrity.

An infolding of the surface membrane creates an internal space, termed the 'flagellar pocket' (Webster and Russell 1993; Landfear and Ignatushchenko 2001). This is so named because a flagellum emerges from the surface membrane and projects into the pocket. This flagellum is not functional in the amastigote and does not extend beyond the cell body. The flagellar pocket is thus topologically external to the cell although contained within it. In addition to anchoring the flagellum, the main function of the pocket is as a site of endocytosis and exocytosis (McConville et al. 2002; Morgan et al.

2002). Microtubules do not underlie the surface of the flagellar pocket membrane thus allowing access to vesicular traffic. Immediately below the origin of the flagellum lies the kinetoplast, a dense mass of mitochondrial DNA. The kinetoplast DNA is composed of several thousand circular DNA molecules linked together in a catenated network (Shlomai 1994; Shapiro and Englund 1995; Lukes et al. 2002). The DNA circles are of two size classes: each kinetoplast contains 25–50 maxicircles of c. 30 kb, and 5000–10 000 minicircles of c. 2 kb. Together these constitute the mitochondrial genome. Branches of the surrounding mitochondrion extend throughout the cell body and contain plate-like cristae.

The cytoplasm contains both rough and smooth endoplasmic reticulum. The Golgi apparatus is typically found in the vicinity of the flagellar pocket, which probably reflects the role of this organelle in the endocytic and exocytic pathways. Lysosomes are also found in the cytoplasm together with an organelle unique to kinetoplastids, the glycosome. This is so named because a

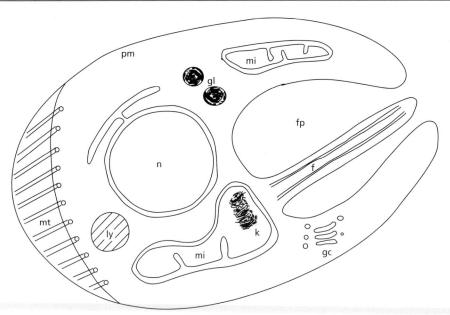

Figure 16.5 *Ultrastructure of a* Leishmania *amastigote. Amastigotes possess a central nucleus (n) and adjacent kinetoplast (k) within a single branching mitochondrion (mi). The flagellum (f) arises from a flagellar pocket (fp) but does not extend beyond the cell body. Lysosomes (ly), glycosomes (gl) and Golgi complex (gc) are found in the cytoplasm. Rows of microtubules (mt) run just below the plasma membrane (pm).*

number of glycolytic enzymes are specifically located in this organelle, together with some others (Michels et al. 1997; Parsons et al. 2001).

Promastigote structure

In the sandfly host, the parasite is found mainly as a promastigote form (Figure 16.6). The structural elements of promastigotes are the same as those described for amastigotes although there may be variations in organelle numbers. The main differences from the amastigotes are that the cell body is elongated, 8–15 μm, and that the flagellum emerges from the cell body and is functional thus making these motile cells. The flagellum is found at the anterior end of the cell, i.e. the cell body is trailed behind the flagellum. Desmosomal plaques

anchor the flagellum to the cell body as it emerges from the flagellar pocket. The promastigote flagellum has a paraxial rod, a paracrystalline structure running parallel to the microtubules of the axoneme, which plays a role in flagellar motility (Maga et al. 1999). Recently, a novel tubular lysosome that is part of the endocytic pathway has been described in promastigotes (Waller and McConville 2002).

Life cycle

DEVELOPMENT IN THE SANDFLY HOST

The phase of the life cycle in female sandflies is of relatively short duration. Flies that acquire a *Leishmania* infection probably remain infected for life, but this is

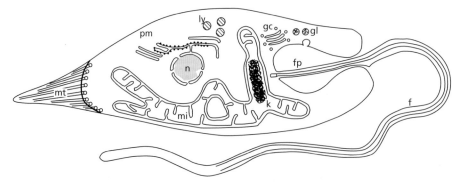

Figure 16.6 *Ultrastructure of a* Leishmania *promastigote. Many of the features found in the amastigote stage are also found in the promastigotes (abbreviations, see Figure 16.5). Some differences are that the cell body is elongated, the flagellum extends beyond the cell body and the kinetoplast has a more anterior location relative to the nucleus.*

usually only a matter of weeks. In most parasite–vector combinations that have been studied experimentally, the development of the parasite is sufficiently rapid for mammal-infective promastigotes to be produced by the time the female is ready to take her next bloodmeal. This may be as little as 5–7 days under optimal conditions. Development occurs exclusively in the sandfly gut (Figure 16.7) and begins with infected macrophages or free amastigotes in the bloodmeal. The macrophages disintegrate over a matter of hours and, as with the mammalian phase, the first event is transformation, in this case to the promastigote form, which takes 24–48 h. Promastigotes of various kinds grow and divide in the sandfly gut. The ultimate products of the developmental cycle are the mammal-infective forms, termed 'metacyclic promastigotes', which accumulate in the anterior midgut and foregut of the sandfly.

The developmental cycle of *Leishmania* in the sandfly host is illustrated in Figure 16.8 and reviewed in Bates and Rogers (2004). Promastigote development has been studied using a variety of combinations of parasite and vector (Killick-Kendrick 1990a; Walters 1993; Bates 1994a) and two patterns have emerged for those species infective to mammals. Certain species usually include a phase of development in the hindgut of the sandfly, for example the New World species *L. braziliensis*, share certain other features, and form a distinct cluster of species according to isoenzyme and DNA sequence analysis. These exclusively New World species are placed in a separate subgenus, (*Viannia*) and are described in detail in Chapter 17, New World leishmaniasis. The majority of *Leishmania* species, including all the Old World parasites, in which there is no hindgut development are placed in another subgenus (*Leishmania*).

There are a variety of different promastigote forms that can be distinguished on morphological grounds and various authors refer to these by different names, often interchanging cultured and sandfly-derived promastigotes. However, the evidence for functional distinction is less complete. In the Old World *Leishmania* species, six developmental forms are recognized during development: procyclic promastigotes, nectomonad promastigotes, leptomonad promastigotes, haptomonad promastigotes, paramastigotes, and metacyclic promastigotes (Rogers et al. 2002; Gossage et al. 2003; Figure 16.9). Other categories and subdivisions can be found in the literature, but at present there is little evidence for any functional distinction between them.

The first major developmental event in the sandfly is the transformation of amastigotes to procyclic promastigotes. However, very few observations on the early events have been made and it is also possible that some amastigotes may divide before transforming to promastigotes. Amastigote to promastigote transformation occurs readily in vitro and is usually completed in 24–48 h when

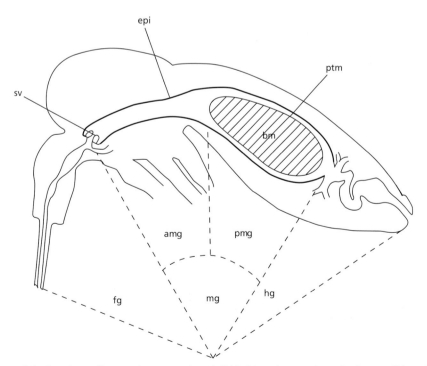

Figure 16.7 *Structure of the female sandfly gut. The gut can be subdivided into three sections: the foregut (fg), midgut (mg), and hindgut (hg). The midgut is lined by epithelial cells (epi) with microvilli that project into the lumen of the gut and is subdivided itself into the anterior or thoracic midgut (amg) and the posterior or abdominal midgut (pmg). The junction between the anterior midgut and the foregut is formed by the stomodeal valve (sv). The foregut, stomodeal valve and hindgut are lined by a chitinous cuticular layer. Initial development of the parasite occurs within the bloodmeal (bm) encased in a peritrophic matrix (ptm) in the posterior midgut.*

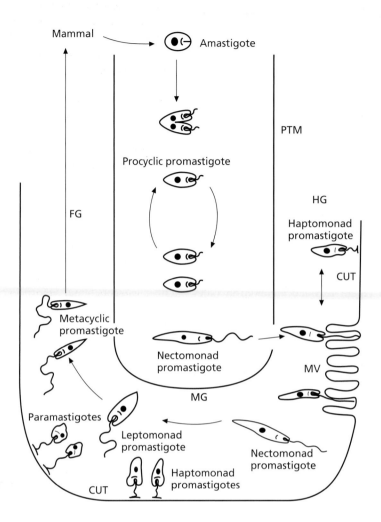

Figure 16.8 *Developmental cycle of* Leishmania *promastigotes. Amastigotes are ingested by a female sandfly and transform to procyclic promastigotes in the bloodmeal, itself encased in a peritrophic matrix (PTM) in the midgut (MG). Procyclic promastigotes are gradually replaced by nectomonad promastigotes that escape from the ptm; some attach to the microvilli (mv) via their flagella. In some species, haptomonad forms attach to the cuticular lining (cut) in the hindgut (HG), in others they attach in only the foregut (FG) and stomodeal valve. Paramastigotes may also be found in the foregut. Leptomonad promastigotes multiply in the anterior midgut and differentiate into metacyclic promastigotes to complete the developmental cycle.*

a suitable culture medium and an appropriate temperature (about 26°C) is used. The media used vary widely, but the optimal pH for promastigote growth, and that used in culture media, is generally neutral to slightly alkaline (pH 7.0–7.5). In vitro transformation to promastigotes and cell division occur coincidentally, such that, of the first forms observed in division, the majority are partially transformed intermediates (Bates 1994b). The signals initiating both processes may be the same and it is possible that they are mechanistically linked.

In the sandfly, these events occur in the posterior midgut in the bloodmeal that is itself encased in a peritrophic matrix (PTM) secreted by the midgut epithelium. The PTM is a lattice of chitin fibrils, proteins and glycoproteins secreted by the midgut epithelium in

response to a bloodmeal (Shao et al. 2001). In phlebotomine sandflies, the PTM is produced at one time and encapsulates the whole bloodmeal offering some protection to the parasite during amastigote to procyclic promastigote transformation by limiting their exposure to trypsin secreted by the midgut epithelium (Pimenta et al. 1997). This temporary vulnerability to trypsin is probably a result of reorganization of the surface glycocalyx during transformation. The procyclics are short, ellipsoid promastigotes generally about 6–8 μm in body length. Multiplication of procyclic promastigotes occurs within the PTM but gradually the promastigotes elongate and transform to nectomonad forms of 15–20 μm body length.

Approximately 3 days after blood feeding, the PTM usually begins to break down and promastigotes begin to

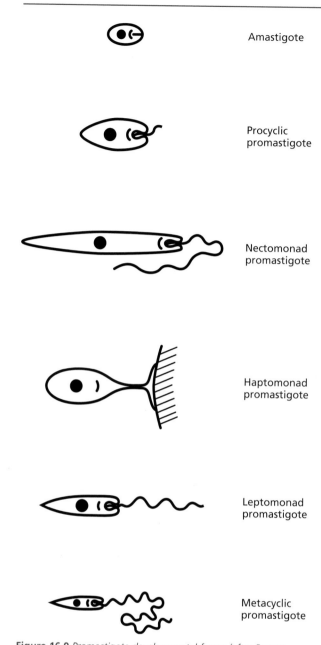

Amastigote

Procyclic promastigote

Nectomonad promastigote

Haptomonad promastigote

Leptomonad promastigote

Metacyclic promastigote

Figure 16.9 *Promastigote developmental forms (after Rogers et al. 2002). The six main morphological stages found in the life-cycle of* Leishmania *drawn to show their relative sizes and shapes.*

migrate forward to the anterior midgut. There is evidence for an accelerated breakdown of the PTM in infections of *L. infantum* in *Lutzomyia longipalpis* (Walters 1993) and *L. major* in *Phlebotomus papatasi* (Schlein et al. 1991). One explanation for this phenomenon is the presence of a chitinase secreted by promastigotes (Schlein et al. 1991; Shakarian and Dwyer 1998, 2000). This is presumed to have survival value to the parasite as it would permit establishment of a midgut infection outside the PTM earlier than would otherwise be the case. In some vectors, it could be essential for escape from the PTM before defecation of the bloodmeal remnants by the sandfly.

The transformation from procyclics to nectomonads can be mimicked in vitro (Bates 1994b). In this study, lesion amastigotes were allowed to transform, procyclics appeared on day 2 and, after 3 days of growth, a population consisting predominantly of nectomonad forms was obtained. The timing of these events is strongly reminiscent of the situation in sandfly infections and occurs in vitro at a particular growth phase (late exponential phase). The latter suggests that transformation to nectomonads is a density-dependent phenomenon and could be triggered, for example, by a factor such as nutrient depletion. This could be a useful adaptation in vivo as the initial source of nutrition, the bloodmeal, becomes exhausted. Charlab and Ribeiro (1993) found that inhibition of growth by exposure of promastigotes to salivary gland homogenate from *Lu. longipalpis* caused transformation to nectomonads in vitro. Whether parasites are exposed to sandfly saliva during their development in the midgut is uncertain, but is a possibility after escape from the PTM. Flies continue to take sugar meals over the several days that the bloodmeal is being digested and as promastigote development occurs.

Some of the nectomonads attach to the midgut epithelium, inserting their flagella between the microvilli (Killick-Kendrick 1990a). The major surface glycolipid of promastigotes, lipophosphoglycan (LPG) (Turco and Descoteaux 1992), mediates this attachment, which is important for establishment of a mature infection, i.e. one that persists beyond the initial multiplicative phase in the bloodmeal (Lang et al. 1991; Pimenta et al. 1992, 1994). Binding of promastigotes to dissected midguts in vitro can be inhibited by free LPG or its constituents and there is a positive correlation between the binding of LPG isolated from a particular parasite to dissected midguts of sandflies, and the ability of the given vector to transmit that parasite (Pimenta et al. 1994; Sacks et al. 1994; Sacks 2001). Mutants specifically unable to express LPG cannot attach to the midgut and maintain an infection in sandflies (Sacks et al. 2000).

LPG covers the entire surface including the flagellar membrane where it could mediate binding (Lang et al. 1991). However, one unexplained feature of midgut attachment is the striking feature of flagellar insertion, a phenomenon restricted to the subgenus *Leishmania*. The reason may be that the juxtaposition of elongated flagella and microvilli provides a good fit that maximizes the number of binding sites and hence, with time, the promastigotes will adopt this orientation. However, it is premature to discount the possibility that other flagellar-specific molecules could be involved and provide additional specificity.

By 5 days, the infection has usually spread to the anterior midgut and the cuticular surface of the stomodeal valve at the junction with the foregut. Nectomonad forms are responsible for this anterior migration and, upon reaching the anterior midgut, give rise to leptomonad promastigotes and haptomonad promastigotes (Rogers et al. 2002). Leptomonad forms, promastigotes of 6-8 μm, accumulate in the lumen of the midgut producing a gel-like substance, promastigote secretory gel (PSG) (Stierhof et al. 1999; Rogers et al. 2002). Haptomonad forms, broad cells of 5–8 μm, are found attached to the stomodeal valve via hemidesmosomes in expansions of their flagellar membranes (reviewed by Vickerman and Tetley 1990). The PSG produced by leptomonad forms and the attached haptomonads both assist with transmission by helping to create a 'blocked fly' (see below). Attached paramastigotes (nucleus adjacent to kinetoplast) and haptomonads may also be found in the foregut. In these attached forms, the flagellar membrane can be seen at the ultrastructural level to be closely apposed to the surface forming a junctional complex. An electron-dense plaque and associated filaments are located below the flagellar membrane, and thus these attachment organelles bear a superficial resemblance to the hemidesmosomes found at the basal surfaces of vertebrate epithelial cells.

From day 5 onwards, increasing numbers of small (5–8 μm), narrow, highly motile, metacyclic promastigotes, the mammal-infective forms, can be observed in the lumen of the anterior midgut and/or foregut in a position that facilitates their transmission on subsequent blood feeding by the female sandfly. Metacyclics differentiate from leptomonad promastigotes (Rogers et al. 2002), a process that can be induced in vitro by culture at low pH (Bates and Tetley 1993; Zakai et al. 1998), under anaerobic conditions (Mendez et al. 1999), or nutrient depletion (Cunningham et al. 2001). Conversely, both metacyclogenesis and chitinase secretion by *L. major* are inhibited by the inclusion of hemoglobin in the culture medium (Schlein and Jacobson 1994a). Unlike the other transformations described so far, this is more properly described as a differentiation process, as the end-product is in a non-dividing state.

TRANSMISSION FROM SANDFLY TO MAMMALIAN HOST

The opportunity for transmission from an infected sandfly to a mammalian host occurs when the fly is ready to take another bloodmeal. There is debate concerning the precise mechanisms that aid transmission and the extent to which the parasites themselves contribute to this process; these may vary with the specific vector–parasite combination. For the parasites to become lodged in the skin of a mammal, they must travel against the predominant flow of blood, which is into the gut of the fly.

In a sandfly containing a mature infection, the majority of promastigotes are found in the anterior midgut with a smaller population in the foregut. The number of promastigotes required to initiate a mammalian infection varies between hosts and parasite/vector combinations; the numbers inoculated are believed to be in the range 10–1 000 metacyclic promastigotes per infective bite (Rogers et al. 2004).

Under experimental conditions, infected flies show differences in feeding behavior when compared to uninfected flies. Frequently, they are seen to probe the skin an increased number of times and appear to find difficulty in taking a bloodmeal, a behavior that may enhance transmission (reviewed by Molyneux and Jeffries 1986; Molyneux and Killick-Kendrick 1987). Massive infections develop in and block the anterior midgut of both experimentally infected and wild-caught flies, with the promastigotes embedded in PSG (Stierhof et al. 1999; Rogers et al. 2002). The main component of PSG is a mucin-like filamentous proteophosphoglycan (Stierhof et al. 1999; Ilg 2000). Haptomonad forms attached to the stomodeal valve and foregut also contribute to the blockage of the gut. The parasite/gel plug appears to form a physical obstruction to feeding, a so-called 'blocked fly', resulting in an increased frequency of multiple probing and failed attempts to engorge.

During sandfly feeding, infective metacyclic promastigotes are carried out along with sandfly saliva and PSG (Rogers et al. 2004). Both saliva and PSG can enhance the infectivity of the promastigotes in the mammalian host (Titus and Ribeiro 1990; Belkaid et al. 1998; Kamhawi 2000; Rogers et al. 2004). Anti-saliva immune responses may be a useful component of anti-leishmanial vaccines (Kamhawi et al. 2000). Much debate has centered on the position and the types of promastigote found in the foregut and midgut: specifically where the mammal-infective forms are found (Killick-Kendrick 1990a). The two main ideas proposed are that metacyclic promastigotes are directly inoculated from the proboscis (the anterior part of the foregut), or regurgitated forward from more distal parts of the foregut and possibly the midgut. The current consensus favors regurgitation as the main mode of transmission, and this has been recently proven in *L. mexicana*-infected sandflies (Rogers et al. 2004). A variation on simple regurgitation comes from observations that the stomodeal valve can be damaged and remain permanently open as a result of the parasite infection, as has been described with infections of *L. major* in *Phlebotomus papatasi* (Schlein et al. 1992). Where such damage does occur, it may permit mixing of the contents of the anterior midgut with an incoming bloodmeal, such that when the pharynx contracts to deposit blood in the midgut, there is also a forward surge of promastigote-contaminated blood into the wound.

DEVELOPMENT IN THE MAMMALIAN HOST

The developmental cycle of *Leishmania* in the mammalian host is illustrated in Figure 16.10. This cycle is initi-

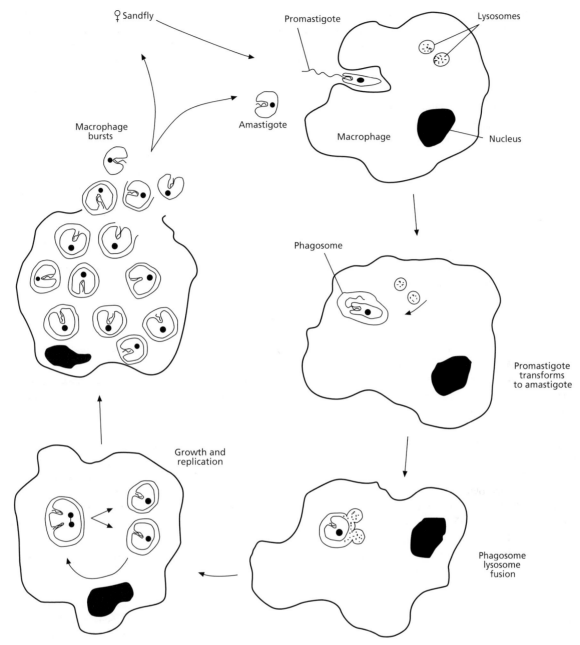

Labels within figure:
♀ Sandfly
Promastigote
Lysosomes
Amastigote
Macrophage bursts
Macrophage
Nucleus
Phagosome
Growth and replication
Promastigote transforms to amastigote
Phagosome lysosome fusion

Figure 16.10 *Developmental cycle of* Leishmania *amastigotes*

ated by the interaction of metacyclic promastigotes with skin macrophages. The precise kinetics of parasite transformation and uptake by skin macrophages is uncertain, but it is generally assumed that uptake occurs soon after inoculation and that the promastigote to amastigote transformation occurs mainly inside the host cell. Direct studies on the binding and uptake of metacyclics by skin macrophages in situ are not possible but this process has been modeled in vitro (Alexander and Russell 1992; Rittig and Bogdan 2000; Handman and Bullen 2002). Two important caveats must be attached to the published reports to date: the types of macrophage used have been either a peritoneal exudate cell or a macrophage-like cell line (inappropriate host cells) and the

parasites used have not been sandfly-derived metacyclic promastigotes but cultured forms that are sometimes not properly characterized. Nevertheless, these studies have shown that complement fixation by metacyclic promastigotes is not only an inevitable consequence of exposure to mammalian blood and tissue fluids but is an important adaptation for gaining entry to the host cell (Mosser and Brittingham 1997). Metacyclic promastigotes are relatively resistant to complement-mediated lysis and use the surface-bound complement components as ligands for binding to macrophage complement receptors. This promotes phagocytosis by the macrophage. After uptake and internalization in a phagosome, fusion with lysosomes occurs and the parasite inhabits a secondary lysosome or phagolysosome (Antoine et al. 1998). During this process, the metacyclic promastigote transforms into an amastigote, a process that takes 12–24 h to complete. Interestingly, the parasite appears able to delay phagosome maturation until this transformation is complete (Swanson and Fernandez-Moreia 2002), a survival strategy mediated by LPG (Descoteaux and Turco 2002). Thereafter, the parasite remains in the amastigote form for the duration of the mammalian phase of the life cycle. Fully transformed amastigotes then continue to grow and divide within the phagolyso-somal compartment of the host cell in a parasitophorous vacuole. Different species of parasite inhabit different sized vacuoles. The species causing Old World leishmaniasis all inhabit small vacuoles in which the vacuole membrane is closely opposed to the surface of the parasite itself. As the parasite divides so does the surrounding vacuole and, consequently, the host cell becomes occupied by multiple parasites and vacuoles. Amastigote division is by simple binary fission. Some of the New World parasites, for example L. mexicana (see Chapter 17, New World leishmaniasis), occupy larger vacuoles, and these have a different effect on host cell endocytic traffic through their vacuoles (Antoine et al. 1998). After division, the parasites remain together in clusters within large vacuoles.

The phagolysosomal compartment presents two threats to amastigotes: the battery of lysosomal enzymes and low pH (4.5–5.5). Defense against the enzymes is mainly passive and relies on the unusual structures of the surface glycolipids and lipopho-sphoglycans rendering the amastigotes indigestible (Ilgoutz and McConville 2001). Low pH is not a problem as amastigotes are acidophiles: they have effective means of regulating their internal pH in an acidic environment; they are metabolically more active at low pH; and some species can only be cultured axenically in vitro as amastigotes at low pH (Bates 1993; Zilberstein and Shapira 1994). There is no known specific mechanism for escape from the host cell, although recent observations suggest the possibility that amastigotes are shed from peripheral vacuoles (Rittig and Bogdan 2000). The alternative is that parasites simply multiply until the host cell bursts open, making them available for uptake by other macrophages. Amastigotes are usually regarded as obligate intracellular parasites but this may not be strictly true: it is now possible to culture certain species axenically and at least in experimental infections it is possible to find extracellular amastigotes in heavily parasitized tissues. Despite this, the vast majority have an intracellular existence. Different Leishmania species infect macrophages at different sites in the body, but it is not known if amastigotes are specifically able to target subpopulations of mononuclear phagocytic cells or if they simply do not have access to or do not survive in macrophages in the wrong site. All amasti-

gotes appear to be equivalent and there is no evidence for functionally distinct subpopulations.

TRANSMISSION FROM MAMMALIAN TO SANDFLY HOST

The opportunity for transmission from an infected mammalian host to a female sandfly occurs when the sandfly feeds. Sandflies are pool feeders, i.e. they possess cutting mouthparts that slice into the skin, and feed from the small pool of blood that seeps into the wound. Thus, despite their small size, 2–3 mm, sandflies give a noticeable and sometimes painful bite. Parasites causing cutaneous infections are acquired by a sandfly along with the bloodmeal if the fly happens to feed on a cutaneous lesion. Tissue damage from the bite releases infected macrophages or free amastigotes into the wound. Visceral parasites are probably acquired with the blood itself. Although the principal pathological events of visceral leishmaniasis occur in the spleen and liver, where the host cells probably remain resident, another site of infection is the bone marrow. This is a major site of hemopoiesis and blood monocytes infected with Leishmania amastigotes are released into the peripheral circulation and are thus available to feeding sandflies. It is also possible that infected monocytes leave the circulation and become resident in the skin to act as a source of parasites especially in the case of post kala-azar dermal leishmaniasis or the generalized leishmaniasis typical of canine infections with L. infantum. Nevertheless, in most endemic foci the acquisition of Leishmania parasites by individual flies is a rare event and the vast majority of flies are uninfected. The few infected flies are, however, efficient vectors.

BIOCHEMISTRY, MOLECULAR BIOLOGY, AND GENETICS

Surface membrane structure

The surface membrane proteins of Leishmania can be subdivided into two groups. Those in the first group are present at high copy numbers per cell and are major structural components of the surface membrane. Members of the second group are present at low copy numbers per cell and have been identified using functional assays, for example, enzymes, transporters, and ion pumps.

Three major surface components have been described: gp63 (or promastigote surface protease (PSP)), lipopho-sphoglycan and glycosylphosphatidylinositols (Ilgoutz and McConville 2001). gp63 is a major surface protein of c. 63 000 kDa found on the surface of cultured promastigotes in all species of Leishmania that have been examined (Chang et al. 1990; Etges and Bouvier 1991) and has also been demonstrated on the surface of sandfly-derived promastigotes. gp63 is a zinc metalloprotease, the active site of which is exposed on the external face of the surface membrane (an ectoenzyme). There is some evidence for low level expression of gp63 by amastigotes in some species, but efforts to demonstrate active protease on the amastigote surface have revealed very low or no activity. Potential functions of gp63 in the life cycle of *Leishmania* are not certain, but are likely to be most important in the sandfly or during the brief existence of metacyclic promastigotes in the mammalian host. In the sandfly, gp63 may fulfil a digestive function, helping to provide a supply of amino acids for the parasite. The most abundant substrate would be hemoglobin and by digesting this molecule gp63 could also help to provide pre-formed heme, an essential nutrient for *Leishmania*. It might also help the parasite to penetrate the peritrophic membrane, which is partially proteinaceous in nature. gp63 can also be released from the parasite surface, so it may be able to exert biological effects away from the surface of the parasite (McGwire et al. 2002).

After transmission to a mammalian host, the proteolytic activity of gp63 may help metacyclic promastigotes to defend themselves against lysosomal enzymes while they are transforming to amastigotes in the macrophage. gp63 has also been identified as a significant site of complement fixation on the promastigote surface (Mosser and Brittingham 1997) and has therefore been proposed to be involved in the binding of promastigotes to macrophages via complement receptors. Another potential role is in facilitation of parasite migration through host tissues (McGwire et al. 2003).

LPG is the major surface molecule found on the promastigotes of *Leishmania* species (Turco and Descoteaux 1992; Beverley and Turco 1998). LPG contains four covalently linked structural components: a phospholipid tail that anchors the molecule in the surface membrane; a glycan core of several saccharide residues that lies immediately above the surface membrane; a repeating phosphodisaccharide backbone comprising 15–30 phosphate-galactose-mannose units; and a cap of 2–3 saccharide residues. LPG has been most extensively studied in *L. major* but is assumed to fulfil similar functions in other species. The number of repeated phosphodisaccharide units varies between life cycle stages, approximately doubling in number between *L. major* multiplicative and metacyclic promastigotes. Modeling suggests that the backbone projects away from the surface; it also carries a number of short saccharide side chains which, in *L. major*, terminate mainly in galactose or arabinose residues. LPG is either absent or expressed at low levels on the surface of amastigotes.

LPG has been proposed to fulfil a variety of functions in both the sandfly and mammalian hosts. The unusual structure of LPG may give the parasites general protection against hydrolytic enzymes in both the sandfly midgut and macrophage phagolysosome. One specific function, that of mediating attachment of promastigotes to microvilli on the sandfly midgut epithelium, has been described above. Detachment from the midgut is correlated with developmental modification of the side chains of LPG in *L. major*. In procyclic promastigotes, the side chains terminate in galactose residues that can bind to the midgut epithelium. Subsequently, these are replaced with LPG-carrying side chains that terminate with arabinose residues. These do not bind to the microvilli, allowing the promastigotes to detach, migrate forward to the anterior midgut, and differentiate into metacyclic promastigotes (Pimenta et al. 1992). At the same time, the number of repeat units increases. This latter change is believed to be responsible for the observed resistance to complement-mediated lysis shown by metacyclic promastigotes, a prerequisite for survival in the mammalian host. LPG is the major site of complement fixation on the metacyclic promastigote. Mechanistically, it is proposed that elongated LPG interferes with complement lysis by steric hindrance, the components of a potential membrane attack complex being too far from and lacking access to the plasma membrane to enable insertion. LPG with bound complement components can be released from the surface, which may represent an additional defense mechanism.

LPG is also proposed to have functions in the establishment of the parasite in the macrophage (Descoteaux and Turco 1993, 2002). As with gp63, fixation of complement by LPG provides ligands that can be bound by macrophage complement receptors, stimulating phagocytic uptake of the metacyclic promastigote. LPG has a strong inhibitory effect on macrophage protein kinase C and, consequently, may have the effect of inhibiting signal transduction pathways that trigger microbicidal responses. It is also an efficient scavenger of toxic oxygen radicals.

Glycosylphosphatidylinositols (GPIs) have been found on the surface of both promastigote and amastigotes stages; in the latter, they are a major surface component (McConville and Ferguson 1993; Ferguson 1997). These glycolipids resemble the lipid anchorglycan core moiety of LPG. The glycocalyx formed by these abundant molecules is assumed to help protect amastigotes against the potentially hostile environment of the phagolysosome.

Surface membrane biochemistry

In addition to gp63, a number of other ectoenzymes have been described on the surface membrane of promastigotes: acid phosphatase, 3′ and 5′ nucleotidases (Bates 1991). Promastigotes, and probably amastigotes, possess a surface membrane proton ATPase (Zilberstein and Shapira 1994) which serves an important function in pH regulation, extruding protons and preventing acidification particularly in the phagolysosomal environment. Transporters for glucose, amino acids, ribose, nucleosides, and folate have also been described (Beck and Ullman 1991; Marr 1991; ter Kuile 1993; Zilberstein and Shapira 1994). The surface membrane has an unusual sterol composition with ergosterol as the major membrane sterol (Haughan and Goad 1991; Chance and Goad 1997). Two enzymatically active secretory molecules have been described in promastigotes; acid phosphatase (Bates and Dwyer 1987) and chitinase (Schlein et al. 1991). In addition, promastigotes and amastigotes secrete various phosphoglycans (Ilg et al. 1994; Ilg 2000).

Molecular biology

Leishmania spp., in common with other kinetoplastids, possesses several unusual features that differentiate them from other eukaryotes described to date, and thus merit brief mention here. The structure of kinetoplast DNA (kDNA) is unique for a mitochondrial genome (Shlomai 1994; Shapiro and Englund 1995; Lukes et al. 2002), but kDNA also exhibits a unique form of post-transcriptional RNA processing termed 'RNA editing' (Stuart and Panigrahi 2002). The 25–50 identical maxi-circle DNA molecules of each kinetoplast encode various mitochondrial proteins but the primary RNA transcripts from these are not directly translatable. When compared to the final transcript, the gene sequences are found to contain insertions and deletions of uridine residues. The maxicircle primary transcripts are corrected in an editing process using short sequences, 'guide RNAs', which are themselves encoded by the other component of kDNA, the more numerous minicircles. This process is unique to kinetoplastids and is probably an ancient trait retained since the early divergence of the group from other eukaryotes.

Leishmania and kinetoplastids also exhibit several unusual features in the expression of their nuclear genes (Graham 1995; Stiles et al. 1999). Such genes are usually grouped together in polycistronic transcription units, an arrangement more commonly found in prokaryote organisms. Promoters have not been precisely defined, but there may be as few as one or two transcription start sites per chromosome (McDonagh et al. 2000; Myler and Stuart 2000). Once transcription is initiated, RNA polymerase then proceeds to transcribe in turn many individual genes of the array. Although transcription appears to be continuous, a giant transcript is not produced because, as it is synthesized, each gene is individually processed. A specific 39-nucleotide leader sequence is spliced in *trans* to the 5′ end and a poly(A) tail added to the 3′ end of each gene transcript. These two processes may be mechanistically linked. There are no intervening sequences in the coding regions of nuclear genes. Finally, gene expression itself is rarely, if at all, regulated at the transcriptional level in *Leishmania* (the general rule for eukaryotes) but occurs at a number of post-transcriptional steps depending on the specific gene concerned. These include the RNA processing steps mentioned above, mRNA stability, translational control and post-translational turnover of proteins.

Genetics

Leishmania appear to be essentially diploid organisms (Bastien et al. 1992; Lighthall and Giannini 1992) although occasional examples of aneuploidy have been found. Diploidy has been confirmed for a number of specific genes in which two rounds of gene disruption are required in order to generate null mutants. Chromosomes do not condense during mitosis but can be separated and enumerated by pulsed-field gel electrophoresis. The use of this technique, the identification of genetic markers for specific chromosomes and an ongoing effort to sequence the *Leishmania major* genome have recently established the karyotype of *Leishmania* (Ivens and Blackwell 1999). There are 36 chromosomes in most species, ranging in size from 285 to 2 800 kb. Species related to *L. mexicana* and *L. braziliensis* (both from the New World) possess 35 and 34 chromosomes respectively, a result of possessing fused versions of certain chromosomes that are separate in other species (Britto et al. 1998). However, synteny is well conserved between species examined to date, variation residing mainly in the spacing of genes rather than their order (Britto et al. 1998; Ravel et al. 1999). There is some circumstantial evidence for a sexual cycle in *Leishmania* (Bastien et al. 1992), but so far no one has succeeding in performing a genetic cross (as in African trypanosomes, Chapter 18). Population genetic data indicate that *Leishmania* has a predominantly clonal structure and reproduces asexually (Tibayrenc et al. 1990; Victoir and Dujardin 2002). Therefore, sexual reproduction, if present, may be more important on an evolutionary scale than in the routine life cycle progression of the parasite.

CLINICAL ASPECTS

Clinical manifestations

One of the most remarkable features of the leishmaniases is the diversity of diseases caused by morphologically similar parasites living in a single series of cells (Figure 16.11). This diversity is by no means fully explained by genetic diversity among the parasites and is generated in part by the variety of host responses to the infection and in part by the (largely unexplained) restriction of the parasites to specific parts or organs of the body. One factor restricting parasites to the skin may simply be temperature, to which some species of

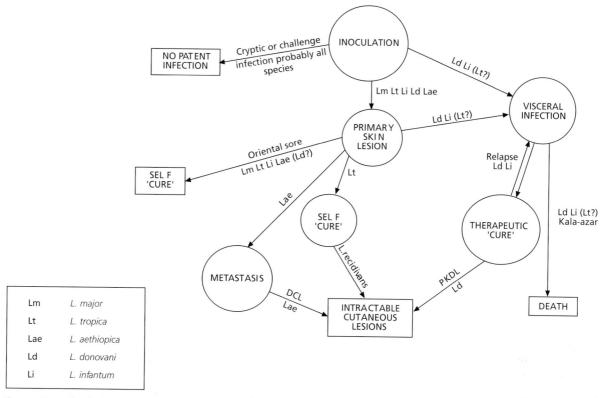

Figure 16.11 The 'leishmaniasis spectrum', showing the potential progress and outcome of infection with Old World species. (Modified from Molyneux and Ashford 1983)

Leishmania are particularly sensitive. In experimental animals, the genetic make-up of the host has been shown to have a great effect on the host response to infection.

There is considerable evidence that even the most virulent *Leishmania* parasites may cause no detectable disease in certain individuals. Wherever transmission occurs, there are numerous people who have positive immunological reactions but no cutaneous scar or history of visceral disease. Sub-clinical infection must be particularly common in foci of *L. infantum*. Here, disease in adults is usually associated with underlying immunosuppressive disease and may then be very common; even among children overt disease may be extremely rare despite intense transmission.

CUTANEOUS LEISHMANIASIS OR ORIENTAL SORE

A single cutaneous lesion, or leishmanioma, at the site of each infective sandfly bite is characteristic of infection with *L. tropica*, *L. major*, *L. aethiopica*, certain strains of *L. infantum* and, rarely, *L. donovani*. Oriental sore, caused by *L. major* or *L. tropica*, is usually painless and self-curing, and is frequently passed off as a trivial matter. Although this may be true of a single lesion in a hidden place, multiple lesions of *L. major* or disfiguring

facial lesions due to *L. tropica* may be physically or psychologically crippling.

Generally, all lesions on any one patient have a similar appearance and progress synchronously. In *L. major* infection, the center of the 'wet' lesion becomes necrotic and exudative forming a loose crust of congealed serum above a granulomatous base that eventually produces the characteristic scar. *L. major* lesions may number more than 100. With *L. tropica* the 'dry' lesion is more swollen and less necrotic; the exudate is less profuse and accumulates as a thicker crust. Numerous lesions are rare, the average number being around two. These differences are by no means consistent and are certainly not diagnostic. *L. aethiopica* lesions are even more swollen and less necrotic than those of *L. tropica*; frequently they are barely exudative, with gradual scaling or exfoliation of the dermis at the center. These lesions often last for years before healing. Cutaneous lesions resulting from *L. infantum* resemble those of *L. tropica* but are generally smaller and more indolent.

Oriental sore is not usually associated with systemic manifestations, although there may be enlargement of the draining lymph nodes. The necrotic centers of lesions are heavily contaminated with bacteria but these are not usually invasive, so inflammation, suppuration, and pain are atypical.

LEISHMANIASIS RECIDIVANS

Leishmaniasis recidivans (LR) is an unusual sequel of oriental sore caused by *L. tropica* infection. In this condition, small and usually non-ulcerating lesions appear, mainly on the margin of an apparently healed scar of oriental sore. These may last indefinitely, gradually extending the limits of the initial scar, and causing severe disfigurement. LR is associated with a strong cell-mediated immune response; as it closely resembles cutaneous tuberculosis, both superficially and histologically, the name 'lupoid leishmaniasis' is sometimes used.

DIFFUSE CUTANEOUS LEISHMANIASIS

Diffuse (more correctly 'disseminated') cutaneous leishmaniasis (DCL) is a rare form of disease, caused by *L. aethiopica* in the Old World; just over 100 cases have been recorded. The parasites are restricted to the skin but become widely distributed over much of the surface in large, swollen plaques and nodules. DCL resembles lepromatous leprosy, for which it was initially mistaken, but the abundant parasites in the nodules provide easy distinction. This is an immunologically anergic condition in which neither humoral nor cell-mediated responses are activated; it is very difficult to treat and may last for the rest of the greatly disrupted life of the patient.

VISCERAL LEISHMANIASIS OR KALA-AZAR

Visceral leishmaniasis is characteristic of infection with *L. donovani* and *L. infantum*. There are also suggestions that occasional cases are caused by *L. tropica*. The two main viscerotropic species cause broadly similar diseases although *L. infantum* mainly affects young children and has a greater tendency to cause lymph node enlargement. The general course of the disease is thought to be closely related to the health status of the patient at the time of infection and longitudinal studies have provided good evidence that malnutrition exacerbates *L. infantum* infection (Badaro et al. 1986).

The infection begins at the site of the infective sandfly bite, where there may be an initial ulcerating lesion. This has been best described in dogs, where its occurrence does not reflect the eventual outcome (Vidor et al. 1991). In humans, a leishmanioma is rarely seen and, if present, it may become a self-limiting lesion resembling oriental sore without producing visceral disease.

The clinical features of visceral leishmaniasis are well described by Rees and Kager (1987). Typically, visceral disease develops after an incubation period of 2–6 months and is accompanied by a persistent, irregular, low-grade fever, but onset may be delayed for many years. Once established, subsequent development of disease is quite variable; onset of severe symptoms may be very acute with rapid progression to life-threa-tening disease within 2 weeks, or progression may be insidious, almost unnoticed, and probably passed off by the patient as malaria, until the abdominal swelling becomes a major concern. Splenomegaly is the most consistent and noticeable sign. The spleen becomes enormously enlarged, extending well below the umbilicus, its size emphasized by an accompanying cachexia. Hepatomegaly is less consistent and less extreme but is usually present in late cases. The hematological picture is greatly altered with anemia and leukopenia being the most manifest changes. The outcome of untreated fully symptomatic visceral leishmaniasis is usually fatal.

POST KALA-AZAR DERMAL LEISHMANIASIS

Post kala-azar dermal leishmaniasis (PKDL) is a relatively common consequence of therapeutic cure from visceral leishmaniasis caused by *L. donovani*. It is not associated with *L. infantum* infection. PKDL is occasionally seen in patients with no history of visceral disease but only in places where *L. donovani* is transmitted; this is probably a sequel to subclinical infection. Sometimes, PKDL develops before the visceral infection has cured but its onset may be delayed for as much as 2 years. The extent to which chemotherapy actually causes PKDL is an interesting conundrum and one that is unlikely to be resolved soon as visceral disease is always treated if drugs are available and untreated cases have a poor prognosis. PKDL has been reported in some 20 percent of cured cases in India and is said to be rare in Africa but a very high incidence has been recorded in a recent Sudanese epidemic (Zijlstra et al. 1994).

PKDL is a variable disease; it may start as a widespread punctate, progressive depigmentation giving the skin a mottled or freckled effect, or may first be noticed as discrete papules, mainly on surfaces exposed to light. It can progress to produce an extensive surface of coalescing papules or large discrete nodules which superficially resemble DCL or lepromatous leprosy. The lesions are delicate, but do not ulcerate unless traumatized. The duration of the untreated condition, the numbers of parasites and the immune responses of patients with PKDL are very variable.

LEISHMANIA AND HUMAN IMMUNODEFICIENCY VIRUS (HIV) COINFECTION

Most of the information on coinfection relates to *L. infantum* acquired in south-western Europe (France, Italy, Portugal, and Spain); few coinfections with other *Leishmania* species have been described to date although they must certainly occur. From 1990–98, 50 percent of visceral leishmaniasis cases presenting in these countries were coinfected with HIV (Dereure et al. 1995; WHO 2000). Visceral leishmaniasis and HIV together are synergistic infections because visceral leishmaniasis accelerates the development of aquired immune defeciency syndrome (AIDS) and the presence of HIV infection enhances the spread of visceral disease (Wolday et al. 1999). The diversity of strains found is much greater than usual and strains that are normally associated with cutaneous disease may become visceral in coinfection (Pratlong et al. 1995). Serological evidence suggests that HIV may both activate subclinical leishmaniasis and make the patient susceptible to a new infection (Gradoni et al. 1993); serology is of little use in diagnosing coinfection. In a group of patients in Spain, the clinical presentation was characteristic of kala-azar and most were at CDC HIV stage 4 with fewer than 200×10^6 CD4+ lymphocytes per l. Visceral leishmaniasis was the first reported severe infection in 10 of 47 cases (Medrano et al. 1992). Other workers have found a wider diversity of presentations, frequently with infection of the upper alimentary tract (Peters et al. 1990). Many authors have suggested that visceral leishmaniasis should be included in the list of opportunistic infections indicative of AIDS (e.g. Altes et al. 1991). The predominance of intravenous drug users among these patients suggests that syringe transmission may occur (Alvar and Jimenez 1994). Cases of coinfection in Europe are expected to fall as a result of availability of new anti-HIV drugs but the problem is increasing in eastern Africa, the Indian subcontinent, and South America. The true extent is unclear, but it is certainly grossly under-reported, as few of these countries implement active surveillance of Leishmania/HIV coinfection.

IMMUNOLOGY

Introduction

The responses of the immune system to Leishmania infection are highly complex (reviewed by Mauel and Behin 1987; Liew and O'Donnell 1993; Bogdan and Rollinghoff 1999). They may accelerate cure or exacerbate the disease depending on the particular circumstances. This is partly due to the effects of genetic variation in the mammalian host, partly due to genetic variation in the parasites between species and strains and partly due to chance factors, such as the location, inoculum size, and number of infective bites received.

Host genetic variability

The evidence that genetic variation in humans contributes to variability in their immune responses and clinical outcome is mainly anecdotal and essentially based on the wide clinical spectrum produced by apparently identical parasites. More definitive evidence is now accumulating (see Blackwell 1999), but even under circumstances in which individuals share the same environment

and differences exist in incidence or severity of disease, it is impossible to be certain whether this reflects human genetic variation or, for example, different numbers of bites received. The main evidence supporting the importance of genetic variation comes from experimental studies conducted with inbred strains of mice (reviewed by Bradley 1987; Blackwell 1992). It is difficult to determine the extent to which these findings can be extrapolated to human patients. Nevertheless, this experimental approach has proved productive and several genetic loci have been identified in the mouse that are linked to resistance phenotypes. The genes encoding some of these have now been cloned and, although the functions of these are not fully resolved, this opens up the possibility of a search for homologs in the human genome. This should lead to progress in assessing the contribution of human genetic variability to the clinical spectrum of disease.

Human cutaneous leishmaniasis

Spontaneous cure of uncomplicated L. major and L. tropica infection usually results in a solid immunity. There is a marked development of cell-mediated reactions but a weak antibody response, although specific antibody can be detected. The cell-mediated reactions are responsible for a marked delayed type hypersensitivity (DTH) response to the inoculation of leishmanin (washed promastigotes in 0.5 percent phenol saline) in active and cured cases. The DTH response can often be detected before healing. In persistent L. tropica infections that result in LR there is still a strong DTH response, indicating that the parasite can survive despite the high immunological reactivity of the host. In contrast, DCL caused by L. aethiopica is characterized by a lack of DTH response but the presence of antibodies thus tending more to resemble response to visceral leishmaniasis.

Human visceral leishmaniasis

Radical cure of L. donovani infection by chemotherapy can also generate protective immunity in some individuals. In contrast to cutaneous leishmaniasis, cell-mediated immunity is impaired in active kala-azar patients who consequently lack a DTH response but this can be demonstrated after cure. There is, however, a marked humoral response to visceral leishmaniasis. Polyclonal B cell activation produces mainly immunoglobulin (Ig)G and IgM antibody, most of which is non-specific, but high titers of specific antibody can also be detected. In PKDL, a curious situation arises in which there seems to be immunity in the viscera but not in the skin. Only a proportion of PKDL patients show a DTH response; antibody levels are elevated but are not as high as in active kala-azar.

Experimental leishmaniasis

The relative paucity of information on the immunology of human leishmaniasis is in contrast to an explosion in the literature relating to experimental infections in inbred strains of mice particularly using *L. major* (Liew and O'Donnell 1993; Milon et al.1995; Solbach and Laskay 1995, 2000; Nabors 1997; Alexander et al. 1999; Sacks and Noben-Trauth 2002). This explosion is due not only to the potential impact that such work may eventually have on human medicine, but also to the role that *Leishmania* has played in fundamental immunological research itself. The relatively recent nature of these findings, together with the difficulties of investigating human immunology, means that the extent to which they can be extrapolated to human leishmaniasis is difficult to predict. In all of these experimental studies, it should be remembered that the infection occurs against a specific genetic background, usually in a highly susceptible strain of mouse, and that the interaction between innate genetic factors and specific immune responses is not well understood. Despite these caveats, some of the major findings are likely to be borne out and thus merit brief discussion. The main effector cells in murine cutaneous leishmaniasis are two subsets of CD4+ T cells, T helper 1 (Th1) and T helper 2 (Th2) cells, each associated with a distinct cytokine profile. Resistance and healing in mice is associated with a Th1 type response whereas susceptibility and disease progression is accompanied by a Th2 response. These polarized Th1 or Th2 responses have not been demonstrated in murine *L. donovani* infections which instead are associated with an absence of a Th1 response in susceptible strains (Kaye 1995). Early events in the dermis and epidermis of the skin may play a crucial role influencing the development of resistance or susceptibility (Moll 1993; Solbach and Laskay 1995; Ritter and Korner 2002). A Th1/Th2-like dichotomy in CD4+ T cell clones can be demonstrated in human leishmaniasis (Kemp et al. 1993, 1994; Ghosh et al. 1995). The main effector cell in protective immunity is the macrophage host cell itself, in which the killing mechanism is a cytokine-induced production of nitric oxide (Liew and Cox 1991).

PATHOLOGY

In visceral leishmaniasis, parasites may be found in practically every organ, whereas they are largely limited to the skin in cutaneous leishmaniasis; nevertheless the histological spectrum of the different diseases is surprisingly uniform. Cells of the mononuclear phagocyte system are infected (Ridley 1987), particularly the more actively phagocytic members of the series. Further phagocytic cells are attracted to the site which, in turn, become infected so a colony of infected cells is produced. The parasites may be restricted to the initial cutaneous site or may, after a variable delay, disseminate in the blood rather than the lymphatics to establish metastatic colonies in almost any organ. The temperature dependence of each species may partly be responsible for the ultimate distribution of the lesions. Dissemination is also limited by the histological immune response.

In cutaneous leishmaniasis, a spectrum of histological response to infection, somewhat analogous to that in leprosy, is based on the amount of lymphocyte, plasma cell, and giant cell infiltration constituting the granulomatous inflammation. The minimal, anergic, (Ridley's 'group I') response is best seen in DCL with abundant heavily infected macrophages, few other cells and no giant cells. There is little necrosis and no ulceration. The other extreme, the 'group V' hypersensitive response, is tuberculoid, with an epithelioid cell granuloma, large Langerhans cells, variable plasma cell numbers and scanty lymphocytes. Few parasites are present and, again, there is no necrosis. Langerhans cells may migrate to the draining lymph nodes where they present antigen to T cells initiating the immune response (Moll 1993). Between these two histologically stable extremes, both of which produce long lasting lesions intractable to treatment, are three less well defined categories which, to some extent, represent a progression associated with lymphocyte infiltration, variable amounts of necrosis, and healing. Necrosis is the main feature of an effective immune response and subsequent healing, whereas small numbers of parasites may be destroyed by macrophages and giant cells.

The histological pattern in deep tissues is less fully described but seems to be similar; again, the main feature of healing is the presence of lymphocytic granulomata and necrotic centers. In contrast to cutaneous leishmaniasis, however, natural healing is not the rule and the lesions are more like the anergic, group I type described above with widely disseminated proliferations of infected macrophages and little infiltration. Most of the pathology is caused by this hyperplasia and damage to the organs caused by associated congestion. According to El Hag and Hashim (1994), the liver shows infected Kupffer cells and macrophages together with chronic mononuclear cell infiltration of the portal tracts and lobules, ballooning degeneration of the hepatocytes and fibrosis of the terminal hepatic venules.

DIAGNOSIS

Relative unfamiliarity with the leishmaniases among practitioners in non-endemic areas occasionally leads to serious misdiagnosis and mistreatment. A good travel history may arouse suspicion in those versed in geographical medicine, but reference to a tropical specialist is usually required to confirm the diagnosis. In areas where the disease is better known, clinical diagnosis is frequently regarded as sufficient. Although this may be justified during epidemics, a number of alternative diagnoses will inevitably be missed if demonstration of the parasites is not attempted on all patients.

Clinical diagnosis

Clinical diagnosis of any of the leishmaniases depends primarily on awareness. None of the signs or symptoms is diagnostic but, when combined with a relevant travel history, or in an endemic area, a good presumptive diagnosis may be made clinically and can be confirmed parasitologically. 'Diagnosis' by therapeutic trial is commonly practiced but is unsatisfactory, particularly in view of the expense and potential toxicity of the drugs.

Visceral leishmaniasis is often mistakenly recorded post-mortem as 'malaria that failed to respond to treatment' (and is probably included in malaria mortality statistics). The typical patient presents at a late stage, with persistent but fluctuating low grade fever, weight loss giving the appearance of severe starvation, and spleno- or hepatosplenomegaly. The patient is usually alert and feels remarkably well, considering his or her condition. The skin is sometimes said to be 'muddy', 'pale' or 'dark', but these are not diagnostically useful descriptions. There is frequently persistent diarrhea. Lymphadenopathy is an irregular feature.

Non-specific laboratory tests will show marked leukopenia (pancytopenia, mainly neutropenia), anemia, raised serum proteins with reversal of the albumin/globulin ratio due to greatly raised IgG levels.

Manson-Bahr (1987) suggests the following differential diagnoses: malaria, relapsing fever, trypanosomiasis, brucellosis, liver abscess, tuberculosis, tropical splenomegaly, myeloid leukemia, lymphoma, cirrhosis of the liver, schistosomiasis, thalassemia, histoplasmosis, and various gammopathies.

Diagnosis of visceral leishmaniasis in the presence of HIV infection is particularly difficult as the presentation may be very atypical and serological tests may be negative.

All forms of cutaneous leishmaniasis usually present without systemic disease. Typical lesions of oriental sore are readily recognized by the raised edges, necrotic, exudative center with granulomatous base, long duration, and painlessness. Less typical lesions take a bewildering array of forms and must be distinguished from basal cell carcinoma, tuberculosis, various mycoses, cheloid and lepromatous leprosy.

LR may be indicated by the presence of a scar and positive skin test, whereas DCL shows no ulceration or scarring and has a negative skin test.

Parasitological diagnosis

A diagnosis of leishmaniasis can be formally confirmed only by the demonstration of the parasites. These are readily seen in smears or touch preparations of infected tissue stained with Giemsa's stain, preferably at pH 7.2 rather than the pH 6.8 normally used in hematology. Sections of tissue stained more conventionally with hematoxylin and eosin are much more difficult to interpret so the diagnosis must be suspected before material is prepared. Preparations from cutaneous lesions may be made from punch biopsies, but it is usually sufficient to take a small quantity of material from the living, but diseased, tissue at the edge of the lesion by a slit smear or other minimally invasive technique.

Deeper tissues may be sampled by needle biopsy. Material may be aspirated from lymph nodes following injection of a small quantity of physiological saline. Bone marrow smears are made using standard hematological techniques. Spleen aspirate is the most reliable material for demonstrating parasites in kala-azar. Although many physicians are reluctant to take spleen aspirates others have no hesitation and even use this method to monitor treatment. It is vital to use the correct technique and equipment with confidence so that the capsule of the spleen is penetrated by a fine needle for only a fraction of a second (see Bryceson 1987a for precise instructions).

Parasites may be scanty and are mostly extracellular in slide preparations so these may have to be examined for at least 15 minutes using a ×100 oil immersion objective before the diagnosis can be confirmed. Amastigotes are identified by their size (3–5 μm) and the possession of a nucleus and kinetoplast; although smaller, the kinetoplast is often easier to detect because the DNA is very densely packed and produces a deep purple spot or rod on staining. It may be possible to provisionally identify the parasites to species level by considering the circumstances and clinical features. *L. major* and *L. tropica* can be separated with some confidence by the abundance of parasites in the latter, their smaller size and the presence of more numerous parasites in each infected cell. Complete identification requires the isolation of parasites in culture.

Whatever material is collected, it is desirable to culture the parasites in blood-agar or insect tissue culture medium so that they can be identified at strain level by isoenzyme analysis. For special purposes, material can be inoculated into a hamster, generally the most susceptible experimental host. Monoclonal antibody and DNA probes, with or without PCR amplification, have been developed for identification of the parasites. These are not yet commercially available but are of great use in epidemiological or ecological investigations.

Two indirect approaches for investigating the presence of parasites in clinical samples are currently under development. There are numerous publications describing the potential application of PCR to *Leishmania* diagnosis (for example, Wiegle et al., 2002; Salotra et al. 2001; Lachaud et al. 2001). There is, as yet, no general consensus about which sequences are the best targets for diagnosis, but repetitive nuclear genes, such as ribosomal and mini-exon genes, or kDNA are often used. Also, there is no one-step PCR diagnostic test that will identify all species. However, these drawbacks are likely to be solved in the near future. The second area of advance is in antigen detection systems, specifically the detection of antigen in the urine of patients with visceral leishmaniasis (Attar et al. 2001; Sarkari et al. 2002). This is particularly welcome as this is a completely non-invasive technique, a significant advantage compared to bone marrow or spleen biopsy. As yet, these methods have not supplanted direct parasitological diagnosis and require further testing and clinical trials but, given their technical advantages, this seems a matter of time.

Immunological diagnosis

Immunological diagnosis (reviewed by Kar 1995) is of value in screening suspected kala-azar patients, other than those coinfected with HIV, and in epidemiological studies.

In the formol-gel test, a drop of full strength formalin is added to 1 ml of serum. A positive result is indicated by the rapid and complete coagulation of the serum. This simple test merely indicates greatly increased serum proteins and thus is non-specific. Nevertheless, it is still widely used; in order to be acceptable in peripheral clinics in endemic areas, this is the level of technology that any new test will have to approach.

More specific tests, which become positive earlier in kala-azar, include the indirect fluorescent antibody test (IFAT), the enzyme-linked immunosorbent assay (ELISA) and a direct agglutination test (DAT). This last technique requires minimal equipment, is robust in the most difficult conditions and is exquisitely sensitive. DAT is fast becoming available for use at peripheral-level clinics and the development of freeze-dried antigen (Meredith et al. 1995) promises even wider availability of this test. For active case detection, numerous positive responses to serological tests are found in people who show no disease. To treat all these people as patients would be a mistake; probably the best thing is to examine them carefully, treat those with signs of disease and to monitor the others carefully.

The specific reagents for the above serological tests are not readily available commercially, but they can be prepared relatively easily in the laboratory.

A delayed hypersensitivity reaction to intradermal crude Leishmania antigen is produced in healing or cured cases of both cutaneous and visceral leishmaniasis. This leishmanin, or Montenegro skin test is of great value in epidemiological studies, but is of little clinical use as the number of positive reactors is usually much greater than the number with a history of disease

suggesting that either the test is non-specific or there is a considerable amount of sub-clinical infection.

EPIDEMIOLOGY

Introduction

Knowledge of the epidemiology of the leishmaniases has benefited in recent years from our growing understanding of the taxonomy of the group and from the possibility of accurate identification of morphologically similar parasites derived from human cases, sandfly vectors or non-human mammalian hosts. Current taxonomy corresponds much better with ecology and epidemiology than it does with clinical manifestations (see Molyneux and Ashford 1983). Anomalies and questions remain and the 'spectral epidemiology' (different diseases caused by ostensibly similar organisms) is largely unexplained.

It has been conservatively estimated that, in the Old World, 150 million people in 40 countries are at risk of cutaneous leishmaniasis and 180 million people in 39 countries are at risk of visceral leishmaniasis. The total number of reported cases has been estimated at 240 000 and 73 000 annually for cutaneous and visceral leishmaniasis respectively but this is doubtless a gross underestimate of the real numbers (Desjeux 1991; Ashford et al. 1992a). The figure of 12 million cases annually (sometimes misquoted as 1.2 million) was estimated some time ago by Walsh and Warren (1979), but although the World Health Organization (WHO 1990) uses the higher figure of 12 million, it is hard to know how it was derived.

Although transmission is possible by contagion, by transfusion or in utero, these routes have little if any known epidemiological significance. Effectively, all transmission is by the bite of a phlebotomine sandfly (Diptera: Phlebotominae) (reviewed by Killick-Kendrick 1990b). The leishmaniases illustrate the entire gamut of zoonotic patterns of transmission.

Epidemiology of *L. major* infection

L. major is widely distributed in the Palaearctic and Aethiopian zoogeographical regions along the northern and southern fringes of the Sahara, extending south into northern Kenya and through Arabia, southern Iraq and Iran, into the central Asian states of Turkmenistan, Uzbekistan, and northern Afghanistan. It also ranges through Iranian and Pakistani Baluchistan to the Indian border in Rajasthan (see Figure 16.1).

Throughout this wide range, the parasite is restricted to arid and semi-arid zones. Its broad distribution seems to be determined by that of its vectors and its detailed distribution is largely determined by that of its natural mammalian hosts.

The three known vectors belong to the subgenus *Phlebotomus* (*Phlebotomus*) of which *P. papatasi* is the most widespread and abundant. *P. papatasi* transmits *L. major* in all but the sub-Saharan and Indian parts of its distribution; it is replaced by *P. duboscqi* in the former and *P. salehi* in the latter. *P. papatasi* has a much wider distribution than *L. major*, especially in northern India, but here it is largely synanthropic and has no access to suitable reservoir hosts. The efficiency of this species as a vector may be affected by the plant juices on which it feeds, some of which may inhibit growth of the parasites (Schlein and Jacobson 1994b).

The known reservoir hosts of *L. major* are rodents. In the central Asian deserts, the great gerbil *Rhombomys opimus* lives in dense colonies with deep and complex burrow systems that provide shelter and possibly a breeding site for *P. papatasi*. The gerbils breed throughout the warmer months and by the end of the transmission season, in autumn when the rodent population is greatest, almost every animal may be infected. The infection is usually restricted to the ears, which become only slightly swollen and eroded, and there is no indication that the rodent is seriously affected.

Great gerbil colonies are best developed in areas where the soil is deep and of a consistency that allows permanent burrows to be constructed. These conditions are found on the alluvial fans in the valleys of the largest rivers draining the great central Asian loess deposits, such as the Amudarya.

The distribution of the rodent extends much further north in central Asia than that of the vector. In these zones, they are infected with other parasites, *L. gerbilli* and *L. turanica*, which behave similarly in the rodent but do not infect humans and are transmitted by different sandfly vectors (Strelkova et al. 1993).

The other rodent reservoir host of *L. major*, which has very narrow habitat requirements, is the fat sand rat *Psammomys obesus*. This host maintains populations of the parasite in Algeria, Tunisia, Libya, north Sinai, the Jordan Valley, southern Syria, and much of Arabia. This rodent is unique in being able to derive its entire nutrient and water requirements from the succulent, but highly saline, stems and leaves of halophilic plants of the family Chenopodiaceae. These plants dominate the internally draining salt pans, sebchet or playas of much of the desert fringes and it is here that the sand rats are concentrated. *P. obesus* burrows are less deep or extensive than those of *R. opimus* and provide a less favorable habitat for *P. papatasi*. Furthermore, the rodents breed mainly in the winter when there is no transmission so the young are not exposed to infection until they are a few months old. Nevertheless, all the evidence supports the conclusion that *P. obesus* is responsible for the maintenance of *L. major* in most of this part of its range.

Some foci of *L. major* in the Palaearctic occur in the absence of both *P. obesus* and *R. opimus* and have been attributed to various rodent hosts such as jirds, *Meriones* spp. (Rioux et al. 1982) or the mole rat *Nesokia indica*. The structure of these foci, notably that on the Iraq–Iran border, remain to be fully understood.

In its Aethiopian range, where *L. major* is usually transmitted by *P. duboscqi*, various rodents have been found infected, notably the grass rats *Arvicanthis* spp. and the multimammate rats *Mastomys* spp., in which the parasites are found in the viscera but do not seem to cause serious pathology. These rodent populations fluctuate wildly and do not have narrow habitat requirements so they are not good candidates for the maintenance of vector-borne infections. Possibly there is no individual rodent species maintaining the parasite but a diversity of hosts may be required.

L. major infection in humans is most conspicuous in epidemics in groups of people entering sparsely inhabited zoonotic foci. Major outbreaks have occurred in military groups and in workers on development projects. Development frequently alters the environment sufficiently so that, following a brief epidemic period, the infection disappears as has been seen particularly in irrigation schemes. Residents in zoonotic foci usually become infected at an early age and, being familiar with the infection, are less concerned.

The dynamics of human infection are largely governed by the rate of exposure to infection and the almost complete resistance to reinfection. Beljaev and Lysenko (1977) have modeled the stable-state system, making minimal, realistic assumptions; they show that at low rates of exposure (fewer than 0.05 exposures per person per year) incidence is proportional to this rate. They term locations where these conditions prevail 'hypoendemic foci'. In contrast, hyperendemic foci are those where the exposure rate is greater than 0.25 exposures per person per year. In such foci, lifetime incidence is largely independent of the exposure rate, change in which alters only the mean age at which people become infected. As most people who live in such hyperendemic foci are infected when young and subsequently become immune, the incidence is dependent on the rate at which non-immune people are recruited into the population, whether by birth or immigration.

In sub-Saharan Africa, outbreaks are much less frequent and human cases occur sporadically in both space and time. In some foci, in Ethiopia and Kenya for example, human infection is extremely rare even in places where infected rodents have been found.

Transmission between humans without a mammalian reservoir host has not been well established, although the possibility cannot be excluded. A series of epidemics in Khartoum and nearby towns in Sudan in the late 1980s occurred in dense, stable human populations that had never previously known the infection (El Safi et al. 1988). The best available explanation is that the parasite was introduced by displaced people from endemic areas to the west and that it was transmitted to residents by the abundant *P. papatasi* living close to the Nile.

Epidemiology of *L. tropica* infection

L. tropica is mainly found in the ancient cities of west and central Asia (see Figure 16.1). Generally, *L. tropica* inhabits slightly wetter areas than *L. major*. The two species are geographically sympatric in places but are ecologically separated. Vernacular names for the infection frequently refer to endemic cities, for example Balkh sore, Baghdad boil, Aleppo evil, and Delhi boil; this probably reflects the perception of visitors rather than indigenous people.

Urban cutaneous leishmaniasis is transmitted by *Phlebotomus* (*Paraphlebotomus*) *sergenti* throughout its range. This sandfly can reach high populations in crowded cities and suburbs but is not restricted to the peridomestic environment. Dogs have been found to be infected but it is generally thought that the main hosts are humans. Unlike *L. major*, the rate of exposure to infection depends on the number of active human lesions as well as other factors. The human lesion normally lasts around 1 year and people with cured lesions are immune. Therefore, in conditions of potentially high transmission, most people are immune, but few are susceptible leaving even fewer actually infected. The maintenance of stable transmission in a stable human population has been estimated to require some 20 000 sandfly bites per person per year which, in the seasonal climates of central Asia, may mean 100 bites per person per night. Such conditions are rarely, if ever, met and the infection must rely either on movements of people (Ashford et al. 1992b) or on spatial heterogeneity in the immunity of the population, moving from place to place in localized epidemics (Ashford et al. 1993). In this way *L. tropica* resembles a viral infection such as measles rather than most other eukaryotic parasites.

There are reports of *L. tropica* infection occurring sporadically in rural settings, which cannot easily be explained by the above patterns. Furthermore, the finding of *L. tropica* causing visceral leishmaniasis in American soldiers who served in the Gulf region during the first Gulf war in 1991 (Magill et al. 1993, 1994) and in Kenyan and Indian villagers (Mehbratu et al. 1989; Sacks et al. 1995) in the heart of areas of *L. donovani* transmission, pose intriguing epidemiological questions for which no sensible answer can yet be proposed.

It was thought at one time that *L. tropica* did not occur in Africa. Parasites from an ostensibly zoonotic outbreak of cutaneous leishmaniasis in Tunisia were identified as a new species, *L. killicki*, which, on the cladograms of Rioux et al. (1990) (see Figure 16.3), is close to *L. tropica*. A wide variety of *L. tropica*-like strains are responsible for another apparently zoonotic focus in Morocco (Pratlong et al. 1991). Parasites from sporadic cutaneous cases, hyraxes *Procavia* spp. and from the sandflies *P. rossi* and *P. guggisbergi*, in Namibia and Kenya were found to be intermediate between the original *L. killicki* and *L. tropica*. It is a matter of opinion whether these parasites should be included with *L. tropica* or not (Sang et al. 1992). Occasional aberrant cutaneous, visceral, or mixed infections have been reported elsewhere in Africa, notably in Tanzania, Zambia, and Malawi (Pharoah et al. 1993). These parasites have not been isolated and their identity is unknown but presumably they originated in zoonotic foci that await discovery and description.

Epidemiology of *L. aethiopica* infection

L. aethiopica is restricted to the highlands of East Africa in Ethiopia and Kenya, between c. 1 800 m and 2 700 m above sea level and within the 800 mm isohyet (see Figure 16.1). In all known foci, it is associated with hyraxes (Hyracoidea) which are the natural reservoir hosts (Ashford et al. 1973). These animals have very poor temperature regulation and live in deep clefts in rocks or trees in which they can find the shelter they require for the maintenance of their body temperature. Despite this restriction, they are found at all altitudes from below sea level in the Danakil desert to above 4 000 m where there is frost every night. Hyraxes are long-lived and live in colonies or family groups so, although their rate of reproduction is low, they are good candidate reservoir hosts. Parasites have been isolated from normal skin and external mucosa showing no sign of the infection which presumably lasts a long time.

The vectors, *P. longipes* and *P. pedifer*, have less specific habitat requirements, being common in houses or associated with cattle, but they are restricted in altitude. The distribution of the parasite is, therefore, governed by the habitat of the reservoir host and the altitude range of the vectors. Wherever they have been adequately studied, human cases can be attributed to the nearby presence of infected hyraxes. Incidence of disease in humans is rarely high and the rare cases of diffuse cutaneous leishmaniasis caused by this parasite have not been explained epidemiologically.

Epidemiology of *L. donovani* infection

L. donovani is surely one of the great scourges of mankind having been responsible for a series of historical deadly epidemics each causing depopulation of affected areas. The geographical distribution of *L. donovani* infection is somewhat labile (see Figure 16.2). In the Indian subcontinent, it occurs in the valleys of the Ganges and Brahmaputra Rivers, in Bihar, lowland Nepal, West Bengal, Bangladesh, and Assam, as well as further south, around Madras (reviewed by Sanyal

1985). It is often presumed that visceral leishmaniasis in south China was caused by *L. donovani* but this has not been confirmed. In Africa, the infection is largely restricted to Sudan, Ethiopia, and Kenya.

The ecological distribution of *L. donovani* is well known but poorly understood. In Asia, where humans are the only known hosts, the infection occurs in heavily populated rural areas in the rich silty flood-plains of the great rivers. Here, the vector *P. argentipes* is abundant and strictly synanthropic, breeding in the organically rich material on the floors of cow byres, almost the only sandfly whose larval habitat is reasonably well described (Ghosh and Bhattacharya 1991). The only known reservoirs of infection here are humans and persistent cases of PKDL may be particularly important in maintaining the parasite between epidemics (Addy and Nandy 1992).

In Africa, there are two types of habitat, each associated with a specific vector. *P. orientalis* occurs in *Acacia seyal* woodland, which is the dominant natural vegetation of the light, alluvial, montmorillonite, silty clays of eastern Sudan between latitudes 9°N and 14°N, extending into the western Ethiopian lowlands. *P. martini* occurs in less well defined habitats but is commonly associated with eroded termitaria in the heavy laterite clays of the southern borders of Sudan and Ethiopia, northern Uganda, and much of Kenya. *L. donovani* infection occurs either in fairly stable endemic conditions or as violent epidemics throughout the distribution of these two vectors. The Nile grass rat, *Arvicanthis niloticus*, has been found infected in substantial numbers in southern Sudan and may be an important reservoir host, but the shifting distribution of the epidemics and persistence of intensely endemic residual foci, too dangerous to even visit in the transmission season, remain largely unexplained.

Human *L. donovani* infection is endemic in parts of southern Ethiopia, Sudan, and northern Kenya, where most people are exposed to infection at some time during their lives but relatively few become sick (e.g. Zijlstra et al. 1994). An Ethiopian study indicated an exposure rate of c. 11 percent per year, but there were only 2 percent cases of overt disease annually per 100 non-immune individuals (Ali and Ashford 1994). Nearly all adult males and more than half the females showed evidence of previous exposure.

This apparent insusceptibility of most of the people cannot be true of the early epidemics in northwest India where the disease swept through the population like a plague killing a large proportion (and, incidentally, giving rise to the dreadful reputation of malaria: kala-azar was thought to be *the* malignant, lethal form of malaria (Rogers 1908)). The epidemic of the late 1980s in southern Sudan was equally virulent, and was estimated to have killed half of the affected population, some 100 000 people, in 5 years (Seaman et al. 1992).

Epidemics of kala-azar are often associated with natural disasters or social upheavals, although causal relationships are difficult to establish. The work of Sati (1962) documents the spread of an epidemic in Sudan with famine-related displacement. The 1980s Sudan epidemic may have resulted from the introduction of the parasite by people coming from endemic areas on the Ethiopian border to settled areas with potential transmission. These movements were related to the civil war, as was the poor condition of the recipient people, but any precise interpretation is largely informed speculation.

Although epidemics in Africa occur in rural, largely pastoral environments, those in India are in towns and villages with settled agriculture. In the north Bihar epidemic, maximum incidence was estimated at 6 per 1 000 per year, with more than 100 000 cases reported in some years. In 1992, there were 80 000 cases (Sacks et al. 1995). Reported cases are surely only a small fraction of the actual number occurring.

Epidemiology of *L. infantum* infection

L. infantum parasites are difficult to separate objectively from *L. donovani* and the distribution of the two forms is largely allopatric, so there is a good case for regarding them as subspecies. However, it is convenient to treat them separately because they are very distinct in both their ecology and epidemiology.

L. infantum is the only *Leishmania* species that occurs in both the Old and New Worlds (in the Americas it is sometimes called by its junior synonym *L. chagasi*; see Chapter 17, New World leishmaniasis for New World features). In the Old World, it is distributed in much of the Mediterranean basin, both in Europe and North Africa (see Figure 16.2). It extends erratically through Arabia, Turkey, and Iran to the Central Asian republics, Pakistan, and Kashmir (Rab and Evans 1995), as well as into northwest China. It is thought to have been much more widely distributed in China but visceral leishmaniasis has been almost eradicated throughout that country.

Throughout this range, the main host is the domestic dog, which develops an acute or chronic disease. Though eventually fatal, canine visceral leishmaniasis presents abundant parasites in the skin, available for transmission. The distribution of foci of infection depends on the presence of vector sandflies, usually species of the subgenus *Phlebotomus* (*Larroussius*). Although we are ignorant of the mechanisms by which the distribution of sandflies is regulated, each species has clearly defined ecological preferences. *P. ariasi*, for example, is largely restricted to hillsides with mixed oak *Quercus* spp. woodland. A more important vector, *P. perniciosus*, occurs at lower altitudes, in drier, limestone country, where it is particularly abundant in gardens and suburbs with outcrops and dry-stone walls. These are the main vectors in the western Mediterranean; they are replaced further east by *P. major* and a wide variety of less well studied probable vectors.

In areas where transmission is intense, almost all dogs become infected and kennels become very difficult to keep. Individual dogs vary greatly in their response to infection and there is some evidence that regionally bred dogs are less susceptible than exotic breeds.

Human infection must be very much more common than overt disease. The latter is classically restricted to children, especially those below the age of 2 years. In the post-war years, the incidence in children, though not in dogs, in most of southern Europe has decreased greatly so the small number of adult cases has become a greater proportion of the total (Marty et al. 1994). In fact, adult cases have now actually increased owing to greater susceptibility to disease in association with HIV and immunosuppression associated with transplant surgery.

CHEMOTHERAPY AND TREATMENT

Treatment of leishmaniasis has a chequered history. Few adequate clinical trials have been carried out, especially with cutaneous leishmaniasis; recommended dosages are frequently based on spurious information and definition of the state of the disease is rarely standardized. Patients frequently present at a late stage, having been treated informally with unknown remedies. Courses of treatment are usually prolonged and so are difficult to monitor or even achieve. Despite these misgivings, the various diseases can generally be cured quite effectively, albeit mainly with drugs that come from archaic pharmacopoeia (Chance 1995; Croft and Yardley 2002). In addition to healing the patient, treatment may also reduce the transmission of anthroponotic *L. tropica* or *L. donovani* infection.

Placebo and physical treatment

The first question is whether or not chemical treatment is justified or even required. Simple, single lesions due to *L. major* will heal naturally in a few months and leave the patient immune for life. It is only with multiple or disfiguring lesions that chemotherapy is justified. The same may be said of lesions due to *L. tropica*, though these are of longer duration. With *L. donovani* infection, sick patients have a high risk of death if left untreated, but active case detection may reveal numerous sub-clinical infections. It may be preferable to keep these under observation if possible, especially if, as is frequently the case, resources are very limited.

Cutaneous lesions, whether or not to be treated, should be disinfected and covered, to prevent secondary infection and to avoid infecting sandflies. Heat treatment, with hot compresses or infra-red radiation may accelerate cure. Hot water, in the form of frequent baths or saunas, is beneficial in diffuse cutaneous leishmaniasis, which responds very poorly to chemotherapy. Removal of lesions by cryotherapy has been recommended, but is controversial. For visceral leishmaniasis associated with massive splenomegaly and unresponsiveness to antimonial drugs, surgical removal of the spleen used to be practiced occasionally and has also been undertaken relatively recently (Magill et al. 1994).

Pentavalent antimonials

Pentavalent antimony, currently used in the form of meglumine antimoniate or sodium stibogluconate, replaced tartar emetic many years ago and is still the first line drug of choice (Croft and Yardley 2002). However, increasing incidence of unresponsive infections, now known to be due to genuine parasite drug resistance, means that this may not remain true for much longer (Sundar 2001). Intralesional infiltration is used for simple, single cutaneous lesions but intramuscular injection is required for all other cases. Despite the prolonged period of repeated treatment required for the cure of visceral leishmaniasis, these drugs are remarkably safe if administered in the correct doses; their reputation for toxicity is unjustified and relates rather to the trivalent antimonials used previously (Bryceson 1987b). Antimony is excreted rapidly from the body so doses are repeated daily and treatment is continuous throughout each course.

For visceral leishmaniasis, the World Health Organization (WHO 1995) recommends intramuscular injection of 20 mg Sb^{5+}/kg daily (to a maximum of 850 mg) for a minimum of 20 days, and until no parasites can be seen in spleen or marrow biopsies taken 14 days apart. Doses may be divided to counteract rapid excretion. Relapsing patients should be treated similarly.

For cutaneous leishmaniasis, the World Health Organization (WHO 1995) recommends 1–3 ml of either of the main antimony preparations, to be infiltrated into the base of the lesion, to be repeated once or twice, at intervals of 1–2 days. For parenteral treatment, 10–20 mg Sb^{5+}/kg i.m. is given daily until a few days after clinical cure.

Aromatic diamidines

Pentamidine, as the isethionate or dimethylsulfonate, is used as a second choice in cases that are unresponsive to antimony. This product may, however, have serious toxic effects, due to sensitivity at the site of intramuscular injection, too rapid intravenous injection, accumulation, or overdosage. The drug may disrupt blood-sugar homeostasis and lead to intractable diabetes after prolonged use. The margin between therapeutic and toxic doses is narrow and the drug is only slowly excreted, so prolonged treatment at well spaced intervals is required.

The World Health Organization (WHO 1995) recommends treatment with 4 mg/kg three times per week for 5–25 weeks, or even longer, for visceral leishmaniasis. For African diffuse cutaneous leishmaniasis, the only other likely use of this drug in Old World leishmaniasis, treatment is once weekly, but continued for at least 4 months.

Monomycin/paromomycin/aminosidine

The antibiotic monomycin was developed in Russia both for intralesional injection and as a topical treatment for cutaneous leishmaniasis. In the latter use, the active ingredient was contained in a collagen pad tightly applied to the lesion. Paromomycin and aminosidine are probably the same as monomycin, and show considerable promise in topical ointments but in current formulations, irritation, and inflammation are too severe for these products to be more than experimental. This antibiotic shows promise as a synergist when used in combination with antimonials, in both visceral and diffuse cutaneous leishmaniasis (Seaman et al. 1993; Teklemariam et al. 1994).

Amphotericin B

In soluble form, this product is too toxic to be used except where all else has failed. It is administered very slowly, by the intravenous route, requires careful monitoring, and the early doses must be reduced in order to check for excessive side-effects. This drug is mainly used to treat cases of mucocutaneous leishmaniasis in South America (see Chapter 17, New World leishmaniasis) and is rarely indicated in Old World leishmaniasis. The recommended dose is 0.5–1.0 mg/kg on alternate days (WHO 1995). More recently, however, there have been important developments in the use of amphotericin, with lipid-associated forms of amphotericin (e.g. liposomal encapsulation) showing much reduced toxicity (Chance 1995). These are now commercially available (AmBisome, Amphocil) and, although relatively expensive, are suitable alternatives to antimonials.

Allopurinol

Allopurinol has the attraction that it is a widely used and well-established drug. Furthermore, biochemical and in vitro studies indicate that it should be effective against Leishmania. Despite these factors, clinical trials have been disappointing, although there are suggestions that the drug may have a synergistic effect when used together with antimony.

Miltefosine

Miltefosine or hexadecylphosphocholine, is the phosphorylcholine ester of hexadecanol and was originally developed as an anticancer agent but showed activity against experimental visceral leishmaniasis in mice (Kuhlencord et al. 1992). Recent clinical trials using miltefosine against Indian visceral leishmaniasis have been very encouraging (Jha et al. 1999; Sundar et al. 2002). If miltefosine proves to be clinically useful after more extensive testing, it may represent the most significant step forward in leishmanial chemotherapy for some time as, unlike the other treatments for visceral leishmaniasis, the drug is delivered orally. If safe and effective it is likely to become the first line drug of choice in the future.

VACCINATION

Two important factors suggest that vaccination or other immunological intervention holds promise for the protection of people from at least some of the leishmaniases. First, many individuals appear innately not to be susceptible. This is demonstrated by the survival of some people in even the most serious epidemics and by the finding that some individuals, who have never been exposed, produce a lymphocyte proliferation response to antigen (Akuffo and Britton 1992; Kurtzhals et al. 1995). Second, the immunity produced naturally in response to L. major infection is usually solid and there is no evidence that it is not sterile. There may even be a cross-immunity between L. major and other species. At present, however, no vaccine is available against Old World leishmaniasis.

Immunization against cutaneous leishmaniasis by deliberate infection with virulent parasites is traditional in parts of west Asia and was carried out on a massive scale during the Iran–Iraq war (Nadim 1988). Both the Russian and Israeli armies abandoned this measure as there were an unacceptable number of serious side-effects. Trials are in progress using various crude and recombinant antigen preparations with acceptable adjuvants and it is hoped that vaccination, at least against L. major infection, may soon become feasible (Modabber 1995; Handman 1997; McMahon-Pratt et al. 1998; Handman 2001).

INTEGRATED CONTROL

The principles of control, with particular reference to vector control, are discussed in Chapter 5, Control of parasites, parasitic infections, and parasitic diseases. Control of infection depends greatly on local information which, in turn, depends on efficient passive case detection and epidemiological surveillance. These are the minimal activities recommended by the World

Health Organization (WHO 1990). Increased knowledge of the structure of foci of leishmaniasis has led to the proposal of numerous potential active control strategies, but few of these have been evaluated or even tried in practice.

Avoidance of areas of risk is the simplest method and could be applied to military or other activities involving entry into semi-arid areas where zoonotic *L. major* can be a serious risk. In North Africa and west Asia, barracks for soldiers or laborers should be situated well away from depressions with halophilic vegetation and *P. obesus* colonies. People infected with HIV should avoid much of southern Europe in summer, although the coast itself is generally free from *L. infantum* infection. The few known residual foci of intense *L. donovani* transmission in Africa should be avoided. Some of the national parks of Sudan are particularly notorious and are rarely visited in the transmission season.

Environmental management, such as the destruction of desert rodent colonies in inhabited areas, has been attempted with great success in central Asian irrigation schemes, the designs of which were modified to prevent reinvasion by *R. opimus*. Pilot schemes to clear land around affected towns in Tunisia (Ben Ismail 1994) and Jordan (Kamhawi et al. 1993) were inconclusive. The incidence of infection can vary greatly from year to year, so experimental controls are hard to establish. In Ethiopia, it has been suggested that villages affected by *L. aethiopica* infection could be protected by shooting the hyraxes in the vicinity, but this intervention has yet to be tried.

In areas where transmission is peridomestic, as with *L. tropica* and *L. donovani* in Asia, sandfly numbers can be greatly reduced by insecticides. In fact, antimalarial campaigns were credited with the near-eradication of kala-azar in India in the 1950s. Active case detection, treatment and insecticide spraying eliminated *L. tropica* in Azerbaijan in the 1960s. A mass campaign in China in the 1950s, using case detection and treatment, insecticides and elimination of dogs, reduced the number of cases from an estimated 500 000 in 1951 to almost zero (Wang 1985; Guan 1991). Experiments with insecticide-impregnated bednets or curtains have given promising results (Maroli and Majori 1991).

Where transmission is sylvatic, as in the *L. donovani* epidemics in southern Sudan, the economic and political problems that prevent effective control elsewhere are compounded by technological problems. There is no realistic way to alter the environment, we know of no way to reduce sandfly numbers, and the reservoir hosts maintaining residual foci have not been identified. Little can be done other than the provision of active case detection and treatment or the promotion of impreg-nated bednets, with no guarantee that the latter will be effective. These measures are almost impracticable in such a poor, isolated, and war-torn country. Much of northern Sudan has undergone massive environmental change with the replacement of woodland by mechanized agricultural projects. This elimination of much of the *P. orientalis* habitat has incidentally greatly reduced the area in which kala-azar epidemics can be expected to occur.

REFERENCES

Addy, M. and Nandy, A. 1992. Ten years of kala azar in West Bengal Part 1.Did post kala azar dermal leishmaniasis initiate the outbreak in 24-Parganas? *Bull World Health Organ*, **70**, 341–6.

Akuffo, H.O. and Britton, S.F. 1992. Contribution of non *Leishmania*-specific immunity to resistance to *Leishmania* infection in humans. *Clin Exp Immunol*, **87**, 58–64.

Alexander, J. and Russell, D.G. 1992. The interaction of *Leishmania* species with macrophages. *Adv Parasitol*, **31**, 175–254.

Alexander, J., Satoskar, A.R. and Russell, D.G. 1999. *Leishmania* species: models of intracellular parasitism. *J Cell Sci*, **112**, 2993–3002.

Ali, A. and Ashford, R.W. 1994. Visceral leishmaniasis in Ethiopia IV. Prevalence, incidence and relation of infection to disease in an endemic area. *Ann Trop Med Parasitol*, **88**, 289–93.

Altes, J., Salas, A., et al. 1991. Visceral leishmaniasis: another HIV-associated opportunistic infection? Report of eight cases and review of the literature. *AIDS*, **5**, 201–7.

Alvar, J. and Jimenez, M. 1994. Could infected drug users be potential *Leishmania* reservoirs? *AIDS*, **8**, 854.

Antoine, J.C., Prina, E., et al. 1998. The biogenesis and properties of the parasitophorous vacuoles that harbour *Leishmania* in murine macrophages. *Trends Microbiol*, **7**, 392–401.

Ashford, R.W., Desjeux, P. and de Raadt, P. 1992a. Estimation of population at risk and numbers of cases of leishmaniasis. *Parasitol Today*, **8**, 104–5.

Ashford, R.W., Kohestany, K.A. and Karimzad, M.A. 1992b. Cutaneous leishmaniasis in Kabul, Afghanistan: observations on a 'prolonged epidemic'. *Ann Trop Med Parasitol*, **86**, 361–71.

Ashford, R.W., Bray, R.S., et al. 1973. The epidemiology of cutaneous leishmaniasis in Ethiopia . *Trans R Soc Trop Med Hyg*, **67**, 568–601.

Ashford, R.W., Rioux, J.A., et al. 1993. Evidence for a long term increase in the incidence of *Leishmania tropica* in Aleppo, Syria. T. *Trans R Soc Trop Med Hyg*, **87**, 247–9.

Attar, Z.J., Chance, M.L., et al. 2001. Latex agglutination test for the detection of urinary antigens in visceral leishmaniasis. *Acta Trop*, **78**, 11–16.

Badaro, R., Jones, T.C., et al. 1986. A prospective study of visceral leishmaniasis in an endemic area of Brazil. *J Infect Dis*, **154**, 639–49.

Bastien, P., Blaineau, C. and Pagès, M. 1992. *Leishmania*: sex, lies and karyotype. *Parasitol Today*, **8**, 174–7.

Bates, P.A. 1991. Phosphomonoesterases of parasitic protozoa. In: Coombs, G.H. and North, M.J. (eds), *Biochemical protozoology*. London: Taylor and Francis, 537–53.

Bates, P.A. 1993. Axenic culture of *Leishmania* amastigotes. *Parasitol Today*, **9**, 143–6.

Bates, P.A. 1994a. The developmental biology of *Leishmania* promastigotes. *Exp Parasitol*, **79**, 215–18.

Bates, P.A. 1994b. Complete developmental cycle of *Leishmania mexicana* in axenic culture. *Parasitology*, **108**, 1–9.

Bates, P.A. and Dwyer, D.M. 1987. Biosynthesis and secretion of acid phosphatase by *Leishmania donovani* promastigotes. *Mol Biochem Parasitol*, **26**, 289–95.

Bates, P.A. and Rogers, M.E. 2004. New insights into the developmental biology and transmission mechanisms of *Leishmania*. *Curr Mol Med*, **4**, 601–9.

Bates, P.A. and Tetley, L. 1993. *Leishmania mexicana*: induction of metacyclogenesis by cultivation at acidic pH. *Exp Parasitol*, **76**, 412–23.

Beck, J.T. and Ullman, B. 1991. Genetic analysis of folate transport and metabolism in *Leishmania donovani*. In: Coombs, G.H. and North, M.J. (eds), *Biochemical protozoology*. London: Taylor and Francis, 554–9.

Beljaev, A.E. and Lysenko, A.J. 1977. Measurements of endemicity of zoonotic cutaneous leishmaniasis. *Ecologie des Leishmanioses. Colloques Internationaux du Centre National de la Reserche Scientifique No. 239*. Paris: CNRS, 271–8.

Belkaid, Y., Kamhawi, S., et al. 1998. Development of a natural model of cutaneous leishmaniasis: powerful effects of vector saliva and saliva preexposure on the long-term outcome of *Leishmania major* infection in the mouse ear dermis. *J Exp Med*, **188**, 1941–53.

Ben Ismail, R. 1994. Rapport de fonctionnement: Laboratoire d'Epidemiologie et d'Ecologie Medicale. *Arch Inst Pasteur Tunis*, **71**, 86–107.

Beverley, S.M. and Turco, S.J. 1998. Lipophosphoglycan (LPG) and the identification of virulence genes in the protozoan parasite *Leishmania*. *Trends Microbiol*, **6**, 35–40.

Blackwell, J.M. 1992. Leishmaniasis epidemiology: all down to the DNA. *Parasitology*, **104**, Supplement, S19–34.

Blackwell, J.M. 1999. Tumour necrosis factor alpha and mucocutaneous leishmaniasis. *Parasitol Today*, **15**, 73–5.

Bogdan, C. and Rollinghoff, M. 1999. How do protozoan parasites survive inside macrophages? *Parasitol Today*, **15**, 22–8.

Bradley, D.J. 1987. Genetics of susceptibility and resistance in the vertebrate host. In: Peters, W. and Killick-Kendrick, R. (eds), *The leishmaniases*. vol. 2. London: Academic Press, 551–81.

Britto, C., Ravel, C.S., et al. 1998. Conserved linkage groups associated with large-scale chromosomal rearrangements between Old World and New World *Leishmania* genomes.. *Gene*, **222**, 107–17.

Bryceson, A. 1987a. Splenic aspiration procedure as performed at the Clinical Research Centre, Nairobi, 1982. In: Peters, W. and Killick-Kendrick, R. (eds), *The leishmaniases*. vol. 2. London: Academic Press, 728–9.

Bryceson, A. 1987b. Therapy in man. In: Peters, W. and Killick-Kendrick, R. (eds), *The leishmaniases*. vol. 2. London: Academic Press, 847–907.

Bulle, B., Millon, L., et al. 2002. Practical approach for typing strains of *Leishmania infantum* by microsatellite analysis. *J Clin Microbiol*, **40**, 3391–7.

Chance, M.L. 1995. New developments in the chemotherapy of leishmaniasis. *Ann Trop Med Parasitol*, **89**, 37–43.

Chance, M.L. and Goad, L.J. 1997. Sterol metabolism of *Leishmania* and trypanosomes: potential for chemotherapeutic exploitation. In: Peters, W. and Killick-Kendrick, R. (eds), *The leishmaniases*. vol. 2. London: Academic Press, 163–76.

Chang, K-P., Chauduri, G. and Fong, D. 1990. Molecular determinants of *Leishmania* virulence. *Annu Rev Microbiol*, **44**, 499–529.

Charlab, R. and Ribeiro, J.M.C. 1993. Cytostatic effect of *Lutzomyia longipalpis* salivary gland homogenates on *Leishmania* parasites. *Am J Trop Med Hyg*, **48**, 831–8.

Courtenay, O., Quinnell, R.J., et al. 2002. Low infectiousness of a wildlife host of *Leishmania infantum*: the crab-eating fox is not important for transmission. *Parasitology*, **125**, 407–14.

Croft, S.L. and Yardley, V. 2002. Chemotherapy of leishmaniasis. *Curr Pharm Des*, **8**, 319–42.

Cunningham, M.L., Titus, R.G., et al. 2001. Regulation of differentiation to the infective stage of the protozoan parasite *Leishmania major* by tetrahydrobiopterin. *Science*, **292**, 285–7.

Dereure, J., Reyes, J., et al. 1995. Visceral leishmaniasis in HIV-infected patients in the south of France. *Bull World Health Organ*, **73**, 245–6.

Descoteaux, A. and Turco, S.J. 1993. The lipophosphoglycan of *Leishmania* and macrophage protein kinase C. *Parasitol Today*, **9**, 468–71.

Descoteaux, A. and Turco, S.J. 2002. Functional aspects of the *Leishmania donovani* lipophosphoglycan during macrophage infection. *Microbes Infect*, **4**, 975–81.

Desjeux, P. 1991. *Information on the epidemiology and control of the leishmaniases by country and territory*. WHO Report WHO/Leish/91.30.

El Hag, I.A. and Hashim, F.A. 1994. Liver morphology and function in visceral leishmaniasis (kala-azar). *J Clin Pathol*, **47**, 547–51.

El Safi, S.H., Peters, W. and Evans, D. 1988. Current situation with regard to leishmaniasis in Sudan with particular reference to the recent outbreak of cutaneous leishmaniasis in Khartoum. *Research on control strategies for the leishmaniases. Proceedings of an International Workshop, Ottawa, Canada, 1-4 June 1987, I.D.R.C. Ottawa*. Ottawa: IDRC, 60–77.

Etges, R. and Bouvier, J. 1991. The promastigote surface proteinase of *Leishmania*. In: Coombs, G.H. and North, M.J. (eds), *Biochemical protozoology*. London: Taylor and Francis, 221–33.

Ferguson, M.A.J. 1997. The structure and biosynthesis of trypanosomatid glycosylphosphatidylinositols. In: Peters, W. and Killick-Kendrick, R. (eds), *The leishmaniases*. vol. 2. London: Academic Press, 65–77.

Garcia-del Portillo, F. and Finlay, B.B. 1995. The varied lifestyles of intracellular pathogens within eukaryotic vacuolar compartments. *Trends Microbiol*, **3**, 373–80.

Ghosh, K.N. and Bhattacharya, A. 1991. Breeding places of *Phlebotomus argentipes* in West Bengal, India. *Parassitologia*, **33**, Supplement, 267–72.

Ghosh, M.K., Nandy, A., et al. 1995. Subpopulations of T lymphocytes in the peripheral blood, dermal lesions and lymph nodes of post kala-azar dermal leishmaniasis patients.. *Scand J Immunol*, **41**, 11–17.

Gossage, S.M., Rogers, M.E. and Bates, P.A. 2003. Two separate growth phases during the development of *Leishmania* in sand flies: implications for understanding the life cycle. *Int J Parasitol*, **33**, 1027–34.

Gradoni, L., Scalone, A. and Gramiccia, M. 1993. HIV-*Leishmania* coinfections in Italy: serological data as an indication of the sequence of acquisition of the two infections. *Trans R Soc Trop Med Hyg*, **87**, 94–6.

Graham, S.V. 1995. Mechanisms of stage-regulated gene expression in Kinetoplastida. *Parasitol Today*, **11**, 217–23.

Guan, L-R. 1991. Current status of kala-azar and vector control in China.. *Bull World Health Organ*, **69**, 595–601.

Handman, E. 1997. *Leishmania* vaccines: old and new. *Parasitol Today*, **13**, 236–8.

Handman, E. 2001. Leishmaniasis: current status of vaccine development. *Clin Microbiol Rev*, **14**, 229–43.

Handman, E. and Bullen, D.V.R. 2002. Interaction of *Leishmania* with the host macrophage. *Trends Parasitol*, **18**, 332–4.

Haughan, P.A. and Goad, L.J. 1991. Lipid biochemistry of trypanosomatids. In: Coombs, G.H. and North, M.J. (eds), *Biochemical protozoology*. London: Taylor and Francis, 312–28.

Ilg, T. 2000. Proteophosphoglycans of *Leishmania*. *Parasitol Today*, **16**, 489–97.

Ilg, T., Stierhof, Y-D., et al. 1994. Characterization of phosphoglycan-containing secretory products of *Leishmania*. *Parasitology*, **108**, Supplement, S63–71.

Ilgoutz, S.C. and McConville, M.J. 2001. Function and assembly of the *Leishmania* surface coat. *Int J Parasitol*, **31**, 899–908.

Ivens, A.C. and Blackwell, J.M. 1999. The *Leishmania* genome comes of age. *Parasitol Today*, **15**, 225–31.

Jamjoom, M.B., Ashford, R.W., et al. 2002. Towards a standard battery of microsatellite markers for the analysis of the *Leishmania donovani* complex . *Ann Trop Med Parasitol*, **96**, 265–70.

Jamjoom, M.B., Ashford, R.W., et al. 2004. Leishmania donovani is the only cause of visceral leishmaniasis in East Africa; previous descriptions of *L. infantum* and *L. archibaldi* from this region are a consequence of convergent evolution in the isoenzyme data. *Parasitology*, **129**, 399–409.

Jha, T.K., Sundar, S., et al. 1999. Miltefosine, an oral agent, for the treatment of Indian visceral leishmaniasis.. *N Engl J Med*, **341**, 1795–800.

Kamhawi, S. 2000. The biological and immunomodulatory properties of sand fly saliva and its role in the establishment of *Leishmania* infections. *Microbes Infect*, **2**, 1765–73.

Kamhawi, S., Arbagi, A., et al. 1993. Environmental manipulation in the control of a zoonotic cutaneous leishmaniasis focus. *Arch Inst Pasteur Tunis*, **70**, 383–90.

Kamhawi, S., Belkaid, Y., et al. 2000. Protection against cutaneous leishmaniasis resulting from bites of uninfected sand flies. *Science*, **290**, 1351–4.

Kar, K. 1995. Serodiagnosis of leishmaniasis. *Crit Rev Microbiol*, **21**, 123–52.

Kaye, P.M. 1995. Costimulation and the regulation of antimicrobial immunity. *Immunol Today*, **16**, 423–7.

Kemp, M., Kurtzhals, J.A., et al. 1993. Interferon gamma and interleukin 4 in human *Leishmania donovani* infection. *Immunol Cell Biol*, **71**, 583–7.

Kemp, M., Kurtzhals, J.A., et al. 1994. Dichotomy in the human CD4+ T-cell response to *Leishmania* infection. *Acta Pathol Microbiol Immunol Scand*, **102**, 81–8.

Killick-Kendrick, R. 1990a. The life-cycle of *Leishmania* in the sandfly with special reference to the form infective to the vertebrate host. *Ann Parasitol Hum Comp*, **65**, Supplement 1, 37–42.

Killick-Kendrick, R. 1990b. Phlebotomine vectors of the leishmaniases: a review. *Med Vet Entomol*, **4**, 1–24.

Kuhlencord, A., Maniera, T., et al. 1992. Hexadecylphosphocholine: oral treatment of visceral leishmaniasis in mice. *Antimicrob Agents Chemother*, **36**, 1630–4.

Kurtzhals, J.A., Kemp, M., et al. 1995. Interleukin 4 and interferon gamma production by *Leishmania*-stimulated peripheral blood mononuclear cells from non-exposed individuals. *Scand J Immunol*, **41**, 343–9.

Lachaud, L., Chabbert, E., et al. 2001. Comparison of various sample preparation methods for PCR diagnosis of visceral leishmaniasis using peripheral blood. *J Clin Microbiol*, **39**, 613–17.

Lainson, R. and Shaw, J.J. 1987. Evolution, classification and geographical distribution. In: Peters, W. and Killick-Kendrick, R. (eds), *The leishmaniases*. vol. 2. London: Academic Press, 1–120.

Landfear, S.M. and Ignatushchenko, M. 2001. The flagellum and flagellar pocket of trypanosomatids. *Mol Biochem Parasitol*, **115**, 1–17.

Lang, T., Warburg, A., et al. 1991. Transmission and scanning EM-immunogold labeling of *Leishmania major* lipophosphoglycan in the sandfly *Phlebotomus papatasi*. *Eur J Cell Biol*, **55**, 362–72.

Liew, F.Y. and Cox, F.E.G. 1991. Nonspecific defence mechanisms: the role of nitric oxide. *Parasitol Today*, **7**, Supplement, A17–21.

Liew, F.Y. and O'Donnell, C.A. 1993. Immunology of leishmaniasis. *Adv Parasitol*, **32**, 161–259.

Lighthall, G.K. and Giannini, S.H. 1992. The chromosomes of *Leishmania*. *Parasitol Today*, **8**, 192–9.

Lukes, J., Guilbride, D.L., et al. 2002. Kinetoplast DNA network: evolution of an improbable structure. *Eukaryot Cell*, **1**, 495–502.

Maga, J.A., Sherwin, T., et al. 1999. Genetic dissection of the *Leishmania* paraflagellar rod, a unique flagellar cytoskeleton structure. *J Cell Sci*, **112**, 2753–63.

Magill, A.J., Grogl, M., et al. 1993. Viscerotropic leishmaniasis caused by *Leishmania tropica* in veterans of Operation Desert Storm. *N Engl J Med*, **328**, 1383–7.

Magill, A.J., Grogl, M., et al. 1994. Visceral leishmaniasis due to *L. tropica* in a veteran of Operation Desert Storm who presented two years after leaving Saudi Arabia. *Clin Infect Dis*, **19**, 805–6.

Manson-Bahr, P.E.C. 1987. Diagnosis. In: Peters, W. and Killick-Kendrick, R. (eds), *The leishmaniases*. vol. 2. London: Academic Press, 703–29.

Maroli, M. and Majori, G. 1991. Permethrin impregnated curtains against phlebotomine sandflies: laboratory studies. *Parassitologia*, **33**, Supplement, 399–404.

Marr, J.J. 1991. Purine metabolism in parasitic protozoa and its relationship to chemotherapy. In: Coombs, G.H. and North, M.J. (eds), *Biochemical protozoology*. London: Taylor and Francis, 524–36.

Marty, P., Le Fichoux, Y., et al. 1994. Human visceral leishmaniasis in Alpes Maritimes, France: epidemiological characteristics for the period 1985-1992. *Trans R Soc Trop Med Hyg*, **88**, 33–4.

Mauel, J. 1996. Intracellular survival of protozoan parasites with special reference to *Leishmania* spp. *Toxoplasma gondii* and *Trypanosoma cruzi*. *Adv Parasitol*, **38**, 1–51.

Mauel, J. and Behin, R. 1987. Immunity: clinical and experimental. In: Peters, W. and Killick-Kendrick, R. (eds), *The leishmaniases*. vol. 2. London: Academic Press,, 731–91.

Mauricio, I.L., Howard, M.K., et al. 1999. Genomic diversity in the *Leishmania donovani* complex. *Parasitology*, **119**, 237–46.

Mauricio, I.L., Stothard, J.R., et al. 2000. The strange case of *Leishmania chagasi*. *Parasitol Today*, **16**, 188–9.

McConville, M.J. and Ferguson, M.A.J. 1993. The structure, biosynthesis and function of glycosylated phosphatidylinositols in the parasitic protozoa and higher eukaryotes. *Biochem J*, **294**, 305–24.

McConville, M.J., Mullin, K.A., et al. 2002. Secretory pathway of trypanosomatid parasites. *MMBR*, **66**, 122–54.

McDonagh, P.D., Myler, P.J. and Stuart, K. 2000. The unusual gene organization of *Leishmania major* chromosome 1 may reflect novel transcription processes. *Nucleic Acids Res*, **28**, 2800–3.

McGwire, B.S., Chang, K.P. and Engman, D.M. 2003. Migration through the extracellular matrix by the parasitic protozoan *Leishmania* is enhanced by surface metalloprotease gp63. *Infect Immun*, **71**, 1008–10.

McGwire, B.S., O'Connell, W.A., et al. 2002. Extracellular release of the glycophosphatidylinositol (GPI) linked *Leishmania* surface metalloprotease, gp63, is independent of GPI phospholipolysis: implications for parasite virulence. *J Biol Chem*, **277**, 8802–9.

McMahon-Pratt, D., Kima, P.E. and Soong, L. 1998. *Leishmania* amastigote target antigens: the challenge of a stealthy intracellular parasite. *Parasitol Today*, **14**, 31–4.

Medrano, F.J., Hernandez-Quero, J., et al. 1992. Visceral leishmaniasis in HIV 1 infected individuals: a common opportunistic infection in Spain? *AIDS*, **6**, 1499–503.

Mehbratu, Y., Lawyer, P., et al. 1989. Visceral leishmaniasis unresponsive to Pentostam caused by *Leishmania tropica* in Kenya. *Am J Trop Med Hyg*, **41**, 289–94.

Mendez, S., Fernandez-Perez, F.J., et al. 1999. Partial anaerobiosis induces infectivity of *Leishmania infantum* promastigotes. *Parasitol Res*, **85**, 507–9.

Meredith, S.E., Kroon, N.C., et al. 1995. Leish-KIT, a stable direct agglutination test based on freeze-dried antigen for serodiagnosis of visceral leishmaniasis. *J Clin Microbiol*, **33**, 1742–5.

Michels, P.A.M., Hannaert, V. and Bakker, B.M. 1997. Glycolysis of Kinetoplastida. In: Peters, W. and Killick-Kendrick, R. (eds), *The leishmaniases*. vol. 2. London: Academic Press, 133–48.

Milon, G., Del Giudice, G. and Louis, J.A. 1995. Immunobiology of experimental cutaneous leishmaniasis. *Parasitol Today*, **11**, 244–7.

Modabber, F. 1995. Vaccines against leishmaniasis. *Ann Trop Med Parasitol*, **89**, 83–8.

Moll, H. 1993. Epidermal Langerhans cells are critical for immunoregulation of cutaneous leishmaniasis. *Immunol Today*, **14**, 383–7.

Molyneux, D.H. and Ashford, R.W. 1983. *The Biology of Trypanosoma and Leishmania parasites of man and domestic animals*. London: Taylor and Francis.

Molyneux, D.H. and Jeffries, D. 1986. Feeding behaviour of pathogen-infected vectors. *Parasitology*, **92**, 721–36.

Molyneux, D.H. and Killick-Kendrick, R. 1987. Morphology, ultrastructure and life cycles. In: Peters, W. and Killick-Kendrick, R. (eds), *The leishmaniases*. vol. 2. London: Academic Press, 121–76.

Momen, H., Pacheco, R.S., et al. 1993. Molecular evidence for the importation of Old World *Leishmania* into the Americas. *Biol Res*, **26**, 249–55.

Morgan, G.W., Hall, B.S., et al. 2002. The kinetoplastida endocyctic apparatus. Part I: a dynamic system for nutrition and evasion of host defences. *Trends Parasitol*, **18**, 491–6.

Mosser, D.M. and Brittingham, A. 1997. *Leishmania*, macrophages and complement: a tale of subversion and exploitation. *Parasitology*, **115**, Supplement, S9–23.

Myler, P.J. and Stuart, K. 2000. Recent developments from the *Leishmania* genome project. *Curr Opin Microbiol*, **3**, 412–16.

Nabors, G.S. 1997. Modulating ongoing Th2-cell responses in experimental leishmaniasis. *Parasitol Today*, **13**, 76–9.

Nadim, A. 1988. Leishmanization in the Islamic Republic of Iran. *Research on Control Strategies for the leishmaniases. Proceedings of an International Workshop, Ottawa, Canada, 1–4 June 1987*. Ottawa: IRDC, 336–9.

Parsons, M., Furuya, T., et al. 2001. Biogenesis and function of peroxisomes and glycosomes. *Mol Biochem Parasitol*, **115**, 19–28.

Peters, B.S., Fish, D., et al. 1990. Visceral leishmaniasis in HIV infection and AIDS: clinical features and response to therapy. *Q J Med*, **77**, 1101–11.

Pharoah, P.D., Ponnighaus, J.M., et al. 1993. Two cases of cutaneous leishmaniasis in Malawi. *Trans R Soc Trop Med Hyg*, **87**, 668–70.

Pimenta, P.F.P., Turco, S.J., et al. 1992. Stage-specific adhesion of *Leishmania* promastigotes to the sandfly midgut. *Science*, **256**, 1812–15.

Pimenta, P.F.P., Saraiva, E.M.B., et al. 1994. Evidence that the vectorial competence of phlebotomine sandflies for different species of *Leishmania* is controlled by structural polymorphisms in the surface lipophosphoglycan. *Proc Natl Acad Sci USA*, **91**, 9155–9.

Pimenta, P.F.P., Modi, G.B., et al. 1997. A novel role for the peritrophic matrix in protecting *Leishmania* from the hydrolytic activities of the sand fly midgut. *Parasitology*, **115**, 359–69.

Pratlong, F., Rioux, J.A., et al. 1991. *Leishmania tropica* in Morocco IV. Intrafocal enzyme diversity. *Ann Parasitol Hum Comp*, **66**, 100–4.

Pratlong, F., Dedet, J.P., et al. 1995. *Leishmania*-Human immunodeficiency virus coinfection in the Mediterranean basin: isoenzymatic characterization of 100 isolates of the *Leishmania infantum* complex. *J Infect Dis*, **172**, 323–7.

Rab, M.A. and Evans, D.A. 1995. *Leishmania infantum* infection in the Himalayas. *Trans R Soc Trop Med Hyg*, **89**, 27–32.

Ravel, C., Dubessay, P., et al. 1999. High conservation of the fine-scale organization of chromosome 5 between two pathogenic *Leishmania* species. *Nucleic Acids Res*, **27**, 2473–7.

Rees, P.H. and Kager, P.A. 1987. Visceral leishmaniasis and post-kala-azar dermal leishmaniasis. In: Peters, W. and Killick-Kendrick, R. (eds), *The leishmaniases*. vol. 2. London: Academic Press, 583–615.

Ridley, D.S. 1987. Pathology. In: Peters, W. and Killick-Kendrick, R. (eds), *The leishmaniases*. vol. 2. London: Academic Press, 665–701.

Rioux, J.A., Petter, F., et al. 1982. *Meriones shawi* reservoir de *Leishmania major* dans le sud maroccain. *C R Acad Sci*, **294**, 515–17.

Rioux, J.A., Lanotte, G., et al. 1990. Taxonomy of *Leishmania*. Use of isoenzymes. Suggestions for a new classification. *Ann Parasitol Hum Comp*, **65**, 111–25.

Ritter, U. and Korner, H. 2002. Divergent expression of inflammatory dermal chemokines in cutaneous leishmaniasis. *Parasite Immunol*, **24**, 295–301.

Rittig, M.G. and Bogdan, C. 2000. *Leishmania*-host cell interaction: complexities and alternative views. *Parasitol Today*, **16**, 292–7.

Rogers, L. 1908. *Fevers in the Tropics*. London: Oxford University Press.

Rogers, M.E., Chance, M.L. and Bates, P.A. 2002. The role of promastigote secretory gel in the origin and transmission of the infective stage of *Leishmania mexicana* by the sandfly *Lutzomyia longipalpis*. *Parasitology*, **124**, 495–507.

Rogers, M.E., Ilg, T., et al. 2004. Transmission of cutaneous leishmaniasis by sand flies is enhanced by regurgitation of fPPG. *Nature*, **430**, 463–7.

Sacks, D.L. 2001. *Leishmania*-sand fly interactions controlling species-specific vector competence. *Cell Microbiol*, **3**, 189–96.

Sacks, D.L. and Kamhawi, S. 2001. Molecular aspects of parasite-vector and vector-host interactions in leishmaniasis. *Ann Rev Microbiol*, **55**, 453–83.

Sacks, D.L. and Noben-Trauth, N. 2002. The immunology of susceptibility and resistance to *Leishmania major* in mice. *Nature Rev Immunol*, **2**, 845–58.

Sacks, D.L., Saraiva, E.M., et al. 1994. The role of lipophosphoglycan of *Leishmania* in vector competence. *Parasitology*, **108**, Supplement, S55–62.

Sacks, D.L., Kenney, R.T., et al. 1995. Indian kala-azar caused by *Leishmania tropica*. *Lancet*, **345**, 959–61.

Sacks, D.L., Modi, G., et al. 2000. The role of phosphoglycans in *Leishmania*-sand fly interactions. *Proc Natl Acad Sci USA*, **97**, 406–11.

Salotra, P., Sreenivas, G., et al. 2001. Development of species-specific PCR assay for detection of *Leishmania donovani* in clinical samples from patients with kala-azar and post-kala-azar-dermal leishmaniasis. *J Clin Microbiol*, **39**, 849–54.

Sang, D.K., Pratlong, F. and Ashford, R.W. 1992. The identity of *Leishmania tropica* in Kenya. *Trans R Soc Trop Med Hyg*, **86**, 621–2.

Sanyal, R.K. 1985. Leishmaniasis in the Indian sub-continent. In: Chang, K.P. and Bray, R.S. (eds), *Leishmaniasis*. Amsterdam: Elsevier, 443–67.

Sarkari, B., Chance, M.L. and Hommel, M. 2002. Antigenuria in visceral leishmaniasis: detection and partial characterization of a carbohydrate antigen. *Acta Trop*, **82**, 339–48.

Sati, M.H. 1962. Early phases of an outbreak of kala azar in the southern Fung. *Sud Med J*, **1**, 98–111.

Schlein, Y. 1993. *Leishmania* and sandflies: interactions in the life cycle and transmission. *Parasitol Today*, **9**, 255–8.

Schlein, Y. and Jacobson, R.L. 1994a. Haemoglobin inhibits the development of infective promastigotes and chitinase secretion in *Leishmania major* cultures. *Parasitology*, **109**, 23–8.

Schlein, Y. and Jacobson, R.L. 1994b. Mortality of *Leishmania major* in *Phlebotomus papatasi* caused by plant feeding of sandflies. *Am J Trop Med Hyg*, **50**, 20–7.

Schlein, Y., Jacobson, R.L. and Messer, G. 1992. *Leishmania* infections damage the feeding mechanism of the sandfly vector and implement parasite transmission by bite. *Proc Natl Acad Sci USA*, **89**, 9944–8.

Schlein, Y., Jacobson, R.L. and Shlomai, J. 1991. Chitinase secreted by *Leishmania* functions in the sandfly vector. *Proc R Soc Lond B Biol Sci*, **245**, 121–6.

Seaman, J., Ashford, R.W., et al. 1992. Visceral leishmaniasis in southern Sudan: status of healthy villagers in epidemic conditions. *Ann Trop Med Parasitol*, **86**, 481–6.

Seaman, J., Pryce, D., et al. 1993. Epidemic visceral leishmaniasis in Sudan: a randomised trial of aminosidine plus sodium stibogluconate versus sodium stibogluconate alone. *J Infect Dis*, **168**, 715–20.

Shakarian, A.M. and Dwyer, D.M. 1998. The *Ld Cht1* gene encodes the secretory chitinase of the human pathogen *Leishmania donovani*. *Gene*, **208**, 315–22.

Shakarian, A. and Dwyer, D.M. 2000. Pathogenic *Leishmania* secrete antigenically related chitinases which are encoded by a highly conserved gene locus. *Exp Parasitol*, **94**, 238–42.

Shao, L., Devenport, M. and Jacobs-Lorena, M. 2001. The peritrophic matrix of hematophagous insects. *Arch Insect Biochem Physiol*, **47**, 119–25.

Shapiro, T.A. and Englund, P.T. 1995. The structure and replication of kinetoplast DNA. *Ann Rev Microbiol*, **49**, 117–43.

Shlomai, J. 1994. The assembly of kinetoplast DNA. *Parasitol Today*, **10**, 341–6.

Solbach, W. and Laskay, T. 1995. *Leishmania major* infection: the overture. *Parasitol Today*, **11**, 394–7.

Solbach, W. and Laskay, T. 2000. The host response to *Leishmania* infection. *Adv Immunol*, **74**, 275–317.

Stiles, J.K., Hicock, P.I., et al. 1999. Genomic organization, transcription, splicing and gene regulation in *Leishmania*. *Ann Trop Med Parasitol*, **93**, 781–807.

Strelkova, M.V., Eliseev, L.N., et al. 1993. The isoenzyme identification of *Leishmania* isolates taken from great gerbils, sandflies and human patients in foci of zoonotic cutaneous leishmaniasis in Turkmenistan. *Med Parazitol Bolezni*, **5**, 34–7.

Stierhof, Y-D., Bates, P.A., et al. 1999. Filamentous proteophosphoglycan secreted by *Leishmania* promastigotes forms gel-like three-dimensional networks that obstruct the digestive tract of infected sandfly vectors. *Eur J Cell Biol*, **78**, 675–89.

Stuart, K. and Panigrahi, A.K. 2002. RNA editing: complexity and complications. *Mol Microbiol*, **45**, 591–6.

Sundar, S. 2001. Drug resistance in Indian visceral leishmaniasis. *Trop Med Int Health*, **6**, 849–54.

Sundar, S., Jha, T.K., et al. 2002. Oral miltefosine for Indian visceral leishmaniasis. *N Engl J Med*, **347**, 1739–46.

Swanson, M.S. and Fernandez-Moreia, E. 2002. A microbial strategy to multiply in macrophages: the pregnant pause. *Traffic*, **3**, 170–7.

Teklemariam, S., Hiwot, A.G., et al. 1994. Aminosidine and its combination with sodium stibogluconate in the treatment of diffuse cutaneous leishmaniasis caused by *Leishmania aethiopica*. *Trans R Soc Trop Med Hyg*, **88**, 334–9.

ter Kuile, B.H. 1993. Glucose and proline transport in kinetoplastids. *Parasitol Today*, **9**, 206–10.

Tibayrenc, M., Kjellberg, F. and Ayala, F.J. 1990. A clonal theory of parasitic protozoa: The population structures of *Entamoeba*, *Giardia*, *Leishmania*, *Naegleria*, *Plasmodium*, *Trichomonas*, and *Trypanosoma* and their medical and taxonomical consequences. *Proc Natl Acad Sci USA*, **87**, 2414–18.

Titus, R.G. and Ribeiro, J.M.C. 1990. The role of vector saliva in transmission of arthropod-borne disease. *Parasitol Today*, **6**, 157–60.

Turco, S.J. and Descoteaux, A. 1992. The lipophosphoglycan of *Leishmania* parasites. *Ann Rev Microbiol*, **46**, 65–94.

Vickerman, K. and Tetley, L. 1990. Flagellar surfaces of parasitic protozoa and their role in attachment. In: Bloodgood, R.A. (ed.), *Ciliary and Flagellar Membranes*. New York and London: Plenum, 267–304.

Victoir, K. and Dujardin, J-C. 2002. How to succeed in parasitic life without sex? Asking *Leishmania*. *Trends Parasitol*, **18**, 81–5.

Vidor, E., Dereure, J., et al. 1991. Le chancre d'inoculation dans la leishmaniose canine à *Leishmania infantum*. Etude d'une cohorte en région cérenole. *Prat Méd Chir Animal Compagnie*, **26**, 133–7.

Waller, R.F. and McConville, M.J. 2002. Developmental changes in lysosome morphology and function in *Leishmania* parasites. *Int J Parasitol*, **32**, 1435–45.

Walsh, J.A. and Warren, K.S. 1979. Selective primary health care. An interim strategy for disease control in developing countries. *N Engl J Med*, **301**, 967–74.

Walters, L.L. 1993. *Leishmania* differentiation in natural and unnatural sand fly hosts. *J Eukaryot Microbiol*, **40**, 196–206.

Wang, C-T. 1985. Leishmaniasis in China: epidemiology and control programme. In: Chang, K-P. and Bray, R.S. (eds), *Leishmaniasis*. Amsterdam: Elsevier, 469–78.

Webster, P. and Russell, D.G. 1993. The flagellar pocket of trypanosomatids. *Parasitol Today*, **9**, 201–6.

WHO, 1990. *Control of the leishmaniases*. WHO Technical Report Series No. 793, 158. World Health Organization, Geneva.

WHO, 1995. *WHO model prescribing information. Drugs used in parasitic diseases*, 2nd edn. Geneva: World Health Organization, 1–146.

WHO, 2000. *Leishmania/HIV co-infection*. Geneva: WHO Department of Communicable Disease Surveillance and Response.

Wiegle, K.A., Labrada, L.A., et al. 2002. PCR-based diagnosis of acute and chronic cutaneous leishmaniasis caused by *Leishmania* (*Viannia*). *J Clin Microbiol*, **40**, 601–6.

Wolday, D., Berhe, N., et al. 1999. *Leishmania*-HIV interaction: immunopathogenic mechanisms. *Parasitol Today*, **15**, 182–7.

Zakai, H., Chance, M.L. and Bates, P.A. 1998. In vitro stimulation of metacyclogenesis in *Leishmania braziliensis*, *L. donovani*, *L. major* and *L. mexicana*. *Parasitology*, **116**, 305–9.

Zijlstra, E.E., El Hassan, A.M., et al. 1994. Endemic kala azar in eastern Sudan: a longitudinal study on the incidence of clinical and subclinical infection and post kala azar dermal leishmaniasis. *Am J Trop Med Hyg*, **51**, 826–36.

Zilberstein, D. and Shapira, M. 1994. The role of pH and temperature in the development of *Leishmania* parasites. *Ann Rev Microbiol*, **48**, 449–70.

17

New World leishmaniasis

RALPH LAINSON AND JEFFREY J. SHAW

INTRODUCTION

Neotropical cutaneous leishmaniasis (NCL) seems to be of great antiquity. Pottery from Peru and Ecuador, dated c. AD 400–900, commonly depicted human faces with mutilations that are remarkably similar to those caused by present day cutaneous and mucosal leishmaniasis (Figure 17.1) and, at the time of the conquistadores, Spanish historians wrote about ugly facial lesions that frequently afflicted the local Amerindians.

The earliest traceable clinical description of the disease is probably that of a certain Dr L. Villar who, in 1859, wrote 'the disease (Peruvian "uta") is very like the Aleppo button' (i.e. oriental sore due to *Leishmania tropica*) (Matta 1918).

The first suspicions that phlebotomine sandflies might be involved in the transmission of NCL seem to have been those of Cosme Bueno who, in 1764, implicated these insects in uta in endemic areas; uta is caused by *L. (Viannia) peruviana* in the Peruvian highlands (Herrer and Christensen 1975). Final proof that NCL was due to infection with *Leishmania* was to await publications by Lindenberg (1909) and Carini and Paranhos (1909), who independently demonstrated 'Leishman-Donovan bodies' in the skin lesions of patients from the State of São Paulo, Brazil. It was left to another Brazilian, Gaspar Vianna (1911) to give the name of *Leishmania braziliense* to the parasite, later corrected to *L. braziliensis* by Matta (1916).

It is quite extraordinary that until 1972, all cases of NCL in Brazil and a number of neighboring countries were attributed solely to this parasite, although specific species names were attributed in different regions: Velez (1913) had given the name *L. peruviana* to the causative agent of Peruvian uta; Biagi (1953) had named the parasite of Mexican chiclero's ulcer as *L. tropica mexicana,* emended to *L. mexicana* by Garnham (1962); and Floch (1954) considered pian-bois in French Guyana to be due to *L. tropica guyanensis,* later emended to *L. braziliensis guyanensis* by the Brazilian parasitologist Pessôa (1961). During the past 30 years, intensive ecological and epidemiological studies in the Americas have revolutionized previous ideas regarding the etiology of NCL and the classification of its causative parasites. No less than 21 species of *Leishmania* are now recognized within the neotropical region, of which 15 are known to cause either cutaneous or mucocutaneous leishmaniasis or both in humans. There is little doubt that others remain to be discovered and described, particularly in the rich sandfly and mammalian faunas of the great South American forests.

Cutaneous leishmaniasis is widespread in Latin America, and the only two countries that seem to be free of the disease are Chile and Uruguay: cases have even been recorded in the extreme south of the USA in Texas, and rodents naturally infected with *L. (L.) mexicana* have been found in Arizona, USA. A reliable figure for the incidence of the human disease in the Americas is virtually impossible to obtain: it is not a notifiable disease in most of the countries where it occurs. Most of these countries also have poor facilities for unequivocal diagnosis so heavy reliance is placed on

Figure 17.1 *Peruvian pottery (huaco) showing facial mutilations thought to represent mucosal leishmaniasis. (From Pessôa and Barretto 1948)*

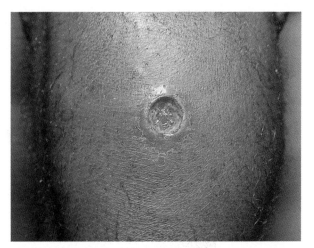

Figure 17.2 *Simple skin lesion of the arm, due to* Leishmania (Viannia) braziliensis; *Pará, Brazil*

clinical aspects, which are often misleading. In Costa Rica, a small country, but with very good communications and medical assistance, the impressive figure has been given of more than 2 000 cases a year in a population of 2 000 000 (Walton 1987). In Brazil, the Ministry of Health reported an increase in the number of cases from 2 856 in 1977 to 4 821 in 1982, (Walton 1987). Lacerda (1994) suggested the considerably higher figure of 154 103 recorded cases between the years 1980 and 1990, with an incidence of c. 5 people in 1 000 infected between the years 1980 and 1984, rising to 25 per 1 000 during 1987–1990.

In all known instances, NCL is a zoonosis: that is, the causative parasites are primarily those of wild animals. When the various sandfly vectors also feed on humans there may be transmission of a number of species of *Leishmania*; in this 'unnatural' host, they usually provoke an intense reaction and the eventual development of a skin lesion at the site of the bite. Some 7–10 days later, a tiny papule appears which, although usually painless, may itch considerably. In most cases, the papule eventually ulcerates, producing a steadily growing and crater-like lesion with a characteristically inflamed, elevated border (Figure 17.2). A lesion may remain single, but in some cases infected macrophages may transport the parasites to other parts of the body and establish secondary lesions. One parasite in particular, *L. (V.) braziliensis*, tends to produce

such metastatic lesions in the nasal, pharyngeal and laryngeal mucosae. These may arise within a few months of the original skin lesion, or years later when the patient has supposedly been cured of the initial infection; they may be extremely mutilating (Figures 17.3 and 17.4).

Other neotropical *Leishmania* species may produce an even more serious and incurable condition in individuals who fail to mount a fully functional cell-mediated immune reaction against the parasite. Such patients, who develop large numbers of nodular lesions scattered over almost the whole skin surface (Figure 17.5), have been referred to as cases of diffuse cutaneous leishmaniasis (DCL). This is a misleading term as other perfectly curable cases of cutaneous leishmaniasis in immunologically competent patients may have numerous lesions

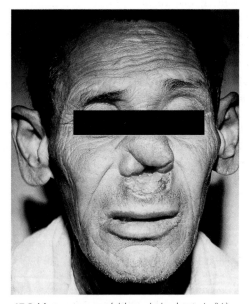

Figure 17.3 *Mucocutaneous leishmaniasis, due to* L. (V.) braziliensis; *Pará, Brazil*

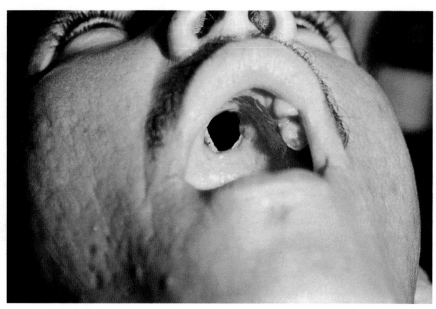

Figure 17.4 *Destruction of the palate, due to* L. (V.) braziliensis; *Pará, Brazil*

scattered over the body surface (Figure 17.6; see also *Leishmania (Viannia) guyanensis* Floch, 1954, below); the term anergic diffuse cutaneous leishmaniasis (ADCL) is more appropriate. In the Americas, this condition has been found associated only with infection by members of the *mexicana* complex, e.g. *L. (Leishmania) mexicana*, *L. (L.) pifanoi*, and *L. (L.) amazonensis*. In the Old World, another member of the subgenus *Leishmania*, *L. (L.) aethiopica*, causes a similar incurable disease in immunologically incompetent patients (see Chapter 16, Old World leishmaniasis).

The fact that a number of *Leishmania* species cause disease in humans in the neotropics is reflected in variations in chemotherapeutic responses. Thus, drugs that work well in one region may not be so efficient in another because the species of the parasite infecting the patients are different. In Guatemala, for example, individuals infected with *L. (L.) mexicana* responded to treatment with ketoconazole better than those infected with *L. (V.) braziliensis* s.l., but the reverse applied when patients were treated with sodium stibogluconate (Navin et al. 1992).

Unlike humans, the natural sylvatic hosts of the various neotropical *Leishmania* species rarely suffer disease from the infection, which is usually of a benign, inapparent nature. Under certain circumstances (see *Leishmania (Viannia) braziliensis* (Vianna, 1911) Matta, 1916), domestic animals such as dogs, mules, and horses may be found with extensive skin ulcers due to *Leishmania*: an indication that they, like humans, are unnatural and unaccustomed hosts.

The known history of American visceral leishmaniasis (AVL) is comparatively short compared with that of the Old World (Chapter 16, Old World leishmaniasis). The first record was probably that of Migone (1913), who saw 'corpuscles', which he was convinced were amastigotes of *Leishmania*, in the blood of a sick man in Paraguay. The patient's symptoms were highly indicative of visceral leishmaniasis and, failing to respond to antimalarial treatment, he died. Prior to his illness, the man had been working on the construction of the São Paulo–Corumbá railway in Brazil, where he most probably became infected. The first undoubted cases to be registered in Latin America were documented by Mazza and Cornejo (1926) in two Argentinian children.

Penna (1934) used the viscerotome to examine liver samples from patients suspected to have died from yellow fever in various parts of Brazil. He diagnosed 41 of them as being cases of visceral leishmaniasis, the largest number coming from the north-east of that country. Sporadic cases began to be recorded in Bolivia, Colombia, Guatemala, Paraguay, and El Salvador, but the full importance of the disease as a public health problem was not realized until as recently as 1953, when a dramatic outbreak was estimated to have been responsible for more than 100 deaths in the small country town of Sobral, in the State of Ceará, north-east Brazil (Deane 1956). Clearly, the history of AVL goes back much further than this and deaths from this highly lethal infection must have long been attributed to other causes, including malaria and yellow fever. To this day, AVL remains second in importance only to malaria among the Latin American tropical diseases, with the very conservative number of over 6 000 cases recorded in Brazil alone, up to 1980 (Deane and Grimaldi 1985).

Although a wide variety of *Leishmania* species have been recorded in humans, the prevalence of each of these will clearly depend on just how anthropophilic their respective sandfly vectors are. Thus, by far the

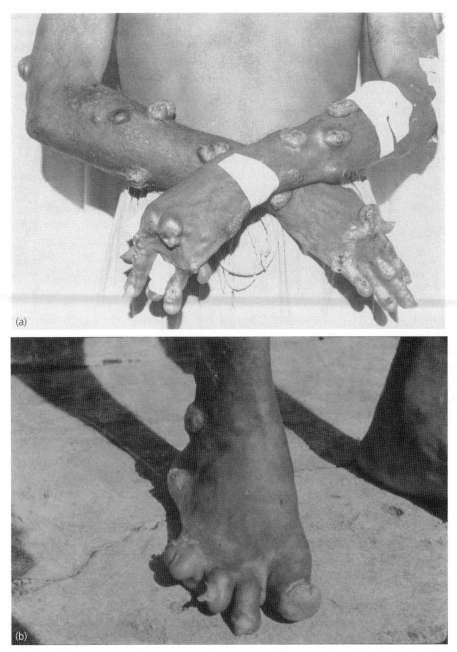

Figure 17.5 *ADCL due to* L. (L.) amazonensis. *Note the amputation of some fingers* **(a)** *and toes* **(b)**, *simulating lepromatous leprosy, with which disease ADCL is often confused; Pará, Brazil. (Figure 17.5a from Lainson 1982b)*

largest proportion of NCL cases in Brazil are due to *L. (V.) braziliensis* and *L. (V.) guyanensis*, both of which have sandfly vectors that feed avidly on humans. On the other hand, human infection with *L. (V.) lainsoni* and *L. (V.) naiffi* is relatively rare, as their vectors are disinclined to bite humans. Some neotropical *Leishmania* species, like *L. (L.) hertigi* and *L. (L.) deanei* of porcupines, are unknown in humans, possibly due to their inability to survive in human tissues, but more probably because their sandfly vectors never feed on humans.

It is likely that all neotropical *Leishmania* species once shared a sylvatic ecology, as is still the case in the remaining areas of extensive primary and secondary forest in the Amazon Region. Following the Iberian colonization and ensuing destruction of the forests, some phlebotomine sandfly species adapted to a peridomestic habitat. The persistence of cutaneous leishmaniasis in such ecologically disturbed, although still essentially rural, areas has been taken to indicate the evolution of a secondary, peridomestic transmission of the causative parasite; the occurrence of leishmanial skin lesions in village dogs, mules, and horses suggests that such domestic animals might have now become secondary reservoir hosts. Their development of extensive skin

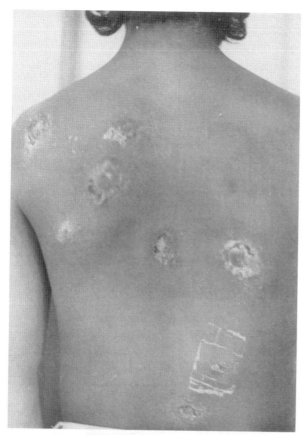

Figure 17.6 *Multiple skin lesions due to* L. (V.) guyanensis. *Similar lesions were present on this man's legs, arms and face; Pará, Brazil. (From Lainson 1982b)*

The role of humans as a source of infection for sandfly vectors of cutaneous leishmaniasis in the Americas remains controversial. The Brazilian parasitologist Aragão (1922) succeeded in infecting *Lu. intermedia* on the skin lesions of patients, and Strangways-Dixon and Lainson (1966) infected six out of eight sandflies (*Lu. cruciata* and *Lu. ylephilator*) by feeding them on the indurated margin of lesions due to *L. (L.) mexicana* in Belize. The fact that sandflies fed on healthy skin of the same patients consistently failed to become infected and that the insects showed a marked reluctance to feed on old, scabbed lesions, led them to conclude that only on rare occasions were humans likely to serve as a reservoir of infection for the sandfly vector. In a similar experiment, Montoya-Lerma et al. (1998) obtained two infected *Lu. longipalpis* out of 103 that took a bloodmeal on the edges of lesions due to *L. (V.) braziliensis*, but none among those that had fed on normal skin of the same patients. Understandably, the systemic nature of visceral leishmaniasis provides a more ready source of amastigotes in the blood of persons with active infections. Costa et al. (2000) fed 3 747 laboratory-bred *Lu. longipalpis* on persons with active, asymptomatic and apparently cured infections, in an area endemic for AVL: 26 flies acquired infection from 11 of 44 persons with active infections, but none of those fed on 137 asymptomatic individuals became infected. This supports the observation of Deane (1956) that man is a poor source of infection compared with the acutely infected dog.

Although *Lutzomyia longipalpis*, the major vector of AVL, is better known as a peridomestic or intra-domiciliary sandfly, its origin is sylvatic (Lainson 1989; Lainson et al. 1990a). In Amazonian Brazil, for example, it has been captured in primary rain forest far from human habitation and when crude roads are cut through such forest, newly constructed houses and animal sheds are soon invaded by this insect. It follows that whether the source of the disease is local foxes (*Cerdocyon thous*), in which the infection rate may be more than 50 percent, or infected dogs brought to the area from distant endemic foci of AVL, transmission of the parasite will ensue and a new focus of the canine and human disease will be established.

Studies on the ecology of the New World leishmanial parasites have strongly suggested the existence of environmental barriers that limit the different species of *Leishmania* to specific sandfly species that transmit to certain mammalian hosts in distinct ecotopes. Extreme care and considerable field research is needed, therefore, before conclusions can be reached as to the principal sandfly vectors of the different parasites. The mere presence of a *Leishmania* in a specimen of a given sandfly species does not necessarily mean that this insect is the vector of that parasite. Before such a conclusion can be reached, the organism must be found with frequency in that particular sandfly and must show

lesions tends to suggest, however, that, like humans, these animals still remain unnatural and unaccustomed 'victim' hosts. Firm evidence is required to show that infected humans, dogs, and equines are capable of infecting sandflies fed on them and thus maintaining the parasite in the absence of another source of infection in wild animals still existing in nearby surviving pockets of woodland. Reithinger and Davies (1999), in a review of the literature, concluded that evidence suggesting the domestic dog as a reservoir for human American cutaneous leishmaniasis was largely circumstantial. This conclusion was supported by Savani et al. (1999) who found no serological evidence of infection in 973 stray dogs examined in an endemic area of cutaneous leishmaniasis in São Paulo State, Brazil. Tolezano et al. (1998) carried out serological surveys of dogs in two endemic areas of the Ribeira River Valley, São Paulo State. It was suggested that dogs played only a minor role in maintaining the parasite population in one area, with only six of 77 (7.8 percent) animals positive to the skin test: in the other area, a positivity of 10 out of 56 (17.9 percent) led to their conclusion that dogs are involved in the maintenance of *L. (V.) braziliensis* to different degrees in different regions of Brazil. Positive skin tests are no indication, however, that the dogs are a source of parasites for the sandfly vector(s).

abundant proliferation in the alimentary tract, with migration to the foregut and mouthparts. Ideally, a definite association with the wild mammalian host should be shown to exist and, if the parasite is a cause of human leishmaniasis, the fly must be shown to bite humans.

Similarly, the efficiency of a given mammal in maintaining a *Leishmania* species in nature and in serving as a source of infection for the sandfly vector must be considered. Determining a significant infection rate in the mammalian host and providing experimental proof that the sandfly vector can be infected by feeding on it are prerequisites in labeling the animal as a reservoir host.

A number of the neotropical *Leishmania* species were first described in their wild mammalian or sandfly hosts, considerable time elapsing before they were incriminated as a cause of human leishmaniasis. For this reason, this chapter discusses all the known New World species, regardless of the fact that some have yet to be found in humans. Chapter 16, Old World leishmaniasis, deals with the systematic position of the genus *Leishmania*, morphology and development of the organism in the mammalian and sandfly hosts, the application of biochemistry and molecular biology to identification of the different species of the parasite, genetics and immunology, clinical features and the treatment of human cutaneous and visceral leishmaniasis. Many features of these topics are common to both Old World and New World leishmanial parasites and this chapter discusses only those that seem to be peculiar to the American leishmaniases and their causative agents. Lengthy discussions on the evolution, ecology, epidemiology, and classification of the parasites have been given elsewhere (Lainson and Shaw 1979, 1987; Lainson 1983; Lainson et al. 1994; Shaw and Lainson 1987; Shaw 1994), and reviews on the history of the neotropical leishmaniases have also been dealt with in other publications (Lainson and Shaw 1992; Lainson 1996). For methods in the laboratory diagnosis of these diseases, see Lainson and Shaw (1981) and Chapter 16, Old World leishmaniasis.

CLASSIFICATION OF THE RECOGNIZED NEOTROPICAL *LEISHMANIA* SPECIES

The genus *Leishmania* has been subdivided into two subgenera, as follows.

The subgenus *Leishmania* Ross, 1903

This subgenus possesses the characteristics of the genus (Chapter 16, Old World leishmaniasis). The life-cycle in the natural sandfly vector is limited to the midgut and foregut of the alimentary tract. Species occur in both the Old World and the New World (type species *Leishmania (Leishmania) donovani* (Laveran and Mesnil, 1903) Ross, 1903 of the Old World). The recorded neotropical species are:

1 *Leishmania (Leishmania) infantum chagasi** Cunha and Chagas, 1937
2 *L. (L.) enriettii* Muniz and Medina, 1948
3 *L. (L.) mexicana** (Biagi, 1953) Garnham, 1962
4 *L. (L.) pifanoi** (Medina and Romero, 1959) Medina and Romero, 1962
5 *L. (L.) hertigi* Herrer, 1971
6 *L. (L.) amazonensis** Lainson and Shaw, 1972
7 *L. (L.) deanei* Lainson and Shaw, 1977
8 *L. (L.) aristidesi* Lainson and Shaw, 1979
9 *L. (L.) garnhami** Scorza et al., 1979
10 *L. (L.) venezuelensis** Bonfante-Garrido, 1980
11 *L. (L.) forattinii* Yoshida et al., 1993.

*Recorded from humans. For distribution, see Table 17.1.

The subgenus *Viannia* Lainson and Shaw, 1987

This subgenus possesses the characteristics of the genus (Chapter 16, Old World leishmaniasis). The developmental cycle in the natural sandfly vector includes a prolific and prolonged phase of division of rounded or ovoid paramastigotes and promastigotes attached to the wall of the hindgut (pylorus and ileum, see Figure 17.7, p. 321) by flagellar hemidesmosomes, followed by migration of free promastigotes to the midgut and foregut. Members of this subgenus are known only in the neotropical region and the recorded species are as follows:

1 *Leishmania (Viannia) braziliensis** (Vianna, 1911) Matta, 1916; type species
2 *L. (V.) peruviana** Velez, 1913
3 *L. (V.) guyanensis** Floch, 1954
4 *L. (V.) panamensis** Lainson and Shaw, 1972
5 *L. (V.) lainsoni** Silveira et al., 1987
6 *L. (V.) shawi** Lainson et al., 1989
7 *L. (V.) naiffi** Lainson and Shaw, 1989
8 *L. (V.) colombiensis** Kreutzer et al., 1991
9 *L. (V.) equatorensis* Grimaldi et al., 1992
10 *L. (V.) lindenbergi** Silveira et al., 2002
11 *L. (V.) utingensis* Braga et al., 2003

*Recorded from humans. For distribution, see Table 17.1.

Species within the subgenus *Leishmania* Ross, 1903

LEISHMANIA (LEISHMANIA) INFANTUM CHAGASI CUNHA AND CHAGAS, 1937

Known geographical distribution

Distribution has been noted throughout most of the Latin American continent: Argentina, Bolivia, Brazil,

Table 17.1 *A country-by-country list of the neotropical* Leishmania *species recorded in humans and the resultant pathologies*

Country	Species	Disease forms recorded[g]
Argentina	*L. (L.) infantum chagasi*	VL
	L. (V.) braziliensis s.l.	CL
Belize	*L. (L.) mexicana*	CL
	L. (V.) braziliensis s.l.	CL
Bolivia	*L. (L.) amazonensis*	CL, ADCL
	L. (L.) infantum chagasi	VL
	L. (L.) sp.	CL
	L. (V.) braziliensis s.l.	CL, MCL
	L.(V.) lainsoni[o]	CL
Brazil	*L. (L.) amazonensis*	CL, ADCL, MCL[h], VL[i]
	L. (L.) infantum chagasi	VL, (CL[j])
	L. (L.) sp.	CL[k]
	L. (V.) braziliensis	CL, MCL
	L. (V.) guyanensis	CL, MCL
	L. (V.) lainsoni	CL
	L. (V.) naiffi	CL
	L.(V) shawi	CL
	L.(V.) lindenbergi[p]	CL
Colombia	*L. (L.) amazonensis*	CL, ADCL
	L. (L.) infantum chagasi	VL
	L. (L.) mexicana	CL, ADCL
	L. (V.) braziliensis s.l.	CL, MCL
	L. (V.) colombiensis	CL
	L. (V.) guyanensis	CL
	L. (V.) panamensis	CL, MCL
Costa Rica	*L. (L.) infantum chagasi*	CL[l], VL[q]
	L. (L.) mexicana	CL
	L. (V.) braziliensis s.l.	CL, MCL
	L. (V.) panamensis	CL
Dominican Republic	*L. (L.) mexicana*-like	ADCL
Ecuador	*L. (L.) sp.*[a]	CL
	L. (L.) mexicana[a]	CL
	L. (V.) braziliensis s.l.	CL, MCL
	L. (V.) braziliensis / L.(V.) panamensis?	CL
	L. (V.) panamensis / L.(V.) guyanensis?	CL
El Salvador	*L. (L.) infantum chagasi*	VL
	L. (L.) mexicana	CL
French Guyana	*L. (L.) amazonensis*	CL, ADCL
	L. (V.) braziliensis s.l.[b]	CL, MCL
	L. (V.) guyanensis	CL
	L. (V.) naiffi[c]	CL
Guadeloupe	*L. (L.) infantum chagasi*	VL
	Leishmania sp.	CL
Guatemala	*L. (L.) chagasi*	VL
	L. (L.) mexicana	CL
	L. (V.) braziliensis s.l.	CL
Guyana	*L. (V.) guyanensis*	CL
	Leishmania sp.	MCL
Honduras	*L. (L.) infantum chagasi*	VL, CL[l]
	L. (L.) mexicana	CL, ADCL
	L. (V.) braziliensis s.l.	CL, MCL
	L. (V.) panamensis	CL, MCL
Martinique	*L. (L.) sp.*	CL

(Continued over)

Table 17.1 *A country-by-country list of the neotropical* Leishmania species recorded in humans and the resultant pathologies *(Continued)*

Country	Species	Disease forms recorded[g]
Mexico	*L. (L.) infantum* chagasi	VL
	L. (L.) mexicana	CL, ADCL
	L. (L.) sp.	CL, ADCL[m]
	L. (V.) braziliensis	CL
Nicaragua	*L. (L.) infantum* chagasi	VL, CL[r]
	L. (V.) braziliensis s.l.	CL, MCL
	L. (V.) panamensis	CL, MCL
	L. (V.) braziliensis / panamensis[d]	CL
Panama	*L. (V.) braziliensis* s.l.	CL
	L. (V.) panamensis	CL
	Leishmania sp.	MCL
Paraguay	*L. (L.) amazonensis*	CL, ADCL
	L. (L.) infantum chagasi	VL
	L. (V.) braziliensis s.l.	CL, MCL
Peru	*L. (V.) braziliensis* s.l.	CL, MCL
	L. (V.) peruviana	CL, MCL
	L. (V.) braziliensis / L. (V.) peruviana[e]	CL
	L. (V.) lainsoni[s]	CL
Surinam	*Leishmania* sp.	CL
USA	*L. (L.) mexicana*	CL, ADCL[m]
Venezuela	*L. (L.) infantum* chagasi	VL
	L. (L.) garnhami	CL
	L. (L.) pifanoi	CL, ADCL
	L. (L.) venezuelensis	CL
	L. (V.) braziliensis s.l.	CL, MCL
	L. (V.) colombiensis	VL[n]
	L. (V.) braziliensis / L. (V.) guyanensis[f]	CL

Data from the review articles of Lainson and Shaw 1979, 1987; Shaw and Lainson 1987; Grimaldi et al. 1989 except where otherwise indicated.
a) Katakura et al. 1993. b) Raccurt et al. 1995. c) Darie et al. 1995. d) Darce et al. 1991. e) Dujardin et al. 1995. f) Bonfante-Garrido et al.1992. g) ADCL, anergic diffuse cutaneous leishmaniasis; CL, cutaneous leishmaniasis; MCL, mucocutaneous leishmaniasis; VL, visceral leishmaniasis. h) In cases of ADCL. i) Only recorded in a small region of Bahia State, BR. j) Only in Rio de Janeiro, BR. k) Localized in Minas Gerais, BR. l) Zelodón et al. 1989; Ponce et al. 1991. m) In Texas, southern USA. n) Bone marrow only. o) Martinez et al. (2001). p) Silveira et al. (2002). q) Carrillo et al. (1999). r) Belli et al. (1999). s) Lucas et al. (1994).

Colombia, Ecuador, El Salvador, Guadeloupe, Guatemala, Honduras, Martinique, Mexico, Nicaragua, Paraguay, Surinam, and Venezuela.

Known mammalian hosts

Known mammalian hosts include humans and the domestic dog, *Canis familiaris*. Among wild animals the fox, *Cerdocyon thous*, appears to be an important natural reservoir in Brazil in the northeast (Ceará), north (Pará), and southeast (Mato Grosso do Sul) (Deane 1956; Lainson et al. 1987; Mello et al. 1988; respectively). Evidence has been presented suggesting that the infected foxes identified by Deane (1956) in Ceará as *Lycalopex vetulus* were most probably *C. thous* (Courtenay et al. 1996).

There are reports of isolates of the parasite from domestic rats in Honduras (Walton, personal communication) and, on rare occasions, from opossums of the genus *Didelphis* in Bahia, northeast Brazil (Sherlock et al. 1984). In Colombia, *D. marsupialis* is considered to be an important reservoir host (Corredor et al. 1989; Travi et al. 1994).

The latter authors (Travi et al. 1998) obtained positive results using the polymerase chain reaction (PCR)-hybridization test for three out of 34 specimens of the 'spiny-rat', *Proechimys canicollis* from forest in Colombia and suggested that these animals had active infections, although no parasites were isolated from them. Continuing their studies on the epidemiology of visceral leishmaniasis in Colombia, Travi et al. (2002) tested the susceptibility of another species of *Proechimys* (*P. semi-spinosus*) to *L. infantum chagasi* following inoculation of promastigotes by the intracardial and intradermal routes. No parasites could be detected in smears of the viscera from all of the 10 animals when they were sacrificed 7 months later, but positive cultures were obtained from splenic material of two of five animals inoculated by the intracardial route and three of five inoculated by the intradermal route. Attempts to infect sandflies (*Lu. longipalpis*) after two or three xenodiagnoses on each animal, during a period of 1–6 months, all failed: PCR-

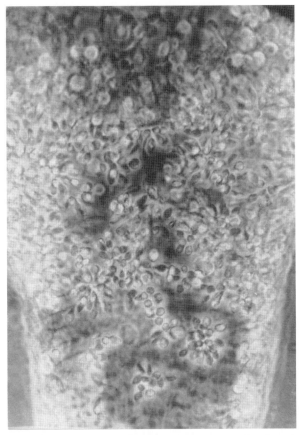

Figure 17.7 *The pylorus ('hindgut triangle') of a sandfly infected with* L. (V.) braziliensis. *Prolific division of rounded or ovoid promastigotes and paramastigotes attached by flagellar hemidesmosomes to the gut wall, a characteristic of* Leishmania *species within the subgenus* Viannia.

host for this parasite and plays no important part in the eco-epidemiology of the disease in north Brazil.

Recorded sandfly hosts

Lutzomyia (Lutzomyia) longipalpis is the major vector of *L. (L.) infantum chagasi* throughout its geographic range (Deane and Grimaldi 1985; Lainson 1989) and the parasite has been transmitted experimentally by the bite of this sandfly, using both laboratory-infected and naturally infected flies (Lainson et al. 1977, 1985). Strong evidence was provided, however, that *Lu. longipalpis* represents a species complex of at least two taxa (Ward et al. 1988), based on slight morphological differences (males with pale spots on terga 3 and 4, versus others with a similar spotted pheromonal gland on only the 4th tergum). Attempts to cross-breed the two forms failed, adding support to this suggestion (Ward et al. 1983). Following a study of population genetics and phylogenetic analyses of *Lu. longipalpis* from Central and South America, Soto et al. (2001) suggested that specific allopatric populations had differentiated to the extent of forming sibling species, with four distinct lineages corresponding to central and northern South America, Brazil and an isolated population in Colombia.

Lanzaro et al. (1993) compared populations of *Lu. longipalpis* from Costa Rica, Colombia and Brazil, using isoenzyme electrophoresis and cross-breeding experiments. They concluded that the three populations represented three different species. This finding has helped to explain why different clinical manifestations of *L. (L.) infantum chagasi* infection in humans occur in some geographic regions (see Clinical features).

In one particular region endemic for visceral leishmaniasis, in the Córdoba Department of Colombia, 87 percent of the sandflies captured were found to be *Lutzomyia evansi* and as one of these flies was infected with *L. (L.) chagasi* it was suggested that this sandfly might be acting as an alternative vector (Travi et al. 1990). In later studies (Travi et al. 1996) infections were found in nine more specimens of *Lu. evansi*, and, taking into account the apparent absence of *Lu. longipalpis* and the peridomestic and intradomiciliary habits of *Lu. evansi*, these authors concluded that this species was the principal vector in that part of Colombia. *Lu. evansi* had long been suspected as an alternative vector of *L. (L.) infantum chagasi* in Venezuela, following failure to find *Lu. longipalpis* in some foci of AVL where *Lu. evansi* was abundant (Potenza and Anduze 1942; Pifano and Romero 1964). The conclusions of these authors have gained strong support from the recent isolation of promastigotes from this sandfly in Venezuela and the finding that k-DNA restriction analysis showed high homologies between the cultures and *L. (L.) infantum chagasi* (Feliciangeli et al. 1999).

In the State of Mato Grosso do Sul, Brazil, suspicions have been raised that *Lu. cruzi* might be another alter-

hybridization tests gave negative results. It was concluded that the infection in *P. canicollis* 'is contained and compartmentalized', but that there may be differing degrees of susceptibility among different species of *Proechimys*.

Lainson et al. (2002) attempted to infect 12 laboratory-bred *Proechimys guyanensis* with a strain of *L. i. chagasi* from Amazonian Brazil, by massive intraperitoneal inoculation of amastigotes or promastigotes. Although control hamsters succumbed to acute infection 6 months p.i., no infection could be registered in any of the *P. guyanensis* when sacrificed at 6 or 12 months p.i., by stained smears of liver and spleen, culture of these tissues and their inoculation into hamsters, and the PCR-hybridization test. Furthermore, during the experiment laboratory-bred *Lu. longipalpis* showed a complete reluctance to feed on the experimental animals. Based on these results, previous unsuccessful attempts to capture *Lu. longipalpis* in Disney-traps baited with *P. guyanensis* and placed in the backyards of sandfly-infested houses, and a failure to isolate *L. i. chagasi* from large numbers of this rodent captured in foci of AVL in northern Brazil (Lainson et al. 1987), the authors concluded that *P. guyanensis* is an unsuitable native vector (dos Santos et al. 1998). Unfortunately, the females of this sandfly are morphologically indistinguishable from those of *Lu. longipalpis* and conclusions were based on an infection in a single female sandfly (*Lu. cruzi* or *Lu. longipalpis?*) and the apparent absence of *Lu. longipalpis* males. The limited geographical distribution of these two 'alternative' vectors compared with that of *Lu. longipalpis*, however, leaves no doubt regarding the overwhelming importance of the latter as the principal sandfly host of *L. (L.) infantum chagasi*.

Disease caused by the parasite in humans

The parasite predominantly causes visceral leishmaniasis, commonly with a fatal outcome unless treated (see Chapter 16, Old World leishmaniasis). On rare occasions, the visceral disease may be preceded by a cutaneous lesion, and in some geographical regions the parasite is responsible for an almost exclusively cutaneous disease (see Clinical features). Until recently, only cutaneous lesions were recorded in Costa Rica, but a single case of visceral leishmaniasis has now been recorded (Carrillo et al. 1999). Common names for the visceral disease include kala-azar or calazar but it is more appropriate to reserve these terms for Indian visceral leishmaniasis and to use the name American visceral leishmaniasis or AVL.

Opinions have been divided as to whether the causative agent of AVL is indigenous to the Americas, or whether it is simply *L. (L.) infantum*, which was introduced into the New World by immigrants from the Iberian peninsula in post-Columbian times. Points in favor of the first hypothesis and retention of the specific name *L. (L.) chagasi*, originally given to the parasite, are as follows:

- There is a high incidence of infection in the native fox, *Cerdocyon thous*, in relatively remote areas of Amazonian Brazil and the benign and inapparent nature of these infections is indicative of an ancient host–parasite relationship.
- Wild canids are considered to be the source from which members of the *L. (L.) donovani* complex originated (Lysenko 1971) and canids were present in the Americas as long ago as the Pleistocene era, some 2–3 million years ago.
- *Lu. longipalpis* is a vector of *L. (L.) chagasi* throughout almost the entire geographic range of the parasite, but is not known to transmit any other neotropical species of *Leishmania*; the vectors of *L. (L.) infantum* belong to a different sandfly genus. The host specificity of *Leishmania* species in nature is most pronounced among their sandfly vectors, thus it seems unlikely that introduced *L. (L.) infantum* could have made the relatively sudden jump from one phlebotomine genus to another.
- There are differences in both the kinetoplast DNA fragment patterns and the radio-respirometry profiles of *L. (L.) infantum* and *L. (L.) chagasi* (Jackson et al.

1982, 1984; Decker-Jackson and Tang 1982) and the two parasites are antigenically different (Santoro et al. 1986). The hypothesis that American visceral leishmaniasis is the result of imported *L. (L.) infantum* is based on the similarity of those isoenzyme profiles of the two parasites that have been studied (Rioux et al. 1990), the dual role of the domestic dog as an amplification host in their respective epidemiologies (Killick-Kendrick 1985) and recent studies on the genomic diversity in the *Leishmania donovani* complex (Maurício et al. 1999). The authors of this chapter have long used the name *L. (L.) chagasi*. From the close biochemical and molecular similarities that have now been shown, however, we feel it best to separate the two parasites only at subspecific level, with the names *L. (L.) infantum infantum* and *L. (L.) infantum chagasi*.

LEISHMANIA (LEISHMANIA) ENRIETTII MUNIZ AND MEDINA, 1948

Known geographical distribution

To date, this parasite has been found only in the States of Paraná and São Paulo, Brazil.

Known mammalian hosts

Natural infections have, until now, only been found in domestic guinea-pigs (*Cavia porcellus*). When this strange parasite was first discovered, producing large tumor-like lesions on the ears of two laboratory guinea-pigs (Medina 1946), the origin of the infected animals was obscure. As phlebotomine sandflies are the only known vectors of *Leishmania* species, however, it is difficult to imagine that domestic guinea-pigs are the principal natural hosts of *L. (L.) enriettii*, and it can only be assumed that the animals had spent some time in or near a rural area where transmission was occurring among the true, wild mammalian hosts. There have been two further spontaneous reappearances of the parasite in domestic guinea-pigs, again in Curitiba, Paraná (Luz et al. 1967) and more recently in a rural district of São Paulo State (Machado et al. 1994). Although the exact locality of the animals was on these occasions well documented, attempts to discover the wild animal source of the parasite were not made.

Recorded sandfly hosts

The natural vector of *L. (L.) enriettii* remains to be discovered. Luz et al. (1967) examined the sandfly population of neighboring forest around the site of isolation, where *Lu. monticola* and *Lu. correalimai* were the only species encountered. *Lu. monticola* was caught on tree trunks, in the nests of opossums (*Didelphis*) and from human bait, and six of 10 specimens fed on the lesions of guinea-pigs became heavily infected.

Absence of infection in humans

Human infection has not yet been reported. Muniz and Medina (1948) attempted to infect human volunteers, rhesus monkeys, dogs, mice, and the wild guinea-pig (preá) by the intradermal inoculation of promastigotes from in vitro cultures, without success. Out of eight hamsters inoculated, only one developed an inconspicuous lesion containing scanty amastigotes. Failure to infect the closely related wild guinea-pig (*Cavia aperea*) at first seems surprising. Lainson and Shaw (1979) suggested, however, that if this animal is the natural host of *L. (L.) enriettii* the inoculated preá may well have developed an inapparent infection that went unnoticed.

LEISHMANIA (LEISHMANIA) MEXICANA (BIAGI, 1953) GARNHAM, 1962

Known geographical distribution

The areas of known geographical distribution are Southern USA (Arizona and Texas), Mexico, Belize, Guatemala, Honduras, and Costa Rica. In view of extensive geographic separation from the type locality of *L. (L.) mexicana* and considerable differences in the mammalian and phlebotomine sandfly faunas, reports of this parasite in Panama and some South American countries must be viewed with caution.

Known mammalian hosts

The known mammalian hosts are humans and the forest rodents *Ototylomys phyllotis* (primary host), *Nyctomys sumichrasti*, *Heteromys desmarestianus* and *Sigmodon hispidus* (secondary hosts). In southern USA, the rodent *Neotoma albigula* has been found infected in Pima County, Arizona.

Recorded sandfly hosts

Lutzomyia olmeca olmeca is the only proven vector, but *Lu. (Lu.) diabolica* has been suspected as a vector in foci of cutaneous leishmaniasis due to *L. (L.) mexicana* in southern Texas and northern Mexico, and *Lu. anthophora* in Arizona.

Disease caused by the parasite in humans

This parasite causes cutaneous leishmaniasis, with a pronounced tendency towards long-lasting and destructive lesions of the external ear (Figure 17.8). Relatively rare cases of ADCL have been recorded, principally from southern Texas and northern Mexico. Local names include chiclero's ulcer, chiclero's ear, and bay-sore.

Parasites clearly related to *L. (L.) mexicana* have been reported from humans and wild animals in Panama, Colombia, Venezuela, Peru, and parts of Brazil. In the absence of the vector, *Lu. olmeca olmeca,* it is perhaps unlikely that any of these are true *L. (L.) mexicana*.

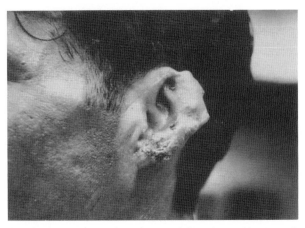

Figure 17.8 *Chiclero's ulcer, due to* L. (L.) mexicana. *Almost total destruction of the the external ear in a chiclero from Belize with an infection of many years' duration. (From Lainson and Strangways-Dixon 1963)*

Sandflies of the *Lu. flaviscutellata* complex, to which subspecies of *Lu. olmeca* belong, range through these South American countries, however, and are probably the vectors of a number of closely related parasites within the *mexicana* complex: at present this includes *L. (L.) pifanoi, L. (L.) amazonensis, L. (L.) aristidesi, L. (L.) garnhami, L. (L.) venezuelensis* and *L. (L.) forattinii.*

LEISHMANIA (LEISHMANIA) PIFANOI (MEDINA AND ROMERO, 1959) MEDINA AND ROMERO, 1962

Known geographical distribution

Known distribution is limited to Venezuela, specifically in the States of Yaracuy, Lara and Miranda.

Known mammalian hosts

Humans are the only known mammalian host: the wild mammalian hosts of the parasite have yet to be discovered. A parasite isolated from the forest rodent *Heteromys anomalus* was shown to behave in hamsters in the same way as *L. (L.) pifanoi* isolated from humans, but its conclusive identification remains in doubt.

Recorded sandfly hosts

A parasite with similar characteristics was also found in a specimen of *Lu. flaviscutellata* but, once again, has not been conclusively identified. This sandfly is the proven vector of *L. (L.) amazonensis,* another member of the *mexicana* complex often found in rodents in South American forests and it remains likely that the Venezuelan workers were dealing with this parasite.

Disease caused by the parasite in humans

All isolates found to date have been from cases of ADCL. It is most likely, however, that simple, curable lesions also occur in immunologically competent individuals. There has been much controversy regarding the validity of this member of the *mexicana* complex, some authors regarding it merely as an enzymic variant of either *L. (L.) amazonensis* or *L. (L.) mexicana*. Much confusion has certainly resulted from the laboratory mix-up of parasites. Although there are relatively few isolates of *L. (L.) pifanoi*, interest in the clinical features of the human infection has resulted in a wide distribution of the parasite to laboratories throughout the world.

LEISHMANIA (LEISHMANIA) AMAZONENSIS LAINSON AND SHAW, 1972

Known geographical distribution

The parasite has been recorded in Bolivia, Brazil, Colombia, French Guyana, and Paraguay. It is also very likely to occur in other South American countries where the sandfly vector is found.

Known mammalian hosts

Known mammalian hosts are the forest rodents *Proechimys* (principal host), *Oryzomys*, *Neacomys*, *Nectomys*, and *Dasyprocta*; the marsupials *Marmosa*, *Metachirus*, *Didelphis*, and *Philander*; and the fox *Cerdocyon*.

Recorded sandfly hosts

The principal vector of the parasite is *Lutzomyia (Nyssomyia) flaviscutellata*. Occasional infections have been found in the closely related flies *Lu. (N.) olmeca nociva* and *Lu. (N.) reducta* but, if these are capable of transmitting the parasite, they probably play a small role in its ecology and epidemiology. In Bolivia, a parasite identified as *L. (L.) amazonensis* has been isolated from 16 of 1 715 *Lu. nuneztovari* dissected (Martinez et al. 1999).

Disease caused by the parasite in humans

The parasite causes cutaneous leishmaniasis, usually of the single sore type and ADCL in individuals with a defective cell-mediated immune system (Figures 17.5 and 17.9). If it does occur, classical mucocutaneous leishmaniasis following metastasis to the nasopharyngeal mucosae from a simple cutaneous lesion is extremely rare. In advanced cases of ADCL, however, the disseminated infection may also include those tissues.

Typical visceral leishmaniasis in patients from one particular region of Bahia State Brazil, has been attributed to *L. (L.) amazonensis* (Barral et al. 1986). Conversely, no records exist to confirm this anywhere else in the geographical range of the parasite and cases of ADCL of very long duration show no signs or symptoms of visceral involvement, in spite of their defective immune system. Silveira examined the tissues from a patient who suffered from ADCL due to *L. (L.) amazonensis* from the age of 5 years until his death at the age of 57 years (Figure 17.5a) (F.T. Silveira, unpublished

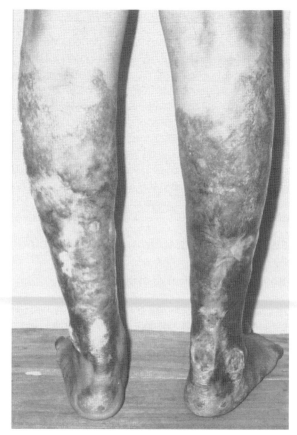

Figure 17.9 *ADCL due to* L. (L.) amazonensis *in a young woman from Pará, Brazil, showing active lesions and extensive scarring of the legs in an infection of some 15 years' duration.*

observations). At the time of his death, scarcely any of his body surface remained unaffected and the cutaneous lesions contained enormous numbers of parasites (Figure 17.10). No macroscopic or microscopic pathological changes were seen in the viscera and no amastigotes were found in stained impression smears prepared from the spleen, liver and lungs, and bone marrow. Finally, hamsters inoculated intradermally with triturates of these visceral tissues failed to become infected.

LEISHMANIA (LEISHMANIA) ARISTIDESI LAINSON AND SHAW, 1979

Known geographical distribution

The parasite is known to occur in the Sasardi forest, San Blas Territory, Eastern Panama.

Known mammalian hosts

Known mammalian hosts are the rodents *Oryzomys capito*, *Proechimys semispinosus*, and *Dasyprocta punctata*; and the marsupial *Marmosa robinsoni*.

Recorded sandfly hosts

The species most suspected is *Lutzomyia (Nyssomyia) olmeca bicolor*. Christensen et al. (1972) showed it to be

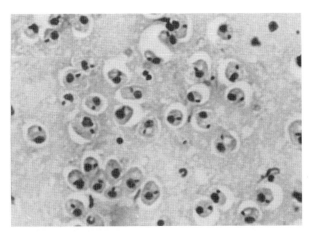

Figure 17.10 *Amastigotes of* L. (L.) amazonensis *in a Giemsa-stained smear from one of the nodular lesions of the patient shown in Figure 17.5a.*

the dominant fly on Disney-traps baited with rodents and opossums in the area where infected animals had been captured and the most common species collected among leaf litter on the forest floor.

Possibility of infection in humans

Human infection has not yet been reported. *Lu. olmeca bicolor* does bite humans on rare occasions and it is likely that the parasite may eventually be found infecting a human. In this respect, it should be remembered that following the discovery of *L. (V.) naiffi* in armadillos in 1979, 11 years were to pass before cases of human cutaneous leishmaniasis due to this parasite were diagnosed (Lainson et al. 1979, 1990b; Naiff et al. 1989).

LEISHMANIA (LEISHMANIA) GARNHAMI SCORZA ET AL., 1979

Difference of opinion exists regarding the validity of *L. (L.) garnhami*, which is indistinguishable from *L. (L.) amazonensis* on isoenzyme profiles (Rioux et al. 1990). Guevara et al. (1992), however, noted clear differences between the nontranscribed rDNA intergenic spacer sequences of these two parasites and, in our own laboratory, they have been separated by monoclonal antibodies (Shaw, Ishikawa, and Lainson, unpublished observations).

Known geographical distribution

The parasite has been found only in the Venezuelan Andes.

Known mammalian hosts

Humans are known hosts and a single infection has been recorded in the marsupial *Didelphis marsupialis*.

Recorded sandfly hosts

Experimental infections in the sandfly *Lu. youngi* (*verrucarum* group) in Venezuela have led some authors to suggest this species to be the vector (Grimaldi et al. 1989; Young and Duncan 1994; Killick-Kendrick 1990). In addition, Scorza (in Màrquez and Scorza 1982) reported that he had found natural infections with flagellates in *Lu. youngi* (at the time identified as *Lu. townsendi*) that, on inoculation into the skin of hamsters, produced 'amastigotes de *Leishmania garnhami*'. Unfortunately, the parasite was not specifically identified and the role of *Lu. youngi* as the vector still remains in doubt.

Disease caused by the parasite in humans

The parasite causes cutaneous leishmaniasis, but there are no recorded cases of mucocutaneous leishmaniasis or ADCL.

LEISHMANIA (LEISHMANIA) VENEZUELENSIS BONFANTE-GARRIDO, 1980
Known geographical distribution

The parasite has been found in the Lara and Yaracuy States, Venezuela.

Known mammalian hosts

The known mammalian hosts are humans, equines, and the domestic cat. These are best regarded as 'victim' hosts and the wild animal source of infection has yet to be ascertained.

Recorded sandfly hosts

Lu. olmeca bicolor and *Lu. rangeliana* are suspected as possible vectors.

Disease caused by the parasite in humans

The parasite causes single or multiple skin lesions, sometimes of a disseminated, nodular type simulating ADCL but curable by the current method of antimonial treatment.

LEISHMANIA (LEISHMANIA) FORATTINII YOSHIDA ET AL., 1993
Known geographical distribution

The parasite has been found in the States of São Paulo, Bahia and Espirito Santo, Brazil.

Known mammalian hosts

The opossum *Didelphis marsupialis aurita* (São Paulo) and the rodent *Proechimys iheringi denigratus* (Bahia and Espirito Santo, Brazil) are the known mammalian hosts.

Recorded sandfly hosts

The sandfly vector is unknown. Barretto et al. (1985) showed experimentally that the parasite was capable of development in the sandflies *Psychodopygus ayrozai* and *Lutzomyia yuilli* from the Três Braços area, Bahia, where the two insects are very common.

Possibility of infection in humans

The parasite has not yet been reported to cause human disease. Both *Ps. ayrozai* and *Lu. yuilli* occasionally feed on humans and if one or other of these sandflies is indeed the vector among the wild animal hosts, human infection with *L. (L.) forattini* may well occur.

Species within the subgenus *Viannia* Lainson and Shaw, 1987

These parasites have been separated by their isoenzymatic profiles, the use of monoclonal antibodies, and their biological characteristics. As is clear from Supplement 1 to volume 96 of *Trans R Soc Trop Med Hyg*, molecular biology techniques are now important tools in the field of parasitology and a set of microsatellite markers has recently been developed that enables separation of all known species of *Leishmania* within the subgenus *Viannia* and also indicates possible species hybrids (Russell et al. 1999). It will be particularly useful in the detection and identification of the more closely related pair *L. (V.) braziliensis* and *L. (V.) peruviana*, and the trio *L. (V.) panamensis*, *L. (V.) guyanensis* and *L. (V.) shawi*. A rapid method for the identification of *Leishmania* species using formalin-fixed biopsy samples and PCR is another promising new development (Mimori et al. 1998), as is the claim that these parasites can be reliably detected and identified, at least to the *L. (V.) braziliensis* complex level, by PCR using boiled dermal scrapings instead of skin biopsies (Belli et al. 1998).

LEISHMANIA (VIANNIA) BRAZILIENSIS (VIANNA, 1911) MATTA, 1916
Known geographical distribution

The distribution of this important parasite is badly defined, due to inadequate methods of identification used in the past. Parasites variously described as '*L. braziliensis*', '*L. braziliensis braziliensis*' or '*L. braziliensis* sensu lato' have been reported from most Latin American countries, including Argentina, Belize, Bolivia, Brazil, Colombia, Costa Rica, Ecuador, French Guyana, Guatemala, Honduras, Mexico, Nicaragua, Panama, Paraguay, Peru, Surinam, and Venezuela.

Environmental factors seem to govern the combinations of sandfly vector and wild mammalian host in the natural history of the different species of *Leishmania*. It seems unlikely, therefore, that *L. (V.) braziliensis* sensu stricto can have such an enormous geographic range and the parasites recorded in many of these regions may represent related, but different, parasites of the *braziliensis* complex. The situation is aggravated by the fact that the type material of *L. (V.) braziliensis* Vianna, 1911 from Além Paraiba, Minas Gerais, Brazil, is no longer available for comparison. It is further complicated by distinctly different ecological and epidemiological features of cutaneous and mucocutaneous leishmaniasis due to *L. (V.) braziliensis* s.l. in different regions, sometimes within the same country, due to human destruction of the sylvatic habitat. Finally, a high degree of enzymic polymorphism has been shown to exist in isolates of *L. (V.) braziliensis* s.l. from Central and South America, with as many as 44 zymodemes obtained in a total of 137 isolates examined by 10 enzyme systems (Chouicha et al. 1997). In a similar manner, a variety of different serodemes have been shown to exist in the parasite from Brazil (Shaw et al. 1986).

Known mammalian hosts

In primary forest in the Serra dos Carajás, Pará State, Brazil, the low-flying habit and high attraction of the principal sandfly vector to both humans and rodent-baited traps led to the conclusion that terrestrial mammals, such as rodents, were probably hosts of *L. (V.) braziliensis* (Ward et al. 1973; Lainson et al. 1973). Subsequent records of parasites considered as this species in the Brazilian rodents *Oryzomys concolor, O. capito, O. nigripes, Akodon arviculoides, Proechimys* spp., *Rattus rattus* and *Rhipidomys leucodactylus*, and the opossum *Didelphis marsupialis* did much to support this suggestion (Forattini et al. 1972, 1973; Lainson and Shaw 1970, 1979; Lainson et al. 1981a; Rocha et al. 1988). In Venezuela, records have been given for the rodents *R. rattus* and *Sigmodon hispidus* (De Lima et al. 2002). In most cases, identification of the parasite was based on only biological features and the isolates are no longer available for confirmation by modern biochemical methods, monoclonal antibodies or molecular techniques. Recently, however, a definitive identification of isolates from the Brazilian rodents *Bolomys lasiurus* and *R. rattus* as *L. (V.) braziliensis* by multilocus enzyme electrophoresis (Brandão-Filho et al. 2003) does suggest that the earlier records of rodent hosts of *L. (V.) braziliensis* were probably correct. It now remains to show that these animals serve as an efficient source of infection for the sandfly vectors.

Domestic animals such as dogs, mules, horses, and (very rarely) cats have been found with skin lesions produced by parasites regarded as *L. (V.) braziliensis* s.l. or simply recorded as leishmanias of the *braziliensis* complex. These reports, in areas of suspected peridomestic transmission of the parasite, come principally from localities of extensive deforestation in Argentina, southern Brazil, Bolivia, Colombia, and Venezuela.

Recorded sandfly hosts

Uncertainties regarding the exact distribution of *L. (V.) braziliensis* make it difficult to indicate its vector (or vectors). An isolate of the parasite responsible for human cutaneous leishmaniasis in primary Amazonian rain forest in the Carajás highlands of Pará State, Brazil, has been used as a reference strain of *L. (V.) braziliensis* (MHOM/BR/1975/M2903) and in this area the vector is undoubtedly *Psychodopygus wellcomei*. This sandfly, referred to by some as *Lutzomyia (Psychodopygus) wellcomei*, has been found heavily infected on numerous occasions (Lainson et al. 1973) and *L. (V.) braziliensis* has been experimentally transmitted to a hamster by the bite of a naturally infected specimen (Ryan et al. 1987).

Ps. wellcomei is essentially sylvatic and avidly feeds on humans, not only at night but also in the daylight hours during overcast weather. It is extremely abundant in the rainy season (November–April), when it may represent c. 65 percent of the total catch of some 25 different species of sandflies taken off human bait. Captures of sandflies from humans stationed at different heights on tree-ladders have shown that *Ps. wellcomei* has a vertical flight range of only 1–2 meters above ground level: this, and the insect's abundance (25.5 percent) in catches of different sandflies taken from rodent-baited traps, suggest that the principal wild mammalian hosts of *L. (V.) braziliensis*, in the area in question, are likely to be terrestrial animals, probably rodents.

In the lowland regions of Pará State, a parasite identified as *L. (V.) braziliensis* was isolated from a single specimen of a closely related and highly anthropophilic sandfly, *Ps. complexus* (de Souza et al. 1996). The females of this species are morphologically indistinguishable from those of *Ps. wellcomei*, but the males of each species are quite distinct.

In the State of Amazonas another highly anthrophilic species, *Ps. carrerai*, has been found infected with *L. (V.) braziliensis* s.l. (Grimaldi et al. 1989). The strain was biochemically similar to *L. (V.) braziliensis* of humans from the same region but antigenically different from those from other areas of Brazil, including the lower Amazon region.

In other parts of Brazil, occasional isolates of *L. (V.) braziliensis* s.l. have been made from the sandfly *Lutzomyia (Nyssomyia) whitmani* sensu stricto, caught in and around houses in rural parts of Bahia and Ceará States, north-east Brazil (Hoch et al. 1986; Ryan et al. 1990; de Queiroz et al. 1994) and in the States of São Paulo and Minas Gerais (south-east Brazil) it is again suspected

as a vector due to its highly anthropophilic feeding habits and its high density in and around human dwelling places and animal sheds in the endemic areas of cutaneous leishmaniasis. In the northern part of Paraná *Lu. whitmani* was shown to form 62 percent of the sandfly population captured, and to be highly anthropophilic. Of 1 628 specimens dissected, three were infected (0.2 percent) and in each case the isolated parasite was identified as *L. (V.) braziliensis* on enzyme profiles using the standard World Health Organization (WHO) reference strains for comparison (Luz et al. 2000).

In the State of Rondônia, Brazil, a parasite identified as *L. (V.) braziliensis* has been isolated from three specimens of *Lutzomyia davisi* (Grimaldi et al. 1991). This sandfly is also highly anthropophilic.

Lutzomyia (Nyssomyia) intermedia is another highly anthropophilic sandfly that, although originally sylvatic, has now adapted well to a peridomestic habitat in deforested, rural areas. It is highly suspected as a vector of *L. (V.) braziliensis* s.l. in the State of Rio de Janeiro and some parts of the State of São Paulo, Brazil and certain regions of Argentina. Flagellates thought to have been promastigotes of *Leishmania* were on one occasion seen in histological sections of the intestines of the sandflies *Lutzomyia migonei* and *Lu. (Pintomyia) pessoai* caught in endemic areas of cutaneous leishmaniasis in São Paulo State (Pessôa and Barretto 1948). Their true nature, however, remains obscure.

In Bolivia *L. (V.) braziliensis* s.l. has been isolated from *Ps. carrerai carrerai* (Le Pont et al. 1988), *Ps. llanosmartini*, and *Ps. yucumensis* (Le Pont and Desjeux 1986), whereas in Colombia (Young et al. 1987) and Venezuela (Young and Duncan 1994) similar parasites have been found in *Lu. spinicrassa*. In north central Venezuela, parasites isolated from *Lu. ovallesi* and *Lu. gomezi* have been typed as *L. (V.) braziliensis* and these sandflies are considered as primary and secondary vectors, respectively, on epidemiological grounds (Feliciangeli et al. 1994). In the Andean region, there is strong evidence that *Lu. youngi* is the vector of local cutaneous leishmaniasis, but the parasites encountered in this sandfly require precise identification. *Lu. spinicrassa* may be a vector throughout areas bordering the endemic Colombian foci.

Disease caused by the parasite in humans

The parasite causes cutaneous leishmaniasis, usually with one or few lesions and also mucocutaneous and mucosal leishmaniasis (Figures 17.2–17.4). Common names include: úlcera de Bauru, ferida brava, ferida sêca, bouba, buba, nariz de anta (tapir nose), and espundia.

The various zymodemes of *L. (V.) braziliensis* s.l. are placed in the *braziliensis* complex, together with the closely related parasite *L. (V.) peruviana*.

LEISHMANIA (VIANNIA) PERUVIANA VELEZ, 1913

Known geographical distribution

The parasite has been noted in Peru, on the western slopes of the Andes and in the inter-Andean valleys. Its range may possibly extend into the Argentinian highlands and it is probably more widely distributed in the Andean countries than previously suspected. Transmission seems to take place in relatively barren, mountainous areas with scant vegetation and a relatively restricted wild mammalian fauna.

Known mammalian hosts

Until recently humans and dogs (*Canis familiaris*) were the only known mammalian hosts (Herrer 1951) The role of the dog in the epidemiology of the human disease has remained obscure, however, and it has long been postulated that both humans and dogs are merely 'victim hosts' of a parasite maintained in wild animals of the Peruvian Andes. Recent studies (Llanos-Cuentas et al. 1999) report that isolations of a parasite from two specimens of the rodent *Phyllotis andinum* and an opossum, *Didelphis marsupialis* have been characterized as *L. (V.) braziliensis* by their isoenzyme profiles. Five other isolates from the rodent *Akodon* sp., were identified only to the subgenus *Viannia*. That the prevalence in the wild animals is much the same as in dogs, supports the existence of enzootic disease in wild animals.

Recorded sandfly hosts

Lutzomyia (Helcocyrtomyia) peruensis and *Lutzomyia verrucarum* have long been suspected as probable vectors, in view of their anthropophilic feeding habits. Isolation of a parasite with the biological characteristics of *L. (V.) peruviana* from the former fly, captured in an endemic area, implies its involvement in transmission (Herrer 1982), but much research in the field is needed before significant evidence can be obtained.

Disease caused by the parasite in humans

The parasite causes simple cutaneous leishmaniasis, not associated with the mucocutaneous disease. It is particularly frequent in school children, commonly resulting in extensive facial scars. Ulcers are usually self-healing and a firm immunity to reinfection with the same parasite is usually imparted. Common names include uta, tiaccaraña and llaga.

LEISHMANIA (VIANNIA) GUYANENSIS FLOCH, 1954

Known geographical distribution

This is an essentially sylvatic species that is an extremely common cause of human cutaneous leishmaniasis, particularly in Brazil, north of the Amazon river and the Guyanas. Its range is reported to also extend into Colombia, Ecuador, Venezuela, and the lowland forests of Peru.

Known mammalian hosts

Known mammalian hosts include humans and, in primary forest, the major reservoir hosts are the sloth *Choloepus didactylus* and the lesser anteater *Tamandua tetradactyla* (Xenarthra) (Lainson et al. 1981b; Gentile et al. 1981) with occasional infections found in rodents and opossums (Lainson et al. 1981a; Gentile et al. 1981) The infection is always inapparent, with parasites located in apparently normal skin and in viscera such as the spleen and liver.

Recorded sandfly hosts

The principal vector among wild animals and to humans is the sandfly *Lutzomyia (Nyssomyia) umbratilis*, with infections relatively infrequently found in a closely related fly, *Lu. (N.) anduzei* and *Lu. (N.) whitmani* s.l. Some early records of the parasite in the latter sandfly in Amazonian Brazil, possibly refer to *Leishmania (V.) shawi*.

Lu. (N.) umbratilis is a sandfly that dwells in the forest canopy and on the larger tree trunks. During the early hours of daylight, it may be found in large numbers resting on the larger tree trunks from which the flies will readily fly off and attack humans when disturbed. Although infection of this sandfly clearly takes place when it feeds on the reservoir hosts at night in the canopy, transmission to humans is principally during the day (early morning), when gangs of forest laborers are engaged in their work, particularly deforestation. Others at risk include the collectors of Brazil nuts and other fruits, topographers, visiting botanists and zoologists, and even the occasional tourist.

The enzootic of *L. (V.) guyanensis*, as studied in primary rain forest, is unlikely to survive in secondary forests or man-made plantations of non-indigenous trees. The small girth of young trees provides a microhabitat that is unsuitable for resting sandflies due to the low surface humidity of the smooth trunks. In addition, such immature trees are an equally unsuitable environment for relatively large and heavy animals, such as sloths and anteaters. Finally, in monoculture plantations (e.g. pine and gmelina) sloths are deprived of their normal diet of indigenous fruits and foliage.

Cutaneous leishmaniasis due to *L. (V.) guyanensis* may reach high prevalence in human communities situated in or very near primary forest, leading to an erroneous impression that the sandfly vector, *Lu. (N.) umbratilis*, has adapted to a peridomestic habitat. There is, as yet, no evidence that this occurs and peridomestic acquisition of the disease is doubtless due to infected flies that have been attracted to the lights of houses at night, from nearby forest. Esterre et al. (1986) showed experimentally that clearing forest to c. 500 meters from a village situated in primary forest in French Guyana completely interrupted the transmission of *L. (V.) guyanensis* among its inhabitants.

The marsupial *Didelphis marsupialis* has rarely been found infected with *L. (V.) guyanensis* in primary forest where there is intensive transmission of the parasite among sloths and anteaters and to humans. Strangely, however, a high rate of infection has been recorded in the abnormally large populations of this opossum that are attracted to human refuse in villages on the borders of virgin forest (Arias and Naiff 1981). Reasons for this are not clear, nor is it certain whether opossums serve as a source of *L. (V.) guyanensis* for the sandfly vector, or if they merely represent dead-ends in the life-cycle of the parasite.

Disease caused by the parasite in humans

The parasite causes cutaneous leishmaniasis, very frequently with multiple skin lesions (Figure 17.6). Cases of mucocutaneous disease appear to be rare. Common names for the disease include pian-bois, bosch-yaws, and forest-yaws.

Multiplicity of the skin lesions arises in two very different ways. First, sloths are rather sedentary animals and an infected animal may remain in a given spot for a considerable period. The infection rate of *Lu. (N.) umbratilis* resting on neighboring tree trunks will thus tend to rise to a high level, with levels as high as 25 percent found among many hundreds of specimens taken from a single tree. It follows that people attacked by sandflies in the area may receive numerous infective bites at the same time on all exposed parts of the body. Forest workers tend to be shirtless and frequently wear shorts: for this reason many patients present with lesions scattered over the face, trunk, arms, and legs (Figure 17.6). The developing multiple lesions of such individuals tend to be of a similar size and evolution. Secondly, there is abundant clinical evidence indicating the formation of metastatic lesions in persons originally presenting with a single skin lesion. These lesions tend to follow a distinct migratory course along the lymphatics and because of this evolution they are frequently nodular and, when ulcerated, of very unequal size.

L. (V.) guyanensis gives its name to the *guyanensis* complex of closely related parasites including *L. (V.) panamensis* and *L. (V.) shawi*.

LEISHMANIA (VIANNIA) PANAMENSIS LAINSON AND SHAW, 1972
Known geographical distribution

As the name suggests, most information on this parasite comes from Panama and the Canal Zone, where the very frequent acquisition of cutaneous leishmaniasis by American military personnel prompted intensive eco-epidemiological studies. It is also recorded in west and central Colombia, Ecuador, Venezuela, Costa Rica, Honduras, and Nicaragua.

Known mammalian hosts

Humans are a host to the parasite. The eco-epidemiology of *L. (V.) panamensis* follows a very similar pattern to that of *L. (V.) guyanensis*, which is not surprising in view of the close biological and biochemical relationship of the two parasites. The major host is the two-toed sloth *Choloepus hoffmanni* with occasional infections reported in the three-toed sloth *Bradypus infuscatus* and *B. griseus* (Herrer et al. 1973). More rarely, infections have been registered in other sylvatic animals such as *Bassaricyon gabbi*, *Nasua nasua* and *Potos flavus* (Carnivora: Procyonidae), *Aotus trivirgatus*, and *Saguinus geoffroyi* (Primates: Cebidae and Callitrichidae) and *Heteromys* (Rodentia). Hunting dogs occasionally develop skin lesions due to *L.(V.) panamensis*: like humans they are 'victim hosts' that rarely, if ever, serve as a source of parasites for the sandfly vector(s), or as a means of maintaining the enzootic.

Recorded sandfly hosts

The major sandfly vector is considered to be *Lu. (N.) trapidoi*, whereas *Lu.(N.) ylephiletor*, *Lu. (Lu.) gomezi*, and *Psychodopygus panamensis* may act as secondary vectors (Johnson et al. 1963; Christensen et al. 1969)

Disease caused by the parasite in humans

The parasite usually causes single or a limited number of skin lesions. Rare cases of mucocutaneous leishmaniasis have been attributed to *L. (V.) panamensis*.

Studies on populations of *Leishmania* in Ecuador have shown a cluster in which the isoenzyme profile of 6-phosphate gluconate dehydrogenase (6PGD) showed considerable variation (Bañuls et al. 1999). As it is this enzyme that has been extensively used to separate *L. (V.) panamensis* and *L. (V.) guyanensis*, these authors suggested that these parasites are not separate species. Within the cluster of these species in eastern Amazonia, however, *L. (V.) guyanensis* strains all grouped together, while isolates from Ecuador grouped with *L. (V.) panamensis* markers. For this reason, and taking eco-epidemiological and clinical data into consideration, we prefer to maintain specific separation of the two parasites.

LEISHMANIA (VIANNIA) LAINSONI SILVEIRA ET AL., 1987
Known geographical distribution

Until recently, this parasite was recorded only in the State of Pará, north Brazil. Its presence has now been noted, however, in forested areas of Peru and Bolivia (Lucas et al. 1994; Martinez et al. 2001). It probably exists in other regions where the known mammalian and sandfly hosts coexist.

Known mammalian hosts

Humans are a known mammalian host. The only known host among wild animals is the rodent *Agouti paca* (Rodentia: Dasyproctidae) (Silveira et al. 1991a).

Recorded sandfly hosts

Lu. (Trichophoromyia) ubiquitalis is a known host and the first representative of the subgenus *Trichophoromyia* to be incriminated as a vector of a *Leishmania* species.

L. (V.) lainsoni was isolated only from this sandfly, among many other species dissected in forested areas where patients had become infected with this parasite (Silveira et al. 1991b). The puzzling fact remained that *Lu. (T.) ubiquitalis* had not been caught biting humans in the forest. It was found, however, that this sandfly would feed avidly on humans if maintained for some hours in the laboratory after capture and this prompted the conclusion that under certain conditions it must also feed on humans in its natural habitat. Continuing field studies confirmed this (Lainson et al. 1992), although the factors influencing the sandfly's biting habits remain obscure. *Lu. (T.) ubiquitalis* is clearly not particularly fond of human blood, which accounts for the relatively low rate of infection with *L. (V.) lainsoni* in humans, compared with other species of *Leishmania*, such as *L. (V.) braziliensis* and *L. (V.) guyanensis*; these have highly anthropophilic sandfly vectors. In Bolivia, promastigote infections have been recorded in the sandfly *Lu. velascoi*, but the parasite was not isolated for characterization. As this fly is the only species of the subgenus *Trichophoromyia* in a region where *L.(V.) lainsoni* was encountered infecting man, it remains highly likely that it is the local vector of this parasite (Martinez et al. 2001).

Disease caused by the parasite in humans

The parasite causes cutaneous leishmaniasis, usually with a single ulcerating skin lesion. Cases of mucocutaneous leishmaniasis due to this parasite have not yet been encountered.

LEISHMANIA (VIANNIA) SHAWI LAINSON ET AL., 1989
Geographical distribution

To date, the parasite has been found in various localities in the Amazon Region of north Brazil, south of the Amazon river.

Known mammalian hosts

Humans are commonly infected. Hosts among the forest animals include the monkeys, *Cebus apella* and *Chiropotes satanas* (Cebidae); the sloths *Choloepus didac-*

tylus and *Bradypus tridactylus* (Xenarthra); and the coatimundi *Nasua nasua* (Procyonidae) (Lainson et al. 1988, 1989). It remains likely that other arboreal animals harbor the parasite.

Recorded sandfly hosts

Infections have so far been recorded in only one species of sandfly, which was provisionally identified as *Lutzomyia (Nyssomyia) whitmani* (Lainson et al. 1989). Morphometric differences have been noted, however, between this insect and the type material of *Lu. (N.) whitmani* sensu stricto from Bahia, northeast Brazil. These, and separation by DNA probes, suggested the vector to be a 'cryptic' species of a *Lu. (N.) whitmani* complex (Rangel et al. 1996). A recent phylogenetic analysis of the described mitochondrial (cytochrome b) haplotypes of *Lu. whitmani*, however, disputes this hypothesis, and suggests the existence of clades of haplotypes and a continuum of interbreeding populations of *Lu. whitmani* in the rain-forest regions of Brazil (Ishikawa et al. 1999). The differences in behavior of the fly in the type locality in Bahia, northeast Brazil, and that in Pará State, north Brazil are nevertheless striking. The former is highly anthropophilic, abundant in human dwelling places, and transmits *L. (V.) braziliensis*; the latter rarely bites man, until now has not been recorded in houses even when these are situated very close to the fly's normal habitat in primary forest, and is a vector of *L. (V.) shawi*.

Disease caused by the parasite in humans

The parasite causes cutaneous leishmaniasis, usually with a single ulcerating skin lesion, but cases of multiple lesions of varying pathologies have been observed (Figure 17.11). Cases of mucocutaneous leishmaniasis due to this parasite have not yet been encountered.

LEISHMANIA (VIANNIA) NAIFFI LAINSON AND SHAW, 1989
Known geographical distribution

Isolates of the parasite are registered in the Brazilian States of Pará and Amazonas and in French Guyana. The range of this parasite will almost certainly extend into other parts of Latin America where the wild animal host and the sandfly vector coexist.

Known mammalian hosts

Humans are a known mammalian host (Lainson et al. 1990b). To date, the only known wild animal host is the nine-banded armadillo, *Dasypus novemcinctus* (Xenarthra: Dasypodidae), in which there is a high infection rate in apparently normal skin and viscera.

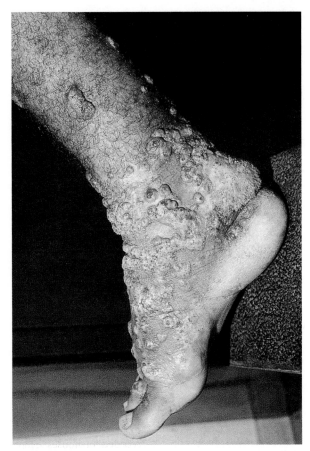

Figure 17.11 *Strange multiple nodules on the foot due to* L. (V.) *shawi, clinically resembling mycotic disease; Pará, Brazil*

Recorded sandfly hosts

Most recorded sandfly infections (Arias et al. 1985; Naiff et al. 1989) involve the sandfly *Psychodopygus ayrozai* which is, therefore, most suspected as the vector among armadillos. This fly is not highly anthropophilic, which possibly accounts for the paucity of human infection with *L. (V.) naiffi*. On the other hand, the parasite has, on rarer occasions, also been found in *Ps. paraensis* and *Ps. s. squamiventris*, both of which are highly anthropophilic and possibly involved in transmission to humans. Further studies are clearly necessary to establish the respective roles of these three sandflies in the eco-epidemiology of *L. (V.) naiffi*.

Disease caused by the parasite in humans

The parasite causes cutaneous leishmaniasis, usually with single, small, ulcerating lesions. No case of mucocutaneous leishmaniasis has yet been attributed to this parasite.

Unlike most species of *Leishmania*, *L. (V.) naiffi* rarely produces a visible lesion when inoculated into the skin of hamsters, although the parasite may be re-isolated following culture of skin from the inoculation site in blood agar media after at least 1 year. For this reason, human infection with this parasite may have remained undiagnosed in the past, when inoculation of hamsters has been the sole method of isolation attempted from human skin lesions. In addition, it may be that *L. (V.) naiffi* can also produce an occult, benign infection in the skin of humans, as it does in the hamster and that transmission to humans is more frequent than has been suspected.

LEISHMANIA (VIANNIA) COLOMBIENSIS KREUTZER ET AL., 1991
Known geographical distribution

The parasite exists in Colombia, Panama, and Venezuela, probably extending into the neighboring forests of Brazil, the Peruvian lowlands, and other Latin American countries where the wild mammalian and sandfly hosts coexist.

Known mammalian hosts

The parasite has been recorded in humans and the sloth, *Choloepus hoffmanni* (Panama).

Recorded sandfly hosts

Lutzomyia (Helcocyrtomyia) hartmanni has been found infected in Colombia and *Lu. (Lutzomyia) gomezi* and *Psychodopygus panamensis* in Panama (Kreutzer et al. 1991).

Disease caused by the parasite in humans

The parasite causes single to multiple ulcerating skin lesions. Cases of mucocutaneous leishmaniasis caused by the parasite have not yet been seen. The strain identified from Venezuela by Delgado et al. (1993) was isolated from a bone marrow aspirate of a patient with visceral leishmaniasis. It is, however, uncertain whether or not this was the parasite responsible for the clinical symptoms observed.

LEISHMANIA (VIANNIA) EQUATORENSIS GRIMALDI ET AL., 1992
Known geographical distribution

The parasite exists along the Pacific coast of Ecuador.

Known mammalian hosts

The sloth, *Choloepus hoffmanni*, and the squirrel, *Sciurus granatensis* have been identified as hosts. In both animals, the parasite was isolated from the liver and spleen but was not found in the skin.

Recorded sandfly hosts

Lutzomyia hartmanni has been found infected in Ecuador (Furuya et al. 1998).

Disease caused by the parasite in humans

As yet, the parasite has not been found in humans.

LEISHMANIA (VIANNIA) LINDENBERGI SILVEIRA ET AL., 2002

Known geographical distribution

At present, this recently described parasite has been recorded only from degraded forest on the outskirts of Belém, Pará State, north Brazil, with isolates made from a number of soldiers who acquired their infections while undertaking maneuvers in this forest at night.

Known mammalian hosts

Humans are the only recorded host so far. From the behavior of the suspected vector and the circumstances in which the soldiers became infected, the wild animal host(s) is (are) most likely terrestrial.

Recorded sandfly hosts

Lutzomyia (Nyssomyia) antunesi is by far the most common sandfly avidly biting man in the type locality. It is of particular interest that the infected soldiers spent much of their time in slit-trenches, so that their heads and resting arms were at ground level: most of their lesions were on these parts of the body. Previous workers recorded heavy promastigote infections in three specimens of *Lu. (N.) antunesi* from Marajó island, near Belém (Ryan et al. 1984), but they failed to identify the parasite. Its development in the sandfly gut was described as being 'suprapylarian in nature' (i.e. characteristic of the subgenus *Leishmania*), whereas Silveira et al. (2002) clearly showed the development of *L. (V.) lindenbergi* to be peripylarian in experimentally infected laboratory-bred *Lu. longipalpis*.

Disease caused by the parasite in humans

Localized cutaneous lesions: as yet, no cases of mucosal lesions have been recorded. *L. (V.) lindenbergi* is distinguished from other members of the subgenus *Viannia* by its enzyme profiles and the use of monoclonal antibodies. It is clearly closely related to *L. (V.) naiffi* and also produces an inapparent infection in the skin of hamsters.

LEISHMANIA (VIANNIA) UTINGENSIS BRAGA ET AL., 2003

Known geographical distribution

As yet, this parasite is recorded only from the Utinga forest, on the outskirts of Belém, Pará State, north Brazil.

Known mammalian hosts

At present, mammalian hosts are unknown, but given the habits of the vector are likely to be arboreal.

Recorded sandfly hosts

The parasite was isolated from a single specimen of the sandfly *Lutzomyia tuberculata*. This sandfly is commonly found on the larger tree-trunks of primary and secondary forests.

Disease caused by the parasite in humans

Not yet found infecting humans. As the vector shows no tendency to bite man, human infections with *L. (V.) utingensis* are unlikely to occur by way of this insect. If any anthropophilic sandflies are also hosts of the parasite, however, human infection may well be found in the future.

'Hybrid' *Leishmania* of the subgenus *Viannia*

Strains with phenotypic and genotypic characters of two species have been recorded in different geographical areas of Latin America. There are two possible interpretations for such strains: either they represent strains that originated directly from a common ancestor or they are the result of genetic exchange.

L. (V.) BRAZILIENSIS/L. (V.) PANAMENSIS HYBRID

Known geographical distribution

The parasite has been found in northern Nicaragua (Darce et al. 1991) close to the border with Honduras. To the south, the strains are *L. (V.) panamensis* and to the north, *L. (V.) braziliensis*. Bañuls et al. (1997) recorded what appears to be the result of hybridization between *L.(V.) braziliensis/L.(V.) panamensis* and *L. (V.) panamensis/L. (V.) guyanensis* in isolates from human infections in Ecuador.

Known mammalian hosts

So far, the parasite is only known in humans.

Recorded sandfly hosts

No sandfly host has been recorded to date.

Disease caused by the parasite in humans

The parasite causes ulcerating skin lesions. Cases of mucocutaneous infections have not yet been seen.

L. (V.) BRAZILIENSIS/L. (V.) GUYANENSIS HYBRIDS

Known geographical distribution

In Venezuela, the parasite has been found in Lara State; Tachira; DF, El Junquito; and Miranda, Guarenas. The isolates from the last three localities were similar, and their hybrid alleles differed from those of the isolate previously made in Lara (Bonfante-Garrido et al. 1992; Delgado et al. 1997).

Known mammalian hosts

So far, the parasite has only been recorded from humans.

Recorded sandfly hosts

No sandfly hosts have yet been found.

Disease caused by the parasite in humans

The parasite causes ulcerating skin lesions. Cases of mucocutaneous leishmaniasis caused by this parasite have not yet been seen.

L. (V.) BRAZILIENSIS/L. (V.) PERUVIANA HYBRID

Known geographical distribution

To date, four strains have been isolated from patients in the Limapampa region of the Huanuco Valley, Peru (Dujardin et al. 1995).

Known mammalian hosts

Humans are the only known mammalian host to date.

Recorded sandfly hosts

No sandfly vectors have been recorded.

Disease caused by the parasite in humans

Ulcerating skin lesions were noted in all patients and one also had a mucosal lesion.

Leishmania-like parasites of uncertain taxonomic position

Molecular studies suggest that the following parasites are more closely related to *Endotrypanum* (an intra-erythrocytic flagellate of sloths) than they are to *Leishmania*. Until more information is available, however, we feel it best to retain their present names.

LEISHMANIA (LEISHMANIA) HERTIGI HERRER, 1971

Known geographical distribution

The known distribution is limited to Panama and Costa Rica.

Known mammalian hosts

The tree-porcupine *Coendou rothschildi* is the only known mammalian host.

Recorded sandfly hosts

The sandfly vector of the parasite has still to be discovered. The remarkably high infection rate of 88 percent found in the porcupines studied suggests a close association of the vector with the mammalian host, possibly in the hollow trees where these animals live.

Absence of infection in humans

Human infection has not been recorded. *L. (L.) hertigi* seems to be specific to *Coendou rothschildi*, as the parasite has not been found in a wide variety of other forest mammals studied in the Panamanian forests. The apparent absence of human infection may be due to the failure of the parasite to survive in human tissues, or because the sandfly vector does not bite humans.

L. (L.) hertigi is included in the *hertigi* complex, together with another closely related parasite of porcupines, *L. (L.) deanei*.

LEISHMANIA (LEISHMANIA) DEANEI LAINSON AND SHAW, 1977

Known geographical distribution

The parasite has been noted only in Amazonian Brazil.

Known mammalian hosts

The known mammalian hosts are the tree-porcupine *Coendou p. prehensilis* and another, as yet unnamed, species of *Coendou*.

Recorded sandfly hosts

The vector of *L. (L.) deanei* remains to be discovered. As with *L. (L.) hertigi*, the infection rate in Brazilian porcupines is very high, again suggesting a close association of the vector with the mammalian host, probably in the animal's home in hollow trees. A tree-inhabiting sandfly, *Lu. (Viannamyia) furcata*, taken from a tree-hole in which an infected porcupine was living, was shown to have promastigotes of *L. (L.) deanei* in its midgut (Miles et al. 1980). There was, however, no evidence of migration of the parasites to the anterior station of the gut and, in subsequent experimental infections of this sandfly, it was shown that the promastigotes disappeared following complete digestion of the blood-meal (Lainson and Shaw 1987).

Absence of infection in humans

Human infection has not been recorded. As with *L. (L.) hertigi*, this could be because the sandfly vector is not attracted to humans or because the parasite cannot survive in human tissues.

L. (L.) deanei, like *L. (L.) hertigi*, seems to be restricted to species of the porcupine *Coendou* and an exhaustive examination of other animals in areas of forest where porcupines are commonly infected failed to indicate any other mammalian hosts (Lainson and Shaw, unpublished observations). Although both parasites seem to be peculiar to porcupines, they are readily distinguishable by their isoenzyme profiles and the morphology of their amastigote stages; those of *L. (L.) hertigi* are strangely elongated and measure from 3.5×1.2 to 4.8×2.5 μm. The amastigotes of *L. (L.) deanei* are the largest of all known species of *Leishmania*, measuring from 5.1×3.1 to 6.8×3.7 μm.

Working with 12 isolates of *L. (L.) deanei* from Pará, north Brazil, Miles et al. (1980) found that they were separable into groups by the enzyme profiles of:

- malate dehydrogenase (oxaloacetate-decarboxylating) (NADP) E.C.1.1.1.40 (ME)
- phosphoglucomutase E.C.2.7.5.1. (PGM)
- malate dehydrogenase E.C.1.1.1.37 (MDH).

Whether or not this is indicative of a third species of *Leishmania* in the *hertigi* complex is debatable and further study is indicated on the biology, biochemistry and molecular biology of the two zymodemes.

Questionable leishmanial parasites

LEISHMANIA HERRERI ZELODÓN ET AL. 1979

This parasite was isolated in Costa Rica from the sloth *Bradypus griseus* and regarded as a new species of *Leishmania*. Mention was made that the DNA buoyant densities and enzyme profiles of the organism were 'totally different from other known hemoflagellates', but no details of the biochemistry were given and, until further isolates of the parasite are available, there must remain considerable doubt as to its true nature.

Infection of human skin by nonleishmanial amastigotes

Amastigotes found in skin lesions of individuals infected with human immunodeficiency virus (HIV) and non-infected individuals on the Island of Martinique have been considered to be those of monoxenous insect flagellates (Boisseau-Garsaud et al. 2000). It is likely that some such lesions have, in the past, been wrongly attributed to *Leishmania*.

GENETICS

The New World *Leishmania* have 20–25 chromosomes that exhibit high degrees of intra- and interspecies size polymorphism. This is thought to reflect a high level of genomic plasticity. Leishmanial genomic DNA is composed of repetitive, moderately repetitive and unique sequences that, respectively, form ca. 25, 13, and 60 percent of the total. Repeated sequences are useful for identification regardless of their distribution, function or copy numbers. Fernandes et al. (1994) found that the mini-exon repeat units of four *Viannia* species were different. Considerable genomic differences have been noted between the two subgenera, but they appear greater for species of *Viannia*. Mendoza-León et al. (1995) found that the β-tubulin gene regions showed a much greater degree of heterogeneity within and between parasites of the *braziliensis* complex than those of the *mexicana* complex. This suggests that there are more random mutations in the former than in the latter. The same gene may be located in more than one chromosome. This may be the result of gene duplication and transposition to another chromosome or chromosomal duplication followed by size divergence (Spithill and Samaras 1985). A useful summary of the recent developments from the *Leishmania* genome project have been given by Myler and Stuart (2000).

Genetic exchange between neotropical *Leishmania* has not been demonstrated experimentally, but the finding of apparent hybrids among *Viannia* species suggests that it may occur. Hybrids only occur in some localities and it is possible that ecological conditions may be important in determining the contact between the species involved. No hybrids have been noted among neotropical species of the subgenus *Leishmania*. Extreme care is needed in the characterization of flagellates from sandflies, as they may sometimes harbor more than one parasite. Monoxenous flagellates, species of *Trypanosoma*, and *Endotrypanum* will contribute to confusion, in addition to the possible coexistence of two species of *Leishmania* in the same insect, as recorded by Barrios et al. (1994). Ideally, all isolates should be carefully cloned, especially if showing unusual characteristics by current methods of identification.

The possible role in humans of genes and chromosome regions suggested by previous studies in mice to be linked to disease resistance or susceptibility to visceral leishmaniasis has been investigated in 638 individuals in 89 families with multiple cases of AVL in northeast Brazil, using complex segregation analyses (Blackwell 1998). Preliminary results indicated that the major histocompatibility complex showed only weak association with AVL; that there is no evidence for linkage between *nramp1* (the positionally cloned candidate for a murine macrophage resistance gene) and susceptibility to AVL; and that the T helper 2 cytokine gene cluster is not linked to human susceptibility for this disease. It is suggested that the mouse model, together with knowledge of human response to infection, may lead to identification of important candidate gene regions in humans. Feitosa et al. (1999), used complex segregation analyses in a study of the intradermic reaction to antigens derived from *L. i. chagasi* in 502 individuals from 94 families in Bahia State, northeast Brazil. They concluded that the results gave evidence of a major genetic mechanism, with a frequency of a recessive susceptibility gene (q) of approximately 0.45, and that a small multifactorial component (H = 0.29), acting in combination with a major recessive gene (q = 0.37) could be a concomitant factor.

Viral infections in *Leishmania*

Virus-like particles have been described in the cytoplasm of cultured promastigotes of five isolates of *L. (L.) hertigi* and three of *L. (L.) deanei* (Molyneux and Killick-Kendrick 1987). Although unidentified, they were shown to be associated with cytopathological changes in the mitochondrion of the promastigote. The number of virus-like particles was drastically reduced when *L. (L.) hertigi* were grown as amastigotes in mouse peritoneal macrophages and dog sarcoma cell lines incubated at 32°C. The particles were not transmissible to other *Leishmania* species, nor to cell lines susceptible to viruses.

RNA virus particles belonging to the family Totaviridae were described by Tarr et al. (1988) in cultured promastigotes of a strain of *L. (V.) guyanensis* from French Guyana. Subsequent studies revealed the presence of similar viruses in promastigotes of 11 *Leishmania* strains isolated from humans, including both *L. (V.) guyanensis* and *L. (V.) braziliensis* from the Amazonian region of Brazil and Peru (Guilbride et al. 1992). All the viral isolates are considered to be different and are classified within a single genus, *Leishmaniavirus*. None, however, has been given specific status within this genus. The virus-bearing promastigotes were those of *Leishmania* isolated from uncomplicated cases of cutaneous leishmaniasis, but so far the viral particles have not been demonstrated in amastigotes from the skin lesions. The effect that such viruses may have on the *Leishmania* is not known but in experimental infections of *Leishmania major* in mice, it was noted that an infected line was less pathogenic than an uninfected one.

When human biopsy material from some cases of cutaneous leishmaniasis from Peru were examined for the presence of *Leishmaniavirus* RNA by the reverse transcription polymerase chain reaction (RT-PCR), two samples showed RT-PCR bands of the expected size. Sequence analysis indicated them to share approximately 90 percent sequence with the neotropical strain LRV 1-4 of that virus (Saiz et al. 1998).

CLINICAL FEATURES

American visceral leishmaniasis

Throughout most of its geographical range, the clinical features of AVL closely resemble those of infantile visceral leishmaniasis caused by *L. (L.) infantum* of the Old World (see Chapter 16, Old World leishmaniasis) and this similarity extends to the fact that the disease is seen mainly in children (Figure 17.12). However, It has been shown that the same parasite may produce almost exclusively nonulcerative cutaneous lesions in Costa Rica (Zelodón et al. 1989), and may cause both visceral and cutaneous disease in the same focus of AVL in Honduras and Nicaragua (Ponce et al. 1991; Belli et al. 1999).

The saliva of *Lu. longipalpis* has been shown to contain a potent vasodilatory peptide, maxadilan (Lerner et al. 1991). It was found that when sandflies from Brazilian, Colombian, and Costa Rican populations were fed on the arms of volunteers, the degree of erythema was highest for the Brazilian flies, slightly less for the Colombian, and very low for the Costa Rican examples (Warburg et al. 1994). These authors considered this difference to be due to a variation in the *potency* of the peptide but similar recent experiments have suggested it to be due to differences in the *amount* of maxadilan in the saliva of the three populations (Yin et al. 2000). This is in agreement with the distribution of atypical cutaneous leishmaniasis due to *L. i. chagasi*, and lends credence to the involvement of maxadilan in visceralization of the parasite.

There is evidence that *L. (L.) infantum chagasi* may produce a benign, inapparent infection in some individuals and that severity of the disease depends to some extent on the nutritional state of the infected person (Badaró et al. 1986). Differential diagnosis has to be made principally from malaria, schistosomiasis, cirrhosis of the liver, visceral syphilis and other causes of hepatosplenomegaly.

Although serological methods such as the indirect fluorescent antibody, dot-enzyme-linked immunosorbent assay (ELISA) and direct agglutination tests are useful indicators, the novel use of a dipstick (InBios International, Inc) based on recombinant RK 39 antigen for the differential diagnosis of AVL from other sympatric endemic diseases, has been shown to be rapid and highly specific (Delgado et al. 2001). Also, new latex agglutination test (KATEX) is claimed to detect leishmanial antigen in the urine of patients with a 100 percent specificity (Attar et al. 2001). Unequivocal diagnosis of the disease, however, depends on the demonstration of amastigotes in stained smears of aspirates from the spleen, bone marrow or lymph glands. Usually, such smears contain abundant parasites; if not, the material can be cultured in a suitable blood agar medium (varieties of Novy, MacNeal, Nicolle (NNN) medium).

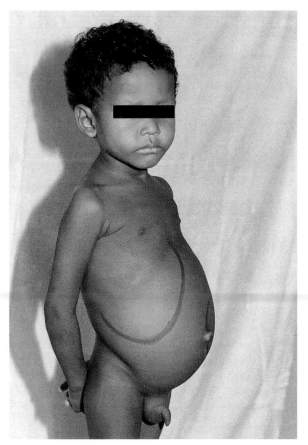

Figure 17.12 *AVL, due to* L. (L.) infantum chagasi, *in a boy from Marajó island, Pará, Brazil. Note the greatly distended abdomen resulting from hepatosplenomegaly.*

However, some difficulty may be experienced in culturing the parasite, especially when the aspirates contain scanty amastigotes. The intraperitoneal inoculation of aspirates into hamsters is by far the most reliable method of isolating *L. (L.) infantum chagasi* for further study but the long delay before parasites are detectable in these animals makes the method impractical for quick diagnosis. The PCR, together with specific hybridization techniques, will probably be the method of choice in the future (Noyes et al. 1996; Rodriguez et al. 1997; Breniere et al. 1999), at least in the larger and better-equipped hospitals, clinics, and research laboratories. Traditional methods of demonstrating the presence of the parasite will remain vitally important for many more years to come, however, particularly in the more remote rural areas where the majority of cases are concentrated.

Cutaneous leishmaniasis

Simple cutaneous lesions are produced by 12 of the 14 neotropical species of *Leishmania* known to infect humans. It is not possible, however, to diagnose the causative species by the appearance of these lesions, which are in many cases indistinguishable from the classical leishmanial lesion of the Old World, forming a rounded, crater-like ulcer with a raised border (see Chapter 16, Old World leishmaniasis, and Figure 17.2). In addition, the simple lesion may vary greatly in appearance (Figures 17.11 and 17.13–17.15), evoking such descriptions as framboesiform, lichenoid, lupoid, nodular, vegetative, verrucose, ulcerative, etc. from clinicians and dermatologists. All of these forms have in common that they are painless; unfortunately, this means that the infected person often fails to seek medical advice until the lesion has reached large proportions. Differential diagnosis needs to consider tropical ulcer (painful and suppurative), sporotrichosis, cutaneous tuberculosis, yaws, blastomycosis, lupus, and tertiary syphilis. Bacterial infections of insect bites or skin abrasions are frequently misdiagnosed as early lesions due to *Leishmania* and are particularly common in children living in rural areas with an abundance of biting flies, such as *Simulium* and *Culicoides*.

Although patients under examination may show a positive Montenegro (leishmanin) skin-test, this may be due to a previous, subsequently eliminated infection with *Leishmania* and may be unrelated to present skin lesions. The demonstration of amastigotes in stained smears prepared from the border of the lesions and

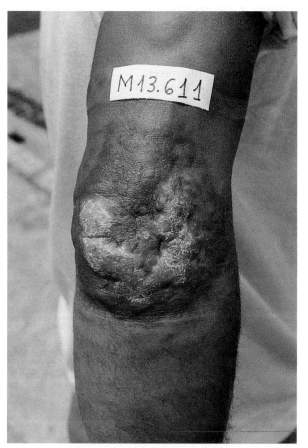

Figure 17.13 *Infection with* L. (L.) amazonensis, *on the elbow; Pará, Brazil*

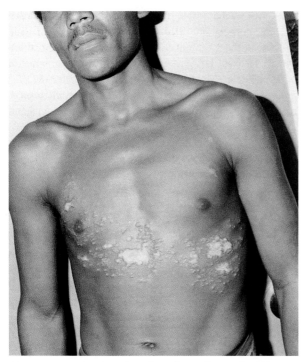

Figure 17.14 *A large number of small, papular satellite lesions surrounding the larger, primary lesions in a case of pian-bois due to L. (V.) guyanensis; Pará, Brazil*

isolation of the parasite in blood agar culture medium are prerequisite for diagnosis.

Vacuum aspiration of amastigotes in material from human lymph nodes, first used in the examination of patients with visceral leishmaniasis, has been used in cases of cutaneous leishmaniasis resulting from *L. (V.) braziliensis* infection in Brazil (Marzochi et al. 1993; Romero et al. 1999). Aspiration directly into the culture medium, through the rubber cap, greatly reduces the risk of bacterial or fungal contamination.

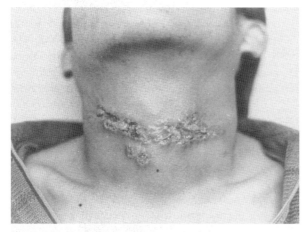

Figure 17.15 *Another atypical lesion, due to the same parasite; Pará, Brazil*

Mucocutaneous and mucosal leishmaniasis

For excellent general reading on this unique form of leishmaniasis, reference should be made to Pessôa and Barretto (1948), Marsden (1986), and Walton (1987).

Although the disease has, on rare occasions, been attributed to *L. (V.) guyanensis* and *L. (V.) panamensis*, the great villain is undoubtedly *L. (V.) braziliensis* s.l. The exact percentage of individuals infected with this parasite who develop mucosal lesions is difficult to calculate. In what was almost certainly a hospital series in São Paulo State, southern Brazil, Pessôa (1941) produced figures indicating that among 171 patients with cutaneous lesions of less than 1 year's duration, 105 lacked mucosal lesions and 66 (38.5 percent) had mucosal involvement. Of 110 individuals with cutaneous lesions of more than 1 year's duration, 21 had no mucosal lesions, and 89 (80.9 percent) had developed them. Marsden (1986), however, described Pessôa's figure of 80.9 percent as 'widely quoted out of text ... giving the impression that the great majority (of cases) will proceed to mucosal metastasis'.

In field studies in an endemic area in Três Braços, Bahia State, north-east Brazil, Marsden and his colleagues found only 2.7 percent of 371 patients examined to have both cutaneous and mucosal lesions. Mucosal infection seems to be much more common in some Latin American countries (e.g. Bolivia and Ecuador) than in others, suggesting that certain parasites referred to as *L. (V.) braziliensis* have a greater propensity for producing mucosal lesions than others (these may, in fact, represent different subspecies, or even species, of the *braziliensis* complex).

In spite of its antiquity, the most detailed study of the mucosal lesions remains that of Klotz and Lindenberg (1923), but a wealth of clinical and pathological data is also available in the works mentioned above, and that of Ridley (1987). It is generally agreed that involvement of the mouth, nose, and throat is always the result of metastases from a simple skin lesion elsewhere on the body (Figure 17.2). Apparently, migration of the parasites to the mucosae, by way of the lymphatics or the bloodstream, may take place quite early in the infection, for amastigotes have been demonstrated in scrapings of the apparently normal nasal mucosae of 12 patients with skin lesions of only 1–11 months' evolution (Villela et al. 1939). The nose is the major site of the metastases (Figure 17.3) and it remains a mystery as to why this occurs in some individuals and not in others. From the observations of Villela et al. and the fact that years may elapse between the disappearance of the primary skin lesion and the onset of mucosal disease, it seems that the parasite remains dormant in the mucosae for a variable period. Exactly what triggers this occult infection into destructive activity is another mystery, although minor injury to the mucosae may be one cause. The following account of the developing pathology of the mucosal disease is summarized from Ridley (1987).

When a nasal lesion develops, it is initiated in the deep mucosa of the nose, an accumulation of plasma cells and lymphocytes forming around the small blood vessels. A few amastigotes may appear in the endothelial cells. The major part of the lesion stays in the deep mucosa, accompanied by congestion and edema, a pronounced infiltration of plasma cells and a characteristic proliferation of the vascular endothelial cells, which contain variable numbers of amastigotes. Inflammatory foci proceed towards the mucosal surface, resulting in a patchy desquamation, followed by hyalinization and necrosis of the exposed tissue, accompanied by polymorph infiltration. It should be stressed that the ulcer is due to the desquamation resulting from the inflammatory process and not to the necrosis. In the area of deep inflammation, the endothelial nodule, with its perivascular inflammatory cells, undergoes central necrosis or, in the case of large nodules, hyalinization. Scanty amastigotes may still be found, but they are absent in the necrotic area. Considerable endarteritis may be associated with thrombosis which, together with subsequent fibrosis, deforms and erodes the nasal septum. Liquefaction of cartilage continues, even some distance from the leishmanial nodule and the vascular supply is so reduced that only coarse fibrous tissue can survive.

Blockage of the nasal passages due to the developing lesion usually results in respiratory distress, mouth breathing, and a high frequency of pulmonary infection that may lead to death. Palatal (Figure 17.4), laryngeal and tracheal lesions are less common; for first-hand accounts of these, reference should be made to Marsden (1986) and Walton (1987).

Differential diagnosis of mucosal leishmaniasis must be made from nasal syphilis (no destruction of the septum), gangosa (ulcerating yaws), blastomycosis, rhinosporidiosis, midline granuloma, carcinoma, and cancrum oris.

Sudanese and Ethiopian oro-nasal (mucosal) leishmaniasis bears a superficial resemblance to American espundia (see El-Hassan et al. 1995, for review). Due largely to *L. (L.) donovani* s.l. and, more rarely, *L. (L.) major*, it is not preceded by a cutaneous lesion. In addition, unlike patients with the South American disease, advanced cases of Sudanese mucosal leishmaniasis respond readily to treatment with pentavalent antimonials and ketoconazole.

Anergic, diffuse cutaneous leishmaniasis

In Venezuela, Convit and Lapenta (1948) described a bizarre form of cutaneous leishmaniasis characterized by nodular lesions scattered all over the body and containing vast numbers of rather large amastigotes; a negative Montenegro skin-test reaction; and almost total resistance to chemotherapy. Further cases were recorded in that country and subsequently in Bolivia, Brazil, Colombia, the Dominican Republic, Honduras, northern Mexico, Texas, USA, and Peru. The causative parasites in Venezuela, Mexico, Texas, and Brazil are *L. (L.) pifanoi*, *L. (L.) mexicana* and *L. (L.) amazonensis* respectively and it is likely that this form of leishmaniasis elsewhere in the Americas is also due to members of the *mexicana* complex within the subgenus *Leishmania*. The disease, as its name suggests, is the outcome of infection by this group of parasites in individuals with little or no cell-mediated immunity. Curiously, parasites of the subgenus *Viannia* do not seem to possess the potential to produce this strange disease.

At one time it appeared that ADCL was an irreversible condition and that the best treatment merely kept the disease in check, without eliminating the parasite. There is some evidence to suggest, however, that the prognosis is not always so grim. One of the several ADCL patients studied in the authors' laboratory acquired her infection with *L. (L.) amazonensis* when she was less than 5 years old and, in spite of constant chemotherapy, her condition had not improved greatly by the age of 28 years. However, the lesions faded away and the patient made a complete recovery following immunochemotherapy. We had noted that, unlike the lesions of other cases of ADCL, those of this young woman had occasionally ulcerated and healed, leaving her with very unsightly scars on her face and legs (Figure 17.9). This and the final recovery, suggests that there is a gradation in the degree of anergic condition, from total to partial. In many patients, the nodular lesions remain unulcerated, whereas in others, there may be ulceration and even amputation of fingers and toes, resembling that seen in leprosy (Figure 17.5).

Usually, ADCL forms a minute proportion of the total number of cases of cutaneous leishmaniasis caused by parasites of the *mexicana* complex but an exception is found in the Dominican Republic, where ADCL occurs in the apparent absence of simple, curable cutaneous leishmaniasis. The causative parasite is clearly a member of the *mexicana* complex, but is, as yet, unnamed and its animal reservoir and sandfly vector are unknown. The occurrence of three cases in a single family in the Dominican Republic suggests involvement of an hereditary component (Walton 1987); Petersen et al. (1982) demonstrated a population of specific suppressor cells in four other patients. The deficient immune response in ADCL in general is considered to be related to a thymus-dependent system in both the New World and the Old World diseases (Convit et al. 1971; Bryceson 1970). The only other disease with which ADCL has frequently been confused is lepromatous leprosy; this mistake is unpardonable if Giemsa-stained smears of the skin lesions have been examined (Figure 17.10).

In a recent paper, Silveira et al. (2004) discussed the clinical and immunopathological spectrum of American cutaneous leishmaniasis, with particular reference to the disease in Amazonian Brazil. They emphasized the necessity of accurately identifying the species of *Leishmania* stimulating the patient's immune response, together with the quality and magnitude of this response. At the diagramatic center of the authors' clinical spectrum are the most commonly seen localized forms of cutaneous leishmaniasis (LCL), which may be caused by members of both subgenera, *Leishmania* and *Viannia*, and which usually respond well to conventional chemotherapy. The two pathogenicity poles of the spectrum generally recognized are the hyposensitivity pole, occupied by cases of ADCL, and the hypersensitivity pole represented by mucocutaneous leishmaniasis. They proposed the term borderline disseminated cutaneous leishmaniasis (BDCL) for the cases showing disseminated lesions due to some parasites of both the subgenera where there is a partial failure of the cellular immune response (unlike the complete failure seen in ADCL), enabling eventual cure of the patient provided that treatment is promptly given. If this is not undertaken, however, these infections may develop into either ADCL (due, for example, to *L. (L.) amazonensis*) or mucocutaneous leishmaniasis (due mainly to *L. (V.) braziliensis*).

Leishmania infections and the immunosuppressed patient

Cutaneous and visceral leishmaniasis are now firmly established in the list of infections that the clinician must consider when dealing with immunosuppressed patients, in particular those with acquired immune deficiency syndrome (AIDS). Visceral leishmaniasis is the clinical form most commonly associated with HIV/AIDS and has become a public health problem of considerable proportions in south-western Europe. Strangely, although visceral leishmaniasis is far more common in many parts of Latin America (e.g. Brazil) than it is in Europe, cases of AVL/HIV infection are, until now, less common (WHO 1998). Possibly this is because, in the Americas, HIV is predominantly found in the larger cities; the situation will doubtless change as HIV spreads to the more rural areas where AVL is highly endemic or with the migration of leishmaniasis carriers into the large cities and their acquisition of HIV. One of the most striking examples of such migration is that of individuals from the rural areas of northeast Brazil, where AVL is very common, into São Paulo city in search of work. Co-infection of HIV and *Leishmania* species responsible for cutaneous leishmaniasis seems to be less frequent in the Americas, but may result in serious complications, ranging from disseminated skin lesions and mucosal lesions (Coura et al. 1987; Da-Cruz et al. 1999) to rectal involvement (Hernández et al. 1995). *L. (V.) guyanensis* has been isolated from the lesions of an HIV patient presenting with mucocutaneous leishma-

niasis in the State of Amazonas, Brazil (de Souza e Souza et al. 1998), and visceral leishmaniasis due to *L. (L.) mexicana* has been recorded in a patient with HIV in Mexico (Ramos-Santos et al. 2000).

PREVENTION AND CONTROL

Visceral leishmaniasis

Campaigns are of fundamental importance, with the distribution of illustrated pamphlets to alert the populations as to the early symptoms of the disease, the signs of infection in the dog and the appearance and habits of the sandfly vector. The staff of small, rural clinics must be trained to recognize visceral leishmaniasis and should have means of reporting suspected cases to centers where more conclusive diagnosis can be made. In view of the impecunious situation of many inhabitants of the rural districts of developing countries, the staff of such centers may need to travel to the patient's village, where the possibility of other cases must be investigated. Periodic surveillance of populations at risk may detect cases of early infection either clinically or, more effectively, by serological methods that can easily be carried out under field conditions, such as the direct agglutination test (DAT) or the dot-ELISA.

In areas of high endemicity, such as the states of Ceará and Bahia in northeast Brazil, past control measures (i.e. destroying infected dogs, regular insecticide spraying of houses and animal sheds, and the early treatment of patients) resulted in a dramatic drop in the number of cases of AVL. A problem arises, however, in maintaining such a control program, which is costly and inevitably meets with considerable opposition on the part of dog owners, who fail to understand why their apparently healthy (but serologically positive) animals must be killed. Many deliberately conceal their dogs, and many strays are never examined. A critical evaluation of the cost-effectiveness of this dog-slaughtering policy (Akhavan 1996) points out the vast amount of work and expense expended in surveying dog populations, and the apparent failure of this method in control programs in some parts of Brazil has led to the question as to what extent the dog population needs to be reduced in order to eliminate AVL or to bring it under control. It has been suggested that it may be preferable to find improved methods of eliminating or controlling the sandfly vector.

Insecticide (deltamethrin)-impregnated plastic dog collars have been found to provide an efficient and prolonged protection of dogs against the bites of sandfly vectors of *L. i. infantum*, thus breaking transmission of the parasite in its major reservoir host (Killick-Kendrick et al. 1997). Such collars have been shown to offer similar protection against the bites of *Lu. longipalpis* in Brazil (David et al. 2001), and preliminary results following field-trials in three regions of Brazil have been discussed in the 2002 *Proceedings of the Second International Cannine Leishmaniasis Forum, Seville, Spain.* Only after the completion of more lengthy field-trials, however, will it be possible to assess both the cost-efficiency of deltamethrin-impregnated dog collars, or other methods of insecticide application, and the effects they may have on the incidence of human AVL. In the meantime, the development and use of an effective and long-lasting vaccine, preferably in a well-supervised, governmental program similar to that of antirabies campaigns, remains of great importance. Encouraging results have been obtained after the inoculation of dogs with a mixture of promastigotes of *L. (V.) braziliensis*, disrupted by ultrasound, and bacille Camette Guerin (BCG). Of 10 vaccinated dogs, only one succumbed to subsequent challenge with 2.3×10^6 infective promastigotes of *L. i. chagasi*, whereas all of nine nonvaccinated dogs became infected (Mayrink et al. 1996). In addition, field trials using a fucose–mannose ligand (FML)-vaccine prepared from *L. donovani* have afforded a 92 percent protection among dogs in an endemic area of AVL in Rio Grande do Norte, Brazil, 2 years after their vaccination. The 8 percent of dogs that did acquire infection showed only mild signs of disease with no deaths and, significantly, the number of human cases of AVL in the area decreased from 15 in 1996 to six in 1997 and none in 1998 (da Silva et al. 2001).

Cutaneous and mucosal leishmaniasis

Most inhabitants of the endemic areas are very familiar with the dermal leishmaniases, under their wide variety of local names, but ignorance and negligence are all too frequently to blame for allowing these diseases to reach debilitating or mutilating proportions. The painless nature of the lesions makes them of no great inconvenience in their early stages and this encourages the tendency to wait and see if they will cure spontaneously. Again, as these diseases are predominantly zoonotic, most infections are acquired by those living in rural areas, often long distances from the simplest of medical attention. As a result, a high proportion of cases only seek help when the infection is well advanced and, in regions where the causal agent is frequently *L. (V.) braziliensis*, this may prove disastrous. As for visceral leishmaniasis, health education campaigns can help to indicate the importance of early treatment.

Personal avoidance of cutaneous leishmaniasis is, at present, limited to the use of insect repellents, protective clothing and the avoidance of danger areas, particularly at night when the sandfly vectors are most active. These are precautions that may be feasible for the visiting tourist, but they are not very practical for the shirtless forest-worker who can ill afford to be constantly purchasing insect repellents, who is most comfortable wearing shorts (far less expensive than trousers) and

short-sleeved shirts and who has to eke out his living by hunting in the forest at night.

The prevention or control of sylvatic leishmaniasis among gangs of laborers, topographers, and other forestry workers can be effective on a small scale by the following measures: placing the encampments of such men in adequate clearings; spraying the bases of the larger, nearby tree trunks with insecticides (e.g. in areas where the vectors are known to be arboreal – see *Leishmania (Viannia) guyanensis* Floch, 1954 above); and prohibiting night-time hunting. Destruction of the wild animal reservoirs of sylvatic leishmaniasis is clearly neither practical nor desirable. Finally, knowledge of the ecology of a vector can sometimes help in preventing acquisition of the disease under certain circumstances. Thus, *Psychodopygus wellcomei*, an important vector of *L. (V.) braziliensis* s.l. in the highland forests of Pará, north Brazil, is highly anthropophilic and attacks humans not only at night but also frequently during the day. Field studies have shown, however, that this sandfly is only active for about 6 months of the year, during the rainy season (November–April) and that it enters into diapause in the dry season during the rest of the year, when adult flies are rarely seen. The area in question is one of intense human activity due to the mining of iron ore and other minerals and the incidence of cutaneous leishmaniasis has been very high among those clearing the primary forest. Planning such work for the dry season clearly avoids contact with the sandfly vector.

Although drastic ecological changes such as deforestation and the planting of non-indigenous pine, gmelina, and eucalyptus trees might lead to unfavorable conditions for the enzootics of some species of *Leishmania*, it can actually encourage others. Thus, the creation of vast monoculture plantations of non-indigenous trees for paper pulp production in north Brazil may eliminate cutaneous leishmaniasis due to *L. (V.) guyanensis* in the immediate areas, as neither the major reservoir host (the two-toed sloth) nor the sandfly vector (*Lu. umbratilis*) find this new environment suitable. On the other hand, the wild rodent and marsupial hosts of *L. (L.) amazonensis* and the sandfly vector, *Lu. flaviscutellata*, find it ideal.

In those parts of Latin America where vector species have adapted to a peridomestic or domiciliary habitat, the use of insecticides is clearly indicated. In the absence of firm evidence that domestic animals with leishmanial skin lesions offer a source of infection to sandflies, it would seem unwise to recommend their destruction, even in the unlikely event of their owners' consent. Equines respond well to antimonial treatment.

Direct transmission risks

Cases of congenital transmission of visceral leishmaniasis have been reported (Bialek and Knobloch 1999; Knobloch et al. 2001) but would appear to be rare.

It has been estimated that in some endemic areas infection with leishmanial parasites may be as high as 70 percent of the population, and data on patients with *Leishmania*/HIV co-infection have indicated that viable parasites may persist in individuals long after initial infection, irrespective of the presence or absence of primary disease (WHO 1998). Leishmanial antibodies have been detected in 9 percent of a group of blood donors in the city of Natal, Rio Grande do Norte, Brazil, and in 37 percent of patients undergoing hemodialysis in the same locality (Luz et al. 1997); the latter figure was significantly higher than in any other group, including inhabitants in AVL foci on the outskirts of the city! The importance of transfusion transmission of visceral leishmaniasis in Europe has been clearly indicated in non-endemic regions where the sandfly vector is absent (Mauny et al. 1993). With increases in HIV infection and organ transplant surgery involving prolonged use of immunosuppressants, occult infections and contaminated blood or organs become life-threatening risks necessitating special precautions (Schulman 1994).

TREATMENT

Of the various drugs discussed for the treatment of leishmaniasis (see Chapter 16, Old World leishmaniasis; Lainson 1982a; Amato et al. 1996; de Carvalho et al. 2000), the pentavalent antimonials are those of first choice in dealing with the neotropical leishmaniases. It is of historical interest that the first use of antimony to treat leishmaniasis was in Brazil. A young clinician, Gaspar Vianna, impressed by the effectiveness of tartar emetic (antimony potassium tartrate) in the treatment of African trypanosomiasis (see Chapter 18, African trypanosomiasis), realized its potential against the related organism *Leishmania* and, in 1912, published his spectacular results obtained with this drug in advanced cases of mucocutaneous leishmaniasis (Vianna 1912). There followed development of the somewhat less toxic trivalent antimonials, which still produced unpleasant side-effects, and then the much better-tolerated pentavalent antimonials which, some five decades later, are still the clinician's principal armaments. Even so, these latter, more refined antimonials may sometimes show some cardiotoxicity (Ribeiro et al. 1999).

American visceral leishmaniasis

Treatment usually follows that given in Chapter 16, Old world leishmaniasis. Unresponsiveness to the recommended antimonial dosage schedule, however, has been noted in some AVL patients from Bahia, Brazil. In such cases, there is little alternative other than elevating the dosage or using second-line drugs such as pentamidine. In this manner, Bryceson et al. (1985) cured four of ten unresponsive patients with Kenyan kala azar, but all ten suffered serious side-effects from both stibogluconate and pentamidine at the dosage levels used. Liposomal amphotericin B (AmBisome®) is recommended for the treatment of visceral leishmaniasis by the US Food and Drug Administration (Meyerhoff 1999).

Cutaneous leishmaniasis

Once again, the treatment is much the same as that for Old World cutaneous leishmaniasis, using the pentavalent antimonials. Many clinicians prefer, however, to use the intravenous route for both pentostam and Glucantime, rather than intramuscular inoculation.

In general, lesions due to the neotropical leishmanias tend to much greater chronicity than those of the Old World and there is not such a ready response to treatment. There is no justification in delaying treatment in anticipation of an early spontaneous cure, as recommended with *L. (L.) major* or *L. (L.) tropica* infections, or limiting treatment to intralesional injection of drugs or the topical application of these in ointments. The following hazards must be considered: subsequent mucosal disease due to *L. (V.) braziliensis*; ADCL due to parasites of the *mexicana* complex; and multiple lesions following lymphatic or hematogenous spread on the part of *L. (V.) guyanensis*. Treatment should, therefore, be systemic and immediate and identification of the causative parasite is most important. If it proves to be *L. (V.) braziliensis*, treatment should be particularly intensive.

Effectiveness of the pentavalents Pentostam® and Glucantime may vary considerably, not only when dealing with different parasites, but in treating different patients infected with the same organism. Some patients with simple lesions due to *L. (V.) guyanensis*, for example, may be cured by a single course of treatment, whereas others may require three or four. In some cases of poor response, recourse must be made to other drugs, or combinations of drugs, in spite of their greater toxicity. The most commonly used second-line drug is pentamidine, which has been used routinely by French workers in treating cases due to *L. (V.) guyanensis* in French Guyana.

Treatment of patients with the antimalarial drug mefloquine has produced conflicting results (Gómez et al. 1996; Hendrickx et al. 1998; Laguna-Torres et al. 1999).

Mucocutaneous and mucosal leishmaniasis

Advanced cases are difficult to treat, with slow response to the pentavalent antimonials. The high dosage recommended (20 mg Sb^{5+} per kg body weight, given in a single daily injection for a mean of 30 days, or until no evidence of activity of the lesion has been noted for a week) may sometimes produce pronounced side-effects, so that treatment has to be suspended (Marsden 1986).

The considerably more toxic amphotericin B and pentamidine are used only when there is failure to respond to the pentavalent antimonials. In some patients, the early treatment provokes a severe inflammation around the lesion, presumed to be caused by antigen released from killed parasites. Although this reaction is regarded as a favorable prognostic sign, it may prove highly dangerous in patients with laryngeal or tracheal lesions. It is recommended that corticosteroids be used as a prophylactic measure when treating such cases and that there should be a gradual elevation of the drug dose (Marsden 1986). Successful treatment of mucosal leishmaniasis in patients unresponsive to Glucantime has been achieved with liposomal amphotericin B (Sampaio and Marsden 1997).

Another difficulty confronting the clinician is in evaluating the effectiveness of treatment. What are the criteria of cure? Serological monitoring has used complement fixation and indirect fluorescent antibody tests, and it has been suggested that cure is indicated by decline and disappearance of antibody titers (Walton 1987). This method is likely to prove the most useful, provided that measurements of leishmanial antibody can be standardized. Both parasitological and histological examination entail biopsy trauma to the treated lesion and may reactivate a lesion that seems to be healed but still contains parasites. Serological follow-up should accompany testing for antigen: a negative antigen test together with reduced titer in serological tests indicates a lower likelihood of relapse (Amato et al. 1998), but an inflammatory process may persist, even in these patients, and could indicate continued presence of amastigotes.

Mutilation following the mucosal lesions may be so extreme that even after cure the patient is ostracized and unable to lead a normal life. Plastic surgery is helpful, but only when there is no doubt of cure. Walton (1987) cites the case of one patient, with apparently complete healing, who underwent cosmetic surgery to reconstruct his nose: 'The results were disastrous, with widespread reactivation along the surgical wounds and, in the patient's words – the new nose fell off!'.

Immunotherapy, utilizing vaccines with or without BCG (Convit et al. 1987, 1989; Mayrink et al. 1991) has proved effective, but has been shown to give much the same cure rate as conventional chemotherapy (Convit et al. 1987). As pointed out by the latter authors, however, it is considerably cheaper and safer.

Anergic diffuse cutaneous leishmaniasis

A smooth, fleshy, unulcerated lesion containing very abundant, large amastigotes in a patient with a negative Montenegro skin-test reaction is a danger signal that should prompt immediate, high-dosage antimonial treatment. In the authors' laboratory in Amazonian Brazil, this is virtually diagnostic of an early lesion due to *L.*

(L.) amazonensis which, in an individual with a defective cell-mediated immune response (suggested by his negative skin-test), will proceed to ADCL unless adequately treated. Fortunately, although simple, curable skin lesions due to this parasite are quite common, ADCL is relatively rare.

Of all the forms of leishmaniasis, this disease undoubtedly represents the clinician's greatest challenge, because the patient is unable to offer the all-important immunological collaboration necessary for successful drug treatment. Advanced cases of ADCL, with multiple nodular lesions, may respond dramatically to the first antimonial treatment, with complete disappearance of the lesions. Unfortunately, this may be taken to indicate cure, but with the cessation of treatment the nodules reappear some time later, usually more abundantly. Repeated treatment with the same drug may again give good results, but the patient will relapse again and the response steadily diminishes until the treatment is virtually ineffective. Similar results may be obtained with a variety of other drugs, until the list is exhausted and the patient's situation becomes desperate.

In patients with ADCL due to *L. (L.) amazonensis*, very hot baths taken daily, together with periodic chemotherapy, have been found to reduce the size of the nodules substantially, giving the patient a much improved appearance over periods of many years (Lainson 1982a). More recently, immunochemotherapy has been used with considerable success in the treatment of persons suffering from ADCL in Venezuela (Convit et al. 1989). 'Marked clinical improvement' was observed in nine out of ten patients given intradermal injections of a mixture of heat-killed promastigotes of '*L. mexicana amazonensis*' (*L. (L.) pifanoi*?) isolated from a Venezuelan case of ADCL, plus 'variable amounts' of BCG, together with the standard Glucantime treatment. A similar treatment of Brazilian ADCL patients infected with *L. (L.) amazonensis*, however, has met with limited success (F.T. Silveira, unpublished observations). Treatment with an association of Glucantime and paramomycin (Gabbrox, via oral) has been found useful (Costa et al. 1999).

VACCINATION

Immunology of infectious diseases in general is discussed in Chapter 4, Immunology and immunopathology of human parasitic infections. With regards to leishmaniasis, this is just as well, because a detailed review of the extensive literature on this subject over the past few years would fill many more pages here than space allows.

The pioneer work of Mayrink and collaborators in Brazil, on the production of a vaccine against American cutaneous and mucocutaneous leishmaniasis (reviewed by Genaro et al. 1996), and that of Convit and coworkers in Venezuela on the use of vaccines in immunotherapy (reviewed by Convit 1996) have triggered a veritable explosion of studies directed towards the production of vaccines

In early trials, the Mayrink vaccine was a mixture of killed, sonicated promastigotes of *L. (L.) amazonensis*, *L. (V.) braziliensis*, and a *Leishmania* of doubtful identity, without the use of an adjuvant. More recent versions have been a cocktail of promastigotes of *L. (L.) mexicana*, *L. (L.) amazonensis*, *L. (V.) guyanensis*, and *L. (V.) braziliensis*, with or without *Corynebacterium parvum* as an adjuvant. While apparently still some way from offering complete protection, the vaccine at least has the distinction of being the only one commercially available (Leishvacin®), and it is to be hoped that continuing studies will improve its efficacy. The protective efficacy of a similar vaccine against cutaneous leishmaniasis in Ecuador is claimed to be 72.9 percent (Armijos et al. 1998).

The immune response of volunteers vaccinated with BCG plus killed promastigotes of *L. (L.) mexicana* in Venezuela indicated the vaccine to be potentially protective for the majority of the vaccinees (Sharples et al. 1994), and follow-up studies showed an immune response in more than 85 percent of the vaccines as indicated either by skin-test conversion, lymphocyte proliferation or interferon-γ production.

Kenney et al. (1999) used rhesus monkeys to assess the safety, immunogenicity, and efficacy of a vaccine prepared from heat-killed *L. (L.) amazonensis* promastigotes combined with recombinant human interleukin-12 (rhIL-12) and aluminum hydroxide gel as adjuvants. Challenge with 10^7 metacyclic *L. (L.) amazonensis* promastigotes 4 weeks after vaccination demonstrated complete protection in 12 monkeys that received 2 μg rhIL-12 with alum/antigen. The safety and efficacy of this vaccine in monkeys suggest a basis for human trials.

Handman (2001) gives a useful update of work on killed and live, attenuated vaccines; recombinant vaccines; synthetic peptides; nonprotein antigens; and 'naked' DNA vaccine. Transmission-blocking vaccines are discussed by Tonui (1999).

CONCLUDING REMARKS

We have discussed the possible origin of the parasitic Kinetoplastida, the family Trypanosomatidae, and the genus *Leishmania* elsewhere (Lainson and Shaw 1987). It is the general opinion that the trypanosomatids have their origin in monogenetic intestinal flagellates of invertebrates and that they subsequently adapted to spend a part of their life-cycle in vertebrates. Thus, it is more correct to consider the phlebotomine sandfly as the primary host of *Leishmania* species, rather than the vertebrate hosts that merely function as reservoirs of infection for the sandfly (Lainson 1997). The finding that promastigotes may undergo a form of conjugation (Lanotte and Rioux 1990), with possible exchange of

nuclear material, supports this hypothesis on the reasonable assumption that such a process is more likely to take place in the definitive or primary host of a heteroxenous parasite.

It is difficult to assess the specificity of the *Leishmania* species in their sandfly hosts and unwise to base one's conclusions on the results of laboratory experiments, when unnaturally large numbers of amastigotes or promastigotes, fed to laboratory-bred flies, may well overwhelm the natural resistance of a nonvector species. In nature, however, there is considerable evidence suggesting the limitation of the life-cycle of most leishmanial parasites to specific sandfly vectors. Thus, *Lu. longipalpis* is the major vector of *L. (L.) chagasi* throughout the whole geographical range of this parasite; the closely related *Lu. olmeca olmeca* and *Lu. flaviscutellata* are the only confirmed vectors of *L. (L.) mexicana* and *L. (L.) amazonensis* respectively, in Central and South America, in spite of the presence of many other species of sandflies known to feed on rodents in the endemic areas of these two parasites. As far as is known, *Ps. wellcomei* is the sole vector of *L. (V.) braziliensis* s.l. in the Carajás highlands of Pará and *Lu. umbratilis* is the major vector of *L. (V.) guyanensis* throughout its geographical distribution, again in spite of the presence of a large number of other species of sandflies. On the other hand, because some sandfly vectors feed on a variety of mammalian hosts in nature, the *Leishmania* of a given species of sandfly may sometimes be isolated from a variety of mammalian hosts sharing the same habitat. *Lu. flaviscutellata*, for example, is a low-flying sandfly and transmits *L. (L.) amazonensis* to a number of predominantly terrestrial rodents and marsupials; canopy and tree trunk-dwelling sandflies such as *Lu. umbratilis* and *Lu. whitmani* s.l. transmit *L. (V.) guyanensis* and *L. (V.) shawi*, respectively, to arboreal animals such as sloths, anteaters, monkeys, and procyonids.

The number of *Leishmania* species in a given locality will largely be governed by the number of sandfly species, although some sandflies seem to be resistant to infection with this parasite. Thus, we (Lainson and Shaw, unpublished observations) have failed to infect the sandfly *Lu. carmelinoi* with any neotropical *Leishmania* species. The multiplicity of species within the genus is a relatively recent realization and much research is still needed to further our knowledge regarding the diversity, ecology, and taxonomy of the neotropical leishmanias. For the clinician, the continued isolation and characterization of parasites from cases of human leishmaniasis is sufficient to indicate the spectrum of *Leishmania* species commonly infecting humans in a given area. The parasitologist's interests, however, are much wider and include the wild mammalian reservoir and sandfly vectors of those parasites infecting humans, as well as the possible existence of other *Leishmania* species that rarely, if ever, infect humans.

The number of *Leishmania* species in the neotropical region is anybody's guess and some idea of it will only be gained when sufficient numbers of mammalian and sandfly species have been examined. Clearly, it is easier and more economical to concentrate simply on the sandfly population, a truly gigantic task in itself considering that nearly 400 different species of these insects have been identified in the Americas (Young and Duncan 1994). The Amazon Region has already provided us with almost half of the recognized species of neotropical leishmanias and doubtless this great forest will continue to provide us with many more!

ACKNOWLEDGMENTS

The authors are indebted to the Wellcome Trust, London, for the financial support of our studies on the ecology, epidemiology, and taxonomy of *Leishmania* in the Amazon Region of Brazil over the past 30 years and to the Instituto Evandro Chagas, Belém, Pará, where this work was carried out.

REFERENCES

Akhavan, D. 1996. Análise de custo-efetividade do componente de leishmaniose no projeto de controle de doenças endêmicas no nordeste do Brasil. *Rev Patol Trop*, **25**, 203–52.

Amato, V.S., de Paula, J.G., et al. 1996. Tratamento de leishmaniose tegumentar americana, com lesão em mucosa, por meio do isotionato de pentamidina. *Rev Soc Brasil Med Trop*, **29**, 477–81.

Amato, V.S., Duarte, M.I.S., et al. 1998. An evaluation of clinical, serologic, anatomopathologic and immunohistochemical findings for fifteen patients with mucosal leishmaniasis before and after treatment. *Rev Inst Med Trop São Paulo*, **40**, 23–30.

Aragão, H.B. 1922. Transmissão de leishmaniose no Brasil pelo Phlebotomus intermedius. *Brasil Méd*, **36**, 129–30.

Arias, J.R. and Naiff, R.D. 1981. The principal reservoir host of cutaneous leishmaniasis in the urban areas of Manaus, Central Amazon of Brazil. *Mem Inst Oswaldo Cruz*, **76**, 279–86.

Arias, J.R., Miles, M.A., et al. 1985. Flagellate infections of Brazilian sandflies (Diptera, Psychodidae), isolation in vitro and biochemical identifications of *Endotrypanum* and *Leishmania*. *Am J Trop Med Hyg*, **34**, 1098–108.

Armijos, R.X., Weigel, M.M., et al. 1998. Field trial of a vaccine against New World cutaneous leishmaniasis in an at-risk child population, safety, immunogenicity, and efficacy during the first 12 months of follow-up. *J Infect Dis*, **177**, 1352–7.

Attar, Z.T., Chance, M.L., et al. 2001. Latex agglutination test for the detection of urinary antigens in visceral leishmaniasis. *Acta Trop*, **78**, 11–16.

Badaró, R., Jones, T.C., et al. 1986. A prospective study of visceral leishmaniasis in an endemic area of Brazil. *J Infect Dis*, **154**, 639–649.

Bañuls, A.L., Guerrini, F., et al. 1997. Evidence for hybridization by multilocus enzyme electrophoresis and random amplified polymorphic DNA between *Leishmania braziliensis* and *Leishmania panamensis/guyanensis* in Ecuador. *J Eukaryot Microbiol*, **44**, 408–11.

Bañuls, A.L., Jonquieres, R., et al. 1999. Genetic analysis of *Leishmania* parasites in Ecuador. Are *Leishmania (Viannia) panamensis* and *Leishmania (V.) guyanensis* distinct taxa? *Am J Trop Med Hyg*, **61**, 838–45.

Barral, A., Badaró, R., et al. 1986. Isolation of *Leishmania mexicana amazonensis* from the bone marrow in a case of American visceral leishmaniasis. *Am J Trop Med Hyg*, **35**, 732–4.

Barrios, M., Rodriguez, N., et al. 1994. Coexistence of two species of *Leishmania* in the digestive tract of the vector *Lutzomyia ovallesi*. *Am J Trop Med Hyg*, **51**, 669–75.

Barretto, A.C., Peterson, N.E., et al. 1985. *Leishmania mexicana* in *Proechimys iheringi denigratus* Moojen (Rodentia, Echimyidae) in a region endemic for American cutaneous leishmaniasis. *Rev Soc Brasil Med Trop*, **18**, 243–6.

Belli, A., Rodriguez, B., et al. 1998. Simplified polymerase chain reaction detection of New World *Leishmania* in clinical specimens of cutaneous leishmaniasis. *Am J Trop Med Hyg*, **58**, 102–9.

Belli, A., Garcia, D., et al. 1999. Widespread atypical cutaneous leishmaniasis caused by *Leishmania (L.) chagasi* in Nicaragua. *Am J Trop Med Hyg*, **61**, 380–5.

Biagi, F.F. 1953. Algunos comentarios sobre las leishmaniasis y sus agentes etiológicos, *Leishmania tropica mexicana*, nueva subespecie. *Med México*, **33**, 401–6.

Bialek, R. and Knobloch, J. 1999. Parasitäre Infektionen in der Schwangerschaft und konnatale Parasitosen. I. Teil, Protozoeninfektionen. *Z Geburtschilfe Neonatol*, **203**, 55–62.

Blackwell, J.M. 1998. Genetics of host resistance and susceptibility to intramacrophage pathogens, a study of multi-case families of tuberculosis, leprosy and leishmaniasis in north eastern Brazil. *Int J Parasitol*, **28**, 21–8.

Boisseau-Garsaud, A.M., Cales-Quist, D., et al. 2000. A new case of cutaneous infection by a presumed monoxenous trypanosomatid in the island of Martinique (French West Indies). *Trans R Soc Trop Med Hyg*, **94**, 51–2.

Bonfante-Garrido, R. 1980. New sub-species of leishmaniasis isolated in Venezuela. *Proceedings of the 10th International Congress of Tropical Medicine and Malaria, Manila*, p. 203.

Bonfante-Garrido, R., Meléndez, E., et al. 1992. Cutaneous leishmaniasis in Western Venezuela caused by infection with *Leishmania venezuelensis* and *L braziliensis* variants. *Trans R Soc Trop Med Hyg*, **86**, 141–8.

Braga, R.R., Lainson, R., et al. 2003. *Leishmania (Viannia) utingensis* n. sp., a parasite from the sandfly *Lutzomyia tuberculata* in amazonian Brazil. *Parasite*, **10**, 111–18.

Brandão-Filho, S.P., Brito, M.E., et al. 2003. Wild and synanthropic hosts of *Leishmania (Viannia) braziliensis* in the endemic cutaneous leishmaniasis locality of Amaraji, Pernambuco State, Brazil. *Trans R Soc Trop Med Hyg*, **97**, 291–6.

Breniere, S.F., Telleria, J., et al. 1999. Polymerase chain reaction-based identification of New World *Leishmania* species complexes by specific kDNA probes. *Acta Trop*, **73**, 283–93.

Bryceson, A.D.M. 1970. Diffuse cutaneous leishmaniasis in Ethiopia III. Immunological studies. *Trans R Soc Trop Med Hyg*, **64**, 380–7.

Bryceson, A.D.M., Chulay, J.D., et al. 1985. Visceral leishmaniasis unresponsive to antimonial drugs 1. Clinical and immunological studies. *Trans R Soc Trop Med Hyg*, **79**, 700–4.

Carini, A. and Paranhos, U. 1909. Identification de l'Úlcera de Bauru avec le bouton d'Orient. *Bull Soc Pathol Exot*, **2**, 225–6.

Carrillo, J., Chinchilla, M., et al. 1999. Visceral leishmaniasis in Costa Rica, first case report. *Clin Infect Dis*, **29**, 678–9.

Chouicha, N., Lanotte, G., et al. 1997. Phylogenetic taxonomy of *Leishmania (Viannia) braziliensis* based on isoenzymatic study of 137 isolates. *Parasitology*, **115**, 343–8.

Christensen, H.A., Herrer, A. and Telford, S.R. 1972. Enzootic cutaneous leishmaniasis in eastern Panama. II. Entomological investigations. *Ann Trop Med Parasitol*, **66**, 55–66.

Christensen, H.A., Herrer, A. and Telford, S.R. 1969. Leishmania braziliensis from Lutzomyia panamensis in Panama. *J Parasitol*, **55**, 1090–1.

Convit, J. 1996. Leishmaniasis, immunological and clinical aspects and vaccines in Venezuela. *Clin Dermatol*, **14**, 479–87.

Convit, J. and Lapenta, P. 1948. Sobre un caso de leishmaniose tegumentaria de forma disseminada. *Rev Policlin, Caracas*, **18**, 153–8.

Convit, J., Pinardi, M.E. and Rondón, A.J. 1971. Diffuse cutaneous leishmaniasis, a disease due to an immunological defect of the host. *Trans R Soc Trop Med Hyg*, **66**, 603–10.

Convit, J., Castellanos, P.L., et al. 1987. Immunotherapy versus chemotherapy in localized cutaneous leishmaniasis. *Lancet*, **i**, 401–4.

Convit, J., Castellanos, P.L., et al. 1989. Immunotherapy of localized, intermediate and diffuse forms of American cutaneous leishmaniasis. *J Infect Dis*, **160**, 104–15.

Corredor, A., Gallego, J.F., et al. 1989. *Didelphis marsupialis*, an apparent wild reservoir of *Leishmania donovani chagasi* in Colombia, South America. *Trans R Soc Trop Med Hyg*, **83**, 195.

Costa, C.H.N., Gomes, R.B.B., et al. 2000. Competence of the human host as a reservoir for *Leishmania chagasi*. *J Infect Dis*, **182**, 997–1000.

Costa, J.M.L., Mendes, S., et al. 1999. Tratamento da leishmaniose cutânea difusa (LCD) com a associação antimoniato-n-metilglucamina (Glucantime) e sulfato de paramomicina(Gabbrox). *An Brasil Dermatol*, **74**, 63–7.

Coura, J.R., Galvão-Castro, B. and Grimaldi, G. Jr. 1987. Disseminated American cutaneous leishmaniasis in a patient with AIDS. *Mem Inst Oswaldo Cruz*, **82**, 581–2.

Courtenay, O., Santana, E.W., et al. 1996. Visceral leishmaniasis in the hoary zorro *Dusicyon vetulus*, a case of mistaken identity. *Trans R Soc Trop Med Hyg*, **90**, 498–502.

Cunha, A.M. and Chagas, E. 1937. Nova espécie de protozoário do gênero *Leishmania* pathogenico para o homem. *Leishmania chagasi*, n. sp. Nota Prévia. *Hospital (Rio de Janeiro)*, **11**, 3–9.

Da-Cruz, A.M., Filgueiras, D.V., et al. 1999. Atypical mucocutaneous leishmaniasis caused by *Leishmania braziliensis* in an acquired immunodeficiency syndrome patient, T-cell responses and remission of lesions associated with antigen immunotherapy. *Mem Inst Oswaldo Cruz*, **94**, 537–42.

Darce, M., Moran, J., et al. 1991. Etiology of human cutaneous leishmaniasis in Nicaragua. *Trans R Soc Trop Med Hyg*, **85**, 58–9.

da Silva, V.O., Borja-Cabrera, G.P., et al. 2001. A phase III trial of efficacy of the FML-vaccine against canine kal-azar in an endemic area of Brazil (São Gonçalo do Amaranto, RN). *Vaccine*, **19**, 1082–92.

David, J.R., Stamm, L.M., et al. 2001. Deltamethrin - impregnated plastic dog collars have a potent antifeeding effect on *Lutzomyia longipalpis* and *Lutzomyia migonei*. *Mem Inst Oswaldo Cruz*, **96**, 839–47.

Darie, H., Deniau, M., et al. 1995. Cutaneous leishmaniasis of humans due to *Leishmania (Viannia) naiffi* outside Brazil. *Trans R Soc Trop Med Hyg*, **89**, 476–7.

Deane, L.M. 1956. *Leishmaniose visceral no Brasil*. Rio de Janerio: Serviço Nacional de Educação Sanitária.

Deane, L.M. and Grimaldi, G. 1985. Leishmaniasis in Brazil. In: Chang, K.P. and Bray, R.S. (eds), *Leishmaniasis*. New York: Elsevier, 247–75.

de Carvalho, P.B., Da Arribas, M.A. and Ferreira, E.I. 2000. Leishmaniasis. What do we know about its chemotherapy? *Rev Brasil Ciênc Farmacêut*, **36**, 69–96.

Decker-Jackson, J.E. and Tang, D.B. 1982. Identification of *Leishmania* spp. by radiorespirometry II. A statistical method of data analysis to evaluate the reproducibility and sensitivity of the technique. In: Chance, M.L. and Walton, B.C. (eds), *Proceedings of a Workshop held at the Pan American Health Organization, 9–11 December 1980*. Biochemical characterization of Leishmania. Geneva: UNDP/WORLD BANK/WHO, 205–45.

Delgado, O., Castes, M., et al. 1993. *Leishmania colombiensis* in Venezuela. *Am J Trop Med Hyg*, **48**, 145–7.

Delgado, O., Cupolillo, E., et al. 1997. Cutaneous leishmaniasis in Venezuela caused by infection with a new hybrid between *Leishmania (Viannia) braziliensis* and *L. (V.) guyanensis*. *Mem Inst Oswaldo Cruz*, **92**, 581–2.

Delgado, O., Feliciangeli, M.D., et al. 2001. Value of a dipstick based on recombinant RK39 antigen for differential diagnosis of American

visceral leishmaniasis from other sympatric endemic diseases in Venezuela. *Parasite*, **8**, 355–7.

De Lima, H., De Gugielmo, Z., et al. 2002. Cotton rats (*Sigmodon hispidus*) and black rats (*Rattus rattus*) as possible reservoirs of *Leishmania* spp. in Lara State, Venezuela. *Mem Inst Oswaldo Cruz*, **97**, 169–74.

de Souza, A., Ishikawa, E., et al. 1996. *Pyschodopygus complexus*, a new vector of *Leishmania braziliensis* to humans in Pará State Brazil. *Trans R Soc Trop Med Hyg*, **90**, 112–13.

de Souza e Souza, I., Naiff, R.D., et al. 1998. American cutaneous leishmaniasis due to *Leishmania (Viannia) guyanensis* as an initial clinical presentation of human immunodeficiency virus infection. *J Eur Acad Dermatol Veneriol*, **10**, 214–17.

de Queiroz, R.G., Vasconcelos, Ide. A., et al. 1994. Cutaneous leishmaniasis in Ceará State in Northeastern Brazil, incrimination of *Lutzomyia whitmani* (Diptera, Psychodidae) as a vector of *Leishmania braziliensis* in Baturite municipality. *Am J Trop Med Hyg*, **50**, 693–8.

dos Santos, S.O., Arias, J., et al. 1998. Incrimination of *Lutzomyia cruzi* as a vector of American visceral leishmaniasis. *Med Vet Entomol*, **12**, 315–17.

Dujardin, J.C., Bañuls, A.L., et al. 1995. Putative *Leishmania* hybrids in the Eastern Andean valley of Huanuco, Peru. *Acta Trop*, **59**, 293–307.

El-Hassan, A.M., Meredith, S.E.O., et al. 1995. Sudanese mucosal leishmaniasis, epidemiology, clinical features, diagnosis, immune responses and treatment. *Trans R Soc Trop Med Hyg*, **89**, 647–52.

Esterre, P., Chippaux, J.P., et al. 1986. Evaluation d'un programme de lutte contre la leishmaniose cutanée dans un village forestier de Guyane française. *Bull World Health Organ*, **64**, 559–65.

Feitosa, M.F., Azevêdo, E., et al. 1999. Genetic causes involved in *Leishmania chagasi* infection in northeastern Brazil. *Genet Mol Biol*, **22**, 1–5.

Feliciangeli, M.D., Rodriguez, N., et al. 1994. Vectors of cutaneous leishmaniasis in north-central Venezuela. *Med Vet Entomol*, **8**, 317–24.

Feliciangeli, M.D., Rodriguez, N., et al. 1999. The re-emergence of American visceral leishmaniasis in an old focus in Venezuela. II. Vectors and parasites. *Parasite*, **6**, 113–20.

Fernandes, O., Murthy, V.K., et al. 1994. Mini-exon gene variation in human pathogenic *Leishmania* species. *Mol Biochem Parasitol*, **66**, 261–77.

Floch, H. 1954. *Leishmania tropica guyanensis* n,ssp., agent de la leishmaniose tégumentaire des Guyanas et de l'Amérique Centrale. *Arch Inst Pasteur Guyane française*, **15**, 1–4.

Forattini, O.P., Pattoli, D.B.G., et al. 1972. Infecções naturais de mamiferos silvestres em área endêmica de leishmaniose tegumentardo Estado de São Paulo, Brasil. *Rev Saúde Publ*, **6**, 255–61.

Forattini, O.P., Pattoli, D.B.G., et al. 1973. Nota sôbre infecção natural de *Oryzomys capito laticeps* em foco enzoótica de leishmaniose tegumentar no Estado de São Paulo, Brasil. *Rev Saúde Publ*, **7**, 181–4.

Furuya, M., Shiraishi, M., et al. 1998. Natural infection of *Lutzomyia hartmanni* with *Leishmania (Viannia) equatorensis* in Ecuador. *Parasitol Int*, **47**, 121–6.

Garnham, P.C.C. 1962. Cutaneous leishmaniasis in the New World with special reference to *Leishmania mexicana*. *Sci Rep Inst Sup Sanita (Rome)*, **2**, 76–82.

Genaro, O., Peixoto de Toledo, V.P.C., et al. 1996. Vaccine for prophylaxis and immunotherapy, Brazil. *Clin Dermatol*, **14**, 503–12.

Gentile, B.F., Le Pont, F., et al. 1981. Dermal leishmaniasis in French Guiana, the sloth (*Choloepus didactylus*) as a reservoir host. *Trans R Soc Trop Med Hyg*, **75**, 612–13.

Gómez, E., Loor, C. and Vaca, G. 1996. Tratamento oral de la leishmaniasis americana en el Ecuador, primer reporte. *Medicina (Quito)*, **2**, 121–8.

Grimaldi, G., Tesh, R.B. and McMahon-Pratt, D. 1989. A review of the geographic distribution and epidemiology of leishmaniasis in the New World. *Am J Trop Med Hyg*, **41**, 687–725.

Grimaldi, G., Momen, H., et al. 1991. Characterization and classification of leishmanial parasites from humans, wild mammals and sandflies in the Amazon region of Brazil. *Am J Trop Med Parasitol*, **44**, 645–61.

Grimaldi, G., Kreutzer, R.D., et al. 1992. Description of *Leishmania equatorensis* sp.n. (Kinetoplastida, Trypanosomatidae), a new parasite infecting arboreal mammals in Ecuador. *Mem Inst Oswaldo Cruz*, **87**, 221–8.

Guevara, P., Alonso, G., et al. 1992. Identification of new world *Leishmania* using ribosomal gene spacer probes. *Mol Biochem Parasitol*, **56**, 15–26.

Guilbride, L., Myler, P.J. and Stuart, K. 1992. Distribution and sequence divergence of LRV1 viruses among different *Leishmania* species. *Mol Biochem Parasitol*, **54**, 101–4.

Handman, E. 2001. Leishmaniasis: current status of vaccine development. *Clin Microbiol Rev*, **14**, 229–43.

Hendrickx, E.P., Agudelo, S.P., et al. 1998. Lack of efficacy of mefloquine in the treatment of New World cutaneous leishmaniasis in Colombia. *Am J Trop Med Hyg*, **59**, 889–92.

Hernández, D.E., Oliver, M., et al. 1995. Visceral leishmaniasis with cutaneous and rectal dissemination due to *Leishmania braziliensis* in acquired immunodeficiency syndrome (AIDS). *Int J Dermatol*, **34**, 114–15.

Herrer, A. 1951. Estudios sobre leishmaniasis tegumentaria en el Perú. V. Leishmaniasis natural en perros procedentes de localidades utógenas. *Rev Med Exp (Lima)*, **8**, 87–117.

Herrer, A. 1971. *Leishmania hertigi* sp. n., from the tropical porcupine, *Coendou rothschildi* Thomas. *J Parasitol*, **57**, 626–9.

Herrer, A. 1982. *Lutzomyia peruensis* (Shannon 1929), possible vector natural de la uta (*Leishmaniasis tegumentaria*). *Rev Inst Med Trop São Paulo*, **24**, 168–72.

Herrer, A. and Christensen, H.A. 1975. Implication of *Phlebotomus* sandflies as vectors of bartonellosis and leishmaniasis as early as 1764. *Science*, **190**, 154–5.

Herrer, A., Christensen, H.A. and Beumer, R.J. 1973. Reservoir hosts of cutaneous leishmaniasis among Panamanian forest mammals. *Am J Trop Med Hyg*, **22**, 585–91.

Hoch, A., Ryan, L., et al. 1986. Isolation of *Leishmania braziliensis braziliensis* and other trypanosomatids from phlebotomines in a mucocutaneous leishmaniasis endemic area, Bahia, Brazil. *Mem Inst Oswaldo Cruz*, **81**, Suppl, 44.

Ishikawa, E.A.Y., Ready, P.D., et al. 1999. A mitochondrial DNA phylogeny indicates close relationships between populations of *Lutzomyia whitmani* (Diptera, Psychodidae, Phlebotominae) from the rain-forest regions of Amazônia and northeast Brazil. *Mem Inst Oswaldo Cruz*, **94**, 339–45.

Jackson, P.R., Wohlhieter, J.A. and Hockmeyer, W.T. 1982. *Leishmania* characterization by restriction endonuclease digestion of kinetoplastic DNA. *Abstracts of the Vth International Congress of Parasitology, 7–14 August 1982, Toronto, Canada*, p. 342.

Jackson, P.R., Stiteler, J.M. et al. 1984. Characterization of *Leishmania* responsible for visceral disease in Brazil by restriction endonuclease digestion and hybridization of kinetoplast DNA. *Proceedings of the 11th International Congress of Tropical Medicine and Malaria, Calgary*, p. 68.

Johnson, P.T., McConnell, E. and Hertig, M. 1963. Natural infections of leptomonal flagellates in Panamanian *Phlebotomus* sandflies. *Exp Parasitol*, **14**, 107–22.

Katakura, K., Matsumoto, Y., et al. 1993. Molecular karyotype characterization of *Leishmania panamensis*, *Leishmania mexicana* and *Leishmania major*-like parasites, agents of cutaneous leishmaniasis in Ecuador. *Am J Trop Med Hyg*, **48**, 707–15.

Kenney, R.T., Sacks, D.L., et al. 1999. Protective immunity using recombinant human IL-12 and alum as adjuvants in a primate model of cutaneous leishmaniasis. *J Immunol*, **163**, 4481–8.

Killick-Kendrick, R. 1985. Some epidemiological consequences of the evolutionary fit between leishmaniae and their phlebotomine vectors. *Bull Soc Pathol Exot*, **78**, 747–55.

Killick-Kendrick, R. 1990. Phlebotomine vectors of the leishmaniases, a review. *Med Vet Entomol*, **4**, 1–24.

Killick-Kendrick, R., Killick-Kendrick, M., et al. 1997. Protection of dogs from bites of phlebotomine sandflies by deltamethrin collars for control of canine leishmaniasis. *Med Vet Entomol*, **11**, 105–11.

Klotz, O. and Lindenberg, H. 1923. The pathology of leishmaniasis of the nose. *Am J Trop Med Hyg*, **3**, 117–41.

Knobloch, J., Hassler, D. and Braun, R. 2001. Wenn die Schmetterlingsmücke auf den Hund kommt, Leishmaniosen. *Dtsch Med Wochenschr*, **126**, A401–2.

Kreutzer, R.D., Corredor, A., et al. 1991. Characterization of *Leishmania colombiensis* sp. n. (Kinetoplastida, Trypanosomatidae), a new parasite infecting humans, animals, and phlebotomine sandflies in Colombia and Panama. *Am J Trop Med Hyg*, **44**, 662–75.

Lacerda, M.M. 1994. The Brazilian Leishmaniasis Control Program. *Mem Inst Oswaldo Cruz*, **89**, 489–95.

Laguna-Torres, V.A., Silva, C.A.C., et al. 1999. Mefloquina no tratamento da leishmaniose cutânea em uma área endêmica de Leishmania (Viannia) braziliensis. *Rev Soc Brasil Med Trop*, **32**, 529–32.

Lainson, R. 1982a. Leishmaniasis. In: Steele, J.H. (ed.), *Handbook series in zoonoses, section C. Parasitic zoonoses*, Vol. 1. . Boca Raton, Florida: CRC Press, 41–103.

Lainson, R. 1982b. Leishmanial parasites of mammals in relation to human disease. In: Edwards, M.A. and McDonnel, U. (eds), *Animal disease in relation to animal conservation*. London: Academic Press, 137–79.

Lainson, R. 1983. The American leishmaniases: some observations on their ecology and epidemiology. *Trans R Soc Trop Med Hyg*, **77**, 569–96.

Lainson, R. 1989. Demographic changes and their influence on the epidemiology of the American leishmaniases. In: Service, M.W. (ed.), *Demography and vector-borne diseases*. Boca Raton, Florida: CRC Press, 85–106.

Lainson, R. 1996. New World leishmaniasis. In: Cox, F.E.G. (ed.), *The Wellcome history of tropical diseases*. London: The Wellcome Trust, 218–29.

Lainson, R. 1997. On *Leishmania enriettii* and other enigmatic *Leishmania* species of the neotropics. *Mem Inst Oswaldo Cruz*, **92**, 377–87.

Lainson, R. and Shaw, J.J. 1970. Leishmaniasis in Brazil: V. Studies on the epidemiology of cutaneous leishmaniasis in Mato Grosso State, and observations on two distinct strains of *Leishmania* isolated from man and forest animals. *Trans R Soc Trop Med Hyg*, **64**, 654–67.

Lainson, R. and Shaw, J.J. 1972. Leishmaniasis of the New World, taxonomic problems. *Br Med Bull*, **28**, 44–8.

Lainson, R. and Shaw, J.J. 1977. Leishmanias of neotropical porcupines, *Leishmania hertigi deanei* nov. subsp. *Acta Amazon*, **7**, 51–7.

Lainson, R. and Shaw, J.J. 1979. The role of animals in the epidemiology of South American leishmaniasis. In: Lumsden, W.H.R. and Evans, D.A. (eds), *Biology of the Kinetoplastida*, Vol. 2. . London: Academic Press, 1–116.

Lainson, R. and Shaw, J.J. 1981. The leishmanial parasites. In: Cheesbrough, M. (ed.), *Medical laboratory manual for tropical countries*. Hertford, UK: S Austin and Sons, 206–17.

Lainson, R. and Shaw, J.J. 1987. Evolution, classification and geographical distribution. In: Peters, W. and Killick-Kendrick, R. (eds), *The leishmaniases in biology and medicine*, Vol. 1. . London: Academic Press, 1–120.

Lainson, R. and Shaw, J.J. 1989. *Leishmania (Viannia) naiffi* sp.n., a parasite of the armadillo, *Dasypus novemcinctus* (L.) in Amazonian Brazil. *Ann Parasitol Hum Comp*, **64**, 3–9.

Lainson, R. and Shaw, J.J. 1992. A brief history of the genus *Leishmania* (Protozoa, Kinetoplastida) in the Americas with particular reference to Amazonian Brazil. *Ciênc Cult*, **44**, 94–106.

Lainson, R. and Strangways-Dixon, J. 1963. *Leishmania mexicana*, The epidemiology of dermal leishmaniasis in British Honduras. *Trans R Soc Trop Med Hyg*, **57**, 242–65.

Lainson, R., Shaw, J.J. and Póvoa, M. 1981b. The importance of edentates (sloths and anteaters) as primary reservoirs of *Leishmania braziliensis guyanensis*, causative agent of pian bois in North Brazil. *Trans R Soc Trop Med Hyg*, **75**, 611–12.

Lainson, R., Ward, R.D. and Shaw, J.J. 1977. Experimental transmission of *Leishmania chagasi*, the causative agent of neotropical visceral leishmaniasis, by the sandfly *Lutzomyia longipalpis*. *Nature*, **226**, 628–30.

Lainson, R., Shaw, J.J., et al. 1973. Leishmaniasis in Brazil, IX. Considerations on the *Leishmania braziliensis* complex, importance of sandflies of the genus *Psychodopygus* (Mangabeira) in the transmission of *L. braziliensis braziliensis* in north Brazil. *Trans R Soc Trop Med Hyg*, **67**, 184–96.

Lainson, R., Shaw, J.J., et al. 1979. Leishmaniasis in Brazil, XIII. Isolation of *Leishmania* from armadillos (*Dasypus novemcinctus*), and observations on the epidemiology of cutaneous leishmaniasis in North Pará State. *Trans R Soc Trop Med Hyg*, **73**, 239–42.

Lainson, R., Shaw, J.J., et al. 1981a. Leishmaniasis in Brazil: XVI. Isolation and identification of *Leishmania* species from sandflies, wild mammals and man in north Pará State, with particular reference to *L. braziliensis guyanensis*, causative agent of 'pian-bois'. *Trans R Soc Trop Med Hyg*, **75**, 530–6.

Lainson, R., Shaw, J.J., et al. 1985. Leishmaniasis in Brazil. XXI, Visceral leishmaniasis in the Amazon Region and further observations on the role of *Lutzomyia longipalpis* (Lutz and Neiva 1912) as the vector. *Trans R Soc Trop Med Hyg*, **79**, 223–6.

Lainson, R., Shaw, J.J., et al. 1987. American visceral leishmaniasis, on the origin of *Leishmania* (*Leishmania*) *chagasi*. *Trans R Soc Trop Med Hyg*, **81**, 517.

Lainson, R., Shaw, J.J., et al. 1988. Isolation of *Leishmania* from monkeys in the Amazon Region of Brazil. *Trans R Soc Trop Med Hyg*, **82**, 231.

Lainson, R., Braga, R.R., et al. 1989. *Leishmania* (*Viannia*) *shawi* n.sp., a parasite of monkeys, sloths and procyonids in Amazonian Brazil. *Ann Parasitol Hum Comp*, **64**, 200–7.

Lainson, R., Dye, C., et al. 1990a. Amazonian visceral leishmaniasis - Distribution of the vector *Lutzomyia longipalpis* (Lutz and Neiva) in relation to the fox *Cerdocyon thous* (Linn.) and the efficiency of this reservoir host as a source of infection. *Mem Inst Oswaldo Cruz*, **85**, 135–7.

Lainson, R., Shaw, J.J., et al. 1990b. Cutaneous leishmaniasis of humans due to *Leishmania* (*Viannia*) *naiffi* Lainson and Shaw 1989. *Ann Parasitol Hum Comp*, **65**, 282–4.

Lainson, R., Shaw, J.J., et al. 1992. Further observations on *Lutzomyia ubiquitalis* (Psychodidae, Phlebotominae), the sandfly vector of *Leishmania* (*Viannia*) *lainson*. *Mem Inst Oswaldo Cruz*, **87**, 437–9.

Lainson, R., Shaw, J.J., et al. 1994. The dermal leishmaniases of Brazil, with special reference to the eco-epidemiology of the disease in Amazonia. *Mem Inst Oswaldo Cruz*, **89**, 435–43.

Lainson, R., Ishikawa, E.A.Y., et al. 2002. American visceral leishmaniasis, wild animal hosts. *Trans R Soc Trop Med Hyg*, **96**, 630–1.

Lanotte, G. and Rioux, J.-A. 1990. Fusion cellulaire chez les *Leishmania* (Kinetoplastida, Trypanosomatidae). *C R Séances Acad Sci*, **310**, 285–8.

Lanzaro, G.C., Ostrovska, K., et al. 1993. *Lutzomyia longipalpis* is a species complex, genetic divergence and interspecific hybrid sterility among three populations. *Am J Trop Med Hyg*, **48**, 839–47.

Laveran, A. and Mesnil, F. 1903. Sur un protozoaire nouveau (*Piroplasma donovani* Lav. et Mesn.). Parasite d'une fièvre de l'Inde. *C R Séances Acad Sci*, **137**, 957–61.

Le Pont, F. and Desjeux, P. 1986. Leishmaniasis in Bolivia. II. The involvement of *Psychodopygus yucumensis* and *Psychodopygus llanosmartinsi* in the sylvatic transmission cycle of *Leishmania braziliensis braziliensis* in a lowland subandean region. *Mem Inst Oswaldo Cruz*, **81**, 311–18.

Le Pont, F., Breniere, F.S., et al. 1988. Leishmaniose en Bolivie. III. *Psychodopygus carrerai carrerai* (Barretto 1946) nouveau vecteur de *Leishmania braziliensis braziliensis* em milieu sylvatique de région subandine basse. *CR Séances Acad Sci*, **307**, 279–82.

Lerner, E.A., Ribeiro, J.M., et al. 1991. Isolation of maxadilan, a potent vasodilatory peptide from the salivary glands of the sandfly *Lutzomyia longipalpis*. *J Biol Chem*, **266**, 11234–6.

Llanos-Cuentas, E.A., Roncal, N., et al. 1999. Natural infections of *Leishmania peruviana* in animals in the Peruvian Andes. *Trans R Soc Trop Med Hyg*, **93**, 15–20.

Lindenberg, A. 1909. A úlcera de Bauru e seu micróbio. *Revi Inst Med Trop São Paulo*, **12**, 116–20.

Lucas, C.M., Franke, E.D., et al. 1994. *Leishmania* (*Viannia*) *lainsoni*, first isolation in Peru. *Am J Trop Med Hyg*, **51**, 533–7.

Luz, E., Giovannoni, M. and Borba, A.M. 1967. Infecção de *Lutzomyia monticola* por *Leishmania enriettii*. *An Fac Med Uni Fed Paraná*, **9-10**, 121–8.

Luz, E., Membrive, N., et al. 2000. *Lutzomyia whitmani* (Diptera, Psychodidae) as vector of *Leishmania* (*V.*) *braziliensis* in Partana State, southern Brazil. *Ann Trop Med Parasitol*, **94**, 623–31.

Luz, K.G., Da Silva, V.O., et al. 1997. Prevalence of anti-*Leishmania donovani* antibodies among Brazilian blood donors and multiple transfused hemodialysis patients. *Am J Trop Med Hyg*, **57**, 168–71.

Lysenko, A.J. 1971. Distribution of leishmaniasis in the Old World. *Bull World Health Organ*, **44**, 515–20.

Machado, M.I., Milder, R.V., et al. 1994. Naturally acquired infections of *Leishmania enriettii* Muniz and Medina 1948 in guinea-pigs from São Paulo, Brazil. *Parasitology*, **109**, 135–8.

Marsden, P.D. 1986. Mucosal leishmaniasis ('espundia' Escomel 1911). *Trans R S Trop Med Hyg*, **80**, 859–76.

Màrquez, M. and Scorza, J.V. 1982. Criterios de nuliparidad y paridad en *Lutzomyia townsendi* (Ortiz 1959) del occidente de Venezuela. *Mem Inst Oswaldo Cruz*, **77**, 229–46.

Martinez, E., Le Pont, F., et al. 1999. *Lutzomyia nuneztovari anglesi* (Le Pont and Desjeux 1984) as a vector of *Leishmania amazonensis* in a sub-Andean leishmaniasis focus of Bolivia. *Am J Trop Med Hyg*, **61**, 846–9.

Martinez, E., Le Pont, F., et al. 2001. A first case of cutaneous leishmaniasis due to *Leishmania* (*Viannia*) *lainsoni* in Bolivia. *Trans R Soc Trop Med Hyg*, **95**, 375–7.

Marzochi, M.C.A., Teixeira, P.C., et al. 1993. Vacuum aspiratory puncture system for *Leishmania* culturing, isolation and transport. Preliminary report. *Rev Inst Med Trop São Paulo*, **35**, 301–3.

Matta, A. 1916. Sur les leishmanioses tégumentaires. Classification générale des leishmanioses. *Bull Soc Pathol Exot*, **9**, 494–503.

Matta, A. 1918. Notas para a historia das leishmanioses da pele e das mucosas. *Amazon Méd*, **1**, 11–17.

Mauny, I., Blanchot, I., et al. 1993. Visceral leishmaniasis in an infant in Brittany, discussion on the modes of transmission outside endemic zones. *Pédiatrie (Lyon)*, **48**, 237–9.

Maurício, I.L., Howard, M.K., et al. 1999. Genomic diversity in the *Leishmania donovani* complex. *Parasitology*, **119**, 237–46.

Mayrink, W., Michalick, M.S.M., et al. 1991. Tratamento da leishmaniose tegumentar Americana utilizando vacina. *An Brasil Dermatol*, **66**, 55–9.

Mayrink, W., Genaro, O., et al. 1996. Phase I and II open clinical trials of a vaccine against *Leishmania chagasi* infections in dogs. *Mem Inst Oswaldo Cruz*, **91**, 695–7.

Mazza, S. and Cornejo, A.J. 1926. Primeros casos autóctonos de kala-azar infantil comprobados en el norte de la República (Tabacal y Orán, Salta). *Bol Inst Clín Quir (Buenos Aires)*, **2**, 140–4.

Medina, H. 1946. Estudos sôbre leishmaniose. I. Primeiros casos de leishmaniose espontânea observados em cobaias. *Arq Biol Tecnol (Curitiba)*, **1**, 39–74.

Medina, R. and Romero, J. 1959. Estudio clinico y parasitologico de una nueva cepa de leishmania. *Arch Venez Patol Trop Parasitol Méd*, **3**, 298–326.

Medina, R. and Romero, J. 1962. *Leishmania pifanoi* n.sp. El agente causal de la leishmaniasis tegumentaria difusa. *Arch Venez Patol Trop Parasitol Méd*, **4**, 349–53.

Mello, D.A., Rego, Fde.A., et al. 1988. *Cerdocyon thous* (L) (Carnivora, Canidae) naturally infected with *Leishmania donovani chagasi* (Cunha and Chagas 1937) in Corumbá (Mato Grosso do Sul State, Brazil). *Mem Inst Oswaldo Cruz*, **83**, 259.

Mendoza-León, A., Havercroft, J.C. and Barker, D.C. 1995. The RFLP analysis of the β-tubulin gene region in New World *Leishmania*. *Parasitology*, **111**, 1–9.

Meyerhoff, A. 1999. U.S. Food and Drug Administration approval of AmBisome (liposomal amphotericin B) for treatment of visceral leishmaniasis. *Clin Infect Dis*, **28**, 42–8.

Migone, L.E. 1913. Un caso de kala-azar a Assuncion (Paraguay). *Bull Soc Pathol Exot*, **6**, 118–20.

Miles, M.A., Póvoa, M.M., et al. 1980. Some methods for the enzymic characterization of Latin-American *Leishmania* with particular reference to *Leishmania mexicana amazonensis* and subspecies of *Leishmania hertigi*. *Trans R Soc Trop Med Hyg*, **74**, 243–52.

Mimori, T., Sasaki, J., et al. 1998. Rapid identification of *Leishmania* species from formalin-fixed biopsy samples by polymorphism-specific polymerase chain reaction. *Gene*, **210**, 179–86.

Molyneux, D.H. and Killick-Kendrick, R. 1987. Morphology, ultrastructure and life cycles. In: Peters, W. and Killick-Kendrick, R. (eds), *The leishmaniases in biology and medicine*, Vol. 1. . London: Academic Press, 121–76.

Montoya-Lerma, J., Palacios, R., et al. 1998. Further evidence of humans as source of *Leishmania Viannia* for sandflies. *Mem Inst Oswaldo Cruz*, **93**, 735–6.

Muniz, J. and Medina, H. 1948. Leishmaniose tegumentar do cobaio (*Leishmania enriettii* n. sp.). *Hospital (Rio de Janeiro)*, **33**, 7–25.

Myler, P.J. and Stuart, K.D. 2000. Recent developments from the *Leishmania* genome project. *Curr Opin Microbiol*, **3**, 412–16.

Naiff, R.D., Freitas, R.A. et al. 1989. Aspectos epidemiológicos de uma *Leishmamia* de tatus (*Dasypus novemcinctus*). *Resumos do XI Congresso Brasileira de Parasitologia, July–August 1989, Rio de Janeiro*, p. 24.

Navin, T.R., Arana, B.A., et al. 1992. Placebo-controlled clinical trial of sodium stibogluconate (Pentostam) versus ketoconazole for treating cutaneous leishmaniasis in Guatemala. *J Infect Dis*, **165**, 528–34.

Noyes, H.A., Belli, A.A. and Maingon, R. 1996. Appraisal of various random amplified polymorphic DNA-polymerase chain reaction primers for *Leishmania* identification. *Am J Trop Med Hyg*, **55**, 98–105.

Oliveira Neto, M.P., Marzochi, M.C.A., et al. 1986. Concurrent human infection with *Leishmania donovani* and *Leishmania braziliensis braziliensis*. *Ann Trop Med Parasitol*, **80**, 587–92.

Penna, H.A. 1934. Leishmaniose visceral no Brasil. *Brasil Méd*, **48**, 949–50.

Pessôa, S.B. 1941. Dados sobre a epidemiologia da leishmaniose tegumentar em São Paulo. *O Hospital*, **19**, 389–409.

Pessôa, S.B. 1961. Classificação das leishmanioses e das espécies do gênero *Leishmania*. *Arq Hig Saúde Pública*, **26**, 41–50.

Pessôa, S.B. and Barretto, M.P. 1948. *Leishmaniose tegumentar Americana*. Rio de Janeiro: Imprensa Nacional.

Petersen, E.A., Neva, F.A., et al. 1982. Specific inhibition of lymphocyte-proliferation response by adherent suppressor cells in diffuse cutaneous leishmaniasis. *N Engl J Med*, **306**, 387–92.

Pifano, C.F. and Romero, M.J. 1964. Comprobación de un nuevo foco de leishmaniasis en Venezuela, Valle de Cumanacoa, Edo. Sucre. *Gac Méd Caracas*, **72**, 473–9.

Ponce, C., Ponce, E., et al. 1991. *Leishmania donovani chagasi*, new clinical variant of cutaneous leishmaniasis in Honduras. *Lancet*, **337**, 67–70.

Potenza, L. and Anduze, J. 1942. Kala-azar en el Estado de Bolivar, Venezuela. *Rev Policlin Caracas*, **11**, 312–17.

Raccurt, C.P., Pratlong, F.M., et al. 1995. French Guiana must be recognized as an endemic area of *Leishmania (Viannia) braziliensis* in South America. *Trans R Soc Trop Med Hyg*, **89**, 372, .

Ramos-Santos, C., Hernandez-Montes, O., et al. 2000. Visceral leishmaniasis caused by *Leishmania (L.) mexicana* in a Mexican patient with human immunodeficiency virus infection. *Mem Inst Oswaldo Cruz*, **95**, 733–7.

Rangel, E.F., Lainson, R., et al. 1996. Variation between geographical populations of *Lutzomyia (Nyssomyia) whitmani* (Antunes and Coutinho 1939) *sensu lato* (Diptera, Psychodidae, Phlebotominae) in Brazil. *Mem Inst Oswaldo Cruz*, **91**, 43–50.

Reithinger, R. and Davies, C.R. 1999. Is the domestic dog (*Canis familiaris*) a reservoir host of American cutaneous leishmaniasis? A critical review of the current evidence. *Am J Trop Med Hyg*, **61**, 530–41.

Ribeiro, A.L.P., Drummond, J.B., et al. 1999. Electrocardiographic changes during low-dose, short-term therapy of cutaneous leishmaniasis with the pentavalent antimonial meglumine. *Braz J Med Biol Res*, **32**, 297–301.

Ridley, D.S. 1987. Pathology. In: Peters, W. and Killick-Kendrick, R. (eds), *The leishmaniases in biology and medicine*, Vol. 2. . London: Academic Press, 665–701.

Rioux, J.A., Lanotte, G., et al. 1990. Taxonomy of *Leishmania*. Use of isoenzymes. Suggestions for a new classification. *Ann Parasitol Hum Comp*, **65**, 111–25.

Rocha, N.M.M., Melo, M.N., et al. 1988. *Leishmania braziliensis braziliensis* isolated from *Akodon arviculoides* captured in Caratinga, Minas Gerais, Brazil. *Trans R Soc Trop Med Hyg*, **82**, 68.

Rodriguez, N., De Lima, H., et al. 1997. Genomic DNA repeat from *Leishmania (Viannia) braziliensis* (Venezuelan strain) containing simple repeats and microsatellites. *Parasitology*, **115**, 349–58.

Romero, G.A.S., Sampaio, R.N.R., et al. 1999. Sensitivity of a vacuum aspiratory culture technique for diagnosis of localized cutaneous leishmaniasis in an endemic area of *Leishmania (Viannia) braziliensis* transmission. *Mem Inst Oswaldo Cruz*, **94**, 505–8.

Ross, R. 1903. (1) Note on the bodies recently described by Leishman and Donovan and (2) further notes on Leishman's bodies. *Br Med J*, **ii**, 1261–1262, 1401.

Russell, R., Iribar, M.P., et al. 1999. Intra- and inter-specific microsatellite variation in the *Leishmania* subgenus *Viannia*. *Mol Biochem Parasitol*, **103**, 71–7.

Ryan, L., Lainson, R. and Shaw, J.J. 1987. Leishmaniasis in Brazil. XXIV. Natural flagellate infections of sandflies (Diptera, Psychodidae) in Pará State, with particular reference to the role of *Psychodopygus wellcomei* as the vector of *Leishmania braziliensis braziliensis* in the Serra dos Carajás. *Trans R Soc Trop Med Hyg*, **81**, 353–9.

Ryan, L., Silveira, F.T., et al. 1984. Leishmanial infections in *Lutzomyia longipalpis* and *Lu. antunesi* (Diptera, Psychodidae) on the island of Marajó, Pará State, Brazil. *Trans R Soc Trop Med Hyg*, **78**, 547–8.

Ryan, L., Vexinat, J.A., et al. 1990. The importance of rapid diagnosis of new cases of cutaneous leishmaniasis in pin-pointing the sandfly vector. *Trans R SocTrop Med Hyg*, **84**, 786.

Saiz, M., Llanos-Cuentas, A., et al. 1998. Detection of *Leishmaniavirus* in human biopsy samples of leishmaniasis from Peru. *Am J Trop Med Hyg*, **58**, 192–4.

Sampaio, R.N.R. and Marsden, P.D. 1997. Mucosal leishmaniasis unresponsive to glucantime therapy successfully treated with AmBisome™. *Trans R Soc Trop Med Hyg*, **91**, 77.

Santoro F., Lemesre, J.L. et al. 1986. Spécificité au niveau des protéines de surface des promastigotes de *Leishmania donovani* (Laveran et Mesnil 1903) *Leishmania infantum* Nicolle 1908 et *Leishmania chagasi* Cunha et Chagas 1937. In Rioux, J.-A. (ed.) *Leishmania. Taxonomie et phylogenèse. Applications éco-épidémiologiques*. Colloque International, 2–6 July 1984, Montpellier: IMEEE, pp. 71–75.

Savani, E.S.M.M., Galati, E.A.B., et al. 1999. Inquérito sorológico sobre leishmaniose tegumentat Americana em cães errantes no Estado de São Paulo, Brasil. *Rev Saúde Pública*, **33**, 629–31.

Schulman, I.A. 1994. Parasitic infections and their impact on blood donor selection and testing. *Arch Pathol Lab Med*, **118**, 366–70.

Scorza, J.V., Valera, M., et al. 1979. A new species of *Leishmania* parasite from the Venezuelan Andes region. *Trans R Soc Trop Med Hyg*, **73**, 293–8.

Sharples, C.E., Shaw, M.A., et al. 1994. Immune response in healthy volunteers vaccinated with BCG plus killed leishmanial promastigotes, antibody responses to mycobacterial and leishmanial antigens. *Vaccine*, **12**, 1402–12.

Shaw, J.J. 1994. Taxonomy of the genus *Leishmania*, Present and future trends and their implications. *Mem Inst Oswaldo Cruz*, **89**, 471–8.

Shaw, J.J. and Lainson, R. 1987. Ecology and epidemiology, New World. In: Peters, W. and Killick-Kendrick, R. (eds), *The leishmaniases in biology and medicine*, Vol. 1. . London: Academic Press, 291–363.

Shaw, J.J., Lainson, R. et al. 1986. Serodemes of the *Leishmania braziliensis* complex. In Rioux, J.-A. (ed.) *Leishmania. Taxonomie et phylogenèse. Applications éco-épidémiologiques*. Colloque International, 2–6 July 1984, Montpellier: IMEEE, pp. 179–183

Sherlock, I.A., Miranda, J.C., et al. 1984. Natural infection of the opossum *Didelphis albiventris* (Marsupialia, Didelphidae) with *Leishmania donovani* in Brazil. *Mem Inst Oswaldo Cruz*, **79**, 511.

Silveira, F.T., Shaw, J.J., et al. 1987. Dermal leishmaniasis in the Amazon Region of Brazil, *Leishmania* (*Viannaia*) *lainsoni* sp.n., a new parasite from the State of Pará. *Mem Inst Oswaldo Cruz*, **82**, 289–92.

Silveira, F.T., Lainson, R., et al. 1991a. Leishmaniose cutânea na Amazonia, isolamento de *Leishmania (Viannia)lainsoni* do roedor *Agouti paca* (Rodentia, Dasyproctidae), no Estado do Pará, Brasil. *Rev Inst Med Trop São Paulo*, **33**, 18–22.

Silveira, F.T., Souza, A.A.A., et al. 1991b. Cutaneous leishmaniasis in the Amazon Region, natural infection of the sandfly *Lutzomyia ubiquitalis* (Psychodidae, Phlebotominae) by *Leishmania (Viannia) lainsoni* in Pará State, Brazil. *Mem Inst Oswaldo Cruz*, **86**, 127–30.

Silveira, F.T., Ishikawa, E.A.Y., et al. 2002. An outbreak of cutaneous leishmaniasis among soldiers in Belém,Pará State, Brazil, caused by *Leishmania (Viannia) lindenbergi* n.sp. A new leishmanial parasite of man in the Amazon region . *Parasite*, **9**, 43–50.

Silveira, F.T., Lainson, R. and Corbett, E.P. 2004. Clinical and immunopathological spectrum of American cutaneous leishmaniasis with special reference to the disease in Amazonian Brazil - a review. *Mem Inst Oswaldo Cruz*, **99**, 239–51.

Soto, S.I.U., Lehmann, T., et al. 2001. Speciation and population structure in the morphospecies *Lutzomyia longipalpis* (Lutz and Neiva) as derived from mitochondrial ND4 gene. *Mol Phylogenet Evol*, **18**, 84–93.

Spithill, T.W. and Samaras, N. 1985. The molecular karyotype of *Leishmania major* and mapping of α and β tubulin gene families to multiple unlinked chromosomal loci. *Nucleic Acids Res*, **13**, 4155–69.

Strangways-Dixon, J. and Lainson, R. 1966. The epidemiology of dermal leishmaniasis in British Hondurus. Part III. The transmission of *Leishmania mexicana* to man by *Phlebotomus pessoanus*, with observations on the development of the parasite in differents species of *Phlebotomus*. *Trans R Soc Trop Med Hyg*, **60**, 192–207.

Tarr, P.I., Aline, R.F., et al. 1988. LR1, a candidate RNA virus of *Leishmania*. *Proc Natl Acad Sci USA*, **85**, 9572–5.

Tolezano, J.E., Taniguchi, H.H., et al. 1998. Epidemiologia da leishmaniose tegumentar americana no Estado de São Paulo, Brasil. II. Utilização de antígeno particulado de *Leishmania* (*Viannia*) *braziliensis* em inquérito canino em regiões endêmicas. *Rev Inst Adolfo Lutz*, **57**, 65–71.

Tonui, W.K. 1999. *Leishmania* transmission-blocking vaccines, a review. *East Afr Med J*, **76**, 93–6.

Travi, B.L., Vélez, I.D., et al. 1990. *Lutzomyia evansi*, an alternate vector of *Leishmania chagasi* in a Colombian focus of visceral leishmaniasis. *Trans R Soc Trop Med Hyg*, **84**, 676–7.

Travi, B.L., Jaramillo, C., et al. 1994. *Didelphis marsupialis*, an important reservoir of *Trypanosoma* (*Schizotrypanum*) *cruzi* and

Leishmania (*Leishmania*) *chagasi* in Colombia. *Am J Trop Med Hyg*, **50**, 557–65.

Travi, B.L., Montoya, J., et al. 1996. Bionomics of *Lutzomyia evansi* (Diptera, Psychodidae), vector of visceral leishmaniasis in northern Colombia. *J Med Entomol*, **33**, 278–85.

Travi, B.L., Osorio, Y., et al. 1998. Dynamics of *Leishmania chagasi* infection in small mammals of the undisturbed and degraded tropical dry forests of northern Colombia. *Trans R Soc Trop Med Hyg*, **92**, 275–8.

Travi, B.L., Arteaga, L.T., et al. 2002. Susceptibility of spiny rats (*Proechimys semispinosus*) to *Leishmania* (*Viannia*) *panamensis* and *Leishmania* (*Leishmania*) *chagasi*. *Mem Inst Oswaldo Cruz*, **97**, 887–92.

Velez, L.R. 1913. Uta e espundia. *Bull Soc Pathol Exot*, **6**, 545.

Vianna, G. 1911. Sôbre uma nova especie de *Leishmania* (Nota Preliminar). *Brasil Méd*, **25**, 411.

Vianna, G. 1912. Tratamento da leishmaniose pelo tartaro emético. *Arch Venez Patol Trop Parasitol Méd*, **2**, 426–8.

Villela, F., Pestana, B.R. and Pessôa, S.B. 1939. Presença de *Leishmania braziliensis* na mucosa nasal sem lesão aparente em casos recentes de leishmaniose cutânea. *O Hospital (Rio de Janeiro)*, **16**, 953–60.

Walton, B.C. 1987. American cutaneous and mucocutaneous leishmaniasis. In: Peters, W. and Killick-Kendrick, R. (eds), *The leishmaniases in biology and medicine*, Vol. 2. . London: Academic Press, 637–64.

Warburg, A., Saraiva, E., et al. 1994. Saliva of *Lutzomyia longipalpis* sibling species differs in its composition and capacity to enhance leishmaniasis. *Philos Trans R Soc Lond B Biol Sci*, **345**, 223–30.

Ward, R.D., Shaw, J.J., et al. 1973. Leishmaniasis in Brazil, VIII. Observations on the phlebotomine fauna of an area highly endemic for cutaneous leishmaniasis, in the Serra dos Carajás, Pará State. *Trans R Soc Trop Med Hyg*, **67**, 174–83.

Ward, R.D., Ribeiro, A.L., et al. 1983. Reproductive isolation between different forms of *Lutzomyia longipalpis* (Lutz and Neiva), (Diptera, Psychodidae), the vector of *Leishmania donovani chagasi* Cunha and Chagas and its significance to kala-azar distribution in South America. *Mem Inst Oswaldo Cruz*, **78**, 269–80.

Ward, R.D., Phillips, A., et al. 1988. The *Lutzomyia longipalpis* complex, reproduction and distribution. In: Service, M.W. (ed.), *Biosystematics of haematophagus insects*, Systematics Association Special Volume 37. . Oxford: Clarendon Press, 257–9.

WHO, 1998. Leishmania *and HIV in gridlock*, WHO/CTD/LEISH/98.9. Desjeux, P. (ed.) Geneva: World Health Organization.

Yin, H., Norris, D.E. and Lanzaro, G.C. 2000. Sibling species in the *Lutzomyia longipalpis* complex differ in levels of mRNA expression for the salivary gland peptide, maxadilan. *Insect Mol Biol*, **9**, 309–14.

Yoshida, E.L.A., Cuba Cuba, C.A., et al. 1993. Description of *Leishmania* (*Leishmania*) *forattinii* sp.n., a new parasite infecting opossums and rodents in Brazil. *Mem Inst Oswaldo Cruz*, **88**, 397–406.

Young, D.G., Morales, R.D., et al. 1987. Isolation of *Leishmania braziliensis* (Kinetoplastida, Trypanosomatidae) from cryopreserved Colombian sandflies (Diptera, Psychodidae). *J Med Entomol*, **24**, 587–589, 703.

Young, D.G. and Duncan, M.A. 1994. *Guide to the identification and geographic distribution of* Lutzomyia *sandflies in Mexico, the West Indies, Central and South America (Diptera, Psychodidae)*. Memoirs of the American Entomological Institute No. 54. Associated Publishers, Gainsville, Florida.

Zelodón, R., Ponce, C. and Murillo, J. 1979. *Leishmania herreri* sp.n. from sloths and sandflies of Costa Rica. *J Parasitol*, **65**, 275–9.

Zelodón, R., Hidalgo, H., et al. 1989. Atypical cutaneous leishmaniasis in a semiarid region of north-west Costa Rica. *Trans R Soc Trop Med Hyg*, **83**, 786.

18

African trypanosomiasis

SAMUEL J. BLACK AND JOHN RICHARD SEED

INTRODUCTION

African trypanosomiasis, or sleeping sickness, is a human disease caused by *Trypanosoma brucei rhodesiense* and *T. b. gambiense* that inhabit the fluid compartments of blood, lymph, and interstitial fluids of their human hosts (Figure 18.1). The parasites belong to:

    Kingdom Protozoa
        Phylum Euglenozoa
            Class Kinetoplastea
                Order Trypanosomatida
                    Genus *Trypanosoma* Gruby, 1843
                        Section *Salivaria*
                        Subgenus *Trypanozoon*
                        Species *Trypanosoma brucei*

This pedigree, distilled from more than one and a half centuries of observation by systematists, signifies that in addition to causing sleeping sickness in people, the parasites are single-celled eukaryotes with a single flagellum, no chloroplast, a conspicuous aggregate of DNA located near the flagellar kinetosomes in a single mitochondrion, and that they are transmitted from the anterior station, i.e. the proboscis, of tsetse flies.

Sleeping sickness is endemic to sub-Saharan Africa and arises within the habitat of tsetse flies (Genus *Glossina*, Order Diptera), which are the primary vectors of African trypanosomes and in which trypanosomes undergo cyclic development. Tsetse flies are constrained by temperature and humidity boundaries that are met within vast swaths of the humid and semi-humid zones of Africa. They infest a land mass of about 10 million square kilometers ranging from the southern boundary

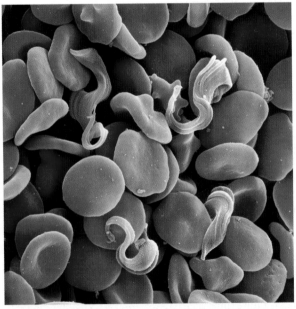

Figure 18.1 Trypanosoma brucei *in a typical habitat within a mammal. The parasites inhabit the extracellular compartment of blood and tissue fluids and are shown here among red blood cells in a smear of infected blood. (Courtesy of Jürgen Berger and Dr Peter Overath, Max Planck Institute for Developmental Biology, Tübingen)*

of the Saharan desert to around 20° south of the Equator. Within this region, the *Morsitans* subgenus inhabits the savanna, the *Palpalis* subgenus inhabits riverine valleys, the *Austeni* subgenus inhabits thickets, and the *Fusca* subgenus inhabits the forests. All subgenera of *Glossina* transmit African trypanosomes but *T. b. gambiense* is mainly transmitted by the *Palpalis* subgenus and *T. b. rhodesiense* by the *Morsitans* subgenus.

Some species of trypanosomes (see below) infect cattle and other domesticated animals causing a wasting disease, nagana, which costs producers about US$3 billion each year in lost productivity. Some environmentalists see nagana as a disease barrier that protects Central Africa from becoming home to monocultures of grassland and cattle. To others, it is a garrote strangling the capacity of sedentary agriculturists in Central Africa to establish profitable integrated agricultural systems. Both points of view have merit. Overgrazing and environmental degradation are prevalent in some grazed regions abutting the Sahara and could spread south if nagana was no longer a problem. On the other hand, the inability to keep cattle in Central Africa because of trypanosomiasis prevents that region from following successful models of agrarian development established elsewhere.

The importance of trypanosomiasis to human and livestock health in Africa was recognized early in the period of European colonial expansion in Africa (Ford 1971). Various historical accounts of the early work on the epidemiology, the control measures, and the social and economic consequences of trypanosomiasis are available (Duggan 1970; Ford 1971). By the late-1950s, human trypanosomiasis had been contained through vector control, vector avoidance, and a combination of diagnosis and treatment through chemotherapy. Despite this early success, the disease is now a serious public health problem in much of sub-Saharan Africa. The breakdown of public health programs and, as a consequence, a reduction of disease surveillance and treatment, appear to be key factors in disease resurgence. The incidence of sleeping sickness has now reached epidemic proportions in Angola and the Democratic Republic of Congo (formerly the Republic of Zaire) and is high and increasing in Uganda and Sudan. Only about 10 percent of people at risk are screened for trypanosomiasis and of these only a small proportion are treated. It is estimated that 300 000–500 000 people have sleeping sickness, and in some areas the incidence of disease exceeds that of acquired immune deficiency syndrome (AIDS). The incidence of trypanosomiasis is unlikely to decline in the near future unless aggressive diagnosis and treatment regimes are re-established and efforts made to reduce the prevalence of tsetse flies and to limit their contact with people. The World Health Organization is working to achieve these goals by re-enacting earlier successful control programs. These include:

- the assembly of mobile medical teams equipped with the latest diagnostic capability and treatment drugs
- survey of the population at risk with subsequent provision of chemotherapy
- establishment of local health centers to provide long-term care and post-therapy follow up
- motivation of rural communities to assume responsibility for vector control programs and health care initiatives.

This is all timely but vulnerable to unstable sociopolitical conditions. Unfortunately, there are no alternatives: neither long-lived chemoprophylactic treatments nor effective vaccines have been developed.

CLASSIFICATION

There are several species of African trypanosomes distributed through four subgenera: subgenus *Duttonella* (*D*), *Nannomonas* (*N*), *Trypanozoon* (*T*) and *Pycnomonas* (*P*).

Order Trypanosomatida
    Family Trypanosomatidae
        Genus *Trypanosoma*
        Section Stercoraria
        Subgenus *Schizotrypanum*
            Species *T. (S) cruzi*
        Section Salivaria
        Subgenus *Duttonella*
            Species *T. (D) vivax*
        Subgenus *Nannomonas*
            Species *T. (N) congolense*
            Species *T. (N) simiae*
        Subgenus *Trypanozoon*
            Species *T. (T) brucei*
            Subspecies *T. (T) b. brucei*
            Subspecies *T. (T) b. rhodesiense*
            Subspecies *T. (T) b. gambiense*
            Subspecies *T. (T) equiperdum*
            Subspecies *T. (T) evansi*
        Subgenus *Pycnomonas*
            Species *T. (P) suis*

The subgenera and species are distinguished on morphologic, isoenzyme, molecular diagnostic, and host-specificity criteria. The sub-genus *Trypanozoon* contains the human infective trypanosomes and is considered to be 'in the midst of a period of rapid evolution at the present time, which makes it difficult for systematists to divide it up neatly into distinct species' (Kreier and Baker 1987). There are two groups of *Trypanozoon* organisms. The first, referred to as the *T. (T) brucei* group, includes the human infective pathogens *T. b. gambiense* and *T. b. rhodesiense*, and a morphologically identical subspecies, *T. b. brucei*, that does not infect humans. All members of the *T. (T) brucei* group are transmitted by tsetse flies. Their taxonomic relationships are further discussed below based on molecular characterization and sensitivity to human serum. Kreier and

Baker (1987) use the term *T. brucei* s.l. (= sensu lato, in the broad sense) to describe the *T. (T) brucei* group. This usage is awkward and in the rest of this chapter when we refer to *T. brucei* without any further qualification it should be taken to mean *T. brucei* s.l.

The second group of *Trypanozoon* species is comprised of *T. evansi* and *T. equiperdum*. These parasites are not cyclically transmitted by tsetse flies. Their taxonomic relationship with each other and with their putative parent species *T. brucei* is unclear. It is thought that they evolved from *T. brucei* organisms that had been carried beyond the northern boundary of the tsetse habitat by infected camels and subsequently 'adapted to noncyclical transmission by biting Diptera and so survived and underwent speciation on their own' (Kreier and Baker 1987). *T. evansi* is widely spread through North Africa, Asia, and South America. Camels, cattle, water buffalo, horses, and dogs are common hosts. *T. equiperdum*, a parasite of horses and donkeys, is morphologically identical to *T. evansi*. It is not dependent on an invertebrate vector but accumulates in inflammatory sites on the skin, including the genitalia that are abraded during coitus thus permitting transfer of the parasites to uninfected breeding partners.

Trypanosoma b. brucei, which is morphologically indistinguishable from the human-infective *T. brucei* subspecies, does not infect humans but does infect cattle and, together with *T. (D) vivax* and *T. (N) congolense*, causes nagana, a fatal disease of these livestock. *T. b. brucei*, *T. congolense*, and *T. vivax* are widely distributed throughout the tsetse fly habitat and their presence excludes cattle from all but the fringes of it.

The *T. (T) brucei* complex and human infectivity

Members of the *T. (T) brucei* complex, *T. b. brucei*, *T. b. rhodesiense*, and *T. b. gambiense*, share phenotypic characteristics including their morphology, life cycle, and major biochemical features. However, there is significant intraspecific variation evidenced by differences in isoenzyme electrophoretic patterns and in restriction fragment length polymorphisms (RFLP). Cluster analysis of data on isoenzyme variation show that the African trypanosomes can be placed into a number of different principal zymodeme groups. *T. b. gambiense* trypanosomes are fairly homogenous by zymodeme analysis and can be distinguished by this method from *T. b. rhodesiense* and *T. b. brucei* (Stevens and Godfrey 1992). *T. b. gambiense*, *T. b. rhodesiense*, and *T. b. brucei* can be distinguished on the basis of RFLP fingerprints (Hide et al. 1994, Agbo et al. 2002) but intra-subspecies variability exists reducing the usefulness of this test. Evaluation of *Trypanozoon* organisms using a repeated coding sequence and microsatellite markers (Biteau et al. 2000) has revealed great heterogeneity in genotype related to

these sequences; however, group-specific genotypes, or alleles, were evident and a *T. b. gambiense* group could be distinguished on the basis of their pattern of reactivity with the probes. Many workers have concluded that the African trypanosomes are evolving rapidly with respect to their population genetics and epidemiology, accounting for the complex variations in isoenzyme and RFLP patterns found among various isolates from tsetse, different host groups, and distinct geographical locations.

T. b. gambiense and *T. b. rhodesiense*, but no other African trypanosomes, are infective for humans. Infections initiated with *T. b. rhodesiense* are more acute than those initiated with *T. b. gambiense* although both are fatal if left untreated. *T. b. rhodesiense* and *T. b. gambiense* also have differing geographic distributions: the former are prevalent in the Eastern half of the African continent within the tsetse habitat, and the latter are prevalent in West and Central Africa. The two groups meet and overlap around the Northern and Eastern shores of Lake Tanganyika.

T. b. gambiense and *T. b. rhodesiense* can be incubated in human serum with no ill effects. In contrast, *T. b. brucei* is lysed by incubation in human serum and is not infective to humans. *T. vivax* and *T. congolense*, which also are not infective for humans, are similarly lysed when incubated in human serum (Black et al. 2001) and it is likely that the other trypanosome species that do not infect humans will be equally sensitive. Lysis in human serum provides a simple, although only partly reliable, method of distinguishing *T. b. brucei* from the human infective *T. brucei* subspecies (Rickman and Robson 1970; Robson and Rickman 1978). The test can record false negatives because human-infective *T. brucei* can reversibly give rise to forms sensitive to human serum when freed from the selective pressure of human serum. Furthermore, individual human serum samples vary with respect to their lytic activity on sensitive *T. brucei* subspecies and a given sample can have higher activity against one or another isolate.

Two species of trypanolytic material have been isolated from human serum, a very high density lipoprotein (VHDL) subfraction containing apoliproteins A-I, A-II, C-I, C-II, C-III, haptoglobin-related protein (hpr), and paraoxonase-arylesterase (Smith et al. 1995), and a non-lipoprotein component that contains apo AI, hpr, and IgM (Raper et al. 1999). *T. b. brucei* that are incubated with human serum undergo characteristic changes in their morphology, starting with pronounced swelling around the flagellar pocket region. It is considered that uptake of the lytic material results in disruption of an endocytic and most likely a lysosomal compartment. The hpr component of the trypanolytic factors has been proposed as the targeting device, allowing the lytic complexes to bind to, and be endocytosed by, the parasites. However, hpr alone, or in combination with mouse serum components, does not harm the parasites (Hatada et al. 2002). In a stunning series of experiments, Pays and his group have now shown that apolipoprotein L-1 is the lytic factor in human serum and that this minor component of high density lipoprotein exerts its effect in the presence or absence of hpr (Vanhamme et al. 2003). The apolipoprotein L-1 is endocytosed by the parasites and processed in a lysosome. The mechanism through which endocytosed apolipoprotein L-1 destabilizes the *T. b. brucei* lysosomal membrane has not been resolved.

T. b. rhodesiense express a human serum resistance associated (SRA) gene (De Greef et al. 1989), which, upon transfection into *T. b. brucei*, confers resistance to lysis by human serum (Xong et al. 1998) and which can be used as a molecular confirmation of *T. b. rhodesiense* (Welburn et al. 2001; Gibson 2001a). SRA has been shown by Pays and colleagues (Vanhamme et al. 2003) to encode a lysosomal protein that most likely confers resistance to the lytic activity of apolipoprotein L-1 by reacting with the protein in the lysosome. Unlike *T. b. rhodesiense*, *T. b. gambiense* do not express SRA and the mechanism used by these parasites to resist lysis in human serum has not been resolved. Although many crucial details remain to be resolved as to how human serum lyses *T. b. brucei* and how this process is prevented in other *T. brucei* subspecies, the basic finding that apolipoprotein L-1 is the lytic factor of human serum is of huge significance. This finding may lead to novel transgenic solutions to nagana as well as to the development of therapeutic peptides for *T. b. rhodesiense*-sleeping sickness.

STRUCTURE AND LIFE CYCLE

Kinetoplast

Trypanosoma brucei are spindle shaped protozoa about 20 μm in length and 1–2 μm in width. The parasites have the common features of eukaryotic cells, namely a lipid bilayer plasma membrane that encloses a number of membrane-bound organelles including the nucleus, mitochondria, Golgi apparatus, endoplasmic reticulum, and endosomal vesicles (Figure 18.2). They have a single flagellum that emerges from an invagination in the plasma membrane called the flagellar pocket. African trypanosomes also have a number of special organelles. The most notable of these is a mass of mitochondrial DNA called the kinetoplast from which the Class Kinetoplastea gains its name and, as a consequence, the group as a whole is frequently referred to as kinetoplastid flagellates. The kinetoplast is a network of DNA arranged as a flask-shaped body. It is located within the organism's single mitochondrion at the posterior end of the parasite, close to its flagellar pocket. The kinetoplast is composed of a few identical maxicircles each of 20–40 kb intertwined with thousands of minicircles of 0.5–3 kb. The interlinked nature of maxicircle and

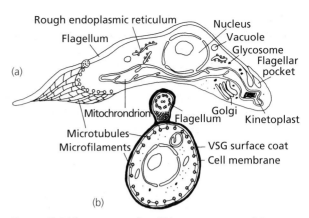

Figure 18.2 *Ultrastructure of an African trypanosome:* **(a)** *longitudinal section of an intermediate-to-short stumpy blood trypanosome;* **(b)** *cross-sectional view (Redrawn from Seed and Hall 1992, with permission of Academic Press)*

minicircle DNA in the kinetoplast presents unique problems with respect to replication of this body and has been the subject of extensive investigation (reviewed in Morris et al. 2001). The kinetoplast maxicircles house mitochondrial genes, some of which require extensive editing at the mRNA level (discussed later) to yield functional proteins.

Glycosome

African trypanosomes and related organisms have glycosomes that are membrane-bound assemblies of enzymes that support glycolysis, purine salvage, ether-lipid synthesis, and beta-oxidation of fatty acids. Sequence similarities between genes involved in glycosome and peroxisome biogenesis suggest that these organelles share a common evolutionary origin (Parsons et al. 2001). Enzyme function in the glycosome is considered later in the text.

Variable surface glycoprotein coat and the VSG–glycosylphosphatidylinositiol-specific phospholipase C

African trypanosomes inhabit the blood plasma and, in the case of *T. brucei*, also the interstitial fluids, the lymph, and cerebrospinal fluid. Their extracellular lifestyle presents two unique challenges. First, the parasites must avoid lysis by serum complement factors, a group of enzymes that can assemble into a donut-shaped protein complex on permissive lipid bilayer membranes, leading to loss of the membrane permeability barrier. Secondly, the parasites must evade immune elimination to establish chronic infections. Protection against complement activation and immune elimination is mediated by the variable surface glycoprotein (VSG) coat. In electron micrographs, the coat appears as an amorphous layer 12–15 nm thick on the outer lipid leaflet of the plasma membrane. On each trypanosome, it comprises about 10^7 VSG molecules encoded by a single VSG gene. The VSG molecules assemble as non-covalent homodimers that stand proud of the membrane (Figure 18.3). Each VSG monomer is attached to the outer lipid leaflet of the parasites plasma membrane by a glycosylphosphatidylinositiol (GPI) group containing C14 fatty acid tails (Mehlert et al. 1998; Ferguson 1999). The GPI is covalently bound to the VSG C-terminal

carboxyl group and the GPI-anchored VSG is free to diffuse in the lateral plain of the membrane. Immune evasion is achieved by antigenic variation of the VSG coat, a process that entails switching among any of a thousand VSG genes and their recombinants of which only one is expressed at a time. Analysis of crystallized VSGs that shared only 16 percent amino acid sequence identity showed that they have highly conserved secondary and tertiary structure (Blum et al. 1993).

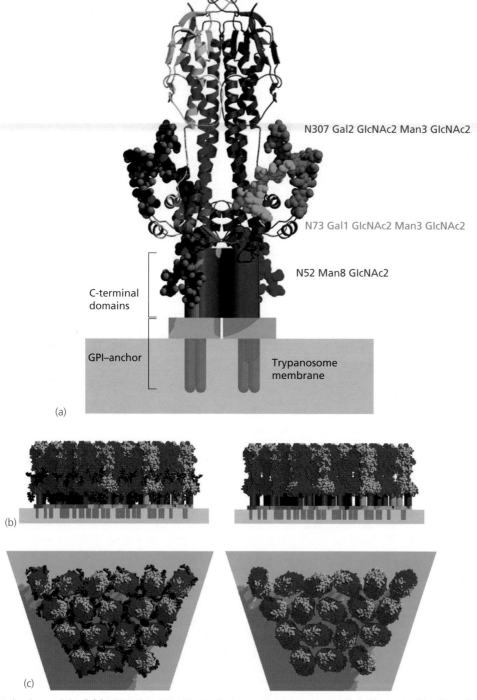

Figure 18.3 *Molecular models of:* **(a)** *VSG MITat 1.5 with attached, energy minimized, N-linked oligosaccharides;* **(b and c)** *the cell surface coat, viewed from the side and above without (right) and with (left) attached N-linked oligosaccharides (Redrawn from an image by Dr Charles Bond, from Mehlert et al. 2002)*

Consequently, antigenically distinct VSGs assemble into very similar surface coats.

Trypanosomes that are subjected to hypotonic and other stresses release their VSG coat. The released coat lacks its GPI anchor and has an exposed conserved carbohydrate side chain called the cross-reacting determinant (CRD). Release of the VSG is mediated by a VSG–GPI-specific phospholipase (PLC), which is reported to be localized on the cytoplasmic face of vesicles distinct from the endoplasmic reticulum and where it does not normally encounter VSG–GPI. The VSG–GPI-PLC encounters VSG–GPI only under appropriate stress conditions, a process that is thought to involve flipping of VSG from the outer to the inner leaflet of the plasma membrane (Cardoso de Almeida et al. 1999). Conservation of the VSG–GPI-PLC among trypanosomes suggests that it plays an important role in trypanosome biology and/or the parasite–host relationship. However, *T. b. brucei* organisms that are genetically engineered to lack the VSG–GPI-PLC infect and grow in laboratory rodents and tsetse flies. Further studies showed that the organisms do not have a duplicate PLC, suggesting that the VSG–GPI-PLC is not essential for normal *T. b. brucei* function and differentiation events (reviewed in Carrington et al. 1998). However, mice that were infected with the mutant trypanosomes lived for twice as long as those infected with control trypanosomes indicating a role for the enzyme, or for soluble VSG with exposed CRD, in the mammal–trypanosome relationship. The laboratory mouse is not a natural host for *T. brucei*, hence, additional insights into the roles of the VSG–GPI-PLC and released VSG might be gained by studying the behavior of mutant parasites lacking the enzyme in trypanosomiasis-resistant African wildlife species, extant representatives of the hosts in which trypanosomes were selected through evolution.

Additional components of the trypanosome surface

The trypanosome plasma membrane is equipped with transporters for the facilitated diffusion of glucose, pyruvate, and purines. It also has a number of invariant surface glycoproteins (ISG) of unknown function. These are present at about 1 percent of the density of VSG. In addition, the parasite has receptors for host-derived holo- (i.e. iron-loaded) transferrin and serum, and low- and high-density lipoproteins (reviewed by Borst and Fairlamb 1998). The transferrin receptor is encoded by VSG gene expression site associated genes (ESAG) 6 and 7 for which there are several alleles (Steverding 2000). As with VSGs, only one form of the transferrin receptor is expressed at a time by trypanosomes. Although isolation of a receptor for serum low-density lipoprotein has been reported, this was most likely an erroneous assignment (reviewed in Black et al. 2001)

and at this time the trypanosome receptors for serum low and high density are not known.

Flagellum and flagellar pocket

African trypanosomes are highly motile. They have a single flagellum, which emerges from an invagination at the posterior end of the cell called the flagellar pocket (Balber 1990) (Figure 18.4). The structural elements that cause the invagination to form have not been defined. The flagellum comprises an axoneme, a paraflagellar rod, and an attachment zone filament that mediates adhesion to the trypanosome plasma membrane. The flagellum lies in a shallow groove on the cell surface. When the flagellum beats, the attached portion of the cell surface is thrown into folds and gives rise to the undulating membrane visible in the light microscope. The body of the trypanosome has an underlying corset of microtubules that prevent any vesicular exchange with the plasma membrane. This microtubular array is open in the region of the flagellar pocket and at the posterior end of the cell. Although vesicles could transit to or from the cell membrane at either of these zones, endocytosis and exocytosis are restricted to the flagellar pocket region. Bloodstream stage *T. brucei* have much more endocytic activity than other differentiative stages. This is associated with massive upregulation of a clathrin-like heavy chain that is involved in the formation of endocytic vesicles (Morgan et al. 2001).

Trypanosome receptors for host transferrin and serum lipoproteins are thought to be restricted to the flagellar pocket, and to bind ligand that diffuses into the pocket. Indeed, the transferrin receptor has been shown to be restricted to the pocket, at least in organisms that were incubated in the presence of high-affinity transferrin (i.e. donor transferrin) that could bind with high affinity to the receptor. When these same organisms were incubated for 48 hours with low-affinity transferrin, which elevates receptor gene expression, the transferrin receptor was detected over the surface of the parasite. A similar result was obtained using organisms that were engineered to overexpress transferrin receptors and that were incubated in the presence of high-affinity transferrin (Mussmann et al. 2003). These observations were taken to show that the transferrin receptor is held in the flagellar pocket by a saturable mechanism. The transferrin receptor has also been shown to be present in the pocket matrix associated with a gel-like substance. The significance of this dual location is not known. Trypanosome endocytic pathways, endocytic compartments, and sorting signals are the subject of on-going study (Pal et al. 2002). Figure 18.5 presents a comparison of secretory and endocytic pathways in *T. brucei* and other trypanosomatids. In the case of *T. brucei*, the flagellar pocket and endosomal glycoproteins have been shown to have an attached linear poly-*N*-acetyllactosamine

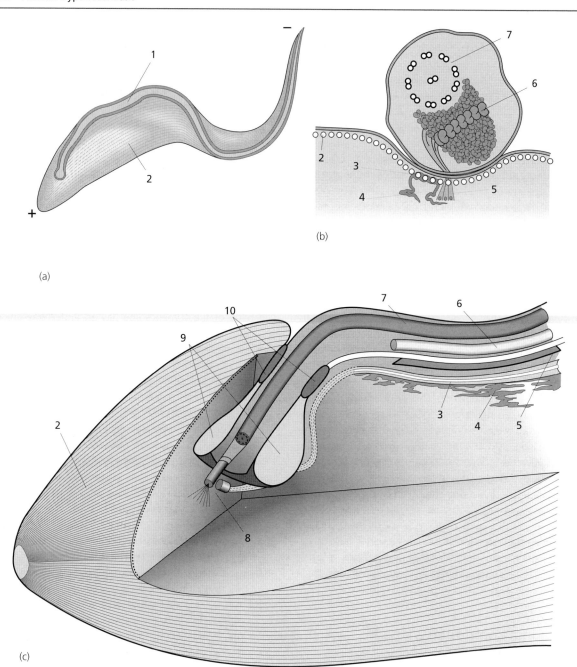

Figure 18.4 *Schematic illustrations of a trypanosome cell emphasizing the structural components of the cytoskeletion:* **(a)** *detergent-extracted cytoskeleton;* **(b)** *idealized transverse section taken through the midpoint of the cell looking from posterior to anterior;* **(c)** *cut-out view of the flagellar pocket area. Flagellum (1), sub-pellicular microtubules (2), special quartet of reticulum-associated microtubules (3), microtubule-associated reticulum (4), flagellum attachment zone filament (5), paraflagellar rod (6), axoneme (7), basal body complex (8), flagellar pocket (9), adhesion zones (10) (From Hill et al. 2000)*

moiety, a unique carbohydrate that is developmentally regulated and may be required for endocytosis (Nolan et al. 1999).

Pleomorphism

Bloodstream stage *T. brucei* undergo morphological and biochemical differentiation to non-replicating stumpy forms on reaching peak parasitemia. Stumpy forms have

a very truncated flagellum and are greater in diameter than the replicating bloodstream forms. This transition plays a significant role in setting the level of parasitemic waves in infected mammals (Black et al. 1985) and is thought to be directed by endogenous quorum-sensing factors and host-derived growth-inhibitors (Murphy and Olijhoek 2001). The slender to stumpy transition occurs via a series of intermediate forms and *T. brucei* strains that undergo this differentiation are referred to as pleomorphic. Strains that are maintained by passage in

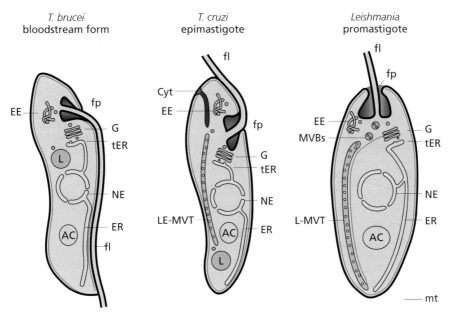

Figure 18.5 *Schematic representations of the secretory and endocytic organelles of* T. brucei *bloodstream form,* T. cruzi *epimastiogote and* Leishmania *promastigote. Endoplasmic reticulum (ER), transitional endoplasmic reticulum (tER), multivesicular bodies (MVBs), nuclear envelope (NE), Golgi apparatus (G), flagellar pocket (fp), early endosomes (EE), multi vesicular tubule (MVT), lysosome (L), late endosome (LE), cytostome (a second invagination that is specialized for endocytosis, Cyt) flagellum (fl), microtubules (mt) (From McConville et al. 2002)*

laboratory rodents or in tissue culture in the absence of high molecular weight agarose supplements often lose the property of pleomorphism at least with respect to infections that they give rise to in laboratory rodents, i.e. they remain as replicating slender forms in these hosts. These so-called monomorphic *T. brucei* are typically very virulent and poorly or non-infective for tsetse flies.

Slender form *T. brucei* derive energy solely from glycolysis. In contrast, stumpy form *T. brucei* express α-ketoglutarate dehydrogenase, a Kreb's cycle enzyme, and obtain additional ATP by converting α-ketoglutarate, if supplied to them, to succinate. The partial acquisition of mitochondrial function by stumpy form *T. brucei* is considered to be in preparation for establishing an infection in the tsetse vector.

Life cycle

When an uninfected tsetse bites an infected vertebrate, trypanosomes acquired with ingested blood pass through the esophagus, the crop, and the proventriculus until they reach the midgut. In the fly's midgut, the short stumpy forms complete their morphological and biochemical differentiation into procyclic stages. They complete the development of their mitochondrion, cease synthesizing VSG and express a new coat glycoprotein called procyclin (Mehlert et al. 1998), which facilitates survival and growth in the tsetse fly by mechanisms that are not understood (Ruepp et al. 1997). The procyclic form has a fully functional mitochondrion and consequently differs from the slender bloodstream forms in

metabolic activity. The procyclic forms are also elongated with respect to the bloodstream-stage organisms and differ from these in the relative position of the kinetoplast and nucleus.

The trypanosomes have a complex life cycle and a low infection rate in their vectors. It has recently been shown that trypanosomes evoke relatively poor protective immune responses in tsetse flies and other factors are being sought with respect to identifying those that influence infectivity (Aksoy et al. 2002). It is known that age, sex, and genotype of the vector, the type of bloodmeal taken by the vector, and many environmental factors can all influence the susceptibility of the tsetse to the trypanosomes (Maudlin and Ellis 1985; Molyneux and Ashford 1983; Otieno et al. 1983; Vickerman et al. 1988).

After a period of growth in the midgut, the trypanosomes undergo a further series of morphological changes, migrate from the midgut into the endoperitrophic space and finally into the tsetse's salivary glands where they develop into epimastigotes. The epimastigote form is not infective for the mammalian host. The epimastigotes attach to the cells of the salivary gland. They divide repeatedly and then transform into non-dividing metacyclic forms, which are small, highly motile, short, and stumpy (Figure 18.6). They have a terminally located kinetoplast but no free flagellum. When mature, the metacyclic forms detach from the salivary gland cells, synthesize a surface coat of the type found on the bloodstream forms, and become infective to the vertebrate host (Vickerman et al. 1988; Vickerman 1989). The signals for these morphological and

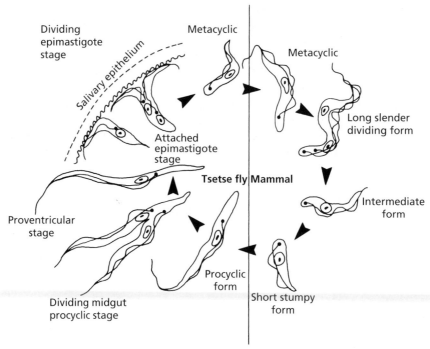

Figure 18.6 *Life cycle of trypanosomes of the* Trypanosoma brucei *group (Redrawn from Seed and Hall 1992, with permission of Academic Press)*

physiological changes in the trypanosome in its vector are not known. The time required for the trypanosomes ingested by the fly to complete their development in the tsetse and to regain infectivity is approximately 3–4 weeks. However, the length of time will vary depending on the external environmental conditions of humidity and temperature and on the age, sex, and other factors in the tsetse (Molyneux and Ashford 1983). The infective metacyclic forms of the trypanosome are injected into the mammalian host by the biting fly. They transform in the vertebrate host into the long slender trypanomastigote form and increase in number reinitiating the cycle.

BIOCHEMISTRY

In vitro cultivation

The in vitro cultivation of any organism is a prerequisite for any biochemical study and trypanosomes are no exception. Cultivation of procyclic-like *T. brucei* organisms in blood agar medium was achieved early in the 20th century. By the 1970s, semi-defined and defined procyclic culture media had been developed, and by the 1980s, methods were developed for the efficient transformation of bloodstream-stage *T. brucei* into procyclic forms in vitro. Typically, incubations are at 27°C in serum-supplemented medium further supplemented with citrate and *cis*-aconitate (Brun and Jenni 1985; Kaminsky et al. 1988). The signal pathways that are regulated by the drop in temperature and presence of

citrate and *cis*-aconitate are not resolved. Procyclic form *T. brucei* can undergo further differentiation in vitro to give rise to mammal-infective metacyclic form trypanosomes (Hirumi et al. 1992). The signaling systems that regulate this differentiation are also not known. Propagation of bloodstream-form *T. brucei* in vitro was achieved in 1977 following a heroic checkerboard comparison involving several media, sera, and fibroblast feeder-layer cells with respect to their *T. b. brucei* growth-supporting capacity (Hirumi et al. 1977). Axenic culture conditions were subsequently established by supplementing culture medium with cysteine or a reducing agent, β2-mercaptoethanol. It is now known that cysteine is an essential growth factor for the parasites (Duszenko et al. 1992). Cysteine is readily oxidized to cystine in the presence of copper, and survival and growth of bloodstream-stage *T. brucei* in vitro is enhanced by inclusion of the copper chelating agent bathocuproine in culture medium. The ability to grow trypanosomes in axenic cultures has greatly facilitated analysis of cell function and biochemistry both by providing controlled experimental conditions, and by expediting the development of mutant organisms through the use of genetic engineering techniques. Gene disruption and addition mutants have been developed, as have organisms expressing inhibitory RNA. These mutant cell lines are a valuable resource in elucidating pathways that regulate endocytosis and exocytosis, VSG switching, and other critical processes.

Not all strains of *T. brucei* grow as bloodstream-form organisms in tissue culture medium, and among those that do there is a high tendency to lose the pleomorphic phenotype. However, it has recently been found that the inclusion of high molecular weight agarose in the medium facilitates both growth of the parasites in vitro and retention of pleomorphism (Vassella and Boshart 1996). The mechanism underlying the effect is unknown. A serum-free culture medium has also been described for *T. b. brucei* (Hirumi et al. 1997). This medium is supplemented with fatty acid-free albumin and serum lipoproteins, components that are required for progression of the parasites through G_1 of their cell division cycle (Morgan et al. 1996) as well as bovine alpha 2 macroglobulin, which has an undefined function with respect to *T. brucei*. Serum lipoproteins are complexes of apolipoproteins, phospholipids of several types, cholesteryl esters, triglycerides, and cholesterol. It is not known which, or how many, of these components are required to drive progression of the parasites through G_1 of their cell division cycle.

Energy metabolism

The blood stage trypanosomes have a high rate of glucose catabolism. As mentioned above, the glycolytic enzymes carrying out this process are present in a unique microbody-like organelle called the glycosome. An outline of glucose metabolism is shown in Figure 18.7. Under aerobic conditions, glycerol-3-phosphate is produced from dihydroxy-acetone-phosphate during which process NADH is oxidized to NAD. The glycerol-3-phosphate moves from the glycosome to the parasite's mitochondrion where it is converted back to dihydroxy-acetone-phosphate by glycerol-3-phosphate oxidase. The dihydroxy-acetone phosphate re-enters the glycosome and is further metabolized. Under aerobic

conditions, glucose is exclusively converted to pyruvate, which is exported by a pyruvate transporter or transaminated and exported as alanine or utilized in protein synthesis. Under anaerobic conditions, glycerol-3-phosphate cannot be converted back to dihydroxy-acetone-phosphate, hence glucose metabolism results in equal portions of pyruvate and glycerol being produced and exported. In bloodstream *T. brucei* the nine enzymes involved in the conversion of glucose into phosphoglycerate or glycerol account for more than 90 percent of the protein in the glycosome (Hannaert and Michels 1994; Opperdoes 1987; Opperdoes et al. 1990).

The compartmentalization of the trypanosome's glycolytic enzymes in the glycosomes is a pattern different from that occuring in other eukaryotic organisms where the enzymes involved in glycolysis are present in the cytoplasm. It has been suggested that the close physical association of the glycolytic enzymes in the glycosome and their high concentration there account for the very high rate of glucose catabolism by the blood-stage parasites (Hannaert and Michels 1994; Sommer and Wang 1994). However, model data obtained using a kinetic computer replica of glycolysis in *T. brucei* did not support this idea. Rather, the model suggests that compartmentalization may serve to prevent glycolytic intermediates accumulating to dangerous levels in the trypanosome cytosol, and to facilitate recovery from glucose deprivation (Bakker et al. 2000).

The procyclic forms are adapted to the environment of the vector, which is entirely different from the environment that the blood-forms inhabit. In the procyclic forms, the mitochondrion becomes functional; it has cristae, and contains the tricarboxylic acid cycle enzymes, cytochromes, and cytochrome-dependent respiratory chain enzymes, all of which are absent from the slender bloodstream-forms. The procyclic form has a much lower rate of glucose catabolism than does the slender bloodstream-form. Phosphoenolpyruvate is carboxylated by procyclics to form oxaloacetic acid with subsequent conversion to malate, fumarate, and succinate. The major end products of glucose utilization by the procyclic forms are succinate, acetate, carbon dioxide, and alanine. ATP is generated primarily by oxidative phosphorylation. Amino acids, particularly proline, are the primary substrates used by the procyclics to obtain energy.

Both bloodstream-stage and procyclic *T. brucei* lack key enzymes necessary for classical glyconeogenesis, and both lack the enzymes of the Entner–Douderoff pathway (Cronin et al. 1989).

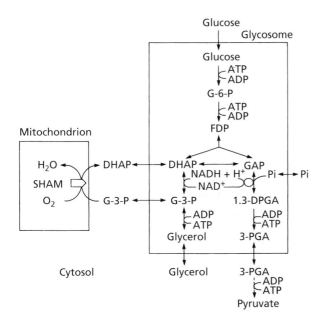

Figure 18.7 *Glucose metabolism in the bloodstream form of African trypanosomes (From Opperdoes 1987)*

Redox-regulation

Trypanosomes, like other eukaryotic cells, use NADPH as their major source of reducing power. This is produced in the pentose–phosphate pathway. All the enzymes of this pathway have been found in *T. b. brucei* (Cronin et al. 1989). Glucose-6-phosphate dehydrogenase, the sentinel that sits at the gateway of the pentose–phosphate pathway has been demonstrated to localize to the glycosome, and the gene encoding the downstream enzyme 6-phospho-gluconolactonase contains a C-terminal peroxisome targeting signal consistent with a likely localization in the glycosome (Duffieux et al. 2000). The trypanosomes possess the enzymes required to produce d-ribose-5-phosphate for nucleic acid synthesis and NADPH for other synthetic reactions. However, there are differences between the procyclic forms and blood-forms in the enzymes of the non-oxidative segment of the pentose–phosphate pathway. Bloodstream-stage *T. brucei* suppress expression of genes encoding ribulose-5-phosphate 3'-epimerase and transketolase (Cronin et al. 1989) and thus terminate their NADPH-yielding pentose–phosphate pathway at ribulose-5-phosphate instead of recycling this product back into the glycolytic pathway. Consequently, the glucose metabolic pathways in bloodstream-stage *T. brucei* appear to be configured to yield ATP or NADPH but not both at the same time. The evolutionary advantage accrued by the trypanosomes through adopting this unusual arrangement is not known.

Trypanosomes spontaneously generate hydrogen peroxide as a result of energy metabolism. They are also exposed to hydrogen peroxide and superoxide anion that are produced by their mammal hosts. The parasites have an Fe-superoxide dismutase that catalizes the reduction of superoxide anion to hydrogen peroxide. However, they lack catalase and rely on coupled redox reactions fueled by the reducing power of NADPH to inactivate hydrogen peroxide (Fairlamb and Cerami 1992; Flohe et al. 1999). The utilization of NADPH deflects glucose-6-phosphate from the glycolytic to the pentose–phosphate pathway by unleashing glucose-6-phosphate dehydrogenase. This may account for the almost instantaneous depletion of the ATP content of bloodstream-stage trypanosomes that occurs when the parasites are placed under oxidative stress (Muranjan et al. 1997). Hydrogen peroxide metabolism by *T. brucei* and other kinetoplastid flagellates involves 'a cascade of three enzymes, trypanothione reductase, tryparedoxin, and tryparedoxin peroxidase... The terminal peroxidase of the system belongs to the protein family of peroxiredoxins, which is also represented in Entamoeba and a variety of metazoan parasites' (Flohe et al. 1999).

Amino acid synthesis

There has been relatively little study of amino acid synthesis and utilization by trypanosomes, but it appears that the organisms have limited ability to synthesize amino acids de novo. A majority of the amino acids

required by trypanosomes must either be formed by a salvage pathway such as that bringing about the transamination of pyruvate, oxaloacetic, and α-ketoglutarate to alanine, aspartic, and glutamic acids, respectively, or be directly supplied by the host (Gutteridge and Coombs 1977).

Some amino acids are used for purposes other than protein synthesis. In the procyclic forms, proline is used as a respiratory substrate, and threonine may serve as the preferred substrate for formation of acetyl coenzyme A, which is required for the elongation of fatty acid chains. The only enzymes involved in amino acid metabolism that have received extensive study are those bringing about aromatic amino acid catabolism and those for the use of ornithine in the synthesis of polyamines. The polyamines are known to be important in the proliferation and differentiation of the blood-stage trypanosomes. Inhibition of putrescine and spermidine synthesis blocks trypanosome division and causes changes in cellular morphology (Pegg and McCann 1988).

Purine synthesis

Neither the blood-form nor the procyclic form of the trypanosomes synthesize purines de novo; they require salvage mechanisms to obtain them (Hammond and Gutteridge 1984). Most of the enzymes of the salvage pathway are located in the cytoplasm, and hypoxanthine–guanine phosphoribosyl transferase is found in the glycosome. The trypanosomes can synthesize pyrimidine by a pathway which appears similar to that used by mammalian cells, although there are two differences: (i) in mammals, the enzyme dihydroorotate dehydrogenase is present in the mitochondria but in trypanosomes it is present in the cytosol; (ii) in the trypanosomes, the last two enzymes of the pathway (orotate phosphoribosyl transferase and orotidine 5-phosphate decarboxylase) are in the glycosome rather than in the cytosol as in mammals (Fairlamb 1989).

Lipids

Bloodstream-stage *T. b. brucei* have a limited ability to synthesize lipids and generally depend on the host for fatty acids, phospholipids, cholesterol, and other lipids (reviewed in Mellors and Samad 1989). They obtain phospholipids, lysolipids, cholesterol, and cholesteryl esters from serum low- and high-density lipoproteins (Vandeweerd and Black 1989) and remodel the phospholipids by removing and replacing fatty acids in the sn1 and sn2 positions (Mellors and Samad 1989). The bloodstream-stage parasites also synthesize myristate (Morita et al. 2000), which they preferentially use to remodel the glycosyl-phosphatidylinositol anchor that holds their glycoprotein coat in the outer leaflet of the plasma membrane.

The trypanosome plasma membrane contains domains with detergent extraction properties shared with the lipid rafts of higher eukaryotic cells. Most of the VSG may be associated with these putative rafts (Denny et al. 2001) and there is evidence that the putative rafts are enriched in the flagellar membrane of the parasites (Toriello et al. 2002).

MOLECULAR BIOLOGY AND GENETICS

The general organization of trypanosome nuclear DNA has been succinctly described by Donelson (2001).

> *Trypanosoma brucei* is a diploid organism with a nuclear haploid DNA content of 35 ± 9 megabase (Mb) pairs depending on the trypanosome isolate. About 15 percent of total cellular DNA is in the kinetoplast... The remaining 85 percent occurs in the nucleus as linear DNA molecules ranging in size from 50 kb to 6 Mb. At least 11 pairs of megabase chromosomes of 1–6 Mb exist that are numbered I–XI from smallest to largest. The two homologues of a megabase chromosome pair can differ in size by as much as 4-fold. Several intermediate-sized chromosomes of 0.2–0.9 Mb and uncertain ploidy are also present... [and there are about 100 minichromosomes]. About 50 percent of the nuclear genome is coding sequence. To date, only one tRNA gene and one protein-encoding gene have been found to contain an intron. [Significant portions of the genome have now been sequenced]... Based on analogy with the *Leishmania* genome, much of the African trypanosome nuclear genome is likely to be arrayed as long transcription units of 50 or more intronless genes.

Chromosomal DNA of all eukaryotes is folded with nucleosomes that are composed of core histones H2A, H2b, H3, and H4. These are highly conserved from yeast to humans but are much more diverse in trypanosomatids. However, there is a degree of sequence conservation suggesting that the overall structure of the *T. brucei* nucleosome is probably similar to that of other eukaryotes (reviewed in Horn 2001). For example, lysine residues are conserved in the N-terminal tails of the core histones of *T. brucei* and higher eukaryotes and may be targets for covalent modifications that affect chromatin structure. Acetylation, phosphorylation, ubiquitination, and methylation of the N-terminal tails of core histones are known to regulate chromatin structure in higher eukaryotes. Nevertheless, chromatin remodeling in *T. brucei* does not fully mimic that in higher eukaryotes. For example, phosphorylation of serine 10 in histone H3 is involved in chromosome condensation in many eukar-

yotes. However, this amino acid residue is absent from the N-terminal tail of *T. brucei* histone, which may account for the absence of chromosome condensation during trypanosome mitosis.

In higher eukaryotes, nucleosomes form bundles as a result of the binding of linker histone H1 to linker DNA flanking the nucleosome core. *T. brucei* organisms also have linker histone H1 but this has a simple repetitive sequence and is not precisely analogous to that of higher eukaryotes (reviewed in Horn 2001). The level of histone H1 is higher in the bloodstream-forms than in procyclic *T. brucei*, a larger number of H1 variants is expressed and the H1−chromatin association is stronger in the bloodstream-forms (Schlimme et al. 1993) possibly associated with a greater need to stabilize chromatin at the higher temperature encountered in the mammal host.

Sequence information in the *T. brucei* genome database indicates that the parasites also have deacetylases, acetyltransferases, and bromodomain proteins (reviewed in Horn 2001). These proteins affect chromatin by removing groups that covalently modify histones and DNA, and there is evidence that they act as transcription regulators in higher eukaryotes. Perhaps they perform a similar function in trypanosomes. There is evidence that differences in chromatin accessibility may account for developmentally regulated gene expression in *T. brucei* (Navarro et al. 1999).

There is considerable evidence that exchange of genetic information can occur in the African trypanosomes, and that this exchange occurs in the vector's salivary glands, but not in the tsetse gut form or in the mammalian host. The evidence suggests that genetic exchange involves a meiotic division. However, a haploid trypanosome stage has not been identified (reviewed in Gibson 2001b). Genetic exchange has been demonstrated between the various *T. brucei* subspecies, but in the case of *T. b. gambiense* was observed with the relatively virulent so-called group 2 parasites that transmit through the tsetse fly. The less virulent group 1 *T. b. gambiense* do not transmit readily through the fly. The relevance of genetic exchange to the survival and function of *Trypanozoon* organisms is far from clear. In this regard, group 1 *T. gambiense* and *T. evansi*, neither of which are tsetse transmitted, survive quite well without it. Nevertheless, it is apparent from the analysis of isoenzyme patterns, as well as RFLP among different *T. brucei* stocks, that sexual recombination does occur among trypanosomes in the field.

There is debate as to the ploidy of the trypanosomes. The results of isoenzyme and RFLP analysis of genetic crosses between trypanosome clones suggest that the trypanosomes are generally diploid. However, there is evidence to suggest that some hybrids may be triploid and that changes in chromosome size may occur during genetic exchange (Kooy et al. 1989; Gibson et al. 1992). Indeed, 'hybrids with a DNA content of either $2n$ or $3n$

(but not intermediate amounts) have been found in five of six crosses for which DNA contents have been measured, suggesting that trypanosome DNA is inherited in packets of *n'* (reviewed in Gibson 2001b).

Cell division

In the African trypanosomes, the nuclear membrane does not break down during cell division and chromosomes do not condense. However, as would be expected, chromosomal reorganization does occur as determined by the location of chromosome ends, i.e. telomeres. During interphase, telomeres are located around the periphery of the nucleus in nondividing stumpy form *T. brucei*, and the insect procyclic form, but in the center of the nucleus in dividing bloodstream-forms. The distinct orientation of telomeres in bloodstream-stage trypanosomes may be required for expression of VSG genes as the expression sites are telomeric. At the onset of mitosis, in all trypanosome life cycle stages the telomeres congregate in a central zone of the nucleus and split into two clusters that migrate to opposite nuclear poles. The double-ended nucleus then divides and the telomeres reorganize to the foci present during interphase. During migration, the telomeres are located at the free end of the mitotic spindle, which is contained within the nuclear membrane. The mechanism of chromosome segregation is discussed by Ogbadoyi et al. (2000) and Perez-Morga et al. (2001) and readers are referred to these papers for further information.

RNA editing

RNA editing (i.e. 'the alteration of RNA sequences by base modifications, substitutions, insertions, and deletions') was first described in trypanosomes and has since been identified in other eukaryotes including humans (reviewed in Horton and Landweber 2002). RNA editing in trypanosomes is mediated by guide RNAs that are encoded in the kinetoplast minicircles. These direct the insertion and deletion of U, thus converting mRNA transcribed from so-called 'cryptogenes' into functional gene products. RNA editing can be a cumbersome process sometimes involving multiple guide RNAs, sequentially applied, to edit a single mRNA, and presumably had selective value to be retained. Possible advantageous properties of the process include the stable maintenance of otherwise deleterious mutations in DNA genomes, improved storage of information in crowded genomes, and novel posttranscriptional modes of genetic regulation.

Transplicing

All trypanosome mRNAs are capped with a 39-nucleotide structure that is added at the 5′ end of pre-mRNA by trans-splicing. The totally conserved 39-nucleotide structure, called the spliced leader or mini-exon, is trimmed from an approximately 100-nucleotide nontranslated RNA that is transcribed from a set of nonidentical genes in the trypanosome genome. The regulation of this process is not known. The spliced leader is subsequently attached to each protein-coding sequence present in long polycistronic precursors. This process is an essential step in the biosynthesis of trypanosome mRNAs.

VSG antigenic variation

Trypanosomes contain within their genome over 1 000 different VSG genes. In addition, new variant antigen types (VAT) are created by VSG gene mutations and recombination. With a few special exceptions, each trypanosome expresses only a single VSG on its surface at any given time. This is encoded by a single VSG gene that is expressed at a telomeric expression site. Unexpressed VSG genes are found throughout the trypanosome genome including on minichromosomes and in unexpressed VSG-gene expression sites. There are 20 expression sites for bloodstream-stage trypanosome VSG genes present in the genome, of which only one is active at a time. The promoter for the expressed VSG gene is upstream of and separated from the expressed VSG gene by at least eight ESAGs. Analyses carried out using trypanosomes expressing the AnTat 1.3 VSG show that the ESAGs and expressed VSG are transcribed as one polycistronic unit (Vanhamme et al. 2001). Although only one of the possible 20 bloodstream-stage trypanosome VSG-gene expression sites is fully expressed at a time, transcripts of ESAG 6 from unexpressed expression sites has been recorded (Ansorge et al. 1999). Switching of the expressed VSG gene requires exchange of VSG genes within an active expression site, or silencing of an active expression site and activation of another. Mechanisms of VSG gene switching and expression site silencing are reviewed by Cross et al. (1998) and Borst and Ulbert (2001) and readers are referred to these reviews for further information.

Metacyclic *T. brucei*, i.e. the infective form transmitted from tsetse to the mammal host, are also coated with VSG (Barry et al. 1998). The genes that encode metacyclic VSGs are distinct from those of bloodstream-form organisms and only 27 have been identified. Each metacyclic VSG gene is located at the end of a large chromosome and expression does not require DNA rearrangements. The diversity of metacyclic VSG genes may have evolved to permit multiple reinfection of hosts. Although there are many fewer metacyclic VSG genes than bloodstream-form VSGs, it is unlikely that a metacyclic VSG vaccine will be effective. The expressed metacyclic VSG repertoire has been observed to change over time (Barry et al. 1983).

Transferrin receptor variants

Transferrin is a required growth factor for trypanosomes and is acquired by receptor-mediated endocytosis using receptors that are distinct from mammalian counterparts and are encoded by genes ESAG 6 and 7, which are upstream of the VSG-gene expression site. ESAG 6 and 7 gene products are structurally similar to the N-terminus of VSGs suggesting that they arose from a VSG gene. ESAG 7 encodes a polypeptide of 340 amino acids while ESAG 6 encodes one of 401 amino acids. Only the latter has a carboxy-terminal GPI-addition signal. Hence, the assembled heterodimeric receptor is held in the membrane by only a single GPI. This dimer binds transferrin. It is about 75 percent of the size of a VSG dimer and may be buried in the VSG layer in such a way that its ligand-binding site is accessible to transferrin but the rest of the receptor is not accessible to antibody. The affinity of binding depends on the donor origin of the transferrin, and on the bloodstream-form trypanosome VSG expression site used. Different transferrin receptor alleles associated with different VSG-gene expression sites encode similar but not identical receptors resulting in variation in affinity for transferrins from different mammal host species. This flexibility may allow trypanosomes to escape binding-site-specific antibodies and to meet the needs of the parasites for transferrin acquisition in a particular host. In this regard, antibody specific for the transferrin receptor selectively affected the in vitro growth of parasites expressing receptors with low affinity for the transferrin type present in the culture medium (Gerrits et al. 2002). While discouraging from the point of view of vaccinology, this result suggests that the receptor is directly accessible to antibody, supporting the view that not all molecules that are critical for trypanosome function are obscured by VSG.

CLINICAL ASPECTS

The cell and molecular biology of African trypanosomes is so fascinating that it is sometimes possible to forget that the organisms are devastating pathogens. However, this is not a luxury afforded to the victims of sleeping sickness. *T. b. rhodesiense* and *T. b. gambiense* both cause fatal disease. However, there are differences in the clinical manifestations of the diseases they cause and in their epidemiology (Table 18.1). The disease produced by *T. b. gambiense* (also called West African trypanosomiasis) is chronic in nature lasting up to 4 years, whereas the disease produced by *T. b. rhodesiense* is more acute, rarely lasting more than 9 months before death occurs. The molecular basis for this difference in virulence is not known.

As noted earlier, one of the key aspects of African trypanosomiasis is the phenomenon of antigenic variation that results in waves of parasitemia occurring in the blood of infected people and animals (Figure 18.8). The clinical aspects of trypanosomiasis and the chronic nature of the disease can be attributed to these waves of parasitemia.

Phases of the disease

African trypanosomiasis in humans can be separated into three phases (Table 18.2).

PHASE 1

The first phase occurs just after the inoculation of the metacyclic trypomastigotes by biting tsetse flies. A chancre develops at the site of the bite, and the trypanosomes remain at this site for a short time. There is inflammation of the subdermal tissue at the site, with edema, erythema, tenderness, and heat. The chancre is

Table 18.1 *A comparison of the biology of the subspecies of African trypanosomes,* T. b. gambiense *and* T. b. rhodesiense [a]

Characteristic	T. b. gambiense	T. b. rhodesiense
Disease	Chronic	Acute
	Low parasitemia	High parasitemia
	Incubation period is months to years	Incubation period is days to weeks
Main tsetse vector	G. palpalis group	G. morsitans group
	G. palpalis	G. morsitans
	G. fuscipes	G. pallidipas
	G. tachinoides	G. swynnertoni
Transmission	Human reservoir	Animal reservoir
	– (primary)	– (primary)
	Riverine tsetse	Savanna and woodland tsetse
	– (secondary)	– (secondary)
	Animal reservoir	Human reservoir
Reservoir hosts	Possibly kob, hartebeest, domestic pigs, dogs	Bushbuck, other antelope, hartebeest, hyena, lion, domestic cattle; possibly warthog and giraffe
Geographical range	West Africa, western, and northern Central Africa	East Africa and north Central Africa

a) Reviewed in Seed and Hall (1992)

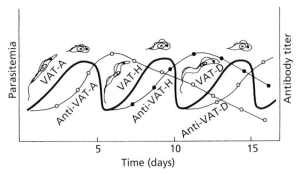

Figure 18.8 *Waves of parasitemia and their relationship to the humoral response in a mammalian host infected with the African trypanosomes. The first wave consists of a trypanosome population in which the major variant antigen type is VAT-A. This wave of parasitemia is followed by an antibody response (antiVAT-A) whose peak titer is observed shortly after the wave of parasitemia is cleared. A minor VAT (VAT-H) present on some trypanosomes from the preceding population gives rise to the next wave of parasitemia. This population is later cleared by host antiVAT-H. The rise and fall of trypanosome populations continues throughout the life of the infected host. In experimentally infected rodents or rabbits, a peak in parasitemia is observed every 5–7 days. Infected rabbits may live for as long as 6 months, during which time they may experience as many as 30 waves of parasitemia. (From Seed and Hall 1992, with permission of Academic Press)*

more readily observed in white people than in black people and is most frequently found in individuals infected with *T. b. rhodesiense*.

PHASE 2

During the second phase the trypanosomes spread throughout the entire body. They move through the blood and lymphatic vessels and multiply rapidly. Clinical symptoms during this stage appear to be associated with the waves of parasitemia. The high parasitemias are accompanied by fever, headache, joint pain, and malaise (Molyneux et al. 1984). These clinical symptoms may decrease in intensity as the disease becomes chronic.

Splenomegaly and lymphadenopathy are major features of this stage of the infection. Histopathological examination of the lesions shows vasculitis with edema and perivascular infiltration of leukocytes.

PHASE 3

Humans infected with the African trypanosomes, if untreated, eventually develop central nervous system (CNS) disorders. They may develop a wide array of behavioral changes ranging from aggressiveness to sleep-like states. The final stage involves complete somnolence (sleeping sickness). The common clinical symptoms of third-phase infection result from damage to the heart, nervous system, and other organs.

IMMUNOLOGY AND PATHOLOGY

Trypanosomiasis-susceptible hosts

Sequential microscopic and serologic analyses of blood from infected patients and experimental animals have shown that African trypanosomiasis is characterized by recurring waves of parasitemia. Each rise in parasitemia is paralleled by an increase in antibody specific for major variant antigenic types in the wave. The variant specific antibody is involved in the removal or clearance of a major portion of the trypanosome population from the blood and other body fluids, a process that ends each wave of parasitemia. Although this specific antibody clears a large majority of the trypanosomes present during a given wave of parasitemia, there pre-exists within that population a minor population of trypanosomes with a different VSG. This minor population gives rise to the next wave of parasitemia. New surface coat variants appear at frequencies of 1 in 10^6 cells in syringe-passaged populations and an estimated 1 in 100 cells in fly-passaged field isolates. Antibody therefore acts as a selecting agent removing the major VSG popu-

Table 18.2 *Some common clinical symptoms of African trypanosomiasis of humans*

Stage of infection	Symptoms [a]
Fly bite, chancre	Localized tenderness and erythema at site of fly bite
Early parasitemic stage	Headache, joint pain, fever, lymphadenopathy, weight erythema, itching, anemia, dizziness
Later stages – symptoms associated with particular organ systems	
Edema	Peripheral ascites, lung edema, pericardial effusion
Nervous system	Waking EEG changes and sleep–wake cycles[b,c], insomnia/somnolence, mental disorders, slurred speech, paralysis, brisk reflexes, epileptiform fits
Cardiac system	ECG changes, congestive heart failure
Endocrine system	Amenorrhea and impotence with hypogonadism[d], altered thyroid function[e], adrenocortical function[f], and changes in the circadian rhythms of plasma cortisol and prolactin[g]
Other	Puffy facial appearance, diarrhea, anorexia, splenomegaly

a) For further details, see Seed and Hall (1992), b) Buguet et al. (1993), c) Hamon et al. (1993), d) Soudan et al. (1993), e) Reincke et al. (1993), f) Reincke et al. (1994), g) Radomski et al. (1995).

lation and allowing for other VATs that exist in the population to increase in number. VSG gene switching and antibody selection of variants results in the appearance of new peaks in parasitemia every 5–10 days (Figure 18.8). There is evidence that tumor necrosis factor, which is elaborated by macrophages in infected animals, can also kill trypanosomes, but this action appears to be restricted to one or a few isolates and may therefore be of limited significance (Kitani et al. 2002).

Trypanosome-infected mammals first make VSG-specific antibodies of the IgM class. These facilitate trypanosome clearance in two ways:

1 by promoting assembly of complement components on the parasite surface to yield a membrane attack complex, which lyses the parasites
2 by causing opsonins, namely, complement fragments C3b and iC3b, to be deposited on the parasite surface resulting in their binding to complement receptors on macrophages and neutrophils and their phagocytosis.

It has been shown experimentally that the VSG-specific IgM immune response is primarily a B-cell response of a type that, in mice, does not require T-cell help. However, T-cell-dependent IgG responses also arise in infected animals. VSG-specific antibodies of the IgG class are thought to play an important role in clearing trypanosomes from tissue fluids because IgG is smaller than IgM and can diffuse more readily into the tissues.

The development of broad, and possibly polyclonal, B-cell activation and macroglobulinemia are among the defining characteristics of trypanosomiasis. Both antihost and antitrypanosome antibodies are present in the sera of infected patients and contribute to the macroglobulinemia. After the disease has persisted for some time, the lymph nodes and spleen become enlarged. There is also CNS involvement. Large numbers of white cells are found in the cerebrospinal fluid, which contains high levels of IgM and other proteins, especially during the late stages of the disease (Haller et al. 1986; Molyneux et al. 1984). The presence in brain tissue of 'Morula cells of Mott', which are plasma cells with highly distended sacs of endoplasmic reticulum, was once considered pathognomic of sleeping sickness.

Many signs of disease appear in infected trypanosomiasis-susceptible hosts. There are inflammatory lesions in the heart and brain with perivascular infiltration by lymphocytes, plasma cells, and monocytes. In chronically ill patients, there may be chronic inflammation of muscle and nerve fibers. In patients with advanced disease, these changes lead to edema and fibrosis. Pathology, visible by microscopic examination, has also been seen in the lungs, spleen, and liver. In the blood, there is a reduction in the numbers of erythrocytes and platelets, and a reduction in hemoglobulin content. Edema and hemorrhage develop, due to changes in the vascular beds of the heart and brain. In the CNS, there is demye-

lination and neuronal damage that may extend into the white matter. The perivascular cuffing and inflammatory cell infiltration extend deep into the choroid plexus. As the disease progresses, chronic meningoencephalitis occurs. Changes in electroencephalograms and diurnal sleep patterns have been detected in chronically infected humans (Bentivoglio et al. 1994; Buguet et al. 1993; Hamon et al. 1993).

Experimental data from studies of infected laboratory animals suggest that the encephalopathy observed in patients with advanced trypanosomiasis has an immunological basis (Jennings et al. 1989). It can be observed, for example, that many aspects of the pathology of African trypanosomiasis are similar to those present in animals undergoing a progressive Arthus-type reaction. For example, high immunoglobulin levels and large quantities of immune complexes are present in the blood and the CNS of experimentally infected animals and infected humans. In addition, high kinin levels with changes in prothrombin activity as well as changes in the levels of fibrin, fibrinogen, and complement have been reported to occur in patients as well as in experimental animals suffering from trypanosomiasis (reviewed in Seed and Hall 1992). These last changes are not only consistent with those occurring during an Arthus reaction, but also with those induced by the activation of the coagulation system as occurs in patients with the disseminated intravascular coagulation often present in patients with the Rhodesian form of this disease. Furthermore, antiCNS components (galactocerebrosides, neurofilaments, tryptophane) have been described in sera and cerebrospinal fluid (CSF) of sleeping sickness patients (Dumas and Bouteille 1996). These may arise as a result of cross reactions with common epitopes between host and trypanosomes or from a breakdown in immune regulation leading to autoimmunity. All of these changes, and others including the changes reported to occur in cytokine levels of patients, are suggestive of an immunologically mediated pathology in trypanosomiasis. Furthermore, the reactive arsenical encephalopathy that occurs in 3–5 percent of late-stage infection patients treated with melarsoprol may also have an immunological basis (Adams et al. 1986; Haller et al. 1986), which is the rationale for the use of anti-inflammatory corticosteroids in the treatment of these patients (Jennings et al. 1989).

Although immunoglobulin levels are increased in humans with trypanosomiasis and their immunological system is very active, it has been found that infected humans respond to secondary infections as if they were immunosuppressed. Such immunosuppression in patients with trypanosomiasis is well documented (Greenwood et al. 1973). The actual mechanisms involved are unknown but there are a number of reasonable suggestions based on data from animal studies. Impaired B-cell responses may result from overactivation and clonal depletion, and from a loss of immune system follicular structure during infections. It is unclear whether unique trypanosome components are the cause of this suppression or if it results from massive antigenic overstimulation. T-cell suppression is thought to result from the development of immunosuppressive macrophages responding to the combined effects of trypanosome products and interferon-γ with production of immunosuppressive levels of nitric oxide and prostaglandins. In this regard, *T. brucei* have been shown to directly stimulate murine T cells to produce interferon-γ. Trypanosomes have also been recently shown to have an endogenous prostaglandin F(2alpha) synthase and generation of this prostaglandin by the parasites may contribute to host pathology (Kubata et al. 2000).

In addition to the microscopic pathology and the immunopathology that the infection induces, there are extensive changes in host physiology in persons with trypanosomiasis. For example, the reproductive capacity of infected individuals is low. Low estradiol levels are reported to occur in 65 percent of infected females and low testosterone levels in 50 percent of infected males (Ikede et al. 1988; Boersma et al. 1989; Soudan et al. 1993). Abnormalities in thyroid hormones have also been observed (Reincke et al. 1993). These observations suggest that all organ systems and physiological functions are affected by the infection. A summary of some of the physiological changes that occur in people with African trypanosomiasis is shown in Table 18.2.

Trypanosomiasis-resistant hosts

Several African wildlife species are found throughout areas where trypanosomiasis is endemic. These animals were presumably selected by co-evolution with trypanosomes and tsetse flies to have developed methods to limit the growth of African trypanosomes and the severity of trypanosome-induced pathology. The little information available on mechanisms of resistance expressed by these animals was made possible by the pioneering work of Jan Grootenhuis, a Dutch veterinarian who established a breeding herd of Cape buffalo and eland in Nairobi, Kenya, outside of the tsetse habitat. There, several generations of the wild bovids were raised in the complete absence of trypanosomiasis. Experimental challenge with trypanosome-infected tsetse flies and needle inoculation with cloned bloodstream-form organisms showed that the bovids could be infected but showed few or no signs of disease. In both cases, only one or a few waves of parasitemia developed after which infections became cryptic being characterized by the presence of between one and ten mammalinfective trypanosome(s) per ml blood. (Reduth et al. 1994). Eventually, the infections were cleared. During the period of cryptic parasitemia, the wild bovids could not be super-infected with different trypanosome isolates.

Infected Cape buffalo were shown to develop broadacting trypanocidal activity in their blood plasma. This was mediated by production of a trypanocidal concentration of hydrogen peroxide as a result of catabolism of endogenous plasma purine by plasma xanthine oxidase (reviewed in Wang et al. 2002), and was contingent on an infection-induced decline in blood catalase (Wang et al. 1999). However, the change in blood chemistry was not responsible for the long-term suppression of parasitemia in the infected Cape buffalo. By a month after infection, blood catalase was restored to preinfection levels but parasitemia remained cryptic (Wang et al. 2002) Consequently, a second mchanism must be responsible for control of parasite population growth at this time. Possible mechanisms include the development of trypanosome growth-inhibitory antibodies directed against conserved receptors for required macromolecules, and the priming of a cell type that can accelerate the development of antibodies specific for trypanosome VSG thus limiting population growth in a VSG-specific manner.

DIAGNOSIS

A primary consideration for diagnosis of African trypanosomiasis should be a history of travel or residence of the patient in an area of Africa where the disease occurs. This said, diagnosis of African trypanosomiasis depends on the microscopic demonstration of trypanosomes in the blood, in lymph node aspirates or in cerebrospinal fluid. The very elevated macroglobulinemia that occurs in patients has been considered diagnostic of African trypanosomiasis but it is only suggestive as other diseases such as leishmaniasis also cause a macroglobulinemia.

Unfortunately, the clinical features of the disease are not sufficiently unique to be diagnostic, except possibly during the late sleeping sickness stage; the number of parasites in the blood and other body fluids is often very low, making detection difficult. Multiple sampling and concentration techniques are often required to find the parasites. In *T. b. gambiense*-infected people in particular, the parasitemias can be extremely low and make infection with this trypanosome particularly difficult to diagnose (DeRaadt and Seed 1977). It should be noted that there have been attempts to develop a list of symptoms (enlarged lymph nodes, CNS-associated signs, etc.) that are common to sleeping sickness patients, which could be used by rural health personnel to make a tentative diagnosis of a *T. b. rhodiense* infection (Boatin et al. 1986).

Examination of stained thin or thick blood films for trypanosomes is still a good diagnostic technique, and repeated daily examination can increase the probability of detecting the parasite. The use of concentration methods such as miniature anion exchange columns (Lanham and Godfrey 1970; Lumsden et al. 1979, 1981), or hematocrit tube centrifugation (Woo 1971; WHO 1986; Woo and Hauck 1987; Levine et al. 1989) coupled with microscopic examination can increase the sensitivity of methods for detection of trypanosomes several fold. A modification of the hematocrit tube centrifugation test is the acridine orange quantitative buffy coat (QBC) technique (Bailey and Smith 1992). This test is simple, quick, and sensitive. Since it detects trypanosomes microscopically, it does not produce false positives. The concentration techniques have been modified for use in field surveys, but they are still difficult to use in field settings because of the need for a centrifuge and electrical power (WHO 1986).

If trypanosomes cannot be detected in the blood of a suspected trypanosomiasis patient, the microscopic examination of a wet preparation from an enlarged cervical lymph node may confirm the diagnosis. Once the trypanosomes are detected in the blood and lymph fluids, the CSF must also be examined in order to determine the stage of the disease. Distinguishing between the acute blood stage of the infection and the later chronic neurologic stage is important in determining the appropriate chemotherapeutic regime. Centrifuged sediments of the CSF should be examined for trypanosomes. The fluid should also be examined to determine if there are elevated white cell counts and increases in protein concentration (DeRaadt and Seed 1977). Both CSF and blood from patients should be examined periodically after completion of chemotherapy in order to ensure that the patient is cured.

The use of in vitro cultivation of trypanosomes or the inoculation of suspected samples into experimental animals is not currently practical for routine diagnosis, although both procedures are used extensively for research purposes. A kit for in vitro isolation (KIVI) of trypanosomes from sleeping sickness patients in the field has been developed (Aerts et al. 1992; Truc et al. 1992). This test is of value in isolating trypanosomes from infected individuals; however, because KIVI requires days for the cultures to produce detectable numbers of trypanosomes, it will probably have limited diagnostic value. Its main importance will be as a standard for evaluating other quicker diagnostic tests.

Immunodiagnostic tests can be especially useful for mass field surveys and for other epidemiological investigations as well as for the detection of latent infections. A number of serological tests have been developed and many are currently being tested in the field. Immunofluorescence, complement fixation test, and card agglutination test for trypanosomes (CATT) have all been utilized in epidemiological studies. Several of these procedures are available in the form of commercial kits for field use. These tests are based on the detection of antibody in the sera of infected individuals and utilize antigens from blood-stage trypanosomes. For example, the CATT test employs a fixed, stained suspension of intact trypanosomes containing a mixture of various antigenic types. The various types selected are based on the frequency with which they appear in the human population in a broad geographic area. This test is easy to perform, inexpensive, and convenient for field work. However, like many indirect diagnostic tests, it is not perfect. For example, it is not equally effective in all geographic areas, nor is it equally effective for detection of the two forms of human trypanosomiasis. This test is also reported to give positive reactions with antibodies to species of trypanosomes infecting animals but not humans (Penchenier et al. 1991).

Tests that use monoclonal antibodies to detect trypanosome antigens in serum are now being developed (Komba et al. 1992; Nantulya et al. 1992). Antigen detection tests would be valuable since they are based on detection of trypanosome antigens in the blood and, therefore, on active infection. Tests that detect antibody cannot distinguish between a past and a present infection, but those that detect antigen can. The antigens produced by modern biotechnology have not yet been used for diagnostic purposes. Although molecular probes have been used successfully in epidemiological studies, hybridization and polymerase chain reaction (PCR) technologies are not currently used for routine diagnosis because tests based on them are still too slow and cumbersome.

In a recent comparison of the CATT, the hematocrit centrifuge tube assay, the minianion exchange column technique, the QBC, the KIVI, and the thick blood film (TBF) procedure, the TBF was found to be as sensitive as any of the others. It also has the advantage of being the simplest and cheapest of all the diagnostic methods. It is rapid and simple, requiring only the staining and examination of the slides (Truc et al. 1994).

EPIDEMIOLOGY AND CONTROL

As mentioned at the start of this chapter, African trypanosomiasis is restricted to Central Africa because of the ecology of the insect vector (Glossina, tsetse flies). The tsetse habitat is limited by the Sahara Desert to the north and the cool, dry areas of Southern Africa to the south. The tsetse habitat is approximately the size of the USA, and it is estimated that within this region about 60 million people are at risk. However, this figure is based on a number of suppositions and there is no way to truly know the number of people at risk. A small portion of people in trypanosomiasis endemic areas has been evaluated for the disease and based on this evaluation it is estimated that 300 000–500 000 people presently have sleeping sickness. Of these, only a few will be detected by the screening systems in place and even less will be treated. Sleeping sickness now occurs in 36 countries south of the Sahara and there are approximately 200 known disease foci (Kuzoe 1991, 1993).

Epidemiology

The epidemiology of infection caused by *T. b. gambiense* differs from that of infection caused by *T. b. rhodesiense* in many aspects. These include the species of tsetse involved, the type of tsetse habitat, and the number and type of reservoir hosts (Table 18.1). *T. b. gambiense* is transmitted by flies of the *Glossina palpalis* group, which live primarily in the vegetation along river banks and in moist forests. Tsetse of this group feed frequently on humans in areas where contact is common, such as at river crossings and water holes. There are animals that serve as reservoir hosts of *T. b. gambiense* but people with chronic infection who have relatively mild symptoms provide the major contribution to the persistence of the Gambian type of infection (Jordan 1986; Molyneux and Ashford 1983). *T. b. rhodesiense* is transmitted by tsetse of the *G. morsitans* group. These flies inhabit low woodlands and thickets on lake shores and much of the savanna of Central and East Africa. This group of flies feeds primarily on nonhuman hosts, and humans are only a secondary food source (Seed and Hall 1992). Fishermen, game wardens, and others who enter areas with high tsetse numbers are particularly at risk (Wyatt et al. 1985). In addition, epidemics of the Rhodesian infection occur when there is an increase in tsetse numbers in close association with villages, a development causing an increase in fly–human contact. The areas in which the West African infection occurs overlap the areas of infection caused by *T. b. rhodesiense* in countries such as Zaire and Uganda.

Control

In the past, control of trypanosomiasis was partly based on the removal of reservoir hosts from areas near human settlements. This involved the fencing out and killing of game animals. Other early attempts at control of trypanosomiasis involved the destruction of tsetse habitats. Both the elimination of reservoir hosts and the destruction of tsetse habitats are now considered ecologically unsound. In addition, game animals are a tourist attraction and consequently an important source of revenue for African countries with managed National Parks and Safari operations including lodges.

The epidemiology of human African trypanosomiasis is shaped by many factors (Table 18.3). They include the type and density of the vector, the type and density of the reservoir host population, and the density of the human population. The densities of the fly population and of the host population are dependent on environmental conditions such as temperature, humidity, water availability, and other factors of the habitat. It is a general truth that any condition that increases the contact between tsetse, the reservoir hosts, and humans, including growth of population, will increase the risk of human infection. The chronic nature of the infection in both the vector and mammalian host makes continued surveillance and treatment of the human population, and vector control necessities if low prevalence rates are to be maintained in the human population. Current control of human African trypanosomiasis is therefore based on three arms (Seed 2000). The first is the continued surveillance in individuals by mobile field teams for the detection of trypanosomiasis. It is believed that every individual in an endemic area should be screened at least once a year. The second arm of a control program is the treatment of the infected individuals. These two phases are essential for a successful control program because African trypanosomiasis is a chronic disease and ambulatory individuals can act as reservoirs of the infection. This is especially important for the West African

Table 18.3 *Some factors involved in the epidemiology of African trypanosomiasis of humans* [a]

Fly	Mammal	Environmental	Parasite
Fly numbers	Host numbers	Temperature	Parasite numbers
Sex	Host species	Humidity	Genetics
Age at infection	Habitat	Water sources	Growth rates, virulence, drug resistance, human serum resistance, sub-species, strain
Symbiotic infections, viral, bacterial	Attractiveness to fly	Vegetation	
Fly behavior, habitat selection, host preference	Genetics	Wind direction and current	
Physiological status	Immune response		
Genetic factors	Non-specific factors		
Species, subspecies	Race		
Stage of infection in fly	Trypanocidal factors		
	Degree of immunosuppression		
	Intercurrent infections		

a) From Molyneux et al. (1984), Jordan (1986), Seed and Hall (1992), Vale (1993)

form of the infection. Since East African trypanoso-miasis is a zoonotic disease, the treatment of infected individuals has less effect on the incidence of human infections. In East Africa, cattle are an important reservoir of the disease. It has been shown that cattle infected with *T. b. rhodesiense* and their movement can significantly contribute to the epidemiology of the human infection especially in the vicinity of cattle markets. It has, therefore, been suggested that the surveillance of cattle for human-infective *T. b. rhodesiense* and the targeted treatment of these cattle would be an important addition to the control of sleeping sickness (Hutchinson et al. 2003). The third arm of current control programs involves vector control.

All species of *Glossina* are blood feeders and as adults are totally dependent on this food source. The different species of *Glossina* have different host preferences, and both sexes take bloodmeals. In addition, tsetse are larviparous and a female produces only 8–12 larvae during her limited life span of 3–5 months. The larvae are deposited on the ground and immediately burrow into patches of moist soil or sand such as occur in shaded areas under fallen logs, rocks or bushes. The need for frequent bloodmeals and the reproduction dynamics of tsetse are important factors in shaping the epidemiology of trypanosomiasis (Molyneux and Ashford 1983).

The environmental requirements of the flies and their low reproductive rates make then vulnerable to control by a variety of techniques. The techniques used are based on our knowledge of tsetse behavior and other aspects of epidemiology of trypanosomiasis. Control programs using insecticide application and traps are in wide use. Traps and insecticides such as the chlorinated hydrocarbons and synthetic pyrethroids, when properly used, cause limited environmental damage (Molyneux and Ashford 1983; Jordan 1986). Depending on the environmental conditions, application may be by spraying with individually carried sprayers or from aircraft. The latter technique is used when broad coverage is needed. Technical factors that are considered important for achieving good results in spraying programs include appropriate formulation of the spray to obtain proper droplet size and the proper selection of insecticide. Either a residual or a nonresidual insecticide may be selected, depending on the retention time desired. Selection of appropriate sites for spraying is also important. Many of the requirements for successful use of insecticides for control of tsetse are reviewed by Jordan (1986) and Molyneux and Ashford (1983). Although control by insecticide application is effective, it is also expensive and repeated application is required if results are to last.

The use of fly traps and screens or rags impregnated with insecticides is more economical than environmental spraying and has been demonstrated to lower tsetse numbers dramatically in limited areas (Molyneux and Ashford 1983; Jordan 1986; Vale 1993). The biconical

trap (Figure 18.9) has been shown to be effective in reducing tsetse numbers in West Africa. Detailed knowledge of fly behavior in the area where control is attempted is required for proper design and placement of traps and screens.

There are chemicals to which tsetse are known to be attracted. These include carbon dioxide and volatile compounds in the breath and urine of cattle (Vale and Hall 1985a, b; Vale et al. 1988). Some of these compounds have been identified: they include acetone, 1-octenol-3-ol, and the phenolic compounds 4-methyl-phenol and 3-n-propylphenol. The baiting of traps with these synthetic compounds and with carbon dioxide has been shown to increase the numbers of flies caught. There are also substances from humans that are repellent to some tsetse (Vale 1993), and these must not be permitted to contaminate traps. In addition, color and design of traps or screens are important factors in attracting and inducing the landing of flies. In the Congo Republic, a modified biconical trap is used for the control of tsetse in areas where trypanosomiasis occurs (Lancien 1981). The traps are relatively inexpensive and can be constructed and maintained by the villagers themselves if the necessary materials are provided (Okoth 1986).

In Zambia, disposable screens impregnated with insecticide have replaced traps (Figure 18.9b). The screens are 2×1 m and have a central black area of 1×1 m surrounded by two 0.5×1 m rectangular blue areas. The screens are supported by wooden frames that can be stuck into the ground. The insecticide used, Deltame-thrin, persists on the screens for months (Vale 1993). The screens are inexpensive, require no maintenance, and can be utilized with or without the use of chemo-attractive odors (Vale 1993). These screens are extremely effective in killing tsetse and in some areas are believed to have reduced or even eradicated tsetse populations. Similar screens have been used to control *G. pallidipes* in Kenya (Opiyo et al. 1990). The use of traps and screens in the control of tsetse has been extensively discussed by Jordan (1986), Laveissiere et al. (1990), and Vale (1993). They hold great promise for future control of tsetse populations and, therefore, African trypanoso-miasis in an environmentally friendly way. However, although both traps and disposable screens have been successfully used, they require continued maintenance and replacement due to environmental damage, theft, and animal damage. They are not cost-free and require a strong infrastructure for maintenance.

An approach related to the use of screens and traps in tsetse control has been the use of cattle dipped in insecticide, or on which insecticides have been poured or sprayed. This technique has been shown to lower tsetse numbers (Fox et al. 1993). This procedure is relatively inexpensive and has the additional advantage that it also reduces the populations of ticks and other biting insects.

There have been attempts to control tsetse by the release of sterile flies. *Glossina austeni* was successfully

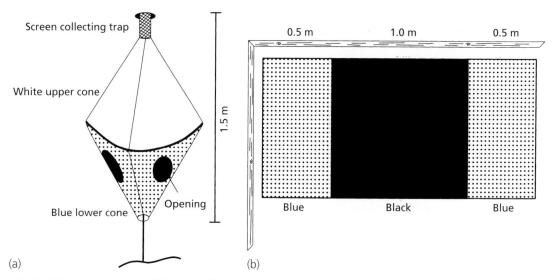

Figure 18.9 (a) *Biconical tsetse trap and* **(b)** *insecticide impregnated screen or target.*

eradicated from the Island of Zanzibar (Vreysen et al. 2000). Based on this success are programs currently being initiated to attempt to eradicate tsetse using sterile fly technology in combination with other control activities. However, there is considerable debate as to the validity of this approach because of the high cost required to eliminate tsetse from Zanzibar (Molyneux 2001). In addition, the fact that Zanzibar is a small, isolated island that tsetse cannot reinvade has suggested that Zanzibar is not a satisfactory model for the tsetse–African trypanosomiasis situation in mainland Africa (Rogers and Randolph 2002). There have also been attempts to find tsetse pathogens, and to identify pheromones for use in programs to reduce fly populations (Molyneux and Ashford 1983; Langley et al. 1988). An approach using biotechnology to develop a transmissible vector that would cause male sterility has been proposed, as has an approach to control tsetse fly susceptibility to infection that is based on the release of antitrypanosome peptides by symbiotic prokaryotes (reviewed by Aksoy 2001; Aksoy et al. 2002). However, although this is promising, control based on these techniques has not reached a stage of development sufficiently advanced to even permit testing.

The lack of economic resources is a major factor limiting effective control of trypanosomiasis. In many of the endemic countries, the per capita gross national product is less than US$500, and there may be an expenditure of less than US$10 per individual for total health care. Several years ago, the cost of surveillance alone was estimated at approximately US$1 per individual at risk, and the cost of treatment at US$35 per patient in the early stage and US$135 per patient in the late stage of disease. These costs are likely to have increased. In addition, there are costs for vector surveillance and control (Kuzoe 1991; WHO 1986). It is obvious that the necessary resources for control of

sleeping sickness are very limited in most areas of Africa, and any change in government priorities, e.g. to control political unrest, will decrease the amount of money available for long-term control activities. Civil unrest also disrupts the flow of external aid required to supplement local funds for control activities.

CHEMOTHERAPY

Chemotherapeutic treatment of African trypanosomiasis is important in reducing the incidence of the infection especially in the West African form of the disease. However, surveillance and early detection of both forms of the disease are essential because the treatment protocol is shorter for patients in the early stages of the disease and drug-induced morbidity and mortality are lower than in patients in the late stages of the disease. Cure rates are also higher when infected individuals are treated early in the disease rather than during the late secondary stages. Control of both forms of sleeping sickness, therefore, works best if early detection and treatment are used together.

Suramin is used to treat patients with primary stage infections that do not involve the CNS. This drug is effective against both the Gambian and Rhodesian form of the disease but because it does not cross the blood–brain barrier, it is not effective against the secondary CNS stages. Suramin is a polysulfonated naphthylamine derivative of trypan red and was first found to have trypanolytic activity by Ehrlich. It is relatively toxic and may cause optic atrophy, blindness, nephrotoxicity, and adrenal insufficiency in some patients (Pepin and Milord 1994). Its exact mode of action is not known, although it inhibits a wide variety of trypanosome enzymes including glycolytic enzymes and the mitochondrial glycerol phosphate oxidase enzymes.

Pentamidine isethionate and other aromatic diamidines were first examined for trypanocidal activity in the 1930s. Like suramin, the dicationic pentamidine isethionate does not cross the blood–brain barrier, and is therefore used in patients with CNS involvement only to clear the blood of trypanosomes prior to treatment with melarsoprol. Pentamidine isethionate may produce nephrotoxicity, hepatotoxicity, and pancreatic toxicity in some patients (Sands et al. 1985; WHO 1986; Goa and Campoli-Richards 1987; Kapusnik and Mills 1988; Pepin and Milord 1994). The exact mode of action of this drug is unknown, but the strongly basic dicationic molecule binds to many cellular components, including a variety of trypanosome enzymes, causing their inhibition. It also binds to DNA, preferentially to adenosine–thymidine-rich regions in the minor groove. It has recently been shown that pentamidine isethionate is actively metabolized by the mammalian cytochrome P450 drug-metabolizing system (Berger et al. 1990).

In contrast to suramin and pentamidine isethionate, melarsoprol does cross the blood–brain barrier and is used to treat patients in the late secondary CNS stages of trypanosomiasis. Until very recently, melarsoprol has been the only drug available for treatment in this stage of the disease. Melarsoprol failures have been reported in several areas of sub-Saharan Africa, where sleeping sickness cases have a high prevalence. At this stage, it is unclear whether the treatment failures are a result of drug metabolism by the patient, or development of resistance in the parasites, or both. melarsoprol is an organic arsenical, first developed by Friedheim (1949). As with pentamidine isethionate and suramin, its exact mechanism of action is unknown, although it has been shown to inhibit a variety of trypanosome enzymes and functions in vitro. It is toxic, and in 1–10 percent of treated patients there is encephalopathy with mortality rates of 1–5 percent. Other adverse side-effects include fever, headache, joint pain, gastrointestinal disturbances, renal damage, and hypertension (Pepin and Milord 1994; WHO 1986). As noted above (see Trypanosomiasis-susceptible hosts) the encephalopathy may have an immunological basis (Jennings et al. 1989). It has been possible to reduce the development of meningoencephalitis and perivascular cuffing in infected mice by treatment with the immunosuppressive drug azathioprine (Jennings et al. 1989).

Suramin, pentamidine isethionate, and melarsoprol are still the primary drugs used to treat patients with African trypanosomiasis. The only new drug that has been developed is difluoromethylornithine (DFMO), also known as eflornithine and Ornidyl) a specific inhibitor of ornithine decarboxylase, which is the first enzyme in the polyamine biosynthetic pathway (Bacchi and McCann 1987). Currently, it is the only drug other than melarsoprol available for treatment of late-stage trypanosomiasis (Schechter et al. 1987). In the trypanosomes, exposure to DFMO leads to rapid depletion of

putrescine and to low spermidine levels. It induces transformation of the replicating long slender form to the nonreplicating short stumpy form and, therefore, causes an inhibition of trypanosome growth. DFMO is relatively non-toxic but adverse reactions can include anemia, thrombocytopenia, and gastrointestinal disturbances, all of which are reversible. It is an extremely valuable chemotherapeutic agent for the treatment of late-stage patients who do not respond to melarsoprol. T. b. gambiense seems to be more susceptible to DFMO than is T. b. rhodesiense, although the reason for this difference is not clear (Bacchi et al. 1990).

Because DFMO is a specific inhibitor of ornithine decarboxylase with no other known inhibitory activity, rapid selection for drug-resistant trypanosomes has been possible in experimental studies. Resistance has also been observed in trypanosomes isolated from humans treated with DFMO. Considerable research is being conducted in an effort to find other compounds that inhibit the polyamine pathway. Treatments using combinations of drugs are also being investigated in an effort to decrease the probability of selecting for drug resistance.

The current treatment regime recommended by the WHO for reducing parasitemia in patients with T. b. gambiense infection is the use of pentamidine isethionate or, where there is resistance to this drug, with suramin. This is done in patients in both early (or primary) and late (or secondary) stages of the disease. In the patients in the late stages of the disease who have CNS involvement, this treatment is followed by treating with multiple injections of melarsoprol. The treatment of patients in the early stages of a T. b. rhodesiense infection is with suramin. In patients with the late stages of a T. b. rhodesiense infection, suramin is used to clear the blood and lymph and then multiple injections of melarsoprol are given (WHO 1986). The details of the treatment regimes are outlined in Table 18.4 (WHO 1986).

Currently, the number of drugs available for the treatment of African trypanosomiasis is limited and all except DFMO are quite toxic to humans. Researchers are currently searching for better analogues of existing drugs, investigating the use of drugs in combinations, and trying to develop compounds that inhibit biochemical pathways unique to trypanosomes. Although there are new leads in this field, further research is critically required.

VACCINATION

Because of the phenomenon of antigenic variation no vaccine is available.

CONCLUSION

Control of trypanosomiasis is possible, but it requires continuous surveillance of the human population and

Table 18.4 *The protocol used in the Côte d'Ivoire for the treatment of* T. b. gambiense *in infected patients*[a]

Time (days)	Drug used	Dose (mg/kg)	Route
1 and 2	Pentamidine isethionate	4.0	IM
4	Melarsoprol	1.2	IV
5	Melarsoprol	2.4	IV
6	Melarsoprol	3.6	IV
17	Melarsoprol	1.2	IV
18	Melarsoprol	2.4	IV
19	Melarsoprol	3.6	IV
20	Melarsoprol	3.6	IV
30	Melarsoprol	1.2	IV
31	Melarsoprol	2.4	IV
32	Melarsoprol	3.6	IV
33	Melarsoprol	3.6	IV

a) Abstracted from WHO (1986). For full details on treatment regimens for *T. b. gambiense* and *T. b. rhodesiense* infections, see Annex 5, pp. 118–21 of that publication.

the treatment of all infected individuals. It also requires control of tsetse flies through the use of traps and insecticide-impregnated screens and the judicious application of insecticides to the environment. It is possible that in the future insecticides will be required only in limited circumstances. The application of any control measure is currently limited in many areas of Africa owing to political instability as well as the lack of resources to initiate and maintain them adequately (Williams et al. 1993). While an enormous amount of information on trypanosome cellular and molecular biology and host immune responses has been gathered using modern research approaches and tools, this knowledge has not yet, with the exception of the introduction of DFMO for chemotherapy, resulted in new or improved methods to control trypanosomiasis. However, the discovery that lysis of *T. b. brucei* by human serum is due to the presence of apolipoprotein L-1 (Vanhamme et al. 2003), and the likelihood that this protein is also responsible for the lysis of *T. congolense* and *T. vivax* in human serum, may result in translational research leading to novel prophylactic regimes against nagana and therapeutic regimes against *T. b. rhodesiense*-sleeping sickness.

REFERENCES

Adams, J.H., Haller, L., et al. 1986. Human African trypanosomiasis (*T. b. gambiense*), A study of 16 fatal cases of sleeping sickness with some observations on acute reactive arsenical encephalopathy. *Neuropathol Appl Neurobiol*, **12**, 81–94.

Aerts, D., Truc, P., et al. 1992. A kit for in vitro isolation of trypanosomes in the field, first trial with sleeping sickness patients in the Congo Republic. *Trans R Soc Trop Med Hyg*, **86**, 394–5.

Agbo, E.E.C., Majiwa, P.A., et al. 2002. Molecular variation of *Trypanosoma brucei* subspecies as revealed by AFLP fingerprinting. *Parasitology*, **124**, 349–58.

Aksoy, S. 2001. Tsetse based strategies for control of African trypanosomes. In Black, S.J. and Seed, J.R. (eds). *World class parasites*, vol. 1, *The African trypanosomes*. Boston: Kluwer Academic Publishers, 39–49.

Aksoy, S., Hao, Z. and Strickler, P.M. 2002. What can we hope to gain for trypanosomiasis control from molecular studies on tsetse biology? *Kinetoplastid Biology and Disease*, **1**, 4, (www.kinetoplastids.com/content/1/1/4).

Ansorge, I., Steverding, D., et al. 1999. Transcription of 'inactive' expression sites in African trypanosomes leads to expression of multiple transferring receptor RNAs in bloodstream forms. *Mol Biochem Parasitol*, **101**, 81–94.

Bacchi, C.J. and McCann, P.P. 1987. Parasitic protozoa and polyamines. In: McCann, P.P., Pegg, A.E. and Sjoerdsma, A. (eds), *Inhibition of polyamine metabolism*. New York: Academic Press, 317–44.

Bacchi, C.J., Nathan, H.C., et al. 1990. Differential susceptibility to DL-alpha-difluoromethylornithine in clinical isolates of *Trypanosoma brucei rhodesiense*. *Antimicrob Agents Chemother*, **34**, 1183–8.

Bailey, J.W. and Smith, D.W. 1992. The use of acridine orange QBC technique in the diagnosis of African trypanosomiasis. *Trans R Soc Trop Med Hyg*, **86**, 630.

Bakker, B.M., Mensonides, F.I., et al. 2000. Compartmentation protects trypanosomes from the dangerous design of gycolysis. *Proc Natl Acad Sci USA*, **97**, 2087–92.

Balber, A.E. 1990. The pellicle and the membrane of the flagellum, flagellar adhesion zone, and flagellar pocket, functionally discrete surface domains of the bloodstream form of African trypanosomes. *Crit Rev Immunol*, **10**, 177–201.

Barry, J.D., Crowe, J.S., et al. 1983. Instability of the *Trypanosoma brucei rhodesiense* metacyclic variable antigen repertoire. *Nature*, **306**, 699–701.

Barry, J.D., Graham, S.V., et al. 1998. VSG gene control and infectivity strategy of metacyclic stage *Trypanosoma brucei*. *Mol Biochem Parasitol*, **91**, 93–105.

Bentivoglio, M., Grassi-Zucconi, G., et al. 1994. *Trypanosoma brucei* and the nervous system. *Trends Neurosci*, **17**, 325–9.

Berger, B.J., Lombardy, R.J., et al. 1990. Metabolic *N*-hydroxylation of pendamidine in vitro. *Antimicrob Agents Chemother*, **34**, 1678–85.

Biteau, N., Bringaud, F., et al. 2000. Characterization of Trypanozoon isolates using a repeated coding sequence and microsatellite markers. *Mol Biochem Parasitol*, **105**, 185–201.

Black, S.J., Sendashonga, C.N., et al. 1985. Regulation of parasitemia in mice infected with *Trypanosoma brucei*. *Curr Top Microbiol Immunol*, **117**, 93–118.

Black, S.J., Seed, J.R., et al. 2001. Innate and acquired resistance to African trypanosomiasis. *J Parasitol*, **87**, 1–9.

Blum, M.L., Down, J.A., et al. 1993. A structural motif in the variant surface glycoproteins of *Trypanosoma brucei*. *Nature*, **362**, 603–9.

Boatin, B.A., Wyatt, G.B., et al. 1986. Use of symptoms and signs for diagnosis of *Trypanosoma brucei rhodesiense* trypanosomiasis by rural health personnel. *Bull World Health Organ*, **64**, 389–95.

Boersma, A., Noireau, F., et al. 1989. Gondadotropic axis and *Trypanosoma brucei gambiense* infection. *Ann Soc Belge Méd Trop*, **69**, 127–35.

Borst, P. and Fairlamb, A.H. 1998. Surface receptors and transporters of *Trypanosoma brucei*. *Ann Rev Microbiol*, **52**, 745–78.

Borst, P. and Ulbert, S. 2001. Control of VSG gene expression sites. *Mol Biochem Parasitol*, **114**, 17–27.

Brun, R. and Jenni, L. 1985. Cultivation of African and South American trypanosomes of medical or veterinary importance. *Br Med Bull*, **41**, 122–9.

Buguet, A., Bert, J., et al. 1993. Sleep–wake cycle in human African trypanosomiasis. *J Clin Neurophysiol*, **10**, 190–6.

Cardosa de Almeida, M.L., Geuskens, M. and Pays, E. 1999. Cell lysis induces redistribution of the GPI-anchored variant surface glycoprotein on both faces of the plasma membrane of *Trypanosoma brucei*. *J Cell Sci*, **112**, 4461–73.

Carrington, M., Carnall, N., et al. 1998. The properties and function of the glycosylphosphatidylinositol-phospholipase C in *Trypanosoma brucei*. *Mol Biochem Parasitol*, **91**, 153–64.

Cronin, C.N., Nolan, D.P. and Voorheis, H.P. 1989. The enzymes of the classical pentose phosphate pathway display differential activities in procyclic and bloodstream forms of *Trypanosoma brucei*. *FEBS Lett*, **244**, 26–30.

Cross, G.A., Wirtz, L.E. and Navarro, M. 1998. Regulation of VSG gene expression site transcription and switching in *Trypanosoma brucei*. *Mol Biochem Parasitol*, **91**, 7–91.

Denny, P.W., Field, M.C. and Smith, D.F. 2001. GPI-anchored proteins and glycoconjugates segregate into lipid rafts in Kinetoplastida. *FEBS Lett*, **491**, 148–53.

De Greef, C.H., Imberechts, G., et al. 1989. A gene expressed only in serum-resistant variants of *Trypanosoma brucei rhodesiense*. *Mol Biochem Parasitol*, **36**, 169–76.

DeRaadt, P. and Seed, J.R. 1977. Trypanosomes causing disease in man in Africa. In Kreier, J.P. (ed), *Parasitic Protozoa*, vol. 1, The African trypanosomes. New York: Academic Press, 176–237.

Donelson, J.E. 2001. The genome of the African trypanosome. In Black, S.J. and Seed, J.R (eds). *World class parasites*, vol. 1, *The African trypanosomes*. Boston: Kluwer Academic Publishers, 143–58.

Duffieux, F., Van Roy, J., et al. 2000. Molecular characterization of the first two enzymes of the pentose phosphate pathway of *Trypanosoma brucei*. Glucose-6-phosphate dehydrogenase and 6-phosphogluconolactonase. *J Biol Chem*, **275**, 27559–65.

Duggan, A.J. 1970. An historical perspective. In: Mulligan, H.W. (ed.), *The African Trypanosomiases*. New York: Wiley-Interscience.

Dumas, M. and Bouteille, B. 1996. Human African trypanosomiasis. *C R Soc Biol*, **190**, 395–408.

Duszenko, M., Muhlstadt, K. and Broder, A. 1992. Cysteine is an essential growth factor for *Trypanosoma brucei* bloodstream forms. *Mol Biochem Parasitol*, **50**, 269–73.

Fairlamb, A.H. 1989. Novel biochemical pathways in parasitic protozoa. *Parasitology*, **99**, Supplement, S93–112.

Fairlamb, A.H. and Cerami, H. 1992. Metabolism and functions of trypanothione in the Kinetoplastida. *Ann Rev Microbiol*, **46**, 695–729.

Ferguson, M.A.J. 1999. The structure, biosynthesis and functions of glycosylphosphatidylinositol anchors, and the contributions of trypanosome research. *J Cell Sci*, **112**, 2799–809.

Flohe, L., Hecht, H.J. and Steinert, P. 1999. Glutathione and trypanothione in parasitic hydroperoxide metabolism. *Free Radic Biol Med*, **27**, 966–84.

Ford, J. 1971. *The role of the Trypanosomiases in African ecology: A study of the tsetse fly problem*. Oxford: Clarendon Press.

Fox, R.G.R., Mmbando, S.O., et al. 1993. Effect on herd health and productivity of controlling tsetse and trypanosomiasis by applying Deltamethrin to cattle. *Trop Anim Health Prod*, **25**, 203–14.

Friedheim, E.A.H. 1949. Mel B in the treatment of human trypanosomiasis. *Am J Trop Med Hyg*, **29**, 173–80.

Gerrits, H., Mussmann, R., et al. 2002. The physiological significance of transferrin receptor variations in *Trypanosoma brucei*. *Mol Biochem Parasitol*, **119**, 237–47.

Gibson, W. 2001a. Molecular characterization of field isolates of human pathogenic trypanosomes. *Trop Med Int Health*, **6**, 401–6.

Gibson, W. 2001b. Sex and evolution in trypanosomes. *Int J Parasitol*, **31**, 642–6.

Gibson, W.G., Garside, L. and Bailey, M. 1992. Trisomy and chromosome size changes in hybrid trypanosomes from a genetic cross between *Trypanosoma brucei rhodesiense* and *T. b. brucei*. *Mol Biochem Parasitol*, **52**, 189–200.

Goa, K.L. and Campoli-Richards, D.M. 1987. Pentamidine isethionate. A review of its antiprotozoal activity, pharmacokinetic properties and therapeutic use in *Pneumocystis carinii* pneumonia. *Drugs*, **33**, 242–58.

Greenwood, B.M., Whittle, H.C. and Molyneux, D.H. 1973. Immunosuppression in Gambian trypanosomiasis. *Trans R Soc Tropl Med Hyg*, **67**, 846–50.

Gutteridge, W.E. and Coombs, G.H. 1977. *Biochemistry of parasitic protozoa*. Baltimore: Baltimore University Park Press, pp. 89–107.

Haller, L., Adams, H., et al. 1986. Clinical and pathological aspects of human African trypanosomiasis (*T. b. gambiense*) with particular reference to reactive arsenical encephalopathy. *Am J Trop Med Hyg*, **35**, 94–9.

Hammond, D.J. and Gutteridge, W.E. 1984. Purine and pyrimidine metabolism in the trypanosomatidae. *Mol Biochem Parasitol*, **13**, 242–61.

Hamon, J.F., Camara, P., et al. 1993. Waking electroencephalograms in blood lymph and encephalitic stages of Gambian trypanosomiasis. *Ann Trop Med Parasitol*, **87**, 149–55.

Hannaert, V. and Michels, P.A.M. 1994. Structure, function, and biogenesis of glycosomes in Kinetoplastida. *J Bioenerg Biomembr*, **26**, 205–12.

Hatada, S., Seed, J.R., et al. 2002. No trypanosome lytic activity in the sera of mice producing human haptoglobin-related protein. *Mol Biochem Parasitol*, **119**, 291–4.

Hide, G., Welburn, S.C., et al. 1994. Epidemiological relationships of *Trypanosoma brucei* stocks from South East Uganda, evidence for different population structures in human infective and non-human infective isolates. *Parasitology*, **109**, 95–111.

Hill, K.L., Hutchings, N.R., et al. 2000. T lymphocyte-triggering factor of African trypanosomes is associated with the flagellar fraction of the cytoskeleton and represents a new family of proteins that are present in several divergent eukaryotes. *J Biol Chem*, **275**, 39369–78.

Hirumi, H., Doyle, J.J. and Hirumi, K. 1977. African trypanosomes, cultivation of animal-infective *Trypanosoma brucei* in vitro. *Science*, **196**, 992–4.

Hirumi, H., Hirumi, K., et al. 1992. *Trypanosoma brucei brucei:* in vitro production of metacyclic forms. *J Protozool*, **39**, 619–27.

Hirumi, H., Martin, S., et al. 1997. Cultivation of bloodstream forms of *Trypanosoma brucei* and *T. evansi* in a serum free medium. *Trop Med Int Health*, **2**, 240–4.

Horn, D. 2001. Nuclear gene transcription and chromatin in *Trypanosoma brucei*. *Int J Parasitol*, **31**, 1157–65.

Horton, T.L. and Landweber, L.F. 2002. Rewriting the information in DNA, RNA editing in kinetoplastids and myxomycetes. *Curr Opin Microbiol*, **5**, 620–6.

Hutchinson, O.C., Fevre, E.M., et al. 2003. Lessons learned from the emergence of a new *Trypanosoma brucei rhodesiense* sleeping sickness focus in Uganda. *Lancet Infect Dis*, **3**, 42–5.

Ikede, B.O., Elhassan, E. and Akpavie, S.O. 1988. Reproductive disorders in African trypanosomiasis. A review. *Acta Trop*, **45**, 5–10.

Jennings, R.W., McNeil, P.E., et al. 1989. Trypanosomiasis and encephalitis, possible aetiology and treatment. *Trans R Soc Trop Med Hyg*, **83**, 578–618.

Jordan, A.M. 1986. *Trypanosomiasis control and African rural development*. New York: Longman.

Kaminsky, R., Beaudoin, E. and Cunningham, I. 1988. Cultivation of the life cycle stage of *Trypanosoma brucei* spp. *Acta Trop*, **45**, 33–43.

Kapusnik, J.E. and Mills, J. 1988. Pentamidine. In: Peterson, P.K. and Verhoef, J. (eds), *The antimicrobial agent annual*. vol. 3. New York: Elsevier, 299–311.

Kitani, H., Black, S.J. and Nakamura, Y. 2002. Recombinant human tumor necrosis factor does not inhibit the growth of trypanosomes in axenic culture. *Infect Immun*, **70**, 2210–14.

Komba, E.K., Odiit, M., et al. 1992. Multicenter evaluation of an antigen-detection ELISA for the diagnosis of *Trypanosoma brucei rhodesiense* sleeping sickness. *Bull World Health Organ*, **70**, 57–61.

Kooy, R.F., Hirumi, H., et al. 1989. Evidence for diploidy in metacyclic forms of African trypanosomes. *Proc Natl Acad Sci USA*, **86**, 5469–72.

Kreier, J.P. and Baker, J.R. 1987. *Parasitic protozoa*. Boston, Massachussetts: Allen and Unwin.

Kubata, B.K., Duszenko, M., et al. 2000. Identification of a novel prostaglandin f(2alpha) synthase in *Trypanosoma brucei*. *J Exp Med*, **192**, 1327–38.

Kuzoe, F.A.S. 1991. Perspectives in research on and control of African trypanosomiasis. *Ann Trop Med Parasitol*, **85**, 33–41.

Kuzoe, F.A.S. 1993. *African trypanosomiasis 6*, Tropical Disease Research Progress 1991–2. Eleventh Programme Report of the UNDP/World Bank/WHO Special Programme for Research and Training in Tropical Diseases, World Health Organization, Geneva, 57–66.

Lancien, J. 1981. Descriptions du piège monoconique utilisé pour l'élimination des glossines en République de Congo. *Cah ORSTROM. Série Entomol Méd Parasitol*, **19**, 235–8.

Langley, P.A., Felton, T. and Doichi, H. 1988. Juvenile hormone mimics as effective sterilants for the tsetse fly *Glossina morsitans morsitans*. *Med Vet Entomol*, **2**, 29–35.

Lanham, S.M. and Godfrey, D.G. 1970. Isolation of salivarian trypanosomes from man and other mammals using DEAE cellulose. *Exp Parasitol*, **28**, 521–34.

Laveissiere, C., Vale, G.A. and Gouteux, J.P. 1990. Bait methods for tsetse control. In: Curtis, C.F. (ed.), *Appropriate technology in vector control*. Boca Raton: CRC Press, 47–74.

Levine, R.A., Wardlaw, S.C. and Patton, C.L. 1989. Detection of haematoparasites using quantitative buffy coat analysis tubes. *Parasitol Today*, **5**, 132–4.

Lumsden, W.H.R., Kimber, C.D., et al. 1979. *Trypanosoma brucei*, miniature anion-exchange centrifugation technique for detecting of low parasitemias; adaptation for field use. *Trans R Soc Trop Med Hyg*, **73**, 312–17.

Lumsden, W.H.R., Kimber, C.D., et al. 1981. Field diagnosis of sleeping sickness in the Ivory Coast. 1. Comparison of the miniature anion-exchange/centrifugation technique with other protozoological methods. *Trans R Soc Trop Med Hyg*, **75**, 242–50.

Maudlin, I. and Ellis, D. 1985. Extrachromasomal inheritance of susceptibility to trypanosome infection in tsetse flies. 1. Selection of susceptible and refractory lines of *Glossina morsitans morsitans*. *Ann Trop Med Parasitol*, **79**, 317–24.

McConville, M.J., Mullin, K.A., et al. 2002. Secretory pathway of trypanosomatid parasites. *MMBR*, **66**, 122–54.

Mehlert, A., Bond, C.S. and Ferguson, M.A.J. 2002. The glycoforms of a *Trypanosoma brucei* variant surface glycoprotein and molecular modelling of a glycosylated surface coat. *Glycobiology*, **12**, 607–12.

Mehlert, A., Zitzmann, N., et al. 1998. The glycosylation of the variant surface glycoproteins and procyclic acidic repetitive proteins of *Trypanosoma brucei*. *Mol Biochem Parasitol*, **91**, 145–52.

Mellors, A. and Samad, A. 1989. The acquisition of lipids by African trypanosomes. *Parasitol Today*, **5**, 239–44.

Molyneux, D.H. 2001. African trypanosomiasis, failure of science and public health. In Black, S.J. and Seed, J.R. (eds). *World class parasites*, vol. 1, *The African trypanosomes*. Boston: Kluwer Academic Publishers, 1–10.

Molyneux, D.H. and Ashford, R.W. 1983. *The biology of Trypanosoma and Leishmania. Parasites of man and domestic animals*. London: Taylor and Francis.

Molyneux, D.H., DeRaadt, P. and Seed, J.R. 1984. African human trypanosomiasis, epidemiological, experimental, and clinical aspects. In: Labno, J. and Gilles, H.M. (eds), *Recent advances in tropical medicine*. New York: Livingstone Churchill, 39–62.

Morgan, G.A., Hamilton, E.A. and Black, S.J. 1996. The requirements for G1 checkpoint progression of *Trypanosoma brucei* S 427 clone 1. *Mol Biochem Parasitol*, **78**, 195–207.

Morgan, G.W., Allen, C.A., et al. 2001. Developmental and morphological regulation of clathrin-mediated endocytosis in *Trypanosoma brucei*. *J Cell Sci*, **114**, 2605–15.

Morita, Y.S., Paul, K.S. and Englund, P.T. 2000. Specialized fatty acid synthesis in African trypanosomes, myristate for GPI anchors. *Science*, **288**, 140–3.

Morris, J.C., Drew, M.E., et al. 2001. Replication of kinetoplast DNA, an update for the new millennium. *Int J Parasitol*, **31**, 453–8.

Muranjan, M., Wang, Q., et al. 1997. The trypanocidal Cape buffalo serum protein is xanthine oxidase. *Infect Immun*, **65**, 3806–14.

Murphy, N.B. and Olijhoek, T. 2001. Trypanosome factors controlling population size and differentiation status, In Black, S.J. and Seed, J.R. (eds). *World class parasites*, vol. 1, *The African trypanosomes*. Boston: Kluwer Academic Publishers, 113–26.

Mussmann, R., Janssen, H., et al. 2003. The expression level determines the surface distribution of the transferrin receptor in *Trypanosoma brucei*. *Mol Microbiol*, **47**, 23–35.

Nantulya, V.M., Doua, F. and Molisho, S. 1992. Diagnosis of *Trypanosoma brucei gambiense* sleeping sickness using an antigen detection enzyme-linked immunosorbent assay. *Trans R Soc Trop Med Hyg*, **86**, 42–5.

Navarro, M., Cross, G.A.M. and Wirtz, E. 1999. *Trypanosoma brucei* variant surface glycoprotein regulation involves coupled activation/inactivation and chromatin remodeling of expression sites. *EMBO J*, **18**, 2265–72.

Nolan, D.P., Geuskens, M. and Pays, E. 1999. Linear poly-n-acetyllactosamine as sorting signals in exo/endocytosis in *Trypanosome brucei*. *Curr Biol*, **9**, 1169–72.

Ogbadoyi, E., Ersfeld, K. and Robinson, D. 2000. Architecture of the *Trypanosoma brucei* nucleus during interphase and mitosis. *Chromosoma*, **108**, 501–13.

Okoth, J.O. 1986. Community participation in tsetse control. *Parasitol Today*, **2**, 88.

Opiyo, E.A., Njogu, A.R. and Omuse, J.K. 1990. Use of impregnated targets for control of *Glossina pallidipes* in Kenya. *Insect Sci Appl*, **11**, 417–25.

Opperdoes, F.R. 1987. Compartmentation of carbohydrate metabolism in trypanosomes. *Ann Rev Microbiol*, **41**, 127–51.

Opperdoes, F.R., Wierenga, R.K., et al. 1990. Unique properties of glycosomal enzymes. In: Agabian, N. and Cerami, A. (eds), *Parasites: molecular biology, drug and vaccine design*. New York: Wiley-Liss, 233–46.

Otieno, L.H., Darji, N., et al. 1983. Some observations of factors associated with the development of *Trypanosoma brucei brucei* infections in *Glossina morsitans morsitans*. *Acta Trop*, **40**, 113–20.

Pal, A., Hall, B.S., et al. 2002. Differential endocytic functions of *Trypanosoma brucei* Rab isoforms reveal a glycosylphosphatidylinositol-specific endosomal pathway. *J Biol Chem*, **277**, 9529–39.

Parsons, M., Furuya, T., et al. 2001. Biogenesis and function of peroxisomes and glycosomes. *Mol Biochem Parasitol*, **115**, 19–28.

Pegg, A.E. and McCann, P.P. 1988. Polyamine metabolism and function in mammalian cells and protozoans. In *ISI Atlas of Science: Biochemistry*. Philadelphia: Institute for Science Information Incorporated, 11–18.

Penchenier, L., Jannin, J., et al. 1991. Le problème de l'interpretation du CATT dans le depistage de la trypanosomiase humaine à *Trypanosoma brucei gambiense*. *Ann Soc Belge Méd Trop*, **71**, 221–8.

Pepin, J. and Milord, F. 1994. The treatment of human African trypanosomiasis. *Adv Parasitol*, **33**, 2–47.

Perez-Morga, D., Amiguet-Vercher, A., et al. 2001. Organization of telomeres during the cell and life cycles of *Trypanosoma brucei*. *J Eukaryot Microbiol*, **48**, 221–6.

Radomski, M.W., Buguet, A., et al. 1995. Twenty-four-hour plasma cortisol and prolactin in human African trypanosomiasis patients and healthy African controls. *Trans R Soc Trop Med Hyg*, **52**, 281–6.

Raper, J., Fung, R., et al. 1999. Characterization of a novel trypanosome lytic factor from human serum. *Infect Immun*, **67**, 1910–16.

Reduth, D., Grootenhuis, J.G., et al. 1994. African buffalo serum contains novel trypanocidal protein. *J Eukaryot Microbiol*, **41**, 95–103.

Reincke, M., Allolio, B., et al. 1993. Thyroid dysfunction in African trypanosomiasis: a possible role for inflammatory cytokines. *Clin Endocrinol*, **39**, 455–61.

Reincke, M., Heppner, C., et al. 1994. Impairment of adrenocortical function associated with increased plasma tumor necrosis factor-alpha and interleukin-6 concentrations in African trypanosomiasis. *Neuroimmunomodulation*, **1**, 14–22.

Rickman, L.R. and Robson, J. 1970. The blood incubation infectivity test, a simple test which may serve to distinguish *Trypanosoma brucei* from *T. rhodesiense. Bull World Health Organ*, **42**, 650–1.

Robson, J. and Rickman, L.R. 1978. The effect of human serum in vitro on *Trypanosoma (Trypanozoon) brucei* species trypanosomes and its relationship to infectivity in the blood incubation infectivity test. *Med J Zambia*, **11**, 156–8.

Rogers, D.J. and Randolph, S.E. 2002. A response to the aim of eradicating tsetse from Africa. *Trends Parasitol*, **18**, 534–6.

Ruepp, S., Furger, A., et al. 1997. Survival of *Trypanosoma brucei* in the tsetse fly is enhanced by the expression of specific forms of procyclin. *J Cell Biol*, **137**, 1369–79.

Sands, M., Kron, M.A. and Brown, R.B. 1985. Pentamidine. A review. *Rev Infect Dis*, **7**, 625–34.

Schechter, P.J., Barlow, J.L.R. and Sjoerdsma, A. 1987. Clinical aspects of inhibition of ornithine decarboxylase with emphasis on therapeutic trials of eflornithine (DFMO) in cancer and protozoan disease. In: McCann, P.P., Pegg, A.E. and Sjoerdsma, A. (eds), *Inhibition of polyamine metabolism*. New York: Academic Press, 345–64.

Schlimme, W., Burri, M., et al. 1993. *Trypanosoma brucei brucei*: differences in the nuclear chromatin of bloodstream forms and procyclic culture forms. *Parasitology*, **107**, 237–47.

Seed, J.R. 2000. Current status of African trypanosomiasis. *Am Soc Microbiol News*, **66**, 395–402.

Seed, J.R. and Hall, J.E. 1992. Trypanosomes causing disease in men in Africa. In: Kreier, J.P. and Baker, J.R. (eds), *Parasitic protozoa*, vol. 2. . New York: Academic Press, 85–155.

Smith, A.B., Esko, J.D. and Hajduk, S.L. 1995. Killing of trypanosomes by the human haptoglobin-related protein. *Science*, **268**, 284–6.

Sommer, J.M. and Wang, C.C. 1994. Targeting proteins to the glycosomes of African trypanosomes. *Ann Rev Microbiol*, **48**, 105–38.

Soudan, B., Boersma, A., et al. 1993. Hypogonadism induced by African trypanosomes in humans and animals. *Comp Biochem Physiol Comp Physiol*, **104**, 757–63.

Steverding, D. 2000. The transferrin receptor of *Trypanosoma brucei. Parasitol Int*, **48**, 191–8.

Stevens, J.R. and Godfrey, D.G. 1992. Numerical taxonomy of *Trypanozoon* based on polymorphisms in a reduced range of enzymes. *Parasitology*, **104**, 75–86.

Toriello, K.M., Buchanan, J.A., et al. 2002. Abstract 10F Vizualization of lipid rafts in vivo: specialization of the trypanosome flagellar membrane. Woods Hole Molecular Parasitology Meeting XIII. http://e2kroos.cis.upenn.edu/mpm-2002/abstracts

Truc, P., Aerts, D., et al. 1992. Direct isolation in vitro of *Trypanosoma brucei* from man and other animals and its potential value for the diagnosis of Gambian trypanosomiasis. *Trans R Soc Trop Med Hyg*, **86**, 627–9.

Truc, P., Bailey, J.W., et al. 1994. A comparison of parasitological methods for the diagnosis of Gambian trypanosomiasis in an area of low endemicity in Cote d'Ivoire. *Trans R SocTrop Med Hyg*, **88**, 419–21.

Vale, G.A. 1993. Development of baits for tsetse flies (Diptera, Glossinidae) in Zimbabwe. *J Med Entomol*, **30**, 831–42.

Vale, G.A. and Hall, D.R. 1985a. The role of-octen-3-ol, acetone, and carbon dioxide in the attraction of tsetse flies, *Glossina* spp. (Diptera, Glossinidae) to ox odour. *Bull Entomol Res*, **75**, 209–17.

Vale, G.A. and Hall, D.R. 1985b. The role of 1-octen-3-ol, acetone and carbon dioxide to improve baits for tsetse flies, *Glossina* spp. (Diptera, Glossinidae). *Bull Entomol Res*, **75**, 219–31.

Vale, G.A., Hall, D.R. and Gough, A.J.E. 1988. The olfactory responses of tsetse flies, *Glossina* spp (Diptera, Glossinidae), to phenols and urine in the field. *Bull Entomol Res*, **78**, 293–300.

Vandeweerd, V. and Black, S.J. 1989. Serum lipoprotein and *Trypanosoma brucei brucei* interactions in vitro. *Mol Biochem Parasitol*, **37**, 201–11.

Vanhamme, L., Lecordier, L. and Pays, E. 2001. Control and function of the bloodstream variant surface glycoprotein expression sites in *Trypanosoma brucei. Int J Parasitol*, **31**, 522–30.

Vanhamme, L., Paturiaux-Hanocq, F., et al. 2003. Apolipoprotein L-1 is the trypanosome lytic factor of human serum. *Nature (London)*, **422**, 83–7.

Vassella, E. and Boshart, M. 1996. High molecular weight agarose matrix supports growth of bloodstream forms of pleomorphic *Trypanosoma brucei* strains in axenic culture. *Mol Biochem Parasitol*, **82**, 91–105.

Vickerman, K. 1989. Trypanosome sociology and antigenic variation. *Parasitology*, **99**, S37–47.

Vickerman, K., Tetley, L., et al. 1988. Biology of African trypanosomes in the tsetse fly. *Biol Cell*, **64**, 109–19.

Vreysen, M.J.B., Saleh, K.M., et al. 2000. *Glossina austeni* (Diptera, Glossinidae) eradicated on the island of Unguja, Zanzibar, using the sterile insect technique. *J Econ Entomol*, **93**, 123–35.

Wang, J., Van Praagh, A., et al. 2002. Serum xanthine oxidase, origin, regulation and contribution to control of trypanosome parasitemia. *Antioxid Redox Signal*, **4**, 161–78.

Wang, Q., Murphy, N. and Black, S.J. 1999. Infection-associated decline of Cape buffalo blood catalase augments serum tryoanocidal activity. *Infect Immun*, **67**, 2797–803.

Williams, B., Dransfield, R., et al. 1993. Where are we now? Trypanosomiasis. *Health Pol Plan*, **8**, 85–93.

Welburn, S.C., Picozzi, K., et al. 2001. Identification of human-infective trypanosomes in animal reservoir of sleeping sickness in Uganda by means of serum-resistance-associated (SRA) gene. *Lancet*, **358**, 2017–19.

WHO, 1986. *Epidemiology and control of African trypanosomiasis.* WHO Technical Report Series No. 739. World Health Organization, Geneva.

Woo, P.T.K. 1971. Evaluation of the hematocrit centrifuge and other techniques for the field diagnosis of human trypanosomiasis and filariasis. *Acta Trop*, **28**, 298–303.

Woo, P.T.K. and Hauck, L. 1987. The haematocrit centrifuge smear technique for the detection of mammalian *Plasmodium. Trans R Soc Trop Med Hyg*, **81**, 727–8.

Wyatt, G.B., Boatin, B.A. and Wurapa, F.K. 1985. Risk factors associated with the acquisition of sleeping sickness in north-east Zambia, a case-control study. *Ann Trop Med Parasitol*, **79**, 385–92.

Xong, H.V., Vanhamme, L., et al. 1998. A VSG expression site-associated gene confers resistance to human serum in *Trypanosoma rhodesiense. Cell*, **95**, 839–46.

19

New World trypanosomiasis

MICHAEL A. MILES

INTRODUCTION

In 1907, the Brazilian scientist Carlos Chagas left the city of Rio de Janeiro to work as a malaria control officer at Lassance in the state of Minas Gerais. Chagas noted that poor houses in the area were infested by a large blood-sucking insect, the triatomine bug (Hemiptera, Reduviidae) (Figure 19.1). He was aware that blood-sucking insects transmitted human diseases such as malaria and immediately suspected that triatomine bugs might also carry infectious agents. He examined bug feces and found a flagellated protozoan parasite. At the Manguinhos Institute in Rio de Janeiro, marmosets exposed to infected bugs developed blood parasitemias of a new trypanosome, which Chagas named *Trypanosoma cruzi* after his mentor Oswaldo Cruz.

Back in Lassance, *T. cruzi* was found in the blood of sick children living in bug-infested houses (Figure 19.2). Chagas and his distinguished colleagues from what is now the Instituto Oswaldo Cruz, went on to describe clinical aspects of the disease, the life cycle, experimental animal models, and to discover natural mammalian hosts such as the armadillo. These discoveries were remarkable, not least because *T. cruzi* was found first in its insect vector. Chagas' early work was initially controversial, and it was not until some years later that the public health importance of Chagas disease became apparent when scientists in other Latin American countries reported its widespread distribution. The true route

of transmission, through contamination with infected bug feces, was conclusively demonstrated in 1912 by Emile Brumpt. The history of the discovery and early investigations of Chagas disease is described by Miles (1996, 2004) and Lewinsohn (2003).

Chagas disease, for those who survive the acute phase of infection, primarily affects the heart and alimentary tract. Its pathogenesis remains enigmatic in that many infected people remain healthy for life and the precise reasons for a poor prognosis are not fully understood. There has been an intense expansion of interest in *T. cruzi*, driven largely by a desire to understand the disease process, how the organism survives for life in the infected mammalian host, and the unique features of trypanosome molecular biology. *T. cruzi* is also now known to be an opportunistic infection: in addition to being transmissible by blood and organ donors, it also relapses in immuno-compromised individuals. South American trypanosomiasis is predominantly a disease of poverty. As such it is amenable to control by public health interventions to manage the insect vectors, supplemented by measures to eliminate transmission by blood transfusion.

T. cruzi is one of many trypanosome species in the New World, but only one other species is known to infect humans. Like *T. cruzi*, *Trypanosoma rangeli* was first found in triatomine bugs and later in children, but it is not considered to be pathogenic and its medical importance lies in the need to distinguish *T. cruzi* and *T. rangeli* infections during diagnosis, by isolation of the

Figure 19.1 *An adult female triatomine bug* (Panstrongylus megistus) *(Courtesy of T.V. Barrett)*

parasite. Unlike *T. cruzi*, *T. rangeli* is transmitted by the bite of the vector, through infection of the triatomine bug salivary glands.

Hoare's classic monograph (1972) still contains an excellent introduction to the New World human trypanosomiases; Tyler and Miles (2003) gives a recent review on research and WHO (2002) and Maudlin et al. (2004) provide texts on clinical aspects, diagnosis, and disease control.

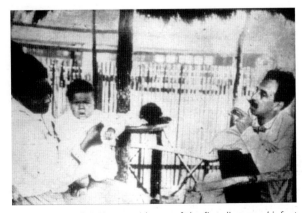

Figure 19.2 *Carlos Chagas with one of the first discovered infant cases of* Trypanosoma cruzi *infection*

CLASSIFICATION

T. cruzi is a kinetoplastid protozoan parasite in the family Trypanosomatidae, which includes the disease agents of both the leishmaniases and the trypanosomiases. The genus *Trypanosoma* includes two species responsible for major public health problems in South America and Africa (*T. cruzi* and *Trypanosoma brucei*, respectively) and others that are not pathogenic (e.g. *T. rangeli*). *T. cruzi* falls into the section Stercoraria (as distinct from the section Salivaria of *T. brucei*; see Chapter 18, African trypanosomiasis). The Stercoraria are characterized as trypomastigotes with a free flagellum, a large kinetoplast that is not terminal, pointed posterior end, discontinuous reproduction in the mammalian host and contaminative transmission (except *T. rangeli*) and nonpathogenic (except *T. cruzi*). The stercorarian trypanosomes, with the notable exception of *T. rangeli*, are transmitted from the vector by contamination with the insect's feces. Reproduction in the mammalian host is typically discontinuous – that is, it is confined to particular life cycle stages or tissue sites. *T. cruzi* is the type species of the subgenus *Schizotrypanum*, an assemblage of morphologically indistinguishable species parasitic in diverse mammals, mostly restricted to New World and comprising *T. cruzi* plus *T. cruzi*-like organisms in the Americas, and a number of cosmopolitan bat trypanosomes. The subgenus *Schizotrypanum* is characterized by trypomastigotes that are small, typically C-shaped with a large kinetoplast near the short, pointed posterior end, and have intracellular reproduction as amastigotes in the mammalian host.

As described below, chemical taxonomic methods, especially phenotypic comparisons by isoenzyme electrophoresis and, more recently, comparative DNA analyses, have demonstrated a remarkable diversity within the *T. cruzi* species and the existence of at least two principal subdivisions named *T. cruzi* I and *T. cruzi* II (Campbell et al. 2004).

STRUCTURE AND LIFE CYCLE

T. cruzi is a eukaryote with a nucleus; it has chromosomes that do not condense during cell division and can only be resolved by electrophoretic methods devised to separate large molecules of DNA. Extranuclear DNA is present in the form of a discrete, visible organelle, the kinetoplast, containing minicircle and maxicircle DNA and associated with a large single-branched cristate mitochondrion. The kinetoplast lies adjacent to a basal body and flagellar pocket from which a flagellum emerges (except in the amastigote stage of the life cycle) having nine peripheral pairs of microtubules, a central doublet and a parallel paraxial rod. The flagellum runs alongside the main body of the organism, in a posterior-to-anterior direction, and adhering to it to form an undulating membrane leading to an anterior free flagellum. Interior organelles include an endoplasmic reticulum and a Golgi apparatus, and a specialized glycolytic organelle (the glycosome). The organism is bound by a complex network of subpellicular microtubules. Hoare (1972) gives ranges of dimensions for proven *T. cruzi* trypanosomes from humans as: length 11.7–30.4 μm; free flagellum 2.0–11.2 μm; breadth 0.7–5.9 μm. Further details of structure can be found in elegant electron microscopical studies of *T. brucei*, with which *T. cruzi* shares many features, and supplementary studies of *T. cruzi* (Tetley and Vickerman 1991).

There are three principal stages in the *T. cruzi* life cycle. The amastigote stage multiplies within nonphagocytic and phagocytic cells by binary fission. The epimastigotes, which have a kinetoplast adjacent to the nucleus and an undulating membrane that runs along approximately the anterior half of the organism, divide by binary fission in the hindgut of the triatomine bug vector. Trypomastigotes, with the kinetoplast at the posterior end, do not divide and are the forms found circulating in the blood of the mammalian host. Trypomastigotes are also the infective (metacyclic) stage that occurs in the rectum of the vector and in bug feces deposited during feeding. There are, therefore, some similarities between the life cycle stages of *T. cruzi* and those of both *Leishmania* and *T. brucei* in that *Leishmania* also has amastigotes that divide intracellularly (see Chapters 16, Old World leishmaniasis and 17, New World leishmaniasis) and *T. brucei* has circulating trypanosomes but, in contrast to *T. cruzi*, the latter divide in the blood by binary fission (see Chapter 18, African trypanosomiasis).

The life cycle is summarized in Figure 19.3. Metacyclic trypomastigotes deposited on the mammalian host in bug feces have the capacity to penetrate abraded skin or the wound made by the bite of the bug. They can also cross the oral and nasal mucosae, or the conjunctiva if bug feces get into the eye. These metacyclic forms are

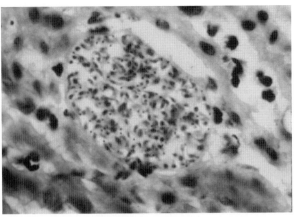

Figure 19.4 *Pseudocyst of* Trypanosoma cruzi *in umbilical cord (Courtesy of Dr Hipolito de Almeida)*

slender, highly motile organisms, which often rapidly traverse the field of view when seen by light microscopy. Once inside the mammalian host, they can penetrate phagocytic or nonphagocytic cells to form a local cutaneous or ocular lesion. Within the cell, the vacuole containing the *T. cruzi* trypomastigote fuses with lysosomes to form a phagolysosome, from which the organism then escapes to lie free in the cytosol. The trypomastigote transforms to an amastigote which divides by binary fission forming a pseudocyst (false cyst), so called because it has no true cyst wall but is simply bounded by the membrane of the host cell. About 5 days later, there may be up to 500 amastigotes transforming to small motile C-shaped trypomastigotes within the pseudocyst (Figure 19.4). The trypomastigotes are released into the surrounding tissue, either to infect other cells and repeat the intracellular cycle or to circulate in the blood (Figure 19.5). When the pseudocyst ruptures, not all the amastigote forms may have transformed to trypomastigotes. Amastigotes released when the cell bursts are thought to be destroyed locally but are occasionally found circulating in the blood of mice with fulminating experimental infections.

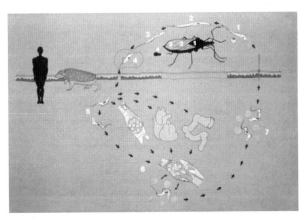

Figure 19.3 *A summary of the life cycle of* Trypanosoma cruzi *(see text) (Courtesy of Meddia)*

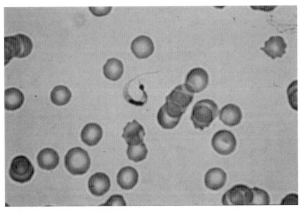

Figure 19.5 Trypanosoma cruzi: *C-shaped trypomastigote in circulating blood (see text)*

Two types of transformation from amastigote to trypomastigote have been described within the pseudocyst. The first, fusiform transformation, involves elongation of the body and migration of the kinetoplast. The second, orbicular cycle transformation is by unrolling of sphaeromastigotes (flagellated amastigote forms with a flagellum that encircles the body without producing an undulating membrane) that have a central vacuole: this orbicular transformation may, however, be an artefact of dry fixation and staining of tissue smears (Hoare 1972).

The released trypomastigotes do not all have the same morphology and behavior. There are two types: slender highly motile trypomastigotes, reminiscent of the infective metacyclic forms found in bug feces, and smaller, broader less motile forms. The slender forms tend to traverse the field rapidly when seen microscopically, whereas the broader forms, although motile, remain for a longer time in the same field of view. The slender motile form has an elongated nucleus, a subterminal kinetoplast and a short free flagellum, whereas the broad form has an oval nucleus, almost terminal kinetoplast and a long free flagellum. The slender forms are generally only seen in the blood of experimental animals, such as mice, during initial acute or fulminating infections (Risso et al. 2004). It has been proposed, although not proven, that the slender forms are pre-adapted to penetrating cells and renewing the intracellular cycle, whereas the shorter, broader forms may persist in the blood to be taken up by the vector with the bloodmeal. Unlike T. brucei, however, both slender and broad forms are thought to have active mitochondria (in T. brucei dividing slender forms depend on glycolysis and the mitochondria are only fully active in the stumpy forms, to prepare them for life in the tsetse fly).

Except in experimental models with T. cruzi strains that are highly virulent to mice, the number of circulating trypomastigotes in the blood is generally quite low (less than one per field at ×400 magnification in a fresh blood film) and the blood parasitemia rapidly becomes subpatent to microscopy. Low levels of circulating organisms, which may remain for the life of the host, can then be detected only by more sensitive methods.

The development of T. cruzi in the triatomine bug is confined to the alimentary tract. Bugs acquire infection by feeding on an infected mammalian host (or rarely by cannibalizing other recently fed bugs) and there is no transovarial transmission from adult female to egg. Once infected, triatomine bugs retain infection throughout the molting cycles. The development of T. cruzi in the vector, dependent on the stage of the insect, in general takes around 10–15 days. Trypomastigotes in the bloodmeal transform to amastigotes and sphaeromastigotes in the foregut, and multiply by binary fission. In the midgut, division is by binary fission in the epimastigote stage. In the rectum, epimastigotes attach to the epithelium and transformation to metacyclic trypomastigotes begins. More detailed descriptions of the life-cycle stages can be found in Hoare (1972) and Brack (1968).

Contamination with infected bug feces during or just after the bloodmeal is not the only route of transmission. Blood transfusion transmission is commonplace in endemic areas where blood is not screened for the presence of T. cruzi antibodies or not pretreated with gentian violet (WHO 2002). All blood donors who have resided in endemic areas and been exposed to triatomine bugs should be screened for antibodies, as should all organ donors or organ recipients with relevant histories. Antibody testing of blood donors is mandatory in some countries. It has been shown experimentally that transmission may also occur by the oral route, by consumption either of triatomine bugs or of food contaminated with bug feces, and possibly by eating uncooked blood and tissues from infected reservoir hosts. Oral transmission may be important in sustaining prevalence rates in insectivorous mammals, particularly those that live in burrows infested with bugs, and several small outbreaks of simultaneous acute cases within families are almost certainly due to oral transmission of T. cruzi (see below). Congenital transmission also occurs in a small proportion of children born to seropositive mothers (WHO 2002). Sexual transmission is thought to be extremely rare, as is transmission through the milk from mother to suckling infant. Accidental transmission in the laboratory is not uncommon, usually through inoculation or ingestion. There are no other important insect vectors, although T. cruzi may survive in cimicid bugs and produce metacyclic trypomastigotes in the hindgut, which are thought to have given rise to infections in primate colonies, possibly by the oral route. Persistent T. cruzi infections may also occur experimentally in ticks (Ornithodorus moubata) (Hoare 1972).

All stages of the life cycle of T. cruzi are reproducible in culture. Thus, epimastigotes are easily grown on blood agar overlays. A wide range of mammalian cell lines can be infected to produce mature pseudocysts with motile trypomastigotes, which are seen as 'boiling cells' by phase microscopy. Emergent trypomastigotes re-enter cells in vitro such that several intracellular cycles can occur on a single cell monolayer maintained for 20 days or more (Miles 1993).

The life cycle of T. rangeli in the arthropod vector is quite distinct from that of T. cruzi in that, in bugs belonging to the genus Rhodnius, the hemolymph is often invaded, usually more than 40 days after the infective bloodmeal. Salivary gland infections become established in a proportion of bugs that have hemolymph infections. T. rangeli is pathogenic to its insect vector and infected bugs may die or not molt successfully. The development of T. rangeli in the lumen of the bug gut and in the hemolymph is indistinguishable. Multiplication occurs in the mammalian host but blood parasitemias are scanty. T. rangeli is capable of infecting an histiocytic cell line in vitro and surviving as nondividing amastigote-like forms, but it is not clear whether such intracellular stages contribute to survival in the mammalian host (Eger-Mangrich et al. 2001). The insect vector stages can be grown in vitro. The importance of *T. rangeli* as a human infective agent is secondary, in that it has to be distinguished from *T. cruzi* during diagnostic procedures.

BIOCHEMISTRY, MOLECULAR BIOLOGY, AND GENETICS

Biochemistry

Drug development is often quoted as the incentive for studying the biochemistry of infectious disease agents such as *T. cruzi*. Thus, the absence of some metabolic pathways may make *T. cruzi* vulnerable to drug action that is circumvented by normal mammalian metabolism and, therefore, not toxic to humans. Alternatively, unique features of trypanosome metabolism may expose the organism to drugs that have no equivalent target in humans. An example of absence of a metabolic pathway is the inactive mitochondrion in the blood-stage form of *T. brucei*, mentioned above (and see Chapter 18, African trypanosomiasis), although this is not the case for *T. cruzi*. Examples of prominent differences between the biochemistry of trypanosomatids and mammals have been found in purine salvage pathways, trypanothione biosynthesis and catabolism, sterol synthesis, and glycosomal metabolism. Several enzymes are reminiscent of those found in plants, suggesting that such enzymes are prospective drug targets, vulnerable to drugs based on herbicides (Martin and Borst 2003; Wilkinson et al. 2002b). In addition, other unusual features of the molecular biology of trypanosomes have excited great interest among molecular biologists and encouraged their use as model organisms. These include the presence of the kinetoplast and radical editing of mitochondrial mRNA, discontinuous transcription and *trans*-splicing, polycistronic transcription and antigenic variation with epigenetic mechanisms of controlling gene expression.

T. cruzi utilizes sugars, especially glucose, through active transport into the cell; nevertheless, it is not dependent on glucose and can survive by catabolism of amino acids and proteins – there appear to be no major reserve polysaccharides. Glucose catabolism, incomplete even in the presence of oxygen, produces succinic acids and acetic acid; anaerobic catabolism leads mainly to succinate and L-alanine. Enzymes of the glycolytic pathway, all of which have been reported, mostly occur in a specialized organelle, the glycosome, related to the peroxisome, which assists fatty acid metabolism in eukaryotes. The pentose phosphate pathway of glucose catabolism is also present. Amastigote, epimastigote, and trypomastigote stages have almost all enzymes of the tricarboxylic acid cycle (Cazzulo 1994).

Glycosomes lack catalase and oxidases, which regulate hydrogen peroxide in peroxisomes, but contain enzymes for conversion of glucose and glycerol to phosphoglycerate, as well as enzymes for β-oxidation of fatty acids and for ether–lipid biosynthesis; other activities include pyrimidine biosynthesis, purine salvage and carbon dioxide fixation. Enzymes found in the glycosome of *T. cruzi* include hexokinase, phosphoglucose isomerase, phosphofructokinase, aldolase, triosephosphate isomerase, glyceraldehyde 3-phosphate dehydrogenase, glycerol 3-phosphate dehydrogenase, glycerol kinase, malate dehydrogenase, adenylate kinase, and phosphoenolpyruvate carboxykinase; some of these enzymes are also present in the cytosol. Transport of enzymes into the glycosome does not involve larger cytosolic precursors and cleavage of signal sequences. Peptide recognition signals govern importation, as is the case for peroxisomes (Cáceres et al. 2003; Wilkinson et al. 2002a).

Glutathione (GSH) is the thiol of low molecular weight that is largely responsible for protection against free radicals and reactive oxygen in most aerobic organisms. In trypanosomatids, GSH is replaced by trypanothione, a glutathione dimer conjugated to spermidine. In *T. cruzi* glutathione reductase (GR) is largely replaced by trypanothione reductase (TR), which reduces trypanothione disulfide (T[S]$_2$) to dihydrotrypanothione (T[SH]$_2$) with NADPH as cofactor. Substrate utilization of TR is distinct from that of GR but catalytic mechanism is conserved (Wilkinson et al. 2002a). The two negatively charged carboxyl groups on the glycyl termini of the GR substrate, glutathione disulfide (GSSG), are replaced by internal amide groups in T[S]$_2$ and an amino group carrying a positive charge on the spermidine. Also, the two glutathione chains in GSSG can rotate freely about the disulfide bridge whereas the T[S$_2$] is more rigid due to the linkage by spermidine. There are distinct distributions of hydrophobic and charged residues in GR and TR adjacent to the glycyl regions of the substrate. The R$_{37}$ and R$_{347}$ amino acids of GR, which are thought to bind the carboxyl groups of GSSG, are replaced by tryptophan and alanine (Taylor et al. 1994). The structural consequences of an engineered change in substrate specificity have been resolved by crystallography (Stoll et al. 1997). Studies of trypanothione metabolism in *T. cruzi* have been an interesting model for the development of rational chemotherapy (Bonse et al. 2000). Other aspects of oxidative defense, such as the presence of an ascorbate hemoperoxidase, may be therapeutically exploitable (Wilkinson et al. 2002b).

Parasitic protozoa such as *T. cruzi* rely on the salvage of free purines or re-use of purines from nucleic acids or nucleotides, whereas mammals can synthesize purines de novo. The hypoxanthine analogue allopurinol has been used as an inhibitor of purine salvage but its value as a therapeutic drug is still greatly in doubt (Stoppani 1999).

There has been much research interest in proteinases of *T. cruzi* with the idea that they may be important for invasion of mammalian host cells, and targets for chemotherapy. Cysteine proteinases are the most prominent of the four main groups (aspartic, cysteine, metallo- and serine) in *T. cruzi* although metalloproteinases and serine proteinase are also present (Cazzulo 2002). There may be several other roles for *T. cruzi* proteinases, such as protection against antibody attack, or adaptation to breakdown of the bloodmeal in the insect vector. Proteinase activities can be demonstrated by using electrophoresis in gels with gelatin or fluorogenic peptides as substrates. A 60 kDa cysteine proteinase (cruzipain) resembles cathepsin and papain, and is also known as the GP57/51 antigen, which is recognized by sera from patients infected with *T. cruzi*. The enzyme produces a C-terminal extension domain by self-proteolysis (Alvarez et al. 2002).

Glycosylphosphatidylinositol (GPI) and GPI-related glycolipids are abundant in trypanosomatids. They serve to anchor surface glycoproteins and, although *T. cruzi* does not have the variant surface glycoprotein (VSG) coat characteristic of the African trypanosomes (see Chapter 18, African trypanosomiasis), several glycoproteins are bound to the surface of *T. cruzi* by GPI anchors, including the Ssp-4 antigen of amastigotes and the 90 kDa antigen of trypomastigotes. Ferguson and his collaborators have devoted much effort to determining the GPI biosynthetic pathway in African trypanosomes, not only for the intrinsic biochemical interest, but also because this pathway might present a target for trypanosome-specific chemotherapy (Crossman et al. 2002; Lillico et al. 2003).

Another biochemical feature of *T. cruzi*, distinct from the vertebrate host, is the requirement for specific endogenous sterols such as egosterol, rather than cholesterol. This has led to promising new inhibitors of sterol biosynthesis in *T. cruzi* (Urbina 2002; Urbina et al. 2003).

Molecular biology

There is a plethora of recent advances in understanding the basic molecular biology of trypanosomes, fuelled by an international interest in their unusual features and their role as models for other studies. Two molecular processes that have received detailed attention are discontinuous transcription, or *trans*-splicing, and RNA editing of transcripts from the maxicircle mitochondrial genome.

Studies of antigenic variation in African trypanosomes revealed that each primary mRNA transcript was processed by addition of a 39-nucleotide spliced leader sequence transcribed from a separate chromosomal location – the mini-exon genes. The spliced leader is added by *trans*-splicing via Y-shaped intermediates that consist of the 3'-leader precursor linked to the 5'-portion of the main transcript. *Trans*-splicing is assisted by small nuclear RNAs (snRNAs) complexed to proteins, as is intron splicing in higher eukaryotes. *Cis*-splicing introns have also been discovered in *T. cruzi* (Mair et al. 2000). Many trypanosome genes are polycistronic, with transcripts of several genes in tandem on an initial primary transcript. The mini-exon derived RNA (medRNA) is thought to be functionally equivalent to the U1-snRNA present in other eukaryotes. Mature mRNAs of *T. cruzi* are polyadenylated. Pyrimidine (C,T)-rich regions of downstream splice acceptor sites apparently participate in the control of polyadenylation of upstream portions of the primary transcript. This may have a role in gene regulation (Matthews et al. 1994).

For many years, the function of trypanosome kinetoplast minicircles was obscure. It was finally discovered that some mitochondrial maxicircle genes are condensed and that guide RNAs (gRNAs) transcribed from minicircles carry information for editing maxicircle transcripts, by the insertion (and deletion) of uridines. It is not clear why such a complex RNA editing process is required by organisms such as *T. cruzi*, but it may be linked to developmental regulation during movement between host, vector and life cycle stages (Simpson et al. 2003).

Other research on the molecular biology of *T. cruzi* has concerned the cloning of genes encoding products involved in cell invasion or in gene expression during transformation to the infective form (metacyclogenesis), the structural and functional analysis of genes encoding antigens recognized by the host immune response, and cloning portions of the genome identified by probing with heterologous sequences from other organisms.

T. cruzi carries a surface *trans*-sialidase which transfers sialic acid from host components to mucin-like glycoproteins on the surface of the parasite. In this way, *trans*-sialidase is thought to influence adhesion and penetration of host cells and may also have a role in trypomastigote escape from the phagolysosome into the cytoplasm of the host cell (Tan and Andrews 2002; Buschiazzo et al. 2002). There is a large family of putative mucin genes in *T. cruzi*, with hundreds of copies per haploid genome, which share a signal peptide on the N-terminus and a presumed GPI anchoring sequence on the C-terminus, with hypervariable central regions (Argibay et al. 2002; Buscaglia et al. 2004). Mucins isolated from non-infective epimastigote and infective metacyclic trypomastigotes differ in lipid structure. Metacyclogenesis may be induced in the triatomine bug gut by products of hemoglobin digestion: synthetic peptides corresponding to amino acids 30–49 and 35–73 of αD-globin have stimulated differentiation of epimastigotes into metacyclic trypomastigotes (Garcia et al. 1995). There is a large heterogeneous *trans*-sialidase multi-gene family in *T. cruzi* which, among other members, includes genes for *T. cruzi* neuraminidase

(TCNA), shared acute phase antigen (SAPA), and the GP85 surface glycoproteins. The gene family has been classified into four groups. It has been proposed that changes in expression of such multigene families may provide a mechanism by which *T. cruzi* evades the host immune response, analogous to but distinct from antigenic variation in African trypanosomes (Frasch 2003).

A major technological breakthrough in research on the molecular biology of *T. cruzi* has been the development of genetic transformation systems. New plasmid and cosmid shuttle vectors allow *T. cruzi* DNA to be propagated in bacteria and reintroduced into *T. cruzi*, either episomally, or integrated into the nuclear genome by homologous recombination. In this way, *T. cruzi* genes can be deleted or overexpressed so that gene structure and function can be analyzed. Functional complementation of deletion mutants allows portions of the genome encoding a particular function to be identified. RNA inhibition (RNAi) (Lillico et al. 2003) and chromosomal fragmentation provide alternative approaches to identifying gene function although, at the time of writing, RNAi has not yet been achieved for *T. cruzi*. The wide interest in *T. cruzi* molecular biology has led to the establishment of a World Health Organization-sponsored international collaboration to sequence the entire *T. cruzi* genome, which is currently underway (Andersson et al. 1998; Aguero et al. 2000).

Genetic diversity

Early experimental work on *T. cruzi*, its wide range of mammalian hosts, and the large number of triatomine bug vector species suggested that *T. cruzi* might be genetically diverse. *T. cruzi* strains were reported to differ in their virulence and histotropism in experimental animals, in their infectivity to triatomine bug species, in their susceptibility to drug treatment, and antigenically. In addition, outcome of chronic Chagas disease appeared to show marked regional variation (Miles 1979).

In the 1970s, isoenzyme electrophoresis began to be applied to study the genetic diversity and population genetics of trypanosomatids (Figure 19.6). Isoenzyme profiles of *T. cruzi* isolates from domestic and silvatic transmission cycles in a single endemic locality (Sao

Felipe, Brazil) showed that domestic and silvatic strains were very different phenotypically and, by implication, genotypically. Thus, more enzyme characters separated these *T. cruzi* strains than distinguished species of *Leishmania*. *T. cruzi* isolates were subsequently characterized from many Latin American countries. Analysis of isoenzyme profiles confirmed the extensive diversity within the species, and initially indicated that there were three main strain groups, which were named principal zymodemes Z1, Z2 and Z3 (Miles 1983). Z1 appeared to be associated with marsupials, especially the genus *Didelphis* and Z3 with the armadillo, *Dasypus novemcinctus*. Host associations for Z2 were less obvious but may be linked to caviamorph rodents in the silvatic cycle in Bolivia, from which *Triatoma infestans* was proposed to have spread to the six southern cone countries of South America (i.e. Argentina, Bolivia, Brazil, Chile, Paraguay, Uruguay) and to southern Peru. The zymodeme concept contributed to the description of *T. cruzi* transmission cycles as:

- non-overlapping or discontinuous, as in Sao Felipe where distinct *T. cruzi* strains were transmitted by domestic bugs (Z2, *Panstrongylus megistus*) and silvatic bugs (Z1, *Triatoma tibiamaculata*)
- overlapping or continuous, as in Venezuela where the same principal zymodeme (Z1) was believed to be transmitted by a single bug species (*Rhodnius prolixus*) considered to be common to domestic and silvatic transmission cycles
- enzootic, as in the Amazon basin and the USA where triatomine bugs do not colonize houses and silvatic *T. cruzi* strains rarely caused human infection (Miles 1979).

One reason for such interest in the heterogeneity of *T. cruzi* is to assess whether different strains are responsible for the diverse clinical outcomes of Chagas disease. The high prevalence of *T. cruzi* Z1 in the north of South America, where mega-esophagus and megacolon are rare, and the abundance of Z2 in central and eastern Brazil, where megasyndromes are common, suggested that some *T. cruzi* strains were more likely to cause chronic Chagas disease (Miles et al. 1981c). Both Z1 and Z2 have, however, been isolated from symptomatic acute phase infections and both can recrudesce after unsuccessful treatment (Luquetti et al. 1986). Although only Z2 was isolated by the same authors from symptomatic chronic Chagas disease, there is no proof that Z1 was not (also) present earlier in the infection and responsible, at least in part, for the poor prognosis. Mixed zymodeme infections have been reported from both mammalian hosts and triatomine bugs.

Identification of *T. cruzi* strains by isoenzyme profiles led to selection of biological clones representing different epidemiologies in Latin America for use as reference strains in experimental studies (together with some traditional uncloned laboratory strains). The

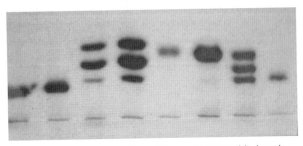

Figure 19.6 *Isoenzyme profiles of* Trypanosoma cruzi *isolates by starch-gel electrophoresis*

World Health Organization *Trypanosoma cruzi* reference strains are as follows (* = derived from clonal populations):

- M/HOM/PE/00/Peru
- M/HOM/BR/00/12 SF
- M/HOM/CO/00/Colombia
- M/HOM/BR/00/Y strain
- M/HOM/BR/00/CL strain
- M/HOM/CH/00/Tulahuen
- M/HOM/AR/74/CA-I
- M/HOM/AR/74/CA-I/72*
- M/HOM/AR/00/CA-I/78*
- M/HOM/AR/00/Miranda 83*
- M/HOM/AR/00/Miranda 88*
- M/HOM/BR/82/Dm 28c*
- M/HOM/BR/78?/Sylvio-X10-CL1*
- M/HOM/BR/Sylvio/X-10-CL4*
- M/HOM/BR/77/Esmeraldo CL3*
- M/HOM/BR/68/CAN III CL1*
- M/HOM/BR/68/CAN III CL2*
- M/HOM/BO/80/CNT/92: 80 CL1*
- I/INF/B0/80/SC43 CL1*
- I/INF/PY/81/P63 CL*

The use of reference strains encouraged detailed characterization by other methods, particularly by Dvorak and his collaborators (Dvorak 1984). It was found that clones of different principal *T. cruzi* zymodemes differed radically in DNA content, in kinetoplast DNA minicircle fragment patterns (schizodemes), in elemental composition of iron, zinc, and potassium, in oxidative metabolism, in antigenic profiles, in extracellular and intracellular growth rates in vitro, in virulence and tropism in experimental animals, and in response to experimental chemotherapy (Nozaki and Dvorak 1993). There is broad agreement between the zymodeme and schizodeme groupings and a correlation with abundance of a particular surface antigen (Carreno et al. 1987; Chapman et al. 1984). Polymerase chain reaction (PCR)-based analyses of polymorphisms in ribosomal RNA genes and in mini-exon genes have confirmed the presence of at least two major lineages, one that corresponds with Z1 and a second that incorporates Z2 (see below) (Stothard et al. 1998). Functional analysis of gene promoters also partitions *T. cruzi* into two major groups (Nunes et al. 1997).

Tibayrenc and colleagues subdivided the Z1, Z2 groups into many zymodemes, performed extensive further studies on their distribution in South America and elaborated a 'clonal hypothesis' which proposes that distinct *T. cruzi* clones are propagated asexually and largely disseminated separately (Tibayrenc et al. 1993; Brisse et al. 2000). PCR has been used to isolate kinetoplast DNA probes specific to widespread 'major clones' of *T. cruzi*. Most allozyme frequencies and measures of linkage disequilibrium have supported clonal propagation, and accord with single or small number of bugs invading houses and introducing a restricted number of populations of *T. cruzi*. Still further methods, including microsatellite analysis (Macedo et al. 2001) confirmed the broad validity of the principal subdivisions of *T. cruzi*.

The two major subdivisions of *T. cruzi* have now been named by international consensus as *T. cruzi* I (Z1) and *T. cruzi* II (incorporating Z2) (Anon. 1999). It has been proposed that *T. cruzi* I evolved in an arboreal palmtree habitat with the triatomine tribe Rhodniini, in association with the opossum *Didelphis* (Gaunt and Miles 2000, 2002). *T. cruzi* II may have evolved in a terrestrial habitat with the triatomine tribe Triatomini, in association with edentates or ground-dwelling marsupials or, alternatively, by more recent host transfer of *T. cruzi* I into primates (Fernandes et al. 1999; Gaunt and Miles 2000). *T. cruzi* II has been further divided by into five subgroups, *T. cruzi* IIa–e, all previously recognized by isoenzyme profiles, and of which *T. cruzi* IIa corresponds with Z3 (Miles et al. 2003a). The affinities of *T. cruzi* IIa (Z3), whether with *T. cruzi* I or *T. cruzi* II, are still somewhat controversial (Brisse et al. 2000; Robello et al. 2000; Sturm et al. 2003; Campbell et al. 2004).

Contemporary studies of allozyme frequencies in African trypanosomes suggested the presence of genetic exchange in what hitherto had been assumed to be an asexual organism. Although the genetic mechanisms are not fully understood, experimental crosses with drug-resistance markers have shown that *T. brucei* undergoes recombination in the tsetse fly vector, probably in the salivary glands. Genetic exchange in *T. cruzi* is most likely to occur, as with African trypanosomes, among undisturbed silvatic cycles of transmission. Allozymes of phosphoglucomutase (PGM) for silvatic isolates of *T. cruzi* from the Amazon basin suggested occurrence of sympatric hybrid and parental strains. This is in accord with random amplification of polymorphic DNA (RAPD), which shows sharing of fragments between putative hybrids and one or other of the putative parents (Carrasco et al. 1996). Further evidence of genetic recombination has come from restriction fragment length polymorphisms (RFLP) studies with gene probes for three *T. cruzi* enzyme genes, and suggests that homozygote and heterozygote loci for these genes occur sympatrically among clinical isolates (Bogliolo et al. 1996).

The presence of an extant capacity of genetic exchange in *T. cruzi* has recently been proven by the production of hybrid clones in the laboratory (Gaunt et al. 2003). Two biological clones of *T. cruzi*, from silvatic transmission in the Amazon basin (Carrasco et al. 1996), were transfected to carry different drug-resistant markers and then passaged through the entire life cycle. Six double-drug-resistant progeny clones were recovered from the mammalian stage of the life cycle. The progeny clones were analyzed by isoenzyme, karyotype, RAPD and microsatellite markers and by sequencing of nuclear and maxicircle DNA targets. The progeny showed fusion of parental genotypes, loss of alleles, apparent uniparental inheritance of maxicircle kDNA and evidence of homologous recombination, although the frequency and precise mechanisms of recombination have not been determined. There were strong genetic parallels between the experimental hybrids and the genotypes among natural isolates of *T. cruzi*. These data (Gaunt et al. 2003) and independent phylogenetic evidence (Machado and Ayala 2001) suggest that *T. cruzi* IId and IIe are hybrid strains, in accord with earlier isoenzyme profiles, and that they are derived from strains similar to *T. cruzi* IIb and IIc. Thus *T. cruzi* IId and IIe haplotypes are split across the IIb and IIc phylogenetic clades. Sequencing of intergenic targets further suggests that *T. cruzi* IIa and IIc may also have some hybrid characters, with *T. cruzi* I and *T. cruzi* IIb being less complicated lineages (Sturm et al. 2003; Campbell et al. 2004). The CL Brewer strain selected for the genome sequence project (www.TcruziDB.org), is a IIe hybrid, which complicates assembly of the sequence.

An absolute link between *T. cruzi* genotypes and clinical outcome of infection remains unproven, and few attempts have yet been made to examine the role of genotypic and phenotypic diversity of the human host in influencing clinical prognosis (Campbell et al. 2004). Multiple human genetic markers can now be used to compare genetic susceptibilities to chronic Chagas disease, although this may have little direct impact on Chagas disease control (Fae et al. 2000). In support of the hypothesis that severity of outcome is determined by the infecting *T. cruzi* strain, Di Noia et al. (2002) report that serological recognition of the *T. cruzi* II specific epitope of a GPI-anchored mucin-like protein is associated with chronic Chagas disease. However, this is not surprising as the sera examined were entirely from the southern cone region of South America, where *T. cruzi* II appears to be overwhelmingly predominant in human populations. Nevertheless, using individual kinetoplast DNA signatures, it has been shown that genetically diverse *T. cruzi* populations may occur in tissues of the same patient, and with a differential distribution between heart and esophagus (Vago et al. 2000).

CLINICAL ASPECTS

Shortly after his pioneering discoveries, Carlos Chagas was once described as a man who searched for diseases that did not exist. Initial acute phase Chagas disease is not frequently noted and is easily confused with other clinical conditions: if present, acute symptoms usually subside within 2 months of their appearance. Patients in certain geographical regions may suffer mega-esophagus and/or megacolon, usually with associated heart disease, but abnormalities in the elecrocardiograph (ECG) and myocardiopathy of chronic Chagas disease may also be mistaken for other forms of heart problem. At least one-

third of the 12 million or so people thought to be infected with *T. cruzi* and surviving the initial stage of infection appear to be clinically compromised.

Clinical manifestations

The initial phase of *T. cruzi* infection may lack specific signs and symptoms. In some cases, there is a lesion at the entry portal of the organism, giving rise, in the case of entry through the skin, to a cutaneous chagoma. If metacyclic trypomastigotes cross the conjunctiva, the resultant painless but inflamed, periopthalmic, unilateral edema and conjunctivitis is known as Romana's sign (Figure 19.7). There may be local infiltration of lymphocytes and monocytes, and regional lymphadenopathy adjacent to these lesions. Rarely, multiple chagomas have been described during acute infections in infants. Acute infections are more common and more severe in children. Less than 10 percent of children die during the acute stage. General clinical changes at this stage may include fever, hepatosplenomegaly, generalized lymphadenopathy, facial or generalized edema, rash, vomiting, diarrhea, and anorexia. Early ECG changes may be sinus tachycardia, increased P–R interval, T-wave changes, and low QRS voltage, and death may be due to

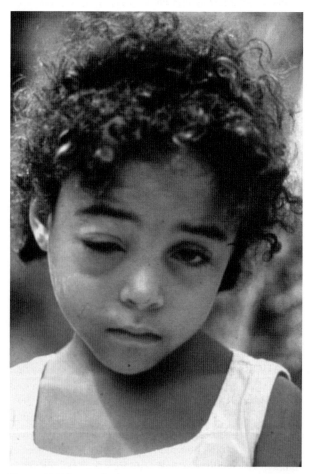

Figure 19.7 *Romana's sign*

acute myocarditis. Meningoencephalitis is infrequent, mainly found in infants and carries a very poor prognosis. In patients suffering recrudescent *T. cruzi* infection associated with acquired immune deficiency syndrome (AIDS), the organism commonly crosses the blood–brain barrier and causes fatal meningoencephalitis (WHO 2002).

Incubation period may be as short as 2 weeks, or several months if infection is acquired by blood transfusion. Prolonged incubation period in recipients given contaminated blood is thought to be due to the poor capacity of circulating broad blood forms to invade cells. Splenomegaly and general lymphadenopathy are frequent in such cases.

In a small proportion of seropositive mothers, *T. cruzi* crosses the placenta, giving rise to abortion or premature birth. Hepatosplenomegaly is common in congenital infections and there may be fever, edema, metastatic chagomas, and neurological signs, such as convulsions, tremors, weak reflexes, and apnea. Signs of cardiac involvement are rare; ECG is usually unchanged but can show low voltage complexes, decreased T-wave height, and increased atrioventricular (AV) conduction time. Premature birth associated with congenital Chagas disease carries a poor prognosis: symptoms are generally less severe if birth is not premature (WHO 2002).

Patients surviving the acute infection enter the indeterminate phase, in which there are no symptoms or signs of Chagas disease. Around 30 percent of these patients, however, develop chronic disease in which cardiac changes are most common, with arrhythmias, palpitations, chest pain, edema, dizziness, syncope, and dyspnea. The most frequent ECG abnormalities are right bundle branch block (RBBB) and left anterior hemiblock (LAH) but there may also be AV conduction abnormalities, sometimes complete AV block. Many different arrhythmias may occur including sinus bradycardia, sinoatrial block, ventricular tachycardia, primary T-wave changes, and abnormal Q-waves. Enlargement of the heart may be seen by chest radiography. Complications include embolism. Death is often sudden (WHO 2002).

The regions of the alimentary tract most often affected in the digestive form of chronic Chagas disease are the esophagus and colon, with loss of peristalsis, regurgitation and dysphagia, or severe constipation, fecaloma, and progressive dilatation of these organs.

Further details of clinical aspects of chagasic heart disease, mega-esophagus and megacolon can be found in a PAHO Expert Committee Report (PAHO 1994).

IMMUNOLOGY

In view of the intense interest in the immune response to *T. cruzi* infection, it is surprising how little is understood about either the protective mechanisms or the autoimmune involvement in the pathogenesis of Chagas disease. Individuals surviving acute infection are able to mount an immune response that suppresses parasitemia but is unable to eradicate infection. In the absence of specific treatment, infection is normally maintained for life. *T. cruzi*-specific antibodies may be detectable within the first 2 weeks, and seropositivity usually remains throughout life in untreated patients. All immunoglobulin (Ig) subclasses are elevated. IgM titers rise early, with increased concentrations, followed by rising IgG titers. Unlike African trypanosomiasis, in which antigenic variation stimulates a sustained increase in IgM, in Chagas disease IgM levels fall as the acute infection subsides. As anticipated for an organism dividing intracellularly, the suppressive immune response also involves cell-mediated immunity. Partial protection by artificial immunization with purified antigens in mice has been shown to be T-cell dependent (Wrightsman et al. 2002).

Numerous studies have been performed on the immune response to *T. cruzi* infection in mice, recently aided by transgenic mouse technology, which allows the course of infection to be observed in knockout mice that are deficient in particular components of the immune response (Tarleton et al. 2000). Hypergammaglobulinemia (of IgG2a, IgM, IgE) persists in mice until 13 weeks post infection and then IgG2a and IgM levels fall. Although mouse strains vary considerably in their antibody response and resistance to *T. cruzi*, IgG1, IgG2a, IgG2b and IgM titers are sustained throughout infection, possibly with a more dominant IgG2b isotope response in a resistant strain of mouse (Rowland et al. 1992). During the acute phase of infection in mice, production of the cytokine interleukin (IL)-2 is severely suppressed, possibly by repression of transcription of the IL-2 gene. Killing in macrophages is dependent on production of interferon-gamma (INF-γ), synergistic with tumor necrosis factor-alpha (TNF-α), triggering macrophage activation, and nitric oxide production (Silva et al. 1995). Nevertheless, studies in mice suggest that *T. cruzi* infection does not necessarily lead to a dominance of T helper (Th)1 patterns of cytokine production. Th1 cells may be a requirement of immunological control of infection and Th2 cells may contribute to parasite persistence and severity of disease (Kumar and Tarleton 2001).

It is not clear how *T. cruzi* survives for life in the infected mammalian host. Immunohistochemical staining suggests that *T. cruzi* antigen can be detected in chronic, chagasic cardiomyopathy, indicating that residual pseudocysts are present. *T. cruzi* amastigotes could be found at autopsy in the hearts of 84 of 556 patients with diffuse, chronic, chagasic myocarditis (Brener 1994). Nests of *T. cruzi* have also been found in the smooth muscle cells of the central vein of the adrenal gland at autopsy in around 50 percent of patients with chronic Chagas disease (Teixeira et al. 1997). Several methods have been proposed by which *T. cruzi* escapes action of antibody, including binding of the third component of complement by a specific trypomastigote surface glyco-protein to inhibit activation of the alternative comple-ment pathway and binding of human C4b to restrict the classical pathway (Beucher et al. 2003). The histopatho-logical picture in chronic Chagas disease suggests some autoimmune involvement: antibodies to host epitopes on specific proteins such as myosin and ribosomal P proteins have been reported.

PATHOLOGY

The pathology at the portal of entry is similar whether the chagoma is cutaneous or a Romana's sign, with infiltration of predominantly lymphocytes and mono-cytes. In the heart, no inflammatory response is said to be directed towards unruptured pseudocysts, but rupture is followed by infiltration of lymphocytes, monocytes and/or polymorphonuclear cells. The degree of parasitism is highly variable in clinical cases and in experimental infections in mice – in the latter, it is dependent on strain of infective organism and strain of mouse (Marinho et al. 2004; Risso et al. 2004). In congenital Chagas disease, amastigotes may be very widespread but are most common in cardiac and skeletal muscle or reticuloendothelial cells. If there is meningoencephalitis, parasites can be found in perivas-cular spaces or in glial and neuronal cells with an histopathology typical of acute meningoencephalitis (WHO 2002).

In patients surviving the acute infection, the inflam-matory response in the heart subsides. Two principal forms of pathology have been described in chronic Chagas disease. In the first or neurogenic form, the heart remains free from any progressive myocarditis, but focal lesions are found in the conducting system of the heart and there is a loss of ganglion cells, particularly from the parasympathetic nervous system. Experimental studies in dogs, in which focal cardiac lesions have been recon-structed from histology of tissue sections taken at autopsy, show a good correlation between ECG abnormalities and histopathology. Longitudinal studies of ECG abnormalities in patients with chronic chagasic heart disease followed up by postmortem histology show a similar correlation between ECG abnormalities and lesions of the conducting system. The right bundle branch is found to be most frequently affected, consis-tent with the frequency of RBBB in chronic Chagas disease. The left conducting system is considered to be less vulnerable to focal lesions because of its more diffuse or bifascicular course. In a proportion of patients, the inflammatory response appears to be reacti-vated resulting in a renewed myogenic form of chronic Chagas disease and a progressive myocarditis with more diffuse lesions, a slower decline in cardiac function, and less sudden death. Lymphocytes and macrophages infiltrate the heart and mycocardial fibers are replaced by interstitial fibrosis (Brener and Andrade 1979).

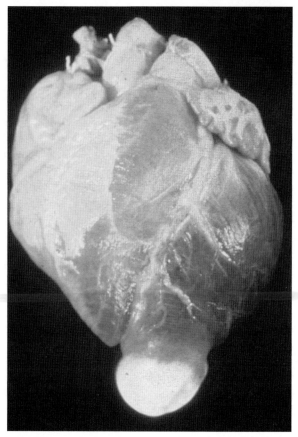

Figure 19.8 *Apical aneurysm of the left ventricle (Courtesy of Dr J.S. de Oliveira)*

At the gross level, the heart may be enlarged (mega-cardia) with focal thinning of the myocardium, which may be particularly frequent and pronounced at the apex of the left ventricle, in the form of an apical aneurysm (Figure 19.8). The presence of apical aneurysm is considered to be a pathognomonic sign of chronic chagasic heart disease. Of the intestinal sequelae chagasic mega-esophagus (Figure 19.9) is generally more common than chagasic megacolon (Figure 19.10); both may occur in the same individual and each is frequently associated with chagasic cardiopathy. A moderate or severe inflammatory response may be seen in the smooth muscle of the esophagus or colon and in the myenteric or Auerbach plexus during acute infection; in chronic Chagas disease the inflammatory response in these organs is usually mild.

Qualitative observations of neurological damage in Chagas disease were recorded as early as the 1930s. Köberle and his colleagues made painstaking quantita-tive studies of the number of ganglion cells present in normal and chagasic sections of heart, esophagus, and colon, and demonstrated widespread and significant ganglion cell loss, especially from the parasympathetic autonomic system (Köberle 1974). Neuron loss was found in many different organs but was most frequent in the heart and alimentary tract. The variable onset of

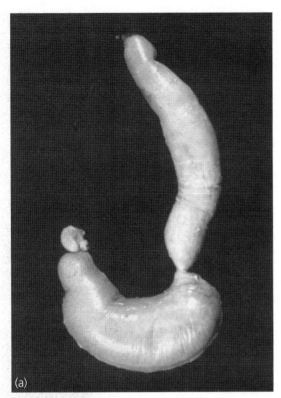
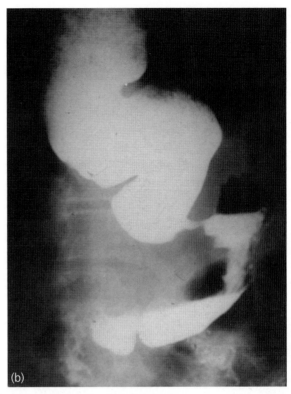

Figure 19.9 *Mega-esophagus:* **(a)** *postmortem;* **(b)** *on radiograph (Courtesy of Dr J.S. de Oliveira)*

ECG abnormalities, mega-esophagus and megacolon was understandable in terms of age-related neuron loss exacerbating the neurological damage attributable to *T. cruzi* infection. In addition, it was proposed that different hollow organs varied in the degree to which they could tolerate neuron loss; beyond these organ-specific thresholds, functional abnormalities and dilatation would begin to appear. Several authors have confirmed neuron cell loss in chronic Chagas disease (Lázzari 1994) and experimental studies in animals have shown that this neurological damage can occur early in the acute infection. Thus, the extent of *T. cruzi* infection in the acute phase and concomitant neurological damage may govern long-term prognosis, at least as far as the neurogenic form of chronic Chagas disease is concerned. Although some nerve cells may be parasitized, it is generally considered that ganglion cells are not destroyed as a direct result of parasitism. Early ganglion cell destruction led Köberle to the conclusion that a neurotoxin was produced by *T. cruzi*. Subsequent studies in experimental animals, however, have showed adsorption of *T. cruzi* antigens to uninfected tissue following pseudocyst rupture, with the implication that a T cell-mediated immune response could give rise to focal destruction of uninfected cells. Some of the physiopathological signs of chronic Chagas disease are compatible with dominance of the sympathetic system, predictable if the parasympathetic system is more exposed to damage from *T. cruzi* infection. Indeed, Oliveira demonstrated

that the pathognomonic sign of apical aneurysm could be induced in rats, in the absence of *T. cruzi* infection, by inoculation of catecholamines (Oliveira 1969; Lázzari 1994). A sympathetic dominance is also apparently associated with sudden heart failure. Histochemical quantification of sympathetic and parasympathetic innervation has shown more severe parasympathetic denervation (Machado et al. 2000).

The trigger that leads some patients to the renewed inflammatory response and progressive decline of chronic myogenic Chagas disease is not clear, but the histopathological picture is suggestive of an autoimmune process. Release of sequestered antigens by destruction of normal cells to which *T. cruzi* antigens are adsorbed could lead to an autoimmune pathology with an unpredictable onset. Much research has been done in an attempt to elucidate the autoimmune pathogenesis of chronic Chagas disease. It seems clear that auto-antibodies appear in experimental animals later than significant neuron destruction. Monoclonal antibodies to rat ganglia cross-react with *T. cruzi*; conversely, some monoclonal antibodies raised to *T. cruzi* cross-react with mammalian tissues. Numerous tissue cross-reactive antibodies have also been reported from patients with chronic Chagas disease. A possible candidate for a cross-reactive epitope between *T. cruzi* and normal host components is the C-terminus of the *T. cruzi* ribosomal P protein (Mahler et al. 2001). IgG fractions of chagasic sera contain antibodies to a motif shared by the C-term-

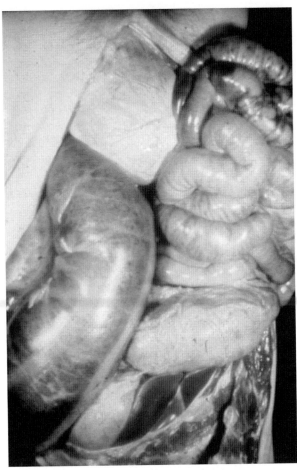

Figure 19.10 *Megacolon (Courtesy of Dr J.S. de Oliveira)*

inal region of the P0 ribosomal protein, the β_1-adrenoreceptor and the M_2 muscarinic receptor, yet are absent from control sera, and antibodies affinity-purified from chagasic sera modulated receptor function (Chiale et al. 2001). Putative cross-reactive epitopes have been described from mammalian proteins such as myosin: recognition of *T. cruzi* B-13 protein antigen and a corresponding epitope in human myosin heavy chain is said to be associated with chronic chagasic cardiopathy (Cunha-Neto and Kalil 2001). Several attempts have been made to implicate cell-mediated autoimmune pathology by studies in vitro, or by adoptive transfer of immunopathology with cells from infected donors to naive recipients: granulomata around the sciatic nerve have been induced experimentally by cell transfer and newborn syngeneic grafts were rejected by *T. cruzi* infected mice, but not by control mice (Ribeiro dos Santos et al. 1992). The CD4 T cell-dependent transplantation model of syngeneic mouse hearts grafted onto chronic chagasic recipients has suggested that anti-laminin antibodies might be used to suppress the process of autoimmune rejection (Silva-Barbosa and Savino 2000). In contrast, there is chronic persistence of the parasite (Schijman et al. 2004), and survival of syngeneic heart transplants in mice, with no signs of autoimmune

rejection in the absence of direct parasitization of heart tissue (Zhang and Tarleton 1999; Tarleton 2001).

The pathogenesis of Chagas disease remains far from understood. A tenuous unifying hypothesis emerges from these diverse considerations: in the acute phase, neuron loss is precipitated by destruction of normal cells that have adsorbed *T. cruzi* antigens, which, dependent on the degree and extent of damage, is followed by a benign or catastrophic neurogenic clinical outcome; a relapse with progressive inflammation, and putative cell-mediated autoimmune involvement, occurs in an unfortunate minority of patients. Girones and Fresno (2003) review evidence for autoimmunity in chronic Chagas cardiomyopathy.

For a review in English of Chagas disease and the nervous system, see PAHO (1994); for an authoritative review on pathogenesis, see Lopes and Chapadeiro (2004).

DIAGNOSIS

In some acute *T. cruzi* infections, trypanosomes may be found transiently in the peripheral blood by direct microscopy of unstained, wet blood films. More scanty parasitemias can be detected by microscopy of Giemsa-stained thick blood films or by microscopy after concentration of trypomastigotes from blood. Concentration methods include:

- hematocrit centrifugation (with caution to avoid tube breakage and exposure to infection) followed by searching for trypomastigotes in or immediately above the Buffy coat layer
- allowing the blood to coagulate and searching for trypomastigotes in centrifuged serum (Strout's method)
- lysis of red cells with 0.85 percent ammonium chloride and centrifugation to sediment trypomastigotes.

In the indeterminate and chronic phases of the infection, *T. cruzi* is present in such low numbers in the peripheral blood that parasitemia is not patent to microscopy even after the above concentration methods. Xenodiagnosis, in which colony-bred uninfected triatomine bugs are fed on a suspect host and dissected about 20–25 days later to detect epimastigotes, or blood culture, are employed for parasitological diagnosis of chronic infections. Xenodiagnosis is generally favored for routine use in Latin America because, unlike blood culture, it does not require stringent aseptic precautions. Colonies of triatomine bugs are usually maintained by feeding on chicken blood, as birds are not susceptible to *T. cruzi* infection. If *T. infestans* is used, colonies must be examined periodically for presence of the mono-genetic kinetoplastid *Blastocrithidia triatomae* as this may be confused with *T. cruzi* when bugs are dissected. Triatomine nymphs can survive for several months without feeding and xenodiagnosis may thus be used in rural locations, as long as the bugs are protected from high temperatures. It is advisable to use a local species of domestic vector; *R. prolixus* and *T. infestans* are favored, *R. prolixus* feeds most avidly but may give rise to hypersensitivity skin reactions in individuals sensitized to bug bites. Precautions must be taken in dissecting infected bugs to avoid exposure to infection. Sterile diluents for bug feces should be used so that results are not confused by motile contaminants, such as ciliates. Under ideal conditions, blood culture on to a blood agar base with physiological saline overlay can be as sensitive as xenodiagnosis; cultures are inoculated 'through the cap' to minimize contamination (Miles 1993).

In central and northern South America *T. cruzi* must be distinguished from nonpathogenic *T. rangeli*, although in practice *T. rangeli* is almost never seen in human blood. *T. rangeli* and *T. cruzi* can both multiply in bugs used for xenodiagnosis. Long epimastigote forms with a small kinetoplast and pointed posterior end indicate the presence of *T. rangeli*, but such forms are not always present. In a proportion of infected bugs *T. rangeli* infections will progress to hemolymph and salivary gland infections (Hoare 1972).

Detection of parasite DNA is theoretically an attractive alternative to demonstration of the whole organism. PCR protocols have been developed that can show the presence of *T. cruzi* infection in the chronic phase. Although extremely sensitive after DNA extraction of blood samples, the PCR approach is not yet simple enough for routine diagnosis. In the long term, if adaptable to a low-cost rapid assay, PCR might be applicable to screening of transfusion blood or monitoring the success of chemotherapy (Castro et al. 2002).

In principle, serology detects exposure to infection rather than an active infection. In *T. cruzi* infection, however, both the infection and seropositivity are generally maintained for life unless specific antiparasite chemotherapy has been given. Only rarely have untreated individuals become seronegative (and presumably lost their infections). A sensitive and specific complement fixation text (CFT) was developed in 1913, but is now rarely used due to its complexity. Favored serological tests are the indirect fluorescent antibody test (IFAT), the enzyme linked immunosorbent assay (ELISA) and the indirect hemagglutination test (IHAT). Care is required with the IFAT to exclude nonspecific antibody binding (usually by selecting a high cut-off serum dilution between positive and negative sera, such as 1 in 80 or above). The ELISA requires careful standardization of conjugate batches, duplicate test wells, and positive and negative serum controls should be run on every plate. The IHAT is commercially available and convenient but said to be somewhat less sensitive than IFAT or ELISA, which is also commercially available (Oelemann et al. 1998). Serum or blood spots collected

on filter paper can be used as the source of antibodies but serum provides a more accurate estimate of antibody titer. Cross-reactions may occur with other infectious diseases, notably visceral leishmaniasis and cutaneous leishmaniasis, which may be sympatric with Chagas disease. The advent of recombinant DNA technology has led to the testing of a series of *T. cruzi*-specific proteins selected from expression libraries (Umezawa et al. 2003). Detection of IgM antibodies (with IgM-specific conjugates) in newborn infants suspected of Chagas disease provides a means of helping to distinguish congenital infection from trans-placental transfer of IgG from mother to child.

In all cases, and especially in travelers returning from Latin America, clinical background should be explored thoroughly for a history of exposure to triatomine bites, blood transfusion or other sources of infection. Parasitological diagnosis is usually performed only if serology is positive. *T. cruzi* is isolated from only about 50 percent of seropositive individuals by xenodiagnosis, with 20 or more triatomine bugs, or by blood culture. Some clinical signs, such as right bundle branch block (RBBB), may be pathognomonic or highly indicative if associated with a supportive history. Hirschsprung's megacolon is seldom confused with chagasic megacolon because the former is rare in adults; when necessary, electromanometry or radiography can help with the differential diagnosis (Miles 1994)

EPIDEMIOLOGY AND CONTROL

Chagas disease is essentially a public health problem that is sustained by poor housing and poverty. The later nymphal stages and adult triatomine bugs are quite large (adults of some species may be more than 4 cm long) and, although they are reclusive insects, bug infestation is a considerable nuisance to householders. Heavy bug infestations would not be tolerated if there were lesser economic and health burdens in the affected communities. It follows that economic development, access to insecticides, improved housing, and health education should interrupt vector-borne transmission of *T. cruzi*, and this has proven to be the case in several successful control campaigns (Wanderley 1993). This is, however, of little comfort to the 11 or 12 million people who are already infected with *T. cruzi*.

A single (unsatisfactory) drug is now readily available for the specific treatment of Chagas disease; access to it may be life-saving in the initial stage of infection. Yet acute cases are seldom seen and recognized, and it is still debated whether the continued presence of the organism in the chronic stage, or repeated infection in endemic areas, has an affect on the long-term prognosis, which might largely be dictated by the severity of early infection. Treatment of chronic cases thus relies on alleviation of symptoms either by drugs or by installation of pacemakers, or by elegant surgical procedures that have been pioneered in Brazil. Although new drugs are highly desirable, their epidemiological impact depends on their ability to eradicate infection from the reservoir of humans already carrying *T. cruzi*: this demands drugs that are inexpensive and without side-effects (Urbina 2002).

The prospects for vaccination against *T. cruzi* are remote, even though immunization has been shown to protect animals against fatal, acute phase infections, and will remain remote until autoimmune involvement in pathogenesis is fully understood and can be excluded as a complication of the antigen administration. Thus, little research is directed towards vaccine development, and it is difficult to justify in the face of successful campaigns to eliminate domestic triatomine bugs.

Elimination of domestic bug populations is an attainable goal, although eradication of *T. cruzi* is beyond reach as it is a widespread zoonosis, with numerous silvatic reservoir and vector species. The distribution of human Chagas disease is somewhat enigmatic, and changing as both the spread of domestic vectors and destruction of forest vector habitats may lead to the emergence of new endemic areas. Much remains to be learned about the complexities of the epidemiology of Chagas disease in the context of parasite, vectors and reservoir hosts.

Epidemiology (including reservoir hosts)

The blood-sucking insect vectors of Chagas disease are a subfamily (Triatominae) of the Reduviidae in the order Hemiptera. Five tribes, 14 genera, and 130 species of triatomine have been described based on morphological characteristics (Lent and Wygodzinsky 1979; Carcavallo et al. 1997). Thirteen species occur in the Old World, and the remaining 105 are confined to the Americas. Seven of the 14 Old World species are related to *Triatoma rubrofasciata*, which occurs worldwide in ports with its host the rat (*Rattus rattus*), to which it transmits *Trypanosoma conorhini*; the remaining six Old World species are grouped into the unusual Indian genus *Linshcosteus* (Galvao et al. 2002).

Most New World triatomine species have silvatic ecotopes, such as palm trees, burrows, hollow trees, rock crevices or caves, where they feed from mammals, birds or reptiles, and many of the New World species have been reported as infected with *T. cruzi*. Five species, of three genera, have adapted to colonize domestic habitats and are responsible for abundant and widespread household infestations; they are: *Triatoma infestans*, *Rhodnius prolixus*, *Panstrongylus megistus*, *Triatoma brasiliensis*, and *Triatoma dimidiata*.

T. infestans is the main vector in the southern cone countries of South America (Argentina, Bolivia, Brazil, Chile, Paraguay, Uruguay) and in southern Peru; it is thought to have spread from the Cochabamba region of Bolivia, where it occurs in rocky silvatic ecotopes: outside Bolivia *T. infestans* is said to be restricted to domestic and peridomestic habitats. *R. prolixus* is the most important vector in northern countries of South America and in central America, with *T. dimidiata* a significant secondary vector. *P. megistus* occurs along the eastern seaboard of Brazil, which was formerly covered by a continuous zone of Atlantic forest, and in central-eastern Brazil. *T. brasiliensis* is the vector in the arid northeastern region of Brazil. *Triatoma sordida*, although extremely widespread in central and eastern South America and often described as an important domestic vector, is usually associated with chickens. Many other species are reported from houses but usually as small, infrequent colonies, or as adult bugs flying into houses attracted to light.

Triatomines are considered to be obligatory blood feeders; most species take mammalian or bird blood but some feed on reptiles. At least one genus (*Eratyrus*) is known to take blood also from large invertebrates that co-inhabit its natural hollow tree ecotope (Miles et al. 1981a). Each nymphal stage must take at least one bloodmeal before molting to the next. Bug pheromones in feces may lead to aggregation of vectors at feeding sites, such as roosting chickens. As there is no transovarial transmission of *T. cruzi*, older bugs are more likely to have fed from an infected host and are, therefore, more likely to be infected.

Adults of all species are winged, with the exception of the Chilean *Triatoma (Mepraia) spinolai*, which has wingless females and either winged or wingless males. Nevertheless, adult triatomines are not frequent flyers; they are almost never observed to fly in laboratory colonies unless artificially induced to do so, but may be seen flying in infested houses at dusk, particularly after heavy rain. Flight is known to be an important method of dispersion since adult bugs of some species commonly fly into houses. Adhesive eggs aid passive dispersion of *Rhodnius*, and nymphal stages can move from host to host by walking. The few natural predators and parasites of triatomines include wasps (*Telonomus*, *Gryon*) that develop in triatomine bug eggs, spiders, lizards, chickens, rodents and various other insectivores, parasitic nematodes, and fungi. Immunological analysis of insect food sources can be used both to determine sources of blood in bugs and to identify their natural predators by detecting ingested bug proteins. When alarmed some triatomines produce the distinctive smell of isobutyric acid, and make sounds by stridulation, in which the proboscis is rubbed in a ridged ventral groove. *Rhodnius* has climbing organs and can scale very smooth surfaces, possibly due to the need to climb smooth-trunked palms and palm-frond stems in its natural arboreal habitats. The distribution of domestic vector species within houses may reflect silvatic habitat preferences: thus *R. prolixus* heavily infests roofs made of palm fronds, and palm crowns are the principal habitat of the genus; *P. megistus* is most frequently found in compressed mud and wooden framed walls, and its most frequent natural habitats are hollow trees or burrows among tree roots; *T. infestans* can be found in quite good quality housing living in the walls or among tiles at the edges of roofs.

More than 150 species of 24 families of mammals have been reported as infected with *T. cruzi*, and all mammal species are considered to be susceptible. Although birds and reptiles are insusceptible to *T. cruzi* infection, domestic chickens are extremely important epidemiologically because they sustain heavy triatomine bug infestations either at roosting sites inside houses, or in neighboring chicken houses, which may be overlooked during spraying campaigns. Important domestic hosts, apart from humans, are dogs, guinea pigs, cats, rats, mice, and any other mammals sleeping inside houses and closely associated with humans. Dogs are particularly important because of high prevalence rates of infection and as they are often abundant in endemic areas (Cohen and Gurtler 2001). Cats, although less frequently studied, also have high prevalence rates and possibly acquire infection via the oral route, by eating either triatomine bugs or house mice that have fed on them. Although smaller hosts, rodents may be abundant and provide an important source of infective bloodmeals. Pigs, goats, cattle, and equines usually have indirect contact with houses and very low prevalence rates of infection, although their enclosures may be infested with bugs.

The most ubiquitous silvatic reservoir host of *T. cruzi* is the opossum, *Didelphis*, which occurs throughout much of the range of *T. cruzi* in the Americas. Multiple nesting or resting sites of *Didelphis* encompass many types of triatomine habitat. High *T. cruzi* prevalence rates may be in part due to the fact that opossums will eat triatomines and might also transmit infection via anal gland secretions, in which infective metacyclic forms typical of the vector stage of the *T. cruzi* life cycle can occur. *Didelphis* also provides an important potential link between silvatic and domestic habitats as it occasionally nests around or inside houses. Edentates (armadillos and anteaters), especially the armadillo *Dasypus novemcinctus*, are commonly infected with *T. cruzi* and their epidemiological importance requires further investigation; they have contact with human households through hunting of armadillos and their consumption for food. *P. geniculatus*, a silvatic triatomine bug frequently found in armadillo burrows, is one species that is attracted to light and often flies into houses. Rodents are thought to have an important epidemiological role as in times of food scarcity or drought they might introduce triatomines and *T. cruzi* strains into new peridomestic nesting sites. Similarly, bats may roost in domestic habitats, encouraging triatomine infestation and bringing in *T. cruzi* strains. The numerous carnivore and primate species that may harbor *T. cruzi* have tenuous links with domestic transmission cycles.

Silvatic cycles of *T. cruzi* transmission extend approximately from southern Argentina and Chile (latitude 46°S) to northern California (latitude 42°N). Human infection is, however, much less widespread, from 44° 45'S in Argentina to the southern states of the USA. Prevalence estimates for human infection are based on rates of seropositivity (Miles 1982). National surveillance by serology indicates that Brazil has the largest number of infected individuals (about 5 million) in accordance with its land mass and population. Bolivia is reported to have the highest prevalence, with seropositivity rates of greater than 70 percent in some localities, and heavy *T. infestans* infestation of rural communities that are difficult to access. The World Health Organization has provided an outline of geographical distribution of *T. cruzi* infection within the affected countries (WHO 2002). As described above, endemic areas can be classed as having either separate domestic and silvatic transmission cycles, with separate vector species (e.g. eastern Brazil) or overlapping domestic and silvatic transmission cycles (e.g. parts of Venezuela and northern Brazil). Enzootic regions have rare human infections with sympatric (e.g. USA) and sometimes abundant (e.g. Amazon basin) silvatic transmission cycles (Miles et al. 1981a). Locally acquired Chagas disease is rare in the USA, partly because of higher living standards and better housing. Although bugs are occasionally found in and around houses with dogs (Beard et al. 2003) they infrequently attack humans, and transmission of *T. cruzi* is less likely as the local bug species tend to defecate after feeding. Occasionally adult bugs are attracted by light into camp sites. Silvatic bugs are widespread in the USA and *T. cruzi* infections are common in opossums, raccoons, and wood rats (*Neotoma*). Nevertheless, transmission might also rarely occur in the USA due to transfusion of blood from infected immigrant Latin American donors (Schmunis et al. 2001).

Around 300 autochthonous human *T. cruzi* infections have been reported from the vast forested Amazon basin, even though there are extensive roadside and riverine communities with houses ideally suited to colonization by triatomine bugs and *T. cruzi* infections abound among species of forest mammals. Tracing of trapped mammals to their nesting sites with a spool-and-line mammal-tracking device (Miles et al. 1981b) has helped to describe the ecotopes of silvatic triatomines, and show that the 13 known species of the central Amazon basin do not readily adapt to household infestation (Miles et al. 1981b). *R. brethesi*, which infests the piacaba palm (*Leopoldinia piacaba*), is reported to attack forest workers harvesting the palm fronds and sleeping nearby, some of whom acquire infection. *P. geniculatus*, normally associated with high humidity within armadillo burrows and difficult to adapt to laboratory colonies, has recently been reported infesting domestic pigsties in Pará state, Brazil and attacking the inhabitants of adjoining houses (Valente et al. 1998).

The destruction of natural habitats around houses in the Amazon may well encourage triatomine infestation, especially as wood and palm fronds are often used in house construction, inevitably bringing some silvatic vectors into houses (Coura et al. 2002).

Nonvectorial transmission of *T. cruzi* that is epidemiologically important includes transmission by blood transfusion, congenital transmission, and oral transmission. Where prevalence is high, seropositivity rates among blood donors may reach 20 percent (endemic localities in Brazil and Argentina) or even higher in Bolivia (60 percent) (WHO 2002). In large cities where there are many migrants from rural areas and many blood transfusions, a significant proportion of recipients will become infected in the absence of serological surveillance of donors or treatment of transfusion blood to destroy *T. cruzi*; there may be thousands of blood-borne infections each year. Control of blood-borne transmission is thus an essential part of campaigns directed primarily at the elimination of domestic triatomine bugs. In endemic areas where prevalence is high, many pregnant women will carry *T. cruzi*. Estimates vary widely but a low percentage (less than 5 percent) of infants born to seropositive mothers may acquire congenital infection (WHO 2002). If resources allow, serological screening of pregnant women, serological and parasitological follow-up of the newborn, and prompt chemotherapy of infected infants, can reduce mortality due to congenital infections (Blanco et al. 2000). Determination of rates of transplacental acquisition of infection are complicated by passive transfer of maternal IgG antibodies to *T. cruzi*, which may cause infants to be seropositive for several months after birth even in the absence of infection. There is convincing evidence of oral transmission and a high proportion of the few human cases in the Amazon basin are thought to have been acquired by this route (Coura et al. 2002). The most likely sources in such outbreaks are contamination of food with triatomine bugs, or with vector-like forms from opossum anal gland secretions, or through consumption of the uncooked meat or blood of an infected reservoir host. An increasing problem is the transfer of *T. cruzi* to organ recipients from infected donors, who must therefore be routinely screened for seropositivity. If already infected with *T. cruzi*, recipients risk a recrudescent acute phase infection in response to immunosuppression. Similarly, human immunodeficiency virus (HIV)-positive patients who develop AIDS may reactivate *T. cruzi* infection and develop meningoencephalitis (Lazo et al. 1998). *T. cruzi* infections in the immunocompromised will become epidemiologically more important as human migration to South American cities continues and urban HIV transmission spreads to rural areas that have endemic Chagas disease (de Gorgolas and Miles 1994).

TREATMENT

Chemotherapy

Two drugs have been used for specific chemotherapy of *T. cruzi* infection, nifurtimox and benznidazole (WHO 2002). Nifurtimox, a synthetic nitrofuran given by mouth and absorbed through the gastrointestinal tract, is no longer readily available. Dose rates employed were 8–10 mg per kg orally in three divided daily doses for 90 days, using tablets of 30, 20, or 250 mg. Higher doses of 15–20 mg per kg in four divided daily doses for the same period were recommended for infected children. Successful treatment thus required either hospitalization or careful monitoring to ensure compliance. Side-effects could include anorexia, nausea, vomiting, gastric pain, insomnia, headache, vertigo, excitability, myalgia, arthralgia, convulsions, and peripheral polyneuritis, and could lead to interruption of treatment. Nifurtimox is thought to act against *T. cruzi* by increasing oxidative stress through the production of free oxygen radicals. Benznidazole is a nitroimidazol which is also absorbed from the alimentary tract. Doses are 5–7 mg per kg orally in two divided doses for adults for 60 days with 100 mg tablets, and 10 mg per kg similarly for children. Adverse affects may include rashes, fever, nausea, peripheral polyneuritis, leukopenia, and, rarely, agranulocytosis. Benznidazole is thought to act by interaction of drug metabolites with DNA but not via oxidative stress.

Cure rates with both nifurtimox and benznidazole are not total and vary regionally. Suppression of parasitemia during acute infection may be life-saving and treatment is, therefore, recommended during the acute phase. The chronic phase is not always treated because the influence of persistent infection on long-term prognosis is uncertain, and there is a significant failure rate, such that up to half of patients may remain infected after treatment. Nevertheless, treatment is better tolerated in children, leading to the suggestion that, as well as acute cases, chronic cases in children might also always be treated (Estani et al. 1998). Treatment is essential for immunocompromised patients, especially if meningoencephalitis is present, in which case double or even higher dose rates may be recommended. Pre-emptive therapy has been described as safe and effective for preventing reactivation of *T. cruzi* infection during bone marrow transplantation (Dictar et al. 1998). In congenital cases, either drug has been used, nifurtimox at 8–25 mg per kg daily for 30 days and benznidazole at 5–10 mg per kg daily for 30–60 days (WHO 2002). Indicators of elimination of the parasite are negative xenodiagnosis together with, in acute cases, serological reversion within a year. However, negative parasitology is not sufficiently sensitive to prove absence of infection and, for chronic cases, reversion of serology may take decades (Rassi and Luquetti 2003).

Allopurinol and itraconazole have been suggested as low cost, nontoxic alternatives for treatment of *T. cruzi* infection, but their efficacy is in doubt. Allopurinol, a structural analogue of hypoxanthine that is metabolized via the purine salvage pathway, prevents formation of ATP, and also interferes with protein synthesis by its incorporation into RNA. Experimental drugs thought to act against the enzyme sterol methyl transferase are under development, have shown promising activities in treatment of animal infection, and may proceed to clinical trial (Urbina 2002; Urbina et al. 2003).

Additional chemotherapy is an important part of supportive treatment in both acute and chronic Chagas disease (WHO 2002). Severe acute infections may require treatment for fever, vomiting, diarrhea, and convulsions; if heart failure is present, sodium intake is restricted; diuretics and digitalis may be recommended; anticonvulsants, sedatives, and intravenous mannitol may assist management of acute meningoencephalitis. Chronic chagasic heart disease may require treatment of heart failure through reduced activity, limitation of sodium, diuretics, vasodilation (angiotensin converting enzyme inhibitors), and maintenance of normal serum potassium; digitalis is only advisable as a last resort for treatment because it may aggravate arrhythmias by inducing premature ventricular beats or impeding AV conduction. A pacemaker may be required to manage bradycardia that does not respond well to atropine, or atrial fibrillation with a slow ventricular response in which vagolytic drugs are not effective, or complete AV block.

Life expectancy of patients with ventricular arrhythmias may be prolonged by treatment to prevent ventricular tachycardia or ventricular fibrillation: lidocaine, mexiletine, propafenone, flecainide, β-adrenoreceptor antagonists, and amiodarone are effective for treatment of ventricular extrasystoles; amiodarone is the most effective drug to treat arrhythmias, but in all cases drug treatment of arrhythmias may have an aggravating effect and patient management may be a complex combination of the use of drugs and a pacemaker. Lidocaine may be used intravenously in emergencies prior to drug administration orally. Surgical resection of arrhythmic endocardial regions may be considered, and surgical resection of ventricular aneurysms has been suggested but not extensively used. Expert reports (WHO 2002) and physicians who have specialized in management of Chagas disease should be consulted before case management can be embarked upon with confidence.

Surgical treatment

Elegant surgical procedures have been developed in Brazil to treat severe chagasic megacolon and mega-esophagus. Laxatives, colonic lavage or manual evacuation may be used to relieve symptoms of megacolon, with laparotomy if the fecaloma cannot be reached. Sigmoid volvulus can be relieved by decompressing intubation or surgery. Sigmoidostomy close to the rectosigmoid junction may simplify subsequent surgery to correct megacolon as it allows more of the colon to be retained.

Surgical treatment of megacolon is often based on the operation of Duhamel. The modified Duhamel–Haddad procedure consists of resection of the sigmoid loop, closing of the rectal stump and insertion of the descending colon through the rear wall of the rectum. The Haddad improvements introduced a two-stage operation, in which the lowered colon is first exteriorized through the retrorectal stump as a perineal colostomy and, secondly, with peridural anesthesia, the stump of the colon is sectioned into anterior and posterior halves, the anterior wall of the colon and posterior wall of the rectum held together in an inverted V, and then cut and sutured together to make a wide join between the colon and rectal stump (Figure 19.11). A series of 624 Duhamel–Haddad operations between 1966 and 1981 had an overall associated mortality of 6.6 percent (Moreira et al. 1985).

Mega-esophagus can be treated by dilatation of the cardiac sphincter or the Heller–Vasconcelos surgical procedure in which a portion of the muscle is removed from the junction of the esophagus and stomach without disturbing the muscular control of the wall of the stomach. More severe mega-esophagus may require partial removal of the distal esophagus and replacement with alternative portion of the alimentary tract such as jejunum. These and other surgical procedures are described in detail in the Brazilian literature (Raia 1983).

VACCINATION

Administration of crude *T. cruzi* lysates to experimental animals has been known for many years to suppress parasitemia, to prevent mortality from a subsequent

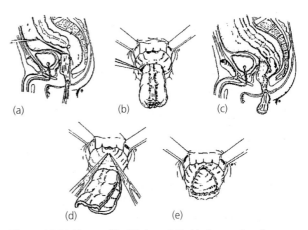

Figure 19.11 *The modified Duhamel–Haddad procedure for surgical correction of megacolon*

challenge infection, and yet not to prevent re-infection. 'Attenuated' *T. cruzi* strains have been used as experimental vaccines in human trials with small numbers of individuals, but organisms were not necessarily entirely non-infective, and are not known to have conferred resistance to infection (or superinfection). Various semipurified or purified antigen preparations have similarly been shown in experimental models to confer protection against mortality on challenge but not to provide a sterile immunity (Taibi et al. 1993). An autoimmune response may have a role in the pathogenesis of Chagas disease, and this does not encourage research on candidate vaccines. Without a full understanding of the potential autoimmune involvement in pathogenesis it is difficult to envisage clinical trials of vaccines: subjects could not be given a live challenge, and would have to be followed for many years for the appearance of vaccine-induced pathology in the absence of infection. These aspects of Chagas disease suggest that resources would be most cost-effectively directed to further improvement of vector control campaigns rather than vaccine development. Immunotherapeutic modulation of progressive autoimmune damage in chronic Chagas disease has been proposed: if the intricacies of immune responses in the pathogenesis of Chagas disease can be convincingly unraveled, this may become possible for a small number of privileged patients.

INTEGRATED CONTROL

In 1948, successful trials of the organochlorine insecticide BHC for control of domestic triatomine bug infestation by Emmanuel Dias and Jose Pellegrino in Minas Gerais, Brazil, prompted their telegram to the Brazilian Ministry of Health suggesting that the end to vector-borne *T. cruzi* transmission was in sight (Miles 1996). The fundamental value of insecticides for control of triatomine bugs has never since been in doubt. Nevertheless, much research has been devoted to developing new methods of eliminating domestic triatomines, including the use of parasitic wasps, nematodes, and fungi; sophisticated traps for catching bugs; and juvenile hormones for preventing development of nymphal instars into adults. These alternative weapons have failed to have significant impact on vector populations.

A combination of insecticide spraying, health education or community participation and housing improvement has been the basis of successful national control programs, a superb example of which is the elimination of domestic *T. infestans* populations from São Paulo State in Brazil (Wanderley 1993). João Carlos Pinto Dias, the son of Emmanuel Dias, has led the elaboration of a systematic approach to triatomine bug control (Dias 1987).

- In the preparatory phase, all houses in the endemic area are mapped and sampled manually with flash-light and forceps for evidence of triatomine infestation: sampling may be assisted by spraying irritant pyrethroids on to suspect wall areas to dislodge bugs hiding in crevices. This mapping and sampling phase allows for the cost of insecticides, personnel, transport and equipment to be calculated, and for detailed planning of operations.

- In the attack phase, all houses and outbuildings are sprayed, irrespective of whether they are known to be bug infested. Failure to spray all infested domestic and peridomestic habitats may lead to rapid local re-emergence of bug populations. The entire spraying may be repeated during the next 3 months if live bugs can still be found in more than 5 percent of houses (except where there is heavy peridomestic infestation and a close link with adjacent silvatic cycles).

- In the vigilance phase, houses are sampled periodically through a surveillance system established with the local community and supported by health education. Thus, householders can report any residual infestation to a local volunteer who mobilizes an immediate insecticide spraying response. Simple surveillance kits consist of a sheet of paper pinned to walls suspected of infestation, and a plastic bag. Periodic examination of the paper may reveal either bug feces or bugs concealed between the paper and wall; the plastic bag is used by householders to collect any live bugs, with care to avoid contamination with bug feces. The vigilance phase is also supported by continuing contact with public health personnel. Although triatomines can sometimes be found in quite good quality housing, for example in tiled roofs (*T. infestans*) or in beds that are not regularly dismantled and cleaned, poor housing is most vulnerable to bug infestation. Encouraging and supporting the community to improve local housing, for example by plastering cracked walls, can contribute to the success of a control campaign. In houses infested by *R. prolixus* it is advisable to replace roofs of palm fronds by alternative materials.

In some localities, for example with domestic and silvatic *Triatoma brasiliensis* in northeastern Brazil, there may be heavy peridomestic infestation and a close link with silvatic habitats, such that rapid re-invasion can occur. In such cases, surveillance (rather than a complete respraying program of all houses and outhouses) may be implemented when the overall infestation rate is greater than 5 percent. Similarly, modification to control strategies, or measures to deal with silvatic foci, may be required at other sites with interaction between silvatic and domestic transmission cycles.

Early spraying campaigns used chlorinated hydrocarbons but their short residual activity (30–180 days) required repeated applications. Carbonates and organophosphates have also been used but the former are expensive and the latter are not well received by communities because of their strong smell. Synthetic pyrethroids have been adopted for triatomine control since 1980 as they have a long residual activity and low toxicity, and several formulations are now available (Schofield 1994). Insecticide resistance (to dieldrin) is known only from small areas of Venezuela and has not interfered with control campaigns.

In 1991, an initiative was launched to use established triatomine control methods in a cooperative international campaign, referred to as the Southern Cone Programme, and designed to eliminate domestic *T. infestans* populations from the Argentina, Bolivia, Brazil, Chile, Paraguay, and Uruguay, with spraying of infested regions that spanned national frontiers intended to prevent re-infestation (Miles 1992). The strategy for this coordinated international program incorporated World Bank recommendations on the need for clear lines of responsibility, regional integration, and pragmatic local autonomy. The Southern Cone Programme has attracted new financial support and stimulated a vigorous and sustained effort. The overall cost of the programme is estimated to be a fraction of that of dealing with the burden of new cases of Chagas disease. An international review of the programme indicates a huge reduction in the geographical area infested by *T. infestans* and a determination to deal with those regions still affected. Miles et al. (2003b) summarize key features governing the success of the programme. Further intergovernmental control collaborations are being established in Central America, the Andean region and Mexico (Schofield and Dias 1998; Dias et al. 2002). Although the Amazon Basin is free of domestic bug populations, it is under threat from migration of species such as *T. infestans* and *R. prolixus* and from adaptation of local forest species to houses (Coura et al. 2002). A network of collaborating centers has recently been established to train local health workers in the recognition of triatomine bugs and acute cases of Chagas disease, and to plan responses to the appearance of domestic bug populations. In the past, lack of funds and poor continuity frequently interrupted control campaigns, in part because resources were diverted to resurgent malaria or outbreaks of dengue. It is essential that campaigns against different arthropod vectors are integrated, at both local and regional levels.

T. cruzi is one of several disease agents, including HIV infection, hepatitis, syphilis, and malaria, that can be transmitted by contaminated transfusion blood. Screening of donors is an essential component of control campaigns and in almost all South American countries serological testing is mandatory. Effective serological tests include the IFAT, which can be used to distinguish maternal transfer of IgG from IgM due to congenital infections, the ELISA, the IHAT, and the CFT, although the latter is now seldom used. A single standardized test may be adequate, but simultaneous use of two tests is often recommended, and some cross-reac-

tions with diseases such as visceral leishmaniasis are inevitable. It has been shown that CFT, IFAT, IHAT, and ELISA can give comparable results. The IFAT has been employed in large scale surveys in Brazil; ELISA has distinct advantages for survey work as it is highly sensitive, relatively low cost, and does not require microscopy. In some countries, such as Chile, cross-reactions with leishmaniasis are not a problem as visceral and cutaneous leishmaniasis are not endemic. Rigorous internal and external quality control of assay performance are required with repetition of a proportion of tests in a central laboratory and in collaborating external laboratories. Sample collection and record systems must be highly efficient and integrated with spraying operations, health education, and community participation aspects of control. The importance of screening blood for HIV infection in major urban areas is likely to improve access to screening for *T. cruzi* infection. Where seropositive rates are very high, positive transfusion blood may be treated with 125 mg of crystal violet per 500 ml and the blood stored at 4°C for 24 hours to destroy trypomastigotes. Side-effects are blue coloration of mucosae and skin in recipients of blood treated with gentian violet.

Serology is a vital and highly effective method of monitoring the success of control programs. In its simplest form, success is demonstrable by zero serological incidence in children born after the initiation of intervention measures. Thus, for a 10-year campaign, 5 years and 10 years after the beginning of vector control no vector-borne infections should occur in children aged 0–5 and 0–10, respectively. Pre-school children and school children obtaining basic or primary and junior schooling (e.g. ages 1–12 years) provide the ideal populations for serological surveillance. Baseline serological data are useful but not essential to planning serological surveillance of control: if interventions are effective, no significant increase in prevalence of seropositivity should occur for any age group after the campaign begins.

Ideally, all children in populations possibly exposed to vector-borne transmission should be tested as serology is a sensitive independent aid for the detection of residual triatomine infestation. Thus, one seropositive case, if congenital transmission and blood transfusion transmission are excluded, may lead to the identification of a single house that has escaped elimination of bugs. In countries such as Chile, which have a relatively small population, good infrastructure, preschool rural health clinics, regional hospitals, demographic data that include individual identification codes, and clearly defined areas of disease transmission, it is feasible to monitor the entire population exposed to transmission. In other countries, such as Brazil, with a very large population in endemic areas, it may not be possible or appropriate to survey the entire population of preschool and school children. There are two obvious alternatives for selection of sub-populations: the first, on a random sample basis, and the second, where effort is concentrated on highly endemic localities. The second alternative is preferred as only in areas with relatively high seroprevalence will a detectable number of the very young acquire infection. Baseline serological data are available for Bolivia, Brazil, Peru, Argentina, Paraguay, and Uruguay to allow such sentinel sites of high endemicity to be selected. It might also be appropriate to survey localities at the margins of endemic areas where explosive outbreaks of disease, transmission might occur in previously unexposed populations.

Social and economic development in Latin America should, in time, eliminate vector-borne and blood-borne transmission of *T. cruzi* as a major public health problem. *T. cruzi* will, however, remain as a widely distributed enzootic infection, with the threat of re-emergence and spread to new areas of human settlement. The burden of management of chronic disease will be a problem for several decades. Further research on immunological evasion and molecular pathogenesis might improve prognosis for those already infected, and a low-cost, nontoxic drug may help eliminate *T. cruzi* from the human reservoir of infection. It is debatable whether other aspects of advanced research will have a significant direct impact on the control of Chagas disease, but they will lead to a fuller understanding of the fascinating epidemiology and biology of trypanosomatid infections.

REFERENCES

Aguero, F., Verdún, R.E., et al. 2000. A random sequencing approach for the analysis of the *Trypanosoma cruzi* genome: general structure, large gene and repetitive DNA families, and gene discovery. *Genome Res*, **10**, 1996–2005.

Alvarez, V., Parussini, F. and Aslund, L. 2002. Expression in insect cells of active mature cruzipain from *Trypanosoma cruzi*, containing its C-terminal domain. *Protein Expr Purif*, **26**, 467–75.

Andersson, B., Aslund, L., et al. 1998. Complete sequence of a 93.4-kb contig from chromosome 3 of *Trypanosoma cruzi* containing a strand-switch region. *Genome Res*, **8**, 4-, 809–15.

Anonymous, 1999. Recommendations from a satellite meeting. *Mem Inst Oswaldo Cruz*, **94**, 429–32.

Argibay, P.F., Di Noia, J.M., et al. 2002. *Trypanosoma cruzi* surface mucin TcMuc-e2 expressed on higher eukaryotic cells induces human T cell anergy, which is reversible. *Glycobiology*, **12**, 25–32.

Beard, C.B., Pye, G., et al. 2003. Chagas disease in a domestic transmission cycle, southern Texas, USA. *Emerg Infect Dis*, **9**, 103–5.

Beucher, M., Meira, W.S., et al. 2003. Expression and purification of functional, recombinant *Trypanosoma cruzi* complement regulatory protein. *Protein Expr Purif*, **27**, 19–26.

Blanco, S.B., Segura, E.L., et al. 2000. Congenital transmission of *Trypanosoma cruzi*: an operational outline for detecting and treating infected infants in north-western Argentina. *Trop Med Int Health*, **5**, 293–301.

Bogliolo, A.R., Lauria-Pires, L. and Gibson, W.C. 1996. Polymorphisms in *Trypanosoma cruzi*: evidence of genetic recombination. *Acta Tropica*, **61**, 31–40.

Bonse, S., Richards, J.M., et al. 2000. (2,2':6',2''-Terpyridine)platinum(II) complexes are irreversible inhibitors of *Trypanosoma cruzi* trypanothione reductase but not of human glutathione reductase. *J Med Chem*, **43**, 4812–21.

Brack, C. 1968. Elektronenmikroskopische untersuchungen zum leben szyklus von *Trypanosoma cruzi*. Unter besonderer Berücksichtigung der entwicklungsformen im ueberträger *Rhodnius prolixus*. *Acta Tropica*, **25**, 289–356.

Brener, Z. 1994. The pathogenesis of Chagas' disease: an overview of current theories. In *Chagas' disease and the nervous system*, Scientific Publication No. 547. Washington, DC: Pan American Health Organization.

Brener, Z. and Andrade, Z. 1979. *Trypanosoma cruzi e doença de Chagas*. Rio de Janeiro: Guanabara Koogan S.A.

Brisse, S., Barnabé, C. and Tibayrenc, M. 2000. Identification of six *Trypanosoma cruzi* phylogenetic lineages by random amplified polymorphic DNA and multilocus enzyme electrophoresis. *Int J Parasitol*, **30**, 35–44.

Buscaglia, C.A., Campo, V.A., et al. 2004. The surface coat of the mammal-dwelling infective trypomastigote stage of *Trypanosoma cruzi* is formed by highly diverse immunogenic mucins. *J Biol Chem*, **279**, 15860–9.

Buschiazzo, A., Amaya, M.F., et al. 2002. The crystal structure and mode of action of trans-sialidase, a key enzyme in *Trypanosoma cruzi* pathogenesis. *Mol Cell*, **10**, 757–68.

Cáceres, A.J., Portillo, R., et al. 2003. Molecular and biochemical characterization of hexokinase from *Trypanosoma cruzi*. *Mol Biochem Parasitol*, **126**, 251–62.

Campbell, D.A., Westenberger, S.J. and Sturm, N.R. 2004. The determinants of Chagas disease: connecting parasite and host genetics. *Curr Mol Med*, **4**, 549–62.

Carcavallo, R.U., Giron, I.G., et al. (eds) 1997. *Atlas of chagas disease vectors in the Americas*, 3 vols. Rio de Janeiro: Fiocruz.

Carrasco, H.J., Frame, I.A., et al. 1996. Genetic exchange as a possible source of genomic diversity in sylvatic populations of *Trypanosoma cruzi*. *Am J Trop Med Hyg*, **54**, 418–24.

Carreno, H., Rojas, C., et al. 1987. Schizodeme analyses of *Trypanosoma cruzi* zymodemes from Chile. *Exp Parasitol*, **64**, 252–60.

Castro, A.M., Luquetti, A.O., et al. 2002. Blood culture and polymerase chain reaction for the diagnosis of the chronic phase of human infection with *Trypanosoma cruzi*. *Parasitol Res*, **88**, 894–900.

Cazzulo, J.J. 1994. Intermediate metabolism in *Trypanosoma cruzi*. *J Bioenerg Biomembr*, **26**, 157–65.

Cazzulo, J.J. 2002. Proteinases of *Trypanosoma cruzi*: potential targets for the chemotherapy of Chagas desease. *Curr Top Med Chem*, **2**, 1261–7.

Chapman, M.D., Snary, D. and Miles, M.A. 1984. Quantitative differences in the expression of a 72,000 molecular weight cell surface glycoprotein (GP72) in *T. cruzi* zymodemes. *J Immunol*, **132**, 3149–53.

Chiale, P.A., Ferrari, I., et al. 2001. Differential profile and biochemical effects of antiautonomic membrane receptor antibodies in ventricular arrhythmias and sinus node dysfunction. *Circulation*, **103**, 1765–71.

Cohen, J.E. and Gurtler, R.E. 2001. Modeling household transmission of American trypanosomiasis. *Science*, **293**, 694–8.

Coura, J.R., Junqueira, A.C., et al. 2002. Emerging Chagas disease in Amazonian Brazil. *Trends Parasitol*, **18**, 171–6.

Crossman, A., Paterson, M.J., et al. 2002. Further probing of the substrate specificities and inhibition of enzymes involved at an early stage of glycosylphosphatidylinositol (GPI) biosynthesis. *Carbohydr Res*, **337**, 2049–59.

Cunha-Neto, E. and Kalil, J. 2001. Heart-infiltrating and peripheral T cells in the pathogenesis of human Chagas' disease cardiomyopathy. *Autoimmunity*, **34**, 187–92.

de Gorgolas, MD and Miles, MA 1994. Visceral leishmaniasis and AIDS. *Nature*, **372**, 374.

Di Noia, J.M., Buscaglia, C.A., et al. 2002. A *Trypanosoma cruzi* small surface molecule provides the first immunological evidence that Chagas' disease is due to a single parasite lineage. *J Exp Med*, **195**, 401–13.

Dias, J.C., Silveira, A.C. and Schofield, C.J. 2002. The impact of Chagas disease control in Latin America: a review. *Mem Inst Oswaldo Cruz*, **97**, 603–12.

Dias, J.C.P. 1987. Control of Chagas disease in Brazil. *Parasitol Today*, **3**, 336–41.

Dictar, M., Sinagra, A., et al. 1998. Recipients and donors of bone marrow transplants suffering from Chagas disease: management and pre-emptive therapy of parasitemia. *Bone Marrow Transplant*, **21**, 391–3.

Dvorak, J.A. 1984. The natural heterogeneity of *Trypanosoma cruzi*: biological and medical implications. *J Cell Biochem*, **24**, 357–71.

Eger-Mangrich, I. and de Oliveira, M.A. 2001. Interaction of *Trypanosoma rangeli* Tejera, 1920 with different cell lines in vitro. *Parasitol Res*, **87**, 505–9.

Estani, S.S., Segura, E.L., et al. 1998. Efficacy of chemotherapy with benznidazole in children in the indeterminate phase Chagas' disease. *Am J Trop Med Hyg*, **59**, 526–9.

Fae, K.C., Drigo, S.A., et al. 2000. HLA and beta-myosin heavy chain do not influence susceptibility to Chagas disease cardiomyopathy. *Microbes Infect*, **2**, 745–51.

Fernandes, O., Mangia, R.H., et al. 1999. The complexity of the sylvatic cycle of *Trypanosoma cruzi* in Rio de Janeiro state (Brazil) revealed by the non-transcribed spacer of the mini-exon gene. *Parasitology*, **118**, 161–6.

Frasch, A.C.C. 2003. Trypanosoma cruzi surface proteins. In: Tyler, K.M. and Miles, M.A. (eds), *World class parasites: American trypanosomiasis*. Boston: Kluwer Academic Publishers, 25–35.

Galvao, C., Patterson, J.S., et al. 2002. A new species of Triatominae from Tamil Nadu, India. *Med Vet Ent*, **16**, 75–82.

Garcia, E.S., Gonzalez, M.S., et al. 1995. Induction of *Trypanosoma cruzi* metacyclogenesis in the gut of the hematophagous insect vector, *Rhodnius prolixus*, by hemoglobin and peptides carrying alpha D-globin sequences. *Exp Parasitol*, **81**, 255–61.

Gaunt, G.W. and Miles, M.A. 2000. The ecotopes and evolution of triatomine bugs (Triatominae) and their associated trypanosomes. *Mem Inst Oswaldo Cruz*, **95**, 557–65.

Gaunt, G.W. and Miles, M.A. 2002. A molecular clock for the insects dates the origin of the insects and accords with paleontological and biogeographic landmarks. *Mol Biol Evol*, **19**, 748–61.

Gaunt, G.W., Yeo, M., et al. 2003. Mechanism of genetic exchange in American trypanosomes. *Nature*, **421**, 936–9.

Girones, N. and Fresno, M. 2003. Etiology of Chagas disease myocarditis: autoimmunity, parasite persistence, or both? *Trends Parasitol*, **19**, 19–22.

Hoare, C.A. 1972. *The trypanosomes of mammals*. Oxford: Blackwell.

Köberle, F. 1974. Pathogenesis of Chagas' disease. In *Trypanosomiasis and leishmaniasis with special reference to Chagas' disease. Ciba Foundation Symposium*, **20**, Amsterdam: Associated Scientific Publishers, 137–58.

Kumar, S. and Tarleton, R.L. 2001. Antigen-specific Th1 but not Th2 cells provide protection from lethal *Trypanosoma cruzi* infection in mice. *J Immunol*, **166**, 4956–603.

Lazo, J.E., Meneses, A.C., et al. 1998. Meningoencefalities toxoplásmica e chagásica em pacientes con infecçao pelo virus de immunodeficiência humana: diagnóstico diferencial anatomopopatológia e tomográfico. *Rev Soc Bras Med Trop*, **31**, 163–71.

Lázzari, J.O. 1994. Autonomic nervous system alterations in Chagas' disease: review of the literature. In *Chagas' disease and the nervous system*. Scientific Publication No. 547. Washington, DC: Pan American Health Organization,

Lent, H. and Wygodzinsky, P. 1979. Revision of the Triatominae (Hemiptera, Reduviidae), and their significance as vectors of Chagas' disease. *Bull Am Mus Nat Hist*, **163**, 123–520.

Lewinsohn, R. 2003. Prophet in his own country: Carlos Chagas and the Nobel Prize. *Perspect Biol Med*, **46**, 532–49.

Lillico, S., Field, M.C., et al. 2003. Essential Roles for GPI-anchored proteins in African trypanosomes revealed using mutants deficient in GPI8. *Mol Biol Cell*, **14**, 1182–94.

Lopes, E.R. and Chapadeiro, E. 2004. Pathogenesis of American trypanosomiasis. In: Maudlin, I., Holmes, P.H. and Miles, M.A. (eds), *The trypanosomiasis*. Wallingford, UK: CABI, 303–30.

Luquetti, A.O., Miles, M.A., et al. 1986. *Trypanosoma cruzi*: zymodemes associated with acute and chronic Chagas' disease in central Brazil. *Trans R Soc Trop Med Hyg*, **80**, 462–70.

Macedo, A.M., Pimenta, J.R., et al. 2001. Usefulness of microsatellite typing in population genetic studies of *Trypanosoma cruzi*. *Mem Inst Oswaldo Cruz*, **96**, 407–13.

Machado, C.A. and Ayala, F.J. 2001. Nucleotide sequences provide evidence of genetic exchange among distantly related lineages of *Trypanosoma cruzi*. *Proc Natl Acad Sci USA*, **98**, 7396–401.

Machado, C.R.S., Camargos, E.R.S., et al. 2000. Cardiac autonomic denervation in congestive heart failure: comparison of Chagas' heart disease with other dilated cardiomyopathy. *Hum Pathol*, **31**, 3–10.

Mahler, E., Sepulveda, P., et al. 2001. A monoclonal antibody against the immunodominant epitope of the ribosomal P2beta protein of *Trypanosoma cruzi* interacts with the human beta 1-adrenergic receptor. *Eur J Immunol*, **31**, 2210–16.

Mair, G., Shi, H., et al. 2000. A new twist in trypanosome RNA metabolism: *cis*-splicing of pre-mRNA. *RNA*, **6**, 163–9.

Marinho, C.R., Bucci, D.Z., et al. 2004. Pathology affects different organs in two mouse strains chronically infected by a *Trypanosoma cruzi* clone: a model for genetic studies of Chagas' disease. *Infect Immun*, **72**, 2350–7.

Martin, W. and Borst, P. 2003. Secondary loss of chloroplasts in trypanosomes. *Proc Natl Acad Sci USA*, **100**, 765–7.

Matthews, K.R., Tshudi, C. and Ullu, E. 1994. A common pyrimidine-rich motif governs *trans*-splicing and polyadenylation of tubulin polycistronic pre-mRNA in trypanosomes. *Genes Dev*, **8**, 491–501.

Maudlin, I., Holmes, P.H. and Miles, M.A. (eds) 2004. *The trypanosomiasis*. Wallingford, UK: CABI.

Miles, M.A. 1979. Transmission cycles and the heterogeneity of *Trypanosoma cruzi*. In: Lumsden, W.H.R. and Evans, D.A. (eds), *Biology of the Kinetoplastida*, Vol. 2. Academic Press: London, 117–196.

Miles, M.A. 1982. *Trypanosoma cruzi*: epidemiology. In: Baker, J.R. (ed.), *Perspectives in trypanosomiasis research*. 21st Trypanosomiasis Seminar: London, 24 September 1981, London: Research Studies Press, 1–15.

Miles, M.A. 1983. The epidemiology of South American trypanosomiasis: biochemical and immunological approaches and their relevance to control. *Trans R Soc Trop Med Hyg*, **77**, 5–23.

Miles, M.A. 1992. Disease control has no frontiers. *Parasitol Today*, **8**, 221–2.

Miles, M.A. 1993. Culturing and biological cloning of *Trypanosoma cruzi*. In: Hyde, J.E. (ed.), *Protocols in molecular parasitology*. Totowa, NJ: Humana Press, 15–28.

Miles, M.A. 1994. Chagas' disease and chagasic megacolon. In: Kamm, M.A. and Lennard-Jones, J.E. (eds), *Constipation*. Petersfield, UK and Bristol, PA: Wrightson Biomedical Publishing, 205–10.

Miles, M.A. 1996. New World trypanosomiasis. In: Cox, F.E.G. (ed.), *The Wellcome Trust illustrated history of tropical diseases*. London: The Wellcome Trust, 192–205.

Miles, M.A. 2004. The discovery of Chagas disease: progress and prejudice. *Infect Dis Clin North Am*, **18**, 247–60.

Miles, M.A., de Souza, A.A. and Póvoa, M. 1981a. Chagas' disease in the Amazon basin. III. Ecotopes of ten triatomine bug species (Hemiptera: Reduviidae) from the vicinity of Belém, Pará, Brazil. *J Med Entomol*, **18**, 266–78.

Miles, M.A., de Souza, A.A. and Póvoa, M. 1981b. Mammal tracking and nest location in Brazilian forest with an improved spool-and-line device. *J Zool*, **195**, 331–47.

Miles, M.A., Yeo, M. and Gaunt, M.W. 2003a. The ecotopes and evolution of *Trypanosoma cruzi* and triatomine bugs. In: Tyler, K.M. and Miles, M.A. (eds), *World class parasites: American trypanosomiasis*. Boston: Kluwer Academic Publishers, 137–45.

Miles, M.A., Feliciangeli, M.D. and de Arias, A.R. 2003b. American trypanosomiasis (Chagas' disease) and the role of molecular epidemiology in guiding control strategies. *Br Med J*, **326**, 1444–8.

Miles, M.A., Cedillos, R.A., et al. 1981c. Do radically dissimilar *Trypanosoma cruzi* strains (zymodemes) cause Venezuelan and Brazilian forms of Chagas disease? *Lancet*, **1**, 1338–40.

Moreira, H., de Rezende, J.M., et al. 1985. Chagasic megacolon. *Colo-Proctology*, **7**, 260–7.

Nozaki, T. and Dvorak, J.A. 1993. Intraspecific diversity in the response of *Trypanosoma cruzi* to environmental stress. *J Parasitol*, **79**, 451–4.

Nunes, L.R., de Carvalho, M.R. and Buck, G.A. 1997. *Trypanosoma cruzi* strains partition into two groups based on the structure and function of the spliced leader RNA and rRNA gene products. *Mol Biochem Parasitol*, **86**, 211–24.

Oelemann, W.M., Teixeira, M.D., et al. 1998. Evaluation of three commerical enzyme-linked immunoabsorbent assays for diagnosis of Chagas disease. *J Clin Microbiol*, **36**, 2423–7.

Oliveira, J.S.M. 1969. Cardiopatia chagásica experimental. *Rev Goiana Med*, **15**, 77–133.

PAHO, 1994. *Chagas' disease and the nervous system*. Washington DC: Pan American Health Organization.

Raia, A.A. 1983. *Manifestações digestivas da moléstia de Chagas*. Sao Paulo: Sarvier.

Rassi, A. and Luquetti, A.O. 2003. Specific treatment for *Trypanosoma cruzi* infection (Chagas disease). In: Tyler, K.M. and Miles, M.A. (eds), *World class parasites: American trypanosomiasis*. Boston: Kluwer Academic Publishers, 118–25.

Ribeiro dos Santos, R., Rossi, M.A., et al. 1992. Anti-CD4 abrogates rejection and reestablishes long-term tolerance to syngeneic newborn hearts grafted in mice chronically infected with *Trypanosoma cruzi*. *J Exp Med*, **175**, 29–39.

Risso, M.G., Garbarino, G.B., et al. 2004. Differential expression of a virulence factor, the trans-sialidase, by the main *Trypanosoma cruzi* phylogenetic lineages. *J Infect Dis*, **189**, 2250–9.

Robello, C., Gamarro, F., et al. 2000. Evolutionary relationships in *Trypanosoma cruzi*: molecular phylogenetics supports the existence of a new major lineage of strains. *Gene*, **246**, 331–8.

Rowland, E.C., Mikhail, K.S. and McCormick, T.S. 1992. Isotype determination of anti-Trypanosma cruzi antibody in murine Chagas' disease. *J Parasitol*, **78**, 557–61.

Schijman, A.G., Vigliano, C.A., et al. 2004. *Trypanosoma cruzi* DNA in cardiac lesions of Argentinean patients with end-stage chronic Chagas heart disease. *Am J Trop Med Hyg*, **70**, 210–20.

Schmunis, G.A., Zicker, F., et al. 2001. Safety of blood supply for infectious diseases in Latin American countries, 1994-1997. *Am J Trop Med Hyg*, **65**, 924–30.

Schofield, C.J. 1994. *Triatominae. Biology and control*. London: Eurocommunica Publications.

Schofield, C.J. and Dias, J.C.P. 1998. The southern cone initiative against Chagas disease. *Adv Parasitol*, **2**, 2–22.

Silva, J.S., Vespa, G.N.R., et al. 1995. Tumor necrosis factor alpha mediates resistance to *Trypanosoma cruzi* infection in mice by inducing nitric oxide production in infected gamma interferon-activated macrophages. *Infect Immun*, **63**, 4862–7.

Silva-Barbosa, S.D. and Savino, W. 2000. The involvement of laminin in anti-myocardial cell autoimmune response in murine Chagas disease. *Dev Immunol*, **7**, 293–301.

Simpson, L., Sbicego, S. and Aphasizhev, R. 2003. Uridine insertion/deletion RNA editing in trypanosome mitochondria: a complex business. *RNA*, **9**, 265–76.

Stoll, V.S., Simpson, S.J., et al. 1997. Glutathione reductase turned into trypanothione reductase: structural analysis of an engineered change in substrate specificity. *Biochemistry*, **36**, 6437–47.

Stoppani, A.O. 1999. The chemotherapy of Chagas disease. *Medicina (Buenos Aires)*, **59**, 147–65.

Stothard, J.R., Frame, I.A., et al. 1998. On the molecular taxonomy of *Trypanosoma cruzi* using riboprinting. *Parasitology*, **117**, 243–7.

Sturm, N.R., Vargas, N.S., et al. 2003. Evidence for multiple hybrid groups in *Trypanosoma cruzi*. *Int J Parasitol*, **33**, 269–79.

Taibi, A., Plumas-Marty, B., et al. 1993. *Trypanosoma cruzi*: immunity-induced in mice and rats by trypomastigote excretory-secretory antigens and identification of a peptide sequence containing a T cell epitope with protective activity. *J Immunol*, **151**, 2676–89.

Tan, H. and Andrews, N.W. 2002. Don't bother to knock – the cell invasion strategy of *Trypanosoma cruzi*. *Trends Parasit*, **18**, 427–8.

Tarleton, R.L. 2001. Parasite persistence in the aetiology of Chagas disease. *Int J Parasitol*, **31**, 550–4.

Tarleton, R.L., Grusby, M.J. and Zhang, L. 2000. Increased susceptibility of Stat4-deficient and enhanced resistance in Stat6-deficient mice to infection with *Trypanosoma cruzi*. *J Immunol*, **165**, 1520–5.

Taylor, M.C., Kelly, J.M., et al. 1994. The structure, organisation and expression of the *Leishmania donovani* gene encoding trypanothione reductase. *Mol Biochem Parasitol*, **64**, 293–301.

Teixeira, Vde P., Hial, V., et al. 1997. Correlation between adrenal central vein parasitism and heart fibrosis in chronic chagasic myocarditis. *Am J Trop Med Hyg*, **56**, 177–80.

Tetley, L. and Vickerman, K. 1991. The glycosomes of trypanosomes: number and distribution as revealed by electron spectroscopic imaging and 3-D reconstruction. *J Microsc*, **162**, 83–90.

Tibayrenc, M., Neubauer, K., et al. 1993. Genetic characterization of six parasitic protozoa: parity between random-primer DNA typing and multilocus enzyme electrophoresis. *Proc Natl Acad Sci USA*, **90**, 1335–9.

Tyler, K.M. and Miles, M.A. (eds) 2003. *World class parasites: American trypanosomiasis*. Boston: Kluwer Academic Publishers.

Umezawa, E.S., Bastos, S.F., et al. 2003. An improved serodiagnostic test for Chagas' disease employing a mixture of *Trypanosoma cruzi* recombinant antigens. *Transfusion*, **43**, 91–7.

Urbina, J.A. 2002. Chemotherapy of Chagas disease. *Curr Pharm Des*, **8**, 287–95.

Urbina, J.A., Payares, G., et al. 2003. Parasitological cure of acute and chronic experimental Chagas disease using the long-acting experimental triazole TAK-187. Activity against drug-resistant *Trypanosoma cruzi* strains. *Int J Antimicrob Agents*, **21**, 39–48.

Vago, A.R., Andrade, L.O., et al. 2000. Genetic characterization of *Trypanosoma cruzi* directly from tissues of patients with chronic Chagas disease: differential distribution of genetic types into diverse organs. *Am J Pathol*, **156**, 1805–9.

Valente, V.C., Valente, S.A.S., et al. 1998. Chagas disease in the Amazon Basin: association of *Panstrongylus geniculatus* (Hemiptera: Reduviidae) with domestic pigs. *J Med Entomol*, **35**, 99–103.

Wanderley, D.M. 1993. Control of *Triatoma infestans* in the State of Sao Paulo. *Rev Soc Bras Med Trop*, **26**, Suppl 3, S17–25.

WHO, 2002. *Control of Chagas disease*, Tech Rep Ser 905. Geneva: World Health Organization.

Wilkinson, S.R., Meyer, D.J., et al. 2002a. The *Trypanosoma cruzi* enzyme TcGPXI is a glycosomal peroxidase and can be linked to trypanothione reduction by glutathione or tryparedoxin. *J Biol Chem*, **277**, 17062–71.

Wilkinson, S., Obado, S.O., et al. 2002b. *Trypanosoma cruzi* expresses a plant-like ascorbate-dependent hemoperoxidase localized to the endoplasmic reticulum. *Proc Natl Acad Sci USA*, **99**, 13453–8.

Wrightsman, R.A., Luhrs, K.A., et al. 2002. Paraflagellar rod protein-specific CD8+ cytotoxic T lymphocytes target *Trypanosoma cruzi*-infected host cells. *Parasite Immunol*, **24**, 401–12.

Zhang, L. and Tarleton, R.L. 1999. Parasite persistence correlates with disease severity and localization in chronic Chagas' disease. *J Infect Dis*, **180**, 480–6.

20

Intestinal coccidia: cryptosporidiosis, isosporiasis, cyclosporiasis

VINCENT MCDONALD AND M. PAUL KELLY

CRYPTOSPORIDIOSIS

Introduction

The first description of the genus *Cryptosporidium* was by the American parasitologist E. E. Tyzzer (1907) who obtained, in considerable detail, the life cycle of *C. muris* present in the gastric glands of the common mouse. An unusual feature of development was that the parasite was located at the apical surface of the epithelium. The only major modification that has been made to Tyzzer's description of the life cycle is that the developing stages are intracellular. However, this observation had to wait until electron microscopy was available. The genus name refers to the absence of sporocysts that normally enclose the sporozoites of coccidia (from Greek 'kruptos' meaning 'hidden'). Tyzzer (1912) later identified the smaller parasite *C. parvum*, in the intestinal villi of mice, that is now recognized as a major cause of mammalian cryptosporidiosis. The same worker subsequently reported cryptosporidial infection of turkeys (Tyzzer 1929). Over many years, *Cryptosporidium* infection was described in different host types and numerous new species were named accordingly. In the early 1980s, however, studies of transmission from one host-type to another indicated host-specificity was less stringent than for some other coccidia, and that there was only a small number of legitimate species of *Cryptosporidium* including *C. parvum* and *C. muris* of mammals plus a few others that infect either birds, reptiles or fish (Fayer et al. 1997). More recent genomic analyses combined with limited animal infectivity studies, however, have suggested that the taxonomy is more complicated and so

the number of species is likely to increase in future (Morgan et al. 1999).

Until the 1950s, it was not appreciated that *Cryptosporidium* could be pathogenic but even then, a report from Slavin (1955) of severe diarrhea and some deaths in a turkey flock infected with the species named *C. meleagridis* attracted little attention. The situation gradually altered starting in the 1970s when a few cases were identified first in cattle (Panciera et al. 1971) and then in humans (Meisel et al. 1976; Nime et al. 1976). With the development of simple diagnostic methods in the 1980s, it became clear that *C. parvum* is a common cause of diarrhea in humans worldwide and also in neonatal domestic mammals such as bovine calves and lambs (Tzipori 1983). In human cryptosporidiosis, diarrhea and other symptoms normally last for 1–2 weeks but infection of hosts with impaired T-cell-mediated immunity, such as individuals with acquired immune deficiency syndrome (AIDS), may be life-threatening and chronic (Current and Garcia 1991). The significance of *Cryptosporidium* as a public health problem has been increased by the lack of an effective chemotherapeutic agent and the transmission of infection through public water supplies (Coombs 1999; Rose et al. 2002).

Classification

The protozoan genus *Cryptosporidium* belongs to the phylum Apicomplexa, which comprises several thousand species that possess an apical complex: a specifalized assembly of organelles believed to be involved in host cell invasion (Levine 1984). *Cryptosporidium* spp. are contained in the suborder Eimeriorina along with other coccidia that cause human or animal disease, *Cyclospora*, *Isospora*, *Toxoplasma*, and *Eimeria*. The oocysts of these latter genera have sporocysts that enclose sporozoites, whereas oocysts of cryptosporidia lack sporocysts and have 'naked' sporozoites; *Cryptosporidium* has, therefore, been placed within a new family, Cryptosporidiidae.

Cryptosporidium spp. were originally designated according to oocyst morphology, site of development, and host specificity. The selection of the latter criterion was presumably based on the assumption that the genus had the restricted host range of related coccidia such as *Eimeria* and *Isospora*. Beginning in the 1980s, the results of cross-transmission studies challenged this view and the number of species was eventually truncated to eight: *C. parvum* and *C. muris* of mammals, *C. wrairi* of guinea-pigs, *C. felis* of cats, *C. baileyi* and *C. meleagridis* of birds, *C. serpentis* of reptiles, and *C. nasorum* of fish (reviewed in Morgan et al. 1999). Cryptosporidiosis of mammals has until recently almost always been associated with *C. parvum*, which has small oocysts (4–5 µm) and infects mainly the intestine, while *C. muris* has larger oocysts and infects the stomach of mice – and perhaps cattle – and is not considered to be a serious pathogen (Upton and Current 1985)

However, evidence from investigations of isoenzyme and antigenic composition of oocysts as well as restriction fragment length polymorphism (RFLP) analysis suggested that *C. parvum* may not be a single species. These studies could clearly distinguish human from animal – mainly bovine – isolates (Awad-El-Kariem et al. 1995; Nichols et al. 1991; Ortega et al. 1991). Furthermore, there are differences between human and bovine isolates in their ability to infect separate host species: bovine isolates are able to infect cattle, mice or humans whereas human isolates usually cannot infect animals (Awad-El-Kariem et al. 1998; Widmer et al. 2000). More recent evidence from gene sequencing (e.g. 18S rDNA, cryptosporidial oocyst wall protein (COWP) and thrombospondin-related adhesion protein (TRAP) and phylogenetic analyses have confirmed this distinctiveness of human and animal isolates, which are now designated Type 1 and Type 2, respectively (Peng et al. 1997; reviewed in Morgan et al. 1999). Additionally, unique *Cryptosporidium* genotypes have been found in oocysts from other mammals, including pigs, dogs, and ferrets (Morgan et al. 1999).

Life cycle

The general features of the life cycle of *C. parvum* are similar to those of other coccidia that infect mammals (Tyzzer 1912; Current and Reese 1986). A unique feature of intracellular development, however, is that the parasite is anchored at the luminal surface and isolated from the host cell cytoplasm. The entire development of the parasite takes place within the same host

and begins following ingestion of oocysts. Sporozoite excystation takes place in the intestine under the influence of numerous factors, including temperature and bile salts (Sundermann et al. 1987) and the parasites attach by the apical pole to the brush border of epithelial cells. The attached sporozoite induces local cytoskeletal rearrangement in the host cell and the brush border membrane envelops the sporozoite forming the parasitophorous vacuole (Chen et al. 1998). Parasite antigens are expressed on the parasitophorous vacuole and on part of the outer host cell membrane surrounding the parasite (McDonald et al. 1995). At the site of attachment, interactions between the parasite and host cell components result in the formation of the so-called feeder organelle, which expands as the parasite grows and acts as a barrier between the host and parasite cytoplasms.

The intracellular sporozoite soon rounds up into a trophozoite and, following a period of growth, the parasite undergoes asexual multiplication by merogony (schizogony). The parasite development induces activation of the transcription factor NF-κB in the host cell which inhibits the cell from entering apoptosis (Chen et al. 2001). Eight merozoites are produced which continue the reproductive cycle following invasion of other host cells (Figure 20.1). Eventually meronts with four merozoites develop and the latter may initiate the development of microgametes (Figure 20.2) and macrogametes (Current and Reese 1986). Fertilization of the gametes follows and two types of oocyst are formed that undergo sporogony within host cells. The first oocyst type, representing about 20 percent of oocysts, has a thin unit membrane and is able to re-initiate the life cycle in the host. This stage may play an important role in maintaining the infection, particularly in immunocompromised hosts who fail to eliminate the parasite. The second type of oocyst, which has a double-layered or thick wall, accounts for about 80 percent of oocysts and is passed in feces (Current and Reese 1986). The development of an autoinfective stage and maturation of oocysts within the host cell are unusual characteristics for coccidia. Oocysts of *Eimeria* and *Isospora* are of a single type and normally have to escape the host before they undergo sporulation, aided by a reduced temperature and increased amount of oxygen.

Parasite cell and molecular studies

At present, there is no consistently effective chemotherapy for cryptosporidiosis and the formulation of novel drugs may depend on rationale design based on targets identified from studies of parasite molecular functions. Studies in cell biology, biochemistry, metabolism, and molecular genetics have been, for the most part, limited to the examination of oocysts and sporozoites, which can be obtained in copious amounts from large animal infections. The biology of intracellular

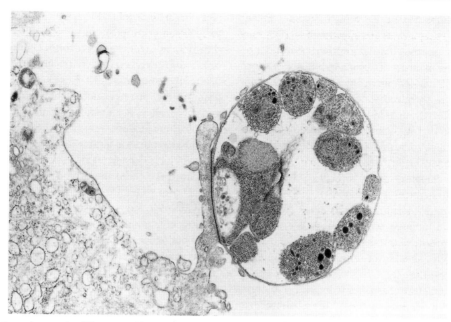

Figure 20.1 *Electronmicrograph of* Cryptosporidium parvum *development in human intestine. The field shows a maturing meront, probably Type 1 with numerous merozoites around the periphery. The parasite's outer membrane and host's cell membrane surrounding the meront are in close apposition. Note the electron-dense band at the interface between the parasite and the host cell. Magnification ×20 000.*

stages has been poorly studied, partly because parasite development in vitro is relatively poor and there is no transfection technology. Although *C. parvum* reproduction may be maintained in vitro for many days (Hijjawi et al. 2001), the development is not yet sufficient for maintenance of parasite isolates. Furthermore, no reliable method for cryopreservation has been described. From the relatively small body of information available on the biology of *Cryptosporidium*, however, it is clear that the parasite has unique features.

CELL BIOLOGY

The robustness of the cryptosporidial oocyst wall is the key to survival of the parasite outside of the host. Little is known about the molecular structure of the wall or how the components gel to provide such an effective barrier against environmental stresses. Ultrastructural studies indicate that the oocyst wall of *Cryptosporidium* comprises of two layers (Reduker et al. 1985) and contains numerous polypeptides (Tilley and Upton

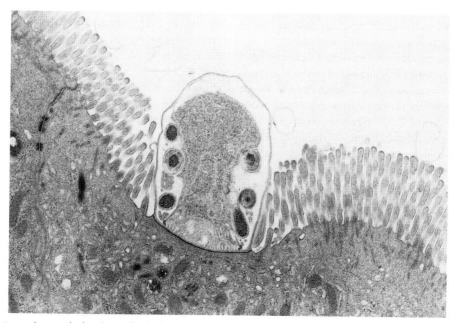

Figure 20.2 *Electron micrograph showing a developing microgametocyte of* Cryptosporidium parvum. *Microgametes are budding from the periphery of the parasite body. Note the absence of an epithelial cell brush border around the parasite. Magnification ×20 000.*

1997). One 190-kDa COWP has a high cysteine content and two types of amino acid repeat motifs that may provide periodic spacing and appropriate conformation to allow a large number of disulfide bridges between the cysteine residues (Ranucci et al. 1993). The outer oocyst wall is highly glycosylated and its polypeptide composition differs from that of the inner wall (Tilley and Upton 1997). The inner surface contains a filamentous array and sonicated protein K-digested walls fragment into small linear sheets indicating there is an integral highly complex lattice structure (Harris and Petry 1999). Ultrastructural studies of *C. parvum* show that the sporozoite morphology is similar to that of other coccidia, including a typical banana-like shape of the organism. Parasite actin helps to maintain the shape of the sporozoite and is also important in motility (Forney et al. 1998).

The apical complex organelles comprising rhoptry, micronemes, and dense granules are believed to discharge contents that are required for invasion and establishment of apicomplexans inside host cells (Lumb et al. 1988). There appears to be only a single rhoptry in *Cryptosporidium*, whereas various other apicomplexan parasites have at least two and sometimes many more (Tetley et al. 1998). Isolated micronemes of *C. parvum* were demonstrated to contain three major proteins of 30, 120, and 200 kDa, and dense granules held five major proteins in the range 120–180 kDa (Petry and Harris 1999). Coccidia normally possess a mitochondrion but the presence of this organelle in *Cryptosporidium* has been disputed. Numerous studies reported the absence of mitochondria (Current and Reese 1986; Tetley et al. 1998) but such a structure was identified in the merozoites of *C. muris* (Uni et al. 1987) and in sporozoites of *C. parvum* (Riordan et al. 1999). A single plastid-like body of unknown function, probably of green algae origin, has been found in a number of apicomplexans and may also be present in *C. parvum* (Tetley et al. 1998).

How the invasive stages attach to and invade epithelial cells is poorly understood. A number of parasite proteins that may be involved have been characterized in recent years, and possible ligands on the host cell surface have also been identified. *Cryptosporidium parvum* micronemes contain a number of these, including TRAP-C1 which may play an important role in attachment or invasion (Naitza et al. 1998). Related proteins of other coccidia (e.g. MIC2 of *Toxoplasma*) have been shown to be involved in zoite locomotion and also attachment and invasion of cells (Soldati et al. 2001). Another micronemal molecule of *C. parvum*, the mucin-like glycoprotein gp900 with a large N-linked oligosaccharide core region and a cysteine-rich region, is shed onto the sporozoite surface during invasion and antibodies to gp900 inhibit infection in vitro (Petersen et al. 1997). Another glycoprotein, gp60, which shares epitopes with gp900 also has mucin-like domains with a terminal peptide specifying attachment of a glycosyl-phosphatidylinositol (GPI) anchor; this protein is cleaved to gp40 and gp15 (Strong et al. 2000). The gp40 was shown to be shed from sporozoites gliding over epithelial cells and is involved in cell attachment. The circumsporozoite-like antigen (CSL) is a 1 300 kDa glycoprotein found in micronemes and on the surface of sporozoites and merozoites. A monoclonal antibody to a repetitive carbohydrate epitope induces a CSL reaction whereby the antigen moves in a posterior direction along the sporozoite pellicle (Langer et al. 2001). The findings that antibody to CSL inhibit infection of cells and that CSL has been shown to bind to an 85 kDa receptor on intestinal epithelial cells suggests that CSL is an important ligand for attachment and invasion. Another protein designated CP47 and associated with the sporozoite membrane has been found to bind to human and animal ileal cells and a putative receptor of 57 kDa was identified (Nesterenko et al. 1999).

During invasion, the host cell cytoskeleton undergoes rearrangement and filamentous actin is assembled into a plaque-like structure at the host–parasite interface (Elliott and Clark 2000). Infection has been demonstrated to be dependent on actin polymerization and the actin polymerizing proteins ARP 2/3, vasodilator-stimulated phosphoprotein, and neural Wiskott Aldrich syndrome protein have been shown to be present at the site of infection (Elliott et al. 2001). This cytoskeletal remodeling appeared to involve host phosphoinositide 3-kinase activity, suggesting that host signaling pathways were induced to elicit actin rearrangement (Forney et al. 1999). As invasion begins, the host cell membrane moves along the length of the sporozoite pellicle to enclose the parasite in a parasitophorous vacuole (Chen et al. 1998). The parasitophorous vacuole membrane is of mixed host and parasite origin as parasite-derived antigens have been located in the membrane by immunogold electron microscopy (McDonald et al. 1995). During invasion, a dense band appears at the interface between the parasite and the host cell (see Figures 20.1 and 20.2) and is retained throughout development (Lumb et al. 1988). Following discharge of the apical organelles, a new parasite vacuole forms at the anterior pole of the sporozoite. This anterior vacuole forms the precursor of the feeder organelle membrane and at the attachment site becomes fused with the parasite pellicular and parasitophorous vacuole membranes (Lumb et al. 1988).

METABOLISM

Years of research have failed to provide adequate chemotherapy for cryptosporidiosis. It may be necessary to adopt a structure-based design approach in which possible molecular targets of the infectious agent are characterized and specific inhibitors constructed (Coombs 1999). Undoubtedly, this process would be facilitated by a better understanding of the biochemical pathways of the parasite, particularly those which are unique or distinct from host equivalents. Little is known about the metabolism of *Cryptosporidium* in comparison with other coccidia such as *Eimeria* or *Toxoplasma*. The significant features of biochemical pathways in *Cryptosporidium* are reviewed elsewhere (Coombs 1999).

Enzymes of the glycolytic pathway have been identified in oocysts of *C. parvum* (Entrala and Mascaro 1997).With the exception of adenosine triphosphate (ATP)-dependent phosphofructokinase, all the classic glycolytic activities leading to the formation of phosphoenolpyruvate were found. A Type I phosphofructokinase specific for inorganic phosphate has been characterized in a number of coccidia, including *C. parvum* (Denton et al. 1996). The use of this enzyme increases the energy efficiency of glycolysis by 50 percent and may represent adaptation towards anaerobiosis. Indeed, two key observations indicate that energy metabolism in *C. parvum* is similar to that of anaerobes, making glycolysis the main energy source. First, sporozoites were not killed by respiratory inhibitors such as potassium cyanide or sodium azide (Brown et al. 1996) and, secondly, enzymes of the Krebs tricarboxylic acid cycle could not be found in oocysts (Denton et al. 1996; Entrala and Mascaro 1997). The reported absence of a mitochondrion in *Cryptosporidium* (Current and Reese 1986; Tetley et al. 1998) is consistent with anaerobic development but later observation of this organelle accompanied by evidence for the presence of various mitochondrial proteins and susceptibility to drugs which target mitochondrial ubiquinones suggested that there are mitochondria (Riordan et al. 1999).

Energy in coccidia may also be obtained via a mannitol cycle. Mannitol synthesis has been shown to occur in *Eimeria* during development of sexual stages (Schmatz 1989) and two of the enzymes involved in this cycle have been found in *Cryptosporidium*. However, the enzyme which initiates mannitol synthesis, mannitol-1-phosphate dehydrogenase, appears to be regulated by a specific inhibitor during asexual development and so mannitol utilization may be relevant only in oocysts (Schmatz 1997). In support of this view, excysted *C. parvum* sporozoites show little hexokinase activity that is necessary for mannitol utilization, suggesting mannitol is not required for establishing infection (Entrala and Mascaro 1997). Furthermore, *Eimeria* hexokinase activity was observed to be greater at environmental temperature than at body temperature (Schmatz 1997).

Polyamines are involved in protein synthesis, stability of ribosomes and cell membranes, and also in regulation of gene expression. Most eukaryotic organisms, including *Eimeria*, initiate polyamine synthesis by decarboxylation of ornithine to give putrescine with ornithine decarboxylase, subsequently leading to the production of spermine. Keithly et al. (1997), however, demonstrated that in *C. parvum*, a quite distinctive route was involved in synthesizing putrescine. Ornithine decarboxylase was undetectable and putrescine was synthesized from arginine via arginine decarboxylase to provide agmatine which was converted to putrescine by agmatine iminohydrolase. This pathway is usually associated with some bacteria and plants, suggesting polyamine synthesis would be a suitable target for development of anticryptosporidial drugs. Another possible target in coccidia is the shikimate pathway which produces chorismate, an essential substrate for the synthesis of *p*-aminobenzoate, folate, ubiquinone, and aromatic amino acids (Roberts et al. 1998). This pathway is absent in mammals but present in plant plastids, algae, bacteria, and fungi. A number of the enzymes involved have also been observed in apicomplexan species which have plastid-like organelles. Also, the herbicide glyphosphate, which inhibits one of the enzymes in the pathway (5-enolpyruvyl shimikate 3-phosphatesynthase), inhibited in vitro development of apicomplexans including *Cryptosporidium* (Roberts et al. 1998). This effect was reduced by addition of *p*-aminobenzoic acid in the study of *T. gondii* and *Plasmodium falciparum*, but development of *C. parvum* was unaffected.

GENETICS

Nucleic acid studies of *Cryptosporidium* have played an important part in building our current understanding of the parasite's molecular cell biology, biochemistry, and taxonomy. In addition, nucleic acid probes have had enormous practical value in molecular epidemiology and as highly sensitive tools for detecting parasite material in tissue.

A difficulty in investigating the karyotypes of *Cryptosporidium* is the absence of chromosome condensation in the course of cell division. However, chromosomal bodies can be separated by pulsed field gel electrophoresis and this method, combined with densitometry analysis, has established that there are at least five, and probably eight, chromosomes in the range 1.03–1.54 Mb and a total genome size of approximately 10.4 Mb (Mead et al. 1988; Blunt et al. 1997). Mead et al. (1988) were able to differentiate between two species, *C. parvum* and *C. baileyi*, each of which demonstrated distinctive banding patterns.

To date, only about 30 genes of *C. parvum* have been fully sequenced and characterized (Spano and Crisanti 2000). One example was the sequencing of the bifunctional enzyme dihydrofolate reductase (DHFR)–thymithymidylate synthase (TS), which was isolated from genomic DNA libraries by hybridization with a probe amplified from *C. parvum* genomic DNA using generic TS primers in the polymerase chain reaction (PCR) (Vasquez et al. 1996). Two significant findings emerged from the sequence data. First, the DHFR active site was found to contain novel residues at several positions where point mutations occurring in other parasites have been demonstrated to produce antifolate resistance in DHFR. This may explain the inherent refractoriness of

C. parvum to treatment with some common antifolates used against other protozoa. Secondly, like other genes that have been characterized, the sequence data for DHFR−TS of different parasite isolates demonstrated polymorphisms that could be employed to examine the genotypic diversity of *C. parvum* isolates.

The adenine, thymine (AT) content for Type 1 and Type 2 *C. parvum* has been estimated at 65 and 68 percent, respectively (Widmer et al. 2002) and there are few introns (e.g. see Vasquez et al. 1996). Telomeres, the repetitive nucleotide sequences at the ends of eukaryotic chromosomes, have been detected in *C. parvum*. A reputed telomeric clone containing the hexanucleotide TTTAGG sequence, similar to a telomeric repeat of *Plasmodium*, was reported to be present in all five separated chromosomal bands of *C. parvum* and a probe containing the repeat hybridized with chromosome VI-specific *Eag*I restriction fragments corresponding to the chromosomal ends (Liu et al. 1998; Putignani et al. 1999). Microsatellite DNAs, which are tandem repeats of short nucleotide sequences have been described in *Cryptosporidium* (reviewed by Spano and Crisanti 2000). Microsatellites are useful genetic tools for intraspecific variation as they are usually highly polymorphic. For example, *C. parvum* Types 1 and 2 have been differentiated by microsatellite analysis (Spano and Crisanti 2000). More than 200 microsatellites have been identified in *C. parvum* DNA and these are usually AT-rich, thus conforming with the overall AT-richness in nucleotide composition of this species. In other eukaryotic species, microsatellites are densely distributed throughout the genome, making them suitable markers for high-resolution genetic mapping (Spano and Crisanti 2000).

Comprehensive genomic maps of *C. parvum* isolates will facilitate identification and characterization of genes that are crucial to the parasite's development, answer questions about the taxonomy of the genus, and provide suitable targets for chemotherapy. Separate approaches have been taken to map the *C. parvum* genome (Spano and Crisanti 2000; Widmer et al. 2002). In one, sequence analysis has been obtained from sporozoite cDNA cloned in a phage vector (see www.ebi.ac.uk/parasites/cparv.html). This method allows only characterization of genes expressed in the sporozoite, however. Other systems have employed sheared genomic DNA inserts derived from a plasmid library or genomic DNA cut with restriction enzymes and cloned into a phage vector. Analysis of sequenced DNA for translated parts of the *C. parvum* genome has indicated that the DNA contains approximately 85 percent coding sequences and about 5 000 genes (Widmer et al. 2002).

A major benefit of molecular genetic studies has been the provision of molecular markers for taxonomic evaluation of the parasite. Use of DNA sequencing and PCR with parasite-specific primers combined with RFLP has been valuable in the differentiation of *Cryptospor-idium* spp. and in configuring *C. parvum* isolates into groups. Early evidence of phenotypic variation between isolates of *C. parvum* relating to parasite virulence, drug sensitivity, isoenzyme analysis, and western blotting has been supported by DNA studies (reviewed by Morgan et al. 1999). Phylogenetic analysis of *Cryptosporidium* at different loci indicates that *Cryptosporidium* is not closely related to other apicomplexans such as *Toxoplasma* (Char et al. 1996). Sequence data indicate that there are several distinct species; for example, the similarity between the 18S rDNA of two bovine isolates of *C. parvum* (99.94 percent) was greater than that between one of these isolates and *C. baileyi* (96.26 percent) (Morgan et al. 1999). Importantly, genomic studies based on gene sequences for proteins, rDNA and microsatellite DNA point to the possibility that human Type 1 and animal Type 2 isolates may represent separate species of *Cryptosporidium*. In support of this, cross-transmission studies have shown that human Type 1 isolates rarely infected animals (Awad-El-Kariem et al. 1998; Widmer et al. 2000) and a multilocus study found no evidence of recombination between Type 1 and Type 2 isolates of *C. parvum* (Spano et al. 1998). The widespread prevalence of identical genotypes implies that there is a clonal population structure (Morgan et al. 1997) as there is in *Toxoplasma* (Howe and Sibley 1995). However, using microsatellite loci as genetic markers, sexual recombination was reported to occur during mixed laboratory infections of two distinct Type 2 isolates in interferon-gamma (IFN-γ) knockout mice (Feng et al. 2002). The whole question of whether there is significant genetic recombination in *Cryptosporidium* is still under debate.

A significant practical benefit of studies in molecular genetics is the development of probes for identifying isolates involved in outbreaks of cryptosporidiosis. For example, a Type 1 parasite was likely to have been the cause of a major waterborne outbreak in Milwaukee in the 1990s (Peng et al. 1997).

Clinical aspects

GENERAL

The commonest manifestation of infection is diarrhea. This may be indistinguishable from any other acute nonbloody diarrheal illness, but cryptosporidiosis is more likely to become persistent and this usually leads to weight loss. Other less common clinical features include abdominal pain, low-grade fever, malaise, and other nonspecific symptoms. As cryptosporidiosis is caused by an intracellular parasite, abnormalities of T-cell immunology seriously impair the host capacity to clear infection, leading to persistent infection. Human immunodeficiency virus (HIV) infection and AIDS lead to a greatly increased risk of cryptosporidiosis, as do other primary and secondary immunodeficiencies. These include severe combined immunodeficiency (SCID), other T-cell deficiencies, deficiencies of mannose-binding lectin (Jacyna et al. 1990; Kelly et al. 2000), and X-linked immunodeficiency with hyper-immunoglobulin (Ig)M caused by mutations in the CD40 ligand gene (Hayward et al. 1997). Children with acute leukemia also have increased susceptibility to cryptosporidiosis (Hunter and Nichols 2002).

There is an important interaction between cryptosporidiosis and nutritional impairment. This is true of children with and without HIV infection (Amadi et al. 2001), probably because persistent cryptosporidiosis causes severe anorexia, and mucosal damage also leads to malabsorption of micronutrients. In children, cryptosporidiosis seems to lead to an increased risk of nutritional problems. Phillips et al. (1992) showed an association between cryptosporidiosis and nutritional impairment in immunocompetent children living in London. A series of studies on children in Guinea-Bissau has revealed that children who were undernourished were not more likely to develop the infection (Molbak et al. 1993), but children with cryptosporidiosis are more likely to go on to lose weight after the infection (Molbak et al. 1997). In a series of studies in Brazil, cryptosporidiosis was associated with persistent diarrhea (Newman et al. 1999) and this is associated with nutritional shortfalls (Lima et al. 2000) and with diminished cognitive function (Niehaus et al. 2002). In adults, the great majority of cases of persistent cryptosporidiosis are related to HIV infection and, again, undernutrition is common but there is no evidence that cryptosporidiosis in this setting is associated with more severe nutritional problems than other causes of HIV-related persistent diarrhea.

AIDS

In many countries, the problem of cryptosporidiosis in AIDS patients is now much less severe than it was because of the widespread use of antiretroviral therapy, which has revolutionized the care of patients with HIV infection. Information on clinical illness is largely based on older studies. However, some cases are still seen because of late presentation or failure of therapy due to resistance of the virus.

The severity of illness in AIDS patients is variable, ranging from asymptomatic infection to fulminant disease in a minority. Fulminant disease is characterized by a high-volume diarrhea and high mortality (Blanshard et al. 1992). Generally, there is a relationship between the severity of symptoms and the CD4 cell count. As the CD4 cell count falls and HIV infection progresses, the severity of the illness associated with cryptosporidial infection increases (Flanigan et al. 1992). The majority of patients in a study reported by Connolly et al. (1988) had stool volumes of 500–1 500 ml per 24 h, with stool frequency of 2–10 per day. The natural history of persistent cryptosporidiosis is a remitting and relapsing diarrhea. In one study, 11 of 38 patients with HIV-related cryptosporidiosis underwent spontaneous remission lasting over 2 months. These patients had higher peripheral blood CD4 cell counts than those who did not undergo remission (McGowan et al. 1993). In a series of patients studied in Lusaka, Zambia (Kelly, unpublished data) the median duration of diarrhea at presentation with cryptosporidiosis was 5 months, with 60 percent admitting to intermittent diarrhea. Among Congolese patients, 89 percent had intermittent diarrhea and the mean duration was 9 months; no difference was demonstrated between patients with cryptosporidiosis and those with other forms of HIV-related diarrhea (Colebunders et al. 1987).

BILIARY DISEASE

The situation here has also changed due to the use of antiretroviral therapy, and this manifestation is now much less common. Some patients with AIDS develop a sclerosing cholangitis, sometimes associated with cholecystitis (Teixidor et al. 1991). This may be associated with cryptosporidiosis, microsporidiosis (see below), cytomegalovirus, or it may be impossible to identify a cause. The disorder usually, but not always, occurs in patients with chronic diarrhea, and it leads to a progressive right-upper-quadrant abdominal pain. Biochemical tests of hepatic damage usually show elevated serum alkaline phosphatase and γ-glutamyl transferase levels in the absence of jaundice. Transaminases may or may not be elevated. Ultrasound examination of the liver may show irregularly dilated intrahepatic bile ducts. The definitive test is endoscopic retrograde cholangio-pancreatography (ERCP), which shows this distortion of the biliary anatomy, with or without papillary stenosis. Forbes et al. (1993) found cryptosporidia in 13 of 20 cases, and estimated that up to one in six of all cases of AIDS-related cryptosporidiosis may also have sclerosing cholangitis.

Immunology

ANIMAL MODELS

As attempts to treat cryptosporidiosis with drugs usually have disappointing results, resolution of infection depends on the ability of the host's immune system to bring the infection under control. When the immune system is defective, as for example in AIDS, the infection may persist, spread to extraintestinal sites, and cause severe morbidity and mortality. The ultimate outcome of infection depends on adaptive immune responses but innate immunity also appears to have a protective role. Mechanisms of parasite killing are still poorly understood, however.

Investigations with animal-infection models have played an important part in the development of our understanding of immune responses to *Cryptosporidium*.

A limitation of animal models is that susceptibility to *C. parvum* infection decreases within a few weeks of birth and adult animals are usually strongly resistant to infection (Sherwood et al. 1982). This age-related alteration in susceptibility to infection appears to be absent in human infection (Current and Garcia 1991). One of the most commonly used models of infection for experimentation involves conventional neonatal mice (Sherwood et al. 1982). Adult immunocompromised mice such as athymic nude mice or SCID mice are susceptible to infection and have often been used in immunological studies (reviewed in McDonald et al. 2000). Drug-immunosuppressed rodents may also be susceptible to *C. parvum* infection and although such a model may be useful in drug studies it is of little value for immunological investigation (Brasseur et al. 1988). The gastric parasite *C. muris* readily produces intense acute infections in adult mice and this model has been used to study mechanisms of T-cell-mediated immunity (McDonald et al. 1992).

Another important tool in the study of host—parasite interactions is in vitro parasite culture in enterocyte cell lines. *C. parvum* infection of human cell lines stimulates the proinflammatory transcription factor NF-κB which activates production of chemokines (Laurent et al. 1997) that may play an important part in initiating an early inflammatory response. Infection also stimulates production of prostaglandins that may regulate immunity (Laurent et al. 1998) and in bovine calf infections upregulation of antimicrobial β-defensin expression by intestinal epithelial cells has been reported (Tarver et al. 1998). Proinflammatory cytokines, particularly IFN-γ, have been shown to activate enterocyte cell lines to inhibit *C. parvum* development (Pollok et al. 2001).

IFN-γ has been demonstrated to be a crucial factor in protection conferred by innate immunity. Infections in nude mice and SCID mice, which are T-cell deficient and T + B-cell deficient, respectively, were exacerbated by administration of anti-IFN-γ-neutralizing antibodies (Ungar et al. 1991). The most likely source of IFN-γ, apart from T cells, are natual killer (NK) cells activated by cytokines including interleukin (IL)-12 and tumor necrosis factor (TNF-α) from macrophages (Tripp et al. 1993) and it was shown in vitro that sporozoites of *C. muris* could stimulate NK cells to produce IFN-γ under regulation by these cytokines (McDonald et al. 2000). Injection of neonatal SCID mice with IL-12 prior to inoculation with *C. parvum* oocysts conferred complete resistance to infection (Urban et al. 1996). However, treatment of SCID mice with either anti-asialo-GM1 antibodies that deplete NK cells or anti-TNF-α antibodies had no effect on *C. parvum* reproduction (McDonald and Bancroft 1994). Nude or SCID mice infected with *C. parvum* survive for several weeks due at least in part to the innate IFN-γ response, but the infection gradually increases in magnitude, spreading to extra-intestinal sites, and the animals usually develop signs of illness and die (reviewed in McDonald et al. 2000).

Clearance of cryptosporidial infection is dependent on T-cell activity and specifically the CD4 T cell subpopulation. Conventional neonatal mice develop acute *C. parvum* infections which disappear around day 14 of infection. The development of chronic cryptosporidial infections in T-cell deficient nude and SCID mice indicates the requirement for T cells to resolve infection (Heine et al. 1984; Mead et al. 1991). Little is known about the role of T-cell costimulatory molecules in immunity to this parasite, but CD40/CD40L interaction, required for T-cell activation and antibody isotype switching, appear to be necessary since both CD40 and CD40L knockout mice develop fulminant infections of *C. parvum* (Cosyns et al. 1998). Mice lacking major histocompatibility complex (MHC) class II and hence CD4 T cells are highly susceptible to cryptosporidial infection and fail to recover, whereas animals deficient in MHC class I and therefore CD8 T cells can control the infection normally (Aguirre et al. 1994). Similarly, reconstitution of the SCID mouse immune system with conventional lymphoid cells confers resistance against cryptosporidial infection, but if the CD4 T cells are depleted this effect is lost; depletion of CD8 T cells affects immunity only marginally (reviewed in Theodos 1998). Intra-epithelial lymphocytes may be important effector cells in controlling infection as the transfer of these cells from *C. muris*-infected mice to SCID mice allowed the recipients to recover from infection (Culshaw et al. 1997).

The protective adaptive response appears to involve a cell-mediated – T helper (Th)1 – response since IFN-γ and IL-12 activity is required. *C. parvum* infection levels increased in mice injected with neutralizing antibodies to these cytokines and IFN-γ knockout mice were more susceptible to infection (Ungar et al. 1991; Urban et al. 1996; Theodos 1998). However, there is also evidence that the allergic or antibody-mediated (Th2) cytokines IL-4 and IL-5 play a part in immunological control. *C. parvum* infection was more intense in mice treated with anti-IL-4 or anti-IL-5 neutralizing antibodies, or in IL-4 knockout mice, than in control animals (Enriquez and Sterling 1993; Aguirre et al. 1998). In one study with C57BL/6 mice, the protective effect of IL-4 did not become apparent until after the peak of infection, suggesting the cytokine was involved in removing parasites surviving the early Th1 response infection (Aguirre et al. 1998). However, in another study, neonatal BALB/c IL-4 knockout mice were more susceptible to *C. parvum* infection than wild-type mice during the acute phase of infection and injection of BALB/c neonates with IL-4 prior to oocyst inoculation increased resistance to infection (McDonald et al., in preparation).

Parasite-specific antibodies develop in the serum and intestine of animals infected with *C. parvum* (Peeters et al. 1992). The peaks of production of intestinal IgA and IgM were found at the time of recovery from infection, suggesting antibodies could have a role in protection. In a rodent study, bile with anticryptosporidial antibodies derived from rats infected with *C. parvum* reduced the level of infection in nude mice (Albert et al. 1994). However, neonatal mice depleted of B cells following anti-μ chain antibody treatment were no more susceptible to infection than control mice (Takhi-Kilani et al. 1990). This implies that although antibodies may have a protective role, they are not essential for elimination of infection. Monoclonal antibodies developed against a variety of sporozoite antigens or polyclonal antibodies produced in bovine colostrum after immunization with oocyst antigens have been shown to passively transfer immunity to *C. parvum* infection either in mice after oral ingestion or in cell culture (Bjorneby et al. 1990).

HUMAN INFECTION

The human immune response to *C. parvum* is not well understood. T-cell activity is probably necessary for early control of infection since cryptosporidiosis may be more severe in individuals undergoing immunosuppressive chemotherapy or who are immunosuppressed as a result of other infections or malnutrition (Current and Garcia 1991). It is also likely that the CD4 T-cell subpopulation is required for control of infection since, in AIDS, the incidence of infection and severity of disease increases as the CD4 T cell count decreases (Blanshard et al. 1992). Treatment of HIV infection with highly active antiretroviral therapy (HAART) has been found to be highly beneficial in the control of cryptosporidiosis, probably as a result of recovery of CD4 cell counts (Miller 1998). There is a requirement for ligation of CD40/CD40L involved in T-cell activation by dendritic cells as boys with hyper-IgM syndrome caused by a mutation in CD40L have a high incidence of infection (Hayward et al. 1997). The precise role of CD4 T cells in immunity is unclear but, as with animals, resistance to infection appears to be associated with IFN-γ. Studies with human volunteers have shown that individuals who clear *C. parvum* infection express IFN-γ mRNA in the intestine while those with chronic infection as a result of AIDS express little or no IFN-γ mRNA (White et al. 2000). Increased intestinal expression of transforming growth factor-beta (TGF-β) has been demonstrated and this anti-inflammatory cytokine may have an important role in regulating the activities of inflammatory cells and repairing epithelial barrier function (Robinson et al. 2000).

Infection has been shown to induce production of parasite-specific antibodies in the serum and secretory IgA in the gastrointestinal tract (Ungar et al. 1986; Cozon et al. 1994). Persistent infection has been reported in individuals with hypogammaglobulinemia, although this condition may be associated with other immunodeficiencies (Tzipori et al. 1986). A number of studies in developing countries showed a lower incidence of cryptosporidiosis in breast-fed children, but environmental factors such as extent of use of contaminated water may have contributed to these findings (reviewed in Fayer and Ungar 1986). High titers of secretory IgA have been measured in AIDS patients with chronic infection (Cozon et al. 1994), which may indicate that antibody in general is unimportant in control of infection or that these individuals are producing the wrong type of antibodies. Alternatively, it is possible that these antibodies are protective but cannot fully compensate for a defective cell-mediated response.

Pathology and pathological physiology

DISTRIBUTION OF INFECTION IN THE GUT

Studies in piglets suggest that the infection moves along the gut in a caudal direction (Vitovec and Koudela 1992). The heaviest infections were seen in the distal jejunum and ileum, moving to the colon as time passed; they were associated with partial villous atrophy, crypt hyperplasia, and an inflammatory infiltrate in the lamina propria.

Human studies on the distribution of infection are rather limited. Blanshard et al. (1992) found the parasite in 40 percent of duodenal biopsies and 52 percent of rectal biopsies from patients with AIDS-related cryptosporidiosis. In children with normal immune function, Phillips et al. (1992) found trophozoites in jejunal biopsies by light or scanning electron microscopy in seven out of nine children with chronic cryptosporidiosis. Goodgame et al. (1993) found parasites in 8 of 12 patients with AIDS-related cryptosporidiosis. Clayton et al. (1994) found two patterns of disease: one characterized by proximal small intestinal infection and one by distal infection. Patients with proximal infection had more severe disease. Using PCR to detect parasites in different anatomical regions of the intestine in 83 AIDS patients, the most frequently colonized site was the ileo–cecal region (Kelly et al. 1998).

MUCOSAL ABNORMALITIES

Electron microscopy of the infection shows that the presence of the trophozoites leads to destruction of the microvilli comprising the brush border (Figure 20.2). It seems likely from the appearance of these biopsies that brush border enzyme activity must be reduced, and this has been verified for disaccharidases in children (Phillips et al. 1992).

In germ-free lambs, the villus architecture was normal in the jejunum and ileum during colonization, until the appearance of sexual stages of the life cycle. At this point, the villi shrank and became fused, and an inflammatory infiltrate appeared in the lamina propria. Simultaneously, diarrhea supervened (Angus 1990). This change, however, probably coincides with the beginning of a heavy parasite burden consisting of both sexual and asexual stages.

In human infections, histopathological abnormalities can be severe even when the patient does not have AIDS (Meisel et al. 1976). In AIDS patients in the USA, the jejunal mucosa from patients with cryptosporidiosis was associated with more severe enteropathy than mucosa from patients in whom no pathogen was detected (Kotler et al. 1990), but this was not found in an African series (Kelly et al. 1997a).

Cryptosporidial infection has been associated with increased numbers of mitotic figures in the crypts and villous atrophy. In the normal human intestine, cell maturation takes place during migration up the crypt and along the length of the villus. This process is accompanied by the change from a crypt cell generating net water and salt secretion to a villus cell with net salt and water absorption. Maturation is also accompanied by the synthesis of a normal complement of brush border enzymes, including disaccharidases, lipases, and alkaline phosphatase. According to the 'enterocyte immaturity hypothesis' (Buret et al. 1990), accelerated cell turnover populates the villus with immature enterocytes, contributing to the maldigestion and malabsorption seen in many protozoal small-bowel parasitoses. An increased apoptosis rate has also been observed in cultured epithelial cells, but anti-apoptotic effects have also been observed (McCole et al. 2000).

DISTURBANCES OF PHYSIOLOGY

It has been assumed that, because some conspicuous individuals have very high salt and water losses in stool, often quoted as up to 20 l per day, the principal mechanism of diarrhea must be a cholera-like jejunal secretory state. However, as noted above, this severe secretory diarrhea is not the commonest manifestation. Also, animal perfusion studies carried out in a neonatal piglet model of cryptosporidiosis do not support this hypothesis (Argenzio et al. 1990). Neither do studies carried out in humans using a jejunal perfusion technique (Kelly et al. 1996). However, there is some preliminary evidence suggesting that C. parvum might elaborate a secretory enterotoxin. A protein extract of supernatant from calves with cryptosporidial diarrhea induced a short circuit current in human jejunum mounted in Ussing chambers (Guarino et al. 1994). The same group later showed that this as yet unidentified enterotoxic factor was only present in stool supernatants from patients with an osmotic gap consistent with secretory diarrhea (Guarino et al. 1995). However, it is not clear whether this enterotoxic factor could be parasite derived or an inflammatory mediator secreted by the host.

There is ample evidence that in cryptosporidiosis there is malabsorption of disaccharides (Phillips et al. 1992), and in AIDS patients there is malabsorption of bile acids, and vitamin B_{12} in patients with cryptosporidiosis and other causes of AIDS-related diarrhea

(Bjarnason et al. 1996). In pigs, cryptosporidiosis is associated with impaired salt and water absorption (Argenzio et al. 1990), partly brought about by prostanoid action on the enteric nervous system (Argenzio et al. 1996). Glutamine can overcome this effect (Blikslager et al. 2001). At a cellular level, C. parvum penetration and intracellular growth causes several signaling effects within the enterocyte, leading to cytoskeletal rearrangements especially of β-actin (Elliott and Clark 2000) and to increased paracellular permeability in vitro (Adams et al. 1994), though there is as yet no direct evidence linking these two events through, for example, β-catenin. In the piglet model, macromolecular permeability was not increased (Moore et al. 1995), but studies in children found increased permeability (Zhang et al. 2000), and in AIDS patients, many of whom had cryptosporidiosis, permeability was increased (Keating et al. 1995). There is evidence for host-derived phosphoinositide 3-kinase participation in the cytoskeletal changes (Forney et al. 1999).

Diagnosis

Cryptosporidiosis is usually diagnosed by the detection of stained oocysts in fecal smears using acid-fast methods such as modified Ziehl-Neelsen (Figure 20.3) and phenol auramine. The oocysts are 4.5–5.0 μm in size, and round or slightly oval in shape. Oocysts may be concentrated beforehand by centrifugation in Sheather's sugar solution if necessary or saturated sodium chloride (Casemore et al. 1985). Monoclonal antibodies against oocyst-wall antigens have been employed in Cryptosporidium-specific immunofluoresence antibody tests with fecal specimens (Arrowood and Sterling 1989). Enzyme-linked immunoassay (ELISA) may also be used for detection of cryptosporidial antibodies in serum and has been used in sero-epidemiological investigations (Ungar et al. 1986). Sensitive ELISA methods for detection of cryptosporidial antigens in fecal samples have been

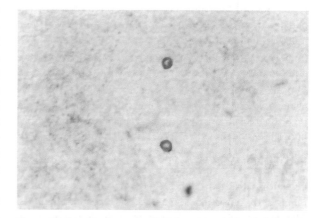

Figure 20.3 *Light microscopy field showing acid-fast staining of two oocysts of* Cryptosporidium parvum *in a fecal smear. These oocysts are almost spherical with a diameter of approximately 5 μm. Magnification ×1 000.*

described and commercial diagnostic ELISA as well as monoclonal antibody kits are available (reviewed by Petry 2000). The immunological techniques are more sensitive than direct observation or histological staining methods but even more sensitive detection of parasites has been achieved using PCR. Many protocols are now available to target *C. parvum* DNA sequences from genes or unknown DNA sequences (Petry 2000). PCR has been shown to be more sensitive and specific than microscopy in detecting oocysts in fecal specimens (Morgan et al. 1998). Ultimately, however, the method used for diagnosis will usually be determined by cost.

Epidemiology

TRANSMISSION

C. parvum is a highly infectious organism. In infectivity studies in healthy volunteers, the median infective dose was 132 oocysts (DuPont et al. 1995). Transmission of *C. parvum* is seasonal in children (Casemore 1990; Perch et al. 2001) and in adults with AIDS (Blanshard et al. 1992). There is some evidence that transmission may relate to waterborne oocyst contamination (Kelly et al. 1997b; Nchito et al. 1998)

Exposure to *C. parvum* oocysts occurs commonly in childhood in developing countries, and as immunity to this organism is strong, cryptosporidiosis is a relatively uncommon cause of diarrhea in adults with a healthy immune system. *C. parvum* causes diarrhea in adults in such populations principally when there is a problem with T-cell function, as in AIDS.

The pathogenicity of human and bovine genotypes may be host-specific in that bovine genotype (Type 2) *C. parvum* produced more severe disease than human genotype (Type 1) infections in pigs (Pereira et al. 2002). As molecular tools (see above) are now beginning to be applied to analysis of transmission, the next few years will see a rapid expansion of our understanding of routes of spread between human populations and animal reservoirs, and directly and indirectly from person to person. In a recent example, application of these markers demonstrated diversity of *C. parvum* genotypes in AIDS patients in Thailand (Gatei et al. 2002). These authors also found other *Cryptosporidium* species in these patients, and this somewhat surprising finding has also emerged in other studies despite uncertainty regarding species definition within the genus (see above under Classification).

EPIDEMIOLOGY IN THE IMMUNOCOMPETENT

Over the last two decades, there has been a large number of outbreaks of waterborne cryptosporidiosis. The largest of these affected over 400 000 people in Milwaukee, Wisconsin (MacKenzie et al. 1994). In response to this and other outbreaks, water supply companies around the world are concerned with how to prevent waterborne cryptosporidiosis and the UK has introduced a statutory requirement to monitor water supplies for oocysts (Fairley et al. 1999). Chlorination of water at usual levels fails to inactivate oocysts, so outbreaks seem to occur when filtration systems fail (MacKenzie et al. 1994).

Cryptosporidiosis is also a major cause of endemic childhood diarrhea among the poor of the Third World, and now in much of Africa and South and South-East Asia cryptosporidiosis occurs in children with and without HIV infection. In the industrialized world, outbreaks of cryptosporidiosis have often originated in day care nurseries. Traveler's diarrhea may also be caused by this protozoan (Katelaris and Farthing 1992).

The prevalence among children with diarrhea in the industrialized world ranges from 1.4 percent in the UK (Hart et al. 1984) to 4 percent in Ireland (Carson 1989), 4.6 percent in Switzerland (Egger et al. 1990), and 6 percent in another study in the UK (Thomson et al. 1987). In underdeveloped countries, prevalences among children with diarrhea are higher, for example ranging from 2.4 percent in Malaysia (Ludin et al. 1991) to 8.4 percent in Rwanda (Bogaerts et al. 1987), and 15 percent in Nigeria (Nwabuisi 2001). In studies around China, the prevalence ranged from 1 percent to 13 percent (Zu et al. 1992). These estimates are, of course, approximate, and the studies vary in their selection of children in hospitals or in the community. In population-based studies in Lusaka, 17 percent of diarrheal episodes in 222 children were associated with this infection (Nchito et al. 1998). In malnourished children with diarrhea admitted to the University Teaching Hospital, Lusaka, the prevalence was 25 percent, and the prevalence in HIV-seropositive children was similar to that in HIV-seronegative children (Amadi et al. 2001).

Serological studies suggest that even in low-transmission countries, a majority of children are exposed before their 21st birthday (Kuhls et al. 1994).

EPIDEMIOLOGY IN IMMUNOCOMPROMISED INDIVIDUALS

An interesting insight into the transmission of the infection was obtained during an outbreak of cryptosporidiosis in Denmark (Ravn et al. 1991). The setting was an infectious disease ward with a mixed population of HIV-seropositive and HIV-seronegative patients; the index case was a demented HIV-seropositive man with cryptosporidial diarrhea who contaminated an ice machine with feces. Among 73 HIV-seronegative inpatients, the attack rate was zero, but among 57 HIV-positives, the attack rate was 18 (32 percent); 17 of these had AIDS. The mean incubation time was at least 13 days.

Cryptosporidiosis was first recognized as a problem in the context of AIDS to the extent that chronic cryptosporidial diarrhea is a case-defining diagnosis for AIDS.

The prevalence of cryptosporidiosis in patients with AIDS-related diarrhea in industrialized countries ranges from 7 percent in Germany (Ullrich et al. 1992) and 8 percent in the USA (Antony et al. 1988) to 21 percent in France (Rene et al. 1989), but its frequency is now much less following the widespread adoption of anti-retroviral therapy. In underdeveloped countries, AIDS-related diarrhea has become a significant public health problem and cryptosporidiosis is still a major contributor to morbidity and mortality. Prevalence ranges from 7 percent in Zambia (Zulu et al. 2002) and 12 percent in Thailand (Saksirisampant et al. 2002) to 22 percent in Zaire (Colebunders et al. 1987) and 46 percent in Haiti (DeHovitz et al. 1986). All of these estimates are, of course, approximate and subject to the same biases as discussed above. Among HIV-2-infected adults in Guinea-Bissau, the prevalence (25 percent) was similar to the studies in HIV-1-infected people (Lebbad et al. 2001).

It is likely that patients with AIDS-related cryptosporidiosis act as reservoirs of the parasite and transmit it to other adults and children in poor communities, as has already been suggested for children with cryptosporidiosis (Newman et al. 1994).

Chemotherapy

C. parvum has proved to be one of the most difficult protozoal infections to treat. Over 100 compounds have been tried without success (O'Donohue 1995). Earlier hopes that spiramycin, letrazuril, and now paromomycin (Hewitt et al. 2000) would be effective have faded. In AIDS patients, effective antiretroviral therapy is a very important component of management (Miao et al. 2000), but for most of the world's AIDS patients this approach is unaffordable.

Nitazoxanide, a nitrothiazolyl–salicylamide derivative, is a broad-spectrum antimicrobial agent with activity against protozoa, nematodes, cestodes, trematodes, and bacteria (Gilles and Hoffman 2002). It has demonstrated activity against *C. parvum* in cell culture and in animal models (Theodos et al. 1998; Gargala et al. 2000). In randomized controlled trials, nitazoxanide was found to have significant benefit in HIV-seronegative children with cryptosporidiosis in Zambia (Amadi et al. 2002), in immunocompetent adults and children in Egypt (Rossignol et al. 2001), and in AIDS patients in Mexico (Rossignol et al. 1998).

Early data suggest intriguingly that rifabutin may prevent cryptosporidiosis in AIDS patients (Fichtenbaum et al. 2000).

IMMUNOTHERAPY

Experimental studies and some case reports suggest that systematic immunotherapy against unrelenting cryptosporidiosis may be an achievable goal. The most studied approach has involved colostral antibodies. The evidence available from animal studies suggests that parasite-specific antibodies developed during infection and present in colostrum are unable to protect offspring from *C. parvum* infection (Moon et al. 1988). However, signs of disease and levels of infection could be shown to be diminished in colostrum from cows immunized mucosally with oocyst antigens and adjuvant (Tzipori et al. 1986). The bovine colostral antibodies that recognize sporozoite antigens were demonstrated to be an important protective element in a murine infection model (Fayer et al. 1990). In a similar approach, antibodies produced in chicken egg yolk following immunization of hens with oocyst antigen provided a degree of protection against infection in mice (Cama and Sterling 1991). Murine monoclonal antibodies directed against particular sporozoite surface or apical complex antigens involved in host cell invasion may potentially have a more potent protective effect. Such antibodies were able to inhibit parasite reproduction in vitro or in neonatal mice (reviewed in Riggs 1997).

A small number of studies have examined the possibility of cytokine therapy. There has been one report of successful treatment of chronic cryptosporidiosis in an immunocompromised child using recombinant human IFN-γ (Gooi 1994) and another demonstrating that treatment of AIDS patients with IL-2 could ameliorate the symptoms of cryptosporidiosis (Connolly et al. 1989).

ISOSPORIASIS

Introduction

Isospora is characterized by the morphology of the oocyst stage and is differentiated from its relative *Eimeria*, a parasite of animals, by the quantity of sporocysts found in the oocyst and number of sporozoites within each sporocyst. Many species of *Isospora* from animals have been described and some hosts are susceptible to more than one species of the parasite. The first identification of human intestinal coccidiosis was by Virchow (1860) and this may have been isosporiasis. Human infection by *Isospora* was first described in a survey of intestinal pathogens found in British troops returning from Turkey during the First World War suffering from dysentery (Ledingham et al. 1915). The parasitologist for the study, H. M. Woodcock, observed large unsporulated and sporulating oocysts with two sporoblasts but no sporulated oocysts were observed. In the same year, Wenyon (1915), with knowledge of Woodcock's findings, described further cases of infection with this parasite and, furthermore, found that oocysts would sporulate when left at room temperature for 3–4 days. The excreted oocysts matured from having a single sporoblast to two sporoblasts from which two oval sporocysts with four sporozoites developed. Recognizing the distinctive features of an *Isospora* sp., Wenyon suggested that as animal species of *Isospora* could cause severe enteritis, this human parasite might be pathogenic (Wenyon 1915). Wenyon later named the parasite *I. belli* and made a comparative study of *I. belli* and other species from domestic animals (Wenyon 1923). Matsubayashi and Nozawa (1948) demonstrated that the oocysts were infectious, using volunteers who developed symptoms including diarrhea. The development of the capability to obtain intestinal biopsies allowed detailed characterization of the endogenous asexual and sexual stages by light and electron microscopy (Brandborg et al. 1970; Trier et al. 1974) and confirmed the parasite taxonomy.

For many years, *I. belli* infection was considered to be uncommon and cases were normally located in tropical countries (Jarpa Gana 1966), although some cases in the USA were reported (Faust et al. 1961); generally, isosporiasis in the western hemisphere was associated with foreign travel (Godiwala and Yaeger 1987). An important development in the early 1980s was the increasing number of cases of chronic, and sometimes life-threatening, diarrhea in AIDS patients associated with *I. belli* infection. In some surveys, particularly in developing countries, the incidence of *I. belli* infection in AIDS-related diarrhea was high (Lindsay et al. 1997). This problem also arose in western countries, particularly the USA, although many of the individuals were from Latin America. Unlike *Cryptosporidium*, *I. belli* infection is readily treatable, although the infection is often recurrent in immunocompromised individuals (Pape et al. 1989). Infection is usually transient in immunocompetent persons but chronic and even fatal cases have been described in these hosts (Brandborg et al. 1970; Liebman et al. 1980).

Classification

Isospora belongs to the Apicomplexa and the suborder Eimeriorina (Levine 1987). It has been placed in the family Eimeriidae, which includes the animal parasite *Eimeria* and also *Cyclospora*. However, a recent phylogenetic study of the sequences of small-subunit ribosomal RNA of *I. belli* and other coccidia suggests that *Isospora* should belong to the family Sarcocystidae (Franzen et al. 2000). An important characteristic of the genus *Isospora* is that, as in *Eimeria*, the entire endogenous development occurs within a single host. One further defining feature of *Isospora* is the presence in mature oocysts of two sporocysts that contain four sporozoites, whereas *Eimeria* has four sporocysts that have two sporozoites. Another member of the suborder, *Toxoplasma*, has a similar type of development and oocyst morphology but this genus normally has two hosts, one of which is the cat, in which sexual develop-

ment occurs. Many organisms that were formerly classified as *Isospora* spp. have been renamed as species of *Sarcocystis* (see Chapter 22, *Sarcocystis*). The development of the latter occurs in two hosts, a prey (asexual) and a predator (sexual), and the oocysts mature within the predator (Levine 1987). Currently, *I. belli* is the only species of *Isospora* known to infect humans, although historically several others were described. One in particular, *I. hominis*, was at one time commonly reported but its morphology suggests it was probably *Sarcocystis suihominis* (Lindsay et al. 1997). Attempts to infect numerous species of domestic animals with *I. belli* have failed (reviewed by Lindsay et al. 1997), so humans may be the only significant reservoir of infection.

Life cycle

I. belli is transmitted by oocysts in a fecal–oral manner. Following ingestion, oocysts release their sporozoites in the intestine. Excystation of *Isospora* in vitro is induced by body temperature in the presence of bile or bile salts and trypsin (McKenna and Charleston 1982). The sporozoites become motile within the sporocyst and eventually emerge through an opening that appears in the sporocyst wall. The oocyst wall also becomes fractured which allows the sporozoites to escape. *I. belli* develops within epithelial cells of the small intestine and intracellular parasites are rarely observed in the large intestine. The zoites of *I. belli* probably multiply by endodyogeny to form meronts and the merozoite progeny will infect and multiply in other enterocytes. Eventually, some merozoites form micro- or macrogametes that, on maturation, fuse to form a zygote and then an immature oocyst. The asexual and sexual stages have been identified by light microscopy (Brandborg et al. 1970), and ultrastructural study by electron microscopy (Figure 20.4) demonstrates coccidian structures such as micronemes, conoid, and subpellicular microtubules in merozoites (Trier et al. 1974). In immunocompromised individuals, extraintestinal development of the parasite has been observed in biliary epithelial cells (Benator et al. 1994).

Parasite cysts comprising undeveloped sporozoites or merozoites within a wall rich in polysaccharide have been observed at extraintestinal sites such as lymph nodes, liver, and spleen (Michiels et al. 1994). Zoites were found in lymphatic channels suggesting a route of transport from the intestine. It is possible that recurrence of infection in AIDS patients following successful chemotherapy is caused by activation of these cyst stages. Some animal species of *Isospora* have paratenic nondefinitive hosts in which sporozoites form cysts in extraintestinal tissues (Lindsay et al. 1997). The sporozoites do not undergo development in this host but infection takes place in the definitive host following

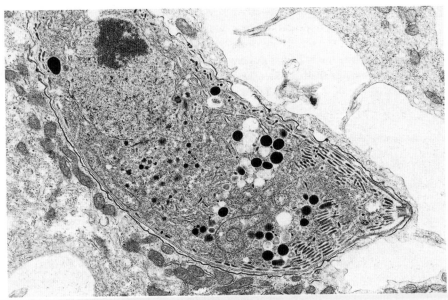

Figure 20.4 *Electronmicrograph with a merozoite of* Isospora belli *demonstrating some typical apicomplexan organelles including conoid at the apical pole, micronemes extending along the periphery from the apical pole and many dense granules in the cytoplasm. Magnification ×20 000.*

ingestion of the paratenic host. It is not known if there are paratenic hosts for *I. belli*.

Cell and molecular biology

Little is known about the biology of *I. belli*, probably in part because infection is not common in developed countries and parasite material is not easily obtained. Some *Isospora* species such as *I. suis* are able to grow in cell culture but development is limited and no sexual stages or oocysts are produced (Lindsay and Blagburn 1987). Complete development from sporozoites to oocysts of *I. suis* has also been obtained in the chick embryo chorioallantoic membrane but the oocysts do not sporulate (Lindsay and Current 1984).

Clinical aspects

I. belli infects both immunocompetent and immunocompromised adults and children. Infection may be asymptomatic in immunocompetent individuals or it may lead to a mild, self-limiting diarrhea lasting from 6 weeks to 6 months. The most extensive experience of human isosporiasis is in Chile. The most common manifestations reported were diarrhea (98 percent), weight loss (86 percent), and abdominal pain (61 percent). The stools contained fat and Charcot–Leyden crystals, and, most interestingly for a protozoal infection, a peripheral eosinophilia was seen in 54 percent of cases (Brandborg et al. 1970). Vomiting, steatorrhea, headache, fever, and malaise may also be present, and dehydration follows when diarrhea is severe. Persistent nonbloody diarrhea, indistinguishable from that caused by microsporidia and

Cryptosporidium parvum, is the major manifestation in immunocompromised individuals. Persistent diarrhea may occur in immunocompetent persons as well, but clinical features are usually less severe (Lindsay et al. 1997). Manifestations in children can sometimes be severe.

In AIDS patients, extra-intestinal infections can occur, though they are rare. Necropsy occasionally reveals infection of mesenteric lymph nodes, liver, and spleen. Biliary disease has been described.

Immunology

There is a paucity of data on mechanisms of immunity against *I. belli* and extrapolation from studies with related parasites give only a possible picture of the main elements of the adaptive immune system involved in clearance of the infection. It is likely that *I. belli* infection induces immunity against reinfection as piglets recovered from *I. suis* infection were shown to exhibit strong resistance to reinfection (Stuart et al. 1982). Colostral parasite-specific antibodies did not protect piglets against *I. suis* infection (Lindsay et al. 1997). Presumably CD4 T cells are involved in immunity to *I. belli* as infection in AIDS patients is more severe than in immunocompetent individuals (Soave and Johnson 1988). Immunity to parasites of the related genus *Eimeria*, which has a similar life cycle to *Isospora*, has been studied extensively in rodents and chickens (reviewed in Lillehoj and Trout 1996; Smith and Hayday 1998). The significant features of immunity are that CD4 αβTCR T cells are involved, developing a Th1-cell-mediated response with IFN-γ production. Although parasite-specific antibody, including secretory IgA, is produced during *Eimeria* infection, it does not appear to play a major role in recovery.

Pathology

There have been few publications on human isosporiasis, possibly fewer than on any other common human infection. Brandborg et al. (1970) observed moderate-to-severe villous atrophy and 'spectacular' collagen deposition in the lamina propria, with eosinophil infiltration which was extreme in two of the six cases analyzed. Ultrastructural work shows vacuolation of the enterocytes, microvillous shortening, and lipid deposits (Trier et al. 1974). Sexual and asexual stages of the parasite life-cycle are seen in the epithelium, enclosed by a parasitophorous vacuole (Comin and Santucci 1994).

There are no data on the disturbed physiology of human isosporiasis, but it is likely that there would be similarities with disturbances found in cryptosporidiosis.

Diagnosis

Examination of a fecal specimen for oocysts would normally allow diagnosis of infection with *I. belli* as the oocysts are large and oval (23−36 × 12−17 μm) (Lindsay et al. 1997). The oocysts may not be as numerous as in cryptosporidiosis and concentration by centrifugal flotation methods are sometimes employed (Whiteside et al. 1984). Oocysts can also be detected in fecal smears following acid-fast staining (Figure 20.5) or

staining with auramine-rhodamine (Forthal and Guest 1984). Use of the latter technique should be employed in conjunction with another method for confirmation, however. Unstained oocysts are autofluorescent, appearing violet under ultraviolet light and green under violet or blue-violet light (Varea et al. 1998). Sometimes infection is diagnosed from stained intestinal biopsy material showing parasites in enterocytes when oocysts are not found by fecal examination (Whiteside et al. 1984). A highly sensitive and specific method for diagnosis has employed PCR with primers for the small-subunit rRNA sequences of *I. belli* (Muller et al. 2000); no amplification was seen with template DNA extracted from other parasites.

Epidemiology

Isospora belli causes human infection in every continent, but transmission is most frequent in tropical zones, predominantly subequatorial Africa, the Caribbean, and some parts of South America. Isosporiasis is emerging as a major problem in populations heavily infected by HIV. The prevalence in AIDS patients with persistent diarrhea is very variable. Isosporiasis is rarely detected in industrialized countries, the highest reported figure being 3 percent in the USA (Antony et al. 1988). Most cases of isosporiasis in Los Angeles were in AIDS patients from Latin America (Sorvillo et al. 1995). This is clearly different from the epidemiology of cryptosporidiosis. Estimates of prevalence in tropical countries range from 3 percent in Mali (Pichard et al. 1990) to 10 percent in Brazil (Sauda

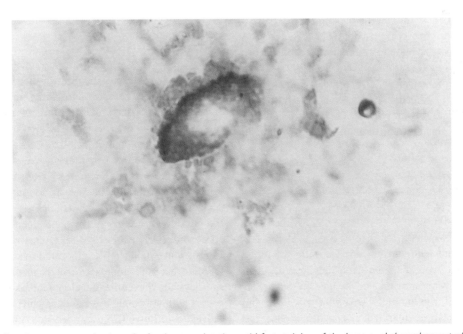

Figure 20.5 *Light microscopy examination of a fecal smear showing acid-fast staining of the large oval-shaped oocyst of* Isospora belli *(approximately 30 μm in length). An oocyst of* Cryptosporidium parvum *is also present which is almost spherical and much smaller (5 μm in diameter). Magnification* ×1 000.

et al. 1993), 20 percent in Burundi (Floch et al. 1989), and 37 percent in Zambia (Zulu et al. 2002). Isosporiasis seems to be relatively uncommon in AIDS patients in East Africa.

Another important difference between the epidemiology of isosporiasis and that of cryptosporidiosis is the rarity of finding *I. belli* in HIV-infected children. In Zambia, the prevalence was 2 percent in children with persistent diarrhea and malnutrition irrespective of HIV status (Amadi et al. 2001). Isosporiasis is rare in Indian children, being detected in seven of 10 126 children with diarrhea (Mirdha et al. 2002). It was found in 2.8 percent of malnourished Egyptian children (Rizk and Soliman 2001). In Zambia, isosporiasis is not seasonal (author, unpublished observations).

Chemotherapy

The first anecdotal report, to our knowledge, of successful treatment was with nitrofurantoin (Brandborg et al. 1970), and the second success was with pyrimethamine–sulfadiazine (Trier et al. 1974). DeHovitz et al. (1986) obtained good responses with trimethoprim–sulfamethoxazole (960 mg four times daily for 10 days, then 960 mg twice daily for 21 days) in all of 20 patients, but 47 percent relapsed within a few weeks. This regimen was further evaluated in Haitian AIDS patients (Pape et al. 1989) and is now the standard therapy, with initial therapy usually given for 10 days. Secondary prophylaxis (to prevent relapse) is usually required, suggesting that true eradication is not achieved. The dose required is 960 mg orally three times per week. Pyrimethamine–sulfonamide combinations (e.g. Fansidar) are also effective (Trier et al. 1974; Pape et al. 1989). In those intolerant of sulfonamides, pyrimethamine alone has been used (Weiss et al. 1988). Other compounds, including primaquine and diclazuril have also been found useful (Lindsay et al. 1997), and a combination of albendazole and ornidazole has been reported to be effective in sulfonamide-intolerant patients (Dionisio et al. 1996). There is conflicting evidence as to the efficacy of nitazoxanide in human isosporiasis.

CYCLOSPORIASIS

Introduction

Cyclospora cayetanensis is the most recently identified member of the coccidia that is a human intestinal pathogen. The genus was created in 1881 by Schneider for a parasite of a myriapod, *Cyclospora glomerica* and other species have been found in rodents, moles, and snakes (reviewed by Ortega et al. 1994) and in three species of monkey (Eberhard et al. 1999). Probably the first report of a human infection was by Ashford (1979)

who described oocysts of a new coccidian from fecal samples of three individuals in Papua New Guinea. From the mid-1980s onwards, similar descriptions were provided in cases of traveler's diarrhea or AIDS-related diarrhea (Soave et al. 1986; Hart et al. 1990; Shlim et al. 1991; Pollok et al. 1992). Cases in the western hemisphere usually involved travelers returning from developing countries in different continents and the symptoms were similar to those associated with cryptosporidiosis, but sometimes lasting longer. The oocysts were approximately 8–10 μm in diameter and stained red with the modified Ziehl-Neelsen method. If the parasite was a coccidian, the oocysts were too small to be from *I. belli* and too large to be from *C. parvum*. Early morphological and ultrastructural studies suggested that the organism was cyanobacterium-like (Long et al. 1990). The parasite was identified as a coccidian and a species of *Cyclospora* following the demonstration of sporulation of the oocysts and sporozoite excystation (Ortega et al. 1993). The mature oocyst had a sporocyst and sporozoite composition characteristic of this genus and distinct from that present in *Cryptosporidium* and *Isospora*. Shortly afterwards, it was reported that *C. cayetanensis* infection was common in Haitians with advanced HIV infection suffering chronic or intermittent diarrhea (Pape et al. 1994). Biliary infection with accompanying symptoms has also been observed in AIDS patients (Sifuentes-Osornio et al. 1995). Cyclosporiasis patients have been found to respond rapidly to antimicrobial therapy but the symptoms often recurred in individuals with AIDS (Wurtz 1994; Pape et al. 1994). There is evidence that infection is transmitted by contaminated water supplies or food, particularly fruit or vegetables (Wright and Collins 1997; Ortega et al. 1997)

Classification

Cyclospora cayetanensis belongs to the phylum Apicomplexa, suborder Eimeriorina, and family Eimeriidae (Ortega et al. 1994). The genus is defined by the morphology of the oocyst which is spheroidal and 8–10 μm in diameter; it matures outside the host and the sporulated oocyst contains two sporocysts, each of which has two sporozoites. Transmission electron microscopy of mature oocysts has demonstrated typical coccidian structures in oocysts such as Stieda body on the sporocyst and rhoptries and micronemes in the sporozoite (Ortega et al. 1998). (Remember, the oocysts of *I. belli* are significantly larger and oval, and the sporulated forms have four sporozoites in each of the two sporocysts.) The entire development of *C. cayetanensis* occurs within a single host and currently humans are the only known hosts (Eberhard et al. 2000). Phylogenetic analysis of the small-subunit ribosomal DNA of *Cyclospora* suggests that it is closely related to members of genus *Eimeria* that infect animals and have oocysts with four sporocysts containing two sporozoites (Relman et al. 1996).

Life cycle

Oocysts of *C. cayetanensis* mature outside the human host over a period of several days. A single sporoblast in the excreted oocyst divides into two and each develops into a sporocyst with two sporozoites (Ortega et al. 1994). *Cyclospora* oocysts have been shown to have maximal sporulation at temperatures of 22–30°C (Smith et al. 1997). It has been demonstrated that after mechanical disruption of oocysts and in the presence of trypsin and sodium taurocholate, sporozoites will excyst. Endogenous development has been observed in small intestinal biopsies and trophozoites, schizonts, merozoites, and sexual stages have been identified at the luminal end of enterocytes (Bendall et al. 1993; Sun et al. 1996; Ortega et al. 1998). Merozoites of different sizes have also been described suggesting that there might be more than one type of meront (Ortega et al. 1998). Parasite development may spread to epithelium of the biliary system in AIDS patients (Zar et al. 2001). The presence of all life cycle stages in the human host indicates that the parasite is monoxenous. Attempts to infect other species including mice, rats, rabbits, girds, hamsters, ferrets, pigs, dogs, and monkeys have failed suggesting that humans may be the only host for *C. cayetanensis* (Eberhard et al. 2000), unless higher primates are susceptible to infection.

Clinical aspects

In common with the other coccidia described in this chapter, *Cyclospora cayetanensis* causes acute and persistent diarrhea in immunocompetent and immunocompromised adults and children. In two studies of Peruvian children, the proportions of episodes of cyclosporiasis which were accompanied by diarrhea were 11 and 28 percent, and the clinical disease produced was indistinguishable from cryptosporidiosis (Ortega et al. 1993).

In Haiti, cyclosporiasis was responsible for 11 percent of AIDS-related diarrhea (Pape et al. 1994). Again, the clinical features were indistinguishable from cryptosporidiosis or isosporiasis.

Pathology

C. cayetanensis occupies an intracellular position in intestinal epithelial cells, and this is accompanied by increased numbers of intra-epithelial lymphocytes and a variable degree of villous blunting (Bendall et al. 1993). There are no significant data on the pathophysiology of human cyclosporiasis.

Diagnosis

Diagnosis of cyclosporiasis is obtained by identification of oocysts in fecal specimens. The 8–10 μm diameter oocysts of *C. cayetanensis* can be readily differentiated microscopically from those of two other human intestinal coccidia, *Cryptosporidium parvum* and *I. belli*, as they are double the size of the former and two to three times smaller than the latter. Under ultraviolet illumination, unstained oocysts of *C. cayetanensis* are autofluorescent, giving a rapid and inexpensive diagnostic method (Eberhard et al. 1997). With normal microscopy, oocysts can be observed by employing acid-fast staining methods such as modified Ziehl-Neelsen or Kinyoun, and also by using safranin (Long et al. 1990; Pape et al. 1994; Wurtz 1994). The organisms can be concentrated by centrifugation in formalin-ethyl acetate or by sucrose flotation. Oocysts may sporulate after several days in potassium dichromate solution at room temperature (Ortega et al. 1998) (Figure 20.6). PCR with primers specific for *C. cayetanensis* have been employed to identify the presence of the parasite in fecal specimens and this approach could provide a highly sensitive diagnostic tool (Relman et al. 1996).

Epidemiology

The first descriptions indicated that travelers (to India, Pakistan, and Morocco) and the immunocompromised are the principal risk groups (Bendall et al. 1993). In expatriates living in Nepal, the annual attack rate was 32 percent (Shlim et al. 1999). In Nepal, *C. cayetanensis* was not found in children with diarrhea under 18 months of age, but was found in 12 percent of episodes in children between 18 and 60 months of age (Hoge et al. 1995a).

Cyclosporiasis can be foodborne or waterborne. A large outbreak of 1 465 cases of cyclosporiasis in 1996 in the USA was traced to raspberries imported from Guatemala (Herwaldt et al. 1997). Exposure to untreated water was associated with a fourfold increased risk of cyclosporiasis in foreign residents in Nepal (Hoge et al. 1993).

In Africa, *C. cayetanensis* infection has not been found in surveys in Zambia or Guinea-Bissau (authors' unpublished data) despite the fact that the organisms should be easily visible in Ziehl-Neelsen stained smears, but small numbers of cases were reported in Tanzania (Cegielski et al. 1999). It would appear that cyclosporiasis is much less common in Africa than in Central America or South Asia.

Chemotherapy

Chemotherapy with trimethoprim–sulfamethoxazole (960 mg twice daily for 7 days) has been shown to be

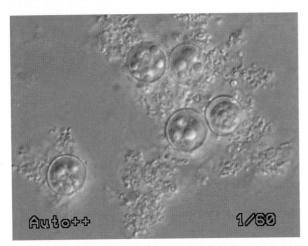

Figure 20.6 *Light microscopy examination of unstained spherical oocysts of* Cyclospora cayetanensis *that have a diameter of 8–10 μm. These are about X2 the size of oocysts of* Cryptosporidium parvum *but are significantly smaller than the oval-shaped oocysts of* Isospora belli. *(Photograph kindly provided by Monika Kettelhut, Hospital for Tropical Diseases, London, UK)*

effective in immunocompetent adults in Nepal (Hoge et al. 1995b). In AIDS patients, relapse is common, and this can be successfully prevented by 960 mg three times weekly indefinitely (Pape et al. 1994). Ciprofloxacin (500 mg twice daily for 7 days) is effective in patients who are intolerant of sulfonamides (Verdier et al. 2000).

REFERENCES

Adams, R.B., Guerrant, R.L., et al. 1994. *Cryptosporidium parvum* infection of intestinal epithelium: morphologic and functional studies in an in vitro model. *J Infect Dis*, **169**, 170–7.

Aguirre, S.A., Mason, P.H. and Perryman, L.E. 1994. Susceptibility of major histocompatibility (MHC) class I- and class II-deficient mice to *Cryptosporidium parvum* infection. *Infect Immun*, **62**, 697–9.

Aguirre, S.A., Perryman, L.E., et al. 1998. IL-4 protects adult C57BL/6 mice from prolonged *Cryptosporidium parvum* infection: analysis of CD4+αβ+IFN-γ+ and CD4+αβ+IL-4+ lymphocytes in gut-associated lymphoid tissue during resolution of infection. *J Immunol*, **161**, 1891–900.

Albert, M.M., Rusnak, J., et al. 1994. Treatment of murine cryptosporidiosis with anticryptosporidial immune rat bile. *Am J Trop Med Hyg*, **50**, 112–19.

Amadi, B.C., Kelly, P., et al. 2001. Intestinal and systemic infection, HIV and mortality in Zambian children with persistent diarrhea and malnutrition. *J Pediatr Gastroenterol Nutr*, **32**, 550–4.

Amadi, B.C., Mwiya, M., et al. 2002. Effect of nitazoxanide on morbidity and mortality in Zambian children with cryptosporidiosis: a randomized controlled trial. *Lancet*, **360**, 1375–80.

Angus, K.W. 1990. Cryptosporidiosis and AIDS. *Bailliere's Clin Gastroenterol*, **4**, 425–41.

Antony, M.A., Brandt, L.J., et al. 1988. Infectious diarrhea in patients with AIDS. *Digest Dis Sci*, **33**, 1141–6.

Argenzio, R.A., Armstrong, M. and Rhoads, J.M. 1996. Role of the enteric nervous system in piglet cryptosporidiosis. *J Pharmacol Exp Ther*, **279**, 1109–15.

Argenzio, R.A., Liacos, J.A., et al. 1990. Villous atrophy, crypt hyperplasia, cellular infiltration, and impaired glucose-Na absorption in enteric cryptosporidiosis of pigs. *Gastroenterology*, **98**, 1129–40.

Arrowood, M.J. and Sterling, C.R. 1989. Comparison of conventional staining methods and monoclonal antibody-based methods for *Cryptosporidium* oocyst detection. *J Clin Microbiol*, **27**, 1490–5.

Ashford, R.W. 1979. Occurrence of an undescribed coccidian in man in Papua New Guinea. *Ann Trop Med Parasitol*, **73**, 497–500.

Awad-El-Kariem, F.M., Robinson, H.A., et al. 1995. Differentiation between human and animal strains of *Cryptosporidium parvum* using isoenzyme typing. *Parasitology*, **110**, 129–32.

Awad-El-Kariem, F.M., Robinson, H.A., et al. 1998. Differentiation between human and animal isolates of *Cryptosporidium parvum* using molecular and biological markers. *Parasitol Res*, **84**, 297–301.

Benator, D.A., French, A.L., et al. 1994. *Isospora belli* infection associated with acalculous cholecystitis in a patient with AIDS. *Ann Intern Med*, **121**, 663–4.

Bendall, R.P., Lucas, S., et al. 1993. Diarrhea associated with *Cyanobacterium*-like bodies: a new coccidian enteritis of man. *Lancet*, **341**, 590–2.

Bjarnason, I., Sharpstone, D.R., et al. 1996. Intestinal inflammation, ileal structure and function in HIV. *AIDS*, **10**, 1385–91.

Bjorneby, J.M., Riggs, M.W. and Perryman, L.E. 1990. *Cryptosporidium parvum* merozoites share neutralization-sensitive epitopes with sporozoites. *J Immunol*, **145**, 298–304.

Blanshard, C., Jackson, A.M., et al. 1992. Cryptosporidiosis in HIV seropositive patients. *Q J Med*, **85**, 813–23.

Blikslager, A., Hunt, E., et al. 2001. Glutamine transporter in crypts compensates for loss of villus absorption in bovine cryptosporidiosis. *Am J Physiolo Gastrointest Liver Physiol*, **281**, G645–653.

Blunt, D.S., Khramtsov, N.V., et al. 1997. Molecular karyotype analysis of *Cryptosporidium parvum*: evidence for eight chromosomes and a low-molecular-size molecule. *Clin Diagn Lab Immunol*, **4**, 11–13.

Bogaerts, J., Lepage, P., et al. 1987. Cryptosporidiosis in Rwanda. Clinical and epidemiological features. *Ann Soc Belgique Med Trop*, **67**, 157–65.

Brandborg, L.L., Goldberg, S.B. and Breidenbach, W.C. 1970. Human coccidiosis – a possible cause of malabsorption. *N Engl J Med*, **283**, 1306–13.

Brasseur, P., Lemeteil, D. and Ballet, J.J. 1988. Rat model for human cryptosporidiosis. *J Clin Microbiol*, **26**, 1037–9.

Brown, S.M.A., McDonald, V., et al. 1996. The use of a new viability assay to determine the susceptibility of *Cryptosporidium* and *Eimeria* sporozoites to respiratory inhibitors and extremes of pH. *FEMS Microbiol Lett*, **142**, 203–8.

Buret, A., Gall, D.G., et al. 1990. Intestinal protozoa and epithelial cell kinetics, structure and function. *Parasitol Today*, **6**, 375–80.

Cama, V.A. and Sterling, C.R. 1991. Hyperimmune hens as a novel source of anti-*Cryptosporidium* antibodies suitable for passive immune transfer. *J Protozool*, **50**, 112–19.

Carson, J.W. 1989. Changing patterns in childhood gastroenteritis. *Ir Med J*, **82**, 66–7.

Casemore, D.P. 1990. Epidemiological aspects of human cryptosporidiosis. *Epidemiol Infect*, **104**, 1–28.

Casemore, D.P., Armstrong, M. and Sands, R.L. 1985. Laboratory diagnosis of cryptosporidiosis. *J Clin Pathol*, **38**, 1337–41.

Cegielski, J.P., Ortega, Y.R., et al. 1999. *Cryptosporidium*, *Enterocytozoon* and *Cyclospora* infections in pediatric and adult patients with diarrhea in Tanzania. *Clin Infect Dis*, **28**, 314–21.

Char, S., Kelly, P., et al. 1996. Codon usage in *Cryptosporidium parvum* differs from that in other Eimeriorina. *Parasitology*, **112**, 357–62.

Chen, X.-M., Levine, S.A., et al. 1998. *Cryptosporidium parvum* is cytopathic for cultured human biliary epithelia via an apoptotic mechanism. *Hepatology*, **28**, 906–13.

Chen, X.-M., Levine, S., et al. 2001. *Cryptosporidium parvum* activates nuclear factor κB in biliary epithelia preventing epithelial cell apoptosis. *Gastroenterology*, **120**, 1774–83.

Clayton, F., Heller, T. and Kotler, D.P. 1994. Variation in the enteric distribution of Cryptosporidia in AIDS. *Am J Clin Pathol*, **102**, 420–5.

Colebunders, R., Francis, H., et al. 1987. Persistent diarrhea, strongly associated with HIV infection in Kinshasa, Zaire. *Am J Gastroenterol*, **82**, 859–64.

Comin, C.E. and Santucci, M. 1994. Submicroscopic profile of *Isospora belli* enteritis in a patient with AIDS. *Ultrastruct Pathol*, **18**, 473–82.

Connolly, G.M., Dryden, M.S., et al. 1988. Cryptosporidial diarrhea in AIDS and its treatment. *Gut*, **29**, 593–7.

Connolly, G.M., Owen, S.L., et al. 1989. Treatment of cryptosporidial infection with recombinant interleukin-2. *5th International Congress on AIDS, Montreal.* WBP47. p. 359

Coombs, G.H. 1999. Biochemical peculiarities and drug targets in *Cryptosporidium parvum*: lessons from other coccidian parasites. *Parasitol Today*, **15**, 333–8.

Cosyns, M., Tsirkin, S., et al. 1998. Requirement for CD40-CD40L interaction for elimination of *Cryptosporidium parvum* infection. *Infect Immun*, **66**, 603–7.

Cozon, G., Biron, F., et al. 1994. Secretory IgA antibodies to *Cryptosporidium parvum* in AIDS patients with chronic cryptosporidiosis. *J Infect Dis*, **169**, 696–9.

Culshaw, R.J., Bancroft, G.J. and McDonald, V. 1997. Gut intraepithelial lymphocytes induce immunity against *Cryptosporidium* infection through a mechanism involving interferon gamma production. *Infect Immun*, **65**, 3074–9.

Current, W.L. and Garcia, L.S. 1991. Cryptosporidiosis. *Clin Microbiol Rev*, **4**, 325–58.

Current, W.L. and Reese, N.C. 1986. A comparison of endogenous development of three isolates of *Cryptosporidium* in suckling mice. *J Protozool*, **33**, 98–108.

DeHovitz, J.A., Pape, J.W., et al. 1986. Clinical manifestations and therapy of *Isospora belli* infection in patients with AIDS. *N Engl J Med*, **315**, 87–90.

Denton, H., Brown, S.M.A., et al. 1996. Comparison of the phosphofructokinase and pyruvate kinase activities of *Cryptosporidium parvum*, *Eimeria tenella* and *Toxoplasma gondii*. *Mol Biochem Parasitol*, **76**, 23–9.

Dionisio, D., Sterrantino, G., et al. 1996. Treatment of isosporiasis with combined albendazole and ornidazole in patients with AIDS. *AIDS*, **10**, 1301–2.

DuPont, H.L., Chappell, C., et al. 1995. The infectivity of *Cryptosporidium parvum* in health volunteers. *N Engl J Med*, **332**, 855–9.

Eberhard, M.L., Pieniazak, N.J. and Arrowood, M.J. 1997. Laboratory diagnosis of *Cyclospora* infections. *Arch Pathol Lab Med*, **121**, 792–7.

Eberhard, M.L., da Silva, A.J., et al. 1999. Morphologic and molecular characterization of new *Cyclospora* species from Ethiopian monkeys: *C. cercopitheci* sp.n., *C. colobi* sp.n., and *C papionis* sp.n.. *Emerg Infect Dis*, **5**, 651–8.

Eberhard, M.L., Ortega, Y.R., et al. 2000. Attempts to establish experimental *Cyclospora cayetanensis* infection in laboratory animals. *J Parasitol*, **86**, 577–82.

Egger, M., Mausezahl, D., et al. 1990. Symptoms and transmission of intestinal cryptosporidiosis. *Arch Dis Child*, **65**, 445–7.

Elliott, D.A. and Clark, D.P. 2000. *Cryptosporidium parvum* induces host cell actin accumulation at the host–parasite interface. *Infect Immun*, **68**, 2315–22.

Elliott, D.A., Coleman, D.J., et al. 2001. *Cryptosporidium parvum* infection requires host cell actin polymerization. *Infect Immun*, **69**, 5940–2.

Enriquez, F.J. and Sterling, C.R. 1993. Role of CD4+ Th1- and Th2-cell-secreted cytokines in cryptosporidiosis. *Folia Parasitol (Praha)*, **40**, 307–11.

Entrala, E. and Mascaro, C. 1997. Glycolytic enzyme activities in *Cryptosporidium parvum* sporozoites. *FEMS Microbiol Lett*, **151**, 51–7.

Fairley, C.K., Sinclair, M.I. and Rizak, S. 1999. Monitoring not the answer to *Cryptosporidium* in water. *Lancet*, **354**, 967–9.

Faust, E.C., Giraldo, L.E., et al. 1961. Human isosporiasis in the Western Hemisphere. *Am J Trop Med Hyg*, **10**, 343–9.

Fayer, R. and Ungar, P.L.B. 1986. *Cryptosporidium* spp. and cryptosporidiosis. *Microbiol Rev*, **50**, 458–83.

Fayer, R., Guidry, A. and Blagburn, B.L. 1990. Immunotherapeutic efficacy of bovine colostral immunoglobulins from a hyperimmunised cow against cryptosporidiosis in neonatal mice. *Infect Immun*, **58**, 2962–5.

Fayer, R., Speer, C.A. and Dubey, J.P. 1997. The general biology of *Cryptosporidum*. In: Fayer, R. (ed.), *Cryptosporidium and cryptosporidiosis*. Boca Raton: CRC Press, 1–42.

Feng, X., Rich, S.M., et al. 2002. Experimental evidence for genetic recombination in the opportunistic pathogen *Cryptosporidium parvum*. *Mol Biochem Parasitol*, **119**, 55–62.

Fichtenbaum, C.J., Zackin, R., et al. 2000. Rifabutin but not clarithromycin prevents cryptosporidiosis in persons with advanced HIV infection. *AIDS*, **14**, 2889–93.

Flanigan, T., Whalen, C., et al. 1992. *Cryptosporidium* infection and CD4 counts. *Ann Intern Med*, **116**, 840–2.

Floch, P.J., Laroche, R., et al. 1989. Parasites, etiologic agents of diarrhea in AIDS. Significance of duodenal aspiration fluid test. *Bull Soc Pathol Exot Filial*, **82**, 316–20.

Forbes, A., Blanshard, C. and Gazzard, B. 1993. Natural history of AIDS related sclerosing cholangitis: a study of 20 cases. *Gut*, **34**, 116–21.

Forney, J.R., Vaughan, D.K., et al. 1998. Actin-dependent motility in *Cryptosporidium parvum* sporozoites. *J Parasitol*, **84**, 908–13.

Forney, J.R. and DeWald, D.B. 1999. A role for host phophoinositide 3-kinase and cytoskeletal remodeling during *Cryptosporidium parvum* infection. *Infect Immun*, **67**, 844–52.

Forthal, D.N. and Guest, S.S. 1984. *Isospora belli* enteritis in three homosexual men. *Am J Trop Med Hyg*, **33**, 1060–4.

Franzen, C., Muller, A., et al. 2000. Taxonomic position of the human intestinal protozoan parasite *Isospora belli* as based on ribosomal RNA sequences. *Parasitol Res*, **86**, 669–76.

Gargala, G., Delaunay, A., et al. 2000. Efficacy of nitazoxanide, tizoxanide and tizoxanide glucuronide against *Cryptosporidium parvum* development in sporozoite-infected HCT-8 enterocytic cells. *J Antimicrob Chemother*, **46**, 57–60.

Gatei, W., Suputtamongkol, Y., et al. 2002. Zoonotic species of *Cryptosporidium* are as prevalent as the anthroponotic in HIV-infected patients in Thailand. *Ann Trop Med Parasitol*, **96**, 797–802.

Gilles, H.M. and Hoffman, P.S. 2002. Treatment of intestinal parasitic infections: a review of nitazoxanide. *Trends Parasitol*, **18**, 95–7.

Godiwala, T. and Yaeger, R. 1987. *Isospora* and traveler's diarrhea. *Ann Inter Med*, **106**, 909–10.

Goodgame, R.W., Genta, R.M., et al. 1993. Intensity of infection in AIDS associated cryptosporidiosis. *J Infect Dis*, **167**, 704–9.

Gooi, H.C. 1994. Gastrointestinal and liver involvement in primary immunodeficiency. In: Heatley, R.V. (ed.), *Gastrointestinal and hepatic immunology*. Cambridge: Cambridge University Press, 169–77.

Guarino, A., Canani, R.B., et al. 1994. Enterotoxic effect of stool supernatant of *Cryptosporidium*-infected calves on human jejunum. *Gastroenterology*, **106**, 28–34.

Guarino, A., Canani, R.B., et al. 1995. Human intestinal cryptosporidiosis: secretory diarrhea and enterotoxic activity in Caco-2 cells. *J Infect Dis*, **171**, 976–83.

Harris, J.R. and Petry, F. 1999. *Cryptosporidium parvum*: structural components of the oocyst wall. *J Parasitol*, **85**, 839–49.

Hart, C.A., Baxby, D. and Blundell, N. 1984. Gastroenteritis due to *Cryptosporidium*: a prospective survey in a children's hospital. *J Infect*, **9**, 264–70.

Hart, A.S., Ridinger, M.T., et al. 1990. Novel organism associated with chronic diarrhea in AIDS. *Lancet*, **335**, 169–70.

Hayward, A.R., Levy, J., et al. 1997. Cholangiopathy and tumors of the pancreas, liver and biliary tree in boys with X-linked immunodeficiency with hyper-IgM (XHIM). *J Immunol*, **158**, 977–83.

Heine, J., Moon, H.W. and Woodmansee, D.B. 1984. Persistent *Cryptosporidium* infection in congenitally athymic (nude) mice. *Infect Immun*, **43**, 856–9.

Herwaldt, B.L. and Ackers, M.L. Cyclospora Working Group, 1997. An outbreak in 1996 of cyclosporiasis associated with imported raspberries. *N Engl J Med*, **336**, 1548–56.

Hewitt, R.G., Yiannoutsos, C.T., et al. 2000. Paromomycin: no more effective than placebo for the treatment of cryptosporidiosis in patients with advanced HIV infection. AIDS Clinical Trial Group. *Clin Infect Dis*, **31**, 1084–92.

Hijjawi, N.S., Meloni, B.P., et al. 2001. Complete development and long-term maintenance of *Cryptosporidium parvum* human and cattle genotypes in cell culture. *Intl J Parasitol*, **31**, 1048–55.

Hoge, C.W., Shlim, D.R., et al. 1993. Epidemiology of diarrheal illness associated with coccidian-like organism among travellers and foreign residents in Nepal. *Lancet*, **341**, 1175–9.

Hoge, C.W., Echeverria, P., et al. 1995a. Prevalence of *Cyclospora* species and other enteric pathogens among children less than 5 years of age in Nepal. *J Clin Microbiol*, **33**, 3058–60.

Hoge, C.W., Shlim, D.R., et al. 1995b. Placebo-controlled trial of co-trimoxazole for *Cyclospora* infections among travellers and foreign residents in Nepal. *Lancet*, **345**, 691–3.

Howe, D.K. and Sibley, L.D. 1995. *Toxoplasma gondii* comprises three clonal lineages: correlation of parasite genotype with human disease. *J Infect Dis*, **172**, 1561–6.

Hunter, P.R. and Nichols, G. 2002. Epidemiology and clinical features of *Cryptosporidium* infection in immunocompromised patients. *Clin Microbiol Rev*, **15**, 145–54.

Jacyna, M.R., Parkin, J., et al. 1990. Protracted enteric cryptosporidial infection in selective immunoglobulin A and saccharomyces opsonin deficiencies. *Gut*, **31**, 714–16.

Jarpa Gana, A. 1966. Coccidiosis humana. *Biologica (Santiago)*, **39**, 3–26.

Katelaris, P. and Farthing, M.J.G. 1992. Cryptosporidiosis – an emerging risk to travellers. *Travel Med Int*, **1**, 10–14.

Keating, J., Bjarnason, I., et al. 1995. Intestinal absorptive capacity, intestinal permeability and jejunal histology in HIV and their relation to diarrhea. *Gut*, **37**, 623–9.

Keithly, J.S., Zhu, G., et al. 1997. Polyamine biosynthesis in *Cryptosporidium parvum* and its implications for chemotherapy. *Mol Biochem Parasitol*, **88**, 35–42.

Kelly, P., Thillainayagam, A.V., et al. 1996. Jejunal water and electrolyte transport in human cryptosporidiosis. *Dig Dis Sci*, **41**, 2095–9.

Kelly, P., Davies, S.E., et al. 1997a. Enteropathy in Zambians with HIV related diarrhea: regression modelling of potential determinants of mucosal damage. *Gut*, **41**, 811–16.

Kelly, P., Nchito, M., et al. 1997b. Cryptosporidiosis in adults in Lusaka, Zambia, and its relationship to oocyst contamination of drinking water. *J Infect Dis*, **176**, 1120–3.

Kelly, P., Makumbi, F.A., et al. 1998. Variable distribution of *Cryptosporidium parvum* in small and large intestine in AIDS revealed by polymerase chain reaction. *Eur J Gastroenterol Hepatol*, **10**, 855–8.

Kelly, P., Jack, D., et al. 2000. Mannose binding lectin is a contributor to mucosal defence against *Cryptosporidium parvum* in AIDS patients. *Gastroenterology*, **119**, 1236–42.

Kotler, D.P., Francisco, A., et al. 1990. Small intestinal injury and parasitic diseases in AIDS. *Ann Intern Med*, **113**, 444–9.

Kuhls, T.L., Mosier, D.A., et al. 1994. Seroprevalence of cryptosporidial antibodies during infancy, childhood, and adolescence. *Clin Infect Dis*, **18**, 731–5.

Langer, R.C., Schaefer, D.A. and Riggs, M.W. 2001. Characterization of an intestinal epithelial cell receptor recognized by the *Cryptosporidium parvum* sporozoite ligand CSL. *Infect Immun*, **69**, 1661–70.

Laurent, F., Eckmann, L., et al. 1997. *Cryptosporidium parvum* infection of human intestinal epithelial cells induces polarized secretion of C-X-C chemokines. *Infect Immun*, **65**, 5067–73.

Laurent, F., Kagnoff, M.F., et al. 1998. Human intestinal epithelial cells respond to *Cryptosporidium parvum* infection with increased prostaglandin H synthase 2 expression and prostaglandin E2 and F2α production. *Infect Immun*, **66**, 1787–90.

Lebbad, M., Norrgren, H., et al. 2001. Intestinal parasites in HIV-2 associated AIDS cases with chronic diarrhea in Guinea-Bissau. *Acta Trop*, **80**, 45–9.

Ledingham, J.C.G., Penfold, W.J. and Woodcock, H.M. 1915. Recent bacteriological experiences with typhoidal disease and dysentery. *Br Med J*, **II**, 704–11.

Levine, N.D. 1984. Taxonomy and review of the coccidian genus *Cryptosporidium* (Protozoa, Apicomplexa). *J Protozool*, **31**, 94–8.

Levine, N.D. 1987. Whatever became of *Isospora bigemina*? *Parsitol Today*, **3**, 101–5.

Liebman, W.M., Thaler, M.M., et al. 1980. Intractable diarrhea of infancy due to intestinal coccidiosis. *Gastroenterology*, **78**, 579–84.

Lillehoj, H.S. and Trout, J.M. 1996. Avian gut-associated lymphoid tissues and intestinal immune responses to *Eimeria* parasites. *Clin Microbiol Rev*, **9**, 349–60.

Lima, A.A., Moore, S.R., et al. 2000. Persistent diarrhea signals a critical period of increased diarrhea burdens and nutritional shortfalls: a prospective cohort study among children in northeastern Brazil. *J Infect Dis*, **181**, 1643–51.

Lindsay, D.S. and Blagburn, B.L. 1987. Development of *Isospora suis* from pigs in primary porcine and bovine cell cultures. *Vet Parasitol*, **24**, 301–4.

Lindsay, D.S. and Current, W.L. 1984. Complete development of *Isospora suis* of swine in chicken embryos. *J Protozool*, **31**, 152–5.

Lindsay, D.S., Dubey, J.P. and Blagburn, B.L. 1997. Biology of *Isospora* spp. from humans, nonhuman primates, and domestic animals. *Clin Microbiol Rev*, **10**, 19–34.

Liu, C., Schroeder, A.A., et al. 1998. Telomeric sequences of *Cryptosporidium parvum*. *Mol Biochem Parasitol*, **94**, 291–6.

Long, E.G., Ebrahimzadeh, A., et al. 1990. Alga associated with diarrhea in patients with acquired immunodeficiency syndrome and in travellers. *J Clin Microbiol*, **28**, 1101–4.

Lumb, R., Smith, K., et al. 1988. Ultrastructure of the attachment of *Cryptosporidium* sporozoites to tissue culture cells. *Parasitol Res*, **74**, 531–6.

Ludin, C.M., Afifi, S.A., et al. 1991. Cryptosporidiosis among children with acute gastroenteritis in a pediatric ward in the General Hospital, Penang. *Southeast Asian J Trop Med Public Health*, **22**, 200–2.

MacKenzie, W.R., Hoxie, N.J., et al. 1994. A massive outbreak in Milwaukee of *Cryptosporidium* infection transmitted through the public water supply. *N Engl J Med*, **331**, 161–7.

Matsubayashi, H. and Nozawa, T. 1948. *Isospora hominis* infections among American personnel in the Southwest Pacific. *Am J Trop Med*, **28**, 639–44.

McCole, D., Eckmann, L., et al. 2000. Intestinal epithelial cell apoptosis following *Cryptosporidium parvum* infection. *Infect Immun*, **68**, 1710–13.

McDonald, V. and Bancroft, G.J. 1994. Mechanisms of innate and acquired immunity in SCID mice infected with *Cryptosporidium parvum*. *Parasite Immunol*, **16**, 315–20.

McDonald, V., McCrossan, M.V. and Petry, F. 1995. Localization of parasite antigens in *Cryptosporidium parvum*-infected cells using monoclonal antibodies. *Parasitology*, **110**, 259–68.

McDonald, V., Deer, R., et al. 1992. Immune responses to *Cryptosporidum muris* and *Cryptosporidium parvum* in adult immunocompetent or immunocompromised (nude and SCID) mice. *Infect Immun*, **60**, 3325–31.

McDonald, V., Smith, R., et al. 2000. Host immune responses against *Cryptosporidium*. *Contrib Microbiol*, **6**, 75–91.

McGowan, I., Hawkins, A.S. and Weller, I.V. 1993. The natural history of cryptosporidial diarrhea in HIV-infected patients. *AIDS*, **7**, 349–54.

McKenna, P.B. and Charleston, W.A.G. 1982. Activation and excystation of *Isospora felis* and *Isospora rivolta* sporozoites. *J Parasitol*, **68**, 276–86.

Mead, J.R., Arrowood, M.J., et al. 1988. Field inversion gel electrophoretic separation of *Cryptosporidium* spp. chromosome-sized DNA. *J Parasitol*, **74**, 366–9.

Mead, J.R., Arrowood, M.J., et al. 1991. Cryptosporidial infections in SCID mice reconstituted with human or murine lymphocytes. *J Protozool*, **38**, Supplement, 59S–61S.

Meisel, J.L., Perera, D.G., et al. 1976. Overwhelming wasting diarrhea associated with a *Cryptosporidium* in an immunodepressed patient. *Gastroenterology*, **70**, 1156–60.

Miao, Y.M., Awad-El-Kariem, F.M., et al. 2000. Eradication of cryptosporidia and microsporidia following successful antiretroviral therapy. *J Acquir Immune Defic Syndr*, **25**, 124–9.

Michiels, J.F., Hofman, P., et al. 1994. Intestinal and extraintestinal *Isospora belli* infection in an AIDS patient. *Pathol Res Pract*, **190**, 1089–93.

Miller, J.R. 1998. Decreasing cryptosporidiosis among HIV-infected persons in New York City, 1995–1997. *J Urban Health*, **75**, 601–2.

Mirdha, B.R., Kabra, S.K. and Samantray, J.C. 2002. Isosporiasis in children. *Indian Pediatr*, **39**, 941–4.

Molbak, K., Hojlyng, N., et al. 1993. Cryptosporidiosis in infancy and childhood mortality in Guinea Bissau. *Br Med J*, **307**, 417–20.

Molbak, K., Andersen, M., et al. 1997. *Cryptosporidium* infection in infancy as a cause of malnutrition: a community study from Guinea-Bissau, West Africa. *Am J Clin Nutr*, **65**, 149–52.

Moon, H.W., Woodmansee, W.B., et al. 1988. Lacteal immunity to enteric cryptosporidiosis in mice: Immune dams do not protect their suckling pups. *Infect Immun*, **56**, 649–53.

Moore, R., Tzipori, S., et al. 1995. Temporal changes in permeability and structure of piglet ileum after site-specific infection by *Cryptosporidium parvum*. *Gastroenterology*, **108**, 1030–9.

Morgan, U.M., Constantine, C.C. and Thompson, R.C.A. 1997. Is *Cryptosporidium* clonal? *Parasitol Today*, **13**, 825–30.

Morgan, U.M., Pallant, L., et al. 1998. Comparison of PCR and microscopy for detection of *Cryptosporidium parvum* in human fecal specimens: clinical trial. *J Clin Microbiol*, **36**, 995–8.

Morgan, U.M., Xiao, L., et al. 1999. Variation in *Cryptosporidium*: towards a taxonomic revision of the genus. *Int J Parasitol*, **29**, 1733–51.

Muller, A., Bialek, R., et al. 2000. Detection of *Isospora belli* by polymerase chain reaction using primers based on small-subunit ribosomal RNA sequences. *Eur J Clin Microbiol Infect Dis*, **19**, 631–4.

Naitza, S., Spano, F., et al. 1998. The thrombospondin-related protein family of apicomplexan parasites: the gears of the cell invasion machinery. *Parasitol Today*, **14**, 479–84.

Nchito, M., Kelly, P., et al. 1998. Cryptosporidiosis in urban Zambian children: an analysis of risk factors. *Am J Trop Med Hyg*, **59**, 435–7.

Nesterenko, M.V., Woods, K. and Upton, S.J. 1999. Receptor/ligand interactions between *Cryptosporidium parvum* and the surface of the host cell. *Biochim Biophys Acta*, **1454**, 165–73.

Newman, R.F.D., Zu, S.-X., et al. 1994. Household epidemiology of *Cryptosporidium parvum* infection in an urban community in northeast Brazil. *Ann Intern Med*, **120**, 500–5.

Newman, R.F.D., Sears, C.L., et al. 1999. Longitudinal study of *Cryptosporidium* infection in children in Northeastern Brazil. *J Infect Dis*, **180**, 167–75.

Nichols, G.L., McLauchlin, J. and Samuel, D. 1991. A technique for typing *Cryptosporidium* isolates. *J Protozool*, **38**, Suppl, 237S–40S.

Niehaus, M.D., Moore, S.R., et al. 2002. Early childhood diarrhea is associated with diminished cognitive function 4 to 7 years later in children in a northeast Brazilian shantytown. *Am J Trop Med Hyg*, **66**, 590–3.

Nime, R.A., Burek, J.D., et al. 1976. Acute enterocolitis in a human being infected with the protozoan *Cryptosporidium*. *Gastroenterology*, **70**, 592–8.

Nwabuisi, C. 2001. Childhood cryptosporidiosis and intestinal parasitosis in association with diarrhea in Kwara Stata, Nigeria. *West Afr J Med*, **20**, 165–8.

O'Donoghue, P.J. 1995. *Cryptosporidium* and cryptosporidiosis in man and animals. *Int J Parasitol*, **25**, 139–95.

Ortega, Y.R., Gilman, R.H. and Sterling, C.R. 1994. A new coccidian parasite (Apicomplexa: Eimeriidae) from humans. *J Parasitol*, **80**, 625–9.

Ortega, Y.R., Sterling, C.R. and Gilman, R.H. 1998. *Cyclospora cayetanensis*. *Adv Parasitol*, **40**, 399–418.

Ortega, Y.R., Sheehy, R.R., et al. 1991. Restriction fragment length polymorphism analysis of *Cryptosporidium parvum* isolates of bovine and human origin. *J Protozool*, **38**, Suppl, 40–1.

Ortega, Y.R., Sterling, C.R., et al. 1993. *Cyclospora* species – a new protozoan pathogen of humans. *N Engl J Med*, **328**, 1308–12.

Ortega, Y.R., Roxas, C.R., et al. 1997. Isolation of *Cryptosporidium parvum* and *Cyclospora cayetanensis* from vegetables collected in markets of an endemic region in Peru. *Am J Trop Med Hyg*, **57**, 683–6.

Panciera, R.J., Thomassen, R.W. and Garner, F.M. 1971. Cryptosporidial infection in a calf. *Vet Pathol*, **8**, 479–84.

Pape, J.W., Verdier, R.-I. and Johnson, W.D. 1989. Treatment and prophylaxis of *Isospora belli* infection in patients with the acquired immunodeficiency syndrome. *N Engl J Med*, **32**, 1044–77.

Pape, J.W., Verdier, R.-I., et al. 1994. *Cyclospora* infection in adults infected with HIV. *Ann Intern Med*, **121**, 654–7.

Peeters, J.E., Villacorta, I., et al. 1992. *Cryptosporidium parvum* in calves: kinetics and immunoblot analysis of specific serum and local antibody responses (immunoglobulin A [IgA], IgG, and IgM) after natural and experimental infections. *Infect Immun*, **60**, 2309–16.

Peng, M.M., Xiao, L., et al. 1997. Genetic polymorphisms among *Cryptosporidium parvum* isolates: evidence of two distinct human transmission cycles. *Emerg Infect Dis*, **3**, 567–73.

Perch, M., Sodemann, M., et al. 2001. Seven years' experience with *Cryptosporidium parvum* in Guinea-Bissau, West Africa. *Ann Trop Paediatr*, **21**, 313–18.

Pereira, S.J., Ramirez, N.E., et al. 2002. Pathogenesis of human and bovine *Cryptosporidium parvum* in gnotobiotic pigs. *J Infect Dis*, **186**, 715–18.

Petersen, C., Barnes, D.A. and Gousset, L. 1997. *Cryptosporidium parvum* GP900, a unique invasion protein. *J Eukaryot Microbiol*, **44**, Suppl, 89S–90S.

Petry, F. 2000. Laboratory diagnosis of *Cryptosporidium parvum* infection. *Contrib Microbiol*, **6**, 33–49.

Petry, F. and Harris, J.R. 1999. Ultrastructure, fractionation, and biochemical analysis of *Cryptosporidium parvum* sporozoites. *Int J Parasitol*, **29**, 1249–60.

Phillips, A.D., Thomas, A.G. and Walker-Smith, J.A. 1992. *Cryptosporidium*, chronic diarrhea, and the proximal small intestinal mucosa. *Gut*, **33**, 1057–61.

Pichard, E., Doumbo, O., et al. 1990. Role of cryptosporidiosis in diarrhea among hospitalized adults in Bamako. *Bull Soc Pathol Exot Filial*, **83**, 473–8.

Pollok, R.C.G., Bendell, R.P., et al. 1992. Traveler's diarrhea associated with cyanobacterium-like bodies. *Lancet*, **340**, 556–7.

Pollok, R.C.G., Farthing, M.J.G., et al. 2001. Interferon gamma induces enterocyte resistance against infection by the intracellular pathogen *Cryptosporidium parvum*. *Gastroenterology*, **120**, 99–107.

Putignani, L., Sallicandro, P., et al. 1999. Chromosome mapping in *Cryptosporidium parvum* and establishment of a long-range restriction map for chromosome VI. *FEMS Microbiol Lett*, **175**, 231–8.

Ranucci, L., Muller, H.-M., et al. 1993. Characterization and immunolocalization of a *Cryptosporidium* protein containing repeated amino acid motifs. *Infect Immun*, **61**, 2347–56.

Ravn, P., Lungren, J.D., et al. 1991. Nosocomial outbreak of cryptosporidiosis in AIDS patients. *Brit Med J*, **302**, 277–80.

Reduker, D.W., Speer, C.A. and Blixt, J.A. 1985. Ultrastructural changes in the oocyst wall during excystation of *Cryptosporidium parvum* (Apicomplexa: Eucoccidiorida). *Can J Zool*, **63**, 1892–6.

Relman, D.A., Schmidt, T.M., et al. 1996. Molecular phylogenetic analysis of *Cyclospora*, the human intestinal pathogen, suggests that it is closely related to *Eimeria* species. *J Infect Dis*, **173**, 440–5.

Rene, E., Marche, C., et al. 1989. Intestinal infections in patients with acquired immunodeficiency syndrome. A prospective study in 132 patients. *Dig Dis Sci*, **34**, 773–80.

Riggs, M.W. 1997. Immunology: host response and development of passive immunotherapy and vaccines. In: Fayer, R. (ed.), *Cryptosporidium and cryptosporidiosis*. Boca Raton: CRC Press, 129–62.

Riordan, C.E., Langreth, S.G., et al. 1999. Preliminary evidence for a mitochondrion in *Cryptosporidium parvum:* phylogenetic and therapeutic implications. *J Eukaryot Microbiol*, **46**, Suppl, 52–5.

Rizk, H. and Soliman, M. 2001. Coccidiosis among malnourished children in Mansoura, Dakar Governorate, Egypt. *J Egypt Soc Parasitol*, **31**, 877–86.

Roberts, F., Roberts, C.W., et al. 1998. Evidence for the shikimate pathway in apicomplexan parasites. *Nature*, **395**, 801–5.

Robinson, P., Okhuysen, P.C., et al. 2000. Transforming growth factor β1 is expressed in the jejunum after experimental *Cryptosporidium parvum* infection. *Infect Immun*, **68**, 5405–7.

Rose, J.B., Huffman, D.E. and Gennaccaro, A. 2002. Risk and control of waterborne cryptosporidiosis. *FEMS Microbiol Rev*, **26**, 113–23.

Rossignol, J.F., Ayoub, A. and Ayers, M.S. 2001. Treatment of diarrhea caused by *Cryptosporidium parvum:* a prospective randomized double-blind placebo-controlled study of nitazoxanide. *J Infect Dis*, **184**, 103–6.

Rossignol, J.F., Hidalgo, H., et al. 1998. A double-blind placebo controlled study of nitazoxanide in the treatment of cryptosporidial diarrhea in AIDS patients in Mexico. *Trans R Soc Trop Med Hyg*, **92**, 663–6.

Saksirisampant, W., Eampokalap, B., et al. 2002. A prevalence of *Cryptosporidium* infections among Thai HIV-infected patients. *J Med Assoc Thai*, **85**, S424–8.

Sauda, F.C., Zamarioli, L.A., et al. 1993. Prevalence of *Cryptosporidium* spp. and *Isospora belli* among AIDS patients attending Santos Reference Center for AIDS, Sao Paulo, Brazil. *J Parasitol*, **79**, 454–6.

Schmatz, D.M. 1989. The mannitol cycle – a new metabolic pathway in the coccidia. *Parasitol Today*, **5**, 205–8.

Schmatz, D.M. 1997. The mannitol cycle in *Eimeria*. *Parasitology*, **114**, S81–9.

Sherwood, D.K., Angus, K.W., et al. 1982. Experimental cryptosporidiosis in laboratory mice. *Infect Immun*, **38**, 471–5.

Shlim, D.R., Cohen, M.T. and Eaton, M. 1991. An alga-like organism associated with an outbreak of prolonged diarrhea among foreigners in Nepal. *Am J Trop Med Hyg*, **45**, 383–9.

Shlim, D.R., Hoge, C.W., et al. 1999. Persistent high risk of diarrhea among foreigners in Nepal during the first 2 years of residence. *Clin Infect Dis*, **29**, 613–16.

Sifuentes-Osornio, J., Porras-Cortes, G., et al. 1995. *Cyclospora cayetanensis* infection in patients with and without AIDS: biliary disease as another clinical manifestation. *Clin Infect Dis*, **21**, 1092–7.

Slavin, D. 1955. *Cryptosporidium meleagridis* sp. nov. *J Comp Pathol*, **65**, 262–6.

Smith, A.L. and Hayday, A.C. 1998. Genetic analysis of the essential components of the immunoprotective response to infection with *Eimeria vermiformis*. *Int J Parasitol*, **28**, 1061–9.

Smith, H.V., Paton, C.A., et al. 1997. Sporulation of *Cyclospora* sp. oocysts. *Appl Environ Microbiol*, **63**, 1631–2.

Soave, R. and Johnson, W.D. 1988. *Cryptosporidium* and *Isospora belli* infections. *J Infect Dis*, **157**, 225–9.

Soave, R., Dubey, J.P., et al. 1986. A new intestinal pathogen? *Clin Res*, **34**, Appendix, 533A.

Soldati, D., Dubremetz, J.F. and Leburn, M. 2001. Microneme proteins: structural and functional requirements to promote adhesion and invasion by the apicomplexan parasite *Toxoplasma gondii*. *Int J Parasitol*, **31**, 1293–303.

Sorvillo, F.J., Lieb, L.E., et al. 1995. Epidemiology of isosporiasis among persons with AIDS in Los Angeles County. *Am J Trop Med Hyg*, **53**, 656–9.

Spano, F. and Crisanti, A. 2000. *Cryptosporidium parvum:* the many secrets of a small genome. *Int J Parasitol*, **30**, 553–65.

Spano, F., Putignani, L., et al. 1998. Multi-locus genotypic analysis of *Cryptosporidium parvum* isolates from different hosts and geographic origins. *J Clin Microbiol*, **36**, 3255–9.

Strong, W.B., Gut, J. and Nelson, R.G. 2000. Cloning and sequence analysis of a highly polymorphic *Cryptosporidium parvum* gene encoding a 60-kilodalton glycoprotein and characterization of its 15- and 45-kilodalton zoite surface antigen products. *Infect Immun*, **68**, 4117–34.

Stuart, B.P., Sisk, D.B., et al. 1982. Demonstration of immunity against *Isospora suis* in swine. *Vet Parasitol*, **9**, 185–91.

Sun, T., Illardi, C.F., et al. 1996. Light and electron microscopic identification of *Cyclospora* species in the small intestine. Evidence of the presence of asexual life cycle in human host. *Am J Clin Pathol*, **105**, 216–20.

Sundermann, C.A., Lindsay, D.S. and Blagburn, B.L. 1987. In vitro excystation of *Cryptosporidium baileyi* from chickens. *J Protozool*, **34**, 28–30.

Takhi-Kilani, R., Sekla, L. and Hayglass, K.T. 1990. The role of humoral immunity in *Cryptosporidum* spp. infection studies with B cell-depleted mice. *J Immunol*, **145**, 1571–6.

Tarver, A.P., Clark, D.P., et al. 1998. Enteric β-defensin: molecular cloning and characterization of a gene with inducible intestinal epithelial cell expression associated with *Cryptosporidium parvum* infection. *Infect Immun*, **66**, 1045–56.

Teixidor, H.S., Godwin, T.A. and Ramirez, E.A. 1991. Cryptosporidiosis of the biliary tract in AIDS. *Radiology*, **180**, 51–6.

Tetley, L., Brown, S.M.A., et al. 1998. Ultrastructural analysis of the sporozoite of *Cryptosporidium parvum*. *Microbiology*, **144**, 3249–55.

Theodos, C.M. 1998. Innate and cell-mediated immune responses to *Cryptosporidium parvum*. *Adv Parasitol*, **40**, 87–119.

Theodos, C.M., Griffiths, J.K., et al. 1998. The efficacy of nitazoxanide against *Cryptosporidium parvum* in cell culture and in animal models. *Antimicrob Agents Chemother*, **42**, 1959–65.

Thomson, M.A., Benson, J.W. and Wright, P.A. 1987. Two year study of *Cryptosporidium* infection. *Arch Dis Child*, **62**, 559–63.

Tilley, M. and Upton, S.J. 1997. Biochemistry of *Cryptosporidium*. In: Fayer, R. (ed.), *Cryptosporidium and cryptosporidiosis*. Boca Raton: CRC Press, 165–81.

Trier, J.S., Moxey, P.C., et al. 1974. Chronic intestinal coccidiosis in man: intestinal morphology and response to treatment. *Gastroenterology*, **66**, 923–35.

Tripp, C.P., Wolf, S.F. and Unanue, E.R. 1993. Interleukin-12 and tumor necrosis factor alpha are costimulators of interferon gamma production by natural killer cells in severe combined immunodeficiency mice with listeriosis, and interleukin-10 is a physiologic antagonist. *Proc Natl Acad Sci*, **90**, 3725–9.

Tyzzer, E.E. 1907. A sporozoan found in the peptic glands of the common mouse. *Proc Soc Exp Biol Med*, **5**, 12–13.

Tyzzer, E.E. 1912. *Cryptosporidium parvum* (sp. nov.), a coccidium found in the small intestine of the common mouse. *Archiv Protist*, **26**, 394–413.

Tyzzer, E.E. 1929. Coccidiosis in gallinaceous birds. *Am J Hyg*, **10**, 269–83.

Tzipori, S. 1983. Cryptosporidiosis in animals and humans. *Microbiol Rev*, **47**, 84–96.

Tzipori, S., Robertson, D. and Chapman, C. 1986. Remission of diarrhea due to cryptosporidiosis in an immunodeficient child treated with hyperimmune colostrum. *Br Med J*, **293**, 1276–7.

Ullrich, R., Heise, W., et al. 1992. Gastrointestinal symptoms in patients infected with human immunodeficiency virus: relevance of infective agents isolated from gastrointestinal tract. *Gut*, **33**, 1080–4.

Ungar, B.L.P., Soave, R., et al. 1986. Enzyme immunoassay detection of immunoglobulin M and G antibodies to *Cryptosporidium* in immunocompetent and immunocompromised persons. *J Infect Dis*, **153**, 570–8.

Ungar, B.L.P., Kao, T.-C., et al. 1991. *Cryptosporidium* infection in an adult mouse model. Independent roles for IFN-γ and CD4+ T lymphocytes in protective immunity. *J Immunol*, **147**, 1014–22.

Uni, S., Iseki, M., et al. 1987. Ultrastructure of *Cryptosporidium muris* (strain RN 66) parasitizing the mouse stomach. *Parasitol Res*, **74**, 123–32.

Upton, S.J. and Current, W.L. 1985. The species of *Cryptosporidium* (Apicomplexa, Cryptosporidiidae) infecting animals. *J Parasitol*, **71**, 625–9.

Urban, J.F. Jr., Fayer, R., et al. 1996. IL-12 protects immunocompetent and immunodeficient neonatal mice against infection with *Cryptosporidium parvum*. *J Immunol*, **156**, 263–8.

Varea, M., Clavel, A., et al. 1998. Fuchsin fluorescence and autofluorescence in *Cryptosporidium*, *Isospora* and *Cyclospora* oocysts. *Int J Parasitol*, **28**, 1881–3.

Vasquez, J.R., Gooze, L., et al. 1996. Potential antifolate resistance determinants and genotypic variation in the bifunctional dihydrofolate reductase-thymidylate synthase gene from human and bovine isolates of *Cryptosporidium parvum*. *Mol Biochem Parasitol*, **79**, 153–65.

Verdier, R.I., Fitzgerald, D.W., et al. 2000. Trimethoprim-sulfamethoxazole compared with ciprofloxacin for treatment and prophylaxis of *Isospora belli* and *Cyclospora cayetanensis* infection in HIV-infected patients. A randomized controlled trial. *Ann Intern Med*, **132**, 885–8.

Virchow, R. 1860. Helminthologische Notizen. 4. Zurkenntniss der Wurmknoten. *Archiv Pathol Anatom Physiol Klin Med*, **18**, 523–36.

Vitovec, J. and Koudela, B. 1992. Pathogenesis of intestinal cryptosporidiosis in conventional and gnotobiotic piglets. *Vet Parasitol*, **43**, 25–36.

Weiss, L.M., Perlman, D.C., et al. 1988. *Isospora belli* infection: treatment with pyrimethamine. *Ann Intern Med*, **109**, 474–5.

Wenyon, C.M. 1915. Observations on the common intestinal protozoa of man: their diagnosis and pathogenicity. *Lancet*, **23**, 1173–83.

Wenyon, C.M. 1923. Coccidiosis of cats and dogs and the status of the *Isospora* of man. *Ann Trop Med Parasitol*, **17**, 231–9.

White, A.C., Robinson, P., et al. 2000. Interferon-γ expression in jejunal biopsies in experimental human cryptosporidiosis correlates with prior sensitization and control of oocyst excretion. *J Infect Dis*, **181**, 701–9.

Whiteside, M.E., Barkin, J.S., et al. 1984. Enteric coccidiosis among patients with the acquired immunodeficiency syndrome. *Am J Trop Med Hyg*, **33**, 1065–72.

Widmer, G., Akiyoshi, D., et al. 2000. Animal propagation and genomic survey of a genotype 1 isolate of *Cryptosporidium parvum*. *Mol Biochem Parasitol*, **108**, 187–97.

Widmer, G., Lin, L., et al. 2002. Genomics and genetics of *Cryptosporidium parvum*: the key to understanding cryptosporidiosis. *Microbes Infect*, **4**, 1081–90.

Wright, M.S. and Collins, P.A. 1997. Waterborne transmission of *Cryptosporidium*, *Cyclospora* and *Giardia*. *Clin Lab Sci*, **10**, 287–90.

Wurtz, R. 1994. *Cyclospora*: a newly identified intestinal pathogen of humans. *Clin Infect Dis*, **18**, 620–3.

Zar, F.A., El-Bayoumi, E. and Yungbluth, M.M. 2001. Histologic proof of acalculous cholecystitis due to *Cyclospora cayetanensis*. *Clin Infect Dis*, **15**, E140–1.

Zhang, Y., Lee, B., et al. 2000. Lactulose-mannitol intestinal permeability test in children with diarrhea caused by rotavirus and cryptosporidium. Diarrhea Working Group, Peru. *J Pediatr Gastroenterol Nutr*, **31**, 16–21.

Zu, S.X., Zhu, S.Y. and Li, J.F. 1992. Human cryptosporidiosis in China. *Trans R Soc Trop Med Hyg*, **86**, 639–40.

Zulu, I., Veitch, A., et al. 2002. Albendazole chemotherapy for AIDS-related diarrhea in Zambia - clinical, parasitological and mucosal responses. *Aliment Pharm Therap*, **16**, 595–601.

Toxoplasmosis

J.P. DUBEY

INTRODUCTION

Infection with *Toxoplasma gondii* is one of the most common parasitic infections of humans and other warm-blooded animals. It is found worldwide from Alaska to Australasia and nearly one-third of humanity has been exposed to this parasite. In most adults, it does not cause serious illness, but it can cause blindness and intellectual deficit in congenitally infected children and severe disease in those with depressed immunity. In animals, it is a common cause of abortion in goats and sheep.

T. gondii was discovered by Nicolle and Manceaux (1908, 1909) in Tunisia in a rodent, *Ctenodactylus gundi* and, independently, in a laboratory rabbit by Splendore (1908) in São Paulo, Brazil. It was recognized as a human pathogen when Wolf et al. (1939) reported a confirmed case of congenital toxoplasmosis in a child (see Table 21.1). The complete life cycle of *T. gondii* was not determined until 1970 (Frenkel et al., 1970; Dubey et al., 1970a, b). The full life cycle clearly shows that this protozoan is a coccidian. Coccidians are very common parasites of animals. Most coccidians are one-host parasites and develop in the intestinal epithelium. An environmentally resistant stage is excreted in feces and the host becomes infected by ingesting feces.

CLASSIFICATION

T. gondii is a coccidian parasite of cats with other warm-blooded animals as intermediate hosts. Coccidiosis is among the most important of parasitic infections of animals. Traditionally, all coccidia of veterinary importance were classified in the family Eimeriidae and were further classified based on the structure of the oocyst. Coccidia having oocysts with four sporocysts each with two sporozoites (total eight sporozoites) are classified as *Eimeria*, and coccidia having oocysts containing two sporocysts each with four sporozoites were classified historically as *Isospora*. After the discovery of the life cycle of *T. gondii*, several other genera (*Sarcocystis*, *Besnoitia*, *Hammondia*, *Frenkelia*, and *Neospora*) were also found to have isosporan oocysts with two sporocysts and eight sporozoites. *T. gondii* and related genera are now classified in the phylum Apicomplexa Levine, 1970, class Coccidea Leuckart, 1879, order Eimeriida Léger, 1911 (see also Chapter 8, Classification and introduction to the parasitic protozoa). Opinions differ regarding the placement of *T. gondii* into a family and subfamily: various authorities have placed it in one or the other of the families Eimeriidae Michin, 1903, Sarcocystidae Poche, 1913, or Toxoplasmatidae Biocca, 1956.

STRUCTURE AND LIFE CYCLE

The name *Toxoplasma* (from 'toxon' meaning 'bow' and 'plasma' meaning 'form') relates to the crescent shape of the tachyzoite stage. There are three infectious stages of *T. gondii*: the tachyzoites (in groups), the bradyzoites (in tissue cysts), and the sporozoites (in oocysts) (Frenkel 1973a).

Table 21.1 *History of* Toxoplasma gondii *and toxoplasmosis*

Contributors and year	Contribution
Nicolle and Manceaux (1908)	Discovered in *gundi*
Splendore (1908)	Discovered in rabbit
Mello (1910)	Disease described in a domestic animal (dog)
Janku (1923)	Identified in human eye at necropsy
Wolf and Cowen (1937)	Congenital transmission documented
Wolf et al. 1939	Congenital transmission coonfirmed
Pinkerton and Weinman (1940)	Fatal disease described in adult humans
Sabin (1942)	Disease characterized in man
Sabin and Feldman (1948)	Dye test described
Siim (1952)	Glandular toxoplasmosis described in man
Weinman and Chandler (1954)	Suggested carnivorous transmission
Hartley and Marshall (1957)	Abortions in sheep recognized
Beverley (1959)	Repeated congenital transmission observed in mice
Jacobs et al. (1960)	Tissue cysts characterized biologically
Hutchison (1965)	Fecal transmission recognized, nematode eggs suspected
Hutchison et al. (1969, 1970, 1971); Frenkel et al. (1970); Dubey et al., (1970a,b); Sheffield and Melton (1970); Overdulve (1970)	Coccidian phase described
Frenkel et al. (1970); Miller et al. (1972)	Definitive and intermediate hosts defined
Dubey and Frenkel (1972)	Five *T. gondii* types described from feline intestinal epithelium
Wallace (1969); Munday (1972)	Confirmation of the epidemiological role of cats from studies on remote islands
Luft et al. (1983)	Toxoplasmosis recognized in AIDS patients
Silveira et al. (1988)	Postnatal opthalmic toxoplasmosis recognized

Reproduced from Dubey 1993; where a complete bibliography may be found.

The tachyzoite is often crescent-shaped and approximately 2×6 μm in smears or globular-to-oval in sections (Figure 21.1). Its anterior (conoidal) end is pointed and its posterior end is round. It has a pellicle (outer covering), a polar ring, a conoid, rhoptries, micronemes, mitochondria, subpellicular microtubules, endoplasmic reticulum, a Golgi apparatus, ribosomes, rough surfaced endoplasmic reticulum, a micropore, and a well defined nucleus (Figure 21.2). The nucleus is usually situated toward the posterior end or in the central area of the cell.

The pellicle consists of three membranes. The inner membrane complex is discontinuous at three points: the anterior end (polar ring), the lateral edge (micropore), and toward the posterior end. The polar ring is an osmiophilic thickening of the inner membrane at the anterior end of the tachyzoite. The polar ring encircles a cylindrical, truncated cone (the conoid) which consists of between six and eight fibrillar elements wound like a compressed spring. There are 22 subpellicular microtubules originating from the anterior end and running longitudinally almost the entire length of the cell. Terminating within the conoid are between four and ten club-shaped organelles called rhoptries (Dubey 1977, 1993; Dubey et al. 1998). The rhoptries are glandlike structures, often labyrinthine, with an anterior narrow neck up to 2.5 μm long. Their saclike posterior end terminates anterior to the nucleus. Micronemes are rice grain-like structures, which occur at the anterior (conoidal) end of the parasite.

The functions of the conoid, rhoptries, and micronemes are not fully known. The conoid is probably associated with the penetration of the tachyzoite through the membrane of the host cell. It can rotate, tilt, extend, and retract as the parasite searches for a host cell. *T. gondii* can move by gliding, undulating and rotating. Rhoptries have a secretory function associated with host-cell penetration, secreting their contents through the conoid to the exterior. The microtubules probably provide the cytoskeleton.

The tachyzoite enters the host cell by active penetration of its membrane. After entering the host cell, the tachyzoite assumes an oval shape and becomes surrounded by a parasitophorous vacuole (PV) which, it has been suggested, is derived from both the parasite and the host. Numerous intravacuolar tubules connect the parasitophorous vacuolar membrane (PVM) to the parasite pellicle (Figure 21.2). *T. gondii* enters host cells by active penetration and the whole event can be completed in 10 seconds, this makes it difficult to study the invasion process. Rhoptries, micronemes, conoid, microtubules, myosin, dense graduals, and other factors are involved in a complex cascade of individual events involving attachment, protrusion of conoid, moving junction formation and exocytosis of contents of rhoptries, micronemes and dense granules (Joiner and

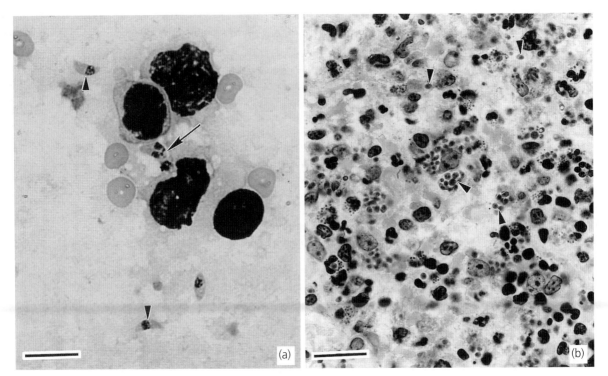

Figure 21.1 *Tachyzoites of* T. gondii: **(a)** *Impression smear. Note individual crescentic (arrowheads) and dividing (arrow) tachyzoites. Giemsa. Bar = 10 μm.* **(b)** *Histological section of mesenteric lymph node. Note numerous oval-to-round tachyzoites (arrowheads). Hematoxylin and eosin (H&E). Bar = 20 μm*

Dubremetz 1993). Parasite proteins P30 and surface lectin-like molecules are involved in finding host cell receptors. Myosin and actin, located at the apical end of the parasite, help in extension of the conoid and forma- tion of a moving junction. The secretions of rhoptries and micronemes probably dissolve the plasma membrane for the parasite invasion. After invasion, the parasite is separated from the host cell cytoplasm by the

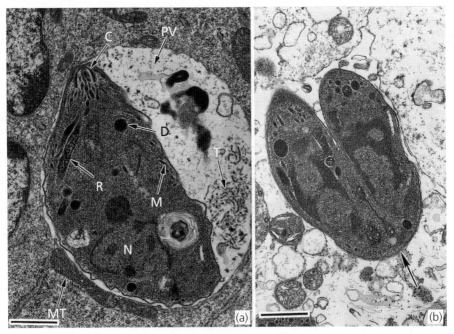

Figure 21.2 *Transmission electron micrographs of* T. gondii *tachyzoites in cell culture:* **(a)** *Tachyzoite in a parasitophorous vacuole (PV) in the cytoplasm of a host cell. Note conoid (C), rhoptries (R), micronemes (M), nucleus (N), dense granules (D), and intravacuolar tubules (T). The host cell mitochondria (MT) are closely associated with the PV. Bar = 2.1 μm.* **(b)** *Dividing tachyzoites with separated apical ends and the undivided posterior end (arrow). Bar = 1.18 μm*

PVM, which is formed by the invagination of host-cell plasma membrane (Sibley 1995). The proteins derived from the host cell are quickly removed from the PV and a complex membranous tubular network (TMN) is formed in the PV; dense granule secretions contribute to TMN formation. These TMN are connected to the PVM and micropores in the PVM allow diffusion of selective host molecules into the PVM, but otherwise the parasite is independent of the host cell.

The tachyzoite multiplies asexually within the host cell by repeated endodyogeny (from 'endon' meaning 'inside', 'dyo' meaning 'two', and 'genesis' meaning 'birth'), a specialized form of reproduction in which two progenies form within the parent parasite, consuming it (Figure 21.2). Tachyzoites continue to divide by endodyogeny until the host cell is filled with parasites.

After a few divisions, T. gondii encysts to form tissue cysts. Tissue cysts grow and remain intracellular (Figure 21.3) as the bradyzoites (encysted zoites) divide by endodyogeny. Tissue cysts vary in size (Figure 21.3). Young tissue cysts may be as small as 5 μm and contain only two bradyzoites, although older ones may contain hundreds of organisms (Figure 21.3a, b). Tissue cysts in

histological sections of brain are often circular and rarely reach a diameter of 60 μm whereas intramuscular cysts are elongated and may reach 100 μm; tissue cysts in unstained live squash preparations vary in size depending on the pressure applied to squash the tissue and the medium of suspension (Figure 21.3b). Although tissue cysts may develop in visceral organs, including lungs, liver and kidneys, they are more prevalent in the neural and muscular tissues, such as the brain, eye, and skeletal and cardiac muscle. Intact tissue cysts probably do not cause any harm and can persist for the life of the host.

The tissue cyst wall is elastic, thin (less than 0.5 μm) and argyrophilic, and may enclose hundreds of crescent-shaped slender bradyzoites approximately 7 × 1.5 μm in size (Figure 21.3). Structurally, bradyzoites differ only slightly from tachyzoites. They have a nucleus situated toward the posterior end, whereas the nucleus in tachyzoites is more centrally located. The contents of rhoptries in bradyzoites in older tissue cysts are electron dense (Figure 21.4) (Ferguson and Hutchison 1987; Dubey 1993). Bradyzoites contain several amylopectin granules, which stain red with periodic acid–Schiff

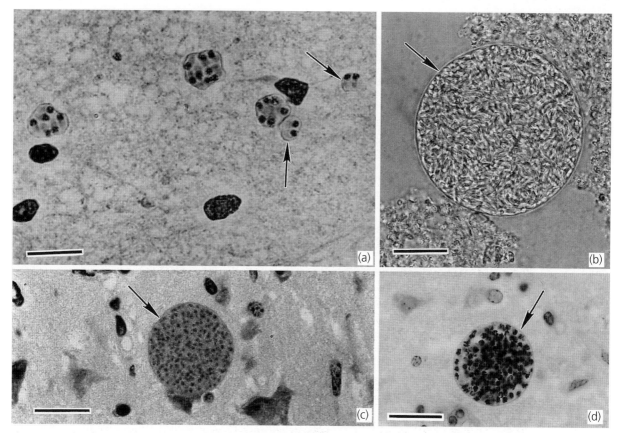

Figure 21.3 *Tissue cysts of* T. gondii *in brain:* **(a)** *Impression smear. Five young tissue cysts with silver positive cyst walls. Two tissue cysts (arrows) each have two bradyzoites with terminal nuclei. Silver stain. Bar = 10 μm.* **(b)** *Impression smear, unstained. This tissue cyst was freed by grinding a piece of brain in a mortar with a pestle. Note thin, elastic cyst wall (arrow) enclosing hundreds of bradyzoites. Bar = 20 μm.* **(c)** *Histological section. Note only nuclei of bradyzoites are visible. H&E. Bar = 20 μm.* **(d)** *Histological section. Note bradyzoites have PAS-positive red granules that appear black in this micrograph. The tissue cyst wall (arrow) is PAS-negative. PAS–hematoxylin. Bar = 20 μm*

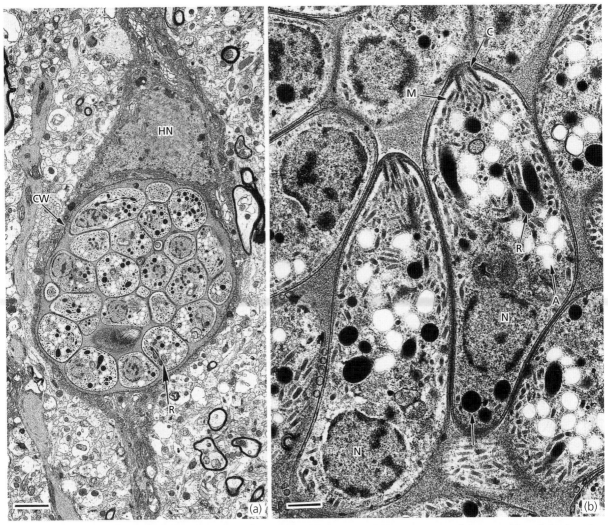

Figure 21.4 *Transmission electron micrographs of tissue cysts of* T. gondii *in brain.* **(a)** *A young intracellular cyst with well developed cyst wall (CW) is visible. The bradyzoites are plump (dividing or preparing to divide) and the contents of the rhoptries (R) in one* T. gondii *have a honeycomb structure (arrow). HN, host nucleus. Bar = 4.4 μm.* **(b)** *Two longitudinally cut bradyzoites from a large cyst. Note electron-dense contents of rhoptries (R), and the terminal nuclei (N), a conoid (C), numerous micronemes (M), and amylopectin granules (A) that appear as empty spaces here. Bar = 0.77 μm.*

(PAS) reagent; such material is either in discrete particles or absent from tachyzoites. Bradyzoites are more slender than tachyzoites and less susceptible to destruction by gastric juice.

Factors influencing tissue cyst formation are not well known. Tissue cysts are more numerous in animals in the chronic stage of infection, after the host has acquired immunity, than during the acute stage of infection. However, tissue cysts have been found in mice infected for only 3 days and in cells in culture systems devoid of known immune factors. It is possible, therefore, that development of functional immunity and the formation of tissue cysts are coincidental.

Cats shed oocysts (Figure 21.5) after ingesting any of the three infectious stages of *T. gondii*, (i.e. tachyzoites, bradyzoites or sporozoites) (Frenkel et al. 1970; Dubey et al. 1970a). Prepatent periods (time to the shedding of oocysts after initial infection) and frequency of oocyst shedding vary according to the stage ingested. Prepatent periods are 3–10 days after ingesting tissue cysts and 18 days or more after ingesting tachyzoites or oocysts (Dubey and Frenkel 1976; Freyre et al. 1989; Dubey 1996, 2002a). Less than 50 percent of cats shed oocysts after ingesting tachyzoites or oocysts, whereas nearly all cats shed oocysts after ingesting tissue cysts.

After the ingestion of tissue cysts by cats, the cyst wall is dissolved by the proteolytic enzymes in the stomach and small intestine. The released bradyzoites penetrate the epithelial cells of the small intestine and initiate development of numerous asexual generations of *T. gondii* (Dubey and Frenkel 1972). Five morphologically distinct types (A–E) of *T. gondii* develop in intestinal epithelial cells before gametogony begins. Types A–E divide asexually by endodyogeny, endopolygeny or schizogony (division into more than two organisms). The origin of gamonts has not been determined; probably the

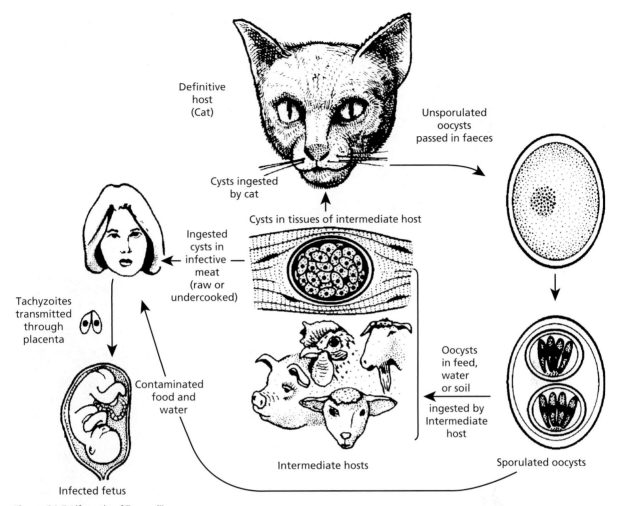

Figure 21.5 *Life cycle of* T. gondii

merozoites released from meronts of types D and E initiate gamete formation. Gamonts occur throughout the small intestine but most commonly in the ileum (Figure 21.6). Gamonts and schizonts are located in surface epithelial cells, usually above the host cell nucleus (Figure 21.7), where they form 3–15 days after infection. The female gamete is subspherical and contains a single centrally located nucleus (Figure 21.7). Mature male gamonts are ovoid to ellipsoidal in shape (Figure 21.6). When microgametogenesis takes place, the nucleus of the male gamont divides to produce 10–21 nuclei which move toward the periphery of the parasite entering protuberances formed in the pellicle of the mother parasite. One or two residual bodies are left in the microgamont after division into microgametes. Each microgamete has two flagella (Figure 21.6) and swims to and penetrates a mature macrogamete. After penetration, oocyst wall formation begins around the fertilized gamete, and, when mature, oocysts are discharged into the intestinal lumen by the rupture of intestinal epithelial cells.

Unsporulated oocysts are subspherical to spherical, 10×12 μm in diameter (Figure 21.8). The oocyst wall contains two layers. The sporont almost fills the oocyst, and sporulation occurs outside the cat within 1–5 days depending on aeration and temperature. Sporulated oocysts are subspherical to ellipsoidal and each sporulated oocyst contains two ellipsoidal sporocysts. These lack a Stieda body (Figure 21.8). Sporocysts measure 6×8 μm. There are four sutures with liplike thickenings in the sporocyst wall (Figure 21.8) which open during excystation of the sporozoites. A sporocyst residuum is present, but no oocyst residuum. Each sporocyst contains four sporozoites, 2×6–8 μm in size with a subterminal to central nucleus and a few PAS-positive granules in the cytoplasm (Dubey et al. 1970a). They have most of the organelles found in other coccidia except a crystalloid body and refractile globules (Figure 21.8).

As the enteroepithelial cycle progresses, bradyzoites penetrate the lamina propria of the feline intestine and multiply as tachyzoites. Within a few hours after infection of cats, *T. gondii* may disseminate to extraintestinal tissues. *T. gondii* persists in intestinal and extraintestinal tissues of cats for at least several months, if not for the life of the cat.

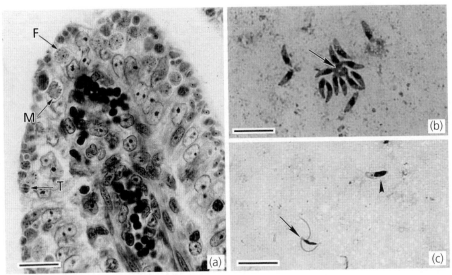

Figure 21.6 *Enteroepithelial stages of* T. gondii, *6 days after feeding tissue cysts to a cat:* **(a)** *Histological section of a villus in small intestine. Note heavy infection of epithelial cells with* T. gondii *types (T), male gamonts (M), and numerous uninucleate female gamonts (F). Cells in the lamina propria are not infected. H&E. Bar = 15 μm.* **(b)** *Impression smear. Note a ruptured type E schizont with a residual body (arrow). Giemsa. Bar = 10 μm.* **(c)** *Impression smear. A biflagellate microgamete (arrow) and a free merozoite (arrowhead). Giemsa. Bar = 10 μm*

CULTIVATION, ANTIGENS, MOLECULAR BIOLOGY, AND GENETICS

T. gondii has not been grown in cell-free media but can be cultivated in laboratory animals, chick embryos and cell cultures. Mice, hamsters, guinea pigs, and rabbits are all susceptible but mice are generally used as hosts because they are more susceptible than the

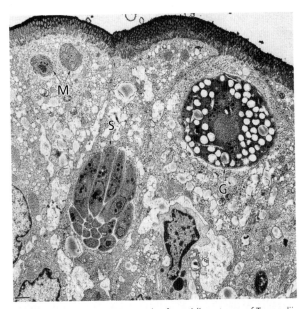

Figure 21.7 *Electron micrograph of coccidian stages of* T. gondii *in epithelial cells of ileum of a cat 6 days after ingesting tissue cysts. Note two merozoites (M) and a female gamont (G) located just below the microvillus border, and a schizont above the host cell nucleus. Bar = 2.5 μm*

others and are not naturally infected when raised in the laboratory on commercial dry food that is free from cat feces.

Tachyzoites of some strains of *T. gondii* grow in the peritoneal cavity of mice, sometimes producing ascites; they also grow in most other tissues after intraperitoneal inoculation with any of the three infectious stages of *T. gondii*. Tissue cysts are prominent in the mouse brain about 8 weeks after infection. Virulent strains usually produce illness in mice and sometimes kill them within 1–2 weeks. Most strains of *T. gondii* do not kill mice.

T. gondii tachyzoites will multiply in many cell lines in cell cultures and although most strains can develop tissue cysts in cell cultures, the yield is lower than that produced in mice. The cysts can develop within 3 days of inoculation of tachyzoites in cell culture. Virulent mouse strains rapidly destroy the cells whereas avirulent strains grow slowly, causing minimal cell damage. The mean generation time of tachyzoites of the virulent RH strain is 5 hours. Feline enteroepithelial stages of *T. gondii* have not yet been cultivated in vitro. Oocysts can be obtained by feeding tissue cysts from infected mice to *T. gondii*-free cats. Although there are stage-specific proteins, all infective stages of *T. gondii* also share common proteins. Most of the proteins that have been characterized are from tachyzoites. *T. gondii* tachyzoites have four major surface proteins (of 22, 30, 35, and 43 kDa) of which p30 (SAG1) is the dominant protein and accounts for 5 percent of the total tachyzoite protein. Information about major *T. gondii* antigens and their localization in organelles of *T. gondii* was summarized by Petersen and Dubey (2001).

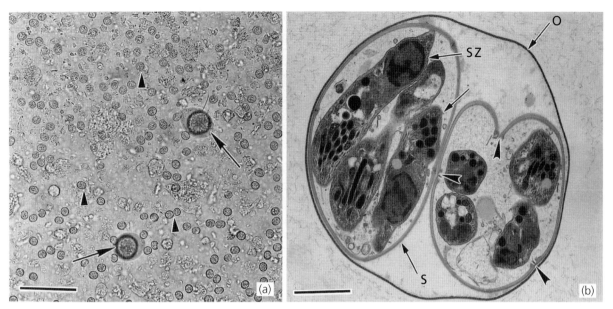

Figure 21.8 *Oocysts of* T. gondii: **(a)** *Unsporulated oocysts (arrowheads) and oocysts of another common feline coccidian,* Isospora felis *(arrows) in a fecal float preparation. Unstained. Bar = 50 μm.* **(b)** *Transmission electron micrograph of a sporulated oocyst. Note thin walled oocyst (O) enclosing the two sporocysts (S) each with four sporozoites (SZ). Each sporocyst has four lip-like thickenings (arrowheads). Bar = 2.8 μm. (Courtesy of Dr D. S. Lindsay)*

The *T. gondii* nucleus is haploid except for the zygote, in the intestine of the cat (Pfefferkorn 1990). Sporozoites result from a miotic division followed by mitotic divisions and genetic segregation seems to follow classical Mendelian laws. The total haploid genome contains approximately 8×10^7 bp and 11 chromosomes. There is also a 36 kb circular mitochondrial DNA that has been partly sequenced. Nine chromosomes have been identified by pulsed field gel electrophoresis. The karyotype has been studied using probes which attach to genes of low copy number (Sibley et al. 1992). Based on a wide host range (all warm blooded animals) and a worldwide distribution, one would expect a great genetic variability among *T. gondii* isolates. Contrary to this expectation, genetic diversity among strains of *T. gondii* so far characterized from humans or animals is low (Howe and Sibley 1995; Lehmann et al. 2000). Evidence suggests that *T. gondii* has a clonal population structure (Howe and Sibley 1995; Grigg et al. 2001a; Su et al. 2002, 2003), which can partly explain the low genetic diversity. Importantly, clonality provides the basis for strain typing using a single locus. Since clonality means that all genes are inherited together, high phenotypic similarity is expected among members of the same type (lineage). *T. gondii* strains have been classified into three genetic types (I, II, III), based on restriction fragment length polymorphism (RFLP). DNA from other sources can be introduced into the *T. gondii* genome (Soldati and Boothroyd 1993).

It has been suggested that type I isolates or recombinants of type I and III are more likely to result in clinical toxoplasmosis (Howe et al. 1997; Grigg et al. 2001a; Fuentes et al. 2001; Aspinall et al. 2003), but genetic characterization has been limited essentially to patients ill with toxoplasmosis. Contrary to the situation in humans, all three types of *T. gondii* have been isolated from asymptomatic animals (Mondragon et al. 1998; Dubey et al., 2002a, b). Little is known of the effect of geography on the genetic variability of *T. gondii* isolates because most of the *T. gondii* isolates genetically typed were from the United States and Europe. Ajzenberg et al. (2002) genotyped 86 isolates of *T. gondii* obtained from suspected or proven cases of congenital toxoplasmosis in Europe. Most of the isolates (73 of 86) were type II, 7 were type I, 2 were type III, and 4 were atypical. Among the 7 type I isolates, 4 were from placenta and the child was not infected. There was no obvious correlation between genotyping and severity of clinical toxoplasmosis. It is noteworthy that in this study all *T. gondii* isolates were genotyped irrespective of the clinical status (Ajzenberg et al. 2002).

T. gondii isolates differ remarkably in their virulence to outbred mice. Type I isolates are usually lethal to mice, irrespective of the dose and virulence is genetically controlled (Howe et al. 1997; Grigg et al. 2001b; Su et al. 2002). Type II and III isolates are less virulent for mice. However, virulence of *T. gondii* in mice should not be equated with virulence in humans or domestic animals. Hundreds of *T. gondii* isolates from aborted ovine fetuses were found to be avirulent for mice.

CLINICAL ASPECTS

T. gondii infection is widespread among humans and its prevalence varies widely from place to place. In the USA and the UK, it is estimated that about

16–40 percent of people are infected whereas in Central and South America and continental Europe infection is estimated to be 50–80 percent (Dubey and Beattie 1988; Tenter et al. 2000; Jones et al. 2001a). Most infections in humans are asymptomatic but at times the parasite can produce devastating disease. Infection may be congenitally or postnatally acquired.

Congenital infection

Congenital infection occurs only when a woman becomes infected during pregnancy; the severity of the disease in the fetus may depend on the stage of pregnancy when the mother becomes infected (Table 21.2) (Desmonts and Couvreur 1974; Daffos et al. 1988). Although the mother rarely has symptoms of infection, she does have a temporary parasitemia. Focal lesions develop in the placenta and the fetus may become infected. In the fetus, there is generalized infection at first, but later it clears from the visceral tissues and may localize in the central nervous system. A wide spectrum of clinical disease occurs in congenitally infected children (Remington et al. 2001; Jones et al. 2001b). Mild disease may consist of slightly diminished vision; severely diseased children may have the full tetrad of signs: retinochoroiditis, hydrocephalus (Figure 21.9), convulsions, and intracerebral calcification (Figure 21.10). Of these, hydrocephalus is the least common but most dramatic lesion of toxoplasmosis. This lesion is unique to congenitally acquired toxoplasmosis in humans and has not been reported in other animals.

By far the most common sequel of congenital toxoplasmosis is ocular disease (Guerina et al. 1994; McAuley et al. 1994; Bosch-Driessen et al. 2002). Except for the occasional involvement of an entire eye, the disease is confined to the posterior chamber (Holland et al. 1996; Lotje et al. 2002) Parasites proliferate in the retina leading to inflammation in the choroid, so the disease is correctly designated as retinochoroiditis. In humans, the characteristic lesions of ocular toxoplasmosis in the acute or subacute stage of inflammation appear as yellowish white, cotton-like patches in the fundus (O'Connor 1975; Dutton 1989). The lesions may be single or multiple and may involve one or both eyes.

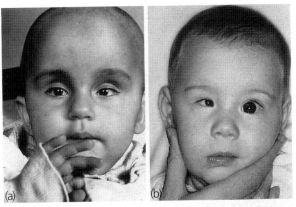

Figure 21.9 *Congenital toxoplasmosis in children.* **(a)***Hydrocephalus with bulging forehead;* **(b)** *microophthalmia of the left eye. (Courtesy of Dr J. Couvreur)*

During the acute stage, inflammatory exudate may cloud the vitreous fluid and may be so dense as to preclude visualization of the fundus by the examiner using an ophthalmoscope. As the inflammation subsides, the vitreous fluid clears and the diseased retina and choroid can be seen through the ophthalmoscope. Retinal lesions may be single or multifocal small gray areas of active retinitis with minimal edema and reaction in the vitreous humor. The punctate lesions are usually harmless unless they are located in a macular area (Figure 21.11). Although severe infections may be detected at birth, milder infections may go undetected until they flare up in adulthood.

The socioeconomic impact of toxoplasmosis in human suffering and the cost of care of sick children, especially those with intellectual deficit and blindness, are enormous (Roberts and Frenkel 1990; Roberts et al. 1994). The testing of all pregnant women for *T. gondii* infection is compulsory in France and Austria, and the cost

Table 21.2 *The relation of clinical toxoplasmosis in children to the time of infection in the mother*

Trimester infected	Children with toxoplasmosis (%)			Total number
	Serious	Mild	Subclinical	
First	40	50	10	10
Second	17.7	45	37	62
Third	2.7	28.7	68.5	108
Undetermined	16.6	20.6	56.6	30

Reproduced from Couvreur et al. 1984 with permission

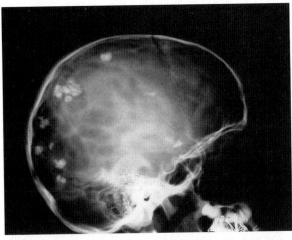

Figure 21.10 *Intracerebral calcification discovered fortuitously in a 10-year-old girl, on a dental panoramic radiograph asked for by a dentist. The girl had unilateral retinochoroiditis and an IQ of 80. (Courtesy of Dr J. Couvreur)*

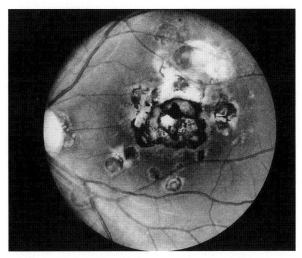

Figure 21.11 *Congenital toxoplasmosis. Retinochoroiditis in the macula of the left eye. (Courtesy of Dr R. Belfort Jnr)*

Table 21.3 *Frequency of symptoms in people with postnatally acquired toxoplasmosis*

Symptom	Patients with symptoms (%)	
	Atlanta outbreak[a] (35 patients)	Panama outbreak[b] (35 patients)
Fever	94	90
Lymphadenopathy	88	77
Headache	88	77
Myalgia	63	68
Stiff neck	57	55
Anorexia	57	NR
Sore throat	46	NR
Artharlgia	26	29
Rash	23	0
Confusion	20	NR
Earache	17	NR
Nausea	17	36
Eye pain	14	26
Abdominal pain	11	55

NR, not reported.
a) From Teutsch et al. (1979).
b) From Benenson et al. (1982).

benefits of such mass screening are being debated in many countries (Remington et al. 2001). Toxoplasmosis is considered to be a leading cause of food-related illness in humans in the US (Mead et al. 1999).

Postnatally acquired infection

Approximately one-third of human populations has been found to have *T. gondii* antibodies. However, serologic prevalence varies a great deal in a geographical region within a country, and within different ethnic groups. Therefore, no generalizations can be made. The type of the serologic test used to measure antibodies and the cut-off values used for serologic surveys also affect the seroprevalence. Dubey and Beattie (1988) and Tenter et al. (2000) summarized the seroprevalence in different countries. In general, seroprevalence is very high in South America and France as compared with North America. Estimates of serologic prevalences are important for control programs. For example, it is important to know if only a minor proportion of a population in a given country is seropositive and thus at a high risk of acquiring toxoplasmosis, particularly during pregnancy. Recent surveys in the US indicate that about 15 percent of women of child-bearing age have antibodies to *T. gondii* and, based on the tests performed under identical conditions, seroprevalence has remained unchanged in the past decade (Jones et al. 2001a, 2003).

The risk of acquiring *T. gondii* infection during pregnancy has been estimated to be less than 1 to 7 per 1 000 pregnancies based on mass screening (Tenter et al. 2000). The risk of intrauterine infection increases with pregnancy and the placenta is thought to be more permeable to *T. gondii* transmission in the later half of pregnancy than in the first half. The chance of fetal infection is about 50 percent of infections acquired during pregnancy.

Postnatally acquired infection may be localized or generalized. Oocyst-transmitted infections may be more severe than tissue cyst-induced infections. Lymphadenitis is the most frequently observed clinical form of toxoplasmosis in humans (Table 21.3). Although any nodes may be involved, the most frequently involved are the deep cervical nodes which, when infected, are tender, discrete but not painful; the infections resolve spontaneously in weeks or months. Lymphadenopathy may be associated with fever, malaise, fatigue, muscle pain, sore throat, and headache (McCabe et al. 1987). Although the condition may be benign, its diagnosis is vital in pregnant women because of the risk to the fetus (Liesenfeld et al. 2001). In tissues from infected nodes examined histologically, reticular cell hyperplasia is usually present whereas necrosis and fibrosis are absent. The node architecture is preserved and usually only a few parasites are present. Diagnosis based on symptoms and histological examination can be confirmed by injection of a homogenate of the lymph node into mice, by immunohistochemical staining of the lymph tissue tagged with *T. gondii* antiserum, or by use of the polymerase chain reaction (PCR) to detect *T. gondii* DNA in tissues.

Until recently, most of toxoplasmic retinochoroiditis was thought to be congenital. Ophthalmologists from Brazil first reported retinochoroiditis in multiple siblings (Silveria et al. 1988). These findings have now been amply confirmed (Couvreur and Thulliez 1996; Nussenblatt and Belfort 1994; Montoya and Remington 1996). The largest outbreak of human toxoplasmosis was epidemiologically linked to drinking water from a municipal water reservoir in Vancouver, British Columbia, Canada; 20 of 100 patients had acquired retinochoroiditis (Bowie et al. 1997; Burnett et al. 1998). Lymphordenitis was found in 51 of the same 100 patients (Bowie et al. 1997).

Encephalitis is the most important manifestation of toxoplasmosis in immunosuppressed patients as it causes the most severe damage to the patient. Infection may occur in any organ. Patients may have headache, disorientation, drowsiness, hemiparesis, reflex changes, and convulsions, and many become comatose. Diagnosis is aided by serological examination. In immunosuppressed patients, both inflammatory signs and antibody production may be suppressed, thus making the diagnosis very difficult. Encephalitis caused by *T. gondii* is now frequently recognized in patients treated with immunosuppressive agents.

Toxoplasmosis ranks high in the list of diseases which lead to the death of patients with acquired immune deficiency syndrome (AIDS); approximately 10 percent of AIDS patients in the USA and up to 30 percent in Europe are estimated to die from toxoplasmosis (Luft and Remington 1992; Luft et al. 1993; Rabaud et al. 1994) Although any organ may be involved including the testis, dermis, and spinal cord, infection of the brain is most frequently reported in AIDS patients. The reason for the different propensity to develop reactivation of latent *T. gondii* infection is unknown (Mariuz and Steigbigel 2001). Most AIDS patients develop toxoplasmic encephalitis when the CD4 lymphocyte count is les than 100 per mm^3. General symptoms in decreasing order of frequency are headache, fever, psychomotor disturbances, meningeal signs, confusion, lethargy, and coma. Focal signs or symptoms include hemiparesis or hemiplegia, convulsions, cranial nerve palsy, visual field defect, cerebellar disturbances, aphasia, and ataxia (Renold et al. 1992). By using computerized axial tomograohy (CAT) scans, lesions were localized in decreasing order of frequency in cortico-medullary, white matter, basal ganglions, cortex, and posterior fossa (Renold et al. 1992). Macroscopically, unilateral or bilateral areas of discoloration indicative of necrosis and hemorrhage were noticed. Microscopically, encephalitis involving many areas is the predominant lesion and the difference may vary depending on whether the patient has received anti-toxoplasma therapy. In untreated patients, the lesion involves a central area of necrosis with degenerating organisms, surrounded by an inflammatory zone with edema, perivascular infiltration of inflammatory cells and hemorrhage (Bertoli et al. 1995). *T. gondii* are more numerous in this peripheral zone surrounding healthy and inflamed tissue. The lesion may be small or the size of a tennis ball and the contents may vary from fluid to solid. Microabscesses are more common in treated patients. In active lesions, numerous tachyzoites are found destroying host tissue. In subacute cases and treated patients, glial nodules predominate.

Patients should be treated empirically for toxoplasmic encephalitis based on clinical and neurological findings and presence of *T. gondii* antibodies because a specific diagnosis may not be possible without a biopsy, which is now rarely indicated.

Transplantation of infected organs or transfusion of infected leukocytes can initiate fatal infection in a seronegative recipient receiving immunosuppressive therapy. Transplantation of non-infected organs and leukocyte transfusion can also activate latent infection in a seropositive recipient receiving immunotherapy. Ordinary blood transfusion is virtually free from danger, but transfusion of packed leukocytes and transplantation of bone marrow have caused toxoplasmosis. It seems that the danger of transplanting an organ from a seropositive donor into a seronegative recipient is greater than that of transplanting an organ from a seronegative donor into a seropositive recipient. Recipients of heart and heart lung are more likely to have symptomatic infection than kidney or liver transplant patients (Wreghitt and Joynson 2001).

Malignancies or immunosuppressive treatment of malignancies can reactivate latent toxoplasmosis. Toxoplasmosis has been reported most commonly in patients treated for Hodgkin's disease. Untreated Hodgkin's disease is rarely associated with clinical toxoplasmosis. A variety of malignancies including lymphoma, leukemia, and myeloma can reactivate toxoplasmosis but there are rare reports of toxoplasmosis associated with solid tumors (Wreghitt and Joynson 2001).

IMMUNOLOGY

In hosts which develop disease, the host may die of acute toxoplasmosis but much more often recovers with the acquisition of immunity (Frenkel 1973b). In the recovering individual, inflammation usually develops where there was initial necrosis. By about the third week after infection, and as recovery develops, *T. gondii* tachyzoites begin to disappear from the visceral tissues and the parasites localize as tissue cysts in neural and muscular tissues. The tachyzoites may persist longer in the spinal cord and brain than in visceral tissues because immunity there is less effective than in visceral organs; they can persist in the placenta for months after the initial infection of the mother.

T. gondii can multiply in virtually any cell of the body, but how it is destroyed in immune cells is not completely known. All extracellular forms of the parasite are directly affected by antibody but intracellular forms are not. It is believed that cellular factors including lymphocytes and lymphokines are more important than humoral ones in immune-mediated destruction of *T. gondii* (Frenkel 1973b). Interferon-γ is considered by some to be the main factor in cell-mediated immunity to *T. gondii* (Gazzinelli et al. 1993). Under experimental conditions, infection with avirulent strains protects the host from damage but does not prevent infection with more virulent strains. In most instances, immunity following a natural *T. gondii* infection persists for the life of the host.

Immunity does not eradicate the infection. *T. gondii* tissue cysts persist for several years after acute infection, and their fate is not fully known. Whether bradyzoites can form new tissue cysts directly without transforming into tachyzoites is not known. However, the finding of new tissue cysts adjacent to old ones suggests that it might happen. It has been proposed that tissue cysts may sometimes rupture during the life of the host and the released bradyzoites may be destroyed by the host's immune responses, possibly causing local necrosis accompanied by inflammation. Hypersensitivity plays a major role in such reactions, which afterwards usually subside with no local renewed multiplication of *T. gondii* in the tissue. However, occasionally there may be formation of new tissue cysts.

In immunosuppressed patients, such as those given large doses of immunosuppressive agents in preparation for organ transplants and in those with AIDS, rupture of a tissue cyst may result in transformation of bradyzoites into tachyzoites and renewed multiplication. The immunosuppressed host may die from toxoplasmosis unless treated. It is not known how corticosteroids cause relapse but it is unlikely that they directly cause rupture of the tissue cysts.

Pathogenicity of *T. gondii* is determined by the virulence of the strain and the susceptibility of the host species. Strains of *T. gondii* may vary in their pathogenicity in a given host; for example, certain strains of mice are more susceptible than others and the severity of infection in individual mice within the same strain may also vary. However, mice of any age are susceptible to clinical *T. gondii* infection, whereas adult rats do not become ill although young rats can die of toxoplasmosis. Adult dogs, like adult rats, are resistant, whereas puppies are fully susceptible to clinical toxoplasmosis. Cattle and horses are among the hosts more resistant to clinical toxoplasmosis and certain marsupials and New World monkeys are the most susceptible. Nothing is known concerning genetic-related susceptibility to clinical toxoplasmosis in higher mammals, including humans.

PATHOLOGY

Humans acquire *T. gondii* mostly by ingestion of tissue cysts in infected meat or by ingestion of oocysts in food or water contaminated with cat feces. The bradyzoites from the tissue cysts or sporozoites from the oocyst penetrate the intestinal epithelial cells where they multiply (Figure 21.12) and spread, first to the mesenteric lymph nodes and then to distant ones by invasion of the lymph and blood. Focal areas of necrosis may develop in many organs. The clinical picture is determined by the extent of injury, especially to vital organs such as the eye, heart, and adrenals. Necrosis is caused

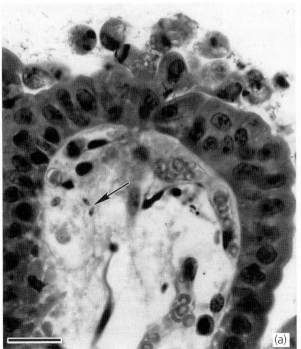 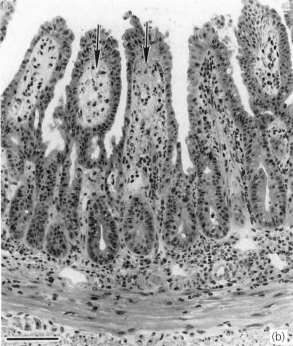

Figure 21.12 *Histological section of small intestine 6 days after feeding* T. gondii *oocysts to a mouse. H&E:* **(a)** *Intestinal villus showing edema and necrosis of the lamina propria cells associated with tachyzoites (arrow), and extrusion of intestinal cells in the lumen. Bar = 20 μm.* **(b)** *Necrosis of lamina propria (arrows). The surface epithelium is not affected. Bar = 100 μm*

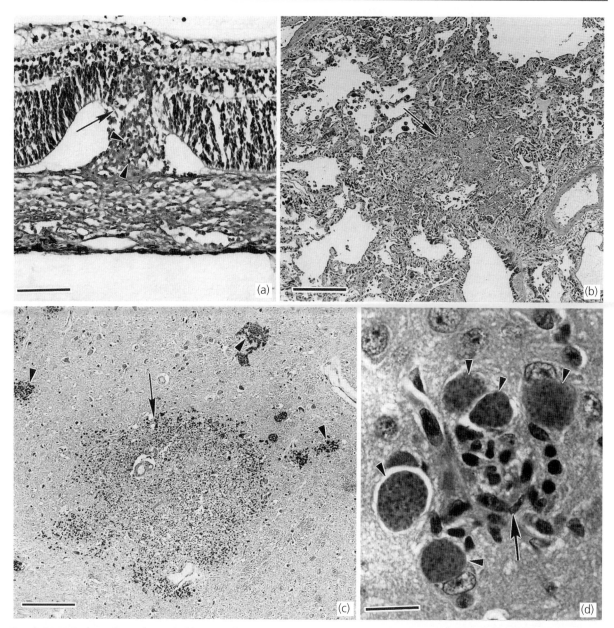

Figure 21.13 *Lesions of toxoplasmosis:* **(a)** *Focus of retinal detachment (arrow) and inflammation associated with tachyzoites (arrowheads). H&E. Bar = 80 μm.* **(b)** *Necrosis in the lung. H&E. Bar = 100 μm.* **(c)** *A large focus of necrosis (arrow) and several satellite small foci (arrowheads). Enormous numbers of T.* gondii *(all black dots) are present in the brain of an AIDS patient. Immunohistochemical stain with anti-T.* gondii *antibody. Bar = 100 μm.* **(d)** *Five tissue cysts (arrowheads) around a glial nodule (arrow) in the brain. H&E. Bar = 20 μm*

by the intracellular growth of tachyzoites (Figure 21.13). *T. gondii* does not produce a toxin.

Necrotic foci are the prominent lesions of toxoplasmosis, varying from microscopic (e.g. in the eye) to macroscopic. Macroscopic areas of necrosis cause encephalitis in AIDS patients. Inflammation usually follows the necrosis and is characterized by infiltration of mononuclear cells. By the time the tissue cysts are forming, the inflammation has begun to subside. However, small inflammatory foci (e.g. glial nodules) may persist for months or even years after the primary infection (Figure 21.13). The hydrocephalus of infected infants is

due to a ventriculitis with a blockage of the aqueduct of Sylvius.

DIAGNOSIS

Diagnosis is made by biological, serological, or histological methods or by some combination of these. Clinical signs are nonspecific and insufficiently characteristic for a definite diagnosis because toxoplasmosis mimics several other infectious diseases.

Numerous serological procedures are available for the detection of humoral antibodies, including the

Sabin–Feldman dye test, indirect hemagglutination assay,indirect fluorescent antibody assay (IFA), direct agglutination test, latex agglutination test, enzyme-linked immunosorbent assay (ELISA), and the immuno-sorbent agglutination assay (ISAGA) (for review see Remington et al. 2001; Roberts et al. 2001; Joynson and Guy 2001; Montoya 2002). The IFA, ISAGA, and ELISA have been modified to detect immunoglobulin (Ig)M antibodies which appear sooner after infection than IgG and disappear faster than IgG after recovery.

The finding of antibodies to *T. gondii* in a single serum sample merely establishes that the host has been infected at some time in the past, so it is best to collect two samples from the same individual, the second 2–4 weeks after the first. A 4–16-fold increase in antibody titer in the second sample indicates an acute infection. A high antibody titer sometimes persists for months after infection. A rise in antibody titer may not be associated with clinical symptoms because most infections in humans are asymptomatic and the fact that titers persist after clinical recovery complicates the interpretation of the results of serological tests. *T. gondii* can be isolated from patients by inoculation into laboratory animals or tissue cultures, of secretions, excretions, body fluids, tissues taken by biopsy, and tissues with macroscopic lesions taken postmortem. Using such specimens, it is possible to attempt isolation of *T. gondii*, to search for *T. gondii* microscopically, and to search for toxoplasmal DNA using PCR (Grover et al. 1990). The PCR and other gene amplification techniques can be designed to target specific regions of the parasite genome. The most commonly used gene of *T. gondii* is B1. By gene amplifi-cation, minute quantities of DNA, representing a single organism, can be detected within a few hours. The main drawbacks of these methods are the inability to differ-entiate between dead and live organisms and the possi-bility of false positives originating during sample collec-tion and processing. It is advisable not to rely entirely on this test for making vital decisions involving the fetus.

As just noted, diagnosis can be made by finding *T. gondii* in host tissue removed by biopsy or at necropsy. A rapid diagnosis may be made by microscopic exam-ination of impression smears of lesions. After drying for 10–30 minutes, the smears are fixed in methyl alcohol and stained with a Romanowsky stain, Giemsa being very satisfactory. Well preserved *T. gondii* are crescent-shaped (Figure 21.1a). In sections, the tachyzoites usually appear round to oval (Figure 21.1b). Electron microscopy can aid diagnosis. *T. gondii* tachyzoites are always located in vacuoles; they have few (usually four) rhoptries and often have a honeycomb structure (Figure 21.3a). Tissue cysts are usually spherical and lack septa, and the cyst wall stains with silver stains. The bradyzoites are strongly PAS positive. The immunohis-tochemical staining of parasites with fluorescent or other types of labeled *T. gondii* antiserum can aid in diagnosis (Figure 21.13).

Serologic screening

Prenatal screening for *T. gondii* infection, to reduce the risk of congenital toxoplasmosis, is routine in Austria and France, and is being considered in several other countries.

Serologic diagnosis in the pregnant woman may be considered under two headings: first, testing of indivi-duals, and second, mass screening. One reason for indi-vidual testing is to detect *T. gondii* infection acquired during pregnancy. To do this, the woman's serum is tested early in pregnancy for specific *T. gondii* IgG anti-bodies by the modified agglutination test (MAT), IgG-ELISA, or Dye test (DT), and for a specific IgM by the double sandwich (DS)-IgM ELISA or IgM-ISAGA tests (Table 21.4).

- If the IgG test is positive, no matter how low the titer, and the IgM test negative, the woman has been infected before pregnancy, there is no risk to her fetus and no treatment is needed (but see Table 21.4, footnote a).
- If both the IgG and IgM tests are positive, the tests should be repeated in 3 weeks time. If the titers have not risen, infection may be assumed to have occurred before pregnancy, and there is no risk to the fetus. If the IgG titer has risen, infection has probably taken place about the time of conception and there is a slight risk to the fetus (but see Table 21.4, footnote b).
- If the IgG and the IgM results are both negative, the woman is susceptible and the tests should be repeated every 4 to 6 weeks to see if either or both become positive. If the woman is infected during pregnancy, the fetus is at high risk, and the woman should be given prophylactic treatment. When she delivers, her child should be examined and, if found to be infected, clinically or subclinically, should be treated.

Another reason for individual testing is the develop-ment of lymphadenopathy during pregnancy. In this event, a high IgG titer (greater than 1 000) accompanied by a positive IgM result makes the diagnosis of active *T. gondii* infection probable. If the IgG titer is less than 1 000, a second test should be carried out 2 to 3 weeks later. If infection is proved, the woman should receive prophylactic treatment and her child should also be treated if found to be infected, clinically or subclinically.

EPIDEMIOLOGY AND CONTROL

Prevalence of infection

T. gondii infection in humans is widespread and occurs throughout the world (Dubey and Beattie 1988; Tenter et al. 2000). Infection rates in humans and other animals differ from one geographical area of a country to another. The causes of these variations are not yet

Table 21.4 *Program for mass screening and prophylactic treatment of pregnant women for* T. gondii

Test 1 (before 2 months of pregnancy)	Test 2 (in second trimester)	Test 3 (in third trimester)	Group
IgG +ve (any titer);	No test;	No test;	I Infection before pregnancy;
IgM –ve	no treatment	No treatment	No risk[a]
IgG +ve;	Repeat IgG after 3 weeks;		II Possible infection soon after conception;
IgM +ve	treat if high or rising titer		slight risk[b]
IgG –ve	Treat if IgG +ve;	Treat if IgG +ve;	III No previous infection;
IgM –ve	Don't treat if IgG –ve	Don't treat if IgG –ve	if seroconversion, high risk[c]

Based on Dubey and Beattie 1988

a) Except in the immunodeficient woman (e.g., with Hodgkin's' disease). In that case, treat the mother if a high or rising IgG titer is observed, and examine the child and treat if found clinically to subclinically infected.

b) In the immunocompetent woman, the slight risk is principally due to infection acquired soon after conception. As only 10 percent of congenital toxoplasmosis arise from maternal infections at this time and as most of these result in abortions or stillbirths, little would be lost by omitting repeated tests.

c) At time of delivery, retest mothers. Examine babies of all mothers who have seroconverted in this pregnancy. Treat the babies if found clinically or subclinically infected.

known. Environmental conditions, cultural habits of the people, and animal fauna are some of the factors that may determine the level of infection, which is more prevalent in hot and humid areas than in dry and cold climates. Only a small proportion (less than 1 percent) of people acquire infection congenitally.

Immunocompetent mothers of congenitally infected children have not been known to give birth to infected children in subsequent pregnancies but repeated congenital infection can occur in mice (Dubey and Beattie 1988). Several litters born to an infected mouse or hamster may be infected even if there is no reinfection from outside sources. A congenitally infected mouse has been shown to be able to produce ten generations of congenitally infected mice. In sheep, congenital infection occurs only when the ewe acquires infection during pregnancy.

Epidemiology

As noted earlier, toxoplasmosis may be acquired by ingestion of oocysts or by ingestion of tissue inhabiting stages of the parasite. The contamination of the environment by oocysts is widespread as oocysts are shed by cats, not only domestic cats but also other members of the Felidae. The domestic cat is probably the major source of contamination as oocyst formation is greatest in this extremely common animal. Widespread natural infection of the environment is possible since a cat can excrete millions of oocysts after ingesting few bradyzoites (Dubey 2001) and these oocysts survive for long periods under most ordinary environmental conditions, for example oocysts in moist soils may survive for months and even years. Oocysts in soil do not always stay there because invertebrates such as flies, cockroaches, dung beetles, and earthworms can spread them mechanically and may even carry them on to food.

Although only a few cats may be shedding *T. gondii* oocysts at any given time (as few as 1 percent) the enor-

mous numbers shed and their resistance to destruction assure widespread contamination. Under experimental conditions, infected cats can shed oocysts after reinoculation with tissue cysts (Dubey 1995) and if this occurs in nature it would greatly facilitate oocyst spread. Congenital infection can also occur in cats, and congenitally infected kittens can excrete oocysts (Dubey and Carpenter 1993), providing another source of contamination. Infection rates in cats are determined by the rate of infection in local avian and rodent populations because cats are thought to become infected by eating these animals. The more oocysts in the environment, the more likely prey animals are to become infected and this, in turn, would increase the infection rate in cats.

Infection in humans is probably most often the result of ingestion of tissue cysts contained in raw or undercooked meat as *T. gondii* is common in many animals used for food, including sheep, pigs, and rabbits. Viable *T. gondii* has not been isolated from beef and the role of cattle in transmission of infection to humans is at best uncertain. Tissue cysts can survive in food animals for years (Dubey and Beattie 1988).

Cultural habits may also affect the acquisition of *T. gondii* infection; for example, in France the prevalence of antibodies to *T. gondii* in humans is very high. In Paris, 84 percent of pregnant women have antibodies to *T. gondii*, comparable figures elsewhere are 32 percent in New York City and 22 percent in London (Dubey and Beattie 1988). The high incidence of *T. gondii* infection in humans in France appears to be related in part to the French habit of eating some of their meat inadequately cooked. In contrast, the high prevalence of the infection in Central and South America is in probably due to high levels of contamination of the environment by oocysts (Teutsch et al. 1979; Bahia-Oliveira et al. 2003). It should be noted, however, that the relative frequency of acquisition of toxoplasmosis from eating raw meat and that due to ingestion of food contaminated by oocysts from cat feces in the general population is unknown and statements on the subject are at best controversial. At present there are no tests to distinguish between oocyst-acquired infection and meat-acquired infection by *T. gondii*.

In addition to infection as a result of ingestion of oocysts or by eating infected raw meat, transmission of toxoplasmosis can be by semen transfusion, by ingestion of milk or saliva, and by eating eggs. The stages most likely to be involved in these transmissions are tachyzoites, which are not environmentally resistant and are killed by water. However, the probability of transmission of *T. gondii* by these means is rare.

There is little, if any, danger of *T. gondii* infection by drinking cows' milk which, in any case, is generally pasteurized or even boiled, but infection has followed drinking unboiled goats' milk. Raw hens' eggs, although an important source of *Salmonella* infection, are extremely unlikely to harbor *T. gondii*. Transmission by sexual activity including kissing is probably rare and epidemiologically unimportant.

Transmission can also occur through blood transfusions and organ transplants, transplantation being the more important; this is a recent development. In people undergoing transplantation, toxoplasmosis may arise from either:

- implantation of an organ or bone marrow from an infected donor into a non-immune immunocompromised recipient, or
- induction of disease in an immunocompromised latently infected recipient.

The tissue cysts in the transplanted tissue or in the latently infected person are probably the source of the infection. In both cases, cytotoxic and immunosuppressive therapy given to the recipient is the cause of induction of the active infection and disease (Wreghitt et al. 1989; Slavin et al. 1994).

CHEMOTHERAPY

Sulfadiazine and pyrimethamine are widely used for therapy of toxoplasmosis (Guerina et al. 1994; St Georgiev 1994). These drugs act synergistically by blocking the metabolic pathway involving *p*-aminobenzoic acid and the folic–folinic acid cycle respectively. The drugs are usually well tolerated; sometimes thrombocytopenia or leukopenia may develop, but these effects can be overcome by administering folinic acid and yeast without interfering with treatment because the vertebrate host can transport presynthesized folinic acid into its cells whereas *Toxoplasma* cannot. Although these drugs have a beneficial action when given in the acute stage of the disease (when there is active multiplication of the parasite), they will not usually eradicate infection. These drugs appear to have little effect on subclinical

infections, but the growth of tissue cysts in mice has been restrained with sulfonamides. Sulfa compounds are excreted within a few hours of administration, so treatment has to be administered in daily divided doses (4 doses of 500 mg each) usually for several weeks or months (WHO 1995). A loading dose (75 mg) of pyrimethamine during the first 3 days has been recommended because it is absorbed slowly and binds to tissues. From the 4th day, the dose of pyrimethamine is reduced to 25 mg, and 2–10 mg of folinic acid plus 5–10 g of baker's yeast are added. Sulfa drugs can cause skin rash and renal toxicity.

The plasma half life of pyrimethamine is 35–139 hours. It has been administered in varying doses and schedules from daily to every 3 or 4 days. Pyrimethamine can cause nausea, headache, dysgeusia, thrombocytopenia, and anemia. Diagnosis and duration of treatment may have to be varied depending on the patient's age and condition (Table 21.5). For example, prenatally infected children, whether with or without clinical manifestations, should be treated for at least 1 year.

Spiramycin, clindamycin, atovaquone, azithromycin, roxithromycin, clarithromycin, dapsone, and several other less commonly used drugs are available for treatment of toxoplasmosis and these were reviewed by McCabe (2001). Spiramycin is relatively nontoxic to mother and the fetus. It is concentrated in placenta and binds to tissues. Thus, it is used to prevent transmission of *T. gondii* from mother to the fetus.

Clindamycin is absorbed quickly and diffuses well into the central nervous system and, therefore, has been used as alternative to sulfadiazine. It is rarely used to treat the primary maternal infection in pregnancy or congenital infection because it enters fetal blood when given to pregnant women. A major side-effect of clindamycin is ulcerative colitis. McCabe (2001) has discussed in detail treatment of toxoplasmosis in patients with various clinical manifestations.

Prophylactic treatment

IMMUNOSUPPRESSED PATIENTS

Before being given immunosuppressive treatments, patients should be serologically tested and, if devoid of *T. gondii* antibodies, be treated with pyrimethamine and sulfadiazine. This is particularly desirable in the case of patients receiving organ transplants. As it is usually impossible to carry out serologic tests on donors, they must all be considered potentially dangerous. There will, however, be time to test the recipients, and if seronegative, they should certainly be given prophylactic treatment. This strategy appears to have been effective in reducing the incidence of latent *T. gondii* infection in patients given marrow transplants from seronegative donors. Perhaps prophylactic treatment should be given to all recipients, irrespective of their sero-status.

Table 21.5 *Treatment schedule for toxoplasmosis*

Condition	Treatment	Dose
General toxoplasmosis	Pyrimethamine + sulfadiazine, 21-day course	Pyrimethamine: 0.5–2 mg per kg per day Sulfadiazine: 50–100 mg per kg per day in two divided doses
	Folic acid (leukovorin calcium)	2–20 mg (or 5–10 g bakers' yeast) twice weekly during pyrimethamine treatment
Toxoplasmosis in pregnancy; start prophylaxis as soon as diagnosis is made	Spiramycin or Spiramycin before week 20 of pregnancy; pyrimethamine + sulfadiazine thereafter	100 mg per kg per day orally in two divided doses; usually 2–4 g daily
Congenital toxoplasmosis	Pyrimethamine + sulfadiazine + folinic acid	Pyrimethamine: 2 mg per kg orally four times per day for 2 days, then 1 mg per kg per day for 6 months, then thrice weekly Sulfadiazine: 100 mg per kg per day orally in two divided doses Folinic acid: 5–10 mg orally thrice weekly
Ocular toxoplasmosis	Pyrimethamine + sulfadiazine	Pyrimethamine: 75 mg per day Sulfadiazine: 2 g per day
	or	
	Clindamycin	300 mg orally, four times daily
	Corticosteroids (only if inflammation present)	Prednisone or methylprednisone: 1–2 mg per kg per day in two divided doses
	Photocoagulation and cryotherapy around active retinochoroiditis lesion to kill *T. gondii* encysted at periphery of lesions	
AIDS patients		
(i) Acute toxoplasmic encephalitis	Pyrimethamine + sulfadiazine	Pyrimethamine: 200 mg oral initially, then 75–100 mg orally four times daily Sulfadiazine: 1–2 g orally, four times daily
	or	
	Pyrimethamine + clindamycin	Pyrimethamine + clindamycin: 60 mg orally or intravenously 6-hourly
	or	
	Pyrimethamine + azithromycin	Pyrimethamine + azithromycin: 1 200–1 500 mg orally four times daily
	or	
	Pyrimethamine + clarithromycin	Pyrimethamine + clarithromycin: 1 g orally two times daily
(ii) Maintenance treatment	Pyrimethamine + sulfadiazine	Dose reduced to half (or less) that given for acute toxoplasmosis and treatment continued for life
	Folic acid	Daily
(iii) Prophylaxis for AIDS patients with *T. gondii* antibodies.	Trimethoprim–sulfamethoxazole	Widely used for the prophylaxis and treatment of *Pneumocystis carinii* infections, thought to reduce onset of toxoplasmic encephalitis

Modified from Dubey and Beattie (1988) and McCabe (2001)

Prophylactic treatment of all AIDS patients with *T. gondii* antibodies is desirable. Fortunately, some drugs used to prophylactically treat *Pneumocystis* pneumonia and bacterial infections may also prevent onset of clinical toxoplasmosis. The combination of trimethoprim (160 mg) and sulfamethoxazole (800 mg), given twice daily and twice per week, is often used because it is inexpensive, convenient and works against *Pneumocystis* (McCabe 2001).

Cutaneous hypersensitivity, however, can be a problem and then alternative therapies are sought. A combination of dapsone (100 mg) and pyrimethamine (25 mg) orally weekly has also been used effectively. Other combinations of pyrimethamine and sulfonamides

(Fansidar, three tablets every 2 weeks) have also been used (Sabauste et al. 1997).

PROPHYLACTIC TREATMENT DURING PREGNANCY

Prevention of infection of the fetus by prophylactic treatment of the mother depends on the delay occuring between maternal infection and its transmission to the fetus. It is also hoped that if infection is already present in the fetus, such treatment may limit its ill effects. Treatment is initiated as soon as possible during the prenatal incubation period. In Austria, it is by spiramycin before week 20 of pregnancy and thereafter by pyrimethamine and sulfonamide; in France treatment is by spiramycin alone. If these measures are begun sufficiently early, they may be expected to reduce the incidence of congenital toxoplasmosis by 50–70 percent.

In places where it has been carried out with thoroughness, persistence, and determination, as in France, education appears to have contributed to a reduction in the incidence of T. gondii infection during pregnancy. Information regarding T. gondii infection should be included with the general instructions given in antenatal clinics and by obstetricians and midwives dealing with individual patients. Personal instruction given by word of mouth is likely to be most effective, and should be supplemented by booklets printed in various languages and by videos in the waiting rooms of antenatal clinics.

VACCINATION

The objectives of vaccines against toxoplasmosis include reducing fetal damage, reducing the number of tissue cysts in animals and preventing the formation of oocysts in cats (Araujo 1994; Dubey 1994; Fishback and Frenkel 2001). None of these objectives can be realized by the use of any currently available single vaccine. At present there are no effective subunit or killed vaccines for immunization against T. gondii but research is under way in many laboratories (Araujo 1994). There is no vaccine to prevent toxoplasmosis in humans, and none is on the horizon.

Prevention of oocyst shedding by cats is the key to controlling the spread of T. gondii and the oral ingestion of live bradyzoites is necessary to induce immunity to oocyst shedding, as the administration of any stage of T. gondii parenterally does not induce an immunity capable of inhibiting oocyst shedding (Frenkel and Smith 1982). A proposed vaccine for use in cats contains live bradyzoites from a mutant strain (T-263) of T. gondii. After oral administration of T-263 bradyzoites, the cycle of development of T. gondii is arrested at the sexual stage because only gamonts of a single sex develop; thus oocysts are not produced (Frenkel et al. 1991). In one trial, 84 percent of cats vaccinated with T-263 bradyzoites failed to shed oocysts following challenge (Frenkel et al. 1991). The duration of immunity in cats induced by the vaccine has not yet been determined. This vaccine is not available commercially.

The objectives of vaccination of farm animals are to reduce the incidence of abortions in sheep and goats resulting from transplacental infection of fetuses, and to reduce the risk of human exposure resulting from ingestion of infected meat (with the subsequent risk of fetal infection). For these purposes, vaccines made of nonpersistent strains of T. gondii are under study. One vaccine that contains a strain (S48) of tachyzoites that does not persist in the tissues of sheep is available in Europe and New Zealand, where it is used to reduce fetal losses attributable to toxoplasmosis (Wilkins et al. 1988; Buxton 1993). Ewes vaccinated with the S48 strain vaccine retain immunity for at least 18 months (Buxton 1993). Another strain of T. gondii (RH) does not persist in tissues of swine but does induce immunity (Dubey et al. 1994). A mutant (ts-4) of the RH strain is being studied as a vaccine for immunizing hosts other than cats. It grows better at 33°C than it does at 37°C (body temperature). The ts-4 stain is nonpathogenic even to suckling pigs (Lindsay et al. 1993). These strains of T. gondii (S48, RH, ts-4) do not induce oocyst shedding in cats.

INTEGRATED CONTROL

To prevent infection of humans by T. gondii, the hands of people handling meat should be washed thoroughly with soap and water before they go to other tasks. All cutting boards, sink tops, knives, and other materials coming in contact with uncooked meat should also be washed with soap and water. Washing is effective because the stages of T. gondii in meat are killed by contact with soap and water.

Parasites in meat can be killed by exposure to extreme cold or heat; tissue cysts are killed by heating the meat throughout to 67°C (Dubey et al. 1990) or by cooling it to −13°C (Kotula et al. 1991). Tissue cysts are also killed by exposure to 0.5 Gy of γ-irradiation (Dubey and Thayer 1994). All meat should be cooked to 67°C before consumption, and tasting meat while cooking or while seasoning should be avoided. Pregnant women, especially, should avoid contact with cats, soil, and raw meat. Pet cats should be fed only dry, canned or cooked food and the cat litter box should be emptied every day, preferably not by a pregnant woman. Gloves should be worn while gardening and vegetables should be washed thoroughly before eating because of possible contamination with cat feces. Expectant mothers should be aware of the dangers of toxoplasmosis (Foulon et al. 2000; Jones et al. 2001b).

Infection in animals other than humans

T. gondii is capable of causing severe disease in animals other than humans (Dubey and Beattie 1988) and is responsible for great losses to the livestock industry. In sheep and goats it may cause embryonic death and resorption, fetal death and mummification, abortion, stillbirth, and neonatal death. Disease is more severe in goats than in sheep. Outbreaks of toxoplasmosis in pigs have been reported from several countries, especially Japan, and mortality is more common in young pigs than in adult pigs. Pneumonia, myocarditis, encephalitis and placental necrosis occur in infected pigs. Cattle and horses are more resistant to clinical toxoplasmosis than are other species of livestock. In cats and dogs, the disease is most severe in young animals. Common clinical manifestations of canine toxoplasmosis are respiratory distress, ataxia, and diarrhea. In most infected dogs, pneumonia is caused by a combination of *T. gondii* and distemper virus as the virus is immunosuppressive. Respiratory distress is a common clinical sign in felines with toxoplasmosis. Sporadic and widespread outbreaks of toxoplasmosis occur in rabbits, mink, birds, and other domesticated and wild animals (Tenter et al. 2000; Dubey 2002b). Toxoplasmosis is severe in many species of Australian marsupials and in New World monkeys (Dubey and Beattie 1988). Free-ranging marine mammals have died of acute toxoplasmosis.

REFERENCES

Ajzenberg, D., Cogné, N., et al. 2002. Genotype of 86 *Toxoplasma gondii* isolates associated with human congenital toxoplasmosis, and correlation with clinical findings. *J Infect Dis*, **186**, 684–9.

Araujo, F.G. 1994. Immunization against *Toxoplasma gondii*. *Parasitol Today*, **10**, 358–60.

Aspinall, T.V., Guy, E.C., et al. 2003. Molecular evidence for multiple *Toxoplasma gondii* infections in individual patients in England and Wales: public health implications. *Int J Parasitol*, **33**, 97–103.

Bahia-Oliveira, L.M.G., Jones, J.L., et al. 2003. Highly endemic, waterborne toxoplasmosis in north Rio de Janeiro State, Brazil. *Emerg Infect Dis*, **9**, 55–62.

Benenson, M.W., Takafuji, E.T., et al. 1982. Oocyst transmitted toxoplasmosis associated with ingestion of contaminated water. *N Engl J Med*, **307**, 666–9.

Bertoli, F., Espino, M., et al. 1995. A spectrum in the pathology of toxoplasmosis in patients with acquired immunodeficiency syndrome. *Arch Pathol Lab Med*, **119**, 214–24.

Beverley, J.K.A. 1959. Congenital transmission of toxoplasmosis through successive generations of mice. *Nature*, **183**, 1348–9.

Bosch-Driessen, L.E.H., Berendschot, T.T.J.M., et al. 2002. Ocular toxoplasmosis. Clinical features and prognosis of 154 patients. *Ophthalmology*, **109**, 869–78.

Bowie, W.R., King, A.S., et al. 1997. Outbreak of toxoplasmosis associated with municipal drinking water. *Lancet*, **350**, 173–7.

Burnett, A.J., Short, S.G., et al. 1998. Multiple cases of acquired toxoplasmosis retinitis presenting in an outbreak. *Ophthalmology*, **105**, 1032–7.

Buxton, D. 1993. Toxoplasmosis: the first commercial vaccine. *Parasitol Today*, **9**, 335–7.

Couvreur, J. and Thulliez, P. 1996. [Acquired toxoplasmosis of ocular or neurologic site: 49 cases]. *Presse Méd*, **25**, 438–42.

Couvreur, J., Desmonts, G., et al. 1984. Etude d'une série homogène de 210 cas de toxoplasmose congénitale chez des nourrissons agés de 0 a 11 mois et dépistés de façon prospective. *Ann Pediatr (Paris)*, **31**, 815–19.

Daffos, F., Forestier, F., et al. 1988. Prenatal management of 746 pregnancies at risk for congenital toxoplasmosis. *N Engl J Med*, **318**, 271–5.

Desmonts, G. and Couvreur, J. 1974. Congenital toxoplasmosis. A propective study of 378 pregnancies. *N Engl J Med*, **290**, 1110–16.

Dubey, J.P. 1977. *Toxoplasma, Hammondia, Besnoitia, Sarcocystis,* and other tissue cyst-forming coccidia of man and animals. In: Kreier, J.P. (ed.), *Parasitic protozoa*, Vol. III, 1st edn. 1st edn. New York: Academic Press, 101–237.

Dubey, J.P. 1993. *Toxoplasma, Neospora, Sarcocystis,* and other tissue cyst-forming coccidia of humans and animals. In: Kreier, J.P. (ed.), *Parasitic protozoa*, Vol. 6, 2nd edn. 2nd edn. New York: Academic Press, 1–158.

Dubey, J.P. 1994. Toxoplasmosis. *J Am Vet Med Assoc*, **205**, 1593–8.

Dubey, J.P. 1995. Duration of immunity to shedding of *Toxoplasma gondii* oocysts by cats. *J Parasitol*, **81**, 410–15.

Dubey, J.P. 1996. Infectivity and pathogenicity of *Toxoplasma gondii* oocysts for cats. *J Parasitol*, **82**, 957–61.

Dubey, J.P. 2001. Oocyst shedding by cats fed isolated bradyzoites and comparision of infectivity of bradyzoites of the VEG strain *Toxoplasma gondii* to cats and mice. *J Parasitol*, **87**, 215–19.

Dubey, J.P. 2002a. Tachyzoite-induced life cycle of *Toxoplasma gondii* in cats. *J Parasitol*, **88**, 4, 713–17.

Dubey, J.P. 2002b. A review of toxoplasmosis in wild birds. *Vet Parasitol*, **106**, 121–53.

Dubey, J.P. and Beattie, C.P. 1988. *Toxoplasmosis of animals and man.* Boca Raton: CRC Press.

Dubey, J.P. and Carpenter, J.L. 1993. Neonatal toxoplasmosis in littermate cats. *J Am Vet Med Assoc*, **203**, 1546–9.

Dubey, J.P. and Frenkel, J.K. 1972. Cyst-induced toxoplasmosis in cats. *J Protozool*, **19**, 155–77.

Dubey, J.P. and Frenkel, J.K. 1976. Feline toxoplasmosis from acutely infected mice and the development of *Toxoplasma* cyst. *J Protozool*, **23**, 537–46.

Dubey, J.P. and Thayer, D.W. 1994. Killing of different strains of *Toxoplasma gondii* tissue cysts by irradiation under defined conditions. *J Parasitol*, **80**, 764–7.

Dubey, J.P., Lindsay, D.S. and Speer, C.A. 1998. Structure of *Toxoplasma gondii* tachyzoites, bradyzoites and sporozoites, and biology and development of tissue cysts. *Clin Microbiol Rev*, **11**, 267–99.

Dubey, J.P., Miller, N.L. and Frenkel, J.K. 1970a. Characterization of the new fecal form of *Toxoplasma gondii*. *J Parasitol*, **56**, 447–56.

Dubey, J.P., Miller, N.L. and Frenkel, J.K. 1970b. The *Toxoplasma gondii* oocyst from cat feces. *J Exp Med*, **132**, 636–62.

Dubey, J.P., Kotula, A.W., et al. 1990. Effect of high temperature on infectivity of *Toxoplasma gondii* tissue cysts in pork. *J Parasitol*, **76**, 201–4.

Dubey, J.P., Baker, D.G., et al. 1994. Persistence of immunity to toxoplasmosis in pigs vaccinated with a nonpersistent strain of *Toxoplasma gondii*. *Am J Vet Res*, **55**, 982–7.

Dubey, J.P., Gamble, H.R., et al. 2002a. High prevalence of viable *Toxoplasma gondii* infection in market weight pigs from a farm in Massachusetts. *J Parasitol*, **88**, 1234–8.

Dubey, J.P., Graham, D.H., et al. 2002b. Biological and genetic characterisation of *Toxoplasma gondii* isolates from chickens (*Gallus domesticus*) from São Paulo, Brazil: Unexpected findings. *Int J Parasitol*, **32**, 99–105.

Dutton, G.N. 1989. Toxoplasmic retinochoroiditis – A historical review and current concepts. *Ann Acad Med*, **18**, 214–21.

Ferguson, D.J.P. and Hutchison, W.M. 1987. An ultrastructural study of the early development and tissue cyst formation of *Toxoplasma gondii* in the brains of mice. *Z Parasitenkd*, **73**, 483–91.

Fishback, J.L. and Frenkel, J.K. 2001. Toxoplasma vaccines. In: Joynson, D.H.M. and Wreghitt, T.G. (eds), *Toxoplasmosis. A comprehensive clinical guide.* Cambridge: Cambridge University Press, 360–76.

Foulon, W., Naessens, A. and Ho-Yen, D. 2000. Prevention of congenital toxoplasmosis. *J Perinat Med*, **28**, 337–45.

Frenkel, J.K. 1973a. *Toxoplasma* in and around us. *BioScience*, **23**, 343–52.

Frenkel, J.K. 1973b. Toxoplasmosis: parasite life cycle, pathology and immunology. In: Hammond, D.M. and Long, P.L. (eds), *The Coccidia. Eimeria, Isospora, Toxoplasma and related genera*. Baltimore, MD: University Park Press, 343–410.

Frenkel, J.K. and Smith, D.D. 1982. Immunization of cats against shedding of *Toxoplasma* oocysts. *J Parasitol*, **68**, 744–8.

Frenkel, J.K., Dubey, J.P. and Miller, N.L. 1970. *Toxoplasma gondii* in cats: fecal stages identified as coccidian oocysts. *Science*, **167**, 893–6.

Frenkel, J.K., Pfefferkorn, E.R., et al. 1991. Prospective vaccine prepared from a new mutant of *Toxoplasma gondii* for use in cats. *Am J Vet Res*, **52**, 759–63.

Freyre, A., Dubey, J.P., et al. 1989. Oocyst-induced *Toxoplasma gondii* infections in cats. *J Parasitol*, **75**, 750–5.

Fuentes, I., Rubio, J.M., et al. 2001. Genotypic characterization of *Toxoplasma gondii* strains associated with human toxoplasmosis in Spain: direct analysis from clinical samples. *J Clin Microbiol*, **39**, 1566–70.

Gazzinelli, R.T., Denkers, E.Y. and Sher, A. 1993. Host resistance to *Toxoplasma gondii*: model for studying the selective induction of cell-mediated immunity by intracellular parasites. *Infect Agent Dis*, **2**, 139–49.

Grigg, M.E., Bonnefoy, S., et al. 2001a. Success and virulence in *Toxoplasma* as the result of sexual recombination between two distinct ancestries. *Science*, **294**, 161–5.

Grigg, M.E., Ganatra, J., et al. 2001b. Unusual abundance of atypical strains associated with human ocular toxoplasmosis. *J Infect Dis*, **184**, 633–9.

Grover, C.M., Thulliez, P., et al. 1990. Rapid prenatal-diagnosis of congenital *Toxoplasma* infection by using polymerase chain-reaction and amniotic-fluid. *J Clin Microbiol*, **28**, 2297–301.

Guerina, N.G., Hsu, H.W., et al. 1994. Neonatal serologic screening and early treatment for congenital *Toxoplasma gondii* infection. *N Engl J Med*, **330**, 1858–63.

Hartley, W.J. and Marshall, S.C. 1957. Toxoplasmosis as a cause of ovine perinatal mortality. *NZ Vet J*, **5**, 119–24.

Holland, G.N., O'Connor, G.R., et al. 1996. Toxoplasmosis. In: Pepose, J.S., Holland, G.S. and Wilhelmus, K.R. (eds), *Ocular infection and immunity*, 1st edn. St. Louis: Mosby, 1183–223.

Howe, D.K. and Sibley, L.D. 1995. *Toxoplasma gondii* comprises three clonal lineages: Correlation of parasite genotype with human disease. *J Infect Dis*, **172**, 1561–6.

Howe, D.K., Honoré, S., et al. 1997. Determination of genotypes of *Toxoplasma gondii* strains isolated from patients with toxoplasmosis. *J Clin Microbiol*, **35**, 1411–14.

Hutchison, W.M. 1965. Experimental transmission of *Toxoplasma gondii*. *Nature*, **206**, 961–2.

Hutchison, W.M., Dunachie, J.F., et al. 1969. Life cycle of *Toxoplasma gondii*. *Br Med J*, **4**, 806.

Hutchison, W.M., Dunachie, J.F., et al. 1970. Coccidian-like nature of *Toxoplasma gondii*. *Br Med J*, **1**, 142–4.

Hutchison, W.M., Dunachie, J.F., et al. 1971. The life cycle of the coccidian parasite, *Toxoplasma gondii*, in the domestic cat. *Trans Roy Soc Trop Med Hyg*, **65**, 380–99.

Jacobs, L., Remington, J.S. and Melton, M.L. 1960. The resistance of the encysted form of *Toxoplasma gondii*. *J Parasitol*, **46**, 11–21.

Janku, J. 1923. Pathogenesis and pathologic anatomy of coloboma of the macula lutea in an eye of normal dimensions and in a microphthalmic eye with parasites in the retina. *Cas Lek cesk*, **62**, 1021, 1052, 1081, 111 and 1138.

Joiner, K.A. and Dubremetz, J.F. 1993. *Toxoplasma gondii*: a protozoan for the nineties. *Infect Immun*, **61**, 1169–72.

Jones, J.L., Kruszon-Moran, D. and Wilson, M. 2003. *Toxoplasma gondii* infection in the United States. 1999–2000. *Emerg Infect Dis*, **9**, 1371–4.

Jones, J.L., Kruszon-Moran, D., et al. 2001a. *Toxoplasma gondii* infection in the United States: seroprevalence and risk factors. *Am J Epidemiol*, **154**, 357–65.

Jones, J.L., Lopez, A., et al. 2001b. Congenital toxoplasosis: A review. *Obstetr Gynecol Surv*, **56**, 296–305.

Joynson, D.H.M. and Guy, E.C. 2001. Laboratory diagnosis of *Toxoplasma* infection. In: Joynson, D.H.M. and Wreghitt, T.G. (eds), *Toxoplasmosis. A comprehensive clinical guide*. Cambridge: Cambridge University Press, 296–318.

Kotula, A.W., Dubey, J.P., et al. 1991. Effect of freezing on infectivity of *Toxoplasma gondii* tissue cysts in pork. *J Food Protection*, **54**, 687–90.

Lehmann, T., Blackston, C.R., et al. 2000. Strain typing of *Toxoplasma gondii*: comparison of antigen-coding and housekeeping genes. *J Parasitol*, **86**, 960–71.

Liesenfeld, O., Montoya, J.G., et al. 2001. Confirmatory serologic testing for acute toxoplasmosis and rate of induced abortion among women reported to have positive *Toxoplasma* immunoglobulin M antibody titers. *Am J Obstet Gynecol*, **184**, 140–5.

Lindsay, D.S., Blagburn, B.L. and Dubey, J.P. 1993. Safety and results of challenge of weaned pigs given a temperature-sensitive mutant of *Toxoplasma gondii*. *J Parasitol*, **79**, 71–6.

Lotje, E., Bosch-Driessen, M.D., et al. 2002. Ocular toxoplasmosis: Clinical features and prognosis of 154 patients. *Opthalmology*, **109**, 869–78.

Luft, B.J. and Remington, J.S. 1992. Toxoplasmic encephalitis in AIDS. *Clin Infect Dis*, **15**, 211–22.

Luft, B.J., Conley, F.K., et al. 1983. Outbreak of central nervous system toxoplasmosis in Western Europe and North America. *Lancet*, **1**, 781–4.

Luft, B.J., Hafner, R., et al. 1993. Toxoplasmic encephalitis in patients with the acquired immunodeficiency syndrome. *N Engl J Med*, **329**, 995–1000.

Mariuz, P. and Steigbigel, R.T. 2001. *Toxoplasma* infection in HIV-infected patients. In: Joynson, D.H.M. and Wreghitt, T.G. (eds), *Toxoplasmosis. A comprehensive clinical guide*. Cambridge: Cambridge University Press, 147–77.

McAuley, J., Boyer, K.M., et al. 1994. Early and longitudinal evaluations of treated infants and children and untreated historical patients with congenital toxoplasmosis – the Chicago collaborative treatment trial. *Clin Infect Dis*, **18**, 38–72.

McCabe, R.E. 2001. Antitoxoplasma chemotherapy. In: Joynson, D.H.M. and Wreghitt, T.G. (eds), *Toxoplasmosis. A comprehensive clinical guide*. Cambridge: Cambridge University Press, 319–59.

McCabe, R.E., Brooks, R.G., et al. 1987. Clinical spectrum in 107 cases of toxoplasmic lymphadenopathy. *Rev Infect Dis*, **9**, 754–74.

Mead, P.S., Slutsker, L., et al. 1999. Food-related illness and death in the United States. *Emerg Infect Dis*, **5**, 607–24.

Mello, U. 1910. Un cas de toxoplasmose du chien observé à Turin (2). *Bull Soc Pathol Exot Filial*, **3**, 359–63.

Miller, N.L., Frenkel, J.K. and Dubey, J.P. 1972. Oral infections with *Toxoplasma* cysts and oocysts in felines, other mammals, and in birds. *J Parasitol*, **58**, 928–37.

Mondragon, R., Howe, D.K., et al. 1998. Genotypic analysis of *Toxoplasma gondii* isolates from pigs. *J Parasitol*, **84**, 639–41.

Montoya, J.G. 2002. Laboratory diagnosis of *Toxoplasma gondii* infection and toxoplasmosis. *J Infect Dis*, **185**, Suppl. S73–82.

Montoya, J.G. and Remington, J.S. 1996. Toxoplasmic chorioretinitis in the setting of acute acquired toxoplasmosis. *Clin Infect Dis*, **23**, 277–82.

Munday, B.L. 1972. Serological evidence of *Toxoplasma* infection in isolated groups of sheep. *Res Vet Sci*, **13**, 100–2.

Nicolle, C. and Manceaux, L. 1908. Sur une infection à corps de Leishman (ou organismes voisins) du gondi. *CR Acad Sci Paris*, **147**, 763–6.

Nicolle, C. and Manceaux, L. 1909. Sur un protozoaire nouveau du gondi. *CR Acad Sci Paris*, **148**, 369–72.

Nussenblatt, R.B. and Belfort, R. 1994. Ocular toxoplasmosis – an old disease revisited. *J Am Med Assoc*, **271**, 304–7.

O'Connor, G.R. 1975. Ocular toxoplasmosis. *Jpn J Opthalmol*, **19**, 1–24.

Overdulve, J.P. 1970. The probable identity of *Toxoplasma* and *Isospora* and the role of the cat in the transmission of toxoplasmosis. *Tijdschr Diergeneesk*, **95**, 149–55.

Petersen, E and Dubey, J.P. 2001. Biology of toxoplasmosis. In: Joynson, D.H.M. and Wreghitt, T.G. (eds), *Toxoplasmosis. A comprehensive clinical guide*. Cambridge: Cambridge University Press, 1–42.

Pfefferkorn, E.R. 1990. *Cell biology of Toxoplasma gondii*. New York: Freedman.

Pinkerton, H. and Weinman, D. 1940. *Toxoplasma* infection in man. *Arch Pathol*, **30**, 374–92.

Rabaud, C., May, T., et al. 1994. Extracerebral toxoplasmosis in patients infected with HIV. A French National Survey. *Medicine*, **73**, 306–14.

Remington, J.S., McLeod, R., et al. 2001. Toxoplasmosis. In: Remington, J.S. and Klein, J.O. (eds), *Infectious diseases of the fetus and newborn infant*, 5th edn. Philadelphia: W.B. Saunders, 205–346.

Renold, C., Sugar, A., et al. 1992. *Toxoplasma* encephalitis in patients with the acquired-immunodeficiency-syndrome. *Medicine*, **71**, 4, 224–39.

Roberts, A., Hedman, K., et al. 2001. Multicenter evaluation of strategies for serodiagnosis of primary infection with *Toxoplasma gondii*. *Eur J Clin Microbiol Infect Dis*, **20**, 467–74.

Roberts, T. and Frenkel, J.K. 1990. Estimating income losses and other preventable costs caused by congenital toxoplasmosis in people in the United States. *J Am Vet Med Assoc*, **196**, 249–56.

Roberts, T., Murrell, K.D. and Marks, S. 1994. Economic losses caused by foodborne parasitic diseases. *Parasitol Today*, **10**, 419–23.

Sabin, A.B. 1942. Toxoplasmosis. A recently recognized disease of human beings. *Adv Pediatr*, **1**, 1–53.

Sabin, A.B. and Feldman, H.A. 1948. Dyes as microchemical indicators of a new immunity phenomenon affecting a protozoon parasite (*Toxoplasma*). *Science*, **108**, 660–3.

Sheffield, H.G. and Melton, M.L. 1970. *Toxoplasma gondii*: The oocyst, sporozoite, and infection of cultured cells. *Science*, **167**, 892–3.

Sibley, L.D. 1995. Invasion of vertebrate cells by *Toxoplasma gondii*. *Trends Cell Biol*, **5**, 129–32.

Sibley, L.D., LeBlanc, A.J., et al. 1992. Generation of a restriction fragment length polymorphism linkage map for *Toxoplasma gondii*. *Genetics*, **132**, 1003–15.

Siim, J.C. 1952. Studies on acquired toxoplasmosis. II. Report of a case with pathological changes in a lymph node removed at biopsy. *Acta Pathol Microbiol Scand B*, **30**, 104–8.

Silveira, C., Helfort, R., et al. 1988. Acquired toxoplasmic infection as the cause of toxoplasmic retinochoroiditis in families. *Am J Ophthalmol*, **106**, 362–3.

Slavin, M.A., Meyers, J.D., et al. 1994. *Toxoplasma gondii* infection in marrow transplant recipients: A 20-year experience. *Bone Marrow Transplant*, **13**, 549–57.

Soldati, D. and Boothroyd, J.C. 1993. Transient transfection and expression in the obligate intracellular parasite *Toxoplasma gondii*. *Science*, **260**, 349–52.

Splendore, A. 1908. Un nuovo protozoa parassita de' conigli, incontrato nelle lesioni anatomiche d'une malattia che ricorda in molti punti il Kala-azar dell'uomo. Nota preliminare pel. *Rev Soc Scient Sao Paulo*, **3**, 109–12.

St Georgiev, V. 1994. Management of toxoplasmosis. *Drugs*, **48**, 179–88.

Su, C., Howe, D.K., et al. 2002. Identification of quantitative trait loci controlling acute virulence in *Toxoplasma gondii*. *Proc Natl Acad Sci*, **99**, 10753–8.

Su, C., Evans, D., et al. 2003. Recent expansion of *Toxoplasma* through enhanced oral transmission. *Science*, **299**, 414–16.

Subauste, C.S., Wong, S.Y., Remington, J.S., et al. 1997. AIDS-associated toxoplasmosis. In: Sande, M.A. and Volberding, P.A. (eds), *The medical management of AIDS*. Philadelphia: W.B. Saunders, 343–62.

Tenter, A.M., Heckeroth, A.R. and Weiss, L.M. 2000. *Toxoplasma gondii*: from animals to humans. *Int J Parasitol*, **30**, 1217–58.

Teutsch, S.M., Juranek, D.D., et al. 1979. Epidemic toxoplasmosis associated with infected cats. *N Engl J Med*, **300**, 695–9.

Wallace, G.D. 1969. Serologic and epidemiologic observations on toxoplasmosis on three Pacific atolls. *Am J Epidemiol*, **90**, 103–11.

Weinman, D. and Chandler, A.H. 1954. Toxoplasmosis in swine and rodents. Reciprocal oral infection and potential human hazard. *Proc Soc Exp Biol Med*, **87**, 211–16.

WHO, 1995. *Drugs used in parasitic diseases*. 2nd edn. Geneva: World Health Organization.

Wilkins, M.F., O'Connell, E. and Te Punga, W.A. 1988. Toxoplasmosis in sheep. III. Further evaluation of the ability of a live *Toxoplasmosis gondii* vaccine to prevent lamb losses and reduce congenital infection following experimental oral challenge. *NZ Vet J*, **36**, 86–9.

Wolf, A. and Cowen, D. 1937. Granulomatous encephalomyelitis due to an encephalitozoon (encephalitozic encephalomyelitis): a new protozoan disease of man. *Bull Neurol Inst NY*, **6**, 306–35.

Wolf, A., Cowen, D. and Paige, B. 1939. Human toxoplasmosis: occurrence in infants as an encephalomyelitis verification by transmission to animals. *Science*, **89**, 226–7.

Wreghitt, T.G. and Joynson, D.H.M. 2001. *Toxoplasma* infection in immunosuppressed (HIV-negative) patients. In: Joynson, D.H.M. and Wreghitt, T.G. (eds), *Toxoplasmosis. A comprehensive clinical guide*. Cambridge: Cambridge University Press, 178–92.

Wreghitt, T.G., Hakim, M., et al. 1989. Toxoplasmosis in heart and lung transplant recipients. *J Clin Pathol*, **42**, 194–9.

22

Sarcocystis

J.P. DUBEY

INTRODUCTION AND CLASSIFICATION

The *Sarcocystis* parasite was first described in the skeletal muscle of a house mouse (*Mus musculus*) in Switzerland in 1843 (Table 22.1). Before 1972, many of these parasites were given names based on the finding of their cysts in the muscles of a variety of hosts. The true nature of these intramuscular cysts remained unknown until the discovery of the life cycle of *Sarcocystis* in 1972 (Table 22.1).

Sarcocystis species are coccidian parasites and are classified as follows.

Phylum Apicomplexa Levine, 1979
Class Coccidea Leuckart, 1879
Order Eimeriida Léger 1911
Family Sarcocystidae Poche, 1913
Genus *Sarcocystis* Lankester, 1882

STRUCTURE AND LIFE CYCLE

Sarcocysts (from Greek 'sarkos' meaning 'flesh' and 'kystis' meaning 'bladder'), the terminal asexual stage of development of these parasites, are found primarily in the striated muscles of mammals, birds, marsupials, and poikilothermic animals (Figure 22.1).

Sarcocystis has an obligatory prey–predator (two-host) life cycle (Figure 22.2). Asexual stages develop only in the intermediate host, which in nature is often a prey animal, and sexual stages develop only in the definitive host, which is carnivorous. There are different intermediate and definitive hosts for each species of *Sarcocystis*; for example, there are three named species in

cattle: *S. cruzi*, *S. hirsuta*, and *S. hominis* (Dubey et al. 1989), the definitive hosts for these species being Canidae, Felidae, and primates, respectively. Species of *Sarcocystis* are generally more specific for their intermediate hosts than for their definitive hosts; for *S. cruzi* for example, ox and bison are the only intermediate hosts whereas dogs, wolves, coyotes, raccoons, jackals, and foxes can act as definitive hosts. In the following description of life cycle and structure, *S. cruzi* serves as the example because its complete life cycle is known.

The intermediate host becomes infected by ingesting sporocysts in food or water. Sporozoites excyst from sporocysts in the small intestine and first generation schizonts are formed in endothelial cells of arteries 7–15 days after inoculation. Second generation schizonts occur 19–46 days after inoculation, predominantly in capillaries throughout the body (Figure 22.3a). Merozoites are found in mononuclear blood cells 24–46 days after inoculation.

The schizonts divide by endopolygeny (from 'endon' meaning 'inside', 'poly' meaning 'many', and 'genesis' meaning 'birth'). The nucleus becomes lobulated and divides into several nuclei. Merozoites form at the periphery of the schizont. Both first and second generation schizonts are located within the host cytoplasm and are not surrounded by a parasitophorous vacuole. *Sarcocystis* merozoites have much the same organelles as *Toxoplasma gondii* tachyzoites including a conoid, micronemes and a plastid body but no rhoptries (Dubey et al. 1989). Rhoptries are present in *Sarcocystis* bradyzoites.

Merozoites liberated from the terminal vascular generation of the developing parasite initiate sarcocyst

Table 22.1 *Historical landmarks concerning* Sarcocystis

Year	Findings	Reference
1843	Sarcocysts found in muscles of a mouse	Miesher (1843)
1882	Genus *Sarcocystis* introduced	Lankaster (1882)
1943	*Sarcocystis* not transmitted from sheep to sheep, role of carnivores suspected but not proven	Scott (1943)
1972	Sexual phase cultured in vitro	Fayer (1972)
1972	Two-host life cycle found	Rommel and Heydorn (1972), Rommel et al. (1972)
1973	Vascular phase recognized and pathogenicity demonstrated	Fayer and Johnson (1975)
1975	Multiple *Sarcocystis* species within a given host recognized	Heydorn et al. (1975b)
1975	Chemotherapy demonstrated	Fayer and Johnson (1973)
1976	Abortion due to sarcocytosis recognized	Fayer et al. (1979b)
1981	Protective immunity demonstrated	Dubey (1980a)
1986	Vascular phase cultured in vitro	Speer and Dubey (1986)
2000	Unusual life cycle of *S. neurona* discovered	Dubey et al. (2000)

Modified from Dubey et al. (1989); where a complete bibliography can be found

formation. These merozoites penetrate appropriate host cells. The intracellular merozoite, which is surrounded by a parasitophorous vacuole (PV), becomes round to ovoid (metrocyte) and undergoes repeated binary division eventually producing banana-shaped zoites called bradyzoites (Figure 22.3b) (also called cystozoites) containing prominent amylopectin granules that stain bright red when treated with the periodic acid–Schiff (PAS) reagent. Some mature sarcocysts may contain some peripherally arranged metrocytes in addition to zoites (Figure 22.1b). Eventually, the sarcocyst is filled with bradyzoites and it is this stage that is infective for the predator definitive host. Sarcocysts generally become infectious about 75 days after infection, but there is considerable variation between species of *Sarcocystis*. The definitive host becomes infected by ingesting tissues containing mature sarcocysts. Bradyzoites liberated from the sarcocyst by digestion in the stomach and intestine penetrate the mucosa of the small intestine and transform into male (micro-) and female (macro-) gamonts. Within 6 h of ingesting infected tissue, gamonts are found within a PV in goblet cells near the tips of the villi (Figure 22.3c). Macrogamonts are 10–20 μm in diameter and contain a single nucleus whereas microgamonts are 7 × 5 μm and contain between 3 and 11 slender gametes (Figure 22.3d). The microgametes,

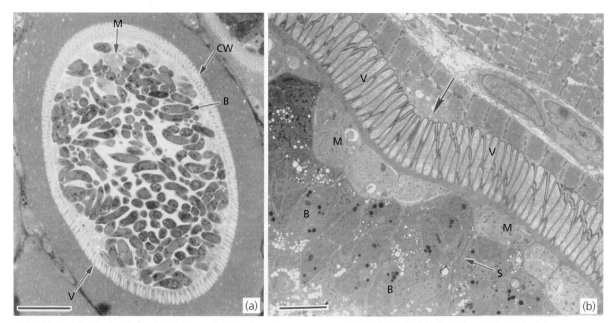

Figure 22.1 *Intramuscular* Sarcocystis hominis *sarcocysts.* **(a)** *Histological section of a mature sarcocyst. Note finger-like villar protrusions (V) on the cyst wall (CW) enclosing numerous bradyzoites (B) and a few metrocytes (M). Toluidine blue. Bar = 20 μm.* **(b)** *Transmission electron micrograph. Note villar projections (V) on the cyst wall, metrocytes (M), bradyzoites (B) and septa (S). Arrow points to border of the parasite and host cell cytoplasm. Bar = 4.3 μm.*

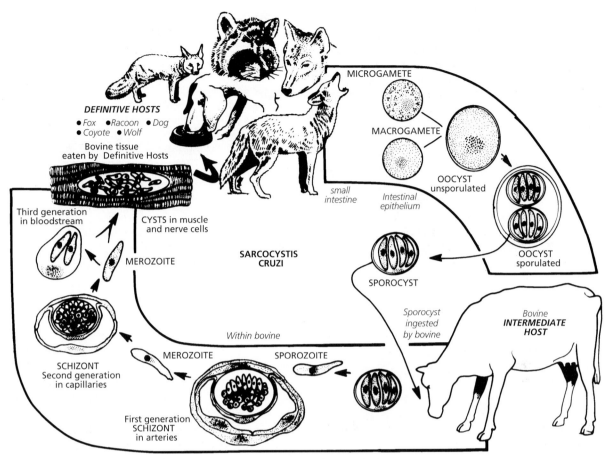

Figure 22.2 *Life cycle of* Sarcocystis cruzi *(from Dubey et al. 1989, with permission).*

which are about 4×0.5 μm, have a compact nucleus, and two flagella. After fertilization of a macrogamete by a microgamete, a wall develops around the zygote and an oocyst is formed. The entire process of gametogony and fertilization can be completed within 24 h. Gamonts and oocysts may be found at the same time in a host (Figure 22.3).

Oocysts of *Sarcocystis* species sporulate in the lamina propria (Figure 22.3e). Sporulated oocysts are generally colorless, thin-walled (less than 1 μm), and contain two elongate sporocysts. There is neither an oocyst residuum nor a micropyle. Each sporocyst contains four elongated sporozoites and a granular sporocyst residuum which may be compact or dispersed. There is no Stieda body (a cap-like thickening at one pole). Each sporozoite has a central-to-terminal nucleus, several cytoplasmic granules, and a crystalloid body, but there is no refractile body. The thin oocyst wall often ruptures, releasing the sporocysts into the intestinal lumen from which they are passed in the feces. The prepatent and patent periods vary, but for most *Sarcocystis* species, oocysts are first shed in feces 7–14 days after ingesting sarcocysts.

The number of generations of schizogony and the type of host cell in which schizogony may occur vary with each species of *Sarcocystis*, but trends are apparent. For example, all species of *Sarcocystis* of large domestic animals (sheep, goats, cattle, and pigs) form first and second generation schizonts in the vascular endothelium, whereas only a single precystic generation of schizogony has been found in *Sarcocystis* species of small mammals (mice and deer mice) and this is generally in hepatocytes (Dubey et al. 1989).

Sarcocysts, which are always located within a PV in the host cell cytoplasm, consist of a cyst wall that surrounds the metrocyte or the bradyzoites. The structure and thickness of the cyst wall differs among species of *Sarcocystis* and within each species as the sarcocyst matures (Tadros and Laarman 1978). Histologically, the sarcocyst wall may be smooth, striated or hirsute, or may possess complex branched protrusions. These protrusions are of taxonomic importance (Figure 22.1). Internally, groups of zoites may be segregated into compartments by septa that originate from the sarcocyst wall or they may not be compartmentalized.

Not all species of *Sarcocystis* cause disease in their host species (Dubey et al. 1989). Generally, species using canids as definitive hosts are more pathogenic than those using felids. For example, of the three species in cattle, *S. cruzi*, for which the dog is the definitive host, is the most pathogenic for cattle, whereas *S. hirsuta* and *S. hominis*, which undergo sexual development in cats and primates respectively, are only mildly pathogenic

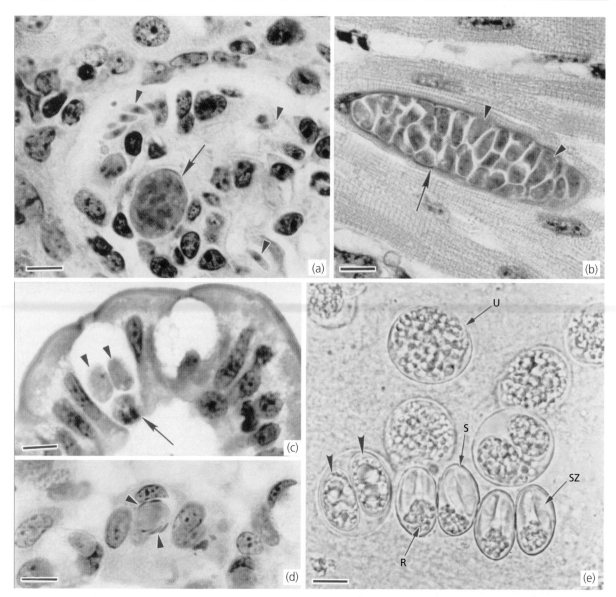

Figure 22.3 *Asexual and sexual stages of* S. cruzi. **(a)** *Second generation schizont (arrow) and released merozoites (arrowheads) in renal glomerulus of a calf. Hematoxylin and eosin (H&E). Bar = 10 μm.* **(b)** *Intramuscular immature sarcocyst (arrow) containing metrocytes (arrowheads). H&E. Bar = 10 μm.* **(c)** *Two female gamonts (arrowheads) and a male gamont (arrow) in a goblet cell, 6 h after infection of a coyote. Iron hematoxylin. Bar = 10 μm.* **(d)** *Mature microgamont with five microgametes (arrowheads) in an epithelial cell 6 h after infection of a coyote. Iron hematoxylin. Bar = 10 μm.* **(e)** *Sporogony of* S. cruzi *in the intestinal lamina propria of a coyote. Note unsporulated oocyst (U), partially sporulated oocysts with immature sporocysts (arrowheads), an oocyst with two sporocysts (S), and two fully sporulated oocysts containing sporozoites (SZ) and residual body (R). Unstained. Bar = 10 μm (from Dubey et al. 1989, with permission).*

(Table 22.2). Pathogenicity is manifested in the intermediate host. *Sarcocystis* generally does not cause illness in definitive hosts.

S. neurona is an unusual species of the genus that does not follow the life-cycle pattern of *S. cruzi* outlined above. It is also one of the most pathogenic species of the genus. *S. neurona* is the most frequent cause of a fatal disease in horses called equine protozoal encephalomyelitis (EPM) in the Americas (Dubey et al. 2001). Horses are considered the aberrant host because only schizonts are found in their tissues. Unlike *S. cruzi*, *S. neurona* schizonts occur in neural cells, not in the

vascular endothelium, and schizonts may persist in the central nervous system for months. Sarcocysts of *S. neurona* occur in domestic cats, striped skunks, raccoons, sea otters, and armadillos. Opossums (*Didelphis virginianus*, *D. abbreventis*) are its definitive hosts. Only the sexual cycle occurs in the definitive host and it is confined to the small intestine. Encephalomyelitis associated with *S. neurona* has been reported in horses, ponies, zebras, skunks, raccoons, cats, lynx, mink, and marine mammals (Pacific harbor seals and sea otters).

S. canis is another unusual species of the genus with an unknown life cycle. Its sarcocysts, sexual phase, and

Table 22.2 Sarcocystis species in livestock

Intermediate host	Sarcocystis sp.	References	Sarcocysts Max length (mm)	Wall type	Pathogenicity	Definitive hosts
Cattle (Bos taurus)	S. cruzi	(Hasselmann, 1926) Wenyon, 1926	<1	7	++	Dog, coyote, red fox, raccoon, wolf
	S. hirsuta	Moulé, 1888	7	10	±	Cat
	S. hominis	(Railliet and Lucet, 1891) Dubey, 1976	7	10	±	Humans and other primates
Sheep (Ovis aries)	S. tenella	(Railliet, 1886) Moulé, 1886	0.7	14	++	Dog, coyote, red fox
	S. arieticanis	Heydorn, 1985	0.9	7	+	Dog
	S. gigantea	(Railliet, 1886) Ashford, 1977	10	21	–	Cat
	S. medusiformis	Collins, Atkinson, and Charleston, 1979	8	20	–	Cat
Goat (Capra hircus)	S. capracanis	Fisher, 1979	1	14	++	Dog, coyote, red fox
	S. hircicanis	Heydorn and Unterhozner, 1983	2.5	7	++	Dog
	S. moule	Nevu-Nemaire, 1912	7.5	7	?	Cat
Pigs (Sus scrofa)	S. miescheriana	(Künn, 1865) Labbé, 1899	1.5	10(?)	+	Dog, raccoon, wolf, red fox, jackal
	S. porcifelis	Dubey, 1976	?	?	?	Cat
	S. suihominis	(Tadros and Laarman, 1976) Heydorn, 1977	1.5	10	+	Humans and other primates
Horses (Equus caballus)	S. fayeri	Dubey, Streitel, Stromberg, and Toussant, 1977	1.0	11	±	Dog
	S. equicanis	Rommel and Geisel, 1975	0.35	?	?	Dog
	S. bertrami	Doflein, 1901	12	?	?	Dog
Water buffalo (Bubalus bubalis)	S. levinei	Dissanike and Kan, 1978; Huong et al; 1997	1.1	7(?)	+	Dog
	S. fusiformis	(Railliet, 1897) Bernard and Bauche, 1912	3	21	–	Cat
	S. dubeyi	Huong and Uggla, 1999	<1 (?)	9	?	?
	S. buffalonis	Huong, Dubey, Nikka and Uggla, 1997				
Came l(Camelus spp.)	S. cameli	Mason, 1910	0.38	?	?	Dog
	Sarcocystis sp	Mason, 1910	?	?	?	?
Chickens (Gallus gallus)	S. horvathi	Ratz, 1908	0.98	?	?	?
	Sarcocystis sp.	Wenzel, Erber, Boch and Schellner, 1982	?	?	?	Dog, cat
	S. rileyi	(Stiles, 1893), Michin, 1903	12	23	?	Skunk
Ducks (Anas spp.)						

Reprinted with permission from Dubey et al. (1989), where a complete bibliography can be found
++) very pathogenic; +) pathogenic, mildly pathogenic; –) nonpathogenic; ?) unknown or unclassified; ±) questionable pathogenicity.

definitive hosts are unknown. The schizont is the only stage that is known. *S. canis* has been found associated with fatal hepatitis in sea lions, dogs, black and grizzly bears, a horse, and a dolphin (Dubey and Speer 1991; Resendes et al. 2002). Congenital infection has been documented in dogs.

CLINICAL ASPECTS

Humans serve as the definitive host for *S. hominis* and *S. suihominis* and also serve as accidental intermediate hosts for several unidentified species of *Sarcocystis*. Symptoms vary with the species of *Sarcocystis* causing the infection.

Intestinal sarcocystosis

SARCOCYSTIS HOMINIS (RAILLIET AND LUCET, 1891) DUBEY, 1976

Infection with *S. hominis* is acquired by ingesting uncooked beef containing sarcocysts. *S. hominis* is only mildly pathogenic. A volunteer who ate raw beef from an experimentally infected calf developed nausea, stomach ache, and diarrhea 3–6 h after ingesting the beef; these symptoms lasted 24–36 h and *S. hominis* sporocysts were excreted between 14–18 days after ingestion of the beef, during which time the volunteer had diarrhea and stomach ache (Heydorn 1977). Somewhat similar but milder symptoms were experienced by other volunteers who ate uncooked naturally infected beef (Rommel and Heydorn 1972; Aryeetey and Piekarski 1976; Hiepe et al. 1979). Recently, six of seven human volunteers that ate 128–260 grams of Kibbe (a preparation made from raw beef and spices) obtained from an Arabian restaurant in São Paulo, Brazil, excreted *S. hominis* sporocysts 10–14 days later. Two of these volunteers became ill. One of them had abdominal pain and diarrhea 1–3 days post ingestion and the other had diarrhea 11 days post ingestion (Pena et al. 2001). A patient in Spain developed abdominal pain and loose stools after eating raw beef; the diagnosis was confirmed by finding *S. hominis* sporocysts in feces (Clavel et al. 2001).

A *Sarcocystis* species similar to *S. hominis* named *S. dubeyi* (Huong and Uggla 1999) has been found in water buffaloes. The definitive host for *S. dubeyi* is unknown but suspected to be humans. Although all *Sarcocystis* species capable of infecting domestic livestock are host-specific, researchers from the People's Republic of China believe that *Sarcocystis* species from cattle and water buffaloes (including *S. hominis*) are the same and even suggest that there may be more of them than one species now called *S. hominis* (for review see Yang et al. 2001a, b). However, in my opinion, there is no critical evidence for the hypothesis proposed by Yang et al. (2001a, b) and until proven otherwise, *Sarcocystis*

species in cattle, although structurally similar to those in water buffaloes, should be regarded as distinct parasites based on transmission experiments (Dubey et al. 1989).

SARCOCYSTIS SUIHOMINIS (TADROS AND LAARMAN, 1976) HEYDORN, 1977

This species, acquired by eating undercooked pork, is more pathogenic than *S. hominis*. Human volunteers developed hypersensitivity-like symptoms; nausea, vomiting, stomach ache, diarrhea, and dyspnea within 24 h of ingestion of uncooked pork from naturally or experimentally infected pigs. Sporocysts were shed 11–13 days after ingesting the infected pork (Rommel and Heydorn 1972l; Piekarski et al. 1978; Hiepe et al. 1979; Kimmig et al. 1979). Zuo (1991) found sporocysts in stools of 123 of 414 (29.7 percent) people from three villages in Xianguan City in China. He reported that these people ate raw pork but not raw beef.

NATURAL PREVALENCE OF INTESTINAL SARCOCYSTOSIS OF HUMANS

Before the discovery of the life cycle of *Sarcocystis* and recognition of cattle and pigs as sources of human infection, *Sarcocystis* sporocysts in human feces were referred to as *Isospora hominis*. Because of structural similarities between *S. hominis* and *S. suihominis* sporocysts, it is not possible to distinguish between them by microscopic examination. Intestinal sarcocystosis is more common in Europe than in other continents. *Sarcocystis* sporocysts were seen in 2 percent of 3 500 fecal samples in France (Deluol et al. 1980), 1.6 percent of 1 518 samples in Germany (Flentje et al. 1975; Janitschke 1975), and 10.4 percent of 125 fecal samples from 7–18-year-old children in Poland (Plotkowiak 1976). Enteritis was associated with shedding of *Sarcocystis* sporocysts in six cases in Thailand and two cases reported from the People's Republic of China (reviewed in Dubey et al. 1989). Straka et al. (1991) found *Sarcocystis* spp. sporocysts in the stools of 14 of 1 228 students from Vietnam who worked in Slovakia.

Muscular sarcocystosis

Sarcocysts have been found in striated muscles of humans, mostly as incidental findings. Judging from the published reports, sarcocysts in humans are rare (Beaver et al. 1979); most reported cases are from Asia. Of the 40 histologically diagnosed reports that Beaver et al. (1979) reviewed and 6 additional cases since 1982 reviewed by Dubey et al. (1989), 15 were from southeast Asia, 11 from India, 5 from Central and South America, 4 from Europe, 4 from Africa, 3 from the USA, and 1 from China; the source of 2 was undetermined. Of the 46 confirmed cases, sarcocysts were found in skeletal muscles of 35 and in the heart of 11. The clinical significance of sarcocysts and the life cycles of the *Sarcocystis* spp. giving rise to these sarcocysts in humans are unknown. Recently, 7 of 15 US military men developed acute illness after an army exercise in rural Malaysia (Arness et al. 1999). The illness was characterized by fever, myalgias, bronchspasm, fleeting pruritic rashes, transient lymphadenopathy, and subcutaneous nodules associated with eosinophilia, elevated erythrocyte sedimentation rate, and elevated levels of muscle creatinine kinase. Sarcocysts of an unidentified *Sarcocystis* species were found in skeletal muscle biopsies of the index case. Symptoms in 5 other men were mild to moderate and self-limited, and 1 team member with laboratory abnormalities were asymptomatic. Of eight team members tested for antibody to *Sarcocystis*, six were positive; of four with the eosinophilic myositis syndrome who were tested, all were positive. The illness was considered to be sarcocystosis. Arness et al. (1999) also reviewed other cases of sarcocystosis after 1990.

DIAGNOSIS

The antemortem diagnosis of muscular sarcocystosis can only be made by histological examination of muscle collected by biopsy. The finding of immature sarcocysts with metrocytes suggests recently acquired infection; the finding of mature sarcocysts indicates only past infection.

The diagnosis of intestinal sarcocystosis is easily made by fecal examination. As has been mentioned, sporocysts or oocysts of sarcocystis are shed fully sporulated in feces whereas those of *Isospora belli* are often shed unsporulated. It is not possible to distinguish one species of *Sarcocystis* from another by the examination of sporocysts.

An inflammatory response associated with sarcocystosis may help to distinguish an active disease process from incidental findings of sarcocysts. *Sarcocystis* schizonts have not yet been identified in humans. Although there are several serologic tests and polymerase chain reaction (PCR) techniques developed experimentally to distinguish *Sarcocystis* species in animals, none have been applied to cases of sarcocystosis in humans. Tenter (1995) has reviewed in detail pitfalls of serologic and molecular diagnosis of sarcocystosis in animals.

EPIDEMIOLOGY AND CONTROL

Sarcocystis infection is common in many species of animals worldwide (Dubey et al. 1989). A variety of conditions permit such high prevalence:

- a host may harbor any of several species of *Sarcocystis*
- many definitive hosts are involved in transmission
- large numbers of sporocysts may be shed
- *Sarcocystis* oocysts and sporocysts develop in the lamina propria, and are discharged over a period of many months
- *Sarcocystis* oocysts and sporocysts are resistant to freezing, so they can overwinter on pasture

- *Sarcocystis* sporocysts and oocysts remain viable for many months in the environment and they may be spread by invertebrate transport hosts
- there is little or no immunity to reshedding of sporocysts and, therefore, each meal of infected meat can initiate a new round of sporocyst production
- *Sarcocystis* oocysts, unlike those of many other species of coccidia, are passed in feces in the infective form, this frees them from dependence on weather conditions for maturation and infectivity.

Poor hygiene during handling of meat between slaughter and cooking can be a source of *Sarcocystis* infection. In one survey in India, *S. suihominis* oocysts were found in feces of 14 of 20 3–12-year-old children (Banerjee et al. 1994) indicating that meat was consumed raw at least by some because *S. suihominis* can be transmitted to humans only by the consumption of raw pork. In another study, 3–5-year-old children from a slum area were found to consume meat scraps virtually raw, and many pigs from that area harbored *S. suihominis* sarcocysts (Solanki et al. 1991). In European countries where the frequency of consumption of raw or undercooked meat is relatively high, humans are likely to have intestinal sarcocystosis. In one survey, 60.7 percent of pigs had *S. suihominis* sarcocysts in their muscles (Boch et al. 1978).

CHEMOTHERAPY

There is no treatment for *Sarcocystis* infection of humans. On the basis of results in experimental animals, it is probable that sulfonamides and pyrimethamine are helpful in treating sarcocystosis (Rommel et al. 1981).

There is no vaccine to protect livestock or humans against sarcocystosis. Shedding of *Sarcocystis* oocysts and sporocysts in feces of the definitive hosts is the key factor in the spread of *Sarcocystis* infection; to interrupt this cycle, carnivores should be excluded from animal houses and from feed, water and bedding for livestock. Uncooked meat or offal should never be fed to carnivores. As freezing can drastically reduce or eliminate infectious sarcocysts, meat should be frozen if not cooked. Exposure to heat at 55°C for 20 min kills sarcocysts so only limited cooking or heating is required to kill sporocysts (Fayer 1975). Dead livestock should be buried or incinerated. Dead animals should never be left in the field for vultures and carnivores to eat.

REFERENCES

Arness, M.K., Brown, J.D., et al. 1999. An outbreak of acute eosinophillic myositis attributed to human *Sarcocystis* parasitism. *Am J Trop Med Hyg*, **61**, 548–53.

Aryeetey, M.E. and Piekarski, G. 1976. Serologische *Sarcocystis*-Studien an Menschen und Ratten. *Z Parasitenkd*, **50**, 109–24.

Banerjee, P.S., Bhatia, B.B. and Pandit, B.A. 1994. *Sarcocystis suihominis* infection in human beings in India. *J Vet Parasitol*, **8**, 57–8.

Beaver, P.C., Gadgil, R.K. and Morera, P. 1979. *Sarcocystis* in man: A review and report of five cases. *AmJ Trop Med Hyg*, **28**, 819–44.

Boch, J., Mannewitz, U. and Erber, M. 1978. Sarkosporidien bei Schlachtschweinen in Süddeutschland. *Berl Münch Tierärztl Wochenschr*, **91**, 106–11.

Clavel, A., Doiz, O., et al. 2001. Molestias abdominales y heces blandas en consumidor habitual de carne de vacuno poco cocinada. *Enferm Infec Microbiol Clin*, **19**, 29–30.

Deluol, A.M., Mechall, D., et al. 1980. Incidence et aspects cliniques des coccidioses intestinales dans une consultation de médecine tropicale. *Bull Soc Pathol Exot*, **73**, 259–66.

Dubey, J.P. and Speer, C.A. 1991. *Sarcocystis canis* n. sp. (Apicomplexa: Sarcocystidae), the etiologic agent of generalized coccidiosis in dogs. *J Parasitol*, **77**, 522–7.

Dubey, J.P., Speer, C.A. and Fayer, R. 1989. *Sarcocystosis of animals and man*. Boca Raton, FL: CRC Press.

Dubey, J.P., Saville, W.J.A., et al. 2000. Completion of the life cycle of *Sarcocystis neurona*. *J Parasitol*, **86**, 1276–80.

Dubey, J.P., Lindsay, D.S., et al. 2001. A review of *Sarcocystis neurona* and equine protozoal myeloencephalitis (EPM). *Vet Parasitol*, **95**, 89–131.

Fayer, R. 1975. Effects of refrigeration, cooking, and freezing on *Sarcocystis* in beef from retail food stores. *Proc Helminthol Soc Wash*, **42**, 138–40.

Flentje, B., Jungman, R. and Hiepe, T. 1975. Vorkommen von *Isospora-hominis*-Sporozysten beim Menschen. *Dt Gesundh-Wesen*, **90**, 523–5.

Heydorn, A.O. 1977. Sarkosporidieninfiziertes Fleisch als mögliche Krankheitsursache für den Menschen. *Arch Lebensmittelhygiene*, **28**, 27–31.

Hiepe, F., Hiepe, T., et al. 1979. Experimentelle Infektion des Menschen und von Tieraffen (Cercopithecus callitrichus) mit Sarkosporidien-Zysten von Rind und Schwein. *Arch Exp Vet Med*, **33**, 819–30.

Huong, L.T.T. and Uggla, A. 1999. *Sarcocystis dubeyi* n. sp. (Protozoa: Sarcocystidae) in the water buffalo (*Bubalus bubalis*). *J Parasitol*, **85**, 102–4.

Janistchke, K. 1975. Neue Erkenntnisse über die Kokzidien-Infektionen des Menschen. II. *Isospora*-Infektion. *Bundesgesundheitsblatt*, **18**, 419–22.

Kimmig, P., Piekarski, G. and Heydorn, A.O. 1979. Zur Sarkosporidiose (*Sarcocystis suihominis*) des Menschen (II). *Immun Infekt*, **7**, 170–7.

Pena, H.F., Ogassawara, S. and Sinhorini, I.L. 2001. Occurence of cattle *Sarcocystis* species in raw kibbe from arabian food establishments in the city of São Paulo, Brazil, and experimental transmission to humans. *J Parasitol*, **87**, 1459–65.

Piekarski, G., Heydorn, A.O., et al. 1978. Klinische, parasitologische und serologische Untersuchungen zur Sarkosporidiose (*Sarcocystis suihominis*) des Menschen. *Immun Infekt*, **6**, 153–9.

Plotkowiak, J. 1976. Wynikl dalszych badan nad wystepowaniem l epidemiologia inwazjl *Isospora hominis* (Raillet and Lucet, 1891). *Wiadomasci Parazytol*, **22**, 137–47.

Resendes, A.R., Juan-Sallés, C., et al. 2002. Hepatic sarcocystosis in a striped dolphin (*Stenella coeruleoalba*) from the Spanish Mediterranean coast. *J Parasitol*, **88**, 206–9.

Rommel, M. and Heydorn, A.O. 1972. Beitrage zum Lebenszyklus der Sarkosporidien. III. *Isospora hominis* (Railliet und Lucet, 1891) Wenyon, 1923, eine Dauerform der Sarkosporidien des Rindes und des Schweins. *Berl Münch Tierärztl Wochenschr*, **85**, 143–5.

Rommel, M., Schwerdfeger, A. and Blewaska, S. 1981. The *Sarcocystis muris*-infection as a model for research on the chemotherapy of acute sarcocystosis of domestic animals. *Zentralbl Bakteriol Hyg I Abt Orig A*, **250**, 268–76.

Solanki, P.K., Shrivastava, H.O.P. and Shah, H.L. 1991. Prevalence of *Sarcocystis* in naturally infected pigs in Madhya-Pradesh with an epidemiological explanation for the higher prevalence of *Sarcocystis suihominis*. *Indian J Anim Sci*, **61**, 820–1.

Straka, S., Skracikova, J., et al. 1991. *Sarcocystis* species in Vietnamese apprentices. *Cesk Epidemiol Mikrobiol Imunol*, **40**, 204–8.

Tadros, W. and Laarman, J.J. 1978. A comparative study of the light and electron microscopic structure of the walls of the muscle cysts of several species of sarcocystid eimeriid coccidia. *Proc Konink Nederl Akad Wetensch*, **81**, 469–91.

Tenter, A.M. 1995. Current research on *Sarcocystis* species of domestic animals. *Int J Parasitol*, **25**, 1311–30.

Yang, Z.Q., Zuo, Y.X., et al. 2001a. Analysis of the 18S rRNA genes of *Sarcocystis* species suggests that the morphologically similar organisms from cattle and water buffalo should be considered the same species. *Mol Biochem Parasitol*, **115**, 283–8.

Yang, Z.Q., Zuo, Y.X., et al. 2001b. Identification of *Sarcocystis hominis*-like (protozoa: Sarcocystidae) cyst in water buffalo (*Bubalus bubalis*) based on 18S rRNA gene sequences. *J Parasitol*, **87**, 934–7.

Zuo, Y.X., Zhou, Z.B., et al. 1991. The prevalence of *Sarcocystis suihominis* in the natives of Yunnan Province, China. *Chin Sci Bull*, **36**, 965–7.

23

Babesiosis of humans

SAM R. TELFORD III AND ANDREW SPIELMAN

INTRODUCTION

Babesiosis, caused by certain tick-borne intra-erythrocytic sporozoans, has affected human affairs since antiquity. The 'grievous murrain' described in Exodus IX was probably redwater fever of cattle, caused by *Babesia bovis*. Babesiosis occupies a prominent place in the history of biomedical sciences. Smith and Kilborne's landmark discovery (1893) of the role of ticks in the transmission of *B. bigemina*, the Texas cattle fever pathogen, stands as the first proof that hematophagous arthropods can transmit pathogens of vertebrates.

With the exception of the trypanosomes, *Babesia* spp. are the most ubiquitous of the mammalian hemoprotozoa, occurring wherever certain ticks proliferate. Numerous types of mammal serve as hosts for these pathogens, certain of which, particularly *Babesia microti*, have taken a prominent place among the emerging infections of humans. Babesiosis, therefore, spans human history and continues to demand our attention.

Babesias are small sporozoans that cycle between ixodid ticks and vertebrates. They undergo sexual reproduction after ingestion by ticks and replicate one to several times before reaching maturity in the vector's salivary glands. The tick, which is the definitive host, then deposits the infectious sporozoite stage of the *Babesia* within the vertebrate host's dermis. The mode of development at the site of deposition and subsequent path of dissemination remain largely unknown.

Ultimately, the parasites appear as characteristic pear-shaped trophozoites (piroplasms) in infected erythrocytes where they multiply, appearing as simple rings, paired or single piroplasms or the pathognomonic tetrads. The parasite's 'accordion-like' intra-erythrocytic gametocyte stage is only identifiable by electron microscopy or staining with DNA-specific stains. Although certain basic elements of the babesial life cycle appear to be well established, much remains to be learned about the biology of this diverse group of parasites.

CLASSIFICATION

The Babesias are protozoa classified within the phylum Sporozoa (= Apicomplexa), order Piroplasmida and family Babesiidae. Those that parasitize mammals have long been considered to form two main lineages, the 'large' and 'small' types of *Babesia*. *Babesia* are divided informally between those with intra-erythrocytic forms that are 1.0–2.5 µm in diameter and those of 2.5–5.0 µm. No pre- or exo-erythrocytic forms have been described for most *Babesia*, which implies that the sporozoites directly invade erythrocytes. Indeed, the absence of a pre-erythrocytic cycle has served as the main feature that distinguishes the Babesiidae from the Theileriidae (Riek 1968), *Theileria* being characterized by a lymphocytic pre-erythrocytic stage whereas *Babesia* lack such a stage in their life cycles. Direct evidence for penetration of erythrocytes by babesial sporozoites, however, is lacking. The lymphocytic exo-erythrocytic forms that have been shown to occur in the life cycle of one small *Babesia*, *B. equi* (Schein et al. 1981; Mehlhorn and Schein 1984) may also occur in the life cycle of *B. microti* (Mehlhorn and Schein 1984). The existence of lymphocytic stages in the life cycle of some of the small *Babesia* raises a question about the validity of the generally accepted separation of *Babesia*, particularly the small types, from *Theileria*.

Although none of the small *Babesia* so far studied passes from generation to generation within the vector tick (transovarial transmission) (Walter and Weber 1981; Mehlhorn and Schein 1984; Gray et al. 2002), all of the large ones appear to do so. The small *Babesia* share this exclusively intragenerational pattern of development with *Theileria*. The absence of transovarial transmission gives additional biological support to the suggestion that the small *Babesia* should be classified with *Theileria*. Whether all of the small *Babesia* produce pre-erythrocytic stages, and whether all lack the capacity for inherited infection in the tick host remain to be determined.

At least 99 *Babesia* species infecting mammals have been described (Levine 1988), 78 of which infect non-ruminants. The vectors and other details of the life cycles of most of these species, except for those of veterinary or public health importance, remain largely undescribed. Virtually all piroplasm species were described on the basis of their microscopical appearance in Giemsa-stained blood smears (the morphology of the blood forms and the occurrence of binary or tetrad forms), as well as their host range and endemicity. Although it had been expected that many of these species would be synonymized with further study, molecular phylogenetic analysis demonstrates unexpected diversity (Ellis et al., 1992; Thomford et al., 1994; Allsopp et al., 1994; Caccio et al., 2000; Kjemtrup et al., 2000a; Zahler et al., 2000; Goethert and Telford, 2003a). Well-characterized, widely distributed 'species' such as *B. canis* (Carret et al. 1999), *B. gibsoni* (Kjemtrup et al. 2000b) and *B. microti* (Goethert and Telford, 2003a) are now considered to comprise groups of genetically diverse entities based on sequencing of 18S rDNA or other genes. Thus, the genus *Babesia* may be more speciose than previously appreciated, may even comprise several distinct genera. Much work, however, remains to be done prior to naming new species, let alone major taxonomic revision. In particular, a consensus must be formed among protozoologists as to the degree of distinction required to interpret a given difference in nucleotide sequences between samples as representative of distinct species. Such a consensus would require some knowledge of the degree of inter- and intrapopulation variability of the gene being used for classifying the piroplasms, as well as concordance with aspects of their biology such as the identity of their vectors.

STRUCTURE AND LIFE CYCLE

Babesias belong to the sporozoans, the group of protozoa also sometimes referred to as the Apicomplexa, which are characterized by having a variety of apical complex organelles, including rhoptries and micronemes. Mature babesial sporozoites have been described as pear-shaped with the anterior end being the broad end (Kakoma and Mehlhorn 1994). They have one anterior rhoptry and several micronemes but lack conoids, polar rings or subpellicular microtubules. Sporozoites appear to contain free ribosomes, a smooth endoplasmic reticulum, mitochondria-like structures and coiled organelles.

Recently invaded erythrocytes contain merozoites that initially lie within a parasitophorous vacuole (Rudzinska 1981) but this invaginated host-derived membrane soon disintegrates, leaving the maturing trophozoite free in the cytoplasm. Perhaps this unique freedom from a constricting parasitophorous vacuole contributes to the frequently noted pleomorphism of the intra-erythrocytic stage, which often is contorted or ameboid.

As the asexually reproducing intra-erythrocytic trophozoites of *B. microti* mature, new organelles appear, including polar rings, micronemes, larger rhoptries, subpellicular microtubules, and double membrane segments (Rudzinska et al. 1979). Interestingly, no cytostome is apparent; their method for deriving nourishment from a host cell remains undescribed.

Following the ingestion of parasitized erythrocytes by feeding *Ixodes dammini* tick larvae (the main vector for *B. microti* in the northern USA), gametocytes emerge and undergo a process of development into gametes (Strahlenkörper) which then fuse within the tick gut (Rudzinska et al. 1983). The existence of the ray-like Strahlenkörper that Koch associated with sexuality was confirmed by a demonstration of the fusion of the ray bodies emerging from neighboring erythrocytes (Rudzinska et al. 1983). The zygotes resulting from this fusion mature and become ookinetes 14–18 hours after the feeding larval tick becomes replete. The ookinetes subsequently penetrate into the hemolymph and elongate to form the kinetes, which penetrate fat body cells and nephrocytes. A series of divisions within the fat body cells and nephrocytes then results in the production of secondary ookinetes which in some *Babesia* may invade the ovaries (Karakashian et al. 1986). This secondary cycle of division and invasion of the ovaries is characteristic of the large *Babesia* and may facilitate transovarial transmission. Primary ookinetes enter the salivary glands directly, where they have been observed as early as 13 days after the larva became replete.

The salivary acini become hypertrophied when ookinetes invade the salivary glands. The ookinetes in the salivary acini become sporoblasts, which then remain dormant until the larval tick becomes a nymph and the nymphal tick attaches to a host. In the case of *B. microti*, this state of dormancy in *I. dammini* (the vector) extends naturally through the winter, a period of 9–10 months (Piesman et al. 1987). After the infected nymphal tick attaches, nuclear division occurs, resulting in the formation of about 10 000 sporozoites from each sporoblast. Many thousands of sporozoites appear to be deposited in the dermis around the tick's mouthparts during the final hours of attachment (Mehlhorn and Schein 1984). Some 10 000–25 000 syringe-injected *B. microti* sporozoites are required to infect white-footed mice and hamsters (Piesman and Spielman 1982), but the anti-inflammatory pharmacological activity of tick saliva (Ribeiro 1987) greatly facilitates infection, suggesting that ticks may transmit the infectious agents far more effectively than does a needle.

The events that follow inoculation of babesial sporozoites by the feeding vector tick are poorly understood. In the absence of evidence to the contrary, early reviews of the biology of *Babesia* by Mehlhorn and Schein (1984), Mahoney (1977), and Young and Morzaria (1986) have generally inferred that in *Babesia* no pre-erythrocytic cycle occurs. But in *B. equi*, an agent of equine babesiosis, sporozoites appear to enter lymphocytes directly, both in vitro and in vivo, where they undergo a cycle of merogony before the resulting merozoites emerge to infect erythrocytes (Schein et al. 1981). The exo-erythrocytic cycle of this small *Babesia* was only recently discovered, unlike the exo-erythrocytic stages of *Theileria* which were discovered decades ago and called lymphocytic 'blue bodies' by Robert Koch.

The intra-erythrocytic parasites replicate when a nucleus and other organelles migrate to a location under particular double membrane areas of the parent. These areas then develop and pinch off from the parental piroplasm in a process of budding (Rudzinska 1981). *Babesia* do not undergo synchronous budding (schizogony) as do the plasmodia, but rather one, two or, rarely, three bud off at one time.

In addition to the asexually reproducing trophozoites and their merozoite progeny, there are also non-reproducing accordion-like gametocytes. These latter forms fail to develop the double membrane segments that become the anlage for merozoite formation and do not acquire a rhoptry; rather they just appear to grow larger, folding or coiling themselves within the confines of the erythrocyte (Mehlhorn and Schein 1984; Rudzinska et al. 1979). These accordion-like forms emerge from erythrocytes within the gut of the tick and differentiate into gametes, pairs of which may fuse in a process of syngamy. Unlike the gametocytes of the plasmodia, *Babesia* gametocytes cannot be distinguished from asexual forms by light microscopy; electron microscopy must be used to make the distinction.

BIOCHEMISTRY, MOLECULAR BIOLOGY, AND GENETICS

The biochemistry, molecular biology, and genetics of the zoonotic *Babesia* have been inadequately explored. Such work as has been reported has generally had as its goal the discovery of diagnostic antigens, the elucidation of diagnostically useful DNA sequences, or the identification of variation among strains of species. Isoenzymes, for example, were examined in an attempt to distinguish various populations of *B. microti* (Momen et al. 1979). High resolution polyacrylamide gel electrophoresis (Moss et al. 1986) was used to study the degree of polymorphism in the glucose phosphate isomerase locus in low- and high-passage Gray strain *B. microti* (the index strain for human infection) and showed that the Gray strain is heterogeneous, at least in terms of polymorphism at the locus. These preliminary results suggest that study of cloned isolates may be required to detect differences between populations.

Babesia microti infection may only slightly alter the physiology of hamster erythrocytes (Roth et al. 1981). Reduced glutathione levels double, but 2,3-diphosphoglycerate levels remain unchanged. Hamsters in which reticulocytosis is chemically induced experience similar changes, however, suggesting that the observed biochemical changes relate to anemia and not necessarily to the presence of parasites.

The existence of sexuality in piroplasms, initially determined by observation of fusing gametes within the tick with the aid of electron microscopy (Rudzinska et al. 1983), was confirmed by fluorescence microscopy of *Babesia* stained with the DNA-specific bisbenzimide Hoechst 33258. These studies revealed the ploidy of the various life-cycle stages of the cattle piroplasm *B. divergens* (Mackenstedt et al. 1990). Merozoites are haploid, two haploid merozoites are formed during budding within an erythrocyte; zygotes within the tick gut are diploid; kinetes are polyploid, and the polyploid sporoblasts divide to give rise to the uninuclear (haploid) infective sporozoites. Meiosis, however, has not been described.

A putative virulence-associated gene, *Bm13*, was cloned from a mung bean nuclease DNA expression library prepared from *B. microti* DNA (Tetzlaff et al. 1990). Mung bean nuclease, under certain conditions, appears to cleave AT-rich DNA at sites flanking intact genes (McCutchan et al. 1984). Using hyperimmune serum from mice infected by the ATCC 30221 strain of *B. microti*, which kills murine hosts, a clone coding a 54 kDa immunodominant antigen was selected. This sequence did not hybridize with DNA prepared from the less virulent Peabody strain of *B. microti*. Further experiments characterizing the protein that is coded for by the gene, and describing the protein's role in pathogenesis have not yet been published.

CLINICAL ASPECTS

The first convincingly demonstrated case of human babesial infection was reported in 1957 in a splenecto-mized resident of what was then Yugoslavia. The man died after an acute illness marked by anemia, fever, hemoglobinuria, and renal failure (Skrabalo and Deanovic 1957). Intra-erythrocytic parasites were detected and tentatively identified as *B. bovis*, a piro-plasm of cattle. In 1969, an elderly resident of Nantucket Island (USA) with no predisposing immune-compro-mising factors became infected with a rodent piroplasm, *B. microti* (Western et al. 1970). Since then, infection by *B. microti* has been documented in hundreds of residents of the northeastern and upper Midwestern USA. Mole-cular phylogenetic methods have demonstrated that several other distinct *Babesia* (designated WA-1, CA-1, M0-1, EU-1) are agents of human babesiosis. Human babesiosis has also been convincingly documented in Taiwan, Japan, and South Africa. The public health significance of zoonotic babesiosis may be greater than previously estimated.

Babesiosis is a malaria-like infection and its pathology, like that of malaria, is caused mainly by the asexually dividing intra-erythrocytic forms. Non-synchro-nous fevers, chills, myalgia, sweats, and prostration may be observed in patients with acute disease. In such patients, a fulminating parasitemia may lead to anemia, hemoglobinuria, renal failure, disseminated intravascular coagulation, and acute respiratory distress syndrome. Human infection, although occasionally severe, is usually subclinical.

Clinical manifestations

Symptoms of babesiosis caused by *B. microti* may commence 1–4 weeks after a human is bitten by an infected tick. In persons who develop clinical disease, a gradual onset of malaise, anorexia, and fatigue ensues, with subsequent development (within a week) of fever with a temperature as high as 40°C, drenching sweats and myalgia (Ruebush et al. 1977). Nausea, vomiting, headache, shaking chills, emotional lability and depres-sion, hemoglobinuria, and hyperesthesia have also been reported in such patients (Golightly et al. 1989). Pulmonary edema may be a frequent complication in symptomatic persons (Gordon et al. 1984; Boustani et al. 1994). Splenomegaly may occur, but other findings on physical examination are unremarkable. Anemia, throm-bocytopenia, and low or generally normal white cell counts may be observed; parasitemias range from 1 to 20 percent in spleen-intact patients and reach 85 percent in asplenic patients (Sun et al. 1983). Lactic dehy-drogenase, bilirubin and transaminases may be elevated (Ruebush et al. 1977). The ambiguity of the symptoms of human babesiosis renders clinical diagnosis difficult, particularly in sites where the agents of Lyme disease and human granulocytic ehrlichiosis are co-transmitted (Telford et al. 1995). The presence of fever, malaise, headache, splenomegaly, anemia, emotional lability and, particularly, profound fatigue provide clues useful to the physician.

As noted before, babesial infection in humans is usually subclinical or very mild. A large asymptomatic to symptomatic ratio is evident from serosurveys. For example, 8 percent of residents of Block Island, Rhode Island (USA) have been infected without clinically diag-nosed illness (Krause et al. 1994). Even though most infections are subclinical in areas with a high level of infection, moderate to severe disease may be common. Nantucket Island reported 21 cases in 1994, which trans-lates to 280 cases per 100 000 population, placing the community burden of this disease in a category with that of gonorrhea (Wilson 1991). The case fatality rate was estimated at 5 percent in a retrospective study of 136 New York cases (Meldrum et al. 1992). In certain sites, in years of particularly high transmission, babesiosis may constitute a public health burden.

Several factors appear to determine the severity of infection. There appears to be a relationship between the severity of illness and the age of the patient. The average age of seven spleen-intact New York patients with clinically apparent disease was 63 years (Benach and Habicht 1981) as was that of five similar patients from Nantucket (Ruebush et al. 1977). Similarly, infec-tion by human immunodeficiency virus (HIV) appears to promote the severe manifestations of babesial infection (Ong et al. 1990; Benezra et al. 1987; Machtinger et al. 1993). Although *B. microti* is enzootic in much of coastal New England, virtually all cases are diagnosed on Nantucket Island, eastern Long Island or in southeastern Connecticut. Parasite strain variation (as yet unchar-acterized) may influence the course of infection, as well as the geographical distribution of cases.

Babesial infections acquired in Europe have been attributed to the cattle piroplasm *B. divergens* and tend to be severe. Virtually all European patients had been splenectomized and the disease followed a severe course with the rapid multiplication of parasites (Gorenflot et al. 1990). Infection is usually fulminant; more than half of the 22 recorded patients died. Acute illness appears suddenly and is characterized by hemoglobinuria, which generally serves as the presenting symptom. Jaundice rapidly ensues and is accompanied by persistent non-periodic high fever (40–41°C), shaking chills, intense sweats, headaches, and myalgia, as well as lumbar and abdominal pain. Vomiting and diarrhea may be present. In severe cases, renal failure ensues rapidly (16 of the recorded 22 cases) and is induced by intravascular hemolysis. In fatal cases, the patient loses consciousness and dies in coma. A recent report suggests that the causative agent for reported cases of 'divergens' babe-siosis may be uncertain. A *Babesia* sp. that is genetically more similar to *B. odocoilei* (a parasite of American deer) than to *B. divergens* caused fulminating infection in two Italian patients (Herwaldt et al. 2003). This agent has been designated EU-1; it is morphologically similar to *B. capreoli* from European deer (Enigk and Friedhoff 1962) but could be mistaken for *B. divergens* by less experienced microscopists. *B. divergens* has been definitively identified as a causative agent of babesiosis in Europe by the use of cattle subinoculation and protection of such animals from challenge (e.g. Entrican et al. 1979). Thus, there are at least two agents of 'divergens' babesiosis.

Two splenectomized American patients with no travel history to Europe have been infected by *B. divergens* or something very closely related (Herwaldt et al. 1996; Beattie et al. 2002); one case terminated fatally. The patients presented with fever, headache, and rigors; thrombocytopenia, hemoglobinuria, and proteinuria were noted. Blood smears demonstrated parasites with typical *B. divergens* morphology (accole forms, paired pyriforms), and 18S rDNA sequencing indicated more than 99 percent sequence similarity with the European cattle agent. The same agent is enzootic within cottontail rabbits and their ticks (Goethert and Telford 2003b). Rabbit sera reacts with *B. divergens* (Purnell strain) antigen, more so than to the closely related *B. odocoilei* that is present in deer from the sites where rabbits are found to be infected. All available data suggest that this agent is *B. divergens* regardless of its American location or unusual enzootic host, but final taxonomic assignment awaits definitive cattle inoculation experiments.

Although only a few cases of babesiosis due to the WA-1 *Babesia* have been described, including two transfusion cases (Herwaldt et al. 1997; Kjemtrup et al. 2002) the course of the infection would appear to be similar to that caused by severe *B. microti*. The index case (Quick et al. 1993) was a 41-year-old spleen-intact man from a rural area in south central Washington (USA) with no history of travel. He was hospitalized with a 1-week history of fever, rigors, anorexia, cough, and headache, which progressed to severe rigors and vomiting. Dark-colored urine was noted, but a urine dipstick assay was only weakly positive for occult blood. Intra-erythrocytic parasites were observed, including characteristic tetrad forms. The patient was treated with clindamycin and quinine and recovered uneventfully. An isolate was made by inoculating hamsters with the patient's blood, but unlike *B. microti*, this agent killed these and other rodents within 10 days of infection (unpublished results 1992). Serum reacted weakly with *B. microti* antigen but more intensely with that of *B. gibsoni*. Subsequent molecular analysis confirmed that WA-1 was more closely related to *B. gibsoni* than to other *Babesia* (Thomford et al. 1994; Kjemtrup et al. 2000b).

Since the index case, similar infections have occurred in splenectomized residents of northern California (Persing et al. 1995). One case terminated fatally and another developed disseminated intravascular coagulation and renal insufficiency but recovered. 18S rDNA sequencing demonstrated that the agent was distinct from but related to WA-1 and it has been designated the CA-type babesia. Recent phylogenetic analyses suggest that mule deer and bighorn sheep are infected by parasites that are virtually identical (Kjemtrup et al. 2000a), but no other information is available regarding vectors and reservoirs.

IMMUNOLOGY

The relatively greater severity of human babesiosis in patients lacking a spleen indicates that this organ plays an important role in limiting parasitemia. In particular, early in the infection, parasitized and otherwise altered erythrocytes may be trapped and phagocytosed in the spleen. Absence of the spleen, however, does not necessarily imply that parasitemia may not be eventually limited.

Treatment of splenectomized and intact hamsters with antilymphocyte serum (ALS) demonstrated the importance of cellular immunity because all ALS recipients died upon challenge with *B. microti*, whereas only 20 percent of splenectomized animals died (Wolf 1974). Persistent intense parasitemias in athymic (nude) mice, in contrast to those of their heterozygous (thymus-intact) littermates, suggests the critical role of T cells in regulating the level of parasitemia (Clark and Allison 1974). CD4 gene knockout mice sustained *B. microti* parasitemias for longer periods of time than did congeneic controls (Hemmer et al. 2000). In contrast, humoral immunity appears to be less important than cellular immunity in resolving *B. microti* infection. Passive transfer of immune serum to severe combined immunodeficiency (SCID) mice and to nude mice fails to protect them from *B. microti* infection (Igarashi et al. 1999; Matsubara et al. 1993). T-cell-receptor deficient mice are more easily infected and maintain greater parasitemias than are those that are B-cell deficient (Clawson et al. 2002). Indeed, the severe babesial disease that has been reported to occur in HIV-positive patients would seem to affirm the critical role of T cells in regulating this infection (Machtinger et al. 1993; Falagas and Klempner 1996).

Experiments using inbred mice and murine-adapted *B. microti* have explored the possibility of a genetic basis for susceptibility, defined as the inability of mice of a given strain to regulate and eventually clear parasitemia. Parasitemia was lower in intact BALB/c mice than in intact mice of four other strains inoculated with *B. microti* but was highest in BALB/c mice when they were splenectomized (Ruebush and Hanson, 1979, 1980). Intact C3H mice were most susceptible to infection. Susceptibility does not appear to be related to H-2 (major histocompatibility complex) haplotype, however, because mice of identical haplotypes develop different maximum parasitemias. The degree of protection of mice against *B. microti* induced by *Propionibacterium acnes* inoculation does not correlate with H-2 haplotype (Wood and Clark 1982). On the other hand, the ability to express this induced protection trait is heritable. Some as yet undefined genetic component appears to influence susceptibility to *B. microti* infection.

Various non-specific factors seem to modify susceptibility to piroplasm infection. The presence of eperythrozoa, for example, may antagonize infection (Peters 1965). Injection of various reagents and coinfection with diverse organisms may non-specifically protect mice against *B. microti* (Clark 1979). The prepatent period is increased and peak parasitemia reduced in mice on a severely restricted protein diet (Tetzlaff et al. 1988). Finally, although old mice experience later and lower peak parasitemias than do younger animals, they fail to clear infection (Habicht et al. 1983). The factors that control susceptibility are complex.

In patients with *B. microti* babesiosis, some antibody responses are generally evident when they are first diagnosed. Seventeen patients with positive blood smears had a median indirect fluorescent antibody test (IFAT) immunoglobulin (Ig)G titer of 1:1 024 (negative, 1:64) with a range of 1:128–1:4 096 in their serum at the time of diagnosis (Telford unpublished). The presence of antibody in the serum when these individuals first became ill indicates that the prepatent period may be prolonged. Titer level is not related to parasitemia, however, consistent with the conclusion from experiments in mice that antibody alone cannot modulate infection.

In one comprehensive study of immunoresponsiveness in humans with *B. microti* babesiosis, lymphocyte responses to non-specific mitogens were markedly suppressed in acutely ill patients (Benach et al. 1982a). The numbers of T cells bearing the IgG Fc receptor were high, as was the relative proportion of B lymphocytes. Diminished levels of complement factors C3 and C4 in the serum of patients with acute babesiosis suggest activation of the classical complement pathway. In addition, the patient's blood contained a high level of circulating immune complexes. These data indicate that *B. microti* infection profoundly alters the cellular immune status of patients.

Reinfection of those who have recovered from babesiosis has not been reported but it would not be easy to prove reinfection because of the difficulty of excluding recrudescence of an earlier infection. The concomitant immunity (or premunition) that characterizes babesiosis in cattle may also occur in humans. Using primers designed from species-specific sequences of the small subunit ribosomal DNA (Persing et al. 1992), the polymerase chain reaction (PCR) assay may facilitate a determination of the prevalence of chronic babesiosis in humans. Detection of non-clinical carriers is important because transfusion-acquired cases are not uncommon (Jacoby et al. 1980; Marcus et al. 1982; Smith et al. 1986).

PATHOLOGY

The development of the pathophysiology of babesial infection is directly related to the development of the parasitemia. Studies on hamsters and laboratory mice infected by syringe with *B. microti* of human origin have provided a wealth of information on host–parasite interactions. Intravascular and extravascular hemolysis develops as the parasitemia rises, resulting in profound anemia; the hematocrit may fall to 20 percent (Lykins et al. 1975). During this acute phase of disease, there is extramedullary hematopoiesis and hyperplasia of the splenic red pulp, giving rise to the virtually pathognomonic gross splenomegaly that accompanies babesiosis in wild rodents (Fay and Rausch 1969). In addition, livers of infected hamsters contain hypertrophied Kupffer cells, many with ingested parasitized erythrocytes but little hemoglobin breakdown products (Cullen and Levine 1987). The proximal convoluted tubules of the kidneys contain abundant hemosiderin, an observation consistent with occurrence of marked intravascular hemolysis (Cullen and Levine 1987).

WA1-infected hamsters die within 2 weeks of infection, and necropsy demonstrates intravascular mononuclear cell aggregates, as well as occlusion of small vessels by parasitised erythrocytes (Dao and Eberhard 1996; Wozniak et al. 1996). Multifocal coagulative necrosis was present in the heart, spleen, lung, and liver, perhaps due to vascular stasis, resultant anoxia, and tissue damage. Mice with a genetic disruption in the tumor necrosis factor-alpha (TNF-α) pathway were less likely to die of fulminating WA1 infection, as were CD8 gene knockout mice, whereas in $\gamma\delta$ T cell knockout mice and control mice WA-1 infection terminated fatally. Thus, CD8 T cells and local expression of TNF-α may contribute to the WA1-induced pathology (Hemmer et al. 2000).

DIAGNOSIS

Diagnosis of babesiosis in human hosts depends mainly on a demonstration of the organism in erythrocytes. Examination of conventional Giemsa-stained thin blood films remains the most generally useful diagnostic procedure. Although the presence of tetrad (Maltese cross) forms is said to be diagnostic, such elements are rarely encountered. Similarly, the absence of parasite hemozoin (malarial pigment) is often considered to be diagnostic for the piroplasms but early ring stages of malaria parasites also lack pigment. Diagnosis of *B. microti* (Figure 23.1) is made by a combination of criteria, including the presence of an intense parasitemia (1–50 percent), erythrocytes infected by multiple parasites, basket-shaped merozoites which are often extracellular

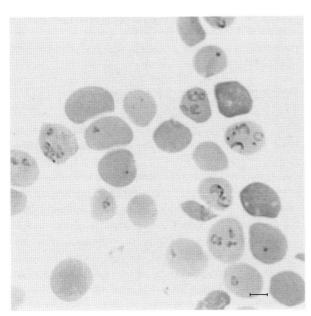

Figure 23.1 Babesia microti, *hamster blood. Note the varied morphology of the parasites, ranging from that of merozoites to that of mature trophiozoites. The occurrence of multiple infection of single erythrocytes is a characteristic which helps to distinguish these ring-shaped parasites from* Plasmodium falciparum. *Bar = 4 μm*

and which often have a light-colored cytoplasm. The small *Babesia* are difficult to recognize with confidence in a thick film, but the presence of tetrads or paired pyriforms might easily be detected. Parasitemias may sometimes also be exceedingly sparse, and persons with infections with low parasitemias may escape diagnosis when a thin blood film is examined. Inoculation into hamsters of a sample of patient's blood facilitates diagnosis by amplifying the parasitemia but the blood of the hamster should be examined microscopically at weekly intervals for at least 6 weeks before the test is declared negative. Demonstration of the characteristic organisms in the hamster blood proves infection.

Serological testing is useful, particularly in diagnosing chronic *B. microti* infection. The IFAT, using antigen derived from infected hamster red cells (Chisholm et al. 1978), is sensitive and specific and is currently the serological method of choice. A serological reaction at a titer above the cutoff point of 1:64 for IgG is generally considered to be diagnostic (Krause et al. 1994). Examination of paired acute and convalescent serum samples is most useful for a diagnosis of *B. microti* infection. Detection of parasite-specific IgM may indicate that the patient has an acute infection even in the absence of a readily demonstrable parasitemia.

In persons with *B. divergens* babesiosis, specific antibodies do not become detectable until at least 1 week after the onset of illness. Because this infection develops rapidly, serological procedures are not practical for the diagnosis of the infection in its acute form, but serological conversion serves as an aid in making a

retrospective diagnosis in survivors. Of course, the high parasitemias usually present in patients with this infection in its acute form are easily detected by examination of blood smears. Paired pyriform and accole parasites are frequently observed in erythrocytes from patients with acute *B. divergens* infection (Figure 23.2).

Infection with *Babesia* of the WA-1 strain may be readily detected by IFAT and this organism may also be detected by in vitro cultivation (Thomford et al. 1994). Its ease of cultivation facilitates production of a standardized antigen. For unknown reasons, a higher titer in the IFAT (more than 1:160) is required for diagnosis of WA-1 infection than for diagnosis of infection with *B. microti*. Tetrad forms are more frequently seen in blood smears from patients and rodents with WA-1 and CA-type infection (Figure 23.3) than in those with *B. microti* infections, but otherwise infection with this agent is difficult to distinguish from that with *B. microti*.

The PCR has largely supplanted the hamster inoculation test. In experienced hands, the PCR test detects *Babesia* DNA accurately and usually within a single working day. Hamster inoculation, long thought to be the gold standard for confirmatory diagnosis, has its problems. For example, it requires at least a week and often more than a month for incubation before results are obtained. Blood samples for hamster inoculation must also be taken prior to initiating treatment. An additional problem with use of hamster inoculation for diagnosis is that in many instances infection is transient in hamsters and may be missed if monitoring of blood smears is done only weekly. Finally, an infection due to a *Babesia* other than *B. microti*, *B. divergens* or WA-1, might be missed entirely if it lacks the ability to infect hamsters. These problems are avoided if diagnosis is made with the aid of a genus-specific DNA amplification assay.

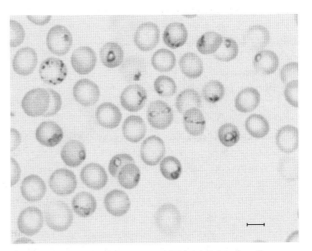

Figure 23.2 Babesia divergens, *gerbil blood. Pairs of pyriform organisms occur frequently and their presence helps to distinguish* B. divergens *from the other* Babesia *infecting humans. Ring-shaped organisms may also occur, as may tetrad forms. Bar = 4.8 μm*

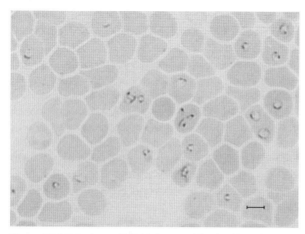

Figure 23.3 Babesia *sp. (WA-1), hamster blood. Although its morphology is very similar to that of* B. microti, *WA-1 may more frequently occur as tetrads than does* B. microti; *multiple infection of erythrocytes occurs less frequently in blood infected with WA-1 than in blood infected with* B. microti. *Bar = 5 μm*

EPIDEMIOLOGY AND CONTROL

The acquisition of babesiosis due to *B. divergens* or *B. microti* depends on contact with the subadult stages of certain *Ixodes* ticks. The vectors of WA-1, CA-type, or the atypical *B. divergens*-like infections, however, have not yet been described. Environmental disturbance, increasing recreational use of the wilderness and suburbanization promote contact between people, ticks, and the pathogens that they maintain (Spielman 1988).

Two epidemiological patterns are apparent in human babesiosis. The first (pattern I) involves splenectomized or otherwise immunologically compromised people and diverse *Babesia* spp. This pattern includes European 'divergens' babesiosis. The risk of acquiring human babesiosis in the northeastern and northern Midwestern USA does not appear to depend on splenectomy. This second epidemiological pattern (pattern II) involves infection by the rodent piroplasm *B. microti* and being bitten by *I. dammini* ticks. The vectors of *B. microti*-like babesiosis in Asia (Shih et al. 1997; Saito-Ito et al. 2000) remain undescribed.

Although one might predict that the increasing prevalence of HIV infection in many tropical areas of the world may allow the emergence of 'tropical' babesiosis, such cases have not yet been reported. Babesiosis is particularly severe in individuals with HIV and several case reports of babesiosis in patients with acquired immune deficiency syndrome (AIDS) already exist in the literature (Benezra et al. 1987; Ong et al. 1990; Machtinger et al. 1993; Falagas and Klempner 1996). The diversity of *Babesia* species and the ubiquity of ticks suggest a high potential for the emergence of *Babesia* as zoonotic agents. It is at least possible that some chloroquine-resistant malaria parasites may eventually be identified as *Babesia* (Young and Morzaria 1986). Whether or not this will be the case will be determined

by careful microscopy, complemented by the use of molecular diagnostic techniques by people aware of the possibility of babesial infection in humans.

Epidemiology and ecology

The ecology of *B. microti* is the best understood of the three known zoonotic piroplasms. The enzootic cycle of *B. microti* clearly depends on the interaction of subadult *I. dammini* and their main host, the white-footed mouse (*Peromyscus leucopus*). Deer (*Odocoileus virginianus*) serve as the host upon which adult ticks commonly feed, but deer are not reservoirs for *B. microti* (Piesman et al. 1979). Adult ticks feed during the autumn, overwinter, feed, and engorge again and lay eggs during the spring (Yuval and Spielman 1990). The eggs hatch synchronously in late July, with larvae feeding on mice mainly during August and September, when they acquire babesial infection; the prevalence of *Babesia* infection in the mice appears to be about 60 percent (Etkind et al. 1980).

Fed larvae overwinter and molt to the nymphal stage during the spring. Thus, non-infected mice resulting from reproduction during May and June are inoculated by nymphs infected as larvae during the preceding fall. About 40 percent of nymphal ticks on Nantucket Island (USA) may contain babesial sporozoites in their salivary glands during June. People become exposed mainly during June (Piesman et al. 1987) when they visit zoonotic sites for recreation or inhabit summer homes located in such sites. The population of Nantucket Island increases from about 7 000 winter residents to over 35 000 summer residents after Memorial Day (at the end of May). The small size of nymphal *I. dammini* makes its detection and prompt removal difficult. Failure to remove a feeding tick promptly determines risk because the probability that the tick will transfer the *Babesia* to the host on which it feeds is directly proportional to the duration of feeding by the tick (Piesman et al. 1986).

The life cycle of the tick is completed when nymphs that have fed on a host molt to the adult stage in the fall. Thus, *I. dammini* ticks develop from egg to egg over a span of at least 2 years (Yuval and Spielman 1990), permitting cohorts of nymphal ticks to overlap, thereby buffering the tick population against years of host scarcity.

The epizootiology of *B. divergens* and the role of its presumed vector, *I. ricinus*, in the spread of infection is poorly understood. In Ireland, a high incidence in cattle is associated with elevated air temperature, which is when there is much tick activity. As in the case of *I. dammini*, nymphs may become active in the spring before the larvae hatch, but incrimination of the vector stage responsible for transmitting infection is confounded by the presence of adult ticks concurrently with nymphs (Gray 1980). Most cases of babesiosis due to *B. divergens* in humans have occurred in farmers or others who are frequently in contact with cattle (Clarke et al. 1989). Interestingly, although *B. microti* is enzootic in many European sites, human babesiosis due to this parasite has not been observed there, possibly because a uniquely mouse-specific tick, *I. trianguliceps*, maintains *B. microti* in Europe, although the aggressively human-biting *I. ricinus* may densely infest the same sites. Perhaps European strains of *B. microti* are not pathogenic to humans. Recent serosurveys in Germany and Switzerland (Hunfeld et al. 2002; Foppa et al. 2002) suggest exposure, but clinically apparent *B. microti* babesiosis has not been recorded from these European sites.

Neither the vector nor the reservoir of the WA-1 and CA-type *Babesia* are known. Three ticks are prevalent in sites where cases of WA-1 or CA-type babesiosis have occurred: *I. pacificus*, *Dermacentor variabilis*, and *Ornithodoros coriaceus*. Phylogenetic studies (Kjemtrup et al. 2000a) suggest that mule deer or bighorn sheep may be infected by CA-type parasites, and a dog isolate serves as the most closely related agent to WA-1. Identification of candidate reservoir hosts may help focus investigations to incriminate the vectors of these *Babesia* spp.

Residents of few countries outside the USA and Europe appear to have experienced clinical babesiosis. One case has been observed in China (Li and Meng 1984) and two in South Africa (Bush et al. 1990). The species of *Babesia* involved in these cases has not been determined. One fatal case due to infection by *B. caucasica* has been observed in the former Soviet Union (Rabinovich et al. 1978). Unidentified *Babesia* were isolated by hamster inoculation from asymptomatic residents of Mexico (Osorno et al. 1976). A serosurvey conducted in Venezuela revealed that there was a low prevalence of antibody against *B. bovis* in the Venezuelan population (Montenegro-James et al. 1990). In Asia, serosurveys using *B. microti* antigen indicated exposure of a few residents of Taiwan (Hsu and Cross 1977) and three cases of human babesiosis, caused by an agent that infected hamsters and whose 16s rDNA sequence closely resembles *B. microti*, were observed (Shih et al. 1997). An authocthonous Japanese case of *B. microti*-like babesiosis, transmitted by transfusion, has been reported (Tsuji et al. 2001); although *B. microti*-like parasites are prevalent in Japanese rodents (Shiota et al. 1984), the genotype recovered from the patient is apparently only found on a small island, suggesting the hypothesis that only this insular strain of parasite may be zooonotic (Tsuji et al. 2001). Asymptomatic *Babesia* infections have been observed in Africa. A survey conducted in Nigeria indicated that 54 percent of 173 men from the northwestern border may have been infected (Leeflang et al. 1976) and a serological survey in Mozambique suggested that *B. bovis* may infect residents there (Rodriguez et al. 1984).

Two human infections have been ascribed to *Entopolypoides macaci* (Wolf et al. 1978). Organisms in this genus, however, are now generally considered to be *Babesia* (Levine 1988; Bronsdon et al. 1999). This piroplasm (which should now be referred to as *Babesia macaci* although the designation PB-1 was applied by Bronsdon et al. 1999) appears to be commonly found in research primate colonies where ectoparasite control is stringent. It may be that this *Babesia* can be maintained in the absence of tick vectors by transplacental or perinatal transmission or by direct blood contact during aggressive interactions.

CHEMOTHERAPY

Because of its use for malaria, orally administered chloroquine was used in the treatment of the first few Americans in whom *Babesia* infections were observed. Symptomatic improvement was described, but parasitemia tended to continue (Ruebush et al. 1977). Indeed, administration of chloroquine, sulfadiazine, or pyrimethamine fails to reduce parasitemia in hamsters (Miller et al. 1978). Although pentamidine appears to have been useful in several patients who were treated with this drug, such therapy fails to eliminate the parasitemia completely (Francioli et al. 1981). The treatment of choice for babesiosis in humans caused by *B. microti* is quinine and clindamycin, administered in combination (Dammin et al. 1983). Quinine should be administered orally in a regimen of 650 mg, three times daily and clindamycin intravenously at 1200 mg, twice a day. Alternatively, clindamycin can be administered in an oral regimen of 600 mg, three times a day. The treatment should be continued for at least 7 days (Anon 1986) or until the parasitemia is eliminated, whichever comes first. This treatment generally is effective, except in those who are immunosuppressed (Smith et al. 1986) or infected by HIV (Benezra et al. 1987; Ong et al. 1990). Atovaquone (750 mg every 12 hours) and azithromycin (500 mg on day 1 and 250 mg per day thereafter) orally for 7 days appears to be as effective and better tolerated than quinine–clindamycin (Krause et al. 2000). In fulminating cases, exchange transfusion is life-saving (Jacoby et al. 1980; Cahill et al. 1981; Sun et al. 1983).

Any person with babesiosis acquired in Europe should be treated on an emergency basis. In addition to supportive treatment, such patients should receive prompt specific therapy designed to reduce parasitemia and to prevent the extensive hemolysis and consequent renal failure that may follow. Massive exchange transfusion (between two and three blood volumes) should be followed by administration of 600 mg clindamycin intravenously, between three and four times daily, together with quinine (600 mg base) administered orally three times a day (Gorenflot et al. 1990, 1987; Uhnoo et al. 1992). Indeed, because of the rapidly increasing parasitemia characteristic of the disease, exchange transfusion should be instituted on the first signs of disease due to *B. divergens* infection. The time element becomes crucial because the request for medical attention generally follows the onset of hemoglobinuria, a manifestation of the disease that signals a fulminating parasitemia. Unless blood exchange is promptly undertaken, the prognosis is poor, as the rapidly increasing intravascular hemolysis leads to renal failure.

Babesiosis due to WA-1, CA-type, or atypical *B. divergens* appears to respond to quinine and clindamycin, although too few cases have been reported to gauge the efficacy of this treatment regimen.

VACCINATION

Strong immunity in cattle against bovine babesiosis has been induced by injection of attenuated parasites (Callow 1971) or exoantigen preparations (Smith and Ristic 1981). Less successful vaccines have been prepared from recombinant antigens. Crude parasite lysates serve to immunize hamsters against infection by *B. microti* (Benach et al. 1982b). Culture-derived parasites, crude culture supernatants, and purified exoantigen preparations protect gerbils and cattle against *B. divergens* (Valentin et al. 1993). Immunoscreening of *B. microti* expression libraries have yielded candidate protective antigens such as bmn1-6 and p58 (Homer et al. 2000; Nishisaka et al. 2001), but development of vaccine candidates for human use remains a long-term objective. Other than for those individuals who are splenectomized or otherwise immunocompromised or living in certain high-risk areas, however, the value of an antibabesial vaccine for prevention of babesiosis in humans is doubtful.

INTEGRATED CONTROL

The accepted public health strategies to protect human populations against zoonotic babesial infection depend on reducing the density of ticks. The spraying of acaricidal emulsions (Stafford 1991) has been used to reduce the abundance of ticks on vegetation. Less environmentally intrusive methods of application of acaricides, one particularly based on fiber-formulated permethrin may be more acceptable (Mather et al. 1987). These methods interrupt transmission of *B. microti* infection by depositing acaricide in the nests and on the coats of rodent reservoir hosts. Tick infestations may also be reduced locally by destroying the animals (i.e. deer) that serve as host to the adult stage of the tick (Wilson et al. 1988). An effort to reduce the abundance of the hosts of subadult ticks might be counterproductive because the absence of these hosts might increase the density of ticks seeking alternative hosts (Spielman et al. 1981). Acaricides applied to the coats of cattle could reduce the possible transmission of cattle babesiosis to humans.

Repellents are useful for personal protection, particularly permethrin-based formulations applied to clothing (Schreck et al. 1986). Even volatile compounds such as diethyltoluamide (DEET) may provide a certain measure of protection. Personal protection, based on daily examination of the body surface of a person who has visited a site where transmission is intense, is the most effective means of reducing the risk of contracting babesiosis. All ticks found should be promptly disposed of to prevent their attachment. Ticks already attached should be removed by means of forceps to take advantage of the initial 50–60 hour 'grace period' during which attached ticks rarely transmit infection.

ACKNOWLEDGMENTS

Our work has been supported by grants from the National Institutes of Health (AI 19693, AI 37993, and AI 39002), SmithKlineBeecham Pharmaceuticals, the Chace Fund, the Gibson Island Corporation, and David Arnold.

REFERENCES

Allsopp, M.T., Cavalier-Smith, T., et al. 1994. Phylogeny and evolution of the piroplasms. *Parasitology*, **108**, 147–52.

Anon, 1986. Drugs for parasitic infections. *Med Let Drug Ther*, **28**, 9–16.

Beattie, J.F., Michelson, M.L. and Holman, P.J. 2002. Acute babesiosis caused by *Babesia divergens* in a resident of Kentucky. *N Eng J Med*, **347**, 697–8.

Benach, J.L. and Habicht, G.S. 1981. Clinical characteristics of human babesiosis. *J Infect Dis*, **144**, 48.

Benach, J.L., Habicht, G.S. and Hamburger, M.I. 1982a. Immunoresponsiveness in acute babesiosis in humans. *J Infect Dis*, **146**, 369–80.

Benach, J.L., Habicht, G.S., et al. 1982b. Glucan as an adjuvant for a murine *Babesia microti* immunisation trial. *Infect Immun*, **35**, 947–51.

Benezra, D., Brown, A.E., et al. 1987. Babesiosis and infection with human immunodeficiency virus (HIV). *Ann Intern Med*, **107**, 944.

Boustani, M.R., Lepore, T.J., et al. 1994. Acute respiratory failure in patients treated for babesiosis. *Am J Resp Crit Care Med*, **149**, 1689–91.

Bronsdon, M.A., Homer, M.J., et al. 1999. Detection of enzootic babesiosis in baboons (*Papio cynocephalus*) and phylogenetic evidence supporting synonymy of the genera *Entopolypoides* and *Babesia*. *J Clin Microbiol*, **37**, 1548–53.

Bush, J.B., Isaacson, M., et al. 1990. Human babesiosis: a preliminary report of 2 suspect cases in southern Africa. *S Afr J Med*, **78**, 699.

Caccio, S., Camma, C., et al. 2000. The beta tubulin gene of *Babesia* and *Theileria* parasites is an informative marker for species discrimination. *Int J Parasitol*, **30**, 1181–5.

Cahill, K.M., Benach, J.L., et al. 1981. Red cell exchange: Treatment of babesiosis in a splenectomized patient. *Transfusion*, **21**, 193–8.

Callow, L.L. 1971. The control of babesiosis with a highly effective attenuated vaccine. *Proc World Vet Congress*, **1**, 357–60.

Carret, C., Walas, F., et al. 1999. *Babesia canis canis, Babesia canis vogeli, Babesia canis rossi*: differentiation of the three subspecies by a restriction fragment length polymorphism analysis on amplified small subunit ribosomal RNA genes. *J Euk Microbiol*, **46**, 298–303.

Chisholm, E.S. and Ruebush II, T.K. 1978. *Babesia microti* infection in man: evaluation of an indirect immunofluorescent antibody test. *Am J Trop Med Hyg*, **27**, 14–19.

Clark, I.A. 1979. Protection of mice against *Babesia microti* with cord factor, COAM, zymosan, glucan, *Salmonella* and *Listeria*. *Parasite Immunol*, **1**, 179–96.

Clark, I.A. and Allison, A.C. 1974. *Babesia microti* and *Plasmodium berghei yoelii* infections in nude mice. *Nature*, **252**, 328–9.

Clarke, C.S., Rogers, E.T. and Egan, E.L. 1989. Babesiosis: under-reporting or case-clustering? *Postgrad Med J*, **65**, 591–3.

Clawson, M., Paciorkowski, N., et al. 2002. Cellular immunity, but not gamma interferon, is essential for resolution of *Babesia microti* infection in BALB/c mice. *Infect Immun*, **70**, 5304–6.

Cullen, J.M. and Levine, J.F. 1987. Pathology of experimental *Babesia microti* infection in the Syrian hamster. *Lab Anim Sci*, **37**, 640–3.

Dammin, G.J., Spielman, A., et al. 1983. Clindamycin and quinine treatment for *Babesia microti* infections. *Morb Mort Rep*, **32**, 65–6.

Dao, A.H. and Eberhard, M.L. 1996. Pathology of acute fatal babesiosis in hamsters experimentally infected with the WA-1 strain of *Babesia*. *Lab Invest*, **74**, 853–9.

Ellis, J., Hefford, C., et al. 1992. Ribosomal DNA sequence comparison of *Babesia* and *Theileria*. *Mol Biochem Parasitol*, **54**, 87–96.

Enigk, K. and Friedhoff, K. 1962. *Babesia capreoli* n.sp. beim Reh (Capreolus capreolus L.). *Z Tropenmed Parasitol*, **13**, 8–20.

Entrican, J.H., Williams, H., et al. 1979. Babesiosis in man: report of a case from Scotland with observations on the infecting strain. *J Infection*, **1**, 227–34.

Etkind, P., Piesman, J., et al. 1980. Methods for detecting *Babesia* microti infection in wild rodents. *J Parasitol*, **66**, 107–10.

Falagas, M.E. and Klempner, M.S. 1996. Babesiosis in patients with AIDS: a chronic infection presenting as fever of unknown origin. *Clin Infect Dis*, **22**, 809–12.

Fay, F.G. and Rausch, R.L. 1969. Parasitic organisms in the blood of arvicoline rodents in Alaska. *J Parasitol*, **55**, 1258–65.

Foppa, I.M., Krause, P.J., et al. 2002. Entomologic and serologic evidence of zoonotic transmission of *Babesia microti*, eastern Switzerland. *Emerg Infect Dis*, **8**, 722–6.

Francioli, P.B., Keithly, J.S., et al. 1981. Response of babesiosis to pentamidine therapy. *Ann Intern Med*, **94**, 326–30.

Goethert, H.K. and Telford III, S.R. 2003a. What is *Babesia microti*? *Parasitology*, **127**, 301–9.

Goethert, H.K. and Telford III, S.R. 2003b. Enzootic transmission of *Babesia divergens* among cottontail rabbits on Nantucket Island, Mass.. *Am J Trop Med Hyg*, **69**, 455–60.

Golightly, L.M., Hirschhorn, L.R. and Weller, P.F. 1989. Fever and headache in a splenectomized woman. *Rev Infect Dis*, **11**, 629–37.

Gordon, S., Cordon, R.A., et al. 1984. Adult respiratory distress syndrome in babesiosis. *Chest*, **86**, 633–4.

Gorenflot, A., Precigout, E., et al. 1987. *Babesia divergens* vaccine. *Mem Inst Oswaldo Cruz*, **87**, Supplement 3, 279–81.

Gorenflot, A., Brasseur, P., et al. 1990. Deux cas de babesiose humaine grave traités avec succes. *Presse Med*, **19**, 335.

Gray, J.S. 1980. Studies on the activity of *Ixodes ricinus* in relation to the epidemiology of babesiosis in County Meath, Ireland. *Br Vet J*, **136**, 427–36.

Gray, J.S., Von Stedingk, L.V., et al. 2002. Transmission studies of *Babesia microti* in *Ixodes ricinus* ticks and gerbils. *J Clin Microbiol*, **40**, 1259–63.

Habicht, G.S., Benach, J.L., et al. 1983. The effect of age on the infection and immunoresponsiveness of mice to *Babesia microti*. *Mech Aging Dev*, **23**, 357–69.

Hemmer, R.M., Ferrick, D.A. and Conrad, P.A. 2000. Role of T cells and cytokines in fatal and resolving experimental babesiosis: protection in TNFRp55-/- mice infected with the human *Babesia* WA1 parasite. *J Parasitol*, **86**, 736–42.

Herwaldt, B.L., Persing, D.H., et al. 1996. A fatal case of babesiosis in Missouri: identification of another piroplasm that infects humans. *Ann Int Med*, **124**, 643–50.

Herwaldt, B.L., Kjemtrup, A.M., et al. 1997. Transfusion-transmitted babesiosis in Washington State: first reported case caused by a WA1-type parasite. *J Infect Dis*, **175**, 1259–62.

Herwaldt, B.L., Cacciò, S., et al. 2003. Molecular characterization of a non-*Babesia divergens* organism causing zoonotic babesiosis in Europe. *Emerg Infect Dis*, **9**, 02–0748, http://www.cdc.gov/ncidod/EID/vol9no8/02-0748.htm.

Homer, M.J., Bruinsma, E.S., et al. 2000. A polymorphic multigene family encoding an immunodominant protein from *Babesia microti*. *J Clin Microbiol*, **38**, 362–8.

Hsu, N.H.M. and Cross, J.H. 1977. Serologic survey for human babesiosis on Taiwan. *J Formosan Med Assoc*, **76**, 950–4.

Hunfeld, K.P., Lambert, A., et al. 2002. Seroprevalence of *Babesia* infections in humans exposed to ticks in midwestern Germany. *J Clin Microbiol*, **40**, 2431–6.

Igarishi, I., Suzuki, R., et al. 1999. Roles of CD4(+) T cells and gamma interferon in protective immunity against *Babesia microti* infection in mice. *Infect Immun*, **67**, 4143–8.

Jacoby, G.A., Hunt, J.V., et al. 1980. Treatment of transfusion-transmitted babesiosis by exchange transfusion. *N Engl J Med*, **303**, 1098–100.

Kakoma, I. and Mehlhorn, H. 1994. *Babesia* of domestic animals. In: Kreier, J.P. (ed.), *Parasitic protozoa*, 2nd edn. San Diego: Academic Press, 141–216.

Karakashian, S.J., Rudzinska, M.A., et al. 1986. Primary and secondary ookinetes of *Babesia microti* in the larval and nymphal stages of the tick, *Ixodes dammini*. *Can J Zool*, **64**, 328–39.

Kjemtrup, A.M., Thomford, J., et al. 2000a. Phylogenetic relationships of human and wildlife piroplasm isolates in the western United States inferred from the 18S nuclear small subunit RNA gene. *Parasitology*, **120**, 487–93.

Kjemtrup, A.M., Kocan, A.A., et al. 2000b. There are at least three genetically distinct small piroplasms from dogs. *Int J Parasitol*, **30**, 1501–5.

Kjemtrup, A.M., Lee, B., et al. 2002. Investigation of transfusion transmission of a WA1-type babesial parasite to a premature infant in California. *Transfusion*, **42**, 1482–7.

Krause, P.J., Telford III, S.R., et al. 1994. Diagnosis of babesiosis: evaluation of a serologic test for the detection of *Babesia microti* antibody. *J Infect Dis*, **169**, 923–6.

Krause, P.J., Lepore, T.J., et al. 2000. Atovaquone and azithromycin for the treatment of babesiosis. *N Engl J Med*, **343**, 1454–8.

Leeflang, P., Oomen, J.M.V., et al. 1976. The presence of *Babesia* antibody in Nigerians. *Int J Parasitol*, **6**, 159–61.

Levine, N.D. *The protozoan phylum Apicomplexa*, Vol. II. . Boca Raton, FL: CRC Press.

Li, J.F. and Meng, D.B. 1984. The discovery of human babesiosis in China. *Chin J Vet Med*, **10**, 19–20.

Lykins, J.D., Ristic, M., et al. 1975. *Babesia microti*: pathogenesis of parasite of human origin in the hamster. *Exp Parasitol*, **37**, 388–97.

Machtinger, L., Telford III, S.R., et al. 1993. Treatment of babesiosis by red blood cell exchange in an HIV positive splenectomised patient. *J Clin Apheresis*, **8**, 78–81.

Mackenstedt, U., Gauer, M., et al. 1990. Sexual cycle of *Babesia divergens* confirmed by DNA measurements. *Parasitol Res*, **76**, 199–206.

Mahoney, D.F. 1977. Babesiosis of domestic animals. In: Kreier, J.P. (ed.), *Parasitic protozoa*, Vol. IV. . New York: Academic Press, 1–52.

Marcus, L.C., Valigorsky, J.M., et al. 1982. A case report of transfusion induced babesiosis. *JAMA*, **248**, 465–7.

Mather, T.N., Ribeiro, J.M.C. and Spielman, A. 1987. Lyme disease and babesiosis: acaricide focused on potentially infected ticks. *Am J Trop Med Hyg*, **36**, 609–14.

Matsubara, J., Koura, M. and Kamiyama, T. 1993. Infection of immunodeficient mice with a mouse-adapted strain of the Gray strain of *Babesia microti*. *J Parasitol*, **79**, 783–6.

McCutchan, J.L., Hansen, T.F., et al. 1984. Mung bean nuclease cleaves *Plasmodium* genomic DNA at sites before and after genes. *Science*, **225**, 625–8.

Mehlhorn, H. and Schein, E. 1984. The piroplasms: life cycle and sexual stages. *Adv Parasitol*, **23**, 37–103.

Meldrum, S.C., Birkhead, G.S., et al. 1992. Human babesiosis in New York State: an epidemiological description of 136 cases. *Clin Infect Dis*, **15**, 1019–23.

Miller, L.H., Neva, F.A. and Gill, F. 1978. Failure of chloroquine in human babesiosis (*Babesia microti*): case report and chemotherapeutic trials in hamsters. *Ann Intern Med*, **88**, 200–2.

Momen, H., Chance, M.L. and Peters, W. 1979. Biochemistry of intraerythrocytic parasites III. Biochemical taxonomy of rodent *Babesia. Ann Trop Med Parasitol*, **73**, 203–12.

Montenegro-James, S., James, M.A., Lopez, R.1990. Seroprevalence of human babesiosis in Venezuela, *Annual Meeting, American Society of Tropical Medicine and Hygiene.* New Orleans, LA.

Moss, D.M., Healy, G.R., et al. 1986. Isoenzyme analysis of *Babesia microti* infections in humans. *J Protozool*, **33**, 213–15.

Nishisaka, M., Yokoyama, N., et al. 2001. Characterisation of the gene encoding a protective antigen from *Babesia microti* identified it as η subunit of chaperonin containing T-complex protein 1. *Int J Parasitol*, **31**, 1673–9.

Ong, K.R., Stavropoulos, C. and Inada, Y. 1990. Babesiosis, asplenia and AIDS. *Lancet*, **336**, 112.

Osorno, B.M., Vega, C., et al. 1976. Isolation of *Babesia* spp. from asymptomatic human beings. *Vet Parasitol*, **2**, 111–20.

Persing, D.H., Mathiesen, D., et al. 1992. Detection of *Babesia microti* by polymerase chain reaction. *J Clin Microbiol*, **30**, 2097–103.

Persing, D.H., Herwaldt, B.L., et al. 1995. Infection with a *Babesia*-like organism in northern California. *N Engl J Med*, **332**, 298–303.

Peters, W. 1965. Competitive relationship between *Eperythrozoon coccoides* and *Plasmodium berghei* in the mouse. *Exp Parasitol*, **16**, 158–66.

Piesman, J. and Spielman, A. 1982. *Babesia microti*: infectivity of parasites from ticks from hamsters and white-footed mice. *Exp Parasitol*, **53**, 242–8.

Piesman, J., Spielman, A., et al. 1979. Role of deer in the epizootiology of *Babesia microti* in Massachusetts, USA. *J Med Entomol*, **15**, 537–40.

Piesman, J., Mather, T.N., et al. 1986. Duration of tick attachment and *Borrelia burgdorferi* transmission. *J Clin Microbiol*, **25**, 557–8.

Piesman, J., Mather, T.N., et al. 1987. Seasonal variation of transmission risk of Lyme disease and human babesiosis. *Am J Epidemiol*, **126**, 1187–9.

Quick, R.E., Herwaldt, B.L., et al. 1993. Babesiosis in Washington state: a new species of Babesia? *Ann Intern Med*, **119**, 284–90.

Rabinovich, S.A., Voronina, Z.K., et al. 1978. First detection of human babesiosis in the USSR and brief analysis of cases described in the literature. *Med Parazitol*, **47**, 97–107.

Ribeiro, J.M.C. 1987. Role of saliva in blood-feeding by arthropods. *Annu Rev Entomol*, **32**, 463–78.

Riek, R.F. 1968. Babesiosis. In: Weinman, D. and Ristic, M. (eds), *Infectious blood diseases of man and animals*, Vol. 2. . New York: Academic Press, 220–68.

Rodriguez, O.N., Dias, M. and Rodriguez, P. 1984. Reporte de la infeccion por *Babesia bovis* (Babes) en la poblacion humana de la Republica popular de Mozambique. *Rev Cubana Cien Vet*, **15**, 41–50.

Roth, E.F. Jnr., Tanowitz, H., et al. 1981. *Babesia microti*: biochemistry and function of hamster erythrocytes infected from a human source. *Exp Parasitol*, **51**, 116–23.

Rudzinska, M.A. 1981. Morphological aspects of host-cell-parasite relationships in babesiosis. In: Ristic, M. and Kreier, J.P. (eds), *Babesiosis*. New York: Academic Press, 87–141.

Rudzinska, M.A., Spielman, A., et al. 1979. Intraerythrocytic 'gametocytes' of *Babesia microti* and their maturation in ticks. *Can J Zool*, **57**, 424–34.

Rudzinska, M.A., Spielman, A., et al. 1983. Sexuality in piroplasms as revealed by electron microscopy in *Babesia microti. Proc Natl Acad Sci USA*, **80**, 2966–70.

Ruebush, M.J. and Hanson, W.L. 1979. Susceptibility of five strains of mice to *Babesia microti* of human origin. *J Parasitol*, **65**, 430–3.

Ruebush, M.J. and Hanson, W.L. 1980. Thymus dependence of resistance to infection with *Babesia microti* of human origin in mice. *Am J Trop Med Hyg*, **29**, 507–15.

Ruebush, T.K., Cassaday, P.B., et al. 1977. Human babesiosis on Nantucket Island: clinical features. *Ann Intern Med*, **86**, 6–9.

Saito-Ito, A., Tsuji, M., et al. 2000. Transfusion-acquired, autochthonous human babesiosis in Japan: isolation of *Babesia microti*-like parasites with hu-RBC-SCID mice. *J Clin Microbiol*, **38**, 4511–16.

Schein, E., Reihbein, G., et al. 1981. *Babesi equi* (Laveran 1901) 1 Development in horses and in lymphocyte culture. *Z Parasitenk*, **34**, 68–94.

Schreck, C.E., Snoddy, E.L. and Spielman, A. 1986. Pressurized sprays of permethrin or deet on military clothing for personal protection against *Ixodes dammini* (Acari: Ixodidae). *J Med Entomol*, **23**, 396–9.

Shih, C.M., Liu, L.P., et al. 1997. Human babesiosis in Taiwan: asymptomatic infection with a *Babesia microti*-like organism in a Taiwanese woman. *J Clin Microbiol*, **35**, 450–4.

Shiota, T., Kurimoto, H., et al. 1984. Studies on babesia first found in murine in Japan: epidemiology, morphology, and experimental infection. *Z Bakt Mikrobiol Hyg*, **256**, 347–55.

Skrabalo, Z. and Deanovic, Z. 1957. Piroplasmosis in man. Report on a case. *Doc Med Geogr Trop*, **9**, 11–16.

Smith, R.D. and Ristic, M. 1981. Immunization against babesiosis with culture-derived antigens. In: Ristic, M. and Kreier, J.P. (eds), *Babesiosis*. New York: Academic Press, 485–507.

Smith, R.P., Evans, A.T., et al. 1986. Tranfusion-acquired babesiosis and failure of antibiotic treatment. *JAMA*, **256**, 2726–7.

Smith, T., Kilbourne, F.L. 1893. *Investigation into the nature, causation, and prevention of Texas or southern cattle fever.* Bureau of Animal Industries Bulletin No. 1. Washington, DC: US Department of Agriculture.

Spielman, A. 1988. Lyme disease and human babesiosis: evidence incriminating vector and reservoir hosts. In: Englund, P.T. and Sher, A. (eds), *The biology of parasitism*. New York: AR Liss, 147–65.

Spielman, A., Etkind, P. and Piesman, J. 1981. Reservoir hosts of human babesiosis on Nantucket Island. *Am J Trop Med Hyg*, **30**, 560–5.

Stafford, K.C. 1991. Effectiveness of carbaryl applications for the control of *Ixodes dammini* (Acari:Ixodidae) nymphs in an endemic residential area. *J Med Entomol*, **28**, 32–6.

Sun, T., Tenenbaum, M.J., et al. 1983. Morphologic and clinical observations in human infection with *Babesia microti. J Infect Dis*, **148**, 239–48.

Telford III, S.R. and Lepore, T.J. 1995. Human granulocytic ehrlichiosis in Massachusetts. *Ann Intern Med*, **123**, 277–9.

Tetzlaff, C.L., Carlomagno, M.A. and McMurray, D.N. 1988. Reduced dietary protein content suppresses infection with *Babesia microti. Med Microbiol Immunol*, **177**, 305–15.

Tetzlaff, C.L., McMurray, D.N. and Rice-Ficht, A.C. 1990. Isolation and characterisation of a gene associated with a virulent strain of *Babesia microti. Mol Biochem Parasitol*, **40**, 183–92.

Thomford, J.W., Conrad, P.A., et al. 1994. Cultivation and characterisation of a newly recognised human pathogenic protozoan. *J Infect Dis*, **169**, 1050–6.

Tsuji, M., Wei, Q., et al. 2001. Human babesiosis in Japan: epizootiologic survey of rodent reservoir and isolation of new type of *Babesia microti*-like parasite. *J Clin Microbiol*, **39**, 4316–22.

Uhnoo, I., Cars, O., et al. 1992. First documented case of human babesiosis in Sweden. *Scan J Infect Dis*, **24**, 541–7.

Valentin, A., Precigout, E., et al. 1993. Cellular and humoral immune responses induced in cattle by vaccination with *Babesiaa divergens* culture-derived exoantigens correlate with protection. *Infect Immun*, **61**, 734–41.

Walter, G. and Weber, G. 1981. Untersuchung zur Ubertragung (transstadial, transovarial) von *Babesia microti* in Stamm 'Hannover I', in *Ixodes ricinus. Tropenmed Parasitol*, **32**, 228–30.

Western, K.A., Benson, G.D. and Gleason, N.N. 1970. Babesiosis in a Massachusetts resident. *N Engl J Med*, **283**, 854–6.

Wilson, M.E. 1991. *A world guide to infections*. New York: Oxford University Press.

Wilson, M.L., Telford III, S.R., et al. 1988. Reduced abundance of immature *Ixodes dammini* (Acari: Ixodidae) following elimination of deer. *J Med Ent*, **25**, 224–8.

Wolf, R.E. 1974. Effects of antilymphocyte serum and splenectomy on resistance to *Babesia microti* infection in hamsters. *Clin Immunol Immunopathol*, **2**, 381–94.

Wolf, R.E., Gleason, N.N., et al. 1978. Intraerythrocytic parasitosis in humans with *Entopolypoides* species (Family Babesiidae). *Ann Intern Med*, **88**, 769–73.

Wood, P.R. and Clark, I.A. 1982. Genetic control of *Propionibacterium acnes*-induced protection of mice against *Babesia microti*. *Infect Immun*, **35**, 52–7.

Wozniak, E.J., Lowenstine, L.J., et al. 1996. Comparative pathogenesis of human WA1 and *Babesia microti* isolates in a Syrian hamster model. *Lab Animal Sci*, **46**, 507–15.

Young, A.S. and Morzaria, S.P. 1986. Biology of *Babesia*. *Parasitol Today*, **2**, 211–19.

Yuval, B. and Spielman, A. 1990. Duration and regulation of the developmental cycle of Ixodes dammini (Acari:Ixodidae). *J Med Entomol*, **27**, 196–201.

Zahler, M., Rinder, H. and Gothe, R. 2000. Genotypic status of *Babesia microti* within the piroplasms. *Parasitol Res*, **86**, 642–6.

24

Malaria

MARCEL HOMMEL AND HERBERT M. GILLES

INTRODUCTION

Malaria is the most important of all the tropical diseases in terms of morbidity and mortality. Worldwide, some two billion individuals are at risk; 100 million develop overt clinical disease and 1.5 to 2.7 million die every year. Nearly 85 percent of the cases and 90 percent of carriers (many asymptomatic) are found in tropical Africa, where in some countries 20–30 percent of deaths in childhood are attributed to the disease (Defo 1995; Greenwood et al. 1987; Snow et al. 1999, 2003). Human malaria is caused by four species of protozoans belonging to the genus *Plasmodium*, *P. falciparum*, *P. vivax*, *P. ovale*, and *P. malariae* causing malignant tertian, benign tertian, ovale tertian, and quartan malaria, respectively.

Drug and insecticide resistance have aggravated the complexity of the malaria problem worldwide. Thus, among the countries where falciparum malaria persists, only those of Central America have not recorded the resistance of *Plasmodium falciparum* to chloroquine. In contrast, in the Mekong region in Southeast Asia, multidrug resistance to chloroquine, sulfadoxine-pyrimethamine, and even mefloquine, now occurs and the sensitivity to quinine is also diminishing (WHO 1994a).

Maps showing the global distribution of malaria and the countries from which *P. falciparum* resistance to chloroquine and other antimalarial drugs has been reported, are given in Figures 24.1 and 24.2. It should be emphasized that degrees of resistance vary within individual countries; for example, in Cambodia multidrug resistance occurs in one part of the country, whereas in another part of the country *P. falciparum* parasites are still sensitive to chloroquine.

CLASSIFICATION

Malaria parasites belong to the genus *Plasmodium* that includes over 125 species infecting reptiles, birds, and mammals. *Plasmodium* is the only genus in the Family Plasmodiidae, Order Haemosporida, Class Coccidea, Phylum Sporozoa (Apicomplexa). The phylum is characterized by the presence, at specific stages of the life cycle, of an apical complex consisting of a number of specialized organelles, which may include a conoid, polar rings, rhoptries, and micronemes.

Morphological characteristics and features of the life cycle, the major criteria used in Garnham's (1966) classification, include the shape of the trophozoite, the gametocyte, and the oocyst, the number of nuclei in the erythrocytic and exo-erythrocytic schizonts, the aspect and distribution of the pigment, and the nature of the damage induced by the parasite in the host cell (e.g. 'Maurer's clefts' or 'Schüffner's dots'). The main biological criteria include the host range, the type of host cell infected, the duration of the different stages in the life cycle, the presence or absence of relapse (with or without recrudescences), the nature of the vector, and

☐ Areas where malaria transmission occurs

☐ Areas with limited risk

☐ No malaria

Figure 24.1 *Global distribution of malaria in 2003 (reproduced by permission of WHO)*

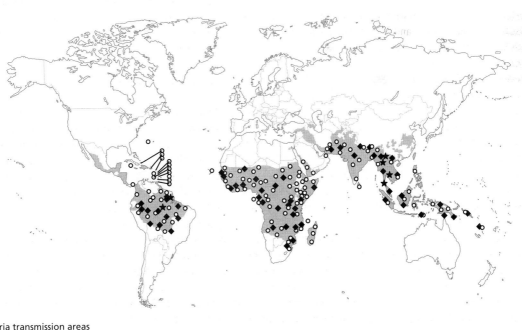

☐ Malaria transmission areas

○ Chloroquine resistance

◆ Sulfadoxine-pyrimethamine resistance

★ Multidrug resistance

Figure 24.2 *Malaria transmission areas and reported* Plasmodium falciparum *drug resistance in 2001 (reproduced by permission of WHO)*

the geographical distribution. While none of these criteria are ideal taxonomic markers, in most cases they have been considered broadly adequate to differentiate and classify species.

LIFE CYCLE, DEVELOPMENTAL STAGES, AND TRANSMISSION

Life cycle

Human malaria parasites undergo a complex cycle of development, including stages in a female *Anopheles* mosquito. The infection starts with the bite of an infected mosquito, when the sporozoite is introduced into the skin together with mosquito saliva. The sporozoite rapidly makes its way to the liver, where it invades a hepatocyte. Within the hepatocyte, the parasite undergoes a period of differentiation and multiplication to produce the pre-erythrocytic schizont, containing a few thousand merozoites. The merozoites, when released from the hepatocyte, enter the bloodstream and invade new erythrocytes to start a characteristic periodic phase of differentiation and asexual multiplication leading to the erythrocytic schizont. This contains a small number of merozoites, which in turn invade new erythrocytes. At some stage, some merozoites undergo sexual differentiation and gametocytogenesis. When a new biting mosquito picks up the male and female gametocytes, gametocytes are released from the erythrocytes in the acidic, low temperature environment of the insect midgut, and transform into gametes, which fertilize to form first a zygote, then an ookinete. The motile ookinete crosses the epithelial wall of the midgut before transforming into an oocyst, which divides by schizogony to produce thousands of sporozoites. These undergo a final differentiation into their infective forms while migrating to the salivary glands of the insect, where they stay until the next bite, thus completing the cycle. Figure 24.3 illustrates the life cycle of *P. falciparum*.

Developmental stages

The various stages of malaria parasites exhibit dramatically different morphological features, which reflect both their very different function and the different microenvironments inhabited. Only the sporozoite, merozoite, and ookinete (which are designed for the invasion of, respectively, the hepatocyte, erythrocyte, or midgut epithelial cell of the mosquito) possess a surface coat and the specialized apical complex. The surface coat is responsible for host cell recognition and adherence, and the invasion organelles allow the parasite to actively enter the target cell. In contrast, other stages of the life cycle, which are designed for growth and development within the host cell, lack these invasion organelles, but possess structures involved in respiration, intake of food or digestion of nutrients (e.g. mitochondrion, cytostome, food vacuoles, and pigment). The sexual stages of the parasite undergo a complex differentiation in order to become equipped for the fertilization process that takes place within the mosquito midgut; the zygote and ookinete are the only parts of the life cycle in which the organism is diploid and where meiosis is observed.

Plasmodium spp. possess a single mitochondrion (Fry 1991), which seems to play a very different function during the vertebrate stage (when it is almost acristate, particularly in mammalian plasmodia) and the invertebrate stage (when it is always fully cristate). A switch of enzymatic activity takes place as the parasite differentiates from one stage to the other. Mitochondria are self-replicating and each stage of the life cycle carries at least one, albeit rudimentary in some cases.

Malaria parasites are unable to ingest particulate food and are entirely dependent upon the intake of soluble cellular macromolecules. Early trophozoites feed by micropinocytosis, pinching off small portions of the erythrocyte cytoplasm and producing small digestive vesicles. Later trophozoites are equipped with a cytostome (i.e. primitive 'mouth'), which becomes functional when micropinocytosis ceases. For most growing stages, the cytostome represents the major route for food intake.

Within the food vacuole, the digestion of hemoglobin leaves an insoluble waste product consisting of hematin and ferriprotoporphyrin coupled to partially degraded globin and plasmodial proteins; this forms malaria pigment or hemozoin. When the mature schizont finally breaks up, liberating merozoites, the pigment is left behind in the 'residual body;' these bodies eventually accumulate and remain within the body, mainly in the liver and spleen, for many years.

The plasma membrane acts as an interface between the parasite and its host and provides both an effective protection against host defense mechanisms (e.g. proteolytic enzymes, cytokines, or antibodies) and the means for a selective permeability and macromolecular exchanges. The parasitophorous vacuole membrane, which separates the intracellular parasite from the host cell cytoplasm, is involved in the complex molecular trafficking events required for macromolecular exchange between the parasite and its host.

Pre-erythrocytic stage

The sporozoite enters the bloodstream and is rapidly carried to the liver, where it invades a hepatocyte and undergoes drastic changes in morphology. It loses its apical complex and surface coat and transforms into a round or oval trophozoite. Development takes place

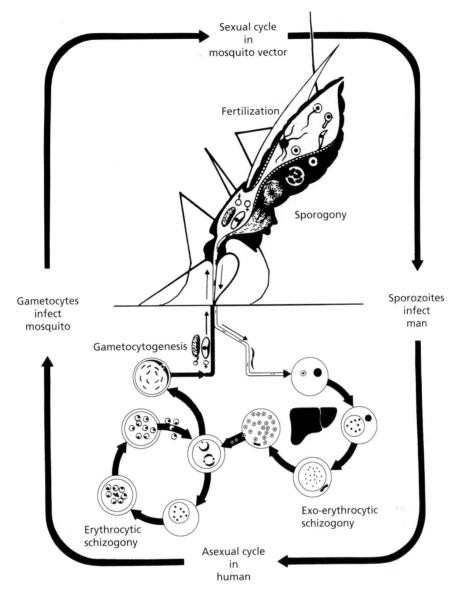

Figure 24.3 *Life cycle of* Plasmodium falciparum

within a parasitophorous vacuole in the hepatocyte. The nucleus divides many times and the cytoplasmic mass grows substantially; the number of nuclear divisions and the duration of schizogony varies considerably from one species to another, as does the ultimate size of the pre-erythrocytic schizont (which reaches a diameter of 60 μm in *P. falciparum*, and contains more than 30 000 nuclei) (Figure 24.4).

After completion of nuclear divisions, the cytoplasm segments and individual merozoites are formed. The duration of the pre-erythrocytic schizogony is characteristic for each species; for instance, minimum maturation times of 5.5 days in *P. falciparum* and 15 days in *P. malariae*. In mammalian plasmodia the merozoites arising from pre-erythrocytic schizogony can only invade red cells (i.e. they cannot invade hepatocytes or start a secondary exo-erythrocytic schizogony).

Erythrocytic stage

After invading the erythrocyte, the parasite loses its specific invasion organelles and de-differentiates into a round trophozoite located within a parasitophorous vacuole in the red cell cytoplasm. The young trophozoite (so-called 'ring' stage, because of its morphology on stained blood films) grows substantially before undergoing several nuclear divisions. The number of nuclear divisions and the duration of schizogony vary from one species to another. The timing of the different stages of the erythrocytic schizogony of *P. falciparum* is illustrated in Figure 24.5

The length of the intra-erythrocytic development is variable; 48 h for *P. falciparum*, *P. vivax*, and *P. ovale* and 72 h for *P. malariae*. At the end of schizogonic development, new merozoites are formed which have a

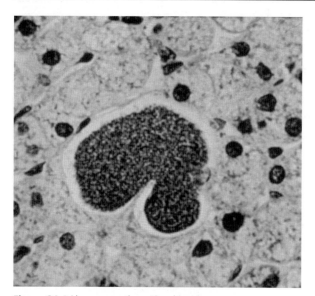

Figure 24.4 *Liver pre-erythrocytic schizont*

very short extracellular viability (probably less than 30 min) and can only invade erythrocytes.

The development of the parasite within the erythrocyte produces a series of alterations of the host cell, including the presence in the cytoplasm of the red cell of vacuoles or clefts and morphological changes of the erythrocyte membrane (e.g. 'knobs' or 'caveolae'). Some of these alterations are visible in stained parasites and represent useful taxonomic features (e.g. Maurer's clefts in *P. falciparum* or Schüffner's dots in *P. vivax*) (Figure 24.6).

The metabolism of the malaria parasite is largely dependent on the digestion of red cell hemoglobin, which is transformed into malaria pigment. Pigment is absent in the ring stage and only becomes detectable in the late trophozoite and the schizont. The aspect and distribution of the pigment in the parasite may vary

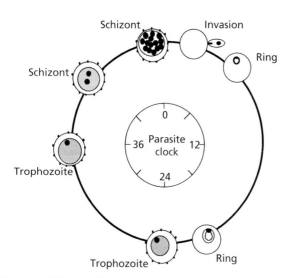

Figure 24.5 *Timing of the intra-erythrocytic cycle of* Plasmodium falciparum

from species to species, and represents another useful distinguishing feature.

Gametocytes and gametes

Some merozoites develop either into male or female gametocytes but the factors which induce sexual differentiation, rather than schizogonic development, are essentially unknown. All merozoites emerging from a sexually committed gametocyte transform into gametocytes of the same sex (Smith et al. 2000) and in most *Plasmodium* species the process of gametocytogenesis leads to a female-biased sex ratio. A variety of factors have been shown to stimulate gametocytogenesis, including hyperparasitemia, anemia, and antimalarial drug treatment and it looks as though the parasite is capable of sensing hostile conditions and, by transforming into gametocytes, prepares to escape into a new host. Chloroquine and sulfadoxine-pyrimethamine both stimulate gametocytogenesis, particularly when used in suboptimal doses (Buckling and Read 1999; Sokhna et al. 2001). This may have important implications in the spread of drug resistance, since resistant parasite populations selected by drug treatment have an increased chance of being transmitted (Robert et al. 2000). The effect of anemia on gametocytogenesis is paradoxical since anemia also induces a strongly male-biased sex ratio (Paul et al. 2002).

Mature gametocytes of *P. vivax*, *P. ovale*, and *P. malariae* are round, but the gametocytes of *P. falciparum* are elongated or crescent-shaped (Figure 24.7). Maturation of *P. falciparum* gametocytes also takes much long than that of other species; 10 days compared to 3 or 4. Although the longevity of mature gametocytes may exceed several weeks, their half-life in the bloodstream may only be 2 or 3 days, while waiting in an arrested state of development to be taken up by a mosquito.

When taken up by a suitable mosquito vector, the gametocytes transform into gametes stimulated by the drop in temperature and the biochemical features of the new environment (Carter and Graves 1988). The maturation of the female gametocyte into a macrogamete takes place without major morphological changes. In contrast, in the male gamete, the nucleus divides three times and each of the eight nuclei formed combines with cytoplasm to form eight thread-like microgametes, a process called 'exflagellation.'

Fertilization and sporogony

Microgametes are motile, highly differentiated stages, which move actively towards the macrogametes to start the process of fertilization and sexual stage of the parasite's life cycle. The haploid nuclei of male and female gametes merge forming the diploid form of the parasite ('zygote'). Meiosis and genetic recombination

Diagram of ultrastructure of malaria-infected red blood cells

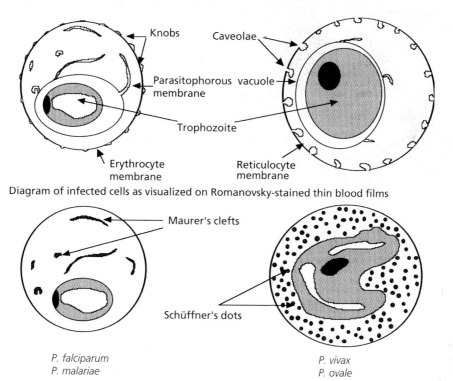

Diagram of infected cells as visualized on Romanovsky-stained thin blood films

P. falciparum
P. malariae

P. vivax
P. ovale

Figure 24.6 *Diagram describing the difference between Maurer's clefts and Schüffner's dots. The ultrastructure shows that the objects identified as Schüffner's dots in a Romanovsky-stained thin film are caveolae on the surface of infected reticulocytes, while Maurer's clefts correspond to expansions of the parasitophorous vacuole membrane and vesicles derived from this membrane.*

take place while the zygote transforms into a motile form, the ookinete, which crosses the midgut epithelium and settles below the basal lamina to form an oocyst; at this stage, the parasite has returned to a haploid stage. The oocyst grows to reach a diameter of 40–50 µm after 4–21 days and each infected mosquito may carry a few hundred oocysts (as many as 1 600 in some laboratory-infected mosquitoes) (Figure 24.8), but normally the number of oocysts per mosquito is <100. The size of the mature oocyst and the distribution of pigment are characteristic for each species and may be used as taxonomic markers.

Within the oocyst, the nucleus divides in a schizogonic mode, where multiple, synchronized nuclear divisions take place before individual daughter cells are formed. At the end of the maturation period, sporozoites are formed, the number varying from species to species (e.g. estimated between 1 000 and 20 000 in *P. falciparum*). The immature sporozoites emerge through the oocyst wall into the hemolymph, and migrate to the salivary glands of the mosquito. Infected salivary glands contain between 10 000 and 200 000 sporozoites.

Biology of transmission

The transmission of *Plasmodium* by mosquitoes is driven by the female mosquito's own biological need for regular blood-meals and successful transmission is regulated by a variety of intrinsic and environmental factors. A sequence of blood-feeding, egg maturation, and oviposition is repeated several times throughout the mosquito's life cycle and is referred to as the 'gonotrophic cycle,' the length of which depends on external temperature. In *Anopheles gambiae*, for example, the cycle takes 48 h when the average day–night temperature is 23°C. The average lifespan of the female has direct relevance to its efficiency as a vector: only females that live long enough for the sporogonic cycle to be completed (i.e. approximately 10 days) can successfully transmit *Plasmodium*. It also follows that any intervention to reduce the lifespan of the female (e.g. exposure to insecticides) will affect transmission efficiency. Infected humans are only infective to mosquitoes if they carry mature gametocytes; infection and infectivity generally coincide in *P. vivax*, *P. ovale*, and *P. malariae*, but not always in *P. falciparum* due to the much longer time required for gametocytes to reach maturity.

Within the mosquito, the maturation of the oocyst takes 4–21 days, depending largely on environmental temperature. The sporozoites that emerge from the oocyst are not yet infective and it is during their migration through the hemolymph towards the salivary glands that final differentiation takes place (including the expression of two key binding molecules on their

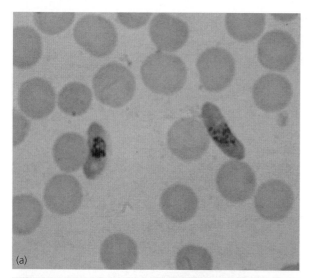

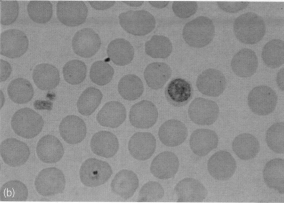

(a)

(b)

Figure 24.7 *Gametocytes of* Plasmodium falciparum **(a)** *and* P. vivax **(b)**.

surface, CSP and TRAP/SSP2). Less than 25 percent of the sporozoites produced successfully reach the glands, bind to the cells by means of CSP, then penetrate into the lumen of the glands, from where they are released into droplets of saliva with each subsequent blood-meal (Pimenta et al. 1994).

Only mosquitoes of the genus *Anopheles* support the sporogonic development of human malaria parasites. Of

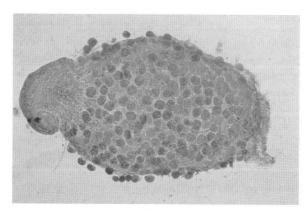

Figure 24.8 *Oocysts in mosquito*

the 460 species of *Anopheles* described worldwide, fewer than 70 are competent vectors and there are substantial differences in vectorial competence, even between subpopulations of a given species. Vectorial competence measures the ability of a mosquito which has ingested mature gametocytes to sustain their development until they again become infectious: this must not be confused with vectorial capacity which measures the ability of a mosquito species to transmit the infection in the field and is a major parameter of malaria epidemiology.

Female *Anopheles* mosquitoes usually take their blood-meal between dusk and dawn and can feed on a variety of warm-blooded animals, including humans. Some species prefer to feed on humans and are termed anthropophilic, while others prefer to feed on animals, particularly cattle, and are termed zoophilic, but these preferences are by no means absolute. Obviously, species that prefer to feed on humans will be better vectors of human malaria parasites than zoophilic species. Females are attracted to their hosts by a variety of stimuli, including exhaled carbon dioxide and host odor (Braks et al. 1999).

Some species of *Anopheles* bite humans mainly outdoors ('exophagic'), for example *A. albimanus*, a major malaria vector in Central and South America, whereas mosquitoes of the *A. gambiae* complex, the world's most effective malaria vectors bite mostly indoors ('endophagic') and at night. After feeding, females either rest inside houses ('endophilic' species) or outdoors in a variety of natural shelters ('exophilic' species). Many species exhibit a mixture of these extreme behaviors. The time of biting, the biting behavior, and the resting behavior following the blood-meal are important determinants of malaria epidemiology and transmission; it would, for instance, serve no useful purpose to spray the inside of houses with insecticides if the local vectors are exophagic and exophilic.

Because of their nocturnal habits, *Anopheles* females usually feed while their human prey is sleeping, and can thus take a single, uninterrupted blood-meal, which usually lasts for 2–3 min. During the bite, blood is ingested either directly from a capillary or from a micro-hemorrhage resulting from the local damage caused by the tearing action of the toothed mandibles and maxillae. The quantity of blood ingested at a single feed varies from 1 to 2.5 μl representing a larger volume of blood, since *Anopheles* concentrate blood by diuresis (release of a few drops of clear liquid from the anus during feeding).

At the time of mosquito bite, each drop of saliva carries with it a small number of sporozoites which are injected into the skin. An infective bite from a single mosquito will characteristically represent between five and a few hundred sporozoites (Rosenberg et al. 1990), sufficient to induce an infection. Interventions that affect the vector-to-human contact are likely to reduce malaria transmission, this includes the use of repellents (such as diethyltoluamide (DEET)), special clothing, awareness of risk, use of bednets, or better, insecticide-impregnated bednets, design of housing, and position of housing relative to breeding sites.

Recrudescences and relapses

Some sporozoites develop immediately after entering the hepatocyte and a few days later, when the pre-erythrocytic schizont is mature, produce merozoites that infect red blood cells. This first wave of parasitemia is responsible for the primary attack of malaria and may, in the absence of specific antimalarial treatment, last from a few weeks to a few months in a nonimmune individual (2–3 weeks in some *P. falciparum* infections to 4–6 months in *P. malariae* infections). Further waves of parasitemia may follow a few weeks, months, or years after the primary attack. In the absence of reinfection, there are two possible causes for such secondary attacks:

1 The presence of dormant parasites or 'hypnozoites.' This is a situation in which some of the sporozoites do not immediately start to grow and divide, but remain in a dormant stage for weeks or months (Krotoski et al. 1980). The duration of latency is variable from one hypnozoite to another and the factors that eventually trigger growth are not known; this explains how a single infection can be responsible for a series of waves of parasitemia or 'relapses.' *P. vivax* and *P. ovale* are the only two human malaria parasites which produce hypnozoites and, thus, relapses. Different strains of parasites have their own characteristic pattern of relapse. Lengthy prepatencies or delayed relapses appear to be a feature of parasites endemic in temperate rather than tropical areas (Brumpt 1949).

2 Parasites which do not form hypnozoites (*P. malariae* and *P. falciparum*) may, nevertheless, produce secondary waves of parasitemia called 'recrudescences,' either by long-term survival of erythrocytic stages (e.g. within a 'sequestration' site) or the continuation, at a low or undetectable level, of erythrocytic schizogony in the peripheral bloodstream. The basis for this latent survival of parasites has not, so far, been elucidated. In the case of *P. malariae*, recrudescences have been described as long as 60 years after the primary attack (Brumpt 1949); note that a recrudescence occurring more than 24 weeks after the original attack is sometimes called a 'recurrence.' Drug resistance creates a situation in which the initial peak of parasitemia is only partially controlled and a recrudescence of resistant parasites occurs shortly thereafter. Figure 24.9 illustrates the differences between reinfection, recrudescence, and relapse.

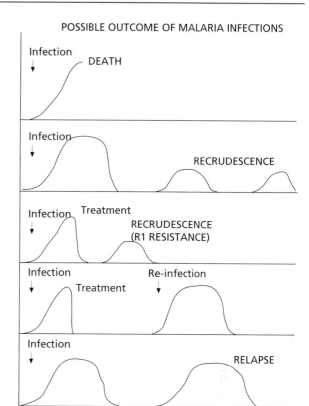

Figure 24.9 *Relapse, recrudescence, reinfection, and drug resistance*

The main practical difference between relapse and recrudescence is that, whereas latent erythrocytic stages are susceptible to standard chemotherapy, hypnozoites (being cells of low metabolic activity) require a specific treatment (e.g. with primaquine). A true relapse is, therefore, always exo-erythrocytic in nature and may be defined as the 'reappearance of parasitemia in a sporozoite-induced infection following adequate blood schizonticidal therapy' (Cogswell 1992). This also means that a blood-induced infection (e.g. by blood transfusion) cannot relapse.

Synchronicity

One of the striking features of the erythrocytic cycle of mammalian malaria parasites is the fact that the parasites tend to grow in synchrony. The consequence is that parasites examined on a patient's blood film are frequently all at the same developmental stage and that clinical symptoms (e.g. fever paroxysms) often tend to occur at regular intervals of time (e.g. 48 h in tertian malaria or 72 h in quartan malaria). This synchronous life cycle of the parasite is determined, at least in part, by host factors. In *P. falciparum*, a parasite in which synchronicity is particularly striking, it has been suggested that it is the paroxysms of fever themselves which sharpen the level of synchronicity (Kwiatkowski and Greenwood 1989), in consequence of which the periodicity of fever becomes more regular as the malaria infection progresses. The host's circadian rhythm is also a determining factor in controlling the timing of parasite development. This very accurate 'biological clock' of the parasite is believed to represent an adaptation designed to facilitate the transmission of the parasite by night-biting mosquitoes (Hawking 1975).

Sequestration

The overall distribution of both asexual forms and gametocytes in the blood is not always random and there are circumstances in which parasites may be removed from the peripheral bloodstream and be retained or 'sequestered' in various host tissues.

In *P. falciparum*, only the early trophozoites ('rings') are present in the peripheral circulation, while later developmental stages are sequestered within the capillaries of various organs. This sequestration is caused by the adherence of infected erythrocytes to capillary endothelial cells by means of a specific interaction between a parasite-derived molecule present at the surface of the malaria-infected red cell and specific host cell receptors (Hommel 1997). Adherent malaria-infected red cells can be found in various tissues, including the heart, the liver, and the brain (Figure 24.10).

Sequestration has important consequences for the diagnosis of falciparum malaria, because it means that parasites may not be found on a blood film at a time when the clinical picture is most suggestive. During pregnancy, infected erythrocytes are preferentially retained in the placenta. This is believed to be of a mechanical nature, due to the sluggish blood flow of the placenta rather than any interaction between infected erythrocytes and specific receptors on the surface of syncytiotrophoblasts (Galbraith et al. 1980).

A radically different form of sequestration is responsible for the removal of immature gametocytes of *P. falciparum* from the peripheral circulation. As early as the first 24 h of gametocytogenesis, the immature gametocytes are retained within the bone marrow and the

spleen (Smalley et al. 1980), and are released into the peripheral circulation only after reaching maturity, 8–10 days later. In contrast to the sequestration of immature gametocytes, which is unique to *P. falciparum*, mature gametocytes of all species may be submitted to intermittent sequestration, controlled by the diurnal rhythm of the host. It is conceivable that this sequestration may take place in the capillaries of the upper layer of the skin, more accessible to the feeding mosquito. This would explain why the number of gametocytes in the blood-meal may be far greater than that present in the peripheral circulation of the host (Carter and Graves 1988).

Interaction of parasite and host cell

INVASION OF THE ERYTHROCYTE

In order to invade the erythrocyte, the merozoite has to recognize and attach to its surface components. The nature of the erythrocyte receptor involved varies from one merozoite to another, because each parasite species has specific host cell preferences (Hadley and Miller 1988; Bannister and Dluzewski 1990). In *P. falciparum*, a parasite which can invade erythrocytes of all ages, the receptor varies from one group of isolates to another (Perkins and Holt 1988) and optional receptors include sialic acid (Facer 1983), glycophorins (Pasvol et al. 1982), and band 3 (Okoye and Bennett 1985). In *P. vivax*, a parasite which is restricted to immature red blood cells (reticulocytes), attachment to two receptors is required: one is reticulocyte-specific, the other has been shown to be associated with a glycoprotein serologically defined by the Duffy (FyFy) blood group determinant (Wertheimer and Barnwell 1989). The red cell receptor preferences of *P. ovale* and *P. malariae* are presumed to be different from those of *P. falciparum* and *P. vivax*. *P. ovale* can infect Duffy-negative red cells, which is compatible with the fact that the species is found in West Africa (where the Duffy-negative phenotype is highly prevalent) while *P. malariae* is said to have a predilection for senescent red cells.

INVASION OF THE HEPATOCYTE

After a sporozoite has entered the bloodstream, it rapidly 'homes in' towards liver sinusoids and establishes itself within a hepatocyte; this is supported by the finding that injected CSP is selectively targeted to the liver (Cerami et al. 1994). The membrane of the sporozoite is entirely covered by two molecules (CSP and TRAP/SSP2, see Table 24.3), which are responsible for the initial attachment to the hepatocyte surface. Heparan sulfate protein glycoconjugates (HSPG) and low-density lipoprotein receptor-related (LPR) protein have both been incriminated as the host receptors for sporozoite invasion (Frevert et al. 1996). Whether sporozoites enter the hepatocytes directly (e.g. into the space

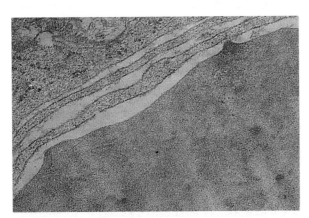

Figure 24.10 *Cytoadherence of* Plasmodium falciparum

of Disse through gaps in the endothelial lining) or whether they first cross a Kupffer cell, has been a controversial issue for years (Verhave and Meis 1984). Recent findings suggest that sporozoites enter Kupffer cells, manage to survive the passage through these phagocytic cells by blocking phago-lysosomal fusion, then exit from the Kupffer cell and enter a hepatocyte; the binding ligands of the surface molecules, TRAP and CSP, have such a broad specificity that recognition of different cell types is possible (Barnwell 2001; Pradel et al. 2002). Furthermore, it appears that the passage through Kupffer cells is an essential step in the sporozoite maturation, without which the apical end organelles cannot exocytose and enable invasion with the formation of a parasitophorous vacuole. It has been suggested that the sporozoite may, in some circumstances, have to cross more than one cell before this maturation is achieved (Mota et al. 2002). Once inside the hepatocyte, the sporozoite de-differentiates, losing its apical organelles, and starts to grow.

ALTERATIONS OF HOST CELLS

After infection by the malaria parasite, the host cell undergoes a variety of structural changes, which may substantially alter its function, appearance or antigenicity. The nature of the alterations induced varies from one species to another. These host cell alterations have been studied most comprehensively in the erythrocytic stage of the parasite (Hommel and Semoff 1988; Sherman et al. 2003), but it is conceivable that they may also exist in the other intracellular stages.

The alterations identified so far in the membrane of malaria-infected erythrocytes include:

- a visible change of shape and reduced deformability;
- the presence of electron-dense protrusions or 'knobs' (in *P. falciparum* and *P. malariae*) (Figure 24.11);
- the presence of small depressions, or 'caveolae,' at the surface of the erythrocyte, connected by a network of small vesicles and clefts (in *P. vivax* or *P. ovale*) (Aikawa 1988);

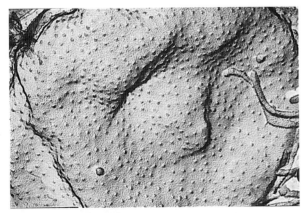

Figure 24.11 *Knobs on* Plasmodium falciparum-*infected erythrocytes*

- the expression of new sugar moieties, particularly galactose (David et al. 1981);
- the cytoadherence to endothelial cells or rosetting with normal erythrocytes (Wahlgren et al. 1987);
- the presence of new metabolic channels (Ginsburg et al. 1985);
- the evidence of new parasite-specific antigens associated with the red cell membrane (Hommel et al. 1983); and
- the reorganization of normal erythrocyte components (e.g. disruption of spectrin or conformational changes of band 3) (Yuthavong et al. 1979; Winograd and Sherman 1989).

Characteristics of human malaria parasites

The main features of human malaria parasites, as observed on Romanovsky-stained thin blood films, are shown in Figure 24.12.

PLASMODIUM FALCIPARUM

Of all the human malaria parasites, *P. falciparum* is the most highly pathogenic and responsible for 'malignant tertian malaria,' a form of the disease which runs an acute course in nonimmune patients and is frequently fatal if untreated. This parasite represents the major cause of malaria in tropical Africa and is also responsible for the great regional epidemics that sometimes occur in north-west India or Sri Lanka; it is generally confined to tropical or subtropical areas because its development in the mosquito is greatly retarded when the temperature falls below 20°C. The most characteristic biological features of *P. falciparum* are as follows:

- it infects mature and young erythrocytes;
- the surface of erythrocytes infected with late-stage trophozoites or schizonts is altered so they stick to endothelial cells in various tissues (this cytoadherence is responsible for the sequestration of the parasite);
- the pre-erythrocytic cycle starts immediately after injection of sporozoites by the mosquito, without the production of hypnozoites (i.e. there are no relapses);
- schizogony is particularly prolific in all stages (pre-erythrocytic schizogony produces up to 30 000 merozoites; erythrocytic schizogony produces 16–18 merozoites and sporogony produces up to 20 000 sporozoites) which may be the cause of its success as a species and its virulence;
- infection in the peripheral blood is characterized by the predominant presence of ring forms and gametocytes, whereas late trophozoites and schizonts are only seen exceptionally;
- the level of parasitemia may be high and multiple infection in a single erythrocyte is common; and
- the gametocytes are characteristically crescent-shaped and, unlike the gametocytes of other species, are very

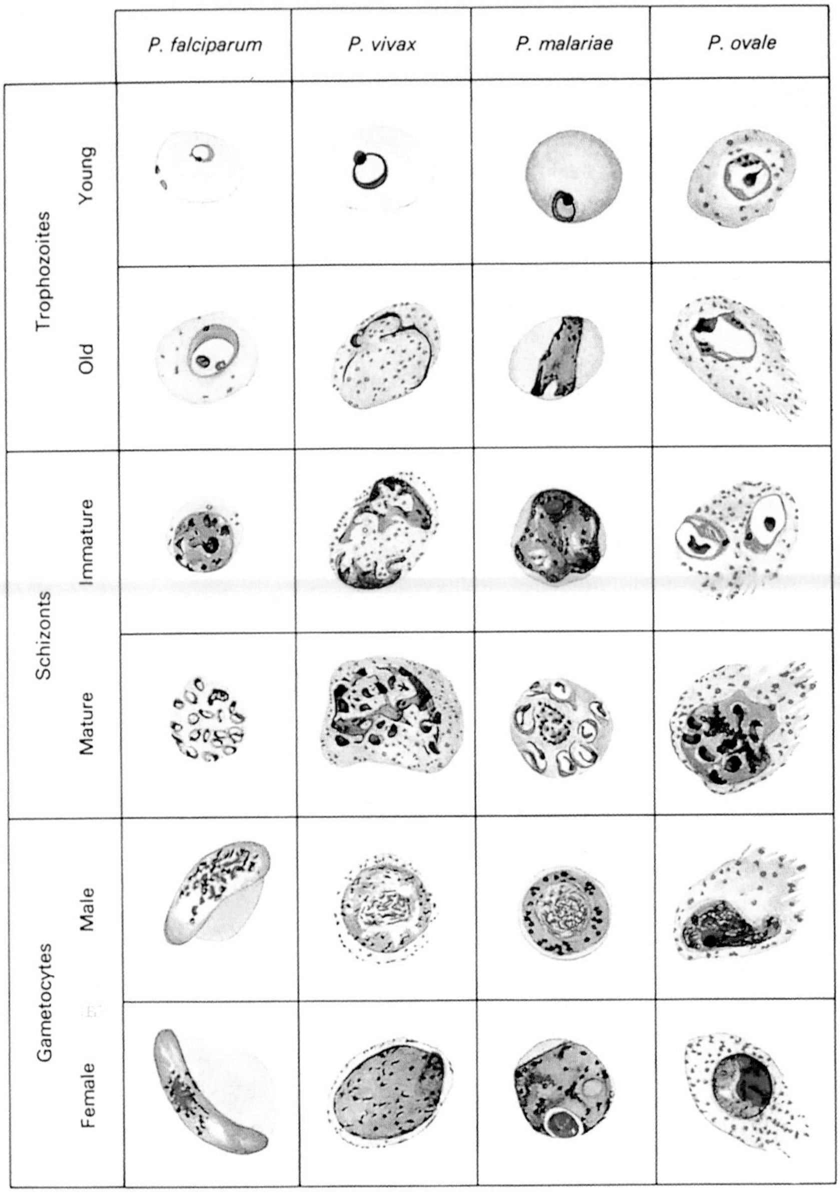

		P. falciparum	P. vivax	P. malariae	P. ovale
Trophozoites	Young				
Trophozoites	Old				
Schizonts	Immature				
Schizonts	Mature				
Gametocytes	Male				
Gametocytes	Female				

Figure 24.12 *Morphological characteristics of human malaria parasites*

slow to reach maturity (up to 10 days) and early forms of gametocytes are sequestered.

PLASMODIUM VIVAX

Plasmodium vivax occurs throughout most of the temperate zones as well as large areas of the tropics (but is mainly absent from tropical West Africa). It causes 'benign' tertian malaria, the pattern of which varies with different parasite strains. As a species, *P. vivax* is particularly polymorphic and the subspecies status proposed for some strains may be justified either on morphological grounds (e.g. *Plasmodium vivax multinucleatum* found in China, appears to have a characteristic oocyst), on biological grounds (e.g. *Plasmodium vivax hibernans* found in Russia, whose sporozoites produce only hypnozoites), or molecular grounds (e.g. a series of *P. vivax*

isolates with substantial differences in their genetic make-up have been identified, including a *P. vivax*-like parasite, found in humans in Papua New Guinea, Brazil, and Madagascar, indistinguishable from the simian parasite *Plasmodium simiovale*) (Qari et al. 1993). The distinctive biological features are: (1) a restriction of erythrocyte invasion to reticulocytes bearing Duffy blood group determinants, a feature which explains why red blood cells infected with trophozoites of *P. vivax* are sometimes described as 'larger' than normal; (2) the presence of caveolar structures on the surface of the infected erythrocyte membrane, which communicate with underlying cytoplasmic vesicles; these structures readily take up the Romanovsky stain and, to the microscopist, look like a reddish cytoplasmic stippling or 'Schüffner's dots;' and (3) after invading the hepatocyte some, or all, of the sporozoites may transform into hypnozoites, then remain latent for months or years and be responsible for subsequent relapses.

PLASMODIUM OVALE

Plasmodium ovale is a species truly distinct from *P. vivax*, not only on minor morphological differences (e.g. oval shape of infected red cell with ragged or fimbriated margin, smaller number of merozoites in schizont, aspect of gametocytes), but also presenting antigenic and molecular differences. Nevertheless, most of the biological and clinical features are identical. The major biological difference is that *P. ovale* can infect Duffy-negative reticulocytes, whereas *P. vivax* cannot.

PLASMODIUM MALARIAE

Plasmodium malariae differs from the other three human malaria parasites in its slow development and its longer asexual cycle. Development is slow in both the vector and the human host, because of less efficient schizogony (pre-erythrocytic schizonts have 6 000 merozoites, erythrocytic schizonts have 6–12 merozoites). The asexual cycle is 72 h instead of 48 h, and this feature has given the clinical form of *P. malariae* its name, quartan malaria, because fever paroxysms occur every fourth day according to the Roman custom of regarding day 0 as day 1. The biological characteristics of the parasite are:

- an apparent preference for old erythrocytes, explaining why infected cells are often described as 'smaller' by microscopists (this feature is poorly documented in view of the absence of an in vitro culture system);
- the presence of 'knobs' at the surface of infected erythrocytes which are similar to *P. falciparum*, but the cells do not exhibit any cytoadherence (and therefore no sequestration);
- the surface of infected cells does not exhibit any caveolar/vesicle complexes and, consequently, Schüffner's dots are absent; and
- sporozoites of *P. malariae* do not transform into hypnozoites and, hence, there are no relapses. However, *P. malariae* can survive for a very long time in the peripheral blood (10 years or more) at a very low level of parasitemia, occasionally producing detectable peaks with a recrudescence of clinical symptoms.

BIOCHEMISTRY, MOLECULAR BIOLOGY, AND GENETICS

Metabolic pathways

Much of what is known about the metabolism of malaria parasites has in the past been derived from studies of blood stages, either isolated from their vertebrate host or produced in culture. However, the availability of the complete genome sequence of *P. falciparum* (Gardner et al. 2002) has generated valuable information

regarding malarial enzymes, which classical biochemical methods had failed to achieve. As a result, many metabolic pathways can now be reconstructed in their entirety using similarity-searching methods. Since many of the enzymes are different from the host enzymes, this can potentially lead to the finding of novel antimalarial compounds, probably the most important achievement of parasite genome sequencing.

ACCESS TO NUTRIENTS

The passage of nutrients from the extracellular space to the parasite involves a transfer across the host cell membrane, the parasitophorous membrane, and the parasite's own membrane. Transfer across membranes may be achieved by a variety of mechanisms including diffusion along a gradient of substrate and carrier-mediated transport. Because of rapid parasite growth and replication, the normal transport mechanisms of the red cell are not sufficient to cope with the increased demand (e.g. an infected red cell utilizes almost 100 times more glucose than a normal red cell) (Roth et al. 1982). Increased traffic across the membrane is achieved not only by an increase in the transmembrane gradient of substrate (created by increased consumption of nutrients by the growing parasite) but also by new permeation pathways introduced by the parasite into the host cells (Sherman 1988; Ginsburg and Kirk 1998). The need for such new pathways is probably greater for parasites restricted to mature erythrocytes, since these cells are relatively impermeable to many of the nutrients needed for parasite growth (e.g. L-glutamine) (Elford 1986) and the presence of such pathways may represent a target for new antimalarials (e.g. the derivatives of the Chinese antimalarial qinhaosu are potent inhibitors of the L-glutamine influx).

Most nutrients required by the parasite originate from the extracellular space, e.g. the host plasma. These include glucose, purines and pyrimidines, amino acids, anions, cations, iron, zinc, vitamins, fatty acids, and phospholipids. Some of these nutrients may also be obtained from the host cytoplasm either by transport across the membranes or by uptake through the cytostome, e.g. the uptake of hemoglobin and its breakdown to form malarial pigment. The uptake of antimalarial drugs follows the same pathways as the transport of nutrients and, to some extent, these pathways may also function in the opposite direction for the disposal of waste products such as lactate. Genome analysis has revealed that *P. falciparum* possesses only a very limited repertoire of membrane transporters compared to other eukaryotes, which probably reflects its obligate intracellular parasite status (Gardner et al. 2002).

Secretion pathways in higher eukaryotes normally include a translocation across the endoplasmic reticulum followed by the transfer, by means of transport vesicles, first to the stacked Golgi cisternae then to the trans-

Golgi network. In *Plasmodium*, the understanding of the fate of proteins synthesized by the parasite and destined for secretion raises interesting questions, since it appears to lack 'classical' Golgi stacks. A comparison between two secreted proteins, the glycophorin-binding protein (GBP) and the serine-rich protein (SERP), proved the existence of secretory pathways and showed that different molecules may behave in different ways. SERP is released into the space between the parasite and the parasitophorous vacuole membrane (PVM), while GBP goes through a two step process, being first released into the vacuolar space then actively translocated through the PVM into the erythrocyte cytosol (Lingelbach 1997).

CARBOHYDRATE METABOLISM

The intra-erythrocytic malaria parasites lack reserves of glycogen and of other polysaccharides and are, consequently, dependent on glucose as their primary source of energy. Almost all the glucose utilized by the parasite is used by anaerobic glycolysis, a relatively inefficient process since oxidization stops at the lactate stage; as a result, this generates a large amount of lactate, which reduces the internal pH and has to be efficiently disposed of. All the malarial enzymes of the glycolytic pathway have been identified and found to be isoenzymes different from host enzymes (Roth 1990). Although the hexose monophosphate shunt pathway is operative and represents an essential source of nicotinamide adenine dinucleotide phosphate (reduced) oxidase (NADPH), the activity of the parasite-encoded glucose-6-phosphate dehydrogenase (G6PD) is only about 5 percent of that of the host cell enzyme (Ling and Wilson 1988). This suggests that the host enzyme normally initiates the pathway and that the parasite enzyme may be up-regulated only when the parasite is in a G6PD-deficient red cell; this would explain why the in vitro growth of *P. falciparum* in G6PD-deficient cells is initially impaired but may be overcome after a few cycles of growth.

There was, until recently, considerable doubt concerning the presence or absence of a functional tricarboxylic acid (TCA) cycle in malaria parasites, since the only TCA cycle enzyme identified with any certainty in blood stage plasmodia was malate dehydrogenase (Sherman 1979). Although all the enzymes of the TCA cycle have been identified in the genome of *P. falciparum*, some doubts remain regarding their functionality. It has been suggested that, rather than using the TCA cycle for glycolysis, the parasite may use it to synthesize succinylCoA, which can in turn be used in the heme biosynthesis pathway. The localization of some enzymes of the TCA pathway in the mitochondrion and the apicoplast, rather than the cytoplasm, support this hypothesis.

All malaria parasites appear capable of fixing CO_2, with the production of α-ketoglutarate and oxaloacetate as first intermediates. It has been suggested that a branched pathway from the oxaloacetate thus produced may be preferred by malaria parasites to the Krebs cycle for the production of aspartate, glutamate, malate, and citrate (Scheibel 1988). Like bacteria and plants, the malaria parasite has to replenish its oxaloacetate pool from phosphoenolpyruvate.

PROTEIN AND POLYAMINE SYNTHESIS

The molecular mechanisms involved in plasmodial protein synthesis are typically eukaryotic. The parasite obtains most of the amino acids it needs either from the digestion of red cell proteins (particularly hemoglobin) or by salvage from the free amino acid pool in the plasma; some amino acids, e.g. glutamate, aspartate, alanine, and leucine, are biosynthesized by the parasite itself from glucose and CO_2. The genome sequence has confirmed that few of the enzymes required for amino acid synthesis are present in the parasite genome, but has identified the enzymes required for the interconversion of glycine-serine, proline-ornithine, cysteine-alanine, and glutamine-glutamate (Gardner et al. 2002). The minimal amino acid requirements of the parasite are still not known.

Hemoglobin is broken down within the parasite food vacuoles, into heme (the major component of hemozoin, the malarial pigment) and globin, which is further hydrolyzed to free amino acids by a series of parasite proteases (including serine, aspartic, and cysteine proteases) (Schrével et al. 1990; Rosenthal and Meshnick 1998); up to 75 percent of the hemoglobin of an infected cell may be degraded by the parasite.

Three steps are critical in the biosynthesis of polyamines: the decarboxylation of ornithine to putrescine via ornithine decarboxylase (ODC), the formation of S-adenosylmethionine (AdoMet) from L-methionine and adenosine triphosphate (ATP), and the decarboxylation of AdoMet which provides the aminopropyl groups required for the synthesis of spermidine and spermine (Assaraf et al. 1984). All three steps represent potential targets for chemotherapy.

NUCLEIC ACID METABOLISM

The amount of DNA in a malaria parasite is approximately 10^{-13} g and there is about two to five times more RNA than DNA (Gutteridge and Trigg 1970). DNA synthesis occurs mostly in the late trophozoite to the early part of schizogony (29–44 h in the *P. falciparum* cycle). The nuclear DNA base composition of malaria parasites is characteristically adenine and thymidine (A+T) rich. Because of its rapid schizogonic development (e.g. a merozoite may generate 8–20 new merozoites in 48 h), the parasite needs to possess an effective nucleic acid synthesis, and this implies access to nucleic acid precursors. The mature human erythrocyte has no requirements for pyrimidine and, lacking the capacity for purine synthesis de novo, relies on the salvage of preformed purines (mostly for its ATP synthesis). Nucleosides are transported into erythrocytes by means of a specific transport protein in the membrane, and adenosine, hypoxanthine, and guanine can thus be salvaged from the plasma. In infected erythrocytes, the influx of purines into the red cell is increased and there is evidence to suggest that the erythrocyte's ATP is broken down to adenosine monophosphate (AMP) and hypoxanthine; although hypoxanthine appears to be the major purine incorporated into the parasite, the parasite has a range of purine salvage pathway enzymes (Reyes et al. 1982).

Malaria parasites must synthesize pyrimidines de novo, because the erythrocyte cannot supply them; the small amount of uridine and thymidine taken up by the erythrocyte does not appear to be incorporated by the parasite. The parasite has all the enzymes required for the synthesis of uridine monophosphate (UMP) from glutamine, ATP, and CO_2 (Gero and O'Sullivan 1990); some of these enzymes have been found to be functionally different from the corresponding host enzymes.

The synthesis of pyrimidines is intimately linked to folate metabolism. All the enzymes of the folate cycle have been studied in malaria parasites, with particular emphasis on dihydrofolate reductase (DHFR) and dihydropteroate synthase (DHPS), which are the targets of two different classes of antifolates used as antimalarials. It seems that the parasite relies more on the de novo folate pathway than folate salvage, which explains the synergy of the antimalarial activity of sulfa drugs (which inhibit de novo synthesis) and pyrimethamine (which inhibits the parasite DHFR, an enzyme which is considerably more sensitive to the drug than the host enzyme). The synthesis of pyrimidines also involves the vitamin *para*-amino benzoic acid (*p*ABA) and several authors have shown that the growth of plasmodia in experimental hosts is affected by the absence of *p*ABA in the diet (Ferone 1977).

LIPID METABOLISM

The growth of the parasite within the host erythrocyte requires a substantial increase of its total membranes; lipids generally constitute about 50 percent of the membrane mass. The lipid content of infected erythrocytes is higher than in normal ones (Homewood and Neame 1980) and the relative distribution of various lipids reflects the separate contributions that are made to erythrocyte membrane, parasitophorous vacuole membrane, external parasite membrane, and parasite organelles membranes.

During the erythrocytic cycle of the parasite, there is a 500–700 percent increase in phospholipid levels and the four major types of phospholipids are found in infected cells (Vial et al. 1990). Although infected cells have the capacity to incorporate some intact phospholipids from the plasma (probably by means of a specific phospholipid transfer molecule), it is thought that de novo synthesis and conversion from one phospholipid species to another represent the two major sources of new phospholipids. Malaria parasites appear to be incapable of producing cholesterol de novo from acetate or mevalonic acid. While malaria parasites do not synthesize their own sterols, they make use of the enzymes of the mevalonate pathway for the synthesis of nonsteroid isoprenoids, which in turn are involved in many different cell functions. The mevalonate pathway has been recognized as an important new target for chemotherapy (Grellier et al. 1994).

As there appears to be no fatty acid synthesis in infected cells, nor the capacity for retailoring available molecules (by elongation or desaturation processes) (Holz 1977), the host plasma must represent the main source of fatty acids. Infected cells incorporate considerable amounts of palmitic, stearic, oleic, arachidonic, and linoleic acid. Parasites have a high acyl-CoA synthetase activity (20 times more active than in normal erythrocytes), which produces fatty acyl-CoA, the usual fatty acid donor molecule in biosynthesis pathways. Choline enters the infected cell by means of a transport-mediated process, similar to that of normal erythrocytes.

ION METABOLISM

The uptake of extracellular calcium (Ca^{2+}) is essential for the growth of malaria parasites. The in vitro growth of *P. falciparum* trophozoites and merozoite invasion is inhibited after depletion of Ca^{2+} by ethyleneglycol-bis(b-aminoethylether)-N,N,N′,N′-tetraacetic acid (EGTA). This is consistent with the finding of calmodulin, one of the Ca^{2+}-binding proteins, both in the parasite cytoplasm and in apical organelles (Scheibel et al. 1987). The Ca^{2+} content of infected cells increases as the parasite matures due to an increased permeability of infected cells to external Ca^{2+} (Tanabe 1990). Since malaria-infected cells actively incorporate extracellular Ca^{2+}, it is not surprising that blockers of Ca^{2+} channels (e.g. verapamil) or antagonists of calmodulin (e.g. diltiazem or calmidazolium) may arrest parasite development.

Malaria parasites, like most eukaryotic cells, are capable of maintaining a high level of K^+ and a low level of Na^+ in their cytoplasm by means of a Na-, K+-ATPase, and at the expense of the host cell ionic environment. It is interesting that the ATPase has been identified in the parasitophorous vacuole membrane (not in the parasite membrane) and this implies that the parasite is living in a low Na^+, high K^+ extracellular environment. This unusual feature is consistent with the observation that it is necessary to use a low Na^+, high K^+ medium to grow trophozoites extracellularly in vitro (Trager and Williams 1992).

The requirement for iron in a number of metabolic pathways, including the synthesis of DNA, explains why the parasite may be inhibited by relatively low amounts of the iron chelator, desferrioxamine (Raventos-Suarez et al. 1982). Although hemoglobin is degraded within its food vacuoles, most of the heme released is transformed into crystalline hemozoin and does not represent a usable source of free iron. Older studies suggested that extracellular iron may be taken up by the binding of ferrotransferrin (from serum transferrin) to a parasite transferrin receptor in the infected erythrocyte membrane but a transferrin-independent mechanism is now thought to be more likely (Sanchez-Lopez and Haldar 1992). The role of iron in the viability of malaria parasites is a matter of controversy, since iron deficiency protects mice against *P. chabaudi*, while people with iron deficiency are still susceptible to malaria, but may have an increased parasitemia if given iron supplementation (Murray et al. 1978a). While the excess of heme produced in the food vacuole by hemoglobin degradation is toxic if not transformed into hemozoin, there is evidence that the parasite has its own de novo pathway for the synthesis of the heme required for the production of its iron-containing proteins in the mitochondrion and the apicoplast (Bonday et al. 2000).

OXYGEN UPTAKE AND REDOX STATUS

Low oxygen levels have a beneficial effect on the in vitro growth of *P. falciparum* and increased oxygen uptake is observed in infected erythrocytes in the presence of glucose, suggesting that malaria parasites are essentially microaerophilic (Scheibel et al. 1979). For many years, there was little tangible evidence of the existence of respiration or the presence of an electron transport chain in the parasite. The isolation of purified mitochondria from *P. falciparum* has allowed an unambiguous measurement of the cytochrome content of these organelles and confirmed the existence of a classical respiratory chain including the presence of cytochromes (Fry 1991). Ubiquinone cycling plays a crucial role in mitochondrial physiology and its selective inhibition by hydroxyquinones has provided a target for one of the most potent antimalarials discovered in recent years (Atovaquone®) (Fry and Pudney 1992).

The status of reduction-oxidation ('redox') reactions of the infected red cell is complex since it involves a combination of the oxidative stress exerted by the parasite on the host cell, the ability of the host cell to mount anti-oxidant defenses to avoid oxidative damage, oxidative stress exerted by the host on the parasite (which may include the effect of immune responses and antimalarial drugs) and, finally, the ability of the parasite to mount its own anti-oxidant defenses (Hunt and Stocker 1990). The generation of reactive oxygen intermediates (ROI) is the major factor of oxidative stress both in the parasite and in the host cell. Parasites have been shown to be susceptible to various ROIs and hydrogen peroxide, t-butyl-hydroperoxide, xanthine-xanthine oxidase, and alloxan all have antiparasitic activity (Clark et al. 1989b). A number of antimalarial drugs (e.g. primaquine and qinghaosu) also seem to act by means of oxidation. The reduced growth of parasites in erythrocytes with certain genetic abnormalities (sickle-cell disease, α- and β-thalassemia, persistence of hemoglobin E) may be explained either by a reduced ability to mount anti-oxidant defenses or by an increased level of oxidative stress. Among the anti-oxidant mechanisms used both by the red cell and the parasite, detoxication by the glutathione cycle (by means of glutathione reductase and NADPH) plays a crucial role (Fritsch et al. 1987). Vitamins C and E are found in increased amounts in infected cells and are thought to have a protective role, particularly in preventing the peroxidation of membrane lipids and in preventing the formation of methemoglobin. Both host cell and parasite superoxide dismutases (SOD) have been identified in *P. falciparum*-infected cells, the parasite SOD being cyanide insensitive (Fairfield et al. 1988).

Molecular biology and genetics

The genome of malaria parasites is complex, presenting a high degree of diversity not only from species to species, but also from one isolate to another. Diversity is generated either by cross-fertilization during meiosis or by genomic reorganization during mitosis. The genome is haploid throughout most of the life cycle, except for a short time after zygote formation; it consists of 14 chromosomes of various sizes.

GENOME STRUCTURE

Measured sizes of the haploid genome of malaria parasites range between 20 and 25 Mb, values comparable to the genome of yeast but three to four times greater than that of *Escherichia coli*. The base composition of the genome of all malaria parasites is always A+T rich (from 70 to 83 percent A+T, depending on the species) (Weber 1988). Codon usage by malaria parasites is not random and it has been shown that for some genes (e.g. the repeat region of the circumsporozoite, CS protein), the usage of synonymous codons could be highly biased. The entire genome of *P. falciparum* (clone 3D7) has been sequenced and it was shown to contain 5 300 protein-encoding genes (Gardner et al. 2002). The genome of other species of *Plasmodium* are in the process of being sequenced.

The parasite genome shows variation at many levels and the genetic polymorphism of parasite isolates may be used as a tool for molecular epidemiological studies. The extreme diversity of the genomic make-up of malaria isolates has been demonstrated by the use of Southern blot fingerprinting which has revealed a different fingerprint pattern for almost every single isolate from a given geographical area (Langsley et al. 1988; Hughes et al. 1989).

As well as nuclear DNA, malaria parasites have two other extrachromosomal forms of DNA: a 6-kb element that is related to the mitochondrial DNA of other species and a 35-kb circular element (or 'apicoplast') which seems to be related to the chloroplast element found in algae (Clough and Wilson 2001). It has been suggested that this organelle is the target of antimalarial antibiotics, such as tetracycline, doxycyclin, and clindamycin, which inhibit prokaryotic translation, as well as rifampicin and quinolones which inhibit prokaryotic transcription (Wilson et al. 1996; Fichera and Roos 1997).

MALARIA GENETICS

Genetic recombination in malaria parasites occurs in the mosquito where fertilization of gametes, zygote formation, and meiosis takes place. For recombination to occur, the parasites picked up by the mosquito must consist of more than one population of parasites of the same species. During recombination, members of pairs of chromosomes segregate randomly into the haploid progeny and crossing-over events take place between homologous chromosomes. Such studies provide information on the relationships between various genes (e.g. the presence of linkages) and the different ways in which recombination can occur. In view of the fact that malarial infections observed in humans are rarely clonal, but more often a mixture of between three and nine different parasite clones, it is surprising that the level of recombination in mosquitoes actually observed is lower than expected (one-and-a-half to two on average). This may be explained by the fact that, although different populations are present at the same time, the different populations are rarely synchronized with regard to gametocytogenesis, with the consequence that homozygote crosses are more likely than heterozygote crosses (Paul et al. 1995).

GENETICS OF CHLOROQUINE RESISTANCE

Forty-five years ago, resistance to chloroquine (CQ) spontaneously appeared, almost at the same time in both Indochina and Brazil, then spread throughout Asia and South America, to eventually reach Africa in the late 1970s (Wernsdorfer and Payne 1988); the current situation is illustrated in Figure 24.2 showing an almost worldwide distribution of resistance, with the need for countries to switch to new first-line antimalarials, when resistance exceeds 25 percent.

The mechanism of action of CQ is complex and remains controversial. CQ, being a lysosomo-tropic drug, accumulates in the parasite food vacuole where it binds to ferriprotoporphyrin IX (Fitch et al. 1982), a toxic metabolite of heme which is normally detoxified by the formation of hemozoin crystals; the accumulation of toxic metabolites of heme kills the parasite. Chloroquine-resistant (CQR) parasites accumulate less CQ than chloroquine-sensitive (CQS) parasites, below the level necessary to inhibit heme polymerization. Different hypotheses have been proposed to explain the reduced accumulation of CQ: (1) an abnormal proton pump in the food vacuole membrane leading to an elevated pH in the normally acid food vacuole would inhibit CQ transport (Ginsburg and Stein 1991); (2) a reduction of the carrier-mediated entry of CQ, in which CQR cells have a constitutively activated Na^+/H^+ exchanger in the vacuole membrane which makes it impossible for CQ to stimulate its own active uptake (Wünsch et al. 1998); or (3) the presence in CQR parasites of a rapid efflux mechanism by means of a P-glycoprotein *mdr* transporter, which would effectively pump CQ out of the cell. The third hypothesis seemed to take into account the observation that verapamil, and other calcium channel blockers, can partially reverse CQ resistance, presumably by blocking the *mdr* transporter, which is overexpressed in resistant parasites (Martin et al. 1987). The finding that the pH of the food vacuole in CQR parasites was actually decreased (rather than increased, as had been assumed in the first two hypotheses) produced yet another possible mechanism of resistance taking into account the fact that the CQ accumulation in the infected red cell relies on binding to heme (Bray et al. 1998; Dzekunov et al. 2000). In this model, the lower pH in the vacuole of CQR parasites is more suitable for hemozoin production, thus leaving less free heme available for forming toxic complexes with CQ.

Linkage analysis experiments suggested that CQ resistance could be associated with a single locus, perhaps a single gene (Wellems et al. 1990). This was an important step forward, since it was believed, until then, that CQ resistance was multigenic. The search for a potential CQR gene identified the *pfcrt* gene as the most likely CQR gene candidate. The *pfcrt* gene product appears to be localized in the food vacuole membrane. Various *pfcrt* alleles are associated with CQR, which is not surprising since multiple mutations are expected, in view of the slow genesis and spread of CQ resistance (Fidock et al. 2000; Djimde et al. 2001).

RESISTANCE TO ANTIFOLATES

The folate metabolic pathway leads to the de novo synthesis of pyrimidines and is, therefore, essential to malaria parasites incapable, as they are, of pyrimidine salvage. In addition, this pathway includes enzymes with no human counterparts. Despite being nearly ideal antimalarial drugs, resistance to antifolates spread very fast after their deployment. When pyrimethamine was introduced to Tanzania in 1957, rates of resistance rose from zero to 37 percent in 5 months (Clyde and Shute 1957).

Genetic crosses have provided crucial information on the linkage between resistance and the *dhfr-ts* gene (Walliker et al. 1987). The molecule involved is a bifunctional enzyme, where different domains of the molecule act as DHFR and thymidylate synthase (TS), both enzymes of the folate pathway. Pyrimethamine, cycloguanil (a metabolite of proguanil), and chlorcycloguanil are DHFR inhibitors. A series of *dhfr* mutations have been reported, each with a single amino acid substitution within the active site of DHFR; in the primary mutation (which is found in all resistant lines), asparagine is substituted for serine at codon 108. Asn-108 occurs first, conferring only modest pyrimethamine resistance, but further mutations at codon 51, 59, and 164 may increase resistance substantially. So far, only triple resistances (Asn-108, Ile-51, Arg-59) have been found in Africa, while quadruple resistances involving Leu-164 in addition have already been found in South East Asia (Wichmann et al. 2003).

Sulfa drugs act on a different enzyme of the folate pathway, dihydropteroate synthase (DHPS): sulfonamides, and sulfones are structural analogs of *p*ABA, the natural substrate of DHPS. A variety of point mutations of DHPS have been described; the triple mutation (Gly-437, Glu-540, Ala-581), which confers most resistance, is prevalent in areas of widespread sulfadoxine-pyrimethamine failure in South America and Asia, but has not yet been reported in Africa (Plowe 2001).

Point mutations affect antifolates selectively: for example, a parasite line with mutation Asn-108 may be resistant to pyrimethamine but remain sensitive to cycloguanil and the triple resistant (Asn-108, Ile-51, Arg-59), which is highly resistant to pyrimethamine, has only moderate resistance to cycloguanil or chlorcycloguanil. This explains why the association of two drugs working in synergy, such as sulfadoxine-pyrimethamine (SP), has remained effective much longer even in areas with high prevalence of DHFR mutations. This also explains why it is conceivable to replace one antifolate combination with another, e.g. to replace SP with the new combination chlorproguanil (Lapudrine®-dapsone (LapDap)) (Winstanley 2001).

Any widespread use of antifolates is likely to exert a selective pressure, which may increase the spread of resistance. Although trimethoprim-sulfamethoxazole (Co-trimoxazole) is not a very good antimalarial, its widespread prophylactic use in areas where human immunodeficiency virus (HIV) and malaria are both highly endemic could shorten the life-span of SP (Jelinek et al. 1999).

MOLECULAR EPIDEMIOLOGY OF DRUG RESISTANCE

Two approaches have been used extensively in the past to test for drug resistance – in vivo tests and in vitro culture assays. Each of these methods has disadvantages. Molecular techniques, which allow the detection of mutations associated with resistance, offer an alternative approach with many advantages. The difficulty of this approach has been the definition of mutations that accurately define resistance or, more specifically, to predict in vivo resistance based on the finding of mutations. The presence of a mutation of the DHFR gene may confer some resistance to pyrimethamine, but this does not necessarily mean that a patient carrying such a parasite can no longer be treated with SP. Only parasites carrying triple mutations for *dhfr* and double mutations for *dhps* are always associated with SP failure and a high prevalence of such parasites in a particular area suggests the need to switch to a new drug for first-line treatment. Such molecular mapping studies are essential for following the spread of resistance at country level and to provide an early warning system for policymakers. Because it is much easier to perform than in vivo and in vitro resistance tests, molecular screening is likely to succeed where other surveillance programs have failed. A number of regional centers have been set up to coordinate information and work towards a rational implementation of coherent policies, e.g. the best example so far is the East African Network for Monitoring Antimalarial Treatment (EANMAT) (Mutabingwa 2001). Figure 24.13 provides an example of how Africa-wide information can be gathered at a distance by screening parasites isolated from European tourists using molecular drug resistance markers (TropNetEurop) (Jelinek et al. 2002).

CLINICAL ASPECTS

Malaria can mimic many diseases and there are no absolute diagnostic clinical features. The classical periodicity of febrile paroxysms every 48 h for *P. falciparum*, *P. vivax*, and *P. ovale* (days 1 and 3) or every 72 h for *P. malariae* (days 1 and 4) are usually absent at the beginning of the disease. Periodicity develops only as a result of a delay in treatment when a sufficient number of schizonts rupture at the same time and the infection becomes synchronized.

Clinical manifestations

FALCIPARUM MALARIA ('MALIGNANT TERTIAN MALARIA')

In nonimmune people, falciparum malaria is a medical emergency, as there are few conditions in tropical medicine that can change dramatically from a relatively benign illness to a catastrophic and fatal one (Gilles and Phillips 1988; Molyneux and Fox 1993; Warrell 2002). The incubation period is generally 9–14 days, but it can be as short as 7 days. The symptoms are nonspecific with headache, pains in the back and limbs, simulating influenza, anorexia, nausea, and a feeling of chill rather than a distinct cold phase (as in vivax malaria). Fever is very common but not invariable; it is continuous or

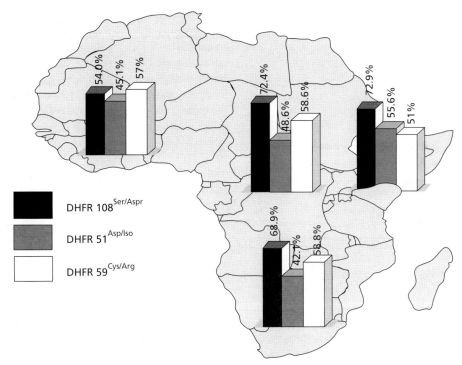

Figure 24.13 *TropNetEurop a network of travel clinics in Europe acts as sentinel for monitoring the spread of drug resistance in Africa– distribution of DHFR mutants in malaria isolates from European tourists.*

remittent, not tertian. The physical findings include prostration, a tinge of jaundice often misdiagnosed as viral hepatitis, *Herpes simplex* of the lips and tender hepatosplenomegaly. An ophthalmoscopic examination should always be carried out, because multiple retinal hemorrhages (predominantly white-centered) and other retinal changes occur in both children and adults. They are best seen, however, by indirect ophthalmoscopy (Beare et al. 2003). Hyponatremia can occur in both uncomplicated and severe malaria (Ustianowski et al. 2004).

MANIFESTATIONS OF SEVERE FALCIPARUM MALARIA

Cerebral malaria is the most common presentation of severe malaria in adults. They have usually been ill for 4–5 days with fever, slowly lapsing into coma, with or without convulsions. The usual neurological picture is of a symmetrical upper motor neurone lesion with increased muscle tone, brisk, tendon reflexes, ankle clonus, and extensor plantar responses. The abdominal reflexes are absent. A number of bizarre neurological signs can occur; dysconjugate gaze is common. A whole variety of other neurological signs have been described (Warrell 2002). Retinal hemorrhages occur in about 15 percent of adults with cerebral malaria and are associated with low hematocrits and high parasitemias (Looareesuwan et al. 1983) (Figure 24.14). Even where good standards of care can be provided, the morbidity is about 15–20 percent.

Anemia is common in young children (see below), in pregnant women (see below), and in adults with severe malaria in whom it correlates with parasitemia.

Biochemical evidence of renal dysfunction progressing to oliguria, anuria, and acute renal failure occurs in about 10 percent of cases.

Hypoglycemia is common in patients with hyperparasitemia or as a result of treatment with quinine. The classic symptoms of hypoglycemia may not be present,

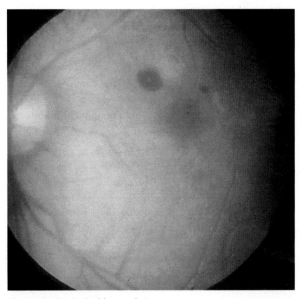

Figure 24.14 *Retinal hemorrhage*

Table 24.1 *Complications of severe falciparum malaria*

Complications
Cerebral malaria
Anemia
Renal failure
Hypoglycemia
Hyperparasitemia
Malarial hemoglobinuria
Metabolic acidosis presenting as 'respiratory distress'
Fluid, electrolyte, and acid base disturbances
Pulmonary edema
Circulatory collapse ('Algid malaria')
Bleeding and clotting disturbances
Hyperpyrexia

the only sign being deterioration in the level of consciousness.

Pulmonary edema is the most dreaded complication of malaria with a mortality of over 50 percent. It is sometimes precipitated by excessive parenteral fluid therapy and, as such, avoidable. The earliest signs are an increase in respiratory rate, dyspnea, and crepitations. In most patients, however, it is not iatrogenically produced and the picture is similar to that of the adult respiratory distress syndrome (ARDS) (Charoenpan et al. 1990). The pathogenesis of pulmonary edema in *P. falciparum* is complex and not fully understood.

Malarial hemoglobinuria is now usually associated with hyperparasitemia resulting in severe intravascular hemolysis. Some patients are mistakenly labeled as such when in fact the hemoglobinuria is due to glucose-6-phosphate dehydrogenase deficiency, triggered by a variety of drugs, including the oxidant antimalarial primaquine (Gilles and Ikeme 1960). The known complications of severe falciparum malaria are given in Table 24.1.

VIVAX AND OVALE MALARIA ('BENIGN TERTIAN MALARIA')

The incubation period of vivax malaria is usually between 12–17 days, but with some strains can be as long as 250–637 days; that of ovale malaria is usually 16–18 days but may also be longer. The clinical manifestations are similar.

After 2–3 days of nonspecific symptoms, similar to those described for falciparum malaria, the classical febrile paroxysm occurs. The cold stage, which lasts for 15–60 min, is characterized by violent rigors, a high core temperature, a cold dry skin due to intense peripheral vasoconstriction, and a rapid, low-volume pulse. The hot stage follows, the patient becoming unbearably hot. The temperature rises to 40–41°C, accompanied by a severe throbbing headache, palpitations, prostration, confusion, and delirium. The skin is flushed, the pulse rapid and full. The patient looks ill, may be anemic, mildly jaun-

diced, and the liver and spleen are enlarged and tender. The hot stage lasts 2–6 h. It is followed by the sweating stage characterized by drenching sweats, defervescence, and exhausted sleep. Unless treated, the febrile paroxysms recur every other day. Thrombocytopenia is common and the only serious (often fatal) complication is splenic rupture. Untreated or inadequately treated cases relapse after a period of quiescence at intervals varying from 8–40 weeks or more, depending on the strain of the parasite. A few cases of reversible pulmonary edema and neurological manifestations have been described in patients with confirmed vivax malaria (Tanios et al. 2001; Beg et al. 2002).

MALARIAE MALARIA ('QUARTAN MALARIA')

The incubation period of quartan malaria is 18–40 days. The clinical picture is similar to that of benign tertian malaria with the febrile paroxysms occurring every 72 h. Symptomatic recrudescences, which can occur up to 52 years after the last exposure to infection are due to persistent undetectable parasitemia, not to hypnozoites. In several parts of the world there is strong epidemiological evidence that *P. malariae* is an important cause of nephrotic syndrome especially in children under 15 years of age (Gilles and Hendrickse 1963).

Malaria in children

The manifestations of the primary attack of malaria in nonimmune children are very variable, consisting of any of the following symptoms, or combinations thereof: restlessness or drowsiness, refusal to eat or suck, headache, vomiting, loose stools, and cough. In the majority, the temperature is high (40°C), the child is flushed, and febrile convulsions are common, lasting only a few minutes. If the child does not regain consciousness within 30 min following the convulsion, cerebral malaria must be suspected. The liver is often enlarged and tender, splenomegaly develops later. Ocular fundus findings such as papilledema indicate a poor prognosis (Lewallen et al. 1993). Other retinal changes, which are best seen by indirect ophthalmoscopy, include patchy whitening of the peripheral retina and discoloration of the retinal vessels to pale orange or white (Lewallen et al. 2000). Retinal hemorrhages are common.

Children living in highly endemic malarious areas are equally vulnerable until the age of about 5 years, after which they achieve a relative immunity to malaria infection. Many, especially in tropical Africa, are either asymptomatic or have a mild illness, despite a parasitemia of 10–30 percent. Hepatosplenomaly occurs in 80 percent of these individuals.

Many of the clinical features of severe malaria (described in Clinical manifestations above) also occur in children. The commonest and most important complications are cerebral malaria, severe anemia, and

Table 24.2 *Differences between severe malaria in adults and children*

Sign or symptom	Adults	Children
Cough	Uncommon	Common
Convulsions	Common	Very common
Jaundice	Common	Uncommon
Duration of antecedent illness	5–7 days	1–2 days
Pretreatment hypoglycemia	Uncommon	Common
Resolution of coma	2–4 days	1–2 days
CSF opening pressure	Usually normal	Often raised
Abnormality of brain stem reflexes(e.g. occulovestibular, occulocervical)	Rare	More common
Neurological sequelae	<5%	>10%
Pulmonary edema	Common	Rare
Renal failure	Common	Rare
Bleeding/clotting disturbances	Up to 10%	Rare

Derived from studies in South-East Asian adults and African children (White et al. 1987; Molyneux et al. 1989).

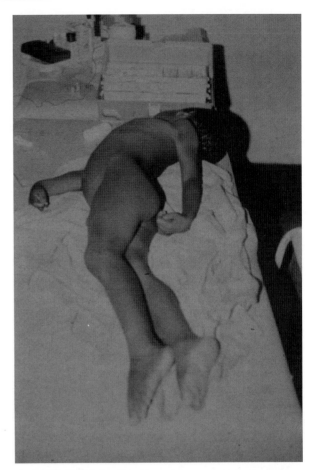

Figure 24.15 *Opisthotonos*

metabolic acidosis. Cerebral malaria is the commonest presentation in the Gambia and in Burkina Faso (Waller et al. 1995; Modiano et al. 1995), whereas anemia is the most frequent presentation in Kenya and in Papua New Guinea (Marsh et al. 1995; Allen et al. 1996).

The differences between severe malaria in adults and in children are given in Table 24.2. The reasons for these differences have yet to be fully elucidated.

CEREBRAL MALARIA IN CHILDREN

In holoendemic areas of malaria, cerebral malaria occurs in children between 6 months and 5 years, most commonly in those 3–4 years old. The earliest symptom is usually fever (37.5–41°C) followed by failure to eat or drink. Vomiting and cough are common; diarrhea is unusual. The history of symptoms preceding coma may be very brief (1 or 2 days). Hypoglycemia is a particularly common presenting feature in children under 3 years and in those with convulsions, hyperparasitemia, or profound coma (White et al. 1987; Taylor et al. 1988). In the latter, corneal and vestibular-ocular reflexes may be absent. Extreme opisthotonos is sometimes seen (Figure 24.15) and is easily mistaken for tetanus or meningitis (Taylor and Molyneux 2002).

Some children are in a state of shock. Cerebrospinal fluid (CSF) opening pressure is often raised, sometimes markedly so. Intracranial hypertension and neurological evidence of cerebral herniation has been documented (Newton et al. 1991). Convulsions are common before or after the onset of coma, they are significantly associated with morbidity and sequelae. Neurological sequelae, which occur in about 10 percent of children,

include: hemiparesis, cerebellar ataxia, cortical blindness, severe hypotonia, mental retardation, generalized spasticity, and aphasia (Molyneux et al. 1989; Newton et al. 1994). Significative cognitive problems affecting school performance may occur in children who appear to have made a complete neurological recovery (Holding et al. 1999).

ANEMIA

This is particularly common in children between 6 months and 2 years. It is often the result of repeated untreated or partially treated episodes of uncomplicated malaria. Parasitemia is often scanty, although numerous pigmented monocytes are seen in the peripheral blood. The anemia is normochromic with prominent dyserythropoietic changes in the bone marrow. In older children the anemia is acute and hemolytic and associated with hyperparasitemia. Children with severe anemia may present with tachycardia, dyspnea, respiratory distress, confusion, restlessness, coma, retinal hemorrhages, cardiac failure, and pulmonary edema. A study in Tanzania has shown that treating infants with SP at 2, 3, and 9 months, to coincide with routine immunization, reduces episodes of clinical malaria by 60 percent and severe anemia by 50 percent (Schellenberg et al. 2001).

METABOLIC ACIDOSIS

Metabolic acidosis is an important feature of severe malaria in children. It may present separately or in combination with cerebral malaria or anemia (Taylor et al. 1993; Marsh et al. 1995). Deep breathing ('Kussmaul') is a good clinical indicator of the presence of acidosis. Lactic acidosis is a major contributor to academia (Krishna et al. 1994). Metabolic acidosis is associated with a poor prognosis. The child often presents with other signs of respiratory distress, but is clinically and radiologically clear.

Malaria in pregnancy

Nonimmune pregnant women are susceptible to all the usual manifestations of malaria. Moreover, they have an increased risk of abortion, stillbirth, premature delivery, and low birth-weight of their infant. Mortality from severe malaria is higher than in nonpregnant patients (Wickramasuriya 1935; Menon 1972; Bray and Anderson 1979; Brabin 1983; Looareesuwan et al. 1985; Brabin et al. 1993).

Pregnant women are particularly prone to hypoglycemia which may occur on admission in severe disease, in otherwise uncomplicated malaria and as a complication of quinine therapy. It may be asymptomatic or manifest itself merely as an alteration in conscious level. It is dangerous for the baby causing fetal bradycardia and other signs of fetal distress. In the quinine-induced hyperinsulinemic hypoglycemia, abnormal behavior, sweating, and sudden loss of consciousness usually occur (White et al. 1983). The prevalence and intensity of malaria infection in pregnancy is higher in women who are HIV positive.

Pulmonary edema may occur in pregnant women on admission, may develop suddenly and unexpectedly several days after admission, or may develop immediately after childbirth (see also Figure 24.16). In holoendemic areas of malaria, partially immune pregnant women, especially primigravidae are susceptible to abor-

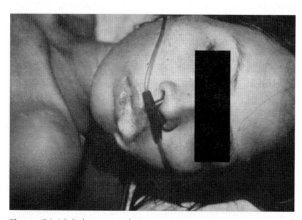

Figure 24.16 *Pulmonary edema*

tion, stillbirth, premature delivery, and low birth-weight of their infants (Archibald 1956; Bray and Anderson 1979; McGregor et al. 1983; Morgan 1994; Greenwood et al. 1994). Significant increases in parasite rates and densities occur particularly in primigravidae. In the second trimester primiparae are prone to develop a severe hemolytic anemia which bears little relation to their peripheral parasitemia (Gilles et al. 1969; Rougemont et al. 1977; Brabin 1983; McGregor 1984; Fleming et al. 1986; Steketee et al. 1996a). The other complications of severe malaria are, however, rarely encountered.

Vertical transmission of malaria parasites across the placenta from mother to fetus occurs and in endemic areas neonates not infrequently have cord-blood and peripheral parasitemia, which disappears within a day or two. Congenital malaria (i.e. symptoms or signs resulting from malarial infection in the neonate) is more common in infants born to nonimmune mothers and during epidemics of malaria. In malaria-endemic regions, the incidence is low, despite a high prevalence of placental infection. A detailed description of malaria in pregnancy is given by Shulman and Dorman (2002).

Hyperactive malarial splenomegaly (tropical splenomegaly syndrome)

In some malaria-endemic regions of the world and particularly in parts of Papua New Guinea and sub-Saharan Africa, adults develop a progressive, sometimes massive enlargement of the spleen (more than 10 cm below the costal margin). The spleen may weigh over 4 kg. The liver is often enlarged with dilated sinuses containing lymphocytes. Hypersplenism can cause a normocytic anemia, leukopenia, and thrombocytopenia. Pregnant women may suffer attacks of acute hemolysis. There is a gradual onset of early lethargy that can progress to an incapacitating weakness. Left upper quadrant pain is common. Physical examination reveals gross splenomegaly with an obvious notch in an afebrile patient, evidence of hypovolemia and moderate anemia (Bates and Bedu-Addo 1997). There are suggestions that hyperactive malarial splenomegaly may, in some patients, lead to the development of a tropical splenic lymphoma (Bates et al. 2001).

The syndrome seems to be an abnormal immune response to recurrent malaria infections with excessive production of IgM. Raised titers of IgM and IgG malaria antibodies and sinusoidal lymphatic infiltration of hepatic sinuses are also often found. Patients respond well, both clinically and immunologically, within 3 months of starting continuous antimalarial chemoprophylaxis (Fakunle 1981; Crane 1986). Recent studies on this syndrome have not been carried out, nor has the response to antimalarial therapy in the light of increasing parasite resistance been recorded. Such a study should be undertaken.

IMMUNOLOGY

Susceptibility to malaria is restricted by a variety of genetic features of the host, generally referred to as 'natural immunity' or 'innate resistance.' When infection does occur, the host immune system responds in a variety of ways, some of which may eventually lead to the clearance of parasites and protection against subsequent infection. In many circumstances, the development of protective immunity is slow and this may be explained by both the extreme antigenic diversity of the parasite and the existence of efficient escape mechanisms.

Innate resistance

The relationship between malaria parasites and their host is a delicately balanced one, which may be affected by a variety of host and parasite features (Miller 1976). When different hosts are exposed to a given malaria parasite, the ensuing infection may range from negligible to fatal. For example, *Plasmodium knowlesi* which produces a chronic but low-level infection in *Macaca fascicularis* (the natural host), a low-level and rapidly self-limiting infection in humans (accidental or experimental host), and a fulminant, fatal infection in *Macaca mulatta* (experimental host). The degree of susceptibility may vary with time, particularly after serial sub-inoculations of the parasite, and it appears that 'adaptation' to an artificial host may occur. When *P. knowlesi* was extensively used for malariotherapy of neurosyphylitic patients in Romania, after a number of sub-inoculations the strain became so highly virulent to humans that it became unmanageable (Coatney et al. 1971).

Although it has often been stated that ethnic groups, which have had a particularly long association with malaria, are less susceptible to infection and disease (e.g. Pygmies in Central Africa, Tharu in Nepal or Orang Asli in Malaysia), the evidence is mostly anecdotal. The best proof of this concept comes from a comparative study of three sympatric ethnic groups in Burkina Faso that showed reduced infection rates and morbidity in Fulanis, a traditionally nomadic tribal group, compared to Mossis and Ribaimas, in a situation where all three groups cohabit in the same villages (Modiano et al. 1996). The follow up of well-characterized family groups seems to implicate a region of chromosome 5 (5q3l-q33) as the locus for susceptibility to *P. falciparum* infection (Flori et al. 2003). Our improved knowledge of human and parasite genomes will open the field for future genetic susceptibility studies.

Susceptibility to infection is primarily determined by the ability of the parasite to invade and survive in appropriate host cells (e.g. hepatocyte and erythrocyte in the mammalian host). With regard to human malaria parasites, specific receptors for hepatocyte invasion are still unknown and there is no evidence to suggest differential susceptibility or resistance in any particular human ethnic group or in association with any specific genetical characteristic of hepatocytes. The maturity of the hepatocyte is crucial, since fetal hepatocytes appear to be resistant to infection.

Susceptibility to erythrocytic infection may be determined at the level of merozoite invasion, intracellular growth, or erythrocytic lysis at the time of merozoite release. *P. vivax* is restricted to reticulocytes and Duffy-positive cells, while *P. falciparum* invasion may be substantially reduced in red cells with glycophorin abnormalities, e.g. the En(a-) or the S-s-U phenotypes (Pasvol and Wilson 1982). Resistance to parasite entry is observed in cells with cytoskeleton abnormalities; ovalocytosis confers resistance to all malarial species while elliptocytosis, due to the absence of band 4.1, may hinder the entry of *P. falciparum* (Nagel 1990). Intra-erythrocytic development of *P. falciparum* is reduced or retarded in red cells presenting various hemoglobinopathies, including HbS, HbE, HbF, HbC, G6PD deficiency or α-thalassemia; the reduced growth of *P. falciparum* in HbC/C cells has been attributed to an increased resistance to red cell lysis and to intracellular degenerescence, compatible with an incapacity of parasites to induce merozoite release. In some cases, reduced growth may be observed only in the initial cycle of development in abnormal cells, as is the case when *P. falciparum* is grown in vitro in G6PD-deficient cells, where the parasite can 'adapt' by switching on its own enzyme.

Resistance to infection may also be caused by the antiparasitic effect of components of the host serum. The serum of adult Sudanese living in an endemic area for malaria may cause intraerythocytic death of *P. falciparum*; the toxic component of serum, named 'crisis forming factor,' being independent of host immunity (Jensen et al. 1983).

In addition to innate resistance to infection, there is considerable diversity in individual susceptibility to severe disease; for instance, it has been estimated that *P. falciparum* leads to cerebral malaria in only a small percentage of cases in an endemic area (estimated at 1:200), and that only a percentage of those are fatal (Greenwood et al. 1991). Many factors can explain this apparently low percentage of disease severity, including the nutritional status and the level of acquired immunity, which is directly related to transmission intensity. A number of genetic susceptibility factors have been identified, the importance of which may vary from one endemic area to another (Hill and Weatherall 1998). For instance, individuals heterozygous for the HLA-B53 haplotype had a significantly decreased susceptibility to severe falciparum disease in the Gambia, but not in Kenya. In contrast, genetic variation in the promoter gene for inducible nitric oxide synthase (iNOS) was found to be responsible for reduced susceptibility to severe malaria in Gabon and increased susceptibility to cerebral malaria in the Gambia (Burgner et al. 2003). The intercellular adhesion molecule-1 (ICAM-1)[Kilifi] allele was found to be associated with increased susceptibility to cerebral malaria in Kenya, but not in the Gambia and was associated with protection in Gabon (Bellamy et al. 1998). Variable susceptibility to severe malaria was also shown to be significantly linked to the polymorphism of the TNF-α promoter gene (McGuire et al. 1994), the interferon- receptor I gene (Aucan et al. 2003), and the CD36 gene (Aitman et al. 2000). As confusing as such contradictory results may appear to be, they provide a clue to the role of host genetic susceptibility in the complex events leading to different forms of severe malaria. Paradoxically, while the carriers of the α-thalassemia gene are less likely to develop severe falciparum malaria, they have been shown to have an increased susceptibility for *P. vivax* infection, probably because of the preference of this parasite for reticulocytes, which are more abundant in thalassemics (Williams et al. 1996).

Acquired immunity

Even in individuals living in an area endemic for malaria and who are, therefore, exposed to frequent reinfection, an effective immunity against *Plasmodium* takes a long time to develop. The different stages of the development of immunity may be described as follows:

- A child born to an immune mother is protected during the first 6 months of life, as the result of the passive transfer of maternal immunity.
- During that period, the infant is exposed to infection and starts to develop its own immunity. It will, however, take many years before this immunity becomes protective. Once maternal immunity has waned, each subsequent infection will result in clinical malaria and a percentage of children will die in situations where early diagnosis and treatment is not available.
- Once the child reaches school age, malaria paroxysms will progressively become less frequent and less severe: this is the stage of anti-disease immunity (or clinical tolerance). At this stage, the child continues to become infected, sometimes with very high levels of parasitemia, but either develops only mild forms of malaria or remains asymptomatic. Age-dependency is a crucial feature of malaria immunity.
- The state of anti-disease immunity progressively develops into a state of premunition, when the child not only exhibits clinical tolerance but also much reduced parasitemia and shorter episodes of infection.
- A state of complete, sterile immunity may never be reached, even in adults who have never left the endemic zone. At this stage, infections remain mostly asymptomatic, with very low parasitemia and short episodes of infection.
- The state of incomplete immunity (or semi-immunity) is solid, but requires frequent reinfection. Any period spent outside the endemic area may result in a loss of premunition.
- During pregnancy, adult women, who had previously reached a state of premunition, appear to lose it and may again present clinical malaria, sometimes in severe, life-threatening forms.

Important *Plasmodium* antigens

Much of what is known of the antigenic structure of malaria parasites has been derived from studies aimed at identifying protective antigens with a view to the development of an antimalarial vaccine. Table 24.3 lists the *P. falciparum* antigens which have been considered important in antimalarial immunity, because they are expressed on the surface of extracellular infective forms, on the surface of malaria-infected cells or because their recognition by patient antibodies appears to be consistent with the development of protective immunity. In many cases, these *P. falciparum* molecules are representative of a family of antigens and similar, if not necessarily biochemically identical, antigens may be found in all malarial species.

The antigens of *Plasmodium* spp. present a number of notable features:

1 A relatively small number of molecules actually expressed on the surface of invasive stages and this reflects the fact that these stages are generally covered by a surface coat with crucial biological functions in terms of recognition and adherence to future host cells (CSP-1 and SSP2/TRAP on sporozoites; MSP1, MSP-2, and MSP-4 on merozoites; Pfs230, Pfs45/48, and Pfs25 on gametes or ookinetes).
2 Extracellular stages all possess the characteristic invasion organelles and many immunodominant antigens appear to be associated with these organelles. Some are found in rhoptries (AMA-1, RAP-1-3, Rhop1-3), some in micronemes (EBA-175), and some in dense granules (RESA/Pf155). Although these molecules are not actually on the surface, they seem, nevertheless, to play an important role at the parasite–host cell interface and immune reactions against them interfere with invasion. It is conceivable that the good immunogenicity of these molecules may reflect more their uniqueness (i.e. no similar antigens in the host) than their true biological importance.
3 Some molecules of parasite origin are expressed at the surface of infected erythrocytes or infected hepatocytes, either directly by the parasite in order to fulfill a fundamental parasite function (as is probably the case of PfEMP-1 and rifins, expressed on the surface of host erythrocyte) or by means of a host processing function related to MHC class I expression (e.g. the exo-erythrocytic antigens at the surface of hepatocytes, including CSP-1, hsp70, LSA-1, and LSA-2, STARP, SALSA); and

Table 24.3 *Important antigens of* Plasmodium falciparum

Antigen	Full name and alternative	Location
Intra-erythrocytic stages		
hsp70	Heat shock protein-1, p75	Parasite cytoplasm
GRP78	Glucose-regulated protein, heat shock protein-2	Parasite cytoplasm
Aldolase	Aldolase	Parasite cytoplasm
HRP-3	Histidine-rich protein-3, SHARP	Parasite cytoplasm
CARP	Clustered asparagine-rich protein	Parasite cytoplasm
RESA	Ring-infected erythrocyte surface antigen, Pf155	RBC internal surface, merozoite dense granule
MESA	Mature-parasite infected surface antigen-2, PfEMP-2	RBC internal surface
AARP	Asparagine- and aspartate rich protein	RBC internal surface
HRP-1	Histidine-rich protein-1, KAHRP, KP	RBC internal surface
HRP-2	Histidine-rich protein-2	RBC internal surface and secreted
Pf332	Pf332, Ag332, giant protein	RBC internal surface
GBP	Glycophorin-binding protein, Pf120-130,	RBC cytoplasm
FIRA	Falciparum interspersed repeat antigen	RBC cytoplasm
S-antigen	S-antigen	Parasitophorous vacuole
SERA	Serine-rich antigen, SERP, p113, p126, Pf140	Parasitophorous vacuole
GLURP	Glutamate-rich protein	Parasitophorous vacuole (erythro, exo-erythro)
ABRA	Acidic basic repeat antigen, p101	Parasitophorous vacuole
SPAM	Secreted polymorphic antigen associated with merozoite, MSP3	Parasitophorous vacuole
PfEMP-1	Erythrocyte membrane-associated malaria protein-1	RBC external surface
Merozoite		
MSP-1	Merozoite surface antigen-1, MSA-1, PMMSA, P190, gp185	Merozoite surface, surface of schizont
MSP-2	Merozoite surface antigen-2, MSA-2, gp56, QF122, GYMMSA	Merozoite surface
MSP-4	Merozoite surface antigen-4, MSA-4	Merozoite surface
MSP-5	Merozoite surface antigen-2, MSA-5	Merozoite surface
AMA-1	Apical membrane antigen-1, Pf83	Apical region, merozoite surface
EBA-175	Erythrocyte-binding antigen, SABP	Micronemes, merozoite surface
RAP-1/-2/-3	Rhoptry-associated protein-1, 2 and 3	Rhoptries
RhopH	Complex of Rhop1, Rhop-2 and Rhop-3	Rhoptries
Ag512	Ag512	Rhoptries
MCP-1	Merozoite cap protein-1	Apical surface
RIMA	Ring-stage membrane antigen	Dense granule
Sporozoite and pre-erythrocytic stages		
CSP-1	Circum sporozoite protein, CSP	Sporozoite surface, infected hepatocyte
SSP-2	Sporozoite surface protein-2, TRAP	Sporozoite surface
LSA-1	Liver-specific antigen-1	Infected hepatocyte
LSA-2	Liver-specific antigen-2	Infected hepatocyte
LSA-3	Liver-specific antigen-3	Infected hepatocyte
STARP	Sporozoite threonine and asparagine-rich protein	Sporozoite surface, infected hepatocyte
SALSA	Sporozoite and liver stage antigen	Sporozoite surface, infected hepatocyte
Sexual stages		
Pfs230	Pfs230	Gametocytes, surface gametes
Pfs45/48	Pfs45/48	Gametocytes, surface gametes
Pfs25	Pfs25	Surface ookinete
Pfs16	Pfs16	Gametes and sporozoite
11.1	11.1	Parasite cytoplasm

4 a large number of antigens are found inside the parasite and, as such, would not be expected to play a role in protective immunity (GLURP, SERA, hsp-70). These molecules are either excreted during schizogony (by means of a transfer from the parasite to the parasitophorous vacuole and then to the outside of the cell) or, more often, released at the time of schizont rupture. They may be highly immunogenic and be responsible for a considerable portion of the total host immune response; some of these molecules are of interest for immunodiagnosis and some may be involved in immunopathological events.

Epidemiological evidence for immunity

The main lesson from epidemiological studies is that malaria is a chronic disease in which immunity is essentially an age-dependent event. Repeated reinfection is necessary to build up protection and this generally takes a few years. The first sign of immunity is a protection against severe forms of the disease, which, according to one study, could occur after only a small number of clinical episodes (Gupta et al. 1999). This is followed by a progressive reduction of the frequency of malarial episodes, without any reduction in the levels of parasitemia; as a result, surveys of malaria prevalence in a malaria-endemic area will show paradoxically high levels of parasitemia in older children in the community (hence the 'clinical threshold' concept discussed under Diagnosis below) (Trape et al. 1994). Even adults living in an area of high endemicity still continue to present asymptomatic infections; in a longitudinal study of adults in a West African community, Bruce-Chwatt (1963) showed that nearly all adults had a patent parasitemia at one time or another during the survey.

This progressive evolution from a state of immunity against 'malaria-disease' to a state of partial immunity against 'malaria-infection' is almost unique among infectious diseases. It has been described as 'premunition' by early malariologists (Sergent et al. 1924) and refers to a form of nonsterile immunity, which has to be maintained by almost continuous exposure to the parasites. This continuous exposure is achieved either as the result of a long-term survival of parasites in the host (P. falciparum can survive for over a year and P. malariae is said to survive for as long as 25 years in the absence of treatment) or as the result of frequent reinfections. Individuals who temporarily leave the endemic area may lose their premunition and develop clinical paroxysms on their return. This observation does not, however, mean that no immunological memory is associated with premunition. The persistence of protection has best been demonstrated during the 1987 epidemic of falciparum malaria on the Hauts Plateaux of Madagascar. Having been eliminated for over 30 years, malaria returned in the form of an epidemic with many fatalities, but it was noticed that individuals over 40 years of age (who had been exposed to malaria prior to elimination) were significantly less susceptible than younger age groups (Deloron and Chougnet 1992).

Immune effector mechanisms

ANTIBODIES

A rapid increase of immunoglobulin levels is observed in the course of malaria, but only a small percentage of these antibodies is directed against the parasite; most of the characteristic hypergammaglobulinemia observed during acute malaria may be attributed mainly to polyclonal activation (Rosenberg 1978). The role of antibodies in protection has best been demonstrated by the passive transfer of serum from immune adult donors to infected Gambian children (Cohen et al. 1961), in whom falciparum infection was completely cleared 48 h after the serum transfer. Similarly, in the newborn child of an immune mother, the passive transfer of IgG across the placenta protects the child during the early months of life (Edozien et al. 1962).

Individuals living in an area endemic for malaria have high levels of antibodies against many malarial antigens as demonstrated by the use of enzyme-linked immunosorbent assay (ELISA) or immunofluorescence antibody tests (IFAT); these antibodies are generally directed against immunodominant epitopes of the parasite, but are not necessarily protective. Protective antibodies may either act directly on the parasite or parasitized cells, or act in synergy with various effector cells. The possible targets of immune effector mechanisms in the life cycle of malaria parasites are illustrated in Figure 24.17.

Antibodies may interfere with parasite development at different sites: (a) by blockage of merozoite dispersion; (b) by inhibition of invasion (whether of merozoites into red cells or of sporozoites into hepatocytes); (c) by intracellular killing of erythrocytic stages (presumably by a form of metabolic poisoning induced by interaction of antibodies with specific 'pumps' introduced by the parasite into the red cell membrane); or, (d) by the inhibition or reversal of cytoadherence, which obliges the infected erythrocyte to undergo, 20 times every hour, the perilous journey around the peripheral bloodstream of the host, including passage through the spleen, instead of remaining safely sequestered in capillary venules. Antibodies to sexual stages of the parasite may be picked up, together with gametocytes, by a feeding mosquito and thus block gamete fertilization, a feature which forms the rational basis for the development of transmission-blocking immunity. Finally, antibodies may act in co-operation with various cells (including macrophages and T cells) to increase cell-mediated killing.

Depending on the expected mode of action of the protective antibodies and the target, various in vitro correlates

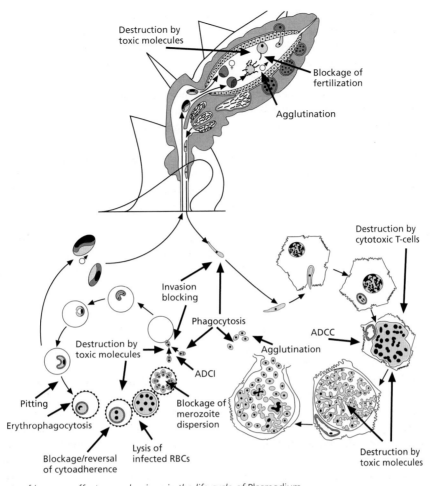

Figure 24.17 *Targets of immune effector mechanisms in the life cycle of* Plasmodium

for protection have been described including circumsporozoite precipitation (Spitalny and Nussenzweig 1973), inhibition of merozoite reinvasion (Butcher et al. 1978), inhibition of merozoite dispersal from schizonts (Green et al. 1981), recognition of ring-infected erythrocyte surface antigen (RESA) (Perlmann et al. 1984), recognition of the surface of infected erythrocytes by surface immunofluorescence, agglutination or immunogold staining (Hommel et al. 1991; Marsh and Howard 1986), inhibition/reversal of cytoadherence (Singh et al. 1988) or inhibition of rosette formation (Carlson et al. 1990). All these in vitro tests may have a degree of correlation with protection but none is perfect. When a number of such tests was compared and related to clinical outcome in a study of Gambian children, it was shown that, while high antibodies titers had a predictable age distribution, none of the individual tests had a significant correlation with protection (Marsh et al. 1989).

The activity of antibodies may be a consequence of opsonization, since the passive transfer of immune serum is more effective in intact than in splenectomized recipients (Brown and Phillips 1974). While it has been convincingly demonstrated that immune phagocytosis is an important mechanism for the removal of dead parasites (Shear et al. 1979), the true role of opsonization in

parasite killing and elimination remains unclear. Cytophilic antibodies (IgG1 and IgG3) may be crucial and this may explain why antibody-dependent cellular inhibition (ADCI), an assay that measures cooperation between antibodies and monocytes, appears to have a better correlation with protection than other in vitro antibody assays (Bouharoun-Tayoun et al. 1995).

MACROPHAGES

The role of macrophages in malaria was recognized by early pathologists, who observed an increased number of macrophages in the spleen, liver, and bone marrow of infected animals (Taliaferro and Cannon 1936). The increased number of macrophages in the spleen has been related to the release of T cell cytokines, which appear to trigger a major cellular influx, leading to splenomegaly (Wozencraft et al. 1984); both macrophage influx and splenomegaly are absent in T-cell-deficient mice (Roberts and Weidanz 1979). During parasitic 'crisis' (when parasitemia starts to decrease under the influence of acquired immunity), there is a considerable increase in phagocytosis of malaria parasites, infected erythrocytes and malarial debris (particularly residual bodies and pigment); in rodent malaria, the phagocytic activity of splenic macrophages has been estimated to be 20–50-fold greater than that of normal macrophages (Zuckerman 1977). This increase in phagocytic activity has been linked both to a specific activation of macrophages and to humoral factors, particularly opsonizing antibodies, which act in synergy with macrophages to enhance the destruction of parasites. Despite this massive deployment of defense mechanisms against the parasite, the process is not very effective in eliminating the infection; this may be explained by an impairment of some macrophage functions during malarial infection. The antigen-presenting function of macrophages, which is crucial to T-cell activation, may be impaired as the result of the suppressive effect of the malarial pigment ingested (Schwarzer et al. 1998; Taramelli et al. 2000). It has been suggested that cytoadherence of *P. falciparum*-infected erythrocytes to cells of the reticulo-endothelial lineage, particularly dendritic cells, may directly switch off their immunoregulatory function (Urban and Roberts 2002).

In addition to their phagocytic activity, macrophages and monocytes can produce a variety of toxic substances, which may damage or destroy malaria parasites; the production of these toxic molecules is generally enhanced in activated macrophages. Effector functions of macrophages include the release of reactive oxygen intermediates (hydrogen peroxide and hydroxyl radicals) (Clark and Hunt 1983), the production of tumor necrosis factor (TNF-α) (Taverne et al. 1990), and the production of many other cytokines and enzymes, which may adversely affect the parasite (e.g. polyamine oxidase which can kill *P. falciparum* in vitro) (Kumaratilake and Ferrante 1994). While activated macrophage-derived toxic products may damage parasites within erythrocytes (producing 'crisis forms'), these products may also damage the parasite in its exo-erythrocytic cycle, either directly or by means of the production of nitric oxide (Green et al. 1990). The infectivity of gametocytes may also be reduced as a result of the presence of macrophage products (particularly TNF and nitric oxide (NO)) (Naotunne et al. 1991).

T CELLS

It is well documented that T cells are crucial for malaria immunity (Weidanz and Long 1988; Ho and Webster 1989). As far as immunity to the erythrocytic stage of the parasite is concerned, the major functions of T cells seem to be to provide help for the production of antibodies and to activate macrophages. Studies of the development of immunity in murine malaria models, as well as a small number of studies in humans, suggest a slow progression from an initial Th1 response to a Th2 response. The Th1 response may be sufficient to maintain the infection in check (avoiding the kind of fulminant, rapidly lethal infection which occurs in T-deficient mice) but not sufficient to eliminate parasites altogether;

clinical protection only occurs when the immune response has switched to an essentially Th2 mode, in which antibody-mediated mechanisms play the major role in the elimination of parasites (Taylor-Robinson et al. 1993; Langhorne et al. 2002).

While neither cytotoxic CD4$^+$ or CD8$^+$ T cells (CTLs) nor antibody-dependent cellular cytotoxicity (ADCC) have ever been conclusively incriminated in immunity to erythrocytic parasites, the situation is very different with exo-erythrocytic parasites since these stages reside within hepatocytes, which bear the MHC class I antigens required for CTL activity. In experimental models, CTLs from mice immunized with x-irradiated sporozoites can kill malaria parasites grown in vitro in mouse hepatocytes and appear to be involved in the destruction of exo-erythrocytic parasites in the livers of immunized mice (Hoffman et al. 1989). C-reactive protein, interferon (IFN)-γ, IL-1, IL-6, TNF, and nitric oxide have all been incriminated in the killing of exo-erythrocytic stages, probably in the form of a cascade of events triggered by T-cell recognition of sporozoite or liver stage antigens and involving both hepatocytes and Küpffer cells (Suhrbier 1991). There is now a reasonable body of evidence to incriminate CTLs in protection against the exo-erythrocytic stages of malaria parasites in humans (Lalvani et al. 1994).

$\gamma\delta$ T CELLS

The $\gamma\delta$ T cell receptor is normally only expressed in a small percentage of peripheral lymphocytes and, while the number of $\gamma\delta$ T cells may be increased in certain infections, the role of these cells in immunity is still imperfectly understood. A substantial increase in the number of $\gamma\delta$ cells has been reported both in experimental mouse malaria (Langhorne et al. 1993) and during the acute stage of infection in humans (Ho et al. 1994). Activated $\gamma\delta$ cells from malaria-naive donors inhibit parasite growth in vitro, which confirms their role in the innate defense against infection (Troye-Blomberg et al. 1999). The presence of high levels of activated $\gamma\delta$ cells contributes to the increased production of cytokines, particularly TNF-α, and may, thus, contribute to the development of a Th1 environment which would, in turn, facilitate the pathophysiological events leading to disease (Grau and Behr 1994).

SOLUBLE MEDIATORS AND CYTOKINES

In malaria-naive individuals, the first contact with the parasites triggers nonspecific defense mechanisms which include the production of IFN-γ by $\gamma\delta$-T cells and the production of interleukin (IL)-1 and TNF-α by macrophages; these will, in turn, stimulate phagocytosis and the production of oxygen-derived reactive species (ODRS) and NO by neutrophils and macrophages. A cascade of reactions leading to the production of NO is responsible for the killing of infected hepatocytes

(Green et al. 1990). As the infection progresses, the host response develops to a typical Th1 response, with the production of pro-inflammatory cytokines (including IFN-γ, TNF-α, lymphotoxin, IL-12, and IL-18) by T cells and activated macrophages) (Torre et al. 2001; Malaguarnera et al. 2002). At this stage, the ability of the host to produce IFN-γ is crucial: peripheral blood mononuclear cells taken from African children with mild malaria are significantly better at producing IFN-γ in response to malarial antigens in vitro, than cells taken from children with severe disease (Luty et al. 1999). It is now also well established that TNF, together with NO, can kill intra-erythrocytic parasites and contribute to the sudden decrease of parasitemia observed during 'crisis' (Naotunne et al. 1991).

As in other infectious diseases, there is only a fine line between the host-beneficial effects and the pathological consequences of immunity. Cytokines may be beneficial and contribute to avoid the development of an overwhelming parasitemia; increased pro-inflammatory cytokine levels also play a crucial role in the pathophysiology of malaria and the evolution towards severe forms of the disease. The outcome hinges on the ability of the host to regulate the inflammatory response.

After repeated infection, the immune response progressively transforms to a Th2 response, with the production of anti-inflammatory cytokines (including IL-4, IL-10, and TNF-β), which enable a more effective production of antibodies and down-regulate the inflammatory events leading to pathology (Troye-Blomberg et al. 1990; Kurtzhals et al. 1998; Omer et al. 2000).

Evasion of host immunity

Protective immunity against malaria parasites is slow to develop and a child born in a malaria endemic zone may need to be exposed to successive infections for years before developing any resistance; complete protection may never be achieved. This 'real-life' situation is in sharp contrast to what is observed in experimental situations where solid, sterile immunity can be achieved. The apparent inefficiency of antimalarial immunity may be explained by the ability of the malaria parasite to evade host immunity by the intrinsic poor immunogenicity of its antigens, by sequestration, by antigenic diversity and variation, or by an alteration of the host immune response (Mercereau-Puijalon et al. 1991).

POOR IMMUNOGENICITY

Crucial antigens involved in protection may be poorly immunogenic either because of their intrinsic molecular structure or their analogy to host molecules or because of immune restriction, which may impair recognition.

The presence of multiple repeat sequences on a number of highly immunogenic erythrocytic stage antigens (e.g. RESA, S-antigens, or FIRA) has been interpreted as an evasion mechanism, since such immunodominant structures may act as a 'smoke screen' preventing the development of effective immunity to more relevant epitopes. The existence of cross-reactivity between repetitive epitopes of different malarial antigens (e.g. between RESA, FIRA, Pf11.1, Ag332, and S-antigens, see Table 24.3) has been interpreted as a further cause for poor immunogenicity since the presence of these epitopes may interfere with the normal maturation of the immune response towards a progressively increasing percentage of high affinity antibodies (Anders 1986).

SEQUESTRATION

The ability of parasites to remain sequestered by cytoadherence to the capillary lining of certain tissues must be regarded as a selected advantage since such parasites can avoid frequent passage through the spleen and thus exposure to immune effector mechanisms. P. falciparum isolates are able to switch rapidly from one endothelial receptor to another and this may be part of a parasite survival strategy, particularly since the change of receptor requirements may be in response to a changing environment of the host (acute malaria infection induces cytokine production which in turn up-regulates certain endothelial cell surface antigens) (Hommel 1997). Sequestration does not exist in the other human malaria parasites and this is considered the main reason for the difference in disease severity.

ALTERATION OF THE IMMUNE RESPONSE

Several reports demonstrate that malaria infection can induce a suppression of host immune responsiveness, including increased severity of Salmonella infections (Bennett and Hook 1959), decreased efficiency of tetanus vaccination (McGregor and Barr 1962), and the strong association between malaria and Burkitt's lymphoma (Marsh and Greenwood 1986). Various mechanisms of immunosuppression have been proposed, including polyclonal activation, macrophage dysfunction, abnormal antigenic presentation, disruption of lymphatic and splenic tissue architecture, activation of suppressor cells, and antigenic competition. While these mechanisms may affect the outcome of concurrent infections, they may also affect the ability of the host to mount an effective immune response to the malaria parasite itself.

ANTIGENIC DIVERSITY

In addition to the marked species- and stage-specificity of immunity to malaria parasites, there are also marked differences between isolates of the same species. Studies of S-antigens were the first to suggest that field isolates of *P. falciparum* may vary in their antigenic structure (Wilson et al. 1975). The use of finer methods of analysis (including isoenzymes, two-dimensional gel electrophoresis, monoclonal antibodies, or restriction fragment length polymorphism (RFLP)) has provided further evidence of considerable diversity of both the phenotypes and genotypes of field isolates. Markers of diversity used for the typing of field isolates are mostly polymerase chain reaction (PCR)-based (or PCR combined with RFLP). The most frequently used primers are those derived from MSP-1, MSP-2, CSP-1, TRAP, GLURP, and the Pf60.1 multigene family (Contamin et al. 1995; Carcy et al. 1995). The polymorphism between isolates is usually considerable. In one African village, the study of isolates collected at a single point in time from 19 children revealed 35 different genotypes. The multiplicity of infection (i.e. when isolates consist of more than one genotype) has been linked to age, parasite density, and clinical severity; in a study in Tanzanian children, a greater multiplicity was observed in asymptomatic than in clinical cases (Beck et al. 1997).

The considerable diversity between field isolates may be the major reason why a long time is required to develop immunity, since an individual living in an endemic area would need to be exposed to a vast repertoire of local strains (Day and Marsh 1991). However, before accepting that antigenic diversity is a likely explanation for the slow development of immunity, it is necessary to assume that strain-specific immunity can ultimately be transcended by a broader species-specific immunity. In the absence of such a transcending immunity, it would be difficult to understand why infections become rarer in adults, since adults continue to be exposed to new isolates. The constant fluctuation of isolates 'in circulation' in a given geographical area has best been demonstrated in a study of parasite genotypes over time in villages of Papua New Guinea (Bruce et al. 2000).

ANTIGENIC VARIATION

Malaria parasites, as many other micro-organisms, are capable of periodically changing the expression of their antigens. This provides the parasite with a powerful means for evading host immunity particularly when antigenic variation occurs in conditions where a selective pressure is exerted. In all examples of antigenic variation, the common feature is the presence of successive peaks of infection with each new peak antigenically distinct from the previous one (the latter having been eliminated by host immunity) (Hommel 1985). Surface antigens, which are most exposed to immune pressure, are obviously most likely to exhibit antigenic variation.

In malaria, antigenic variation was first described in the *P. knowlesi*/rhesus monkey model (Brown and Brown 1965) and has since been shown to exist in *P. falciparum*. Antigenic variation has not yet been

demonstrated for *P. malariae* or *P. vivax* but the long chronicity of the former and the presence of successive waves of recrudescence in the latter suggest the existence of either a fluctuating immunity or antigenic variation.

Many of the studies on antigenic variation in *P. falciparum* have concentrated on the molecules expressed on the surface of infected erythrocytes (e.g. PfEMP-1). In one study, a single isolate of *P. falciparum* was shown to be able to express ten different variant surface antigens when sub-inoculated in squirrel monkeys. For reasons that remain unclear, the set of variant antigens observed was different in intact and splenectomized animals (Hommel et al. 1991). The rate of switching from one set of antigens to another is thought to be fast; in an in vitro study of *P. falciparum*, 2 percent of parasites switched to a new antigen in every erythrocytic cycle (Roberts et al. 1992).

Antigenic variation of *P. falciparum* has been linked to a large gene family (the *var* genes), which appear to be differentially expressed in different individual parasites (Su et al. 1995; Smith et al. 1995). It has been calculated that the genome of a given clone of malaria parasites has between 50 and 60 different *var* genes (i.e. approximately 5 percent of the total genome), that *var* genes are scattered over most of the malarial chromosomes, and that only one *var* gene may be expressed at a time in a given parasite. Mathematical modeling of switching rates, based on the number of *var* genes that can be transcribed in vitro in one cycle of division, suggests that switching may be as fast as 18 percent per generation (i.e. nine times higher than estimates based on biological observations) (Gatton et al. 2003). A group of similar molecules exists in other *Plamodium* species, e.g. *vir* genes in *P. vivax* (Del Portillo et al. 2001) or *SICAvar* in *P. knowlesi* (encoding for what was previously known as the schizont-infected cell agglutination antigen (SICA), the first variant malarial variant antigen described) (Al-Khedery et al. 1999).

ANTIDISEASE IMMUNITY

Acquired immunity to malaria in humans is a mixture of 'antidisease' immunity, which results in decreased clinical manifestations despite infection, and of 'antiparasite' immunity, which results in a control of the infection itself. The mechanisms of these two types of immunity are clearly distinct since antidisease immunity usually occurs much before antiparasite immunity. The consequence of this is that an individual living in a highly endemic area of the world may harbor very heavy parasite loads and be clinically well (McGregor et al. 1956; Miller 1958). Little is known of the mechanisms leading to antidisease immunity but the most attractive hypothesis so far is that host immunity interrupts a cascade of events, initially triggered by the production of a series of cytokines (including TNF and IL-6) in response to malarial 'endotoxins' (Playfair et al. 1990); the nature of malarial endotoxins may be either lipidic or carbohydrate in nature, which suggests that anti-disease immunity may be controled by T-cell-independent mechanisms and consequently requires frequent boosting to be sustained.

The term 'premunition' refers to a situation where malarial infection is only partially controled by the host (i.e. a state of 'relative immunity') with elements of both antidisease immunity (e.g. the existence of asymptomatic infections) and antiparasite immunity (e.g. an age-dependent reduction of parasitemia), hence the confusion the use of the term has generated in the literature. Recent studies in Indonesia have suggested that, in contrast to established dogma, a state of premunition may be more easily reached by adult migrants to an endemic area rather than by young children with life-long residence in such an area (Baird 1995).

Whether antidisease immunity is specifically linked to a response to the so-called malarial toxins has not yet been conclusively established. *Plasmodium* glycosylphosphatidylinositol (GPI), which serves as a membrane anchor for a number of parasite surface antigens, is the most likely candidate molecule, as it has been shown to trigger a variety of pathways involved in the pathophysiology of malaria, including the induction of TNF-α and NO (Schofield et al. 1996). In a study in Papua New Guinea, antibodies to GPI were inversely associated with increased tolerance to high parasite loads in local residents (Boutlis et al. 2002), but the same association between anti-GPI antibodies and protection against mild or severe disease was not found in a study in the Gambia, showing that the acquisition of anti-GPI antibodies was age- and exposure-dependent (De Souza et al. 2002).

Immunity and nutrition

The nutritional status of the host plays an important role in parasite growth and the outcome of malarial infections. This was first demonstrated in mice fed on a milk diet (lacking the crucial vitamin pABA), which were partially protected from infection (Ferone 1977). Similarly, riboflavin (vitamin B2) deficiency confers a degree of protection against malaria infection (Das et al. 1988). Malnutrition and kwashiorkor are also said to reduce susceptibility and, in famine-relief situations, outbreaks of clinical malaria frequently occur during refeeding (termed 'refeeding malaria') where the improved nutritional status allows previously suppressed malaria to develop (Edington 1967; Murray et al. 1978b). This classical concept has recently been challenged after a careful re-analysis of published studies (reviewed by Shankar 2000), which showed that malnutrition was actually more often associated with an increased risk of malaria, suggesting a more complex relationship between nutrition and malaria than previously supposed. Malnutrition in Kenyan children has been shown to be an important risk factor of severe malaria (Marsh et al. 1995) and a cohort study in Colombia showed that malnourished children had lower antimalarial antibody levels, suggesting an impaired immune response. A variety of specific deficiencies (including vitamin A, thiamin, and zinc deficiencies) has been found to be more frequent in individuals with severe malaria than individuals with mild forms of the disease (Gibson and Huddle 1998; Krishna et al. 1999). Micronutrient deficiencies are of particular interest since their mode of action is more often on host response to infection and immunity than direct effect on parasite development. Both zinc and vitamin A are, for instance, essential for effective immune responses. Intervention campaigns using micronutrient supplementation (particularly iron, vitamin A, and zinc) have had beneficial effects on the whole, showing reduced risk and reduced intensity of infection, but their impact was variable from one population to another. For example, a study in Papua New Guinea has shown that vitamin A supplementation (200 000 IU every 3 months) significantly reduced the number of clinical episodes and that zinc supplementation reduced both the number and the severity of clinical episodes (Shankar et al. 1999). By contrast, a randomized trial of zinc supplementation in children in Burkina Faso did not show any significant effect on malaria (Müller et al. 2001).

Immunity and chemoprophylaxis

It seems well established that appropriate chemoprophylaxis, taken regularly (as is done by expatriates residing in an endemic area) interferes with the development of immunity. The situation is more confused when the effects of targeted mass chemoprophylaxis are studied, as in the Garki Project in Nigeria or, more recently, in the Gambia. Young children receiving chemoprophylaxis are not fully protected from malaria infections, but infections are less heavy, less frequent, and there is a substantially reduced mortality and morbidity (Greenwood 1991). In these children, antibody levels are generally significantly reduced, but reduction is less marked the longer the chemoprophylaxis is given; in contrast, lymphocyte proliferation or IFN-γ responses are increased suggesting that reduced infection levels generate more effective cell-mediated responses. Once chemoprophylaxis is interrupted, there does not appear to be a 'rebound' effect in clinical attacks, morbidity, or mortality, which suggests that children, having received some form of chemoprophylaxis over the first few years of life and thus suffered less from their malaria, have nevertheless developed an antimalarial immunity comparable to that of children without chemoprophylaxis. In any situation where mass chemoprophylaxis is associated with other effective means of prophylaxis (e.g. impregnated bednets), the chance of infection may be so substantially reduced that the development of immunity can be impaired (Alonso et al. 1993). Intermittent presumptive treatment (IPT), whether in pregnant women or in infants, is theoretically less likely to interfere with the development of immunity, but extensive studies of the immune response have not yet been performed during IPT interventions.

Antimalarial drugs may act much more effectively in semi-immune than in naive individuals, due to a synergy between the effect of drugs and immunity. A consequence of this is, in areas where there is resistance to a given drug, the clinical response may be very different in different age groups. For example, in a holoendemic area, there may appear to be a lower degree of drug resistance in school age children than in infants. Another consequence of this synergy is that, in hyper- or holoendemic areas, where most of the population reaches a state of semi-immunity from a relatively early age, drugs like CQ may continue to be of benefit to the local population even when a relatively high degree of parasite resistance has been reported. This has implications in the interpretation of in vivo drug susceptibility tests and explains some of the discrepancies observed between in vitro and in vivo tests.

Malaria immunity and acquired immunodeficiency syndrome

The question of whether the two major causes of mortality in sub-Saharan Africa, malaria, and acquired immunodeficiency syndrome (AIDS), interact with each other still remains controversial. This is surprising since even minor synergy between the two would be considered of public health importance. While none of the early hospital-based studies in Zaire, Rwanda, or Zambia showed any obvious interaction during co-infections in adults (Nguyen-Dinh et al. 1987; Simooya et al. 1988; Allen et al. 1991), a cohort study did show increased levels of parasitemia in Ugandan adults with HIV, with a progressively increasing incidence of clinical malaria associated with falling CD4 counts (Whitworth et al. 2000). The situation is much clearer during pregnancy. A number of studies in Malawi and Kenya have demonstrated a synergy between HIV infection and malaria, leading to higher maternal parasitemia and increased fetal mortality in co-infections (Bloland et al. 1995; Steketee et al. 1996b; Verhoeff et al. 1999; Van Eijk et al. 2003).

Does HIV-induced immunosuppression affect immunity to malaria? HIV is well known to influence infection by a variety of intracellular parasites (e.g. *Leishmania*, *Toxoplasma gondii*, or *Cryptosporidium parvum*) but this is always explained by a decrease in the Th1 response and would, therefore, not be expected to have a substantial impact on protection against malaria in semi-immune adults, since immunity is believed to be mostly Th2/antibody-dependent. Antimalarial antibody levels are not significantly reduced in African adults with HIV and only a few other parameters of malaria immunity have so far been examined in HIV-co-infected patients. The Th1 cytokine response to MSP1 was, for instance, predictably reduced in HIV-infected adults in Burkina Faso (Migot et al. 1996). A loss of pathogen-specific $CD4^+$ T cell immunity would mostly affect immunity to exo-erythrocytic parasites and the increase in parasitemia, observed in the Ugandan study, could thus be explained. However, increased levels of parasitemia were not observed in another Ugandan cohort study, in which the incidence of malarial fever was significantly higher in HIV-positive individuals (French et al. 2001), suggesting that HIV-positives may be more responsive to parasites or parasite-derived toxins. This is surprising since fever is usually induced by the pro-inflammatory cytokines produced in response to parasite toxins and this would be less effective in a Th1-depressed environment. Clearly, HIV infection raises unresolved issues in our understanding of acquired immunity to malaria. Even the reports which indicate increased clinical malaria in HIV-co-infected individuals do not suggest a significant increase in malaria mortality and, in particular, no increased incidence of cerebral malaria. This is in keeping with existing hypotheses regarding the pathophysiology of cerebral malaria, which involve the up-regulation of endothelial cell cytoadherence receptors (e.g. ICAM-1) which occurs only in a Th1 environment; in support of this are the results obtained in murine malaria models, where the immunosuppression induced by murine AIDS or in IFN-γ knock-out mice, prevents the development of cerebral malaria in the *P. berghei* ANKA model (Eckwalanga et al. 1994).

In pregnancy, HIV infection clearly acts in synergy with malaria to increase both maternal and fetal morbidity/mortality (e.g. a 3.4-fold increased infant mortality has been reported) (Bloland et al. 1995). The most striking finding is the fact that this increased severity of malaria in HIV-co-infected women has been reported to equally affect primi- and multigravidae. This observation contradicts current hypotheses regarding the pathophysiology of malaria in pregnancy, where the crucial event is the difference between primi- and multigravidae, explained by the acquisition of immunity against novel, placenta-specific malaria isolates rather than by the previously accepted belief in pregnancy-associated immunosuppression. It is probable that cytokine dysregulation plays an important part in this. Protection against placental malaria relies on a Th1 environment in the placenta, with increased IFN-γ response to malarial antigens, and this response is impaired in HIV-infected women (Moore et al. 2000).

Does malaria affect the course of HIV infection? Enhanced viral replication and higher viremia have been reported in association with malaria, which suggest a potentially more rapid evolution and an increased risk of transmission (see review by Rowland-Jones and Lohman 2002). The fact that severe anemia may require blood transfusion is in itself a risk of HIV infection, particularly in children. The suggestion that malariotherapy may be reintroduced in order to treat patients with AIDS is highly controversial, despite the impressive increase in CD4 levels observed during pilot studies performed by the Heimlich Institute in China (Nierengarten 2003).

PATHOLOGY AND PATHOPHYSIOLOGY

Infection by malarial parasites may lead to a variety of clinical syndromes, depending on a combination of different elements, including the 'virulence' of the parasite isolate and a variety of host-related factors such as the status of host immunity and its genetic make up. The development of asexual parasites in the blood plays the central role in disease pathophysiology and, for most malarial species, it is the rupture of the schizont which triggers off the major events leading to the characteristic symptoms of the malarial paroxysm.

Direct effects of the parasite on the host

The rupture of the mature schizont has three consequences:

1 the liberation of merozoites, which leads to further invasion and a sharp increase in parasitemia;
2 the destruction of infected erythrocytes;
3 the liberation of malarial antigens, pigment, and malarial toxins.

While erythrocyte destruction may contribute to the anemia observed in malaria, this is only of importance in chronic infections of semi-immune individuals where high parasitemia (e.g. 5–15 percent) is relatively well tolerated. In nonimmunes, malarial paroxysms occur at very low parasitemia (less than 1 percent) and the effects of massive erythrocyte destruction are negligible at this stage. In contrast, it is the release of malarial antigens, pigment and toxins which are responsible for most of the pathology. Most of these effects are indirect (by means of a cascade of pathological events in the host) (see Indirect effects of the parasite on the host). However, among the parasite products released, two main ones have direct pathological effects: (1) Malarial pigment, which is taken by monocytes, inhibits some of their phagocytic and immunological functions and contributes to the immunosuppression observed (particularly since pigment remains in the reticulo-endothelial system for many years after infection) (Arese et al. 1991); and (2) malarial 'toxins,' including GPI and lipid antigens, may be responsible for the induction of hypo-

glycemia by acting directly on glucose uptake by adipocytes in synergy with insulin (Taylor et al. 1992; Schofield et al. 1966; Zakeri et al. 2000).

Indirect effects of the parasite on the host

Indirect effects are by far the most important and the production of cytokines, particularly TNF-α induced by the release of parasite products after schizont rupture, appears to play the central role in the pathophysiology of malaria.

TUMOR NECROSIS FACTOR

Several arguments have been put forward to support the hypothesis that TNF-α plays a central role: the experimental injection of TNF-α in humans produces symptoms that closely resemble a malarial paroxysm (Clark et al. 1989a); the rise in temperature during P. vivax paroxysms closely follows the rise in circulating levels of TNF-α (Karunaweera et al. 1992); the levels of TNF-α are higher in severe than in mild forms of falciparum malaria and particularly high in fatal cases (Grau et al. 1989); the injection of anti-TNF-α monoclonal antibodies to children with severe malaria abolishes some of the symptoms, particularly fever (Kwiatkowski et al. 1993); there are many similarities between severe malaria and sepsis, and TNF-α and phospholipase-A2 are substantially increased in both (Vadas et al. 1993). The multifaceted effects of TNF-α are illustrated in Figure 24.18.

At physiological levels, TNF-α may destroy intracellular parasites or reduce the infectivity of gametocytes to mosquitoes. At increased levels, TNF-α induces fever, depresses erythropoiesis, and increases erythro-phagocytosis, thereby contributing to anemia, and directly causes many of the nonspecific symptoms of malaria including nausea, vomiting, and diarrhea.

Experiments in mice suggest that cerebral malaria (CM) may be due to an overproduction of lymphotoxin-α (LT-α) (a cytokine which was previously known as TNF-β) rather than TNF-α. Mice deficient in LT-α were resistant to CM, dying from hyperparasitemia and severe anemia instead, while mice deficient in TNF-α behaved like controls (Engwerda et al. 2002). This study suggests that caution may required in the interpretation of TNF data, since the two molecules share many features and most immunological assays for TNF do not discriminate between TNF-α and LT-α.

Although the circulating level of other 'endogenous pyrogens,' such as IL-1 and IL-6, is also increased during malaria (Kern et al. 1989), they appear to play only a modest pathophysiological role in severe malaria, probably in synergy with TNF-α, as suggested by Rockett and colleagues (1994). IL-6 may play a role in the induction of hypergammaglobulinemia and contribute to some

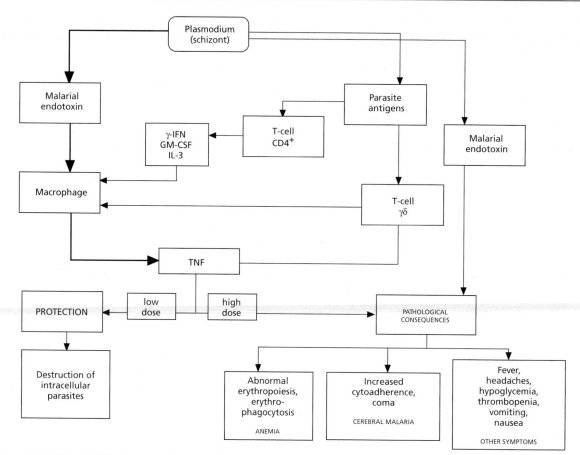

Figure 24.18 *Role of TNF in the pathophysiology of malaria*

of the malaria complication (e.g. glomerulonephritis and increased frequency of Burkitt's lymphoma) (Grau et al. 1990). One of the major consequences of the inflammatory events is the production of NO as the result of the up-regulation of the synthesis of iNOS. NO is a short-lived, highly reactive molecule with a wide spectrum of biological activities.

INDUCTION OF CYTOKINE PRODUCTION

The induction of inflammatory cytokines is determined by two independent pathways: (1) the direct action of malarial toxins on host macrophages and of malarial superantigens on γδ T cells; and (2) the immune response against malarial antigens and the production of Th1 cytokines (particularly IFN-γ), which up-regulate the production of inflammatory cytokines and act in synergy with them (Clark and Rockett 1994).

The nature of malarial toxins is still controversial, but whether it is hemozoin (Sherry et al. 1995), a proteolipid (Bate and Kwiatkowski 1994), GPI (Schofield et al. 1993), or polar lipids (Zakeri et al. 2000), most authors agree that the toxin is released at the time of schizont rupture, that it is at least partly lipidic in nature, and that it probably is associated with the pigment-containing 'residual body.' Such a toxin can induce TNF-α production by host cells in vitro and this

production may be blocked by antitoxic antibodies (Bate et al. 1992). Different isolates of *P. falciparum* can vary considerably in their ability to induce TNF-α and this may be one explanation for the observed differences in the clinical response (Allan et al. 1993). The reader needs to be aware that, since *Mycoplasma* contamination of *P. falciparum* cultures is common (Turrini et al. 1997), the reported results on so-called cytokine-inducing malaria toxins must be carefully disentangled from the effects of potent cytokine-inducing molecules of mycoplasmal origin; this is not always possible from published data. The malarial 'superantigens' responsible for the proliferation of γδ T cells are phosphorylated molecules, similar to isopentenol pyrophosphate from *Mycobacterium tuberculosis* (Behr et al. 1996).

UP-REGULATION OF CYTOADHERENCE RECEPTORS

Cytoadherence of infected erythrocytes to endothelial cells plays a central role in pathophysiology, particularly in the induction of cerebral malaria (MacPherson et al. 1985). Rosetting, which is another form of cytoadherence where each infected erythrocyte binds a number of uninfected erythrocytes, may contribute to this effect. Different isolates of *P. falciparum* are more or less cytoadherent to a variety of receptor molecules, including CD36, ICAM-1, thrombospondin, E-selectin, vascular cell adhesion molecule-1 (VCAM-1), chondroitin sulfate (see reviews by Hommel 1993; Sherman et al. 2003); differences in cytoadherence characteristics may represent a virulence factor and there is some evidence of a correlation between high cytoadherence/rosetting and cerebral malaria (Carlson et al. 1990; Ho et al. 1991).

One of the features of inflammatory cytokines in malaria is to up-regulate the expression of cytoadherence receptors (e.g. ICAM-1 or VCAM-1) in certain tissues; this very localized effect results in an increased sequestration in specific tissues. For example, cerebral malaria may occur when increased cytoadherence takes place preferentially in the brain. There is evidence that receptor up-regulation may also require a Th1 immune response (particularly IFN-γ) (Grau and Behr 1994) and that the development of cerebral complications may be avoided by the injection of Th2 mediators such as IL-10 (Ho et al. 1995).

One of the effects of cytokine-induced up-regulation of cytoadherence receptors is an increased level of the soluble forms of these molecules in the serum of patients. High levels of sICMA-1, sVCAM-1, sELAM-1, or thrombomodulin have not, however, been correlated with increased malaria severity (Hemmer et al. 1994; McGuire et al. 1996).

OXYGEN-DERIVED FREE RADICALS AND NO

The stimulation of pro-inflammatory cytokines leads to a cascade of events, including the release of various mediators, including oxygen-derived free radicals (ODFR) and NO, which are responsible for the pathology observed. The role of ODFR and NO has been extensively reviewed by Clark and Cowden (2003), who suggest that a better understanding of the mode of action of these mediators may open up new avenues for the treatment of severe malaria. For instance, ethyl pyruvate has been shown to be a potent scavenger of ODFRs and can dramatically reduce the effects of septic shock and systemic inflammation in mice (Ulloa et al. 2002); also ethyl pyruvate has not yet been used in severe malaria, desferrioxamine, another scavenger of reactive oxygen, has been shown to reduce the duration of deep coma in African children with cerebral malaria (Gordeuk et al. 1992). Unfortunately, research in this field is difficult because of the evanescent nature of molecules such as NO, as well as the complex interactions between cytokines, secondary mediators, host cells, and parasites. Suggestive experimental data in murine models have often proved hard to confirm in humans, particularly in falciparum malaria where the parasites are sequestered and the interactions occur in inaccessible tissues. In order to measure increased NO synthesis as the result of iNOS stimulation in the brain, the circulating levels of nitrates

and nitrites, both NO metabolites, have been measured; the confounding effect of diet and renal impairment often make these measurements difficult to interpret and contradictory results are found in the literature. Al-Yaman et al. (1996) reported an elevation in reactive nitrogen metabolites in children with cerebral malaria in Papua New Guinea, while such an increase was not observed in comparable studies in Tanzania (Anstey et al. 1996) or Ghana (Agbenyega et al. 1997). More direct evidence of the role of NO in pathogenesis is the detection of local iNOS production. This has been found in a small number of autopsy studies by immunocytochemical staining of iNOS in cerebral vascular walls (Maneerat et al. 2000; Clark et al. 2003).

Pathophysiology of different forms of malaria

The cascade of events leading to pathology is a complex one: it is influenced by both parasite and host factors (including host immunity and genetic 'susceptibility' factors). Why do apparently similar events lead to mild malarial paroxysms in some cases and to cerebral malaria or severe anemia in others?

MALARIAL PAROXYSM

The clinical features of uncomplicated malaria, whatever the parasite species responsible, include fever, headache and muscle ache, lethargy, sweating, and shivering; in its more typical form, the 'malarial paroxysm' resembles the endotoxin reactions described in other infections and has a characteristic periodicity (e.g. recurring every 48 h. Classical malarial paroxysms are more often observed during *P. vivax* than *P. falciparum* infections (see review by Karunaweera et al. 2003). The following sequence of events: rupture of schizont, with release of malarial antigens, hemozoin and pigment, which in turn trigger the release of TNF and other pyrogens by host cells, is generally accepted as the starting point of the paroxysm.

The pivotal role of TNF has best been demonstrated by the injection of TNF into humans, which reproduces all the symptoms of a malarial paroxysm. In clinical studies of vivax paroxysms, the rise of body temperature follows the rise of circulating TNF after an interval of 30–45 min, suggesting a direct causal relationship (Karunaweera et al. 1992). The level of parasitemia needs to reach a critical threshold before the rupture of schizonts is able to trigger a clinical paroxysm and, hence, the first few erythrocytic cycles are generally asymptomatic. The threshold of parasitemia is, however, variable from one individual to another: a malaria-naive individual is likely to experience a malarial parasitemia at a barely detectable parasitemia (0.001 percent), while school-age children in a malaria-endemic area may remain asymptomatic at very high parasitemia (5–15 percent). Fever threshold and antidisease immunity are closely related concepts.

CEREBRAL MALARIA

Cerebral malaria is a severe complication of falciparum malaria and frequently leads to death, even when appropriate therapy has been given. Various hypotheses for the pathophysiology of CM have been proposed over time. While the theories implicating microvascular obstruction by thrombus formation, disseminated coagulation, or immunopathological events due to the deposition of immune complexes have now been generally abandoned, the currently favored hypothesis assumes a central role for intracapillary sequestration of infected erythrocytes by cytoadherence to various endothelial receptors (MacPherson et al. 1985).

Although it has been established that different isolates are more or less cytoadherent to a variety of receptors, it is becoming clear that if 'stickiness' is a necessary requirement, it cannot solely explain the evolution of infection towards CM. The effects of TNF (in synergy with other pro-inflammatory cytokines) in the up-regulation of endothelial cell adherence receptors is another necessary feature in the events that lead to CM in susceptible individuals; immune status is as important as host genetic make up, since receptor up-regulation works best in a Th1 (i.e. during the early stages of malarial immunity) and is reversed in a Th2 environment (for example, by injection of IL-10) (Ho et al. 1995). Genetic susceptibility to CM has been described in various situations (e.g. in certain alleles of iNOS promoter gene, the TNF-α promoter gene or in individuals with the ICAM-1[Kilifi] allele). It is conceivable that, as a consequence of the presence of Th1 cytokines, other cascades of mediators may be generated (e.g. prostaglandins or tryptophan pathways) which aggravate the situation; it has been shown, in mice, that animals with CM had increased levels of the neurotoxic quinolinic acid, a by-product of tryptophan metabolism (Sanni et al. 1998).

This classical description of the pathophysiology of CM is based on studies of CM in Thailand. More recent studies of CM in children in Malawi and Kenya suggest that this classical picture may not always be found and that CM is, in reality, a wide spectrum of clinical entities. In an autopsy study in Malawi, only 14/32 children had a histological picture of brain tissue compatible with the cascade described above, i.e. significant sequestered parasites, microhemorrhages, intravascular accumulation of monocytes, fibrin, and presence of iNOS staining in vascular walls (Clark et al. 2003). The other cases described in the same study were histologically atypical, many with little or no tissue damage, no detectable iNOS in vascular walls and only minimal sequestration of parasites. The authors suggested that such atypical cases of CM could be the result of concomitant sepsis or metabolic disorders, which had led to coma before the level of parasitemia was sufficient to fill capillaries with sequestered infected erythrocytes. It had previously been established that the risk of mortality was four times greater when metabolic acidosis was associated with CM (Marsh et al. 1996). Histological features, like the disruption of endothelial intercellular junctions (Brown et al. 2001), suggesting an impaired blood–brain barrier, have only rarely been observed and it is generally believed that the blood–brain barrier remains intact in CM (Warrell et al. 1986; Badibanga et al. 1986). This is confirmed by the absence of leakage of radioactive albumin from the serum to the CSF and the lack of detectable cerebral edema on computed tomography (CT) scans.

In view of the different possible pathways leading to CM, coma may be caused by poor oxygen delivery to the brain, due to sluggish blood flow and partial mechanical obstruction or reduced oxygen utilization by brain tissue resulting from an inhibition of mitochondrial functions by the inflammatory process. In both cases, local hypoxia is the crucial factor leading to coma. When hypoxia occurs at the same time as hypoglycemia, or the accumulation of lactate, due either to the high parasitemia in capillaries resulting from sequestration or to the accumulation of glutamate (as a consequence of a reduction of glutamate uptake by astrocytes induced by NO), the resulting neurological damage is always greater than with hypoxia alone.

Neurological damage in CM is usually reversible, with less than 10 percent of African children suffering from neurological sequelae. Raised intracranial pressure is a major cause of poor outcome in encephalopathies, because of the risk of transtentorial herniation or a reduction in cerebral perfusion pressure. This may happen during severe forms of CM and is considered an indicator of poor outcome (i.e. death or neurological sequelae) (Newton and Krishna 1998). Seizures also reflect intense neurological damage when seen in children with CM; multiple and prolonged seizures during CM are always a sign of bad prognosis (Molyneux et al. 1989).

Figure 24.19 shows the sequence of events leading to CM, taking into account the relevant features of the parasite, the ability to cytoadhere, and the ability of parasite toxins to induce inflammatory cytokines, as well as the various points in the cascade where host factors may intervene.

ANEMIA

The pathophysiology of anemia is still poorly understood (Menendez et al. 2000). In nonimmunes, the destruction of erythrocytes by growing parasites cannot alone explain the degree of anemia observed, because malarial paroxysms occur at low parasitemia (1 percent or less), insufficient to produce anemia. By contrast, individuals having lived their life in malarious areas

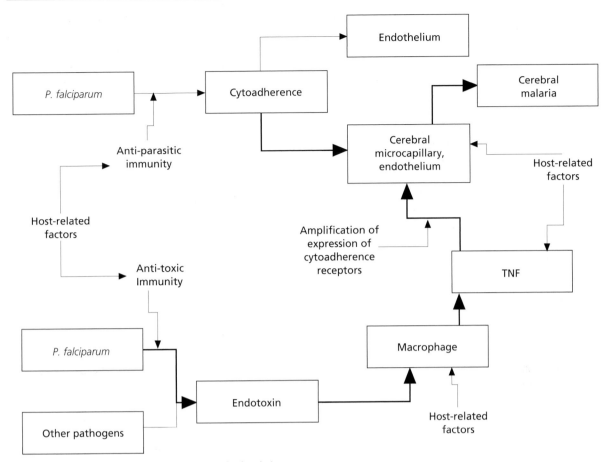

Figure 24.19 *Sequence of events leading to cerebral malaria*

have acquired a degree of tolerance to high parasitemia and, while such individuals suffer from mild anemia as the direct consequence of important erythrocyte destruction, anemia in older children or adults is not considered a serious health problem.

Severe anemia, such as occurs in early childhood before immunity is effective, is due to the combination of two abnormalities: (1) an increased destruction of normal erythrocytes by erythro-phagocytosis, particularly in the spleen; and (2) impaired production of new erythrocytes in the bone marrow. Dyserythropoiesis, with its accompanying low reticulocyte counts has long been known to be a major cause of malarial anemia (Abdalla et al. 1980). TNF is known to stimulate erythrophagocytosis and depress erythropoiesis in bone marrow (Clark and Chaudri 1988) and an association between high levels of pro-inflammatory cytokines and severe anemia has been observed in African children (Biemba et al. 2000). In addition to TNF, another T-cell cytokine, macrophage inhibitory factor (MIF), has recently been shown to play a role in bone marrow failure in a murine malaria model (Martiney et al. 2000), but this has not yet been confirmed in human malaria. Some authors believe that cytokine-mediated suppression alone could not explain the gross morphological bone marrow abnormalities described in malaria and

that other contributing factors need to be looked for. The observation that a high proportion of hemozoin-containing monocytes in the peripheral blood (over 1/500 monocytes) correlates well with the severity of anemia, suggests that hemozoin, or toxic metabolites, induced by the presence of hemozoin, may contribute to malarial dyserythropoiesis. In addition, nutritional factors, particularly severe iron deficiency, may contribute to the pathophysiology of severe anemia, as was observed in adults in Thailand (Phillips et al. 1986), but not confirmed in children in Kenya (Newton et al. 1997).

Of interest is the reported difference in age prevalence of severe anemia (in early childhood) and CM (in children over 2 years old). This is consistent with the view that the pathogenesis of CM has a greater immunological component than anemia, such as a synergy between TNF and IFN-γ. It has been suggested that the priming of T cells in early life may lead to excessive IFN-γ production on reinfection, predisposing the individual to over-production of TNF and resulting in an increased risk of severe pathology (Riley 1999).

MALARIA IN PREGNANCY

A paradoxical situation prevails during pregnancy in women with lifelong residence in a malarious area where a state of semi-immunity has been reached. Despite an apparent loss of immunity during pregnancy, these women are able to transfer a fully effective protective immunity to their infant. This phenomenon is particularly pronounced in primigravidae and becomes less of a problem during later pregnancies.

The features of falciparum malaria in a primigravida are the following: infection and parasite development take place in the placenta, while remaining at a low level in the peripheral circulation. In the absence of treatment, an infection acquired during the early stages of pregnancy will be maintained throughout pregnancy in the placenta, often with a heavy parasite load. Severe anemia in the mother and a low birth weight in the infant are the two major consequences. In extreme cases, there may be maternal death or abortion of the fetus. It is noteworthy that other forms of severe malaria (such as CM) do not occur in this situation.

The classical explanation of what happens during malaria in first pregnancies is that a reduction of acquired immunity takes place, consistent with the immunosuppression generally observed in all pregnancies. This is in line with the observation of reduced lymphoproliferative responses in pregnant women as compared to age-matched nonpregnant individuals (Rasheed et al. 1993). However, crucial studies, such as the longitudinal follow up of a cohort of women throughout their first pregnancy, have never been performed. The suggested loss of immunity does not account for the ability to transfer immunity to the infant, even if we accept that the described reduction of immune response is mostly in the cell-mediated response, while passive transfer refers mostly to antibodies.

Two alternative explanations for malaria in pregnancy have been proposed. The first takes account of the fact that the placenta is a 'new' organ in primigravidae, which could by-pass existing immune responses in the host, without actual loss of immunity, or allowing for the development of unusual phenotypes of *P. falciparum*. This would be compatible with the observation that parasites which cytoadhere to chondroitin sulfate are unusually found in the general population, but may be common in the placenta (Fried and Duffy 1996). The acquisition of a placenta-specific immunity or previous exposure to new malaria phenotypes in an earlier pregnancy would explain decreased susceptibility to infection in multigravidae. The second alternative suggests that what happens in pregnancy is a change in the balance of the local placental environment from an essentially Th2 environment, as normally prevails in the placenta, to a Th1 environment during acute malaria, a feature which is consistent with the presence of a large number of monocytes in an infected placenta.

Further details on the pathogenesis of malaria may be found in one of the many excellent reviews recently published on the subject (White 2003b; Newton and Krishna 1998; Miller et al. 2002; Clark and Cowden 2003; Clark et al. 2004). The histopathological changes of falciparum malaria have been described in detail on numerous previous occasions (Edington 1967; Edington and Gilles 1976; Lucas 1992; Warrell et al. 2002).

Immunopathology

The fact that the host immune system appears to have difficulty controlling the malarial infection is probably the reason why so many aberrant responses have been described. Aberrant responses are generally caused by an over-reaction of the host, where an immunologically inappropriate response has no effect on the parasite but may have immunopathological consequences of various degrees of severity. Table 24.4 gives a list of immunopathological complications of malaria, some of which have already been briefly described; detailed descriptions of these complications may be found in reviews by Marsh and Greenwood (1986) and by Ho and Sexton (1995).

DIAGNOSIS

Clinical diagnosis

The most important aspect in clinical diagnosis of malaria is a high index of suspicion (Doherty et al. 1995). In these days of widespread travel, a history of travel should be elicited in all patients, particularly febrile ones. It is also important to remember that population movements are common within countries where malaria is not uniformly endemic.

Malaria can mimic many diseases and the differential diagnosis is an extensive one. The golden rule is to always exclude malaria, irrespective of the clinical presentation, if a history of exposure is elicited, even if the patient has been on chemoprophylaxis. The commonest misdiagnoses in nonimmune subjects have been influenza, viral hepatitis, viral encephalitis, meningitis, psychosis, and viral hemorrhagic fever.

Clinical diagnosis can be further guided by epidemiological considerations, e.g. acute febrile illness in high risk groups, such as pregnant women or children under 5 years of age in holoendemic areas. Association with particular occupations, such as those working in forest or forest fringe areas, may be strongly correlated with malaria infection and may serve as a pointer to diagnosis.

The discovery of parasitemia provides an explanation for symptoms and signs in a nonimmune patient, this is not necessarily the case in semi-immune individuals in whom parasitemia may be incidental. In these patients, the diagnosis is more difficult and no 'gold standard' exists. A reasonable set of guidelines is: (1) a clinical illness compatible with malaria, including objective evidence of fever; (2) exclusion of any other likely cause

Table 24.4 *Immunopathology of malaria*

Pathology	Possible mechanisms
Hypergammaglobulinemia	Antigen induced cytokine production (IL-6)
	Antigenic variation
	Polyclonal activation
Immunosuppression	Antigenic competition
	Structural disruption of germinal centers
	Disruption of spleen function
	Macrophage dysfunction
	Polyclonal activation and immune 'exhaustion'
Nephrotic syndrome	Immune complex deposition
	Auto-immunity
Autoimmunity	Auto-antibodies
	Anti-nuclear antibodies
Anemia	Anti-erythrocyte antibodies
	Dyserythropoiesis (e.g. effect of TNF)
	Excessive erythro-phagocytosis
Thrombocytopenia	Excessive removal of platelets
	Coating of platelets with malaria antigen
Hyperreactive malarial splenomegaly	Genetic predisposition
	Hypergammaglobulinemia
	Chronic increase of lymphocyte proliferation
Burkitt's lymphoma	Co-endemicity with Epstein–Barr virus
	Polyclonal activation
	Antigen induced cytokine production

of fever following physical examination; (3) malaria parasitemia above 10 000/µl; and, (4) appropriate response to treatment.

Parasitological methods

Despite the wide range of serological, immunological, and molecular techniques currently available, the only certain means of diagnosing all four species of human malaria is the detection of the *Plasmodium* spp. by microscopic examination of the blood (see Figure 24.20). Both thick and thin blood films should be made. The thick film method, which concentrates layers of red

blood cells on a small surface by a factor of 20–30, is the most sensitive and by far the best for clinical use. Diagnostic characteristics of human malaria parasites, as seen in a well-stained thick or thin film are given in Table 24.5.

Various methods have been designed to enhance the examination of blood films in order to reduce the time spent reading the slides or to enable less well-trained personnel to achieve equally reliable results. Staining with acridine orange, which can be read either on a fluorescence microscope or a microscope equipped with an interference filter system (Kawamoto 1991) allows a quicker screening of films, because parasites are more readily recognized and a lower power microscope objective may be used. The quantitative buffy coat (QBC®) method is also based on acridine-orange staining, but in this case the blood is centrifuged in a specially designed and patented microcapillary tube fitted with a plastic float. The float spreads the buffy coat against the edge of the tube; parasites and leukocytes take up the dye, which is fluorescent when examined under ultraviolet light (Figure 24.21). This elegant method is easy to perform, fast, and easy to read, but requires specialized equipment (a microcentrifuge and a fluorescence microscope) and the purchase of expensive QBC capillary tubes (Spielman and Perrone 1989; Petersen and Marbiah 1994). The sensitivity of QBC, when used in field conditions, is comparable to, or marginally better than that of thick films. Both acridine-orange staining techniques are inferior to Romanovsky staining for the

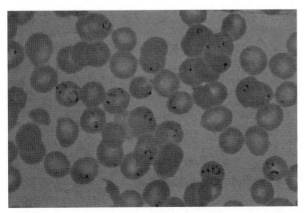

Figure 24.20 *Blood film of* P. falciparum-*infected blood showing the presence of ring-stage parasites*

Table 24.5 *Diagnostic characteristics of human malaria*

	P. falciparum	P. malariae	P. vivax	P. ovale
Appearance of infected red blood cells (size and shape)	Both normal	Normal shape; size normal or smaller	1.5–2 times larger than normal; shape normal or oval	As for *P. vivax*, but some have irregular frayed edges
Schüffner's dots (eosinophilic stippling)	None (but presence of occasional comma-like red Maurer's dots)	None	Present in all stages, except early ring form	As for *P. vivax*
Red cells with multiple parasites per cell	Common	Rare	Occasional	As for *P. vivax*
Stages present in peripheral blood	Rings and gametocytes Schizonts rarely seen	All stages	All stages	All stages
Ring form (young trophozoite)	Delicate, small ring; scanty cytoplasm; sometimes at edge of cell ('accolé' form)	Ring 1/3 diameter of cell; heavy chromatin dots sometimes 'filled in'	Ring 1/3 to 1/2 diameter of red cell; heavy chromatin dots	As for *P. vivax*
Schizont	Rarely seen in peripheral blood; 16–18 merozoites	6–12 merozoites in rosette; coarse pigment clump in center	12–24 merozoites in rosette filling the entire RBC; central pigment	8–12 merozoites in rosette
Gametocyte	'Crescent' shape characteristic	Round or oval; coarse pigment	Round or oval	Round or oval (smaller than *P. vivax*)
Main criteria	Only rings and crescent shaped gametocytes in blood; multiple infection; level of infection high; normal RBC s xhape; no Schüffner dots	All stages in blood; trophozoites compact and intensly stained; band forms suggestive; coarse pigment; normal/small RBC; no Schüffner dots	Large pale RBC; presence of Schüffner dots; round gametocytes; large amoeboid trophozoite with pale pigment	Generally like *P. vivax*, oval RBC with fimbriated edges characteristic but not always present

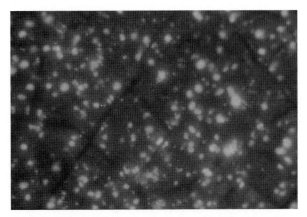

Figure 24.21 *Quantitative buffy coat (QBC). Blood infected with P. falciparum as seen in QBC capillary tubes. The large fluorescent cells are white blood cells, the small fluorescent dots in the dark area are ring-stage parasites within non-fluorescing erythrocytes (courtesy of Becton-Dickinson Tropical Disease Diagnostics, USA).*

precise identification of malaria species or their accurate enumeration. Where it is necessary to identify parasite species with precision, the use of blood films stained by immunocytochemical methods using species monoclonal antibodies may be envisaged (Lindergard 1995; Perez et al. 1995).

Despite its shortcomings, microscopy has qualitative and quantitative features that are not associated with most other techniques. For example, when carefully examined by an experienced microscopist, a thin film with *P. falciparum* can provide clues regarding the degree of disease severity, which include not only the high level of parasitemia but also the presence of more 'mature' ring-stage parasites (linked to the existence of a greater sequestered biomass of parasites) (Silamut and White 1993) or the presence of an unusually high number of circulating schizonts. In addition to the observation of the parasites themselves, the overall examination of a blood film may provide other clues. The presence of visible malarial pigment in neutrophils or monocytes may be used as a criterion of poor prognosis; a fatal outcome being associated with pigment in more than 5 percent of neutrophils (Nguyen et al. 1995). One must, however, be careful not to overinterpret such observations, since pigmented monocytes are not uncommon in asymptomatic African children.

Immunological methods

Instead of identifying the parasite itself, immunological methods provide the means for detecting either the parasite antigens or the host antibodies directed against the parasite. The detection of antigens may be an acceptable alternative to parasite detection, particularly if the assay is robust, inexpensive, easy to use in field conditions, and does not require a microscope, but the detection of antibodies merely provides information on past malaria experience and is of limited use for individual diagnosis.

SEROLOGY

Serological methods have been in use since the early 1960s, when IFAT and indirect hemagglutination assays (IHA) were described. Since such tests detect antimalarial antibodies they cannot distinguish between current or past infection and are therefore of limited value as a guide to the treatment or management of the disease. At best, a negative serological assay may help to eliminate the possibility of malaria, since it has been shown that antibody levels become detectable a few days after the blood is invaded. A detailed description of the various methods may be found elsewhere (Voller 1988; Hommel 2002).

IFAT is the main method for routine serodiagnosis, because it is relatively easy to make antigen slides for all human malarial parasites (commercial IFAT slides for malaria are also available from various sources) (Figure 24.22). The disadvantages are the need for a fluorescence microscope, the subjectivity of the reading, and the labor-intensive nature of the method, which limits its application to specialized centers with a relatively small through-put of samples.

ELISA uses a soluble malarial antigen, generally prepared from asexual stages of *P. falciparum*, which can be readily grown in vitro (Trager and Jensen 1976), coated on the wells of a microtiter plate. A large number of samples can thus be processed at the same time and produce quantitative results, when the assay is read on a spectrophotometer (portable, battery-operated ELISA readers are available for field use). In addition to crude extracts of malarial antigens, ELISA has been applied to a variety of defined, synthetic, or recombinant malarial antigens (e.g. MSP-1, RESA, or CSP) (Zavala et al. 1986; Del Giudice et al. 1987; Riley et al. 1992); such studies have been useful in elucidating the role of target malarial antigen in immunity and protection.

A variety of other serological test formats have been explored, including radioimmunoassays, latex agglutination, indirect hemagglutination, solid-phase dipstick, and membrane dot-blot, which may have specific advantages in given situations.

Figure 24.22 *Positive immunofluorescence antibody test with malaria antigen (courtesy of Dr. W. Bailey, Liverpool School of Tropical Medicine)*

ANTIGEN DETECTION

In contrast to serology, a positive antigen detection assay should only detect a current infection. Experimental tests for detecting malarial antigens are based on either an antigen-capture or an antigen-competition format, and often use ELISA or the radioimmunoassay (RIA) methodology. Once optimal reagents have been identified (i.e. monoclonal or polyclonal antibodies to specific malarial antigens), the assay may be simplified to a straightforward agglutination or immunochromatography ('dipstick') format. The best antigen detection assays described have a maximum sensitivity of 0.01 to 0.001 percent parasitemia and are five to ten times inferior to good quality microscopy (Mackey et al. 1982; Khusmith et al. 1987; Fortier et al. 1987; Taylor and Voller 1993).

ParaSight[TM]-F (Shiff et al. 1993) was the first commercial detection test in dipstick format (Figure 24.23), in which a monoclonal antibody captures a specific antigen of *P. falciparum*, PfHRP-2, present in the parasite throughout the erythrocytic cycle. If antigen is present, the positivity of the test is visualized by a second anti-HRP2 antibody labeled with a colored marker, which produces a visible line on the dipstick. The whole test takes only 10 min and gives a sensitivity almost comparable to that of thick films. This simple, robust assay requires no equipment and could be taught to village health workers, since its reading was a straightforward positive/negative assessment (Premij et al. 1994). Its high cost and the fact that the assay was not quantitative are serious limitations to its use in the very field situations for which it was designed. Many other versions of the test are now commercially available.

The detection of parasite lactate dehydrogenase (pLDH) was originally developed as a way to monitor in vitro drug susceptibility assays (Makler and Hinrichs 1993). The principle of the assay is that pLDH has different biochemical characteristics from human lactate dehydrogenase (LDH) and may, therefore, be differen-

tially measured using a simple colorimetric assay. The principle of OptiMAL[®] (Diamed, Switzerland) is based on the detection of pLDH using a series of monoclonal and polyclonal antibodies; it is the most recently developed immunochromatographic rapid malaria strip tests (Piper et al. 1999). Differentiation of malaria species in the OptiMAL test is based on antigenic differences between pLDH isoforms. Unlike HRP-2, pLDH does not persist in the blood but clears at about the same time as the parasites, following successful treatment. The test is, therefore, useful for monitoring responses to drug therapy and detecting drug-resistant malaria because the level of pLDH correlates with the number of viable malaria parasites in the blood. If less sensitive than HPR-2 tests (with a threshold of 100–200 parasites/μl), the test produces fewer false-positives in patients with rheumatoid factor. Evaluations published so far indicate that the test is sensitive and able to distinguish between *P. vivax* and *P. falciparum* (Palmer et al. 1998).

MOLECULAR METHODS

The application of DNA or RNA hybridization to malaria diagnosis has several advantages over traditional methods as a research tool to monitor malaria control programs, to perform quality control checks on microscopic diagnosis or to determine the distribution of important genes (e.g. genes associated with drug resistance).

Various methods based on the principle of nucleic acid hybridization have been developed in order to detect parasite DNA and RNA. A known sequence of nucleic acid (oligonucleotide) is synthesized and labeled either with radioactive ^{32}P or a nonradioactive colorimetric reagent and this 'probe' is used to detect parasite nucleic acid taking advantage of the fact that complementary sequences will hybridize. The simplest version of this technique is the use of DNA probes to detect parasites directly in a drop of patient's blood immobilized on filter paper (Franzen et al. 1984). In this test format, the sensitivity and specificity of the technique depends largely upon the choice of nucleic acid sequence. In order to achieve a higher degree of sensitivity, most of the first-generation DNA probes were directed towards repetitive sequences of parasite genes. Since the availability of amplification methods, such as the PCR, the sensitivity of nucleic acid probes has increased exponentially. This has made it possible to use probes against nonrepetitive sequences, which have been useful, for example, to examine whether *pfmdr* genes are or are not associated with CQ resistance (Foote et al. 1989; Wellems et al. 1990) or to use species-specific small subunit ribosomal RNA sequences to differentiate between *P. falciparum*, *P. vivax*, *P. ovale*, or *P. malariae* (Snounou et al. 1993) (Figure 24.24).

Many technical refinements of the PCR method have been described, some of which have been applied to

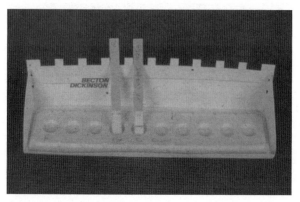

Figure 24.23 *Positive and negative malarial antigenemia using the ParaSight F dipstick (courtesy of Dr. W. Bailey, Liverpool School of Tropical Medicine)*

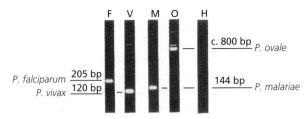

Figure 24.24 *Polymerase chain reaction (PCR) detection of human* Plasmodium *species (courtesy of Dr. G. Snounou, Institut Pasteur, Paris)*

malaria diagnosis. The use of methods for the detection of immobilized amplified nucleic acid or liquid phase hybridization provides a test format akin to ELISA methods (Oliveira et al. 1995). Various ways have also been described to make the PCR technique more user-friendly under field conditions by reducing the manipulation of the samples and the likely contamination of specimens (e.g. by collection of blood on filter paper) (Long et al. 1995).

The PCR technique has been reported to be capable of detecting parasitemias of less than 0.00002 percent, when used in the best possible conditions. It is theoretically capable of detecting the presence of a single parasite in the sample, although this is rarely achieved, partly because the blood contains poorly defined products which may inhibit the PCR reaction. The detection of a parasitemia of 0.00002 percent, which corresponds to one parasite per mm^3 or five parasites per 5-µl sample of blood, is a detection threshold at least five times lower than the detection threshold achieved by means of a thick film performed in optimal conditions (i.e. 0.0001). A sensitivity of one parasite per 20 µl of blood, as reported by Tirasophon and colleagues (1991) would, therefore, correspond to a sensitivity 100 times higher than a thick film. In reality, the sensitivity of PCR is rarely as good, particularly when performed in field conditions, and may be considered comparable or only marginally better than that of the microscopic examination of a thick film (Laserson et al. 1994). In contrast, the specificity of PCR is generally considered to be better than microscopy.

Appraisal of the relative value of diagnostic methods

Whatever the method used, a diagnostic test should be able to correctly differentiate between individuals that are infected and those who are not; consequently, the validity of a test is usually determined by its sensitivity (i.e. the test with the highest sensitivity has the lowest number of false-negatives) and its specificity (i.e. the test with the highest specificity has the lowest number of false-positives). In reality, it is generally more important to know the capability of a positive assay to predict the probability of infection, i.e. its positive predictive value.

Individual diagnosis

In most situations, the 'gold standard' for individual diagnosis is the microscopic examination of thick and thin films. There are, however, situations where this may not apply. In areas of high endemicity, clinical diagnosis alone is usually the only feasible and cost-effective method for recommending the first-line treatment (CQ, Fansidar® or Mefloquine®). When microscopy is not available, a variety of clinical features may be used as predictive markers of disease (Genton et al. 1994).

The ideal use of methods like QBC, PCR, or Para-Sight is in sophisticated laboratories in developed countries that can afford the cost of the assays and have expertise in complex technologies but have been unable to maintain competency in malaria microscopy. Ironically, in these situations, the need for a quantitative evaluation of the level of parasitemia to monitor treatment efficacy may make it necessary to perform a microscopic examination of blood films as a second-line test. In contrast, in hyperendemic areas, in situations where microscopy is not available, one of the most important applications of ParaSight or OptiMal is to identify clinical cases that are not attributable to malaria, since such cases would otherwise be wrongly assumed to be malaria and given antimalarials when the real bacteriological etiology would, for example, require antibiotics. PCR is generally considered unsuitable for routine diagnosis in field conditions (Hänscheid and Grobusch 2002).

When microscopy is used in areas of high endemicity, it is usually necessary to redefine the critical threshold for each situation, i.e. a level above which parasitemia may have a clinical significance. A parasitemia above 5 000 to 10 000 parasites/µl is usually suggested as a guideline, but precise counting may not always be feasible; Coosemans et al. (1994) have proposed the sensible 'rule-of-thumb' that a 100 percent field positivity on a thick film may be used as a reliable morbidity indicator.

Epidemiological surveillance

Active or passive surveillance of malaria prevalence is an important tool for malaria control, particularly when the efficacy of control measures is being evaluated. Microscopy and, by extension, antigen detection or PCR, have serious limitations for such an evaluation, because these techniques only measure a 'point prevalence' of the infection at the time of the survey and is particularly unhelpful in situations where transmission is seasonal. In contrast, sero-epidemiology may help to delineate those areas where there is transmission, may also provide information on species-prevalence or age-related prevalence, and may chart the changes that are taking place as a result of the control intervention.

Another issue in diagnosis for epidemiological purposes is the follow up of the distribution in a community of isolates with specific features, such as antigens of interest, in the context of a vaccination program or drug resistance markers, for the definition of treatment policies. PCR, whose specificity can be changed at will by changing the primer sequences used, is ideally suited for this purpose and is increasingly being used for isolate-specific surveys (Contamin et al. 1995).

Transfusion malaria

The transmission of malaria parasites by blood transfusion is a serious risk, since the diagnosis of malaria in the recipient, being unexpected, is often missed. Microscopic examination of donor blood is highly unsatisfactory since most donor infections are at a sub-microscopic level. Outside endemic areas, the policy for screening donor history for known episodes of clinical malaria or for tropical travel in the past 5 years is generally sufficient. With increasing frequency of tropical travel, the policy in some countries is to reject only donors whose malaria serology is positive, generally using IFAT with homologous antigens for *P. falciparum*, *P. vivax*, and *P. malariae* for maximum assay sensitivity. This screening method is not perfect, but a negative serology gives a high probability of freedom from infection. The use of PCR (recently explored by the Blood Transfusion Center of Ho Chi Minh City in Vietnam) is not, despite its much increased sensitivity, a complete guarantee of safe blood, since the absence of parasites in a 20 μl sample does not exclude the possibility of infection in the remaining 450 ml of the blood unit (Vu thi Ty Hang et al. 1995). In endemic areas, the only safe prevention of transfusion malaria is appropriate preventive antimalarial therapy of the recipient. Malaria transmission has occurred as a result of organ transplantation, e.g. kidney, heart, and bone marrow.

CHEMOTHERAPY AND CHEMOPROPHYLAXIS

Chemotherapy

Although chemotherapy is mandatory in the treatment of malaria (White 2003a; Greenwood 2004), correct management of the patient and the complications that occur with severe malaria is also vital (Warrell 2002; Molyneux and Fox 1993; Wilkinson et al. 1994). A summary of the management of severe and complicated malaria is given in Table 24.6 (WHO 2000).

For uncomplicated falciparum malaria, oral therapy is the administration of choice and parenteral therapy should only be considered if the patient presents with severe nausea and vomiting. For severe and complicated falciparum malaria parenteral therapy is always indi-

cated, but reversal to oral therapy should be instituted as soon as the patient's condition has improved sufficiently to take the drugs by mouth.

The antimalarial drugs in general use are chloroquine (CQ), amodiaquine, quinine, mefloquine, halofantrine, sulfadoxine-pyrimethamine, atovaquone-proguanil (Malarone®), lapudrine-dapsone (LapDap®), pyronaridine, and artemisinin derivatives singly or in combination, e.g. artesunate, artemether, artemether-benflumetol (Riamet®, Coartem®) and primaquine. Tafenoquine is likely to be on the market soon.

When choosing a treatment it is important to be aware of the regimen recommended for the specific geographical area, both by the national malaria control program and by WHO.

The treatment of uncomplicated *P. falciparum* malaria is given in Table 24.7; severe malaria in Table 24.8; *P. vivax* and *P. ovale* malaria in Table 24.9, and *P. malariae* in Table 24.10.

Although the proposed regimen for *P. malariae* is that which is usually recommended, the authors feel that a 4-day treatment would be more logical for this parasite, in view of its longer intra-erythrocytic cycle of 72 h instead of 48 h for the other human malaria species. This approach may help to deal with a 'lesser susceptibility' of this parasite to CQ treatment which has sometimes been reported but has not been thoroughly documented.

In the greater Mekong region (i.e. the region including Yunam Province, Thailand, Vietnam, Laos PDR, Cambodia, and Myanmar), where multidrug resistance has been recognized, the treatment of choice is with artemisinin and its analogs (Hien and White 1993; Jing-Bo-Jiang et al. 1982; Looareesuwan et al. 1992; Karbwang and Harinasuta 1992; Bunnag et al. 1995).

In pregnancy, artemisinins should only be used in the second and third trimesters when other treatments, like quinine with clindamycin, are not available. They are not recommended for treatment in the first trimester, unless considered life-saving and when no other suitable antimalarial, e.g. quinine, is available (Nosten et al. 2003).

Combination therapy with antimalarial drugs

The concept of combination therapy is based on the synergistic or additive potential of two or more antimalarial drugs (WHO 2001). In Southeast Asia, artemisinin-based combinations have been successfully used to improve efficiency of treatment and control resistance (White and Olliaro 1996).

The following combinations are possible options:

Amodiaquine + sulfadoxine-pyrimethamine (SP)
Artesunate + amodiaquine
Artesunate + SP (providing levels of resistance to SP less than 15%)
Artesunate + LapDap

Table 24.6 *Summary of the management of severe falciparum malaria (WHO 2000)*

Summary of management

In all cases, an appropriate antimalarial drug should be started immediately and complications managed appropriately as below.

1. Coma (cerebral malaria)

Maintain airway; nurse on side; exclude other treatable causes of coma (e.g. hypoglycemia, bacterial meningoencephalitis). Avoid harmful adjuvant treatments such as corticosteroids, heparin, and epinephrine (adrenaline).

2. Convulsions

Maintain airway; treat with diazepam given intravenously (0.15 mg/kg of body weight) or intrarectally (0.5 mg/kg of body weight), or intramuscular paraldehyde injection (0.1 ml/kg of body weight). Paraldehyde should, if possible, be given from a glass syringe. A disposable plastic syringe may be used, provided that the injection is given immediately the paraldehyde is drawn up and that the syringe is never reused.

3. Severe anemia

Transfuse screened fresh whole blood or packed cells.

4. Acute renal failure

Exclude dehydration; maintain strict fluid balance; carry out dialysis if indicated.

5. Hypoglycemia

Measure blood glucose, give 50% dextrose injection 50 ml (1 ml/kg of body weight for children) followed by 5% or 10% dextrose infusion.

6. Metabolic acidosis

Exclude or treat hypoglycemia, hypovolemia, and gram-negative septicemia. Give isotonic saline 10 ml/kg of body weight rapidly or screened whole blood 10 ml/kg of body weight over 30 min if hemoglobin is <5 g/dl.

7. Acute pulmonary edema

Prevent by avoiding excessive rehydration. Prop patient up; give oxygen. If pulmonary edema is due to overhydration, stop intravenous fluids, give a diuretic (furosemide (frusemide) 40 mg intravenously), and withdraw 3 ml/kg of blood by venesection into a donor bag.

8. Shock, algid malaria

Suspect gram-negative septicemia; take blood samples for culture. Give parenteral antimicrobials; correct hemodynamic disturbances.

9. Spontaneous bleeding and coagulopathy

Transfuse screened fresh whole blood or clotting factors; give vitamin K, 10 mg intravenously.

10. Hyperpyrexia

Give antipyretic (paracetamol 15 mg/kg of body weight) and use tepid sponging and fanning.

11. Hyperparasitemia

Give initial dose of parenteral antimalarial therapy; consider exchange transfusion if there are other signs of severity.

12. Malarial hemoglobinuria

Continue antimalarial treatment; transfuse screened fresh blood if needed.

13. Aspiration pneumonia

Give parenteral antimicrobials; change position of patient; give physiotherapy; give oxygen.

14. Fluid and electrolyte balance

Replacement of water and salt with isotonic saline. Monitor fluid therapy by measurement of jugular or central venous pressure.

Table 24.7 *Antimalarial chemotherapy for adults and children who can swallow tablets (Warrell et al. 2002)*

Chloroquine-resistant *P. falciparum* or origin of species unknown	Chloroquine-sensitive *P. falciparum* or *P. vivax, P. ovale, P. malariae*
1. Mefloquine Adults: 15–25 mg base/kg[a] given as 2 doses 6–8 h apart Children: 25 mg base/kg given as 2 doses 6–8 h apart or	1. Chloroquine[b] Adults: 600 mg base on the 1st and 2nd days; 300 mg on the 3rd day Children: approximately 10 mg base/kg on the 1st and 2nd days; 5 mg base/kg on the 3rd day For radical cure of vivax/ovale add
2. Proguanil *with* atovaquone (Malarone) Adults: 4 tablets (each containing 100 mg proguanil and 250 mg atovaquone) once daily for 3 days Children: 11–20 kg, 1 tablet; 21–30 kg, 2 tablets; 31–40 kg, 3 tablets, all once daily for 3 days or	2. Primaquine Adults (except pregnant and lactating women and G6PD-deficient patients): 15 mg base/day on days 4–17 *or* 45 mg/week for 8 weeks[c] Children: 0.25 mg/kg per day on days 4–17 *or* 0.75 mg/kg per week for 8 weeks[c]
3. Artemether *with* lumefantrine (Riamet) Adults: 4 tablets (each containing 20 mg artemether and 120 mg lumefantrine) twice daily for 3 days Children: <15 kg, 1 tablet; 15–<25 kg, 2 tablets; 25–<35 kg, 3 tablets, all twice daily for 3 days or	
4. *Quinine* Adults: 600 mg salt 3 times daily for 7 days[d] Children: approximately 10 mg salt/kg 3 times daily for 7 days or	
5. Chloroproguanil *with* dapsone (Lapdap) Adults and children: chlorproguanil 2.0 mg/kg with dapsone 2.5 mg/kg once daily for 3 days or	
6. Sulphonamide-pyrimethamine[e] Sulfadoxine (500 mg per tablet) or sulfalene (500 mg) plus pyrimethamine (25 mg) Adults: 3 tablets as single dose Children: <5 years, 1/2 tablet; <9 years, 1 tablet; <15 years, 2 tablets, all as single doses	

a) For chloroquine-resistant *P. vivax*, repeat the course.
b) For Chesson-type strains (SE Asia, W. Pacific), use double dose or double duration up to a total dose of 6 mg base/kg in daily doses of 15–22.5 mg in adults.
c) Depending on geographical area and presumed immunity.
d) In areas where 7 days of quinine is not curative (e.g. Thailand), add tetracycline 250 mg four times each day or doxycycline 100 mg daily for 7 days except for children under 8 years and pregnant women or add clindamycin 1 mg/kg twice daily for 3–7 days.
e) Sulfadoxine + pyrimethamine (Fansidar); sulfalene + pyrimethamine (Metakelfin). Contraindicated if patient has known sulfonamide hypersensitivity.
Note (1) Combination therapy is now advocated by WHO; (2) Suppositories of artemisinin and its derivatives can be used if severe vomiting prevents oral therapy and parenteral therapy is unavailable, e.g. in peripheral health centers.

Artesunate + pyronaridine
Artesunate + lumefantrine (Coartem®)

Constraints to the use of combination therapy, especially in Africa, are:

● scanty information on the safety and efficacy of the proposed combinations, particularly in pregnant women and children;
● the present high cost of artemisinin-based compounds. In this context, however, WHO has negotiated an agreement with a major pharmaceutical company to make a co-formulated artemisinin-containing combination therapy for public sector use in malaria-endemic countries at cost price, resulting in a reduction of the cost for a full treatment course for a child to US$ 1.

The optimal time to use combination therapy is before resistance has developed to either of the individual drugs chosen, or when resistance is at a low level – below 15 percent (Adjuik et al. 2004).

Table 24.8 *Severe falciparum malaria*[a]: *antimalarial chemotherapy in adults and children (WHO 2000)*

Chloroquine-resistant, or sensitivity unknown	Chloroquine-sensitive
1. Quinine	1. Chloroquine
Adults: 20 mg dihydrochloride salt/kg (loading dose)[b,c] diluted in 10 ml/kg isotonic fluid by intravenous infusion over 4 h then, 8 h after the start of the loading dose, 10 mg/kg[d] over 4 h every 8 h until the patient can swallow, then quinine tablets approximately 10 mg quinine salt/kg (maximum 600 mg) every 8–12 h to complete 7 d treatment[e] or give a single dose of 25 mg/kg sulfadoxine and 1.25 mg/kg pyrimethamine (maximum 1500 mg sulfadoxine/75 mg pyrimethamine). Children: 20 mg dihydrochloride salt/kg (loading dose)[b,c] diluted in 10 ml/kg isotonic fluid by intravenous infusion over 4 h then, 12 h after the start of the loading dose, 10 mg/kg[d] over 2 h every 12 h until the patient can swallow, then quinine tablets approximately 10 mg quinine salt/kg (maximum 600 mg) every 8 h to complete 7 d treatment[e] or give a single dose of 25 mg/kg sufadoxine and 1.25 mg/kg pyrimethamine	10 mg base/kg in isotonic fluid by constant rate intravenous infusion over 8 h, followed by 15 mg/kg over 24 h
or	or
2. Artesunate[f]	2. Chloroquine
2.4 mg/kg intravenously on the first day followed by 1.2 mg/kg daily for a minimum of 3 d until the patient can take oral therapy or another effective antimalarial	5 mg base/kg in isotonic fluid by constant rate infusion over 6 h, every 6 h, to a total dose of 25 mg/kg over 30 h
or	or
3. Artemether	3. Quinine, artemether or quinidine
1.2 mg/kg intramuscularly on the first day, followed by 1.6 mg/kg daily for a minimum of 3 d until the patient can take oral reatment or another effective antimalarial. In children the use of a 1 ml tuberculin syringe is advisable since the injection volumes will be small.	(see left-hand column)

a) For definition see text.
b) Loading dose should not be used if the patient received quinine, quinidine, or mefloquine within the preceding 12 h.
c) Alternatively, the loading dose can be administered as 7 mg salt/kg by intravenous infusion (or pump) over 30 min followed immediately by 10 mg/kg diluted in 10 ml/kg isotonic fluid by intravenous infusion over 4 h.
d) If there is no clinical improvement after 48 h of parenteral therapy, reduce the maintenance dose to 5–7 mg quinine dihydrochloride/kg or to 3.75–5 mg quinidine base/kg every 8 h.
e) In areas where a 7 d course of quinine is not curative (e.g Thailand) add a course of oral tetracycline 4 mg/kg daily or doxycycline 3 mg/kg once daily except for children under 8 years of age and pregnant women, or clindamycin 10 mg/kg twice daily for 3–7 d.
f) Artesunic acid 60 mg is dissolved in 0.6 ml of 5% sodium bicarbonate diluted to 3–5 ml with 5% (W/v) dextrose and given immediately by intravenous ('push') bolus injection. Note (1) Combination therapy is now advocated by WHO; (2) Suppositories of artemisinin and its derivatives can be used if severe vomiting prevents oral therapy and parenteral therapy is unavailable, e.g. in peripheral health centers.

Drug resistance

Monitoring the spread of drug resistance is one of the most important functions of malaria control programs and such monitoring provides the basis for a national antimalarial treatment policy on evidence-based recommendations (WHO 1994a). The protocol for the assessment of therapeutic efficacy of antimalarial drugs has recently been updated (WHO 2001; White 2002). It takes into consideration the important role of immunity in high transmission areas with more emphasis on parasitological clearance and different transmission intensity (Table 24.11).

The mechanism of drug resistance of malaria parasites is only well understood for antifolate drugs and CQ. The four groups of antifolates, sulfonamides, sulfones, pyrimethamine, and proguanil, act at different points of the parasite pathway for folate production, with two major target enzymes DHPS and DHFR. Resistance to antifolates is generally induced by one or more mutations on either of these genes and these mutations generally occur in a limited number of sites on the molecules (Foote et al. 1990). Resistance to CQ is associated with mutations in the *pfcrt* gene, but may also correlate with mutations on the pfmdr genes (Fidock et al. 1999). It is now feasible to use molecular techniques, such as PCR, as an alternative to in vitro assays for monitoring the spread of such mutations in an area where antifolate or CQ drug resistance has been reported. The higher the prevalence of the mutations, the lower the efficacy of the drug (Watkins et al. 2005).

Whereas CQ resistance of *P. falciparum* (see Figure 24.2) is widespread, resistance to CQ by *P. vivax*

Table 24.9 *Antimalarial chemotherapy for vivax- and ovale-malaria*

Treatment	
1. Chloroquinine	10 mg base/kg on days 1 and 25 mg base/kg on day 3
Plus for radical cure	
2. Primaquinine[a]	0.25–0.33 mg[b] base/kg daily on days 4-17

a) Contraindicated for pregnant and lactating women. For G6PD-deficient patients daily doses of primaquine are also contraindicated, but radical cure can be obtained using 0.75 mg/kg once weekly for 8 weeks, with minimal risk of hemolysis.
b) For Oceania and South East Asia strains.

Table 24.10 *Antimalarial chemotherapy for malariae-malaria*

Antimalarial chemotherapy	
Chloroquine	10 mg base/kg on days 1 and 2 5 mg base/kg on day 3 and 4

is still limited to Papua New Guinea, parts of India, Indonesia, Myanmar, Brazil, and Guatemala. For *P. vivax*, there is only a single definition of treatment failure namely:

- clinical deterioration requiring hospitalization in the presence of parasitemia;
- presence of parasitemia and axillary temperature >37.5°C anytime between days 3 and 28;
- presence of parasitemia on any day between days 7 and 28, irrespective of clinical condition.

Four basic methods have been routinely used to assess antimalarial drug resistance: in vivo, in vitro, animal models, and molecular characterization. In general, resistance occurs through spontaneous mutations – single or multiple – that result in reduced sensitivity to drugs. The biochemical mechanism of resistance has been described for CQ, the antifolate combination drugs and atovaquone (Bloland 2001).

Chemoprophylaxis of malaria

Drug resistance of *P. falciparum* to CQ and other antimalarial drugs continues to increase in both intensity and geographical distribution. The situation is a dynamic one and advice on the current chemoprophylaxis recommendations for travelers should be sought from the various specialized centers in respective countries (Steffen et al. 1993; Lobel et al. 1995; Bradley and Bannister 2001). To date, resistance of *P. vivax* to CQ has been reported only occasionally from Indonesia, Myanmar, parts of India, Brazil, Guatemala, Papua New Guinea, and Vanuatu, and these reports do not yet justify a recommendation for alternative chemoprophylaxis for these areas.

The principles of prevention that should be borne in mind are:

- Awareness of the malaria risk coupled with a high index of suspicion if fever occurs while in a malarious country or within 3–12 months of return even if all recommended precautions have been taken.
- Personal protection to diminish contact with the mosquito vector is very important. This includes (1) sleeping in screened rooms and spraying the room with a knock-down insecticide before sleeping to kill any mosquitoes that have entered during the day; (2) using nets impregnated with pyrethroids; (3) using an electric mat to vaporize synthetic pyrethroids, or burning mosquito coils; (4) wearing long-sleeved clothing and trousers after sunset; and (5) application of repellents containing diethyltaluanide to exposed skin.
- Strict compliance with the chemoprophylaxis regimens recommended (Bradley and Bannister 2001).

Chemoprophylaxis at community level is a contentious issue, which is not recommended by WHO at present for a variety of reasons: potential dangers of drug resistance; poor compliance in areas of perennial transmission; possible delay in the development of natural immunity; cost; toxicity of drugs; regularity of supply and logistics of delivery. Nevertheless, there may be a place for community chemoprophylaxis targeted to young children and pregnant women in areas where transmission is seasonal and short and where the population is co-operative and well-educated with regard to malaria (as is the case in the Gambia) (Greenwood 1991).

Chemoprophylaxis in pregnancy has long been recommended, particularly in primigravidae (Gilles et al. 1969; Fleming et al. 1986; Greenwood et al. 1989; WHO 1994b, Steketee et al. 1996a). The advent of drug resistance, increasing evidence of poor compliance, unacceptable adverse drug reactions, and economic considerations have resulted in alternative strategies being sought. For instance, in Malawi and Kenya, a two-dose regimen of sulfadoxine-pyrimethamine given at first attendance in the second trimester of pregnancy and repeated at the beginning of the third trimester (IPT), was found to be a cost-effective intervention to reduce the incidence of low birth-weight infants, placental parasitemia, and anemia (Schultz et al. 1995; Parise et al. 2003). This regimen is generally recommended. Meanwhile, a curative treatment with CQ given at first attendance, followed by a prophylaxis continued throughout pregnancy is recommended for Central America where *P. falciparum* is still predominantly sensitive to CQ (WHO 1994a). The spread of SP resistance presents a challenge in areas of high transmission for which no clear cut solution is readily available. In areas of low to moderate transmission, prompt treatment with quinine and clindamycin is an option.

Table 24.11 *WHO definitions of treatment failures*

Adequate clinical response (ACR)	Early treatment failure (ETF)	Late treatment failure (LTF)
Absence of parasitemia on day 14, irrespective of axillary temperature, without previously meeting any of the criteria of early or late treatment failure	Development of danger signs or severe malaria on day 1, 2, or 3 after drug treatment in the presence of parasitemia	Development of danger signs or severe malaria in the presence of parasitemia on any day from day 4 to 14 after treatment, without previously meeting any of the criteria of ETF
Axillary temperature <37.5°C, irrespective of the presence of parasitemia, without previously meeting any of the criteria of early or late treatment failure	Axillary temperature ⩾37.5°C on day 2 with parasitemia greater than that of day 0 count, irrespective of temperature	In areas of intense transmission, axillary temperature ⩾37.5°C in the presence of parasitemia on any day from day 4 to 14, without previously meeting any of the criteria of ETF
	Axillary temperature ⩾37.5°C on day 3 in the presence of parasitemia. In areas with low to moderate transmission, there must be a measured increase in axillary temperature on day 3	In areas of low to moderate transmission, presence of parasitemia on any day from Day 4 to 28, and a measured axillary temperature ⩾37.5°C, without previously meeting any of the criteria of ETF. If a history of fever, rather than measured fever, was accepted as an entry criterion, then parasitemia with history of fever suffices for LTF

The 2002 WHO Consultancy has modified the definition of cure to adequate clinical and parasitological response (ACPR). In a high-transmission setting, ACPR is defined as an adequate clinical response, but in low to moderate transmission settings, the assessment is made on day 28. The consultancy has also included the definition of late parasitological failure (LPF). In high-transmission settings, LPF is defined as 'presence of parasitemia on day 14 and a measured axillary temperature of <37.5°C, without previously meeting any of the criteria of ETF or LTF.' In low to moderate transmission settings, the definition is the same, except that parasitemia at any time from day 7 to 28 qualifies.

EPIDEMIOLOGY

The epidemiology of malaria can be described in at least four different ways, each of which is complementary: (1) the classical approach; (2) the ecological classification with eight paradigms; (3) a clinico-epidemiological approach; and (4) a conceptual diagram which merely elaborates on the above three aspects.

In addition to transmission by various species of *Anopheles* mosquitoes, malaria parasites can also be transmitted by blood transfusion, by infected needles among drug users, and by organ transplantation. Mosquitoes surviving an aircraft journey or taxi journey from an endemic country may infect persons in a nonendemic area ('airport malaria,' 'baggage malaria,' or 'taxi-rank malaria'). Malaria can be contracted during a short refueling stop 'runway malaria.' Animal reservoirs are of no importance in the epidemiology of human malaria.

The classical macro-epidemiology approach

Two epidemiological extremes are described: stable and unstable malaria. The salient differences are shown in Table 24.12. In between these two extremes, variable degrees of transmission occur. Moreover, the prevalence of malaria varies considerably within the

same country. In Bangladesh, holoendemic falciparum malaria occurs in one area, whilst unstable malaria (or epidemic prone) occurs in another, whereas in most of the country transmission is low and predominantly due to *P. vivax*.

Four major determinants relevant to the dynamics of malaria transmission are: (1) the parasite, (2) the vector, (3) the human host, and (4) the environment.

THE PARASITE

Differences in the biology of the different species of *Plasmodium* are important factors in the epidemiology of malaria. Thus, the prepatency period is shortest in *P. falciparum* (6–25 days) and longest in *P. malariae* (18–27 days) while the incubation period for *P. falciparum* is 7–27 days, in contrast to that for *P. malariae* which is 23–69 days. The parasite life span is usually only 1 year for *P. falciparum* while for *P. malariae* it is many decades. The appearance of gametocytes occurs simultaneously with the asexual parasitemia in *P. vivax*, *P. ovale*, and *P. malariae* is in contrast to *P. falciparum* in which viable gametocytes appear in the circulation only 8–15 days after blood infection. Genetic diversity in *P. falciparum* antigens is well documented and has already been discussed. The potential for multiplication and the ability to relapse in the human host also varies with the different species of parasite.

Table 24.12 *Epidemiological features of stable and unstable malaria*

	Stable malaria	Unstable malaria
Transmission pattern	Transmission occurs throughout the year. Fairly uniform intensity. Pattern repeats annually	Seasonal transmission variable in intensity. Liable to flare up in dramatic epidemics
Immunity	Potent resistance in the community due to prevailing intense transmission	General lack of communal immunity due to low level and variable intensity of transmission
Impact	Mainly young children	All age groups
Control	Difficult	Much easier than stable malaria
Occurrence	Rural West Africa and East Africa, coastal areas of PNG	Plateau of Ethiopia, Plateau of Madagascar, Sri Lanka, north west India

THE VECTOR

Many factors affect the susceptibility of anopheline mosquitoes (of which around 70 species can transmit malaria) to specific *Plasmodium* species. Air temperature, relative humidity, and types of breeding places can affect the longevity of the adult mosquito and the development of the aquatic stages. The density of the vectors in relation to humans, their parous rates, the frequency of mosquitoes feeding on humans (anthropophilic or zoophilic), the duration of sporogony, the sporozoite rates, the peak biting time, the preference for biting indoors or outdoors (endophagy, exophagy), and the choice of resting place (endophily, exophily), are important determinants which affect the transmission dynamics and are usually measured to quantify the malaria risk (Bockarie et al. 1994; Service and Townson 2002).

THE HOST

Human biological factors, human–parasite interactions, and human behavior all influence the epidemiology of malaria.

Several genetic factors prevent parasite development in the red cell and conclusive evidence of protection against *P. falciparum* malaria is now available for the inherited genes for hemoglobin S, alpha- and β-thalassemia, HbC, G6PD deficiency, and ovalocytosis (Allison 1954; Gilles et al. 1967; Edington 1967; Willcox 1975; Cattani 1987; Ruwende et al. 1995; Agarwal et al. 2000).

Miller et al. (1979), demonstrated natural selection to *P. vivax* malaria in individuals in whom the Duffy blood group antigen is absent (Fy^{a-b-} phenotype). More recently, it has been shown that two HLA types, HLA-BW53 and a haplotype bearing the DRW13.02 antigen, and R111 gene confer a degree of protection against cerebral malaria and malarial anemia in the Gambia (Hill et al. 1991). In contrast, in the same population, homozygotes for the TNF2 allele, a variant of the TNF-γh gene promoter region and H121 phenotype have a relatively higher risk of death or severe neurological sequelae due to cerebral malaria (McGuire et al. 1994; Cooke et al. 2004). Acquired immunity has already been discussed. Other human variables such as pregnancy (especially in primigravidae) modify malaria prevalence and density, as well as the birth-weight of infants and the incidence of anemia (Brabin et al. 1993). Nutritional factors seem to have a paradoxical effect. Severe malaria is seldom evident at necropsy in children who died with marasmus or Kwashiorkor in West Africa (Edington 1967). In Kenya, however, poor nutritional status was found to be a risk factor for severe malaria (Marsh et al. 1995). Human behavior, e.g. sleeping habits, types of housing, occupational activities, and political and social factors can all be responsible for changes in the epidemiology of malaria.

THE ENVIRONMENT

The physical environment has a direct effect on both the parasite and on the vector. Thus, altitude, temperature, rainfall, the rate of flow of rivers and streams, collections of water, and availability of animals all influence transmission.

The effect of global warming could present a risk of epidemics in the highland areas of tropical Africa, by pushing malaria transmission uphill into these populated areas (Bradley 1995). It is, however, quite another matter to attribute recent resurgences of malaria in the highlands of East Africa to climatic changes (Hay et al. 2002). The following can all have a significant impact on malaria endemicity: increases in agricultural colonization; the construction of large economic projects such as dams, irrigation schemes, and highways; mining; deforestation; drug resistance; poor vector control; deterioration of health services; and mobilization of large migrant labor forces. The cultivation of rice and cotton are often associated with increased malaria risk. The creation of new microclimates, following deforestation in urban areas of West Africa, seems to have favored the establishment of *A. arabiensis*, a savannah species, at the expense of *A. gambiae* (Coluzzi et al. 1979). Environmental factors may interact with social, political, and economic pressures, e.g. movements of thousands of refugees, urbanization, smuggling, illegal logging and gem mining, and the existence of ethnic and marginalized communities. The Garki Project is probably the best example of a study where many of the parameters of malaria epidemiology were critically examined (Molineaux and Grammivcia 1980).

Geographic information systems (GIS) provide a powerful means of capturing large amounts of information and modeling spatial risks. Applications of GIS using meteorological data have been used to develop provisional malaria risk maps in Africa (Craig et al. 1999).

The ecological classification

A pragmatic approach to the epidemiological stratification of malaria has been developed by Najera et al. (1991), defining eight major ecological prototypes:

1 African savannah;
2 plains and valleys outside Africa (areas of traditional agriculture);
3 forest and forest fringe;
4 desert and highland fringe;
5 coastal and marshland;
6 urban slums;
7 agricultural developments;
8 sociopolitical disturbances.

For each paradigm, a description of the circumstances that are responsible for the transmission of malaria is given and the salient features of the impact of the disease are described.

A clinico-epidemiological approach

The natural history of 'stable' malaria infection has been extensively documented. The salient features are as follows: infants born from semi-immune mothers are protected for the first 3–6 months of life, predominantly by passive immunity acquired from their mothers and possibly other factors such as a high concentration of fetal hemoglobin. Although parasitemia occurs during this period, severe and complicated malaria is seldom encountered. From 6 months to 5 years or more (depending on the level of transmission and other factors such as the availability of antimalarial drugs, impregnated bednets, etc.), the child is susceptible to severe attacks of malaria resulting in death in a proportion of children exposed. In the Gambia it has been estimated that 25 percent of all childhood deaths under 5 years were due to malaria (Greenwood et al. 1987). After this danger period, when the cause of death is usually either malarial anemia or cerebral malaria, immunity is gradually acquired and clinical disease is rare except in primigravidae. This sequence of events is graphically presented in Figure 24.25.

A conceptual diagram of the epidemiology of malaria

A conceptual diagram of the epidemiology of malaria is given in Figure 24.26. Various factors are known to determine the relative sizes of the circles in various communities. For example, in 'stable' areas of malaria, a substantial proportion of the population will have asymptomatic parasitemia at any one period in time, whereas comparatively few will have clinical malaria and fewer still will have severe malaria (approximately 0.3 percent); this will occur mainly in children and a small proportion (estimated at 0.1 percent) will die. Nevertheless, this amounts to the alarming figure of one million children under 5 years of age in Africa alone. In contrast, in 'unstable' areas the largest circle will be that for uncomplicated malaria, since most of those who get infected usually develop clinical illness.

We have only just begun to identify the factors that might determine which circle an individual patient falls into. For example, some genetic factors that protect against developing severe disease have already been identified but many questions remain unanswered. Some genetic traits that increase the susceptibility to severe disease have also been identified (Knight et al. 1999).

HIV and malaria

The prevalence and intensity of malaria in pregnancy are higher in women who are HIV-positive. In areas of moderate or high transmission, HIV renders multigravidae as susceptible to malaria as primigravidae (Steketee et al. 1996b). Both HIV infection and malaria are independent risk factors. In lower birth weight and maternal anemia, a common occurrence especially in sub-Saharan Africa, the risks for both mother and baby are high.

While most studies have stated that, paradoxically, HIV does not seem to increase the severity of the disease nor significantly enhance the intensity of parasitemia in non-pregnant women and all others exposed to malaria, a study in Uganda (French et al. 2001) has shown a marked inverse relationship between malarial fever and CD4 T-cell counts. In situations where co-infection is a problem, there is some evidence to suggest that antimalarial drugs might be less effective and carry a greater risk of adverse drug reactions (Rowland-Jones and Lohman 2002).

Epidemic malaria

Severe epidemics have occurred in Africa and elsewhere as recently as 2002. They are often the result of exceptional meteorological conditions or massive destruction as a consequence of war or natural disasters followed by population movements. The following risk factors have been identified: (1) abnormally prolonged rains; (2) extensive floods; (3) global warming; (4) colonization of tropical jungle areas by agricultural settlers; (5) explosive growth of urban areas; (6) open-cast mining; (7) arrival of a nonimmune population into a malarious area (e.g. refugees); (8) the introduction of a number of infected individuals into a malaria-free area where both

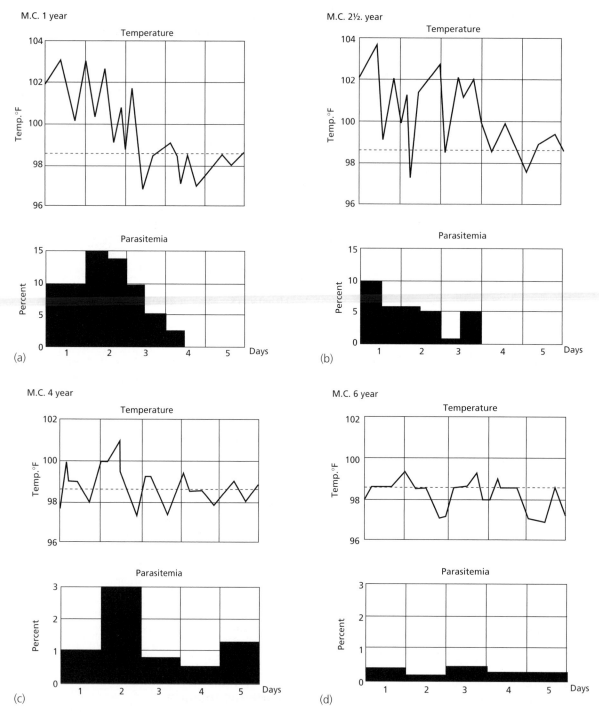

Figure 24.25 *Natural history of* Plasmodium falciparum *malaria in a 'stable' area showing how clinical immunity develops in a child. With each subsequent phase there is first a decline in the febrile response to infection and later in parasitemia.* **(a)** *Child looks ill, has high temperature and heavy parasitemia.* **(b)** *Child does not look ill, despite high temperature and heavy parasitemia.* **(c)** *Child does not look ill, has mild temperature despite appreciable parasitemia, but may develop anemia insidiously.* **(d)** *Child does not look ill, has no temperature and light parasitemia.*

the *Anopheles* vector and conditions for transmission exist; (9) admixtures of large numbers of immunes and nonimmunes living under primitive conditions (e.g. labor camps); (10) sudden increase in *Anopheles* densities; (11) agricultural development schemes; (12) failure to maintain previous control; (13) breakdown of health services (e.g. Afghanistan); and (14) progressive spread

of drug resistance to antimalarials (Snow and Gilles 2002).

Imported malaria

International travel has grown exponentially in the last 50 years and with it imported malaria. This growth has

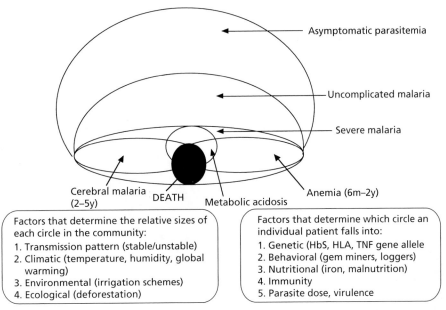

Factors that determine the relative sizes of each circle in the community:
1. Transmission pattern (stable/unstable)
2. Climatic (temperature, humidity, global warming)
3. Environmental (irrigation schemes)
4. Ecological (deforestation)

Factors that determine which circle an individual patient falls into:
1. Genetic (HbS, HLA, TNF gene allele
2. Behavioral (gem miners, loggers)
3. Nutritional (iron, malnutrition)
4. Immunity
5. Parasite dose, virulence

Figure 24.26 *Conceptual diagram of epidemiology of malaria*

been compounded by the emergence and spread of drug resistance. Labels such as 'airport malaria,' 'baggage malaria' or 'taxi rank malaria' are the result of infected vectors transported from endemic areas in aircraft, suitcases, and taxis, biting nonimmune individuals who had not journeyed to malarious areas.

In the UK, around 1 200 cases of *P. falciparum* are notified yearly; in 2003, the number of cases of *P. falciparum* was 847. Deaths have varied from four in 1989 to 15 in 2000 (PHLS Malaria Reference Laboratory, London School of Hygiene and Tropical Medicine) (Figure 24.27).

VACCINATION

The prevention of malaria by vaccination is a conceivable approach, which may eventually have a place in malaria control. Unfortunately, no effective malaria

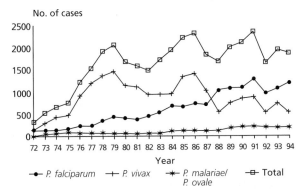

Figure 24.27 *Annual number of imported cases of malaria in the United Kingdom (based on data from the Malaria Reference Laboratory)*

vaccine is as yet available and none is likely to be available at an operational level for many years. The topic of malaria vaccines is vast and complex and has been reviewed at length elsewhere (Mendis 1991; Phillips 1994; Hoffman 1996; Facer and Tanner 1997; Engers and Godal 1998; Richie and Saul 2002; Greenwood and Alonso 2002).

The aim of a vaccine is to reduce morbidity/mortality due to malaria and this can be achieved in one of two different ways: either by interrupting the infection at one or other of the stages of the parasite life cycle, or by means of an 'anti-disease' vaccine, which would reduce the pathophysiological effects triggered by the release of malarial toxins as proposed by Playfair et al. (1990). *Plasmodium* GPI has been suggested as a potential toxin candidate and the first results of vaccination with this molecule have been reported (Schofield 2002; Schofield et al. 2002).

The interruption of the life cycle may take place at the time of entry of the sporozoite into the host, at the level of schizogony in the hepatocyte, at the time of erythrocyte invasion and intra-erythrocytic parasite development, or by interruption of the sexual development within the mosquito. Figure 24.28 illustrates potential vaccine targets and indicates which possible parasite molecules may be used as a vaccine in each situation.

The methods by which host immunity will control the infection are variable depending on the target and this will have an implication on the choice of immunogen. For example, a pre-erythrocytic vaccine is likely to rely more on the production of CTLs than antibodies, while a transmission-blocking vaccine is likely to rely essentially on antibodies to sexual stages. It follows that new methods for immunization using DNA injection are likely to be particularly useful for pre-erythrocytic

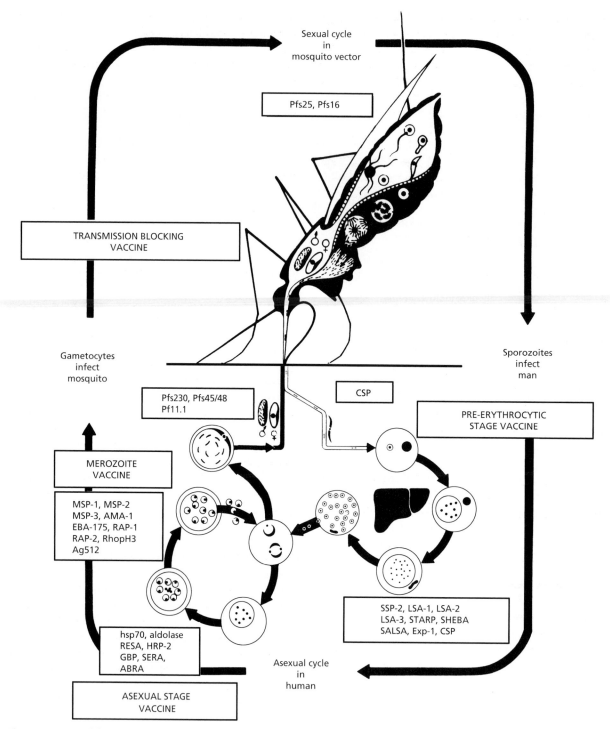

Figure 24.28 *Possible targets of a malaria vaccine*

stages, since they are good for producing CTLs, and perhaps less useful for transmission-blocking vaccines.

A number of vaccine candidates have been tested in human volunteers with variable success. The best results were obtained with x-irradiated sporozoites (Clyde et al. 1973), while synthetic and recombinant antigens have so far given disappointing results: 1/34 volunteers were protected in a trial using a recombinant CS epitope (Ballou et al. 1987); 0/20 volunteers vaccinated with a recombinant CS co-expressed in yeast with hepatitis B surface antigen were protected (Gordon et al. 1995); a 31 percent efficacy was reported for the synthetic 'Spf66' vaccine, when tested in Tanzania (Alonso et al. 1994), but only a 3 percent efficacy for the same molecule, when tested in the Gambia (D'Alessandro et al. 1995). Current research trends in vaccine development are geared towards 'cocktail' vaccines, which include a combination of multiple epitopes from different malarial antigens and from different stages of the parasite life cycle. These may include a combination of CSP, LSA-1/3, MSP-1/2, SERA, AMA-1, Pfs48/45, Pfs25, STARP, SSP-2. A synthetic vaccine consisting of part of the CSP fused to the S-antigen of hepatitis B (RTS, S/ASOZA) has produced encouraging results in Mozambique in which 57.5 percent of children were protected (Alonso et al. 2004; Doolan and Hoffman 1997).

The predicted impact of a malaria vaccine in controling the disease has been discussed and analyzed with a variety of mathematical models and with various degrees of optimism (Halloran et al. 1989; Saul 1992).

INTEGRATED CONTROL

Between 1898 and 1940, malaria control was effected by a combination of the use of quinine (prophylactically as well as therapeutically) and mosquito larval control.

The Second World War (1940–45) revolutionized malaria control through the development of the synthetic antimalarials (atebrin, CQ, and proguanil) for effective prophylaxis and treatment, and the first residual insecticide (dichlorodiphenyltrichloroethane (DDT)) which made possible the control of adult mosquitoes by house spraying. Between 1945 and 1955, house spraying with DDT combined with CQ treatment of cases, eliminated malaria from several countries in Europe and gave rise to the concept of malaria eradication as opposed to malaria control. Thus in 1956, the WHO World Health Assembly recommended that WHO should implement a program that would eliminate malaria from the world and this policy was pursued until 1975 when it was abandoned. The reasons for failure can be listed as follows:

1 eradication was an unrealistic objective, particularly in tropical Africa;
2 inflationary price increases of equipment, fuel, and insecticides;
3 political instability;
4 financial and administrative shortcomings;
5 vector resistance;
6 vector-exophily;
7 resistance to antimalarial drugs;
8 household objections to regular spraying; and
9 uncontroled population movements.

In 1976, the WHO World Health Assembly adopted an integrated malaria control strategy, the elements of which are shown in Figure 24.29. Even this reduced goal proved to be overambitious and unachievable in most of the endemic malaria countries, especially those in tropical Africa, and in 1985 the World Health Assembly recommended that malaria control should be developed as an integrated part of national primary health care systems (see Chapter 5, Control of parasites, parasitic infections, and parasitic diseases). The malaria control

strategy now being promoted by WHO (Ministerial Conference on Malaria 1992) recognizes four basic technical elements: (1) to provide early diagnosis and prompt treatment; (2) to plan and implement selective and sustainable preventive measures against the parasite, as well as the vector; (3) to detect early, contain or prevent epidemics; and, (4) to reassess regularly a country's malaria situation, in particular the ecological, social, and economic determinants of the disease. This strategy involves a radical shift from the primary emphasis on transmission control of the parasite to the present focus of reduction of malarial disease.

In this context, the use of impregnated bednets is of particular interest. Pyrethroid-treated bednets have been shown to reduce malarial morbidity in certain areas, e.g. China and the Gambia. In areas of intense perennial malarial transmission, there is evidence of reduction of *P. falciparum* parasitemia, reduction in transmission, and reduction in rate of reinfection. In some trials but not in others, widespread community use of insecticide-treated bednets protect even non-users of nets (Greenwood and Baker 1993; Smith et al. 1993; Bockarie et al. 1994; Somboon et al. 1995; WHO 1995). Sustained utilization of nets is clearly important and remains an unsolved challenge (Meek 1995); sustainability is likely to be improved by the development of nets that do not have to be reimpregnated frequently or possibly not at all for many years despite frequent washing (Kroeger et al. 2004), e.g. Olyset® and PermaNet®. The existing vector control tools can still be effective, but they must be selectively deployed, cost-effective, and sustainable.

In the long term, the impact of any malaria control measure (including early diagnosis and treatment, indoor residual spraying, personal protection, or environmental management) depends on the importance given to these measures by the community at risk and on their understanding of and involvement in their application (WHO 1995). The long-neglected social, behavioral, and economic aspects of malaria and its control are being increasingly recognized (Combie 1994; Ruiz and Kroeger 1994; Manderson 1994; Agyepong et al. 1995; Heggenhougen et al. 2004).

It would be wrong to assume that the involvement of primary health care in malaria control in endemic countries is an easy way of solving the difficulties related to administrative, social, or economic obstacles. The following must all be achieved: a proper degree of cooperation between the different levels of the health care system; inter-sectoral coordination; substantial political commitment; adequate human and financial resources; and the full partnership of communities. In addition to these obstacles, the appropriate choice of antimalarial in areas of increasing parasite resistance must also be made if the relatively modest goal of reducing malaria morbidity and mortality is to be achieved (Bryce et al. 1994; Steketee et al. 1996a). Treatment uptake needs to be increased (von Seidlein et al.

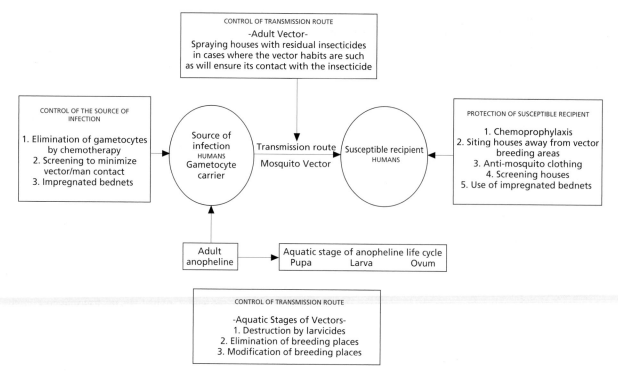

Figure 24.29 *Integrated malaria control strategy*

2002) and the private sector needs to be controled since it cannot easily be controled especially as it is thegreater provider of antimalarials in most endemic countries.

Roll back malaria

In June 1997, African heads of state met in Harare (Zimbabwe) and issued a declaration on malaria prevention and control in the context of African economic recovery and development. This political commitment was endorsed by the leaders of the industrialized G8 nations. In 1978, this led to a new WHO initiative to roll back malaria (RBM). The goals of RBM include:

1 support to endemic countries in developing their national health systems as a major strategy for controling malaria;
2 developing the broader health sector, i.e. all providers of health care to the community, including the private sector (drug vendors and traditional healers, pharmacists and others); and
3 encouraging the human and financial investments necessary for health system development, nationally and internationally.

The elements of the RBM strategy are:

1 early diagnosis;
2 prompt treatment;
3 multiple prevention;
4 well-coordinated action;
5 dynamic global movement;
6 focussed research.

A functioning partnership with a range of organizations at global, regional, and country levels is being established to ensure sustained capacity to address malaria and other priority health problems. The partners in RBM include malaria-endemic countries, other UN organizations, development banks including the World Bank, bilateral development agencies, nongovernmental organizations, and the private sector.

The RBM campaign focuses first on Africa, where the impact of malaria morbidity and mortality is greatest. It will build upon existing initiatives, such as the African Initiative on Malaria (AIM), the WHO Special Fund for Africa, Medicines for Malaria Venture (MMV), and the Multilateral Initiative on Malaria (MIM). Although the need for new drugs, insecticides, and vaccines still remains, major gains can still be made through better application of current knowledge and best practice.

There are a number of evidence-based, cost-effective interventions available to RBM. These are:

1 insecticide-treated nets (ITN) that are long-lasting (effective up to 4 years) and wash-resistant;
2 intermittent preventive treatment;
3 antimalarial drug combinations;
4 improved access to antimalarial drugs and improving compliance by the use of blister packs and simple dose combination;
5 emergency treatment in the home, at village level, or in rural health centers using either artemisinin suppositories or intra-rectal quinine.

The most difficult hurdle facing RBM is to ensure that sufficient effective drugs are available at the right time, in the right place, and where they are most needed. New strategies to ensure this is the case must be developed if RBM is to reach its goal of significantly reducing mortality within the next 10 years. A promising strategy entitled 'home management of malaria' is vigorously being pursued (WHO 2004), and may turn out to be the breakthrough we have been awaiting for years, combined with ACT or artesunate suppositories.

REFERENCES

Abdalla, S., Weatherall, D.J., et al. 1980. The anaemia of *Plasmodium falciparum* malaria. *Br J Haematol*, **46**, 171–83.

Adjuik, M., Babiker, A., et al. 2004. Artesunate combinations for treatment of malaria: meta-analysis. *Lancet*, **3**, 9–17.

Agarwal, A., Guindo, A., et al. 2000. Hemoglobin C associated with protection from severe malaria in the Dogon of Mali, a West African population with a low prevalence of hemoglobin S. *Blood*, **96**, 2358–63.

Agbenyega, T., Angus, B., et al. 1997. Plasma nitrogen oxides and blood lactate concentrations in Ghanaian children with malaria. *Trans R Soc Trop Med Hyg*, **91**, 298–302.

Agyepong, I.A., Aryee, B. et al. 1995. *The malaria manual*. Geneva: WHO TDR/SER/MSR/95.1, pp. 17

Aikawa, M. 1988. Morphological changes in erythrocytes induced by malarial parasites. *Biol Cell*, **64**, 169–77.

Aitman, T.J., Cooper, L.D., et al. 2000. Malaria susceptibility and CD36 mutation. *Nature*, **405**, 1015–16.

Al-Khedery, B., Barnwell, J. and Galinski, M. 1999. Antigenic variation in malaria: a 3′ genomic alteration associated with the expression of a *P. knowlesi* variant antigen. *Mol Cell*, **3**, 131–41.

Allan, R.J., Rowe, A. and Kwiatkowski, D. 1993. *Plasmodium falciparum* varies in its ability to induce tumor necrosis factor. *Infect Immun*, **61**, 4772–6.

Allen, S., Vandeperre, P., et al. 1991. Human-immunodeficiency-virus and malaria in a representative sample of childbearing women in Kigali, Rwanda. *J Infect Dis*, **164**, 67–71.

Allen, S.J., O'Donnell, A., et al. 1996. Severe malaria in children in Papua New Guinea. *Quart J Med*, **89**, 779–88.

Allison, A.C. 1954. Protection afforded by the sickle cell trait against subtertian malarial infection. *Br Med J*, **i**, 290–4.

Alonso, P., Lindsay, S.W., et al. 1993. A malaria control trial using insecticide-treated nets and targeted chemoprophylaxis in a rural area of The Gambia, West Africa. 6. The impact of the interventions on mortality and morbidity from malaria. *Trans R Soc Trop Med Hyg*, **87**, Suppl. 2, 37–44.

Alonso, P., Smith, T., et al. 1994. Randomised trial of efficacy of SPf66 vaccine against *Plasmodium falciparum* in children in southern Tanzania. *Lancet*, **344**, 1175–81.

Alonso, P., Sacarlal, J., et al. 2004. Efficacy of the RTS, S/ASOZA vaccine against *Plasmodium falciparum* infection and disease in young African children: randomised control trial. *Lancet*, **364**, 1411–20.

Al-Yaman, F.M., Mokela, D., et al. 1996. Association between serum levels of reactive nitrogen intermediates and coma in children with cerebral malaria in Papua New Guinea. *Trans R Soc Trop Med Hyg*, **90**, 270–3.

Anders, R.F. 1986. Multiple cross-reactivities among antigens of *Plasmodium falciparum* impair the development of protective immunity against malaria. *Parasite Immunol*, **8**, 529–39.

Anstey, N.M., Weinberg, J.B., et al. 1996. Nitric oxide in Tanzanian children with malaria: inverse relationship between malaria severity and nitric oxide production nitric oxide synthase type 2 expression. *J Exp Med*, **184**, 557–67.

Archibald, H.M. 1956. The influence of malarial infection of the placenta on the incidence of prematurity. *Bull World Health Organ*, **15**, 842–5.

Arese, P., Turini, F. and Ginsburg, H. 1991. Erythrophagocytosis in malaria: host defence or menace to the macrophage. *Parasitol Today*, **7**, 25–8.

Assaraf, Y.G., Golenser, J., et al. 1984. Polyamine levels and the activity of their biosynthetic enzymes in human, erythrocytes infected with the malarial parasite, *Plasmodium falciparum*. *Biochem J*, **222**, 815–19.

Aucan, C., Walley, A.J., et al. 2003. Interferon-alpha receptor-1 (IFNAR1) variants are associated with protection against cerebral malaria in the Gambia. *Genes Immun*, **4**, 275–82.

Badibanga, B., Dayal, R., et al. 1986. Etude des principaux facteurs immunologiques et de la barrière hémato-méningée au cours de la malaria cérébrale chez l'enfant en pays d'endémie (Zaire). *Ann Soc Belg Med Trop*, **66**, 23–37.

Baird, J.K. 1995. Host age as a determinant of naturally acquired-immunity to *Plasmodium-falciparum*. *Parasitol Today*, **11**, 105–11.

Ballou, W.R., Hoffman, S.L., et al. 1987. Safety and efficacy of a recombinant DNA *Plasmodium falciparum* sporozoite, vaccine. *Lancet*, **i**, 1277–81.

Bannister, L.H. and Dluzewski, A.R. 1990. The ultrastructure of red cell invasion in malaria infection: a review. *Blood Cells*, **16**, 257–92.

Barnwell, J.W. 2001. Hepatic Kupffer cells: the portal that permits infection of hepatocytes by malarial sporozoites? *Hepatology*, **33**, 1331–3.

Bate, C.A.W. and Kwiatkowski, D. 1994. A monoclonal antibody that recognizes phosphatidylinositol inhibits, induction of tumor necrosis factor alpha by different strains of *Plasmodium falciparum*. *Infect Immun*, **62**, 5261–6.

Bate, C.A., Taverne, J., et al. 1992. Serological relationship of tumor necrosis factor-inducing exoantigens of *Plasmodium falciparum* and *Plasmodium vivax*. *Infect Immun*, **60**, 1241–3.

Bates, I. and Bedu-Addo, G. 1997. Review of diagnostic criteria of hyper-reactive malarial splenomegaly. *Lancet*, **349**, 1178.

Bates, I., Bedu-Addo, G., et al. 2001. B-lymphotropic viruses in a novel tropical splenic lymphoma. *Br J Haematol*, **112**, 161–6.

Beare, N.A.V., Lewis, D.K., et al. 2003. Retinal changes in adults with cerebral malaria. *Ann Trop Med Parasitol*, **97**, 313–15.

Beck, H.P., Felger, I., et al. 1997. Analysis of multiple *Plasmodium falciparum* infections in Tanzanian children during the phase III trial of the malaria vaccine SPf66. *J Infect Dis*, **175**, 921–6.

Beg, M.A., Khan, R., et al. 2002. Cerebral involvement in benign tertian malaria. *Am J Trop Med Hyg*, **67**, 230–2.

Behr, C., Poupot, R., et al. 1996. *Plasmodium falciparum* stimuli for human gamma delta T cells are related to phosphorylated antigens of mycobacteria. *Infect Immun*, **64**, 8, 2892–6.

Bellamy, R., Kwiatkowski, D. and Hill, A.V.S. 1998. Absence of an association between intercellular adhesion molecule 1, complement receptor 1 and interleukin 1 receptor antagonist gene polymorphisms and severe malaria in a West African population. *Trans R Soc Trop Med Hyg*, **92**, 312–16.

Bennett, I.L. and Hook, E.W. 1959. Infectious diseases (some aspects of salmonellosis). *Annu Rev Med*, **16**, 1–19.

Biemba, G., Gordeuk, V.R., et al. 2000. Markers of inflammation in children with severe malarial anaemia. *Trop Med Int Health*, **5**, 256–62.

Bloland, P.B. 2001. *Drug resistance in malaria*. WHO/CDS/CSR/DSR/2001.4, 32 pp.

Bloland, P.B., Wirima, J.J., et al. 1995. Maternal HIV-infection and infant-mortality in Malawi – evidence for increased mortality due to placental malaria infection. *AIDS*, **9**, 721–6.

Bockarie, M.J., Service, M.W., et al. 1994. Malaria in a rural area of Sierra Leone. III Vector ecology and disease, transmission. *Ann Trop Med Parasitol*, **88**, 251–62.

Bonday, Z.Q., Dhanasekaran, S., et al. 2000. Import of host δ-aminolevulinate dehydratase into the malarial parasite: identification of a new drug target. *Nature Med*, **6**, 898–903.

Bouharoun-Tayoun, H., Oeuvray, C. and Druilhe, P. 1995. Mechanisms underlying the monocytes-mediated antibody-dependent killing of *Plasmodium falciparum* asexual blood stages. *J Exp Med*, **182**, 409–15.

Boutlis, C.S., Gowda, D.C., et al. 2002. Antibodies to *Plasmodium falciparum* glycosylphosphatidylinositols: inverse associations with tolerance of parasitaemia in Papua New Guinean children and adults. *Infect Immun*, **70**, 5052–7.

Brabin, B.J. 1983. An analysis of malaria in pregnancy in Africa. *Bull World Health Organ*, **61**, 1005–16.

Brabin, B.J., Maxwell, S., et al. 1993. A study of the consequences of malarial infection in pregnant women and their infants. *Parassitologia*, **35**, Suppl, 9–11.

Bradley, D.J. 1995. The epidemiology of malaria in the tropics and in travellers. In: Pasvol, G. (ed.), *Bailliere's clinical infectious diseases – malaria*. London: Bailliere Tindall, 211–26.

Bradley, D.J. and Bannister, B. 2001. Guidelines for malaria prevention in travellers from the United Kingdom for 2001. *Comm Dis Publ Health*, **4**, 84–101.

Braks, M.A.H., Anderson, R.A. and Knols, B.G.J. 1999. Infochemicals in mosquito host selection: human skin microflora and *Plasmodium* parasites. *Parasitol Today*, **15**, 409–13.

Bray, P.G., Mungthin, M., et al. 1998. Access to hematin: the basis of chloroquine resistance. *Mol Pharmacol*, **54**, 170–9.

Bray, R.S. and Anderson, M.J. 1979. Falciparum malaria in pregnancy. *Trans R Soc Trop Med Hyg*, **73**, 427–31.

Brown, H., Rogerson, S., et al. 2001. Blood–brain barrier function in cerebral malaria in Malawian children. *Am J Trop Med Hyg*, **64**, 207–13.

Brown, K.N. and Brown, I.N. 1965. Immunity to malaria: antigenic variation in chronic infections of *Plasmodium knowlesi*. *Nature*, **208**, 1286–90.

Brown, K.N. and Phillips, R.S. 1974. Immunity to *Plasmodium berghei* in rats: passive serum transfer and role of the spleen. *Infect Immun*, **10**, 1213–18.

Bruce, M.C., Galinski, M.R., et al. 2000. Genetic diversity and dynamics of *Plasmodium falciparum* and *P. vivax* populations in multiply infected children with asymptomatic malaria infections in Papua New Guinea. *Parasitology*, **121**, 257–72.

Bruce-Chwatt, L.J. 1963. A longitudinal survey of natural malaria infection in a group of West African adults. *West Afr Med J*, **12**, 141–73.

Brumpt, E. 1949. The human parasites of the genus *Plasmodium*. In: Boyd, M.F. (ed.), *Malariology*. vol. 1. Philadelphia: W.B. Saunders, 65–121.

Bryce, J., Roungou, J.B., et al. 1994. Evaluation of national malaria control programmes in Africa. *Bull World Health Organ*, **72**, 371–81.

Bunnag, D., Kanda, T., et al. 1995. Artemether-mefloquine combination in multidrug-resistant falciparum malaria. *Trans R Soc Trop Med Hyg*, **89**, 213–15.

Buckling, A.G.J. and Read, A.F. 1999. The effect of chloroquine treatment on the infectivity of *Plasmodium chabaudi* gametocytes. *Int J Parasitol*, **29**, 6190626.

Burgner, D., Usen, S., et al. 2003. Nucleotide and haplotypic diversity of the NOS2A promoter region and its relationship to cerebral malaria. *Hum Genet*, **112**, 379–86.

Butcher, G.A., Mitchell, G.H. and Cohen, S. 1978. Antibody mediated mechanisms of immunity to malaria induced by vaccination with *Plasmodium knowlesi* merozoites. *Immunology*, **34**, 77–86.

Carcy, B., Bonnefoy, S., et al. 1995. *Plasmodium falciparum*: typing of malaria parasites based on polymorphism of a novel multigene family. *Exp Parasitol*, **80**, 463–70.

Carlson, J., Helmby, H., et al. 1990. Human cerebral malaria: association with erythrocyte rosetting and lack of anti-rosette antibodies. *Lancet*, **336**, 1457–60.

Carter, R. and Graves, P.M. 1988. Gametocytes. In: Wernsdorfer, W.H. and McGregor, I. (eds), *Malaria. Principles and practice of malariology*. vol. 1. Edinburgh: Churchill Livingstone, 253–305.

Cattani, J.A. 1987. Hereditary ovalocytosis and reduced susceptibility to malaria in Papua New Guinea. *Trans R Soc Trop Med Hyg*, **81**, 705–9.

Cerami, C., Frevert, U., et al. 1994. Rapid clearance of malaria circumsporozoite protein (CS) by hepatocytes. *J Exp Med*, **179**, 695–701.

Charoenpan, P., Indraprasit, S., et al. 1990. Pulmonary edema in severe falciparum malaria. Haemodynamic study and clinicophysiological correlation. *Chest*, **97**, 1190–7.

Clark, I.A. and Chaudri, G. 1988. Tumour necrosis factor may contribute to the anaemia of malaria by causing dyserythropoiesis and erythrophagocytosis. *Br J Haematol*, **70**, 99–103.

Clark, I.A. and Cowden, W.B. 2003. The pathophysiology of falciparum malaria. *Pharm Ther*, **99**, 221–61.

Clark, I.A. and Hunt, N.H. 1983. Evidence for reactive oxygen intermediates causing hemolysis and parasite death in malaria. *Infect Immun*, **39**, 1–6.

Clark, I.A. and Rockett, K.A. 1994. T cells and malarial pathology. *Res Immunol*, **145**, 437–41.

Clark, I.A., Chaudri, G. and Cowden, W.B. 1989a. Role of tumour necrosis factor in the illness and pathology of malaria. *Trans R Soc Trop Med Hyg*, **83**, 436–40.

Clark, I.A., Cowden, W.B. and Chaudhri, G. 1989b. Possible roles for oxidants through tumor necrosis factor in malarial anemia. In: Eaton, J.W. and Meshnick, S.R. (eds), *Malaria and the red cell*. New York: Alan R. Liss Inc, 73–82.

Clark, I.A., Awburn, M.M., et al. 2003. Tissue distribution of migration inhibitory factor and inducible nitric oxid synthase in falciparum malaria and sepsis in African children. *Malaria J*, **2**, 6.

Clark, I.A., Alleva, L.M., et al. 2004. Pathogenesis of malaria and clinically similar conditions. *Clin Microbiol Rev*, **17**, 509–39.

Clough, B. and Wilson, R.J.M. 2001. Antibiotics and the plasmodial plastid organelle. In: Rosenthal, P.J. (ed.), *Antimalarial chemotherapy: mechanisms of action resistance and new directions in drug discovery*. Totowa, NJ: Human Press, 265–86.

Clyde, D.F. and Shute, G.T. 1957. Resistance of *Plasmodium falciparum* in Tanganyika to pyrimethamine administered at weekly intervals. *Trans R Soc Trop Med Hyg*, **51**, 505–13.

Clyde, D.F., Most, H., et al. 1973. Immunization of man against sporozoite-induced falciparum malaria. *Am J Med Sci*, **266**, 169–77.

Coatney, G.R., Collins, W. et al. 1971. *The primate malarias*. US Department of Health Education and Welfare, 366 pp.

Cogswell, F.B. 1992. The hypnozoite and relapse in primate malaria. *Clin Microbiol Rev*, **5**, 26–35.

Cohen, S., McGregor, I.A. and Carrington, S. 1961. Gamma globulin and acquired immunity to human malaria. *Nature*, **192**, 733–7.

Coluzzi, M., Sabatini, A., et al. 1979. Chromosomal differentiation and adaptation to human environments in the *Anopheles gambiae* complex. *Trans R Soc Trop Med Hyg*, **73**, 483–97.

Combie, S.C. 1994. *Treatment seeking for malaria*. TDR/SER/RP/94.1, reference paper no. 2. TDR Geneva, pp. 29.

Contamin, H., Fandeur, T., et al. 1995. PCR typing of field isolates of *Plasmodium falciparum*. *J Clin Microbiol*, **33**, 944–51.

Cooke, G.S., Aucan, C., et al. 2003. Association of Fcγ receptor IIa (CD32). Polymorphism with severe malaria in West Africa. *Am J Trop Med Hyg*, **69**, 565–8.

Coosemans, M., Van der Stuyft, P. and Delacollette, C. 1994. A hundred per cent of fields positive in a thick film: a useful indicator of relative changes in morbidity in areas with seasonal malaria. *Ann Trop Med Parasitol*, **88**, 581–6.

Craig, M.H., Snow, R.W. and le Sueur, D. 1999. A climate-based distribution model of malaria transmission in sub-Saharan Africa. *Parasitol Today*, **15**, 105–11.

Crane, G.C. 1986. Hyperreactive malarious splenomegaly (tropical splenomegaly syndrome). *Parasitol Today*, **2**, 4–9.

D'Alessandro, U., Leach, A., et al. 1995. Efficacy trial of malaria vaccine Spf66 in Gambian infants. *Lancet*, **346**, 462–7.

Das, B.S., Das, D.B., et al. 1988. Riboflavin deficiency and severity of malaria. *Eur J Clin Nutr*, **42**, 277–83.

David, P.H., Hommel, M. and Oligino, L.D. 1981. Interactions of *Plasmodium falciparum*-infected erythrocytes with ligand, coated agarose beads. *Mol Biochem Parasitol*, **4**, 195–204.

Day, K.P. and Marsh, K. 1991. Naturally acquired immunity to *Plasmodium falciparum*. *Parasitol Today*, **7**, 68–71.

Defo, B.K. 1995. Epidemiology and control of infant and early childhood malaria: A competing risks analysis. *Int J Epid*, **24**, 204–17.

Del Giudice, G., Engers, H.D., et al. 1987. Antibodies to the repetitive epitope of *Plasmodium falciparum*, circumsporozoite protein in a rural Tanzanian community. A longitudinal study of 132 children. *Am J Trop Med Hyg*, **36**, 203–12.

Deloron, P. and Chougnet, C. 1992. Is immunity to malaria really short-lived? *Parasitol Today*, **8**, 375–8.

Del Portillo, H.A., Fernandez-Becerra, C., et al. 2001. A superfamily of variant genes encoded in the subtelomeric region of *Plasmodium vivax*. *Nature*, **410**, 839–42.

De Souza, J.B., Todd, J., et al. 2002. Prevalence and boosting of antibodies to *Plasmodium falciparum* glycosylphosphatidylinositols and evaluation of their association with protection from mild and severe clinical malaria. *Infect Immun*, **70**, 5045–51.

Djimde, A., Doumbo, O.K., et al. 2001. Application of a molecular marker for surveillance of chloroquine-resistant falciparum malaria. *Lancet*, **358**, 890–1.

Doherty, J.F., Grant, A.D. and Bryceson, A.D.M. 1995. Fever as the presenting complaint of travellers returning from the tropics. *Quart J Med*, **88**, 277–81.

Doolan, D.L. and Hoffman, S.L. 1997. Multi-gene vaccination against malaria: a multistage, multi-immune response approach. *Parasitol Today*, **13**, 171–8.

Dzekunov, S., Ursos, L. and Roepe, P. 2000. Digestive vacuolar pH of intact erythrocytic *P. falciparum* either sensitive or resistant to chloroquine. *Mol Biochem Parasitol*, **110**, 107–24.

Eckwalanga, M., Marussig, M., et al. 1994. Murine AIDS protects mice against experimental cerebral malaria: down-regulation by interleukin-10 of a T-helper type 1 CD4+ cell-mediated pathology. *Proc Natl Acad Sci USA*, **91**, 8097–101.

Edington, G.M. 1967. Pathology of malaria in West Africa. *Br Med J*, **i**, 715–18.

Edington, G.M. and Gilles, H.M. 1976. Malaria. *Pathology in the tropics*, 2nd edn. London: Edward Arnold, 10–33.

Edozien, J.C., Gilles, H.M. and Udeozo, I.O. 1962. Adult and cord blood gammaglobulins and immunity to malaria in Nigerians. *Lancet*, **ii**, 951–5.

Elford, B.C. 1986. L-glutamine influx in malaria-infected erythrocytes: a target for antimalarials? *Parasitol Today*, **2**, 309–11.

Engers, H.D. and Godal, T. 1998. Malaria vaccine development: current status. *Parasitol Today*, **14**, 56–64.

Engwerda, C.R., Mynott, T.L., et al. 2002. Locally up-regulated lymphotoxin alpha, not systemic tumor necrosis factor alpha, is the principle mediator of murine cerebral malaria. *J Exp Med*, **195**, 1371–7.

Facer, C.A. 1983. Erythrocyte sialoglycoproteins and *Plasmodium falciparum* isolates. *Trans R Soc Trop Med Hyg*, **77**, 524–30.

Facer, C.A. and Tanner, M. 1997. Clinical trials of malaria vaccines: progress and prospects. *Adv Parasitol*, **39**, 1–68.

Fairfield, A.S., Abosch, A., et al. 1988. Oxidant defense enzymes of *Plasmodium falciparum*. *Mol Biochem Parasitol*, **30**, 77–82.

Fakunle, Y.M. 1981. Tropical splenomegaly. Part 1. Tropical Africa. *Clin Haematol*, **10**, 963–75.

Ferone, R. 1977. Folate metabolism in malaria. *Bull World Health Organ*, **55**, 291–8.

Fichera, M. and Roos, D. 1997. A plastid organelle as a drug target in apicomplexan parasites. *Nature*, **390**, 407–9.

Fidock, D.A., Su, X.Z., et al. 1999. Genetic approaches to the determinants of drug response, pathogenesis and infectivity in *Plasmodium falciparum* malaria. In: Wahlgren, M. and Perlmann, P. (eds), *Malaria: molecular and clinical aspects*. Amsterdam: Harwood Academic Publishers, 217–48.

Fidock, D.A., Nomura, A.K., et al. 2000. Mutations in the *Plasmodium falciparum* digestive vacuole transmembrane PfCRT and evidence for their role in chloroquine resistance. *Mol Cell*, **6**, 861–71.

Fitch, C.D., Chevli, R., et al. 1982. Lysis of *Plasmodium falciparum* by ferriprotoporphyrin IX and a chloroquine-ferriprotoporphyrin IX complex. *Antimicrob Agents Chemother*, **21**, 819–22.

Fleming, A.F., Ghatoura, G.B., et al. 1986. The prevention of anaemia in pregnancy in primigravidae in the guinea savanna of Nigeria. *Ann Trop Med Parasitol*, **80**, 211–33.

Flori, L., Kumulungui, B., et al. 2003. Linkage and association between *Plasmodium falciparum* blood infection levels and chromosome 5q31-q33. *Genes Immun*, **4**, 265–8.

Foote, S.J., Thompson, J.K., et al. 1989. Amplification of the multidrug-resistance gene in some chloroquine resistant isolates of *Plasmodium falciparum*. *Cell*, **57**, 921–30.

Foote, S.J., Galatis, D. and Cowman, A.F. 1990. Amino acids in the dihydrofolate reductase-thymidylate synthase gene of *Plasmodium falciparum* involved in cycloguanil resistance differ from those involved in pyrimethamine resistance. *Proc Natl Acad Sci USA*, **87**, 3014–17.

Fortier, B., Delplace, J.F., et al. 1987. Enzyme immunoassay for detection of antigen in acute *Plasmodium falciparum* malaria. *Eur J Clin Microbiol*, **6**, 596–8.

Franzen, L., Westin, G., et al. 1984. Analysis of clinical specimens by hybridization with a probe containing repetitive DNA for *Plasmodium falciparum* malaria. *Lancet*, **i**, 525–7.

French, N., Nakiyingi, J., et al. 2001. Increasing rates of malarial fever with deteriorating immune status in HIV-1-infected Ugandan adults. *AIDS*, **15**, 899–906.

Frevert, U.P., Sinnis, P., et al. 1996. Cell surface glycosaminoglycans are not obligatory for *Plasmodium berghei* invasion *in vitro*. *Mol Biochem Parasitol*, **76**, 257–66.

Fried, M. and Duffy, P.E. 1996. Adherence of *Plasmodium falciparum* to chondroitin sulfate A in the human placenta. *Science*, **272**, 1502–4.

Fritsch, B., Dieckmann, A., et al. 1987. Glutathione and peroxide metabolism in malaria-parasitized erythrocytes. *Parasitol Res*, **73**, 515–17.

Fry, M. 1991. Mitochondria of *Plasmodium*. In: Coombs, G.H. and North, M.J. (eds), *Biochemical protozoology*. London: Taylor & Francis, 154–67.

Fry, M. and Pudney, M. 1992. Site of action of the antimalarial hydroxynaphthoquinone, 2-[trans-4-(4′-chlorophenyl) cyclohexyl]-3-hydroxy-1,4-naphthoquinone (566C80). *Biochem Pharmacol*, **43**, 1545–53.

Galbraith, R.M., Fox, H., et al. 1980. The human materno–foetal relationship in malaria. II. Histological, ultrastructural and immunopathological studies of placenta. *Trans R Soc Trop Med Hyg*, **74**, 61–72.

Gardner, M.J., Hall, N., et al. 2002. Genome sequence of the human malaria parasite *Plasmodium falciparum*. *Nature*, **419**, 498–511.

Garnham, P.C.C. 1966. *Malaria parasites and other Haemosporidia*. Oxford: Blackwell Scientific Publications.

Gatton, M.L., Peters, J.M., et al. 2003. Switching rates of *Plasmodium falciparum var* genes: faster than we thought? *Trends Parasit*, **19**, 2002–208.

Genton, B., Smith, T., et al. 1994. Malaria: how useful are clinical criteria for improving the diagnosis in a highly endemic area. *Trans R Soc Trop Med Hyg*, **88**, 537–41.

Gero, A.M. and O'Sullivan, W.J. 1990. Purines and pyrimidines in malarial parasites. *Blood Cells*, **16**, 467–84.

Gibson, R.S. and Huddle, J.M. 1998. Suboptimal zinc status in pregnant Malawian women: its association with low intakes of poorly available zinc, frequent reproductive cycling, and malaria. *Am J Clin Nutr*, **67**, 702–9.

Gilles, H.M. and Hendrickse, R.G. 1963. Nephrosis in Nigerian children. Role of *P. malariae* and effect of antimalarial treatment. *Br Med J*, **2**, 27–9.

Gilles, H.M. and Ikeme, A.C. 1960. Haemoglobinuria among adult Nigerians due to glucose phosphate dehydrogenease deficiency with drug sensitivity. *Lancet*, **ii**, 889–91.

Gilles, H.M. and Phillips, R.E. 1988. Malaria. In: Gilles, H.M. and Warrell, D.A. (eds), *Medicine International – infections*. Part 4. Oxford: Medical Education (International), 2220–5.

Gilles, H.M., Fletcher, K.A., et al. 1967. Glucose-6-phosphate dehydrogenase deficiency sickling and malaria in African children in South Western Nigeria. *Lancet*, **i**, 138–40.

Gilles, H.M., Lawson, J.B., et al. 1969. Malaria, anaemia and pregnancy. *Ann Trop Med Parasitol*, **63**, 245–63.

Ginsburg, H. and Kirk, K. 1998. Membrane transport in the malaria-infected erythrocyte. In: Sherman, I.W. (ed.), *Malaria: parasite biology, pathogenesis and protection*. Washington DC: ASM Press, 219–32.

Ginsburg, H. and Stein, W.D. 1991. Kinetic modelling of chloroquine uptake by malaria-infected erythrocytes. Assessment of the factors that may determine drug resistance. *Biochem Pharmacol*, **41**, 1463–70.

Ginsburg, H., Krugliak, M., et al. 1985. New permeability pathways induced in membranes of *Plasmodium falciparum*-infected erythrocytes. *Mol Biochem Parasitol*, **8**, 177–90.

Gordeuk, V., Thuma, P., et al. 1992. Effect of iron chelation therapy on recovery from deep coma in children with cerebral malaria. *N Engl J Med*, **327**, 1473–7.

Gordon, D.M., McGovern, T.W., et al. 1995. Safety, immunogenicity, and efficacy of a recombinantly produced *Plasmodium falciparum* circumsporozoite protein-hepatitis B surface antigen subunit vaccine. *J Infect Dis*, **171**, 1576–85.

Grau, G.E. and Behr, C. 1994. T cells and malaria: is Th1 cell activation a prerequisite for pathology? *Res Immunol*, **145**, 441–54.

Grau, G.E., Taylor, T.E., et al. 1989. Tumour necrosis factor and disease severity in children with falciparum malaria. *N Engl J Med*, **320**, 1586–91.

Grau, G.E., Frei, K., et al. 1990. Interleukin-6 production in experimental cerebral malaria. Modulation by anti-cytokine antibodies and possible role in hyper gammaglobulinemia. *J Exp Med*, **172**, 1505–8.

Green, S.J., Mellouk, S., et al. 1990. Cellular mechanisms of nonspecific immunity to intracellular infection: cytokine-induced synthesis of toxic nitrogen oxides from L-arginine by macrophages and hepatocytes. *Immun Lett*, **25**, 15–20.

Green, T.J., Morhardt, M., et al. 1981. Serum inhibition of merozoite dispersal from *Plasmodium falciparum* schizonts; indicator of immune status. *Infect Immun*, **31**, 1203–8.

Greenwood, B.M. 1991. Malaria chemoprophylaxis in endemic regions. In: Targett, G.A.T. (ed.), *Malaria: waiting for the vaccine*. Chichester: John Wiley & Sons, 83–102.

Greenwood, B. 2004. Use of antimalarial drugs to prevent malaria in the population of malaria endemic areas. *Am J Trop Med Hyg*, **70**, 1–7.

Greenwood, B.M. and Alonso, P. 2002. Malaria vaccine trials. *Chem Immunol*, **80**, 366–95.

Greenwood, B.M. and Baker, J.R. 1993. A malaria control trial using insecticide-treated bed nets and targeted chemoprophylaxis in a rural area of the Gambia, West Africa. *Trans R Soc Trop Med Hyg*, **87**, suppl 2, 60.

Greenwood, B.M., Bradley, A.K., et al. 1987. Mortality and morbidity from malaria among children in a rural area of the Gambia, West Africa. *Trans R Soc Trop Med Hyg*, **81**, 478–86.

Greenwood, B.M., Greenwood, A.M., et al. 1989. The effects of malaria chemoprophylaxis given by traditional birth attendants on the course and outcome of pregnancy. *Trans R Soc Trop Med Hyg*, **83**, 589–94.

Greenwood, B.M., Marsh, K. and Snow, R. 1991. Why do some African cildren develop severe malaria? *Parasitol Today*, **7**, 277–81.

Greenwood, A.M., Menendez, C., et al. 1994. The distribution of birthweight in Gambian women who received malaria chemoprophylaxis during their first pregnancy and in control women. *Trans R Soc Trop Med Hyg*, **88**, 311–12.

Grellier, P., Valentin, A., et al. 1994. 3-Hydroxy-methylglutaryl co-enzyme A reductase inhibitors lovastatin and simvastatin inhibit *in vitro* development of *Plasmodium falciparum* and *Babesia divergens* in human erythrocytes. *Antimicrob Agents Chemother*, **38**, 1144–8.

Gupta, S., Snow, R.W., et al. 1999. Immunity to non-cerebral severe malaria is acquired after one or two infections. *Nature Med*, **5**, 340–3.

Gutteridge, W.E. and Trigg, P.I. 1970. Incorporation of radioactive precursors into DNA and RNA of *Plasmodium knowlesi in vitro*. *J Protozool*, **17**, 89–96.

Hadley, T.J. and Miller, L.H. 1988. Invasion of erythrocytes by malaria parasites; erythrocyte ligands and parasite receptors. *Prog Allerg*, **41**, 49–71.

Halloran, M.E., Struchiner, C.J. and Spielman, A. 1989. Modelling malaria vaccines II. Population effects of stage-specific malaria vaccines dependent on natural boosting. *Math Biosci*, **94**, 115–49.

Hänscheid, T. and Grobusch, M.P. 2002. How useful is PCR in the diagnosis of malaria? *Trends Parasitol*, **18**, 395–400.

Hawking, F. 1975. Circadian and other rythms of parasites. *Adv Parasit*, **13**, 123–82.

Hay, S.I., Rogers, D.J., et al. 2002. Hot topic or hot air? Climate change and malaria resurgence in East African highlands. *Trends Parasitol*, **18**, 530–4.

Heggenhougen, H.K., Hackethal, V. and Vivek, P. 2004. The behavioural and social aspect of malaria and its control. An introduction and annotated bibliography. WHO on behalf of the special programme for research and training in tropical diseases, 2003.

Hemmer, C.J., Bierhaus, A., et al. 1994. Elevated thrombomodulin plasma levels as a result of endothelial involvement in *Plasmodium falciparum* malaria. *Thromb Haemostasis*, **72**, 457–64.

Hien, T.T. and White, N.J. 1993. Qinghaosu. *Lancet*, **341**, 603–8.

Hill, A.V.S. and Weatherall, D.J. 1998. Host genetic factors in resistance to malaria. In: Sherman, I.W. (ed.), *Malaria: parasite biology, pathogenesis and protection*. Washington, DC: ASM Press, 445–55.

Hill, A.V.S., Allsopp, C.E.M., et al. 1991. Common West African HLA antigens are associated with protection from severe malaria. *Nature*, **352**, 595–600.

Ho, M. and Sexton, M.M. 1995. Clinical immunology of malaria. In: Pasvol, G. (ed.), *Bailliere's clinical infectious diseases – malaria*. London: Baillière Tindall, 227–24.

Ho, M. and Webster, H.K. 1989. Immunology of human malaria. A cellular perspective. *Parasite Immunol*, **11**, 105–16.

Ho, M., Singh, B., et al. 1991. Clinical correlates of in vitro *Plasmodium falciparum* cytoadherence. *Infect Immun*, **59**, 873–8.

Ho, M., Tongtawe, P., et al. 1994. Polyclonal expansion of peripheral γδ T cells in human *Plasmodium falciparum* malaria. *Infect Immun*, **62**, 855–62.

Ho, M., Sexton, M.M., et al. 1995. Interleukin-10 inhibits tumor necrosis factor production but not antigen-specific lymphoproliferation in acute *Plasmodium falciparum* malaria. *J Infect Dis*, **172**, 838–44.

Hoffman, S.L. 1996. *Malaria vaccine development, A multi-immune response approach*. Washington: ASM Press, 310 pp.

Hoffman, S.L., Isenbarger, D., et al. 1989. Sporozoite vaccine induces genetically restricted T cell elimination of malaria from hepatocytes. *Science*, **244**, 1078–81.

Holding, P.A., Stevenson, J., et al. 1999. Cognitive sequelae of severe malaria with impaired consciousness. *Trans R Soc Trop Med Hyg*, **93**, 529–34.

Holz, G.G. 1977. Lipids and the malaria parasite. *Bull World Health Organ*, **55**, 237–48.

Homewood, C.A. and Neame, K.D. 1980. Biochemistry of malarial parasites. In: Kreier, J.P. (ed.), *Malaria*. vol. 1. New York: Academic Press, 345–405.

Hommel, M. 1985. Antigenic variation in malaria parasites. *Immun Today*, **6**, 28–33.

Hommel, M. 1993. Amplification of cytoadherence in cerebral malaria: towards a more rational explanation of disease pathophysiology. *Ann Trop Med Parasitol*, **87**, 627–35.

Hommel, M. 1997. Modulation of host receptors: a mechanism for the survival of malaria parasites. *Parasitology*, **115**, S45–54.

Hommel, M. 2002. Diagnostic methods in malaria. In: Warrell, D.A. and Gilles, H.M. (eds), *Essential malariology*, 4th edn. London: Arnold, 35–58.

Hommel, M. and Semoff, S. 1988. Expression and function of erythrocyte-associated surface antigens in malaria. *Biol Cell*, **64**, 183–204.

Hommel, M., David, P.H. and Oligino, L.D. 1983. Surface alterations of erythrocytes in *Plasmodium falciparum* malaria. *J Exp Med*, **157**, 1137–48.

Hommel, M., Hughes, M., et al. 1991. Antibody and DNA probes used to analyse variant populations of the Indochina-1 strain of *Plasmodium falciparum*. *Infect Immun*, **59**, 3975–81.

Hughes, M.A., Hommel, M. and Crampton, J.M. 1989. The use of biotin-labelled oligomers for the detection and identification of *Plasmodium falciparum*. *Parasitology*, **100**, 382–7.

Hunt, N.H. and Stocker, R. 1990. Oxidative stress and redox status of malaria-infected erythrocytes. *Blood Cells*, **16**, 499–526.

Jelinek, T., Kilian, A.H.D., et al. 1999. *Plasmodium falciparum* resistance to sulfadoxine/pyrimethamine in Uganda: correlation with polymorphisms in the dihydrofolate reductase and dihydropteroate synthetase genes. *Am J Trop Med Hyg*, **61**, 463–70.

Jelinek, T., Peyerl-Hoffmann, G., et al. 2002. Molecular surveillance of drug resistance through imported isolates of *Plasmodium falciparum* in Europe. *Malaria J*, **1**, 11.

Jensen, J.B., Boland, M.T., et al. 1983. Association between human serum induced crisis forms in cultured *Plasmodium falciparum* and clinical immunity to malaria in Sudan. *Infect Immun*, **41**, 1302–11.

Jing-Bo-Jiang, Xing-Bo-Go et al. 1982. Antimalarial activity of mefloquine and qinghaosu. *Lancet*, **2**, 295–288

Karbwang, J. and Harinasuta, T. 1992. *Chemotherapy of malaria in Southeast Asia*. Bangkok: Ruantasan Co Ltd, 125.

Karunaweera, N.D., Grau, G.E., et al. 1992. Dynamics of fever and serum levels of tumour necrosis factor are closely associated during clinical paroxysms in *Plasmodium vivax* malaria. *Proc Natl Acad Sci USA*, **89**, 3200–3.

Karunaweera, N.D., Wijesekera, S.K., et al. 2003. The paroxysm of *Plasmodium vivax* malaria. *Trends Parasitol*, **19**, 188–93.

Kawamoto, F. 1991. Rapid detection of *Plasmodium* by a new thick smear method using transmission fluorescence microscopy: direct staining with acridine orange. *J Protozool Res*, **1**, 27–34.

Kern, P., Hemmer, C.J., et al. 1989. Elevated tumour necrosis factor alpha and interleukin-6 serum levels as markers for complicated *Plasmodium falciparum* malaria. *Am J Med*, **57**, 139–43.

Khusmith, S., Tharavanij, S., et al. 1987. Two-site immunoradiometric assay for detection of *Plasmodium falciparum* antigen in blood using monoclonal and polyclonal antibodies. *J Clin Microbiol*, **25**, 1467–71.

Knight, J.C., Udalova, I., et al. 1999. A polymorphism that affects OCT-1 binding to the TNF promoter region is associated with severe malaria. *Nat Genet*, **22**, 145–50.

Kroeger, A., Skovmand, O., et al. 2004. Combined field and laboratory evaluation of a long-term impregnated bednet, PermaNet. *Trans R Soc Trop Med Hyg*, **98**, 152–5.

Krishna, S., Waller, D.W., et al. 1994. Lactic acidosis and hypoglycaemia in children with severe malaria: pathophysiological and prognostic significance. *Trans R Soc Trop Med Hyg*, **88**, 67–73.

Krishna, S., Taylor, A.M., et al. 1999. Thiamine deficiency and malaria in adults from southeast Asia. *Lancet*, **353**, 546–9.

Krotoski, W.A., Krotoski, D.M., et al. 1980. Relapses in primate malaria: discovery of two populations of exoerythrocytic stages. Preliminary note. *Br Med J*, **1**, 153–4.

Kumaratilake, L.M. and Ferrante, A. 1994. T-cell cytokines in malaria: their role in the regulation of neutrophil- and macrophage-mediated killing of *Plasmodium falciparum* asexual blood forms. *Res Immunol*, **145**, 423–9.

Kurtzhals, J.A.L., Adabayeri, V., et al. 1998. Low plasma concentrations of interleukin 10 in severe malarial anaemia compared with cerebral and uncomplicated malaria. *Lancet*, **351**, 1768–72.

Kwiatkowski, D. and Greenwood, B.M. 1989. Why is malaria fever periodic? A hypothesis. *Parasitol Today*, **5**, 264–8.

Kwiatkowski, D., Molyneux, M., et al. 1993. Anti-TNF therapy inhibits fever in cerebral malaria. *Quart J Med*, **86**, 91–8.

Lalvani, A., Aidoo, M., et al. 1994. An HLA-based approach to the design of a CTL-inducing vaccine against *Plasmodium falciparum*. *Res Immunol*, **145**, 461–8.

Langhorne, J., Pells, S. and Eichmann, K. 1993. Phenotypic characterization of splenic T cells from mice infected with *Plasmodium chabaudi chabaudi*. *Scand J Immunol*, **38**, 521–8.

Langhorne, J., Quin, S.J. and Sanni, L.A. 2002. Mouse models of blood-stage malaria infections: immune responses and cytokines involved in protection and pathology. *Chem Immunol*, **80**, 97–124.

Langsley, G., Patarapotikul, J., et al. 1988. *Plasmodium vivax*: karyotype polymorphism of field isolates. *Exp Parasitol*, **67**, 301–6.

Laserson, K.F., Petralanda, I., et al. 1994. Use of polymerase chain reaction to directly detect malaria parasites in blood samples from the Venezuelan Amazon. *Am J Trop Med Hyg*, **50**, 169–80.

Lewallen, S., Taylor, T.E., et al. 1993. Ocular fundus filings in Malawian children with cerebral malaria. *Opthalmology*, **100**, 857–61.

Lewallen, S., White, V.A., et al. 2000. Clinical-histopathological correlation of the abnormal retinal vessels in cerebral malaria. *Arch Opthalmol*, **118**, 924–8.

Lindergard, G. 1995. Tools for the evaluation of *Plasmodium malariae* endemicity. MSc dissertation, University of Liverpool.

Ling, I.T. and Wilson, R.J.M. 1988. Glucose-6-phosphate dehydrogenase activity of the malarial parasite *Plasmodium falciparum*. *Mol Biochem Parasitol*, **31**, 47–56.

Lingelbach, K. 1997. Protein trafficking in the *Plasmodium falciparum*-infected erythrocyte – from models to mechanisms. *Ann Trop Med Parasit*, **91**, 543–9.

Lobel, H.O., Miani, M., et al. 1995. Long-term malaria prophylaxis with weekly mefloquine. *Lancet*, **341**, 848–51.

Long, G.W., Fries, L., et al. 1995. Polymerase chain reaction amplification from *Plasmodium falciparum* on dried blood spots. *Am J Trop Med Hyg*, **52**, 344–6.

Looareesuwan, S., Warrell, D.A., et al. 1983. Retinal hemorrhage, a common sign of prognostic significance in cerebral malaria. *Am J Trop Med Hyg*, **32**, 911–15.

Looareesuwan, S., Phillips, R.E., et al. 1985. Quinine and severe falciparum malaria in late pregnancy. *Lancet*, **ii**, 4–8.

Looareesuwan, S., Viravan, C., et al. 1992. Randomised trial of artesunate and mefloquine alone and in sequence for acute uncomplicated falciparum malaria. *Lancet*, **339**, 821–4.

Lucas, S. 1992. Malaria. In: Macsween, R.N.M. and Luhaley, K. (eds), *Muir's textbook of pathology*, 13th edn. London: Edward Arnold, 1144–8.

Luty, A.J.F., Lell, B., et al. 1999. Interferon-gamma responses are associated with resistance to reinfection with *Plasmodium falciparum* in young African children. *J Infect Dis*, **179**, 980–8.

Mackey, L.J., McGregor, I.A., et al. 1982. Diagnosis of *Plasmodium falciparum* infection in man: detection of parasite antigens by ELISA. *Bull World Health Organ*, **60**, 69–75.

MacPherson, G., Warrell, M.J., et al. 1985. Human cerebral malaria: a quantitative ultrastructural analysis of parasitized erythrocyte sequestration. *Am J Pathol*, **119**, 385–401.

Makler, M.T. and Hinrichs, D.J. 1993. Measurement of the lactate dehydrogenase activity of *Plasmodium falciparum* as an assessment of parasitaemia. *Am J Trop Med Hyg*, **48**, 205–10.

Malaguarnera, L., Imbesi, R.M., et al. 2002. Increased levels of interleukin-12 in *Plasmodium falciparum* malaria: correlation with the severity of disease. *Parasite Immunol*, **24**, 387–9.

Manderson, L. 1994. Community participation and malaria control in Southeast Asia: Defining the principles of involvement. *South East Asia J Trop Med Publ Hlth*, **23**, 9–17.

Maneerat, Y., Viriyavejakul, P., et al. 2000. Inducible nitric oxide synthase expression is increased in the brain in fatal cerebral malaria. *Histopathol*, **37**, 269–77.

Marsh, K. and Greenwood, B.M. 1986. The immunopathology of malaria. In: Strickland, G.T. (ed.), *Malaria, clinics in tropical medicine and communicable diseases*, vol. 1. . London: W.B. Saunders, 91–125.

Marsh, K. and Howard, R.J. 1986. Antigens induced on erythrocytes by *Plasmodium falciparum*: expression of diverse and conserved determinants. *Science*, **231**, 150–3.

Marsh, K., Otoo, L., et al. 1989. Antibodies to blood stage antigens of *Plasmodium falciparum* in rural Gambians and their relation to protection against infection. *Trans R Soc Trop Med Hyg*, **83**, 293–303.

Marsh, K., Forster, D., et al. 1995. Indicators of life-threatening malaria in African children. *N Engl J Med*, **332**, 13-99-140, 4.

Marsh, K., English, M., et al. 1996. Clinical algorithm for malaria in Africa. *Lancet*, **347**, 1327–8.

Martin, S.K., Oduola, A.M. and Milhous, W.K. 1987. Reversal of chloroquine resistance in *Plasmodium falciparum* by verapamil. *Science*, **235**, 899–901.

Martiney, J.A., Sherry, B., et al. 2000. Macrophage migration inhibitory factor release by macrophages after ingestion of *Plasmodium chabaudi*-infected erythrocytes: Possible role in the pathogenesis of malarial anemia. *Infect Immun*, **68**, 2259–67.

McGregor, I.A. 1984. Epidemiology, malaria and pregnancy. *Am J Trop Med Hyg*, **33**, 517–25.

McGregor, I.A. and Barr, M. 1962. Antibody response to tetanus toxoid inoculation in malarious and non-malarious Gambian children. *Trans R Soc Trop Med Hyg*, **56**, 364–7.

McGregor, I.A., Gilles, H.M. and Walters, J.H. 1956. Effects of heavy and repeated malarial infections on Gambian infants and children. *Br Med J*, **ii**, 686–92.

McGregor, I.A., Wilson, M.E. and Billewicz, W.Z. 1983. Malaria infection of the placenta in the Gambia. Its incidence and relationship to stillbirth and placental weight. *Trans R Soc Trop Med Hyg*, **77**, 232–44.

McGuire, W., Hill, A.V.S., et al. 1994. Variation in the TNF-alpha promoter region associated with susceptibility to cerebral malaria. *Nature*, **371**, 508–10.

McGuire, W., Hill, A.V.S., et al. 1996. Circulating ICAM-1 levels in falciparum malaria are high but unrelated to disease severity. *Trans R Soc Trop Med Hyg*, **90**, 274–6.

Meek, S.R. 1995. Vector control in some countries of South East Asia: Comparing the vectors and the strategies. *Ann Trop Med Parasitol*, **89**, 135–47.

Mendis, K.N. 1991. Malaria vaccine research – a game of chess. In: Targett, G.A.T. (ed.), *Malaria: waiting for the vaccine*. Chichester: John Wiley & Sons, 183–97.

Menendez, C., Fleming, A.F. and Alonso, P.L. 2000. Malaria-related anaemia. *Parasitol Today*, **16**, 469–76.

Menon, R. 1972. Pregnancy and malaria. *Med J Malaysia*, **27**, 115–19.

Mercereau-Puijalon, O., Fandeur, T., et al. 1991. Parasite features impeding malaria immunity: antigenic diversity, antigenic, variation and poor immunogenicity. *Res Immunol*, **142**, 690–7.

Migot, F., Ouedraogo, J.B., et al. 1996. Selected *Plasmodium falciparum* specific immune responses are maintained in AIDS adults in Burkina Faso. *Parasite Immunol*, **18**, 333–9.

Miller, L.H. 1976. Innate resistance in malaria. *Exp Parasitol*, **40**, 132–46.

Miller, L.H., Mason, S.J., et al. 1979. The resistance factor to *Plasmodium vivax* in blacks. The Duffy blood group genotype. *N Engl J Med*, **295**, 302–4.

Miller, L.H., Baruch, D.I., et al. 2002. The pathogenic basis of malaria. *Nature*, **415**, 673–9.

Miller, M.F. 1958. Observations on the natural history of malaria in semi-resistant West Africans. *Trans R Soc Trop Med Hyg*, **52**, 152–68.

Modiano, D., Sawadogo, A. and Pagnoni, F. 1995. Indicators of life-threatening malaria. *New Engl J Med*, **333**, 1011–11.

Modiano, D., Petrarca, V., et al. 1996. Different response to *Plasmodium falciparum* malaria in West African sympatric ethnic groups. *Proc Natl Acad Sci USA*, **93**, 13206–11.

Molineaux, L. and Grammiccia, G. 1980. *The Garki project*. Geneva: World Health Organisation.

Molyneux, M.E. and Fox, R. 1993. Diagnosis and treatment of malaria in Britain. *Br Med J*, **306**, 1175–80.

Molyneux, M.E., Taylor, T.E., et al. 1989. Clinical features and prognostic indicators in paediatric cerebral malaria: a study of 131 comatose Malawian children. *Quart J Med*, **71**, 441–59.

Moore, J.M., Ayisi, J., et al. 2000. Immunity to placental malaria. II. Placental antigen-specific cytokine responses are impaired in human immunodeficiency virus-infected women. *J Infect Dis*, **182**, 960–4.

Morgan, H.G. 1994. Placental malaria and low birthweight neonates in urban Sierra Leone. *Ann Trop Med Parasit*, **88**, 575–80.

Mota, M.M., Hafalla, J.C.R. and Rodriguez, A. 2002. Migration through host cells activates *Plasmodium* sporozoites for infection. *Nature Med*, **8**, 1318–22.

Müller, O., Becher, H., et al. 2001. Effect of zinc supplementation on malaria and other causes of morbidity in West African children: randomized, double bind placebo controlled trial. *Br Med J*, **322**, 1–6.

Murray, M.J., Murray, A.B., et al. 1978a. The adverse effect of iron repletion on the course of certain infections. *Br Med J*, **ii**, 1113–15.

Murray, M.J., Murray, A.B., et al. 1978b. Diet and cerebral malaria: the effect of famine and refeeding. *Am J Clin Nutr*, **31**, 57–61.

Mutabingwa, T.K. 2001. Monitoring antimalarial drug resistance within National Malaria Control Programmes: the EANMAT experience. *Trop Med Int Health*, **6**, 891–8.

Nagel, R.L. 1990. Innate resistance to malaria: the intraerythrocytic cycle. *Blood Cells*, **16**, 321–39.

Najera, J., Liese, B. and Hamer, J.S. 1991. Malaria. In: Jameson, D.T. and Mosley, W.H. (eds), *Disease control priorities in developing countries*. Oxford: Oxford University Press for the World Bank, 200–10.

Naotunne, T.D., Karunaweera, N.D., et al. 1991. Cytokines kill malaria parasites during infection crisis – extracellular complementary factors are essential. *J Exp Med*, **173**, 523–9.

Newton, C.R. and Krishna, S. 1998. Severe malaria in children: current understanding of pathophysiology and supportive treatment. *Pharmacol Ther*, **79**, 1–53.

Newton, C.R., Kirkham, F.J., et al. 1991. Intracranial pressure in African children with cerebral malaria. *Lancet*, **337**, 573–6.

Newton, C.R., Peshu, N., et al. 1994. Brain swelling and ischaemia in Kenyans with cerebral malaria. *Arch Dis Child*, **70**, 281–7.

Newton, C.R., Warn, P.A., et al. 1997. Severe anaemia in children living in a malaria endemic area of Kenya. *Trop Med Int Health*, **2**, 165–78.

Nguyen-Dinh, P., Greenberg, A.E., et al. 1987. Absence of association between *Plasmodium falciparum* malaria and human immunodeficiency virus infection in children in Kinshasa, Zaire. *Bull World Health Organ*, **65**, 607–13.

Nguyen, P.H., Day, N., et al. 1995. Intraleucocytic malaria pigment and prognosis in severe malaria. *Trans R Soc Trop Med Hyg*, **89**, 200–4.

Nierengarten, M.B. 2003. Malariotherapy to treat HIV patients? *Lancet Inf Dis*, **3**, 321.

Nosten, F., McGready, R., et al. 2003. Editorial: Maternal malaria: time for action. *Trop Med Int Health*, **8**, 485–7.

Okoye, V.C. and Bennett, V. 1985. *Plasmodium falciparum* malaria: band 3 as a possible receptor during invasion of human erythrocytes. *Science*, **227**, 169–71.

Oliveira, D.A., Holloway, B.P., et al. 1995. Polymerase chain reaction and a liquid-phase, nonisotopic hybridization for species-specific and sensitive detection of malaria infection. *Am J Trop Med Hyg*, **52**, 139–44.

Omer, F.M., Kurtzhals, J.A. and Riley, E.M. 2000. Maintaining the immunological balance in parasitic infections: a role for TGF-beta? *Parasitol Today*, **16**, 18–23.

Palmer, C.J., Lindo, J.F., et al. 1998. Evaluation of the OptiMal test for rapid diagnosis of *Plasmodium falciparum* and *Plasmodium vivax*. *J Clin Microbiol*, **36**, 203–6.

Parise, M.R., Lewis, L.S., et al. 2003. A rapid assessment approach for public health decision-making related to the prevention of malaria during pregnancy. *Bull World Health Organ*, **81**, 315–22.

Pasvol, G. and Wilson, R.J.M. 1982. The interaction of malaria parasites with red blood cells. *Br Med Bull*, **38**, 133–40.

Pasvol, G., Wainscoat, J.S. and Weatherall, D.J. 1982. Erythrocytes deficiency in glycophorin resist invasion by the malarial parasite *Plasmodium falciparum*. *Nature*, **297**, 64–6.

Paul, R.E.L., Packer, M.J., et al. 1995. Mating patterns in malaria parasite populations of Papua New Guinea. *Science*, **269**, 1709–11.

Paul, R.E.L., Brey, P. and Robert, V. 2002. Plasmodium sex determination and transmission to mosquitoes. *Trends Parasitol*, **18**, 32–8.

Perez, H.A., Wide, A., et al. 1995. *Plasmodium vivax*: detection of blood parasites using fluorochrome labelled monoclonal antibodies. *Parasite Immunol*, **17**, 305–12.

Perkins, M.E. and Holt, E.H. 1988. Erythrocyte receptor varies in *Plasmodium falciparum* isolates. *Mol Biochem Parasitol*, **27**, 23–34.

Perlmann, H., Berzins, K., et al. 1984. Antibodies in malaria sera to parasite antigens in the membrane of erythrocytes infected with early asexual stages of *Plasmodium falciparum*. *J Exp Med*, **159**, 1686–704.

Petersen, E. and Marbiah, N.T. 1994. QBC[®] and thick fims for malaria diagnosis under field conditions. *Trans R Soc Trop Med Hyg*, **88**, 416–17.

Phillips, R.E., Looareesuwan, S., et al. 1986. The importance of anaemia in cerebral and uncomplicated falciparum malaria: role of complications, dyserythropoiesis and iron sequestration. *Q J Med*, **58**, 305–23.

Pimenta, O.F., Touray, M. and Miller, L. 1994. The journey of malaria sporozoites in the mosquito salivary gland. *J Eukaryot Microbiol*, **41**, 608–24.

Piper, R., Lebras, J., et al. 1999. Immunocapture diagnostic assays for malaria using *Plasmodium* lactate dehydrogenase (pLDH). *Am J Trop Med Hyg*, **60**, 109–18.

Phillips, R.S. 1994. Malaria vaccines – a problem solved or simply a promising start? *Protozool Abstr*, **18**, 459–86.

Playfair, J.H.L., Taverne, J., et al. 1990. The malaria vaccine: anti-parasite or anti-disease? *Immun Today*, **11**, 25–7.

Plowe, C.V. 2001. Folate antagonists and mechanisms of resistance. In: Rosenthal, P.J. (ed.), *Antimalarial chemotherapy: mechanisms of action, resistance and new directions in drug discovery*. Toronto: Humana Press Inc, 173–90.

Pradel, G., Garapaty, S. and Frevert, U. 2002. Proteoglycans mediate malaria sporozoite targeting to the liver. *Mol Microbiol*, **45**, 637–51.

Premij, Z., Minjas, J.N. and Shiff, C.J. 1994. Laboratory diagnosis of malaria by village health-workers using the rapid manual ParaSight[®]-F test. *Trans R Soc Trop Med Hyg*, **88**, 418.

Qari, S.H., Shi, Y.P., et al. 1993. Identification of *Plasmodium vivax*-like human malaria parasites. *Lancet*, **341**, 780–3.

Rasheed, F.N., Bulmer, J.N., et al. 1993. Suppressed peripheral and placental blood lymphoproliferative responses in 1st pregnancies – relevance to malaria. *Am J Trop Med Hyg*, **48**, 154–60.

Raventos-Suarez, C., Pollack, S. and Nagel, R.L. 1982. *Plasmodium falciparum*: inhibition of *in vitro* growth by desferrioxamine. *Am J Trop Med Hyg*, **31**, 919–22.

Reyes, P., Rathod, P., et al. 1982. Enzymes of purine and pyrimidine metabolism from the human malaria parasite *Plasmodium falciparum*. *Mol Biochem Parasitol*, **5**, 275–90.

Richie, T.L. and Saul, A. 2002. Progress and challenges for malaria vaccines. *Nature*, **415**, 694–701.

Riley, E.M. 1999. Is T-cell priming required for initiation of pathology in malaria infections? *Immunol Today*, **20**, 228–33.

Riley, E.M., Allen, S.J., et al. 1992. Naturally acquired cellular and humoral immune responses to the major merozoite surface antigens (Pf MSP-1) of *Plasmodium falciparum* are associated with reduced malaria morbidity. *Parasite Immunol*, **14**, 321–37.

Robert, V., Awono-Ambene, H.P., et al. 2000. Gametocytemia and infectivity to mosquitoes of patients with uncomplicated *Plasmodium falciparum* malaria attacks treated with chloroquine or sulfadoxine plus pyrimethamine. *Am J Trop Med Hyg*, **62**, 210–16.

Roberts, D.J., Craig, A.G., et al. 1992. Rapid switching to multiple antigenic and adhesive phenotypes in malaria. *Nature*, **357**, 689–91.

Roberts, D.W. and Weidanz, W.P. 1979. T-cell immunity to malaria in the B-cell deficient mouse. *Am J Trop Med Hyg*, **28**, 1–3.

Rockett, K.A., Awburn, M.M., et al. 1994. Tumor necrosis factor and interleukin-1 synergy in the context of malaria pathology. *Am J Trop Med Hyg*, **50**, 735–42.

Rosenberg, Y.J. 1978. Autoimmune and polyclonal B cell responses during murine malaria. *Nature*, **274**, 170–2.

Rosenberg, R., Wirtz, R.A., et al. 1990. An estimation of the number of malaria sporozoites ejected by a feeding mosquito. *Trans R Soc Trop Med Hyg*, **84**, 209–12.

Rosenthal, P.J. and Meshnick, S.R. 1998. Hemoglobin processing and the metabolism of amino acids, heme and iron. In: Sherman, I.W. (ed.), *Malaria: parasite biology, pathogenesis and protection*. Washington DC: ASM Press, 145–58.

Roth, E. 1990. Plasmodium falciparum carbohydrate metabolism: a connection between host cell and parasite. *Blood Cells*, **16**, 453–60.

Roth, E.F., Raventos-Suarez, C., et al. 1982. Glutathione stability and oxidative stress in *Plasmodium falciparum* infection in vitro: response of normal and G6PD deficient cells. *Biochem Biophys Res Commun*, **109**, 355–62.

Rougemont, A., Boisson, M.E., et al. 1977. Paludisme et anémie de la grossesse en zone de savane africaine. *Bull Soc Pathol Exot*, **70**, 265–73.

Rowland-Jones, S.L. and Lohman, B. 2002. Interactions between malaria and HIV infection – an emerging public health problem? *Microbes Infect*, **4**, 1265–70.

Ruiz, W. and Kroeger, A. 1994. The socioeconomic impact of malaria in Colombia and Ecuador. *Health Policy Plan*, **9**, 144–54.

Ruwende, C., Khoo, S.C., et al. 1995. Natural selection of hemi- and heterozygotes for glucose 6-phosphate dehydrogenase deficiency in Africa by resistance to severe malaria. *Nature*, **376**, 246–9.

Sanchez-Lopez, R. and Haldar, K. 1992. A transferrin-independent iron uptake activity in *Plasmodium falciparum*-infected and uninfected erythrocytes. *Mol Biochem Parasitol*, **55**, 9–20.

Sanni, L.A., Thomas, S.R., et al. 1998. Dramatic changes in oxidative tryptophan metabolism along the kynurenine pathway in experimental cerebral and noncerebral malaria. *Am J Pathol*, **152**, 611–19.

Saul, A. 1992. Towards a malaria vaccine: riding the rollercoaster between unrealistic optimism and lethal pessimism. *South East Asian J Trop Med Publ Hlth*, **23**, 656–71.

Scheibel, L.W. 1988. Plasmodial metabolism and related organellar function during various stages of the life-cycle: carbohydrates. In: Wernsdorfer, W.H. and McGregor, I. (eds), *Malaria. Principles and practice of malariology*. vol. 1. Edinburgh: Churchill Livingstone, 171–217.

Scheibel, L.W., Ashton, S.H. and Trager, W. 1979. *Plasmodium falciparum*: microaerophilic requirements in human red cells. *Exp Parasitol*, **47**, 410–18.

Scheibel, L.W., Colombani, P.M., et al. 1987. Calcium and calmodulin antagonists inhibit human malaria parasites (*Plasmodium falciparum*): implications for drug design. *Proc Natl Acad Sci USA*, **84**, 7310–14.

Schellenberg, J.R.M.A., Smith, T., et al. 2001. What is clinical malaria? Finding case definitions for field-research in highly endemic areas. *Parasitol Today*, **10**, 439–42.

Schofield, L. 2002. Antidisease vaccines. *Chem Immunol*, **80**, 322–42.

Schofield, L., Vivas, L., et al. 1993. Neutralizing monoclonal antibodies to glycosylphosphatidylinositol, the dominant TNF-a inducing toxin of *Plasmodium falciparum*: prospects for the immunotherapy of severe malaria. *Ann Trop Med Parasitol*, **87**, 617–26.

Schofield, L., Novakovic, S., et al. 1996. Glycosylphosphatidylinositol toxin of *Plasmodium* up-regulates intercellular adhesion molecule-1, vascular adhesion molecule-1 and E-selectin expression in vascular endothelial cells and increases leukocyte and parasite cytoadherence via tyrosine kinase-dependent signal transduction. *J Immunol*, **156**, 1886–96.

Schofield, L., Hewitt, M.C., et al. 2002. Synthetic GPI as a candidate anti-toxic vaccine in a model of malaria. *Nature*, **418**, 785–9.

Schrével, J., Deguercy, A., et al. 1990. Proteases in malaria-infected red blood cells. *Blood Cells*, **16**, 563–84.

Schultz, L.J., Steketee, R.W., et al. 1995. Antimalarials during pregnancy: a cost effectiveness analysis. *Bull World Health Organ*, **73**, 207–14.

Schwarzer, E., Alessio, M., et al. 1998. Phagocytosis of the malarial pigment, hemozoin, impairs expression of major histocompatibility complex class II antigen, CD54, and CD11c in human monocytes. *Infect Immun*, **66**, 1601–6.

Sergent, E., Parrot, L. and Donatien, A. 1924. Une question de terminologie: immuniser et prémunir. *Bull Word Health Organ*, **17**, 37–8.

Service, M. and Townson, H. 2002. The Anopheles vector. In: Warrell, D.A. and Gilles, H.M. (eds), *Essential malariology*, 4th edn. London: Arnold, 59–84.

Shankar, A.H. 2000. Nutritional modulation of malaria morbidity and mortality. *J Infect Dis*, **182**, Suppl, S37–53.

Shankar, A.H., Genton, B., et al. 1999. Effect of vitamin A supplementation on morbidity due to *Plasmodium falciparum* in young children in Papua New Guinea: a randomised trial. *Lancet*, **354**, 203–9.

Shear, H.L., Nussenzweig, R.S. and Bianco, C. 1979. Immune phagocytosis in murine malaria. *J Exp Med*, **149**, 1288–93.

Sherman, I.W. 1979. Biochemistry of malarial parasites. *Microbiol Rev*, **43**, 453–95.

Sherman, I.W. 1988. Mechanisms of molecular trafficking in malaria. *Parasitology*, **96**, 857–81.

Sherman, I.W., Eda, S. and Winograd, E. 2003. Cytoadherence and sequestration in *Plasmodium falciparum*: defining the ties that bind. *Microbes Infect*, **5**, 897–909.

Sherry, B.A., Alava, G., et al. 1995. Malaria-specific metabolite hemozoin mediates the release of several potent endogenous pyrogens (TNF, MIP-1α and MIP-1β) in vitro and alters thermoregulation in vivo. *J Inflamm*, **45**, 85–96.

Shiff, C.J., Premij, Z. and Minjas, J.N. 1993. The rapid ParaSight™-F test. A new diagnostic tool for *Plasmodium falciparum* infection. *Trans R Soc Trop Med Hyg*, **87**, 29–31.

Shulman, C. and Dorman, E. 2002. Clinical features of malaria in pregnancy. In: Warrell, D.A. and Gilles, H.M. (eds), *Essential malariology*. London: Arnold, 217–35.

Silamut, K. and White, N.J. 1993. Relation of the stage of parasite development in the peripheral blood to prognosis in severe falciparum malaria. *Trans R Soc Trop Med Hyg*, **87**, 436–43.

Simooya, O.O., Mwendapole, R.M., et al. 1988. Relation between falciparum-malaria and HIV seropositivity in Ndola, Zambia. *Br Med J*, **297**, 30–1.

Singh, B., Ho, M., et al. 1988. *Plasmodium falciparum*: Inhibition/reversal of cytoadherence of Thai isolates to melanoma cells by local immune sera. *Clin Exp Immunol*, **72**, 145–50.

Smalley, M.E., Abdalla, S. and Brown, J. 1980. The distribution of *Plasmodium falciparum* in the peripheral blood and bone marrow of Gambian children. *Trans R Soc Trop Med Hyg*, **75**, 103–5.

Smith, J.D., Chitnis, C.E., et al. 1995. Switches in expression of *Plasmodium falciparum var* genes correlate with changes in antigenic and cytoadherent phenotypes of infected erythrocytes. *Cell*, **82**, 101–10.

Smith, T., Charlwood, J.D., et al. 1993. Absence of seasonal variation in malarial parasitaemia in an area of intense seasonal transmission. *Acta Trop*, **54**, 55–72.

Smith, T.G., Lourenco, P., et al. 2000. Commitment to sexual differentiation in the human malaria parasite, *Plasmodium falciparum*. *Parasitology*, **121**, 127–33.

Snounou, G., Viriyakosol, S., et al. 1993. Identification of the four human malaria parasite species in field samples by the polymerase chain reaction and detection of a high prevalence of mixed infections. *Mol Biochem Parasitol*, **58**, 283–92.

Snow, R.W. and Gilles, H.M. 2002. The epidemiology of malaria. In: Warrell, D.A. and Gilles, H.M. (eds), *Essential malariology*, 4th edn. London: Arnold, 85–106.

Snow, R.W., Craig, M., et al. 1999. Estimating mortality, morbidity and disability due to malaria among Africa's non-pregnant population. *Bull World Health Organ*, **77**, 624–40.

Snow, R.W., Newton, C.R.J.C., et al. 2003. The public health burden of *Plasmodium falciparum* malaria in Africa: deriving the numbers. Disease Control Priorities Project Working Paper No. 111.

Sokhna, C.S., Trape, J.F. and Robert, V. 2001. Gametocytes in Senegalese children with uncomplicated falciparum malaria treated with chloroquine, amodiaquine or sulfadoxine plus pyrimethamine. *Parasite*, **8**, 243–50.

Somboon, P., Lines, J., et al. 1995. Entomological evaluation of community-wide use of lambdacyhalothian-impregnated bed nets against malaria in a border area of north west Thailand. *Trans R Soc Trop Med Hyg*, **89**, 248–54.

Spielman, A. and Perrone, J.B. 1989. Rapid diagnosis of malaria. *Lancet*, **1**, 727.

Spitalny, G.L. and Nussenzweig, R.S. 1973. Plasmodium berghei: relationship between protective immunity and antisporozoite (CSP) antibody in mice. *Exp Parasitol*, **33**, 168–78.

Steffen, R., Fuchs, E., et al. 1993. Mefloquine compared with other chemoprophylactic regimens in tourists visiting East Africa. *Lancet*, **341**, 1299–303.

Steketee, R.W., Wirima, J.J., et al. 1996a. The effect of malaria and malaria prevention in pregnancy on offspring birthweight, prematurity and intrauterine growth retardation in rural Malawi. *Am J Trop Med Hyg*, **55**, Suppl 1, 33–41.

Steketee, R.W., Wirima, J.J., et al. 1996b. Impairment of a pregnant woman's acquired ability to limit *Plasmodium falciparum* by infection with human immunodeficiency virus type-1. *Am J Trop Med Hyg*, **55**, Suppl., 42–9.

Su, X.Z., Heatwole, V.M., et al. 1995. The large diverse gene family *var* encodes proteins involved in cytoadherence and antigenic variation of *Plasmodium falciparum*-infected erythrocytes. *Cell*, **82**, 89–100.

Suhrbier, A. 1991. Immunity to the liver stage of malaria. *Parasitol Today*, **7**, 160–3.

Taliaferro, W.H. and Cannon, P.R. 1936. The cellular reactions during primary infections and superinfections of *Plasmodium brasilianum* in Panamian monkeys. *J Infect Dis*, **59**, 72–125.

Tanabe, K. 1990. Ion metabolism in malaria-infected erythrocytes. *Blood Cells*, **16**, 437–49.

Tanios, M.A., Kogelman, L., et al. 2001. Acute respiratory distress syndrome complicating *Plasmodium vivax* malaria. *Crit Care Med*, **29**, 665–7.

Taramelli, D., Recalcati, S., et al. 2000. Macrophage preconditioning with synthetic malaria pigment reduces cytokine production via heme iron-dependent oxidative stress. *Lab Invest*, **80**, 1781–8.

Taverne, J., Bate, C.A. and Playfair, J.H. 1990. Malaria exoantigens induce TNF are toxic and are blocked by T-independent antibody. *Immunol Lett*, **25**, 207–12.

Taylor, T.E. and Molyneux, M.E. 2002. Clinical features of malaria in children. In: Warrell, D.A. and Gilles, H.M. (eds), *Essential malariology*, 4th edn. London: Arnold, 206–18.

Taylor, D.W. and Voller, A. 1993. The development and validation of a simple antigen detection ELISA for *Plasmodium falciparum* malaria. *Trans R Soc Trop Med Hyg*, **87**, 29–31.

Taylor, K., Bate, C.A.W., et al. 1992. Phospholipid containing toxic malaria antigens induce hypoglycaemia. *Clin Exp Immunol*, **90**, 1–5.

Taylor, T.E., Molyneux, M.E., et al. 1988. Blood-glucose levels in Malawian children before and during the administration of intravenous quinine for severe falciparum-malaria. *New Engl J Med*, **319**, 1040–7.

Taylor, T.E., Borgstein, A. and Molyneux, M.E. 1993. Acid-base status in paediatric *Plasmodium falciparum* malaria. *Quart J Med*, **86**, 99–109.

Taylor-Robinson, A.W., Phillips, R.S., et al. 1993. The role of Th1 and Th2 cells in a rodent malaria infection. *Science*, **260**, 1931–4.

Tirasophon, W., Ponglikitmongkol, M., et al. 1991. A novel detection of a single *Plasmodium falciparum* in infected blood. *Biochem Biophys Res Comm*, **175**, 179–84.

Torre, D., Giola, M., et al. 2001. Serum levels of interleukin-18 in patients with uncomplicated *Plasmodium falciparum* malaria. *Eur Cytokine Netw*, **2**, 361–4.

Trager, W. and Jensen, J.B. 1976. Human malaria parasites in continuous culture. *Science*, **193**, 125–9.

Trager, W. and Williams, J. 1992. Extracellular (axenic) development in vitro of the erythrocytic cycle of *Plasmodium falciparum*. *Proc Natl Acad Sci USA*, **89**, 5351–5.

Trape, J.F., Rogier, C., et al. 1994. The Dielmo project – a longitudinal study of natural malaria infection and the mechanisms of protective immunity in a community living in a holoendemic area of Senegal. *Am J Trop Med Hyg*, **51**, 123–37.

Troye-Blomberg, M., Riley, E.M., et al. 1990. Production by activated T cells of interleukin 4 but not interferon gamma is associated with elevated levels of serum antibodies to activating malaria antigens. *Proc Natl Acad Sci USA*, **87**, 5484–8.

Troye-Blomberg, M., Worku, S., et al. 1999. Human gamma delta T cells that inhibit the in vitro growth of the asexual blood stages of the *Plasmodium falciparum* parasite express cytolytic proinflammatory molecules. *Scand J Immunol*, **50**, 642–50.

Turrini, F., Giribaldi, G., et al. 1997. Mycoplasma contamination of *Plasmodium* cultures – a case of parasite parasitism. *Parasitol Today*, **13**, 367–8.

Ulloa, L., Ochani, M., et al. 2002. Ethyl pyruvate prevents lethality in mice with established lethal sepsis and systemic inflammation. *Proc Natl Acad Sci USA*, **99**, 12351–6.

Urban, B.C. and Roberts, D.J. 2002. Malaria, monocytes, macrophages and myeloid dendritic cells: sticking of infected erythrocytes switches off host cells. *Curr Opinion Immunol*, **14**, 458–65.

Ustianowski, A., Schwab, U. and Pasvol, G. 2004. Case report: severe acute symptomatic hyponatraemia in falciparum malaria. *Trans R Soc Trop Med Hyg*, **96**, 647–8.

Vadas, P., Taylor, T.E., et al. 1993. Increased serum phospholipase A2 activity in Malawian children with Falciparum malaria. *Am J Trop Med Hyg*, **49**, 455–9.

Van Eijk, A.M., Ayisi, J.G., et al. 2003. HIV increases the risk of malaria in women of all gravidities in Kisumu, Kenya. *AIDS*, **17**, 595–603.

Verhave, J.P. and Meis, J.F.G. 1984. The biology of tissue forms and other sexual stages in mammalian plasmodia. *Experientia*, **40**, 1317–29.

Verhoeff, F.H., Brabin, B.J., et al. 1999. Increased prevalence of malaria in HIV-infected pregnant women and its implications for malaria control. *Trop Med Int Health*, **4**, 5–12.

Vial, H.J., Ancelin, M.L., et al. 1990. Biosynthesis and dynamics of lipids in *Plasmodium falciparum*-infected mature mammalian erythrocytes. *Blood Cells*, **16**, 531–55.

Voller, A. 1988. The immunodiagnosis of malaria. In: Wernsdorfer, W.H. and McGregor, I. (eds), *Malaria. Principles and practice of malariology*. vol. 1. Edinburgh: Churchill Livingstone, 815–25.

von Seidlein, L., Clarke, S., et al. 2002. Treatment uptake by individuals infected with *Plasmodium falciparum* in rural Gambia, West Africa. *Bull Wld Hlth Organ*, **80**, 790–6.

Vu thi Ty Hang, Tran Van Be, et al. 1995. Screening donor blood for malaria by polymerase chain reaction. *Trans R Soc Trop Med Hyg*, **89**, 44–7.

Wahlgren, M., Carlson, J. and Udomsangpetch, R. 1987. Why do *Plasmodium falciparum*-infected erythrocytes form spontaneous rosettes? *Parasitol Today*, **5**, 183–5.

Waller, D., Krishna, S., et al. 1995. Clinical-features and outcome of severe malaria in Gambian children. *Clin Infect Dis*, **21**, 577–87.

Walliker, D., Quakyi, I.A., et al. 1987. Genetic analysis of the human malaria parasite *Plasmodium falciparum*. *Science*, **236**, 1661–6.

Warrell, D.A. 2002. Clinical features of malaria. In: Warrell, D.A. and Gilles, H.M. (eds), *Essential malariology*, 4th edn. London: Arnold, 191–205.

Warrell, D.A., Looareesuwan, S., et al. 1986. Function of the blood–cerebrospinal fluid barrier in human cerebral malaria – rejection of the permeability hypothesis. *Am J Trop Med Hyg*, **35**, 882–9.

Warrell, D.A., Turner, G.D.H. and Francis, N. 2002. Pathology and pathophysiology of human malaria. In: Warrell, D.A. and Gilles, H.M. (eds), *Essential malariology*, 4th edn. London: Arnold, 236–51.

Watkins et al., 2005. The search for effective and sustainable treatments for Plasmodium falciparum malaria in Africa: a model of the selection of resistance by antifolate drugs and their combinations. *Am J Trop Med Hyg*, **72**, in press.

Weber, J.L. 1988. Molecular biology of malaria parasites. *Exp Parasitol*, **66**, 143–70.

Weidanz, W.P. and Long, C.A. 1988. The role of T-cells in immunity to malaria. In: Perlmann, P. and Wigzell, H. (eds), *Malaria immunology*. Basel: Karger, 215–22.

Wellems, T.E., Panton, L.J., et al. 1990. Chloroquine resistance not linked to mdr-like genes in a *Plasmodium falciparum* cross. *Science*, **345**, 253–8.

Wernsdorfer, W.H. and Payne, D. 1988. Drug sensitivity tests in malaria parasites. In: Wernsdorfer, W.H. and McGregor, I. (eds), *Malaria. Principles and practice of malariology*. vol. 2. Edinburgh: Churchill Livingstone, 1765–800.

Wertheimer, S.P. and Barnwell, J.W. 1989. Plasmodium vivax interaction with the human blood group glycoprotein: identification of a parasite receptor-like protein. *Exp Parasitol*, **69**, 340–50.

White, N.J. 2002. The assessment of antimalarial drug efficacy. *Trends Parasitol*, **18**, 458–64.

White, N.J. 2003a. The management of severe falciparum malaria. *Am J Resp Crit Care*, **167**, 673–4.

White, N.J. 2003b. Malaria. In: Cook, G. and Zumla, A. (eds), *Manson's tropical diseases*, 21st edn. London: Elsevier Saunders, 1205–95.

White, N.J. and Olliaro, P.L. 1996. Strategies for the prevention of antimalarial drug resistance: rationale for combination chemotherapy for malaria. *Parasitol Today*, **12**, 399–401.

White, N.J., Warrell, D.A., et al. 1983. Severe hypoglycemia and hyperinsulinemia in falciparum-malaria. *New Engl J Med*, **309**, 61–6.

White, N.J., Miller, K.D., et al. 1987. Hypoglycaemia in African children with severe malaria. *Lancet*, **i**, 708–11.

Whitworth, J., Morgen, D., et al. 2000. Effect of HIV-1 and increasing immunosuppression on malaria parasitaemia and clinical episodes in adults in rural Uganda: a cohort study. *Lancet*, **356**, 1051–6.

Wichmann, O., Jelinek, T., et al. 2003. Molecular surveillance of ntifolate-resistant mutation I164L in imported African isolates of

Plasmodium falciparum in Europe: sentinel data from Trop Net Europ. *Malaria J*, **2**, 17.

Wickramasuriya, G.W.A. 1935. Some observations on malaria occurring in association with pregnancy. *J Obs Gynaceol Br Empire*, 816–833

Wilkinson, R.J., Brown, J.L., et al. 1994. Severe falciparum malaria: Predicting the effect of exchange transfusion. *Quart J Med*, **87**, 553–5.

Willcox, M.C. 1975. Thalassaemia in Northern Liberia: a survey in the Mount Nimba area. *J Med Genetics*, **12**, 55–63.

Williams, T.N., Maitland, K., et al. 1996. High incidence of malaria in alpha-thalassaemic children. *Nature*, **383**, 522–5.

Wilson, R.J.M., McGregor, I.A. and Williams, K. 1975. Occurrence of S-antigens in serum in *Plasmodium falciparum* infections in man. *Trans R Soc Trop Med Hyg*, **69**, 453–9.

Wilson, R.J.M., Denny, P.W., et al. 1996. Complete gene map of the plastid-like DNA of the malaria parasite *Plasmodium falciparum*. *J Mol Biol*, **261**, 155–72.

Winograd, E. and Sherman, I.W. 1989. Characterization of a modified red cell membrane protein expressed on erythrocytes infected with the human malaria parasite *Plasmodium falciparum*: possible role as a cytoadherence mediating protein. *J Cell Biol*, **108**, 23–30.

Winstanley, P. 2001. Chlorproguanil-dapsone (LAPDAP) for uncomplicated falciparum malaria. *Trop Med Int Heath*, **6**, 952–4.

World Health Organization. 1994a. *Antimalarial drug policies*. Report of an informal consultation, WHO/Mal/94.1070, Geneva, 14–18 March 1994, pp. 67

World Health Organization. 1994b. *World malaria situation in 1992*. Wkly Epidemic Rec. No. 42, 309–14: No. 43, 317–21: No. 44, 325–30.

World Health Organization. 1995. *Vector control for malaria and other mosquito-borne diseases*. WHO Technical Report Series 857, 99 pp.

World Health Organization, 2000. *Management of severe malaria. A practical handbook*, 2nd edn. Geneva: WHO, 84 pp.

World Health Organization. 2001. *The use of antimalarial drugs*. Report of a WHO informal consultation, WHO/CDS/RBM/2001.33.

World Health Organization. 2004. Scaling up home-based management of malaria from research to implementation. WHO/HTM/MAL/2004.1096. TDR/IDE/HMM/04.1 Roll-back Malaria Department/Unicef/UNDP/World Bank. WHO Special Programme for Research and Training in Tropical Diseases.

Wozencraft, A.O., Dockrell, H.M., et al. 1984. Killing of human malaria parasites by macrophage secretory products. *Infect Immun*, **43**, 664–9.

Wünsch, S., Sanchez, C.P., et al. 1998. Differential stimulation of the Na+/H+ exchanger determines chloroquine uptake in *Plasmodium falciparum*. *Cell*, **140**, 335–45.

Yuthavong, Y., Wilairat, P., et al. 1979. Alterations in membrane proteins of mouse erythrocytes infected with different species and strains of malaria parasites. *Comp Biochem Physiol*, **63B**, 83–5.

Zakeri, S., Taylor, K., et al. 2000. Polar *Plasmodium falciparum* lipids induce lipogenesis in rat adipocytes in vitro. *Microbes Infect*, **2**, 1789–98.

Zavala, F., Tam, J.P. and Masuda, A. 1986. Synthetic peptides as antigens for the detection of humoral immunity to *Plasmodium falciparum* sporozoites. *J Immunol Methods*, **93**, 55–61.

Zuckerman, A. 1977. Current status of the immunology of blood and tissue protozoa. II. Plasmodium. *Exp Parasitol*, **42**, 374–446.

Microsporidiosis

ALAN CURRY

INTRODUCTION

Microsporidians are all obligate intracellular parasites with a unique mode of entering host cells via a polar tube within a spore. Microsporidians are among the most successful and widespread groups of intracellular parasites in animals (Sprague and Vávra 1977; Canning and Lom 1986; Weidner 1991), infecting other protozoa (Cali 1991; Foissner and Foissner 1995), bryozoans, arthropods, fish, amphibians, reptiles, birds, and mammals. Some are hyperparasites (Hussey 1971). Microsporidians are eukaryotes of ancient origin (Canning and Lom 1986; Vossbrinck et al. 1987); about 100 genera and about 1 000 species are currently recognized worldwide (Sprague et al. 1992). Microsporidians have had a chequered taxonomic history and have nearly always been classified with or alongside the protozoa. At one time, they were classified with the Sporozoa, later they were grouped with the myxosporidians as cnidosporidians and, more recently, elevated to the status of phylum of their own, the Microspora. The Microspora are now thought to be more closely related to fungi than to protozoa (Cavalier-Smith 2001; Hirt et al. 1999), but are retained in this volume because they are traditionally considered together with the protozoa and are more likely to be of interest to parasitologists than mycologists. Relatively few microsporidian species

infect homeothermic vertebrates (birds and mammals) and before the recognition of human immunodeficiency virus (HIV) infection and the acquired immunodeficiency syndrome (AIDS), only a handful of human infections were reported in the literature. However, with the spread of HIV infection and AIDS, microsporidia are becoming increasingly recognized as important causes of opportunistic disease in infected individuals.

STRUCTURE AND LIFE CYCLE

The most familiar stage of microsporidians is the small, highly resistant, gram-positive staining spore (Figure 25.1). Spores contain a coiled filament (polar filament or polar tube) and an infective sporoplasm. The polar tube and its associated organelles are responsible for the unique mode of entering host cells exhibited by microsporidia. Several reviews of the ultrastructure and general biology of the Microsporidia have been published (Canning and Lom 1986; Cali 1991; Perkins 1991; Canning 1993; Bigliardi and Sacchi 2001).

Microsporidian spores are small, more or less ovoid, with a double-layered spore wall. The outer layer, or exospore, is proteinaceous and electron-dense and, according to Weidner and Halonen (1993), is partially stabilized by keratins. The inner layer, or endospore, is chitinous and electron-lucent. The plasma membrane

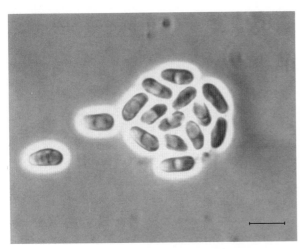

Figure 25.1 *Light micrograph of spores of* Trachipleistophora hominis *showing shape and posterior vacuole. Bar = 4 μm. (Reproduced with permission from W.S. Hollister and E.U. Canning)*

lines the inside of the spore wall. Organelles within the cytoplasm are the polar sac, polar tube, polaroplast, nucleus, and posterior vacuole (Figures 25.2 and 25.3). All forms lack mitochondria, centrioles, peroxisomes, and a classical Golgi apparatus (Canning 1988; Cali 1991). However, there is molecular evidence that microsporidians possessed mitochondria in their early evolutionary history (Hirt et al. 1997) and that a newly recognized organelle (the multilayered interlaced network) found in the spore and sporoplasm of *Brachiola algerae* may be of Golgi origin (Cali et al. 2002). Ribosomes are

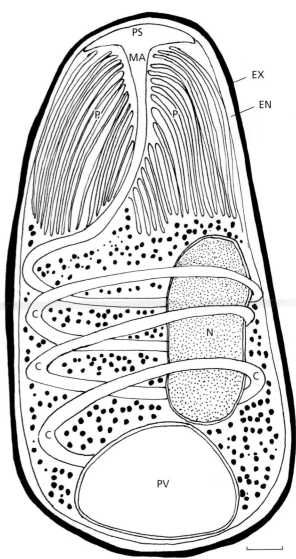

Figure 25.3 *Diagram of the general features of a microsporidian spore showing coils of polar tube (C), endospore (EN), exospore (EX), manubrium (MA), nucleus (N), polaroplast (P), polar sac (PS), posterior vacuole (PV)*

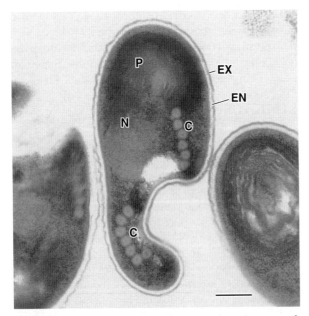

Figure 25.2 *Electron micrograph of a section through a spore of* Encephalitozoon hellem *showing the coiled polar tube (C), electron-dense exospore (EX), electron-lucent endospore (EN), nucleus (N), polaroplast (P). The indentation in the spore marks the position of the posterior vacuole which collapses during processing for electron microscopy. Bar = 0.18 μm.*

of prokaryotic size. The polar sac (or anchoring disc), shaped like the cap of a mushroom, is continuous with the polar tube, the base of which appears like the stalk of the mushroom. A thin layer of cytoplasm separates the polar sac from the spore wall. The endospore layer immediately above the polar sac is thinned. From its insertion in the polar sac, the polar tube is initially straight (the manubrium) before coiling into loops around the inside of the spore wall (Figure 25.4). This polar tube is a complex organelle with intricate anatomical relationships to the other structures of the mature spore (Jensen and Wellings 1972). The structure of the polar tube in the ungerminated spore appears to be tubular, with a filled lumen and concentric rings (Weidner 1982), and was thought to be composed of a single polypeptide of low molecular weight (Weidner 1976) but is now considered to be composed of several

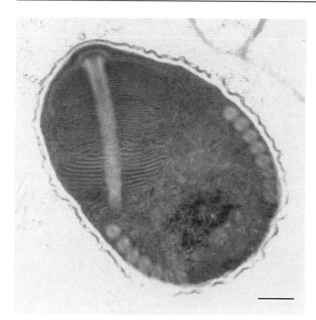

Figure 25.4 *Electron micrograph of a mature spore of* Trachipleistophora hominis *showing manubrium, polaroplast and coiled polar tube. Bar = 0.48 μm. (Reproduced with permission from Andrew S. Field)*

proteins (Keohane et al. 1999). It is surrounded by a membrane. Around the manubrium is located the polaroplast, which is a component of the extrusion apparatus. Ultrastructurally, the polaroplast is composed of a stack of closely apposed membranes. By light microscopy, the polaroplast region appears transparent and is often referred to as the anterior vacuole. The nucleus is surrounded by a nuclear envelope. In some microsporidial species, nuclei are paired with apposed membranes flattened against one another in a diplokaryon arrangement (Figure 25.16). The posterior region of the spore is occupied by a membrane-bound posterior vacuole.

On entering a suitable host, the emergence of the polar tube is initiated within a fraction of a second (Weidner 1976). The extruded tube can be very long compared with the size of the spore and is a flexible structure, the purpose of which is to penetrate a host cell and allow entry of the infective sporoplasm. The extrusion process appears to be initiated by swelling of the spore stimulated by a rise in pH and the presence of calcium ions (Weidner and Byrd 1982). Swelling of the polar sac and polaroplast causes the thin anterior area of the spore wall to rupture, a prerequisite for polar tube eversion. The polar tube is believed to be extruded as a solid cylinder, the interior of which flows outward at the growing tip to form a hollow cylinder (Weidner 1982). Once fully everted, the migration of the sporoplasm commences, driven by processes associated with the posterior vacuole. The multilamellar nature of the polaroplast diminishes during extrusion and provides the plasma membrane surrounding the sporoplasm as it arrives within the host cell cytoplasm (Weidner 1982).

The migration of the sporoplasm through the lumen of the extruded polar tube occurs within 5–30 s (Weidner 1976) and causes some slight distension of the tube (Figure 25.5).

Discharged spores, devoid of the sporoplasm, appear empty and show that the proximal part of the polar tube is funnel-shaped, perhaps to guide the sporoplasm cytoplasm into the tube lumen. Little remains of the polaroplast membranes and the externalized polar tube has a flaccid appearance after passage of the sporoplasm cytoplasm.

In some species, penetration into a host cell occurs by the polar tube punching a hole in the host-cell plasma membrane without loss of cytoplasm: an example is *Enterocytozoon bieneusi*. In other species, penetration appears to induce host-cell plasma membrane expansion to cover the emerging sporoplasm, which is itself covered by a membrane thought to be derived from the polaroplast (Figure 25.6): an example is *Encephalitozoon* (Canning et al. 1992).

Inside the host cell, proliferation begins and two major phases are recognized: merogony and sporogony (Figure 25.7). In both phases, the parasite nuclei divide without breakdown of the nuclear envelope. Nutrients are absorbed from the host cell to fuel parasite development (Canning and Lom 1986).

Meronts are rounded, irregular or elongated cells with little differentiation of the cytoplasm and are surrounded by a plasma membrane (Figure 25.8). In most species, the meronts are in direct contact with the host cell cytoplasm (e.g. *Enterocytozoon bieneusi*). However, meronts of *Encephalitozoon* are surrounded by a membrane derived from the host cell that

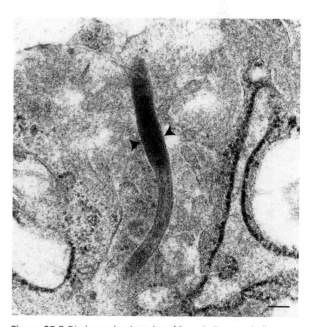

Figure 25.5 *Discharged polar tube of* Encephalitozoon hellem *showing slight expansion (arrowheads) due to passage of dense sporoplasm cytoplasm. Bar = 0.13 μm.*

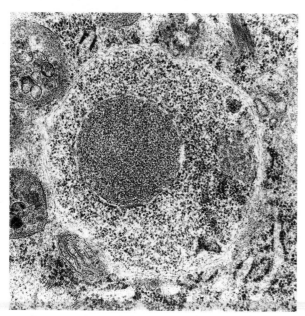

Figure 25.6 *Possible sporoplasm of* Encephalitozoon hellem *in a nasal epithelial cell. The parasite is surrounded by two membranes: the inner one is the plasma membrane of the parasite, and the outer one a membrane derived from the plasma membrane of the host cell. This latter membrane will ultimately form the margin of the parasitophorous vacuolar membrane seen in the later stages of development. Parasite is about 1 μm in diameter.*

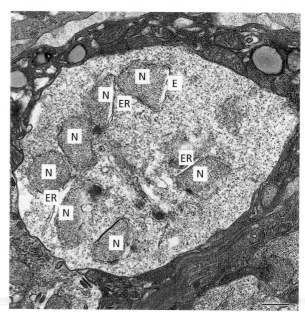

Figure 25.8 *Meront of* Enterocytozoon bieneusi *showing numerous nuclei (N) and associated cisternae of endoplasmic reticulum (ER). The expanded endoplasmic reticulum can sometimes be seen by optical microscopy as 'cytoplasmic slits' within the host small intestinal epithelial cells. Bar = 0.48 μm*

ultimately forms the margin of a parasitophorous vacuole seen in the later stages of parasite development. Meronts may divide by binary fission, multiple fission of a multinucleate meront, or plasmotomy (fission of multinucleate parasite to form multinucleate offspring by division of the cytoplasm without relation to that of the nuclei).

Meronts develop into sporonts, characterized by the presence of an electron-dense surface coat, on the outside of the plasma membrane, which will ultimately become the exospore layer of the mature spore wall (Figure 25.9). Sporonts may divide by binary fission directly into sporoblasts (cells that differentiate into spores without further division) or may become multi-

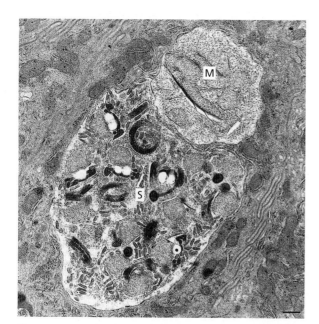

Figure 25.7 *Duodenal enterocyte infected with* Enterocytozoon bieneusi. *Here, a merogonic (M) and a sporogonic (S) stage lie adjacent to one another within the same host cell. Bar = 0.5 μm*

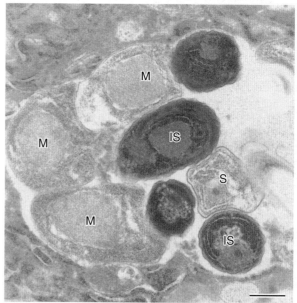

Figure 25.9 Encephalitozoon hellem *from nasal epithelium. Electron micrograph showing parasitophorous vacuole containing meronts (M), a sporont (S) and immature spores (IS). Bar = 0.37 μm*

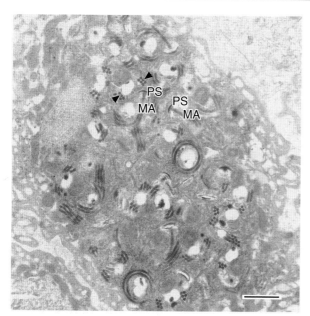

Figure 25.10 *Late sporogonial sporoplasm of* Enterocytozoon bieneusi *showing many copies of spore organelles. Multiple fission will produce sporoblasts that will mature into spores. Note manubrium (MA), polar sac (PS) and six coils in two rows of a polar tube (arrowheads). Bar = 0.64 μm*

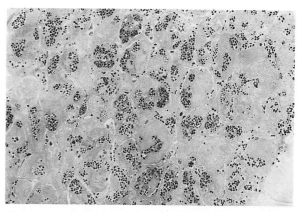

Figure 25.12 *Light micrograph of* Trachipleistophora hominis *in skeletal muscle of an AIDS patient. Note that the spores are in packets (sporophorous vesicles).*

nucleate and form sporogonial plasmodia (Figures 25.10 and 25.11). Sporogonial plasmodia undergo multiple (sequential) fission to produce sporoblasts that ultimately develop into spores. In some genera, such as *Trachipleistophora* and *Pleistophora*, spores are packaged within sporophorous vesicles (SPOV) (Figure 25.12). The endospore layer of the spore wall is synthesized between the dense exospore layer and the plasma membrane. Spores may be liberated by lysis of the host cell, although in some species, such as *Encephalitozoon hellem*, mature spores may germinate and infect other neighboring cells without lysis of the host cell (Canning et al. 1992) (Figure 25.13).

An intranuclear spindle facilitates chromosomal separation during mitosis. The mitotic apparatus consists of two centriolar plaques (electron-dense regions associated with nuclear pores) at the spindle apices (Figure 25.14), on which the spindle microtubules converge (Canning 1988). Sexual processes, as indicated by synaptonemal complexes, have been reported in some genera parasitizing invertebrates (Andreadis and Vossbrinck 2002). Among the genera that infect vertebrates, meiosis is either unknown or unconfirmed.

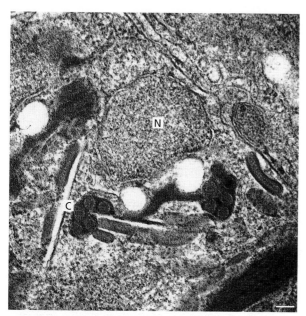

Figure 25.11 *High magnification micrograph of part of the cytoplasm of a sporogonial plasmodium of* Enterocytozoon bieneusi *showing individual spore organelles. Note nucleus (N) and associated profiles of the coiled polar tube (C). Bar = 0.12 μm*

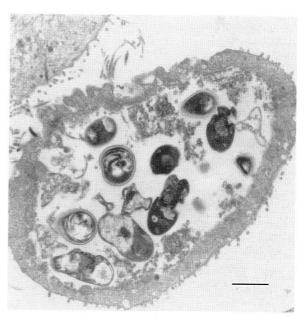

Figure 25.13 Encephalitozoon hellem *from conjunctival epithelium. Parasitophorous vacuole containing spores. Note that some have germinated as indicated by the presence of profiles of polar tubes between and around the spores. Bar = 1 μm*

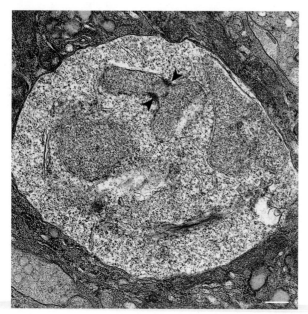

Figure 25.14 *A meront of* Enterocytozoon bieneusi. *Note dense nuclear plaques (arrowheads) indicating nuclear division. Bar = 0.33 μm*

GENERA, SPECIES, AND TAXONOMIC CHARACTERISTICS

Much of the current taxonomy is based on light microscopy, ultrastructural features, host species and organ or tissue specificity. Important ultrastructural features include nuclear arrangement (mono- or diplokaryotic), mode of division, development of proliferative forms in direct contact with host cell cytoplasm or within a parasitophorous vacuole, sporogony producing spores dispersed or aggregated in a SPOV and spore structure. Many amendments and revisions to the current classification are likely as extensive molecular studies are undertaken.

Encephalitozoon

See Canning and Lom (1986). Unpaired (isolated) nuclei are seen in all stages of development. Development occurs in parasitophorous vacuoles, which contain a finely granular matrix. Meronts lie attached to the vacuolar membrane, whereas sporogony, which is generally disporoblastic, occurs within the lumen of the parasitophorous vacuole. Merogony, typically by binary fission, supplies abundant potential sporonts and the vacuole concomitantly expands to accommodate the growing numbers of parasites. Vacuoles contain mature and maturing spores. Spores are small with a thick endospore layer and a rugose exospore.

Three species of *Encephalitozoon* have been recognized in humans: *E. cuniculi* (type species), *E. hellem*, and *E. intestinalis* (synonym for *Septata intestinalis*). Spores of *E. cuniculi* measure about 2.5–3.2 × 1.2–

1.6 μm, with 4–6 coils of the polar tube (Canning and Lom 1986). Spores of *E. hellem* measure 2–2.5 × 1–1.5 μm, with 6–8 coils of the polar tube (Didier E.S. et al. 1991). Spores of *E. intestinalis* measure 2.2 × 1.2 μm, with 5–7 coils of the polar tube (Canning et al. 1994).

Enterocytozoon

See Desportes et al. (1985); Cali and Owen (1990). Unpaired nuclei are seen in all stages of development. Merogonic and sporogonic stages lie in direct contact with the host cell cytoplasm. Proliferative stages develop into sporogonial plasmodia, which produce sporoblasts by multiple fission. Membrane-bound electron-lucent clefts are conspicuous in both meronts and sporonts. In the multinucleated sporogonial plasmodia, polar tube precursors form as electron-dense discs adjacent to the clefts. A unique characteristic of this genus is the complete differentiation of the polar tube within the sporogonial plasmodium, prior to fission (Wongtavatchai et al. 1995). On division into sporoblasts and maturation of these into spores, each becomes dispersed within the host cell cytoplasm.

One species has been recorded from humans, the type species *Enterocytozoon bieneusi*. Spores are very small (about 1.5 × 0.5 μm), with 4–6 coils of the polar tube in two rows. The endospore layer is thin.

Trachipleistophora

See Hollister et al. (1996). Unpaired nuclei are seen in all stages of development. The plasma membrane of meronts is overlain by a dense surface coat, which extends outwards as branched and anastomosing processes. Meronts divide by binary fission or plasmotomy. At the onset of sporogony, the surface coat separates and becomes the thick envelope of the SPOV. Sporonts divide repeatedly by binary fission of binucleate stages to give a variable number of uninucleate sporoblasts and spores. Multinucleate sporogonial plasmodia are not formed.

Two species have been recorded from humans, the type species *T. hominis* (Hollister et al. 1996) and a newly recognized species, *T. anthropophthera* (Vávra et al. 1998). Spores of *T. hominis* measure 5.2 × 2.4 μm, with about 11 coils of the polar tube. The polar tube shows an anisofilar arrangement with 8–11 larger diameter coils and 1–3 smaller diameter coils (Canning et al. 1998). The endospore is thick. *T. anthropophthera* has two types of SPOV: type I SPOV contains usually eight spores measuring 3.7 × 2 μm, with an anisofilar polar tube arrangement (7 larger diameter coils in a single row with 2 posterior coils of smaller diameter shifted inwards towards the centre of the spore); type II SPOV contains two spores, each measuring 2.5 × 1.4 μm, with about 4–5 coils of the polar tube.

Pleistophora

See Canning and Lom (1986). Unpaired nuclei are seen in all stages of development. Meronts form as multi-nucleate plasmodia that possess a thick amorphous electron-dense surface coat external to the plasma membrane. Division into smaller segments occurs. The surface coat becomes the wall of the SPOV (pansporoblast membranes). Sporogony is polysporoblastic. Division of the sporogonial plasmodium is by repeated segmentation, finally producing uninucleate sporoblasts, which mature into spores. The number of spores produced within the sporophorous vesicles is large and variable (polysporoblastic).

A few cases of human infection have been ascribed to parasites belonging to the genus *Pleistophora*, but only in one case has the organism been specifically named (as *P. ronneafiei* – see Cali and Takvorian (2003). Some of these parasites may not be true species of Pleistophora as their full development has not been adequately described and some may have to be reclassified within the genus *Trachipleistophora* (Hollister et al. 1996). Spores from human infections measure 3.2–3.4 × 2.8 µm, with 11 coils of the polar tube (Ledford et al. 1985) or about 4 × 2 µm, with 9–12 coils of the polar tube (Chupp et al. 1993).

Vittaforma

See Silveira and Canning (1995). Nuclei are seen in diplokaryotic arrangement throughout the life cycle. All stages are individually enveloped by cisternae of the host cell rough endoplasmic reticulum. Merogony is by binary fission and sporogony is polysporoblastic, giving rise to 4–8 linearly arranged sporoblasts.

One species has been recorded from humans, the type species *V. corneae*. Spores, still enveloped by host endoplasmic reticulum, measure 3.8 × 1.2 µm, with 5–7 coils of the polar tube.

Nosema

See Canning and Lom (1986). Nuclei are found in diplokaryotic arrangement throughout the life cycle. Parasites are in direct contact with the host cell cytoplasm. Merogony is by binary fission and sporogony is disporoblastic giving rise to spores, which are dispersed in host cell cytoplasm.

One species is currently recorded from humans, *N. ocularum* (Cali et al. 1991a). Spores measures 5 × 3 µm, with between 9 and 12 coils of the polar tube. Three former species of *Nosema* from humans have now been reclassified: *Nosema corneum* (Shadduck et al. 1990) is a synonym for *V. corneae*; *N. algerae*, a mosquite parasite (Visvesvara et al. 1999), has been renamed *Brachiola algerae* (Lowman et al. 2000) and *N. connori* has been reclassified as *B. connori* (Cali et al. 1998).

Brachiola

See Cali et al. (1998). Nuclei are found in diplokaryotic arrangement throughout the life cycle. Merogonic and sporogonic stages lie in direct contact with the host cell cytoplasm. No plasmodial stages develop. All stages possess electron-dense secretions external to the plasma membrane. Abundant tubulovesicular structures embedded in the secreted coat often extend like appendages into surrounding host cell cytoplasm. Merogony is by binary fission and sporogony is disporoblastic.

Three species have been recorded from humans. The type species is *Brachiola vesicularum* with spores which measure 2.5–2.9 × 1.9–2.0 µm and contain 7–10 coils of the polar tube arranged in two rows. The last 2–3 coils of the polar tube are smaller in diameter than the rest (anisofilar arrangement). The other species are *B. connori*, formerly *Nosema connori* (see Margileth et al. 1973; Sprague 1974) and *B. algerae*, formerly *N. algerae* (Visvesvara et al. 1999; Lowman et al. 2000).

'Microsporidium'

See Canning and Lom (1986). This is an assemblage of identifiable species for which the generic positions are uncertain because of insufficient taxonomic detail. They include *M. ceylonensis* (Ashton and Wirasinha 1973) in which the spores measure 3.5 × 1.5 µm and with about 9 coils of the polar tube and *M. africanum* (Pinnolis et al. 1981) 4.5–5 × 2.5–3 µm, with 11–13 coils of the polar tube.

HISTORICAL ASPECTS OF HUMAN INFECTION

The first authenticated report of human microsporidial infection was by Matsubayashi et al. (1959) who described a case of microsporidial infection in a 9-year-old Japanese boy suffering a severe convulsive illness admitted to hospital unconscious and with a fever. Organisms resembling *Encephalitozoon cuniculi* were isolated on day 5 of the illness from the cerebrospinal fluid, and on days 13–15 spores were identified from the urine. On regaining consciousness, headache and vomiting were recurrent symptoms, but the boy is reported to have made a full recovery.

A Colombian child suffered a similar convulsive illness (Bergquist et al. 1984) and spores of *E. cuniculi* were isolated from the urine. After anticonvulsive therapy, the child made a full recovery. Ashton and Wirasinha (1973) described a case of corneal infection in an 11-year-old Sri Lankan boy. Although the original authors did not specifically identify it, Canning and Lom

(1986) transferred this organism to the collective group *Microsporidium* and named it *Microsporidium ceylonensis*. In 1981, another microsporidian was described from the eye of a Botswanan woman (Pinnolis et al. 1981) and this was later also transferred to the collective group *Microsporidium* as *M. africanum* (Canning and Lom 1986). Margileth et al. (1973) described a case history of an overwhelming disseminated infection in a 4-month-old infant with a defective lymphoid system. After an illness lasting 4 months, characterized by diarrhea and malabsorption and concurrent infection with *Pneumocystis carinii*, the child died. Microsporidian spores were seen in cardiac and smooth muscle, diaphragm, myocardium, kidney tubules, liver, lungs, and adrenal cortex and within the walls of arteries in many organs. On the basis of the cytoplasmic and ultrastructural features of the sporoblasts, immature and mature spores, particularly the diplokarya and about 11 coils of the polar tube, the organism was tentatively classified as *Nosema connori* (Sprague 1974; Shadduck et al. 1979; Canning and Lom 1986). This organism has since been allocated to a newly established genus *Brachiola*, as *Brachiola connori* (Cali et al. 1998).

Since the recognition of HIV and AIDS in the late 1970s and early 1980s, several previously rare or unknown human parasitic protozoa have been recognized in this immunocompromised group (Curry et al. 1991). In 1985, a new microsporidian, *Enterocytozoon bieneusi*, was identified in the duodeno-jejunal enterocytes of an AIDS patient with a history of diarrhea and weight loss (Desportes et al. 1985; Modigliani et al. 1985). Since this description, more microsporidian species have been described. Fifteen species are currently recognized from humans and many of these were previously unknown.

SEROLOGICAL SURVEYS OF MICROSPORIDIAL INFECTION IN HUMANS

Before our recent awareness of human microsporidial infections, one of the few species known to infect homeothermic vertebrates was *E. cuniculi* (Levine 1985; Canning and Lom 1986). It has a worldwide distribution in many mammals including rodents, lagomorphs, carnivores and primates. It is a multiorgan pathogen that can be found in the brain, kidneys, liver, spleen, and other organs. In the past *E. cuniculi* caused many problems in animal houses, and serological tests are now available to test for such infections (Canning and Lom 1986). Considering its wide host range in homeothermic vertebrates, it is not surprising that encephalitozoonosis has also been reported in humans. Bergquist et al. (1984) reported the results of a serological survey for *E. cuniculi* in 22 serum samples from human patients with disorders of the central nervous system (all of uncertain origin); one sample gave clearly positive results. The patient was

an apparently healthy Colombian boy, adopted when just over 1 year old by a Swedish family. He was admitted to hospital a year later with generalized convulsive seizures. Based on an indirect immunofluorescence technique, a serum sample taken from the boy about the time of adoption (a year before initial admission to hospital) showed a positive immunoglobulin IgM antibody response to *E. cuniculi*. This original serum sample and those taken during his subsequent admissions to hospital also showed high titers of IgG against *E. cuniculi*. Sedimented urine, obtained during his first stay in hospital, contained gram-positive spores measuring 1.5×2.5 μm, which reacted with a fluorescent anti-*E. cuniculi* conjugate. Urine-derived organisms were injected intraperitoneally into five mice known to be free of encephalitozoonosis and, after 3 weeks, two became infected with parasites indistinguishable from *E. cuniculi*. Hematological testing revealed that the child had a lymphocyte abnormality. The child was given prophylactic anticonvulsive therapy and recovered.

A later serological survey for antibodies to *E. cuniculi* using enzyme-linked immunosorbent assay (ELISA) and utilizing spores harvested from cell culture, demonstrated antibodies to *E. cuniculi* in some patients with psychiatric disorders and also some of those previously exposed to schistosomiasis and malaria (Hollister and Canning 1987).

A more comprehensive survey of human sera (Hollister et al. 1991) utilizing ELISA and also immunofluorescence, immunoperoxidase and Western blots of sodium dodecylsulfate–polyacrylamide gel electrophoresis (SDS–PAGE) protein profiles of *E. cuniculi* spores concluded that human *E. cuniculi* infections were common in the tropics and that reactivation of such infections could account for the occurrence of some infections in AIDS patients. Hollister et al. (1993b), in a retrospective analysis of stored serum samples from an AIDS patient, detected antibodies to *Encephalitozoon* sp. 32 months before signs of keratoconjunctivitis developed and 3 years before nasal obstruction became a problem (Hollister et al. 1993b).

Another serological survey in Europe has shown a high seroprevalence to *Encephalitozoon* species in the immunocompetent population (Van Gool et al. 1997): 8 percent of Dutch blood donors and 5 percent of pregnant French women had antibodies. It was concluded that subclinical infection, or infection involving mild transient symptoms may be common. Such transient episodes may be followed by latency (Hollister et al. 1991), which only becomes problematical if the individual subsequently becomes immunosuppressed.

HUMAN MICROSPORIDIAL INFECTIONS AND IMMUNE STATUS

Current information indicates that patients with severe cellular immunodeficiency are at the greatest risk of developing microsporidial disease (Weber and Bryan 1994). Experimental evidence in mice supports this view (Schmidt and Shadduck 1983) and that resistance to lethal disease appears to be T-cell dependent. *E. cuniculi* causes chronic but not life-threatening infections in euthymic BALB mice, whereas athymic mice succumb to infection. Serum antibodies are not protective.

With the increase in knowledge about microsporidial infections, we now know that they are not restricted to those infected with HIV. Other immunodeficient states, such as those associated with organ transplantation, can make such patients susceptible to microsporidial infection. Indeed, microsporidial infection should be considered for any individual with symptoms who is known to be immunodeficient from any cause. In addition, microsporidial infections are also being reported from individuals with normal immune function and from the elderly in whom the immune system goes into decline.

The clinical spectrum of microsporidial infection includes ocular, intestinal, muscular, and systemic aspects. Most, but by no means all, are associated with HIV infection. The most common microsporidian-associated disease is chronic non-bloody diarrhea and wasting in HIV-infected patients.

Enteric infections

Diarrhea, malabsorption, and wasting are frequently encountered in AIDS patients (Modigliani et al. 1985) and many microbiological causes have been identified for these enteric symptoms; among the parasitic protozoa, both *Cryptosporidium parvum* (see Chapter 20, Intestinal coccidia: cryptosporidiosis, isosporiasis, cyclosporiasis) and microsporidial infections are significant causes of morbidity.

ENTEROCYTOZOON BIENEUSI INFECTIONS

The first reported case of enteric microsporidial infection in an AIDS patient was from France by Desportes et al. (1985). The patient was a 29-year-old nonhomosexual Haitian patient with AIDS, who had severe diarrhea and *Giardia* in stool specimens. Electron microscopic investigations of duodeno–jejunal and ileal biopsies revealed the presence of developmental stages of a microsporidian parasite in the enterocytes (Figures 25.8, 25.10, and 25.11). The ultrastructural details of the development of this microsporidian supported the view that the microsporidial species involved was distinctive enough to merit classification as a new species within a new genus. The name allocated was *Enterocytozoon bieneusi* (in order to avoid confusion with species of *Encephalitozoon*, *Enterocytozoon bieneusi* will subsequently be referred to as *Ent. bieneusi*, as used in Canning et al. 1993). Further details of the intracellular development of this

species were reported from cases in the USA (Cali and Owen 1990). Fever is not associated with such infection, but abdominal cramping is (Asmuth et al. 1994). Malabsorption of fat and D-xylose, and low levels of zinc have been noted in such patients (Asmuth et al. 1994).

Ent. bieneusi is likely to have a worldwide distribution as it has been reported from many countries around the world including the UK (Curry et al. 1988), the Netherlands (Rijpstra et al. 1988), Africa (Lucas et al. 1989), USA (Cali and Owen 1990), and Australia (Field et al. 1993b). It is now recognized that the majority of microsporidial infections found in patients infected with HIV are attributable to this species (Weber and Bryan 1994). *Ent. bieneusi* infection in the small bowel can show histological changes such as villous blunting or ballooning (Orenstein et al. 1990a; Kotler et al. 1993; Asmuth et al. 1994), although some biopsies show more or less normal villous architecture (Eeftinck Schattenkerk et al. 1991). Crypt hyperplasia may also be found (Field et al. 1993c). Inflammation in infected tissues is often minimal (Lucas 1989; Orenstein 1991). Such a variability of changes seen in biopsies may reflect the degree of infection, with heavy infections causing the most cellular and histological changes. *Ent. bieneusi* infection is generally restricted to the enterocytes of the small bowel but infection in other sites has been reported. The biliary tract can be infected, leading to cholangitis (Beaugerie et al. 1992). Such infection may be caused by spread along the intestinal–bile duct epithelium. The gall bladder can also be infected causing acalculous cholecystitis. In addition, there are reports of *Ent. bieneusi* infection in nasal (Hollister and Canning unpublished results, quoted in Canning et al. 1993) and bronchial (Weber et al. 1992b) epithelia, respiratory samples (del Aguila et al. 1997) and the liver (Pol et al. 1993). Such non-small-intestinal sites of infection involving *Ent. bieneusi* appear to be uncommon, and the mechanism of spread is uncertain.

Association of *Enterocytozoon bieneusi* with symptoms of diarrhea

A significant number of patients with intestinal microsporidiosis are co-infected with other potential enteric pathogens (Weber and Bryan 1994) making the pathogenic role of microsporidia difficult to assess. However, most prevalence studies of intestinal microsporidiosis in HIV-positive patients have shown a strong association with diarrhea (Greenson et al. 1991; Weber et al. 1992a; Field et al. 1993b; Eeftinck Schattenkerk et al. 1991; Van Gool et al. 1993). In contrast, Rabeneck and colleagues (1993) found that the association between enteric microsporidial infection and diarrhea was not as strong as previous studies had suggested but this is the only study not to link microsporidial infection to symptoms of chronic diarrhea in HIV disease.

Prevalence of *Enterocytozoon bieneusi* infection in Africa

In developing countries, such as those in central Africa, diarrhea and weight loss are important indicators for the clinical diagnosis of AIDS (Kelly et al. 1994). These common manifestations are often referred to as entero-pathic AIDS or 'slim disease' (Serwadda et al. 1985; Lucas et al. 1993). Lucas et al. (1989) identified a lower prevalence of microsporidial infection from Uganda and Zambia than in developed countries; 6.5 percent (5 of 77 patients with chronic diarrhea and wasting) had *Ent. bieneusi* infection. This low prevalence suggested that patients were dying from other causes (particularly tuberculosis) before their CD4 lymphocyte count had dropped to the critical level of about 100×10^6/l, where microsporidial infection becomes significant (Canning et al. 1993).

Bretagne et al. (1993), in a survey of children of unknown HIV status from an area of Africa (Niamey, Niger) with a low occurrence of HIV, found that in 990 stool samples tested, 6 of 593 (1 percent) with diarrhea and 2 of 397 (0.5 percent) without diarrhea contained *Ent. bieneusi* spores.

ENCEPHALITOZOON (=SEPTATA) INTESTINALIS INFECTIONS

Although originally described as resembling *E. cuniculi* (Orenstein et al. 1992a,b), this microsporidian was subsequently assigned to a new genus and named *Septata intestinalis* (Cali et al. 1993). Unlike *Ent. bien-eusi*, intracellular development takes place within a lobed parasitophorous vacuole bounded by a membrane (Figure 25.15). Within the vacuole, parasites are sepa-rated by granular septa secreted by the parasite (Cali et al. 1993). However, Hartskeerl et al. (1995) consid-ered that this organism should be classified as *Encephali-tozoon intestinalis*, on the basis of restriction fragment length polymorphism analysis of amplified ribosomal RNA sequences which showed about 90 percent identity with *E. cuniculi* and *E. hellem*. In addition, Western blots reveal a significant cross-reactivity between these three species. Hartskeerl and colleagues concluded that *S. intestinalis* should be regarded as a species within the genus *Encephalitozoon* and this is what is accepted in this chapter.

E. intestinalis is not confined to epithelial cells as it is also found in macrophages in the lamina propria. This species appears to have a primary infection site in the small bowel, but can disseminate to the viscera (see below) and has been found in epithelial enterocytes and the lamina propria of the duodenum, jejunum (Oren-stein et al. 1992b), ileum, colon (Field et al. 1993a), kidney (Orenstein et al. 1992a; Field et al. 1993a), liver, and gallbladder (Orenstein et al. 1992a) and other sites, including the lower airways (Schwartz et al. 1993b). The prevalence of *E. intestinalis* in small-intestinal biopsies

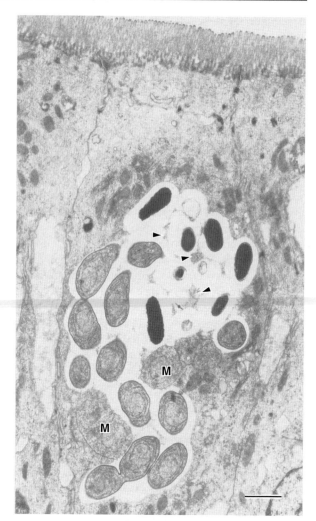

Figure 25.15 *Electron micrograph of a duodenal enterocyte with a parasitophorous vacuole containing* Encephalitozoon intestinalis. *Note meronts (M) apposed to vacuolar membrane and granular septal material between parasites (arrowheads). Bar = 0.53 μm. (Reproduced with permission from Canning et al. 1994)*

appears to be significantly less than *Ent. bieneusi* infec-tion to which there are now numerous references (Canning et al. 1993). Kelly et al. (1994) found 2 of 75 (3 percent) patients with HIV-related diarrhea infected with *E. intestinalis* in an African (Zambia) study and the prevalence is similar to that found in other studies else-where.

Concurrent infections involving both *E. intestinalis* and *Ent. bieneusi* have been reported (Blanshard et al. 1992b; Orenstein et al. 1992a; Cali et al. 1991c). Van Gool et al. (1994), in an attempt to culture *Ent. bieneusi* from stool samples from four AIDS patients with biopsy-proven *Ent. bieneusi* infections, established cultures, but the organism grown was identified as *E. intestinalis*. A possible explanation for this somewhat unexpected result is that *Ent. bieneusi* and *E. intestinalis* are dimorphic forms of the same species. As no evidence of *E. intestinalis* infection had been found in the original biopsies, Van Gool and colleagues dismissed this explanation and suggested that all the patients from whom stool samples were obtained had heavy infections of *Ent. bieneusi* and light infections of *E. intestinalis*. Critical re-examination of the stool samples revealed abundant *Ent. bieneusi* spores and rarely seen, larger spores, which could possibly have been *E. intestinalis*. These results suggest that *E. intestinalis* can occur as an unapparent infection and that its true prevalence may be much higher than previously thought.

ENCEPHALITOZOON CUNICULI INFECTIONS

It is also possible that *E. cuniculi* infects the gut as it has also been reported from the intestinal tract of an AIDS patient with disseminated *E. cuniculi* infection (Franzen et al. 1995). However, the authors point out that the intestinal infection was not confirmed and was presumed to be *E. cuniculi*, as *E. cuniculi* infection had been confirmed from other sites. In this case, the possibility exists that the enteric infection involved *E. intestinalis* rather than *E. cuniculi*.

CLINICAL FEATURES OF ENTERIC MICROSPORIDIAL INFECTIONS IN AIDS

Asmuth et al. (1994) retrospectively reviewed 20 patients who had small-intestinal microsporidiosis. Fever was not associated with enteric microsporidial infection, but abdominal cramping was. Mean CD4 counts were $35 \pm 29 \times 10^6/l$ and most had been lower than $100 \times 10^6/l$ for at least 16 months. Mean duration of diarrheal symptoms was 8.5 ± 6.9 months, with stool frequency at the time of diagnosis 5.7 ± 2 per day. Within this group of 20 patients, 18 had their microsporidial parasites identified to species level: 14 (78 percent) were infected with *Ent. bieneusi* and 4 (22 percent) with *E. intestinalis*. Asmuth et al. (1994) also showed malabsorption of fat and D-xylose and low levels of zinc, and suggested that the enteric microsporidia have the ability to turn on secretory processes within the small intestine and to induce diarrhea.

Fecal tumor necrosis factor-alpha (TNF-α) has been found to be elevated in cases of HIV-related diarrhea due to enteric microsporidial infection (Sharpstone et al. 1997).

Normally, *Ent. bieneusi* infects enterocytes of the small bowel, but this species can occasionally be found in the lamina propria. Less commonly, it can invade the biliary tree causing cholangitis, and the gall bladder causing acalculous cholecystitis. *E. intestinalis* can also cause cholecystitis (French et al. 1995) and cholangitis (Orenstein et al. 1992a).

ENT. BIENEUSI AND E. INTESTINALIS IN IMMUNOCOMPETENT INDIVIDUALS (TRAVELER'S DIARRHEA)

Microsporidia are increasingly recognized among individuals not infected with HIV. *Ent. bieneusi* and *E. intestinalis* have been reported as a cause of traveler's diarrhea (Sandfort et al. 1994; Sobottka et al. 1995a;

Wanke et al. 1996; Fournier et al. 1998; Gainzarain et al. 1998; Raynaud et al. 1998; Muller et al. 2001). In such cases, without underlying immunodeficiency, microsporidial infection appears to be self-limiting.

A report from Zimbabwe suggests that *Ent. bieneusi* infection may be a more common cause of diarrhea than previously suspected and that larger prospective studies are needed to assess the true incidence of this infection in patients without evidence of immunosuppression (Gumbo et al. 2000).

ENT. BIENEUSI INFECTION IN THE ELDERLY

Chronically ill, elderly individuals may be an additional group at risk of acquiring microsporidial infection as a result of the diminishment in immune response. A Spanish study has shown significant rates of *Ent. bieneusi* infection in elderly individuals (mean age 75 years) with chronic and nonchronic diarrhea (Lores et al. 2002).

MICROSPORIDIAL INFECTION IN ORGAN TRANSPLANT RECIPIENTS

There are now several reports of microsporidial infection in organ transplant recipients undergoing immunosuppressive therapy (Sax et al. 1995; Rabodonirina et al. 1996; Kelkar et al. 1997). *Ent. bieneusi* infection was diagnosed from diarrheal samples from a heart–lung recipient (Rabodonirina et al. 1996) and an allogenic marrow recipient (Kelkar et al. 1997). In a liver recipient (Sax et al. 1995), the species involved was not identified but, as the spores were measured at 1 μm in length, it is likely that the species involved was *Ent. bieneusi*. Clinical presentation and evolution of the infection were identical to those observed in patients with AIDS.

Ocular microsporidioses

Two types of ocular infection have been described: the first involves the conjunctival and corneal epithelium and occurs concomitantly with HIV. The second involves the corneal stroma and leads to ulceration and suppurative keratitis, and occurs in immunocompetent individuals, free of HIV infection (Shadduck et al. 1990).

CORNEAL STROMAL INFECTIONS

Pinnolis et al. (1981) published a report of a 26-year-old Botswanan woman with a perforated corneal ulcer, who became blind and underwent enucleation. Microsporidial spores were identified mainly in the cytoplasm of corneal histiocytes. Canning and Lom (1986) placed this species in the collective group *Microsporidium*, as *M. africanum*. Canning et al. (1998) drew attention to the similarities between *M. ceylonensis* and *M. africanum*, but it is unlikely that further taxonomically significant details of these microsporidians will become available. Shadduck et al. (1990) reported the case of a

45-year-old HIV-seronegative man with an 18-month history of central disciform keratitis, recurrent patchy infiltration of the anterior stroma and iritis. At biopsy, microsporidia were seen in the corneal stroma by light and electron microscopy. He was treated with topical steroids and broad spectrum antibiotics but ultimately required a corneal transplant. During the transplantation, part of the explanted cornea was inoculated into cell cultures and an infection was established. The organism contained diplokarya, division was by binary fission and the parasite was in intimate contact with the host cell cytoplasm for all stages of the life cycle. Because of these features, it was named *Nosema corneum* and differentiated from *N. connori* (now *B. conneri*) (Margileth et al. 1973) by having 6 coils of the polar tube, compared with 10–11 coils in *B. connori*.

Cali et al. (1991a) described the case of a 39-year-old man with irritation and blurred vision of his left eye. A foreign body was removed but visual problems and a corneal ulcer persisted. A biopsy revealed microsporidia in direct contact with host cell cytoplasm. The nuclei of the parasites were in a diplokaryotic arrangement. The spore size was 3 × 5 μm, with 9–12 coils of the polar tube. The organism was placed in the genus *Nosema* and named *Nosema ocularum*.

Silveira et al. (1993) established an infection of *N. corneum* in athymic mice. Subsequently, Silveira and Canning (1995) revised the classification of this microsporidian using the ultrastructural development of the organism in the mouse hepatocytes, and concluded that it was not a species of *Nosema*, as only the diplokaryotic arrangement of the nuclei was consistent with that genus (Figure 25.16). Sporogony was polysporoblastic and sporonts were ribbon-shaped, dividing to produce linear arrays of sporoblasts. Each parasite was also completely enveloped by endoplasmic reticulum derived from the host cell (Figure 25.16). Based on these new taxonomically significant features, the organism was placed in a new genus, *Vittaforma*, and named *Vittaforma corneae*.

PREDOMINANTLY OCULAR INFECTIONS IN HIV DISEASE

In the USA in 1990, five homosexual men with AIDS were diagnosed as having bilateral microsporidial keratoconjunctivitis (Friedberg et al. 1990; Lowder et al. 1990; Orenstein et al. 1990b; Yee et al. 1991). Histological sections of corneal or conjunctival scrapings contained numerous oval spores, confirmed as being microsporidial by electron microscopy. Further cases of ocular microsporidiosis have subsequently been reported

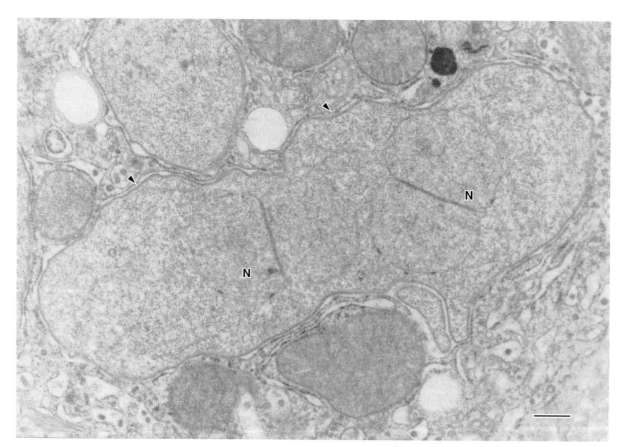

Figure 25.16 *Electron micrograph of a meront of* Vittaforma corneae *with two diplokaryotic nuclei (N). Note that parasite is enveloped by host endoplasmic reticulum (arrowheads) Bar = 0.26 μm. (Reproduced with permission of the authors and publisher from Silveira and Canning 1995)*

from the USA (Schwartz et al. 1993a). An *Encephalitozoon*-like species was involved in all of the above cases.

Additional cases of ocular infection from other parts of the world have followed these initial reports. Metcalfe et al. (1992) reported a case of keratoconjunctivitis from the UK and McCluskey et al. (1993) reported one from Australia. An *Encephalitozoon*-like species was thought likely to be the microsporidian involved.

Isolates from conjunctival or corneal scrapings obtained from three American AIDS patients, grown in cultures of Madin–Darby canine kidney cells (MDCK cells) by Didier E.S et al. (1991), were all identical by SDS–PAGE analysis but were different from a well-characterized isolate of *E. cuniculi*. Identical banding patterns on Western immunoblotting were obtained from each patient's serum, whereas murine antisera to *E. cuniculi* reacted to only some antigens from the tissue culture-grown isolates. The differences were sufficient for Didier E.S. et al. (1991) to propose that the ocular isolates from the AIDS patients studied should be redesignated as a new species, *E. hellem*. Subsequently, the UK isolate was also found to be *E. hellem* (Hollister et al. 1993b).

In addition to *E. hellem* another microsporidian, *Trachipleistophora hominis* can infect the cornea and was first described from an AIDS patient with muscle, ocular, and respiratory symptoms (Field et al. 1996). Although *V. corneae* was originally described from the eyes of a non HIV infected patient (see above), this species has now been identified in the urinary tract of a patient with AIDS (Deplazes et al. 1998) and must, therefore, be considered as having the potential to infect ocular sites in patients infected with HIV.

CLINICAL FEATURES OF OCULAR INFECTIONS IN HIV DISEASE

Clinical manifestations of ocular infections include conjunctivitis, foreign-body sensation, blurred vision, and photophobia. Concomitant unilateral cytomegalovirus retinitis was noted in two patients. Ophthalmic examination revealed a diffuse punctate keratopathy (Friedberg et al. 1990; Lowder et al. 1990; Orenstein et al. 1990b; Yee et al. 1991). In patients for whom data were available, CD4 lymphocytes were profoundly decreased (mean 26×10^6/l; range 2–50) (Schwartz et al. 1993a). The Australian case was a homosexual male with a CD4 count of 83×10^6/l (McCluskey et al. 1993).

The case reported from the UK involved a married man who had had homosexual activity before marriage. In contrast to the other published cases, his symptoms of ocular infection had been noted 2 years before diagnosis of microsporidial infection, when his CD4 count was of the order of 200×10^6/l. In this individual, both corneas were diffusely covered with fine punctate epithelial opacities and there was also some punctate staining of the interpalpetral bulbar conjunctiva of both eyes. Conjunctival scrapings established the microsporidial

nature of the infection (Figure 25.13). The taxonomic features of this parasite seen by electron microscopy were consistent with a species of the genus *Encephalitozoon*. Further study confirmed the species involved to be *E. hellem* (Hollister et al. 1993b). The same patient also had nasal obstruction and discharge (Lacey et al. 1992). Ear, nose and throat (ENT) examination showed multiple nasal polyps and a computed tomography (CT) scan showed extensive opacities in the maxillary antra, and ethmoid and sphenoid sinuses, as well as minor cerebral atrophy. A nasal polypectomy was performed and on both histological and electron microscopical examination, many superficial epithelial cells were found to contain microsporidial spores, identical to those found in the corneal and conjunctival scrapings. An interesting feature of the ultrastructural examination was the presence of discharged (or germinated) spores within intact parasitophorous vacuoles (Canning et al. 1992). These would probably have contributed to the chronic nature of this infection (Lacey et al. 1992) if spores discharged their sporogonial contents into adjacent epithelial cells without release from the cell in which they developed. Cali and colleagues have also noted this premature germination (Cali et al. 1991b).

It is possible that in the UK case, infection was initially caused by inoculation of spores into corneal abrasions and infection could have spread from the corneal epithelium through the lacrimal canaliculi and nasolacrimal ducts that drain secretions from the eyes into the nasal sinuses (Curry and Canning 1993). However, ocular infection may possibly be acquired by reverse passage from a respiratory source, as microsporidial infection has been identified within the tracheobronchial epithelium (Weber et al. 1992b; Schwartz et al. 1993b).

Other sites of infection and other species

Encephalitozoon cuniculi has been reported from a small number of AIDS cases. A case of peritonitis was reported by Zender et al. (1989) and a case of hepatitis by Terada et al. (1987). Several cases of renal infection involving species of *Encephalitozoon* have been reported (Orenstein et al. 1992a; Cali et al. 1993; Aarons et al. 1994; Dore et al. 1995), as have respiratory tract (Lacey et al. 1992; Schwartz et al. 1993b; Molina et al. 1995) and urinary tract infections (Corcoran et al. 1996; Birthistle et al. 1996).

The genus *Pleistophora* is normally associated with infections of fish (Canning and Lom 1986) but a few cases of myositis caused by *Pleistophora* sp. have been reported in humans (Chupp et al. 1993; Ledford et al. 1985; Macher et al. 1988). In the case reported by Ledford and colleagues, the patient was immunodeficient but HIV seronegative. Further work on this organism has classified it as a new species, *P. ronneafiei*

(Cali and Takvorian 2003). The other reported cases were from AIDS patients. In the case described by Chupp and colleagues, the patient was a 33-year-old Haitian man with AIDS who had pain and weakness spreading to the posterior thighs and upper extremities and daily fevers. Electromyography findings were consistent with a diffuse, active myopathic process including denervation characteristic of inflammatory myopathy. The patient died without the cause of death being immediately known but a muscle biopsy taken before death showed, in one area, that all the fibers were atrophic and contained intracellular basophilic organisms. Electron microscopy revealed all the developmental stages of a microsporidian, which was ascribed to the genus *Pleistophora*.

In a case originating from Australia, *Pleistophora*-like organisms were diagnosed in a muscle biopsy but were also present in corneal epithelium, urine, and nasopharyngeal washings, suggesting disseminated infection (Hollister et al. 1996). Studies of this microsporidian in culture and in athymic mice (Figure 25.17), indicated that the parasite differed from established species of *Pleistophora* and it has therefore been reclassified as a new species, *Trachipleistophora hominis*.

A second species of *Trachipleistophora*, *Trachipleistophora anthropophthera*, has been reported from the brain and other organs of two AIDS patients (Vávra et al. 1998). This species has two sporogonial development sequences which give rise to two types of SPOV, containing either approximately eight spores (type I SPOV) or containing two thin-walled spores (type II SPOV).

Another species of microsporidian, *Brachiola vesicularum*, has been described from skeletal muscle of an AIDS patient (Cali et al. 1998). Key ultrastructural features of this organism are:

- a diplokaryotic arrangement of nuclei
- all stages in intimate contact with the host cell cytoplasm
- surface appendages that reach into the surrounding host cell cytoplasm
- an anisofilar polar tube arrangement within the spore (posterior coils of a smaller diameter compared to anterior coils of the polar tube).

Disseminated infections

There are now several reports of multiorgan infection (Gunnarsson et al. 1995) or systemic dissemination (Orenstein et al. 1992a; Cali et al. 1993; Doultree et al. 1995). In patients presenting with symptoms localized to one organ system, consideration must be given to the possibility of systemic infection, particularly if species of *Encephalitozoon* are involved (Schwartz et al. 1993b). Dore et al. (1995) suggest that a distinguishing feature of disseminated infection with *E. intestinalis* is the prevalence of symptoms of chronic rhinosinusitis. De Groote and colleagues (1995) describe a case of disseminated *E. cuniculi* infection in which the patient had advanced renal failure. The organism was found in urine and sputum and many tests, including the polymerase chain reaction (PCR) and indirect immunofluorescence, confirmed the presence of *E. cuniculi*. Treatment with albendazole, a broad-spectrum antiparasitic agent with activity against protozoa resulted in improvement of symptoms.

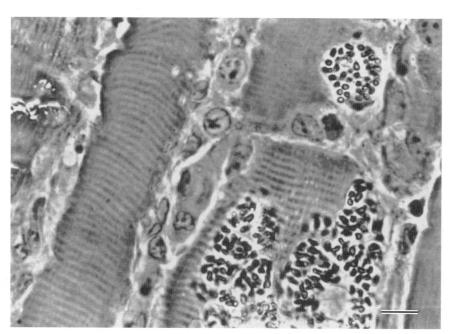

Figure 25.17 *Light micrograph of* Trachipleistophora hominis *in skeletal muscle fibers of an athymic mouse. Note that muscle fibre is packed with spores. Bar = 10 μm. (Reproduced with permission from W.S. Hollister and E.U. Canning)*

Schwartz et al. (1993b) describe a case of a patient previously diagnosed as having keratoconjunctivitis involving *E. hellem* infection, but who subsequently presented with respiratory problems and fever. A chest radiograph revealed a left lower lobe interstitial infiltrate. Bronchoscopy with biopsy and bronchoalveolar lavage revealed abundant *E. hellem* spores within epithelial cells. Sputum and urine samples also contained the same microsporidian.

Molina et al. (1995) describe five cases of disseminated infection due to *E. intestinalis*. Symptoms included chronic diarrhea, fever, cholangitis, sinusitis, and bronchitis. Albendazole cleared the stool samples of all five patients but the urine of only three patients. Other disseminated infections involving *E. intestinalis* have been reported (Orenstein et al. 1992a; Cali et al. 1993).

The reports outlined above strongly suggest that AIDS patients with low CD4 counts presenting with infections involving any species of *Encephalitozoon* may have unsuspected disseminated infection. Early treatment with albendazole can result in significant improvement of symptoms in most cases. The mechanism of dissemination of microsporidia is uncertain, but infected macrophages may play a significant role in this process (Doultree et al. 1995).

Although only known from a few patients, other species are capable of dissemination and include *T. hominis* (Field et al. 1996) and *T. anthropophthera* (Vávra et al. 1998). *T. hominis* has been demonstrated in corneal scrapings, skeletal muscle and nasal discharge (Field et al. 1996) and *T. anthropophthera* in brain, kidneys, heart, pancreas, thyroid, parathyroid, liver, spleen, and bone marrow (Vávra et al. 1998).

V. corneae was first described from a localized corneal infection in an immunocompetent individual (Shadduck et al. 1990), but experimental infection of athymic mice suggested that it could spread systemically in immunocompromised humans and not be restricted to the eyes (Silveira et al. 1993). *V. corneae* has subsequently been identified from the urinary tract of an AIDS patient (Deplazes et al. 1998).

Dissemination should be seriously considered particularly if the patient is diagnosed with *Encephalitozoon*, *Trachipleistophora* or *V. corneae* infection.

Progression of infection

In an experimental investigation of the route and progression of infection with *E. cuniculi* in adult rabbits, it was found that administration of spores either orally or intratracheally induced serum antibodies within 3 weeks and excretion of spores in urine by 6 weeks (Cox et al. 1979). Kondova et al. (1998) challenged gnotobiotic piglets with *Ent. bieneusi* from human and primate sources. All challenged animals became infected, but immunosuppressed piglets had an earlier onset of excretion and considerably increased spore shedding in feces. Immunosuppressed animals started to excrete spores 5–8 days after challenge compared to 9–12 days for nonimmunosuppressed control piglets. The piglets showed little or no symptoms of diarrhea or wasting and this, the researchers believed, was due to the relatively short period of immunosuppression and was comparable to individuals with AIDS who usually only show symptoms when their CD4 T-cell count drops below $100 \times 10^6/l$. *Ent. bieneusi* spores from a human AIDS patient were orally transmitted to two simian immunodeficiency-infected rhesus monkeys and the animals started to excrete spores within 7 and 8 days of inoculation respectively (Tzipori et al. 1997). Shedding of spores continued until the rhesus monkeys were euthanatized 7 and 8 months later, respectively. Jejunal infection was found to be sparse and this was thought to be due to either the number of circulating CD4 cells or that rhesus monkeys are less susceptible than humans to *Ent. bieneusi* infection. Similar information is not available for human microsporidial infections, but the incubation period may be similar to that seen in primate infection.

EPIDEMIOLOGY AND POSSIBLE ANIMAL RESERVOIRS

Microsporidial infections have only become well recognized in humans since the advent of HIV and AIDS. The open question is whether the organisms involved are solely human and primate parasites or may, in some cases, have a zoonotic origin. Any explanation of the current situation, where microsporidial infections are becoming more commonly recognized, must consider possible zoonotic transmission (Glaser et al. 1994). Curry (1999) has reviewed human microsporidial infection and possible animal sources.

Most human microsporidial infections are transmitted (horizontally) by spores (either by ingestion, inhalation, inoculation into abrasions or sexually), although there is some evidence of vertical transmission in other vertebrates such as lagomorphs and rodents (Hunt et al. 1972; Innes et al. 1962). As microsporidia are ubiquitous in the wild, environmental contamination must be common. Spores from infected animals must be widespread in the environment because of urination, defecation or decay after death (Canning 2001). Human exposure to microsporidia is inevitable. A French analysis of risk factors for microsporidiosis showed that swimming in pools was significant, but, surprisingly, that pet contact was not (Hutin et al. 1998). Microsporidian spores (*Ent. bieneusi*, *E. intestinalis* and *V. corneae*) have been identified in water, but the source or sources of this contamination remain unknown (Sparfel et al. 1997; Dowd et al. 1998).

High prevalence rates of *Encephalitozoon* and *Enterocytozoon* in humans may indicate natural human-to-human transmission, but several studies have found these parasites in animals that are associated with or farmed by humans, thus allowing the possibility that some human infections may be zoonotic.

Domestic animals, such as dogs, could be a source of some human infections. *E. cuniculi*, in particular, is known to occur in dogs (Didier et al. 1996) and a wide range of mammals. Three strains of *E. cuniculi* have been identified so far (Biderre et al. 1999). A culture of *E. cuniculi* from an AIDS patient has been found to be identical to *E. cuniculi* strain III, which had previously only been isolated from domestic dogs (Didier et al. 1996; Snowden et al. 1999). Small mammals such as rats and mice, which are pests of human habitation, can also harbor *E. cuniculi* (Canning and Lom 1986) and may contaminate the environment with spores. Ingestion of spores by severely immunocompromised individuals could result in disease. However, even with our improved diagnostic capabilities, only a few cases of *E. cuniculi* infection have been reported in humans (Franzen et al. 1995; Canning 1998).

The other two species of *Encephalitozoon*, *E. intestinalis* and *E. hellem*, are more commonly reported from human infections and these species are now being identified from animals. Stool samples from domestic animals were collected in two rural villages in Mexico and examined by microscopy (Gram-chromotrope and specific immunofluorescence) and PCR (Bornay-Llinares et al. 1998). *E. intestinalis* was identified in dogs and donkeys and in some animals reared for food (cow, pig, and goat). Bornay-Llinares and colleagues discuss the possibility of transmission to humans via contaminated food or water or via an aerosol route, but such routes of transmission remain to be demonstrated.

The finding of *E. hellem* in birds has highlighted possible avian sources of this parasite. Black et al. (1997) and Pulparampil et al. (1998) have identified *E. hellem* in psittacine birds and Snowden et al. (2000) have also identified *E. hellem* in lovebirds. As birds (particularly psittacine birds) are often kept as pets, this raises the possibility that these may be sources of human disease. In the above studies, the intestinal epithelium, kidneys, cloaca or droppings were found to contain spores. Droppings are dry or become dry quickly and can produce dust (Curry 1999). Inhalation into the respiratory tract of dust containing viable spores may initiate infection in humans, particularly if the pet owner is immunosuppressed.

Infected fish muscle is also emerging as a possible source of human infection. Species of *Pleistophora* are common fish and crustacean pathogens (Cheney et al. 2000). A small number of cases of *Pleistophora* infection in humans have now been described, all involving skeletal muscle infection. Inadequately cooked infected fish or crustaceans may have a role in these infections, particularly in the immunosuppressed. Some evidence for this comes from the incidental finding of microsporidial spores associated with muscle fibers in a human

stool sample from an AIDS patient with diarrhea (McDougall et al. 1993) suggesting that infected fish had been consumed and that spores had passed through the gut largely intact.

A molecular study by Cheney et al. (2000) has shown that *T. hominis* has similarities to organisms within the microsporidian genus *Vavraia*, which are mosquito parasites, thus raising the possibility of transmission to humans from biting insects. The likelihood of microsporidial species from poikilothermic hosts being able to infect homeothermic humans, particularly if severely immunosuppressed (as in HIV disease) needs further study. Evidence for this possibility is emerging. *Nosema algerae*, a mosquito parasite was successfully transmitted to mice (Trammer et al. 1997) and has been diagnosed from an ocular lesion in an immunocompetent human (Visvesvara et al. 1999). This mosquito parasite has now been renamed *Brachiola algerae* (Lowman et al. 2000).

Enterocytozoon bieneusi, the commonest microsporidian species to infect humans, has currently been identified in macaques, pigs, rabbits, dogs, cattle, a llama, and a cat (Kondova et al. 1998; Breitenmoser et al. 1999; del Aguila et al. 1999; Mathis et al. 1999; Dengjel et al. 2001). Molecular methods have shown a number of genotypes (Rinder et al. 1997; Dengjel et al. 2001) or genetically distinct strains (Liguory et al. 1998) of *Ent. bieneusi* in humans. In the study of Liguory et al. (1998), four strains were identified, one of which (type IV strain) was similar to a porcine isolate (Deplazes et al. 1996). A study from Sadler et al. (2002) has shown that of 13 human *Ent. bieneusi*-positive samples tested, 11 were of a genotype (genotype B) so far only found in humans, one had been reported from humans and a laboratory macaque (genotype D) and one (genotype K) had previously only been identified from a cat. Sadler et al. concluded that a possible zoonotic link with cats had been demonstrated, but that the origin of the majority of human cases was unknown and this may indicate that genotype B is solely a human pathogen. Dengjel et al. (2001) examined fecal samples from humans and animals and identified 14 different genotypes of *Ent. bieneusi*, 6 of them previously undescribed. Phylogenetic analysis of these genotypes revealed the lack of a transmission barrier between *Ent. bieneusi* from humans and cats, pigs and cattle and they conclude that *Ent. bieneusi* appears to be a zoonotic pathogen.

Mansfield et al. (1998) found simian immunodeficiency virus (SIV)-infected and non-SIV-infected (normal) rhesus macaques to be infected with *Ent bieneusi*, but infection rates were significantly higher in the SIV-infected group (33.9 percent compared to 16.7 percent). The SIV-infected animals also showed more symptoms of diarrhea and *Ent. bieneusi* was detected in the small intestine. In normal animals, *Ent. bieneusi* was not detectable in the small intestine, but was found in bile. *Ent. bieneusi* has been successfully transmitted from humans and rhesus macaques to gnotobiotic piglets and propagated serially (Kondova et al. 1998). More studies are needed to determine whether *Ent. bieneusi* is a primate parasite or a widespread pathogen of mammals, particularly primates.

Whether animal infections with *Ent. bieneusi* and *E. intestinalis* are natural or from human handlers remains to be determined. Although still unproven and with firm evidence of transmission currently lacking, clues are emerging as to possible animal links with human microsporidial disease. We must remain open-minded to the possibility that some human microsporidial infection is from animal sources.

DIAGNOSIS

Microsporidiosis in AIDS patients is probably significantly under-reported because of diagnostic difficulties (Aldras et al. 1994). Diagnosing microsporidial infections depends on recognizing that the patient could be infected with microsporidia, particularly when the CD4 lymphocyte count drops below $100 \times 10^6/l$, thus allowing the most appropriate specimens to be taken and the laboratory to choose the most appropriate methods of testing. Laboratory diagnosis of microsporidial infections in tissues is not easily accomplished because of the intracellular nature of these parasites, and their small size and poor staining properties (particularly of proliferative stages) with routinely used histological stains. In addition, the histopathologist requires experience in detecting such organisms. To date, 15 microsporidian species have been identified from humans (see Genera, species and taxonomic characteristics). Given the number of microsporidian species now detected, electron microscopy should be considered, particularly if the patient is immunocompromised and a biopsy is to be taken as an infection is suspected.

Microbiological examination of a fecal sample must include appropriate stains for identifying microsporidial spores. Laboratory diagnosis of microsporidial infections initially requires light microscopy (either bright field or epifluorescence, depending on stains used) and putative positives should, ideally, be confirmed by either electron microscopy or molecular techniques.

Small-intestinal biopsies

Enteric infection is the most common type of microsporidial infection. Until about 1990, diagnosis required both light-microscopic and electron-microscopic examination of small-intestinal biopsy sections (Orenstein et al. 1990a) or touch preparations of biopsies (Rijpstra et al. 1988). Microsporidia stain poorly with conventional histopathology stains, but diagnosis can still be accomplished with experience (Peacock et al. 1991). Appropriate special stains should be chosen for the tissue sections, although these are not always considered. Diagnostic workers interested in microsporidial

detection have tried many alternative staining methods and often advocate their own particular favorite. Stains used include Giemsa, tissue Gram stain (Rabeneck et al. 1993) and periodic acid–Schiff reagent (Field et al. 1993b). A tissue stain that shows great promise is Warthin–Starry (Field et al. 1993 a, b, c), which stains some sporogonic stages and spores (Figure 25.18).

Microsporidial infections can be misdiagnosed in tissues. Furuta et al. (1991) thought that pulmonary lesions found in autopsy material from leprosy patients involved microsporidia, but after immunohistochemical staining it was concluded that these lesions were, in fact, caused by *Cryptococcus neoformans*. Resin-embedded tissue stained with toluidine blue can also be useful, but mucus granules in goblet cells also take up stain and can cause confusion. The excellent preservation and thin tissue sections (1 μm) enhance the resolution of the cellular detail compared with wax-embedded material. This intermediate step of resin embedding naturally leads to possible electron-microscopic examination of ultrathin tissue sections, and electron microscopy remains the gold standard for confirmation of microsporidial infection in tissues (Weber et al. 1994; Rabeneck et al. 1993; Asmuth et al. 1994; Field et al. 1993b). Demonstration of the coiled polar tube within spores is pathognomic of microsporidial infection (Figure 25.2) and identification to generic and sometimes species level can be made from ultrastructural features (Cali 1991). The identification of the species involved in new patients is important, as effective drug therapies can then be instigated. Some workers regard electron microscopy as cumbersome and time-consuming (Visvesvara et al. 1994), but this technique has been involved in the detection and description of most new human microsporidial infections. Molecular detection has usually followed after the detection and description by electron microscopy. In the expanding field of human microsporidial infection, both electron microscopy and molecular methods have their merits.

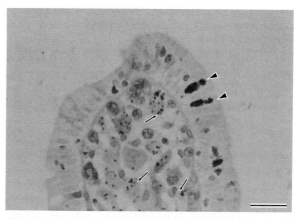

Figure 25.18 *Light micrograph of a tip of a duodenal villus showing aggregates of* Encephalitozoon intestinalis *in enterocytes (arrowheads) and macrophages in the lamina propria (arrows). Warthin–Starry stain. (Reproduced with permission of the authors and publisher from Field et al. 1993b)*

Immunofluorescent or immunoperoxidase antibody staining of microsporidia in tissue sections would be useful diagnostically if such antibodies to species or genera were commercially available. Visvesvara et al. (1994) produced both polyclonal and monoclonal antibodies to *E. hellem* that successfully differentiated this organism from other species in clinical specimens and in established cultures. Other studies (Schwartz et al. 1993a) have shown that, using the indirect fluorescent antibody test (IFAT) with species-specific antibodies, it is possible to identify microsporidial infection in sections of corneal biopsies.

Because *E. cuniculi*, *E. hellem*, and *E. intestinalis* can be grown in culture (Schwartz et al. 1993a; Hollister et al. 1993b; Beauvais et al. 1994; Van Gool et al. 1994; Visvesvara et al. 1995; Doultree et al. 1995), purified spore preparations can be harvested for inoculation into rabbits or mice. Aldras et al. (1994) describe the production of monoclonal antibodies raised against *E. hellem* and polyclonal antibodies raised against *E. cuniculi* and *E. hellem* which, when used in IFAT, were able to identify specifically the presence of microsporidial spores, including those of *Ent. bieneusi*, in stool samples.

The use of a commercially available fluorescein-labeled monoclonal antibody against species of *Encephalitozoon* has now been reported (Enriquez et al. 1998) and was used to demonstrate the prevalence of *Encephalitozoon* (7.84 percent) in two rural villages in Mexico.

Microsporidial spores in stool samples

Noninvasive techniques have been developed for the detection of microsporidial spores in, for example, stool samples. The contents of the duodenum–jejunum can also be examined for microsporidial spores using the Entero-test (HDC Corporation) method, which is often used to obtain luminal contents for detection of *Giardia* (Hamour et al. personal communication 1995). The staining properties of the microsporidian spore wall differentiate microsporidians from most other microorganisms. Spore-staining methods utilizing a modified Trichrome stain (chromotrope 2R stain) (Weber et al. 1992a), Giemsa (Van Gool et al. 1990) or fluorescent dyes (optical brightening agents) (Van Gool et al. 1993; Vávra et al. 1993) have been described; all these microscopical methods require adequate illumination, high magnification objectives (oil immersion) (Weber et al. 1994), examination of multiple stool samples and an adequate examination time (minimum 10 minutes) (Wuhib et al. 1994) or examination of at least 100 high power fields (Bendall and Chiodini 1993) for reliable diagnosis.

Weber's method (Weber et al. 1992a) utilizing chromotrope 2R has become a routine diagnostic staining method in many laboratories, particularly in the USA.

Microsporidial spores stain pink and surrounding bacteria light green in thin stool smears. A number of modifications to Weber's method have been published (Ryan et al. 1993; Kokoskin et al. 1994; Sianongo et al. 2001).

Microsporidial spores are of the same order of size as bacteria and any diagnostic staining method must allow differentiation between spores and fecal bacteria. In addition, the microscopist must be able to differentiate between microsporidial spores and those of fungi and yeasts, as well as oocysts of coccidian protozoal parasites. Of the fluorescent dyes, Uvitex 2B (Ciba-Geigy 48) and Calcofluor (Sigma) are useful for fecal screening as they both stain the chitin in the endospore layer of microsporidial spores. They are quick to use, but require an epi-illumination, fluorescence microscope fitted with a 350–380 nm excitation filter and a light source that emits such wavelengths (e.g. a mercury vapor lamp). Spores are identified by their size, shape, and brilliant blue-white fluorescence staining properties (Figures 25.19 and 25.20). As fungal and yeast spores also contain chitin, adding Formalin fixative to a fresh fecal sample as quickly as possible reduces contamination due to continued fungal growth. In addition, putative positive microsporidial samples should ideally be confirmed and the species determined by another method such as PCR or electron microscopy (Corcoran et al. 1995). If such confirmatory techniques are not readily available, it is sometimes possible to determine the species involved by careful measurement of any spores detected. *Ent. bieneusi* has very small spores that measure about 1.5 μm in length, compared to spores of *Encephalitozoon* species that are about 2.5 μm in length. Experience is required for reliable results.

Some caution may be needed in ascribing the finding of microsporidial spores in stool samples to patient symptoms. McDougall et al. (1993) report the finding of microsporidian spores in a stool sample from an AIDS patient with chronic diarrhea, anorexia, and lethargy. Further investigations suggested that the microsporidia found had been ingested as food in heavily infected muscle and were, therefore, an incidental finding. This situation is, perhaps, not uncommon. Levine (1985) has reported that oocysts of a number of animal coccidian parasites ingested in food and found in human feces have been mistaken for parasites of humans.

Concurrent infections involving both microsporidia and *Cryptosporidium parvum* may also be common. Garcia et al. (1994) showed that 17 of 60 (28 percent) mainly immunocompromised patients with diarrhea had simultaneous *Cryptosporidium* and microsporidial infections. The finding of one parasite should not preclude further examination by a different diagnostic method for other parasites. In addition, enteric infection with both *Ent. bieneusi* and *E. intestinalis* may not be unusual (Blanshard et al. 1992b; Van Gool et al. 1994) and this may have important implications for treatment, as *E.*

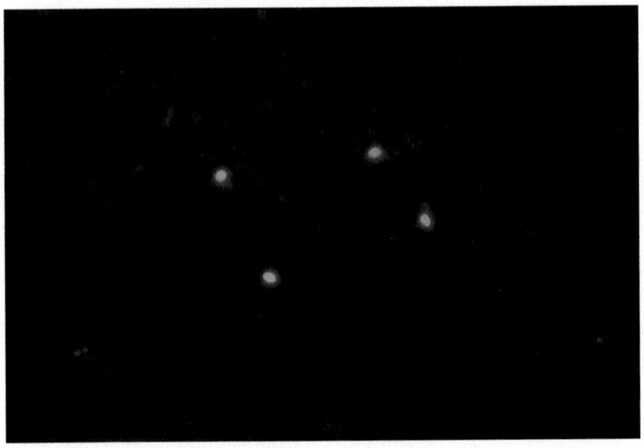

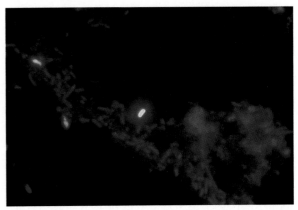

Figure 25.19 *Spores of* Enterocytozoon bieneusi *from a stool sample stained with calcofluor and examined with 350–380 nm wavelength illumination under an epi-illumination fluorescence microscope. Note ovoid shape of spores, which are shorter than those of* E. intestinalis *(c.f. Figure 25.20).*

Figure 25.20 *Spores of* Encephalitozoon intestinalis *from a stool sample stained with calcofluor. Note that the spores of this species are longer than those of* Enterocytozoon bieneusi *(c.f. Figure 25.19)*

intestinalis appears to be fully susceptible to treatment with albendazole, whereas the response of *Ent. bieneusi* is variable.

Molecular methods of detection

It is now possible to generate sequence data rapidly by direct sequencing of amplified ribosomal DNA using the PCR technique. This technique has identified sequence variation and shows considerable promise for the specific identification of microsporidian species as well as for increasing our knowledge of the epidemiology of these organisms. Molecular techniques for identification of the commoner species of microsporidian infecting humans using either stool or intestinal biopsy specimens are available (Kock et al. 1997).

Sequences have been produced mainly from the small subunit ribosomal RNA gene of microsporidia and from these it is possible to identify genus or species.

Microsporidian DNA can be extracted fairly easily from infected tissues or cells by procedures such as proteinase K digestion followed by phenol-chloroform extraction and ethanol precipitation, but DNA extraction from spores is more difficult. Mechanical disruption of the spores in addition to digestion is required (Franzen and Müller 1999; da Silva et al. 1999).

Vossbrinck et al. (1993) sequenced a segment of ribosomal DNA about 1350 bp long and demonstrated that both *E. cuniculi* and *E. hellem* could be specifically identified. A comparison of the sequence data showed relatively high sequence homology, justifying the same generic status of these organisms but also confirmed that the isolates of *E. hellem* and *E. cuniculi* were not of the same species. Restriction digests of the amplified region of ribosomal DNA provides the potential for a rapid method of distinguishing between *E. hellem* and *E. cuniculi*. Visvesvara et al. (1994) using direct sequencing of PCR amplified small-subunit ribosomal RNA (SSU-

rRNA) demonstrated similar specific differences between these two species. Initially, PCR techniques were largely based on cultured organisms, but amplification directly from clinical specimens has improved the speed of diagnosis (Visvesvara et al. 1994; Zhu et al. 1994).

A PCR assay of stool samples from AIDS patients known to be infected with microsporidia (identified by chromotrope 2R stain) not only detected microsporidial infection but also identified the enteric species involved (Fedorko et al. 1995). However, this assay required a laborious 4-day extraction procedure and was strongly inhibited by fresh stools, although this could be overcome with the addition of sodium hypochlorite. Current extraction methods are much faster (Franzen and Müller 1999; da Silva et al. 1999). Cultured spores of *E. intestinalis* were added to a stool specimen positive for *Ent. bieneusi* to simulate a dual infection. A single primer (PMP2) complementary to conserved sequences of the SSU-rRNA enabled amplification of DNA from both species. An important observation was that whereas *Ent. bieneusi* lacks a Pst1 restriction site in the amplicon, that of *E. intestinalis* cut into two distinct fragments, thus enabling differentiation of the two species. The authors suggest that PCR may become the method of choice for identification of microsporidia in clinical specimens.

Canning et al. (1993) have also demonstrated differences between *V. corneae*, *E. cuniculi* and *E. hellem* using a random amplified polymorphic DNA (RAPD) polymerase chain reaction.

Vossbrink et al. (1993) suggested that sequence information appropriate for comparison with other microsporidial species should be included in future species descriptions for the specific identification and phylogenetic comparisons of microsporidia.

There is a comprehensive review of molecular techniques applied to microsporidia by (Franzen and Müller (1999).

Culture of microsporidia from clinical samples

Microsporidial spores from clinical samples such as urine (Visvesvara et al. 1991; Bocket et al. 1992; Hollister et al. 1993a; Visvesvara et al. 1995; Hollister et al. 1995; Deplazes et al. 1998), nasal polyps (Hollister et al. 1993b), nasopharyngeal aspirates (Doultree et al. 1995), feces (Van Gool et al. 1994), and muscle biopsies (Hollister et al. 1995) can sometimes be cultured in various cell lines. Once established in culture, enough spores can be produced from the original human sample to allow further tests, such as SDS–PAGE or PCR, to be carried out, which can help to establish the species involved. Ultrastructural studies of the organisms in culture can also be undertaken (Didier P.J. et al. 1991) Of the microsporidia that infect humans, *E. cuniculi*, *E. intestinalis*, *E. hellem*, *T. hominis*, *V. corneae* and *B. algerae* have been successfully cultured (Visvesvara 2002). Once established, activity of anti-microsporidial agents, such as albendazole, can be examined using the culture system (Colbourn et al. 1994; Lafranchi-Tristem et al. 2001). However, *Ent. bieneusi*, the commonest human microsporidial infection, has yet to be successfully cultured and this has hampered research into effective treatments against this species.

CHEMOTHERAPY

Problems of treatment

Microsporidial infections are difficult to treat because of their intracellular habitat and the resistant nature of the spores. Enteric infections have been treated with varying degrees of success with several drugs, but carefully controlled comparative treatment trials are lacking.

Treatment of enteric microsporidial infections

The most frequent microsporidial infections of humans are enteric, *Ent bieneusi* being more commonly detected than *E. intestinalis*. *Encephalitozoon intestinalis* appears to be more susceptible to treatment than *Ent. bieneusi*.

OCTREOTIDE

Octreotide produced a partial remission in one patient infected by *Ent. bieneusi* (Simon et al. 1991) and metronidazole caused some remission in 10 of 13 HIV seropositive patients with mild microsporidial diarrhea caused by *Ent. bieneusi* (Eeftinck Schattenkerk et al. 1991). Blanshard et al. (1992a) treated six AIDS patients with diarrhea and biopsy confirmed infection with *Ent. bieneusi*, but no other identified cause of symptoms.

ALBENDAZOLE

Albendazole was found to be well tolerated. Within 7 days of treatment, there was a marked reduction in stool frequency, volume, and incontinence, with all patients either gaining weight or weight loss being arrested. Four patients relapsed within a month of cessation of treatment, three of whom responded to a further course of albendazole treatment. It was concluded that albendazole had a partial parasitostatic effect, rather than a parasiticidal action. Further studies by Blanshard et al. (1993) confirmed this conclusion. In biopsy material taken from patients being treated with albendazole, damage to developmental stages of *Ent. bieneusi*, causing a partial inhibition of parasite replication, was apparent. Dieterich et al. (1994) showed a similar reduction of symptoms in patients treated with albendazole, but again concluded that *Ent. bieneusi* was not eliminated.

In contrast to *Ent. bieneusi*, *E. intestinalis* appears to be more susceptible to treatment with albendazole, as demonstrated by the disappearance of parasites from the intestine of a case of disseminated infection (Orenstein et al. 1993). Gunnarsson et al. (1995) used albendazole to treat five patients with evidence of multiorgan infection with *E. intestinalis*. After treatment, most clinical samples were negative for microsporidia and there was a significant improvement of symptoms.

The direct effect of albendazole on cell cultures infected with *E. cuniculi* has been examined by Colbourn et al. (1994). Many of the proliferative stages of the parasite became grossly enlarged and devoid of nuclei. As albendazole prevents the polymerization of microtubules, which are known to occur only within the intranuclear spindles of dividing nuclei in microsporidia, parasite growth continues in the absence of nuclear division.

Albendazole was administered to five AIDS patients (400 mg orally twice a day) with evidence of disseminated microsporidiosis caused by *E. intestinalis* (Molina et al. 1995). Spores of *E. intestinalis* were detected in stool and urine samples and confirmed by transmission electron microscopy of duodenal biopsies. There was a rapid clinical response to therapy, with spores being cleared from feces of all patients and from the urine of three. However, during follow-up, spores were detected in feces and mild diarrhea recurred in two patients. It was concluded that albendazole has a significant effect on *E. intestinalis* infection, but that its effects were transient.

Not all AIDS patients relapse. Parasitological cure following albendazole therapy of disseminated *E. intestinalis* infection has been verified at autopsy in two AIDS patients (Sobottka et al. 1995b; Joste et al. 1996). Both patients died of bacterial pneumonia and thorough examination of tissue samples taken at autopsy failed to reveal the presence of microsporidia.

A randomized, double-blind, placebo-controlled trial of albendazole treatment of microsporidiosis due to *E. intestinalis* has been reported by Molina et al. (1998). All four patients in the albendazole treatment group showed clearance of infection compared to none of the four in the control group. The control patients were then given albendazole treatment and the infection cleared. Subsequently, the effect of albendazole prophylaxis was investigated in these eight patients. They were randomly assigned to receive maintenance albendazole therapy (three patients) or no treatment (five patients) for a 12-month period. None of the three patients receiving prophylaxis had a recurrence, but three relapses were recorded among the patients not receiving albendazole. It was concluded that albendazole has both parasitological and clinical efficacy and reduces the risk of relapse.

NITAZOXANIDE

One AIDS patient with chronic diarrhea caused by *Ent. bieneusi* was successfully treated with nitazoxanide twice a day for 60 consecutive days (Bicart-See et al. 2000). At the time of treatment, the patient was not receiving any antiretroviral therapy and his CD4 count was $85 \times 10^6/l$. Fecal examination by microscopy was negative for microsporidian spores after 11 days and post-treatment examination (including PCR) did not detect any microsporidia.

FUMAGILLIN

Although reported as toxic to humans (see below), purified fumagillin (98 percent pure) has been administered to a group of four patients infected with *Ent. bieneusi* to test its efficacy (Molina et al. 1997). Other drug regimens were also used on small groups of patients infected with *Ent. bieneusi* in the same study, but only those treated with the purified fumagillin showed clearance of this parasite (mean follow-up period of 10 months) but the drug also induced thrombocytopenia that was reversed when treatment was discontinued. A semi-synthetic analog of fumagillin, TNP-470, has also shown its effectiveness against the microsporidian *E. cuniculi* in infected rabbit kidney (RK) 13 cells and an athymic nude mouse model (Coyle et al. 1998).

THALIDOMIDE

Thalidomide also appears to have anti-microsporidial activity (Sharpstone et al. 1997). Fecal TNF-α was found to be elevated in cases of HIV-related diarrhea due to enteric microsporidial infection. Thalidomide, an anti-TNF-α agent, was given to 18 patients with AIDS and chronic diarrhea due to *Ent. bieneusi*. Seven of these patients had a complete clinical response and three a partial response to thalidomide therapy. These new treatments may prove to be an effective therapy against *Ent. bieneusi* infection.

NUTRITIONAL THERAPY

Asmuth et al. (1994) also suggested that benefit could be derived from nutritional therapy directed at minimizing malabsorption. Low-fat diets and introduction of simple carbohydrates could also induce an improvement of symptoms.

Treatment of keratoconjunctivitis

Various topical antimicrobial, lubricating and anti-inflammatory agents have been used to treat eyes infected with microsporidia. Itraconazole (Yee et al. 1991) was responsible, at least in part, for the resolution of a corneal infection with *Encephalitozoon* sp. (later identified as *E. hellem* by Didier E.S. et al. (1991). Treatment with propamidine isethionate resolved symptoms in a UK patient infected with *Encephalitozoon* sp. (Metcalfe et al. 1992) (also later identified as *E. hellem* by Hollister et al. 1993b). The symptoms returned when treatment was discontinued (Metcalfe et al. 1992) but albendazole treatment improved the nasal infection, with significant regression of sinus opacification. The patient remained free from nasal symptoms until his death from AIDS dementia complex (Lacey et al. 1992). An Australian case of keratoconjunctivitis was also treated with propamidine isethionate eye ointment, with similar resolution of symptoms (McCluskey et al. 1993).

Fumagillin has been known for some time to control *Nosema* infection in honeybees, where infection causes severe economic loss (Katznelson and Jamieson 1952). Shadduck (1980) observed the in vitro effects of fumagillin on *E. cuniculi* and noted that it protected noninfected cells but did not eliminate the parasite from infected ones. This anti-microsporidial agent is toxic to humans (McCowen et al. 1951), but has been used topically to resolve ocular infection by *E. hellem*. Fumagillin is insoluble in water, but a derivative, Fumidil B, is soluble. A Fumidil B preparation applied topically reduced the keratoconjunctivitis in two AIDS patients (Diesenhouse et al. 1993). No toxic side-effects were observed but infection recurred when treatment was stopped. Both Fumidil B and propamidine isethionate, when used topically to treat microsporidial infections of the eye, appear to be inhibitory rather than parasiticidal in action. Rosberger et al. (1993) has also reported successful treatment of keratoconjunctivitis with topical fumagillin.

Treatment of muscle and other microsporidial infections

Albendazole, sometimes in combination with other chemotherapeutic agents, appears to be the treatment of choice for many microsporidial infections.

In one patient with AIDS, myositis due to infection with *T. hominis* was successfully treated with albendazole, sulfadiazine, and pyrimethamine, with resolution of clinical symptoms (Field et al. 1996). Another AIDS patient with myositis due to *B. vesicularum* infection was treated with albendazole and itraconazole, which successfully cleared the infection both clinically and histologically (Cali et al. 1998).

An AIDS patient with respiratory, abdominal, and urogenital symptoms was found to be infected with two different species of microsporidian, *E. hellem* and *V. corneae*. Treatment with albendazole resulted in abatement of symptoms. Microsporidian spores disappeared from respiratory secretions and there was a reduction in the urinary excretion of spores (Deplazes et al. 1998).

Dunand et al. (1997) reported five cases of parasitic sinusitis and otitis, three of which had *E. hellem* infection of the sinuses. Two of these patients were treated with albendazole. In one patient, the sinus symptoms decreased moderately, whereas in the second, the sinus symptoms initially decreased but because the patient did not receive maintenance (suppressive) therapy, the symptoms recurred within 5 months.

Aarons et al. (1994) describe a case of renal failure in an HIV-seropositive patient, caused by a species of *Encephalitozoon*, which was reversible on treatment with albendazole and correlated with the disappearance of the parasite from the urine. In another AIDS patient with disseminated *E. intestinalis* infection and evidence of impairment of renal function, renal function returned to normal after a course of treatment with albendazole (Dore et al. 1995).

Birthistle et al. (1996) have reported a case of urethritis caused by microsporidial infection, which was successfully cleared after albendazole treatment. The authors believe that this patient had acquired microsporidial infection sexually from his partner who, at the same time, had diarrhea due to intestinal microsporidiosis.

Reconstitution of the immune system

In developed countries, highly active antiretroviral therapy (HAART) is available to AIDS patients and this induces progressive reconstitution of the immune system (Montaner et al. 1998). Remission of intestinal microsporidiosis occurs in patients given HAART and this appears to correlate with an increase in CD4 and CD8 lymphocyte counts (Goguel et al. 1997; Kartalija and Sande 1999).

DISINFECTION AND SUSCEPTIBILITY TO PHYSICAL AND CHEMICAL AGENTS

Disinfection of surfaces contaminated with microsporidia has received little attention. That spores are highly resistant has been demonstrated in the laboratory, where they have been found to be viable after storage

for up to 10 years in distilled water (Sprague and Vávra 1977). In the experimental infection of gnotobiotic piglets, Kondova et al. (1998) found that spores of *Ent bieneusi* could remain infectious for at least 2 years when stored at $+4°C$, although the extent, if any, of the loss of viability over this period remains unknown. Shadduck and Polley (1978) used a rabbit isolate of *E. cuniculi* grown in a rabbit choroid plexus cell line to evaluate various factors influencing the infectivity and replication of this organism. *E. cuniculi* was not affected by penicillin, streptomycin or gentamicin, nor was it affected by sonication, freezing and thawing or distilled water. Organisms survived for 60 min but not 120 min at $56°C$. They were, however, killed after 10 min of autoclaving at $120°C$, or exposure to 2 percent (v/v) lysol, 10 percent (v/v) formalin or 70 percent (v/v) ethyl alcohol for 10 min. Whether all microsporidian species are affected by these physical conditions to the same degree as *E. cuniculi* is not known.

Undeen and Vander Meer (1990) subjected spores of *N. algerae* to ultraviolet radiation and found that very high dosages (3.8 J/cm^2) were required to inhibit germination. Similar results had been obtained with γ-radiation (Undeen et al. 1984) and it was concluded that spore germination did not involve nuclear function. Significantly, after receiving γ-radiation above 0.5 kGy, this microsporidian was unable to infect its mosquito host, but spore germination rate remained unaffected up to 10 kGy.

In work undertaken by Leitch et al. (1993), on the inhibition of polar tube extrusion in purified spores derived from cell cultures infected with *E. hellem*, germination was found to be pH dependent and was enhanced in the presence of calcium in the medium. Under experimental conditions, four agents inhibited polar tube extrusion: cytochalasin D disrupts microfilaments, demecolcine disrupts microtubules, nifedipine is a calcium channel blocker, and itraconazole an antifungal agent. As a result of this work, it was concluded that future clinical trials employing drug combinations might be useful in treating microsporidiosis.

CONCLUSION

Much remains to be learned about microsporidians, particularly in relation to human infection (Current and Blagburn 1991) and little is known about the epidemiology or pathogenesis of these parasites in humans. Whether or not there are animal sources of human infection will become clearer as epidemiological and experimental studies become better established. Molecular techniques are becoming more commonly available and may provide incontrovertible evidence of zoonotic transmission and also establish the phylogenetic and taxonomic relationships of the microsporidian species involved. Further developments in treatment are needed, particularly in relation to *Ent. bieneusi* infection. Developments in the culture and of animal models of this species would aid improvements in treatment and understanding.

In the developed world, because of improved treatment against HIV infection and corresponding partial recovery of cellular immunity, human microsporidiosis is now relatively infrequently diagnosed. This contrasts with the situation in the developing world, where microsporidial infections are much more commonly involved in human disease, particularly in those suffering from AIDS. Because of such inequalities in treatment, further study of these parasites is warranted to further improve our knowledge of these emerging pathogens.

REFERENCES

Aarons, E.J., Woodrow, D., et al. 1994. Reversible renal failure caused by a microsporidian infection. *AIDS*, **8**, 1119–21.

Aldras, A.M., Orenstein, J.M., et al. 1994. Detection of microsporidia by indirect immunofluorescence antibody test using polyclonal and monoclonal antibodies. *J Clin Microbiol*, **32**, 608–12.

Andreadis, T.G. and Vossbrinck, C.R. 2002. Life cycle, ultrastructure and molecular phylogeny of *Hyalinocysta chapmani* (Microsporidia: Thelohaniidae), a parasite of *Culiseta melanura* (Diptera: Culicidae) and *Orthocyclops modestus* (Copepoda: Cyclopidae). *J Eukaryot Microbiol*, **49**, 350–64.

Ashton, N. and Wirasinha, P.A. 1973. Encephalitozoonosis (nosematosis) of the cornea. *Br J Ophthalmol*, **57**, 669–74.

Asmuth, D.M., DeGirolami, P.C., et al. 1994. Clinical features of microsporidiosis in patients with AIDS. *Clin Infect Dis*, **18**, 819–25.

Beaugerie, L., Teilhac, M.F., et al. 1992. Cholangiopathy associated with Microsporidia infection of the common bile duct mucosa in a patient with HIV infection. *Ann Intern Med*, **117**, 401–2.

Beauvais, B., Sarfati, C., et al. 1994. In vitro model to assess effect of antimicrobial agents on *Encephalitozoon cuniculi*. *Antimicrob Agents Chemother*, **38**, 2440–8.

Bendall, R.P. and Chiodini, P.L. 1993. New diagnostic methods for parasitic infections. *Curr Opin Infect Dis*, **6**, 318–23.

Bergquist, N.R., Stintzing, G., et al. 1984. Diagnosis of encephalitozoonosis in man by serological tests. *Br Med J*, **288**, 902.

Bicart-See, A., Massip, P., et al. 2000. Successful treatment with nitazoxanide of *Enterocytozoon bieneusi* microsporidiosis in a patient with AIDS. *Antimicrob Agents Chemother*, **44**, 167–8.

Biderre, C., Mathis, A., et al. 1999. Molecular karyotype diversity in the microsporidian *Encephalitozoon cuniculi*. *Parasitology*, **118**, 439–45.

Bigliardi, E. and Sacchi, L. 2001. Cell biology and invasion of the microsporidia. *Microbes Infect*, **3**, 373–9.

Birthistle, K., Moore, P. and Hay, P. 1996. Microsporidia: a new sexually transmissible cause of urethritis. *Genitourin Med*, **72**, 445.

Black, S.S., Steinohrt, L.A., et al. 1997. *Encephalitozoon hellem* in budgerigars (*Melopsittacus undulatus*). *Vet Pathol*, **34**, 189–98.

Blanshard, C., Ellis, D.S., et al. 1992a. Treatment of intestinal microsporidiosis with albendazole in patients with AIDS. *AIDS*, **6**, 311–13.

Blanshard, C., Hollister, W.S., et al. 1992b. Simultaneous infection with two types of intestinal microsporidia in a patient with AIDS. *Gut*, **33**, 418–20.

Blanshard, C., Ellis, D.S., et al. 1993. Electron microscopic changes in *Enterocytozoon bieneusi* following treatment with albendazole. *J Clin Pathol*, **46**, 898–902.

Bocket, L., Marquette, C.H., et al. 1992. Isolation and replication in human fibroblast cells (MRC-5) of a microsporidian from an AIDS patient. *Microb Pathog*, **12**, 187–91.

Bornay-Llinares, F.J., da Silva, A.J., et al. 1998. Immunologic, microscopic, and molecular evidence of *Encephalitozoon intestinalis*

(*Septata intestinalis*) infection in mammals other than humans. *J Infect Dis*, **178**, 820–6.

Breitenmoser, A.C., Mathis, A., et al. 1999. High prevalence of *Enterocytozoon bieneusi* in swine with four genotypes that differ from those identified in humans. *Parasitology*, **118**, 447–53.

Bretagne, S., Foulet, F., et al. 1993. Prevalence of microsporidial spores in stools from children in Niamey, Niger. *AIDS*, **7**, Suppl, S34–5.

Cali, A. 1991. General microsporidian features and recent findings on AIDS isolates. *J Protozool*, **38**, 625–30.

Cali, A. and Owen, R.L. 1990. Intracellular development of *Enterocytozoon*, a unique microsporidian found in the intestine of AIDS patients. *J Protozool*, **37**, 145–55.

Cali, A. and Takvoriah, P.M. 2003. Ultrastructure and development of *Pleistophora ronneafiei* n. sp., a microsporidium (Protista) in the skeletal muscle of an immune-compromised individual. *J Eukaryot Microbiol*, **50**, 77–85.

Cali, A., Kotler, D.P. and Orenstein, J.M. 1993. *Septata intestinalis* n. g., n. sp., an intestinal microsporidian associated with chronic diarrhea and dissemination in AIDS patients. *J Eukaryot Microbiol*, **40**, 101–12.

Cali, A., Weiss, L.M. and Takvorian, P.M. 2002. *Brachiola algerae* spore membrane systems, their activity during extrusion, and a new structural entity, the multilayered interlaced network, associated with the polar tube and the sporoplasm. *J Eukaryot Microbiol*, **49**, 164–74.

Cali, A., Meisler, D.M., et al. 1991a. Corneal microsporidioses: characterization and identification. *J Protozool*, **38**, Suppl, S215–7.

Cali, A., Meisler, D.M., et al. 1991b. Corneal microsporidiosis in a patient with AIDS. *Am J Trop Med Hyg*, **44**, 463–8.

Cali, A., Orenstein, J.M., et al. 1991c. A comparison of two microsporidian parasites in enterocytes of AIDS patients with chronic diarrhea. *J Protozool*, **38**, Suppl, S96–8.

Cali, A., Takvorian, P.M., et al. 1998. *Brachiola vesicularum*, n. g., n. sp., a new microsporidium associated with AIDS and myositis. *J Eukaryot Microbiol*, **45**, 240–51.

Canning, E.U. 1988. Nuclear division and chromosome cycle in microsporidia. *Biosystems*, **21**, 333–40.

Canning, E.U. 1993. Microsporidia. In: Kreier, J.P. and Baker, J.R. (eds), *Parasitic Protozoa*. New York: Academic Press, 299–370.

Canning, E.U. 1998. Microsporidiosis. In: Palmer, S.R., Soulsby, L. and Simpson, D.I.H. (eds), *Zoonoses*. Oxford: Oxford Medical Publications, 609–23.

Canning, E.U. 2001. Microsporidia. In: Gillespie, S.H. and Pearson, R.D. (eds), *Principles and practice of clinical parasitology*. Chichester: John Wiley and Sons, 171–95.

Canning, E.U. and Lom, J. 1986. *The microsporidia of vertebrates*. London: Academic Press.

Canning, E.U., Curry, A., et al. 1992. Ultrastructure of *Encephalitozoon* sp. infecting the conjunctival, corneal and nasal epithelia of a patient with AIDS. *Eur J Protistol*, **28**, 226–37.

Canning, E.U., Hollister, W.S., et al. 1993. Human microsporidioses: site specificity, prevalence and species identification. *AIDS*, **7**, Suppl 3, S3–7.

Canning, E.U., Field, A.S., et al. 1994. Further observations on the ultrastructure of *Septata intestinalis* Cali, Kotler and Orenstein, 1993. *Eur J Protistol*, **30**, 414–22.

Canning, E.U., Curry, A., et al. 1998. Some ultrastructural data on *Microsporidium ceylonensis*, a cause of corneal microsporidiosis. *Parasite*, **5**, 247–54.

Cavalier-Smith, T. 2001. What are fungi? In: McLauchlin, D.J., McLauchlin, E.G. and Lemke, P.A. (eds), *Mycota. Part A. Systematics and evolution*, VII. . Berlin: Springer-Verlag, 1–37.

Cheney, S.A., Lafranchi-Tristem, N.J. and Canning, E.U. 2000. Phylogenetic relationships of *Pleistophora*-like microsporidia based on small subunit ribosomal DNA sequences and implications for the source of *Trachipleistophora hominis* infections. *J Eukaryot Microbiol*, **47**, 280–7.

Chupp, G.L., Alroy, J., et al. 1993. Myositis due to *Pleistophora* (Microsporidia) in a patient with AIDS. *Clin Infect Dis*, **16**, 15–21.

Colbourn, N.I., Hollister, W.S., et al. 1994. Activity of albendazole against *Encephalitozoon cuniculi* in vitro. *Eur J Protistol*, **30**, 211–20.

Corcoran, G.D., Tovey, D.G., et al. 1995. Detection and identification of gastrointestinal microsporidia using non-invasive techniques. *J Clin Pathol*, **48**, 725–7.

Corcoran, G.D., Isaacson, J.R., et al. 1996. Urethritis associated with disseminated microsporidiosis: clinical response to albendazole. *Clin Infect Dis*, **22**, 592–3.

Cox, J.C., Hamilton, R.C. and Attwood, H.D. 1979. An investigation of the route and progression of *Encephalitozoon cuniculi* infection in adult rabbits. *J Protozool*, **26**, 260–5.

Coyle, C., Kent, M., et al. 1998. TNP-470 is an effective antimicrosporidial agent. *J Infect Dis*, **177**, 515–18.

Current, W.L. and Blagburn, B.L. 1991. *Cryptosporidium* and microsporidia: some closing comments. *J Protozool*, **38**, Suppl, S244–5.

Curry, A. 1999. Human microsporidial infection and possible animal sources. *Curr Opin Infect Dis*, **12**, 473–80.

Curry, A. and Canning, E.U. 1993. Human microsporidiosis. *J Infect*, **27**, 229–36.

Curry, A., Turner, A.J. and Lucas, S. 1991. Opportunistic protozoan infections in human immunodeficiency virus disease: review highlighting diagnostic and therapeutic aspects. *J Clin Pathol*, **44**, 182–93.

Curry, A., McWilliam, L.J., et al. 1988. Microsporidiosis in a British patient with AIDS. *J Clin Pathol*, **41**, 477–8.

da Silva, A.J., Bornay-Llinares, F.J., et al. 1999. Fast and reliable extraction of protozoan parasite DNA from faecal specimens. *Mol Diagn*, **4**, 57–64.

De Groote, M.A., Visvesvara, G., et al. 1995. Polymerase chain reaction and culture confirmation of disseminated *Encephalitozoon cuniculi* in a patient with AIDS: successful therapy with albendazole. *J Infect Dis*, **171**, 1375–8.

del Aguila, C., Lopez-Velez, R., et al. 1997. Identification of *Enterocytozoon bieneusi* spores in respiratory samples from an AIDS patient with a 2-year history of intestinal microsporidiosis. *J Clin Microbiol*, **35**, 1862–6.

del Aguila, C., Izquierdo, F., et al. 1999. *Enterocytozoon bieneusi* in animals: rabbits and dogs as new hosts. *J Eukaryot Microbiol*, **46**, Suppl, 8S–9S.

Dengjel, B., Zahler, M., et al. 2001. Zoonotic potential of *Enterocytozoon bieneusi*. *J Clin Microbiol*, **39**, 4495–9.

Deplazes, P., Mathis, A., et al. 1996. Molecular epidemiology of *Encephalitozoon cuniculi* and first detection of *Enterocytozoon bieneusi* in faecal samples of pigs. *J Eukaryot Microbiol*, **43**, Suppl, 93S.

Deplazes, P., Mathis, A., et al. 1998. Dual microsporidial infection due to *Vittaforma corneae* and *Encephalitozoon hellem* in a patient with AIDS. *Clin Infect Dis*, **27**, 1521–4.

Desportes, I., Le Charpentier, Y., et al. 1985. Occurrence of a new microsporidan: *Enterocytozoon bieneusi* n. g., n. sp., in the enterocytes of a human patient with AIDS. *J Protozool*, **32**, 250–4.

Didier, E.S., Didier, P.J., et al. 1991. Isolation and characterization of a new human microsporidian, *Encephalitozoon hellem* (n. sp.), from three AIDS patients with keratoconjunctivitis. *J Infect Dis*, **163**, 617–21.

Didier, E.S., Visvesvara, G.S., et al. 1996. A microsporidian isolated from an AIDS patient corresponds to *Encephalitozoon cuniculi* III, originally isolated from domestic dogs. *J Clin Microbiol*, **34**, 2835–7.

Didier, P.J., Didier, E.S., et al. 1991. Fine structure of a new human microsporidian, *Encephalitozoon hellem*, in culture. *J Protozool*, **38**, 502–7.

Diesenhouse, M.C., Wilson, L.A., et al. 1993. Treatment of microsporidial keratoconjunctivitis with topical fumagillin. *Am J Ophthalmol*, **115**, 293–8.

Dieterich, D.T., Lew, E.A., et al. 1994. Treatment with albendazole for intestinal disease due to *Enterocytozoon bieneusi* in patients with AIDS. *J Infect Dis*, **169**, 178–83.

Dore, G.J., Marriott, D.J., et al. 1995. Disseminated microsporidiosis due to *Septata intestinalis* in nine patients infected with the human immunodeficiency virus: response to therapy with albendazole. *Clin Infect Dis*, **21**, 70–6.

Doultree, J.C., Maerz, A.L., et al. 1995. In vitro growth of the microsporidian *Septata intestinalis* from an AIDS patient with disseminated illness. *J Clin Microbiol*, **33**, 463–70.

Dowd, S.E., Gerba, C.P. and Pepper, I.L. 1998. Confirmation of the human-pathogenic microsporidia *Enterocytozoon bieneusi*, *Encephalitozoon intestinalis* and *Vittaforma corneae* in water. *Appl Environ Microbiol*, **64**, 3332–5.

Dunand, V.A., Hammer, S.M., et al. 1997. Parasitic sinusitis and otitis in patients infected with human immunodeficiency virus: report of five cases and review. *Clin Infect Dis*, **25**, 267–72.

Eeftinck Schattenkerk, J.K., Van Gool, T., et al. 1991. Clinical significance of small-intestinal microsporidiosis in HIV-1-infected individuals. *Lancet*, **337**, 895–8.

Enriquez, F.J., Taren, D., et al. 1998. Prevalence of intestinal encephalitozoonosis in Mexico. *Clin Infect Dis*, **26**, 1227–9.

Fedorko, D.P., Nelson, N.A. and Cartwright, C.P. 1995. Identification of microsporidia in stool specimens by using PCR and restriction endonucleases. *J Clin Microbiol*, **33**, 1739–41.

Field, A.S., Canning, E.U., et al. 1993a. Microsporidia in HIV-infected patients in Sydney, Australia: a report of 37 cases, a new diagnostic technique and the light microscopy and ultrastructure of a disseminated species. *AIDS*, **7**, Suppl 3, S27–33.

Field, A.S., Hing, M.C., et al. 1993b. Microsporidia in the small intestine of HIV-infected patients. A new diagnostic technique and a new species. *Med J Aust*, **158**, 390–4.

Field, A.S., Marriott, D.J. and Hing, M.C. 1993c. The Warthin-Starry stain in the diagnosis of small intestinal microsporidiosis in HIV-infected patients. *Folia Parasitol (Praha)*, **40**, 261–6.

Field, A.S., Marriott, D.J., et al. 1996. Myositis associated with a newly described microsporidian, *Trachipleistophora hominis*, in a patient with AIDS. *J Clin Microbiol*, **34**, 2803–11.

Foissner, I. and Foissner, W. 1995. Ciliatosporidium platyophryae nov. gen., nov. spec. (Microspora incerta sedis), a parasite of *Platyophrya terricola* (Ciliophora, Colpodea). *Eur J Protistol*, **31**, 248–59.

Fournier, S., Liguory, O., et al. 1998. Microsporidiosis due to *Enterocytozoon bieneusi* infection as a possible cause of traveler's diarrhea. *Eur J Clin Microbiol Infect Dise*, **17**, 743–4.

Franzen, C. and Müller, A. 1999. Molecular techniques for detection, species differentiation, and phylogenetic analysis of microsporidia. *Clin Microbiol Rev*, **12**, 243–85.

Franzen, C., Schwartz, D.A., et al. 1995. Immunologically confirmed disseminated, asymptomatic *Encephalitozoon cuniculi* infection of the gastrointestinal tract in a patient with AIDS. *Clin Infect Dis*, **21**, 1480–4.

French, A.L., Beaudet, L.M., et al. 1995. Cholecystectomy in patients with AIDS: clinopathologic correlations in 107 cases. *Clin Infect Dis*, **21**, 852–8.

Friedberg, D.N., Stenson, S.M., et al. 1990. Microsporidial keratoconjunctivitis in acquired immunodeficiency syndrome. *Arch Ophthalmol*, **108**, 504–8.

Furuta, M., Obara, A., et al. 1991. *Cryptococcus neoformans* can be misidentified as a microsporidian: studies of lung lesions in leprosy patients. *J Protozool*, **38**, Suppl, S95–6.

Gainzarain, J.C., Canut, A., et al. 1998. Detection of *Enterocytozoon bieneusi* in two human immunodeficiency virus-negative patients with chronic diarrhea by polymerase chain reaction in duodenal biopsy specimens and review. *Clin Infect Dis*, **27**, 394–8.

Garcia, L.S., Shimizu, R.Y. and Bruckner, D.A. 1994. Detection of microsporidial spores in fecal specimens from patients diagnosed with cryptosporidiosis. *J Clin Microbiol*, **32**, 1739–41.

Glaser, C.A., Angulo, F.J. and Rooney, J.A. 1994. Animal-associated opportunistic infections among persons infected with the human immunodeficiency virus. *Clin Infect Dis*, **18**, 14–24.

Goguel, J., Katlama, C., et al. 1997. Remission of AIDS-associated intestinal microsporidiosis with highly active antiretroviral therapy. *AIDS*, **11**, 1658–9.

Greenson, J.K., Belitsos, P.C., et al. 1991. AIDS enteropathy: occult enteric infections and duodenal mucosal alterations in chronic diarrhea. *Ann Intern Med*, **114**, 366–72.

Gumbo, T., Gangaidzo, I.T., et al. 2000. *Enterocytozoon bieneusi* infection in patients without evidence of immunosuppression: two cases from Zimbabwe found to have positive stools by PCR. *AnnTrop Med Parasitol*, **94**, 699–702.

Gunnarsson, G., Hurlbut, D., et al. 1995. Multiorgan microsporidiosis: report of five cases and review. *Clin Infect Dis*, **21**, 37–44.

Hartskeerl, R.A., Van Gool, T., et al. 1995. Genetic and immunological characterization of the microsporidian *Septata intestinalis* Cali, Kotler and Orenstein, 1993: reclassification to *Encephalitozoon intestinalis*. *Parasitology*, **110**, 277–85.

Hirt, R.P., Healy, B., et al. 1997. A mitochondrial Hsp70 orthologue in *Vairimorpha necatrix*: molecular evidence that microsporidia once contained mitochondria. *Curr Biol*, **7**, 995–8.

Hirt, R.P., Logsdon, J.M., et al. 1999. Microsporidia are related to Fungi: evidence from the largest subunit of RNA polymerase II and other proteins. *Proc Natl Acad Sci USA*, **96**, 580–5.

Hollister, W.S. and Canning, E.U. 1987. An enzyme-linked immunosorbent assay (ELISA) for detection of antibodies to *Encephalitozoon cuniculi* and its use in determination of infections in man. *Parasitology*, **94**, 209–19.

Hollister, W.S., Canning, E.U. and Willcox, A. 1991. Evidence for widespread occurrence of antibodies to *Encephalitozoon cuniculi* (Microspora) in man provided by ELISA and other serological tests. *Parasitology*, **102**, 33–43.

Hollister, W.S., Canning, E.U. and Colbourn, N.I. 1993a. A species of *Encephalitozoon* isolated from an AIDS patient: criteria for species differentiation. *Folia Parasitol (Praha)*, **40**, 293–5.

Hollister, W.S., Canning, E.U., et al. 1993b. Characterization of *Encephalitozoon hellem* (Microspora) isolated from the nasal mucosa of a patient with AIDS. *Parasitology*, **107**, 351–8.

Hollister, W.S., Canning, E.U., et al. 1995. *Encephalitozoon cuniculi* isolated from the urine of an AIDS patient, which differs from canine and murine isolates. *J Eukaryot Microbiol*, **42**, 367–72.

Hollister, W.S., Canning, E.U., et al. 1996. Development and ultrastructure of *Trachipleistophora hominis* n.g., n.sp. after in vitro isolation from an AIDS patient and inoculation into athymic mice. *Parasitology*, **112**, 143–54.

Hunt, R.D., King, N.W. and Foster, H.L. 1972. Encephalitozoonosis: evidence for vertical transmission. *J Infect Dis*, **126**, 212–14.

Hussey, K.L. 1971. A microsporidan hyperparasite of strigeoid trematodes, *Nosema strigeoideae* sp. n.. *J Protozool*, **18**, 676–9.

Hutin, Y.J., Sombardier, M.N., et al. 1998. Risk factors for intestinal microsporidiosis in patients with human immunodeficiency virus infection: a case-control study. *J Infect Dis*, **178**, 904–7.

Innes, J.R.M., Zeman, W., et al. 1962. Occult endemic encephalitozoonosis of the central nervous system in mice (Swiss Bagg-O'Grady strain). *J Neuropathol Exp Neurol*, **21**, 519–33.

Jensen, H.M. and Wellings, S.R. 1972. Development of the polar filament-polaroplast complex in a microsporidian parasite. *J Protozool*, **19**, 297–305.

Joste, N.E., Rich, J.D., et al. 1996. Autopsy verification of *Encephalitozoon intestinalis* (microsporidiosis) eradication following albendazole therapy. *Arch Pathol Lab Med*, **120**, 199–203.

Kartalija, M. and Sande, M.A. 1999. Diarrhea and AIDS in the era of highly active antiretroviral therapy. *Clin Infect Dis*, **28**, 701–5.

Katznelson, H. and Jamieson, C.A. 1952. Control of nosema disease of honey bees with fumagillin. *Science*, **115**, 70–1.

Kelkar, R., Sastry, P.S., et al. 1997. Pulmonary microsporidial infection in a patient with CML undergoing allogeneic marrow transplant. *Bone Marrow Transplant*, **19**, 179–82.

Kelly, P., McPhail, G., et al. 1994. *Septata intestinalis*: a new microsporidian in Africa. *Lancet*, **344**, 271–2.

Keohane, E.M., Orr, G.A., et al. 1999. Polar tube proteins of microsporidia of the family Encephalitozoonidae. *J Eukaryot Microbiol*, **46**, 1–5.

Kock, N.P., Petersen, H., et al. 1997. Species-specific identification of microsporidia in stool and intestinal biopsy specimens by the polymerase chain reaction. *Eur J Clin Microbiol Infect Dis*, **16**, 369–76.

Kokoskin, E., Gyorkos, T.W., et al. 1994. Modified technique for efficient detection of microsporidia. *J Clin Microbiol*, **32**, 1074–5.

Kondova, I., Mansfield, K., et al. 1998. Transmission and serial propagation of *Enterocytozoon bieneusi* from humans and Rhesus macaques in gnotobiotic piglets. *Infect Immun*, **66**, 5515–19.

Kotler, D.P., Reka, S., et al. 1993. Effects of enteric parasitoses and HIV infection upon small intestinal structure and function in patients with AIDS. *J Clin Gastroenterol*, **16**, 10–15.

Lacey, C.J., Clarke, A.M., et al. 1992. Chronic microsporidian infection of the nasal mucosae, sinuses and conjunctivae in HIV disease. *Genitourin Med*, **68**, 179–81.

Lafranchi-Tristem, N.J., Curry, A., et al. 2001. Growth of *Trachipleistophora hominis* (Microsporidia: Pleistophoridae) in C_2,C_{12} mouse myoblast cells and response to treatment with albendazole. *Folia Parasitol (Praha)*, **48**, 192–200.

Ledford, D.K., Overman, M.D., et al. 1985. Microsporidiosis myositis in a patient with the acquired immunodeficiency syndrome. *Ann Intern Med*, **102**, 628–30.

Leitch, G.J., He, Q., et al. 1993. Inhibition of the spore polar filament extrusion of the microsporidium, *Encephalitozoon hellem*, isolated from an AIDS patient. *J Eukaryot Microbiol*, **40**, 711–17.

Levine, N.D. 1985. *Veterinary protozoology*. Ames: Iowa State University Press.

Liguory, O., David, F., et al. 1998. Determination of types of *Enterocytozoon bieneusi* strains isolated from patients with intestinal microsporidiosis. *J Clin Microbiol*, **36**, 1882–5.

Lores, B., Lopez-Miragaya, I., et al. 2002. Intestinal microsporidiosis due to *Enterocytozoon bieneusi* in elderly human immunodeficiency virus-negative patients from Vigo, Spain. *Clin Infect Dis*, **34**, 918–21.

Lowder, C.Y., Meisler, D.M., et al. 1990. Microsporidia infection of the cornea in a man seropositive for human immunodeficiency virus. *Am J Ophthalmol*, **109**, 242–4.

Lowman, P.M., Takvorian, P.M. and Cali, A. 2000. The effects of elevated temperatures and various time-temperature combinations on the development of *Brachiola (Nosema) algerae* N. Comb. in mammalian cell culture. *J Eukaryot Microbiol*, **47**, 221–34.

Lucas, S.B. 1989. Aspects of infectious disease. In: Anthony, P.P. and MacSween, R.N.M. (eds), *Recent advances in histopathology*. Edinburgh: Churchill Livingstone,, 281–302.

Lucas, S.B., Papadaki, L., et al. 1989. Diagnosis of intestinal microsporidiosis in patients with AIDS. *J Clin Pathol*, **42**, 885–7.

Lucas, S.B., Hounnou, A., et al. 1993. The mortality and pathology of HIV infection in a West African city. *AIDS*, **7**, 1569–79.

McCluskey, P.J., Goonan, P.V., et al. 1993. Microsporidial keratoconjunctivitis in AIDS. *Eye*, **7**, 80–3.

McCowen, M.C., Callender, M.E. and Lawlis Jnr, J.F. 1951. Fumagillin(H-3), a new antibiotic with amebicidal properties. *Science*, **113**, 202–3.

McDougall, R.J., Tandy, M.W., et al. 1993. Incidental finding of a microsporidian parasite from an AIDS patient. *J Clin Microbiol*, **31**, 436–9.

Macher, A.M., Neafie, R., et al. 1988. Microsporidial myositis and the acquired immunodeficiency syndrome (AIDS): a four-year follow-up. *Ann Intern Med*, **109**, 343.

Mansfield, K.G., Carville, A., et al. 1998. Localization of persistent *Enterocytozoon bieneusi* infection in normal rhesus macaques (*Macaca mulatta*) to the hepatobiliary tree. *J Clin Microbiol*, **36**, 2336–8.

Margileth, A.M., Strano, A.J., et al. 1973. Disseminated nosematosis in an immunologically compromised infant. *Arch Pathol*, **95**, 145–50.

Mathis, A., Breitenmoser, A.C. and Deplazes, P. 1999. Detection of new *Enterocytozoon* genotypes in faecal samples of farm dogs and a cat. *Parasite*, **6**, 189–93.

Matsubayashi, H., Koike, T., et al. 1959. A case of *Encephalitozoon*-like body infection in man. *Arch Pathol*, **67**, 181–7.

Metcalfe, T.W., Doran, R.M., et al. 1992. Microsporidial keratoconjunctivitis in a patient with AIDS. *Br J Ophthalmol*, **76**, 177–8.

Modigliani, R., Bories, C., et al. 1985. Diarrhoea and malabsorption in acquired immune deficiency syndrome: a study of four cases with special emphasis on opportunistic protozoan infestations. *Gut*, **26**, 179–87.

Molina, J.M., Oksenhendler, E., et al. 1995. Disseminated microsporidiosis due to *Septata intestinalis* in patients with AIDS: clinical features and response to albendazole therapy. *J Infect Dis*, **171**, 245–9.

Molina, J.M., Goguel, J., et al. 1997. Potential efficacy of fumagillin in intestinal microsporidiosis due to *Enterocytozoon bieneusi* in patients with HIV infection: results of a drug screening study. The French Microsporidiosis Study Group. *AIDS*, **11**, 1603–10.

Molina, J.M., Chastang, C., et al. 1998. Albendazole for treatment and prophylaxis of microsporidiosis due to *Encephalitozoon intestinalis* in patients with AIDS: a randomized double-blind controlled trial. *J Infect Dis*, **177**, 1373–7.

Montaner, J.S., Hogg, R., et al. 1998. Antiretroviral treatment in 1998. *Lancet*, **352**, 1919–22.

Muller, A., Bialek, R., et al. 2001. Detection of microsporidia in travelers with diarrhea. *J Clin Microbiol*, **39**, 1630–2.

Orenstein, J.M. 1991. Microsporidiosis in the acquired immunodeficiency syndrome. *J Parasitol*, **77**, 843–64.

Orenstein, J.M., Dieterich, D.T. and Kotler, D.P. 1993. Albendazole as a treatment of disseminated microsporidiosis due to *Septata intestinalis* in AIDS patients: a report of four patients. *AIDS*, **7**, Suppl 3, S40–2.

Orenstein, J.M., Chiang, J., et al. 1990a. Intestinal microsporidiosis as a cause of diarrhea in human immunodeficiency virus-infected patients: a report of 20 cases. *Hum Pathol*, **21**, 475–81.

Orenstein, J.M., Seedor, J., et al. 1990b. Microsporidian keratoconjunctivitis in patients with AIDS. *Morb Mortal Weekly Rep*, **39**, 188–9.

Orenstein, J.M., Dieterich, D.T. and Kotler, D.P. 1992a. Systemic dissemination by a newly recognized intestinal microsporidia species in AIDS. *AIDS*, **6**, 1143–50.

Orenstein, J.M., Tenner, M., et al. 1992b. A microsporidian previously undescribed in humans, infecting enterocytes and macrophages, and associated with diarrhea in an acquired immunodeficiency syndrome patient. *Hum Pathol*, **23**, 722–8.

Peacock, C.S., Blanshard, C., et al. 1991. Histological diagnosis of intestinal microsporidiosis in patients with AIDS. *J Clin Pathol*, **44**, 558–63.

Perkins, F.O. 1991. 'Sporozoa' Apicomplexa, Microsporidia, Haplosporidia, Paramyxea, Myxosporidia and Actinosporidia. In: Harrison, F.W. and Corliss, J.O. (eds), *Microscopic anatomy of invertebrates*, 1. . New York: Wiley-Liss, 288–302.

Pinnolis, M., Egbert, P.R., et al. 1981. Nosematosis of the cornea. Case report, including electron microscopic studies. *Arch Ophthalmol*, **99**, 1044–7.

Pol, S., Romana, C.A., et al. 1993. Microsporidia infection in patients with the human immunodeficiency virus and unexplained cholangitis.. *N Engl J Med*, **328**, 95–9.

Pulparampil, N., Graham, D., et al. 1998. *Encephalitozoon hellem* in two eclectus parrots (*Eclectus roratus*): identification from archival tissues. *J Eukaryot Microbiol*, **45**, 651–5.

Rabeneck, L., Gyorkey, F., et al. 1993. The role of Microsporidia in the pathogenesis of HIV-related chronic diarrhea. *Ann Intern Med*, **119**, 895–9.

Rabodonirina, M., Bertocchi, M., et al. 1996. *Enterocytozoon bieneusi* as a cause of chronic diarrhea in a heart-lung transplant recipient who was seronegative for human immunodeficiency virus. *Clin Infect Dis*, **23**, 114–17.

Raynaud, L., Delbac, F., et al. 1998. Identification of *Encephalitozoon intestinalis* in travelers with chronic diarrhea by specific PCR amplification. *J Clin Microbiol*, **36**, 37–40.

Rijpstra, A.C., Canning, E.U., et al. 1988. Use of light microscopy to diagnose small-intestinal microsporidiosis in patients with AIDS. *J Infect Dis*, **157**, 827–31.

Rinder, H., Katzwinkel-Wladarsch, S. and Loscher, T. 1997. Evidence for the existence of genetically distinct strains of *Enterocytozoon bieneusi*. *Parasitol Res*, **83**, 670–2.

Rosberger, D.F., Serdarevic, O.N., et al. 1993. Successful treatment of microsporidial keratoconjunctivitis with topical fumagillin in a patient with AIDS. *Cornea*, **12**, 261–5.

Ryan, N.J., Sutherland, G., et al. 1993. A new trichrome-blue stain for detection of microsporidial species in urine, stool, and nasopharyngeal specimens. *J Clin Microbiol*, **31**, 3264–9.

Sadler, F., Peake, N., et al. 2002. Genotyping of *Enterocytozoon bieneusi* in AIDS patients from the north west of England. *J Infect*, **44**, 39–42.

Sandfort, J., Hannemann, A., et al. 1994. *Enterocytozoon bieneusi* infection in an immunocompetent patient who had acute diarrhea and who was not infected with the human immunodeficiency virus. *Clin Infect Dis*, **19**, 514–16.

Sax, P.E., Rich, J.D., et al. 1995. Intestinal microsporidiosis occurring in a liver transplant recipient. *Transplant*, **60**, 617–18.

Schmidt, E.C. and Shadduck, J.A. 1983. Murine encephalitozoonosis model for studying the host–parasite relationship of a chronic infection. *Infect Immun*, **40**, 936–42.

Schwartz, D.A., Visvesvara, G.S., et al. 1993a. Pathologic features and immunofluorescent antibody demonstration of ocular microsporidiosis (*Encephalitozoon hellem*) in seven patients with acquired immunodeficiency syndrome. *Am J Ophthalmol*, **115**, 285–92.

Schwartz, D.A., Visvesvara, G.S., et al. 1993b. Pathology of symptomatic microsporidial (*Encephalitozoon hellem*) bronchiolitis in the acquired immunodeficiency syndrome: a new respiratory pathogen diagnosed from lung biopsy, bronchoalveolar lavage, sputum, and tissue culture. *Hum Pathol*, **24**, 937–43.

Serwadda, D., Mugerwa, R.D., et al. 1985. Slim disease: a new disease in Uganda and its association with HTLV-III infection. *Lancet*, **2**, 849–52.

Shadduck, J.A. 1980. Effect of fumagillin on in vitro multiplication of *Encephalitozoon cuniculi*. *J Protozool*, **27**, 202–8.

Shadduck, J.A. and Polley, M.B. 1978. Some factors influencing the in vitro infectivity and replication of *Encephalitozoon cuniculi*. *J Protozool*, **25**, 491–6.

Shadduck, J.A., Kelsoe, G. and Helmke, R.J. 1979. A microsporidan contaminant of a non-human primate cell culture: ultrastructural comparison with *Nosema connori*. *J Parasitol*, **65**, 185–8.

Shadduck, J.A., Meccoli, R.A., et al. 1990. Isolation of a microsporidian from a human patient. *J Infect Dis*, **162**, 773–6.

Sharpstone, D., Rowbottom, A., et al. 1997. Thalidomide: a novel therapy for microsporidiosis. *Gastroenterology*, **112**, 1823–9.

Sianongo, S., McDonald, V. and Kelly, P. 2001. A method for diagnosis of microsporidiosis adapted for use in developing countries. *Trans R Soc Trop Med Hyg*, **95**, 605–7.

Silveira, H. and Canning, E.U. 1995. *Vittaforma corneae* n. comb. for the human microsporidium *Nosema corneum* Shadduck, Meccoli, Davis & Font, 1990, based on its ultrastructure in the liver of experimentally infected athymic mice. *J Eukaryot Microbiol*, **42**, 158–65.

Silveira, H., Canning, E.U. and Shadduck, J.A. 1993. Experimental infection of athymic mice with the human microsporidian *Nosema corneum*. *Parasitology*, **107**, 489–96.

Simon, D., Weiss, L.M., et al. 1991. Light microscopic diagnosis of human microsporidiosis and variable response to octreotide. *Gastroenterology*, **100**, 271–3.

Snowden, K., Logan, K. and Didier, E.S. 1999. *Encephalitozoon cuniculi* strain III is a cause of encephalitozoonosis in both humans and dogs. *J Infect Dis*, **180**, 2086–8.

Snowden, K.F., Logan, K. and Phalen, D.N. 2000. Isolation and characterization of an avian isolate of *Encephalitozoon hellem*. *Parasitology*, **121**, 9–14.

Sobottka, I., Albrecht, H., et al. 1995a. Self-limited traveler's diarrhea due to a dual infection with *Enterocytozoon bieneusi* and *Cryptosporidium parvum* in an immunocompetent HIV-negative child. *Eur J Clin Microbiol Infect Dis*, **14**, 919–20.

Sobottka, I., Albrecht, H., et al. 1995b. Disseminated *Encephalitozoon* (*Septata*) *intestinalis* infection in a patient with AIDS: novel diagnostic approaches and autopsy-confirmed parasitological cure following treatment with albendazole. *J Clin Microbiol*, **33**, 2948–52.

Sparfel, J.M., Sarfati, C., et al. 1997. Detection of microsporidia and identification of *Enterocytozoon bieneusi* in surface water by filtration followed by specific PCR. *J Eukaryot Microbiol*, **44**, Suppl, S78.

Sprague, V. 1974. *Nosema connori* n. sp., a microsporidian parasite of man. *Trans Am Microsc Soc*, **93**, 400–3.

Sprague, V. and Vávra, J. 1977. Systematics of the microsporidia. In: Bulla, L.A.J. and Cheng, T.C. (eds), *Comparative pathobiology*, vol. 2. . New York: Plenum Press.

Sprague, V., Becnel, J.J. and Hazard, E.I. 1992. Taxonomy of phylum Microspora. *Crit Rev Microbiol*, **18**, 285–395.

Terada, S., Reddy, K.R., et al. 1987. Microsporidan hepatitis in the acquired immunodeficiency syndrome. *Ann Intern Med*, **107**, 61–2.

Trammer, T., Dombrowski, F., et al. 1997. Opportunistic properties of *Nosema algerae* (Microspora), a mosquito parasite, in immunocompromised mice. *J Eukaryot Microbiol*, **44**, 258–62.

Tzipori, S., Carville, A., et al. 1997. Transmission and establishment of a persistent infection of *E. bieneusi* derived from a human with AIDS in SIV-infected rhesus monkeys. *J Infect Dis*, **175**, 1016–20.

Undeen, A.H. and Vander Meer, R.K. 1990. The effect of ultraviolet radiation on the germination of *Nosema algerae* Vávra and Undeen (Microsporida: Nosematidae) spores. *J Protozool*, **37**, 194–9.

Undeen, A.H., Vander Meer, R.K., et al. 1984. The effect of gamma radiation on *Nosema algerae* (Microspora: Nosematidae) spore viability and germination. *J Protozool*, **31**, 479–82.

Van Gool, T., Canning, E.U., et al. 1994. *Septata intestinalis* frequently isolated from stool of AIDS patients with a new cultivation method. *Parasitology*, **109**, 281–9.

Van Gool, T., Hollister, W.S., et al. 1990. Diagnosis of *Enterocytozoon bieneusi* microsporidiosis in AIDS patients by recovery of spores from faeces. *Lancet*, **336**, 697–8.

Van Gool, T., Snijders, F., et al. 1993. Diagnosis of intestinal and disseminated microsporidial infections in patients with HIV by a new rapid fluorescence technique. *J Clin Pathol*, **46**, 694–9.

Van Gool, T., Vetter, J.C., et al. 1997. High seroprevalence of *Encephalitozoon* species in immunocompetent subjects. *J Infect Dis*, **175**, 1020–4.

Vávra, J., Dahbiova, R., et al. 1993. Staining of microsporidian spores by optical brighteners with remarks on the use of brighteners for the diagnosis of AIDS associated human microsporidioses. *Folia Parasitol (Praha)*, **40**, 267–72.

Vávra, J., Yachnis, A.T., et al. 1998. Microsporidia of the genus *Trachipleistophora* – causative agents of human microsporidiosis: description of *Trachipleistophora anthropophthera* n. sp. (Protozoa: Microsporidia). *J Eukaryot Microbiol*, **45**, 273–83.

Visvesvara, G.S. 2002. In vitro cultivation of microsporidia of clinical importance. *Clin Microbiol Rev*, **15**, 401–13.

Visvesvara, G.S., Leitch, G.J., et al. 1991. Culture, electron microscopy, and immunoblot studies on a microsporidian parasite isolated from the urine of a patient with AIDS. *J Protozool*, **38**, Suppl, S105–11.

Visvesvara, G.S., Leitch, G.J., et al. 1994. Polyclonal and monoclonal antibody and PCR-amplified small-subunit rRNA identification of a microsporidian *Encephalitozoon hellem*, isolated from an AIDS patient with disseminated infection. *J Clin Microbiol*, **32**, 2760–8.

Visvesvara, G.S., da Silva, A.J., et al. 1995. In vitro culture and serologic and molecular identification of *Septata intestinalis* isolated from urine of a patient with AIDS. *J Clin Microbiol*, **33**, 930–6.

Visvesvara, G.S., Belloso, M., et al. 1999. Isolation of *Nosema algerae* from the cornea of an immunocompetent patient. *J Eukaryot Microbiol*, **46**, Suppl, S10.

Vossbrinck, C.R., Maddox, J.V., et al. 1987. Ribosomal RNA sequence suggests microsporidia are extremely ancient eukaryotes. *Nature*, **326**, 411–14.

Vossbrinck, C.R., Baker, M.D., et al. 1993. Ribosomal DNA sequences of *Encephalitozoon hellem* and *Encephalitozoon cuniculi*: species identification and phylogenetic construction. *J Eukaryot Microbiol*, **40**, 354–62.

Wanke, C.A., DeGirolami, P. and Federman, M. 1996. *Enterocytozoon bieneusi* infection and diarrheal disease in patients who were not infected with human immunodeficiency virus: case report and review. *Clin Infect Dis*, **23**, 816–18.

Weber, R. and Bryan, R.T. 1994. Microsporidial infections in immunodeficient and immunocompetent patients. *Clin Infect Dis*, **19**, 517–21.

Weber, R., Bryan, R.T., The Enteric Opportunistic Infections Working Group, et al. 1992a. Improved light-microscopical detection of microsporidia spores in stool and duodenal aspirates. *N Engl J Med*, **326**, 161–6.

Weber, R., Kuster, H., et al. 1992b. Pulmonary and intestinal microsporidiosis in a patient with the acquired immunodeficiency syndrome. *Am Rev Resp Dis*, **146**, 1603–5.

Weber, R., Bryan, R.T., et al. 1994. Human microsporidial infections. *Clin Microbiol Rev*, **7**, 426–61.

Weidner, E. 1976. The microsporidian spore invasion tube. The ultrastructure, isolation, and characterization of the protein comprising the tube. *J Cell Biol*, **71**, 23–34.

Weidner, E. 1982. The microsporidian spore invasion tube. III. Tube extrusion and assembly. *J Cell Biol*, **93**, 976–9.

Weidner, E. 1991. Closing remarks on opportunistic microsporidians in humans. *J Protozool*, **38**, 638.

Weidner, E. and Byrd, W. 1982. The microsporidian spore invasion tube. II. Role of calcium in the activation of invasion tube discharge. *J Cell Biol*, **93**, 970–5.

Weidner, E. and Halonen, S.K. 1993. Microsporidian spore envelope keratins phosphorylate and disassemble during spore activation. *J Eukaryot Microbiol*, **40**, 783–8.

Wongtavatchai, J., Conrad, P.A. and Hedrick, R.P. 1995. In vitro characteristics of the microsporidian: *Enterocytozoon salmonis*. *J Eukaryot Microbiol*, **42**, 401–5.

Wuhib, T., Silva, T.M., et al. 1994. Cryptosporidial and microsporidial infections in human immunodeficiency virus-infected patients in northeastern Brazil. *J Infect Dis*, **170**, 494–7.

Yee, R.W., Tio, F.O., et al. 1991. Resolution of microsporidial epithelial keratopathy in a patient with AIDS. *Ophthalmology*, **98**, 196–201.

Zender, H.O., Arrigoni, E., et al. 1989. A case of *Encephalitozoon cuniculi* peritonitis in a patient with AIDS. *Am J Clin Pathol*, **92**, 352–6.

Zhu, X., Wittner, M., et al. 1994. Ribosomal RNA sequences of *Enterocytozoon bieneusi*, *Septata intestinalis* and *Ameson michaelis*: phylogenetic construction and structural correspondence. *J Eukaryot Microbiol*, **41**, 204–9.

26

Blastocystis hominis

JEFFREY J. WINDSOR

INTRODUCTION

Blastocystis hominis is an enigmatic human parasite that has puzzled microbiologists since the early part of the last century. The genus *Blastocystis* was first described by Alexeieff (1911). The following year Brumpt (1912) proposed *B. hominis* for organisms found in human feces. Controversy has surrounded both the taxonomy and pathogenicity of *B. hominis* since its discovery. Although the taxonomy of *B. hominis* has only recently been resolved (Silberman et al. 1996), debate continues on the pathogenic potential of this organism (Zierdt 1991; Stenzel and Boreham 1996; Tan et al. 2002). Indeed, conflicting scientific evidence continues to be published, and it has been reported in both symptomatic and asymptomatic individuals. Although a wealth of information is available on the morphology of *B. hominis* (Boreham and Stenzel 1993; Stenzel and Boreham 1996), relatively little is currently known regarding the mode of transmission and life cycle of this fascinating parasite. *B. hominis* has a worldwide distribution and has been described as probably the most common human gut protozoan (Clark 2000). Prevalence rates of more than 50 percent have been reported in developing countries (Stenzel and Boreham 1996).

CLASSIFICATION

See Zierdt (1991) and Stenzel and Boreham (1996) for a complete history of *B. hominis* classification. The earliest workers regarded *B. hominis* as a harmless intestinal yeast (Alexeieff 1911; Brumpt 1912). Barret (1921) regarded *B. hominis* as of plant origin. Subsequent workers thought *B. hominis* to be a degenerate or cyst form of a flagellate, while others regarded it as a sporozoan or an ameba (Swellengrebel 1917; Zierdt 1991; Stenzel and Boreham 1996). There is no doubt that the presence of multiple morphological forms aided the misclassification of *B. hominis* (Zierdt 1991; Boreham and Stenzel 1993; Stenzel and Boreham 1996). Zierdt and colleagues (1967) described many protozoan-like characteristics of *B. hominis*, which resulted in the organism being reclassified and placed in the class Sporozoa (Zierdt 1978). Silberman and colleagues (1996), using phylogenetic analysis of ribosomal RNAs from *B. hominis*, classified it as the only member of the Stramenopiles found to infect humans. This diverse group of eukaryotes also includes diatoms, kelp, water molds, and slime nets. In Cavalier-Smith's (1998) six-kingdom classification of life, *B. hominis* is placed in Kingdom Chromista, Subkingdom Chromobiota, Infrakingdom Heterokonta (synonymous with Stramenopiles), Subphylum Opalinata, and (new) Class Blastocystea. More recently, Arisue and colleagues (2002) examined the phylogenetic position of *B. hominis* using individual phylogenies based on four molecules and confirmed its position within the Stramenopiles. Furthermore, *Blastocystis*/Stramenopiles were found to be the closest relatives of alveolates. Alveolates are another diverse group of protists and include ciliates, sporozoa, and dinoflagellates.

Blastocystis is a ubiquitous parasite that is found throughout the animal kingdom (Tan et al. 2002), and has been described in many different species including, mammals, birds, reptiles, and insects (Boreham and Stenzel 1993; Abe et al. 2002; Tan et al. 2002). Molecular typing has revealed that some isolates of *Blastocystis* from animal sources share very similar typing patterns to *B. hominis* (Yoshikawa et al. 1996; Clark 1997a). In addition, Snowden and colleagues (2000) described restriction fragment length polymorphism (RFLP) analysis of *Blastocystis* small subunit (SSU) rRNA genes isolated from animal species and found that different animal species shared a single-genotype of *Blastocystis*. Therefore, the potential for possible zoonotic transmission exists and, although not proven conclusively to date, it is plausible that some strains of *Blastocystis* may cross the species barrier.

STRUCTURE

B. hominis shows great morphological diversity. Four main morphological types have been described: vacuolar, granular, ameboid, and cystic (Zierdt 1991; Stenzel and Boreham 1996). *B. hominis* is a strict anaerobe (Zierdt 1983) and exposure to air may adversely affect its morphology (Zierdt 1991). *B. hominis* morphology also differs between fresh fecal specimens and cultures. Discussion continues on whether these different morphologic types are all involved in the life cycle or whether some are merely artifactual or degenerative forms.

Vacuolar form

This form is also termed the central body form and is perhaps the most common form recognized in diagnostic laboratories (Figure 26.1). The vacuolar form is seen in

fecal specimens and is the predominant form in culture material (Stenzel and Boreham 1996). Vdovenko (2000) considered vacuolar and granular forms of *B. hominis* to be degenerative and due to environmental exposure. Vacuolar forms are usually spherical with a large size variation from 2 μm to more than 200 μm (average size between 4 and 15 μm) (Zierdt 1991). Transmission electron microscopy (TEM) studies of the vacuolar form reveal a thin band of cytoplasm around a large vacuole that occupies at least 90 percent of the cell volume (Zierdt 1983) (Figures 26.2 and 26.3). The central vacuole is composed of unevenly distributed, finely granular material, the function of which is unknown (Stenzel and Boreham 1996). Mitochondrion-like organelles are found in expanded areas of the cytoplasm and are surrounded by a bilaminar outer membrane, with cristae-like structures on the inner membrane (Stenzel and Boreham 1996) (Figure 26.4). As *B. hominis* is anaerobic, it has been hypothesized that these mitochondrion-like structures may, in fact, be hydrogenosomes (Boreham and Stenzel 1993; Stenzel and Boreham 1996). The nucleus shows a typical crescent-shaped band of condensed chromatin on one side. *B. hominis* often has a surface slime layer or capsule surrounding the cell membrane (Figure 26.4).

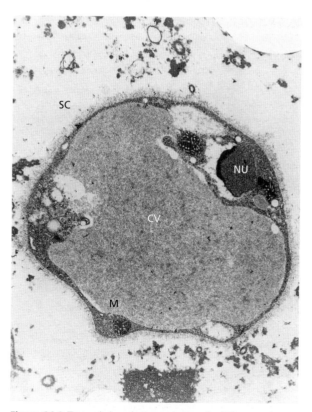

Figure 26.2 *Transmission electron micrograph of* Blastocystis hominis *from culture. Vacuolated form showing nucleus (NU), central vacuole (CV) surrounded by a thin cytoplasm, mitochondrion-like organelles (M), and a surface coat (SC) (original magnification ×5 200). (Reproduced with permission from A. Curry and J. J. Windsor)*

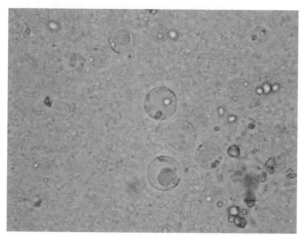

Figure 26.1 *Vacuolar form of* Blastocystis hominis *(original magnification ×400)*

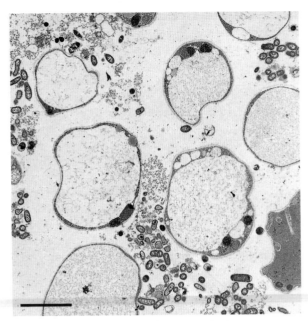

Figure 26.3 *Transmission electron micrograph of numerous vacuolar forms of* Blastocystis hominis *from culture. Bar = 5 μm. (Reproduced with permission from D. J. Stenzel)*

Granular form

This morphological form is similar to the vacuolar type, and can be regarded simply as a vacuolar form with

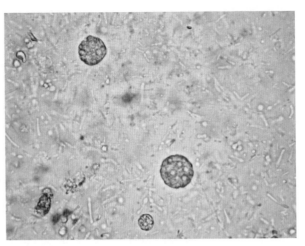

Figure 26.5 *Granular form of* Blastocystis hominis *(original magnification ×400)*

granules in the central vacuole (Stenzel and Boreham 1996). The size range is similar and the granules are visible in unstained preparations (Figures 26.5 and 26.6). Many functions have been proposed for the different types of granule, but it is likely that they are involved in metabolism and storage (Stenzel and Boreham 1996). In older cultures, the granular form is often the most frequently found morphotype (Zierdt et al. 1967). Granular forms can be induced in culture by the use of antibiotics, axenic culture, increased serum concentration, and transfer to a different culture medium (Zierdt 1973; Stenzel and Boreham 1996).

Ameboid form

The ameboid form of *B. hominis* is only rarely encountered and there have been inconsistencies in the published data describing this form (Tan and Zierdt 1973;

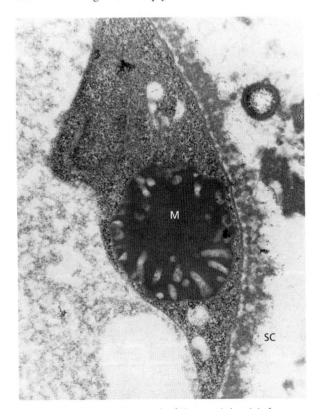

Figure 26.4 *Electron micrograph of* Blastocystis hominis *from cultured material. Vacuolar form showing mitochondrion-like organelle (M), and a surface coat (SC) (original magnification ×21 000). (Reproduced with permission from A. Curry and J. J. Windsor)*

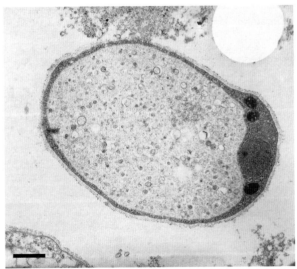

Figure 26.6 *Granular form of* Blastocystis hominis. *Note the presence of clumps of granules in the central vacuole. Bar = 1 μm. (Reproduced with permission from D. J. Stenzel)*

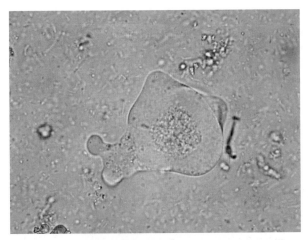

Figure 26.7 *Ameboid form of* Blastocystis hominis *from culture. Note the irregular shape (original magnification ×400)*

Zierdt 1991; Stenzel and Boreham 1996). Extended pseudopodia are often seen in ameboid forms of *B. hominis* (Tan and Zierdt 1973; Zierdt 1973) (Figures 26.7 and 26.10b). The ameboid form has been reported to feed on bacteria (Zierdt 1973; Boreham and Stenzel 1993) and cellular debris from dead cells (Tan et al. 2002). It has been postulated that this might provide nutrition for subsequent encystations (Singh et al. 1995). Tan and colleagues (2002) agreed, and concluded that the ameboid form of *B. hominis* is likely to be involved in a nutritional or a regulatory role in programmed cell death.

Cyst form

Mehlhorn (1988) is credited with the first report of a cyst-like form of *B. hominis* in the stool of an HIV-positive patient. Three years later, Stenzel and Boreham (1991) provided a detailed description of the cyst stage. The cysts were often found in stored fecal samples and were small (3.7–5.0 μm), round or ovoid. TEM of the cysts showed a thick wall and a bi-layered membrane between the wall and the cytoplasm (Figures 26.8 and 26.9). An external capsular layer was present on some cells. Only one nucleus per cell was noted, though in a subsequent publication, the same group described multinucleated cysts (Boreham et al. 1996). Cysts were also occasionally found in long-term cultures. Zaman and colleagues (1997) reported variations in *B. hominis* cyst morphology. They described two types of cyst, with or without the capsular layer. The surface coat was shed as the cyst matured. Although Mehlhorn (1988) did not see mitochondria in his original study, they have been reported by other workers (Stenzel and Boreham 1991; Boreham et al. 1996; Moe et al. 1996; Zaman and Zaki 1996).

Avacuolar and multivacuolar forms

There is evidence that the morphology of *B. hominis* in vivo may be different than that normally found in stool

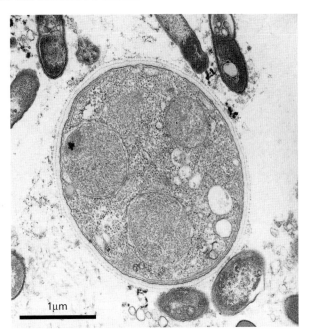

Figure 26.8 *Cyst of* Blastocystis hominis *showing the multilayered wall. Bar = 1 μm. (Reproduced with permission from D. J. Stenzel)*

samples and fecal cultures. Zierdt and Tan (1976) reported an unusual form of *B. hominis* in a patient with severe gastrointestinal symptoms, producing large volumes of diarrheal fluid. This form did not have a vacuole with numerous mitochondria, and they termed it the trophozoite form. In a later report, Stenzel and colleagues (1991) described a similar form from a colono-

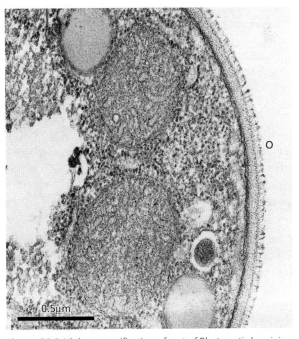

Figure 26.9 *Higher magnification of cyst of* Blastocystis hominis *showing inner layer (I), electron-dense layer (D) and outer layer (O). Bar = 0.5 μm. (Reproduced with permission from D. J. Stenzel)*

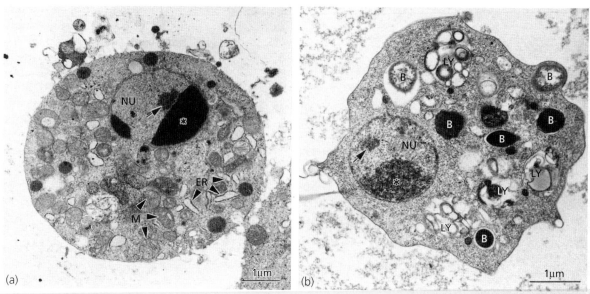

Figure 26.10 *Transmission electron micrographs of* B. hominis. **(a)** *Avacuolar form obtained at colonoscopy. In addition to a crescentic band (asterisk), the nucleus (NU) contains electron-opaque material (arrow). Mitochondrion-like organelles (M) and rough endoplasmic reticulum (ER) are abundant in the cytoplasm.* **(b)** *Ameboid form showing numerous lysosome-like bodies (LY) in the cytoplasm. Bacteria are seen within these bodies and in other membrane-bounded structures. The nucleus contains a compact band (asterisk) and clumps (arrow) of electron-opaque material. (Reproduced from Stenzel and Boreham 1996, with permission from author and publisher)*

scopy specimen and called it avacuolar. Avacuolar forms lack a central vacuole and contain one or two nuclei (Figure 26.10a). They are smaller than vacuolar forms (approximately 5 μm as compared to 4–15 μm) and, like the ameboid form, are not surrounded by a surface coat (Stenzel and Boreham 1996).

Multivacuolar forms of *B. hominis* have many small vacuoles of different sizes within the cytoplasm, and are found in fecal material. They are smaller (5–8 μm) than the vacuolar form of *B. hominis* seen in feces and have a surface coat (Figures 26.11 and 26.12). Many studies of *B. hominis* concentrated on the forms seen in culture (Boreham and Stenzel 1993). This might explain why the multivacuolar form has not been widely reported. However, Stenzel and colleagues (1991) found the multivacuolar form to be the predominant form in human fecal material, when using TEM. Indeed, they considered the multivacuolar and avacuolar forms more likely to be representative of the parasite in vivo.

LIFE CYCLE

Conflicting reports have been published regarding the life cycle of *B. hominis* and these may reflect the use of light microscopy alone, which has limitations and can lead to inaccurate interpretations (Tan et al. 2002). A number of different life cycles have been proposed, dependent on the division process of *B. hominis*. Some workers support the theory that *B. hominis* divides by either plasmotomy, endodgeny or schizogony (Zierdt and Tan 1976; Zierdt 1988, 1991; Jiang and He 1993; Singh et al. 1995; Zaman et al. 1999), in addition to

binary fission. However, TEM has failed to support these hypotheses and it is possible that the schizonts filled with progeny described by light microscopy (Zierdt 1991; Singh et al. 1995) might be granules or inclusions within the cells (Stenzel and Boreham 1996). The most plausible life cycle proposed to date is by Stenzel and Boreham (1996), based on binary fission as the only

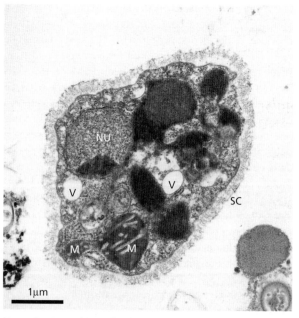

Figure 26.11 *Multivacuolar form of* Blastocystis hominis *from fresh fecal sample showing nucleus (NU), vacuoles (V), mitochondrion-like oroganelles (M), and a thick fibrillar surface coat (SC). Bar = 1 μm. (Reproduced with permission from D. J. Stenzel)*

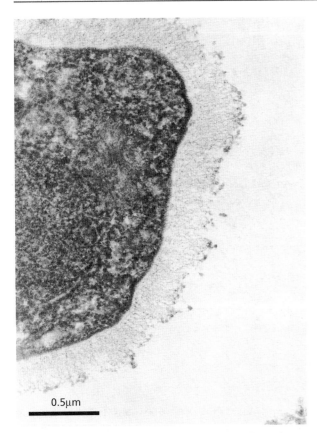

0.5μm

Figure 26.12 *Higher magnification of multivacuolar form of* Blastocystis hominis *showing surface coat. Bar = 0.5 μm. (Reproduced with permission from D. J. Stenzel)*

method of division (Figure 26.13). The cyst is regarded as the infective stage, although the role of the ameboid and granular forms remains poorly defined. There is still much to be elucidated, and the factors responsible for excystation and encystation are currently unknown (Stenzel and Boreham 2001).

BIOCHEMISTRY, MOLECULAR BIOLOGY, AND GENETICS

Data regarding the biochemical cycles and cytochemistry of *B. hominis* is somewhat limited and comprehensive studies have not been performed. Conflicting data has been published on many aspects of basic *B. hominis* biochemistry. For a more detailed review see Stenzel and Boreham (1996).

Although early studies indicated otherwise (Zierdt et al. 1967; Zierdt 1973), carbohydrates have been detected in *B. hominis* using periodic acid-Schiff (PAS) and alcian blue (AB) stains (Yoshikawa et al. 1995). Positive reactions were found in the central vacuole and the cytoplasm; however, the intensity varied suggesting the distribution of carbohydrates may be different in individual cells. At the electron microscopic level, using periodic acid methenamine silver, positive reactions

were found in the Golgi apparatus, cytoplasmic vesicles, external filamentous layer, and the central vacuole. Fluorescein isothiocyanate (FITC)-labeled lectins have been used to elucidate the surface-coat carbohydrates of *B. hominis* (Lanuza et al. 1996). Although the surface coat was found to be complex, they all contained α-D-mannose, α-D-glucose, *N*-acetyl-α-D-glucosamine, α-L-fucose, chitin and sialic acid. By contrast, other workers, using lectin localization at the ultrastructural level, have found that carbohydrates are not a major component of the surface coat (Stenzel and Boreham 1996).

B. hominis may accumulate lipid in the central vacuole and cytoplasm, as demonstrated with Nile blue and Sudan III staining (Dunn et al. 1989). Lipid analysis of axenic *B. hominis* strains has revealed a variety of neutral lipids, phospholipids, and polar lipids (Keenan et al. 1992). Cardiolipin was also detected, which is almost exclusively found in the inner mitochondrial membrane. Paradoxically, although cardiolipin plays a pivotal role in the cytochrome oxidase complex in mammalian cells, cytochrome oxidase has not been found in *B. hominis* (Zierdt 1988). Other enzymes normally associated with mitochondrial activity were also absent from *B. hominis* isolates (Zierdt 1986; Zierdt 1988). Boreham and Stenzel (1993) suggested that these organelles may be hydrogenosomes. Hydrogenosomes lack catalase and cytochromes and were first described in trichomonads (Landmark and Müller 1973). The mitochondrion-like organelles of *B. hominis* stain positively with the vital stains rhodamine 123 and Janus green, indicating that they are functional and not vestigial remnants (Zierdt 1988). Chitin has been demonstrated in the cyst stage of *B. hominis* (Stenzel and Boreham 1996).

Isoenzyme electrophoretic analysis of *B. hominis* isolates revealed malic enzyme, phospholucomatase, glucose phosphate isomerase, 6-phosphogluconate dehydrogenase, and hexokinase (Mansour et al. 1995). Two variants or zymodemes were found, although the study size was small and only 11 isolates were investigated. Analysis of the polypeptide patterns and immunological characteristics of axenic strains of *B. hominis* revealed at least two different variants (Kukoschke and Müller 1991; Mansour et al. 1995). The existence of biochemically and immunologically different strains of *B. hominis* supported the theory that some types of *B. hominis* may have differing pathogenic potential (Kukoschke et al. 1990). Protein and DNA analysis of ten Australian *B. hominis* isolates also revealed two different demes (Boreham et al. 1992). Lanuza and colleagues (1999) described the soluble-protein profile and antigenic cross-reactivity of *B. hominis* isolates and found three protein groups (I–III) and two antigenic groups (1 and 2). The isolates from patients with chronic diarrhea were in protein patterns I and II and antigenic group 1. Isolates from patients with acute diarrhea were clustered in protein pattern III and antigenic group 2.

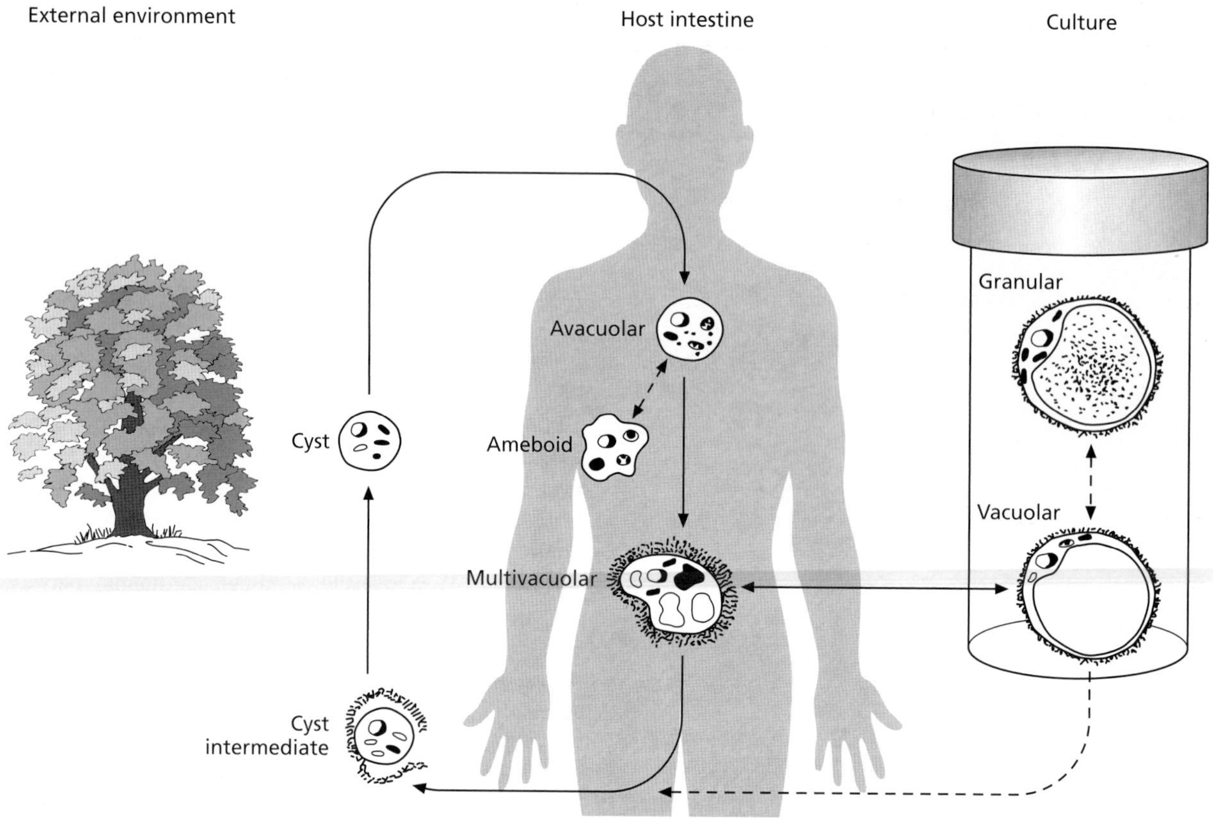

External environment

Host intestine

Culture

Figure 26.13 *Proposed life cycle for* Blastocystis hominis *(Redrawn from Stenzel and Boreham 1996, with permission from author and publisher)*

Conflicting data has been reported when karyotyping *B. hominis* isolates. Some workers (Upcroft et al. 1989; Carbajal et al. 1997) found genetic heterogeneity whereas others have reported significant homology (Teow et al. 1991; Ho et al. 1994). Discrepancies may be due to different investigation procedures or may reflect genetic changes in axenic cultures or laboratory stock cultures (Ho et al. 1994).

Various molecular methods have confirmed that *B. hominis* exhibits extensive genetic diversity. The methods used include RFLP analysis (Böhm-Gloning et al. 1997; Clark 1997a; Hoevers et al. 2000; Yoshikawa et al. 2000; Kaneda et al. 2001) and random amplified polymorphic DNA (RAPD) (Yoshikawa et al. 1998).

Clark (1997a) amplified SSU rDNA and analyzed the products using restriction enzyme analysis. Some areas of rDNA are highly conserved in all eukaryotes, but are distinct from bacterial rDNA. An SSU rRNA gene amplicon of approximately 1800 bp was produced. He termed this method riboprinting (Clark 1997b) and obtained seven different ribodemes (distinct ribotypes) after using 11 restriction enzymes. The restriction enzymes *Hin*fI and *Rsa*I, however, could be used to identify all seven ribodemes. Three human *B. hominis* isolates were identical to a guinea-pig isolate, indicating the possibility of cross-species infection. Phylogenetic analysis of the seven ribodemes revealed two distinct

lineages; ribodemes 1 and 6 were closely related, but very distant from the others. Using the same methodology, other workers subsequently described a further three ribodemes: 8, 9 and 10 (Yoshikawa et al. 2000; Kaneda et al. 2001). In one of these studies, the results provided the first molecular evidence of possible human-to-human transmission of *B. hominis* (Yoshikawa et al. 2000). Restriction site analysis of polymerase chain reaction (PCR)-amplified 16S-like rDNA of *B. hominis* from symptomatic and asymptomatic individuals revealed five subgroups (Böhm-Gloning et al. 1997). However, not all the SSU rDNA product was examined (approx. 850 bp) and only three restriction enzymes were used. Although it appeared that subgroup II was more prevalent in symptomatic patients, none of the groups were statistically correlated with disease. Analysis of DNA polymorphism of *Blastocystis* strains from humans, a chicken, and a reptile, using an arbitrary primer PCR method, found similar products with human and avian isolates, suggesting that the chicken isolate was a zoonotic strain (Yoshikawa et al. 1996).

Ho and colleagues (2001) used a PCR-based RFLP technique to analyze the elongation factor gene (EF-1α) of *B. hominis*. They termed this technique elfaprinting and the different subtypes elfatypes. Again, extensive genetic diversity was found with *Blastocystis* species. As the EF-1α is highly conserved, elfaprinting has potential as an epidemiological tool. However, it was used only on axenic isolates of *B. hominis* and the authors concede that, in future, it would be necessary to amplify directly from direct (xenic) cultures.

There is a need to optimize and standardize the molecular methods and terms used to characterize *B. hominis* strains (Hoevers et al. 2000). The terms subgroup, genotype, ribodeme, and elfatype have all been used to describe *B. hominis* strains. Similar problems are found with the use of different primers and restriction enzymes. It has been suggested that future studies should use reference strains and a standard panel of restriction enzymes (Tan et al. 2002). See Tan et al. (2002) for a detailed review of *B. hominis* typing methods.

CLINICAL ASPECTS

Before describing the clinical spectrum associated with *B. hominis*, it must be stressed that the organism has also been frequently found in asymptomatic individuals (Udkow and Markell 1993; Hellard et al. 2000; Tan et al. 2002). In addition, many of the reported cases of *B. hominis* linked with symptomatology are anecdotal and are uncontrolled, making evaluation of its pathogenicity problematical. Moreover, some studies have failed to eliminate all other causes and infections that may produce intestinal symptoms.

The most common symptoms reported with *B. hominis* infection include diarrhea, nausea, abdominal pain, cramps and discomfort, and flatulence (Sheehan et al. 1986; Diaczok and Rival 1987; Guirges and Al-Waili 1987; Kain et al. 1987; Stenzel and Boreham 1996). Other, less frequently reported, symptoms include fever, constipation, vomiting, eosinophilia, anorexia, fatigue, and nonspecific gastrointestinal effects (Sheehan et al. 1986; Stenzel and Boreham 1996).

B. hominis has been reported in patients who had an intestinal blockage due to carcinoma (Horiki et al. 1999). *B. hominis* was considered coincidental rather than related to the neoplastic growth, although it appeared to take advantage of the altered physiology of the intestine. Two studies have found an association between *B. hominis* and irritable bowel syndrome (IBS) (Hussain et al. 1997; Giacometti et al. 1999). In addition, an earlier study by Markell and Udkow (1986) described five patients with pure *B. hominis* infection who fulfilled the medical criteria for IBS. Giacometti et al. (1999) detected *B. hominis* in 18.1 percent of IBS patients versus 6.9 percent of IBS-negative patients. They speculated that *B. hominis* might be an indicator of intestinal dysfunction. Significantly increased immunoglobulin (Ig)G2 antibody titers against *B. hominis* have been described in IBS patients when compared with asymptomatic controls (Hussain et al. 1997). Increased levels of IgG2 were also found in IBS patients without *B. hominis* in their stools,

although this might be explained by prior treatment of *B. hominis*, resulting in elimination of the parasite. Indeed, 18 out of 22 (85 percent) patients in one center had received previous antibiotic treatment. It is presently unknown whether *B. hominis* is actively involved in the pathogenesis of IBS or simply takes advantage of the disruption in the microbial flora.

B. hominis has been reported in traveler's diarrhea in some studies (Sheehan et al. 1986; Jelinek et al. 1997), but not in others (Shlim et al. 1995). Jelinek et al. (1997) reported the association of *B. hominis* with diarrhea in German travelers returning from the tropics. In contrast, Shlim et al. (1995) conducted a prospective controlled study of traveler's diarrhea in Nepal and concluded that *B. hominis* did not cause diarrhea. However, this study would seem to be weakened by design flaws as, once other known pathogens were excluded, only 19 patients with diarrhea and 26 controls were infected with *B. hominis* (Keystone 1995). These relatively small numbers reduced the power of the study. Furthermore, only one stool per person was examined and stains to exclude cryptosporidia and *Cyclospora* were not performed.

In 1989, Rolston et al. reported nine cases of symptomatic gastroenteritis in patients with pure *B. hominis* infection and underlying pathology. Three patients had acquired immune deficiency syndrome (AIDS) and the other conditions included Hodgkin's disease, leukemia, bone marrow transplant (BMT), lung and bladder cancer, and pelvic inflammatory disease (PID). In 1990, Garavelli et al. described five cases of *B. hominis* infection in symptomatic human immunodeficiency virus (HIV)-positive patients that responded to treatment with metronidazole. In both of these anecdotal studies, little data is available on laboratory methodology and neither states which pathogens were excluded. Cirioni et al. (1999) described the prevalence and relevance of *B. hominis* in different patient cohorts (immunocompromised patients, elderly subjects, psychiatric in-patients, immigrants from developing countries, travelers to tropical countries, and controls) and found it to be the most common parasite in all groups. Symptoms were found only in patients with *B. hominis* when severe immunosuppression was present and these workers concluded that immunodepression was an important factor in pathogenicity of the organism. Other studies have failed to find a link between *B. hominis* and clinical symptoms in AIDS patients, despite a high carriage rate (Albrecht et al. 1995).

IMMUNOLOGY

Relatively little is known regarding the host immune response to *B. hominis* and serological techniques have been used with limited success. Chen et al. (1987) failed to detect a humoral response in four patients using an IgG immunoblot assay. In contrast, Zierdt et al. (1995)

detected IgG antibodies to *B. hominis* in 28 out of 30 patients using an enzyme-linked immunosorbent assay (ELISA). The patients in this latter study were symptomatic and other protozoa and bacterial pathogens were excluded. Although the cut-off was low (1:50), nineteen titers were 1:100 or greater and all the controls (blood bank sera) were negative. In another study, levels of *B. hominis* IgG antibodies were significantly raised in patients with IBS when compared to asymptomatic controls (Hussain et al. 1997). When the IgG antibodies were further subdivided, a selective increase was found in the IgG2 subclass. IgG2 antibodies are primarily directed against carbohydrate antigens and it was thought that these antibodies might be produced in response to the fibrillar surface coat surrounding *B. hominis*. The discrepancies between studies might be due to differences in antigen preparation, one study used a sonicated antigen that was likely to contain carbohydrates (Zierdt et al. 1995), whereas the other used a protein antigen (Chen et al. 1987). The usefulness of serological testing may be limited however, as antibodies have been reported in 70 percent of asymptomatic individuals infected with *B. hominis* (Kaneda et al. 2000). Using immunoblotting analysis, it was found that the antibody response was directed against a 12 kDa surface protein of *B. hominis*.

PATHOLOGY

The general consensus is that *B. hominis* does not usually invade the intestinal mucosa, but may be associated with inflammation and edema (Stenzel and Boreham 1996). Kain et al. (1987) reported that endoscopic, histopathologic, and radiologic findings were normal in symptomatic patients with *B. hominis* infection. However, Russo et al. (1988) described a case of biopsy-proven colitis with large numbers of *B. hominis* present in the stool sample. The patient in this case report became asymptomatic following treatment for *B. hominis* infection with metronidazole. An invasive *B. hominis* infection has been reported in a child, presenting as diffuse superficial ulcers and pseudomembranes on endoscopy (Al-Tawil et al. 1994). Histological examination revealed numerous *B. hominis* present over the surface and within the superficial lamina propria in the region of ulceration. All known viruses, bacteria, and parasites were excluded, including *Clostridium difficile* toxins. A case of hemorrhagic proctosigmoiditis has also been described in association with *B. hominis* (Carrascosa et al. 1996). Acute inflammation was found in rectal and sigmoid biopsy specimens, and sigmoidoscopy revealed marked diffuse erythema. These findings should be interpreted with caution however, as the stool sample was not examined for viruses or toxins. From the above studies, it would seem that in certain circumstances *B. hominis* might be responsible for causing atypical pathology.

Pathogenicity

Difficulties in proving the pathogenicity of *B. hominis* include the lack of a suitable animal model, failure to exclude all other possible causes of symptoms, and the inability to fulfill even the extended Koch's postulates (Stenzel and Boreham 1996; Anon. 1991). Some workers have proposed that *B. hominis* is only pathogenic when present in large numbers ($\geqslant 5$ organisms per $\times 400$ or $\times 100$ field) (Zierdt 1983; Miller and Minshew 1988; Sheehan et al. 1986). This hypothesis has been disputed by others (Shlim et al. 1995; Stenzel and Boreham 1996), and it seems somewhat implausible given that no other human intestinal protozoan is only pathogenic when found in high numbers.

Phillips and Zierdt (1976) inoculated *B. hominis* together with the intestinal bacterial flora from humans into gnotobiotic guinea pigs. Of the 43 animals inoculated orally, 14 developed *B. hominis* infection. Microscopic examination showed penetration of *B. hominis* into the epithelium, but not the lamina propria. When a monoxenic (with *Proteus vulgaris*) culture was used, only one of eight animals was infected. Axenic cultures failed to produce any infection. This failure of an axenic culture of *B. hominis* to cause disease suggests that the symptoms may have been caused by the presence of the original bacterial flora, the bacteria provide essential cofactors for *B. hominis* pathogenicity, or that some strains of *B. hominis* are more pathogenic than others. Xenic *B. hominis* cultures have been shown to produce cytopathic effects in cell lines, although filtrates in which *B. hominis* failed to grow produced similar effects (Walderich et al. 1998). This would seem to suggest that the cytopathic effect was caused by toxins produced by the co-cultivated bacteria and may also explain the earlier findings of Zierdt (1991).

An axenic culture of *B. hominis* did not produce any cytopathic effect on colonic epithelium cell lines (HT-29 and T84) (Long et al. 2001). However, significant increases in the cytokines interleukin (IL)-8 and granulocyte–macrophage colony-stimulating factor (GM-CSF) were discovered. IL-8 activates neutrophils, monocytes, and T lymphocytes, and GM-CSF is a strong chemoattractant for neutrophils and eosinophils. Consequently, Long et al. (2001) concluded that *B. hominis* induces and modulates the immune response in intestinal epithelial cells. A cytopathic effect was produced when *B. hominis* was inoculated into laboratory mice intramuscularly (Moe et al. 1998). Histological examination revealed that *B. hominis* produced a severe inflammatory reaction and myonecrosis. In addition, *B. hominis* was shown to have a strong chemoattractant activity for neutrophils.

Dagci et al. (2002) described an increase in intestinal permeability in patients infected with *Giardia lamblia* and *B. hominis*, but not in patients harboring nonpathogenic *Entamoeba coli*. Intestinal permeability can be affected by various factors and demonstrates the intactness of the gut.

DIAGNOSIS

Light microscopy remains the mainstay of *B. hominis* detection (MacPherson and MacQueen 1994), with culture techniques usually reserved for research purposes. However, the variable forms of *B. hominis* can present difficulties in reliable detection of this parasite. Indeed, laboratory awareness and the use of suitable methodologies are of paramount importance when detecting *B. hominis* (Windsor et al. 2001a). Irregular shedding of *B. hominis* has been reported and, therefore, more than one fecal specimen must be taken to exclude this parasite in patients with clinical symptoms (Vennila et al. 1999).

Detection in unstained preparations

The vacuolar form of *B. hominis* is usually easily recognizable in either saline or iodine wet preparations (Figure 26.1). *B. hominis* can be seen in ethylene-acetate fecal concentrations, but reports on the reliability of this method vary considerably (Stenzel and Boreham 1996). Care must be taken to distinguish *B. hominis* from leukocytes and parasites such as *Dientamoeba fragilis*, *Entamoeba* spp, and *Endolimax nana* (Windsor et al. 2001b). Lee (1991) recommended the use of phase-contrast microscopy for the detection of *B. hominis*. Putative *B. hominis* positives should be confirmed with a suitable staining method (see below), as false-positive detections have occurred when using unstained preparations (Windsor et al. 2002).

Permanent staining techniques

Although the use of permanently stained smears is recommended by parasitologists in North America for the detection and identification of fecal protozoan parasites (Garcia 1990), relatively few laboratories in the UK employ such techniques (Windsor and Johnson 1999). Indeed, such permanently stained smears are considered by many to be the gold standard for detection of *B. hominis* (Garcia et al. 1984; Lee 1991; Stenzel and Boreham 1996; Windsor et al. 2001b). Suitable staining methods include trichrome (Figure 26.14) (Garcia 1990; Windsor et al. 2002) and iron-hematoxylin (MacPherson and MacQueen 1994), which are usually used in conjunction with a specialized fixative such as polyvinyl alcohol (PVA) (Goldman and Brooke 1953) or sodium-acetate acetic-acid formalin (SAF) (Yang and Scholten 1977). Giemsa stain (Ricci et al. 1984) and Fields stain (Moody and Fleck 1985) can also be used after fixing fecal smears in methanol. Gram's stain has been used

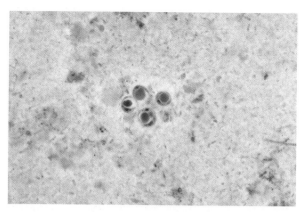

Figure 26.14 *Trichrome-stained smear of vacuolar forms of* Blastocystis hominis *(original magnification ×600) (Redrawn from Windsor et al. 2002, with permission from author and publisher)*

(Miller and Minshew 1988), but is of limited use as *B. hominis* cells often lyse when using this method (Zierdt 1991). Although *B. hominis* is not acid-fast, it may stain with the counterstain (malachite green or methylene blue) when the modified Ziehl-Neelsen stain is used for the detection of *Cryptosporidium* (Windsor et al. 2001b).

Studies comparing unstained microscopy and permanent stains have shown the latter to be much more sensitive and specific (Windsor et al. 2002). Furthermore, it is likely that some of the smaller multivacuolar and cystic forms of *B. hominis* would only be detected using stained smears. Much variability occurs when using permanent stains, especially with the appearance of the central vacuole (MacPherson and MacQueen 1994; Stenzel and Boreham 1996). However, this should not pose a significant problem to experienced laboratory personnel.

Culture

Zierdt (1991) does not recommend culture of *B. hominis* as a routine procedure, but acknowledges its usefulness when the microscopic diagnosis is uncertain. Clark and Diamond (2002) concur and state that cultivation of intestinal protists plays only a minor role in diagnosis. Kukoschke et al. (1990) compared detection of *B. hominis* by direct microscopy and culture and found almost complete agreement. Zaman and Khan (1994) however, disagreed and found culture to be more sensitive than direct microscopy. They found only 18 positives by microscopy out of 100 patients, whereas 45 were positive for *B. hominis* using culture. Despite these conflicting reports there can be no doubt that culture of *B. hominis* has been of great importance in the study of this organism in the past and will continue to be a useful research tool in the future.

B. hominis will grow in a wide variety of different culture media. In 1921, Barret first described the growth of *B. hominis* in a simple medium of human serum and 0.5 percent salt solution. He was attempting to grow the ciliate parasite *Balantidium coli*, but found that *Blastocystis hominis* grew very well in this medium. Many of the media designed specifically for culture of *Entamoeba* species also support the growth of *B. hominis* (Jones 1946; Robinson 1968). Indeed, Clark and Diamond (2002) comment that *B. hominis* grows luxuriantly in all xenic media formulated for both *Entamoeba* species and *Dientamoeba fragilis*. Some of the media are monophasic (Jones 1946), whereas others have a solid and a liquid phase (Robinson 1968; Zierdt 1988). Zierdt (1988) favored the use of a modified egg slant with Locke solution and 30 percent horse serum. Antibiotics, such as erythromycin, are often used to suppress unwanted bacterial species (Robinson 1968).

Antibiotics can also be used to produce an axenic culture, which is grown in the absence of bacteria (Zierdt 1991). *B. hominis* was grown axenically for the first time by Zierdt and Williams in 1974. Ho et al. (1993) subsequently produced an axenic growth of *B. hominis* in monophasic Iscove's modified Dulbecco's medium (IMDM). Tan et al. (1996, 2000) took the culture of *B. hominis* to the next level by establishing colonies from single cells, initially in soft agar (Tan et al. 1996) and then on solid agar (Tan et al. 2000). Production of an axenic culture of *B. hominis* is an essential first step for the production of antibodies. However, commercial diagnostic antibodies against vacuolar and cystic forms of *B. hominis* are not currently available (Tan et al. 2002).

Fluorescent staining

Very little has been published on the use of fluorescent staining techniques for the detection of *B. hominis*. Zierdt (1991) reported previously unpublished data on the use of rabbit antisera to unheated whole-cell *B. hominis* antigens. Reactions were clear-cut at a dilution of 1:200, and immunostaining detected vacuolar, ameboid, and granular forms of *B. hominis* from both feces and culture. Antigenic diversity is extensive among isolates of *B. hominis*, and may be one reason why appropriate antibodies are still not available for diagnostic work (Kukoschke and Müller 1991; Lanuza et al. 1999; Tan et al. 2001).

EPIDEMIOLOGY

Stenzel and Boreham (1996) commented that the epidemiology of *B. hominis* is totally unknown. They attributed this to the confusion over the status of the organism. When suitable methodologies are used, *B. hominis* is often the most frequent intestinal protozoan found in humans worldwide (Stenzel and Boreham 1996; Clark 2000). Unfortunately, some studies of intestinal infection do not report potential pathogens such as *B. hominis* and *Dientamoeba fragilis* (Tompkins et al.

1999). Nevertheless, in developed countries, prevalences of *B. hominis* range from 1.5 to 23 percent (Garcia et al. 1984; Kain et al. 1987; Doyle et al. 1990; Senay and MacPherson 1990; O'Gorman et al. 1993; Logar et al. 1994; Amin 2002; Windsor et al. 2002). Higher rates are seen in developing countries, ranging from 25 to 50 percent (Nimri 1993; Nimri and Batchoun 1994; Stenzel and Boreham 1996). Hellard et al. (2000) reported *B. hominis* in 6 percent of asymptomatic individuals in Australia. Martín-Sánchez et al. (1992) also described *B. hominis* in a high number of asymptomatic cases in children from different population groups in Spain (5.3–19.4 percent). Increased prevalences of *B. hominis* may be found in lower socioeconomic groups and in individuals with poor standards of hygiene (Stenzel and Boreham 1996).

In the UK, little has been published on the prevalence of *B. hominis*. Pakianathan and McMillan (1999) described an incidence rate of 26 percent in homosexual men in Edinburgh, though they did not correlate carriage of *B. hominis* with diarrhea. This population subgroup usually has a high carriage rate of intestinal protozoa per se and, therefore, is not representative of the general population. Casemore et al. (1984), while examining 2 000 specimens in North Wales for *Cryptosporidium* species, only found one sample containing large numbers of *B. hominis*. They concluded that *B. hominis* does not contribute significantly to the etiology of gastroenteritis in North Wales. However, small numbers of *B. hominis* may have been overlooked. Furthermore, optimal staining techniques for *B. hominis* detection were not employed. When trichrome-stained smears were examined, *B. hominis* was found in 6.9 percent of unselected fecal samples in mid-Wales (Windsor et al. 2002).

CHEMOTHERAPY

Anecdotal reports of successful treatment of *B. hominis* infection abound in the published literature. As with proving pathogenicity, a similar question arises when assessing successful treatment of *B. hominis* infection – are all other causes of gastrointestinal symptoms excluded? Furthermore, many of the drugs used for treatment of *B. hominis* are also active against a wide range of gram-positive and gram-negative bacteria (Stenzel and Boreham 2001). This makes it difficult to accurately interpret treatment regimens. There is a dearth of experimental data to confirm the efficacy of treatment for *B. hominis* infections.

Guirges and Al-Waili (1987) used metronidazole therapy to treat 103 patients with pure *B. hominis* infection (i.e. they were not knowingly co-infected with another enteropathogen). The main symptoms were diarrhea with abdominal pain and flatulence with abdominal discomfort. They found metronidazole to be effective and concluded that *B. hominis* was a pathogenic parasite that could be successfully eradicated with this antibiotic. O'Gorman et al. (1993) studied *B. hominis* infections in children and found that gastrointestinal symptoms (abdominal pain, diarrhea, vomiting, and weight loss) resolved in 90 percent of patients receiving a month of antimicrobial therapy, but only in 58 percent of those who received no treatment. Metronidazole was the recommended drug, although furazolidone, mebendazole, and iodoquinol were used for treatment of concurrent *B. hominis* infection with other parasites.

Dunn and Boreham (1991) tested the in vitro activity of various antimicrobials against *B. hominis*. Ketaconazole and iodoquinol were less active than metronidazole, three times and twenty-five times less respectively. Metronidazole resistance was reported by Haresh et al. (1999), with Malaysian strains of *B. hominis* being resistant to 0.01 mg per ml and Indonesian isolates surviving 1.0 mg per ml. Metronidazole has also been shown to have no effect on *B. hominis* cysts (Zaman and Zaki 1996).

Cabello et al. (1997) evaluated nitazoxanide, a 5-nitrothiazol derivative, for the treatment of both intestinal protozoan and helminthic infections in Mexico. It was well tolerated and had a 'cure rate' of 100 percent in ten cases of *B. hominis* infection.

Ok et al. (1999) evaluated the use of trimethoprim–sulfamethaxazole for the treatment of *B. hominis* infection in children and adults. Fifty-three patients (38 children and 15 adults) were treated. Children received 6 mg per kg trimethoprim and 30 mg per kg sulfamethaxazole, and adults received 320 mg per kg trimethoprim, and 1600 mg per kg sulfamethaxazole daily. *B. hominis* was eradicated in 36 of 38 (94.7 percent) children and 14 of 15 (93.3 percent) adults. Clinical symptoms disappeared in 39 (73.6 percent) patients, decreased in 10 (18.9 percent), with one patient showing no change. In 3 (5.7 percent) patients, symptoms persisted and *B. hominis* could not be eradicated.

Crude extracts of traditional Chinese medicinal herbs have been evaluated for activity against *B. hominis* in vitro, and *Coptis chinensis* and *Brucea javanica* were active in concentrations of 100 μg per ml and 500 μg per ml respectively (Yang et al. 1996). These concentrations were not as effective as metronidazole, which was used as a standard and was active at 10 μg per ml. The active ingredients may prove to be more active if identified and purified before evaluating (Stenzel and Boreham 2001). Interestingly, emulsified oil of oregano has been used with good effect to treat 11 patients infected with *B. hominis* (Force et al. 2000). In eight cases *B. hominis* was eradicated and in the other three cases the parasite load declined.

Treatment is not indicated in either asymptomatic individuals or patients with mild or transient symptoms (Stenzel and Boreham 1996). In addition, there is evidence that some *B. hominis* infections may be self-limiting (Boreham and Stenzel 1993). As doubt still remains with regard to its pathogenic potential, *B. hominis* should be treated as a pathogen by exclusion. Consequently, treatment should only be considered when all other causes of gastrointestinal symptoms have been excluded.

CONCLUSION

Ninety years after its first description in humans, many questions remain unanswered regarding the pathogenicity of *B. hominis*. Despite growing anecdotal evidence that *B. hominis* may be pathogenic under certain conditions, scientific studies have largely failed to elucidate its pathogenesis. However, it is difficult to properly evaluate the pathogenic potential of *B. hominis* as elimination of all causes of gastrointestinal symptoms (infectious and non-infectious) in case studies is problematic. Diagnosis has been hampered by the wide variation in morphology, and this is often compounded by unsuitable methodologies and lack of awareness. Consequently, the epidemiology of *B. hominis* is unknown. Molecular techniques have revealed extensive genetic diversity, suggesting that some strains of *B. hominis* may be more pathogenic than others. Some animal isolates of *Blastocystis* have been indistinguishable from human isolates, raising the possibility of zoonotic transmission. Future molecular studies are needed to compare *B. hominis* isolates from differing clinical groups, including symptomatic and asymptomatic individuals.

REFERENCES

Abe, N., Nagoshi, M., et al. 2002. A survey of *Blastocystis* sp. In livestock, pets, and zoo animals in Japan. *Vet Parasitol*, **106**, 203–12.

Albrecht, H., Stellbrink, H.J., et al. 1995. *Blastocystis hominis* in human immunodeficiency virus-related diarrhea. *Scand J Gastroenterol*, **30**, 909–14.

Alexeieff, A. 1911. Sur la nature des formations dites 'kystes de *Trichomonas intestinalis'*. *Compt Rend Soc Biol*, **71**, 296–8, (cited in Stenzel and Boreham 1996).

Al-Tawil, Y.S., Gilger, M.A., et al. 1994. Invasive *Blastocystis hominis* infection in a child. *Arch Paediatr Adolesc Med*, **148**, 882–5.

Amin, O.M. 2002. Seasonal prevalence of intestinal parasites in the United States during 2000. *Am J Trop Med Hyg*, **66**, 799–803.

Anon., 1991. *Blastocystis hominis*: commensal or pathogen? *Lancet*, **337**, 521–2.

Arisue, N., Hashimoto, T., et al. 2002. Phylogenetic position of *Blastocystis hominis* and of stramenopiles inferred from multiple molecular sequence data. *J Eukaryot Microbiol*, **49**, 42–53.

Barret, H.P. 1921. A method for the cultivation of *Blastocystis*. *Ann Trop Med Parasitol*, **15**, 113–16.

Böhm-Gloning, B., Knobloch, J. and Walderich, B. 1997. Five subgroups of *Blastocystis hominis* isolates from symptomatic and asymptomatic patients revealed by restriction site analysis of PCR-amplified 16S-like rDNA. *Trop Med Int Health*, **2**, 771–8.

Boreham, P.F.L. and Stenzel, D.J. 1993. *Blastocystis* in humans and animals: morphology, biology, epizootiology. *Adv Parasitol*, **32**, 1–70.

Boreham, P.F.L., Upcroft, J.A. and Dunn, L.A. 1992. Protein and DNA evidence for two demes of *Blastocystis hominis* from humans. *Int J Parasitol*, **22**, 49–53.

Boreham, R.E., Benson, S., et al. 1996. *Blastocystis hominis* infection. *Lancet*, **348**, 272–3.

Brumpt, E. 1912. *Blastocystis hominis* N. sp. Et formes voisines. *Bull Soc Pathol Exot*, **5**, 725–30, (cited in Zierdt 1991).

Cabello, R.C., Guerrero, L.R., et al. 1997. Nitazoxanide for the treatment of intestinal protozoan and helminthic infections in Mexico. *Trans R Soc Trop Med Hyg*, **91**, 701–3.

Carbajal, J.A., Del Castillo, L., et al. 1997. Karyotypic diversity among *Blastocystis hominis* isolates. *Int J Parasitol*, **27**, 941–5.

Carrascosa, M., Martinez, J. and Perez-Castrillon, J.L. 1996. Hemorrhagic proctosigmoiditis and *Blastocystis hominis* infection. *Ann Intern Med*, **124**, 278–9.

Cavalier-Smith, T. 1998. A revised six-kingdom system of life. *Biol Rev Camb Philos Soc*, **73**, 203–66.

Casemore, D.P., Armstrong, M. and Bruce Jackson, F. 1984. Clinical relevance of *Blastocystis hominis*. *Lancet*, **i**, 1234.

Chen, J., Vaudry, W.L., et al. 1987. Lack of serum immune response to *Blastocystis hominis*. *Lancet*, **ii**, 1021.

Cirioni, O., Giacometti, A., et al. 1999. Prevalence and clinical relevance of *Blastocystis hominis* in diverse patient cohorts. *Eur J Epidemiol*, **15**, 389–93.

Clark, C.G. 1997a. Extensive genetic diversity in *Blastocystis hominis*. *Mol Biochem Parasitol*, **87**, 79–83.

Clark, C.G. 1997b. Riboprinting: a tool for the study of genetic diversity in microorganisms. *J Eukaryot Microbiol*, **44**, 277–83.

Clark, C.G. 2000. Cryptic genetic variation in parasitic protozoa. *J Med Microbiol*, **49**, 489–91.

Clark, C.G. and Diamond, L.S. 2002. Methods for the cultivation of luminal parasitic protests of clinical importance. *Clin Microbiol Rev*, **15**, 329–41.

Dagci, H., Ustun, S., et al. 2002. Protozoon infections and intestinal permeability. *Acta Trop*, **81**, 1–5.

Diaczok, B.J. and Rival, J. 1987. Diarrhea due to *Blastocystis hominis*: an old organism revisited. *South Med J*, **80**, 931–2.

Doyle, P.W., Helgason, M.M., et al. 1990. Epidemiology and pathogenicity of *Blastocystis hominis*. *J Clin Microbiol*, **28**, 116–21.

Dunn, L.A. and Boreham, P.F.L. 1991. The in-vitro activity of drugs against *Blastocystis hominis*. *J Antimicrob Chemother*, **27**, 507–16.

Dunn, L.A., Boreham, P.F.L. and Stenzel, D.J. 1989. Ultrastructural variation of *Blastocystis hominis* stocks in culture. *Int J Parasitol*, **19**, 43–56.

Force, M., Sparks, W.S. and Ronzio, R.A. 2000. Inhibition of enteric parasites by emulsified oil of oregano in vivo. *Phytother Res*, **14**, 213–214.

Garavelli, P.L., Scaglione, L., et al. 1990. Blastocystosis: a new disease in the acquired immunodeficiency syndrome. *Int J STD AIDS*, **1**, 134–5.

Garcia, L.S., Bruckner, D.A. and Clancy, M.N. 1984. Clinical relevance of *Blastocystis hominis*. *Lancet*, **i**, 1233–4.

Garcia, L.S. 1990. Laboratory method for diagnosis of parasitic infections. In: Barron, E.J. and Finegold, S. M. (eds), *Bailey and Scott's diagnostic microbiology*. St Louis: Mosby, 776–861.

Giacometti, A., Cirioni, O., et al. 1999. Irritable bowel syndrome in patients with *Blastocystis hominis* infection. *Eur J Clin Microbiol Infect Dis*, **18**, 436–9.

Goldman, M. and Brooke, M.M. 1953. Protozoans in stools unpreserved and preserved in PVA-fixative. *Public Health Rep*, **68**, 703–6.

Guirges, S.Y. and Al-Waili, N.S. 1987. *Blastocystis hominis*: evidence for human pathogenicity and effectiveness of metronidazole therapy. *Clin Exp Pharmacol Physiol*, **14**, 333–5.

Haresh, K., Suresh, K., et al. 1999. Isolate resistant of *Blastocystis hominis* to metronidazole. *Trop Med Int Health*, **4**, 274–7.

Hellard, M.E., Sinclair, M.I., et al. 2000. Prevalence of enteric pathogens among community based asymptomatic individuals. *J Gastroenterol Hepatol*, **15**, 290–3.

Ho, L.C., Singh, M., et al. 1993. Axenic culture of *Blastocystis hominis* in Iscove's modified Dulbecco's medium. *Parasitol Res*, **79**, 614–16.

Ho, L.C., Singh, M., et al. 1994. A study of the karotypic patterns of *Blastocystis hominis* by pulsed-field gradient electrophoresis. *Parasitol Res*, **80**, 620–2.

Ho, L.C., Jeyaseelam, K. and Singh, M. 2001. Use of elongation-1 alpha gene in a polymerase chain reaction-based restriction-fragment-length polymorphism analysis of genetic heterogeneity among *Blastocystis* species. *Mol Biochem Parasitol*, **112**, 287–91.

Hoevers, J., Holman, P., et al. 2000. Restriction-fragment-length polymorphism analysis of small-subunit rRNA genes of *Blastocystis hominis* isolates from geographically diverse human hosts. *Parasitol Res*, **86**, 57–61.

Horiki, N., Kaneda, Y., et al. 1999. Intestinal blockage by carcinoma and *Blastocystis hominis* infection. *Am J Trop Med Hyg*, **60**, 400–2.

Hussain, R., Jaferi, W., et al. 1997. Significantly increased IgG2 subclass antibody levels to *Blastocystis hominis* in patients with irritable bowel syndrome. *Am J Trop Med Hyg*, **56**, 301–6.

Jelinek, T., Peyerl, G., et al. 1997. The role of *Blastocystis hominis* as a possible intestinal pathogen in traveler's. *J Infect*, **35**, 63–6.

Jiang, J.-B. and He, J.-G. 1993. Taxonomic status of *Blastocystis hominis*. *Parasitol Today*, **9**, 2–3.

Jones, W.R. 1946. The experimental infection of rats with *Entamoeba histolytica*; with a method for evaluating the anti-amoebic properties of new compounds. *Ann Trop Med Parasitol*, **40**, 130–40.

Kain, K.C., Noble, M.A., et al. 1987. Epidemiology and clinical features associated with *Blastocystis hominis* infection. *Diagn Microbiol Infect Dis*, **8**, 235–44.

Kaneda, Y., Horiki, N., et al. 2000. Serological response to *Blastocystis hominis* infection in asymptomatic individuals. *Tokai J Exp Clin Med*, **25**, 51–6.

Kaneda, Y., Horiki, N., et al. 2001. Ribodemes of *Blastocystis hominis* isolated in Japan. *Am J Trop Med Hyg*, **65**, 393–6.

Keenan, T.W., Huang, C.M. and Zierdt, C.H. 1992. Comparative analysis of lipid composition in axenic strains of *Blastocystis hominis*. *Comp Biochem Physiol*, **102B**, 611–15.

Keystone, J.S. 1995. *Blastocystis hominis* and traveler's diarrhea. *Clin Infect Dis*, **21**, 102–3.

Kukoschke, K.G., Necker, A. and Muller, H.E. 1990. Detection of *Blastocystis hominis* by direct microscopy and culture. *Eur J Clin Microbiol Infect Dis*, **9**, 305–7.

Kukoschke, K.G. and Müller, H.E. 1991. SDS-PAGE and immunological analysis of different axenic *Blastocystis hominis* strains. *J Med Microbiol*, **35**, 35–9.

Landmark, D.G. and Müller, M. 1973. Hydrogenosome, a cytoplasmic organelle of the anaerobic flagellate *Tritrichomonas foetus*, and its role in pyruvate metabolism. *J Biol Chem*, **248**, 7724–8.

Lanuza, M.D., Carbajal, J.A. and Borrás, R. 1996. Identification of surface coat carbohydrates in *Blastocystis hominis* by lectin probes. *Int J Parasitol*, **26**, 527–32.

Lanuza, M.D., Carbajal, J.A., et al. 1999. Soluble-protein and antigenic heterogeneity in axenic *Blastocystis hominis* isolates: pathogenic implications. *Parasitol Res*, **85**, 93–7.

Lee, M.J. 1991. Pathogenicity of *Blastocystis hominis*. *J Clin Microbiol*, **29**, 2089.

Logar, J., Andlovic, A. and Poljšak-Prijate, M. 1994. Incidence of *Blastocystis hominis* in patients with diarrhea. *J Infect*, **28**, 151–4.

Long, H.Y., Handschack, A., et al. 2001. *Blastocystis hominis* modulates immune responses and cytokine release in colonic epithelial cells. *Parasitol Res*, **87**, 1029–30.

MacPherson, D.W. and MacQueen, W.M. 1994. Morphological diversity of *Blastocystis hominis* in sodium acetate-acetic acid-formalin-preserved stool samples stained with iron hematoxylin. *J Clin Microbiol*, **32**, 267–8.

Mansour, N.S., Mikhail, E.M., et al. 1995. Biochemical characterisation of human isolates of *Blastocystis hominis*. *J Med Microbiol*, **42**, 304–7.

Markell, E.K. and Udkow, M.P. 1986. *Blastocystis hominis*: pathogen or fellow traveller? *Am J Trop Med Hyg*, **35**, 1023–6.

Martín-Sánchez, A.M., Canut-Blasco, A., et al. 1992. Epidemiology and clinical significance of *Blastocystis hominis* in different population groups in Salamanca (Spain). *Eur J Epidemiol*, **8**, 553–9.

Mehlhorn, H. 1988. *Blastocystis hominis*, Brumpt 1912: are there different stages or species? *Parasitol Res*, **74**, 393–5.

Miller, R.A. and Minshew, B.H. 1988. *Blastocystis hominis*: an organism in search of a disease. *Rev Infect Dis*, **10**, 930–8.

Moe, K.T., Singh, M., et al. 1996. Observations on the ultrastructure and viability of the cystic stage of *Blastocystis hominis* from human feces. *Parasitol Res*, **82**, 439–44.

Moe, K.T., Singh, M., et al. 1998. Cytopathic effect of *Blastocystis hominis* after intramuscular inoculation into laboratory mice. *Parasitol Res*, **84**, 450–4.

Moody, A.H. and Fleck, S.L. 1985. Versatile Field's stain. *J Clin Pathol*, **38**, 842–3.

Nimri, L.F. 1993. Evidence of an epidemic of *Blastocystis hominis* infection in preschool children in northern Jordan. *J Clin Microbiol*, **31**, 2706–8.

Nimri, L. and Batchoun, R. 1994. Intestinal colonisation of symptomatic and asymptomatic school children with *Blastocystis hominis*. *J Clin Microbiol*, **32**, 2865–6.

O'Gorman, M.A., Orenstein, S.R., et al. 1993. Prevalence and characteristics of *Blastocystis hominis* infection in children. *Clin Paediatr (Phila)*, **32**, 91–6.

Ok, U.Z., Girginkardesler, N., et al. 1999. Effect of trimethoprim–sullfamethaxaxole in *Blastocystisd hominis* infection. *Am J Gastroenterol*, **94**, 3245–7.

Pakianathan, M.R. and McMillan, A. 1999. Intestinal protozoa in homosexual men in Edinburgh. *Int J STD AIDS*, **10**, 780–4.

Phillips, B.P. and Zierdt, C.H. 1976. *Blastocystis hominis*: pathogenic potential in human patients and in gnotobiotes. *Exp Parasitol*, **39**, 358–64.

Ricci, N., Toma, P., et al. 1984. *Blastocystis hominis*: a neglected cause of diarrhea? *Lancet*, **i**, 966.

Robinson, G.L. 1968. The laboratory diagnosis of human parasitic amoeba. *Trans R Soc Trop Med Hyg*, **62**, 285–94.

Rolston, K.V.I., Winans, R. and Rodriguez, S. 1989. *Blastocystis hominis*: Pathogen or not? *Rev Infect Dis*, **11**, 661–2.

Russo, A.R., Stone, S.L., et al. 1988. Presumptive evidence for *Blastocystis hominis* as a cause of colitis. *Arch Intern Med*, **148**, 1064.

Senay, H. and MacPherson, D. 1990. *Blastocystis hominis*: epidemiology and natural history. *J Infect Dis*, **162**, 987–9.

Sheehan, D.J., Raucher, B.G. and McKitrick, J.C. 1986. Association of *Blastocystis hominis* with signs and symptoms of human disease. *J Clin Microbiol*, **24**, 548–50.

Shlim, D.R., Hoge, C.W., et al. 1995. Is *Blastocystis hominis* a cause of diarrhea in travelers? A prospective controlled study in Nepal. *Clin Infect Dis*, **21**, 97–101.

Silberman, J.D., Sogin, M.L., et al. 1996. Human parasite finds taxonomic home. *Nature*, **380**, 398.

Singh, M., Suresh, K., et al. 1995. Elucidation of the life cycle of the intestinal protozoan *Blastocystis hominis*. *Parasitol Res*, **81**, 446–50.

Snowden, K., Logan, K., et al. 2000. Restriction-fragment-length polymorphism analysis of small-subunit rRNA genes of *Blastocystis* isolates from animal hosts. *Parasitol Res*, **86**, 62–6.

Stenzel, D.J. and Boreham, P.F.L. 1991. A cyst-like stage of *Blastocystis hominis*. *Int J Parasitol*, **21**, 613–15.

Stenzel, D.J. and Boreham, P.F.L. 1996. *Blastocystis hominis* revisited. *Clin Microbiol Rev*, **9**, 563–84.

Stenzel, D.J. and Boreham, R.E. 2001. Blastocystis. In: Gillespie, S.H. and Pearson, R.D. (eds), *Principles and practice of clinical parasitology*. Chichester, England: John Willey and Sons, 355–68.

Stenzel, D.J., Boreham, P.F.L. and McDougall, R. 1991. Ultrastructure of *Blastocystis hominis* in human stool samples. *Int J Parasitol*, **21**, 807–12.

Swellengrebel, N.H. 1917. Observations on *Blastocystis hominis*. *Parasitology*, **9**, 451–9.

Tan, H.K. and Zierdt, C.H. 1973. Ultrastructure of *Blastocystis hominis*. *Z Parasitenkd/Parasitol Res*, **42**, 315–24.

Tan, K.S., Singh, M. and Yap, E.H. 2002. Recent advances in *Blastocystis hominis* research: hot spots in terra incognita. *Int J Parasitol*, **32**, 789–804.

Tan, K.S.W., Ng, G.C., et al. 2000. *Blastocystis hominis*: A simplified, high-efficiency method for clonal growth on solid agar. *Exp Parasitol*, **96**, 9–15.

Tan, K.S.W., Ibrahim, M., et al. 2001. Exposure of *Blastocystis* species to a cytotoxic monoclonal antibody. *Parasitol Res*, **87**, 534–8.

Tan, S.W., Singh, M., et al. 1996. Clonal growth of *Blastocystis hominis* in soft agar with sodium thioglycollate. *Parasitol Res*, **82**, 737–9.

Teow, W.L., Zaman, V., et al. 1991. A *Blastocystis* species from the sea-snake, *Lapemis hardwickii* (Serpentes: hydrophiidae). *Int J Parasitol*, **21**, 723–6.

Tompkins, D.S., Hudson, M.J., et al. 1999. A study of infectious intestinal disease in England: Microbiological findings in cases and controls. *Comm Dis Public Health*, **2**, 108–13.

Udkow, M.P. and Markell, E.K. 1993. *Blastocystis hominis*: prevalence in asymptomatic versus symptomatic hosts. *J Infect Dis*, **168**, 242–4.

Upcroft, J.A., Dunn, L.A., et al. 1989. Chromosomes of *Blastocystis hominis*. *Int J Parasitol*, **19**, 879–83.

Vdovenko, A.A. 2000. *Blastocystis hominis*: origin and significance of vacuolar and granular forms. *Parasitol Res*, **86**, 8–10.

Vennila, G.D., Suresh Kumar, G., et al. 1999. Irregular shedding of *Blastocystis hominis*. *Parasitol Res*, **85**, 162–4.

Walderich, B., Bernauer, S., et al. 1998. Cytopathic effects of *Blastocystis hominis* on Chinese hamster ovary (CHO) and adeno carcinoma HT29 cell cultures. *Trop Med Int Health*, **3**, 385–90.

Windsor, J.J. and Johnson, E.H. 1999. More laboratories should test for *Dientamoeba fragilis*. *Br Med J*, **318**, 735.

Windsor, J.J., Macfarlane, L., et al. 2001a. *Blastocystis hominis*: a common yet neglected human parasite. *Br J Biomed Sci*, **58**, 129.

Windsor, J.J., Macfarlane, L. and Whiteside, T.M. 2001b. *Blastocystis hominis*. *Br J Biomed Sci*, **58**, 253.

Windsor, J.J., Macfarlane, L., et al. 2002. The incidence of *Blastocystis hominis* in fecal samples submitted for routine microbiological analysis. *Br J Biomed Sci*, **59**, 154–7.

Yang, J. and Scholten, T. 1977. A fixative for intestinal parasites permitting the use of concentration and permanent staining procedures. *Am J Clin Pathol*, **67**, 300–4.

Yang, L.Q., Singh, M., et al. 1996. In vitro response of *Blastocystis hominis* against traditional Chinese medicine. *J Ethnopharmacol*, **55**, 35–42.

Yoshikawa, H., Kuwayama, N. and Enose, Y. 1995. Histochemical detection of carbohydrates of *Blastocystis hominis*. *J Eukaryot Microbiol*, **42**, 70–4.

Yoshikawa, H., Nagano, I., et al. 1996. DNA polymorphism revealed by arbitrary primers polymerase chain reaction among *Blastocystis* strains isolated from humans, a chicken, and a reptile. *J Eukaryot Microbiol*, **43**, 127–30.

Yoshikawa, H., Nagano, I., et al. 1998. Genomic polymorphism among *Blastocystis hominis* strains and development of subtype-specific diagnostic primers. *Mol Cell Probes*, **12**, 153–9.

Yoshikawa, H., Abe, N., et al. 2000. Genomic analysis of *Blastocystis hominis* strains isolated from two long-term health care facilities. *J Clin Microbiol*, **38**, 1324–30.

Zaman, V. and Khan, K.Z. 1994. A comparison of direct microscopy with culture for the diagnosis of *Blastocystis hominis*. *Southeast Asian J Trop Med Public Health*, **25**, 792–3.

Zaman, V. and Zaki, M. 1996. Resistance of *Blastocystis hominis* cysts to metronidazole. *Trop Med Int Health*, **1**, 677–8.

Zaman, V., Howe, J. and Ng, M. 1997. Variation in the cyst morphology of *Blastocystis hominis*. *Parasitol Res*, **83**, 306–8.

Zaman, V., Zaki, M., et al. 1999. Postcystic development of *Blastocystis hominis*. *Parasitol Res*, **85**, 437–40.

Zierdt, C.H. 1973. Studies of *Blastocystis hominis*. *J Protozool*, **20**, 114–21.

Zierdt, C.H. 1978. *Blastocystis hominis*, an intestinal protozoan parasite of man. *Public Health Lab*, **36**, 147–60.

Zierdt, C.H. 1983. *Blastocystis hominis*, a protozoan parasite and intestinal pathogen of human beings. *Clin Microbiol News*, **5**, 57–9.

Zierdt, C.H. 1986. Cytochrome-free mitochondria of an anaerobic protozoan - *Blastocystis hominis*. *J Protozool*, **33**, 67–9.

Zierdt, C.H. 1988. *Blastocystis hominis*, a long-misunderstood intestinal parasite. *Parasitol Today*, **4**, 15–17.

Zierdt, C.H. 1991. *Blastocystis hominis* – past and future. *Clin Microbiol Rev*, **4**, 61–79.

Zierdt, C.H. and Tan, H.K. 1976. Ultrastructure and light microscope appearance of *Blastocystis hominis* in a patient with enteric disease. *Z Parasitenkd/Parasitol Res*, **50**, 277–83.

Zierdt, C.H. and Wiliams, R.L. 1974. *Blastocystis hominis*: axenic cultivation. *Exp Parasitol*, **36**, 233–43.

Zierdt, C.H., Rude, W.S. and Bull, B.S. 1967. Protozoan characteristics of *Blastocystis hominis*. *Am J Clin Pathol*, **48**, 495–501.

Zierdt, C.H., Zierdt, W.S. and Nagy, B. 1995. Enzyme-linked immunosorbent assay for detection of serum antibody to *Blastocystis hominis* in symptomatic infections. *J Parasitol*, **81**, 127–9.

PART III

HELMINTHS

27

Nature and classification of parasitic helminths

DAVID I. GIBSON

GENERAL INTRODUCTION

The parasitic worms or helminths, which occur in humans, comprise four major groups:

- two groups of flatworms (phylum Platyhelminthes), the trematodes or digeneans (flukes) and the cestodes (tapeworms)
- the nematodes (roundworms, phylum Nematoda)
- the acanthocephalans (thorny-headed worms; phylum Acanthocephala).

Other groups have been reported as parasites of humans, for example free-living flatworms (turbellarians), hairworms (gordiids, Phylum Nematomorpha) free-living nematodes (such as mermithids) and even earthworms (oligochaete annelids). Although these groups may be seen at the site of a toilet (in WCs, after defecation or urination on the ground, etc.) they are very unlikely to have passed through the human digestive system.

Nomenclature

The more important levels used in the classification of parasites can often be recognized by commonly used suffixes:

Phylum (e.g. Platyhelminthes) suffix varies
    Class (e.g. Trematoda) suffix commonly -a
      Subclass (e.g. Digenea) suffixes -ea, -a
Order (e.g. Ascaridida) suffixes -ida, -idea, -iformes
    Superfamily (e.g. Diplostomoidea) suffix -oidea
    Family (e.g. Diplostomidae) suffix -idae
    Subfamily (e.g. Diplostominae) suffix -inae
      Genus (e.g. *Schistosoma*) suffix varies, if used
      Species (e.g. *mansoni*) suffix varies but depends on derivation and certain rules.

Levels above the family level vary because they are not covered by the International Code for Zoological Nomenclature. Vernacular versions are derived from some groups by the addition of -s or -es to the root, e.g. platyhelminths, diplostomids, diplostomines. The genus and species together, e.g. *Schistosoma mansoni*, form the scientific name of an organism. Any personal name and date following this name indicate the original authority for the name.

A direct life cycle is one in which the final host is re-infected directly without the involvement of intermediate hosts. An indirect life cycle is one in which intermediate hosts or paratenic hosts harbor one or more life-history stages. The final or definitive host harbors the adult (sexual adult in the case of the Digenea). Intermediate hosts are those in which one or more larval stages develop as a necessary part of the life cycle. A paratenic host is one which larval stages may survive but do not normally develop; they are often not a necessary part of the life cycle.

Identification and morphometric analysis

The identification of parasitic worms is possible in certain cases using a variety of techniques, such as:

- the identification of eggs found during coprological examination (e.g. Thienpont et al. 1986)
- the recognition of worm sections in histological investigations (e.g. Chitwood and Lichtenfels 1972)
- immunodiagnostic technique (e.g. Lightowlers and Gottstein 1995)
- enzyme electrophoresis (e.g. Wright and Ross 1980)
- molecular biological approaches (e.g. de Clercq et al. 1994).

However, in the majority of cases the most effective method is still morphometrics. Although this may well change in the future, helminth classification, whether it involves the intuitive approach of the specialist or the more rigorous protocols of cladistics (a method based on the use of derived characters as indicators of evolutionary events), still relies mainly on morphological features.

FIXATION

Morphometric analysis relies on having material in good condition; in order to achieve this, it is best if specimens are fixed live. In many cases, for example when worms are voided after the administration of anthelmintics, this is not possible, and any conventional fixative, such as 10 percent formalin or 70–80 percent alcohol, may be used. Live material is best fixed in Berland's fluid (19 parts glacial acetic acid to 1 part pure formalin) or (especially in the case of cestodes) a very hot conventional fixative or boiling water. After fixation for a minute or so, the specimens are best stored in 80 percent alcohol. Specimens should not be fixed under the pressure (e.g. a coverslip, glass slide or other device) except in the case of large tapeworms and acanthocephalans; some pressure or prior soaking in tap water may help evert the proboscis of live acanthocephalans.

MOUNTING

Nematodes and acanthocephalans are best examined as temporary mounts between a glass slide and a coverslip in a clearing agent, such as beechwood creosote, lactophenol or glycerine; the latter is really only useful for smaller worms (shorter than 15 mm in length). Trematodes and cestodes (or fragments of the latter, usually the head and mature segments) need to be mounted on slides permanently. This requires staining of the worms in a good carmine-based stain such as Mayer's paracarmine for 1–20 min (depending on size), destaining in acid alcohol until the worm is a pale pink color, dehydrating, and clearing and mounting in Canada balsam or some other mountant.

HISTOLOGY

In some cases, serial histological sections of specimens are required for identification. In this procedure, orientation is usually critical, so it must be done manually. The technique is given in detail by Cooper (1988).

SPECIALIST HELP

Help may be obtained from specialist laboratories, which are usually in museums or similar institutes and where there are helminth taxonomists. Contact should be made before sending material. Any material sent should be very well packed: well dried in the case of slidemounts, and in strong plastic (not glass) containers in the case of wet material.

The helminth groups

An account of the helminth groups that might occur as parasites in humans is given below. The higher taxa dealt with tend to be those generally considered important in the systematics of the group, and most information is given at the family level. The account is presented in the form of a summary of the gross morphological features by which a group is recognized, detailed information on many organs and organ systems is omitted. This is followed by a section listing the genera involved and some idea of the basic recognition features of the group; information on the life history, how humans become infected and any control measures relating to the life history; and information on the pathogenicity of the worm which might relate to its morphology, mode of existence or life history. The information was gathered from a variety of sources, but Muller (2001), Coombs and Crompton (1991), and Garcia and Bruckner (1993) were useful references for nonsystematic data.

PHYLUM PLATYHELMINTHES

The platyhelminths or flatworms are bilaterally symmetrical, dorsoventrally flattened worms with a definite head end and lacking a body cavity. They include a variety of free-living turbellarians, which occur in aquatic and terrestrial conditions (a small number are parasitic), and a number of entirely parasitic groups. There are three classes:

1 the Monogenea (mainly ectoparasites of fishes)
2 the Cestoidea (tapeworms; endoparasites)
3 the Trematoda (endoparasitic flukes; mainly digeneans).

Only the latter two classes infect humans.

Class Cestoidea Rudolphi, 1808

Main features Platyhelminthes; primarily intestinal parasites of vertebrates; usually single generation in life cycle and sexual adult in vertebrate, rarely with asexual reproduction in intermediate host; life cycle indirect (one exception), wide variety of invertebrates and vertebrates used as intermediate hosts; adult segmented or not with duplication of reproductive organs along body (polyzoic) or not (monozoic); segmented forms polyzoic, unsegmented forms monozoic or polyzoic; distinct scolex (head) present or absent; syncytial tegument usually but not always unarmed (at light microscope level); gut absent (Figure 27.1).

The cestodes, or tapeworms, occur as intestinal parasites of all groups of vertebrates. Their closest relatives are the monogeneans, ectoparasitic flukes mainly of fishes, from which they are likely to have been derived. There are two subclasses, the Cestodaria, which are monozoic forms lacking a scolex and parasitic in fishes and turtles, and the Eucestoda, the majority of which are segmented and polyzoic. Only the latter subclass parasitizes humans.

The higher classification of the cestodes is controversial and complex (see, for example, Brooks 1989).

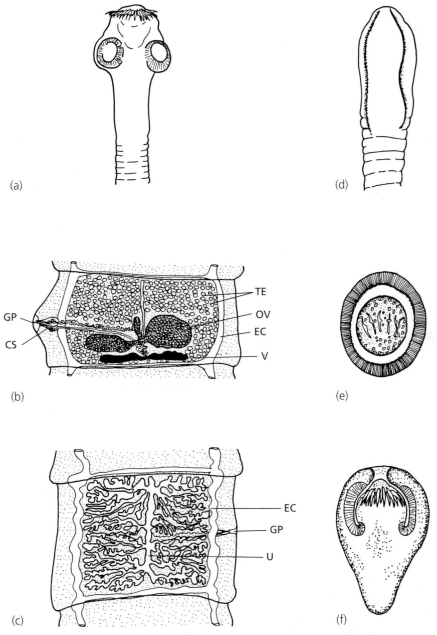

Figure 27.1 Cestodes: **(a)** scolex (head) of Taenia (Taeniidae) with armed rostellum and four suckers; **(b)** mature segment of Taenia; **(c)** gravid segment of Taenia; **(d)** scolex of Diphyllobothrium (Diphyllobothriidae); **(e)** egg of Taenia containing hexacanth larva; **(f)** taeniid cysticercus larva (these can occur singly or, in certain genera, in large numbers within a cyst following asexual multiplication). Cirrus sac (CS), excretory canal (EC), genital pore (GP), ovary (OV), testes (TE), uterus (U), vitellarium (V). Adapted from (a–d) Khalil et al. 1994; **(e)** Abuladze 1964; **(f)** Fuhrmann 1931.

The classification of cestodes found in humans used here at ordinal and subordinal taxonomic levels follows that of Khalil et al. (1994):

Order Pseudophyllidea
    Family Diphyllobothriidae
Order Cyclophyllidea
    Family Anoplocephalidae
    Family Davaineidae
    Family Dipylidiidae
    Family Hymenolepididae
    Family Mesocestoididae
    Family Taeniidae.

Subclass Eucestoda Southwell, 1930

Main features Cestoidea; usually segmented, usually polyzoic with one or more sets of reproductive organs per segment; distinct scolex normally present; invariably parasitic in intestine of vertebrates.

The tapeworms of this subclass that infect humans are all segmented as adults (Figure 27.1) with one or more copies of the reproductive system in each segment (proglottid) along the body (strobila). There are two orders involved:

- the Pseudophyllidea, recognizable by the fact that the scolex lacks suckers and hooks, the attachment organ being a pair of bothria (dorsal and ventral longitudinal grooves on the scolex)
- the Cyclophyllidea, where the scolex is armed with a ring of four suckers and sometimes one or more rings of apical hooks.

In some cases, humans can become infected with the larval stages of tapeworms; these normally reside in the tissues (see below, sections on Diphyllobothriidae and Taeniidae).

Cestodes generally have a life cycle involving one or two intermediate hosts. Only one species (*Hymenolepis*) is capable of autoinfection without the use of intermediate hosts. Since adult cestodes are intestinal parasites of vertebrates, the eggs or gravid segments containing eggs pass out with the feces. In the cyclophyllideans, the eggs must be eaten by a suitable intermediate host, which may be a terrestrial invertebrate (commonly an arthropod) or a vertebrate, in order to hatch and release a hexacanth (six-hooked) larva, called an onchosphere. In the case of the pseudophyllideans, the eggs hatch in water to release a ciliated, motile hexacanth, called a coracidium, which is eaten by an aquatic arthropod, such as a copepod. Once in the intermediate host, the hexacanth usually penetrates the gut wall and develops in the body cavity into a procercoid. This develops further, either in the same host or in a second intermediate host if the first host is eaten, into a resting, normally encysted, stage that takes on a variety of names depending on its form (e.g. cysticercus, cysticercoid, plerocercoid). The final host acquires the parasite when it feeds on the host harboring the encysted stage. In a few instances (see below, Taeniidae) some asexual multiplication of larval heads (protoscoleces) can occur when vertebrates act as an intermediate host.

The form of the attachment organ on the scolex differs markedly in different forms and can be readily used to distinguish the various orders. The form and armament of the scolex distinguishes genera, and the number and morphometrics of the hooks that make up the armature are useful at the specific level. Other important characters relate to the shape of the segments and the arrangement and form of the reproductive system(s) within the segments, e.g. the position of the genital pore, the nature of the vitellarium, the size of the cirrus sac, the shape of the ovary, and the nature of the uterus.

Since adult tapeworms are intestinal parasites that absorb nutrients though their tegument rather than browse, except perhaps at the point of attachment, the worms tend to do little physical damage to their host, apart from the possibility of bowel obstruction in the case of large infections. Their presence in humans, when apparent, is often accompanied by diarrhea and loss of appetite. In some cases, the absorption of certain nutrients is a factor (see below, *Diphyllobothrium* for selective absorption), especially in the malnourished. The presence of encysted larval tapeworms in the tissues is much more serious, notably in certain organs, such as the brain, liver, and lungs, since the cysts can, in some cases, reach a large size. Control is dependent on the life cycle of the worm, but it in most cases it can be effected by the cooking or freezing of meat (especially that of game, fish or other wild animals) and insects intended for ingestion, or the removal of insects from salad and other plant material to be eaten.

ORDER PSEUDOPHYLLIDEA CARUS, 1863

Main features Eucestoda; without suckers or hooks on scolex but normally with pair of longitudinal grooves (bothria) which aid attachment; polyzoic (segmented); first larval stage in crustaceans, second usually in fish; adults in all vertebrate groups, especially fish.

Family Diphyllobothriidae Lühe, 1910

Main features Pseudophyllidea; strobila medium to large; scolex occasionally poorly developed; bothria usually well developed; set of reproductive organs in each segment usually single, occasionally double or multiple; second larval stage (plerocercoid) usually in fish, occasionally in reptiles and mammals; adults in reptiles, birds, and mammals.

Diphyllobothriids are usually parasitic in piscivorous higher vertebrates. One genus, *Diphyllobothrium* Cobbold, 1858, regularly occurs as an adult in humans, *Diplogonoporus* Lönnberg, 1892 occurs occasionally, and other genera on rare occasions. These may be huge worms, reaching many meters in length. Diphyllobothriids in humans are recognizable by the absence of suckers and hooks on the scolex (head). One genus, *Spirometra* Faust, Campbell and Kellog, 1929, occurs in humans as a plerocercoid larva (called a sparganum).

Adult diphyllobothriids live in the intestine of their final host, producing huge numbers of eggs that pass out with the feces and hatch in water to release a ciliated larva (coracidium). If the coracidium is eaten by a suitable copepod, it develops in the hemocoel as a procercoid larva. If the infected copepod is eaten by a suitable vertebrate host, usually a fish, the larva penetrates the tissues and develops (sometimes encysting) in the body cavity, muscles or other tissues as a plerocercoid larva. If this host is then eaten by a suitable higher vertebrate, the plerocercoid develops to an adult in the intestine. Humans become infected with adult diphyllobothriids by feeding on raw, lightly marinated or inadequately cooked freshwater or, less frequently, marine fishes. Since these parasites are more usually parasitic in other mammals, little can be done to control the presence of larvae in fishes, but the freezing or cooking of fish and fish products kills the worms. Larval forms (*Spirometra*) are acquired by humans through drinking water containing infected copepods, by eating raw or poorly cooked amphibians, reptiles or mammals and, apparently, by using the flesh of reptiles or other species as a poultice for wounds, etc. Control is effected by filtering water and avoiding consumption of or contact with raw flesh that might harbor the plerocercoid larva.

The adult worms are large and absorb nutrients from the gut, notably vitamin B_{12}, absorption of which results in pernicious anemia. Large infections can also result in obstruction of the bowel. Infections with plerocercoids (spargana) are more of a problem, as these larvae wander through the tissues before becoming encysted in a fibrous nodule reaching about 2 cm in diameter. These can be painful in subcutaneous regions, but can be much more serious if they end up in the eye, lymphatic system, brain, etc. (see Chapter 32, Intestinal tapeworms).

ORDER CYCLOPHYLLIDEA VAN BENEDEN IN BRAUN, 1900

Main features Eucestoda; scolex normally with four suckers; rostellum (apical protrusion on scolex) usually present but sometimes absent, armed with hooks or not; polyzoic (segmented); parasitic as adults in all vertebrate groups except fish; normally in intestine.

Family Anoplocephalidae Cholodkowsky, 1902

Main features Cyclophyllidea; small to large worms; scolex without rostellum; suckers unarmed; segments craspedote (posterior border overlaps anterior border of next segment) or not; single or double set of reproductive organs per segment; in mammals, birds, and reptiles.

Several genera have been recorded as occasional or accidental parasites of humans, including *Bertiella* Stiles and Hassall, 1902 (species normally occurring in primates), *Inermicapsifer* Janicki, 1910 and *Mathevotaenia* Akhumyan, 1946 (species normally occurring in rodents) and *Moniezia* Blanchard, 1891 (species normally parasitic in domesticated ruminants). A general, but not infallible, recognition feature of anoplocephalids is the absence of both hooks and a rostellum (muscular apical organ) on the scolex.

The eggs of anoplocephalids leave with the feces either freely or in the form of gravid segments. If eaten by a suitable terrestrial arthropod, such as a soil mite, the egg hatches and the embryo (onchosphere) develops into a cysticercoid larva in the hemocoel. Herbivorous vertebrates become infected when they feed on infected arthropods or accidentally ingest them with vegetation. Humans acquire infections by accidentally ingesting mites with raw vegetation (*Bertiella* and *Moniezia*) or by eating larger insects, e.g. beetles, either as food or as medical remedies (*Mathevotaenia* and probably *Inermicapsifer*). Control measures include the careful washing of salad and other items of vegetation intended to be eaten raw and the avoidance or cooking or freezing of insects intended to be eaten.

Little is known regarding the pathogenicity of anoplocephalid infections in humans (diarrhea has been reported), but it is likely that these unarmed worms cause little harm.

Family Davaineidae Braun, 1900

Main features Cyclophyllidea; body small to large; rostellum usually present, rarely rudimentary, armed with crown of hooks usually but not always in two rows; crown of hooks round, oval or undulating, interrupted or not; hooks characteristically numerous, small, hammer-shaped; suckers normally present, armed with small spines or not; segments normally numerous; one or two sets of reproductive organs per segment; adults in birds and mammals.

Several species of *Raillietina* Furhmann, 1920, normally parasites of rodents, have been recorded as occasional parasites of humans. Davaineids are usually recognizable by the presence of a rostellum on the scolex armed with two rings of many minute, hammer-shaped hooks.

The adult worms release gravid segments which pass out in the feces. Eggs eaten by suitable terrestrial insects develop in the hemocoel into a cysticercoid larva. The vertebrate host picks up the parasite by feeding on the infected intermediate host. Humans are thought to acquire *Raillietina* by ingesting insects, such as ants, beetles or cockroaches, either accidentally with food or as part of the diet. Control can be effected by the careful washing of raw and cold food, keeping cold food in a meat safe or refrigerator, and cooking or freezing of insects intended for ingestion.

Family Dipylidiidae Stiles, 1896

Main features Cyclophyllidea; body small to medium-sized; scolex with protrusible rostellum armed with several rows of hooks; hooks usually rose-thorn-shaped; rostellar sac absent; suckers unarmed; segments numerous, mature and gravid segments longer than wide; two sets of reproductive organs per segment; larval stages in insects, amphibians or reptiles; adults in carnivorous mammals.

One genus, *Dipylidium* Leuckart, 1863, occurs in humans. Dipylidiids are recognizable by the absence of a rostellar sac and the presence of two sets of reproductive organs in each segment.

Dipylidiid tapeworms normally occur in the intestine of carnivorous mammals. The gravid segments break off and leave with the feces, releasing large numbers of eggs. This group of tapeworms appears to have two life-cycle strategies. It seems likely that all use insects as intermediate hosts in which the larval cysticercoid stage develops. In some cases, the mammalian host becomes infected when these insects are ingested; in other cases, insectivorous amphibians or reptiles act as second intermediate or paratenic hosts. In the case of *Dipylidium*, dogs, cats, humans (usually children), etc. become infected by feeding on fleas harboring the cysticercoid larvae. Control relates to improvements in hygiene and the regular de-worming and de-fleaing of pets.

Although the rostellum of these worms is armed with hooks, they appear not to be very pathogenic, causing only slight indigestion and loss of appetite. Infection may only become apparent when segments are observed in stools.

Family Hymenolepididae Perrier, 1897

Main features Cyclophyllidea; rostellum usually present but occasionally lacking or rudimentary, unarmed or armed with single, rarely double, crown of hooks (hooks may be very small); suckers armed or not; segments broader than long, craspedote; single (rarely double) set of reproductive organs in each segment; normally one to three (usually three) testes per segment; adults in birds and mammals.

Two genera, *Hymenolepis* Weinland, 1858 and *Rodentolepis* Spasskii, 1954, occur relatively commonly in humans; both normally parasitize rodents. Although previously considered as *Hymenolepis*, species of *Rodentolepis* differ from *Hymenolepis* in that the rostellum is armed with a crown of hooks. Another genus, *Drepanidotaenia* Railliet, 1892, a parasite of anseriform birds, has also been reported in humans. Hymenolepidids are relatively small tapeworms generally recognizable by the fact that each segment has only three testes.

Hymenolepidids are intestinal parasites. Their eggs leave with the feces and are ingested by arthropods, within which develops a cysticercoid larva. The vertebrate host normally becomes infected by swallowing arthropods harboring this stage. In terrestrial forms (e.g. *Hymenolepis* and *Rodentolepis*) the usual intermediate hosts are beetles and fleas; whereas in aquatic forms (e.g. *Drepanidotaenia*) they are copepods. One species of *Rodentolepis* (*R. nana*), which occurs in humans, is unique for a tapeworm in that direct infection is possible, either by reinfection (autoinfection) of the intestinal mucosa by eggs in the gut or by swallowing eggs that have passed in the feces. Humans also become infected with hymenolepidids by accidentally eating beetles and fleas and, in the rare cases of *Depanidotaenia*, by drinking water containing infected copepods. Control is effected by rodent control, personal hygiene, care when handling laboratory animals, especially where these parasites are used as laboratory models, de-fleaing laboratory rodents, and filtering drinking water.

These worms cause few symptoms, although heavy infections of *R. nana* caused by autoinfection may result in enteritis. This is thought to be caused more by the waste-products of these worms than damage at the site of attachment resulting from the armed scolex (see Chapter 32, Intestinal tapeworms.)

Family Mesocestoididae Lühe, 1894

Main features Cyclophyllidea; body narrow; with numerous segments; rostellum absent; mature segments wider than long, gravid segments may be longer than wide; single set of reproductive organs in each segment with median genital pore; larval stage (tetrathyridium) in amphibians, reptiles, birds, and mammals; adults in mammals, rarely birds.

One genus, *Mesocestoides* Lühe, 1894, has been reported from humans, the species concerned being natural parasites of carnivorous mammals, such as foxes, racoons, skunks, etc. Mesocestoidids are recognizable by the absence of a rostellum on the scolex and the median position of the genital pore on the segments.

Eggs pass out with the feces in gravid segments. If these are eaten by a suitable arthropod, probably an insect, the onchosphere develops in the hemocoel as a procercoid larva. If the arthropod is eaten by a vertebrate such as an amphibian, reptile, bird or mammal, the worm develops in the tissue as a type of plerocercoid referred to as a tetrathyridium. The final host acquires the parasite by feeding on the vertebrate intermediate host. Infections in humans appear to have been caused by people feeding on poorly cooked game, raw birds and, especially, raw snake livers. Control is by the proper cooking or freezing of meat of all sorts obtained from the wild.

These unarmed worms appear not to be very pathogenic, but diarrhea, poor appetite, and some anemia has been reported in relation to human infections.

Family Taeniidae Ludwig, 1886

Main features Cyclophyllidea; strobila ribbon-like, usually with many segments; rostellum usually well developed and usually bears two rows of hooks; anterior row of hooks usually larger and alternating with second row; single set of reproductive organs in each segment; larval stage either cysticercus or large cystic structure producing multiple protoscoleces asexually within brood capsules; larval stage and adult in mammals.

Taeniid tapeworms may occur in humans in both larval and adult form. Two genera frequently parasitize humans, *Taenia* Linnaeus, 1758 (now includes *Multiceps* Goeze, 1782 and *Taeniarhynchus* Weinland, 1858) and *Echinococcus* Rudolphi, 1801, the latter only in its larval form.

Taeniids normally occur as adults in the intestine of carnivorous mammals, such as canids and felids. Eggs or segments containing eggs leave with the feces. If the eggs fall on grass or other vegetation and are eaten by herbivorous mammals, the egg hatches in the gut and releases a larva (the onchosphere), which penetrates the gut wall and enters the blood system. These larvae are transported via the circulation until they reach the body muscles, where they encyst and develop as cysticercus larve which have invaginated scoleces (protoscoleces). Carnivores become infected with the adult parasite when they feed on herbivorous mammals. Similarly, humans become infected by feeding on raw or poorly cooked meat, especially pork and beef. Humans can also harbor the larval stage (cysticercus), when the eggs passed by infected humans or carnivores are accidentally ingested. In some species this larva (a coenurus) produces within its body many small protoscoleces (heads), each one of which can develop into an adult worm if eaten by a suitable host; whereas the hydatid cyst of *Echinococcus* contains huge number of 'brood capsules' within each of which develop numerous protoscoleces. In the latter case, the cysts may be huge, reaching 30 cm in diameter. A consequence of this asexual reproduction is that the adult worms are much smaller, have fewer segments, and produce fewer eggs than more conventional taeniids. Control includes the proper cooking or freezing of meat, regular de-worming of dogs especially farm dogs, not feeding raw offal and other parts of slaughtered animals to dogs, keeping dog (and human) feces away from farm animals and vegetable crops, and the proper washing of raw vegetation intended as food.

Adult worms in the intestine cause few symptoms, although diarrhea, bowel discomfort, weight loss, irritation at the site of attachment in the case of species armed with hooks, and bowel obstruction have been reported. The presence of segments in the stools or crawling out of the anus may be the only evidence of infection. The presence of larval stages (cysticercosis, hydatid disease, coenurosis) in the tissues is another matter as, depending on size and location, the cysts can be very dangerous. These larvae may live for many years and may have to be removed surgically. Hydatid cysts can occur in many parts of the body; the commonest site is the liver and another frequent site is the lungs. Symptoms are similar to those of a slow-growing tumor. The cysticerci and coenurus larvae of *Taenia* spp. (including *Multiceps*) frequently occur in the brain in humans, sometimes in the eye or under the skin. Symptoms of brain infections are similar to those caused by any space-filling lesion, including epileptic seizures, paraplegia, etc. (see Chapter 32, Intestinal tapeworms and Chapter 33, Larval cestodes).

Class Trematoda Rudolphi, 1808

Main features Platyhelminthes; primarily permanent parasites of tissues of mollusks and have single or multiple generations in life history, life cycle may or may not involve other additional hosts; mollusk normally harbors asexual generations or single sexual generation; additional host usually a vertebrate, harbors adult sexual stage, especially in gut and associated organs; sexual adult usually with two organs of attachment organs (normally suckers or sucker-like structures), one generally anterior and one ventral or posterior; syncytial tegument armed with spines or smooth; gut or its vestige always present.

The trematodes or 'flukes' are flatworms that originally evolved as parasites of mollusks; virtually all species retain a molluskan element in the their life history. There are two subgroups, the Digenea and the Aspidogastrea, but only the former occurs in humans and other mammals.

There is no modern accepted classification of the Trematoda or the Digenea. Brooks et al. (1985, 1989) presented a version based on a cladistic analysis, but this has been heavily criticized by, among others, Pearson (1992). For keys to genera, see Gibson et al. (2002) and subsequent volumes. The classification below for forms occurring in humans follows the general outline indicated by Gibson and Bray (1994):

Order Strigeida
    Superfamily Diplostomoidea
        Family Diplostomidae
    Superfamily Gymnophalloidea
        Family Gymnophallidae
    Superfamily Schistosomatoidea
        Family Schistosomatidae
Order Echinostomida
    Superfamily Echinostomatoidea
        Family Echinostomatidae
        Family Fasciolidae

Superfamily Paramphistomatoidea
    Family Zygocotylidae
Order Plagiorchiida
    Superfamily Dicrocoelioidea
        Family Dicrocoeliidea
    Superfamily Opsithorchioidea
        Family Heterophyidae
        Family Opsithorchiidae
    Superfamily Plagiorchioidea
        Family Lecithodendriidae
        Family Paragonimidae
        Family Plagiorchiidae
        Family Troglotrematidae.

Subclass Digenea Carus, 1863

Main features Trematoda; alternation of generations, normally two (sometimes more) asexual generations in mollusk and single sexual generation in vertebrate (occasionally invertebrate); normally hermaphroditic, occasionally partly or entirely dioecious; generally small elongate-oval to tubular worms, but various other forms, including round and filamentous, occur; body smooth or armed with spines, usually with two muscular suckers, sometimes one, occasionally none; form of testes variable, one to many, often two; muscular sac (cirrus sac) often envelopes all or part of male terminal genitalia; genital atrium (chamber into which male and female ducts open, opens via genital pore) present or absent; common genital pore normally present, position variable, usually ventral on anterior body, occasionally terminal; ovary normally single, occasionally multiple, position variable; uterus variable in size, form and distribution; eggs normally oval and operculate, normally tanned, occasionally with spine or filament(s); vitellarium (yolk gland) variable in shape and distribution, exhibiting all forms between follicular and single compact mass; parasitic as sexual adult in all vertebrate groups, occasionally in invertebrates, usually present in gut or other body cavities, occasionally in blood or other tissues, rarely ectoparasitic.

Those trematodes (i.e. digeneans) that occur in humans are readily recognized by the fact that they possess two suckers, which act as attachment organs; they have only a single set of reproductive organs and they have a functional gut (Figure 27.2). They are endoparasites occurring in humans mainly in the gut, liver, bile ducts, lungs or blood system.

Digeneans have a unique life cycle that usually involves three generations: two asexual generations in a mollusk and a sexual generation in a vertebrate. The eggs leave the vertebrate, normally with the feces, and hatch to release a ciliated larva, the miracidium, which penetrates a particular molluskan host; sometimes the egg is eaten by this host prior to release of the miracidium. Within the mollusk, the miracidium develops into a sac-like parthenogenetic adult, the mother sporocyst.

Within this stage develop daughter stages called daughter sporocysts or, if a sac-like gut is present, rediae. Within the second parthenogenetic generation develop short-lived, usually tailed, larvae, the cercariae, often in huge numbers (Figure 27.2). These larvae usually escape from the mollusk and swim or crawl to the next host. Some cercariae penetrate the final vertebrate host directly, but the majority penetrate or settle on its food as a long-lived resting stage called the metacercaria. Metacercariae are normally encysted within invertebrates or other vertebrates, which act as intermediate hosts, or on vegetation. The final vertebrate hosts become infected by feeding on prey or vegetation harboring metacercariae. Digeneans tend to be very specific in relation to their molluskan host, relatively unspecific for intermediate hosts and exhibit various levels of specificity for their final vertebrate host.

Some digeneans differ considerably in shape and morphology, but the majority are similar in overall appearance. Features used to distinguish groups include the presence or absence of spines on the tegument, the position and number of suckers, the form and arrangement of the reproductive organs, and the shape of the gut and excretory vesicle. At the specific level, details such as sucker ratio and egg size are important.

Although, like tapeworms, digeneans can take in nutrients through their body surface, they also tend to feed by browsing. In the gut, this is usually not a great problem especially in light infestations, but in other sites the damage, irritation, and toxins produced can result in inflammation, tissue reactions, fibrosis, obstruction, etc. The body spines of some worms, which help them maintain position, may also cause irritation. In view of the nature of the life cycle, obvious targets for the control of digenean infections are the molluskan host and the development of sewage systems and standards of hygiene that prevent contact between worm eggs and mollusks. Similarly, transmission to humans can be pre-empted by avoidance of contact between bare skin and water containing cercariae, where these penetrate the skin directly, or by changes in food preparation and eating habits.

SUPERFAMILY DIPLOSTOMOIDEA POIRIER, 1886
Family Diplostomidae Poirier, 1886

Main features Diplostomoidea; body usually divided into two regions; anterior region wide, foliate with ventral depression; hindbody cylindrical; oral and ventral suckers well developed, sometimes small; ventral suckers in base of ventral concavity; pseudosuckers often present lateral to oral sucker; cirrus sac absent; metacercaria in fishes or amphibians; adults in intestine of birds and mammals.

One species of the genus, *Neodiplostomum* Railliet, 1919, which is more commonly a parasite of rats, has been recorded in the intestine of humans in Korea on several occasions. Diplostomids are recognizable by the bipartite nature of the body and wide, ventrally concave forebody.

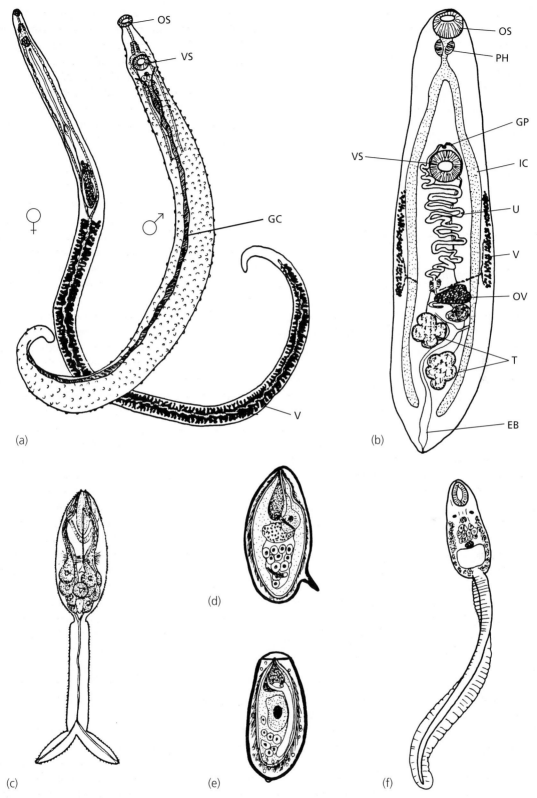

Figure 27.2 *Trematodes:* **(a)** *male and female* Schistosoma *(Schistosomatidae) (the presence of separate sexes is very unusual for a flatworm);* **(b)** Opisthorchis *(Opisthorchiidae) with more conventional trematode morphology;* **(c)** *cercaria (larva) of* Schistosoma *with bifid tail;* **(d)** *egg of* Schistosoma *containing miracidium (ciliated larva);* **(e)** *egg of* Opisthorchis *with lid (operculum) through which miracidium escapes;* **(f)** *cercaria of* Opisthorchis. *Excretory bladder (EB), gynecophoric canal within which female lies (GC), genital pore (GP), intestinal cecum (IC), oral sucker (OS), ovary (OV), pharynx (PH), testes (T), uterus (U), vitellarium (V), ventral sucker (VS). Adapted from* **(a)** *Schell (1985);* **(b, e, f)** *Vogel (1934);* **(c, d)** *Cort (1919)*

Eggs of diplostomids leave with the feces. If they enter water, the miracidium is released; after penetrating an aquatic snail, it transforms into a mother sporocyst. The mother sporocyst produces daughter sporocysts that release fork-tailed cercariae. The latter leave the snail and encyst as metacercariae, usually in fishes but in the case of *Neodiplostomum*, in frogs. Snakes, which acquire the parasites by feeding on frogs, act as paratenic hosts. Humans may become infected by feeding on raw or poorly cooked frogs or snakes. Control can be effected by the proper cooking of amphibians or reptiles used as food. Little is known about the pathogenicity of these worms in humans.

SUPERFAMILY GYMNOPHALLOIDEA ODHNER, 1905

Family Gymnophallidae Odhner, 1905

Main features Gymnophalloidea; body small, oval to pear-shaped, flattened; tegument armed with small spines; oral sucker large, often very large, may support pair of lateral papillae; ventral sucker small, in anterior or posterior half of body (posterior for *Gymnophalloides*); cirrus sac absent; in intestine, bursa of Fabricius or gallbladder of birds (mainly) and mammals.

Gymnophallids are usually parasites of birds. Only one genus, *Gymnophalloides* Fujita, 1925, is parasitic in humans; this is a very small worm that can occur in the intestine, or possibly the biliary system, in huge numbers in Korea. An obvious recognition feature is that the small ventral sucker occurs in the posterior quarter of the body. It is readily distinguished from paramphistomoids by its smaller size, spiny tegument, and smaller, nonterminal ventral sucker.

The full life cycle is not known, but the intermediate host is presumed to be a marine oyster or clam. The asexual generations of gymnophallids usually occur in bivalves; but there is no free-living cercarial stage, as the metacercariae encyst directly inside the daughter sporocyst. Consequently, the final host acquires the parasite by feeding on bivalve mollusks. Control can, therefore, be implemented by the avoidance of feeding on raw bivalves or at least those which have not been treated by freezing to kill the parasite.

Little is known about this parasite, but it appears to cause gastric discomfort, indigestion, and diarrhea, and possibly pancreatitis or cholecystitis. Presumably, when they are present in large numbers, the spined tegument of these worms can cause local irritation to the tissues.

SUPERFAMILY SCHISTOSOMATOIDEA STILES AND HASSALL, 1898

Family Schistosomatidae Stiles and Hassall, 1898

Main features Schistosomatoidea; dioecious; elongate worms, females narrower and longer than males; female clasped in gynecophoric canal (longitudinal groove) of male; oral and ventral suckers normally present; eggs nonoperculate with or without spine; parasitic in blood system of birds and mammals.

Only one genus of schistosome, i.e. *Schistosoma* Weinland, 1858, occurs as an adult in humans. The fact that the sexes are separate and their presence in the blood system makes the group readily recognizable (Figure 27.2).

Eggs leave the body with the feces or urine. The miracidium hatches from the egg, penetrates an aquatic snail, and transforms into a mother sporocyst. This gives rise to daughter sporocysts, which produce cercariae. The life cycle in this group is different from the majority of digeneans in that there is no metacercarial stage and no intermediate host between the molluskan and vertebrate hosts: the fork-tailed cercariae leave the molluskan host, swim and penetrate the skin of the final host, and enter the blood system. The mode of transmission means that avoidance of contact between the skin and water containing cercariae is important in preventing infection, as are control of the relevant species of mollusks, general sanitation, and hygiene.

Since schistosomes occur in the blood system, there is no direct route for the voiding of eggs, which become lodged in small blood vessels in various parts of the body, including the liver and lungs, causing blockages, local granulomata, and abscesses that form around the eggs and aid their passage into the lumen of the gut or bladder. These worms are, therefore, very pathogenic and cause a variety of symptoms. In addition, the fact that the cercariae penetrate the skin means that local immune responses are elicited, with a resulting skin reaction known as cercarial dermatitis: this reaction is caused not only by species of *Schistosoma* of humans and other mammals but also by avian schistosomes of the genera *Trichobilharzia* Skrjabin and Zakharov, 1920 and *Ornithobilharzia* Odhner, 1912 where the reaction is often more pronounced (see Chapters 28–30, Schistosomes: general, African, and Asian).

SUPERFAMILY ECHINOSTOMATOIDEA LOOSS, 1899

Family Echinostomatidae Looss, 1899

Main features Echinostomatoidea; body oval to elongate; tegument armed with small spines; head collar present, usually armed with single or double crown of spines; suckers well developed; encysted as metacercariae in or on mollusks, annelids, fishes, tadpoles, etc.; eggs relatively large; adults mainly in intestine, occasionally other organs, of birds, mammals, and occasionally reptiles.

Several echinostomatid genera, the most common of which is *Echinostoma* Rudolphi, 1809, occur in humans especially in parts of southeast Asia. Other genera include *Echinochasmus* Dietz, 1909, *Echinoparyphium*

Dietz, 1909 and *Hypoderaeum* Dietz, 1909; these are primarily parasites of birds and mammals and infect humans only accidentally. Echinostomatids, which occur in the intestine, are readily recognized by the single or double crown of large circumoral spines; but worms must be in good condition when they are fixed, since the spines are readily lost in frozen or poorly preserved material.

The life cycle involves two intermediate hosts, both of which may be snails. The eggs pass out from the vertebrate host with the feces and hatch to release the ciliated miracidium. This penetrates a snail and develops into a mother sporocyst. Within this develops a brood of rediae. When these are released into the mollusk's tissues, they produce cercariae that leave the mollusk and then encyst in or on the same snail or neighboring snails. Therefore, the metacercarial stage of echinostomatids tends to occur in or on snails but some species also encyst in or on annelids, fish, and tadpoles. Humans acquire this parasite by feeding on uncooked snails or, more likely, on poorly washed salad or other raw vegetation harboring the snails. Control is, therefore, possible by avoiding the ingestion of snails, by the careful washing or cooking of vegetation intended as food, and by limiting the use of human feces as a fertilizer. *Echinostoma* is not specific to humans and occurs in other mammals; rats and dogs are especially noted for carrying infections.

These worms attach to the wall of the intestine and cause local inflammation, presumably due in part to the large oral spines and general body spination. The oral spines are used by the worm for abrading the gut wall. Heavy infections can cause ulceration of the bowel, diarrhea, and abdominal pain.

Family Fasciolidae Railliet, 1895

Main features Echinostomatoidea; body large, flattened, often broad, sometimes narrow; forebody short; tegument normally armed with small spines; head collar and circumoral crown of spines absent; intestinal ceca with or without dendritic lateral branches; testes usually deeply lobed, sometimes unlobed; in intestine and biliary system of mammals.

Two genera occur in humans, *Fasciola* Linnaeus, 1758 and *Fasciolopsis* Looss, 1899. These are large, broad, leaf-like worms, recognizable by the presence of deeply lobed gonads, the lobes being so deep that the gonads appear dendritic. *Fasciola*, the common liver fluke of sheep, occurs in the biliary system, whereas *Fasciolopisis* occurs in the intestine and is restricted to southeast Asia. The two genera are readily distinguished as *Fasciola* has branched intestinal ceca.

The eggs of fasciolids, whether they be in the intestine or biliary system, pass out with the feces. The miracidia infect freshwater snails, within which the mother sporocyst develops and produces rediae. Accounts of the life history within the molluskan host vary but, apparently, within the rediae may develop a second or even third generation of rediae and any generation of rediae may produce cercariae. The latter leave the snail and encyst as metacercariae usually on vegetation. Encysted metacercariae are especially common on grass close to water and on plants such as watercress and water chestnut. The final hosts, normally herbivorores but occasionally humans, become infected by feeding on vegetation harboring the encysted metacercariae. Metacercariae of *Fasciola* migrate through the liver en route for the bile ducts, where the adults live. These parasites are very common in domesticated animals, such as sheep, pigs, and cattle, which form huge reservoirs of infection. Control is, therefore, facilitated by the avoidance of raw (or unpeeled) vegetables or other plant material which has been associated with water. Similarly, control can be assisted by avoidance where possible of the links between human and animal feces and bodies of water, e.g. by the use of drinking troughs and the draining of pasture land.

Fasciolids cause inflammation at the site of attachment, which may lead to ulceration and hemorrhage, as these spine-covered worms browse on the tissues. The amount of damage caused depends on the worm burden. In the liver, the presence of these worms can cause traumatic damage resulting in necrosis ('liver rot'); in the bile ducts, mechanical irritation can cause hyperplasia and fibrosis. When flukes are present in large numbers, they can result in blockages of the bile ducts or even the intestine (see Chapter 31, Lung and liver flukes).

SUPERFAMILY PARAMPHISTOMOIDEA FISCHOEDER, 1901

Family Zygocotylidae Ward, 1917

Main features Paramphistomoidea; body stout, often large, pear-shaped to elongate-oval, ventrally flattened; tegument unarmed, tegumentary papillae present or absent; appears to have suckers at each end of body; oral sucker absent; pharynx resembles oral sucker, terminal at anterior extremity, opening via mouth, with one or two posterolateral sacs; posterior (ventral) sucker normally occurs ventroterminally at posterior end of body, well developed, often very large; eggs very large, numerous; in intestine of mammals and birds (mainly herbivores).

Two genera of zygocotylid paramphistomoids occur in humans, *Gastrodiscoides* Leiper, 1913 and *Watsonius* Näsmark, 1937. These relatively large worms are readily recognizable by the apparent presence of a sucker at each end of the body, hence their vernacular name, 'amphistomes.' The two genera can be distinguished by shape; in *Gastrodiscoides* the body consists of a posterior discoid region and a narrower anterior region.

The zygocotylid life cycle commences with eggs passed out with the feces. When an egg reaches water, it hatches and the released larva, the miracidium, penetrates an aquatic gastropod mollusk. Within the mollusk, the miracidium transforms into a mother sporocyst, within which develop rediae (some authors maintain that only a single mother redia derives from the mother sporocyst and this produces a second generation of daughter rediae). Cercariae, which develop within the rediae, are released from the mollusk and encyst as metacercariae on the substratum, especially on vegetation and the snail host. The final host, a bird or mammal, normally becomes infected by feeding on vegetation or snails harboring the metacercarial stage.

The general similarity in transmission between the zygocotylids and the fasciolids means that the method of human infection and control measures linked to the worms' life cycles are essentially the same, i.e. avoiding the consumption of uncooked vegetation which has originated from or close to water. There is a difference, however, in that primates rather than domesticated herbivores tend to act as reservoir hosts, although pigs, rats, and other rodents may also be involved. Human feces should be kept away from bodies of water.

Little pathogenicity is observed in light infections, but inflammation of the intestinal mucosa may occur where the large posterior sucker attaches. Heavy infestations may cause edema of the intestinal wall and diarrhea.

SUPERFAMILY OPISTHORCHIOIDEA LOOSS, 1899

Family Heterophyidae Leiper, 1909

Main features Opisthorchioidea; body small, oval to elongate; tegument spined; oral and ventral suckers present; ventral sucker occasionally small or submedian, often associated with genital atrium forming ventrogenital complex; cirrus sac absent; gonotyl(s) (genital sucker, a muscular structure associated with genital atrium) may or may not be developed, armed or unarmed; eggs small; metacercariae on skin and fins and in tissues of fishes; adults in piscivorous birds and mammals.

Two heterophyid genera occur relatively commonly in humans, *Heterophyes* Cobbold, 1886 in lands bordering the western Mediterranean and both *Heterophyes* and *Metagonimus* Katsurada, 1913 in China, Korea, Japan, and neighboring regions. Other genera, such as *Centrocestus* Looss, 1899, *Stellantchasmus* Onji and Nishio, 1915, and *Haplorchis* Looss, 1899, may also occur on rare occasions. *Heterophyes* and *Metagonimus* can be distinguished by the fact that in the latter the mouth of the ventrogenital sac is directed anteriorly, there is no muscular gonotyl and the genital pore is anterior to the ventral sucker. Heterophyids are small worms recognizable by the fact that the genital atrium is associated with the ventral sucker to form a ventrogenital complex.

Heterophyids in humans are intestinal parasites. Their eggs pass out with the feces and hatch to release

miracidia which penetrate aquatic snails. Within the snail, there are two generations of rediae. Cercariae produced by the rediae leave the mollusk and encyst under the scales and in the superficial muscles of freshwater fishes. Humans and other piscivorous mammals obtain the infection by feeding on raw, pickled or poorly cooked fish harboring the metacercaria. Infection can be avoided by ensuring that fish intended as food is well-cooked or frozen prior to ingestion. A reduction in the likelihood of infection may also occur if fish in the food of reservoir hosts, such as cats, dogs, and pigs, is treated in the same way. Improvements in sanitation also help disrupt the life cycle.

In small numbers, these small, spined worms cause little damage to the intestine apart from local inflammation at the site of attachment; but more serious damage, including ulceration, may occur when large numbers are present. Occasionally the worms may penetrate the gut wall and eggs may end up in various parts of the body, including the heart and brain.

Family Opisthorchiidae Looss, 1899

Main features Opisthorchioidea; body small to medium-sized, flattened, elongate-oval to fusiform, often translucent; tegument spined; suckers well developed; cirrus sac absent; gonotyl and ventrogenital sac (see above, Heterophyidae) absent; metacercariae present in tissues of freshwater fishes; adults in piscivorous mammals and birds.

Two genera of opisthorchiids occur in humans, *Opisthorchis* Blanchard, 1895 and *Clonorchis* Looss, 1907. They can be distinguished by the fact that in *Clonorchis* the testes are tandem and have long, branched lobes that overlap the intestinal ceca. Both are parasitic in the biliary system and are most common in southeast Asia. They can be distinguished from heterophyids by their greater size, more elongate shape, and the absence of a ventrogenital complex. Body spines are very small and commonly lost in preserved specimens.

Worms in the bile ducts release eggs that pass into the intestine to be voided with the feces. If an egg finds its way to water and is ingested by a suitable aquatic gastropod snail, it hatches in the gut and releases the miracidium, which penetrates the mollusk and develops into a mother sporocyst. Rediae develop within the latter stage, which in turn give rise to cercariae. These leave the snail and swim and encyst as metacercariae under the scales or in the superficial tissues of freshwater fishes. Piscivorous mammals and birds become infected by feeding on fishes harboring the metacercarial stage. Control measures are similar to those outlined for heterophyids, i.e. primarily the avoidance of raw, pickled or poorly cooked fishes. Opisthorchiids are not primarily parasites of humans, as those infecting humans also occur commonly in many piscivorous mammals, including cats, foxes, dogs, pigs, etc., which act as reservoir hosts. Avoidance of the link between both human and animal feces and bodies of water is needed to break the life cycle.

Light infections are not pathogenic, but heavier infections of these spine-covered worms cause abdominal discomfort, diarrhea, and even acute pain. This results from inflammation of the biliary epithelium and thickening of the bile ducts as a consequence of mechanical irritation and their browsing activity. The presence of worms may cause gallstones, cirrhosis of the liver, and perhaps even carcinoma of the bile ducts (see Chapter 31, Lung and liver flukes).

SUPERFAMILY DICROCOELIOIDEA LOOSS, 1899

Family Dicrocoeliidae Looss, 1899

Main features Dicrocoelioidea; body oval, lanceolate or filiform, flattened, often translucent; tegument usually spined; oral sucker subterminal; ventral sucker in anterior third of body, occasionally very reduced or absent; cirrus sac present close to ventral sucker; metacercaria occurs in terrestrial arthropods; in biliary system or pancreatic ducts of reptiles, birds, and mammals.

Only one dicrocoeliid genus, *Dicrocoelium* Dujardin, 1845, occurs in humans and then only rarely. It normally parasitizes a wide range of mammals and even birds, but is very common in domesticated ruminants, especially sheep. Dicrocoeliids can be distinguished from other digeneans occurring in the biliary system of humans in that the testes occur close to the ventral sucker and the vitelline fields are in the post-testicular region of the hindbody.

As in the case of other parasites of the biliary system, the eggs pass out with the feces, but, in contrast to other digeneans occurring in humans, the snail host is a terrestrial gastropod. Inside the mollusk, the miracidium transforms into a mother sporocyst within which form daughter sporocysts. The latter give rise to cercariae which end up in slime balls produced by the snail. When the slime balls are eaten by ants, the parasite encysts as a metacercaria in the insect. Herbivores, and humans, become infected with the parasite when ants are eaten accidentally with vegetation. Human infection can, therefore, be avoided by careful washing of salad material and other items of vegetation eaten raw.

Little is known regarding the pathogenicity of these worms in humans, but, since infections are likely to be light, they are presumably similar to mild infections of fasciolids or opisthorchiids.

SUPERFAMILY PLAGIORCHIOIDEA LÜHE, 1901

Family Lecithodendriidae Lühe, 1901

Main features Plagiorchioidea; body spherical to elongate; tegument usually spined; oral sucker well developed; ventral sucker often small, near middle of body; testes usually symmetrical and widely separated near middle of body; cirrus sac usually present; vitelline follicles in forebody, occasionally hindbody; eggs small; metacercariae in aquatic insects; in gut of amphibians, reptiles, birds, and mammals (especially bats).

Lecithodendriids occur in a wide variety of insectivorous vertebrates and are noted parasites of bats. Only one genus, *Phaneropsolus* Looss, 1899, regularly parasitizes humans, although others, such as *Paralecithodendrium* Odhner, 1910 and *Prosthodendrium* Dollfus, 1931 have also been reported. These minute parasites are recognizable by the fact that the gonads and vitelline follicles are in the forebody or at least level with the ventral sucker.

Lecithodendriid eggs pass into water with the host's feces. The miracidium enters the tissue of an aquatic snail and transforms into a mother sporocyst. It is likely that these produce daughter sporocysts, which in turn produce cercariae. The cercariae leave the snail and some are eaten by or penetrate the larvae of aquatic insects, within which they encyst as cercariae. The final vertebrate host acquires the parasite by feeding on larval or adult insects harboring the parasite. It is not clear how people in southeast Asia (Thailand and Indonesia) become infected with lecithodendriids, but it is presumably by feeding on larval or adult insects in regions where these are local delicacies or by accidentally ingesting insect larvae with aquatic vegetation or water. Control measures include the filtering or heating of water, the cleaning of vegetable material, and the cooking or freezing of insects intended as food.

Virtually nothing is known concerning the pathogenicity of lecithodendriids in humans, but it seems likely that some local irritation occurs, similar to that caused by mild infections of other small, intestinal digeneans armed with spines, such as heterophyids.

Family Paragonimidae Dollfus, 1939

Main features Plagiorchioidea; body oval to fusiform; tegument spinose; suckers relatively small, ventral sucker near middle of body; testes symmetrical, lobed, in hindbody; cirrus sac absent; genital pore closely posterior to ventral sucker; metacercariae in tissues of crustaceans; in lungs of mammals.

Numerous species of the genus *Paragonimus* Braun, 1899 occur in the lungs of humans in parts of Africa, South America and especially southeast Asia. They are medium-sized worms readily recognizable by their site in the lungs and the position of the genital pore posterior to the ventral sucker. They can be distinguished from the closely related troglotrematids by the absence of a cirrus sac. Paragonimids are usually parasitic in carnivores, especially felids and others likely to feed on freshwater crabs and other large crustaceans.

The eggs pass out of the lungs with mucus, are swallowed, and then voided with the feces or in sputum. The hatched miracidium penetrates a freshwater snail and develops into a mother sporocyst. Two generations of rediae are then produced. The second generation gives rise to cercariae, which leave the snail and encyst in the tissues (especially muscles and gills) of crustaceans (usually crabs or crayfish); the latter may apparently also become infected by feeding directly on infected snails. Humans acquire the parasite by eating raw or inadequately cooked crustaceans; the young flukes penetrate the gut wall and find their way to the lungs via the abdominal cavity, diaphragm, and pleural cavity. Control is effected by the cooking of crustaceans intended as food and sanitary measures which break the link between human and, where possible, carnivore feces and bodies of water.

Paragonimids are covered with small spines and presumably browse on the walls of the bronchi, so their presence in the lungs causes local inflammation, resulting in a tissue reaction, cyst formation, and fibrosis. The net result is coughing, bloodstained sputum, chest pains, and bronchitis, which occur when the cysts rupture. The migration of the young worms through the body tissues does not cause any serious problems; but occasionally worms reach other parts of the body where problems may arise, especially when worms reach the brain where they may cause paralysis or even death (see Chapter 31, Lung and liver flukes).

Family Plagiorchiidae Lühe, 1901

Main features Plagiorchioidea; body oval, pyriform, lanceolate or elongate; tegument armed with spines; suckers well developed; ventral sucker normally in anterior half of body; testes in hindbody, usually diagonal; cirrus sac present, usually long; genital pore anterior to ventral sucker; ovary pretesticular; uterus in hindbod; in intestine, occasionally biliary system, of vertebrates, especially birds and mammals.

Species of the genus *Plagiorchis* Lühe, 1899 have occasionally been recorded from the intestine of humans in parts of southeast Asia. Plagiorchiids are parasites of insectivorous vertebrates, especially mammals. Although some nominal species are known only from humans, it is not known whether they are zoonotic infections or specifically human parasites. The worms differ from other plagiorchioids found in humans in that the testes tend to be diagonal to tandem rather than symmetrical.

Plagiorchiid eggs leave with the feces and the miracidium enters an aquatic snail and develops into a mother sporocyst. These give rise to daughter sporocysts from which cercariae emerge. These leave the snail and find their way into the larval stages of aquatic insects which act as intermediate hosts. The vertebrate host becomes infected by feeding on larval or adult insects harboring cercariae. Humans are thought to become infected by eating insect grubs.

Little is known regarding the pathogenicity of these worms, but infections are likely to be light and effect

similar local responses to those of other small, spined digeneans parasitic in the gut, such as heterophyids.

Family Troglotrematidae Odhner, 1914

Main features Plagiorchioidea; body oval; tegument spinose; suckers well developed, often similar in size; ventral sucker in middle or anterior half of body; testes symmetrical; cirrus sac present; genital pore usually closely posterior to ventral sucker; metacercaria in tissues of fish or crustaceans; in intestine, body sinuses, kidney, liver, etc. of mammals.

One genus of troglotrematid, *Nanophyetus* Chapin, 1927, occasionally occurs in humans. These worms normally occur as parasites of piscivorous carnivores in North America. This group appears closely related to the paragonimids and, like the latter group, tends not to be parasitic in the alimentary system (though *Nanophyetus* is an intestinal parasite) and be recognizable by the position of the genital pore posterior to the ventral sucker.

Eggs leave with the feces. The miracidium enters aquatic snails within which a generation of rediae produce more rediae and cercariae. The cercariae leave the snail and become attached to fish encysting under the scales as metacercariae (some troglotrematid cercariae encyst in crustaceans). Carnivores become infected by feeding on fish carrying the metacercariae. Similarly, humans become infected by eating raw, smoked or poorly cooked freshwater fish. Obvious control measures involve the skinning, cooking or freezing of fish prior to consumption.

Little is known concerning the pathogenicity of these worms, but, as they have body spines and may be acquired in some numbers, some irritation and inflammation of the bowel is to be expected.

PHYLUM NEMATODA

Nematodes are probably the most abundant and widespread animal group, often occurring in huge numbers in environments ranging from the polar regions to hot springs. In addition to free-living marine and freshwater forms, there are free-living forms in the soil and parasitic forms in both animals and plants. Nematodes are symmetrically bilateral, unsegmented, normally dioecious worms that are usually filiform; they have a body cavity with a high hydrostatic pressure, a straight digestive tract with an anteriorly terminal mouth and posteriorly subterminal anus, no circulatory system, a simple excretory system, and a body wall consisting of an outer layer of cuticle and an inner layer of longitudinal muscles (Figure 27.3). As animal parasites, they occur in virtually all groups, both invertebrate and vertebrate. The phylum is divided into two classes, the Adenophora and the Secernentea; both groups have evolved parasitic members, although the majority of

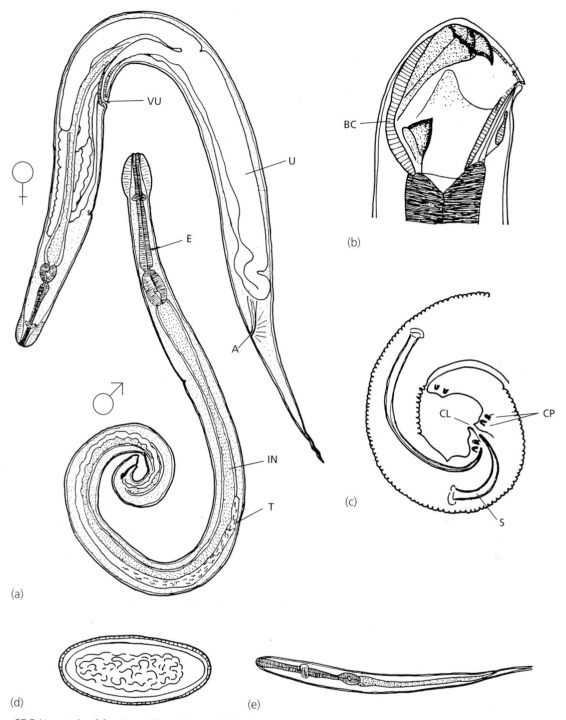

Figure 27.3 Nematodes: **(a)** male and female Strongyloides (Strongyloididae); **(b)** head of Ancylostoma (Ancylostomatidae) (hookworm) with cuticularized buccal capsule armed with teeth; **(c)** caudal end of male Onchocerca (Onchocercidae); **(d)** egg of Strongyloides; **(e)** third-stage larva of Strongyloides. Anus (A), buccal capsule (BC), cloaca (CL) (in the male, the gut and reproductive system open through the same aperture), caudal papillae (CP), esophagus (E), intestine (IN), spicule (S), testis (T), uterus (U), vulva (VU). Adapted from (a, d, e) Hugot and Tourte-Schaefer (1985); **(b)** Yorke and Maplestone (1926); **(c)** Caballero y Caballero 1944)

animal parasites belong to the latter group. The differences between the groups reflect the presence and absence of small sensory structures (phasmids) on the tail and the nature of the excretory system; but there is a fundamental biological difference in the parasitic members, since in the Adenophora the first-stage larva is infective to the definitive (final) host, whereas in the Secernentea it is the third-stage larva.

All nematodes have five life history stages, four larval and one adult, which are separated by a molt of the cuticle. It is common for the first one or two molts to occur within the egg. The life cycle may be direct or indirect. Except in the case of the Adenophora, the stage infective to the final host is the third-stage larva. Direct life cycles can involve the ingestion of eggs or larvae with food or, in some cases, direct penetration of larvae through the skin. Indirect life cycles usually utilize invertebrate intermediate hosts but sometimes vertebrates (or other invertebrates) may act as intermediate or paratenic hosts. The larvae normally reside, often encysted, in the tissues of intermediate hosts. The majority of nematode parasites of vertebrates occur in the alimentary canal; those in other parts of the body often require the migration of larvae through the body to reach these sites. Some groups also have a larval migration from the gut and into the tissues and back to the gut; this represents the vestige of an indirect life cycle in its evolutionary past. Whichever mode of transmission is utilized, the chance of a particular egg or larva developing into an adult worm is very slight; this is compensated for in many cases by a huge output of eggs, which in some cases reaches 200 000 per female per day.

The morphological features by which nematodes are recognized vary from group to group, but at higher taxonomic levels, the nature of the anterior regions of the alimentary canal, e.g. the esophagus, the form of the head (presence, number of lips, teeth, etc.), and the form of the male tail are usually important. At the specific level, details of the male tail, such as the arrangement of caudal papillae (sensory structures used during copulation) and the length and shape of the spicule(s) (sclerotized copulatory aids) are important. In the majority of cases, males carry more taxonomically useful information than females, the latter often being unidentifiable at the specific level in the absence of males.

The pathogenicity of nematodes in their final host varies considerably, usually being dependent on the size of the infection. Those such as hookworm, which are heavily armed with teeth or other sclerotized mouthparts and browse on the gut wall, can cause considerable damage. Similarly, forms that migrate around the body, both adults in the tissues and larvae (termed larva migrans), can cause serious problems, especially if they reach sensitive regions such as the brain, liver or eyes (see Chapters 36, *Toxocara*, and 37, *Trichinella*).

The classification of the nematodes below is based on that of Anderson et al. (1974–83); it includes most of the forms that have been recorded from humans regularly, but accidental infections of other forms do occur infrequently:

Class Adenophorea
    Superfamily Trichinelloidea
        Family Trichinellidae
        Family Trichuridae
Class Secernentea
    Superfamily Ancylostomatoidea
        Family Ancylostomatidae
    Superfamily Ascaridoidea

        Family Anisakidae
        Family Ascarididae
    Superfamily Dracunculoidea
        Family Dracunculidae
    Superfamily Filarioidea
        Family Onchocercidae
    Superfamily Gnathostomatoidea
        Family Gnathostomatidae
    Superfamily Metastrongyloidea
        Family Angiostrongylidae
    Superfamily Oxyuroidea
        Family Oxyuridae
    Superfamily Physalopteroidea
        Family Physalopteridae
    Superfamily Rhabditoidea
        Family Rhabditidae
        Family Strongyloididae
    Superfamily Spiruroidea
        Family Gongylonematidae
    Superfamily Strongyloidea
        Family Chabertiidae
        Family Syngamidae
    Superfamily Thelazioidea
        Family Thelaziidae
    Superfamily Trichostrongyloidea
        Family Trichostrongylidae.

Class Adenophorea Dougherty, 1958

Main features Nematoda; tail occasionally modified to form sucker; esophagus with normal appearance or modified, with esophageal gland cells forming row of large gland cells (stichocytes) or reserve organ (trophosome); oviparous or viviparous, eggs with plug at either end; excretory system without lateral canals; caudal papillae absent or few; phasmids absent; first larval stage usually infective to final host.

SUPERFAMILY TRICHINELLOIDEA WARD, 1907

Family Trichinellidae Ward, 1907

Main features Trichinelloidea; modified esophagus with one row of stichocytes; intestine unmodified, anus present; vulva of female close to middle of esophagus; viviparous; spicule in male absent; larval stage in muscles of mammals; adults usually in intestine of carnivorous mammals.

One genus, *Trichinella* Railliet, 1895, occurs in humans both as an adult in the intestine (usually gut wall) and as a larval stage in the tissues (usually muscles). The adult worms are very small, thread-like, and recognizable by the row of large gland cells (stichocytes) surrounding the esophagus and the absence of spicules in the male.

Adult worms occur in the intestine and intestinal mucosa. Females are viviparous and release first stage larvae into the intestinal mucosa. These larvae migrate via the blood and lymphatic system to muscles where they encyst in a coiled form. Transmission to another host occurs when the host is eaten by a carnivore or omnivore; here the larvae molt four times in the gut prior to reaching the adult stage. The same animal may serve, therefore, as both intermediate and definitive host. Humans usually become infected by feeding on raw or inadequately cooked meat, especially pork, but also wild carnivores, such as bears. Control is effected by the cooking or freezing of meat, preventing pigs eating raw meat and offal, and getting rid of rats (which may harbor the infection) on pig farms.

The effects of *Trichinella* infection range from diarrhea and nausea during the intestinal phase to eosinophilia, general weakness due to muscle damage, and difficulties in breathing and myocarditis caused by larvae in the muscles. Heavy infection can result in death, especially in cases where the heart or brain is involved (see Chapter 37, *Trichinella*).

Family Trichuridae Ransom, 1911

Main features Trichinelloidea; modified esophagus with one to three rows of stichocytes; intestine unmodified; vulva of female close to posterior end of esophagus; oviparous, eggs with plug at each end; single spicule and spicule sheath usually present; in alimentary canal, tissues, and various organs of vertebrates.

Several trichurid genera are known to infect humans, including *Trichuris* Roederer, 1761, *Aonchotheca* Lopez-Neyra, 1947, *Calodium* Dujardin, 1845, and *Eucoleus* Dujardin, 1845; the latter three genera are still commonly referred to as *Capillaria* Zeder, 1800, when the latter is used in its wide sense (see Moravec 1982 for generic differences). *Trichuris* is a common parasite of humans, whereas the capillariines are normally parasites of wild animals which occasionally infect humans. Like trichinellids, these worms have a row of stichocytes around the esophagus, but they can be distinguished by the presence of thick-shelled eggs with terminal plugs in the females and a spicule in the male.

Trichurids inhabit the intestine, intestinal mucosa, liver, lungs, and other tissues of vertebrates, depending on the species involved. Eggs pass out with the feces, or remain in the body until the host dies or is eaten by a predator or scavenger and are voided with its feces. Again, depending on the species involved, the life cycle may be direct (*Trichuris, Aonchotheca, Eucoleus*), in which case the definitive host becomes infected by feeding on eggs, or indirect (e.g. *Calodium*) and involve an intermediate host that has swallowed the eggs. Humans become infected by ingesting eggs in soil (especially children) or on vegetation or, in the case of *Calodium*, by ingesting raw fish harboring the larva. Control is effected by modern sewage disposal and improvements in personal hygiene, the proper cooking of fish (*Calodium*), and the control of rats and other rodents which act as reservoirs (*Aonchotheca*).

Symptoms of infection vary according to the species but in the case of *Aonchotheca* in the liver, they resemble hepatitis as eggs are deposited in the liver parenchyma. Intestinal forms may cause diarrhea, dysentery, and weight loss as the worms penetrate the mucosa, but the severity depends on the size of the infection. In extreme cases rectal prolapse, muscle wasting, and death can occur (see Chapter 34, Gastrointestinal nematodes and Chapter 35, *Strongyloides* and *Capillaria*).

Class Secernentea Dougherty, 1958

Main features Nematoda; esophagus without stichocytes or reserve organ (trophosome); oviparous or viviparous, eggs without plug at either end; excretory system with lateral canals; caudal papillae numerous; phasmids present; third larval stage infective to final host.

SUPERFAMILY ANCYLOSTOMATOIDEA LOOSS, 1905

Family Ancylostomatidae Looss, 1905

Main features Ancylostomatoidea; buccal cavity large; cuticularized; subglobular; mouth not surrounded by lips or comb-like leaf-crown(s), unarmed or armed with teeth or cutting plates; esophagus claviform, not divided into two regions; male with caudal bursa normally supported by rays; in alimentary canal of mammals.

Two genera of ancylostomatids ('hookworms'), *Ancylostoma* Dubini, 1843 and *Necator* Stiles, 1903, occur in humans mainly in the tropics. These are intestinal parasites recognizable by their large globular, cuticularized buccal cavity, mouth armed with teeth or cutting plates, and dorsally curved anterior extremity. Other genera may occur as accidental parasites on rare occasions.

The life cycle is direct. Eggs laid by female worms in the intestine are voided with the feces and hatch in the soil. The larva feeds on bacteria in the soil and molts twice. The third-stage larvae penetrate the skin of a suitable mammalian host and then migrate via the blood system to the lungs where they pass from the blood system into the alveoli, ascend the bronchi and trachea, and are then swallowed. Once in the intestine, they molt twice and mature. Humans become infected via their bare feet, the skin of which is penetrated by the larvae. Human infection is also possible from larvae swallowed with salad or other raw vegetation. Control is effected by sewage management, the wearing of shoes, and the careful washing of salad material.

The penetration of the skin by larvae can cause an itching and the migrating larvae can result in coughing and bronchitis. The heavy armature of the mouth parts of adults, as they feed on the tissues and blood of the host, damages the intestinal wall, resulting in blood loss, abdominal pain, general weakness, and sometimes even death (see Chapter 34, Gastrointestinal nematodes).

SUPERFAMILY ASCARIDOIDEA BAIRD, 1853

Family Ascarididae Baird, 1853

Main features Ascaridoidea; head with three well-defined lips; lips sometimes separated by interlabia, unadorned by spines or cuticular embellishments posteriorly, single row of denticles on anterior margin present or absent; buccal cavity absent or weakly developed; esophagus cylindrical, occasionally divided into two regions (when posterior glandular ventriculus is present), without posterior muscular bulb; excretory system symmetrical, 2-sided and tubular, excretory pore at level of nerve ring; in all vertebrate groups, mainly but not always terrestrial forms.

Three genera of ascaridid nematode, *Ascaris* Linnaeus, 1758, *Toxocara* Stiles, 1905, and *Toxascaris* Leiper, 1907, regularly occur as parasites of humans although only *Ascaris* is a true human parasite, the other two being common parasites of dogs or cats. Other genera occurring in wild animals, such as *Baylisascaris* Sprent, 1968 and *Lagochilascaris* Leiper, 1909, have also been recorded from humans on rare occasions. Only *Ascaris* normally occurs in humans as an adult; the others are present in the tissues in larval form only. Ascaridids are related to anisakids, and like that group have three distinct lips; they differ in that (with the exception of the subfamily Heterocheilinae) their life cycle is linked to terrestrial rather than aquatic conditions.

Ascaridids occur in all vertebrate groups and have a variety of life cycles. Most involve invertebrate or vertebrate intermediate hosts or both, but those occurring in humans normally have a direct life cycle. Nevertheless, even in these forms the vestige of an indirect life cycle is apparent in the form of a larval migration through the tissues of the host prior to maturation in the intestine. In these cases, eggs pass out with the feces and the first two larval molts occur within the egg. If the eggs are eaten by a suitable host, the third-stage larvae hatch in the gut, penetrate the gut wall, and are transported by the blood to the lungs. After a short period in the lungs where they molt twice, they escape into the alveoli, pass up the bronchi and trachea with mucus and are swallowed. Once they reach the intestine, they mature. There are variations: in some genera developing in mammals, migrating larvae in the blood can result in transplacental or transmammary infections of young; and occasionally insects and rodents may act as intermediate hosts. Humans, especially children, become infected by accidentally ingesting soil or food contaminated with eggs. Control is effected by careful sewage management, avoiding the use of human feces as manure, improvements in personal hygiene, keeping children away from the feces of domesticated animals, especially dogs, the regular de-worming of pets, and the careful washing of salad material, fruit, etc.

Despite the fact that *Ascaris* is a large worm, its presence in the intestine does not cause many symptoms unless the worm burden is high. Light infections can cause nausea and other bowel disorders; heavy infections can result in bowel obstruction or peritonitis and even death. The migration of the larval stage of *Ascaris* and especially the other ascaridids mentioned above can be more serious. Larvae passing through the lungs cause some pneumonitis, but larva migrans that lodge in other tissues, such as the liver, spleen, kidneys, placenta, and especially the eye and CNS (e.g. *Toxocara*), can represent a danger, producing inflammation and granulomata in these sensitive areas (see Chapter 34, Gastrointestinal nematodes and Chapter 36, *Toxocara*).

Family Anisakidae Railliet and Henry, 1912

Main features Ascaridoidea; head with three well-defined lips; lips sometimes separated by interlabia, unadorned by spines or cuticular embellishments posteriorly, single row of denticles on anterior margin present or absent; buccal cavity absent or weakly developed; esophagus with posterior ventriculus; gut ceca present or absent; excretory system asymmetrical, entirely or mainly one-sided, tubular or ribbon-like, excretory pore at level of nerve ring or ventral interlabium; larval stages in fishes or invertebrates or both; in piscivorous vertebrates.

Anisakids capable of accidentally infecting humans are normally parasites in the stomach of marine mammals. The group occurs in all vertebrate groups, but those parasitic as adults in other hosts, such as fishes, are not capable of infecting humans. There are two main genera involved, *Anisakis* Dujardin, 1845 and *Pseudoterranova* Mozgovoi, 1950; the former is the more widely reported. Since humans do not provide a suitable environment for the development of these parasites, they do not reach maturity and usually attempt to penetrate the gut wall in order to re-encyst as they would in any other unsuitable host. Like ascaridids, anisakids have three distinct lips at the anterior end, but differ in the configuration of the excretory system.

All anisakids have an indirect life cycle involving aquatic intermediate hosts. Eggs leave with the feces. The first one or two (accounts vary) larval molts occur within the egg. If eggs released into water are eaten by small planktonic invertebrates, such as copepods, the third-stage larvae develop in the hemocoel. If this host is eaten by a larger crustacean, squid or bony fish, the larvae are transferred and re-encyst in the body cavity or muscles of the new host. The larvae can be transferred to new hosts several times. If a third-stage larva within its intermediate host is consumed by a suitable (usually piscivorous) vertebrate host, the larva molts twice and matures in the stomach or intestine. Humans usually acquire infections of anisakids by feeding on raw, lightly marinated or inadequately cooked marine bony fishes (infection by consuming raw squid and raw marine crustaceans is also possible). Infection can be prevented by cooking fish properly or by the freezing of fish or fish products intended for consumption raw or marinated. Fish inspection may eliminate heavily infected fish and evisceration immediately following netting may prevent the migration of larvae to the muscles, but it is not possible to prevent the infection of wild marine fishes without a heavy culling of marine mammals.

Although the worms may molt once in humans, they tend to penetrate the gut wall in an attempt to re-encyst; whether this is in the stomach or intestine may depend on the size of the food bolus. Symptoms of infection can mimic gastric or duodenal ulcers. Repeated infections are more serious, and worms may have to be removed by surgery. In Japan, where this has been an important parasite of humans because of local fish-eating habits, the presence of these worms has been linked to cases of gastric and intestinal cancer (see Chapter 40, *Angiostrongylus* (*Parastrongylus*) and less common nematodes).

SUPERFAMILY DRACUNCULOIDEA STILES, 1907

Family Dracunculidae Stiles, 1907

Main features Dracunculoidea; considerable sexual dimorphism; head-end rounded, bilaterally symmetrical; lips and pseudolabia absent; buccal capsule weakly developed, reduced to cuticular ring; esophagus apparently undivided; esophageal glands usually uninucleate; anus atrophied in adult; male tail without caudal bursa; body of female long, cylindrical; vulva normally in middle of body; viviparous; transmission indirect, copepods act as intermediate hosts; usually in tissues and tissue spaces of reptiles, birds, and mammals.

One genus, *Dracunculus* Reichard, 1759, infects the subcutaneous tissues of humans, the species involved also occurs in a variety of mammals, especially dogs. These worms resemble filarioids in being tissue parasites, but they have a different life cycle, which involves an aquatic intermediate host, and they do not release microfilariae into the blood.

Fully developed female worms migrate to the skin and produce a blister that bursts, permitting the rupture of the worm and its uterus when contact is made with water. First-stage larvae eaten by copepods enter the hemocoel and molt twice. Vertebrates become infected with these parasites when they drink water containing copepods harboring the third-stage larva. The larvae penetrate the gut wall and enter connective tissue within the body, where they molt twice. After mating, the females develop in the subcutaneous regions. Humans become infected by drinking water containing infected copepods. The parasite is readily controlled by the filtering of drinking water or by using water sources, such as springs or deep wells, where contact with animal skin is not possible and copepods are absent.

Migrating worms in the subcutaneous regions may cause some tenderness, but the presence of the parasite may not be apparent until the blister forms. The blister may be painful and cause considerable inflammation, a large ulcer and, when secondary infections are involved, an abscess (see Chapter 41, Dracunculiasis).

SUPERFAMILY FILARIOIDEA WEINLAND, 1858

Family Onchocercidae Leiper, 1911

Main features Filarioidea; body long, cylindrical; head bilaterally symmetrical; lips and pseudolabia absent; buccal cavity usually relatively small, not greatly cuticularized; esophagus usually with anterior muscular region and posterior glandular region, sometimes regions difficult or impossible to differentiate; male tail without caudal bursa, caudal alae usually narrow or apparently absent; vulva in anterior half of body but posterior to nerve ring; ovoviviparous or viviparous, eggs thin-shelled, poorly differentiated larvae (microfilariae); microfilariae appear in blood, lymph or skin of host; transmission indirect, with blood-feeding insects acting as intermediate hosts; in tissues and tissue spaces of amphibians, reptiles, birds, and mammals.

Several genera of filarioids are parasitic in humans, the more important of which are *Brugia* Buckley, 1958, *Dipetalonema* Diesing, 1861, *Dirofilaria* Railliet and Henry, 1910, *Loa* Stiles, 1905, *Mansonella* Faust, 1929, *Onchocerca* Diesing, 1841, and *Wuchereria* Silva Araujo, 1877. These tissue parasites are usually found in the lymphatic system, subcutaneous connective tissue or other connective tissues. They differ from dracunculids in that their life cycle is terrestrial and they produce larvae (microfilariae) that are found in the blood, skin, etc. A generic key to microfilariae, which are more likely to be seen than the adult worms, was produced in French by Bain and Chabaud (1986).

The life cycle is indirect, involving a biting insect as an intermediate host. Mature female worms release first- or second-stage larvae surrounded by their first-stage cuticle into the blood. These often exhibit a diurnal periodicity that brings them into the peripheral circulation during the period when blood-feeding insects, such as mosquitoes and blackfly, are likely to feed. When taken in by a suitable insect, they pass into the hemocoel, grow, molt into the infective third stage and migrate to the region of the insect's mouthparts. The worm is then injected into a new host when the insect next feeds. After entering a suitable host, the larvae migrate to the lymphatic system or connective tissue, molt twice, and mature. Humans, therefore, become infected when they are bitten by insects, and some control can be achieved by reducing the numbers of vectors by the drainage of marshes, use of insecticides, etc.

Human infection (filariasis) causes various symptoms, depending on the species and site of the parasite; elephantiasis caused by inflammation in the lymphatic system in various parts of the body is the most extreme. In some cases, symptoms are few; in others, subcutaneous swellings or nodules occur that can result in abscesses if the worm dies. Such infections are often accompanied by rashes and intense itching. Microfilariae may also cause problems when they become lodged in regions such as the central nervous system or eyes (see Chapter 38, Lymphatic filariasis and Chapter 39, Onchocerciasis).

SUPERFAMILY GNATHOSTOMATOIDEA RAILLIET, 1895

Family Gnathostomatidae Railliet, 1895

Main features Gnathostomatoidea; body long, cylindrical or stout often armed with spines; head bilaterally symmetrical with two large, trilobed, lateral pseudolabia; collar region may be enlarged to form cephalic bulb, which may be covered with transverse rows of recurved hooks; inner face of pseudolabia folded into rounded, tooth-like formations that interlock with those on opposite interlabium; buccal cavity normally relatively uncuticularized; esophagus with anterior muscular region and longer posterior glandular region; male tail without caudal bursa; transmission indirect with copepods and vertebrates acting as intermediate hosts; in anterior gut of vertebrates.

Several species of the genus, *Gnathostoma* Owen, 1836, occur occasionally as accidental parasites of humans in their larval form. These worms normally occur as adults in the stomach, usually in nodules in the stomach wall, of carnivorous mammals and pigs. They are recognizable by the presence of regular rows of spines on the body; on an inflated cephalic bulb these spines are larger and flattened.

The life cycle normally involves two intermediate hosts. Eggs leave the body in the feces. In water, these develop and hatch to release a larva that has molted once. When eaten by a suitable copepod, the larva penetrates the hemocoel; if the copepod is then eaten by a suitable freshwater fish or frog, the larva molts into the infective third stage. Fish, snakes, rodents, chickens, and humans may eat these hosts and act as paratenic hosts. The definitive host becomes infected by feeding on the second intermediate or a paratenic host. Worms found in humans, normally the third-stage larva, are obtained through eating raw or poorly cooked fish (occasionally, perhaps, chicken). Prevention involves the cooking or freezing of freshwater fish and fish products.

In humans, larvae penetrate the gut wall and migrate via the liver to the subcutaneous regions. Their penetration of the gut wall can cause some pain and presence in the subcutaneous regions which results in swelling and symptoms similar to the cuticular larva migrans of some ascaridids. More serious problems and even death can be caused if worms reach more sensitive areas of the body, such as the eyes or central nervous system (see Chapter 40, *Angiostrongylus* (*Parastrongylus*) and less common nematodes).

SUPERFAMILY METASTRONGYLOIDEA LEIPER, 1908

Family Angiostrongylidae Böhm and Gebauer, 1934

Main features Metastrongyloidea; longitudinal cuticular ridges on body absent; cephalic vesicle (cuticular inflation) absent; buccal cavity absent or reduced and weakly cuticularized; mouth unarmed, not surrounded by lips or comb-like leaf-crown(s); esophagus cylindrical and not divided into two regions; male with caudal bursa supported by rays, bursa often reduced; vulva near anus; larvae in mollusks; adults in respiratory system and, occasionally, blood system and other organs of marsupials, insectivores, rodents, and carnivores.

One metastrongyloid genus, *Angiostrongylus* Kamensky, 1905 (includes *Parastrongylus* Baylis, 1928) occurs in humans. Metastrongyloids are commonly called lungworms. Angiostrongylids capable of infecting humans normally parasitize the lungs or blood system of rodents. They are recognizable by the presence in the male of a caudal bursa supported by rays, in conjunction with their site of infection, i.e. the respiratory or blood systems.

Eggs produced by worms in the pulmonary or mesenteric vessels of rodents, etc., release first-stage larvae into the intestine (those in the lungs are swallowed with mucus), which are voided with the feces. If these larvae penetrate or are eaten by certain snails, slugs or, less frequently, crabs and land planarians, they penetrate the tissues and molt twice. When an infected mollusk is eaten by a suitable mammal, the third-stage larvae penetrate the gut wall and migrate in the blood system or lymphatic system, molting twice (sometimes in the central nervous system) prior to maturing in the pulmonary or mesenteric arteries. Humans presumably become infected with these worms by eating raw or poorly cooked snails, or accidentally consuming mollusks on salad and other raw plant material. Prevention involves the cooking or freezing of snails, the careful washing of salad material and rodent control.

The migration of these worms to the central nervous system is potentially serious, producing symptoms resembling meningitis, paralysis, and blindness; occasionally this results in death. Pulmonary symptoms may not be apparent, but worms in vessels associated with the intestine can cause inflammation, granulomata, thrombosis, and necrosis, especially in the region of the appendix (see Chapter 40, *Angiostrongylus* (*Parastrongylus*) and less common nematodes).

SUPERFAMILY OXYUROIDEA COBBOLD, 1864

Family Oxyuridae Cobbold, 1864

Main features Oxyuroidea; body short and stout; buccal cavity absent or weakly developed; esophagus relatively short, not divided into anterior muscular and posterior glandular regions but with posterior muscular bulb; male tail with wide caudal alae which may resemble caudal bursa but supported by arrangement of usually similar pedunculate papillae rather than assortment of rays; usually one spicule; eggs often flattened on one side with relatively thick shells, usually embryonated on deposition, two larval molts occur within egg; life cycle direct; in intestine of mammals.

Only one genus, *Enterobius* Leach, 1853 (commonly known as 'threadworm' or 'pinworm'), regularly occurs in humans and other primates, although *Syphacia* Seurat, 1916, a rodent parasite, has been reported as an accidental infection in humans. Oxyurids are small, intestinal parasites with a direct life cycle and recognizable by the presence of a large posterior bulb on the esophagus.

Gravid female worms leave the body through the anus, depositing eggs in the perianal region. As their bodies dehydrate and burst, they eject eggs in huge numbers, which cause an itch. Two larval molts occur within the egg. Humans are reinfected by swallowing the eggs, which become accidentally attached to food or lodged under fingernails following scratching of the perianal region. Once swallowed, the eggs hatch and the third-stage larvae molt and develop into adults in the intestine. Autoinfection of humans is common. Prevention involves improvements in personal hygiene, in terms of both individuals and the entire population, and especial care in relation to food preparation.

The presence of these worms generally does little harm, apart from discomfort related to itching of the perianal region, although penetration of the gut wall has been observed, resulting in nonproven links with appendicitis and other complaints. Gravid females that migrate into the vagina may cause a mucoid discharge (see Chapter 34, Gastrointestinal nematodes).

SUPERFAMILY PHYSALOPTEROIDEA RAILLIET, 1893

Family Physalopteridae Railliet, 1893

Main features Physalopteroidea; body cylindrical without lateral rows of spines; head bilaterally symmetrical; two pseudolabia, unlobed, armed with denticles (teeth) on free border; cuticle posterior to pseudolabia often expanded to form cephalic collar; buccal cavity usually relatively small, not greatly cuticularized; esophagus with anterior muscular region and longer posterior glandular region; male tail without caudal bursa, but caudal alae often wide; transmission indirect with arthropods acting as intermediate hosts; adults in gut of all vertebrate groups and occasionally crustaceans.

Only one physalopteroid, a species of the genus *Physaloptera* Rudolphi, 1819, which normally occurs in the intestine of primates, has been reported from humans. Other species of *Physaloptera* occur in birds of prey and insectivorous or carnivorous mammals. Important diagnostic features of physalopterids include the presence of two large, symmetrical, unlobed pseudolips armed with teeth and a cephalic collar (an expansion of the cuticle immediately posterior to the pseudolips).

Physalopterids have an indirect life cycle, normally involving an arthropod intermediate host and often a vertebrate paratenic host. Eggs of these gut parasites leave with the feces; if eaten by insects, such as cockroaches, other beetles, and crickets, the third-stage larva develops. The final host becomes infected by feeding on insects or, especially in the case of carnivores, on insectivorous vertebrates such as frogs and rodents. Humans presumably become infected by feeding (usually accidentally) on insects. Accidental infection can presumably be avoided by keeping food in a meat safe, refrigerator, etc.

Little is known about the pathogenicity of the worm, but infections are likely to be light. Some damage to the intestinal wall may be caused by the armed mouthparts, as species in animals do appear to feed on the gut wall and erode the mucosa, leaving bleeding wounds behind them as they browse.

SUPERFAMILY RHABDITOIDEA ÖRLEY, 1880

Family Rhabditidae Örley, 1880

Main features Rhabditoidea; small, slender worms; head with six lips (or three double lips); buccal cavity long, cylindrical with chitinized thickenings in wall; esophagus long, narrow, with prominent posterior valved bulb; caudal alae of male form bursa; mainly free-living saprophagous forms, occasionally accidental parasites of vertebrates.

Rhabditids are free-living worms that are often swallowed accidentally or become associated with human feces; the most frequently recorded examples are from the genus *Rhabditis* Dujardin, 1844. Some species of the genus *Pelodera* Schneider, 1866 are known to infect human skin in rare cases, producing symptoms similar to cutaneous larva migrans. Humans presumably become infected by soil contamination of existing wounds. The occurrence of these parasites in the skin is so rare that preventative measures are not worthwhile.

Family Strongyloididae Chitwood and McIntosh, 1928

Main features Rhabditoidea; parasitic stage usually a parthenogenetic female (alternates with free-living sexual generation with esophagus bearing two swellings); small, slender worms; buccal cavity reduced or well developed, but not long; esophagus long, narrow, lacking prominent valved bulb posteriorly; vulva in posterior quarter of body; in alimentary canal of terrestrial vertebrates.

This is a group with an irregular alternation of generations, a parasitic parthenogenetic female alternating with a free-living sexual generation. One genus, *Strongyloides* Grassi, 1879, occurs in humans. The parasitic females found in humans are small, slender worms that live in the intestinal mucosa. They can be confused with trichinellids in gross morphology, but the long, cylindrical esophagus is quite different, lacking the large gland cells of the latter.

Several life cycle pathways are possible, both direct and indirect, the latter involving alternation of generations. Eggs laid in the mucosa or submucosa of the intestine release first-stage larvae that escape into the intestine and are voided with the feces. These larva feed in the soil and molt twice. The third-stage larvae *either* continue to develop into male and female worms capable of continuing to produce free-living generations (and thus more third-stage larvae), especially in warmer climates *or* penetrate the skin of a terrestrial vertebrate, migrate via the blood stream to the lungs, escape into the alveoli where they molt twice to become a parthenogenetic female, pass up the trachea, and are swallowed. Variations on the pulmonary migration route may be possible.

Once in the intestine, the worms enter the mucosa and submucosa. Humans become infected by contamination with infected soil or feces; autoinfection is also possible as, in some cases, larvae can reinfect the same host be reinvading the gut wall and migrating via the pulmonary route or penetrating the skin in the perianal region. Prevention involves improvements in sewage disposal and personal hygiene.

Symptoms of intestinal infection vary from none to severe enteritis, although diarrhea is most common. Very heavy infections may result in damage to the mucosa. Penetration of the skin by larvae may cause a rash and symptoms similar to a cutaneous larva migrans, and the passage through the lungs may result in a cough or pneumonia. Autoinfection may cause the disease to last for many years. Hyperinfestation can occur in immunosuppressed patients (see Chapter 35, *Strongyloides* and *Capillaria*).

SUPERFAMILY SPIRUROIDEA ÖRLEY, 1885

Family Gongylonematidae Hall, 1916

Main features Spiruroidea; body covered with large verruciform thickenings; head bilaterally symmetrical; mouth octagonal; pseudolabia absent; buccal cavity cuticularized, short; esophagus divided into anterior muscular and posterior glandular regions; caudal alae present in male; adult in mucosa of anterior alimentary canal of birds and mammals.

A species of *Gongylonema* Molin, 1957, the 'gullet worm' of domestic ruminants, has been recorded in humans as an accidental parasite. These worms usually occur in the esophagus or stomach of various mammals

embedded in the mucosa or submucosa. They are readily recognized by the verruciform cuticular thickenings on the surface of the body.

The life cycle is indirect. Eggs of these worms are voided with the feces and hatch once they have been swallowed by an insect, such as a coprophagous beetle. The final host becomes infected by feeding on beetles harboring the third-stage larva. Once in the new host, the worms molt twice and penetrate the surface layers of the anterior parts of the alimentary canal, where they reside. Humans, in whom infections are usually sited in the mucosa and submucosa of the buccal cavity, presumably become infected by accidentally eating beetles. Prevention can be helped by regular de-worming of farm animals and careful washing of food items.

Although the worms may be quite large, their presence tends to cause little in the way of symptoms apart from some local irritation and inflammation (see Chapter 40, *Angiostrongylus* (*Parastrongylus*) and less common nematodes).

SUPERFAMILY STRONGYLOIDEA BAIRD, 1853

Family Chabertiidae Popova, 1952

Main features Strongyloidea; mouth round or oval, usually surrounded by well developed comb-like leaf-crown (single or double) or lips; buccal cavity cuticularized; esophagus claviform, not divided into two regions; male with caudal bursa supported by rays, dorsal ray usually with two branches on each side of median fissure; in alimentary canal of mammals.

Strongyloid nematodes are occasional accidental parasites of humans. Chabertiids usually occur in the intestine of various herbivorous mammals, including farm animals, where they often cause nodules in the intestinal wall. Species of the genera *Oesophagostomum* Molin, 1861 and, less frequently, *Ternidens* Railliet and Henry, 1909 that parasitize primates, have been reported in humans. They are recognized by the presence of a double crown of leaf-like spines which surround the mouth, a cuticular buccal capsule and by the wide caudal bursa of the male which is supported by rays.

The life cycle is direct. Eggs laid in the intestine pass out with the feces; the first-stage larva emerges from the egg and feeds, developing into an infective third-stage larva. The host, usually a herbivore, acquires the worms when feeding on vegetation; the larvae penetrate the intestinal mucosa, grow and molt, prior to returning to the intestine, molting once again, and maturing. Humans presumably become infected by accidentally ingesting larvae with vegetation from the vicinity of primate populations. In some situations, the parasite can be passed between humans.

When the larvae penetrate the gut wall, they become encapsulated in a cyst that forms a fibrous nodule. Rupture of the nodule, which occurs when the worm returns to the lumen, may result in dysentery or even peritonitis. These nodules are sometimes misinterpreted as other diseases, such as carcinomas. The worms can also occur in subcutaneous cysts in humans, but how they arrive in this position is not known (see Chapter 40, *Angiostrongylus* (*Parastrongylus*) and less common nematodes).

Family Syngamidae Leiper, 1912

Main features Strongyloidea; mouth hexagonal, leaf-crown and lips rudimentary or absent; buccal cavity cuticularized, hexagonal in cross-section, with teeth at base; esophagus claviform, not divided into two regions; male with caudal bursa supported by rays; vulva usually near mid-body; usually in respiratory system of birds and mammals or urinary system of pigs.

One genus of syngamid, *Mammomonogamus* Ryzhikov, 1948, is occasionally reported from humans in the tropics. Syngamids are usually parasitic in the respiratory system (trachea and bronchi) of domesticated ruminants and infect humans only accidentally. Syngamids are recognizable in that the male and larger female are usually joined together in a Y formation; they differ considerably from other lungworms (the metastrongyloids) in the presence of a large, cuticular buccal cavity.

The life cycle of syngamids may be direct, similar to that of the chabertiid strongyloids, or indirect and involve paratenic hosts. Eggs produced in the respiratory tract are swallowed with mucus and voided with the feces. Development in the soil is similar to that of chabertiids, except that the larvae may be eaten by a variety of invertebrate hosts, such as earthworms, snails, slugs, and arthropods, which act as paratenic hosts. If the third-stage larva or its paratenic host is eaten by a suitable vertebrate, the larvae penetrate the gut lining and migrate to the lungs via the blood system. Here they molt and develop to maturity in the respiratory tract. Humans become infected by accidentally ingesting the larvae or its paratenic host. Some prevention is possible by the careful washing of salad material and other vegetation.

Worms in the trachea, larynx, and bronchi may cause irritation and a cough.

SUPERFAMILY THELAZIOIDEA SKRJABIN, 1915
Family Thelaziidae Skrjabin, 1915

Main features Thelazioidea; body often small, cylindrical, unspined; head bilaterally symmetrical; pseudolabia absent; mouth hexagonal or oval; buccal cavity cuticularized, variable, sometimes long, and cylindrical; esophagus may or may not be visibly divided into anterior muscular region and longer posterior glandular region; male tail without caudal bursa, caudal alae usually absent; transmission indirect with invertebrates (but not copepods) and vertebrates acting as intermediate hosts; in orbit of birds and mammals.

One thelaziid genus, *Thelazia* Travassos, 1918, normally parasitic in domesticated animals such as dogs and farm animals, has occasionally been reported in the eye of humans. Thelaziids can be recognized by their location in the eye, the possession of a cuticularized buccal cavity, and the absence of pseudolabia.

Thelaziid worms usually occur in the orbit, conjunctival sac or lachrymal glands. First-stage larvae occur in eye secretions. If ingested by a fly, they molt twice in its tissues, migrate to the mouth parts, and are transmitted to the mammalian host, and presumably to humans, when the fly feeds on eye secretions. Two final molts occur in the eye. There seem to be no practical control measures.

Symptoms caused by the presence of these worms include excessive tear formation, itching, and some discomfort. In animals, conjunctivitis and some damage to the eye tissues may occur, but the worms are not generally very pathogenic.

SUPERFAMILY TRICHOSTRONGYLOIDEA LEIPER, 1908
Family Trichostrongylidae Leiper, 1908

Main features Trichostrongyloidea; body usually not coiled; longitudinal cuticular ridges present or absent; cephalic vesicle absent; male with caudal bursa supported by rays, dorsal ray usually short and not deeply divided; life cycle direct; in mammals such as lagomorphs and ruminants, occasionally in birds.

Numerous trichostrongyloid genera, common stomach parasites of domesticated ruminants, have been reported as accidental parasites of humans. These include *Ostertagia* Ransom, 1907, *Nematodirus* Ransom, 1907, *Haemonchus* Cobb, 1898, *Marshallagia* Orloff, 1933, and, by far the most frequently, *Trichostrongylus* Looss, 1905; with the exception of the molineid *Nematodirus*, these are all trichostrongylids. Trichostrongylids are generally recognizable by the presence of a series of longitudinal cuticular ridges which run along the body, a caudal bursa supported by rays in the male and their presence in the stomach or intestine.

The life cycle is direct. Eggs are passed out with the feces, hatch on the ground, and release a first-stage larva that feeds and develops in the soil. After two molts, the infective third-stage larva can be transmitted to a suitable herbivore on vegetation. Within the gut, the worms molt twice prior to maturing as adults. Humans presumably become infected by feeding on raw vegetation harboring the larvae. Some level of prevention is possible by the cooking or careful washing of vegetables and other plant material.

Light infections pose no problem, but diarrhea and some abdominal pain have been reported. Limited damage at the site of attachment may occur as the head of the worms tend to penetrate the mucosa (see Chapter 40, *Angiostrongylus* (*Parastrongylus*) and less common nematodes).

PHYLUM ACANTHOCEPHALA

Acanthocephalans, or 'thorny-headed worms', represent a phylum of parasites of uncertain affinities. Like the tapeworms, members of this group are so well adapted to parasitism that they have lost their digestive system and reduced their muscular, nervous, and excretory systems, but they differ in that they are dioecious worms with a body cavity. The body is composed of a large trunk and a retractable anterior proboscis armed with a regular array of hooks used for attachment to the gut wall (Figure 27.4). Like tapeworms, they are intestinal parasites and absorb nutrients through their body-wall. Movement of the proboscis and its armature is complex, involving retractor muscles, a muscular proboscis receptacle, and fluid-filled sacs (lemnisci). The body wall is a syncytium with large, often fragmented nuclei; it comprises a complex outer tegument and an inner layer of muscles, and contains a lacunar system of interconnecting canals that may serve as a circulatory system. The males contain two testes and a series of accessory glands called cement glands. In females, the ovary breaks down early in development to form ovarian balls; the body cavity of mature females is usually full of eggs.

The life history of acanthocephalans is remarkably constant relative to that of other helminth groups. All have the same larval stages and all utilize arthropods as intermediate hosts. Essentially, female worms in the intestine produce eggs that are voided with the feces. If eaten by a suitable arthropod, the egg releases a larva armed with hooks, called an acanthor, which enters the hemocoel. This develops into a stage called an acanthella within which the internal organs of the adult begin to develop. When the proboscis develops, it usually invaginates into the body and the larva becomes encysted; this resting stage is called a cystacanth (Figure 27.4). The final host becomes infected either by feeding directly on an arthropod harboring the cystacanth larva or a vertebrate paratenic host harboring a juvenile form which is basically a re-encysted cystacanth. Once in the intestine of a suitable final host, the parasites attach to the gut wall and develop to maturity. In some cases, where the final host does not normally feed on arthropods, the paratenic host becomes a necessary part of the life history.

The different classes of acanthocephalans are not readily distinguished using gross morphology, their systematics are instead based on internal details that are often difficult to elucidate. At the specific level, the armature of the proboscis, in terms of the number of rows of hooks and the number of hooks per row, is the most important criterion.

No acanthocephalans are primarily human parasites. Data on their pathogenicity are limited, as records in humans tend to be scarce. However, since the proboscis with its armature of hooks does penetrate the gut wall,

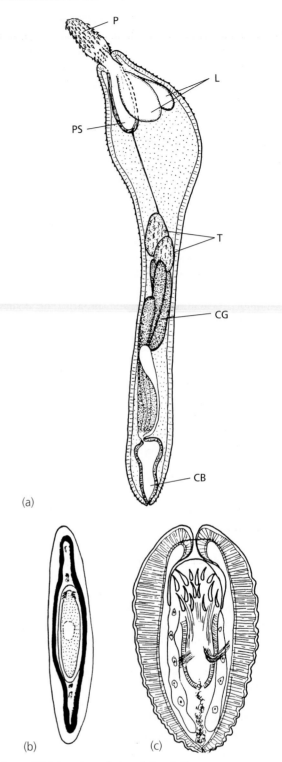

(a)

(b)   (c)

Figure 27.4 *Acanthocephalans:* **(a)** *male* Corynosoma *(Polymorphidae) (the body cavity of the female is usually full of eggs, apart from which it contains no other useful taxonomic features not present in the male);* **(b)** *egg of* Bolbosoma *(Polymorphidae) containing acanthor larva;* **(c)** *cystacanth larva of* Macracanthorhynchus *(Oligacanthorhynchidae) with withdrawn proboscis. Copulatory bursa (CB), cement glands (CG), lemnisci (L), proboscis (P), proboscis sac (PS), testes (T). Adapted from* **(a)***Yamaguti 1963;* **(b)** *Meyer 1933;* **(c)** *Van Cleave 1947.*

some local inflammation and abdominal pain is to be expected; but, as the worms do not feed on the gut wall and their location tends to be semi-permanent, this may not persist. Nevertheless, in some cases, especially when the juvenile parasite finds the human gut unsuitable for development, it may pass right through and perforate the intestine with a consequent possibility of causing peritonitis.

The classification of forms recorded from humans below is based on Schmidt (1972) and Amin (1985):

Class Archaeacanthocephala
    Order Moniliformida
        Family Moniliformidae
    Order Oligacanthorhynchida
        Family Oligacanthorhynchidae
Class Palaeacanthocephala
    Order Echinorhynchida
        Family Echinorhynchidae
    Order Polymorphida
        Family Polymorphidae.

Class Archiacanthocephala Meyer, 1931

Main features Acanthocephala; relatively large worms; spines on body trunk absent; main longitudinal vessels of lacunar system dorsal or dorsal and ventral; subcuticular nuclei few, elongate, branched or with residual fragments situated close together; cement glands of male usually eight, uninucleate; insects (and millipedes) act as intermediate hosts; adults in intestine of terrestrial vertebrates.

ORDER MONILIFORMIDA SCHMIDT, 1972

Family Moniliformidae Van Cleave, 1924

Main features Moniliformida; body medium-sized to long, usually with pseudosegmented wall; proboscis cylindrical with long, straight rows of hooks; proboscis receptacle with double-layered wall, retractor muscles piercing its posterior end and cerebral ganglion positioned posteriorly; protonephridial organs absent; eggs oval, with sculptured surface; in birds and mammals.

The order Moniliformida has only one family containing a single genus, *Moniliformis* Travassos, 1915. This genus, which normally occurs in rodents, has been reported on rare occasions in humans. It is recognizable by a pseudosegmentation of the body (i.e. a series of annular thickenings of the body wall) and its cylindrical proboscis. The larger females can reach more than 20 cm in length.

Eggs of these intestinal worms are passed in the feces. When eaten by an insect, normally a beetle (especially a grain beetle) or a cockroach, the cystacanth larva develops in the hemocoel. If the arthropod host is eaten by a rodent, the worms excyst in the gut, the proboscis evaginates, and the worm attaches to the intestinal wall and develops to maturity. Humans become infected by accidentally ingesting beetles with food. Prevention can

presumably be effected by rodent control and keeping food intended to be eaten cold in beetle-proof containers.

Since the proboscis of these worms does penetrate the gut lining, it is likely to produce some local inflammation. Symptoms such as abdominal pain and diarrhea have been reported, but data on the effect of these parasites in humans are few.

ORDER OLIGACANTHORHYNCHIDA PETROCHENKO, 1956

Family Oligacanthorhynchidae Southwell and MacFie, 1925

Main features Oligacanthorhynchida; Body medium-sized to very long; proboscis subspherical with short, straight rows of small numbers of hooks; sensory papilla present at apex of proboscis and each side of neck; proboscis receptacle with thick, single-layered wall, retractor muscles piercing it dorsally and cerebral ganglion positioned on ventral inner surface; protonephridial organs present; in birds and mammals.

These parasites are especially common in pigs, but also occur in a wide range of terrestrial mammals. They are commonly covered in transverse wrinkles but lack any regular pseudosegmentation; the females especially reach a large size, sometimes more than 40 cm. One genus, *Macracanthorhynchus* Travassos, 1917, occurs on rare occasions in humans.

The life cycle is similar to that of moniliformids except that, although a range of beetles, cockroaches, etc. can harbor the parasite, the main intermediate hosts would appear to be dung beetles that become infected when they feed on manure. Humans acquire the parasite by ingesting beetles as food, accidentally with food or as a cure for ailments. The keeping of food intended to be eaten cold in beetle-proof containers should help prevent accidental infection.

The worms cause inflammation and a granuloma at the site of attachment; but abdominal pain and perforation of the bowel have been reported and, in pigs, death from peritonitis does occasionally occur.

Class Palaeacanthocephala Meyer, 1931

Main features Acanthocephala; small to medium-sized worms; spines present or absent on trunk; proboscis shape variable; main longitudinal canals of lacunar system lateral; subcuticular nuclei numerous, fragmented, occasionally limited to anterior region of trunk; cement glands of male separate, tubular to globular; in intestine of aquatic vertebrates.

ORDER ECHINORHYNCHIDA SOUTHWELL AND MACFIE, 1925

Family Echinorhynchidae Cobbold, 1876

Main features Echinorhynchida; body not spined; normally in intestine of bony fish and amphibians.

Two echinorhynchid genera, *Acanthocephalus* Koelreuter, 1771 and *Pseudoacanthocephalus* Petrochenko, 1956, have been reported as occurring accidentally in humans. These genera normally occur in the intestine of fishes or amphibians. Humans presumably acquire the parasite by feeding on freshwater crustacean intermediate hosts harboring the cystacanth larvae or a vertebrate host that had ingested the cystacanth. They are unlikely to have any great pathogenic effect, although there is one report of a young worm having penetrated the gut wall and ended up on the peritoneum.

ORDER POLYMORPHIDA PETROCHENKO, 1956
Family Polymorphidae Meyer, 1931

Main features Polymorphida; body spined; normally in intestine of aquatic birds and marine mammals.

Two polymorphid genera, *Corynosoma* Lühe, 1904 and *Bolbosoma* Porta, 1908, are occasionally reported from humans. These are normally parasitic in marine mammals, usually whales and dolphins in the case of *Bolbosoma* and seals in the case of *Corynosoma*. In addition to their marine link, these worms are recognizable by the presence of spines on the trunk of the body and, in the case of *Bolbosoma*, by a large expansion of the neck region posterior to the proboscis.

The basic life history of these marine polymorphids is similar to that of the terrestrial forms, but the eggs are voided with the feces into water and the arthropod host harboring the infective cystacanth larva is a crustacean rather than an insect. Furthermore, bony fishes, which feed on infected crustaceans, act as paratenic hosts, the cystacanth larva passing through the fish's gut wall and re-encysting as a juvenile worm. Marine mammals become infected by feeding on these fishes, which, in cases where the definitive host does not feed on crustaceans, represent a necessary step in the life cycle. Humans become infected by feeding on raw or poorly cooked marine fish. It seems unlikely that the worms can mature in humans. The proper cooking or freezing of fish would kill the parasite.

There is little information on the pathogenicity of these worms; but they are known to begin to develop in humans, so penetration of the gut wall by the proboscis will elicit local inflammation and symptoms of abdominal pain (which is sometimes acute). There is one reported case of perforation of the intestine.

NOTE ADDED BY THE EDITORS FOR THIS EDITION

The classification of helminths as described in the 9th edition has remained essentially unchanged, and therefore that chapter has been repeated in this edition. It still provides an excellent and systematic introduction to the variety of helminths and provides the basis for accurate identification and taxonomic work.

During the last five years there has been an increasing interest in molecular methods of identification and classification to supplement the more traditional morphological approach. An important element of this work has been the development of genetic markers, such as mitochondrial DNA, that are valuable not only in taxonomy, but also for studies relevant to epidemiology, population biology, and parasite control. Although genetic studies with helminths are less complete than those carried out with some of the protozoan parasites, there are a number of programmes designed to provide genomic maps of species of major interest in human medicine, and extensive computer databases of sequences are available. The further reading section at the end of this chapter lists some recent references that will be useful for readers wishing to extend their knowledge of these areas of classification and parasite biology.

REFERENCES

Abuladze, K.I. 1964. Taeniata of animals and man and diseases caused by them (in Russian). In: Skrjabin, K.I. (ed.), *Osnovy Tsestodologii*. vol. 4. Moscow: Nauka.

Amin, O. 1985. Classification. In: Crompton, D.W.T. and Nickol, B.B. (eds), *Biology of the Acanthocephala*. Cambridge: Cambridge University Press, 27–72.

Anderson, R.C., Chabaud, A.G. and Willmott, S. 1974–83. *CIH keys to the nematode parasites of vertebrates*, vols 1–10. Commonwealth Agricultural Bureaux, Farnham Royal.

Bain, O. and Chabaud, A.G. 1986. Atlas des larves infestantes de Filaires. *Trop Med Parasitol*, **37**, 301–40.

Brooks, D.R. 1989. The phylogeny of the Cercomeria (Platyhelminthes: Rhabdocoela) and general evolutionary principles. *J Parasitol*, **75**, 606–16.

Brooks, D.R., O'Grady, R.T. and Glen, D.R. 1985. Phylogenetic analysis of the Digenea (Platyhelminthes: Cercomeria) with comments on their adaptive radiation. *Can J Zool*, **63**, 411–43.

Brooks, D.R., Bandoni, S., et al. 1989. Aspects of the phylogeny of the Trematoda Rudolphi, 1808 (Platyhelminthes: Cercomeria). *Can J Zool*, **67**, 2609–24.

Caballero y Caballero, E. 1944. Estudios helminthologicos de la region oncecercosa de Mexico y la Republica de Guatemala. Nematoda. Prima Parta. Filarioidea. I. *An Inst Biol Univ Mex*, **15**, 87–105.

Chitwood, M. and Lichtenfels, J.R. 1972. Identification of parasitic Metazoa in tissue sections. *Exp Parasitol*, **32**, 407–519.

Coombs, I. and Crompton, D.W.T. 1991. *A guide to human helminths*. London: Taylor and Francis.

Cooper, D.W. 1988. The preparation of serial sections of platyhelminth parasites, with details of the materials and facilities required. *Syst Parasitol*, **12**, 211–29.

Cort, W.W. 1919. The cercaria of the Japanese blood fluke, *Schistosoma japonicum Katsurada*. *Univ Calif Publs Zool*, **18**, 17, 485–507.

de Clercq, D., Rollinson, D., et al. 1994. Schistosomiasis in Dogon country, Mali: identification and prevalence of the species responsible for infection in the local community. *Trans R Soc Trop Med Hyg*, **88**, 653–6.

Fuhrmann, O. 1927. Dritte Klasse des Cladus Plathelminthes. In: Kükenthal, W. and Krumbach, T. (eds), *Handbuch der Zoologie, Cestoidea*. vol. II(2). Berlin: Walter de Gruyter, 141–416.

Garcia, L.S. and Bruckner, D.A. 1993. *Diagnostic Medical Parasitology*, 2nd edn. Washington DC: American Society for Parasitology.

Gibson, D.I. and Bray, R.A. 1994. The evolutionary expansion and host-parasite relationships of the Digenea. *Int J Parasitol*, **24**, 1213–26.

Gibson, D.I., Jones, A. and Bray, R.A. 2002. *Keys to the trematoda*, vol. 1. Wallingford: CABI.

Hugot, J.P. and Tourte-Schaefer, C. 1985. Morphological study of the two pinworms parasitic in man: *Enterobius vermicularis* and *E. gregorii*. *Ann Parasit Hum Comp*, **60**, 57–64.

Khalil, L.F., Jones, A. and Bray, R.A. 1994. *Keys to the Cestode Parasites of Vertebrates*. Wallingford, Oxon: CAB International.

Lightowlers, M.W. and Gottstein, B. 1995. Echinococcosis/hydatidosis: antigens, immunological and molecular diagnosis. In: Thompson, R.C.A. and Lymbery, A.J. (eds), *Echinococcus and Hydatid Disease*. Wallingford, Oxon: CAB International, 355–410.

Meyer, A. 1933. Acanthocephala. In: Bronn, H.G. (ed.), *Klassen und ordnungen teirreichs*. vol. IV(2,2). Leipzig: Akademisches Verlagsgesellschaft, 333–582.

Moravec, F. 1982. Proposal for a new systematic arrangement of nematodes of the family Capillariidae. *Folia Parasitol*, **29**, 119–32.

Muller, R. 2001. *Worms and human diseases*, 2nd edn. Wallingford: CABI.

Pearson, J.C. 1992. On the position of the digenean family Heronimidae: an inquiry into a cladistic classification of the Digenea. *Syst Parasitol*, **21**, 81–166.

Schell, S.C. 1985. *Handbook of Trematodes of North America*. Dubuque, Idaho: Idaho University Press.

Schmidt, G.D. 1972. Revision of the class Archiacanthocephala Meyer, 1931 (Phylum Acanthocephala), with emphasis on Oligacanthorhynchidae Southwell et Macfie, 1925. *J Parasitol*, **58**, 290–7.

Thienpont, D., Rochette, F. and Vanparijs, O.F.J. 1986. *Diagnosing Helminthiasis by Coprological Examination*. Beerse, Belgium: Janssen Research Foundation.

Van Cleave, H.J. 1947. A critical review of the terminology for immature stages in acanthocephalan life histories. *J Parasitol*, **33**, 118–25.

Vogel, H. 1934. Der Entwicklungzyklus von *Opisthorchis felineus* (Riv.) nebst Berkungen über die Systematik und Epidemiologie. *Zoologica (Stuttgart)*, **33**, 86, 1–103.

Wright, C.A. and Ross, G.C. 1980. Hybrids between *Schistosoma haematobium* and *S.mattheei* and their identification by isoelectric focusing of enzymes. *Trans R Soc Trop Med Hyg*, **74**, 326–32.

Yamaguti, S. 1963. *Systema Helminthum*, vol. III. New York: Interscience.

Yorke, W. and Maplestone, P.A. 1926. *The Nematode Parasites of Vertebrates*. London: Churchill.

FURTHER READING

Anderson, T.J.C., Blouin, M.S. and Beech, R.N. 1998. Population biology of parasitic nematodes: applications of genetic markers. *Adv Parasitol*, **41**, 219–83.

Blouin, M.S., Yowell, C.A., et al. 1998. Substitution bias, rapid saturation and the use of mtDNA for nematode systematics. *Mol Biol Evol*, **15**, 1719–27.

Brindley, P. (ed.). 2000. Parasite genomes. *Int J Parasitol*, **30**, 327.

Brooks, D.R. and Isaac, R.E. 2002. Functional genomics of parasitic worms – the dawn of a new era. *Parasitol Int*, **51**, 319–25.

Curtis, J. and Minchella, D.J. 2000. Schistosome population genetic structure: when clumping worms is not just splitting hairs. *Parasitol Today*, **16**, 68–71.

Dias Nieto, E., Souza, et al. 1993. The random amplification of polymorphic DNA allows the identification of strains and species of schistosome. *Mol Biochem Parasitol*, **57**, 83–8.

Franco, G.R. and Simpson, A.J.G. 2001. The structure and expression of the schistosome genome. In: Mahmoud, A.A.F. (ed.), *Schistosomiasis*. London: Imperial College Press, 85–113.

Gasser, R.B. 2001. Identification of parasitic nematodes and study of genetic variability using PCR approaches. In: Kennedy, M.W. and Harnett, W. (eds), *Parasitic nematodes. Molecular biology, biochemistry and immunology*. Wallingford, Oxon: CABI Publishing, 53–82.

Hoberg, E.P., Jones, A., et al. 2000. A phylogenetic hypothesis for species of the genus *Taenia* (Eucestoda: Taeniidae). *J Parasitol*, **86**, 89–98.

Hu, M., Chilton, N.B. and Gasser, R.B. 2002. The mitochondrial genomes of the human hookworms, *Ancylostoma duodenale* and *Necator americanus* (Nematoda: Sercenentea). *Int J Parasitol*, **32**, 145–58.

Le, T.H., Blair, D. and McManus, D.P. 2000. Mitochondrial geneomes of human helminths and their use as markers in population genetics and phylogeny. *Acta Tropica*, **77**, 243–56.

Le, T.H., Humar, P.F., et al. 2001. Mitochondrial gene content, arrangement and composition compared in African and Asian schistosomes. *Mol Biochem Parasitol*, **28**, 61–71.

Le, T.H., Pearson, M.S., et al. 2002. Complete mitochondrial genomes confirm the distinctiveness of the horse-dog and sheep-dog strains of *Echinococcus granulosus*. *Parasitology*, **124**, 97–112.

Littlewood, D.T.J. and Bray, R.A. (eds) 2001. *Interrelationships of the Platyhelminthes*. New York: Taylor & Francis Publishing.

Nadler, S.A. and Hudspeth, D.S.S. 2000. Phylogeny of the Ascaridoidea (Nematoda: Ascaridida) based on three genes and morphology: hypotheses of structural and sequence evolution. *J Parasitol*, **86**, 380–93.

Rollinson, D., Kaukas, A., et al. 1997. Some molecular insights into schistosome evolution. *Int J Parasitol*, **27**, 11–28.

Thompson, R.C. and McManus, D.P. 2002. Towards the taxonomic revision of the genus *Echinococcus*. *Trends Parasitol*, **18**, 452–7.

Williams, S.A. and Laney, S.J. 2002. Filaria genomes: gene discovery and gene expression. In: Klei, T.R. and Rajan, T.V. (eds), *World Class Parasites. The Filaria*. Boston: Kluwer Academic Publishing, 31–42.

Zhu, X.Q., Gasser, R.B., et al. 2000. Relationships among some ascaridoid nematodes based on ribosomal DNA sequence data. *Parasitol Res*, **86**, 738–44.

28

Schistosomes: general

SOMEI KOJIMA AND ANDREW S. MACDONALD

THE PARASITES

Schistosomes are the causative agents of the disease schistosomiasis. Over 190 million people in 76 countries and territories in the world are affected and about 652 million are at risk, according to recent calculations applied to the 1995 world population estimates (Engels et al. 2002). In terms of global clinical impact, an estimated 85 percent of people infected with schistosomes are thought to be located in the African continent (Engels et al. 2002).

Classification

The human schistosomes or blood flukes are digenetic trematodes belonging to the superfamily Schistosomatoidea of the suborder Strigeata. They differ from other trematodes in that (1) the adults are dioecious, being either male or female; (2) the adult worms parasitize blood vessels; (3) they lack a muscular pharynx; (4) they produce nonoperculate eggs; and (5) the cercaria, with a bifurcated tail, invades the final host percutaneously. Several species can infect humans (Table 28.1). Four, *S. japonicum*, *S. mekongi*, *S. mansoni*, and *S. haematobium*, are important agents of human schistosomiasis. *S. intercalatum* has a very restricted distribution in humans; *S. bovis* and *S. mattheei*, although primarily parasitic in other mammalian hosts, may produce infections in humans. In addition, human cases with *S. malayensis* have been described among aborigines in Peninsular Malaysia (Murugasu and Por 1973; Murugasu et al. 1978). The advent of modern molecular techniques

should make it possible to construct a detailed map of genetic diversity both between species, and within populations, of schistosomes (Rollinson et al. 1997; Oliveira and Johnston 2001).

Morphology and structure

The human-infective schistosome species appear similar morphologically. Near the anterior end of the body there are two suckers, oral and ventral. Following a short esophagus, the intestine divides at the level of the ventral sucker into two parallel gut ceca that rejoin behind the gonads to form a blind cecum ending near the posterior end of the body. In the male, between four and nine testes (depending upon the species) are usually present on the dorsal side just below the ventral sucker. The vas efferens from each of the testes leads to the common vas deferens which runs via a small seminal vesicle to the genital pore situated posterior to the ventral sucker. On the ventral side, the body is flattened and folded to form the gynecophoric canal (Figure 28.1), in which the longer and more slender female is held firmly (Figure 28.2). In the females, the ovary is located near the posterior union of the lateral gut ceca about halfway along the body. The vitelline glands surround the cecum from the extreme end of the body up to the level of the posterior end of the ovary. The oviduct runs forward from the end of the ovary and joins the vitelline duct to form the ootype, where ova are fertilized, provided with vitelline materials and shells, and then passed into the uterus. The uterus is a long straight tube and opens behind the ventral sucker; 50–150 eggs

Table 28.1 *Schistosomes that may parasitize humans*

Schistosome species	Final hosts	Distribution
S. haematobium	Human, monkey, baboon, chimpanzee	Africa, Middle East
S. japonicum	Human, dog, cat, rodent, pig, sheep, goat, water buffalo, horse	China, Philippines, Celebes
S. malayensis	Human, rodent	Malaysia
S. mansoni	Human, monkey, baboon, chimpanzee, dog, rodent	Africa, South America, Caribbean
S. mekongi	Human, dog	Laos, Cambodia, Thailand
S. intercalatum	Human	West and Central Africa
S. bovis	Cattle, sheep, goat, equine, baboon, human	Southern Europe, Africa, Iraq
S. mattheei	Horse, sheep, cow, zebra, antelopes, baboon, human	Southern Africa

develop in utero at one time. Ova deposited in tissues develop to the larval stage, the miracidium, within a week. Passage of the eggs through the vein wall and tissues to the lumen of the intestine is aided by the release through micropores in the shell of histolytic enzymes secreted by the miracidium.

In comparison to both *S. mansoni* and *S. haematobium*, the *S. japonicum* tegumental surface appears smooth under light microscopy, rather than overtly tuberculated. Although the structure of adult worms has been studied mainly using *S. mansoni*, it is considered basically common among other schistosomes in that the worm body is covered by a syncytial tegument approxi-

mately 4 µm in thickness, in which surface pits form deep, tortuous channels which may be branched and interconnected, thereby increasing the effective surface by a factor of as much as 10 (Hockley 1973). The physiological role of the pits is as yet unknown. The external surface of the tegument is a plasma membrane composed of two lipid bilayers (Hockley and McLaren 1973), of which the outer may fuse with foreign lipid membranes of host neutrophils and eosinophils (Caulfield et al. 1980), probably as a result of the production of monopalmitoylphosphatidylcholine by the parasite (Furlong and Caulfield 1989). Since the parasite cannot synthesize essential lipids de novo, lipids, particularly free fatty acids, bound to serum proteins such as albumin, provide potential sources for the parasite. The parasite also has receptors for serum lipoproteins on the outer bilipid membrane, by which the parasite may acquire another source of lipids from the host (Rumjanek et al. 1983; Rogers et al. 1990). The outer bilipid membrane is shed continuously and replaced by products of organelles transported through a channel originated from the tegumental cells that are located under the circular and longitudinal muscle layers.

Schistosomes absorb glucose from the host plasma through the tegument (Fripps 1967; Rogers and Beuding

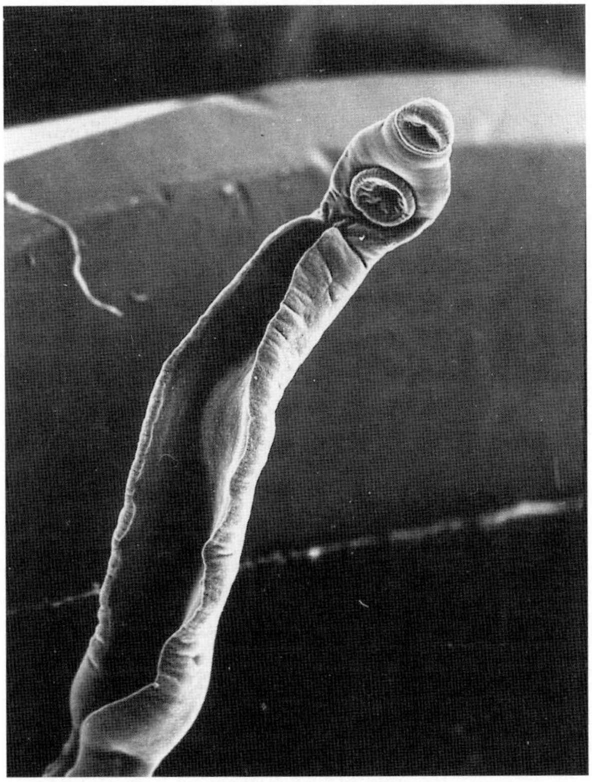

Figure 28.1 *Scanning electron micrograph of a male worm of* Schistosoma japonicum, *showing the gynecophoric canal*

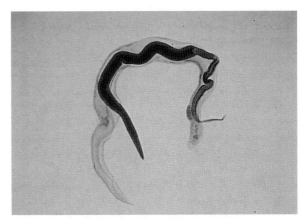

Figure 28.2 *Adult worms of* Schistosoma mansoni

1975; Uglem and Read 1975). Glucose must be transported to internal cells across several membranes such as the two lipid bilayers and the basal membrane of the tegument, as well as the plasma membranes of internal cells, a transport system that is unique to schistosomes. Indeed, two cDNA clones encoding the proteins SGTP1 and SGTP4, with features conserved in the facilitated diffusion glucose transporter family, have been identified by screening an adult *S. mansoni* cDNA library (Skelly et al. 1994). Immunolocalization studies have revealed the presence of SGTP4 in both of the lipid bilayers that cover the tegumental surface of adults and schistosomula, and also in the discoid and multilamellar bodies in adults, as well as in the membranous bodies in schistosomula (Jiang et al. 1996), while SGTP1 is present in the basal membrane of the tegument (Zhong et al. 1995). These results suggest that SGTP4 may be responsible for transporting glucose from plasma into the tegument, whereas SGTP1 may function to transport free glucose from the tegument into the extracellular matrix beneath the tegument, as well as into the muscle (Jiang et al. 1996; Zhong et al. 1995). These proteins are homologs of GLUT-1, the human glucose transporter protein (Mueckler et al. 1985).

Absorbed glucose is metabolized through the Embden–Meyerhof pathway and the Krebs cycle. Studies on the mode of action of anti-schistosomal drugs have demonstrated the importance of the key enzymes involved in these pathways as the target of chemotherapy. For example, compounds of antimony, though no longer used for treatment because of their side effects, inhibit phosphofructokinase activity at concentrations 70–80 times lower than those inhibiting the isozyme of the host, while praziquantel, the most effective drug currently available, affects the transport of calcium ions across cell membranes, which may result in tegumental vacuolization and disruption (Mehlhorn et al. 1981, 1983).

Adult worms also ingest red blood cells through the mouth and utilize hemoglobins, as well as serum globulins as nutrients after cleaving by proteolytic enzymes. Binding of the proteolytic enzymes to hemoglobin is optimal at pH 4.0 in the lumen of the gut. Since their digestive system ends in a blind cecum, residual black hematin is regurgitated together with the proteolytic enzyme to the circulation, from where it is taken up by phagocytic cells in the liver and spleen of the host. Meanwhile, the enzyme–hemoglobin complex dissociates owing to the change of pH, resulting in release of the enzyme in the plasma as a circulating antigen that in turn sensitizes hosts, inducing immediate-type hypersensitivity reactions to the enzyme (Senft and Maddison 1975).

It is likely that the female may obtain supplementary nutrients through the intimate contact with the male ventral tegument. This may be related to the fact that unisexual infections usually fail to induce sexual maturation of females, although certain lipid metabolites, such as ecdysteroids and dolichols, have been suggested to act as promotors for sexual development and egg production by adult worms, respectively (Furlong 1991; Fried and Haseeb 1991).

Life cycle

The life cycle of all species follows a common pathway from the sexual generation of adult schistosomes within the vascular system of the definitive host to asexual generations in the intermediate host of freshwater snails, and a return to the definitive host by skin penetration of infective larval stages (Figure 28.3).

When embryonated eggs discharged in feces or urine enter fresh water, miracidia hatch from the eggs, rapidly swim by means of cilia that cover the body surface except the apical papilla, and penetrate the exposed surface of the body of snail intermediate hosts. The attachment and penetration of the miracidium are assisted by secretions from the anterior penetration glands which open anterolaterally via ducts to the base of the apical papilla. The genera of snails that act as intermediate hosts differ from species to species of schistosome (Table 28.2). Moreover, differences in host specificity may exist among species or strains of the snail intermediate hosts in terms of infectivity of geographically different strains of schistosomes.

After penetration into the intermediate host, the miracidium loses the ciliated surface and develops into mother and daughter sporocysts, and then produces fork-tailed cercariae in the course of several weeks (4 weeks in the case of *S. mansoni* under favorable conditions but longer in other species, especially in *S. japonicum*) (Figure 28.4). As a result of asexual multiplication within mother and daughter sporocysts, thousands of cercariae, all of the same sex, are produced from a single miracidium. The cercariae measure around 400 μm in length including the tail, the terminal one-third of which is bifurcated. Their surface and internal structures are essentially identical among schistosome species. They have a muscular oral sucker occupying

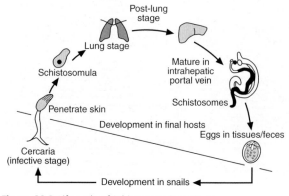

Figure 28.3 *Life cycle of* Schistosoma japonicum

Table 28.2 *Genera of snails acting as intermediate hosts for the human schistosomes*

Schistosome species	Intermediate snail hosts
S. haematobium	*Bulinus* and *Physopsis* spp.
S. japonicum	*Oncomelania* spp.
S. malayensis	*Robertsiella*
S. mansoni	*Biomphalaria* spp.
S. mekongi	*Neotricula aperta*

about one-third of the body and a small ventral sucker or acetabulum. The cercariae infect the definitive host by penetrating the skin, aided by proteolytic enzymes secreted from the penetration glands, losing their tails and transforming to the next larval stage, the schistoso-mulum. During the whole penetration process, drastic morphological and physiological changes take place in the larvae. Thus, the cercarial glycocalyx is lost and replaced by a double bilayer; this is marked by a change from a trilaminate appearance to a heptalaminate membrane.

Schistosomula migrate into the lungs via the venous circulation. This migration route seems to be essential for their development and is associated with the acquisition of biological differences, since schistosomula recovered from the lungs are not only elongated but are more resistant than newly transformed schistosomula to antibody-dependent cell-mediated cytotoxicity, although the mechanisms are as yet unknown. The schistosomula break out from the lungs through the pulmonary capillaries and are carried through the left heart into the systemic circulation, finally reaching the portal system. It takes 3–4 days for *S. japonicum* schistosomula to migrate into the lungs after skin penetration, followed by another 3–4 days stay in the lungs; *S. mansoni* reaches the lungs and leaves for the final destination at least 2 days earlier than *S. japonicum* (Usawattanakul et al. 1982). Some schistosomula may take another route through the diaphragm to reach the liver. In the intrahepatic portal circulation, feeding begins and further

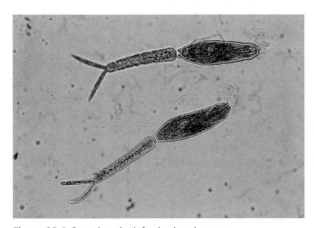

Figure 28.4 *Cercariae, the infective larval stage*

growth occurs. The lateral gut ceca grow posteriorly and eventually unite behind the ventral sucker and the developing gonads. Within several weeks, pairing of the worms takes place on sexual maturation, and they migrate to the mesenteric veins, where the females lay eggs. For *S. mansoni* and *S. japonicum*, the prepatent period between penetration by cercariae and the first appearance of eggs in the excreta is around 35 days.

Once the eggs have been released by the female within the vasculature, they cross the endothelium and basement membrane of the vein, traverse the intervening tissue, basement membrane, and epithelium of the intestine (*S. mansoni* and *S. japonicum*) or bladder (*S. haematobium*) en route to the exterior. While it is unclear precisely how this process occurs, there may be an immunological component since egg excretion is reduced in immunocompromised mice, but can be increased by transfer of sera or lymphocytes from infected animals (Doenhoff 1997). Further supporting this idea, *S. mansoni*-infected Human immunodeficiency virus (HIV)-positive individuals, who have significantly reduced CD4[+] T cell counts, display significantly diminished egg excretion in comparison to HIV-negative controls (Karanja et al. 1997).

Transmission and epidemiology

As is obvious from the life cycle of schistosomes, human infections depend absolutely on the presence of intermediate snail hosts in bodies of water which may be contaminated with human feces and excreta as a result of insanitary habits or polluted with excreta from reservoir hosts including livestock, which have to be considered in disease control, especially in the case of *S. japonicum* infection. Thus, transmission is influenced by biological, economic, and sociocultural factors, such as (1) the distribution, biology and population dynamics of the intermediate hosts; (2) the patterns and extent of environmental contamination with human excreta through unsanitary conditions and utilization as fertilizer, or with excreta from domestic animals in the case of *S. japonicum*; (3) human water contact activities, patterns, and duration; (4) water development projects including extension of irrigation systems and construction of dams; and (5) most seriously, people's knowledge of the disease itself and their attitudes towards, or practices for, avoiding the infection.

Despite its public health importance in many tropical and subtropical areas, which is second only to that of malaria, schistosomiasis has been largely ignored and/or given a low priority in health policy mainly because of the low socioeconomic status of developing countries, and the tendency to give economic development priority over health and welfare. Thus, water development projects, particularly construction of dams, artificial lakes, and irrigation canals, often associated with dynamic population resettlement, have become a major factor in the spread and intensification of the disease.

Schistosomiasis may also be categorized as one of the emerging and re-emerging diseases. A typical example is an outbreak of intestinal schistosomiasis in the delta of the Senegal river basin, where the presence of the disease had not been reported before the mid-1980s. In the 1980s the Diama dam was constructed at the mouth of the river to avoid a backward flow from the sea. A year and a half after it became operational (August 1986), a first case of *S. mansoni* infection was detected at Richard-Toll. The prevalence of *S. mansoni* infection rapidly increased from 1.9 percent in 1988 to 71.5 percent in late 1989, with an accumulative total of 1935 positive cases out of 3 926 stool examinations. *Biomphalaria pfeifferi* was found to be the major intermediate host in this area (Talla et al. 1990). Another example of schistosomiasis as a re-emerging disease occurred in Japan where endemic schistosomiasis was observed among dairy cows pastured along the bank of the Tone River in Chiba Prefecture, and also involved some inhabitants in the infection. This was 15 years after the disease was no longer considered to be endemic (Yokogawa et al. 1971, 1973).

Community-based epidemiological studies of schistosomiasis demonstrate that prevalence (expressed as the percentage of the population found infected at a given point of time) and intensity of infection (expressed in terms of egg output in excreta) are usually higher among younger age groups than in older age groups. A typical example obtained from a study of urinary schistosomiasis in a defined area of southern Ghana is shown in Figure 28.5, indicating that the peak prevalence and intensity occur in children aged 10–14 years, accompanied by a gradual decline in prevalence and intensity in older age groups (Aryeetey et al. unpublished data). This phenomenon is more marked in *S. haematobium* than in *S. mansoni*, but in *S. japonicum* infection the

reduction in prevalence and intensity in older age groups may not be observed. The decline in egg output in community-based studies may reflect the development of immunity to re-infection, or alternatively the possibility of reduced water contact in the older age groups (see Immune responses).

Thus, the control of schistosomiasis apparently involves many factors and requires a long-term effort with a complex of strategies. So far, there are only three countries – Japan, Montserrat, and Tunisia – where the transmission of the disease is considered to have ceased (WHO 1993).

CLINICAL AND PATHOLOGICAL ASPECTS OF INFECTION

Clinical manifestations, pathogenesis, and pathology

Clinical symptoms and pathological changes caused by schistosome infections may reflect migratory routes and host responses to toxic or antigenic substances derived from developmental stages of the parasites after their penetration of the skin. By far the predominant cause of pathology resulting from infection is the host reaction to parasite eggs that are laid in the vasculature, and subsequently trapped within the host tissues. The specific pathological changes associated with the African and the Asian species are described in detail in Chapters 29 and 30.

In general, acute infections are not accompanied by symptoms among people living in endemic areas, unless they are exposed to a massive infection for a specific reason, such as flooding. In these individuals, pathological changes are caused mainly by schistosome eggs deposited in various tissues. The host immune response to antigens excreted from embryonated eggs, which is

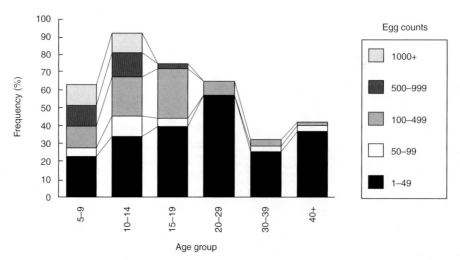

Figure 28.5 *Prevalence and intensity of* Schistosoma haematobium *infection in southern Ghana (Aryeetey et al. unpublished data)*

now the focus of cytokine studies, results in formation of localized inflammatory reactions to the eggs, which are termed granulomas. These granulomas appear to be simultaneously beneficial and detrimental for both host and parasite. In terms of the parasite, the granuloma may promote the process of egg excretion (Doenhoff 1997). In terms of the host, by effectively isolating the inflammatory reaction to eggs, they may protect host tissue adjacent to the granuloma (Sandor et al. 2003), while also sequestering potentially damaging toxins secreted by the eggs (von Lichtenberg 1964). However, the granulomas themselves are ultimately pathogenic, as they induce fibrosis, increased portal blood pressure, and development of portal systemic shunts.

In primary infections, the granuloma is composed of aggregations of mononuclear phagocytes, neutrophils, lymphocytes, plasma cells, and fibroblasts (Figure 28.6). Giant cells are also frequently observed in the granulomas. Granulomas may vary in size and cellular components with the immune status of the host; in experimental infections in immunized animals, a dominant cellular infiltration of eosinophils and lymphocytes is observed around the eggs, and the egg granuloma is smaller (Satoh et al. 1997).

Granuloma formation around schistosome eggs has been considered to be the result of delayed-type hypersensitivity reactions mediated through a T-cell mediated immune response to soluble egg antigens (Warren et al. 1967). However, recent studies have demonstrated that there exist at least two subsets of T helper cells with a $CD4^+$ phenotype, termed Th1 and Th2 cells, which can be distinguished from each other by their cytokine productions (Mosmann and Coffman 1989). The cytokines derived from Th1 cells, such as IL-2, interferon-γ, and tumor necrosis factor (TNF)-β, may be responsible for activation of macrophages and cell-mediated immunity, whereas IL-4 and IL-5, cytokines produced by Th2 cells, stimulate IgE production and eosinophilia, respectively (Mosmann and Coffman 1989; Takatsu et al. 1994) (see Chapter 4, Immunology and immunopathology of human parasitic infections). The most detailed studies of patho-

genesis have been carried out with *S. mansoni*, and these are described in Chapter 29, Schistosomes: African.

Perhaps a less appreciated consequence of schistosomiasis is that the physical effects of infection can result in attention deficits, learning difficulties, and absenteeism. Unfortunately, such 'downstream' effects can particularly affect children, who often harbor the highest levels of infection.

IMMUNE RESPONSES

It has been assumed from epidemiological studies in endemic areas that age-dependent immunity may develop against infection, or against re-infection after treatment, with *S. mansoni* (Butterworth et al. 1985; Dessein et al. 1988) and *S. haematobium* (Hagan et al. 1991) (for review, see Butterworth 1994; Gryseels 1994). Prevalence and intensity of infection are usually higher among younger age groups than in older age groups. Using a mathematical model, it has also been shown that predicted patterns of variation in age-related changes in the intensity and prevalence of *S. haematobium* infection are consistent with the epidemiological effects of acquired immunity (Woolhouse et al. 1991).

In *S. japonicum* infection, however, the reduction in prevalence and intensity in older age groups may not be observed. Prevalence by age and sex of *S. japonicum* infection in China based on national surveys carried out in eight endemic provinces in 1989 and 1995 revealed that the infection rates peaked at around the age of 35 years and remained high thereafter, especially in males, and that males were twice as likely to become infected as females (Ross et al. 2001), which might be due to male-dominated exposure to cercariae-contaminated water while fishing during the time coinciding with peak cercarial shedding, at least in the Dongting Lake region (Ross et al. 1998). Despite these observations, another study carried out in a well-defined population in Leyte, Philippines, was able to identify individuals resistant and susceptible to *S. japonicum* infection and to demonstrate that the latter individuals, who were most vulnerable to rapid re-infection, belonged to a younger age group of 5–14-year-old children, whereas the former, who belonged to the 15–19 year age group or older, showed decreased incidence and intensity (Acosta et al. 2002a). Cellular immune responses of these two groups to defined vaccine candidate antigens (see below) suggested that a Th1 type of response appeared to be important in predicting resistance in this population (Acosta et al. 2002b).

To better understand effector mechanisms in immunity to schistosomiasis, animal models may be useful at least for the initial steps of analysis, although results observed in animals do not always parallel those in humans (Cheever et al. 2002). There are two experimental models for the induction of immunity to schistosome infections in vivo: the concomitant immunity

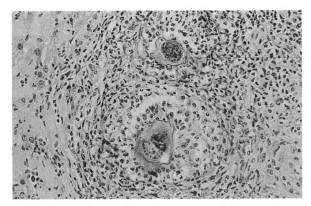

Figure 28.6 *Granuloma formation around newly deposited eggs of Schistosoma japonicum*

model and the attenuated vaccine model. Concomitant immunity is so called after the immunity developed to a second tumor graft in animals that already carry the same tumor; thus animals (originally rhesus monkeys) carrying an initial infection of adult schistosomes show partial but significant protection to a cercarial challenge infection (Smithers and Terry 1967). Although this model may reflect more closely the immunity observed in humans, it has been replaced by the attenuated vaccine model, in which animals are immunized with irradiated cercariae that can migrate into the lungs or liver, but cannot mature. In the former model much of the parasite attrition appears to be closely linked to the pathology induced by the previous infection (Dean et al. 1978; Wilson et al. 1983; von Lichtenberg 1985) rather than to immunologically specific anti-parasite mechanisms.

As in other parasitic infections, cellular and humoral immune responses are observed in schistosomiasis against antigens derived from various developmental stages of schistosomes. Responses may be observed against stage-specific antigens or alternatively cross-reactive antigens common to various stages. Immune responses may vary due to host genetic differences, and a number of studies have identified particular genes regulating both protective immunity and pathology (Marquet et al. 1996; Dessein et al. 1999). In general, $CD4^+$ T helper cells (Th cells) play a major role in responses to schistosome infection through their release of cytokines (Wynn and Cheever 1995; Pearce and MacDonald 2002). Recent work using murine models to address the involvement of Th cells and their cytokines in immune response development during infection is covered in the following chapters.

Worm development and the immune response

Evidence is accumulating that host immune factors may actually act as developmental cues for schistosomes (Davies and McKerrow 2003). While it has been known for some time that parasite fecundity can be severely reduced in immunocompromised animals, there is continued debate about the identity of the host immune factors that are responsible for influencing schistosome development, with IL-7 (Wolowczuk et al. 1999), TNF (Amiri et al. 1992), and TGFβ (Beall and Pearce 2001) all being possible candidates.

Vaccination

Reflecting the complexity of the host–parasite relationship and underscoring the need to more fully understand both immune-response development and parasite biology, an effective rationally designed anti-schistosome vaccine has yet to be produced (Bergquist and Colley 1998; Wilson and Coulson 1998; Capron et al. 2001; Pearce 2003). Because there is a complex of cytokine networks and pathways of signal transduction, the development of vaccines against schistosomiasis must be carefully designed in order to focus particular effectors against specific target stages or molecules. In addition, further studies are urgently required in humans to elucidate immune effector mechanisms. It should be noted that chemotherapy with praziquantel may induce immunological changes (by burst destruction of adult worms and miracidia in eggs) in terms of antigen recognition, antibody classes, and cytokine responses, and that age- and sex-difference of the host may affect immune responses after treatment (Mutapi 2001). Vaccines that can reduce schistosomiasis morbidity and mortality by lowering intensity of the infection should be adopted for practical use, when they become available, even if they are not effective in inducing the complete elimination of parasites. Factors such as the technical feasibility of vaccine production, the prospects of passage through existing regulatory bodies, and the ease of incorporation into existing immunization programs, must also be taken into account (Bergquist et al. 2002). Compared to the other major schistosome infections, schistosomiasis japonica has a characteristic nature as a zoonosis, giving serious economic problems for livestock. However, this characteristic may assist the development of a schistosomiasis vaccine for livestock as the first step, then followed by further application to humans.

Details of vaccines for African and Asian schistosomes are given in Chapters 29 and 30, respectively.

DIAGNOSIS

Specific diagnosis of schistosomiasis can be made by detection of the characteristic eggs in the stools or urine under microscopic examination. Biopsy at rectoscopy or cystoscopy may reveal eggs in the mucosa. For quantitative determination, which is necessary for evaluation of the success of chemotherapy or control operations, the Kato-Katz smear technique (though semiquantitative) or urine filtration technique may be used for intestinal or urinary schistosomiasis, respectively.

Indirect methods for diagnosis depend on clinical symptoms and signs, and biochemical or immunological analyses. Immunodiagnosis may be useful for demonstration of active or chronic schistosomiasis. A unique immunological method for the diagnosis of schistosomiasis is the circum oval precipitin (COP) test in which precipitate is formed around the eggs containing live miracidia after incubation in the serum of infected individuals (Figure 28.7). The enzyme-linked immunosorbent assay (ELISA) is also widely used in diagnosis.

The selection and application of any diagnostic method used for field studies must correspond to the type of information sought by the public health officer or epidemiologist, based on operational and financial

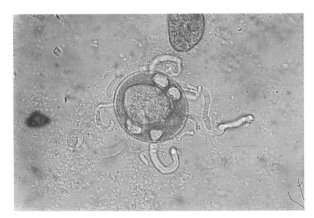

Figure 28.7 *Circum oval precipitin (COP) test*

constraints, and such constraints and drawbacks must be taken into consideration for interpretation of test results (Feldmeier and Poggensee 1993).

CHEMOTHERAPY, PREVENTION, AND CONTROL

The drug of choice is praziquantel, which is effective against all species of schistosomes, although the precise mechanism by which it kills the parasites remains to be clarified (Redman et al. 1996). In vitro treatment of *S. japonicum* with praziquantel resulted in contraction of the parasite and the vacuolization of the tegument within 5 min after exposure to 1 µg/ml praziquantel in medium TC199 (Mehlhorn et al. 1983). Praziquantel induces an influx of calcium ions across the tegument, causing an immediate muscular contraction (Mehlhorn et al. 1981). Praziquantel-induced contraction was found to be biphasic in normal parasites when incubated in high-magnesium medium, although only the larger, second contraction was observed in worms of which the tegument had been removed (Blair et al. 1994). These observations may suggest the presence of praziquantel-sensitive sites in the tegument (Redman et al. 1996).

Since immature (2–4 week old) worms are less susceptible to praziquantel than are larval (1–2 week old) worms or adult (5 or more weeks old) worms (Fallon et al. 1996) and some isolates of *S. mansoni* have been shown to be resistant to praziquantel, it is important to find new drugs for schistosomiasis (Doenhoff et al. 2002). Quite recently, it has been reported that artemether, the methyl ether derivative of artemisinin, is not only effective against malaria but also therapeutically effective against schistosomes, and is more effective against 7-day-old schistosomules than other developmental stages of schistosomes (Xiao and Catto 1989).

Thus, artemether was shown to be effective for the early phase of infection with *S. mansoni* and *S. haematobium* and morphological alterations, such as swelling and fusion of tegumental ridges, vesiculation, peeling, and erosion were observed for schistsomula, and also for adult worms, with this compound (Utzinger et al. 2001b). Field trials carried out in endemic areas with a single oral dose of 6 mg/kg of artemether administered repeatedly between four and ten times every 15 days resulted in successful prophylactic effects on at least *S. japonicum* and *S. mansoni* infections (Xiao et al. 2002) and this was also confirmed later for *S. haematobium* infection by a randomized, double-blind, placebo-controlled trial, although the protective efficacy was considerably lower (N'Goran et al. 2003). Combination therapy with praziquantel plus arthemether may be recommended since it is more effective than each drug given separately (Xiao et al. 2000, 2002; Utzinger et al. 2001a). Long-term toxicity tests carried out in the rat with up to 11.7-fold the dose recommended for humans revealed no adverse effect on any organs so far examined, except rapidly reversible reduction of the reticulocyte count and rapidly reversible increase in hemoglobin levels (Xiao et al. 2002).

It is clear that a number of factors must be addressed to enable effective national control programs for schistosomiasis. Long-term provision of sufficient clinical care facilities, community-based treatment programs, adequate health education to reduce the chances of transmission and encourage local recognition of schistosomiasis as a public health problem, implementation of environmental management measures, and development of enhanced levels of sanitation and hygiene, are all required to sustain prevention of the emergence or spread of schistosomiasis in endemic areas (Engels et al. 2002).

REFERENCES

Acosta, L.P., Aligui, G.D.L., et al. 2002a. Immune correlate study on human *Schistosoma japonicum* in a well-defined population in Leyte, Philippines: I. Assessment of 'resistance' versus 'susceptibility' to *S. japonicum* infection. *Acta Trop*, **84**, 127–36.

Acosta, L.P., Waine, G., et al. 2002b. Immune correlate study on human *Schistosoma japonicum* in a well-defined population in Leyte, Philippines: II. Cellular immune responses to *S. japonicum* recombinant and native antigens. *Acta Trop*, **84**, 137–49.

Amiri, P., Locksley, R.M., et al. 1992. Tumour necrosis factor alpha restores granulomas and induces parasite egg-laying in schistosome-infected SCID mice. *Nature*, **356**, 604–7.

Beall, M.J. and Pearce, E.J. 2001. Human transforming growth factor-beta activates a receptor serine/threonine kinase from the intravascular parasite *Schistosoma mansoni*. *J Biol Chem*, **276**, 31613–19.

Bergquist, N.R. and Colley, D.G. 1998. Schistosomiasis vaccines: research to development. *Parasitol Today*, **14**, 99–104.

Bergquist, N.R., Al-Sherbiny, M., et al. 2002. Blueprint for schistosomiasis vaccine development. *Acta Trop*, **82**, 183–92.

Blair, K.L., Bennett, J.L. and Pax, R.A. 1994. *Schistosoma mansoni*: myogenic characteristics of phorbol ester-induced muscle contraction. *Exp Parasitol*, **78**, 302–16.

Butterworth, A.E. 1994. Human immunity to schistosomes: some questions. *Parasitol Today*, **10**, 378–80.

Butterworth, A.E., Capron, M., et al. 1985. Immunity after treatment of human schistosomiasis mansoni. II. Identification of resistant individuals, and analysis of their immune responses. *Trans R Soc Trop Med Hyg*, **79**, 393–408.

Capron, A., Capron, M., et al. 2001. Vaccine strategies against schistosomiasis: from concepts to clinical trials. *Int Arch Allergy Immunol*, **124**, 9–15.

Caulfield, J.P., Korman, G., et al. 1980. The adherence of human neutrophils and eosinophils to schistosomula: evidence for membrane fusion between cells and parasites. *J Cell Biol*, **86**, 46–63.

Cheever, A.W., Lenzi, J.A., et al. 2002. Experimental models of *Schistosoma mansoni* infection. *Mem Inst Oswaldo Cruz*, **97**, 917–40.

Davies, S.J. and McKerrow, J.H. 2003. Developmental plasticity in schistosomes and other helminths. *Int J Parasitol*, **33**, 1277–84.

Dean, D.A., Minard, P., et al. 1978. Resistance of mice to secondary infection with *Schistosoma mansoni*. II. Evidence for a correlation between egg deposition and worm elimination. *Am J Trop Med Hyg*, **27**, 957–65.

Dessein, A.J., Begley, M., et al. 1988. Human resistance to *Schistosoma mansoni* is associated with IgG reactivity to a 37-kDa larval surface antigen. *J Immunol*, **140**, 2727–36.

Dessein, A.J., Hillaire, D., et al. 1999. Severe hepatic fibrosis in *Schistosoma mansoni* infection is controlled by a major locus that is closely linked to the interferon-gamma receptor gene. *Am J Hum Genet*, **65**, 709–21.

Doenhoff, M.J. 1997. A role for granulomatous inflammation in the transmission of infectious disease: schistosomiasis and tuberculosis. *Parasitology*, **115**, Suppl., S113–25.

Doenhoff, M.J., Kusel, J.R., et al. 2002. Resistance of *Schistosoma mansoni* to praziquantel: is there a problem? *Trans R Soc Trop Med Hyg*, **96**, 465–9.

Engels, D., Chitsulo, L., et al. 2002. The global epidemiological situation of schistosomiasis and new approaches to control and research. *Acta Trop*, **82**, 139–46.

Fallon, P.G., Tao, L.-F., et al. 1996. Schistosome resistance to praziquantel: fact or artifact? *Parasitol Today*, **12**, 316–20.

Feldmeier, H. and Poggensee, G. 1993. Diagnostic techniques in schistosomiasis control. A review. *Acta Trop*, **52**, 205–20.

Fried, B. and Haseeb, M.A. 1991. Role of lipids in *Schistosoma mansoni*. *Parasitol Today*, **7**, 204.

Fripps, P.J. 1967. The site of (1-14C) glucose assimilation in *Schistosoma haematobium*. *Comp Biochem Physiol*, **23**, 893–8.

Furlong, S.T. 1991. Unique roles for lipids in *Schistosoma mansoni*. *Parasitol Today*, **7**, 59–62.

Furlong, S.T. and Caulfield, J.P. 1989. *Schistosoma mansoni*: synthesis and release of phospholipids, lysophospholipids, and neutral lipids by schistosomula. *Exp Parasitol*, **69**, 65–77.

Gryseels, B. 1994. Human resistance to *Schistosoma* infections: age or experience? *Parasitol Today*, **10**, 380–4.

Hagan, P., Blumenthal, U.J., et al. 1991. Human IgE, IgG4 and resistance to reinfection with *Schistosoma haematobium*. *Nature*, **349**, 243–5.

Hockley, D.J. 1973. Ultrastructure of the tegument of *Schistosoma*. *Adv Parasitol*, **11**, 233–305.

Hockley, D.J. and McLaren, J. 1973. *Schistosoma mansoni*: changes in the outer membrane of the tegument during development from cercariae to adult worm. *Int J Parasitol*, **3**, 13–25.

Jiang, J., Skelly, P.J., et al. 1996. *Schistosoma mansoni*: the glucose transport protein SGTP4 is present in tegumental multilamellar bodies, discoid bodies, and the surface lipid bilayers. *Exp Parasitol*, **82**, 201–10.

Karanja, D.M., Colley, D.G., et al. 1997. Studies on schistosomiasis in western Kenya: I. Evidence for immune-facilitated excretion of schistosome eggs from patients with *Schistosoma mansoni* and human immunodeficiency virus coinfections. *Am J Trop Med Hyg*, **56**, 515–21.

Marquet, S., Abel, L., et al. 1996. Genetic localization of a locus controlling the intensity of infection by *Schistosoma mansoni* on chromosome 5q31-q33. *Nat Genet*, **14**, 181–4.

Mehlhorn, H., Becker, B., et al. 1981. In vivo and in vitro experiments on the effects of praziquantel on *Schistosoma mansoni*. A light and electron microscopic study. *Arzneimittel Forsch*, **31**, 544–54.

Mehlhorn, H., Kojima, S., et al. 1983. Ultrastructural investigations on the effects of praziquantel on human trematodes from Asia: *Clonorchis sinensis*, *Metagonimus yokogawai*, *Opisthorchis viverrini*, *Paragonimus westermani* and *Schistosoma japonicum*. *Arzneimittel Forschung*, **33**, 91–8.

Mosmann, T.R. and Coffman, R.L. 1989. Heterogeneity of cytokine secretion patterns and function of helper T cells. *Adv Immunol*, **46**, 111–47.

Mueckler, M., Caruso, C., et al. 1985. Sequence and structure of a human glucose transporter. *Science*, **229**, 941–5.

Murugasu, R. and Por, P. 1973. First case of schistosomiasis in Malaysia. *Southeast Asian J Trop Med Pub Hlth*, **4**, 519–23.

Murugasu, R., Wang, F. and Dissanaike, A.S. 1978. *Schistosoma japonicum*-type infection in Malaysia: report of the first living case. *Trans R Soc Trop Med Hyg*, **72**, 389–91.

Mutapi, F. 2001. Heterogeneities in anti-schistosome humoral responses following chemotherapy. *Trends Parasitol*, **17**, 518–24.

N'Goran, E.K., Utzinger, J., et al. 2003. Randomized, double-blind, placebo-controlled trial of oral artemether for the prevention of patent *Schistosoma haematobium* infections. *Am J Trop Med Hyg*, **68**, 24–32.

Oliveira, G. and Johnston, D.A. 2001. Mining the schistosome DNA sequence database. *Trends Parasitol*, **17**, 501–3.

Pearce, E.J. 2003. Progress towards a vaccine for schistosomiasis. *Acta Trop*, **86**, 309–13.

Pearce, E.J. and MacDonald, A.S. 2002. The immunobiology of schistosomiasis. *Nat Rev Immunol*, **2**, 499–511.

Redman, C.A., Robertson, A., et al. 1996. Praziquantel: an urgent and exciting challenge. *Parasitol Today*, **12**, 14–20.

Rogers, M.V., Quilici, D., et al. 1990. Purification of a putative receptor from *Schistosoma japonicum* adult worms. *Mol Biochem Parasitol*, **41**, 93–100.

Rogers, S.H. and Beuding, E. 1975. Anatomical localization of glucose uptake by *Schistosoma mansoni* adults. *Int J Parasitol*, **5**, 369–71.

Rollinson, D., Kaukas, A., et al. 1997. Some molecular insights into schistosome evolution. *Int J Parasitol*, **27**, 11–28.

Ross, A.G.P., Li, Y., et al. 1998. Measuring exposure to *S. japonicum* in China. I. Activity diaries to assess water contact and comparison to other measures. *Acta Trop*, **71**, 213–28.

Ross, A.G.P., Sleigh, A.C., et al. 2001. Schistosomiasis in the People's Republic of China: prospects and challenges for the 21st century. *Clin Microbiol Rev*, **14**, 270–95.

Rumjanek, F.D., McLaren, D.J. and Smithers, S.R. 1983. Serum-induced expression of a surface protein in schistosomula of *Schistosoma mansoni*: A possible receptor for lipid uptake. *Mol Biochem Parasitol*, **4**, 337–50.

Sandor, M., Weinstock, J.V., et al. 2003. Granulomas in schistosome and mycobacterial infections: a model of local immune responses. *Trends Immunol*, **24**, 44–52.

Satoh, M., Nara, T., et al. 1997. Squirrel monkey as a useful vaccine model for *Schistosoma japonicum* infection. *Parasitol Int*, **46**, 31–9.

Senft, A.W. and Maddison, S.E. 1975. Hypersensitivity to parasite proteolytic enzyme in schistosomiasis. *Am J Trop Med Hyg*, **24**, 83–9.

Skelly, P.J., Kim, J.W., et al. 1994. Cloning, characterization and functional expression of cDNAs encoding glucose transporter proteins from the human parasite. *Schistosoma mansoni*. *J Biol Chem*, **269**, 4247–53.

Smithers, S.R. and Terry, R.J. 1967. Resistance to experimental infection with *Schistosoma mansoni* in rhesus monkeys induced by the transfer of adult worms. *Trans R Soc Trop Med Hyg*, **61**, 517–33.

Takatsu, K., Takaki, S. and Hitoshi, Y. 1994. Interleukin-5 and its receptor system: implications in the immune system and inflammation. *Adv Immunol*, **57**, 145–90.

Talla, I., Kongs, A., et al. 1990. Outbreak of intestinal schistosomiasis in the Senegal river basin. *Ann Soc Belge Med Trop*, **70**, 173–80.

Uglem, G.L. and Read, C.P. 1975. Sugar transport and metabolism in *Schistosoma mansoni*. *J Parasitol*, **61**, 390–7.

Usawattanakul, W., Kamijo, T. and Kojima, S. 1982. Comparison of recovery of schistosomula of *Schistosoma japonicum* from lungs of mice and rats. *J Parasitol*, **68**, 783–90.

Utzinger, J., Chollet, J., et al. 2001a. Effect of combined treatment with praziquantel and artemether on *Schistosoma japonicum* and *Schistosoma mansoni* in experimentally infected animals. *Acta Trop*, **80**, 9–18.

Utzinger, J., Xiao, S.H., et al. 2001b. The potential of artemether for the control of schistosomiasis. *Int J Parasitol*, **31**, 1549–62.

von Lichtenberg, F. 1964. Studies on granuloma formation. III. Antigen sequestration and destruction in the schistosome pseudotubercle. *Am J Pathol*, **45**, 711–31.

von Lichtenberg, F. 1985. Conference on contended issues of immunity to schistosomes. *AmJ Top Med Hyg*, **34**, 78–85.

Warren, K.S., Domingo, E.O. and Cowan, R.B.T. 1967. Granuloma formation around schistosome eggs as a manifestation of delayed hypersensitivity. *Am J Pathol*, **51**, 735–56.

WHO. 1993. *The control of schistosomiasis* (WHO Expert Committee). Geneva: World Health Organization, pp. 15, 18 and 79.

Wilson, R.A. and Coulson, P.S. 1998. Why don't we have a schistosomiasis vaccine? *Parasitol Today*, **14**, 97–8.

Wilson, R.A., Coulson, P.S. and McHugh, S.M. 1983. A significant part of the concomitant immunity of mice to *Schistosoma mansoni* is a consequence of a leaky hepatic portal system, not immune killing. *Parasite Immunol*, **5**, 595–601.

Wolowczuk, I., Nutten, S., et al. 1999. Infection of mice lacking interleukin-7 (IL-7) reveals an unexpected role for IL-7 in the development of the parasite *Schistosoma mansoni*. *Infect Immun*, **67**, 4183–90.

Woolhouse, M.E., Taylor, P., et al. 1991. Acquired immunity and epidemiology of *Schistosoma haematobium*. *Nature (Lond)*, **351**, 757–759.

Wynn, T.A. and Cheever, A.W. 1995. Cytokine regulation of granuloma formation in schistosomiasis. *Curr Opin Immunol*, **7**, 505–11.

Xiao, S.H. and Catto, B.A. 1989. In vitro and in vivo studies of the effect of artemether on *Schistosoma mansoni*. *Antimicrobial Agent Chemother*, **33**, 1557–62.

Xiao, S.H., You, J.Q., et al. 2000. Effect of praziquantel together with artemether on *Schistosoma japonicum* parasites of different ages in rabbits. *Parasitol Int*, **49**, 25–30.

Xiao, S.H., Tanner, M., et al. 2002. Recent investigations of artemether, a novel agent for the prevention of schistosomiasis japonica, mansoni and haematobia. *Acta Trop*, **82**, 175–81.

Yokogawa, M., Sano, M., et al. 1971. An outbreak of *Schistosoma* infection among dairy-cows in the Tone river basin in Chiba Prefecture (1). *Jpn J Parasitol*, **20**, 507–11.

Yokogawa, M., Sano, M., et al. 1973. Epidemiological survey for schistosomiasis among the inhabitants in Tone river basin, Chiba Prefecture and snail control by burning. *Jpn J Parasitol*, **22**, 116–25.

Zhong, C., Skelly, P.J., et al. 1995. Immunolocalization of a *Schistosoma mansoni* facilitated diffusion glucose transporter to the basal, but not the apical, membranes of the surface syncytium. *Parasitology*, **110**, 383–94.

29

Schistosomes: African

SOMEI KOJIMA AND ANDREW S. MACDONALD

THE PARASITES

Approximately 85 percent of the nearly 200 million individuals estimated to be infected with schistosomes worldwide are thought to live on the African continent (Engels et al. 2002). Three species occur there, *S. intercalatum*, *S. haematobium*, and *S. mansoni* (see Table 29.1). *S. intercalatum* has a restricted distribution, being found only in parts of West Africa (Cameroon, Equatorial Guinea, Gabon, Nigeria, Democratic Republic of Congo, and Sao Tomé – Tchuente et al. 2003). *S. mansoni* and *S. haematobium* are the main species that cause human schistosomiasis in Africa, although both species can also be found in the Middle East, and *S. mansoni* can be found in regions of South America and the Caribbean. Schistosomiasis is an important public health concern in these regions, with the estimated mortality rates attributed to *S. mansoni* and *S. haematobium* in sub-Saharan Africa alone being some 280 000 per year (van der Werf et al. 2003). For this reason, this chapter will focus on these two species.

Since ancient times there has been an awareness of schistosome infection in endemic areas of Africa. Hierographics in the papyrus of Kahun refer to hematuria (bloody urine), a typical sign of urinary schistosomiasis, which is mentioned 50 times in various medical papyri (Contis and David 1996). In Cairo, in 1851, Theodor Bilharz found a trematode at autopsy in the blood of mesenteric veins of a young man. It was not hermaphrodite, as were other flukes known at that time, but had separate sexes. He named the worm *Distomum haematobium* and described this as the cause of hematuria, based on the presence of terminal-spined eggs in the urine, although he considered lateral-spined eggs to belong to the same species. In 1859, Cobbold used a new generic name *Bilharzia* for the worm he found in the portal vein of a sooty monkey, although the name *Schistosoma* (meaning split body) had already been used by Weinland in 1858. In 1864, Harley found eggs with a terminal spine, but not those with a lateral spine, in the urine of patients with hematuria living in South Africa. In 1902, Manson found lateral-spined ova in a stool sample of a British patient with anemia instead of the hookworm eggs that he had expected, but no ova were detected in the urine of this patient, who had spent 15 years in the West Indies. He suggested the possible existence of two species of *Bilharzia*, one with terminal-spined eggs which were found in the urine, the other depositing lateral-spined eggs in the rectum; in 1907 Sambon designated the latter as *Schistosoma mansoni*.

Classification

Gene discovery and gene mapping projects, which have so far mainly focused on *S. mansoni*, generally agree with original classifications of schistosome species based on morphological or life history characteristics, with a high level of identity observed within species (Rollinson et al. 1997; Oliveira and Johnston 2001). It is hoped that continued work in this important area should provide illuminating information about the evolution and biology of the parasites and prove invaluable for the rational design of control measures.

Table 29.1 *Intermediate hosts of* S. mansoni *and* S. haematobium[a]

Schistosome species	Intermediate hosts	Geographical distribution
S. mansoni	BIOMPHALARIA	
	B. glabrata	Caribbean islands, Venezuela, Brazil, Argentina
	B. straminea	Caribbean islands, Central and South America
	B. tenagophila	Southeast Brazil
	B. pfeifferi group	
	B. pfeifferi	All sub-Saharan Africa
	B. rhodesiensis	East/Central to South Africa
	B. ruepelli	Yemen to East Africa
	B. arabica	Arabian peninsula
	B. choanomphala group	
	B. choanomphala	Central African lakes
	B. alexandrina group	
	B. alexandrina	Nile valley from North Sudan
	B. angulosa	Mountain areas from East to South Africa
	B. sudanica group	
	B. sudanica	Eastern tropical Africa
	B. camerunensis	Western tropical Africa
	B. salinarum	Southwest Africa
S. haematobium	BULINUS	
	B. africanus group	
	B. abyssinicus	Somalia, Ethiopia
	B. africanus	East, Central and Southern Africa
	B. globosus	Most of subsaharan Africa
	B. jousseaumei	West Africa
	B. nasutus	East Africa
	B. obtusispira	Madagascar
	B. forskali group	
	B. beccarii	Aden
	B. camerunensis	West Cameroon crater lakes
	B. cernicus	Mauritius
	B. forskali	Most of tropical Africa
	B. senegalensis	Mauritania, Senegal, Gambia
	B. reticulatus group	
	B. reticulatus[b]	Focal from Ethiopia to South Africa
	B. wrighti	Saudi Arabia, Oman, Yemen
	B. truncatus/tropics complex	
	B. guernei	West Africa
	B. liratus	Madagascar
	B. rohlfsi	West Africa
	B. truncatus	Mediterranean, south to Mauritania and Malawi, east to Iran

a) Adapted from Sturrock (1993)
b) Experimentally infected with a strain of *S. haematobium*

Morphology and structure

SCHISTOSOMA MANSONI

The surface of adult male *S. mansoni* is covered with characteristic coarse tubercles. Elliptical eggs (110–175 × 45–70 μm), hundreds of which are produced each day by the sexually mature *S. mansoni* female, are discharged mainly in the feces. They are distinguished from the other species of schistosome by the presence of a lateral spine (Figure 29.1).

SCHISTOSOMA HAEMATOBIUM

Adult *S. haematobium* worms resemble *S. mansoni* in that their surface tegument is coated in tubercles, rather than appearing smooth as in *S. japonicum*, although these tubercles are smaller than those present on the surface of *S. mansoni*. The elliptical eggs (110–170 × 40–70 μm) produced by sexually mature females are discharged mainly in the urine, and are characterized by the presence of a terminal spine (Figure 29.2). As with *S. mansoni*, hundreds of eggs can be produced daily by each female.

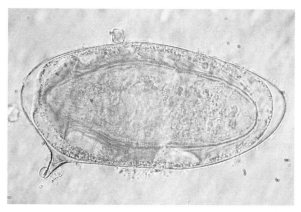

Figure 29.1 Schistosoma mansoni *ovum*

Location in the host

Adult worms of *S. mansoni* inhabit the mesenteric veins of the portal circulation, whereas *S. haematobium* adults generally remain in the systemic circulation, maturing in the blood vessels of the ureteric and vesical plexus surrounding the urinary bladder.

Life cycle

The life cycles of both *S. mansoni* and *S. haematobium* follow the common pathway outlined in Chapter 28, Schistosomes: general. Intermediate hosts for *S. haematobium* are provided by several species of *Bulinus* (Figure 29.3) and *Physopsis* snail, and possibly also *Planorbarius*. The main intermediate hosts for *S. mansoni* in Africa and the western hemisphere are *Biomphalaria* snails: *B. alexandrina* in northern Africa, Saudi Arabia, and Yemen; *B. sudanica*, *B. pfeifferi*, and *B. rupellii* in other parts of Africa; *B. glabrata* (Figure 29.4), *B. tenagophila*, and *B. straminea* in the Americas (Paraense 2001). Intermediate hosts are listed in Table 29.1.

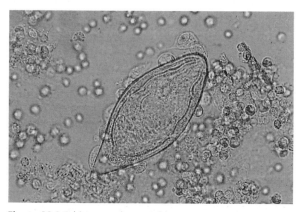

Figure 29.2 Schistosoma haematobium *ovum*

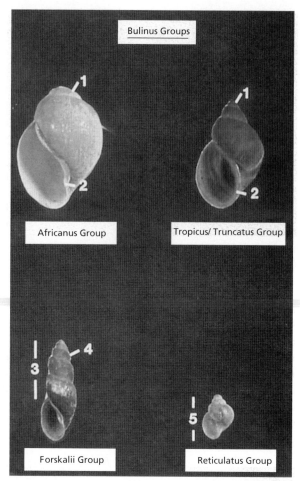

Figure 29.3 *Four species groups of the intermediate snail hosts of* Schistosoma haematobium *(WHO Slide Set Series; snail hosts, Schistosomiasis)*

CLINICAL AND PATHOLOGICAL ASPECTS OF INFECTION

Clinical manifestations, pathogenesis, and pathology

Symptoms characteristic of chronic infection with *S. mansoni* include abdominal pain, diarrhea, blood in the stool, and, more severely, hematemesis. Mortality caused by *S. mansoni* is generally associated with the development of liver fibrosis and hepatosplenomegaly, with many individuals dying each year from hematemesis or liver failure as a result of portal hypertension (van der Werf et al. 2003). However, it is not yet clear whether the development of hepatosplenic disease actually requires the development of fibrosis (Dessein et al. 1999b). During *S. haematobium* infection, the most common symptom is hematuria, as well as polyuria and dysuria. A significant proportion of those affected will additionally suffer from major bladder wall pathology and hydronephrosis (Figure 29.5). Mortality is primarily due to bladder cancer and kidney failure (van der Werf

Figure 29.4 Biomphalaria glabrata, *an important intermediate host of* Schistosoma mansoni

et al. 2003). In addition to the fact that lesions due to *S. haematobium* eggs are found in the vulva, vagina, cervix, and less commonly the ovaries, Fallopian tubes, or uterus (Wright et al. 1982), *S. haematobium* may migrate through the network of female pelvic vasculature, and adaptive changes in the vasculature during puberty and especially during pregnancy make 'ectopic' localization of the parasites possible (Feldmeier et al. 1995). Thus, it

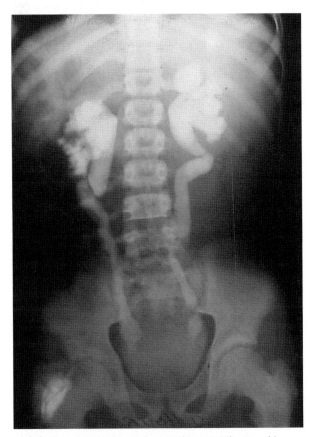

Figure 29.5 *Hydronephrosis due to* Schistosoma haematobium *infection (courtesy of Dr Yukiko Wagatsuma).*

has been pointed out that female genital schistosomiasis may be an important risk factor for transmission of Human immunodeficiency virus (HIV), as it is known that lesions in the genital mucosa, associated with sexually-transmitted diseases, increase the probability of viral uptake (Feldmeier et al. 1994). From published data there seems to be pathophysiological, immunological, and epidemiological evidence for an association between genital ulcer due to *S. haematobium* and HIV infection in women (Feldmeier et al. 1995). If so, control of female genital schistosomiasis should be tackled seriously to prevent further widespread dissemination of HIV infection.

Immunopathology

In recent years, cytokine production from T helper subsets has been intensively studied to gain new insights into granuloma formation (see Chapter 28, Schistosomes: general, Figure 28.6) and development of immunopathology, with the focus of research primarily being murine schistosomiasis mansoni (Hoffmann et al. 2002; Pearce and MacDonald 2002). Initial studies revealed that the production of Th2 cytokines was found to correspond to the onset of egg deposition (Grzych et al. 1991; Pearce et al. 1991). Lymphocytes from mice vaccinated with irradiated cercariae or from infected animals were compared for their ability to produce interferon-γ and IL-2, or IL-4 and IL-5. After stimulation with specific antigen or mitogen, T cells from vaccinated mice or prepatently infected animals responded primarily with Th1 cytokines, whereas lymphocytes from patently infected mice instead produced Th2 cytokines. The Th2 response in infected animals was shown to be induced by schistosome eggs and directed largely against egg antigens, whereas the Th1 reactivity in vaccinated mice was triggered primarily by larval antigens. Interestingly, Th1 responses in mice carrying egg-producing infections were found to be profoundly down-regulated. Moreover, the injection of eggs into vaccinated mice resulted in a reduction of antigen and mitogen-stimulated Th1 function accompanied by a coincident expression of Th2 responses (Pearce et al. 1991). Such studies were instrumental in establishing that it is the egg stage of the schistosome that is the major Th2 stimulus during infection, and in defining those components of egg antigens that are responsible. Th2 response development is an area of active research, with current evidence suggesting that both lipid (Faveeuw et al. 2002) and carbohydrate (Okano et al. 1999) may be involved. Interestingly, IL-4 production has recently been found to be initiated in Kupffer cells in the liver by adolescent *S. mansoni* worms before the females become mature and produce eggs, although the physiological relevance of the levels of cytokine produced has yet to be determined (Hayashi et al. 1999).

Cheever et al. (1992a) were able to demonstrate that in vivo treatment of infected mice with anti-IL-2 antibodies significantly diminished the size of circumoval granulomas in the liver, resulting in a decrease of hepatic fibrosis to half of that in untreated mice. Antibody-treated animals also displayed a marked reduction in both peripheral blood and tissue eosinophilia, while IgE levels were unchanged or increased. Spleen cell cytokine production in response to antigen or mitogen stimulation was selectively altered by in vivo anti-IL-2 administration. IL-5 responses were dramatically reduced, whereas IL-4, IL-2, and interferon-γ responses were not consistently changed. These findings suggest a role for IL-2 in egg-induced pathology, but indicate that the primary function of this cytokine in schistosome-infected mice may be in the generation of Th2- rather than Th1-associated responses (Cheever et al. 1992a).

By using a synchronized granuloma development model in the lung (von Lichtenberg 1962), it has been demonstrated that although the effects of neutralization of interferon-γ or IL-2 were variable, in vivo treatment of egg-injected mice with either anti-IL-2 or anti-IL-4 antibodies dramatically diminished the size of egg granulomas in the lungs (Chensue et al. 1992; Wynn et al. 1993). Both groups of antibody-treated animals displayed a marked reduction in IL-4, as well as IL-5 mRNA expression, although interferon-γ and IL-2 mRNA levels were unchanged or slightly increased. The up-regulation of mRNA expression of the IL-4 and IL-5 genes was also shown to be greater than that of Th1 cytokine genes in the granulomatous liver at 8 weeks after infection (Henderson et al. 1991; Wynn et al. 1993).

Further studies, in which mice were treated with anti-IL-4 before egg deposition, demonstrated decreased IL-4, IL-5, and IL-10 production in response to in vitro antigenic stimulation, as well as decreased IL-5 and IL-13 mRNA levels in the liver (Cheever et al. 1994). It was found that non-B, non-T cells were a major source of IL-4 in infected mice treated with control monoclonal antibody, and that the diminished IL-4 response in anti-IL-4-treated animals was caused, at least in part, by a reduction in the number of these cells, as well as by decreased secretion of IL-4 per cell. In contrast, production of the Th1 cytokines, IL-2 and interferon-γ was elevated in anti-IL-4-treated infected mice in vitro, and the corresponding mRNAs in the liver increased. Anti-IL-4 treatment did not consistently reduce the size of hepatic granulomas around *S. mansoni* eggs, but markedly inhibited granuloma formation in the lungs of the same animals after intravenous egg injection. Nevertheless, anti-IL-4-treatment showed consistent and marked reductions in hepatic collagen deposition at 8 weeks in *S. mansoni*-infected mice (Cheever et al. 1994). Similarly, hepatic fibrosis was markedly diminished in anti-IL-4-treated mice at 10 weeks after infection with *S. japonicum* (Cheever et al. 1995). These findings suggested that IL-4 plays a major role in the development of the Th2 response in murine schistosomiasis and contributes to the pathogenesis of hepatic fibrosis. Moreover, the administration of cytokines themselves into infected mice has also been shown to modulate granuloma formation. According to Yamashita and Boros (1992), chronically infected mice treated with 10–1 000 U of recombinant IL-4 showed significantly enhanced liver granulomatous responses compared with untreated animals and the augmented granulomas contained more enlarged macrophages and connective tissue matrix. However, when mice in which the IL-4 gene had been deleted (IL-4$^{-/-}$ mice) were infected with *S. mansoni* at 8 and 16 weeks after infection, liver pathology was similar to that in wild-type animals in terms of the size, cellularity, cellular composition, and collagen content of granulomas. When compared to normal mice, smaller granulomatous responses were observed in IL-4$^{-/-}$ mice only when eggs were injected intravenously to form granulomas in the lungs, despite the fact that Th1 cytokine production seemed to be dominant in lymphoid cells of IL-4-deficient mice stimulated with soluble egg antigen (Pearce et al. 1996). These results suggest that IL-4 is not an absolute requirement for hepatic granuloma formation. Nevertheless, subsequent investigation of disease progression in Th2-impaired (IL-4$^{-/-}$) mice has shown that a failure to develop a Th2 response has lethal consequences, resulting from a failure to regulate the initial pro-inflammatory response associated with acute infection (Brunet et al. 1997; Fallon et al. 2000). Increased disease severity and mortality in these studies was not due to increased parasite burden, but rather appeared to be a direct consequence of the failure to regulate the production and effector function of inflammatory mediators during infection. Thus, somewhat paradoxically, the dominant polarized Th2 immune response that is induced by schistosomes is intimately involved in many of the pathological changes that accompany infection, but is also required to allow host survival while infected. Whether such a clear requirement for Th2 development to avoid overt pathology is also the case in active human infection remains debatable.

Following these initial studies investigating the role of IL-4 in Th2 induction during schistosome infection, it has been shown that neither IL-6 nor IL-13 cytokines, that have both been implicated in Th2 development in other systems, appear to be critical for the development of this type of response during schistosomiasis. Although it may be involved at some level for regulation of IFN-γ and IL-12 production in the initial response to schistosome eggs (La Flamme et al. 2000), IL-6 is not an absolute requirement for Th2 development to the same stimulus (La Flamme and Pearce 1999). IL-13, a cytokine which is in many ways closely related to IL-4, is not required for Th2 development, but appears to play a key role in granuloma formation and fibrosis during infection (Chiaramonte et al. 1999; Fallon et al. 2000). Indeed, it appears that the ability of both IL-4 and IL-13 to promote fibrosis is linked to the ability of these cytokines to 'alternatively' activate macrophages (Hesse et al. 2001). An interesting recent study suggests that liver macrophages (Kupffer cells) can provide a source of both IL-4 and IL-13 during the early stages of Th2 development (Hayashi et al. 1999), highlighting the fact that these key cytokines can originate from a number of sources. Additional possible sources of early IL-4 during schistosomiasis include naive T cells (Schmitz et al. 1994; Reiner 2001), basophils (Falcone et al. 1996), the so-called non-B, non-T cells (Williams et al. 1993) (which are most likely mast cells or basophils), and eosinophils (Sabin et al. 1996).

Administration of recombinant interferon-γ or IL-12 can also inhibit granuloma formation (Wynn et al. 1994). IL-12 is a key cytokine that promotes NK cell activity and Th1 responses. It has been demonstrated that the Th1-related cytokine, interferon-γ, augments endotoxin-stimulated IL-12 production in oil-elicited macrophages, whereas IL-4 and especially IL-10, are profoundly inhibitory (Wynn et al. 1994). Further experiments have revealed that sensitization with eggs plus IL-12 partly inhibits granuloma formation and dramatically reduces the tissue fibrosis induced by natural infection with *S. mansoni*. These results may provide an example of a vaccine against parasites which acts by preventing pathology rather than infection (Wynn et al. 1995). However, it has been shown recently that immunization with soluble egg antigens (SEA) plus complete Freund's adjuvant induces a pronounced Th1-shifted response in the low-pathology C57BL/6 mice, which results in a dramatic enhancement of hepatic egg-induced immunopathology, manifested by a marked increase in granuloma size and parenchymal inflammation, thereby leading to early death (Rutitzky et al. 2001). This raises the concern that predisposing individuals to make an extreme Th1 response during schistosome infection may in fact exacerbate disease, and so caution must be exercised in following this kind of immune deviation-based therapeutic approach.

IL-10, a cytokine often associated with Th2 response development and which is produced by a range of cell types, including Th1 cells (Sornasse et al. 1996), has been shown to down-regulate Th1 responses by working synergistically with IL-4 (Oswald et al. 1992). It has long been known that IL-10 is produced rapidly after schistosome egg production, with levels remaining high through the chronic stages of infection (Sher et al. 1991). In the past few years, renewed interest in the immuno-regulatory role of IL-10 has revealed a number of interesting functions for this cytokine during infection. IL-10 decreases egg antigen-specific Th cell responses by down-regulating MHC class II, as well as B7 costimulatory molecule expression, on accessory cells (Flores-Villanueva et al. 1994). Systemic administration of IL-10

significantly inhibited delayed-type hypersensitivity reactions to egg antigen as well as primary and secondary granuloma formation to eggs embolized in the lung. However, significant inhibition of hepatic granuloma formation associated with the natural infection required the use of an IL-10/Fc fusion protein with a prolonged in vivo half-life. Lymph node cells from IL-10/Fc-treated mice produced less IL-2 and interferon-γ and more IL-4 and IL-10 than control cells, suggesting that reduced egg granuloma formation resulted primarily from down-regulation of Th1 responses. These results indicate that suitable administration of exogenous IL-10 can be effective in ameliorating immunopathologic damage associated with schistosomiasis (Flores-Villanueva et al. 1996). An elegant series of experiments recently highlighted the importance of IL-10 in maintaining a balanced immune response during schistosome infection. Mice that were doubly deficient for IL-4 and IL-10, IL-12 and IL-10, or IFNγ and IL-10, all developed severe disease (Hoffmann et al. 2000; Vaillant et al. 2001). Infected 'Th1 polarized' IL-4/10-deficient animals, while displaying minimal signs of fibrosis, developed an exaggerated form of the hepatotoxic disease that is seen in IL-4$^{-/-}$ animals which resulted in their death by around 8 weeks post-infection. Interestingly, in contrast to this, infected 'Th2-polarized' IL-12/IL-10 or IFN-γ/IL-10 doubly deficient mice displayed elevated levels of fibrosis, that led to the death of around 50 percent of the IL-12/IL-10-deficient animals by 12 weeks post-infection. These important studies indicate that extreme polarization of the immune response towards either Th1 or Th2 can have fatal consequences during infection with *S. mansoni*.

An additional unexpected role for IL-10 produced in response to schistosomes has become apparent as a result of several recent studies, which have reported reduced allergen responsiveness during *S. mansoni* (Araujo et al. 2000) and *S. haematobium* (van den Biggelaar et al. 2000) infections. The suggestion is that production of IL-10 (and perhaps TGFβ) during infection with schistosomes, and perhaps other helminths, may help prevent or reduce inflammation associated with conditions such as allergy and atopy, providing a new interpretation of the so-called 'hygiene hypothesis' (Yazdanbakhsh et al. 2002).

Granulomatous inflammation in schistosomiasis mansoni seems to be a complexly regulated consequence of T cell-mediated hypersensitivity to egg antigens. Three consecutive independent chromatographic procedures have been carried out to fractionate and identify soluble egg antigens (SEA) recognized by schistosome-specific cloned murine, CD4$^+$ Th1-type lymphocytes, which had previously been shown to be capable of mediating granuloma formation in vivo when adoptively transferred to normal syngeneic hosts challenged with an intravenous injection of eggs. The stimulatory activity resided in two acidic egg molecules, with apparent molecular masses of 64–68 kDa and 38–42 kDa, each of which ran as a single band on SDS-PAGE after purification. Fast performance liquid chromatography and SDS-PAGE performed under reducing conditions suggested that the two molecules are related and that the 38–42 kDa molecule is a subunit of the 64–68 kDa molecule. Polyclonal lymphoid cells from schistosome-infected mice were stimulated by both the purified 64–68 kDa and 38–42 kDa molecules, implying that these are sensitizing antigens in the natural disease (Chikunguwo et al. 1993). This kind of identification of the molecules among crude SEA seems to be important for further studies on the identification of vaccine candidates that might regulate granuloma formation.

Indeed, further studies using SEA-specific monoclonal or hybridoma T cells have revealed that at least three important molecules in SEA are involved in the granuloma formation mediated by MHC class-II-restricted CD4$^+$ Th lymphocytes (Stadecker et al. 2001). Firstly, the most abundant egg component of 40 kDa protein was sequenced as a molecule of 354 amino acids with substantial homology to alpha-crystallins and heat shock proteins (Nene et al. 1986) and this Sm-p40 was found remarkably immunodominant in the large granuloma-forming H-2^k mouse strains, such as C3H and CBA, whereas none of the SEA-specific T cell hybridomas derived from the small granuloma-forming strain (C57BL/6) responded to Sm-p40 (Hernandez et al. 1997a). Within the immunodominant 13mer peptide 234PKSDNQIKAVPAS246, D237 has been revealed as the MHC class II (I-A^k)-anchoring residue, while N238, Q239, and K241 have been shown as the principal T cell receptor (TCR)-contact residues (Hernandez and Stadecker 1999). Second, with the aid of a specific C57BL/6 T-cell hybridoma (4E6), a 62 kDa egg component was isolated and identified as S. mansoni phosphoenolpyruvate carboxykinase (Sm-PEPCK) (Asahi et al. 1999). The recombinant Sm-PEPCK of 626 amino acids stimulates CD4$^+$ Th cells from schistosome-infected or SEA-immunized mice (Asahi et al. 2000), although the response is distinctly polarized towards the Th1 type despite the fact that the native Sm-PEPCK induces a more balanced Th1 and Th2 type response in infected C57BL/6 mice (Asahi et al. 1999). A 12-amino-acid region of 398DKSKDPKAHPNS409 contained a T cell epitope (Asahi et al. 2000). Third, another immunogenic component of 26 kDa was identified as thioredoxin peroxidase-1 (TPx-1) (Williams et al. 2001). Interestingly, TPx-1 is secreted by the eggs and has been localized to the region between the miracidium and the eggshell (von Lichtenberg's envelop), stimulating antibody responses in all mouse strains so far examined (Williams et al. 2001), and thereby producing the circum oval precipitin (COP) reaction (Alger et al. 2002).

Antibody-mediated modulation of granulomas may occur in cases of schistosomiasis japonica, in which IgG1 antibodies are involved (Olds and Stavitsky 1986). Mitchell et al. (1991) have discussed similarities and differences between S. mansoni and S. japonicum infections from various aspects of biology, as well as immune responses, and also suggested that differences may exist even among geographical isolates or strains of the latter species.

Another important aspect of granuloma formation is the nature of the cellular components. Infiltration of eosinophils is one of the conspicuous cellular responses observed. IL-5 is essential for granuloma eosinophilia, since treatment of mice with anti-IL-5 monoclonal antibody suppressed the eosinophil response (Cheever et al. 1992b). The granuloma eosinophils make substance P, a cytokine with immunoregulatory properties, which belongs to a family of hormones called tachykinins. This substance may modulate interferon-γ production through interaction with a substance P-like receptor expressed on CD4$^+$ granuloma T lymphocytes (Cook et al. 1994). Other prominent cellular components in granulomas are macrophages and fibroblasts. Macrophages produce macrophage inflammatory protein 1α (MIP-1α), a 6–8 kDa protein which is lipopolysaccharide-inducible and monocyte- and neutrophil-chemotactic, and it has been shown that this protein contributes to cellular recruitment during schistosome egg granuloma formation (Lukacs et al. 1993). Moreover, IL-1, interferon-γ, and IL-10, cytokines found within the granuloma were able to induce significant production of MIP-1 by granuloma fibroblasts, whereas the constitutive expression of monocyte chemoattractant protein-1 (MCP-1) was demonstrated in both unstimulated and cytokine-stimulated granuloma fibroblasts. Interestingly, normal noninflammatory fibroblasts from uninfected mice showed no significant production of MIP-1 or MCP-1 in response to these cytokines. These results suggest that granuloma fibroblasts may be phenotypically altered compared with normal fibroblasts and have a significant role in leukocyte recruitment, granuloma growth, and maintenance of the egg-induced lesions (Lukacs et al. 1994a).

Further studies have demonstrated a crucial role for tumor necrosis factor (TNF) during inflammatory granuloma formation. In addition, TNF has been shown to up-regulate adhesion molecules that participate in cellular recruitment and lymphocyte activation. The mechanism of TNF activation during S. mansoni egg granuloma formation and its relationship to the expression of ICAM-1 have been studied in some detail. Firstly, high affinity human soluble TNF receptor (TNFR) coupled to the Fc portion of an Ig (sTNFR:Fc construct) could effectively diminish granuloma formation and lymphocyte activation in vivo. Second, increased steady state ICAM-1 mRNA expression was observed in primary egg granulomas when compared with normal lung and foreign body (Sephadex bead) granulomas, which suggests a role for ICAM-1 in antigen-induced lesion formation. Subsequent studies have demonstrated that sTNFR:Fc treatment down-regulated granuloma formation and ICAM-1 expression, whereas anti-ICAM-1 decreased SEA-specific T cell proliferation in vitro. In addition, passive immunization of mice with anti-ICAM-1 monoclonal antibody during primary granuloma formation resulted in attenuation of lesion development as compared with lesion development in a control antibody-treated group. The proliferative response to SEA was also significantly reduced in ex vivo experiments that used spleen cells from the anti-ICAM-1 treated mice. These data demonstrate that both TNF and ICAM-1 participate in lymphocyte activation and granuloma formation and suggest that one mechanism of TNF in granuloma development is through TNF-induced ICAM-1 expression (Lukacs et al. 1994b).

In addition to cytokines, it is becoming increasingly clear that antigen presenting cells expressing costimulatory molecules participate in the induction of the Th2 response. Both macrophages and dendritic cells have been shown to have the ability potently to activate T cells during schistosomiasis (Hernandez et al. 1997b; Hayashi et al. 1999; MacDonald et al. 2001), although such a role for B cells remains less clear (Hernandez et al. 1997b; Jankovic et al. 1998). It seems that schistosomes might affect Th2 response development by influencing dendritic cell activation (MacDonald et al. 2001; Sher et al. 2003). Schistosome antigens, in contrast to most microbial pathogens, do not seem to activate DC conventionally. In particular, they fail to stimulate IL-12 production, while also failing to up-regulate the expression of surface markers normally associated with the activation or maturation of dendritic cells (MacDonald et al. 2001; Zaccone et al. 2003). Expression of the costimulatory molecules CD40 (MacDonald et al. 2002b) and OX-40L (de Jong et al. 2002) by dendritic cells exposed to schistosomal antigens seems to be important for Th2 skewing. In relation to this, several costimulatory partnerships (including CD40/CD40L (MacDonald et al. 2002a), B7/CD28 (Hernandez et al. 1999), and B7RP-1/ICOS (Rutitzky et al. 2003)) have been implicated in Th2 development and appropriate granuloma formation during active murine infection with *S. mansoni*.

Thus, it is evident from the observations described above that cell–cell interactions through cytokine production and their responses to these cytokines, including expression of costimulatory and adhesion molecules, are essential for appropriate immune response and egg granuloma development during schistosome infection. In future, it might therefore be possible that immunomodulation, using cytokines or antibodies to influence key molecules involved in response development and granuloma formation, may be applied for clinical use in the prevention or treatment of tissue damage due to schistosome eggs.

IMMUNE RESPONSES

In humans, epidemiological surveys carried out in areas endemic for schistosomiasis mansoni or urinary schistosomiasis have suggested that there is a good correlation between the development of IgE antibodies, resulting from Th2 responses, and age-dependent resistance to reinfection (Hagan et al. 1991; Hagan 1992; Butterworth 1994).

Effector cells

In human and rat systems, eosinophils have been shown to be the effector cells involved in antibody- or complement-dependent damage to various parasites including *S. mansoni* (Mackenzie et al. 1977; Capron et al. 1978; Ramalho-Pinto et al. 1978; Anwar et al. 1979). Similar results have not been obtained during murine infection, perhaps as a result of the fact that murine eosinophils, in contrast to human and rat, express no IgE receptor (Dombrowicz and Capron 2001). In the rat system, Capron et al. (1978) demonstrated that IgG2a is the isotype responsible for promoting the adherence of these cells. Adherence of eosinophils to schistosomula, followed by movement of secretion granules toward the basal region of the cell interacting with the parasite surface, is a critical step for mediating the killing activity, the fusion of granules forming small vacuoles, the contents of which are released on to the surface of schistosomula (McLaren 1980). Using a slow-motion movie camera, these steps were also observed in antibody-dependent cell-mediated cytotoxicity (ADCC) against schistosomula of *S. japonicum* by a human eosinophilic leukemia cell line, EoL-3, when the activity of the cells was enhanced by pretreatment with recombinant TNF (Figure 29.6) (Janecharut et al. 1992). This was consistent with previous results demonstrating that

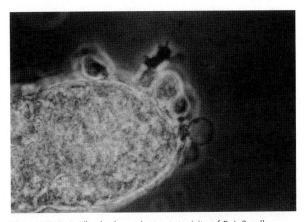

Figure 29.6 *Antibody-dependent cytotoxicity of EoL-3 cells against schistosomula of* Schistosoma japonicum. *An observation with a slow-motion camera, showing that granules and other contents of the cells are concentrated toward the surface of the schistosomula (Janecharut et al. 1992).*

TNF has no direct effect on the parasite, but enhances human eosinophil cytotoxicity to *S. mansoni* larvae in a dose-dependent fashion (Silberstein and David 1986). In another experiment, TNF was reported to exhibit direct toxicity to schistosomula at high concentrations or at lower concentrations in the presence of interferon-γ (James et al. 1990).

Antigens

Paramyosin is a muscle protein of invertebrates, and Pearce et al. (1988) have demonstrated that vaccination of mice with this molecule is effective in induction of resistance to *S. mansoni* infection. When administered intradermally with BCG at total doses of only 4–40 μg per mouse, both the native molecule and a recombinant expression product containing approximately 50 percent of the whole paramyosin were found to confer significant resistance (26–33 percent) against challenge infection, whereas 2 mg of an unfractionated complex soluble worm antigen preparation (SWAP) was required to induce similar levels of protection. In addition, paramyosin was shown to stimulate T lymphocytes from vaccinated mice to produce lymphokines such as interferon-γ that activate macrophages to kill schistosomula. Neither schistosome myosin nor a heterologous paramyosin from a different invertebrate genus was protective, indicating a requirement for specific epitopes in the immunization. That the protection induced by paramyosin involves a T-cell-mediated mechanism was supported by the failure of anti-paramyosin antibodies to transfer significant resistance to infection in recipient mice. Lymphocytes from mice vaccinated with paramyosin were found to produce interferon-γ in response to living schistosomula, suggesting that, during challenge infection of vaccinated hosts, paramyosin (a non-surface antigen according to Pearce et al. (1988), but now known as the molecule either localized in the tegument or even secreted from larval schistosomes) may elicit a protective T-cell response as a consequence of its release from migrating parasite larvae. These results suggest that the induction of T-cell-dependent, cell-mediated immunity against soluble non-surface antigens may be an effective strategy for immunization against multicellular parasites and, in the case of schistosomes, identify paramyosin as a candidate vaccine immunogen (Pearce et al. 1988). However, it is known that paramyosin is recognized by IgG antibodies in sera from patients with chronic schistosomiasis japonica (Kojima et al. 1987) and elevated levels of antibody responses to paramyosin have been reported in *S. mansoni* infection among cured individuals after treatment with praziquantel (Correa-Oliveira et al. 1989). Therefore, although it may be possible to induce an antibody-dependent cell-mediated immunity to human schistosomiasis by immunization with paramyosin, as is shown by resistance to infection to recipient mice

(Kojima et al. 1998), at present, the critical role of induction of Th1 responses in protection to human schistosomiasis cannot be excluded.

GENETIC CONTROL

Genetic segregation analyses carried out on *S. mansoni* infected individuals in the Sudan have associated hepatic fibrosis and portal hypertension with a major gene located in the genetic region 6q22-q23, referred to as *SM2* (Dessein et al. 1999a). A similar approach in Brazil has associated infection intensity with a different major codominant gene located in the genetic region 5q31-q33, termed *SM1* (Marquet et al. 1996). These studies provide compelling evidence that human disease progression and infection intensity may be under distinct major gene control (*SM2* and *SM1*, respectively).

Vaccination

In addition to structural molecules, such as paramyosin, parasite enzymes have also been identified as vaccine candidates (see Figure 29.7 and Table 29.2). The triosephosphate isomerase (TPI) of *S. mansoni* was first described as a 28 kDa target antigen recognized by a monoclonal antibody (M.1) generated from mice immunized with membrane enriched extracts of mechanically transformed schistosomula (Harn et al. 1985). The monoclonal antibody M.1 passively transfers partial resistance (41–49 percent) to cercarial challenge in naive mice. Thus, the 28 kDa antigen recognized by M.1 is a putative vaccine candidate. Purified native 28 kDa antigen from adult parasites was shown to function enzymatically in a manner analogous to yeast and mammalian TPI. Addition of M.1 antibody to the enzyme reaction altered the catalytic activity of schistosome TPI. To determine the immunologic cross-reactivity of this vaccine candidate with mammalian TPI, Western blot analysis was performed and demonstrated that M.1 was immunologically specific for the schistosome enzyme (Harn et al. 1992). Amino acid sequence analysis of the purified native antigen revealed a significant homology to the human glycolytic enzyme, triose-phosphate isomerase (D-glyceraldehyde-3-phosphate ketol-isomerase, EC 5.3.1.1). The complete coding DNA for *S. mansoni* TPI was isolated and it was confirmed that this cDNA encodes the 28 kDa antigen recognized by M.1. The complete cDNA has been expressed within *Escherichia coli* to produce high levels of soluble recombinant *S. mansoni* TPI protein. The product was purified by the M.1 antibody and a functional TPI was obtained with an intrinsic specific activity comparable to that of rabbit and yeast TPI (Shoemaker et al. 1992). Further studies by Reynolds et al. (1994) have demonstrated that schistosome TPI is a potent inducer of IL-2 and interferon-γ production, driving production of these cytokines in the

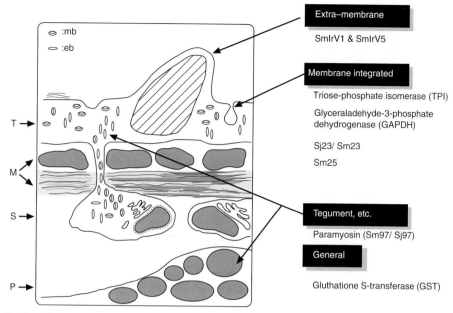

Figure 29.7 *Localization of vaccine candidates for schistosomiasis. M, muscle layers; P, postacetabular gland; S, subtegumental tissues; T, tegument.*

same cell populations of infected animals that have high Th2 responses directed at other SEA. With the goal of synthetic peptide vaccine design, recombinant TPI was used to determine specific T-cell and B-cell epitopes recognized by two strains of mice representing high and moderate responders (C57Bl/6J and CBA/J). All epitopes were selected from non-conserved regions of TPI and were thus parasite-specific. The investigators defined minimal size immunoreactive epitopes and synthesized four-armed multiple antigenic peptides (MAP) consisting of T-cell and B-cell epitopes that could be recognized by both strains of mice in the same molecule. Characterization of the immunoreactivity of the MAP showed that higher antibody recognition of the MAP was attained when the B-cell epitope was placed on the N-termini relative to the T-cell epitope, whereas T-cell immunoreactivity was equivalent in either position. Most interesting was the finding that one of the minimal T-cell epitopes, when incorporated into the MAP, required enlargement to retain immunoreactivity. Finally, both the full-length TPI molecule and the final version of the MAP were found to be immunogenic to T cells in naive animals, and these molecules induced crossrecognition reflected in IL-2 and interferon-γ production (Reynolds et al. 1994).

Table 29.2 *Vaccine candidates for schistosomiasis*

Identity of antigen	Size (kDa)	Stage	Function/localization	Protection (%) (mouse)	Reference
Paramyosin (Sm97)	97	Schistosomula Adult worms	Muscle protein	30	Pearce et al. (1988)
(Sj97)		Schistosomula adult worms cercariae	Muscle, tegument, postacetabular gland	20–60	Kojima et al. (1987, 1998), Nara et al. (1994)
Sj23/Sm23	23	All stages	Tegument		Reynolds et al. (1992)
TPI	28	All stages	Enzyme	30–60	Shoemaker et al. (1992), Reynolds et al. (1994)
GAPDH	37	Schistosomula Adult worms	Enzyme		Dessein et al. (1988)
GST (Sj26/Sj28) (Sm26/Sm28)	26/28	Schistosomula Adult worms	Enzyme	30–70	Balloul et al. (1987b)
IrV-5	200	Schistosomula	Muscle	32–75	Soisson et al. (1992)
IrV-1	90	All stages	Tegument		Hawn et al. (1993)
FABP (Sm14)	14	Schistosomula	Tegument	67	Tendler et al. (1996)

Glyceraldehyde-3-phosphate dehydrogenase (GAPDH) was first identified as a 37 kDa molecule of schistosomula recognized by Brazilian children resistant to *S. mansoni* infection (Dessein et al. 1988). The cDNA for this antigen was cloned by screening a schistosome cDNA expression library with antibodies against the purified protein. The amino acid sequence of the encoded polypeptide shows 72.5 percent positional identity with human GAPDH. Antibodies against the recombinant protein identified the 37 kDa molecule on the larvae (Goudot-Crozel et al. 1989). Since a number of conserved proteins have been found to be major targets of host-protective immunity against *S. mansoni*, it is interesting to note, as suggested by these authors, that genetic restriction of the immune response to these antigens may occur in heterogeneous human populations because of the limited number of T-cell epitopes carried by these host-like proteins and that such genetic effects might allow parasite transmission through nonresponder (susceptible) individuals.

Other vaccine candidates of an enzymatic nature are the glutathione *S*-transferases (GST), which consist of at least two isoenzymes of 26 and 28 kDa for both *S. japonicum* and *S. mansoni* (Sj26, Sj28, Sm26, and Sm28) (Tiu et al. 1988). Immunization of rats and hamsters with Sm28 resulted in significant protection against a natural challenge infection with live cercariae (Balloul et al. 1987a). Moreover, vaccination of Fisher rats and BALB/c mice with purified 28 kDa protein resulted in a marked decrease (up to 70 percent) in the parasite burden in both experimental infection models, and the antibody raised against this protein was able to kill *S. mansoni* larvae in the eosinophil-mediated cytotoxicity assay (Balloul et al. 1987b). *S. haematobium* GST has successfully passed the industrial scale-up and safety testing and is now undergoing phase II clinical trials (Capron et al. 2001); phase III field trials of this vaccine candidate are being planned (Bergquist et al. 2002).

An interesting molecule has been identified as a schistosome vaccine candidate by Strand and colleagues (Soisson et al. 1992). Mice exposed to radiation-attenuated cercariae of *S. mansoni* were highly resistant to challenge infection, and sera from these mice could confer partial resistance when transferred to naive recipients. These sera recognized antigens present in schistosomula and adult worms, among them particularly a 200 kDa antigen. A cDNA encoding a 62 kDa portion of this antigen was cloned and the deduced amino acid sequence of this cDNA clone was found to share homology with myosins of other species. To assess the immunoprophylactic potential, vaccination trials were carried out in mice using the recombinant polypeptide expressed as a fusion protein with β-galactosidase presented in the form of proteosome complexes with the outer membrane protein of meningococcus. The level of protection achieved was 32 percent, and this could be increased to 75 percent by removal of those amino acids included in the fusion protein that were derived from the vector to yield a polypeptide, designated rIrV-5. A similar level of protection was achieved when mice were immunized with the same dose of rIrV-5 in the form of protein complexes but without outer membrane protein, suggesting that protection did not require the use of adjuvant. However, at least three immunizations were necessary to achieve protection. Using monoclonal antibodies and sera from mice vaccinated with rIrV-5, the native protein recognized by antibodies against rIrV-5 was found to be a 200 kDa protein that was expressed on the surface of newly transformed schistosomula (Soisson et al. 1992). A further vaccination trial has been carried out in baboons with the rIrV-5 or radiation-attenuated cercariae. rIrV-5 was presented either in the form of protein micelles or complexed with the outer membrane protein of meningococcus to form proteosomes. The level of protection achieved in these groups ranged from 0 to 54 percent, with a mean of 27.7 percent, whereas in baboons exposed to radiation-attenuated cercariae the level of protection was very high, with a mean of 84 percent. The resistance observed after vaccination with rIrV-5 or radiation-attenuated cercariae was reflected in the overall histopathology. Vaccination of baboons with rIrV-5 or radiation-attenuated cercariae elicited an antibody response against epitopes exposed on the surface of newly transformed schistosomula. Analysis of individual baboon sera by enzyme linked immunosorbent assay (ELISA) demonstrated that there was a direct correlation between the anti-rIrV-5 titer and resistance to challenge worm burden, suggesting that the immunoprotective mechanism is antibody-dependent (Soisson et al. 1993).

Furthermore, in relation to SmIrV5, Hawn et al. (1993) have carried out molecular cloning and sequencing of SmIrV1, another candidate antigen recognized by immune sera raised in mice vaccinated with irradiated cercariae. SmIrV1 contained a deduced amino acid sequence of 582 residues with similarity to three proteins: calnexin, calreticulin, and OvRal1 (a surface antigen of the filarial nematode *Onchocerca volvulus*). SmIrV1 was divided into three regions: a neutral N-terminal region with a putative signal sequence, followed by a proline- and tryptophan-rich P region in which two sets of sequences are repeated four times and a C-terminal region which is highly acidic with an isoelectric point of 4.7. The P and C regions of SmIrV1 reacted with sera of immunized, as well as chronically infected, mice.

Immunoprecipitation studies with antibodies raised against a portion of recombinant IrV1 demonstrated its presence in cercariae, schistosomula, and adult worms with an apparent molecular mass on SDS-PAGE of 90 kDa. There was an approximate six-fold increase in protein expression level during the transformation from cercariae to schistosomula. Consistent with a potential role as a molecular chaperone, IrV1 was associated with several metabolically labeled proteins in co-immunoprecipitation studies with the adult worm tegumental fraction. Similar to calnexin, IrV1 was metabolically labeled with phosphorus-32 on serine and threonine residues in adult worms and was one of the major phosphoproteins of this stage. This phosphorylation was developmentally regulated and coincided with the transformation of cercariae into schistosomula. The localization was also stage-specific, as IrV1 was transported from internal regions of cercariae to the outer tegumental layer of schistosomula (Hawn and Strand 1994). Cloning of a cDNA encoding SjIrV1 has also been carried out and shown that the molecule was 83 percent identical to SmIrV1 at the predicted amino acid level. Recombinant, full-length SjIrV1 was expressed with a hexahistidine tag in *E. coli* and the recombinant protein isolated by nickel-chelate chromatography. The recombinant molecule was shown to exhibit calcium-dependent, differential electrophoretic migration and to bind ruthenium red in the absence, but not in the presence, of calcium ions. That SjIrV1 was a functional calcium-binding protein was confirmed by the presence of conserved Ca^{2+}-binding motifs predicted from the primary sequence, together with the Ca^{2+}-dependent electrophoretic mobility of recombinant SjIrV1 (Hooker and Brindley 1999).

A new approach has been carried out to identify a vaccine candidate (Jankovic et al. 1996). A CD4[+] clone (clone B) was established from C57BL/6 mice protectively immunized with SWAP plus BCG and this T cell clone was characterized as Th1, based on its selective production of interferon-γ and IL-2. Transfer of the clone to syngeneic mice could activate antigen-elicited peritoneal macrophages to kill schistosomula in vitro. Recipients of the clone also displayed siginificant resistance against cercarial challenge. By screening a battery of λgt11 clones from an adult worm cDNA library, one recombinant (25B) was identified that stimulated clone B specifically. Analysis of the 25B cDNA insert revealed a nucleotide sequence identical with that of the large subunit of schistosome calpain, a Ca^{2+}-activated neutral proteinase. By expressing the products of polymerase chain reaction (PCR) subcloning, the authors identified a 146-amino acid region of the 25B gene containing immunologic activity equivalent to the whole polypeptide. Overlapping peptides spanning this region were synthesized, and a core epitope was identified with sequence EWKGAWCDGS. Since clone B responded to supernatants from cultured schistosomula, it was postulated that the recognition of calpain released by invading larvae and the resulting induction of Th1 cytokines accounted for the protection mediated by the adoptively transferred clone. Thus, calpain was implicated as a target of protective immunity in schistosomes. This report provided the first example of a candidate vaccine antigen for this parasite identified on the basis of T cell reactivity.

A fatty acid binding protein (Sm14) was shown to form the basis of the protective immune crossreactivity between the parasitic trematode worms *Fasciola hepatica* and *S. mansoni* (Tendler et al. 1995). A recombinant form of the *S. mansoni* antigen, rSm14, protected outbred Swiss mice by up to 67 percent against challenge with *S. mansoni* cercariae in the absence of adjuvant and without provoking any observable autoimmune response. The same antigen also provided complete protection against challenge with *F. hepatica* metacercariae in the same animal model. The results suggest that it is possible to produce a single vaccine that would be effective against at least two parasites, *F. hepatica* and *S. mansoni*, which are of veterinary and human importance, respectively (Tendler et al. 1996). Moreover, this vaccine candidate (recombinant Sm14) has been shown to induce proliferative responses of peripheral blood mononuclear cells (PBMC) of individuals living in an endemic area for schistsosomiasis (Brito et al. 2000). The highest proliferation index to rSm14 was detected in uninfected endemic normal (EN) individuals who are naturally resistant to schistosomiasis. Regarding the cytokine produced, the levels of Th2 cytokines IL-5 and IL-10 were not statistically different between chronic, treated patients and EN individuals. In contrast, IFN-γ and TNF-α were produced in significantly higher amounts by PBMC of EN individuals following rSm14 stimulation. Flow cytometry demonstrated that CD4[+] T cells from these individuals were the main lymphocyte subpopulation producing IFN-γ and TNF-α. Exogenous rIL-10 suppressed T-cell proliferation, neutralization of endogenous IL-10 with a monoclonal antibody to this cytokine restored lymphocyte activation, and enhanced IFN-γ and TNF-α production in chronically infected patients. In contrast, the addition of anti-IFN-γ totally abrogated the PMNC proliferation within the EN group. These results suggest that Th1 type of immune response induced in EN individuals to a specific schistosome antigen might be associated with resistance to infection and also highlighted the importance of Sm14 as a potential vaccine candidate (Brito et al. 2000).

DIAGNOSIS

In addition to the use of classical parasitological techniques for detection of eggs in urine (*S. haematobium*) or stools (*S. mansoni*), ELISAs are widely used for diagnosis of infection. ELISAs for the detection of circulating anodic antigen (CAA) and circulating cathodic antigen (CCA) in serum and urine have been developed and applied as an epidemiologic tool in a recent, intense focus of *S. mansoni* in Senegal (Polman et al. 1995). CAA and CCA in serum and CCA in urine were found in 94, 83, and 95 percent, respectively, of the population, of which 91 percent were positive on stool examination. Circulating antigens were also detectable in sera and urine of most egg-negative individuals. The sensitivies of the urine CCA and serum CAA ELISA were substantially higher than that of a single egg count, and increased with egg output. The CAA and CCA levels correlated well with egg counts and with each other. The age-related evolution of antigen levels followed a similar pattern to egg counts, providing supplementary evidence for a genuine reduction of worm burden in adults, in spite of the supposed absence of acquired immunity in this recently exposed community.

CHEMOTHERAPY, PREVENTION, AND CONTROL

Praziquantel is the drug of choice, being safe and effective against all of the major human-infective schistosomes (Harder 2002). Further, there is currently little evidence of emergence of drug resistant parasites against this treatment, although this of course remains a potential future problem (Kusel and Hagan 1999). A single oral dose of praziquantel (40 mg/kg body weight) is prescribed for the treatment of *S. mansoni* and *S. haematobium*. Control programs based on praziquantel treatment to target the parasites, coupled with use of chemical molluskicides to control snail populations, have been successful in regions such as South America, Egypt, and the Phillipines (Engels et al. 2002; Fenwick et al. 2003). However, in areas of socio-economic deprivation, these programs have faltered and proved to be difficult to sustain, even though the cost of praziquantel has significantly reduced in recent years. New initiatives, aimed primarily at providing resources to enable sustained control strategies to be implemented in the most needy areas (particularly sub-Saharan Africa) are currently underway (Engels et al. 2002; Fenwick et al. 2003). Prior to the discovery of praziquantel, oxamniquine was used to treat *S. mansoni*, and metrifonate for *S. haematobium*. Oxamniquine is a miracil derivative that has an anticholinergic effect, which is thought to increase parasite motility and inhibit nucleic acid synthesis. It does not appear to affect any of the other schistosome species that cause human disease. Metrifonate is an organophosphate, the mode of action of which may be to target acetylcholinesterase on the parasite surface (Camacho and Agnew 1995). Neither oxamniquine nor metrifonate are currently in widespread use, probably due to a combination of their species-specificity, toxicity (in comparison to praziquantel), lack of availability, and the fact that they are less effective than praziquantel (Ferrari et al. 2003). As has been found for *S. japonicum* (see Chapter 30, Schistosomes: Asian), there is potential in the use of artemether, a derivative of the antimalarial artemesinin, for the treatment of *S. mansoni* and *S. haematobium* infection (Utzinger et al. 2001). It is thought that artemether acts primarily on the juvenile stages of schistosomes. This limitation, together with potential long-term toxicity and the possibility that treat-ment in areas where schistosomes and malaria are co-endemic might select for drug-resistant malaria parasites, somewhat restricts its potential for schistosome control, particularly in sub-Saharan Africa.

REFERENCES

Alger, H.M., Sayed, A.A., et al. 2002. Molecular and enzymatic characterisation of *Schistosoma mansoni* thioredoxin. *Int J Parasitol*, **32**, 1285–92.

Anwar, A.R.E., Smithers, S.R. and Kay, A.B. 1979. Killing of schistosomula of *Schistosoma mansoni* coated with antibody and/or complement in preferential killing by eosinophil. *J Immunol*, **122**, 628–37.

Araujo, M.I., Lopes, A.A., et al. 2000. Inverse association between skin response to aeroallergens and *Schistosoma mansoni* infection. *Int Arch Allergy Immunol*, **123**, 145–8.

Asahi, H., Hernandez, H.J. and Stadecker, M.J. 1999. A novel 62-kilodalton egg antigen from *Schistosoma mansoni* induces a potent CD4+ T helper cell response in the C57BL/6 mouse. *Infect Immun*, **67**, 1729–35.

Asahi, H., Osman, A., et al. 2000. *Schistosoma mansoni* phosphoenolpyruvate carboxykinase, a novel egg antigen: immunological properties of the recombinant protein and identification of a T-cell epitope. *Infect Immun*, **68**, 3385–93.

Balloul, J.M., Sondermeyer, P., et al. 1987a. Molecular cloning of a protective antigen of schistosomes. *Nature (Lond)*, **326**, 149–53.

Balloul, J.M., Grzych, J.M., et al. 1987b. A purified 28,000 dalton protein from *Schistosoma mansoni* adult worms protects rats and mice against experimental schistosomiasis. *J Immunol*, **138**, 3448–53.

Bergquist, N.R., Al-Sherbiny, M., et al. 2002. Blueprint for schistosomiasis vaccine development. *Acta Trop*, **82**, 183–92.

Brito, C.F., Caldas, I.R., et al. 2000. CD4+ T cells of schistosomiasis naturally resistant individuals living in an endemic area produce interferon-γ and tumour necrosis factor-α in response to the recombinant 14 kDa *Schistosoma mansoni* fatty acid-binding protein. *Scand J Immunol*, **51**, 595–601.

Brunet, L.R., Finkelman, F.D., et al. 1997. IL-4 protects against TNF-alpha-mediated cachexia and death during acute schistosomiasis. *J Immunol*, **159**, 777–85.

Butterworth, A.E. 1994. Human immunity to schistosomes: some questions. *Parasitol Today*, **10**, 378–80.

Camacho, M. and Agnew, A. 1995. *Schistosoma*: rate of glucose import is altered by acetylcholine interaction with tegumental acetylcholine receptors and acetylcholinesterase. *Exp Parasitol*, **81**, 584–91.

Capron, M., Capron, A., et al. 1978. Eosinophil-dependent cytotoxicity in rat schistosomiasis. Involvement of IgG2a antibody and role of mast cells. *Eur J Immunol*, **8**, 127–33.

Capron, A., Capron, M., et al. 2001. Vaccine strategies against schistosomiasis: from concepts to clinical trials. *Int Arch Allergy Immunol*, **124**, 9–15.

Cheever, A., Finkelman, F.D., et al. 1992a. Treatment with anti-IL-2 antibodies reduces hepatic pathology and eosinophilia in *Schistosoma mansoni*-infected mice while selectively inhibiting T cell IL-5 production. *J Immunol*, **148**, 3244–8.

Cheever, A.W., Xu, Y., et al. 1992b. The role of cytokines in the pathogenesis of hepatic granulomatous disease in *Schistosoma mansoni* infected mice. *Mem Inst Oswaldo Cruz*, **4**, 81–5.

Cheever, A.W., Williams, M.E., et al. 1994. Anti-IL-4 treatment of *Schistosoma mansoni*-infected mice inhibits development of T cells and non-B, non-T cells expressing Th2 cytokines while decreasing egg-induced hepatic fibrosis. *J Immunol*, **153**, 753–9.

Cheever, A.W., Finkelman, F.D. and Cox, T.M. 1995. Anti-interleukin-4 treatment diminishes secretion of Th2 cytokines and inhibits hepatic fibrosis in murine schistosomiasis japonica. *Parasite Immunol*, **17**, 103–9.

Chensue, S.W., Terebuh, P.D., et al. 1992. Role of IL-4 and IFN-γ in *Schistosoma mansoni* egg-induced hypersensitivity granuloma formation. Orchestration, relative contribution, and relationship to macrophage function. *J Immunol*, **148**, 900–6.

Chiaramonte, M.G., Donaldson, D.D., et al. 1999. An IL-13 inhibitor blocks the development of hepatic fibrosis during a T-helper type 2-dominated inflammatory response. *J Clin Invest*, **104**, 777–85.

Chikunguwo, S.M., Quinn, J.J., et al. 1993. The cell-mediated response to schistosomal antigens at the clonal level. III. Identification of soluble egg antigens recognized by cloned specific granulomagenic murine CD4+ Th1-type lymphocytes. *J Immunol*, **150**, 1413–21.

Contis, G. and David, A.R. 1996. The epidemiology of bilharzia in ancient Egypt: 5000 years of schistosomiasis. *Parasitol Today*, **12**, 253–5.

Cook, G.A., Elliott, D., et al. 1994. Molecular evidence that granuloma T lymphocytes in murine schistosomisis mansoni express an authentic substance P (NK-1) receptor. *J Immunol*, **152**, 1830–5.

Correa-Oliveira, R., Pearce, E.J., et al. 1989. The human immune response to defined immunogens of *Schistosoma mansoni*: Elevated antibody levels to paramyosin in stool-negative individuals from two endemic areas in Brazil. *Trans R Soc Trop Med Hyg*, **83**, 798–804.

de Jong, E.C., Vieira, P.L., et al. 2002. Microbial compounds selectively induce Th1 cell-promoting or Th2 cell-promoting dendritic cells in vitro with diverse Th cell-polarizing signals. *J Immunol*, **168**, 1704–9.

Dessein, A.J., Begley, M., et al. 1988. Human resistance to *Schistosoma mansoni* is associated with IgG reactivity to a 37-kDa larval surface antigen. *J Immunol*, **140**, 2727–36.

Dessein, A.J., Hillaire, D., et al. 1999a. Severe hepatic fibrosis in *Schistosoma mansoni* infection is controlled by a major locus that is closely linked to the interferon-gamma receptor gene. *Am J Hum Genet*, **65**, 709–21.

Dessein, A.J., Marquet, S., et al. 1999b. Infection and disease in human schistosomiasis mansoni are under distinct major gene control. *Microbes Infect*, **1**, 561–7.

Dombrowicz, D. and Capron, M. 2001. Eosinophils, allergy and parasites. *Curr Opin Immunol*, **13**, 716–20.

Engels, D., Chitsulo, L., et al. 2002. The global epidemiological situation of schistosomiasis and new approaches to control and research. *Acta Trop*, **82**, 139–46.

Falcone, F.H., Dahinden, C.A., et al. 1996. Human basophils release interleukin-4 after stimulation with *Schistosoma mansoni* egg antigen. *Eur J Immunol*, **26**, 1147–55.

Fallon, P.G., Richardson, E.J., et al. 2000. Schistosome infection of transgenic mice defines distinct and contrasting pathogenic roles for IL-4 and IL-13: IL-13 is a profibrotic agent. *J Immunol*, **164**, 2585–91.

Faveeuw, C., Angeli, V., et al. 2002. Antigen presentation by CD1d contributes to the amplification of Th2 responses to *Schistosoma mansoni* glycoconjugates in mice. *J Immunol*, **169**, 906–12.

Feldmeier, H., Krantz, I. and Poggensee, G. 1994. Female genital schistosomiasis as a risk-factor for the transmission of HIV. *Int J Std Aids*, **5**, 368–72.

Feldmeier, H., Poggensee, G., et al. 1995. Female genital schistosomiasis. New challenges from a gender perspective. *Trop Geogr Med*, **47**, Suppl. 2, 2–15.

Fenwick, A., Savioli, L., et al. 2003. Drugs for the control of parasitic diseases: current status and development in schistosomiasis. *Trends Parasitol*, **19**, 509–15.

Ferrari, M.L., Coelho, P.M., et al. 2003. Efficacy of oxamniquine and praziquantel in the treatment of *Schistosoma mansoni* infection: a controlled trial. *Bull WHO*, **81**, 3, 190–6.

Flores-Villanueva, P.O., Reiser, H. and Stadecker, M.J. 1994. Regulation of T helper cell responses in experimental murine schistosomiasis by IL-10. Effect on expression of B7 and B7-2 costimulatory molecules by macrophages. *J Immunol*, **153**, 5190–9.

Flores-Villanueva, P.O., Zheng, X.X., et al. 1996. Recombinant IL-10 and IL-10/Fc treatment down-regulate egg antigen-specific delayed hypersensitivity reactions and egg granuloma formation in schistosomiasis. *J Immunol*, **156**, 3315–20.

Goudot-Crozel, V., Caillol, D., et al. 1989. The major parasite surface antigen associated with human resistance to schistosomiasis is a 37-kD glyceraldehyde-3P-dehydrogenase. *J Exp Med*, **170**, 2065–80.

Grzych, J.M., Pearce, E., et al. 1991. Egg deposition is the major stimulus for the production of Th2 cytokines in murine schistosomiasis mansoni. *J Immunol*, **146**, 1322–7.

Hagan, P. 1992. Reinfection, exposure and immunity in human schistosomiasis. *Parasitol Today*, **8**, 12–16.

Hagan, P., Blumenthal, U.J., et al. 1991. Human IgE, IgG4 and resistance to reinfection with *Schistosoma haematobium*. *Nature (Lond)*, **349**, 243–5.

Harder, A. 2002. Chemotherapeutic approaches to schistosomes: current knowledge and outlook. *Parasitol Res*, **88**, 395–7.

Harn, D.A., Mitsuyama, M., et al. 1985. Identification by monoclonal antibody of a major (28 kDa) surface membrane antigen of *Schistosoma mansoni*. *Mol Biochem Parasitol*, **16**, 345–54.

Harn, D.A., Gu, W., et al. 1992. A protective monoclonal antibody specifically recognizes and alters the catalytic activity of schistosome triose-phosphate isomerase. *J Immunol*, **148**, 562–7.

Hawn, T.R. and Strand, M. 1994. Developmentally regulated localization and phosphorylation of SmIrV1, a *Schistosoma mansoni* antigen with similarity to calnexin. *J Biol Chem*, **269**, 20083–9.

Hawn, T.R., Tom, T.D. and Strand, M. 1993. Molecular cloning and expression of SmIrV1, a *Schistosoma mansoni* antigen with similarity to calnexin, calreticulin and OvRal1. *J Biol Chem*, **268**, 7692–8.

Hayashi, N., Matsui, K., et al. 1999. Kupffer cells from *Schistosoma mansoni*-infected mice participate in the prompt type 2 differentiation of hepatic T cells in response to worm antigens. *J Immunol*, **163**, 6702–11.

Henderson, G.S., Conary, J.T., et al. 1991. In vivo molecular analysis of lymphokines involved in the murine immune response during *Schistosoma mansoni* infection. I. IL-4 mRNA, not IL-2 mRNA, is abundant in the granulomatous livers, mesenteric lymph nodes, and spleens of infected mice. *J Immunol*, **147**, 992–7.

Hernandez, H.J. and Stadecker, M.J. 1999. Elucidation and role of critical residues of immunodominant peptide associated with T cell-mediated parasitic disease. *J Immunol*, **163**, 3877–82.

Hernandez, H.J., Wang, Y., et al. 1997a. Expression of class II, but not class I, major histocompatibility complex molecules is required for granuloma formation in infection with *Schistosoma mansoni*. *Eur J Immunol*, **27**, 1170–6.

Hernandez, H.J., Wang, Y., et al. 1997b. In infection with *Schistosoma mansoni*, B cells are required for T helper type 2 cell responses but not for granuloma formation. *J Immunol*, **158**, 4832–7.

Hernandez, H.J., Sharpe, A.H., et al. 1999. Experimental murine schistosomiasis in the absence of B7 costimulatory molecules: reversal of elicited T cell cytokine profile and partial inhibition of egg granuloma formation. *J Immunol*, **162**, 2884–9.

Hesse, M., Modolell, M., et al. 2001. Differential regulation of nitric oxide synthase-2 and arginase-1 by type 1/type 2 cytokines in vivo: granulomatous pathology is shaped by the pattern of L-arginine metabolism. *J Immunol*, **167**, 6533–44.

Hoffmann, K.F., Cheever, A.W., et al. 2000. IL-10 and the dangers of immune polarization: excessive type 1 and type 2 cytokine responses induce distinct forms of lethal immunopathology in murine schistosomiasis. *J Immunol*, **164**, 6406–16.

Hoffmann, K.F., Wynn, T.A., et al. 2002. Cytokine-mediated host responses during schistosome infections; walking the fine line between immunological control and immunopathology. *Adv Parasitol*, **52**, 265–307.

Hooker, C.W. and Brindley, P.J. 1999. Cloning of a cDNA encoding SjIrV1, a *Schistosoma japonicum* calcium-binding protein similar to calnexin, and expression of the recombinant protein in *Escherichia coli*. *Biochim Biophys Acta*, **1429**, 331–41.

James, S.L., Glaven, J., et al. 1990. Tumour necrosis factor (TNF) as a mediator of macrophage helminthotoxic activity. *Parasite Immunol*, **12**, 1–13.

Janecharut, T., Hata, H., et al. 1992. Effects of recombinant tumour necrosis factor on antibody-dependent eosinophil-mediated damage to *Schistosoma japonicum* larvae. *Parasite Immunol*, **14**, 605–16.

Jankovic, D., Aslund, L., et al. 1996. Calpain is the target antigen of a Th1 clone that transfers protective immunity against *Schistosoma mansoni*. *J Immunol*, **157**, 806–14.

Jankovic, D., Cheever, A.W., et al. 1998. CD4[+] T cell-mediated granulomatous pathology in schistosomiasis is downregulated by a B cell-dependent mechanism requiring Fc receptor signaling. *J Exp Med*, **187**, 619–29.

Kojima, S., Niimura, M. and Kanazawa, T. 1987. Production and properties of a mouse monoclonal IgE antibody to *Schistosoma japonicum*. *J Immunol*, **139**, 2044–9.

Kojima, S., Nara, T., et al. 1998. A vaccine trial for controlling reservoir livestock against schistosomiasis japonica. In Tada, I., Kojima, S. and Tsuji, M. (eds), *Proceedings of the 9th International Congress of Parasitology*. Bologna: Monduzzi Editore, 489–94.

Kusel, J. and Hagan, P. 1999. Praziquantel – its use, cost and possible development of resistance. *Parasitol Today*, **15**, 352–4.

La Flamme, A.C. and Pearce, E.J. 1999. The absence of IL-6 does not affect Th2 cell development in vivo, but does lead to impaired proliferation, IL-2 receptor expression and B cell responses. *J Immunol*, **162**, 5829–37.

La Flamme, A.C., MacDonald, A.S., et al. 2000. Role of IL-6 in directing the initial immune response to schistosome eggs. *J Immunol*, **164**, 2419–26.

Lukacs, N.W., Kunkel, S.L., et al. 1993. The role of macrophage inflammatory protein 1 α in *Schistosoma mansoni* egg-induced granulomatous inflammation. *J Exp Med*, **177**, 1551–9.

Lukacs, N.W., Chensue, S.W., et al. 1994a. Production of monocyte chemoattractant protein-1 and macrophage inflammatory protein-1α by inflammatory granuloma fibroblasts. *Am J Pathol*, **144**, 711–18.

Lukacs, N.W., Chensue, S.W., et al. 1994b. Inflammatory granuloma formation is mediated by TNF-α inducible intercellular adhesion molecule-1. *J Immunol*, **152**, 5883–9.

MacDonald, A.S., Straw, A.D., et al. 2001. CD8-dendritic cell activation status plays an integral role in influencing Th2 response development. *J Immunol*, **167**, 1982–8.

MacDonald, A.S., Patton, E.A., et al. 2002a. Impaired Th2 development and increased mortality during *Schistosoma mansoni* infection in the absence of CD40/CD154 interaction. *J Immunol*, **168**, 4643–9.

MacDonald, A.S., Straw, A.D., et al. 2002b. Cutting edge: Th2 response induction by dendritic cells: a role for CD40. *J Immunol*, **168**, 537–40.

Mackenzie, C.D., Ramalho-Pinto, F.J., et al. 1977. Antibody-mediated adherence of rat eosinophils to schistosomula of *Schistosoma mansoni* in vitro. *Clin Exp Immunol*, **30**, 97–104.

Marquet, S., Abel, L., et al. 1996. Genetic localization of a locus controlling the intensity of infection by *Schistosoma mansoni* on chromosome 5q31-q33. *Nat Genet*, **14**, 181–4.

McLaren, D.J. 1980. *Schistosoma mansoni: the parasite surface in relation to host immunity*. Chichester: Research Studies Press, 1–229.

Mitchell, G.F., Tiu, W.U. and Garcia, E.G. 1991. Infection characteristics of *Schistosoma japonicum* in mice and relevance to the assessment of schistosome vaccines. *Adv Parasitol*, **30**, 167–200.

Nara, T., Matsumoto, N., et al. 1994. Demonstration of the target molecule of a protective IgE antibody in secretory glands of *Schistosoma japonicum* larvae. *Int Immunol*, **6**, 963–71.

Nene, V., Dunne, D.W., et al. 1986. Sequence and expression of a major egg antigen from *Schistosoma mansoni*. Homologies to heat shock proteins and alph-crystallins. *Mol Biochem Parasitol*, **21**, 179–88.

Okano, M., Satoskar, A.R., et al. 1999. Induction of Th2 responses and IgE is largely due to carbohydrates functioning as adjuvants on *Schistosoma mansoni* egg antigens. *J Immunol*, **163**, 6712–17.

Olds, G.R. and Stavitsky, A.B. 1986. Mechanisms of in vivo modulation of granulomatous inflammation in murine schistosomiasis japonica. *Infect Immun*, **52**, 513–18.

Oliveira, G. and Johnston, D.A. 2001. Mining the schistosome DNA sequence database. *Trends Parasitol*, **17**, 501–3.

Oswald, I.P., Gazzinelli, R.T., et al. 1992. IL-10 synergizes with IL-4 and transforming growth factor-β to inhibit macrophage cytotoxic activity. *J Immunol*, **148**, 3578–82.

Paraense, W.L. 2001. The schistosome vectors in the Americas. *Mem Inst Oswaldo Cruz*, **96**, Suppl., 7–16.

Pearce, E.J. and MacDonald, A.S. 2002. The immunobiology of schistosomiasis. *Nat Rev Immunol*, **2**, 499–511.

Pearce, E.J., Casper, P., et al. 1991. Downregulation of Th1 cytokine production accompanies induction of Th2 responses by a parasitic helminth, *Schistosoma mansoni*. *J Exp Med*, **173**, 159–66.

Pearce, E.J., James, S.L., et al. 1988. Induction of protective immunity against *Schistosoma mansoni* by vaccination with schistosome paramyosin (Sm97), a nonsurface parasite antigen. *Proc Natl Acad Sci USA*, **85**, 5678–82.

Pearce, E.J., Cheever, A., et al. 1996. *Schistosoma mansoni* in IL-4-deficient mice. *Int J Immunol*, **8**, 435–44.

Polman, K., Stelma, F.F., et al. 1995. Epidemiologic application of circulating antigen detection in a recent *Schistosoma mansoni* focus in northern Senegal. *Am J Trop Med Hyg*, **53**, 152–7.

Ramalho-Pinto, F.J., McLaren, D.J. and Smithers, S.R. 1978. Complement-mediated killing of schistosomula of *Schistosoma mansoni* by rat eosinophils in vitro. *J Exp Med*, **147**, 147–56.

Reiner, S.L. 2001. Helper T cell differentiation, inside and out. *Curr Opin Immunol*, **13**, 351–5.

Reynolds, S.R., Shoemaker, C.B. and Harn, D.A. 1992. T and B cell epitope mapping of Sm23, an integral membrane protein of *Schistosoma mansoni*. *J Immunol*, **149**, 3995–4001.

Reynolds, S.R., Dahl, C.E. and Harn, D.A. 1994. T and B epitope determination and analysis of multiple antigenic peptides for the *Schistosoma mansoni* experimental vaccine triose-phosphate isomerase. *J Immunol*, **152**, 193–200.

Rollinson, D., Kaukas, A., et al. 1997. Some molecular insights into schistosome evolution. *Int J Parasitol*, **27**, 11–28.

Rutitzky, L.I., Ozkaynak, E., et al. 2003. Disruption of the ICOS-B7RP-1 costimulatory pathway leads to enhanced hepatic immunopathology and increased gamma interferon production by CD4 T cells in murine schistosomiasis. *Infect Immun*, **71**, 4040–4.

Rutitzky, L.I., Hernandez, H.J. and Stadecker, M.J. 2001. Th1-polarizing immunization with egg antigens correlates with severe exacerbation of immunopathology and death in schistosome infection. *Proc Natl Acad Sci USA*, **98**, 13243–8.

Sabin, E.A., Kopf, M.A., et al. 1996. *Schistosoma mansoni* egg-induced early IL-4 production is dependent upon IL-5 and eosinophils. *J Exp Med*, **184**, 1871–8.

Schmitz, J., Thiel, A., et al. 1994. Induction of interleukin 4 (IL-4) expression in T helper (Th) cells is not dependent on IL-4 from non-Th cells. *J Exp Med*, **179**, 1349–53.

Sher, A., Fiorentino, D., et al. 1991. Production of IL-10 by CD4[+] T lymphocytes correlates with down-regulation of Th1 cytokine synthesis in helmith infection. *J Immunol*, **147**, 2713–16.

Sher, A., Pearce, E., et al. 2003. Shaping the immune response to parasites: role of dendritic cells. *Curr Opin Immunol*, **15**, 421–9.

Shoemaker, C., Gross, A., et al. 1992. cDNA cloning and functional expression of the *Schistosoma mansoni* protective antigen triose-phosphate isomerase. *Proc Natl Acad Sci USA*, **89**, 1842–6.

Silberstein, D.S. and David, J.R. 1986. Tumor necrosis factor enhances eosinophil toxicity to *Schistosoma mansoni* larvae. *Proc Natl Acad Sci USA*, **83**, 1055–9.

Soisson, L.M., Masterson, P., et al. 1992. Induction of protective immunity in mice using a 62-kDa recombinant fragment of a *Schistosoma mansoni* surface antigen. *J Immunol*, **149**, 3612–20.

Soisson, L.A., Reid, G.D., et al. 1993. Protective immunity in baboons vaccinated with a recombinant antigen or radiation-attenuated cercariae of *Schistosoma mansoni* is antibody-dependent. *J Immunol*, **151**, 4782–9.

Sornasse, T., Larenas, P.V., et al. 1996. Differentiation and stability of T helper 1 and 2 cells derived from naive human neonatal CD4+ T cells, analyzed at the single-cell level. *J Exp Med*, **184**, 473–83.

Stadecker, M.J., Hernandez, H.J. and Asahi, H. 2001. The identification and characterization of new immunogenic egg components: implications for evaluation and control of the immunopathogenic T cell response in schistosomiasis. *Mem Inst Oswald Cruz*, **96**, Suppl., 29–33.

Sturrock, R.F. 1993. The intermediate hosts and host-parasite relationships. In: Jordan, P., Webbe, G. and Sturrock, R.F. (eds), *Human schistosomiasis*. Oxon: CAB International, 33–85.

Tchuente, L.A.T., Southgate, V.R., et al. 2003. *Schistosoma intercalatum*: an endangered species in Cameroon? *Trends Parasitol*, **19**, 389–93.

Tendler, M., Brito, C.A., et al. 1996. A *Schistosoma mansoni* fatty acid-binding protein, Sm14, is the potential basis of a dual-purpose anti-helminth vaccine. *Proc Natl Acad Sci USA*, **93**, 269–73.

Tiu, W.U., Davern, K.M., et al. 1988. Molecular and serological characteristics of the glutathione S-transferases of *Schistosoma japonicum* and *Schistosoma mansoni*. *Parasite Immunol*, **10**, 693–706.

Utzinger, J., Xiao, S.H., et al. 2001. The potential of artemether for the control of schistosomiasis. *Int J Parasitol*, **31**, 1549–62.

Vaillant, B., Chiaramonte, M.G., et al. 2001. Regulation of hepatic fibrosis and extracellular matrix genes by the Th response: new insight into the role of tissue inhibitors of matrix metalloproteinases. *J Immunol*, **167**, 7017–26.

van den Biggelaar, A.H., van Ree, R., et al. 2000. Decreased atopy in children infected with *Schistosoma haematobium*: a role for parasite-induced interleukin-10. *Lancet*, **356**, 9243, 1723–7.

van der Werf, M.J., de Vlas, S.J., et al. 2003. Quantification of clinical morbidity associated with schistosome infection in sub-Saharan Africa. *Acta Trop*, **86**, 125–39.

von Lichtenberg, F. 1962. Host response to eggs of *S. mansoni*. I. Granuloma formation in the unsensitized laboratory mouse. *Am J Pathol*, **41**, 711–31.

Williams, D.L., Asahi, H., et al. 2001. Schistosome infection stimulates host CD4+ T helper cell and B-cell responses against a novel egg antigen, thioredoxin peroxidase. *Infect Immun*, **69**, 1134–41.

Williams, M.E., Kullberg, M.C., et al. 1993. Fc epsilon receptor-positive cells are a major source of antigen-induced interleukin-4 in spleens of mice infected with *Schistosoma mansoni*. *Eur J Immunol*, **23**, 1910–16.

Wright, E.D., Chiphangwi, J. and Hutt, M. 1982. Schistosomiasis of the female genital tract. A histopathological study of 176 cases from Malawi. *Trans R Soc Trop Med Hyg*, **76**, 822–9.

Wynn, T.A., Cheever, A.W., et al. 1995. An IL-12-based vaccination method for preventing fibrosis induced by schistosome infection. *Nature (Lond)*, **376**, 594–6.

Wynn, T.A., Eltoum, I., et al. 1993. Analysis of cytokine mRNA expression during primary granuloma formation induced by eggs of *Schistosoma mansoni*. *J Immunol*, **151**, 1430–40.

Wynn, T.A., Eltoum, I., et al. 1994. Endogenous interleukin 12 (IL-12) regulates granuloma formation induced by eggs of *Schistosoma mansoni* and exogenous IL-12 both inhibits and prophylactically immunizes against egg pathology. *J Exp Med*, **179**, 1551–61.

Yamashita, T. and Boros, D.L. 1992. IL-4 influences IL-2 production and granulomatous inflammation in murine schistosomiasis mansoni. *J Immunol*, **149**, 3659–64.

Yazdanbakhsh, M., Kremsner, P.G., et al. 2002. Allergy, parasites, and the hygiene hypothesis. *Science*, **296**, 5567, 490–4.

Zaccone, P., Fehervari, Z., et al. 2003. *Schistosoma mansoni* antigens modulate the activity of the innate immune response and prevent onset of type 1 diabetes. *Eur J Immunol*, **33**, 1439–49.

30

Schistosomes: Asian

SOMEI KOJIMA

THE PARASITES

In Asia, human schistosomiasis is caused by *Schistosoma japonicum*, *S. mekongi*, or *S. malayensis* (Table 30.1), while zoophilic species such as *S. sinensium*, *S. incognitum*, and *S. spindale* have also been described from this region. *S. japonicum* affects over 800 000 people in the area along the Yangtze River basin in China, 200 000 people in various islands in the Philippines, and a rather limited population around Lake Lindo in Sulawesi, Indonesia. Control measures have been successfully introduced in some endemic areas, thereby resulting in a significant decrease in the prevalence and intensity of the disease. In other areas, however, the endemic situation has changed little, with severe infections even today, as was the case half a century ago.

Although schistosomiasis has a long history of more than 2 100 years in China, which is evident from the fact that eggs of *S. japonicum* were identified in two ancient corpses excavated from Hunan and Hubei (Chen and Feng 1999), it was in Japan that clinical symptoms were clearly described. In the middle of the nineteenth century Daijiro (Yoshinao) Fujii, a physician of the late Edo era in Japan, gave precise descriptions of clinical symptoms, which clearly refer to schistosomiasis, observed among peasants in Katayama District, although the disease was then of unknown etiology. In 1888, Majima detected a peculiar parasite ovum at autopsy in the liver of a patient who had ascites and systemic edema. On 26 May 1904, Fujiro Katsurada discovered the anterior half of a male worm in the portal vein of a female cat kept for more than 10 years

as a pet of a physician in Kofu Valley, Yamanashi Prefecture, where a serious endemic disease with abdominal swelling or abdominal fluid had been noted. Two months later, Katsurada found 24 males and eight females from another cat and described the worms as *Schistosomum japonicum* (Katsurada 1904). Following autopsy of a farmer from Katayama District carried out on 30 May of the same year, Akira Fujinami confirmed the fact that this was a human parasite (Fujinami 1904). It was also in 1904, in Singapore, that Catto independently discovered worms of the same type in a Chinese man from Fukien who had died from cholera, and Blanchard named the parasite *Schistosoma cattoi*. This work was published in 1905, but later the parasite was found to be identical with one described by Katsurada. Using cows for experimental infections, Fujinami and Nakamura (1909) found that the infection was acquired through the percutaneous route. Miyairi and Suzuki (1913) were able to determine that an amphibious snail, *Oncomelania nosophora*, was the intermediate host of this parasite. This discovery had a strong influence on the work of Leiper (1916), who was able to find the snail intermediate hosts of schistosomiasis in Egypt and to differentiate definitively between *S. haematobium* and *S. mansoni* on the infectivity to snails of miracidia derived, respectively, from terminal-spined or lateral-spined eggs.

Meanwhile, the presence of human schistosomiasis has been known in the lower Mekong river basin since the late 1950s, and direct evidence of endemicity in Khong Island, Laos, was obtained by Iijima and Garcia (1967) (cited in Iijima et al. 1971) who demonstrated that 8.6 percent of 547 inhabitants of the island were positive

Table 30.1 *Human schistosomes distributed in Asia*

Schistosome species	Final hosts	Intermediate hosts	Geographical distribution
S. japonicum	Human, dog, cat, rodent, cow, pig, sheep, goat, horse	Oncomelania hupensis hupensis	China
		O. h. nosophora	Japan
		O. h. quadrasi	Philippines
		O. h. lindoensis	Sulawesi
		(O. formosana)[a]	(Taiwan)[a]
S. mekongi	Human, dog, pig	Neotricula aperta	Laos, Cambodia, Thailand
S. malayensis	Human, rodent	Robertsiella	Malaysia

a) Zoophilic strain, not infective to humans.

for *S. japonicum*-like eggs. Iijima et al. (1971) reported that the Mekong schistosome differed from the Japanese strain of *S. japonicum*: the ovary was larger, the variation in number of testes was greater, and the eggs were smaller. Harinasuta et al. (1972) identified an aquatic snail, now known as *Tricula* (currently *Neotricula*) *aperta* (Davis and Greer 1980), as the intermediate host. Voge et al. (1978) concluded that the Mekong schistosome was a new species, *S. mekongi*. In addition, human cases with *S. malayensis* infection have been reported from Malaysia.

Classification

Recent studies on genetic variation among Asian schistosomes have revealed that all of the *S. japonicum* isolates obtained from different localities in mainland China (two from Sichuan and one from Hunan) and the Philippines (Sorsogon) showed limited variation (less than 1 percent) in both the nucleotide and amino acid sequences in mitochondrial DNA (4.9 kbp), whereas *S. mekongi* (obtained from Khong Island, Laos) and *S. malayensis* (Baling, Malaysia), both being recognized as separate but closely related, differed from each other by 10 percent, and differed by about 25 percent from *S. japonicum* (Le et al. 2002). Similarly, a high level of identity (87–94 percent at both the nucleotide and amino acid levels) was observed for all protein-encoding mitochondrial genes of *S. mekongi* and *S. malayensis*, while lesser identity was observed between genes of these two species and those of *S. japonicum* (Le et al. 2001).

Morphology and structure

The adult forms of all schistosomes infective to humans basically resemble one another. In particular, schistosomes distributed in Asia belong to the *S. japonicum* species complex because of their morphological similarities.

SCHISTOSOMA JAPONICUM

The male of *S. japonicum* measures about 15 mm in length by 0.5 mm in breadth, and possesses seven testes;

the female is long and slender like a nematode, measuring about 22 × 0.3 mm. The body surface of the male looks smooth under light microscopy, but scanning electron microscopy demonstrates folded protrusions with semi-spherical sensory organs that possess a small process in the center (Sakamoto and Ishii 1977). The surface of the female is covered with minute spines. The eggs discharged in feces are ovoid (70–100 × 50–65 μm) with a small spinose process near one end, but without an operculum (Figure 30.1).

SCHISTOSOMA MEKONGI

Adult worms of this schistosome closely resemble those of *S. japonicum* in all aspects of size, shape, internal structure, and body surface, except that the ovary is larger (Iijima et al. 1971). Embryonated eggs of *S. mekongi* are also quite similar in shape and structure to those of *S. japonicum*, but are smaller, measuring 61.7 × 51.2 μm (Iijima et al. 1971).

SCHISTOSOMA MALAYENSIS

Eggs of this species are similar to those of *S. mekongi*, with both of them being smaller than those of *S. japonicum*, although only minor morphometric differences were observed among adult worms (Greer et al. 1988).

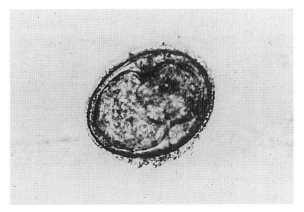

Figure 30.1 Schistosoma japonicum *ovum*

Location in host and general biology

Adult worms of *S. japonicum* and *S. mekongi* inhabit the mesenteric veins.

Life cycle

The life cycle of all Asian species follows the common pathway described in Chapter 28, Schistosomes: general, Figure 28.3. The intermediate hosts of *S. japonicum* are a complex of amphibious snails of *Oncomelania hupensis* (Figure 30.2), of which there are six subspecies: *O. h. hupensis* in mainland China; *O. h. quadrasi* in the Philippines; *O. h. nosophora* in Japan; *O. h. lindoensis* in Sulawesi, Indonesia; *O. h. formosana* in Taiwan where a zoophilic strain of *S. japonicum* exists; finally, experimental infection with the strain has been demonstrated in *O. h. chiui*, another subspecies found in Taiwan. Recent studies suggest that these subspecies should be elevated to specific status like *O. hupensis* or *O. quadrasi* (Sturrock 1993). These oncomelanid snails have conical or subconical dextral shells of four to eight whorls with a height of less than 10 mm. *O. quadrasi* is more aquatic than the other species, which may crawl out of water on nearby vegetation and mud surfaces and can survive for at least a few months of a dry season in natural conditions. *S. mekongi* is transmitted by the α, β, and γ races of *Neotricula aperta* (Sornmani 1976), of which the γ race is the most susceptible to infection under laboratory conditions (Liang and Kitikoon 1980), although only the γ race was found to be infected naturally (Sornmani 1976). The intermediate host of *S. malayensis* was described as *Robertsiella*, a sister genus to *Neotricula* (Davis and Greer 1980).

Transmission and epidemiology

From 1947, in Japan, a program to control the disease in previously known endemic areas was undertaken by the Japanese Ministry of Health and Welfare and other institutions in collaboration with the United States 406th Medical General Laboratory. Voluntary organizations,

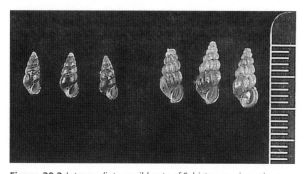

Figure 30.2 *Intermediate snail hosts of* Schistosoma japonicum: *left,* Oncomelania hupensis nosophora *collected from Yamanashi, Japan; right,* Oncomelania hupensis hupensis *collected from Jian-xi Province, China*

composed of staff from local governments, physicians' associations, teachers' associations, and representatives from endemic communities, made a major contribution to the intensive control program, which included snail control and improvement of environmental and sanitary conditions, since they had been established before the Second World War. Based on these activities, the government carried out massive surveys in 1973 and 1978 on the inhabitants, *Oncomelania* snails and the reservoir hosts throughout endemic areas. Although two snail colonies still exist, including a new habitat found recently in a limited area that has no relation to former endemic areas (Kojima et al. 1988), schistosomiasis japonica is considered to have been eradicated from Japan since 1978 (Hunter and Yokogawa 1984; Tanaka and Tsuji 1997).

In China, around the time of the founding of the People's Republic in 1949, schistosomiasis was raging in many areas. For example, in 1954 in Jiangxi Province, only two persons were left alive in one particular township which had had about 1 000 households in the mid-nineteenth century. In another township in Jiansu Province, 4 019 people out of about 7 000 villagers acquired acute infection during a flood season in 1950 and 1 335 patients died of schistosomiasis (Chen and Feng 1999). Large-scale national epidemiological surveys carried out in the mid-1950s revealed that schistosomiasis was endemic in 380 counties in 12 provinces along the Yangtze River, with 12 million infected people among approximately 100 million at risk of infection (Li et al. 2000). Thus, schistosomiasis control was given high priority by the Chinese Government and various measures were taken in an integrated approach, including health education through schools, hand-outs, and mass media, as well as principal measures such as snail control, environmental modification, and large-scale chemotherapy, both for infected persons and cattle, molluskiciding activities, sanitation, etc., to control the disease over the last four decades. A World Bank loan was also implemented to strengthen operational reseach capabilities from 1992 to 1998. As a result, the number of infected persons reduced from 1 638 103 in 1989 to 865 084 in 1995, and the prevalence of infected bovines also reduced from 13.3 to 9.1 percent (100 251 cattle and buffalo are estimated to be infected) (Chen and Feng 1999). According to another report, the number of reported cases of schistosomiasis further reduced to 694 788 in the year of 2000, which included 664 acute cases, 671 400 chronic cases and 22 786 advanced cases (Jiang et al. 2002). However, most of the 118 counties and cities still endemic for schistosomiasis are distributed in the swamp and lake regions and also in mountainous regions, where snail elimination is quite difficult, and therefore the only control strategy is morbidity control by repeated chemotherapy (Chen and Feng 1999).

Similarly, the government of the Philippines has adopted the objective of schistosomiasis control from a comprehensive approach that includes treatment of cases, environmental sanitation, snail control, and health education to morbidity or disease reduction by chemotherapy supplemented with environmental sanitation, health education, and molluskiciding (Leonardo et al. 2002). However, schistosomiasis continues to remain as a public health problem affecting ten out of 16 regions, with 6.7 million people at risk. Many factors may account for this problem: (1) detrimental integration of the schistosomiasis control program with other health services causing confusion in coordination of services, implementation of plans, leadership among related personnel, etc.; (2) low disease awareness among people at risk because of their poverty, resulting in avoidance of participation in control activities, refusal of treatment for fear of drug toxicity if treatment is done without prior examination and laboratory evidence of infection, and refusal of water-sealed toilets; and (3) aggravated security problems in some areas that diminish access of people to the health delivery system (Leonardo et al. 2002). Thus, prevalence of schistosomiasis remains at 4.7 percent in the whole country, with an infection rate nearly as high as 19 percent in some provinces.

In Indonesia, schistosomiasis is limited to two isolated areas in central Sulawesi. A recent survey carried out in 1999 revealed that 11 (0.46 percent) out of 2 387 villagers examined were positive for schistosome eggs in Lindu valley, while 67 (1.55 percent) out of 4 312 persons were positive in Napu valley, indicating a great decrease in the prevalence if compared to previous data in the 1970s (Izhar et al. 2002). However, it seems likely that the life cycle of S. japonicum is still maintained among snails (about 7 percent positive for cercariae) and wild rats (5.6 percent infected) in the natural environment.

Meanwhile, transmission of mekongi schistosomiasis occurs in rocky banks of the Mekong River due to the characteristics of the snail host, which is aquatic and usually attaches to small rocks, stones, or old tree branches under slow-running shallow water. According to a recent report, about 140 000 people are estimated to be at risk of infection in Laos and Cambodia, despite universal treatment campaigns (Urbani et al. 2002). Daily water contact activities such as fishing, laundering, bathing, swimming, etc., are risk factors, although the role of dogs and pigs, which harbor the parasite, in transmission (Sornmani et al. 1971; Strandgaard et al. 2001) is not clear.

CLINICAL AND PATHOLOGICAL ASPECTS OF INFECTION

Clinical manifestations, pathogenesis, and pathology

The initial manifestations may appear following penetration of cercariae and migration of schistosomula from the skin to lungs. Table 30.2 summarizes the symptoms

Table 30.2 Symptoms observed among people with acute schistomiasis japonica (figures are percentages)

Symptoms	Chinese cases (200 cases)[a]	Japanese cases (75 cases)[b]
Initial phase		
Dermatitis	44.0	67.0
Prepatent phase		
Malaise/fatigue		67.0
Cough	63.0	
Bloody sputum	5.5	
Anorexia	88.0	38.6
Abdominal pain	41.0	(+)
Anaemia	23.0	24.0
Oviposition phase		
Chill	78.0	77.3
Fever	100.0	77.3
Sweat	72.0	73.3
Diarrhea	51.0	5.3
Dysentery	10.0	(−)
Hepatomegaly	96.5	85.3
Splenomegaly	62.5	22.7
Ascites	8.5	9.3

a) Ling et al. (1949)
b) Kashiwado et al. (1927)

of Chinese and Japanese patients who were infected with S. japonicum in the floods that occurred in the Yangtze river in 1945 and the Tone river in 1926, indicating that cercariae may induce dermatitis with slight exanthema and itch upon their penetration. Dermatitis usually disappears within a week and migration of schistosomula into the lungs provokes a cough with association of a mild fever and dullness. Anemia may be observed among one-fifth of patients (Table 30.2).

After a latent period of about a month, the acute phase of disease may suddenly start with a high fever accompanied by rigors, termed Katayama fever after the Katayama District in Hiroshima Prefecture, Japan, where Yoshinao Fujii originally described the acute onset of the illness (see The parasites). Among the Japanese patients described in Table 30.2, the majority had such an onset 31–50 days after the infection. The onset corresponds to the start of oviposition by mature female worms. S. japonicum female worms lay eggs in the mesenteric branches of the portal vein along the intestinal wall and although a relatively large proportion of the eggs are carried into the liver and other organs by the blood flow, the remainder of them may stay in the small venules until the embryos they contain develop to miracidia within 10 days. Antigenic substances excreted from miracidia diffuse out through submicroscopic pores in the eggshell and elicit an acute inflammation in the surrounding tissues, resulting in the rupture of the vascular wall and escape of the eggs from the venules through the intestinal submucosa and mucosa into the intestinal lumen. Typical egg granulomas formed in the liver are shown in Chapter 28, Schistosomes: general, Figure 28.6. The inflammation causes recurrent daily fever, abdominal pain, and enlarged tender liver and spleen, and discharge of eggs into the intestinal canal is accompanied by dysentery or diarrhea (Table 30.2). Blood chemistry may reveal a transient elevation of glutamic pyruvic transaminase, glutamic oxaloacetic transaminase, and alkaline phosphatase 5–6 weeks after infection. Eosinophilia may be observed in most of the patients with or without increase of leukocyte counts. Serum levels of IgE may increase as observed in other helminth infections (Kojima et al. 1972). These symptoms characterize intestinal schistosomiasis.

The chronic phase of manifestations in *S. japonicum* infection is characterized by hepatosplenomegaly and may therefore be called hepatosplenic schistosomiasis, although development of polyps or mucosal proliferation of the intestine may also be observed in most cases. Egg granulomas are replaced by fibrotic tissues, which are prominent in the periportal areas and lead to the development of pipestem fibrosis (Figure 30.3) similarly to that described in cases with *S. mansoni* infection (von Lichtenberg et al. 1971). In association with liver dysfunction, diagnostic techniques such as computed

tomography or ultrasonography may reveal pathological changes with characteristic reticulate or hexagonal patterns in patients with chronic schistosomiasis (Figure 30.4a, b). The liver gradually decreases in size, but increases in hardness as fibrosis is gradually extended into the parenchyma, resulting eventually in liver cirrhosis in severe cases (Figure 30.5). However, it has been pointed out that the use of the term 'schistosomal cirrhosis' may not be appropriate because of the lack of the essential features of cirrhosis, such as disruption of the lobular architecture (Ross et al. 2001). Even though schistosomiasis cases with cirrhosis are observed, it might be difficult to determine the pathogenesis unless there is evidence that alcoholic and/or postviral cirrhosis are not involved. Despite the possible presence of such coincident factors, post-schistosomal hepatic fibrosis has been suggested to be under the control of HLA class II genes, since associations of HLA-DR-DQ alleles or HLA-DP alleles were observed with protection from the early or late phase of fibrosis, respectively, in a study performed on patients with schistosomiasis japonica in China (Hirayama et al. 1999). However, further studies, using defined antigens, for example, will be necessary to clarify the immunogenetic regulation of fibrosis in humans as a result of responses to soluble egg antigens

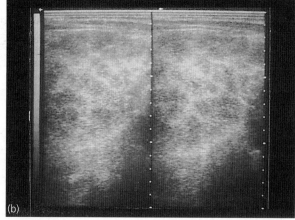

Figure 30.3 *Pipestem fibrosis in a Japanese patient with schistosomiasis japonica*

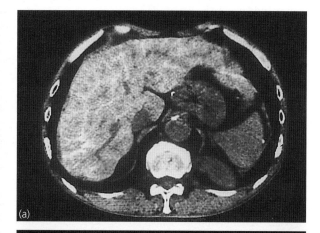

Figure 30.4 *Hepatic fibrosis in a patient with chronic schistosomiasis japonica:* **(a)** *computed tomography;* **(b)***ultrasonography. Note the characteristic reticulate or hexagonal patterns.*

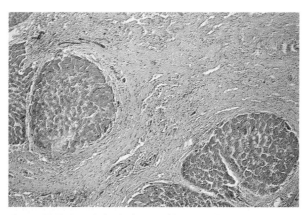

Figure 30.5 *Liver cirrhosis due to* Schistosoma japonicum *infection. A case found in Yamanashi Prefecture, Japan*

(SEA) (see below). In experimental schistosomiasis japonica, it was shown that killing the eggs deposited in the liver by chemotherapy might result in recovery of fibrotic tissues to normal (Kojima 1970). The enlarged spleen may reach the level of the umbilicus or even at times expand to fill most of the abdomen. Portal hypertension and hypoalbuminemia may induce ascites (Figure 30.6). Dilatation of abdominal collateral veins (Figure 30.6) and esophagogastric varices are observed in advanced cases, and bleeding from the varices may cause sudden death. Anemia is more noticeable than in the previous stages of the disease. Acute or chronic cerebral involvement may occur in schistosomiasis japonica due to embolism of eggs or heterotopic parasitism of female worms, causing headache, Jacksonian epileptic seizures, paresthesia, and poor vision (Ariizumi 1963). Hepatic coma may occur as a complication, although its frequency is far less than that of liver cirrhosis.

Similar chronic symptoms and signs, including cachexia, hepatosplenomegaly, stunting, retardation of puberty, and decompensation of portal hypertension with ascites, may be observed in cases of mekongi schistosomiasis and rupture of esophageal varices is usually the cause of death (Urbani et al. 2002).

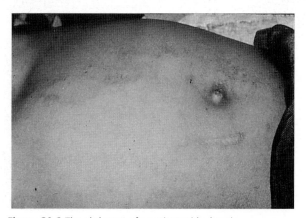

Figure 30.6 *The abdomen of a patient with chronic schistosomiasis japonica, showing ascites and dilatation of abdominal collateral veins*

In addition, it has been reported that female genital tract diseases may occur in schistosomiasis japonica, although the involvement of gynecological organs is more usually observed in *S. haematobium* infection. A case-control study carried out in 2 106 women aged 18–50 years revealed that cases with chronic cervicitis and uterine enlargement were observed more frequently in the experimental group of 244 women with schistosomiasis japonica than in the control group of 236 age-matched healthy women (Liu et al. 2000). Menstrual disorder and shorter stature and lighter weight of the first newborn also occurred more frequently in the experimental group. It has been pointed out that female genital schistosomiasis may be an important risk factor for transmission of Human immunodeficiency virus (HIV), since sexually transmitted diseases increase the probability of HIV transmission, presumably through lesions in the genital mucosa (Feldmeier et al. 1994). Further studies are necessary for cases with schistosomiasis japonica.

IMMUNE RESPONSES

A hybridoma that produces a monoclonal IgE antibody to this species was established (Kojima et al. 1987) to examine the roles of IgE antibodies in protection against *S. japonicum*. This monoclonal antibody (SJ18ε.1) recognized a 97 kDa antigen and the antibody was shown to be protective when it was passively transferred into mice during the early phase of the infection or used in antibody dependent cell-mediated cytotoxicity (ADCC) in the presence of macrophages or rat eosinophils (Kojima et al. 1987; Janecharut et al. 1991, 1992). The antibody reacted with a single protein of the same size extracted from *S. mansoni* that had been identified as paramyosin (Figure 30.7), a muscle protein unique for invertebrates, by gene cloning and analysis of the deduced amino acid sequence (Laclette et al. 1991). By using SJ18ε.1 to screen a cDNA library obtained from adult *S. japonicum*, the gene coding for the target molecule has been cloned to confirm that the target of this monoclonal antibody is paramyosin (Nara et al. 1994). It has also been demonstrated that paramyosin localizes in the tegument and postacetabular glands, as well as in the muscle tissues of schistosomula (Nara et al. 1994) (Figure 30.8), although the localization of the molecule in tissues other than the muscles, even in isolated teguments, is questionable (Schmidt et al. 1996). However, a comparative immunolocalization study using electron microscopy among schistosome adults, cercariae, and lung schistosomula has revealed that paramyosin is localized within the muscle layer of all three developmental stages, within granules of the post-acetabular glands of cercariae, and within the tegument matrix and surface of lung schistosomula (Gobert et al. 1997). *S. japonicum* paramyosin is composed of some 866 amino acids and is thought to have an α-helical conformation. If compared with the amino acid sequences of paramyosin demonstrated for other helminths,

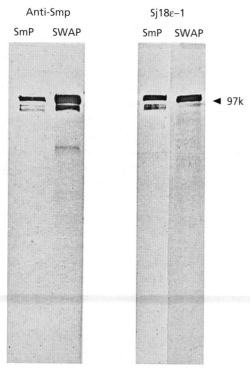

Figure 30.7 *SJ18ε.1 recognizes paramyosin of* Schistosoma mansoni. *SWAP, soluble worm antigen preparation of* S. mansoni; *SmP, paramyosin purified from SWAP; Anti-SmP, rabbit antiserum specific to SmP; SJ18ε.1, a mouse monoclonal IgE antibody (in collaboration with Dr Isabella Oswald and Dr Alan Sher)*

S. japonicum shares 96 percent homology with *S. mansoni* (Laclette et al. 1991), and 72 percent with *Echinococcus granulosus* (Muhlschlegel et al. 1993), but only around 35 percent with a parasitic or non-parasitic nematode, *Dirofilaria immitis* (Limberger and McReynolds 1990) or *Caenorhabditis elegans* (Kagawa et al. 1989), indicating that the sequence of paramyosin is well conserved among platyhelminths.

Since ADCC in the presence of IgE antibodies has been shown to be involved in immunity to schistosome infections (Kojima et al. 1987; Janecharut et al. 1991, 1992; Capron and Capron 1994), determination of the

B-cell epitope of paramyosin responsible for protection is important for the development of a peptide vaccine for schistosomiasis. The epitope recognized by SJ18ε.1 was examined using a series of deletion mutants expressed in *Escherichia coli* and it was found that the antibody reacted with recombinant paramyosin containing 113 amino acids (Glu[301]–Ala[413]), but not with a shorter peptide (Glu[301]–Asp[343]). Thus, the epitope recognized by this antibody was expected to exist within 71 amino acid residues (Asp[343]–Ala[413]). Further analysis was carried out by a multi-pin system using heptameric peptides synthesized sequentially from these 71 amino acids of paramyosin and the result demonstrated significant binding of SJ18ε.1 to a sequence consisting of four amino acid residues (Ile-Arg-Arg-Ala[359]). The target epitope of SJ18ε.1 was common to paramyosins of *S. mansoni*, *Taenia solium*, and *Echinococcus granulosus*, but not to nematode paramyosins, suggesting that the epitope is specific for platyhelminths (Nara et al. 1997).

It is known that paramyosin is recognized by IgG antibodies in sera from patients with chronic schistosomiasis japonica (Kojima et al. 1987). Therefore, it may be possible to induce an antibody-dependent cell-mediated immunity to human schistosomiasis by immunization with this molecule. With regard to this, Hernandez et al. (1999) reported that titers of IgA antibodies to soluble worm antigen preparation (SWAP) correlated with age-dependent resistance to *S. japonicum* infection among Filipinos and that the antibodies from the majority of high IgA responders recognized 97 kDa molecules. Based on these observations, the authors proposed that IgA antibodies to paramyosin might contribute to parasite attrition during migration in the lungs.

Vaccination

The efficacy of vaccination with native or recombinant paramyosin has been examined in various animal models with different regimes including adjuvants and route of immunization. Promising results were obtained against *S. japonicum* infections in mice (Kojima et al. 1998;

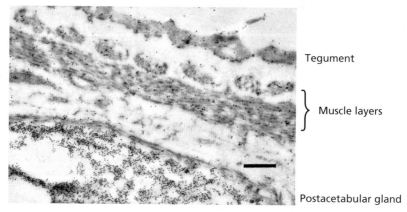

Figure 30.8 *Localization of paramyosin shown by immunoelectromicroscopy (Nara et al. 1994)*

McManus et al. 1998), sheep (Taylor et al. 1998), pigs (Chen et al. 2000), and water buffalo (McManus et al. 2001). Domestic pigs were used in China for evaluation of vaccination with recombinant *S. japonicum* paramyosin (rSJPM) (Chen et al. 2000). As a positive control, 16-week-old pigs were vaccinated with 400 UV attenuated cercariae and they showed a significant reduction of worm recovery (53 percent, $p < 0.001$, if compared with non-immunized controls). The experimental groups were immunized intradermally with rSJPM and alum or TiterMax and they were partially but significantly protected against the challenge infection (32–35 percent reduction) (Chen et al. 2000). Among three water buffalo immunized with rSJPM plus Quil A adjuvant, a significant reduction in liver eggs was recorded in two buffalo and all of them showed high levels of specific anti-paramyosin IgG antibodies. There was no evidence of any toxic effects and the vaccine preparations and Quil A were well tolerated (McManus et al. 2001). Thus, the unique character of schistosomiasis japonica as a zoonosis may provide a useful tool applicable to development of vaccines for the other major human schistosomiasis.

The other vaccine candidates for schistosomiasis are summarized in Table 30.3 and Figure 30.9). To avoid duplication with the preceding chapter, however, only molecules related to Asian schistosomes will be discussed below. Among them, the 23 kDa molecules of *S. japonicum* (Sj23) (or *S. mansoni*, Sm23) are integral membrane proteins (Rogers et al. 1988) and the molecules share an 84 percent amino acid homology (Davern et al. 1991). Hydrophobicity analyses for both molecules suggest the presence of transmembrane regions arranged as three consecutive regions at the N-terminus followed by a relatively hydrophilic domain with a fourth domain at the C-terminus (Wright et al. 1990). Using antibodies, this protein has been detected in all stages of the parasite found in the human host, notably the lung stage, and therefore is of interest as a vaccine candidate. It is of particular significance that Sm23 (composed of 218 amino acids) was strikingly similar, with respect to both amino acid sequence (36 percent identity) and putative domain structure, to ME491, a human stage-specific melanoma-associated antigen (Wright et al. 1990). Sm23 and Sj23 also show a strong homology with a family of membrane proteins including the lymphocyte-expressed surface molecules CD37 and TAPA-1 and have therefore been suggested to play a role in cellular proliferation and parasite growth, because these membrane proteins are considered to be involved in cell proliferation (Davern et al. 1991). Thus, Sm23 and Sj23 have been shown to be members of a proposed new superfamily of membrane proteins whose structures do not conform to the previously known classifications. To date, the other members include CD9 (p23), CD53, MRC OX-44, CO-029, MRP-1, L6, the gene product of TI-1, the target of monoclonal antibody AD-1. Most of these molecules, except for Sm23 and Sj23, are found in membranes of hemopoietic and/or malignant cells, although their function is not yet clear. With regard to schistosomes dwelling in the blood, however, this homology may be considered as an example of molecular mimicry, in which the parasite might have evolved a host-like molecule in an attempt to escape from host immune responses. Nevertheless, Sm23 has been shown to be a major antigen recognized by sera from animals vaccinated with irradiated cercariae (Richter and Harn 1993). The two predicted external hydrophilic domains were found to be highly immunogenic and contained several B-cell epitopes. There were at least four T-cell

Table 30.3 *Vaccine candidates for schistosomiasis*

Identity of antigen	Size (kDa)	Stage	Function/ localization	Protection (%) (mouse)	Reference
Paramyosin (Sm97)	97	Schistosomula Adult worms	Muscle protein	30	Pearce et al. (1988)
(Sj97)		Schistosomula adult worms cercariae	Muscle, tegument, postacetabular gland	20–60	Kojima et al. (1987, 1998), Nara et al. (1994)
Sj23/Sm23	23	All stages	Tegument		Reynolds et al. (1992)
TPI	28	All stages	Enzyme	30–60	Shoemaker et al. (1992), Reynolds et al. (1994)
GAPDH	37	Schistosomula Adult worms	Enzyme		Dessein et al. (1988)
GST (Sj26/Sj28) (Sm26/Sm28)	26/28	Schistosomula Adult worms	Enzyme	30–70	Balloul et al. (1987b)
IrV-5	200	Schistosomula	Muscle	32–75	Soisson et al. (1992)
IrV-1	90	All stages	Tegument		Hawn et al. (1993)
FABP (Sm14)	14	Schistosomula	Tegument	67	Tendler et al. (1996)

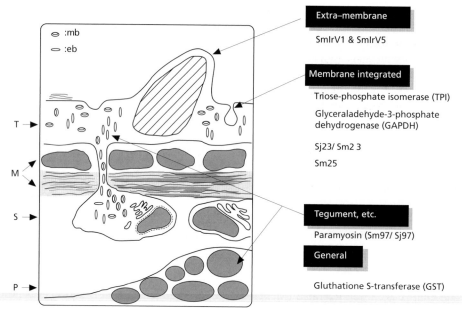

Figure 30.9 *Localization of vaccine candidates for schistosomiasis. M, muscle layers; P, postacetabular gland; S, subtegumental tissues; T, tegument*

epitopes in the large hydrophilic domain. One segment of 23 amino acids contained both a T-cell and a B-cell epitope, as well as the putative glycosylation site. This particular segment was recognized by immune sera and cells of every mouse strain tested. The elucidation of these epitopes demonstrates the immunogenic nature of this molecule and raises questions as to the role of Sm23 in the host–parasite relationship (Reynolds et al. 1992). However, it is still obscure whether these molecules are able to induce protection to human schistosomiasis, although a monoclonal antibody to Sj23 has been successfully used in immunodiagnostic assays to detect *S. japonicum* infection in Filipino patients (Davern et al. 1991). By using full-length Sm23 prepared with the baculovirus expression system, and the N-terminal 133 amino acids and the 85 C-terminal amino acids expressed by means of the prokaryotic vector pGEX, a total of 70 sera from patients from Sudan and Egypt were examined by enzyme linked immunosorbent assay (ELISA) or Western blot analysis (Koster et al. 1993). The anti-Sm23 antibody titers in infected patients varied widely and were not correlated with egg counts or age of the individuals. Most of the seroreactivity was directed against the C-terminal polypeptide.

Parasite enzymes have also been identified as vaccine candidates. Triose-phosphate isomerase (TPI) was isolated from a Chinese strain of *S. japonicum* and this native protein was shown to be functional with enzyme activity of 177 IU/mg of protein. Interestingly, when examined for protective activity, the number of eggs in the liver of immunized mice was significantly reduced compared with adjuvant control animals, whereas a modest reduction in worm burdens was observed in immunized animals (Miao et al. 1998). Further studies

made it possible to obtain recombinant TPI with a specific enzyme activity as high as 7687 IU/mg of protein (Hooker and Brindley 1996; Sun et al. 1999).

Other candidates of an enzymatic nature are the glutathione S-transferases (GST) that consist of at least two isoenzymes of 26 and 28 kDa for both *S. japonicum* and *S. mansoni* (Sj26, Sj28, Sm26, and Sm28) (Tiu et al. 1988). Mice of the inbred strain 129/J (WEHI 129/J) relatively resistant to chronic infection with *S. japonicum* were shown to be high responders to Sj26 (Smith et al. 1986). The cDNA sequence encoding Sj26, Sj28, Sm26, and Sm28 has been isolated and expressed in *E. coli* (Smith et al. 1986; Balloul et al. 1987a, b; Henkle et al. 1990; Trottein et al. 1990). Despite their immunological cross-reactivity using rabbit antisera, Sj28 is weakly immunogenic relative to Sm28 in mouse immunization experiments using GSTs purified from adult worms. The difference in immunogenicity is also observed during schistosome infection in mice. Using surface-labeled living *S. japonicum* worms, evidence was obtained for a surface location of Sj28 comparable to that reported for the *S. mansoni* molecule. The nucleotide and deduced amino acid sequences of cDNA clones corresponding to Sj28 and Sm28 have been compared. Despite obvious homology (77 percent identity), differences were found in regions known to contain T-cell epitopes in the *S. mansoni* protein, which may be an explanation for the striking differences in immunogenicity in regard to antibody production in mice. The 26 kDa GSTs (Sj26 and Sm26) are also closely related on the basis of nucleotide and deduced amino acid sequences, with 82 percent identity in the putative coding regions. When the amino acid sequences of Sj28 and Sm28 were compared with those of Sj26 and Sm26, the overall sequence identity was approximately 20 percent. However, a relatively conserved region was identified in otherwise structurally different molecules which may participate in common properties of these enzymes (Henkle et al. 1990). A study of the tissue distribution of the cloned Sm26 by immunoelectron microscopy demonstrated similarities to Sm28 in that they are present in the tegument and in subtegumentary parenchymal cells. However, a major difference was observed in the protonephridial region in which Sm26 was present in the cytoplasmic digitations localized in the apical chamber, delineated by the flame cell body, suggesting that Sm26 may be actively excreted by adult worms (Trottein et al. 1990). Experimental vaccination of rats, hamsters, and monkeys with a recombinant fusion protein induced a strongly cytotoxic antibody response. Immunization of rats and hamsters with Sm28 resulted in significant protection against a natural challenge infection with live cercariae (Balloul et al. 1987a, b). Moreover, the antibody response raised against this protein was able to kill *S. mansoni* schistosomula in in vitro cytotoxicity assays in the presence of rat eosinophils. The inhibition of this cytotoxic activity by an aggregated myeloma IgG2a indicated that one of the major isotypes involved in this in vitro model is IgG2a. The passive transfer of Sm28 antisera induced a significant level of protection against experimental infection. However, immunization with recombinant Sj26 was not sufficient to induce consistent immunity and maximum resistance was approximately 50 percent, while no protection was obtained in BALB/c mice, known low responders to Sj26. Although only Freund's complete adjuvant was used, the data indicated that satisfactory levels of resistance to *S. japonicum* might not be attained by vaccination with Sj26 alone. The requirement for other antigens, including the additional GST isoenzyme of Sj28, has been suggested to establish whether Sj26 will be an important component of a defined multivalent vaccine against schistosomiasis japonica (Mitchell et al. 1988). Complete cloning of Sj28 has been carried out recently and GST active recombinant proteins were used for examination of vaccine efficacy in various strains of mice. Unfortunately, however, consistent results were not obtained in terms of reduction in the worm burden, liver eggs, or hatched miracidia numbers (Scott and McManus 2000).

A cDNA coding for *S. japonicum* calpain has been cloned from two isolates, Yamanashi and Hunan strains (Zhang et al. 2000). Nucleotide sequence and amino acid sequence revealed 99.1 and 99.8 percent identity between the two calpains, respectively. Both calpains were considered to be translated as a preproenzyme, and a 746-amino acid mature enzyme contained eight motifs without a signal peptide at the N-terminal based on the deduced amino acid sequences. Although mRNA for calpain was detectable in different developmental stages, sera obtained from mice immunized with recombinant

calpain showed enhanced binding to cercarial antigen. Human sera from *S. japonicum*-individuals recognized the large subunit, and sera of individuals with light infection showed stronger reactivity than those obtained from moderate or high infection cases. When tested with synthetic peptides, human B-cell epitopes on *S. japonicum* calpain were observed to be common with those of *S. mansoni* calpain. From these results, the authors suggested that calpain of *S. japonicum* has a diagnostic value in addition to being a vaccine candidate.

DIAGNOSIS

Detection of circulating antigens with ELISA kits has been developed for *S. japonicum* infection with high sensitivity and specificity (Wang et al. 1999). Interestingly, it has been reported that keyhole limpet hemocyanin (KLH), extracted from the marine mollusk *Megathura crenulata*, may be utilized as a diagnostic antigen just as *S. japonicum* SEA is used for the dot ELISA (Li et al. 1997). Moreover, if serum samples were diluted at 1:500, the sensitivity of the KLH dot–ELISA for diagnosis of acute and chronic schistosomiasis was 82 and 15 percent, respectively, suggesting possible serological differentiation of both forms.

CHEMOTHERAPY, PREVENTION, AND CONTROL

For schistosomiasis japonica, oral administration of praziquantel in three 20 mg/kg^{-1} doses at 4-h intervals may result in 80 percent cure. At this dosage, about a half of the individuals receiving the treatment may complain of side effects, such as abdominal discomfort, drowsiness, headache, backache, fever, sweating, and dizziness, although these are usually mild and transient. Recent observations using mice heavily infected with *S. japonicum* demonstrated that praziquantel administration induced typical signs of systemic anaphylaxis until half of them died shortly after the treatment (Matsumoto 2002). Autopsy of these mice revealed remarkable intestinal alterations characterized by increased mucosal permeability, mucosal edema, petechial hemorrhage, and histologically, degranulation of intestinal mast cells, which was particularly remarkable around eggs hatched as an effect of praziquantel. Thus, the author suggested that the adverse effects of praziquantel treatment characterized by abdominal signs might result from anaphylactic reactions to antigens released from hatched eggs (Matsumoto 2002).

Artemether, the methyl ether derivative of artemisinin has been tested against *S. japonicum*. When administered intragastrically to mice on day 7 after infection with *S. japonicum* cercariae at a single dose of 300 mg/kg body weight, and the same dose of artemether was repeated at 1–3-week intervals between one and four times after the first dosing, most of the female worms were killed before oviposition, with female worm reduction rates of 70–90 percent, resulting in protection of the host from damage induced by schistosome eggs. When rabbits and dogs were treated intragastrically with artemether at 10 mg/kg on day 7 after infection, followed by between one and four repeated weekly doses, the worm reduction rates were 85–99 percent. Moreover, some parameters related to acute schistosomiasis, such as temperature, eosinophil count, and eggs in the feces, were negative, and low specific antigen and antibody levels in serum were seen. Further study showed that the appropriate regimens of artemether were also effective in early treatment of re-infection with cercariae. Histopathological examination of the livers showed that early treatment with artemether exhibited a promising protective effect in dogs and rabbits. Based on these results, these authors concluded that early treatment with artemether could be recommended for field trials for controlling acute schistosomiasis, reducing infection rate and intensity of infection (Xiao et al. 1995). Control of schistosomiasis is not an easy task. Even after successful treatment, re-infection easily takes place in most endemic areas, unless transmission is cut off somewhere between the intermediate hosts and the final hosts in the life cycle of the parasites. Control of schistosomiasis japonica is complicated by the existence of reservoir hosts, mainly domestic animals such as cows, pigs, sheep, and goats in China, and dogs, cats, and water buffalo in the Philippines, in addition to wild rats. Use of human night soil for fertilizer should be avoided in endemic areas. The snail hosts constitute a key factor in the control of schistosomiasis. In Japan, a periodical distribution of molluskicides, such as sodium pentachlorophenate, Yurimin and later B-2 (sodium 2,5-dichloro-4-bromophenol), was applied in combination with cement lining of irrigation ditches in endemic areas for control of *Oncomelania* snails (Figure 30.10). However, these measures can hardly be applied in such huge endemic areas as in China, and molluskicides are

not suitable for application in water sources where aquatic snail hosts are abundant and on which people's daily life depends. Indeed, even if limited only to the Dongting Lake region, snail habitats are estimated at 1 768 km^2 in 1996, increasing at a rate of 34.7 km^2 annually due to high silt deposition from the Yangtze River and from connecting rivers in that area, and construction of embankments in the lake. In addition, it is anticipated that the construction of the Three Gorges superdam may result in substantial expansion of snail habitats and increase the number of new schistosomiasis cases (Li et al. 2000). The positive impact on the control of schistosomiasis may be that floods in the Yangtze River will occur to a lesser degree, thereby decreasing the chances of snail dispersal and infection of humans and animals, whereas the negative impact should also be taken into consideration in that flushed beaches and migratory settlements may create snail habitats at certain altitudes (Zheng et al. 2002). In addition, the chances of infection may increase in the Dongting Lake beach because of an increase in the likelihood of human and reservoir animal water exposure due to water regression that occurs ahead of time in the autumn, while the impact of the dam construction on the transmission of schistosomiasis was not significant in the Poyang Lake and other areas on the lower reach of the Lake (Zheng et al. 2002). Although remarkable progress has been made in the control of the disease after intensive efforts over four decades in China (Li et al. 2000), the construction of the Three Gorges superdam may make it possible to utilize riverbeds for pasture for a whole year, otherwise impossible during the rainy season, increasing the risk of schistosome infection among livestock and local people.

Thus, improvements in environmental sanitation and safety of supply water are essential for control. The successful control accomplished in Japan by 1978 indicates that to achieve the goal of control, the organization of voluntary associations including mothers' groups, and groups of medical doctors and workers, in cooperation with national and local government, may be as important as the education and motivation of the people themselves. Japan has undergone rapid socioeconomic development resulting in drastic changes in agriculture or living styles, land reclamation projects, and increased awareness of schistosomiasis (Hunter and Yokogawa 1984). It should be noted that to overcome a complex of issues involved in the control of schistosomiasis, strengthening of operational research on cost-effective methods of disease control proved to be useful from the experiences in China under the World Bank Schistosomiasis Research Initiative (Yuan et al. 2000).

To control mekongi schistosomiasis, universal treatment campaigns were organized in 1989 in Laos and later in 1996 in Cambodia under the guidance and support of the World Health Organization, in collaboration with several agencies and international organizations. A single

Figure 30.10 *Cement lining of irrigation ditches for control of* Oncomelania *snails (courtesy of Yamanashi Prefectural Institute of Public Health)*

dose of praziquantel of 40 mg/kg was distributed to whole populations in endemic areas where the prevalence was above 50 percent, except children under 2 years of age and pregnant women, while below this prevalence only children between 2 and 14 years were treated. The impact of this campaign was apparent in a remarkable reduction in hepatosplenomegaly in association with decreasing infection rates and improvement of liver pathology monitored with abdominal ultrasonography in a cohort study of an endemic village in north-eastern Cambodia (Hatz 2001). In addition, a decrease in the number of in- and out-patients suffering from schistosomiasis in hospitals in the endemic province of Kratie, Cambodia, was noted, although further ultrasound surveys suggested that repeated treatment might not prevent or reverse advanced periportal fibrosis and development of portal hypertension (Urbani et al. 2002). The ecology of the intermediate host makes it difficult to complete the control of mekongi schistosomiasis transmission by taking measures such as snail control and environmental modification that have been applied to other endemic areas of schistosomiasis japonica as mentioned above. Health education and people's motivation to better manage biological waste are apparently important, but it seems difficult for local people to construct latrines and washing facilities without external financial support (Urbani et al. 2002). Due to the low sensitivity of the Kato-Katz stool examination and high costs of available immunological tests, focusing on school-aged children, in combination with the soil-transmitted helminthiases control program, may be the cost-effective way to control mekongi schistosomiasis.

REFERENCES

Ariizumi, M. 1963. Cerebral schistosomiasis japonica. Report of one operated case and fifty clinical cases. *Am J Trop Med Hyg*, **12**, 40–55.

Balloul, J.M., Gryzch, J.M., et al. 1987a. A purified 28,000 dalton protein from *Schistosoma mansoni* adult worms protects rats and mice against experimental schistosomiasis. *J Immunol*, **138**, 3448–53.

Balloul, J.M., Sondermeyer, P., et al. 1987b. Molecular cloning of a protective antigen of schistosomes. *Nature (London)*, **326**, 149–53.

Capron, M. and Capron, A. 1994. Immunoglobulin E and effector cells in schistosomiasis. *Science*, **264**, 1876–7.

Chen, H., Nara, T., et al. 2000. Vaccination of domestic pig with recombinant paramyosin against *Schistosoma japonicum* in China. *Vaccine*, **18**, 2142–6.

Chen, M. and Feng, Z. 1999. Schistosomiasis control in China. *Parasitol Int*, **48**, 11–19.

Davern, K.M., Wright, M.D., et al. 1991. Further characterisation of the *Schistosoma japonicum* protein Sj23, a target antigen of an immunodiagnostic monoclonal antibody. *Mol Biochem Parasitol*, **48**, 67–75.

Davis, G.M. and Greer, G.J. 1980. A new genus and two new species of Triculinae (Gastropoda: Prosobranchia) and the transmission of a Malaysian mammalian *Schistosoma* sp. *Proc Acad Nat Sci Phil*, **132**, 245–76.

Dessein, A.J., Begley, M., et al. 1988. Human resistance to *Schistosoma mansoni* is associated with IgG reactivity to a 37-kDa larval surface antigen. *J Immunol*, **140**, 2727–36.

Feldmeier, H., Krantz, I. and Poggensee, G. 1994. Female genital schistosomiasis as a risk-factor for the transmission of HIV. *Int J Std Aids*, **5**, 368–72.

Fujii, Y. 1847. Katayama-ki. *Chugai Iji Shimpo*, **691**, 55–6, re-description by Fujinami A., Chinese.

Fujinami, A. 1904. Wetere Mitteilung uber die pathologische Anatomie der sog 'Katayama-Krankheit' und der Krankheitserreger derselben. *Kyoto Igaku Zassi*, **1**, 201–13, Japanese, with German abstract.

Fujinami, A. and Nakamura, H. 1909. Infection route. *Infection route, development of the parasite, and infectivity of animals in Katayama disease (schistosomiasis japonica) in Hiroshima Prefecture*, **6**, 224–52, in Japanese.

Gobert, G.N., Stenzel, D.J., et al. 1997. *Schistosoma japonicum*: immunolocalization of paramyosin during development. *Parasitology*, **114**, 45–52.

Greer, G.J., Ow-Yang, C.K. and Yong, H.S. 1988. *Schistosoma malayensis* n.sp.: a *Schistosoma japonicum*-complex schistosome from Peninsular Malaysia. *J Parasitol*, **74**, 471–80.

Harinasuta, C., Sornmani, S., et al. 1972. Infection of aquatic hydrobiid snails and animals with *Schistosoma japonicum*-like parasites from Khong Island, Southern Laos. *Trans R Soc Trop Med Hyg*, **66**, 184–5.

Hatz, C. 2001. The use of ultrasounds in schistosomiasis. *Adv Parasitol*, **48**, 225–84.

Hawn, T.R., Tom, T.D. and Strand, M. 1993. Molecular cloning and expression of SmIrV1, a *Schistosoma mansoni* antigen with similarity to calnexin, calreticulin and OvRal1. *J Biol Chem*, **268**, 7692–8.

Henkle, K.J., Davern, K.M., et al. 1990. Comparison of the cloned genes of the 26- and 28-kilodalton glutathione S-transferases of *Schistosoma japonicum* and *Schistosoma mansoni*. *Mol Biochem Parasitol*, **40**, 23–34.

Hernandez, M.G., Hafalla, J.C., et al. 1999. Paramyosin is a major target of the human IgA response against *Schistosoma japonicum*. *Parasite Immunol*, **21**, 641–7.

Hirayama, K., Chen, H., et al. 1999. HLA-DR-DQ alleles and HLA-DP alleles are independently associated with susceptibility to different stages of post-schistosomal hepatic fibrosis in the Chinese population. *Tissue Antigens*, **53**, 269–74.

Hooker, C.W. and Brindley, P.J. 1996. Cloning and characterization of strain-specific transcripts encoding triosephosphate isomerase, a candidate vaccine antigen from *Schistosoma japonicum*. *Mol Biochem Parasitol*, **82**, 265–9.

Hunter III, G.W. and Yokogawa, M. 1984. Control of schistosomiasis japonica in Japan – a review 1950–1978. *Jpn J Parasitol*, **33**, 341–51.

Iijima, T., Lo, C.-T. and Ito, Y. 1971. Studies on schistosomiasis in the Mekong Basin. I. Morphological observation of the schistosomes and detection of their reservoir hosts. *Jpn J Parasitol*, **20**, 24–33.

Izhar, A., Sinaga, R.M., et al. 2002. Recent situation of schistosomiasis in Indonesia. *Acta Tropica*, **82**, 283–8.

Janecharut, T., Hata, H. and Kojima, S. 1991. Effects of heterologous helminth infections on passive transfer of immunity using a mouse monoclonal IgE antibody against *Schistosoma japonicum*. *Parasitol Res*, **77**, 668–74.

Janecharut, T., Hata, H., et al. 1992. Effects of recombinant tumour necrosis factor on antibody-dependent eosinophil-mediated damage to *Schistosoma japonicum* larvae. *Parasite Immunol*, **14**, 605–16.

Jiang, Q.-W., Wang, L.-Y., et al. 2002. Morbidity control of schistosomiasis in China. *Acta Trop*, **82**, 115–25.

Kagawa, H., Gengyo, K., et al. 1989. Paramyosin gene (*unc-15*) of *Caenorhabditis elegans*. Molecular cloning, nucleotide sequence and models for thick filament structure. *J Mol Biol*, **207**, 311–33.

Kashiwado, T., Ishijima, F., et al. 1927. Clinical report on the outbreak of schistosomiasis japonica in the Sakura region. *J Chiba Med Soc*, **5**, 1473–528.

Katsurada, F. 1904. *Schistosomum japonicum*, ein neuer menschlicher Parasit, durch welchen eine endemische Krankheit in verschiedenen Gegenden Japans verursacht wird. *Annot Zool Japan*, **5**, 147–60.

Kojima, S. 1970. Histopathological and imunological studies on the experimental schistosomiasis japonica, with special reference to the chemotherapeutic effects on the involving tissues. *Jpn J Parasitol*, **19**, 54–75.

Kojima, S., Yokogawa, M. and Tada, T. 1972. Raised levels of serum IgE in human helminthiases. *Am J Trop Med Hyg*, **21**, 913–18.

Kojima, S., Niimura, M. and Kanazawa, T. 1987. Production and properties of a mouse monoclonal IgE antibody to *Schistosoma japonicum*. *J Immunol*, **139**, 2044–9.

Kojima, S., Kanazawa, T., et al. 1988. Epidemiologic studies on schistosomiasis japonica in a newly found habitat of *Oncomelania* snails in Japan. *Am J Trop Med Hyg*, **38**, 92–6.

Kojima, S., Nara, T., et al. 1998. A vaccine trial for controlling reservoir livestock against schistosomiasis japonica. In Tada, I., Kojima, S. and Tsuji, M. (eds), *Proceedings of the 9th International Congress of Parasitology*. Bologna: Monduzzi Editore, 489–94.

Koster, B., Hall, M.R. and Strand, M. 1993. *Schistosoma mansoni*: immunoreactivity of human sera with the surface antigen Sm23. *Exp Parasitol*, **77**, 282–94.

Laclette, J.P., Landa, A., et al. 1991. Paramyosin is the *Schistosoma mansoni* (Trematoda) homologue of antigen B from *Taenia solium* (Cestoda). *Mol Biochem Parasitol*, **44**, 287–95.

Le, T.H., Humair, P.F., et al. 2001. Mitochondrial gene content, arrangement and composition compared in Africa and Asian schistosomes. *Mol Biochem Parasitol*, **117**, 61–71.

Le, T.H., Blair, D. and McManus, D.P. 2002. Revisiting the question of limited genetic variation within *Schistosoma japonicum*. *Ann Trop Med Parasitol*, **96**, 155–64.

Leiper, R.T. 1916. On the relation between the terminal-spined and lateral-spined eggs of *Bilharzia*. *Br Med J*, **i**, 411.

Leonardo, L.R., Acosta, L.P., et al. 2002. Difficulties and strategies in the control of schistosomiasis in the Philippines. *Acta Trop*, **82**, 295–9.

Li, Y.S., Ross, A.G.S., et al. 1997. Serological diagnosis of *Schistosoma japonicum* infections in China. *Trans R Soc Trop Med Hyg*, **91**, 19–21.

Li, Y.S., Sleigh, A.C., et al. 2000. Epidemiology of *Schistosoma japonicum* in China: morbidity and strategies for control in the Dongting Lake region. *Int J Parasitol*, **30**, 273–81.

Liang, Y.-S., Kitikoon, V. 1980. Susceptibility of *Lithoglyphopsis aperta* to *Schistosoma mekongi* and *Schistosoma japonicum*. The Mekong Schistosome. *Malacological Rev*, 53–60.

Limberger, R.J. and McReynolds, L.A. 1990. Filarial paramyosin: cDNA sequences from *Dirofilaria immitis* and *Onchocerca volvulus*. *Mol Biochem Parasitol*, **38**, 271–80.

Ling, C.C., Cheng, W.J. and Chung, H.L. 1949. Clinical and diagnostic features of schistosomiasis japonica. A review of 200 cases. *Chin Med J*, **67**, 347–66.

Liu, Q.H., Zhang, J.W., et al. 2000. Investigation of association between female genises and schistosomiasis japonica infection. *Acta Trop*, **77**, 179–83.

Matsumoto, J. 2002. Adverse effects of praziquantel treatment of *Schistosoma japonicum* infection: involvement of host anaphylactic reactions induced by parasite antigen release. *Int J Parasitol*, **32**, 461–71.

McManus, D.P., Liu, S., et al. 1998. The vaccine efficacy of native paramyosin (Sj-97) against Chinese *Schistosoma japonicum*. *Int J Parasitol*, **28**, 1739–42.

McManus, D.P., Wong, J.Y.M., et al. 2001. Recombinant paramyosin (rec-Sj-97) tested for immunogenicity and vaccine efficacy against *Schistosoma japonicum* in mice and water buffaloes. *Vaccine*, **20**, 870–8.

Miao, Y.X., Liu, S.X. and McManus, D.P. 1998. Isolation of native, biochemically purified triosephosphate isomerase from a Chinese strain of *Schistosoma japonicum* and its protective efficacy in mice. *Parasitol Int*, **47**, 195–9.

Mitchell, G.F., Garcia, E.G., et al. 1988. Sensitization against the parasite antigen Sj26 is not sufficient for consistent expression of resistance to *Schistosoma japonicum* in mice. *Trans R Soc Trop Med Hyg*, **82**, 885–9.

Miyairi, K. and Suzuki, M. 1913. On the development of *Schistosoma japonicum*. *Tokyo Iji Shinshi*, **1836**, 1–5, in Japanese.

Muhlschlegel, F., Sygulla, L., et al. 1993. Paramyosin of *Echinococcus granulosus*: cDNA sequence and characterization of a tegumental antigen. *Parasitol Res*, **79**, 660–6.

Nara, T., Matsumoto, N., et al. 1994. Demonstration of the target molecule of a protective IgE antibody in secretory glands of *Schistosoma japonicum* larvae. *Int Immunol*, **6**, 963–71.

Nara, T., Tanabe, K., et al. 1997. The B cell epitope of paramyosin recognized by a protective monoclonal IgE antibody to *Schistosoma japonicum*. *Vaccine*, **15**, 79–84.

Pearce, E.J., James, S.L., et al. 1988. Induction of protective immunity against *Schistosoma mansoni* by vaccination with schistosome paramyosin (Sm97), a nonsurface parasite antigen. *Proc Natl Acad Sci USA*, **85**, 5678–82.

Reynolds, S.R., Shoemaker, C.B. and Harn, D.A. 1992. T and B cell epitope mapping of SM23, an integral membrane protein of *Schistosoma mansoni*. *J Immunol*, **149**, 3995–4001.

Reynolds, S.R., Dahl, C.E. and Harn, D.A. 1994. T and B epitope determination and analysis of multiple antigenic peptides for the *Schistosoma mansoni* experimental vaccine triose-phosphate isomerase. *J Immunol*, **152**, 193–200.

Richter, D. and Harn, D.A. 1993. Candidate vaccine antigens identified by antibodies from mice vaccinated with 15- or 50-kilorad-irradiated cercariae of *Schistosoma mansoni*. *Infect Immun*, **61**, 146–54.

Rogers, M.V., Davern, K.M., et al. 1988. Immunoblotting analysis of the major integral membrane protein antigens of *Schistosoma japonicum*. *Mol Biochem Parasitol*, **29**, 77–87.

Ross, A.G.P., Sleigh, A.C., et al. 2001. Schistosomiasis in the People's Republic of China: prospects and challenges for the 21st century. *Clin Microbiol Rev*, **14**, 270–95.

Sakamoto, K. and Ishii, Y. 1977. Scanning electron microscope observations on adult *Schistosoma japonicum*. *J Parasitol*, **63**, 407–12.

Schmidt, J., Bodor, O., et al. 1996. Paramyosin isoforms of *Schistosoma mansoni* are phosphorylated and localized in a large variety of muscle types. *Parasitology*, **112**, 459–67.

Scott, J.C. and McManus, D.P. 2000. Molecular cloning and enzymatic expression of the 28-kDa glutathione S-transferase of *Schistosoma japonicum*: evidence for sequence variation but lack of consistent vaccine efficacy in the murine host. *Parasitol Int*, **49**, 289–300.

Shoemaker, C., Gross, A., et al. 1992. cDNA cloning and functional expression of the *Schistosoma mansoni* protective antigen triose-phosphate isomerase. *Proc Natl Acad Sci USA*, **89**, 1842–6.

Smith, D.B., Davern, K.M., et al. 1986. Mr 26,000 antigen of *Schistosoma japonicum* recognized by resistant WEHI 129/J mice is a parasite glutathione S-transferase. *Proc Natl Acad Sci USA*, **83**, 8703–7.

Soisson, L.M., Masterson, C.P., et al. 1992. Induction of protective immunity in mice using a 62-kDa recombinant fragment of a *Schistosoma mansoni* surface antigen. *J Immunol*, **149**, 3612–20.

Sornmani, S. 1976. Current status of research on the biology of Mekong *Schistosoma*. *SE Asian J Trop Med Pub Hlth*, **7**, 208–13.

Sornmani, S., Kitikoon, V., et al. 1971. Epidemiological study of schistosomiasis japonica on Khong Island, southern Laos. *SE Asian J Trop Med Pub Hlth*, **2**, 365–74.

Strandgaard, H., Johansen, M.V., et al. 2001. The pig as a host for *Schistosoma mekongi* in Laos. *J Parasitol*, **87**, 708–9.

Sturrock, R.F. 1993. The intermediate hosts and host-parasite relationships. In: Jordan, P., Webbe, G. and Sturrock, R.F. (eds), *Human schistosomiasis*. Oxon: CAB International, 33–85.

Sun, W.Y., Liu, S.X., et al. 1999. Bacterial expression and characterization of functional recombinant triosephosphate isomerase from *Schistosoma japonicum*. *Protein Expr Purif*, **17**, 410–13.

Tanaka, H. and Tsuji, M. 1997. From discovery to eradication of schistosomiasis in Japan: 1847–1996. *Int J Parasitol*, **27**, 1465–80.

Taylor, M.G., Huggins, M.C., et al. 1998. Production and testing of *Schistosoma japonicum* candidate vaccine antigens in the natural ovine host. *Vaccine*, **16**, 1290–8.

Tendler, M., Brito, C.A., et al. 1996. A *Schistosoma mansoni* fatty acid-binding protein, Sm14, is the potential basis of a dual-purpose anti-helminth vaccine. *Proc Natl Acad Sci USA*, **93**, 269–73.

Tiu, W.U., Davern, K.M., et al. 1988. Molecular and serological characteristics of the glutathione S-transferases of *Schistosoma japonicum* and *Schistosoma mansoni*. *Parasite Immunol*, **10**, 693–706.

Trottein, F., Kieny, M.P., et al. 1990. Molecular cloning and tissue distribution of a 26-kilodalton *Schistosoma mansoni* glutathione S-transferase. *Mol Biochem Parasitol*, **41**, 35–44.

Urbani, C., Sinoun, M., et al. 2002. Epidemiology and control of menkongi schistosomiasis. *Acta Tropica*, **82**, 157–68.

Voge, M., Bruckner, D. and Bruce, J.I. 1978. *Schistosoma mekongi* sp. n. from man and animals, compared with four geographic strains of *Schistosoma japonicum*. *J Parasitol*, **64**, 577–84.

von Lichtenberg, F., Sadun, E.H., et al. 1971. Experimental infection with *Schistosoma japonicum* in chimpanzees. Parasitologic, clinical, serologic, and pathological observations. *Am J Trop Med Hyg*, **20**, 850–93.

Wang, X.Z., Li, S.T. and Zhou, Z.X. 1999. A rapid one-step method of EIA for detection of circulating antigen of *Schistosoma japonicum*. *Chin Med J (Engl)*, **112**, 124–8.

Wright, M.D., Henkle, K.J. and Mitchell, G.F. 1990. An immunogenic Mr 23,000 integral membrane protein of *Schistosoma mansoni* worms that closely resembles a human tumor-associated antigen. *J Immunol*, **144**, 3195–200.

Xiao, S.H., You, J.Q., et al. 1995. Experimental studies on early treatment of schistosomal infection with artemether. *SE Asian J Trop Med Pub Hlth*, **26**, 306–18.

Yuan, H.C., Guo, J.G., et al. 2000. The 1992–1999 World Bank Schistosomiasis Research Initiative in China: outcome and perspectives. *Parasitol Int*, **49**, 195–200.

Zhang, R., Suzuki, T., et al. 2000. Cloning and molecular characterization of calpain, a calcium-activated neutral proteinase, from different strains of *Schistosoma japonicum*. *Parasitol Int*, **48**, 232–42.

Zheng, J.A., Gu, X.G., et al. 2002. Relationship between the transmission of schistosomiasis japonica and the construction of the Three Gorge Reservoir. *Acta Trop*, **82**, 147–56.

31

Lung and liver flukes

MELISSA R. HASWELL-ELKINS

THE PARASITES

An estimated 40 million people harbor food-borne trematodes globally (WHO 1995). Although previous estimates of numbers infected may have been inaccurately high, there is evidence that the pattern of human infections with these parasites is in a state of change regarding number, geographic range, and economic and clinical significance (Xu et al. 1995; Eckert 1996). In many areas, their prevalence has decreased as a result of control efforts using drug treatment and health education (Jongsuksuntigul and Imsomboon 1997; Ooi et al. 1997). However, in other places, infections have persisted or increased because of environmental changes increasing snail habitats, the free-market economy with wider distribution networks for food, lack of food inspections, poor sanitation, human migration, and declining economic conditions (Maurice 1994; Mas-Coma et al. 1999; Anantaphruti 2001). Many people also continue to enjoy raw or undercooked foods, often in spite of knowledge about its potential health hazards.

This large group of parasites includes the liver flukes, *Opisthorchis viverrini*, *Clonorchis sinensis*, *O. felineus*, *Fasciola hepatica*, and *F. gigantica*, and nine or more species of the lung fluke, *Paragonimus* (Beaver et al. 1984; Toscano et al. 1995). A third group of food-borne trematodes found in humans, not detailed here, are the intestinal flukes (e.g. *Fasciolopsis*, *Echinostoma*, and *Heterophyes*). Schistosomes, or blood flukes, are also trematodes, but are acquired by skin penetration, not via contaminated food.

Fasciola hepatica (Figure 31.1a) was first referred to by de Brie in 1379 in sheep with 'liver rot,' 500 years

before the discovery and formal description of trematodes that infect humans. The parasite was named by Linnaeus in 1758, just before its discovery in a human by Pallas in 1760. *F. gigantica* was found in a giraffe and named by Cobbold in 1855. The life cycle of *Fasciola* was the first to be elucidated for a trematode; this was achieved concurrently but independently by Leuckart

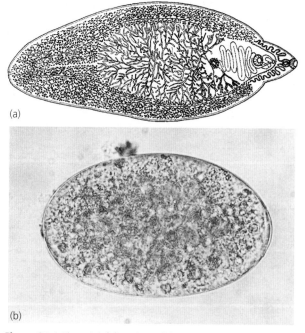

Figure 31.1 *The adult* **(a)** *and egg* **(b)** *of the liver fluke,* Fasciola hepatica. *(Drawn by S. Kaewkes, reproduced with permission)*

and Thomas in the 1880s and Lutz in 1892 (Beaver et al. 1984).

F. hepatica and *F. gigantica* are parasites of global veterinary importance, causing huge productivity losses in sheep, goats, and cattle. The two species are difficult to differentiate morphologically, although they utilize different snail hosts. Their relative importance to veterinary and human health is debated. The less studied *F. gigantica* may be the more important species in the tropics, while *F. hepatica* predominates in cooler regions. Human fascioliasis is uncommon but distributed globally (Chen and Mott 1990).

Paragonimus westermani (Figure 31.2a) was first discovered in the lungs of Bengal tigers by Kerbert in 1878. Shortly after, the worm was found at human autopsy and eggs were found in sputum by Ringer, Manson, and Baelz in Taiwan and Japan. Kobayashi, S. Yokogawa, and other Japanese investigators revealed the complete life cycle. There is debate regarding the number and relative importance of individual *Paragonimus* species infecting humans. Until recently, *P. westermani* was thought to be the only agent of human infection, but at least nine species are now known to be involved (Miyazaki 1982; Yokogawa 1982; Toscano et al. 1995; Velez et al. 2002). Considerable work has been carried out to attempt to characterize similarities and differences in their karyotypes and nuclear and mitochondrial genomes (see Blair et al. 1999; Blair 2000).

C. sinensis was discovered by McConnel in 1874 at the autopsy of a Chinese carpenter in Calcutta, and the parasite was subsequently named by Cobbold in 1875. *O. felineus* was first reported in a cat, while in 1892 Winogradoff reported the first human infection in Siberia. *O. viverrini* (Figure 31.3a) was first described in cats in 1886 by Poirier, then infection was discovered in a resident of Chiangmai, northern Thailand by Kerr. Leiper identified the worms as a species distinct from *O. felineus* in 1915. These parasites have lived with humans for millennia; *Clonorchis* eggs have been found in a corpse from the West Han dynasty in China dated 278BC (cited in Chen et al. 1994).

The liver flukes have attracted considerable interest because of their close association with bile duct cancer. *O. viverrini* infection was classified as a human carcinogen by the International Agency for Research on Cancer, and *C. sinensis* infection was judged a probable carcinogen (IARC 1994). The current understanding of the link between liver fluke infection and cancer has recently been reviewed (Okuda et al. 2002; Watanapa and Watanapa 2002).

Classification

The liver and lung flukes belong to the Phylum Platyhelminths (flatworms), Class Trematoda, and Subclass

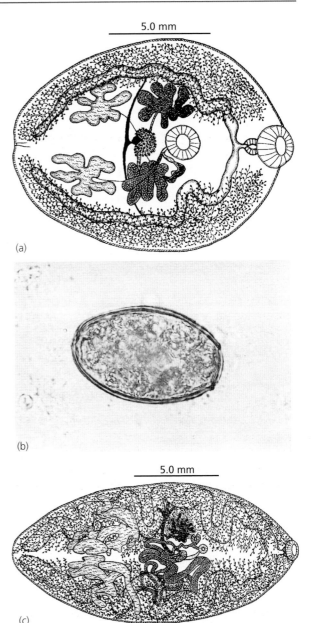

Figure 31.2 *The adult* **(a)** *and egg* **(b)** *of the lung fluke,* Paragonimus westermani *and adult worm of* P. heterotremus **(c)** *(Drawn by S. Kaewkes, reproduced with permission)*

Digenea (Beaver et al. 1984). The latter term (digenetic trematodes) is often used to indicate the indirect life cycle involving several morphological stages and at least one intermediate host. The medically important foodborne flukes were once classified under a single genus, *Distoma*, referring to their two conspicuous suckers. Later it became clear that distomate flukes comprised a large and complex group of parasites, as indicated by today's classification into several superfamilies.

Fasciola, along with intestinal flukes *Fasciolopsis* and *Echinostoma*, belongs to the superfamily Echinostomatoidea. *Paragonimus* is a member of the superfamily Plagiorchioidea; whereas *Opisthorchis* and *Clonorchis* fall within the superfamily Opisthorchioidea. These

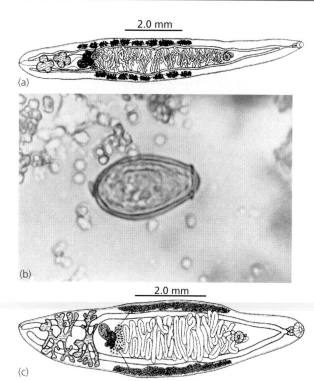

Figure 31.3 *The adult* **(a)** *and egg* **(b)** *of the liver fluke,* Opisthorchis viverrini *and adult worm of* Clonorchis sinensis **(c)** *(Drawn by S. Kaewkes, reproduced with permission)*

orders are differentiated on the basis of life cycle (intermediate hosts), morphological structure (especially of the excretory bladder), the suckers in the adult worm and spines in intermediate stages, and size and maturity of eggs when laid.

Most food-borne trematodes are zoonotic, i.e. usual parasites of nonhuman animals which 'accidentally' infect people. Several species, for example of *Dicrocelium*, *Opisthorchis*, and *Eurytrema*, infect birds, livestock, and wildlife and are reported rarely in humans (Beaver et al. 1984).

Morphology and structure

The adult flukes are elongated, often described as 'leaf shaped' and bilaterally symmetrical with three body layers, but have no true body cavity (Schmidt and Roberts 1994; Rim 1986). They are covered with a nonciliated integument which is not simply a secreted sheath: it is composed of complex, living tissue often containing invaginations, spines, and lectin-binding carbohydrate residues (Apinhasmit et al. 2000). *Fasciola* and *Paragonimus* species are covered with large spines. Muscles lie subcutaneously below the integument enveloping the body, while specialized structures, such as the esophagus and suckers, possess radial muscle fibers.

As mentioned above, the oral (surrounding the mouth) and ventral suckers are located in the anterior end, while the digestive system ends blindly with no rectum or anus. A specialized group of cells called solenocytes, or flame cells, join into an excretory bladder that exits through an excretory pore at the posterior end. This system probably carries out excretory functions and controls water balance. There is no circulatory system. The nervous system comprises an esophageal commissure with ladder-like pairs of nerve trunks running in each plane of the body ending in sensory processes. Eyespots may occur on free-swimming larval stages, but not on the adults.

A special feature of food-borne trematodes is their hermaphroditic (monoecious) reproductive system, that is, each individual worm produces both eggs and sperm. This contrasts with the schistosome trematodes which are dioecious (the sexes are separate) but in very close physical association. Most of the reproductive organs are positioned posteriorly to the ventral sucker, with their external opening (genital pore) often in close proximity to the ventral sucker.

The largest of the human liver and lung flukes is *Fasciola gigantica* (up to 75 mm long by 12 mm wide), followed by *F. hepatica* (30 mm × 13 mm) (see Figures 31.1a and 31.4). Both possess 'shoulders' and a conical anterior end. No single morphological feature definitively separates the two species, although there are genetic differences (Blair et al. 1999). *F. gigantica* tends to be more oblong with a longer, rounded posterior end, as compared to the short and more angular posterior end of *F. hepatica*. The testes of *F. gigantica* are located closer toward the anterior end, and its ventral sucker is larger than that of *F. hepatica*.

The intestinal caeca, testes, and vitelline follicles of *Fasciola* are extensively branched. The eggs (Figure 31.1b) are large, ovoid, yellowish brown with a small operculum, and contain an immature larva, the miracidium. The eggs of *F. gigantica* (190 μm) tend to be consistently larger than those of *F. hepatica*, which are 140–150 μm long (Beaver et al. 1984).

Paragonimus (see Figure 31.2a) is roughly half the size of *Fasciola*, measuring up to 16 mm long and 8 mm wide, oval-shaped, and reddish brown in color with an integument covered with scale-like spines (Miyazaki 1982). The ventral sucker is located toward the middle of the body and is of similar size to the oral sucker on the anterior end. The fluke possesses a large excretory bladder. Various morphological and life cycle differences in the adult worms distinguish the many species capable of infecting humans (e.g. *P. heterotremus*, Figure 31.2c). The yellowish brown *Paragonimus* eggs are oval, measure 90 × 55 μm, and have a flattened operculum (Figure 31.2b). They are unembryonated when laid.

In humans, *Clonorchis* measures up to 20 mm long and 5 mm wide (see Figure 31.3c), whereas the *Opisthorchis* species are somewhat smaller (see Figures 31.3a and 31.5; Komiya 1966; Rim 1986; Sadun 1955). The

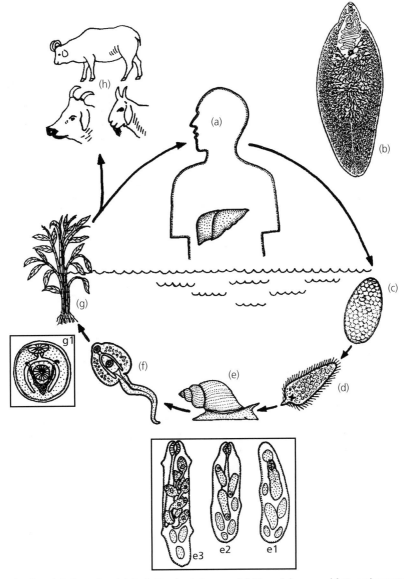

Figure 31.4 *The life cycle of* Fasciola hepatica. **(a)** *Definitive host: human;* **(b)** *The adult worm;* **(c)** *Unembryonated egg;* **(d)** *Miracidium;* **(e)** *First intermediate host:* Lymnaea sp. *Intramolluscan stages: e1, sporocyst; e2, mother redia; e3, daughter redia;* **(f)** *Cercaria;* **(g)** *Second intermediate host: aquatic plants. Aquatic plant stage: g1, metacercaria;* **(h)** *Reservoir hosts: sheep, cattle, goat. (Drawn by S. Kaewkes, reproduced with permission)*

ventral sucker is smaller than the oral and located within the anterior half of the worm. Like *Fasciola*, the adult worm morphologies do not provide definitive separation, but the flame cell patterns on cercariae are distinct (Wykoff et al. 1965). The number and shape of testicular lobes, their location and the appearance of vitelline glands vary between, but also within, the species.

The genital pore through which eggs pass is located near the ventral sucker, with the large, coiled uterine glands filling most of the body. Eggs (Figure 31.3b) are small (approximately 25 × 15 μm), yellowish brown, contain fully developed miracidia when laid, and have a knob at the posterior end and a distinct operculum on the anterior end. Measurements and shape vary within

and between species (Sadun 1955; Kaewkes et al. 1991; Ditrich et al. 1992).

General biology

Movement is a very important feature of trematodes. Their complex life cycle requires finding and entering appropriate hosts and extensive migration from the duodenum to the appropriate site of maturation within the final host (Schmidt and Roberts 1994). Flukes also undergo remarkable physiological adjustment in order to tolerate and function in their highly varied environments, from fresh water to invertebrate host to mammalian tissue. The fact that trematodes successfully utilize

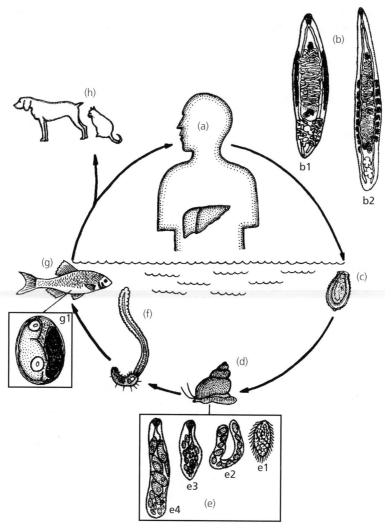

Figure 31.5 *The life cycle of human liver flukes:* **(a)** *Definitive host: human;* **(b)** *The adult worm. b1,* Clonorchis sinensis; *b2,* Opisthorchis viverrini; **(c)** *Embryonated egg;* **(d)** *First intermediate host:* Bithynia sp.; **(e)** *Intramolluscan stages. e1, miracidium; e2, sporocyst; e3, mother redia; e4, daughter redia;* **(f)** *Cercaria;* **(g)** *Second intermediate host: cyprinoid fish. Stage in fish: g1, metacercaria;* **(h)** *Reservoir hosts: dog, cat, and other mammals. (Drawn by S. Kaewkes, reproduced with permission)*

these extremely complex routes to achieve adulthood and reproduction reflects complex evolutionary development to maximize transmission.

Hou (1955) described the movement of *Clonorchis* by means of attachment and detachment of its two suckers combined with extension and contraction of the body. The two suckers and a collar of spines on the immature worm are presumably used to migrate up the biliary tract, against the flow of bile (Apinhasmit et al. 1993). Mature worms probably move short distances within the ducts. Attachment is secured by the ventral sucker adhering to the biliary epithelium, leaving the oral sucker free for feeding.

The nutritional supply of human trematodes is not precisely understood. Initially it was postulated that *Clonorchis* feeds on red blood cells, bile, and epithelial cells. Hou (1955), however, suggested that the parasite lives on protein, oxygen, and glycogen derived from mucin and tissue fluid entering the lumen following

desquamation. *Fasciola* feeds on blood and hepatic tissue as it transverses the parenchyma during the migration phase, and its food supply in the bile duct may be similar to that of *Clonorchis*. The biliary flukes live in conditions of very low oxygen tension, and anaerobic glycolysis of glycogen to glucose is the major energy pathway (Schmidt and Roberts 1994). A glycolytic enzyme, phosphoglycerate kinase, is found in the tegument of *Clonorchis sinensis* and probably assists in the utilization of external glucose for its metabolism (Hong et al. 2000). Ammonia, urea, and amino acids are excreted. Among the amino acids released by *Fasciola* are remarkably high levels of proline (Wolf-Spengler and Isseroff 1983).

Life cycle

The life cycles of food-borne trematodes are complex, involving one or more intermediate hosts (the first always a snail), several morphological stages, and distinct generations. These stages are similar between liver and lung flukes, and although intermediate host species vary, all are freshwater not marine.

DETAILS OF THE PARASITES IN HUMANS AND OTHER DEFINITIVE HOSTS

The infective stages of these flukes are called metacercariae. Humans become infected with *Opisthorchis* and *Clonorchis* by consuming raw or undercooked fish containing encysted metacercariae (Figure 31.5). *Paragonimus* is acquired by eating metacercariae in raw crabs (Figure 31.6), whereas *Fasciola* metacercariae encyst on aquatic plants (see Figure 31.4). Ingested metacercariae excyst in the duodenum, releasing larvae.

Newly excysted *Opisthorchis* and *Clonorchis* larvae migrate through the ampulla of Vater and the extra-

hepatic bile ducts to the smaller, intrahepatic ducts where they mature. Within 1 month, adult worms begin producing an average of 10 000 eggs per day which exit the bile ducts and are excreted in the feces (Sithithaworn et al. 1991).

Paragonimus larvae undergo systemic migration. Upon ingestion, the larvae are released in the duodenum, penetrate the intestine, and enter into the abdominal cavity. The parasites enter and remain in the abdominal wall for some days, then continue migration through the diaphragm into the pleural cavity and lungs. When they reach the terminal alveoli or under the pleura, they become encapsulated by the host's inflammatory response and produce eggs. Eggs are expelled in the sputum or may be dislodged by coughing, are then swallowed and excreted in the feces.

Fasciola metacercariae excyst in the duodenum, and the immature worms migrate through the duodenal wall,

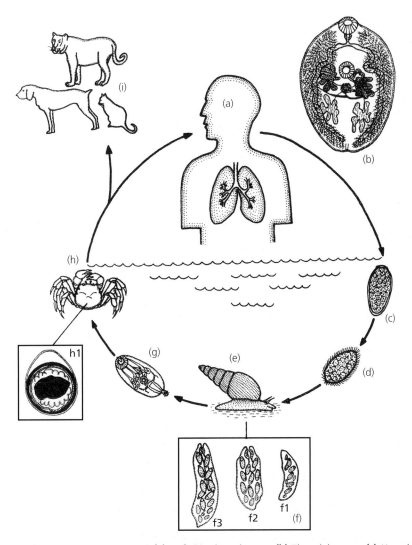

Figure 31.6 *The life cycle of* Paragonimus westermani: **(a)** *Definitive host: human;* **(b)** *The adult worm;* **(c)** *Unembryonated egg;* **(d)** *Miracidium;* **(e)** *First intermediate host: Thiarid snail;* **(f)** *Intramolluscan stages. f1, sporocyst; f2, mother redia; f3, daughter redia;* **(g)** *Cercaria;* **(h)** *Second intermediate host: crustaceans. Stage in crustacean: h1, metacercaria;* **(i)** *Reservoir hosts: dog, cat, tiger. (Drawn by S. Kaewkes, reproduced with permission)*

into the body cavity. They then burrow through Glisson's capsule and across the hepatic parenchyma to the proximal bile ducts and gallbladder where they mature. This migration takes 6–7 weeks. The adult worms begin producing eggs that exit in the feces 3–4 months after ingestion of metacercariae.

The worms may live many years. The maximum reported life span of *Clonorchis* in the absence of reinfection is 26 years from a Chinese emigrant, and average life span in endemic areas is about 10 years (Attwood and Chou 1978; Chen et al. 1994). Similarly *Paragonimus* is thought to live a maximum of 20 years and an average of 6 years; and *Fasciola* probably lives around 10 years (Chen and Mott 1990).

O. viverrini, *C. sinensis*, and *O. felineus* infect pigs, cats, civets, rats, dogs, foxes, and other mammals. *Fasciola* infections are maintained in domestic livestock, e.g. sheep, goats, camels, buffaloes, and cattle. Wild rabbits, hares, and rodents are thought to maintain infection in uncultivated watercress beds in France (Rondelaud et al. 2001). *Paragonimus* infection is found in a range of reservoir hosts such as tigers and cats, rats, wolves, dogs, and foxes. Although infection cannot occur by consuming adult worms or eggs within livers of infected animals, *Paragonimus* may establish through consumption of incompletely cooked pig or wild boar meat containing lung fluke larvae (Miyazaki 1982).

Patterns of infection in reservoir hosts are not always closely linked with those in humans, since these are determined by distributions of intermediate hosts, whereas human infection is further limited by eating behavior (Sadun 1955; Rim 1986). However, reservoir hosts may be particularly important where human infections are uncommon or sporadic, and human egg excretion is inefficient (e.g. *Fasciola*, *Paragonimus*), or blocked by sanitation or anthelmintic treatment (*Opisthorchis* and *Clonorchis*).

DEVELOPMENTAL STAGES OF THE PARASITES

Freshwater bodies (small ponds, streams and rivers, flooded rice fields, and large reservoirs) or (for *Fasciola hepatica*) submerged grass become contaminated with eggs from the feces and sputum (for *Paragonimus*) of infected people or reservoir hosts. The parasites must enter an appropriate snail. Species of *Bithynia*, *Melanoides*, *Parafossarulus*, and *Assiminea* are important first intermediate hosts of *Opisthorchis* and *Clonorchis*. These snails become infected by ingesting eggs which contain fully mature miracidial larvae. In contrast, *Paragonimus* eggs develop and hatch in water after approximately 3 weeks. The miracidia must find and penetrate a permissible species of snail; the major ones are of the genera *Semisulcospira*, *Thiara*, and *Oncomelania*. *F. hepatica* miracidia develop and hatch within 2 weeks and swim until finding an appropriate snail of the genus *Lymnaea*. *F. gigantica* requires aquatic species of *Lymnaea*. Both the snails and the developmental stages of the fluke appear to have made adaptations that provide a greater survival capacity in the high altitudes of the Altiplano of Bolivia and Peru (Mas-Coma et al. 2001).

Inside the snail, the miracidia develop into sporocysts and rediae, which undergo parthenogenetic reproduction, finally giving birth to cercariae. Cercariae emerge and swim to find an appropriate site for transformation into metacercariae. Nearly 100 species of cyprinoid fish serve as the second intermediate host of *Opisthorchis* or *Clonorchis* (Komiya 1966; Waikagul 1998; Fattakhov 2002). *Paragonimus* cercariae penetrate gills and muscles of freshwater mitten crabs (*Potamon*, *Sesarma*, *Eriocher*) or crayfish (*Astacus*). Fifty three crustacean species have been reported to serve as second intermediate hosts of *Paragonimus* spp. (Toscano et al. 1995). The species of *Paragonimus* present appears to be related to water flow and suitability of the aquatic environment for various crab and snail species (Cheng 1999). The cercariae transform into metacercariae in the viscera, muscles, or gills. *Fasciola* encysts on green water vegetation, e.g. watercress; it does not enter a second intermediate host.

These life cycles require at least 4 months to complete and may be prolonged by winter hibernation of snails (Rim 1986). Seasonal variation in transmission, water temperature, densities of intermediate and reservoir hosts, and human sanitation and eating behavior govern life cycle completion and the prevalence of human infection.

TRANSMISSION AND EPIDEMIOLOGY

Geographic distribution

A crude estimate of the global number of human liver fluke infections is in the order of 17 million, made up of 7 million with *Clonorchis*, 9 million with *O. viverrini*, and 1.2 million with *O. felineus* infections (Iarotski and Be'er 1993; WHO 1995). Countries with endemic human liver fluke infection are Thailand, Laos, and possibly Cambodia (*O. viverrini*) (Lee et al. 2002), Korea, China, Taiwan, (formerly Japan), Hong Kong, Macao, and Vietnam (*C. sinensis*), and Russia, Kazakhstan, Ukraine, and possibly some eastern Europe countries (*O. felineus*). Encouraging news has come from northeast Thailand, where the regional prevalence of infection appears to have dropped from 34 percent in 1981, to 24–30 percent in 1992 to 18.6 percent in 1994 (Jongsuksuntigul and Imsomboon 1997). There also seems to be a decline in the frequency of *Opisthorchis* metacercariae found in fish over the same time period (Waikagul 1998). Prevalences remain high in neighboring Laos (Vannachone et al. 1998; Kobayashi et al. 2000), however, and may be increasing in central and northern Thailand. Research in Northern Thailand collecting adult worms post-treatment and examining metacercariae in fish has suggested this increase may be partly a result of misdiagnosis of minute intestinal fluke infections (Radomyos et al. 1998; Sukontason et al. 2001), which are much less dangerous than liver flukes. Infections in China are also ethnically and geographically associated. Cantonese, notably Hakka people, are most frequently infected in the southern provinces of Guangdong and Guangxi, while infections in northeast China occur among the Korean national minority who migrated there (Chen et al. 1994). Infection in Hong Kong is probably acquired from eating fish imported alive from southern China. A study conducted in Ninh Binh Province in Vietnam revealed a high prevalence of *Clonorchis* metacercariae among fish farmed in ponds (Kino et al. 1998).

The human lung flukes, which comprise approximately nine species of *Paragonimus*, infect an estimated 22 million people globally (Miyazaki 1982; Yokogawa 1982; Toscano et al. 1995; Velez et al. 2002); with approximately 10 million in China alone. China harbors three species infecting humans, *P. westermani*, *P. heterotremus*, and *P. skrjabini*. Other Asian countries with endemic *Paragonimus* infection include Philippines, Korea, India, Thailand, and Laos. Peru, Ecuador, and Colombia report large areas endemic for *P. mexicanus*, while Cameroon and Nigeria report *P. africanus* and *P. uterobilateralis*. Scattered reports have come from parts of India, Japan, Latin America, Liberia, Guinea, and some Pacific countries. In the United States, there are occasional human infections with the lung fluke, *Paragonimus kellicotti*, which normally parasitize wildlife (DeFrain and Hooker 2002).

Fasciola infections number approximately 2.4 million and are more globally widespread than *Paragonimus*. The World Health Organization identified 2 594 human cases of *Fasciola hepatica* in 42 countries from the scientific literature and public health reports between 1970 and 1990 (Chen and Mott 1990). Countries that reported the most cases were France, the UK, Portugal, Spain, the former USSR (Tadzhikstan), Peru, Cuba, Ecuador, Egypt, and Iran. Community-based studies of *Fasciola* infection using improved diagnostic methods have demonstrated a very high prevalence and intensity of infection in the high altitude regions of Bolivia and Peru (Esteban et al. 1997; Mas-Coma et al. 2001; Haseeb et al. 2002) and Egypt (Hassan et al. 1995). Some argue that *Fasciola* should no longer be considered an accidental zoonosis because it is an important human parasite (Mas-Coma et al. 1999). Less is known about the extent of *F. gigantica* infection, except that it has been reported in Africa, Asia, Hawaii, the former USSR, and Iraq. Although *Fasciola* was previously considered rare in Vietnam, 500 cases have been reported since 1997 (Tran et al. 2001).

In addition to the endemic areas for all of these flukes, regional and global migration of people has sometimes broadened their distribution and returned travelers seek medical treatment in areas well outside the endemic zones. Since life cycles usually do not become established, this may have limited epidemiological relevance. Given the potential severity of resulting disease, however, it is of clinical importance that physicians are better informed to recognize these infections (Chan and Lam 1987; LaPook et al. 2000). The World Health Organization is concerned about the spreading of human trematode infections via the increasing international trade in fresh aquatic foods from countries in which such infections are endemic (Maurice 1994).

Social factors related to transmission

Raw or undercooked freshwater fish and crab are prepared in many different ways and these dishes are often of considerable cultural, medicinal, and nutritional significance, making change difficult. This is aggravated by frequently held beliefs that marination in various sauces or consumption with alcohol kills the parasites. Transmission can also occur via contamination of utensils, hands, and surfaces used first to prepare crabs, fish or vegetables for cooking, and then for other foods taken raw. *Fasciola* and *Paragonimus* infection through contaminated water may be possible, but is generally of lesser importance than through food.

Traditionally in southern China and among the Cantonese of Hong Kong, raw fish containing *Clonorchis* metacercariae is eaten by dipping in rice porridge or kongee (Chen et al. 1994). Large fish may be sliced and eaten raw with ginger and garlic. In some areas, children become infected by catching and incompletely roasting fish during play. Koreans eat raw fish soaked in vinegar, red-pepper mash or hot bean paste with rice wine at social gatherings (Choi 1984; Rim 1986). Vietnamese people, particularly men, reportedly eat raw fish from contaminated fish ponds in salads in the hot season (Chen et al. 1994; Kino et al. 1998). Infection in Japan, which is now very rare, came from eating slices of large raw fish with vinegar or soya bean paste (Komiya 1966). Sushi and other uncooked sea fish eaten in Japan today do not carry *Clonorchis*.

In northeast Thailand and Laos, preparations may contain fresh (*koi pla*), partially fermented (*som pla*) or fully fermented (*pla ra*) uncooked fish which contain *Opisthorchis* metacercariae. *Koi pla* is probably the most important source of infection (Bunnag and Harinasuta 1984). People in western Siberia and other areas of the former USSR, endemic for *O. felineus*, enjoy eating uncooked fish frozen, salted, or smoked with condiments. In addition to some groups of aboriginal people, migrants to endemic areas also become infected (Iarotski and Be'er 1993).

Raw crabs which may transmit *Paragonimus* infection also carry important social significance in some areas. For example in Korea, raw crayfish juice was used as a treatment for childhood measles (Choi 1990; Cho 1994), and young Bakossi women in Cameroon ate raw crabs to increase fertility (Kum and Nchinda 1982). Both practices have apparently become rare. Crabs and crayfish are also caught and eaten raw or incompletely roasted by children and workers in rice fields (Cabrera 1984). Traditional dishes often involve raw crab or crayfish meat or juice mixed with lemon juice (*ceviche* in Peru), brine, soya sauce (Korean *ke-jang*), alcohol (drunken crabs, China), or vinegar (Laos).

The consumption of watercress has been implicated in most human *F. hepatica* infections, especially in Europe (Hardman et al. 1970; Rondelaud et al. 2000). In the high altitudes of the Bolivian Altiplano region, people (especially children) consume kjosco (raw water plant salad) while tending their animals in the fields. However, inhabitants of the Peruvian Altiplano who also have a high prevalence and intensity of *Fasciola* infection do not consume aquatic vegetables. Exposure in this group appears to be through contaminated irrigation water used for drinking (Esteban et al. 2002). Morning glory and other water plants are important vehicles of *F. gigantica* infection in Asia (Tesana et al. 1989).

Distribution of infection in human communities

Local patterns of infection within endemic communities are partially determined by social customs and attitudes toward raw foods. This is especially true where these foods are thought to have medicinal or fertility-enhancing properties, where catching and eating raw foods is part of children's play, or where the foods accompany certain social activities, such as drinking parties. Infection patterns may also reflect differential effectiveness of health education efforts in endemic countries.

Levels of *Opisthorchis* and *Clonorchis* infection vary greatly between communities, but age-related patterns are generally similar. The youngest ages show low prevalence and intensity, while these increase in the pre- and early teens and often reach a plateau in the late teen age groups (e.g. 15–19 years) or continue to rise. Prevalence and intensity either do not differ, or are higher, among males compared to females; sex differences tend to be greater in areas endemic for *Clonorchis* (Upatham et al., 1984; Rim 1986; Haswell-Elkins et al. 1991; Kino et al. 1998). The population of *O. viverrini*, and probably of all the liver and lung flukes, is highly aggregated within a small minority of heavily infected people.

There have been very few community-based studies of *Paragonimus* infection, so that its distribution is not well characterized. Studies in the 1950s and 1960s in Korea revealed prevalences of *Paragonimus* infection of up to 45 percent (according to skin testing) in endemic communities and as high as 13 percent nationwide. Recently, however, the occurrence of areas of high prevalence has dropped dramatically (Choi 1990; Cho 1994). In other endemic countries, e.g. China, Thailand,

Laos, Cameroon, Nigeria, Peru, and China, infections tend to occur at low prevalences (under 10 percent) in defined areas. Mountainous areas with unpolluted water are most favorable to *Paragonimus* transmission.

Fasciola infection occurs mainly in rural areas, with sheep and cattle herders and people living in small villages typically being most at risk of infection. In France, however, an increase in cases from people living in larger towns within cattle-rearing areas has been observed (Rondelaud et al. 2000). Very high prevalences (up to 75 percent in children) and intensities of infection (>1 000 eggs/g of feces) with *Fasciola hepatica* are found in communities of the Bolivian Altiplano region, where transmission occurs year round, especially from December to March (Esteban et al. 1997; Mas-Coma et al. 1999). A study of schoolchildren in Egypt revealed 11 percent prevalence of infection with *Fasciola* (Hassan et al. 1995). In addition to chronic infection, fascioliasis sometimes occurs in acute outbreaks within communities and households where food from a common source of contaminated aquatic plants is consumed.

Location in host

Opisthorchis and *Clonorchis* locate within the small, intrahepatic bile ducts; in heavy infections, adult worms may also be found in the extrahepatic bile ducts, pancreatic ducts, and gallbladder (Hou 1955; Sithitha-worn et al. 1991). Infection is confined to the hepato-biliary tract lumen; there is no tissue migration phase (Sun et al. 1968). *Fasciola* also finally dwells in the bile duct, but its migration out of the duodenum and through the hepatic parenchyma causes much more tissue trauma and acute inflammation than that of *Clonorchis* and *Opisthorchis*. *Paragonimus* flukes are found within fibrotic capsules in the terminal alveoli or under the pleura in the lungs after going through a similarly complex migratory process.

In addition to these usual migratory routes, some worms of *Fasciola* and *Paragonimus* may become lodged in ectopic sites. Ectopic sites of *Paragonimus* include the liver, brain, mesenteric lymph nodes, intestinal wall, and subcutaneous tissue of the groin. *Fasciola* may locate in the liver parenchyma, blood vessels, lungs, subcutaneous tissue, and the brain. Thus, as discussed below, clinical manifestations of this infection may be due to their accumulation in their 'normal' site or due to one worm that becomes encapsulated in a sensitive, ectopic site.

CLINICAL AND PATHOLOGICAL ASPECTS OF INFECTION

Although these infections are not considered numerous enough to be among the top tropical parasitic diseases according to the World Health Organization, there is no question that they are highly pathogenic and cause significant human disease. The very important clinical aspect of *Opisthorchis viverrini*, and to a lesser extent of *Clonorchis* infection, is the extreme susceptibility of infected people to bile duct cancer (IARC 1994). This cancer has a very poor prognosis, and few patients live longer than 1 or 2 years after diagnosis and/or surgery (Watanapa and Watanapa 2002). Thus, mortality due to complications of this fluke infection is very high in the endemic area of northeast Thailand, which reports the world's highest incidence of liver cancer (Vatanasapt et al. 1993).

Very importantly, in endemic areas, these infections are usually asymptomatic or may contribute to a high background level of nonspecific illness that is difficult to distinguish from other causes. For example, the hemoptysis of *Paragonimus* requires parasitological and bacteriological differentiation from that of the more prevalent tuberculosis in co-endemic areas (Toscano et al. 1995). Similarly, the abdominal pain reportedly occurring in *Opisthorchis* infection is difficult to differentiate from that due to other causes. However, particularly in the case of the liver flukes associated with cholangiocarcinoma, the absence of symptoms does not indicate a medically unimportant infection. In fact, the typical lack of specific clinical signs and symptoms (or pain and discomfort) makes treatment and prevention even more difficult.

Clinical manifestations

The likelihood and types of clinical manifestations associated with these fluke infections are dependent on the location of the parasite in the liver, lungs, or ectopic site, the species of fluke, number of worms harbored, duration of infection, and, possibly, nutritional status and the individual's pattern of immune responses to parasite antigens. Chronic trematode infection may also play a role in promoting chronic renal disease in tropical countries (Boonpucknavig and Soontornniyomkij 2003).

O. VIVERRINI, C. SINENSIS, AND *O. FELINEUS*

The frequency and types of clinical disease seem to differ between the three closely related species. Most notably, there are many reports in the *O. felineus* and *C. sinensis* literature detailing specific signs and symptoms accompanying well-defined clinical stages of opisthorchiasis, from acute to chronic (Bronshtein 1986; Chen et al. 1994). Acute infection, characterized by chills and high fever, abdominal pain and distension, hepatitis-like symptoms, and eosinophilia, is frequently reported in *O. felineus*. In contrast, there are fewer reports of acute clonorchiasis (Koenigstein 1949; Chen et al. 1994) or opisthorchiasis viverrini. This difference may be due to the large number of migrants entering endemic areas of *O. felineus* who first become infected as adults (Iarotski

and Be'er 1993), which may be rare for the other two species.

Studies in two communities with very high levels of *O. viverrini* infection reported significantly increased frequencies of hepatomegaly, abdominal pain in the right upper quadrant, flatulence, dyspepsia, and weakness associated with increasing intensity of infection (Upatham et al. 1984). An estimated 5–10 percent of the community had mild symptoms attributable to the infection. Hospital-based studies on patients with *Opisthorchis* and *Clonorchis* infection (lacking comparison with uninfected controls) have reported much higher frequencies of signs and symptoms. In addition to the above-mentioned symptoms, eosinophilia, anorexia, dizziness, weight loss, diarrhea, anemia, edema, neuropsychiatric symptoms, and retardation of growth and sexual maturity have been reported (Bunnag and Harinasuta 1984; Chen et al. 1994).

Stones in the gallbladder, liver, and bile ducts have been linked to liver fluke infection. Eggs or worm fragments are often found in the nidus (Teoh 1963; Riganti et al. 1988). Hepatolithiasis is considered a major clinical manifestation of *Clonorchis* infection (Leung and Yu 1997). An increase in gallbladder stone frequency with increasing intensity of *Clonorchis* infection was found by ultrasonography among Hakka people in Taiwan, from 4.2 percent in uninfected subjects to over 14 percent of those with heavy infections (Hou et al. 1989). Epidemiological associations with stones, however, are weaker for *O. viverrini* infection.

Ascending cholangitis, obstructive jaundice, portal hypertension, ascites, and gastrointestinal bleeding are severe complications of liver fluke infection. However, only 88 cases of severe disease were reported among 15 243 infected people who presented to a Bangkok hospital for praziquantel treatment (Pungpak et al. 1985). Furthermore, these disease presentations are also typical of cholangiocarcinoma that may have been present in some patients.

Studies using ultrasonography have shown very strong relationships between gallbladder enlargement, wall irregularities, and sludge and intensity of infection (Lim 1990; Mairiang et al. 1992). These abnormalities are not accompanied by clinical signs and symptoms and are reversible following praziquantel treatment (Mairiang et al. 1993; Pungpak et al. 1997; Richter et al. 1999).

As mentioned above, the most important clinical manifestation of opisthorchiasis, also occurring in clonorchiasis, is bile duct cancer. People with heavy *O. viverrini* infection face at least a 14-fold increased risk of cholangiocarcinoma over uninfected people from the same communities (Haswell-Elkins et al. 1994a). This cancer is a major cause of death in the endemic area of northeast Thailand, where age-standardized incidences are 84.6 and 36.8 per 100 000 males and females, respectively (Vatanasapt et al. 1993). This compares to two to four cases per 100 000 typically arising in non-endemic areas. Areas endemic for *Clonorchis* infection also report higher frequencies of bile duct cancer (Rim 1986; IARC 1994). In a study of liver cancer in Korea, *Clonorchis* infection was found to carry a relative risk of 2.7 (95 percent confidence intervals 1.1–6.3) for cholangiocarcinoma (Shin et al. 1996). A number of case studies have recently been published describing extrahepatic cholangiocarcinoma and gall bladder carcinoma associated with *Clonorchis* infection (Kim 2003a, b). The average age of patients is 50 years, with a higher frequency of males and, due to an absence of early symptoms, those affected present to hospital in the late stages of the disease. Prognosis with or without surgery is poor.

FASCIOLA HEPATICA AND F. GIGANTICA

As with opisthorchiasis and clonorchiasis, fascioliasis is often asymptomatic, especially in light or chronic infection. Reported symptoms are grouped into three phases, depending on where the parasites are in their migration (Chen and Mott 1990).

First, the acute or invasive stage occurring when the flukes are in migration through the liver parenchyma may be accompanied by fever and night sweats, mild to severe abdominal pain, gastrointestinal disturbances, urticaria, malaise, weight loss, dermatographia, and cough (Hardman et al. 1970; Chen and Mott 1990; Patrick and Isaac-Renton 1992). Clinical signs may include hepatomegaly and splenomegaly, ascites containing eosinophils and other leukocytes, mild to moderate anemia with its accompanying symptoms, pulmonary signs, eosinophilia, and jaundice. Biochemical abnormalities are also observed, for example, elevation of acute phase proteins (Basha et al. 1998), raised serum bile acids (Osman et al. 1999), and sometimes increased liver enzymes (e.g. alkaline phosphatase, although not GGT and SAP).

The latent phase of infection, during which the flukes are in the bile ducts, is generally asymptomatic, perhaps involving gastrointestinal symptoms, and can last for years. Some individuals progress to the obstructive phase of infection with chronic cholecystitis and cholangitis, which may be accompanied by biliary colic, epigastric pain, jaundice, nausea, pruritis, and right upper quadrant pain precipitated by fatty foods. Gallbladder enlargement and stones are frequently found in these patients. Anemia and ascites may occur as a result of blood loss and deaths, although rare, can occur through hemorrhaging in the bile duct.

Ultrasonography, ERCP, and MRI may reveal the worms themselves in the liver or on the gallbladder wall, bile duct dilatation and filling defects. *Fasciola* has been implicated as a carrier for *Salmonella typhimurium* as the bacteria preferentially bind to the metacercariae, thus treatment for both infections may be required (El Zawawy et al. 2002).

Ectopic *Fasciola* infections leading to pulmonary, cardiac, gastric, cecal, cerebral, and neurological disorders have all been reported. These appear to be rare, although the frequency with which parasitic origins of such diseases are missed is unknown.

PARAGONIMUS WESTERMANI AND OTHER LUNG SPECIES

Because of interspecies differences in migratory behavior and extent of pulmonary pathology, clinical manifestations vary between the various *Paragonimus* species infecting humans (Miyazaki 1982).

In general, however, the clinical manifestations of *Paragonimus* infection are often misdiagnosed as tuberculosis or lung cancer (Beaver et al. 1984; Toscano et al. 1995; Nagakura et al. 2002). Nonspecific symptoms, e.g. diarrhea, abdominal and chest pain, allergic reactions, fever and chills, may be present during the migration phase. Once the worms establish, the most common symptoms are cough and hemoptysis which may be accompanied by night sweats and general malaise (Yokogawa 1982). Up to 50 ml of gelatinous, rusty brown sputum containing traces of blood and parasite eggs may be expectorated daily during paroxysmal coughing. Severe infections might progress to pleurisy, persistent rales, clubbed fingers, and pneumothorax. The severity of pulmonary lesions viewed by chest roentgenograms is associated with intensity of egg output and duration of infection with *P. heterotremus* in Thailand (Vanijanonta et al. 1984).

Ectopic infection with *Paragonimus* can occur in the brain, omentum, skin, and liver (Miyazaki 1982; Sasaki et al. 2002; Hughes and Biggs 2002). If parasites lodge in the brain or spinal cord, which happens more frequently in children, severe disease may result (Miyazaki 1982; Bunnag and Harinasuta 1984; Jaroonvesama 1988). Symptoms may include headache, fever, paralysis, visual disturbances, and (sometimes fatal) convulsive seizures. Recovery may be spontaneous, but symptoms may recur. Lesions are visualized by computed tomography or cerebral angiography.

Migration of worms under the skin leads to the formation of migratory swellings containing the parasites and eggs. Intestinal paragonimiasis which manifests as multiple granulomatous nodules or ulcers can be mistaken for gastric or abdominal tumors.

Pathogenesis and pathology

Although the role of immune responses in protection against infection is debated in all of these parasites, it is clear that these play an important role in pathogenesis and may determine whether clinically significant disease results.

PARAGONIMUS INFECTIONS

Juvenile *Paragonimus* which migrate successfully to the lungs elicit insignificant pathology. However, adult worms and trapped eggs in the pulmonary parenchyma stimulate inflammation and granuloma formation, consisting mainly of eosinophils and neutrophils (Miyazaki 1982; Matsumoto et al. 2002). These granulomas develop into fibrous capsules, 1.5–5 cm in diameter, usually surrounding a pair or triplet of worms, eggs, and blood-streaked fluid. The pleural fluid of patients recently infected with *Paragonimus westermani* may contain large numbers of eosinophils and IgM and high concentrations of IL-5, thymus, and activation-related chemokine and eotaxin which may be involved in the acute pathogenesis of infection (Matsumoto et al. 2002). Leakage of fluid into the bronchioles causes paroxysmal coughing, hemorrhage, and blood in the sputum. Later in infection, parasite-specific IgG dominates and nodular, cavitating lesions develop in the lung parenchyma (Nakamura-Uchiyama et al. 2001). Flukes which lodge in ectopic sites, e.g. the brain, spinal cord, intestine, and heart, invoke similar inflammatory responses leading to ulcerations and abscesses which cause severe damage in sensitive sites.

O. VIVERRINI , O. FELINEUS, AND C. *SINENSIS*

The acute phase of disease appears to result from acute inflammatory reactions to parasite antigens met for the first time (Chen et al. 1994). Typically light, chronic infections show minimal pathological change. In heavier infections, the liver may be enlarged with localized dilatation of slightly thickened, fibrosed peripheral bile ducts (Hou 1955; Rim 1986). Histopathological changes (see Figure 31.7) include proliferation and desquamation of bile duct epithelial cells, glandular formation, goblet cell metaplasia, inflammatory infiltration by lymphocytes, monocytes and eosinophils and severe fibrosis (Pairojkul et al. 1991). Although pathological changes are most frequent in the small ducts where the flukes reside, the gallbladder is commonly enlarged and its

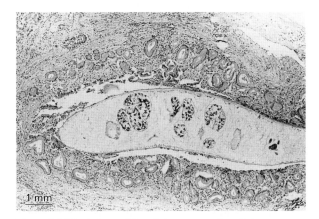

Figure 31.7 *Pathology of liver flukes: mature liver fluke,* Opisthorchis viverrini, *residing in the lumen of the human intrahepatic bile duct. Epithelial desquamation, adenomatous hyperplasia, periductal fibrosis and inflammatory infiltration are seen in the bile duct wall. Bar = 1 mm. (Photograph by B. Sripa, reproduced with permission)*

function affected, leading to bile stasis. A 24 kDa antigen of *Clonorchis* is thought to be a major cytotoxic protease causing direct pathology (Park et al. 1995), while Sripa and Kaewkes (2000, 2002) found inflammatory changes associated with excretory and secretory antigens throughout intrahepatic and extrahepatic bile ducts in experimental *O. viverrini* infection. Eosinophils were found to be the major inflammatory cell in both *Opisthorchis*- and *Clonorchis*-associated cholecystitis (Kim 1999; Sripa and Kaewkes 2000). In some cases, the extrahepatic bile ducts may also become inflamed and fibrosed, leading to strictures and stagnant bile. This might facilitate bacterial infection, leading to abscess formation and acute cholangitis (Chen et al. 1994). Increased levels of various biochemical indices of hepatic function suggest that chronic liver fluke infection enhances cytolysis of hepatocytes, cholestasis, and hepatic cell insufficiency (Bakshtanovskaia et al. 2002).

Parasite products, mechanical or obstructive changes, and the immune response, all appear to play a role in chronic pathogenesis, as well as in the carcinogenesis associated with infection. Based on experimental evidence, some authors suggest that oval cell or bile duct epithelial cell proliferation during chronic infection increases their susceptibility to carcinogens, especially *N*-nitrosodimethylamine, in the diet (Thamavit et al. 1996; Migasena 1982; Lee et al. 1997). High levels of nitrosamines have been found in some traditional Thai foods and Thai brand cigarettes (Migasena 1982; Mitacek et al. 1999a, b). The work of others has suggested that chronic inflammatory responses to *O. viverrini* infection increase the generation of nitric oxide, which combines with amines (Haswell-Elkins et al. 1994b; Satarug et al. 1996a) to form nitrosamines endogenously within the inflamed tissue (Srivatanakul et al. 1991; Satarug and Haswell-Elkins 1998). Specific T-cell responses to particular *Opisthorchis* antigens are closely associated with increased levels of endogenous nitrate and nitrosamine production in humans. Infection and chronic fibrosis of the small bile ducts also appear to stimulate the levels of cytochrome P450 2A6, which can activate dimethylnitrosamine to its highly toxic form (Kirby et al. 1994; Satarug et al. 1996b). All of these reactions are reversed after elimination of the fluke with praziquantel treatment.

FASCIOLA HEPATICA AND *F. GIGANTICA*

Although *Fasciola* inhabits the bile duct and stimulates chronic inflammation and fibrosis, like *Opisthorchis* and *Clonorchis*, it has not been associated with human biliary carcinoma. The pathological lesions are more similar to those of *Paragonimus*. This includes granulomatous inflammation encompassing eggs and abscess formation around the adult worms appearing grossly as multiple yellow nodules. Track-like inflammatory lesions appear along the migration route of the worms (Chen and Mott 1990). The cells initially involved are mainly polymorphonuclear leukocytes, eosinophils, histiocytes, and lymphocytes. Later lesions contain lymphocytes and plasmocytes with fibrosis and calcification. Similarly, ectopically located parasites, mostly found in subcutaneous tissue, may form abscesses.

Monocytes tested during the acute phase of infection in humans actively produce higher levels of pro-inflammatory cytokines (GM-CSF, interleukin 8, and interleukin 6) in vitro, compared to those taken during the chronic phase (Khalil et al. 1999). Immunoproliferative responses and nitric oxide production appears to be suppressed within a week of infection in experimental animals through production of interleukin 4, gamma interferon, and interleukin 10 (Osman and Abo-El-Nazar 1999; Cervi et al. 2001). Suppression of inflammatory responses may limit the degree of immunopathology during chronic infection but may also inhibit otherwise effective parasiticidal responses.

The pathogenesis of *Fasciola*, therefore, involves acute and chronic immune responses leading to granulomas, abscess, and hepatic fibrosis. Direct parasite products and movement play a well-defined role in pathology as the parasites eat their way through the liver to the bile ducts undergoing considerable growth. In addition, large amounts of proline produced and excreted by the flukes may be directly responsible for bile duct hyperplasia and dilatation and collagen deposition (Wolf-Spengler and Isseroff 1983).

Immune responses, nature, and source of antigen

Whether the immune responses which develop in humans in response to liver and lung fluke infections confer protective immunity against the currently held parasites or newly acquired flukes remains unclear. Little direct progress has been made towards vaccine development for humans, however, a great deal of work has been carried out to characterize certain enzymes shared by these flukes, particularly the cysteine proteases, cathepsin L-like proteases, and glutathione S-transferases. These enzymes are abundant in the tegument and/or excretory and secretory products and play important roles in parasite survival (Kofta et al. 2000). While glutathione S-transferases protect flukes from bioreactive compounds (Sobhon et al. 1998; Hong and Yun Kim 2002), the proteases appear to play several roles in immune evasion (Trap and Boireau 2000; Shin et al. 2001; Park et al. 2002).

Because of the economic and veterinary importance of fascioliasis, the immunology of *Fasciola* has been extensively studied in sheep, cattle, and rodents. Differing degrees of acquired resistance are seen among these animals. Cattle and most rodents develop resistance to reinfection, but sheep and goats do not and may die in the face of continued exposure and accumulation of worms. Human *Fasciola* infections are usually self-limiting and only few migrating larvae achieve establishment. However, this may not be related to immune responses.

Experimental vaccine trials against *Fasciola* in livestock have demonstrated that significant protection (up to 98 percent) can be conferred with irradiated larval vaccines (Spithill 1992). Vaccines targeting important individual antigens have been partially successful. For example, immunization with a recombinant glutathione S-transferase of *F. hepatica* has conferred over 50 percent protection against worm establishment in sheep. A detailed review of the elegant and promising work on understanding the immune response in animals and the progress towards an effective vaccine for animals is available (Spithill et al. 1997).

Epidemiological patterns reveal little evidence of, but also do not rule out, protective immunity in humans exposed to *Opisthorchis* and *Clonorchis*. There is no decline in prevalence or intensity of infection among individuals exposed to decades of infection, and reinfection may occur rapidly following treatment in areas of heavy infection (Upatham et al. 1984, 1988). The parasites clearly survive in the face of high levels of parasite-specific IgG, IgA, and IgE in both serum and bile (Wongratanacheewin et al. 1988), and of T-cell reactivity to parasite antigens as demonstrated by delayed-type hypersensitivity responses following skin testing (Rim 1986). Recent studies by Tesana et al. (2000) has suggested that immunization of experimental animals with *Opisthorchis viverrini* antigens prior to live infection and administration of carcinogen, dimethylnitrosamine, may accelerate carcinogenesis.

Even less is known about acquired immunity in human *Paragonimus* infection. The use of skin testing and antibody tests as diagnostic tools (see below) indicates that people mount strong cellular and humoral immune responses to parasite antigens. However, these responses do not appear to be protective, possibly due to the action of secreted cysteine proteases which appear to destroy IgG bound to the parasite surface, reduce eosinophil survival, IL-8 production and degranulation, and inhibit superoxide production by granulocytes (Shin et al. 2001).

Diagnosis

Similarities in egg morphology and cross-reactive antigens between intestinal flukes and liver flukes complicate both parasitological and immunological diagnosis. *Fasciola* eggs are indistinguishable from those of human *Fasciolopsis* and *Echinostoma* species, and *Opisthorchis* and *Clonorchis* eggs look similar to minute intestinal fluke eggs. Food-borne parasites often present in mixed infections and have overlapping endemic areas (Lee et al. 1984; Radomyos et al. 1984; Kaewkes et al. 1991). Nevertheless, diagnosis of lung and liver fluke infections is reliably and frequently based on the observation of eggs in the stool (all of the parasites) or sputum (*Paragonimus* only). Stoll's dilution method, formalin-ether (or formalin-ethyl acetate) concentration method (FECT) and AMS III are the most sensitive methods for processing specimens for egg examination (Chen and Mott 1990; Sithithaworn et al. 1991; Toscano et al. 1995). Rapid staining with potassium permanganate can help differentiate *O. viverrini* eggs from those of common minute intestinal flukes (Sukontason et al. 2001). *Fasciola* and *Paragonimus* infection pose a problem of diagnosis in the migratory phase, or in ectopic infection, when no eggs are passed.

Peripheral blood eosinophilia and elevated serum IgE levels often accompany parasite infections, including those of lung and liver flukes. Specific serological diagnostic methods, particularly for fascioliasis and paragonimiasis, have improved. Many authors report very good results with enzyme linked immunosorbent assay (ELISA) detecting circulating parasite antigens (Shehab et al. 1999) or specific antibodies to adult worm, excretory, and secretory or metacercarial preparations (Osman et al. 1995; Silva et al. 1996; Maher et al. 1999). Various modifications can enhance assay performance, for example IgM for early infections (Osman et al. 1995) or IgG4 subclass (O'Neill et al. 1999; Hong et al. 1999). The use of single purified antigens (Maleewong et al. 1997; Mansour et al. 1998) or using cystatin or Protein A-treated plates for capture ELISA and the detection of cysteine protease activity (Ikeda 2001) can increase specificity. It has become very clear that serological and antigen detection ELISAs are far superior to egg detection in *Fasciola*, especially in the prepatent, acute phase of infection and are also very useful in assessing treatment efficacy (Hassan et al. 1995; Hammouda et al. 1997; Chaithirayanon et al. 2002).

Potentially promising diagnostic antigens for *Clonorchis* and *Opisthorchis* are being explored, some in isotype-specific assays. An IgG4-specific assay to detect 7–8 kDa antigen (Kim 1998) and an IgE-detection assay detecting antibodies to a 28 kDa glutathione *S*-transferase (Yong et al. 1999; Na et al. 2002) of *Clonorchis* have been shown to be highly specific. However, a mix of antigens appears to be needed to increase sensitivity (Hong and Yun Kim 2002). Excretory and secretory antigens have been the most studied and generate the dominant antibody responses to infection with *Opisthorchis*. Most antigens and antibodies that have been studied have been nonspecific and persistent after treatment (Akai et al. 1995); however, high sensitivity and specificity has been reported using affinity purified oval antigens from adult worms (Wongsaroj et al. 2001) and Bithynia snail antigens (Waikagul et al. 2002). Sirisinha et al. (1995) and Wongratanacheewin et al. (2001) have developed sensitive and specific diagnostic tests that

detect *Opisthorchis* material in feces using monoclonal-antibody capture ELISAs and polymerase chain reaction (PCR)-based assay. The PCR assay is also able to detect cercaria and metacercaria in artificially inoculated snails and fish (Maleewong and Intapan 2003). These tests can be easily used in large-scale field studies and control programs.

Clinical, pathological, radiological, and potential exposure information clearly plays an important contributing role in addition to parasitological and immunological tests in diagnosis as well as in determining the extent of pathological change that has occurred. Evidence of *Paragonimus* infection may be shown on radiographs as patchy foci of fibrotic change, with a characteristic 'ring shadow' in the lung (Bunnag and Harinasuta 1984; Toscano et al. 1995) which may be mistaken for tuberculosis (Nagakura et al. 2002). Ultrasound diagnosis reveals abnormalities associated with liver flukes, e.g. gallbladder enlargement, stones and bile duct tumors, and sometimes the parasites themselves (van Beers et al. 1990; Lim 1990; Mairiang et al. 1992; Richter et al. 1999). Endoscopic retrograde cholangiopancreatography (ECRP) may reveal filamentous or elliptic filling defects of the biliary tract associated with clonorchiasis and fascioliasis (Chan et al. 2002). Parasite fragments and eggs can sometimes be seen in surgical tissue or biopsies, revealing the unsuspected cause of the disease.

CONTROL

The main tools for control of liver and lung flukes have been anthelmintic treatment, sanitation improvement, and health education. The rationale is that treatment is required to eliminate the long-lived parasites immediately, sanitation interrupts transmission from human feces to snails, and health education stops people from eating raw foods and becoming reinfected after treatment. Alternative risk reduction strategies to kill metacercaria, such as freezing or irradiating fish and soaking leafy greens in vinegar and potassium permanganate, may be effective if widely promoted (Song 1987; Iarotski and Be'er 1993). Improving sanitation through provision of latrines, stopping night-soil use on fields and in fish ponds, and moving pig farming from lakesides have been widely implemented and useful in preventing liver fluke infection (Jongsuksuntigul and Imsomboon 1997; Ooi et al. 1997). Hazard analysis critical control point procedures are increasingly being used in food safety programs and have been found to be helpful in decreasing contamination of farmed fish (Khamboonruang et al. 1997). Alternative strategies, such as freezing and irradiating fish (Song 1987; Iarotski and Be'er 1993; Park and Yong 2003), biological agents to destroy cercariae (Intapan et al. 1992), and treating reservoir hosts have been suggested but not widely implemented. Improving sanitation through latrines and stopping night-soil use on fields and fish ponds has been widely implemented to prevent liver fluke infection. Furthermore, environmental changes, the use of pesticides and pollution of river systems by industrial effluents, have also reduced intermediate host populations (Komiya 1966; Cho 1994).

A single dose of praziquantel is generally used at 40 mg/kg body weight for *O. viverrini* and *C. sinensis* in Korea, while higher, multiple doses (3 × 25 mg/kg for 1–3 days) are used in China. Praziquantel appears to induce an influx of calcium causing depolymerization of the tubular networks and detachment of the parasite tegument (Apinhasmit and Sobhon 1996). *Paragonimus* is also effectively killed by praziquantel at a dose of 25 mg/kg, three times a day for 3 days. Side effects are transient and relatively minor. Published efficacy of praziquantel at these doses is over 90 percent (Chen et al. 1983; Sui et al. 1988). Treatment with *Fasciola* is more problematic as it is not particularly sensitive to praziquantel, unless with multiple treatments at very high doses (Queneau et al. 1997). Recent literature reports good outcomes and low side-effects with the use of triclabendazole at 10 or 20 mg/kg (Apt et al. 1995; Lecaillon et al. 1998). Repeated treatment is frequently required, but generally effective the second time. Recent reports of triclabendazole resistance developing in livestock animals has led to the search for new drugs against *Fasciola* infection with nitazoxanide looking promising in a randomized control trial in Peru (Favennec et al. 2003).

Despite the value of treatment, reinfection can and does occur in the case of *Opisthorchis* and *Clonorchis*. Complete success in control has been hampered by the difficulty in changing human behavior. Health education campaigns have focused primarily on providing information about the disease, rather than on prevention. Since raw food consumption is often culturally specific, it often holds great importance for the identity of minorities, e.g. the Laos-descendent Thais of the northeastern region, the Korean minority of northeastern China, the aboriginal people of Siberia. Community participation and culturally appropriate health education messages are a vital element of long-term success (Sornmani 1987; Jongsuksuntigul and Imsomboon 1997).

REFERENCES

Akai, P.S., Pungpak, S., et al. 1995. Serum antibody responses in opisthorchiasis. *Int J Parasitol*, **25**, 971–3.

Anantaphruti, M.T. 2001. Parasitic contaminants in food. *Southeast Asian J Trop Med Pub Health*, **32**, Suppl 2, 218–28.

Apinhasmit, W., Sobhon, P., et al. 1993. *Opisthorchis viverrini*: changes of the tegumental surface in newly excysted juvenile, first week and adult flukes. *Int J Parasitol*, **23**, 829–39.

Apinhasmit, W. and Sobhon, P. 1996. *Opisthorchis viverrini*: effect of praziquantel on the adult tegument. *Southeast J Trop Med Pub Health*, **27**, 304–11.

Apinhasmit, W., Sobhon, P., et al. 2000. *Opisthorchis viverrini*: ultrastructure and cytochemistry of the glycocalyx of the tegument. *J Helminth*, **74**, 23–9.

Apt, W., Aguilera, X., et al. 1995. Treatment of human chronic fascioliasis with triclabendazole: Drug efficacy and serologic response. *Am J Trop Med Hyg*, **52**, 532–5.

Attwood, H.D. and Chou, S.T. 1978. The longevity of *Clonorchis sinensis. Pathology*, **10**, 153–6.

Bakhtanovskaia, I.V., Stepanova, T.F., et al. 2002. [Biochemical characteristics of hepatic functions in different clinical forms of chronic opisthorchiasis]. *Meditsinskaia Parazitologiia I Parazitarnye Bolezni*, **1**, 12–16, in Russian.

Basha, L.A., Abaza, M., et al. 1998. Study of some acute phase reactant proteins, in acute human fascioliasis. *J Med Research Inst*, **19**, 147–51.

Beaver, P.C., Jung, R.C. and Cupp, E.W. 1984. *Clinical parasitology*, 9th edn. Philadelphia, PA: Lea and Febiger, 406–81.

Blair, D. 2000. Genomes of *Paragonimus westermani* and related species: current state of knowledge. *Int J Parasitol*, **30**, 421–6.

Blair, D., Xu, Z.B., et al. 1999. Paragonimiasis and the genus *Paragonimus. Adv Parasitol*, **42**, 113–222.

Boonpucknavig, V. and Soontornniyomkij, V. 2003. Pathology of renal diseases in the tropics. *Semin Nephrol*, **23**, 88–106, review.

Bronshtein, A.M. 1986. Morbidity from opisthorchiasis and diphyllobothriasis in the aboriginal population of the Kyshik village in the Khanty-Mansy Autonomous Region. *Med Parazitol (Mosk)*, **3**, 44–8, in Russian.

Bunnag, D. and Harinasuta, T. 1984. Opisthorchiasis, clonorchiasis, and paragonimiasis. In: McGraw, R.P. and McIvor, D. (eds), *Tropical and geographic medicine*. New York: McGraw-Hill, 461–9.

Cabrera, B.D. 1984. Paragonimiasis in the Phillipines: current status. *Arzneimittel Forsch*, **34**, 1188–92.

Cervi, L., Cejas, H., et al. 2001. Cytokines involved in the immunosuppressor period in experimental fasciolosis in rats. *Int J Parasitol*, **31**, 1467–73.

Chaithirayanon, K., Wanichanon, C., et al. 2002. Production and characterization of a monoclonal antibody against 28.5 kDa tegument antigen of *Fasciola gigantica. Acta Tropica*, **84**, 1–8.

Chan, C.W. and Lam, S.K. 1987. Diseases caused by liver flukes and cholangiocarcinoma. *Baill Clin Gastroenterol*, **1**, 297–318.

Chan, H.H., Lai, K.H., et al. 2002. The clinical and cholangiographic picture of hepatic *clonorchiasis. J Clin Gastroenterol*, **34**, 183–6.

Chen, M.G. and Mott, K.E. 1990. Progress in assessment of morbidity due to *Fasciola hepatica* infection: a review of recent literature. *Trop Dis Bull*, **87**, R1–R38.

Chen, M.G., Hua, X.J., et al. 1983. Praziquantel in 237 cases of clonorchiasis sinensis. *Chin Med J*, **96**, 935–40.

Chen, M.G., Lu, Y., et al. 1994. Progress in assessment of morbidity due to *Clonorchis sinensis* infection: a review of recent literature. *Trop Dis Bull*, **91**, R7–R65.

Cheng, Y. 1999. Studies on the relationship between the infection of intermediate hosts of *Paragonimus* and ecological environment. *Chin J Parasitol Parasitic Dis*, **17**, 212–14, in Chinese.

Cho, S.-Y. 1994. Epidemiology of paragonimiasis in Korea. In: Chai, J.-Y. and Cho, S.-Y. (eds), *Collected papers on parasite control in Korea*. Seoul: The Korean Association of Health, 51–7.

Choi, D.W. 1984. *Clonorchis sinensis*: life cycle, intermediate hosts, transmission to man and geographical distribution in Korea. *Arzneimittel Forsch*, **34**, 1145–51.

Choi, D.W. 1990. *Paragonimus* and paragonimiasis in Korea. *Korean J Parasitol*, **28**, 79–102.

DeFrain, M. and Hooker, R. 2002. North American paragonimiasis: case report of a severe clinical infection. *Chest*, **121**, 1368–72.

Ditrich, O., Giboda, M., et al. 1992. Comparative morphology of eggs of the Haplorchiinae (Trematoda: Heterophyidae) and some other medically important heterophyid and Opisthorchiid flukes. *Folia Parasitol*, **39**, 123–32.

Eckert, J. 1996. Workshop summary: food safety: meat- and fish-borne zoonoses. *Vet Parasitol*, **64**, 143–7.

El Zawawy, L.A. and Ali, S.M. 2002. Bacterial-parasite interaction between *Salmonella* and each of *Fasciola gigantica* and *Trichinella spiralis*. *J Egyptian Society Parasitol*, **32**, 745–54.

Esteban, J.G., Flores, A., et al. 1997. A population-based coprological study of human fascioliasis in a hyperendemic area of the Bolivian Altiplano. *Trop Med Int Health*, **2**, 695–9.

Esteban, J., Gonzalez, C., et al. 2002. High fascioliasis infection in children linked to a man-made irrigation zone in Peru. *Trop Med Int Health*, **7**, 339–48.

Fattakhov, R.G. 2002. [Fish infection with *Opisthorchis* larvae in Russia and some contiguous countries (by the materials of the 'Cadaster of *opisthorchis* infection foci in Russia in 1994')]. *Meditsinskaia Parazitologiia I Parazitarnye Bolezni*, **1**, 25–7, in Russian.

Favennec, L., Ortiz, J.J., et al. 2003. Double-blind, randomised, placebo-controlled study of nitazoxanide in the treatment of fascioliasis in adults and children from northern Peru. *Aliment Pharmacol Ther*, **17**, 265–70.

Hammouda, N.A. and El Mansoury, S.T. 1997. Detection of circulating antigens in blood to evaluate treatment of fascioliasis. *J Egyptian Soc Parasitol*, **27**, 365–71.

Hardman, E.W., Jones, R.L.H. and Davies, A.H. 1970. Fascioliasis – a large outbreak. *Br Med J*, **3**, 502–5.

Haseeb, A.N., el-Shazly, A.M., et al. 2002. A review on fascioliasis in Egypt. *J Egypt Soc Parasitol*, **32**, 317–54.

Hassan, M.M., Moustafa, N.E., et al. 1995. Prevalence of *Fasciola* infection among school children in Sharkia Governorate, Egypt. *J Egyptian Soc Parasitol*, **25**, 543–9.

Haswell-Elkins, M.R., Elkins, D.B., et al. 1991. Distribution patterns of *Opisthorchis viverrini* within a human community. *Parasitology*, **103**, 97–101.

Haswell-Elkins, M.R., Mairiang, E., et al. 1994a. Cross-sectional study of *Opisthorchis viverrini* infection and cholangiocarcinoma in communities within a high-risk area in Northeast Thailand. *Int J Cancer*, **59**, 505–9.

Haswell-Elkins, M.R., Satarug, S., et al. 1994b. Liver fluke infection and cholangiocarcinoma: model of endogenous nitric oxide and extragastric nitrosation in human carcinogenesis. *Mutat Res*, **305**, 241–52.

Hong, S.J. and Yun Kim, T. 2002. *Clonorchis sinensis*: glutathione S-transferase as a serodiagnostic antigen for detecting IgG and IgE antibodies. *Exp Parasitol*, **101**, 231–3.

Hong, S.J., Seong, K.Y., et al. 2000. Molecular cloning and immunological characterization of phosphoglycerate kinase from *Clonorchis sinensis*. *Mol Biochem Parasitol*, **108**, 207–16.

Hong, S.T., Lee, M., et al. 1999. Usefulness of IgG4 subclass antibodies for diagnosis of human clonorchiasis. *Korean J Parasitol*, **37**, 243–8.

Hou, M.F., Ker, C.G., et al. 1989. The ultrasound survey of gallstone diseases of patients infected with *Clonorchis sinensis* in Southern Taiwan. *J Trop Med Hyg*, **92**, 108–11.

Hou, P.C. 1955. The pathology of *Clonorchis sinensis* infestation of the liver. *J Pathol Bacteriol*, **70**, 53–68.

Hughes, A.J. and Biggs, B.A. 2002. Parasitic worms of the central nervous system: an Australian perspective. *Intern Med J*, **32**, 11, 541–53, review.

IARC Working Group. 1994. *Schistosomes, liver flukes and* Helicobacter pylori. *IARC monographs on the evaluation of carcinogenic risks to humans*, vol. 6. Lyon: International Agency for Research on Cancer, 121–75.

Iarotski, L.S., Be'er, S.A. 1993. *Epidemiology and control of opisthorchiasis in the former USSR*, Unpublished document SCH/SG/93/WP.12. Geneva: WHO.

Ikeda, T. 2001. Protein A immunocapture assay detecting antibodies to fluke cyteine proteinases for immunodiagnosis of human paragonimiasis and fascioliasis. *J Helminth*, **75**, 3, 245–9.

Intapan, P., Kaewkes, S. and Maleewong, W. 1992. Control of *Opisthorchis viverrini* cercariae using the copepod *Mesocyclops leuckarti*. *Southeast Asian J Trop Med Public Health*, **23**, 348–9.

Jaroonvesama, N. 1988. Differential diagnosis of eosinophilic meningitis. *Parasitol Today*, **88**, 262–6.

Jongsuksuntigul, P. and Imsomboon, T. 1997. The impact of a decade long opisthorchiasis control program in northeastern Thailand. *Southeast Asian J Trop Med Public Health*, **28**, 551–7.

Kaewkes, S., Elkins, D.B., et al. 1991. Comparative studies on the morphology of the eggs of *Opisthorchis viverrini* and lecithodendriid trematodes. *Southeast Asian J Trop Med Public Health*, **22**, 623–30.

Khalil, S.S., Abou Shousha, S., et al. 1999. Production of pro-inflammatory cytokines (GM-CSF, IL-8 and IL-6) by monocytes from fasciolosis patients. *J Egypt Soc Parasitol*, **29**, 1007–15.

Khamboonruang, C., Keawvichit, R., et al. 1997. Application of hazard analysis critical control point (HACCP) as a possible control measure for *Opisthorchis viverrini* infection in cultured carp (*Puntius gonionotus*). *Southeast Asian J Trop Med Public Health*, **28**, 65–72.

Kim, S.I. 1998. A *Clonorchis sinensis*-specific antigen that detects active human clonorchiasis. *Korean J Parasitol*, **36**, 37–45.

Kim, Y.H. 1999. Eosinophilic cholecystitis in association with *Clonorchis sinensis* infestation in the common bile duct. *Clin Radiol*, **54**, 552–4.

Kim, Y.H. 2003a. Carcinoma of the gallbladder associated with *clonorchiasis*: clinicopathologic and CT evaluation. *Abdom Imaging*, **28**, 83–6.

Kim, Y.H. 2003b. Extrahepatic cholangiocarcinoma associated with *clonorchiasis*: CT evaluation. *Abdom Imaging*, **28**, 68–71.

Kino, H., Inaba, H., et al. 1998. Epidemiology of clonorchiasis in Ninh Binh Province, Vietnam. *Southeast Asian J Trop Med Public Health*, **29**, 250–4.

Kirby, G.M., Pelkonen, P., et al. 1994. Association of liver fluke (*Opisthorchis viverrini*) infestation with increased expression of CYP2A and carcinogen metabolism in male hamster liver. *Mol Carcinog*, **11**, 81–9.

Kobayashi, J., Vannachone, B., et al. 2000. An epidemiological study on *Opisthorchis viverrini* infection in Lao villages. *Southeast Asian J Trop Med Pub Health*, **31**, 128–32.

Koenigstein, R.P. 1949. Observations on the epidemiology of infections with *Clonorchissinensis*. *Trans R Soc Trop Med Hyg*, **42**, 503–6.

Kofta, W., Mieszczanek, J., et al. 2000. Successful DNA immunisation of rats against fasciolosis. *Vaccine*, **18**, 2985–90.

Komiya, Y. 1966. *Clonorchis* and clonorchiasis. *Adv Parasitol*, **4**, 53–106.

Kum, P.N. and Nchinda, T.C. 1982. Pulmonary paragonimiasis in Cameroon. *Trans R Soc Trop Med Hyg*, **76**, 768–72.

LaPook, J.D., Magun, A.M., et al. 2000. Sheep, watercress, and the Internet. *Lancet*, **356**, 218–19.

Lecaillon, J.B., Godbillon, J., et al. 1998. Effect of food on the bioavailability of triclabendazole in patients with fascioliasis. *Br J Clin Pharmacol*, **45**, 601–4.

Lee, J.H., Rim, H.J. and Sell, S. 1997. Heterogeneity of the 'oval-cell' response in the hamster liver during cholangiocarcinogenesis following *Clonorchis sinensis* infection and dimethylnitrosamine treatment. *J Hepatol*, **26**, 1313–23.

Lee, K.J., Bae, Y.T., et al. 2002. Status of intestinal parasites infection among primary school children in Kampongcham, Cambodia. *Korean J Parasitol*, **40**, 153–5.

Lee, S.H., Hwang, S.W., et al. 1984. Comparative morphology of eggs of heterophyids and *Clonorchissinensis* causing human infections in Korea. *Korean J Parasitol*, **22**, 171–80.

Leung, J.W. and Yu, A.S. 1997. Hepatolithiasis and biliary parasites. *Baill Clin Gastroenterol*, **11**, 681–706, review.

Lim, J.H. 1990. Radiologic findings in clonorchiasis. *Am J Radiol*, **155**, 1001–8.

Maher, K. and El Ridi, R. 1999. Parasite-specific antibody profile in human fascioliasis: Application for immunodiagnosis of infection. *Am J Trop Med Hyg*, **61**, 738–42.

Mairiang, E., Elkins, D.B., et al. 1992. Relationship between intensity of *Opisthorchis viverrini* infection and hepatobiliary disease detected by ultrasonography. *J Gastroenterol Hepatol*, **7**, 17–21.

Mairiang, E., Haswell-Elkins, M.R., et al. 1993. Reversal of biliary tract abnormalities associated with *Opisthorchis viverrini* infection following praziquantel treatment. *Trans R Soc Trop Med Hyg*, **87**, 194–7.

Maleewong, W. and Intapan, P.M. 2003. Detection of *Opisthorchis viverrini* in experimentally infected bithynid snails and cyprinoid fishes by a PCR-based method. *Parasitology*, **126**, 63–7.

Maleewong, W., Intapan, P.M., et al. 1997. Antigenic components of somatic extract from adult *Fasciola gigantica* recognized by infected human sera. *Asian Pacific J Allergy Immunol*, **15**, 213–18.

Mansour, W.A., Kaddah, M.A., et al. 1998. A monoclonal antibody diagnoses active *Fasciola* infection in humans. *J Egyptian Soc Parasitol*, **28**, 711–27.

Mas-Coma, M.S., Esteban, J.G. and Bargues, M.D. 1999. Epidemiology of human fascioliasis: A review and proposed classification. *Bull WHO*, **77**, 340–6.

Mas-Coma, S., Funatsa, I.R., et al. 2001. *Fasciola hepatica* and lymnaeid snails occurring at very high altitude in South America. *Parasitol*, **123**, Suppl, S115–27.

Matsumoto, N., Mukae, H., et al. 2002. Elevated levels of thymus and activation-regulated chemokine (TARC) in pleural effusion samples from patients infested with *Paragonimus westermani*. *Clin Exp Immunol*, **130**, 314–18.

Maurice, J. 1994. Is something lurking in your liver? *New Scientist*, **19 March**, 26–31.

Migasena, P. 1982. Liver flukes relationship to dietary habits and development programs in Thailand. In: Patrice Jellife, E.F. and Jellife, D.B. (eds), *Adverse effects of foods*. New York: Plenum, 307–11.

Mitacek, E.J., Brunnemann, K.D., et al. 1999a. Volatile nitrosamines and tobacco-specific nitrosamines in the smoke of Thai cigarettes: a risk factor for lung cancer and a suspected risk factor for liver cancer in Thailand. *Carcinogenesis*, **20**, 133–7.

Mitacek, E.J., Brunnemann, K.D., et al. 1999b. Exposure to N-nitroso compounds in a population of high liver cancer regions in Thailand: volatile nitrosamine levels in Thai food. *Food Chem Toxicol*, **37**, 297–305.

Miyazaki, I. 1982. Paragonimiasis. In: Hillyer, G.V. and Hopla, C.E. (eds), *Parasitic zoonoses. , CRC handbook series in zoonoses*, vol. III, section C. . Boca Raton, FL: CRC Press, 143–64.

Na, K. and Lee, H.J. 2002. Expression of cysteine proteinase of *Clonorchis sinensis* and its use in serodiagnosis of clonorchiasis. *J Parasitol*, **88**, 1000–6.

Nakamura-Uchiyama, F., Onah, D.N., et al. 2001. Clinical features and parasite-specific IgM/IgG antibodies of paragonimiasis patients recently found in Japan. *Southeast Asian J Trop Med Pub Health*, **32**, Suppl 2, 55–8.

Nagakura, K., Oouchi, M., et al. 2002. Pulmonary paragonimiasis misdiagnosed as tuberculous: with specific references on paragonimiasis. *Tokai J Exp Clin Med*, **27**, 97–100.

Okuda, K., Nakanuma, Y., et al. 2002. Cholangiocarcinoma: recent progress. Part 1: Epidemiology and etiology. *J Gastroenterol Hepatol*, **17**, 1049–55, review.

O'Neill, S.M., Parkinson, M., et al. 1999. Short report: Immunodiagnosis of human fascioliasis using *Fasciola hepatica* cathepsin L1 cysteine proteinase. *Am J Trop Med Hyg*, **60**, 749–51.

Ooi, H.K., Chen, C.I., et al. 1997. Metacercariae in fishes of Sun Moon lake which is an endemic area for Clonorchis sinensis in Taiwan. *Southeast Asian J Trop Med Public Health*, **28**, 1, 222–3.

Osman, M.M. and Abo-El-Nazar, S.Y. 1999. IL-10, IFN-gamma and TNF-alpha in acute and chronic human fascioliasis. *J Egyptian Soc Parasitol*, **29**, 13–20.

Osman, M.M., Shehab, A.Y., et al. 1995. Evaluation of *Fasciola* excretory-secretory (E/S) product in diagnosis of acute human fasciolosis by IgM ELISA. *Trop Med Parasitol*, **46**, 115–18.

Osman, M.M., Ismail, Y. and Aref, T.Y. 1999. Human fasciolosis: a study on the relation of infection intensity and treatment to hepatobiliary affection. *J Egyptian Soc Parasitol*, **29**, 353–63.

Pairojkul, C., Sithithaworn, P., et al. 1991. Risk groups for opisthorchiasis-associated cholangiocarcinoma indicated by a study of worm burden-related biliary pathology. *Kan-Tan-Sui*, **22**, 111–20.

Park, G.M. and Yong, T.S. 2003. Effects of gamma-irradiation on the infectivity and chromosome aberration of *Clonorchis sisensis*. *Korean J Parasitol*, **42**, 41–5.

Park, H., Ko, M.Y., et al. 1995. Cytotoxicity of a cysteine proteinase of adult *Clonorchis sinensis*. *Korean J Parasitol*, **33**, 211–18.

Park, H., Kim, S.I., et al. 2002. Characterization and classification of five cysteine proteinases expressed by *Paragonimus westermani* adult worm. *Exp Parasitol*, **102**, 143–9.

Patrick, K.M. and Isaac-Renton, J. 1992. Praziquantel failure in treatment of *Fasciola hepatica*. *Can J Infect Dis*, **3**, 33–6.

Pungpak, S., Riganti, M., et al. 1985. Clinical features in severe opisthorchiasis viverrini. *Southeast Asian J Trop Med Pub Health*, **16**, 405–9.

Pungpak, S., Viravan, C., et al. 1997. *Opisthorchis viverrini* infection in Thailand: studies on the morbidity of the infection and resolution following praziquantel treatment. *Am J Trop Med Hyg*, **56**, 311–14.

Queneau, P.-E., Koch, S., et al. 1997. Nodular intra-hepatic lesions of liver distomatosis: Efficacy of praziquantel. *Gastroenterol Clin Biol*, **21**, 511–13.

Radomyos, P., Bunnag, D. and Harinasuta, T. 1984. Worms recovered in stools following praziquantel treatment. *Arzneimittel Forsch*, **34**, 1215–17.

Radomyos, B., Wongsaroj, T., et al. 1998. Opisthorchiasis and intestinal fluke infections in northern Thailand. *Southeast Asian J Trop Med Public Health*, **29**, 123–7.

Richter, J., Freise, S., et al. 1999. Fascioliasis: Sonographic abnormalities of the biliary tract and evolution after treatment with triclabendazole. *Trop Med Int Health*, **4**, 774–81.

Riganti, M., Pungpak, S., et al. 1988. *Opisthorchis viverrini* eggs and adult flukes as nidus and composition of gallstones. *Southeast Asian J Trop Med Pub Health*, **19**, 633–6.

Rim, H.J. 1986. The current pathobiology and chemotherapy of clonorchiasis. *Korean J Parasitol*, **24**, Suppl, 1–141.

Rondelaud, D., Dreyfuss, G., et al. 2000. Changes in human fasciolosis in a temperate area: About some observations over a 28-year period in central France. *Parasitol Res*, **86**, 753–7.

Rondelaud, D., Vignoles, P., et al. 2001. The definitive and intermediate hosts of *Fasciola hepatica* in the natural watercress beds in central France. *Parasitol Res*, **87**, 475–8.

Sadun, E.H. 1955. Studies on *Opisthorchis viverrini* in Thailand. *Am J Hyg*, **2**, 81–115.

Sasaki, M., Kamiyama, T., et al. 2002. Active hepatic capsulitis caused by *Paragonimus westermani* infection. *Intern Med*, **41**, 661–3.

Satarug, S. and Haswell-Elkins, M.R. 1998. Relationships between the synthesis of N-nitrosodimethylamine and immune responses to chronic infection with the carcinogenic parasite, *Opisthorchis viverrini*, in men. *Carcinogenesis*, **19**, 485–91.

Satarug, S., Haswell-Elkins, M.R., et al. 1996a. Thiocyanate-independent nitrosation in humans with carcinogenic parasite infection. *Carcinogenesis*, **17**, 1075–81.

Satarug, S., Lang, M.A., et al. 1996b. Induction of cytochrome P450 2A6 expression in humans by the carcinogenic parasite infection, *Opisthorchiasis viverrini*. *Cancer Epidemiol Biomarkers Prev*, **5**, 795–800.

Schmidt, G.D. and Roberts, L.S. 1994. *Foundations of parasitology*, 5th edn. St. Louis, MO: Mosby Year Book, 227–59.

Shehab, A.Y., Abou Basha, L.M., et al. 1999. Circulating antibodies and antigens correlate with egg counts in human fascioliasis. *Trop Med Int Health*, **4**, 691–4.

Shin, H.R., Lee, C.U., et al. 1996. Hepatitis B and C virus, *Clonorchis sinensis* for the risk of liver cancer: a case-control study in Pusan, Korea. *Int J Epidemiol*, **25**, 933–40.

Shin, M.H., Kita, H., et al. 2001. Cysteine protease secreted by *Paragonimus westermani* attenuates effector functions of human eosinophils stimulated with immunoglobulin G. *Infect Immun*, **69**, 1599–604.

Silva, M.L., Sampaio, C., et al. 1996. Antigenic components of excretory-secretory products of adult *Fasciola hepatica* recognized in human infections. *Am J Trop Med Hyg*, **54**, 146–8.

Sirisinha, S., Chawengkirttikul, R., et al. 1995. Evaluation of a monoclonal antibody-based enzyme linked immunosorbent assay for the detection of *Opisthorchis viverrini* infection in an endemic area. *Am J Trop Med Hyg*, **52**, 521–4.

Sithithaworn, P., Tesana, S., et al. 1991. Relationship between faecal egg count and worm burden of *Opisthorchis viverrini* in human autopsy cases. *Parasitology*, **102**, 277–81.

Sobhon, P., Anantavara, S., et al. 1998. *Fasciola gigantica*: studies of the tegument as a basis for the developments of immunodiagnosis and vaccine. *Southeast Asian J Trop Med Pub Health*, **29**, 387–400.

Song, S.B. 1987. Larvicidal action of liquid nitrogen against metacercariae of *Clonorchis sinensis. Korean J Parasitol*, **25**, 129–40.

Sornmani, S. 1987. Control of opisthorchiasis through community participation. *Parasitol Today*, **3**, 31–3.

Spithill, T.W. 1992. Control of tissue parasites. III. Trematodes. In: Yong, W.K. (ed.), *Animal parasite control utilizing biotechnology*. Boca Raton, FL: CRC Press, 199–219.

Spithill, T.W., Piedrafita, D. and Smooker, P.M. 1997. Immunological approaches for the control of fasciolosis. *Int J Parasitol*, **27**, 1221–35.

Sripa, B. and Kaewkes, S. 2000. Localisation of parasite antigens and inflammatory responses in experimental opisthorchiasis. *Int J Parasitol*, **30**, 735–40.

Sripa, B. and Kaewkes, S. 2002. Gall bladder and extrahepatic bile duct changes in *Opisthorchis viverrini*-infected hamsters. *Acta Tropica*, **83**, 29–36.

Srivatanakul, P., Ohshima, H., et al. 1991. Endogenous nitrosamines and liver fluke as risk factors for cholangiocarcinoma in Thailand. *Int J Cancer*, **48**, 821–5.

Sui, F., Shu-hua, X. and Catto, B.A. 1988. Clinical use of praziquantel in China. *Parasitol Today*, **4**, 312–15.

Sukontason, K.L., Sukontason, K., et al. 2001. Prevalence of *Opisthorchis viverrini* infection among villagers harbouring *Opisthorchis*-like eggs. *Southeast Asian J Trop Med Pub Health*, **32**, Suppl 2, 23–6.

Sun, T., Chou, S.T. and Gibson, J.B. 1968. Route of entry of *Clonorchis sinensis* to the mammalian liver. *Exp Parasitol*, **22**, 346–51.

Teoh, T.B. 1963. A study of gall-stones and included worms in recurrent pyogenic cholangitis. *J Pathol Bacteriol*, **86**, 123–9.

Tesana, S., Pamarapa, A. and Sae Sio, O.-T. 1989. Acute cholecystitis and *Fasciola* sp. infection in Thailand: report of two cases. *Southeast Asian J Trop Med Pub Health*, **20**, 447–52.

Tesana, S., Takahashi, Y., et al. 2000. Ultrastructural and immunohistochemical analysis of cholangiocarcinoma in immunized Syrian golden hamsters infected with *Opisthorchis viverrini* and administered with dimethylnitrosamine. *Parasitol Int*, **49**, 239–51.

Thamavit, W., Tiwawech, D., et al. 1996. Equivocal evidence of complete carcinogenicity after repeated infection of Syrian hamsters with *Opisthorchis viverrini. Toxicol Pathol*, **24**, 493–7.

Toscano, C., Hai, Y.S., et al. 1995. Paragonimiasis and tuberculosis – diagnostic confusion: a review of the literature. *Trop Dis Bull*, **92**, R1–R27.

Tran, V.H., Tran, T.K., et al. 2001. Fascioliasis in Vietnam. *Southeast Asian J Trop Med Pub Health*, **32**, Suppl 2, 48–50.

Trap, C. and Boireau, P. 2000. [Proteases in helminthic parasites]. *Vet Res*, **31**, 461–71, in French.

Upatham, E.S., Viyanant, V., et al. 1984. Relationship between prevalence and intensity of *Opisthorchis viverrini* infection, and clinical symptoms and signs in a rural community in northeast Thailand. *Bull WHO*, **62**, 451–61.

Upatham, E.S., Viyanant, V., et al. 1988. Rate of re-infection by *Opisthorchis viverrini* in an endemic Northeast Thai community after chemotherapy. *Int J Parasitol*, **18**, 643–9.

van Beers, B., Pringot, J., et al. 1990. Hepatobiliary fascioliasis: noninvasive imaging findings. *Radiology*, **174**, 809–10.

Vannachone, B., Kobayashi, J., et al. 1998. An epidemiological survey on intestinal parasite infection in Khammouane Province, Lao PDR, with special reference to *Strongyloides* infection. *Southeast Asian J Trop Med Public Health*, **29**, 717–22.

Vanijanonta, S., Bunnag, D. and Harinasuta, T. 1984. Radiological findings in pulmonary paragonimiasis heterotremus. *Southeast Asian J Trop Med Pub Health*, **15**, 122–8.

Vatanasapt, V., Martin, N. et al. 1993. *Cancer in Thailand 1988–1991*, IARC Technical Report No. 16. Lyon: International Agency for Research on Cancer, 57–92.

Velez, I.D., Ortega, J.E., et al. 2002. Paragonimiasis: a view from Columbia. *Clin Chest Med*, **23**, 421–31.

Waikagul, J. 1998. *Opisthorchis viverrini* metacercaria in Thai freshwater fish. *Southeast Asian J Trop Med Public Health*, **29**, 324–6.

Waikagul, J., Dekumroy, P., et al. 2002. Serodiagnosis of human opisthorchiasis using cocktail and electroeluted Bithynia snail antigens. *Parasitol Int*, **51**, 237–47.

Watanapa, P. and Watanapa, W.B. 2002. Liver fluke-associated cholangiocarcinoma. *Br J Surg*, **89**, 962–70, review.

WHO. 1995. Control of food-borne trematode infections. *WHO Tech Rev Serv*, **849**.

Wolf-Spengler, M.L. and Isseroff, H. 1983. Fasciolasis: bile duct collagen induced by proline from the worm. *J Parasitol*, **69**, 290–4.

Wongratanacheewin, S., Bunnag, D., et al. 1988. Characterization of humoral immune response in the serum and bile of patients with opisthorchiasis and its application in immunodiagnosis. *Am J Trop Med Hyg*, **38**, 356–62.

Wongratanacheewin, S., Pumidonming, W., et al. 2001. Development of a PCR-based method for the detection of *Opisthorchis viverrini* in experimentally infected hamsters. *Parasitology*, **122**, 175–80.

Wongsaroj, T., Sakolvaree, Y., et al. 2001. Affinity purified oval antigen for diagnosis of *Opisthorchiasis viverrini. Asian Pacific J Allergy Immunol*, **19**, 245–58.

Wykoff, D.E., Harinasuta, C., et al. 1965. *Opisthorchis viverrini* in Thailand – the life cycle and comparison with *O. felineus. J Parasitol*, **51**, 207–14.

Xu, L., Jiang, Z., et al. 1995. [Characteristics and recent trends in endemicity of human parasitic diseases in China]. *Chung-Kuo-Chi-Sheng-Chung-Hsueh-Yu-Chi-Sheng-Chung-Ping-Tsa-Chih*, **13**, 214–17, Chinese.

Yokogawa, M. 1982. Paragonimiasis. In: Hillyer, G.V. and Hopla, C.E. (eds), *Parasitic Zoonoses.* , *CRC handbook series in zoonoses*, vol. III, Section C. Boca Raton, FL: CRC Press, 123–42.

Yong, T.S., Park, S.J., et al. 1999. Identification of IgE-reacting *Clonorchis sinensis* antigens. *Yonsei Med J*, **40**, 178–83.

32

Intestinal tapeworms

JØRN ANDREASSEN

THE PARASITES

Introduction and historical perspective

Intestinal tapeworms are among the earliest known human parasites. They have been known since prehistoric times and are referred to in the Papyrus Ebers from about 1500BC (Hoeppli 1956). As early as the second century, Galen recognized *Taenia*, but it was not until 1592 that Dunas discovered *Diphyllobothrium*, the broad tapeworm. The first illustration of a tapeworm (*Taenia*) was the *pittoresce* picture made by Andry in 1700. In 1782, the connection between tapeworms and hydatids was found by Goeze, who also found differences between the two human species of *Taenia*. In 1790, Abildgaard demonstrated that plerocercoid larvae from sticklebacks became adult tapeworms in ducks to which they were fed, but it was not until 1845 that Dujardin showed adult *Taenia* in humans to arise from cysticerci in meat. Six years later, the life cycle of *Taenia saginata* was established by Küchenmeister, but it was not until 1917 that the complex life cycle of *D. latum* was found by Janicki (Grove 1990; Smyth 1990; Saklatvala 1993). As late as 1993, a new human species, *Taenia asiatica*, was described (Eom and Rim 1993), although it is now questioned whether this is a separate species or a subspecies of *T. saginata* (McManus and Bowles 1994) (see section on *Taenia saginata saginata* (beef tapeworm) below).

Until the early 1960s, tapeworms were said to have a more or less inert outer cuticle thought to prevent the worm being digested by the host's intestinal enzymes. However, at the same time, the worms had to absorb all nutrients through this surface because tapeworms have no gut. By electron microscopy, it was then shown that the surface was a living structure, a tegument (see section on Morphology and structure), which at the same time could absorb nutrients (see section on Feeding) and have mechanisms to inhibit host enzymes from digesting the worms (see section on Physiology).

The exact prevalence of human intestinal tapeworms is not known, but it has been estimated that as many as 100 million people worldwide may be infected with either *T. solium* or *T. saginata* (FAO 1991), and about 50–70 million with the dwarf tapeworm, *Hymenolepis nana* (Pawlowski 1984; Crompton 1999). None of the intestinal tapeworm infections are really life-threatening, but the presence of the adult pig tapeworm, *T. solium*, increases the risk of acquiring cysticercosis by the uptake of infective eggs, and an infection with the broad tapeworm, *D. latum*, may give rise to pernicious anemia (see section on Clinical and pathological aspects of infection).

The geographical distribution of intestinal tapeworm infections in humans corresponds to the distribution of their intermediate hosts. Since the cyclophyllid tapeworms, such as species of *Taenia* and *Hymenolepis*, have intermediate hosts living in close contact to humans, their distribution is worldwide, while the pseudophyllid species of *Diphyllobothrium* are more restricted, because they need certain freshwater copepods and fishes as intermediate hosts.

Classification

All tapeworms belong to the class Cestoda and the subclass Eucestoda, which is made up of a number of orders. Tapeworm species which may infect humans belong to six families from two orders (Khalil et al. 1994). (See Table 32.1.)

Morphology and structure

Human adult tapeworms are flat worms (Platy-helminthes) with a length ranging from a few centi-meters (*Hymenolepis nana*) up to several meters (e.g. *Taenia* species). They are white or grayish to yellow. Calcareous corpuscles are present in tissues filling the spaces between internal organs. Yellow coloration comes from eggs in the posterior part of the worms.

Tapeworms found in the small intestine of humans are adults in different developmental stages. The adult worm always consists of a head (scolex) and a neck region (the growth region). If time and conditions allow, the neck generates a series of proglottides (segments), which together form a strobila. The scolex has muscular suckers which function as holdfast organs and are used in locomotion. There are either four round suckers (acetabula), as in the cyclophyllids, or two elongated grooves (bothria), as in the pseudophyllids. On the scolex, some species also have small hooks, which can be inserted into the intestinal epithelium and thereby help the worm to maintain its position in the gut of the host. The hooks are often in one or two circles (two in *T. solium* and one in *H. nana*) on a retractable rostellum (cone) between the suckers on the anterior end of the scolex. From the narrow neck region, the width of the worm increases and folds of the body wall are made at regular intervals along its length, forming the proglottides, giving the tapeworm the appearance of a segmented worm (Figure 32.1) (Mehlhorn et al. 1981).

The body wall (see Figure 32.2) is called a tegument. This consists of a distal syncytial cytoplasm, with cyto-plasmic connections to a perinuclear cytoplasm lying beneath the basal plasma membrane, and circular and longitudinal muscles. The surface of the tegument is made up of microvillus-like structures called micro-triches, microthrix in the singular (see Arme 1984). A microthrix consists of a shaft and an electron-dense bent spine and is, like a microvillus on most other absorptive cells, covered by a thin layer of glycoproteins and muco-polysaccharides called a glycocalyx.

Beneath the tegumental basal plasma membrane lie two layers of muscles: an outer circular layer and an inner longitudinal one, allowing the worms to move. Special muscles run to the suckers, where bands of criss-cross fibers help improve the function of the suckers. Radial muscles may go to inner organs such as the cirrus, uterus, and vagina.

The nervous system consists of a brain-like structure in the scolex built up of a central ganglion, and lateral and rostellar ganglia connected by a central nerve ring. From the ganglia, nerves run to the suckers and rostellum. Lateral and median longitudinal nerves run from the ring all the way back to the posterior end of the tapeworm. There are transverse commissures between the longitudinal nerves and connections to inner organs and sensory organs in the body wall.

The excretory system is a typical protonephridial system, as in other platyhelminths, consisting of two lateral canals in each side (dorsal and ventral) connected in the scolex. The two ventral, lateral canals are connected with a transverse canal in the posterior of each proglottid. The excretory canals are built up of terminal cells (flame cells) connected to canal cells; flow through the system is generated by a bundle of cilia in these terminal cells. The excretory fluid is carried to the scolex in the dorsal canals and back through the ventral canals that open to the exterior at the terminal proglottis. Around the cilia in the flame cells is a lattice-like system of projections from the flame cell and the

Table 32.1 Classification of human tapeworms

Order	Family	Species
Cyclophyllidea Scolex with four round suckers and mostly armed with hooks. Development through two larval stages	Taeniidae	*Taenia saginata saginata, T. saginata asiatica, T. solium*
	Hymenolepididae	*Hymenolepis nana, H. diminuta*
	Dipylidiidae	*Dipylidium caninum*
	Mesocestoididae	*Mesocestoides corti*
	Anoplocephalidae	*Bertiella* e.g. *studeri, Inermicapsifer madagascariensis*
Pseudophyllidea Scolex has two elongated slit-like suckers. Development through three larval stages	Diphyllobothridae	*Diphyllobothrium latum.* A number of zoonotic animal species of *Diphyllobothrium* may sporadically infect humans, e.g. *D. dendriticum, D. nihonkaiense, D. pacificum, D. cordatum, D. hians, D. houghtoni, D. orcini, D. scoticum, D. ursi, D. yonagoense*

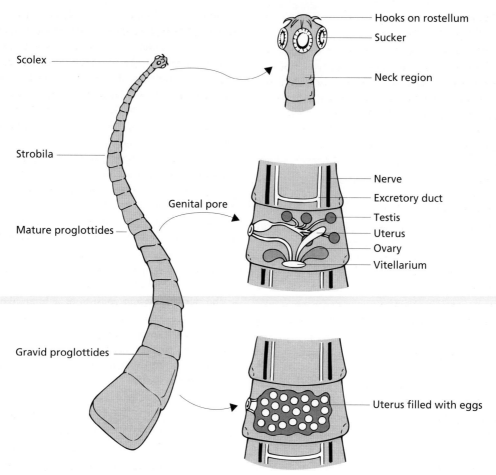

Figure 32.1 *The organization of an adult tapeworm. (Based on* Taenia*)*

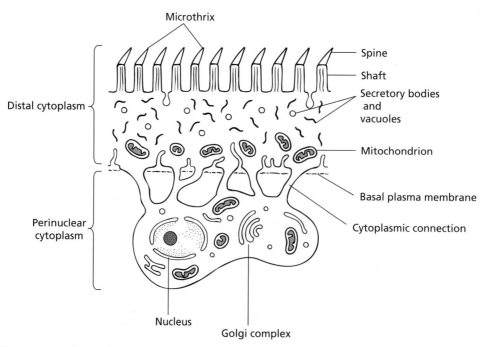

Figure 32.2 *The tegument of an adult tapeworm*

canal cells. Ultrafiltration takes place at the basal lamina through gaps in the lattice-like system.

Tapeworms are acoelomate, lacking not only a body cavity (coelom) but also a circulatory system.

The reproductive system is a typical hermaphroditic system, where the male and female organs differentiate progressively from the anterior region. The reproductive organs are repeated, with one set in each of the proglottides. Proglottides are considered mature when the reproductive organs found in them are sexually mature. Copulation then takes place and fertilized eggs are formed. When the eggs are mature, i.e. contain the first stage larva, the proglottid is said to be gravid. Male organs typically consist of a few large, or many small, testes from which spermatozoa are passed via vasa efferentia to a common vas deferens and then via a seminal vesicle to a long muscular and protrusible cirrus surrounded by a cirrus sac. The female organs normally consist of a single bilobed ovary, an ootype, where egg composition and wall formation start, a vagina leading to a common genital pore, and a compact or dispersed vitellarium from which vitelline cells are brought to the ootype to participate in the formation of a capsule (normally called an eggshell), as in the case of pseudophyllids, or in nourishment of the fertilized egg cell. In the cyclophyllids, the capsule is absent or poorly developed, while the embryophore (the inner envelope) forms a shell. In the cyclophyllid *Hymenolepis*, only one vitelline cell takes part in egg formation. Eggs are then passed into the uterus, which may or may not extend to the common genital pore. In some species, such as *Diphyllobothrium*, the eggs are operculate, as in most digeneans.

A fertilized egg cell develops to the first larval stage, called either an oncosphere (ball with hooks) or hexacanth (larva with six hooks), inside an eggshell (see Figure 32.3).

General biology

Usually, only one or a few of the large human tapeworms are present in a patient, but up to 28 worms (*T. saginata*) have been recorded, indicating an aggregated distribution of the parasite in the human population. In infections of the dwarf tapeworm (*H. nana*), many worms may be present in a human intestine. Individuals heavily infected with intestinal nematodes, often called wormy people, have been shown to be predisposed to this condition (Anderson 1986). Whether this situation also exists in humans infected with tapeworms is not known.

The large human tapeworms are capable of living for many years, in some cases up to 20 years (Wright 1984), but perhaps as long as the host. It has been shown that the rat tapeworm, *H. diminuta*, can live as long as the host, and, if repeatedly transplanted into the duodenum of young rats, can survive for at least 14 years (Read 1967).

When only a single tapeworm is present, it has been shown that self-insemination, i.e. transfer of sperm within the same segment or between segments in the same strobila, takes place. When multiple tapeworms are present, both self-insemination and cross-insemination take place (Nollen 1983).

MOVEMENT

The tapeworm utilizes its suckers and, if present, the hooks found on the scolex, to hold on to the villi of the intestine. Movement about the intestine is carried out by means of the suckers and the subtegumental musculature. Prepatent and diurnal movements are discussed below (see section entitled Location in host).

FEEDING

Since tapeworms have no mouth, all food has to be absorbed through the tegument. The microtriches help to increase the absorptive area of the tegument. A functional amplification factor for tapeworms has been shown to vary from slightly less than two to about 12, compared to about 26 for a mouse small intestine (Threadgold and Robinson 1984).

Tapeworms are normally in close contact to the intestinal mucosa (see section Location in host), which may be important for absorption of food. Small molecules,

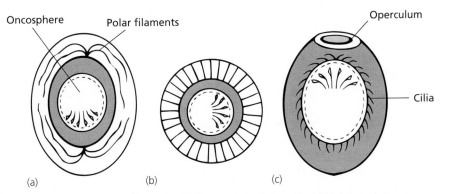

Figure 32.3 *Representative tapeworm eggs (not to scale):* **(a)** *hymenolepid;* **(b)** *taeniid;* **(c)** Diphyllobothrium

such as amino acids and glucose, have been shown to be taken up through the tegument via diffusion or active transport, but pinocytosis (endocytosis) of larger molecules such as proteins is still controversial (see Siles-Lucas and Hemphill 2002).

Although the energy metabolism of tapeworms mainly involves carbohydrates, amino acids play a major role in the high protein synthesis necessary for the large egg production. Since the worms cannot synthesize long chain fatty acids and lipids such as cholesterol, the absorption of these compounds is essential. It has been shown that bile salts are required for this transport across the tegument. The uptake of purine and pyrimidine is highly complex. For further details see Pappas and Read (1975).

PHYSIOLOGY

Excretion of waste products must take place through the tegument by exocytosis or through the excretory system.

Tapeworms have no special breathing organs and no vascular system: all exchange of oxygen and carbon dioxide has to take place over the tegument. The mitochondria in the tegument have few cristae, indicating that the metabolism of tapeworms is largely anaerobic.

There were few investigations of the physiology and biochemistry of the nervous system of cestodes until the development of immunocytochemical techniques. These techniques have demonstrated a number of neuropeptides which probably function as either neurotransmitters or neuromodulators and are responsible for coordinating behavioral responses (see reviews by Halton et al. 1990 and Siles-Lucas and Hemphill 2002).

BIOCHEMISTRY

Cestodes normally show a rather high level of carbohydrates (especially glycogen) and lipids and a relatively low level of proteins.

Mucopolysaccharides or complexes with proteins, such as mucoproteins or glycoproteins, are the main components of the surface glycocalyx. It has been suggested that this surface glycocalyx protects the worms from being digested by the host's proteolytic enzymes (Robertson and Cain 1984). On the other hand, digestion of nutrients to a state available for absorption can be made by enzymes bound to the glycocalyx, which has a turnover time of about 6 h. These enzymes may be of parasitic origin (intrinsic) or derived from their host (extrinsic).

Several neurotransmitters have been found in cestodes, e.g. acetylcholine, serotonin (5-hydroxytryptamine), noradrenaline (norepinephrine), dopamine, and octopamine. Serotonin is regarded as the main excitatory neurotransmitter of motor activity and acetylcholine as the main inhibitory neurotransmitter.

Life cycle

Not only a final host (a human being) but also at least one intermediate host is necessary for the completion of the life cycle of a cestode. For pseudophyllids, at least two intermediate hosts are necessary. Only one human tapeworm, *H. nana*, can complete its life cycle without an intermediate host. All larval stages are passed on to the next host via the oral route (for details see under the individual species). Humans become infected with an adult tapeworm by eating an infected intermediate host or infected organs from an intermediate host. When an infective larva is eaten by a human, excystation occurs if the larva is encysted, and, if the scolex is 'retracted' it will be 'inverted;' growth then commences from the neck region. In pseudophyllids, growth does not commence until everything except the scolex and neck has been shed from the plerocercoid larva. This active process takes about 60 h in golden hamsters.

Transmission and epidemiology

Human tapeworm infections caused by *T. saginata* and *T. solium*, commonly called taeniosis or taeniasis, are cosmopolitan. Prevalence is normally low but very much dependent on cultural factors and the level of hygiene and veterinary control. In some African countries, prevalences above 10 percent have been reported. In contrast to these low levels in the final host, the prevalence of cysticercus infections in the intermediate hosts may be very high in some regions.

The total production of tapeworm eggs per infected human is an important epidemiological factor, which, for the large human tapeworms, is probably negatively correlated with the total number of tapeworms present per host, so that the fewer tapeworms per person, the more eggs are produced.

Location in host

All adult human tapeworms live in the lumen of the small intestine, but, dependent on the species, in different parts of the small intestine. The two species of *Taenia* have their scoleces in the upper part of the jejunum, while those of *Diphyllobothrium* and *H. nana* normally lie in the upper two-thirds of the ileum. The scolex is found deeply embedded in the villi, while the strobila lies with its flat side close to the mucosa, sometimes in a spiral along the intestine.

The rare human tapeworm, *Hymenolepis diminuta*, which normally has rats as final hosts, has been shown to migrate in the small intestine of rats both during its prepatent period and diurnally. The newly excysted worm establishes itself about 30–40 percent down the small intestine. During the second week, the worm migrates anteriorly, stopping in the upper 10–20 percent of the small intestine. If many worms are present, they can be found spread over a larger part of the small intestine. Circadian migratory behavior has also been found for *H. diminuta* in rats, where the worms are found more anteriorly in the small intestine in the morning and more posteriorly in the evening. This migration of the worms has been shown to be dependent on the host's feeding rhythm (see review by Arai 1980). Whether these migrations take place for the large human tapeworms is not known, but it is likely.

CLINICAL AND PATHOLOGICAL ASPECTS OF INFECTION

Generally, intestinal tapeworms are considered to be minimally pathogenic, causing no or very slight symptoms as compared to many other human parasitic diseases. However, they do utilize some of the food consumed by the host. Normally, sufficient food supplies are available and the host is not affected. In cases of malnutrition, tapeworm infections may have a deleterious effect on the infected person. The amount of energy taken from the host by tapeworms has been investigated for *Hymenolepis diminuta* in rats and shown to be 0.8 percent, which is comparable to carnivore predation on a prey population (Bailey 1975).

The effect of the broad tapeworm (*Diphyllobothrium latum*) on vitamin B_{12} and cobalamin absorption is discussed later (see section entitled Clinical and pathological aspects of infection).

Clinical manifestations

The large human tapeworms may disturb gastrointestinal motility. Although no specific symptoms are known for any of the human intestinal tapeworms, patients may experience abdominal pain, nausea, weakness, loss of weight, increased appetite, headache, constipation, dizziness, diarrhea, pruritis ani, excitation, and even vomiting and anorexia. Both nausea and abdominal pain may be explained as a result of extension or spasm of the intestine caused by the movement of the tapeworms. The mechanisms of the other symptoms mentioned are obscure.

Pathogenesis and pathology

Our knowledge of specific changes in the human intestinal environment and intestinal absorption caused by tapeworms is very limited (see review by Rees 1967). However, in rats infected with *H. diminuta*, the worms cause a decrease in pH (perhaps because they excrete fatty acids), an increase in pCO_2 and pO_2, and a reduction to about half the number of intestinal microorganisms, when compared to uninfected rats (Mettrick 1971).

Blood eosinophilia, commonly found in many other human helminth infections, has also been described in some cases, as have inflammatory reactions in the intestinal mucosa. However, only in experimental models with rats and mice have changes in the small intestine in the number of eosinophils, mast cells, and goblet cells been followed during a tapeworm infection. Peripheral eosinophilia, combined with pruritis, urticaria, or asthma, has been described in human taeniasis and is indicative of an allergic reaction stimulated by the tapeworms. Human self-infections with the tapeworm *H. diminuta* also showed pronounced eosinophilia, along with an increase in plasma viscosity, but circulating IgE was not detected (Turton et al. 1975).

Complications may occur when proglottides move into new locations, such as the appendix or the uterine cavity or, after vomiting, into the cavities connecting with the oral cavity.

Immune responses

Although antibodies in sera from humans infected with adult tapeworms have been demonstrated in many cases, almost nothing is known about functional immunity to adult tapeworms in humans; our knowledge is mainly confined to animal models. Turton et al. (1975) made self-infections with *H. diminuta* and found increases in specific IgG and IgM, but not IgE. However, IgE is produced against intestinal tapeworm infections in dogs (Williams and Perez-Esandi 1971), mice (Moss 1971), and rats (Harris and Turton 1973), as in human intestinal nematode infections. It has also been shown, using the macrophage migration inhibition test, that humans infected with adult *Taenia saginata* develop a delayed-type hypersensitivity reaction (Boro'n-Kaczmarska et al. 1978).

In few cases, spontaneous cure has been observed in *T. saginata* infections, but it is not known whether this is caused by a direct immunological reaction or through a nonspecific cross-reaction caused by another parasite, e.g. as a result of increased plasma leakage to the gut lumen caused by hookworms.

The reason why only one or a few of the large human tapeworms are found at a time is not known, but the possibility exists that it is a result of concomitant immunity (previously called premunition), i.e. the host cannot get rid of existing worms but at the same time is in some way immune to new infections. The resistance to new infections is not a result of crowding effects from the present worms, as has been shown in rats infected with *H. diminuta*. The greater the biomass of primary worms present in the intestine, the greater the biomass of the superimposed worms. This indicates that the adverse effect on the superimposed worms comes from the host, i.e. it is immunologically mediated and not just an effect caused by the present primary worms (Andreassen 1991).

Human taeniid tapeworms are found not only together with other tapeworm species, but also with other intestinal worms, especially nematodes (large roundworms and hookworms) and protozoans. The interspecific interactions between these species are not known, but it is obvious that the presence of other parasites in the intestine will interfere with the presence of tapeworms and with the host's reactions against them. It is known from experimental infections in mice and rats with the tapeworm *H. diminuta* that the immunologically mediated expulsion of existing intestinal nematodes will, at the same time, expel a new infection of the tapeworm and destrobilate an existing tapeworm population (Behnke et al. 1977; Christie et al. 1979). These experiments indicate that adult tapeworms are more resistant to nonspecific host reactions than newly excysted worms. In vitro experiments have indeed shown that newly excysted cysticercoids of *H. diminuta* are completely lysed in fresh normal rat serum, while the scolex and neck region of older worms are more resistant (Christensen et al. 1986).

Protozoan infections, such as trypanosomes in the blood and flagellates such as *Giardia* in the gut, are often immunosuppressive and may increase the survival time of tapeworms in humans. Other immunosuppressive infections, special situations such as pregnancy and lactation, and immunosuppressive treatments will probably be favorable to a human tapeworm infection.

The antigens stimulating immune reactions in the final host probably come from the surface of the tapeworm and from gland secretions. It has been suggested (Elowni 1982) that the scolex and the germinative (neck) region of intestinal tapeworms are the main source of protective antigens. This is very likely, because these parts of the tapeworm are in the closest contact with the host's mucosa. Furthermore, the scolex has a rostellar gland, and antibodies have been shown to be produced in the host against secretions from this gland (Hoole et al. 1994).

Diagnosis

An intestinal tapeworm infection as such can be diagnosed by finding proglottides or cestode eggs in the patient's feces. If only eggs are present, it is not possible to diagnose *T. saginata* from *T. solium*. To diagnose the correct infective species, the scolex or those stages of proglottides showing species characteristics (see sections dealing with individual parasites) are needed. Several techniques for detecting eggs in feces are available, e.g. thick and thin smears and concentration techniques using sedimentation or flotation. Anal swabs may also be effective. However, free eggs in feces may be difficult to find and often multiple samples are needed. To help species identification, patients should therefore be asked to look for and collect proglottides.

Serological diagnosis of human intestinal tapeworms has been carried out using different antigen preparations in precipitation, complement fixation, skin hypersensitivity, agglutination, and enzyme-linked immunosorbent assay (ELISA) tests, but with variable results. A single serological test can determine the presence of antibodies, but cannot differentiate between a present and a past infection.

Methods have been developed to detect coproantigens of *Taenia* species in human fecal samples by means of a coproantigen ELISA (Allan et al. 1992). Diagnostic identification of a tapeworm proglottid or a fraction of it has been carried out with *T. saginata* using the polymerase chain reaction to amplify extracted genomic DNA and electrophoresis to identify a 0.55-kb DNA fragment (Gottstein et al. 1991). This is a highly sensitive, specific, and easy technique for identification of *T. saginata* proglottides and eggs; a single egg should be enough to make a species identification (Gottstein and Mowatt 1991). Recently, mitochondrial DNA analyses of human taeniids have shown unique profiles of *T. saginata*, the subspecies *T. saginata asiatica*, and two types of *T. solium* (Yamasaki et al. 2002).

Control

From a human health point of view, the tapeworms *T. solium* and *D. latum* are particularly important: *T. solium* because of the risk of cysticercosis (see Chapter 33, Larval cestodes) and *D. latum* because of the risk of pernicious anemia due to vitamin B_{12} deficiency.

CHEMOTHERAPY

Two compounds, the older niclosamide and the newer praziquantel, act against adult intestinal worms (WHO 1995), but other drugs have been or still are in use, including extracts of certain seeds from locally grown plants. Niclosamide is a very safe drug because minimal amounts are absorbed in the gastrointestinal tract; in contrast, praziquantel is extensively absorbed by the host. Praziquantel has been shown also to have some effects against cysticerci of *T. solium*. Normally, a single dose is highly effective (at least 80 percent). Niclosamide functions by blocking glucose absorption by the adult tapeworm. Adult worms treated with praziquantel rapidly contract and disintegrate in the intestine.

PROPHYLAXIS

For human consumers, prophylaxis depends on the tapeworm species concerned (see descriptions of individual parasites and the section entitled Sporadic zoonotic animal tapeworms that may infect humans). Generally speaking, humans can avoid tapeworm contamination by avoiding food that has not been properly prepared. Meat or fish should be heated to a minimum of 57°C at the center, deep frozen at −10°C for 10 days, or pickled in a

25 percent salt solution for 5 days before being eaten. To reduce the risk of infected meat being eaten by humans, it is important to have an adequate veterinary meat inspection before meat is released for consumption, although a veterinary inspection can never be 100 percent effective. The dwarf tapeworm, *H. nana*, needs special control measures (see section entitled *Hymenolepis diminuta*).

TRANSMISSION

To avoid infection of intermediate hosts, it is important that human fecal material is kept away from these animals. Safe disposal of sewage is important in minimizing the transmission; water from sewage plants should not be used to water grasslands grazed by cattle or pigs. Viable tapeworm eggs can also reach grasslands and cattle grazing along rivers and lakes via effluent from sewage plants. Viable eggs in proglottides taken up by birds can pass through them and be dispersed on to fields, where the intermediate host can be infected or its feed contaminated. Under natural conditions, desiccation is normally the most important factor reducing the survival of the eggs: hot, dry summers will kill more eggs than cold, wet winters. Health education, not only of consumers, but also of people involved in every step in the life cycle, is important to prevent transmission of human tapeworms.

TAENIA SAGINATA SAGINATA

T. saginata (beef tapeworm) has a cosmopolitan distribution wherever inadequately cooked beef is eaten. The name 'beef tapeworm' comes from the fact that beef is the main source of infection.

Stoll (1947) estimated that about 39 million people were infected. Although present data are not sufficient to make a new estimate, it seems justified to multiply Stoll's number by the increase in the total global population, giving about 60 million people currently infected with *T. saginata*.

The economic loss due to taeniasis or cysticercosis in cattle is very high in some countries. The cost due to medical treatment of people is only a small proportion of the loss due to cysticercosis, which varies from region to region, not only in total loss, but also in loss per animal.

Morphology and structure

The mature tapeworm, which can reach a length of 4–6 m or more, has 1 000–2 000 proglottides. In the proglottides the genital pore is seen laterally but irregularly alternating from one side to the other. The scolex has four suckers but no hooks. Because of the lack of hooks, the species has also been named *Taeniarhynchus saginatus*.

General biology

The maximum size of the worm is reached when only one tapeworm is present. When two or more are present the individual length decreases to about 2 m when eight worms are present (Tesfa-Yohannes 1990), and as little as 50–80 cm when 16 worms are present (Altmann and Bubis 1959). An increasing number of worms present in the small intestine of humans coincides with a decrease in the number of proglottides; an increase from one to eight worms shows a decrease in the number of proglottides from about 600 to 350. This is a typical example of the effect of crowding, but, in parasites (unlike freeliving animals), not only do the worms themselves interfere with each other and compete for space and food, but also the immunological reactions from the host have an influence on the growth of the worms.

LIFE CYCLE

In the life cycle (see Figure 32.4a) humans are the only final host and infections are acquired by eating raw or undercooked meat, e.g. by eating a 'rare' steak containing the cysticercus stage in a host capsule. When the cysticercus or bladder worm (previously named as a separate species, *Cysticercus bovis*) reaches the stomach, proteolytic enzymes start dissolving the capsule. In the small intestine, the cysticercus is stimulated to evaginate. The scolex attaches to the intestinal mucosa by means of the four suckers and starts growing into a mature tapeworm. Maturity is reached after about 10–14 weeks. The gravid proglottides contain up to about 80 000 mature eggs each and an average of six to nine proglottides may be expelled in 24 h, giving a total daily egg output of about 600 000 eggs. Meat from domestic cattle is the main source, but water buffaloes are also known as intermediate hosts. Unhooked cysticerci have also been reported in meat and livers from wild ruminants in Africa, but the significance of these observations needs further investigation. Llamas in South America and reindeer in the northern hemisphere have also been reported as intermediate hosts. A dozen cases of hookless cysticerci found in humans and described as *T. saginata* cysticercosis have been reported, but these require confirmation. The intermediate hosts become infected by eating mature eggs or gravid proglottides filled with eggs. The proglottides, which are quite mobile after being deposited, may migrate out of the fecal pat and thereby be more easily eaten by cattle, although this is not proven to have any significance for transmission. During the migrations, most eggs leave the proglottid. The eggs hatch in the small intestine and the free oncospheres then penetrate the gut wall and reach the muscles via the blood. In the intramuscular connective tissue they develop to a cysticercus, which, in the mature stage after about 10–15 weeks, is a grayish white bladder about 5–9 mm with an opaque invaginated scolex and neck.

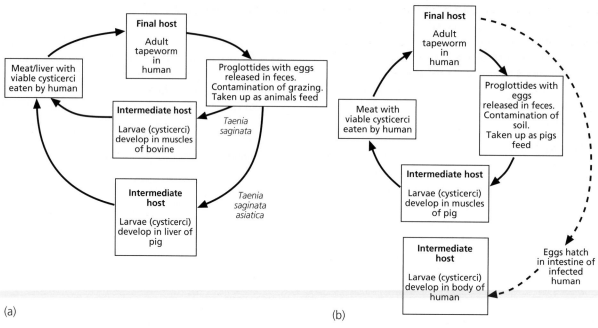

(a)   (b)

Figure 32.4 *Life cycles of the human tapeworms:* **(a)** T. saginata *and* T. saginata asiatica; **(b)** T. solium

TRANSMISSION AND EPIDEMIOLOGY

The beef tapeworm is an important infection of both humans and cattle throughout the beef-eating world, but especially in the tropics, where cattle-grazing areas are more often contaminated with human feces. It is important from the point of view both of human health and of economics, especially in potentially beef-exporting African countries with a high incidence of cysticercosis in cattle (see section entitled Control above). Under ideal conditions, the eggs may survive in the field for up to 6 months.

Gravid proglottides from infected humans living in areas with water closets accumulate in sewage plants. When sewage sludge is used as fertilizer on grazing pastures, cattle may become infected. Pastures may also be contaminated with eggs from sewage plants when treated effluent (still containing viable cestode eggs) is released into rivers running through grazing areas. Cattle may then become infected by drinking or eating along the rivers or by farmers watering pastures with water from the contaminated rivers. A more 'sophisticated' way to contaminate pastures is through droppings from gulls which have eaten proglottides at open sewage plants or at places where human feces are deposited directly in water. Cattle may also be infected through human activities such as camping, where human feces are deposited on grassland. A farmer may also risk a cysticercus 'storm' in his cattle if he empties septic tanks, mixes the content with slurry, and sprays it over the pastures as a fertilizer.

Another parameter of special epidemiological interest is the longevity of infective cysticerci. This is quite variable, even in the same host. Cysts in some organs are killed, yet remain alive in others. Cysticerci have been shown to be viable for a maximum of 3 years in their intermediate host.

LOCATION IN HOST

The mature tapeworm has been shown to attach mainly to the upper jejunum, with the posterior part of the body down in the ileum. However, the tapeworm is mobile in the intestinal lumen (Prévot et al. 1952).

CLINICAL AND PATHOLOGICAL ASPECTS OF INFECTION

Clinical manifestations

Infected humans are not always aware of having a beef tapeworm, but the observation of moving proglottides in the anal region, in the underclothing, or on the stool normally brings the patient to the doctor. Infected persons often complain of having abdominal discomfort (epigastric pain), nausea, weakness, loss of weight, increased or decreased appetite, and headache. Constipation, dizziness, diarrhea, pruritis ani, and excitation are also noted, while vomiting is an infrequent symptom.

Pathogenesis and pathology

Occasionally, obstructive appendicitis or cholangitis occurs in *T. saginata* infections due to aberrant migration of segments, which have also been found in the uterine cavity. After vomiting, proglottides may, in very rare cases, obstruct the respiratory tract, enter the middle ear through the eustachian tube or localize in the adenoid tissue of the nasopharynx. Symptoms such as abdominal pain and nausea suggest that the tapeworm has an irritative action, and biopsies of the intestinal mucosa taken from infected patients often show slight inflammatory reactions. Moderate eosinophilia has been reported in varying percentages of infected persons, but an increase in IgE has not yet been demonstrated.

Immunity

Acquired immunity against *T. saginata* infections in humans after elimination of an infection has not been demonstrated and is probably of little importance; in contrast, concomitant immunity may be important.

Diagnosis

If only gravid proglottides are present for diagnosis, normally as a result of apolytic discharge in feces, the presence of a vaginal sphincter muscle, which can be seen on cleared specimens, can identify the specimens as *T. saginata* and distinguish them from *T. solium* (Figure 32.5a).

The presence of two ovarian lobes is also species specific for *T. saginata*. If the scolex is recovered after therapy, the absence of hooks is a characteristic of *T. saginata*, while the pork tapeworm (*T. solium*) has hooks on the rostellum (see Verster 1967). The two species overlap in the number of uterine branches in gravid proglottides, which therefore can only be used when the number on each side of the stem is greater than 16 (*T. saginata*) or less than ten (*T. solium*).

For detecting eggs (Figure 32.3), anal swabs are superior to methods using feces (thick smears and concentration or flotation methods). However, eggs cannot be used for species determination, and more feces samples are often necessary to find eggs.

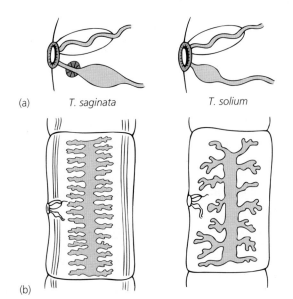

(a)   *T. saginata*   *T. solium*

(b)

Figure 32.5 *Diagnostic feature of* T. saginata *and* T. solium*: **(a)** genital atria; **(b)** gravid proglottides*

Control

Details of prophylaxis and chemotherapy are given in Table 32.2 (see also section entitled Chemotherapy). Although it seems simple to break the life cycle of *T. saginata* (humans being the only final host), both human and animal infections are in fact very difficult to control. To control cattle infections, contact with human feces should be avoided. Cattle should be kept away from pastures where there is risk of contamination. Humans should not eat undercooked or raw meat unless it has been properly frozen, meat production should be controlled by veterinarians and infected people should be treated in order to break the parasite life cycle.

Eggs of *T. saginata* can survive for a long period of time, depending on temperature and moisture. Higher temperatures and lower humidity result in shorter egg survival times. To prevent establishment of viable cysticerci in cattle, immunization may be a possibility. Although an effective vaccine has been produced, no commercial vaccine is available as yet.

TAENIA SAGINATA ASIATICA

This human tapeworm was described by Eom and Rim (1993) as a new species, *T. asiatica* (Asian tapeworm). Some authors still regard it as a separate species (Galán-Puchades and Mas-Coma, 1996; Queiroz et al. 1998; Hoberg et al. 2000; Eom et al. 2002), while others consider it as a subspecies of *T. saginata* (McManus and Bowles 1994; Fan and Chung 1998; Ito et al. 2002a). It has been characterized genetically by Bowles and McManus (1994). According to Fan (1988), it has been present in Taiwan since at least 1915. It is now found in Korea, Taiwan, Indonesia, China, Malaysia, Philippines, and Thailand, but probably also occurs in other Asian countries (Ito et al. 2002a).

Morphology and structure

The morphology of the adult tapeworm is very similar to that of *T. saginata*, but the scolex generally has a larger diameter and the cysticercus – named *C. viscerotropica* – is different in morphology, development, and species of intermediate host; for further details see Fan (1988), Eom and Rim (1993), Fan and Chung (1998), and Ito et al. (2002a).

LIFE CYCLE

The life cycle differs from that of *T. saginata* in that the intermediate hosts are pigs rather than cattle, and the preferred location in the pig is the liver, rarely the omentum, serosa or muscles (Fan and Chung 1998). Pigs, calves, goats, and monkeys have been successfully infected experimentally with eggs of *T. saginata asiatica*. Fully mature cysticerci are developed within 4 weeks, whereas it takes 10–12 weeks for *T. saginata* in cattle

Table 32.2 *Control measures against taeniasis (taeniosis)*

Measure	Details
Prophylaxis	
Reduction of human consumption of raw infected meat	Cook properly and/or freeze meat from cattle and pigs and other intermediate hosts
	Veterinary meat inspection
Reduction of intermediate hosts becoming infected	Keep human feces away from cattle and pigs and other intermediate hosts
Therapy	
Chemotherapy (drugs)	
Against intestinal tapeworms in humans	Praziquantel, niclosamide
Against larval stages in the intermediate host	(see Chapter 33, Larval cestodes)
Vaccination	
Against adult intestinal tapeworms	Not yet possible
Against larval stages in the tapeworms' intermediate hosts	(see Chapter 33, Larval cestodes)

(Figure 32.4a). Normal and immunosuppressed mice have been shown experimentally to function as intermediate hosts (Wang et al. 1999).

TRANSMISSION AND EPIDEMIOLOGY

To understand the transmission and epidemiology of this new strain of *T. saginata* we require more knowledge of the prevalence and egg production of the adult tapeworms in the human population. Furthermore, it is important to know not only which animals are confirmed intermediate hosts, along with the corresponding prevalence and infection intensity, but also the potential intermediate hosts in the area concerned (see Galan-Puchades and Fuentes (2000) and Eom and Rim (2001)).

CLINICAL AND PATHOLOGICAL ASPECTS OF INFECTION

See under *T. saginata*.

DIAGNOSIS

When a human tapeworm infection is diagnosed as *T. saginata* (see section on *Taenia saginata saginata*), only identification of the source of infection, i.e. whether the patient has eaten undercooked beef or raw pig liver, may indicate if the infection is caused by the Asian subspecies *T. saginata asiatica*.

TAENIA SOLIUM

Taenia solium is known as pork tapeworm because pork is the main source of human infection. It has a world-wide distribution except among Muslims and Jews, who do not eat pork, or where strict veterinary inspection and treatment of human cases have caused its eradication. *T. solium* is especially common in Latin America.

Morphology and structure

The mature tapeworm normally reaches a length of about 2–4 m (sometimes more), and contains about 800–1 000 proglottides. The scolex has four suckers and a small rostellum with a double crown of 25–30 small hooks (see Chapter 33, Larval cestodes, Figure 33.1).

LIFE CYCLE

Humans can act not only as the final host, as for *T. saginata*, but also as an intermediate host (Figure 32.4b), which is the more serious situation (see Chapter 33, Larval cestodes). When humans act as final hosts the source of infection is meat from domestic pigs, wild boar, or dogs containing the cysticercus stage, previously named as a separate species (*Cysticercus cellulosae*). When this is eaten, development to the adult stage is as described for *T. saginata* (see section on *Taenia saginata saginata*). Pigs, wild boar, and occasionally other mammals become infected by eating mature eggs or gravid proglottides filled with up to 40 000 eggs. The proglottides, which, unlike those of *T. saginata*, are only slightly mobile after being deposited in human feces, are in certain places easily eaten by pigs, which will quite happily eat human feces. In pigs, the development to an infectious cysticercus is essentially as for *T. saginata* in cattle, except that the invaginated scolex has hooks on the rostellum and the developmental time is shorter, 7–9 weeks. Chinchillas have been shown to function as definitive hosts for *T. solium* (Maravilla et al. 1998).

TRANSMISSION AND EPIDEMIOLOGY

In some areas, the prevalence in pigs may reach 25 percent, and in heavily infected pigs (so-called measly pork) the infection can be detected by the presence of cysts in the tongue of the living pig. Such levels of infection are not uncommon in countries where pigs live in close proximity to humans who defecate in places where the pigs can eat the feces. Dogs are intermediate hosts in Irian Jaya (Ito et al. 2002b), where prevalences of human intestinal *T. solium* ranges from 8 to 51 percent (Singh et al. 2002).

LOCATION IN HOST

The mature tapeworm has been shown to attach mainly to the upper jejunum, as in *T. saginata*.

CLINICAL AND PATHOLOGICAL ASPECTS OF INFECTION

Clinical manifestations

Unlike those of *T. saginata*, the gravid proglottides of *T. solium* are released in groups and are not as active as those of *T. saginata*. Therefore, infected persons do not recognize the infection as easily. Otherwise the clinical manifestations are as described for *T. saginata* (see section on *Taenia saginata saginata*).

Pathogenesis and pathology

See Pathogenesis and pathology section and *T. saginata*.

Immunity

See Immune responses section and *T. saginata*.

Diagnosis

Adult *T. solium* worms can be identified by the presence of hooks on the scolex, unlike *T. saginata*. For species characteristics in the proglottides, see *T. saginata*. An immunoblot assay that is 95 percent sensitive and 100 percent specific has been developed for serological diagnosis of *T. solium* from *T. saginata* (Wilkins et al. 1999). Two genotypes of *T. solium* have been found, one from Asia and one from Latin America and Africa (Nakao et al. 2002)

CONTROL

See Table 32.2 and Pawlowski (1990) for more details and references.

Prophylaxis

To prevent infection, veterinary inspection and thorough cooking of pork are essential. In most parts of the world these methods are easier and cheaper than the measures necessary for prohibiting porcine access to human feces or food contaminated with eggs of *T. solium*.

Chemotherapy

Praziquantel, at a minimum dose of 5 mg/kg body weight, is the drug of choice for *T. solium* because it not only kills the adult tapeworm in a single dose, but, when taken in high doses over 3–7 days, also kills the cysti-cerci. Other drugs may also help cure human cysti-cercosis (see Chapter 33, Larval cestodes). As for *T. saginata*, treatment with a single dose of four tablets (each of 500 mg) of niclosamide chewed thoroughly after a small meal is also effective against the adult worm in the intestine.

When treating patients, it is important that nausea, and especially vomiting, is avoided; should proglottides with eggs or eggs alone enter the stomach and later the small intestine, the patient may acquire cysticercosis. A purgative should therefore be given 1–2 h after the anthelmintic treatment.

Transmission

Theoretically, transmission to humans may also be inhibited by vaccination of pigs against oncospheres from *T. solium* eggs. However, a commercial vaccine is not yet available (see Chapter 33, Larval cestodes).

HYMENOLEPIS NANA

Hymenolepis nana (dwarf tapeworm) infects not only humans but also rodents such as mice and rats. Pawlowski (1984) and Crompton (1999) estimated that about 50–75 million people worldwide are infected. In rodents, the tapeworm is regarded by some as a special strain (*H. nana* var. *fraterna*), but cross-infections in both directions are possible. This means that *H. nana* is a zoonosis, where the infection can be maintained in animals and accidentally infect humans. However, Macnish et al. (2002) failed to infect laboratory rodents with human isolates of *H. nana* from Australia. Some taxonomists (e.g. Czaplinski and Vaucher (1994)) consider *H. nana* to belong to a separate genus *Rodentolepis*.

Morphology and structure

The adult tapeworm is small, as the name indicates, reaching only 4–5 cm in length. The scolex has four cup-shaped suckers and a retractable, prominent, and movable rostellum with a ring of small hooks.

LIFE CYCLE

This tapeworm has an indirect life cycle with an insect as the intermediate host. It is exceptional among tapeworms in also being able to use the final host for development directly from the egg containing the onco-sphere larva (Figure 32.6).

Intermediate hosts include grain- and flour-eating beetles such as species of *Tribolium* and *Tenebrio*, fleas such as *Pulex irritans*, *Xenopsylla cheopis*, and *Ctenocephalides canis*, and moths. The indirect life cycle begins when these insects or their larvae eat an infective *H. nana* egg. They crush the eggshell, and enzymes in the gut then stimulate the oncosphere to free itself from the enclosing membranes. When free in the gut lumen,

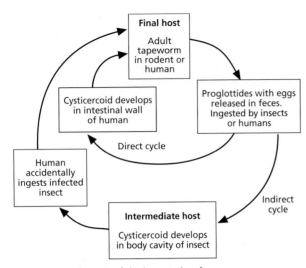

Figure 32.6 *Life cycle of the human dwarf tapeworm,* Hymenolepis nana

the oncosphere penetrates the gut wall by means of its six hooks and glandular secretions. In the body cavity of the insect, the oncospheres transform and grow into the second larval stage, the cysticercoid, which is infective to the final host. After an infected insect is eaten by humans (chiefly children), bile salts in the small intestine stimulate the emergence and liberation of the young tapeworm from the cysticercoid. In humans the tapeworm grows to patency within 3–4 weeks. The direct life cycle of *H. nana* takes place when infective eggs are ingested. In the lumen of the small intestine a free oncosphere penetrates an intestinal villus and grows to the cysticercoid stage in about 96 h. Thereafter, the villus ruptures, the cysticercoid becomes free in the lumen of the small intestine and grows to an adult tapeworm. It is said that in heavy infections eggs may hatch in the intestine before passing out with the feces resulting in autoinfection, but the importance of this is questionable.

TRANSMISSION AND EPIDEMIOLOGY

Transmission takes place directly from hand to mouth, by contaminated food and drinking water, or in rare cases, by infected insects. Infection is most prevalent where hygiene is insufficient. e.g. in young children and inhabitants of institutions. Although humans are the main source of infection, infected rodents are always a potential source. This tapeworm is present throughout the world, with overall prevalences ranging from nearly zero to about 4 percent in some countries and as high as 16 percent in children.

LOCATION IN HOST

In humans the adult tapeworms are said to be found in the upper two-thirds of the ileum, whereas in mice they are found in the posterior part of the ileum.

CLINICAL AND PATHOLOGICAL ASPECTS OF INFECTION

Clinical manifestations

The presence of low numbers of these tapeworms causes no symptoms, but, if as many as 1 000–2 000 worms are present in children, there may be abdominal pain, lack of appetite, eventual diarrhea or dizziness.

Pathogenesis, pathology, and immunity

In heavy infections, enteritis may be produced. Nothing seems to be known about immune reactions in human infections, although there is considerable knowledge of experimental infections in mice and rats (see review by Ito and Smyth 1987).

Diagnosis

Free eggs or proglottides filled with eggs may be found in feces. The characteristic two groups of polar filaments in the egg is species-specific.

CONTROL

Chemotherapy

The drug of choice is praziquantel, of which a single dose is highly effective. A second choice is niclosamide. When a person is found to be infected, treatment should include the entire household.

Prophylaxis

Sanitary improvements, uncontaminated food supplies, and rodent control in houses and nearby surroundings are important in prevention of this infection, but the hygienic status of children and inhabitants of institutions is the most important factor in control. Because of the direct life cycle and the zoonotic status of this infection, this tapeworm is difficult to control and impossible to eradicate.

HYMENOLEPIS DIMINUTA

Morphology and structure

The scolex of this species has no hooks on its rostellum, unlike *H. nana*. Furthermore, *H. diminuta* is larger than *H. nana*, reaching a length of at least 1 m in single infections in rats, and perhaps more in human infections.

LIFE CYCLE

The life cycle of this tapeworm is identical to the indirect life cycle of *H. nana* (see section Hymenolepis nana) and always needs an intermediate host. The prepatent period in man is about 3 weeks. It has been shown in rats that, when only a few tapeworms are present, they may live as long as the host (see section entitled General biology) In a self-induced infection with 30 cysticercoids and a superimposed infection with 100 cysticercoids on day 22, Turton et al. (1975) did not find any eggs in feces and concluded that the worms were expelled before patency.

TRANSMISSION AND EPIDEMIOLOGY

In order to be infected with an adult *H. diminuta*, humans have to ingest larvae (cysticercoids) from the body cavity of an insect. Since the tapeworm is very rare in humans, the source of infection must come from infected rats expelling gravid proglottides in their feces, from where the insects eat infective eggs. This infection is therefore a true zoonosis: infected rats must be present in order to infect insects, which in turn are consumed by humans. Poor sanitary conditions are a requirement for occurrence of this parasite.

CLINICAL AND PATHOLOGICAL ASPECTS OF INFECTION

Since infected rats and insects, functioning as final and intermediate hosts, respectively, may live worldwide, this is a cosmopolitan zoonosis. However, few (200–300) cases have been reported (Levi et al. 1987; Millon et al. 1994). It is found mainly in children under 3 years of age, but surveys have also found infected adults. Because of the few human cases reported and minor pathological effects of the adult tapeworms, this parasite is of little medical importance.

Clinical manifestations

Although diarrhea, anorexia, nausea, headache, and dizziness have been reported, no clinical symptoms were observed in the only experimental human infection (Turton et al. 1975). However, this was in a healthy adult, and a different picture might possibly be detected in children naturally infected with this tapeworm.

Pathogenesis, pathology, and immune responses

In the experimental infection referred to above, eosinophilia was observed, but there was no evidence of anemia or change in serum transaminases, indicating an absence of tissue damage, although there was an increase in plasma viscosity (Turton et al. 1975). Parasite-specific antibodies were detected using a fluorescent antibody technique at a titer of 1:160 on day 21 after an infection with 30 cysticercoids. On day 22, another 100 cysticercoids were ingested. The antibody titer remained at 160 after the second infection and, since no tapeworm eggs were found using saline flotation techniques, the authors concluded that the infection was established but the tapeworms were expelled before reaching maturity. The reason for this spontaneous cure could be host immune reactions caused by the rather high infection

doses. A year later, the same person infected himself with 200 cysticercoids on days 0, 6, 13, 19, and 21 and was treated with niclosamide seven times between days 24 and 31. These infections caused pronounced eosinophilia and an increase in parasite-specific IgM and IgG, but not IgE (see Andreassen et al. 1998).

Diagnosis

In fecal samples, this species can easily be differentiated from *H. nana* by the larger size of proglottides or the absence of polar filaments on the inner membrane of the eggs.

CONTROL
Chemotherapy and prophylaxis

Treatment is as for *H. nana*. Good sanitary conditions, reduced rodent populations (especially rats), and avoidance of close rodent contact with humans or human food supplies will reduce the risk of infection.

DIPHYLLOBOTHRIUM LATUM

Diphyllobothrium latum is also known as 'human broad' or 'fish' tapeworm. The name fish tapeworm relates to the fact that a fish is the source of infection to humans, and the name broad (*latum*) to the fact that the proglottides are much wider than they are long, unlike the *Taenia* species.

This tapeworm is found globally, mainly in temperate zones. It has been very common in parts of Europe where fresh water fishes are a common part of the diet; especially where certain lightly salted fish are eaten as delicacies. In Europe, the Baltic countries, Russia, and especially Finland have been pointed out as having a high prevalence of human fish tapeworms. However, it is also present in tropical Africa and parts of Asia. In North and South America it is believed to have been introduced by immigrants. *Diphyllobothrium latum* is the subject of a monograph by Von Bonsdorff (1977).

Morphology and structure

D. latum is the longest tapeworm found in man, reaching up to 10 m or more, with over 3 000 proglottides.

LIFE CYCLE

Fertilized eggs (50×60–70 μm in size) are released from mature proglottides through the uterus pore and can be found free in the feces. After the liberation of the eggs, groups of proglottides then detach and degenerate; this is known as pseudoapolysis. The fertilized egg starts development in a suitable environment outside the host and time of embryonation is dependent on the temperature. When the first larva, the coracidium, has developed (after 8–12 days at 16–20°C), it is ciliated, containing an oncosphere. It hatches from the operculated egg only when exposed to light. When freely

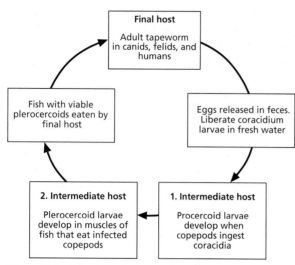

Figure 32.7 *Life cycle of the human fish tapeworm,* Diphyllobothrium latum

suspended in fresh water, its food reserves enable it to swim around for about 12 h or until eaten. In order to complete its life cycle (Figure 32.7), the coracidium must be eaten by suitable species and stages of small copepods, mainly of the genera *Diaptomus*, *Eudiaptomus*, and *Cyclops*. In the intestine of the copepod, the ciliated embryophore is shed and the oncosphere (hexacanth) penetrates the intestine into the hemocoele. Here it develops to an infective procercoid larva, where the six hooks of the hexacanth are confined to a small constricted posterior part called a cercomer. When copepods with infective procercoid larvae about 0.5 mm in length are eaten by a suitable second intermediate host (a freshwater fish) the procercoid penetrates the intestine of the fish and grows to the final larval stage, the plerocercoid. This lies freely in the body cavity or in the gut wall, or migrates to the muscles or the ovaries. The elongated plerocercoid (10–20 × 2–3 mm), which is not encysted or encapsulated, has a well-developed and normally contracted, partly invaginated scolex. Only certain species of freshwater fishes are suitable intermediate hosts, and they vary from location to location. If a small infected fish is eaten by a larger suitable fish host, the plerocercoids are, to some degree, able to penetrate the intestine of this second host and survive without further development; i.e. this host is a paratenic host in which the infective stage may accumulate. Secretions of enzymes from glands in the scolex are probably assisting the penetration of the intestinal wall. Humans become infected when they eat undercooked, raw, or lightly salted meat or roe from infected freshwater fishes. The adult, egg-producing stage is reached after about 4 weeks.

TRANSMISSION AND EPIDEMIOLOGY

Humans are the main final host and contribute most to the spread of the infection, but dogs, cats, and foxes and,

in some places, bears and pigs, may be possible final hosts. Depending on the geographical location, several different species of both copepods and freshwater fishes may act as first and second intermediate hosts, respectively. In Europe, the most important second intermediate hosts for human infections are pike (*Esox lucius*), perch (*Perca fluviatilis*), and burbot (*Lota lota*). Ruffs (*Acerina cernua*) and small specimens of the other three species are important prey for the larger specimens of the first three species. The pike is also an important intermediate host in North America, where another species of burbot (*Lota maculosa*), yellow perch (*Perca flavescens*), wall-eyed pike (*Stizostedeon vitreum*), and sand pike (*Stizostedeon canadense*) are important species. On the Pacific coast of the Russian Far East, some Russian authors claim that *D. latum* is not present and a previous synonym of *D. latum*, namely *D. luxi*, is a valid species and the principal agent of human diphyllobothriasis, with *D. giljacicum*, *D. nihonkaiense*, and *D. klebanovskii* as synonyms (Dovgalev and Valovaya 1996). Investigations using DNA techniques seem to be necessary to solve this problem.

LOCATION IN HOST

The tapeworm is usually found in the ileum or jejunum.

CLINICAL AND PATHOLOGICAL ASPECTS OF INFECTION

The number of human carriers of *D. latum*, which was estimated as about 10 million by Stoll (1947), has decreased in countries like Finland, where it is endemic and has caused many cases of pernicious anemia. Although there are no recent data on the prevalence of *D. latum* infections, this infection is probably still frequent in countries where much freshwater fish is eaten.

Clinical manifestations

Most human infections are with only one tapeworm and cause no, or very vague, ill effects. Minor clinical manifestations such as fatigue, weakness, diarrhea, and numbness of the extremities may occur. It has been shown that worm carriers have a significantly higher frequency of these symptoms than controls (Saarni et al. 1963). In a few cases, pernicious anemia may develop due to manifest vitamin B_{12} deficiency (deficiency of cobalamins in general). Clinical diagnosis of this tapeworm-induced pernicious anemia is made not only by a decreased vitamin B_{12} level in serum (around 100 pg/ml), but also by the presence of macrocytic megaloblastic anemia in conjunction with leukopenia, thrombocytopenia, and increased hemolysis. Furthermore, symptoms involving the central nervous system, such as paraesthesia, disturbances of motility and co-ordination, and impairment of deep sensibility, are also clinical criteria of tapeworm-induced pernicious anemia.

Pathogenesis and pathology

D. latum, which contains more than 50 times as much vitamin B_{12} as *T. saginata*, has been shown to absorb as much as 80–100 percent of a single oral dose of vitamin B_{12}, thereby competing with the host for this important vitamin. The size of the tapeworm and its proximity to the stomach will influence the amount of B_{12} absorption by the parasite: large worms in close proximity to the stomach reduce B_{12} availability to the host, greatly enhancing the probability of parasite-induced pernicious anemia. In fact, it has been indicated that the anemia may be relieved, if the tapeworm is forced further back in the intestine. However, other factors such as a relative deficiency in intrinsic factor due to endogenous or exogenous damage to the gastric mucosa, an inadequate supply of vitamin B_{12} in the diet or an increased requirement of vitamin B_{12} may contribute to the development of tapeworm-induced pernicious anemia.

Immunity

Practically nothing is known about immune reactions in humans against *D. latum*, and there is no evidence of acquired immunity.

Diagnosis

Diagnosis of diphyllobothriasis is impossible from purely clinical symptoms. Confirmation of pernicious anemia in a patient coming from an endemic area and known to have a diet of raw fish are relatively good indications. Evidence of the characteristic eggs of *D. latum* in fecal examinations is a reliable diagnosis, although eggs may be overlooked at routine examinations in about 5 percent of worm carriers. In suspected cases, several fecal specimens must be examined.

CONTROL

Chemotherapy

The broad tapeworm is more easily expelled than the *Taenia* species. Several drugs have been used, but niclosamide or praziquantel are the drugs of choice. Since the cure is not always 100 percent, examination for eggs in fecal samples should be performed 3 weeks after treatment.

Prophylaxis

It is very difficult, and in practice impossible, for sewage plants to remove 100 percent of eggs present in waste water. Adequate heating of fish dishes for a minimum of 10 min at 50°C (bone deep), or deep freezing (at least −10°C for 24 h) of fish or roe intended to be eaten raw or semi-raw is another prophylactic control measure. Brining of infected fish or roe should be very strong (10–12 percent NaCl) in order to kill plerocercoid larvae.

However, this does not generally suit the taste of people eating the fish. In certain endemic areas, education of the public about the danger, and how to avoid it, should be given high priority.

Transmission

Because of uncontrolled defecation, many *D. latum* eggs are often transferred directly to freshwater lakes, where first and second intermediate hosts are present. Therefore, control of this infection must cover chemotherapy of infected humans (removal of adult worms) and control of human feces (reducing risk of freshwater contamination). Even if infected persons are found and treated and the egg-removal efficacy of sewage plants is improved to 100 percent, fish infections may still be maintained through infection of wild mammals.

SPORADIC ZOONOTIC ANIMAL TAPEWORMS THAT MAY INFECT HUMANS

Species are listed in alphabetical order of genus and species name.

Bertiella spp.

Since the first report of human infection in 1913, about 60 cases of *Bertiella* in man have been published. Species of *Bertiella* (*mucronata* and *studeri*) are found in non-human primates and have been described as incidental zoonotic infections primarily in children and young people in the tropics. The life cycle of these tapeworms is not yet completely known but mites are suspected to be the infective intermediate host. Human cases of *B. studeri* have been described in Sri Lanka (Karunaweera et al. 2001), India (Panda and Panda 1994), and Indonesia (Kagei et al. 1992). A review of human cases has been published by Denegri and Perez-Serrano (1997).

Diphyllobothrium dendriticum

This tapeworm, which is found in adult form in the intestine of many fish-eating birds, has also been shown to be infective to humans. The plerocercoid is found encysted on the entrails of many salmonid fishes. Because these are normally removed from the fish before preparation for human consumption, it is a rare human infection.

Diphyllobothrium nihonkaiense

This tapeworm, a close relative of *D. latum*, is described in Japan (Ohnishi and Murata 1994), where the number of cases has increased (Nishiyama 1994). Its epidemiology in Japan has been described by Yamane et al. (1998).

Diphyllobothrium pacificum

This species, which is adult in sea lions, has been described as a human tapeworm acquired from marine fishes. It has been found in Peru (Baer et al. 1967; Baer 1969), Japan (Tsuboi et al. 1993), Chile (Gonzalez et al. 1999), and Equador (Gallegos and Brousselle 1991).

Other Diphyllobothrium species

A number of *Diphyllobothrium* species naturally occurring in free-living animals have been reported in a single or very few human cases. *Diphyllobothrium cordatum*, which occurs in seals, walruses, and dogs, has been reported in humans in Greenland and Iceland; *D. hians* has been reported from Japan; *D. houghtoni*, a canine and feline tapeworm, has been found in humans in China; *D. orcini* and *D. scoticum* in Japan; *D. ursi*, a tapeworm of North American bears, has infected one person in British Columbia, Canada; *D. yonagoense* has been described from Japan. Some of these species may not be identified correctly or may be synonyms.

Dipylidium caninum (the double-pored tapeworm)

This is a very common tapeworm in dogs and cats. Human infections are rare and normally restricted to young children, because infected fleas from cats or dogs must be ingested in order to acquire the infective cysticercoid stage (Gleason 1962). However, it has been suggested that infected fleas may be crushed in the dog's mouth, thereby releasing the cysticercoids, and these may be transmitted to children in the dog's saliva (Chappell 1991). This infection, which seldom causes any symptoms, can easily be diagnosed by the presence of active barrel-shaped proglottides. Furthermore, packets of about 15 eggs can be seen microscopically in the proglottides. Cases have been described in Austria (Brandstetter and Auer 1994), USA (Raitiere 1992), and Japan (Watanabe et al. 1993).

Mesocestoides spp.

Species of this tapeworm are common in carnivorous mammals, such as foxes. Mites are probably the infective intermediate host. Incidental intake of infected mites has resulted in a few human cases published in Japan and the United States (Schultz et al. 1992). *Mesocestoides lineatus* has also been found in humans in China (Jin et al. 1991) and Korea (Eom et al. 1992).

Inermicapsifer madagascariensis

This small (about 5 cm long) tapeworm, has been reported in humans from the eastern hemisphere, but also in children in Cuba and South America. It is common in rodents and hyraxes, and human cases have been reported from Madagascar and Mauritius and several countries in South and East Africa, where it is suspected to be more common than records indicate (Nelson et al. 1965). It was found in a 10-month-old girl in Kenya (Chunge et al. 1987).

REFERENCES

Allan, J.C., Craig, P.S., et al. 1992. Coproantigen detection for immunodiagnosis of echinococcosis and taeniasis in dogs and humans. *Parasitology*, **104**, 347–55.

Altmann, G. and Bubis, J.J. 1959. A case of multiple infection with *Taenia saginata*. *Israel Med J*, **18**, 35.

Anderson, R.M. 1986. The population dynamics and epidemiology of intestinal nematode infections. *Trans R Soc Trop Med Hyg*, **80**, 686–96.

Andreassen, J. 1991. Immunity to adult cestodes: basic knowledge and vaccination problems. A review. *Parassitologia*, **33**, 45–53.

Andreassen, J., Bennet-Jenkins, E.M. and Bryant, C. 1998. Immunology and biochemistry of *Hymenolepis diminuta*. *Adv in Parasitol*, **42**, 223–75.

Arai, H.P. 1980. Migratory activity and related phenomena. *Biology of the tapeworm, Hymenolepis diminuta*. New York: Academic Press, 615–37.

Arme, C. 1984. The terminology of parasitology: the need for uniformity. *Int J Parasitol*, **14**, 539–40.

Baer, J.G. 1969. *Diphyllobothrium pacificum*, a tapeworm from sea lions endemic in man along the coastal area of Peru. *J Fish Res Bd (Canada)*, **26**, 717–23.

Baer, J.G., Miranda, C.H., et al. 1967. Human diphyllobothriasis in Peru. *Z Parasitenkd*, **28**, 277–89.

Bailey, G.N.A. 1975. Energetics of a host-parasite system: a preliminary report. *Int J Parasitol*, **5**, 609–13.

Behnke, J.M., Bland, P.W. and Wakelin, D. 1977. Effect of the expulsion phase of *Trichinella spiralis* on *Hymenolepis diminuta* infection in mice. *Parasitology*, **75**, 79–88.

Boro'n-Kaczmarska, A., Machicka-Roguska, B., et al. 1978. Untersuchungen über das Verhalten des Macrophagenmigrations-hemmtestes mit dem Polysaccharidantigen 'C' von *Taenia saginata* in der menschlichen Täniase. *Zbl Bakt Hyg, IAbt Orig A*, **240**, 538–41.

Bowles, J. and McManus, D.P. 1994. Genetic characterization of the Asian *Taenia*, a newly described taeniid cestode of humans. *Am J Trop Med Hyg*, **50**, 33–44.

Brandstetter, W. and Auer, H. 1994. *Dipylidium caninum*, ein seltener Parasit des Menschen. *Wiener Klin Wochen*, **106**, 115–16.

Chappell, C.L. 1991. Misinformation about *Dipylidium*. In reply. *Ped Infect Dis J*, **10**, 169.

Christensen, J.P.B., Bøgh, H.O. and Andreassen, J. 1986. *Hymenolepis diminuta*: the effect of serum on different ages of worms in vitro. *Int J Parasitol*, **16**, 447–53.

Christie, P.R., Wakelin, D. and Wilson, M.M. 1979. The effect of the expulsion phase of *Trichinella spiralis* on *Hymenolepis diminuta* infection in rats. *Parasitology*, **78**, 323–30.

Chunge, R.N., Kabiru, E.W. and Mugo, B.M. 1987. A human case of infection with a rodent cestode (*Inermicapsifer*) in Kenya. *East African Med J*, **64**, 424–7.

Crompton, D.W.T. 1999. How much human helminthiasis is there in the world? *J Parasitol*, **85**, 397–403.

Czaplinski, B. and Vaucher, C. 1994. Family Hymenolepididae Ariola, 1899. In: Khalil, L.F., Jones, A. and Bray, R.A. (eds), *Key to the cestode parasites of vertebrates*. Wallingford: CAB International, 595–663.

Denegri, G.M. and Perez-Serrano, J. 1997. Bertiellosis in man: a review of cases. *Rev Inst Med Trop Sao Paulo*, **39**, 123–7.

Dovgalev, A.S., Valovaya, M.A. 1996. [Species that cause human diphyllobothriasis in the Pacific zone of Russia] (in Russian), *Med Parazitol Parazit Bolezni*, No. 3 (probably part 3 of that year), 31–4. Published in *Helminthological Abstracts*, 1997, **66**, 529 (abstr no. 3410).

Elowni, E.E. 1982. *Hymenolepis diminuta*: the origin of protective antigens. *Exp Parasitol*, **53**, 157–63.

Eom, K.S. and Rim, H.J. 1993. Morphologic description of *Taenia asiatica* sp. n. *Korean J Parasitol*, **31**, 1–6.

Eom, K.S. and Rim, H.J. 2001. Epidemiological understanding of *Taenia* tapeworm infections with special reference to *Taenia asiatica* in Korea. *Korean J Parasitol*, **39**, 267–83.

Eom, K., Kim, S.H. and Rim, H.J. 1992. Second case of human infection with *Mesocestoides lineatus* in Korea. *Korean J Parasitol*, **30**, 147–50.

Eom, K.S., Jeon, H.K., et al. 2002. Identification of *Taenia asiatica* in China: molecular, morphological, and epidemiological analysis of a Luzhai isolate. *J Parasitol*, **88**, 758–64.

Fan, P.C. 1988. Taiwan *Taenia* and taeniasis. *Parasitol Today*, **4**, 86–8.

Fan, P.C. and Chung, W.C. 1998. *Taenia saginata asiatica*: epidemiology, infection, immunological and molecular studies. *Chinese J Microbiol Immunol*, **31**, 84–9.

FAO. 1991. *Report of the FAO expert consultation on helminth infections of livestock in developing countries (AGA, 815)*. Rome: FAO, 16–17.

Galan-Puchades, M.T. and Fuentes, M.V. 2000. Human cysticercosis and larval tropism of *Taenia asiatica*. *Parasitol Today*, **16**.

Galán-Puchades, M.T. and Mas-Coma, S. 1996. Considering *Taenia asiatica* at species level. *Parasitol Today*, **12**, 123.

Gallegos, R. and Brousselle, C. 1991. Parasitoses intestinales chez les habitants d'un archipel Equatorien. *Bull Soc Franç Parasitol*, **9**, 219–23.

Gleason, N.N. 1962. Records of *Dipylidium caninum*, the double-pored tapeworm. *J Parasitol*, **48**, 812.

Gonzalez, B.A., Sagua, F.H., et al. 1999. Difilobotriasis humana por *Diphyllobothrium pacificum*. Un nuevo caso en Antofagasta, norte de Chile. *Rev Med Chile*, **127**, 75–7.

Gottstein, B. and Mowatt, M.R. 1991. Sequencing and characterization of an *Echinococcus multilocularis* DNA probe and its use in the polymerase chain reaction (PCR). *Mol Biochem Parasitol*, **44**, 183–94.

Gottstein, B., Deplazes, P., et al. 1991. Diagnostic identification of *Taenia saginata* with the polymerase chain reaction. *Trans R Soc Trop Med Hyg*, **85**, 248–9.

Grove, D.I. 1990. *A history of human helminthology*. Wallingford, Oxon: CAB International, 355–438.

Halton, D.W., Fairweather, L., et al. 1990. Regulatory peptides in parasitic platyhelminths. *Parasitol Today*, **6**, 284–90.

Harris, W.G. and Turton, J.A. 1973. Antibody response to tapeworm (*Hymenolepis diminuta*) in the rat. *Nature (London)*, **246**, 521–2.

Hoberg, E.P., Jones, A., et al. 2000. A phylogenetic hypothesis for species of the genus *Taenia* (Eucestoda: Taeniidae). *J Parasitol*, **86**, 89–98.

Hoeppli, R. 1956. The knowledge of parasites and parasitic infections from ancient times to the 17th century. *Exp Parasitol*, **5**, 398–419.

Hoole, D., Andreassen, J. and Birklund, D. 1994. Microscopical observations on immune precipitates formed in vitro on the surface of hymenolepid tapeworms. *Parasitology*, **109**, 243–8.

Ito, A. and Smyth, J.D. 1987. Adult cestodes - Immunology of the lumen-dwelling cestode infections. In: Soulsby, E.J.L. (ed.), *Immune responses in parasitic infections: immunology, immunopathology and immunoprophylaxis. , Trematodes and cestodes*. vol. II. Boca Raton, FL: CRC Press, 115–63.

Ito, A., Nakao, M., et al. 2002a. Mitochondrial DNA of *Taenia solium*: From basic to applied science. In: Singh, G. and Prabhakar, S. (eds),

Taenia solium Cysticercosis. Wallingford, Oxon: CAB International, 47–55.

Ito, A., Putra, M.I., et al. 2002b. Dogs as alternative intermediate hosts of *Taenia solium* in Papua (Irian Jaya), Indonesia confirmed by highly specific ELISA and immunoblot using native and recombinant antigens and mitochondrial DNA analysis. *J Helminth*, **72**, 311–14.

Jin, L.G., Yi, S.H. and Liu, Z. 1991. (The first case of human infection with *Mesocestoides lineatus* (Goeze, 2) in Jilin Province). In Chinese with English summary. *J Norman Bethune Univ Med Sci*, **178**, 4, 360–1.

Kagei, N., Purba, Y. and Sakamoto, O. 1992. Two cases of human infection with *Bertiella studeri* in North Sumatra, Indonesia. *Jpn J Trop Med Hyg*, **20**, 166–8.

Karunaweera, N.D., Ihalamulla, R.L., et al. 2001. *Bertiella studeri*: a case of human infection. *Ceylon J Med Sci*, **44**, 23–4.

Khalil, L.F., Jones, A. and Bray, R.A. 1994. *Keys to the cestode parasites of vertebrates*. Wallingford, Oxon: CAB International.

Levi, M.H., Raucher, B.G., et al. 1987. *Hymenolepis diminuta*: one of three enteric pathogens isolated from a child. *Diagnostic Microbiol Infect Dis*, **7**, 255–9.

Macnish, M.G., Morgan, U.M., et al. 2002. Failure to infect laboratory rodent hosts with human isolates of *Rodentolepis* (=*Hymenolepis*) *nana*. *J Helminth*, **76**, 37–43.

Maravilla, P., Avila, G., et al. 1998. Comparative development of *Taenia solium* in experimental models. *J Parasitol*, **84**, 882–6.

McManus, D.P. and Bowles, J. 1994. Asian (Taiwan) *Taenia*: species or strain? *Parasitol Today*, **10**, 273–5.

Mehlhorn, H., Becker, B., et al. 1981. On the nature of the proglottids of cestodes: light and electron microscope study of *Taenia*, *Hymenolepis* and *Echinococcus*. *Z Parasitenkde*, **65**, 243–59.

Mettrick, D.F. 1971. Effect of host dietary constituents on intestinal pH and the migratory behaviour of the rat tapeworm, *Hymenolepis diminuta*. *Can J Zool*, **49**, 1513–25.

Millon, L., Berbineau, L. and Barale, T. 1994. *Hymenolepis diminuta* chez l'enfant. À propos de 2 cas. *Bull Soc Franç Parasitol*, **12**, 157–60.

Moss, G.D. 1971. The nature of the immune response of the mouse to the bile duct cestode. *Parasitology*, **62**, 285–94.

Nakao, M., Okamoto, M., et al. 2002. A phylogenetic hypothesis for the distribution of two genotypes of the pig tapeworm *Taenia solium* worldwide. *Parasitology*, **124**, 657–62.

Nelson, G.S., Pester, F.R.N. and Rickman, R. 1965. The significance of wild animals in the transmission of cestodes of medical importance in Kenya. *Trans R Soc Trop Med Hyg*, **59**, 507–24.

Nishiyama, T. 1994. Environmental changes and tapeworm diseases in Japan – special reference to diphyllobothriasis nihonkaiense (diphyllobothriasis latum) (in Japanese with English summary). *Jpn J Parasitol*, **43**, 471–6.

Nollen, P.M. 1983. Patterns of sexual reproduction among parasitic platyhelminths. *Parasitology*, **86**, 99–120.

Ohnishi, K. and Murata, M. 1994. Praziquantel for the treatment of *Diphyllobothrium nihonkaiense* infections in humans. *Trans R Soc Trop Med*, **88**, 580.

Panda, D.N. and Panda, M.R. 1994. Record of *Bertiella studeri* (Blanchard, 1891), an anoplocephalid tapeworm, from a child. *Ann Trop Med Parasitol*, **88**, 451–2.

Pappas, P.W. and Read, C.P. 1975. Membrane transport in helminth parasites: a review. *Exp Parasitol*, **37**, 469–530.

Pawlowski, P.W. 1984. Cestodiases: taeniasis, diphyllobothriasis, hymenolepiasis and others. In: Warren, K.S. and Mahmoud, A.A.F. (eds), *Tropical and geographical medicine*. NY: McGraw-Hill Book Company, 471–86.

Pawlowski, Z.S. 1990. Perspectives on the control of *Taenia solium*. *Parasitol Today*, **6**, 371–3.

Prévot, R., Hornbostel, H. and Dorken, H. 1952. Lokalisationsstudien bei *Taenia saginata*. *Klin Wochenschr*, **30**, 78–80.

Queiroz, A. de, Alkire, N.L. and de-Queiroz, A. 1998. The phylogenetic placement of *Taenia* cestodes that parasitize humans. *J Parasitol*, **84**, 379–83.

Raitiere, C.R. 1992. Dog tapeworm (*Dipylidium caninum*) infestation in a 6-month-old infant. *J Fam Prac*, **34**, 101–2.

Read, C.P. 1967. Longevity of the tapeworm, *Hymenolepis diminuta*. *J Parasitol*, **53**, 1055–6.

Rees, G. 1967. Pathogenesis of adult cestodes. *Helminth Abstr*, **36**, 1–23.

Robertson, N.P. and Cain, G.D. 1984. Glycosaminoglycans of tegumental fractions of *Hymenolepis diminuta*. *Mol Biochem Parasitol*, **12**, 173–83.

Saklatvala, T. 1993. Milestones in parasitology. *Parasitol Today*, **9**, 347–8.

Saarni, M., Nyberg, W., et al. 1963. Symptoms in carriers of *Diphyllobothrium latum* and in non-infected controls. *Acta Med Scand*, **173**, 147–54.

Schultz, L.J., Roberto, R.R., et al. 1992. *Mesocestoides* (Cestoda) infection in a California child. *Ped Infect Dis J*, **11**, 332–4.

Siles-Lucas, M. and Hemphill, A. 2002. Cestode parasites: application of in vivo and in vitro models for studies on the host–parasite relationship. *Adv Parasitol*, **51**, 134–230.

Singh, G., Prabhakar, S., et al. 2002. *Taenia solium* Taeniasis and Cysticercosis in Asia. In: Singh, G. and Prabhakar, S. (eds), *Taenia solium Cysticercosis*. Wallingford, Oxon: CAB International, 111–27.

Smyth, J.D. 1990. Peter Abildgaard: forgotten pioneer of parasitology. *Parasitol Today*, **6**, 337–9.

Stoll, N. 1947. This wormy world. *J Parasitol*, **33**, 1–18.

Tesfa-Yohannes, T.-M. 1990. Effectiveness of praziquantel against *Taenia saginata* infections in Ethiopia. *Ann Trop Med Parasitol*, **84**, 581–5.

Threadgold, L.T. and Robinson, A. 1984. Amplification of the cestode surface: a stereological analysis. *Parasitology*, **89**, 523–35.

Tsuboi, T., Torii, M. and Hirai, K. 1993. Light and scanning electron microscopy of *Diphyllobothrium pacificum* expelled from a man. *Jpn J Parasitol*, **42**, 422–8.

Turton, J.A., Williamson, J.R. and Harris, W.G. 1975. Haematological and immunological responses to the tapeworm. *Tropenmed Parasitol*, **26**, 196–200.

Verster, A.J.M. 1967. Redescription of *Taenia solium* Linnaeus, 1758 and *Taenia saginata* Goeze, 1782. *Z Parasitenkde*, **29**, 313–28.

Von Bonsdorff, B. 1977. *Diphyllobothriasis in man*. London: Academic Press.

Wang, I.C., Ma, Y.X., et al. 1999. Oncospheres of *Taenia solium* and *T. saginata asiatica* develop into metacestodes in normal and immunosuppressed mice. *J Helminth*, **73**, 183–6.

Watanabe, T., Horii, Y. and Nawa, Y. 1993. A case of *Dipylidium caninum* in an infant – first case found in Miyazaki prefecture, Japan. *Jpn J Parasitol*, **42**, 234–6.

Wilkins, P.P., Allan, J.C., et al. 1999. Development of a serological assay to detect *Taenia solium* taeniasis. *Am J Trop Med Hyg*, **60**, 199–204.

Williams, J.F. and Perez-Esandi, M.V. 1971. Reaginic antibodies in dogs infected with *Echinococcus granulosus*. *Immunology*, **20**, 451–5.

WHO, 1995. *Model prescribing information – drugs used in parasitic diseases*, 2nd edn. Geneva: WHO, 91–8.

Wright, E.P. 1984. Human infestation by *Taenia saginata* lasting over 20 years. *Postgrad Med J*, **60**, 495–6.

Yamane, Y., Shiwaku, K., et al. 1998. The taxonomic study of diphyllobothriid cestodes with special reference to *Diphyllobothrium nihonkaiense* in Japan. In: Ishikura, H., Aikawa, M., et al. (eds), *Host response to international parasitic zoonoses*. Tokyo: Springer, 25–38.

Yamasaki, H., Nakao, M., et al. 2002. DNA differential diagnosis of human taeniid cestodes by base excision sequence scanning thymine-base reader analysis with mitochondrial genes. *J Clin Microbiol*, **40**, 3818–21.

33

Larval cestodes

ANA FLISSER AND PHILIP S. CRAIG

THE PARASITES

Introduction, classification, and history

Larval cestodes belong to the phylum Platyhelminthes, which are acoelomate metazoa with an elongated dorso-ventrally flattened body in their adult stage and a vesicular bladder in their larval stage. The most important human parasites of this group belong to the class Cestoda, subclass Eucestoda, order Cyclophyllidea, families Taeniidae, Hymenolepididae, Diphylobothriidae and genera *Taenia*, *Hymenolepis*, *Diphyllobothrium*, and *Echinococcus*. Except for *Echinococcus*, these parasites live in the human intestine throughout the adult stage. *Taenia solium*, *Echinococcus granulosus*, *E. multilocularis*, and *E. vogeli* also parasitize humans in their larval stage. The metacestode of *T. solium* is known as a cysticercus and that of *Echinococcus* as a hydatid cyst. The general term for the disease caused by the cysticercus is cysticercosis and, depending upon the site of infection, is also known as neuro, ocular, muscular or subcutaneous cysticercosis. In the case of hydatid cyst, the disease used to be called hydatidosis, but now it is usually termed echinococcosis: there are three types: cystic echinococcosis due to *E. granulosus*, alveolar echinococcosis due to *E. multilocularis*, and polycystic echinococcosis due to *E. vogeli* and *E. oligarthrus*. The first is the most prevalent of the four species, the second one is the most pathogenic, and the last two species are rare zoonoses restricted to Latin America.

Human cysticercosis is considered a public health problem in many developing countries of Latin America, non-Islamic Asia, and Africa, and it is clearly associated with lack of public health education and proper sanitary conditions (Flisser 1988, 1994, 2002a). In addition, immigration from Mexico and other Latin American countries to the United States has increased markedly in each of the last three decades. It is estimated that movement in both directions across the US–Mexico border exceeds 200 million persons per year. Opportunities for acquiring and transporting *T. solium* infections are abundant (Schantz et al. 1998). Cystic echinococcosis is cosmopolitan in its distribution but concentrated in the major sheep-raising and pastoral areas especially of South America, the Mediterranean, the Middle and Near East, East Africa, Russia, Central Asia, and China. Alveolar echinococcosis has been reported in Alaska, China, Japan, parts of Europe, the Near East, and Russia, but since the life cycle involves sylvatic animals such as foxes, rodents, and other small mammals, exposure to humans is uncommon. Polycystic echinococcosis has been identified in a total of 100 cases from Brazil, Colombia, Ecuador, Panama, and Venezuela, because *E. vogeli* is native to the humid tropical forests of South America (D'Alessandro et al. 1979; Kammerer and Schantz 1993; Rausch 1997). To date, only three human infections with *E. oligarthrus* have been reported from Venezuela, Brazil, and Suriname (Basset et al. 1998).

Adult tapeworms in human feces were first recognized by the Egyptians. The species was probably *Taenia saginata* since Egyptians did not eat pork. Hippocrates, Aristotle, and Threophrastus called them 'flatworms' meaning band or ribbon worm, whereas Romans, such as Celsus, Pliny the Elder, and Galen, named them 'lumbricus latus', meaning broad or wide worm. Only many centuries later were cysticerci discovered in human beings. They were first reported by Rumler in 1558, but the disease was not identified as parasitic until in 1697 Malpighi discovered the animal nature of cysticerci. Goeze recognized their helminthic essence in 1784 and Leuckart (1886) defined them in great detail. Tyson, in 1683, detailed the head of tapeworms and later in 1691 described the hydatides that he found in 'rotten sheep'. Goeze in 1782 described the scolices of echinococcal cysts and indicated their similarity to the heads of tapeworms; he concluded that echinococci were alive, verminous in nature, and related to tapeworms.

The life cycle of *T. solium* was defined by van Beneden in 1854 who fed a pig with eggs from a human *T. solium* and found numerous cysts in muscles, and by Kuchenmeister, who in 1855 discovered adult tapeworms in the intestines of convicts who ate pork infected with cysticerci some time before execution (Grove 1990). In 1933, Yoshino ingested swine cysticerci in order to obtain a 'reliable source' of eggs to infect pigs and studied, with great histological detail, the early development of cysticerci. He also reported that, following his own infection, between one and five gravid proglottids were expelled per day for 2 years (Yoshino 1933). The name 'Cysticercus cellulosae' was given by Zeder and by Rudolphi at the beginning of last century, but was abolished when cysticerci were shown to be the larval stage of *Taenia*.

The life cycle of *Echinococcus* was demonstrated by von Siebold in 1853 when he infected dogs with hydatid cysts from sheep and 27 days later discovered intestinal worms, consisting of a 'head' and three segments containing reproductive organs and eggs, which he called *Taenia echinococcus*. Leuckart, in 1867, infected suckling pigs with ova of *T. echinococcus* and was able to study the resulting cysts at intervals after infection. Naunyn, in 1863, infected dogs with hundreds of scolices obtained after autopsy of a patient with a large hepatic cyst. The dog, when killed after 5 weeks, had mature small tapeworms identical to *Taenia echinococcus*. Naunyn not only confirmed that hydatid cysts of human origin undergo the same development as cysts of animal origin but, in demonstrating that the adult worms obtained from the two sources were similar, also showed that there was only a single species. Naunyn concluded that the *Echinococcus* from man is the bladder worm stage of *T. echinococcus* living in the intestine of the dog. These results were confirmed soon afterwards by Krabbe and Finsen in Iceland in 1866 and by Thomas in Australia in 1883, all of whom used hydatid cysts obtained from humans (Grove 1990). Until the 1950s however, controversy, surrounded whether human *Echinococcus* infection was due to one species, *E. granulosus*, or two, the second being the cause of multi-vesicular or multi-locular lesions. Parasitological and

ecological studies in Alaska by Rausch showed that human infections in Eskimos were the same as those described in Europe, and that similar larval cestode infections occurred in tundra voles. Experimental infections of dogs by Vogel using human isolates confirmed that it was a separate species and it was designated *Echinococcus multilocularis* (Rausch 1954; Vogel 1957).

Morphology and structure

The body of an adult tapeworm has a head (scolex), a neck, and a chain of segments (strobila). The scolex is the size of a pinhead, while the strobila may be several meters long, as in *Taenia*, containing more than 1 000 proglottids. In *Echinococcus* there are only between two and five segments and the strobila measures only a few millimeters. The scolex has four suckers and a rostellum with a double crown of hooks (Figure 33.1a–c). The development of adult *Echinococcus* may not be a limiting factor for its study, since several canids are definitive hosts, but in the case of *Taenia solium* only humans are definitive hosts. Experimental rodent models for *Taenia solium* have been developed (Maravilla et al. 1998) in which the characteristics of the implantation sites have been described (Merchant et al. 1998) and other studies can now be performed. Adult *E. multilocularis* tapeworms have also been grown successfully in immunosuppressed rodents (Kamiya and Sato 1990).

The most conspicuous features of tapeworms are the lack of a mouth or digestive cavity and the presence of repeated units or proglottids at different developmental stages. The proximal proglottids are immature (i.e. have not yet developed sexual organs), the mature proglottids contain functional hermaphroditic sexual organs, where fertilization occurs, and the terminal proglottids are gravid, being little more than sacs full of eggs (see Chapter 32, Intestinal tapeworms).

Eggs are spherical and range in size from 20 to 50 mm. They have a radial appearance under light microcopy because the embryophore that surrounds the oncosphere is formed by contiguous keratinaceous blocks. The eggs of *Taenia* and *Echinococcus* are morphologically indistinguishable at the light microscope level. The embryophore protects the oncosphere, while the egg is in the external environment, making eggs extremely resistant and enabling them to disperse and withstand a wide range of environmental temperatures (Gemmell and Lawson 1986; Torgerson et al. 1995; Veit et al. 1995). When the eggs are released from the definitive host they are, or quickly become, fully embryonated and infective to a suitable host. When eggs are ingested by the intermediate host, the cementing substance that joins the embryophoric blocks is digested and the oncosphere is released in the small intestine (Flisser 1994). Experimental hatching methods include pepsin/trypsin

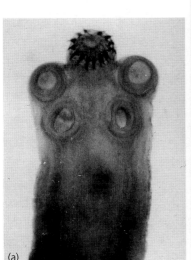

Figure 33.1 (a) Taenia solium *scolex showing the double crown of hooks and one sucker;* **(b)** Taenia solium *strobila with immature proglottids (upper left) and mature proglottids;* **(c)** *Multiple* Echinoccocus granulosus *adult parasites in the intestine of a dog*

or sodium hypochlorite, the latter being more efficient (Heath 1982; Wang et al. 1997).

The liberated and activated oncosphere is mobile and uses its six hooks to penetrate into the mucosa, reaching the lamina propria within 3–120 min after hatching. Penetration is presumably assisted by secretions from the paired penetration glands, but may be purely mechanical, involving hook and body movements (Thompson 1995; Pawlowski 2002). After penetration, larvae are transported through blood vessels into the tissues, where they develop into the metacestode, specifically named cysticercus (*Taenia*) or hydatid cyst (*Echinococcus*). Macroscopic *T. solium* cysticerci measuring around 0.3 mm were seen in liver, brain, and skeletal muscles of pigs 6 days after infection, and after 60–70 days cysticerci had fully developed scolices and measured 6–9 mm. Full development of cysticerci (to a size of 8–15 mm) takes 2–3 months in pigs (Yoshino 1933). Cysticerci are formed of two chambers; the inner one contains the scolex and the spiral canal and is surrounded by the outer compartment that contains the vesicular fluid, usually less than 0.5 ml. When a living cysticercus is ingested by the definitive host, the first event that takes place is the widening of the pore of the bladder wall for the scolex and neck to emerge, leaving the bladder wall and vesicular fluid to disintegrate in the digestive tract (Rabiela et al. 2000). Several in vivo and in vitro studies have demonstrated that *Echinococcus* oncospheres rapidly undergo a series of re-organizational events during the first 14 days, involving cellular proliferation, degeneration of oncospheral hooks, muscular atrophy, vesicularization, central cavity formation, and development of both germinal and laminated layers (Rausch 1954; Heath and Lawrence 1976). The minimum time required for the development of protoscolices of *E. granulosus* inside cysts within the human

host is not known, in pigs it takes 10–12 months, and in sheep 10 months to 4 years, although protoscolices can be formed in hydatid cysts that measure 0.5–2 cm. For *E. multilocularis* in rodent intermediate hosts protoscoleces may develop within 2–4 months. The cysticercus of *T. solium* is formed by one scolex, while all species of hydatid cyst contain multiple protoscolices (often within brood capsules) formed by asexual budding processes of the metacestode germinal layer. All metacestodes and adult worm stages are macroscopic and have a similar scolex, but eggs and oncospheres are microscopic.

There are two morphological types of cysticerci: cellulose and racemose (Figure 33.2). Cellulose cysticerci are small, spherical or oval, white or yellow, with vesicles that measure 0.5–1.5 cm and have a translucent bladder wall, through which the scolex can be seen as a small solid eccentric granule. Usually this stage grows no further. This type of cysticercus is frequently separated from the host tissue by a thin collagenous capsule, within which it remains alive (Rabiela et al. 1982). The racemose cysticercus either appears as a large, round or lobulated bladder circumscribed by a delicate wall, or resembles a cluster of grapes. It measures up to 10 or even 20 cm and may contain 60 ml fluid. The scolex cannot be seen; in some cases only detailed histological studies will reveal its remains. Neuropathological observations of human brain suggest that parasites lodged in spacious areas are able to grow and transform into racemose cysticerci, since a parasite with a bilobulated aspect was found (Rabiela et al. 1989). In support of this hypothesis is the finding that experimental infections with *T. serialis* in mice demonstrated an aberrant large cysticercus devoid of scolex in the peritoneal cavity (Lachberg et al. 1990). Alternatively, growth of cysticerci might be controlled by the immune response of the host, as suggested by in vitro experiments that showed

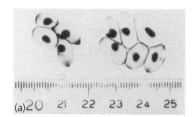  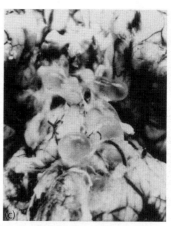

Figure 33.2 *Cellulose type* Taenia solium *cysticerci from pig muscle (by courtesy of Dr D. Correa);* **(b)** *Histological section of a cellulose type cysticercus in host muscle showing the spiral canal and the bladder wall (by courtesy of Dr A.S. de Aluja);* **(c)** *Racemose type* Taenia solium *cysticerci in the base of a human brain (by courtesy of Dr M.T. Rabiela)*

that the presence of the inflammatory capsule that surrounds *T. solium* cysticerci inhibits their evagination (Ostrosky et al. 1991). A recent study documented the events of evagination employing light and electron microscopy. The authors show that the scolex uncoils during evagination, but does not turn inside out, and that the scolex and the neck comprise a different structure from the bladder wall, although they are contiguous (Rabiela et al. 2000). An autoradiographic analysis of the germinative tissue in evaginated *T. solium* cysticerci identified stem cells that proliferate continuously, differentiate, and migrate to the tegument, constituting the main process by which these worms develop from metacestode to adult stage (Merchant et al. 1997).

The fully developed *E. granulosus* hydatid cyst is typically unilocular, subspherical in shape and fluid filled. It may range in size from a few mm to >30 cm. The cyst consists of an inner germinal nucleated layer supported externally by a tough, elastic, glycan rich, acellular laminated layer of variable thickness, surrounded by a host-produced fibrous adventitial layer. Asexual proliferation takes place in the germinal layer and brood capsule formation is entirely endogenous; within each capsule several protoscolices are found. Occasionally cysts may adjoin and coalesce, forming groups of clusters of small cysts of different size. In humans, the slowly growing hydatid cysts may attain a volume of many liters and contain many tens of thousands of protoscolices. With time, internal septae and daughter cysts may form within the primary cyst, most probably from vesiculating protoscolices, disrupting the unilocular pattern (Schantz 1994; Thompson 1995) (Figure 33.3).

The metacestode of *E. multilocularis* is the most complex and develops quite differently from that of *E. granulosus*. The size of the parasitic focus varies from less than 1 mm to >20 cm, as lesions caused by the metacestodes vary from minor foci (a few millimeters in diameter) up to large areas of infiltration. The

metacestode is a multi-vesicular infiltrating structure with no limiting host–parasite barrier (adventitial layer). In human infections the larval mass usually contains a semisolid matrix rather than fluid and is formed by numerous small vesicles embedded in a dense stroma of connective tissue. The metacestode consists of a network of filamentous solid cellular protrusions of the germinal layer that are responsible for infiltrating growth, transforming into tube-like and cystic structures where proliferation occurs both endogenously and exogenously (Figure 33.4).

Detachment of germinal cells or microvesicles from infiltrating cellular protrusions and their subsequent distribution via the lymph or blood can give rise to the distant metastatic foci characteristic of alveolar echinococcosis. Protoscolices of *E. multilocularis* appear not to be responsible for secondary spread of the metacestode. In rodents, the natural intermediate hosts, the larval mass proliferates rapidly by exogenous budding of germinative tissue and produces an alveolar-like pattern of microvesicles filled with protoscolices, but in humans the larval mass resembles a malignancy in appearance and behavior because it proliferates continuously by exogenous budding and invades the surrounding tissues. Protoscolices are rarely observed in infections of humans (Wilson and Rausch 1980; Ali-Khan et al. 1983).

The metacestode of *E. vogeli* exhibits developmental and structural characteristics considered intermediate to those of *E. granulosus* and *E. multilocularis*, with sizes that range from 2 to 80 mm. It may occur singly, in small groups, or occasionally in dense aggregations in which each cyst is enclosed by its separate adventitial layer (polycystic development). It has internal division of fluid-filled cysts to form multi-chambered growths due to endogenous proliferation and convolution of both germinal and laminated layers. This leads to the formation of secondary subdivisions of the primary vesicle

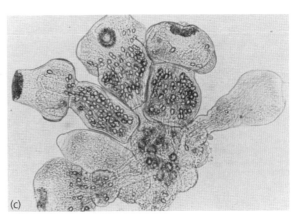

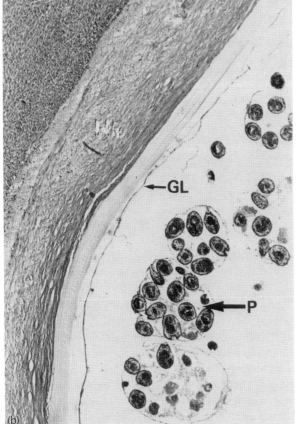

Figure 33.3 (a) Echinococcus *cyst from the liver of a patient in Urumqi, China;* **(b)** *Histologic section of an* Echinococcus granulosus *cyst with many protoscolices (P) and germinal layer (GL) (by courtesy of Dr P. Schantz);* **(c)** Echinococcus granulosus *protoscolices seen under light microscopy, partly evaginated*

with production of brood capsules and protoscolices in the resultant chambers, which are often interconnected (D'Alessandro et al. 1979; Rausch et al. 1981).

Hydatid cysts due to *E. granulosus* develop in 6 months in mice, 10–18 months in sheep, and 1 year in swine. Alveolar cysts due to *E. multilocularis* develop in 2–4 months in laboratory mice, but faster in experimental voles. Development of any of these metacestodes in humans is not known because it is impossible to define the time of infection. The life span of these para-

sites can be as long as 4 or more years in sheep, up to 16 years in horses and, impressively, several decades in humans (Dixon and Lipscomb 1961; Spruance 1974).

General biology

Studies on the biochemistry of Platyhelminthes are sparse. The first contributions dealing with tapeworms, from Read (1952), related to anaerobic metabolism in the rat tapeworm and the role of carbohydrates in cestode biology, and these stimulated work on the biochemistry of protoscolices of *E. granulosus*. More recently it was shown that cysticerci use either aerobic or anaerobic pathways according to oxygen availability in the environment (Cervantes-Vazquez et al. 1990), oxygen uptake, and evagination are useful measurements of their viability (Correa et al. 1987; Flisser et al. 1990a; Maravilla et al. 1998).

Larval cestodes obtain nutrients by diffusion through their bladder wall that bears plasma-membrane bounded microvilli covered by a loose glycocalyx. Two facilitated diffusion glucose transporters (TGPT1 and TGTP2) have been identified, the first being abundant in structures underlying the tegument in adult and larval parasites, whereas the latter appears to be localized only in the tegumentary surface of the larvae. None of the glucose transporters were recognized by antibodies in the sera from patients and pigs with cysticercosis (Rodriguez-Contreras et al. 1998, 2002). The syncytial tegument is filled with ellipsoidal vesicles of different sizes and is connected by cytoplasmic processes to the underlying nucleated cell bodies. The excretory or protonephridial system consists of flame cells which form a dense network attached to excretory ducts. The bladder wall also contains a network of nerve-like cells. Immunohistochemical studies on protoscolices of *E. granulosus* have identified serotonin in nerve cell bodies in the lateral ganglia and in association with the lateral longitudinal nerve cords, as well as in the central nerve ring, the rostellar nerves and the nerve plexus of the suckers; some peptidergic nerve elements were also detected

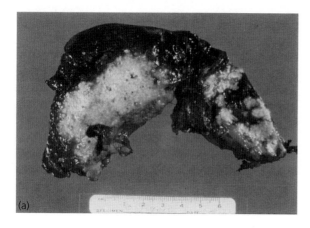

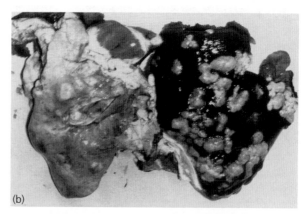

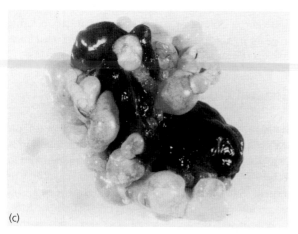

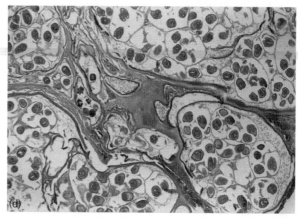

Figure 33.4 Echinococcus multilocularis *in the liver of a 61-year-old female from Minnesota, USA (by courtesy of Dr P. Schantz);* **(b)** Echinococcus multilocularis *in the liver and lungs of a sheep (by courtesy of Professor R.C.A. Thompson);* **(c)** Echinococcus multilocularis *in the liver of a cotton rat 4 months after infection;* **(d)** *Histological section of* Echinococcus multilocularis *showing multiple vesicle filled with protoscolices (by courtesy of Dr P. Schantz)*

(Fairweather et al. 1994). Platyhelminthes produce mineral concretions which in cestodes are referred to as calcareous corpuscles. In contrast to the intracellular origin formation described for other cestodes, in cysticerci of *Taenia solium*, calcareous corpuscles occur extra-cellularly in the lumen of protonephridial ducts in a way similar to that proposed for trematodes (Vargas-Parada et al. 1999). Their role in larval physiology has been reviewed (Vargas-Parada and Laclette 1999). Formation of calcareous corpuscles in *E. multilocularis* has been reported on the brood capsule and on the germinal layer (Ohnishii and Kutsumi 1991).

Several glycoproteins have been detected on the tegumentary surface of the bladder wall of *T. solium* cysticerci: the most abundant (of 55 kDa) was found to be the heavy chain of the host's IgG and one of 180 kDa was also present on proglottid surface of several taeniids (Landa et al. 1994). Surfaces also contain cytoskeleton proteins, such as myosin, which in *T. solium* cysticerci and tapeworms is of the conventional type II and can also be found in supernatants of parasite cultures and in feces of experimentally infected rodents (Ambrosio et al.

1997). The excretory-secretory molecule designated *T. solium* antigen B, initially identified as an immunodominant antigen (Flisser et al. 1980), has been purified and characterized. It is a glycoprotein (Guerra et al. 1982) with significant homology to the paramyosin of *Schistosoma mansoni* (Laclette et al. 1991). Functionally, it has been shown to inhibit the C1 component of the complement system (Laclette et al. 1992). Recent studies are focused in defining the immunological reactivity of the different regions of paramyosin (Gazarian et al., 2000; Vazquez-Talavera et al. 2001a, b). The gene for paramyosin has been expressed and sequenced. The entire gene is 6 106 bp (from its start to the stop translation codons) and contains 13 introns that have higher A + T content than its exons. A tendency towards decreasing the number of introns during evolution was inferred from gene comparison with the other two genes for which the complete structure is known: *Caenorhabditis elegans* and *Drosophila melanogaster*. No evidence of alternative splicing sites was found, excluding the possibility that *T. solium* expresses a mini-paramyosin, as is the case for *D. melangaster* (Vargas-Parada and Laclette 2003). Cholinesterases, cysteine, and aspartic, but not serine, protease activities have been identified in *T. solium* cysticerci (White et al. 1992), as well as type II myosin (Ambrosio et al. 1997). Recently, actin was identified and characterized in *T. solium* cysticerci and seven isoforms were resolved, pointing to a family of actin genes (Ambrosio et al. 2003). *Taenia solium* cysticerci respond to temperature stress through two major stress proteins: HSP80 and HSP70; stress proteins were also found in the excretory-secretory products. Patients with neurocysticercosis recognized at least two of the excreted-secreted products, one of which appears to belong to the HSP60 family (Vargas-Parada et al. 2001). Paramyosin from *E. granulosus* shows 71 percent identity with *S. mansoni* paramyosin and a significant homology with *T. solium* paramyosin (Mühlschlegel et al. 1993). In *E. granulosus* several enzymes have been demonstrated in the tegument that may have a digestive or absorptive function (McManus and Bryant 1995). At least five proteolytic enzymes have been detected in the cyst fluid (Marco and Nieto 1991) and gluthathione S-transferase has been isolated from protoscolices (Fernandez and Hormaeche 1994). A characteristic feature of the metacestodes of *Echinococcus* spp. is the formation of an acellular laminated layer which in *E. multilocularis* consists of mucin type glycosylated proteins that appear to have structural and immunoregulatory roles (Gottstein et al. 2002; Hulsmeier et al. 2002). The chemical composition of hydatid cysts from different geographic and host origins has been studied in detail (Sanchez and Sanchez 1971; Frayha and Haddad 1980) and has revealed basic differences between the UK horse and sheep strains of *E. granulosus* and between these and *E. multilocularis* (McManus and Smyth 1978). A subsequent study of the biochemical composition of *E. granulosus* in Kenya indicates differences between protoscolices of cattle, goat, camel, and sheep origin, but similarities between the sheep and human parasites (McManus 1981).

An interesting recent finding in a rural village of central Mexico, where natural conditions favor *Taenia solium* transmission, was that castration of male pigs increased prevalence of cysticercosis from 23 to 50 percent, and pregnancy in sows also increased prevalence from 28 to 59 percent. Endocrinological conditions characterized by low levels of androgens or high levels of female hormones probably influence the susceptibility of pigs to cysticercosis. Therefore, delaying castration of male pigs and confinement of sows during pregnancy might significantly decrease the prevalence of the infection and help curb transmission without much cost or difficulty (Morales et al. 2002).

The use of various molecular techniques for clarification of the taxonomic status within cestodes generated knowledge related to the evolution, ecology, and population genetics of these parasites. Intra-specific DNA variability occurs in *T. solium* isolates from India, Mexico, and Zimbabwe (McManus et al. 1989), while the analysis of the sequence variation within a 366 nucleotide for cytochrome C oxidase I (COI) from *T. solium* isolates recovered from India, China, and Zimbabwe showed only slight differences in three conservative changes in the third base position of codons of the isolate from Zimbabwe (Bowles and McManus 1994). Also, low variability in *T. solium* DNA coding for COI, ITS1, and the diagnostic antigen Ts 14 has been found (Hancock et al. 2001). In contrast, the sequence of the entire COI and cytochrome b from 13 isolates of *T. solium* from various regions showed 1.7 and 2.9 percent variant nucleotide positions for COI and cytochrome b among all isolates and the phylogenies obtained show that the isolates from Asia (China, India, Irian Jaya, and Thailand) formed a single cluster, whereas the isolates from Latin America (Bolivia, Brazil, Ecuador, Mexico, and Peru) combined with those from Africa (Cameroon, Mozambique, and Tanzania) form a second cluster (Nakao et al. 2002). Furthermore, recent studies on the variability of individual *T. solium* cysticerci from Mexico, Honduras, and Africa by random amplified polymorphic DNA analysis (RAPD) showed several alleles fixed and the presence of a clonal structure, suggesting the existence of local lineages with events of genetic recombination within them (Maravilla et al. 2003; Vega et al. 2003).

Also, many different molecular biological techniques have been applied with great success to the characterization of *E. granulosus* host isolates or strains. Biological criteria indicate that distinct horse-dog and sheep-dog forms of *E. granulosus* occur in UK (reviewed by Thompson and Lymbery 1988) and this has been confirmed by a variety of molecular techniques (Bowles and McManus 1993). These studies also indicate that the sheep strain is genetically uniform and cosmopolitan in its geographical distribution and that the horse strain or genotype is genetically similar to the one that infects equines in other countries. DNA sequence data show that the sheep and horse genotypes do not interbreed despite the fact that they use the same definitive host and occur sympatrically. Based on this and complete mitochondrial sequence data, it has been suggested that the sheep and horse genotypes of *E. granulosus* could probably be regarded as different species (McManus and Bryant 1995; Le et al. 2002). In total, nine distinct genotypes or strains of *E. granulosus* have been described using DNA sequence data and have been designated G1–G9. Apart from the sheep and horse strains, genotypes characteristic for cattle, buffalo, camel, pig, and cervid (deer) hosts have been described, as well as a second sheep strain in Australia and a pig-related isolate from Poland (Bowles et al. 1995). The majority of human isolates of *E. granulosus* have been shown to be the common sheep strain, however surgical biopsy material has also indicated that infection with cattle, pig, or camel genotypes is possible (Rozenzvit et al. 1999; Bardonnet et al. 2002).

Life cycle

The life cycle of taeniid larval cestodes involves two mammalian hosts, one definitive and one intermediate, and three developmental stages: the adult tapeworm in the definitive host, eggs in the environment, and the metacestode in the intermediate host (Figures 33.5 and 33.6). The cycle is simple in principle: the adult stage develops in the intestine after the host ingests the metacestode lodged in the tissue of the intermediate host. The metacestode develops in muscle, liver, lungs, or elsewhere after the intermediate host ingests eggs from the environment. Eggs or proglottids are released into the environment from the adult intestinal parasite. The adult stage of *T. solium* occurs only in human beings, whereas canids and felids are the hosts of adult *Echinococcus* spp. Intermediate hosts for *T. solium* are mainly pigs; however, hydatid cysts can occur in a great variety of mammals. Cysticerci can develop in other mammals but have no epidemiological relevance. The importance of *T. solium* and *Echinococcus* spp. is that humans can also harbor the larval stage that may cause severe disease (see Figures 33.5 and 33.6). A recent cladistic analysis of the genus *Taenia* indicates that there has been an extensive host-switching among definitive carnivorous hosts, in contrast there has been a more pervasive co-evolution with intermediate herbivorous hosts, and rodents are postulated as ancestral for both *Taenia* and *Echinococcus* (Hoberg et al. 2000).

Depending on the geographical location and the intermediate hosts, two biological forms of *E. granulosus* have been recognized, the so-called northern and the European biotypes. The northern biotype is maintained in the tundra and taiga biomes by a predator–prey relationship between the wolf (*Canis lupus*) and large deer (moose, reindeer, and other cervids), but may also occur in humans where reindeer are domesticated, as in parts of western Alaska and Eurasia, being acquired from dogs used in herding or from eggs dispersed by wolves. Coyotes (*Canis latrans*) may also become infected from scavenging. The intermediate hosts of the European biotype include camels, cattle, goat, horses, pigs, and sheep; while buffalo, wild boar, llama, kangaroo, wallaby, gazelle, giraffe, and impala, among others, have also been reported to have hydatid cysts. The European biotype of *E. granulosus* is almost cosmopolitan, as a result of the introduction of domestic animals and their helminths by Europeans in colonizing other regions from the early sixteenth century onwards. The adult stage occurs mainly in dogs but also in foxes, dingoes, jackals, hyenas, and lions (Rausch 1995).

The adult stage of *E. multilocularis* is found mainly in foxes (principally the red fox *Vulpes vulpes*), rarely in wolves, regionally in coyotes, and also in dogs and cats. At least eight families of rodents and small mammals have been reported to be infected with the metacestodes, these include voles and lemmings, hamsters and gerbils, rats and mice, squirrels, shrews, muskrats, and pikas. Cattle and swine, as well as captive non-human primates, may also acquire the infection accidentally but are not important epidemiologically.

Echinococcus vogeli is a neotropical species maintained in the bush dog (*Speothus venaticus*) and forest rodents like the paca (*Cunniculus paca*). Other

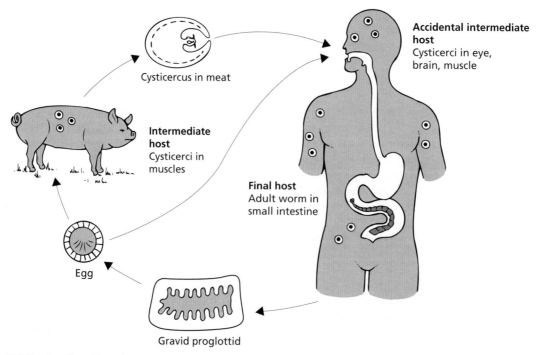

Figure 33.5 Taenia solium *life cycle*

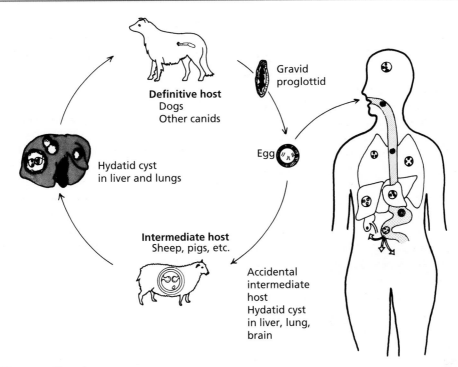

Figure 33.6 Echinococcus *life cycle*

mammals can be easily infected, as was seen in Los Angeles Zoo where several young animals were housed near a bush dog captured in Colombia. During the following 10 years, seven orangutans, three chimpanzees, two gibbons, and a siamang died or were killed with terminal disease; thereafter three gorillas died and two survivors had advanced disease (Howard and Gendron 1980). The second neotropical species *E. oligarthrus* has been recorded in six species of indigenous felids including the jaguar (*Panthera onca*) in South America and also the bobcat (*Lynx rufus*) in northern Mexico, and in several rodent intermediate hosts including the agouti (*Dasyprocta* spp.) (Rausch and D'Alessandro 2002).

Location in the host

Characteristically, *T. solium* cysticerci lodge in the central nervous system, the eye, striated muscle, and subcutaneous tissue (Figure 33.7). It is known that cysticerci are able to evaginate in the eye (Figure 33.7b). Unexpectedly, an evaginated cysticercus was seen by magnetic resonance in the fourth ventricle of the brain (Flisser and Madrazo 1996). These data support experimental ones that indicate that the host's inflammatory capsule, and not physical restraint, inhibits spontaneous evagination of cysticerci (Ostrosky et al. 1991). Parasites lodged in eye chambers or brain ventricles, as opposed to those found in muscle or brain parenchyma, are usually not surrounded by the host's capsule (Rabiela et al. 1982). It is not known why cysticerci develop

primarily in these tissues; cysticerci have exceptionally been found in other organs and tissues, most probably as a result of severe immunosuppression. The location of parasites in 2 188 cases of cysticercosis collected from several Latin American countries was: central nervous system 82 percent, eye and annexes 17 percent, subcutaneous tissue 7 percent, muscle 5 percent, other organs 6 percent, generalized 1 percent. In the central nervous system cysticerci are more frequently found in subarachnoidal spaces, and less often in parenchyma and ventricles (Rabiela et al. 1982; Escobar 1983). From 161 neurosurgical cases, 89 had cysticerci in the fourth ventricle, 20 in the third ventricle, 17 in the lateral ventricle, 27 were subarachnoidal, and eight were spinal (Madrazo and Flisser 1992). Cysticerci found in basal ganglia are related to the total number of parasites, ranging from 5 percent in patients with a single lesion to 60 percent in patients with more than five parasites; putamen and caudate nuclei were the most frequent sites of lesions (Cosentino et al. 2002). The cellulose type is the most frequent cysticercus in human brain, although in 9–13 percent of studied cases, cysticerci of both types coexisted in the same brain (Rabiela et al. 1982). Racemose cysticerci are only found in spacious areas of the brain such as ventricular cavities and some subarachnoidal spaces, mainly the basal meningeal cisternae (Rabiela et al. 1989); they are found only in human brain and not in swine. Studies of ocular infections indicate that 46 percent of cysticerci are found in vitreous humor, 37 percent are subretinal, 12 percent subconjuntival, 2 percent occur in the orbit, 2 percent are subcutaneous, and 1 percent occurs in the anterior

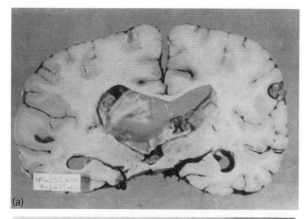

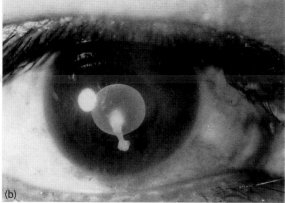

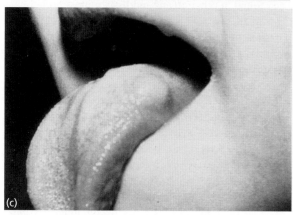

Figure 33.7 (a) *Brain section showing a huge ventricular cysticercus and a small subarachnoidal parasite (by courtesy of Dr J. Olvera);* **(b)** *Taenia solium cysticercus in the anterior chamber of the eye: interestingly, it is evaginated (by courtesy of Dr D. Lozano);* **(c)** *Tongue of child with a cysticercus in the muscle (by courtesy of Dr A. Martuscelli)*

chamber (Puig-Solanes 1974; Gomez-Leal 1989). In ocular, muscular, and subcutaneous cysticercosis only cellulose cysticerci are found.

Most primary infections of *E. granulosus* in humans consist of a single hydatid cyst; however, up to 20–40 percent of patients have multiple cysts or multiple organ involvement. The liver is the most common site, followed by the lungs. The frequency of *E. granulosus* cysts recorded in the Australian Hydatid Registry was: liver 63 percent, lungs 25 percent, muscles 5 percent,

bones 3 percent, kidney 2 percent, spleen 1 percent, brain 1 percent, and exceptionally other organs. Similar figures were recorded in Switzerland. In the Xinjiang region of China, of a series of 15 289 surgical cases, 70 percent were in the liver, 19 percent in the lungs, 3 percent liver and lungs, 3 percent in the abdominal cavity, 0.4 percent spleen, 0.4 percent brain, 0.3 percent kidney, 0.3 percent pelvic cavity, and 0.3 percent bone (National Hydatid Disease Center of China 1993). Cysts may be very big (one reported in 1928 contained 48 liters of fluid), but at the time of diagnosis most cysts measure 1–10 cm. The average size of 304 hepatic hydatid cysts in 212 patients treated in China was 10.1 cm with a range of 4–22 cm, while in 65 hepatic cases detected by ultrasound based mass screening (therefore largely asymptomatic), the average cyst size was 7.2 cm, with a range of 2–13 cm (Wang et al. 2003). In contrast to cysticercosis, where parasites do not change in size, except when racemose cysts are formed, *E. granulosus* hydatid cysts can continue to grow. A study performed in 265 Turkana patients showed that average growth was 9 cm per year, 16 percent of cysts did not expand, 30 percent grew slowly (1–5 mm per year), 42 percent showed a moderate growth rate (6–15 mm per year), and 11 percent increased rapidly in size (31 mm on average per year), with a maximum growth in one case of 160 mm per year (Romig et al. 1986).

The primary location of alveolar echinococcosis is the liver (approximately 98 percent), but the capacity for tumor-like proliferation and the potential for metastasis means that development can also occur in adjacent tissues and structures, and in lung, brain, bone, or other organs (Eckert et al. 1983; Ammann and Eckert 1995). Distant metastases were found in 13 percent of 70 Swiss patients and in 10 percent of 152 Japanese patients (Mesarina-Wicki 1991; Sato et al. 1993). Polycystic echinococcosis has characteristics intermediate between the cystic and alveolar forms (D'Alessandro et al. 1979; Meneghelli et al. 1992). The relatively large cysts are filled with liquid and contain brood capsules with numerous protoscolices. The primary location is the liver, but cysts may spread to contiguous sites.

THE DISEASES

Clinical manifestations and pathology

Neurocysticercosis is characterized by its great diversity of signs and symptoms. It is a complex disease with major manifestations depending on the number, location, and type of parasites lodged in the central nervous system (CNS) and meninges, as well as by the extent of the inflammatory response (Earnest et al. 1987; Takayanagui and Jardim 1983; del Brutto and Sotelo 1988, Sotelo and Del Brutto 2000). Seizures is the most

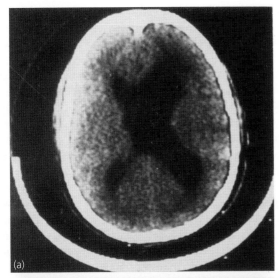

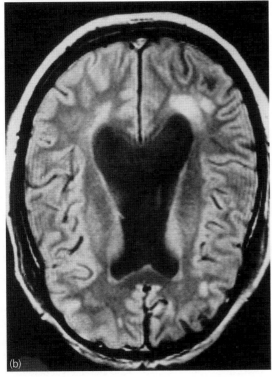

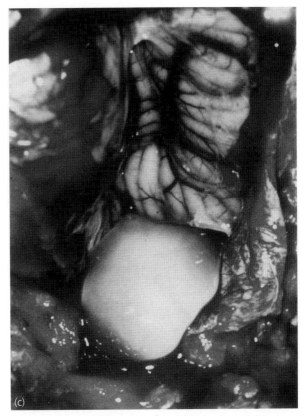

Figure 33.8 *Hydrocephalus due to CSF blockage caused by a racemose type cysticercus seen in CT* **(a)**, *MR* **(b)**, *and during surgery* **(c)** *(by courtesy of Dr I. Madrazo)*

frequent in enhancing calcified lesions (Nash et al. 2001). The glial fibrillary acidic protein was shown to be elevated in the cerebrospinal fluid (CSF) of patients with neurocysticercosis, thus it can be used as a marker of glial damage (Quintanar et al. 2003).

Racemose cysticerci elicit a more intense inflammatory reaction and frequently produce severe mass effects. Progressive inflammation is elicited when the parasite is in contact with the meninges and causes scar tissue proliferation that gives rise to mechanical obstruction of CSF circulation and intracranial hypertension due to hydrocephalus. Acute stages of the disease produce intracranial hypertension due to intense edema induced by numerous developing cysticerci (most commonly found in young patients), ependymitis or arachnoiditis (Rabiela et al. 1982; Escobar 1983; Rabiela et al. 1989; Puri et al. 1991; Madrazo and Flisser 1992; Sotelo and Del Brutto 2000). Idiopathic epilepsy, multiple intracranial space-occupying lesions, chronic meningitis, some bacterial and parasitic infections (including cystic echinococcosis), as well as primary and secondary malignancies affecting the central nervous system can mimic neurocysticercosis (Rodriguez-Carbajal and Boleaga-Duran 1982; Almeida-Pinto et al. 1988; Puri et al. 1991). A similar picture of clinical heterogeneity is also found in ocular cysticercosis (Puig-Solanes 1974; Gomez-Leal 1989). Cysticerci in the

frequent clinical manifestation, occurring on average in 70 percent of cases (Flisser 1994; Garcia et al. 1995; Palacio et al. 1998) and late onset epilepsy shows a strong association with cysticercosis (Vazquez and Sotelo 1992; Garcia et al. 1993; Sanchez et al. 1999). Patients with seizures caused by parenchymal calcified or cellulose single cysticerci have a mild disease as compared to those that have hydrocephalus as a consequence of meningeal arachnoiditis, which is frequently fatal (Estanol et al. 1986; Sotelo et al., 1985a; Figure 33.8). Perilesional edema has been observed in patients with only calcified cysticerci and is associated with seizures and neurological morbidity, being more

subcutaneous tissue and muscles are usually asymptomatic or well tolerated (Botero et al. 1993; Cruz et al. 1994), probably because they are found in small numbers. Muscular pseudo-hypertrophy due to cysticercosis has been reported in cases that have many parasites and marked muscle enlargement and pain (Zhipiao et al. 1980; Rim and Joo 1989). In 2 188 patients with neurocysticercosis from several Latin American countries, 68 percent were between 20 and 49 years of age, only 19 percent were detected below 20 years, and 13 percent were above 50 years. In contrast, in 30 Mexican cases of ocular cysticercosis, 38 percent were under 20 years, 58 percent were between 20 and 49 years and 4 percent were above 50 years (Cardenas et al. 1989), probably indicating that cysticerci in the eyes are easily seen and thus diagnosed at a younger age, and that neurological symptoms start several years after infection.

Symptoms due to cystic echinococcosis (CE) is correlated with the site of infection and parasite burden or biomass. Because of the slowly growing nature of echinococcal cysts, most cases of liver and lung cysts are diagnosed in adult patients and symptoms are related to space occupying or tumor-like lesions. Only 10–20 percent of CE cases are diagnosed at hospital in patients less than 16 years, primarily where cysts are located in the brain or eye, the latter being similar to cysticercosis. Clinical manifestations are variable and determined by the site, size, and condition of the cysts, for example, 67 percent of 297 CE patients with cysts >5 cm diameter had symptoms, as compared to 48 percent of 123 cases with cysts <5 cm; overall, 38 percent were asymptomatic. Mass screening programs using portable ultrasound scanners have shed light on the natural history of pathology of hepatic cystic echinococcosis in endemic communities (Craig et al. 1996). Studies in North Africa and South America found that simple univesicular or hyaline hydatid cysts, called type 1, occurred in 17–28 percent of asymptomatic persons, and were considered to be an early or evolutive developmental stage. Cysts with a clear laminated membrane and/or presence of daughter cysts (types 2 and 3) comprised 11–55 percent of community detected cases, while hydatid cysts showing signs of involution, such as infiltration and calcification (types 4 and 5) made up 18–45 percent of cases (Cohen et al. 1998; Shambesh et al. 1999). These and other similar studies indicate that many cases of cystic echinococcosis are asymptomatic and that spontaneous regression and healing occurs in a significant proportion of infected people in endemic communities. These features are also similar to cysticercosis. The signs and symptoms of hepatic echinococcosis may include hepatic enlargement, with or without a palpable mass, liver abscesses, calcified lesions, epigastric pain, nausea and vomiting, portal hypertension, inferior vena cava compression or thrombosis, secondary biliary cirrhosis, biliary colic-like

symptoms, biliary peritonitis or fistula formation. If a cyst ruptures, the sudden release of its contents may precipitate allergic reactions ranging from mild to fatal anaphylaxis, bacterial infection may occur and there is spread of protoscolices, which may result in a multiple secondary echinococcosis disease (Kammerer and Schantz 1993; Ammann and Eckert 1995). Clinical symptoms of pulmonary cystic echinococcosis are tumor-like chest pain, chronic cough, fever, hemoptysis pneumothorax, pleuritis, lung abscess, eosinophilic pneumonitis, lung embolism, and sometimes expectoration of cyst content (Schantz and Okelo 1990). Nearly 40 percent of patients with pulmonary hydatidosis have liver involvement as well (Little 1976). Cysts in other locations such as the heart, spine, brain, and even bone also generate tumor-like symptomatology (Ammann and Eckert 1995). In that respect bone infection may also resemble an alveolar echinococcosis lesion (Bonifacino et al. 1997). Brain hydatid cysts may be confused with neurocysticercosis because of similar images in tomography. In these cases, the epidemiological history is the best way of confirming diagnosis, together with specific immunological tests (Flisser 1988).

Infection of the liver with *E. multilocularis*, the primary location in humans, closely mimics hepatic carcinoma or cirrhosis. Clinical cases are characterized by a chronic course of the disease lasting for weeks, months, or years. Advanced lesions often consist of a central necrotic cavity surrounded by a white amorphous material that is covered with a thin peripheral layer of dense fibrous tissue (Wilson and Rausch 1980; Rausch et al. 1987). Focal areas of calcification exist, as does extensive infiltration by proliferating vesicles. The initial symptoms of alveolar echinococcosis (AE) are usually vague. Mild upper quadrant and epigastric pain with hepatomegaly may progress to obstructive jaundice. Occasionally, the initial manifestations are related to metastases to the lungs or brain. Symptoms at diagnosis in 40 Turkish AE patients included cholestatic jaundice (45 percent), dyspepsia (40 percent), and epigastric pain (35 percent), the disease was detected incidentally during a check-up in 38 percent of the cases (Polat et al. 2002). The mortality in progressive, clinically manifest AE cases may be 50 to 75 percent (Wilson and Rausch 1980; Kammerer and Schantz 1993) and much higher figures (up to 100 percent) have been recorded in untreated or inadequately treated patients. Radiographic or serologic screening of high-risk patients results in detection of early-stage disease, improving prognosis through earlier application of therapy (Kammerer and Schantz 1993; WHO 1996). Spontaneous death of alveolar hydatid cysts was clearly documented in five asymptomatic individuals in Alaska; lesions were circumscribed and calcified with a mineralized wall and a cavity filled with amorphous necrotic material, in some cases also with folded parasite membranes, suggesting that death at early stages is probably not uncommon

(Rausch et al. 1987). Community mass screening studies in France and China also identified seropositive individuals with calcified lesions >2 cm which were considered to be abortive infections (Bresson-Hadni et al. 1994; Bartholomot et al. 2002). Alveolar echinococcosis is typically seen clinically in persons of advanced age, the average age being >50 years as compared to <40 in cystic echinococcosis.

The clinical characteristics of polycystic echinococcosis, like that for cystic echinococcosis, depend on the location and larval biomass, but, as for alveolar echinococcosis, also on the degree of tissue infiltration by exogenous proliferation. It is less organ specific than *E. multilocularis* and proliferates more rapidly. Up to 78 percent of cases due to *E. vogeli* infection involved the liver alone or in association with the lungs and abdominal organs, while pulmonary, mesentery, or stomach locations have also been described (D'Alessandro 1997). Three cases of human infection with *E. oligarthrus* involved the orbit or the heart. Mean age of 51 cases of polycystic echinococcosis was 44 years (range 6–78 years) of which 40 percent survived at least 1.5 years after initial diagnosis.

IMMUNE RESPONSE

Immune responses in patients with neurocysticercosis have been studied mainly because of the need to standardize immunological methods for diagnosis (Flisser et al. 1980; Flisser and Larralde 1986; Richards and Schantz 1991; Dorny et al. 2003). Anti-cysticercus antibodies of several immunoglobulin isotypes have been detected, although the most frequently found are IgG antibodies in serum, cerebrospinal fluid, and saliva. Interestingly, IgG anti-cysticercus antibodies are not found in all compartments of the same patient (Cho et al. 1986; Espinoza et al. 1986; Feldman et al. 1990; Wilson et al. 1991). The presence of IgG confirms that the disease is usually chronic and long-term. A few reports indicate an increase in total or specific IgE (Goldberg et al. 1981; Gorodezky et al. 1987; Short et al. 1991), IgM is less frequently found in CSF than IgG, and IgA antibodies have been reported in few cases, although not in saliva. Antibodies have been detected in 94 percent of cases with two or more parasites, but only 28 percent of cases with single lesions, and in most cases with undamaged parasites, but in only 44 percent of cases with calficied cysticerci (Espinoza et al. 1986; Baily et al. 1988; Chang et al. 1988; Michault et al. 1990; Michel et al. 1990; Wilson et al. 1991). The humoral immune response in patients with neurocysticercosis is quite heterogeneous as evidenced by the number of antigens recognized: patients' antibodies may react with one to eight antigens in immunoelectrophoresis and up to 30 antigens in Western blot (Flisser and Larralde 1986; Correa et al. 1985, 1989a; Larralde et al. 1989; Proaño-Narvaez et al. 2002). Differences have also been found

between benign and malignant cysticercosis, the later state being more antigenic. It has also been shown that the immune response may be transient in households of neurocysticercosis patients and in apparently healthy individuals in the open population (Garcia et al. 2001; Meza-Lucas et al. 2003).

As in cysticercosis, the humoral immune response in patients with hydatid disease has been used widely to diagnose infection or disease and the main emphasis has been characterization of parasite antigens (Lightowlers et al. 1993; Lightowlers and Gottstein 1995). Experimental studies indicate that antibodies to antigen 5 are among the first detected following infection with *E. granulosus* (Conder et al. 1980), but are also found in patients with neurocysticercosis or alveolar echinococcosis (Varela-Díaz et al. 1978; Schantz and Gottstein 1986; Moro et al. 1992). ELISA and Western blot are usually performed with anti-human IgG as this is the main isotype in human hydatid disease. Antibodies of the subclass IgG4 are particularly evident in the serum of advanced CE and AE patients (Wen and Craig 1994; Shambesh et al., 1997; Ortona et al. 2002). The demonstration of parasite-specific IgE in cystic echinococcosis has attracted particular attention but, as in human neurocysticercosis, appears to have no significant diagnostic advantage. However, high levels of IgE antibody may correlate with allergic reactions to *E. granulosus* antigens with homology to cyclophilin, and low levels with beneficial post-treatment chemotherapy (Ortona et al. 2002; Rigano et al. 2002). In contrast, 68 percent of patients with alveolar echinococcosis tested positive for parasite-specific IgE by ELISA and between 50 and 88 percent by the radioallergosorbent test (Vuitton et al. 1988; Gottstein 1992; Liu et al. 1992; Moro et al. 1992; Lightowlers and Gottstein 1995). Strain variation in the parasite, geographical origin of the patient, and differences in the host–parasite relationship may significantly affect the humoral responses (Rafiei and Craig 2002).

The main antigen recognized by patients in both diseases, though different, has unfortunately been designated the same, i.e. antigen B. However, *T. solium* antigen B is a different molecule from *E. granulosus* antigen B. The first is a glycoprotein of 110 kDa (Guerra et al. 1982), while that of *E. granulosus* is a lipoprotein of 120–160 kDa (Oriol and Oriol 1975). Both are immunodominant; 85 percent of patients with neurocysticercosis react with *T. solium* antigen B (Flisser et al. 1980; Espinoza et al. 1986) and 80–90 percent of patients with advanced cystic hydatid disease recognize *E. granulosus* antigen B (Williams et al. 1971; Wen and Craig 1994). Recognition of antigen B in asymptomatic cystic hydatid patients is, however, variable with greater responses in patients with evolutive disease (Shambesh et al. 1999; Daeki et al. 2000; Rigano et al. 2002). Immunological and biochemical assays have demonstrated that for echinococcus immunodominant antigen B is highly genus specific. *T. solium* antigen B, or as recently termed, paramyosin, (Laclette et al. 1991) is non-specific; it is found in many platyhelminthes, and sera from patients with hydatid disease cross-react with this antigen (Olivo et al. 1988). *E. granulosus* antigen B, and antigen 5, another large lipoprotein, have been extensively characterized (reviewed in Lightowlers and Gottstein 1995), and, as these authors suggest, 'it would be valuable for these antigens to be compared between the different laboratories involved, in order to clarify whether the molecules are indeed one and the same'. It is known that antigen B of *E. granulosus* consists of at least three components with molecular weights of 8, 16, and 24 kDa, suggesting that this antigen is a polymer of an 8 kDa subunit (Lightowlers et al. 1989). The 8 kDa subunit is considered to be the most specific for *E. granulosus* and has been cloned (Shepherd et al. 1991; Frosch et al. 1994a) and more recently shown in fact to comprise two 8 kDa subunits (AgB8/1 and AgB8/2) with approximately 50 percent amino acid homology between them (Gonzalez et al. 1996). Both native and recombinant antigen B of *E. granulosus* preferentially bind IgG4 subclass antibodies which is most pronounced in the active or evolutive phase of cystic hydatidosis (Wen and Craig 1994; Rigano et al. 2002).

For the diagnosis of cysticercosis, antigenic extracts are usually prepared from *T. solium* cysticerci from infected pigs (Flisser and Larralde 1986). Alternatively, antigens from other sources such as *T. crassiceps*, *T. hydatigena*, and *T. saginata*, or those obtained by genetic engineering (McManus et al. 1989) have also been used. Cyst fluid, protoscolices, and cyst membranes of *E. granulosus* from a variety of hosts are the sources of antigen for diagnostic tests of hydatid disease. Fluid from fertile cysts contains higher concentrations of antigens than from sterile cysts, although viable non-fertile cysts also contain suitable antigens, while fluid from sheep and human liver hydatid cysts also contain higher concentration of the two major antigens (antigen B and antigen 5) than that of cattle and pig cysts. In general, hydatid cyst fluid has been a better source of *E. granulosus* antigens than protoscolices or cyst membranes; however, both these also contain dominant antigens (Rogan and Craig 1997; Rafiei and Craig 2002). *T. hydatigena* and *T. ovis* have also been evaluated as source of antigens (Yong et al. 1984), enabling successful antigen preparation from heterologous taeniid parasite species. Oncospheres have also been used as a source of antigens: antibodies to high molecular weight antigens were detected in pig sera after 1 month of infection (Garcia-Allan et al. 1996), in 20 percent of serum samples from patients with active neurocysticercosis, and in 95 percent of tapeworm carriers (Verastegui et al. 2003).

Because crude antigen extracts generate cross-reactions, the use of purified or semi-purified antigens will increase specificity. Characterization of *T. solium* and *E. granulosus* antigens has changed from gel diffusion and immunoelectrophoresis to Western blot, lentil-lectin chromatography, immunoprecipitation, monoclonal antibodies, chromatofocusing, and recombinant DNA technology (reviewed in Flisser (1994) for cysticercosis and in Lightowlers and Gottstein (1995) for hydatid disease). In cysticercosis the use of a crude electrophoretic extract gave no clear specific band patterns in Western blots (Espinoza and Flisser 1986; Larralde et al. 1989). However, a similar assay using a semi-purified antigenic fraction identified seven specific glycoproteins (Gp50, Gp42-39, GP24, GP21, GP18, GP14, and GP13), the three with highest molecular weights reacting most frequently with sera from clinical cases (Tsang et al. 1989; Feldman et al. 1990; Wilson et al. 1991) and population samples (Schantz et al. 1994; Thais et al. 1994). In contrast, low molecular weight antigens have also been identified as immunodominant in human cysticercosis (Chung et al. 1999; Plancarte et al., 1999b; Restrepo et al. 2000, 2001; Sako et al. 2000; Dorny et al. 2003). Further progress has been made by recombinant DNA techniques, using *E. granulosus* clones that react with sera from patients with cystic hydatid disease (see for example, Ferreira and Zaha 1994), and clones from *E. multilocularis* immunoreactive with sera from patients with alveolar hydatid disease (see for example, Hemmings and McManus (1989) and Muller et al. (1989), and review by Siles-Lucas and Gottstein (2001)).

Data on cellular immune responses are limited (Flisser et al. 2003), but, interestingly, show similarities in both larval cestode diseases: an increase in $CD8^+$ lymphocytes (Allan et al. 1981; Flisser et al. 1986; Vuitton et al. 1989; Molinari et al. 1990; Craig 1994), polyclonal activation of B lymphocytes (Craig 1994), and a decrease in γ-interferon (IFN) (Suntsov et al. 1990a, b; Ostrosky-Zeichner et al. 1996), although γ-IFN was increased in a group of patients who had not received previous treatment of any kind (Medina-Escutia et al. 2001), and tumor necrosis factor α was increased in children with active neurocysticercosis (Aguilar-Rebolledo et al. 2001). Parasite molecules with immunosuppressive effects have also been reported (Molinari et al. 1990). Nevertheless, a systematic and focused analysis of the relationship between cytokine effects and the outcome of larval cestode diseases is still needed. The role of Th1/Th2 balance upon resolution of the disease is still not clear. Increased levels of IL-1 and IL-6 in CSF of patients with inflammatory neurocysticercosis were found (Ostrosky-Zeichner et al. 1996). Serum levels of IL-5 and eotaxin were also increased in patients as compared to controls in one study (Evans et al. 1998), but were not detected in another (Aguilar-Rebolledo et al. 2001). IL-2 was synthesized by peripheral blood cells in half of patients with neurocysticercosis (Medina-Escutia et al. 2001) and a Th1/Th2 mixed pattern of interleukins were found at the host–parasite interface (Restrepo et al. 1998) and in the CSF (Rolfs et al. 1995).

The mechanisms by which established larval cestodes survive in hosts resistant to re-infection remain unknown. Evidence exists for a variety of immune-evasion mechanisms. Several have been described in cysticercosis (Flisser 1989; White et al. 1997; Flisser et al. 2003): survival of parasites lodged in 'immunologically privileged sites', masking of cysticerci by host immunoglobulins, suppression of host responses by taeniaestatin, paramyosin, sulfated polysaccharides, enzymes that detoxify reactive oxygen intermediates, low molecular RNA, cysteine proteinases, and PGE2. In hydatid disease, studies in experimental models have demonstrated the influence of the parasite on the specific and non-specific immune responses of the host, such as pathological alterations in the architecture of lymphoid organs, inhibition of cellular immune responses to specific antigens and of leukocyte chemotaxis, reduction in macrophage activity, and reduced blastoid transformation and autoantibody production (Dixon and Jenkins 1995; Rogan and Craig 1997; Vuitton 2003). The impact of acquired immune deficiency syndrome (AIDS) on the susceptibility of endemic populations for cysticercosis and hydatid disease could result in a significant terminal disease (Heath 1995). As an example, huge brain cysticerci have been found in two HIV patients (Soto-Hernandez et al. 1996). Few studies have been undertaken to determine factors that regulate innate susceptibility to larval cestode infections. Host age, sex, strain, and physiological state have marked influences in determining innate resistance to infection (reviewed by Flisser 1994). Intermediate hosts develop specific humoral and cellular responses to the parasites that confer a significant level of resistance to reinfection. The contribution of these immune responses to the destruction of the parasite after initial establishment is less clear. A proportion of cysticerci and of hydatid cysts die some time after initial establishment, so that calcified lesions can be observed together with viable and with dead parasites. Although it is not known whether immunity is responsible for the death of these parasites, there is evidence, for example, that the inflammatory response is associated with restricting the growth and metastasis of the cyst mass in hosts refractory to infection with *E. multilocularis* (Ali-Khan and Siboo 1980) and that in experimental treatment of pigs with cysticercosis, cestocidal drugs damage cysticerci and eosinophils, lymphocytes, and macrophages destroy them (see Figure 33.9) (Flisser et al. 1990b; Torres et al. 1992). Nevertheless, despite the development of specific immune responses, viable cysts frequently persist for long periods (Rogan and Craig 1997). Production of a T-helper cell type 2 (Th2) response in human cystic echinococcosis, was characterized by low γ-IFN, and high IL4 and IL10 when patients had a poor prognosis after albendazole chemotherapy. In contrast, a Th1 type response characterized by production of γ-IFN and low IL4 and IL10 levels was correlated with beneficial therapeutic response to chemotherapy (Rigano et al. 1995). High counts of CD4$^+$ T cells in hepatic granuloma was correlated in persons with abortive AE lesions, while CD8$^+$ T cells were invariably present in chronic active AE disease (Vuitton 2003). Evidence for a genetic susceptibility to human AE has also been demonstrated to be associated with the HLA DR3/4 gene with a possible phenotypic outcome of increased IL10 production (Godot et al. 2000).

Several aspects of the immunobiology of larval cestode infections in the intermediate host have

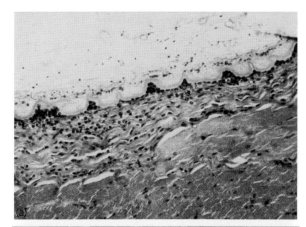

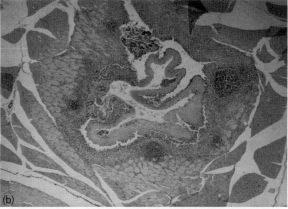

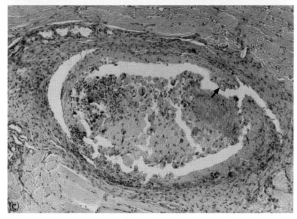

Figure 33.9 *Effect of treatment of pigs with praziquantel on cysticerci: eosinophils* **(a)** *and lymphocytes* **(b)** *surround the parasite and macrophages phagocytose cell debris* **(c)**. *The arrow in* **(c)** *indicates a larval hook (by courtesy of Dr A.S. de Aluja)*

provided the framework for research on vaccination. Many studies have been performed in order to evaluate protection against cysticercosis and against hydatid disease (Mitchell 1990; Rickard et al. 1995; Lightowlers 1996; Lightowlers et al. 2000). The immune response against larval cestodes may be divided into two phases, the first directed against recently hatched oncospheres attempting to penetrate the gut mucosa and establish themselves in host tissues, and the second aimed at the established metacestode. Immune effector mechanisms of the first phase are more successful in destroying the parasites than those of the second phase because established metacestodes have evolved highly effective physical and biochemical mechanisms to evade the host's defenses, some mentioned above. Early oncospheres are highly vulnerable to antibody-dependent complement mediated immune attack, but only briefly, and thus there is a race between parasite development and the generation of the host immune response in the critical first few days of infection (Ito 1997). Antigens from various sources have been used as vaccines (Flisser 1994) and immunological studies are being undertaken (Toledo et al. 1999; Flisser and Lightowlers 2001; Verastegui et al. 2002). Recombinant oncosphere antigen-based vaccines were first used successfully against ovine cysticercosis caused by *Taenia ovis* (Johnson et al. 1989) and most recently against *T. solium* in pigs and *E. granulosus* in sheep (Manoutcharian et al. 1996; Lightowlers et al. 1996; Plancarte et al. 1999a). A successful field trial with a synthetic peptide vaccine in pigs raised and kept in their original rural environment was reported (Huerta et al. 2001). Furthermore, experimental vaccination of rustically raised pigs was more effective if performed when pigs were 70 days of age as compared to 40 days (Huerta et al. 2000), and cestocidal treatment of pigs protected them for at least 3 months against cysticercosis acquired naturally in hyperendemic areas (Gonzalez et al. 2001). The 45W protective antigen from *Taenia ovis* shows high homology to *T. solium* in studies of genes and transcripts of this antigen in both species, making it a vaccine candidate for the latter species (Gauci and Lightowlers 2001). A cDNA library immunization, based on the use of a large number of pathogens' cDNA clones, induced a cellular protective immune response (Manoutcharian et al. 1998, 1999, 2000). A high degree of immune protection in sheep (96–100 percent) against cystic echinococcosis was obtained with an *E. granulosus* oncosphere sub-unit recombinant vaccine (EG95) (Lightowlers et al. 1996), which was antibody mediated and directed at conformational epitopes of a 16.9 kDa protein expressing a fibronectin type domain (Chow et al. 2001; Woolard et al. 2001). The EG95 homolog was also identified in an *E. multilocularis* cDNA library and purified as an experimental vaccine which gave 78–83 percent protection in mice against an *E. multilocularis* egg challenge (Gauci et al. 2002). An unrelated metacestode protein of *E. multilocularis* called

14-3-3, has also been cloned and provided >95 percent protection of mice against an egg challenge, but not secondary infection and therefore must have shared epitopes with the oncosphere (Siles-Lucas et al. 2003).

DIAGNOSIS

In the past, the pleomorphic symptomatology of neurocysticercosis has required the use of a variety of diagnostic procedures. Most of these have now been replaced by imaging techniques, such as computed tomography (CT) and magnetic resonance (MR) (Suss et al. 1986; Almeida-Pinto et al. 1988; Jena et al. 1988; Rodacki et al. 1989; Teitelbaum et al. 1989, Flisser et al. 1988; Rajshekhar 1991; Rajshekhar and Chandy 1997; Sotelo and Del Brutto 2000; Garcia and Del Brutto 2003). As alternative or complementary diagnostic procedures, immunological assays are used to detect anti-cysticercus antibodies in CSF or serum. Enzyme-linked immunosorbent assays (ELISA) and Western blot (WB) are currently used for clinical support of symptomatic patients in several countries (Espinoza et al. 1986; Gottstein et al. 1986; Pammenter et al. 1987; Michault et al. 1989; Tsang et al. 1989; Zini et al. 1990; Wilson et al. 1991; Sanchez et al. 1999; Dorny et al. 2003). Their high positivity facilitates diagnosis of neurocysticercosis, especially when CT or MR is not available or not conclusive. Diagnostic critera for neurocysticercosis have been proposed recently by a panel of experts that are based on objective clinical, imaging, immunologic, and epidemiologic data. These include four categories of criteria stratified on the basis of their diagnostic strength: absolute, major, minor, and epidemiologic. Interpretation of these criteria permits two degrees of diagnostic certainty: definitive and probable diagnosis (Del Brutto et al. 2001).

CT and MR are used to identify cysticerci in the brain and thus confirm the etiology of the disease and define the number, stage, location, and extent of lesions. Because of their vesicular structure, living cysticerci are seen as hypodense or low signal intensity images, the scolex giving a hyperdense or high signal intensity, usually eccentric, inside the vesicle (see Figure 33.10).

Isolated cysticerci detected in brain parenchyma by imaging techniques are frequently seronegative, whereas multiple or subarachnoidal cysticerci are usually seropositive (Wilson et al. 1991). Cysticerci in ventricles can also be detected, especially by MR. Imaging techniques also show vasculitis, inflammation, and edema surrounding the parasite, ring-like enhancement being associated with an acute infection (cysticercotic encephalitis) and indicating an active inflammatory process. Hydrocephalus is a common finding associated with intraventricular and/or basal cysticerci, and imaging techniques are very helpful non-invasive techniques for following the outcome, especially after a shunt has been introduced. In long-term disease, parasites of different

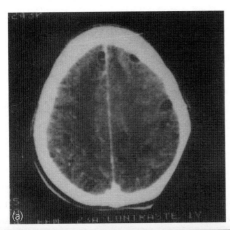

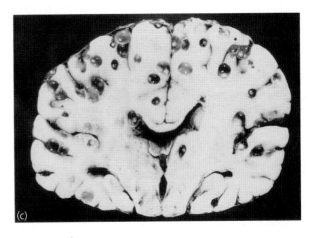

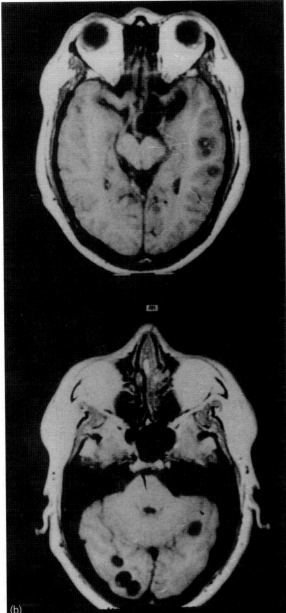

Figure 33.10 *Multiple small cysticerci seen by* **(a)** *CT,* **(b)** *MR and* **(c)** *in anatomical section of the brain. The scolex can be seen in some parasites (by courtesy of Drs I. Madrazo and J. Olvera)*

sizes can be seen with diverse tomographic pictures of brain response (Flisser et al. 1988). Calcified cysticerci are seen in CT as hyperdense dots, but are usually not detected by MR because of the lack of signal from calcifications. In countries where cysticercosis is endemic, clinicians usually consider this etiology in the initial diagnosis of cases that have mass occupying lesions, enhancing nodules, hydrocephalus, or other images that might be associated with neurocysticercosis. Solitary cerebral granulomas causing seizures can be due to cysticerci or to tuberculomas, and are not only common in developing countries but also in developed ones. Diagnostic criteria have been defined for accurately and specifically identifying solitary cerebral cysticercus granulomas (Rajshekhar and Chandy 1997). More consideration should be given to it in developed countries, where cases of neurocysticercosis are currently being diagnosed or misdiagnosed (Michael et al. 1990). Cysticerci in eyes are easily diagnosed when parasites are viable, but can easily be misdiagnosed when there are inflammatory reactions or involution of the parasite (Gomez-Leal 1989). Subcutaneous and muscle cysticerci may be detected by palpation and diagnosed by biopsy.

Immunodiagnosis has the great advantage of lower cost than CT or MR, and the presence of specific anti-cysticercus antibodies may confirm the disease. ELISA has clearly improved immunodiagnosis and is being used routinely for confirmation of clinical diagnosis when imaging techniques are not available or are not conclusive. It has a high sensitivity, between 75 and 90 percent with serum, and, seemingly, lumbar CSF gives better results. The main disadvantage of ELISA is cross-reactivity with other helminth infections since the antigen employed is a crude homogenate of cysticerci or vesicular fluid. Western blot overcomes cross-reactivity because it employs an enriched fraction of glycoproteins (Gp). In addition the technique separates the antigens present in the Gp fraction during electrophoresis prior to the antigen–antibody reaction. The presence of between one and seven specific Gp bands is considered diagnostic of *T. solium* infections. When two or more cysticerci are present, the blot is 100 percent specific and has 100 percent sensitivity with serum samples and 95 percent with CSF. Antigens can be detected in CSF by ELISA using monoclonal antibodies (Correa et al. 1989b; Choromanski et al. 1990; Chen et al. 1991; Garcia et al. 1998a, 2000, 2002c). Although the sensitivity of antigen detection is lower than that of antibody detection, positive cases confirm the presence of the parasite without the need for other diagnostic procedures.

Improvement of immunodiagnosis of human neurocysticercosis, specifically regarding the characterization and purification of antigens (Garcia-Allan et al. 1996; Rodriguez-Canul et al. 1999; Vaz et al. 1997; Ito et al. 1998; Ko and Ng 1998; Yang et al. 1998; Greene et al. 1999; Plancarte et al. 1999a, b; Casagranda et al. 2000; Hernandez et al. 2000; Chung et al. 1999; Restrepo et al. 2000, 2001; Sako et al. 2000; Gazarian et al. 2001; Quintanar et al. 2003; Dorny et al. 2003), the follow-up of patients (Rajshekhar 1991; García et al. 1997a, 2000) and the diagnosis of swine cysticercosis (Pinto et al. 2000a, b; Rodriguez-Canul et al. 1998; Sciutto et al. 1998a, b) has obvious priority in cysticercosis, others relate to taeniasis regarding the distinction of *T. solium* from *T. saginata* by DNA analysis (Chapman et al. 1995; Gasser et al. 1995; Mayta et al. 2000; Gonzalez et al. 2002; Yamasaki et al. 2002) and the identification of serum antibodies (Wilkins et al. 1999).

The presence of a cyst-like mass in a person with a history of exposure to dogs in areas in which *E. granulosus* is endemic supports the diagnosis of cystic echinococcosis. However, hydatid cysts must be differentiated from benign cysts, cavitary tuberculosis, mycoses, abscesses (microbial or parasitic), racemose cysticercosis, and benign or malignant neoplasms. A non-invasive confirmation of the diagnosis can usually be accomplished by the use of ultrasonography (US), CT, MR, or radiography (Ammann and Eckert 1995). The liver is the most frequently involved organ and 60–85 percent of the cysts are located in the right lobe. Ultrasonography, CT, and MR are useful in diagnosing deep-seated lesions in all organs and defining the extent and condition of avascular fluid-filled cysts (Figure 33.11) (von Sinner 1991; Choji et al. 1992; Ammann and Eckert 1995). Images of *E. granulosus* hydatid disease typically show round, solitary or multiple, sharply contoured cysts, measuring from 1 to over 15 cm, these represent about 38–48 percent of all cysts. In 29–46 percent of all cysts the presence of internal daughter cysts produces structures comparable to a cartwheel. Thin, crescent or ring-shaped calcifications of variable degree located in the cyst wall occur in about 10 percent of the cysts, giving an eggshell pattern. These require about 5–10 years to develop (Figure 33.11a, b). Since Gharbi used ultrasound images to classify *E. granulosus* cyst presentation into five basic types (types 1–5), several similar hydatid cyst pathology profiles have been described (Gharbi

et al. 1981; Perdomo et al. 1997; Larrieu et al. 2000). Recently, a revised US classification for human cystic echinococcosis (CE) was published by WHO in order to try and simplify and combine elements of several similar classifications (WHO/OIE 2001). CT allows measurement of the size of the parasites, which is useful for chemotherapeutic follow-up and gives a correct diagnosis in a high proportion of cases (60–90 percent). Serology for cystic hydatid disease is useful in confirmation of radiographic images, but may be variable dependent on site, type, number, and condition of the cysts.

The usual CT image of *E. multilocularis* infection is that of indistinct solid tumors seen as heterogeneous hypodense masses often associated with central necrotic areas and calcifications, lesion contours are irregular without a well-defined wall (Figure 33.11c, d). Frequently, the lesion extends beyond the liver, which can cause compression or obstruction of other structures, such as the inferior vena cava, hepatic veins, and portal branches (Ammann and Eckert 1995). Clusters of micro-calcifications or irregular plaque-like calcified foci are often found in central or peripheral parts of the lesions. In a study of 30 AE patients, CT was considered the method of choice for hepatic lesions followed by MR and US; MR was less effective for calcified lesions though superior for extra-hepatic sites (Reuter et al. 2001). Ultrasound, as for CE, has been demonstrated to be effective in mass screening for hepatic AE in endemic communities (Bartholomot et al. 2002). The lungs may be involved either by direct extension of the liver process or by parasite metastases seen as multiple small solid foci located usually eccentrically at the periphery of the lobes. Serologic tests are usually positive at high titers; purified *E. multilocularis* antigens are highly specific, and comparing a patient's titers to both purified-specific and shared antigens allows serologic discrimination between patients infected with *E. multilocularis* and those infected with *E. granulosus* (Ito et al. 2002).

Diagnostic puncture should usually be avoided because of the risk of secondary echinococcosis due to spillage of viable protoscolices or of metastases in the site of injection from germinative tissue adhering to the needle. There is also the risk of anaphylactic reactions from leakage of cyst fluid especially from *E. granulosus* cysts (Ammann and Eckert 1995). The availability of immunodiagnostic tests for confirmation of US, CT, or MR images reduces the need of puncture for diagnosis, although puncture is presently being used successfully as an alternative therapeutic treatment for cystic hydatidosis (PAIR; see section on Therapy), and has been used in a diagnostic biopsy capacity for specific antigen detection (Paul and Stefaniak 1997).

Virtually every serodiagnostic technique devised for any disease has been evaluated for diagnosis of cystic echinococcosis (CE), often with considerable

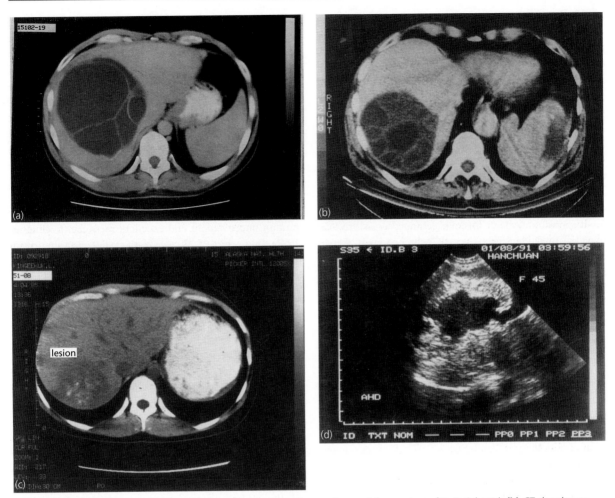

Figure 33.11 (a) *CT showing a solitary Echinococcus granulosus human liver cyst (by courtesy of Dr P. Schantz);* **(b)** *CT showing an Echinococcus granulosus human liver cyst with a cartwheel appearance (by courtesy of Dr P Schantz);* **(c)** *CT showing an Echinococcus multilocularis liver cyst showing plaque-like calicified foci (by courtesy of Dr P. Schantz);* **(d)** *Ultrasonography of a liver from a patient from Gansu, China with alveolar echinococcosis (by courtesy of Professor P. Craig)*

discrepancies between laboratories (Lightowlers and Gottstein 1995). As Schantz and Gottstein (1986) discuss, selection of a particular immunodiagnostic test involves consideration of the sensitivity and specificity of the techniques, the purpose for which they will be used, available technical expertise, and cost. Sensitivity and specificity vary according to the technique used, the quality of antigens and the characteristics of the groups of CE patients and controls used in the study. ELISA, indirect hemagglutination, latex agglutination, and indirect immunofluorescence have a high sensitivity (60–90 percent), although some patients do not develop a detectable immune response. Hepatic CE cysts are more likely to elicit an immune response than pulmonary cysts, but about 10 percent of patients with hepatic cysts and 40 percent with pulmonary cysts do not produce detectable serum antibodies and give false-negative results, which may result in a dangerous puncture of a hydatid cyst (Ammann and Eckert 1995; Paul and Stefaniak 1997). ELISA, which is now used routinely, has a similar sensitivity to the indirect hemagglutination test

and thus both are procedures of choice for the initial screening of serum. Specific confirmation of reactivity for CE can be obtained by demonstrating antibodies to antigen 5 (arc 5) by immunodiffusion or immunoelectrophoresis or more specifically to the 8–12 kDa bands (= antigen B) in Western blot (Madisson et al. 1989). Antigen B of *E. granulosus* has been cloned by several groups (Siles-Lucas and Gottstein 2001), and both native and recombinant antigens have the propensity to bind specific IgG4 antibodies in CE patients with advanced disease (Wen and Craig 1994; McVie et al. 1997). Antigen B is an antigen of choice for seroepidemiological studies and in support of ultrasound for CE mass screening surveys (Cohen et al. 1998; Wang et al., 2001). Eosinophilia is present in fewer than 25 percent of infected patients and hypergammaglobulinemia in about 30 percent of the patients with CE (Ammann and Eckert 1995). Sero-conversion is usually higher in human alveolar echinococcosis (AE) compared to human CE. About 95 percent or more of patients suffering from advanced AE exhibit antibodies against the laminated layer derived glycan Em2 antigen (Siles-Lucas and Gottstein 2001) and also the recombinant EM10 antigen, a 65 kDa molecule which is 99 percent specific in *E. multilocularis* (Helbig et al. 1993; Frosch et al. 1994b). An *E. multilocularis* 18 kDa native antigen (Em18) was identified using immunoblotting to be highly species specific and sensitive for AE, and a recombinant Em18 was also useful in differential immunodiagnosis of alveolar and cystic echinococcosis (Ito et al. 1999, 2002). Serological tests have also proved useful in confirmation of AE cases in screening surveys and for epidemiological studies in endemic areas (Romig et al. 1999; Bartholomot et al. 2002). Techniques useful for diagnosing cystic or alveolar echinococcosis are also of value in diagnosing polycystic echinococcosis due to *E. vogeli* infection, since these parasites have common antigens (Gottstein et al. 1995).

THERAPY

Treatment is based on palliative drugs to control symptoms and inflammation, placement of ventricular shunts in patients with neurocysticercosis that drain CSF to the peritoneal cavity, bile shunts in AE and CE, or surgery to remove brain cysticerci or hydatid cysts/lesions from liver, lungs or other organs; the advent of cestocidal drugs has improved prognosis of many cases (Earnest et al. 1987; Madrazo and Flisser 1992; Kammerer and Schantz, 1993; Ammann and Eckert 1995; Escobedo 2000; Sotelo and Del Brutto 2000; White 2000, Garcia and Del Brutto 2000; Garcia et al. 2002a). Recently a panel of experts published current consensus guidelines for treatment of neurocysticercosis, where a basic set of principles to follow is clearly described. They advise that treatment of neurocysticercosis should not be generalized; instead, each case should be approached and assessed individually. Furthermore, the selection of the treatment option must include in the consideration of risks and benefits, the economic situation of the patient. Also, the current controversy of the benefits of anti-parasitic therapy is thoroughly discussed (Garcia et al. 2002b).

Anti-epileptic drugs are the treatment of choice in patients with neurocysticercosis that have calcifications, in which seizures are the only manifestation of the disease and there is no imaging evidence of living parasites. Ventricular shunting is the most common neurosurgical approach to control hydrocephalus caused by cysticerci in the ventricles, periventricular or basal cisternae, or by ependimitis, inflammatory scarring, arachnoiditis or adhesive basilar meningitis. Patients with hydrocephalus always require a ventricular shunt before other measures are attempted. CSF shunting is a relatively simple surgical technique used to treat intracranial hypertension due to blockage of CSF circulation, but complications may arise from the high protein levels in CSF (which can cause blockage), from bacterial infections or from surgical inexperience (Colli et al. 1986; Madrazo and Flisser 1992). A ventriculoperitoneal shunt with a new design based on the rate of CSF production and shunt resistance, rather than on ventricular pressure, has recently been evaluated with promising results (Sotelo et al. 1995). In a hospital study of 632 cases subjected to ventriculoperitoneal shunting, 77 percent were resolved with one operation, but 11 percent needed three or more shunt placements (Mateos and Zenteno 1987). In another study, 68 percent of 69 cases were re-admitted to the hospital one or more times for shunt revision (Colli et al. 1986). Removal of solitary brain cysticerci is often followed by prompt improvement and excellent recovery (Colli et al. 1994). In a series of 27 cases with cysticerci in the fourth ventricle submitted for direct surgical removal of the parasite, 81 percent had excellent results, but the remaining cases had poor results owing to severe arachnoiditis found at surgery (Loyo et al. 1980) (Figure 33.12). Recently, endoscopic excision of a cysticercus in the fourth ventricle via a frontal burr hole was performed. The patient had an uneventful recovery and remained asymptomatic and shunt free 1 year after the procedure. Follow-up MRI showed no signs of hydrocephalus and a normal fourth ventricle (Zymberg et al. 2003). The neuroendoscopic management of intra-ventricular cysticercosis should be considered as the primary treatment whenever possible since it is safe, effective, and provides a definitive treatment for this disorder. In addition to avoiding a CSF shunt in many cases, removal of the cyst(s) offers a reduced risk of inflammatory sequelae. Neurosurgeons with familiarity and experience with flexible neuroendoscopes should find these cases straightforward and highly gratifying (Bergsneider and Nieto 2003).

Parasites in the intraspinal subarachnoidal space require surgery because the compression causes spinal cord dysfunction. Neurocysticercosis can involve any cranial nerve; surgery is the most frequent indication for optic nerve involvement and removal of arachnoiditic tissue surrounding nerves will produce different degrees of clinical benefit (Madrazo and Flisser 1992; Escobedo 2000). Treatment of ocular cysticercosis usually involves removal of parasites from vitreous or subretinal locations. In some cases photocoagulation with a laser beam is used and in severe inflammation extrusion of the eye is recommended (Santos et al. 1979; Kruger-Leite et al. 1985; Cardenas et al. 1992). Muscle and subcutaneous cysticerci are efficaciously eliminated by cestocidal treatment (Rim and Joo 1989).

Surgical removal of single hydatid cysts due to *E. granulosus*, which can be performed in about 90 percent of patients, may have few complications and the best prognosis (WHO 1996). It is the preferred treatment when cysts are large (>10 cm diameter), secondarily infected, or located in the brain or the heart. The aim of surgery is total removal of the cyst carefully avoiding the

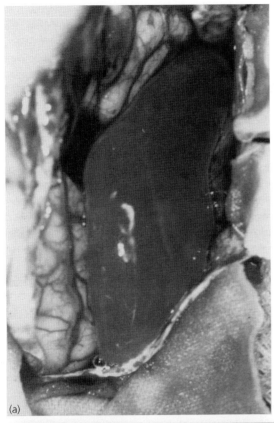

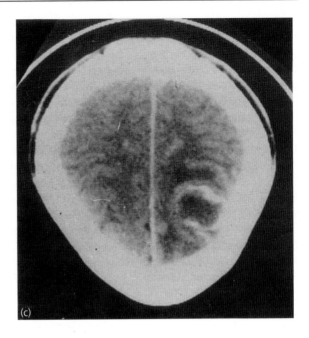

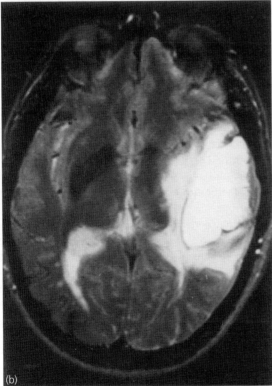

Figure 33.12 *Surgical removal of a huge brain cysticercus* **(a)** *that can be seen to have a similar size in MR* **(b)** *and as the end of the vesicle surrounded by an inflammatory reaction in CT* **(c)***. (by courtesy of Dr G. Zenteno)*

adverse consequences of spilling its contents. Pericystectomy is the usual procedure, but resection of the involved organ may be used depending on the location and condition of the cyst. Although surgery provides cure in a high percentage of cases there may be 2–25 percent recurrence (Kammerer and Schantz 1993; Ammann and Eckert 1995). Postoperative complications may occur in 10 to 25 percent of cases and mortality is around 2 percent, but may increase considerably with further operations. Surgical resection of the infected liver segment and of alveolar echinococcosis lesions from other affected organs is indicated in all operable cases even though it is impossible to know whether an operation has removed all parasite tissue. Liver transplantation has been performed in French AE cases with end stage liver failure; however, re-colonization of the donor liver in immunosuppressed patients was a significant problem (WHO 1996). It is now common practice for postoperative chemotherapy to be routinely carried out for at least 2 years after radical surgery, with careful monitoring of the patient over a minimum of 10 years for possible recurrence (Ammann and Eckert 1995). Because alveolar hydatid disease is often not diagnosed until the disease is advanced, the lesion is then inoperable and albendazole chemotherapy is the recommended treatment; however, efficacy may be variable (Polat et al. 2002).

Cestocidal treatment of cysticercosis is based on praziquantel (a synthetic acylated isoquinoline-pyrazine) and albendazole (a benzimidazole derivative). Pharmacokinetic and toxicological studies of both drugs indicate rapid absorption and, in general, no toxicological effects (Marriner et al. 1986; Frohberg 1989). Both drugs are useful for eliminating living cysticerci located in brain parenchyma and the subarachnoidal space (Cruz and Cruz 1991; Takayanagui and Jardim 1992), but the proper dosages have not yet been precisely defined. Praziquantel is usually used at 50 mg/kg body weight in three daily doses given with meals over 15 days with improvement in 70–96 percent of patients with cysticerci in brain parenchyma (Earnest et al. 1987; Madrazo and Flisser 1992; Kammerer and Schantz 1993; Garcia and Del Brutto 2000; Sotelo and Del Brutto 2000; White 2000). One-day treatment has also been successful (Bittencourt et al. 1990; Corona et al. 1996; Sotelo and Flisser 1997; Pretell et al. 2001a), especially in patients with a single or few viable brain cysticerci, but is poorly effective for multiple cysts, it can be optimized with a high carbohydrate diet (Lopez-Gómez et al. 2001; Pretell et al. 2001b). Albendazole was initially used at 15 mg/kg for 1 month, but in later protocols this was reduced to 7–8 days (García et al. 1997b). Efficacy of cestocidal treatment is measured by the reduction in number and/or size of cysticerci seen by imaging techniques, clinical improvement, withdrawal of corticoid or anti-epileptic treatment, and disappearance of ventricular dilatation. A case of myositis in disseminated muscular cysticercosis induced by praziquantel therapy (Takayanagui and Chimelli 1998) demonstrates that these effects are seemingly due to a strong inflammatory reaction and can be considered as a reliable indicator of drug effectiveness. Steroids should be used in this type of patients who show exacerbation of neurological symptoms or develop adverse reactions not related to drug toxicity. Headache, nausea, and seizures are common during treatment, but are usually transient and can be ameliorated with analgesic, anti-emetic, or antiepileptic drugs when reactions are mild, or with steroids when severe (Sotelo et al. 1985b; Del Brutto and Sotelo 1988). Pharmacokinetics of albendazole, in combination with dexamethasone or cimetidine, indicates that their presence decreases the rate of elimination of albendazole (Jung et al. 1990; Takayanagui et al. 1997). Albendazole therapy results in significantly faster and increased resolution of solitary cysticercus lesions and appears to reduce the risk of late onset epilepsy recurrences (Baranwal et al. 1998).

The need for cestocidal drugs has been questioned mainly because children appear not to warrant antiparasitic drug therapy (Mitchell and Snodgrass 1985; Mitchell and Crawford 1988). The controversy for treatment depends on the associated inflammatory reaction, as well as clinical and pathological features (Evans et al. 1997). As Evans and colleagues (1997) state: 'Although recommendations cannot yet be definitive, available evidence suggests that viable, intact cysticerci which cause epilepsy or other symptoms can be treated with cestocidal therapy, especially if they are causing mass effect.' Furthermore, there are several reasons why cestocidal treatment should always be considered: intraparenchymal single lesions can mimic neoplastic or other

infectious brain diseases; drugs are efficacious and well tolerated, side effects can be controlled, and some cases are seronegative even with the highly sensitive Western blot. Subarachnoidal cysticerci in the cerebral convexity or in the cisternas of the skull base should be surgically treated only when cestocidal or anti-inflammatory treatment has failed, especially when there is secondary hydrocephalus or diagnostic doubt (Madrazo and Flisser 1992; Sotelo and Del Brutto 2000). A recent study showed that children with neurocysticercosis who were only under antiepileptic treatment and had more than five lesions in active or transitional stages, showed a high seizure frequency, predicting a worse short-term prognosis (Ferreira et al. 2002). Even today, there are several controversies with regard to the medical treatment of cysticercosis. Most importantly, debate continues over the usefulness of anticysticercal drugs. Other issues, where opinion varies, include the specific drug – prazqiuantel versus albendazole – to be used; the drug dosage and duration of treatment; the specific role, indications, and duration of corticosteroid co-medication and antiepileptic drugs (Singhal and Salinas 2003).

Albendazole and mebendazole, in relatively high doses and over prolonged periods, can severely damage or kill protoscolices and cysts of E. granulosus in animals (Pérez-Serrano et al. 1994; Taylor and Morris 1988). When surgery is not recommended in CE because of the patient's condition and the extent and location of the cysts, cestocidal treatment should be used. Both albendazole (10 mg/kg body weight per day) and mebendazole (40–50 mg/kg) are effective. However, because of its superior intestinal absorption and penetration into the cysts, albendazole is slightly more efficacious. Adverse reactions (neutropenia, liver toxicity, alopecia, and others), reversible upon cessation of treatment, have been noted with both drugs (WHO 1996). A minimum period of treatment is 3 months, but the long-term prognosis in individual patients is difficult to predict, therefore, prolonged follow-up with ultrasound or other imaging procedures is needed to determine the eventual outcome. Recent trials have shown that, with both drugs, success rates for CE are variable. Between 20 and 70 percent of patients improved, but this included high proportions of inoperable and severe cases (Davis et al. 1989; Todorov et al. 1992; WHO 1996). In general, approximately one-third of CE patients treated with benzimidazole drugs have been cured of their disease, as seen by complete and permanent disappearance of cysts, and an even higher proportion have responded with significant regression of cyst size and alleviation of symptoms. Small (<7 mm diameter) isolated cysts, surrounded by minimal adventitial reaction, respond best, whereas complicated cysts with multiple compartments and daughter cysts, or with thick or calcified surrounding adventitial reactions, are relatively refractory to treatment. Difficult presentations such as bone cysts, however, may also respond favorably for individual patients (Bonifacino et al. 1997). CE disease responds more readily to chemotherapy in children than it does in adults (Goccmen et al. 1993; Kammerer and Schantz 1993; Ammann and Eckert 1995). The few available results of praziquantel treatment of cystic echinococcosis do not yet allow conclusions to be drawn (Craig 1994; Ammann and Eckert 1995), but the potential efficacy of praziquantel in association with albendazole needs critical assessment for cystic hydatid disease (Bonifacino et al. 1997). The use of cyclosporin A for postoperative control of secondary cystic echinococcosis has been suggested (Hurd et al. 1993). Recently hydatid cysts in infected gerbils were injected by a laparotomic procedure with ivermectin and found severely damaged and with no viable protoscolices between 44 and 58 days after injection (Ochieng-Mitula and Burt 1996). Co-administration of albendazole with cimetidine in Chinese CE patients significantly increased bile and cyst fluid concentrations of the active metabolite, albendazole sulfoxide, and resulted in improved therapeutic benefit (Wen et al. 1994).

Long-term treatment with mebendazole (50 mg/kg per day) or albendazole (10 mg/kg) inhibits growth of larval *E. multilocularis*, reduces metastasis, and enhances both the quality and length of survival; prolonged therapy may eventually be larvicidal in some patients, but recurrence of the disease also occurs (Ammann et al. 1993). In one study of 23 inoperable Turkish AE patients, a beneficial outcome was observed in eight cases, while nine others showed lesion progression, and none were considered cured (Polat et al. 2002). Liver transplantation has been employed successfully on otherwise terminal AE cases (WHO 1996). Because the lesions due to *E. vogeli* are so extensive, surgical resection is always difficult and usually incomplete and a combination of surgery with albendazole is most likely to be successful (Kammerer and Schantz 1993). The principles of management of cystic and alveolar echinococcosis also apply to polycystic echinococcosis.

Combinations of cyst puncture, aspiration, and drainage, with or without injection of chemicals (called percutaneous aspiration–injection–re-aspiration (PAIR)), have also been evaluated. Percutaneous puncture is performed under sonographic guidance, followed by aspiration of the liquid contents, instillation of a protoscolicidal agent (95 percent ethanol or 20 percent sodium chloride solution) and re-aspiration (Filice et al. 1990; Khuroo et al. 1993). To avoid sclerosing cholangitis, this procedure must not be performed in patients whose cysts have biliary communication. The presence of the latter can be determined by testing the cyst fluid for presence of bilirubin or by intraoperative cholangiogram. Mild anaphylaxis has been reported using PAIR, and care is required to prevent secondary microbial contamination when drainage tubes are temporarily left in place (WHO 1996; Wang et al. 2003). The

possibility of secondary echinococcosis resulting from accidental spillage during this procedure can be minimized by concurrent treatment with albendazole. Indeed, combining PAIR and chemotherapy may improve the results of either treatment alone. PAIR may have advantages over surgery particularly in less well equipped hospitals (WHO 1996; Filice and Brunetti 1997).

TRANSMISSION, EPIDEMIOLOGY, AND CONTROL

Human cysticercosis is a disease related to underdevelopment. It is present in countries that lack proper sanitary infrastructure and hygiene, as well as insufficient health education (Gemmell et al. 1983; Schantz et al. 1998), as exemplified by the emergence of neurocysticercosis in 1978 in West New Guinea where it became a disaster among the Ekari population, to whom the disease was unknown prior to the entrance of cysticercotic pigs as official gifts. Some 18–20 percent of the population acquired cysticercosis. The disease was detected by an epidemic of severe burns resulting from convulsions manifested while the people were sleeping around house fires. Individuals also had subcutaneous nodules (Muller et al. 1987). Between 1991 and 1995, a rapid increase in the number of cases of epileptic seizures and burns in Irian Jaya was found, 67 and 65 percent of persons with epileptic seizures and with subcutaneous nodules, respectively, showed antibody responses highly specific to cysticercosis, indicating that most cases of epilepsy and burns were associated with cysticercosis (Wandra et al. 2000). Furthermore, serological analysis of pigs in this region appears to be a good indicator of infection (Subahar et al. 2001). Human cysticercosis is still highly endemic in Brazil, Colombia, Ecuador, Guatemala, Mexico, Peru, China, India, Indonesia, New Guinea, Benin, Mozambique, South Africa, West Africa, Zimbabwe, Ile de la Reunion, and Madagascar. Cases are also reported in the USA, Honduras, Panama, Venezuela, Nepal, Portugal, Spain, and the UK, though these include imported cases (Flisser 1994; Chimelli et al. 1998; Ferrer et al. 2002; García et al. 1995, 1999; Garcia-Noval et al. 1996; Garcia and Del Brutto 2000; Takayanagui et al. 1996; Houinato et al. 1998; Schantz et al. 1998, Vilhena et al. 1999; Wandra et al. 2000).

Many epidemiological studies have shown a correlation between human cysticercosis, taeniasis, and epilepsy and between seropositive people, infected pigs, and disposal of feces (Diaz-Camacho et al. 1990; Michault et al. 1990; Sarti et al. 1992, 1994; Aranda-Alvarez et al. 1995; Cruz et al. 1995; García et al. 1995, 1997c, 1998b; Correa et al. 1999; Garcia-Garcia et al. 1999; Rodriguez-Canul et al. 1997; Sakai et al. 2001; Widdowson et al. 2000). The results of surveys have identified community, behavioral, and environmental practices that must be modified to prevent continued transmission of cysticercosis and taeniasis. Most importantly, these studies have shown that the main risk factor is the presence of a Taenia carrier in the immediate environment. Strawberries and flies were considered of high importance in the transmission of Taenia eggs, but recent publications demonstrated that both potential vectors did not carry Taenia eggs (Spindola-Felix et al. 1996; Martinez et al. 2000). A proposal to declare neurocysticercosis an international reportable disease has been published (Roman et al. 2000). This proposal, if taken into account, could be helpful in the control of cysticercosis, since, if cases of cysticercosis and of taeniosis are reported in all countries, it will provide accurate quantification of the incidence and prevalence of neurocysticercosis at regional level, thus permitting the rational use of resources in eradication campaigns.

Recent studies suggest that neurocysticercosis may be a risk factor for human cancer, 1 271 autopsy files were reviewed, 113 of which had a malignant hematological disease, of these 6 percent also had neurocysticercosis, while only 1 percent of those with other neoplasias ($n = 448$) and 2 percent of those that did not have any malignant neoplasia ($n = 710$) had neurocysticercosis (Herrera et al. 1999). The authors conclude that 'while it will be difficult to demonstrate that neurocysticercosis is a causal agent of malignant hematological diseases, it should be considered a potential risk factor for cancer induction in countries where cysticercosis remains a public health problem'. Several data support this hypothesis: chromosome aberrations induced in peripheral blood lymphocytes were higher in human patients with neurocysticercosis and in cysticercotic pigs as compared to those observed in the same cases after cestocidal treatment and in healthy controls (Flisser et al. 1990a; Montero et al. 1994; Herrera et al. 2000). Also, 17 percent of 43 patients with glioma, but only 3 percent of 172 controls had neurocysticercosis (Del Brutto et al. 1997). The apparent association of neurocysticercosis and oncogenesis was reviewed recently (Del Brutto and Dolezail 2000; Herrera and Ostrosky-Wegman 2000), that the parasite is capable of immunosuppressing the host's immune response has long ago been proposed, and recently data have been compiled (Flisser 1989; White et al. 1997).

Taeniosis, the infection with the adult stage of a tapeworm, is the main element that maintains the life cycle of Taenia in the environment, since tapeworms release permanently thousands of eggs inside gravid proglottids. Two tapeworms infect human beings: Taenia saginata and Taenia solium. It is known that only the second species causes human cysticercosis. The first one is, even nowadays, a cosmopolitan parasite. Its biological characteristics and life cycle have been thoroughly studied (Pawlowski and Schultz 1972), many of these studies referring to T. solium. Nevertheless, it is remarkably difficult to find human T. solium carriers, while those of T. saginata are easily detected. Coproparasitoscopic studies are generally used to identify carriers in epidemiological studies, but unless gravid proglottids or the scolex are found, it is impossible to differentiate between the species, unless molecular biology techniques are used (Chapman et al. 1995; Gasser et al. 1995; Gonzalez et al. 2002; Mayta et al. 2000). The establishment of experimental models (Maravilla et al. 1998) will hopefully support the knowledge of the host–parasite relationship in T. solium for example the understanding of the characteristics of the cellular and humoral immune response in the intestinal mucosa (Avila et al. 2002, 2003), but of special relevance is the understanding the life span of adult T. solium in field conditions for its control. Field characteristics of this parasite were studied in rural Guatemala showing a high prevalence of tapeworm carriers (Allan et al. 1996). A higher prevalence of taeniasis than previously reported, among patients with neurocysticercosis, was found in Peru. In addition, a clear association between the presence of taeniasis and the severity of neurocysticercosis was found. Most massive cerebral infections probably resulted from a maintained infective source in patients harboring the adult tapeworm in the intestine. Therefore, the perception that Taenia are silent guests causing no harm to humans is erroneous, and tapeworm carriers should be regarded as potential sources of contagion to themselves and to those living in their close environment (Gilman et al. 2000). A recent study performed in 68 asymptomatic individuals from Amerindians living in the Venezuelan state of Amazonas showed that 65 percent had circulating T. solium antigens and 77 percent had specific IgM antibodies, suggesting that recent exposure has occurred and active recent infections are present (Ferrer et al. 2002), therefore a control program adapted to local conditions, knowledge, and practices should be implemented.

The knowledge of the main risk factors has led to the evaluation of control measures (Flisser 2002b, 2003). Mass treatment against intestinal tapeworms provides quite good results (Cruz et al. 1989; Pawlowski 1990; Diaz-Camacho et al. 1991; Allan et al. 1997; Sarti et al. 2000), although in one study, a case of occult neurocysticercosis was exacerbated (Flisser et al. 1994). Targeted treatment, instead of mass drug administration, can be used since potential tapeworm carriers are identified by direct questioning or by detection of parasite antigens or eggs, as well as by association with late onset epilepsy (Sarti et al. 1992, 1994, 2000; Garcia-Noval et al. 1996). One intervention study based on a specially prepared health education strategy was very successful since 4 years and 6 months after the project ended no pigs with cysticercosis were identified in the whole community (Sarti et al. 1997). Apparently an economic factor facilitated the success, since people learned that by having pigs restrained in certain areas without access to human feces or garbage they would not acquire the disease and thus could be sold at a higher price. A third alternative for the control of cysticercosis is vaccination, especially because pigs are the only intermediate hosts that participate in the maintenance of the parasite in the environment. Many studies have been performed to evaluate the effectiveness of vaccination against larval cestodes (Lightowlers 1994, 1996; Lightowlers et al. 2000; Huerta et al. 2000; Flisser and Lightowlers 2001). Results show that it is possible to protect pigs from acquiring cysticercosis. The present challenge is to identify the optimal vaccination protocol and carriers for the vaccine. Meanwhile cysticercotic pigs can receive 1-day treatment with cestocidal drugs (Torres et al. 1992; Gonzalez et al. 1998) to interrupt transmission of *T. solium*.

Neurocysticercosis has been recognized in recent years as a significant public health problem due to the fact that the tapeworm carrier can migrate to countries where *Taenia solium* was previously absent. This has generated various review articles and books of interest for the international scientific community and governments (Del Brutto 1997; Flisser 1995; Flisser et al. 1998; San Esteban et al. 1997; Del Brutto and Sotelo 1988; Del Brutto et al. 1998; Schantz et al. 1998; Garcia and Martinez 1999; Garcia and Del Brutto 2000; Passos-Barbosa et al. 2000; Roman et al. 2000; Sciutto et al. 2000; Sotelo and Del Brutto 2000; White 2000; Singh and Prabhakar 2002; Craig and Pawlowski 2002), as well as special issues of scientific journals (Van der Kaay and Overbosch 1989; Murrell 2003).

The greatest prevalence of cystic echinococcosis is in temperate countries, including southern South America, the entire Mediterranean littoral, Eastern Europe, the Middle East, southern and central parts of the former Soviet Union, Central Asia, India, Nepal, China including Tibet, Australia, and parts of Africa. In the USA, most infections are seen in immigrants from countries in which the disease is highly endemic. Sporadic autochthonous transmission is currently recognized in Alaska, California, Utah, Arizona, and New Mexico. Geographic strains or genotypes of *E. granulosus* exist with different host affinities. The northern or sylvatic strain is maintained in wolves and wild cervids (moose and reindeer) in northern Alaska, Canada, Scandinavia, and Eurasia. Pastoral strains are maintained in dogs and domestic ungulates throughout the world. Populations of *E. granulosus* in different assemblages of domestic hosts (dog-sheep, dog-horse, dog-pig, dog-camel and dog-cattle) may differ morphologically, developmentally, biochemically, and possibly in infectivity and pathogenicity to humans. For example, the dog-sheep strain, the most widespread of the variants, is relatively pathogenic in humans in comparison to the northern sylvatic strain that occurs in wolf-wild cervid hosts. There is much evidence that the strain adapted to the dog-horse cycle, occurring in parts of Europe and the Middle East, rarely, if ever, infects humans. Probes characterizing the

nuclear and mitochondrial DNA of the variant populations provide reliable genetic markers to distinguish them; nine genotypes (G1–G9) have been identified. Globally, sheep are the most important intermediate hosts, but swine, cattle, buffalo, horses, and camels are more important in certain regions. Apart from the sheep strain (G1), human CE cases that have been DNA tested after surgical biopsy have been identified infected with *E. granulosus* of bovid, camel, or cervid genotypes (Pearson et al. 2002). Certain human activities, such as the widespread rural practice of feeding the viscera of home-butchered sheep to dogs, facilitate transmission of the sheep strain and consequently increase the risk that humans will become infected. People who practice seasonal movement of sheep between summer and winter pastures and other nomadic lifestyles appear to be at greater risk of transmission of *E. granulosus* (Macpherson and Craig 2000; Wang et al. 2001).

Genetic heterogeneity is much less evident in *E. multilocularis*, although some differences in nuclear and mitochondrial genes have suggested separation into the traditional view of Alaskan and Eurasian types (Rinder et al. 1997). Despite this, it is known that a great variety of small mammal species are susceptible to metacestode development (WHO/OIE 2001). Those small mammal species that are prone to rapid population increases either seasonally and/or pluri-annually (e.g. the vole species *Arvicola terrestris*, and *Microtus arvalis* in Europe, and *Microtus oeconomus* in Alaska, or the lagomorph *Ochotona curzonae* on the Tibetan plateau) appear to contribute most significantly to the parasitic life-cycle of *E. multilocularis*. That is because they become preferential prey for the fox definitive host (red fox in Europe, arctic fox in Alaska, and the Tibetan fox in China), and also may be eaten by domestic dogs if they occur in endemic communities (WHO/OIE 2001; Giraudoux et al. 2002). The availability of suitable habitats for small mammal hosts within upland temperate and holarctic landscapes also appears to be important in the ecology of transmission of *E. multilocularis* in sylvatic host cycles. Application of remote sensing and geographic information system (GIS) to quantify landscape cover appears to be useful in development of risk models for human AE at village level in China (Danson et al. 2002).

Dogs infected with *Echinococcus* spp. tapeworms pass proglottids and eggs in their feces, and humans become infected through fecal–oral contact, particularly in the course of playful and intimate contact between people and dogs. Eggs adhere to hairs around the infected dog's anus and are also found on the muzzle and paws. Indirect transfer of eggs, either through contaminated water and uncooked food or through the intermediary of flies and other arthropods, may also result in infection of humans. In *E. granulosus* endemic areas, personal preventive measures include careful hygiene, strict dietary regulation of pet dogs to preclude ingestion of livestock offal, and avoidance of dogs that are not so regulated. Periodic prophylactic treatment of pet dogs for intestinal echinococcosis may sometimes be necessary. Control measures applicable in communities include health education, regulation of livestock slaughtering in abattoirs and on farms, control of dogs, and periodic or regular mass treatments of dogs with praziquantel (5 mg/kg) to reduce the prevalence of *E. granulosus* below levels necessary for continued transmission (reviewed in Gemmell and Roberts 1998). A targeted campaign of education and surveillance based on specific antibody detection in dogs as an indicator of exposure to *E. granulosus* has been used in Western Australia (Thompson et al. 1993). The most successful national or provincial control programs for *E. granulosus* have, however, focused on dog registration and regular dosing of dogs at 6–12-week intervals using arecoline or preferably praziquantel in the 'attack phase', followed by slaughter-house surveillance of transmission using sheep hydatid prevalence rates. This approach at a national or regional level in Chile, Argentina, Falkland Islands (Islas Malvinas), New Zealand, and Tasmania has resulted in significant reduction in the incidence of human cystic echinococcosis within 10 years in some programs (Gemmell and Roberts 1998).

The life cycle of *E. multilocularis* involves foxes (principally the red fox and the arctic fox) and their small mammal prey, particularly microtine voles, in ecosystems generally separate from humans. There is often, however, an ecological overlap with humans, because domestic dogs or cats may become infected when they eat infected wild rodents. As a result, exposure of humans to *E. multilocularis* is relatively less common than exposure to *E. granulosus*. Alveolar echinococcosis has been reported in parts of central Europe, Turkey, and Iran, much of eastern Russia, the Central Asian Republics, and western China, Japan, the northwestern portion of Canada, and western Alaska. Wildlife transmission, involving red foxes, coyotes, and prairie voles, appears to be increasing in central North America. Hunters, trappers, and persons who work with fox fur may be exposed to AE, but the main risk factors appear to be farming or agricultural occupations and a history of dog ownership (Stehr-Green et al. 1988; Eckert and Deplazes 1999; Craig et al. 2000; Kern et al. 2003). Hyperendemic foci have been described in some Inuit villages of the Alaskan tundra and also in China, where both communities owned dogs regularly fed on infected commensal rodents. The epidemiology of alveolar echinococcosis in China remains one of the major challenges in temperate areas with human prevalences of 2–15 percent in some communities (Craig et al. 1992, 2000). Eliminating *E. multilocularis* from its wild animal hosts is virtually impossible; therefore, contact with dogs and foxes in areas where the infection is endemic should be avoided. Preventing infections in humans depends on education to improve hygiene and sanitation. Infections in dogs and cats prone to eat infected rodents can be prevented by monthly treatments with praziquantel (Kammerer and Schantz 1993; Rausch 1995). Use of fox baits containing praziquantel has had some success in reducing fox prevalence rates of *E. multilocularis* in focal endemic areas of Germany (Tackmann et al. 2001).

Very little is known about the circumstances associated with polycystic hydatid disease due to *E. vogeli*. Bush dogs (*Speothus venaticus*) are rare and avoid man, and therefore probably play little role in transmission to humans. In endemic areas, infections are probably acquired from the feces of domestic dogs that have been fed on rodent viscera of infected bush rodents (e.g. pacas), a practice that has been reported commonly by patients (Meneghelli et al. 1992; D'Alessandro 1997).

REFERENCES

Aguilar-Rebolledo, F., Cedillo-Rivera, R., et al. 2001. Interleukin levels in cerebrospinal fluid from children with neurocysticercosis. *Am J Trop Med Hyg*, **64**, 35–40.

Ali-Khan, Z. and Siboo, R. 1980. Pathogenesis and host response in subcutaneous alveolar hydatidosis. I. Histogenesis of alveolar cyst and a qualitative analysis of the inflammatory infiltrates. *Z Parasitenk*, **62**, 241–4.

Ali-Khan, Z., Siboo, R., et al. 1983. Cystolytic events and the possible role of germinal cells in metastasis in chronic alveolar hydatidosis. *Ann Trop Med Parasitol*, **77**, 497–512.

Allan, D., Jenkins, P., et al. 1981. A study of immunoregulation of BALB/c mice by *Echinococcus granulosus equinus* during prolonged infection. *Parasite Immunol*, **3**, 137–42.

Allan, J.C., Velasquez-Tohom, M., et al. 1996. Epidemiology of intestinal taeniasis in four, rural, Guatemalan communities. *Ann Trop Med Parasitol*, **90**, 157–65.

Allan, J.C., Velasquez-Tohom, M., et al. 1997. Mass chemotherapy for intestinal *Taenia solium* infection: effect of prevalence in humans and pigs. *Trans R Soc Trop Med Hyg*, **91**, 595–8.

Almeida-Pinto, J., Veiga-Pires, J.A., et al. 1988. Cysticercosis of the brain. The value of computed tomography. *Acta Radiol*, **29**, 625–8.

Ammann, R. and Eckert, R. 1995. Clinical diagnosis and treatment of echinococcosis in humans. In: Thompson, R.C.A. and Lymbery, A.J. (eds), *Echinococcus and hydatid disease*. Wallingford, UK: CAB International, 411–51.

Ammann, R., Ilitsch, N., et al. 1993. Effect of chemotherapy on the larval mass and on the long-term course of alveolar echinococcosis. *Hepatology*, **19**, 735–42.

Ambrosio, J., Cruz-Rivera, M., et al. 1997. Identification and partial characterization of a myosin-like protein from cysticerci and adults of *Taenia solium* using a monoclonal antibody. *Parasitology*, **114**, 545–53.

Ambrosio, J., Reynoso-Ducoing, O., et al. 2003. Actin expression in *Taenia solium* cysticerci (cestoda): tisular distribution and detection of isoforms. *Cell Biol Int*, **27**, 727–33.

Aranda-Alvarez, J.G., Tapia-Romero, R., et al. 1995. Human cysticercosis: risk factors associated with circulating serum antigens in an open community of San Luis Potosi, Mexico. *Ann Trop Med Parasitol*, **89**, 689–92.

Avila, G., Aguilar, L., et al. 2002. Inflammatory response in the intestinal mucosa of gerbils and hamsters experimentally infected with the adult stage of *Taenia solium*. *Int J Parasitol*, **32**, 1301–8.

Avila, G., Benítez, M., et al. 2003. Kinetics of *Taenia solium* antibodies and antigens in experimental taeniosis. *Parasitol Res*, **89**, 284–9.

Baily, G.G., Mason, P.R., et al. 1988. Serological diagnosis of neurocysticercosis: evaluation of ELISA tests using cyst fluid and other components of *Taenia solium* cysticerci as antigens. *Trans R Soc Trop Med Hyg*, **82**, 295–9.

Baranwal, A.K., Singhi, P.D., et al. 1998. Albendazole therapy in children with focal seizures and single small enhancing computerized tomographic lesions: a randomised, placebo-controlled, double blind trial. *Pediatr Infect Dis J*, **17**, 696–700.

Bardonnet, K., Piarroux, R., et al. 2002. Combined eco-epidemiological and molecular biology approaches to assess *Echinococcus granulosus* transmission to humans in Mauritania: occurrence of the 'camel strain' and human cystic echinococcosis. *Trans R Soc Trop Med Hyg*, **96**, 383–6.

Bartholomot, B., Vuitton, A., et al. 2002. Combined ultrasound and serologic screening for hepatic alveolar echinococcosis in central China. *Am J Trop Med Hyg*, **66**, 23–9.

Basset, D., Girou, I.P., et al. 1998. Neotropical echinococcosis in Suriname: *Echinococcus oligarthrus* in the orbit and *Echinococcus vogeli* in the abdomen. *Am J Trop Med Hyg*, **59**, 787–90.

Bergsneider, M. and Nieto, J.H. 2003. Endoscopic management of intraventricular cysticercosis. In: Singh, G. and Prabhakar, S. (eds), *Taenia solium cysticercosis: from basic to clinical science*. Oxon, UK: CABI Publishing, 399–410.

Bittencourt, P.R.M., Gracia, C.M., et al. 1990. High-dose praziquantel for neurocysticercosis: efficacy and tolerability. *Eur Neurol*, **30**, 229–34.

Bonifacino, R., Dogliani, E. and Craig, P.S. 1997. Albendazole treatment and serological follow-up in hydatid disease of bone. *Int Orthop*, **21**, 127–32.

Botero, D., Tanowitx, H.B., et al. 1993. Taeniasis and cysticercosis. *Parasit Dis*, **7**, 683–97.

Bowles, J. and McManus, D.P. 1993. Molecular variation in *Echinococcus*. *Acta Trop*, **53**, 291–305.

Bowles, J. and McManus, D.P. 1994. Genetic characterization of the Asian *Taenia*, a newly described taeniid of humans. *Am J Trop Med Hyg*, **50**, 33–44.

Bowles, J., Blair, D. and McManus, D.P. 1995. A molecular phylogeny of the genus *Echinococcus*. *Parasitology*, **110**, 317–28.

Bresson-Hadni, S., Laplante, J.J., et al. 1994. Seroepidemiological screening of *Echinococcus multilocularis* infection in a European area endemic for alveolar echinococcosis. *Am J Trop Med Hyg*, **51**, 837–46.

Cardenas, F., Plancarte, A., et al. 1989. *Taenia crassiceps*: experimental model of intraocular cysticercosis. *Exp Parasitol*, **69**, 324–9.

Cardenas, F., Quiroz, H., et al. 1992. *Taenia solium* ocular cysticercosis: findings in 30 cases. *Ann Ophthalmol*, **24**, 25–8.

Casagranda, B.E., Vaz, A.J., et al. 2000. Specific *Taenia crassiceps* and *Taenia solium* antigenic peptides for neurocysticercosis immunodiagnosis using serum samples. *J Clin Microbiol*, **38**, 146–51.

Cervantes-Vazquez, M., Correa, D., et al. 1990. Respiratory changes associated with the in vitro evagination of *Taenia solium* cysterci. *J Parasitol*, **76**, 108–12.

Chang, K.H., Kim, W.S., et al. 1988. Comparative evaluation of brain CT and ELISA in the diagnosis of neurocysticercosis. *AJNR*, **9**, 125–30.

Chapman, A., Vallejo, V., et al. 1995. Isolation and characterization of species-specific DNA probes from *Taenia solium* and *Taenia saginata* and their use in an egg detection assay. *J Clin Microbiol*, **33**, 1283–8.

Chen, J.P., Zhang, X.Y., et al. 1991. Determination of circulating antigen in cysticercosis patients using McAb-based ELISA. *Chung Kuo Chin Sheng Chung Hsueh Yu Chi Sheng Chung Ping Tsa Chih*, **9**, 122–5.

Chimelli, L., Lovalho, A.F. and Takayanagui, O.M. 1998. Neurocisticercose. Contribuicao da necropsia na consolidacao da notificao compulsoria em ribeirao preto-sp. *Arq Neuro Psiq*, **56**, 577–84.

Cho, S.Y., Kim, S.I., et al. 1986. Evaluation of enzyme-linked immunosorbent assay in serological diagnosis of human neurocysticercosis using paired samples of serum and cerebrospinal fluid. *Korean J Parasitol*, **24**, 25–41.

Choji, K., Fujita, N., et al. 1992. Alveolar hydatid disease of the liver: computed tomography and transabdominal ultrasound with histopathological correlation. *Clin Radiol*, **46**, 97–103.

Choromanski, L., Estrada, J.J. and Kuhn, R.E. 1990. Detection of antigens of larval *Taenia solium* in the cerebrospinal fluid of patients with the use of HPLC and ELISA. *J Parasitol*, **76**, 69–73.

Chow, C., Gauchi, C.G., et al. 2001. A gene family expressing a host-protective antigen of *Echinococcus granulosus*. *Mol Biochem Parasitol*, **118**, 83–8.

Chung, J.Y., Balik, Y.Y., et al. 1999. A recombinant 10 kDa protein of *Taenia solium* metacestodes specific to active neurocysticercosis. *J Infect Dis*, **180**, 1307–15.

Cohen, H., Paolillo, E., et al. 1998. Human cystic echinococcosis in a Uruguayan community: A sonographic, serologic and epidemiologic study. *Am J Trop Med Hyg*, **59**, 620–7.

Colli, B.O., Martelli, N., et al. 1986. Results of surgical treatment of neurocysticercosis in 69 cases. *J Neurosurg*, **65**, 309–15.

Colli, B.O., Martelli, N., et al. 1994. Cysticercosis in the central nervous system. *Arq Neuropsiquiatr*, **52**, 166–86.

Conder, G.A., Andersen, F.L. and Schantz, P.M. 1980. Immunodiagnostic tests for hydatidosis in sheep: an evaluation of double diffusion, immunoelectrophoresis, indirect hemagglutination and intradermal tests. *J Parasitol*, **66**, 577–84.

Corona, T., Lugo, R., et al. 1996. Single day praziquantel therapy for neurocysticercosis. *N Engl J Med*, **334**, 125.

Correa, D., Dalma, D., et al. 1985. Heterogeneity of humoral immune components in human cysticercosis. *J Parasitol*, **71**, 533–41.

Correa, D., Laclette, J.P., et al. 1987. Heterogeneity of *Taenia solium* cysticerci obtained from different naturally infected pigs. *J Parasitol*, **73**, 443–5.

Correa, D., Plancarte, A., et al. 1989a. Immunodiagnosis of human and porcine cysticercosis. Detection of antibodies and parasite products. *Acta Leidensia*, **57**, 93–100.

Correa, D., Sandoval, M.A., et al. 1989b. Human neurocysticercosis: comparison of enzyme-immunoassay capture techniques based on monoclonal and polyclonal antibodies for the detection of parasite products in cerebrospinal fluid. *Trans R Soc Trop Med Hyg*, **83**, 814–16.

Correa, D., Sarti, E., et al. 1999. Antigens and antibodies in sera from human cases of epilepsy or taeniasis from an area of México where *Taenia solium* cysticercosis is endemic. *Ann Trop Med Parasitol*, **93**, 69–74.

Cosentino, C., Velez, M., et al. 2002. Cysticercosis lesions in basal ganglia are common but clinically silent. *Clin Neurol Neurosurg*, **104**, 57–60.

Craig, P.S. 1994. Current research in echinococcosis. *Parasitol Today*, **10**, 209–11.

Craig, P. and Pawlowski, Z. (eds) 2002. *Cestode zoonoses: Echinococcosis and cysticercosis, an emergent and global problem*. 2002. *NATO Science Series*. vol. 341. Amsterdam: IOS Press, 393.

Craig, P.S., Liu, D., et al. 1992. A large focus of human alveolar echinococcosis in central China. *Lancet*, **340**, 826–31.

Craig, P.S., Rogan, M.T. and Allan, J.C. 1996. Detection, screening and community epidemiology of taeniid cestode zoonoses: cystic echinococcosis, alveolar echinococcosis and neurocysticercosis. *Adv Parasitol*, **38**, 169–250.

Craig, P.S., Giraudoux, P., et al. 2000. An epidemiological and ecological study of human alveolar echinococcosis transmission in south Gansu. *Acta Trop*, **77**, 167–77.

Cruz, M. and Cruz, I. 1991. Albendazole versus praziquantel in the treatment of cerebral cysticercosis: clinical evaluation. *Trans R Soc Trop Med Hyg*, **85**, 224–47.

Cruz, M., Davis, A., et al. 1989. Operational studies on the control of *Taenia solium* taeniasis/cysticercosis in Ecuador. *Bull WHO*, **67**, 401–7.

Cruz, I., Cruz, M.E., et al. 1994. Human subcutaneous *Taenia solium* cysticercosis in an Andean population with neurocysticercosis. *Am J Trop Med Hyg*, **51**, 405–7.

Cruz, M., Cruz, I., et al. 1995. Headache and cysticercosis in Ecuador, South America. *Headache J*, **35**, 93–7.

Daeki, A.O., Craig, P.S. and Shambesh, M.K. 2000. IgG-subclass antibody responses and the natural history of hepatic cystic echinococcosis in asymptomatic patients. *Ann Trop Med Parasit*, **94**, 319–28.

D'Alessandro, A. 1997. Polycystic echinococcosis in tropical America: *Echinococcus vogeli* and *E. oligarthrus*. *Acta Tropica*, **67**, 43–65.

D'Alessandro, A., Rausch, R.L., et al. 1979. *Echinococcus vogeli* in man, with a review of polycystic hydatid disease in Colombia and neighbouring countries. *Am J Trop Med Hyg*, **28**, 303–17.

Danson, F.M., Craig, P.S., et al. 2002. Satellite remote sensing and geographical information systems for risk modelling of alveolar echinococcosis. In: Craig, P.S. and Pawlowski, Z. (eds), *Cestode zoonoses: Echinococcosis and cysticercosis an emergent and global problem.* , *NATO Science Series.* vol. 341. Amsterdam: IOS Press, 237–48.

Davis, A., Dixon, H. and Pawlowski, Z.S. 1989. Multicentre clinical trial of benzimidazole-carbamates in human cystic echinococcosis. *Bull WHO*, **67**, 503–8.

Del Brutto, O.H. 1997. Neurocysticercosis. *Curr Opin Neurol*, **10**, 268–74.

Del Bruto, O.H. and Dolezail, M. 2000. Neurocysticercosis and oncogenesis. *Arch Med Res*, **31**, 151–5.

Del Brutto, O.H. and Sotelo, J. 1988. Neurocysticercosis: an update. *Rev Infect Dis*, **6**, 1075–87.

Del Brutto, O.H., Castillo, P.R., et al. 1997. Neurocysticercosis among patients with cerebral glioma. *Arch Neurol*, **54**, 1125–8.

Del Brutto, O.H., Sotelo, J. and Roman, G. 1998. *Neurocysticercosis: a clinical handbook.* The Netherlands: Lisser, Swets & Zeitlinger, 207 pp.

DelBrutto, O.H., Rajshekhar, V., et al. 2001. Proposed diagnostic criteria for neurocysticercosis. *Neurology*, **57**, 177–83.

Dixon, J.B. and Jenkins, P. 1995. Immunology of mammalian metacestode infections II. Immune recognition and effector function. *Helminth Abstr*, **64**, 559–661.

Dixon, H.B.F. and Lipscomb, F.M. 1961. Cysticercosis: an analysis and follow up of 450 cases. *Privy Council Med Res Special Rep Ser*, **229**, 1–58.

Diaz-Camacho, S., Candil, A., et al. 1990. Serology as an indicator of *Taenia solium* tapeworm infections in a rural community in Mexico. *Trans R Soc Trop Med Parasitol*, **84**, 563–6.

Diaz-Camacho, S., Candil, A., et al. 1991. Epidemiologic study and control of *Taenia solium* infections with praziquantel in a rural village of Mexico. *Am J Trop Med Hyg*, **45**, 522–31.

Dorny, P., Brandt, J., et al. 2003. Immunodiagnostic tools for human and porcine cysticercosis. *Acta Trop*, **87**, 79–86.

Earnest, M.P., Reller, L.B., et al. 1987. Neurocysticercosis in the United States: 35 cases and a review. *Rev Infect Dis*, **9**, 961–79.

Eckert, J. and Deplazes, P. 1999. Alveolar echinococcosis in humans: The current situation in Central Europe and need for counter-measures. *Parasitol Today*, **15**, 315–19.

Eckert, J., Thompson, R.C.A. and Mehlhorn, H. 1983. Proliferation and metastases formation of larval *Echinococcus multilocularis*. I. Animal model, macroscopical and histological findings. *Z Parasitenk*, **69**, 737–48.

Escobar, A. 1983. The pathology of neurocysticercosis. In: Palacios, E., Rodriguez-Carbajal, J. and Taveras, J.M. (eds), *Cysticercosis of the central nervous system.* Springfield IL: Thomas, 27–54.

Escobedo, F. 2000. Neurosurgical aspects of neurocysticercosis. In: Schmidek, H.H. (ed.), *Operative neurosurgical techniques*, 4th edn. Philadelphia: WB Saunders Co, 1756–68.

Espinoza, B. and Flisser, A. 1986. Antigenos especificos y de reaccion cruzada de helmintos parasitos. *Arch Invest Med (Mex)*, **17**, 299–311.

Espinoza, B., Ruiz-Palacios, G., et al. 1986. Characterization by enzyme linked immunosorbent assay of the humoral immune response in patients with neurocysticercosis and its application in immunodiagnosis. *J Clin Microbiol*, **24**, 536–41.

Estanol, B., Corona, T. and Abad, P. 1986. A prognostic classification of cerebral cysticercosis: therapeutic implications. *J Neurol Neurosurg Psychiat*, **49**, 1131–4.

Evans, C.A.W., Garcia, H.H., et al. 1997. Controversies in the management of cysticercosis. *Emerg Infect Dis*, **3**, 403–5.

Evans, C.A.W., García, H.H., et al. 1998. Elevated concentration of eotaxin and interleukin-5 in human neurocysticercosis. *Infect Immun*, **66**, 4522–5.

Fairweather, Y., McMullan, M.T., et al. 1994. Serotoninergic and peptidergic nerve elements in the protoscolex of *Echinococcus granulosus* (Cestoda, Cyclophyllidea). *Parasitol Res*, **80**, 649–56.

Feldman, M., Plancarte, A., et al. 1990. Comparison of two assays (EIA and EITB) and two samples (saliva and serum) for the diagnosis of neurocysticercosis. *Trans R Soc Trop Med Hyg*, **84**, 559–62.

Fernandez, C. and Hormaeche, C.E. 1994. Isolation and biochemical characterisation of a glutathione S-transferase from *Echinococcus granulosus* protoscoleces. *Int J Parasitol*, **24**, 1063–6.

Ferreira, H.B. and Zaha, A. 1994. Expression and analysis of the diagnostic value of an *Echinococcus granulosus* antigen gene clone. *Int J Parasitol*, **24**, 863–70.

Ferreira, L.S., Min, L.F., et al. 2002. Number and viability of parasite influence seizure frequency in children with neurocysticercosis. *Arq Neuropsiquiatr*, **60**, 909–11.

Ferrer, E., Cortez, M.M., et al. 2002. Serological evidence for recent exposure to *Taenia solium* in Venezuelan Amerindians. *Am J Trop Med Hyg*, **66**, 170–4.

Filice, C. and Brunetti, E. 1997. Use of PAIR in human cystic echinococcosis. *Acta Trop*, **64**, 95–107.

Filice, C., Pirola, F., et al. 1990. A new therapeutic approach for hydatid liver cysts. Aspiration and alcohol injection under sonographic guidance. *Gastroenterology*, **98**, 1366–8.

Flisser, A. 1988. Neurocysticercosis in Mexico. *Parasitol Today*, **4**, 131–7.

Flisser, A. 1989. *Taenia solium* cysticercosis: some mechanisms of parasite survival in immunocompetent hosts. *Acta Leiden*, **57**, 259–63.

Flisser, A. 1994. Taeniasis and cysticercosis due to *Taenia solium*. In: Sun, T. (ed.), *Progress in clinical parasitology*. Boca Raton, FL: CRC Press, 77–116.

Flisser, A. 1995. *Taenia solium, Taenia saginata* and *Hymenolepis nana*. In: Farthing, M.J.G., Keusch, G.T. and Wakelin, D. (eds), *Enteric infection.* , *Intestinal helminths.* vol. 2. London: Chapman and Hall, 173–89.

Flisser, A. 2002a. Epidemiological studies of taeniosis and cysticercosis in Latin America. In: Craig, P. and Pawlowski, Z. (eds), *Cestode zoonoses: echinococcosis and cysticercosis.* , *NATO Science Series.* vol. 341. Amsterdam: IOS Press, 3–11.

Flisser, A. 2002b. Risk factors and control measures for taeniosis/cysticercosis. In: Craig, P. and Pawlowski, Z. (eds), *Cestode zoonoses: echinococcosis and cysticercosis.* , *NATO Science Series.* vol. 341. Amsterdam: NATO Science Series, 335–42.

Flisser, A. 2003. Neurocysticercosis: regional status, epidemiology, impact and control measures in the Americas. *Acta Tropica*, **87**, 43–51.

Flisser, A. and Larralde, C. 1986. Cysticercosis. In: Walls, K.W. and Schantz, P.M. (eds), *Immunodiagnosis of parasitic diseases*. Orlando, FL: Academic Press, 109–61.

Flisser, A. and Lightowlers, M.W. 2001. Vaccination against *Taenia solium* cysticercosis. *Mem Inst Oswaldo Cruz*, **96**, 353–6.

Flisser, A. and Madrazo, I. 1996. Evagination of *Taenia solium* in the fourth ventricle. *N Engl J Med*, **335**, 753–4.

Flisser, A., Woodhouse, E. and Larralde, E. 1980. Human cysticercosis: antigens, antibodies and non-responders. *Clin Exp Immunol*, **39**, 27–37.

Flisser, A., Espinoza, B., et al. 1986. Host–parasite relationship in cysticercosis: immunologic study in different compartments of the host. *Vet Parasitol*, **20**, 95–202.

Flisser, A., Madrazo, I., et al. 1988. Comparative analysis of human and porcine neurocysticercosis by computed tomography. *Trans R Soc Trop Med Hyg*, **82**, 739–42.

Flisser, A., Gonzalez, D., et al. 1990a. Praziquantel treatment of brain and muscle porcine *Taenia solium* cysticercosis. 2. Immunological and cytogenetic studies. *Parasitol Res*, **76**, 640–2.

Flisser, A., Gonzalez, D., et al. 1990b. Praziquantel treatment of porcine brain and muscle *Taenia solium* cysticercosis. 1. Radiological, physiological and histopathological studies. *Parasitol Res*, **76**, 263–9.

Flisser, A., Madrazo, I., et al. 1994. Neurological symptoms in occult neurocysticercosis after a single taenicidal dose of praziquantel. *Lancet*, **342**, 748.

Flisser, A., Madrazo, I. and Delgado, H. 1998. *Cisticercosis humana*. México DF: El Manual Moderno, 176 pp.

Flisser, A., Correa, D. and Evans, C.A.W. 2003. *Taenia solium* cysticercosis: new and revisited immunological aspects. In: Singh, G. and Prabhakar, S. (eds), *Taenia solium cysticercosis: from basic to clinical science*. Oxon, UK: CABI Publishing, 15–24.

Frayha, G.J. and Haddad, R. 1980. Comparative chemical composition of protoscoleces and hydatid cyst fluid of *Echinococcus granulosus* (Cestoda). *Int J Parasitol*, **10**, 359–64.

Frohberg, H. 1989. The toxicological profile of praziquantel in comparison to other anthelminthic drugs. *Acta Leiden*, **57**, 201–15.

Frosch, P.M., Muhlschlegel, F. and Sygulla, L. 1994a. Identification of a cDNA clone from the larval stage of *Echinococcus granulosus* with homologies to the *E. multilocularis* antigen EM10-expressing cDNA clone. *Parasitol Res*, **80**, 703–5.

Frosch, P., Hartmann, M., et al. 1994b. Sequence heterogeneity of echinococcal antigen B. *Mol Biochem Parasitol*, **64**, 171–5.

Garcia-Allan, C., Martinez, N., et al. 1996. Immunocharacterization of *Taenia solium* oncosphere and metacestode antigens. *J Helminthol*, **70**, 271–80.

Garcia-Garcia, M.L., Torres, M., et al. 1999. Prevalence and risk of cysticercosis and taeniasis in an urban population of soldiers and their relatives. *Am J Trop Med Hyg*, **61**, 386–9.

Garcia, H.H. and Del Brutto, O.H. 2000. *Taenia solium* cysticercosis. *Infect Dis Clin N Am*, **14**, 97–119.

Garcia, H.H. and Del Brutto, O.H. 2003. Imaging findings in neurocysticercosis. *Acta Trop*, **87**, 71–8.

Garcia, H.H. and Martinez, S.M. (eds) 1999. Taenia solium *taeniasis/cisticercosis*. Lima, Peru: Editorial Universo, 346 pp.

Garcia, H.H., Gilman, R., et al. 1993. Cysticercosis as a major cause of epilepsy in Peru. *Lancet*, **341**, 197–9.

García, H.H., Gilman, R.H., et al. 1995. Factors associated with *Taenia solium* cysticercosis: Analysis of 946 Peruvian neurologic patients. *Am J Trop Med Hyg*, **52**, 145–8.

García, H.H., Gilman, R.H., et al. 1997a. Clinical significance of neurocysticercosis in endemic villages. *Trans R Soc Trop Med Hyg*, **91**, 176–8.

García, H.H., Gilman, R.H., et al. 1997b. Serologic evolution of neurocysticercosis patients after antiparasitic therapy. *J Infect Dis*, **175**, 486–9.

García, H.H., Gilman, R.H., et al. 1997c. Albendazole therapy for neurocysticercosis: a prospective double-blind trial comparing 7 versus 14 days of treatment. *Neurology*, **48**, 1421–7.

Garcia, H.H., Harrison, L.J.S., et al. 1998a. A specific antigen-detection ELISA for the diagnosis of human neurocysticercosis. *Trans R Soc Trop Med Hyg*, **92**, 411–14.

Garcia, H.H., Araoz, R., et al. 1998b. Increased prevalence of cysticercosis and taeniasis among professional fried pork vendors and the general population of a village in the Peruvian highlands. *Am J Trop Med Hyg*, **59**, 902–5.

Garcia, H.H., Gilman, R.H., et al. 1999. Human and porcine *Taenia solium* infection in a village in the highlands of Cusco. *Peru Acta Trop*, **73**, 31–6.

Garcia, H.H., Parkhouse, R.M.E., et al. 2000. Serum antigen detection in the diagnosis, treatment and follow-up of neurocysticercosis patients. *Trans R Soc Trop Med Hyg*, **94**, 673–6.

Garcia, H.H., Gonzalez, A.E., et al. 2001. Transient antibody response in *Taenia solium* infection in field conditions – a major contributor to high seroprevalence. *Am J Trop Med Hyg*, **65**, 31–2.

Garcia, H.H., Evans, C.A.W., et al. 2002a. Current consensus guidelines for treatment of neurocysticercosis. *Clin Microbiol Rev*, **15**, 747–56.

Garcia, H.H., Gonzalez, A.E., et al. 2002b. Circulating parasite antigens in patients with hydrocephalus secondary to neurocysticercosis. *Am J Trop Med Hyg*, **66**, 427–30.

Garcia, H.H., Pretell, J. and Gilman, R.H. 2002c. Neurocysticercosis and the global world. *J Neurol*, **249**, 1107–8.

Garcia-Noval, J., Allan, J.C., et al. 1996. Epidemiology of *Taenia solium* taeniasis and cysticercosis in two rural Guatemalan communities. *Am J Trop Med Hyg*, **55**, 282–9.

Gasser, R.B., Chilton, N.B., et al. 1995. Characterisation of taeniid cestode species by PCR-RFLP of ITS2 ribosomal DNA. *Acta Trop*, **59**, 31–40.

Gauci, C.G. and Lightowlers, M.W. 2001. Alternative splicing and sequence diversity of transcripts from the oncosphere stage of *Taenia solium* with homology to the 45W antigen of *Taenia ovis*. *Mol Bioch Parasitol*, **112**, 173–81.

Gauci, C., Merli, M., et al. 2002. Molecular cloning of a vaccine antigen against infection with the larval stage of *Echinococcus multilocularis*. *Infect Immun*, **70**, 3969–72.

Gazarian, K.G., Gazarian, T.G., et al. 2000. Epitope mapping on N-terminal region of *Taenia solium* paramyosin. *Immunol Lett*, **72**, 191–5.

Gazarian, K.G., Rowley, M.J., et al. 2001. Post-panning computer-aided analysis of phagotope collections selected with neurocysticercosis patient polyclonal antibodies: separation of disease-relevant and irrelevant peptide sequences. *Comb Chem High Throughput Screen*, **4**, 221–35.

Gemmell, M.A. and Lawson, J.R. 1986. Epidemiology and control of hydatid disease. In: Thompson, R.C.A. (ed.), *The biology of echinococcus and hydatid disease*. London: Allen and Unwin, 189–216.

Gemmell, M.A. and Roberts, M.G. 1998. Cystic echinococcosis (*Echinococcus granulosus*). In: Palmer, S.R., Soulsby, L. and Simpson, I.H. (eds), *Zoonoses*. Oxford: Oxford University Press, 665–88.

Gemmell, M., Matyas, Z., et al. 1983. *Guidelines for surveillance, prevention and control of taeniasis/cysticercosis. VPH/83.49*. Geneva: World Health Organization.

Gharbi, H.A., Hassine, W. and Brauner, M.W. 1981. Ultrasound examination of the hydatic liver. *Radiology*, **139**, 459–63.

Giraudoux, P., Delattre, P., et al. 2002. Transmission ecology of Echinococcus multilocularis in wildlife: what can be learned from comparative studies and multiscale approaches? In: Craig, P.S. and Pawlowski, Z. (eds), *Cestode zoonoses: echinococcosis and cysticercosis an emergent and global problem.*, *NATO Science Series*. 341. Amsterdam: IOS Press, 251–66.

Goccmen, A., Toppare, M.F. and Kiper, N. 1993. Treatment of hydatid disease in childhood with mebendazole. *Eur Resp J*, **6**, 253–7.

Godot, V., Harraga, S., et al. 2000. Resistance/susceptibility to *Echinococcus multilocularis* infection and cytokine profile in humans II. Influence of the HLA B8 DR3 haplotype. *Clin Exp Immunol*, **121**, 491–8.

Goldberg, A.S., Heiner, D.C., et al. 1981. Cerebrospinal fluid IgE and the diagnosis of cerebral cysticercosis. *Bull Los Angeles Neurol Soc*, **46**, 21–5.

Gomez-Leal, A. 1989. Cisticercosis del globo ocular y sus anexos. In: Flisser, A. and Malagon, F. (eds), *Cisticercosis humana y porcina, su conocimiento e investigacion en Mexico*. Mexico DF: Limusa-Noriega, 129–39.

Gonzalez, A.E., Falcon, N., et al. 1998. Time–response curve of oxfendazole in the therapy of swine cysticercosis. *Am J Trop Med Hyg*, **59**, 832–6.

Gonzalez, A.E., Gavidia, C., et al. 2001. Protection of pigs with cysticercosis from further infections after treatment with oxfendazole. *Am J Trop Med Hyg*, **65**, 15–18.

Gonzalez, G., Nieto, A., et al. 1996. Two different 8 kDa monomers are involved in the oligomeric organisation of the native *Echinococcus granulosus* antigen B. *Parasite Immunol*, **18**, 587–96.

Gonzalez, L.M., Montero, E., et al. 2002. Differential diagnosis of *Taenia saginata* and *Taenia solium* infections: from DNA probes to polymerase chain fraction. *Trans R Soc Trop Med Hyg*, **96**, S1/243–50.

Gorodezky, C., Diaz, M.L., et al. 1987. IgE concentration in sera of patients with neurocysticercosis. *Arch Invest Med (Mex)*, **18**, 225–7.

Gottstein, B. 1992. Molecular and immunological diagnosis of echinococcosis. *Clin Microbiol Rec*, **5**, 248–61.

Gottstein, B., Zinni, D. and Schantz, P.M. 1986. Species specific immunodiagnosis of *Taenia solium* cysticercosis by ELISA and immunoblotting. *Trop Med Parasitol*, **38**, 299–303.

Gottstein, B., D'Alessandro, A. and Rausch, R. 1995. Immunodiagnosis of polycystic hydatid disease/polycystic echinococcosis due to *Echinococcus vogeli*. *Am J Trop Med Hyg*, **53**, 558–63.

Gottstein, B., Dai, W.J., et al. 2002. An intact laminated layer is important for the establishment of secondary *Echinococcus multilocularis* infection. *Parasitol Res*, **88**, 822–8.

Grove, D.I. 1990. *A history of human helminthology*. Wallingford, Oxon: CAB International, pp. 355–383.

Greene, R.M., Wilkins, P.P. and Tsang, V.C.W. 1999. Diagnostic glycoproteins of *Taenia solium* cysts share homologous 14- and 18-kDa subunits. *Mol Bioch Parasitol*, **99**, 257–64.

Guerra, G., Flisser, A., et al. 1982. Biochemical and immunological characterization of antigen B purified from cysticerci of *Taenia solium*. In: Flisser, A., Willms, K., et al. (eds), *Cysticercosis: Present state of knowledge and perspectives*. New York: Academic Press, 437–51.

Gilman, R.H., Del Brutto, O.H., et al. 2000. Prevalence of taeniosis among patients with neurocysticercosis is related to severity of infection. *Neurology*, **55**, 1062.

Hancock, K., Broughel, D.E., et al. 2001. Sequence variation in the cytochrome oxidase I, internal transcribed spacer 1, and Ts 14 diagnostic antigen sequences of *Taenia solium* isolates from South and Central America, India, and Asia. *Int J Parasitol*, **31**, 1601–7.

Heath, D.D. 1982. In vitro culture of cysticerci and aid to investigations of morphological development ad host–parasite relationships. In: Flisser, A., Willms, K., et al. (eds), *Cysticercosis: Present state of knowledge and perspectives*. New York: Academic Press, 477–93.

Heath, D.D. 1995. Immunology of Echinococcus infections. In: Thompson, R.C.A. and Lymbery, A.J. (eds), *Echinococcus and hydatid disease*. Wallingford, UK: CAB International, 183–200.

Heath, D.D. and Lawrence, S.B. 1976. *Echinococcus granulosus*: development in vitro from oncosphere to immature hydatid cyst. *J Parasitol*, **73**, 417–23.

Helbig, M., Frosch, P., et al. 1993. Serological differentiation between cystic and alveolar echinococcosis by use of recombinant larval antigens. *J Clin Microbiol*, **31**, 3211–15.

Hemmings, L. and McManus, D.P. 1989. The isolation and differential antibody screening of *Echinococcus multilocularis* antigen gene clones with potential for immunodiagnosis. *Mol Biochem Parasitol*, **33**, 171–82.

Hernandez, M., Beltran, C., et al. 2000. Cysticercosis: towards the design of a diagnostic kit based on synthetic peptides. *Immunol Lett*, **71**, 13–17.

Herrera, L.A., Benita-Bordes, A., et al. 1999. Possible relationship between neurocysticercosis and hematological malignancies. *Arch Med Res*, **30**, 154–8.

Herrera, L.A., Ramírez, T., et al. 2000. Possible association between *Taenia solium* cysticercosis and cancer: Increased frequency of DNA damage in peripheral lymphocytes from neurocysticercosis patients. *Trans R Soc Trop Med Hyg*, **94**, 1–5.

Herrera, L.A. and Ostrosky-Wegman, P. 2000. Do helminths play a role in carcinogenesis? *Trends Parasitol*, **17**, 172–5.

Hoberg, E.P., Jones, A., et al. 2000. A phylogenetic hypothesis for species of the genus *Taenia* (Eucestoda: Taeniidae). *J Parasitol*, **86**, 89–98.

Houinato, D., Ramanankandrasana, B., et al. 1998. Seroprevalence of cysticercosis in Benin. *Trans R Soc Trop Med Hyg*, **92**, 621–4.

Howard, E.B. and Gendron, A.P. 1980. *Echinococcus vogeli* infection in higher primates at the Los Angeles Zoo. In: Montali, R.J. and Migaki,

G. (eds), *The comparative pathology of zoonosis animals*. Washington, DC: Smithsonian Institution Press, 379–82.

Huerta, M., Sciutto, E., et al. 2000. Vaccination against *Taenia solium* cysticercosis in underfed rustic pigs of México: roles of age, genetic background and antibody response. *Vet Parasitol*, **90**, 209–19.

Huerta, M., de Aluja, A.S., et al. 2001. Synthetic peptide vaccine against *Taenia solium* pig cysticercosis: successful vaccination in a controlled field trial in rural Mexico. *Vaccine*, **30**, 262–6.

Hulsmeier, A.J., Gehrig, P.M., et al. 2002. A major *Echinococcus multilocularis* antigen is a mucin-type glycoprotein. *J Biol Chem*, **277**, 5742–8.

Hurd, H., Mackenzie, K.S. and Chappell, L.H. 1993. Anthelmintic effects of cyclosporin A on protoscoleces and secondary hydatid cysts of *Echinococcus granulosis* in the mouse. *Int J Parasitol*, **23**, 315–20.

Ito, A. 1997. Basic and applied immunology in cestode infections: from Hymenolepis to Taenia and Echinococcus. *Int J Parasitol*, **27**, 1203–11.

Ito, A., Plancarte, A., et al. 1998. Novel antigens for neurocysticercosis: simple method for preparation and evaluation for serodiagnosis. *Am J Trop Med Hyg*, **59**, 291–4.

Ito, A., Ma, L., et al. 1999. Differential serodiagnosis for cystic and alveolar echinoccocosis using fractions of *Echinoccocus granulosus* cyst fluid (antigen B) and *E. multilocularis* protoscolex (Em18). *Am J Trop Med Hyg*, **60**, 188–92.

Ito, A., Ning, X., et al. 2002. Evaluation of an enzyme-linked immunosorbent assay (ELISA) with affinity purified Em18 and an ELISA with recombinant Em18 for differential diagnosis of alveolar echinococcosis: results of a blind test. *J Clin Microbiol*, **40**, 4161–5.

Jena, A., Sanchetee, P.C., et al. 1988. Cysticercosis of the brain shown by magnetic resonance imaging. *Clin Radiol*, **39**, 542–6.

Johnson, K.S., Harrison, G.B.L., et al. 1989. Vaccination against ovine cysticercosis using a defined recombinant antigen. *Nature (London)*, **338**, 585–7.

Jung, H., Hurtado, M., et al. 1990. Dexamethasone increases plasma levels of albendazole. *J Neurol*, **237**, 279–80.

Kammerer, W.S. and Schantz, P.M. 1993. Echinococcal disease. *Parasit Dis*, **7**, 605–18.

Kamiya, M. and Sato, H. 1990. Complete lifecycle of the canid tapeworm, *Echinococcus multilocularis* in laboratory rodents. *FASEB J*, **4**, 334–9.

Kern, P., Bardonnet, K., et al. 2003. European Echinococcosis Registry: human alveolar echinococcosis, Europe, 1982–2000. *Emerg Infect Dis*, **9**, 343–9.

Khuroo, M.S., Dar, M.Y., et al. 1993. Percutaneous drainage versus albendazole therapy in hepatic hydatidosis: A prospective, randomised study. *Gastroenterology*, **104**, 1452–9.

Ko, R.C. and Ng, T.F. 1998. Specificity of isoelectric focusing-purified antigens in the diagnosis of human cysticercosis. *Parasitol Res*, **84**, 565–9.

Kruger-Leite, E., Jalkh, A.E., et al. 1985. Intraocular cysticercosis. *Am J Ophthalmol*, **99**, 252–7.

Kuchenmeister, F. 1855. Offeness sendschreiben an die k.k. Gesellschaft der Aertze zu Wien. Experimenteller nachweis, dass Cysticercus cellulosae innerhalb des menschlichen darmkanales sich in Taenia solium unwandelt (translated in Kean, B.H., Mott, J.E., Russell, A.J. (eds) 1978. *Tropical medicine and parasitology: classic investigations*. Ithaca, NY: Cornell University Press, 677). *Wien Med Wochenschr*, **5**, 1–4.

Lachberg, S., Thompson, R.C.A. and Lymbery, A.J. 1990. A contribution to the etiology of racemose cysticercosis. *J Parasitol*, **76**, 592–4.

Laclette, J.P., Landa, A., et al. 1991. Paramyosin is the *Schistosoma mansoni* (trematoda) homologue of antigen B from *Taenia solium* (cestoda). *Mol Biochem Parasitol*, **44**, 287–96.

Laclette, J.P., Shoemaker, C., et al. 1992. Paramyosin inhibits complement C1. *J Immunol*, **148**, 124–8.

Landa, A., Merchant, M.T., et al. 1994. Purification and ultrastructural localization of surface glycoproteins of *Taenia solium* (Cestoda) cysticerci. *Int J Parasitol*, **24**, 265–9.

Larralde, C., Montoya, R.M., et al. 1989. Deciphering western blots of tapeworm antigens (*Taenia solium, Echinococcus granulosus* and *Taenia crassiceps*) reacting with sera from neurocysticercosis and hydatid disease patients. *Am J Trop Med Hyg*, **40**, 282–90.

Larrieu, E., Frider, B., et al. 2000. Asymptomatic carriers of hydatidosis: epidemiology, diagnosis and treatment. *Rev Panam Salud Publica*, **8**, 250–7.

Le, T.H., Pearson, M.S., et al. 2002. Complete mitochondrial genomes confirm the distinctiveness of the horse-dog and sheep-dog strains of *Echinococcus granulosus. Parasitology*, **124**, 97–112.

Leuckart, R. 1879. Die Parasiten des Menschen und die von ihnen herruhrenden Krankheiten. In *Ein Hand- und Lehrbuch fur Naturforscher und Aertze*, vol. 1. Leipzig: CF Winter'sche Verlangshandlung, 1009. The parasites of man and the diseases which proceed from them. In *A textbook for students and practitioners*, translated by Hoyle, W.E., Young, J. 1886. Edinburgh: Pentland, 771.

Lightowlers, M.W. 1994. Vaccination against animal parasites. *Vet Parasitol*, **54**, 177–204.

Lightowlers, M.W. 1996. Vaccination against cestode parasites. *Int J Parasitol*, **26**, 819–24.

Lightowlers, M.W. and Gottstein, B. 1995. Echinococcosis/hydatidosis: antigens, immunological and molecular diagnosis. In: Thompson, R.C.A. and Lymbery, A.J. (eds), *Echinococcus and hydatid disease*. Wallingford, UK: CAB International, 355–93.

Lightowlers, M.W., Liu, D., et al. 1989. Subunit composition and specificity of the major cyst fluid antigens of *Echinococcus granulosus. Mol Biochem*, **37**, 171–82.

Lightowlers, M.W., Mitchell, G.F. and Rickard, M.D. 1993. Cestodes. In: Cestodes, I. (ed.), *Immunology and molecular biology of parasitic infections*, 3rd edn. Oxford: Blackwell Scientific, 436–70.

Lightowlers, M.W., Lawrence, S.B., et al. 1996. Vaccination against hydatidosis using a defined recombinant antigen. *Parasite Immunol*, **18**, 457–62.

Lightowlers, M.W., Flisser, A., et al. 2000. Vaccination against cysticercosis and hydatid disease. *Parasitol Today*, **16**, 191–6.

Little, J.M. 1976. Hydatid disease at Royal Prince Alfred Hospital. 1964 to 1974. *Med J Aust*, **1**, 903–8.

Liu, D., Lightowlers, M.W. and Rickard, M.D. 1992. Evaluation of a monoclonal antibody-based competition ELISA for the diagnosis of human hydatidosis. *Parasitology*, **104**, 357–61.

Lopez-Gómez, M., Castro, N., et al. 2001. Optimization of the signle-day praziquantel therapy for neurocysticercosis. *Neurology*, **57**, 1929–30.

Loyo, M., Kleriga, E. and Estanol, B. 1980. Fourth ventricular cysticercosis. *Neurosurgery*, **7**, 456–8.

Macpherson, C.N.L. and Craig, P.S. 2000. Dogs and cestode zoonoses. In: Macpherson, C.N.L., Meslin, F.X. and Wandeler, A.I. (eds), *Dogs, zoonoses and public health*. Wallingford, UK: CABI Publishing, 177–211.

Madisson, S.E., Slemenda, S.B., et al. 1989. A specific diagnostic antigen of *Echinococcus granulosus* with an apparent molecular weight of 8 Kda. *Am J Trop Med Hyg*, **40**, 377–83.

Madrazo, I. and Flisser, A. 1992. Parasitic infestations of the cerebrum. In: Appuzo, J.M.L. (ed.), *Cysticercosis, brain surgery. Complication avoidance and management*. Edinburgh: Churchill Livingston, 1419–30.

Manoutcharian, K., Rosas, G., et al. 1996. Cysticercosis: identification and cloning ofprotective recombinant antigens. *J Parasitol*, **82**, 250–4.

Manoutcharian, K., Terrazas, L.I., et al. 1998. Protection against murine cysticercosis using cDNA expression library immunization. *Immunol Lett*, **62**, 131–6.

Manoutcharian, K., Terrazas, L.I., et al. 1999. Phage-displayed T-cell epitope grafted into immunoglobulin heavy-chain complementarity-determining regions: an effective vaccine design tested in murine cysticercosis. *Infect Immun*, **67**, 4764–70.

Manoutcharian, K., Terrazas, L.I., et al. 2000. DNA pulsed macrophage-mediated cDNA expression library immunization in vaccine development. *Vaccine*, **18**, 389–91.

Maravilla, P., Avila, G., et al. 1998. Comparative development of *Taenia solium* in experimental models. *J Parasitol*, **84**, 882–6.

Maravilla, P., Souza, V., et al. 2003. Genetic variation in *Taenia solium* isolates. *J. Parasitol*, **89**, 1250–4.

Marco, M. and Nieto, A. 1991. Metalloproteinases in the larvae of *Echinococcus granulosus. Int J Parasitol*, **21**, 743–6.

Marriner, S.E., Morris, D.L., et al. 1986. Pharmacokinetics of albendazole in man. *Eur J Clin Pharmacol*, **30**, 705–8, 413–22.

Martinez, M.J., de Aluja, A.S. and Gemmell, M. 2000. Failure to incriminate domestic flies (Diptera: Muscidae) as mechanical vectors of *Taenia* eggs (Cyclophillidea: Taeniidae) in rural Mexico. *J Med Entomol*, **37**, 489–91.

Mateos, J.H. and Zenteno, G.H. 1987. Neurocisticercosis. Analisis de mil casos consecutivos. *Neurol Neuropsiq Psiquiat (Mex)*, **27**, 53–5.

Mayta, H., Talley, A., et al. 2000. Differenatiating *Taenia solium* and *Taenia saginata* infections by simple hematoxylin-eosin staining and PCR-restriction enzyme analysis. *J Clin Microbiol*, **38**, 133–7.

McManus, D.P. 1981. A biochemical study of adult and cystic stages of *Echinococcus granulosus* of human and animal origin from Kenya. *J Helminthol*, **55**, 21–7.

McManus, D.P. and Bryant, C. 1995. Biochemistry, physiology and molecular biology of *Echinococcus*. In: Thompson, R.C.A. and Lymbery, A.J. (eds), *Echinococcus and hydatid disease*. Wallingford, UK: CAB International, 135–71.

McManus, D.P. and Smyth, J.D. 1978. Differences in the chemical composition and carbohydrate metabolism of *Echinococcus granulosus* (horse and sheep strains) and *E. multilocularis. Parasitology*, **77**, 103–9.

McManus, D.P., Garcia-Zepeda, E., et al. 1989. Human cysticercosis and taeniasis: molecular approaches for specific diagnosis and parasite identification. *Acta Leiden*, **57**, 81–91.

McVie, A., Ersfeld, K., et al. 1997. Expression and immunological characterisation of *Echinococcus granulosus* recombinant antigen B for IgG4 subclass detection in human cystic echinococcosis. *Acta Trop*, **67**, 19–35.

Medina-Escutia, E., Morales-López, Z., et al. 2001. Cellular immune response and Th1/Th2 cytokines in human neurocysticercosis: lack of immune supression. *J Parasitol*, **87**, 587–90.

Meneghelli, V.G., Martinelli, A.L.C., et al. 1992. Polycystic hydatid dusease (*Echinococcus vogeli*): Clinical laboratory, and morphological findings in nine Brazilian patients. *J Hepatol*, **14**, 203–10.

Merchant, M.T., Corella, C. and Willms, K. 1997. Autoradiographic analysis of the germinative tissue in evaginated *Taenia solium* metacestodes. *J Parasitol*, **83**, 363–7.

Merchant, M.T., Aguilar, L., et al. 1998. *Taenia solium*: description of the intestinal implantation sites in experimental hamster infections. *J Parasitol*, **84**, 681–5.

Mesarina-Wicki, B. 1991. Long-term course of alveolar echinoccosis in 70 patients treated by benzimidazole derivatives (mebendazole and albendazole) (1976–1989). Medical dissertation, University of Zurich.

Meza-Lucas, A., Carmona-Miranda, L., et al. 2003. Limited and short lasting humoral response in *Taenia solium* seropositive households compared with patients with neurocysticercosis. *Am J Trop Med Hyg*, **69**, 223–7.

Michault, A., Leroy, D., et al. 1989. Diagnostic immunologique dans le liquide cephalo-rachidien et le serum de la cysticercose encephalique evolutive. *Path Biol (Paris)*, **37**, 249–53.

Michault, A., Duval, G., et al. 1990. Etude seroepidemiologique de la cysticercose a l'ile de la Réunion. *Bull Soc Path Exot Filiales*, **83**, 82–92.

Michael, A.S., Levy, J.M. and Paige, M.L. 1990. Cysticercosis mimicking brain neoplasm: MR and CT appearance. *J Comput Assist Tomogr*, **14**, 708–11.

Michel, P., Michault, A., et al. 1990. Le serodiagnostic de la cysticercose par ELISA et western blot. Son interet et ses limites a Madagascar. *Arch Inst Pasteur Madagascar*, **57**, 115–42.

Mitchel, G.F. 1990. Vaccines and vaccination strategies against helminths. In: Agabian, N. and Cerami, A. (eds), *Parasites. Molecular biology, drug and vaccine design*. New York: Wiley-Liss, 349–63.

Mitchell, G.W. and Crawford, R.O. 1988. Intraparenchymal cerebral cysticercosis in children: a benign prognosis. *Pediatrics*, **82**, 76–82.

Mitchell, G.W. and Snodgrass, S.R. 1985. Intraparenchymal cerebral cysticercosis in children: diagnosis and treatment. *Pediatr Neurol*, **1**, 151–6.

Molinari, J.L., Tato, P., et al. 1990. Depresive effect of a *Taenia solium* cysticercus factor on cultured human lymphocytes stimulated with phytohemaglutinin. *Ann Trop Med Parasitol*, **84**, 205–8.

Montero, R., Flisser, A., et al. 1994. Mutation at the HPRT locus in patients with neurocysticercosis treated with praziquantel. *Mut Res*, **305**, 181–8.

Morales, J., Velasco, T., et al. 2002. Castration and pregnancy of rural pigs significantly increase the prevalence of naturally acquired *Taenia solium* cysticercosis. *Vet Parasitol*, **108**, 41–8.

Moro, P.L., Gilman, R.H., et al. 1992. Immunoblot (western blot) and double immunodiffusion (DD5) tests for hydatid disease cross-react with sera from patients with cysticercosis. *Trans R Soc Trop Med Hyg*, **86**, 422–3.

Mühlschlegel, F., Sygulla, L., et al. 1993. Paramyosin of *Echinococcus granulosus*: CDNA sequence and characterization of a tegumental antigen. *Parasitol Res*, **79**, 660–6.

Muller, N., Gottstein, B., et al. 1989. Application of a recombinant *Echinococcus multilocularis* antigen in an enzyme-linked immunosorbant assay for immunodiagnosis of human alveolar echinococcosis. *Mol Biochem Parasitol*, **36**, 151–60.

Muller, R., Lillywhite, J., et al. 1987. Human cysticercosis and intestinal parasitism amongst the Ekari people of Irian Jaya. *J Trop Med Hyg*, **90**, 291–6.

Murrell, K.D. (ed.) 2003. International action planning workshop on *Taenia solium* cysticercosis/taeniosis with special focus on Eastern and Southern Africa. *Acta Tropica*, **87**, 1–191 (Special issue).

Nakao, M., Okamoto, M., et al. 2002. A phylogenetic hypothesis for the distribution of two genotypes of the pig tapeworm *Taenia solium* worldwide. *Parasitology*, **124**, 657–62.

Nash, T.E., Pretall, J. and Garica, H.H. 2001. Calcified cysticerci provoke perilesional edema and seizures. *Clin Infect Dis*, **33**, 1649–53.

National Hydatid Disease Center of China, 1993. A retrospective survey for surgical cases of cystic echinococcosis in the Xinjiang Uygur Autonomous Region, PRC (1951–90). In Anderson, F.L., Chai, J.J. and Liu, F.J. (eds), *Compendium on cystic echinococcosis with special reference to the Xinjiang Uygur Autonomous Region, P.R. China*. Provo, UT: Brigham Youn University Print Services.

Ochieng-Mitula, P.J. and Burt, M.D.B. 1996. The effects of ivermectin on the hydatid cyst of *Echinococcus granulosus* after direct injection at laparotomy. *J Parasitol*, **82**, 155–7.

Ohnishii, K. and Kutsumi, K. 1991. Possible formation of calcareous corpuscles by the brood capsule in secondary hepatic metacestodes of *Echinococcus multilocularis*. *Parasitol Res*, **77**, 600–1.

Olivo, A., Plancarte, A. and Flisser, A. 1988. Presence of antigen B from *Taenia solium* cysticercus in other platyhelminthes. *Int J Parasitol*, **18**, 543–5.

Oriol, C. and Oriol, R. 1975. Physicochemical properties of a lipoprotein antigen of *Echinococcus granulosus*. *Am J Trop Med Hyg*, **24**, 96–100.

Ortona, E., Vaccari, S., et al. 2002. Immunological characterisation of *Echinococcus granulosus* cyclophilin, an allergen reactive with IgE from patients with cystic echinococcosis. *Clin Exp Immunol*, **128**, 124–30.

Ostrosky, L., Correa, D., et al. 1991. *Taenia solium*: inhibition of spontaneous evagination of cysticerci by the host inflammatory capsule. *Int J Parasitol*, **21**, 603–4.

Ostrosky-Zeichner, L., García-Mendoza, E., et al. 1996. Humoral and cellular immune response within the subarachnoid space of patients with neurocysticercosis. *Arch Med Res*, **27**, 513–17.

Palacio, L.G., Jimenez, I., et al. 1998. Neurocysticercosis in persons with epilepsy in Medellín, Colombia. *Epilepsia*, **39**, 1334–9.

Pammenter, M.D., Rossouw, E.J. and Epstein, S.R. 1987. Diagnosis of neurocysticercosis by enzyme-linked immunosorbent assay. *S Afr Med J*, **71**, 512–14.

Passos-Barbosa, A., Costa-Cruz, J.M., et al. 2000. Cisticercose: fatores relacionados a interacao parasito-hospedeiro, diagnostico e soroprevalencia. *Rev Patol Trop*, **29**, 17–34.

Paul, M. and Stefaniak, J. 1997. Detection of specific *Echinococcus granulosus* antigen 5 in liver cyst bioptate from human patients. *Acta Trop*, **64**, 65–77.

Pawlowski, Z.S. 1990. Perspectives on the control of *Taenia solium*. *Parasitol Today*, **6**, 371–3.

Pawlowski, Z.S. 2002. *Taenia solium*: Basic biology and transmission in *Taenia solium* cysticercosis. In: Singh, G. and Prabhakar, S. (eds), *From basic to clinical science*. Oxon, UK: CABI Publishing, 1–13.

Pawlowski, Z.S. and Schultz, M.G. 1972. Taeniasis and cysticercosis. *Adv Parasitol*, **10**, 269–343.

Pearson, M., Le, T.H., et al. 2002. Molecular taxonomy and strain analysis in *Echinococcus*. In: Craig, P.S. and Pawlowski, Z. (eds), *Cestode zoonoses: Echinococosis and cysticercosis an emergent and global problem.*, *NATO Science Series*. vol. 341. Amsterdam: IOS Press, 205–19.

Perdomo, R., Alvarez, C., et al. 1997. Principles of the surgical approach in human liver cystic echinococcosis. *Acta Trop*, **64**, 109–22.

Pérez-Serrano, J., Casado, N., et al. 1994. The effects of albendazole and albendazole sulphoxide combination-therapy on *Echinococcus granulosus* in vitro. *Int J Parasitol*, **24**, 219–24.

Pinto, P.S.A., Vaz, A.J., et al. 2000a. ELISA test for the diagnosis of cysticercosis in pigs using antigens of *Taenia solium* and *Taenia crassiceps* cysticerci. *Rev Inst Med Trop S Paulo*, **42**, 71–9.

Pinto, P.S.A., Vaz, A.J., et al. 2000b. Performance of the ELISA test for swine cysticercosis using antigens of *Taenia solium* and *Taenia crassiceps* cysticerci. *Vet Parasitol*, **88**, 127–30.

Plancarte, A., Flisser, A., et al. 1999a. Vaccination against *Taenia solium* cysticercosis in pigs using native and recombinant oncosphere antigens. *Intl J Parasitol*, **29**, 643–7.

Plancarte, A., Hirota, C., et al. 1999b. Characterization of GP39-42 and GP24 antigens from *Taenia solium* cysticerci and of their antigenic GP10 subunit. *Parasitol Res*, **85**, 680–4.

Polat, K., Balik, A.A., et al. 2002. Hepatic alveolar echinococcosis: clinical report from an endemic region. *Can J Surg*, **45**, 415–19.

Pretell, E.J., Garcia, H.H., et al. 2001a. Short regimen of praziquantel in the treatment of single brain enhancing lesions. *Clin Neurol Neurosurg*, **102**, 215–18.

Pretell, E.J., Garcia, H.H., et al. 2001b. Failure of one-day praziquantel treatment in patients with multiple neurocysticercosis lesions. *Clin Neurol Neurosurg*, **103**, 175–7.

Proaño-Narvaez, J.V., Meza-Lucas, A., et al. 2002. Laboratory diagnosis of human neurocysticercosis: double blind comparison of ELISA and EITB. *J Clin Microbiol*, **40**, 2115–18.

Puig-Solanes, M. 1974. Consideraciones clinico-patologicas acerca de la cisticercosis intraocular: cisticerco viable y cisticerco en involucion. *Arch Soc Española Oftalmol*, **34**, 341–64.

Puri, V., Sharma, D.K., et al. 1991. Neurocysticercosis in children. *Indian J Pediatr*, **28**, 1309–17.

Quintanar, J.L., Franco, L.M. and Salinas, E. 2003. Detection of glial fibrillary acidic protein and neurofilaments in the cerebrospinal fluid of patients with neurocysticercosis. *Parasitol Res*, **90**, 261–3.

Rabiela, M.T., Rivas, A., et al. 1982. Anatomopathological aspects of human brain cysticercosis. In: Flisser, A., Willms, K., et al. (eds), *Cysticercosis: present state of knowledge and perspectives*. New York: Academic Press, 179–200.

Rabiela, M.T., Rivas, A. and Flisser, A. 1989. Morphological types of *Taenia solium* cysticerci. *Parasitol Today*, **5**, 357–9.

Rabiela, M.T., Hornelas, Y., et al. 2000. Evagination of *Taenia solium* cysticerci: a histologic and electrón microscopy study. *Arch Med Res*, **31**, 605–7.

Rafiei, A. and Craig, P.S. 2002. The immunodiagnostic potential of protoscolex antigens in human cystic echinococcosis and the possible influence of parasite strain. *Ann Trop Med Parasitol*, **96**, 383–9.

Rajshekhar, V. 1991. Etiology and management of single small CT lesions in patients with seizures: understanding a controversy. *Acta Neurol Scand*, **84**, 465–70.

Rajshekhar, V. and Chandy, M.J. 1997. Validation of diagnostic criteria for solitary cerebral cysticercus granuloma in patients presenting with seizures. *Acta Neurol Scand*, **96**, 76–81.

Rausch, R.L. 1954. Studies on the helminth fauna of Alaska. XX. The histogenesis of the alveolar larva of *Echinococcus* species. *J Infect Dis*, **94**, 178–86.

Rausch, R.L. 1995. Life cycle patterns and geographic distribution of *Echinococcus* species. In: Thompson, R.C.A. and Lymbery, A.J. (eds), *Echinococcus and hydatid disease*. Wallingford, UK: CAB International, 89–119.

Rausch, R.L. 1997. *Echinococcus granulosus*: biology and ecology. In: Andersen, F.L., Ouhelli, H. and Kachani, M. (eds), *Compendium on echinococcosis in Africa and in Middle Eastern countries with special reference to Morocco*. Provo, UT: Brigham Young University Print Services, 18–53.

Rausch, R.L. and D'Alessandro, A. 2002. The epidemiology of echinococcosis caused by *Echinococcus oligarthrus* and *E. vogeli* in the neotropics. In: Craig, P.S. and Pawlowski, Z. (eds), *Cestode zoonoses: echinococcosis and cysticercosis an emergent and global problem.*, *NATO Science Series*. vol. 341. Amsterdam: IOS Press, 107–13.

Rausch, R.L., D'Alessandro, A. and Rausch, V.R. 1981. Characteristics of the larval *Echinococcus vogeli* Rausch and Bernstein, 1972 in the natural intermediate host, the paca, *Cuniculus paca* L (Rodentia: Dasyproctidae). *Am J Trop Med Hyg*, **30**, 1043–52.

Rausch, R.L., Wilson, J.F., et al. 1987. Spontaneous death of *Echinococcus multilocularis* cases diagnosed serologically (by Em2 ELISA) and clinical significance. *Am J Trop Med Hyg*, **36**, 576–85.

Read, C.P. 1952. Contributions to cestode enzymology. 1. The cytochrome system and succinic dehydrogenase in *Hymenolepis diminuta*. *Exp Parasitol*, **1**, 353–62.

Restrepo, B., Llaguno, P., et al. 1998. Analysis of immune lesions in neurocysticercosis patients: central nervous system response to helminth appears Th1-like instead of Th2. *J Neuroimmunol*, **89**, 64–72.

Restrepo, B.I., Obregon-Henao, A., et al. 2000. Characterization of the carbohydrate components of *Taenia solium* metacestode glycoprotein antigens. *Int J Parasitol*, **30**, 689–96.

Restrepo, B.I., Aguilar, M.I., et al. 2001. Analysis of the peripheral immune response in patients with neurocysticercosis: evidence for T cell reactivity to parasite glycoprotein and vesicular fluir antigens. *Am J Trop Med Hyg*, **65**, 366–70.

Reuter, S., Nussle, K., et al. 2001. Alveolar liver echinococcosis: a comparative study of three imaging techniques. *Infection*, **29**, 119–25.

Rim, H.J. and Joo, K.H. 1989. Clinical evaluation of the therapeutic efficacy of praziquantel against human cysticercosis. *Acta Leiden*, **57**, 235–45.

Richards, F. and Schantz, P.M. 1991. Laboratory diagnosis of cysticercosis. *Clin Lab Med*, **11**, 1011–28.

Rickard, M.D., Harrison, G.B., et al. 1995. *Taenia ovis* recombinant vaccine – 'quo vadit'. *Parasitology*, **110**, S5–9.

Rigano, R., Profumo, E., et al. 1995. Immunological markers indicating the effectiveness of pharmacological treatment in human hydatid disease. *Clin Exp Immunol*, **102**, 281–5.

Rigano, R., Ioppolo, S., et al. 2002. Long term serological evaluation of patients with cystic echinococcosis treated with benzimidazole carbamates. *Clin Exp Immunol*, **129**, 485–92.

Rinder, H., Rausch, R.L., et al. 1997. Limited range of genetic variation in *Echinococcus multilocularis*. *J Parasitol*, **83**, 1045–50.

Rodacki, M.A., Detoni, X.A., et al. 1989. CT features of cellulosae and racemosus neurocysticercosis. *J Comput Assist Tomogr*, **13**, 1013–16.

Rodriguez-Canul, R., Allan, J.C., et al. 1997. Comparative evaluation of purified *Taenia solium* glycoproteins and crude metacestode extracts by immunoblotting for the serodiagnosis of human *T. solium* cysticercosis. *Clin Diagn Lab Immunol*, **4**, 579–82.

Rodriguez-Canul, R., Allan, J.C., et al. 1998. Application of an immunoassay to determine risk factors associated with porcine cysticercosis in rural areas of Yucatan, Mexico. *Vet Parasitol*, **79**, 165–80.

Rodriguez-Canul, R., Fraser, R., et al. 1999. Epidemiological study of *Taenia solium* taeniasis/cysticercosis in a rural village in Yucatan state, Mexico. *Ann Trop Med Parasitol*, **93**, 57–67.

Rodriguez-Carbajal, J. and Boleaga-Duran, B. 1982. Neuroradiology of human cysticercosis. In: Flisser, A., Willms, K., et al. (eds), *Cysticercosis: present state of knowledge and perspectives*. New York: Academic Press, 139–62.

Rodriguez-Contreras, D., Skelly, P.J., et al. 1998. Molecular and functional characterization and tissue localization of 2 glucose transporter homologues (TGTP1 and TGTP2) from the tapeworm *Taenia solium*. *Parasitology*, **117**, 579–88.

Rodriguez-Contreras, D., de, T.P., et al. 2002. The *Taenia solium* glucose transporters TGTP1 and TGTP2 are not immunologically recognized by cysticercotic humans and swine. *Parasitol Res*, **88**, 280–2.

Rogan, M.T. and Craig, P.S. 1997. Immunology of *Echinococcus granulosus* infections. *Acta Trop*, **67**, 7–17.

Roman, G., Sotelo, J., et al. 2000. A proposal to declare neurocysticercosis an international reportable disease. *Bull WHO*, **78**, 399–406.

Romig, T., Zeyhle, E., et al. 1986. Cyst growth and spontaneous cure in hydatid disease. *Lancet*, **ii**, 861.

Romig, T., Kratzer, W., et al. 1999. An epidemiologic survey of human alveolar echinococcosis in southwestern Germany. *Am J Trop Med Hyg*, **61**, 566–73.

Rolfs, A., Muhlschlegel, F., et al. 1995. Clinical and immunological follow-up study of patients with neurocysticercosis after treatment with praziquantel. *Neurology*, **45**, 532–8.

Rozenzvit, M.C., Zhang, L.H., et al. 1999. Genetic variation and epidemiology of *Echinococcus granulosus* in Argentina. *Parasitology*, **118**, 523–30.

Sakai, H., Vilela-Barbosa, H., et al. 2001. Short report: seroprevalence of Taenia solium cysticercosis in pigs in Bahia state, northeastern Brazil. *Am J Trop Med Hyg*, **64**, 268–9.

San Esteban, E., Flisser, A. and Gonzalez-Astiazaran, A. (eds) 1997. *Neurocisticercosis en la infancia*. México, DF: Miguel Angel Porrua, PUIS-UNAM, 317pp.

Sanchez, A.L., Ljungstrom, I. and Medina, M.T. 1999. Diagnosis of human neurocysticercosis in endemic countries: a clinical study in Honduras. *Parasitol Intl*, **48**, 81–9.

Sanchez, F.A. and Sanchez, A.C. 1971. Estudio de algunas propiedades físicas y componentes químicos del líquido y pared germinativa de quistes hidatídicos de diversas especies y de diferente localización. *Rev Iber Parasitol*, **31**, 347–66.

Santos, R., Dalma, A. and Ortiz, E. 1979. Management of subretinal and viteous cysticercosis: Role of photocoagulation and surgery. *Ophthalmology*, **86**, 1501–4.

Sarti, E., Schantz, P.M., et al. 1992. Prevalence and risk factors for *Taeniasolium* taeniasis and cysticercosis in humans and pigs in a village in Morelos, Mexico. *Am J Trop Med Hyg*, **46**, 677–84.

Sarti, E., Schantz, P.M., et al. 1994. Epidemiologic investigation of *Taenia solium* taeniasis and cysticercosis in a rural village of Michoacan State, Mexico. *Trans R Soc Trop Med Hyg*, **88**, 49–52.

Sarti, E., Flisser, A., et al. 1997. Development and evaluation of a health education intervention against *Taenia solium* in a rural community in Mexico. *Am J Trop Med Hyg*, **56**, 127–32.

Sarti, E., Schantz, P.M., et al. 2000. Mass treatment against human taeniasis for the control of cysticercosis: a population-based intervention study. *Trans R Soc Trop Med Hyg*, **94**, 85–9.

Sako, Y., Nakao, M., et al. 2000. Molecular characterization and diagnostic value of *Taenia solium* low- molecular-weight antigen genes. *J Clin Microbiol*, **38**, 4439–44.

Sato, N., Aoki, S., et al. 1993. Clinical features. In: Uchino, J. and Sato, N. (eds), *Alveolar echinococcosis of the liver*. Sapporo: Hokkaido University School of Medicine, 63–8.

Schantz, P.M. 1994. Larval cestodiases. In: Hoeprich, P.D., Jordan, M.C. and Ronald, A.R. (eds), *Infectious diseases. A treatise of infectious processes*, 4th edn. Philadelphia: JB Lippincott, 850–60.

Schantz, P.M. and Gottstein, B. 1986. Echinococcosis (hydatidosis). In: Walls, K.W. and Schantz, P.M. (eds), *Immunodiagnosis of parasitic diseases. , Heminthic diseases.* vol. 1. Orlando, FL: Academic Press, 69–107.

Schantz, P.M. and Okelo, G.B.A. 1990. Echinococcosis (Hydatidosis). In: Warren, K.S. and Mahmoud, A.A.F. (eds), *Tropical and geographical medicine*. Cleveland: University Hospitals of Cleveland, 504–18.

Schantz, P.M., Sarti, E., et al. 1994. Community-based epidemiological investigations of cysticercosis due to *Taenia solium*: comparison of serological screening tests and clinical findings in two populations in Mexico. *Clin Infect Dis*, **18**, 879–85.

Schantz, P.M., Wilkins, P.P. and Tsang, V.C.W. 1998. Immigrants, imaging and immunoblots: the emergence of neurocysticercosis as a significant public health problem. In: Scheld, W.M., Craig, W.A. and Hughes, W.A. (eds), *Emerging infections 2*. Washington DC: ASM Press, 213–42.

Sciutto, E., Hernandez, M., et al. 1998a. Diagnosis of porcine cisticercosis: a comparative study of serological tests for detection of circulating antibody and viable parasites. *Vet Parasitol*, **78**, 185–94.

Sciutto, E., Martinez, J.J., et al. 1998b. Limitations of current diagnostic procedures for the diagnosis of *Taenia solium* cysticercosis in rural pigs. *Vet Parasitol*, **79**, 299–313.

Sciutto, E., Fragoso, G., et al. 2000. *Taenia solium* disease in humans and pigs: an ancient parasitosis disease rooted in developing countries and emerging as a major health problem of global dimensions. *Microb Infect*, **2**, 1875–90.

Shambesh, M.K., Craig, P.S., et al. 1997. IgGl and IgG4 serum antibody responses in asymptomatic and clinically expressed cystic echinococcis patients. *Acta Trop*, **64**, 53–63.

Shambesh, M.K., Craig, P.S., et al. 1999. An extensive ultrasound and serologic study to investigate the prevalence of human cystic echinococcosis in northern Libya. *Am J Trop Med Hyg*, **60**, 462–8.

Shepherd, J.C., Aitken, A. and McManus, D.P. 1991. A protein secreted in vivo by *Echinococcus granulosus* inhibits elastase activity and neutrophil chemotaxis. *Mol Biochem Parasitol*, **44**, 81–90.

Short, J.A., Heiner, D.C., et al. 1991. Immunoglobulin E and G4 antibodies in cysticercosis. *J Clin Microbiol*, **28**, 1635–9.

Siles-Lucas, M. and Gottstein, B. 2001. Molecular tools for the diagnosis of cystic and alveolar echinococcosis. *Trop Med Int Hlth*, **6**, 463–75.

Siles-Lucas, M., Merli, M., et al. 2003. The *Echinococcus multilocularis* 14-3-3 protein protects mice against primary but not secondary alveolar echinococcosis. *Vaccine*, **21**, 431–9.

Singh, G. and Prabhakar, S. (eds) 2002. Taenia solium *cysticercosis: from basic to clinical science*. Oxon UK: CABI Publishers, 457 pp.

Singhal, B.S. and Salinas, R.A. 2003. Controversies in the drug treatment of neurocysticercosis. In: Singh, G. and Prabhakar, S. (eds), *Taenia solium cysticercosis: from basic to clinical science*. Oxon, UK: CABI Publishers, 375–85.

Sotelo, J. and Del Brutto, O.H. 2000. Brain cysticercosis. *Arch Med Res*, **31**, 3–14.

Sotelo, J. and Flisser, A. 1997. Neurocysticercosis, practical treatment guidelines. *CNS Drugs*, **7**, 17–25.

Sotelo, J., Guerrero, V. and Rubio, F. 1985a. Neurocysticercosis: a new clasification based on active and inactive forms. A study of 753 cases. *Arch Intern Med*, **145**, 442–5.

Sotelo, J., Torres, B., et al. 1985b. Praziquantel in the treatment of neurocysticercosis: long-term follow-up. *Neurology*, **35**, 752–5.

Sotelo, J., Rubalcava, M.A. and Gomez-Llata, S. 1995. A new shunt for hydrocephalus that relies on CSF production rather than on ventricular pressure: initial clinical experiences. *Surg Neurol*, **43**, 324–32.

Soto-Hernandez, J.L., Ostrosky-Zeichner, L., et al. 1996. Neurocysticercosis and HIV infection, report of two cases and review. *Surg Neurol*, **45**, 57–61.

Spindola-Felix, N., Rojas-Wastavino, G., et al. 1996. Parasite search in strawberries from Irapuato, Guanajuato and Zamora, Michoacan (Mexico). *Arch Med Res*, **27**, 229–31.

Spruance, S.L. 1974. Latent period of 53 years in a case of hydatid cyst disease. *Arch Intern Med*, **134**, 741–2.

Stehr-Green, J.A., Stehr-Green, P.A. and Schantz, P.M. 1988. Risk factors for infection with *Echinococcus multilocularis* in Alaska. *Am J Trop Med Hyg*, **38**, 380–5.

Subahar, R., Hamid, A., et al. 2001. *Taenia solium* infection in Irian Jaya (West Papua), Indonesia: a pilot serological survey of human and porcine cysticercosis in Jayawijaya District. *Trans R Soc Trop Med Hyg*, **95**, 388–90.

Suntsov, S., Ozeretskovskaya, N., et al. 1990a. Status of interferon during helminthiases. Communication 1. Human unilocular hydatidosis. *Med Parazitol (Mosk)*, **6**, 27–9.

Suntsov, S., Ozeretskovskaya, N., et al. 1990b. Interferon status during helminthiases. Communication 2. Multilocular hydatidosis. *Med Parazitol (Mosk)*, **2**, 43–4.

Suss, R.A., Maravilla, K.R. and Thompson, J. 1986. MR imaging of intracranial cysticercosis: comparison with CT and anatomopathologic features. *AJNR*, **7**, 235–41.

Tackmann, T., Loscher, U., et al. 2001. A field study to control *Echinococcus multilocularis* infections of the red fox (*Vulpes vulpes*) in an endemic area. *Epidemiol Infect*, **127**, 577–87.

Takayanagui, O.M. and Chimelli, L. 1998. Disseminated muscular cysticercosis with myositis induced by praziquantel therapy. *Am J Trop Med Hyg*, **59**, 1002–3.

Takayanagui, O.M. and Jardim, E. 1983. Aspectos clinicos da neurocisticercose. Analise de 500 casos. *Arq Neuro-Psiquiatr*, **41**, 50–63.

Takayanagui, O.M. and Jardim, E. 1992. Therapy for neurocysticercosis. Comparison between albendazole and praziquantel. *Arch Neurol*, **49**, 290–4.

Takayanagui, O.M., Castro e Silva, A.A.M.C., et al. 1996. Notificacao compulsoria da cisticercose em Ribeirao Preto-SP. *Arq Neuro Psiq*, **54**, 557–64.

Takayanagui, O.M., Lanchote, V.L., et al. 1997. Therapy for neurocysticercosis: pharmacokinetic interaction of albendazole sulfoxide with dexamethasone. *Ther Drug Monitor*, **19**, 51–5.

Taylor, D.H. and Morris, D.L. 1988. In vitro culture of *Echinococcus multilocularis*: protoscolicidal action of praziquantel and albendazole sulphoxide. *Trans R Soc Trop Med Hyg*, **82**, 265–7.

Teitelbaum, G.P., Otto, R.J., et al. 1989. MR imaging of neurocysticercosis. *Am J Roentgenol*, **153**, 857–66.

Thais, J.H., Goldsmith, R.S., et al. 1994. Detection by immunoblot assay of antibodies to *Taenia solium* cysticerci in sera from residents of rural communities and from epileptic patients in Bali, Indonesia. *Southeast Asian J Trop Med Pub Hlth*, **25**, 464–8.

Thompson, R.C.A. 1995. Biology and systematics of *Echinococcus*. In: Thompson, R.C.A. and Lymbery, A.J. (eds), *Echinococcus and hydatid disease*. Wallingford, UK: CAB International, 1–37.

Thompson, R.C.A. and Lymbery, A.J. 1988. The nature, extent and significance of variation within the genus *Echinococcus*. *Adv Parasitol*, **27**, 210–63.

Thompson, R.C.A., Robertson, I.D., et al. 1993. Hydatid disease in Western Australia: a novel approach to education and surveillance. *Parasitol Today*, **9**, 431–3.

Todorov, T., Mechkov, G., et al. 1992. Factors influencing the response to chemotherapy in human cystic echinococcosis. *Bull WHO*, **70**, 347–58.

Toledo, A., Larralde, C., et al. 1999. Towards a *Taenia solium* cysticercosis vaccine: an epitope shared by *Taenia crassiceps* and *Taenia solium* protects mice against experimental cysticercosis. *Infect Immun*, **67**, 2522–30.

Torgerson, P.R., Pilkington, J., et al. 1995. Further evidence for the long distance dispersal of taeniid eggs. *Int J Parasitol*, **25**, 265–7.

Torres, A., Plancarte, A., et al. 1992. Praziquantel treatment of porcine brain and muscle *Taenia solium* cysticercosis. 3. Effect of 1-day treatment. *Parasitol Res*, **78**, 161–4.

Tsang, V.C.W., Brand, A.J. and Boyer, A.E. 1989. An enzyme-linked immunoelectrotransfer blot assay by glycoprotein antigens for diagnosing human cysticercosis (*Taenia solium*). *J Infect Dis*, **159**, 50–9.

van Beneden, P.J. 1854. Note sur des experiences relatives au developpement des cysticerques. *Ann Sci Nat*, **1**, 104.

Van der Kaay, H.J. and Overbosch, D. 1989. Proceedings of the symposium on neurocysticercosis. *Acta Leid*, **57**, 79–275.

Varela-Díaz, V.M., Coltorti, E.A. and D'Alessandro, A. 1978. Immunoelectrophoresis tests showing *Echinococcus granulosus* arc 5 in human cases of *Echinococcus vogeli* and cysticercosis-multiple myeloma. *Am J Trop Med Hyg*, **27**, 554–7.

Vargas-Parada, L. and Laclette, J.P. 1999. Role of calcareous corpuscles in cestode physiology: a review. *Rev Latamer Microbiol*, **41**, 303–7.

Vargas-Parada, L. and Laclette, J.P. 2003. Gene structure of *Taenia solium* paramyosin. *Parasitol Res*, **89**, 375–8.

Vargas-Parada, L., Merchant, M.T., et al. 1999. Formation of calcareous corpuscles in the lumen of excretory canals of *Taenia solium* cysticerci. *Parasitol Res*, **85**, 88–92.

Vargas-Parada, L., Solís, C.F. and Laclette, J.P. 2001. Heat shock and stress response of *Taenia solium* and *T. crassiceps* (Cestoda). *Parasitology*, **122**, 583–8.

Vaz, A.J., Nunes, C.M., et al. 1997. Immunoblot with cerebrospinal fluid from patients with neurocysticercosis using antigen from cysticerci of *Taenia solium* and *Taenia crassiceps*. *Am J Trop Med Hyg*, **57**, 354–7.

Vazquez, V. and Sotelo, J. 1992. The course of seizures after treatment of cerebral cysticercosis. *N Engl J Med*, **327**, 696–701.

Vazquez-Talavera, J., Solis, C.F., et al. 2001a. Characterization and protective potential of the immune response to *Taenia solium* paramyosin in a murine model of cysticercosis. *Infect Immun*, **69**, 5412–16.

Vazquez-Talavera, J., Solis, C.F., et al. 2001b. Human T and B cell epitope mapping of *Taenia solium*. *Parasite Immunol*, **23**, 575–9.

Vega, R., Piñero, D., et al. 2003. Population genetic structure of *Taenia solium* from Madagascar and Mexico: implications for clinical profile diversity and immunological technology. *Int J Parasitol*, **33**, 1479–85.

Veit, P., Bilger, B., et al. 1995. Influence of environmental factors on the infectivity of *Echinococcus multilocularis* eggs. *Parasitology*, **110**, 79–86.

Verastegui, M., Gilman, R.H., et al. 2002. *Taenia solium* oncosphere antigens induce immunity in pigs against experimental cysticercosis. *Vet Parasitol*, **108**, 49–62.

Verastegui, M., Gilman, R.H., et al. 2003. Prevalence of antibodies to unique *Taenia solium* oncosphere antigens in taeniasis and human and porcine cysticercosis. *Am J Trop Med Hyg*, **69**, 438–44.

Vilhena, M., Santos, M. and Torgal, J. 1999. Seroprevalence of human cysticercosis in Maputo, Mozambique. *Am J Trop Med Hyg*, **61**, 59–62.

Vogel, H. 1957. Uber den *Echinococcus multilocularis* Suddeutschlands. Das Bandwurmstadium von Stammen Menschlicher und tierischer Herkunft. *Zeit fur Tropenmed Parasitkd*, **8**, 404–56.

von Sinner, W.N. 1991. New diagnostic signs in hydatid disease: radiography, ultrasound, CT and MRI correlated to pathology. *Eur J Radiol*, **12**, 150–9.

Vuitton, D.A. 2003. The ambiguous role of immunity in echinococcosis: protection of the host or the parasite? *Acta Trop*, **85**, 119–32.

Vuitton, D.A., Bresson, H.S., et al. 1988. IgE-dependent humoral immune response in *Echinococcus multilocularis* infection: circulating and basophil-bound specific IgE against *Echinococcus* antigens in patients with alveolar echinococcosis. *Clin Exp Immunol*, **71**, 247–52.

Vuitton, D., Bresson, H.S., et al. 1989. Cellular immune response in *Echinococcus multilocularis* infection in humans. II Natural killer cell activity and cell subpopulations in the blood and in the periparasitic granuloma of patients with alveolar echinococcosis. *Clin Exp Immunol*, **78**, 67–74.

Wang, I.C., Ma, Y.X., et al. 1997. A comparative study on egg hatching methods and oncosphere viability determination for *Taenia solium* eggs. *Intl J Parasitol*, **27**, 1311–14.

Wang, Y.H., Rogan, M.T. and Vuitton, D.A. 2001. Cystic echinococcosis in semi-nomadic pastoral communities in northwest China. *Trans R Soc Trop Med Hyg*, **95**, 153–8.

Wang, Y.H., Zhang, X., et al. 2003. Classification, follow-up and recurrence of hepatic cystic echinococcosis using ultrasound images. *Trans R Soc Trop Med Hyg*, **97**, 203–11.

Wandra, T., Subahar, R., et al. 2000. Resurgence of cases of epileptic seizures and burns associated with cysticercosis in Assologaima, Jayawijaya, Irian Jaya, Indonesia, 1991–95. *Trans R Soc Trop Med Hyg*, **94**, 46–50.

Wen, H. and Craig, P.S. 1994. IgG subclass responses in human cystic and alveolar echinococcosis. *Am J Trop Med Hyg*, **51**, 479–86.

Wen, H., Zhang, H.W., et al. 1994. Initial observations on albendazole in combination with cimetidine for the treatment of human cystic echinococcosis. *Ann Trop Med Parasitol*, **88**, 49–52.

White, A.C. 2000. Neurocysticercosis: updates on epidemiology, pathogenesis, diagnosis and management. *Ann Rev Med*, **51**, 187–206.

White, A.C., Robinson, P. and Kuhn, R. 1997. *Taenia solium* cysticercosis: host–parasite interactions and the immune response. In: Freedman, D.O. (ed.), *Chemical immunology.* , *Immunopathogenic aspects of diseases induced by helminth parasites*. vol. 66. Basel: Karger, 209–30.

White, A.C., Molinari, J.L., et al. 1992. Detection and preliminary characterization of *Taenia solium* metacestode proteases. *J Parasitol*, **78**, 281–7.

WHO. 1996. Informal Working Group on Echinococcosis. Guidelines for treatment of cystic and alveolar echinococcosis in humans. *Bull WHO*, **74**, 231–42.

WHO/OIE. 2001. *Manual on echinococcosis in humans and animals: a public health problem of global concern*. Eckert, J. et al. (eds). Geneva: WHO, pp 265.

Wilkins, P.P., Allan, J.C., et al. 1999. Development of a serologic assay to detect *Taenia solium* taeniasis. *Am J Trop Med Hyg*, **60**, 199–204.

Widdowson, M.A., Cook, A.J.C., et al. 2000. Investigation of risk factors for porcine *Taenia solium* cysticercosis: a multiple regression analysis of a cross-sectional study in the Yucatan peninsula, Mexico. *Trans R Soc Trop Med Hyg*, **94**, 620–4.

Williams, J.F., Perez-Escandi, M.V. and Oriol, R. 1971. Evaluation of purified lipoprotein antigens of *Echinococcus granulosus* in the immunodiagnosis of human infection. *Am J Trop Med Hyg*, **20**, 575–9.

Wilson, J.F. and Rausch, R.L. 1980. Alveolar hydatid disease. A review of clinical features of 33 indigenous cases of *Echinococcus multilocularis* infection in Alaskan Eskimos. *Am J Trop Med Hyg*, **29**, 1340–55.

Wilson, M., Bryan, R.T., et al. 1991. Clinical evaluation of the cysticercosis enzyme-linked immunoelectrotransfer blot in patients with neurocysticercosis. *J Infect Dis*, **164**, 1007–9.

Woolard, D.J., Gauci, C.G., et al. 2001. Protection against hydatid disease induced with the EG95 vaccine is associated with conformational epitopes. *Vaccine*, **19**, 498–507.

Yamasaki, H., Nakao, M., et al. 2002. DNA differential diagnosis of human taenidd cestodes by base excision sequence scanning thymine-base reader analysis with mitochondrial genes. *J Clin Microbiol*, **40**, 3818–21.

Yang, H.J., Chung, J.Y., et al. 1998. Immunoblot of a 10 kDa antign in cyst fluid of *Taenia solium* metacestodes. *Parasite Immunol*, **20**, 483–8.

Yong, W.K., Heath, D.D. and van Knapen, F. 1984. Comparison of cestode antigen in an enzyme-linked immunosorbent assay for the diagnosis of *Echinococcus granulosus*, *Taenia hydatigena* and *T. ovis* infections in sheep. *Res Vet Sci*, **36**, 24–31.

Yoshino, K. 1933. Studies on the post-embryonal development of *Taenia solium*. Part III. On the development of cysticercus cellulosae within the definite intermediate host. *J Med Ass Formosa*, **32**, 166–9.

Zhipiao, X.B., Yvequing, Z., et al. 1980. Muscular pseudohypertrophy due to cysticercosis cellulosae. Report of 3 cases. *Chin Med J*, **93**, 4853.

Zini, D., Farrell, V.J.R. and Wadee, A.A. 1990. The relationship of antibody levels to the clinical spectrum of human neurocysticercosis. *J Neurol Neurosurg Psych*, **53**, 656–61.

Zymberg, S.T., Palva-Nieto, M.A., et al. 2003. Endoscopic approach to fourth ventricle cysticercosis. *Arq Neuro-Psiquiatr*, **61**, 1–7.

34

Gastrointestinal nematodes – *Ascaris*, hookworm, *Trichuris*, and *Enterobius*

CELIA V. HOLLAND

Nearly 200 species of helminth have been found associated with the human alimentary tract (Coombs and Crompton 1991), many of these probably being the result of accidental or spurious infections. Of the habitual parasites, the pinworm or threadworm *Enterobius vermicularis* (L, 1758) and four species of soil-transmitted nematodes, the roundworm *Ascaris lumbricoides* L, 1758, hookworms *Ancylostoma duodenale* (Dubini, 1843) Creplin, 1845 and *Necator americanus* (Stiles, 1902) Stiles, 1903 and the whipworm *Trichuris trichiura* (L, 1771) Stiles, 1901 are some of the commonest infections on earth (Table 34.1).

The origins of these human–helminth relations are intriguing and might perhaps be traced to human activities leading to the domestication of animals. A qualitative survey indicates that humans share surprisingly few intestinal helminth infections with the other 180 or so species of primate (see Coombs and Crompton 1991; Macdonald 1984). *E. vermicularis* may well have evolved with humans from our primate ancestors, but the soil-transmitted species might perhaps be traced to human activities leading to the domestication of animals. Humans do, however, seem to have the same or similar types of infection to those found in domesticated animals, especially dogs (Beaver 1954) and pigs (Bell et al. 1988). Perhaps human *A. lumbricoides*, if it really is a species in its own right (see Crompton 1989), might have been acquired from porcine *A. suum* or its ancestor after the last ice age. Pigs were probably domesticated in several places (Clutton-Brock 1987), but China is likely to have been one of the first and there remains an intimate relationship today between Chinese people and their pigs (Peng et al. 1995).

The domestication of the very few species of mammal that humans now rely on for food, clothing, and transport began about 9 000 years BC (Clutton-Brock 1987) when the human population was about 10 million (Lewin 1989). Evidence from early writings (Bryan 1930; Ebbell 1937), from mummified bodies (see Cockburn and Cockburn 1980) and from archaeological digs (Fry and Moore 1969; Bundy and Cooper 1989) indicates that pinworm, roundworm, hookworm, and whipworm were

Table 34.1 *Features of* Ascaris, hookworms *and* Trichuris *in human hosts*

Feature	Ascaris	Ancylostoma	Necator	Trichuris
No. species/subspecies in genus[a]	16	23	8	71
No. species infecting humans[b]	2	7	3	3
Global no. of infections[c]	1470×10^6	135×10^6	735×10^6	1300×10^6
Estimate of cases of morbidity[d]	$120–215 \times 10^6$	$27–39 \times 10^6$	$63–91 \times 10^6$	$90–130 \times 10^6$

a) Yamaguti (1961).
b) See Table 34.2 and Coombs and Crompton (1991).
c) Chan (1997).
d) Chan et al. (1994). The estimates for the separate species of hookworm are based on the assumption that 30 percent are due to *A. duodenale* and 70 percent to *N. americanus* (Crompton and Stephenson 1990).

well established in humans a few thousand years ago. Today, the human population is over 6.3 billion and by the year 2010 is likely to be 7 billion with around 80 percent of the people living in the countries currently identified as less developed (Bulatao et al. 1990). Pinworms are the only species common in temperate countries; pinworms and the soil-transmitted species are widely distributed in countries with warmer climates, often occurring concurrently in individuals. It has been estimated that around a billion people are infected with *A. lumbricoides*, *A. duodenale*, *N. americanus*, and *T. trichiura*, often concurrently (Crompton 1989); this means that four out of every ten people in much of Africa, Asia, and South America harbor worms.

The notion that humans may have acquired the soil-transmitted nematodes during the domestication of a few species of wild mammal over a period of a few thousand years should not be discarded too lightly without further consideration. The human host has undergone about 400 generations and a vast social and technical evolution during the last 10 000 years, while *A. lumbricoides* has undergone about 20 000 generations which, coupled with its prodigious fecundity, could have created the circumstances for an explosive evolution.

The history of human helminth infections has been carefully documented by Grove (1990). Although three of these intestinal species (*Ascaris*, *Enterobius*, and *Trichuris*) were known to Linnaeus in the 18th century, and *Ancylostoma* was identified in the early 19th century, details of their life histories did not become known until much later. A fundamental observation in the case of *A. lumbricoides* was made by Koino (1922) who courageously swallowed about 2 000 eggs and subsequently passed 667 worms after taking an anthelmintic drug. He was extremely ill during the course of this experimental infection at a time when the larvae would have been occupying his lungs. Looss (1901) published the results of an experiment which proved that hookworm larvae penetrate skin. He placed a drop of water containing infective larvae of *A. duodenale* on the skin of a leg awaiting amputation. An hour later, when the surgery had been completed, Looss retrieved samples of skin and found the larvae beneath the skin. Calandrucci in 1886, having confirmed over a period of 6

months that he was free from infection with *T. trichiura*, swallowed infected eggs of the helminth and found eggs in his stools 27 days later. This result was published by Grassi (1887) who may not have given Calandrucci sufficient credit for his effort (Grove 1990). These studies are of great importance because they established the nature of the helminths' life cycles and the routes of transmission to human hosts.

CLASSIFICATION

In this chapter, the five species of helminth are assumed to belong to the phylum Nematoda, as advocated by Maggenti (1982), rather than to a class of the phylum Aschelminthes, as adhered to by Hyman (1951). Much work needs to be done before we have a secure understanding of the phylogeny of these ubiquitous animals, which thrive as free-living, plant-parasitic and animal-parasitic organisms (Andrassy 1976; Maggenti 1982; Poinar 1983). Here is a simple classification for soil-transmitted nematodes, based mainly on Anderson et al. (1974):

> Phylum Nematoda
>> Class Secernentea
>>> Order Ascaridida
>>>> Family Ascarididae
>>>>> Genus *Ascaris*
>>>>>> Species *lumbricoides*
>>>>>> Trivial name roundworm
>>> Order Strongylida
>>>> Family Ancylostomatidae
>>>>> Genus *Ancylostoma*
>>>>>> Species *duodenale*
>>>>>> Trivial name hookworm
>>>> Family Ancylostomatidae
>>>>> Genus *Necator*
>>>>>> Species *americanus*
>>>>>> Trivial name hookworm
>> Class Adenophorea
>>> Order Enoplida
>>>> Family Trichuridae
>>>>> Genus *Trichuris*
>>>>>> Species *trichiura*
>>>>>> Trivial name whipworm

Table 34.2 *Zoonotic infections of humans with species of* Ascaris, Ancylostoma, Necator *and* Trichuris

Genus and species	Nonhuman hosts	Observations on infection in human hosts	References
Ascaris suum	Pig	Sexually mature; in small intestine	Davies and Goldsmid (1978)
Ancylostoma brasiliense	Cat, dog	Cutaneous larva migrans (CLM)[a]	Yoshida et al. (1974)
A. caninum	Dog	CLM; adult in small intestine	Croese et al. (1994)
A. ceylanicum	Cat, dog	Adult in small intestine	Carroll and Grove (1986)
A. japonica	Unresolved	Reported once from human in Japan	Barriga (1982a)
A. malayanum	Bear	No information	Yorke and Maplestone (1926)
A. tubaeforme	Cat	CLM	Muller (1975)
Necator argentinus	Unresolved	Reported once from human	Yoshida (1973)
N. suillus	Pig	Reported at least once from human	Barriga (1982a)
Trichuris suis	Pig	Adult in large intestine	Barriga (1982b)
T. vulpis	Dog	Sexually mature adult in appendix	Kenney and Eveland (1978)

a) For a full discussion of CLM, see Beaver (1956).

Although the ecology and general biology of *A. lumbricoides*, *A. duodenale*, *E. vermicularis*, *N. americanus* and *T. trichiura* have much in common, the latter species has major morphological differences from the others. Most noticeably, it lacks phasmids (paired, glandular sensory organs) at its caudal end and is equipped with a stichosome esophagus. Taxonomists have used the presence or absence of phasmids as a basic feature in the systematic organization of nematodes.

If adult worms (either alive or in a good state of preservation) are available, an investigator should have no difficulty in identifying the genera under discussion. The keys prepared by Yamaguti (1961) and Anderson et al. (1974) are recommended together with the biological and clinical descriptions published by Beaver and Jung (1985). If only morphological evidence is available, problems are likely to be encountered in trying to distinguish from each other, for example, all the seven species of *Ancylostoma* reported from humans (Tables 34.1 and 34.2). Adult *A. lumbricoides* and adult *A. suum* remain virtually indistinguishable on morphological grounds (see Crompton 1989) as are *T. trichiura* and the other species of *Trichuris* found in non-human primates and pigs.

MORPHOLOGY AND STRUCTURE

Detailed accounts of the anatomy, functional morphology, and structure of animal-parasitic nematodes, including the species considered here, *Trichuris*, have been published by Chitwood and Chitwood (1974), Gibbons (1986), and Bird and Bird (1991) among others. Each of these treatises deals comparatively with different features such as the cephalic region and the nervous system rather than presenting descriptions of individual species. The brief accounts given below for the four species of soil-transmitted nematode under review in this chapter cover features not highlighted in the definitive works cited above. Inevitably, some of the details are based on extrapolation from the results of studies on *A. suum*, *Anc. caninum*, and *T. suis* from animal hosts.

Ascaris lumbricoides

The eggs of *A. lumbricoides* are seen in human stool samples in two general forms: unfertilized eggs (Figure 34.1a) and fertilized eggs which may or may not possess a cortical layer (Figure 34.1b, c). Typically eggs in stools appear to be brown in color, probably because they become stained by bile pigments while in the gastrointestinal tract. The eggshell of a fertilized egg ($50–70 \times 40–50$ μm) consists of an inner lipid layer responsible for selective permeability (Perry and Clark 1982), a chitin–protein layer responsible for structural strength and an outer vitelline layer (Foor 1967; Wharton 1980). The inner layer contains a remarkably resilient lipoprotein, known as ascaroside, which explains how the eggs with enclosed infective larvae can survive formaldehyde, disinfectants and other destructive chemicals. The fertilized egg is frequently observed to have an uneven deposit of mucopolysaccharide (the cortical layer) on its outer surface (Figure 34.1b). This deposit is obtained when the egg is passing through the uterus of the female worm (Foor 1967) and is responsible for its adhesive properties (Kagei 1983).

The first-stage larva of *A. lumbricoides* develops inside the eggshell, molts there, and forms the second-stage larva. Soon after hatching, second-stage larvae have a typical filariform appearance and measure about 250×14 μm (Nichols 1956). Just before the second molt in the lungs, they measure about 560×28 μm in length. After the fourth (final) molt in the small intestine, growth is rapid. Freshly recovered adult male and female *A. lumbricoides* are often observed to be a pinkish cream in color and to measure 200–300 mm in length. A male, which is usually smaller than a female of

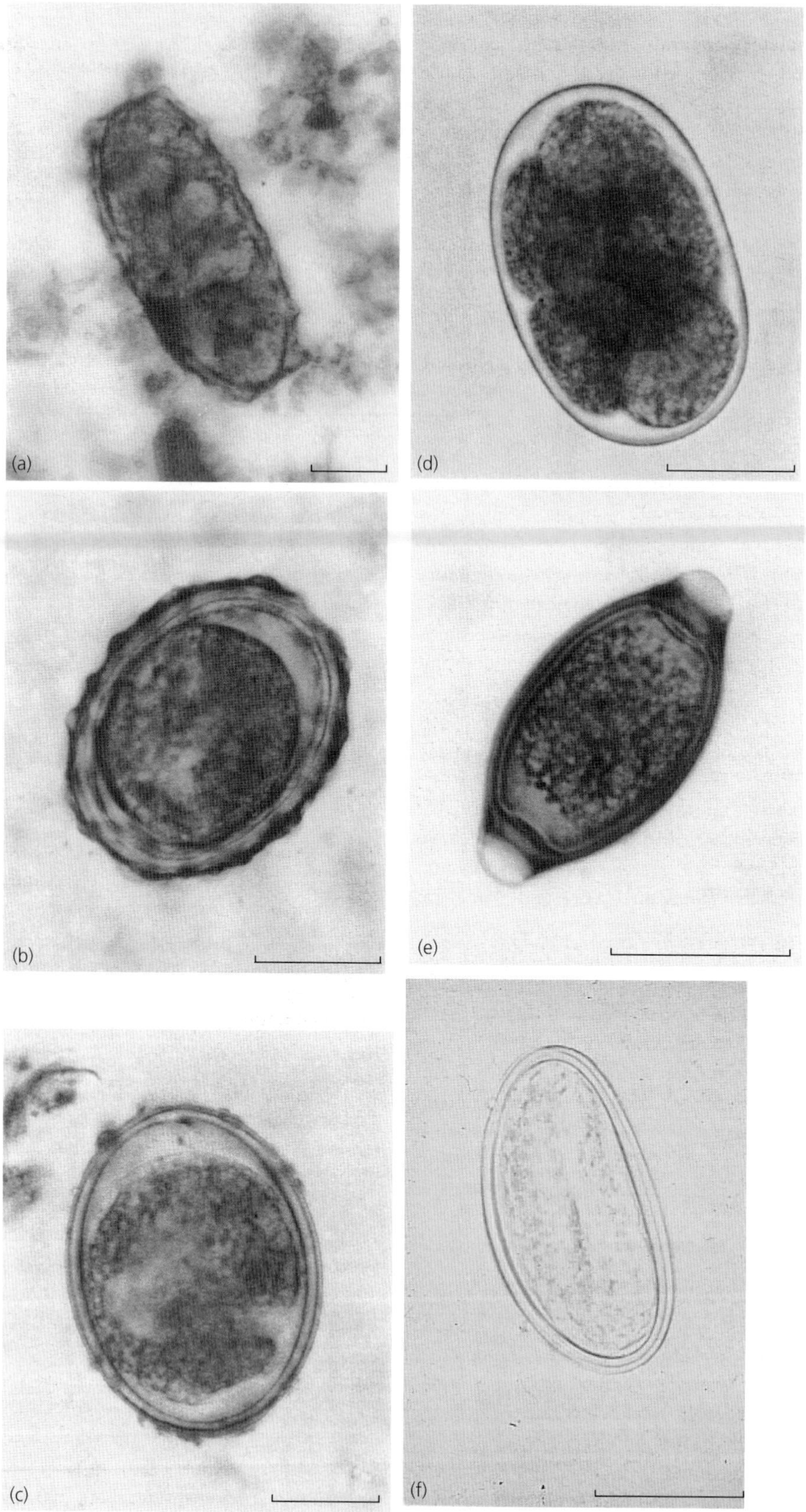

Figure 34.1 *Photomicrographs of eggs of soil-transmitted nematodes as seen in human stool samples:* **(a)** *unfertilized egg of* Ascaris lumbricoides; **(b)** *fertilized egg of* A. lumbricoides *with cortical layer;* **(c)** *fertilized egg of* A. lumbricoides *without cortical layer;* **(d)** *egg of a hookworm;* **(e)** *egg of* Trichuris trichiura; **(f)** *egg of* Enterobius vermicularis. *Bar = 25 μm. (Photomicrographs (a–e) from WHO 1994;* **(f)** *courtesy of R. Muller)*

the same age, has a curved posterior tail that accommodates the copulatory apparatus. Skryabin et al. (1991) describe adult *A. lumbricoides* as having a mouth surrounded by three well-developed lips, rows of small 'teeth' (i.e. denticles of Sprent (1952)) on the lips, a cuticle with transverse striations, a didelphic uterus and the vulva in the anterior half of the body. The morphology of these denticles is a character widely used by workers who seek to distinguish *A. lumbricoides* from *A. suum* (Maung 1973).

The cuticle of *A. lumbricoides* under natural circumstances is only ever exposed to the conditions prevailing in the host. Collagen, stabilized by disulfide links, forms the main structural protein of the adult cuticle, which becomes more complex with each molt (Bird and Bird 1991). Collagen is now known to be the principal protein in the cuticles of all adult nematode species investigated to date. An investigation into the composition of the cuticles of developmental stages of *A. suum* from pigs revealed that the proteins extracted with 2-mercaptoethanol from third- and fourth-stage larvae and from adult worms were readily digested on incubation with bacterial collagenase (Fetterer et al. 1990). A significant amount of the cuticle from second-stage larvae was not affected by 2-mercaptoethanol and was not digested by bacterial collagenase. Presumably, the same results may apply to *A. lumbricoides*, with the differences detected in the second-stage larval cuticle being an adaptation either to survival outside the host or the infection process or both events.

Ancylostoma duodenale and Necator americanus

When seen in human stools (Figure 34.1d), the eggs of these two species are indistinguishable, being characteristically barrel-shaped with a thin shell and measuring about $60 \times 75 \times 36-40$ μm (WHO 1994). If hookworm eggs alone are available, identification of the species cannot be made without first using the Harada–Mori technique to culture the newly-hatched larvae to the third stage when morphological differences become apparent (WHO 1981).

Descriptions of the morphology of the larval stages of hookworms are available through the experimental studies of Nichols (1956) and observations summarized by Pawlowski et al. (1991). The first-stage larvae, which escape from the eggs, measure about 200 μm in length. They feed on organic debris, then molt and the resulting second-stage larvae measure up to 500 μm. After another molt, the infective third-stage larvae measure from 500–700 μm, those of *Ancylostoma* being generally longer than those of *Necator* (WHO 1981).

The most detailed description of the morphology of adult *A. duodenale* is to be found in the classic account by Looss (1905). For both species, male worms

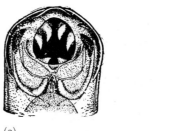 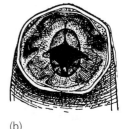

(a)   (b)

Figure 34.2 *Buccal capsules:* **(a)** *adult* Ancylostoma duodenale; **(b)** *adult* Necator americanus

(5–11 mm) are shorter than females (9–13 mm) and *A. duodenale* is generally longer and more sturdily built than *N. americanus*. Proof of identity is obtained by comparing the morphologies of the buccal capsule (Figure 34.2). The term 'hookworm' derives from the prominent curve to the anterior end of specimens of *N. americanus* (Pawlowski et al. 1991).

Enterobius vermicularis

Pinworm eggs are ovoid, measuring 50–54 μm × 20–27 μm. They are asymmetrically flattened on one side, and appear colorless when recovered from the perianal skin (Figure 34.1f). The outer layer of the eggshell is albuminous and sticky, enabling the egg to adhere readily. When laid, the egg contains an immature first-stage larva, but this develops rapidly. When swallowed, the eggs hatch in the small bowel and the larvae pass down into the large bowel, where they complete their molting and development. Adult worms, which measure about 10×0.5 mm (female) and 2.5×0.2 mm (male), live primarily in the cecum, in close contact with the mucosa. Males are rarely seen, but the gravid females can sometimes be seen as white or yellowish worms around the anus. Such worms have prominent uteri filled with eggs.

Trichuris trichiura

The eggs of *T. trichiura* are lemon-shaped with a characteristic plug at each end (Figure 34.1e). They are usually brown in a human stool sample and measure from 57–58 × 26–30 μm. As with eggs of *Ascaris* spp. and hookworms, it is virtually impossible to distinguish eggs of *T. trichiura* from those of *T. suis* and *T. vulpis* (Kenney and Yermakov 1980; Yoshikawa et al. 1989). According to studies by Wharton and Jenkins (1978), the eggs of *T. suis* have a shell similar in structure to that of *A. lumbricoides*, an observation that explains the survival of the enclosed infective larva under adverse environmental conditions (see Bundy and Cooper 1989). Apparently, there is a higher proportion of chitin in the plugs at the poles of an egg of *T. trichiura* than elsewhere in the eggshell. Wharton and Jenkins have suggested that this may facilitate the hatching process if the activated larva releases chitinase (see Rogers 1960). On hatching in the intestine, the second-stage larval *T. trichiura* measures about 260 × 15 μm in length (Beck and Beverley-Burton 1968). These larvae, if they behave as do those of other species of *Trichuris*, burrow into the intestinal mucosa, and it is Bundy and Cooper's view that *Trichuris* spp. are essentially tissue parasites. There is some confusion about where the intestinal penetration takes place and whether or not the larvae migrate along the intestinal tissues to reach the large intestine. During this prepatent phase (Table 34.3), *T. trichiura* must complete the series of four molts and is found as an adult intimately associated with the wall of the large intestine.

Adult whipworms have a highly characteristic shape from which the trivial name is derived. The long, thin anterior end lies in a burrow in the mucosa while the thicker end, which contains the reproductive tract, extends into the intestinal lumen. The worms are white in color and the males (30–45 mm) are not only shorter than the females (30–50 mm), but also have a coiled posterior end when viewed in vitro. The most characteristic morphological feature of species of *Trichuris* and their relatives is the stichosome, which is a glandular structure encircling the slender esophagus in the thin, anterior half of the worms (Beck and Beverley-Burton 1968).

Although *Enterobius* occurs throughout the world, it is usually perceived as important only in temperate countries, primarily because it is the only common intestinal nematode there. Many characteristics of pinworms (e.g. their relatively low pathogenicity and their biology and mode of transmission, which is directly contaminative and not dependent on initial fecal dissemination into the environment) differentiate

them from the other parasites covered in this chapter. *Ascaris*, hookworms and *Trichuris* are more pathogenic, they often occur together, their modes of transmission are facilitated by similar environmental and socioeconomic factors, and they are studied epidemiologically as a coherent group. For these reasons, the soil-transmitted species will be considered both together and separately.

SOIL-TRANSMITTED NEMATODES: BIOLOGY AND FEEDING

Comprehensive summaries of current knowledge and assumed knowledge of specific aspects of the biology of *A. lumbricoides*, *A. duodenale*, *N. americanus*, and *T. trichiura* are available in reviews by Crompton (1989, 1994, 2001), Bundy and Cooper (1989), Schad and Warren (1990) and Misegna and Gilles (1987). We may assume that molting, energy metabolism, excretion and reproduction occur as described for animal-parasitic nematodes generally by Bird and Bird (1991), Barrett (1981), Lee and Atkinson (1976), and Adiyodi and Adiyodi (1983). Their feeding activities merit special consideration, partly because little is understood about how they obtain nutrients and energy while living as larval parasites and partly because of the role of the feeding of the adult stages in pathogenesis and chronic disease. Feeding is based on the development of the muscular esophagus to pump fluids or materials in suspension along a simple gut (Bennet-Clark 1976).

Ascaris lumbricoides

To ensure development of the eggs, the parent worm must have supplied the zygote with nutrients sufficient to

Table 34.3 *Some features of the life histories of* Ascaris, *hookworms and* Trichuris[a]

Feature	Ascaris lumbricoides	Ancylostoma duodenale	Necator americanus	Trichuris trichiura
Longevity in host (years)	1–2	1	3–5	1–3
Location of adult worms	Jejunal lumen	Jejunal mucosa	Jejunal mucosa	Tissue of large intestine
Reproductive strategy	Dioecious	Dioecious	Dioecious	Dioecious
Fecundity (eggs/female/day)	c. 240 000	10 000–25 000	5 000–10 000	14 000–20 000
Prepatent period (days)	67–76	53	49–56	60–90
Patent period (days)	c. 300–650	c. 300	c. 1 050–1 775	c. 300–1 000
Embryonation (days)	12 ± 2, at 31°C			28, at 25°C
Infective stage	L2 in eggshell[b]	Free L3	Free L3	L2 in eggshell
Max. survival of infective stage	Up to 15 years			Up to 6 years[c]
Usual transmission route	Oral	Skin/oral	Skin	Oral
Tissue migration	Liver and lungs	Not necessary	Lungs	
Arrested development		+[d]		

a) Abstracted mainly from Hoagland and Schad (1978); Bundy and Cooper (1989); Crompton (1994).

b) L, larva; L1, L2, L3, L4 refer to the first, second, third and fourth larval (juvenile) stages, there being a molt after each stage has completed its contribution to the life history.

c) Based on observations made on *T. suis* by Hill (1957).

d) Discovered by Schad et al. (1973).

ensure progression to the infective second-stage larva and subsequent survival. After hatching in the human gut, the larvae must either carry with them or obtain, through direct feeding on host tissues and metabolites, enough food and energy to ensure a successful migration back to the gut and a series of development molts. The tissue migrations of parasitic larvae have remained an enigma: are the parasites reliving their life cycles in a once intermediate but now definitive host? (Smyth 1994). Given that *A. lumbricoides* starts its association with its host in the gut and then returns there, perhaps natural selection would have been expected to have eliminated the migratory phase, with its exposure to the host's full array of immunological mechanisms, unless there were some compensatory benefits for the parasite (Read and Skorping 1995). Read and Skorping found that helminths that migrated grew relatively larger than those that did not. Nematode fecundity is linked to worm size, so a benefit for the survivors of the tissue migration becomes apparent.

Once *A. lumbricoides* is established in the lumen of the host's small intestine, it feeds on digestion products by pumping chyme through its own gut. In doing so, adult *A. lumbricoides* are likely to swallow bacteria including some that may be pathogenic to the worms, such as *E. coli* and *S. aureus* (Adedeja and Ogunba 1991). Wardlaw et al. (1994) found that the pseudocoelomic fluid of *A. suum* contained potent bactericidal activity that was not due to lysozyme. Antibacterial defense mechanisms would be expected to be present in an organism adapted to live in the intestine and feed on its intestinal contents.

Davey (1964), after examining *A. suum* obtained within half an hour from freshly slaughtered pigs, observed many pig intestinal epithelial cells in the parasites' guts. He estimated that whereas the concentration of free epithelial cells in the pig's gut would be expected to be about 6×10^4/mm^3, the concentration in the anterior part of the gut of the worms was about 4×10^6/mm^3. This finding suggests that *A. suum* may either selectively feed on discharged epithelial cells or even browse on the mucosa in addition to ingesting chyme. There is no doubt that the mucosal surface of a pig harboring intestinal stages of *A. suum* shows histological damage (Martin et al. 1984), a finding commensurate with the observed maldigestion of lactose in *Ascaris*-infected pigs (Forsum et al. 1981) and children (Carrera et al. 1984).

Ancylostoma duodenale and Necator americanus

Hatched larval hookworms feed voraciously on bacterial and organic matter until the non-feeding third-stage larvae are formed. Presumably, these contain sufficient stored energy to support skin penetration. It is not known whether newly penetrated third-stage larvae feed in the host or whether the next molt must occur and then the fourth-stage larva will feed actively. Third-stage larvae of *A. duodenale* entering the host by the oral route appear not to undergo a tissue migration (Pawlowski et al. 1991). Those larvae of *A. duodenale* that penetrate skin and migrate through tissue must need much more food and energy than those entering the oral route. The third-stage larvae of *N. americanus* must always penetrate skin and undergo tissue migration to reach the gut.

The intestinal stages of hookworms feed on blood obtained by puncturing the capillary network in the mucosa. Host blood is then pumped along the worm's gut and nutrient molecules absorbed from the host's plasma. Feeding by both *A. caninum* and *A. duodenale* involves the release of an anticoagulant secreted at the time of feeding (Hotez and Cerami 1983). Probably much of the blood taken by a hookworm as it feeds passes through its body into the host's gut lumen (Crompton and Whitehead 1993). When the worm bites into the mucosa at a second site, the lacerations at the first site, together with residual anticoagulant activity, mean that further blood loss occurs. From an experimental study of *A. caninum* in dogs, Wang et al. (1983) concluded that an individual female worm pumped about 0.043 ± 0.04 ml of host blood per day and that a further 0.046 ± 0.019 ml was lost as a hemorrhage from the abandoned feeding site. Female worms cause more blood loss than males.

Trichuris trichiura

Despite current interest in the biology of whipworms, largely stimulated by the work of Bundy and colleagues (see Bundy and Cooper 1989), there remains some controversy about their feeding activities. Do the worms suck blood, a view supported by Beck and Beverley-Burton (1968), or do they digest host tissues by releasing proteolytic enzymes (see Watson 1960)? Burrows and Lillis (1964) clearly demonstrated that *T. vulpis* feeds on dog blood. They found dog blood in the gut of *T. vulpis* and by direct observation described how the worms, through a series of rapier-like slashing movements, could lacerate capillaries and then pump released blood into their guts. They also found blood in the gut of *T. trichiura*. The structure of the buccal apparatus with its piercing stylet also suggests that blood-feeding is its probable purpose. Furthermore, rectal bleeding and iron-deficiency anemia are characteristics of some patients suffering from trichiuriasis (Fisher and Cremin 1970; Lotero et al. 1974; Bundy and Cooper 1989). The etiology of iron deficiency anemia is complex and may not arise directly from the feeding activity of *T. trichiura*; the nutritional status of the hosts and polyparasitism are two of the factors which must be accounted for. Nevertheless, whipworms undoubtedly ingest host blood obtained from the mucosa of the large intestine.

SOIL-TRANSMITTED NEMATODES: LIFE HISTORIES

The direct life-history patterns of *A. lumbricoides*, *A. duodenale*, *N. americanus* and *T. trichiura* are well understood in general terms and can be represented diagrammatically in the form of a flow chart (Figure 34.3). This scheme is important because it focuses attention on the fact that a soil-transmitted nematode exists in two populations: a free-living cohort based on the eggs and related early larval stages, and an endoparasitic cohort based on the later larval stages and the adult worms in the gut. More detailed information about the life histories is summarized in Table 34.3, and much recent information is available in articles edited by Crompton et al. (1985, 1989) and Schad and Warren (1990), and in the review by Bundy and Cooper (1989). Aspects of the life histories are treated in some detail in the sections dealing with transmission and with epidemiology.

SOIL-TRANSMITTED NEMATODES: TRANSMISSION AND ESTABLISHMENT OF INFECTION

Ascaris lumbricoides

Infective second-stage larvae of *A. lumbricoides* must usually be swallowed by the human host in order to initiate the establishment of an infection. The larvae escape from the egg, presumably in response to the same stimuli identified by Rogers (1960) for the hatching in vitro of eggs of *A. suum*. Experimental work by Murrell et al. (1997) has shown that *A. suum* L_2 in the pig host almost exclusively invade the wall of the cecum and colon en route to the liver, rather than doing so into the intestine as was generally thought. The findings of

these authors raise important questions as to whether *A. suum* and *A. lumbricoides* share a similar migratory pathway and may suggest a greater potential for liver damage in human hosts. If the time course for *A. lumbricoides* in humans is the same as that for *A. suum* in pigs, mice, and rabbits, the larval stages spend about 4 days in the liver and about 14 days in the lungs and then begin to re-enter the gut via the bronchi and trachea. At least 65 days must pass after infection before eggs are first detected in the stools (Table 34.3; Takata 1951).

The eggs of *A. lumbricoides* are produced in vast numbers (Table 34.3) and, although many will either perish during embryonation due to exposure to ultraviolet radiation or become inaccessible to humans through the activities of earthworms or pass into the soil during rainfall, enough survive to ensure the persistence of the infections. In a recent experimental study carried out with eggs of *A. suum* contaminating soil in Poland, Mizgajska (1993) found that 0.1–4.5 percent of the eggs retained infectivity (observed larval movements) for 17 months and most of these had remained within the top 50 mm of the soil. During this period, some eggs had reached a depth of 210 mm in the soil. Any community lacking facilities for the safe disposal of human feces will remain vulnerable to infection with *A. lumbricoides*, and use of untreated night-soil as a fertilizer for vegetables will probably increase the risk of infection. The complex eggshells of *A. lumbricoides* (Wharton 1980) protect the larvae from mechanical, chemical, and physical damage, and the external mucopolysaccharide ensures that the eggs adhere to objects. In areas where infection is endemic, infective eggs have been found adhering to cooking and eating utensils, money, furniture, door handles, fruit, vegetables, and fingers (Kagei 1983), as well as contaminating the soil in households, gardens, and public parks (Morishita 1972; Yadav and Tandon 1989; Wong and Bundy 1990). Some evidence suggests that the eggs can become airborne and be inhaled in dust, thus gaining access to the alimentary tract (Bidinger et al. 1981; Kroeger et al. 1992).

Although there need be no doubt that transmission nearly always depends on swallowing eggs from a contaminated environment, other routes of transmission may also exist. For example, three case studies indicate that *A. lumbricoides* may occasionally pass the placenta (Chu et al. 1972; Rathi et al. 1981; da Costa-Macedo and Rey 1990) and, since larval *Ascaris* can be transplanted between hosts experimentally, there is even the slight chance that infections may be established inadvertently through organ-transplant surgery.

Ancylostoma duodenale and Necator americanus

The third-stage larvae of hookworms are responsible for transmission and the establishment of infections.

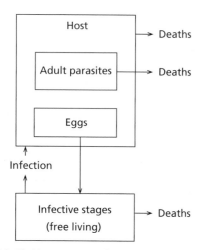

Figure 34.3 *Life-history pattern of soil-transmitted nematodes*

Hoagland and Schad (1978) have concluded that *A. duodenale* is an opportunistic species, partly because the third larval stages (Table 34.3) can penetrate skin or enter via the oral route. Recent evidence from China indicates that larval *A. duodenale* may pass from the mother to infect the fetus in utero; transmammary infection with *A. duodenale* cannot be ignored (see Banwell and Schad 1978).

Under favorable conditions, first-stage larvae hatch from hookworm eggs within about 24 hours of feces being deposited on the soil. One of the factors which explains why *A. duodenale* is not confined to tropical and subtropical countries is the ability of its eggs to survive at 7°C even if embryonation is not occurring (Beaver and Jung 1985). Hookworms flourish particularly in rural communities where there are long-established traditions of the daily use of defecation fields. The rhabditiform larvae feed on bacteria and organic debris, molt and continue to feed as second-stage larvae. These molt in turn to form third-stage filariform larvae, which no longer feed, but wait for the conditions facilitating skin contact and penetration. Hookworm larvae that fail to make contact with a susceptible host probably die from desiccation within a few days. According to Schad, even under the most favorable survival conditions of shade, moisture, and soil texture, probably less than 1 percent of a batch of infective larvae will live for more than a month (see Banwell and Schad 1978).

Skin penetration by hookworms has been studied in most detail using a hamster-adapted strain of *N. americanus*. Based on earlier work which showed that larval enzymes were involved (Matthews 1982), Salafsky et al. (1990) developed an artificial membrane that the larvae would attack and penetrate. They found that the presence of human essential fatty acids in the membrane significantly increased the ability of the larvae to penetrate in vitro and that these acids, especially linoleate, influenced the secretion of eicosanoid compounds by the larvae. Eicosanoids affect T and B cells and so may modulate the host's immune response during skin penetration (Salafsky et al. 1990).

Hoagland and Schad (1978) took the view that *N. americanus* was less of an opportunist than *A. duodenale* with less flexibility in the infection process. *Ancylostoma duodenale* can function as a food-borne infection (Schad et al. 1983).

Trichuris trichiura

The transmission and establishment of an infection of *T. trichiura* has much in common with that of *A. lumbricoides*, being dependent on the ingestion of an infective egg from a contaminated environment. The overall success of the process depends on the numbers of infective eggs, their accessibility to humans and their survival characteristics, which are related to climatic and soil conditions (see Bundy and Cooper 1989). Geophagia was demonstrated to be a significant risk factor for exposure to *Trichuris* among 10- to 18-year-old Kenyans, with a threefold higher intensity of reinfection with *T. trichiura* 11 months after treatment in geophagic individuals compared to those who did not exhibit geophagia (Geissler et al. 1998).

SOIL-TRANSMITTED NEMATODES: EPIDEMIOLOGY AND POPULATION BIOLOGY

Knowledge of the epidemiology and population biology of soil-transmitted nematodes has advanced rapidly during the last decade. Recognition of the need to quantify the public health significance of the infections and the development of a mathematical approach to study how these infections survive have given impetus to this work. The theoretical aspects of the population biology of the helminths have also led to the development of models for:

- the control of soil-transmitted helminth infections by means of chemotherapy (Anderson 1989)
- evaluating the effects of control measures (Medley et al. 1993)
- estimating the morbidity attributable to the infections (Guyatt and Bundy 1991; Lwambo et al. 1992; Chan et al. 1994)
- simulating host immune responses (Anderson 1994).

Some workers have expressed concern about the lack of validation of some of the models and the fact that they are generally based on limited data (Goodman 1994). This view does, however, not detract from their value in helping to understand the relationships between humans, roundworms, hookworms, and whipworms.

The four common species of soil-transmitted nematode are particularly amenable to study and the following epidemiological generalizations apply.

- Accurate diagnosis can be made by the detection of eggs in the stools.
- Counts of eggs in defined quantities of stool can give a useful comparative indirect measure of the intensity of infection, assuming that the egg count increases as the number of female worms in the gut increases.
- In practice, density-dependent constraints apply to the individual fecundity of female worms; after a certain worm burden has been reached, individual egg production will decline.
- The life histories are direct and the population of worms in a host does not increase unless new infective stages are acquired.
- The distribution of numbers of worms per host is over-dispersed or aggregated so that in a given population of hosts a few will tend to harbor most of the worms.

- Individual hosts appear to be predisposed to a particular infection intensity.

Detailed treatments of the principles underlying the epidemiology and population biology of soil-transmitted nematodes are documented and developed by Bundy (1986, 1988), Anderson and May (1985, 1991), and Scott and Smith (1994). Perhaps Croll and Ghadirian's (1981) research on 'wormy persons', which dealt with infections on *A. lumbricoides*, *A. duodenale*, *N. americanus* and *T. trichiura* was the catalyst for much of the endeavor that has been applied to understanding soil-transmitted nematodes. In this section of this chapter, the four species are treated together since their population biology has so much in common and they often occur together in the same community or individual.

Global distribution and abundance

Crompton (1989) reckoned that human infections with *A. lumbricoides* were occurring in 150 of the world's countries; some national situations were trivial but others were probably of considerable public health significance. The same global pattern of infection probably applies to the hookworms and to *T. trichiura*. Of the hookworms, *N. americanus* prevails in tropical and subtropical regions, whereas *A. duodenale* tends to occur in the cooler and somewhat drier regions (Pawlowski et al. 1991). Mixed infections of hookworms also exist, especially in northern India and the middle of China. Bundy and Cooper (1989) pointed out that while *T. trichiura* overlaps with *A. lumbricoides* in its distribution, it is also still a problem in some temperate countries. Estimates of the numbers of cases of infection with the four species of soil-transmitted helminth are given in Table 34.1. Overall, at least a fifth of the world's population is infected. The recent collaboration of scientists between China and other countries has enabled new estimates to be made of the numbers of cases of *A. lumbricoides* in that country. Chan et al. (1994) estimated 568 million cases and Peng et al. (1995) have calculated the number to be 538 million, indicating that the national prevalence of *A. lumbricoides* infection in China is about 47 percent. A national prevalence figure, however, belies the fact that infections like *A. lumbricoides* in humans are not distributed evenly between communities (Crompton 1989; Yu 1994; Peng et al. 1995). Climate, local economies, social customs, and ethnicity all influence the distribution.

Prevalence

The results of surveys to estimate the prevalence of these infections are of limited usefulness. They identify locations where infections occur and provide information about the numbers of cases; they contribute little to

assessments of the extent of morbidity. Apart from providing information about the number of cases, the most useful aspect of prevalence data is to examine the relationship between prevalence and host age. Typical patterns are shown diagrammatically for *A. lumbricoides*, hookworm, and *T. trichiura* (Figure 34.4). These relationships have been detected in many surveys carried out in many countries (Bundy and Cooper 1989; Anderson and May 1991; Crompton 1994). The data are generally obtained by examining stool samples for the presence of helminth eggs, the stools having been obtained during cross-sectional surveys. The conduct of a survey is often complex and the reader is recommended to consult the protocol described by Thein Hlaing (1989).

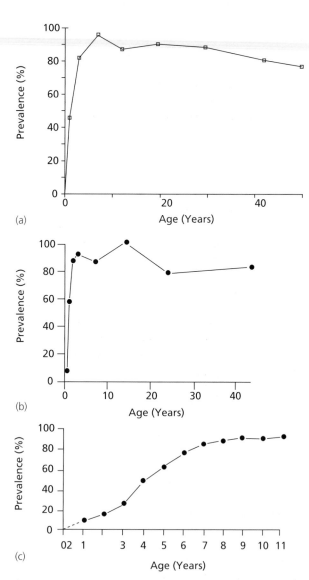

Figure 34.4 *Diagrammatic representations of the age–prevalence pattern between human hosts and soil-transmitted nematodes:* **(a)** Ascaris lumbricoides; **(b)** Trichuris trichiura; **(c)** hookworm *(Redrawn from:* **(a)** *Elkins et al. 1986;* **(b)** *Bundy 1986;* **(c)** *Nawalinski et al. 1978)*

The relationship between prevalence and age reveals that *A. lumbricoides* and *T. trichiura* become established in infants, soon after weaning (Figure 34.4a, b). Maximum prevalence values are attained in most cases when children reach the age of 10 years and then a gradual decline with age is observed. The age–prevalence pattern is somewhat different in the case of hookworm infections, there being a more gradual rise in the number of cases with plateau values observed as people reach young adulthood (Figure 34.4c).

Intensity

The population biology and public health significance of soil-transmitted nematodes cannot be understood, and strategies for their control cannot be developed, without reliable information about the intensities of infection. Intensity in this context is defined as the number of worms per infected host and it can be measured directly by counting worms expelled with stools after anthelmintic chemotherapy or indirectly by counting eggs in known amounts of stool.

Provided there is good compliance, expulsion chemotherapy is the preferred method for measuring intensity, but it relies on highly effective anthelmintic drugs. Ideally, all the worms present in the host's intestine should be expelled within 24 hours so that relatively few stools have to be examined. This is generally the case for *A. lumbricoides*. Hookworms and *T. trichiura* are much smaller than *A. lumbricoides* and are much harder to find in the stools. Furthermore, more than one dose of drug and more than one stool collection may be required to release and recover all the *T. trichiura* from the large intestine (Bundy and Cooper 1989). Although quantitative egg counts are much easier to perform and many more subjects can be investigated in an intensity survey as a result, the information is inevitably less accurate than that obtained from the expulsion method. Egg counts may fail to give information about intensity if the infection consists of male or immature worms only or if there are density-dependent constraints on female worm fecundity (Thein Hlaing et al. 1984). Generally, egg counts using the Kato Katz method (WHO 1994) do provide a useful estimate of intensity with the number of eggs per gram (epg) of stool being observed to increase as the worm burden increases (Forrester and Scott 1990).

The typical relationship between infection intensity and host age is shown diagrammatically in Figure 34.5 for *A. lumbricoides*, hookworms, and *T. trichiura*. During childhood, the age–intensity profile for *A. lumbricoides* and *T. trichiura* mirrors that for the age–prevalence profile, there being a rapid acquisition of worms. In later life, the intensity of these two species declines (Figure 34.5a, b); due to the development of some degree of immunity or to changes in behavior or

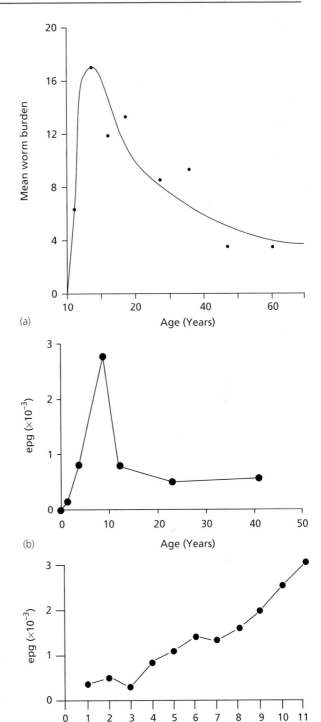

Figure 34.5 *Diagrammatic representations of the age–intensity patterns between human hosts and soil-transmitted nematodes:* **(a)** Ascaris lumbricoides; **(b)** Trichuris trichiura; **(c)** *hookworm (Redrawn from:* **(a)** *Thein-Hlang 1985;* **(b)** *Bundy 1986;* **(c)** *Schad and Anderson 1985)*

to a combination of these factors. The intensity of hookworm infections follows the same trend as the age–prevalence relationships (Figure 34.5b) with adults tending to carry the greater worm burdens. These age–intensity relationships determine which sections of the population will be most at risk of morbidity, which will be responsible for most contamination of the environment with transmission stages, and which will be most in need of treatment if resources become available.

Frequency distribution of numbers of worm per host

Prevalence and intensity data reveal that soil-transmitted nematodes are not distributed randomly in host populations; most infected hosts harbor a few worms each, while a few hosts harbor most of the worms (Anderson 1986; Bundy 1986; Crompton 1994). The observed pattern (Figure 34.6) is best described by the negative binomial distribution and the worms are said to be aggregated or over-dispersed. When the negative binomial model is fitted to worm frequency distribution data, the aggregation constant (k) is usually found to be <1, indicating aggregation (Anderson 1986). Similarly, if the ratio of the variance to mean number of worms in the population under study ($S^2/\bar{x}$) is >1, aggregation or over-dispersion is also confirmed (Anderson and Gordon 1982).

Predisposition to intensity of infection

Predisposition, in the context of soil-transmitted nematodes, is the term used to describe the observation that after anthelmintic chemotherapy, individuals seem to acquire worm burdens similar to those they harbored before treatment. The observation appears to be secure whether expelled numbers of worms or egg counts are used as measures of intensity.

Predisposition is now a well-established epidemiological feature of *Ascaris* (Elkins et al. 1986), *Trichuris* (Bundy et al. 1987), and hookworm infection (Schad and Anderson 1985). Evidence for multiple species predisposition (*Ascaris*, *Trichuris*, hookworm, and *Enterobius*) was also provided by Haswell-Elkins et al. (1987a). The factors that contribute to the observed predisposition and their relative importance have been the subject of review and debate by a number of authors (Bundy 1988; Keymer and Pagal 1990; Bundy and Medley 1992; Holland and Boes 2002). Whether predisposition is a feature of long-term causal factors, such as host genetics and host socioeconomic status, or short-term factors, such as the host acquired immune response, is obviously important for the design of appropriate control strategies. Recently, Quinnell et al. (2001) assessed reinfection and predisposition to *N. americanus* in a rural village in Papua New Guinea over an 8-year period. Interestingly, predisposition could be detected 6–8 years after a single round of chemotherapy but was not detectable after repeated chemotherapy. The authors concluded that differences in susceptibility are likely to influence predisposition but that longer-term variation in either expo-

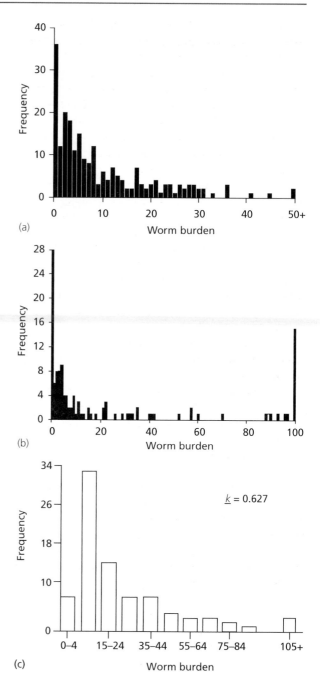

Figure 34.6 *Diagrammatic representations of the frequency distribution of numbers of worms per host:* **(a)** *Ascaris lumbricoides;* **(b)** *Trichuris trichiura;* **(c)** *hookworm (Redrawn from:* **(a)** *Elkins et al. 1986;* **(b)** *Bundy 1986;* **(c)** *Schad and Anderson 1985)*

sure or susceptibility limits the period over which significant predisposition can be detected. Holland and Boes (2002) reviewed the evidence available from field-based studies concerning the relationship between measures of host exposure or susceptibility and reinfection or predisposition to geohelminth infection. Two studies concerning the involvement of the immune response illustrate the difficulties experienced in trying to unravel this complex phenomenon.

Palmer et al. (1995), in a carefully designed case-control study, compared consistently lightly infected subjects with those consistently heavily infected. A range of antibody isotypes were measured, including total immunoglobulin (Ig)G, IgG1, IgG2, IgG3, IgG4, IgA, total IgE, and parasite-specific IgE. Children who were predisposed to heavy infection showed higher concentrations of antibody isotypes compared to children predisposed to light infections. In contrast to the findings of Hagel et al. (1993), the concentrations of total IgE and parasite-specific IgE in this study mirrored the infection intensity of the subjects. The authors do not rule out an effector role for these antibodies, but suggest that the utilization of more specific antigens may rule out polyspecific responses to numerous antigens that may mask any epitope-specific protective responses.

This point was borne out in a study by McSharry et al. (1999) who compared a range of serum factors in children predisposed to remain uninfected, lightly infected, and heavily infected. These groups of children showed few differences in measures of socioeconomic status and lived in environments where samples of soil contained eggs of *Ascaris*, assumed to be those of *A. lumbricoides*. Three different sources of *Ascaris* antigen were used but only the most defined allergen, *Ascaris* rABA-1, provided evidence for a significant relationship between predisposition status and parasite-specific IgE. A subgroup of children, who responded to the ABA-1 allergen, was selected and a relationship between reduced rABA-1 specific IgE titer and increasing parasite load was detected. Subjects were further divided according to high or low levels of IgE antibody using a threshold median value, and a distinct pattern emerged. The putatively immune group tended to have higher levels of rABA-1 specific IgE and the susceptible groups had low levels. Significantly higher levels of inflammatory indicators – such as serum ferritin, eosinophil cationic protein and C-reactive protein – were detected in the putatively immune group. The authors concluded that IgE responses, in conjunction with innate inflammatory responses, are associated with natural immunity to ascariasis.

Remarkably little work has been performed on the relationship between susceptibility to human helminths and host genetics (in contrast to genetic studies in laboratory and other animals), but recent contributions by Williams-Blangero and colleagues have provided important evidence for the role of host genetics in helminth aggregation and predisposition. In a large-scale study, which involved 1 261 subjects, all of whom belonged to the same pedigree, these authors demonstrated a strong genetic component accounting for 30–50 percent variation in *Ascaris* worm burden. Sharing a household accounted for only 3–13 percent of the total phenotypic variance (Williams-Blangero et al. 1999). Furthermore, a variance components linkage analysis resulted in the localization of two genes (one on chromosome 1 and another on chromosome 13) with clear, significant effects on susceptibility to *Ascaris* infection (Williams-Blangero et al. 2002).

SOIL-TRANSMITTED NEMATODES: CLINICAL MANIFESTATIONS, MORTALITY, AND MORBIDITY

Many of the millions of people infected with soil-transmitted nematodes may show no signs or symptoms of any ill health. Clinical problems are generally related to the intensity of infection and so only the relatively few hosts who carry the heavy worm burdens will be expected to be ill. Nevertheless, diseased individuals will not be numerically rare (Bundy and Cooper 1989). Chan et al. (1994) have estimated that there may be 120–215 million cases of morbidity due to *A. lumbricoides*, 90–130 million due to hookworm, and 60–100 million due to *T. trichiura*.

The difficulties encountered in estimating morbidity rates include:

- the problem of deciding what constitutes a clinical case (see Pawlowski 1982)
- the possibility that the infections exacerbate existing problems rather than cause them
- the scarcity of accurate records and reporting, particularly in the regions where the infections are highly endemic.

In such regions, growth stunting should be considered as a feature of chronic ascariasis, hookworm disease, and trichuriasis; iron-deficiency anemia as a feature of hookworm disease and sometimes of trichuriasis; and rectal prolapse and chronic dysentery as features of trichuriasis. These conditions, the mechanisms by which they arise and the related pathology have been summarized by Holland (1987a, b), Stephenson (1987), Bundy and Cooper (1989), and Tomkins and Watson (1989). The impact of each infection invariably has its main effect on a different section of the population and the threshold burden responsible for causing disease may vary depending on local conditions.

Mortality rates are even more difficult to estimate although people regularly die as a result of soil-transmitted nematode infections. Pawlowski and Davis (1989) considered that perhaps 100 000 people die annually from ascariasis and Pawlowski et al. (1991) suggested that 60 000 die from hookworm disease.

Ascariasis

The illness associated with *A. lumbricoides* infection occurs in acute and chronic forms (Table 34.4). The acute forms may arise from the larval migration through the lungs, from allergic responses or from complications involving the adult worms.

Table 34.4 *Features of morbidity due to ascariasis (based on Stephenson 1987)*

Stage	Event	Clinical features	Outcome
Larval migration	Migration of larvae through liver and lungs	Pneumonitis, asthma, dyspnea, cough, substernal pain	?Decrease food intake
Maturation, oviposition	Presence of juveniles and patent adult worms in small intestine	Abdominal pain, abdominal distention, colic, nausea, vomiting, intermittent diarrhea, anorexia, restlessness, anal itching, enterocolitis; disordered small bowel pattern, jejunal mucosal abnormalities	Decrease food intake; increase nutrient loss; malabsorption of protein, fat, d-xylose, lactose, and vitamin A
Allergic reaction	Exposure to *Ascaris* allergen at any stage of life cycle[a]	Hypersensitivity reactions including asthma, conjunctivitis, facial edema, urticaria abdominal pain, heartburn, diarrhea	Decrease food intake; increase nutrient loss
Complications	Migration or aggregation of adult *Ascaris* in intestine	Intestinal obstruction, intussusception, volvulus; invasion of bile duct (producing obstructive jaundice, gallstones, cholangitis, or liver abscesses); acute pancreatitis, acute appendicitis, intestinal perforation, peritonitis, upper respiratory tract obstruction	Life-threatening illnesses that all decrease food intake and may increase nutrient requirements (due to fever) and nutrient losses (due to diarrhea)

a) Uninfected individuals can also develop *Ascaris* allergy.

Larval migration leading to acute pulmonary ascariasis is associated with fever, skin rash, pneumonitis, substernal pain, and eosinophilia (Stephenson 1987). The problem appears to be more serious in regions like Saudi Arabia where climatic conditions ensure that *A. lumbricoides* is transmitted seasonally (Gelphi and Mustafa 1967). Stephenson pointed out that larval *A. lumbricoides* will remain associated with respiratory complications rather than being identified as a causative agent. However, the deliberate self-infection by Koino (1922) and the malicious contamination of a festive meal with infective eggs of *A. suum* described by Warren (1972) indicate that the larval stages initiate acute pulmonary disease.

Larval and adult *A. lumbricoides* and *A. suum* secrete allergens that elicit the production of IgE by the host leading to hypersensitivity and histamine release (Pawlowski 1982; Coles 1985). Hypersensitivity may be induced in uninfected laboratory workers studying the worms. With its most serious form, a sensitized person will collapse with bronchospasms. There is little quantitative information about the extent of this aspect of the biology of *A. lumbricoides*. In both infected and unin-

fected people, the presence of *Ascaris* appears to either stimulate or complicate asthma but more research is needed to clarify this problem. Lynch and colleagues have undertaken a number of studies on the relationship between *Ascaris* and asthma in Venezuelan patients (Lynch et al. 1992a, 1992b, 1997) and provided some evidence for the amelioration of asthmatic symptoms for a period of up to 2 years following anthelminthic treatment (Lynch et al. 1997). Experimental infections of rhesus monkeys with eggs of *A. suum* induces an IgE-mediated pathology in the lungs (Patterson and Harris 1985).

Acute ascariasis involving adults presents as a variety of obstructions and surgical complications (Table 34.5). Among children, intestinal obstruction caused by a bolus of worms appears to be the commonest problem and may lead to tissue necrosis and perforation of the intestinal wall. From the data summarized in Table 34.5, it would appear that girls suffer more intestinal obstructions than boys. Louw (1966) drew attention to the importance of *A. lumbricoides* as a cause of abdominal complications by recording that 100 (12.8 percent) of the 731 abdominal emergencies in children (aged 1–12

Table 34.5 *Complications of acute ascariasis related to adult* Ascaris lumbricoides *in relation to patient age and sex (Crompton 1994)*

Complication	All cases	Age		Sex		Deaths	
		Child	Adult	Male	Female	Child	Adult
Biliary system	1 124	380	374	92	117	4	23
Gastrointestinal tract	3 408	2 149	73	572	716	85	9
Hepatic abscess	100	52	6	17	15	13	1
Pancreatitis	67	28	26	11	15	1	2
Miscellaneous complications	94	77	15	50	41	9	3
Totals	4 793	2 686	494	742	964	112	38

The cases recorded in this table were abstracted from 230 reports published between 1971 and 1992. The cases described were from 54 countries. Often reports were found to be incomplete, with no information given about the age or gender of the patients.

years) admitted between 1958 and 1962 to a hospital in Cape Town, South Africa, were due to *A. lumbricoides*. Of the 100 cases, 68 proved to be intestinal obstructions. Among adults, complications involving the biliary system appear to be the commonest acute complication due to *A. lumbricoides* (Table 34.5). Presumably, the adult worms migrate from small intestine up the common bile duct where they may remain or may move on to the liver or to the pancreas. It is not known why adult worms leave the intestine; in some cases, larval worms may have reached unusual sites during the natural tissue migratory phase. That might explain how *A. lumbricoides* have been recovered from sites such as the middle ear (Berkowitz et al. 1980). Details of the diagnosis and management of cases of acute ascariasis due to adult worms have been given by Pinus (1985) and Erdener et al. (1992) among others.

Chronic ascariasis (Table 34.4) has been extensively studied in children in many locations during the past 20 years (Crompton 1992; Thein Hlaing 1993). Given that the study design is adequate and there is good compliance in the study population (Crompton and Stephenson 1985), convincing circumstantial evidence can be obtained to show that infection with *A. lumbricoides* interferes with the growth and development of children, especially during the period of 2–10 years of age. This conclusion is based on a study design in which the nutritional status (height and weight for age) was compared before and after intervention with an effective anthelmintic drug. The adverse contribution of chronic ascariasis to childhood nutritional status has been demonstrated in various countries worldwide (Gupta et al. 1977; Stephenson et al. 1980a; Thein Hlaing et al. 1992; see Tomkins and Watson 1989), although not every investigation has detected this effect.

Various experimental studies of ascariasis in pigs and clinical investigations of ascariasis in humans have shown that the presence of adult worms in the small intestine is associated with some degree of crypt hyperplasia and villous atrophy (Tripathy et al. 1972; Stephenson et al. 1980b; Martin et al. 1984). Under certain conditions, maldigestion and malabsorption occur (Forsum et al. 1981; Carrera et al. 1984) with recovery being observed following anthelmintic treatment (Taren et al. 1987). Probably the most serious aspect of an *Ascaris* infection is a reduction in food intake, which has been well documented for *A. suum* infections in pigs (see Nesheim 1985) and has been demonstrated more recently by Jalal (1991) in rural children in Indonesia.

Currently, ascariasis (and hookworm disease and trichuriasis) should be considered as contributors to different degrees of childhood malnutrition. Infections with soil-transmitted nematodes have also been convincingly linked to impaired cognitive performance in young children (Nokes et al. 1992; Nokes and Bundy 1994; Simeon et al. 1994). It may be wiser to view these infections as one of the many interacting socioeconomic determinants that impair cognitive performance (see Connolly and Kvalsvig 1993).

Hookworm disease

Not all the people infected with hookworms suffer from iron-deficiency anemia, but many do and many of them are at risk because the intestinal stages feed on blood and cause a loss of blood into the intestinal lumen (Roche and Layrisse 1966). Anemia is defined as a reduction in blood hemoglobin concentration below expected values for age and sex (WHO 1972). In physiological terms, in an anemic individual, the circulating number of red blood cells becomes insufficient to meet the oxygen needs of the body and the symptoms of a severe hookworm infection mimic those of anemia: breathlessness, lassitude, headache, palpitations, apathy and depression. Often occult blood is found in the stools (Holland 1987a). Studies from many countries including India, Kenya, Thailand, and Venezuela have detected the relationship depicted in Figure 34.7 (Hill and Andrews 1942). Blood hemoglobin concentration in adults is observed to fall in a nonlinear manner in relation to increasing hookworm infection intensity. A value of 100 mg/ml in an adult is judged to indicate anemia (WHO 1972).

Whether anemia develops or not during the course of a hookworm infection depends not only on the intensity of the infection and species of hookworm (*A. duodenale* takes more blood than *N. americanus*), but also on the iron status and physiological needs of the host, the

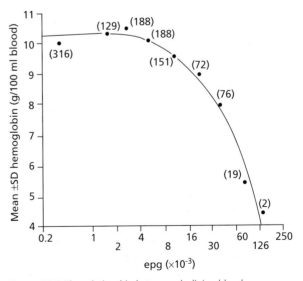

Figure 34.7 *The relationship between declining blood hemoglobin concentration and increasing hookworm intensity, expressed as egg-count classes, obtained from a study of 1 141 residents of South Georgia, USA. The egg counts have been corrected to a formed-stool basis (Data from Hill and Andrews 1942)*

quality and quantity of the daily iron intake and the bioavailability of the iron for absorption from the small intestine. This complex relationship has been discussed in detail by Holland (1987a), Crompton and Stephenson (1990), and Crompton and Whitehead (1993). Maintaining the required number of circulating red blood cells probably has a high priority for iron metabolism, so hookworm-infected individuals who are not clinically anemic may have a deteriorating iron status and latent iron deficiency. Measurements of plasma ferritin concentration, transferrin saturation, and erythrocyte protoporphyin would probably indicate that clinically nonanemic individuals with hookworm infections are under iron stress, as has been demonstrated by Mansour et al. (1985) for people suffering from schistosomiasis mansoni. Stoltzfus and colleagues studied the impact of hookworms on the iron status of Zanzibar schoolchildren (Stoltzfus et al. 1997, 1998). They demonstrated that in the absence of iron supplementation, anthelminthic treatment of varying frequency significantly reduced the risk of these children developing both moderate-to-severe and severe anemia (Stoltzfus et al. 1998).

Since many features of hookworm disease indicate iron-deficiency anemia, it is to be expected that women and girls of child-bearing age, pregnant women, and people engaged in heavy manual labor will be most at risk from the adverse effects of infection. On average, menstruating females lose about 1.5 mg iron per day, about the same amount as is absorbed from the gut under optimal conditions (Crompton and Whitehead 1993). Although there is no menstrual loss during pregnancy and lactation, the mother must meet all the iron requirements of the developing fetus and the suckling infant. H. Torlesse (personal communication 1994) has recently estimated that at any given time about 30 million women are currently pregnant and infected with hookworm. Many of these women must be experiencing acute anemia exacerbated by the demands of pregnancy and the blood loss due to hookworms. The birth of underweight children is a major consequence of anemia during pregnancy, and, in developing countries, the deaths of about 200 000 women can be ascribed to anemia as a complication of childbirth or the early postpartum period (Viteri 1994). As many as 1 in 25 pregnant women die in some developing countries due to complications associated with the pregnancy and birth (WHO 1995).

Worker productivity in relation to anemia has attracted considerable research attention. Study designs have sought to examine the cost–benefit ratios in financial terms of expenditure to relieve anemia (iron supplements and antiparasite drugs) compared with the increased revenue from increased productivity by the nonanemic workers. The results have shown that anemic sugar cane cutters (Viteri and Torun 1974), road builders (Brookes et al. 1979; Wolgemuth et al. 1982),

rubber tappers (Basta et al. 1979), and tea pickers (Gardner et al. 1977) do not have the same work output as either nonanemic controls or those given treatment to relieve the anemia. Vigorous physical work is difficult to carry out and sustain once blood hemoglobin concentration has fallen to a value of 7 mg/ml (see Figure 34.7). In developing countries, where productivity depends largely on manual labor, widespread hookworm infections may make a substantial contribution to a depressed economy (Holland 1987a).

Iron deficiency anemia is a factor that reduces cognitive performance: on measures of achievement at school, anemic children usually show lower scores than their nonanemic counterparts (see Connolly and Kvalsvig 1993). According to Pollitt (1990) there is sufficient evidence to conclude that iron-deficiency anemia is linked to impaired educational performance. Again hookworm infections, with their contribution to iron loss, must contribute to this problem. It should be noted that reduced appetite and loss of nutrients including zinc (Migasena et al. 1984) through plasma leakage into the gut, together with malabsorption, which also occurs in some cases, are likely to affect school performance, as well as growth (Holland 1987a) and physical fitness (Stephenson et al. 1993).

Trichuriasis

The pathology of an infection with *T. trichiura* involves chronic inflammation of the mucosa of the large intestine associated with the intimate contact that the worms make with the mucosa. The degree of inflammation may extend from the distal part of the small intestine to the rectum depending on the intensity of the infection (Bundy and Cooper 1989). Since the infection is confined largely to the large intestine, malabsorption of digestion products is not likely to be a serious consequence. However, the lacerations caused by the feeding activities of the worms may enable secondary bacterial infections to become established. Secondary bacterial establishment has been demonstrated in experimental infections of *T. suis* in pigs (Hall et al. 1976), which developed dysentery. It is frequently observed that children with heavy infections of *T. trichiura* often suffer from chronic dysentery (see Holland 1987b).

Rectal prolapse is the most striking lesion associated with trichuriasis. Apparently, the surface tissue of the rectum becomes extremely edematous and the prolapsing occurs as the patient strains to defecate. Bundy and Cooper (1989) suggest that the mechanism of the prolapse is similar to that of intussusception and that remission often occurs within a few days of effective anthelmintic chemotherapy.

Rectal bleeding occurs during trichuriasis (Fisher and Cremin 1970; Lotero et al. 1974) and this has prompted a vigorous debate about the contribution of the worms to the development of iron-deficiency anemia. Infected people undoubtedly lose blood during an infection from both the feeding activities of the worms and the extensive damage to the mucosa of the large intestine. Iron lost into the gut lumen from this tissue would essentially represent a net loss to the host. After a critical review of an extensive literature, Holland (1987b) and Bundy and Cooper (1989) independently concluded that trichuriasis contributed significantly to iron-deficiency anemia. Recently, Ramdath et al. (1995) compared the blood picture of 264 *Trichuris*-infected children with that of 157 matched, uninfected children. Children judged to have heavy infections (more than 10 000 eggs/g) had significantly lower ($p < 0.05$) blood hemoglobin concentrations and mean cell volumes than uninfected children or those with lower infection intensities. Also, 33 percent of the heavily infected children were diagnosed as anemic compared with 11 percent of the rest of those in the study.

Several studies have shown that trichuriasis, like ascariasis and often in combination with it, is a factor involved in growth stunting in children (Bowie et al. 1978; Holland 1987b; Simeon et al. 1994). For example, Gilman et al. (1983) found that the nutritional status of children increased significantly following anthelmintic expulsion of *T. trichiura* ($p < 0.05$). The children in this study who were judged to be heavily infected also showed blood in their stools and some degree of anemia and dysentery. Twelve percent of the children presented with finger clubbing; this lesion is a curious thickening at the ends of the fingers and toes and is associated with other illnesses besides trichuriasis. In a cross-sectional survey of 260 children in St Lucia, Cooper and Bundy (1987) found a strong association between measures of impaired rates of growth and infection intensities greater than 20 000 eggs/g. More recently, Cooper et al. (1990) have linked the observed growth depression in children with heavy infections to the dysentery syndrome that is a feature of trichuriasis. Nineteen Jamaican children, ranging in age from 2 to 10 years and known to be suffering from heavy *T. trichiura* infections, were studied. Nutritional status was assessed and blood hemoglobin concentrations ($\bar{x} = 7.0 \pm 2.5$ mg/ml) measured. The growth rates of 11 of these children were calculated, following anthelmintic treatment, for a period of 6 months, and a remarkable rate of catch-up growth was detected despite the absence of any food supplementation other than the provision of oral ferrous sulfate to correct the iron-deficiency anemia. Cooper et al. (1990) proposed that the chronic inflammation of the large intestine which persists during *T. trichiura* infections was in some way linked to the growth deficits detected in these heavily infected children. Recent developments concerning the pathophysiology of *Trichuris* dysentery syndrome are reviewed by Stephenson et al. (2001).

SOIL-TRANSMITTED NEMATODES: IMMUNITY

The human response to *A. lumbricoides* is characterized by prominent antibody and cellular responses that are directed at the larval stages of infection (Cooper 2002). The most consistent finding relates to the production of large amounts of total and parasite-specific IgE, which is most likely to be stimulated by the tissue migratory stage of the life cycle (Hagel et al. 1993; Lynch et al. 1993; McSharry et al. 1999). There is a paucity of data on cellular responses to geohelminths in humans in general, but a recent paper by Cooper et al. (2000) demonstrated a polarized T helper 2 (Th2) response accompanied by interleukin (IL)-4 and IL-5 production in *Ascaris*-infected subjects. Furthermore, for this parasite there is more convincing evidence that differential immune responses both humoral and cellular play a part in the observed predisposition as discussed in the previous section.

In contrast to *Ascaris*, studies on immunity to trichuriasis have been greatly assisted by the provision of an entirely appropriate animal model – *T. muris* in the mouse (Faulkner and Bradley 2002). Evidence for the importance of host genetic background has been provided from the identification of resistant and susceptible strains of mice (Else and Wakelin 1988) and cytokines produced by Th2 cells such as IL-4, Il-3, and Il-9 have been identified as important for worm expulsion. Studies of immune responses in humans have produced less clear-cut results and evidence for protective responses remains uncertain. *Trichuris*-specific IgG appears to reflect current infection status, whereas IgA persists at high levels in adulthood (Needham et al. 1992). Faulkner and Bradley (2002) advocate the use of more defined antigens in order to detect key responses mirroring the experience of McSharry et al. (1999) working with *Ascaris*.

In a series of field-based studies based in Papua New Guinea, Pritchard and co-workers (Pritchard et al. 1992, 1995; Quinnell et al. 1995) reported the relationship between *N. americanus* infection and humoral antibody responses in subjects who experienced reinfection after chemotherapy. After controlling for the effects of age, the authors demonstrated that correlation coefficients between levels of anti-adult worm ES IgG and worm burden declined significantly with age and did not persist after reinfection. The trend was the same for antilarval IgG response, but the pattern persisted after reinfection. The switch from positive to negative correlation in adults appears consistent with a protective role (Quinnell et al. 1995). Furthermore, the effect of the humoral immune response on the weight and fecundity of the parasite was investigated in the same subjects. After controlling for the effects of age and parasite burden, a significant negative correlation between total and specific IgE and the weight and fecundity of *Necator* worms was detected at initial treatment and after reinfection (Pritchard et al. 1995), which the authors suggest may reflect a Th2 response. For a recent update on approaches to vaccination for hookworm including the identification of novel vaccine candidates, see Pritchard et al. (2002).

Important evidence is now accumulating that helminth infections can modify the host response to secondary infections or vaccination (see Bentwich et al. 1995, 1999; reviewed in Borkow and Bentwich 2002). This may be particularly important with regard to populations in developing countries severely affected by epidemics of human immunodeficiency virus (HIV) and tuberculosis (TB). Disease progression and susceptibility appears to accelerate in individuals co-infected with helminths; one explanation for this may be an influence on the cellular immune response. Helminth infections are associated with a strong Th2-type responses that can down-regulate production of T helper 1 (Th1)-type cytokines important in the control of microparasite infections (Urban et al. 1992).

SOIL-TRANSMITTED NEMATODES: DIAGNOSIS

For epidemiological surveys, operational research programs and control activities, the diagnosis of soil-transmitted nematode infections continues to rely on the collection and examination of stool samples for helminth eggs or larvae (Figure 34.1) and the method to be used depends on the purpose of the study or survey (Thein Hlaing 1989). A useful summary of those available is given by Theinpont et al. (1986). The procedure now recommended by the World Health Organization (WHO 1994) is that known universally as the Kato Katz method, which gives reliable information about infection intensities. Despite the efforts to develop serological tests to aid diagnosis and techniques to detect antigens and metabolites in feces, coprological examination remains the method best suited to the resources and skills available in developing countries where the infections are endemic (Wakelin et al. 1993).

In modern hospitals and clinics, other diagnostic techniques are available, but not for mass application. For example, in the case of acute ascariasis, real-time ultrasound (Cremlin 1982), radiography (Choudhuri et al. 1986), and endoscopic retrograde cholangiopancreatography (Khan et al. 1993) are regularly employed. Fiber optics may be used; Cooper and Bundy (1987) describe the use of a flexible fiber optic colonoscope to study infections of *T. trichiura*.

SOIL-TRANSMITTED NEMATODES: PROSPECTS OF PREVENTION AND CONTROL

Geohelminth infections represent a serious public health problem in countries where sanitation is poor and access to safe, effective anthelmintic drugs is minimal (Savioli et al. 2002). The World Bank has ranked intestinal helminth infections first as the main cause of morbidity in children aged 5 to 14 years in developing countries and has also identified these infections as being targets for efficient control by cost-effective intervention (World Bank 1993). It is now agreed that one of the best available means of reducing morbidity and mortality due to these infections is to treat high-risk groups, particularly school-age children, through the regular administration of single dose, oral anthelmintic drugs (WHO 2001). Schools can serve as a gathering point for drug delivery and administration, which can be undertaken by teachers; schools can further act as a focus for education in the community so that health education can be reinforced. A successful example of this approach is taking place in Zanzibar where, in addition to the treatment of school-enrolled pupils, an outreach program for non-enrolled but school-age children was also put in place (Savioli et al. 2002; Montresor et al. 2001). Recently, attention has been focused on preschool children and the impact that soil-transmitted helminths may be having on their health and well being (Stoltzfus et al. 2001). Furthermore, in a recent informal consultation the WHO has been considering the issues surrounding treatment of these younger children (Montresor et al. 2002; WHO, 2003). For a recent review of the economics of worm control see Guyatt (2002).

Geohelminth infections are also very suitable diseases to incorporate into integrated control efforts. This has been successfully employed with family planning and nutrition programs by the Japanese Organization for International Co-operation in Family planning (Kunii 1983) and is now the focus of an important initiative concerning the eradication of lymphatic filariasis. The provision of two-drug regimens, namely albendazole-plus-ivermectin or albendazole-plus-diethycarbamazine designed principally to interrupt the transmission of lymphatic filariasis through mass drug administration to at-risk populations, will also have a major impact on intestinal helminth infections (Stephenson et al. 2000).

ENTEROBIUS: BIOLOGY, TRANSMISSION, AND EPIDEMIOLOGY

Biology

Infection occurs as a result of ingesting fully developed, infective eggs. After hatching in the small bowel, subsequent development takes place entirely within the intestine, there being no systemic migration similar to that undertaken by *Ascaris*, and worms mature within about 5 weeks. The adults are found primarily in the large bowel, lying closely applied to the mucosal surface. On the basis of studies made with rodent pinworm, it can be assumed that the worms feed on material present in the lumen and possibly on the cellular debris that accumulates on the epithelium and in the crypts. The worms are mobile and move within the large bowel, the females also migrate at night down the rectum to the anus to release eggs. Males live for some 7 weeks, females 5–13 weeks (Cook 1995).

Unlike the soil-transmitted species considered above, the eggs of *Enterobius* are not released into the intestinal lumen, but are released on the perianal skin after the females have migrated to the anal region. Hence, diagnosis of pinworm infection cannot be made reliably through fecal smears or similar techniques. Egg release occurs through the female reproductive opening or when the worm dies and disintegrates on the surface of the skin, each female releasing an estimated 10 000 eggs (Faust et al. 1970). Movement to the anal region causes local irritation and itching. When laid, the eggs already contain immature larvae and these complete development very rapidly, maturing within 6 hours at body temperature (Cook 1995). The eggs are 'sticky' and readily adhere but, being light, are also easily dispersed. If the local environmental conditions are not too dry, they can survive for considerable periods.

Transmission

The characteristics of the eggs, and the way in which they are released from females favors several routes of transmission. In children, the itching associated with the presence of female worms on the perianal skin prompts scratching, as a result of which eggs adhere to the hands and fingers, from where they are easily transferred to the mouth. Contamination of nightclothes or bedding can also result in eggs being transferred to the hands. The lightness of the eggs results in a wide dispersal in bedrooms and houses, and accidental transmission to other family members can occur from these sources. In one school, counts of eggs present on the wall of the lavatory showed some 5 000 per square foot (0.093 m^2) (Br Med J 1974). Eggs can become airborne when clothes, bedding or dust are disturbed. They may then contaminate food or the hands of people in the vicinity, but it has also been suggested that airborne eggs can be inhaled and then swallowed.

Epidemiology

Pinworm infections are most frequent in school-age children, but can occur in any age group. They are the commonest worm parasites in temperate, developed countries, where prevalence figures in young children can reach 80–90 percent, but infections are distributed globally, and *Enterobius* often accompanies the soil-transmitted nematodes in the warmer, less developed parts of the world (reviewed by Haswell-Elkins et al. 1987b). Data on prevalence can be misleading, as some figures have been based on detection of eggs in stool samples, which frequently fails to detect infections. Haswell-Elkins et al. (1987b) determined prevalence in an Indian village by recovering worms after chemotherapy and recorded overall figures of 70.8 percent (74.3 percent in males, 67.5 percent in females) with a mean intensity of 25.8 worms. By the age of 4 years, 60 percent of children were infected. Eleven months after initial treatment, the population was sampled again and prevalence and intensity were found to be significantly higher (87.1 percent and 40.2 worms). Infections were markedly aggregated in the host population: for example, in the second study the heaviest infection recorded was 2 152 worms.

ENTEROBIUS: CLINICAL MANIFESTATIONS

In most cases, *Enterobius* infections are asymptomatic. The commonest symptom is the itching associated with movement of the female worms to the anal regions (anal and perineal pruritus), although it has been suggested that these symptoms reflect the existence of other dermatological problems (Ganor 1987). Infected children may show irritability, enuresis, and weight loss. The worms normally cause only minor pathology in the intestine. Unlike the other intestinal nematodes, *Enterobius* infection is not associated with a pronounced eosinophilia. or with elevated IgE (Jarrett and Kerr 1973). In girls, worms may migrate into the vagina, causing a vaginitis. Urinary infections in children have been associated with such migratory behavior and adult women may also be affected in this way. The presence of *Enterobius* in the appendix (2.7 percent of 1 419 appendices removed during a 5-year period in Bristol, UK (Budd and Armstrong 1987), 2.5 percent of 2 921 appendices in India (Gupta et al. 1989)) has suggested a link with appendicitis, but this is considered rare.

Infections can lead occasionally to more severe complications, if the worms find their way into the peritoneum, either via intestinal perforations or following migration through the female reproductive tract. Dead worms and eggs have been identified in granulomata in several sites (vagina, cervix, fallopian tubes, omentum, peritoneum, even the liver, kidneys, and lungs (Cook 1995).

ENTEROBIUS: DIAGNOSIS AND CONTROL

Infection can be diagnosed directly by observing worms in the perianal region or on the surface of the stool. More commonly, diagnosis requires the detection of eggs. These occur on the perianal skin and can be collected on to the surface of an adhesive transparent tape, pressed on to the skin. The tape can then be transferred to a slide, with a clearing agent (e.g. toluene, cedar oil) for microscopic examination. Such examination is best performed in the morning when the patient first wakes, and repeated examination (3–6 tests) may be necessary to ensure positive diagnosis. Commercial detection kits are available. Eggs can sometimes be found in stools, but diagnosis is very unreliable. If infection is confirmed in one individual in a household, other family members should also be examined and treated.

Pinworms are easily removed by anthelmintic treatment. The use of piperazine compounds has been standard for many years but requires treatment daily for 7 days. Pyrantel compounds are effective in a single (10 mg/kg) dose, as are the benzimidazoles mebendazole (100 mg tablet) and albendazole (400 mg tablet). Where *Enterobius* occurs with other intestinal nematodes, use of the benzimadazoles has the additional benefit of eliminating the other species as well.

Although treatment is effective, reinfection occurs readily unless steps are taken to minimize contact with infective eggs. Personal hygiene is important, particularly hand washing, fingernail cleaning, and regular bathing. Infected children can be made to wear gloves in bed to prevent contamination of their hands. Heavily infected areas such as bedrooms must be kept scrupulously clean and dust free, and bed linen and night clothes should be changed and laundered frequently. In practice, elimination of eggs in this way is very difficult and to be effective these approaches must be accompanied by repeated chemotherapy to remove the adult worms.

ACKNOWLEDGMENTS

This chapter is adapted and expanded from Crompton, D.W.T. 1998. Gastrointestinal nematodes – Ascaris, Hookworm, Trichuris and Enterobius. In: Cox, F.E.G., Kreier, J.P. and Wakelin, D. (eds) *Topley & Wilson's Microbiology and Microbial Infections*, vol. 5, *Parasitology*, 9th edn. London: Edward Arnold, pp. 561–584.

REFERENCES

Adedeja, S.O. and Ogunba, E.O. 1991. Bacterial flora of *Ascaris suum* (Goeze 1782) and its relationship to the host flora. *Trop Vet*, **9**, 123–9.

Adiyodi, K.G. and Adiyodi, R.G. 1983. *Reproductive Biology of Invertebrates*. 6 vols. . Chichester: John Wiley.

Anderson, R.C., Chabaud, A.G. and Willmott, S. 1974. *CIH Keys to the Nematode Parasites of Vertebrates*. Farnham Royal, UK: Commonwealth Agricultural Bureaux.

Anderson, R.M. 1986. The population dynamics and epidemiology of intestinal nematode infections. *Trans R Soc Trop Med Hyg*, **80**, 686–96.

Anderson, R.M. 1989. Transmission dynamics of *Ascaris lumbricoides* and the impact of chemotherapy. In: Crompton, D.W.T., Nesheim, M.C. and Pawlowski, Z.S. (eds), *Ascariasis and its Prevention and Control*. London: Taylor and Francis, 253–73.

Anderson, R.M. 1994. Mathematical studies of parasitic infection and immunity. *Science*, **264**, 1884–6.

Anderson, R.M. and Gordon, D.M. 1982. Processes influencing the distribution of parasite numbers within host populations with special emphasis on parasite-induced host mortalities. *Parasitology*, **85**, 373–98.

Anderson, R.M. and May, R.M. 1985. Helminth infections of humans: mathematical models, population dynamics, and control. *Adv Parasitol*, **24**, 1–101.

Anderson, R.M. and May, R.M. 1991. *Infectious Diseases of Humans. Dynamics and Control*. Oxford: Oxford University Press.

Andrassy, I. 1976. *Evolution as a Basis for the Systematization of Nematodes*. London: Pitman.

Banwell, J.G. and Schad, G.A. 1978. Hookworms. *Clin Gastroenterol*, **7**, 129–56.

Barrett, J. 1981. *Biochemistry of Parasitic Helminths*. London: Macmillan.

Barriga, O.O. 1982a. Ancylostomiasis. In: Schulz, M.G. (ed.), *Section C: Parasitic Zoonoses. CRC Handbook Series in Zoonoses* , vol. II. Boca Raton, FL: CRC Press, 3–24.

Barriga, O.O. 1982b. Trichuriasis. In: Schulz, M.G. (ed.), *Section C: Parasitic Zoonoses. CRC Handbook Series in Zoonoses* , vol. II. Boca Raton, FL: CRC Press, 339–45.

Basta, S.S., Soekirman, K.D. and Scrimshaw, N.S. 1979. Iron deficiency anaemia and productivity of adult males in Indonesia. *Am J Clin Nutr*, **32**, 916–25.

Beaver, P.C. 1954. Parasitic diseases of animals and their relation to public health. *Small Anim Pract*, **49**, 199–205.

Beaver, P.C. 1956. Larva migrans. *Exp Parasitol*, **6**, 587–621.

Beaver, P.C. and Jung, R.C. 1985. *Animal Agents and Vectors of Human Disease*, 5th edn. Philadelphia, PA: Lea and Febiger.

Beck, J.W. and Beverley-Burton, M. 1968. The pathology of *Trichuris, Capillaria* and *Trichinella* infections. *Helminth Abstr*, **37**, 1–26.

Bell, J.C., Palmer, S.R. and Payne, J.M. 1988. *The Zoonoses. Infections Transmitted from Animals to Man*. London: Edward Arnold.

Bennet-Clark, H.C. 1976. Mechanics of nematode feeding. In: Croll, N.A. (ed.), *The Organization of Nematodes*. London: Academic Press, 313–42.

Bentwich, Z., Kalinkovich, A. and Weisman, Z. 1995. Immune activation is a dominant factor in the pathogenesis of African AIDS. *Immunol Today*, **16**, 187–91.

Bentwich, Z., Kalinkovich, A., et al. 1999. Can eradication of helminthic infections change the face of AIDS and tuberculosis? *Immunol Today*, **20**, 485–7.

Berkowitz, F.E., Sochet, E. and Packirisamy, G. 1980. *Ascaris* in the middle ear. *S Afr Med J*, **58**, 680.

Bidinger, P.D., Crompton, D.W.T. and Arnold, S.E. 1981. Aspects of intestinal parasitism in villages from rural peninsular India. *Parasitology*, **83**, 373–80.

Bird, A.F. and Bird, J. 1991. *The Structure of Nematodes*, 2nd edn. London: Academic Press.

Br Med J 1974. Children's worms (Editorial). *Br Med J*, **4**, 3.

Borkow, G. and Bentwich, Z. 2002. Geohelminths, HIV/AIDS and TB. In: Holland, C.V. and Kennedy, M.W. (eds), *The geohelminths: Ascaris, Trichuris and hookworm*. Boston/Dordrecht/London: Kluwer Academic Publishers, 302–17.

Bowie, M.D., Morison, A., et al. 1978. Clubbing and whipworm infestation. *Arch Dis Child*, **53**, 411–13.

Brookes, R.M., Latham, M.C. and Crompton, D.W.T. 1979. The relationship of nutrition and health to worker productivity in Kenya. *East Afr Med J*, **56**, 413–21.

Bryan, C.P. 1930. *The Papyrus Ebers*. London: Geoffrey Bles.

Budd, J.S. and Armstrong, C. 1987. Role of *Enterobius vermicularis* in the aetiology of appendicitis. *Br J Surg*, **74**, 748–9.

Bulatao, R.A., Bos, E., et al. 1990. *World Population Projections*. Baltimore, MD: Johns Hopkins University Press.

Bundy, D.A.P. 1986. Epidemiological aspects of *Trichuris* and trichuriasis in Caribbean commuities. *Trans R Soc Trop Med Hyg*, **80**, 706–18.

Bundy, D.A.P. 1988. Population ecology of intestinal helminth infections in human communities. *Phil Trans R Soc B*, **321**, 405–20.

Bundy, D.A.P. and Cooper, E.S. 1989. *Trichuris* and trichuriasis in humans. *Adv Parasitol*, **28**, 107–73.

Bundy, D.A.P. and Medley, G.F. 1992. Immuno-epidemiology of human geohelminthiasis: ecological and immunological determinants of worm burden. *Parasitology*, **104**, Supplement, S105–19.

Bundy, D.A.P., Cooper, E.S., et al. 1987. Predisposition to *Trichuris trichiura* in humans. *Epid Infect*, **98**, 65–71.

Burrows, R.B. and Lillis, W.G. 1964. The whipworm as a blood sucker. *J Parasitol*, **50**, 675–80.

Carrera, E., Nesheim, M.C. and Crompton, D.W.T. 1984. Lactose maldigestion in *Ascaris*-infected pre-school children. *Am J Clin Nutr*, **39**, 255–64.

Carroll, S.M. and Grove, D.I. 1986. Experimental infections of humans with *Ancylostoma ceylanicum*: clinical, parasitological, haematological and immunological findings. *Trop Geogr Med*, **38**, 38–45.

Chan, M.-S. 1997. The global burden of intestinal nematode infections – fifty years on. *Parasitol Today*, **13**, 439–43.

Chan, M.-S., Medley, G.F., et al. 1994. The evaluation of potential global morbidity attributable to intestinal nematode infections. *Parasitology*, **109**, 373–87.

Chitwood, B.G. and Chitwood, M.B. 1974. *Introduction to Nematology*. Baltimore, MD: University Park Press.

Choudhuri, G., Saha, S.S. and Tandon, R.K. 1986. Gastric ascariasis. *Am J Gastroenterol*, **81**, 788–90.

Chu, W.-G., Pen, P.-M., et al. 1972. Neonatal ascariasis. *J Pediatr*, **81**, 783–5.

Clutton-Brock, J. 1987. *A Natural History of Domesticated Mammals*. Cambridge: Cambridge University Press.

Cockburn, A. and Cockburn, E. 1980. *Mummies, Disease and Ancient Cultures*. Cambridge: Cambridge University Press.

Coles, G.C. 1985. Allergy and immunopathology of ascariasis. In: Crompton, D.W.T., Nesheim, M. and Pawlowski, Z.S. (eds), *Ascariasis and its public health significance*. London: Taylor and Francis, 167–84.

Connolly, K.J. and Kvalsvig, J.D. 1993. Infection, nutrition and cognitive performance in children. *Parasitology*, **107**, Supplement, S187–200.

Cook, G.C. 1995. *Enterobius vermicularis* infection. In: Farthing, M.J.G, Keusch, G.T. and Wakelin, D. (eds), *Intestinal Helminthes. , Enteric infection*. vol. 2. London: Chapman and Hall, 213–23.

Coombs, I. and Crompton, D.W.T. 1991. *A Guide to Human Helminths*. London: Taylor and Francis.

Cooper, E.S. and Bundy, D.A.P. 1987. Trichuriasis. In: Pawlowski, Z.S. (ed.), *Intestinal helminthic infections*. London: Baillière Tindall.

Cooper, E.S., Bundy, D.A.P., et al. 1990. Growth suppression in the *Trichuris* dysentery syndrome. *Eur J Clin Nutr*, **44**, 285–91.

Cooper, P.J. 2002. Immune response in humans. In: Holland, C.V. and Kennedy, M.W. (eds), *The Geohelminths: Ascaris, Trichuris and hookworm*. Boston/ Dordrecht/ London: Kluwer Academic Publishers, 89–104.

Cooper, P.J., Chico, M.E., et al. 2000. Human infection with *Ascaris lumbricoides* is associated with a polarized cytokine response. *JID*, **182**, 1207–13.

Cremlin, B.J. 1982. Real-time ultrasound in paediatric biliary ascariasis. *S Afr Med J*, **61**, 914–16.

Croese, J., Loukas, A., et al. 1994. Occult enteric infection by *Ancylostoma caninum*: a previously unrecognized zoonosis. *Gastroenterology*, **106**, 3–12.

Croll, N.A. and Ghadirian, F. 1981. Wormy persons: contributions to understanding the nature and patterns of overdispersion with *Ascaris lumbricoides, Ancylostoma duodenale, Necator americanus* and *Trichuris trichiura*. *Top Geogr Med*, **33**, 241–8.

Crompton, D.W.T. 1989. Biology of *Ascaris lumbricoides*. In: Crompton, D.W.T., Nesheim, M.C. and Pawlowski, Z.S. (eds), *Ascariasis and its Prevention and Control*. London: Taylor and Francis, 9–44.

Crompton, D.W.T. 1992. *Ascaris* and childhood malnutrition. *Trans R Soc Trop Med Hyg*, **86**, 577–9.

Crompton, D.W.T. 1994. *Ascaris lumbricoides*. In: Scott, M.E. and Smith, G. (eds), *Parasitic and Infectious Diseases*. London: Academic Press, 175–96.

Crompton, D.W.T. 2001. *Ascaris* and ascariasis. *Adv Parasitol*, **48**, 286–375.

Crompton, D.W.T. and Stephenson, L.S. 1985. Ascariasis in Africa. In: Crompton, D.W.T., Nesheim, M.C. and Pawlowski, Z.S. (eds), *Ascariasis and its Public Health Significance*. London: Taylor and Francis, 185–202.

Crompton, D.W.T. and Stephenson, L.S. 1990. Hookworm infection, nutritional status and productivity. In: Schad, G.A. and Warren, K.S. (eds), *Hookworm Disease*. London: Taylor and Francis, 231–64.

Crompton, D.W.T. and Whitehead, R.R. 1993. Hookworm infections and human iron metabolism. *Parasitology*, **107**, Supplement, S137–45.

Crompton, D.W.T., Nesheim, M.C. and Pawlowski, Z.S. (eds) 1985. *Ascaris and its Public Health Significance*. London: Taylor and Francis.

Crompton, D.W.T., Nesheim, M.C. and Pawlowski, Z.S. (eds) 1989. *Ascariasis and its Prevention and Control*. London: Taylor and Francis.

da Costa-Macedo, L.M. and Rey, L. 1990. *Ascaris lumbricoides* in neonate evidence of congenital transmission of intestinal nematodes. *Rev Inst Med Trop Sao Paulo*, **32**, 351–4.

Davey, K. 1964. The food of *Ascaris. Canad J Zool*, **42**, 1160–1.

Davies, N.J. and Goldsmid, J.M. 1978. Intestinal obstruction due to *Ascaris suum* infection. *Trans R Soc Trop Med Hyg*, **72**, 107.

Ebbell, B. 1937. *The Papyrus Ebers: The Greatest Egyptian Medical document*. Copenhagen: Levin and Munksgaard.

Elkins, D.B., Haswell-Elkins, M. and Anderson, R.M. 1986. The epidemiology and control of intestinal helminths in the Pulicat Lake region of Southern India. 1. Study design and pre- and post-treatment observations on *Ascaris lumbricoides* infection. *Trans R Soc Trop Med Hyg*, **80**, 774–92.

Else, K.J. and Wakelin, D. 1988. The effects of H-2 and non-H-2 genes on the expulsion of the nematode *Trichuris muris* from inbred congenic mice. *Parasitology*, **96**, 543–50.

Erdener, A., Ozok, G., et al. 1992. Abdominal complications of *Ascaris lumbricoides* in children. *JAMA*, **42**, 73–4.

Faulkner, H. and Bradley, J.E. 2002. Immune response in humans. In: Holland, C.V. and Kennedy, M.W. (eds), *The Geohelminths: Ascaris, Trichuris and hookworm*. Boston/Dordrecht/London: Kluwer Academic Publishers, 126–42.

Faust, E.C., Russell, P.F. and Jung, R.C. 1970. *Craig and Faust's Clinical Parasitology*, 8th edn. Philadelphia, PA: Lea and Fabinger.

Fetterer, R.H., Hill, D.E. and Urban, J.F. 1990. The cuticular biology in developmental stages of *Ascaris suum. Acta Trop (Basel)*, **47**, 289–95.

Fisher, R.M. and Cremin, B.J. 1970. Rectal bleeding due to *Trichuris trichiura. Br J Radiol*, **43**, 214–15.

Foor, W.E. 1967. Ultrastructural aspects of oocyte development and shell formation in *Ascaris lumbricoides. J Parasitol*, **53**, 1245–61.

Forrester, J.E. and Scott, M.E. 1990. Measurement of *Ascaris lumbricoides* infection intensity and the dynamics of expulsion following treatment with mebendazole. *Parasitology*, **100**, 303–8.

Forsum, E., Nesheim, M.C. and Crompton, D.W.T. 1981. Nutritional aspects of *Ascaris* infection in young protein-deficient pigs. *Parasitology*, **83**, 497–512.

Fry, G.H. and Moore, J.G. 1969. *Enterobius vermicularis*: 10,000 year-old human infection. *Science*, **166**, 1620.

Ganor, S. 1987. In whom does pinworm infection itch? *Int J Dermatol*, **26**, 667.

Gardner, G.W., Edgerton, R.V., et al. 1977. Physical work capacity and metabolic stress in subjects with iron deficiency anaemia. *Am J Clin Nutr*, **30**, 910–17.

Geissler, P.W., Mwaniki, D.L., et al. 1998. Geophagy as a risk factor for geohelminth infections: a longitudinal study of Kenyan primary schoolchildren. *Trans R Soc Trop Med Hyg*, **92**, 7–11.

Gelphi, A.P. and Mustafa, A. 1967. Seasonal pneumonitis with eosinophilia: a study of larval ascariasis in Saudi Arabs. *Am J Trop Med Hyg*, **16**, 646–57.

Gibbons, L.M. 1986. *SEM Guide to the Morphology of Nematode Parasites of Vertebrates*. Farnham Royal, Slough: CAB International.

Gilman, R.H., Chong, Y.H., et al. 1983. The adverse consequences of heavy *Trichuris* infection. *Trans R Soc Trop Med Hyg*, **77**, 432–8.

Goodman, B. 1994. Models aid understanding, help control parasites. *Science*, **264**, 1862–3.

Grassi, B. 1887. *Trichocephalus* und Ascarisentwicklung. Preliminarnot. *Zentralbl Bakteriol Parasitenkunde*, **1**, 131–2.

Grove, D.J. 1990. *A History of Human Helminthology*. Wallingford, Oxon: CAB International.

Gupta, A.K., Gupta, S.C. and Keswani, N.K. 1989. Pathology of tropical appendicitis. *J Clin Pathol*, **42**, 1169–72.

Gupta, M.C., Mithal, S., et al. 1977. Effect of periodic deworming on nutritional status of *Ascaris*-infected pre-school children receiving supplementary food. *Lancet*, **2**, 108–10.

Guyatt, H. 2002. The economics of worm control. In: Holland, C.V. and Kennedy, M.W. (eds), *The Geohelminths: Ascaris, Trichuris and hookworm*. Boston/ Dordrecht/ London: Kluwer Academic Publishers, 75–87.

Guyatt, H.L. and Bundy, D.A.P. 1991. Estimating the prevalence of morbidity due to intestinal helminths: prevalence of infection as an indicator of prevalence of disease. *Trans R Soc Trop Med Hyg*, **85**, 778–82.

Hall, G.A., Rutter, J.M. and Beer, R.J. 1976. A comparative study of the histopathology of the large intestine of convenionally reared, specific pathogen free and gnotobiotic pigs infected with *Trichuris suis*. *J Comp Pathol*, **86**, 285–92.

Hagel, I., Lynch, N.R., et al. 1993. *Ascaris* reinfection of slum children: relation with the IgE response. *Clin Exp Immunol*, **94**, 80–3.

Haswell-Elkins, M.R., Elkins, D.B. and Anderson, R.M. 1987a. Evidence for predisposition in humans to infection with *Ascaris*, hookworm, *Enterobius* and *Trichuris* in a South Indian fishing community. *Parasitology*, **95**, 323–37.

Haswell-Elkins, M.R., Elkins, D.B., et al. 1987b. The distribution and abundance of *Enterobius vermicularis* in a South Indian fishing community. *Parasitology*, **95**, 339–54.

Hill, A.W. and Andrews, J. 1942. Relation of hookworm burden to physical status in Georgia. *Am J Trop Med*, **22**, 499–506.

Hill, C.H. 1957. The survival of swine whipworm eggs in hog lots. *J Parasitol*, **43**, 104.

Hoagland, K.E. and Schad, G.A. 1978. *Necator americanus* and *Ancylostoma duodenale*: life history parameters and epidemiological implications of two sympatric hookworms of humans. *Exp Parasitol*, **44**, 36–49.

Holland, C.V. 1987a. Hookworm infection. In: Stephenson, L.S. (ed.), *Impact of Helminth Infections on Human Nutrition*. London: Taylor and Francis, 128–60.

Holland, C.V. 1987b. Neglected infections-trichuriasis and strongyloidiasis. In: Stephenson, L.S. (ed.), *Impact of Helminth Infections on Human Nutrition*. London: Taylor and Francis, 161–201.

Holland, C.V. and Boes, J. 2002. Distributions and predisposition: people and pigs. In: Holland, C.V. and Kennedy, M.W. (eds), *The Geohelminths: Ascaris, Trichuris and hookworm*. Boston/ Dordrecht/ London: Kluwer Academic Publishers, 1–24.

Hotez, P.J. and Cerami, A. 1983. Secretion of a proteolytic anticoagulant by *Ancylostoma duodenale* hookworms. *J Exp Med*, **157**, 1594–603.

Hyman, L.H. 1951. *The Invertebrates: Acanthocephala, Aschelminthes and Entoprocta*. vol. III. New York: McGraw-Hill.

Jalal, F. 1991. Effects of deworming, dietary fat, and carotenoid-rich diets on vitamin A status of preschool children infected with *Ascaris lumbricoides* in West Sumatera Province, Indonesia. PhD Dissertation, Cornell University, New York.

Jarrett, E.E.E. and Kerr, J.W. 1973. Threadworms and IgE in allergic asthma. *Clin Allergy*, **3**, 203–7.

Kagei, N. 1983. Techniques for the measurement of environmental pollution by infective stages of soil-transmitted helminths. In: Yokogawa, M., et al. (eds), *Collected Papers on the Control of Soil-Transmitted Helminthiases*. vol. 2. Tokyo: Asian Parasite Control Organization, 27–46.

Kenney, M. and Eveland, L.R. 1978. Infection of man with *Trichuris vulpis*, the whipworm of dogs. *Am J Clin Pathol*, **69**, 199.

Kenney, M. and Yermakov, V. 1980. Infection of man with *Trichuris vulpis*, the whipworm of dogs. *Am J Trop Med Hyg*, **29**, 1205–8.

Keymer, A. and Pagal, M. 1990. Predisposition to helminth infection. In: Schad, G.A. and Warren, K.S. (eds), *Hookworm Disease*. London: Taylor and Francis, 77–209.

Khan, T.T.F., Raj, S.M. and Visvanathan, R. 1993. Spectrum of cholangitis in a rural setting in North-eastern Peninsular Malaysia. *Trop Doct*, **23**, 117–18.

Koino, S. 1922. Experimental infections on the human body with ascarides. *Jpn Med World*, **15**, 317–20.

Kroeger, A., Schulz, B., et al. 1992. Helminthiasis and cultural change in the Peruvian rainforest. *J Trop Med Hyg*, **95**, 104–13.

Kunii, C. 1983. *Humanistic Family Planning Approaches: The Integration of Family Planning and Health Goals*. New York: United Nations Fund for Population Activities.

Lee, D.L. and Atkinson, H. 1976. *Physiology of Nematodes*, 2nd edn. London: Macmillan.

Lewin, R. 1989. *Human Evolution*, 2nd edn. Oxford: Blackwell Scientific Publications.

Looss, A. 1901. Über das Eindrigen der Ankylostomalarven in die menschliche Haut. *Zentralbl Bakteriol Parasitenkunde*, **29**, 733–9.

Looss, A. 1905. The anatomy and life history of *Ankylostoma duodenale* Dub. Pt I. The anatomy of the adult worm. *Records of the Egyptian Government, School of Medicine*, vol. III. Cairo: National Printing Department, 1–158.

Lotero, H., Tripathy, K. and Bolanos, O. 1974. Gastrointestinal blood loss in *Trichuris* infection. *Am J Trop Med Hyg*, **23**, 1203–4.

Louw, J.H. 1966. Abdominal complications of *Ascaris lumbricoides* infestation in children. *Br J Surg*, **53**, 510–21.

Lwambo, N.J.S., Bundy, D.A.P. and Medley, G.F.H. 1992. A new approach to morbidity assessment in hookworm endemic countries. *Epidemiol Infect*, **108**, 469–81.

Lynch, N.R., Isturiz, G., et al. 1992a. Bronchial challenge of tropical asthmatics with *Ascaris lumbricoides*. *J Invest Allergy Clin Immun*, **2**, 97–105.

Lynch, N.R., Isturiz, G., et al. 1992b. Bronchoconstriction in helminthic infection. *Int Arch Allergy Immunol*, **98**, 77–9.

Lynch, N.R., Hagel, I., et al. 1993. Effect of age and helminthic infection on IgE levels in slum children. *J Investig Allergol Clin Immunol*, **3**, 96–9.

Lynch, N.R., Palenque, M., et al. 1997. Clinical improvement of asthma after anthelmintic treatment in a tropical situation. *Am J Respir Crit Care Med*, **156**, 50–4.

Macdonald, D. 1984. *The Encyclopaedia of Mammals*. London: Unwin Hyman.

Maggenti, A. 1982. Nemata (nematodes). In: Parker , S.P. (ed.), *Synopsis and Classification of Living Organisms*. vol. 1. New York: McGraw-Hill.

Mansour, M.M., Francis, W.M. and Farid, Z. 1985. Prevalence of latent iron deficiency in patients with chronic *S. mansoni* infection. *Trop Geogr Med*, **37**, 124–8.

Martin, J., Crompton, D.W.T., et al. 1984. Mucosal surface lesions in young protein-deficient pigs infected with *Ascaris suum* (Nematoda). *Parasitology*, **88**, 333–40.

Matthews, B.E. 1982. Skin penetration by *Necator americanus* larvae. *Z Parasitenkunde*, **68**, 81–6.

Maung, M. 1973. *Ascaris lumbricoides* Linne, 1758 and *Ascaris suum* Goeze, 1782: morphological differences between specimens obtained from man and pig. *Southeast Asian J Trop Med Public Health*, **4**, 41–5.

McSharry, C., Xia, Y., et al. 1999. Natural immunity to *Ascaris lumbricoides* associated with Immunoglobulin E antibody to ABA-1 allergen and inflammation indicators in children. *Infect Immun*, **67**, 1–6.

Medley, G.F.H., Guyatt, H.L. and Bundy, D.A.P. 1993. A quantitative framework for evaluating the effect of community treatment on the morbidity due to *Ascaris*. *Parasitology*, **106**, 211–21.

Migasena, S., Sumetchotimaytha, J., et al. 1984. Zinc in the pathogenesis of hookworm anaemia. *Southeast Asian J Trop Med Public Health*, **15**, 206–8.

Misegna, S. and Gilles, H.M 1987. Hookworm infection. In: Pawlowski, Z.S. (ed.), *Intestinal Helminthic Infections*. London: Baillière Tindall, 617–27.

Mizgajska, H. 1993. The distribution and survival of eggs of *Ascaris suum* in six different natural soil profiles. *Acta Parasitol*, **38**, 170–4.

Montresor, A., Ramsen, M., et al. 2001. Extending anthelmintic coverage to non-enrolled school children using a simple and low-cost school-based method. *Trop Med Int Health*, **6**, 535–7.

Montresor, A., Stoltzfus, R.J., et al. 2002. Is the exclusion of children under 24 months from anthelmintic treatment justifiable? *Trans R Soc Trop Med Hyg*, **96**, 197–9.

Morishita, K. 1972. Studies on the epidemiological aspects of ascariasis in Japan and basic knowledge concerning its control. In: Morishita, K., et al. (eds), *Progress of Medical Parasitology in Japan*. 4. Tokyo: Meguro Parasitological Museum, 3–153.

Muller, R. 1975. *Worms and Disease*. London: Heinemanns Medical.

Murrell, K.D., Eriksen, L., et al. 1997. *Ascaris suum*: A revision of its early migratory path and implications for human ascariasis. *J Parasitol*, **83**, 255–60.

Nawalinski, T.A., Schad, G.A. and Chowdhury, A.B. 1978. Population biology of hookworm in children in rural Bengal, 1. General parasitological observations. *Am J Trop Med Hyg*, **27**, 1152–61.

Needham, C.S., Bundy, D.A.P., et al. 1992. The relationship between *Trichuris trichiura* transmission intensity and the age-profiles of parasite-specific antibody isotypes in two endemic communities. *Parasitol*, **105**, 273–83.

Nesheim, M.C. 1985. Nutritional aspects of *Ascaris suum* and *A. lumbricoides* infection. In: Crompton, D.W.T., Nesheim, M.C. and Pawlowski, Z.S. (eds), *Ascariasis and its public health significance*. London: Taylor and Francis, 147–60.

Nichols, R.L. 1956. The aetiology of visceral larva migrans, II. Comparative larval morphology of *Ascaris lumbricoides*, *Necator americanus*, *Strongyloides stercoralis* and *Ancylostoma caninum*. *J Parasitol*, **42**, 363–99.

Nokes, C. and Bundy, D.A.P. 1994. Does helminth infection affect mental processing and educational achievement? *Parasitol Today*, **10**, 14–18.

Nokes, C., Grantham-McGregor, S.M., et al. 1992. Moderate to heavy infections of *Trichuris trichiura* affect cognitive function in Jamaican schoolchildren. *Parasitology*, **104**, 539–47.

Palmer, D.R., Hall, A., et al. 1995. Antibody isotype responses to antigens of *Ascaris lumbricoides* in a case-control study of persistently heavily infected Bangladeshi children. *Parasitology*, **111**, 385–93.

Patterson, R. and Harris, K.E. 1985. Parallel induction of immunoglobulin E-mediated *Ascaris* antigen airway responses and increased carbachol reactivity in rhesus monkeys by infection with *Ascaris suum*. *J Lab Clin Medicine*, **106**, 293–7.

Pawlowski, Z.S. 1982. Ascariasis: host-pathogen biology. *Rev Infect Dis*, **4**, 806–14.

Pawlowski, Z.S. and Davis, A. 1989. Morbidity and mortality in ascariasis. In: Crompton, D.W.T., Nesheim, M.C. and and Pawlowski, Z.S. (eds), *Ascariasis and its Prevention and Control*. London: Taylor and Francis, 71–86.

Pawlowski, Z.S., Schad, G.A. and Stott, G.J. 1991. *Hookworm Infection and Anaemia*. Geneva: World Health Organization.

Peng, W., Zhou, X. and Crompton, D.W.T. 1995. Aspects of ascariasis in China. *Helminthologia*, **32**, 97–100.

Perry, R.N. and Clark, A.J. 1982. Hatching mechanisms of nematodes. *Parasitology*, **83**, 435–49.

Pinus, J. 1985. Surgical complications of ascariasis in Brazil. In: Crompton, D.W.T., Nesheim, M.C. and Pawlowski, Z.S. (eds), *Ascariasis and its Public Health Significance*. London: Taylor and Francis, 161–6.

Poinar, G.O. 1983. *The Natural History of Nematodes*. Englewood Cliffs, NJ: Prentice-Hall.

Pollitt, E. 1990. *Malnutrition and Infection in the Classroom*. Paris: UNESCO.

Pritchard, D.I., Quinnell, R.J. and Walsh, E.A. 1995. Immunity in humans to *Necator americanus*: IgE, parasite weight and fecundity. *Parasite Immunol*, **17**, 71–5.

Pritchard, D.I., Walsh, E.A., et al. 1992. Isotypic variation in antibody responses in a community in Papua New Guinea to larval and adult antigens during infection and following reinfection, with the hookworm *Necator americanus*. *Parasite Immunol*, **14**, 617–31.

Pritchard, D.I., Quinnell, R.J., et al. 2002. The immunobiology of hookworm infection. In: Holland, C.V. and Kennedy, M.W. (eds), *The Geohelminths: Ascaris, Trichuris and hookworm*. Boston/Dordrecht/London: Kluwer Academic Publishers, 143–65.

Quinnell, R.J., Woolhouse, M.E.J., et al. 1995. Immunoepidemiology of human necatoriasis: correlations between antibody responses and parasite burdens. *Parasite Immunol*, **17**, 313–18.

Quinnell, R.J., Griffin, J., et al. 2001. Predisposition to hookworm infection in Papua New Guinea. *Trans R Soc Trop Med Hyg*, **95**, 139–42.

Ramdath, D.D., Simeon, D.T., et al. 1995. Iron status of schoolchildren with varying intensities of *Trichuris trichiura* infection. *Parasitology*, **110**, 347–51.

Rathi, A.K., Batra, S., et al. 1981. Ascariasis causing intestinal obstruction in a 45-day-old infant. *Indian Pediatr*, **18**, 751–2.

Read, A.F. and Skorping, A. 1995. The evolution of tissue migration by parasitic nematode larvae. *Parasitology*, **111**, 359–71.

Roche, M. and Layrisse, M. 1966. The nature and causes of 'hookworm anaemia'. *Am J Trop Med Hyg*, **15**, 1030–100.

Rogers, W.P. 1960. The physiology of infective processes of nematode parasites: the stimulus from the animal host. *Proc R Soc London B*, **152**, 367–86.

Salafsky, B., Fusco, A.C. and Siddiqui, A. 1990. *Necator americanus*: factors influencing skin penetration by larvae. In: Schad, G.A. and Warren, K.S. (eds), *Hookworm Disease*. London: Taylor and Francis, 329–39.

Savioli, L., Montresor, M. and Albonico, M. 2002. Control strategies. In: Holland, C.V. and Kennedy, M.W. (eds), *The Geohelminths: Ascaris, Trichuris and hookworm*. Boston/Dordrecht/London: Kluwer Academic Publishers, 25–37.

Schad, G.A. and Anderson, R.M. 1985. Predisposition to hookworm infection in humans. *Science*, **228**, 1537–40.

Schad, G.A. and Warren, K.S. 1990. *Hookworm Disease: Current Status and New Directions*. London: Taylor and Francis.

Schad, G.A., Nawalinski, T.A. and Kochar, V. 1983. Human ecology and the distribution and abundance of hookworm populations. In: Croll, N.A. and Cross, J.H. (eds), *Human Ecology and Infectious Diseases*. London: Academic Press, 187–223.

Schad, G.A., Chowdhury, A.B., et al. 1973. Arrested development in human hookworm infections. *Science*, **180**, 502–4.

Scott, M.E. and Smith, G. 1994. *Parasitic and Infectious Diseases*. London: Academic Press.

Simeon, D., Callender, J., et al. 1994. School performance, nutritional status and trichuriasis in Jamaican school children. *Acta Paediatr*, **83**, 1188–93.

Skryabin, K.I., Shikhobalova, N.P. and Mozgovoi, A.A. 1991. Oxyurata and Ascaridata. In: Skryabin, K.I. (ed.), *Key to Parasitic Nematodes*. vol. 2. Leiden: E.J. Brill.

Smyth, J.D. 1994. *Introduction to animal parasitology*, 3rd edn. Cambridge: Cambridge University Press.

Sprent, J.F.A. 1952. Anatomical distinction between human and pig strains of *Ascaris*. *Nature (London)*, **170**, 627–8.

Stephenson, L.S. 1987. *Impact of Helminth Infections on Human Nutrition*. London: Taylor and Francis.

Stephenson, L.S., Holland, C.V. and Cooper, E.S. 2001. The public health significance of *Trichuris trichiura*. *Parasitology*, **121**, Suppl., S573–95.

Stephenson, L.S., Holland, C.V. and Ottesen, E.A. 2000. Controlling intestinal helminths while eliminating lymphatic filariasis. *Parasitology*, **121**, Suppl S173.

Stephenson, L.S., Crompton, D.W.T., et al. 1980a. Relationships between *Ascaris* infection and growth of malnourished pre-school children in Kenya. *Am J Clin Nutr*, **33**, 1165–72.

Stephenson, L.S., Pond, W.G., et al. 1980b. *Ascaris suum*: nutrient absorption, growth, and intestinal pathology in young pigs experimentally infected with 15-day-old larvae. *Exp Parasitol*, **49**, 15–25.

Stephenson, L.S., Latham, M.C., et al. 1993. Physical fitness, growth and appetite of Kenyan school boys with hookworm, *Trichuris trichiura* and *Ascaris lumbricoides* are improved four months after a single dose of albendazole. *J Nutr*, **123**, 1036–46.

Stoltzfus, R.J., Chwaya, H.M., et al. 1997. Epidemiology of iron deficiency anaemia in Zanzibari schoolchildren: the importance of hookworms. *Am J Clin Nutr*, **65**, 153–9.

Stoltzfus, R.J., Albonico, M., et al. 1998. Effects of the Zanzibar school-based deworming program on iron status in children. *Am J Clin Nutr*, **68**, 179–86.

Stoltzfus, R.J., Kvalsig, J.D., et al. 2001. Effects of iron supplementation and anthelminthic treatment on motor and language development of pre-school children in Zanzibar: double blind placebo controlled study. *Br Med J*, **323**, 1389–93.

Takata, I. 1951. Experimental infection of man with *Ascaris* of man and the pig. *Kitasato Arch Exp Med*, **23**, 49–59.

Taren, D.L., Nesheim, M.C., et al. 1987. Contributions of ascariasis to poor nutritional status of children from Chiriqui Province, Republic of Panama. *Parasitology*, **95**, 603–13.

Thein-Hlaing 1985. *Ascaris lumbricoides* infections in Burma. In: Crompton, D.W.T., Nesheim, M.C. and Pawlowski, Z.S. (eds), *Ascariasis and its Public Health Significance*. London: Taylor and Francis, 83–112.

Thein-Hlaing 1989. Epidemiological basis of survey design, methodology and data analysis for ascariasis. In: Crompton, D.W.T., Nesheim, M.C. and Pawlowski, Z.S. (eds), *Ascariasis and its Prevention and Control*. London: Taylor and Francis, 351–68.

Thein-Hlaing 1993. Ascariasis and childhood malnutrition. *Parasitology*, **107**, Supplement, S125–36.

Thein-Hlaing, Than-Saw, et al. 1984. Epidemiology and transmission dynamics of *Ascaris lumbricoides* in Okpo village, rural Burma. *Trans R Soc Trop Med Hyg*, **78**, 497–504.

Thein-Hlaing, Thane-Toe, et al. 1992. A controlled chemotherapeutic intervention trial on the relationship between *Ascaris lumbricoides* infection and malnutrition in children. *Trans R Soc Trop Med Hyg*, **85**, 523–8.

Theinpont, D., Rochette, F. and Vanparijs, O.F.J. 1980. *Diagnosing Helminthiasis through Coprological Examination*. Beerse, Belgium: Janssen Research Foundation.

Tomkins, A. and Watson, F. 1989. *Malnutrition and infection: a review*. Geneva: World Health Organization ACC/SCN.

Tripathy, K., Duque, E., et al. 1972. Malabsorption syndrome in ascariasis. *Am J Clin Nutr*, **25**, 1276–87.

Urban, J.F., Madden, K.B., et al. 1992. The importance of Th2 cytokines in the protective immunity to nematodes. *Immunol Rev*, **12**, 205–20.

Viteri, F.E. 1994. The consequences of iron deficiency and anaemia in pregnancy on maternal health, the foetus and the infant. *SCN News*, **11**, 14–18.

Viteri, F.E. and Torun, B. 1974. Anaemia and physical work capacity. *Clin Haematol*, **3**, 609–26.

Wakelin, D., Harnett, W. and Parkhouse, R.M.E. 1983. Nematodes. In: Warren, K.S. (ed.), *Immunology and Molecular Biology of*

Parasitic Infections, 3rd edn. Oxford: Blackwell Scientific Publications, 496–526.

Wang, Z.Y., Wang, X.Z., et al. 1983. Blood sucking activities of hookworms. *Chin Med J*, **96**, 281–6.

Wardlaw, A.C., Forsyth, L.M.G. and Crompton, D.W.T. 1994. Bactericidal activity in the pig roundworm *Ascaris suum*. *J Appl Bacteriol*, **76**, 36–41.

Warren, K. 1972. *Ascaris* – a practical joke? *N Engl J Med*, **286**, 999–1000.

Watson, J.M. 1960. *Medical Helminthology*. London: Ballière, Tindall and Cox.

Wharton, D. 1980. Nematode egg-shells. *Parasitology*, **81**, 447–63.

Wharton, D. and Jenkins, T. 1978. Structure and chemistry of the egg-shell of a nematode (*Trichuris suis*). *Tissue Cell*, **10**, 427–40.

WHO, 1972. *Nutritional anaemias*. Technical Report Series 503. Geneva: World Health Organization.

WHO, 1981. *Intestinal protozoan and helminthic infections*. Technical Report Series 666. Geneva: World Health Organization.

WHO, 1994. *Bench Aids for the Diagnosis of Intestinal Parasites*. Geneva: World Health Organization.

WHO, 1995. *Hookworm infection and anaemia in girls and women* (WHO/CDS/IPI/95.1). Geneva: World Health Organization.

WHO, 2001. Schistosomiasis and soil-transmitted helminth infections. Agenda item 13.3. Fifty-fourth World Health Assembly, 22 May 2001. Geneva: World Health Organization.

WHO, 2003. *Report of the informal consultation on the use of praziquantel during pregnancy/lactation and albendazole/mebendazole in children under 24 months*. Geneva: World Health Organization.

Williams-Blangero, S., Subedi, J., et al. 1999. Genetic analysis of susceptibility to infection with *Ascaris lumbricoides*. *Am J Trop Med Hyg*, **60**, 921–6.

Williams-Blangero, S., VandeBerg, J.L., et al. 2002. Genes on chromosomes 1 and 13 have significant effects on *Ascaris* infection. *PNAS*, **99**, 5533–8.

Wolgemuth, J.C., Latham, M.C., et al. 1982. Worker productivity and the nutritional status of Kenyan road construction laborers. *Am J Clin Nutr*, **32**, 68–78.

Wong, M.S. and Bundy, D.A.P. 1990. Quantitative assessment of contamination of soil by eggs of *Ascaris lumbricoides* and Trich*uris trichiura*. *Trans R Soc Trop Med Hyg*, **84**, 567–70.

World Bank, 1993. *World Development Report 1993: Investing in Health*. New York: Oxford University Press.

Yadav, A.K. and Tandon, V. 1989. Prevalence of nematode eggs in the urban area of the city of Shilong, India – a public health problem. *Health Hyg*, **10**, 158–61.

Yamaguti, S. 1961. The nematodes of vertebrates. *Systema helminthum*. vol. III. London: Interscience Publishers.

Yorke, W. and Maplestone, R.A. 1926. *The Nematode Parasites of Vertebrates*. London: J and A Churchill.

Yoshida, Y. 1973. Species of hookworms infecting man: patterns of development. *Abstracts of the 9th International Congress of Tropical Medicine and Malaria*, Athens, 174.

Yoshida, Y., Kondo, K., et al. 1974. Comparative studies on *Ancylostoma braziliense* and *Ancylostoma ceylanicum*. III. Life history in the definitive host. *J Parasitol*, **60**, 636–41.

Yoshikawa, H., Yamada, M., et al. 1989. Variations in egg size of *Trichuris trichiura*. *Parasitol Res*, **75**, 649–54.

Yu, S. 1994. Report on the first nationwide survey on the distribution of human parasites in China. 1. Regional distribution of parasite species. *Chin J Parasitol Parasitic Dis*, **12**, 241–7.

35

Stronglyloides and *Capillaria*

MICHAEL CAPPELLO AND PETER J. HOTEZ

STRONGYLOIDES

THE PARASITE

Nematodes belonging to the genus *Strongyloides* were first identified as a cause of intestinal disease by Normand in 1876, who discovered these worms in the feces of French troops in Cochin-China. At autopsy, morphologically distinct worms were identified within the intestinal mucosa of those suffering from overwhelming disease. Bavay (1876) subsequently named the intestinal forms *Anguillula intestinalis* and the fecal forms *A. stercoralis*. Soon after, it was determined that these represented different stages of development of the same parasite. The existence of an autoinfective cycle was suggested by Fuelleborn (1914), and later work by Nishigori (1928) demonstrated that first stage larvae could indeed develop into infective filariform larvae and invade the colonic mucosa, thereby initiating development within the same host.

Classification

Strongyloides belongs to the phylum Nematoda and the class Phasmidia, i.e. nematodes containing phasmids or caudal chemoreceptors. Within this class, they are assigned to the order Rhabditida and the superfamily Rhabditoidea, which is composed of both free-living and parasitic worms. They are distinguished from the superfamily Rhabdiasoidea, also within the suborder Rhabditina, by the presence of a distinct esophageal bulb. Two species cause disease in humans, *S. stercoralis* and *S. fuelleborni*.

Morphology and structure

Parasitic females, which reside in the mucosa of the small intestine, are small and thin, measuring approximately 2–3 mm in length and 30–50 μm in width (Figure 35.1). The anterior portion is thicker than the posterior, and contains the esophagus. The triradiate pharynx contains two subventral and one dorsal gland, which deposit their secretions into the intestinal lumen. The reproductive system contains a midline vulva and paired uteri leading anteriorly and posteriorly from the vagina. The spirally wound ovaries of *S. fuelleborni* help

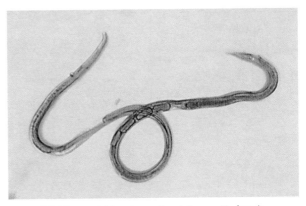

Figure 35.1 Strongyloides stercoralis *adult parasitic female (courtesy of Ash and Orihel (1997))*

to distinguish it morphologically from *S. stercoralis*, whose ovaries are linearly aligned within the body of the worm.

The existence of parasitic males has been debated over many years. Early observations by Faust (1933) that parasitic males exist, and that sexual reproduction may represent one route by which *Strongyloides* reproduces within the host, have not been confirmed by others. Part of the reason for the skepticism surrounding his findings may be the close morphological resemblance of the putative adult males to the free-living rhabditiform stage (Schad 1990).

The free-living adult female is approximately half the length of its parasitic counterpart (approximately 1.0 mm), although it is nearly twice as thick (80 μm). While the reproductive systems are morphologically similar, the uteri in the free-living adult female contain significantly more eggs. The free-living male is slightly smaller than the female, measuring approximately 50 μm in width. The reproductive system is composed of a tubular structure containing the testis, vas deferens, and seminal vesicle. The copulatory spicules, which penetrate the female during copulation, are located on each side of the gubernaculum.

First-stage rhabditiform larvae of *S. stercoralis* measure approximately 250 × 15 μm (Figure 35.2). They are characterized by a muscular esophagus, which comprises the anterior third of the body, and a short buccal cavity that distinguishes them from other nematodes, such as hookworms, which possess a more developed buccal tube. Third-stage filariform, or infective, larvae are long and thin compared to the rhabditiform stages (Figure 35.3). Paired alae on both sides of the worm offer structural stability during its movements along the ground. The mouth opening is small and closed to the external environment, as the filariform stage is probably non-feeding. Recent work using electron microscopy with three-dimensional reconstruction has delineated the location and structure of the amphidial neurons, highly specialized organs that may represent a means through which larvae sense environmental stimuli (Ashton et al. 1995; Lopez et al. 2000).

The eggs of the parasitic female are deposited within the mucosa of the small intestine and usually hatch before reaching the lumen. As a result, they are rarely excreted in the feces. Eggs from the free-living adult female are partially embryonated, and measure approximately 50–70 μm in length with an oval shape.

General biology

Perhaps the most intriguing aspect of the biology of *Strongyloides* is its ability to alternate parasitic (homogonic) and free-living (heterogonic) life cycles. The factors that determine which route will be chosen have been studied for some species, and it appears that both internal, i.e. host specific, and external, i.e. environmental, conditions play a role in influencing developmental outcome (Moncol and Triantaphyllou 1978). In addition, certain geographic strains of *Strongyloides* may have a particular predilection for developing into one generation or the other (Neva 1986). It has also been observed that the free-living cycle develops most frequently in tropical climates with moist soil, while the parasitic life cycle may predominate in more temperate regions.

Rhabditiform first-stage larvae deposited with feces feed on bacteria and organic debris found in the

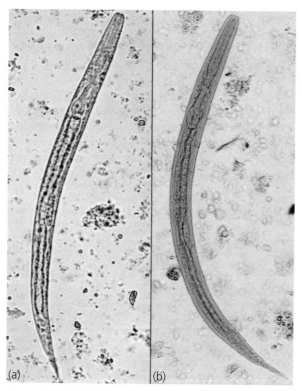

Figure 35.2 Strongyloides stercoralis *first stage (L1) rhabditiform larva, with* **(a)** *and without* **(b)** *iodine stain (courtesy of Ash and Orihel (1997))*

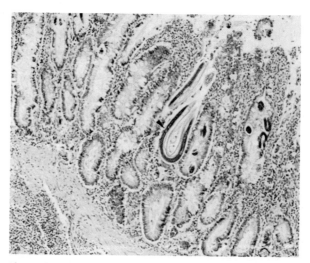

Figure 35.3 *Adult female* Strongyloides stercoralis *worm (see arrow) in biopsy of small intestine (hematoxylin and eosin stain) (courtesy of Orihel and Ash (1995))*

surrounding soil. Those that develop into free-living adults continue to survive in this environment, whereas those that become filariform (i.e. infective third-stage larvae) will ultimately cease to feed, while engaging in activities that might facilitate contact with a susceptible host. Filariform larvae have been shown to cluster in groups and stand on their tails, a behavior called 'questing,' which is characteristic of the infective larval stage of parasitic nematodes.

Life cycle

Adult parasitic females, which reside within the crypts of the small intestine, deposit eggs that hatch prior to migrating into the lumen. The resulting first-stage larvae are transported through the bowel, and excreted with the feces. Larvae that are deposited on to moist, warm soil will develop either directly into infective filariform larvae or into free-living adult males and females, whose progeny will ultimately molt to become infective larvae. These infect their host by penetrating the skin, a process associated with the release of hydrolytic enzymes (Brindley et al. 1995; Gallego et al. 1998) that may function to degrade the extracellular dermal matrix. The traditionally accepted view of the life cycle of *S. stercoralis* maintains that, once within the dermis, larvae invade the venous circulation and are carried passively to the lungs, where they become trapped within capillaries and break through to the alveolar space. After migrating up the respiratory tree into the pharynx, the larvae are swallowed and travel to the small intestine. During this process, they molt to the fourth stage, probably during passage through the lungs. Work by Schad et al. (1989), however, has raised the possibility that this 'canonical route' represents only one of many potential pathways by which larvae may reach the small intestine.

Only females develop into adults within the intestine, where they reside and deposit eggs that have been produced by parthenogenesis. These eggs normally hatch within the epithelium, liberating first-stage larvae into the lumen of the intestine. As a result, *Strongyloides* eggs are not routinely detected in the feces of infected individuals. If the first-stage larvae molt to the filariform stage during passage through the bowel, they may penetrate colonic mucosa or perianal skin, where they once again locate the venous circulation and begin the infective process anew. This autoinfection provides a means by which the parasite can multiply within the host, and explains how infections can be maintained for years after an individual leaves an endemic area. What regulates larval development in the intestine is not known, nor are the factors that sometimes trigger the explosive amplification of *Strongyloides* within the host called hyperinfection (see section entitled Clinical manifestations), which can lead to the dissemination of parasites throughout various tissues.

Transmission and epidemiology

The predominant route of transmission of *S. stercoralis* is by skin contact with fecally contaminated soil. As a result, communities where close living conditions and poor sanitation facilities exist are frequently characterized by high prevalence rates of strongyloidiasis. These include both rural and urban areas in the developing world, as well as closed communities, such as institutions for the mentally handicapped (Braun et al. 1989). Zoonotic transmission of strongyloidiasis from asymptomatic dogs to animal workers has also been reported (Georgi and Sprinkle 1974). Although transmammary transmission of *S. stercoralis* has been demonstrated in laboratory animals infected experimentally (Shoop et al. 2002), this has not been confirmed to occur in humans.

The geographic distribution of *Strongyloides* is quite broad, and includes temperate as well as tropical climates. This undoubtedly is attributable to the remarkable phenotypic flexibility of the parasite, as it is able to adapt its life cycle and development to suit environmental conditions. Moreover, internal autoinfection allows for maintenance of the parasite within the host for years following initial exposure. Surveys of prisoners of war from World War II conducted more than 40 years later succeeded in identifying patent infections in up to 37 percent of those soldiers who were held captive in various parts of the Pacific (Hill 1988; Genta 1989). Since all infected individuals are at risk of hyperinfection and disseminated disease, the identification of those who may have acquired their infection in the distant past is of significant clinical importance (Link and Orenstein 1999). This group includes not just former military personnel, but also immigrants and refugees from highly endemic areas (De Silva et al. 2002; Gyorkos et al. 1992). Strongyloidiasis remains a chronic infection among people from certain parts of the United States, particularly Appalachia (Kitchen et al. 2000; Siddiqui and Berk 2001), as well as throughout Europe (Sanchez et al. 2001).

Strongyloides hyperinfection is most frequently seen in those with underlying immune defects such as hematological malignancy, autoimmune disease or severe protein malnutrition. In addition, immunosuppressive therapy, particularly glucocorticoids, associated with any of the above conditions, as well as organ transplantation, may also precipitate hyperinfection and dissemination. Whether this is due to immunosuppression or a direct effect of steroids on the reproductive and developmental patterns of *Strongyloides* remains to be determined (Genta 1992). A cDNA encoding a putative steroid hormone receptor has recently been cloned from *S. stercoralis*, suggestive of the possibility that host hormones may, in fact, influence the behavior of the parasite in vivo (Siddiqui et al. 2000). Coinfection with human immunodeficiency virus (HIV), although reported in association with severe strongyloidiasis, is less common than might have been expected, given the significant overlap in the geographic distribution of these two diseases (Celedon et al. 1994; Gompels et al. 1991). In contrast to HIV, there is evidence that coinfection with human T-lymphotropic virus type 1 (HTLV-1) may be associated with more severe strongyloidiasis than that which is seen in HTLV-1 seronegative individuals (Robinson et al. 1994; Gotuzzo et al. 1999; Satoh et al. 2002).

S. fuelleborni is also transmitted primarily by skin contact. However, epidemiological data from parts of Africa and Papua New Guinea suggest that this may not be the only route of infection. Investigations into the high prevalence of *S. fuelleborni* in African children led to the postpartum identification of larvae in the breast milk of a woman from Zaire (Brown and Girardeau 1977). In Papua New Guinea, high rates of infection (up to 65 percent) in children up to 3 months of age suggest that transmammary infection might also occur in this endemic area (Barnish and Ashford 1989; Ashford and Barnish 1990). In addition, *S. fuelleborni* has been identified as the causative agent of 'swollen belly syndrome,' a fulminant and fatal enteritis that has been well documented among infants less than 6 months of age born in the highly endemic areas of Papua New Guinea (see Clinical manifestations). It has been suggested that transmammary infection followed by early external autoinfection from soiled nappies may result in the rapid multiplication of worms in these very young children (Ashford and Barnish 1990).

CLINICAL AND PATHOLOGICAL ASPECTS OF INFECTION

Strongyloidiasis can be divided into the following clinical categories: (1) asymptomatic carriage; (2) intestinal disease; and (3) hyperinfection with or without dissemination. Symptomatology is somewhat dependent on worm burden, which appears to be regulated by a delicate balance between the immune status of the host and the developmental tendencies of the parasite. In chronic strongyloidiasis, egg output remains relatively low and constant, proportionally few larvae undergoing internal autoinfection. In contrast, hyperinfection is associated with both an increase in the number of eggs produced as well as the proportion of larvae that develop internally to the filariform stage and reinvade the host.

Clinical manifestations

Chronic infection with *S. stercoralis* is most frequently asymptomatic (Milder et al. 1981; Neva 1986; Berk et al. 1987). When symptoms are present, they usually involve the gastrointestinal tract. Intermittent abdominal pain, distention, bloating, and diarrhea alternating with constipation are characteristic of intestinal disease. Children

may develop a malabsorption syndrome caused by infection with *S. stercoralis*, characterized by growth stunting and failure to thrive (O'Brien 1975; Burke 1978). Pruritus and urticaria, particularly involving the perianal skin and buttocks, are also common symptoms of chronic *Strongyloides* infection. A serpigenous, pruritic eruption usually found on the trunk or buttocks, named larva currens (racing larva) by Arthur and Shelley (1958), may represent skin penetration by autoinfective filariform larvae in this area. This pathognomonic rash tends to progress and subside over a period of hours, and its intermittent nature often hinders the diagnosis of strongyloidiasis. Other dermatoses, including prurigo and lichen simplex chronicus, have also been associated with chronic strongyloidiasis (Albanese et al. 2001; Jacob and Patten 1999).

The most serious clinical manifestations of strongyloidiasis occur in the setting of hyperinfection, where the dramatic increase in the number of parasites within the host results in dissemination of larvae to various tissues. The increase in worm burden results from an increase in both the number of eggs produced by the female parasites in the intestine and the proportion of those larvae that undergo autoinfection. Usually associated with some underlying host immune defect, hyperinfection is most frequently associated with immunosuppressive therapy, particularly glucocorticoids, which are often used as part of the treatment of such conditions as hematological malignancies and autoimmune diseases. Coinfection with HTLV-1 may also predispose to *Strongyloides* hyperinfection, perhaps due to a defect in effector IgE responses (Dixon et al. 1989; Robinson et al. 1994; Satoh et al. 2002).

Patients with *Strongyloides* hyperinfection generally experience a worsening in abdominal symptoms, often accompanied by paralytic ileus (Cookson et al. 1990), gastrointestinal bleeding (Bhatt et al. 1990; Dees et al. 1990; Daubenton et al. 1998), and even perforation. In addition, the increase in the number of worms migrating through the lungs often results in wheezing, dyspnea, and occasionally pulmonary hemorrhage. Patchy infiltrates, diffuse interstitial pneumonitis, and bronchopneumonia may be detected by chest radiography or CT scan (Meltzer et al. 1979; Berenson et al. 1987; Chu et al. 1990; Kramer et al. 1990; Upadhyay et al. 2001). This can occasionally be interpreted as a relapse in underlying malignancy (Sandlund et al. 1997).

During hyperinfection, filariform larvae may gain access to the arterial circulation, thereby allowing dissemination to numerous tissues and solid organs. In fact, filariform (and rarely rhabditiform) larvae have been recovered from biopsies of lymph nodes (Adam et al. 1973), endocardium (Neefe et al. 1973), pancreas (Kuberski et al. 1975), liver (Neefe et al. 1973), kidney (Civantos and Robinson 1969), and brain (Neefe et al. 1976), and have even been isolated from peripheral blood (Onuigbo and Ibeachum 1991). Involvement of the central nervous system can be associated with seizures and mental status changes (Igra-Siegman et al. 1981; Genta and Walzer 1989; Cappello and Hotez 1993), and is frequently accompanied by pyogenic abscess or meningitis caused by gram-negative bacteria, which may be carried from the gut by migrating larvae (Link and Orenstein 1999). Cutaneous manifestations of disseminated strongyloidiasis include petechiae and sometimes severe purpura (Kalb and Grossman 1986; Berenson et al. 1987; Genta and von Kuster 1988; Ronan et al. 1989). Unlike the rash of chronic strongyloidiasis (larva currens), skin biopsies in disseminated disease often reveal filariform larvae.

Children suffering from swollen belly syndrome caused by *S. fuelleborni* usually present early in life (8–10 weeks of age) with protein-losing enteropathy, abdominal distention, and respiratory distress (Ashford and Barnish 1990). Diarrhea is generally not pronounced and fever is absent. Severe hypoproteinemia (without proteinuria) is a characteristic finding and leads to the development of peripheral edema and ascites. Egg counts are often greater than 100 000/g feces. Untreated, the condition is associated with significant mortality.

Pathogenesis and pathology

The intestinal pathology associated with *Strongyloides* infection is indicated by a variety of histological findings. The spectrum of damage, ranging from scattered petechiae with mild edema to ulcerated, atrophic, and fibrotic mucosa can occasionally coexist in a single patient. Parasites may be found in all layers of the intestinal wall, and infiltrates of mononuclear cells and neutrophils may be identified (Genta and Walzer 1989). A recent histologic evaluation of small bowel biopsy specimens demonstrated no significant inflammatory response in 19 individuals with mild chronic infections, compared to uninfected controls (Trajman et al. 2000).

In disseminated disease, migrating larvae cause extensive tissue damage to the colon, lungs, and other organs. Because of the association with immunosuppression, biopsy specimens containing these aberrantly migrating larvae may reveal only a mild inflammatory response surrounding the parasite. Invasion of the central nervous system can be associated with meningitis characterized by a predominance of either neutrophils or lymphocytes.

Immune responses

The immune responses that accompany *Strongyloides* infection are not well understood. To date, there have been reports of elevated serum IgE in chronic strongyloidiasis. However, the diagnostic significance of this observation is unclear, since at least one study reported elevated serum IgE levels in less than 10 percent of chronically infected individuals (Willis and Nwokolo 1966). Moreover, it appears that IgE levels may wane over time, thus allowing for the maintenance of a chronically infected state (Atkins et al. 1997). An association between HTLV-1 infection and fecal excretion of *Strongyloides* larvae has been reported from Jamaica (Robinson et al. 1994). The authors found that HTLV-1-infected individuals with strongyloidiasis showed an age-related depression in serum IgE levels when compared to those seronegative for HTLV-1. These data suggest that serum IgE may confer some degree of protection against heavy infection with *Strongyloides*, and that this immune response is somehow inhibited by HTLV-1. More recent studies have shown that HTLV-1-infected patients with strongyloidiasis produce elevated levels of the cytokines IFN-γ and TGF-β1 compared to those infected with *S. stercoralis* alone. Moreover, coinfected individuals had reduced circulating levels of IgE and IgG4, which was associated with lower cure rates following anthelminthic treatment (Satoh et al. 2002).

Diagnosis

The diagnosis of strongyloidiasis, particularly chronic or asymptomatic infection, can be extremely challenging (Siddiqui and Berk 2001). A definitive diagnosis is most reliably made by identifying rhabditiform (or rarely filariform) larvae in the feces of an infected individual. However, the excretion of eggs in uncomplicated intestinal infection can be quite erratic, and multiple stool examinations may be required. Various stool concentration techniques and use of the Baermann method (by which motile larvae are extracted from stool specimens) may increase the diagnostic yield of fecal examination (Genta et al. 1987). An agar plate method that relies on the use of inverted microscopy to identify tracks made by motile *S. stercoralis* larvae may be more sensitive than some of these other techniques (Sukhavat et al. 1994; Iwamoto et al. 1998).

Strongyloides larvae can also be detected in duodenal aspirates, either by endoscopy or by the use of the string test (Beal et al. 1970; Goldsmid and Davies 1978). The latter involves swallowing a gelatin capsule containing a coiled piece of string. The proximal portion of the string is attached externally, and approximately 4 h later, the string is removed. The duodenal fluid adhering to the string is then examined microscopically for the presence of larvae. The sensitivity of the string test is probably slightly higher than that of routine stool examination. Adult worms may be identified in small bowel biopsy specimens (Figure 35.3).

Serological tests for strongyloidiasis are available through certain reference laboratories. Multiple enzyme-linked immunosorbent assays (ELISA) have been developed, some of which are sensitive and relatively specific (Loutfy et al. 2002; Ramachandran et al. 1998; Carroll et al. 1981; Genta et al. 1987). An indirect immunofluorescence antibody test (IFAT), which utilizes antigens from the rodent species *S. venezuelensis* has also been reported to be sensitive for diagnosing human infections (Machado et al. 2001). Recent identification of immunoreactive recombinant antigens from *S. stercoralis* should provide for standardization of testing for strongyloidiasis, particularly as a means of screening patients who are at risk of hyperinfection and disseminated disease (Ravi et al. 2002). Anyone with a positive serological test for *Strongyloides* should have a thorough stool examination to identify the presence of larvae and confirm the diagnosis.

During hyperinfection with dissemination, *Strongyloides* larvae can be detected in a variety of body fluids and tissues. The massive migration of parasites through the pulmonary circulation and into the lungs results in the frequent identification of larvae in sputum or bronchoalveolar lavage specimens (Berenson et al. 1987; Chu et al. 1990; Kramer et al. 1990) (Figure 35.4). Strongyloides larvae have also been identified in pericardial fluid (Lai et al. 2002), cerebrospinal fluid (Cappello and Hotez 1993), and brain biopsies (Neefe et al. 1976). *Strongyloides* eggs have even been isolated from a urethral smear from a patient following bone marrow transplant (Steiner et al. 2002). Eosinophilia, which is a frequent finding in chronic strongyloidiasis, is often absent in hyperinfection, in many cases because of exogenous corticosteroid therapy.

In contrast to *S. stercoralis*, the diagnosis of *S. fuelleborni* infection is made by identifying eggs in the feces, rather than larvae. Children with swollen belly syndrome tend to have extremely heavy infections, and excrete large numbers of eggs daily.

Control

For many years, thiabendazole was the drug of choice for the treatment of chronic strongyloidiasis. Although

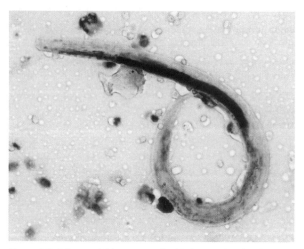

Figure 35.4 Strongyloides stercoralis *third-stage (L3) filariform larvae detected in sputum specimen (courtesy of Orihel and Ash (1995))*

cure rates of 60–100 percent have been reported (Schaffel et al. 2000), its use is associated with an extremely high occurrence of adverse side effects. These include nausea, dizziness, and neuropsychiatric symptoms, which have been reported in up to 89 percent of patients taking this drug (Grove 1982). It has recently been shown that ivermectin, a broad spectrum anthelmintic, is at least as effective and better tolerated than thiabendazole for the treatment of chronic strongyloidiasis (Naquira et al. 1989; Lyagoubi et al. 1992; Gann et al. 1994; Adenusi 1997; Zaha et al. 2002). Ivermectin, therefore, is now considered the drug of choice for chronic strongyloidiasis, with a two-dose regimen of 200 µg/kg given on successive days or at an interval of 2 weeks demonstrated to cure nearly all chronic infections (Gann et al. 1994; Zaha et al. 2002; Medical Letter 2002). In addition, albendazole has also been suggested as an alternative therapy (Datry et al. 1994; Sato et al. 2000), although this agent appears to be less efficacious than ivermectin.

Follow up of patients after treatment for chronic strongyloidiasis should include a follow-up eosinophil count and serologies. It has been reported that both eosinophil counts and *Strongyloides* specific IgG should return to baseline within 6–9 months following effective treatment (Lindo et al. 1996; Loutfy et al. 2002). Because the cure rate in patients coinfected with HTLV-1 appears to be lower, these patients should be monitored closely for evidence of treatment failure (Satoh et al. 2002; Terashima et al. 2002).

Successful treatment of *Strongyloides* hyperinfection requires prompt diagnosis and initiation of anthelmintic therapy. Treatment with ivermectin (Torres et al. 1994) or thiabendazole (25 mg/kg twice a day) should be continued indefinitely, pending both clinical response and the eradication of larvae from stool, sputum, and other body fluids. Although no intravenous formulations are currently available, veterinary preparations of ivermectin have been successfully used to treat severe human infections (Chiodini et al. 2000). Concurrent bacterial infections, e.g. bronchopneumonia and meningitis, should be treated aggressively with antibiotics. Even with appropriate therapy, the mortality from disseminated strongyloidiasis approaches 50–75 percent (Genta and Walzer 1989). In light of this, appropriate screening of high risk individuals should be carried out prior to the initiation of therapy that is associated with immunosuppression, particularly glucocorticoids (Avery 2002).

Thiabendazole treatment of children with swollen belly syndrome caused by *S. fuelleborni* appears to be effective, although reinfection is likely to occur in endemic areas (Ashford and Barnish 1990; Barnish and Barker 1987).

Because *Strongyloides* larvae can survive in warm, moist environments for days, control of transmission within an endemic area requires proper disposal of human waste. In addition, outbreaks documented among animal care workers underline the importance of exercising caution when handling specimens from laboratory animals. Despite the overwhelming numbers of larvae excreted by individuals experiencing hyperinfection syndrome, the risk of person to person transmission appears to be low (Maraha et al. 2001).

CAPILLARIA

THE PARASITE

Introduction and historical perspective

Human parasitic zoonoses caused by members of the genus *Capillaria* are uncommon, except in the Philippines and Thailand, where epidemics caused by the intestinal species *C. philippinensis* have been described. Since 1964, intestinal capillariasis has been recognized as an important cause of morbidity and mortality in these areas, with impressive outbreaks in Northern Luzon, Southern Leyte (Cross 1995), and the Compestela Valley (Belizario et al. 2000). Outbreaks have also been described elsewhere in Asia, and more recently, case reports of *C. philippinensis* infection have been reported in Egypt (Anis et al. 1998; Ahmed et al. 1999) and Colombia, South America (Dronda et al. 1993). Unlike most other intestinal nematodes, *C. philippinensis* is associated with substantial mortality during epidemics (Detels et al. 1969), and shares with *Strongyloides stercoralis* the unusual ability to undergo autoinfection and hyperinfection in humans. Much of our existing knowledge about human intestinal capillariasis derives from the earlier investigative work of Professor John H. Cross during his tenure at the US Naval Medical Research Unit in the Philippines (Cross 1992).

In contrast to intestinal capillariasis, extraintestinal zoonotic capillariasis, caused by other members of the genus *Capillaria* (*C. hepatica*, *C. aerophila*, and *C. plica*) is probably rare. However, with improved detection methods (Juncker-Voss et al. 2000), the number of actual human cases may greatly exceed those reported to date.

Classification

The Capillariinae is one of three subfamilies belonging to the superfamily Trichinelloidea. The Trichurinae (whipworms) and the Trichosomoidinae comprise the other two subfamilies. All members of the Trichinelloidea have a unique esophageal structure that distinguishes them from other nematodes (see below). Anderson (1992) points out that the 'classification of the Capillariinae is one of the most difficult and unsatisfactory in the Nematoda.' Many investigators including Skrjabin et al. (1957) have assigned the four human parasites of the genus *Capillaria*, namely *C. philippi-*

nensis, *C. hepatica*, *C. aerophilus*, and *C. plica*, to other genera, including *Calodium*, *Eucoleus*, *Pearsonema* (Anderson 1992), *Paracapillarias*, and *Crossicapillaria* (Moravec 2001). Because these terms are usually unfamiliar to medical parasitologists and clinicians we will adopt the common usage term *Capillaria*.

Morphology and structure

Like all members of the Trichinelloidea, the adult *Capillaria* have a modified esophageal gland called a stichosome comprised of unique gland cells called stichocytes. Each stichocyte communicates with the lumen of the esophagus by a single pore (Wright et al. 1985; Anderson 1992). Generally speaking, the Trichinelloidea have a long narrow anterior end containing the stichosome, which broadens into a thickened posterior end. In the *Capillaria*, however, the differences in width between the anterior and posterior regions are much less exaggerated. The adult Capillariinae are also more delicate and smaller than the other Trichinelloidea subfamilies. Adult males of *C. philippinensis* measure 1.5–3.9 mm in length and females measure 2.3–5.3 mm in length (Chitwood et al. 1968; Cross 1995). Because it is difficult to recover and measure intact *C. hepatica* and *C. aerophila* from human tissue, there are no precise estimates of their length. From nonhuman hosts, lengths of 4–12 mm have been reported for *C. hepatica*, and 30–40 mm for *C. aerophila*.

The uterus of the female adult *C. philippinensis* contains thick-shelled eggs, thin-shelled eggs, and larvae (Cross 1992). Eggs passed in the feces are bipolar 'lemon-shaped' or 'barrel-shaped' and bear a superficial resemblance to those of the human whipworm *Trichuris trichiura* (see Figure 35.5). They are, however, smaller (36–45 × 20 μm for *C. philippinensis* vs 50–54 × 22–23 μm for *T. trichiura*), more cylindrical in shape and have less prominent bipolar plugs which are somewhat flattened (Cross 1992). In contrast, the eggs of *C. aerophila* are large, measuring 59–80 × 30–40 μm (Soulsby 1982).

General biology

Humans are accidental hosts for all members of the genus *Capillaria*.

C. PHILIPPINENSIS

There is evidence to suggest that fish-eating birds are the natural definitive hosts of *C. philippinensis* (Cross and Basaca-Sevilla 1983). Humans become infected when they accidentally interrupt the fish–bird life cycle by eating infected, uncooked freshwater fish. *C. philippinensis* and *Strongyloides stercoralis* are the two major nematodes that can undergo autoinfection and hyperinfection in the human host.

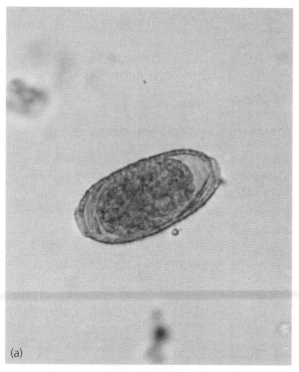

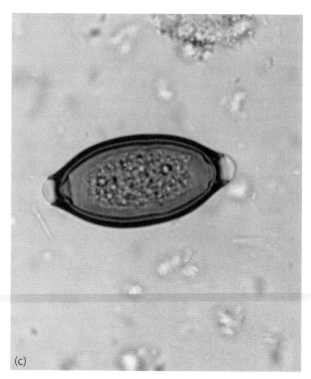

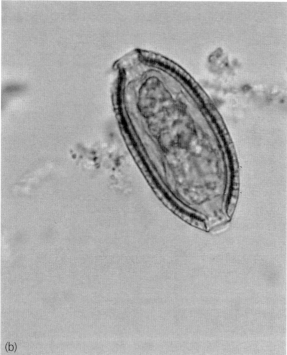

Figure 35.5 *Eggs of* Capillaria philippinensis **(a)** *and* C. hepatica **(b)**. *Included for comparison is an egg from* Trichuris trichiura **(c)** *(courtesy of Dickson D. Despommier, Columbia University School of Public Health)*

C. HEPATICA

C. hepatica occurs in the liver of numerous rodent species worldwide, but will also infect at least 20 other mammals, including squirrel, muskrat, opossum, porcupines, and, rarely, humans (Borucinska and

Nielsen 1993; Hamir and Rupprecht 2000). *C. hepatica* has the unusual feature that all of its mammalian hosts are 'dead end' in that the eggs must be released from the liver by a predation (including cannibalism and scavenging).

C. AEROPHILA

C. aerophila is a worldwide parasite of wild carnivores where it lives in the upper and lower respiratory tree and parenchyma.

Life cycle

C. PHILIPPINENSIS

As Cross (1995) points out, the *C. philippinensis* life cycle is unique among human nematode parasites and was elucidated primarily by laboratory experimentation. Eggs from the feces of infected patients embryonate in water at ambient temperatures in 5–10 days and develop further only if swallowed by small freshwater and brackish water fish. The bagsit (*Hypseleotris bipartita*) is a major intermediate fish host in the Philippines (Cross 1992). Upon ingestion the eggs hatch, giving rise to larval stages that grow in size. Humans become infected when they ingest larvae contained within the uncooked fish. The larvae become adults in the lumen and mucosa of the intestine. The length of time during which ingested larvae develop into adult male and oviparous adult female worms in humans is not known, although in experimental gerbil infections, the larvae develop into adults in 10–11 days (Cross 1992, 1995).

An unusual feature of *C. philippinensis* is its ability to cause autoinfection and hyperinfection in humans. The initial evidence for hyperinfection arose from studies with monkeys (*Macaca* spp.), in which 10 000– 30 000 worms were recovered from monkeys infected with 30– 50 larvae contained within fish (Cross et al. 1972; Cross 1992). This observation explains the extreme morbidity and relatively high mortality of intestinal capillariasis in humans. Autoinfection arises because the female adult worm is sometimes larviparous or can produce thin-walled eggs that hatch within the intestinal tract; these newborn larvae presumably can molt and reinvade the intestinal mucosa. Very heavy infections can develop as a result, as many as 200 000 worms having been recovered from 1 l of human bowel fluid at autopsy (Cross 1992).

C. HEPATICA

The adult female resides in the liver, where it deposits groups of eggs. These develop larvae but remain encapsulated in the liver (Wright 1974). Therefore, for transmission to occur, the embryonated eggs contained within infected livers must be eaten by animals or humans during cannibalism, predation, or scavenging (Anderson 1992). In contrast, the unembryonated eggs will pass through the gut of so-called 'disseminator animals' and are then dispersed into the environment with the feces (Anderson 1992). Ingestion of nonembryonated eggs does not lead to hepatic capillariasis (Juncker-Voss et al. 2000). When infective eggs are ingested accidentally by humans, they hatch in the small intestine and produce larvae that migrate through the portal system to the liver (Gutierrez 1990). After approximately 4 weeks, the female adult worm begins to release eggs.

C. AEROPHILA

The adult female lives in the upper and lower respiratory tree (under the mucosa), where it produces eggs that are coughed and swallowed. Eggs passing out with the feces will embryonate and enter a new host via contaminated food or water (Gutierrez 1990).

Transmission and epidemiology

C. PHILIPPINENSIS

The diet and eating habits of indigenous populations living in areas such as the Philippines and Thailand have been implicated in the transmission of intestinal capillariasis. For example, the Ilocanos of Northern Luzon use the intestinal juices from animals to season rice and typically consume whole freshwater fish uncooked (Cross 1992). Since the original reports of widespread intestinal capillariasis in the Philippines and in Thailand (Pradatsundarasar et al. 1973), infection with *C. philippinensis*

has been subsequently reported from other parts of Asia, including Japan (Nawa 1988), Indonesia (Chichino et al. 1992; Bangs et al. 1994) and India (Kang et al. 1994), as well as from nonendemic areas such as Iran, Egypt, and South America (Dronda et al. 1993).

C. HEPATICA

Probably because of its broad host specificity, there is a wide distribution of animal and human hepatic capillariasis. Cases have been reported from the Americas, Europe, Asia, and Africa. Humans may be at high risk for *C. hepatica* infection in Zaire, Nigeria, and other parts of West Africa where they consume the Gambian rat or cricetoma (Malekani et al. 1994).

C. AEROPHILA

Like *C. hepatica*, *C. aerophila* is a worldwide parasite of wild carnivores. Human infections have been reported primarily from Russia, although other cases have been reported from Morocco and Iran.

Location in host

C. PHILIPPINENSIS

In heavy infections, all stages of *C. philippinensis* can be found in the small intestine, particularly the jejunum. Extraintestinal *C. philippinensis* infection is a rare occurrence.

C. HEPATICA

The adult worms live in a host-derived capsule within the liver where they feed on cytoplasmic debris.

C. AEROPHILA

The adult worms live beneath the mucosa of the upper and lower respiratory tree.

CLINICAL AND PATHOLOGICAL ASPECTS OF INFECTION

C. philippinensis can be one of the most virulent helminthic pathogens of humans, with extremely high mortality rates during epidemics of intestinal capillariasis. Parasite virulence can be largely ascribed to the ability of the worm to cause hyperinfection.

Extraintestinal infections caused by other members of the genus *Capillaria* are relatively rare. All together, only 37 cases of human hepatic capillariasis caused by *C. hepatica* have been reported between 1924 and 1996 (Juncker-Voss et al. 2000), although the potential for transmission of this zoonosis may be high in some areas (Malekani et al. 1994). It has been postulated that many subclinical cases of *C. hepatica* may also go undetected (Juncker-Voss et al. 2000). Zoonotic transmissions of *C. aerophila* and *C. plica* are probably very rare, as reflected by only a handful of case reports in the medical literature.

Clinical manifestations

C. PHILIPPINENSIS (INTESTINAL CAPILLARIASIS)

There is no evidence for asymptomatic intestinal capillariasis. Patients usually develop diarrhea and abdominal pain that may progress in intensity. Borborygmi is a common presenting sign of intestinal capillariasis. Dehydration resulting from intestinal fluid losses is exacerbated by vomiting and anorexia. Infections lasting 2–3 months result in a severe protein-losing enteropathy characterized by cachexia, dehydration, and anasarca. Death is often the outcome in patients who do not receive adequate fluid resuscitation, nutritional supplementation, and anthelmintic chemotherapy. These patients often die from electrolyte disturbances that contribute to heart malfunction, as well as secondary bacterial superinfections and sepsis (Cross 1995).

C. HEPATICA (HEPATIC CAPILLARIASIS)

Heavy infections may cause an acute or subacute hepatitis accompanied by peritonitis, ascites, and eosinophilia. A triad of persistent fever, hepatomegaly, and leukocytosis with eosinophilia has been described (Juncker-Voss et al. 2000). Anemia, pulmonary symptoms, splenomegaly, kidney enlargement, and cachexia may also occur (Juncker-Voss et al. 2000).

C. AEROPHILA (PULMONARY CAPILLARIASIS)

Heavy infections cause a tracheobronchitis accompanied by dyspnea, dry cough, mild hemoptysis, and pulmonary infiltrates (Gutierrez 1990).

Pathogenesis and pathology

C. PHILIPPINENSIS

The protein-losing enteropathy of intestinal capillariasis is a result of direct mucosal invasion by the adult worms, which is sometimes exacerbated by the host inflammatory response. Although the intestinal tract frequently has a gross normal appearance, a histopathological examination of the intestine from human autopsies reveals villus flattening and dilated crypts, with occasional inflammatory infiltrates (Cross 1995); numerous adult and larval worms can be identified in the intestinal mucosa where they are associated with ulcerative and degenerative changes. Nothing is known about the molecular and cellular events associated with *C. philippinensis* invasion, although the parasite may secrete bioactive molecules similar to those of *Trichuris*, which releases a proteolytic enzyme and a pore-forming protein from their stichosome (Drake et al. 1994).

C. HEPATICA

In hepatic capillariasis there is a remarkable degree of fibrosis in the liver (Attah et al. 1983). Histopathological examination reveals areas of focal destruction and numerous granulomata consisting of mononuclear cells and eosinophils. Many of these inflammatory changes occur at the site of worms and eggs. The adult worm is frequently identified by the characteristic presence of the stichosome on histological section. The eggs frequently appear in clusters.

C. AEROPHILA

A single case report of a biopsy in pulmonary capillariasis demonstrated numerous granulomas within the bronchiolar wall, producing marked airway destruction (Aftandelians et al. 1977). The parasite was identified in tissue sections (Gutierrez 1990).

Immune responses

Very little has been published on the immune responses to human capillariasis. Patients infected with *C. philippinensis* acquire humoral antibodies, including circulating immunoglobulin E (Rosenberg et al. 1970).

Diagnosis

C. PHILIPPINENSIS

In epidemic situations, patients who present with diarrhea, borborygmi, and abdominal pain are presumed to have intestinal capillariasis. The definitive diagnosis is usually made by identifying the characteristic-shaped eggs in the stool, although in severe hyperinfection it is not uncommon to also find larvae, or even adult worms (Cross 1995). For lightly infected patients a number of concentration techniques are available in order to increase the sensitivity of fecal examination.

C. HEPATICA

The clinical manifestations of human *C. hepatica* infection can resemble those of visceral larva migrans caused by *Toxocara canis*. A definitive diagnosis can be made by demonstrating the presence of eggs, larvae or adults in liver biopsy specimens. Usually, there is extensive accompanying hepatic fibrosis (Gutierrez 1990). More recently, an indirect immunofluorescence assay based on human serum samples has been developed (Juncker-Voss et al. 2000).

C. AEROPHILA

Conceivably, parasite eggs may be demonstrated in sputum or feces, although a pulmonary biopsy may be required to establish the diagnosis definitively (Aftandelians et al. 1977).

Control

C. PHILIPPINENSIS

Anthelmintics of the benzimidazole class, namely albendazole and mebendazole, are the treatments of choice for intestinal capillariasis. Mebendazole treatment usually requires 200 mg twice daily for 20 days, or albendazole 400 mg daily for 10 days (Medical Letter 2002). Relapses can occur with either agent, so that the patient must be followed and retreated for longer periods if necessary. Mebendazole is thought to be less active against the larval stages of the parasite and therefore may require longer treatment regimens (Cross 1995). Outpatient management of patients with intestinal capillariasis is often unsatisfactory because of the need for supportive therapy and the possibility of relapse following treatment. Cross (1995) points out that patients receiving specific anthelmintic therapy are often poorly compliant as outpatients, particularly with regard to receiving a full treatment course. Therapy must therefore be closely monitored, preferably in an in-patient setting. Patients need to be educated on the dangers of eating uncooked small fish, especially in endemic areas. The sanitary disposal of feces will also reduce parasite transmission.

C. HEPATICA AND C. AEROPHILA

Little is known about the efficacy of specific anthelmintic therapy for human infections with either of these two parasites. Because mebendazole is poorly absorbed outside the gastrointestinal tract, high doses would presumably be needed to achieve a therapeutic effect. Albendazole or thiabendazole may also be effective.

REFERENCES

Adam, M., Morgan, O., et al. 1973. Hyperinfection syndrome with *Strongyloides stercoralis* in malignant lymphoma. *Br Med J*, **1**, 264–6.

Adenusi, A.A. 1997. Cure by ivermectin of a chronic, persistent, intestinal strongyloidosis. *Acta Tropica*, **66**, 163–7.

Aftandelians, R., Raafat, F., et al. 1977. Pulmonary capillariasis in a child in Iran. *Am J Trop Med Hyg*, **26**, 64–71.

Ahmed, L., el-Dib, N.A., et al. 1999. Capillaria philippinensis: an emerging parasite causing severe diarrhoea in Egypt. *J Egypt Soc Parasitol*, **29**, 483–93.

Albanese, G., Venturi, C., et al. 2001. Prurigo in a patient with intestinal strongyloidiasis. *Int J Dermatol*, **40**, 52–4.

Anderson, R.C. 1992. *Nematode parasites of vertebrates, their development and transmission*. Wallingford, Oxon: CAB International, 544–50.

Anis, M.H., Shafeek, H., et al. 1998. Intestinal capillariasis as a cause of chronic diarrhoea in Egypt. *J Egypt Soc Parasitol*, **28**, 143–7.

Arthur, R.P. and Shelley, W.B. 1958. Larva currens: a distinct variant of cutaneous larva migrans due to *Strongyloides stercoralis*. *Arch Dermatol*, **78**, 186–90.

Ash, L. and Orihel, T. 1997. *Atlas of human parasitology*, 4th edn. Chicago, IL: ACSP Press.

Ashford, R.W. and Barnish, G. 1990. *Strongyloides fuelleborni* and similar parasites in animals and man. In: Grove, D.I. (ed.), *Strongyloidiasis: a major roundworm infection of man*. London: Taylor and Francis, 271–86.

Ashton, F.T., Bhopale, V.M., et al. 1995. Sensory neuroanatomy of a skin-penetrating nematode parasite: *Strongyloides stercoralis*. I. Amphidial neurons. *J Comp Neurol*, **357**, 281–95.

Atkins, N.S., Lindo, F.J., et al. 1997. Humoral responses in human strongyloidiasis: correlations with infection chronicity. *Trans R Soc Trop Med Hyg*, **91**, 609–13.

Attah, E.B., Nagarajan, S., et al. 1983. Hepatic capillariasis. *Am J Clin Pathol*, **79**, 127–30.

Avery, R.K. 2002. Recipient screening prior to solid-organ transplantation. *Clin Infect Dis*, **35**, 1513–19.

Bangs, M.J., Purnomo and Andersen, E.M. 1994. A case of capillariasis in a highland community of Irian Jaya, Indonesia. *Ann Trop Med Parasitol*, **88**, 685–7.

Barnish, G. and Ashford, R.W. 1989. *Strongyloides fuelleborni* and hookworm in Papua New Guinea: patterns of infection within the community. *Trans R Soc Trop Med Hyg*, **83**, 684–8.

Barnish, G. and Barker, J. 1987. An intervention study using thiabendazole suspension against *Strongyloides fuelleborni*-like infections in Papua New Guinea. *Trans R Soc Trop Med Hyg*, **81**, 60–3.

Bavay, A. 1876. Sur l'anguillule stercorale. *C R Acad Sci Paris*, **83**, 694–6.

Beal, C.B., Viens, P., et al. 1970. A new technique for sampling duodenal contents. *Am J Trop Med Hyg*, **19**, 349–52.

Belizario, V.Y., de Leon, W.U., et al. 2000. Compestela Valley: a new endemic focus for *Capillariasis philippinensis*. *Southeast Asian J Trop Med Publ Health*, **31**, 478–81.

Berenson, C.S., Dobuler, K.J. and Bia, F.J. 1987. Fever, petechiae, and pulmonary infiltrates in an immunocompromised Peruvian man. *Yale J Biol Med*, **60**, 437–45.

Berk, S.L., Verghese, A., et al. 1987. Clinical and epidemiologic features of strongyloidiasis. *Arch Intern Med*, **147**, 1257–61.

Bhatt, B.D., Cappell, M.S., et al. 1990. Recurrent massive upper gastrointestinal hemorrhage due to *Strongyloides stercoralis* infection. *Am J Gastroenterol*, **85**, 1034–6.

Borucinska, J.D. and Nielsen, S.W. 1993. Hepatic capillariasis in muskrats (*Ondatra zibethicus*). *J Wildlife Dis*, **29**, 518–20.

Braun, T.I., Fekete, T. and Lynch, A. 1989. Strongyloidiasis in an institution for mentally retarded adults. *Arch Intern Med 1*, **48**, 634.

Brindley, P.J., Gam, A.A., et al. 1995. Ss40: the zinc endopeptidase secreted by infective larvae of *Strongyloides stercoralis*. *Exp Parasitol*, **80**, 1–7.

Brown, R.C. and Girardeau, M.H.F. 1977. Transmammary passage of *Strongyloides* sp. larvae in the human host. *Am J Trop Med Hyg*, **26**, 215–19.

Burke, J.A. 1978. Strongyloides in childhood. *Am J Dis Child*, **132**, 1130–6.

Cappello, M. and Hotez, P.J. 1993. Disseminated strongyloidiasis. *Semin Neurol*, **13**, 169–74.

Carroll, S.M., Karthigasu, D.T., et al. 1981. Serodiagnosis of human strongyloidiasis by an enzyme-linked immunosorbent assay. *Trans R Soc Trop Med Hyg*, **75**, 706–9.

Celedon, J.C., Mathur-Waugh, U., et al. 1994. Systemic strongyloidiasis in patients infected with the human immunodeficiency virus. *Medicine*, **73**, 256–63.

Chichino, G., Bernuzzi, A.M., et al. 1992. Intestinal capillariasis (*Capillaria philippinensis*) acquired in Indonesia: a case report. *Am J Trop Med Hyg*, **47**, 10–12.

Chiodini, P.L., Reid, A.J.C., et al. 2000. Parenteral ivermectin in *Strongyloides* hyperinfection. *Lancet*, **355**, 43–4.

Chitwood, M.B., Valasquez, C. and Salazar, N.G. 1968. *Capillaria philippinensis* sp. n. (Nematoda: Trichinellida) from intestine of man in the Philippines. *J Parasitol*, **54**, 368–71.

Chu, E.C., Whitlock, W.L. and Dietrich, R.A. 1990. Pulmonary hyperinfection syndrome with *Strongyloides stercoralis*. *Chest*, **97**, 1475–7.

Civantos, F. and Robinson, M.J. 1969. Fatal strongyloidiasis following corticosteroid therapy. *Am J Dis Dis*, **14**, 643–51.

Cookson, J.B., Montgomery, R.D., et al. 1990. Fatal paralytic ileus due to strongyloidiasis. *Br Med J*, **4**, 771–2.

Cross, J.H. 1992. Intestinal capillariasis. *Clin Microbiol Rev*, **5**, 120–9.

Cross, J.H. 1995. *Capillaria philippinensis* and *Trichostrongylus orientalis*. In: Farthing, M.J.G., Keusch, G.T. and Wakelin, G.T. (eds), *Enteric infection 2, intestinal helminths*. London: Chapman & Hall Medical, 151–64.

Cross, J.H. and Basaca-Sevilla, V. 1983. Experimental transmission of *Capillaria philippinensis* to birds. *Trans R Soc Trop Med Hyg*, **77**, 511–14.

Cross, J.H., Banzon, T.C., et al. 1972. Studies on the experimental transmission of *Capillaria philippinensis* in monkeys. *Trans Roy Soc Trop Med Hyg*, **66**, 819–27.

Datry, A., Hilmarsdottir, I., et al. 1994. Treatment of *Strongyloides stercoralis* infection with ivermectin compared with albendazole: results of an open study of 60 cases. *Trans R Soc Trop Med Hyg*, **88**, 344–5.

Daubenton, J.D., Buys, H.A. and Hartley, P.S. 1998. Disseminated strongyloidiasis in a child with lymphoblastic lymphoma. *J Pediatr Hematol Oncol*, **20**, 260–3.

Dees, A., Batenburg, P.L., et al. 1990. *Strongyloides stercoralis* associated with a bleeding gastric ulcer. *Gut*, **31**, 1414–15.

De Silva, S., Saykao, P., et al. 2002. Chronic *Strongyloides stercoralis* infection in Laotian immigrants and refugees 7–20 years after resettlement in Australia. *Epidemiol Infect*, **128**, 439–44.

Detels, R.L., Gutman, L., et al. 1969. An epidemic of human intestinal capillariasis: a study in a barrio in North Luzon. *Am J Trop Med Hyg*, **18**, 676–82.

Dixon, A.C., Yanaghihara, E.T., et al. 1989. Strongyloidiasis associated with human T-cell lymphotropic virus type 1 infection in a nonendemic area. *West J Med*, **151**, 410–13.

Drake, L., Korchev, Y., et al. 1994. The major secreted product of the whipworm *Trichuris*, is a pore-forming protein. *Proc R Soc London B*, **257**, 255–61.

Dronda, F., Chaves, F., et al. 1993. Human intestinal capillariasis in an area of nonendemicity: case report and review. *Clin Infect Dis*, **17**, 909–12.

Faust, E.C. 1933. The development of *Strongyloides* in the experimental host. *Am J Hyg*, **18**, 114–32.

Fuelleborn, F. 1914. Untersuchungen uber den Infektionsweg bei *Strongyloides* und *Ankylostomum* und die biologie dieser parasiten. *Arch Schiffs-u Tropenhyg*, **18**, 80.

Gallego, S.G., Slade, R.W. and Brindley, P.J. 1998. A cDNA encoding a pepsinogen-like aspartic protease from the human roundworm parasite *Strongyloides stercoralis*. *Acta Tropica*, **71**, 17–26.

Gann, P.H., Neva, F.A. and Gam, A.A. 1994. A randomized trial of single and two-dose ivermectin versus thiabendazole for treatment of strongyloidiasis. *J Infect Dis*, **169**, 1076–9.

Genta, R.M. 1989. Global prevalence of strongyloidiasis: critical review with epidemiologic insights into the prevention of disseminated disease. *Rev Infect Dis*, **11**, 755–67.

Genta, R.M. 1992. Dysregulation of strongyloidiasis: a new hypothesis. *Clin Microbiol Rev*, **5**, 345–55.

Genta, R.M. and von Kuster, L.C. 1988. Cutaneous manifestations of strongyloidiasis. *Arch Dermatol*, **124**, 1826–30.

Genta, R.M. and Walzer, P.D. 1989. Strongyloidiasis. In: Walzer, P.D. and Genta, R.M. (eds), *Parasitic infections in the compromised host*. New York: Marcel Dekker, 463–525.

Genta, R.M., Weesner, R., et al. 1987. Strongyloidiasis in US veterans of the Vietnam and other wars. *J Am Med Assoc*, **258**, 49–52.

Georgi, J.R. and Sprinkle, C.L. 1974. A case of human strongyloidiasis apparently contracted from asymptomatic colony dogs. *Am J Trop Med Hyg*, **23**, 899–901.

Goldsmid, J.M. and Davies, N. 1978. Diagnosis of parasitic infections of the small intestine by the enterotest duodenal capsule. *Med J Aust*, **1**, 519–20.

Gompels, M.M., Todd, J., et al. 1991. Disseminated strongyloidiasis in AIDS: uncommon but important. *AIDS*, **5**, 329–32.

Gotuzzo, E., Terashima, A., et al. 1999. *Strongyloides stercoralis* hyperinfection associated with human T-cell lymphotropic virus type-1 infection in Peru. *Am J Trop Med Hyg*, **60**, 146–9.

Grove, D.I. 1982. Treatment of strongyloidiasis with thiabendazole: an analysis of toxicity and effectiveness. *Trans R Soc Trop Med Hyg*, **76**, 114–18.

Gutierrez, Y. 1990. *Diagnostic pathology of parasitic infections with clinical correlations*. Philadelphia, PA: Lea and Febiger.

Gyorkos, T.W., MacLean, J.D., et al. 1992. Intestinal parasite infection in the Kampuchean refugee population 6 years after resettlement in Canada. *J Infect Dis*, **166**, 413–17.

Hamir, A.N. and Rupprecht, C.E. 2000. Hepatic capillariasis (*Capillaria hepatica*) in porcupines (*Erethizon dorsatum*) in Pennsylvania. *J Vet Diagn Invest*, **12**, 463–5.

Hill, J.A. 1988. Strongyloidiasis in ex-Far East prisoners of war. *Br Med J*, **296**, 753.

Igra-Siegman, Y., Kapila, R., et al. 1981. Syndrome of hyperinfection with *Strongyloides stercoralis*. *Rev Infect Dis*, **3**, 397–407.

Iwamoto, T., Kitoh, M., et al. 1998. Larva currens: The usefulness of the agar plate method. *Dermatology*, **196**, 343–5.

Jacob, C.I. and Patten, S.F. 1999. *Strongyloides stercoralis* infection presenting as generalized prurigo nodularis and lichen simplex chronicus. *J Am Acad Dermatol*, **41**, 357–61.

Juncker-Voss, M., Prosl, H., et al. 2000. Serological detection of Capillaria hepatica by indirect immunofluorescence assay. *J Clin Microbiol*, **38**, 431–3.

Kalb, R.E. and Grossman, M.E. 1986. Periumbilical purpura in disseminated strongyloidiasis. *J Am Med Assoc*, **256**, 1170–1.

Kang, G., Mathan, M., et al. 1994. Human intestinal capillariasis: first report from India. *Trans R Soc Trop Med Hyg*, **88**, 204–5.

Kitchen, L.W., Tu, K.K. and Kerns, F.T. 2000. *Strongyloides*-infected patients at Charleston Area Medical Center, West Virginia, 1997–1998. *CID*, **31**, e5–6.

Kramer, M.R., Gregg, P.A., et al. 1990. Disseminated strongyloidiasis in AIDS and non-AIDS immunocompromised hosts: diagnosis by sputum and bronchoalveolar lavage. *South Med J*, **83**, 1226–9.

Kuberski, T.T., Gabor, E.P. and Boudreaux, D. 1975. Disseminated strongyloidiasis: a complication of the immunosuppressed host. *West J Med*, **122**, 504–8.

Lai, C.P., Hsu, Y.H., et al. 2002. *Strongyloides stercoralis* infection with bloody pericardial effusion in a non-immunocompromised patient. *Circ J*, **66**, 613–14.

Lindo, J.F., Atkins, N.S., et al. 1996. Parasite-specific serum IgG following successful treatment of endemic strongyloidiasis using ivermectin. *Trans R Soc Trop Med Hyg*, **90**, 702–3.

Link, K. and Orenstein, R. 1999. Bacterial complications of strongyloidiasis: *Streptococcus bovis* meningitis. *South Med J*, **92**, 728–32.

Lopez, P.M., Boston, R., et al. 2000. The neurons of class ALD mediate thermotaxis in the parasitic nematode, *Strongyloides stercoralis*. *Int J Parasitol*, **30**, 1115–21.

Loutfy, M.R., Wilson, M., et al. 2002. Serology and eosinophil count in the diagnosis and management of strongyloidiasis in a non-endemic area. *Am J Trop Med Hyg*, **66**, 749–52.

Lyagoubi, M., Datry, A., et al. 1992. Chronic persistent strongyloidiasis cured by ivermectin. *Trans R Soc Trop Med Hyg*, **86**, 541.

Machado, E.R., Ueta, M.T., et al. 2001. Diagnosis of human strongyloidiasis using particulate antigen of two strains of *Strongyloides venezuelensis* in indirect immunofluorescence antibody test. *Exp Parasitol*, **99**, 52–5.

Malekani, M., Kumar, V. and Pandey, V.S. 1994. Hepatic capillariasis in edible *Cricetomys* spp. (Rodentia: Cricetidae) in Zaire and its possible public health implications. *Ann Trop Med Parasitol*, **88**, 569–72.

Maraha, B., Buiting, A.G.M., et al. 2001. The risk of *Strongyloides stercoralis* transmission from patients with disseminated strongyloidiasis to the medical staff. *J Hosp Infect*, **49**, 222–4.

The Medical Letter, 2002. Drugs for parasitic infections. *The Medical Letter on Drugs and Therapeutics*, **April**, 1–12.

Meltzer, R.S., Singer, C., et al. 1979. Antemortem diagnosis of central nervous system strongyloidiasis. *Am J Med Sci*, **277**, 91–8.

Milder, J.E., Walzer, P.D., et al. 1981. Clinical features of *Strongyloides stercoralis* infection in an endemic area of the United States. *Gastroenterology*, **80**, 1481–8.

Moncol, D.J. and Triantaphyllou, A.C. 1978. *Strongyloides ransomi*: factors influencing the in vitro development of the free living generation. *J Parasitol*, **64**, 220–5.

Moravec, F. 2001. Redescription and systematic status of *Capillaria pilippinensis*, and intestinal parasite of human beings. *J Parasitol*, **87**, 161–4.

Naquira, C., Jimenez, G., et al. 1989. Ivermectin for human strongyloidiasis and other intestinal helminths. *Am J Trop Med Hyg*, **403**, 304–9.

Nawa, Y. 1988. A case report of intestinal capillariasis. The second case found in Japan. *Jpn J Parasitol*, **37**, 113–18.

Neefe, L.I., Pinilla, O., et al. 1973. Disseminated strongyloidiasis with cerebral involvement. *Am J Med*, **55**, 832–8.

Neefe, L.I., Owor, R. and Wamukota, W.M. 1976. A fatal case of strongyloidiasis with strongyloides larvae in the meninges. *Trans R Soc Trop Med Hyg*, **70**, 497–9.

Neva, F. 1986. Biology and immunology of human strongyloidiasis. *J Infect Dis*, **153**, 397–406.

Nishigori, M. 1928. The factors which influence the external development of *Strongyloides stercoralis* and on autoinfection with this parasite. *J Form Med Assn*, **276**, 1–56.

Normand, A. 1876. Sur la maladie dite diarrhée de Cochin-Chine. *C R Acad Sci Paris*, **83**, 316–18.

O'Brien, W. 1975. Intestinal malabsorption in acute infection with *Strongyloides stercoralis*. *Trans R Soc Trop Med Hyg*, **69**, 69–77.

Onuigbo, M.A.C. and Ibeachum, G.I. 1991. *Strongyloides stercoralis* larvae in peripheral blood. *Trans R Soc Trop Med Hyg*, **85**, 97.

Orihel, T. and Ash, L. 1995. *Parasites in human tissues*. Chicago, IL: ACSP Press.

Pradatsundarasar, A., Pecharanond, K., et al. 1973. The first case of intestinal capillariasis in Thailand. *Southeast Asian J Trop Med Public Health*, **4**, 131–4.

Ramachandran, S., Thompson, R.W., et al. 1998. Recombinant cDNA clones for immunodiagnosis of strongyloidiasis. *J Infect Dis*, **177**, 196–203.

Ravi, V., Ramachandran, S., et al. 2002. Characterization of a recombinant immunodiagnostic antigen (NIE) from *Strongyloides stercoralis* L3-stage larvae. *Mol Biochem Parasitol*, **125**, 73–81.

Robinson, R.D., Lindo, J.F., et al. 1994. Immunoepidemiologic studies of *Strongyloides stercoralis* and human T lymphotropic virus type I infections in Jamaica. *J Infect Dis*, **169**, 692–6.

Ronan, S.G., Reddy, R.L., et al. 1989. Disseminated strongyloidiasis presenting as purpura. *J Am Acad Dermatol*, **21**, 1123–5.

Rosenberg, E.B., Whalen, G.E., et al. 1970. Increased circulating IgE in a new parasitic disease – human intestinal capillariasis. *N Engl J Med*, **283**, 1148–9.

Sanchez, P.R., Guzman, A.P., et al. 2001. Endemic strongyloidiasis on the Spanish Mediterranean coast. *Q J Med*, **94**, 357–63.

Sandlund, J.T., Kauffman, W. and Flynn, P.M. 1997. *Strongyloides stercoralis* infection mimicking relapse in a child with small non-cleaved cell lymphoma. *Am J Clin Oncol*, **20**, 215–16.

Sato, T.H., Shiroma, Y., et al. 2000. Comparative studies on the efficacy of three anthelminthics on the treatment of human strongyloidiasis in Okinawa, Japan. *Southeast Asian J Trop Med Public Health*, **31**, 147–51.

Satoh, M., Toma, H., et al. 2002. Reduced efficacy of treatment of strongyloidiasis in HTLV-1 carriers related to enhanced expression of IFN-γ and TGF-β1. *Clin Exp Immunol*, **127**, 354–9.

Schad, G.A. 1990. Morphology and life history of *Strongyloides stercoralis*. In: Grove, D.I. (ed.), *Strongyloidiasis: a major roundworm infection of man*. London: Taylor and Francis, 85–104.

Schad, G.A., Aikens, L.M. and Smith, G. 1989. *Strongyloides stercoralis*: is there a canonical migratory route through the host? *J Parasitol*, **75**, 740–9.

Schaffel, R., Nucci, M., et al. 2000. Thiabendazole for the treatment of strongyloidiasis in patients with hematologic malignancies. *CID*, **31**, 821–2.

Shoop, W.L., Michael, B.F., et al. 2002. Transmammary transmission of *Strongyloides stercoralis* in dogs. *J Parasitol*, **88**, 536–9.

Siddiqui, A.A. and Berk, S.L. 2001. Diagnosis of *Strongyloides stercoralis* infection. *CID*, **33**, 1040–7.

Siddiqui, A.A., Stanley, C., et al. 2000. A cDNA encoding a nuclear hormone receptor of the steroid/thyroid hormone receptor family from the human parasitic nematode *Strongyloides stercoralis*. *Parasitol Res*, **86**, 24–9.

Skrjabin, K.E., Shikhobalova, N.P. and Orlov, I.V. 1957 [1970]. *Essentials of nematology, Trichocephalidae and Capillariidae of animals and man and the diseases caused by them*, vol. VI. Jerusalem: Academy of Sciences of the USSR, Israel Program for Scientific Translations.

Soulsby, E.J.L. 1982. *Helminths, arthropods and protozoa of domesticated animals*, 7th edn. Philadelphia, PA: Lea and Febiger.

Steiner, B., Riebold, D., et al. 2002. *Strongyloides stercoralis* eggs in a urethral smear after bone marrow transplant. *CID*, **34**, 1280–1.

Sukhavat, K., Morakote, N., et al. 1994. Comparative efficacy of four methods for the detection of *Strongyloides stercoralis* in human stool specimens. *Ann Trop Med Parasitol*, **88**, 95–6.

Terashima, A., Alvarez, H., et al. 2002. Treatment failure in intestinal strongyloidiasis: an indicator of HTLV-1 infection. *Int J Infect Dis*, **6**, 28–30.

Torres, J.R., Isturiz, R., et al. 1994. Efficacy of ivermectin in the treatment of strongyloidiasis complicating AIDS. *Clin Infect Dis*, **17**, 900–2.

Trajman, A., MacDonald, T.T. and Elia, C.C.S. 2000. Intestinal immune cells in *Strongyloides stercoralis* infection. *J Clin Path*, **50**, 991–5.

Upadhyay, D., Corbridge, T., et al. 2001. Pulmonary hyperinfection syndrome with *Strongyloides stercoralis*. *Am J Med*, **111**, 167–9.

Willis, A.J. and Nwokolo, C. 1966. Steroid therapy and strongyloidiasis. *Lancet*, **1**, 1396–8.

Wright, K.A. 1974. The feeding site and probable feeding mechanism of the parasitic nematode *Capillaria hepatica* (Bancroft, 1893). *Can J Zool*, **52**, 1215–20.

Wright, K.A., Lee, D.L. and Shivers, R.R. 1985. A freeze-fracture study of the digestive tract of the parasitic nematode *Trichinella*. *Tissue Cell*, **17**, 189–98.

Zaha, O., Hirata, T., et al. 2002. Efficacy of ivermectin for chronic strongyloidiasis: two single doses given 2 weeks apart. *J Infect Chemother*, **8**, 94–8.

36

Toxocara

ELENA PINELLI, LAETITIA M. KORTBEEK, AND JOKE W.B. VAN DER GIESSEN

THE PARASITES

The discovery of a larval nematode within the retinal granuloma of a child by Wilder in 1950 (Wilder 1950) led to the description of the condition known today as visceral larva migrans (VLM). Two years later, Beaver and colleagues (Beaver 1952) described a number of infections, again in children, in which eosinophilia and severe multisystem disease were caused by migrating larvae of *Toxocara canis* and *Toxocara cati*, parasites whose definitive hosts are dogs and cats, respectively.

MORPHOLOGY AND STRUCTURE

Toxocara adult parasites are large, pink roundworms, measuring 7–10 cm × 5–6 mm (Figure 36.1). The male has a curved posterior end, which distinguishes it from the straight-tailed female (Figure 36.2a). Females lay large quantities of unembryonated eggs (Figure 36.2b) into the lumen of the small intestine. These eggs need to pass into the external environment in order to embryonate (Figure 36.2c). The larvae of *T. canis* are 290–350 × 18–21 μm (Figure 36.2d). The diameter of *T. cati* larvae is somewhat smaller (65–75 μm) (Ash and Orihel 1984). The eggs of *T. canis* measure approximately 85 × 75 μm and the eggs of *T. cati* are approximately 75 × 65 μm (Beaver 1984).

GENERAL BIOLOGY

Adult worms occupy the lumen of the small intestine and live unattached to the host. They maintain their position in the lumen by serpentine movement against the flow of peristalsis. Worms ingest partially digested food of the host and process it through their gut. Little is known about the digestive physiology, but it is likely to be similar to that of *Ascaris lumbricoides*, i.e. to have the capability of digesting most foodstuffs and absorbing the resulting small molecular weight nutrients across the midgut microvilli. *Toxocara* is a facultative anaerobe and derives most of its energy from glycogen degradation (Gopinath and Keystone 1995). The pseudocoelom

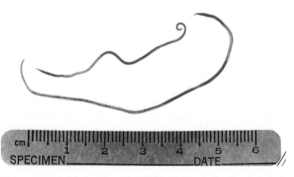

Figure 36.1 *Adult male and female* Toxocara canis. *The female is larger than the male and has a straight tail. Unstained, formalin-fixed specimens*

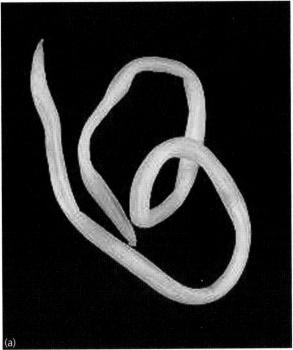

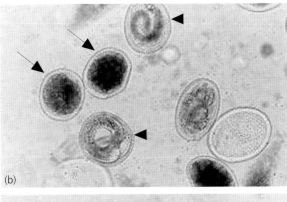

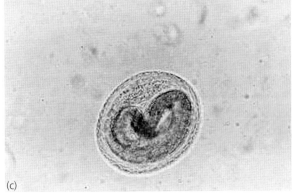

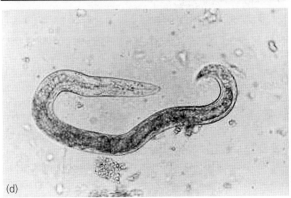

Figure 36.2 *Adult worm, eggs and larvae of* Toxocara canis: **(a)** *female worms measuring 7–10 cm shed* **(b)** *unembryonated eggs (arrows) which are passed with the feces into the environment. Eggs embryonate (arrow heads) within a period of 3–6 weeks (×200);* **(c)** *embryonated eggs containing the L2 larvae (×400).* **(d)** *L2 larvae (×100) hatch in the intestine, invade the intestinal mucosa, and migrate through the body causing the visceral larval migrans (VLM) syndrome.*

contains a special hemoglobin, which presumably binds oxygen thus augmenting energy metabolism. The worms' pink color is due to the presence of this substance.

MOLECULAR BIOLOGY

Ribosomal DNA (rDNA) markers and, in particular, the internal transcribed spacer (ITS) regions can be used to characterize and to identify the different *Toxocara* species. Polymerase chain reaction (PCR) derived strategies based on ribosomal DNA can be used as diagnostic tools where animal fecal material, as well as animal and human tissues, could be used (Jacobs et al. 1997; Zhu et al. 2001). Recently, a *Toxocara* variant in cats,

originating from Malaysia and morphologically resembling *T. canis*, appeared to be genetically different when using ITS 1 and ITS 2 regions and was characterized as the new species *Toxocara malaysiensis* (Gibbons et al. 2001; Zhu et al. 1998)

LIFE CYCLE

In their definitive hosts, *T. canis* and *T. cati* have similar life cycles, which resemble that of *Ascaris lumbricoides* in humans (Figure 36.3). Nonembryonated eggs are passed with the feces into the environment and incubate in soil. Within 2 weeks, the eggs contain an infectious second-stage larva that, if eaten by a dog (*T. canis*) or cat (*T. cati*, *T. malaysiensis*), initiates infection by hatching in the small intestine. The freed larvae penetrate the small intestine and enter the general circulation. In the liver, they lodge in the presinusoidal capillaries because of their large diameter, and are stimulated to enter the parenchymal tissue where they feed on liver cells. They molt in the liver to the third stage, re-enter the general circulation, and are carried to the lungs. Trapped by the alveolar capillaries, again because of their size, the larvae penetrate into the alveolar space, crawl up the bronchioles into the trachea, bypass the epiglottis and are swallowed. In the small intestine, larvae molt for a fourth time, transforming into adult worms. Mating ensues soon after and the females begin to pass unembryonated eggs, thus completing the life cycle in their definitive hosts.

Toxocara spp. cannot complete their life cycle in humans. Infection is initiated, as in the dogs or cats, by the ingestion of embryonated eggs. Larvae hatch in the small intestine but, as they do not receive the proper environmental signals (Castro 1982), begin their odyssey, wandering throughout the body, invading all organs and damaging tissue wherever they go. The larvae never mature and eventually die in situ. Death may occur soon after infection, but many larvae can survive for several months, even up to years (Smith and Beaver 1953). In experimentally infected rhesus monkeys, *Toxocara* larvae have been reported to remain viable and infective in tissues for at least 9 years (Schantz 1989).

TRANSMISSION AND EPIDEMIOLOGY

Transmission of *Toxocara* occurs by ingestion of embryonated eggs present in soil either directly by geophagia or indirectly by consumption of unwashed contaminated fresh vegetables. Human infections with larvae have also been described and can take place by eating undercooked tissues of a paratenic host (Schantz 1989).

Toxocara spp. are found throughout the world but the recorded seroprevalence varies among countries or even within countries. In Ireland, *Toxocara* seroprevalence in schoolchildren is high (31 percent) and rises with age (Holland et al. 1995). The seroprevalence in children in the state of Connecticut, USA, varied from 6.1 percent in New Haven to 27.9 percent in Bridgeport (Sharghi et al. 2001). Children are most at risk of infection, especially when there is a history of pica. Several risk factors have been reported, which differ between countries. In the USA, an association with living in an urban area, race, and family income has been described. In this study, no association was found with owning dogs or puppies and gender (Sharghi et al. 2001). In Ireland however, males were significantly more infected than females, as well as children attending rural schools versus urban schools (Holland et al. 1995).

The prevalence of ocular toxocariasis is still largely unknown. In a 1-year survey in Alabama, USA, the estimated prevalence was 1 per 1 000 population (Maetz et al. 1987). There are no accurate data on the number of patients with ocular lava migrans (OLM) who seek medical attention. Patients with OLM are usually older than VLM patients, probably because they lack systemic signs and do not have a history of pica (Schantz 1989).

Concerning the definitive hosts, infections with *Toxocara* spp. are not only prevalent in dog and cat populations but several epidemiological studies around the world showed a high prevalence of *Toxocara* spp. in wild carnivores like foxes and wolves. In the Netherlands, the high prevalence of *Toxocara* spp. in domestic and wild carnivores ranges from 4.7 percent in domestic cats and 2.9 percent in domestic dogs to 21 percent in stray cats (Overgaauw 1997). In Poland, the prevalence has been reported to be 39 percent in cats, 32 percent in dogs, and 16 percent in foxes (Luty 2001); in Spain, 6.4 percent has been reported in wolves (Segovia et al. 2001). Because of the reservoir in wild carnivores, eggs are not only found in high densities in public and private places but also in more remote areas around the world. The numbers of eggs detected in soil depends on season, presence of earthworms, and depth of the soil (Weller et al. 1983). Children's sandpits can be contaminated by *T. canis* and *T. cati*, depending on their maintenance (Jansen et al. 1993).

CLINICAL SYNDROMES AND MANIFESTATIONS

Three clinical syndromes of human toxocariasis are recognized: VLM, OLM, and covert toxocariasis (CT).

VLM is mainly a disease of young children and its symptoms include fever, respiratory distress, and hepatosplenomegaly (Gutierrez 1990; reviewed in Despommier 2003). Other symptoms include an eosinophilic pneumonitis (Loeffler's pneumonia) that bears a clinical resemblance to the pulmonary inflammatory responses observed in asthmatic patients. Myocarditis, nephritis, and central nervous system (CNS) symptoms are not uncommon. CNS involvement may lead to seizures, psychiatric manifestations, or encephalopathy (Fortenberry et al. 1991). The different manifestations of VLM and OLM have been related to the infective dose. This has been clearly shown in murine models for toxocariasis. After infection of mice with large numbers of *T. canis* embryonated eggs, larvae can be recovered from brain tissue. The presence of larvae in the brain has been associated with behavioral changes observed in these infected animals (Holland and Cox 2001).

OLM occurs primarily in older children and usually manifests as a unilateral vision disorder often accompanied by strabismus (Dinning et al. 1988). However, the age of onset of ocular toxocariasis has been reported to range from 2 to 50 years (Taylor 2001). Invasion of the retina leading to granuloma formation is the most serious consequence of the infection (Gillespie et al. 1993) and occurs peripherally or in the posterior pole. These granulomata distort the retina creating heteropia, or macular detachment (Small et al. 1989). Diffuse uniocular endophthalmitis or papillitis with secondary glaucoma are other complications of larval migration and death. Blindness is common, but the degree of damage depends on the area affected. It is still not clear what role *T. cati* might play in causing ocular lesions (Taylor 2001).

Patients with toxocariasis whose symptoms do not fit into the categories of VLM or OLM are described as having CT. The symptoms are nonspecific but when grouped together form a recognizable syndrome. These

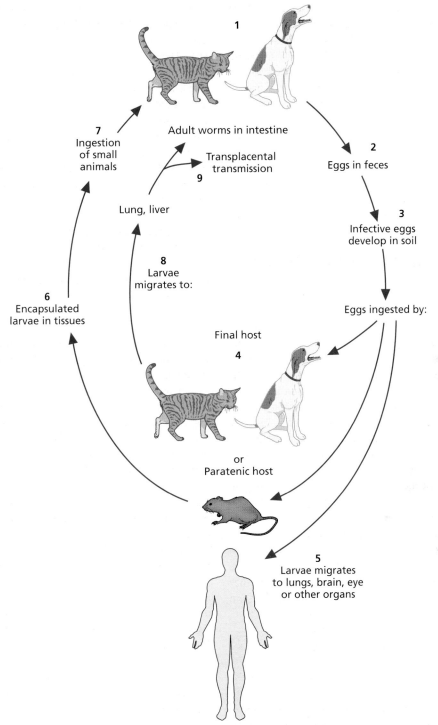

1

Adult worms in intestine

7
Ingestion
of small
animals

9
Transplacental
transmission

2
Eggs in feces

Lung, liver

3
Infective eggs
develop in soil

8
Larvae
migrates to:

6
Encapsulated
larvae in tissues

Eggs ingested by:

Final host
4

or
Paratenic host

5
Larvae migrates
to lungs, brain, eye
or other organs

Figure 36.3 *Life cycle of* Toxocara canis *and* Toxocara cati. *Adult worms present in the intestinal lumen of dogs or cats (1) shed a large number of unembryonated eggs that are passed into the environment via the feces (2). These eggs are usually found in soil contaminated with feces of these animals. In the soil, the first and second larval stages (L1, L2) are formed within the eggshell under favorable conditions (3). The infective eggs can be ingested by final (4) or paratenic (5) hosts. In the intestinal lumen of the paratenic host, the L2 larvae hatch from the egg, penetrate through the intestinal wall and migrate through the body, but there is no further development. Such larvae may become included in granulomas, remain in tissues (6) for a long time and are infective to final hosts (7). In the final host, the L2 larvae hatch from the egg inside the intestinal lumen, penetrate the intestinal wall and are transported via the bloodstream to the liver and lung (8). The larvae molt twice (L3, L4) in the lungs. The L4 larvae migrate up the trachea and are swallowed, reaching the intestinal lumen where they mature. After mating, females release fertilized eggs and the cycle starts again. In the majority of adult final hosts, the larvae do not complete the lung migration and remain 'dormant'. During pregnancy, larvae are activated, re-enter the circulatory system, are carried to the placenta, and penetrate the fetus (9).*

symptoms include abdominal pain, anorexia, behavior disturbances, cervical adenitis, wheezing, limb pains, and fever (Schantz 1989; Taylor et al. 1988).

Asymptomatic cases, only detected by positive serology, do occur. However, there are no data available on the extent of these cases (Schantz 1989).

PATHOGENESIS AND PATHOLOGY

Damage depends on the tissue invaded; the lungs, eyes, and CNS are the most seriously affected (Figure 36.4). Although tissue damage due to the migration of worms is a major cause of cell death, it is less harmful than the eosinophilic granulomata that are the hallmark of worm death. When worm death occurs in the eye, the resulting damage often resembles retinoblastoma (Despommier et al. 1994). In the past, enucleation was performed in some unfortunate cases because of the unavailability of modern diagnostic tests. Granulomata induced by dying worms can be sufficiently intense to be responsible for the loss of sight in the affected eye.

Infection with this parasite can also result in asthmatic manifestations associated with the hypereosinophilic syndrome (Feldman and Parker 1992). Studies in the murine model for toxocariasis have shown that *T. canis* larvae migrate to the lungs resulting in pulmonary inflammation. The cells that infiltrate the lung include eosinophils, macrophages, and lymphocytes (Buijs et al. 1995). Additional changes after infection include hypertrophy of goblet cells, indicative of mucus production, and increased total immunoglobulin (Ig)E in serum (Pinelli et al. 2001).

Epidemiological studies have suggested that allergic manifestations occur more often in *Toxocara*-seropositive patients (Buijs et al. 1997; Chan et al. 2001). However, this remains a point of discussion since other authors have not found any significant association

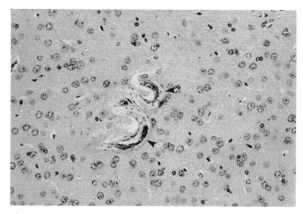

Figure 36.4 *A larva of* Toxocara canis *in the brain tissue of an experimentally infected mouse. Note the lack of any inflammatory response around the area of damage. The worm is approximately 20 μm in diameter. Hematoxylin and eosin-stained section of paraffin-embedded tissue.*

between *Toxocara* infection and asthma (Sharghi et al. 2001).

IMMUNE RESPONSES

Immunological features of infection with *Toxocara* spp. are increased serum IgE levels and eosinophilia (Brunello et al. 1983; Kayes 1997; Sugane and Ohshima 1984). The death of migrating larvae is accompanied by striking delayed-type and immediate-type hypersensitivity. Humoral responses include an elevation in levels of IgG, IgM, and IgE only some of which is specific for parasite components (Schantz et al. 1979). Most antibody responses are against a unique cuticular-associated trisaccharide epitope that is shed by the worm during its migration throughout the tissues. This antigen, which can be collected in vitro, is useful in the immunodiagnosis of the infection. However, immune responses against this antigen appear not to be protective. Nothing is known regarding the immune mechanisms responsible for the death of the worm.

Infection with helminths is generally associated with the induction of a T-helper 2 (Th2) type of immune response. Th2 cells produce cytokines such as interleukin (IL)-4/IL-13 and IL-5 that are required for the production of IgE and the maturation of eosinophils respectively (Mosmann and Coffman 1989). Infection with *Toxocara* spp. results in elevated circulating levels of eosinophils, which target dying worms in tissues. Granulomata consist of collections of eosinophils, lymphocytes, fibroblasts, epithelioid cells, and giant cells. Remnants of larval tissue may be present in the center of each granuloma. The nature of the role of eosinophils during infection is not clear. There is a limited number of studies in vivo to support a protective role for eosinophils against helminth infections (Weiss 2000). Experiments in vitro, however, have demonstrated that eosinophils can adhere in an antibody-dependent manner and kill larvae of *Schistosoma mansoni* (Butterworth et al. 1979), newborn larvae of *Trichinella spiralis* (Kazura and Aikawa 1980), and microfilariae of *Onchocerca volvulus* (Greene et al. 1981). For *Toxocara canis* however, eosinophils do not appear to play a major role in the killing of larvae (Fattah et al. 1986; Badley et al. 1987).

DIAGNOSIS

Biopsy of liver or other tissues in which larvae are suspected is rarely positive, hence diagnosis depends primarily on indirect measures of the presence of the worm, such as immunological tests, particularly enzyme-linked immunosorbent assay (ELISA) and Western blotting (Speiser and Gottstein 1984; Taylor et al. 1988). Using antigens secreted by the second-stage larvae and ELISA, sufficient specificity (91–93 percent) and sensitivity (78–91 percent) has been reported (Jacquier et al.

1991; Speiser and Gottstein 1984; Taylor et al. 1988). The specificity depends on the threshold used. A titer greater than 1:32 is considered to be indicative for the diagnosis (Smith et al. 1983; Taylor et al. 1988). Higher titers, with optical denisty (OD) values greater than 1.8, are common in well expressed toxocariasis (Luzna-Lyskov et al. 2000). Elevated gammaglobulin and an elevated isohemagglutinin titer (Despommier et al. 1994) can also help the clinician to confirm toxocariasis. Extreme eosinophilia is rarely seen in toxocariasis cases, although it can be an important indicator (Taylor et al. 1988).

OLM is diagnosed on clinical criteria during an ophthalmologic examination. Serological tests for antibodies are not as reliable for OLM as they are for VLM. In a study with OLM patients, only 45 percent of the patients had antibody levels that correlated with the infection (Schantz et al. 1979). A positive titer may be a diagnostic aid, but a negative titer cannot exclude the diagnosis. Western blotting with specific anti-*Toxocara* IgE detection appears to be an accurate procedure for the immunodiagnosis of OLM, provided that the testing is performed simultaneously on serum and ocular fluid (Magnaval et al. 2002).

CONTROL (THERAPY, PROPHYLAXIS, AND TRANSMISSION)

Although diethylcarbamazine (DEC) is listed in some references as an effective drug, its use is not recommended. Patients treated with DEC reported a significantly higher rate of adverse reactions than patients treated with mebendazole. Mebendazole in a dose of 20–25 mg/kg for 21 days was effective in treating an adult case of VLM (Magnaval 1995). Albendazole (15 mg/kg, 5 days) may also be effective (Pawlowski 2001). Corticosteroids suppress intense inflammation and thus relieve some of the more serious symptoms associated with worm migration and worm death. Ivermectin given as a single oral dose of 12 mg was less effective in clinical improvement and reduction of eosinophilia (Magnaval 1998). Liposome-incorporated benzimidazole carbamates have been shown to be effective in a murine model. Fenbendazole was more effective in muscles, whereas albendazole was more effective in the brain (Hrckova and Velebny 2001; Okada et al. 1996).

Treatments for OLM include surgery (vitrectomy), anthelmintic chemotherapy, and corticosteroids. In an Austrian study, albendazole (adults: 800 mg b.i.d.; children: 400 mg b.i.d.), in combination with systemic steroids, was shown to be a useful regimen to treat OLM syndrome (Barisani-Asenbauer et al. 2001).

Education of patients and healthcare workers on the importance of avoiding contact with contaminated environments is essential. The most important preventive measures are:

- regular de-worming of dogs and cats, beginning at 2 weeks of age (pups), 4, 6, and 8 weeks (pups and cats), followed by treatment every 2 months until 6 months of age and thereafter twice a year
- preventing contamination of soil by removing cat and dog feces in places immediately adjacent to houses and children's playgrounds
- covering children's sandpits
- regular washing of hands after handling soil and before eating
- teaching children not to put dirty objects into their mouths (Chin and Asher 2000).

REFERENCES

Ash, L.R. and Orihel, T.C. 1984. *Atlas of human parasitology. Toxocara canis*. Chicago: Am. Soc. of Clin. Pathologists Press.

Badley, J.E., Grieve, R.B., et al. 1987. Immune-mediated adherence of eosinophils to *Toxocara canis* infective larvae: the role of excretory-secretory antigens. *Parasite Immunol*, **9**, 133–43.

Barisani-Asenbauer, T., Maca, S.M., et al. 2001. Treatment of ocular toxocariasis with albendazole. *J Ocul Pharmacol Ther*, **17**, 287–94.

Beaver, P.C. 1984. Oxyuroidea and ascaridoidea. In: Beaver, P.C., Jung, R.C. and Cupp, E.W. (eds), *Clinical parasitology*. Philadelphia: Lea and Febiger, 320–2.

Beaver, P.C.S.C.H. 1952. Chronic eosinophilia due to visceral larva migrans. *Pediatrics*, **9**, 7–19.

Brunello, F., Genchi, C. and Falagiani, P. 1983. Detection of larva-specific IgE in human toxocariasis. *Trans R Soc Trop Med Hyg*, **77**, 279–89.

Buijs, J., Egbers, M.W. and Nijkamp, F.P. 1995. *Toxocara canis*-induced airway eosinophilia and tracheal hyporeactivity in guinea pigs and mice. *Eur J Pharmacol*, **293**, 207–15.

Buijs, J., Borsboom, G., et al. 1997. Relationship between allergic manifestations and *Toxocara* seropositivity: a cross-sectional study among elementary school children. *Eur Respir J*, **10**, 1467–75.

Butterworth, A.E., Vadas, M.A., et al. 1979. Interactions between human eosinophils and schistosomula of *Schistosoma mansoni*. II. The mechanism of irreversible eosinophil adherence. *J Exp Med*, **150**, 1456–71.

Castro, G.A. 1982. Gastrointestinal physiology: environmental factors influencing infection and pathogenicity. In: Bailey, W.S. (ed.), *Cues that influence behavior of internal parasites*. New Orleans: US Department of Agriculture, 1–21.

Chan, P.W., Anuar, A.K., et al. 2001. *Toxocara* seroprevalence and childhood asthma among Malaysian children. *Pediatr Int*, **43**, 350–3.

Chin, J. and Asher, M.S. 2000. Toxocariasis. In: Chin, J. and Asher, M.S. (eds), *Control of communicable diseases manual*. Washington, DC: American Public Health Association, 497–9.

Despommier, D.D. 2003. Toxocariasis: clinical aspects, epidemiology, medical ecology, and molecular aspects. *Clin Microbiol Rev*, **16**, 265–72.

Despommier, D.D., Gwadz, R.G. and Hotez, P.J. 1994. *Parasitic diseases*, 3rd edn. New York: Springer-Verlag.

Dinning, W.J., Gillespie, S.H., et al. 1988. Toxocariasis: a practical approach to management of ocular disease. *Eye*, **2**, 580–2.

Fattah, D.I., Maizels, R.M., et al. 1986. *Toxocara canis*: interaction of human blood eosinophils with the infective larvae. *Exp Parasitol*, **61**, 421–31.

Feldman, G.J. and Parker, H.W. 1992. Visceral larva migrans associated with the hypereosinophilic syndrome and the onset of severe asthma. *Ann Intern Med*, **116**, 838–40.

Fortenberry, J.D., Kenney, R.D. and Younger, J. 1991. Visceral larva migrans producing static encephalopathy in an infant. *Pediatr Infect Dis J*, **10**, 403–6.

Gibbons, L.M., Jacobs, D.E. and Sani, R.A. 2001. *Toxocara malaysiensis* n. sp. (Nematoda: Ascaridoidea) from the domestic cat (*Felis catus* Linnaeus, 1758). *J Parasitol*, **87**, 660–5.

Gillespie, S.H., Dinning, W.J., et al. 1993. The spectrum of ocular toxocariasis. *Eye*, **7**, 415–18.

Gopinath, R. and Keystone, J.S. 1995. Ascariasis, trichuriasis, and enterobiasis. In: Blaser, M.J. and Smith, P.D. (eds), *Infections of the gastrointestinal tract*. New York: Raven Press, 1167–88.

Greene, B.M., Taylor, H.R. and Aikawa, M. 1981. Cellular killing of microfilariae of Onchocerca volvulus: eosinophil and neutrophil-mediated immune serum-dependent destruction. *J Immunol*, **127**, 1611–18.

Gutierrez, Y. 1990. *Toxocara* – visceral larva migrans. In: Gutierrez, Y. (ed.), *Diagnostic pathology of parasitic infections with clinical correlations*. Philadelphia: Lea and Febiger, 262–72.

Holland, C.V. and Cox, D.M. 2001. *Toxocara* in the mouse: a model for parasite-altered host behaviour? *J Helminthol*, **75**, 125–35.

Holland, C.V., O'Lorcain, P., et al. 1995. Sero-epidemiology of toxocariasis in school children. *Parasitology*, **110**, 535–45.

Hrckova, G. and Velebny, S. 2001. Treatment of *Toxocara canis* infections in mice with liposome-incorporated benzimidazole carbamates and immunomodulator glucan. *J Helminthol*, **75**, 141–6.

Jacobs, D.E., Zhu, X., et al. 1997. PCR-based methods for identification of potentially zoonotic ascaridoid parasites of the dog, fox and cat. *Acta Trop*, **68**, 191–200.

Jacquier, P., Gottstein, B., et al. 1991. Immunodiagnosis of toxocarosis in humans: evaluation of a new enzyme-linked immunosorbent assay kit. *J Clin Microbiol*, **29**, 1831–5.

Jansen, J., van Knapen, F., et al. 1993. *Toxocara* ova in parks and sand-boxes in the city of Utrecht. *Tijdschr Diergeneeskd*, **118**, 611–14.

Kayes, S.G. 1997. Human toxocariasis and the visceral larva migrans syndrome: correlative immunopathology. *Chem Immunol*, **66**, 99–124.

Kazura, J.W. and Aikawa, M. 1980. Host defense mechanisms against *Trichinella spiralis* infection in the mouse: eosinophil-mediated destruction of newborn larvae in vitro. *J Immunol*, **124**, 355–61.

Luty, T. 2001. Prevalence of species of *Toxocara* in dogs, cats and red foxes from the Poznan region, Poland. *J Helminthol*, **75**, 153–6.

Luzna-Lyskov, A., Andrzejewska, L., et al. 2000. Clinical interpretation of eosinophilia and ELISA values (OD) in toxocariasis. *Acta Parasiologica*, **45**, 35–9.

Maetz, H.M., Kleinstein, R.N., et al. 1987. Estimated prevalence of ocular toxoplasmosis and toxocariasis in Alabama. *J Infect Dis*, **156**, 414.

Magnaval, J.F. 1995. Comparative efficacy of diethylcarbamazine and mebendazole for the treatment of human toxocariasis. *Parasitology*, **110**, 529–33.

Magnaval, J.F. 1998. Apparent weak efficacy of ivermectin for treatment of human toxocariasis. *Antimicro Agents Chemother*, **42**, 2770.

Magnaval, J.F., Malard, L., et al. 2002. Immunodiagnosis of ocular toxocariasis using western-blot for the detection of specific anti-*Toxocara* IgG and CAP for the measurement of specific anti-*Toxocara* IgE. *J Helminthol*, **76**, 335–9.

Mosmann, T.R. and Coffman, R.L. 1989. TH1 and TH2 cells: different patterns of lymphokine secretion lead to different functional properties. *Ann Rev Immunol*, **7**, 145–73.

Okada, K., Fujimoto, K., et al. 1996. Eosinophil chemotactic activity in bronchoalveolar lavage fluid obtained from *Toxocara canis*-infected rats. *Clin Immunol Immunopathol*, **78**, 256–62.

Overgaauw, P.A. 1997. Aspects of *Toxocara* epidemiology: toxocarosis in dogs and cats. *Crit Rev Microbiol*, **23**, 233–51.

Pawlowski, Z. 2001. Toxocariasis in humans: clinical expression and treatment dilemma. *J Helminthol*, **75**, 299–305.

Pinelli, E., Dormans, J., et al. 2001. A comparative study of toxocariasis and allergic asthma in murine models. *J Helminthol*, **75**, 137–40.

Schantz, P.M. 1989. *Toxocara* larva migrans now. *Am J Trop Med Hyg*, **41**, 21–34.

Schantz, P.M., Meyer, D. and Glickman, L.T. 1979. Clinical, serologic, and epidemiologic characteristics of ocular toxocariasis. *Am J Trop Med Hyg*, **28**, 24–8.

Segovia, J.M., Torres, J., et al. 2001. Helminths in the wolf, *Canis lupus*, from north-western Spain. *J Helminthol*, **75**, 183–92.

Sharghi, N., Schantz, P.M., et al. 2001. Environmental exposure to *Toxocara* as a possible risk factor for asthma: a clinic-based case-control study. *Clin Infect Dis*, **32**, E111–16.

Small, K.W., McCuen, B.W., et al. 1989. Surgical management of retinal traction caused by toxocariasis. *Am J Ophthalmol*, **108**, 10–14.

Smith, H.V., Kusel, J.R. and Girdwood, R.W. 1983. The production of human A and B blood group like substances by in vitro maintained second stage *Toxocara canis* larvae: their presence on the outer larval surfaces and in their excretions/secretions. *Clin Exp Immunol*, **54**, 625–33.

Smith, M.H.D. and Beaver, P.C. 1953. Persistence and distribution of *Toxocara* larvae in the tissue of children and mice. *Pediatrics,*, **12**, 491–7.

Speiser, F. and Gottstein, B. 1984. A collaborative study on larval excretory/secretory antigens of *Toxocara canis* for the immunodiagnosis of human toxocariasis with ELISA. *Acta Trop*, **41**, 361–72.

Sugane, K. and Ohshima, T. 1984. Interrelationship of eosinophilia and IgE antibody production to larval ES antigen in *Toxocara canis*-infected mice. *Parasite Immunol*, **6**, 409–20.

Taylor, M.R. 2001. The epidemiology of ocular toxocariasis. *J Helminthol*, **75**, 109–18.

Taylor, M.R., Keane, C.T., et al. 1988. The expanded spectrum of toxocaral disease. *Lancet*, **1**, 692–5.

Weiss, S.T. 2000. Parasites and asthma/allergy: what is the relationship? *J Allergy Clin Immunol*, **105**, 205–10.

Weller, P.F., Lee, C.W., et al. 1983. Generation and metabolism of 5-lipoxygenase pathway leukotrienes by human eosinophils: predominant production of leukotriene C4. *Proc Natl Acad Sci USA*, **80**, 7626–30.

Wilder, H. 1950. Nematode endophthalmitis. *Trans Am Acad Ophthalmol Otolaryngol*, **55**, 99–109.

Zhu, X.Q., Jacobs, D.E., et al. 1998. Molecular characterization of a *Toxocara* variant from cats in Kuala Lumpur, Malaysia. *Parasitology*, **117**, 155–64.

Zhu, X.Q., Gasser, R.B., et al. 2001. Molecular approaches for studying ascaridoid nematodes with zoonotic potential, with an emphasis on *Toxocara* species. *J Helminthol*, **75**, 101–8.

37

Trichinella

DICKSON D. DESPOMMIER

THE PARASITE

Trichinella spiralis was described in 1835, first by James Paget, then by Richard Owen from independent observations on muscle tissue from the same cadaver. Paget viewed the encapsulated worms through a compound microscope borrowed from Robert Brawn, a botanist at the British Museum (Campbell 1983a). The infected muscle tissue which harbored the parasites was from a 51-year-old male who had died that day from the ravages of tuberculosis. He had apparently been coincidentally infected with *T. spiralis* some years prior to his death. Following this landmark discovery, work began on deciphering its life cycle and the worm's relationship to various clinical manifestations. By 1859, the major components of its life cycle had been described (Virchow 1859). Much of its natural history and basic biological features are now known (see Campbell 1983b).

All nematodes within the genus *Trichinella* are dependent upon carnivorism in order to move from one host to another, and they infect an extremely wide range of vertebrate host species (birds, reptiles, and mammals). All mammals exhibit some degree of susceptibility to infection with *T. spiralis* and its relatives, ensuring that the parasite's geographic distribution is both widespread and its incidence of infection is extensive. Several new species have been described which will undoubtedly lead to a better understanding regarding the evolutionary history of the genus (Pozio 2001b). There are now eight recognized species, with more surely to be discovered over the next few years. They are: *Trichinella spiralis*, *T. britovi*, *T. nelsoni*, *T. nativa*, *T. murrelli* (Pozio and La Rosa 2000), *T. pseudospiralis*, *T. papuae* (Pozio et al. 1999), and *T. zimbabwensis* (Pozio 2001a; Pozio et al. 2002). Clinically, *Trichinella spiralis* still remains the dominant cause of trichinellosis in humans throughout the world. In addition, the vast majority of experimental work employs *T. spiralis*. Thus, what follows will pertain mainly to this single species.

MORPHOLOGY AND STRUCTURE

T. spiralis adult male and female worms are dimorphic (Figures 37.1 and 37.2).

The female measures 3 mm in length by 36 μm in diameter, and the male measures 1.5 mm in length by 36 μm in diameter. Besides its length, the male can be distinguished from the female by its obvious reproductive organs at its tail end, termed claspers, that it employs to hold on to the female worm while mating. The newborn larva (Figure 37.3) measures 80 μm in length by 7–8 μm in diameter. Its esophagus has a stylet, a spear-like organ (Figure 37.4), which it uses to enter cells (Figure 37.5). The infective muscle larva (Figure 37.6) measures 1 mm in length by 36 μm in diameter, and is an obligate intracellular parasite (Figure 37.7). This enables all species of trichinella to share in the world title as holders of the record for being the largest examples of this kind of parasitism.

GENERAL BIOLOGY

Adult worms of both sexes also live as intracellular parasites (more accurately, intramulticellular), occupying

Figure 37.1 *The adult male* Trichinella spiralis *measures 1.5 mm × 36 μm.*

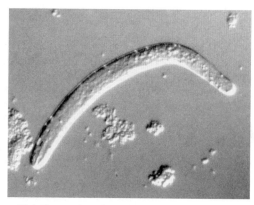

Figure 37.3 *The newborn larva of* Trichinella spiralis *measures 80 × 7 μm. Nomarski interference.*

a row of columnar cells in the upper part of the small intestine (Figure 37.8). In contrast to the larva, their metabolism is aerobic, deriving energy from the surrounding cytoplasmic milieu by an unknown mechanism(s). Host tissue is not directly ingested as evidenced by electron microscopy (Wright 1979). Most likely, the worms actively transport small molecular weight substances across their cuticular surfaces, or perhaps employ the pores of the hypodermal gland cells as an entry point for nutrients (Figure 37.9). Apparently, all stages have the ability to metabolize low molecular weight substrates, synthesizing protein, lipid, DNA, and RNA from them.

The infectious stage is the muscle larva (L1). It occupies a special niche which it creates during its development: the Nurse cell (Despommier 1998) (see Figure 37.10). Its metabolism is strictly anaerobic, despite the fact that the Nurse cell–parasite complex is surrounded by a circulatory rete consisting of a branching network of sinusoidal-like blood vessels (Figure 37.11). Some details of its intermediary metabolism are known, as the larvae are easy to obtain in large numbers (Stewart 1983).

As with all other nematodes, this worm's developmental cycle includes four larval stages. However, trichinella progresses rapidly through them – L1 to L4 in just 28 h after ingestion (another world record?) – all the while remaining in situ within the columnar epithelium (Figure 37.12). Adults can be demonstrated in the gut of experimentally infected rodents as early as 30 h after oral infection. Also, unlike most other nematode species, they do not change their overall dimensions until they achieve adulthood, when they grow longer, but thereafter remain at the same diameter.

The newborn larva migrates from the small intestine to the muscle tissue via the lymphatic circulation and/or the bloodstream, invading tissues, aided by its stylet. This stage is thus the most pathogenic form of trichinella (Capo and Despommier 1996), killing all cells it enters, with one notable exception: the striated skeletal muscle cell. Like the full-grown muscle larva, its metabolism is anaerobic, but few studies on this aspect have been carried out, probably owing to the difficulty and expense in obtaining sufficiently large numbers of this stage on which to work.

LIFE CYCLE

Ingestion by the host of the Nurse cell–parasite complex within infected muscle tissue initiates infection (Figure 37.13). The larvae are immediately freed from their intracellular niche by digestive enzymes and are transported by peristalsis to the upper two-thirds of the small intestinal tract. A small portion of epicuticular material (Figure 37.14) becomes altered by trypsin and perhaps by other proteases as well (Despommier 1983), allowing the parasite to receive external environmental cues within the lumen of the small intestine, resulting in alterations of its behavior (Stewart et al. 1987). Ultimately, these stimuli cause it to penetrate the columnar epithelium, perhaps aided in part by its own complement of proteases (Lun et al. 2003). After undergoing four molts in rapid succession during a 30-h period, the parasite matures to adulthood. Mating

Figure 37.2 *The adult female* Trichinella spiralis *measures 3 mm × 36 μm.*

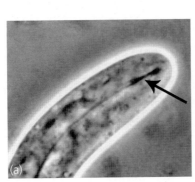 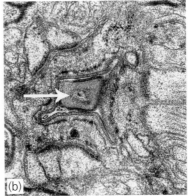

Figure 37.4 (a) *Stylet of the newborn larva of* Trichinella spiralis *(arrow).* **(b)** *Stylet of the newborn larva of* Trichinella spiralis *(arrow). Transmission electron micrograph*

ensues over the next few hours. The gestation period is 5 days, after which females begin depositing live newborn larvae into the lamina propria. Larvae are born continuously over the next several days, depending upon a variety of factors including the development of host immunity. Thus, the total number of newborns produced varies greatly from host to host, and also differs between males and females, and young and old members of the same host species (Despommier 1983). The intestinal phase of the infection in most mammalian hosts, including humans, routinely lasts up to 2–3 weeks, but ultimately adult worms are expelled by immune responses.

The route taken by the newborn larva from the gut to the muscle cell involves traversing the lymphatics as well as the general circulation. As a result, newborns invade all tissues. The migration phase is associated with bacteremia caused by the larvae dragging in members of the enteric flora along with themselves, and deaths in humans due to sepsis may occur during heavy infection. Larvae not finding striated skeletal muscle cells emerge from capillaries and penetrate other cell types, killing them in the process. Only skeletal muscle cells offer the prospect of a suitable niche in which the parasite is able to grow and develop (Despommier 1990; Jasmer 1995).

When the newborn larva penetrates this cell type, it remains in place and begins the next phase of its life, now as an intracellular organism. Within 20 days, the larva has re-engineered that environment, forcing host cytoplasm to transform from that of a contractile cell to one that aids the worm in its quest to achieve infectivity. The mechanisms by which this intriguing outcome of infection occurs are currently only hinted at from an extensive set of data extracted from the experimental literature. The Nurse cell thus becomes the life support system for the parasite, and can sustain the parasite for long periods of time – months to years (Despommier 1993). This ecologically sound strategy enables the parasite to travel great distances in some cases, ensuring the wide distribution of this most unique group of nematode parasites.

Nurse cell formation is thought to be the direct result of host–parasite interactions between the stichosome-derived secretions of the developing worm and host cell DNA. To date, no other nematode group has been shown to be this intimately connected metabolically with its mammalian host (Despommier et al. 1990). The vast majority of larvae eventually die and the remnant Nurse cell–parasite complexes calcify (Gerwel et al. 1970).

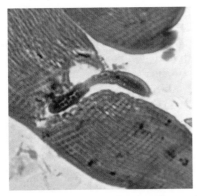 

Figure 37.5 *Newborn larva of* Trichinella spiralis *entering a muscle cell. Note collateral damage to surrounding cytoplasm.*

Figure 37.6 *The infective muscle larva of* Trichinella spiralis *measures 1 mm × 37 μm. Phase contrast*

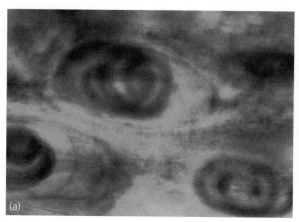
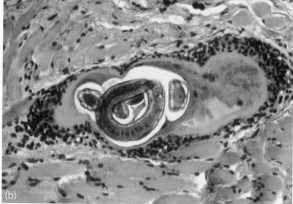

Figure 37.7 (a) *Whole mounted, unstained preparation of infected muscle tissue harboring numerous Nurse cell–parasite complexes.* **(b)** *Histological section of infected muscle tissue shown in* **(a)** *with Nurse cell–parasite complex of* Trichinella spiralis. *Note the intense cellular infiltrate surrounding the Nurse cell capsule.*

TRANSMISSION AND EPIDEMIOLOGY

As already stated, eating meat is the typical way of acquiring infection with *Trichinella spiralis* (Campbell 1983b). From an ecological standpoint, meat is an energy-rich, valued commodity in nature and rarely goes unconsumed when available, either as prey or as a scavenged carcass. Therefore, a wide variety of animals harbor one or more species of trichinella, many of which most would consider to be strict herbivores. The current world prevalence of infection with *T. spiralis* cannot be estimated because there have been no recent global surveys conducted. Several new reports document the prevalence and incidence of infection in some parts of Europe and China. The loss of infrastructure in the war-torn regions of Central Europe (i.e. lack of government-sponsored meat inspection programs, etc.) (Pozio and Marrucci 2003) have ushered in a significant increase in the number of new human cases (Geerts et al. 2002) and a dramatic increase in prevalence in pigs, particularly in those raised on small farms. In China, the incidence of infection in some domestic animals (e.g. dog, cat, pig) (Cui and Wang 2001) and subsequent increase in the rate of human infection in some southern provinces (e.g. Hunan, Szechwan) is substantially higher than previous estimates (Liu and Boireau 2002). Sporadic epidemics continue to occur with regularity (Capo and Despommier 1996; Bruschi and Murrell 2002). Within the last 10–15 years, in addition to those referenced above, outbreaks have occurred in Japan (Takahashi et al. 2000) and the Middle East (Olaison and Ljungstrom 1992). Apparently, Puerto Rico and mainland Australia (but not Tasmania) are among the few places in the world that have remained trichinella-free. *Trichinella pseudospiralis*, a non-capsule forming species, is thought mainly to infect birds, but can also occasionally infect mammals. It has been isolated from the Tasmanian Devil (Obendorf et al. 1990). Human cases have occurred in Tasmania (Andrews et al. 1994) and Thailand (Jonwutiwes et al. 1998). Apparently, the clinical manifestations of infection with this species are more prolonged than for *T. spiralis* (Bruschi and Murrell 2002). Four additional species of trichinella are potentially of medical importance. *T. nativa* occurs sylvatically in the Arctic where the reservoir hosts are polar bears and walruses. Eskimo populations are at high risk for infection due to their penchant for eating raw meats. *T. nelsoni* is also sylvatic and causes trichinellosis in equatorial Africa. There, hyenas and large cats serve as reservoirs. Infection in both wild and domestic pigs with *T. papuae*, another non-encapsulated species, occurs in Western Province, New Guinea (Owen et al. 2000). Human infections may also occur there, as indicated by the presence of *Taenia solium* infection (Margono et al. 2003), since native populations continue the habit of eating undercooked pork. *T. zimbabwensis*, the third species that does not form a capsule as a larva, has recently been described from Central Africa as an infection of domestic crocodiles. Infection in humans with this species has yet to be detected.

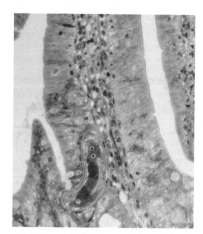

Figure 37.8 *A single adult worm of* Trichinella spiralis *in situ in a row of enterocytes in the small intestine of an experimentally infected mouse*

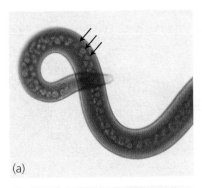
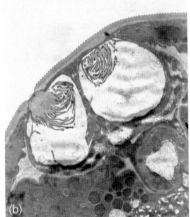
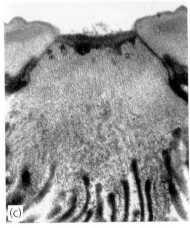

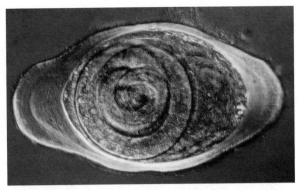

Figure 37.10 *Nurse cell–parasite complex of* Trichinella spiralis. *Nomarski-phase interference. Photo courtesy of Eric Grave*

the lack of physician awareness of this worldwide pathogen, keeping in mind specific signs and symptoms can greatly enhance the chances for making a correct diagnosis prior to running a battery of expensive assays, even in the early stages of the disease. For the duration of the 3-week period of the intestinal infection, patients can suffer many effects related to the damage that occurs there. Development of larvae to adults in the columnar epithelium is associated with enteritis. As newborn larvae are shed (i.e. 5–21 days after infection), mucosal inflammation intensifies, with inflammation consisting of eosinophils, neutrophils, and lymphocytes. Antigen–antibody complexes develop in the surrounding tissues and probably also contribute to intestinal disease experienced by patients who have ingested large numbers of larvae.

PARENTERAL STAGES

Damage caused by larvae penetrating cells becomes serious when this occurs in cardiac and central nervous system (CNS) tissues. Myocarditis, sometimes severe enough to cause death, is transient as Nurse cells cannot form in heart tissue. In the CNS, larvae tend to stay and wander about, frequently causing significant damage even in mild infection.

In skeletal muscle fibers, parasite-secreted proteins and other antigens induce progressive infiltration of inflammatory cells. Myositis and tissue edema develop on or about 14 days after penetration of the fiber. The amount of pathological change caused by the larvae is directly proportional to the number shed in the small intestine. However, there is much variation in the clinical presentation, related mostly to the type of muscle affected and the species of trichinella causing the damage. Lethal infection for most adults is of the order of 15 *T. spiralis* larvae per gram of diaphragmatic muscle, whereas as many as 1 000 *T. nelsoni* parasites per gram of muscle are still well tolerated.

Moderate to heavy trichinellosis presents signs and symptoms in the early (intestinal), middle (systemic and tissue invasion), and late (convalescent) phases (Murrell

Figure 37.9 (a) *Whole mount of an adult female* Trichinella spiralis. *Note double row of hypodermal gland cells (arrows).* **(b)** *Two hypodermal gland cells of an adult female* Trichinella spiralis. *Transmission electron micrograph.* **(c)** *A portion of a hypodermal gland cell, high magnification. Transmission electron micrograph*

CLINICAL AND PATHOLOGICAL ASPECTS OF INFECTION

Clinical manifestations

ENTERIC STAGES

Symptoms resulting from infection with *Trichinella spiralis* are often first ascribed to other illnesses until laboratory tests come back negative for them. Then physician attention shifts to the more 'exotic' diagnoses. Despite

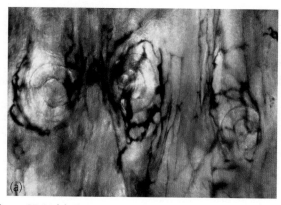
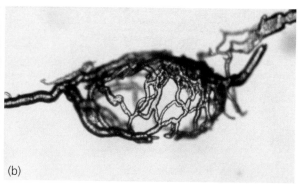

Figure 37.11 (a) *Three circulatory rete surrounding Nurse cell–parasite complexes in a whole mount preparation of an experimentally infected mouse. India ink was injected into the circulatory system immediately after administration of a lethal injection of Nembutal to reveal the blood vessels.* **(b)** *Anatomic plastic cast of the circulatory rete of an experimentally infected mouse*

and Bruschi 1994; Capo and Despommier 1996). Early intestinal distress (in weeks 1–3) is associated with diarrhea, abdominal pain, and vomiting. The systemic and tissue invasion phase (weeks 3–8) is accompanied by fever and myalgia, periorbital edema, and hemorrhages in the nail beds and sclera of the eye, but can also seen in the mucous membranes. Edema leads to muscle tenderness and pain. Photophobia is another consequence of heavy infection.

Progressive infection, in which larvae penetrate a variety of tissues, gives rise to even more serious and often confusing sequelae. Myocarditis occurs in nearly 20 percent of seriously ill patients and electrocardiographic changes are frequently noted during this phase. Dyspnea is another complication of myositis. Neurological signs and symptoms vary greatly, but, coupled with other laboratory findings and a history of raw meat eating, should lead the alert physician to suspect infection with trichinella.

There is no evidence that long-term sequelae occur following even severe infection, at least with *Trichinella*

spiralis (Feldmeier et al. 1991). However, from the worm's perspective there are long-term consequences, since numerous Nurse cells calcify and, as a result, the larvae inside them perish. This process usually takes place over several months to years following infection. Whereas disease caused by other species of *Trichinella* present somewhat differently (Murrell and Bruschi 1994; Capo and Despommier 1996), details are unwarranted because of their low prevalence.

Infection with *T. pseudospiralis* apparently runs a longer course as compared to that typically reported from outbreaks due to infection with *T. spiralis* (Bruschi and Murrell 2002).

Pathogenesis and pathology

The migrating newborn larvae cause the vast majority of pathological consequences which occur mainly during the first 3–5 weeks after ingesting infected meat. The use of their stylet to enter and leave cells accounts for most

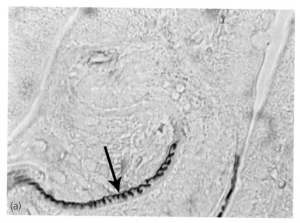
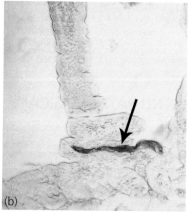

Figure 37.12 (a) *Immunoperoxidase-stained infected small intestine of an experimentally infected mouse. The L2 larva of* Trichinella spiralis *is in the process of molting and is crawling out of its L1 cuticle (arrow). The monoclonal antibody employed recognized only tyvelose epitopes.* **(b)** *Remnant of L1 cuticle (arrow) in enterocytes of a mouse experimentally infected with* Trichinella spiralis. *The same monoclonal antibody was used here as described in* **(a)**.

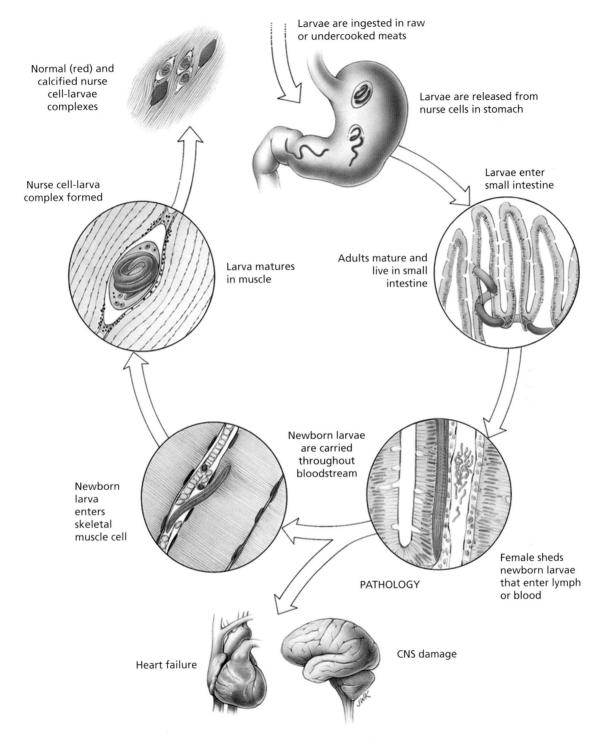

Larvae are ingested in raw or undercooked meats

Larvae are released from nurse cells in stomach

Larvae enter small intestine

Adults mature and live in small intestine

Normal (red) and calcified nurse cell-larvae complexes

Nurse cell-larva complex formed

Larva matures in muscle

Newborn larvae are carried throughout bloodstream

Newborn larva enters skeletal muscle cell

Female sheds newborn larvae that enter lymph or blood

PATHOLOGY

Heart failure

CNS damage

Figure 37.13 *Life cycle of* Trichinella spiralis. *Illustration by John Karapelou*

of the cell death and is associated with inflammation and subsequent granuloma formation. These are the most significant features of clinical trichinellosis. Granulomata are composed of macrophages, neutrophils, and eosinophils, and lymphocytes of various subsets (Wakelin and Denham 1983). The presence of these cells concentrated in a given region of a histological section of tissue should lead the pathologist to the parasite, even if the

offending organism is not seen in that particular section of muscle (Figure 37.15).

Symptoms attributable to the enteric stages (L1–L4 and the adult worm) result from exposure of the host to the secretions of those stages and presumably to the damage they induce in the invaded enterocytes (Castro and Powell 1994). The pharmacological properties of the stichosomal secretions of the enteral stages remain

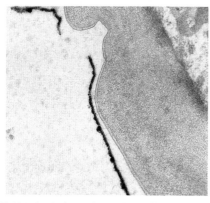

Figure 37.14 *Epicuticular and cuticular surfaces of the L1 larva of* Trichinella spiralis *after exposure for 1 h to trypsin digestion at pH 8.2. Note the break in the epicuticle. This treatment resulted in a marked change in behavior – from coiling–uncoiling to serpentine-like movement – when treated worms were subsequently placed in a mammalian tissue culture medium at pH 7.0. Transmission electron micrograph*

largely undefined, while a few have been characterized in terms of their putative biochemical functions (e.g. metalloproteases, nucleotidases) (Gounaris 2002; Lun et al. 2003). This area of research continues to represent a fertile area for investigation, and eventually may lead to a biochemical description of how this parasite re-models its niche into a Nurse cell. Damage to the enterocytes of the small intestine lasts only for the time the parasites occupy this niche, yet some effects last well beyond the presence of the parasite, at least in experimental infections. Experimentally infected animals lose their wheat germ agglutinin binding sites along the brush border of the entire epithelium and never regain them (Harari and Castro 1988), whereas other changes reverse after the parasites have been eliminated by host immune responses (e.g. myenteric electric potential, peristalsis, and secretory dysfunction (Castro and Powell 1994)). Hypermotility of the small intestine induced by the

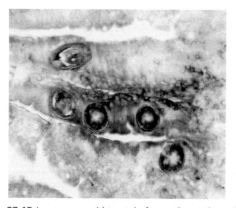

Figure 37.15 *Immunoperoxidase stain for tyvelose epitope in small intestine of an experimentally infected mouse. Note the presence of the antigenic determinant in surrounding non-infected enterocytes. Numerous sections of the parasite are visible in this section.*

enteral stages is well documented, and is ablated by inhibiting nitric oxide synthase (Torrents et al. 2003). The enteral phase of the infection also induces nerve growth factor. Increased numbers of nerve cells in the area of infection accelerates the gut transit time in response to parasite products (Torrents et al. 2002). Ablating the growth of new nerve cells with anti-NGF antibodies eliminates hypermotility, but not the inflammation.

Immune responses and protection

Most immune responses elicited during infection with trichinella, regardless of the host species, are non-protective in nature, and some even result in pathological changes and tissue damage to the host (Garside and Grencis 1992; Wakelin 1997). The vast majority of antigens are thought largely to be contained within the secretions of the parasite's stiochosome cells (Despommier 1998; Wu et al. 1998). Each stage possesses a unique set of these secretory proteins (2-D gel analysis indicates there may be hundreds), most of which are antigenic to the host during the infection (Silberstein 1983; Wakelin and Denham 1983; Gold et al. 1990). This is due to the presence of an unusual sugar residue on most of them, namely tyvelose (Wisenewski et al. 1993). This antigen is also found on the cuticle as well (Despommier et al. 1967; Philipp et al. 1980). Interestingly, although it is one of the most immunodominant of the antigenic determinants the worm secretes (Appleton and Romaris 2001), tyvelose does not elicit protection by itself (Goyal et al. 2002). Thus, the role of tyvelose in the development of protective responses and the epitopes that elicit them remains undefined (Romaris et al. 2002).

Protective immune responses are of the Th2 type (Wakelin and Denham 1983; Despommier 1988; Curman et al. 1992; Murrell and Bruschi 1994; Capo and Despommier 1996). In consequence, many valid immunodiagnostic tests exist for this pathogen (Ljungstrom 1983). Recent evidence suggests that IL-4, IL-5, IL-9, IL-10, IL-13, and IL-18 are important interleukins for mediating protection (Vallance and Matthaei 2000; Urban et al. 2001; Akiho et al. 2002; Helmby and Grencis 2002, 2003; Khan et al. 2003). Inflammation plays a central role in worm expulsion during a primary infection, with IL-4 and IL-13 being the most significant contributors to it (Urban et al. 2001). Reducing tissue mast cells by depletion of IL-10 enables the parasite to establish larger than normal numbers in the small intestine on first challenge due to reduced mast cell production (Helmby and Grencis 2003). IL-4 and IL-10 are down-regulated by IL-18, so when IL-18 is knocked out, mast cell production is increased during infection. The reverse is true when rIL-18 is given, inhibiting the production of mast cells and prolonging the intestinal phase of the infection. However, reducing the IL-10 response allowed the host to act against the larval stages in muscle tissue much more efficiently, by producing larger amounts of INF gamma (Helmby and Grencis 2002).

A monoclonal antibody of the IgA class reacted with various tissues of the infective L1 larva, but not with any other stage of the infection, and conferred high levels of protection (95 percent reduction) when administered prior to primary challenge with L1 larvae (Inaba et al. 2003). The same monoclonal antibody did not interfere with infection if given at the same time as oral challenge with L1, indicating that the epitope(s) of interest were specific to the L1 larva only. Identification of this parasite antigen(s) will undoubtedly shed much needed light on the overall subject of acquired immunity to trichinella.

Diagnosis

If the physician suspects that their patient is infected with any species of trichinella, a positive diagnosis can be made as follows:

- Muscle biopsy (Figure 37.15). This is the most direct means of detecting the infection, but larvae that are newly arrived in muscle tissue are hard to distinguish from muscle fibers and may be missed on biopsy. Low numbers of worms can also escape detection by this method (Figure 37.16). Removal of muscle tissue by digestion in 1 percent pepsin–1 percent HCl makes larvae easier to identify, but young ones may become digested, as well.
- Eliciting a history of eating raw or undercooked meat of any kind, but particularly pork or pork products.
- Laboratory findings consistent with the diagnosis of trichinellosis, such as an elevated circulating eosinophilia (Murrell and Bruschi 1994; Capo and Despommier 1996; Bruschi and Murrell 2002) and antibody detected by enzyme-linked immunosorbent assay (ELISA) (Ljungstrom 1983).

Immunological tests may be positive within 2 weeks after infection. Counterimmunoelectrophoresis and ELISA occasionally detect antibodies within 12 days after infection (Ljungstrom 1983). Both tests stay positive for long periods of time following recovery from infection (i.e. months to years), so a positive test does not necessarily rule in this parasite as the cause of the present illness. In the case of moderate to severe disease, muscle enzymes such as creatine phosphokinase and lactic dehydrogenase are released into the circulation causing an increase in serum levels, and their presence may be another clue to the presence of trichinella (Murrell and Bruschi 1994; Capo and Despommier 1996).

A laboratory finding that is nearly always associated with the infection is circulating eosinophilia (Gould 1970) that shows a pattern of increasing (week 2–5),

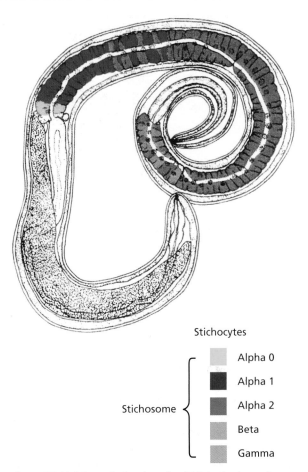

Stichocytes

Alpha 0
Alpha 1
Alpha 2
Beta
Gamma

Stichosome

Figure 37.16 *Schematic drawing of an L1 infective larva of* Trichinella spiralis *color-coded for the five types of stichocytes (Wu 1955)*

plateauing (week 4–8), then decreasing (week 9–12) levels. Eosinophilia may reach levels of 80–95 percent in severe cases (Kaljus 1936). Total white blood cell count is slightly elevated (i.e. 12 000–15 000 cells/mm^3). Admission of having eaten raw or undercooked pork, or meat derived from game animals (Schellenberg et al. 2003) 2 weeks or so earlier, accompanied by a recent bout of gastroenteritis or 'flu-like' illness gives additional useful hints to the diagnosis, and should alert the physician as to the possibility of trichinellosis (Figures 37.17–37.21).

Control (therapy, prophylaxis, transmission)

Clinically, there are no recommended specific therapies once the diagnosis is made. Removal of adult worms from the small intestine with thiabendazole (Gerwel et al. 1974) or mebendazole may reduce the number of newborn larvae that reach the skeletal muscle, but diagnosis is almost invariably determined after most of the larvae have been produced (i.e. weeks 3–6 after infection). Mebendazole and albendazole reduce the worm burden in the muscle tissue in experimental animals, but

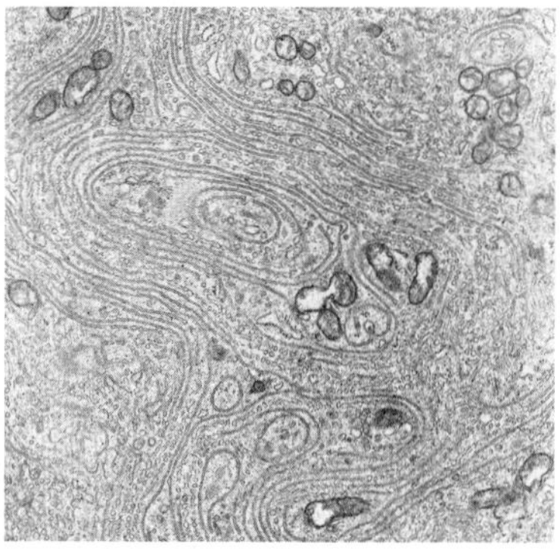

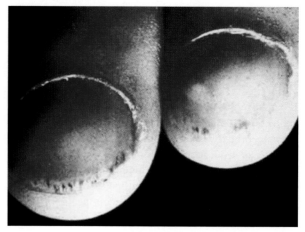

Figure 37.17 *Cytoplasm of a portion of Nurse cell 30 days after infection. Transmission electron micrograph*

Figure 37.19 *Splinter hemorrhages due to infection with* Trichinella spiralis

this stage of the infection does not cause illness once it is in its intracellular niche.

Prevention of infection with any species of trichinella involves cooking all meat products at 59°C (137°F) for 10 min or freezing them at −20°C for 3 days. The one exception is *T. nativa* that survives freezing (Dick 1983; Kapel et al. 1999). Most outbreaks due to infection with *Trichinella spiralis* occur as the result of eating commercially available pork products (Noeckler et al. 2001). Inspection of meat products at the slaughter house is a proven strategy for controlling the incidence of infection in human populations (Blancou 2001), and the breakdown of such programs due to civil unrest and war has led to increases of this infection in those regions of Europe which recently experienced such activity (Djordjevic et al. 2003). In other parts of the world, the United States and Asia, for example, meat is not inspected for *Trichinella*. Spread of infection into commercial meat supplies is by contamination of animal food with infected meat scraps. Outbreaks within the USA are infrequent, and are usually associated with the ingestion

of infected game animal meats. These outbreaks are more difficult to control because few hunters are aware of the risk of acquiring this infection from their kill.

Herbivores are usually not infected with trichinella but occasionally may acquire it by accidentally eating a meal contaminated with meat from an infected carcass (Bellani et al. 1978). An outbreak of trichinellosis in

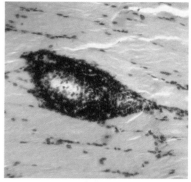

Figure 37.20 *'Conjuntivitis' due to infection with* Trichinella spiralis

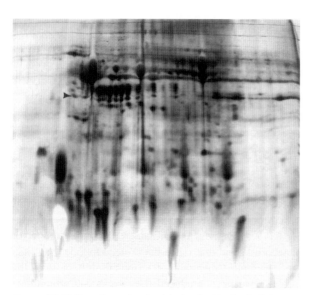

Figure 37.18 *Two-dimensional gel (stained with silver) of the secretions of the L1 larva of* Trichinella spiralis

Figure 37.21 *Histological section of muscle tissue infected with* Trichinella spiralis. *No parasite is visible in this section. Note the intense infiltration of cells surrounding the capsule of the Nurse cell.*

France, involving thousands of people (Ancelle et al. 1985), was traced back to a single horse imported from the United States and sold in Paris as 'steak' tartare. Horsemeat continues to be a source of infection in Europe (Tamburrini et al. 2001).

REFERENCES

Akiho, H., Blennerhassett, P., et al. 2002. Role of IL-4, IL-13, and STAT6 in inflammation-induced hypercontractility of murine smooth muscle cells. *Am J Physiol Gastrointest Liver Physiol*, **282**, R226–32.

Ancelle, T., Dupouy-Camet, J., et al. 1985. Outbreak of trichinosis due to horse meat in the Paris area. *Lancet*, **ii**, 660, 21 September.

Andrews, J.R.H., Ainsworth, R. and Abernathy, D. 1994. *Trichinella pseudospiralis* in humans: description of a case and its treatment. *Trans R Soc Trop Med Hyg*, **88**, 200–3.

Appleton, J.A. and Romaris, F.A. 2001. Pivotal role for glycans at the interface between *Trichinella spiralis* and its host. *Vet Parasitol*, **101**, 249–60.

Bellani, L., Mantovani, A. and Filippini, I. 1978. Observations on an outbreak of human trichinellosis in northern Italy. In: Kim, C.W. and Pawlowski, Z. (eds), *Trichinellosis*. Hanover, NH: University Press of New England, 535–9.

Blancou, J. 2001. History of trichinellosis surveillance. *Parasite*, **8**, Suppl. 2, S16–19.

Bruschi, F. and Murrell, K.D. 2002. New aspects of human trichinellosis: the impact of new Trichinella species. *Postgrad Med J*, **78**, 15–22.

Campbell, W.C. 1983a. Historical introduction. In: Campbell, W.C. (ed.), *Trichinella and trichinosis*. New York: Plenum Press, 1–28.

Campbell, W.C. 1983b. Epidemiology I. Modes of transmission. In: Campbell, W.C. (ed.), *Trichinella and trichinosis*. New York: Plenum Press, 425–44.

Capo, V. and Despommier, D.D. 1996. Clinical aspects of infection with *Trichinella* spp. *Clin Microbiol Rev*, **Jan**, 47–54.

Castro, G.A. and Powell, D.W. 1994. The physiology of the mucosal immune system and immune-mediated responses in the gasrtointestinal tract. In: Johnson, L.R. (ed.), *Physiology of the gastrointestinal tract*, 3rd edn. New York: Raven Press, 709–49.

Cui, J. and Wang, Z.Q. 2001. Outbreaks of human trichinellosis caused by consumption of dog meat in China. *Parasite*, **8**, Suppl. 2, S74–7.

Curman, J.A., Pond, L. and Nashold, F. 1992. Immunity to *Trichinella spiralis* infection in vitamin A deficient mice. *J Exp Med*, **175**, 111–20.

Despommier, D.D. 1983. Biology. In: Campbell, W.C. (ed.), *Trichinella and trichinellosis*. New York: Plenum Press, 75–153.

Despommier, D.D. 1988. The immunobiology of *Trichinella spiralis*. In: Soulsby, E.J.L. (ed.), *Immune responses in parasitic infections: immunology, immunopathology, and immunoprophylaxis. , Nematodes*. vol. 1. Boca Raton, FL: CRC Press, 43–60.

Despommier, D.D. 1990. The worm that would be virus. *Parasitol Today*, **6**, 193–5.

Despommier, D.D. 1993. *Trichinella spiralis* and the concept of niche. *J Parasitol*, **79**, 472–82.

Despommier, D.D. 1998. How *Trichinella spiralis* makes itself at home. *Parasitol Today*, **14**, 318–23.

Despommier, D.D., Kajima, M. and Wostmann, B.S. 1967. Ferritin-conjugated antibody studies on the larva of *Trichinella spiralis*. *J Parasitol*, **53**, 618–24.

Despommier, D.D., Gold, A., et al. 1990. *Trichinella spiralis*: A secreted antigen of the infective L1 larva localizes to the cytoplasm and nucleoplasm of infected host cells. *Exp Parasitol*, **72**, 27–38.

Dick, T.A. 1983. Infectivity of isolates of Trichinella and the ability of the arctic strain to survive freezing temperatures in the raccoon,

Procyon lotor, under experimental conditions. *J Wildlife Dis*, **19**, 333–6.

Djordjevic, M., Bacic, M., et al. 2003. Social, political, and economic factors responsible for the reemergence of trichinellosis in Serbia: a case study. *J Parasitol*, **89**, 226–31.

Feldmeier, H., Biensle, U., et al. 1991. Sequelae after infection with *Trichinella spiralis*: a prospective cohort study. *Wien Klin Wochens*, **103**, 111–16.

Garside, P. and Grencis, R.K. 1992. T lymphocyte dependent enteropathy in murine *Trichinella spiralis* infection. *Parasite Immunol*, **14**, 217–55.

Geerts, S., de Borchgrave, J., et al. 2002. Trichinellosis: old facts and new developments. *Acad Geneeskd Belg*, **64**, 233–48.

Gerwel, C., Kociecka, W. and Pawlowski, Z. 1970. Parasitological examination of muscle several years after trichinosis. *Przegl Epidemiol*, **24**, 381–8.

Gerwel, C., Pawlowski, Z. and Kocieka, W. 1974. Probable sterilization of *Trichinella spiralis* by thiabendazole: further clinical observation of human infection. In: Kim, C.W. (ed.), *Trichinellosis*. New York: Intext Educational, 471–5.

Gold, A.M., Despommier, D.D. and Buck, S.W. 1990. Partial characterization of two antigens secreted by the larva of *Trichinella spiralis*. *Mol Biochem Parasitol*, **41**, 187–96.

Gould, S.E. 1970. B. Clinical pathology: diagnostic laboratory procedures. In: Gould, S.E. (ed.), *Trichinosis in man and animals*. Springfield, IL: Thomas, 191–221.

Gounaris, K. 2002. Nucleotidase cascades are catalyzed by secreted proteins of the parasitic nematode *Trichinella spiralis*. *Infect Immun*, **70**, 4917–24.

Goyal, P.K., Wheatcroft, J. and Wakelin, D. 2002. Tyvelose and protective responses to the intestinal stages of *Trichinella spiralis*. *Parasitol Int*, **51**, 91–8.

Harari, Y. and Castro, G.A. 1988. Evaluation of a possible functional relationship between chemical structure of the intestinal brush border and immunity to *Trichinella spiralis* in the rat. *J Parasitol*, **74**, 244–8.

Helmby, H. and Grencis, R.K. 2002. IL-18 regulates intestinal mastocytosis and Th2 cytokine production independently of IFN-gamma during *Trichinella spiralis* infection. *J Immunol*, **169**, 2553–60.

Helmby, H. and Grencis, R.K. 2003. Contrasting roles for IL-10 in protective immunity to different life cycle stages of intestinal nematode parasites. *Eur J Immunol*, **33**, 2382–90.

Inaba, T., Sato, H. and Kamiya, H. 2003. Monoclonal IgA antibody-mediated expulsion of Trichinella from the intestine of mice. *Parasitology*, **126**, 591–8.

Jasmer, D.P. 1995. *Trichinella spiralis*: subversion of differentiated mammalian skeletal muscle cells. *Parasitol Today*, **11**, 185–8.

Jonwutiwes, S., Chantachum, N., et al. 1998. First outbreak of human trichinellosis caused by *Trichinella pseudospiralis*. *Clin Infect Dis*, **26**, 111–15.

Kaljus, W.A. 1936. On the practical value of the intradermal reaction with the trichinelliasis antigen for the diagnosis of trichinelliasis in man. *Puerto Rico J Public Health*, **11**, 768–90.

Kapel, C.M., Pozio, E., et al. 1999. Freeze tolerance, morphology and RAPD-PCR identification of *Trichinella nativa* in naturally infected arctic foxes. *J Parasitol*, **85**, 144–7.

Khan, W.I., Richard, M., et al. 2003. Modulation of intestinal muscle contraction by interleukin-9 (IL-9) or IL-9 neutralization: correlation with worm expulsion in murine nematode infections. *Infect Immun*, **71**, 2430–8.

Liu, M. and Boireau, P. 2002. Trichinellosis in China: epidemiology and control. *Trends Parasitol*, **18**, 553–6.

Ljungstrom, I. 1983. Immunodiagnosis in man. In: Campbell, W.C. (ed.), *Trichinella and trichinosis*. New York: Plenum Press, 403–24.

Lun, H.M., Mak, C.H. and Ko, R.C. 2003. Characterization and cloning of metallo-proteinase in the excretory/secretory products of the infective-stage larva of *Trichinella spiralis*. *Parasitol Res*, **90**, 27–37.

Margono, S.S., Ito, A., et al. 2003. *Taenia solium* taeniasis/cysticercosis in Papua, Indonesia in 2001: detection of human worm carriers. *J Helminthol*, **77**, 39–42.

Murrell, D. and Bruschi, F. 1994. Clinical trichinellosis. In: Tsien, S. (ed.), *Progress in clinical parasitology*. Boca Raton, FL: CRC Press, 117–50.

Noeckler, K., Reiter-Owona, I., et al. 2001. Aspects of clinical features, diagnosis, notification and tracing back referring to Trichinella outbreaks in north Rhine-Westphalia, Germany, 1998. *Parasite*, **8**, S183–5.

Obendorf, D.L., Handlinger, J.H., et al. 1990. *Trichinella pseudospiralis* in Tasmanian wildlife. *Austr Vet J*, **67**, 108–10.

Olaison, L. and Ljungstrom, I. 1992. An outbreak of trichinosis in Lebanon. *Trans R Soc Trop Med Hyg*, **86**, 658–60.

Owen, I.L., Sims, L.D., et al. 2000. Trichinellosis in Papua New Guinea. *Aust Vet J*, **78**, 698–701.

Philipp, M., Parkhouse, R.M.E. and Ogilvie, B.M. 1980. Changing proteins on the surface of a parasitic nematode. *Nature (Lond)*, **287**, 538.

Pozio, E.G. 2001a. New patterns of Trichinella infection. *Vet Parasitol*, **98**, 133–48.

Pozio, E. 2001b. Taxonomy of Trichinella and the epidemiology of infection in the Southeast Asia and Australian regions. *Southeast Asian J Trop Med Public Health*, **32**, Suppl. 2, 129–32.

Pozio, E., Owen, I.L., et al. 1999. *Trichinella papuae* n.sp. (Nematoda), a new non-encapsulated species from domestic and sylvatic swine of Papua New Guinea. *Int J Parasitol*, **29**, 1825–39.

Pozio, E. and La Rosa, L.K. 2000. *Trichinella murrelli* n. sp: etiological agent of sylvatic trichinellosis in temperate areas of North America. *Parasitol*, **86**, 134–9.

Pozio, E., Foggin, C.M., et al. 2002. *Trichinella zimbabwensis* n.sp. (Nematoda), a new non-encapsulated species from crocodiles (*Crocodylus niloticus*) in Zimbabwe also infecting mammals. *Int J Parasitol*, **32**, 1787–99.

Pozio, E. and Marrucci, G. 2003. Trichinella-infected pork products: a dangerous gift. *Trends Parasitol*, **19**, 338.

Romaris, F., Escalante, M., et al. 2002. Monoclonal antibodies raised in Btk(xid) mice reveal new antigenic relationships and molecular interactions among gp53 and other Trichinella glycoproteins. *Mol Biochem Parasitol*, **125**, 173–83.

Schellenberg, R.S., Tan, B.J., et al. 2003. An outbreak of Trichinellosis due to consumption of bear meat infected with *Trichinella nativa*, in 2 northern Saskatchewan communities. *Infect Dis*, **188**, 835–43.

Silberstein, D.S. 1983. Antigens. In: Campbell, W.C. (ed.), *Trichinella and trichinosis*. New York: Plenum Press, 309–34.

Stewart, G.L. 1983. Biochemistry. In: Campbell, W.C. (ed.), *Trichinella and trichinosis*. New York: Plenum Press, 153–72.

Stewart, G.L., Despommier, D.D., et al. 1987. *Trichinella spiralis*: behavior, structure, and biochemistry of larvae following exposure to components of the host enteric environment. *Exp Parasitol*, **63**, 195–204.

Takahashi, Y., Mingyuan, L. and Waikagul, J. 2000. Epidemiology of trichinellosis in Asia and the Pacific Rim. *Vet Parasitol*, **93**, 227–39.

Tamburrini, A., Sacchini, D. and Pozio, E. 2001. An expected outbreak of human trichinellosis for the consumption of horsemeat. *Parasite*, **8**, S186–7.

Torrents, D., Torres, R., et al. 2002. Antinerve growth factor treatment prevents intestinal dysmotility in *Trichinella spiralis*-infected rats. *J Pharmacol Exp Ther*, **302**, 659–65.

Torrents, D., Prats, N. and Vergara, P. 2003. Inducible nitric oxide synthase inhibitors ameliorate hypermotility observed after *T. spiralis* infection in the rat. *Dig Dis Sci*, **48**, 1035–49.

Urban, J.F. Jr., Noben-Trauth, N., et al. 2001. IL-4 receptor expression by non-bone marrow-derived cells is required to expel gastrointestinal nematode parasites. *J Immunol*, **67**, 6078–81.

Vallance, B.A. and Matthaei, K.I. 2000. Interleukin-5 deficient mice exhibit impaired host defence against challenge *Trichinella spiralis* infections. *Parasite Immunol*, **22**, 487–92.

Virchow, R. 1859. Recherches sur le development du *Trichina spiralis*. *CR Acad Sci*, 660–2.

Wakelin, D. 1997. Parasites and the immune system: conflict or compromise. *Bioscience*, **47**, 32–40.

Wakelin, D. and Denham, D.A. 1983. The immune response. In: Campbell, W.C. (ed.), *Trichinella and trichinosis*. New York: Plenum Press, 265–308.

Wisenewski, N., McNeil, M. and Grieve, R.B. 1993. Characterization of novel fucosyl-containing and tyvelosyl-containing glycoconjugates from *Trichinella spiralis* muscle stage larvae. *Mol Biochem Parasitol*, **61**, 25–36.

Wright, K. 1979. *Trichinella spiralis*: an intracellular parasite in the intestinal phase. *J Parasitol*, **65**, 441–5.

Wu, L.-Y. 1955. Studies on *Trichinella spiralis*, 1. Male and female reproductive systems. *J Parasitol*, **41**, 40–7.

Wu, Z., Nagano, I. and Takahashi, Y. 1998. Differences and similarities between *Trichinella spiralis* and *T. pseudospiralis* in morphology of stichocyte granules, peptide maps of excretory and secretory (E–S) products and messenger RNA of stichosomal glycoproteins. *Parasitology*, **116**, 61–6.

38

Lymphatic filariasis

SUBASH BABU AND THOMAS B. NUTMAN

OVERVIEW

The term 'lymphatic filariasis' encompasses infection with three closely related nematode worms – *Wuchereria bancrofti*, *Brugia malayi*, and *Brugia timori*. All three parasites are transmitted by the bites of infective mosquitoes and have quite similar life cycles in humans (Figure 38.1) with the adult worms living in the afferent lymphatic vessels while their offspring, the microfilariae, circulate in the peripheral blood and are available to infect mosquito vectors when they feed. The majority of infected individuals develop patent infection, with circulating microfilariae, that is subclinical (or clinically asymptomatic). The disease manifestations vary from acute adenolymphangitis to those associated with chronic infection such as hydrocele, lymphatic obstruction, and elephantiasis. Filarial infection can rarely manifest as tropical pulmonary eosinophilia, or more atypical manifestations such as filarial arthritis can also occur (WHO 1992). Though not fatal, the disease is responsible for considerable suffering, deformity, and disability and is the second leading parasitic cause of disability (Ottesen et al. 1997) with disability-adjusted life years (DALY) estimated to be 5.644 million (WHO 2003).

Lymphatic filariasis is a global health problem. At a recent estimate, it has been determined that over two billion people are at risk and at least 129 million people are actually infected (Michael et al. 1996). *W. bancrofti*

accounts for nearly 90 percent of these cases. *W. bancrofti* has the widest geographical distribution and is present in Africa, Asia, the Caribbean, Latin America, and many islands of the western and south Pacific Ocean (Figure 38.2). *B. malayi* is geographically more restricted, being found in southwest India, China, Indonesia, Malaysia, Korea, the Philippines, and Vietnam (Michael et al. 1996). *B. timori* is found in Timor, Flores, Alor, Roti, and southeast Indonesia (WHO 1992) (Table 38.1).

HISTORICAL PERSPECTIVE

The existence of lymphatic filariasis has been recorded in ancient Chinese, Indian, Persian, and Arabic writings. The ancient Indian physician/surgeon Sushruta was the first to describe cases of elephantiasis, which he called *slipada* (*sli* elephant; *pada* leg). The microfilariae of *W. bancrofti* were first discovered by Demarquay in hydrocele fluid from a patient in Cuba in 1863 (Demarquay 1863). Wucherer in 1868 (Wucherer 1868) and Lewis in 1872 (Lewis 1872) were the first to discover microfilariae in the urine and blood of infected patients, respectively. Manson described the periodicity of microfilariae in the peripheral blood (Manson 1899) and also demonstrated that mosquitoes transmit the parasite (Manson 1877). Bancroft was the first to describe the adult female worm (Cobbold 1877), and this was followed by the discovery of the adult male worm by Bourne (Bourne 1888).

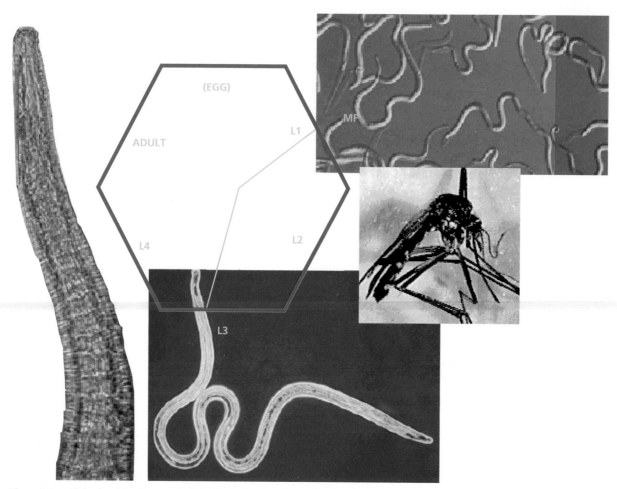

Figure 38.1 *Life cycle of the filarial parasites showing the microfilarial stages (MF), the mosquito, the infective stage (L3) larvae, and adult worm*

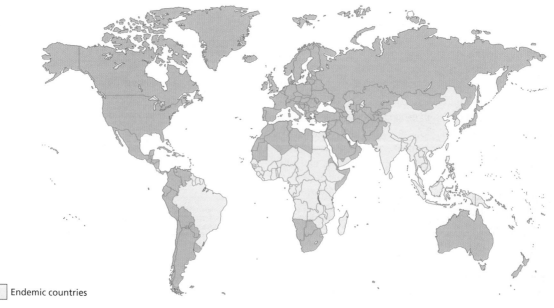

Endemic countries

Figure 38.2 *Geographical distribution of lymphatic filariasis. Lymphatic filariasis is endemic in Asia, Africa, Latin America, and Pacific islands (courtesy of World Health Organization, June 2002)*

Table 38.1 *Lymphatic-dwelling filarial parasites infecting humans*

	Mosquito vectors	Periodicity	Distribution	Primary pathology	Morphology
Wuchereria bancroft	Culex, Anopheles, Aedes	Periodic and non-periodic	Asia, Africa, Australia, Pacific, South America	Lymphatic and lung	Sheathed Mf with tail containing no nuclei
Brugia malayi	Mansonia, Anopheles, Aedes	Periodic and non-periodic	South-East Asia	Lymphatic and lung	Sheathed Mf with tail containing two nuclei
Brugia timori	Anopheles	Periodic	Indonesia and East Timor	Lymphatic and lung	Sheathed Mf with tail containing two nuclei

Bancroft in 1899 and Low in 1900 (Low 1900) established the mode of transmission of the parasite. Lichenstein, Brug, and Buckley were primarily responsible for the identification of the Brugian parasites (Buckley 1960).

MORPHOLOGY

Filarial nematodes belong to the phylum Nematoda, class Secernentea, and superfamily Filarioidea. Adult *W. bancrofti* and *B. malayi* worms are long, slender, tapered, and cylindrical. The males (4 cm × 0.1 mm for *W. bancrofti* and 3.5 cm × 0.1 mm for *B. malayi*) are strikingly smaller than the females (6–10 cm × 0.2–0.3 mm for *W. bancrofti* and 5–6 cm × 0.1 mm for *B. malayi*). The males can be distinguished from the females by their smaller size, as well as the presence of a corkscrew-like tail and the presence of two spicules that serve as organs of copulation (Ash and Orihel 1987). The microfilariae of *W. bancrofti* are ensheathed and measure about 245–300 μm by 7.5–10 μm (Figure 38.3). The tail is pointed and the terminal nuclei do not extend to the tip. The microfilariae of *B. malayi* are ensheathed and measure 175–230 μm by 5–6 μm. The tail is pointed with two terminal nuclei. Finally, the microfilariae of *B. timori* are also ensheathed and have terminal nuclei in the tip but are longer than those of *B. malayi* (265–325 × 4.5–7 μm) (Partono et al. 1977; Eberhard and Lammie 1991). In the mosquito, the filarial larvae

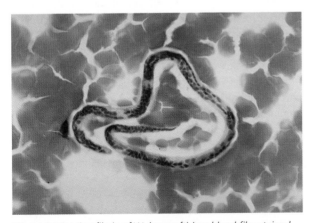

Figure 38.3 *Microfilaria of* W. bancrofti *in a blood film stained with Giemsa. The stain reveals the nuclei and the sheath.*

measure between 0.8 and 1.5 mm. Infective L3 larvae are between 1.2 and 1.6 mm long and newly formed L4 larvae are 3–6 mm in length (Ash and Schacher 1971).

LIFE CYCLE

Humans are the definitive host and mosquitoes the intermediate hosts of *W. bancrofti* and *Brugia* spp. The life cycle of filarial parasites involves four larval stages and an adult stage. Infection begins with the deposition of infective stage larvae (L3) on the skin near the site of puncture during a mosquito bite. The larvae then pass through the puncture wound and reach the lymphatic system. The host or parasite factors dictating this tropism are entirely unknown. The minimum number of L3 larvae required to successfully transmit infection is also not known. However, it is estimated that an average resident in an endemic area is exposed to 50–300 L3 larvae per year (Hati et al. 1989). Within the lymphatics and lymph nodes, the L3 larvae undergo molting and development to form L4 larvae. This takes about 7–10 days for both *W. bancrofti* and *B. malayi*. The L4 larvae undergo a subsequent molting/developmental step to form adult worms. This occurs about 4–6 weeks after L3 entry in the case of *B. malayi* and after several months in the case of *W. bancrofti* (Cross et al. 1979; Ash and Schacher 1971). The adult worms take permanent residence in afferent lymphatics or the cortical sinuses of lymph nodes and generate microscopic live progeny called 'microfilariae.' The female worms can give birth to as many as 50 000 microfilariae per day, which find their way into the blood circulation from the lymphatics. The adult worms are estimated to survive for a period of 5–10 years (Vanamail et al. 1989), although longer durations have been recorded (Carme and Laigret 1979). The time interval between inoculation of infective larvae and the initial appearance of detectable microfilariae in the blood is known as the 'prepatent period' and the development of microfilariae from adult worms is referred to as 'patency.' Information on the duration of the prepatent period in human infection based on experimental infection of human volunteers suggest that it is between 80 and 150 days (Nutman 1991) depending on the species.

The major vectors of *W. bancrofti* are culicine mosquitoes in most urban and semi-urban areas, anophelines in the more rural areas of Africa and elsewhere, and *Aedes* species in many of the endemic Pacific islands. For the Brugian parasites, *Mansonia* species serve as the main vector, but in some areas, anopheline mosquitoes can transmit infection as well. *Culex quinquefasciatus* is the most important vector of *W. bancrofti* and is responsible for more than half of all lymphatic filarial infections. The microfilariae of *W. bancrofti* and *B. malayi*, for the large part, exhibit a phenomenon called nocturnal periodicity (i.e. they appear in larger numbers in the peripheral circulation at night and retreat during the day). This may be a biological and evolutionary co-adaptation to the feeding habits of the mosquito vectors (Manson 1899). Subperiodic or nonperiodic *W. bancrofti* and *B. malayi* are also found in certain parts of the world.

The life cycle in the mosquito begins when the vector ingests microfilariae during a blood meal. The following stages of development have been described in the mosquito (Fuhrman et al. 1987; Devaney and Howells 1979; Agudelo-Silva and Spielman 1985):

1 Exsheathing: the larva comes out of the sheath in which it is enclosed, within 1–2 h of ingestion;
2 First stage (L1) larva: after exsheathing, the larva penetrates the stomach wall within 6–12 h and migrates to the thoracic muscles wherein it grows and develops into a sausage-shaped (short, thick) form;
3 Second stage (L2) larva: the larva molts and increases in length (long, thick form) and is relatively inactive;
4 Third stage (L3) larva: a final molt results in the third stage or infective larva (long, thin form), which may be found in any part of the insect. The L3 is highly active and motile, and, when it migrates to the proboscis of the mosquito, it is ready to be transmitted to a new host.

Under optimum conditions, the mosquito cycle (otherwise known as extrinsic incubation period) is between 10 and 14 days. Although filarial infections occur in animals, human filariasis is not usually a zoonosis (subperiodic form of *B. malayi* may be an exception). Animal reservoirs of Brugia are present in monkeys, cats, and dogs, but these animals are not regarded as important sources of human infection.

GENOMICS AND THE ENDOSYMBIONT

The filarial genome is estimated to be about 85–95 Mb based on work done on *B. malayi*. The genome is AT rich (78 percent) and is organized in five pairs of chromosomes (4 chromosome pairs and one XY or XX sex chromosome pair). Within the parasite, two other genomes have been discovered – the mitochondrial genome (14Kb) and the genome of a bacterial endosymbiont (Williams et al. 2000). Most pathogenic human filarial parasites are infected with a bacterial endosymbiont called *Wolbachia*. It is an alphaproteobacteria, related to *Rickettsia*, *Erlichia*, and *Anaplasma* and is maternally inherited. It is widespread in arthropods, and, in filarial parasites, the genome is 1.1 Mb in length and the *Brugia* endosymbiont has been fully sequenced (Ware et al. 2002). It has been detected in all life cycle stages of the parasite and found to be essential for adult worm viability, normal fertility, and larval development (Taylor and Hoerauf 1999).

An expressed sequence tag (EST) sequencing of the *B. malayi* genome was initiated in 1995 under the auspices of a WHO-sponsored Filarial Genome Project (1995–2000); genomic sequencing of *B. malayi* has now been performed and completion is expected in 2004. The filarial specific databases can be accessed at NCBI (dbEST) (www.ncbi.nlm.nih.gov/dbEST/index.html), TIGR (*B. malayi* Genomic Database) (www.tigr.org/tdb/e2k1/bma1/) and Nembase (Bm cluster database) (www.nematodes.org/nematodeESTs/nembase.html).

EPIDEMIOLOGY AND OCCURRENCE

Lymphatic filariasis is endemic in 80 countries, and more than two billion people worldwide are estimated to be at risk. Approximately 129 million people in tropical and subtropical areas of the world are infected. Of these infections, 90 percent are caused by *Wuchereria bancrofti*, and 10 percent by *Brugia malayi* (limited to Asia and parts of the Pacific). The largest numbers of infected people are present in India (45.5 million) and subSaharan Africa (40 million) with prevalences of 5 and 8 percent, respectively. Another region of importance is the Pacific Islands where high prevalences are found (Papua New Guinea (72 percent) and Republic of Tonga (48 percent)). Some of the low prevalence areas are Egypt and Latin America (Michael et al. 1996). Almost 25 million men suffer from genital disease (most commonly hydrocele); an estimated 15 million people have lymphedema or elephantiasis. Another 76 million have preclinical damage to the lymphatic and renal systems (Michael et al. 1996). In addition, millions of people suffer from acute attacks of adenolymphangitis that can be quite debilitating (WHO 1992) and extract a great socioeconomic and psychological burden.

Men are more susceptible to infection with filarial parasites than women (Brabin 1990). The gender of the individual has an important influence on the incidence of chronic pathology in lymphatic filariasis; males are more commonly susceptible to chronic sequelae than women (Brabin 1990). Interestingly, the incidence of acute filarial infection is not significantly different between males and females. The basis for this gender difference could be both environmental (men are more exposed to mosquito bites than women) and genetic factors (sex hormonal influence on innate and adaptive immunity).

The age dynamics of filarial infection reveals that the frequency of infection increases as a function of age up to about 20–30 years of life, after which it reaches a plateau (Dondero et al. 1976; Day et al. 1991). All infected individuals tend to acquire infection by the third decade of life. Previously, lymphatic filariasis had always been regarded as a debilitating disease of adults, but recent studies using sensitive antigen detection techniques have revealed that infection is first acquired in childhood, with as many as one-third of children infected before 5 years of age (Lammie et al. 1994; Simonsen et al. 1996; Steel et al. 2001). Lymphatic damage begins early in childhood but remains asymptomatic until the development of pathology later in life (Witt and Ottesen 2001). The earliest age when bancroftian microfilariae has been detected in the blood of children is 7–10 months, and the earliest reported case of brugian microfilaremia is 3.5 months.

CLINICAL MANIFESTATIONS

Lymphatic filariasis can manifest itself in a variety of clinical and subclinical conditions.

Subclinical (or asymptomatic) microfilaremia

In areas endemic for lymphatic filariasis, many individuals exhibit no symptoms of filarial infection and yet, on routine blood examinations, demonstrate the presence of significant numbers of parasites. These individuals are carriers of infection (and the reservoir for ongoing transmission) and have commonly been referred to as asymptomatic microfilaremics. The parasite burdens in these individuals can reach dramatically high numbers, exceeding 10 000 microfilariae in 1 ml of blood. With the advent of newer imaging techniques, it has become apparent that virtually all persons with microfilaremia have some degree of subclinical disease. Lymphoscintigraphy has revealed that the lymphatics of these individuals show profound changes including marked dilatation and tortuosity of lymph vessels with collateral channeling, increased flow, and abnormal patterns of lymph flow (Freedman et al. 1994, 1995). Ultrasound examination of scrotal lymphatics in microfilaremic men has revealed that approximately half of the men have nests of motile adult worms in their lymphatics: commonly known as the 'filarial dance sign' (Noroes et al. 1996). In parallel, a considerable degree of scrotal lymphangiectasia has been noted in these individuals. Mirroring the imaging studies, the examination of superficial skin punch biopsies reveal abnormally dilated lymphatic vessels in the limbs of patients with asymptomatic microfilaremia. In addition, approximately 40 percent of microfilaremic individuals suffer from microscopic hematuria and/or proteinuria that are indicative of low-grade renal damage (Dreyer et al. 1992). Thus, while apparently free of overt symptomatology,

the asymptomatic microfilaremic individuals clearly are subject to subtle pathological changes.

Acute clinical disease

The acute manifestations of lymphatic filariasis are characterized by recurrent attacks of fever associated with the inflammation of lymph nodes (lymphadenitis) and lymphatics (lymphangitis) (Partono 1987). In brugian filariasis, episodes of fever, lymphadenitis, and lymphangitis are common, while bancroftian filariasis presents more insidiously with fewer overt acute episodes (Pani et al. 1990). The lymph nodes commonly involved are the inguinal, axillary, and epitrochlear nodes and, in addition, the lymphatic system of the male genitals are frequently affected in *W. bancrofti* infection leading to funiculitis, epididymitis, and/or orchitis (Pani and Srividya 1995).

It has been proposed that there are at least two distinct mechanisms involved in the pathogenesis of acute attacks. The more classical is acute filarial adenolymphangitis, which is felt to reflect an immune-mediated inflammatory response to dead or dying adult worms. The striking manifestation is a distinct well-circumscribed nodule or cord along with lymphadenitis and retrograde lymphangitis. Funiculo-epididymo-orchitis is the usual presenting feature when the attacks involve the male genitalia. Fever is not usually present, but pain and tenderness at the affected site is common (Dreyer et al. 1999a).

The other has been termed acute dermatotolymphangitis, a process characterized by development of a plaque-like lesion of cutaneous or subcutaneous inflammation and accompanied by ascending lymphangitis and regional lymphadenitis. There may or may not be edema of the affected limbs. These pathological features are accompanied by systemic signs of inflammation including fever and chills. This manifestation is thought to result primarily from bacterial and fungal super-infections of the affected limbs (Dreyer et al. 1999a).

Acute filarial attacks have been described in infants as young as 3 months of age, but are usually more common in older children and continue throughout life (Dasgupta 1984; Nanduri and Kazura 1989). They can occur in both Mf+ and Mf– individuals and are frequently associated with chronic pathology (Dreyer et al. 1999a). Acute filariasis is not confined to residents in an endemic area but has also been observed in filarial-infected expatriates (Wartman 1947; McQuay 1967; Melrose et al. 2000), although this may reflect early, prepatent infection and responses to L3 and L4 larvae.

Manifestations of chronic infection

The chronic sequelae of filariasis are postulated to develop approximately 10–15 years after initial infection

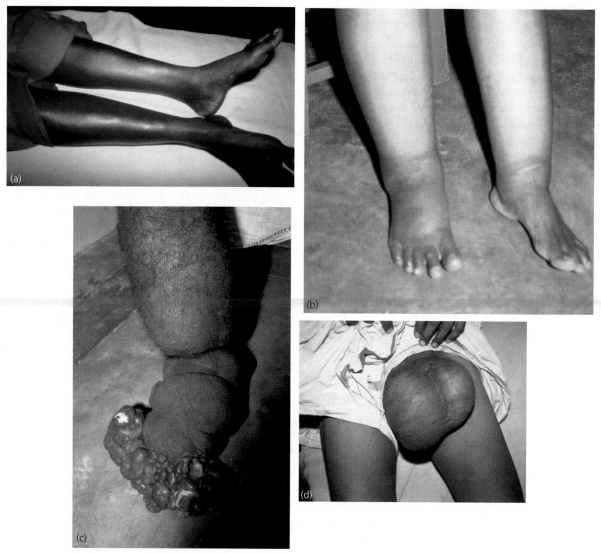

Figure 38.4 *Chronic manifestations of lymphatic filariasis.* **(a)** *Early lymphedema (grade 1) of both the lower limbs;* **(b)** *Advanced lymphedema (grade 3) of both the lower limbs with mild skin changes;* **(c)** *Elephantiasis of the lower limb with verrucous and fibrotic skin changes;* **(d)** *Hydrocele in a male patient.*

(Partono 1987). In bancroftian filariasis, the main clinical features are hydrocele, lymphedema, elephantiasis, and chyluria. The manifestations in descending order of occurrence are hydrocele and swelling of the testis, followed by elephantiasis of the entire lower limb, the scrotum, the entire arm, the vulva, and the breast (Pani and Srividya 1995). In Brugian filariasis, the leg below the knee and the arm below the elbow are commonly involved but rarely the genitals (Pani et al. 1990). Lymphedema can be classified or graded, a scheme proven very useful in clinical trials:

Grade 1: Pitting edema reversible on limb elevation (Figure 38.4a)

Grade 2: Pitting/non-pitting edema not reversible on limb elevation and normal skin

Grade 3: Non-pitting edema of the limb, not reversible on elevation with skin thickening (Figure 38.4b)

Grade 4: Non-pitting edema with fibrotic and verrucous skin changes (elephantiasis) (Figure 38.4c).

In men, scrotal hydrocele is the most common chronic clinical manifestation of bancroftian filariasis (Wijers 1977; Gyapong et al. 1994). Hydroceles are due to accumulation of edematous fluid in the cavity of the tunica vaginalis testis (Figure 38.4d). Chronic epididymitis and funiculitis can also occur. Chyloceles can also occur. The prevalence of chyluria (excretion of chyle, a milky white fluid in the urine) is very low.

Tropical pulmonary eosinophilia

Tropical pulmonary eosinophilia (TPE) is a distinct syndrome that develops in some individuals infected with *W. bancrofti* and *B. malayi* (Ottesen and Nutman

1992; Ong and Doyle 1998). This syndrome affects males and females at a ratio of 4:1, often during the third decade of life. The majority of cases have been reported from India, Pakistan, Sri Lanka, Brazil, Guyana, and Southeast Asia. The main clinical features include paroxysmal cough and wheezing that are usually nocturnal (and probably related to the nocturnal periodicity of microfilariae), weight loss, low-grade fever, adenopathy, and pronounced blood eosinophilia (>3 000 eosinophils/ml). Chest X-rays may be normal but generally show increased bronchovascular markings; diffuse miliary lesions or mottled opacities may be present in the middle and lower lung fields (Khoo and Danaraj 1960). Tests of pulmonary function show restrictive abnormalities in most cases and obstructive defects in half (Nesarajah 1972). Total serum IgE levels (10 000–100 000 ng/ml) and antifilarial antibody titers are characteristically elevated (Neva and Ottesen 1978). In TPE there is rapid clearance of microfilariae and parasite antigens from the bloodstream by the lungs, and the clinical symptoms result from allergic and inflammatory reactions elicited by the cleared parasites (Ottesen et al. 1979; Pinkston et al. 1987; O'Bryan et al. 2003). In some subjects, trapping of microfilariae in other reticuloendothelial organs can cause hepatomegaly, splenomegaly, or lymphadenopathy (Neva and Ottesen 1978). In the absence of successful treatment, interstitial fibrosis can lead to progressive pulmonary damage. Fortunately, TPE responds rapidly to anti-filarial chemotherapy (diethylcarbamazine (DEC)).

Other manifestations

Lymphatic filariasis has been associated with a variety of renal abnormalities including hematuria, proteinuria, nephrotic syndrome, and glomerulonephritis (Melrose 2002). Circulating immune complexes containing filarial antigens have been implicated in the renal damage. Lymphatic filariasis may also present as a mono-arthritis of the knee or ankle joint (Adebajo 1996).

Uninfected, but exposed individuals (asymptomatic amicrofilaremia or endemic normals)

In endemic areas, a proportion of the population remains uninfected despite exposure to the parasite to the same degree as the rest of the population (WHO 1992). This group has been termed endemic normal. The prevalence of endemic normals in a population ranges from zero (Day et al. 1991) to 50 percent (Weil et al. 1987; Ramzy et al. 1991; Steel et al. 2001) in different endemic areas. However, the true prevalence of endemic normal individuals in any population needs to be re-examined with the advent of newer techniques to diagnose subclinical infection. There could be several

mechanisms by which an endemic normal phenotype is achieved: insufficient exposure, low-grade (occult) infections not detected by current techniques, prepatent infections at the time of study, 'burning out' of infections naturally, and finally, genuine immunological resistance to the parasite (Beaver 1970).

PATHOGENESIS AND PATHOLOGY

The main pathological features in chronic lymphatic filariasis are the consequence of inflammatory damage to the lymphatics. The damage to lymphatics is probably the summation of the effect of tissue alterations related to live adult parasites, tissue alterations related to dead adult parasites, the host inflammatory response to living and dead parasites, the host inflammatory response to the endosymbiont Wolbachia, and tissue alterations due to secondary bacterial or fungal infections.

The adult worms reside in the lymphatics and lymph nodes and induce changes that result in dilatation of the lymphatics and thickening of the vessel walls. This is characterized by infiltration of plasma cells, eosinophils, and macrophages in the lymphatic vessels. This is often accompanied by proliferation of endothelial and connective tissue and tortuosity of the lymphatics and damaged lymph valves, lymphedema, and chronic stasis with brawny edema of the overlying skin (Figueredo-Silva et al. 2002).

It is postulated that as long as the adult worms are alive, the lymphatic vessels, though damaged, still remain patent. Death of the worm, however, leads to progressive fibrosis, obliteration of vessels by granuloma, and thrombi formation, extensive perilymphangitis resulting in irreversible lymphatic damage. In addition, secondary bacterial or fungal infections of limbs of affected patients exacerbate the chronic obstructive changes (Dreyer et al. 2000). One of the newer concepts to emerge in the field of pathogenesis is the contribution of the filarial endosymbiont Wolbachia. Several studies have shown that the potent inflammatory response engendered by filarial parasites is due to the endotoxin-like activity of Wolbachia. This response resembles a classical endotoxin response and involves the pattern recognition receptors, CD14 and TLR4 (Taylor et al. 2000, 2001).

IMMUNOLOGY

It has been postulated that the intensity and type of host immune response to the parasite or parasite products reflect the range of clinical manifestations of filariasis. Prepatent (or early) filarial infections are characterized by the presence of parasite-specific delayed-type hypersensitivity responses, parasite-specific T cell proliferation, cytokine production with a mixed phenotype (both Th1 cytokines (IL-2, IFNγ) and Th2 cytokines (IL-4 and IL-5)), profound eosinophilia and high titers of IgE antibody. When the infection progresses to patency (development of microfilaremia), the immune responses are profoundly altered. The marked change is the diminished parasite-specific T cell proliferation, IL-2 and IFNγ production, and an increase in IL-10 production, and class switching to IgG4 antibody production (Mahanty et al. 1992, 1993; Dimock et al. 1996). The mechanisms for this selective immune tolerance have been postulated to include diminished frequency of T and B cell precursors (King et al. 1996), suppressor T cells (Piessens et al. 1982), altered immune responses due to in utero exposure (Steel et al. 1994), increased expression of down-regulatory molecules like CTLA-4 (Steel and Nutman 2003), altered antigen-presenting cell function or apoptosis (Semnani et al. 2003), and the presence of regulatory cytokines such as IL-10 and TGFβ (King et al. 1993; Maizels and Yazdanbakhsh 2003).

One of the hallmarks of chronic filarial infection is the classical Th2 response characterized by the production of IL-4, IL-5, and IL-13 to parasite stimulation (Mahanty et al. 1993, de Boer et al. 1998a). In addition, two other key cytokines produced in chronic infection are IL-10 and TGFβ, which have regulatory functions (King et al. 1993). The Th2 cytokine profile is responsible for the high levels of IgE and eosinophilia that are associated with chronic helminth infections (King et al. 1990, 1991). However, the production of IgG4 in filarial infections is independent of IL-4 and IL-13 (de Boer et al. 1998b) but surprisingly influenced by IL-12 (de Boer et al. 1997). The intensity of transmission in endemic areas (King et al. 2001) and the in utero exposure to filarial antigens (Malhotra et al. 2003) also plays a role in influencing this cytokine bias in filarial infections.

In contrast to Mf-positive patients, patients with chronic pathologic conditions such as elephantiasis exhibit a different type of immune response. They display IgG1, IgG2, and IgG3 antibody responses to parasite antigens (Maizels et al. 1987), mount vigorous IL-2 and IFNγ responses, and good T cell proliferation to parasite antigens (Mahanty et al. 1996). Finally, patients with tropical pulmonary eosinophilia mount vigorous systemic and local immune responses characterized by high levels of IgE and IgG as well as T cells producing IL-4 and IL-5 (Ottesen and Nutman 1992; Nutman and Kumaraswami 2001).

The regulation of the human immune response to filarial parasites is complex with effects on both innate and adaptive immunity. Filarial parasites have been shown to induce apoptosis in dendritic cells, impair their capacity to produce IL-12 and IL-10 and to activate T cells (Semnani et al. 2003). In addition, filarial parasites have also been shown to downregulate the activity of macrophages (Chaussabel et al. 2003). Filarial parasites secrete certain immunoregulatory molecules including homologs of human macrophage migration inhibitory factor (MIF) (Pastrana et al. 1998) and human transforming growth factor-beta (TGFβ) (Gomez-Escobar et al. 1997), which may be partly responsible for the downregulatory effect on professional antigen-presenting cells. Filarial parasites are also capable of inducing pro-inflammatory cytokine expression in T cells early following exposure (Babu and Nutman 2003).

Experimental animal models of filariasis have been used for identifying the kinetics and type of immune response as well as the factors necessary for resistance to the parasite. L3 larvae, as well as adult worms, evoke predominantly Th2 responses (IL-4 and IL-5) in mice, while microfilariae evoke Th1 responses (Pearlman et al. 1993; Lawrence et al. 1994). Wild-type mice with an intact immune system are resistant to infection (Chong and Wong 1967; Carlow and Philipp 1987) with *B. malayi* (*W. bancrofti* has never been observed to survive in any small animal model), whereas mice lacking T cells (nude mice) and mice lacking T and B cells (SCID mice) are susceptible (Vincent et al. 1980; Suswillo et al. 1980; Nelson et al. 1991). Thus, the adaptive immune system is absolutely necessary for the elimination of the parasite. The exact mediators in this resistance phenotype remain to be determined although IL-4 (Babu et al. 2000), B cells (Babu et al. 1999; Paciorkowski et al. 2000), IgE (Spencer et al. 2003), and eosinophils (Martin et al. 2000) have all been postulated to be involved.

DIAGNOSIS

The traditional method of diagnosing lymphatic filarial infections has been the detection of microfilariae in the peripheral blood collected during the night in areas of nocturnal periodicity and during the day in areas of subperiodic lymphatic filariae. The simplest method is a thick blood film of capillary blood stained with Giemsa stain (reviewed in Melrose (2002)), its disadvantage being poor sensitivity. The sensitivity of detection can be augmented by the use of concentration techniques such as the Knott's concentration method, in which 1 ml of whole blood is added to 9 ml of a 2 percent formalin solution, centrifuged, and the sediment examined for microfilariae. Another widely used concentration technique is the membrane filtration technique whereby 1–5 ml of blood is passed through a 3 or 5 µM polycarbonate membrane which retains the microfilariae (Bell 1967).

For bancroftian filariasis, assays for the detection of circulating parasite antigens have been developed based on one of two well-characterized monoclonal antibodies, Og4C3 or AD12 (Weil et al. 1987; More and Copeman 1990). The commercial Og4C3 ELISA (Trop Bio Og4C3 Antigen test, produced by Trop Bio) has a sensitivity approaching 100 percent and specificity of 99–100 percent. The ICT filarial antigen test (Binax) is a rapid format card test with a sensitivity of 96–100 percent and specificity of 95–100 percent. It utilizes capillary or venous blood and is simple enough for field use (Weil et al. 1997). There are currently, however, no antigen tests for Brugian filariasis.

Antibody-based assays for diagnosing filarial infection have typically used crude parasite extracts and have suffered from poor specificity. Improvements have been made by the use of detection of anti-filarial IgG4 antibodies in that they are produced in relative abundance during chronic infection. IgG4 antibodies correlate well with the intensity and duration of filarial exposure and the level of microfilaremia; these IgG4 antibodies also have very little cross-reactivity to non-filarial helminths (Lal and Ottesen 1988; Mahanty et al. 1994). In addition, IgG4 antibodies are also useful in the diagnosis of Brugian infections (Rahmah et al. 1998a; Haarbrink et al. 1999); indeed a diagnostic dipstick test has been used in areas endemic for Brugian filariasis based on a recombinant Brugian antigen (Rahmah et al. 2001).

Polymerase chain reaction (PCR)-based methods have been developed for the detection of *W. bancrofti* DNA in blood, plasma, paraffin-embedded tissue sections (McCarthy et al. 1996), sputum (Abbasi et al. 1996), urine (Lucena et al. 1998), and in infected mosquitoes (Chanteau et al. 1994). For *B. malayi*, DNA in blood (Lizotte et al. 1994; Rahmah et al. 1998b) and in mosquitoes (Vythilingam et al. 1998) has been detected.

Finally, the examination of scrotum (Amaral et al. 1994) and breast (Dreyer et al. 1996) using ultrasonography in conjunction with pulse wave Doppler techniques can identify motile adult worms within the lymphatics. The adult worms exhibit a characteristic pattern of movement known as the filarial dance sign (Amaral et al. 1994) and the location of these adult worm nests remains remarkably stable (Dreyer et al. 1994). On rare occasions, living adult worms reside in the lymphatics of inguinal crural, axillary, and epitrochlear lymph nodes (Dreyer et al. 1999b).

TREATMENT

With better definitions of clinical syndromes in lymphatic filariasis and new tools to assess clinical status (e.g. ultrasound, lymphoscintigraphy, circulating filarial antigen assays, PCR), approaches to treatment based on infection status can be considered. DEC (6 mg/kg daily for 12 days), which has both macro- and microfilaricidal properties, remains the treatment of choice for the individual with active lymphatic filariasis (microfilaremia, antigen positivity, or adult worms on ultrasound), although albendazole (400 mg twice daily for 21 days) has also demonstrated macrofilaricidal efficacy. Ivermectine (400 µg/kg) is another drug with efficacy against filarial parasites.

A growing body of evidence indicates that, although they may be asymptomatic, virtually all people with *W. bancrofti* or *B. malayi* microfilaremia have some degree of subclinical disease (hematuria, proteinuria, abnormalities on lymphoscintigraphy). Thus, early treatment of asymptomatic persons is recommended to prevent further lymphatic damage. Because lymphatic disease is associated with the presence of adult worms, treatment with DEC is recommended for microfilaria-negative adult-worm carriers.

In persons with chronic manifestations of lymphatic filariasis, treatment regimens which emphasize hygiene, prevention of secondary bacterial infections, and physiotherapy have gained wide acceptance for morbidity control. Hydroceles can be drained repeatedly or managed surgically. In patients with chronic manifestations of lymphatic filariasis, drug treatment should be reserved for individuals with evidence of active infection as therapy has been associated with clinical improvement and, in some, reversal of lymphedema.

The recommended course of DEC treatment (12 days; total dose, 72 mg/kg) has remained standard for many years; however, data indicate that single-dose DEC treatment with 6 mg/kg may be equally efficacious (Kimura et al. 1992; Dreyer et al. 1995; Kazura et al. 1993). The 12-day course provides more rapid short-term microfilarial suppression. Regimens that utilize single-dose DEC or ivermectin or combinations of single doses of albendazole and either DEC or ivermectin have all been demonstrated to have a sustained microfilaricidal effect (Horton et al. 2000).

In the past few years, the use of antibiotics such as doxycycline, which have an effect on *Wolbachia*, to treat lymphatic filariasis has been investigated. Preliminary data suggest that 200 mg of doxycycline for 6 weeks is effective in eliminating microfilaremia (Hoerauf et al. 2003).

PREVENTION AND CONTROL

DEC has the ability to kill developing forms of filarial parasites (Ewert and Emerson 1975; Fujimaki et al. 1988) and has been shown to be useful as a prophylactic agent in humans (Lagraulet 1973). DEC has also been shown to reduce community levels of microfilaria so as to interrupt transmission. To this end, DEC has even been used as an additive to common table salt. The combination of community-wide distribution of a yearly single dose of DEC/ivermectin or DEC/albendazole is being touted as a major mechanism of control of the infection in that there is a theoretical basis for interruption of transmission providing the community drug coverage is large.

Like the integrated control programs used for onchocerciasis, long-term microfilarial suppression using mass, annual distribution of single-dose combinations of albendazole with either DEC or ivermectin is underway in many parts of the world. These strategies have as their basis the microfilarial suppression of >1 year using single-dose combinations of albendazole/ivermectin or albendazole/DEC. An added benefit of these combinations is their secondary salutary effects on gastrointestinal helminth infections.

Vector control measures such as the use of insecticides, the use of polystyrene beads in infested pits, the use of *Bacillus sphaericus* as a larvicide, the use of larvivorous fish, and the use of insecticide-treated bed nets have all been advocated as adjunct measures for control of filariasis (Maxwell et al. 1999).

Lymphatic filariasis is one of six potentially eradicable diseases (WHO International Task Force 1993) and the development of a global program to eliminate filariasis (GPELF) came about following a resolution by the WHO Assembly in 1997 (Molyneux et al. 2000; Ottesen 2000). The principal aims of the program are to interrupt transmission of infection and to alleviate and/or prevent disability. The recent advances in immunology, molecular biology, and imaging technology have helped pave the way for the realization of the goal to eliminate lymphatic filariasis.

REFERENCES

Abbasi, I., Hamburger, J., et al. 1996. Detection of *Wuchereria bancrofti* DNA in patients' sputum by the polymerase chain reaction. *Trans R Soc Trop Med Hyg*, **90**, 531–2.

Adebajo, A.O. 1996. Rheumatic manifestations of tropical diseases. *Curr Opin Rheumatol*, **8**, 85–9.

Agudelo-Silva, F. and Spielman, A. 1985. Penetration of mosquito midgut wall by sheathed microfilariae. *J Invertebr Pathol*, **45**, 117–19.

Amaral, F., Dreyer, G., et al. 1994. Live adult worms detected by ultrasonography in human Bancroftian filariasis. *Am J Trop Med Hyg*, **50**, 753–7.

Ash, L.R. and Orihel, T.C. 1987. *Parasites: a guide to laboratory procedures and identification*. Chicago: ASCP Press.

Ash, L.R. and Schacher, J.F. 1971. Early life cycle and larval morphogenesis of *Wuchereria bancrofti* in the jird, *Meriones unguiculatus*. *J Parasitol*, **57**, 1043–51.

Babu, S. and Nutman, T.B. 2003. Proinflammatory cytokines dominate the early immune response to filarial parasites. *J Immunol*, **171**, 6723–32.

Babu, S., Shultz, L.D., et al. 1999. Immunity in experimental murine filariasis: roles of T and B cells revisited. *Infect Immun*, **67**, 3166–7.

Babu, S., Ganley, L.M., et al. 2000. Role of gamma interferon and interleukin-4 in host defense against the human filarial parasite *Brugia malayi*. *Infect Immun*, **68**, 3034–5.

Beaver, P.C. 1970. Filariasis without microfilaremia. *Am J Trop Med Hyg*, **19**, 181–9.

Bell, D. 1967. Membrane filters and microfilariae: a new diagnostic technique. *Ann Trop Med Parasitol*, **61**, 220–3.

Bourne, A.G. 1888. *Br Med J*, **1**, 1050–1.

Brabin, L. 1990. Sex differentials in susceptibility to lymphatic filariasis and implications for maternal child immunity. *Epidemiol Infect*, **105**, 335–53.

Buckley, J.J. 1960. On *Brugia* gen. nov. for *Wuchereria* spp. of the 'malayi' group, i.e., *W. malayi* (Brug, 1927), *W. pahangi* Buckley and Edeson, 1956, and *W. patei* Buckley, Nelson and Heisch, 1958. *Ann Trop Med Parasitol*, **54**, 75–7.

Carlow, C.K. and Philipp, M. 1987. Protective immunity to *Brugia malayi* larvae in BALB/c mice: potential of this model for the identification of protective antigens. *Am J Trop Med Hyg*, **37**, 597–604.

Carme, B. and Laigret, J. 1979. Longevity of *Wuchereria bancrofti* var. pacifica and mosquito infection acquired from a patient with low level parasitemia. *Am J Trop Med Hyg*, **28**, 53–5.

Chanteau, S., Luquiaud, P., et al. 1994. Detection of *Wuchereria bancrofti* larvae in pools of mosquitoes by the polymerase chain reaction. *Trans R Soc Trop Med Hyg*, **88**, 665–6.

Chaussabel, D., Semnani, R.T., et al. 2003. Unique gene expression profiles of human macrophages and dendritic cells to phylogenetically distinct parasites. *Blood*, **102**, 672–81.

Chong, L.K. and Wong, M.M. 1967. Experimental infection of laboratory mice with *Brugia malayi*. *Med J Malaya*, **21**, 382–5.

Cobbold, T.S. 1877. *Lancet*, **2**, 70–1.

Cross, J.H., Partono, F., et al. 1979. Experimental transmission of *Wuchereria bancrofti* to monkeys. *Am J Trop Med Hyg*, **28**, 56–66.

Dasgupta, A. 1984. Cursory survey of lymphatic filariasis. An overview. *Indian J Pathol Microbiol*, **27**, 273–80.

Day, K.P., Gregory, W.F. and Maizels, R.M. 1991. Age-specific acquisition of immunity to infective larvae in a bancroftian filariasis endemic area of Papua New Guinea. *Parasite Immunol*, **13**, 277–90.

de Boer, B.A., Kruize, Y.C., et al. 1997. Interleukin-12 suppresses immunoglobulin E production but enhances immunoglobulin G4 production by human peripheral blood mononuclear cells. *Infect Immun*, **65**, 1122–5.

de Boer, B.A., Fillie, Y.E., et al. 1998a. Antigen-stimulated IL-4, IL-13 and IFN-gamma production by human T cells at a single-cell level. *Eur J Immunol*, **28**, 3154–60.

de Boer, B.A., Kruize, Y.C. and Yazdanbakhsh, M. 1998b. In vitro production of IgG4 by peripheral blood mononuclear cells (PBMC): the contribution of committed B cells. *Clin Exp Immunol*, **114**, 252–7.

Demarquay, J.-N. 1863. Helminthologie. *Gazette Medicale de Paris*, **18**, 665–7.

Devaney, E. and Howells, R.E. 1979. The exsheathment of *Brugia pahangi* microfilariae under controlled conditions in vitro. *Ann Trop Med Parasitol*, **73**, 227–33.

Dimock, K.A., Eberhard, M.L. and Lammie, P.J. 1996. Th1-like antifilarial immune responses predominate in antigen-negative persons. *Infect Immun*, **64**, 2962–7.

Dondero, T.J. Jr., Bhattacharya, N.C., et al. 1976. Clinical manifestations of Bancroftian filariasis in a suburb of Calcutta, India. *Am J Trop Med Hyg*, **25**, 64–73.

Dreyer, G., Ottesen, E.A., et al. 1992. Renal abnormalities in microfilaremic patients with Bancroftian filariasis. *Am J Trop Med Hyg*, **46**, 745–51.

Dreyer, G., Amaral, F., et al. 1994. Ultrasonographic evidence for stability of adult worm location in bancroftian filariasis. *Trans R Soc Trop Med Hyg*, **88**, 558.

Dreyer, G., Coutinho, A., et al. 1995. Treatment of bancroftian filariasis in Recife, Brazil: a two-year comparative study of the efficacy of single treatments with ivermectin or diethylcarbamazine. *Trans R Soc Trop Med Hyg*, **89**, 98–102.

Dreyer, G., Brandao, A.C., et al. 1996. Detection by ultrasound of living adult *Wuchereria bancrofti* in the female breast. *Mem Inst Oswaldo Cruz*, **91**, 1, 95–6.

Dreyer, G., Medeiros, Z., et al. 1999a. Acute attacks in the extremities of persons living in an area endemic for bancroftian filariasis: differentiation of two syndromes. *Trans R Soc Trop Med Hyg*, **93**, 413–17.

Dreyer, G., Santos, A., et al. 1999b. Proposed panel of diagnostic criteria, including the use of ultrasound, to refine the concept of 'endemic normals' in lymphatic filariasis. *Trop Med Int Health*, **4**, 575–9.

Dreyer, G., Noroes, J., et al. 2000. Pathogenesis of lymphatic disease in bancroftian filariasis: a clinical perspective. *Parasitol Today*, **16**, 544–8.

Eberhard, M.L. and Lammie, P.J. 1991. Laboratory diagnosis of filariasis. *Clin Lab Med*, **11**, 977–1010.

Ewert, A. and Emerson, G.A. 1975. Effect of diethylcarbamazine on third stage *Brugia malayi* larvae in cats. *Am J Trop Med Hyg*, **24**, 71–3.

Figueredo-Silva, J., Noroes, J., et al. 2002. The histopathology of bancroftian filariasis revisited: the role of the adult worm in the lymphatic-vessel disease. *Ann Trop Med Parasitol*, **96**, 531–41.

Freedman, D.O., de Almeida Filho, P.J., et al. 1994. Lymphoscintigraphic analysis of lymphatic abnormalities in symptomatic and asymptomatic human filariasis. *J Infect Dis*, **170**, 927–33.

Freedman, D.O., de Almeido Filho, P.J., et al. 1995. Abnormal lymphatic function in presymptomatic bancroftian filariasis. *J Infect Dis*, **171**, 997–1001.

Fuhrman, J.A., Urioste, S.S., et al. 1987. Functional and antigenic maturation of *Brugia malayi* microfilariae. *Am J Trop Med Hyg*, **36**, 70–4.

Fujimaki, Y., Shimada, M., et al. 1988. DEC-inhibited development of third-stage *Brugia pahangi* in vitro. *Parasitol Res*, **74**, 299–300.

Gomez-Escobar, N., van den Biggelaar, A. and Maizels, R. 1997. A member of the TGF-beta receptor gene family in the parasitic nematode *Brugia pahangi*. *Gene*, **199**, 101–9.

Gyapong, J.O., Magnussen, P. and Binka, F.N. 1994. Parasitological and clinical aspects of bancroftian filariasis in Kassena-Nankana District, upper east region, Ghana. *Trans R Soc Trop Med Hyg*, **88**, 555–7.

Haarbrink, M., Terhell, A.J., et al. 1999. Anti-filarial IgG4 in men and women living in *Brugia malayi*-endemic areas. *Trop Med Int Health*, **4**, 93–7.

Hati, A.K., Chandra, G., et al. 1989. Annual transmission potential of bancroftian filariasis in an urban and a rural area of West Bengal, India. *Am J Trop Med Hyg*, **40**, 365–7.

Hoerauf, A., Mand, S., et al. 2003. Doxycycline as a novel strategy against bancroftian filariasis-depletion of *Wolbachia* endosymbionts from *Wuchereria bancrofti* and stop of microfilaria production. *Med Microbiol Immunol (Berl)*, **192**, 211–16.

Horton, J., Witt, C., et al. 2000. An analysis of the safety of the single dose, two drug regimens used in programmes to eliminate lymphatic filariasis. *Parasitology*, **121**, Suppl, S147–60.

Kazura, J., Greenberg, J., et al. 1993. Comparison of single-dose diethylcarbamazine and ivermectin for treatment of bancroftian filariasis in Papua New Guinea. *Am J Trop Med Hyg*, **49**, 804–11.

Khoo, F.Y. and Danaraj, T.J. 1960. The roentgenographic appearance of eosinophilic lung (tropical eosinophilia). *Am J Roentgenol Radium Ther Nucl Med*, **83**, 251–9.

Kimura, E., Spears, G.F., et al. 1992. Long-term efficacy of single-dose mass treatment with diethylcarbamazine citrate against diurnally subperiodic *Wuchereria bancrofti*: eight years' experience in Samoa. *Bull World Health Organ*, **70**, 769–76.

King, C.L., Ottesen, E.A. and Nutman, T.B. 1990. Cytokine regulation of antigen-driven immunoglobulin production in filarial parasite infections in humans. *J Clin Invest*, **85**, 1810–15.

King, C.L., Poindexter, R.W., et al. 1991. Frequency analysis of IgE-secreting B lymphocytes in persons with normal or elevated serum IgE levels. *J Immunol*, **146**, 1478–83.

King, C.L., Mahanty, S., et al. 1993. Cytokine control of parasite-specific anergy in human lymphatic filariasis. Preferential induction of a regulatory T helper type 2 lymphocyte subset. *J Clin Invest*, **92**, 1667–73.

King, C.L., Medhat, A., et al. 1996. Cytokine control of parasite-specific anergy in human urinary schistosomiasis. IL-10 modulates lymphocyte reactivity. *J Immunol*, **156**, 4715–21.

King, C.L., Connelly, M., et al. 2001. Transmission intensity determines lymphocyte responsiveness and cytokine bias in human lymphatic filariasis. *J Immunol*, **166**, 7427–36.

Lagraulet, J. 1973. [Prophylaxis and treatment of lymphatic filariasis in French Polynesia]. *Bull Soc Pathol Exot Filiales*, **66**, 311–20.

Lal, R.B. and Ottesen, E.A. 1988. Characterization of stage-specific antigens of infective larvae of the filarial parasite *Brugia malayi*. *J Immunol*, **140**, 2032–8.

Lammie, P.J., Hightower, A.W. and Eberhard, M.L. 1994. Age-specific prevalence of antigenemia in a *Wuchereria bancrofti*-exposed population. *Am J Trop Med Hyg*, **51**, 348–55.

Lawrence, R.A., Allen, J.E., et al. 1994. Adult and microfilarial stages of the filarial parasite *Brugia malayi* stimulate contrasting cytokine and Ig isotype responses in BALB/c mice. *J Immunol*, **153**, 1216–24.

Lewis, T.R. 1872. *Government of India 8th annual report*, **8**, 241–66.

Lizotte, M.R., Supali, T., et al. 1994. A polymerase chain reaction assay for the detection of *Brugia malayi* in blood. *Am J Trop Med Hyg*, **51**, 314–21.

Low, G.C. 1900. *Br Med J*, **1**, 1456–7.

Lucena, W.A., Dhalia, R., et al. 1998. Diagnosis of *Wuchereria bancrofti* infection by the polymerase chain reaction using urine and day blood samples from amicrofilaraemic patients. *Trans R Soc Trop Med Hyg*, **92**, 290–3.

Mahanty, S., Abrams, J.S., et al. 1992. Parallel regulation of IL-4 and IL-5 in human helminth infections. *J Immunol*, **148**, 3567–71.

Mahanty, S., King, C.L., et al. 1993. IL-4- and IL-5-secreting lymphocyte populations are preferentially stimulated by parasite-derived antigens in human tissue invasive nematode infections. *J Immunol*, **151**, 3704–11.

Mahanty, S., Day, K.P., et al. 1994. Antifilarial IgG4 antibodies in children from filaria-endemic areas correlate with duration of infection and are dissociated from antifilarial IgE antibodies. *J Infect Dis*, **170**, 1339–43.

Mahanty, S., Luke, H.E., et al. 1996. Stage-specific induction of cytokines regulates the immune response in lymphatic filariasis. *Exp Parasitol*, **84**, 282–90.

Maizels, R.M. and Yazdanbakhsh, M. 2003. Immune regulation by helminth parasites: cellular and molecular mechanisms. *Nat Rev Immunol*, **3**, 733–44.

Maizels, R.M., Selkirk, M.E., et al. 1987. Antibody responses to human lymphatic filarial parasites. *Ciba Found Symp*, **127**, 189–202.

Malhotra, I., Ouma, J.H., et al. 2003. Influence of maternal filariasis on childhood infection and immunity to *Wuchereria bancrofti* in Kenya. *Infect Immun*, **71**, 5231–7.

Manson, P. 1877. *China Imperial Maritime Customs*.

Manson, P. 1899. *Br Med J*, **2**, 644–6.

Martin, C., Le Goff, L., et al. 2000. Drastic reduction of a filarial infection in eosinophilic interleukin-5 transgenic mice. *Infect Immun*, **68**, 3651–6.

Maxwell, C.A., Mohammed, K., et al. 1999. Can vector control play a useful supplementary role against bancroftian filariasis? *Bull World Health Organ*, **77**, 138–43.

McCarthy, J.S., Zhong, M., et al. 1996. Evaluation of a polymerase chain reaction-based assay for diagnosis of *Wuchereria bancrofti* infection. *J Infect Dis*, **173**, 1510–14.

McQuay, R.M. 1967. Parasitologic studies in a group of furloughed missionaries. II. Helminth findings. *Am J Trop Med Hyg*, **16**, 161–6.

Melrose, W.D. 2002. Lymphatic filariasis: new insights into an old disease. *Int J Parasitol*, **32**, 947–60.

Melrose, W., Usurup, J., et al. 2000. Development of anti-filarial antibodies in a group of expatriate mine-site workers with varying exposure to the disease. *Trans R Soc Trop Med Hyg*, **94**, 706–707.

Michael, E., Bundy, D.A. and Grenfell, B.T. 1996. Re-assessing the global prevalence and distribution of lymphatic filariasis. *Parasitology*, **112**, Pt 4, 409–28.

Molyneux, D.H., Neira, M., et al. 2000. Lymphatic filariasis: setting the scene for elimination. *Trans R Soc Trop Med Hyg*, **94**, 589–91.

More, S.J. and Copeman, D.B. 1990. A highly specific and sensitive monoclonal antibody-based ELISA for the detection of circulating antigen in bancroftian filariasis. *Trop Med Parasitol*, **41**, 403–6.

Nanduri, J. and Kazura, J.W. 1989. Clinical and laboratory aspects of filariasis. *Clin Microbiol Rev*, **2**, 39–50.

Nelson, F.K., Greiner, D.L., et al. 1991. The immunodeficient scid mouse as a model for human lymphatic filariasis. *J Exp Med*, **173**, 659–63.

Nesarajah, M.S. 1972. Pulmonary function in tropical eosinophilia. *Thorax*, **27**, 185–7.

Neva, F.A. and Ottesen, E.A. 1978. Tropical (filarial) eosinophilia. *N Engl J Med*, **298**, 1129–31.

Noroes, J., Addiss, D., et al. 1996. Ultrasonographic evidence of abnormal lymphatic vessels in young men with adult *Wuchereria bancrofti* infection in the scrotal area. *J Urol*, **156**, 409–12.

Nutman, T.B. 1991. Experimental infection of humans with filariae. *Rev Infect Dis*, **13**, 1018–22.

Nutman, T.B. and Kumaraswami, V. 2001. Regulation of the immune response in lymphatic filariasis: perspectives on acute and chronic infection with *Wuchereria bancrofti* in South India. *Parasite Immunol*, **7**, 389–99.

O'Bryan, L., Pinkston, P., et al. 2003. Localized eosinophil degranulation mediates disease in tropical pulmonary eosinophilia. *Infect Immun*, **71**, 1337–42.

Ong, R.K. and Doyle, R.L. 1998. Tropical pulmonary eosinophilia. *Chest*, **113**, 1673–9.

Ottesen, E.A. 2000. The global programme to eliminate lymphatic filariasis. *Trop Med Int Health*, **5**, 591–4.

Ottesen, E.A. and Nutman, T.B. 1992. Tropical pulmonary eosinophilia. *Annu Rev Med*, **43**, 417–24.

Ottesen, E.A., Neva, F.A., et al. 1979. Specific allergic sensitsation to filarial antigens in tropical eosinophilia syndrome. *Lancet*, **1**, 1158–61.

Ottesen, E.A., Duke, B.O., et al. 1997. Strategies and tools for the control/elimination of lymphatic filariasis. *Bull World Health Organ*, **75**, 491–503.

Paciorkowski, N., Porte, P., et al. 2000. B1 B lymphocytes play a critical role in host protection against lymphatic filarial parasites. *J Exp Med*, **191**, 731–6.

Pani, S.P. and Srividya, A. 1995. Clinical manifestations of bancroftian filariasis with special reference to lymphoedema grading. *Indian J Med Res*, **102**, 114–18.

Pani, S.P., Krishnamoorthy, K., et al. 1990. Clinical manifestations in malayan filariasis infection with special reference to lymphoedema grading. *Indian J Med Res*, **91**, 200–7.

Partono, F. 1987. The spectrum of disease in lymphatic filariasis. *Ciba Found Symp*, **127**, 15–31.

Partono, F., Dennis, D.T., et al. 1977. *Brugia timori* sp. n. (nematoda: filarioidea) from Flores Island, Indonesia. *J Parasitol*, **63**, 540–6.

Pastrana, D.V., Raghavan, N., et al. 1998. Filarial nematode parasites secrete a homologue of the human cytokine macrophage migration inhibitory factor. *Infect Immun*, **66**, 5955–63.

Pearlman, E., Hazlett, F.E. Jr., et al. 1993. Induction of murine T-helper-cell responses to the filarial nematode *Brugia malayi*. *Infect Immun*, **61**, 1105–12.

Piessens, W.F., Partono, F., et al. 1982. Antigen-specific suppressor T lymphocytes in human lymphatic filariasis. *N Engl J Med*, **307**, 144–8.

Pinkston, P., Vijayan, V.K., et al. 1987. Acute tropical pulmonary eosinophilia. Characterization of the lower respiratory tract inflammation and its response to therapy. *J Clin Invest*, **80**, 216–25.

Rahmah, N., Anuar, A.K., et al. 1998a. Use of antifilarial IgG4-ELISA to detect *Brugia malayi* infection in an endemic area of Malaysia. *Trop Med Int Health*, **3**, 184–8.

Rahmah, N., Ashikin, A.N., et al. 1998b. PCR-ELISA for the detection of *Brugia malayi* infection using finger-prick blood. *Trans R Soc Trop Med Hyg*, **92**, 404–6.

Rahmah, N., Taniawati, S., et al. 2001. Specificity and sensitivity of a rapid dipstick test (Brugia Rapid) in the detection of *Brugia malayi* infection. *Trans R Soc Trop Med Hyg*, **95**, 601–4.

Ramzy, R.M., Gad, A.M., et al. 1991. Evaluation of a monoclonal-antibody based antigen assay for diagnosis of *Wuchereria bancrofti* infection in Egypt. *Am J Trop Med Hyg*, **44**, 691–5.

Semnani, R.T., Liu, A.Y., et al. 2003. *Brugia malayi* microfilariae induce cell death in human dendritic cells, inhibit their ability to make IL-12 and IL-10, and reduce their capacity to activate CD4+ T cells. *J Immunol*, **171**, 1950–60.

Simonsen, P.E., Lemnge, M.M., et al. 1996. Bancroftian filariasis: the patterns of filarial-specific immunoglobulin G1 (IgG1), IgG4, and circulating antigens in an endemic community of northeastern Tanzania. *Am J Trop Med Hyg*, **55**, 69–75.

Spencer, L.A., Porte, P., et al. 2003. Mice genetically deficient in immunoglobulin E are more permissive hosts than wild-type mice to a primary, but not secondary, infection with the filarial nematode *Brugia malayi*. *Infect Immun*, **71**, 2462–7.

Steel, C. and Nutman, T.B. 2003. CTLA-4 in filarial infections: implications for a role in diminished T cell reactivity. *J Immunol*, **170**, 1930–8.

Steel, C., Guinea, A., et al. 1994. Long-term effect of prenatal exposure to maternal microfilaraemia on immune responsiveness to filarial parasite antigens. *Lancet*, **343**, 890–3.

Steel, C., Ottesen, E.A., et al. 2001. Worm burden and host responsiveness in *Wuchereria bancrofti* infection: use of antigen detection to refine earlier assessments from the South Pacific. *Am J Trop Med Hyg*, **65**, 498–503.

Suswillo, R.R., Owen, D.G. and Denham, D.A. 1980. Infections of *Brugia pahangi* in conventional and nude (athymic) mice. *Acta Trop*, **37**, 327–35.

Taylor, M.J. and Hoerauf, A. 1999. Wolbachia bacteria of filarial nematodes. *Parasitol Today*, **15**, 11, 437–42.

Taylor, M.J., Cross, H.F. and Bilo, K. 2000. Inflammatory responses induced by the filarial nematode *Brugia malayi* are mediated by lipopolysaccharide-like activity from endosymbiotic *Wolbachia* bacteria. *J Exp Med*, **191**, 1429–36.

Taylor, M.J., Cross, H.F., et al. 2001. *Wolbachia* bacteria in filarial immunity and disease. *Parasite Immunol*, **23**, 401–9.

Vanamail, P., Subramaniam, S., et al. 1989. Estimation of age-specific rates of acquisition and loss of *Wuchereria bancrofti* infection. *Trans R Soc Trop Med Hyg*, **83**, 689–93.

Vincent, A.L., Sodeman, W.A. and Winters, A. 1980. Development of *Brugia pahangi* in normal and nude mice. *J Parasitol*, **66**, 448.

Vythilingam, I., Boaz, L. and Wa, N. 1998. Detection of *Brugia malayi* in mosquitoes by the polymerase chain reaction. *J Am Mosq Control Assoc*, **14**, 243–7.

Ware, J., Moran, L., et al. 2002. Sequencing and analysis of a 63 kb bacterial artificial chromosome insert from the *Wolbachia* endosymbiont of the human filarial parasite *Brugia malayi*. *Int J Parasitol*, **32**, 159–66.

Wartman, W. 1947. Filariasis in American armed in World War II. *Medicine*, **26**, 333–96.

Weil, G.J., Jain, D.C., et al. 1987. A monoclonal antibody-based enzyme immunoassay for detecting parasite antigenemia in bancroftian filariasis. *J Infect Dis*, **156**, 350–5.

Weil, G.J., Lammie, P.J. and Weiss, N. 1997. The ICT Filariasis Test: a rapid-format antigen test for diagnosis of bancroftian filariasis. *Parasitol Today*, **13**, 10, 401–4.

Wijers, D.J. 1977. Bancroftian filariasis in Kenya. IV. Disease distribution and transmission dynamics. *Ann Trop Med Parasitol*, **71**, 452–63.

Williams, S.A., Lizotte-Waniewski, M.R., et al. 2000. The filarial genome project: analysis of the nuclear, mitochondrial and endosymbiont genomes of *Brugia malayi*. *Int J Parasitol*, **30**, 411–19.

Witt, C. and Ottesen, E.A. 2001. Lymphatic filariasis: an infection of childhood. *Trop Med Int Health*, **6**, 582–606.

WHO, 1992. Lymphatic filariasis: the disease and its control. Fifth report of the WHO Expert Committee on Filariasis. *World Health Organ Tech Rep Ser*, **821**, 1–71.

WHO, 2003. Lymphatic filariasis. *Wkly Epidemiol Rec*, **78**, 171–9.

Wucherer, O.E.H. 1868. *Gaz Med Bahia*, **3**, 97–99.

39

Onchocerciasis

JANETTE E. BRADLEY, JAMES WHITWORTH, AND MARIA-GLORIA BASÁÑEZ

THE PARASITE

Introduction and historical perspective

Onchocerciasis is commonly called river blindness and is caused by infection with the parasitic nematode *Onchocerca volvulus*, which occurs naturally mainly in humans and occasionally in gorillas, but can be transmitted experimentally to chimpanzees. Other species of *Onchocerca* are found in ungulates, particularly bovines. Human infections with the cattle parasite *O. gutturosa* have occasionally been documented. These have presented as fibrous nodules containing infertile adult worms.

Adult worms, which have an average estimated lifespan of 10 years, are typically found in subcutaneous nodules where they are usually benign. Female worms produce millions of embryos (microfilariae) throughout their lives and a heavily infected carrier may harbor 50–200 million of them. Microfilariae live for about 1 year in the skin and must be ingested by the correct species of the insect vector (the *Simulium* blackfly) if they are to develop further. Microfilariae that are taken up by a blackfly migrate to its flight muscles, where they mature over the course of about 1 week into infective larvae. These enter another human via the bite wound made when the insect next feeds.

Those microfilariae not taken up by a vector die within skin and eye tissues and provoke an immune response which, it is thought, leads to the inflammatory lesions and subsequent fibrosis that underlie the disease of onchocerciasis. Recent studies have shown that endotoxin-like molecules released from the endosymbiotic rickettsial bacteria *Wolbachia* living in onchocercal worms and many other filarial nematodes generate inflammatory responses that may cause the adverse reactions after chemotherapy and be involved in the pathogenesis of the disease (Saint André et al. 2002). This is a chronic disease characterized by itchy and unsightly skin lesions and progressive eye damage, sometimes leading to blindness.

O. volvulus can only be transmitted by blackflies of the genus *Simulium*. *Onchocerca* species that parasitize other animals may also be transmitted by blackflies or by biting midges of the genus *Culicoides*. It is only the female blackflies that bite humans, in order to obtain the blood necessary for egg development. They use their short, broad, rasping mouthparts, which cause a painful bite but which are unable to penetrate clothing. The flies only bite during daylight hours. Because of the feeding and metabolic requirements of the larvae, blackfly eggs are laid in well-oxygenated waters. Transmission of infection, therefore, usually occurs close to blackfly breeding sites in fast flowing rivers, giving rise to the apt term 'river blindness' for this disease.

The first account of the microfilariae of *O. volvulus* was probably made by O'Neill (1875), who found them in the skin of six West African natives with papular skin rashes. He described the microfilaria as 'easily detectable ... by its violent contortions. Thread-like in form, at one time undulating, and now twisted as if into an inexplicable knot, then, having rapidly untwined itself, it curls and coils into many loops'. This accurate and evocative description is readily recognized by those who have examined microfilariae emerging from skin fragments under a microscope. O'Neill also described the pointed tail, and reported the size of the microfilariae to be about 250 × 12 μm. As this is too long for *Mansonella perstans* and too wide for *M. streptocerca*, there can be little doubt that O'Neill was describing *O. volvulus*.

Prout (1901) described worms from a subcutaneous nodule removed from a Sierra Leone frontier policeman with vague rheumatic pains. He described the adult male and female worms and microfilariae (unsheathed with a sharp tail, central granular appearance, and a size of 250 × 5 μm), which he tentatively called *Filaria volvulus* after Leuckart's 1893 description. Parsons (1908) also gave a description of nodules and commented that he suspected that the disease was more common than generally recognized and probably spread by a blood-sucking insect.

Robles (1917), working in Guatemala, demonstrated an association between nodules and the skin lesions and anterior ocular features of onchocerciasis, and suggested that blackflies could transmit the infection. However, it was Blacklock, working in Sierra Leone, who provided definite evidence of transmission of onchocerciasis by *Simulium damnosum* (Blacklock 1926a, 1926b). He found microfilariae in 45 percent of skin snips taken from subjects with nodules. No definite evidence of skin or eye disease was noted. Blacklock showed that onchocercal larvae developed first in the gut of simuliids, then the thorax, subsequently invading the head and escaping from the proboscis through the membranous labrum.

The use of skin snips for diagnosis was first recorded in 1922 (Macfie and Corson 1922) when a needle and scissors were used to remove a 0.25 cm piece of skin from the lower back. The recognition of ocular onchocerciasis in Africa came from the work of Hissette (1932) in Zaire, who described punctate keratitis, sclerosing keratitis, iritis, retrobulbar neuritis, and retinal lesions, whereas Bryant (1935) noted diffuse retinochoroiditis associated with optic atrophy in the Sudan. The definitive description of ocular onchocerciasis was published as a monograph in 1945 by Ridley, who was working in Ghana (Ridley 1945).

Classification

O. volvulus is a spirurid nematode within the superfamily Filarioidea, and family Onchocercidae. Adult female *O. volvulus* worms, at more than 35 cm, are exceptionally long for a member of the Filarioidea, which usually measure only 2–10 cm in length. Adult *O. volvulus* are normally found in characteristic fibrous nodules, distinguishing them from *Dracunculus medinensis* (superfamily Dracunculoidea), another long (about 100 cm) filarial human parasite found free in subcutaneous tissues (see Chapter 41, Dracunculiasis)

Morphology and structure

The adult worms are long and thin, tapering at both ends, although rounded at the anterior end. Both sexes exhibit sluggish movements. Microscopically, the cuticle is raised in prominent transverse ridges and annular and oblique thickenings, which are more distinct posteriorly. In the male, the cuticular annulations are more closely spaced and less conspicuous, and the cuticle contains more layers and is thinner than in females. The presence of two striae in the inner layer of the cuticle is a helpful diagnostic feature of *Onchocerca* in tissue sections. In females, the prominence of the cuticular ridges diminishes with age as the thickness of the cuticle increases. As worms become older, they become thicker and discolored, changing from transparent white to yellowish to brown. They also exhibit more patches of calcification and cytoplasmic inclusions of iron granules, vacuoles and lipid droplets as they age.

Adult male worms lengthen throughout their life and usually measure 2–4 × 0.15–0.20 cm. Morphologically, male worms are typical nematodes and considerably more mobile than females. Within nodules, males are usually found on the outside of the worm bundle, coiled around the anterior end of a female. The nerve ring is 140 μm from the anterior end. The alimentary canal is straight, ending close to the tip of the tail at the cloaca. The tail ends in a spiral and has a bulbous tip. There are two pairs of preanal and two postanal papillae, an intermediate large papilla and two unequal copulatory spicules that can protrude from the cloaca. The single reproductive tract runs centrally for almost the entire length of the worm. It consists of an anterior testis, a vas deferens, and an ejaculatory duct leading to the cloaca. Spermatogenesis is continuous and synchronized, so that only a single developmental stage is found in a particular transverse section.

Adult female worms are highly modified nematodes, notable for their extreme length, wide lateral hypodermal chords, and their reduced muscle cells and intestinal lumen. They measure 35–70 cm in length by about 400 μm in breadth. The nerve ring is 170 μm from the anterior end, and the anus is about 210 μm from the posterior end of the worm. The female reproductive tract runs from the posterior to the anterior end of the worm and consists of:

- paired ovaries, one lying more posteriorly than the other
- oviducts
- seminal receptacles and uteri
- a single short vagina and vulva, found close to the anterior extremity.

O. volvulus eggs are 30–50 μm in diameter and have a striated shell with a pointed process at each end.

The unsheathed, nonperiodic microfilariae occur mainly in the skin (90 percent) but are also found in nodules. They are occasionally found in blood specimens, in most cases probably by dislodgement from the skin. They are also found in the tissues and chambers of the eye in heavy infections, and have been reported in urine, sputum, cerebrospinal fluid, and ascitic fluid, particularly after treatment with diethylcarbamazine.

Microfilariae are $200-360 \times 5-9$ μm. They have a clear, swollen head with a cephalic space 7–13 μm long. The nuclei are well marked, with the anterior nuclei in rows and elongated terminal nuclei. The tail tapers to a sharp point with no nuclei in the caudal space, which measures 9–15 μm from the terminal nucleus to the tip of the tail. These features distinguish microfilariae of *O. volvulus* from those of the following unsheathed microfilariae that are possible sources of misdiagnosis of onchocerciasis in endemic areas:

- *M. perstans* (<200 μm long and no caudal space)
- *M. streptocerca* (180–240 μm long, a curved tail and no caudal space)
- *M. ozzardi* (200–230 μm long and a fine attenuated hooked tail).

General biology

The nodules (onchocercomata) containing the adult worms are usually less than 2 cm in diameter, firm, mobile, and well defined, and neither tender nor painful. Each nodule typically contains one or two male worms and one or two female worms coiled in a mass, lying within a rim composed of vascularized and hyalinized scar tissue derived from the host. Invading infective larvae seem to be attracted to existing nodules (Duerr et al. 2001) and nodules may occasionally contain ten or more adult worms. Nodules are also found that contain liquefied or calcified necrotic worms. Nodules are well vascularized with fine vessels in close contact with worm coils. Interspersed among the worm bundle is material indicative of chronic inflammation: fibrin, plasma cells, neutrophilic and eosinophilic granulocytes, lymphocytes, and giant cells. Nodules lie subcutaneously in association with bony prominences, sometimes attached to the skin or, occasionally, more deeply. The majority of nodules in African patients occur on the pelvic girdle (Albiez et al. 1988). In young children in Africa, relatively more nodules are found on the upper part of the body, especially the head. In Guatemala and Mexico, nodules are more frequent on the head in all age groups, (Figure 39.1) while in South America nodules are mostly found on the lower half of the body with the exception of the Amazonian focus. It is recognized that not all nodules are palpable and that not all adult worms are found within such nodules (Albiez 1983).

From the evidence that nodules normally contain more female worms than males (ratio 1:1.1 or 1.2) and that some nodules contain fertilized females, but no male worms, it is thought that the males are migratory,

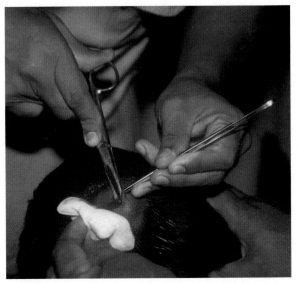

Figure 39.1 *Nodules being excised from the head of a young child in Guatemala (Courtesy of J. Bradley)*

moving from nodule to nodule (Schulz-Key and Albiez 1977), whereas the female worms are generally sessile. Females have a cyclical reproductive pattern, requiring fertilization by males for each successive brood of embryos (Schulz-Key and Karam 1986). Primary oocytes mature as they pass down the length of the ovary and are released as individual cells into the oviduct and seminal receptacles where they are fertilized and begin dividing. Multicellular stages develop along the length of the uteri until, finally, they escape from the vulva as elongated microfilariae. Each reproductive cycle lasts about 2–4 months, development of an oocyte into a microfilaria taking about 3 weeks.

It has been suggested that the majority of microfilariae normally reside in the lymph vessels of dermal papillae, at which site they usually cause no host reaction, and that it is only extralymphatic microfilariae, particularly those that are dead or dying, which elicit the host immune response thought to be responsible for the pathological features of onchocerciasis (Vuong et al. 1988). Adult worms do not directly contribute to the pathological features although cephalic location of nodules is a risk factor for ocular disease (Anderson et al. 1975).

Life-cycle

The life-cycle of *O. volvulus* is shown in Figure 39.2. Infective third-stage (L3) larvae are introduced into the human host by biting female *Simulium damnosum* or other blackfly vectors. There is no significant zoonotic cycle. Larvae undergo two molts before developing into adult worms. The first molt occurs close to the point of entry after 3–10 days and the second molt probably occurs about 1–2 months after the infective bite

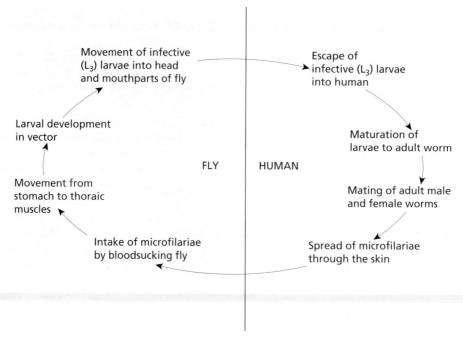

Figure 39.2 *Life cycle of* O. volvulus

(Lok et al. 1984; Strote 1987; Bianco et al. 1989; Duke 1991). The prepatent period from infection to production of detectable microfilariae is usually 12–15 months, with a range of 7–34 months (Prost 1980). Adult worms are long-lived: the average life span of female worms is estimated to be about 10 years, but can be as long as 15 years (Plaisier et al. 1991; Roberts et al. 1967).

Microfilariae are produced by the female worms in large numbers (500–1 500 per day) and migrate to the skin and eyes of infected subjects. They can survive in the body for 1–2 years (Eberhard 1986). When microfilariae are taken up with the bloodmeal by biting simuliids, some migrate from the gut of the vector into the thoracic muscles. There, they molt twice and develop into infective larvae over a period of 6–8 days, increasing in length to 440–700 μm. These larvae then migrate to the head of the blackfly where they may be transmitted to humans at the next bloodmeal by emerging through the membranous labrum of the mouthparts and penetrating into the wound.

TRANSMISSION, EPIDEMIOLOGY, AND CONTROL

Vectors of *O. volvulus*

Human onchocerciasis is transmitted by biting flies of the genus *Simulium*, commonly known as blackflies. Adult blackflies are small (under 4 mm long) Diptera found in all parts of the world, with the exception of a few islands. They are squat, heavy-bodied flies with a pronounced humped thorax. They are usually black and

may have black, white or silvery hairs on the body. The wings are short and broad, with well developed anterior veins but otherwise membranous. In males, the compound eyes occupy almost all of the head and meet anteriorly and dorsally (holoptic), whereas in females, the eyes are separated on top (dichoptic). Only females feed on blood (hematophagic) and can transmit the parasite (Crosskey 1990).

The species of blackflies acting as the main vectors of onchocerciasis in Africa and Arabia belong to the *Simulium damnosum sensu lato* (*s.l.*) species complex. A species complex comprises closely related species that are nearly identical morphologically (sibling species), but that can be distinguished by chromosomal (cytogenetic) techniques (Rothfels 1981). The diversity of simuliid species complexes acting as vectors of human onchocerciasis in the Americas is far greater than in Africa. In the Americas, the main vectors in each of the geographical areas are:

- *S. ochraceum s.l.* in southern Mexico and Guatemala (Dalmat 1955; Ortega and Oliver 1985)
- *S. metallicum s.l.* in northern Venezuela (Duke 1970; Grillet et al. 1994)
- *S. exiguum s.l.* in Colombia and Ecuador (Procunier et al. 1986; Shelley et al. 1990)
- *S. guianense s.l.* in the Amazonian focus between southern Venezuela and northwestern Brazil (Takaoka et al. 1984; Basáñez et al. 1988).

Blackflies lay their eggs in batches on trailing vegetation in fast-flowing water. Each batch may contain 100–900 eggs, which are laid with a secretion of mucus that is immediately wetted, thus cementing the eggs to the substrate. Eggs hatch after 1–2 days; the emerging larvae remain attached to the vegetation and filter the water for food. Larvae develop into pupae after 5–10 days, and adult flies emerge to surface in a gas bubble 2 days later. Swarming and mating usually occur soon after emergence. Female blackflies need mate only once during their life.

All blackflies feed on plant juices and sugar solutions. Only the female feeds on blood and this process has to be repeated for each ovarian cycle, so the female fly lays a batch of eggs as a result of each bloodmeal. Host-seeking and blackfly feeding occurs in daylight, and in *S. damnosum* biting activity peaks soon after sunrise, followed by a lesser peak in the late afternoon (Garms 1973), a pattern not uncommon among other species. Host location is probably related to odor, CO_2 output, movement, color or outline (Wenk 1981). In West Africa, most blackfly bites on humans are on the ankles and calves, with few bites more than 50 cm above the ground (Renz and Wenk 1983). In Central America, most bites by *S. ochraceum* are on the head and shoulders (Dalmat 1955), and in South America, *S. guianense* bites are most frequently on the lower half of the body (Basáñez et al. 1988). If these areas of the body are not exposed, the blackfly may seek an alternative host. The relative efficiency of a species of blackfly as a vector for *O. volvulus* thus depends partly on the degree to which a blackfly will seek a bloodmeal from a human, rather than from an animal. *S. damnosum* normally takes 5 min to feed to full engorgement on humans (Crosskey 1962). The fly's rasp-like mouthparts tear a ragged hole in the host's skin, to a depth of about 400 μm, blood from ruptured capillaries is then sucked up. The egg-laying cycle of the fly takes usually about 3–6 days (gonotrophic cycle), but development of *O. volvulus* to the infective (L3) stage takes 6–8 days under tropical conditions. Thus, a female blackfly infected with microfilariae at her first bloodmeal cannot transmit infective larvae until her third meal. The longevity of blackflies in the wild was originally estimated at about 15 days (Duke 1968a), with a theoretical maximum of four infective bites, but subsequent reports estimate the maximum life span at 4 weeks (WHO 1987) and the average life expectancy at about a week (Basáñez et al. 1996). Studies of the movement of waves of migrating flies suggest that survival of individual blackflies is possible for up to 7–10 weeks (Baker et al. 1990). Dalmat (1955) reported a maximum longevity of up to 85 days for *S. metallicum* in Guatemala.

There are normally major differences in the sizes of blackfly populations in the dry and wet seasons related to the height of rivers and availability of breeding sites (Renz 1987; Grillet et al. 2001). Dispersal of blackflies away from their riverine breeding sites does occur, especially in forested areas but, in general, the risk of infection is highest close to rivers (Renz and Wenk 1987). Migration, as distinct from dispersal, of blackflies is probably a mainly windborne phenomenon. In West Africa, this may occur over long distances, up to 500 km (Baker et al. 1990), and may be crucial for the survival of *S. damnosum* populations during the dry season. Under field conditions, only a small percentage of blackfly populations is infected. Near to breeding sites, the figure averages 3–5 percent of the total population and 6–10 percent of the parous population.

The blackflies belonging to each species complex differ in their ability to transmit the parasite – such differences in transmission ability are referred to as differences in competence. A good illustration is found in the very different vectorial roles played by members of *S. exiguum s.l.* along parts of its distributional range in South America. Members of this complex include the highly anthropophagic and efficient *O. volvulus* vectors in the Esmeraldas province of Ecuador (Santiago focus); blackflies that yield lower numbers of L3 larvae in the San Antonio (River Micay) focus of Colombia; blackflies that are mainly zoophagic in the Amazonian focus (Shelley 1988a, b; Shelley et al. 1990); and even those that are refractory to parasite development in the Colombian–Venezuelan (Orinoco river) border (Basáñez et al. 2000).

Epidemiology

Onchocerciasis is generally a cumulative infection, the severity of clinical features depending on the length of exposure to blackfly bites and the density of microfilariae in the skin. Because of this, the disease tends to affect mainly the rural poor in sub-Saharan Africa, particularly West Africa. It is also endemic in Yemen and in parts of Central and South America. Onchocerciasis is an unpleasant disease, lasting many years and causing ever-increasing disability to those affected. It has serious socioeconomic consequences for the most heavily affected communities in endemic areas. In hyperendemic communities, more than 60 percent of the population are infected with microfilariae, and about half of these will have symptoms. About 15 percent will have serious skin or eye disease, and up to about 5 percent will be blind. Onchocerciasis is, therefore, one of the major causes of infectious blindness in the world, second only to trachoma (Figure 39.3a and b).

The most important determinants of the burden of infection in a community are the infective density of the vector and, more crucially, the number of L3 larvae inoculated per person per year by the biting blackfly vectors. The former is known as the infective biting rate and the latter as the annual transmission potential (ATP). The biting density of blackflies can be measured by regular dawn-to-dusk catches using human collectors at selected sites. This allows the calculation of the monthly biting rate (MBR), a theoretical estimate of the total number of bites an individual could receive if maximally exposed to blackfly bites at that site during

Figure 39.3 (a) *Anterior eye damage caused by onchocerciasis;* **(b)** *a blind man being led to work in West Africa (Courtesy of J. Bradley)*

a month (Duke 1968b; Walsh et al. 1978).

$$MBR = \frac{number\ of\ blackflies\ caught \times number\ of\ days\ in\ month}{number\ of\ catching\ days\ in\ month}$$

The annual biting rate (ABR) is the sum of 12 consecutive MBRs.

The number of third-stage infective larvae can be estimated by fly dissection. This permits the calculation of the monthly transmission potential (MTP), a theoretical estimate of the number of infective larvae that could be transmitted to a maximally exposed individual during a month (Duke 1968b; Walsh et al. 1978).

$$MTP = \frac{MBR \times number\ of\ O.\ volvulus\ L3\ larvae\ detected}{number\ of\ flies\ dissected}$$

The ATP is the sum of 12 consecutive MTPs.

More recently, the proportion of infective flies has been ascertained in pools of flies, rather than by dissecting each individual fly, by a combination of *O. volvulus*-specific DNA probes and the polymerase chain reaction (PCR) assay (Yaméogo et al. 1999; Rodríguez-Pérez et al. 1999).

In practice, these indices give only a rough guide to the level of transmission because:

- not all third stage larvae are truly infective
- only about 50–80 percent of L3 larvae are transmitted during a bloodmeal
- no single individual is likely to be maximally exposed to bites
- the infective larvae of other *Onchocerca* species present in manually dissected blackflies may be hard to distinguish from *O. volvulus*
- there is considerable heterogeneity in individual exposure to vector bites (Renz et al. 1987).

Vector biology is also an important determinant of transmission. Vectorial capacity depends on the density of vectors in relation to that of hosts (vector to human ratio), the biting frequency, and the proportion of bites on humans (Dietz 1982; Basáñez and Boussinesq 1999; Basáñez et al. 2002). Vector species is also of importance. For example, the critical biting rates necessary for the infection to become endemic are roughly ten times higher for *S. ochraceum s.l.* than for *S. damnosum s.l.* In Latin America, altitudinal variation in vector composition and abundance is clearly related to the prevalence and endemicity of onchocerciasis, with lower competence vectors (*S. oyapockense s.l.*) being associated with lower altitudes and with areas of lower endemicity, while highly competent vectors (*S. guianense s.l.*) are mostly responsible for hyperendemic transmission in the highlands (Vivas-Martínez et al. 1998; Grillet et al. 2001). Therefore, distance from breeding sites and blackfly species composition and abundance (Vivas-Martínez et al. 1998) have been used in Africa and Latin America, respectively, as entomological indicators for rapid epidemiological mapping of onchocerciasis (REMO) to set priorities for mass treatment with ivermectin (see Control, below).

Nevertheless, when averaged over long time-periods, it has been shown that the prevalence and intensity of infection and clinical features in humans are related to the numbers of biting flies and measured transmission potentials in both West Africa (Duke et al, 1972, 1975), and Latin America (Basáñez et al. 2002).

It is thought that microfilariae show some geographical strain differences throughout their range. The pattern of clinical features of the disease shows geographical variation, which may be related to differences in parasite strain and vector parasite relationships. In West Africa, there are three major strains of *O. volvulus*: a forest strain with low ocular pathogenicity associated with high nodule numbers and severe skin disease; a dry savannah strain with high ocular pathogenesis and an associated high rate of blindness; and a humid savannah strain with an intermediate pattern (Anderson et al. 1974b; Dadzie et al. 1989). Microfilariae are generally morphologically indistinguishable throughout their range, although significant differences have been reported between savannah and forest microfilariae from West Africa, and between these and microfilariae from South America. In West Africa, at least four different patterns of acid phosphatase staining are found, supporting the hypothesis that a number of biological strains exist (Omar 1978). Differences in the relative frequency of such patterns have also been found between microfilarial populations from West Africa, South America, and the Yemen (Yarzábal et al. 1983). It has been reported that strain-specific DNA probes can distinguish *O. volvulus* parasites originating from forest and savannah areas of West Africa (Erttmann et al. 1987; Zimmerman et al. 1992). A possible mechanism to explain the evolution of these strains is that local populations of microfilariae became adapted to the specific (homologous) sibling species of vector in their area. This could then lead to incompatibility between parasite strains and vectors from different areas. In fact, various degrees of incompatibility have been shown between different strains of *O. volvulus* and heterologous sibling species of *S. damnosum s.l.* from different areas (Duke et al. 1966; WHO 1981). Strong local vector adaptation has also been demonstrated in the Americas (Basáñez et al. 2000). However, recent field studies using the West African forest-specific and savannah-specific DNA probes on vector species from different areas showed that savannah-dwelling vectors carried forest-strain parasites, and vice versa (Toé et al. 1997).

Control

The control of onchocerciasis can be approached either by treating infected patients or by reducing transmission of infection and thereby preventing new cases (Table 39.1). Currently, control is *either* by reducing morbidity in the human through decreasing transmission from human to vector with the drug ivermectin and, to a lesser extent, nodulectomy, *or* by interrupting transmission from vector to human by killing the larval blackflies. Other possibilities for control remain at the development stage; they include the production of a vaccine and chemotherapy by treating the obligate endosymbiotic bacteria *Wolbachia*. An integral part of control has to be effective diagnosis both in the human and the blackfly.

Table 39.1 *Methods of control of onchocerciasis*

Control method	Means of achievement
Reduce human/vector contact	Environmental means Protective clothing Insect repellents
Vector control	Regular larviciding at blackfly breeding sites
Vaccine	A possibility for the future
Nodulectomy	Especially in central America (possibly for head nodules in African children)
Drugs	
(a) Microfilaricides	Ivermectin Diethylcarbamazine (obsolete) Doxycycline (anti-embryogenic)
(b) Macrofilaricides	Suramin (obsolete)
(c) Anti-*Wolbachia* therapy	A future possibility

There are three control programs that have been set up to organize regional control strategies for onchocerciasis: The Onchocerciais Control Programme (OCP) in West Africa, The African Programme for Onchocerciasis Control (APOC) and the Onchocerciasis Elimination Program for the Americas (OEPA). All have employed slightly different control strategies, reviewed in Richards et al. (2001) and Hoerauf et al. (2003), based on the timing of available tools at the time of instigation and local logistical considerations. The OCP primarily concentrated on vector control, APOC on chemotherapy, and OEPA employed nodulectomy in addition to chemotherapy.

The OCP

The first report of successful vector control was made by Buckley (1951) who achieved control of *S. naevei* in a small focus in Kenya. Garnham and McMahon (1947) successfully eliminated *S. naevei*, and eventually eliminated onchocerciasis from four isolated foci in Kenya, by using large doses of DDT at a few selected sites at 10–14 daily intervals. The insecticide was carried downstream, killing simuliid larvae over long stretches of river.

Several attempts to control *S. damnosum* were made in West Africa during the 1950s and 1960s (Taufflieb 1955; Davies et al. 1962; Le Berre 1968). The promising results obtained led to the inception of the OCP in 1974 by the World Health Organization and the World Bank. The original area included Benin, Burkina Faso, Côte D'Ivoire, Ghana, Mali, Niger, and Togo. This area was expanded to include Guinea, Guinea Bissau, Senegal, and Sierra Leone by 1986. The two objectives were to combat a disease that was widespread and severe in West Africa, and to remove a major obstacle to economic development (Walsh et al. 1981). The OCP employed vector control as its primary strategy. This was achieved by helicopter distribution of the larvicides

Figure 39.4 *Ground application of larvicide to a river system (Courtesy of J. Bradley)*

at weekly intervals with supplementary ground applications wherever possible (Figure 39.4), The success of the programs was monitored by teams visiting the treated breeding sites weekly and recording the presence of nulliparous females, pupae, and late instar larvae. This strategy proved highly successful with *S. damnosum* virtually eliminated from large areas, and transmission generally being reduced to very low levels, almost always to less than 10 percent of pre-control figures. However, the program has faced problems. First, the development of widespread insecticide resistance has necessitated the introduction of more expensive and toxic insecticides. Five insecticidal compounds were used in rotation to combat resistance, including the biocide *Bacillus thuringensis*, which has no side-effects on other invertebrates and fish. In addition, the OCP benefited from the use of chemotherapy, as ivermectin was used in conjunction for vector control (see Chemotherapy, below).

The intention of the OCP was to devolve responsibility for maintaining onchocerciasis control to the participating countries. The program was officially closed in December 2002 by which time, it was estimated, between 100 000 and 200 000 people have been prevented from going blind and 30 million people have been protected (Molyneux 1995). The strategy for using vector control in this area has been proved valid (Hougard et al. 2001) and the aim of 'eliminating onchocerciasis as a public health problem in West Africa' (WHO 2002) has been achieved.

APOC

Although vector control proved extremely effective in controlling onchocerciasis in much of West Africa, large parts of Africa remained where vector control was neither cost-effective nor technically feasible. The provision of ivermectin at no cost from the manufacturers provoked the institution of the APOC, which has as its objective 'to establish, within a period of 12 years, effective and self-sustainable community-based ivermectin treatment throughout the remaining endemic areas in Africa' (Remme 1995). It did aim for total elimination of transmission but its objective was to control serious onchocerciasis and reduce transmission. It was, however, suggested that increasing ivermectin distribution from annually to six monthly might increase the possibility of long-term elimination of the parasite (Richards et al. 2000). Simulations using the onchocerciasis simulation model (ONCHOSIM) indicated that if the dosing regime were to be increased from annually to six monthly in areas with medium to high levels of infection, the time required to eliminate infections would be halved from around 25–30 years; however, the simulations also concluded that it was unrealistic to expect the high level of coverage required to be sustained (Winnen et al. 2002). Additional funds for the accelerated elimination of river blindness in Africa were approved at the G8 summit meeting in Kanaskis in 2002.

OEPA

There are six endemic countries in the Americas: Brazil, Colombia, Ecuador, Guatemala, Mexico, and Venezuela. The characteristics of the foci are very different: although ATPs in some of these areas can be as high as those in West African savanna (Basáñez et al., 2002), they lead to less blindness, and transmission appears to occur only in restricted areas that are not rapidly spreading. These factors make elimination of onchocerciasis a viable aim in the Americas. The major strategy employed for the control of onchocerciasis in the Americas is the use of ivermectin. The strategy is for the total elimination of the disease spurred on by the success achieved in isolated foci in Guatemala and Mexico (Cupp 1992; Cupp et al. 1992; Collins et al. 1992; Guderian et al. 1997) when coverage levels were very high. The strategy was originally based on systematic nodulectomy campaigns stimulated by the work of Robles in Guatemala and Mexico. In Guatemala, nodulectomy campaigns have been carried out since 1935 with over 250 000 nodules being removed. During this time, the nodule carrier rate has fallen from 24 to 9 percent. It is not clear if this reduction has been wholly due to the campaign or if other factors have contributed. Nevertheless, in one area with hyperendemic onchocerciasis, the prevalence of blindness fell from 7 percent in 1934 to 0.5 percent in 1979 (WHO 1987).

Nodulectomy has been less popular in Africa with no systematic campaigns. However, since nodules on the head are associated with ocular complications (Anderson et al. 1975; Fuglsang and Anderson 1977), it may be reasonable to remove head nodules, particularly from children.

CHEMOTHERAPY

The drug of choice for the treatment of onchocerciasis is ivermectin. Mass treatment campaigns using this drug are also the cornerstone of the current control programs. Prior to the discovery of ivermectin, diethylcarbazine (DEC) was the only realistic drug available. The filaricidal action of DEC was first reported by Hewitt et al. (1947) and it remains the drug of choice to treat lymphatic filariasis. There was much hope that the drug would be useful for onchocerciasis, but, while DEC is an effective microfilaricidal drug, it does not kill adult O. volvulus and produces serious side-effects (Mazzotti 1948). In the past, it was often used in combination with corticosteroids, aspirin or antihistamines to control the side-effects. Treatment with DEC has now been superseded by ivermectin.

Ivermectin is a semi-synthetic product, being an 80:20 mixture of avermectins B_{1a} and B_{1b}, which are macrocyclic lactones synthesized by the actinomycete Streptomyces avermectilis. It is formulated as 6 mg tablets and given orally. The recommended dose is 150 µg/kg bodyweight taken annually. The manufacturer's exclusion criteria are as follows: children under 5 years of age or under 15 kg bodyweight; pregnant women; breastfeeding mothers within 1 week of delivery; and individuals with neurological disorders or severe intercurrent disease. The drug reaches a peak plasma concentration about 4 hours after administration, is highly bound to albumin, and is reported to have a wide tissue distribution. The precise mechanism of action is unknown, but the drug can act as an agonist of the neurotransmitter γ-aminobutyric acid (GABA), resulting in paralysis of microfilariae (Goa et al. 1991).

Ivermectin was developed initially as a wide-spectrum veterinary anthelminthic agent. It was first studied in humans in Senegal by Aziz et al. (1982) and subsequent studies have found that ivermectin was as effective as DEC in reducing microfilarial counts, but had significantly fewer side-effects (Greene et al. 1985b; Larivière et al. 1985; Diallo et al. 1986). The suppression of microfilarial levels lasts longer after ivermectin than DEC although it is not complete; 20 percent of pretreatment burdens will have reappeared a year after treatment (Awadzi et al. 1986, 1995; Taylor et al. 1986). Ivermectin does not kill adult worms but inhibits microfilarial release from female worms, with a consequent intrauterine accumulation of degenerate microfilariae (Schulz-Key et al. 1985, 1986). Since ivermectin is only microfilaricidal, treatment is suppressive rather than curative and it may be necessary to take the drug for at least the life span of adult worms, i.e. up to 15 years (Whitworth 1992).

Community trials have shown that ivermectin is safe and well accepted and has important effects on the clinical morbidity of the disease, particularly on the ocular lesions (Whitworth et al. 1991b; Abiose et al. 1993). Ivermectin also has some useful activity against other nematode parasites including Wuchereria bancrofti, Ascaris lumbricoides, Strongyloides stercoralis (see Chapters 34, Gastrointestinal nematodes; 35, Strongyloides and Capillaria; and 38, Lymphatic Filariasis), and ectoparasites including head lice (Dunne et al. 1991; Whitworth et al. 1991a). Care is required where Loa loa and O. volvulus coexist because ivermectin has been shown to cause serious side-effects including encephalopathy in heavy infections of loiasis (Gardon et al. 1997; Boussinesq et al. 1998). The risk of adverse reactions to ivermectin is related to the severity of infection. These have been reported in approximately 20 percent of individuals in hyperendemic areas, but the vast majority are mild and self-limiting. About 5 percent of patients have more severe reactions that affect the ability to work for a day or more. Commonly seen adverse reactions include increased skin rash and itching, nontender soft tissue swelling, musculoskeletal pains, fever, lymphgland pain, and swelling. More serious reactions, which include severe postural hypotension, bronchospasm in asthmatics, abscesses, and bullous eruptions, are rare. All of these respond to symptomatic treatment. Adverse reactions are much less common after subsequent doses of treatment, presumably because the microfilarial load has been reduced (Whitworth et al. 1991c).

Ivermectin is provided free by the manufacturers, and the cost of population-based distribution in endemic countries ranges from US$0.29 per person to more than US$2 per person, depending on the precise system that is used. There are now about 20 million people being treated under community-directed treatment schemes, which are likely to be necessary for at least 10 years (www.who.int/tds/about/products/comdt.htm). In nonendemic areas, where infections are usually light and patients are not exposed to further infection, ivermectin may be given at 3–6 monthly intervals as necessary, depending on the reappearance of skin microfilariae or recurrence of symptoms. About two-thirds of patients can be expected to relapse within 6 months of each dose of treatment (Churchill et al. 1994).

For morbidity control, ivermectin is a good drug but it has limitations in its use as a tool for control mostly due to the fact it has little effect on adult worms. The reappearance of microfilariae after treatment appears to be at a sufficient density for transmission to continue (Alley et al. 2001). Studies using albendazole in addition to ivermectin were performed in the hope that it might have additive effects in reducing embryogenesis but these proved to be transient and insufficient to stop transmission (Awadzi et al. 1995). Recent studies using higher doses of ivermectin at more frequent periods has shown some effects on the embryogenesis of microfilariae (Gardon et al. 2002), but the need for an effective macrofilaricidal drug or one that has a sustained effect on the production of microfilariae remains an urgent requirement.

Suramin was the first successful drug for the treatment of onchocerciasis and, although no longer used, it remains the only true macrofilaricide. A dose of 6 g is sufficient to kill all adult worms, but significant side-effects were reported in 10–30 percent of patients. This drug had to be given in a course of intravenous injections over several weeks and so treatment was restricted to hospital in-patients. It is clearly too toxic for general use, with treatment being occasionally fatal.

Recently, it has been realized that *Wolbachia*, the endosymbiotic bacteria of most filarial parasites (see Pathology and pathogenesis of clinical features, below and Figure 39.5a and b), are a good target for chemotherapy. They appear to be essential for worm fertility and reproduction as depletion by chemotherapy disrupts embryogenesis (Taylor and Hoerauf 2001; Bandi et al. 2001) and may even cause worm death (Langworthy et al. 2000). These rickettsial bacteria are sensitive to a range of antibiotics such as tetracyclines, rifampicin, and chloramphenicol (Taylor and Hoerauf 2001; Bandi et al. 2001). Phase II drug trials have been carried out using doxycycline at 100 mg/day for 6 weeks with excellent results. This regime resulted in depletion of the *Wolbachia* and interruption of embryogenesis for the 18-month period of the trial (Hoerauf et al. 2000, 2001). Preliminary data on other treatment regimes have suggested that the treatment period required to deplete the endosymbionts is between 2 and 4 weeks (Hoerauf et al. 2003). The requirement for a long treatment regime is a disadvantage for control strategies because of the logistics of ensuring compliance, but it is a viable treatment for anyone with an imported infection.

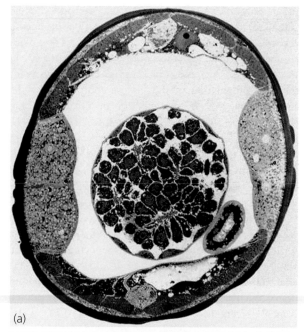

(a)

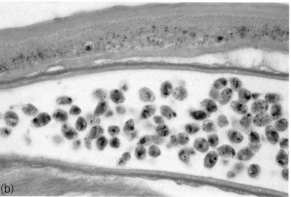

(b)

Figure 39.5 (a) Wolbachia *(yellow) in adult male* Onchocerca ochengi *(Courtesy of M. Taylor);* **(b)** Wolbachia *(red) in* O. volvulus *(Courtesy of D.W. Büttner, A. Hoerauf, M. Taylor)*

The search continues for an effective and safe macrofilaricide to complement the effects of ivermectin and larviciding on onchocerciasis control. Suramin, the only currently available effective macrofilaricide has serious side-effects and is too toxic for general use. Amocarzine shows some success, but its macrofilaricidal properties are not ideal. Moxidectin, which is widely used as a wide-spectrum anthelminthic in veterinary practice, shows promise as a single dose and either kills or sterilizes adult worms in animal models but it has not been evaluated in humans (Alley et al. 2001).

DIAGNOSIS

Parasitological

The gold standard diagnostic test in the human is the demonstration of microfilariae in skin snips. Bloodless

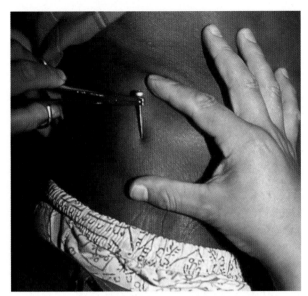

Figure 39.6 *A skin snip being taken from the iliac crest using a corneoscleral punch (Courtesy of J. Bradley)*

skin snips (typically, two to six) 2–3 mm in diameter and 0.5–1 mm deep are taken from both iliac crests, and sometimes the calves in African, Yemeni, and South American cases and the shoulders in Mesoamerican cases. The skin can be raised with a needle and the tip sliced off with a razor blade, or a corneoscleral punch (an obsolete ophthalmologic surgical instrument) can be used (Figure 39.6). These punches enable snips to be taken easily and rapidly, but they are expensive, need regular setting and sharpening, and should be sterilized in cold glutaraldehyde between patients. Microfilariae will emerge from the piece of skin after incubation in normal saline. About 60 percent of microfilariae will emerge after 30 min, rising to over 75 percent after 24 hours. Live microfilariae can be seen moving vigorously in the medium by direct microscopy. They can be distinguished from other species of microfilariae after staining with Giemsa or Mayer's hemalum.

In detailed community-based surveys in endemic areas, the numbers of microfilariae are counted and the skin snip weighed. The community microfilarial load (CMFL) can then be derived as the geometric mean microfilarial count (per skin snip or per mg of skin) for a cohort of adults aged 20 years or older. This is an accurate measure of the endemicity of onchocerciasis in a community (Remme et al. 1986).

For monitoring control programs, the main requirement is to monitor for recrudescence of infection. The requirements of a diagnostic assay for such purposes are totally different from those of an assay used on an individual to determine treatment. The former assay is necessarily specific, but need not be 100 percent sensitive as results will be assessed at a community level. Parasitological diagnosis is not ideal for this purpose, particularly where the microfilaricidal drug ivermectin is employed.

Serological diagnosis

Serological diagnosis can be by the detection of circulating antigen or by the detection of specific antibodies to the parasite. Tests based on the detection of circulating antigen are more desirable as they define the current infection status of an individual. However, in contrast to the success in developing the rapid format card test for the detection of antigen for lymphatic filariasis (Weil et al. 1997), it has not been possible to develop a reliable test for the detection of antigen for onchocerciasis. This could be because the anatomical locations of the microfilariae and adults in onchocerciasis (in the skin and subcutaneous tissues, respectively) do not allow reliable release of antigen into the circulation. Tests based on the detection of antibody have the problem that it is difficult to distinguish between current and historical infection; moreover, early tests using parasite preparations suffered from poor specificity due to the extensive cross-reacting antigens between nematodes (Karam and Weiss 1985). However, recombinant antigen-based tests have been developed that are able to specifically detect exposure to *O. volvulus* (Lucius et al. 1992; Lobos et al. 1991; Bradley et al. 1998). The use of individual recombinant antigens was shown to have poor sensitivity, but this could be overcome by using a cocktail of antigens (Ramachandran 1993; Bradley et al. 1998). The problem of detecting current infection has at least been partially overcome by using anti-immunoglobuin (Ig)G4 reagents to detect the reactivity in the serum because this antibody isotype has been shown to be related to active infection (Weil et al. 1990; Lucius et al. 1992). As mentioned above, to monitor control programs it is important to have a rapid test where specificity is critical but sensitivity is less so. For this purpose, a dipstick is ideal, and such a test based on the use of one of the specific diagnostic antigens used in the cocktail described above that detects IgG4 antibodies in the serum of infected individuals has been developed (Weil et al. 2000)

DNA-based detection

DNA-based detection methods have been developed that are able to distinguish between *Onchocerca* species, which is of particular importance for monitoring transmission in the blackfly vector, and strains of *O. volvulus*, which will define if an individual is infected with blinding or nonblinding parasites (reviewed in Harnett 2002).

DIAGNOSIS IN FLIES

There have been a number of DNA probes developed that can distinguish *O. volvulus* from the other species of *Onchocerca* that infect the blackfly vector (Perler and Karam 1986; Shah et al. 1987; Harnett et al. 1989;

Meredith et al. 1989). These probes are all members of a 150 bp repeat that makes up about 1 percent of the genome (Meredith et al. 1989). Early attempts to detect *O. volvulus* larvae in the blackflies was by DNA hybridization in slot blots but this proved to be a technically demanding procedure and PCR-based assays were developed in which the 150 bp repeat was amplified and then probed with appropriate labeled probes (Meredith et al. 1991). The technique has subsequently been developed to have the sensitivity to detect a single larva in a pool of around 100 blackflies (Katholi et al. 1995; Oskam et al. 1996; Rodríguez-Pérez et al. 1999). It is possible, by the use of DNA probes, to detect the strain of parasite infecting the blackfly (Toé et al. 1997; Fischer et al. 1997).

DIAGNOSIS IN HUMANS

PCR-based diagnostic tests on skin snips (Zimmerman et al. 1994) and, more recently, superficial skin scratches (Toé et al. 1998) have provided an alternative to the skin-snipping-and-microscopy technique and have also proved to be more sensitive (Elson et al. 1994; Freedman et al. 1994). The use of DNA probes can distinguish between forest and savannah (or blinding/nonblinding) isolates of the parasite strains (Erttmann et al. 1987; Zimmerman et al. 1992) and the use of such probes has shown good correlation with the epidemiology and pathogenicity of a strain in a given area (Erttmann et al. 1987; Meredith et al. 1989; Ogunrinade et al. 1999; Zimmerman et al. 1992).

Other tests

None of the available tests is totally ideal for the purposes of monitoring the recrudescence of infection in the post-OCP era because they are either insensitive or technically demanding. The use of a noninvasive patch skin test using DEC has shown promising results (Stingle et al. 1984). This provokes a local reaction due to the death of the microfilariae in the skin and has shown to have better sensitivity than conventional skin snipping and the PCR skin-scratch test (Boatin et al. 2002).

Rapid mapping techniques

A different challenge was posed to APOC in deciding where the mass distribution of ivermectin was required. The approach taken was to use the rapid field assessment process (REMO) to define areas of high risk (Ngoumou et al. 1993). The information provided from REMO was integrated into a geographical information system (GIS) which has allowed delineation of the areas of endemicity and definition of their thresholds for community treatment programs (Noma et al. 2002)

CLINICAL, PATHOLOGICAL, AND IMMUNOLOGICAL ASPECTS OF INFECTION

Introduction

Onchocerciasis causes ocular disease, including blindness in long-standing severe cases, and dermatological pathology. The disease is found in 36 countries of West, East and Central Africa, the Arabian Peninsula and parts of South and Central America (Figure 39.7). It is predominantly a disease of Africa; of 120 million people at risk of infection, 96 percent are in Africa; of the 18 million infected, 99 percent are Africans and of these 270 000 people are blind (WHO 1995).

The communities affected often suffer from socio-economic depression, with the effects of the disease further exacerbating poverty. Visual damage is also significantly associated with blindness, poor nutritional status, and increased risk of premature death (Prost and Vaugelade 1981; Kirkwood et al. 1983b; Pion et al. 2002). Studies from villages in West African savanna

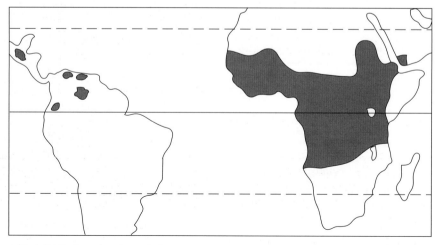

Figure 39.7 *Geographical distribution of onchocerciasis*

regions prior to the instigation of the OCP suggested that blindness from onchocerciasis reduced life expectancy by at least 13 years, and that if the prevalence of blindness in a community is 5 percent, almost half of adult males and one-third of females would become blind before they died (WHO 1987). It has recently been shown that onchocerciasis infection per se is associated with excess human mortality, i.e. not only through its effect on blindness (Little et al. 2004).

The symptoms of onchocerciasis are associated with the presence of the microfilariae in the eye and the skin. Ocular disease is characterized by slowly progressive inflammatory change in most parts of the eye leading to visual impairment and blindness in those most severely affected. Skin disease often starts as an itchy papular dermatitis leading to irreversible skin changes in long-term infection. The pathogenesis has long been thought to be associated with the death of microfilariae (Ottesen 1992; Pearlman 1996). Recently, however, it has been suggested that the release of products from the symbiotic bacteria (*Wolbachia* sp.) of many filarial nematodes may be involved in the pathogenesis of both eye and skin lesions (Saint André et al. 2002).

Clinical features of onchocerciasis

The pattern and frequency of the clinical features of onchocerciasis vary according to duration and frequency of exposure, geographical location, and individual variation (Cook and Zumla 2003). The most frequent symptom is itching. This may be of any degree of severity and is sometimes incapacitating. It is usually generalized and often associated with excoriations. Other acute reactive dermal lesions include papular eruptions anywhere on the body (reflecting intraepithelial abscesses), transient localized intradermal edema, and lymphadenopathy (particularly of the inguinal and femoral glands). These lesions are typically firm and nontender.

In a small minority of individuals, there may be an acute hyperreactive form of the disease known as sowda, in which it appears that there is killing of the microfilariae and severe inflammatory skin lesions occur. Later skin lesions give the appearance of premature aging associated with:

- lichenoid change, hyperkeratosis, and exaggerated wrinkling of the skin
- atrophy of the epidermis, with loose, redundant, thin, and shiny skin (Anderson et al. 1974a)
- depigmentation, often initially hyperpigmented macules, but more typically a spotty depigmentation of the shins that represents islands of repigmentation around hair follicles in areas of depigmentation; this is often termed 'leopard skin' (Buck 1974).

Chronically enlarged lymph nodes and the surrounding fluid may become dependent leading to 'hanging groin' (Nelson 1958) and predisposing to hernia formation, lymphatic obstruction, and mild elephantiasis. Some of these features are shown in Figures 39.8–39.10. A careful characterization of onchocercal skin pathology has been created for Africa where skin lesions are defined as acute or chronic and graded according to severity (Murdoch et al. 1993).

The subcutaneous nodules containing adult worms are often visible and palpable, but they rarely cause any symptoms. Occasionally they may spontaneously rupture through the skin, cause local pressure symptoms or the contents may necrose, leading to abscess formation.

Visual damage is the most serious clinical feature of onchocerciasis and may affect all tissues of the eye. Ocular lesions are usually seen only in those with moderate or heavy microfilarial loads. Two types of lesion affect the cornea: punctate keratitis and sclerosing keratitis (Anderson et al. 1974a). Punctate keratitis is caused by an acute inflammatory exudate surrounding dead and dying microfilariae in the cornea. These give rise to 'snowflake opacities' and resolve without sequelae. Sclerosing keratitis is a progressive exudative process with fibrovascular pannus formation that starts at the inferior or medial and lateral margins of the cornea and slowly becomes confluent. This may lead to irreversible visual damage and blindness if it encroaches on the visual axis. Anterior uveitis and iridocyclitis are usually mild, chronic, and nongranulomatous conditions, but may cause anterior and posterior synechiae, leading to seclusio or oclusio pupillae and serious, potentially blinding, complications such as secondary cataract and glaucoma. Characteristic chorioretinal lesions are atrophy or hyperplasia of the retinal pigment epithelium,

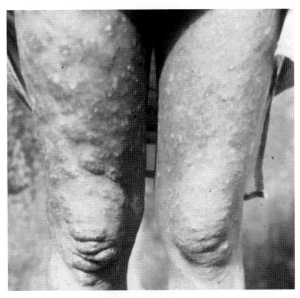

Figure 39.8 *Papular eruption (onchodermatitis) on the thighs of an adolescent girl in West Africa; note the asymmetrical distribution (Courtesy of Dr D. Morgan)*

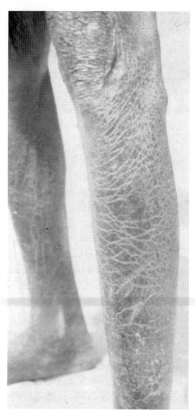

Figure 39.9 *Presbydermia (atrophic skin changes) on the leg of a young man (Courtesy of Dr D. Morgan)*

chronic nongranulomatous chorioretinitis and chorioretinal atrophy. These are typically widespread, unlike toxoplasmosis, and are usually first seen temporal to the macula. Postneuritic optic atrophy with constriction

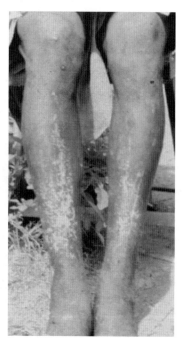

Figure 39.10 *Dispigmentation (leopard skin) on the shins (Courtesy of Dr D. Morgan)*

of visual fields is common in advanced disease, often associated with dense sheathing of retinal vessels.

Other clinical features such as weight loss, musculoskeletal pains and dizziness have been associated with onchocerciasis (Lamp 1967; Pearson et al. 1985; Burnham 1991), but the main clinical burden is due to the skin and eye lesions. In West Africa, the prevalence of infection, as measured by the presence of skin microfilariae, increases with age up to 25 years and is generally higher in males than in females (Kirkwood et al. 1983a). The degree of visual damage has a linear relationship to microfilarial density in savanna regions and the prevalence of visual loss increases with age (Remme et al. 1989). The relationship of blindness to intensity of infection does not hold true for parasites from forest regions, which may be explained by the microfilaria from the nonblinding strains being less eye invasive (Dadzie et al 1989, 1990). Males are 1.5 times more likely to be blind than females of the same age and microfilarial density. Blindness strikes mainly at economically active adults in the prime of life. The development of onchocercal eye disease in an individual is, therefore, both a catastrophe for the individual and a burden for the family. At the community level, visual impairment can have serious socioeconomic consequences for the community, especially if villages are in an area of high transmission. Under these conditions, heavy infections are acquired relatively early in life, so that there may be a considerable proportion of young, productive individuals, especially men, with visual loss. It was in such situations that villages in fertile river valleys were abandoned in Burkina Faso and northern Ghana prior to the start of the OCP in West Africa.

Pathology and pathogenesis of clinical features

Although nodules cause very few symptoms, there are cellular infiltrates around the adult worms in the nodules. There appears to be little or no reaction around live microfilariae, which are characteristically found in the upper dermis but lesions are found around dying microfilariae. Such lesions are characterized by an infiltration of eosinophils and neutrophils and these cells, along with macrophages, have been shown to attack microfilariae (Greene et al. 1981; Medina de la Garza et al. 1990). Early skin changes include:

- perivascular inflammatory infiltrates of eosinophils, neutrophils, plasma cells, histiocytes, and lymphocytes
- hyperkeratosis
- acanthosis with increased melanin in the upper dermis
- dilated tortuous lymph and blood capillaries.

Prolonged and heavy infections of the skin eventually lead to loss of elasticity with fibrosis, scarring of the papillae, replacement of dermal collagen by hyalinized scar tissue and, eventually, atrophy of the epidermis. Although the recruitment and activation of eosinophils has been shown to be a response to parasite products, recent evidence has indicated that neutrophils are attracted to the nodule by products of the endosymbiotic bacteria *Wolbachia* (Brattig et al. 2001). The hyperreactive form of the disease (sowda) is associated with high IgE levels, and a much increased infiltration of lymphocytes and eosonophils around the parasites (Brattig et al. 1987, 1994). Recently, a polymorphism in the T helper 2 (Th2) cytokine interleukin (IL)-13 has been associated with this form of onchocerciasis (Hoerauf et al. 2002).

Lymph node pathology in Africa tends to show scarring of lymphoid tissue with histological evidence of atrophy and fibrosis, chronic inflammatory infiltration and sinus histiocytosis (Gibson and Connor 1978). In the Yemen, follicular hyperplasia is a more common sign of the disease.

Ocular lesions are also thought to be related to the local death of microfilariae. The snowflake opacities of punctate keratitis are focal collections of lymphocytes, eosinophils and neutrophils. Sclerosing keratitis, in contrast, consists of an inflammatory exudate mainly of lymphocytes and eosinophils, scarring, and a fibrovascular pannus formation. Uveitis and chorioretinitis are usually low-grade indolent inflammatory processes. Studies in experimental models have shown that there is an immunological basis of the eye lesion pathology with a role for T cells, neutrophils, parasite-specific antibody, and immune complexes (Pearlman 1997; Pearlman et al. 1995; Hall et al. 1999; Kaifi et al. 2001). It has also been shown that a component of the pathology is regulated by Th2 cytokines: IL-4 is required, as lesions do not develop after immunization in IL-4 gene knockout mice (Hall et al. 2002) and IL-13 is required for eosinophil recruitment into the eye (Berger et al. 2002). Recently, endotoxin-like products of *Wolbachia* have also been shown to provide a major pro-inflammatory stimulus for the induction of eye pathology in the mouse model. Parasites containing *Wolbachia* induced the characteristic infiltrate of neutrophils and keratitis when they were injected, whereas such effects were absent when worms that had been depleted of their *Wolbachia* by chemotherapy were used (Saint André et al. 2002).

Molecular mimicry has been suggested as having a role in the development of onchocercal eye pathology, indeed an *O. volvulus* antigen termed Ov39 was found to cross-react with an antigen of retinal pigment epithelium (Braun et al. 1991). Autoantibodies to this antigen were found in patients with both onchocerciasis and lymphatic filariasis, so it was suggested that intraocular presentation of this antigen by microfilariae is required for the development of disease (McKechnie et al. 1993). Subsequent studies in a rat model showed that immunization with Ov39 can induce ocular inflammation that can be transferred by antigen specific T cell lines (McKechnie et al. 1997, 2002).

Post-treatment adverse reactions can also be attributed to inflammatory responses to *Wolbachia* bacteria. The Mazzotti reaction after treatment with DEC often gave very severe side-effects, and this was originally thought to be due to parasite products. However, recent studies have indicated that there is a link between the side-effects and the concentration of *Wolbachia* DNA (Keiser et al. 2002).

Immunity and immune responses

There is no direct evidence for acquired immunity in individuals exposed to or infected with *O. volvulus*. The fact that worm burdens plateau or decline with age has been suggested as evidence for concomitant immunity, i.e. protection from reinfection despite having a current infection. This phenomenon could equally be explained by exposure levels declining with age. However, in many endemic areas, microfilarial levels appear to continue to rise throughout life. The occurrence of uninfected individuals within endemic areas despite evidence of exposure, has also been cited as evidence for resistance to infection, but as diagnostic techniques are not 100 percent sensitive, it is difficult to prove absence of infection.

Onchocerciasis is a spectral disease with many individuals in a hyperendemic area having high numbers of microfilariae in the skin. Such individuals are often asymptomatic and have suppressed cellular responses to parasite antigens, although they have significant antibody responses (Greene et al. 1985a; King and Nutman 1991; Elson et al. 1994, 1995; Ottesen 1985; Gallin et al. 1988). The characteristics of the cytokines and antibodies produced by these individuals to parasite antigens suggest they are polarized towards type 2 responses as they have high levels of IL-4 and IL-5, and produce significant amounts of IgG4 and IgE.

Increased titers of specific and nonspecific antibodies, particularly IgG, IgM, and IgE, are found in symptomatic subjects. Some are specific for microfilariae, infective larvae and adult worms (Greene et al. 1985a; Ottesen 1985). There are also increased levels of circulating immune complexes of unknown significance. Cell-mediated immune responses are generally suppressed, at least in adults with patent infection (Gallin et al. 1988). Immune responsiveness appears to change with time of exposure to infection (Karam and Weiss 1985), with the IgG–IgE balance being an important determinant of microfilarial levels and clinical disease and, like the suppression of cell-mediated immunity, altering between the ages of 10 and 19 years.

In populations exposed to infection, there are always individuals who appear to be free from infection despite evidence of exposure; these have been suggested to be immune to infection. Such putative immune individuals have immune responses that confirm exposure to the parasite. They have vigorous proliferative responses to parasite antigen (Ward et al. 1988; Elson et al. 1995; Lüder et al. 1996; Brattig et al. 2002), but there are conflicting reports about the types of cytokines produced. Several reports have shown cytokine production indicative of a type 1 response with high interferon gamma (IFN-γ) levels (Ward et al. 1988; Elson et al. 1995; Lüder et al. 1996). Other studies, however, have shown either high Th2 cytokines (IL-5 and IL-13) levels to parasite antigens in individuals with either no parasites or low parasite burdens (Brattig et al. 2002) or heterogeneity in cytokine production particularly to larval and male antigens (IFN-γ and IL-5) (Turaga et al. 2000).

Cross-sectional studies such as those described above do not accurately reflect the fact that onchocerciasis is a dynamic, progressive disease, amicrofilaridermic individuals may in fact be prepatent or postpatent and symptoms may appear late in infection. Longitudinal studies are difficult because of the time scale of infection but age-profile studies can be performed to reflect the accumulated years of exposure in older subjects (Faulkner et al. 2001; MacDonald et al. 2002). If concomitant immunity does exist, it appears to be related to increasing levels of Th2 cytokines to larval antigens and IgG3 and IgE to infective larvae (MacDonald et al. 2002). Studies on migrant populations can also give an indication of the early response to infection and it seems that early infection is associated with a vigorous immune response with elevated IFN-γ and IL-5. In comparison with chronically infected individuals, it seems that this early response is down-regulated partly through an IL-10 dependent mechanism (Cooper et al. 2001).

Immunoregulation

The intriguing features of immunoregulation in onchocerciasis are marked increases in polyclonal IgE and eosinophils, but the persistence of skin microfilariae with minimal host responsiveness (immunotolerance). There is some evidence of active and specific immune suppression by parasite antigens. Most people in endemic areas, although microfilaridermic, are asymptomatic, show reduced specific antibody levels, and have depressed B- and T-cell responses when these are tested in vitro (King and Nutman 1991). One suggestion is that these observations can be explained by a balance between T helper 1 (Th1) and Th2 phenotypes, the asymptomatic microfilaridermics having elevated Th2 cytokine production and Th2 controlled effectors, but depressed proliferative responses and IFN-γ production, the converse being true in exposed uninfected individuals (Soboslay et al. 1999). However, more recent data suggest that the situation is more complex. There is strong evidence that the cellular response is down-regulated by IL-10 and

transforming growth factor beta (TGF-β) which are produced by T regulatory (Tr) cells (Soboslay et al. 1999; Cooper et al. 2001; Doetze et al. 2000). Cells that bear characteristics of Tr1 cells have been cloned from individuals with chronic onchocerciasis. These cells produce substantial amounts of IL-10 and are able to inhibit other T cells in coculture (Satoguina et al. 2002) The presence of the microfilariae is essential for the maintenance of the down-regulation in immune responses as chemotherapy restores responsiveness, at least until microfilarial levels return (Steel et al. 1994).

FUTURE DIRECTIONS

Between 1985 and 1999, a targeted approach to develop a vaccine for onchocerciasis was funded by the Edna McConnell Clark foundation (Cook et al. 2001). Although to date the development of a product to protect against infection by the parasite remains in the future, our understanding of the mechanisms of acquired immunity to parasitic helminths such as *Onchocerca* have been enormously extended because of this program. Major advances have been made in the development of animal model systems for onchocerciasis (Abraham et al. 2002) and a large number of potentially protective antigens have been identified about 14 of which reduce larval survival (Lustigman et al. 2002).

Although there is considerable progress in controlling the worst pathological consequences of onchocerciasis, it is clear that elimination of infection is not currently an achievable goal. Without additional tools such as a vaccine or effective macrofilaricides to combat transmission, it will be necessary to have long-term control and surveillance programs.

With advances in genomic technologies, and as the techniques of transfection and gene knock out become as available for parasitic nematodes as they are for the free-living nematode *Caenorhabditis elegans*, we will be able to increase our knowledge of the biology of these organisms exponentially. Already there are a growing number of *Onchocerca volvulus* expression sequence tag (EST) sequences (Williams et al. 2002) that can be exploited, and with the complete genome sequences for *C. elegans* and the proposed sequencing of the *B. malayi* genome, it is certain that many more vaccine targets will be identified.

REFERENCES

Abiose, A., Jones, B.R., et al. 1993. A randomized, controlled trial of ivermectin for onchocerciasis: evidence for a reduction in the incidence of optic nerve disease. *Lancet*, **341**, 130–4.

Abraham, D., Lucius, R. and Trees, A.J. 2002. Immunity to *Onchocerca* spp. in animal hosts. *Trends Parasitol*, **18**, 86–90.

Albiez, E.J. 1983. Studies on nodules and adult *Onchocerca volvulus* during a nodulectomy trial in hyperendemic villages in Liberia and Upper Volta. I. Palpable and impalpable onchocercomata. *Tropenmed Parasitol*, **34**, 54–60.

Albiez, E.J., Büttner, D.W. and Duke, B.O.L. 1988. Diagnosis and extirpation of nodules in human onchocerciasis. *Trop Med Parasitol*, **39**, 331–46.

Alley, W.S., van Oortmarssen, G.J., et al. 2001. Macrofilaricides and onchocerciasis control, mathematical modeling of the prospects for elimination. *BMC Public Health*, **1**, 12.

Anderson, J., Fuglsang, H., et al. 1974a. Studies on onchocerciasis in the United Cameroon Republic. I. Comparison of populations with and without *Onchocerca volvulus*. *Trans R Soc Trop Med Hyg*, **68**, 190–208.

Anderson, J., Fuglsang, H., et al. 1974b. Studies on onchocerciasis in the United Cameroon Republic. II. Comparison of onchocerciasis in the rainforest and sudan-savanna. *Trans R Soc Trop Med Hyg*, **68**, 209–22.

Anderson, J., Fuglsang, H., et al. 1975. The prognostic value of head nodules and microfilariae in the skin in relation to ocular onchocerciasis. *Tropenmed Parasitol*, **26**, 191–5.

Awadzi, K., Dadzie, K.Y., et al. 1986. The chemotherapy of onchocerciasis XI. A double-blind comparative study of ivermectin, diethylcarbamazine and placebo in human onchocerciasis in Northern Ghana. *Ann Trop Med Parasitol*, **80**, 433–42.

Awadzi, K., Addy, E.T., et al. 1995. The chemotherapy of onchocerciasis XX: ivermectin in combination with albendazole. *Trop Med Parasitol*, **46**, 213–20.

Aziz, M.A., Diallo, S., et al. 1982. Efficacy and tolerance of ivermectin in human onchocerciasis. *Lancet*, **2**, 171–3.

Baker, R.H.A., Guillet, P., et al. 1990. Progress in controlling the reinvasion of windborne vectors into the western area of the Onchocerciasis Control Programme in West Africa. *Philos Trans R Soc Lond B Biol Sci*, **328**, 731–50.

Bandi, C., Trees, A.J. and Brattig, N.W. 2001. *Wolbachia* in filarial nematodes: evolutionary aspects and implications for the pathogenesis and treatment of filarial diseases. *Vet Parasitol*, **98**, 215–38.

Basáñez, M.-G. and Boussinesq, M. 1999. Population biology of human onchocerciasis. *Philos Trans R Soc London B Biol Sci*, **354**, 809–26.

Basáñez, M.-G., Yarzábal, L., et al. 1988. The vectorial role of several blackfly species (Diptera: Simuliidae) in relation to human onchocerciasis in the Sierra Parima and Upper Orinoco regions of Venezuela. *Ann Trop Med Parasitol*, **82**, 597–611.

Basáñez, M.-G., Townson, H., et al. 1996. Density-dependent processes in the transmission of human onchocerciasis: relationship between microfilarial intake and mortality of the simuliid vector. *Parasitology*, **113**, 331–55.

Basáñez, M.-G., Yarzábal, L., et al. 2000. *Onchocerca-Simulium* complexes in Venezuela: can human onchocerciasis spread outside its present endemic areas? *Parasitology*, **120**, 143–60.

Basáñez, M.-G., Collins, R.C., et al. 2002. Transmission intensity and the patterns of *Onchocerca volvulus* infection in human communities. *Am J Trop Med Hyg*, **67**, 669–79.

Berger, R.B., Berger, N.M., et al. 2002. Il-4 and Il-13 regulation of ICAM-1 expression and eosinophil recruitment in *Onchocerca volvulus* keratitis. *Invest Ophthalmol Vis Sci*, **43**, 2992–7.

Bianco, A.E., Mustapha, M.B. and Ham, P.J. 1989. Fate of developing larvae of *Onchocerca lienalis* and *O. volvulus* in micropore chambers implanted into laboratory hosts. *J Helminthol*, **63**, 218–26.

Blacklock, D.B. 1926a. The development of *Onchocerca volvulus* in *Simulium damnosum*. *Ann Trop Med Parasitol*, **20**, 1–48.

Blacklock, D.B. 1926b. The further development of *Onchocerca volvulus* Leuckart in *Simulium damnosum* Theobald. *Ann Trop Med Parasitol*, **20**, 203–18.

Boatin, B.A., Toé, L., et al. 2002. Detection of *Onchocerca volvulus* infection in low prevalence areas: a comparison of three diagnostic methods. *Parasitology*, **125**, 545–52.

Boussinesq, M., Gardon, J., et al. 1998. Three probable cases of *Loa loa* encephalopathy following ivermectin treatment for onchocerciasis. *Am J Trop Med Hyg*, **58**, 461–9.

Bradley, J.E., Atogho, B., et al. 1998. A cocktail of recombinant *Onchocerca volvulus* antigens for serologic diagnosis with the

potential to predict the endemicity of onchocerciasis infection. *Am J Trop Med Hyg*, **59**, 887–92.

Brattig, N.W., Tischendorf, F.W., et al. 1987. Distribution pattern of peripheral lymphocyte subsets in localized and generalized forms of onchocerciasis. *J Inf Dis*, **44**, 149–59.

Brattig, N.W., Krawietz, I., et al. 1994. Strong IgG isotypic antibody response in sowdah type onchocerciasis. *J Inf Dis*, **170**, 955–61.

Brattig, N.W., Büttner, D.W. and Hoerauf, A. 2001. Neutrophil accumulation around *Onchocerca* worms and chemotaxis of neutrophils are dependent on *Wolbachia* endobacteria. *Microbes Infect*, **3**, 439–46.

Brattig, N.W., Leeping, B., et al. 2002. *Onchocerca volvulus*-exposed persons fail to produce interferon (in response to *O.volvulus* antigens but mount proliferative responses with interleukin-5 and IL-13 production that decrease with increasing microfilarial density. *J Inf Dis*, **185**, 1148–54.

Braun, G., McKechnie, N.M., et al. 1991. Immunological crossreactivity between a cloned antigen of *Onchocerca volvulus* and a component of the retinal pigment epithelium. *J Exp Med*, **174**, 169–77.

Bryant, J. 1935. Endemic retino-choroiditis in the Anglo-Egyptian Sudan and its possible relationship to *Onchocerca volvulus*. *Trans R Soc Trop Med Hyg*, **28**, 523–32.

Buck, A.A. 1974. *Onchocerciasis. symptomatology, pathology, diagnosis*. Geneva: World Health Organization.

Buckley, J.J.C. 1951. Studies on human onchocerciasis and *Simulium* in Nyanza province, Kenya. II. The disappearance of *Simulium naevei* from a bush-cleared focus. *J Helminthol*, **25**, 213–22.

Burnham, G.M. 1991. Onchocerciasis in Malawi. 2. Subjective complaints and decreased weight in persons infected with *Onchocerca volvulus* in the Thyolo highland. *Trans R Soc Trop Med Hyg*, **85**, 497–500.

Churchill, D.R., Godfrey-Faussett, P., et al. 1994. A trial of a three-dose regimen of ivermectin for the treatment of patients with onchocerciasis in the UK. *Trans R Soc Trop Med Hyg*, **88**, 242.

Collins, R.C., Gonzalez-Peralta, C., et al. 1992. Ivermectin: reduction in prevalence and intensity of *Onchocerca volvulus* follwing bi-annual treatments in five Guatemalan communities. *Am J Trop Med Hyg*, **47**, 170–80.

Cook, G.C. and Zumla, A.I. (eds) 2003. *Manson's tropical diseases*, 21st edn. Edinburgh: Elsevier Science.

Cook, J.A., Steel, C. and Ottesen, E.A. 2001. Towards a vaccine for onchocerciasis. *Trends Parasitol*, **17**, 555–7.

Cooper, P.J., Mancero, T., et al. 2001. Early human infection with *Onchocerca volvulus* is associated with an enhanced parasite specific cellular immune response. *J Infect Dis*, **183**, 1662–8.

Crosskey, R.W. 1962. Observations on the uptake of human blood by *Simulium damnosum*: the engorgement time and size of the blood meal. *Ann Trop Med Parasitol*, **56**, 141–8.

Crosskey, R.W. 1990. *The natural history of blackflies*. Chichester: John Wiley & Sons Ltd.

Cupp, E.W. 1992. Treatment of onchocerciasis with ivermectin in Central America. *Parasitol Today*, **8**, 212–14.

Cupp, E.W., Ochoa, O., et al. 1992. The effect of repetitive community-wide ivermectin treatment on transmission of *Onchocerca volvulus* in Guatemela. *Am J Trop Med Hyg*, **47**, 170–80.

Dadzie, K.Y., Remme, J., et al. 1989. Ocular onchocerciasis and intensity of infection in the community. II. West African rainforest foci of the vector *Simulium yahense*. *Trop Med Parasitol*, **40**, 348–54.

Dadzie, K.Y., Remme, J., et al. 1990. Ocular onchocerciasis and intensity of infection in the community. III. West Africa rainforest foci of the vector *Simulium sanctipauli*. *Trop Med Parasitol*, **41**, 376–82.

Dalmat H.T. 1955. *The blackflies (Diptera: Simuliidae) of Guatemala and their role as vectors of onchocerciasis*. Smithsonian Miscellaneous Collection No. 125. Smithsonian Institute, Washington DC.

Davies, J.B., Crosskey, R.W. and Johnston, M.R.L. 1962. The control of *Simulium damnosum* at Abuja, Northern Nigeria. *Bull World Health Organ*, **27**, 491–510.

Diallo, S., Aziz, M.A., et al. 1986. A double blind comparison of the efficacy and safety of ivermectin and diethylcarbamazine in a placebo controlled study of Senegalese patients with onchocerciasis. *Trans R Soc Trop Med Hyg*, **80**, 927–34.

Dietz, K. 1982. The population dynamics of onchocerciasis. In: Anderson, R.M. (ed.), *Population dynamics of infectious diseases*. London: Chapman and Hall, 209–41.

Doetze, A., Satoguina, J., et al. 2000. Antigen-specific cellular hyporesponsiveness in a chronic human helminth infection is mediated by T(h)3/T(r)1-type cytokines IL-10 and transforming growth factor-beta but not by a T(h)1 to T(h)2 shift. *Int Immunol*, **12**, 623–30.

Duerr, H.P., Dietz, K., et al. 2001. A stochastic model for the aggregation of *Onchocerca volvulus* in nodules. *Parasitology*, **123**, 193–201.

Duke, B.O.L. 1968a. Studies on factors influencing the transmission of onchocerciasis. V. The stages of *Onchocerca volvulus* in wild 'forest'; *Simulium damnosum*, the fate of the parasite in the fly and the age distribution of the biting population. *Ann Trop Med Parasitol*, **62**, 107–16.

Duke, B.O.L. 1968b. Studies on factors influencing the transmission of onchocerciasis. VI. The infective biting potential of *Simulium damnosum* in different bioclimatic zones and its influence on the transmission potential. *Ann Trop Med Parasitol*, **62**, 164–70.

Duke, B.O.L. 1970. *Onchocerca–Simulium* complexes. VI. Experimental studies on the transmission of Venezuelan and West African strains of *Onchocerca volvulus* by *Simulium metallicum* and *S. exiguum* in Venezuela. *Ann Trop Med Parasitol*, **64**, 421–31.

Duke, B.O.L. 1991. Observations and reflections on the immature stages of *Onchocerca volvulus* in the human host. *Ann Trop Med Parasitol*, **85**, 103–10.

Duke, B.O.L., Lewis, D.J. and Moore, P.J. 1966. *Onchocerca–Simulium* complexes. I. Transmission of forest and Sudan-savanna strains of *Onchocerca volvulus*, from Cameroon, by *Simulium damnosum* from various West African bioclimatic zones. *Ann Trop Med Parasitol*, **60**, 317–36.

Duke, B.O.L., Moore, P.J. and Anderson, J. 1972. Studies on factors influencing the transmission of onchocerciasis. VII. A comparison of the *Onchocerca volvulus* transmission potentials of *Simulium damnosum* populations in four Cameroon rain-forest villages and the pattern of onchocerciasis associated therewith. *Ann Trop Med Parasitol*, **66**, 219–34.

Duke, B.O., Anderson, J. and Fuglsang, H. 1975. The *Onchocerca volvulus* transmission potentials and associated patterns of onchocerciasis at four Cameroon Sudan-savanna villages. *Tropenmed Parasitol*, **26**, 143–54.

Dunne, C.L., Malone, C.J. and Whitworth, J.A.G. 1991. A field study of the effects of ivermectin on ectoparasites in man. *Trans R Soc Trop Med Hyg*, **85**, 550–1.

Eberhard, M.L. 1986. Longevity of microfilariae following removal of the adult worms. *Trop Med Parasitol*, **37**, 361–3.

Elson, L., Guderian, R., et al. 1994. Immunity to onchocerciasis: Identification of a putative immune population in a hyperendemic area of Ecuador. *J Infect Dis*, **169**, 588–94.

Elson, L., Guderian, R., et al. 1995. Immunity to onchocerciasis: Putative immune individuals produce a Th1-like response to *Onchocerca volvulus*. *J Infect Dis*, **171**, 652–8.

Erttmann, K.D., Unnasch, T.R., et al. 1987. A DNA sequence specific for forest form *Onchocerca volvulus*. *Nature*, **327**, 415–17.

Faulkner, H., Gardon, J., et al. 2001. Antibody responses in onchocerciasis as a function of age and infection intensity. *Parasite Immunol*, **23**, 509–16.

Fischer, P., Yocha, J., et al. 1997. PCR and DNA hybridization indicate the absence of animal filariae from vectors of *Onchocerca volvulus* in Uganda. *J Parasitol*, **83**, 1030–4.

Freedman, D.O., Unnasch, T.R., et al. 1994. Truly infection-free persons are rare in areas hyperendemic for African onchocerciasis. *J Inf Dis*, **170**, 1054–5.

Fuglsang, H. and Anderson, J. 1977. The concentration of microfilariae in the skin near the eye as a simple measure of the severity of onchocerciasis in a community and as an indicator of danger to the eye. *Tropenmed Parasitol*, **28**, 63–7.

Gallin, M., Edmonds, K., et al. 1988. Cell-mediated immune responses in human infection with *Onchocerca volvulus*. *J Immunol*, **140**, 1999–2007.

Gardon, J., Gardon-Wendel, N., et al. 1997. Serious reactions after mass ivermectin treatment of onchocerciasis with ivermectin in an area endemic for *Loa loa* infections. *Lancet*, **350**, 18–22.

Gardon, J., Boussineq, M., et al. 2002. Effects of standard and high doses of ivermectin on adult worms of *Onchocerca volvulus*: a randomized control trial. *Lancet*, **360**, 203–10.

Garms, R. 1973. Quantitative studies of the transmission of *Onchocerca volvulus* by *Simulium damnosum* in the Bong Range, Liberia. *Tropenmed Parasitol*, **24**, 358–72.

Garnham, P.C.C. and McMahon, J.P. 1947. The eradication of *Simulium damnosum* Roubaud from an onchocerciasis area in Kenya colony. *Bull Entomol Res*, **37**, 619–28.

Gibson, D.W. and Connor, D.H. 1978. Onchocercal lymphadenitis: clinicopathologic study of 34 patients. *Trans R Soc Trop Med Hyg*, **72**, 137–54.

Goa, K.L., McTavish, D. and Clissold, S.P. 1991. Ivermectin: a review of its antifilarial activity, pharmacokinetic properties and clinical efficacy in onchocerciasis. *Drugs*, **42**, 640–58.

Greene, B.M., Taylor, H.R. and Aikawa, M. 1981. Cellular killing of microfilariae of *Onchocerca volvulus* eosinophil and neutrophil mediated immune serum dependent destruction. *J Immunol*, **127**, 1611–18.

Greene, B.M., Gbakima, A.A., et al. 1985a. Humoral and cellular immune responses to *Onchocerca volvulus* infection in humans. *Rev Infect Dis*, **7**, 789–95.

Greene, B.M., Taylor, H.R., et al. 1985b. Comparison of ivermectin and DEC in the treatment of onchocerciasis. *N Engl J Med*, **313**, 133–8.

Grillet, M.E., Botto, C., et al. 1994. Vector competence of *Simulium metallicum s.l.* (Diptera: Simuliidae) in two endemic areas of human onchocerciasis in northern Venezuela. *Ann Trop Med Parasitol*, **88**, 65–75.

Grillet, M.E., Basáñez, M.-G., et al. 2001. Human onchocerciasis in the Amazonian area of southern Venezuela: spatial and temporal variations in biting and parity rates of black fly (Diptera: Simuliidae) vectors. *J Med Entomol*, **38**, 520–30.

Guderian, R.H., Anselmi, M., et al. 1997. Successful control of onchocerciasis with community-based ivermectin distribution in the Rio Santiago focus in Ecuador. *Trop Med Int Health*, **2**, 982–8.

Hall, L., Lass, J., et al. 1999. An essential role for antibody in neutrophil and eosinophil recruitment to the cornea: B cell deficient (MT) mice fail to develop Th2-dependent, helminth-mediated keratitis. *J Immunol*, **163**, 4970–5.

Hall, L.R., Berger, R.B., et al. 2002. *Onchocerca volvulus* keratitis (river blindness) is exacerbated in BALB/c IL-4 gene knockout mice. *Cell Immunol*, **216**, 1–5.

Harnett, W. 2002. DNA-based detection of *Onchocerca volvulus*. *Trans R Soc Trop Med Hyg*, **96**, Supplement 1, S1/231–4.

Harnett, W., Chambers, A.E., et al. 1989. An oligonucleotide probe specific for *Onchocerca volvulus*. *Mol Biochem Parasitol*, **35**, 119–25.

Hewitt, R.I., Kushner, S., et al. 1947. Experimental chemotherapy of filariasis. III. Effect of 1-diethylcarbamyl-1, 4-methylpiperazine hydrochloride against naturally acquired infections in cotton rats and dogs. *J Lab Clin Med*, **32**, 1314–29.

Hissette, J. 1932. Mèmoire sur l'*Onchocerca volvulus* 'Leuckart' et ses manifestations oculaires au Congo belge. *Ann Soc Belge Med Trop*, **12**, 433–529.

Hoerauf, A., Volkmann, L., et al. 2000. Endosymbiotic bacteria in worms as targets for a novel chemotherapy in filariasis. *Lancet*, **355**, 1242–3.

Hoerauf, A., Mand, S., et al. 2001. Depletion of *Wolbachia* endobacteria in *Onchocerca volvuus* by doxycycline and microfilardermia after ivermectin treatment. *Lancet*, **357**, 1415–16.

Hoerauf, A., Kruse, S., et al. 2002. The variant Arg110Gln of human Il-13 is associated with an immunologically hyper-reactive form of onchocerciasis (sowda). *Microbes Infect*, **4**, 37–42.

Hoerauf, A., Büttner, D., et al. 2003. Onchocerciasis. *Br Med J*, **326**, 207–10.

Hougard, J.M., Alley, E.S., et al. 2001. Eliminating onchocerciasis after 14 years of vector control: A proved strategy. *J Inf Dis*, **814**, 497–503.

Kaifi, J.T., Diaconu, E. and Pearlman, E.A. 2001. Distinct roles for PECAM-1, ICAM-1 and VCAM-1 in recruitment of neutrophils and eosinophils to the cornea in ocular onchoceracisis (river blindness). *J immunol*, **166**, 6795–801.

Karam, M. and Weiss, N. 1985. Seroepidemiological investigations of onchocerciasis in a hyperendemic area of West Africa. *Am J Trop Med Hyg*, **34**, 907–17.

Katholi, C.R., Toé, L., et al. 1995. Determining the prevalence of *Onchocerca volvulus* infection in vector populations by PCR screening of pools of blackflies. *J Inf Dis*, **172**, 1414–17.

Keiser, P.B., Reynolds, S.M., et al. 2002. Bacterial endosymbionts of *Onchocerca volvulus* in the pathogenesis of posttreatment reactions. *J Inf Dis*, **185**, 805–11.

King, T.B. and Nutman, C.L. 1991. Regulation of the immune response in lymphatic filariasis and onchocerciasis. *Parasitol Today*, **7**, Special edition, A54–58.

Kirkwood, B., Smith, P., et al. 1983a. Variations in the prevalence and intensity of microfilarial infections by age, sex, place and time in the area of the Onchocerciasis Control Programme. *Trans R Soc Trop Med Hyg*, **77**, 857–61.

Kirkwood, B., Smith, P., et al. 1983b. Relationships between mortality, visual acuity and microfilarial load in the area of the Onchocerciasis Control Programme. *Trans R Soc Trop Med Hyg*, **77**, 862–8.

Lamp, H.C. 1967. Musculoskeletal pain in onchocerciasis. *W Afr Med J*, **16**, 60–2.

Langworthy, S., Renz, A., et al. 2000. Macrofilaricidal activity of tetracycline against the filarial nematode *Onchocerca ochengi*: elimination of *Wolbachia* precedes worm death and suggests a dependent relationship. *Proc R Soc Lond B Biol Sci*, **267**, 1063–9.

Larivière, M., Aziz, M., et al. 1985. Double-blind study of ivermectin and diethylcarbamazine in African onchocerciasis patients with ocular involvement. *Lancet*, **2**, 174–7.

Le Berre, R. 1968. Bilan sommaire pour 1967 de lutte contre le vecteur de l'onchocercose. *Méd Afr Noire*, **15**, 71–2.

Little, M.P., Breitling, L.P., et al. 2004. Association between microfilarial load and excess mortality in human onchocerciasis. *Lancet*, **363**, 1514–21.

Lobos, E., Weiss, N., et al. 1991. An immunogenic *Onchocerca volvulus* antigen: a specific and early marker of infection. *Science*, **25**, 1603–5.

Lok, J.B., Pollack, R.J., et al. 1984. Development of *Onchocerca lienalis* and *O. volvulus* from the third to fourth larval stage in vitro. *Tropenmed Parasitol*, **35**, 209–11.

Lucius, R., Kern, A., et al. 1992. Specific and sensitive IgG4 immunodiagnosis of onchocerciasis with a recombinant 33 kD *Onchocerca volvulus* protein (Ov33). *Trop Med Parasitol*, **43**, 139–45.

Lüder, C.G., Schulz-Key, H., et al. 1996. Immuno-regulation in onchocerciasis: predominance of Th-1 type responsiveness to low molecular weight antigens of *Onchocerca volvulus* in exposed individuals without microfiladermia and clinical disease. *Clin Exp Immunol*, **105**, 245–53.

Lustigman, S., James, E.R., et al. 2002. Towards a recombinant antigen vaccine against *Onchocerca volvulus*. *Trends Parasitol*, **18**, 135–71.

MacDonald, A.J., Turaga, P.S.D., et al. 2002. Differential cytokine and antibody responses to adult and larval stages of *Onchocerca volvulus* consistent with the development of concomitant immunity. *Infect Immun*, **70**, 2796–804.

Macfie, J.W.S. and Corson, J.F. 1922. Observations on *Onchocerca volvulus*. *Ann Trop Med Parasitol*, **16**, 459–64.

Mazzotti, L. 1948. Possibilidad de utilizar como medio diagnóstico en la oncocercosis, las reacciones alérgicas consecutivas a la administración de 'Hetrazan'. *Rev Inst Salubr Enferm Trop (Mex)*, **9**, 235–7.

McKechnie, N.M., Gurr, W. and Braun, G. 1997. Immunization with the cross-reactive antigens Ov39 from *Onchocerca volvulus* and hr44 from human retinal tissue induces ocular pathology and activates retinal microglia. *J Inf Dis*, **176**, 1334–43.

McKechnie, N.M., Gurr, W., et al. 2002. Antigenic mimicry: *Onchocerca volvulus* antigen-specific T cells and ocular inflammation. *Invest Ophthalmol Vis Sci*, **43**, 411–18.

McKechnie, N.M., Braun, G., et al. 1993. Immunologic cross-reactivity in the pathogenesis of ocular onchocerciasis. *Invest Ophthalmol Vis Sci*, **34**, 2888–902.

Medina de la Garza, C.E., Brattig, N.W., et al. 1990. Serum-dependent interaction of granulocytes with *Onchocerca volvulus* microfilariae in generalized and chronic hyper-reactive onchoceciasis and its modulation by diethylcarbamazine. *Trans R Soc Trop Med Hyg*, **84**, 701–6.

Meredith, S.E.O., Unnasch, T.R., et al. 1989. Cloning and characterization of an *Onchocerca volvulus*-specific DNA sequence. *Mol Biochem Parasitol*, **36**, 1–10.

Meredith, S.E.O., Lando, G., et al. 1991. *Onchocerca volvulus*: application of the polymerase chain reaction to identification and strain differentiation of the parasite. *Exp Parasitol*, **73**, 335–44.

Molyneux, D.H. 1995. Onchocerciasis control in West Africa: Current status and future of the onchocerciasis control programme. *Parasitol Today*, **11**, 399–402.

Murdoch, M.E., Hay, R.J., et al. 1993. A clinical classification and grading system of the cutaneous changes in onchocerciasis. *Br J Dermatol*, **129**, 260–9.

Nelson, G.S. 1958. 'Hanging groin' and hernia complications of onchocerciasis. *Trans R Soc Trop Med Hyg*, **52**, 272–5.

Ngoumou, P., Walsh, J.F., et al. 1993. *A manual for rapid epidemiological mapping of onchocerciasis*. Geneva: World Health Organization, TDR/TDE/ONCHO/93.4..

Noma, M., Nwoke, B.E., et al. 2002. Rapid epidemiological mapping of onchocerciasis (REMO): its application by the African Programme for Onchocerciasis Control (APOC). *Ann Trop Med Parasitol*, **96**, Supplement, S29–39.

Ogunrinade, A., Boakye, D., et al. 1999. Distribution of the blinding and nonblinding strains of *Onchocerca volvulus* In Nigeria. *J Infect Dis*, **179**, 1577–9.

Omar, M.S. 1978. Histochemical enzyme staining patterns of *Onchocerca volvulus* microfilariae and their occurrence in different onchocerciasis areas. *Tropenmed Parasitol*, **29**, 462–72.

O'Neill, J. 1875. On the presence of a filaria in craw-craw. *Lancet*, **1**, 265–6.

Ortega, M. and Oliver, M. 1985. Entomología de la oncocercosis en el Soconusco, Chiapas. II. Estudios sobre dinámica de población de las tres especies de simúlidos considerados transmisores de oncocercosis en el foco sur de Chiapas. *Folia Entomol (Mex)*, **66**, 119–36.

Oskam, L., Schoone, G., et al. 1996. Polymerase chain reaction for detecting *Onchocerca volvulus* in pools of blackflies. *Trop Med Int Health*, **1**, 522–7.

Ottesen, E.A. 1985. Immediate hypersensitivity responses in the immunopathogenesis of human onchocerciasis. *Rev Infect Dis*, **7**, 796–801.

Ottesen, E.A. 1992. Infection and disease in lymphatic filariasis: an immunological perspective. *Parasitology*, **104**, 571–9.

Parsons, A.C. 1908. *Filaria volvulus* Leuckart, its distribution, structure and pathological effects. *Parasitology*, **1**, 359–68.

Pearlman, E. 1996. Experimental onchocercal keratitis. *Parasitol Today*, **14**, 229–34.

Pearlman, E. 1997. Immunopathology of onchocerciasis: a role for eosinophils in onchocercal dermatitis and keratitis. *Chem Immunol*, **66**, 26–40.

Pearlman, E.A., Lass, J.H., et al. 1995. Interleukin 4 and T helper 2 cells are required for the development of experimental onchocercal keratitis (river blindness). *J Exp Med*, **182**, 931–40.

Pearson, C.A., Brieger, W.R., et al. 1985. Improving recognition of onchocerciasis in primary care – 1: non-classical symptoms. *Trop Doct*, **15**, 160–3.

Perler, F.B. and Karam, M. 1986. Cloning and characterization of two *Onchocerca volvulus* repeated DNA sequences. *Mol Biochem Parasitol*, **21**, 171–8.

Pion, S.D., Kamgno, J., et al. 2002. Excess mortality associated with blindness in the onchocerciasis focus of the Mbam valley, Cameroon. *Ann Trop Med Parasitol*, **96**, 181–9.

Plaisier, A.P., van Ortmarssen, G.J., et al. 1991. The reproductive lifespan of *Onchocerca volvulus* in West African savana. *Acta Trop*, **48**, 271–4.

Procunier, W.S., Shelley, A.J. and Arzube, M. 1986. Sibling species of *Simulium exiguum* (Diptera: Simuliidae), the primary vector of onchocerciasis in Ecuador. *Rev Ecuator Higiene Med Trop*, **35**, 49–59.

Prost, A. 1980. Latence parasitaire dans l'onchocercose. *Bull World Health Organ*, **58**, 923–5.

Prost, A. and Vaugelade, J. 1981. La surmortalité des aveugles en zone de savane ouest-africaine. *Bull World Health Organ*, **59**, 773–6.

Prout, W.T. 1901. A filaria found in Sierra Leone? *Filaria volvulus* Leuckart. *Br Med J*, **1**, 209–11.

Ramachandran, C.P. 1993. Improved immunodiagnostic tests to monitor onchocerciasis control programmes-A multicenter effort. *Parasitol Today*, **9**, 76–9.

Remme, J.H.F. 1995. The African Programme for Onchocerciasis Control: preparing to launch. *Parasitol Today*, **11**, 403–6.

Remme, J., Ba, O., et al. 1986. A force of infection model for onchocerciasis and its applications in the epidemiological evaluation of the Onchocerciasis Control Programme in the Volta River basin area. *Bull World Health Organ*, **64**, 667–81.

Remme, J., Dadzie, K.Y., et al. 1989. Ocular onchocerciasis and intensity of infection in the community. I. West African savanna. *Trop Med Parasitol*, **40**, 340–7.

Renz, A. 1987. Studies on the dynamics of transmission of onchocerciasis in a Sudan-savanna area of North Cameroon. II. Seasonal and diurnal changes in the biting densities and in the age-composition of the vector population. *Ann Trop Med Parasitol*, **81**, 229–37.

Renz, A. and Wenk, P. 1983. The distribution of the microfilariae of *Onchocerca volvulus* in the different body regions in relation to the attacking behaviour of *Simulium damnosum s.l.* in the Sudan-savanna of northern Cameroon. *Trans R Soc Trop Med Hyg*, **77**, 748–52.

Renz, A. and Wenk, P. 1987. Studies on the dynamics of transmission of onchocerciasis in a Sudan-savanna area of North Cameroon. I. Prevailing *Simulium* vectors, their biting rates and age-composition at different distances from their breeding sites. *Ann Trop Med Parasitol*, **81**, 215–28.

Renz, A., Fuglsang, H. and Anderson, J. 1987. Studies on the dynamics of transmission of onchocerciasis in a Sudan-savanna area of North Cameroon. IV. The different exposure to *Simulium* bites and transmission of boys and girls and men and women, and the resulting manifestations of onchocerciasis. *Ann Trop Med Parasitol*, **81**, 253–62.

Richards, F.O., Hopkins, D. and Cupp, E. 2000. Programmatic goals and approaches to onchocerciasis. *Lancet*, **355**, 1663–4.

Richards, F.O., Boatin, B., et al. 2001. Control of onchocerciasis today: status and challenges. *Trends Parasitol*, **17**, 558–63.

Ridley, H. 1945. Ocular onchocerciasis, including an investigation in the Gold Coast. *Br J Ophthalmol*, **10**, Supplement, 58.

Roberts, J.M.D., Neumann, E., et al. 1967. Onchocerciasis in Kenya 9 11 and 18 years after elimination of the vector. *Bull World Health Organ*, **37**, 195–212.

Robles, R. 1917. Enfermedad nueva en Guatemala. *La Juventud Medica (Guatemala)*, **17**, 97–115.

Rodríguez-Pérez, M.A., Danis-Lozano, R., et al. 1999. Detection of *Onchocerca volvulus* infection in *Simulium ochraceum sensu lato*: comparison of a PCR assay and fly dissection in a Mexican hypoendemic community. *Parasitology*, **119**, 613–19.

Rothfels, K.H. 1981. Cytotaxonomy: principles and their application to some northern species complexes in *Simulium*. In: Laird, M. (ed.), *Blackflies. The future for biological methods in integrated control*. London: Academic Press, 19–29.

Saint André, A.V., Blackwell, N.M., et al. 2002. A crtical role for endosymbiotic *Wolbachia* bacteria and Toll-4 signalling in the pathogenesis of river blindness. *Science*, **295**, 1892–5.

Satoguina, J., Mepel, M., et al. 2002. Antigen-specific T regulatory-1 cells are associated with immunosupression in a chronic helminth infection (onchocerciasis). *Microbes Infect*, **4**, 1291–300.

Schulz-Key, H. and Albiez, E.J. 1977. Worm burden of *Onchocerca volvulu*s in a hyperendemic village of the rain forest in West Africa. *Tropenmed Parasitol*, **28**, 431–8.

Schulz-Key, H. and Karam, M. 1986. Periodic reproduction of *Onchocerca volvulus*. *Parasit Today*, **2**, 284–6.

Schulz-Key, H., Kläger, S., et al. 1985. Treatment of human onchocerciasis: The efficacy of ivermectin on the parasite. *Trop Med Parasitol*, **36**, Supplement II., 20.

Schulz-Key, H., Greene, B.M., et al. 1986. Efficacy of ivermectin on the reproductivity of female *Onchocerca volvulus*. *Trop Med Parasitol*, **37**, 89.

Shah, J.S., Karam, M.C., et al. 1987. Characterisation of an *Onchocerca*-specific DNA clone from *Onchocerca volvulus*. *Am J Trop Med Hyg*, **37**, 376–84.

Shelley, A.J. 1988a. Vector aspects of the epidemiology of onchocerciasis in Latin America. *Ann Rev Entomol*, **30**, 337–66.

Shelley, A.J. 1988. Biosystematics and medical importance of the *Simulium amazonicum* group and the *S. exiguum* complex in Latin America. In: Service, M.W. (ed.), *Biosystematics of haematophagous insects*. , The Systematics Association Special Volume, 37. . Oxford: Clarendon Press, 203–20.

Shelley, A.J., Charalambous, M. and Arzube, M. 1990. *Onchocerca volvulus* development in four *Simulium exiguum* cytospecies in Ecuador. *Bull Soc Française Parasitol*, **8**, 1145.

Soboslay, P.T., Lüder, C.G., et al. 1999. Regulatory effects of Th1-type (IFN-(, Il-12) and Th-2-type cytokines (IL-10, Il-13) on parasite specific cellular responsiveness in *Onchocerca vovlulus*-infected humans and exposed endemic controls. *Immunol*, **97**, 219–25.

Steel, C.S., Lujan-Trangay, R., et al. 1994. Transient changes in cytokine profiles following ivermectin treatment of onchocerciasis. *J Infect Dis*, **170**, 962–70.

Stingle, P., Ross, M., et al. 1984. A diagnostic patch test for onchocerciasis using topical diethylcarbamazine. *Trans R Soc Trop Med Hyg*, **78**, 254–8.

Strote, G. 1987. Morphology of third and fourth stage larvae of *Onchocerca volvulus*. *Trop Med Parasitol*, **38**, 73–4.

Takaoka, H., Suzuki, H., et al. 1984. Development of *Onchocerca volvulus* larvae in *Simulium pintoi* in the Amazonas region of Venezuela. *Am J Trop Med Hyg*, **33**, 414–19.

Taylor, H.R., Murphy, R.P., et al. 1986. Treatment of onchocerciasis. The ocular effects of ivermectin and diethylcarbamazine. *Arch Ophthalmol*, **104**, 863–70.

Taylor, M.J. and Hoerauf, A. 2001. A new approach to the treatment of filariasis. *Curr Opin Inf Dis*, **14**, 727–31.

Taufflieb, R. 1955. Une campagne de lutte contre *Simulium damnosum* au Mayo Kebbi. *Bull Soc Pathol Exot*, **48**, 564–76.

Toé, L., Tang, J., et al. 1997. Vector-parasite transmission complexes for onchocerciasis in West Africa. *Lancet*, **349**, 163–6.

Toé, L., Boatin, B.A., et al. 1998. Detection of *Onchocerca volvulus* infection by 0-150 polymerase chain reaction analysis of skin scratches. *J Inf Dis*, **178**, 282–5.

Turaga, P.S.D., Tierney, T.J., et al. 2000. Immunity to onchocerciasis: Cells from putatively immune individuals produce enhanced levels of interleukin-5, gamma interferon and granulocyte-macrophage colony-stimulating factor in response to *Onchocerca volvulus* larval and male worm antigens. *Infect Immun*, **68**, 1905–11.

Vivas-Martínez, S., Basáñez, M.-G., et al. 1998. Onchocerciasis in the Amazonian focus of southern Venezuela: parasitological findings in relation to the selection of communities for ivermectin control programmes. *Trans R Soc Trop Med Hyg*, **92**, 371–2.

Vuong, P.N., Bain, O., et al. 1988. Forest and savanna onchocerciasis: comparative morphometric histopathology of skin lesions. *Trop Med Parasitol*, **39**, 105–10.

Walsh, J.F., Davies, J.B. and Cliff, B. 1981. World Health Organization Onchocerciasis Control Programme in the Volta River Basin. In: Laird, M. (ed.), *Blackflies: the future for biological methods in integrated control*. London: Academic Press, 85–103.

Walsh, J.F., Davies, J.B., et al. 1978. Standardization of criteria for assessing the effect of *Simulium* control in onchocerciasis control programmes. *Trans R Soc Trop Med Hyg*, **72**, 675–6.

Ward, D.J., Nutman, T.B., et al. 1988. Onchocerciasis and immunity in humans: Enhanced T cell responsiveness to parasite antigen in putatively immune individuals. *J Infect Dis*, **57**, 536–43.

Weil, G.J., Lammie, P.J. and Weiss, N. 1997. The ICT filariasis test: a rapid format antigen test for the diagnosis of bancroftian filariasis. *Parasitol Today*, **13**, 401–4.

Weil, G.J., Ogunrinade, A.F., et al. 1990. IgG4 subclass antibody serology for onchocerciasis. *J Inf Dis*, **161**, 549–54.

Weil, G.J., Steel, C., et al. 2000. A rapid format antibody card test for diagnosis of onchocercaisis. *J Infect Dis*, **182**, 1796–9.

Wenk, P. 1981. Bionomics of adult blackflies. In: Laird, M. (ed.), *Blackflies: the future for biological methods in integrated control*. London: Academic Press, 259–79.

Whitworth, J. 1992. Drug of the month: ivermectin. *Trop Doct*, **22**, 163–4.

Whitworth, J.A.G., Gilbert, C.E., et al. 1991a. The effects of repeated doses of ivermectin on ocular onchocerciasis – results from a community based trial in Sierra Leone. *Lancet*, **338**, 1100–3.

Whitworth, J.A.G., Morgan, D., et al. 1991b. A field study of the effect of ivermectin on intestinal helminths in man. *Trans R Soc Trop Med Hyg*, **85**, 232–4.

Whitworth, J.A.G., Morgan, D., et al. 1991c. A community trial of ivermectin for onchocerciasis in Sierra Leone: adverse reactions after the first five treatment rounds. *Trans R Soc Trop Med Hyg*, **85**, 501–5.

WHO, 1981. *Report of the WHO Independent Commission on the Long-term Prospects of the Onchocerciasis Control Programme*, 77. Geneva: World Health Organization.

WHO, 1987. *WHO Expert Committee on Onchocerciasis, Third Report*. WHO Technical Report Series 752, 167. Geneva: World Health Organization.

WHO, 1995. *Onchocercaisis and its control*. WHO Technical Report Series 852. Geneva: World Health Organization.

WHO, 2002. *Success in Africa: The Onchocerciasis Control Programme in West Africa 1974–2002*. Geneva: World Health Organization.

Williams, S.A., Laney, S.J., et al. 2002. The river blindness Genome project. *Trends Parasitol*, **18**, 86–90.

Winnen, M., Plaisier, A.P., et al. 2002. Can ivermectin treatments eliminate onchocerciasis in Africa? *Bull World Health Organ*, **80**, 384–91.

Yaméogo, L., Toé, L., et al. 1999. Pool screen polymerase chain reaction for estimating the prevalence of *Onchocerca volvulus* infection in *Simulium damnosum sensu lato*: results of a field trial in an area subject to successful vector control. *Am J Trop Med Hyg*, **60**, 124–8.

Yarzábal, L., Petralanda, I., et al. 1983. Acid phosphatase patterns in microfilariae of *Onchocerca volvulus* from the Upper Orinoco basin, Venezuela. *Tropenmed Parasitol*, **34**, 109–12.

Zimmerman, P.A., Dadzie, K.Y., et al. 1992. *Onchocerca volvulus* DNA probe classification correlates with epidemiological pattern of blindness. *J Infect Dis*, **165**, 964–8.

Zimmerman, P.A., Guderian, R.H., et al. 1994. Polymerase chain reaction-based diagnosis of *Onchocerca volvulus* infection: improved detection of patients with onchocerciasis. *J Infect Dis*, **169**, 686–9.

40

Angiostrongylus (*Parastrongylus*) and less common nematodes

KENTARO YOSHIMURA

ANGIOSTRONGYLUS CANTONENSIS DOUGHERTY, 1946

Introduction

The genus *Angiostrongylus* includes approximately 20 species infective to small mammals, of which two are definitively known to be pathogenic to humans. *A. cantonensis* is inherently neurotropic, causing eosinophilic meningoencephalitis, whereas *A. costaricensis* causes abdominal angiostrongyliasis. *A. mackerrase* may also cause eosinophilic meningitis in Australia. *A. malaysiensis* occurs in South East Asia and three cases of eosinophilic meningoencephalitis might have involved this species (Cross 1987). A comparative study of *A. cantonensis* and *A. malaysiensis* in monkeys revealed that the former was more pathogenic causing fatal infections.

A. cantonensis was first found in rats, a discovery made in China in 1935 (Chen 1935) and in Taiwan 2 years later. Human infection was first confirmed in Taiwan, but its public health importance was not recognized until the parasite was found in the brain of a Filipino patient in Hawaii (Rosen et al. 1962).

Classification and morphology

The genus *Angiostrongylus* belongs to the order Strongylida and the superfamily Metastrongyloidea. Although the genus has now been divided into five, and *A. cantonensis*, *A. costaricensis*, *A. malaysiensis*, and *A. mackerrase* placed in the new genus *Parastrongylus*, *Angiostrongylus* is used for classification here, because most parasitologists still use this terminology.

Adult males measure $20–25 \times 0.32–0.42$ mm. The buccal cavity has two lateral teeth, but no lips. The esophagus is club-shaped, and measures 0.31–0.32 mm long (Figure 40.1). There is an excretory pore just posterior to the esophagus. Spicules are slightly subequal in length, and range between 1.02 and 1.25 mm long. The gubernaculum is 0.095 mm long. The bursa and rays are well developed (Figure 40.1).

Adult females measure $22–34 \times 0.34–0.56$ mm. The esophagus is 0.35–0.46 mm long. The blood-filled intestine and white uterine tubules are spirally wound in a characteristic 'barber's pole' pattern. The vulva and anus open at distances of 0.19–0.27 and 0.04–0.06 mm, respectively, from the tip of the tail (Figure 40.1). Eggs measure approximately 70×30 µm.

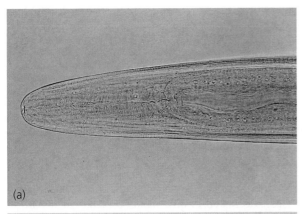

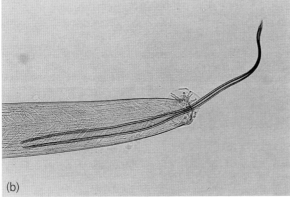

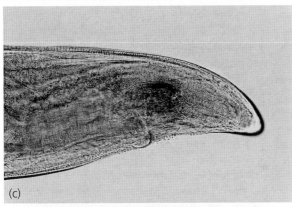

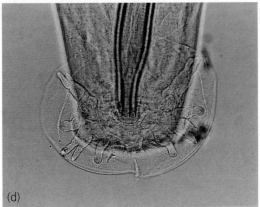

Figure 40.1 Angiostrongylus cantonensis. **(a)** *Anterior end of male (×123).* **(b)** *Posterior end of male, showing copulatory bursa and spicules (×61).* **(c)** *Posterior end of female, showing rectum and vulva (×123).* **(d)** *Copulatory bursa of male, dorsal view (×248).*

Life cycle

The life cycle in the rat was first described by Mackerras and Sandars (1955). Adult worms live and lay eggs mainly in branches of the pulmonary artery, but sometimes in the right ventricle. Eggs are carried to the lung capillaries where they form emboli. They embryonate and hatch after about 1 week. First-stage (L1) larvae enter alveoli, then migrate up the trachea. They are subsequently swallowed and passed out with feces. In the intermediate host, larvae undergo two molts to the second (L2) and third (L3) stage, after 7–9 and 12–16 days respectively. The L3 larvae are 460–510 μm long and, in molluscan intermediate hosts, remain enclosed in the L1 and L2 sheaths.

Many terrestrial and aquatic snails and slugs can serve as intermediate hosts (Chen 1979; Otsuru 1979). Important snails for transmission are the giant African snail, *Achatina fulica* and *Pila* spp. More recently, another edible snail, *Ampullarium canaliculatus*, was found to be naturally infected in Japan and Taiwan.

Animals that acquire L3 larvae by eating infected mollusks may act as paratenic (transport) hosts and may transmit *A. cantonensis* to humans. Known paratenic hosts are toads, frogs, freshwater prawns, land planarians, and land crabs. The yellow tree monitor has also been described as an epidemiologically important paratenic host in Thailand.

When ingested by rats, L3 larvae exsheathe and penetrate into the intestinal wall, reaching the heart via the portal system and the inferior vena cava. Larvae then reach the left ventricle through the pulmonary circulation. Some migrate to the brain directly, whereas others migrate first through other organs or tissues. In either case, L3 larvae reach the central nervous system (CNS) within 2–3 days, after which they migrate within the brain and grow. The larvae molt twice, to become young adults that migrate through brain tissue to the subarachnoid space. Approximately 26–29 days after infection, the worms, now 11.5–13.0 mm long, penetrate cerebral veins and return to the pulmonary arteries via the heart. The worms now grow rapidly and females lay eggs from around day 35. L1 larvae appear in the feces 40–42 days after infection.

In non-permissive hosts, including man, L3 larvae reach the CNS and molt twice, developing into young adults. They die in the brain without returning to the heart and lungs, causing serious eosinophilic meningoencephalitis.

Although *A. cantonensis* has been recovered from the lungs of four patients, it seems unlikely that it can develop to sexual maturity in immunocompetent humans. Among these four, a fatal case reported by Cooke-Yarborough et al. (1999), was an 11-month-old boy who did not show any cerebrospinal fluid (CSF) eosinophilia and harbored adult *A. cantonensis* in the lung after having been treated with intravenous methyl-predonisolone. Experimentally, young adult worms have been recovered from the heart and lungs of immunosuppressed mice (Yoshimura et al. 1982; Sugaya and Yoshimura 1988; Sasaki et al. 1993), but it is still unclear whether full maturity is reached. The pulmonary migration and maturation of *A. cantonensis* in non-permissive hosts may be associated with the immunocompromised state of the host.

Transmission and epidemiology

A. cantonensis infection can occur at any age, even in young infants (Shih et al. 1992). Humans are infected by the following routes: (1) ingestion of L3 larvae in raw or undercooked intermediate or paratenic hosts; (2) by drinking infected water; and (3) oral contact with hands contaminated with larvae released from mollusks. Infections can be established in rodents by various routes including lacerated, abraded, or intact skin. Theoretically, therefore, humans may also acquire infections through the skin.

In Taiwan, most infections occur in aboriginal children, and are acquired by eating infected giant African snails; 80 percent of cases are in children under 14 years old (Chen 1979). The overall snail infection rate is 28 percent, but snails of size >60 mm show 65 percent infection. Infections also occur in infants in the islands of Reunion and Mayotte (Graber et al. 1997; Edmar et al. 1999). In contrast, infections in Thailand are commonly seen in adult males, who often eat chopped raw *Pila* snails with vegetables as an accompaniment to alcoholic drinks. The overall snail infection rate has been calculated as 20 percent, but it was as high as 73 percent in some areas (Teekhasaenee et al. 1986). Infection generally occurs during the rainy months.

Most of the cases reported in Japan have originated from the Ryukyu Islands, where patients have become infected by eating fresh toad liver or raw slugs as medicines (Otsuru 1979). Freshwater prawns, or sauces containing them, are responsible for infections acquired in the Pacific Islands, while in New Caledonia, infections are linked to accidental ingestion of infected small planarians (or fragments of these) on vegetables (Ash 1976). In Micronesia, land crabs and mangrove crabs act as paratenic hosts (Alicata 1965). L3 larvae shed in slug mucus on lettuce may be a source of human infection in Malaysia (Heyneman and Lim 1967). Initial reports of *A. cantonensis* in rodents and mollusks were restricted to tropical and subtropical areas (Alicata and Jindrak 1970), but it has since been found in many other countries.

Cerebral and ocular infections in humans have been confirmed in Hawaii, Japan, China, Taiwan, Thailand, Vietnam, Malaysia, Indonesia, Vanuatu, American Samoa, and the Ivory Coast (Cross 1987). Confirmed or suspected angiostrongyliasis has also been reported in Australia, Fiji, New Orleans, Sri Lanka, Egypt, Cuba, and Madagascar. Cases are common in Taiwan, Ponape island, Tahiti island, and New Caledonia and thousands of cases occur every year in Thailand. Recently, an outbreak of *A. cantonensis*-associated eosinophilic meningitis was recorded in US tourists returning from Jamaica (Slom et al. 2002) and the enzootic *A. cantonensis* infection in rats and snails was noted in Jamaica (Lindo et al. 2002).

It was previously thought that the spread of *A. cantonensis* was linked to dissemination of the giant African snail (Alicata and Jindrak 1970), but this snail does not occur in many of the new endemic areas. This suggests that if infected rats are present, the life cycle of this parasite may easily become adapted to indigenous mollusks (Cross 1987).

Clinical manifestations

These are primarily associated with meningeal irritation and cerebral damage (Table 40.1). Symptoms develop 3–36 days after infection.

Severe headache is the most common symptom (occurring in 86–99 percent of patients). It initially occurs intermittently, but later becomes more frequent. Nausea or vomiting, fever (<38°C), constipation, malaise, and anorexia are frequent (Yii 1976). Less common symptoms are cough, neck stiffness, paresthesia, weakness of extremities, muscle twitching, diplopia, strabismus, and facial paralysis. Patients often present with abnormal (usually decreased) tendon reflexes, Kernig's sign, Brudzinski's sign, sometimes with absence of abdominal reflexes and eye muscle paralysis.

The disease is self-limiting and recovery usually occurs within 4 weeks. Serious cases sometimes result in coma, paralysis of the extremities, urinary incontinence or retention, profuse salivation, convulsions, and the persistence of symptoms (Prociv and Tiernan 1987). Fatality ranges from 0.5 percent (Thailand) to 3 percent (Taiwan).

Eosinophilic radiculomyeloencephalitis has been reported, characterized by peripheral and spinal fluid eosinophilia, severe radicular pain, weakness, and hyporeflexia of the legs and dysfunction of the bladder and bowels (Kliks et al. 1982). Patients may exhibit transient hypertension and lethargy and some become comatose. In eosinophilic myelomeningoencephalitis, patients develop a generalized maculopapular rash followed by myalgia, marked paresthesia, fever, and headache, with progressive weakness, particularly of the legs rather than the arms. Some patients may progress to coma and die (Witoonpanich et al. 1991). A long-lasting, chronic intractable pain, refractory to usual pain management, has also been noted in some patients, the pain of whom

Table 40.1 *Frequency of symptoms/signs among patients with eosinophilic meningitis or meningoencephalitis due to* A. cantonensis *infection*

Symptom/sign	Punyagupta et al. (1975) (484 Thailand cases) (%)	Yii (1976) (114 Taiwanese cases) (%)
Headache	477 (99)	98 (86)
Sensory impairment (lethargy, coma, and confusion)	30 (6)	104 (91)
Vomiting or nausea	425 (88)	94 (83)
Fever	177 (37)	91 (80)
Constipation	–	87 (76)
Malaise	–	81 (71)
Anorexia	–	73 (64)
Neck stiffness	312 (64)	45 (40)
Cough	–	62 (54)
Paraesthesia	181 (37)	32 (28)
Abdominal pain	–	39 (34)
Weakness or paralysis of extremity	4 (1)	26 (23)
Diplopia	184 (38)	11 (10)
Strabismus	–	11 (10)
Muscle twitching	–	15 (13)
Irritability	–	9 (8)
Aching of body and extremities	30 (6)	–
Urinary incontinence or retention	2 (1)	7 (6)
Facial paralysis	20 (4)	–
Convulsion	17 (4)	3 (3)

was successfully relieved by an implantable spinal cord stimulator (Crump et al. 1999).

Infected mice usually develop paralysis of the extremities, ataxia, circling, and torticollis at around day 20 after infection. They develop degenerative or necrotic Purkinje cells that subsequently disappear from the cerebellum, suggesting a Gordon-like phenomenon (Figure 40.2) (Yoshimura et al. 1988, 1994). Eosinophils in the spinal fluid and those around degenerating worms exhibit degranulative changes (Figure 40.3). Eosinophil peroxidase has been detected in the spinal fluid of infected guinea pigs (Perez et al. 1989). Therefore, neurological disorders in eosinophilic meningoencephalitis are probably due not only to mechanical damage caused by migrating worms, but also to the neurotoxicity of eosinophil-derived basic proteins.

Ocular angiostrongyliasis has been reported from Thailand, Taiwan, Vietnam, Indonesia, Japan, Papua New Guinea, and Sri Lanka, worms being found in the anterior chamber, retina, and other sites. This form is rarely accompanied by meningoencephalitis (Nelson et al. 1988). L3 larvae migrate to the base of the brain, move anteriorly into the orbit and finally penetrate the eye via the cribriform plate (Teekhasaenee et al. 1986). Intracerebral migration may also cause optic nerve injury. Worms measuring 500 μm–18.6 mm have been found in the anterior chamber, vitreous humor, and posterior pole of the eye, provoking a variety of symptoms (Scrimgeour et al. 1982; Teekhasaenee et al. 1986). Blindness may develop in rare cases.

Pathogenesis and pathology

Parasites, usually young adult worms, are found mostly in the medulla, pons, and cerebellum and in the adjacent leptomeninges (Jindrak 1975) (Figure 40.4). Some worms are alive at post-mortem, whereas others are disintegrating. The number of worms varies, but may reach several hundreds. Reactions are rarely seen around live worms, but dead worms are often surrounded by infiltrates of polymorphs and eosinophils, or by a zone of suppurative necrosis or granulomatous tissues. Migrating tracks (infiltrates of macrophages, foreign-body giant cells, plasma cells, eosinophils) are found near to worms and at some distance away. Nerve cells near tracks show central chromatolysis and cytoplasmic axonal swelling and blood vessels show perivascular cuffs of eosinophils, lymphocytes, and plasma cells.

The leptomeninges are infiltrated with leukocytes, infiltration being marked over the cerebellum, pons, and medulla. Worms are commonly seen within the underlying nerve tissue and the vessels, particularly veins, are dilated. The subarachnoid space is widened. Worms are found on the surface of the pia mater and foreign-body giant cells are found in the subarachnoid space near to parasites, as well as more remotely. Cases of eosinophilic radiculomyeloencephalitis show a markedly swollen brain, intensely engorged meningeal vessels, and diffuse subarachnoid hemorrhage. Numerous worms are found in sections of the spinal cord, but not the brain

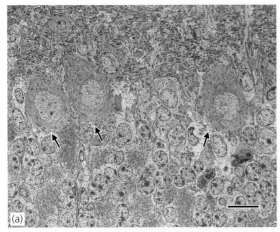

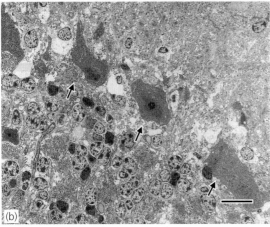

Figure 40.2 *Purkinjé cells (arrows) of the cerebellum from a normal ddY mouse **(a)** and from an infected ddY mouse at 25 days after infection **(b)**. Purkinjé cells in the infected mouse are small and irregular shaped, with pyknotic nuclei. Bar, 10 μm*

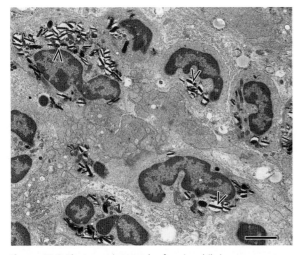

Figure 40.3 *Electron micrograph of eosinophils in a granulomatous reaction around degenerative A. cantonensis worms in a cerebellar fissure of a ddY mouse at 25 days after infection. Eosinophils show marked degranulative changes, i.e. loss of matrix material from specific granules (arrowheads). Bar, 2 μm*

(Kliks et al. 1982). The most striking findings are multiple hemorrhage tracks and cavities caused by worms migrating in the brain parenchyma and spinal cord (Witoonpanich et al. 1991).

Diagnosis

Eosinophilic pleocytosis (Table 40.2) and a history of eating mollusks are indicative. Spinal fluid eosinophilia develops at around day 12 after infection, peaking at days 25–30 (Punyagupta et al. 1975). Opening pressure is raised to >200 mm H_2O. Spinal fluid is usually clear or slightly cloudy, but not xanthochromic (Table 40.2). The white blood cell count in the fluid range from 150 to 2 000/μl, of which eosinophils exceed 10 percent in 95 percent of patients and usually represent 20–70 percent of the total white cells (Weller 1993). Protein is slightly elevated, but glucose is normal (Table 40.2).

Diagnosis should not depend on the identification of larvae in spinal fluid; among 257 cases in Taiwan, worms were found in only 25 patients (Chen 1979). Worm recovery is improved if patients sit upright for some time before sampling (Cross 1987).

According to Punyagupta et al. (1975), most patients had peripheral eosinophilia >10 percent and 43 percent showed eosinophilia of 21–50 percent. Blood eosinophilia did not correlate with cerebrospinal fluid eosinophilia or with the clinical course. Over half of patients (56 percent) showed leukocytosis >10 000/mm². Chest X-ray findings are normal (Koo et al. 1988). Recently, Graber et al. (1999) noticed the presence of enlarged ventricles and cerebral subarachnoid spaces in computed tomography findings of severe eosinophilic meningitis in infants. Meanwhile, Kanpittaya et al. (2000) suggest a diagnostic value of magnetic resonance imaging (MRI); abnormal findings on MRI of eosinophilic meningoencephalitis include dilated perivascular space, subcortical enhancing lesions, and abnormal high T2 signal lesions in the periventricular regions. Another MRI study showed multiple small high intensity areas on contrast-enhanced T1-weighted images, suggesting tissue reactions to dead or dying worms, and local vasodilatation associated with minimal thrombus formation (Ogawa et al. 1998).

Many serological tests have been used to support the diagnosis (Koo et al. 1988). With partially purified antigens, enzyme-linked immunosorbent assay (ELISA) detects significantly higher IgM and IgE levels in serum than in the spinal fluid (Yen and Chen 1991). Immunoblot analyses for detecting serum antibodies to 31 and 29 kDa female worm antigens are also useful (Akao et al. 1992). More recently, the detection of IgG4 antibody to 29 kDa antigen from young adult worms by immunoblotting was found to be most reliable for discriminating between human angiostrongyliasis cantonensis, gnathostomiasis, and cysticercosis (Intapan et al. 2003). Tests are sometimes complicated by possible cross-reactions between *A. cantonensis* and other nematodes. A

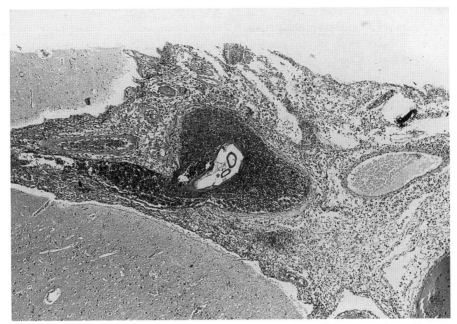

Figure 40.4 *Cross-section of* Angiostrongylus cantonensis *in a blood vessel of the subarachnoid space of a human patient (×47).*

purified 204-kDa (Chye et al. 2000) and 31-kDa (Eamsobhana et al. 2001) antigen with high specificity and sensitivity improved the accuracy of specific immunodiagnosis by ELISA. Shih and Chen (1991) described two monoclonal antibodies that recognize a 91 kDa antigen in excretory-secretory (E-S) products of L3 larvae, and they used an enzyme-linked fluorescent assay to detect circulating antigen in the serum and cerebrospinal fluid. A simple agglutination test using gelatin particles as an antigen carrier is also useful for detecting *A. cantonensis* antibodies in routine immunodiagnosis (Eamsobhana et al. 1999).

Cerebrospinal fluid eosinophilia often occurs in other helminth infections. In Thailand, it is especially important to distinguish between cerebral and ocular angiostrongyliasis, gnathostomiasis, and cysticercosis, all of which are endemic (Teekhasaenee et al. 1986).

Control

Thiabendazole and levamisole have been used to treat eosinophilic meningitis, but their effectiveness is still doubtful. Symptoms may worsen during treatment (Bowden 1981) and anthelminthics should be used with caution, as most of the pathogenesis is ascribed to dead or dying worms (Cross 1987). Corticosteroids may alleviate some symptoms of increased intracranial pressure and relieve allergic reactions. Lumbar puncture may

Table 40.2 *Differential diagnosis of cerebrospinal fluid from patients with angiostrongyliasis cantonensis or gnathostomiasis spinigera*

	Angiostrongyliasis (five cases)	**Gnathostomiasis (39 cases)**
Color	Clear or slightly cloudy	Slightly turbid, xanthochromic, and bloody
Leukocytes/μl	Increased 840–3750 (2 039)[a]	Increased 110–3 000 (920)
Eosinophils (%)	Very high 35–85 (71)	High 15–90 (54)
Protein (g/l)	Slightly increased 0.45–1.25 (0.68)	Increased Not bloody, *n* = 29: 0.43–1.80 (0.80) Bloody, *n* = 6: 3.45–9.00 (4.96) Xanthochromic, *n* = 4: 1.15–3.60 (2.17)
Glucose (g/l)	Normal 0.32–0.62 (0.49)	Normal 0.018–1.00 (0.51)
Opening pressure (mm)	Increased 210–360 (280)	Increased 90–350 (200)

Data from Schmutzhard et al. (1988)
a) Range (mean)

relieve severe headache. Treatment of ocular angiostrongyliasis depends on surgical removal of the parasite. A 2 percent isoptocarpin solution may prevent the worm from entering the deeper parts of the eye (Widagdo et al. 1977).

In order to prevent disease, ingestion of infected hosts and water or vegetables contaminated with L3 larvae must be avoided. Larvae are killed by freezing at −15°C for >12 h or by boiling for 2–3 min (Alicata 1967). The practice of eating toads or slugs for medical purposes (Otsuru 1979) should be prohibited. Controlling the spread of infected rats and mollusks to areas currently free of *A. cantonensis* will restrict further spread of the parasite. Trends towards increased travel, importation of food, and movement of refugees from endemic areas are all factors which suggest that awareness of this infection must be raised.

ANGIOSTRONGYLUS COSTARICENSIS MORERA AND CÉSPEDES, 1971

Introduction

A. costaricensis was discovered in the mesenteric arteries of patients in Costa Rica by Morera and Céspedes (1971), who established it as the cause of abdominal angiostrongyliasis. It has subsequently been reported from Latin America, USA and, more recently, from Africa (Baird et al. 1987). In humans, adult worms lay eggs that can embryonate, but not hatch. *A. costaricensis* therefore seems to be more adapted to humans than *A. cantonensis*, because it attains sexual maturity (Morera and Céspedes 1971).

Morphology

A. costaricensis lacks a buccal capsule and the excretory pore is located slightly posterior to the esophago-intestinal junction (Morera 1973). Adult males measure 17.4–22.2 × 0.28–0.31 mm. The club-shaped esophagus is 0.182–0.225 mm long. Spicules are of equal length, measuring 0.318–0.330 mm. The gubernaculum has two branches, the bursa is symmetrical and well developed. Behind the cloaca are three papillae (Figure 40.5). Adult females measure 28.2–42.0 × 0.322–0.35 mm and the esophagus is 0.23–0.26 mm long. The anus measures 0.060–0.065 mm and the vulva is situated 0.24–0.29 mm from the tip of the tail. The uterine tubules arise near the esophago-intestinal junction and spiral around the intestine.

Life cycle

Although the cotton rat, *Sigmodon hispidus*, is the most important natural final host in Costa Rica, 11 additional rodents can also carry infection. Coati-mundi, marmosets, and dogs can also serve as final hosts (Morera 1985; Brack and Schröpel 1995). In Brazil, small rodents other than cotton rats are important final hosts (Graeff-Teixeira et al. 1990). L1 larvae from the final hosts enter slugs, develop to the L3 stage, and are frequently shed into the mucus (Morera 1973). The most important intermediate host in Costa Rica, Ecuador, and Nicaragua is the slug *Vaginulus plebeius*, with infection rates reaching 85 percent (Morera 1985; Duarte et al. 1992). Other mollusks are also susceptible; for instance, in southern Brazil, *Sarasinula linguaeformis* shows a high

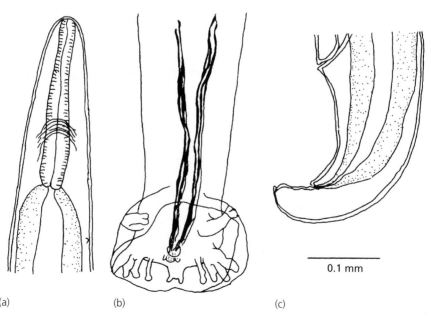

(a)   (b)   (c)

0.1 mm

Figure 40.5 Angiostrongylus costaricensis. **(a)** *Anterior end of the body.* **(b)** *Posterior end of male, showing spicules and bursa.* **(c)** *Posterior end of female, showing rectum and vulva (adapted from Ohbayashi 1979)*

prevalence (86 percent) and individual parasitic burdens (Laitano et al. 2001).

When cotton rats ingest infected slugs, the larvae penetrate the intestinal wall and migrate to the lymphatics. They complete their molts within 7 days. Young adult worms (measuring 3.8–4.4 mm long) migrate to the mesenteric arteries by day 10 and reach maturity and release eggs from day 18. Eggs deposited in the intestinal wall develop and hatch, the L1 larvae appearing in feces 24 days after infection (Morera 1973).

More recently, a new migratory pathway of *A. costaricensis* has been proposed, using the mouse as a definitive host (Mota and Lenzi 1995). The pathway consists of two vascular routes: one is the lymphatic/venous/arterial pathway and the other is the venous portal pathway. The presence of larvae in the lung and liver, especially in the portal veins, of human patients can be accounted for by this new pathway. Moreover, the dog may be a potential reservoir host for *A. costaricensis*, since an experimentally infected dog shed L1 larvae into the feces (Rodriguez et al. 2002).

Transmission and epidemiology

Transmission to humans occurs by ingesting infected slugs, either intentionally or by the accidental ingestion of slugs hidden on vegetables. Larvae shed by slugs can also contaminate hands and food items. L3 larvae can enter abraded (but not unabraded) skin of the cotton rat (Ubelaker et al. 1981).

A. costaricensis is widespread in the American continent, from the USA to northern Argentina. Human infections have been reported from southern Mexico to Argentina. Incidence is particularly high in Costa Rica, numbering some 650 cases annually, or 18 cases/100 000 inhabitants (Morera 1994). Infection is twice as frequent in males as females and occurs predominantly in children under 13 years (99 percent of cases), especially in wet months (Loría-Cortés and Lobo-Sanahuja 1980). In Brazil, the incidence of infection is still increasing.

Clinical manifestations

Symptoms are localized in the abdominal regions, the parasite usually being present in the ileo-cecocolic branches of the anterior mesenteric artery (Morera 1985). Clinical manifestations appear approximately 14 days after infection. Most patients complain of abdominal pain in the right iliac fossa and right flank, and palpation and rectal examinations are painful. Fever lasting for 2–4 weeks is common. Occasionally anorexia, vomiting, diarrhea, and constipation are seen. Patients usually show palpable tumor-like masses in the lower right quadrant and these are often confused with malignant tumors. Hematologically, there is significant leukocytosis and eosinophilia (Table 40.3). The fatality rate ranges from 1.7 to 7.4 percent (Loría-Cortés and Lobo-Sanahuja 1980; Graeff-Teixeira et al. 1991).

Radiology usually reveals abnormalities in the terminal ileum, cecum, and ascending colon. Contrast medium demonstrates spasticity, filling defects, and irritability at the cecum and ascending colon. Fluoroscopy shows Stierlin's sign. The intestinal lumen becomes narrow. When the liver is affected, pain is localized in the upper right quadrant and hepatomegaly is noticeable. Leukocytosis and eosinophilia are usually higher when there is liver involvement than in the intestinal form of the disease, and liver enzymes may be elevated.

Table 40.3 *Frequency of clinical symptoms and signs in patients with abdominal angiostrongyliasis*

Symptom/sign	No. of cases (%)	
	(116 Costa-Rican cases) Loría-Cortés and Lobo-Sanahuja (1980)	(13 Brazilian cases) Graeff-Teixeira et al. (1991)
Abdominal pain	98 (85)	13 (100)
Fever	93 (80)	10 (77)
Abdominal tenderness	79 (68)	–
Anorexia	71 (61)	8 (61)
Malaise	–	7 (58)
Abdominal tumor	59 (51)	4 (31)
Vomiting	52 (45)	5 (38)
Abdominal rigidity	51 (44)	–
Painful rectal examination	50 (43)	–
Weight loss	–	5 (38)
Diarrhea	40 (34)	1 (8)
Nausea	–	4 (31)
Constipation	16 (13)	5 (38)
Hepatomegaly	3 (3)	–
Testicular tumor	1 (1)	–
Jaundice	1 (1)	–

Pathogenesis and pathology

Pathogenesis is attributed either to adult worms in the mesenteric arteries or eggs in the intestinal wall (Morera 1985). Worms damage arterial endothelia, causing thrombosis and distal necrosis. In humans, eggs fail to hatch, but provoke local inflammatory reactions with numerous eosinophilic granulomas and ulceration, causing necrotic lesions and thickening of the intestinal wall. The intestinal lumen becomes narrow and restricts passage of fecal material. Additionally, gangrenous ischemic enterocolitis (Vazquez et al. 1993), eosinophilic ileitis with perforation (Wu et al. 1997), and more rarely, jejunal perforation (Waisberg et al. 1999) have been described.

Hepatic lesions are produced by larvae that invade the peritoneal cavity and migrate to reach branches of the hepatic artery. Larvae cause thromboses and the subsequent inflammation destroys the arterial walls (Vazquez et al. 1994). Ectopic infection in the liver may also occur (Morera 1985). Adult worms may block the testicular arteries and cause extensive necrosis (Ruiz and Morera 1983).

Diagnosis

Diagnosis is difficult, because L1 larvae are not discharged in stools. In endemic areas, physicians must pay special attention to appendicitis-like symptoms in children with prominent eosinophilia. Serological tests have been employed, including ELISA and latex agglutination. More recently, a parasite-specific IgG ELISA has been shown to be useful for the diagnosis of the acute phase of the disease (Geiger et al. 2001). Radiological findings of a mass in the right lower quadrant, thickening of the intestinal wall at the ileo-cecal region, and a narrowed intestinal lumen are also suggestive (Loría-Cortés and Lobo-Sanahuja 1980). Definitive diagnosis depends on the examination of biopsy specimens or surgical resections.

Control

Surgical treatments are generally required for definitive cure, but mild disease may resolve spontaneously (Graeff-Teixeira et al. 1987). The following chemotherapeutic regimens are recommended: (1) thiabendazole in combination with diethylcarbamazine; (2) thiabendazole alone; or (3) high-dose mebendazole.

Prevention is the best management. Children in endemic areas must wash their hands before eating. Inspection and thorough washing of vegetables and related foods before consumption are recommended.

In Guatemala, raw mint, frequently eaten separately or as an ingredient of ceviche, is the most likely vehicle of infection (Kramer et al. 1998). Meanwhile, infective larvae become significantly inactive when they are incubated at 5°C for 12 h in solutions of saturated sodium chloride, vinegar or 1.5 percent sodium hypochlorite (Zanini and Graeff-Teixeira 1995). When the larvae are incubated in tap water in a refrigerator at 5°C, however, it will take 80 days to inactivate all the larvae (Richinitti et al. 1999).

TERNIDENS DEMINUTUS RAILLIET AND HENRY, 1909

Introduction

Ternidens deminutus is found in southern Africa, where it infects the large intestines of primates (baboon, vervet monkey, humans), and in parts of Asia where it has been reported only in monkeys.

Classification and morphology

This genus belongs to the order Strongylida, and the superfamily Strongyloidea. Adult males measure 6–13 mm long, and adult females are 9–17 mm. There is a buccal capsule with three deep-set teeth and a mouth with a mouth-collar. The anterior end has four submedian papillae and two lateral amphids. Males have cup-shaped copulatory bursa, two spicules, and a gubernaculum. Females have protuberant vulva slightly anterior to the anus. The eggs are frequently misdiagnosed as hookworm eggs (and hence are known as 'false hookworm'), but their measurements (81.5×51.8 μm) are slightly larger (Goldsmid 1968). The L3 larva has paired sphincter cells at the esophageal-intestinal junction.

Life cycle

Eggs hatch into L1 larvae in 2–3 days. Subsequently, they molt to L2 larvae in 2–3 days and to L3 filariform larvae in 8–10 days (Goldsmid 1971).

Transmission and epidemiology

Infection is by oral ingestion of L3 larvae. Cases have been noted in Comoros, Mauritius, and southern Africa, from the Congo to the Republic of South Africa. A maximum prevalence of 87 percent was recorded in Rhodesia (Rogers and Goldsmid 1977). Fecocultures of Bush-negroes in Suriname revealed that two (0.2 percent) of 431 positive fecal cultures were positive for *Ternidens* sp. L3 (Jozefzoon and Oostburg 1994).

Clinical manifestations and pathology

L3 larvae penetrate the mucosa of the large intestine and molt, the L4 larvae producing nodules or ulceration. L4 larvae then molt to the adult stage. If worm loads are heavy, anemia might result, although *T. deminutus* was not considered to be a blood sucker by von Boch (1956).

Diagnosis

T. deminutus and hookworm L3 can be differentiated morphologically, or by egg volume (Goldsmid 1968). An indirect fluorescent antibody test, using frozen sections of adult worms has been used to distinguish *T. deminutus* from hookworm infections (Rogers and Goldsmid 1977).

Control

Thiabendazole and albendazole give cure rates of >90 percent (Goldsmid 1972; Bradley 1990). Accidental ingestion of the third stage infective larvae should be avoided.

OESOPHAGOSTOMUM SPECIES

Introduction

Although principally parasitic in the large intestines of ruminants and swine, some species of *Oesophagostomum* parasitize monkeys and apes. Larvae produce nodules in the intestinal wall and cause diarrhea, anemia, and intussusception. Human esophagostomiasis occurs predominantly in Africa, with some cases reported in Indonesia, China, and South America. Five species are known in humans (Polderman et al. 1991), *O. bifurcum* being the most common in Africa and Asia (Blotkamp et al. 1993). *O. aculeatum* occurs in South East Asia.

Classification and morphology

This genus belongs to the order Strongylida and the superfamily Strongyloidea. *O. bifurcum* has a mouth surrounded by an oral collar. The cephalic end has a ventral groove and an excretory pore and the club-shaped esophagus is 0.58–0.63 mm long. Adult males measure 8.6–15.1 mm long. The copulatory bursa is trilobate, the dorsal lobe small. Spicules are 1.02 mm long, and a gubernaculum is present. Adult females are 11.0–16.8 mm long. The vulva and anus are located 0.36–0.62 mm and 0.17–0.26 mm, respectively, from the tip of the tail. Eggs measure $58–69 \times 39–47$ μm, and are indistinguishable from *Necator americanus*.

Life cycle

Eggs develop to the L1 stage in 1 or 2 days at 30°C. The larvae molt twice in 7 days to produce the L3 stage (712–950 μm) which has triangular intestinal cells and a very long, finely tapered tail of the sheath.

Infection occurs by oral ingestion of L3 larvae, which produce nodules in the large intestine wall. After further development, they return to the intestinal lumen, attach to the mucosa, and develop into adult worms. Eggs are discharged in the feces.

Transmission and epidemiology

Oral infection is the most likely route, but infection through the skin is also possible (Ross et al. 1989). *O. bifurcum* L3 are resistant to desiccation and freezing (−15°C for 24 h), implicating the possibility of dust-borne infections in sub-Sahelian areas like northern Togo and Ghana (Pit et al. 2000). Nevertheless, the transmission of infection appears to be limited to the rainy season.

The prevalence of *O. bifurcum* in northern Togo and Ghana has been calculated as 14.2 percent in males and 20.1 percent in females (Polderman et al. 1991). Infection was epidemic in 38/43 villages and prevalence was high (up to 59 percent) in small isolated villages. Infection was rare in children <5 years old, but high levels were found in individuals >5 years old (Krepel et al. 1992). Pit et al. (1999) estimated that more than 100 000 people in Togo were infected with *O. bifurcum*, and actually, infections were more prevalent in women than in men. Likewise, in northern Ghana and Togo, *O. bifurcum* infects an estimated 250 000 people, as determined by cultures of stool samples (Storey et al. 2001).

Clinical manifestations and pathology

L3 larvae penetrate the intestinal wall and produce a solitary, tumor-like inflammatory mass or abscess (helminthoma), measuring one to 2 cm. This is usually located in the ileo-cecal region, but can be present in other organs and in the abdominal wall. The presence of a helminthoma and abdominal pain are major symptoms of the disease. Multiple abdominal masses or nodules containing grayish-green pus and a single immature worm may be present. Intense rectal bleeding and subsequent anemia have been recognized but most cases are totally asymptomatic (Ross et al. 1989).

The clinicopathology has two distinct forms. The multinodular form comprises hundreds of small nodules within a thickened, edematous wall of the large intestine, presenting with weight loss, persistent mucus diarrhea, diffuse abdominal pain, and occasionally, rectal bleeding. Urgent surgery may be required for luminal narrowing of the colon, or bowel obstruction secondary to inflammatory adhesions. On the other hand, the uninodular form, called the Dapaong tumor (Figure 40.6), presents as a painful 30–60 mm granulomatous mass in the abdominal wall or within the abdominal cavity, often associated with fever (Polderman and Blotkamp 1995; Storey et al. 2000a). Ultrasonography is helpful for diagnosis of these two forms; multinodular disease shows nodular 'target' and 'pseudokidney' colonic lesions,

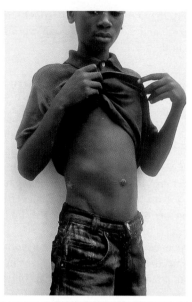

Figure 40.6 *Dapaong tumor resulting from* Oesophagostomum bifurcum *infection (courtesy of Mr P.A. Storey and* Oesophagostomum *research group of Parasitology Department, Leiden University Medical Center)*

whereas the Dapaong tumor is an echo-free ovoid lumen enveloped within a well-defined, poorly reflective wall (Storey et al. 2000a). An epidemiological study in northern Ghana indicates that 13 percent of the cases were multinodular disease, evenly distributed throughout all age-groups, while 87 percent were Dapaong tumor, especially predominant in 5–9-year-old children (Storey et al. 2000b). Oesophagostomes occasionally produce subcutaneous nodules caused by direct skin penetration of L3 larvae or by vascular dissemination of larvae from the bowel (Ross et al. 1989).

Diagnosis

Clinical diagnosis is not necessarily easy and oesophagostomiasis has frequently been misdiagnosed as carcinoma, appendicitis, amoeboma, and other etiologies. Although principally a self-limiting disease, unnecessary radical surgery has frequently been performed. Leukocytosis and blood eosinophilia are not very useful and diagnosis can often be made only by laparotomy (Barrowclough and Crome 1979).

Barium enema examination is useful for detecting a filling defect on the mucosal surface of the large intestine. Harada-Mori test-tube culture technique is used to obtain L3 larvae for morphological identification. A two-step, semi-nested polymerase chain reaction (PCR) method for the specific amplification of minute amounts (fg) of *O. bifurcum* DNA from human fecal samples has been established using genetic markers in the second internal transcribed spacer of ribosomal DNA (Verweij et al. 2000).

ELISA for worm-specific IgG4 antibody has a specificity of >95 percent (Polderman et al. 1993). Body wall

and internal structures help to identify worms in histological sections.

Control

Albendazole and pyrantel pamoate give cure rates >80 percent (Krepel et al. 1993). The effect of the drugs on the tissue-dwelling stages is unknown. Accidental oral ingestion of L3 larvae should be avoided.

TRICHOSTRONGYLUS SPECIES

Introduction

At least ten trichostrongylid species are capable of parasitizing humans. Most species are normally found in herbivores, human infections being accidental, but most *T. orientalis* infections have been recorded in humans. Human infections are prevalent in Iran.

Classification and morphology

This genus belongs to the order Strongylida, superfamily Trichostrongyloidea. The worms are minute and delicate, measuring approximately 4–8 mm long. The head is unarmed and lacks a distinct buccal capsule. The male bursa has long lateral dorsal rays that are poorly developed. Spicules are yellowish-brown, stout, and their chrysanthemum-flower-leaf shape is useful for species differentiation. A gubernaculum is present. The female vulva is situated midway along the body. Eggs resemble hookworm ova, but are longer and narrower (measuring 75–91 × 39–47 μm), with one end more pointed than the other. They are discharged at the 16–32 cell stage (Figure 40.7).

Life cycle

Eggs develop in 1–2 days and then hatch. L1 larvae molt to L2 larvae in 2–3 days and to the ensheathed filariform L3 larvae in 7–8 days. L3 larvae are resistant to desiccation. When ingested by humans with vegetables or drinking water the larvae develop in the intestinal mucosa, molting twice and developing into adult worms in approximately 4 weeks. Percutaneous infection may occur.

Transmission and epidemiology

Trichostrongylus infections have been reported from every continent. Infective larvae are relatively resistant to environmental pressures, and infections are thus prevalent even in relatively cold areas. In Iran, nine species have been recorded from humans; in central Iran, the incidence of *T. orientalis* and *T. colubriformis* has been recorded as 67 percent (Ghadirian and Arfaa 1975). In this region of the world, it is common for

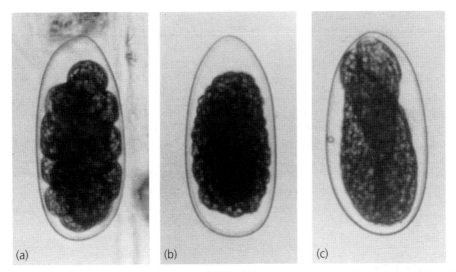

Figure 40.7 Trichostrongylus orientalis *eggs* **(a)** *as seen in fresh feces;* **(b)** *at the morula stage of embryonation;* **(c)** *with developed rhabdtiform larva (×600)*

women to mold animal dung into solid masses, to be dried and burnt as fuel. This practice may be important in the transmission of animal *Trichostrongylus* to humans. Human infections with *T. orientalis* are principally acquired by eating vegetables contaminated with night soil. In Japan, *T. orientalis* infection was previously common in cold areas, the prevalence in Aomori Prefecture in 1975 being approximately 19 percent, but the current figure is estimated to be very low (unpublished data).

Clinical manifestations and pathology

Trichostrongylids penetrate into the intestinal mucosa, producing redness, erosion, and local bleeding. Patients present with epigastric pain, diarrhea, anorexia, nausea, dizziness, and generalized fatigue or malaise (Otsuru and Ito 1972). If worm loads are light patients may be asymptomatic, but heavy worm loads are associated with the development of anemia, and the entry of worms into the biliary tract, provoking cholecystitis.

Diagnosis

Stool examinations are used to detect ova. Harada-Mori cultures are used to obtain L3 larvae for morphological differentiation. Several methods for detecting ova or larvae should be used, because females discharge only a small number of eggs daily.

Control

Pyrantel pamoate, bephenium hydroxynaphthoate, and 1-bromo-naphthol are effective. Ingestion of fresh vegetables or pickles contaminated with L3 larvae should be avoided. Caution must be used when handling animal dung to prevent infection with animal trichostrongylids.

ANISAKIS AND RELATED SPECIES

Introduction

Anisakids live in the stomach of marine mammals. *Anisakis simplex* was first discovered as a cause of larva migrans by van Thiel et al. (1960) who noted a nematode larva in the intestinal wall of a patient suffering from acute abdomimal pain after eating raw herrings. The larva was identified as *Eustoma rotundatum*, but the disease was later named 'anisakiasis' (van Thiel 1962). Prior to this discovery, cross-sections of a nematode-like parasite had frequently been recognized in Japanese patients suffering from acute terminal ileal enteritis, with prominent eosinophilic infiltration and severe allergic tissue reactions. These specimens were initially identified as *Ascaris* larvae. Later, *Anisakis* larvae were removed under gastric endoscopy (Namiki et al. 1970) and since then, many cases of gastric anisakiasis have been reported in Japan, linked to the habit of eating raw fish (sashimi or sushi). Anisakiasis has most frequently been found in Japan, followed by the Netherlands (where infections are acquired from ingesting raw herrings). The increasing popularity of 'sushi bars' has contributed to the spread of infections in the USA and Europe. Anisakid larvae parasitic to humans known to date are *Anisakis simplex*, *A. physeteris*, *Pseudoterranova decipiens*, *Contracaecum osculatum*, *Hysterothylacium aduncum*, and *Porrocaecum reticulatum*; *A. simplex*, and *P. decipiens* being the most important.

Classification and morphology

The genus *Anisakis* belongs to the order Ascaridida, family Heterocheilidae and subfamily Anisakinae. *P. decipiens* was initially called *Ascaris decipiens*, was subsequently transferred to the genus *Porrocaecum*, *Terranova* or *Phocanema* and finally to the genus *Pseudoterranova* (Gibson 1983).

A. simplex L3 were initially called *Anisakis* type-I larva. The larvae have a dorsal lip and two subventral lips orally, with a boring tooth on the dorsal lip. The excretory pore opens at the base of the subventral lips. The esophagus is composed of two parts; the preventriculus and a relatively long ventriculus, with an oblique ventriculo-intestinal junction. The tail is short and rounded, with a mucron (Figure 40.8). Larvae measure 28.40 × 0.45 mm on average. *A. physeteris* was initially called *Anisakis* type-II larva. These measure 27.80 × 0.61 mm on average, have a short ventriculus, a horizontal ventriculo-intestinal junction, and a long tapering tail without a mucron (Figure 40.8).

L3 larvae of *P. decipiens* measure 32.6 × 0.8 mm on average and are yellowish-brown. The ventriculus partly overlaps the intestinal cecum posterior and there is a mucron at the tail end (Figure 40.8). In Japan, *P. decipiens* was previously called *Terranova* type A larva.

Anisakis species have a unique morphology in cross-sections; clear Y-shaped lateral cords (Figure 40.12),

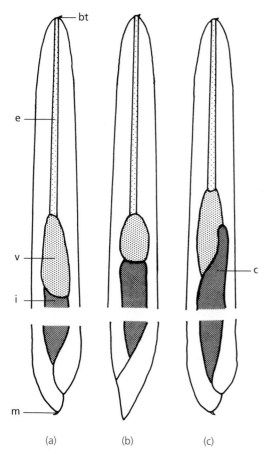

Figure 40.8 *Diagram of the anterior and posterior end of* **(a)** Anisakis simplex; **(b)** A. physeteris; *and* **(c)** Pseudoterranova decipiens. bt, boring tooth; c, cecum; e, esophagus; i, intestine; m, mucron; v, ventriculus (adapted from Koyama et al. 1969)

cuticular lateral alae absent, 60–90 muscle cells/quadrant, and 60–83 columnar intestinal cells. *P. decipiens* larvae have >100 intestinal cells and butterfly-like lateral cords (Oshima 1972).

Life cycle

Figure 40.10 summarizes the life cycles of *A. simplex* and *P. decipiens*. Adult worms live in the stomachs of marine mammals. Eggs (40 × 50 μm) are discharged with the feces, then the larvae molt once into the L2 stage, which hatch when ingested by euphausiid krill and develop into the L3 stage. When infected krill are eaten by squid or marine teleosts, e.g. salmon, cod, herring, and mackerel, the larvae encyst in the viscera or muscles (Figure 40.9), but do not develop further. When these paratenic hosts are eaten by marine mammals, the L3 larvae molt twice in the stomach and develop into adult worms. In Japan, *A. simplex* larvae have been found in 164 species of fish and one squid species, and adult worms have been found in 26 cetacean species and 12 species of pinnipeds. Marine mammals may be infected by ingesting crustacean intermediate hosts. Humans are not suitable hosts; L3 or L4 larvae are found, but never adult worms.

Isopods and mysids serve as intermediate hosts for *P. decipiens*. Cod, haddock, halibut, greenling, pollack, and other fish serve as paratenic hosts (Figure 40.10). L3 larvae have been found in nine species of fish in Japanese waters and 11 species in Californian waters (Ishikura et al. 1992). Seals, sea lions, and walruses act as definitive hosts (Figure 40.10). Adult worms have rarely been recovered from humans.

Transmission and epidemiology

Anisakiasis is linked to the ingestion of raw or undercooked marine fish or squid. Many fish are involved in *A. simplex* transmission in Japan (Figure 40.9) (Oshima and Kliks 1987; Kino et al. 1993). Pacific cod and halibut are important for transmission of *P. decipiens* in Japan. Other important sources of infection include rockfish, salmon, and red snapper in the USA (Kliks 1983; Oshima 1987) and herring in Europe (van Thiel 1976).

Human anisakiasis has been reported from many countries in Asia, Europe, North and South America, and New Zealand. In Japan, a total of 20 582 cases were reported between 1980 and 1993 (Ishikura et al. 1993), but only 660 cases were reported from other countries by the end of 1992. In Japan, 93 percent of cases involved the stomach and only 4.4 percent were intestinal. In other countries intestinal anisakiasis accounts for a much larger percentage of cases. The annual incidence in Japan is >2 000 cases. Changes in dietary habits may account for the escalation of anisakiasis in the USA, where >50 cases have been reported.

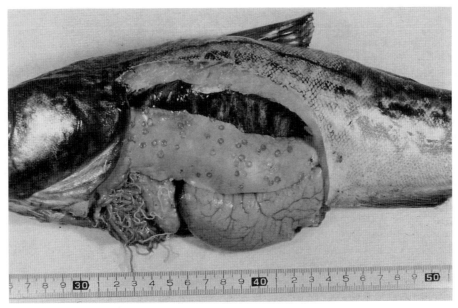

Figure 40.9 Anisakis simplex *third-stage larvae encapsulated in the liver of the pollack,* Theragra chalcogramma *(courtesy of Dr K. Ishida)*

Clinical manifestations

Clinical manifestations can be severe if a secondary infection induces a strong immediate hypersensitivity reaction around the site of worm penetration. Mild clinical symptoms result from local foreign body formation during a primary infection. The disease is also classified into gastric, intestinal, and extra-gastrointestinal (ectopic) anisakiasis. The last form is caused by larvae that penetrate into the abdominal or pleural cavity, entering abdominal or pleural organs or tissues (e.g. omentum, mesentery, peritoneum, liver, pancreas, ovary,

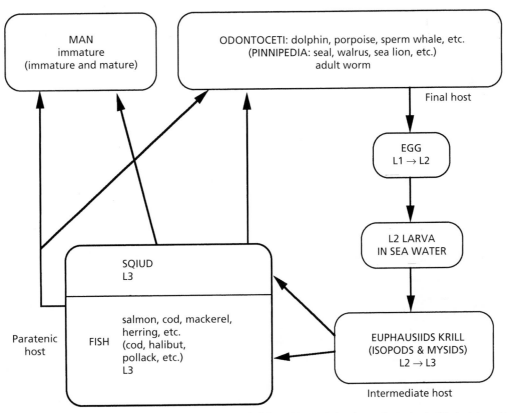

Figure 40.10 *Life-cycle of* Anisakis simplex *and* Pseudoterranova decipiens. *Names in brackets refer to host of* P. decipiens. *L1, L2, and L3: the first-, second- and third-stage larvae (adapted from Oshima 1987).*

oviduct, utero-cervix, lymph node, lung, and even in the subcutaneous tissue), and provoking inflammatory foci.

Acute gastric anisakiasis occurs approximately 6 h after the ingestion of raw seafood (Deardorff et al. 1991). Larvae normally parasitize the fundus of the stomach (rarely tongue, tonsil, uvula, and esophagus), especially the greater curvature (Figure 40.11). Symptoms are summarized in Table 40.4. Rarely, gastric anisakiasis may cause severe chest pain or angina-like pain (Sugano et al. 1993) that is often misdiagnosed as gastric ulcer, cancer, or polyp.

Acute gastric anisakiasis is rare in persons over 60 years of age and in patients who have undergone gastric surgery and this is a result of the low gastric acidity in these people (Muraoka et al. 1996).

Intestinal anisakiasis usually develops within 2 days after infection and occurs in any region between the duodenum and the rectum, most frequently in the ileal region (Matsui et al. 1985). Symptoms (Table 40.4) last for 1–5 days and then lead to ileus. Acute intestinal anisakiasis may cause ascites. The disease is frequently misdiagnosed. Patients usually have no pyrexia, but develop moderate leukocytosis and, in some cases, eosinophilia (4–41 percent) (Ohtaki and Ohtaki 1989; Ishikura 1990; Deardorff et al. 1991).

P. decipiens larvae are less invasive than *A. simplex*, are frequently expelled by vomiting, and rarely cause intestinal anisakiasis (Oshima 1987). Instead, they often provoke a tingling throat syndrome, associated with larval migration from the stomach to the mouth. In contrast, *A. simplex* larvae readily penetrate the gastrointestinal wall and invade the abdominal cavity, provoking peritonitis and intruding into various organs and tissues. Ectopic anisakiasis is usually mild and most cases are discovered accidentally during surgery (Ishikura et al. 1992).

Single worm infection is common, but multiple infections occur, for example approximately 50 *A. simplex* larvae were recovered from a Japanese patient (Takamuku et al. 1994).

A. simplex present in parasitized fish sometimes cause urticaria/angioedema, anaphylaxis, atopic dermatitis, arthralgias/arthritis associated with urticaria, and

Table 40.4 *Clinical symptoms of patients with gastric or intestinal anisakiasis*

Symptom/sign	No. of cases (%)	
	Gastric (363 cases)	Intestinal (124 cases)
Epigastric pain	228 (62.8)	16 (12.9)
Generalized abdominal pain	7 (1.9)	12 (9.7)
Lower abdominal pain	14 (3.9)	56 (45.2)
Epigastric distention	19 (5.2)	1 (0.8)
Abdominal bloating	3 (0.8)	13 (10.5)
Nausea	63 (17.4)	26 (21.0)
Vomiting	34 (9.4)	45 (36.3)
Heart burn	9 (2.5)	–
Anorexia	8 (2.2)	–
Tumor	8 (2.2)	8 (6.5)
General fatigue or malaise	7 (1.9)	1 (0.8)
Blood vomitus	7 (1.9)	–
Diarrhea	5 (1.4)	14 (11.3)
Other peritoneal irritation symptoms	1 (0.3)	18 (14.5)
Urticaria	4 (1.1)	–
Asymptomatic	21 (5.8)	6 (4.8)

Data from Totsuka (1974)

occupational asthma or conjunctivitis in persons handling fish (fishermen/fishmongers) or fish flour (Kasuya et al. 1990; de Corres et al. 1996; Anibarro and Seoane 1998; Armentia et al. 1998; Cuende et al. 1998). Patients presenting with dermatitis after handling fish should be suspected of having contracted protein-contact dermatitis caused by *A. simplex* (Anibarro et al. 1997). Patients with these allergic diseases show positive on skin prick test to *A. simplex* antigen, are positive for histamine release, and have both IgE antibody specific to the worm and a higher total IgE level (del Pozo et al. 1996; Cuende et al. 1998). *Anisakis* allergen is considered to be thermostable and thus, the ingestion of safely cooked but parasitized seafood causes an allergy. The high frequency of *Anisakis*-allergy and the severity of some cases are more important than gastrointestinal anisakiasis, and are becoming a public health problem especially in Spain. Asymptomatic gastro-intestinal anisakiasis sometimes accompanies allergic or anaphylactic reactions (Garcia-Labairu et al. 1999).

Hypersensitivity to *A. simplex* is associated with HLA class II DRB1*1502-DQB1*0601 haplotype, suggesting this haplotype is a susceptibility factor for hypersensitivity to *A. simplex* antigens (Sanches-Velasco et al. 2000).

Pathology

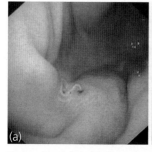 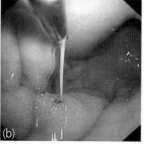

Figure 40.11 Anisakis *larvae just penetrating into the stomach mucosa* **(a)**. *The larva being removed with biopsy forceps of the endoscope* **(b)**. *(Courtesy of Dr I. Sato)*

Initially, lesions are infiltrated mainly by neutrophils, with a few eosinophils and foreign body giant cells.

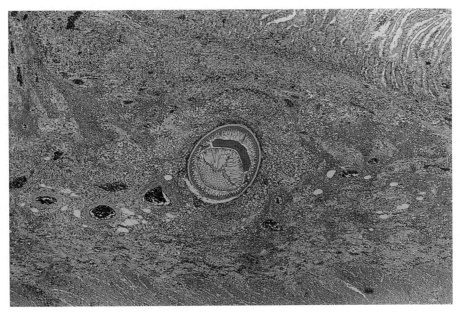

Figure 40.12 *Cross-section of an* Anisakis *larva invading the submucosa of the ileum. Note the characteristic 'Y-shaped' lateral cords and the rennette cell, and intense cellular infiltration around the larva (×47).*

There is little edema, fibrinous exudation, hemorrhage, or vascular damage. Within the first week there may be edematous thickening of the submucosa and infiltration by numerous eosinophils and other cells (Figure 40.12). In chronic gastric and intestinal anisakiasis, necrotic and hemorrhagic abscesses with eosinophilic infiltration may be present. Eosinophilic infiltration becomes less extensive after 6 months and lymphocyte infiltration predominates. Degenerative worms may be surrounded by foreign body giant cells. In more advanced stages (6 months to 1 or more years) the abscess or granulomatous inflammation may be replaced by granulation tissue with some eosinophil infiltration. Fragments of the degenerated worm are usually, but not always, present.

Diagnosis

Clinical manifestations are not specific and therefore diagnosis is difficult, especially with intestinal or extra-gastrointestinal anisakiasis. When patients complain of abdominal pain, nausea, and vomiting, their dietary history immediately prior to the onset of the symptoms should be determined. If raw seafood has been eaten, diagnostic measures must be performed.

ENDOSCOPY

Endoscopy is useful for gastric anisakiasis and sometimes for duodenal anisakiasis (Kliks 1986). The site of larval penetration is edematous, with various degrees of hyperemia, erosion, and bleeding. The penetrating *Anisakis* larva can be easily removed endoscopically with biopsy forceps (Figure 40.11).

RADIOLOGY

Radiology is effective for diagnosing intestinal anisakiasis. Double-contrast, barium-filled roentgenographic images reveal coiled or thread-like filling defects. Edematous, thickened, narrowed, and obstructed areas in the gastrointestinal walls are present (Sugimachi et al. 1985; Matsui et al. 1985). *Anisakis* infection may also cause a vanishing tumor of the stomach.

Sonographic examination is useful for the diagnosis of intestinal anisakiasis when the combined findings of ascitic fluid, dilation of the small intestine, and localized edema of Kerckring's fold is observed in the patients with acute abdomen after recent ingestion of seafood (Ido et al. 1998).

IMMUNODIAGNOSIS

Although several serological tests are used, their application has limitations because of cross-reactions with other nematodes and because low levels of specific antibody may be present in acute anisakiasis. Nevertheless, immunodiagnosis is essential for chronic infections, intestinal, and extra-gastrointestinal anisakiasis. Takahashi et al. (1986) established two monoclonal antibodies, one of which (An2) recognized a specific epitope in the intestine, muscle cells, and E-S products of the larvae. A micro-ELISA assay using this antibody has been developed (Yagihashi et al. 1990). Antilarval *A. simplex*-specific IgG, IgA, and IgM antibodies are detectable 4–5 weeks after the onset of disease; specific IgE antibody can be detected as early as 1–7 days. An immunoblot assay, using E-S antigens of *A. simplex* larvae and detecting IgA or IgE antibodies specific for E-S antigens of the larvae, is also useful (Akao et al. 1990).

IgE immunoblotting with *A. simplex* antigen, positive skin prick test to *A. simplex* but not to fish, and antigen-capture ELISA using O-deglycosylated antigen bound by the monoclonal antibody UA3 are useful approaches to *Anisakis*-allergy diagnosis (del Pozo et al. 1996; Garcia et al. 1997; Lorenzo et al. 2000). More recently, determination of specific IgE directed to *A. simplex* major allergen (24 kDa and named Ani s 1) by immuno-blotting has been shown to be a useful tool for the diag-nosis of hypersensitivity and intestinal anisakiasis, whereas measurement of specific IgG to Ani s 1 is only valid for *A. simplex* allergy (Caballero and Moneo 2002).

Control

Gastric endoscopy is effective for locating and removing penetrating larvae. If diagnosis is confirmed and there is no ileus, surgical treatment is not necessary. Instead, supportive treatment is recommended until the larvae die and are absorbed. In experimental animals this occurs within 3 weeks (Jones et al. 1990).

Thorough cooking and adequate freezing of seafood are easy and practical preventive measures. In the Neth-erlands, legislation (The Green Herring Law) requires fresh herring to be frozen to at least −20°C for 24 h before release to the public. Fish should be cooked so that the internal temperature reaches 60°C or higher for 10 min. Seafood should be frozen at −20°C for 3–5 days (Sakanari and McKerrow 1989; Deardorff et al. 1991). At room temperature, *Anisakis* larvae can survive for many days in vinegar, and for 1 day in soy sauce or Worcester sauce. The larvae may be harmed, but not killed, by gastric fluid. Microwave processing of seafood, for example, arrowtooth flounder fillet, to kill *A. simplex* larvae is effective only when the internal temperature of the fillet reaches higher than 77°C (Adams et al. 1999).

GNATHOSTOMA SPECIES

Owen (1836) collected worms from a stomach wall tumor of a tiger in the London Zoological Gardens and named them *G. spinigerum*. The genus now contains 12 distinct species, four of which (*G. spinigerum*, *G. hispidum*, *G. doloresi*, and *G.nipponicum*) are zoonotic in South East Asia and the Far East.

The number of patients in Mexico has been increasing since the first recorded case in 1970, and the estimated number of patients was more than 1 380 even between 1980 and 1996 (Ogata et al. 1998). Although the causa-tive agent of the human infections in Mexico and Ecuador was identified as *Gnathostoma binucleatum* (Almeyda-Artigas 1991), the validity of *G. binucleatum* and its involvement in human gnathostomiasis was ques-tioned (Akahane et al. 1994; Diaz-Camacho et al. 1998; Ogata et al. 1998). A recent comparative study on ITS-2

rDNA sequences of Mexican *Gnathostoma* spp. indi-cates that *G. binucleatum* is a valid species and incrimi-nated as an etiologic agent of human gnathostomiasis in Mexico and Ecuador (Almeyda-Artigas et al. 2000). In Sinaloa, Mexico, *Gnathostoma* advanced third-stage larvae have been recovered from ichthyophagous birds, e.g. *Egretta alba*, and several fishes, e.g. blue sea catfish (*Arius guatemalensis*), Pacific fat sleeper (*Dormitator latifrons*), fat sleeper (*Gobiomorus* sp.), Nile tilapia (*Oreochromis* sp.), Sinaloan cichlid or green guapote (*Cichlasoma beani*), and spotted sleeper (*Eleotris picta*) (Diaz-Camacho et al. 2002). Moreover, the first human infection of *Gnathostoma malaysiae* was recently found in two Japanese travelers returning from Myanmar (Nomura et al. 2000). These patients ate raw freshwater shrimp in Myanmar and then showed a creeping erup-tion and Quincke's edema.

G. spinigerum is the most important species in Asia, causing cutaneous, cerebrospinal, and ocular lesions. *G. hispidum* was first found to cause human cutaneous gnathostomiasis in Japan (Tsushima et al. 1980), and many cases have subsequently been reported. *G. doloresi*, and *G. nipponicum* have also been shown to provoke creeping eruption in humans in Japan (Ogata et al. 1988; Ando et al. 1988). *G. hispidum*, *G. doloresi*, and *G. nipponicum* usually provoke creeping eruption and rarely affect deeper tissues. They differ from *G. spinigerum* in not causing long-term (>10 years) recur-rent cutaneous lesions and they have not been recog-nized outside Japan.

GNATHOSTOMA SPINIGERUM OWEN, 1836

Introduction

Gnathostoma spinigerum was first discovered in 1889 in Thailand and named *Cheiracanthus siamensis* n. sp., later identified as *G. spinigerum*. Subsequently, many human cases have been reported from various South East Asian countries, Japan, and China. In Thailand, eosinophilic-encephalitis, myelitis, radiculitis, and subarachnoid hemorrhage caused by *G. spinigerum* are often confused with eosinophilic meningoencephalitis caused by *Angios-trongylus cantonensis*; differential diagnosis is therefore important in this country.

Classification and morphology

This genus belongs to the order Spirurida, superfamily Gnathostomatoidea. The adult males and females are 12–30 and 15–33 mm long, respectively. The head-bulb bears between eight and 11 rows of cuticular hooklets. The anterior body has toothed spines. The posterior body is naked apart from the presence of minute term-inal spines. Males have four pairs of large papillae in the

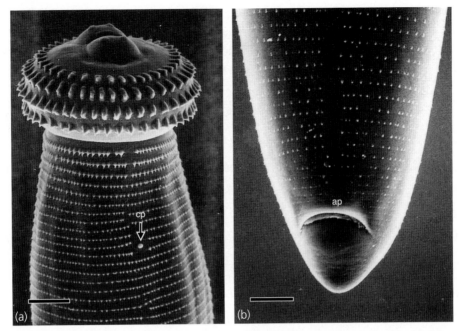

Figure 40.13 *Advanced third-stage larva of* Gnathostoma spinigerum. **(a)** *Anterior end. cp, cervical papilla. Bar, 30 μm;* **(b)** *ventral surface of the terminal end. ap, Anal pore. Bar, 20 μm (courtesy of Dr M. Koga)*

caudal alae and four pairs of small ventral papillae around the cloaca. Spicules are unequal in length, one being between three and four times longer than the other. The vulva is situated just behind the middle of the body. Eggs (measuring 69.3 × 38.5 μm on average) are oval, brownish, and unsegmented with a transparent knob-like thickening at one pole.

Advanced L3 larvae are 3–4 mm long (Figure 40.13 and Table 40.5). The body is entirely covered with cuticular spines. The morphology of the intestinal epithelial cells of the larvae is shown in Table 40.6.

Life cycle

G. spinigerum adult worms occur naturally in the cat, dog, and many other carnivores. In northern Thailand, 4.1 percent of 2 940 dogs examined had adult worms in stomach nodules (Maleewong et al. 1992). Eggs laid in these nodules pass into the gastric lumen through a small aperture and are discharged into the feces. L1 larvae develop in the eggs and molt to the L2 stage (Figure 40.14) after which the eggs hatch. L2 larvae are ingested by an aquatic crustacean (*Cyclops*), after which they develop in the hemocoele and molt into the early L3 stage in 6–10 days.

When early L3 larvae are ingested by freshwater fish, they penetrate into the gut, reaching the muscle and developing into the advanced L3 larvae, measuring 2.8–5.2 × 0.3–0.8 mm. These lie coiled and encysted by connective tissues (Daengsvang 1982). If hosts with L3 larvae are eaten by paratenic hosts (reptiles, birds, or mammals), the larvae encyst again. Advanced L3 larvae have been found in 44 species of vertebrates in Thailand and in 36 species of animals in Japan. Rusnak and Lucey (1993) have listed the many species that can serve as natural and experimental hosts.

When advanced L3 larvae are ingested by a final host they excyst in the stomach, then penetrate the gastric wall to the peritoneal cavity to reach the liver. Subsequently, they migrate through the host and 4 weeks later, they return to the stomach, penetrating the gastric wall from the outside to form a tumor that connects with the stomach lumen through a small aperture. The adult worm develops within 4.5–6 months after infection. Although the adult stage can develop in humans, the worms cannot return to the stomach (Miyazaki 1991).

Transmission and epidemiology

Infections are normally acquired by ingestion of second intermediate or paratenic hosts or, possibly, by drinking water containing infected *Cyclops* or free L3 larvae. Although the most important source of infections is snake-headed fish in both Thailand and Japan, many other species of fish, amphibia, reptiles, birds, and mammals are possible sources. People from Thailand often become infected by eating raw or fermented fish flesh, or raw or undercooked chicken flesh. Freshwater fish and chickens are major sources of human infections in Japan. The incidence of infection in snake-headed fish varies, but overall ranges from 44 to 99 percent (Daengsvang 1982). In Thailand, the incidence in freshwater eel has been reported to reach 80–100 percent, thus both the eel and snake-headed fish are epidemiologically important. Three possible human cases of prenatal infections have been documented (Rusnak and Lucey 1993).

Table 40.5 *Differential diagnosis of advanced third-stage larvae of four species of* Gnathostoma

Species	No. of hooklets on head-bulb				Remarks	Reference
	1st row	2nd row	3rd row	4th row		
G. spinigerum	43.2	44.8	46.7	52.3	The number of hooklets in each row is more than 40, increasing posteriorly	Miyazaki (1960)
G. hispidum	38.3	40.5	41.8	46.0	The number of hooklets gradually increases posteriorly. The first row of hooklets are smaller than the others	Akahane et al. (1982)
G. doloresi	38.3	37.9	35.6	35.7	The number of hooklets in each row is fewer than 40, decreasing posteriorly. The hooklets in the first row are especially small	Mako and Akahane (1985)
G. nipponicum	37.0	37.1	41.0	–	Lack the fourth row of hooklets	Ando et al. (1988)

Table 40.6 *Differential diagnosis in cross-section of the abdominal region of advanced third-stage larvae of four species of* Gnathostoma

Species	No. of muscle cells/quadrant	Number	Intestinal cell shape	No. of nuclei
G. spinigerum	10–15	21–29	Columnar	0–7 (mostly 3–7)
G. hispidum	11–15	19–31	Spherical	0–2 (mostly 1)
G. doloresi	11–15	18–28	Spherical	0–3 (mostly 2)
G. nipponicum	10–14	10–14	Columnar	0–4 (1 in 50% of cells)

After Ando et al. (1991).

Advanced L3 larvae can penetrate the intact skin of mice, rats, and cats, with complete penetration occurring within 30 min. Human infections may also be acquired by this route (Rusnak and Lucey 1993).

Human gnathostomiasis has been recorded from 19 countries, predominantly in Asia, especially Thailand and Japan. In Thailand, infections mainly occur in the rainy season (Punyagupta et al. 1990) and are most prevalent in females, aged 20–25 years, whereas in Japan, they are most prevalent in males in their twenties (Daengsvang 1982). The youngest recorded case in Japan was an infant less than 2 years old.

Clinical manifestations and pathology

Gnathostomiasis spinigera has cutaneous and visceral forms and the latter is divided into pulmonary, gastrointestinal, urogenital, ocular, otorhinolaryngeal, and cerebral forms (Rusnak and Lucey 1993). Patients develop various prodromes, e.g. general malaise, anorexia, urticaria, vomiting, epigastric pain, diarrhea, and fever in about the first 5 days after infection, i.e. when L3 larvae excyst, penetrate into the peritoneal cavity, and migrate through the liver. Right upper quadrant pain may be experienced. The larvae then migrate to various parts of the body, most frequently to muscle and subcutaneous tissues. A specific cutaneous symptom, i.e. migrating intermittent swelling, usually develops within

1 month after infection, but may sometimes occur 1 or 2 years later. During this process, tissues or organs are mechanically injured by the migrating worms and probably also by the toxins they release (Miyazaki 1960). The lesions thus become enlarged, with allergic inflammation around infected foci. Worms in human tissues are not encysted.

In cutaneous gnathostomiasis, migrating swelling frequently appears first in the abdominal area, after which the lesion migrates at random. Worms move at about 1 cm/h in human skin. Swellings are rather hard, variable in size, and mostly erythematous. Skin lesions are normally painless apart from light or moderate itch. Swellings last 1–2 weeks and then disappear, another swelling appearing elsewhere. Migrating swellings may recur intermittently for up to 10–12 years. Worms migrating near the body surface may cause creeping eruption, cutaneous abscesses, or cutaneous nodules, from which the worms can often be resected. Inflammatory foci with profound eosinophilic infiltration are commonly seen in the skin lesions.

Migrating worms in deeper tissues produce visceral gnathostomiasis and serious symptoms may develop, depending on the organs or tissues affected (Rusnak and Lucey 1993). An acute or chronic inflammatory reaction occurs around invading worms and local hemorrhage, necrosis, edema, fibrosis, and tumor formation may occur. When the lungs are affected,

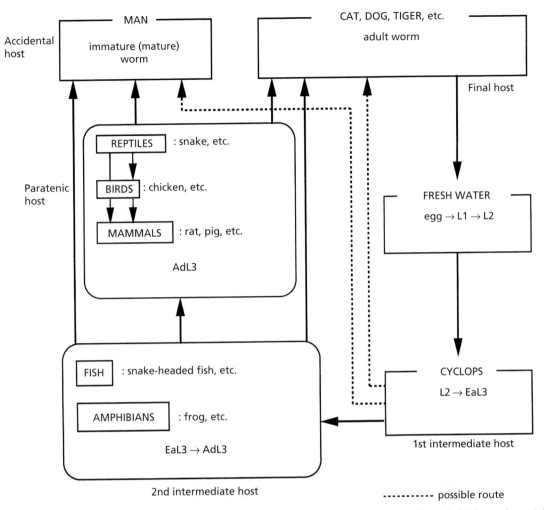

Figure 40.14 *Life-cycle of* Gnathostoma spinigerum. *L1 and L2, the first- and second-stage larvae; EaL3, early third-stage larva; AdL3, advanced third-stage larva.*

respiratory symptoms occur. Involvement of the gastro-intestinal tract is rare and usually asymptomatic, only being found incidentally at surgery. Urogenital involvement is also rare. Ocular involvement presents as pain and visual impairment, with a variety of signs (Teekhasaenee et al. 1986). Iris holes, uveitis, and subretinal hemorrhage with subretinal tract are characteristic features of intraocular gnathostomiasis (Biswas et al. 1994). Involvement of the ear, nose, and throat is always preceded by facial swellings; worms may extrude from tissues and there may be dyspnea and difficulty in swallowing. The prognosis of gnathostomiasis with CNS involvement is poor and the infection is sometimes fatal (Punyagupta et al. 1990). A common clinical symptom in this form is radiculomyelitis, typically characterized by excruciating radicular pain associated with a burning sensation, which may last for 1–5 days as the worms migrate from the nerve root. When a cranial nerve or a cervical nerve is affected, the patient develops severe headache, convulsion, vomiting, paraesthesia, and unconsciousness, followed by paralysis or weakness of the extremities. In Thailand, differential diagnosis

between cerebral angiostrongyliasis and gnathostomiasis is required, because both involve marked eosinophilia in the spinal fluid (Table 40.2).

Initially, patients show peripheral leukocytosis and 5–96 percent eosinophilia, but both these signs later gradually decrease. Patients with ocular gnathostomiasis show normal hematology. Cerebral gnathostomiasis usually shows xanthochromic cerebrospinal fluid with numerous leukocytes, especially lymphocytes and eosinophils (8–98 percent), and some erythrocytes. Protein in the spinal fluid is unchanged or slightly elevated, whereas glucose is normal (Table 40.2). Computed tomography of the head is useful for diagnosis of intracranial hemorrhage or obstructive hydrocephalus. Death is usually due to direct involvement of the brain stem and massive hemorrhage in this area; mortality ranges from 8 to 25 percent (Rusnak and Lucey 1993).

Diagnosis

If biopsy specimens or sections containing worms are available, diagnosis can be made morphologically

(Tables 40.5 and 40.6). Recovery of worms is not always possible and therefore serological tests have also been employed. An ELISA test using crude extract of advanced L3 larvae or their E-S products is used for detecting antigen specific IgE (Soesatyo et al. 1987) or IgG antibody (Tuntipopipat et al. 1989) in sera or spinal fluid. A 24-kDa glycoprotein antigen from advanced L3 larvae is specific to *G. spinigerum* and is thus useful for immunodiagnosis (Nopparatana et al. 1991; Tapchaisri et al. 1991). E-S antigens from advanced L3 larvae give a better diagnostic result than crude somatic extracts. Furthermore, an ELISA assay with low molecular weight E-S or somatic antigens (less than 29 kDa), purified by a simple two-step fractionation process, is sensitive and specific (Tuntipopipat et al. 1993). Detection of circulating antigens seems to be unreliable (Tuntipopipat et al. 1989).

Control

Until recently, there were no effective routine therapeutic measures other than surgical removal of worms, which is not always feasible; supportive, symptomatic, and anti-inflammatory treatments were recommended. Recently, however, it has been shown that cutaneous gnathostomiasis can be treated successfully with a single 3-week course of albendazole (400 mg) either once or twice daily (Kraivichian et al. 1992; Crowley and Kim 1995). Ivermectin is also effective and more convenient in a single-dose treatment (Nontasut et al. 2000). Preventive measures include avoidance of raw or inadequately cooked hosts that may contain L3 larvae and avoidance of drinking water that may contain infected *Cyclops*. L3 larvae are killed by boiling for 5 min or freezing at −20°C for 3–5 days (Schantz 1989; Rusnak and Lucey 1993). Use of gloves or frequent washing of hands while handling food is recommended to prevent possible larval penetration of the skin. Relatively recent increases in world travel and importation of food mean that greater awareness of gnathostomiasis spinigera is required, even in the USA and European countries.

GNATHOSTOMA HISPIDUM FEDTSCHENKO, 1872

Introduction

Gnathostoma hispidum was discovered in 1872 in wild pigs in Turkestan and in domestic pigs in Hungary. Since then it has been found in pigs in Germany and many Asian countries (Daengsvang 1982). Two doubtful human cases were described in 1924 and 1949, and four cases of creeping disease were more recently found in Japan (Tsushima et al. 1980). Subsequently, a further 79 cases of cutaneous gnathostomiasis due to *G. hispidum* have been recognized in Japan (Ando 1992).

Morphology

Adult worms live in the stomach wall, singly or in groups of a few worms, without producing a tumor. Males are 20 mm long and females are 25 mm long. Fresh specimens are red. The head-bulb bears between nine and 12 rows of cephalic hooklets and the entire body surface is covered with cuticular spines. Eight pairs of papillae are located caudally in the male worm, whereas the female has one pair of papillae. Eggs resemble those of *G. spinigerum*, but are slightly larger (72 × 40 µm). Advanced L3 larvae from humans measure approximately 900 µm (Tables 40.5 and 40.6).

Life cycle

Eggs shed in swine feces release L2 larvae in approximately 2 weeks. If ingested by *Cyclops* they molt into early L3 larvae in approximately 8 days. These develop into advanced L3 larvae in second intermediate or paratenic hosts (e.g. fish, amphibians, and rodents) and become encysted in their muscles. Prevalence in loaches from the Beijing and Nanjing areas has been estimated at 12 and 6 percent, respectively (Akahane et al. 1982). In China, advanced L3 larvae have been found in 14 species of freshwater fish, frogs, snakes, chickens, and rats, the incidence in snakes being 100 percent (Chen and Lin 1991). The prepatent period in pigs varies from 3 to 6 months. Unlike *G. spinigerum*, *G. hispidum* cannot mature in humans. Experimentally, cats and rats can be infected by skin penetration of advanced L3 larvae.

Transmission and epidemiology

Humans acquire infections by eating raw loaches containing early L3 larvae or by consuming various vertebrates containing advanced L3 larvae. All human infections in Japan have come from loaches imported from China, Korea, and Taiwan.

Clinical manifestations and pathology

Clinical symptoms develop 3 weeks to 3 months after infection. Initially, patients complain of anorexia, diarrhea, and abdominal pain. All cases in Japan showed creeping eruption or mobile speckled erythema and a few also showed migratory intermittent swellings. Creeping eruption with erythema and itch normally occurs on the trunk or extremities. Morita et al. (1984) classified the disease into five types: (1) skin creeping eruption, on trunk and extremities; (2) migrating tracks on the liver; (3) migrating speckled erythema; (4) cerebrospinal form; and (5) Löffler's syndrome revealed by transient pulmonary infiltration on X-ray. Blood leukocytosis and eosinophilia (up to 84 percent) commonly occur, peaking 1–2 months after infection. Serum IgE levels are highly elevated. Clinical manifestations normally disappear within 3 months and never recur.

Monkeys experimentally infected with larvae from loaches yielded 64 percent encysted advanced L3 larvae after 466 days (Koga et al. 1988). This suggests that encysted larvae may survive in human musculature for long periods.

Diagnosis

Morphological diagnosis (Table 40.6) is possible using resected whole worm specimens or sections. Patients with creeping eruption and peripheral eosinophilia, and a history of eating raw loaches from endemic areas, should be examined for antignathostome antibody. Crude *G. spinigerum* antigens are commonly used in Japan, because *G. hispidum* antigens are unavailable.

Control

No effective drugs are available. Ingestion of raw loaches from endemic areas or flesh from chickens fed with loaches should be avoided.

GNATHOSTOMA DOLORESI TUBANGUI, 1925

Introduction

Gnathostoma doloresi was first discovered by Tubangui (1925) in the stomach wall of a pig in the Philippines. Later, it was also noted in pigs in India by Maplestone (1930) and it has subsequently been recovered from domestic pigs and wild boars in many Asian countries. The first human infections (with creeping eruption) were found by Ogata et al. (1988) in Japan. All eight patients had a history of eating local freshwater fish. Larvae were found in the skin of three out of eight patients. As of 2001, 58 human cases have been noted from Japan.

Morphology

Adult worms occur singly or in groups of two to three worms in the gastric walls of domestic pigs or wild boar, inserting the anterior body into the thickened gastric wall. Males are 20 mm long and females are 34 mm long. The posterior half of the body is thicker than the rest. Worms are covered with cuticular spines, except for a small area at the tail end of the male. Males have eight pairs of caudal papillae. Eggs are 61.3×30.9 μm on average, with transparent knob-like bulges at both ends.

Advanced L3 larvae are approximately 3 mm long. The head-bulb has four rows of cephalic hooklets, those in the first row being extremely small. The average number in each row is <40 (Table 40.5). The morphological characteristics of cross-sections are presented in Table 40.6.

Life cycle

The recent incidence of infection in wild boars examined at two locations in Miyazaki Prefecture, Japan was 95 and 100 percent with an average of 20 and 57 worms per boar, respectively (Ishiwata et al. 1998); in Thailand the incidence in pigs was 2.1–8.7 percent (Daengsvang 1982).

Eggs shed into the feces develop to the L2 stage. These develop in *Cyclops* to the early L3 stage. If *Cyclops* are eaten by fish or amphibians, advanced L3 larvae develop. Snakes that feed on these second intermediate hosts can act as paratenic hosts. The prevalence in the poisonous snake *Agkistrodon halys* has been recorded as 100 percent (Imai et al. 1988). A pig infected with advanced L3 larvae from *A. halys* shed eggs 58 days after infection (Imai et al. 1989).

Transmission and epidemiology

In Japan, freshwater fish are probably the major source of human infections, although several infections occurred by ingesting of the flesh of *A. halys*. *G. doloresi* advanced L3 have been found in a freshwater fish, *Lepomis macrochirus* (common name: blue-gill), which had been ingested by some past patients (Nawa et al. 1993).

Clinical manifestations and pathology

Before cutaneous symptoms develop, patients may experience prodromes (such as fever, 'cold-like' symptoms, vomiting, abdominal pain, weakness, and malaise). Infections normally provoke creeping eruption with local pain and itch on the skin, particularly the trunk, and migratory swellings sometimes develop (Quincke's edema). Creeping eruption leaves pigmentation in the skin. Histopathologically, there is prominent eosinophil infiltration. Total serum IgE may be elevated and patients may show peripheral eosinophilia (6–67 percent).

Miyamoto et al. (1994) reported a case of pulmonary involvement in a patient with right chest pain and high fever. Clinical examination showed eosinophilia (18 percent), elevated IgE, a nodular lesion on r-S4 as revealed by plain X-ray and computed tomography, and anti-*G. doloresi* antibody. No cutaneous lesions developed. A case with an ileus due to an eosinophilic nodular lesion around migrating larvae in the colonic subserosa (Seguchi et al. 1995) and an intraocular case (uveitis) (Sasano et al. 1994) have also been reported. *G. doloresi* larvae can migrate to almost any location in the human body, e.g. the liver and CNS, and give rise to a variety of clinical manifestations.

Diagnosis

Various serological tests employ *G. doloresi* crude extracts as an antigen. If excised skin containing migrating worms is available, the species may be identified by the morphology of whole worms or cross-sections (Tables 40.5 and 40.6).

Control

No effective drugs are available. The best treatment is to excise worms from the skin lesions, if this is possible. No fatal cases have been reported. It is advisable to avoid eating raw, freshwater fish from local rivers in endemic areas.

GNATHOSTOMA NIPPONICUM YAMAGUTI, 1941

Introduction

This species was first found in the esophagus of weasels in Japan, and was mistakenly identified as *G. spinigerum*. It was subsequently renamed *G. nipponicum* by Yamaguti (1941) and has since been found in weasels in various regions of Japan (incidence of 40 percent). The species may also exist in China.

Ando et al. (1988) reported patients in Japan with creeping eruption after eating raw loaches; a *G. nipponicum* larva was found in a skin biopsy specimen. Since then, more than 15 patients with similar gnathostomiasis have been noted.

Morphology

Adult worms live singly or in groups in a hard tumor of the esophageal wall of the weasel, located 2–3 cm from the stomach. The anterior body is embedded and the posterior lies free in the lumen. Adult males are 20–23 mm long and females are 29–34 mm long. The anterior body is covered with cuticular spines, as in *G. spinigerum*, but there are no spines caudally in the female. There are also differences in the shape of the spines of *G. nipponicum* and *G. spinigerum* (Miyazaki 1960, 1991). The caudal alae of the male have small cuticular spines and eight pairs of papillae. Eggs measure 72.3 × 42.1 μm on average, with a transparent knob-like bulge at one end.

Advanced L3 larvae are found encysted in the muscles of the second intermediate or paratenic hosts. They have only three rows of hooklets on the head-bulb (Table 40.5). Morphological features of the cross-sections of the larvae are presented in Table 40.6.

Life cycle

L2 larvae released from fully developed eggs are eaten by *Cyclops* and molt to early L3 larvae. When cyclops are eaten by loaches, the larvae develop to the advanced L3 stage and encyst. Several species of freshwater fish, amphibians, and mammals can act as second intermediate hosts and several species of reptiles, birds, and mammals act as paratenic hosts (Ando et al. 1992). Weasels experimentally infected with advanced L3 larvae shed eggs 3 months after infection. Ferret and mink also serve as experimental hosts.

Transmission and epidemiology

Humans acquire infections by ingesting raw or inadequately cooked loaches or catfish. Recently, a human infection occurred through ingestion of raw, largemouth, black bass (*Micropterus salmoides*) in Akita, Japan (Ishida et al. 2003).

Clinical manifestations and pathology

All 15 cases studied to date developed creeping eruption on the abdomen, waist, and hip areas. The patients had all ingested loaches or various freshwater fish 3 to 4 weeks before the onset of symptoms. Some had peripheral eosinophilia (maximum 29 percent); leukocytosis was not always seen.

Diagnosis

Excised worms or resected skin specimens can be used for morphological identification (Tables 40.5 and 40.6). Some patients show a positive serological reactions, but data are still scanty.

Control

Excision of worms is the best therapeutic measure if feasible. Mebendazole was effective for stopping the creeping eruption (Taniguchi et al. 1991). Eating raw freshwater fish must be avoided in endemic areas.

SPIRUROID TYPE X LARVA

Since 1991, many human cases have been reported from Japan of a creeping disease (most often with blister formations) caused by the type X larva of suborder Spirurina. This larva was first described by Hasegawa (1978) and has subsequently been recovered from human skin, the intestinal wall (especially ileum), and the anterior chamber of the eye. Sometimes, the larvae show intrahepatic erratic parasitism and provoke intestinal perforation. The adult stage has not been identified. Kagei (1991) suggested that the larvae of Spiruroid nematodes other than the Spiruroid type X larva might cause cutaneous or visceral larva migrans.

Spiruroid type X larvae measure 6.7–8.0 × 0.08–0.10 mm, with two tubercles at the tail and two pseudolabia at the head. Larvae detected in sections of resected skin are 80–100 μm in diameter with characteristic cuticular annulations and intracuticular striae. The muscle layer is polymyarian-coelomyarian. Lateral cords are large and Y-shaped. The esophagus has muscular and glandular parts, the latter filled with minute granules that resemble secretory granules or vacuoles. The intestine has between five and six columnar intestinal cells.

These larvae have been found in marine fish which are therefore possible sources of human infections. Some species of squid also harbor type X larvae. Firefly squid are probably the most important source of human infections, because they are the type most frequently eaten raw in Japan. Spiruroid type X infections in humans occur predominantly, especially during March to June, and as of 2001, there are more than 70 cases in Japan.

The disease is classified into ocular, cutaneous, and intestinal (ileus) forms. Cutaneous symptoms develop as early as 2–9 days after eating raw squid (Shinozaki et al. 1993). There is usually only a transient, linear, or meandering erythema, without significant leukocytosis or eosinophilia. Some cases showed a significant rise of serum total IgE levels. Testing with immunofluorescence assay (IFA) or ELISA is recommended for the immunodiagnosis of the disease.

Effective therapeutic measures are unavailable. Most patients with ileus undergo surgical resection of the intestinal lesions and confirmed diagnosis is then made. It is recommended that firefly squid are frozen at −32°C for 30 min or more (in practice at −40°C for more than 40 min) to prevent infection (Akao et al. 1995).

REFERENCES

Adams, A.M., Miller, K.S., et al. 1999. Survival of *Anisakis simplex* in microwave-processed arrowtooth flounder (*Atheresthes stomias*). *J Food Protec*, **62**, 403–9.

Akahane, H., Iwata, K. and Miyazaki, I. 1982. Studies on *Gnathostoma hispidum* Fedchenko, 1872 parasitic in loaches imported from China. *Jpn J Parasitol*, **31**, 507–16, in Japanese with English abstract.

Akahane, H., Lamothe-Argumedo, R., et al. 1994. A morphological observation of the advanced third-stage larvae of Mexican *Gnathostoma. Jpn J Parasitol*, **43**, 18–22.

Akao, N., Ohyama, T.A. and Kondo, K. 1990. Immunoblot analysis of serum IgG, IgA and IgE responses against larval excretory-secretory antigens of *Anisakis simplex* in patients with gastric anisakiasis. *J Helminthol*, **64**, 310–18.

Akao, N., Kondo, K., et al. 1992. Antigens of adult female worm of *Angiostrongylus cantonensis* recognized by infected humans. *Jpn J Parasitol*, **41**, 225–31.

Akao, N., Tsukidate, S., et al. 1995. The lethal effect of freezing on spirurid nematode larvae in firefly squids, *Watasenia scintillans. Jpn J Parasitol*, **44**, 321–4.

Alicata, J.E. 1965. Biology and distribution of the rat lungworm, *Angiostrongylus cantonensis*, and its relationship to eosinophilic meningoencephalitis and other neurological disorders of man and animals. *Adv Parasitol*, **3**, 223–48.

Alicata, J.E. 1967. Effect of freezing and boiling on the infectivity of third-stage larvae of *Angiostrongylus cantonensis* present in land snails and freshwater prawns. *J Parasitol*, **53**, 1064–6.

Alicata, J.E. and Jindrak, K. 1970. *Angiostrongylosis in the Pacific and Southeast Asia*. Springfield, IL: Charles C Thomas.

Almeyda-Artigas, R.J. 1991. Hallazgo de *Gnathostoma binucleatum* n. sp. (Nematoda: Spirurida) en felinos silvestres y el papel de peces dulceacuicolas y oligohalinos como vectores de la gnathostomiasis humana en la cuenca baja del Rio Papaloapan, Oaxaca-Veracruz, Mexico. *Anales del Instituto de Ciencias del Mary Limnologia de la Universidad Nacional Autonoma de Mecixo*, **18**, 295–313.

Almeyda-Artigas, R.J., Bargues, M.D. and Mas-Coma, S. 2000. ITS-2 rDNA sequencing of *Gnathostoma* species (Nematoda) and elucidation of the species causing human gnathostomiasis in the Americas. *J Parasitol*, **86**, 537–44.

Ando, K. 1992. Gnathostomiasis in Japan. *Rinsho Derma (Tokyo)*, **34**, 517–26, in Japanese.

Ando, K., Tanaka, H., et al. 1988. Two human cases of gnathostomiasis and discovery of a second intermediate host of *Gnathostoma nipponicum* in Japan. *J Parasitol*, **74**, 623–7.

Ando, K., Hatsushika, R., et al. 1991. *Gnathostoma nipponicum* infection in the past human cases in Japan. *Jpn J Parasitol*, **40**, 184–6.

Ando, K., Tokura, H., et al. 1992. Life cycle of *Gnathostoma nipponicum* Yamaguti, 1941. *J Helminthol*, **66**, 53–61.

Anibarro, B. and Seoane, F.J. 1998. Occupational conjunctivitis caused by sensitization to *Anisakis simplex. J Allergy Clin Immunol*, **102**, 331–3.

Anibarro, P.C., Carmona, J.B., et al. 1997. Protein contact dermatitis caused by *Anisakis simplex. Contact Derm*, **37**, 247.

Armentia, A., Lombardero, M., et al. 1998. Occupational asthma by *Anisakis simplex. J Allergy Clin Immunol*, **102**, 831–4.

Ash, L.R. 1976. Observations on the role of mollusks and planarians in the transmission of *Angiostrongylus cantonensis* infection to man in New Caledonia. *Rev Biol Trop*, **24**, 163–74.

Baird, J.K., Neafie, R.C., et al. 1987. Abdominal angiostrongylosis in an African man: Case study. *Am J Trop Med Hyg*, **37**, 353–6.

Barrowclough, H. and Crome, L. 1979. Oesophagostomiasis in man. *Trop Geogr Med*, **31**, 133–8.

Biswas, J., Gopal, L., et al. 1994. Intraocular *Gnathostoma spinigerum*. Clinicopathologic study of two cases with review of literature. *Retina*, **14**, 438–44.

Blotkamp, J., Krepel, H.P., et al. 1993. Observations on the morphology of adults and larval stages of *Oesophagostomum* sp. isolated from man in northern Togo and Ghana. *J Helminthol*, **67**, 49–61.

Bowden, D.K. 1981. Eosinophilic meningitis in the New Hebrides: Two outbreaks and two deaths. *Am J Trop Med Hyg*, **30**, 1141–3.

Brack, M. and Schröpel, M. 1995. *Angiostrongylus costaricensis* in a black-eared marmoset. *Trop Geogr Med*, **47**, 136–8.

Bradley, M. 1990. Rate of expulsion of *Necator americanus* and the false hookworm *Ternidens deminutus* Railliet and Henry, 1909 (Nematoda) from humans following albendazole treatment. *Trans R Soc Trop Med Hyg*, **84**, 720.

Caballero, M.L. and Moneo, I. 2002. Specific IgE determination to Ani s 1, a major allergen from *Anisakis simplex*, is a useful tool for diagnosis. *Ann Allergy Asthma Immunol*, **89**, 74–7.

Chen, E.R. 1979. Angiostrongyliasis and eosinophilic meningitis on Taiwan: A review. *NAMRU-2-SP-44*, 57–73.

Chen, H.T. 1935. Un nouveau nematode pulmonaire, *Pulmonema cantonensis* n.g.n.sp., des rats de Canton. *Ann Parasitol*, **13**, 312–17.

Chen, Q.Q. and Lin, X.M. 1991. A survey of epidemiology of *Gnathostoma hispidum* and experimental studies of its larvae in animals. *Southeast Asian J Trop Med Pub Health*, **22**, 611–17.

Chye, S.M., Chang, J.H. and Yen, C.M. 2000. Immunodiagnosis of human eosinophilic meningitis using an antigen of *Angiostrongylus cantonensis* L5 with molecular weight 204 kD. *Acta Tropica*, **75**, 9–17.

Cooke-Yarborough, C.M., Kornberg, A.J., et al. 1999. A fatal case of angiostrongyliasis in an 11-month-old infant. *Med J Aust*, **170**, 541–3.

Cross, J.H. 1987. Public health importance of *Angiostrongylus cantonensis* and its relatives. *Parasitol Today*, **3**, 367–9.

Crowley, J.J. and Kim, Y.H. 1995. Cutaneous gnathostomiasis. *J Am Acad Dermatol*, **33**, 825–8.

Crump, J.A., Chambers, S.T., et al. 1999. Successful management of pain syndrome due to *Angiostrongylus cantonensis* by implantable spinal cord stimulator. *Aust NZ J Med*, **29**, 565.

Cuende, E., Audicana, M.T., et al. 1998. Rheumatic manifestations in the course of anaphylaxis caused by *Anisakis simplex*. *Clin Exp Rheumatol*, **16**, 303–4.

Daengsvang, S. 1982. Gnathostomiasis. In: Steele, J.H. (ed.), *CRC handbook series in zoonoses, section C. , Parasitic zoonoses*. vol. 2. Cleveland: CRC Press, 147–80.

Deardorff, T.L., Kayes, S.G. and Fukumura, T. 1991. Human anisakiasis transmitted by marine food products. *Hawaii Med J*, **50**, 9–16.

de Corres, L.F., Audicana, M., et al. 1996. *Anisakis simplex* induces not only anisakiasis: Report on 28 cases of allergy caused by this nematode. *J Invest Allerg Clin Immunol*, **6**, 315–19.

del Pozo, M.D., Moneo, I., et al. 1996. Laboratory determinations in *Anisakis simplex* allergy. *J Allergy Clin Immunol*, **97**, 977–84.

Diaz-Camacho, S.P., Zazueta-Ramos, M., et al. 1998. Clinical manifestations and immunodiagnosis of gnathostomiasis in Culiacan, Mexico. *Am J Trop Med Hyg*, **59**, 908–15.

Diaz-Camacho, S.P., Willms, K., et al. 2002. Morphology of *Gnathostoma* spp. isolated from natural hosts in Sinaloa, Mexico. *Parasitol Res*, **88**, 639–45.

Duarte, Z., Morera, P., et al. 1992. *Angiostrongylus costaricensis* natural infection in *Vaginulus plebeius* in Nicaragua. *Ann Parasitol Hum Comp*, **67**, 94–6.

Eamsobhana, P., Watthanakulpanich, D., et al. 1999. Detection of antibodies to *Parastrongylus cantonensis* in human sera by gelatin particle indirect agglutination test. *Jpn J Trop Med Hyg*, **27**, 1–5.

Eamsobhana, P., Yoolek, A., et al. 2001. Purification of a specific immunodiagnostic *Parastrongylus cantonensis* antigen by electroelution from SDS-polyacrylamide gels. *Southeast Asian J Trop Med Pub Hlth*, **32**, 308–13.

Edmar, A., Slim, G., et al. 1999. Meningites à eosinophiles et anigostrongylose chez le nourrisson à l'Ile de la Reunion: à propos de deux cas. *Med Mal Infect*, **29**, 480–2.

Garcia, M., Moneo, I., et al. 1997. The use of IgE immunoblotting as a diagnostic tool in *Anisakis simplex* allergy. *J Allergy Clin Immunol*, **99**, 497–501.

Garcia-Labairu, C., Alonso-Martinez, J.L., et al. 1999. Asymptomatic gastroduodenal anisakiasis as the cause of anaphylaxis. *Eur J Gastroenterol Hepatol*, **11**, 785–7.

Geiger, S.M., Laitano, A.C., et al. 2001. Detection of the acute phase of abdominal angiostrongyliasis with a parasite-specific IgG enzyme linked immunosorbent assay. *Mem Inst Oswaldo Cruz*, **96**, 515–18.

Ghadirian, E. and Arfaa, F. 1975. Present status of trichostrongyliasis in Iran. *Am J Trop Med Hyg*, **24**, 935–41.

Gibson, D.I. 1983. The systematics of ascaridoid nematodes: a current assessment. In: Stone, A.F., Platt, H.M. and Khalil, L.E. (eds), *Nematode Systematics Association special volume*. vol. 22. London: Academic Press, 321–38.

Goldsmid, J.M. 1968. The differentiation of *Ternidens deminutus* and hookworm ova in human infections. *Trans R Soc Trop Med Hyg*, **62**, 109–16.

Goldsmid, J.M. 1971. Studies on the life cycle and biology of *Ternidens deminutus* (Railliet & Henry, 1909), (Nematoda: Strongylidae). *J Helminthol*, **45**, 341–52.

Goldsmid, J.M. 1972. Thiabendazole in the treatment of human infections with *Ternidens deminutus* (Nematoda). *S Afr Med J*, **46**, 1046–7.

Graber, D., Jaffar-Bandjee, M.C., et al. 1997. L'angiostrongylose chez le nourrisson à la Reunion et à Mayotte. A propos de trois meningites à eosinophiles dont une radiculomyeloencephalite fatale avec hydrocephalie. *Arch Pediatr*, **4**, 424–9.

Graber, D., Hebert, J.C., et al. 1999. Formes graves de meningites à eosinophiles chez le nourrisson à Mayotte. A propos de 3 observations. *Bull Soc Pathol Exot*, **92**, 3, 1–3.

Graeff-Teixeira, C., Camillo-Coura, L. and Lenzi, H.L. 1987. Abdominal angiostrongyliasis – an under-diagnosed disease. *Mem Inst Oswaldo Cruz*, **82**, 353–4.

Graeff-Teixeira, C., de Avila-Pires, F.D., et al. 1990. Identificação de roedores silvestres como hospedeiros de *Angiostrongylus costaricensis* no sul do Brasil. *Rev Inst Med Trop Sao Paulo*, **32**, 147–50, in Portuguese.

Graeff-Teixeira, C., Camillo-Coura, L. and Lenzi, H.L. 1991. Clinical and epidemiological aspects of abdominal angiostrongyliasis in southern Brazil. *Rev Inst Med Trop São Paulo*, **33**, 373–8.

Hasegawa, H. 1978. Larval nematodes of the superfamily Spiruroidea. A description, identification and examination of their pathogenicity. *Acta Med Biol*, **26**, 79–116.

Heyneman, D. and Lim, B.L. 1967. *Angiostrongylus cantonensis*: Proof of direct transmission with its epidemiological implications. *Science*, **158**, 1057–8.

Ido, K., Yuasa, H., et al. 1998. Sonographic diagnosis of small intestinal anisakiasis. *J Clin Ultrasound*, **26**, 125–30.

Imai, J.I., Asada, Y., et al. 1988. *Gnathostoma doloresi* larvae found in snakes, *Agkistrodon halys*, captured in the central part of Miyazaki Prefecture. *Jpn J Parasitol*, **37**, 444–50.

Imai, J.I., Akahane, H., et al. 1989. *Gnathostoma doloresi*: Development of the larvae obtained from snakes, *Agkistrodon halys*, to adult worms in a pig. *Jpn J Parasitol*, **38**, 221–5.

Intapan, P.M., Maleewong, W., et al. 2003. Evaluation of human IgG subclass antibodies in the serodiagnosis of angiostrongyliasis. *Parasitol Res*, **89**, 425–9.

Ishida, K., Kubota, T., et al. 2003. A human case of gnathostomiasis nipponica confirmed indirectly by finding infective larvae in leftover large mouth bass meat. *J Parasitol*, **89**, 407–9.

Ishikura, H. 1990. Clinical features of intestinal anisakiasis. In: Ishikura, H. and Kikuchi, K. (eds), *Intestinal anisakiasis in Japan. Infected fish, sero-immunological diagnosis, and prevention*. Tokyo: Springer-Verlag, 89–100.

Ishikura, H., Kikuchi, K., et al. 1992. Anisakidae and anisakidosis. In: Sun, T. (ed.), *Progress in clinical parasitology*. vol. 3. New York: Springer-Verlag, 43–102.

Ishikura, H., Sato, S., et al. 1993. Anisakiasis – its outbreak and present status. *Clin Parasitol*, **4**, 152–5, in Japanese.

Ishiwata, K., Diaz-Camacho, S.P., et al. 1998. Gnathostomiasis in wild boars from Japan. *J Wildlife Dis*, **34**, 155–7.

Jindrak, K. 1975. Angiostrongyliasis cantonensis (eosinophilic meningitis, Alicata's disease). In: Hornabrook, R.W. (ed.), *Topics on tropical neurology*. Philadelphia: FA Davis, 133–64.

Jones, R.E., Deardorff, T.L. and Kayes, S.G. 1990. *Anisakis simplex*: histopathological changes in experimentally infected CBA/J mice. *Exp Parasitol*, **70**, 305–13.

Jozefzoon, L.M.E. and Oostburg, B.F.J. 1994. Detection of hookworm and hookworm-like larvae in human fecocultures in Suriname. *Am J Trop Med Hyg*, **51**, 501–5.

Kagei, N. 1991. Morphological identification of parasites in biopsied specimens from creeping disease lesions. *Jpn J Parasitol*, **40**, 437–45.

Kanpittaya, J., Jitpimolmard, S., et al. 2000. MR findings of eosinophilic meningoencephalitis attributed to *Angiostrongylus cantonensis*. *Am J Neuroradiol*, **21**, 1090–4.

Kasuya, S., Hamano, H. and Izumi, S. 1990. Mackerel-induced urticaria and *Anisakis*. *Lancet*, **335**, 665.

Kino, H., Watanabe, K., et al. 1993. Occurrence of anisakiasis in the western part of Shizuoka Prefecture, with special reference to the prevalence of anisakid infections in sardine, *Engraulis japonica*. *Jpn J Parasitol*, **42**, 308–12.

Kliks, M.M. 1983. Anisakiasis in the western United States: Four new case reports from California. *Am J Trop Med Hyg*, **32**, 526–32.

Kliks, M.M. 1986. Human anisakiasis: an update. *JAMA*, **255**, 2605.

Kliks, M.M., Kroenke, K. and Hardman, J.M. 1982. Eosinophilic radiculomyeloencephalitis: An angiostrongyliasis outbreak in American Samoa related to ingestion of *Achatina fulica* snails. *Am J Trop Med Hyg*, **31**, 1114–22.

Koga, M., Ishibashi, J. and Ishii, Y. 1988. Experimental infection in a monkey with *Gnathostoma hispidum* larvae obtained from loaches. *Ann Trop Med Parasitol*, **82**, 383–8.

Koo, J., Pien, F. and Kliks, M.M. 1988. *Angiostrongylus (Parastrongylus)* eosinophilic meningitis. *Rev Infect Dis*, **10**, 1155–62.

Koyama, T., Kobayashi, A., et al. 1969. Morphological and taxonomical studies on Anisakidae larvae found in marine fishes and squids. *Jpn J Parasitol*, **18**, 466–87, in Japanese with English abstract.

Kraivichian, P., Kulkumthorn, M., et al. 1992. Albendazole for the treatment of human gnathostomiasis. *Trans R Soc Trop Med Hyg*, **86**, 418–21.

Kramer, M.H., Greer, G.J., et al. 1998. First reported outbreak of abdominal angiostrongyliasis. *Clin Infect Dis*, **26**, 365–72.

Krepel, H.P., Baeta, S. and Polderman, A.M. 1992. Human *Oesophagostomum* infection in northern Togo and Ghana: epidemiological aspects. *Ann Trop Med Parasitol*, **86**, 289–300.

Krepel, H.P., Haring, T., et al. 1993. Treatment of mixed *Oesophagostomum* and hookworm infection: effect of albendazole, pyrantel pamoate, levamisole and thiabendazole. *Trans R Soc Trop Med Hyg*, **87**, 87–9.

Laitano, A.C., Genro, J.P., et al. 2001. Report on the occurrence of *Angiostrongylus costaricensis* in southern Brazil, in a new intermediate host from the genus *Sarasinula* (Veronicellidae, Gastropoda). *Rev Soc Bras Med Trop*, **34**, 95–7.

Lindo, J.F., Waugh, C., et al. 2002. Enzootic *Angiostrongylus cantonensis* in rats and snails after an outbreak of human eosinophilic meningitis, in Jamaica. *Emerg Inf Dis*, **8**, 324–6.

Lorenzo, S., Iglesias, R., et al. 2000. Usefulness of currently available methods for the diagnosis of *Anisakis simplex* allergy. *Allergy*, **55**, 627–33.

Loría-Cortés, R. and Lobo-Sanahuja, J.F. 1980. Clinical abdominal angiostrongylosis. A study of 116 children with intestinal eosinophilic granuloma caused by *Angiostrongylus costaricensis*. *Am J Trop Med Hyg*, **29**, 538–44.

Mackerras, M.J. and Sandars, D.F. 1955. The life history of the rat lungworm, *Angiostrongylus cantonensis* (Chen) (Nematoda: Metastrongylidae). *Aust J Zool*, **3**, 1–25.

Mako, T. and Akahane, H. 1985. On the larval *Gnathostoma doloresi* found in a snake, *Dinodon semicarinatus* from Amami-Oshima Is., Japan. *Jpn J Parasitol*, **34**, 493–9, in Japanese with English abstract.

Maleewong, W., Wongkham, C., et al. 1992. Detection of circulating parasite antigens in murine gnathostomiasis by a two-site enzyme-linked immunosorbent assay. *Am J Trop Med Hyg*, **46**, 80–4.

Maplestone, P.A. 1930. Nematode parasites of pigs in Bengal. *Records Ind Mus*, **32**, 77–105.

Matsui, T., Iida, M., et al. 1985. Intestinal anisakiasis: Clinical and radiologic features. *Radiology*, **157**, 299–302.

Miyamoto, N., Mishima, K., et al. 1994. A case report of serologically diagnosed pulmonary gnathostomiasis. *Jpn J Parasitol*, **43**, 397–400.

Miyazaki, I. 1960. On the genus *Gnathostoma* and human gnathostomiasis, with special reference to Japan. *Exp Parasitol*, **9**, 338–70.

Miyazaki, I. 1991. *An illustrated book of helminthic zoonoses*. Tokyo: International Medical Foundation of Japan, 368–402.

Morera, P. 1973. Life history and redescription of *Angiostrongylus costaricensis* Morera and Céspedes, 1971. *Am J Trop Med Hyg*, **22**, 613–21.

Morera, P. 1985. Abdominal angiostrongyliasis: A problem of public health. *Parasitol Today*, **1**, 173–5.

Morera, P. 1994. Importance of abdominal angiostrongylosis in the Americas. *8th International Congress on Parasitology* (Izmir, Turkey, October 1994). Abstract, 1: 34.

Morera, P. and Céspedes, R. 1971. *Angiostrongylus costaricensis* n. sp. (Nematoda: Metastrongyloidea), a new lungworm occurring in man in Costa Rica. *Rev Biol Trop*, **18**, 173–85.

Morita, H., Segawa, T., et al. 1984. Gnathostomiasis cases caused by imported loaches. *J Nara Med Assoc*, **35**, 607–19, in Japanese with English abstract.

Mota, E.M. and Lenzi, H.L. 1995. *Angiostrongylus costaricensis* life cycle: a new proposal. *Mem Inst Oswaldo Cruz*, **90**, 707–9.

Muraoka, A., Suehiro, I., et al. 1996. Acute gastric anisakiasis. 28 cases during the last 10 years. *Digest Dis Sci*, **41**, 2362–5.

Namiki, M., Morooka, T., et al. 1970. Diagnosis of acute gastric anisakiasis. *Stomach and Intestine*, **5**, 1437–40, in Japanese.

Nawa, Y., Imai, J., et al. 1993. *Gnathostoma doloresi* larvae found in *Lepomis macrochirus* Rafinesque, a freshwater fish (common name: blue-gill), captured in the central part of Miyazaki Prefecture, Japan. *Jpn J Parasitol*, **42**, 40–3.

Nelson, R.G., Warren, R.C., et al. 1988. Ocular angiostrongyliasis in Japan: A case report. *Am J Trop Med Hyg*, **38**, 130–2.

Nomura, Y., Nagakura, K., et al. 2000. Gnathostomiasis possibly caused by *Gnathostoma malaysiae*. *Tokai J Exp Clin Med*, **25**, 1–6.

Nontasut, P., Bussaratid, V., et al. 2000. Comparison of ivermection and albendazole treatment for gnathostomiasis. *Southeast Asian J Trop Med Pub Hlth*, **31**, 374–7.

Nopparatana, C., Setasuban, P., et al. 1991. Purification of *Gnathostoma spinigerum* specific antigen and immunodiagnosis of human gnathostomiasis. *Int J Parasitol*, **21**, 677–87.

Ogata, K., Imai, J.I. and Nawa, Y. 1988. Three confirmed and five suspected human cases of *Gnathostoma doloresi* infection found in Miyazaki Prefecture, Kyushu. *Jpn J Parasitol*, **37**, 358–64.

Ogata, K., Nawa, Y., et al. 1998. Short report: Gnathostomiasis in Mexico. *Am J Trop Med Hyg*, **58**, 316–18.

Ogawa, K., Kishi, M., et al. 1998. A case of eosinophilic meningoencephalitis caused by *Angiostrongylus cantonensis* with unique brain MRI findings. *Clin Neurol*, **38**, 22–6, in Japanese with English abstract.

Ohbayashi, M. 1979. Hunting parasites in Thailand. *J Hokkaido Vet Assoc*, **23**, 3–15, in Japanese.

Ohtaki, H. and Ohtaki, R. 1989. Clinical manifestation of gastric anisakiasis. In: Ishikura, H. and Namiki, M. (eds), *Gastric anisakiasis in Japan. Epidemiology, diagnosis, treatment*. Tokyo: Springer-Verlag, 37–46.

Oshima, T. 1972. *Anisakis* and anisakiasis in Japan and adjacent area. In: Morishita, K., Komiya, Y. and Matsubayashi, H. (eds), *Progress of medical parasitology in Japan*. vol. 4. Tokyo: Meguro Parasitol Museum, 301–93.

Oshima, T. 1987. Anisakiasis – Is the sushi bar guilty? *Parasitol Today*, **3**, 44–8.

Oshima, T. and Kliks, M. 1987. Effects of marine mammal parasites on human health. *Int J Parasitol*, **17**, 415–21.

Otsuru, M. 1979. *Angiostrongylus cantonensis* and angiostrongyliasis in Japan. *NAMRU-2-SP-44*, 74–117.

Otsuru, M. and Ito, J. 1972. Genus *Trichostrongylus* in Japan. In: Morishita, K., Komiya, Y. and Matsubayashi, H. (eds), *Progress of medical parasitology in Japan*. Tokyo: Meguro Parasitol Museum, 421–63.

Owen, R. 1836. Anatomical description of two species of *Entozoa* from the stomach of a tiger (*Felis tigris* Linn.) one of which forms a new genus of Nematoidea, *Gnathostoma*. *Proc Zool Soc London*, **47**, 123–6.

Perez, O., Capron, M., et al. 1989. *Angiostrongylus cantonensis*: Role of eosinophils in the neurotoxic syndrome (Gordon-like phenomenon). *Exp Parasitol*, **68**, 403–13.

Pit, D.S.S., Rijcken, F.E.M., et al. 1999. Geographic distribution and epidemiology of *Oesophagostomum bifurcum* and hookworm infections in humans in Togo. *Am J Trop Med Hyg*, **61**, 951–5.

Pit, D.S.S., Blotkamp, J., et al. 2000. The capacity of the third-stage larvae of *Oesophagostomum bifurcum* to survive adverse conditions. *Ann Trop Med Parasit*, **94**, 165–71.

Polderman, A.M. and Blotkamp, J. 1995. *Oesophagostomum* infections in humans. *Parasitol Today*, **11**, 451–6.

Polderman, A.M., Krepel, H.P., et al. 1991. Oesophagostomiasis, a common infection of man in northern Togo and Ghana. *Am J Trop Med Hyg*, **44**, 336–44.

Polderman, A.M., Krepel, H.P., et al. 1993. Serological diagnosis of *Oesophagostomum* infections. *Trans R Soc Trop Med Hyg*, **87**, 433–5.

Prociv, P. and Tiernan, J.R. 1987. Eosinophilic meningoencephalitis with permanent sequelae. *Med J Aust*, **147**, 294–5.

Punyagupta, S., Juttijudata, P. and Bunnag, T. 1975. Eosinophilic meningitis in Thailand. Clinical studies of 484 typical cases probably caused by *Angiostrongylus cantonensis*. *Am J Trop Med Hyg*, **24**, 921–31.

Punyagupta, S., Bunnag, T. and Juttijudata, P. 1990. Eosinophilic meningitis in Thailand. Clinical and epidemiological characteristics of 162 patients with myeloencephalitis probably caused by *Gnathostoma spinigerum*. *J Neurol Sci*, **96**, 241–56.

Richinitti, L.M.Z., Fonseca, N.A. and Graeff-Teixeira, C. 1999. The effect of temperature on morbility of *Angiostrongylus costaricensis'* third stage larvae. *Rev Inst Med Trop Sao Paulo*, **41**, 225–8.

Rodriguez, R., Agostini, A.A., et al. 2002. Dogs may be a reservoir host for *Angiostrongylus costaricensis*. *Rev Inst Med Trop Sao Paulo*, **44**, 55–6.

Rogers, S. and Goldsmid, J.M. 1977. Preliminary studies using the indirect fluorescent antibody test for the serological diagnosis of *Ternidens deminutus* infection in man. *Ann Trop Med Parasitol*, **71**, 503–4.

Rosen, L., Chappell, R., et al. 1962. Eosinophilic meningoencephalitis caused by a metastrongylid lung-worm of rats. *JAMA*, **179**, 620–4.

Ross, R.A., Gibson, D.I. and Harris, E.A. 1989. Cutaneous oesophagostomiasis in man. *J Helminthol*, **63**, 261–5.

Ruiz, P.J. and Morera, P. 1983. Spermatic artery obstruction caused by *Angiostrongylus costaricensis* Morera and Céspedes, 1971. *Am J Trop Med Hyg*, **32**, 1458–9.

Rusnak, J.M. and Lucey, D.R. 1993. Clinical gnathostomiasis: Case report and review of the English-language literature. *Clin Infect Dis*, **16**, 33–50.

Sakanari, J.A. and McKerrow, J.H. 1989. Anisakiasis. *Clin Microbiol Rev*, **2**, 278–84.

Sanchez-Velasco, P., Mendizabal, L., et al. 2000. Association of hypersensitivity to the nematode *Anisakis simplex* with HLA class II DRB1*1502-DQB1*0601 haplotype. *Hum Immunol*, **61**, 314–19.

Sasaki, O., Sugaya, H., et al. 1993. Ablation of eosinophils with anti-IL-5 antibody enhances the survival of intracranial worms of *Angiostrongylus cantonensis* in the mouse. *Parasite Immunol*, **15**, 349–54.

Sasano, K., Ando, F., et al. 1994. A case of uveitis due to *Gnathostoma* migration into the vitreous cavity. *J Jpn Ophthalmol Soc*, **98**, 1136–40, in Japanese with English abstract.

Schantz, P.M. 1989. The dangers of eating raw fish. *N Engl J Med*, **320**, 1143–5.

Schmutzhard, E., Boongird, P. and Vejjajiva, A. 1988. Eosinophilic meningitis and radiculomyelitis in Thailand, caused by CNS invasion of *Gnathostoma spinigerum* and *Angiostrongylus cantonensis*. *J Neurol Neurosurg Psych*, **51**, 80–7.

Scrimgeour, E.M., Chambers, B.R. and Kaven, J. 1982. A probable case of ocular angiostrongyliasis in New Britain, Papua New Guinea. *Trans R Soc Trop Med Hyg*, **76**, 538–40.

Seguchi, K., Matsuno, M., et al. 1995. A case report of colonic ileus due to eosinophilic nodular lesions caused by *Gnathostoma doloresi* infection. *Am J Trop Med Hyg*, **53**, 263–6.

Shih, H.H. and Chen, S.N. 1991. Immunodiagnosis of angiostrongyliasis with monoclonal antibodies recognizing a circulating antigen of mol. wt 91,000 from *Angiostrongylus cantonensis*. *Int J Parasitol*, **21**, 171–7.

Shih, S.L., Hsu, C.H., et al. 1992. *Angiostrongylus cantonensis* infection in infants and young children. *Pediatr Infect Dis J*, **11**, 1064–6.

Shinozaki, M., Akao, N., et al. 1993. Detection of type X larva of the suborder Spirurina from a patient with a creeping eruption. *Jpn J Parasitol*, **42**, 51–3.

Slom, T.J., Cortese, M.M., et al. 2002. An outbreak of eosinophilic meningitis caused by *Angiostrongylus cantonensis* in travelers returning from the Caribbean. *N Engl J Med*, **346**, 668–75.

Soesatyo, M.H.N.E., Rattanasiriwilai, W., et al. 1987. IgE responses in human gnathostomiasis. *Trans R Soc Trop Med Hyg*, **81**, 799–801.

Storey, P.A., Anemana, S., et al. 2000a. Ultrasound diagnosis of oesophagostomiasis. *Br J Radiol*, **73**, 328–32.

Storey, P.A., Faile, G., et al. 2000b. Clinical epidemiology and classification of human oesophagostomiasis. *Trans R Soc Trop Med Hyg*, **94**, 177–82.

Storey, P.A., Spannbrucker, N., et al. 2001. Ultrasonographic detection and assessment of preclinical *Oesophagostomum bifurcum*-induced colonic pathology. *Clin Infect Dis*, **33**, 166–70.

Sugano, S., Suzuki, T., et al. 1993. Noncardiac chest pain due to acute gastric anisakiasis. *Dig Dis Sci*, **38**, 1354–6.

Sugaya, H. and Yoshimura, K. 1988. T-cell-dependent eosinophilia in the cerebrospinal fluid of the mouse infected with *Angiostrongylus cantonensis*. *Parasite Immunol*, **10**, 127–38.

Sugimachi, K., Inokuchi, K., et al. 1985. Acute gastric anisakiasis. Analysis of 178 cases. *JAMA*, **253**, 1012–13.

Takahashi, S., Sato, N. and Ishikura, H. 1986. Establishment of monoclonal antibodies that discriminate the antigen distribution specifically found in *Anisakis* larvae (type I). *J Parasitol*, **72**, 960–2.

Takamuku, M., Iino, H. and Fujino, T. 1994. A case of gastric anisakiasis showing multiple infection of larvae in a large gastric ulcer. *Clin Parasitol*, **5**, 74–5, in Japanese.

Taniguchi, Y., Hashimoto, K., et al. 1991. Human gnathostomiasis. *J Cutan Pathol*, **18**, 112–15.

Tapchaisri, P., Nopparatana, C., et al. 1991. Specific antigen of *Gnathostoma spinigerum* for immunodiagnosis of human gnathostomiasis. *Int J Parasitol*, **21**, 315–19.

Teekhasaenee, C., Ritch, R. and Kanchanaranya, C. 1986. Ocular parasitic infection in Thailand. *Rev Infect Dis*, **8**, 350–6.

Totsuka, M. 1974. Human anisakiasis 3. Epidemiology. In The Japanese Society of Scientific Fishery (ed.), *Fish and anisakis*. Tokyo: Kohsei-sha-Kohsei-kaku, 44–57 (in Japanese).

Tsushima, H., Numata, T., et al. 1980. Gnathostomiasis cutis probably infected in Hiroshima city. *J Hiroshima Med Assoc*, **33**, 1183–7, in Japanese.

Tubangui, M.A. 1925. Metazoan parasites of Philippine domesticated animals. *Philippine J Sci*, **28**, 11–37.

Tuntipopipat, S., Chawengkirttikul, R., et al. 1989. Antigens, antibodies and immune complexes in cerebrospinal fluid of patients with cerebral gnathostomiasis. *Southeast Asian J Trop Med Pub Hlth*, **20**, 439–46.

Tuntipopipat, S., Chawengkirttikul, R. and Sirisinha, S. 1993. A simplified method for the fractionation of *Gnathostoma*-specific antigens for serodiagnosis of human gnathostomiasis. *J Helminthol*, **67**, 297–304.

Ubelaker, J.E., Caruso, J. and Peña, A. 1981. Experimental infection of *Sigmodon hispidus* with third-stage larvae of *Angiostrongylus costaricensis*. *J Parasitol*, **67**, 219–21.

van Thiel, P.H. 1962. Anisakiasis. *Parasitology*, **52**, Suppl., 16–17.

van Thiel, P.H. 1976. The present state of anisakiasis and causative worms. *Trop Geogr Med*, **28**, 75–85.

van Thiel, P.H., Kuipers, F.C. and Roskam, R.T. 1960. A nematode parasitic to herring, causing acute abdominal syndromes in man. *Trop Geogr Med*, **2**, 97–113.

Vazquez, J.J., Boils, P.L., et al. 1993. Angiostrongyliasis in a European patient: a rare cause of gangrenous ischemic enterocolitis. *Gastroenterology*, **105**, 1544–9.

Vazquez, J.J., Sola, J.J. and Boils, P.L. 1994. Hepatic lesions induced by *Angiostrongylus costaricensis*. *Histopathology*, **25**, 489–91.

Verweij, J.J., Polderman, A.M., et al. 2000. PCR assay for the specific amplification of *Oesophagostomum bifurcum* DNA from human faeces. *Int J Parasitol*, **30**, 137–42.

von Boch, H. 1956. Knötchenwurmbefall (*Ternidens deminutu*s) bei Rhesusaffen. *Z Angew Zool*, **2**, 207–14.

Waisberg, J., Corsi, C.E., et al. 1999. Jejunal perforation caused by abdominal angiostrongyliasis. *Rev Inst Med Trop Sao Paulo*, **41**, 325–8.

Weller, P.F. 1993. Eosinophilic meningitis. *Am J Med*, **95**, 250–3.

Widagdo, Sunardi, et al. 1977. Ocular angiostrongyliasis in Semarang, Central Java. *Am J Trop Med Hyg*, **26**, 72–4.

Witoonpanich, R., Chuahirun, S., et al. 1991. Eosinophilic myelomeningoencephalitis caused by *Angiostrongylus cantonensis*: A report of three cases. *Southeast Asian J Trop Med Pub Hlth*, **22**, 262–7.

Wu, S.S., French, S.W. and Turner, J.A. 1997. Eosinophilic ileitis with perforation caused by *Angiostrongylus* (*Parastrongylus*) *costaricensis*. A case study and review. *Arch Pathol Lab Med*, **121**, 989–91.

Yagihashi, A., Sato, N., et al. 1990. A serodiagnostic assay by microenzyme-linked immunosorbent assay for human anisakiasis using a monoclonal antibody specific for *Anisakis* larvae antigen. *J Infect Dis*, **161**, 995–8.

Yamaguti, S. 1941. Studies on the helminth fauna of Japan. Part 35. Mammalian nematodes (2). *Jpn J Zool*, **9**, 409–38.

Yen, C.M. and Chen, E.R. 1991. Detection of antibodies to *Angiostrongylus cantonensis* in serum and cerebrospinal fluid of patients with eosinophilic meningitis. *Int J Parasitol*, **21**, 17–21.

Yii, C.Y. 1976. Clinical observations on eosinophilic meningitis and meningoencephalitis caused by *Angiostrongylus cantonensis* on Taiwan. *Am J Trop Med Hyg*, **25**, 233–49.

Yoshimura, K., Sato, K., et al. 1982. The course of *Angiostrongylus cantonensis* infection in athymic nude and neonatally thymectomized mice. *Z Parasitenkd*, **67**, 217–26.

Yoshimura, K., Sugaya, H., et al. 1988. Ultrastructural and morphometric analyses of eosinophils from the cerebrospinal fluid of the mouse and guinea pig infected with *Angiostrongylus cantonensis*. *Parasite Immunol*, **10**, 411–23.

Yoshimura, K., Sugaya, H. and Ishida, K. 1994. The role of eosinophils in *Angiostrongylus cantonensis* infection. *Parasitol Today*, **10**, 231–3.

Zanini, G.M. and Graeff-Teixeira, C. 1995. Angiostrongilose abdominal: profilaxia pela destruicao das larvas infectantes em alimentos tratados com sal, vinagre ou hipoclorito de sodio. *Rev Soc Brasil Med Trop*, **28**, 389–92.

Dracunculiasis

RALPH MULLER

Dracunculiasis is an ancient disease quoted by many classical authors and possibly mentioned in the Old Testament. For the last few years, there has been a world eradication campaign and this chapter has been written in the light of this campaign (Chippaux 1994; Cairncross et al. 2002). It is hoped that it will be superfluous in the next edition.

MORPHOLOGY

The mature female of the dracunculoid nematode *Dracunculus medinensis* (Linnaeus 1758) Gallandant, 1773 measures 500–800 × 1.0–2.0 mm. The mouth has a triangular oval opening surrounded by a quadrangular cuticularized plate, and an internal circle of four double papillae. The vulva opens halfway down the body but is nonfunctional in the mature worm. The uterus has an anterior and a posterior branch; it is filled with 1–3 million embryos and occupies the entire body cavity (pseudocoel), the gut being completely flattened (Cairncross et al. 2002).

Males recovered from experimental infections in animals measure 15–40 × 0.4 mm. The tail has 4(3–6) pairs of preanal and 4–6 pairs of postanal papillae; the subequal spicules are 490–750 μm long with a gubernaculum measuring about 117 μm. The males probably die in the tissues before the females mature but have only doubtfully been seen in human infections.

LIFE CYCLE

In the human body, each female worm takes about one year to mature and moves to the surface of the skin. It provokes a blister at the anterior end which bursts and leaves an ulcer through which about 5 cm of the worm is extruded, particularly after immersion in water (Figure 41.1a). Many thousands of larvae are released into the water where they can live for only a few days; for further development they have to be ingested by small freshwater crustacea (cyclops). The larvae molt twice in the body cavity and are infective in about 2 weeks. If cyclops are ingested in drinking water, released larvae penetrate the human intestinal wall and male and female worms mate in the connective tissues in about 3 months.

TRANSMISSION AND EPIDEMIOLOGY

Dracunculus infection still occurs in 12 countries of West and Central Africa: principally in Sudan, Ghana, and Nigeria but also in Benin, Burkina Faso, Cote d'Ivoire, Ethiopia, Mali, Mauritania, Niger, Togo, and Uganda (Cairncross et al. 2002) and with a recent isolated outbreak in Libya. The disease has been eliminated or has died out in Cameroon, Central African Republic, Chad, Guinea Bissau, India, Iran, Pakistan, Saudi Arabia, Senegal, and Yemen in recent years.

Dracunculiasis is typically a disease of rural communities that obtain their drinking water from ponds (or, formerly, in India, from large step wells) where cyclops can breed (Figure 41.2). In all areas, transmission is markedly seasonal and, in most parts of Africa, the maximum incidence coincides with the planting season, resulting in great economic hardship. In semi-desert (Sahel) areas of Africa (Burkina Faso, Mauritania, Niger, northern Nigeria, and Sudan), drinking water is

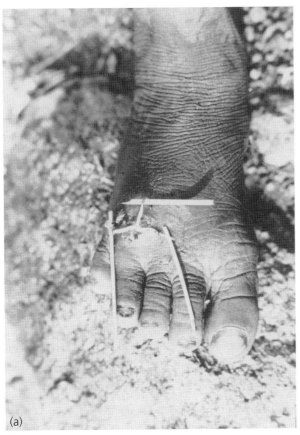
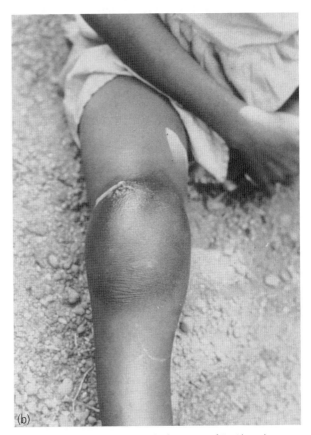

Figure 41.1 *Dracunculiasis:* **(a)** *foot of child with three worms emerging and being wound out on sticks (Courtesy of Dr Ahmed Tayeh);* **(b)** *secondarily infected and swollen knee of girl with guinea worm emerging*

obtained from ponds during the rainy season but from deep (safe) wells for the rest of the year when the ponds are dry. However, in the humid (Guinea) savanna regions of West Africa where rainfall exceeds 150 cm per year (Benin, Cote d'Ivoire, Ghana, southern Nigeria, and Togo) there is almost no transmission during the rainy season (July–September) when ponds turn into streams and cyclops densities are low because of the

large volume and turbidity of the water. The life history of the parasite is well adapted to provide the maximum chances of transmission as the female takes almost exactly a year to mature and release its larvae.

LOCATION IN HOST

One to many adult female worms emerge from the subcutaneous tissues, usually of the foot or lower limbs, but sometimes from any part of the body.

CLINICAL MANIFESTATIONS AND PATHOLOGY

In most patients, the first physical sign is the local lesion, accompanied by an intense burning pain usually relieved by immersion of the affected limb in water. The blister fluid is bacteriologically sterile and contains numerous white cells and larvae. In uncomplicated cases, the worm emerges from the subsequent ulcer over a few weeks and then the lesion rapidly heals. Thus, if there is only one worm present, patency will last for only 4–6 weeks. Unfortunately, secondary infection along the track of the worm in the tissues is very common, often with spreading cellulitis, and approximately 40 percent of patients will be totally incapacitated for an average of 6

Figure 41.2 *One of many large artificial ponds constructed in Nigeria for drinking water and agriculture and which are transmitting both dracunculiasis and schistosomiasis*

weeks (Figure 41.1b). More serious and permanent damage can follow the bursting of a worm in the tissues or as the result of bacterial infection, e.g. ankylosis of joints in about 1 percent of cases or, occasionally, tetanus.

Dracunculiasis is unusual among infectious diseases in that the parasite does not appear to stimulate a protective response, so that the same individual can be reinfected year after year.

DIAGNOSIS

Clinical and parasitological diagnosis

Patients in an endemic area usually have no doubt of the diagnosis as soon as or even before the first signs appear. Local itching, urticaria, and a burning pain at the site of a small blister are usually the first signs of infection. The blister bursts in about 4 days and active larvae, obtained by placing cold water on the resulting small ulcer, can be recognized under a low-powered microscope.

Immunological diagnosis

Immunological methods are not useful in practice. Enzyme-linked immunosorbent assay (ELISA) and sodium dodecylsulfate-polyacrylamide gel electrophoresis (SDS PAGE)/western blotting worked well in one trial for patent infections. Detection of parasite-specific immunoglobulin (Ig) G4 might be able to diagnose prepatent infections possibly up to 6 months before emergence (Bapna and Renapurkar 1996).

TREATMENT

Surgery

Guinea worms have been wound out on sticks since antiquity (e.g. in the Rig Veda of about 1350 BC). Provided that bacterial infection or other complications have not occurred, regular winding out of the worm on a small stick, combined with sterile dressing and acriflavine cream, usually results in complete removal in about 4 weeks with little loss of mobility. Treatment should be commenced as soon after emergence as possible (Magnussen et al. 1994). Sometimes, worms can be seen and surgically removed before emergence while there is no tissue reaction against them.

Chemotherapy

There is no evidence that any chemotherapeutic agent has a direct action against guinea worms. However, many compounds, including thiabendazole, niridazole, metronidazole, mebendazole, and albendazole, have been reported as hastening the expulsion of worms and may act as anti-inflammatory agents (Muller 1971). Ivermectin had no action against pre-emergent worms (Issaka-Tinorgah et al. 1994).

THE ERADICATION CAMPAIGN

The eradication of dracunculiasis from the world was adopted as a sub-goal of the Clean Drinking Water Supply and Sanitation Decade (1981–90) and was formally endorsed by the United Nations World Health Assembly in 1986 and 1989. If the original aim of complete eradication by the end of 1995 had been achieved, this chapter would have been superfluous. However, the task is likely to take quite a while longer, particularly in the Sudan. Nonetheless, the prevalence and distribution of infection have diminished markedly in the last few years: worldwide there were about 3.5 million cases in 1986, 1 million in 1989 but only 46 000 in 2002. Disease has been eliminated from Asia with no cases reported from Pakistan since 1993, nor from India since 1997 (Cairncross et al. 2002). The situation in African countries is not so far advanced but active campaigns are being carried out in almost all endemic countries and some are at the stage of case containment (Figure 41.3). In Nigeria and Ghana for instance, new cases have reduced markedly since 1988, and in all African countries on average 56 percent of cases were contained in 2002 (i.e. all larvae were prevented from reaching ponds). In 2002, only 3 000 cases were reported from Nigeria and 4 900 from Ghana. In 37 countries where the disease used to be present in historical times, applications for certification of absence have been made.

The first priority has been to have accurate figures on the distribution and prevalence in each endemic country. Because infected persons rarely report to health clinics, passive surveillance is almost useless (identifying less than 3 percent of cases) and trained teams of community workers are required to visit each village and obtain the necessary information. These teams have to be carefully monitored and, when trained, can also be associated with other primary healthcare initiatives, such as control of malaria and diarrhea diseases and immunization campaigns.

Once active surveillance measures have been initiated, there are various possible interventions based on knowledge of the life cycle of the parasite.

The provision by governments and aid agencies of safe drinking water from tubewells is a priority in rural areas of Africa and Asia. The United Nations (International) Children's (Emergency) Fund (UNICEF) aimed to provide one borehole for every 200 persons in suitable geological regions by the year 2000. In some areas, traditional (usually hand built) draw wells can be equally effective and in all areas,

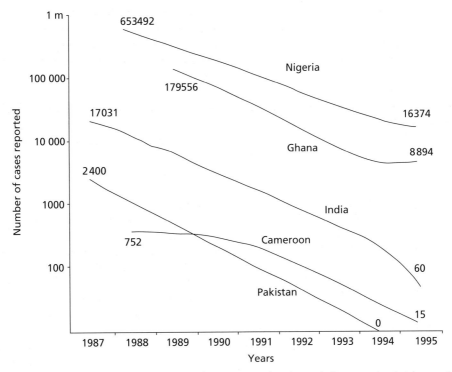

Figure 41.3 *Decrease in numbers of cases of guinea worm infection reported each year in five countries that have active campaigns (note log scale). The campaign in India started in 1980 (Sudan is not included).*

cisterns can be constructed for storing rainwater (Figure 41.4).

Health education interventions by local health workers form a very important component of any control or eradication campaign. Interventions being made include the following:

● Filtering or boiling all drinking water. Boiling water is not usually feasible because fuel is a scarce commodity but filters are playing an important role in all countries now mounting campaigns. The donation by the manufacturer of monofilament nylon nets, which are long lasting, have a regular pore size, are

Figure 41.4 *Cistern built at a school in Nigeria to store rainwater for use during the dry season.*

easily washed and dry quickly, is proving of great help in sieving out cyclops. Health educators in all countries have been very successful in promoting the use of filters.

- Persuading or preventing infected persons with an emerging worm from entering the water source (containment). This is particularly important when prevalence has been greatly reduced (the current motto is 'detect every case, contain every worm'). Bandaging very early lesions can help to prevent subsequent immersion in the water.

- Treating water sources. The chemical treatment of ponds can prove a useful adjunct to other measures, particularly when there is only a low level of transmission remaining. The insecticide temephos added to ponds at a concentration of 1 ppm will kill cyclops for 5–6 weeks and has low toxicity to mammals and fishes. In areas with a long transmission season, it has to be added a few times a year. The amount of temephos estimated to be needed for total eradication in Africa has been donated by the manufacturer. The main problem is the amount of time that needs to be expended by trained technicians.

The disease was eliminated from Kenya and Senegal in 1996, and from Cameroon, Chad, and Yemen in 1999. There are very few cases left in all African countries apart from Ghana, Nigeria, and Sudan (49 000 cases in 2001 and provisionally 40 000 in 2002). The number of cases in Ghana and Nigeria is decreasing greatly each year but the future in Sudan is more uncertain (target eradication date of 2009).

ZOONOTIC ASPECTS

Female worms of the genus *Dracunculus* have been reported as emerging from a wide range of mammals and reptiles from many parts of the world, both endemic and nonendemic for the human disease. Those found in reptiles clearly belong to other species but the situation in regard to those in mammals is not clear. For instance, guinea worm is common in wild carnivores in North America and the species was named *D. insignis* by Leidy in 1858 but there is very little morphological evidence for separating this from the human species.

There have been two documented cases of clearly zoonotic infections, from Japan in 1986 and from Korea in 1926. In both cases, the patients had eaten raw freshwater fish that have been proved experimentally to be capable of acting as paratenic hosts.

In most highly endemic areas, occasional infections in dogs and donkeys with what is presumably the human parasite have been reported but there is no evidence that they have any part in maintaining transmission. The parasite can still be found in dogs in the formerly endemic areas of Tamil Nadu in India and the central Asian republics of the former Soviet Union but no new human cases have been reported, so it is not thought that this will be a problem once world eradication has been achieved.

REFERENCES

Bapna, S. and Renapurkar, D.M. 1996. Immunodiagnosis of early dracunculiasis. *J Commun Dis*, **28**, 33–7.

Cairncross, S., Muller, R. and Zagaria, N. 2002. Dracunculiasis (Guinea Worm Disease) and the eradication initiative. *Clin Microbiol Rev*, **15**, 223–46.

Chippaux, J.-P. 1994. Le Ver de Guinée en Afrique: Methodes de Lutte pour l'Eradication. Paris: ORSTOM.

Issaka-Tinorgah, A., Magnussen, P., et al. 1994. Lack of effect of ivermectin on prepatent guinea-worm: a single-blind, placebo-controlled trial. *Trans R Soc Trop Med Hyg*, **88**, 346–8.

Magnussen, P., Yakubu, A. and Bloch, P. 1994. The effect of antibiotic- and hydrocortisone-containing ointments in preventing secondary infections in Guinea worm disease. *Am J Trop Med Hyg*, **51**, 797–9.

Muller, R. 1971. *Dracunculus* and dracunculiasis. *Adv Parasitol*, **9**, 73–151.

Index

Notes
(Fig.) and (Tab.) refer to figures and tables respectively. *vs.* indicates a comparison or differential diagnosis.
To save space in the index, the following abbreviations have been used
CNS - central nervous system
ELISA - enzyme-linked immunosorbent assay
Ig - immunoglobulin

Complete table of contents for *Topley & Wilson's Microbiology and Microbial Infections*

VIROLOGY, VOLUMES 1 AND 2

BACTERIOLOGY, VOLUMES 1 AND 2

MEDICAL MYCOLOGY

PARASITOLOGY

IMMUNOLOGY